D0146406

ENGINEERING MECHANICS

DYNAMICS

SECOND EDITION

INDIANA
PURDUE
LIBRARY
FORT WAYNE
WITHDRAWN

ANTHONY BEDFORD ▪ WALLACE FOWLER

UNIVERSITY OF TEXAS AT AUSTIN

TA
352
.B382
1999

ADDISON-WESLEY

An imprint of Addison Wesley Longman, Inc.

Menlo Park, California • Reading, Massachusetts • Harlow, England
Berkeley, California • Don Mills, Ontario • Sydney • Bonn • Amsterdam • Tokyo • Mexico City

Senior Acquisitions Editor: Michael Slaughter
Associate Editor: Susan Slater
Editor in Chief: Chuck Iossi
Editorial Assistants: Colleen Kelly, Chandrika Madhavan
Production Manager: Pattie Myers
Production Editor: Caroline Jumper
Art and Design Supervisor: Kevin Berry
Art Coordinator and Production Assistant: Kamila Storr
Copy Editor: Bruce Emmer
Proofreader: Brian Jones
Cover Design and Cover Illustrator: Yvo Riezebos
Text Design: Wilson Graphics & Design (Kenneth J. Wilson)
Frontmatter Design: Vargas Williams Design (Juan Vargas)
Layout: Vargas Williams Design (Edie Williams)
Illustrator: James A. Bryant
Technical Artists: Precision Graphics
Compositor: American Composition & Graphics, Inc.
Printer and Binder: R.R. Donnelley and Sons, Willard
Cover Printer and Separator: Phoenix Color Corporation

Photo Credits
Chapter 1: 1.4, Dennis Mitchell/Allsport Photography, Inc.; P1.22, NASA.
Chapter 2: 2.13, ©Kim Vandiver and Harold Edgerton, courtesy Palm Press, Inc.; 2.38, Courtesy of Sandia National Laboratories.
Chapter 4: P4.107, Courtesy of Victor Austin and David Goldstein.
Chapter 5: 5.5 and P5.70, The Harold Edgerton 1992 Trust, courtesy Palm Press, Inc.
Chapter 6: 6.46 (a & b), NASA.
Chapter 10: 10.22, U.S. Geological Survey.

Copyright © 1999 by Addison Wesley Longman, Inc.

All rights reserved. No part of this publication may be reproduced, stored in a retrieval system, or transmitted in any form or by any means, electronic, mechanical, photocopying, recording, or any other media embodiments now known or hereafter to become known, without the prior written permission of the publisher. Manufactured in the United States of America.

Library of Congress Cataloging-in-Publication Data
Bedford, A.
Engineering mechanics. Dynamics / Anthony Bedford and Wallace Fowler. —2nd ed.
p. cm.
Includes index.
ISBN 0-201-18071-5
1. Mechanics. I. Fowler, Wallace II. Title.
TA352.B382 1998
620.1'04—dc21 98-34437
CIP

1 2 3 4 5 6 7 8 9 10—DOW—02 01 00 99 98

Addison Wesley Longman, Inc.
2725 Sand Hill Road
Menlo Park, CA 94025

About the Authors

Anthony Bedford is Professor of Aerospace Engineering and Engineering Mechanics at the University of Texas at Austin. He received his B.S. degree at the University of Texas at Austin, his M.S. degree at the California Institute of Technology, and his Ph.D. degree at Rice University in 1967. He has industrial experience at the Douglas Aircraft Company and TRW, and has been on the faculty of the University of Texas at Austin since 1968.

Dr. Bedford's main professional activity has been education and research in engineering mechanics. He is the author or coauthor of many technical papers on the mechanics of composite materials and mixtures and of two books, *Hamilton's Principle in Continuum Mechanics* and *Introduction to Elastic Wave Propagation*. From 1973 until 1983 he was a consultant to Sandia National Laboratories, Albuquerque, New Mexico.

He is a licensed professional engineer and a member of the Acoustical Society of America, the American Society for Engineering Education, the American Academy of Mechanics, and the Society for Natural Philosophy.

Wallace Fowler is Paul D. and Betty Robertson Meek Centennial Professor in Engineering at the University of Texas at Austin. He also holds the title of University Distinguished Teaching Professor in Aerospace Engineering and Engineering Mechanics. Dr. Fowler received the B.A. in mathematics and the M.S. and Ph.D. in engineering mechanics at the University of Texas at Austin and has been on the faculty there since 1965. In 1981–1982, he was Distinguished Visiting Professor in Astronautics and Computer Science at the United States Air Force Academy. Since 1991, he has been Associate Director of the Texas Space Grant Consortium.

Dr. Fowler's areas of teaching and research are dynamics, orbital mechanics, and spacecraft/mission analysis and design. He is author or coauthor of many technical papers on spacecraft dynamics and on engineering education. He has received numerous teaching awards at the local, regional, and national levels, including the AIAA-ASEE Distinguished Aerospace Educator (Leland Atwood) Award, the ASEE Fred Merryfield Award for Excellence in Teaching Engineering Design, and the University of Texas Chancellor's Council Outstanding Teaching Award. He is a registered professional engineer. He is a fellow of the American Institute of Aeronautics and Astronautics (AIAA) and the American Society for Engineering Education (ASEE). He served two terms (four years) as Vice President for Member Affairs of ASEE.

Preface

Our original objective in writing this book was to present the foundations and applications of dynamics as we do in the classroom. We used many sequences of figures, emulating the gradual development of a figure by a teacher explaining a concept. We stressed the importance of visual analysis in gaining understanding, especially through the use of free-body diagrams. Because inspiration is so conducive to learning, we based many of our examples and problems on a variety of modern engineering applications. With encouragement and help from many students and fellow teachers who have used the book, we continue and expand upon these themes in this edition.

Goals and Themes

Enhanced Visualization

We help students visualize the developments of solutions, especially the drawing of free-body diagrams:

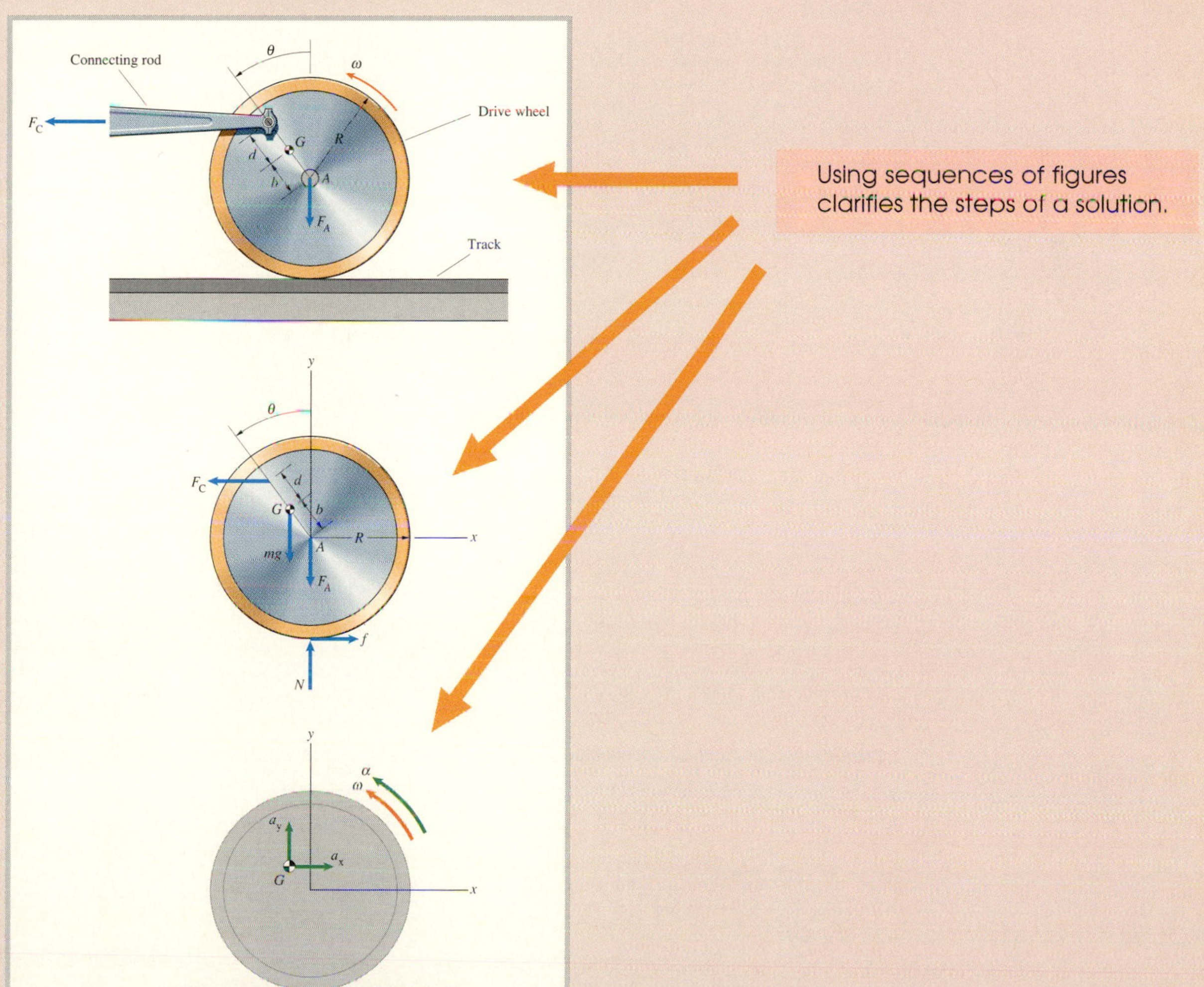

Examples that Teach

The design of our examples helps students learn how to approach problems, solve them, and critically judge the results:

Many examples contain "Strategy" sections showing the preliminary planning needed to begin a solution. What principles and equations apply? What must be determined, and in what order?

The solution is then described in detail, using sequences of figures when needed to clarify the steps.

Many examples conclude with "Discussion" sections pointing out properties of the solution, or commenting on alternative solution methods, or pointing out ways to check answers.

Example 5.1

A 1200-kg helicopter starts from rest at $t = 0$ (Fig. 5.3).
(a) The components of the total force on the helicopter from $t = 0$ to $t = 10$ s are given by

$$\Sigma F_x = 720t \text{ N},$$
$$\Sigma F_y = 2160 - 360t \text{ N},$$
$$\Sigma F_z = 0.$$

Determine the helicopter's velocity at $t = 10$ s.
(b) At $t = 20$ s, the helicopter's velocity is $36\mathbf{i} + 8\mathbf{j}$ (m/s). What is the average of the total force acting on it from $t = 10$ s to $t = 20$ s?

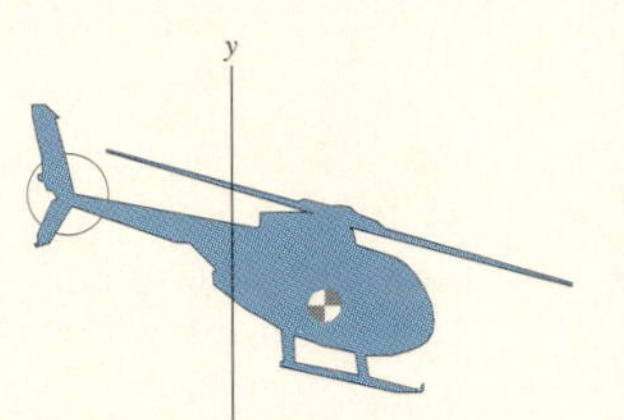

Fig. 5.3

STRATEGY

(a) Since we know the helicopter's velocity at $t = 0$ and the components of the total force acting on it as functions of time, we can use the principle of impulse and momentum, Eq. (5.1), to determine its velocity at $t = 10$ s. (b) Knowing the velocity at $t = 10$ s and at $t = 20$ s, we can determine the average of the total force from Eq. (5.2).

SOLUTION

(a) Applying the principle of impulse and momentum from $t = 0$ to $t = 10$ s,

$$\int_{t_1}^{t_2} \Sigma \mathbf{F}\, dt = m\mathbf{v}_2 - m\mathbf{v}_1:$$
$$\int_0^{10} [720t\,\mathbf{i} + (2160 - 360t)\mathbf{j}]\, dt = (1200)\mathbf{v}_2 - (1200)(0),$$
$$36{,}000\mathbf{i} + 3600\mathbf{j} = 1200\mathbf{v}_2,$$

we find that the velocity at $t = 10$ s is $30\mathbf{i} + 3\mathbf{j}$ (m/s).
(b) To determine the average total force from $t = 10$ s to $t = 20$ s, we apply Eq. (5.2) to this interval of time,

$$(t_2 - t_1)\Sigma \mathbf{F}_{av} = m\mathbf{v}_2 - m\mathbf{v}_1:$$
$$(20 - 10)\Sigma \mathbf{F}_{av} = 1200(36\mathbf{i} + 8\mathbf{j}) - 1200(30\mathbf{i} + 3\mathbf{j}),$$
$$\Sigma \mathbf{F}_{av} = 720\mathbf{i} + 600\mathbf{j} \text{ (N)}.$$

DISCUSSION

Although we did not know the total force acting on the helicopter during the interval from $t = 10$ s to $t = 20$ s, we were able to determine the average of the total force because we knew the velocities at the beginning and end of the interval.

Engineering Design

We include simple design considerations in many examples and problems without compromising emphasis on fundamental mechanics. Optional examples titled "Application to Engineering" provide more detailed discussions of the uses of dynamics in engineering design:

Example 3.6

Application to Engineering

Motor Vehicle Dynamics

A civil engineer's preliminary design for a freeway off-ramp is circular with radius $R = 300$ ft (Fig. 3.12). If she assumes that the coefficient of static friction between tires and road is at least $\mu_s = 0.4$, what is the maximum speed at which vehicles can enter the ramp without losing traction?

A specific engineering application is first described and analyzed.

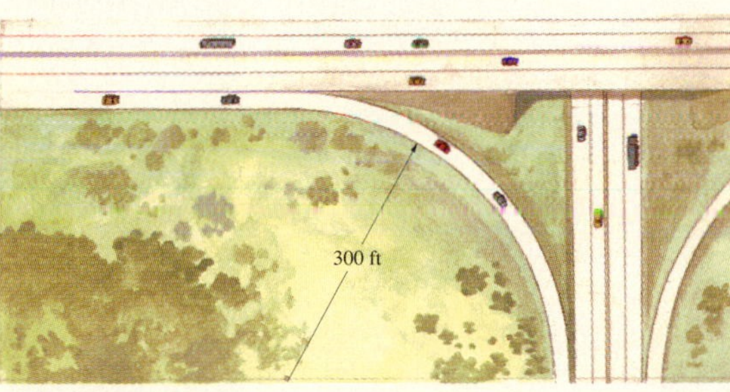

Fig. 3.12

STRATEGY

Since a vehicle on the off-ramp moves in a circular path, it has a normal component of acceleration that depends on its velocity. The necessary normal component of

(a) Top view of the free-body diagram.

A "Design Issues" section then discusses design implications of the application and places it in a broader engineering context.

DESIGN ISSUES

Automotive engineers, civil engineers who design highways, and engineers who study traffic accidents and their prevention must analyze and measure the motions of vehicles under different conditions. By using the methods discussed in this chapter, they can relate the forces acting on vehicles to their motions and study, for example, the factors influencing the distance necessary for a car to be brought to a stop in an emergency, or the effects of banking and curvature on the velocity at which a car can safely be driven on a curved road (Fig. 3.13).

In Example 3.6, the analysis indicates that vehicles will lose traction if they enter the freeway off-ramp at speeds greater than 42.4 mi/hr. This result can be used as an indication of the speed limit that must be posted in order for vehicles to enter the ramp safely, or the off-ramp could be designed for a greater speed by increasing the radius of curvature. Or, if a larger safe speed is desired but space limitations forbid a larger radius of curvature, the off-ramp could be designed to incorporate banking (see Problem 3.93).

Fig. 3.13
Tests of the capabilities of vehicles to negotiate curves influence the design of both vehicles and highways.

Relevant Applications

We place dynamics within the context of engineering practice by including applications from many fields of engineering:

4.52 An astronaut in an excursion module approaches a space station docking collar. The designer of the collar incorporated a spring to attenuate the shock due to docking. The spring constant is $k = 4800$ N/m. The mass of the astronaut and module is 780 kg. If the module contacts the docking collar moving at 0.1 m/s relative to the collar, what distance is required for the spring to decrease its relative velocity to zero? What is the module's maximum relative deceleration?

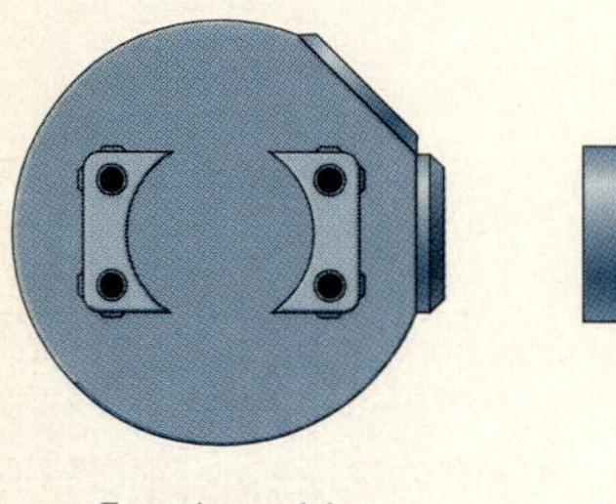
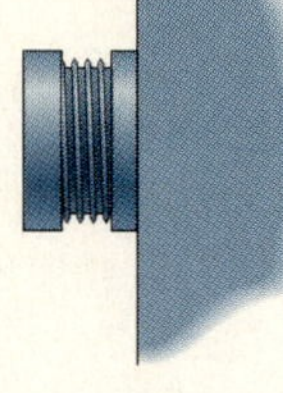

P4.52

5.35 A bioengineer, using an instrumented dummy to test a protective mask for a hockey goalie, launches the 170-g puck so that it strikes the mask moving horizontally at 40 m/s. From photographs of the impact, she estimates its duration to be 0.02 s and observes that the puck rebounds at 5 m/s.
(a) What linear impulse does the puck exert?
(b) What is the average value of the impulsive force exerted on the mask by the puck?

P5.35

5.19 In a cathode-ray tube, an electron (mass = 9.11×10^{-31} kg) is projected at O with velocity $\mathbf{v} = (2.2 \times 10^7)\mathbf{i}$ (m/s). While it is between the charged plates, the electric field generated by the plates subjects it to a force $\mathbf{F} = -eE\,\mathbf{j}$. The charge of the electron is $e = 1.6 \times 10^{-19}$ C (coulombs), and the electric field strength is $E = 15 \sin(\omega t)$ kN/C, where the frequency $\omega = 2 \times 10^9\ \text{s}^{-1}$.
(a) What impulse does the electric field exert on the electron while it is between the plates?
(b) What is the velocity of the electron as it leaves the region between the plates?

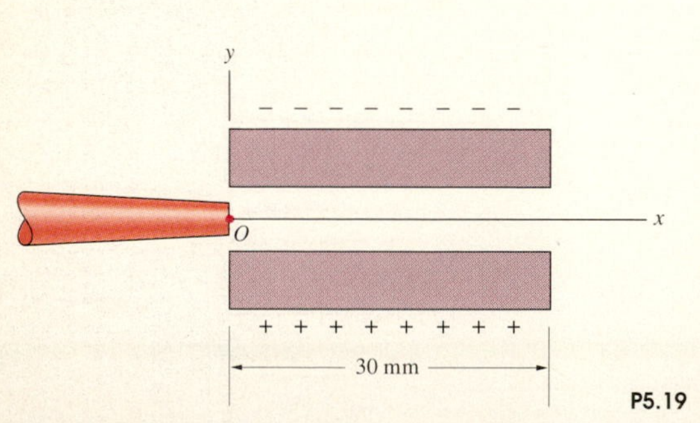

P5.19

6.35 If the crankshaft AB rotates at 6000 rpm in the counterclockwise direction, what is the velocity of the piston at the instant shown?

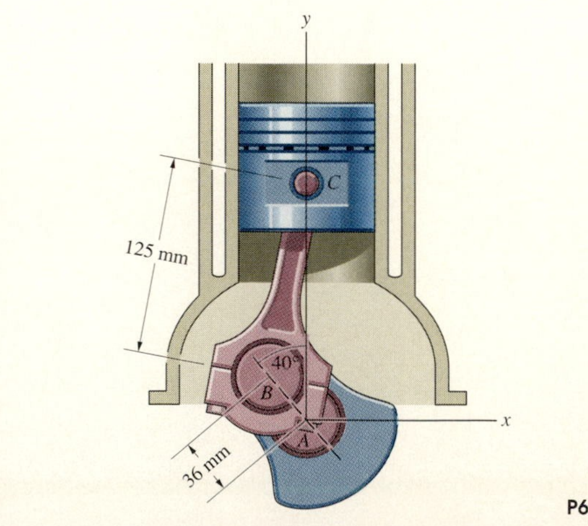

P6.35

Consistent Use of Color

To help students recognize and interpret elements of figures, we use consistent identifying colors:

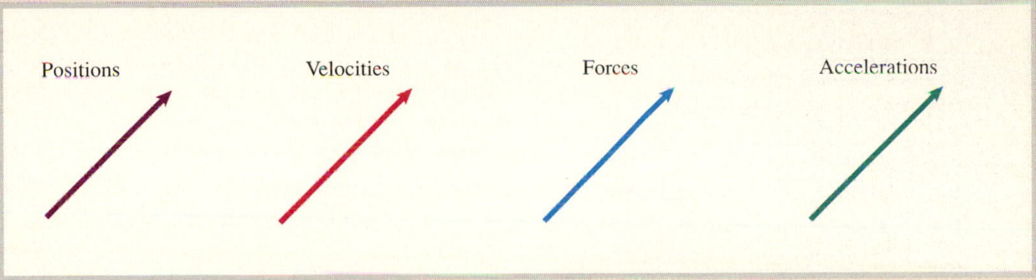

Computational Mechanics

Some instructors prefer to teach dynamics without requiring the use of a computer. Others use dynamics as an opportunity to introduce students to the use of computers in engineering, having them either write their own programs in a lower level language or use higher level problem-solving software. Our book is suitable for each of these approaches. We provide optional, self-contained "Computational Mechanics" sections with examples and problems designed for solution by a programmable calculator or computer. See Examples 3.9 on p. 148 and 4.13 on p. 202.

New to the Second Edition

Positive responses from users and reviewers have led us to retain the basic organization, content, and features of the first edition. During our preparation of this edition, we examined how we presented each concept, example, figure, summary statement, and problem. Where necessary, we made changes, additions, or deletions to simplify and clarify the presentation. In response to requests, we made the following notable changes:

- We have revised and expanded our discussions of reference frames throughout the book, especially in Chapters 2 and 6.
- We have included more free-body diagrams in examples where appropriate, and in most instances use separate diagrams to show velocities and accelerations.
- Our treatments of rigid-body kinematics in Chapters 6 and 9 have been revised.
- Our treatment of D'Alembert's principle in Chapter 7 has been revised.
- We have added new examples where users indicated more were needed. Many of the new examples continue our emphasis on realistic and motivational applications and engineering design.
- We have revised many of the existing problems and have added more than 200 new ones. As with the examples, many of the new problems focus on placing dynamics within the context of engineering practice.

Commitment to Students and Instructors

In revising the textbook and solutions manual, we have taken precautions to ensure accuracy to the best of our ability. Reviewers examined each stage of the manuscript for errors. We have each solved the new problems in an effort to be sure that their answers are correct and that they are of an appropriate level of difficulty. David Hartman of Northern Arizona University also checked the new text, examples, and problems. Any errors that remain are the responsibility of the authors. We welcome communication from students and instructors concerning errors or areas for improvement. Our mailing address is Department of Aerospace Engineering and Engineering Mechanics, University of Texas at Austin, Austin, Texas 78712. Our electronic mail address is abedford@mail.utexas.edu.

Supplements

PRINTED SUPPLEMENTS

Instructor's Solutions Manual The manual for the instructor, revised by Wallace Fowler, contains complete step-by-step solutions to all problems. Each solution includes the problem statement and the associated art.

Design Problems Supplement This is a paperback supplement of approximately 65 pages and over 100 design problems that are based on problems from the textbook. This product makes it possible to introduce students to the concepts of engineering design in the context of dynamics.

Additional Problems Set As a service to aid instructors and students in having the greatest variety and quantity of problems, we have developed an additional problem set. This supplement consists of approximately 500 problems (roughly 200 from *Statics* and 300 from *Dynamics*) using the art from the texts. The problems were written and checked by the authors and are intended to provide a fresh look at the concepts presented.

SOFTWARE SUPPLEMENT

Working Model® Simulations Approximately 100 problems and examples from the text have been recreated on disk as Working Model simulations. These simulations have been constructed to allow the student to change the values of variables and observe the results. The student can explore physical situations in a "what if" manner and thereby develop deeper conceptual insights than possible through quantitative problem solving alone. A site license for these simulations is available free to adopters, or they can be purchased by students bundled with the text for a nominal additional charge.

Acknowledgments

Many students and teachers have given us insightful comments on the first edition. The following colleagues reviewed the first edition or the manuscript of the second edition and made many valuable suggestions.

Nick Altiero
Michigan State University

James G. Andrews
University of Iowa

Gautam Batra
University of Nebraska, Lincoln

Rathi Bhatacharya
Bradley University

Clarence Calder
Oregon State University

Peter Dashner
California Polytechnic University, Pomona

Anthony DeLuzio
Merrimack College

Xioamin Deng
University of South Carolina

James Dent
Montana State University

Bruce R. Dewey
University of Wyoming

Jerry Fine
Rose-Hulman Institute of Technology

Robert W. Fitzgerald
Worcester Polytechnic Institute

David Q. Fletcher
University of the Pacific

Mark Frisina
Wentworth Institute

Robert W. Fuessle
Bradley University

John Giger
Rose State College

Peter J. Gorder
Kansas State University

Hamid Nayeb-Hashemi
Northeastern University

R. Craig Henderson
Tennessee Technological University

Robert A. Howland
University of Notre Dame

Raouf A. Ibrahim
Wayne State University

Robert Johanson
University of the Pacific

David B. Johnson
Southern Methodist University

Charles M. Krousgrill
Purdue University

Kristine M. Larson
University of Colorado, Boulder

Donald G. Lemke
University of Illinois, Chicago

Patrick Lenahan
Pennsylvania State University

Richard Lewis
Louisiana Technological University

Brad S. Liebst
University of Minnesota

Bertram Long
Northeastern University

V. J. Lopardo
U.S. Naval Academy

Frank K. Lu
University of Texas, Arlington

Donald L. Margolis
University of California, Davis

Joseph M. Mansour
Case Western Reserve University

George Mase
Michigan State University

John McPhee
University of Waterloo

Koorosh Naghshineh
Western Michigan University

Satish S. Nair
University of Missouri

Vojin Nikolic
Rose-Hulman Institute of Technology

Karim Nohra
University of South Florida

Colin P. Ratcliffe
United States Naval Academy

Michael J. Rider
Ohio Northern University

George Rosborough
University of Colorado, Boulder

William W. Seto
San Jose State University

Geoffrey Shiflett
University of Southern California

Leo Smith
University of Hartford

Francis M. Thomas
University of Kansas

John Valasek
Texas A&M University

Mark R. Virkler
University of Missouri, Columbia

William H. Walston, Jr.
University of Maryland

Jonathan Wickert
Carnegie Mellon University

Julius Wong
University of Louisville

We also thank the many colleagues who generously provided us with information for examples and problems, particularly Herb Sutherland for specifications and photographs of the 8-m centrifuge facility at Sandia National Laboratories, Chris Juras for data on devices used to polish silicon wafers, and Victor Austin and David Goldstein for information on volcanic eruptions on Io.

The people of Addison Wesley Longman and their colleagues helped us at every stage with professionalism, patience, humor, and friendship, especially Kevin Berry, Chuck Iossi, Dan Joraanstad, Caroline Jumper, Colleen Kelly, Chandrika Madhavan, Rob Merino, Pattie Myers, Susan Slater, Michael Slaughter, Kamila Storr, Janet Weaver, Diane Williams, and Edie Williams.

Finally we thank our wives, Nancy and Marcia, for cheerfully accepting the sacrifices inherent in such a project.

Anthony Bedford and Wallace Fowler
January 1998
Austin, Texas

Contents

1 Introduction 1

1.1 Engineering and Mechanics 2

1.2 Learning Mechanics 2

Problem Solving 2
Calculators and Computers 3
Engineering Applications 3
Subsequent Use of This Text 3

1.3 Fundamental Concepts 4

Space and Time 4
Newton's Laws 4
Newtonian Gravitation 6
Numbers 7

1.4 Units 8

International System of Units 8
U.S. Customary Units 8
Angular Units 9
Conversion of Units 9

2 Motion of a Point 17

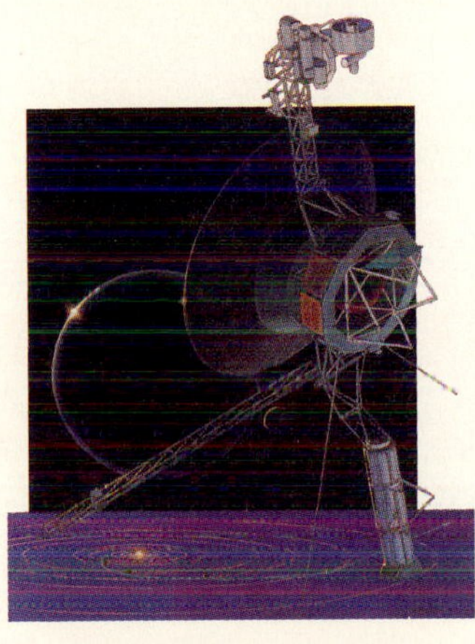

2.1 Position, Velocity, and Acceleration 18

2.2 Straight-Line Motion 21

Description of the Motion 21
Analysis of the Motion 22

2.3 Curvilinear Motion 43

Cartesian Coordinates 43
Angular Motion 53
Normal and Tangential Components 58
Application to Engineering: Centrifuge Design 68
Polar and Cylindrical Coordinates 73

2.4 Relative Motion 84

Computational Mechanics 92

Chapter Summary 98

Review Problems 101

3 Force, Mass, and Acceleration 105

3.1 Newton's Second Law 106

3.2 Equation of Motion for the Center of Mass 106

3.3 Inertial Reference Frames 108

3.4 Applications 110

Cartesian Coordinates and Straight-Line Motion 110
Normal and Tangential Components 126
Application to Engineering: Motor Vehicle Dynamics 130
Polar and Cylindrical Coordinates 136

3.5 Orbital Mechanics 141

Determination of the Orbit 141
Types of Orbits 144
Computational Mechanics 146

Chapter Summary 151

Review Problems 152

4 Energy Methods 157

Work and Kinetic Energy 157

4.1 Principle of Work and Energy 158

4.2 Work and Power 159

Evaluating the Work 159
Power 160

4.3 Work Done by Particular Forces 168

Weight 168
Springs 170
Application to Engineering: Automated Machining 173

Potential Energy 182

4.4 Conservation of Energy 182

4.5 Conservative Forces 183

Potential Energies of Particular Forces 184
Relationships Between Force and Potential Energy 194
Computational Mechanics 202

Chapter Summary 205

Review Problems 208

10 Vibrations 525

10.1 *Conservative Systems* 526
- Examples 526
- Solutions 528

10.2 *Damped Vibrations* 540
- Subcritical Damping 541
- Critical and Supercritical Damping 542

10.3 *Forced Vibrations* 547
- Oscillatory Forcing Function 548
- Polynomial Forcing Function 550
- Application to Engineering: Displacement Transducers 556
- Chapter Summary 560
- Review Problems 563

Appendixes 566

- **A Review of Mathematics** 566
- **B Properties of Areas and Lines** 569
- **C Properties of Volumes and Homogeneous Objects** 571
- **D Spherical Coordinates** 573
- **Answers to Even-Numbered Problems** 574
- **Index** 585

5 Momentum Methods 213

5.1 *Principle of Impulse and Momentum* 214

5.2 *Conservation of Linear Momentum* 224

5.3 *Impacts* 227
- Direct Central Impacts 228
- Oblique Central Impacts 230

5.4 *Angular Momentum* 238
- Principle of Angular Impulse and Momentum 238
- Central-Force Motion 239

5.5 *Mass Flows* 244
- Application to Engineering: Jet Engines 248
- Chapter Summary 253
- Review Problems 255

6 Planar Kinematics of Rigid Bodies 261

6.1 *Rigid Bodies and Types of Motion* 262

6.2 *Rotation About a Fixed Axis* 263

6.3 *General Motions: Velocities* 269

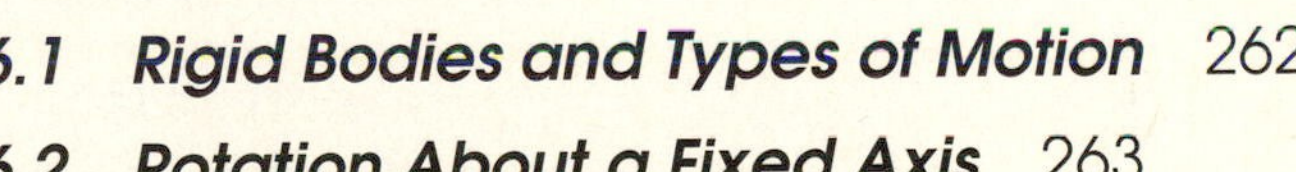

- Relative Velocities 269
- The Angular Velocity Vector 271
- Instantaneous Centers 284

6.4 *General Motions: Accelerations* 290

6.5 *Sliding Contacts* 301

6.6 *Moving Reference Frames* 312
- Motion of a Point Relative to a Moving Reference Frame 313
- Inertial Reference Frames 319
- Chapter Summary 329
- Review Problems 331

7 Planar Dynamics of Rigid Bodies 337

7.1 Preview of the Equations of Motion 338

7.2 Momentum Principles for a System of Particles 339

Force–Linear Momentum Principle 339
Moment–Angular Momentum Principles 340

7.3 Derivation of the Equations of Motion 342

Rotation About a Fixed Axis 343
General Planar Motion 344

7.4 Applications 345

Translation 345
Rotation About a Fixed Axis 347
General Planar Motion 352

7.5 D'Alembert's Principle 358

Application to Engineering:
Internal Forces and Moments in Beams 373
Computational Mechanics 377

Appendix: Moments of Inertia 381

Simple Objects 382
Parallel-Axis Theorem 386
Chapter Summary 395
Review Problems 398

8 Energy and Momentum in Rigid-Body Dynamics 403

8.1 Principle of Work and Energy 404

System of Particles 404
Rigid Body in Planar Motion 405

8.2 Work and Potential Energy 408

8.3 Power 410

8.4 Principles of Impulse and Momentum 425

Linear Momentum 425
Angular Momentum 426

8.5 Impacts 434

Conservation of Momentum 434
Coefficient of Restitution 435
Chapter Summary 448
Review Problems 451

9 Three-Dimensional Kinematics and Dynamics of Rigid Bodies 457

9.1 Kinematics 458

Velocities and Accelerations 458
Moving Reference Frames 459

9.2 Angular Momentum 472

Rotation About a Fixed Point 472
General Motion 473

9.3 Moments and Products of Inertia 474

Simple Objects 475
Parallel-Axis Theorems 477
Moment of Inertia About an Arbitrary Axis 478
Principal Axes 479

9.4 Euler's Equations 490

Rotation About a Fixed Point 490
General Motion 491

9.5 Eulerian Angles 505

Objects with an Axis of Symmetry 505
Arbitrary Objects 510
Chapter Summary 518
Review Problems 520

ENGINEERING MECHANICS

DYNAMICS

In September 1996, the space shuttle Atlantis lifted off and went into orbit 300 kilometers above the earth. By a series of brief firings of its maneuvering rockets, it matched its orbit to that of the Mir space station. While orbiting at more than 8 kilometers per second, Atlantis docked with the Mir to bring astronaut Shannon Lucid home after her 188 days in space.

Chapter 1

Introduction

THE Space Shuttle was conceived as an economical method to transport personnel and equipment to orbit. Throughout its development, engineers used the principles of dynamics to predict its motion during boost, in orbit, and while landing. These predictions were essential for the design of its aerodynamic configuration, structure, rocket engines, and control system. Dynamics is one of the sciences underlying the design of all vehicles and machines.

1.1 *Engineering and Mechanics*

How do engineers design complex systems and predict their characteristics before they are constructed? Engineers have always relied on their knowledge of previous designs, experiments, ingenuity, and creativity to develop new designs. Modern engineers add a powerful technique: They develop mathematical equations based on the physical characteristics of the devices they design. With these mathematical models, engineers predict the behavior of their designs, modify them, and test them prior to their actual construction. Aerospace engineers used mathematical models to predict the paths the space shuttle would follow in flight. Civil engineers used mathematical models to analyze the response to loads of the steel frame of the 1454-ft Sears Tower in Chicago.

Engineers are responsible for the design, construction, and testing of the devices we use, from simple things such as chairs and pencil sharpeners to complicated ones such as dams, cars, airplanes, and spacecraft. They must have a deep understanding of the physics underlying these devices and must be familiar with the use of mathematical models to predict system behavior. Students of engineering begin to learn how to analyze and predict the behavior of physical systems by studying mechanics.

At its most basic level, mechanics is the study of forces and their effects. Elementary mechanics is divided into **statics**, the study of objects in equilibrium, and **dynamics**, the study of objects in motion. The results obtained in elementary mechanics apply directly to many fields of engineering. Mechanical and civil engineers who design structures use the equilibrium equations derived in statics. Civil engineers who analyze the responses of buildings to earthquakes and aerospace engineers who determine the trajectories of satellites use the equations of motion derived in dynamics.

Mechanics was the first analytical science; consequently fundamental concepts, analytical methods, and analogies from mechanics are found in virtually every field of engineering. For example, students of chemical and electrical engineering gain a deeper appreciation for basic concepts in their fields such as equilibrium, energy, and stability by learning them in their original mechanical contexts. In fact, by studying mechanics they retrace the historical development of these ideas.

1.2 *Learning Mechanics*

Mechanics consists of broad principles that govern the behavior of objects. In this book we describe these principles and provide examples that demonstrate some of their applications. Although it is essential that you practice working problems similar to these examples, and we include many problems of this kind, our objective is to help you understand the principles well enough to apply them to situations that are new to you. Each generation of engineers confronts new problems.

Problem Solving

In the study of mechanics you learn problem-solving procedures you will use in succeeding courses and throughout your career. Although different types of

5 Momentum Methods 213

5.1 Principle of Impulse and Momentum 214

5.2 Conservation of Linear Momentum 224

5.3 Impacts 227

Direct Central Impacts 228
Oblique Central Impacts 230

5.4 Angular Momentum 238

Principle of Angular Impulse and Momentum 238
Central-Force Motion 239

5.5 Mass Flows 244

Application to Engineering: Jet Engines 248

Chapter Summary 253

Review Problems 255

6 Planar Kinematics of Rigid Bodies 261

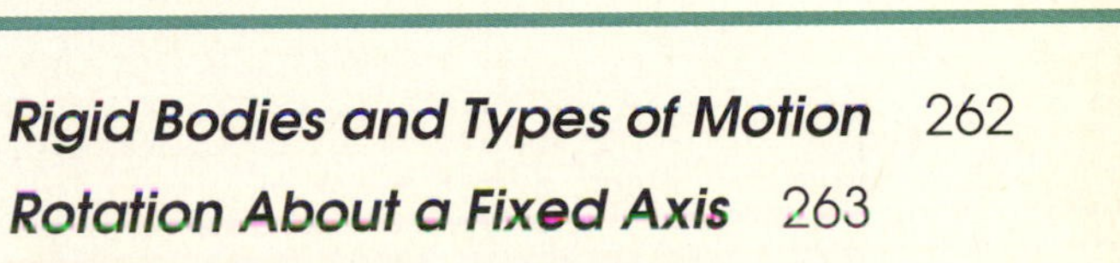

6.1 Rigid Bodies and Types of Motion 262

6.2 Rotation About a Fixed Axis 263

6.3 General Motions: Velocities 269

Relative Velocities 269
The Angular Velocity Vector 271
Instantaneous Centers 284

6.4 General Motions: Accelerations 290

6.5 Sliding Contacts 301

6.6 Moving Reference Frames 312

Motion of a Point Relative to a Moving Reference Frame 313
Inertial Reference Frames 319

Chapter Summary 329

Review Problems 331

7 Planar Dynamics of Rigid Bodies 337

7.1 Preview of the Equations of Motion 338

7.2 Momentum Principles for a System of Particles 339

Force–Linear Momentum Principle 339

Moment–Angular Momentum Principles 340

7.3 Derivation of the Equations of Motion 342

Rotation About a Fixed Axis 343

General Planar Motion 344

7.4 Applications 345

Translation 345

Rotation About a Fixed Axis 347

General Planar Motion 352

7.5 D'Alembert's Principle 358

Application to Engineering:
Internal Forces and Moments in Beams 373

Computational Mechanics 377

Appendix: Moments of Inertia 381

Simple Objects 382

Parallel-Axis Theorem 386

Chapter Summary 395

Review Problems 398

8 Energy and Momentum in Rigid-Body Dynamics 403

8.1 Principle of Work and Energy 404

System of Particles 404

Rigid Body in Planar Motion 405

8.2 Work and Potential Energy 408

8.3 Power 410

8.4 Principles of Impulse and Momentum 425

Linear Momentum 425
Angular Momentum 426

8.5 Impacts 434

Conservation of Momentum 434
Coefficient of Restitution 435

Chapter Summary 448

Review Problems 451

9 Three-Dimensional Kinematics and Dynamics of Rigid Bodies 457

9.1 Kinematics 458

Velocities and Accelerations 458
Moving Reference Frames 459

9.2 Angular Momentum 472

Rotation About a Fixed Point 472
General Motion 473

9.3 Moments and Products of Inertia 474

Simple Objects 475
Parallel-Axis Theorems 477
Moment of Inertia About an Arbitrary Axis 478
Principal Axes 479

9.4 Euler's Equations 490

Rotation About a Fixed Point 490
General Motion 491

9.5 Eulerian Angles 505

Objects with an Axis of Symmetry 505
Arbitrary Objects 510

Chapter Summary 518

Review Problems 520

10 Vibrations 525

10.1 *Conservative Systems* 526

Examples 526

Solutions 528

10.2 *Damped Vibrations* 540

Subcritical Damping 541

Critical and Supercritical Damping 542

10.3 *Forced Vibrations* 547

Oscillatory Forcing Function 548

Polynomial Forcing Function 550

Application to Engineering: Displacement Transducers 556

Chapter Summary 560

Review Problems 563

Appendixes 566

A Review of Mathematics 566

B Properties of Areas and Lines 569

C Properties of Volumes and Homogeneous Objects 571

D Spherical Coordinates 573

Answers to Even-Numbered Problems 574

Index 585

ENGINEERING MECHANICS

DYNAMICS

In September 1996, the space shuttle Atlantis lifted off and went into orbit 300 kilometers above the earth. By a series of brief firings of its maneuvering rockets, it matched its orbit to that of the Mir space station. While orbiting at more than 8 kilometers per second, Atlantis docked with the Mir to bring astronaut Shannon Lucid home after her 188 days in space.

Chapter 1

Introduction

THE Space Shuttle was conceived as an economical method to transport personnel and equipment to orbit. Throughout its development, engineers used the principles of dynamics to predict its motion during boost, in orbit, and while landing. These predictions were essential for the design of its aerodynamic configuration, structure, rocket engines, and control system. Dynamics is one of the sciences underlying the design of all vehicles and machines.

1.1 *Engineering and Mechanics*

How do engineers design complex systems and predict their characteristics before they are constructed? Engineers have always relied on their knowledge of previous designs, experiments, ingenuity, and creativity to develop new designs. Modern engineers add a powerful technique: They develop mathematical equations based on the physical characteristics of the devices they design. With these mathematical models, engineers predict the behavior of their designs, modify them, and test them prior to their actual construction. Aerospace engineers used mathematical models to predict the paths the space shuttle would follow in flight. Civil engineers used mathematical models to analyze the response to loads of the steel frame of the 1454-ft Sears Tower in Chicago.

Engineers are responsible for the design, construction, and testing of the devices we use, from simple things such as chairs and pencil sharpeners to complicated ones such as dams, cars, airplanes, and spacecraft. They must have a deep understanding of the physics underlying these devices and must be familiar with the use of mathematical models to predict system behavior. Students of engineering begin to learn how to analyze and predict the behavior of physical systems by studying mechanics.

At its most basic level, mechanics is the study of forces and their effects. Elementary mechanics is divided into **statics**, the study of objects in equilibrium, and **dynamics**, the study of objects in motion. The results obtained in elementary mechanics apply directly to many fields of engineering. Mechanical and civil engineers who design structures use the equilibrium equations derived in statics. Civil engineers who analyze the responses of buildings to earthquakes and aerospace engineers who determine the trajectories of satellites use the equations of motion derived in dynamics.

Mechanics was the first analytical science; consequently fundamental concepts, analytical methods, and analogies from mechanics are found in virtually every field of engineering. For example, students of chemical and electrical engineering gain a deeper appreciation for basic concepts in their fields such as equilibrium, energy, and stability by learning them in their original mechanical contexts. In fact, by studying mechanics they retrace the historical development of these ideas.

1.2 *Learning Mechanics*

Mechanics consists of broad principles that govern the behavior of objects. In this book we describe these principles and provide examples that demonstrate some of their applications. Although it is essential that you practice working problems similar to these examples, and we include many problems of this kind, our objective is to help you understand the principles well enough to apply them to situations that are new to you. Each generation of engineers confronts new problems.

Problem Solving

In the study of mechanics you learn problem-solving procedures you will use in succeeding courses and throughout your career. Although different types of problems require different approaches, the following steps apply to many of them:

- Identify the information that is given and the information, or answer, you must determine. It's often helpful to restate the problem in your own words. When appropriate, make sure you understand the physical system or model involved.
- Develop a *strategy* for the problem. This means identifying the principles and equations that apply and deciding how you will use them to solve the problem. Whenever possible, draw diagrams to help visualize and solve the problem.
- Whenever you can, try to predict the answer. This will develop your intuition and will often help you recognize an incorrect answer.
- Solve the equations and, whenever possible, interpret your results and compare them with your prediction. This last step is a *reality check.* Is your answer reasonable?

Calculators and Computers

Most of the problems in this book are designed to lead to an algebraic expression with which to calculate the answer in terms of given quantities. A calculator with trigonometric and logarithmic functions is sufficient to determine the numerical value of such answers. The use of a programmable calculator or a computer with problem-solving software such as *Mathcad* or *TK! Solver* is convenient, but be careful not to become too reliant on tools you will not have during tests.

Sections headed "Computational Mechanics" contain examples and problems that are suitable for solution with a programmable calculator or a computer.

Engineering Applications

Although the problems are designed primarily to help you learn mechanics, many of them illustrate uses of mechanics in engineering. Sections headed "Application to Engineering" describe how mechanics is applied in various fields of engineering.

We also include problems that emphasize two essential aspects of engineering:

- *Design.* Some problems ask you to choose values of parameters to satisfy stated design criteria.
- *Safety.* Some problems ask you to evaluate the safety of devices and choose values of parameters to satisfy stated safety requirements.

Subsequent Use of This Text

This book contains tables and information you will find useful in subsequent engineering courses and throughout your engineering career. In addition, you will often want to review fundamental engineering subjects, both during the remainder of your formal education and when you are a practicing engineer.

The most efficient way to do so is by using the textbooks with which you are familiar. Your engineering textbooks will form the core of your professional library.

1.3 *Fundamental Concepts*

Some topics in mechanics will be familiar to you from everyday experience or from previous exposure to them in physics courses. In this section we briefly review the foundations of elementary mechanics.

Space and Time

Space simply refers to the three-dimensional universe in which we live. Our daily experiences give us an intuitive notion of space and the locations, or positions, of points in space. The distance between two points in space is the length of the straight line joining them.

Measuring the distance between points in space requires a unit of length. We use both the International System of units, or SI units, and U.S. Customary units. In SI units, the unit of length is the meter (m). In U.S. Customary units, the unit of length is the foot (ft).

Time is, of course, familiar—our lives are measured by it. The daily cycles of light and darkness and the hours, minutes, and seconds measured by our clocks and watches give us an intuitive notion of time. Time is measured by the intervals between repeatable events, such as the swings of a clock pendulum or the vibrations of a quartz crystal in a watch. In both SI units and U.S. Customary units, the unit of time is the second (s). The minute (min), hour (hr), and day are also frequently used.

If the position of a point in space relative to some reference point changes with time, the rate of change of its position is called its *velocity*, and the rate of change of its velocity is called its *acceleration*. In SI units, the velocity is expressed in meters per second (m/s) and the acceleration is expressed in meters per second per second, or meters per second squared (m/s^2). In U.S. Customary units, the velocity is expressed in feet per second (ft/s) and the acceleration is expressed in feet per second squared (ft/s^2).

Newton's Laws

Elementary mechanics was established on a firm basis with the publication in 1687 of *Philosophiae naturalis principia mathematica*, by Isaac Newton. Although highly original, it built on fundamental concepts developed by many others during a long and difficult struggle toward understanding (Fig. 1.1). Newton stated three "laws" of motion, which we express in modern terms:

1. *When the sum of the forces acting on a particle is zero, its velocity is constant. In particular, if the particle is initially stationary, it will remain stationary.*
2. *When the sum of the forces acting on a particle is not zero, the sum of the forces is equal to the rate of change of the linear momentum of the*

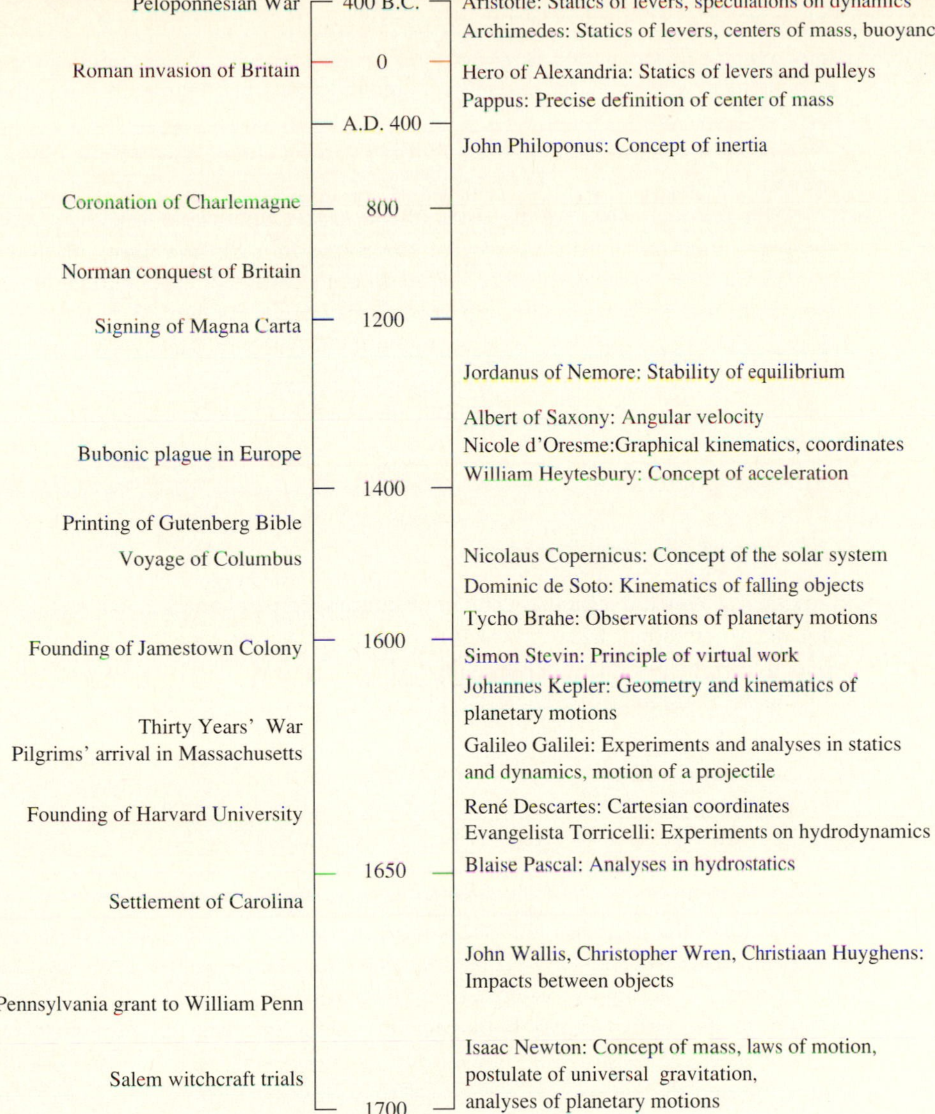

Figure 1.1

Chronology of developments in mechanics up to the publication of Newton's *Principia* in relation to other events in history.

particle. If the mass is constant, the sum of the forces is equal to the product of the mass of the particle and its acceleration.

3. *The forces exerted by two particles on each other are equal in magnitude and opposite in direction.*

Notice that we did not define force and mass before stating Newton's laws. The modern view is that these terms are defined by the second law. To demonstrate, suppose that we choose an arbitrary object and define it to have unit mass. Then we define a unit of force to be the force that gives our unit mass an acceleration of unit magnitude. In principle, we can then determine the mass of any object: We apply a unit force to it, measure the resulting acceleration, and use the second law to determine the mass. We can also determine the magnitude of any force: We apply it to our unit mass, measure the resulting acceleration, and use the second law to determine the force.

Thus Newton's second law gives precise meanings to the terms **mass** and **force**. In SI units, the unit of mass is the kilogram (kg). The unit of force is the newton (N), which is the force required to give a mass of one kilogram an acceleration of one meter per second squared. In U.S. Customary units, the unit of force is the pound (lb). The unit of mass is the slug, which is the amount of mass accelerated at one foot per second squared by a force of one pound.

Although the results we discuss in this book are applicable to many of the problems met in engineering practice, there are limits to the validity of Newton's laws. For example, they don't give accurate results if a problem involves velocities that are not small compared to the velocity of light (3×10^8 m/s). Einstein's special theory of relativity applies to such problems. Elementary mechanics also fails in problems involving dimensions that are not large compared to atomic dimensions. Quantum mechanics must be used to describe phenomena on the atomic scale.

Newtonian Gravitation

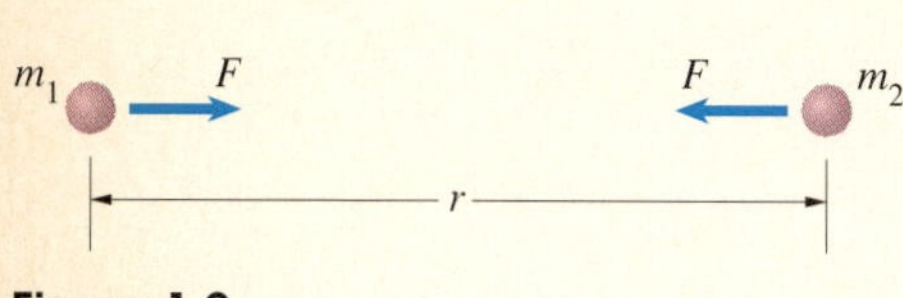

Figure 1.2
The gravitational forces between two particles are equal in magnitude and directed along the line between them.

Another of Newton's fundamental contributions to mechanics is his postulate for the gravitational force between two particles in terms of their masses m_1 and m_2 and the distance r between them (Fig. 1.2). His expression for the magnitude of the force is

$$F = \frac{Gm_1m_2}{r^2}, \tag{1.1}$$

where G is called the universal gravitational constant.

Newton calculated the gravitational force between a particle of mass m_1 and a homogeneous sphere of mass m_2 and found that it is also given by Eq. (1.1), with r denoting the distance from the particle to the center of the sphere. Although the earth is not a homogeneous sphere, we can use this result to approximate the weight of an object of mass m due to the gravitational attraction of the earth,

$$W = \frac{Gmm_E}{r^2}, \tag{1.2}$$

where m_E is the mass of the earth and r is the distance from the center of the earth to the object. Notice that the weight of an object depends on its location relative to the center of the earth, whereas the mass of the object is a measure of the amount of matter it contains and doesn't depend on its position.

When an object's weight is the only force acting on it, the resulting acceleration is called the acceleration due to gravity. In this case, Newton's second law states that $W = ma$, and from Eq. (1.2) we see that the acceleration due to gravity is

$$a = \frac{Gm_E}{r^2}. \tag{1.3}$$

The **acceleration due to gravity at sea level** is denoted by g. Denoting the radius of the earth by R_E, we see from Eq. (1.3) that $Gm_E = gR_E^2$. Substituting this result into Eq. (1.3), we obtain an expression for the acceleration due to gravity at a distance r from the center of the earth in terms of the acceleration due to gravity at sea level:

$$a = g\frac{R_E^2}{r^2}. \tag{1.4}$$

Since the weight of the object $W = ma$, the weight of an object at a distance r from the center of the earth is

$$W = mg\frac{R_E^2}{r^2}. \tag{1.5}$$

At sea level, the weight of an object is given in terms of its mass by the simple relation

$$W = mg. \tag{1.6}$$

The value of g varies from location to location on the surface of the earth. The values we use in examples and problems are $g = 9.81 \text{ m/s}^2$ in SI units and $g = 32.2 \text{ ft/s}^2$ in U.S. Customary units.

Numbers

Engineering measurements, calculations, and results are expressed in numbers. You need to know how we express numbers in the examples and problems and how to express the results of your own calculations.

Significant Digits This term refers to the number of meaningful (that is, accurate) digits in a number, counting to the right starting with the first nonzero digit. The two numbers 7.630 and 0.007630 are each stated to four significant digits. If only the first four digits in the number 7,630,000 are known to be accurate, this can be indicated by writing the number in scientific notation as 7.630×10^6.

If a number is the result of a measurement, the significant digits it contains are limited by the accuracy of the measurement. If the result of a measurement is stated to be 2.43, this means that the actual value is believed to be closer to 2.43 than to 2.42 or 2.44.

Numbers may be rounded off to a certain number of significant digits. For example, we can express the value of π to three significant digits, 3.14, or we can express it to six significant digits, 3.14159. When you use a calculator or computer, the number of significant digits is limited by the number of digits the machine is designed to carry.

Use of Numbers in This Book You should treat numbers given in problems as exact values and not be concerned about how many significant digits they contain. If a problem states that a quantity equals 32.2, you can assume its value is 32.200. . . . We express intermediate results and answers in the examples and the answers to the problems to at least three significant digits. If you use a calculator, your results should be that accurate. Be sure to avoid round-off errors that occur if you round off intermediate results when making a series of calculations. Instead, carry through your calculations with as much accuracy as you can by retaining values in your calculator.

1.4 *Units*

The SI system of units has become nearly standard throughout the world. In the United States, U.S. Customary units are also used. In this section we summarize these two systems of units and explain how to convert units from one system to another.

International System of Units

In SI units, length is measured in meters (m) and mass in kilograms (kg). Time is measured in seconds (s), although other familiar measures such as minutes (min), hours (hr), and days are also used when convenient. Meters, kilograms, and seconds are called the **base units** of the SI system. Force is measured in newtons (N). Recall that these units are related by Newton's second law: One newton is the force required to give an object of one kilogram mass an acceleration of one meter per second squared:

$$1 \text{ N} = (1 \text{ kg})(1 \text{m/s}^2) = 1 \text{ kg-m/s}^2.$$

Because the newton can be expressed in terms of the base units, it is called a **derived unit**.

To express quantities by numbers of convenient size, multiples of units are indicated by prefixes. The most common prefixes, their abbreviations, and the multiples they represent are shown in Table 1.1. For example, 1 km is 1 kilometer, which is 1000 m, and 1 Mg is 1 megagram, which is 10^6 g, or 1000 kg. We frequently use kilonewtons (kN).

Table 1.1 The common prefixes used in SI units and the multiples they represent.

Prefix	Abbreviation	Multiple
nano-	n	10^{-9}
micro-	μ	10^{-6}
milli-	m	10^{-3}
kilo-	k	10^3
mega-	M	10^6
giga-	G	10^9

U.S. Customary Units

In U.S. Customary units, length is measured in feet (ft) and force is measured in pounds (lb). Time is measured in seconds (s). These are the base units of the U.S. Customary system. In this system of units, mass is a derived unit. The unit of mass is the slug, which is the mass of material accelerated at one foot per second squared by a force of one pound. Newton's second law states that

$$1 \text{ lb} = (1 \text{ slug})(1 \text{ ft/s}^2).$$

From this expression we obtain

$$1 \text{ slug} = 1 \text{ lb-s}^2/\text{ft}.$$

We use other U.S. Customary units such as the mile (1 mi = 5280 ft) and the inch (1 ft = 12 in.). We also use the kilopound (kip), which is 1000 lb.

In some engineering applications, an alternative unit of mass called the pound mass (lbm) is used, which is the mass of material having a weight of one pound at sea level. The weight at sea level of an object that has a mass of one slug is

$$W = mg = (1 \text{ slug})(32.2 \text{ ft/s}^2) = 32.2 \text{ lb},$$

so 1 lbm = (1/32.2) slug. When the pound mass is used, a pound of force is usually denoted by the abbreviation lbf.

Angular Units

In both SI and U.S. Customary units, angles are normally expressed in radians (rad). We show the value of an angle θ in radians in Fig. 1.3. It is defined to be the ratio of the part of the circumference subtended by θ to the radius of the circle. Angles are also expressed in degrees. Since there are 360 degrees (360°) in a complete circle, and the complete circumference of the circle is $2\pi R$, 360° equals 2π rad.

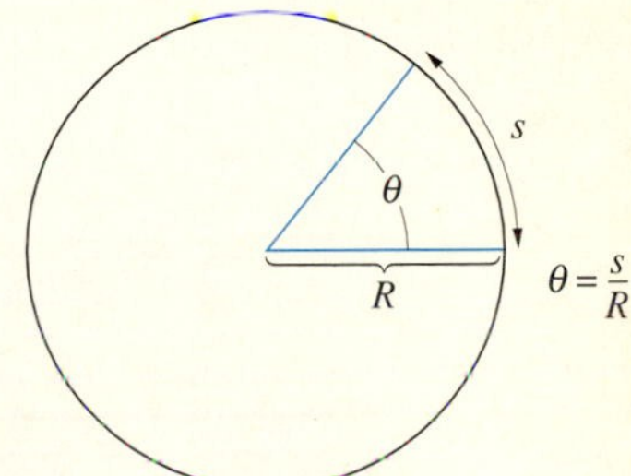

Figure 1.3
Definition of an angle in radians.

Equations containing angles are nearly always derived under the assumption that angles are expressed in radians. Therefore when you want to substitute the value of an angle expressed in degrees into an equation, you should first convert it into radians. A notable exception to this rule is that many calculators are designed to accept angles expressed in either degrees or radians when you use them to evaluate functions such as $\sin \theta$.

Conversion of Units

Many situations arise in engineering practice that require you to convert values expressed in units of one kind into values in other units. If some data in a problem are given in terms of SI units and some are given in terms of U.S. Customary units, you must express all of the data in terms of one system of units. In problems expressed in terms of SI units, you will occasionally be given data in terms of units other than the base units of seconds, meters, kilograms, and newtons. You should convert these data into the base units before working the problem. Similarly, in problems involving U.S. Customary units, you should convert terms into the base units of seconds, feet, slugs, and pounds. After you gain some experience, you will recognize situations in which these rules can be relaxed, but for now the procedure we propose is the safest.

Converting units is straightforward, although you must do it with care. Suppose that we want to express 1 mi/hr in terms of ft/s. Since one mile equals 5280 ft and one hour equals 3600 seconds, we can treat the expressions

$$\left(\frac{5280 \text{ ft}}{1 \text{ mi}}\right) \quad \text{and} \quad \left(\frac{1 \text{ hr}}{3600 \text{ s}}\right)$$

as ratios whose values are 1. In this way we obtain

$$1 \text{ mi/hr} = 1 \text{ mi/hr} \times \left(\frac{5280 \text{ ft}}{1 \text{ mi}}\right) \times \left(\frac{1 \text{ hr}}{3600 \text{ s}}\right) = 1.47 \text{ ft/s}.$$

We give some useful unit conversions in Table 1.2.

Table 1.2 Unit conversions.

Time	1 minute	=	60 seconds
	1 hour	=	60 minutes
	1 day	=	24 hours
Length	1 foot	=	12 inches
	1 mile	=	5280 feet
	1 inch	=	25.4 millimeters
	1 foot	=	0.3048 meter
Angle	2π radians	=	360 degrees
Mass	1 slug	=	14.59 kilograms
Force	1 pound	=	4.448 newtons

Example 1.1

Figure 1.4

If an Olympic sprinter (Fig. 1.4) runs 100 meters in 10 seconds, his average velocity is 10 m/s. What is his average velocity in mi/hr?

SOLUTION

$$10 \text{ m/s} = 10 \text{ m/s} \times \left(\frac{1 \text{ ft}}{0.3048 \text{ m}}\right) \times \left(\frac{1 \text{ mi}}{5280 \text{ ft}}\right) \times \left(\frac{3600 \text{ s}}{1 \text{ hr}}\right)$$

$$= 22.4 \text{ mi/hr}.$$

Example 1.2

Suppose that in Einstein's equation

$$E = mc^2$$

the mass m is in kilograms and the velocity of light c is in meters per second.
(a) What are the SI units of E?
(b) If the value of E in SI units is 20, what is its value in U.S. Customary base units?

STRATEGY

(a) Since we know the units of the terms m and c, we can deduce the units of E from the given equation.
(b) We can use the unit conversions for mass and length from Table 1.2 to convert E from SI units to U.S. Customary units.

SOLUTION

(a) From the equation for E,

$$E = (m \text{ kg})(c \text{ m/s})^2,$$

the SI units of E are kg-m^2/s^2.

(b) From Table 1.2, 1 slug = 14.59 kg and 1 ft = 0.3048 m. Therefore

$$1 \text{ kg-m}^2/\text{s}^2 = 1 \text{ kg-m}^2/\text{s}^2 \times \left(\frac{1 \text{ slug}}{14.59 \text{ kg}}\right) \times \left(\frac{1 \text{ ft}}{0.3048 \text{ m}}\right)^2$$

$$= 0.738 \text{ slug-ft}^2/\text{s}^2.$$

The value of E in U.S. Customary units is

$$E = (20)(0.738) = 14.8 \text{ slug-ft}^2/\text{s}^2.$$

Example 1.3

George Stephenson's *Rocket* (Fig. 1.5), an early steam locomotive, weighed about 7 tons with its tender. (A ton is 2000 lb.) What was its approximate mass in kilograms?

Figure 1.5

STRATEGY

We can use Eq. (1.6) to obtain the mass in slugs and then use the conversion given in Table 1.2 to determine the mass in kilograms.

SOLUTION

The mass in slugs is

$$m = \frac{W}{g} = \frac{14{,}000\ \text{lb}}{32.2\ \text{ft/s}^2} = 434.8\ \text{slugs}.$$

From Table 1.2, 1 slug equals 14.59 kg, so the mass in kilograms is (to three significant digits)

$$m = (434.8)(14.59) = 6340\ \text{kg}.$$

Problems

1.1 The value of π is 3.141592654. . . . What is its value to seven significant digits?

1.2 What is the value of e (the base of natural logarithms) to five significant digits?

1.3 Determine the value of the expression $1/(2 - \pi)$ to three significant digits.

1.4 The opening in a soccer goal is 24 ft wide and 8 ft high. Use these values to determine its dimensions in meters to three significant digits.

1.5 The dimensions of the Boeing 777-200 aircraft are length = 209 ft 1 in., wingspan = 199 ft 11 in., and height (bottom of wheels to top of vertical stabilizer) = 60 ft 6 in. Express these dimensions in meters to three significant digits.

1.6 Suppose that you have just purchased a Ferrari Dino 246GT coupe and you want to know whether you can use your set of SAE (U.S. Customary unit) wrenches to work on it. You have wrenches with widths w = 1/4 in., 1/2 in., 3/4 in., and 1 in., and the car has nuts with dimensions n = 5 mm, 10 mm, 15 mm, 20 mm, and 25 mm. Defining a wrench to fit if w is no more than 2% larger than n, which of your wrenches can you use?

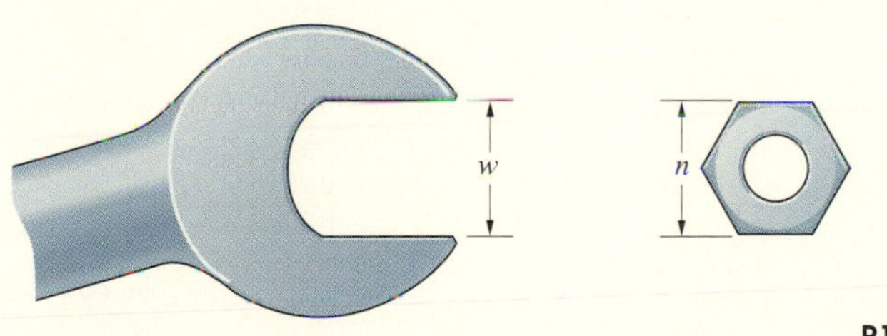

P1.6

1.7 The 1829 *Rocket*, shown in Figure 1.5, could draw a carriage with 30 passengers at 25 mi/hr. Determine its velocity to three significant digits: (a) in ft/s; (b) in km/hr.

1.8 High-speed "bullet trains" began running between Tokyo and Osaka, Japan, in 1964. If a bullet train travels at 240 km/hr, what is its velocity in mi/hr to three significant digits?

1.9 In December 1986, Dick Rutan and Jeana Yeager flew the *Voyager* aircraft around the world nonstop. They flew a distance of 40,212 km in 9 days, 3 minutes, and 44 seconds.
(a) Determine the distance they flew in miles to three significant digits.
(b) Determine their average speed (the distance flown divided by the time required) in kilometers per hour, miles per hour, and knots (nautical miles per hour) to three significant digits.

1.10 Engineers who study shock waves sometimes express velocity in millimeters per microsecond (mm/μs). Suppose the velocity of a wavefront is measured and determined to be 5 mm/μs. Determine its velocity: (a) in m/s; (b) in mi/s.

1.11 Geophysicists measure the motion of a glacier and discover it is moving at 80 mm/year. What is its velocity in m/s?

1.12 The acceleration due to gravity at sea level in SI units is $g = 9.81$ m/s^2. By converting units, use this value to determine the acceleration due to gravity at sea level in U.S. Customary units.

1.13 A *furlong per fortnight* is a facetious unit of velocity, perhaps made up by a student as a satirical comment on the bewildering variety of units engineers must deal with. A furlong is 660 ft (1/8 mile). A fortnight is 2 weeks (14 nights). If you walk to class at 2 m/s, what is your speed in furlongs per fortnight to three significant digits?

1.14 The cross-sectional area of a beam is 480 in^2. What is its cross-sectional area in m^2?

1.15 At sea level, the weight density (weight per unit volume) of water is approximately 62.4 lb/ft^3. Use this value to determine the mass density of water in kg/m^3.

1.16 A pressure transducer measures a value of 300 lb/in^2. Determine the value of the pressure in pascals. A pascal (Pa) is one newton per meter squared.

1.17 A horsepower is 550 ft-lb/s. A watt is 1 N-m/s. Determine the number of watts generated by (a) the Wright brothers' 1903 airplane, which had a 12-horsepower engine; (b) a modern passenger jet with a power of 100,000 horsepower at cruising speed.

P1.17

1.18 In SI units, the universal gravitational constant $G = 6.67 \times 10^{-11}$ N-m^2/kg^2. Determine the value of G in U.S. Customary base units.

1.19 If the earth is modeled as a homogeneous sphere, the velocity of a satellite in a circular orbit is

$$v = \sqrt{\frac{gR_E^2}{r}},$$

where R_E is the radius of the earth and r is the radius of the orbit.
(a) If g is in m/s^2 and R_E and r are in meters, what are the units of v?
(b) If $R_E = 6370$ km and $r = 6670$ km, what is the value of v to three significant digits?
(c) For the orbit described in (b), what is the value of v in mi/s to three significant digits?

1.20 In the equation

$$T = \frac{1}{2}I\omega^2,$$

the term I is in kg-m^2 and ω is in s^{-1}.
(a) What are the SI units of T?
(b) If the value of T is 100 when I is in kg-m^2 and ω is in s^{-1}, what is the value of T when it is expressed in terms of U.S. Customary base units?

1.21 The aerodynamic drag force D exerted on a moving object by a gas is given by the expression

$$D = C_D S \frac{1}{2}\rho v^2,$$

where the drag coefficient C_D is dimensionless, S is a reference area, ρ is the mass per unit volume of the gas, and v is the velocity of the object relative to the gas.
(a) Suppose that the value of D is 800 when S, ρ, and v are expressed in SI base units. By converting units, determine the value of D when S, ρ, and v are expressed in U.S. Customary base units.
(b) The drag force D is in newtons when the expression is evaluated using SI base units and is in pounds when the expression is evaluated using U.S. Customary base units. Using your result from (a), determine the conversion factor from newtons to pounds.

1.22 The Lockheed-Martin X-33 reusable launch test vehicle, when fully fueled, will weigh 273,300 lb at sea level. Its weight at sea level with its fuel expended will be 62,700 lb.
(a) Determine its mass in slugs when it is fully fueled and with its fuel expended.
(b) Determine its weight in meganewtons at sea level when it is fully fueled and with its fuel expended.

P1.22

1.23 The acceleration due to gravity is 13.2 ft/s^2 on the surface of Mars and 32.2 ft/s^2 on the surface of the earth. A woman weighs 125 lb on earth. To survive and work on the surface of Mars, she must wear life-support equipment and carry tools. What is the maximum allowable weight on earth of the woman's clothing, equipment, and tools if the engineers don't want the total weight on Mars of the woman and her clothing, equipment, and tools to exceed 125 lb?

1.24 A person has a mass of 50 kg.
(a) The acceleration due to gravity at sea level is $g = 9.81$ m/s^2. What is the person's weight at sea level?
(b) The acceleration due to gravity on the surface of the moon is 1.62 m/s^2. What would the person weigh on the moon?

1.25 The acceleration due to gravity at sea level is $g = 9.81$ m/s^2. The radius of the earth is 6370 km. The universal gravitational constant $G = 6.67 \times 10^{-11}$ N-m^2/kg^2. Use this information to determine the mass of the earth.

1.26 A person weighs 180 lb at sea level. The radius of the earth is 3960 mi. What force is exerted on the person by the gravitational attraction of the earth if he is in a space station in orbit 200 mi above the surface of the earth?

1.27 The acceleration due to gravity on the surface of the moon is 1.62 m/s^2. The radius of the moon is R_M = 1738 km. Determine the acceleration due to gravity of the moon at a point 1738 km above its surface.

Strategy: Write an equation equivalent to Eq. (1.4) for the acceleration due to gravity of the moon.

1.28 If an object is near the surface of the earth, the variation of its weight with distance from the center of the earth can often be neglected. The acceleration due to gravity at sea level is g = 9.81 m/s^2. The radius of the earth is 6370 km. The weight of an object at sea level is mg, where m is its mass. At what height above the surface of the earth does the weight of the object decrease to 0.99mg?

1.29 The centers of two oranges are 1 m apart. The mass of each orange is 0.2 kg. What gravitational force do they exert on each other? (The universal gravitational constant $G = 6.67 \times 10^{-11}$ N-m^2/kg^2.)

1.30 At a point between the earth and the moon, the magnitude of the earth's gravitational acceleration equals the magnitude of the moon's gravitational acceleration. What is the distance from the center of the earth to that point to three significant digits? The distance from the center of the earth to the center of the moon is 383,000 km, and the radius of the earth is 6370 km. The radius of the moon is 1738 km, and the acceleration due to gravity at its surface is 1.62 m/s^2.

The position and velocity of the *Voyager 2* space probe at the time of its release near earth determined the trajectory (path) it followed to reach the planet Jupiter. The gravitational field of Jupiter altered the trajectory of *Voyager 2* so that it could pass near Saturn, which altered its trajectory again so that it could pass near Uranus, and so on to Neptune. In this chapter we determine trajectories of objects and analyze their positions, velocities, and accelerations using different types of coordinate systems.

Chapter 2

Motion of a Point

ENGINEERS designing a vehicle, whether a bicycle or a spacecraft, must be able to analyze and predict its motion. To design an engine, they must analyze the motions of each of its moving parts. Even when designing "static" structures such as buildings, bridges, and dams, they often must analyze motions resulting from wind loads and potential earthquakes.

In this chapter we begin the study of motion. We are not yet concerned with the properties of objects or the causes of their motions—our objective is to describe and analyze the motion of a point in space. However, keep in mind that the point can represent some point (such as the center of mass) of a moving object. After defining the position, velocity, and acceleration of a point, we consider the simplest example: motion along a straight line. We then show how motion of a point along an arbitrary path, or **trajectory**, is expressed and analyzed in various coordinate systems.

2.1 *Position, Velocity, and Acceleration*

If you observe people in a room, such as a group at a party, you perceive their positions relative to the room. For example, some people may be in the back of the room, some in the middle of the room, and so forth. The colloquial expression is that the room is your "frame of reference." To make this idea precise, we can introduce a cartesian coordinate system with its axes aligned with the walls of the room as in Fig. 2.1(a), and specify the position of a person (actually the position of some point of the person, such as his or her center of mass) by specifying the components of the position vector **r** relative to the origin of the coordinate system. This coordinate system is a convenient reference frame for objects in the room. If you are sitting in an airplane, you perceive the positions of objects within the airplane relative to the airplane. In this case the interior of the airplane is your frame of reference. To precisely specify the position of a person within the airplane, we can introduce a cartesian coordinate system that is fixed relative to the airplane and measure the position of the person's center of mass by specifying the components of the position vector **r** relative to the origin (Fig. 2.1b). A **reference frame** is simply a coordinate system suitable for specifying positions of points. You may be familiar only with cartesian coordinates. We discuss other examples in this chapter and continue our discussion of reference frames throughout the book.

We can describe the position of a point P relative to a given reference frame with origin O by the **position vector r** from O to P (Fig. 2.2a). Suppose

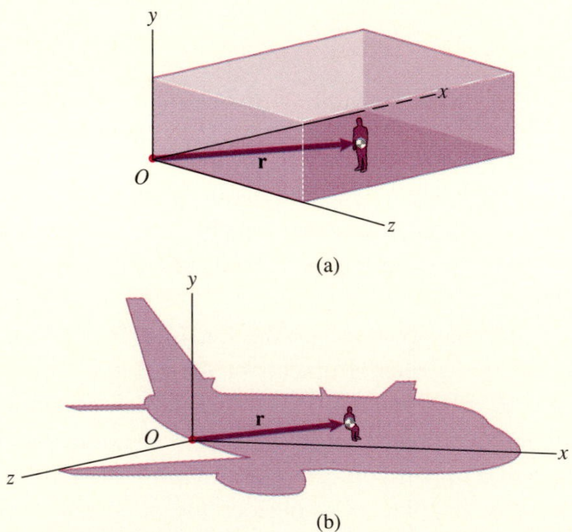

Fig. 2.1

Convenient reference frames for specifying positions of objects:
(a) in a room;
(b) in an airplane.

that P is in motion relative to the chosen reference frame, so that $\mathbf{r}$ is a function of time t (Fig. 2.2b). We express this by the notation

$$\mathbf{r} = \mathbf{r}(t).$$

The **velocity** of P relative to the given reference frame at time t is defined by

$$\mathbf{v} = \frac{d\mathbf{r}}{dt} = \lim_{\Delta t \to 0} \frac{\mathbf{r}(t + \Delta t) - \mathbf{r}(t)}{\Delta t}, \tag{2.1}$$

where the vector $\mathbf{r}(t + \Delta t) - \mathbf{r}(t)$ is the change in position, or **displacement** of P, during the interval of time Δt (Fig. 2.2c). Thus the velocity is the rate of change of the position of P.

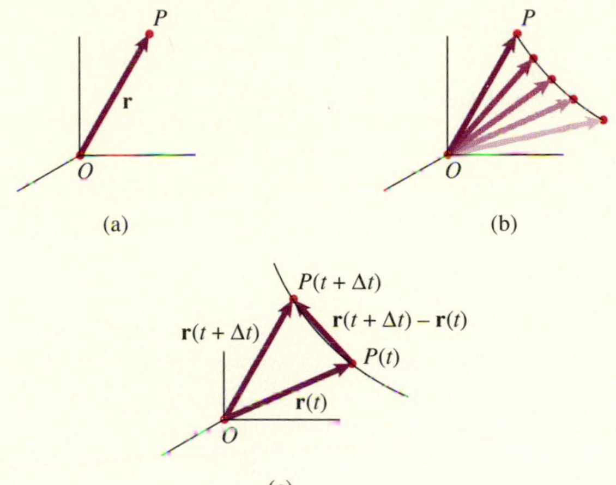

Fig. 2.2
(a) The position vector $\mathbf{r}$ of P relative to O.
(b) Motion of P relative to the reference frame.
(c) Change in position of P from t to $t + \Delta t$.

The dimensions of a derivative are determined just as if it was a ratio, so the dimensions of $\mathbf{v}$ are (distance)/(time). The reference frame being used is often obvious, and we simply call $\mathbf{v}$ the velocity of P. However, you must remember that the position and velocity of a point can be specified only relative to some reference frame.

Notice in Eq. (2.1) that the derivative of a vector with respect to time is defined in exactly the same way as is the derivative of a scalar function. As a result, it shares some of the properties of the derivative of a scalar function. We will use two of these properties: The time derivative of the sum of two vector functions $\mathbf{u}$ and $\mathbf{w}$ is

$$\frac{d}{dt}(\mathbf{u} + \mathbf{w}) = \frac{d\mathbf{u}}{dt} + \frac{d\mathbf{w}}{dt},$$

and the time derivative of the product of a scalar function f and a vector function $\mathbf{u}$ is

$$\frac{d(f\mathbf{u})}{dt} = \frac{df}{dt}\mathbf{u} + f\frac{d\mathbf{u}}{dt}.$$

The **acceleration** of P relative to the given reference frame at time t is defined by

$$\mathbf{a} = \frac{d\mathbf{v}}{dt} = \lim_{\Delta t \to 0} \frac{\mathbf{v}(t + \Delta t) - \mathbf{v}(t)}{\Delta t}, \tag{2.2}$$

where $\mathbf{v}(t + \Delta t) - \mathbf{v}(t)$ is the change in the velocity of P during the interval of time Δt (Fig. 2.3). The acceleration is the rate of change of the velocity of P at time t (the second time derivative of the displacement), and its dimensions are (distance)/(time)^2.

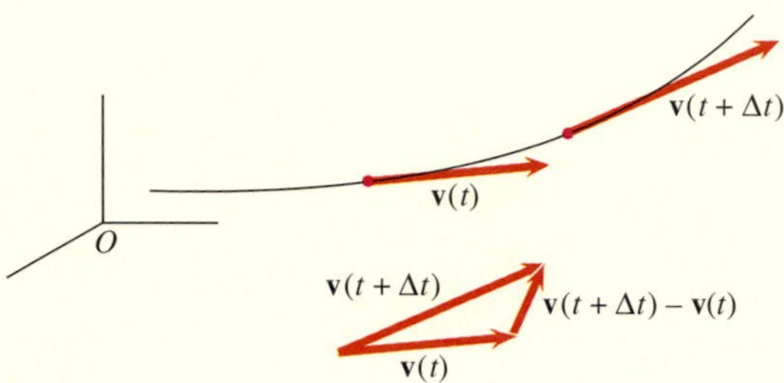

Fig. 2.3
Change in the velocity of P from t to $t + \Delta t$.

We have defined the velocity and acceleration of P relative to the origin O of the reference frame. We can show that *a point has the same velocity and the same acceleration relative to any fixed point in a given reference frame*. Let O' be an arbitrary fixed point, and let $\mathbf{r}'$ be the position vector from O' to P (Fig. 2.4a). The velocity of P relative to O' is $\mathbf{v}' = d\mathbf{r}'/dt$. The velocity of P relative to the origin O is $\mathbf{v} = d\mathbf{r}/dt$. We wish to show that $\mathbf{v}' = \mathbf{v}$. Let $\mathbf{R}$ be the vector from O to O' (Fig. 2.4b), so that

$$\mathbf{r}' = \mathbf{r} - \mathbf{R}.$$

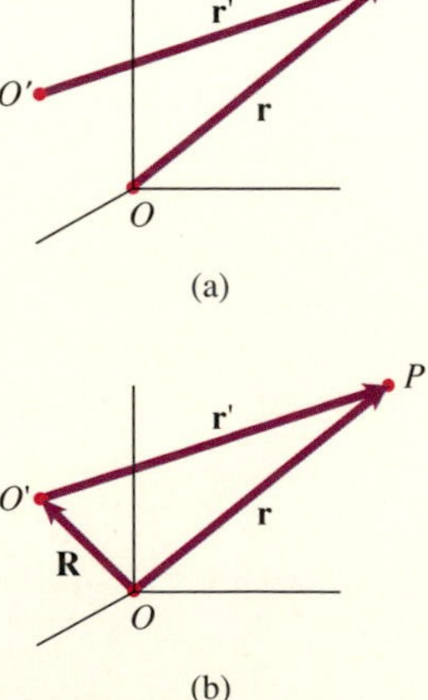

Fig. 2.4
(a) Position vectors of P relative to O and O'.
(b) Position vector of O' relative to O.

Since the vector **R** is constant, the velocity of P relative to O' is

$$\mathbf{v}' = \frac{d\mathbf{r}'}{dt} = \frac{d\mathbf{r}}{dt} - \frac{d\mathbf{R}}{dt} = \frac{d\mathbf{r}}{dt} = \mathbf{v}.$$

The acceleration of P relative to O' is $\mathbf{a}' = d\mathbf{v}'/dt$, and the acceleration of P relative to O is $\mathbf{a} = d\mathbf{v}/dt$. Since $\mathbf{v}' = \mathbf{v}$, $\mathbf{a}' = \mathbf{a}$. The velocity and acceleration of a point P relative to a given reference frame do not depend on the location of the fixed reference point used to specify the position of P.

2.2 Straight-Line Motion

We discuss this simple type of motion primarily so that you can gain experience and insight before proceeding to the general case of motion of a point. But engineers must analyze straight-line motions in many practical situations, such as the motion of a vehicle on a straight road or track or the motion of a piston in an internal combustion engine.

Description of the Motion

We can specify the position of a point P on a straight line relative to a reference point O by the coordinate s measured along the line from O to P (Fig. 2.5a). In this case the straight line is the reference frame we use to describe the position of P, and O is its origin. In Fig. 2.5(a) we define s to be positive to the right, so s is positive when P is to the right of O and negative when P is to the left of O. The **displacement** Δs relative to O during an interval of time from t_0 to t is the change in the position, $\Delta s = s(t) - s(t_0)$.

By introducing a unit vector **e** parallel to the line and pointing in the positive s direction (Fig. 2.5b), we can write the position vector of P relative to O as

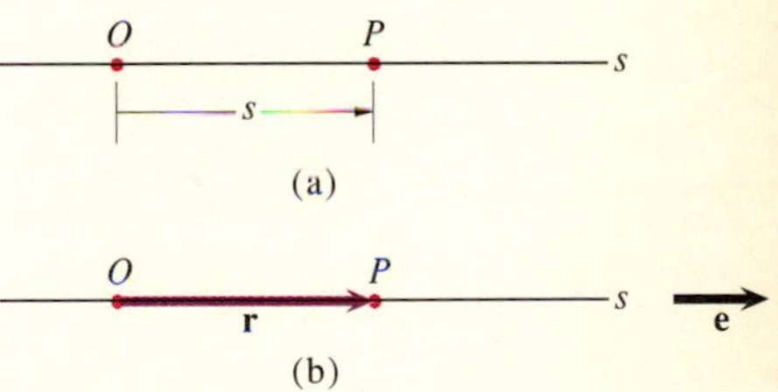

Fig. 2.5
(a) The coordinate s from O to P.
(b) The unit vector **e** and position vector **r**.

$$\mathbf{r} = s\mathbf{e}.$$

The velocity of P relative to O is

$$\mathbf{v} = \frac{d\mathbf{r}}{dt} = \frac{ds}{dt}\mathbf{e}.$$

We can write the velocity vector as $\mathbf{v} = v\mathbf{e}$, obtaining the scalar equation

$$v = \frac{ds}{dt}.$$

The velocity v of point P along the straight line is the rate of change of its position s. Notice that v is equal to the slope at time t of the line tangent to the graph of s as a function of time (Fig. 2.6).

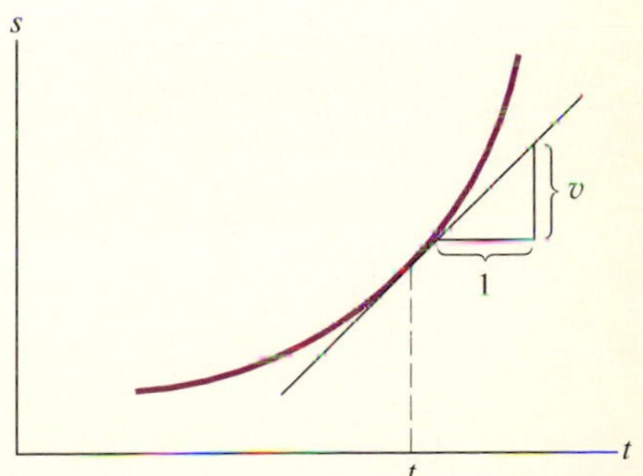

Fig. 2.6
The slope of the straight line tangent to the graph of s versus t is the velocity at time t.

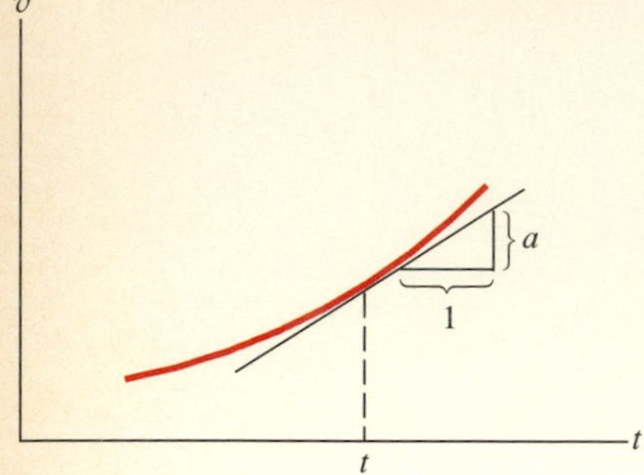

Fig. 2.7
The slope of the straight line tangent to the graph of v versus t is the acceleration at time t.

The acceleration of P relative to O is

$$\mathbf{a} = \frac{d\mathbf{v}}{dt} = \frac{d}{dt}(v\mathbf{e}) = \frac{dv}{dt}\mathbf{e}.$$

Writing the acceleration vector as $\mathbf{a} = a\mathbf{e}$, we obtain the scalar equation

$$a = \frac{dv}{dt} = \frac{d^2s}{dt^2}.$$

The acceleration a is equal to the slope at time t of the line tangent to the graph of v as a function of time (Fig. 2.7).

By introducing the unit vector $\mathbf{e}$, we have obtained scalar equations describing the motion of P. The position is specified by the coordinate s, and the velocity and acceleration are governed by the equations

$$v = \frac{ds}{dt}, \tag{2.3}$$

$$a = \frac{dv}{dt}. \tag{2.4}$$

Analysis of the Motion

In some situations, you will know the position s of some point of an object as a function of time. Engineers use methods such as radar and laser-doppler interferometry to measure positions as functions of time. In this case, you can obtain the velocity and acceleration as functions of time from Eqs. (2.3) and (2.4) by differentiation. For example, if the position of the truck in Fig. 2.8 during the interval of time from $t = 2$ s to $t = 4$ s is given by the equation

$$s = 6 + \frac{1}{3}t^3 \text{ m},$$

its velocity and acceleration during that interval of time are

$$v = \frac{ds}{dt} = t^2 \text{ m/s},$$

$$a = \frac{dv}{dt} = 2t \text{ m/s}^2.$$

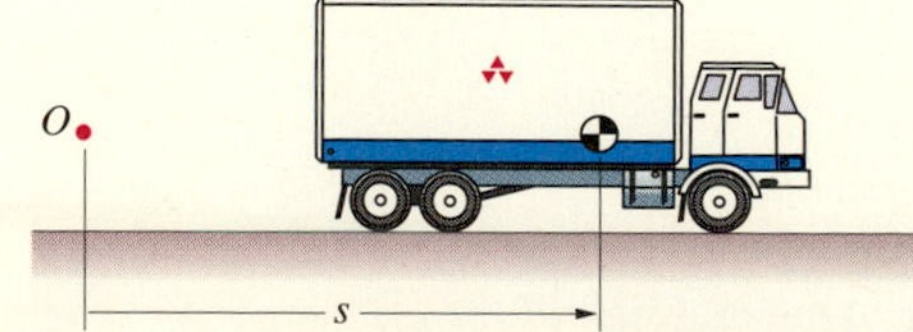

Fig. 2.8
The coordinate s measures the position of the center of mass of the truck relative to a reference point.

However, it is more common to know an object's acceleration than to know its position, because the acceleration of an object can be determined by Newton's second law when the forces acting on it are known. When the acceleration is known, you can determine the velocity and position from Eqs. (2.3) and (2.4) by integration. We discuss three important cases in the following sections.

Acceleration Specified as a Function of Time If the acceleration is a known function of time $a(t)$, we can integrate the relation

$$\frac{dv}{dt} = a(t) \tag{2.5}$$

with respect to time to determine the velocity as a function of time:

$$v = \int a(t)\, dt + A, \tag{2.6}$$

where A is an integration constant. Then we can integrate the relation

$$\frac{ds}{dt} = v \tag{2.7}$$

to determine the position as a function of time:

$$s = \int v\, dt + B, \tag{2.8}$$

where B is another integration constant. We would need additional information about the motion, such as the values of v and s at a given time, to determine the constants A and B.

Instead of using indefinite integrals, we can write Eq. (2.5) as

$$dv = a(t)\, dt$$

and integrate in terms of definite integrals:

$$\int_{v_0}^{v} dv = \int_{t_0}^{t} a(t)\, dt.$$

The lower limit v_0 is the velocity at time t_0, and the upper limit v is the velocity at an arbitrary time t. Evaluating the left integral, we obtain an expression for the velocity as a function of time:

$$v = v_0 + \int_{t_0}^{t} a(t)\, dt. \tag{2.9}$$

We can then write Eq. (2.7) as

$$ds = v\, dt$$

and integrate in terms of definite integrals,

$$\int_{s_0}^{s} ds = \int_{t_0}^{t} v\, dt,$$

where the lower limit s_0 is the position at time t_0 and the upper limit s is the position at an arbitrary time t. Evaluating the left integral, we obtain the position as a function of time:

$$s = s_0 + \int_{t_0}^{t} v\,dt. \tag{2.10}$$

Although we have shown how to determine the velocity and position when you know the acceleration as a function of time, you shouldn't try to remember results such as Eqs. (2.9) and (2.10). As we will demonstrate in the examples, we recommend that you solve straight-line motion problems by beginning with Eqs. (2.3) and (2.4).

We can make some useful observations from Eqs. (2.9) and (2.10):

- The area defined by the graph of the acceleration of P as a function of time from t_0 to t is equal to the change in the velocity from t_0 to t (Fig. 2.9a).
- The area defined by the graph of the velocity of P as a function of time from t_0 to t is equal to the displacement, or change in position, from t_0 to t (Fig. 2.9b).

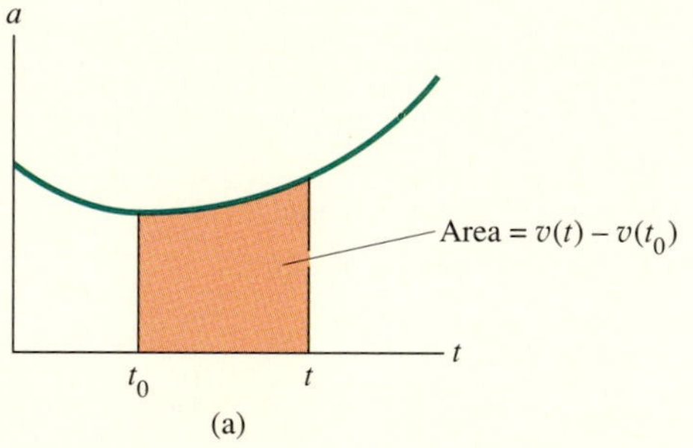

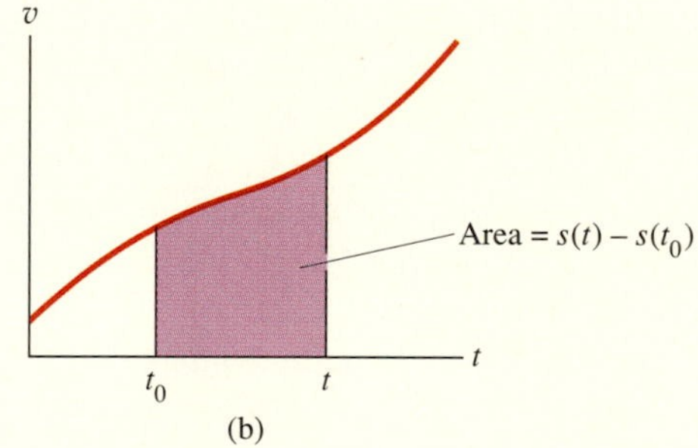

Fig. 2.9
Relations between areas defined by the graphs of the acceleration and velocity of P and changes in its velocity and position.

You often can use these relationships to obtain a qualitative understanding of an object's motion, and in some cases you can even use them to determine its motion.

In some situations, the acceleration of an object is constant, or nearly constant. For example, if you drop a dense object such as a golf ball or rock and it doesn't fall too far, you can neglect aerodynamic drag and assume that its acceleration is equal to the acceleration of gravity at sea level.

Let the acceleration be a known constant a_0. From Eqs. (2.9) and (2.10), the velocity and position as functions of time are

$$v = v_0 + a_0(t - t_0), \tag{2.11}$$

$$s = s_0 + v_0(t - t_0) + \frac{1}{2}a_0(t - t_0)^2, \tag{2.12}$$

where s_0 and v_0 are the position and velocity, respectively, at time t_0. Notice that *if the acceleration is constant, the velocity is a linear function of time.*

We can use the **chain rule** to express the acceleration in terms of a derivative with respect to s:

$$a_0 = \frac{dv}{dt} = \frac{dv}{ds}\frac{ds}{dt} = \frac{dv}{ds}v.$$

Writing this expression as $v\,dv = a_0\,ds$ and integrating,

$$\int_{v_0}^{v} v\,dv = \int_{s_0}^{s} a_0\,ds,$$

we obtain an equation for the velocity as a function of position:

$$v^2 = v_0^2 + 2a_0(s - s_0). \tag{2.13}$$

You are probably familiar with Eqs. (2.11)–(2.13). Although these results can be useful *when you know that the acceleration is constant*, you must be careful not to use them otherwise.

The following examples illustrate how you can use Eqs. (2.3) and (2.4) to obtain information about straight-line motions of objects. You may need to choose the reference point and the positive direction for s. When you know the acceleration as a function of time, you can integrate Eq. (2.4) to determine the velocity and then integrate Eq. (2.3) to determine the position.

Example 2.1

Engineers testing a vehicle that will be dropped by parachute estimate that its vertical velocity when it reaches the ground will be 20 ft/s. If they drop the vehicle from the test rig in Fig. 2.10, from what height h should they drop it to simulate the parachute drop?

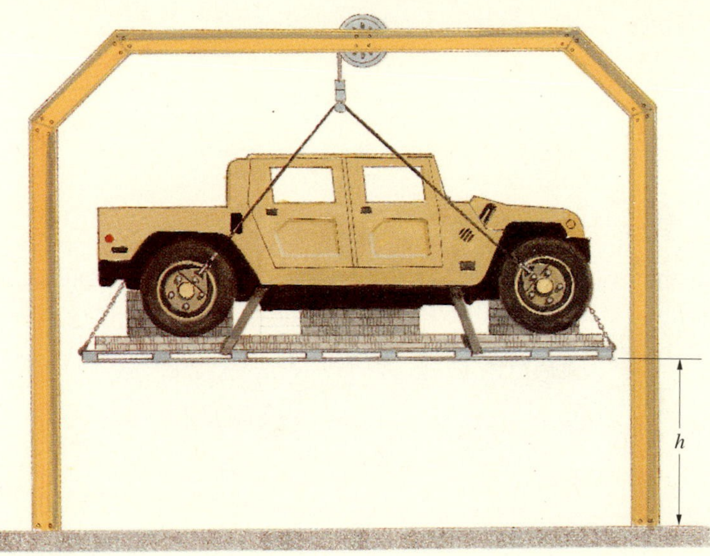

Fig. 2.10

STRATEGY

If the only significant force acting on an object near the earth's surface is its weight, it accelerates downward with the acceleration due to gravity. Therefore we can assume that the vehicle's acceleration during its short fall is $g = 32.2\ \text{ft/s}^2$. We will determine the height h in two ways:

- *First method.* We can integrate Eqs. (2.3) and (2.4) to determine the vehicle's motion.
- *Second method.* We can use Eq. (2.13), which relates the velocity and position when the acceleration is constant.

SOLUTION

We let s be the position of the bottom of the platform supporting the vehicle relative to its initial position (Fig. a). The vehicle's acceleration is $a = 32.2\ \text{ft/s}^2$.

First Method From Eq. (2.4),

$$\frac{dv}{dt} = a = 32.2\ \text{ft/s}^2.$$

Integrating, we obtain

$$v = 32.2t + A,$$

where A is an integration constant. If we let $t = 0$ be the instant the vehicle is dropped, $v = 0$ when $t = 0$, so $A = 0$ and the velocity as a function of time is

(a) The coordinate s measures the position of the bottom of the platform relative to its initial position.

$$v = 32.2t \text{ ft/s}.$$

Then by integrating Eq. (2.3),

$$\frac{ds}{dt} = v = 32.2t,$$

we obtain

$$s = 16.1t^2 + B,$$

where B is a second integration constant. The position $s = 0$ when $t = 0$, so $B = 0$ and the position as a function of time is

$$s = 16.1t^2.$$

From the equation for the velocity, the time of fall necessary for the vehicle to reach 20 ft/s is $t = 20/32.2 = 0.621$ s. Substituting this time into the equation for the position, the height h needed to simulate the parachute drop is

$$h = 16.1(0.621)^2 = 6.21 \text{ ft}.$$

Second Method Because the acceleration is constant, we can use Eq. (2.13) to determine the distance necessary for the velocity to increase to 20 ft/s:

$$v^2 = v_0^2 + 2a_0(s - s_0),$$

$$(20)^2 = 0 + 2(32.2)(s - 0).$$

Solving for s, we obtain $h = 6.21$ ft.

DISCUSSION

In this example we integrated the vehicle's acceleration to determine its velocity as a function of time and then integrated its velocity to determine its position as a function of time. As one check of our procedure, we should differentiate our solution for the position to make sure we obtain the correct velocity and differentiate the velocity to make sure we obtain the correct acceleration. The velocity is

$$v = \frac{ds}{dt} = \frac{d}{dt}(16.1t^2) = 32.2t \text{ ft/s},$$

and the acceleration is

$$a = \frac{dv}{dt} = \frac{d}{dt}(32.2t) = 32.2 \text{ ft/s}^2.$$

Example 2.2

The cheetah, *Acinonyx jubatus* (Fig. 2.11), can run as fast as 75 mi/hr. If you assume that the animal's acceleration is constant and that it reaches top speed in 4 s, what distance can it cover in 10 s?

Fig. 2.11

STRATEGY

The acceleration has a constant value for the first 4 s and is then zero. We can determine the distance traveled during each of these "phases" of the motion and sum them to obtain the total distance covered. We do so both analytically and graphically.

SOLUTION

The top speed in terms of feet per second is

$$75 \text{ mi/hr} = 75 \text{ mi/hr} \times \left(\frac{5280 \text{ ft}}{1 \text{ mi}}\right) \times \left(\frac{1 \text{ hr}}{3600 \text{ s}}\right) = 110 \text{ ft/s}.$$

First Method Let a_0 be the acceleration during the first 4 s. We integrate Eq. (2.4),

$$\int_0^v dv = \int_0^t a_0 \, dt,$$

$$[v]_0^v = [a_0 t]_0^t,$$

obtaining the velocity as a function of time during the first 4 s:

$$v = a_0 t \text{ ft/s}.$$

When $t = 4$ s, $v = 110$ ft/s, so $a_0 = 110/4 = 27.5 \text{ ft/s}^2$. Therefore the velocity during the first 4 s is $v = 27.5t$ ft/s. Now we integrate Eq. (2.3),

$$\int_0^s ds = \int_0^t 27.5t \, dt,$$

$$[s]_0^s = 27.5 \left[\frac{t^2}{2}\right]_0^t,$$

obtaining the position as a function of time during the first 4 s:

$$s = 13.75t^2 \text{ m}.$$

At $t = 4$ s, the position is $s = 13.75(4)^2 = 220$ ft.

From $t = 4$ to $t = 10$ s, the velocity is constant. The distance traveled is

$$(110 \text{ ft/s})(6 \text{ s}) = 660 \text{ ft}.$$

The total distance the animal travels is 220 + 660 = 880 ft, or 293 yd, in 10 s.

Second Method We draw a graph of the animal's velocity as a function of time in Fig. (a). The acceleration is constant during the first 4 s of motion, so the velocity is a linear function of time from $v = 0$ at $t = 0$ to $v = 110$ ft/s at $t = 4$ s. The velocity is constant during the last 6 s. The total distance covered is the sum of the areas during the two phases of motion:

$$\frac{1}{2}(4 \text{ s})(110 \text{ ft/s}) + (6 \text{ s})(110 \text{ ft/s}) = 220 \text{ ft} + 660 \text{ ft} = 880 \text{ ft}.$$

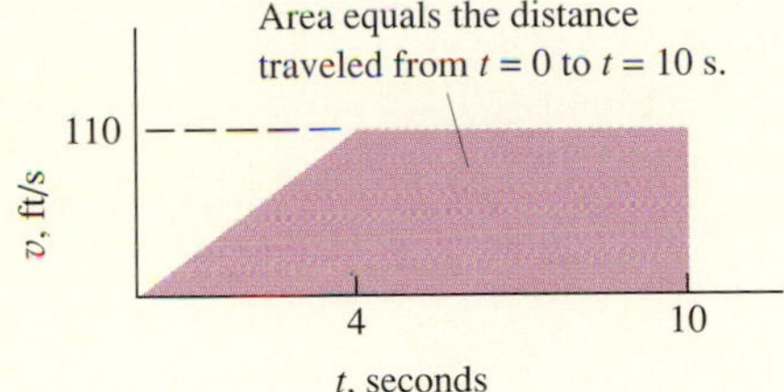

(a) The cheetah's velocity as a function of time.

DISCUSSION

Notice that in the first method we used definite, rather than indefinite, integrals to determine the cheetah's velocity and position as functions of time. You should rework the example using indefinite integrals and compare your results to ours. Whether to use definite or indefinite integrals is primarily a matter of taste, but you need to be familiar with both procedures.

Example 2.3

Fig. 2.12

Suppose that the acceleration of the train in Fig. 2.12 during the interval of time from $t = 2$ s to $t = 4$ s is $a = 2t$ m/s^2, and at $t = 2$ s its velocity is $v = 180$ km/hr. What is the train's velocity at $t = 4$ s, and what is its displacement (change in position) from $t = 2$ s to $t = 4$ s?

STRATEGY

We can integrate Eqs. (2.3) and (2.4) to determine the train's velocity and position as functions of time.

SOLUTION

The velocity at $t = 2$ s in terms of m/s is

$$180 \text{ km/hr} \times \left(\frac{1000 \text{ m}}{1 \text{ km}}\right) \times \left(\frac{1 \text{ hr}}{3600 \text{ s}}\right) = 50 \text{ m/s}.$$

We write Eq. (2.4) as

$$dv = a\,dt = 2t\,dt$$

and integrate, introducing the condition $v = 50$ m/s at $t = 2$ s:

$$\int_{50}^{v} dv = \int_{2}^{t} 2t\,dt,$$

$$[v]_{50}^{v} = [t^2]_2^t,$$

$$v - 50 = t^2 - 4.$$

Solving for v, we obtain

$$v = t^2 + 46 \text{ m/s}.$$

Now that we know the velocity as a function of time, we write Eq. (2.3) as

$$ds = v\,dt = (t^2 + 46)\,dt$$

and integrate, defining the position of the train at $t = 2$ s to be $s = 0$:

$$\int_0^s ds = \int_2^t (t^2 + 46)\,dt,$$

$$[s]_0^s = \left[\frac{t^3}{3} + 46t\right]_2^t,$$

$$s = \frac{t^3}{3} + 46t - \frac{2^3}{3} - 46(2).$$

The position as a function of time is

$$s = \frac{1}{3}t^3 + 46t - 94.7 \text{ m}.$$

Using our equations for the velocity and position, the velocity at $t = 4$ s is

$$v = (4)^2 + 46 = 62 \text{ m/s},$$

and the displacement from $t = 2$ s to $t = 4$ s is

$$\Delta s = \left[\frac{1}{3}(4)^3 + 46(4) - 94.7\right] - 0 = 111 \text{ m}.$$

DISCUSSION

The acceleration in this example is not constant. You must not try to solve such problems by using equations that are valid only when the acceleration is constant. To convince yourself, try applying Eq. (2.11) to this example: Set $a_0 = 2t$ m/s^2, $t_0 = 2$ s, and $v_0 = 50$ m/s, and solve for the velocity at $t = 4$ s.

Problems

The following problems involve straight-line motion. The time *t* is in seconds unless otherwise stated.

2.1 The graph of the position s of a point as a function of time is a straight line. When $t = 4$ s, $s = 24$ m, and when $t = 20$ s, $s = 72$ m.
(a) Determine the velocity of the point by calculating the slope of the straight line.
(b) Obtain the equation for s as a function of time and use it to determine the velocity of the point.

2.2 The graph of the position s of a point of a milling machine as a function of time is a straight line. When $t = 0.2$ s, $s = 90$ mm. During the interval of time from $t = 0.6$ s to $t = 1.2$ s, the displacement of the point is $\Delta s = -180$ mm.
(a) Determine the equation for s as a function of time.
(b) What is the velocity of the point?

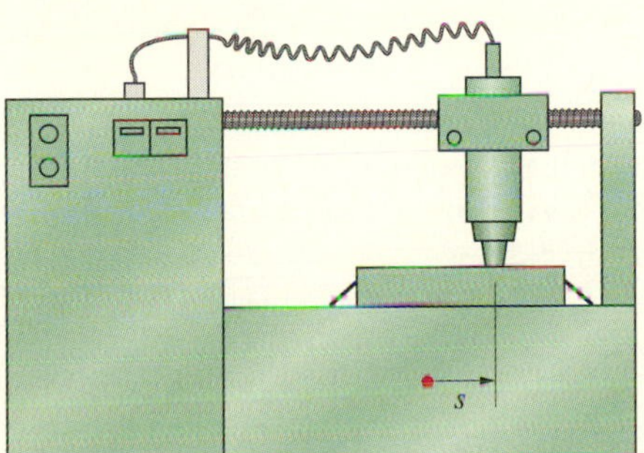

P2.2

2.3 The graph of the velocity v of a point as a function of time is a straight line. When $t = 2$ s, $v = 4$ ft/s, and when $t = 4$ s, $v = -10$ ft/s.
(a) Determine the acceleration of the point by calculating the slope of the straight line.
(b) Obtain the equation for v as a function of time and use it to determine the acceleration of the point.

2.4 The position of a point is $s = 2t^2 - 10$ ft.
(a) What is the displacement of the point from $t = 0$ to $t = 4$ s?
(b) What are the velocity and acceleration at $t = 0$?
(c) What are the velocity and acceleration at $t = 4$ s?

2.5 A rocket starts from rest and travels straight up. Its height above the ground is measured by radar from $t = 0$ to $t = 4$ s and is found to be approximated by the function $s = 10t^2$ m.
(a) What is the displacement during this interval of time?
(b) What is the velocity at $t = 4$ s?
(c) What is the acceleration during the first 4 s?

P2.5

2.6 The position of a point during the interval of time from $t = 0$ to $t = 6$ s is $s = -\frac{1}{2}t^3 + 6t^2 + 4t$ m.
(a) What is the displacement of the point during this interval of time?
(b) What is the maximum velocity during this interval of time, and at what time does it occur?
(c) What is the acceleration when the velocity is a maximum?

2.7 The position of a point during the interval of time from $t = 0$ to $t = 3$ s is $s = 12 + 5t^2 - t^3$ ft.
(a) What is the maximum velocity during this interval of time, and at what time does it occur?
(b) What is the acceleration when the velocity is a maximum?

2.8 The mechanism causes the displacement of point P to be $s = 0.2 \sin(\pi t)$ m, where t is in seconds and the argument πt of the sine function is in radians. Determine the velocity and acceleration of point P at $t = 3.8$ s.

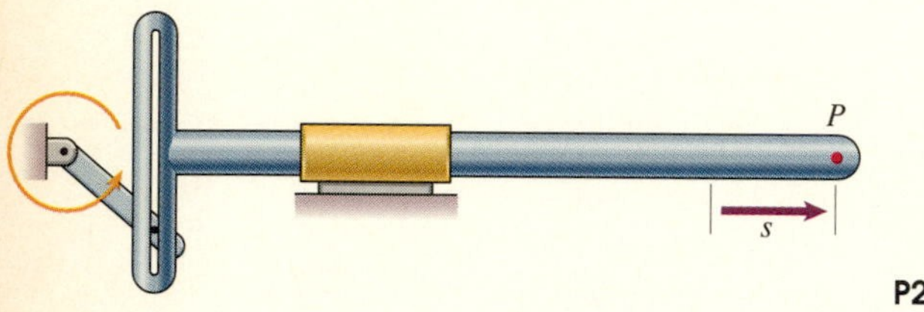

P2.8

2.9 For the mechanism in Problem 2.8, draw graphs of the position s, velocity v, and acceleration a of point P as functions of time for $0 < t < 4$ s. Using your graphs, confirm that the slope of the graph of the position s is zero at times for which the velocity $v = ds/dt$ is zero, and that the slope of the graph of the velocity v is zero at times for which the acceleration $a = dv/dt$ is zero.

2.10 A seismograph measures the horizontal motion of the ground during an earthquake. An engineer analyzing the data determines that for a 10-s interval of time beginning at $t = 0$, the position is approximated by $s = 100 \cos(2\pi t)$ mm. What are the (a) maximum velocity and (b) maximum acceleration of the ground during the 10-s interval?

2.11 During an assembly operation, a robot's arm moves along a straight line. During an interval of time from $t = 0$ to $t = 1$ s, its position is given by $s = 3t^2 - 2t^3$ in. Determine, during this 1-s interval: (a) the displacement of the arm; (b) the maximum and minimum values of the velocity; (c) the maximum and minimum values of the acceleration.

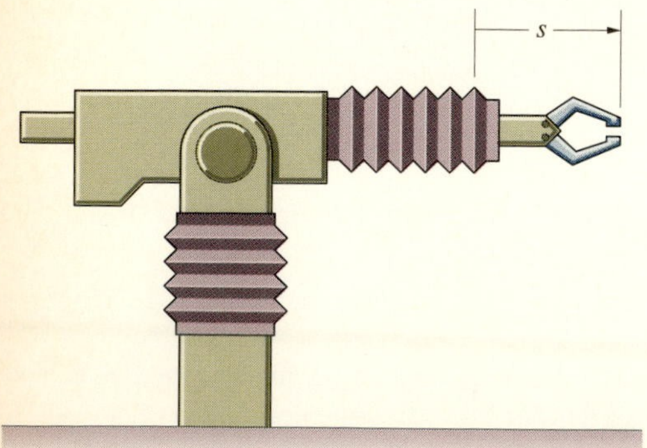

P2.11

2.12 In a test of a prototype car, the driver starts the car from rest at $t = 0$, accelerates, and then applies the brakes. Engineers measuring the position of the car find that from $t = 0$ to $t = 18$ s it is approximated by $s = 5t^2 + \frac{1}{3}t^3 - \frac{1}{50}t^4$ ft.
(a) What is the maximum velocity, and at what time does it occur?
(b) What is the maximum acceleration, and at what time does it occur?

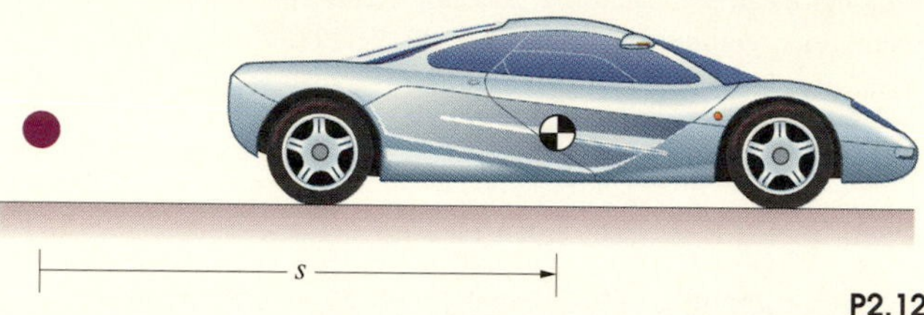

P2.12

2.13 Suppose you want to approximate the position of a vehicle you are testing by the power series $s = A + Bt + Ct^2 + Dt^3$, where A, B, C, and D are constants. The vehicle starts from rest at $t = 0$ and $s = 0$. At $t = 4$ s, $s = 176$ ft, and at $t = 8$ s, $s = 448$ ft.
(a) Determine A, B, C, and D.
(b) What are the approximate velocity and acceleration of the vehicle at $t = 8$ s?

2.14 The acceleration of a point is $a = 20t$ m/s^2. When $t = 0$, $s = 40$ m and $v = -10$ m/s. What are the position and velocity at $t = 3$ s?

2.15 The acceleration of a point is $a = 60t - 36t^2$ ft/s^2. When $t = 0$, $s = 0$ and $v = 20$ ft/s. What are the position and velocity as functions of time?

2.16 Suppose that during the preliminary design of a car, you assume its maximum acceleration is approximately constant. What constant acceleration is necessary if you want the car to be able to accelerate from rest to a velocity of 55 mi/hr in 10 s? What distance would the car travel during that time?

2.17 An entomologist estimates that a flea 1 mm in length attains a velocity of 1.3 m/s in a distance of one body length when jumping. What constant acceleration is necessary to achieve that velocity?

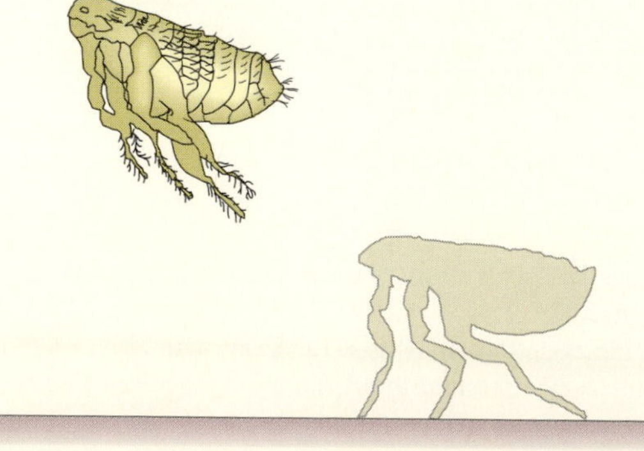

P2.17

2.18 Missiles designed for defense against ballistic missiles achieve accelerations in excess of 100 g's, or 100 times the acceleration of gravity. If a missile has a constant acceleration of 100 g's, how long does it take to go from rest to 60 mi/hr? What is its displacement during that time?

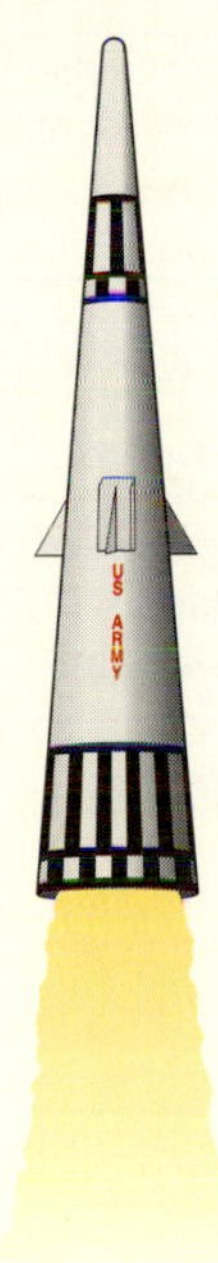

P2.18

2.19 Suppose you want to throw some keys to a friend standing on a second-floor balcony. If you release the keys at 1.5 m above the ground, what vertical velocity is necessary for them to just reach your friend's hand 6 m above the ground?

2.20 The lunar module descends toward the surface of the moon at 1 m/s when its landing probes, which extend 2 m below the landing gear, touch the surface, automatically shutting off the engines. Determine the velocity with which the landing gear contact the surface. (The acceleration due to gravity at the surface of the moon is 1.62 m/s^2.)

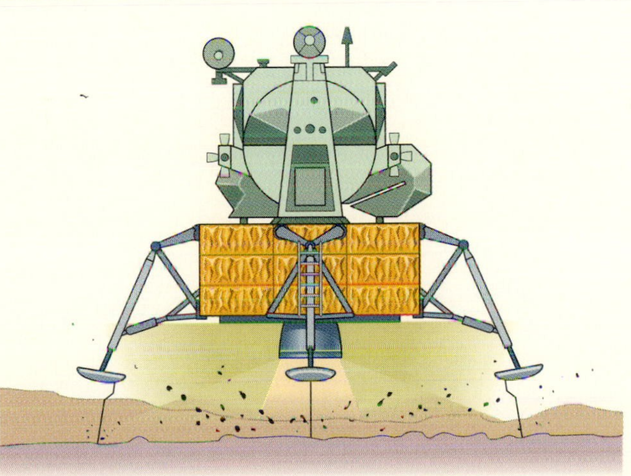

P2.20

2.21 In 1960 R. C. Owens of the Baltimore Colts blocked a Washington Redskins field goal attempt by jumping and knocking the ball away in front of the cross bar at a point 11 ft above the field. If he was 6 ft 3 in. tall and could reach 1 ft 11 in. above his head, what was his vertical velocity as he left the ground?

2.22 The velocity of a bobsled is $v = 10t$ ft/s. When $t = 2$ s, its position is $s = 25$ ft. What is its position when $t = 10$ s?

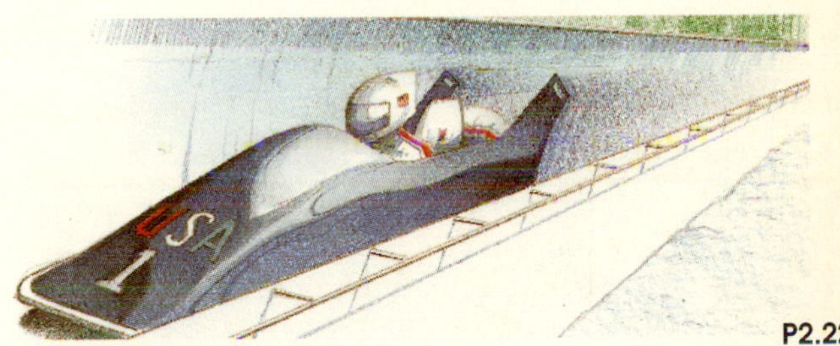

P2.22

2.23 The acceleration of an object is $a = 30 - 6t$ ft/s^2. When $t = 0$, $s = 0$ and $v = 0$. What is its maximum velocity during the interval of time from $t = 0$ to $t = 10$ s?

2.24 The velocity of an object is $v = 200 - 2t^2$ m/s. When $t = 3$ s, its position is $s = 600$ m. What are the position and acceleration of the object at $t = 6$ s?

2.25 The acceleration of a part undergoing a machining operation is measured and is determined to be $a = 12 - 6t$ mm/s^2. When $t = 0$, $v = 0$. For the interval of time from $t = 0$ to $t = 4$ s, determine: (a) the maximum velocity; (b) the displacement.

2.26 The missile shown in Problem 2.18 starts from rest and accelerates straight up for 3 s at 100 g's. After 3 s, its weight and aerodynamic drag cause it to have a nearly constant deceleration of 4 g's. How long does it take the missile to go from the ground to an altitude of 50,000 ft?

2.27 The graph describes an airplane's acceleration during its takeoff run. Determine the airplane's velocity when it rotates (lifts off) at $t = 20$ s.

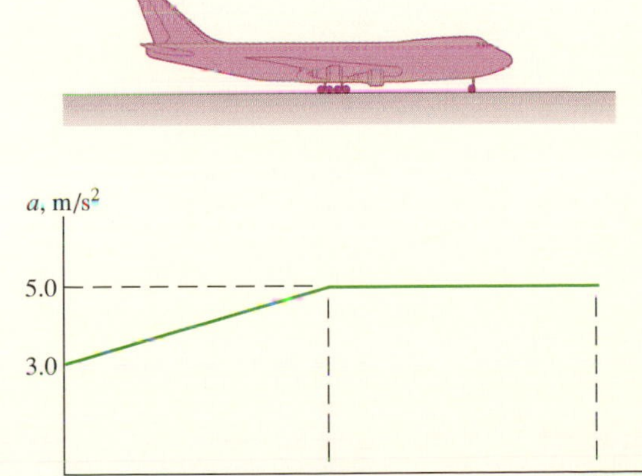

P2.27

2.28 In Problem 2.27, determine the distance the airplane has traveled when it rotates at $t = 20$ s.

2.29 A car is traveling at 30 mi/hr when a traffic light 295 ft ahead turns yellow. The light will remain yellow for 5 s before turning red.
(a) What constant acceleration will cause the car to reach the light at the instant it turns red, and what will the velocity of the car be when it reaches the light?
(b) If the driver decides not to try to make the light, what constant rate of acceleration will cause the car to come to a stop just as it reaches the light?

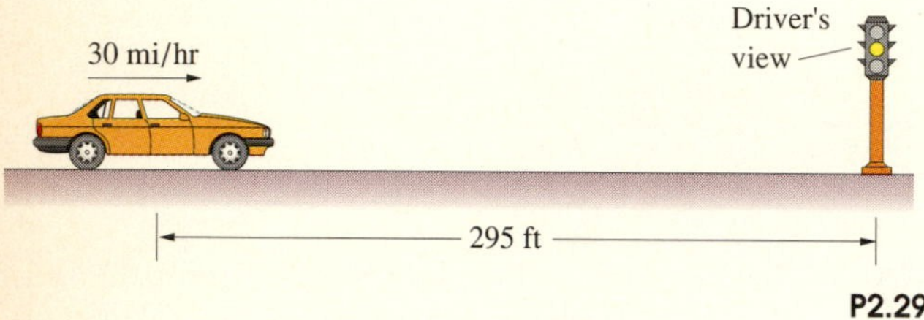

P2.29

2.30 At $t = 0$, a motorist traveling at 100 km/hr sees a deer standing in the road 100 m ahead. After a reaction time of 0.3 s, he applies the brakes and decelerates at a constant rate of 4 m/s^2. If the deer takes 5 s from $t = 0$ to react and leave the road, does the motorist miss him?

2.31 A high-speed rail transportation system has a top speed of 100 m/s. For the comfort of the passengers, the magnitude of the acceleration and deceleration is limited to 2 m/s^2. Determine the minimum time required for a trip of 100 km.

Strategy: A graphical approach can help you solve this problem. Recall that the change in the position from an initial time t_0 to a time t is equal to the area defined by the graph of the velocity as a function of time from t_0 to t.

P2.31

2.32 The nearest star, Proxima Centauri, is 4.22 light years from the earth. Ignoring relative motion between the solar system and Proxima Centauri, suppose that a spacecraft accelerates from the vicinity of the earth at 0.01 g (0.01 times the acceleration due to gravity at sea level) until it reaches one-tenth the speed of light, coasts until time to decelerate, than decelerates at 0.01 g until it comes to rest in the vicinity of Proxima Centauri. How long does the trip take? (Light travels at 3×10^8 m/s; a solar year is 365.2422 solar days.)

2.33 A race car starts from rest and accelerates at $a = 5 + 2t$ ft/s^2 for 10 s. The brakes are then applied, and the car has a constant acceleration $a = -30$ ft/s^2 until it comes to rest. Determine: (a) the maximum velocity; (b) the total distance traveled; (c) the total time of travel.

2.34 When $t = 0$, the position of a point is $s = 6$ m and its velocity is $v = 2$ m/s. From $t = 0$ to $t = 6$ s, its acceleration is $a = 2 + 2t^2$ m/s^2. From $t = 6$ s until it comes to rest, its acceleration is $a = -4$ m/s^2.
(a) What is the total time of travel?
(b) What total distance does it move?

2.35 Zoologists studying the ecology of the Serengeti Plain estimate that the average adult cheetah can run 100 km/hr and the average springbuck can run 65 km/hr. If the animals run along the same straight line, start at the same time, and are each assumed to have constant acceleration and reach top speed in 4 s, how close must a cheetah be when the chase begins to catch a springbuck in 15 s?

2.36 Suppose that a person unwisely drives 75 mi/hr in a 55 mi/hr zone and passes a police car going 55 mi/hr in the same direction. If the police officers begin constant acceleration at the instant they are passed and increase their velocity to 80 mi/hr in 4 s, how long does it take them to be even with the pursued car?

2.37 If $\theta = 1$ rad and $d\theta/dt = 1$ rad/s, what is the velocity of P relative to O?

Strategy: You can write the position of P relative to O as

$$s = (2 \text{ ft}) \cos \theta + (2 \text{ ft}) \cos \theta,$$

then take the derivative of this expression with respect to time to determine the velocity.

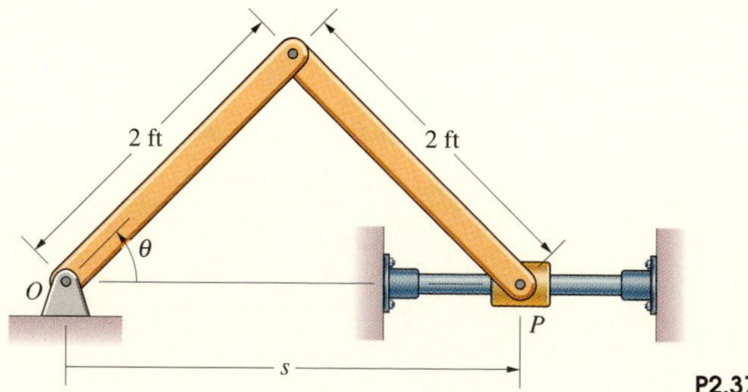

P2.37

2.38 In Problem 2.37, if $\theta = 1$ rad, $d\theta/dt = -2$ rad/s, and $d^2\theta/dt^2 = 0$, what are the velocity and acceleration of P relative to O?

2.39 If $\theta = 1$ rad and $d\theta/dt = 1$ rad/s, what is the velocity of P relative to O?

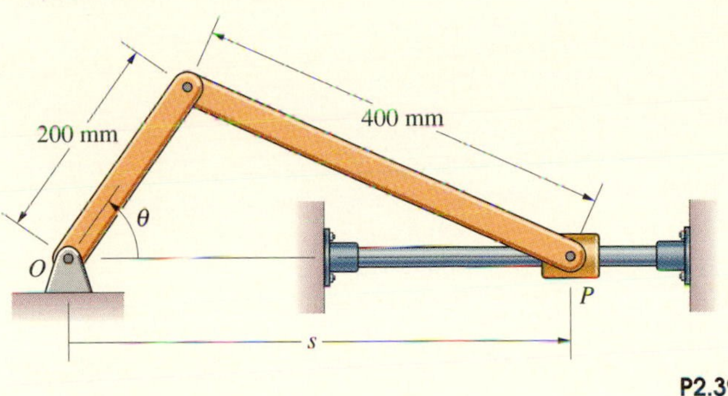

P2.39

Acceleration Specified as a Function of Velocity Aerodynamic and hydrodynamic forces can cause an object's acceleration to depend on its velocity (Fig. 2.13). Suppose that the acceleration is a known function of velocity $a(v)$:

$$\frac{dv}{dt} = a(v). \tag{2.14}$$

We cannot integrate this equation with respect to time to determine the velocity, because $a(v)$ is not known as a function of time. But we can **separate variables**, putting terms involving v on one side of the equation and terms involving t on the other side:

$$\frac{dv}{a(v)} = dt. \tag{2.15}$$

We can now integrate,

$$\int_{v_0}^{v} \frac{dv}{a(v)} = \int_{t_0}^{t} dt, \tag{2.16}$$

where v_0 is the velocity at time t_0. In principle, we can solve this equation for the velocity as a function of time, then integrate the relation

$$\frac{ds}{dt} = v$$

to determine the position as a function of time.

By using the chain rule, we can also determine the velocity as a function of the position. Writing the acceleration as

$$\frac{dv}{dt} = \frac{dv}{ds}\frac{ds}{dt} = \frac{dv}{ds}v$$

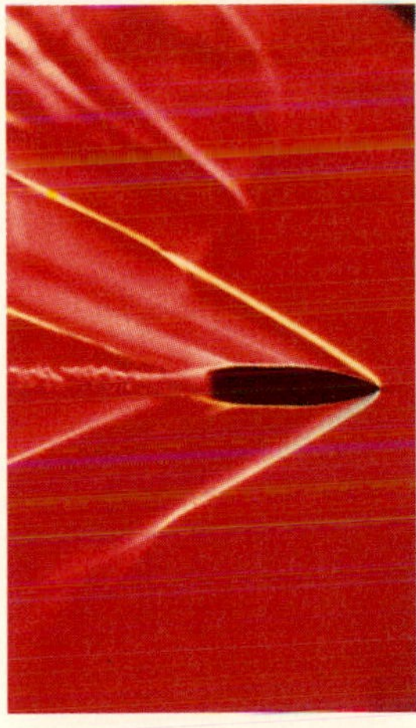

Fig. 2.13

Aerodynamic and hydrodynamic forces depend on an object's velocity. As the bullet slows, the aerodynamic drag force resisting its motion decreases.

and substituting it into Eq. (2.14), we obtain

$$\frac{dv}{ds}v = a(v).$$

Separating variables,

$$\frac{v\,dv}{a(v)} = ds,$$

and integrating,

$$\int_{v_0}^{v} \frac{v\,dv}{a(v)} = \int_{s_0}^{s} ds,$$

we can obtain a relation between the velocity and the position. (See Example 2.4.)

Acceleration Specified as a Function of Position Gravitational forces and forces exerted by springs can cause an object's acceleration to depend on its position. If the acceleration is a known function of position,

$$\frac{dv}{dt} = a(s), \tag{2.17}$$

we cannot integrate with respect to time to determine the velocity, because $a(s)$ is not known as a function of time. Moreover, we cannot separate variables, because the equation contains three variables: v, t, and s. However, by using the chain rule,

$$\frac{dv}{dt} = \frac{dv}{ds}\frac{ds}{dt} = \frac{dv}{ds}v,$$

we can write Eq. (2.17) as

$$\frac{dv}{ds}v = a(s).$$

Now we can separate variables,

$$v\,dv = a(s)\,ds, \tag{2.18}$$

and integrate:

$$\int_{v_0}^{v} v\,dv = \int_{s_0}^{s} a(s)\,ds. \tag{2.19}$$

In principle, we can solve this equation for the velocity as a function of the position:

$$v = \frac{ds}{dt} = v(s). \tag{2.20}$$

Then we can separate variables in this equation and integrate to determine the position as a function of time:

$$\int_{s_0}^{s} \frac{ds}{v(s)} = \int_{t_0}^{t} dt.$$

The next two examples show how you can analyze the motion of an object when its acceleration is a function of velocity or position. The initial steps are summarized in Table 2.1.

Table 2.1 Determining the velocity when you know the acceleration as a function of velocity or position.

If you know $a = a(v)$:	Separate variables, $\frac{dv}{dt} = a(v)$, $\frac{dv}{a(v)} = dt$; or apply the chain rule, $\frac{dv}{dt} = \frac{dv}{ds}\frac{ds}{dt} = \frac{dv}{ds}v = a(v)$, then separate variables, $\frac{v\,dv}{a(v)} = ds$.
If you know $a = a(s)$:	Apply the chain rule, $\frac{dv}{dt} = \frac{dv}{ds}\frac{ds}{dt} = \frac{dv}{ds}v = a(s)$, then separate variables, $v\,dv = a(s)\,ds$.

Example 2.4

After deploying its drag parachute, the airplane in Fig. 2.14 has an acceleration $a = -0.004v^2$ m/s^2.

(a) Determine the time required for the velocity to decrease from 80 m/s to 10 m/s.

(b) What distance does the plane cover during that time?

Fig. 2.14

STRATEGY

In part (b), we will use the chain rule to express the acceleration in terms of a derivative with respect to position and integrate to obtain a relation between the velocity and the position.

SOLUTION

(a) The acceleration is

$$a = \frac{dv}{dt} = -0.004v^2.$$

We separate variables,

$$\frac{dv}{v^2} = -0.004\,dt,$$

and integrate, defining $t = 0$ to be the time at which $v = 80$ m/s:

$$\int_{80}^{v} \frac{dv}{v^2} = \int_0^t -0.004\,dt,$$

$$\left[\frac{-1}{v}\right]_{80}^{v} = -0.004[t]_0^t,$$

$$\left(\frac{1}{v} - \frac{1}{80}\right) = 0.004t.$$

Solving for t, we obtain

$$t = 250\left(\frac{1}{v} - \frac{1}{80}\right).$$

From this equation we find that the time required for the plane to slow to $v = 10$ m/s is 21.9 s. We show the velocity of the airplane as a function of time in Fig. 2.15.

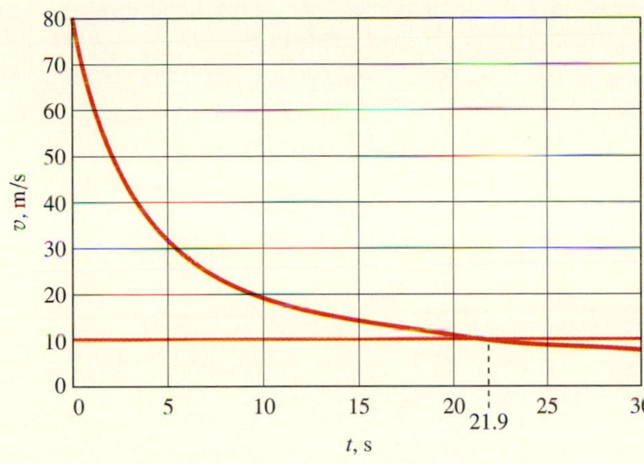

Fig. 2.15
Graph of the airplane's velocity as a function of time.

(b) We write the acceleration as

$$a = \frac{dv}{dt} = \frac{dv}{ds}\frac{ds}{dt} = \frac{dv}{ds}v = -0.004v^2,$$

separate variables,

$$\frac{dv}{v} = -0.004\, ds,$$

and integrate, defining $s = 0$ to be the position at which $v = 80$ m/s:

$$\int_{80}^{v} \frac{dv}{v} = \int_{0}^{s} -0.004\, ds,$$

$$[\ln v]_{80}^{v} = -0.004[s]_{0}^{s},$$

$$\ln v - \ln 80 = -\ln\left(\frac{80}{v}\right) = -0.004s.$$

Solving for s, we obtain

$$s = 250 \ln\left(\frac{80}{v}\right).$$

The distance required for the plane to slow to $v = 10$ m/s is 520 m.

DISCUSSION

Notice that our results predict that the time elapsed and distance traveled continue to increase without bound as the airplane's velocity decreases. The reason is that the modeling is incomplete. The equation for the acceleration includes only aerodynamic drag and does not account for other forces, such as friction in the airplane's wheels.

Example 2.5

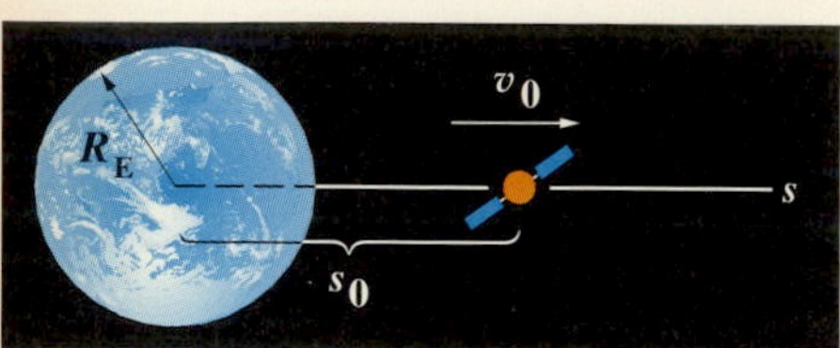

Fig. 2.16

In terms of distance s from the center of the earth, the magnitude of the acceleration due to gravity is gR_E^2/s^2, where R_E is the radius of the earth. (See the discussion of gravity in Section 1.3.) If a spacecraft is a distance s_0 from the center of the earth (Fig. 2.16), what outward velocity v_0 must it be given to reach a specified distance h from the center of the earth?

SOLUTION

The acceleration due to gravity is *toward* the center of the earth:

$$a = -\frac{gR_E^2}{s^2}.$$

Applying the chain rule,

$$a = \frac{dv}{dt} = \frac{dv}{ds}\frac{ds}{dt} = \frac{dv}{ds}v = -\frac{gR_E^2}{s^2},$$

and separating variables, we obtain

$$v\,dv = -\frac{gR_E^2}{s^2}\,ds.$$

We integrate this equation using the initial condition ($v = v_0$ when $s = s_0$) as the lower limits and the final condition ($v = 0$ when $s = h$) as the upper limits:

$$\int_{v_0}^{0} v\,dv = -\int_{s_0}^{h} \frac{gR_E^2}{s^2}\,ds,$$

$$\left[\frac{v^2}{2}\right]_{v_0}^{0} = gR_E^2\left[\frac{1}{s}\right]_{s_0}^{h},$$

$$-\frac{v_0^2}{2} = gR_E^2\left(\frac{1}{h} - \frac{1}{s_0}\right).$$

Solving for v_0, we obtain the initial velocity v_0 necessary for the spacecraft to reach a distance h:

$$v_0 = \sqrt{2gR_E^2\left(\frac{1}{s_0} - \frac{1}{h}\right)}.$$

DISCUSSION

We can make an interesting and important observation from the result of this example. Notice that as the distance h increases, the necessary initial velocity v_0 approaches a finite limit. This limit,

$$v_{\text{esc}} = \lim_{h\to\infty} v_0 = \sqrt{\frac{2gR_E^2}{s_0}},$$

is called the **escape velocity**. In the absence of other effects, an object with this initial velocity will continue moving outward indefinitely. The existence of an escape velocity makes it feasible to send probes and persons to other planets. Once escape velocity is attained, it isn't necessary to expend additional fuel to keep going.

Problems

2.40 An engineer designing a system to control a router for a machining process models it so that the router's acceleration during an interval of time is $a = -2v\,\text{m/s}^2$. When $t = 0$, its position is $s = 0$ and its velocity is $v = 2$ m/s. Determine the router's velocity as a function of time.

2.41 In Problem 2.40, determine the router's position as a function of time.

2.42 The boat is moving at 20 ft/s when its engine is shut down. Due to hydrodynamic drag, its acceleration is $a = -0.1v^2$ ft/s^2. What is the boat's velocity 2 s later?

P2.42

2.43 In Problem 2.42, what distance does the boat move in the 2 s following the shutdown of its engine?

2.44 A steel ball is released from rest in a container of oil. Its downward acceleration is $a = 0.9g - cv$, where g is the acceleration due to gravity at sea level and c is a constant. What is the velocity of the ball as a function of time?

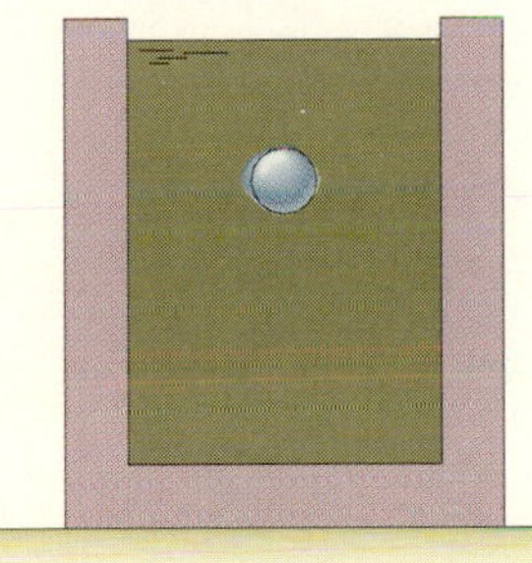

P2.44

2.45 In Problem 2.44, determine the position of the ball relative to its initial position as a function of time.

2.46 The greatest ocean depth yet discovered is in the Marianas Trench in the western Pacific Ocean. A steel ball released at the surface requires 64 min to reach the bottom. The ball's downward acceleration is $a = 0.9g - cv$, where $g = 32.2$ ft/s^2 is the acceleration due to gravity at sea level and the constant $c = 3.02\ s^{-1}$. What is the depth of the Marianas Trench in miles?

2.47 To study the effects of meteor impacts on satellites, engineers use a rail gun to accelerate a plastic pellet to a high velocity. They determine that when the pellet has traveled 1 m from the gun, its velocity is 2.25 km/s, and when it has traveled 2 m from the gun, its velocity is 1.00 km/s. Assume that the acceleration of the pellet after it leaves the gun is given by $a = -cv^2$, where c is a constant.

(a) What is the value of c, and what are its SI units?

(b) What was the velocity of the pellet as it left the rail gun?

P2.47

2.48 If aerodynamic drag is taken into account, the acceleration of a falling object can be approximated by $a = g - cv^2$, where g is the acceleration due to gravity at sea level and c is a constant.

(a) If an object is released from rest, what is its velocity as a function of the distance s from the point of release?

(b) Determine the limit of your answer to part (a) as $c \to 0$, and show that it agrees with the solution you obtain by assuming that the acceleration $a = g$.

2.49 A sky diver jumps from a helicopter and is falling straight down at 30 m/s when her parachute opens. From then on, her downward acceleration is approximately $a = g - cv^2$, where $g = 9.81$ m/s^2 and c is a constant. After an initial "transient" period, she descends at a nearly constant velocity of 5 m/s.

(a) What is the value of c, and what are its SI units?

(b) What maximum deceleration is she subjected to?

(c) What is her downward velocity when she has fallen 2 m from the point where her parachute opens?

P2.49

2.50 A rocket sled starts from rest and accelerates at $a = 3t^2$ m/s^2 until its velocity is 1000 m/s. It then hits a water brake, and its acceleration is $a = -0.001v^2$ m/s until its velocity decreases to 500 m/s. What total distance does the sled travel?

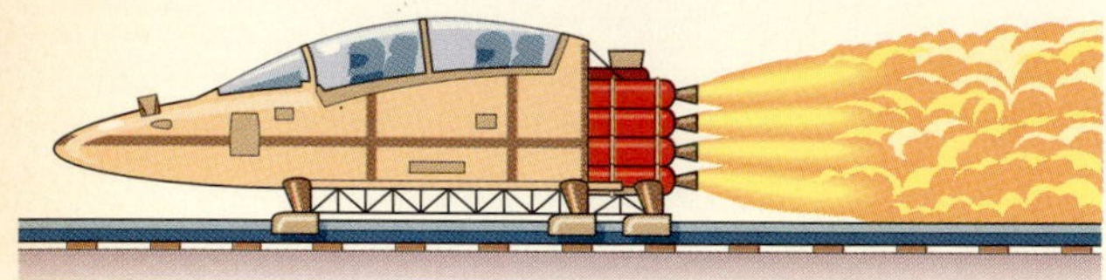

P2.50

2.51 The velocity of a point is given by the equation

$$v = (24 - 2s^2)^{1/2} \text{ m/s}.$$

What is its acceleration when $s = 2$ m?

2.52 The velocity of an object subjected to the earth's gravitational field is

$$v = \left[v_0^2 + 2gR_E^2\left(\frac{1}{s} - \frac{1}{s_0}\right)\right]^{1/2},$$

where v_0 is the velocity at position s_0 and R_E is the radius of the earth. Using this equation, show that the object's acceleration is $a = -gR_E^2/s^2$.

2.53 Engineers analyzing the motion of a linkage determine that the velocity of an attachment point is given by $v = A + 4s^2$ ft/s, where A is a constant. When $s = 2$ ft, its acceleration is measured and determined to be $a = 320$ ft/s^2. What is its velocity when $s = 2$ ft?

2.54 The acceleration of an object is given by the function $a = 2s$ ft/s^2. When $t = 0$, $v = 1$ ft/s. What is the velocity when the object has moved 2 ft from its initial position?

2.55 An object's acceleration is given by $a = 3s^2$ ft/s^2. At $s = 0$, its velocity is $v = 10$ ft/s. What is its velocity when $s = 4$ ft?

2.56 A gas gun used by engineers to investigate the properties of materials subjected to high pressure subjects a projectile to an acceleration $a = 3500/s^{1.4}$ m/s^2, where s is the projectile's position in meters. If the projectile starts from rest at $s = 1.5$ m, what is its velocity when it reaches the position $s = 3$ m?

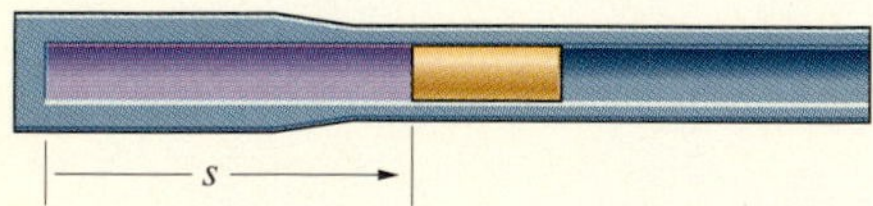

P2.56

2.57 A spring-mass oscillator consists of a mass and a spring connected as shown. The coordinate s measures the displacement of the mass relative to its position when the spring is unstretched. If the spring is linear, the mass is subjected to a deceleration proportional to s. Suppose that $a = -4s$ m/s^2 and that you give the mass a velocity $v = 1$ m/s in the position $s = 0$.
(a) How far will the mass move to the right before the spring brings it to a stop?
(b) What will be the velocity of the mass when it has returned to the position $s = 0$?

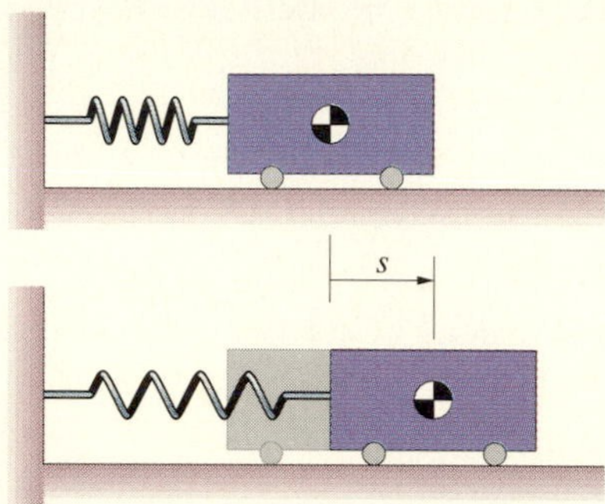

P2.57

2.58 In Problem 2.57, suppose that at $t = 0$ you release the mass from rest in the position $s = 1$ m. Determine the velocity of the mass as a function of s as it moves from the initial position to $s = 0$.

2.59 In Problem 2.57, suppose that at $t = 0$ you release the mass from rest in the position $s = 1$ m. Determine the position of the mass as a function of time as it moves from its initial position to $s = 0$.

2.60 The mass is released from rest with the springs unstretched. Its downward acceleration is $a = 32.2 - 50s$ ft/s^2, where s is the position of the mass measured from the position in which it is released.
(a) How far does the mass fall?
(b) What is the maximum velocity of the mass as it falls?

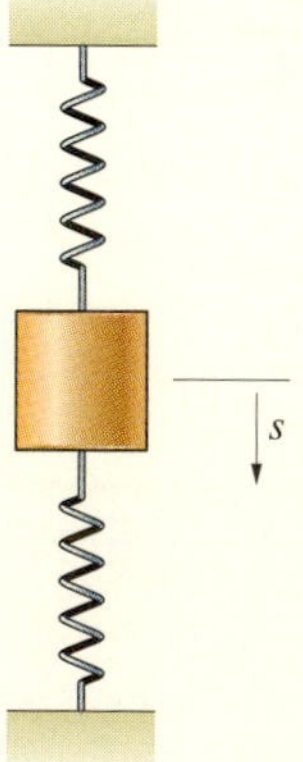

P2.60

2.61 The position of the mass in Problem 2.60 as a function of time is

$$s = \left(\frac{32.2}{50}\right)\left[1 - \cos\left(\sqrt{50}t\right)\right].$$

Prove that this is the correct solution for s; that is, prove that it satisfies the initial conditions and gives the correct acceleration.

2.62 If a spacecraft is 100 mi above the surface of the earth, what initial velocity v_0 straight away from the earth would be required for it to reach the moon's orbit 238,000 mi from the center of the earth? The radius of the earth is 3960 mi. Neglect the effect of the moon's gravity. (See Example 2.5.)

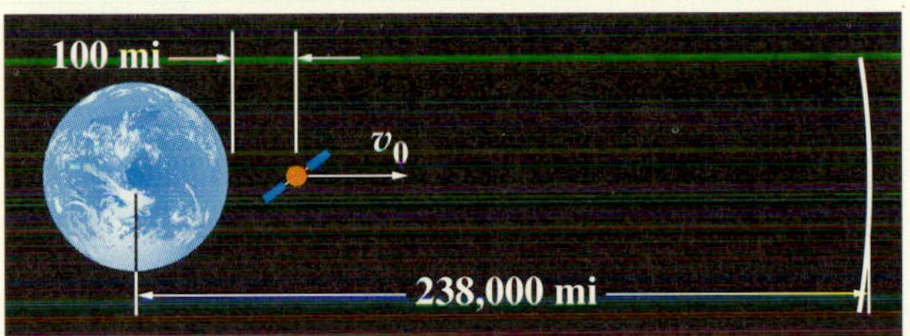

P2.62

2.63 The radius of the moon is $R_M = 1738$ km. The acceleration of gravity at its surface is 1.62 m/s^2. If an object is released from rest 1738 km above the surface of the moon, what is the magnitude of its velocity just before it impacts the surface?

2.64 Using the data in Problem 2.63, determine the escape velocity from the surface of the moon. (See Example 2.5.)

2.65 Suppose that a tunnel could be drilled straight through the earth from the North Pole to the South Pole and the air evacuated. An object dropped from the surface would fall with acceleration $a = -gs/R_E$, where g is the acceleration of gravity at sea level. R_E is the radius of the earth, and s is the distance of the object from the center of the earth. (Gravitational acceleration is equal to zero at the center of the earth and increases linearly with distance from the center.) What is the magnitude of the velocity of the dropped object when it reaches the center of the earth?

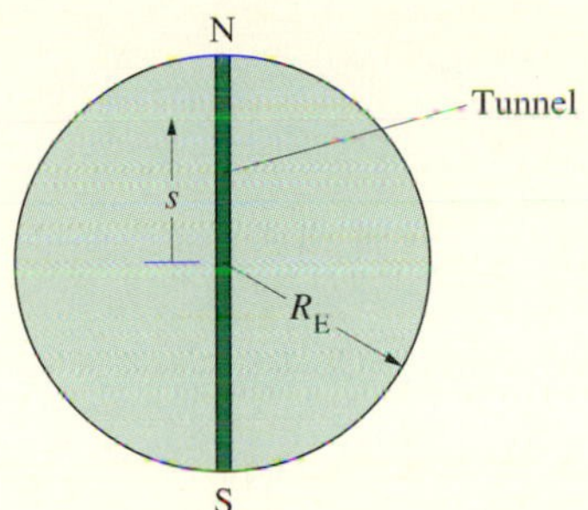

P2.65

2.66 The acceleration of gravity of a hypothetical two-dimensional planet would depend upon the distance s from the center of the planet according to the relation $a = -k/s$, where k is a constant. Let the radius of the planet be R_T and let the magnitude of the acceleration due to gravity at its surface be g_T.

(a) If an object is given an initial outward velocity v_0 at a distance s_0 from the center of the planet, determine its velocity as a function of s.

(b) Show that there is no escape velocity from a two-dimensional planet, thereby explaining why we have never been visited by any two-dimensional beings.

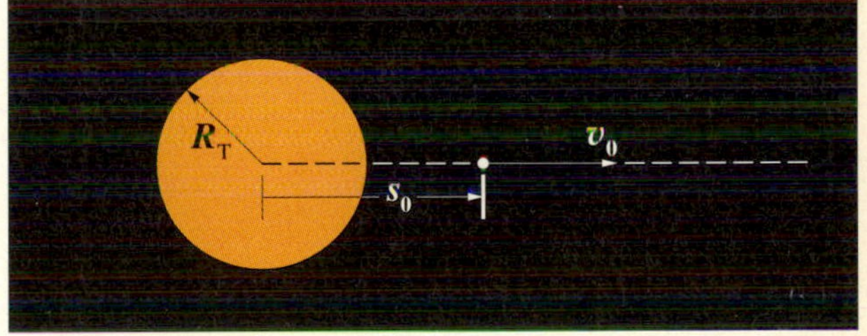

P2.66

2.3 Curvilinear Motion

You have seen that the motion of a point along a straight line is described by the scalars s, v, and a. But if a point describes a **curvilinear** path relative to some reference frame, we must specify its motion in terms of its position, velocity, and acceleration *vectors*. Although the directions and magnitudes of these vectors do not depend on the particular coordinate system used to express them, we will show that the *representations* of these vectors are different in different coordinate systems. We can express many problems in terms of cartesian coordinates, but some situations, including the motions of satellites and rotating machines, can be expressed more naturally using other coordinate systems. In the following sections we show how curvilinear motions of points are analyzed in terms of various coordinate systems.

Cartesian Coordinates

Let **r** be the position vector of a point P relative to the origin O of a cartesian reference frame (Fig. 2.17). The components of **r** are the x, y, z coordinates of P:

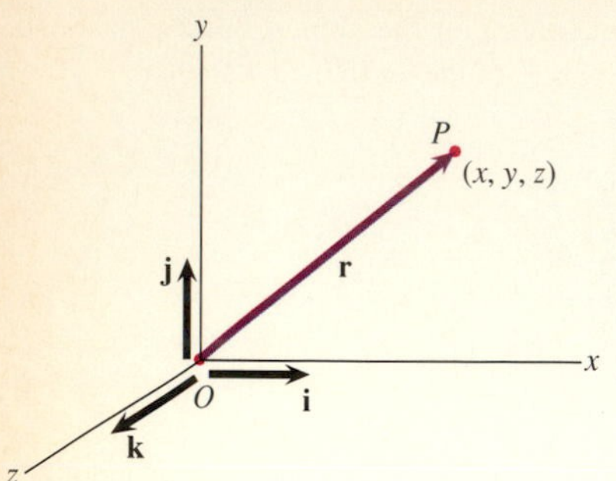

Fig. 2.17
A cartesian coordinate system with origin O.

$$\mathbf{r} = x\mathbf{i} + y\mathbf{j} + z\mathbf{k}.$$

The velocity of P relative to the reference frame is

$$\mathbf{v} = \frac{d\mathbf{r}}{dt} = \frac{dx}{dt}\mathbf{i} + \frac{dy}{dt}\mathbf{j} + \frac{dz}{dt}\mathbf{k}. \tag{2.21}$$

Expressing the velocity in terms of scalar components,

$$\mathbf{v} = v_x\mathbf{i} + v_y\mathbf{j} + v_z\mathbf{k}, \tag{2.22}$$

we obtain scalar equations relating the components of the velocity to the coordinates of P:

$$\boxed{v_x = \frac{dx}{dt}, \qquad v_y = \frac{dy}{dt}, \qquad v_z = \frac{dz}{dt}.} \tag{2.23}$$

The acceleration of P is

$$\mathbf{a} = \frac{d\mathbf{v}}{dt} = \frac{dv_x}{dt}\mathbf{i} + \frac{dv_y}{dt}\mathbf{j} + \frac{dv_z}{dt}\mathbf{k},$$

and by expressing the acceleration in terms of scalar components,

$$\mathbf{a} = a_x\mathbf{i} + a_y\mathbf{j} + a_z\mathbf{k}, \tag{2.24}$$

we obtain the scalar equations

$$\boxed{a_x = \frac{dv_x}{dt}, \qquad a_y = \frac{dv_y}{dt}, \qquad a_z = \frac{dv_z}{dt}.} \tag{2.25}$$

Equations (2.23) and (2.25) describe the motion of a point relative to a cartesian coordinate system. Notice that the equations describing the motion in each coordinate direction are identical in form to the equations that describe the motion of a point along a straight line. As a consequence, you often can analyze the motion in each coordinate direction using the methods you applied to straight-line motion.

The **projectile problem** is the classic example of this kind. If an object is thrown through the air and aerodynamic drag is negligible, it accelerates downward with the acceleration due to gravity. In terms of a fixed cartesian coordinate system with its y axis upward, the acceleration is $a_x = 0$, $a_y = -g$, $a_z = 0$. Suppose that at $t = 0$, the projectile is located at the origin and has velocity v_0 in the x-y plane at an angle θ_0 above the horizontal (Fig. 2.18). At $t = 0$, $x = 0$ and $v_x = v_0 \cos\theta_0$. The acceleration in the x direction is zero,

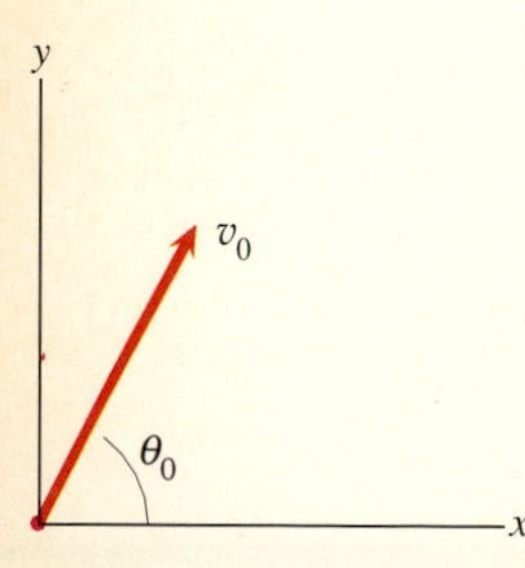

Fig. 2.18
Initial conditions for a projectile problem.

$$a_x = \frac{dv_x}{dt} = 0,$$

so v_x is constant and remains equal to its initial value:

$$v_x = \frac{dx}{dt} = v_0 \cos\theta_0. \tag{2.26}$$

(This result may seem unrealistic. The reason is that your intuition, based upon everyday experience, accounts for drag, whereas this analysis does not.) Integrating this equation,

$$\int_0^x dx = \int_0^t v_0 \cos \theta_0 \, dt,$$

we obtain the x coordinate of the object as a function of time:

$$x = (v_0 \cos \theta_0)t. \tag{2.27}$$

Thus we have determined the position and velocity in the x direction as functions of time without considering the motion in the y or z directions.

At $t = 0$, $y = 0$ and $v_y = v_0 \sin \theta_0$. The acceleration in the y direction is

$$a_y = \frac{dv_y}{dt} = -g.$$

Integrating,

$$\int_{v_0 \sin \theta_0}^{v_y} dv_y = \int_0^t -g \, dt,$$

we obtain

$$v_y = \frac{dy}{dt} = v_0 \sin \theta_0 - gt. \tag{2.28}$$

Integrating this equation,

$$\int_0^y dy = \int_0^t (v_0 \sin \theta_0 - gt) \, dt,$$

we find that the y coordinate as a function of time is

$$y = (v_0 \sin \theta_0)t - \frac{1}{2}gt^2. \tag{2.29}$$

You can see from this analysis that the same vertical velocity and position are obtained by throwing the projectile straight up with initial velocity $v_0 \sin \theta_0$ (Figs. 2.19a, b). The vertical motion is completely independent of the horizontal motion.

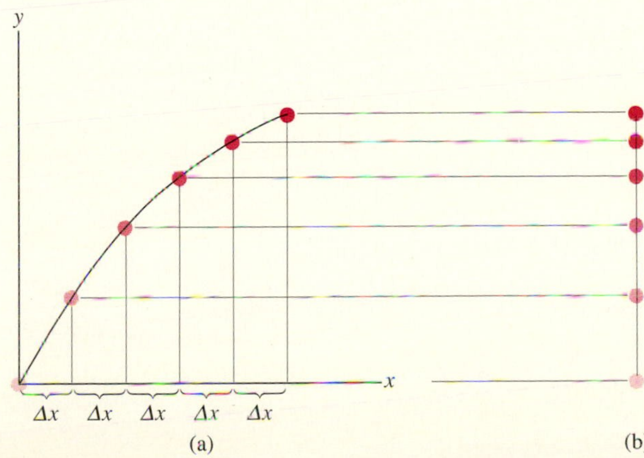

Fig. 2.19

(a) Positions of the projectile at equal time intervals Δt. The distance $\Delta x = v_0 (\cos \theta_0) \Delta t$.
(b) Positions at equal time intervals Δt of a projectile given an initial vertical velocity equal to $v_0 \sin \theta_0$.

By solving Eq. (2.27) for t and substituting the result into Eq. (2.29), we obtain an equation describing the parabolic trajectory of the projectile:

$$y = (\tan \theta_0)x - \frac{g}{2v_0^2 \cos^2 \theta_0} x^2. \qquad (2.30)$$

In the following examples we discuss situations in which you can use Eqs. (2.23) and (2.25) to determine the motions of objects by using cartesian coordinates and analyzing each coordinate direction independently.

Example 2.6

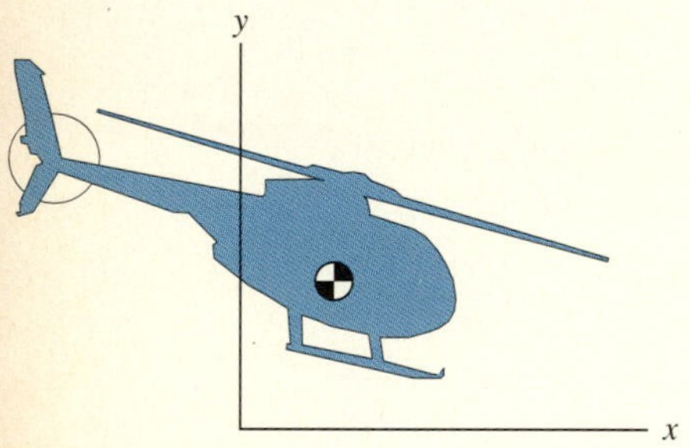

Fig. 2.20

During a test flight in which a helicopter starts from rest at $t = 0$ (Fig. 2.20), accelerometers mounted on board indicate that its components of acceleration from $t = 0$ to $t = 10$ s are closely approximated by

$$a_x = 0.6t \text{ m/s}^2,$$
$$a_y = 1.8 - 0.36t \text{ m/s}^2,$$
$$a_z = 0.$$

Determine the helicopter's velocity and position as functions of time.

STRATEGY

We can analyze the motion in each coordinate direction independently, integrating the acceleration to determine the velocity and then integrating the velocity to determine the position.

SOLUTION

The velocity is zero at $t = 0$, and we assume that $x = y = z = 0$ at $t = 0$. The acceleration in the x direction is

$$a_x = \frac{dv_x}{dt} = 0.6t \text{ m/s}^2.$$

Integrating with respect to time,

$$\int_0^{v_x} dv_x = \int_0^t 0.6t\, dt,$$

we obtain the velocity component v_x as a function of time:

$$v_x = \frac{dx}{dt} = 0.3t^2 \; m/s.$$

Integrating again,

$$\int_0^x dx = \int_0^t 0.3t^2\,dt,$$

we obtain x as a function of time:

$$x = 0.1t^3 \text{ m}.$$

Now we analyze the motion in the y direction in the same way. The acceleration is

$$a_y = \frac{dv_y}{dt} = 1.8 - 0.36t \text{ m/s}^2.$$

Integrating,

$$\int_0^{v_y} dv_y = \int_0^t (1.8 - 0.36t)\,dt,$$

we obtain the velocity,

$$v_y = \frac{dy}{dt} = 1.8t - 0.18t^2 \text{ m/s}.$$

Integrating again,

$$\int_0^y dy = \int_0^t (1.8t - 0.18t^2)\,dt,$$

we determine the position:

$$y = 0.9t^2 - 0.06t^3 \text{ m}.$$

You can easily show that the z components of the velocity and position are $v_z = 0$ and $z = 0$. We show the position of the helicopter as a function of time in Fig. (a).

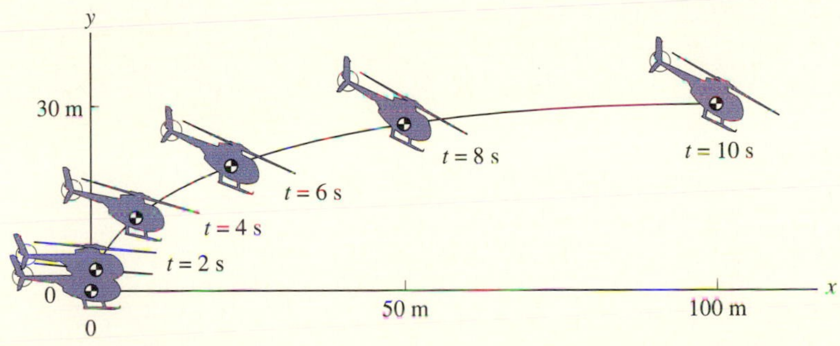

(a) Position of the helicopter at 2-s intervals.

DISCUSSION

This example demonstrates how inertial navigation systems work. They contain accelerometers that measure the x, y, and z components of acceleration. (Gyroscopes maintain the alignments of the accelerometers.) By integrating the acceleration components twice with respect to time, the systems compute changes in the x, y, and z coordinates of the airplane or ship.

Example 2.7

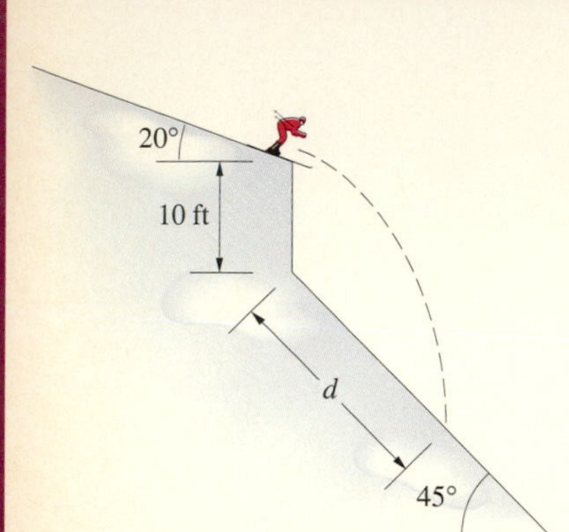

Fig. 2.21

The skier in Fig. 2.21 leaves the 20° surface at 30 ft/s.
(a) Determine the distance d to the point where he lands.
(b) What are the magnitudes of his components of velocity parallel and perpendicular to the 45° surface just before he lands?

STRATEGY

If we neglect aerodynamic drag and treat the skier as a projectile, we can determine his velocity and position as functions of time. By using the equation describing the straight surface on which he lands, we can relate his horizontal and vertical coordinates and thereby obtain an equation for the time at which he lands. Knowing the time, we can determine his position. We can also determine the two components of his velocity just before landing. By using the result that the component of a vector **U** in the direction of a unit vector **e** is $(\mathbf{e} \cdot \mathbf{U})\,\mathbf{e}$, we will determine the components of his velocity parallel and perpendicular to the 45° surface.

SOLUTION

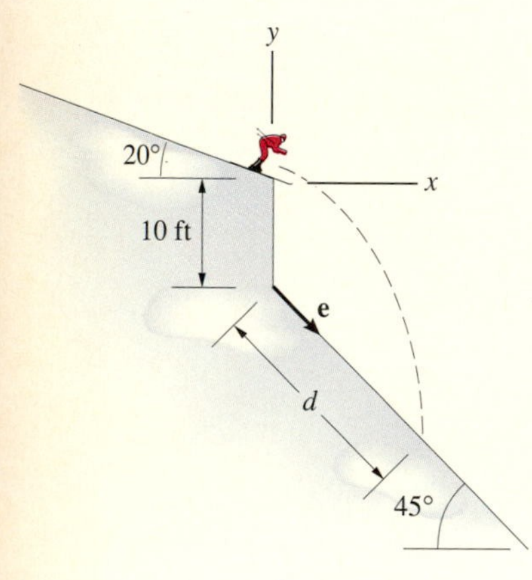

(a)

(a) In Fig. (a) we introduce a coordinate system with its origin at the skier's initial position. The components of his initial velocity are

$$v_x = (30)\cos 20^\circ = 28.2 \text{ ft/s},$$
$$v_y = -(30)\sin 20^\circ = -10.3 \text{ ft/s}.$$

The x component of his acceleration is zero, so the x component of his velocity is constant and his x coordinate as a function of time is

$$x = 28.2t \text{ ft}.$$

The y component of his acceleration is

$$a_y = \frac{dv_y}{dt} = -32.2 \text{ ft/s}^2.$$

We integrate to determine the y component of his velocity as a function of time,

$$\int_{-10.3}^{v_y} dv_y = \int_0^t -32.2\, dt,$$

obtaining

$$v_y = \frac{dy}{dt} = -10.3 - 32.2t \text{ ft/s}.$$

Then we integrate to determine his y coordinate as a function of time,

$$\int_0^y dy = \int_0^t (-10.3 - 32.2t)\, dt,$$

obtaining

$$y = -10.3t - 16.1t^2 \text{ ft}.$$

The slope of the flat surface on which he lands is -1, so we can write the linear equation describing it in the form $y = (-1)x + A$, where A is a constant. At $x = 0$, the y coordinate of the surface is -10 ft, so the constant $A = -10$ and the equation describing the surface is

$$y = -x - 10 \text{ ft}.$$

By substituting our equations for x and y as functions of time into this equation, we obtain an equation for the time at which he lands:

$$-10.3t - 16.1t^2 = -28.2t - 10.$$

Solving this quadratic equation, we find the two roots $t = -0.409$ s and $t = 1.52$ s. The positive root is the time at which he lands. Substituting $t = 1.52$ s into our equations for x and y as functions of time, we find that his x and y coordinates when he lands are

$$x = 28.2(1.52) = 42.9 \text{ ft},$$
$$y = -10.3(1.52) - 16.1(1.52)^2 = -52.8 \text{ ft}.$$

Then the distance d is

$$d = \sqrt{(42.9)^2 + (52.8 - 10)^2} = 60.6 \text{ ft}.$$

(b) His components of velocity just before he lands are

$$v_x = 28.2 \text{ ft/s},$$
$$v_y = -10.3 - 32.2(1.52) = -59.2 \text{ ft/s}.$$

Let $\mathbf{e}$ be a unit vector parallel to the slope on which he lands and pointed down the slope (Fig. a):

$$\mathbf{e} = \cos 45^\circ\, \mathbf{i} - \sin 45^\circ\, \mathbf{j}.$$

The component of his velocity vector parallel to the surface is

$$\begin{aligned}(\mathbf{e} \cdot \mathbf{v})\mathbf{e} &= [(\cos 45^\circ\, \mathbf{i} - \sin 45^\circ\, \mathbf{j})\cdot(28.2\mathbf{i} - 59.2\mathbf{j})]\mathbf{e} \\ &= [(\cos 45^\circ)(28.2) + (-\sin 45^\circ)(-59.2)]\mathbf{e} \\ &= 61.8\mathbf{e} \text{ (ft/s)}.\end{aligned}$$

The magnitude of his velocity parallel to the surface is 61.8 ft/s. The component of his velocity vector perpendicular to the surface is

$$\begin{aligned}\mathbf{v} - (\mathbf{e} \cdot \mathbf{v})\mathbf{e} &= (28.2\mathbf{i} - 59.2\mathbf{j}) - 61.8(\cos 45^\circ\, \mathbf{i} - \sin 45^\circ\, \mathbf{j}) \\ &= -15.5\mathbf{i} - 15.5\mathbf{j} \text{ (ft/s)}.\end{aligned}$$

The magnitude of his velocity perpendicular to the surface is

$$\sqrt{(-15.5)^2 + (-15.5)^2} = 21.9 \text{ ft/s}.$$

Problems

2.67 The cartesian coordinates of a point (in meters) are $x = 2t + 4$, $y = t^3 - 2t$, $z = 4t^2 - 4$, where t is in seconds. What are its velocity and acceleration at $t = 4$ s?

Strategy: Since the cartesian coordinates are given as functions of time, you can use Eqs. (2.23) to determine the components of the velocity as functions of time and then use Eqs. (2.25) to determine the components of the acceleration as functions of time.

2.68 The velocity of a point is $\mathbf{v} = 2\mathbf{i} + 3t^2\mathbf{j}$ (ft/s). At $t = 0$ its position is $\mathbf{r} = -\mathbf{i} + 2\mathbf{j}$ (ft). What is its position at $t = 2$ s?

2.69 The acceleration components of a point (in ft/s^2) are $a_x = 3t^2$, $a_y = 6t$, and $a_z = 0$. At $t = 0$, $x = 5$ ft, $v_x = 3$ ft/s, $y = 1$ ft, $v_y = -2$ ft/s, $z = 0$, and $v_z = 0$. What are its position vector and velocity vector at $t = 3$ s?

2.70 The acceleration components of an object (in m/s^2) are $a_x = 2t$, $a_y = 4t^2 - 2$, and $a_z = -6$. At $t = 0$ the position of the object is $\mathbf{r} = 10\mathbf{j} - 10\mathbf{k}$ (m) and its velocity is $\mathbf{v} = 2\mathbf{i} - 4\mathbf{j}$ (m/s). Determine its position when $t = 4$ s.

2.71 Suppose you are designing a mortar to send a rescue line from a Coast Guard boat to ships in distress. The line is attached to a weight that is fired by the mortar. The mortar is to be mounted so that it fires at 45° above the horizontal. If you neglect aerodynamic drag and the weight of the line for your preliminary design and assume a muzzle velocity of 100 ft/s at $t = 0$, what are the x and y coordinates of the weight as functions of time?

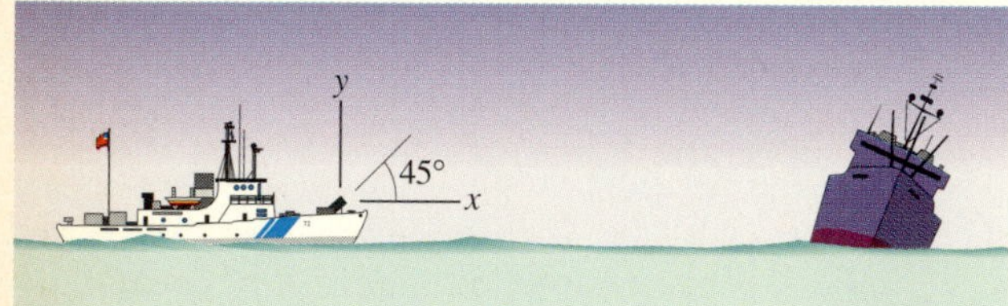

P2.71

2.72 In Problem 2.71, what must the mortar's muzzle velocity be to reach ships 1000 ft away?

2.73 If a stone is thrown horizontally from the top of a 100-ft-tall building at 50 ft/s, at what horizontal distance from the point at which it is thrown does it hit the ground? (Assume level ground.) What is the magnitude of its velocity just before it hits?

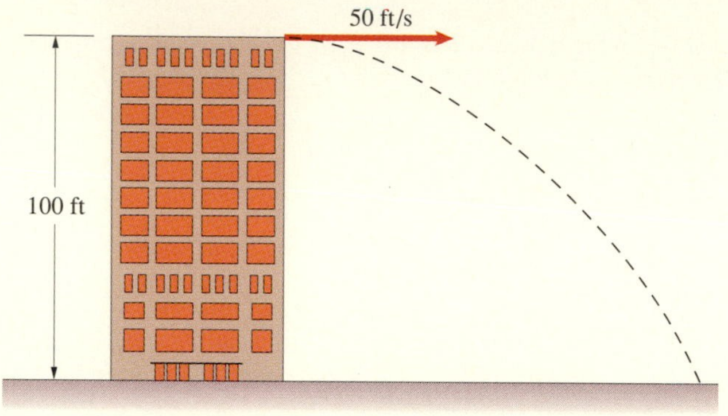

P2.73

2.74 A projectile is launched from ground level with an initial velocity v_0. What initial angle θ_0 above the horizontal causes the range R to be a maximum, and what is the maximum range?

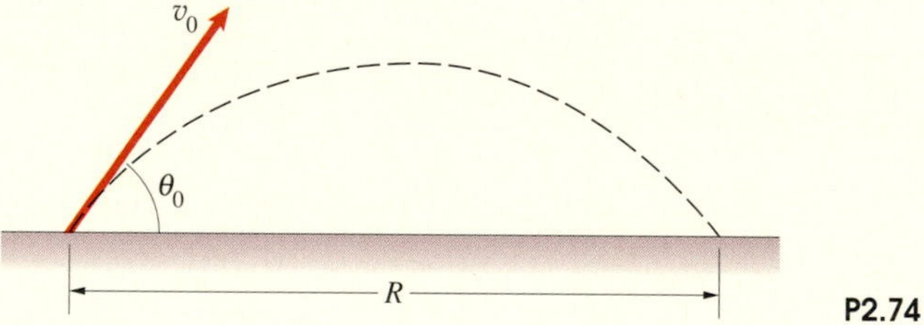

P2.74

2.75 A pilot wants to drop supplies to remote locations in the Australian outback. He intends to fly horizontally and release the packages with no vertical velocity. Derive an equation for the horizontal distance d at which he should release the package in terms of the airplane's velocity v_0 and altitude h.

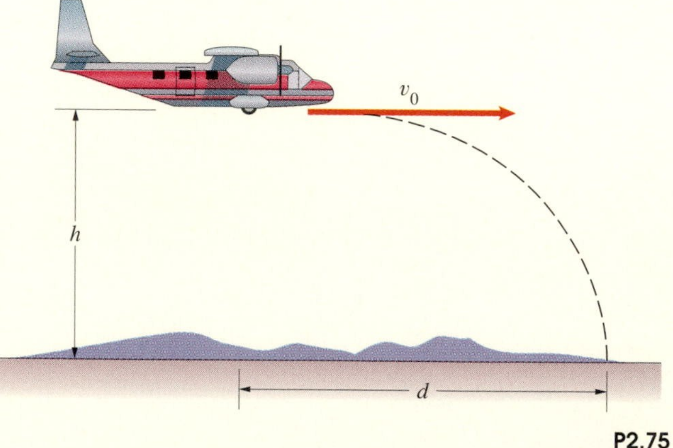

P2.75

2.76 If the pitching wedge the golfer is using gives the ball an initial angle $\theta_0 = 50°$, what range of velocities v_0 will cause the ball to land within 3 ft of the hole? (Assume that the hole lies in the plane of the ball's trajectory.)

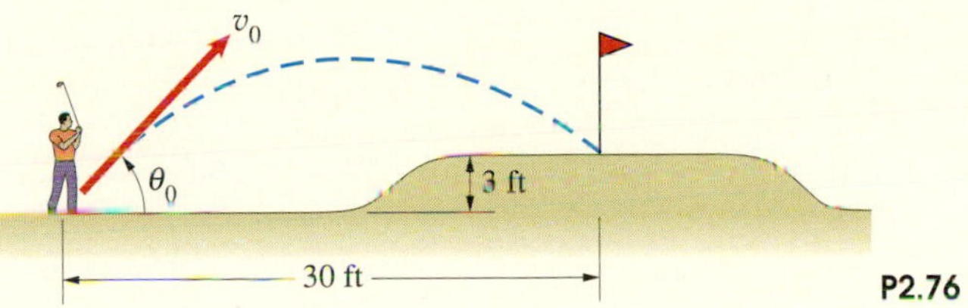

P2.76

2.77 A batter strikes a baseball at 3 ft above home plate and pops it up at an angle of 60° above the horizontal. The second baseman catches it at 6 ft above second base. What was the ball's initial velocity?

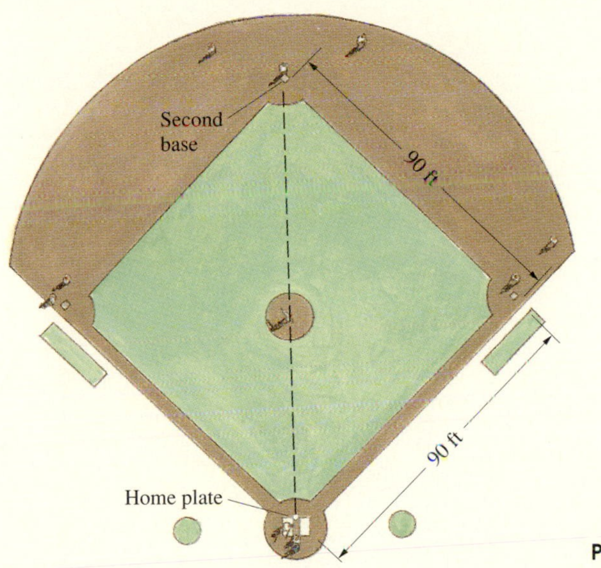

P2.77

2.78 A baseball pitcher releases a fastball with an initial velocity $v_0 = 90$ mi/hr. Let θ be the initial angle of the ball's velocity vector above the horizontal. When it is released, the ball is 6 ft above the ground and 58 ft from the batter's plate. The batter's strike zone extends from 1 ft 10 in. above the ground to 4 ft 6 in. above the ground. Neglecting aerodynamic effects, determine whether the ball will hit the strike zone: (a) if $\theta = 1°$; (b) if $\theta = 2°$.

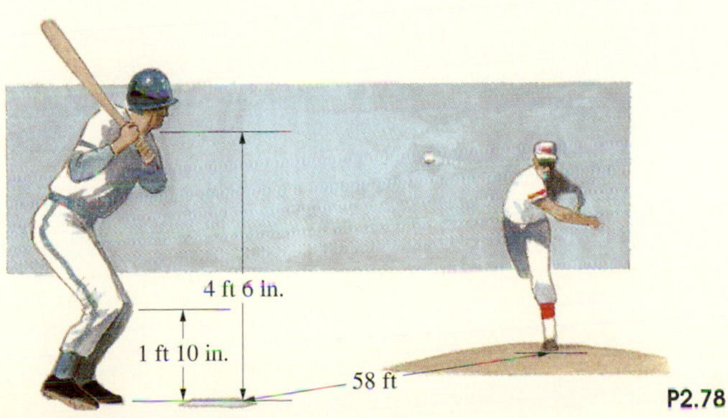

P2.78

2.79 In Problem 2.78, assume that the pitcher releases the ball at an angle $\theta = 1°$ above the horizontal and determine the range of velocities v_0 (in ft/s) within which he must release the ball to hit the strike zone.

2.80 A zoology graduate student is provided with a bow and an arrow tipped with a syringe of sedative and is assigned to measure the temperature of a black rhinoceros (*Diceros bicornis*). The range of his bow when it is fully drawn and aimed 45° above the horizontal is 100 m. A truculent rhino suddenly charges straight toward him at 30 km/hr. If he fully draws his bow and aims 20° above the horizontal, how far away should the rhino be when he releases the arrow?

P2.80

2.81 The crossbar of the goalposts in American football is $y_c = 10$ ft above the ground. To kick a field goal, the ball must go between the two uprights supporting the crossbar and be above the crossbar when it does so. Suppose that the kicker attempts a 40-yd field goal ($x_c = 120$ ft) and kicks the ball with initial velocity $v_0 = 70$ ft/s and angle $\theta_0 = 40°$. By what vertical distance does the ball clear the crossbar?

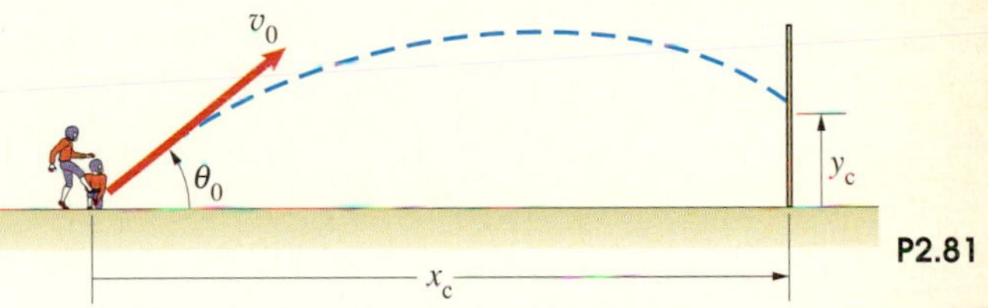

P2.81

2.82 In Problem 2.81, suppose that you want to determine the minimum initial velocity of the football and the corresponding initial angle needed to make a field goal.

(a) Show that the initial angle satisfies the equation

$$\tan\theta_0 = \frac{y_c}{x_c} + \sqrt{\left(\frac{y_c}{x_c}\right)^2 + 1}.$$

(b) Determine the minimum velocity and initial angle needed to make a 40-yd field goal.

2.83 The cliff divers of Acapulco, Mexico, must time their dives so that they enter the water at the crest (high point) of a wave. The crests of the waves are 2 ft above the mean water depth $h = 12$ ft, and the horizontal velocity of the waves is $\sqrt{gh}$. The diver's aiming point is 6 ft out from the base of the cliff. Assume that his velocity is horizontal when he begins the dive.
(a) What is the magnitude of his velocity in miles per hour when he enters the water?
(b) How far from his aiming point should a wave crest be when he dives in order for him to enter the water at the crest?

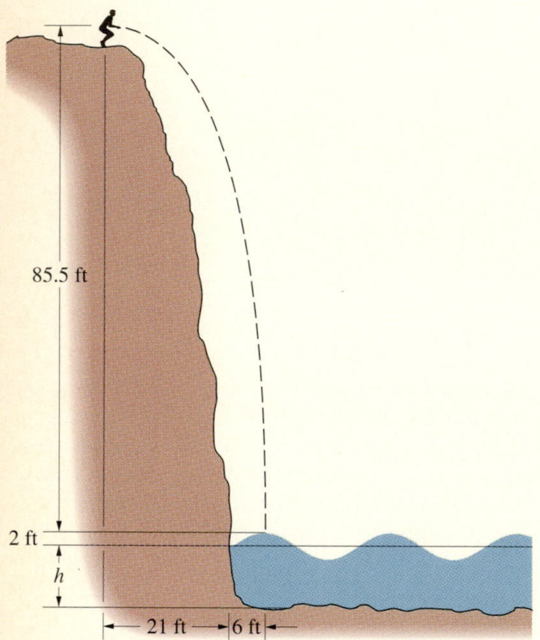

P2.83

2.84 A projectile is launched at 10 m/s from a sloping surface. Determine the range R.

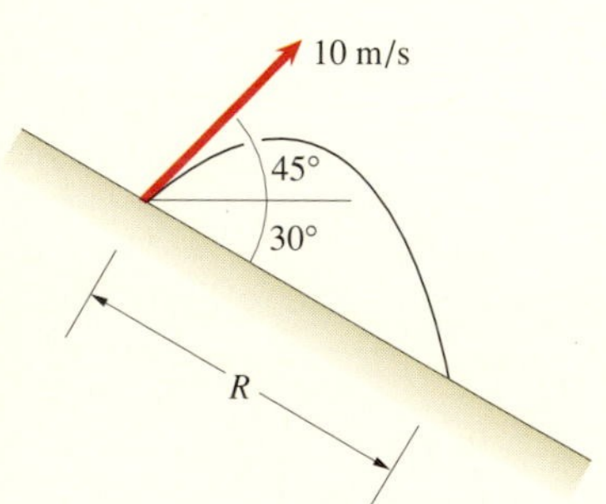

P2.84

2.85 A projectile is launched at 100 ft/s at 60° above the horizontal. The surface on which it lands is described by the equation shown. Determine the x coordinate of the point of impact.

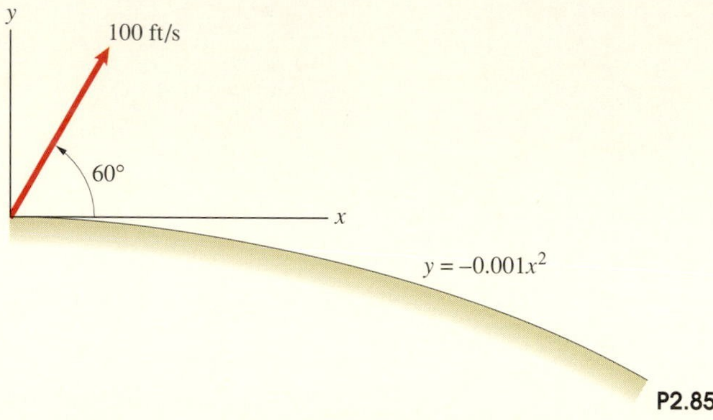

P2.85

2.86 At $t = 0$, a steel ball in a tank of oil is given a horizontal velocity $\mathbf{v} = 2\mathbf{i}$ (m/s). The components of its acceleration in m/s^2 are $a_x = -1.2v_x$, $a_y = -8 - 1.2v_y$, $a_z = -1.2v_z$. What is the velocity of the ball at $t = 1$ s?

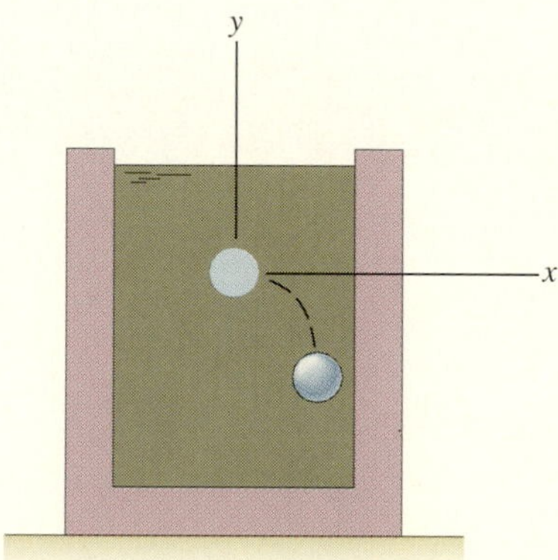

P2.86

2.87 In Problem 2.86, what is the position of the ball at $t = 1$ s relative to its position at $t = 0$?

2.88 You must design a device for an assembly line that launches small parts through the air into a bin. The launch point is $x = 200$ mm, $y = -50$ mm, $z = -100$ mm. (The y axis is vertical and positive upward.) To land in the bin, the parts must pass through the point $x = 600$ mm, $y = 200$ mm, $z = 100$ mm *moving horizontally*. Determine the components of velocity the launcher must give the parts.

2.89 If $y = 150$ mm, $dy/dt = 300$ mm/s, and $d^2y/dt^2 = 0$, what are the magnitudes of the velocity and acceleration of point P?

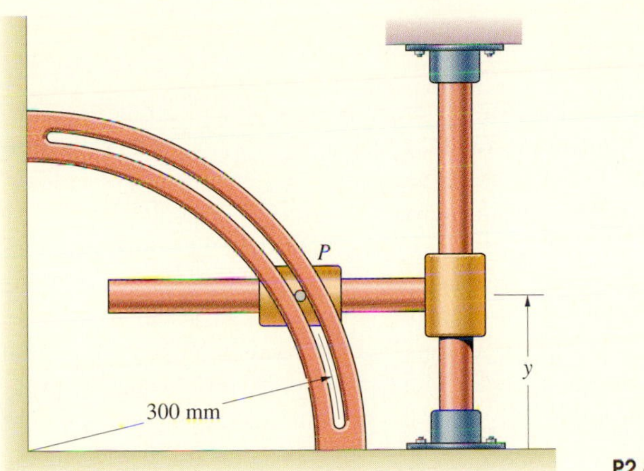

P2.89

2.90 A car travels at a constant speed of 100 km/hr on a straight road of increasing grade whose vertical profile can be approximated by the equation shown. When the car's horizontal coordinate is $x = 400$ m, what is its acceleration?

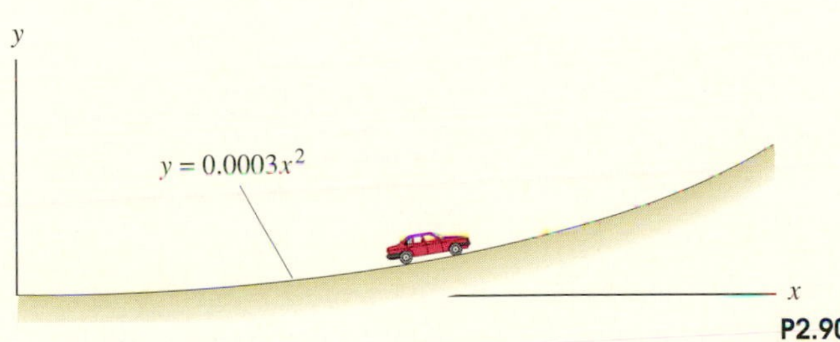

P2.90

2.91 Suppose that a projectile has the initial conditions shown in Fig. 2.18. Show that in terms of the $x'y'$ coordinate system with its origin at the highest point of the trajectory, the equation describing the trajectory is

$$y' = -\frac{g}{2v_0^2 \cos^2 \theta_0}(x')^2.$$

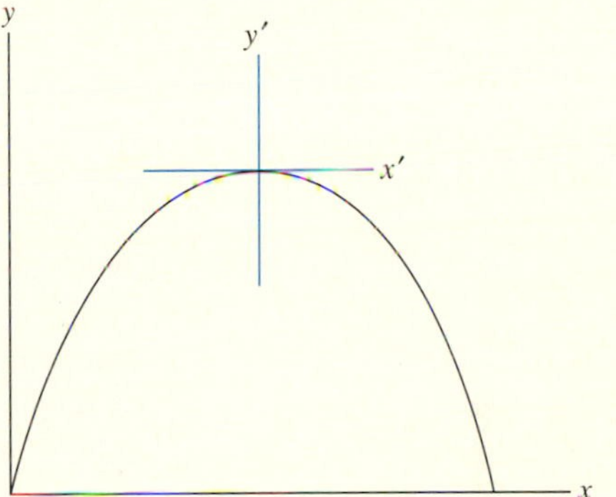

P2.91

2.92 The acceleration components of a point are $a_x = -4 \cos 2t$, $a_y = -4 \sin 2t$, $a_z = 0$. At $t = 0$ its position and velocity are $\mathbf{r} = \mathbf{i}$, $\mathbf{v} = 2\mathbf{j}$. Show that: (a) the magnitude of the velocity is constant; (b) the velocity and acceleration vectors are perpendicular; (c) the magnitude of the acceleration is constant and points toward the origin; (d) the trajectory of the point is a circle with its center at the origin.

Angular Motion

We have seen that in some cases the curvilinear motion of a point can be analyzed using cartesian coordinates. In the following sections we describe problems that can be analyzed more simply in terms of other coordinate systems. To help you understand our discussion of these alternative coordinate systems, we introduce two preliminary topics in this section: the angular motion of a line in a plane and the time derivative of a unit vector rotating in a plane.

Angular Motion of a Line

We can specify the angular position of a line L in a particular plane relative to a reference line L_0 in the plane by an angle θ (Fig. 2.22). The **angular velocity** of L relative to L_0 is defined by

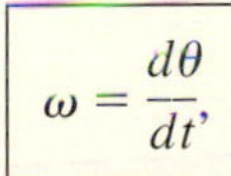

$$\omega = \frac{d\theta}{dt}, \qquad (2.31)$$

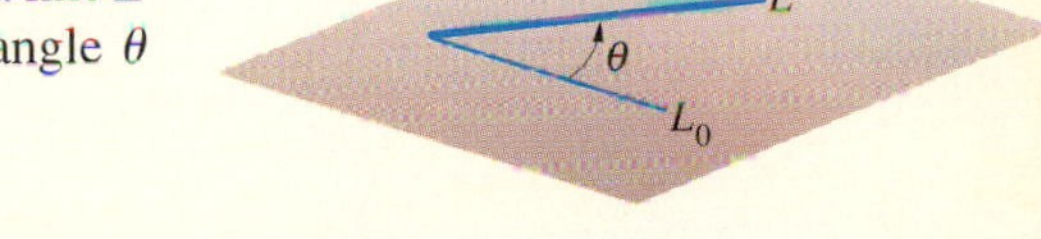

Fig. 2.22
A line L and a reference line L_0 in a plane.

and the **angular acceleration** of L relative to L_0 is defined by

$$\boxed{\alpha = \frac{d\omega}{dt} = \frac{d^2\theta}{dt^2}.} \qquad (2.32)$$

The dimensions of the angular position, angular velocity, and angular acceleration are radians (rad), rad/s, and rad/s^2, respectively. Although these quantities are often expressed in terms of degrees or revolutions instead of radians, you should convert them into radians before using them in calculations.

Notice the analogy between Eqs. (2.31) and (2.32) and the equations relating the position, velocity, and acceleration of a point along a straight line (Table 2.2). In each case the position is specified by a single scalar coordinate, which can be positive or negative. (In Fig. 2.22 the counterclockwise direction is positive.) Because the equations are identical in form you can analyze problems involving angular motion of a line by the same methods you applied to straight-line motion.

Table 2.2 The equations governing straight-line motion and the equations governing the angular motion of a line are identical in form.

Straight-Line Motion	Angular Motion
$v = \frac{ds}{dt}$	$\omega = \frac{d\theta}{dt}$
$a = \frac{dv}{dt} = \frac{d^2s}{dt^2}$	$\alpha = \frac{d\omega}{dt} = \frac{d^2\theta}{dt^2}$

Rotating Unit Vector The directions of the unit vectors **i**, **j**, and **k** relative to the cartesian reference frame are constant. However, in other coordinate systems the unit vectors used to describe the motion of a point rotate as the point moves. To obtain expressions for the velocity and acceleration in such coordinate systems, we must know the time derivative of a rotating unit vector.

We can describe the angular motion of a unit vector **e** in a plane just as we described the angular motion of a line. The direction of **e** relative to a reference line L_0 is specified by the angle θ in Fig. 2.23(a), and the rate of rotation of **e** relative to L_0 is specified by the angular velocity

$$\omega = \frac{d\theta}{dt}.$$

The time derivative of **e** is defined by

$$\frac{d\mathbf{e}}{dt} = \lim_{\Delta t \to 0} \frac{\mathbf{e}(t + \Delta t) - \mathbf{e}(t)}{\Delta t}.$$

Figure 2.23(b) shows the vector **e** at time t and at time $t + \Delta t$. The change in **e** during this interval is $\Delta\mathbf{e} = \mathbf{e}(t + \Delta t) - \mathbf{e}(t)$, and the angle through which **e** rotates is $\Delta\theta = \theta(t + \Delta t) - \theta(t)$. The triangle in Fig. 2.23(b) is isosceles, so the magnitude of $\Delta\mathbf{e}$ is

$$|\Delta\mathbf{e}| = 2|\mathbf{e}|\sin(\Delta\theta/2) = 2\sin(\Delta\theta/2).$$

To write the vector $\Delta\mathbf{e}$ in terms of this expression, we introduce a unit vector $\mathbf{n}$ that points in the direction of $\Delta\mathbf{e}$ (Fig. 2.23b):

$$\Delta\mathbf{e} = |\Delta\mathbf{e}|\mathbf{n} = 2\sin(\Delta\theta/2)\,\mathbf{n}.$$

In terms of this expression, the time derivative of $\mathbf{e}$ is

$$\frac{d\mathbf{e}}{dt} = \lim_{\Delta t\to 0}\frac{\Delta\mathbf{e}}{\Delta t} = \lim_{\Delta t\to 0}\frac{2\sin(\Delta\theta/2)\,\mathbf{n}}{\Delta t}.$$

To evaluate this limit, we write it in the form

$$\frac{d\mathbf{e}}{dt} = \lim_{\Delta t\to 0}\frac{\sin(\Delta\theta/2)}{\Delta\theta/2}\frac{\Delta\theta}{\Delta t}\mathbf{n}.$$

In the limit as Δt approaches zero, $\sin(\Delta\theta/2)/(\Delta\theta/2)$ equals one, $\Delta\theta/\Delta t$ equals $d\theta/dt$, and the unit vector $\mathbf{n}$ is perpendicular to $\mathbf{e}(t)$ (Fig. 2.23c). Therefore the time derivative of $\mathbf{e}$ is

$$\frac{d\mathbf{e}}{dt} = \frac{d\theta}{dt}\mathbf{n} = \omega\mathbf{n}, \tag{2.33}$$

where $\mathbf{n}$ is a unit vector that is perpendicular to $\mathbf{e}$ and points in the positive θ direction (Fig. 2.23d). In the following sections we use this result in deriving expressions for the velocity and acceleration of a point in different coordinate systems.

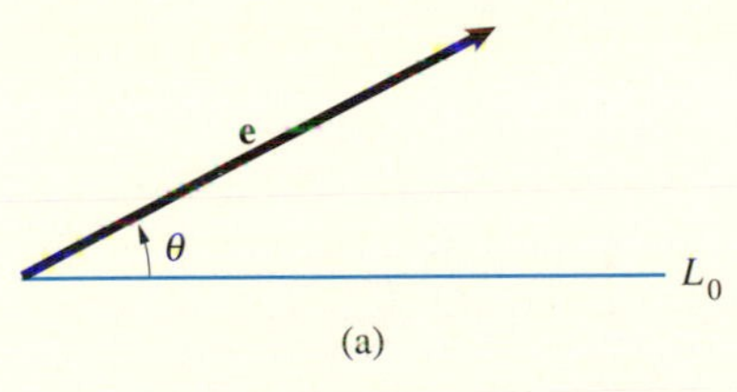

(a)

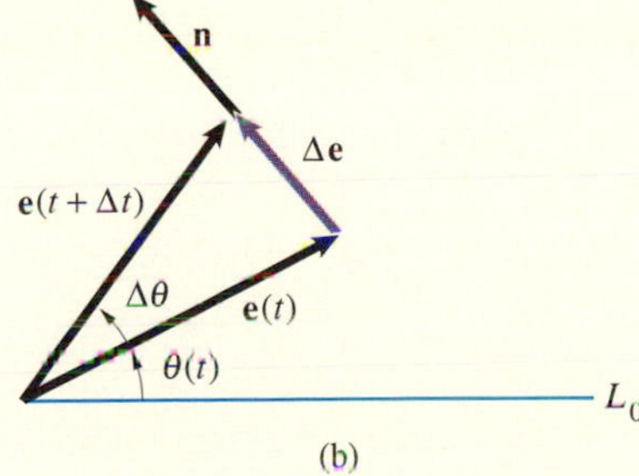

(b)

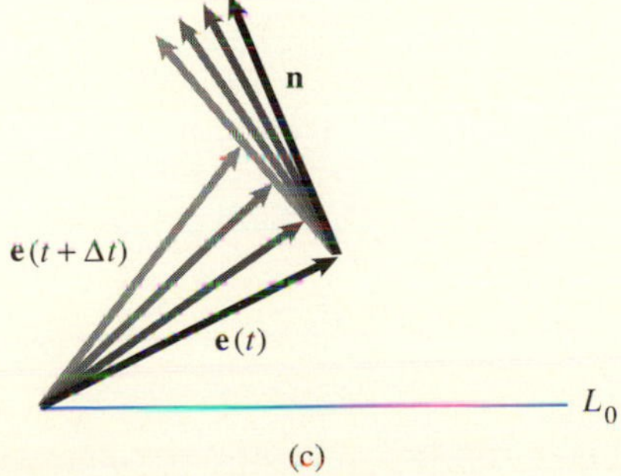

(c)

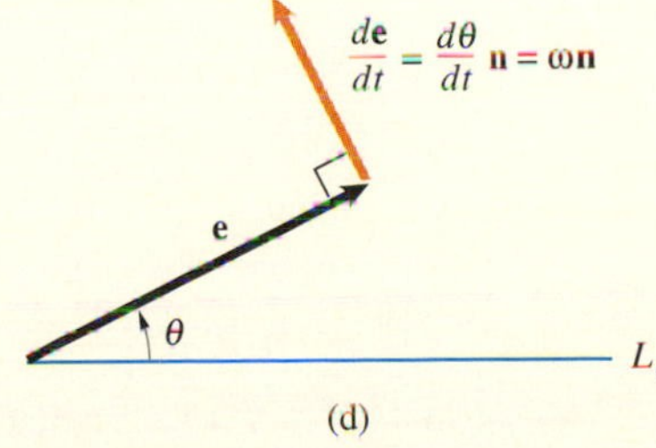

(d)

Fig. 2.23
(a) A unit vector $\mathbf{e}$ and reference line L_0.
(b) The change $\Delta\mathbf{e}$ in $\mathbf{e}$ from t to $t + \Delta t$.
(c) As Δt goes to zero, $\mathbf{n}$ becomes perpendicular to $\mathbf{e}(t)$.
(d) The time derivative of $\mathbf{e}$.

Example 2.8

The rotor of a jet engine is rotating at 10,000 rpm (revolutions per minute) when the fuel is shut off. The ensuing angular acceleration is $\alpha = -0.02\omega$, where ω is the angular velocity in rad/s.

(a) How long does it take the rotor to slow to 1000 rpm?

(b) How many revolutions does the rotor turn while decelerating to 1000 rpm?

STRATEGY

To analyze the angular motion of the rotor, we define a line L that is fixed to the rotor and perpendicular to its axis (Fig. 2.24). Then we examine the motion of L relative to the reference line L_0. The angular position, velocity, and acceleration of L describe the angular motion of the rotor.

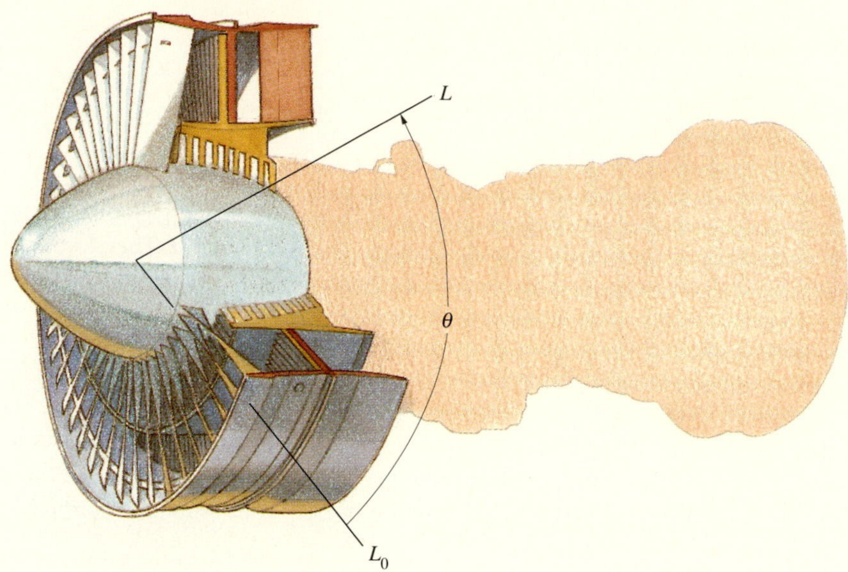

Fig. 2.24

Introducing a line L and reference line L_0 to specify the angular position of the rotor.

SOLUTION

The conversion from rpm to rad/s is

$$1 \text{ rpm} = 1 \text{ revolution/min} \times \left(\frac{2\pi \text{ rad}}{1 \text{ revolution}}\right) \times \left(\frac{1 \text{ min}}{60 \text{ s}}\right)$$

$$= \frac{\pi}{30} \text{ rad/s}.$$

(a) The angular acceleration is

$$\alpha = \frac{d\omega}{dt} = -0.02\omega.$$

We separate variables,

$$\frac{d\omega}{\omega} = -0.02\, dt,$$

and integrate, defining $t = 0$ to be the time at which the fuel is turned off:

$$\int_{10,000\pi/30}^{1000\pi/30} \frac{d\omega}{\omega} = \int_0^t -0.02\, dt.$$

Evaluating the integrals and solving for t, we obtain

$$t = \left(\frac{1}{0.02}\right) \ln\left(\frac{10,000\pi/30}{1000\pi/30}\right) = 115 \text{ s}.$$

(b) We write the angular acceleration as

$$\alpha = \frac{d\omega}{dt} = \frac{d\omega}{d\theta}\frac{d\theta}{dt} = \frac{d\omega}{d\theta}\omega = -0.02\omega,$$

separate variables,

$$d\omega = -0.02\, d\theta,$$

and integrate, defining $\theta = 0$ to be the angular position at which the fuel is turned off:

$$\int_{10,000\pi/30}^{1000\pi/30} d\omega = \int_0^\theta -0.02\, d\theta.$$

Solving for θ, we obtain

$$\theta = \left(\frac{1}{0.02}\right)[(10,000\pi/30) - (1000\pi/30)]$$

$$= 15,000\pi \text{ rad} = 7500 \text{ revolutions}.$$

Problems

2.93 Suppose that the jet engine in Example 2.8 starts from rest and has a constant angular acceleration $\alpha = 5$ rad/s^2.
(a) How long does it take to reach an angular velocity of 10,000 rpm?
(b) How many revolutions does it turn in that time?

2.94 Let L be a line from the center of the earth to a fixed point on the equator and let L_0 denote a fixed reference direction. The figure shows the earth seen from above the North Pole.
(a) Is $d\theta/dt$ positive or negative?
(b) What is the magnitude of $d\theta/dt$ in rad/s?

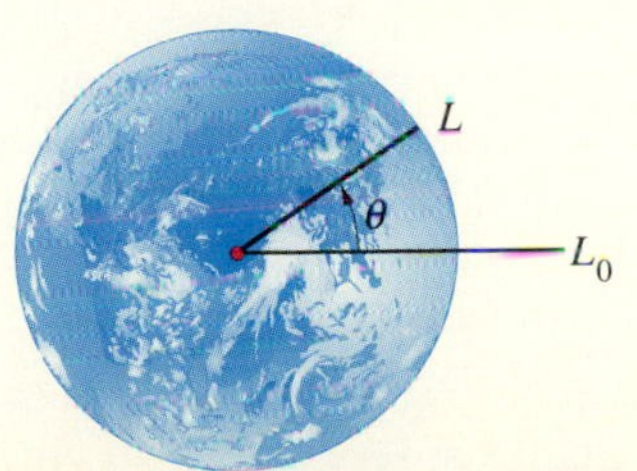

P2.94

2.95 The angle between a line L and a reference line L_0 is $\theta = 2t^2$ rad.
(a) What are the angular velocity and angular acceleration of L relative to L_0 at $t = 6$ s?
(b) How many revolutions does L rotate relative to L_0 during the interval of time from $t = 0$ to $t = 6$ s?

Strategy: Use Eqs. (2.31) and (2.32) to determine the angular velocity and angular acceleration as functions of time.

2.96 The angle θ between the bar and the horizontal line is $\theta = t^3 - 2t^2 + 4$ degrees. Determine the angular velocity and angular acceleration of the bar at $t = 10$ s.

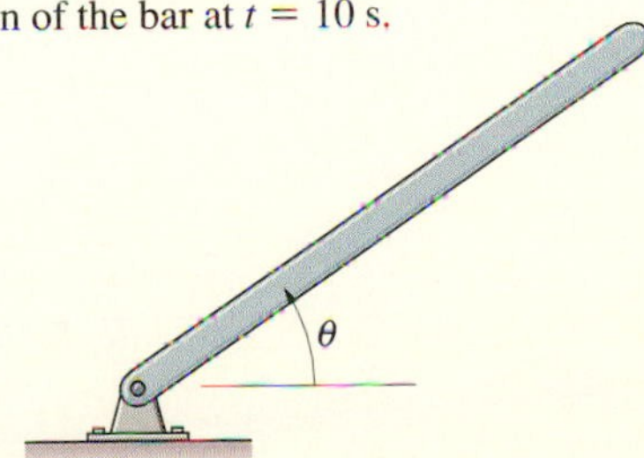

P2.96

2.97 The angular acceleration of a line L relative to a reference line L_0 is $\alpha = 30 - 6t$ rad/s^2. When $t = 0$, $\theta = 0$ and $\omega = 0$. What is the maximum angular velocity of L relative to L_0 during the interval of time from $t = 0$ to $t = 10$ s?

2.98 A gas turbine starts rotating from rest at $t = 0$ and has angular acceleration $\alpha = 6t$ rad/s^2 for 3 s. It then slows down with constant angular deceleration $\alpha = -3$ rad/s^2 until it stops.
(a) What maximum angular velocity does it attain?
(b) Through what total angle does it turn?

2.99 The rotor of an electric generator is rotating at 200 rpm when the motor is turned off. Due to frictional effects, the angular deceleration of the rotor after it is turned off is $\alpha = -0.01\omega$ rad/s^2, where ω is the angular velocity in rad/s. How many revolutions does the rotor turn after the motor is turned off?

2.100 The needle of a measuring instrument is connected to a torsional spring that subjects it to an angular acceleration $\alpha = -4\theta$ rad/s^2, where θ is the needle's angular position in radians relative to a reference direction. If the needle is released from rest at $\theta = 1$ rad, what is its angular velocity at $\theta = 0$?

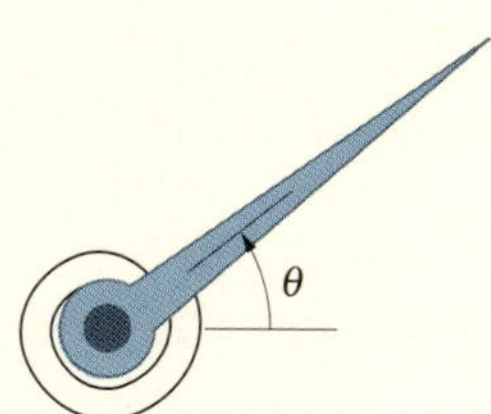

P2.100

2.101 The angle θ measures the direction of the unit vector **e** relative to the x axis. Given that $\omega = d\theta/dt = 2$ rad/s, determine the vector $d\mathbf{e}/dt$: (a) when $\theta = 0$; (b) when $\theta = 90°$; (c) when $\theta = 180°$.

Strategy: You can obtain these results either by using Eq. (2.33) or by expressing **e** in terms of its x and y components and taking its time derivative.

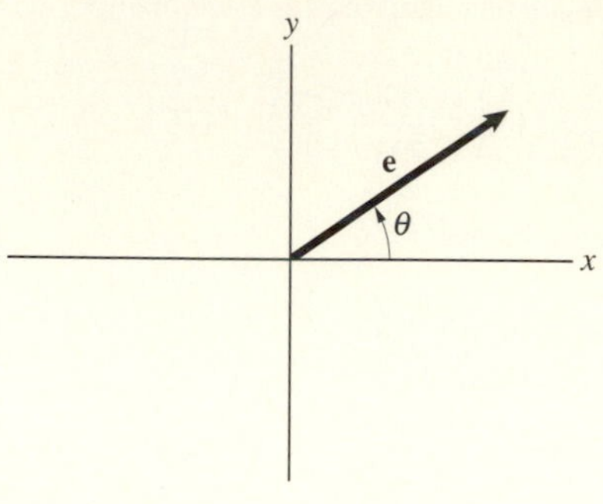

P2.101

2.102 In Problem 2.101, suppose that the angle $\theta = 2t^2$ rad. What is the vector $d\mathbf{e}/dt$ at $t = 4$ s?

2.103 The line OP is of constant length R. The angle $\theta = \omega_0 t$, where ω_0 is a constant.
(a) Use the relations

$$v_x = \frac{dx}{dt}, \qquad v_y = \frac{dy}{dt}$$

to determine the velocity of point P relative to O.
(b) Use Eq. (2.33) to determine the velocity of point P relative to O, and confirm that your result agrees with the result of (a).

Strategy: In part (b), write the position vector of P relative to O as $\mathbf{r} = R\mathbf{e}$, where **e** is a unit vector that points from O toward P.

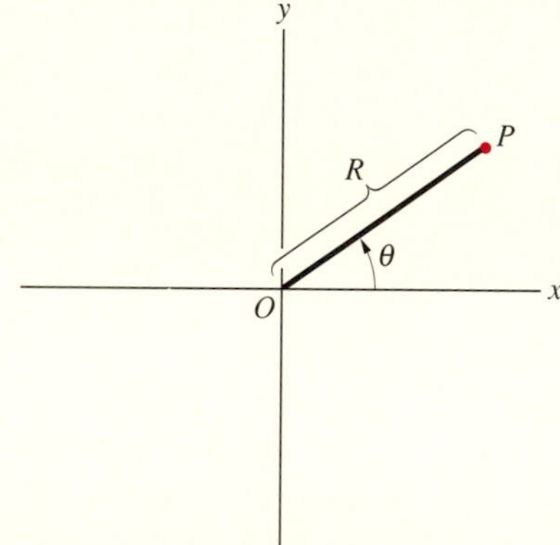

P2.103

Normal and Tangential Components

In this method of describing curvilinear motion, we specify the position of a point by a coordinate measured *along its path*, and express the velocity and acceleration in terms of their components tangential and normal (perpendicular) to the path. Normal and tangential components are particularly useful when a point moves along a circular path. Furthermore, they provide unique insight into the character of the velocity and acceleration in curvilinear motion. We first discuss motion in a plane path because of its conceptual simplicity and also because the examples and problems are limited to planar motion.

Planar Motion Consider a point P moving along a plane, curvilinear path relative to some reference frame (Fig. 2.25a). The position vector $\mathbf{r}$ specifies the position of P relative to the reference point O, and the coordinate s measures its position along the path relative to a point O' on the path. The velocity of P relative to O is

$$\mathbf{v} = \frac{d\mathbf{r}}{dt} = \lim_{\Delta t \to 0} \frac{\mathbf{r}(t + \Delta t) - \mathbf{r}(t)}{\Delta t} = \lim_{\Delta t \to 0} \frac{\Delta \mathbf{r}}{\Delta t}, \tag{2.34}$$

where $\Delta\mathbf{r} = \mathbf{r}(t + \Delta t) - \mathbf{r}(t)$ (Fig. 2.25b). We denote the distance traveled along the path from t to $t + \Delta t$ by Δs. By introducing a unit vector $\mathbf{e}$ defined to point in the direction of $\Delta\mathbf{r}$, we can write Eq. (2.34) as

$$\mathbf{v} = \lim_{\Delta t \to 0} \frac{\Delta \mathrm{s}}{\Delta \mathrm{t}} \mathbf{e}.$$

As Δt approaches zero, $\Delta s/\Delta t$ becomes ds/dt and $\mathbf{e}$ becomes a unit vector tangent to the path at the position of P at time t, which we denote by $\mathbf{e}_t$ (Fig. 2.25c):

$$\mathbf{v} = v\mathbf{e}_t = \frac{ds}{dt}\mathbf{e}_t. \tag{2.35}$$

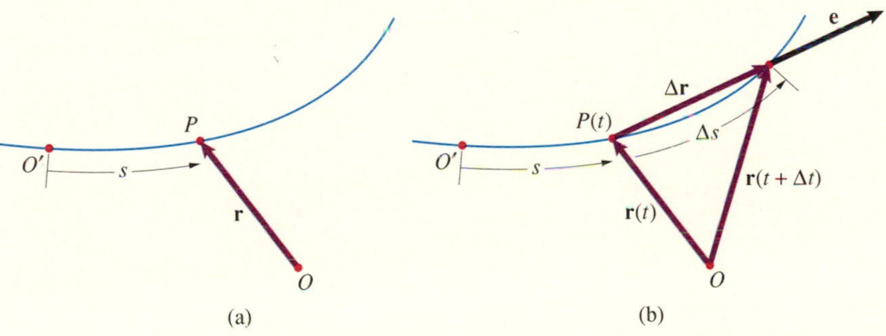

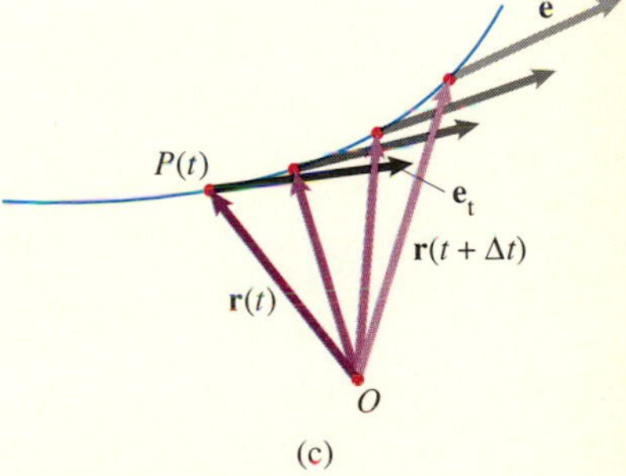

Fig. 2.25

(a) The position of P along its path is specified by the coordinate s.
(b) Position of P at time t and at time $t + \Delta t$.
(c) The limit of $\mathbf{e}$ as $\Delta t \to 0$ is a unit vector tangent to the path.

The velocity of a point in curvilinear motion is a vector whose magnitude equals the rate of change of distance traveled along the path and whose direction is tangent to the path.

To determine the acceleration of P, we take the time derivative of Eq. (2.35):

$$\mathbf{a} = \frac{d\mathbf{v}}{dt} = \frac{dv}{dt}\mathbf{e}_t + v\frac{d\mathbf{e}_t}{dt}. \tag{2.36}$$

If the path is not a straight line, the unit vector $\mathbf{e}_t$ rotates as P moves. As a consequence, the time derivative of $\mathbf{e}_t$ is not zero. In the previous section we derived an expression for the time derivative of a rotating unit vector in terms of the unit vector's angular velocity, Eq. (2.33). To use that result, we define

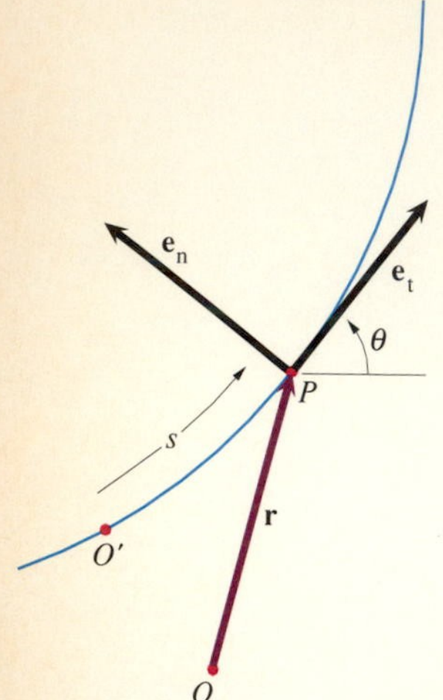

Fig. 2.26
The path angle θ.

the **path angle** θ specifying the direction of $\mathbf{e}_t$ relative to a reference line (Fig. 2.26). Then from Eq. (2.33), the time derivative of $\mathbf{e}_t$ is

$$\frac{d\mathbf{e}_t}{dt} = \frac{d\theta}{dt}\mathbf{e}_n,$$

where $\mathbf{e}_n$ is a unit vector that is normal to $\mathbf{e}_t$ and points in the positive θ direction if $d\theta/dt$ is positive (Fig. 2.26). Substituting this expression into Eq. (2.36), we obtain the acceleration of P:

$$\mathbf{a} = \frac{dv}{dt}\mathbf{e}_t + v\frac{d\theta}{dt}\mathbf{e}_n. \tag{2.37}$$

We can derive this result in another way that is less rigorous but gives additional insight into the meanings of the tangential and normal components of the acceleration. Figure 2.27(a) shows the velocity of P at times t and $t + \Delta t$. In Fig. 2.27(b), you can see that the change in the velocity, $\mathbf{v}(t + \Delta t) - \mathbf{v}(t)$, consists of two components. The component Δv, which is tangent to the path at time t, is due to the change in the *magnitude* of the velocity. The component $v\,\Delta\theta$, which is perpendicular to the path at time t, is due to the change in the *direction* of the velocity vector. Thus the change in the velocity is (approximately)

$$\mathbf{v}(t + \Delta t) - \mathbf{v}(t) = \Delta v\,\mathbf{e}_t + v\Delta\theta\,\mathbf{e}_n.$$

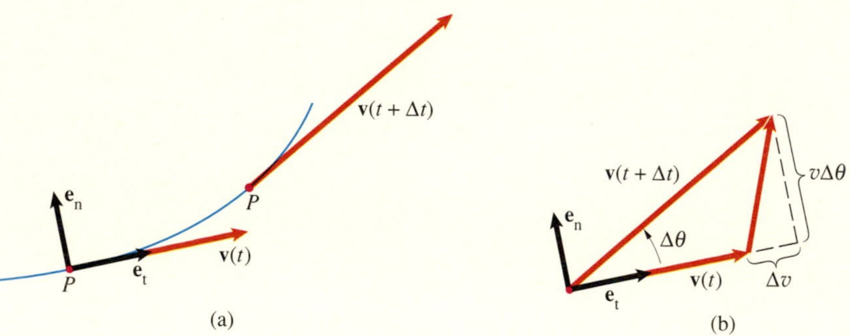

Fig. 2.27
(a) Velocity of P at t and at $t + \Delta t$.
(b) The tangential and normal components of the change in the velocity.

To obtain the acceleration, we divide this expression by Δt and take the limit as $\Delta t \to 0$:

$$\begin{aligned}\mathbf{a} &= \lim_{\Delta t\to 0}\frac{\Delta\mathbf{v}}{\Delta t} = \lim_{\Delta t\to 0}\left(\frac{\Delta v}{\Delta t}\mathbf{e}_t + v\frac{\Delta\theta}{\Delta t}\mathbf{e}_n\right)\\ &= \frac{dv}{dt}\mathbf{e}_t + v\frac{d\theta}{dt}\mathbf{e}_n.\end{aligned}$$

Thus we again obtain Eq. (2.37). However, this derivation clearly points out that the tangential component of the acceleration arises from the rate of change of the magnitude of the velocity, whereas the normal component arises from the rate of change in the direction of the velocity vector. Notice that if the path is a straight line at time t, the normal component of the acceleration equals zero, because in that case $d\theta/dt$ is zero.

We can express the acceleration in another form that often is more convenient to use. Figure 2.28 shows the positions on the path reached by P at times t and $t + dt$. If the path is curved, straight lines extended from these points perpendicular to the path will intersect as shown. The distance ρ from the path to the point where these two lines intersect is called the **instantaneous radius of curvature** of the path. (If the path is circular, ρ is simply the radius of the path.) The angle $d\theta$ is the change in the path angle, and ds is the distance traveled, from t to $t + \Delta t$. You can see from the figure that ρ is related to ds by

$$ds = \rho\, d\theta.$$

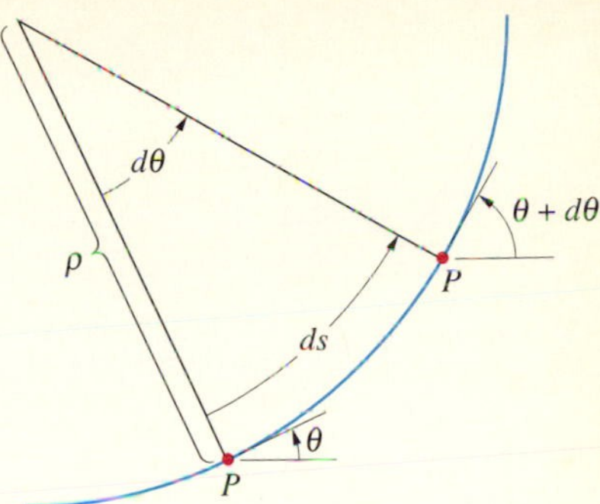

Fig. 2.28
The instantaneous radius of curvature ρ.

Dividing by dt, we obtain

$$\frac{ds}{dt} = v = \rho\frac{d\theta}{dt}.$$

Using this relation, we can write Eq. (2.37) as

$$\mathbf{a} = \frac{dv}{dt}\mathbf{e}_{\mathrm{t}} + \frac{v^2}{\rho}\mathbf{e}_{\mathrm{n}}.$$

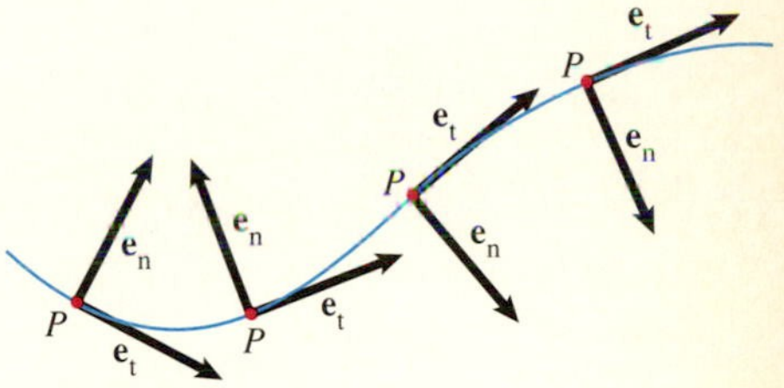

Fig. 2.29
The unit vector normal to the path points toward the concave side.

For a given value of v, the normal component of the acceleration depends on the instantaneous radius of curvature. The greater the curvature of the path, the greater the normal component of acceleration. When the acceleration is expressed in this way, the unit vector $\mathbf{e}_{\mathrm{n}}$ must be defined to point toward the *concave* side of the path (Fig. 2.29).

Thus the velocity and acceleration in terms of normal and tangential components are (Fig. 2.30)

$$\mathbf{v} = v\mathbf{e}_{\mathrm{t}} = \frac{ds}{dt}\mathbf{e}_{\mathrm{t}}, \quad (2.38)$$

$$\mathbf{a} = a_{\mathrm{t}}\mathbf{e}_{\mathrm{t}} + a_{\mathrm{n}}\mathbf{e}_{\mathrm{n}}, \quad (2.39)$$

where

$$a_{\mathrm{t}} = \frac{dv}{dt}, \qquad a_{\mathrm{n}} = v\frac{d\theta}{dt} = \frac{v^2}{\rho}. \quad (2.40)$$

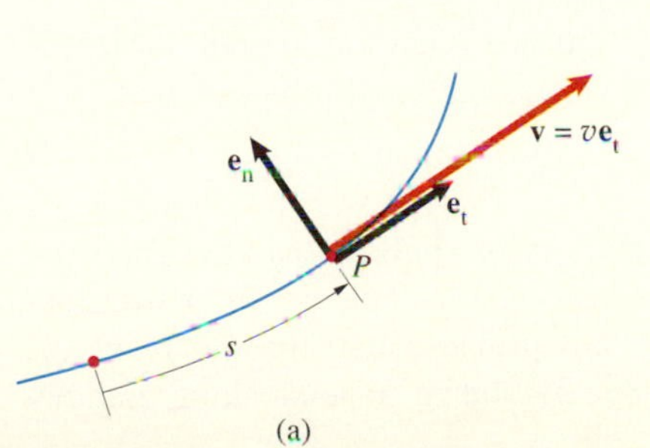

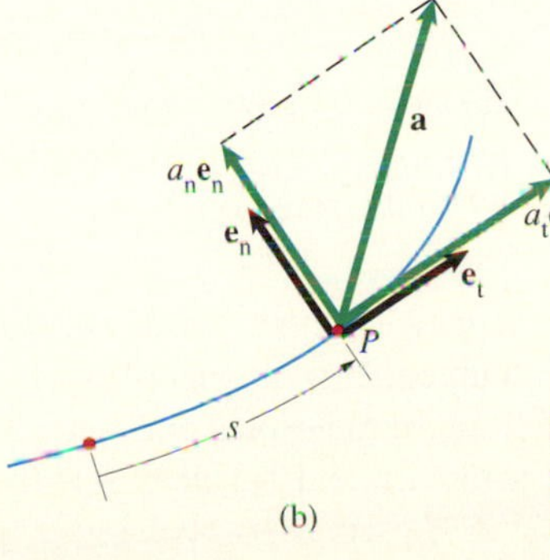

Fig. 2.30
Normal and tangential components of the velocity (a) and acceleration (b).

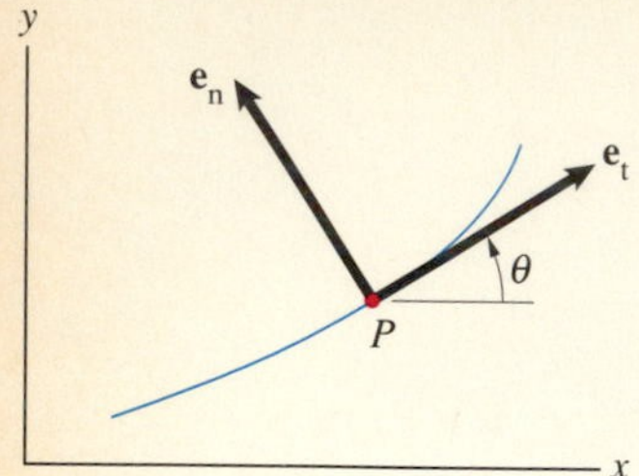

Fig. 2.31
A point P moving in the x-y plane.

If the motion occurs in the x-y plane of a cartesian reference frame (Fig. 2.31) and θ is the angle between the x axis and the unit vector $\mathbf{e}_t$, the unit vectors $\mathbf{e}_t$ and $\mathbf{e}_n$ are related to the cartesian unit vectors by

$$\begin{aligned} \mathbf{e}_t &= \cos\theta\,\mathbf{i} + \sin\theta\,\mathbf{j}, \\ \mathbf{e}_n &= -\sin\theta\,\mathbf{i} + \cos\theta\,\mathbf{j}. \end{aligned} \tag{2.41}$$

If the path in the x-y plane is described by a function $y = y(x)$, it can be shown that the instantaneous radius of curvature is given by

$$\rho = \frac{\left[1 + \left(\dfrac{dy}{dx}\right)^2\right]^{3/2}}{\left|\dfrac{d^2y}{dx^2}\right|}. \tag{2.42}$$

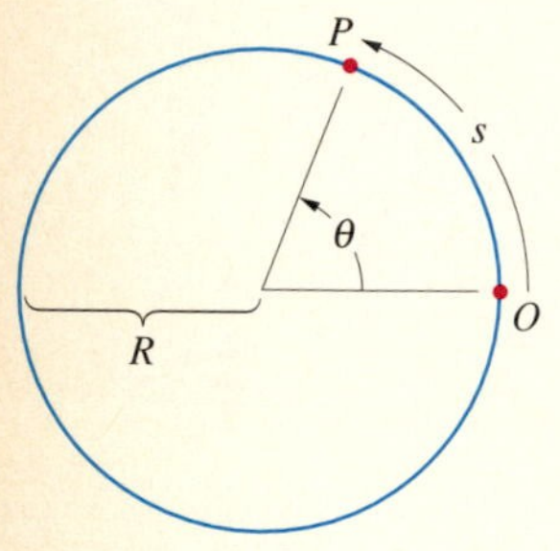

Fig. 2.32
A point moving in a circular path.

Circular Motion If a point P moves in a plane circular path of radius R (Fig. 2.32), the distance s is related to the angle θ by

$$s = R\theta. \qquad \textbf{Circular path}$$

This relation means we can specify the position of P along the circular path either by s or θ. Taking the time derivative of this equation, we obtain a relation between $v = ds/dt$ and the angular velocity of the line from the center of the path to P:

$$\boxed{v = R\frac{d\theta}{dt} = R\omega. \qquad \textbf{Circular path}} \tag{2.43}$$

Taking another time derivative, we obtain a relation between the tangential component of the acceleration $a_t = dv/dt$ and the angular acceleration:

$$\boxed{a_t = R\frac{d\omega}{dt} = R\alpha. \qquad \textbf{Circular path}} \tag{2.44}$$

For this circular path the instantaneous radius of curvature $\rho = R$, so the normal component of the acceleration is

$$\boxed{a_n = \frac{v^2}{R} = R\omega^2. \qquad \textbf{Circular path}} \tag{2.45}$$

Because problems involving circular motion of a point are so common, these relations are worth remembering. But you must be careful to use them *only* when the path is circular.

Three-Dimensional Motion Although most applications of normal and tangential components involve motion of a point in a plane, we briefly discuss three-dimensional motion for the insight it provides into the nature of the velocity and acceleration. If we consider motion of a point along a three-dimensional path relative to some reference frame, the steps leading to Eq. (2.38) are unaltered. The velocity is

$$\mathbf{v} = v\mathbf{e}_t = \frac{ds}{dt}\mathbf{e}_t, \tag{2.46}$$

where $v = ds/dt$ is the rate of change of distance along the path and the unit vector $\mathbf{e}_t$ is tangent to the path and points in the direction of motion. We take the time derivative of this equation to obtain the acceleration:

$$\mathbf{a} = \frac{d\mathbf{v}}{dt} = \frac{dv}{dt}\mathbf{e}_t + v\frac{d\mathbf{e}_t}{dt}.$$

As the point moves along its three-dimensional path, the direction of the unit vector $\mathbf{e}_t$ changes. In the case of motion of a point in a plane, this unit vector rotates in the plane, but in three-dimensional motion the picture is more complicated. Figure 2.33(a) shows the path seen from a viewpoint perpendicular to the plane containing the vector $\mathbf{e}_t$ at times t and $t + dt$. This plane is called the **osculating plane**. It can be thought of as the instantaneous plane of rotation of the unit vector $\mathbf{e}_t$, and its orientation will generally change as P moves along its path. Since $\mathbf{e}_t$ is rotating in the osculating plane at time t, its time derivative is

$$\frac{d\mathbf{e}_t}{dt} = \frac{d\theta}{dt}\mathbf{e}_n, \tag{2.47}$$

where $d\theta/dt$ is the angular velocity of $\mathbf{e}_t$ in the osculating plane and the unit vector $\mathbf{e}_n$ is defined as shown in Fig. 2.33(b). The vector $\mathbf{e}_n$ is perpendicular to $\mathbf{e}_t$, parallel to the osculating plane, and directed toward the concave side of the path. Therefore the acceleration is

$$\mathbf{a} = \frac{dv}{dt}\mathbf{e}_t + v\frac{d\theta}{dt}\mathbf{e}_n. \tag{2.48}$$

In the same way as in the case of motion in a plane, we can also express the acceleration in terms of the instantaneous radius of curvature of the path (Fig. 2.33c):

$$\mathbf{a} = \frac{dv}{dt}\mathbf{e}_t + \frac{v^2}{\rho}\mathbf{e}_n. \tag{2.49}$$

We see that the expressions for the velocity and acceleration in normal and tangential components for three-dimensional motion are identical in form to the expressions for planar motion. The velocity is a vector whose magnitude equals the rate of change of distance traveled along the path, and whose direction is tangent to the path. The acceleration has a component tangential to the path equal to the rate of change of the magnitude of the velocity, and a component perpendicular to the path that depends on the magnitude of the velocity and the instantaneous radius of curvature of the path. In planar motion, the unit vector $\mathbf{e}_n$ is parallel to the plane of the motion. In three-dimensional

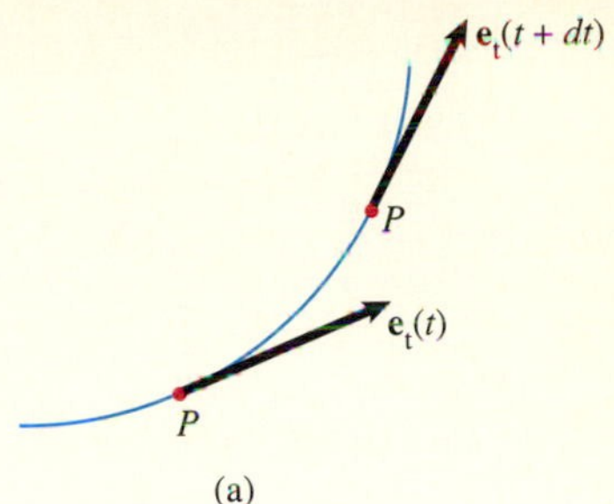

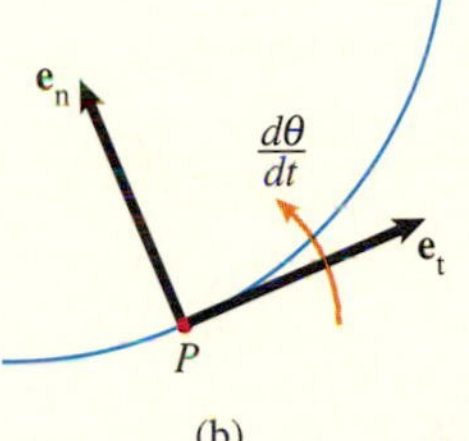

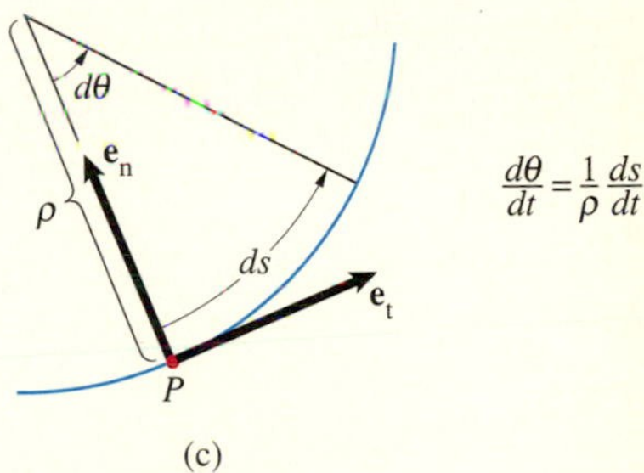

Fig. 2.33
(a) Defining the osculating plane.
(b) Definition of the unit vector $\mathbf{e}_n$.
(c) The instantaneous radius of curvature.

motion $\mathbf{e}_n$ is parallel to the osculating plane, whose orientation depends on the nature of the path. Notice from Eq. (2.47) that $\mathbf{e}_n$ can be expressed in terms of $\mathbf{e}_t$ by

$$\mathbf{e}_n = \frac{\dfrac{d\mathbf{e}_t}{dt}}{\left|\dfrac{d\mathbf{e}_t}{dt}\right|}.$$

As the final step necessary to establish a three-dimensional coordinate system, we introduce a third unit vector that is perpendicular to $\mathbf{e}_t$ and $\mathbf{e}_n$ by the definition

$$\mathbf{e}_p = \mathbf{e}_t \times \mathbf{e}_n.$$

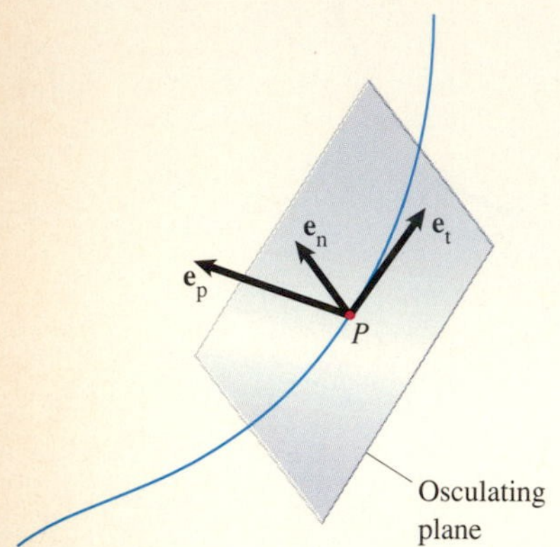

Fig. 2.34
Defining the third unit vector $\mathbf{e}_p$.

The unit vector $\mathbf{e}_p$ is perpendicular to the osculating plane (Fig. 2.34), and therefore defines its orientation.

The following examples demonstrate the use of Eqs. (2.38) and (2.39) to analyze curvilinear motions of objects. Because the equations relating s, y, and the tangential component of the acceleration,

$$v = \frac{ds}{dt},$$

$$a_t = \frac{dv}{dt},$$

are identical in form to the equations that govern the motion of a point along a straight line, you often can solve them using the same methods you applied to straight-line motion.

Example 2.9

Fig. 2.35

The motorcycle in Fig. 2.35 starts from rest at $t = 0$ on a circular track of 400-m radius. The tangential component of its acceleration is $a_t = 2 + 0.2t$ m/s^2. At $t = 10$ s, determine: (a) the distance it has moved along the track; (b) the magnitude of its acceleration.

STRATEGY

Let s be the distance along the track from the initial position O of the motorcycle to its position at time t (Fig. a). Knowing the tangential acceleration as a function of time, we can integrate to determine v and s as functions of time.

SOLUTION

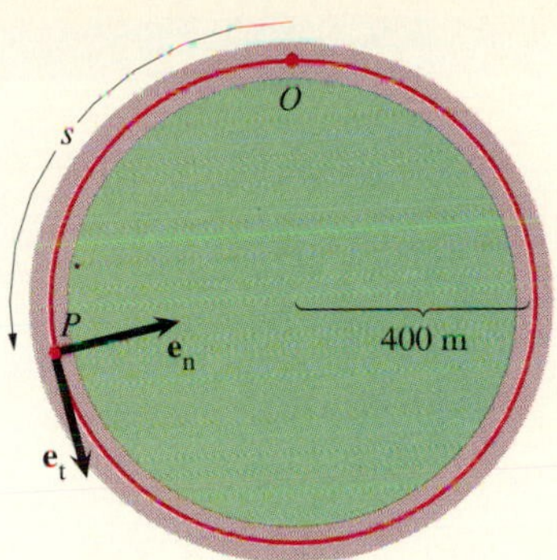

(a) The coordinate s measures the distance along the track.

(a) The tangential acceleration is

$$a_t = \frac{dv}{dt} = 2 + 0.2t \text{ m/s}^2.$$

Integrating,

$$\int_0^v dv = \int_0^t (2 + 0.2t)\, dt,$$

we obtain v as a function of time:

$$v = \frac{ds}{dt} = 2t + 0.1t^2 \text{ m/s}.$$

Integrating this equation,

$$\int_0^s ds = \int_0^t (2t + 0.1t^2)\, dt,$$

the coordinate s as a function of time is

$$s = t^2 + \frac{0.1}{3}t^3 \text{ m}.$$

At $t = 10$ s, the distance moved along the track is

$$s = (10)^2 + \frac{0.1}{3}(10)^3 = 133 \text{ m}.$$

(b) At $t = 10$ s, the tangential component of the acceleration is

$$a_t = 2 + 0.2(10) = 4 \text{ m/s}^2.$$

We must also determine the normal component of acceleration. The instantaneous radius of curvature of the path is the radius of the circular track, $\rho = 400$ m. The magnitude of the velocity at $t = 10$ s is

$$v = 2(10) + 0.1(10)^2 = 30 \text{ m/s}.$$

Therefore the normal acceleration is

$$a_n = \frac{v^2}{\rho} = \frac{(30)^2}{400} = 2.25 \text{ m/s}^2.$$

The magnitude of the acceleration at $t = 10$ s is

$$|\mathbf{a}| = \sqrt{a_t^2 + a_n^2} = \sqrt{(4)^2 + (2.25)^2} = 4.59 \text{ m/s}^2.$$

Example 2.10

A satellite is in a circular orbit of radius R around the earth. What is its velocity?

STRATEGY

The acceleration due to gravity at a distance R from the center of the earth is gR_E^2/R^2, where R_E is the radius of the earth (see Eq. 1.4). By using this expression together with the equation for the acceleration in terms of normal and tangential components, we can obtain an equation for the satellite's velocity.

SOLUTION

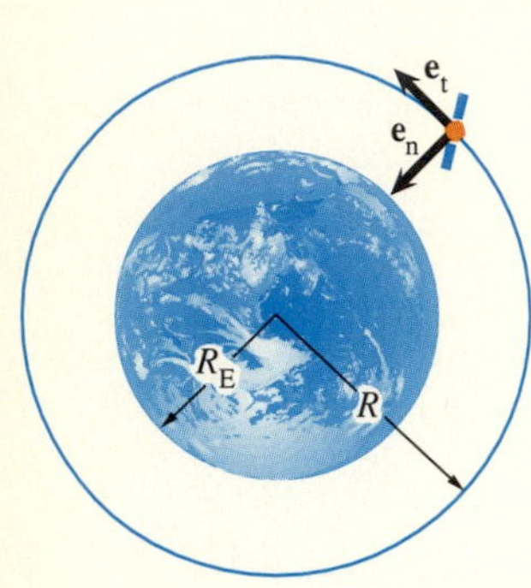

(a) Describing the satellite's motion in terms of normal and tangential components.

In terms of normal and tangential components (Fig. a), the acceleration of the satellite is

$$\mathbf{a} = \frac{dv}{dt}\mathbf{e}_t + \frac{v^2}{R}\mathbf{e}_n.$$

This expression must equal the acceleration due to gravity toward the center of the earth:

$$\frac{dv}{dt}\mathbf{e}_t + \frac{v^2}{R}\mathbf{e}_n = \frac{gR_E^2}{R^2}\mathbf{e}_n.$$

Because there is no $\mathbf{e}_t$ component on the right side, we conclude that the magnitude of the satellite's velocity is constant:

$$\frac{dv}{dt} = 0.$$

Equating the $\mathbf{e}_n$ components and solving for v, we obtain

$$v = \sqrt{\frac{gR_E^2}{R}}.$$

DISCUSSION

In Example 2.5 we determined the escape velocity of an object traveling straight away from the earth in terms of its initial distance from the center of the earth. The escape velocity for an object a distance R from the center of the earth, $v_{esc} = \sqrt{2gR_E^2/\mathrm{R}}$, is only $\sqrt{2}$ times the velocity of an object in a circular orbit of radius R. This explains why it was possible to begin launching probes to other planets not long after the first satellites were placed in orbit.

Example 2.11

During a flight in which a helicopter starts from rest at $t = 0$, the cartesian components of its acceleration are

$$a_x = 0.6t \text{ m/s}^2,$$

$$a_y = 1.8 - 0.36t \text{ m/s}^2.$$

What are the normal and tangential components of its acceleration and the instantaneous radius of curvature of its path at $t = 4$ s?

STRATEGY

We can integrate the cartesian components of acceleration to determine the cartesian components of the velocity at $t = 4$ s, and determine the components of the tangential unit vector $\mathbf{e}_t$ by dividing the velocity vector by its magnitude: $\mathbf{e}_t = \mathbf{v}/|\mathbf{v}|$. We can then determine the tangential component of the acceleration by evaluating the dot product of the acceleration vector with $\mathbf{e}_t$. Knowing the tangential component of the acceleration, we can evaluate the normal component. We can determine the radius of curvature of the path from the relation $a_n = v^2/\rho$.

SOLUTION

Integrating the components of acceleration with respect to time (see Example 2.6), the cartesian components of the velocity are

$$v_x = 0.3t^2 \text{ m/s},$$

$$v_y = 1.8t - 0.18t^2 \text{ m/s}.$$

At $t = 4$ s, $v_x = 4.80$ m/s and $v_y = 4.32$ m/s. The tangential unit vector $\mathbf{e}_t$ at $t = 4$ s is (Fig. a)

$$\mathbf{e}_t = \frac{\mathbf{v}}{|\mathbf{v}|} = \frac{4.80\mathbf{i} + 4.32\mathbf{j}}{\sqrt{(4.80)^2 + (4.32)^2}} = 0.743\mathbf{i} + 0.669\mathbf{j}.$$

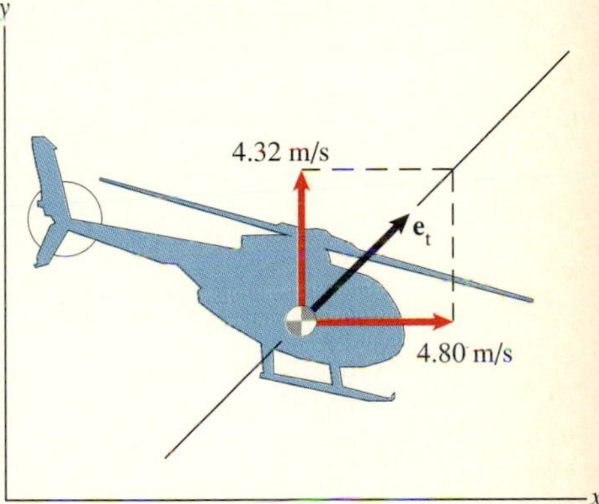

(a) Cartesian components of the velocity and the vector $\mathbf{e}_t$.

The components of the acceleration at $t = 4$ s are

$$a_x = 0.6(4) = 2.4 \text{ m/s}^2,$$

$$a_y = 1.8 - 0.36(4) = 0.36 \text{ m/s}^2,$$

so the tangential component of the acceleration at $t = 4$ s is

$$\begin{aligned} a_t &= \mathbf{e}_t \cdot \mathbf{a} \\ &= (0.743\mathbf{i} + 0.669\mathbf{j}) \cdot (2.4\mathbf{i} + 0.36\mathbf{j}) \\ &= 2.02 \text{ m/s}^2. \end{aligned}$$

The magnitude of the acceleration is $\sqrt{(2.4)^2 + (0.36)^2} = 2.43$ m/s^2, so the magnitude of the normal component of the acceleration is

$$a_n = \sqrt{|\mathbf{a}|^2 - a_t^2} = \sqrt{(2.43)^2 - (2.02)^2} = 1.34 \text{ m/s}^2.$$

The radius of curvature of the path is

$$\rho = \frac{|\mathbf{v}|^2}{a_n} = \frac{(4.80)^2 + (4.32)^2}{1.34} = 31.2 \text{ m}.$$

Example 2.12

Application to Engineering

Centrifuge Design

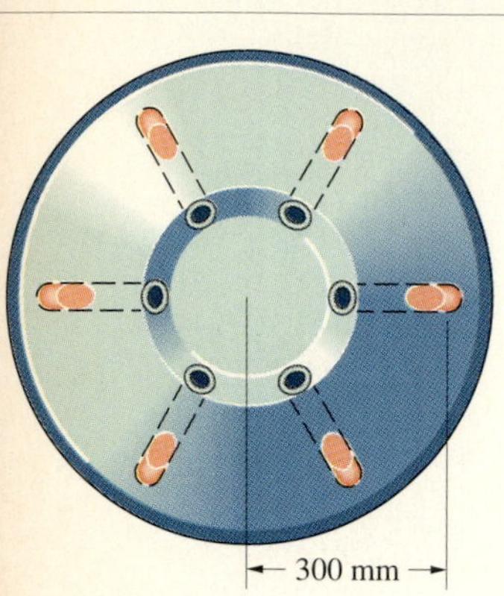

Fig. 2.36

The distance from the center of the medical centrifuge in Fig. 2.36 to its samples is 300 mm. When the centrifuge is turned on, its motor and control system give it an angular acceleration $\alpha = A - B\omega^2$. Choose the constants A and B so that the samples will be subjected to a maximum horizontal acceleration of 12,000 g's and the centrifuge will reach 90% of its maximum operating speed in 2 min.

STRATEGY

Since we know the radius of the circular path in which the samples move and the horizontal acceleration to which they are to be subjected, we can solve for the operating angular velocity of the centrifuge. We will use the given angular acceleration to determine the centrifuge's angular velocity as a function of time in terms of the constants A and B. We can then use the operating angular velocity and the condition that the centrifuge reach 90% of the operating angular velocity in 2 min to determine the constants A and B.

SOLUTION

From Eq. (2.45), the samples are subjected to a normal acceleration

$$a_{\mathrm{n}} = R\omega^2.$$

Setting $a_{\mathrm{n}} = (12{,}000)(9.81)\ \mathrm{m/s^2}$ and $R = 0.3$ m and solving for the angular velocity, we find that the desired maximum operating speed is $\omega_{\mathrm{max}} = 626$ rad/s.

The angular acceleration is

$$\alpha = \frac{d\omega}{dt} = A - B\omega^2.$$

We separate variables,

$$\frac{d\omega}{A - B\omega^2} = dt,$$

and integrate to determine ω as a function of time, assuming the centrifuge starts from rest at $t = 0$:

$$\int_0^{\omega} \frac{d\omega}{A - B\omega^2} = \int_0^t dt.$$

Evaluating the integrals, we obtain

$$\frac{1}{2\sqrt{AB}} \ln\left(\frac{A + \sqrt{AB}\,\omega}{A - \sqrt{AB}\,\omega}\right) = t.$$

The solution of this equation for ω is

$$\omega = \sqrt{\frac{A}{B}}\left(\frac{e^{2\sqrt{AB}t} - 1}{e^{2\sqrt{AB}t} + 1}\right).$$

As t becomes large, ω approaches $\sqrt{A/B}$, so we have the condition that

$$\sqrt{\frac{A}{B}} = \omega_{\text{max}} = 626 \text{ rad/s}, \tag{2.50}$$

and we can write the equation for ω as

$$\omega = \omega_{\text{max}}\left(\frac{e^{2\sqrt{AB}t} - 1}{e^{2\sqrt{AB}t} + 1}\right). \tag{2.51}$$

We also have the condition that $\omega = 0.9\omega_{\text{max}}$ after 2 min. Setting $\omega = 0.9\omega_{\text{max}}$ and $t = 120$ s in Eq. (2.51) and solving for $\sqrt{AB}$, we obtain

$$\sqrt{AB} = \frac{\ln(19)}{240}.$$

We solve this equation together with Eq. (2.50), obtaining $A = 7.69$ rad/s^2 and $B = 1.96 \times 10^{-5}$ rad^{-1}. Figure 2.37 shows the angular velocity of the centrifuge as a function of time.

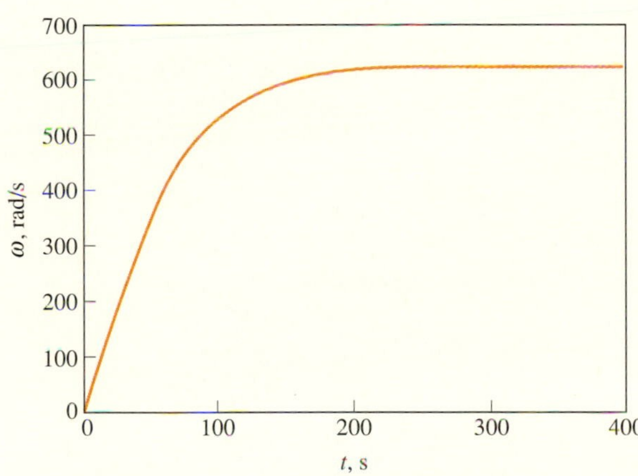

Fig. 2.37
Graph of the angular velocity as a function of time.

DESIGN ISSUES

A centrifuge is a device designed to rotate and thereby subject objects to large normal accelerations. The common spin dryer used to dry clothing is an example. Centrifuges are used extensively in biological and medical testing and research and also in the chemical processing industry to separate constituents of different densities. By using a centrifuge to separate the solid matter in blood (primarily red cells) from the liquid plasma, diagnostic tests can be carried out on the plasma, and the patient's *hematocrit*, the percentage of solid matter, can be determined. During the early days of manned space flight there was concern that the large accelerations to which the astronauts would be subjected during boost to orbit might injure them or affect their ability to carry out essential tasks. Large centrifuges consisting of rotating beams with "gondolas" at the end capable of carrying a person were built and used to study the effects of acceleration on human subjects. Centrifuges have also been designed to investigate the effects of large accelerations on mechanisms and to

Fig. 2.38

A centrifuge designed to subject test articles to large accelerations. (Photograph courtesy of Sandia National Laboratories.)

facilitate modeling studies of geophysical phenomena such as slope stability and cratering. One such centrifuge is shown in Fig. 2.38. Consisting of a large beam that rotates samples in a horizontal circular path with a radius of 8 m, it can subject samples to 150 g's of acceleration.

Centrifuges intended for commercial use, such as medical testing, must be designed so that samples can be installed easily, and they must have an electric motor and control system that permits specified accelerations to be achieved quickly and accurately. The bearing must be designed to sustain the very high angular velocities and have a long lifetime. (The *ultracentrifuge*, used in biological research, reaches angular velocities of millions of revolutions per minute and is usually designed with an air bearing so that there is no solid contact between the spinning centrifuge and its support.) The structure of a centrifuge must be designed to support the large forces resulting from its rapid rotational motion, and the objects being accelerated must be adequately supported.

An essential consideration in the design of a centrifuge is the possibility of structural failure. Such a failure could result in projectiles moving outward from the center of rotation at high velocities. A centrifuge must be designed with a surrounding structure strong enough that such projectiles would be safely contained and not cause additional damage or injury. This design requirement was achieved in the case of the large centrifuge in Fig. 2.38 by locating it underground.

Problems

2.104 The armature of an electric motor rotates at a constant rate. The magnitude of the velocity of point P relative to O is 4 m/s.
(a) What are the normal and tangential components of the acceleration of P relative to O?
(b) What is the angular velocity of the armature?

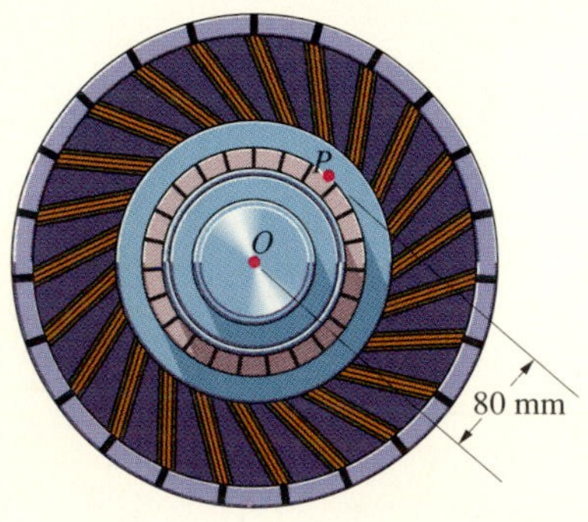

P2.104

2.105 The armature in Problem 2.104 starts from rest and has constant angular acceleration $\alpha = 10\ \text{rad/s}^2$. What are the velocity and acceleration of P relative to O in terms of normal and tangential components after 10 s?

2.106 Suppose that you want to design a medical centrifuge to subject samples to normal accelerations of 1000 g's.
(a) If the distance from the center of the centrifuge to the sample is 300 mm, what speed of rotation in rpm is necessary?
(b) If you want the centrifuge to reach its design rpm in 1 min, what constant angular acceleration is necessary?

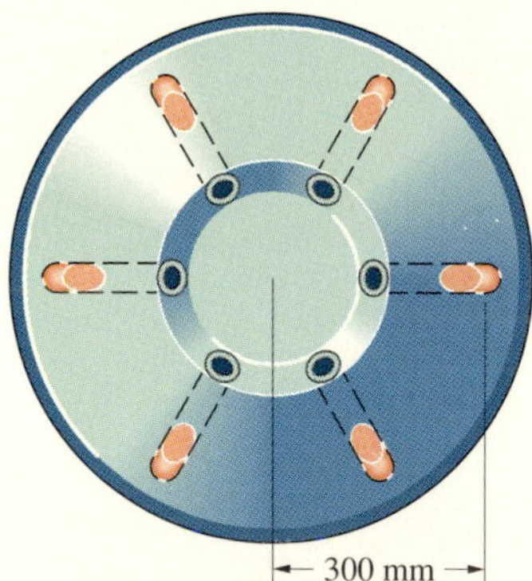

P2.106

2.107 In Example 2.12, what are the magnitudes of the tangential and normal components of acceleration to which the samples are subjected at the instant the centrifuge is turned on?

2.108 In Example 2.12, what are the magnitudes of the tangential and normal components of acceleration to which the samples are subjected 10 s after the centrifuge is turned on?

2.109 A powerboat being tested for maneuverability is started from rest and driven in a circular path of 40-ft radius. The magnitude of its velocity is increased at a constant rate of 2 ft/s^2. In terms of normal and tangential components, determine: (a) the velocity as a function of time; (b) the acceleration as a function of time.

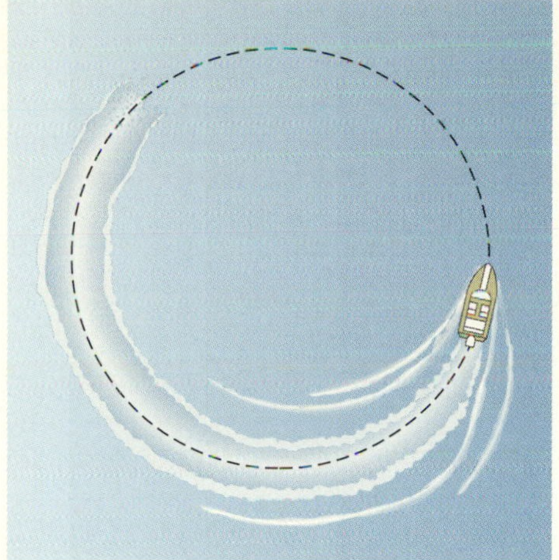

P2.109

2.110 The angle $\theta = 2t^2$ rad.
(a) What are the magnitudes of the velocity and acceleration of P relative to O at $t = 1$ s?
(b) What distance along the circular path does P move from $t = 0$ to $t = 1$ s?

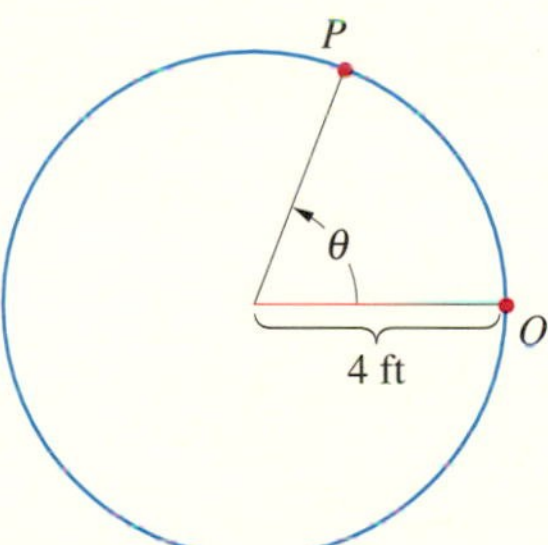

P2.110

2.111 In Problem 2.110, what are the magnitudes of the velocity and acceleration of P relative to O when P has gone one revolution around the circular path starting from $t = 0$?

2.112 The radius of the earth is 3960 miles. If you are standing at the equator, what is the magnitude of your velocity relative to a nonrotating reference frame with its origin at the center of the earth?

2.113 The radius of the earth is 6370 km. If you are standing at the equator, what is the magnitude of your acceleration relative to a nonrotating reference frame with its origin at the center of the earth?

2.114 Suppose that you are standing at point P at 30° north latitude (that is, a point 30° north of the equator). The radius of the earth is $R_E = 3960$ miles. What are the magnitudes of your velocity and acceleration relative to a nonrotating reference frame with its origin at the center of the earth?

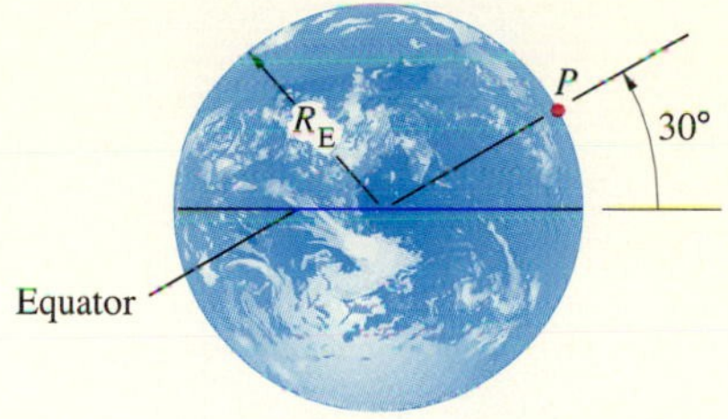

P2.114

2.115 The magnitude of the velocity of the airplane is constant and equal to 400 m/s. The rate of change of the path angle θ is constant and equal to 5°/s.
(a) What are the velocity and acceleration of the airplane in terms of normal and tangential components?
(b) What is the instantaneous radius of curvature of the airplane's path?

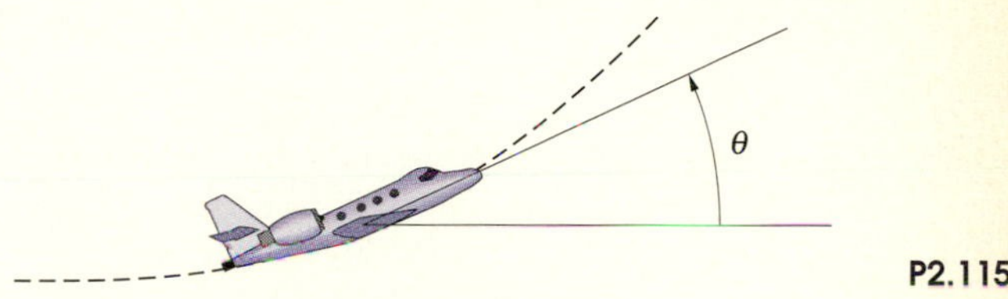

P2.115

2.116 At $t = 0$, a car starts from rest at point A. It moves toward the right, and the tangential component of its acceleration is $a_t = 0.4t$ m/s^2. What is the magnitude of the car's acceleration when it reaches point B?

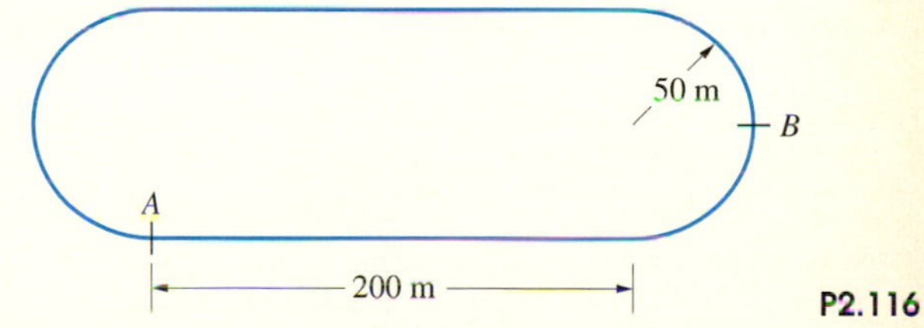

P2.116

2.117 A group of engineering students constructs a sun-powered car and tests it on a circular track of 1000-ft radius. The car starts from rest, and the tangential component of its acceleration is given in terms of the car's velocity by $a_t = 2 - 0.1v$ ft/s^2. Determine v and the magnitude of the car's acceleration 15 s after it starts.

2.118 Suppose that the tangential component of acceleration of the car in Problem 2.117 is $a_t = 2 - 0.008s$ ft/s^2, where s is the distance the car travels along the track from the point where it starts from rest. Determine the velocity v and the magnitude of the car's acceleration when it has traveled a distance $s = 100$ ft.

2.119 A car increases its speed at a constant rate from 40 mi/hr at A to 60 mi/hr at B. What is the magnitude of its acceleration 2 s after it passes point A?

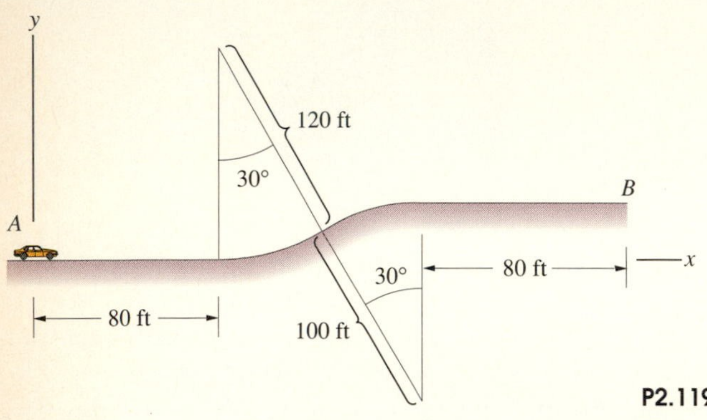

P2.119

2.120 Determine the magnitude of the acceleration of the car in Problem 2.119 when it has traveled along the road a distance (a) 120 ft from A; (b) 160 ft from A.

2.121 An astronaut candidate is to be tested in a centrifuge with a radius of 10 m. He will lose consciousness if his total horizontal acceleration reaches 14 g's. What is the maximum constant angular acceleration of the centrifuge, starting from rest, if he is not to lose consciousness within 1 min?

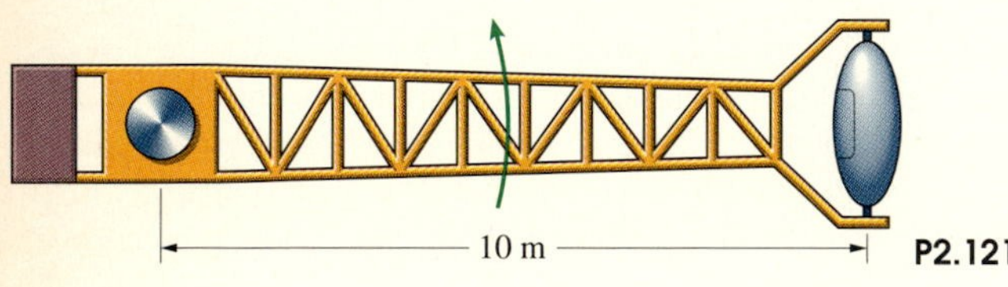

P2.121

2.122 After first-stage separation and before the second-stage engines have fired, a rocket is moving at $v = 3000$ m/s and the angle between its velocity vector and the vertical is 60°. Because aerodynamic forces are negligible, the rocket's acceleration is that due to gravity, which is 9.50 m/s^2 at the rocket's altitude. Determine: (a) the normal and tangential components of the rocket's acceleration; (b) the instantaneous radius of curvature of the rocket's path.

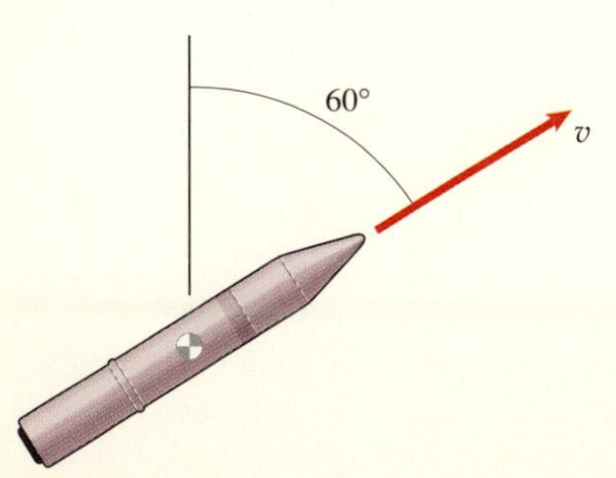

P2.122

2.123 A projectile has an initial velocity of 20 ft/s at 30° above the horizontal.
(a) What are the velocity and acceleration of the projectile in terms of normal and tangential components when it is at the highest point of its trajectory?
(b) What is the instantaneous radius of curvature of the projectile's path when it is at the highest point of its trajectory?

Strategy: In part (b), you can determine the instantaneous radius of curvature from the relation $a_n = v^2/\rho$.

P2.123

2.124 In Problem 2.123, let $t = 0$ be the instant at which the projectile is launched.
(a) What are the velocity and acceleration in terms of normal and tangential components at $t = 0.2$ s?
(b) Use the relation $a_n = v^2/\rho$ to determine the instantaneous radius of curvature of the path at $t = 0.2$ s.

2.125 In Problem 2.123, let $t = 0$ be the instant at which the projectile is launched. Use Eq. (2.42) to determine the instantaneous radius of curvature of the path at $t = 0.2$ s.

2.126 The cartesian coordinates of a point moving in the x-y plane are

$$x = 20 + 4t^2 \text{ m}, \qquad y = 10 - t^3 \text{ m}.$$

What is the instantaneous radius of curvature of the path at $t = 3$ s?

2.127 In Example 2.11, determine the tangential and normal components of the helicopter's acceleration at $t = 6$ s.

2.128 In Example 2.11, use Eq. (2.42) to determine the instantaneous radius of curvature of the helicopter's path at $t = 6$ s.

2.129 For astronaut training, the airplane shown is to achieve "weightlessness" for a short period of time by flying along a path such that its acceleration is $a_x = 0$, $a_y = -g$. If its velocity at O at time $t = 0$ is $\mathbf{v} = v_0\mathbf{i}$, show that the autopilot must fly the airplane so that its tangential component of acceleration as a function of time is

$$a_t = g\frac{(gt/v_0)}{\sqrt{1 + (gt/v_0)^2}}.$$

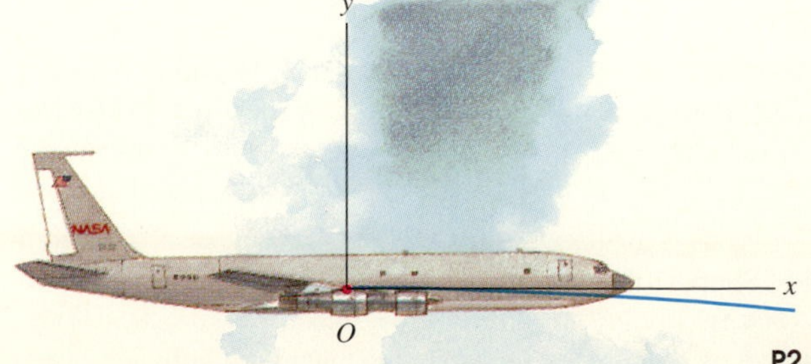

P2.129

2.130 In Problem 2.129, what is the airplane's normal component of acceleration as a function of time?

2.131 If $y = 100$ mm, $dy/dt = 200$ mm/s, and $d^2y/dt^2 = 0$, what are the velocity and acceleration of P in terms of normal and tangential components?

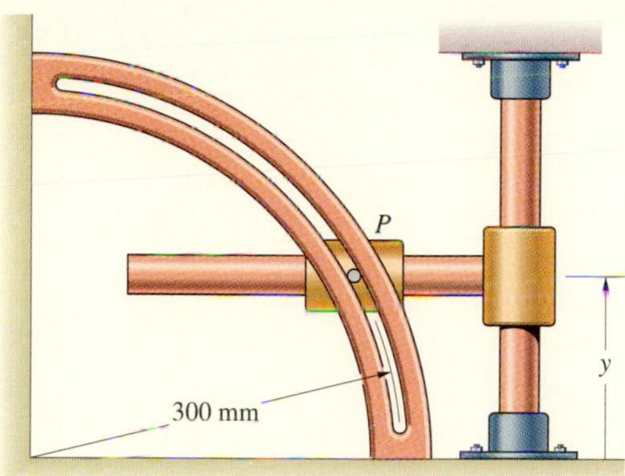

P2.131

2.132 Suppose that the point P in Problem 2.131 moves upward in the slot with velocity $\mathbf{v} = 300\mathbf{e}_t$ (mm/s). When $y = 150$ mm, what are dy/dt and d^2y/dt^2?

2.133 A car travels at 100 km/hr on a straight road of increasing grade whose vertical profile can be approximated by the equation shown. When the car's horizontal coordinate is $x = 400$ m, what are the tangential and normal components of its acceleration?

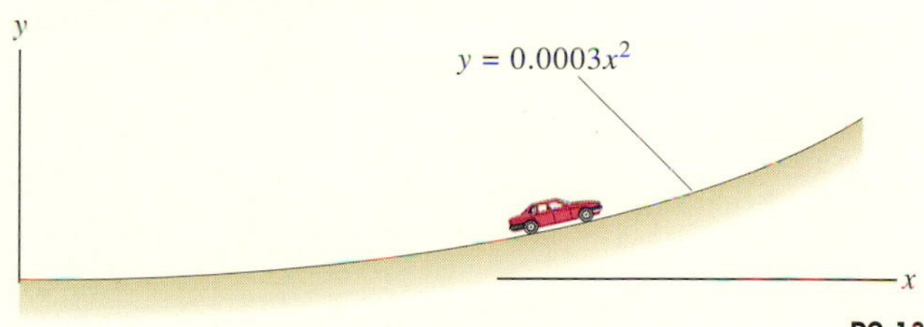

P2.133

2.134 A boy rides a skateboard on the concrete surface of an empty drainage canal described by the equation shown. He starts at $y = 20$ ft, and the magnitude of his velocity is approximated by $v = \sqrt{2(32.2)(20 - y)}$ ft/s.
(a) Use Eq. (2.42) to determine the instantaneous radius of curvature of his path when he reaches the bottom.
(b) What is the normal component of his acceleration when he reaches the bottom?

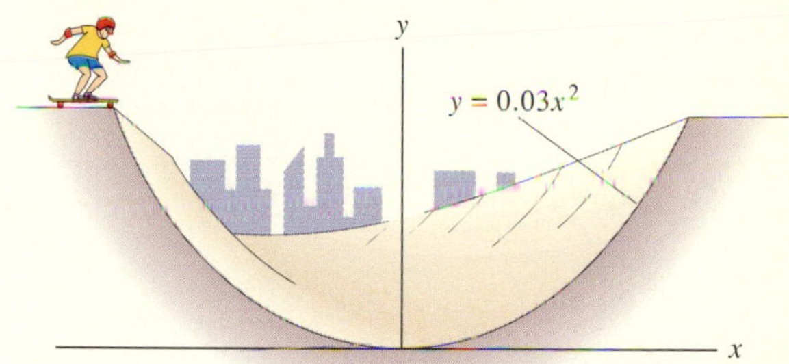

P2.134

2.135 In Problem 2.134, what is the normal component of the boy's acceleration when he has passed the bottom and reached $y = 10$ ft?

2.136 By using Eqs. (2.41): (a) Show that the relations between the cartesian unit vectors and the unit vectors $\mathbf{e}_t$ and $\mathbf{e}_n$ are

$$\mathbf{i} = \cos\theta\,\mathbf{e}_t - \sin\theta\,\mathbf{e}_n,$$

$$\mathbf{j} = \sin\theta\,\mathbf{e}_t + \cos\theta\,\mathbf{e}_n.$$

(b) Show that

$$\frac{d\mathbf{e}_t}{dt} = \frac{d\theta}{dt}\mathbf{e}_n \quad \text{and} \quad \frac{d\mathbf{e}_n}{dt} = -\frac{d\theta}{dt}\mathbf{e}_t.$$

Polar and Cylindrical Coordinates

Polar coordinates are often used to describe the curvilinear motion of a point. Circular motion, certain orbit problems, and more generally, **central force** problems, in which the acceleration of a point is directed toward a given point, can be expressed conveniently in polar coordinates.

Consider a point P in the x-y plane of a cartesian coordinate system. We can specify the position of P relative to the origin O either by its cartesian coordinates x,y or by its polar coordinates r,θ (Fig. 2.39a). To express vectors in terms of polar coordinates, we define a unit vector $\mathbf{e}_r$ that points in the direction of the radial line from the origin to P and a unit vector $\mathbf{e}_\theta$ that is perpendicular to $\mathbf{e}_r$ and points in the direction of increasing θ (Fig. 2.39b). In terms of these vectors, the position vector $\mathbf{r}$ from O to P is

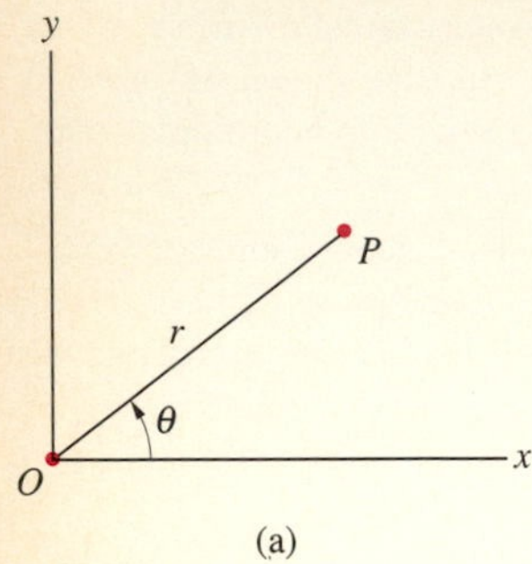

(a)

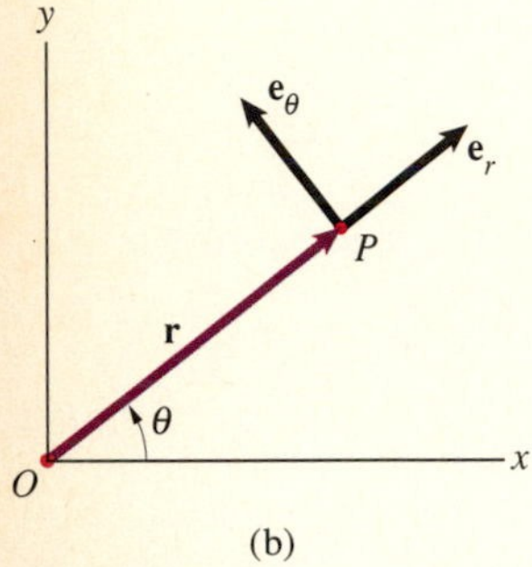

(b)

Fig. 2.39

(a) The polar coordinates of P.
(b) The unit vectors $\mathbf{e}_r$ and $\mathbf{e}_\theta$ and position vector $\mathbf{r}$.

$$\mathbf{r} = r\mathbf{e}_r. \tag{2.52}$$

(Notice that $\mathbf{r}$ has no component in the direction of $\mathbf{e}_\theta$.)

We can determine the velocity of P in terms of polar coordinates by taking the time derivative of Eq. (2.52):

$$\mathbf{v} = \frac{d\mathbf{r}}{dt} = \frac{dr}{dt}\mathbf{e}_r + r\frac{d\mathbf{e}_r}{dt}. \tag{2.53}$$

As P moves along a curvilinear path, the unit vector $\mathbf{e}_r$ rotates with angular velocity $\omega = d\theta/dt$. Therefore, from Eq. (2.33), we can express the time derivative of $\mathbf{e}_r$ in terms of $\mathbf{e}_\theta$ as

$$\frac{d\mathbf{e}_r}{dt} = \frac{d\theta}{dt}\mathbf{e}_\theta. \tag{2.54}$$

Substituting this result into Eq. (2.53), we obtain the velocity of P:

$$\mathbf{v} = \frac{dr}{dt}\mathbf{e}_r + r\frac{d\theta}{dt}\mathbf{e}_\theta = \frac{dr}{dt}\mathbf{e}_r + r\omega\mathbf{e}_\theta. \tag{2.55}$$

We can obtain this result in another way that is less rigorous but more direct and intuitive. Figure 2.40 shows the position vector of P at times t and $t + \Delta t$. The change in the position vector, $\mathbf{r}(t + \Delta t) - \mathbf{r}(t)$, consists of two components. The component Δr is due to the change in the radial position r and is in the $\mathbf{e}_r$ direction. The component $r\Delta\theta$ is due to the change in θ and is in the $\mathbf{e}_\theta$ direction. Thus the change in the position of P is (approximately)

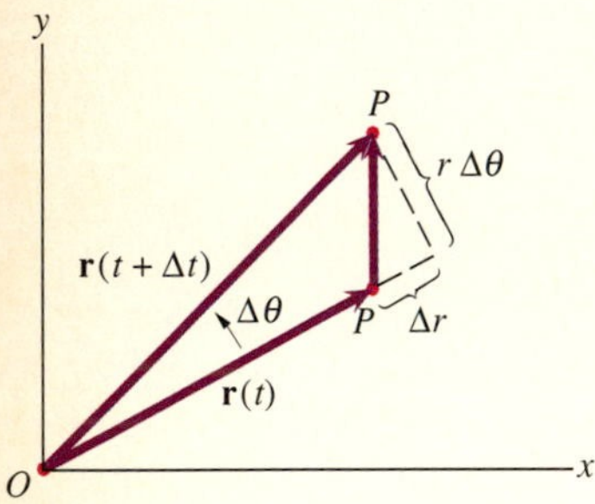

Fig. 2.40

The position vector of P at t and $t + \Delta t$.

$$\mathbf{r}(t + \Delta t) - \mathbf{r}(t) = \Delta r\,\mathbf{e}_r + r\Delta\theta\,\mathbf{e}_\theta.$$

Dividing this expression by Δt and taking the limit as $\Delta t \to 0$, we obtain the velocity of P:

$$\mathbf{v} = \lim_{\Delta t \to 0}\left(\frac{\Delta r}{\Delta t}\mathbf{e}_r + r\frac{\Delta\theta}{\Delta t}\mathbf{e}_\theta\right)$$
$$= \frac{dr}{dt}\mathbf{e}_r + r\omega\mathbf{e}_\theta.$$

One component of the velocity is in the radial direction and is equal to the rate of change of the radial position r. The other component is normal, or *transverse* to the radial direction, and is proportional to the radial distance and to the rate of change of θ.

We obtain the acceleration of P by taking the time derivative of Eq. (2.55):

$$\mathbf{a} = \frac{d\mathbf{v}}{dt} = \frac{d^2r}{dt^2}\mathbf{e}_r + \frac{dr}{dt}\frac{d\mathbf{e}_r}{dt} + \frac{dr}{dt}\frac{d\theta}{dt}\mathbf{e}_\theta$$
$$+ r\frac{d^2\theta}{dt^2}\mathbf{e}_\theta + r\frac{d\theta}{dt}\frac{d\mathbf{e}_\theta}{dt}. \tag{2.56}$$

The time derivative of the unit vector $\mathbf{e}_r$ due to the rate of change of θ is given by Eq. (2.54). As P moves, $\mathbf{e}_\theta$ also rotates with angular velocity $d\theta/dt$ (Fig. 2.41). You can see from this figure that the time derivative of $\mathbf{e}_\theta$ is in the $-\mathbf{e}_r$ direction if $d\theta/dt$ is positive:

$$\frac{d\mathbf{e}_\theta}{dt} = -\frac{d\theta}{dt}\mathbf{e}_r.$$

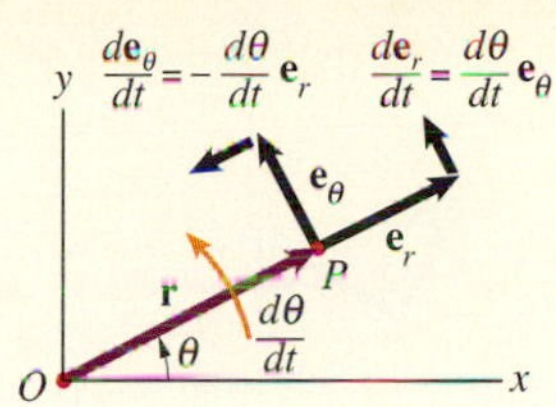

Fig. 2.41
Time derivatives of $\mathbf{e}_r$ and $\mathbf{e}_\theta$.

Substituting this expression and Eq. (2.54) into Eq. (2.56), we obtain the acceleration of P:

$$\mathbf{a} = \left[\frac{d^2r}{dt^2} - r\left(\frac{d\theta}{dt}\right)^2\right]\mathbf{e}_r + \left[r\frac{d^2\theta}{dt^2} + 2\frac{dr}{dt}\frac{d\theta}{dt}\right]\mathbf{e}_\theta.$$

Thus the velocity and acceleration are (Fig. 2.42)

$$\boxed{\mathbf{v} = v_r\mathbf{e}_r + v_\theta\mathbf{e}_\theta = \frac{dr}{dt}\mathbf{e}_r + r\omega\mathbf{e}_\theta} \qquad (2.57)$$

and

$$\boxed{\mathbf{a} = a_r\mathbf{e}_r + a_\theta\mathbf{e}_\theta,} \qquad (2.58)$$

where

$$\boxed{\begin{aligned} a_r &= \frac{d^2r}{dt^2} - r\left(\frac{d\theta}{dt}\right)^2 = \frac{d^2r}{dt^2} - r\omega^2, \\ a_\theta &= r\frac{d^2\theta}{dt^2} + 2\frac{dr}{dt}\frac{d\theta}{dt} = r\alpha + 2\frac{dr}{dt}\omega. \end{aligned}} \qquad (2.59)$$

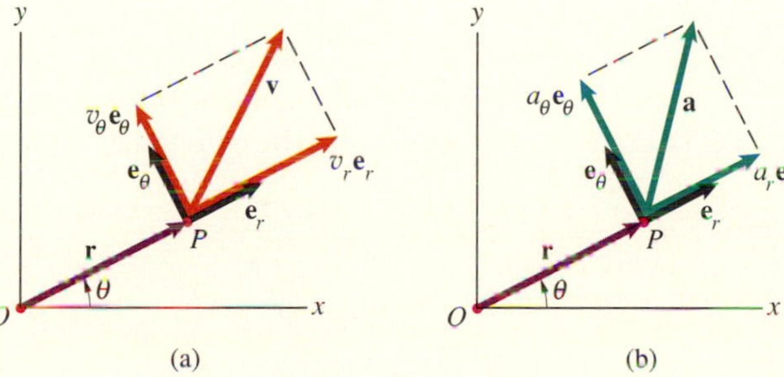

Fig. 2.42
Radial and transverse components of the velocity (a) and acceleration (b).

The term $-r\omega^2$ in the radial component of the acceleration is called the **centripetal acceleration**, and the term $2(dr/dt)\omega$ in the transverse component is called the **Coriolis acceleration**.

The unit vectors $\mathbf{e}_r$ and $\mathbf{e}_\theta$ are related to the cartesian unit vectors by

$$\begin{aligned} \mathbf{e}_r &= \cos\theta\,\mathbf{i} + \sin\theta\,\mathbf{j}, \\ \mathbf{e}_\theta &= -\sin\theta\,\mathbf{i} + \cos\theta\,\mathbf{j}. \end{aligned} \qquad (2.60)$$

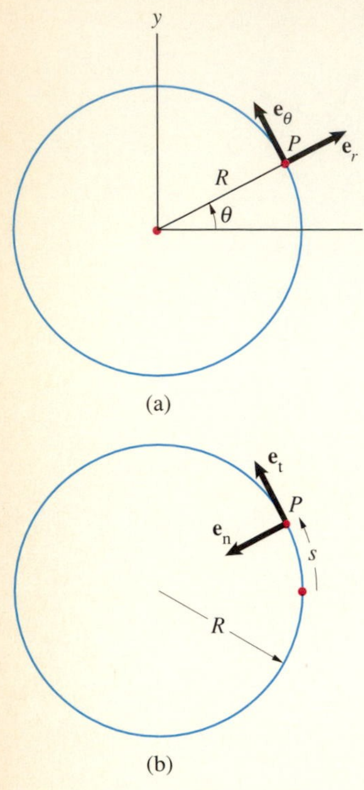

Fig. 2.43
A point P moving in a circular path.
(a) Polar coordinates.
(b) Normal and tangential components.

Circular Motion Circular motion can be conveniently described using either radial and transverse or normal and tangential components. Let us compare these two methods of expressing the velocity and acceleration of a point P moving in a circular path of radius R (Fig. 2.43). Because the polar coordinate $r = R$ is constant, Eq. (2.57) for the velocity reduces to

$$\mathbf{v} = R\omega\mathbf{e}_\theta.$$

In terms of normal and tangential components, the velocity is

$$\mathbf{v} = v\mathbf{e}_t.$$

Notice in Fig. 2.43 that $\mathbf{e}_\theta = \mathbf{e}_t$. Comparing these two expressions for the velocity, we obtain the relation between the velocity and the angular velocity in circular motion:

$$v = R\omega.$$

From Eqs. (2.58) and (2.59), the acceleration in terms of polar coordinates for a circular path of radius R is

$$\mathbf{a} = -R\omega^2\mathbf{e}_r + R\alpha\mathbf{e}_\theta,$$

and the acceleration in terms of normal and tangential components is

$$\mathbf{a} = \frac{dv}{dt}\mathbf{e}_t + \frac{v^2}{R}\mathbf{e}_n.$$

The unit vector $\mathbf{e}_r = -\mathbf{e}_n$. Because of the relation $v = R\omega$, the normal components of acceleration are equal: $v^2/R = R\omega^2$. Equating the transverse and tangential components, we obtain the relation

$$\frac{dv}{dt} = a_t = R\alpha.$$

Cylindrical Coordinates Polar coordinates describe the motion of a point P in the x-y plane. We can describe three-dimensional motion by using **cylindrical coordinates** r, θ, z (Fig. 2.44). The cylindrical coordinates r and θ

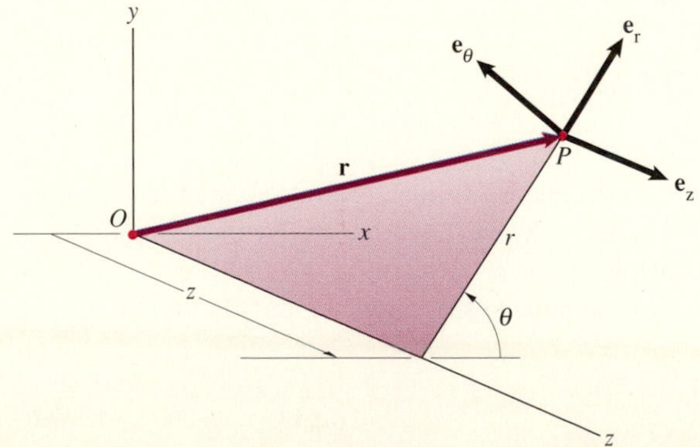

Fig. 2.44
Cylindrical coordinates r, θ, z of point P and the unit vectors $\mathbf{e}_r$, $\mathbf{e}_\theta$, $\mathbf{e}_z$.

are the polar coordinates of P measured in the plane parallel to the x-y plane, and the definitions of the unit vectors $\mathbf{e}_r$ and $\mathbf{e}_\theta$ are unchanged. The position of P perpendicular to the x-y plane is measured by the coordinate z, and the unit vector $\mathbf{e}_z$ points in the positive z-axis direction.

The position vector $\mathbf{r}$ in terms of cylindrical coordinates is the sum of the expression for the position vector in polar coordinates and the z component:

$$\mathbf{r} = r\mathbf{e}_r + z\mathbf{e}_z, \tag{2.61}$$

(The polar coordinate r does not equal the magnitude of $\mathbf{r}$ except when P lies in the x-y plane.) By taking time derivatives, we obtain the velocity,

$$\begin{aligned} \mathbf{v} &= \frac{d\mathbf{r}}{dt} = v_r\mathbf{e}_r + v_\theta\mathbf{e}_\theta + v_z\mathbf{e}_z \\ &= \frac{dr}{dt}\mathbf{e}_r + r\omega\mathbf{e}_\theta + \frac{dz}{dt}\mathbf{e}_z, \end{aligned} \tag{2.62}$$

and the acceleration,

$$\mathbf{a} = \frac{d\mathbf{v}}{dt} = a_r\mathbf{e}_r + a_\theta\mathbf{e}_\theta + a_z\mathbf{e}_z, \tag{2.63}$$

where

$$a_r = \frac{d^2r}{dt^2} - r\omega^2, \qquad a_\theta = r\alpha + 2\frac{dr}{dt}\omega, \qquad a_z = \frac{d^2z}{dt^2}. \tag{2.64}$$

Notice that Eqs. (2.62) and (2.63) reduce to the polar coordinate expressions for the velocity and acceleration, Eqs. (2.57) and (2.58), when P moves along a path in the x-y plane.

The following examples demonstrate the use of Eqs. (2.57) and (2.58) to analyze curvilinear motions of objects in terms of polar coordinates.

Example 2.13

Fig. 2.45

Suppose that you are standing on a large disk (a merry-go-round) rotating with constant angular velocity ω_0 and you start walking at constant speed v_0 along a straight radial line painted on the disk (Fig. 2.45). What are your velocity and acceleration when you are a distance r from the center of the disk?

STRATEGY

We can describe your motion in terms of polar coordinates (Fig. a). By using the information given about your motion and the motion of the disk, we can evaluate the terms in the expressions for the velocity and acceleration in terms of polar coordinates.

SOLUTION

The speed with which you walk along the radial line is the rate of change of r, $dr/dt = v_0$, and the angular velocity of the disk is the rate of change of θ, $\omega = \omega_0$. Your velocity is

$$\mathbf{v} = \frac{dr}{dt}\mathbf{e}_r + r\omega\mathbf{e}_\theta = v_0\mathbf{e}_r + r\omega_0\mathbf{e}_\theta.$$

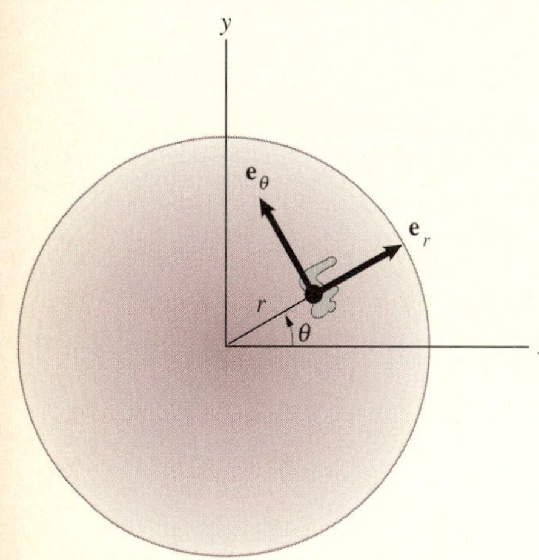

(a) Your position in terms of polar coordinates.

Your velocity consists of two components: the radial component due to the speed at which you are walking and a transverse component due to the disk's rate of rotation. The transverse component increases as your distance from the center of the disk increases.

Your walking speed $v_0 = dr/dt$ is constant, so $d^2r/dt^2 = 0$. Also, the disk's angular velocity $\omega_0 = d\theta/dt$ is constant, so $d^2\theta/dt^2 = 0$. The radial component of your acceleration is

$$a_r = \frac{d^2r}{dt^2} - r\omega^2 = -r\omega_0^2,$$

and the transverse component is

$$a_\theta = r\alpha + 2\frac{dr}{dt}\omega = 2v_0\omega_0.$$

DISCUSSION

If you have ever tried walking on a merry-go-round, you know it is a difficult proposition. This example indicates why: Subjectively, you are walking along a straight line with constant velocity, but you are actually experiencing the centripetal acceleration a_r and the Coriolis acceleration a_θ due to the disk's rotation.

Example 2.14

The robot arm in Fig. 2.46 is programmed so that point P describes the path

$$r = 1 - 0.5 \cos 2\pi t \text{ m},$$

$$\theta = 0.5 - 0.2 \sin 2\pi t \text{ rad}.$$

At $t = 0.8$ s, determine: (a) the velocity of P in terms of radial and transverse components; (b) the cartesian components of the velocity of P.

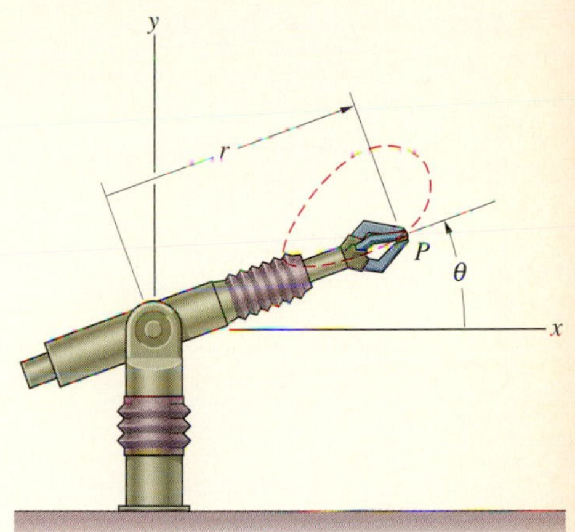

Fig. 2.46

STRATEGY

(a) Since we are given r and θ as functions of time, we can calculate the derivatives in the expression for the velocity in terms of polar coordinates and obtain the velocity as a function of time.
(b) By determining the value of θ at $t = 0.8$ s, we can use trigonometry to determine the cartesian components in terms of the radial and transverse components.

SOLUTION

(a) From Eq. (2.57), the velocity is

$$\begin{aligned}\mathbf{v} &= \frac{dr}{dt}\mathbf{e}_r + r\frac{d\theta}{dt}\mathbf{e}_\theta \\ &= (\pi \sin 2\pi t)\,\mathbf{e}_r + (1 - 0.5\cos 2\pi t)(-0.4\pi \cos 2\pi t)\,\mathbf{e}_\theta.\end{aligned}$$

At $t = 0.8$ s,

$$\mathbf{v} = -2.99\mathbf{e}_r - 0.328\mathbf{e}_\theta \text{ (m/s)}.$$

(b) At $t = 0.8$ s, $\theta = 0.690$ rad $= 39.5°$ (Fig. a). The x component of the velocity of P is

$$\begin{aligned}v_x &= v_r \cos 39.5° - v_\theta \sin 39.5° \\ &= (-2.99)\cos 39.5° - (-0.328)\sin 39.5° = -2.09 \text{ m/s},\end{aligned}$$

and the y component is

$$\begin{aligned}v_y &= v_r \sin 39.5° + v_\theta \cos 39.5° \\ &= (-2.99)\sin 39.5° + (-0.328)\cos 39.5° = -2.16 \text{ m/s}.\end{aligned}$$

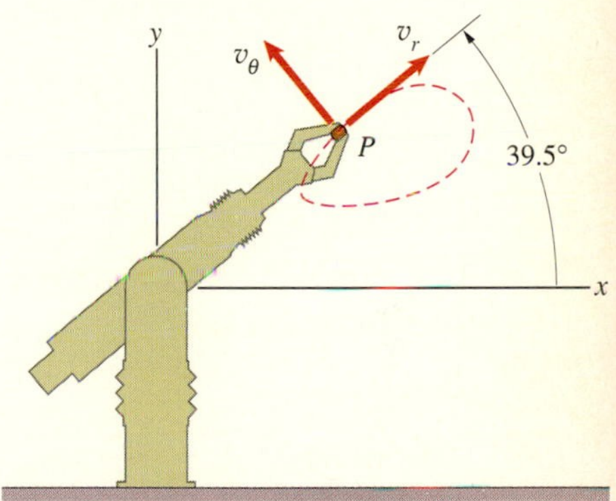

(a) Position at $t = 0.8$ s.

DISCUSSION

When you determine components of a vector in terms of different coordinate systems, you should always check them to make sure they give the same magnitude. In this example,

$$|\mathbf{v}| = \sqrt{(-2.99)^2 + (-0.328)^2} = \sqrt{(-2.09)^2 + (-2.16)^2} = 3.01 \text{ m/s}.$$

Remember that although the components of the velocity are different in the two coordinate systems, those components describe the same velocity vector.

Example 2.15

In the cam-follower mechanism shown in Fig. 2.47, the slotted bar rotates with constant angular velocity $\omega = 4$ rad/s and the radial position of the follower is determined by the elliptic profile of the stationary cam. The path of the follower is described by the polar equation

$$r = \frac{0.15}{1 + 0.5 \cos\theta} \text{ m.}$$

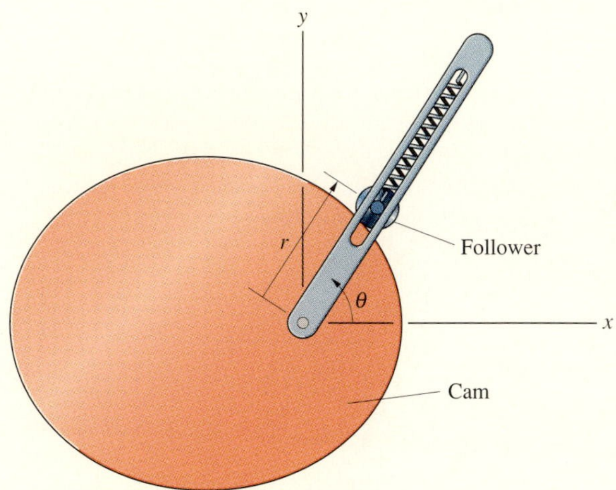

Fig. 2.47

Determine the velocity of the follower when $\theta = 45^\circ$ in terms of (a) polar coordinates; (b) cartesian coordinates.

STRATEGY

By taking the time derivative of the polar equation for the cam profile, we can obtain a relation between the known angular velocity and the radial component of velocity, which permits us to evaluate the velocity in terms of polar coordinates. Then by using Eqs. (2.60) we can obtain the velocity in terms of cartesian coordinates.

SOLUTION

(a) The polar equation for the cam profile is of the form $r = r(\theta)$. Taking its derivative with respect to time, we obtain

$$\begin{aligned}\frac{dr}{dt} &= \frac{dr(\theta)}{d\theta}\frac{d\theta}{dt}\\ &= \frac{d}{d\theta}\left(\frac{0.15}{1 + 0.5\cos\theta}\right)\frac{d\theta}{dt}\\ &= \left[\frac{0.075\sin\theta}{(1 + 0.5\cos\theta)^2}\right]\frac{d\theta}{dt}.\end{aligned}$$

The velocity of the follower in polar coordinates is therefore

$$\mathbf{v} = \frac{dr}{dt}\mathbf{e}_r + r\frac{d\theta}{dt}\mathbf{e}_\theta$$

$$= \left[\frac{0.075 \sin\theta}{(1 + 0.5\cos\theta)^2}\right]\frac{d\theta}{dt}\mathbf{e}_r + \left(\frac{0.15}{1 + 0.5\cos\theta}\right)\frac{d\theta}{dt}\mathbf{e}_\theta.$$

The angular velocity $\omega = d\theta/dt = 4$ rad/s, so we can evaluate the polar components of the velocity when $\theta = 45°$, obtaining

$$\mathbf{v} = 0.116\mathbf{e}_r + 0.443\mathbf{e}_\theta \text{ (m/s)}.$$

(b) Substituting Eqs. (2.60) with $\theta = 45°$ into the polar coordinate expression for the velocity, we obtain the velocity in terms of cartesian coordinates,

$$\mathbf{v} = 0.116\mathbf{e}_r + 0.443\mathbf{e}_\theta$$

$$= 0.116(\cos 45°\,\mathbf{i} + \sin 45°\,\mathbf{j}) + 0.443(-\sin 45°\,\mathbf{i} + \cos 45°\,\mathbf{j})$$

$$= -0.232\mathbf{i} + 0.395\mathbf{j} \text{ (m/s)}.$$

Problems

2.137 At a particular time, the polar coordinates of a point P moving in the x-y plane are $r = 4$ ft, $\theta = 0.5$ rad, and their time derivatives are $dr/dt = 8$ ft/s and $d\theta/dt = -2$ rad/s.
(a) What is the magnitude of the velocity of P?
(b) What are the cartesian components of the velocity of P?

2.138 In Problem 2.137, suppose that $d^2r/dt^2 = 6$ ft/s^2 and $d^2\theta/dt^2 = 3$ rad/s^2. At the instant described, determine: (a) the magnitude of the acceleration of P; (b) the instantaneous radius of curvature of the path.

2.139 The polar coordinates of a point P moving in the x-y plane are $r = t^3 - 4t$ m, $\theta = t^2 - t$ rad. Determine the velocity of P in terms of radial and transverse components at $t = 1$ s.

2.140 In Problem 2.139, what is the acceleration of P in terms of radial and transverse components at $t = 1$ s?

2.141 The radial line rotates with a constant angular velocity of 2 rad/s. Point P moves along the line at a constant speed of 4 m/s. Determine the magnitudes of the velocity and acceleration of P when $r = 2$ m.

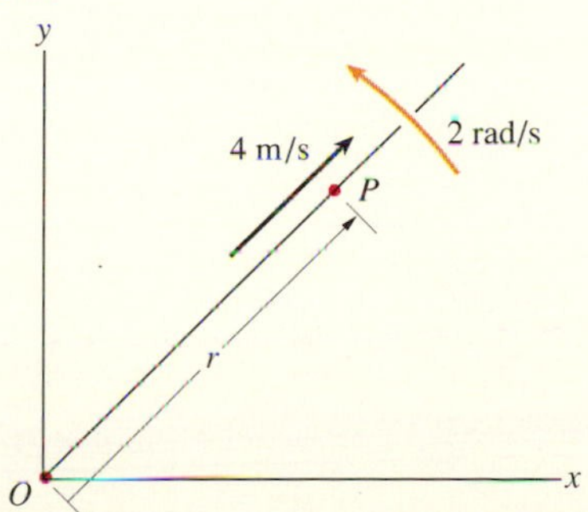

P2.141

2.142 At the instant shown, the coordinates of the slider A are $x = 1.6$ ft, $y = 1.0$ ft, and its velocity and acceleration are $\mathbf{v} = 10\mathbf{j}$ (ft/s), $\mathbf{a} = -32.2\mathbf{j}$ (ft/s^2). Determine the slider's velocity and acceleration in terms of polar coordinates.

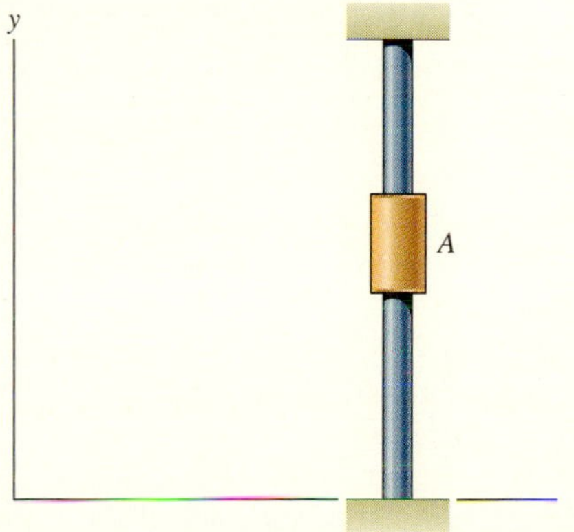

P2.142

2.143 In Problem 2.142, determine d^2r/dt^2 and $d^2\theta/dt^2$ at the instant shown, where r and θ are the slider's polar coordinates.

2.144 A boat searching for underwater archaeological sites in the Aegean Sea moves at 4 knots and follows the path $r = 10\theta$ m, where θ is in radians. (A knot is one nautical mile, or 1852 meters, per hour.) When $\theta = 2\pi$ rad, determine the boat's velocity (a) in terms of polar coordinates; (b) in terms of cartesian coordinates.

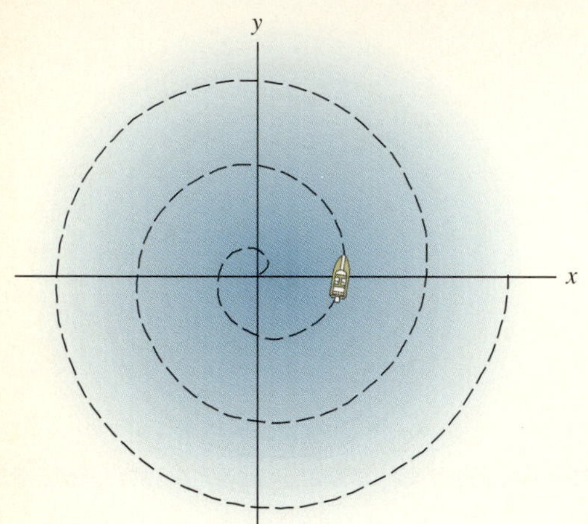

P2.144

2.145 In Problem 2.144, what is the boat's acceleration in terms of polar coordinates?

Strategy: The magnitude of the boat's velocity is constant, so you know that the tangential component of its acceleration equals zero.

2.146 A point P moves in the x-y plane along the path described by the equation $r = e^{\theta}$, where θ is in radians. The angular velocity $d\theta/dt = \omega_0$ = constant, and $\theta = 0$ at $t = 0$.
(a) Draw a polar graph of the path for values of θ from zero to 2π.
(b) Show that the velocity and acceleration as functions of time are $\mathbf{v} = \omega_0 \mathbf{e}^{\omega_0 t}(\mathbf{e}_r + \mathbf{e}_\theta)$, $\mathbf{a} = 2\omega_0^2 \mathbf{e}^{\omega_0 t}\,\mathbf{e}_\theta$.

2.147 In Problem 2.146, show that the instantaneous radius of curvature of the path as a function of time is $\rho = \sqrt{2}\mathbf{e}^{\omega_0 t}$.

2.148 In Example 2.14, determine the acceleration of point P at $t = 0.8$ s (a) in terms of radial and transverse components; (b) in terms of cartesian components.

2.149 A bead slides along a wire that rotates in the x-y plane with constant angular velocity ω_0. The radial component of the bead's acceleration is zero. The radial component of its velocity is v_0 when $r = r_0$. Determine the radial and transverse components of the bead's velocity as a function of r.

Strategy: The radial component of the bead's velocity is

$$v_r = \frac{dr}{dt},$$

and the radial component of its acceleration is

$$a_r = \frac{d^2r}{dt^2} - r\left(\frac{d\theta}{dt}\right)^2 = \frac{dv_r}{dt} - r\omega_0^2.$$

By using the chain rule,

$$\frac{dv_r}{dt} = \frac{dv_r}{dr}\frac{dr}{dt} = \frac{dv_r}{dr}v_r,$$

you can express the radial component of the acceleration in the form

$$a_r = \frac{dv_r}{dr}v_r - r\omega_0^2.$$

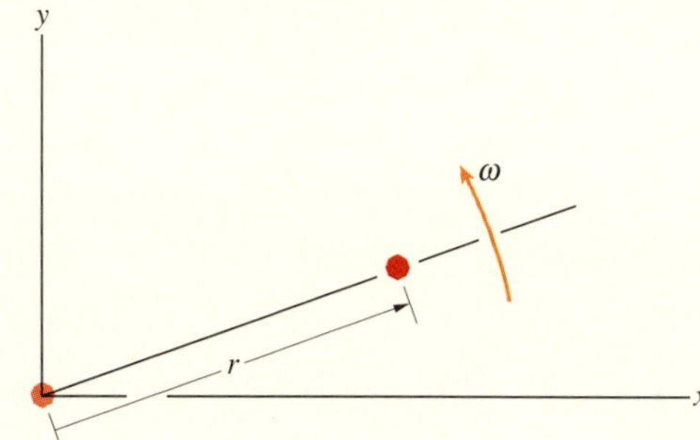

P2.149

2.150 If the motion of a point in the x-y plane is such that its transverse component of acceleration a_θ is zero, show that the product of its radial position and its transverse velocity is constant:

$$rv_\theta = \text{constant}.$$

2.151 From astronomical data, Kepler deduced that the line from the sun to a planet traces out equal areas in equal times (Fig. a). Show that this result follows from the fact that the transverse component a_θ of the planet's acceleration is zero. [When r changes by an amount dr and θ changes by an amount $d\theta$ (Fig. b), the resulting differential element of area is $dA = \frac{1}{2}r(r\,d\theta)$.]

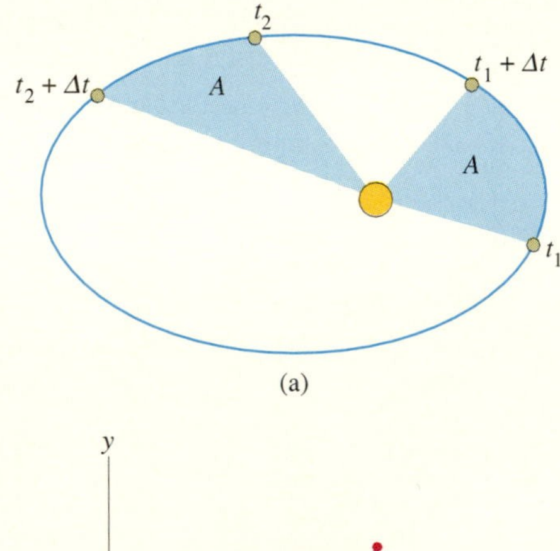

(a)

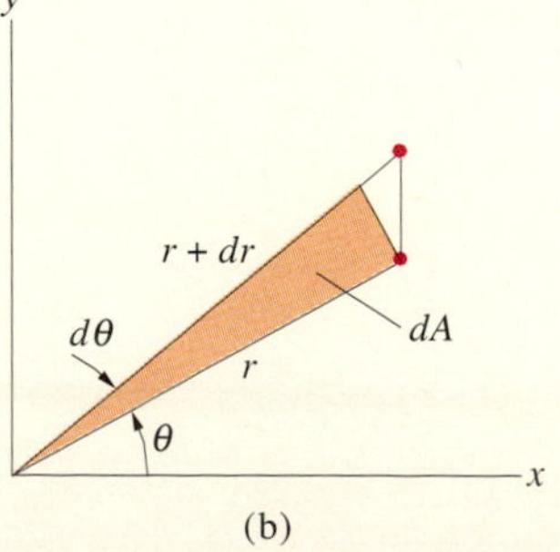

(b)

P2.151

2.152 The bar rotates in the x-y plane with constant angular velocity ω_0. The radial component of acceleration of the collar C is $a_r = -Kr$, where K is a constant. When $r = r_0$, the radial component of velocity of C is v_0. Determine the radial and transverse components of the velocity of C as functions of r.

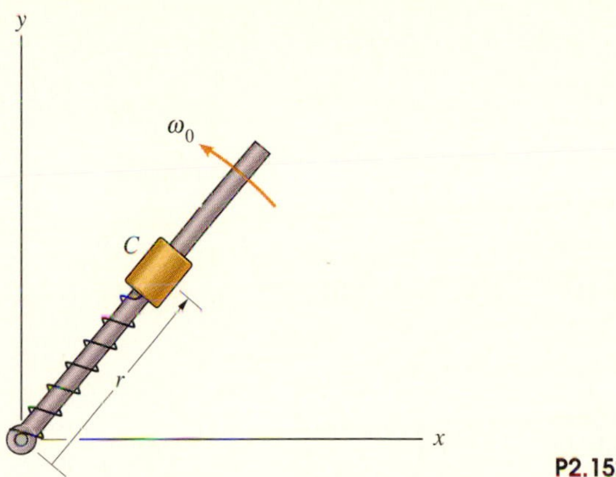

P2.152

2.153 The hydraulic actuator moves the pin P upward with constant velocity $\mathbf{v} = 0.2\mathbf{j}$ (ft/s). Determine the velocity components v_r and v_θ of the pin and the angular velocity of the slotted bar when $\theta = 30°$.

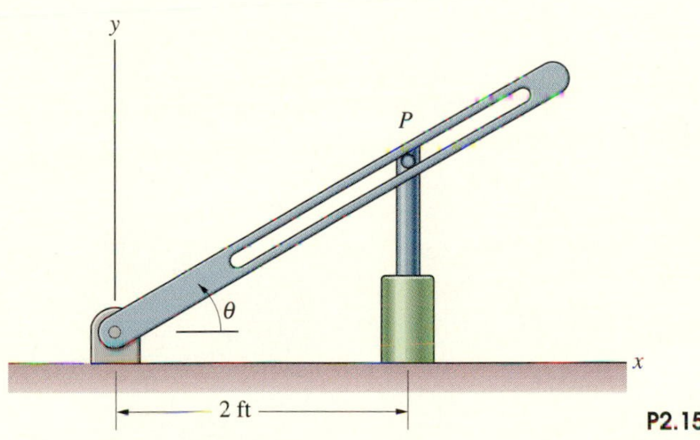

P2.153

2.154 In Problem 2.153, determine the acceleration components a_r and a_θ of the pin and the angular acceleration of the slotted bar when $\theta = 30°$.

2.155 In Example 2.15, determine the velocity of the cam follower when $\theta = 135°$: (a) in terms of polar coordinates; (b) in terms of cartesian coordinates.

2.156 In Example 2.15, determine the acceleration of the cam follower when $\theta = 135°$: (a) in terms of polar coordinates; (b) in terms of cartesian coordinates.

2.157 In the cam-follower mechanism, the slotted bar rotates with constant angular velocity $\omega = 10$ rad/s and the radial position of the follower A is determined by the profile of the stationary cam. The path of the follower is described by the polar equation

$$r = 1 + 0.5 \cos 2\theta \text{ ft}.$$

Determine the velocity of the cam follower when $\theta = 30°$: (a) in terms of polar coordinates; (b) in terms of cartesian coordinates.

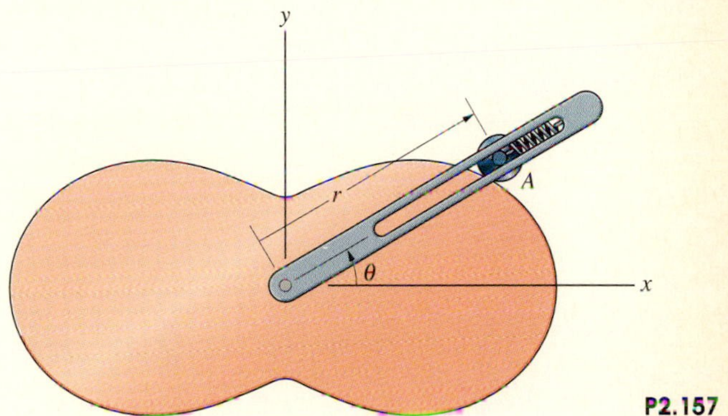

P2.157

2.158 In Problem 2.157, determine the acceleration of the cam follower when $\theta = 30°$: (a) in terms of polar coordinates; (b) in terms of cartesian coordinates.

2.159 The cartesian coordinates of a point P in the x-y plane are related to its polar coordinates by the relations $x = r\cos\theta$, $y = r\sin\theta$.

(a) Show that the unit vectors $\mathbf{i}$ and $\mathbf{j}$ are related to the unit vectors $\mathbf{e}_r$ and $\mathbf{e}_\theta$ by

$$\mathbf{i} = \mathbf{e}_r \cos\theta - \mathbf{e}_\theta \sin\theta,$$

$$\mathbf{j} = \mathbf{e}_r \sin\theta + \mathbf{e}_\theta \cos\theta.$$

(b) Beginning with the expression for the position vector of P in terms of cartesian coordinates, $\mathbf{r} = x\mathbf{i} + y\mathbf{j}$, derive Eq. (2.52) for the position vector in terms of polar coordinates.

(c) By taking the time derivative of the position vector of point P expressed in terms of cartesian coordinates, derive Eq. (2.55) for the velocity in terms of polar coordinates.

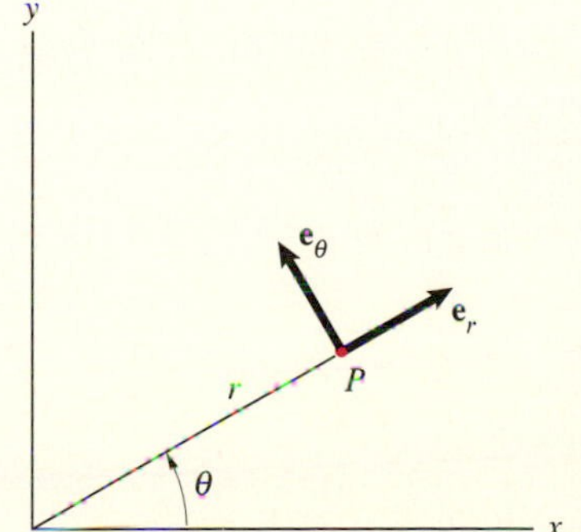

P2.159

2.160 The airplane flies in a straight line at 400 mi/hr. The radius of its propeller is 5 ft, and it turns at 2000 rpm in the counterclockwise direction when seen from the front of the airplane. Determine the velocity and acceleration of a point on the tip of the propeller in terms of cylindrical coordinates. (Let the z axis be oriented as shown in the figure.)

P2.160

2.161 A charged particle P in a magnetic field moves along the spiral path described by $r = 1$ m, $\theta = 2z$ rad, where z is in meters. The particle moves along the path in the direction shown with constant speed $|\mathbf{v}| = 1$ km/s. What is the velocity of the particle in terms of cylindrical coordinates?

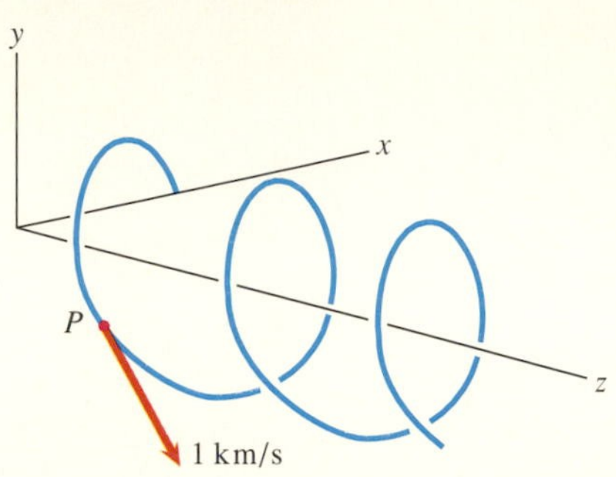

P2.161

2.4 Relative Motion

Our discussion so far has been limited to the motion of a single point relative to a given reference frame. However, often it is not the motion of an individual point, but motions of two or more points *relative to each other* that we must consider. For example, if a pilot wants to land on an aircraft carrier (Fig. 2.48), the individual motions of the carrier and his plane relative to the earth concern him less than the motion of his plane *relative to the carrier*. Pairs skaters (Fig. 2.49) must carefully control both their individual motions relative to the ice and their motion *relative to each other* to successfully complete their moves. In this section we discuss the analysis of the relative motions of points.

(a)

Fig. 2.48
The pilot's primary concern is the motion of his plane relative to the carrier.

Fig. 2.49
Pairs skaters must control their motions relative to each other.

Suppose that A and B are two points whose individual motions we measure relative to a reference frame with origin O, and let's consider how to describe the motion of A relative to B. Let $\mathbf{r}_A$ and $\mathbf{r}_B$ be the position vectors of points A and B relative to O (Fig. 2.50). The vector $\mathbf{r}_{A/B}$ is the position vector of point A relative to point B. These vectors are related by

$$\mathbf{r}_A = \mathbf{r}_B + \mathbf{r}_{A/B}. \tag{2.65}$$

Taking the time derivative of this relation, we obtain

$$\mathbf{v}_A = \mathbf{v}_B + \mathbf{v}_{A/B}, \tag{2.66}$$

where $\mathbf{v}_A$ is the velocity of A relative to O, $\mathbf{v}_B$ is the velocity of B relative to O, and $\mathbf{v}_{A/B} = d\mathbf{r}_{A/B}/dt$ is the velocity of A relative to B.

In our example of an airplane approaching an aircraft carrier, the plane could be point A and the carrier point B. The individual motions of the carrier and the plane would be measured (for example, by using on-board inertial navigation systems) relative to a reference frame fixed with respect to the earth. Knowing the velocities of the plane, $\mathbf{v}_A$, and the carrier, $\mathbf{v}_B$, the pilot could use Eq. (2.66) to determine his velocity relative to the carrier.

Taking the time derivative of Eq. (2.66), we obtain

$$\mathbf{a}_A = \mathbf{a}_B + \mathbf{a}_{A/B}, \tag{2.67}$$

where $\mathbf{a}_A$ and $\mathbf{a}_B$ are the accelerations of A and B relative to O and $\mathbf{a}_{A/B} = d\mathbf{v}_{A/B}/dt$ is the acceleration of A relative to B.

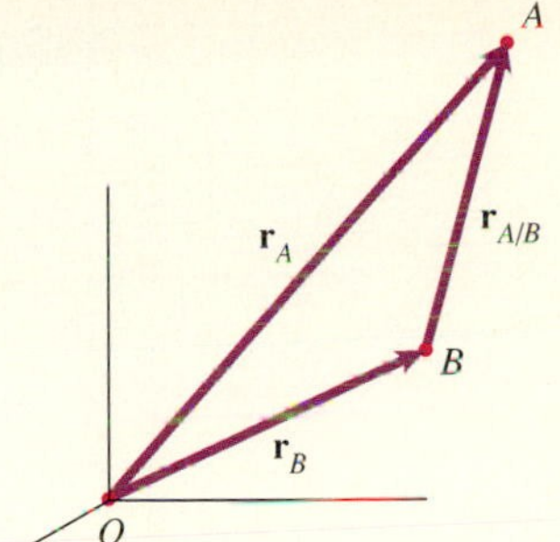

Fig. 2.50
Two points A and B and a reference frame with origin O. The vectors $\mathbf{r}_A$ and $\mathbf{r}_B$ specify the positions of A and B relative to O, and $\mathbf{r}_{A/B}$ specifies the position of A relative to B.

The following examples show how you can use Eqs. (2.65)–(2.67) to analyze relative motions of objects.

Example 2.16

The aircraft carrier in Fig. 2.51 travels north at 15 knots relative to the earth. With its radar, the carrier determines that the velocity of an airplane relative to the carrier is horizontal and of magnitude 300 knots toward the northeast. What are the magnitude and direction of the plane's velocity relative to the earth?

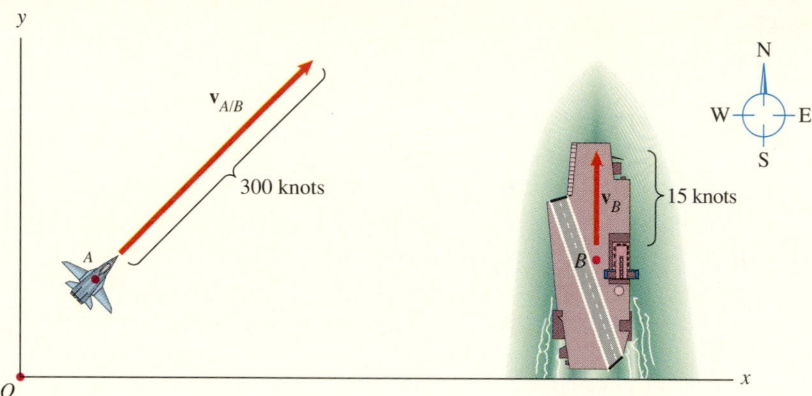

Fig. 2.51

The airplane (A), carrier (B), and a point O fixed relative to the earth.

STRATEGY

Since we know the carrier's velocity relative to the earth and the velocity of the plane relative to the carrier, we can use Eq. (2.66) to determine the plane's velocity relative to the earth.

SOLUTION

Let the airplane be point A and let the aircraft carrier be point B (Fig. 2.51). The xy coordinate system is fixed relative to the earth. The velocity of the carrier relative to the earth and the velocity of the plane *relative to the carrier* are shown. The velocity of the carrier is

$$\mathbf{v}_B = 15\mathbf{j} \text{ (knots)},$$

and the velocity of the plane relative to the carrier is

$$\mathbf{v}_{A/B} = 300 \cos 45^\circ \, \mathbf{i} + 300 \sin 45^\circ \, \mathbf{j} \text{ (knots)}.$$

Therefore the velocity of the plane relative to the earth is

$$\begin{aligned} \mathbf{v}_A = \mathbf{v}_B + \mathbf{v}_{A/B} &= 300 \cos 45^\circ \, \mathbf{i} + (15 + 300 \sin 45^\circ) \, \mathbf{j} \\ &= 212\mathbf{i} + 227\mathbf{j} \text{ (knots)}. \end{aligned}$$

The magnitude of the airplane's velocity relative to the earth is

$$\sqrt{(212)^2 + (227)^2} = 311 \text{ knots},$$

and its direction is arctan (212/227) = 43.0° east of north.

Example 2.17

A ship moving at 5 m/s relative to the water is in a uniform current flowing east at 2 m/s. If the captain wants to sail northwest relative to the earth, what direction should she point the ship? What will be the resulting magnitude of the ship's velocity relative to the earth?

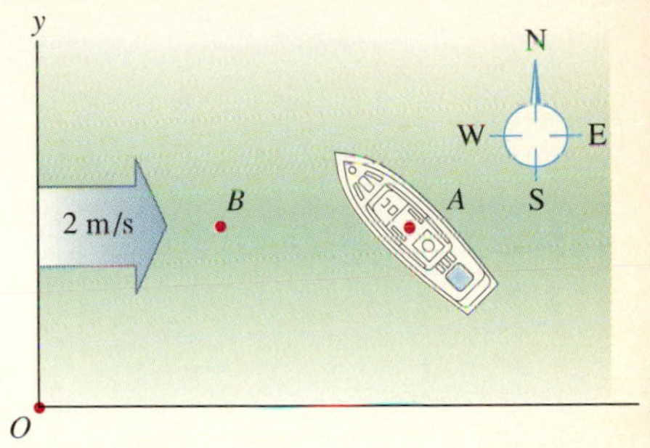

(a) The ship A and a point B moving with the water.

STRATEGY

Let the ship be point A and let B be a point moving with the water (Fig. a). The xy coordinate system is fixed relative to the earth. We know $\mathbf{v}_B$, the desired direction of $\mathbf{v}_A$, and the magnitude of $\mathbf{v}_{A/B}$. We can use Eq. (2.66) to determine the magnitude of $\mathbf{v}_A$ and the direction of $\mathbf{v}_{A/B}$.

SOLUTION

The velocity of the ship relative to the earth is equal to the velocity of the water relative to the earth plus the velocity of the ship relative to the water:

$$\mathbf{v}_A = \mathbf{v}_B + \mathbf{v}_{A/B}.$$

In Fig. (b) we show this relationship together with the information we know about these velocities: The velocity of the current is 2 m/s toward the east, the magnitude of the velocity of the ship relative to the water is 5 m/s, and the direction of the velocity of the ship relative to the earth is northwest. In terms of the coordinate system shown, the velocity of the current is $\mathbf{v}_B = 2\mathbf{i}$ (m/s). We don't know the magnitude of $\mathbf{v}_A$, but because we know its direction, we can write it in terms of components as

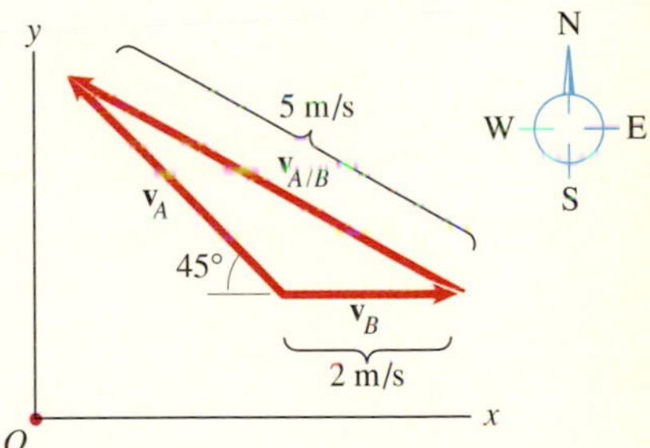

(b) Diagram of the velocity vectors.

$$\mathbf{v}_A = -|\mathbf{v}_A| \cos 45° \, \mathbf{i} + |\mathbf{v}_A| \sin 45° \, \mathbf{j}.$$

The velocity of the ship relative to the water is

$$\mathbf{v}_{A/B} = \mathbf{v}_A - \mathbf{v}_B = -(|\mathbf{v}_A| \cos 45° + 2)\mathbf{i} + |\mathbf{v}_A| \sin 45° \, \mathbf{j}.$$

The magnitude of this vector is

$$|\mathbf{v}_{A/B}| = \sqrt{(|\mathbf{v}_A| \cos 45° + 2)^2 + (|\mathbf{v}_A| \sin 45°)^2} = 5 \text{ m/s}.$$

Solving this equation, we obtain $|\mathbf{v}_A| = 3.38$ m/s, so the velocity of the ship relative to the water is

$$\mathbf{v}_{A/B} = -4.39\mathbf{i} + 2.39\mathbf{j} \text{ (m/s)}.$$

The captain must point her ship at arctan (4.39/2.39) = 61.4° west of north to cause the ship to travel in the northwest direction relative to the earth.

DISCUSSION

The problem described in this example must be solved whenever a ship travels in a current or an airplane flies in a wind that is not parallel to its desired course.

Example 2.18

The bars OP and PQ in Fig. 2.52 rotate with constant angular velocities relative to the fixed coordinate system shown. What is the acceleration of point Q relative to the coordinate system?

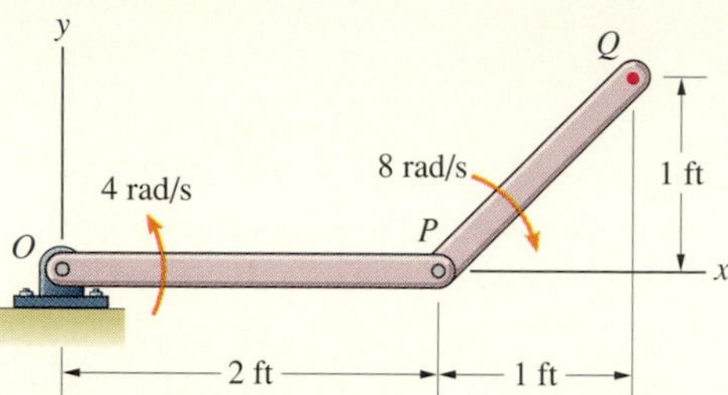

Fig. 2.52

STRATEGY

Relative to point P, point Q moves in a circular path around P with constant angular velocity. We can use polar coordinates to determine $\mathbf{a}_{Q/P}$ and then express it in terms of components in the xy coordinate system. Point P moves in a circular path about O with constant angular velocity, so we can also use polar coordinates to determine the acceleration $\mathbf{a}_P$ and then express it in terms of components in the xy coordinate system. Then the acceleration of Q relative to O is $\mathbf{a}_Q = \mathbf{a}_P + \mathbf{a}_{Q/P}$.

SOLUTION

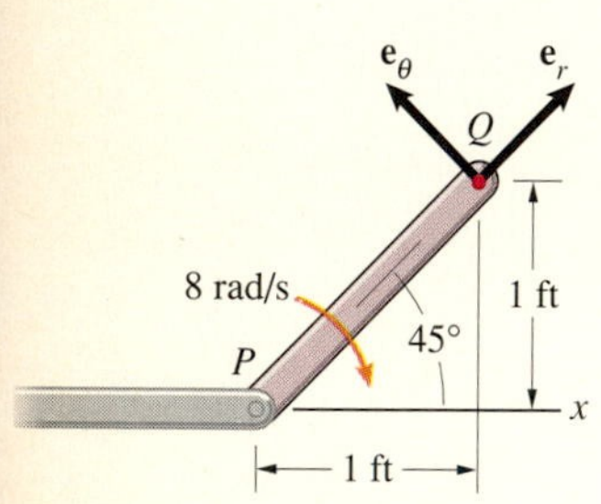

(a) Determining the acceleration of Q relative to P.

Expressing the motion of Q relative to P in terms of polar coordinates (Fig. a), we obtain the radial component of the acceleration,

$$a_r = \frac{d^2r}{dt^2} - r\omega^2 = 0 - (\sqrt{2})(-8)^2 = -90.5 \text{ ft/s}^2,$$

and the transverse component,

$$a_\theta = r\alpha + 2\frac{dr}{dt}\omega = 0.$$

Therefore, the acceleration of Q relative to P in terms of the xy coordinate system is

$$\mathbf{a}_{Q/P} = a_r \cos 45° \, \mathbf{i} + a_r \sin 45° \, \mathbf{j} = -64\mathbf{i} - 64\mathbf{j} \text{ (ft/s}^2\text{)}.$$

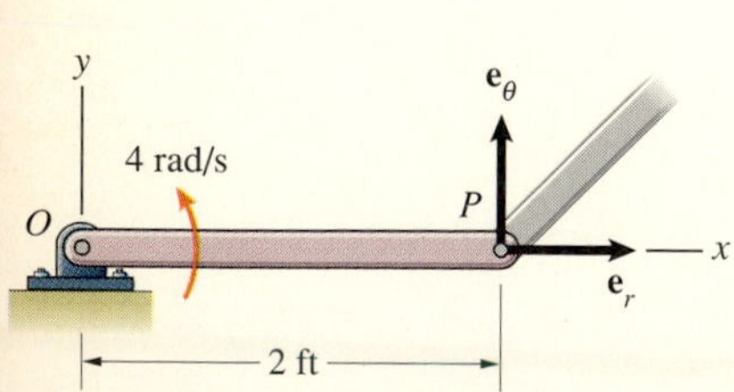

(b) Determining the acceleration of P relative to O.

We also express the acceleration of P relative to O in terms of polar coordinates (Fig. b). The radial component is

$$a_r = \frac{d^2r}{dt^2} - r\omega^2 = 0 - (2)(4)^2 = -32 \text{ ft/s}^2,$$

and the transverse component is

$$a_\theta = r\alpha + 2\frac{dr}{dt}\omega = 0.$$

The acceleration of P relative to O in terms of the xy coordinate system is

$$\mathbf{a}_P = a_r\mathbf{i} = -32\mathbf{i}\ (\text{ft/s}^2).$$

Therefore the acceleration of Q relative to O is

$$\mathbf{a}_Q = \mathbf{a}_P + \mathbf{a}_{Q/P} = -32\mathbf{i} - 64\mathbf{i} - 64\mathbf{j}$$
$$= -96\mathbf{i} - 64\mathbf{j}\ (\text{ft/s}^2).$$

DISCUSSION

This example demonstrates an important use of the concept of relative motion. The motion of point Q relative to point O is quite complicated. But because the motion of Q relative to P and the motion of P relative to O are comparatively simple, we can take advantage of the equations describing relative motion to obtain information about the motion of Q relative to O.

Problems

2.162 Two cars A and B approach an intersection. Car A is going 20 m/s and is decelerating at 2 m/s^2, and car B is going 10 m/s and is decelerating at 3 m/s^2. In terms of the earth-fixed coordinate system shown, determine the velocity of car A relative to car B and the velocity of car B relative to car A.

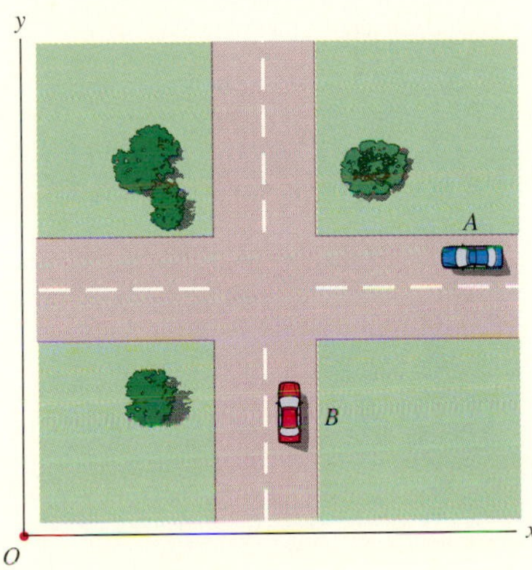

P2.162

2.163 In Problem 2.162, determine the acceleration of car A relative to car B and the acceleration of car B relative to car A.

2.164 Suppose that the two cars in Problem 2.162 approach the intersection with constant velocities. Prove that the cars will reach the intersection at the same time if the velocity of car A relative to car B points from car A toward car B.

2.165 Two sailboats have constant velocities $\mathbf{v}_A$ and $\mathbf{v}_B$ relative to the earth. The skipper of boat A sights a point on the horizon behind boat B. Seeing that boat B appears stationary relative to that point, he knows he must change course to avoid a collision. Use Eq. (2.66) to explain why.

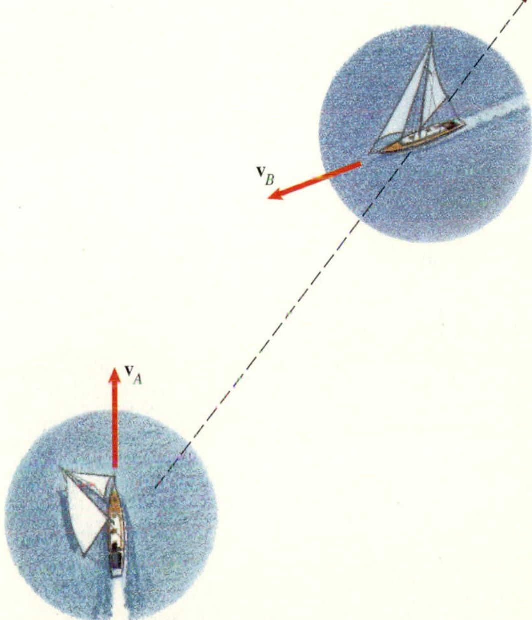

P2.165

2.166 Two projectiles A and B are launched from O at the same time with the initial velocities and elevation angles shown relative to the earth-fixed coordinate system. At the instant B reaches its highest point, determine: (a) the acceleration of A relative to B; (b) the velocity of A relative to B; (c) the position vector of A relative to B.

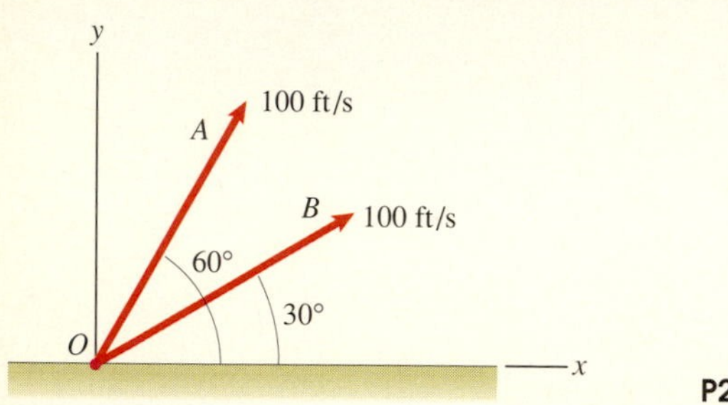

P2.166

2.167 In a machining process, the disk rotates about the fixed point O with a constant angular velocity of 10 rad/s. In terms of the nonrotating coordinate system shown, what is the magnitude of the velocity of A relative to B?

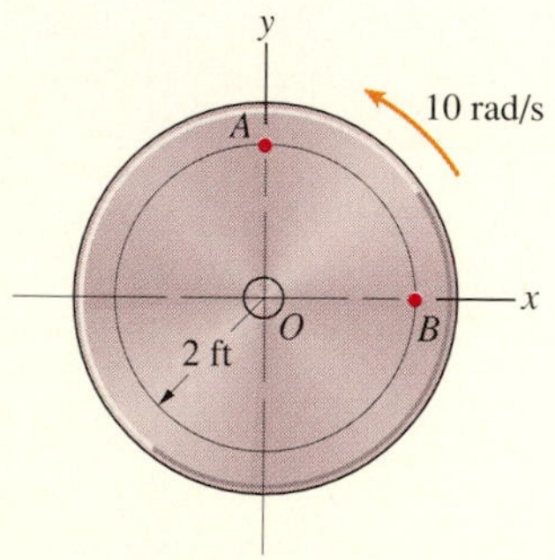

P2.167

2.168 In Problem 2.167, what is the magnitude of the acceleration of A relative to B?

2.169 The bar rotates about the fixed point O with a constant angular velocity of 2 rad/s. Point A moves outward along the bar at a constant rate of 100 mm/s. Point B is a fixed point on the bar. In terms of the nonrotating reference frame with origin O, what is the magnitude of the velocity of point A relative to point B?

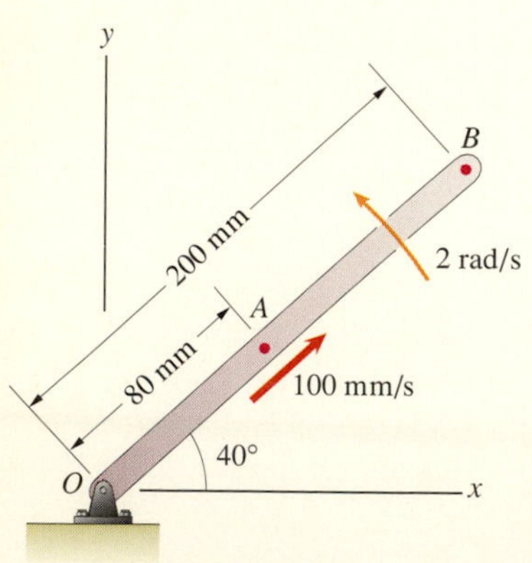

P2.169

2.170 In Problem 2.169, what is the magnitude of the acceleration of point B relative to point A at the instant shown?

2.171 The bars OA and AB are each 400 mm long and rotate in the x-y plane. OA has a counterclockwise angular velocity of 10 rad/s and a counterclockwise angular acceleration of 2 rad/s^2. AB has a constant counterclockwise angular velocity of 5 rad/s relative to the nonrotating coordinate system. What is the velocity of point B relative to point A?

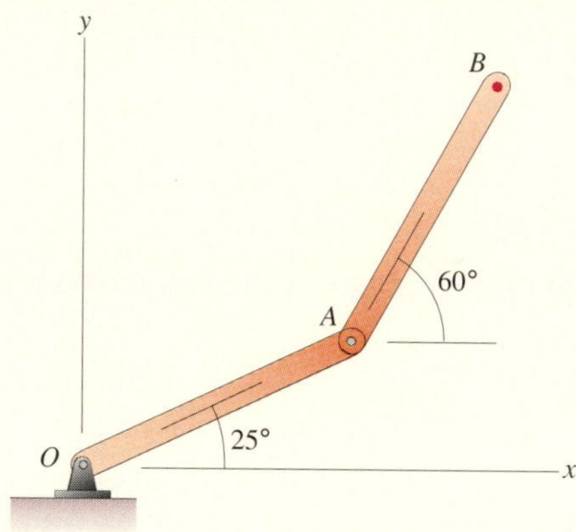

P2.171

2.172 In Problem 2.171, what is the acceleration of point B relative to point A?

2.173 In Problem 2.171, what is the velocity of point B relative to the fixed point O?

2.174 In Problem 2.171, what is the acceleration of point B relative to the fixed point O?

2.175 Points O and O' are fixed relative to the x-y reference frame. At the instant shown, $r_A = 1.8$ m, $\theta_A = 50°$, $dr_A/dt = 12$ m/s, $d\theta_A/dt = 4$ rad/s, $r_B = 1.8$ m, $\theta_B = 40°$, $dr_B/dt = -8$ m/s, and $d\theta_B/dt = 6$ rad/s. What is the magnitude of the velocity of point A relative to point B?

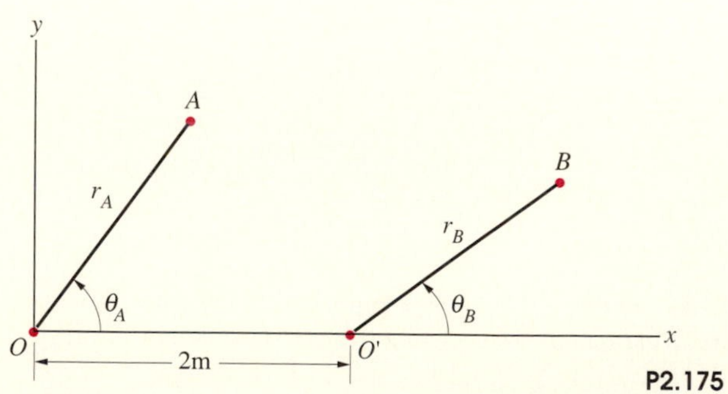

P2.175

2.176 At the instant shown in Problem 2.175, $d^2r_A/dt^2 = 20$ m/s^2, $d^2\theta_A/dt^2 = 0$, $d^2r_B/dt^2 = 10$ m/s^2, and $d^2\theta_B/dt^2 = 0$. What is the magnitude of the acceleration of point A relative to point B?

2.177 The train on the circular track is traveling at a constant speed of 50 ft/s. The train on the straight track is traveling at 20 ft/s and is increasing its speed at 2 ft/s^2. In terms of the earth-fixed coordinate system shown, what is the velocity of passenger A relative to passenger B?

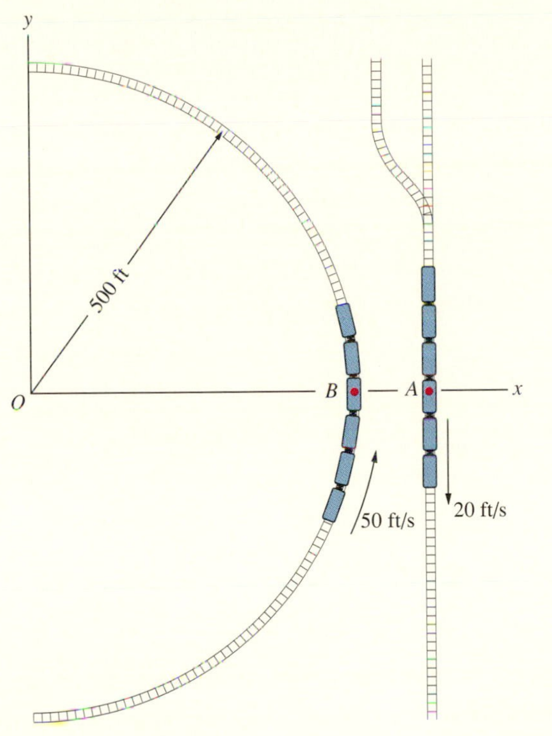

P2.177

2.178 In Problem 2.177, what is the acceleration of passenger A relative to passenger B?

2.179 The velocity of the boat relative to the earth-fixed coordinate system is 40**i** ft/s and is constant. The length of the tow rope is 50 ft. The angle θ is 30° and is increasing at a constant rate of 10°/s. What are the velocity and acceleration of the skier relative to the boat?

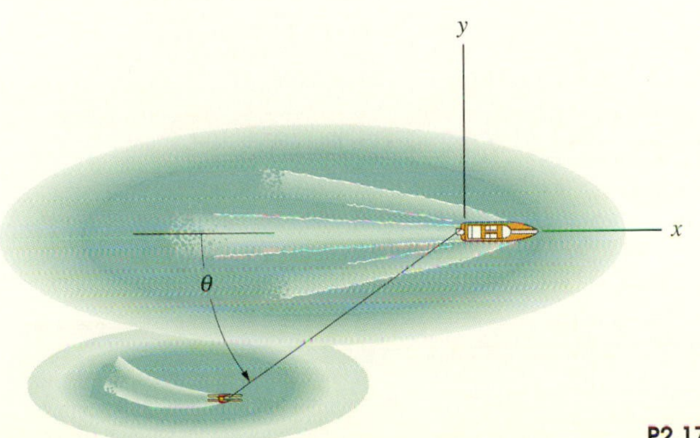

P2.179

2.180 In Problem 2.179, what are the velocity and acceleration of the skier relative to the earth?

2.181 The hockey player is skating with velocity components $v_x = 4$ ft/s, $v_z = -20$ ft/s when he hits a slap shot with a velocity of magnitude 100 ft/s *relative to him.* The position of the puck when he hits it is $x = 12$ ft, $z = 12$ ft. If he hits the puck so that its velocity vector *relative to him* is directed toward the center of the goal, where will be puck intersect the x axis? Will it enter the 6-ft wide goal?

P2.181

2.182 In Problem 2.181, at what point on the x axis should the player aim the puck's velocity vector relative to him so that it enters the center of the goal?

2.183 An airplane flies in a jet stream flowing east at 100 mi/hr. The airplane's airspeed (its velocity relative to the air) is 500 mi/hr toward the northwest. What are the magnitude and direction of the airplane's velocity relative to the earth?

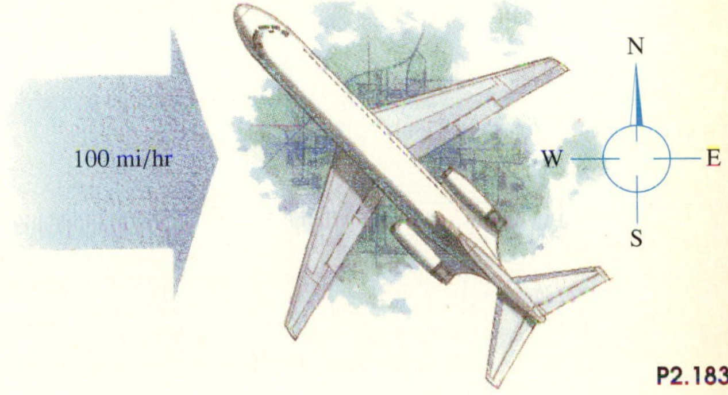

P2.183

2.184 In Problem 2.183, if the pilot wants to fly toward a city that is northwest of his current position, in which direction should he point the airplane, and what will be the magnitude of his velocity relative to the earth?

2.185 A river flows north at 3 m/s. (Assume that the current is uniform.) If you want to travel in a straight line from point C to point D in a boat that moves at a constant speed of 10 m/s relative to the water, in what direction do you point the boat? How long does it take to make the crossing?

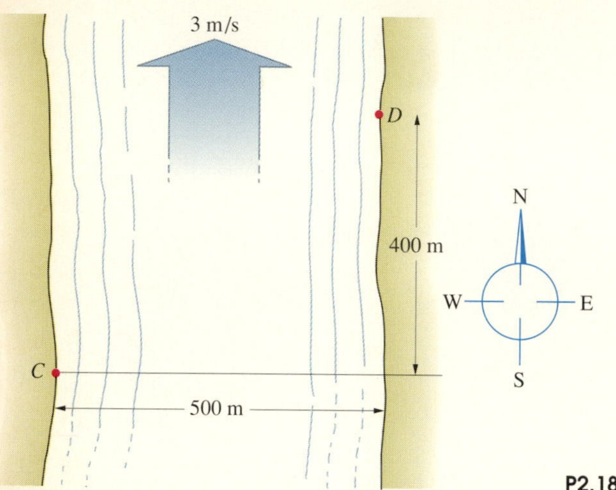

P2.185

2.186 In Problem 2.185, what is the minimum boat speed relative to the water necessary to make the trip from point C to point D?

2.187 Relative to the earth, a sailboat sails north at velocity v_0 and then sails east at the same velocity. The velocity of the wind is uniform and constant. A "tell-tale" on the boat points in the direction of the velocity of the wind relative to the boat. What are the direction and magnitude of the wind's velocity relative to the earth? (Your answer for the magnitude of the velocity of the wind will be in terms of v_0.)

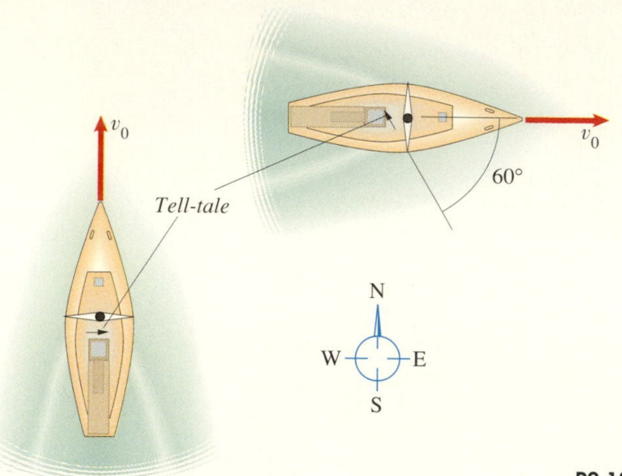

P2.187

2.188 The origin O of the nonrotating coordinate system is at the center of the earth, and the y axis points north. The satellite A on the x axis is in a circular polar orbit of radius R, and its velocity is $v_A\mathbf{j}$. Let ω be the angular velocity of the earth. What is the satellite's velocity relative to the point B on the earth directly below the satellite?

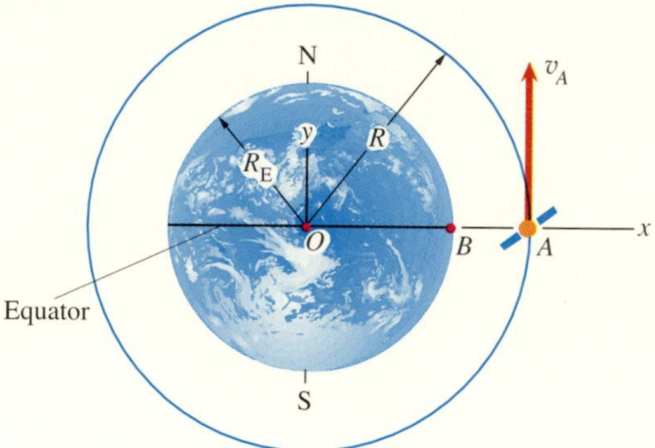

P2.188

2.189 In Problem 2.188, what is the satellite's acceleration relative to the point B on the earth directly below the satellite?

Computational Mechanics

The following examples and problems are designed for the use of a programmable calculator or computer.

Example 2.19

With buoyancy accounted for, the downward acceleration of a steel ball falling in the container of liquid in Fig. 2.53 is $a = 0.9g - cv$, where c is a constant that is proportional to the viscosity of the liquid. To determine the viscosity, a rheologist releases the ball from rest at the surface of the liquid. If the ball requires 2 s to fall the 2 m to the bottom, what is the value of c?

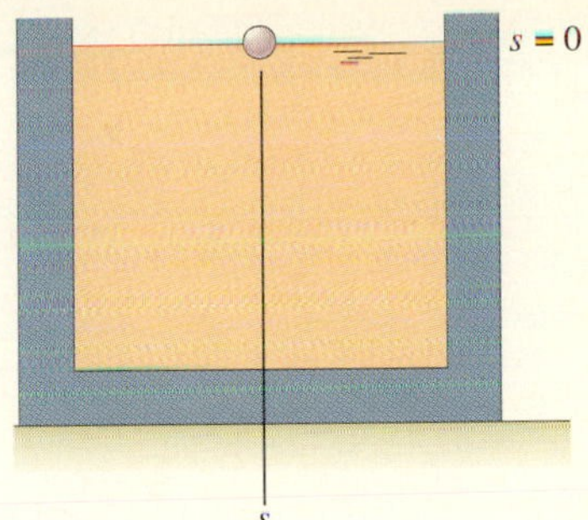

Fig. 2.53

STRATEGY

We can obtain an equation for c by determining the distance the ball falls as a function of time.

SOLUTION

We measure the ball's position s downward from the point of release and let $t = 0$ be the time of release.

The acceleration is

$$a = \frac{dv}{dt} = 0.9g - cv.$$

Separating variables and integrating gives

$$\int_0^v \frac{dv}{0.9g - cv} = \int_0^t dt,$$

$$\left[\frac{1}{(-c)}\ln(0.9g - cv)\right]_0^v = [t]_0^t,$$

$$\frac{1}{(-c)}[\ln(0.9g - cv) - \ln(0.9g)] = \frac{1}{(-c)}\ln\left(\frac{0.9g - cv}{0.9g}\right) = t.$$

Solving for v, we obtain

$$v = \frac{ds}{dt} = \frac{0.9g}{c}(1 - e^{-ct}).$$

Integrating this equation,

$$\int_0^s ds = \int_0^t \frac{0.9g}{c}(1 - e^{-ct})\,dt,$$

$$[s]_0^s = \frac{0.9g}{c}\left[t - \frac{e^{-ct}}{(-c)}\right]_0^t,$$

we obtain the distance the ball has fallen as a function of the time from its release:

$$s = \frac{0.9g}{c^2}(ct - 1 + e^{-ct}).$$

We know that $s = 2$ m when $t = 2$ s, so determining c requires solving the equation

$$f(c) = \frac{(0.9)(9.81)}{c^2}(2c - 1 + e^{-2c}) - 2 = 0.$$

We can't solve this transcendental equation in closed form to determine c. Problem-solving programs such as *Maple*, *MathCad*, *Mathematica*, and *TK! Solver* are designed to obtain roots of such equations. Another approach is to compute the value of $f(c)$ for a range of values of c and plot the results, as we have done in Fig. 2.54. From the graph we estimate that $c = 8.3\ s^{-1}$.

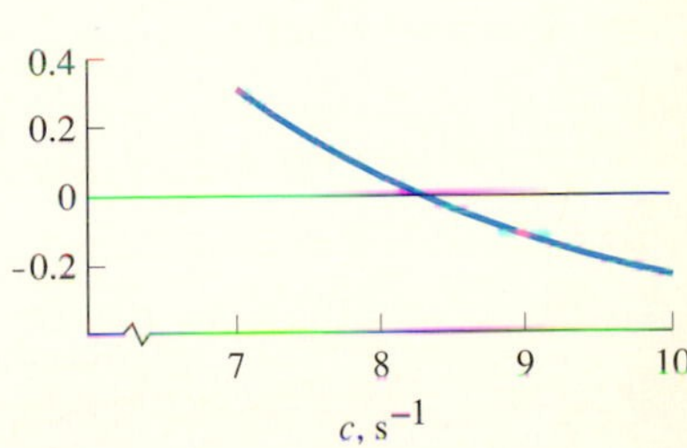

Fig. 2.54
Graph of the function $f(c)$.

Example 2.20

In an industrial process, the sprayer in Fig. 2.55 projects a stream of liquid onto the horizontal surface. An engineer wants to place the sprayer at a horizontal position x_e that maximizes its coverage of the horizontal surface. That is, he wants to maximize the distance $x_s - x_e$ reached by the spray. The angle θ_0 can be varied, but the spray velocity $v_0 = 3$ m/s is fixed. The dimensions $H = 0.2$ m, $L = 0.12$ m, and $h = 0.06$ m. If the droplets of the spray are treated as projectiles, what is the desired position x_e?

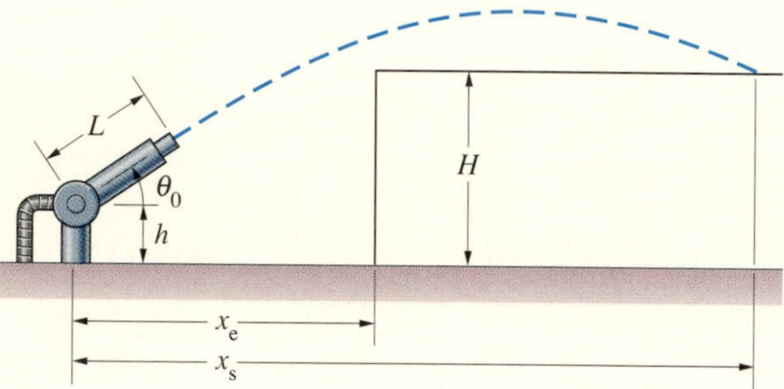

Fig. 2.55

STRATEGY

We first observe that the trajectory that maximizes the coverage of the horizontal surface will be one that just clears the edge of the horizontal surface. For if a trajectory clears the edge by some horizontal distance, as in Fig. 2.55, the sprayer can be moved that distance to the right, increasing its coverage by that amount. Our procedure will be to choose a distance x_e, determine the angle θ_0 that causes the trajectory to just clear the edge, and calculate $x_s - x_e$. By doing this for a range of values of x_e, we can determine the horizontal placement that maximizes $x_s - x_e$.

SOLUTION

Let $t = 0$ be the time at which a droplet leaves the nozzle. In terms of the coordinate system shown in Fig. (a), the x and y coordinates of the droplet are

$$x = L\cos\theta_0 + v_0\cos\theta_0 t,$$

$$y = h + L\sin\theta_0 + v_0\sin\theta_0 t - \frac{1}{2}gt^2.$$

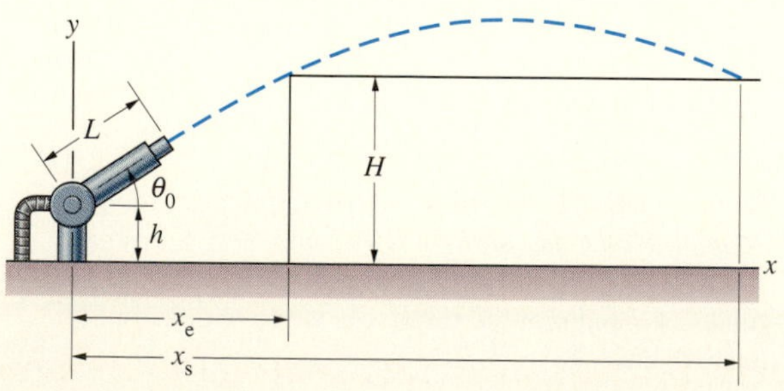

(a)

By setting $y = H$, we can solve for the times at which the droplet is at the height of the horizontal surface. The two resulting solutions for t are

$$t_1, t_2 = \frac{v_0 \sin\theta_0 \pm \sqrt{v_0^2 \sin^2\theta_0 - 2g(H - h - L\sin\theta_0)}}{g}.$$

We wish the smaller root t_1 to be the time at which the droplet passes the edge of the horizontal surface. Then the root t_2 will be the time at which the droplet lands on the surface. Setting $x = x_e$ when $t = t_1$,

$$x_e = L\cos\theta_0 + v_0 \cos\theta_0\, t_1,$$

gives us an equation we can use to determine θ_0 numerically for a given value of x_e. That is, we obtain the angle of the nozzle for which the spray just clears the edge of the surface. Once we know θ_0, we can determine x_s by substituting t_2 into the equation for x:

$$x_s = L\cos\theta_0 + v_0 \cos\theta_0\, t_2.$$

Figure 2.56 shows the values of $x_s - x_e$ we obtained for a range of values of x_e. From the graph we estimate that a maximum coverage of $x_s - x_e = 0.82$ m is obtained at $x_e = 0.12$ m. The trajectory that gives the maximum coverage, obtained with a nozzle angle $\theta_0 = 50°$, is shown in Fig. 2.57.

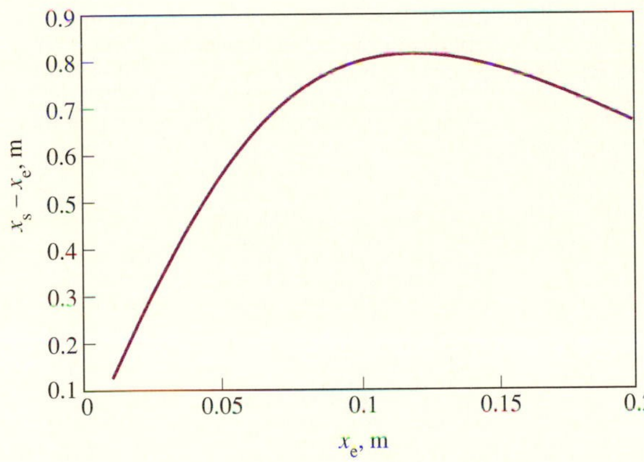

Fig. 2.56
Graph of the extent of coverage as a function of the nozzle placement.

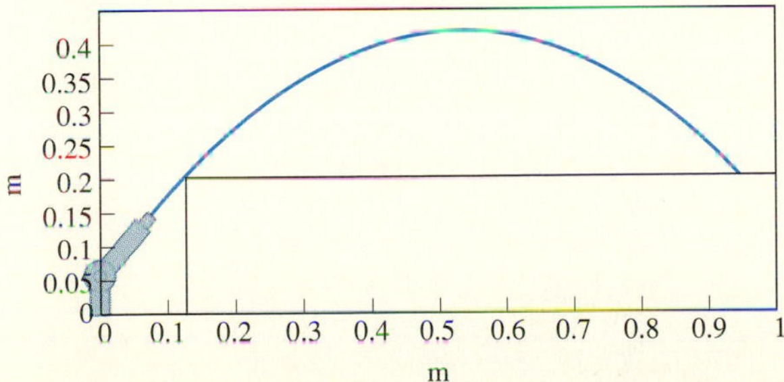

Fig. 2.57
Placement of the nozzle and the trajectory that maximizes coverage of the surface.

Problems

2.190 An engineer analyzing a machining process determines that from $t = 0$ to $t = 4$ s the workpiece starts from rest and moves in a straight line with acceleration

$$a = 2 + t^{0.5} - t^{1.5} \text{ ft/s}^2.$$

(a) Draw a graph of the position of the workpiece relative to its position at $t = 0$ for values of time from $t = 0$ to $t = 4$ s.
(b) Estimate the maximum velocity during this time interval and the time at which it occurs.

2.191 In Problem 2.78, determine the range of angles θ within which the pitcher must release the ball to hit the strike zone.

2.192 A catapult designed to throw a line to ships in distress throws a projectile with initial velocity $v_0(1 - 0.4 \sin \theta_0)$, where θ_0 is the angle above the horizontal. Determine the value of θ_0 for which the distance the projectile is thrown is a maximum, and show that the maximum distance is $0.559 v_0^2/g$.

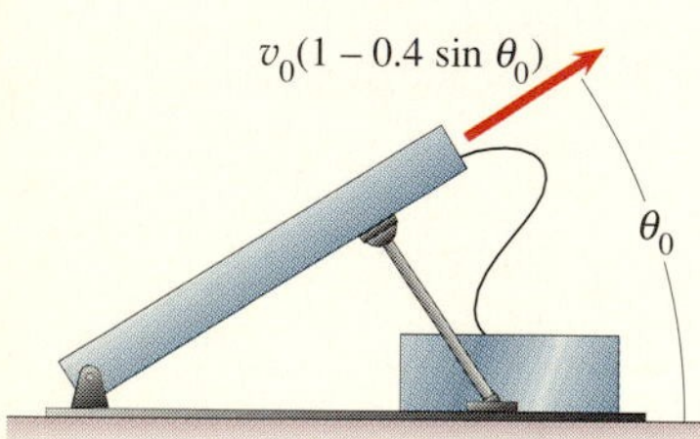

P2.192

2.193 At $t = 0$, a projectile is located at the origin and has a velocity of 20 m/s at 40° above the horizontal. The profile of the ground surface it strikes can be approximated by the equation $y = 0.4x - 0.006x^2$, where x and y are in meters. Determine the approximate coordinates of the point where it hits the ground.

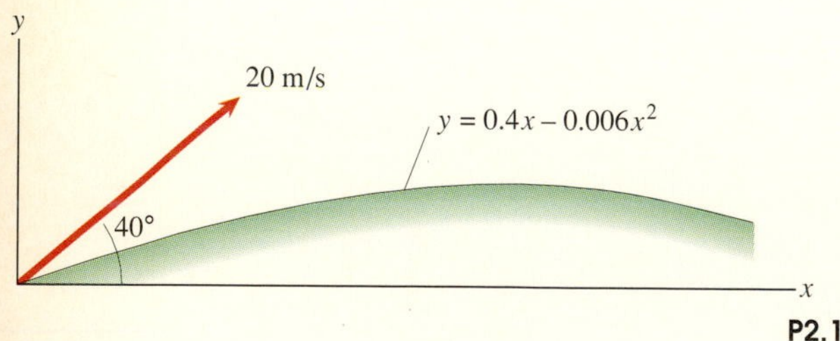

P2.193

2.194 A carpenter working on a house asks his apprentice to throw him an apple. The apple is thrown at 32 ft/s. What two values of θ_0 will cause the apple to land in the carpenter's hand, 12 ft horizontally and 12 ft vertically from the point where it is thrown?

P2.194

2.195 A motorcycle starts from rest at $t = 0$ and moves along a circular track with 400-m radius. The tangential component of its acceleration is $a_t = 2 + 0.2t$ m/s^2. When the magnitude of its total acceleration reaches 6 m/s^2, friction can no longer keep it on the circular track and it spins out. How long after it starts does it spin out, and how fast is it going?

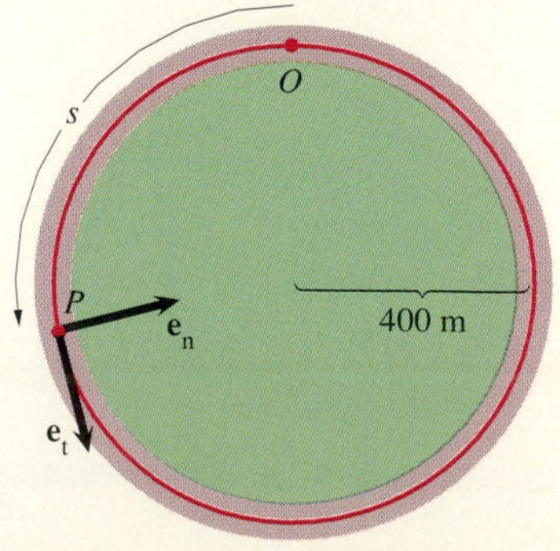

P2.195

2.196 At $t = 0$, a steel ball in a tank of oil is given a horizontal velocity $\mathbf{v} = 2\mathbf{i}$ m/s. The components of its acceleration are $a_x = -cv_x$, $a_y = -0.8g - cv_y$, $a_z = -cv_z$, where c is a constant. When the ball hits the bottom of the tank, its position relative its position at $t = 0$ is $\mathbf{r} = 0.8\mathbf{i} - \mathbf{j}$ (m). What is the value of c?

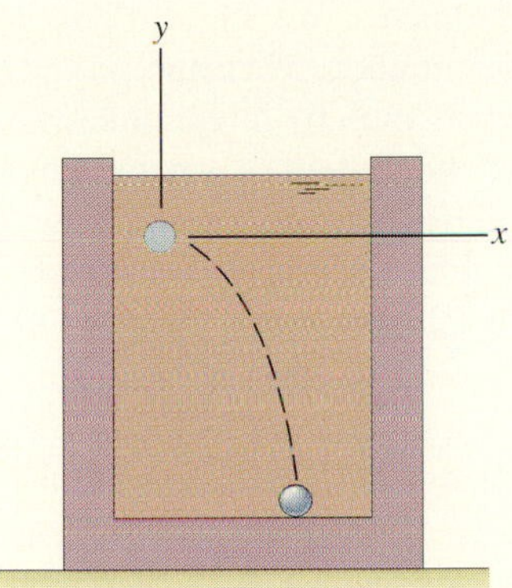

P2.196

2.197 The polar coordinates of a point P moving in the x-y plane are $r = t^3 - 4t$ m, $\theta = t^2 - t$ rad.
(a) Draw a graph of the magnitude of the velocity of P from $t = 0$ to $t = 2$ s.
(b) Estimate the minimum magnitude of the velocity and the time at which it occurs.

2.198 (a) Draw a graph of the magnitude of the acceleration of the point P in Problem 2.197 from $t = 0$ to $t = 2$ s.
(b) Estimate the minimum magnitude of the acceleration and the time at which it occurs.

2.199 The robot is programmed so that point P describes the path

$$r = 1 - 0.5 \cos 2\pi t \text{ m},$$

$$\theta = 0.5 - 0.2 \sin [2\pi(t - 0.1)] \text{ rad}.$$

Determine the values of r and θ at which the magnitude of the velocity of P attains its maximum value.

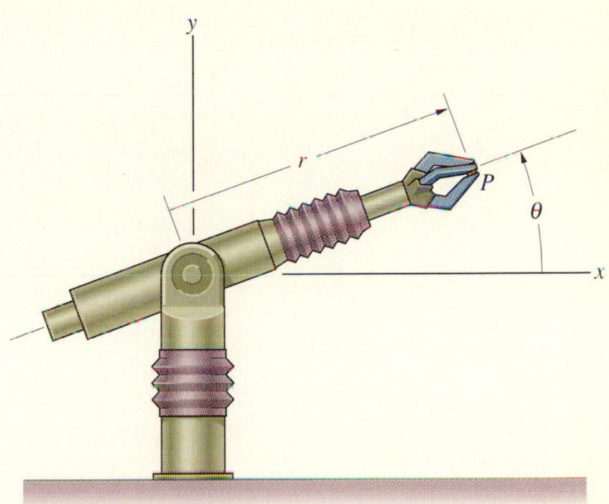

P2.199

2.200 In Problem 2.199, determine the values of r and θ at which the magnitude of the acceleration of P attains its maximum value.

2.201 In the cam-follower mechanism, the slotted bar rotates with constant angular velocity $\omega = 10$ rad/s, and the radial position of the follower A is determined by the profile of the stationary cam. The path of the follower is described by the polar equation

$$r = 1 + 0.5 \cos 2\theta \text{ ft}.$$

(a) Draw a graph of the magnitude of the follower's acceleration as a function of θ for $0 < \theta < 360°$.
(b) Use your graph to estimate the maximum magnitude of the follower's acceleration and the angle(s) at which it occurs.

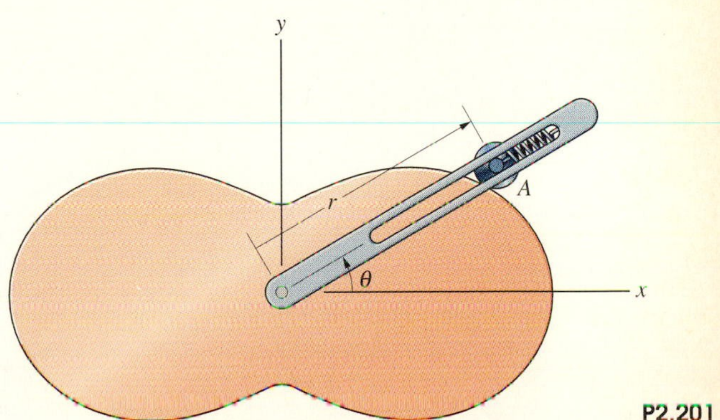

P2.201

Chapter Summary

In this chapter we were concerned with the motion of a point relative to a given reference frame. We defined the position, velocity, and acceleration, and showed how to express them in terms of cartesian coordinates, normal and tangential components, and polar coordinates. We showed how to determine the velocity and acceleration by differentiation when the position is known, and how to determine the velocity and position by integration when the acceleration is known. Finally, we analyzed the motion of a point relative to a second moving point. In Chapter 3 we will use Newton's second law to determine the acceleration of an object when the forces acting on it are known. Once the acceleration is known, the methods developed in this chapter will be applied to obtain information about the velocity and position of the object.

The position of a point P relative to a given reference frame with origin O can be specified by the **position vector r** from O to P. The **velocity** of P relative to the reference frame is

$$\mathbf{v} = \frac{d\mathbf{r}}{dt}, \qquad \textbf{Eq. (2.1)}$$

and the **acceleration** of P relative to the reference frame is

$$\mathbf{a} = \frac{d\mathbf{v}}{dt}. \qquad \textbf{Eq. (2.2)}$$

Straight-Line Motion

The position of a point P on a straight line relative to a reference point O is specified by a coordinate s measured along the line from O to P. The coordinate s and the velocity and acceleration of P along the line are related by

$$v = \frac{ds}{dt}, \qquad \textbf{Eq. (2.3)}$$

$$a = \frac{dv}{dt}. \qquad \textbf{Eq. (2.4)}$$

If the acceleration is specified as a function of time, the velocity and position can be determined as functions of time by integration. If the acceleration is specified as a function of velocity, $dv/dt = a(v)$, the velocity can be determined as a function of time by separating variables:

$$\int_{v_0}^{v} \frac{dv}{a(v)} = \int_{t_0}^{t} dt. \qquad \textbf{Eq. (2.16)}$$

If the acceleration is specified as a function of position, $dv/dt = a(s)$, the chain rule can be used to express the acceleration in terms of a derivative with respect to position:

$$\frac{dv}{dt} = \frac{dv}{ds}\frac{ds}{dt} = \frac{dv}{ds}v = a(s).$$

Separating variables, the velocity can be determined as a function of position:

$$\int_{v_0}^{v} v\,dv = \int_{s_0}^{s} a(s)\,ds. \qquad \textbf{Eq. (2.19)}$$

Cartesian Coordinates

The position, velocity, and acceleration relative to the cartesian coordinate system in Fig. (a) are [Eqs. (2.21)–(2.25)]

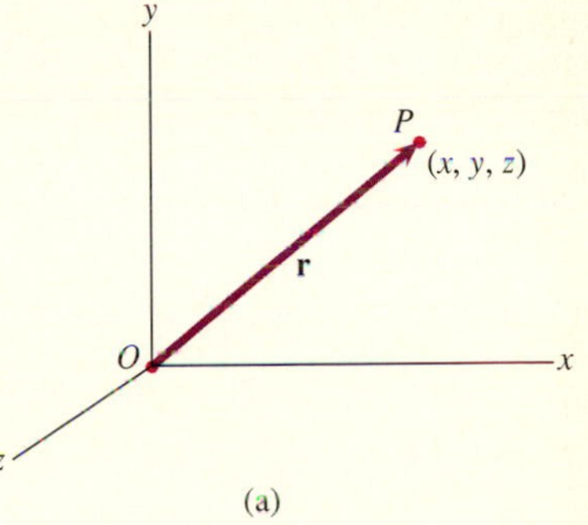

(a)

$$\mathbf{r} = x\mathbf{i} + y\mathbf{j} + z\mathbf{k},$$

$$\mathbf{v} = v_x\mathbf{i} + v_y\mathbf{j} + v_z\mathbf{k} = \frac{dx}{dt}\mathbf{i} + \frac{dy}{dt}\mathbf{j} + \frac{dz}{dt}\mathbf{k},$$

$$\mathbf{a} = a_x\mathbf{i} + a_y\mathbf{j} + a_z\mathbf{k} = \frac{dv_x}{dt}\mathbf{i} + \frac{dv_y}{dt}\mathbf{j} + \frac{dv_z}{dt}\mathbf{k}.$$

The equations describing the motion in each coordinate direction are identical in form to the equations that describe the motion of a point along a straight line.

Angular Motion

The angular velocity ω and angular acceleration α of L relative to L_0 are (Fig. b)

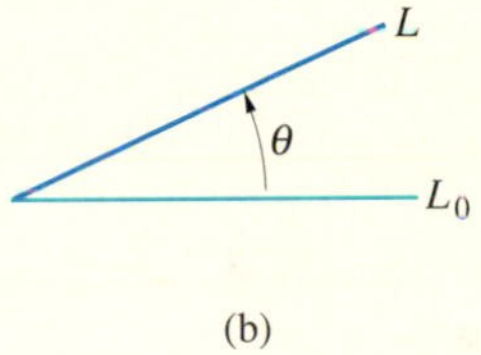

(b)

$$\omega = \frac{d\theta}{dt}, \qquad \textbf{Eq. (2.31)}$$

$$\alpha = \frac{d\omega}{dt} = \frac{d^2\theta}{dt^2}. \qquad \textbf{Eq. (2.32)}$$

Normal and Tangential Components

The velocity and acceleration are (Fig. c)

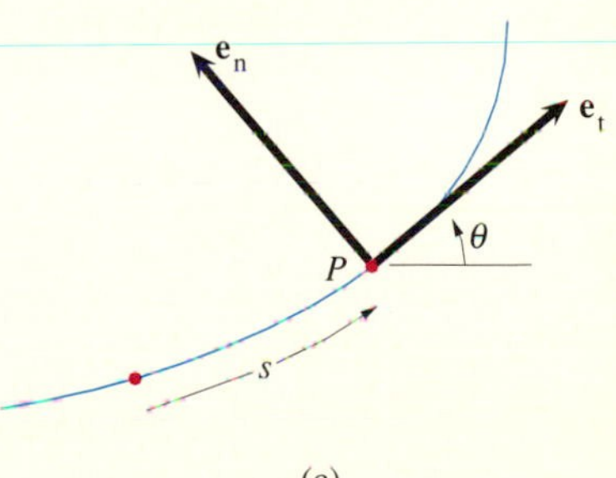

(c)

$$\mathbf{v} = v\mathbf{e}_t = \frac{ds}{dt}\mathbf{e}_t, \qquad \textbf{Eq. (2.38)}$$

$$\mathbf{a} = a_t\mathbf{e}_t + a_n\mathbf{e}_n, \qquad \textbf{Eq. (2.39)}$$

where

$$a_t = \frac{dv}{dt}, \qquad a_n = v\frac{d\theta}{dt} = \frac{v^2}{\rho}. \qquad \textbf{Eq. (2.40)}$$

The unit vector $\mathbf{e}_n$ points toward the concave side of the path. The term ρ is the instantaneous radius of curvature of the path.

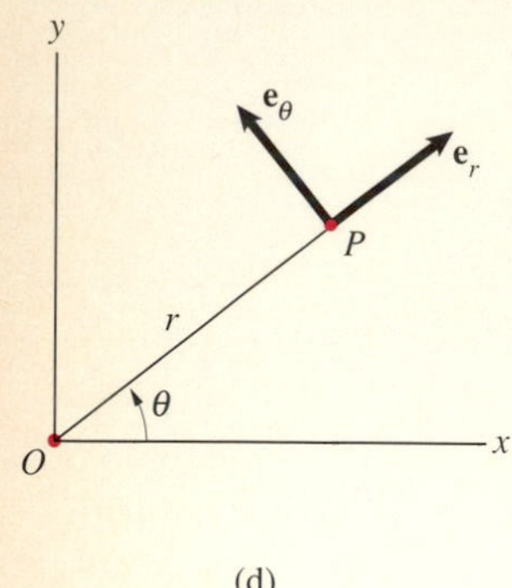

(d)

Polar Coordinates

The position, velocity, and acceleration are (Fig. d)

$$\mathbf{r} = r\mathbf{e}_r, \quad \text{Eq. (2.52)}$$

$$\mathbf{v} = v_r\mathbf{e}_r + v_\theta\mathbf{e}_\theta = \frac{dr}{dt}\mathbf{e}_r + r\omega\mathbf{e}_\theta, \quad \text{Eq. (2.57)}$$

$$\mathbf{a} = a_r\mathbf{e}_r + a_\theta\mathbf{e}_\theta, \quad \text{Eq. (2.58)}$$

where

$$a_r = \frac{d^2r}{dt^2} - r\left(\frac{d\theta}{dt}\right)^2 = \frac{d^2r}{dt^2} - r\omega^2,$$
$$a_\theta = r\frac{d^2\theta}{dt^2} + 2\frac{dr}{dt}\frac{d\theta}{dt} = r\alpha + 2\frac{dr}{dt}\omega. \quad \text{Eq. (2.59)}$$

Relative Motion

The vectors $\mathbf{r}_A$ and $\mathbf{r}_B$ in Fig. (e) specify the positions of A and B relative to O, and $\mathbf{r}_{A/B}$ specifies the position of A relative to B:

$$\mathbf{r}_A = \mathbf{r}_B + \mathbf{r}_{A/B}. \quad \text{Eq. (2.65)}$$

Taking time derivatives of this equation gives the relations

$$\mathbf{v}_A = \mathbf{v}_B + \mathbf{v}_{A/B}, \quad \text{Eq. (2.66)}$$

where $\mathbf{v}_A$ and $\mathbf{v}_B$ are the velocities of A and B relative to O and $\mathbf{v}_{A/B} = d\mathbf{r}_{A/B}/dt$ is the velocity of A relative to B, and

$$\mathbf{a}_A = \mathbf{a}_B + \mathbf{a}_{A/B}, \quad \text{Eq. (2.67)}$$

where $\mathbf{a}_A$ and $\mathbf{a}_B$ are the accelerations of A and B relative to O and $\mathbf{a}_{A/B} = d\mathbf{v}_{A/B}/dt$ is the acceleration of A relative to B.

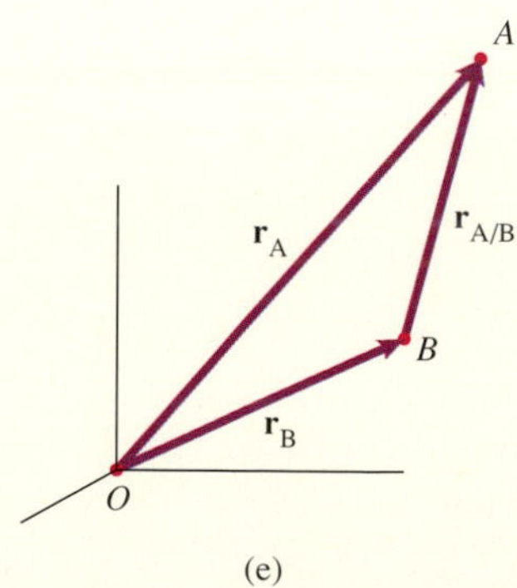

(e)

Review Problems

2.202 Suppose that you throw a ball straight up at 10 m/s and release it at 2 m above the ground.
(a) What maximum height above the ground does it reach?
(b) How long after you release it does it hit the ground?
(c) What is the magnitude of its velocity just before it hits the ground?

2.203 Suppose that you must determine the duration of the yellow light at a highway intersection. Assume that cars will be approaching the intersection traveling as fast as 65 mi/hr, that drivers' reaction times are as long as 0.5 s, and that cars can safely achieve a deceleration of at least $0.4g$.
(a) How long must the light remain yellow to allow drivers to come to a stop safely before the light turns red?
(b) What is the minimum distance cars must be from the intersection when the light turns yellow to come to a stop safely at the intersection?

2.204 The acceleration of a point moving along a straight line is $a = 4t + 2\ \text{m/s}^2$. When $t = 2$ s, its position is $s = 36$ m, and when $t = 4$ s, its position is $s = 90$ m. What is its velocity when $t = 4$ s?

2.205 A model rocket takes off straight up. Its acceleration during the 2 s its motor burns is 25 m/s^2. Neglect aerodynamic drag. Determine: (a) the maximum velocity during the flight; (b) the maximum altitude reached.

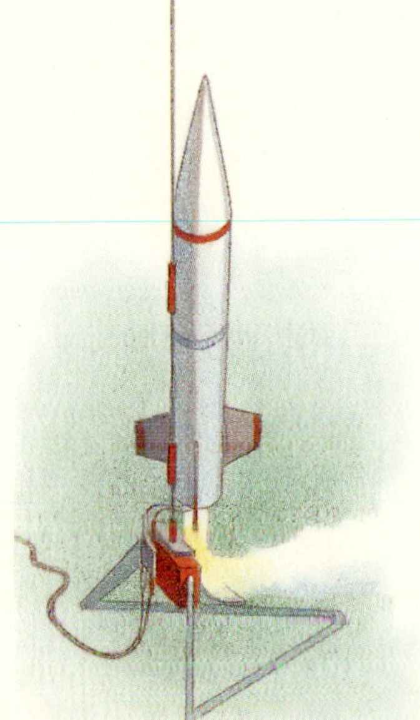

P2.205

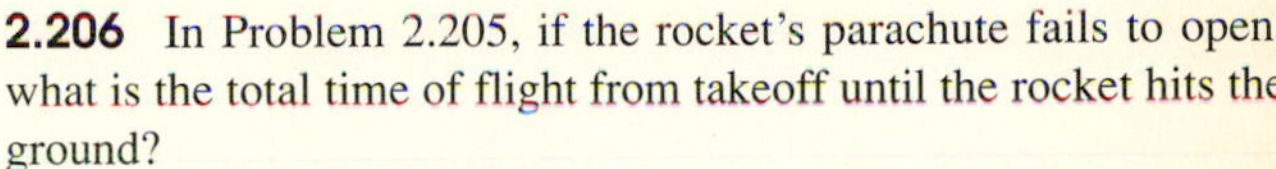

2.206 In Problem 2.205, if the rocket's parachute fails to open, what is the total time of flight from takeoff until the rocket hits the ground?

2.207 The acceleration of a point moving along a straight line is $a = -cv^3$, where c is a constant. If the velocity of the point is v_0, what distance does it move before its velocity decreases to $v_0/2$?

2.208 Water leaves the nozzle at 20° above the horizontal and strikes the wall at the point indicated. What was the velocity of the water as it left the nozzle?

Strategy: Determine the motion of the water by treating each particle of water as a projectile.

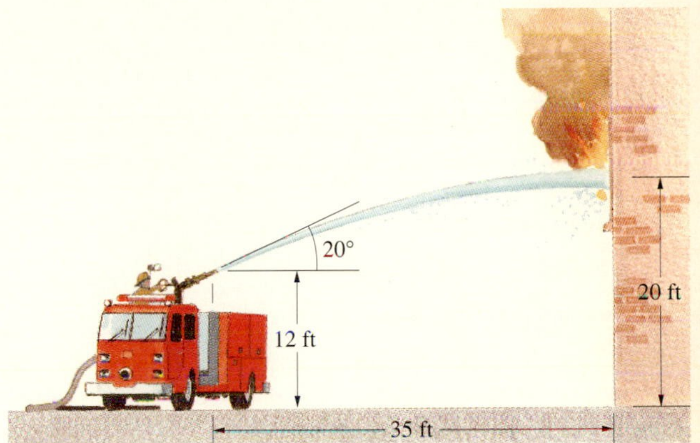

P2.208

2.209 In practice, a quarterback throws the football with velocity v_0 at 45° above the horizontal. At the same instant, the receiver standing 20 ft in front of him starts running straight downfield at 10 ft/s and catches the ball. Assume that the ball is thrown and caught at the same height above the ground. What is the velocity v_0?

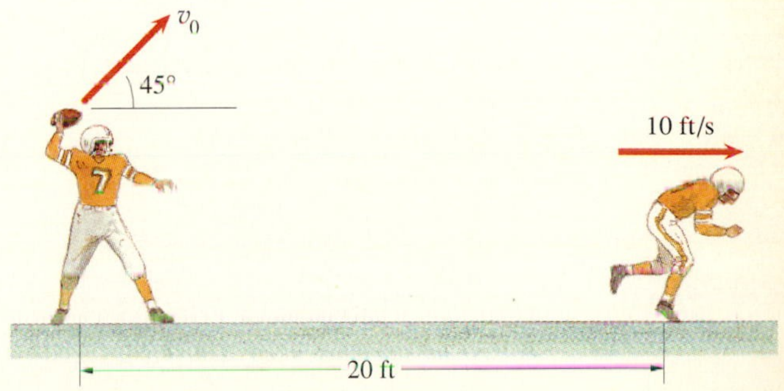

P2.209

2.210 The constant velocity $v = 2$ m/s. What are the magnitudes of the velocity and acceleration of point P when $x = 0.25$ m?

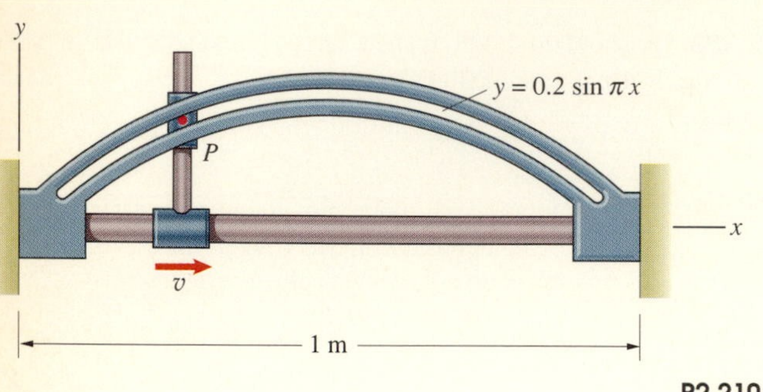

P2.210

2.211 In Problem 2.210, what is the acceleration of point P in terms of normal and tangential components when $x = 0.25$ m? What is the instantaneous radius of curvature of the path?

2.212 In Problem 2.210, what is the acceleration of point P in terms of radial and transverse components (polar coordinates) when $x = 0.25$ m?

2.213 A point P moves along the spiral path $r = (0.1)\theta$ ft, where θ is in radians. The angular position $\theta = 2t$ rad, where t is in seconds, and $r = 0$ at $t = 0$. Determine the magnitudes of the velocity and acceleration of P at $t = 1$ s.

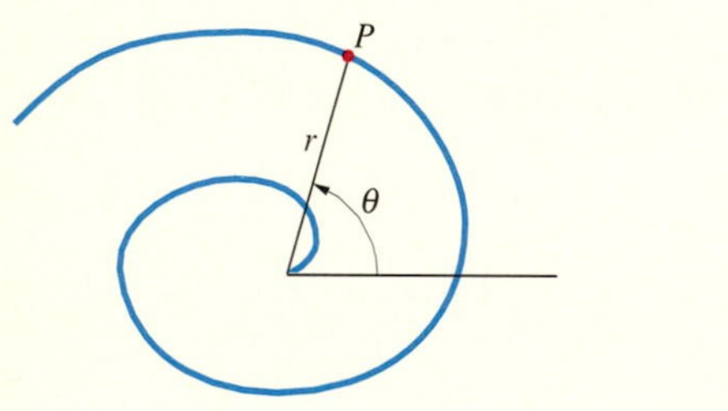

P2.213

2.214 In the cam-follower mechanism, the slotted bar rotates with constant angular velocity $\omega = 12$ rad/s, and the radial position of the follower A is determined by the profile of the stationary cam. The slotted bar is pinned a distance $h = 0.2$ m to the left of the center of the circular cam. The follower moves in a circular path of 0.42-m radius. Determine the velocity of the cam follower when $\theta = 40°$: (a) in terms of polar coordinates; (b) in terms of cartesian coordinates.

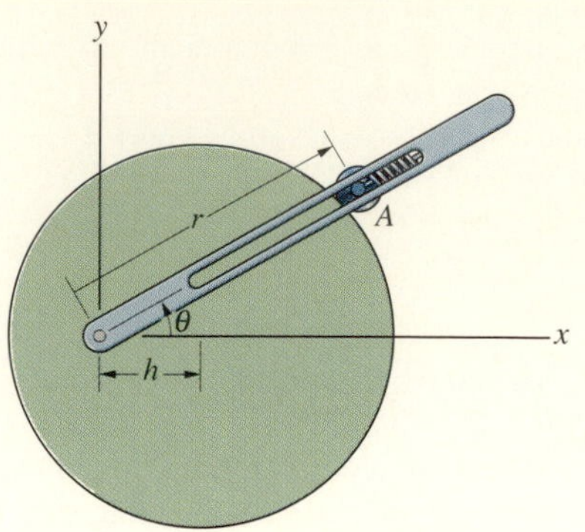

P2.214

2.215 In Problem 2.214, determine the acceleration of the cam follower when $\theta = 40°$: (a) in terms of polar coordinates; (b) in terms of cartesian coordinates.

2.216 A manned vehicle (M) attempts to rendezvous with a satellite (S) to repair it. (They are not shown to scale.) The magnitude of the satellite's velocity is $|\mathbf{v}_s| = 6$ km/s, and a sighting determines that the angle $\beta = 40°$. If you assume that their velocities remain constant and that the vehicles move along the straight lines shown, what should be the magnitude of $\mathbf{v}_M$ to achieve rendezvous?

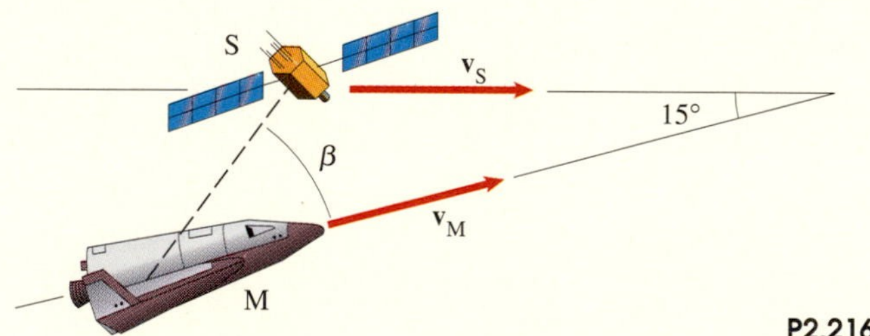

P2.216

2.217 In Problem 2.216, what is the magnitude of the velocity of the manned vehicle relative to the spacecraft once the magnitude of $\mathbf{v}_M$ has been adjusted to achieve rendezvous?

2.218 Each of the three 1-ft bars rotates in the x-y plane with constant angular velocity ω relative to the nonrotating coordinate system. If $\omega = 20$ rad/s, what is the magnitude of the velocity of point C relative to point A?

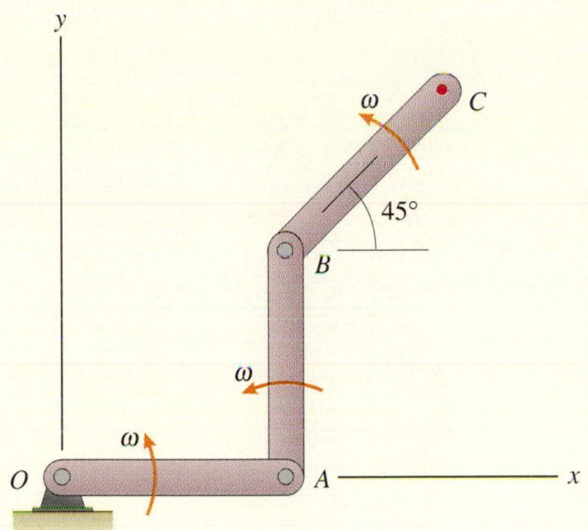

P2.218

2.219 In Problem 2.218, what is the velocity of point C relative to the fixed point O?

2.220 In Problem 2.218, accelerometers mounted at C indicate that the acceleration of point C relative to the fixed point O is $\mathbf{a}_C = -1500\mathbf{i} - 1500\mathbf{j}$ (ft/s^2). What is the angular velocity ω? Can you determine from this information whether ω is counterclockwise or clockwise?

A racing motorcycle can accelerate from rest to 60 mi/hr (96.6 km/hr) in 3 s. Its acceleration is related by Newton's second law to the combined mass of the motorcycle and rider and the external forces acting on them. In this chapter we will use free-body diagrams and Newton's second law to determine the motions that result from the forces acting on objects.

Chapter 3

Force, Mass, and Acceleration

UNTIL now we have analyzed motions of objects without considering the forces causing them. Here we relate cause and effect: By drawing the free-body diagram of an object to identify the forces acting on it, we can use Newton's second law to determine its acceleration. Alternatively, when we know an object's acceleration we can use Newton's second law to obtain information about the forces acting on it.

3.1 *Newton's Second Law*

Newton stated that the total force on a particle is equal to the rate of change of its **linear momentum**, which is the product of its mass and velocity:

$$\mathbf{f} = \frac{d}{dt}(m\mathbf{v}).$$

If the particle's mass is constant, the total force equals the product of its mass and acceleration:

$$\boxed{\mathbf{f} = m\frac{d\mathbf{v}}{dt} = m\mathbf{a}.} \tag{3.1}$$

We pointed out in Chapter 1 that the second law gives precise meanings to the terms *force* and *mass*. Once a unit of mass is chosen, a unit of force is defined to be the force necessary to give one unit of mass an acceleration of unit magnitude. For example, the unit of force in SI units, the newton, is the force necessary to give a mass of one kilogram an acceleration of one meter per second squared. In principle, the second law then gives the value of any force and the mass of any object. By subjecting a one-kilogram mass to an arbitrary force and measuring the acceleration, we can solve the second law for the direction of the force and its magnitude in newtons. By subjecting an arbitrary mass to a one-newton force and measuring the acceleration, we can solve the second law for the value of the mass in kilograms.

If you know a particle's mass and the total force acting on it, you can use Newton's second law to determine its acceleration. In Chapter 2 you learned how to determine the velocity, position, and path, or trajectory, of a point when you know its acceleration. Therefore, *with the second law you can determine a particle's motion when you know the total force acting on it, or if you know its acceleration you can determine the total force.*

3.2 *Equation of Motion for the Center of Mass*

Newton's second law is postulated for a particle, or small element of matter, but an equation of precisely the same form describes the motion of the *center of mass* of an *arbitrary* object. We can show that the total external force on an arbitrary object is equal to the product of its mass and the acceleration of its center of mass.

To do so, we *conceptually* divide an arbitrary object into N particles. Let m_i be the mass of the ith particle, and let $\mathbf{r}_i$ be its position vector (Fig. 3.1a). The object's mass m is the sum of the masses of the particles,

$$m = \sum_i m_i,$$

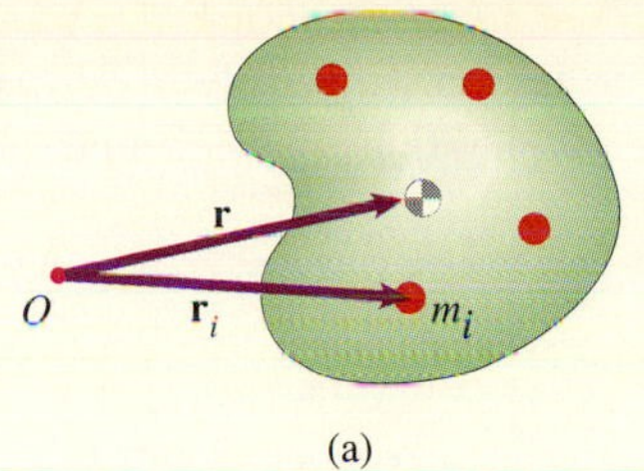

(a)

(b)

Fig. 3.1

(a) Dividing an object into particles. The vector $\mathbf{r}_i$ is the position vector of the ith particle and $\mathbf{r}$ is the position vector of the object's center of mass.
(b) Forces on the ith particle.

where the summation sign with subscript i means "the sum over i from 1 to N." The position of the object's center of mass is

$$\mathbf{r} = \frac{\sum_i m_i \mathbf{r}_i}{m}.$$

By taking two time derivatives of this expression, we obtain

$$\sum_i m_i \frac{d^2\mathbf{r}_i}{dt^2} = m\frac{d^2\mathbf{r}}{dt^2} = m\mathbf{a}, \tag{3.2}$$

where $\mathbf{a}$ is the acceleration of the object's center of mass.

The ith particle of the object may be subjected to forces by the other particles of the object. Let $\mathbf{f}_{ij}$ be the force exerted on the ith particle by the jth particle. Newton's third law states that the ith particle exerts a force on the jth particle of equal magnitude and opposite direction: $\mathbf{f}_{ji} = -\mathbf{f}_{ij}$. Denoting the external force on the ith particle (that is, the total force exerted on the ith particle by objects other than the object we are considering) by $\mathbf{f}_i^{\mathrm{E}}$, Newton's second law for the ith particle is (Fig. 3.1b)

$$\sum_j \mathbf{f}_{ij} + \mathbf{f}_i^{\mathrm{E}} = m_i \frac{d^2\mathbf{r}_i}{dt^2}.$$

We can write this equation for each particle of the object. Summing the resulting equations from $i = 1$ to N, we obtain

$$\sum_i \sum_j \mathbf{f}_{ij} + \sum_i \mathbf{f}_i^{\mathrm{E}} = m\mathbf{a}, \tag{3.3}$$

where we have used Eq. (3.2). The first term on the left side, the sum of the internal forces on the object, is zero due to Newton's third law:

$$\sum_i \sum_j \mathbf{f}_{ij} = \mathbf{f}_{12} + \mathbf{f}_{21} + \mathbf{f}_{13} + \mathbf{f}_{31} + \cdots = \mathbf{0}.$$

The second term on the left side of Eq. (3.3) is the sum of the external forces on the object. Denoting it by $\Sigma \mathbf{F}$, we conclude that the sum of the external forces equals the product of the mass and the acceleration of the center of mass:

$$\boxed{\Sigma \mathbf{F} = m\mathbf{a}.} \tag{3.4}$$

Because this equation is identical in form to Newton's postulate for a particle, for convenience we also refer to it as Newton's second law. Notice that we made no assumptions restricting the nature of the "object" or its state of motion in obtaining this result. The sum of the external forces on any object or collection of objects—solid, liquid, or gas—equals the product of the total mass and the acceleration of the center of mass. For example, suppose that the space shuttle is in orbit and has fuel remaining in its tanks. If its engines are turned on, the fuel sloshes in a complicated manner, affecting the shuttle's motion due to internal forces between the fuel and the shuttle. Nevertheless, we can use Eq. (3.4) to determine the *exact* acceleration of the center of mass of the shuttle, including the fuel it contains, and thereby determine the velocity, position, and trajectory of the center of mass.

3.3 *Inertial Reference Frames*

When we discussed the motion of a point in Chapter 2, we specified its position, velocity, and acceleration relative to an arbitrary reference frame. But Newton's second law cannot be expressed in terms of just any reference frame. Suppose that no force acts on a particle, and that we measure the particle's motion relative to a particular reference frame and determine that its acceleration is zero. In terms of this reference frame, Newton's second law agrees with our observation. But if we then measure the particle's motion relative to a second reference frame that is accelerating or rotating with respect to the first one, we would determine that the particle's acceleration is not zero. In terms of the second reference frame, Newton's second law, at least in the form given by Eq. (3.4), does not predict the correct result.

A well-known example is a person riding on an elevator. Suppose that you conduct an experiment in which you ride an elevator while standing on a set of scales (Fig. 3.2a). The forces acting on you are your weight W and the force N exerted on you by the scales (Fig. 3.2b). You exert an equal and opposite force N on the scales, which is the force they measure. If the elevator is stationary, you observe that the scales read your weight, $N = W$. The sum of the forces on you is zero, and Newton's second law correctly states that your acceleration relative to the elevator is zero. If the elevator accelerates upward with an acceleration a (Fig. 3.2c), you know you will feel heavier, and indeed you observe that the scales read a force greater than your weight, $N > W$. In terms of an earth-fixed reference frame, Newton's second law correctly relates the forces acting on you to your acceleration: $\Sigma F = N - W = ma$. But suppose that you use the elevator as your frame of reference. The sum of the forces acting on you is not zero, so Newton's second law states that you are accelerating relative to the elevator. But you are stationary relative to the elevator.

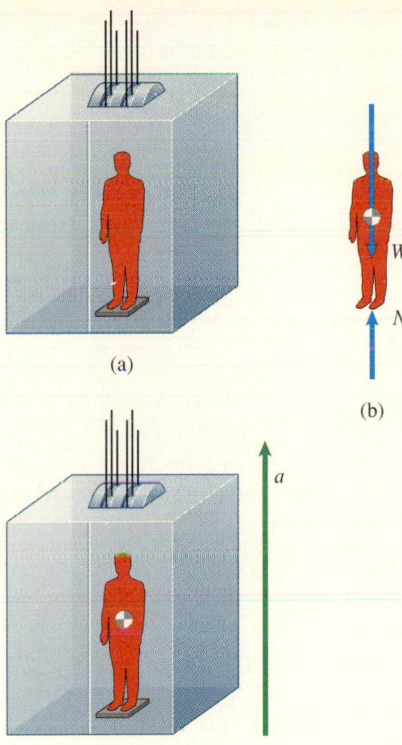

Fig. 3.2

(a) Riding an elevator while standing on scales.
(b) Your free-body diagram.
(c) Upward acceleration of the elevator.

Expressed in terms of this accelerating reference frame, Newton's second law gives an erroneous result.

Newton stated that the second law should be expressed in terms of a reference frame at rest with respect to the "fixed stars." Even if the stars were fixed, that would not be practical advice, because virtually every convenient reference frame accelerates, rotates, or both. Newton's second law *can* be applied rigorously using reference frames that accelerate and rotate by properly accounting for the acceleration and rotation. We explain how in Chapter 6. But for now, we need to indicate when you can apply Newton's second law and when you cannot.

Fortunately, in nearly all "down to earth" situations, you can express Newton's second law in the form given by Eq. (3.4) in terms of a reference frame that is fixed relative to the earth and obtain sufficiently accurate answers. For example, if you throw a piece of chalk across a room, you can use a reference frame that is fixed relative to the room to predict the chalk's motion. While the chalk is in motion, the earth rotates, and therefore the reference frame rotates. But *because the chalk's flight is brief,* the effect on your prediction is very small. (The earth rotates slowly—its angular velocity is one-half that of a clock's hour hand.) You can also usually apply Eq. (3.4) using a reference frame that translates (moves without rotating) at constant velocity relative to the earth. For example, if you and a friend play tennis on the deck of a cruise ship moving with constant velocity relative to the earth, you can apply Eq. (3.4) in terms of a reference frame fixed relative to the ship to analyze the ball's motion. But you cannot if the ship is turning, or changing its speed. In situations that are not "down to earth," such as the motions of earth satellites and spacecraft near the earth, you can apply Eq. (3.4) using a nonrotating reference frame with its origin at the center of the earth.

If a reference frame can be used to apply Eq. (3.4), it is said to be **Newtonian**, or **inertial**. We discuss inertial reference frames in greater detail in Chapter 6. For now, you should assume that examples and problems are expressed in terms of inertial reference frames.

3.4 *Applications*

When you studied statics you became familiar with different types of forces, including the weights of objects, the normal and friction forces exerted by contacting surfaces, and forces exerted by linear springs. You showed these forces on free-body diagrams to obtain information about the systems of forces acting on objects in equilibrium. In this section you will see that in dynamics you deal with the same types of forces. Furthermore, the same techniques you learned for drawing free-body diagrams in statics apply to objects that are not in equilibrium.

To apply Newton's second law in a particular situation, you must choose a coordinate system. Often you will find that you can resolve the forces into components most conveniently in terms of a particular coordinate system, or your choice may be determined by the object's path. In the following sections we use different types of coordinate systems to determine the motions of objects.

Cartesian Coordinates and Straight-Line Motion

If we express the sum of the forces acting on an object of mass m and the acceleration of its center of mass in terms of their components in a cartesian reference frame (Fig. 3.3), Newton's second law states that

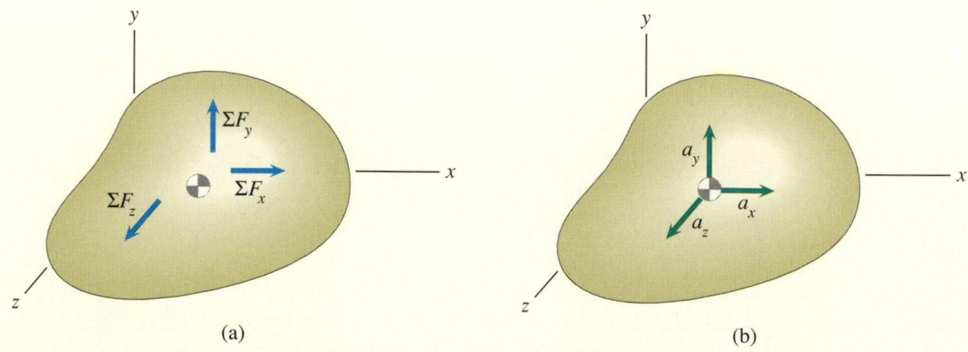

Fig. 3.3

(a) Cartesian components of the sum of the forces on an object.
(b) Components of the acceleration of the center of mass.

$$\Sigma\,\mathbf{F} = m\mathbf{a}:$$

$$(\Sigma\,F_x\mathbf{i} + \Sigma\,F_y\mathbf{j} + \Sigma\,F_z\mathbf{k}) = m(a_x\mathbf{i} + a_y\mathbf{j} + a_z\mathbf{k}).$$

Equating x, y, and z components, we obtain three scalar equations of motion:

$$\boxed{\Sigma\,F_x = ma_x, \qquad \Sigma\,F_y = ma_y, \qquad \Sigma\,F_z = ma_z.} \tag{3.5}$$

The total force in each coordinate direction equals the product of the mass and the component of the acceleration in that direction.

An important example is the projectile problem, in which an object is launched through the air and aerodynamic forces are neglected, so that the only force on the object is its weight. If we describe its motion using an earth-fixed coordinate system with the y axis upward (Fig. 3.4a), the sum of the forces is $\Sigma\,\mathbf{F} = -mg\mathbf{j}$. Therefore, $\Sigma\,F_x = 0$, $\Sigma\,F_y = -mg$, and $\Sigma\,F_z = 0$, and from Eqs. (3.5) we obtain $a_x = 0$, $a_y = -g$, and $a_z = 0$. This was the basis for our assumption in Chapter 2 that a projectile accelerates downward with the acceleration due to gravity and has no horizontal acceleration (Fig. 3.4b).

If an object's motion is confined to the x-y plane, $a_z = 0$, so the sum of the forces in the z direction is zero. Thus when the motion is confined to a fixed plane, the component of the total force normal to that plane equals zero. For straight-line motion along the x axis (Fig. 3.5a), Eqs. (3.5) are

$$\Sigma\,F_x = ma_x, \qquad \Sigma\,F_y = 0, \qquad \Sigma\,F_z = 0.$$

We see that in straight-line motion, the components of the total force perpendicular to the line equal zero, and the component of the total force tangent to the line equals the product of the mass and the acceleration along the line (Fig. 3.5b).

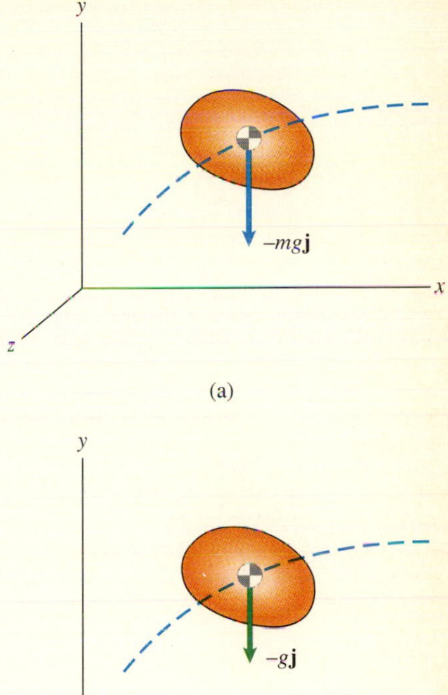

Fig. 3.4
(a) Free-body diagram of a projectile.
(b) The resulting acceleration of the center of mass.

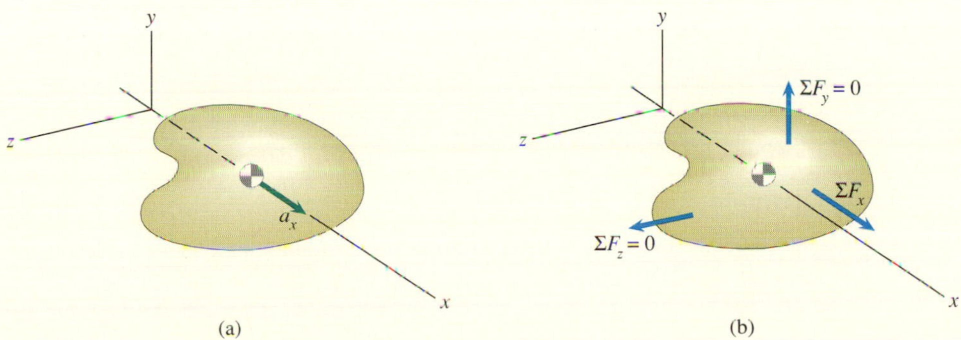

Fig. 3.5
(a) Acceleration of an object in straight-line motion along the x axis.
(b) The y and z components of the total force equal zero.

The following examples demonstrate the use of Newton's second law to analyze motions of objects. By drawing the free-body diagram of an object, you can identify the external forces acting on it and use Newton's second law to determine its acceleration. Conversely, if you know the motion of an object, you can use Newton's second law to determine the external forces acting on it. In particular, if you know that an object's acceleration in a specific direction is zero, the sum of the external forces in that direction also must equal zero.

Example 3.1

The airplane in Fig. 3.6 touches down on the aircraft carrier with a horizontal velocity of 50 m/s relative to the carrier. The arresting gear exerts a horizontal force of magnitude $T_x = 10{,}000v$ newtons (N), where v is the plane's velocity in meters per second. The plane's mass is 6500 kg.

(a) What maximum horizontal force does the arresting gear exert on the plane?

(b) If other horizontal forces can be neglected, what distance does the plane travel before coming to rest?

Fig. 3.6

STRATEGY

(a) Since the plane begins to decelerate when it contacts the arresting gear, the maximum force occurs at first contact when $v = 50$ m/s.

(b) The horizontal force exerted by the arresting gear equals the product of the plane's mass and its acceleration. Once we know the acceleration, we can integrate to determine the distance required for the plane to come to rest.

SOLUTION

(a) We draw the free-body diagram of the airplane and introduce a coordinate system in Fig. (a). The forces T_x and T_y are the horizontal and vertical components of force exerted by the arresting gear, and N is the vertical force on the landing gear. The horizontal force on the plane is $\Sigma F_x = -T_x = -10{,}000v$ N. The magnitude of the maximum force is

$$10{,}000v = (10{,}000)(50) = 500{,}000 \text{ N},$$

or 112,400 lb.

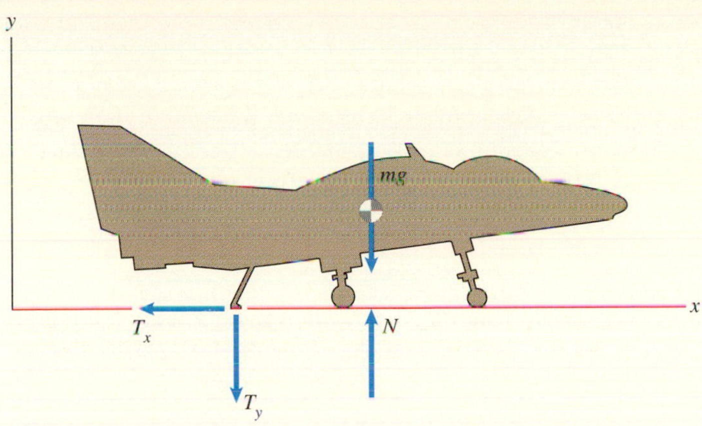

(a) Introducing a coordinate system with the x axis parallel to the horizontal force.

(b) In terms of the plane's horizontal component of acceleration (Fig. b), we obtain the equation of motion

$$\Sigma F_x = ma_x:$$

$$-10{,}000v_x = ma_x.$$

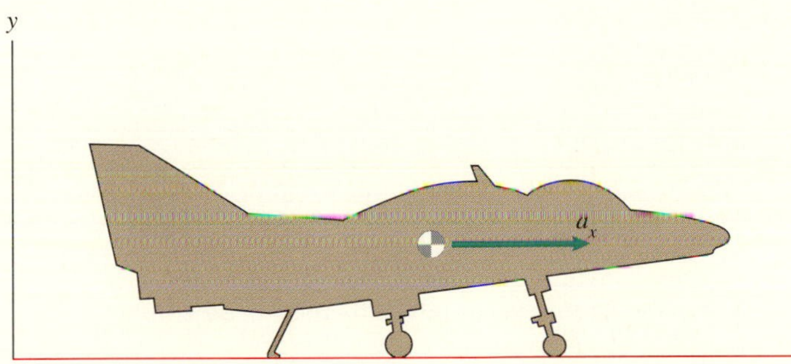

(b) The airplane's horizontal acceleration.

The airplane's acceleration is a function of its velocity. We use the chain rule to express the acceleration in terms of a derivative with respect to x:

$$ma_x = m\frac{dv_x}{dt} = m\frac{dv_x}{dx}\frac{dx}{dt} = m\frac{dv_x}{dx}v_x = -10{,}000v_x.$$

Now we separate variables and integrate, defining $x = 0$ to be the position at which the plane contacts the arresting gear:

$$\int_{50}^{0} m\, dv_x = -\int_0^x 10{,}000\, dx.$$

Evaluating the integrals and solving for x, we obtain

$$x = \frac{50m}{10{,}000} = \frac{(50)(6500)}{10{,}000} = 32.5 \text{ m}.$$

DISCUSSION

As we demonstrate in this example, once you have used Newton's second law to determine an object's acceleration, you can apply the methods developed in Chapter 2 to determine its position and velocity.

Example 3.2

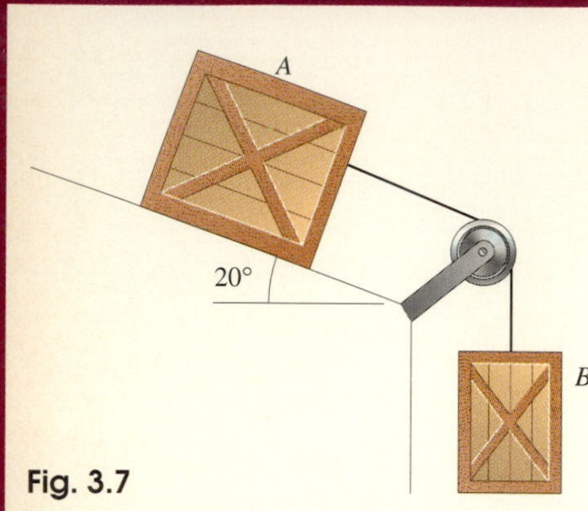

Fig. 3.7

The two crates in Fig. 3.7 are released from rest. Their masses are $m_A = 40$ kg and $m_B = 30$ kg, and the coefficients of friction between crate A and the inclined surface are $\mu_s = 0.2$, $\mu_k = 0.15$. What is their acceleration?

STRATEGY

We must first determine whether A slips. We will assume the crates remain stationary and see whether the friction force necessary for equilibrium exceeds the maximum friction force. If slip occurs, we can determine the resulting acceleration by drawing free-body diagrams of the crates and applying Newton's second law to them individually.

SOLUTION

We draw the free-body diagram of crate A and introduce a coordinate system in Fig. (a). If we assume it does not slip, the equilibrium equations apply,

$$\Sigma F_x = T + m_A g \sin 20° - f = 0,$$

$$\Sigma F_y = N - m_A g \cos 20° = 0,$$

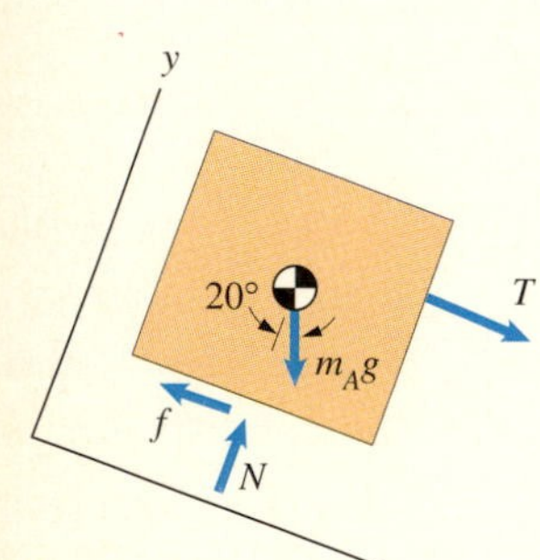

(a) Free-body diagram of crate A.

and the tension T equals the weight of crate B. Therefore the friction force necessary for equilibrium is

$$f = m_B g + m_A g \sin 20° = (30 + 40 \sin 20°)(9.81) = 429 \text{ N}.$$

The normal force $N = m_A g \cos 20°$, so the maximum friction force the surface will support is

$$f_{\max} = \mu_s N = (0.2)[(40)(9.81) \cos 20°] = 73.7 \text{ N}.$$

Crate A will therefore slip, and the friction force is $f = \mu_k N$. We show the crate's acceleration down the plane in Fig. (b). Its acceleration perpendicular to the plane is zero: $a_y = 0$. Applying Newton's second law,

$$\Sigma F_x = T + m_A g \sin 20° - \mu_k N = m_A a_x,$$

$$\Sigma F_y = N - m_A g \cos 20° = 0.$$

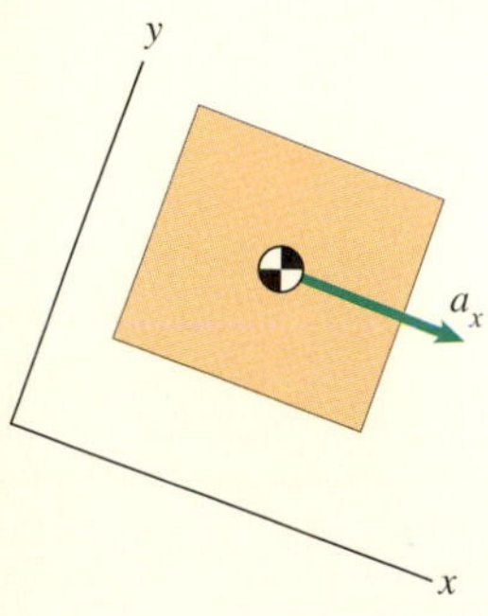

(b) The crate's acceleration parallel to the plane.

In this case *we do not know* the tension T, because crate B is not in equilibrium. We show the free-body diagram of crate B and its vertical acceleration in Figs. (c) and (d). We obtain the equation of motion

$$\Sigma F_x = m_B g - T = m_B a_x.$$

(In terms of the two coordinate systems we use, the two crates have the same acceleration a_x.) By applying Newton's second law to both crates, we have obtained three equations in terms of the unknowns T, N, and a_x. Solving for a_x, we obtain $a_x = 5.33 \text{ m/s}^2$.

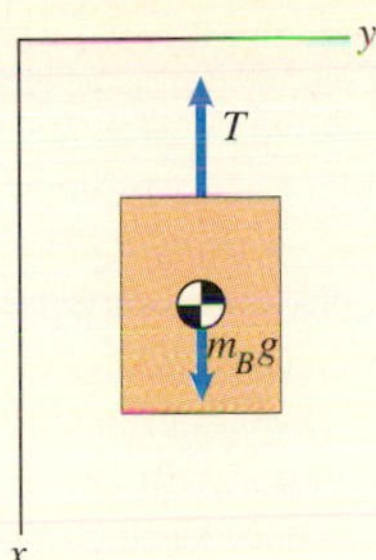

(c) Free-body diagram of crate B.

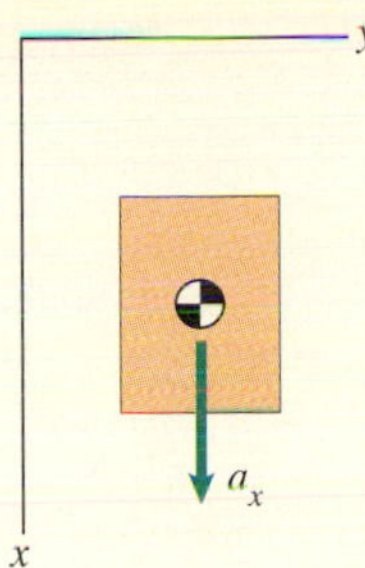

(d) Vertical acceleration of crate B.

DISCUSSION

Notice that we assumed the tension in the cable to be the same on each side of the pulley (Fig. e). In fact, the tensions must be different because a moment is necessary to cause angular acceleration of the pulley. For now, our only recourse is to assume that the pulley is light enough that the moment necessary to accelerate it is negligible. In Chapter 7, we include the analysis of the angular motion of the pulley in problems of this type and obtain more realistic solutions.

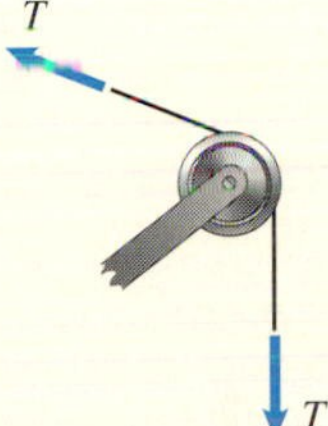

(e) The tension is assumed to be the same on both sides of the pulley.

Example 3.3

The sport utility vehicle in Fig. 3.8, which weighs 3000 lb with its driver, has left the ground after driving over a rise. At the instant shown, it is moving horizontally at 30 mi/hr and the bottoms of its tires are 24 in. above the (approximately) level ground. The earth-fixed coordinate system is placed with its origin 30 in. above the ground, at the height of the vehicle's center of mass when the tires first contact the ground. (Assume the vehicle remains horizontal.) When that occurs, the vehicle's center of mass initially continues moving downward and then rebounds upward due to the flexure of the suspension system. While the tires are in contact with the ground, the force exerted on them by the road is

$$-R\mathbf{i} - N\mathbf{j} = -2400\mathbf{i} - 18{,}000y\mathbf{j}\ (\text{lb}),$$

where y is the vertical position of the center of mass in feet. (The vertical component N is a function of y because of the flexure of the suspension.) When the vehicle hits the ground, what is the magnitude of the maximum acceleration to which it is subjected? If the 160-lb driver is subjected to the same acceleration, what maximum force is exerted on him by the vehicle?

Fig. 3.8

STRATEGY

We will analyze the vehicle's motion in two phases: before and after its wheels contact the ground. If we neglect aerodynamic forces, the only force on the vehicle before its wheels contact the ground is its weight. As a result it accelerates downward with the acceleration due to gravity and has no horizontal acceleration, so we can determine the vehicle's velocity when its wheels first touch the ground. We can then use Newton's second law to determine the vehicle's acceleration while its wheels are in contact with the ground. The maximum acceleration will occur when the suspension system has reached its maximum flexure, which means the center of mass has reached its minimum height. By integrating the y component of the acceleration, we can determine the minimum height reached by the center of mass and evaluate the maximum acceleration.

SOLUTION

Before the Wheels Contact the Ground The components of the vehicle's velocity at the instant shown in Fig. 3.8 are $v_x = (88/60)(30 \text{ mi/hr}) = 44$ ft/s, $v_y = 0$. Before the tires come into contact with the ground, the components of acceleration are $a_x = 0$, $a_y = g$. The horizontal velocity therefore remains constant. We integrate to determine the y component of the velocity as a function of time:

$$a_y = \frac{dv_y}{dt} = g,$$

$$\int_0^{v_y} dv_y = \int_0^t g\, dt,$$

$$v_y = \frac{dy}{dt} = gt.$$

The initial height of the center of mass is $y = -24$ in. $= -2$ ft. Integrating to determine its height as a function of time,

$$\int_{-2}^{y} dy = \int_0^t gt\, dt,$$

$$y = -2 + \frac{1}{2}gt^2.$$

Setting $y = 0$, we determine the time at which the tires contact the ground:

$$t = \sqrt{\frac{4}{g}} = \sqrt{\frac{4}{32.2}} = 0.352 \text{ s}.$$

The vehicle's components of velocity when the tires contact the ground are (Fig. a)

$$v_x = 44 \text{ ft/s}, \qquad v_y = gt = (32.2)(0.352) = 11.3 \text{ ft/s}.$$

(a) Velocity of the center of mass when the tires contact the ground.

After the Wheels Contact the Ground In Figs. (b) and (c) we show the free-body diagram of the vehicle and its components of acceleration while its wheels are in contact with the ground. From the y component of Newton's second law,

$$\Sigma F_y = W - N = ma_y = \left(\frac{W}{g}\right) a_y,$$

we determine the vehicle's vertical acceleration:

$$\begin{aligned} a_y = \frac{dv_y}{dt} &= g\left(1 - \frac{N}{W}\right) \\ &= 32.2\left(1 - \frac{18{,}000y}{3000}\right) \\ &= 32.2(1 - 6y). \end{aligned}$$

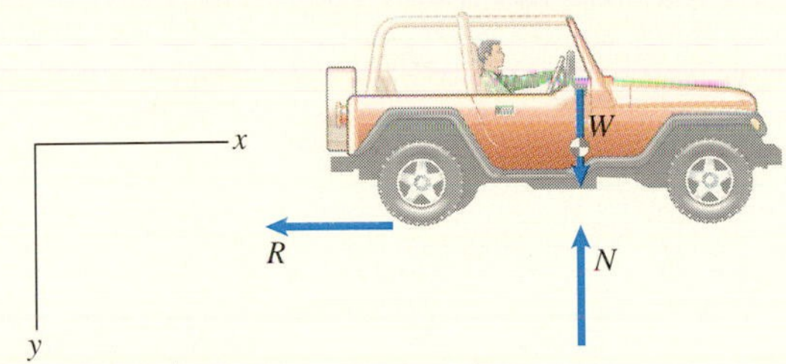

(b) Free-body diagram of the vehicle.

(c) Acceleration of the center of mass.

Because the vehicle's vertical acceleration is a function of y, we use the chain rule to express the acceleration in terms of y instead of t:

$$\frac{dv_y}{dt} = \frac{dv_y}{dy}\frac{dy}{dt} = \frac{dv_y}{dy}v_y = 32.2(1 - 6y).$$

Separating variables and integrating,

$$\int_{11.3}^{v_y} v_y\, dv_y = \int_0^y 32.2(1 - 6y)\, dy,$$

$$\left[\frac{v_y^2}{2}\right]_{11.3}^{v_y} = \left[32.2(y - 3y^2)\right]_0^y,$$

we obtain an equation for the vertical velocity as a function of y:

$$v_y = \sqrt{(11.3)^2 + 64.4(y - 3y^2)}.$$

To determine the maximum flexure of the suspension, we set $v_y = 0$ in this equation and solve for y, obtaining $y = 1$ ft. The maximum force exerted on the vehicle is therefore

$$\begin{aligned} \Sigma \mathbf{F} &= -R\mathbf{i} + (W - N)\mathbf{j} \\ &= -2400\mathbf{i} + [3000 - 18{,}000(1)]\mathbf{j} \\ &= -2400\mathbf{i} - 15{,}000\mathbf{j}\ \text{(lb)}. \end{aligned}$$

The vehicle's maximum acceleration is

$$\mathbf{a} = \frac{1}{m}\Sigma\,\mathbf{F} = \frac{g}{W}\Sigma\,\mathbf{F} = \frac{32.2}{3000}(-2400\mathbf{i} - 15{,}000\mathbf{j})$$

$$= -25.8\mathbf{i} - 161.0\mathbf{j}\ (\text{ft/s}^2).$$

Its magnitude is

$$|\mathbf{a}| = \sqrt{(-25.8)^2 + (-161.0)^2} = 163\ \text{ft/s}^2,$$

which is 5 g's.

Let $\mathbf{F}_\text{D}$ be the force exerted by the vehicle on the 160-lb driver when the vehicle's acceleration is a maximum. Newton's second law for the driver is

$$\Sigma\,\mathbf{F} = m\mathbf{a} = \left(\frac{W}{g}\right)\mathbf{a}:$$

$$\mathbf{F}_\text{D} + 160\mathbf{j} = \left(\frac{160}{32.2}\right)(-25.8\mathbf{i} - 161.0\mathbf{j}).$$

Solving, we determine that the maximum force is

$$\mathbf{F}_\text{D} = -128\mathbf{i} - 960\mathbf{j}\ (\text{lb}).$$

DISCUSSION

Engineers can use calculations of this kind in the design of vehicles and passenger restraint systems, but our modeling is incomplete in important ways. For example, we did not consider the vehicle's shock absorbers, which would affect the dynamics of the impact. To accurately model the vehicle's interaction with the ground, the dynamics of the vehicle's wheels and suspension would need to be included. Also, the vehicle's seat would help cushion the driver from the impact, decreasing the maximum force exerted on him.

Problems

3.1 The 2-kg collar A is initially at rest on the smooth horizontal bar. At $t = 0$ it is subjected to a constant force $F = 4$ N.
(a) How fast is the collar moving at $t = 1$ s?
(b) What distance has the collar moved at $t = 1$ s?

Strategy: Use a cartesian coordinate system with the x axis parallel to the bar. Draw the free-body diagram of the collar and apply Eqs. (3.5) to determine the collar's acceleration.

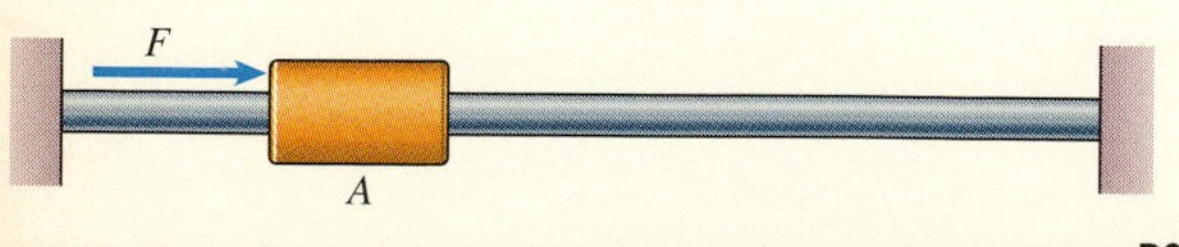

P3.1

3.2 Solve Problem 3.1 if the coefficient of kinetic friction between the collar and the bar is $\mu_\text{k} = 0.1$.

3.3 The 20-lb collar A starts at rest on the smooth bar at $t = 0$ and is subjected to a constant force $F = 10$ lb.
(a) How fast is the collar moving at $t = 1$ s?
(b) What distance has the collar moved along the bar at $t = 1$ s?

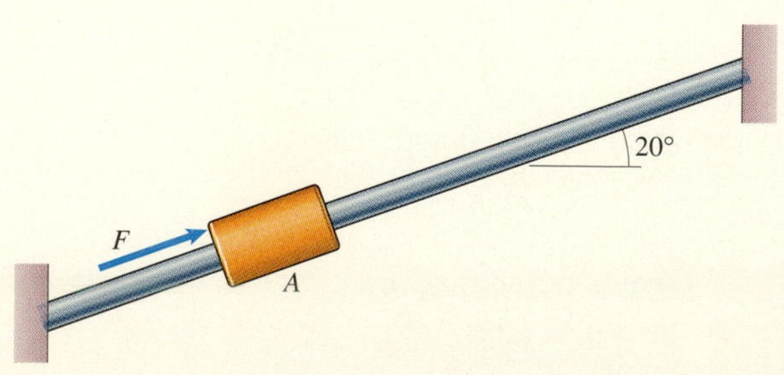

P3.3

3.4 Solve Problem 3.3 if the coefficients of static and kinetic friction between the collar and the bar are $\mu_s = \mu_k = 0.1$.

3.5 Suppose that a person conducts an experiment in which he rides an elevator while standing on a set of scales (Fig. a). Let a be the elevator's upward acceleration relative to the earth. The forces acting on the person are his weight $W = 150$ lb and the force N exerted on him by the scales (Fig. b).
(a) If the scales read 155 lb, what is the elevator's acceleration a?
(b) If the elevator's acceleration is $a = -2$ ft/s^2, what do the scales read?

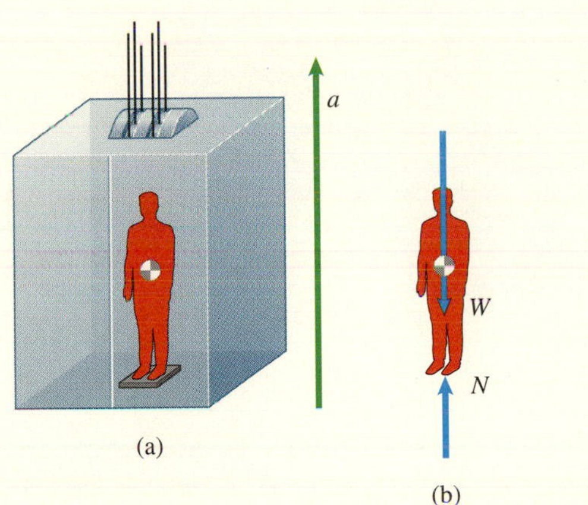

P3.5

3.6 Suppose that the person in Problem 3.5 observes that the scales he is standing on read zero. Use the free-body diagram of the person and Newton's second law to explain what is happening.

3.7 Suppose that the person in Problem 3.5 is standing on a European scale that displays his mass in kilograms, and that his actual mass is 68 kg. (Notice that to determine the force exerted on the scale, you must convert its reading in kilograms into newtons by multiplying by 9.81 m/s^2.)
(a) If the scales read 70 kg, what is the elevator's acceleration a in m/s^2?
(b) If the elevator's acceleration is $a = -0.5$ m/s^2, what do the scales read?

3.8 The total external force on a 10-kg object is $90\mathbf{i} - 60\mathbf{j} + 20\mathbf{k}$ (N). What is the magnitude of its acceleration relative to an inertial reference frame?

3.9 The total external force acting on a 20-lb object is $10\mathbf{i} + 20\mathbf{j}$ (lb). When $t = 0$, its position vector relative to an inertial reference frame is $\mathbf{r} = 0$ and its velocity is $\mathbf{v} = 20\mathbf{i} - 10\mathbf{j}$ (ft/s). Determine the position and velocity of the object when $t = 2$ s.

3.10 The total external force on an object is $10t\mathbf{i} + 60\mathbf{j}$ (lb). When $t = 0$, its position vector relative to an inertial reference frame is $\mathbf{r} = \mathbf{0}$ and its velocity is $\mathbf{v} = 20\mathbf{j}$ ft/s. When $t = 5$ s, the magnitude of its position vector is measured and determined to be 8.75 ft. What is the mass of the object?

3.11 The position of a 10-kg object relative to an inertial reference frame is $\mathbf{r} = \frac{1}{3}t^3\mathbf{i} + 4t\mathbf{j} - 30t^2\mathbf{k}$ (m). What are the components of the total external force acting on the object at $t = 10$ s?

3.12 If the 15,000-lb helicopter starts from rest and its rotor exerts a constant 20,000-lb vertical force, how high does it rise in 2 s?

P3.12

3.13 The 1-lb collar A is initially at rest in the position shown on the smooth horizontal bar. At $t = 0$, a force $\mathbf{F} = \frac{1}{20}t^2\mathbf{i} + \frac{1}{10}t\mathbf{j} - \frac{1}{30}t^3\mathbf{k}$ (lb) is applied to the collar, causing it to slide along the bar. What is the velocity of the collar when it reaches the right end of the bar?

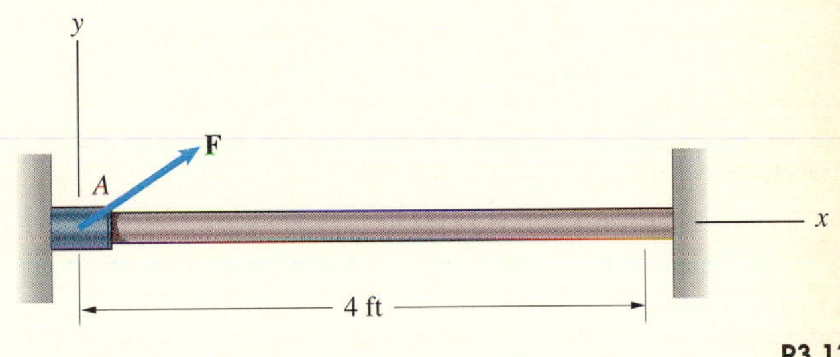

P3.13

3.14 The airplane weighs 20,000 lb. At the instant shown, the pilot increases the thrust T of the engine by 5000 lb. The horizontal component of the airplane's acceleration the instant before the thrust is increased is 20 ft/s^2. What is the horizontal component of the airplane's acceleration the instant after the thrust is increased?

P3.14

3.15 The rocket travels straight up at low altitude. Its weight at the present time is 200 kip, and the thrust of its engine is 270 kip. An on-board accelerometer indicates that its acceleration is 10 ft/s^2 upward. What is the magnitude of the aerodynamic drag force on the rocket?

P3.15

3.16 A cart partially filled with water is initially stationary (Fig. a). The total mass of the cart and water is m. The cart is subjected to a time-dependent force (Fig. b). If the horizontal forces exerted on the wheels by the floor are negligible and no water sloshes out, what is the x coordinate of the center of the cart after the motion of the water has subsided?

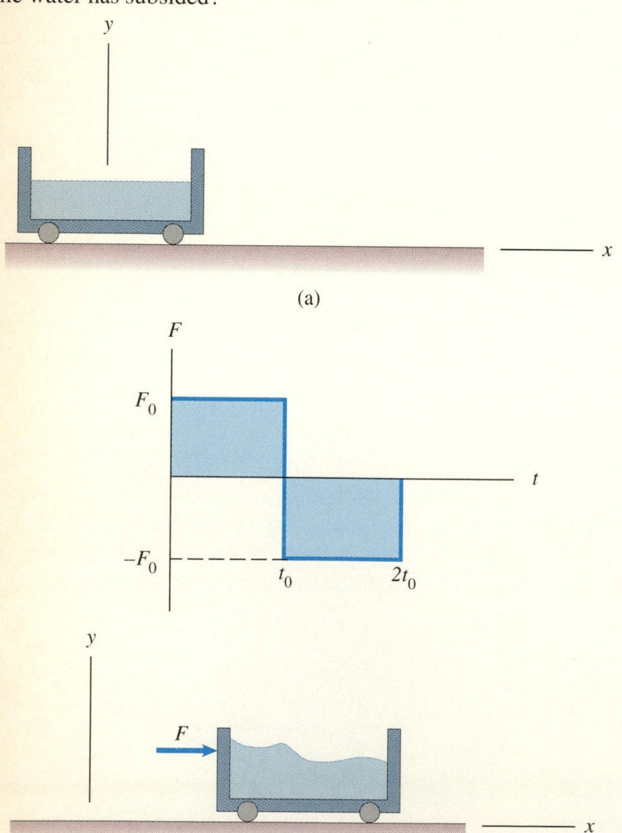

P3.16

3.17 The combined weight of the motorcycle and rider is 360 lb. The coefficient of kinetic friction between the motorcycle's tires and the road is $\mu_k = 0.8$. If he spins the rear (drive) wheel, the normal force between the rear wheel and the road is 250 lb, and the horizontal force exerted on the front wheel by the road is negligible, what is the resulting horizontal acceleration?

P3.17

3.18 The bucket B weighs 400 lb, and the acceleration of its center of mass is $\mathbf{a} = -30\mathbf{i} - 10\mathbf{j}$ (ft/s^2). Determine the x and y components of the total force exerted on the bucket by its supports.

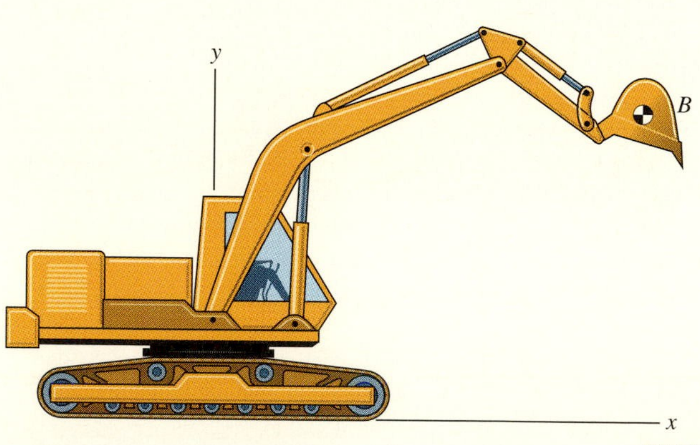

P3.18

3.19 During a test flight in which a 9000-kg helicopter starts from rest at $t = 0$, the acceleration of its center of mass from $t = 0$ to $t = 10$ s is

$$\mathbf{a} = 0.6t\mathbf{i} + (1.8 - 0.36t)\mathbf{j} \text{ (m/s}^2\text{)}.$$

What is the magnitude of the total external force on the helicopter (including its weight) at $t = 6$ s?

3.20 The engineers conducting the test described in Problem 3.19 want to express the total force on the helicopter at $t = 6$ s in terms of three forces: the weight W, a component T tangent to the path, and a component L normal to the path. What are the values of W, T, and L?

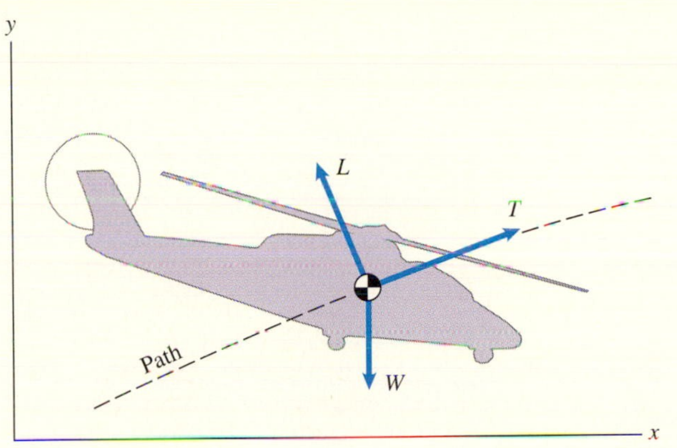

P3.20

3.21 At the instant shown, the 11,000-kg airplane's velocity is $\mathbf{v} = 270\mathbf{i}$ (m/s). The forces acting on it are its weight, the thrust $T = 110$ kN, the lift $L = 260$ kN, and the drag $D = 34$ kN. (The x axis is parallel to the airplane's path.) Determine the magnitude of the airplane's acceleration.

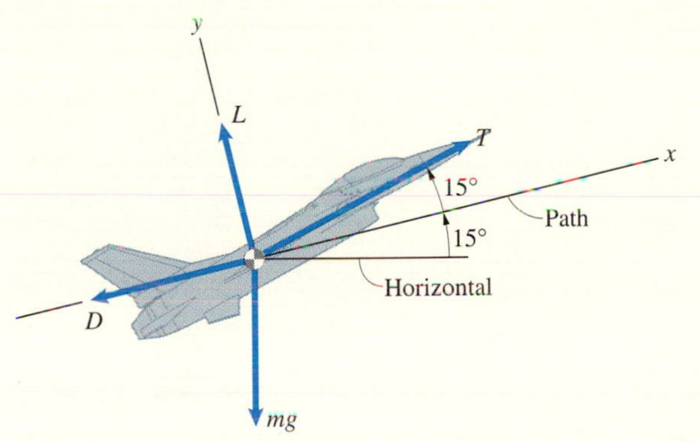

P3.21

3.22 At the instant shown in Problem 3.21, determine: (a) the rate of change of the magnitude of the airplane's velocity dv/dt; (b) the instantaneous radius of curvature of the airplane's path.

3.23 The coordinates in meters of the 360-kg sport plane's center of mass relative to an earth-fixed reference frame during an interval of time are

$$x = 20t - 1.63t^2,$$

$$y = 35t - 0.15t^3,$$

$$z = -20t + 1.38t^2,$$

where t is the time in seconds. The y axis points upward. The forces exerted on the plane are its weight, the thrust vector $\mathbf{T}$ exerted by its engine, the lift force vector $\mathbf{L}$, and the drag force vector $\mathbf{D}$. At $t = 4$ s, determine $\mathbf{T} + \mathbf{L} + \mathbf{D}$.

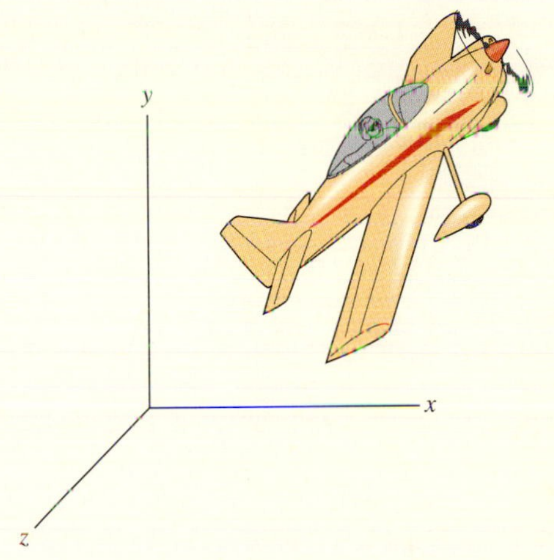

P3.23

3.24 The force in newtons exerted on the 360-kg sport plane in Problem 3.23 by its engine, the lift force, and the drag force during an interval of time is

$$\mathbf{T} + \mathbf{L} + \mathbf{D} = (-1000 + 280t)\mathbf{i} + (4000 - 430t)\mathbf{j} + (720 + 200t)\mathbf{k},$$

where t is the time in seconds. If the coordinates of the plane's center of mass are (0, 0, 0) and its velocity is $20\mathbf{i} + 35\mathbf{j} - 20\mathbf{k}$ (m/s) at $t = 0$, what are the coordinates of the center of mass at $t = 4$ s?

3.25 The robot manipulator is programmed so that $x = 4 + t^2$ in., $y = \frac{1}{4}x^2$ in., $z = 0$ during the interval of time from $t = 0$ to $t = 4$ s. What are the x and y components of the total force exerted by the jaws of the manipulator on the 10-lb widget A at $t = 2$ s?

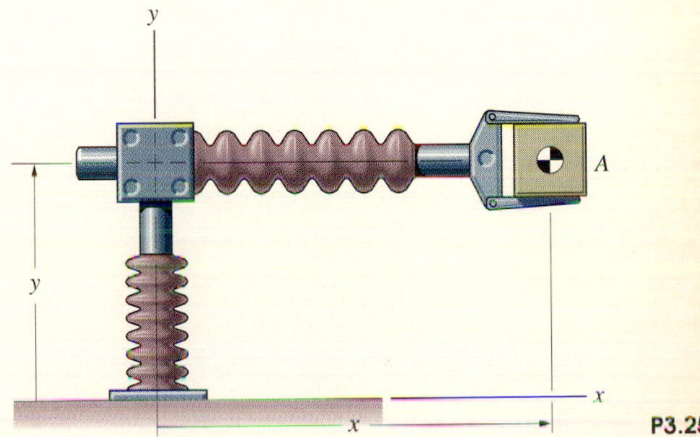

P3.25

3.26 The robot manipulator in Problem 3.25 is stationary at $t = 0$ and is programmed so that $a_x = 2 - 0.4v_x$ in./s^2, $a_y = 1 - 0.2v_y$ in./s^2, $a_z = 0$ during the interval of time from $t = 0$ to $t = 4$ s. What are the x and y components of the total force exerted by the jaws of the manipulator on the 10-lb widget A at $t = 2$ s?

3.27 In the sport of curling, the object is to slide a "stone" weighing 44 lb onto the center of a target located 31 yards from the point of release. If $\mu_k = 0.01$ and the stone is thrown directly toward the target, what initial velocity would result in a perfect shot?

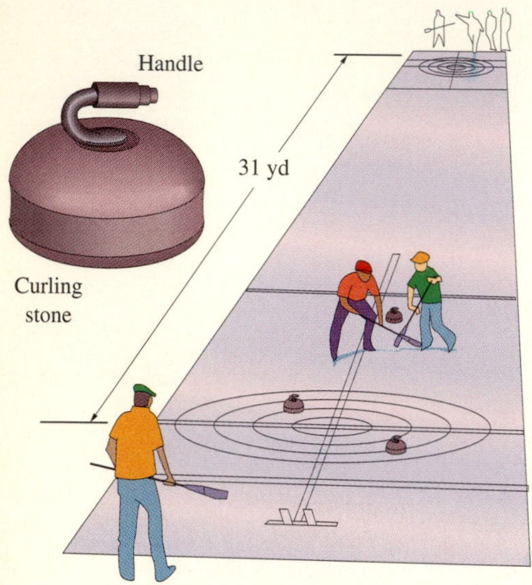

P3.27

3.28 The two weights are released from rest. How far does the 50-lb weight fall in 0.5 s? (See the discussion at the end of Example 3.2.)

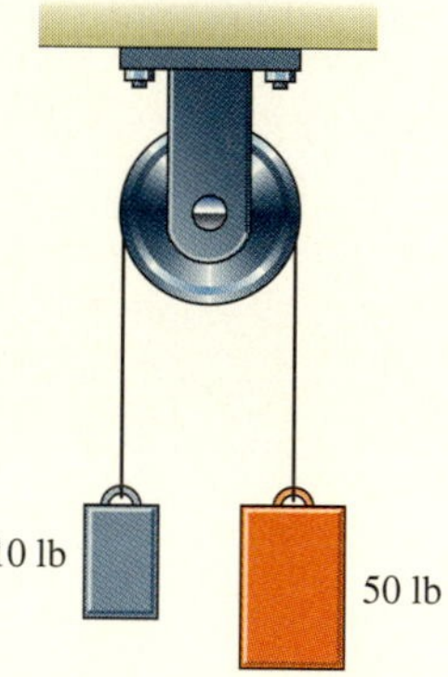

P3.28

3.29 In Example 3.2, what is the ratio of the tension in the cable to the weight of crate B after the crates are released from rest?

3.30 Each box weighs 50 lb, and friction can be neglected. If the boxes start from rest at $t = 0$, determine the magnitude of their velocity and the distance they have moved from their initial position at $t = 1$ s. (See the discussion at the end of Example 3.2.)

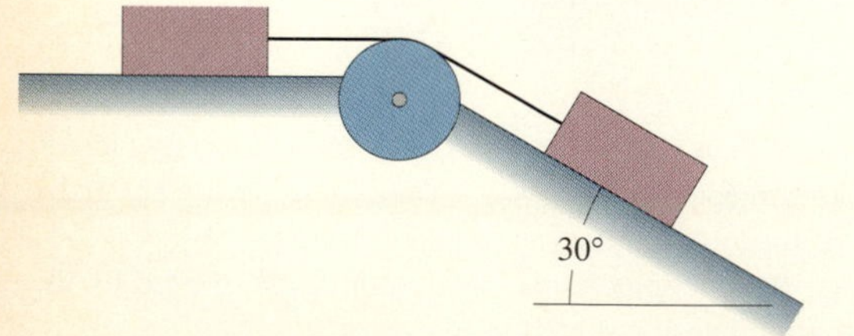

P3.30

3.31 In Problem 3.30, determine the magnitude of the velocity of the boxes and the distance they have moved from their initial position at $t = 1$ s if the coefficient of kinetic friction between the boxes and the surface is $\mu_k = 0.15$.

3.32 The masses $m_A = 15$ kg, $m_B = 30$ kg, and the coefficients of friction between all of the surfaces are $\mu_s = 0.4$, $\mu_k = 0.35$. What is the largest force F that can be applied without causing A to slip relative to B? What is the resulting acceleration?

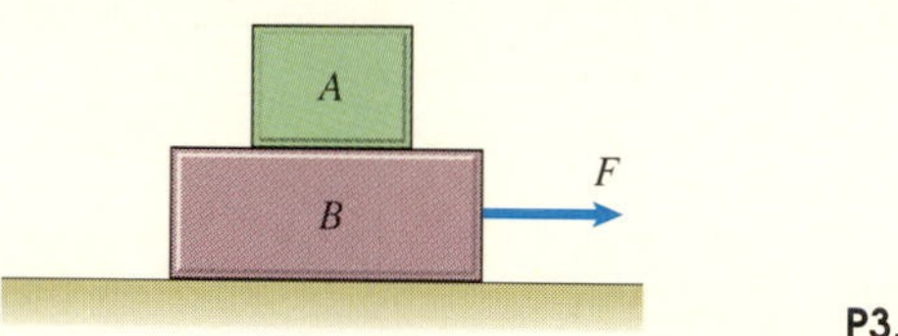

P3.32

3.33 The crane's trolley at A moves to the right with constant acceleration, and the 800-kg load moves without swinging.
(a) What is the acceleration of the trolley and load?
(b) What is the sum of the tensions in the parallel cables supporting the load?

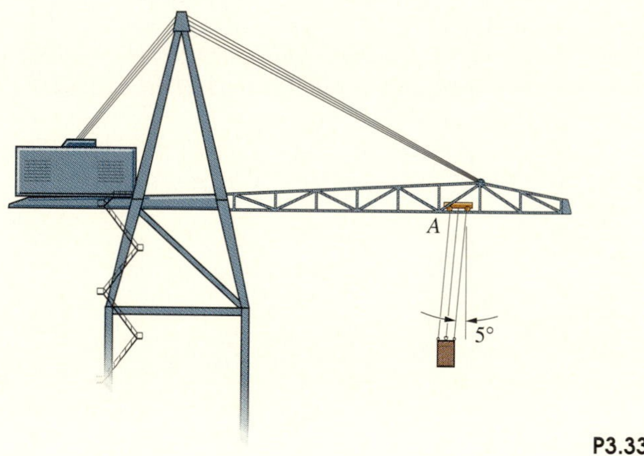

P3.33

3.34 The mass of A is 30 kg and the mass of B is 5 kg. A slides on the smooth surface and the angle $\theta = 20°$ is constant. What is the force F?

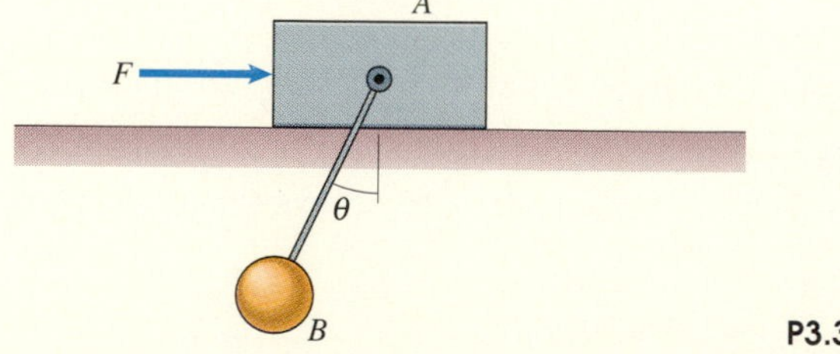

P3.34

3.35 In Problem 3.34, what is the force F if the coefficient of kinetic friction between A and the surface is $\mu_k = 0.24$?

3.36 The 100-lb crate is initially stationary. The coefficients of friction between the crate and the inclined surface are $\mu_s = 0.2$,

$\mu_k = 0.16$. Determine how far the crate moves from its initial position in 2 s if the horizontal force $F = 90$ lb.

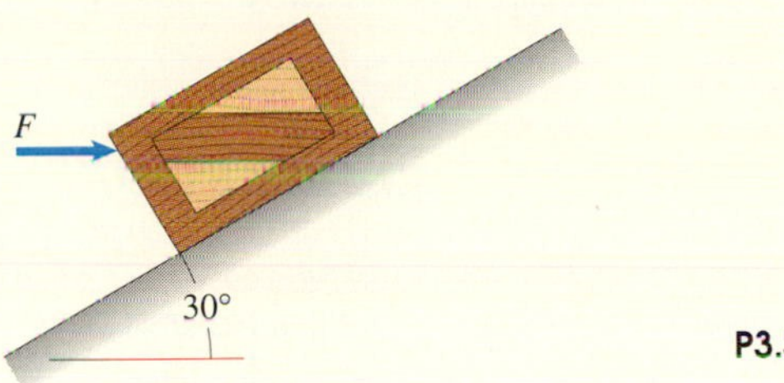

P3.36

3.37 In Problem 3.36, determine how far the crate moves from its initial position in 2 s if the horizontal force $F = 30$ lb.

3.38 The crate has a mass of 120 kg, and the coefficients of friction between it and the sloping dock are $\mu_s = 0.6$, $\mu_k = 0.5$.
(a) What tension must the winch exert on the cable to start the stationary crate sliding up the dock?
(b) If the tension is maintained at the value determined in part (a), what is the magnitude of the crate's velocity when it has moved 2 m up the dock?

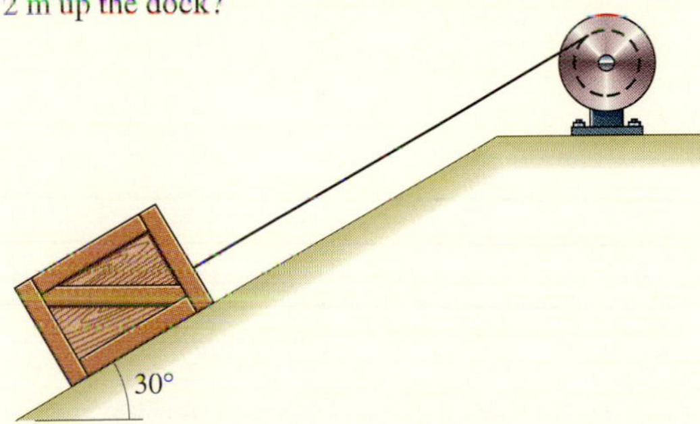

P3.38

3.39 The utility vehicle is moving forward at 10 ft/s. The coefficients of friction between its load A and the bed of the vehicle are $\mu_s = 0.5$, $\mu_k = 0.45$. If $\alpha = 0$, determine the shortest distance in which the vehicle can be brought to a stop without causing the load to slide on the bed.

P3.39

3.40 In Problem 3.39, determine the shortest distance if the angle α is (a) 15°; (b) −15°.

3.41 In an assembly-line process, the 20-kg package A starts from rest and slides down the smooth ramp. Suppose that you want to design the hydraulic device B to exert a constant force of magnitude F on the package and bring it to rest in a distance of 100 mm. What is the required force F?

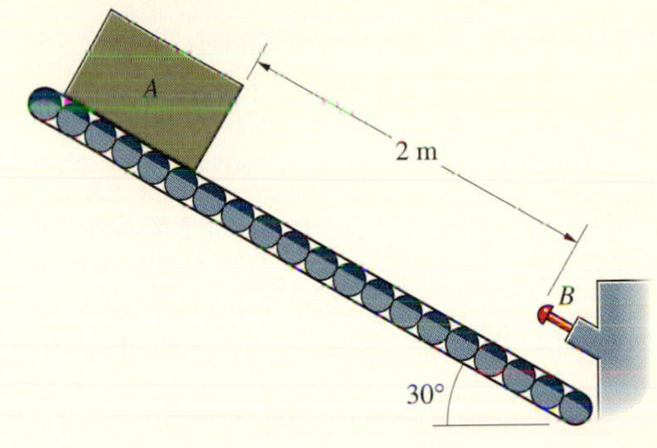

P3.41

3.42 The force exerted on the 10-kg mass by the linear spring is $F = -ks$, where k is the spring constant and s is the displacement of the mass from its position when the spring is unstretched. The value of k is 50 N/m. The mass is released from rest in the position $s = 1$ m.
(a) What is the acceleration of the mass at the instant it is released?
(b) What is the velocity of the mass when it reaches the position $s = 0$?

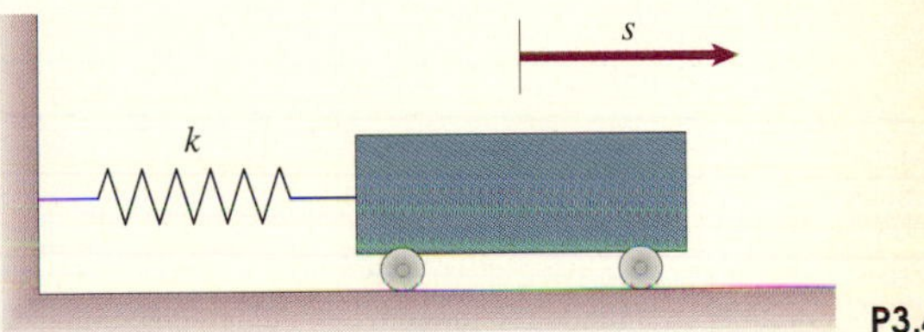

P3.42

3.43 A sky diver and his parachute weigh 200 lb. He is falling vertically at 100 ft/s when his parachute opens. With the parachute open, the magnitude of the drag force (in pounds) is $0.5v^2$.
(a) What is the magnitude of his acceleration at the instant the parachute opens?
(b) What is the magnitude of his velocity when he has descended 20 ft from the point where his parachute opens?

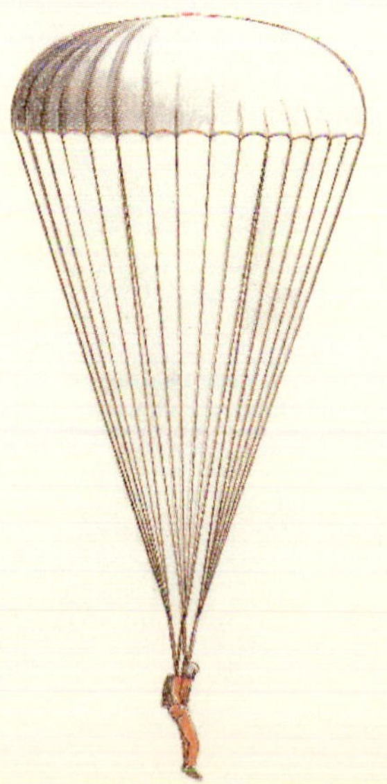

P3.43

3.44 The horizontal force exerted on the C-17 cargo plane during its takeoff run is $T - 127v^2$ N, where $T = 740{,}000$ N is the thrust of its engines and v is the plane's velocity in meters per second. The plane's mass is 265,000 kg. How long does it take the plane to reach its takeoff velocity of 72 m/s?

3.45 In Problem 3.44, what runway length is required for the C-17 to take off?

3.46 A 200-lb "bungee jumper" jumps from a bridge 130 ft above a river. The bungee cord has an unstretched length of 60 ft and has a spring constant $k = 14$ lb/ft.

(a) How far above the river is he when the cord brings him to a stop?

(b) What maximum force does the cord exert on him?

P3.46

3.47 In Problem 3.46, what maximum velocity does the jumper reach, and at what height above the river does it occur?

3.48 In a cathode-ray tube, an electron (mass = 9.11×10^{-31} kg) is projected at O with velocity $\mathbf{v} = (2.2 \times 10^7)\mathbf{i}$ (m/s). While it is between the charged plates, the electric field generated by the plates subjects it to a force $\mathbf{F} = -eE\mathbf{j}$, where the charge of the electron $e = 1.6 \times 10^{-19}$ C (coulombs) and the electric field strength $E = 15$ kN/C. External forces on the electron are negligible when it is not between the plates. Where does it strike the screen?

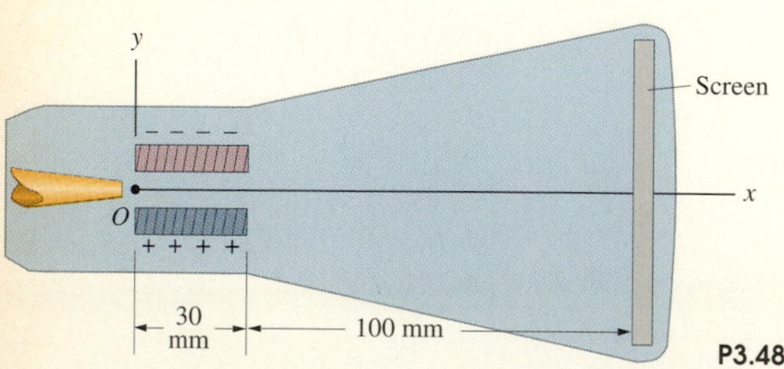

P3.48

3.49 In Problem 3.48, determine where the electron strikes the screen if the electric field strength is $E = 15 \sin(\omega t)$ kN/C, where the circular frequency $\omega = 2 \times 10^9\ \mathrm{s}^{-1}$.

3.50 An astronaut wants to travel from a space station to a satellite that needs repair. He departs the space station at O. A spring-loaded launching device gives his maneuvering unit an initial velocity of 1 m/s (relative to the space station) in the y direction. At that instant, the position of the satellite is $x = 70$ m, $y = 50$ m, $z = 0$, and it is drifting at 2 m/s (relative to the station) in the x direction. The astronaut intercepts the satellite by applying a constant thrust parallel to the x axis. The total mass of the astronaut and his maneuvering unit is 300 kg.

(a) How long does it take him to reach the satellite?

(b) What is the magnitude of the thrust he must apply to make the intercept?

(c) What is his velocity *relative to the satellite* when he reaches it?

P3.50

3.51 What is the acceleration of the 8-kg collar A relative to the smooth bar?

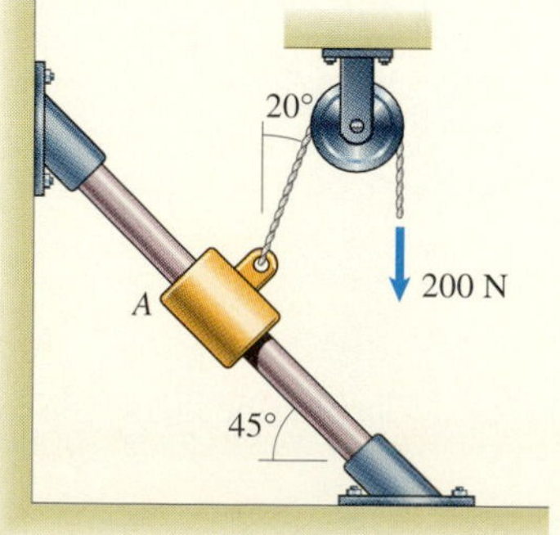

P3.51

3.52 In Problem 3.51, determine the acceleration of the collar A relative to the bar if the coefficient of kinetic friction between the collar and the bar is $\mu_k = 0.1$.

3.53 The acceleration of the 20-lb collar A is $2\mathbf{i} + 3\mathbf{j} - 3\mathbf{k}$ (ft/s^2). What is the force F?

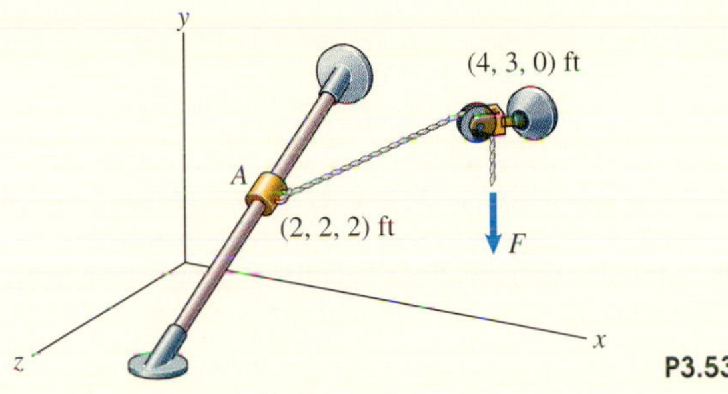

P3.53

3.54 In Problem 3.53, determine the force F if the coefficient of kinetic friction between the collar and the bar is $\mu_k = 0.1$.

3.55 The 6-kg collar starts from rest at position A, where the coordinates of its center of mass are (400, 200, 200) mm, and slides up the smooth bar to position B, where the coordinates of its center of mass are (500, 400, 0) mm, under the action of a constant force $\mathbf{F} = -40\mathbf{i} + 70\mathbf{j} - 40\mathbf{k}$ (N). How long does it take to go from A to B?

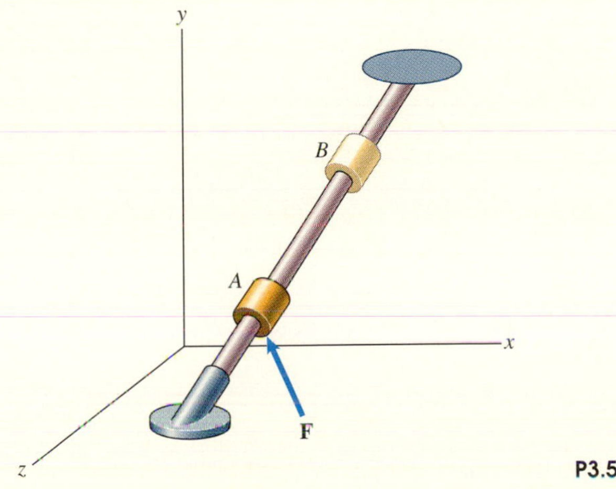

P3.55

3.56 In Problem 3.55, how long does it take the collar to go from A to B if the coefficient of kinetic friction between the collar and the bar is $\mu_k = 0.2$?

3.57 The crate is drawn across the floor by a winch that retracts the cable at a constant rate of 0.2 m/s. The crate's mass is 120 kg, and the coefficient of kinetic friction between the crate and the floor is $\mu_k = 0.24$.
(a) At the instant shown, what is the tension in the cable?
(b) Obtain a "quasi-static" solution for the tension in the cable by ignoring the crate's acceleration. Compare it to your result in (a).

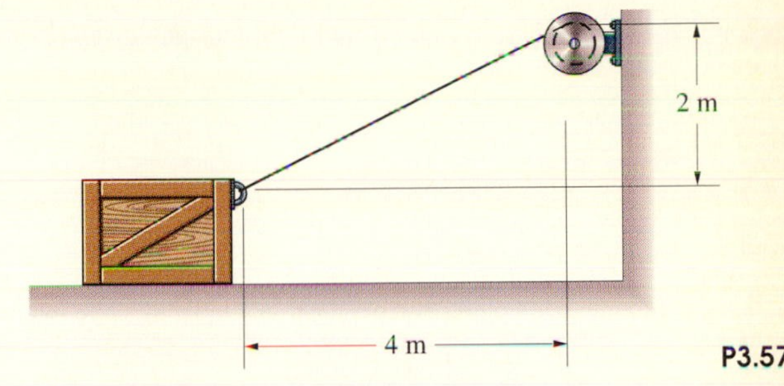

P3.57

3.58 If $y = 100$ mm, $dy/dt = 600$ mm/s, and $d^2y/dt^2 = -200$ mm/s^2, what horizontal force is exerted on the 0.4-kg slider A by the smooth circular slot?

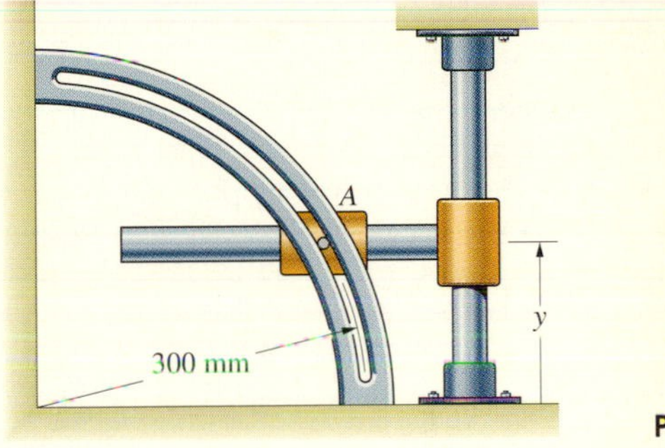

P3.58

3.59 A 3000-lb car travels at a constant speed of 60 mi/hr on a straight road of increasing grade whose vertical profile can be approximated by the equation shown. When $x = 400$ ft, what are the x and y components of the total force acting on the car (including its weight)?

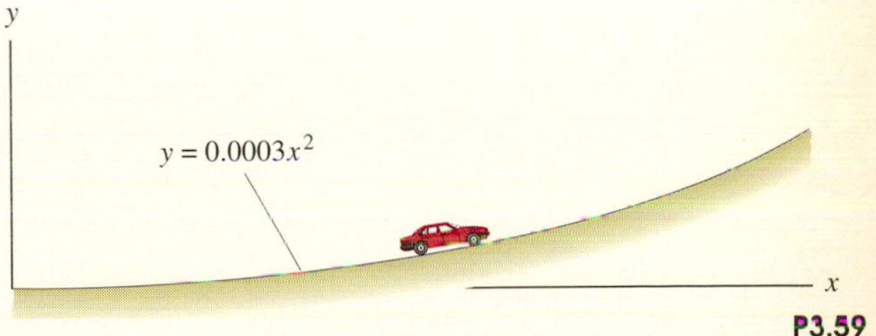

P3.59

3.60 If the car in Problem 3.59 is moving at 60 mi/hr and is increasing its speed at 10 ft/s^2 when $x = 400$ ft, what are the x and y components of the total force acting on the car (including its weight)?

3.61 The two 100-lb blocks are released from rest. Determine the magnitudes of their accelerations if friction at all the contacting surfaces is negligible.

Strategy: Use the fact that the components of the accelerations of the blocks perpendicular to their mutual interface must be equal.

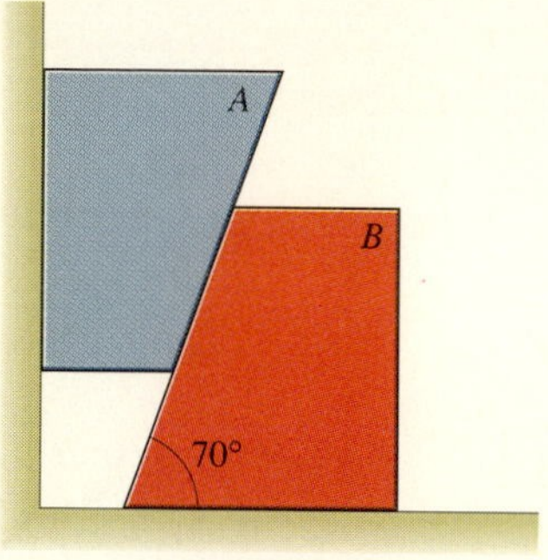

P3.61

3.62 In Problem 3.61, determine how long it takes block A to fall 1 ft if $\mu_k = 0.1$ at all the contacting surfaces.

3.63 In Example 3.3, a sport utility vehicle moves through the air and impacts the ground. After the impact, the vehicle would rebound and leave the ground again. Determine the vertical component of velocity of the vehicle's center of mass at the instant its wheels leave the ground. (The wheels leave the ground when the center of mass is at $y = 0$.)

3.64 In Example 3.3, the duration of the vehicle's impact with the ground is 0.255 s and its wheels leave the ground when the center of mass is at $y = 0$. Determine the horizontal distance the vehicle travels in the air when it rebounds and leaves the ground again. (Assume that the vehicle remains horizontal.)

3.65 The sum of the forces acting on an object of mass m is $\Sigma\,\mathbf{F}$, and the xyz reference frame is inertial, or Newtonian; that is,

$$\Sigma\,\mathbf{F} = m\frac{d^2\mathbf{r}}{dt^2} = m\left(\frac{d^2x}{dt^2}\mathbf{i} + \frac{d^2y}{dt^2}\mathbf{j} + \frac{d^2z}{dt^2}\mathbf{k}\right),$$

where x, y, and z are the coordinates of the object's center of mass. The parallel $x'y'z'$ reference frame is moving at constant velocity relative to the xyz reference frame; that is, $d\mathbf{R}/dt$ is constant. Prove that the $x'y'z'$ reference frame is inertial: A reference frame that translates at constant velocity relative to an inertial reference frame is also inertial.

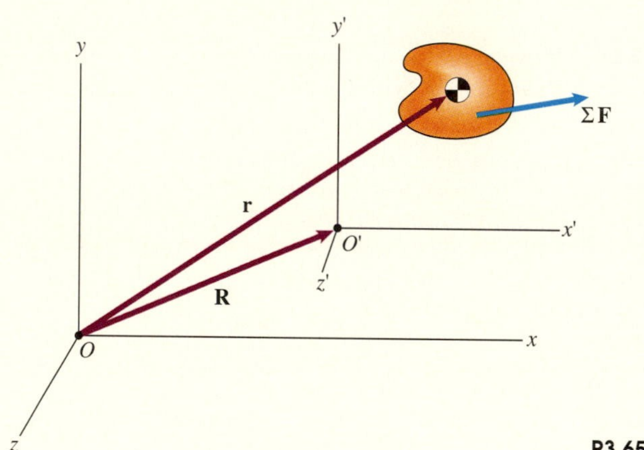

P3.65

Normal and Tangential Components

When an object moves in a curved path, we can resolve the sum of the forces acting on it into normal and tangential components (Fig. 3.9a). We can also express the object's acceleration in terms of normal and tangential components (Fig. 3.9b) and write Newton's second law in the form

$$\Sigma\,\mathbf{F} = m\mathbf{a}:$$

$$(\Sigma\,F_t\mathbf{e}_t + \Sigma\,F_n\mathbf{e}_n) = m(a_t\mathbf{e}_t + a_n\mathbf{e}_n), \tag{3.6}$$

where

$$a_t = \frac{dv}{dt}, \qquad a_n = \frac{v^2}{\rho}.$$

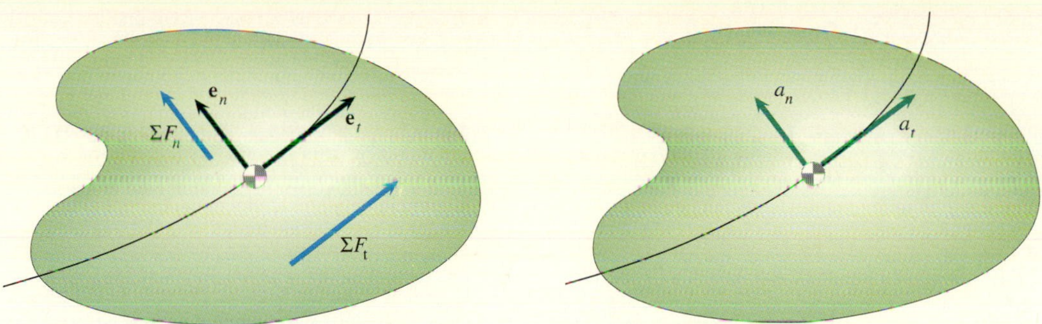

Fig 3.9
(a) Normal and tangential components of the sum of the forces on an object.
(b) Normal and tangential components of the acceleration of the center of mass.

Equating the normal and tangential components in Eq. (3.6), we obtain two scalar equations of motion:

$$\Sigma F_t = ma_t = m\frac{dv}{dt}, \qquad \Sigma F_n = ma_n = m\frac{v^2}{\rho}. \qquad (3.7)$$

The sum of the forces in the tangential direction equals the product of the mass and the rate of change of the magnitude of the velocity, and the sum of the forces in the normal direction equals the product of the mass and the normal component of acceleration. If the path of the object's center of mass lies in a plane, the acceleration of the center of mass perpendicular to the plane is zero and so the sum of the forces perpendicular to the plane is zero.

In the following examples we use Newton's second law expressed in terms of normal and tangential components to analyze motions of objects. By drawing the free-body diagram of an object, you can identify the components of the forces acting on it and use Newton's second law to determine the components of its acceleration. Or, if you know the components of the acceleration, you can use Newton's second law to determine the external forces. When an object follows a circular path, normal and tangential components are usually the simplest choice for analyzing its motion.

Example 3.4

Future space stations may be designed to rotate in order to provide simulated gravity for their inhabitants (Fig. 3.10). If the distance from the axis of rotation of the station to the occupied outer ring is $R = 100$ m, what rotation rate is necessary to simulate one-half of earth's gravity?

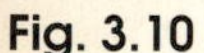

Fig. 3.10

STRATEGY

By drawing the free-body diagram of a person and expressing Newton's second law in terms of normal and tangential components, we can relate the force exerted on the person by the floor to the angular velocity of the station. The person exerts an equal and opposite force on the floor, which is his effective weight.

SOLUTION

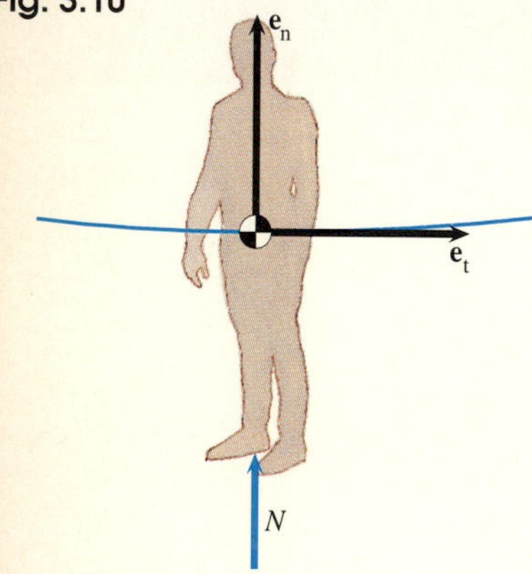

(a) Free-body diagram of a person standing in the occupied ring.

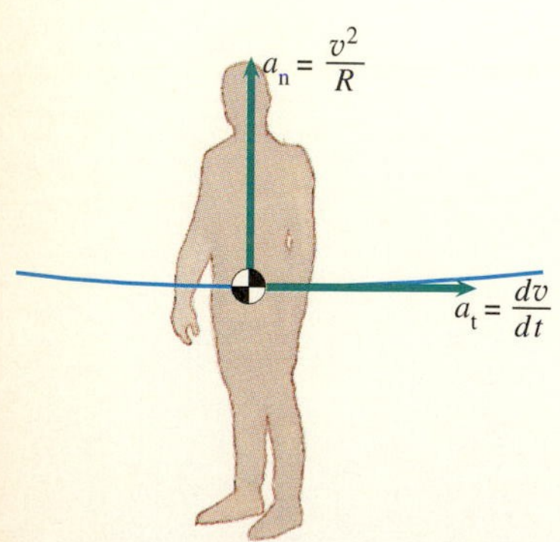

(b) The person's normal and tangential components of acceleration.

We draw the free-body diagram of a person standing in the outer ring in Fig. (a), where N is the force exerted on him by the floor. Relative to a nonrotating reference frame with its origin at the center of the station, he moves in a circular path of radius R. His normal and tangential components of acceleration are shown in Fig. (b). Applying Eqs. (3.7), we obtain

$$\Sigma F_t = 0 = m\frac{dv}{dt},$$

$$\Sigma F_n = N = m\frac{v^2}{R}.$$

The first equation simply indicates that the magnitude of his velocity is constant. The second equation tells us the force N. The magnitude of his velocity is $v = R\omega$, where ω is the angular velocity of the station. If one-half of earth's gravity is simulated, $N = \frac{1}{2}mg$. Therefore

$$N = \frac{1}{2}mg = m\frac{(R\omega)^2}{R}.$$

Solving for ω, we obtain the necessary angular velocity of the station,

$$\omega = \sqrt{\frac{g}{2R}} = \sqrt{\frac{9.81}{(2)(100)}} = 0.221 \text{ rad/s},$$

which is one revolution every 28.4 s.

Example 3.5

The experimental magnetically levitated train in Fig. 3.11 is supported by magnetic repulsion forces exerted normal to the tracks. Motion of the train transverse to the tracks is prevented by lateral supports. The 20-Mg (megagram) train is traveling at 30 m/s on a circular segment of track of radius $R = 150$ m, and the bank angle of the track is 40°. What force must the magnetic levitation system exert to support the train, and what total force is exerted by the lateral supports?

Fig. 3.11

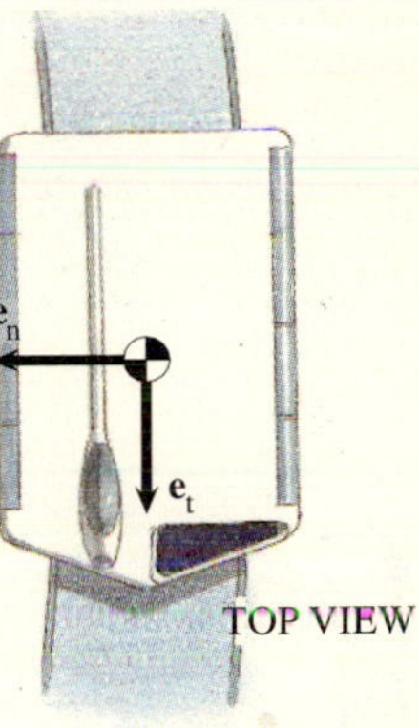

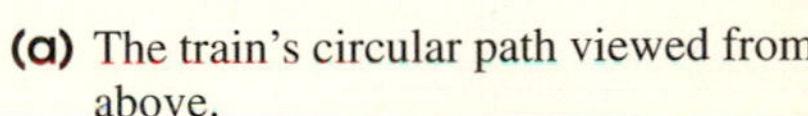

(a) The train's circular path viewed from above.

STRATEGY

We know the train's velocity and the radius of its circular path, so we can determine its normal component of acceleration. By expressing Newton's second law in terms of normal and tangential components, we can determine the components of force normal and transverse to the track.

SOLUTION

Figure (a) shows the train viewed from above. The unit vector $\mathbf{e}_n$ is horizontal and points toward the center of the train's circular path, and $\mathbf{e}_t$ is tangential to the path. In Fig. (b) we draw the free-body diagram of the train seen from the front, where M is the magnetic force normal to the tracks and S is the transverse force. In Fig. (c) we show the train's acceleration, which is perpendicular to its circular path and toward the center of the path. The sum of the forces in the vertical direction (perpendicular to the train's circular path) must equal zero:

$$M \cos 40° + S \sin 40° - mg = 0.$$

The sum of the forces in the $\mathbf{e}_n$ direction equals the product of the mass and the normal component of the acceleration:

$$\Sigma F_n = m\frac{v^2}{\rho}:$$

$$M \sin 40° - S \cos 40° = m\frac{v^2}{R}.$$

Solving these two equations for M and S, we obtain $M = 227.4$ kN, $S = 34.2$ kN.

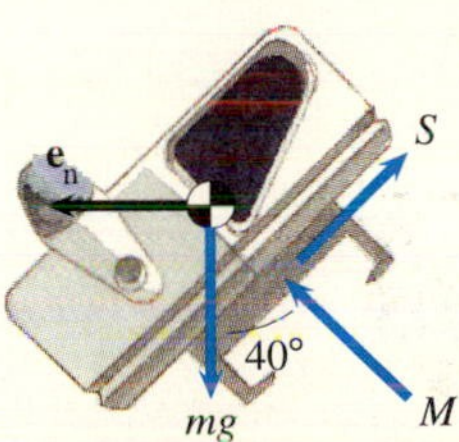

(b) Free-body diagram of the train.

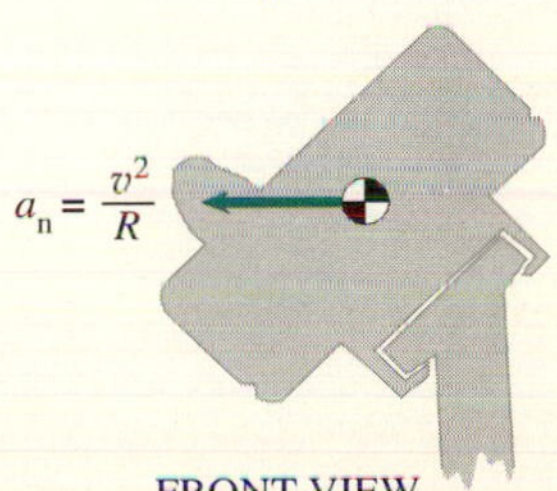

(c) The train's acceleration.

Example 3.6

Application to Engineering

Motor Vehicle Dynamics

A civil engineer's preliminary design for a freeway off-ramp is circular with radius $R = 300$ ft (Fig. 3.12). If she assumes that the coefficient of static friction between tires and road is at least $\mu_s = 0.4$, what is the maximum speed at which vehicles can enter the ramp without losing traction?

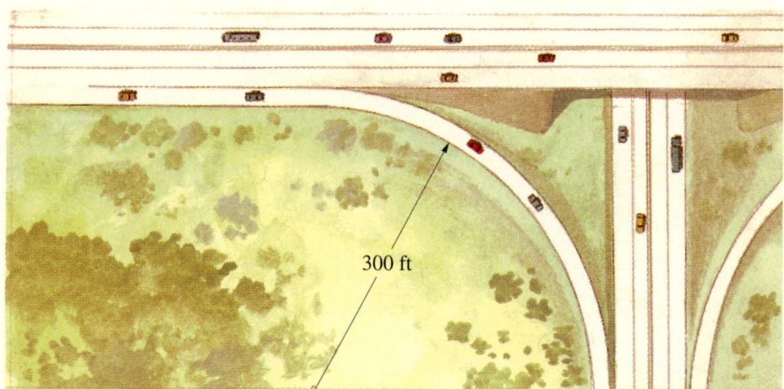

Fig. 3.12

STRATEGY

Since a vehicle on the off-ramp moves in a circular path, it has a normal component of acceleration that depends on its velocity. The necessary normal component of force is exerted by friction between the tires and the road, and the friction force cannot be greater than the product of μ_s and the normal force. By assuming that the friction force is equal to this value, we can determine the maximum velocity for which slipping will not occur.

SOLUTION

We view the free-body diagram of a car on the off-ramp from above the car in Fig. (a) and from the front of the car in Fig. (b). In Fig. (c) we show the car's acceleration, which is perpendicular to its circular path and toward the center of the path. The sum of the forces in the $\mathbf{e}_n$ direction equals the product of the mass and the normal component of the acceleration,

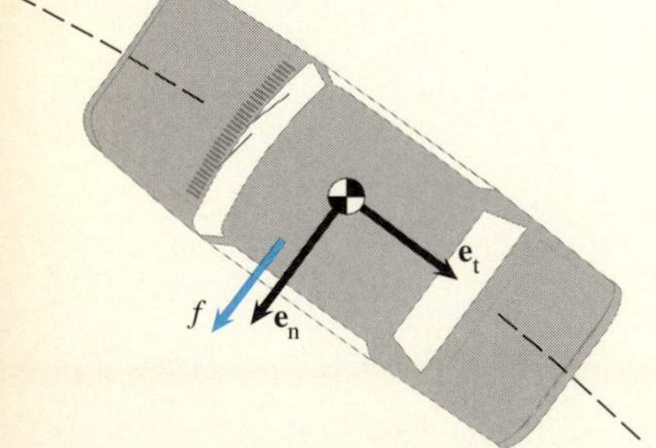

(a) Top view of the free-body diagram.

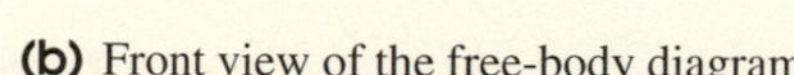

(b) Front view of the free-body diagram.

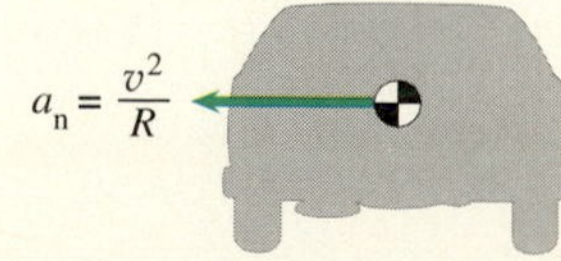

(c) The acceleration seen in the front view.

$$\Sigma F_n = ma_n = m\frac{v^2}{R}:$$

$$f = m\frac{v^2}{R}.$$

The required friction force increases as v increases. The maximum friction force the surfaces will support is $f_{max} = \mu_s N = \mu_s mg$. Therefore the maximum velocity for which slipping does not occur is

$$v = \sqrt{\mu_s gR} = \sqrt{(0.4)(32.2)(300)} = 62.2 \text{ ft/s},$$

or 42.4 mi/hr.

DESIGN ISSUES

Automotive engineers, civil engineers who design highways, and engineers who study traffic accidents and their prevention must analyze and measure the motions of vehicles under different conditions. By using the methods discussed in this chapter, they can relate the forces acting on vehicles to their motions and study, for example, the factors influencing the distance necessary for a car to be brought to a stop in an emergency, or the effects of banking and curvature on the velocity at which a car can safely be driven on a curved road (Fig. 3.13).

In Example 3.6, the analysis indicates that vehicles will lose traction if they enter the freeway off-ramp at speeds greater than 42.4 mi/hr. This result can be used as an indication of the speed limit that must be posted in order for vehicles to enter the ramp safely, or the off-ramp could be designed for a greater speed by increasing the radius of curvature. Or, if a larger safe speed is desired but space limitations forbid a larger radius of curvature, the off-ramp could be designed to incorporate banking (see Problem 3.93).

Fig. 3.13

Tests of the capabilities of vehicles to negotiate curves influence the design of both vehicles and highways.

Problems

3.66 The boat weighs 2600 lb with its passengers. It is moving in a circular path of radius $R = 80$ ft at a constant speed of 15 mi/hr. Determine the magnitudes of: (a) the total horizontal force on the boat in the direction tangent to its path; (b) the total horizontal force on the boat in the direction perpendicular to its path; (c) the total vertical force acting on the boat (including its weight).

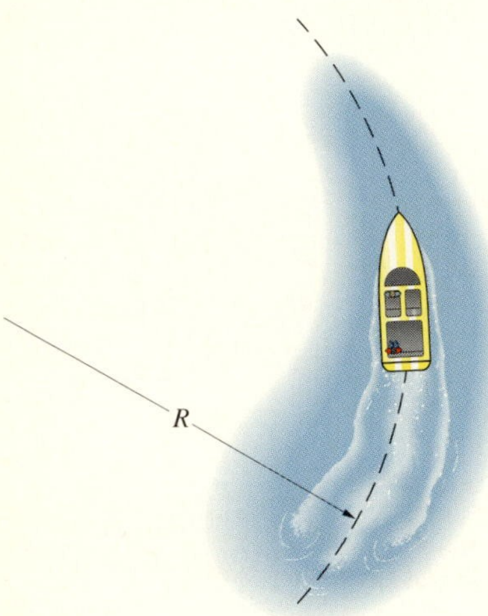

P3.66

3.67 Suppose that the mass of the boat in Problem 3.66 is 1200 kg with its passengers. At the present instant, it is moving in a circular path of radius $R = 30$ m at 4 m/s and is increasing its speed at 2 m/s^2. Determine the magnitudes of: (a) the total horizontal force on the boat in the direction tangent to its path; (b) the total horizontal force on the boat in the direction perpendicular to its path.

3.68 If you choose the velocity of the train in Example 3.5 properly, the lateral force S exerted on it as it travels along the circular track is zero. What is the necessary velocity?

3.69 An astronaut candidate is tested in a centrifuge with a radius of 10 m. If the centrifuge rotates in the horizontal plane with a constant angular velocity of one revolution every two seconds, what horizontal force is exerted on the astronaut? His mass is 72 kg.

Strategy: If a point moves in a circular path of radius R, the magnitude of its velocity v is related to its angular velocity ω by $v = R\omega$, so the normal component of its acceleration can be written $a_n = v^2/R = \omega^2 R$ (see Eq. 2.45). You can use this expression to determine the normal component of force acting on the astronaut.

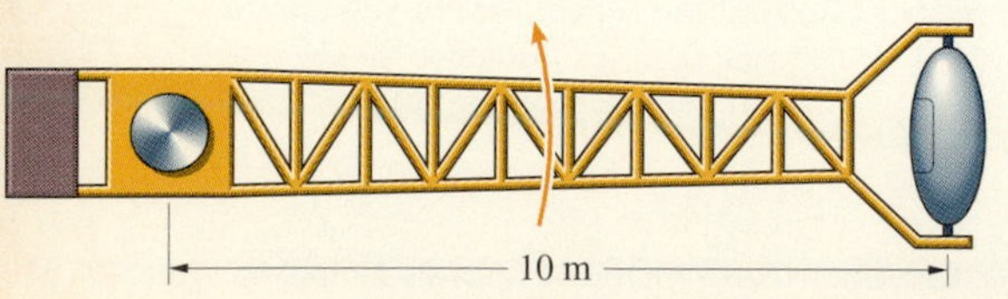

P3.69

3.70 The centrifuge in Problem 3.69 starts from rest at $t = 0$ and is subjected to a constant angular acceleration $\alpha = 0.1$ rad/s^2. What is the magnitude of the total horizontal force exerted on the astronaut at $t = 30$ s?

3.71 The circular disk lies *in the horizontal plane* and rotates with a constant counterclockwise angular velocity of 6 rad/s. The 4-lb slider A is supported horizontally by the smooth slot and the string attached at B. What is the tension in the string?

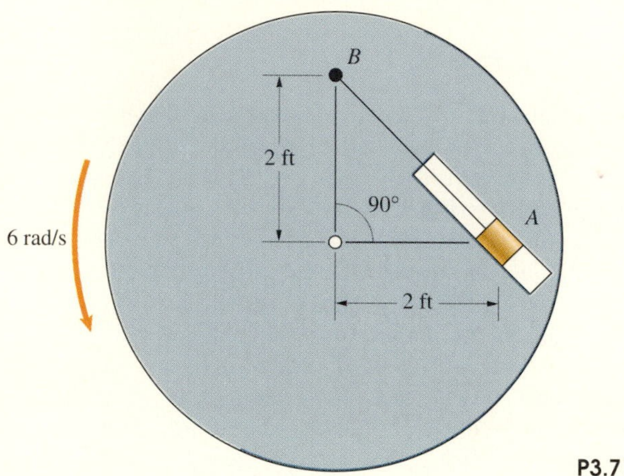

P3.71

3.72 In Problem 3.71, determine the tension in the string if the circular disk has a counterclockwise angular acceleration of 10 rad/s^2 at the instant shown.

3.73 The 2-kg slider A starts from rest and slides *in the horizontal plane* along the smooth circular bar under the action of a tangential force $F_t = 4t$ N. At $t = 4$ s, determine (a) the magnitude of the velocity of the slider; (b) the magnitude of the horizontal force exerted on the slider by the bar.

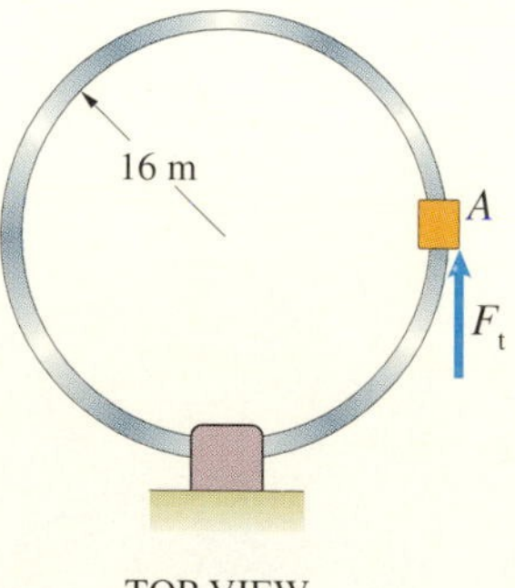

P3.73

3.74 Small parts on a conveyer belt moving with constant velocity v are allowed to drop into a bin. Show that the angle α at which the parts start sliding on the belt satisfies the equation

$$\cos\alpha - \frac{1}{\mu_s}\sin\alpha = \frac{v^2}{gR},$$

where μ_s is the coefficient of static friction between the parts and the belt.

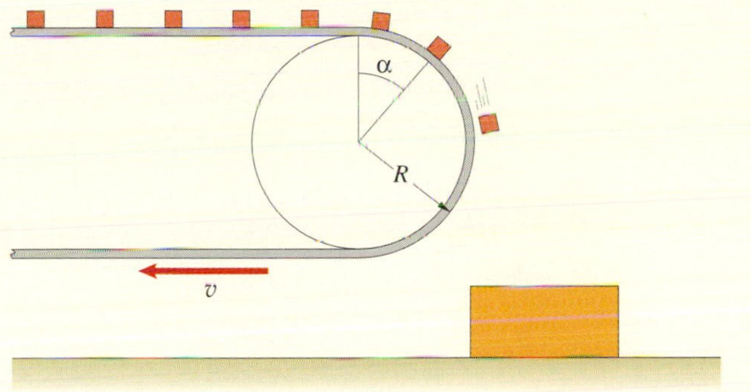

P3.74

3.75 The mass m rotates around the vertical pole in a horizontal circular path. Determine the magnitude of its velocity in terms of θ and L.

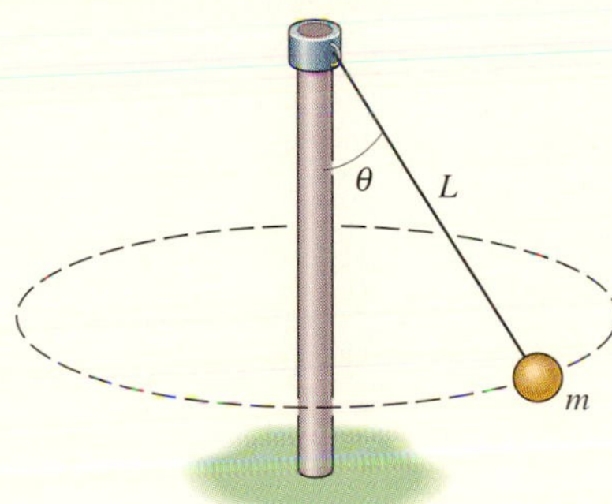

P3.75

3.76 In Problem 3.75, if $m = 1$ slug, $L = 4$ ft, and the mass is moving in its circular path at $v = 15$ ft/s, what is the tension in the string?

3.77 The 10-kg mass m rotates around the vertical pole in a horizontal circular path of radius $R = 1$ m. If the magnitude of its velocity is $v = 3$ m/s, what are the tensions in the strings A and B?

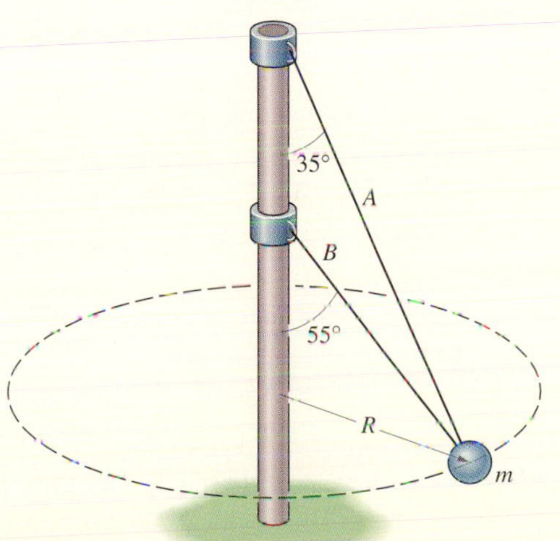

P3.77

3.78 In Problem 3.77, what is the range of values of v for which the mass will remain in the circular path described?

3.79 Suppose you are designing a monorail transportation system that will travel at 50 m/s, and you decide that the angle α that the cars swing out from the vertical when they go through a turn must not be larger than 20°. If the turns in the track consist of circular arcs of constant radius R, what is the minimum allowable value of R?

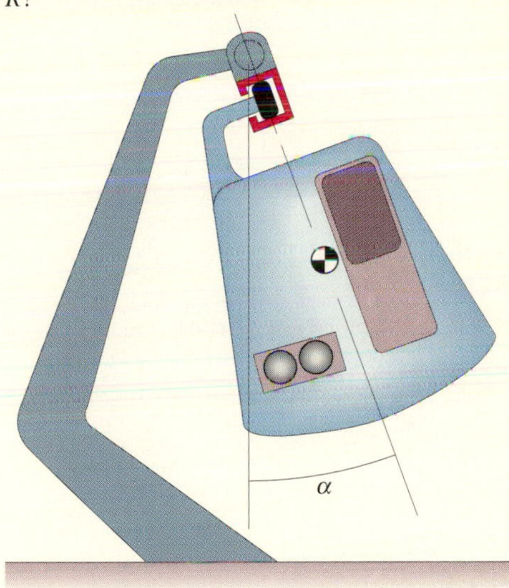

P3.79

3.80 An airplane of weight $W = 200{,}000$ lb makes a turn at constant altitude and at constant velocity $v = 600$ ft/s. The bank angle is 15°.
(a) Determine the lift force L.
(b) What is the radius of curvature of the plane's path?

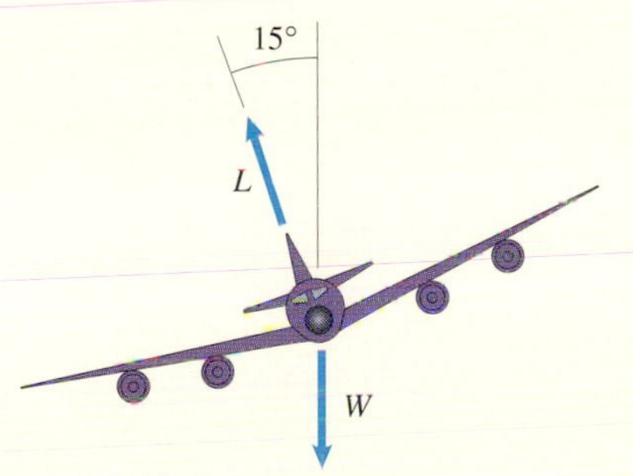

P3.80

3.81 The suspended mass m is stationary.
(a) What are the tensions in the strings?
(b) If string A is cut, what is the tension in string B immediately afterward?

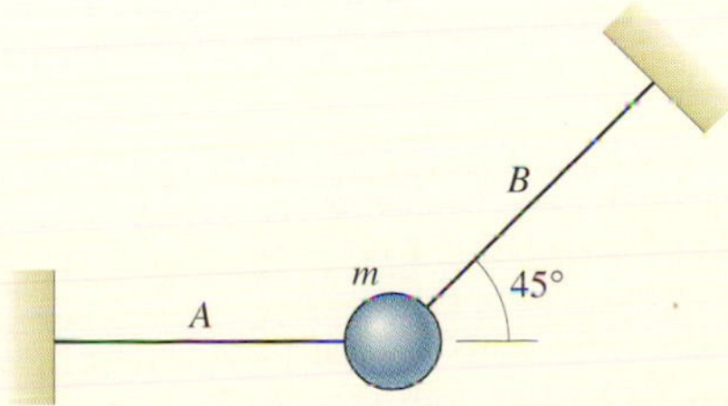

P3.81

3.82 An airplane flies with constant velocity v along a circular path in the vertical plane. The radius of its circular path is 5000 ft. The pilot weighs 150 lb.
(a) The pilot will experience "weightlessness" at the top of the circular path if the airplane exerts no net force on him at that point. Draw a free-body diagram of the pilot, and use it to determine the velocity v necessary to achieve this condition.
(b) Determine the force exerted on the pilot by the airplane at the top of the circular path if the airplane is traveling at twice the velocity determined in part (a).

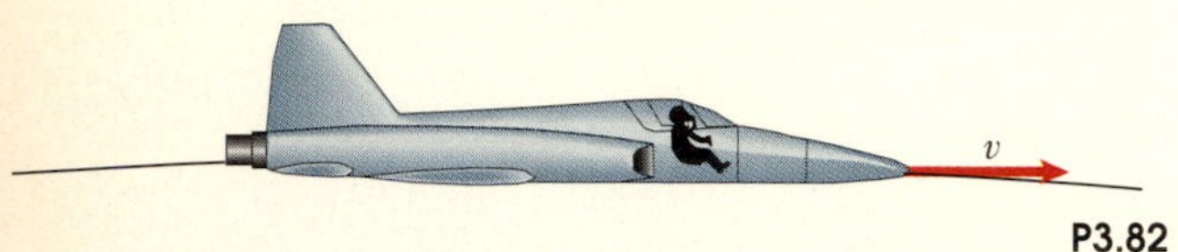

P3.82

3.83 The smooth circular bar rotates with constant angular velocity ω_0 about the vertical axis AB. Determine the angle β at which the slider of mass m will remain stationary relative to the circular bar.

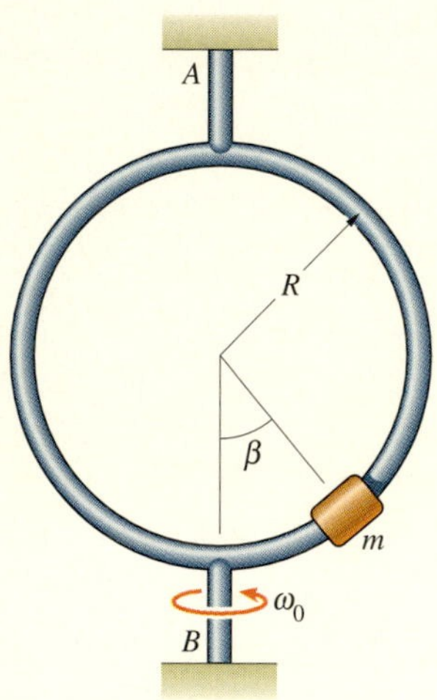

P3.83

3.84 The force exerted on a charged particle by a magnetic field is

$$\mathbf{F} = q\mathbf{v} \times \mathbf{B},$$

where q and $\mathbf{v}$ are the charge and velocity vector of the particle and $\mathbf{B}$ is the magnetic field vector. A particle of mass m and positive charge q is projected at O with velocity $\mathbf{v} = v_0\mathbf{i}$ into a uniform magnetic field $\mathbf{B} = B_0\mathbf{k}$. Using normal and tangential components, show that: (a) the magnitude of the particle's velocity is constant; (b) the particle's path is a circle with radius mv_0/qB_0.

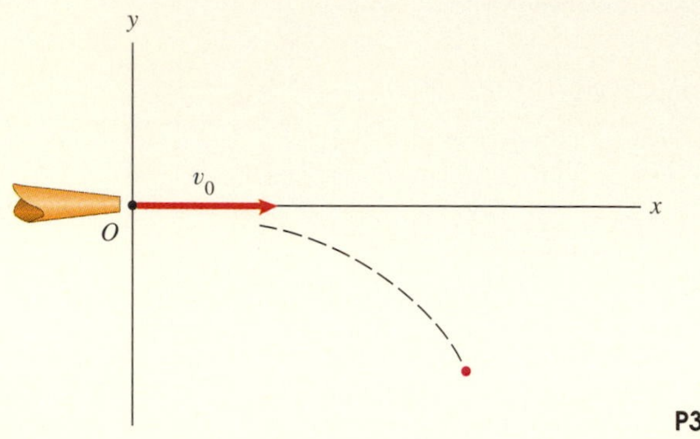

P3.84

3.85 A mass m is attached to a string that is wrapped around a fixed post of radius R. At $t = 0$, the object is given a velocity v_0 as shown. Neglect external forces on m other than the force exerted by the string. Determine the tension in the string as a function of the angle θ.

Strategy: The velocity vector of the mass is perpendicular to the string. Express Newton's second law in terms of normal and tangential components.

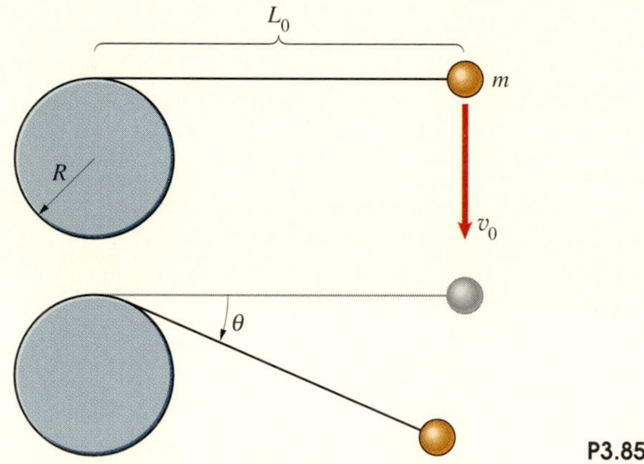

P3.85

3.86 In Problem 3.85, determine the angle θ as a function of time.

3.87 The sum of the forces in newtons exerted on the 360-kg sport plane (including its weight) during an interval of time is

$$(-1000 + 280t)\mathbf{i} + (480 - 430t)\mathbf{j} + (720 + 200t)\mathbf{k},$$

where t is the time in seconds. At $t = 0$, the velocity of the plane's center of mass relative to the earth-fixed reference frame is $20\mathbf{i} + 35\mathbf{j} - 20\mathbf{k}$ (m/s). If you resolve the sum of the forces on the plane into components tangent and normal to the plane's path at $t = 2$ s, what are their values ΣF_t and ΣF_n?

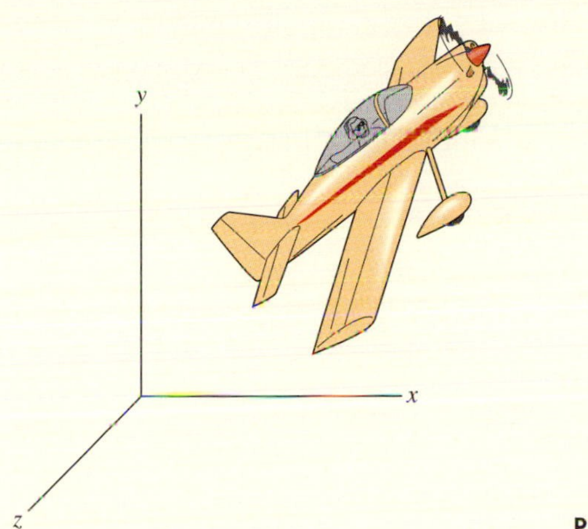

P3.87

3.88 In Problem 3.87, what is the instantaneous radius of curvature of the plane's path at $t = 2$ s? The vector components of the sum of the forces in the directions tangential and normal to the path lie in the osculating plane. Determine the components of a unit vector perpendicular to the osculating plane at $t = 2$ s.

Problems 3.89–3.93 are related to Example 3.6.

3.89 A car is traveling on a straight, level road when the driver perceives a hazard ahead. After a reaction time of 0.5 s, he applies the brakes, locking the wheels. The coefficient of kinetic friction between the tires and the road is $\mu_k = 0.6$. Determine the total distance the car travels before coming to rest, including the distance traveled before the brakes are applied, if it is traveling at (a) 55 mi/hr; (b) 65 mi/hr.

3.90 If the car in Problem 3.89 is traveling at 65 mi/hr and rain decreases the value of μ_k to 0.4, what total distance does the car travel before coming to rest?

3.91 A car traveling at 30 m/s is at the top of a hill. The coefficient of kinetic friction between the tires and the road is $\mu_k = 0.8$ and the instantaneous radius of curvature of the car's path is 200 m. If the driver applies the brakes and the car's wheels lock, what is the resulting deceleration of the car in the direction tangent to its path?

P3.91

3.92 Suppose that the car in Problem 3.91 is at the bottom of a depression whose radius of curvature is 200 m when the driver applies the brakes. What is the resulting deceleration of the car in the direction tangent to its path?

P3.92

3.93 A freeway off-ramp is circular with radius R (Fig. a), and the roadway is banked at an angle β (Fig. b). Show that the maximum constant velocity at which a car can travel the off-ramp without losing traction is

$$v = \sqrt{gR\left(\frac{\sin\beta + \mu_s \cos\beta}{\cos\beta - \mu_s \sin\beta}\right)}.$$

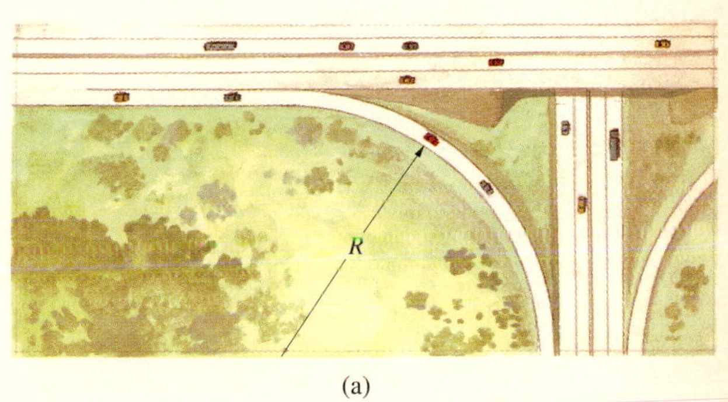

(a)

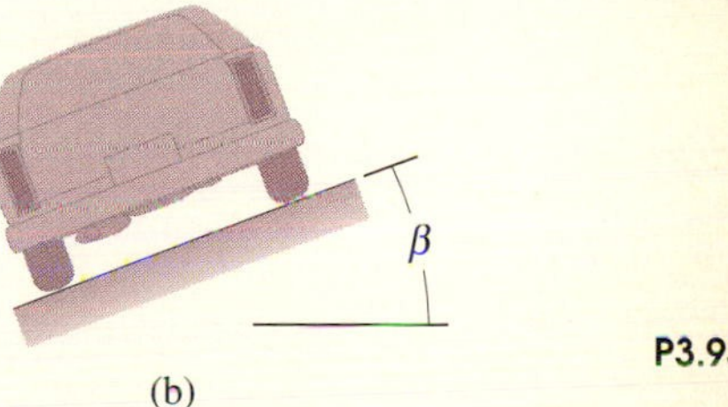

(b)

P3.93

Polar and Cylindrical Coordinates

When an object moves in a plane curved path, we can describe the motion of its center of mass in terms of polar coordinates. Resolving the sum of the forces parallel to the plane into radial and transverse components (Fig. 3.14a) and expressing the acceleration of the center of mass in terms of radial and transverse components (Fig. 3.14b), we can write Newton's second law in the form

$$\Sigma \mathbf{F} = m\mathbf{a}:$$
$$(\Sigma F_r \mathbf{e}_r + \Sigma F_\theta \mathbf{e}_\theta) = m(a_r \mathbf{e}_r + a_\theta \mathbf{e}_\theta), \tag{3.8}$$

where

$$a_r = \frac{d^2r}{dt^2} - r\left(\frac{d\theta}{dt}\right)^2 = \frac{d^2r}{dt^2} - r\omega^2,$$

$$a_\theta = r\frac{d^2\theta}{dt^2} + 2\frac{dr}{dt}\frac{d\theta}{dt} = r\alpha + 2\frac{dr}{dt}\omega.$$

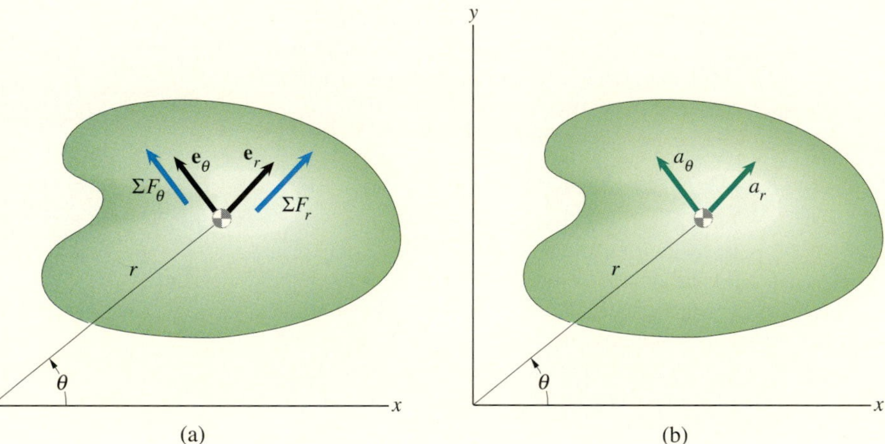

Fig. 3.14
Radial and transverse components of the sum of the forces (a) and the acceleration of the center of mass (b).

Equating the $\mathbf{e}_r$ and $\mathbf{e}_\theta$ components in Eq. (3.8), we obtain the scalar equations

$$\Sigma F_r = ma_r = m\left(\frac{d^2r}{dt^2} - r\omega^2\right), \tag{3.9}$$

$$\Sigma F_\theta = ma_\theta = m\left(r\alpha + 2\frac{dr}{dt}\omega\right). \tag{3.10}$$

The sum of the forces in the radial direction equals the product of the mass and the radial component of the acceleration, and the sum of the forces in the transverse direction equals the product of the mass and the transverse component of the acceleration. Since the object's acceleration perpendicular to the plane in which the motion takes place is zero, the sum of the forces perpendicular to the plane is zero.

We can describe three-dimensional motion of an object using cylindrical coordinates, in which the position of the center of mass perpendicular to the

x-y plane is measured by the coordinate z and the unit vector $\mathbf{e}_z$ points in the positive z axis direction. We resolve the sum of the forces into radial, transverse, and z components (Fig. 3.15a) and express the acceleration of the center of mass in terms of radial, transverse, and z components (Fig. 3.15b). The three scalar equations of motion are the radial and transverse equations (3.9) and (3.10) and the equation of motion in the z direction,

$$\Sigma F_z = ma_z = m\frac{dv_z}{dt} = m\frac{d^2z}{dt^2}. \qquad (3.11)$$

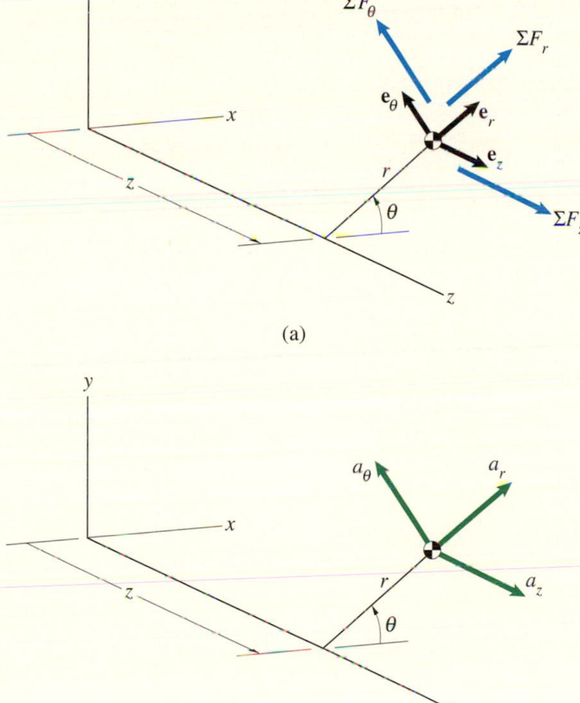

Fig. 3.15

(a) Components of the sum of the forces on an object in cylindrical coordinates.
(b) Components of the acceleration of the center of mass.

In the following example we use Newton's second law expressed in terms of polar coordinates, or radial and transverse components, to analyze the motion of an object.

Example 3.7

Fig. 3.16

The smooth bar in Fig. 3.16 rotates *in the horizontal plane* with constant angular velocity ω_0. The unstretched length of the linear spring is r_0. The collar A has mass m and is released at $r = r_0$ with no radial velocity.
(a) Determine the radial velocity of the collar as a function of r.
(b) Determine the horizontal force exerted on the collar by the bar as a function of r.

STRATEGY

(a) The only force on the collar in the radial direction is the spring force, which we can express in polar coordinates in terms of r. By integrating Eq. (3.9), we can determine the radial velocity v_r as a function of r.
(b) Once $v_r = dr/dt$ is known in terms of r, we can use Eq. (3.10) to determine the transverse force exerted on the collar by the bar.

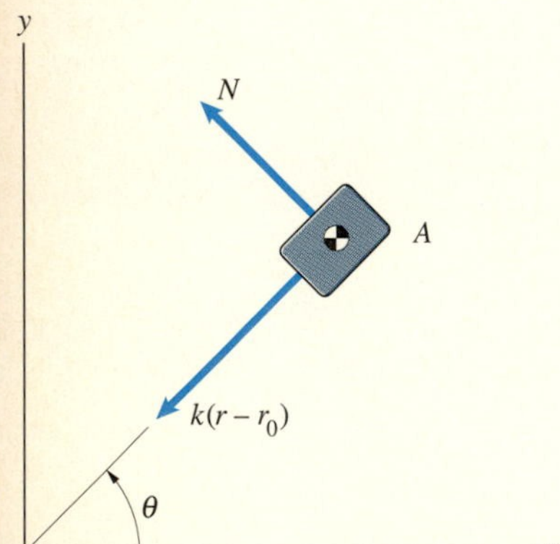

(a) Radial and transverse forces on A.

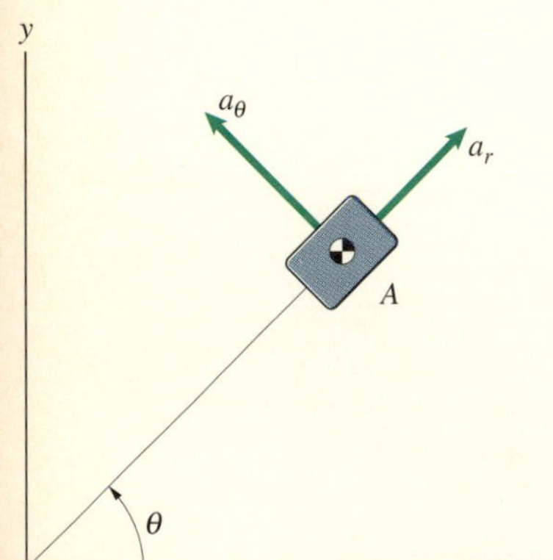

(b) Radial and transverse components of the acceleration of the center of mass.

SOLUTION

(a) The spring exerts a radial force $k(r - r_0)$ in the negative r direction (Fig. a). Since the bar is smooth, it exerts no radial force on A, but may exert a transverse force N. Figure (b) shows the radial and transverse components of the collar's acceleration. Newton's second law in the radial direction is

$$\Sigma F_r = ma_r:$$

$$-k(r - r_0) = m\left(\frac{d^2r}{dt^2} - r\omega^2\right) = m\left(\frac{dv_r}{dt} - r\omega_0^2\right).$$

We solve this equation for the time derivative of v_r:

$$\frac{dv_r}{dt} = r\omega_0^2 - \frac{k}{m}(r - r_0).$$

Applying the chain rule,

$$\frac{dv_r}{dt} = \frac{dv_r}{dr}\frac{dr}{dt} = \frac{dv_r}{dr}v_r,$$

we obtain

$$v_r\,dv_r = \left[\left(\omega_0^2 - \frac{k}{m}\right)r + \frac{k}{m}r_0\right]dr.$$

Integrating,

$$\int_0^{v_r} v_r\,dv_r = \int_{r_0}^{r}\left[\left(\omega_0^2 - \frac{k}{m}\right)r + \frac{k}{m}r_0\right]dr,$$

we obtain the radial velocity as a function of r:

$$v_r = \sqrt{\left(\omega_0^2 - \frac{k}{m}\right)(r^2 - r_0^2) + \frac{2k}{m}r_0(r - r_0)}.$$

(b) To determine the force N, we use Newton's second law in the transverse direction,

$$\Sigma F_\theta = ma_\theta:$$

$$N = m\left(r\alpha + 2\frac{dr}{dt}\omega\right) = 2m\omega_0 v_r.$$

Substituting our expression for v_r as a function of r, we obtain the transverse force exerted by the bar as a function of r:

$$N = 2m\omega_0\sqrt{\left(\omega_0^2 - \frac{k}{m}\right)(r^2 - r_0^2) + \frac{2k}{m}r_0(r - r_0)}.$$

Problems

3.94 The polar coordinates of the center of mass of an object are $r = t^2 + 2$ ft, $\theta = 2t^3 - t^2$ rad, and its mass is 3 slugs. What are the radial and transverse components of the total external force on the object at $t = 1$ s?

3.95 The polar coordinates of the center of mass of an object are $r = 2t^3 + 4t$ m, $\theta = t^2 - t$ rad, and its mass is 20 kg. What are the radial and transverse components of the total external force on the object at $t = 1$ s?

3.96 The robot is programmed so that the 0.4-kg part A describes the path

$$r = 1 - 0.5\cos 2\pi t \text{ m},$$

$$\theta = 0.5 - 0.2\sin 2\pi t \text{ rad}.$$

At $t = 2$ s, determine the radial and transverse components of force exerted on A by the robot's jaws.

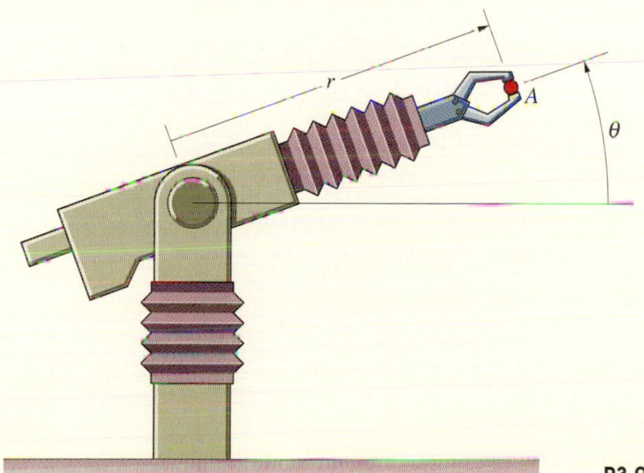

P3.96

3.97 In Example 3.7, what is the maximum radial distance reached by the collar A?

3.98 The smooth bar rotates *in the horizontal plane* with constant angular velocity $\omega_0 = 60$ rpm. If the 2-lb collar A is released at $r = 1$ ft with no radial velocity, what is the magnitude of its velocity when it reaches the end of the bar?

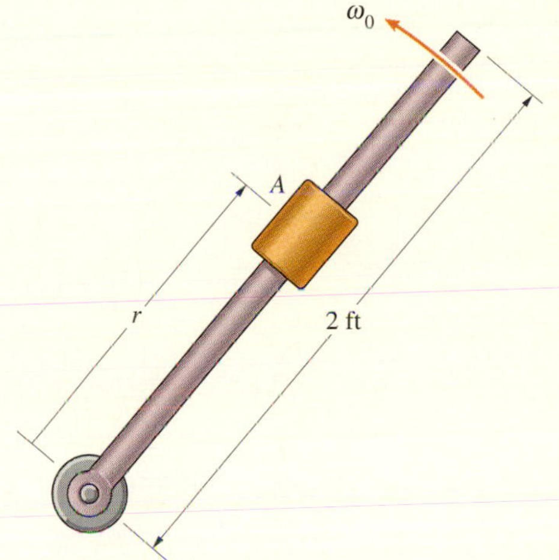

P3.98

3.99 In Problem 3.98, what is the maximum horizontal force exerted on the collar by the bar?

3.100 The mass m is released from rest with the string horizontal. By using Newton's second law in terms of polar coordinates, determine the magnitude of the velocity of the mass and the tension in the string as functions of θ.

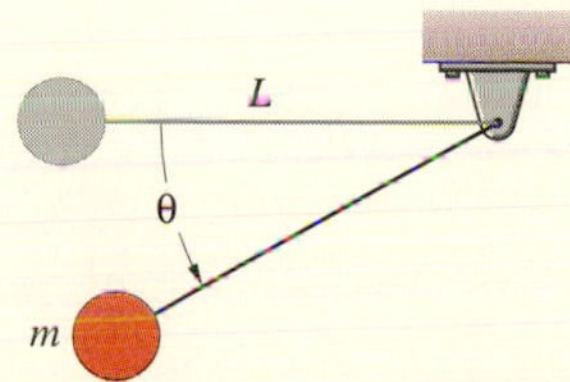

P3.100

3.101 The 1-lb block A is given an initial velocity $v_0 = 14$ ft/s to the right when it is in the position $\theta = 0$, causing it to slide up the smooth circular surface. By using Newton's second law in terms of polar coordinates, determine the magnitude of the velocity of the block when $\theta = 60°$.

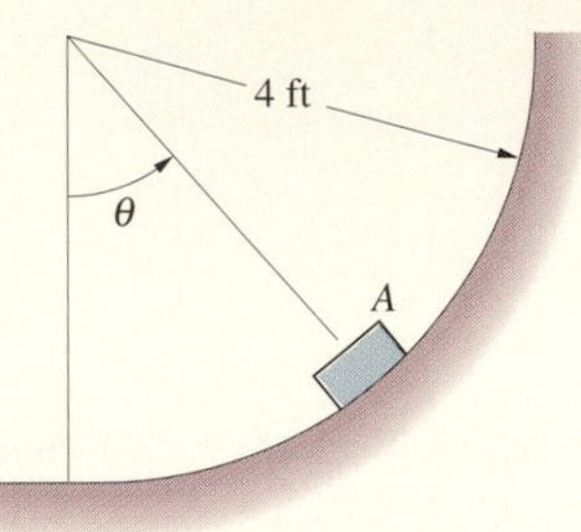

P3.101

3.102 In Problem 3.101, determine the normal force exerted on the block by the smooth surface when $\theta = 60°$.

3.103 The skier passes point A going 17 m/s. From A to B, the radius of his circular path is 6 m. By using Newton's second law in terms of polar coordinates, determine the magnitude of his velocity as he leaves the jump at B. Neglect tangential forces other than the tangential component of his weight.

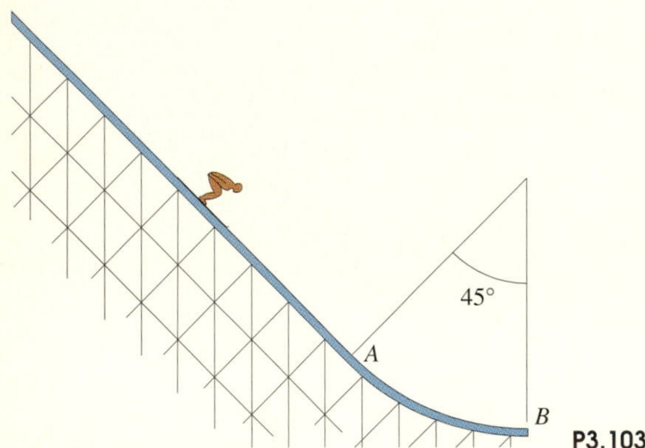

P3.103

3.104 A 2-kg mass rests on a flat horizontal bar. The bar begins rotating *in the vertical plane* about O with a constant angular acceleration of 1 rad/s^2. The mass is observed to slip relative to the bar when the bar is 30° above the horizontal. What is the static coefficient of friction between the mass and the bar? Does the mass slip toward or away from O?

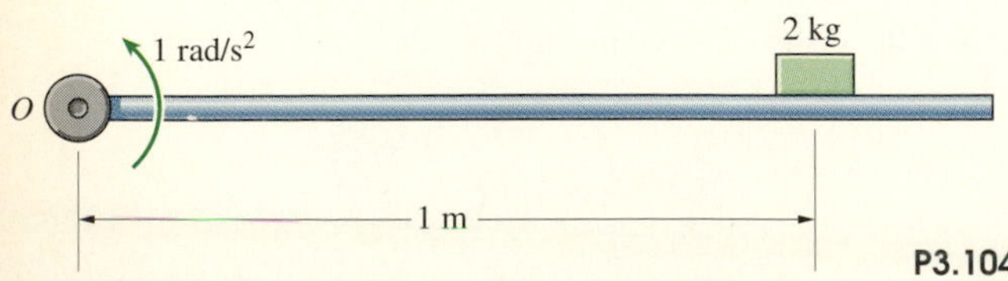

P3.104

3.105 The $\frac{1}{4}$-lb slider A is pushed along the circular bar by the slotted bar. The circular bar lies *in the horizontal plane.* The angular position of the slotted bar is $\theta = 10t^2$ rad. Determine the radial and transverse components of the total external force exerted on the slider at $t = 0.2$ s.

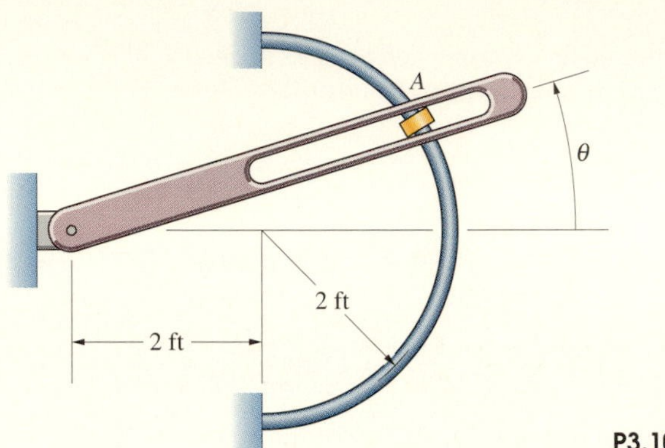

P3.105

3.106 In Problem 3.105, suppose that the circular bar lies *in the vertical plane.* Determine the radial and transverse components of the total force exerted on the slider by the circular and slotted bars at $t = 0.25$ s.

3.107 The slotted bar rotates *in the horizontal plane* with constant angular velocity ω_0. The mass m has a pin that fits in the slot of the bar. A spring holds the pin against the surface of the fixed cam. The surface of the cam is described by $r = r_0(2 - \cos\theta)$. Determine the radial and transverse components of the total external force exerted on the pin as functions of θ.

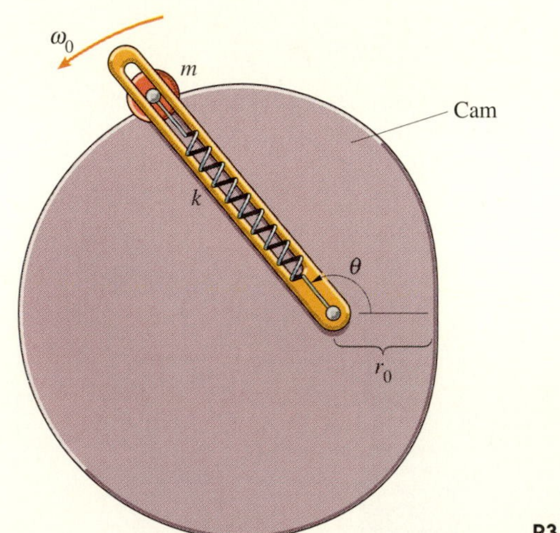

P3.107

3.108 In Problem 3.107, suppose that the unstretched length of the spring is r_0. Determine the smallest value of the spring constant k for which the pin will remain on the surface of the cam.

3.109 A charged particle P in a magnetic field moves along the spiral path described by $r = 1$ m, $\theta = 2z$ rad, where z is in meters. The particle moves along the path in the direction shown with constant speed $|\mathbf{v}| = 1$ km/s. The mass of the particle is 1.67×10^{-27} kg. Determine the sum of the forces on the particle in terms of cylindrical coordinates.

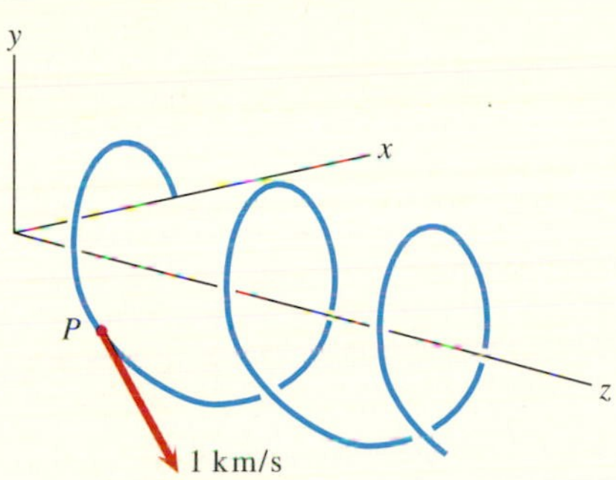

P3.109

3.110 At the instant shown, the cylindrical coordinates of the 4-kg part A held by the robotic manipulator are $r = 0.6$ m, $\theta = 25°$, and $z = 0.8$ m. (The coordinate system is earth-fixed and the y axis points upward.) A's radial position is increasing at $dr/dt = 0.2$ m/s and $d^2r/dt^2 = -0.4$ m/s^2. The angle θ is increasing at $d\theta/dt = 1.2$ rad/s and $d^2\theta/dt^2 = 2.8$ rad/s^2. The base of the manipulator arm is accelerating in the z direction at $d^2z/dt^2 = 2.5$ m/s^2. Determine the force vector exerted on A by the manipulator in terms of cylindrical coordinates.

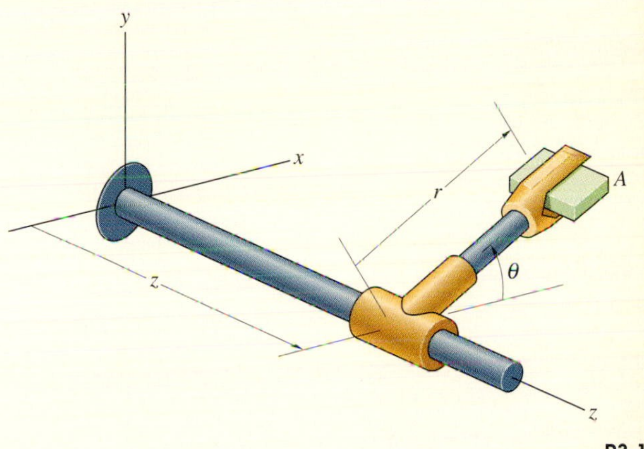

P3.110

3.111 Suppose that the robotic manipulator in Problem 3.110 is used in a space station to investigate zero-g manufacturing techniques. During an interval of time it is programmed so that the cylindrical coordinates of the 4-kg part A are $\theta = 0.15t^2$ rad, $r = 0.5(1 + \sin\theta)$ m, and $z = 0.8(1 + \theta)$ m. Determine the force vector exerted on A by the manipulator at $t = 2$ s in terms of cylindrical coordinates.

3.112 In Problem 3.111, draw a graph of the magnitude of the force exerted on part A by the manipulator as a function of time from $t = 0$ to $t = 5$ s and use it to estimate the maximum force during that interval of time.

3.5 Orbital Mechanics

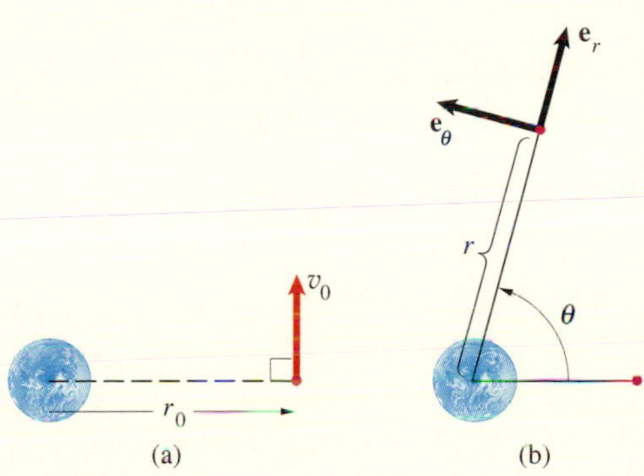

Fig. 3.17
(a) Initial position and velocity of an earth satellite.
(b) Specifying the subsequent path in terms of polar coordinates.

It is appropriate to conclude our chapter on applications of Newton's second law with a discussion of orbital mechanics. Newton's analytical determination of the elliptical orbits of the planets, which had been deduced from observational data by Johannes Kepler, was a triumph for Newtonian mechanics and confirmation of the inverse-square relation for gravitational force.

We can use Newton's second law expressed in polar coordinates to determine the orbit of an earth satellite or planet. Suppose that at $t = 0$ a satellite has an initial velocity v_0 at a distance r_0 from the center of the earth (Fig. 3.17a). We assume that the initial velocity is perpendicular to the line from the center of the earth to the satellite. The satellite's position during its subsequent motion is specified by its polar coordinates (r, θ), where θ is measured from its position at $t = 0$ (Fig. 3.17b). Our objective is to determine r as a function of θ.

Determination of the Orbit

If we model the earth as a homogeneous sphere, the force exerted on the satellite by gravity at a distance r from the center of the earth is mgR_E^2/r^2, where R_E is the earth's radius. (See Eq. 1.5.) From Eqs. (3.9) and (3.10), the equation of motion in the radial direction is

$$\Sigma F_r = ma_r:$$

$$-\frac{mgR_E^2}{r^2} = m\left[\frac{d^2r}{dt^2} - r\left(\frac{d\theta}{dt}\right)^2\right],$$

and the equation of motion in the transverse direction is

$$\Sigma F_\theta = ma_\theta:$$

$$0 = m\left(r\frac{d^2\theta}{dt^2} + 2\frac{dr}{dt}\frac{d\theta}{dt}\right).$$

We therefore obtain the two equations

$$\frac{d^2r}{dt^2} - r\left(\frac{d\theta}{dt}\right)^2 = -\frac{gR_E^2}{r^2}, \tag{3.12}$$

$$r\frac{d^2\theta}{dt^2} + 2\frac{dr}{dt}\frac{d\theta}{dt} = 0. \tag{3.13}$$

We can write Eq. (3.13) in the form

$$\frac{1}{r}\frac{d}{dt}\left(r^2\frac{d\theta}{dt}\right) = 0,$$

which indicates that

$$r^2\frac{d\theta}{dt} = rv_\theta = \text{constant}. \tag{3.14}$$

At $t = 0$ the components of the velocity are $v_r = 0$, $v_\theta = v_0$, and the radial position is $r = r_0$. We can therefore write the constant in Eq. (3.14) in terms of the initial conditions:

$$r^2\frac{d\theta}{dt} = rv_\theta = r_0v_0. \tag{3.15}$$

Using this equation to eliminate $d\theta/dt$ from Eq. (3.12), we obtain

$$\frac{d^2r}{dt^2} - \frac{r_0^2v_0^2}{r^3} = -\frac{gR_E^2}{r^2}. \tag{3.16}$$

We can solve this differential equation by introducing the change of variable

$$u = \frac{1}{r}. \tag{3.17}$$

In doing so, we also change the independent variable from t to θ because we want to determine r as a function of the angle θ instead of time. To express Eq. (3.16) in terms of u, we must determine d^2r/dt^2 in terms of u. Using the chain rule, we write the derivative of r with respect to time as

$$\frac{dr}{dt} = \frac{d}{dt}\left(\frac{1}{u}\right) = -\frac{1}{u^2}\frac{du}{dt} = -\frac{1}{u^2}\frac{du}{d\theta}\frac{d\theta}{dt}. \tag{3.18}$$

Notice from Eq. (3.15) that

$$\frac{d\theta}{dt} = \frac{r_0 v_0}{r^2} = r_0 v_0 u^2. \tag{3.19}$$

Substituting this expression into Eq. (3.18), we obtain

$$\frac{dr}{dt} = -r_0 v_0 \frac{du}{d\theta}. \tag{3.20}$$

We differentiate this expression with respect to time and apply the chain rule again:

$$\frac{d^2r}{dt^2} = \frac{d}{dt}\left(-r_0 v_0 \frac{du}{d\theta}\right) = -r_0 v_0 \frac{d\theta}{dt}\frac{d}{d\theta}\left(\frac{du}{d\theta}\right) = -r_0 v_0 \frac{d\theta}{dt}\frac{d^2u}{d\theta^2}.$$

Using Eq. (3.19) to eliminate $d\theta/dt$ from this expression, we obtain the second time derivative of r in terms of u:

$$\frac{d^2r}{dt^2} = -r_0^2 v_0^2 u^2 \frac{d^2u}{d\theta^2}.$$

Substituting this result into Eq. (3.16) yields a linear differential equation for u as a function of θ:

$$\frac{d^2u}{d\theta^2} + u = \frac{gR_E^2}{r_0^2 v_0^2}.$$

The general solution of this equation is

$$u = A \sin\theta + B\cos\theta + \frac{gR_E^2}{r_0^2 v_0^2}, \tag{3.21}$$

where A and B are constants. We can use the initial conditions to determine A and B. When $\theta = 0$, $u = 1/r_0$. Also, when $\theta = 0$, the radial component of velocity $v_r = dr/dt = 0$, so from Eq. (3.20) we see that $du/d\theta = 0$. From these two conditions, we obtain

$$A = 0, \qquad B = \frac{1}{r_0} - \frac{gR_E^2}{r_0^2 v_0^2}.$$

Substituting these results into Eq. (3.21), we can write the resulting solution for $r = 1/u$ as

$$\boxed{\frac{r}{r_0} = \frac{1+\varepsilon}{1+\varepsilon\cos\theta},} \tag{3.22}$$

where

$$\varepsilon = \frac{r_0 v_0^2}{gR_E^2} - 1. \tag{3.23}$$

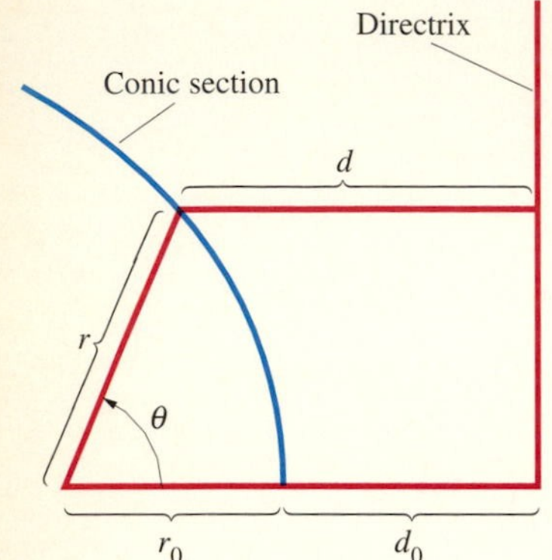

Fig. 3.18
If the ratio r/d is constant, the curve describes a conic section.

Types of Orbits

The curve called a **conic section** (Fig. 3.18) has the property that the ratio of r to the perpendicular distance d to a straight line, called the *directrix*, is constant. This ratio, $r/d = r_0/d_0$, is called the **eccentricity** of the curve. From Fig. 3.18 we see that

$$r \cos \theta + d = r_0 + d_0,$$

which we can write in terms of the eccentricity as

$$\frac{r}{r_0} = \frac{1 + (r_0/d_0)}{1 + (r_0/d_0) \cos \theta}.$$

Comparing this expression to Eq. (3.22), we see that *the satellite's orbit describes a conic section with eccentricity* ε. The value of the eccentricity determines the character of the orbit.

Circular Orbit If the initial velocity v_0 is chosen so that $\varepsilon = 0$, Eq. (3.22) reduces to $r = r_0$ and the orbit is circular (Fig. 3.19). Setting $\varepsilon = 0$ in Eq. (3.23) and solving for v_0, we obtain

$$v_0 = \sqrt{\frac{gR_E^2}{r_0}}, \tag{3.24}$$

which agrees with the velocity for a circular orbit we obtained by a different method in Example 2.10.

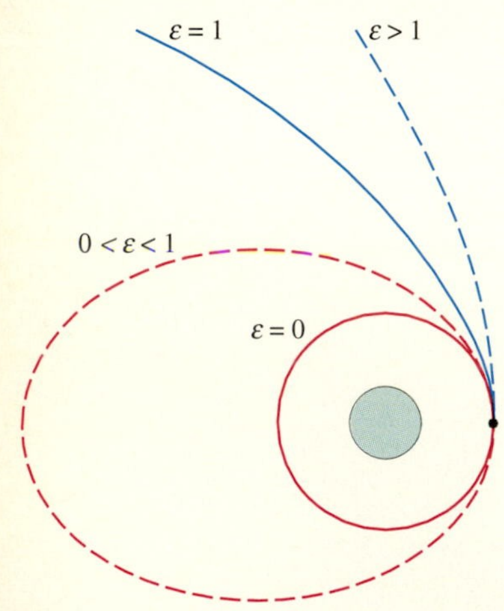

Fig. 3.19
Orbits for different values of eccentricity.

Elliptic Orbit If $0 < \varepsilon < 1$, the orbit is an ellipse (Fig. 3.19). The maximum radius of the ellipse occurs when $\theta = 180°$. Setting θ equal to 180° in Eq. (3.22), we obtain an expression for the maximum radius of the ellipse in terms of the initial radius and ε:

$$r_{max} = r_0\left(\frac{1 + \varepsilon}{1 - \varepsilon}\right). \tag{3.25}$$

Parabolic Orbit Notice from Eq. (3.25) that the maximum radius of the elliptic orbit increases without limit as $\varepsilon \to 1$. When $\varepsilon = 1$, the orbit is a parabola (Fig. 3.19). The corresponding velocity v_0 is the minimum initial velocity for which the radius r increases without limit, which is the escape velocity. Setting $\varepsilon = 1$ in Eq. (3.23) and solving for v_0, we obtain

$$v_0 = \sqrt{\frac{2gR_E^2}{r_0}}.$$

This is the same value for the escape velocity we obtained in Example 2.5 for the case of an object moving in a straight path directly away from the center of the earth.

Hyperbolic Orbit If $\varepsilon > 1$, the orbit is a hyperbola (Fig. 3.19).

The solution we have presented, based on the assumption that the earth is a homogeneous sphere, approximates the orbit of an earth satellite. Determining the orbit accurately requires taking into account the variations in the earth's gravitational field due to its actual mass distribution. Similarly, depending on the accuracy required, determining the orbit of a planet around the sun may require accounting for perturbations due to the gravitational attractions of the other planets.

Example 3.8

An earth satellite is in an elliptic orbit with a minimum radius of 4160 mi and a maximum radius of 10,000 mi. The radius of the earth is 3960 mi.
(a) Determine the satellite's velocity when it is at perigee (its minimum radius) and when it is at apogee (its maximum radius).
(b) Draw a graph of the orbit.

STRATEGY

We can regard the radius and velocity of the satellite at perigee as the initial conditions r_0 and v_0 used in obtaining Eq. (3.22). Since we also know the maximum radius of the orbit, we can solve Eq. (3.25) for the eccentricity of the orbit and then use Eq. (3.23) to determine v_0. From Eq. (3.14), the product of r and the transverse component of the velocity is constant. We can use this condition to determine the velocity at apogee.

SOLUTION

(a) Solving Eq. (3.25) for ε, the eccentricity of the orbit is

$$\varepsilon = \frac{r_{\text{max}}/r_0 - 1}{r_{\text{max}}/r_0 + 1} = \frac{10{,}000/4160 - 1}{10{,}000/4160 + 1} = 0.412.$$

Now from Eq. (3.23), the velocity at perigee is

$$v_0 = \sqrt{\frac{(\varepsilon + 1)gR_E^2}{r_0}} = \sqrt{\frac{(0.412 + 1)(32.2)[(3960)(5280)]^2}{(4160)(5280)}}$$
$$= 30{,}100 \text{ ft/s}.$$

At perigee and apogee, the velocity has only a transverse component. From Eq. (3.14), the velocity at apogee, v_a, is related to the velocity v_0 at perigee by

$$r_0 v_0 = r_{\text{max}} v_a.$$

We solve this equation for the velocity at apogee:

$$v_a = \frac{r_0}{r_{\text{max}}} v_0 = \left(\frac{4160}{10{,}000}\right)(30{,}100) = 12{,}500 \text{ ft/s}.$$

(b) By plotting Eq. (3.22) with $\varepsilon = 0.412$, we obtain the graph of the orbit (Fig. 3.20).

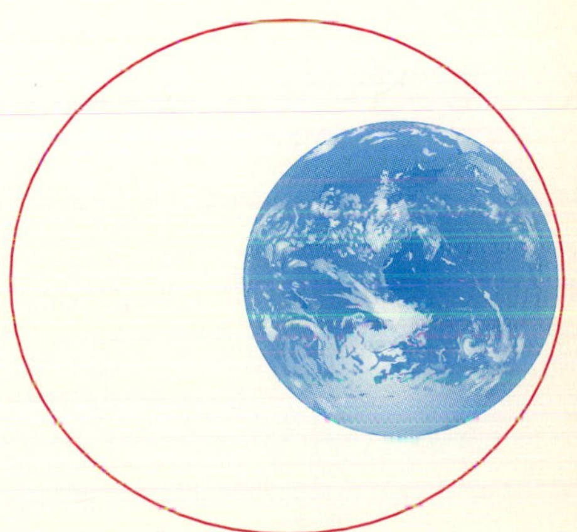

Fig. 3.20
Orbit of an earth satellite with a perigee of 4160 mi and an apogee of 10,000 mi.

Problems

Use the values $R_E = 6730$ km = 3960 mi for the radius of the earth.

3.113 A satellite is in a circular orbit 200 mi above the earth's surface.
(a) What is the magnitude of its velocity?
(b) How long does it take to complete one revolution?

3.114 The moon is approximately 238,000 mi from the earth. Assuming that the moon's orbit around the earth is circular with velocity given by Eq. (3.24), determine the time required for the moon to make one revolution around the earth.

3.115 A satellite is given an initial velocity $v_0 = 22{,}000$ ft/s at a distance $r_0 = 2R_E$ from the center of the earth as shown in Fig. 3.17(a).
(a) What is the maximum radius of the resulting elliptic orbit?
(b) What is the magnitude of the velocity of the satellite when it is at its maximum radius?

3.116 Draw a graph of the elliptic orbit described in Problem 3.115.

3.117 A satellite is given an initial velocity v_0 at a distance $r_0 =$ 6800 km from the center of the earth as shown in Fig. 3.17(a). The resulting elliptic orbit has a maximum radius of 20,000 km. What is v_0?

3.118 In Problem 3.117, what velocity v_0 would be necessary to put the satellite into a parabolic escape orbit?

3.119 At $t = 0$, an earth satellite is a distance r_0 from the center of the earth and has an initial velocity v_0 in the direction shown. Show that the polar equation for the resulting orbit is

$$\frac{r}{r_0} = \frac{(\varepsilon + 1)\cos^2\beta}{[(\varepsilon + 1)\cos^2\beta - 1]\cos\theta - (\varepsilon + 1)\sin\beta\cos\beta\sin\theta + 1},$$

where $\varepsilon = (r_0 v_0^2 / g R_E^2) - 1$.

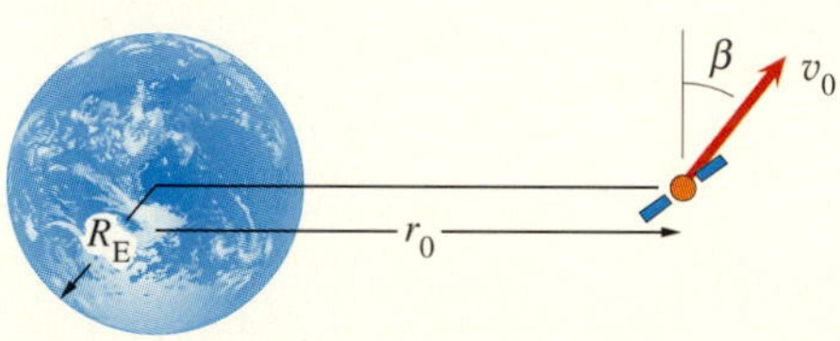

P3.119

3.120 Draw the graphs of the orbits given by the polar equation obtained in Problem 3.119 for $\varepsilon = 0$ and $\beta = 0$, 30°, and 60°.

Computational Mechanics

The material in this section is designed for the use of a programmable calculator or computer.

So far in this chapter you have seen many situations in which you were able to determine the motion of an object by a simple procedure: After using Newton's second law to determine the acceleration, you integrated to obtain analytical, or **closed-form**, expressions for the object's velocity and position. These examples are very valuable—they teach you to use free-body diagrams and express problems in different coordinate systems, and they develop your intuitive understanding of forces and motions. But you would be misled if we presented examples of this kind only, because most problems that must be dealt with in engineering cannot be solved in this way. The functions describing the forces, and therefore the acceleration, are often too complicated for you to integrate and obtain closed-form solutions. In other situations, you will not know the forces in terms of functions but instead will know them in terms of data, either as a continuous recording of force as a function of time (analog data) or as values of force measured at discrete times (digital data).

You can obtain approximate solutions to such problems by using numerical integration. Let's consider an object of mass m in straight-line motion along the x axis (Fig. 3.21) and assume that the x component of the total force may depend on the time, position, and velocity:

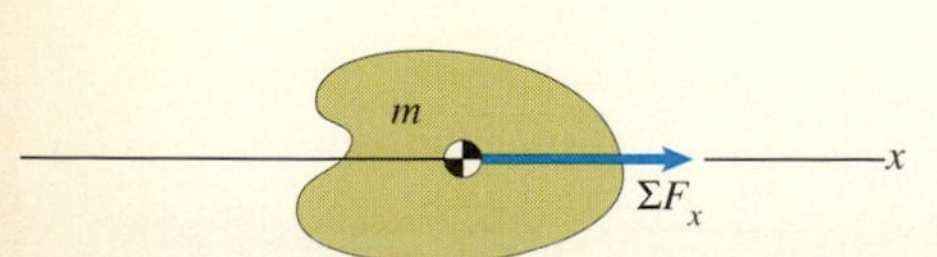

Fig. 3.21
An object moving along the x axis.

$$\Sigma F_x = \Sigma F_x(t, x, v_x). \tag{3.26}$$

Suppose that at a particular time t_0, we know the position $x(t_0)$ and velocity $v_x(t_0)$. The acceleration of the object at t_0 is

$$\frac{dv_x}{dt}(t_0) = \frac{1}{m}\Sigma F_x(t_0, x(t_0), v_x(t_0)). \tag{3.27}$$

To determine the velocity at a time $t_0 + \Delta t$, we express it as a Taylor series:

$$v_x(t_0 + \Delta t) = v_x(t_0) + \frac{dv_x}{dt}(t_0)\,\Delta t + \frac{1}{2}\frac{d^2 v_x}{dt^2}(t_0)(\Delta t)^2 + \cdots$$

By choosing a sufficiently small value of Δt, we can neglect terms in this equation of second and higher order in Δt and substitute Eq. (3.27) to obtain an approximation for the velocity at $t_0 + \Delta t$:

$$v_x(t_0 + \Delta t) = v_x(t_0) + \frac{1}{m}\Sigma F_x(t_0, x(t_0), v_x(t_0))\Delta t. \tag{3.28}$$

We approximate the position at $t_0 + \Delta t$ in the same way. Expressing it as a Taylor series,

$$x(t_0 + \Delta t) = x(t_0) + \frac{dx}{dt}(t_0)\,\Delta t + \frac{1}{2}\frac{d^2 x}{dt^2}(t_0)(\Delta t)^2 + \cdots$$

and neglecting higher-order terms in Δt, we obtain

$$x(t_0 + \Delta t) = x(t_0) + v_x(t_0)\,\Delta t. \tag{3.29}$$

Thus, if we know the position and velocity at a time t_0, we can approximate their values at $t_0 + \Delta t$ by using Eqs. (3.28) and (3.29). We can then repeat the procedure, using $x(t_0 + \Delta t)$ and $v_x(t_0 + \Delta t)$ as initial conditions to determine the approximate position and velocity at $t_0 + 2\Delta t$. By continuing in this way, we obtain approximate solutions for the position and velocity in terms of time. This procedure is easy to carry out using a calculator or computer. It is called a **finite-difference method** because it determines changes in the dependent variables over finite intervals of time. The particular method we describe, due to Leonhard Euler (1707–1783), is called **forward differencing**: The value of the derivative of a function at t_0 is approximated by using its value at t_0 and its value forward in time, at $t_0 + \Delta t$.

More elaborate finite-difference methods, based on retaining more terms in the Taylor series, produce smaller errors in each time step. For example, in the fourth-order Runge-Kutta method, terms through fourth order in Δt are retained. But Euler's method is adequate to introduce you to numerical solutions of problems in dynamics.

Notice that Eq. (3.26) does not need to be a functional expression to carry out the process we have described. The values of the total force must be known at times $t_0, t_0 + \Delta t, \ldots$, and can be determined either from a function or from analog or digital data.

You can determine the velocity and position of an object in curvilinear motion by the same approach. Suppose that an object moves in the x-y plane and that the components of force may depend on the time, position, and velocity:

$$\Sigma F_x = \Sigma F_x(t, x, y, v_x, v_y), \qquad \Sigma F_y = \Sigma F_y(t, x, y, v_x, v_y).$$

If the position and velocity are known at a time t_0, we can use the same steps leading to Eqs. (3.28) and (3.29) to obtain approximate expressions for the components of position and velocity at $t_0 + \Delta t$:

$$
\begin{aligned}
x(t_0 + \Delta t) &= x(t_0) + v_x(t_0)\,\Delta t, \\
y(t_0 + \Delta t) &= y(t_0) + v_y(t_0)\,\Delta t, \\
v_x(t_0 + \Delta t) &= v_x(t_0) + \frac{1}{m}\Sigma F_x(t_0, x(t_0), y(t_0), v_x(t_0), v_y(t_0))\,\Delta t, \\
v_y(t_0 + \Delta t) &= v_y(t_0) + \frac{1}{m}\Sigma F_y(t_0, x(t_0), y(t_0), v_x(t_0), v_y(t_0))\,\Delta t.
\end{aligned}
\tag{3.30}
$$

Example 3.9

A 100-slug projectile is launched from $x = 0$, $y = 0$ with initial velocity $v_x = 400$ ft/s, $v_y = 400$ ft/s. (The y axis is positive upward.) The aerodynamic drag force is of magnitude $C|\mathbf{v}|^2$, where C is a constant. Determine the trajectory for values of C of 0.002, 0.004, and 0.006.

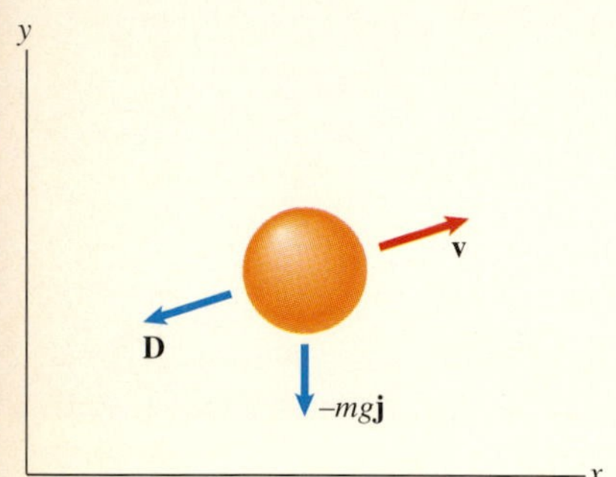

Fig. 3.22
The forces on the projectile are its weight and the drag force **D**.

SOLUTION

To apply Eqs. (3.30), we must determine the x and y components of the total force on the projectile. Let **D** be the drag force (Fig. 3.22). Since $\mathbf{v}/|\mathbf{v}|$ is a unit vector in the direction of **v**, we can write **D** as

$$\mathbf{D} = -C|\mathbf{v}|^2 \frac{\mathbf{v}}{|\mathbf{v}|} = -C|\mathbf{v}|\mathbf{v}.$$

The external forces on the projectile are its weight and the drag,

$$\Sigma \mathbf{F} = -mg\mathbf{j} - C|\mathbf{v}|\mathbf{v},$$

so the components of the total force are

$$\Sigma F_x = -C\sqrt{v_x^2 + v_y^2}\,v_x, \qquad \Sigma F_y = -mg - C\sqrt{v_x^2 + v_y^2}\,v_y. \tag{3.31}$$

Consider the case $C = 0.002$, and let $\Delta t = 0.1$. At the initial time $t_0 = 0$, $x(t_0)$ and $y(t_0)$ are zero, $v_x(t_0) = 400$ ft/s, and $v_y(t_0) = 400$ ft/s. The components of the position and velocity after the first time step are

$$
\begin{aligned}
x(t_0 + \Delta t) &= x(t_0) + v_x(t_0)\,\Delta t: \\
x(0.1) &= x(0) + v_x(0)\,\Delta t \\
&= 0 + (400)(0.1) = 40 \text{ ft},
\end{aligned}
$$

$$
\begin{aligned}
y(t_0 + \Delta t) &= y(t_0) + v_y(t_0)\,\Delta t: \\
y(0.1) &= y(0) + v_y(0)\,\Delta t \\
&= 0 + (400)(0.1) = 40 \text{ ft},
\end{aligned}
$$

$$
\begin{aligned}
v_x(t_0 + \Delta t) &= v_x(t_0) + \frac{1}{m}\Sigma F_x(t_0, x(t_0), y(t_0), v_x(t_0), v_y(t_0))\,\Delta t: \\
v_x(0.1) &= v_x(0) + \left\{-\frac{C}{m}\sqrt{[v_x(0)]^2 + [v_y(0)]^2}\,v_x(0)\right\}\Delta t \\
&= 400 + \left[-\frac{0.002}{100}\sqrt{(400)^2 + (400)^2}\,(400)\right](0.1) \\
&= 399.55 \text{ ft/s},
\end{aligned}
$$

$$v_y(t_0 + \Delta t) = v_y(t_0) + \frac{1}{m}\Sigma F_y(t_0, x(t_0), y(t_0), v_x(t_0), v_y(t_0))\,\Delta t;$$

$$v_y(0.1) = v_y(0) + \left\{-g - \frac{C}{m}\sqrt{[v_x(0)]^2 + [v_y(0)]^2}\, v_y(0)\right\}\Delta t$$

$$= 400 + \left[-32.2 - \frac{0.002}{100}\sqrt{(400)^2 + (400)^2}\,(400)\right](0.1)$$

$$= 396.33 \text{ ft/s}.$$

Continuing in this way, we obtain the following results for the first five time steps:

Time, s	x, ft	y, ft	v_x, ft/s	v_y, ft/s
0.0	0.00	0.00	400.00	400.00
0.1	40.00	40.00	399.55	396.33
0.2	79.95	79.63	399.10	392.66
0.3	119.86	118.90	398.65	389.00
0.4	159.73	157.80	398.21	385.35
0.5	199.55	196.33	397.77	381.70

When there is no drag ($C = 0$), we can obtain the closed-form solution for the trajectory and compare it with numerical solutions. In Fig. 3.23, we present this comparison using $\Delta t = 3.5$ s, 1.0 s, and 0.1 s. Notice that the numerical solution with $\Delta t = 0.1$ s closely approximates the closed-form solution.

In Fig. 3.24, we show the numerical solutions for the various values of C obtained using $\Delta t = 0.1$ s. As expected, the range of the projectile decreases as C increases. Also, when drag is present, the shape of the trajectory is changed. The projectile descends at an angle steeper than the angle at which it was launched.

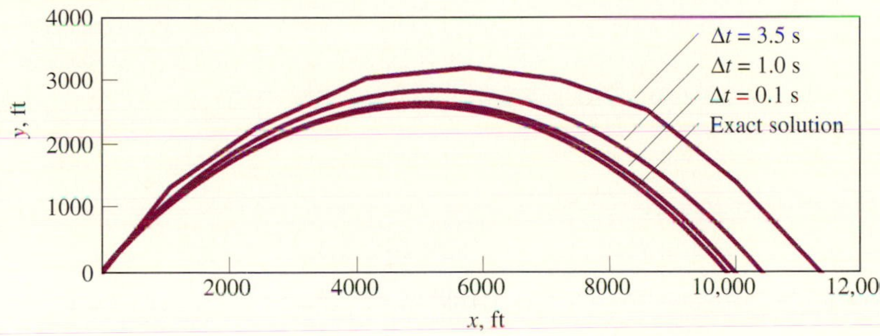

Fig. 3.23
The closed-form solution for the trajectory when $C = 0$ compared with numerical solutions.

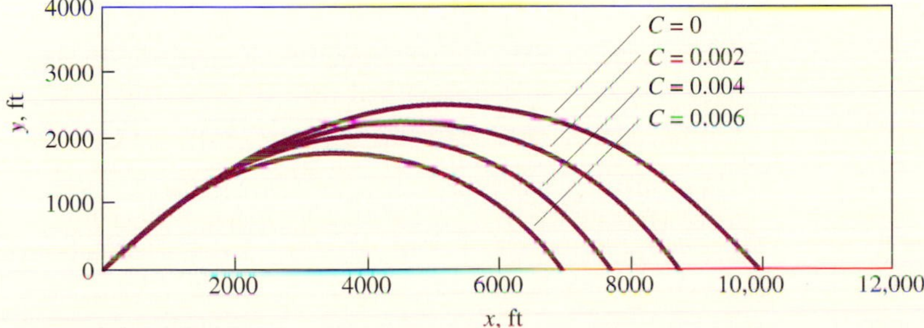

Fig. 3.24
Trajectories for various values of C.

DISCUSSION

The development of the first completely electronic digital computer, the ENIAC (Electronic Numerical Integrator and Computer), built at the University of Pennsylvania between 1943 and 1945, was motivated in part by the need to calculate trajectories of projectiles. A room-sized machine with 18,000 vacuum tubes, it had 20 bytes of random-access memory and 450 bytes of read-only memory.

Problems

3.121 A 1-kg object moves along the x axis under the action of the force $F_x = 6t$ N. At $t = 0$, its position and velocity are $x = 0$ and $v_x = 10$ m/s. Using numerical integration with $\Delta t = 0.1$ s, determine the position and velocity of the object for the first five time steps.

Strategy: At the initial time $t_0 = 0$, $x(t_0) = 0$ and $v_x(t_0) = 10$ m/s. You can use Eqs. (3.28) and (3.29) to determine the velocity and position at time $t_0 + \Delta t = 0.1$ s. The position is

$$x(t_0 + \Delta t) = x(t_0) + v_x(t_0)\,\Delta t:$$

$$x(0.1) = x(0) + v_x(0)\,\Delta t$$

$$= 0 + (10)(0.1) = 1 \text{ m},$$

and the velocity is

$$v_x(t_0 + \Delta t) = v_x(t_0) + \frac{1}{m}F_x(t_0)\,\Delta t:$$

$$v_x(0.1) = 10 + \frac{1}{(1)}6(0)(0.1) = 10 \text{ m/s}.$$

Use these values of the position and velocity as the initial conditions for the next time step.

3.122 For the 1-kg object described in Problem 3.121, draw a graph comparing the closed-form solution from $t = 0$ to $t = 10$ s with the solutions obtained using numerical integration with $\Delta t = 2$ s, $\Delta t = 0.5$ s, and $\Delta t = 0.1$ s.

3.123 At $t = 0$, an object released from rest falls with constant acceleration $g = 9.81$ m/s^2.
(a) Using the closed-form solution, determine the velocity of the object and the distance it has fallen at $t = 2$ s.
(b) Approximate the answers to part (a) by using numerical integration with $\Delta t = 0.2$ s.

3.124 In Problem 3.123, draw a graph of the distance the object falls as a function of time from $t = 0$ to $t = 4$ s, comparing the closed-form solution, the numerical solution using $\Delta t = 0.5$ s, and the numerical solution using $\Delta t = 0.05$ s.

3.125 A 1000-slug rocket starts from rest and travels straight up. The total force exerted on it is $F = 100{,}000 + 10{,}000t - v^2$ lb. Using numerical integration with $\Delta t = 0.1$ s, determine the rocket's height and velocity for the first five time steps. (Assume that the change in the rocket's mass is negligible over this time interval.)

P3.125

3.126 The force exerted on the 50-kg mass by the linear spring is $F = -kx$, where x is the displacement of the mass from its position when the spring is unstretched. The spring constant k is 50 N/m. The mass is released from rest in the position $x = 1$ m. Use numerical integration with $\Delta t = 0.01$ s to determine the position and velocity of the mass for the first five time steps.

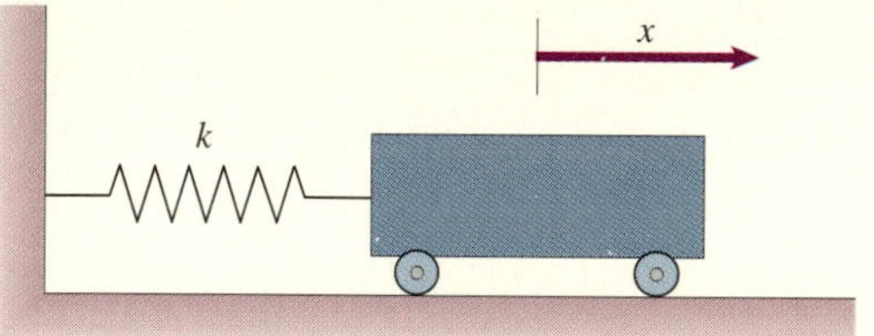

P3.126

3.127 In Problem 3.126, use numerical integration with $\Delta t = 0.01$ s to determine the position and velocity of the mass in terms of time from $t = 0$ to $t = 10$ s. Draw graphs of your results.

3.128 At $t = 0$, the velocity of a 50-slug machine element that moves along the x axis is $v_x = 22$ ft/s. Measurements of the total force ΣF_x acting on the element at 0.1-s intervals from $t = 0$ to $t = 0.9$ s give the following values:

Time, s	Force, lb	Time, s	Force, lb
0.0	50.0	0.5	58.8
0.1	51.1	0.6	57.6
0.2	56.0	0.7	55.4
0.3	57.2	0.8	52.1
0.4	58.5	0.9	49.9

Determine approximately how far the element moves from $t = 0$ to $t = 1$ s and its approximate velocity at $t = 1$ s.

3.129 The lateral supports of a 100-kg structural element exert the horizontal force components

$$F_x = -2000x, \qquad F_y = -2000y,$$

where x and y are the coordinates of the center of mass in meters. At $t = 0$, the coordinates and components of velocity of the center of mass are $x = 0.1$ m, $y = 0$, $v_x = 0$, and $v_y = 1$ m/s. Using $\Delta t = 0.1$ s, determine the approximate position and velocity of the center of mass for the first five time steps.

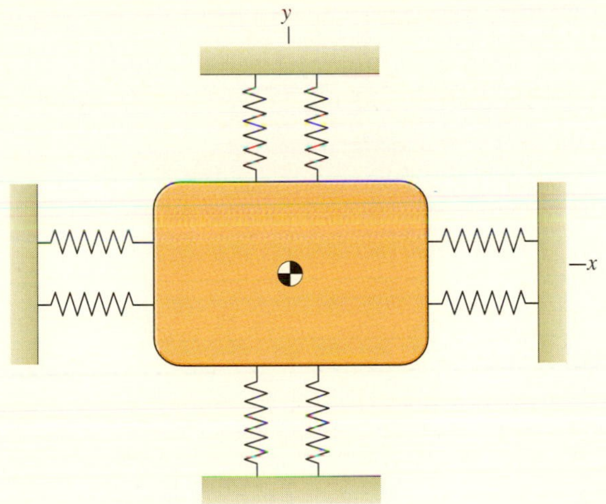

P3.129

3.130 In Problem 3.129, use numerical integration with $\Delta t = 0.001$ s to determine the elliptical path described by the center of mass, and draw a graph of the path.

3.131 A car starts from rest at $t = 0$. Its acceleration is

$$a = 10 + 2t - 0.0185t^3 \text{ ft/s}^2.$$

(a) Using the closed-form solution, determine the distance the car has traveled and its velocity at $t = 6$ s.
(b) Use numerical integration with $\Delta t = 0.1$ s to approximate the answers obtained in (a).
(c) Use numerical integration with $\Delta t = 0.01$ s to approximate the answers obtained in (a).

3.132 A 20-kg projectile is launched from the ground with velocity components $v_x = 100$ m/s, $v_y = 49$ m/s. The magnitude of the aerodynamic drag force is $C|\mathbf{v}|^2$, where C is a constant. If the range of the projectile is 600 m, what is the constant C? (Use numerical integration with $\Delta t = 0.01$ s to compute the trajectory.)

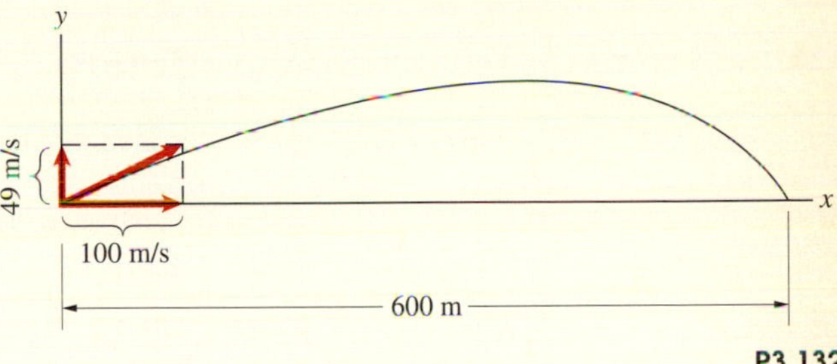

P3.132

3.133 The sport utility vehicle, which weighs 3000 lb with its driver, has left the ground after driving over a rise. At the instant shown, it is moving horizontally at 30 mi/hr and the bottoms of its tires are 24 in. above the (approximately) level ground. The earth-fixed coordinate system is placed with its origin 30 in. above the ground, at the height of the vehicle's center of mass when the tires first contact the ground. (Assume the vehicle remains horizontal.) When that occurs, the vehicle's center of mass initially continues moving downward and then rebounds upward due to the flexure of the suspension system. While the tires are in contact with the ground, the force exerted on them by the road is

$$-R\mathbf{i} - N\mathbf{j} = -2400\mathbf{i} - (18{,}000y + 6000y^3 + 600v_y)\mathbf{j} \text{ (lb)},$$

where y is the vertical position of the center of mass in feet. Letting $t = 0$ be the instant the tires contact the ground, use numerical integration with $\Delta t = 0.001$ to determine the vertical position y and velocity v_y for the first five time steps.

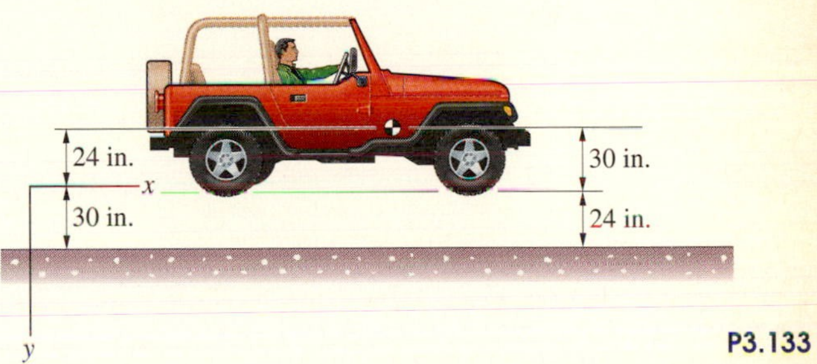

P3.133

3.134 In Problem 3.133, use numerical integration with $\Delta t = 0.001$ to estimate the magnitude of the maximum acceleration to which the vehicle is subjected during its impact with the ground.

3.135 In Problem 3.133, use numerical integration with $\Delta t = 0.001$ to estimate the components of velocity of the vehicle's center of mass at the instant its wheels leave the ground. (The wheels leave the ground when the center of mass is at $y = 0$.)

Chapter Summary

We have used Newton's second law to determine the acceleration of an object when the sum of the forces acting on it is known, and to determine the sum of the forces when the acceleration is known. Once the acceleration of an object was known, we used the methods developed in Chapter 2 to obtain information about the object's velocity and position. In applying Newton's second law, we expressed it in terms of different coordinate systems. The choice of coordinate system was sometimes dictated by the nature of the forces acting on an object. When an object's path is known, especially when it is constrained to move in a circle, normal and tangential components are often advantageous. In Chapter 4 we will use Newton's second law to derive a technique called the method of work and energy, which can greatly simplify the solution of particular types of problems in dynamics.

The total external force on an object is equal to the product of its mass and the acceleration *of its center of mass* relative to an inertial reference frame:

$$\Sigma \mathbf{F} = m\mathbf{a}. \qquad \text{Eq. (3.4)}$$

A reference frame is said to be inertial if it is one in which the second law can be applied in this form. A reference frame translating at constant velocity relative to an inertial reference frame is also inertial.

Expressing Newton's second law in terms of a coordinate system yields scalar equations of motion:

Cartesian Coordinates

$$\Sigma F_x = ma_x, \quad \Sigma F_y = ma_y, \quad \Sigma F_z = ma_z. \qquad \text{Eq. (3.5)}$$

Normal and Tangential Components

$$\Sigma F_{\mathrm{t}} = m\frac{dv}{dt}, \quad \Sigma F_{\mathrm{n}} = m\frac{v^2}{\rho}. \qquad \text{Eq. (3.7)}$$

Polar Coordinates

$$\Sigma F_r = m\left(\frac{d^2r}{dt^2} - r\omega^2\right), \qquad \text{Eq. (3.9)}$$

$$\Sigma F_\theta = m\left(r\alpha + 2\frac{dr}{dt}\omega\right). \qquad \text{Eq. (3.10)}$$

If the motion of an object is confined to a fixed plane, the component of the total force normal to the plane equals zero. In straight-line motion, the components of the total force perpendicular to the line equal zero and the component of the total force tangent to the line equals the product of the mass and the acceleration of the object along the line.

Review Problems

3.136 The Acura NSX, which weighs 3250 lb with its driver, can brake from 60 mi/hr to a stop in a distance of 112 ft. (a) If you assume its deceleration is constant, what are its deceleration and the magnitude of the horizontal force its tires exert on the road? (b) If its tires are at the limit of adhesion (slip is impending), and the normal force exerted on the car by the road equals the car's weight, what is the coefficient of friction μ_s? (This analysis neglects the effects of horizontal and vertical aerodynamic forces.)

3.137 Using the coefficient of friction obtained in Problem 3.136, determine the highest speed at which the NSX could drive on a flat, circular track of 600-ft radius without skidding.

3.138 A "cog" engine hauls three cars of sightseers to a mountaintop in Bavaria. The mass of each car including its passengers is 10 Mg, and the friction forces exerted by the wheels of the cars are negligible. Determine the forces in the couplings 1, 2, and 3 if (a) the engine is moving at constant velocity; (b) the engine is accelerating up the mountain at 1.2 m/s^2.

P3.138

3.139 In a future mission, a spacecraft approaches the surface of an asteroid passing near the earth. Just before it touches down, the spacecraft is moving downward at constant velocity relative to the surface of the asteroid and its downward thrust is 0.01 N. The computer decreases the downward thrust to 0.005 N, and an on-board laser interferometer determines that the acceleration of the spacecraft relative to the surface becomes 5×10^{-6} m/s^2 downward. What is the gravitational acceleration of the asteroid near its surface?

P3.139

3.140 A car, with a mass of 1470 kg with its driver, is driven at 130 km/hr over a slight rise in the road. At the top of the rise the driver applies the brakes. The coefficient of static friction between the tires and the road is $\mu_s = 0.9$, and the radius of curvature of the rise is 160 m. Determine the car's deceleration at the instant the brakes are applied and compare it to the deceleration on a level road.

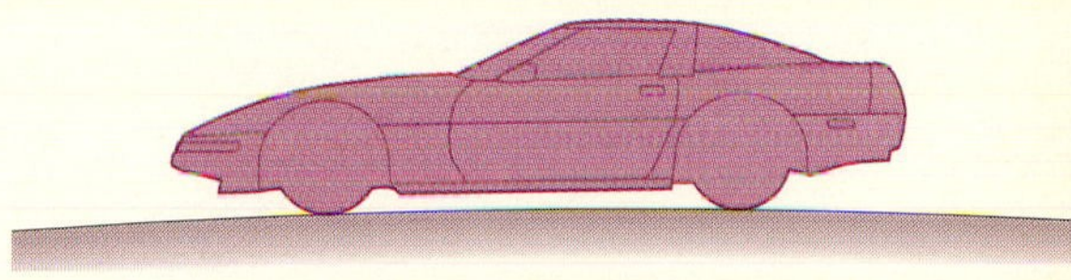

P3.140

3.141 The car drives at constant velocity up the straight segment of road on the left. If the car's tires continue to exert the same tangential force on the road after the car has gone over the crest of the hill and is on the straight segment of road on the right, what will be the car's acceleration?

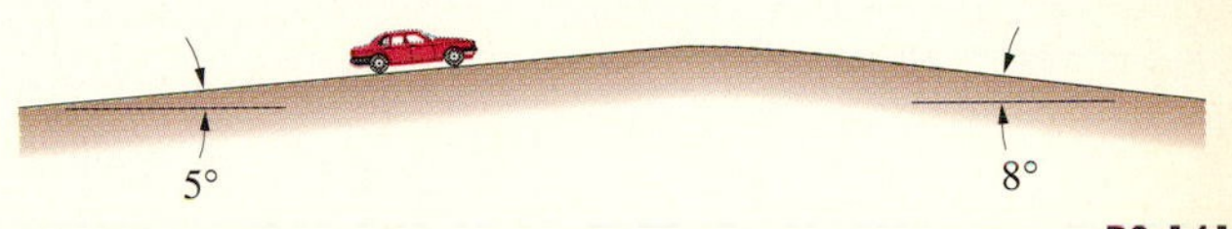

P3.141

3.142 The aircraft carrier *Nimitz* weighs 91,000 tons. (A ton is 2000 lb.) Suppose that it is traveling at its top speed of approximately 30 knots (a knot is 6076 ft/hr) when its engines are shut down. If the water exerts a drag force of magnitude $20{,}000v$ lb, where v is the carrier's velocity in feet per second, what distance does the carrier move before coming to rest?

3.143 If $m_A = 10$ kg, $m_B = 40$ kg, and the coefficient of kinetic friction between all surfaces is $\mu_k = 0.11$, what is the acceleration of B down the inclined surface?

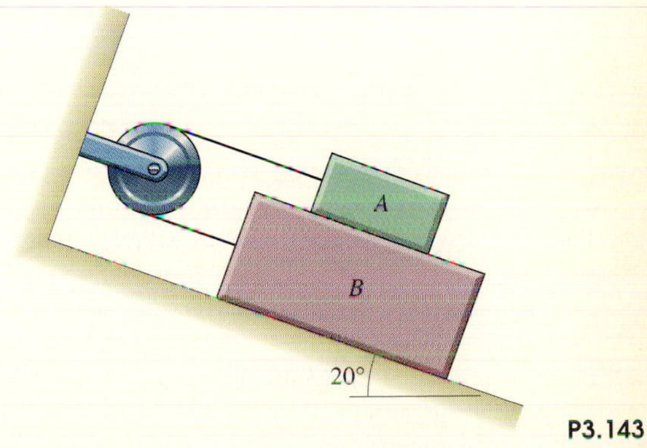

P3.143

3.144 In Problem 3.143, if A weighs 20 lb, B weighs 100 lb, and the coefficient of kinetic friction between all surfaces is $\mu_k = 0.15$, what is the tension in the cord as B slides down the inclined surface?

3.145 A gas gun is used to accelerate projectiles to high velocities for research on material properties. The projectile is held in place while gas is pumped into the tube to a high pressure p_0 on the left and the tube is evacuated on the right. The projectile is then released and is accelerated by the expanding gas. Assume that the pressure p of the gas is related to the volume V it occupies by pV^γ = constant, where γ is a constant. If friction can be neglected, show that the velocity of the projectile at the position x is

$$v = \sqrt{\frac{2p_0 A x_0^\gamma}{m(\gamma - 1)}\left(\frac{1}{x_0^{\gamma-1}} - \frac{1}{x^{\gamma-1}}\right)},$$

where m is the mass of the projectile and A is the cross-sectional area of the tube.

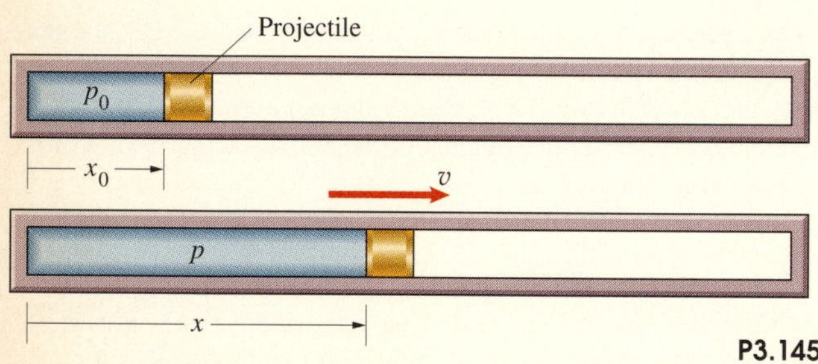

P3.145

3.146 The weights of the blocks are $W_A = 120$ lb and $W_B = 20$ lb, and the surfaces are smooth. Determine the acceleration of block A and the tension in the cord.

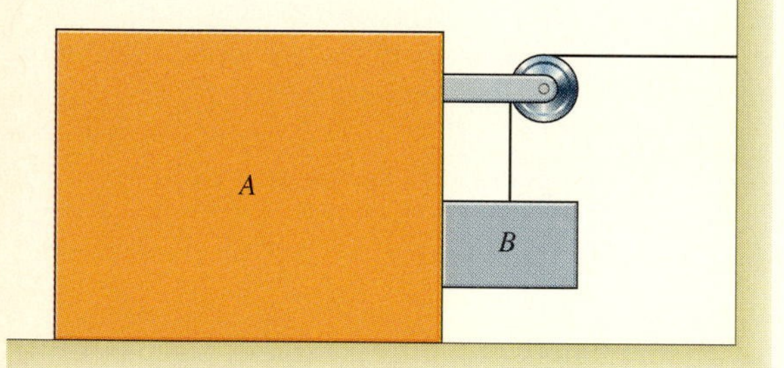

P3.146

3.147 The 100-Mg space shuttle is in orbit when its engines are turned on, exerting a thrust force $\mathbf{T} = 10\mathbf{i} - 20\mathbf{j} + 10\mathbf{k}$ (kN) for 2 s. Neglect the resulting change in its mass. At the end of the 2-s burn, fuel is still sloshing back and forth in the shuttle's tanks. What is the change in the velocity of the center of mass of the shuttle (including the fuel it contains) due to the 2-s burn?

3.148 The water skier contacts the ramp with a velocity of 25 mi/hr parallel to the surface of the ramp. Neglecting friction and assuming that the tow rope exerts no force on him once he touches the ramp, estimate the horizontal length of his jump from the end of the ramp.

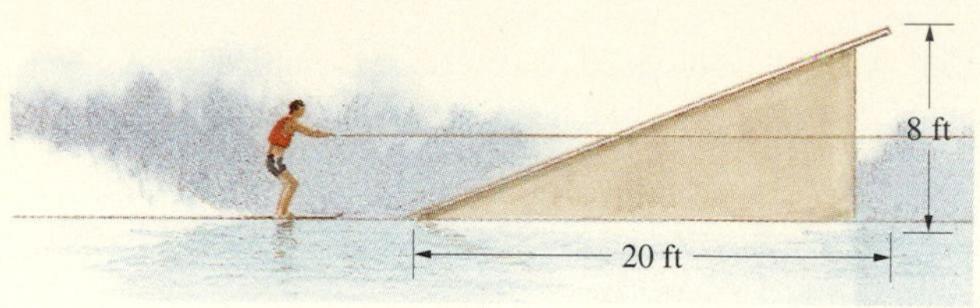

P3.148

3.149 Suppose you are designing a roller coaster track that will take the cars through a vertical loop of 40-ft radius. If you decide that, for safety, the downward force exerted on a passenger by his seat at the top of the loop should be at least one-half his weight, what is the minimum safe velocity of the cars at the top of the loop?

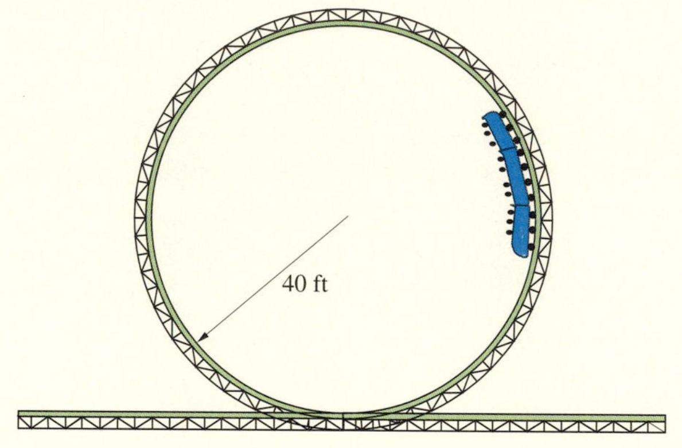

P3.149

3.150 As the smooth bar rotates *in the horizontal plane*, the string winds up on the fixed cylinder and draws the 1-kg collar A inward. The bar starts from rest at $t = 0$ in the position shown and rotates with constant angular acceleration. What is the tension in the string at $t = 1$ s?

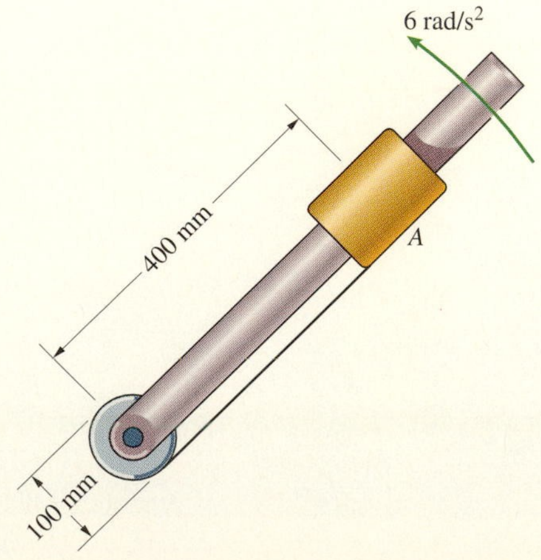

P3.150

3.151 In Problem 3.150, suppose that the coefficient of kinetic friction between the collar and the bar is $\mu_k = 0.2$. What is the tension in the string at $t = 1$ s?

3.152 If you want to design the cars of a train to tilt as the train goes around curves to achieve maximum passenger comfort, what is the relationship between the desired tilt angle α, the velocity v of the train, and the instantaneous radius of curvature ρ of the track?

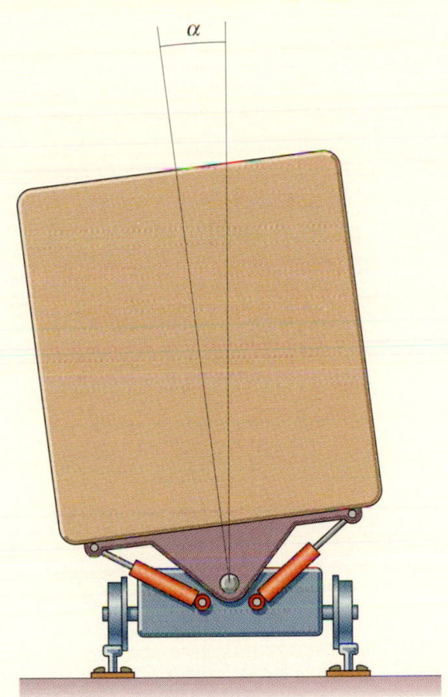

P3.152

3.153 To determine the coefficient of static friction between two materials, an engineer at the U.S. Bureau of Standards places a small sample of one material on a horizontal disk surfaced with the other one, then rotates the disk from rest with a constant angular acceleration of 0.4 rad/s^2. If she determines that the small sample slips on the disk after 9.903 s, what is the coefficient of friction?

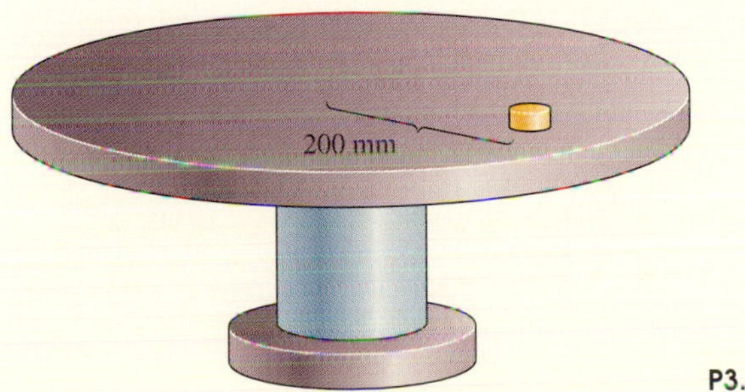

P3.153

3.154 The 1-kg slider A is pushed along the curved bar by the slotted bar. The curved bar lies *in the horizontal plane*, and its profile is described by $r = 2(\theta/2\pi + 1)$ m, where θ is in radians. The angular position of the slotted bar is $\theta = 2t$ rad. Determine the radial and transverse components of the total external force exerted on the slider when $\theta = 120°$.

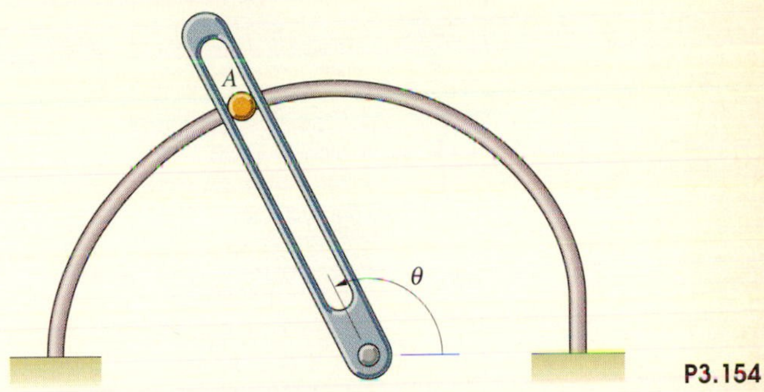

P3.154

3.155 In Problem 3.154, suppose that the curved bar lies *in the vertical plane*. Determine the radial and transverse components of the total force exerted on A by the curved and slotted bars at $t = 0.5$ s.

3.156 The ski boat moves relative to the water with a constant velocity of magnitude $|\mathbf{v}_B| = 30$ ft/s. The magnitude of the 170-lb skier's velocity relative to the boat is $|\mathbf{v}_{S/B}| = 10$ ft/s. The tension in the 36-ft tow rope is 40 lb, and the horizontal force exerted on the skier by the water is perpendicular to the direction of his motion relative to the water. If you can neglect other horizontal forces, what is the skier's acceleration in the direction of his motion relative to the water?

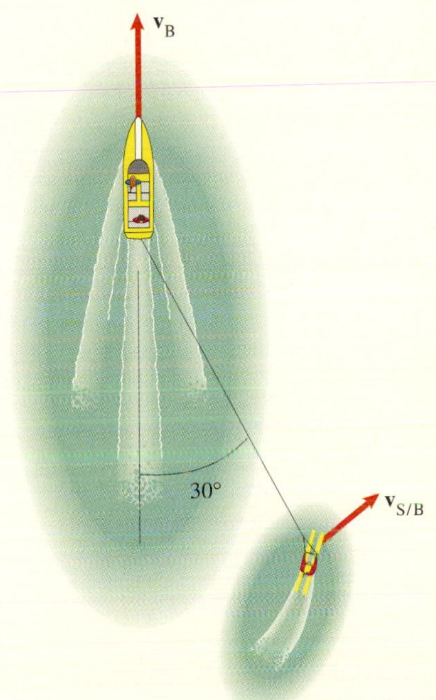

P3.156

3.157 In Problem 3.156, what is the magnitude of the horizontal force exerted on the skier by the water?

The ski lift performs work on the skiers, increasing their gravitational potential energy. Going down the hill, the skiers trade their gravitational potential energy for kinetic energy. To avoid going too fast, they must ski so that the snow performs negative work on them, decreasing their kinetic energy. In this chapter we use the concepts of work and energy to analyze motions of objects.

Chapter 4

Energy Methods

Changes in energy must be considered in the design of any device that moves, including ski lifts as well as skis. The concepts of energy and conservation of energy originated in large part from the study of classical mechanics. A simple transformation of Newton's second law results in an equation that motivates the definitions of work, kinetic energy (energy due to an object's motion), and potential energy (energy due to an object's position). This equation relates the work done by the external forces acting on an object to the change in magnitude of its velocity. This relationship can greatly simplify the solution of problems involving forces that depend on an object's position, including gravitational forces and forces exerted by springs. In addition, studying the derivations and applications in this chapter will develop your intuition concerning energy and its transformations and give you insight into applications of these concepts in other fields.

Work and Kinetic Energy

4.1 *Principle of Work and Energy*

We have used Newton's second law to relate the acceleration of an object's center of mass relative to an inertial reference frame to the mass of the object and the external forces acting on it. We will now show how this vector equation can be expressed in a scalar form that is extremely useful in particular circumstances. We begin with Newton's second law in the form

$$\Sigma \mathbf{F} = m\frac{d\mathbf{v}}{dt}, \tag{4.1}$$

and take the dot product of both sides with the velocity:

$$\Sigma \mathbf{F} \cdot \mathbf{v} = m\frac{d\mathbf{v}}{dt} \cdot \mathbf{v}. \tag{4.2}$$

By expressing the velocity on the left side of this equation as $d\mathbf{r}/dt$ and using the result

$$\frac{d}{dt}(\mathbf{v} \cdot \mathbf{v}) = \frac{d\mathbf{v}}{dt} \cdot \mathbf{v} + \mathbf{v} \cdot \frac{d\mathbf{v}}{dt} = 2\frac{d\mathbf{v}}{dt} \cdot \mathbf{v},$$

we can write Eq. (4.2) as

$$\Sigma \mathbf{F} \cdot d\mathbf{r} = \frac{1}{2}m\, d(\mathbf{v} \cdot \mathbf{v}) = \frac{1}{2}m\, d(v^2), \tag{4.3}$$

where $v^2 = \mathbf{v} \cdot \mathbf{v}$ is the square of the magnitude of $\mathbf{v}$. The term on the left is the **work** expressed in terms of the total external force acting on the object and an infinitesimal displacement $d\mathbf{r}$ of its center of mass. We integrate this equation, obtaining

$$\int_{\mathbf{r}_1}^{\mathbf{r}_2} \Sigma \mathbf{F} \cdot d\mathbf{r} = \int_{v_1^2}^{v_2^2} \frac{1}{2}m\, d(v^2) = \frac{1}{2}mv_2^2 - \frac{1}{2}mv_1^2, \tag{4.4}$$

where v_1 and v_2 are the magnitudes of the velocity at the positions $\mathbf{r}_1$ and $\mathbf{r}_2$. The term $\frac{1}{2}mv^2$ is called the **kinetic energy** associated with the motion of the center of mass. Denoting the work done as the center of mass moves from position $\mathbf{r}_1$ to position $\mathbf{r}_2$ by

$$\boxed{U_{12} = \int_{\mathbf{r}_1}^{\mathbf{r}_2} \Sigma \mathbf{F} \cdot d\mathbf{r},} \tag{4.5}$$

we obtain the **principle of work and energy**:

$$U_{12} = \frac{1}{2}mv_2^2 - \frac{1}{2}mv_1^2. \quad (4.6)$$

The work done on an object as it moves between two positions equals the change in its kinetic energy. The dimensions of work, and therefore the dimensions of kinetic energy, are (force) × (length). In U.S. Customary units, work is usually expressed in ft-lb. In SI units, work is usually expressed in N-m, or joules (J).

If you can evaluate the work done on an object as it moves between two positions, the principle of work and energy allows you to determine the change in the magnitude of its velocity. You can also equate the total work done by external forces on a system of objects to the change in the total kinetic energy of the system *if no net work is done by internal forces.* For example, internal friction forces can do net work on a system. (See Example 4.3.)

Although the principle of work and energy relates changes in position to changes in velocity, you cannot use it to obtain other information about the motion, such as the time required to move from one position to another. Furthermore, since the work is an integral with respect to position, you can usually evaluate it only when the forces doing work are known as functions of position. Despite these limitations, this principle is extremely useful for certain problems because the work can be determined very easily.

4.2 Work and Power

In this section we discuss how to determine the work done on an object. We also define the power transferred to or from an object by the forces acting on it and show how the power is calculated.

Evaluating the Work

Let's consider an object in curvilinear motion relative to an inertial reference frame (Fig. 4.1a) and specify its position by the coordinate s measured along its path from a reference point O. In terms of the tangential unit vector $\mathbf{e}_t$, the object's velocity is

$$\mathbf{v} = \frac{ds}{dt}\mathbf{e}_t.$$

Because $\mathbf{v} = d\mathbf{r}/dt$, we can multiply the velocity by dt to obtain an expression for the vector $d\mathbf{r}$ describing an infinitesimal displacement along the path (Fig. 4.1b):

$$d\mathbf{r} = \mathbf{v}\,dt = ds\,\mathbf{e}_t.$$

The work done by the external forces acting on the object as a result of the displacement $d\mathbf{r}$ is

$$\Sigma\mathbf{F}\cdot d\mathbf{r} = (\Sigma\mathbf{F}\cdot\mathbf{e}_t)\,ds = \Sigma F_t\,ds,$$

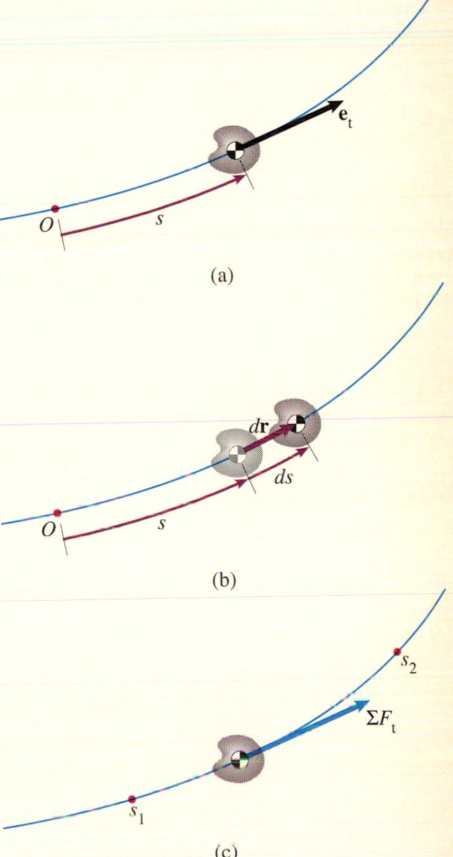

Fig. 4.1

(a) The coordinate s and tangential unit vector.
(b) An infinitesimal displacement $d\mathbf{r}$.
(c) The work done from s_1 to s_2 is determined by the tangential component of the external forces.

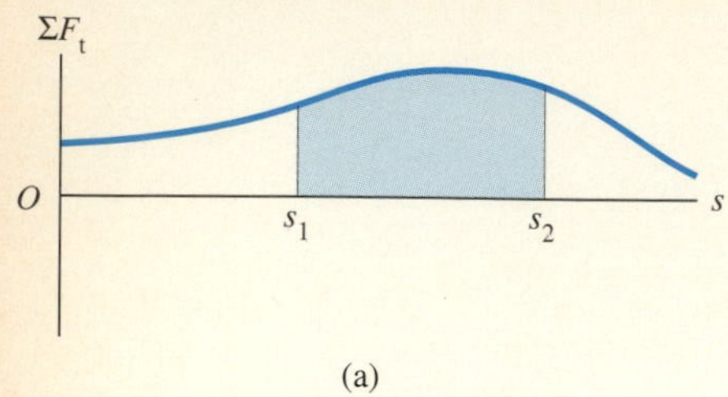

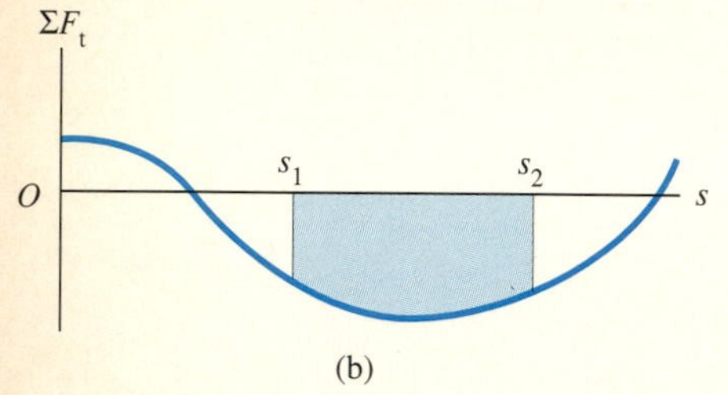

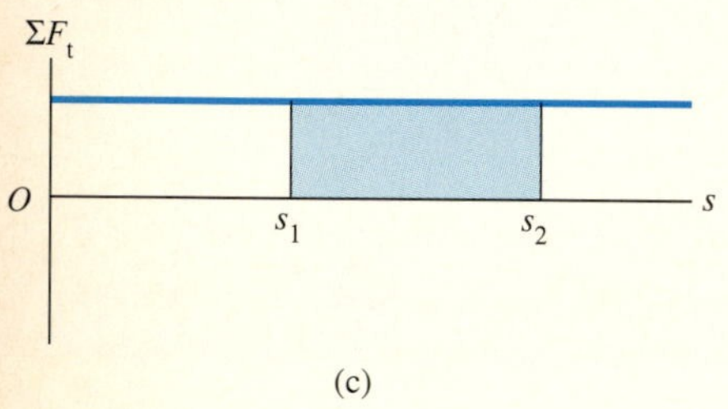

Fig. 4.2

(a) The work equals the area defined by the graph of the tangential force as a function of the distance along the path.
(b) Negative work is done if the tangential force is opposite to the direction of the motion.
(c) The work done by a constant tangential force equals the product of the force and the distance.

where ΣF_t is the tangential component of the total force. Therefore, as the object moves from a position s_1 to a position s_2 (Fig. 4.1c), the work is

$$U_{12} = \int_{s_1}^{s_2} \Sigma F_t \, ds. \tag{4.7}$$

The work is equal to the integral of the tangential component of the total force with respect to distance along the path. Thus the work done is equal to the area defined by the graph of the tangential force from s_1 to s_2 (Fig. 4.2a). *Components of force perpendicular to the path do no work.* Notice that if ΣF_t is opposite to the direction of motion over some part of the path, which means the object is decelerating, the work is negative (Fig. 4.2b). If ΣF_t is constant between s_1 and s_2, the work is simply the product of the total tangential force and the displacement (Fig. 4.2c):

$$U_{12} = \Sigma F_t (s_2 - s_1). \quad \textbf{Constant tangential force} \tag{4.8}$$

Power

Power is the rate at which work is done. The work done by the external forces acting on an object during an infinitesimal displacement $d\mathbf{r}$ is

$$\Sigma \mathbf{F} \cdot d\mathbf{r}.$$

We obtain the power P by dividing this expression by the interval of time dt during which the displacement takes place:

$$P = \Sigma \mathbf{F} \cdot \mathbf{v}. \tag{4.9}$$

This is the power transferred to or from the object, depending on whether P is positive or negative. In SI units, power is expressed in newton-meters per second, which is joules per second (J/s) or watts (W). In U.S. Customary units, power is expressed in foot-pounds per second or in the anachronistic horsepower (hp), which is 746 W or approximately 550 ft-lb/s.

Notice from Eq. (4.3) that the power equals the rate of change of the kinetic energy of the object:

$$P = \frac{d}{dt}\left(\frac{1}{2}mv^2\right).$$

Transferring power to or from an object causes its kinetic energy to increase or decrease. Using this relation, we can write the average with respect to time of the power during an interval of time from t_1 to t_2 as

$$P_{av} = \frac{1}{t_2 - t_1}\int_{t_1}^{t_2} P \, dt = \frac{1}{t_2 - t_1}\int_{v_1^2}^{v_2^2} \frac{1}{2}m \, d(v^2).$$

This result states that the average power transferred to or from an object during an interval of time is equal to the change in its kinetic energy, or the work done, divided by the interval of time:

$$P_{av} = \frac{\frac{1}{2}mv_2^2 - \frac{1}{2}mv_1^2}{t_2 - t_1} = \frac{U_{12}}{t_2 - t_1}. \tag{4.10}$$

In the following examples we apply the principle of work and energy and use Eqs. (4.7) and (4.8) to evaluate the work. You should consider using work and energy when you want to relate the change in velocity of an object to a change in its position. This typically involves two steps:

1. Identify the forces that do work—*By drawing a free-body diagram, you must determine which external forces do work on the object.*
2. Apply work and energy—*Equate the total work done during a change in position to the change in the object's kinetic energy.*

Example 4.1

The 400-lb container A in Fig. 4.3 starts from rest at position $s = 0$ and is subjected to a horizontal force $F = 160 - 10s$ lb by the hydraulic cylinder. The coefficient of kinetic friction between the container and the floor is $\mu_k = 0.26$. What is the velocity of the container when it has reached the position $s = 4$ ft?

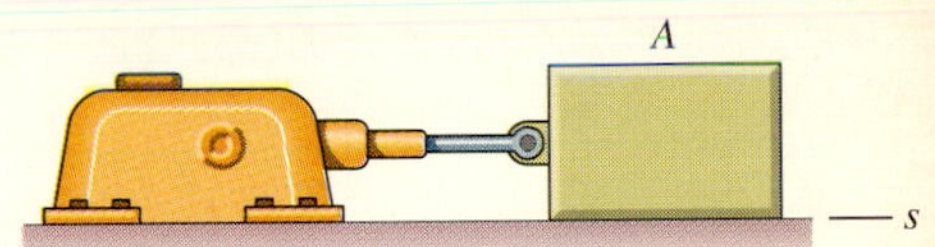

Fig. 4.3

STRATEGY

We are asked to determine the change in the velocity of the container given a change in its position, so we can apply the method of work and energy. We will determine the forces on the container in the direction tangent to its path and use Eq. (4.7) to evaluate the work.

SOLUTION

Identify the Forces That Do Work We draw the free-body diagram of the container in Fig. (a). The forces tangent to its path are the force exerted by the hydraulic cylinder and the friction force. The container's acceleration in the vertical direction is zero, so $N = 400$ lb.

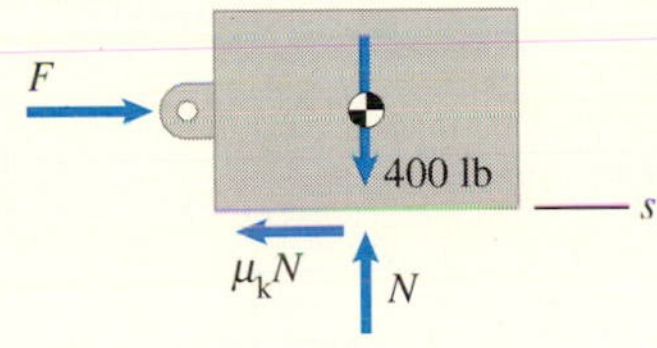

(a) Free-body diagram of the container.

Apply Work and Energy Let v be the magnitude of the container's velocity (Fig. b). At the initial position $s_1 = 0$, its velocity is $v_1 = 0$. We use the principle of work and energy to determine the container's kinetic energy at $s_2 = 4$ ft, applying Eq. (4.7) to evaluate the work.

$$U_{12} = \int_{s_1}^{s_2} \Sigma F_t \, ds = \frac{1}{2}mv_2^2 - \frac{1}{2}mv_1^2:$$

$$\int_0^4 (F - \mu_k N) \, ds = \frac{1}{2}mv_2^2 - 0,$$

$$\int_0^4 [(160 - 10s) - (0.26)(400)] \, ds = \frac{1}{2}\left(\frac{400}{32.2}\right)v_2^2.$$

Evaluating the integral and solving for the velocity, we obtain $v_2 = 4.81$ ft/s.

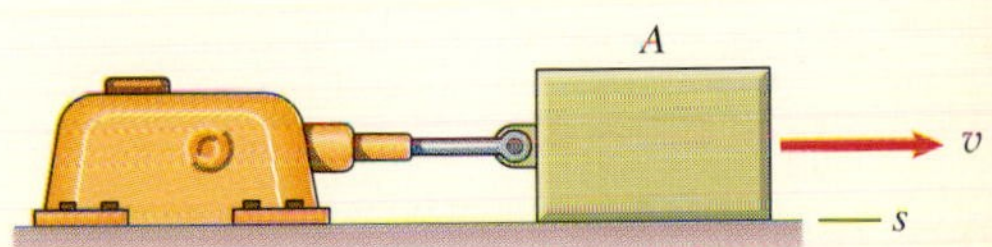

(b) Magnitude v of the container's velocity.

Example 4.2

The two crates in Fig. 4.4 are released from rest. Their masses are $m_A = 40$ kg and $m_B = 30$ kg, and the kinetic coefficient of friction between crate A and the inclined surface is $\mu_k = 0.15$. What is the magnitude of their velocity when they have moved 400 mm?

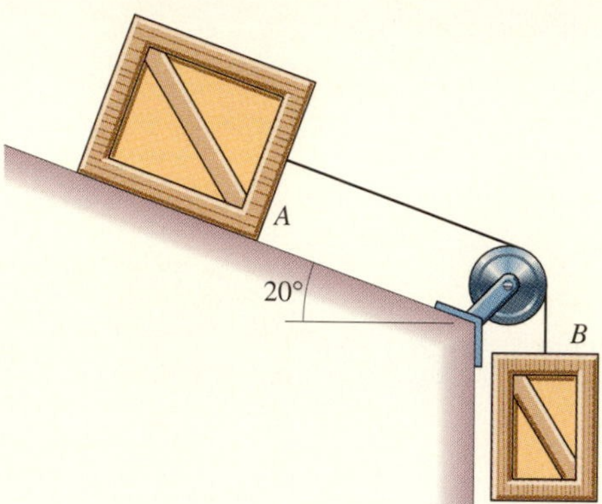

Fig. 4.4

STRATEGY

We will determine the velocity in two ways.

First Method By drawing free-body diagrams of each of the crates and applying the principle of work and energy to them individually, we can obtain two equations in terms of the magnitude of the velocity and the tension in the cable.

Second Method We can draw a single free-body diagram of the two crates, the cable, and the pulley and apply the principle of work and energy to the entire system.

SOLUTION

First Method We draw the free-body diagram of crate A in Fig. (a). The forces that do work as the crate moves down the plane are the forces tangential to its path: the tension T, the tangential component of the weight $m_A g \sin 20°$, and the friction force $\mu_k N$. Because the acceleration of the crate normal to the surface is zero, $N = m_A g \cos 20°$. The magnitude v of the velocity at which A moves parallel to the surface equals the magnitude of the velocity at which B falls (Fig. b). Using Eq. (4.7)

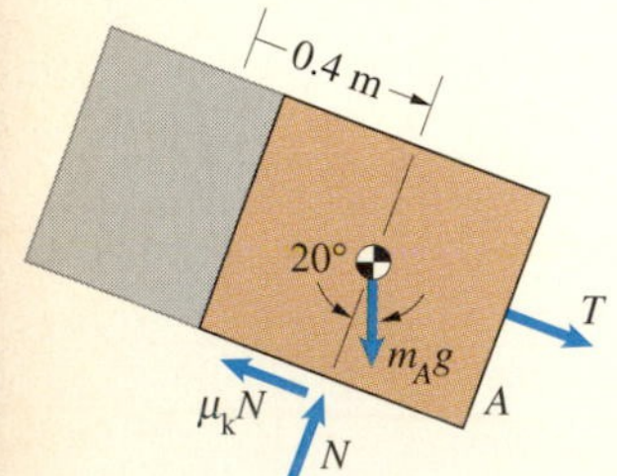

(a) Free-body diagram of A.

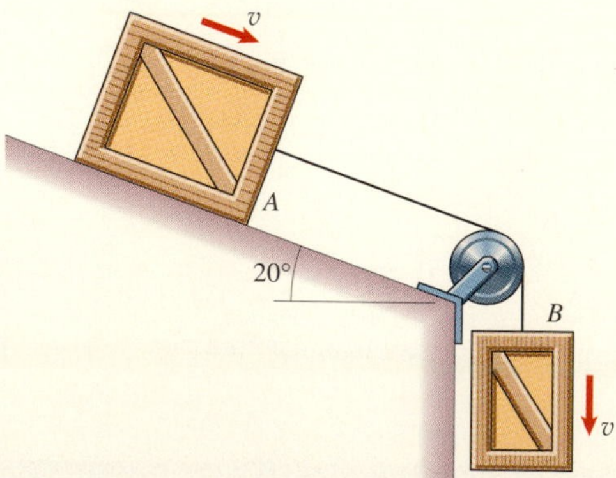

(b) Magnitude of the velocity of each crate is the same.

to determine the work, we equate the work done on A as it moves from $s_1 = 0$ to $s_2 = 0.4$ m to the change in its kinetic energy,

$$\int_{s_1}^{s_2} \Sigma F_t \, ds = \frac{1}{2}mv_2^2 - \frac{1}{2}mv_1^2:$$

$$\int_0^{0.4} [T + m_A g \sin 20° - \mu_k(m_A g \cos 20°)]\, ds = \frac{1}{2}m_A v_2^2 - 0. \quad (4.11)$$

The forces that do work on crate B are its weight $m_B g$ and the tension T (Fig. c). The magnitude of its velocity is the same as that of crate A. The work done on B equals the change in its kinetic energy:

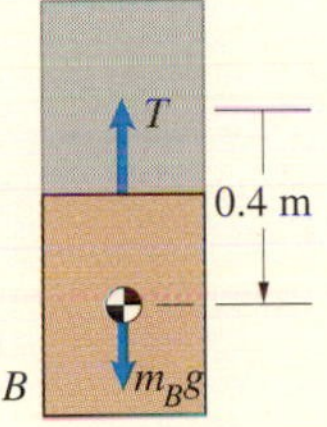

(c) Free-body diagram of B.

$$\int_{s_1}^{s_2} \Sigma F_t \, ds = \frac{1}{2}mv_2^2 - \frac{1}{2}mv_1^2:$$

$$\int_0^{0.4} (m_B g - T)\, ds = \frac{1}{2}m_B v_2^2 - 0. \quad (4.12)$$

By summing Eqs. (4.11) and (4.12), we eliminate T, obtaining

$$\int_0^{0.4} (m_A g \sin 20° - \mu_k m_A g \cos 20° + m_B g) ds = \frac{1}{2}(m_A + m_B)v_2^2:$$

$$[40 \sin 20° - (0.15)(40) \cos 20° + 30](9.81)(0.4) = \frac{1}{2}(40 + 30)v_2^2.$$

Solving for the velocity, we obtain $v_2 = 2.07$ m/s.

Second Method We draw the free-body diagram of the system consisting of the crates, cable, and pulley in Fig. (d). Notice that the cable tension does not appear in this free-body diagram. The reactions at the pin support of the pulley do no work, because the support does not move. The total work done by external forces on the system as the boxes move 400 mm is equal to the change in the total kinetic energy of the system:

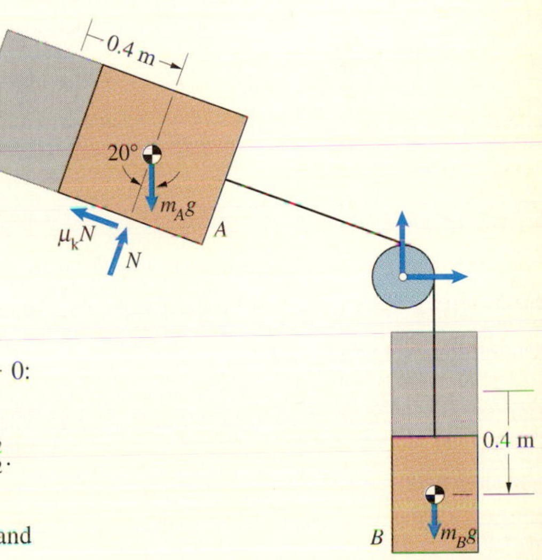

(d) Free-body diagram of the system.

$$\int_0^{0.4} [m_A g \sin 20° - \mu_k(m_A g \cos 20°)] ds + \int_0^{0.4} m_B g \, ds$$

$$= \left(\frac{1}{2}m_A v_2^2 + \frac{1}{2}m_B v_2^2\right) - 0:$$

$$[40 \sin 20° - (0.15)(40) \cos 20° + 30](9.81)(0.4) = \frac{1}{2}(40 + 30)v_2^2.$$

This equation is identical to that we obtained by applying the principle of work and energy to the individual crates.

DISCUSSION

You will often find it simpler to apply the principle of work and energy to an entire system instead of its separate parts. However, as we demonstrate in the next example, you need to be aware that internal forces in a system can do net work.

Example 4.3

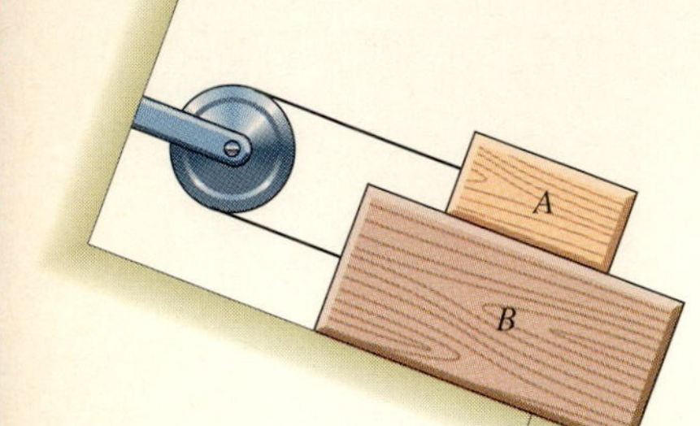

Fig. 4.5

Crates A and B in Fig. 4.5 are released from rest. The coefficient of kinetic friction between A and B is μ_k, and friction between B and the inclined surface can be neglected. What is their velocity when they have moved a distance b?

STRATEGY

By applying the principle of work and energy to each crate, we can obtain two equations in terms of the tension in the cable and the velocity.

SOLUTION

We draw the free-body diagrams of the crates in Figs. (a) and (b). The acceleration of A normal to the inclined surface is zero, so $N = m_A g \cos\theta$. The magnitudes of the velocities of A and B are equal (Fig. c). The work done on A equals the change in its kinetic energy,

$$U_{12} = \frac{1}{2}m_A v_2^2 - \frac{1}{2}m_A v_1^2:$$

$$\int_0^b (T - m_A g \sin\theta - \mu_k m_A g \cos\theta)\, ds = \frac{1}{2}m_A v_2^2, \tag{4.13}$$

and the work done on B equals the change in its kinetic energy,

$$U_{12} = \frac{1}{2}m_B v_2^2 - \frac{1}{2}m_B v_1^2:$$

$$\int_0^b (-T + m_B g \sin\theta - \mu_k m_A g \cos\theta)\, ds = \frac{1}{2}m_B v_2^2. \tag{4.14}$$

Summing these equations to eliminate T and solving for v_2, we obtain

$$v_2 = \sqrt{2gb[(m_B - m_A)\sin\theta - 2\mu_k m_A \cos\theta]/(m_A + m_B)}.$$

(a) Free-body diagram of A.

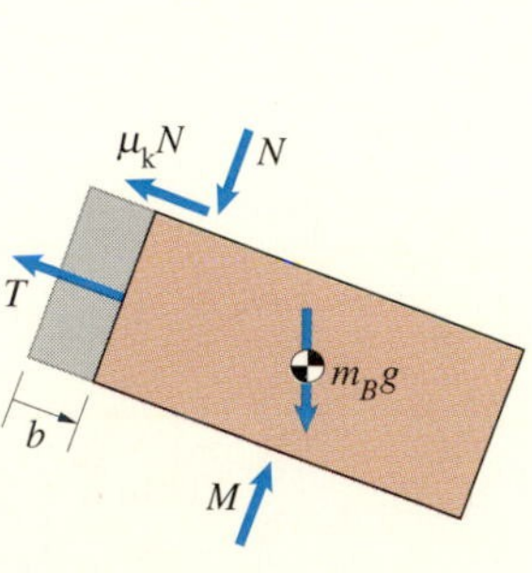

(b) Free-body diagram of B.

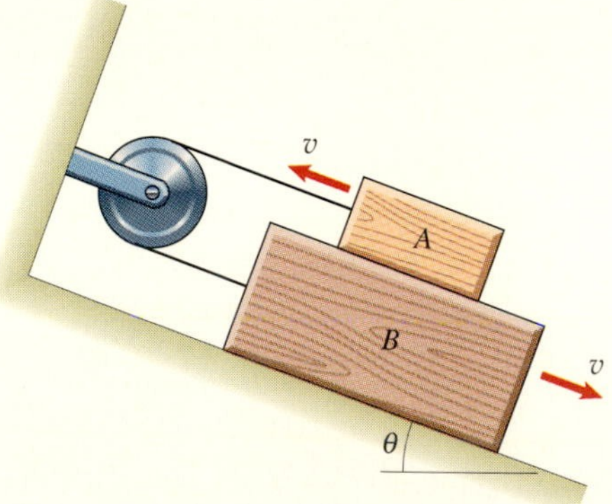

(c) Magnitude of the velocity of each crate is the same.

DISCUSSION

If we attempt to solve this example by applying the principle of work and energy to the system consisting of the crates, the cable, and the pulley (Fig. d), we obtain an incorrect result. Equating the work done by external forces to the change in the total kinetic energy of the system, we obtain

$$\int_0^b m_B g \sin\theta\, ds - \int_0^b m_A g \sin\theta\, ds = \frac{1}{2}m_A v_2^2 + \frac{1}{2}m_B v_2^2:$$

$$(m_B g \sin\theta)b - (m_A g \sin\theta)b = \frac{1}{2}m_A v_2^2 + \frac{1}{2}m_B v_2^2.$$

But if we sum our work and energy equations for the individual crates—Eqs. (4.13) and (4.14)—we obtain the correct equation:

$$\underbrace{[(m_B g \sin\theta)b - (m_A g \sin\theta)b]}_{\text{Work by external forces}} + \underbrace{[-(2\mu_k m_A g \cos\theta)b]}_{\text{Work by internal forces}} = \frac{1}{2}m_A v_2^2 + \frac{1}{2}m_B v_2^2.$$

The internal friction forces the crates exert on each other do net work on the system. We did not account for this work in applying the principle of work and energy to the free-body diagram of the system.

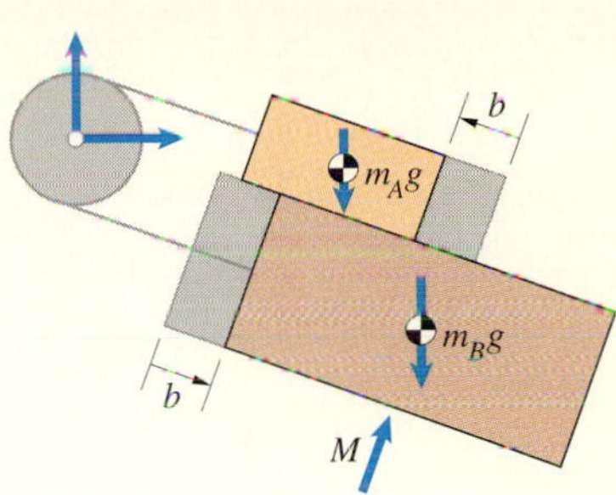

(d) Free-body diagram of the system.

Problems

4.1 The 20-lb box A is at rest on the smooth surface when it is subjected to a constant 10-lb force. (a) By using Newton's second law to determine the acceleration of the box, determine how fast it is traveling when it has moved 4 ft to the right. (b) Use Eq. (4.7) to determine the work done on the box as it moves 4 ft to the right. (c) Use the principle of work and energy to determine how fast the box is traveling when it has moved 4 ft to the right.

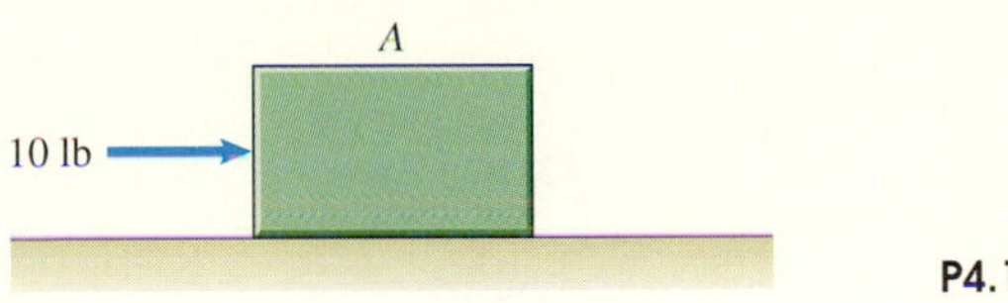

P4.1

4.2 Solve Problem 4.1 if the coefficient of kinetic friction between the box and the surface is $\mu_k = 0.2$.

4.3 The 4-kg box A is released from rest on the smooth inclined surface. Determine how fast the box is traveling when it has moved 2 m in the direction parallel to the surface: (a) by using Newton's second law to determine the acceleration of the box; (b) by using Eq. (4.7) and the principle of work and energy.

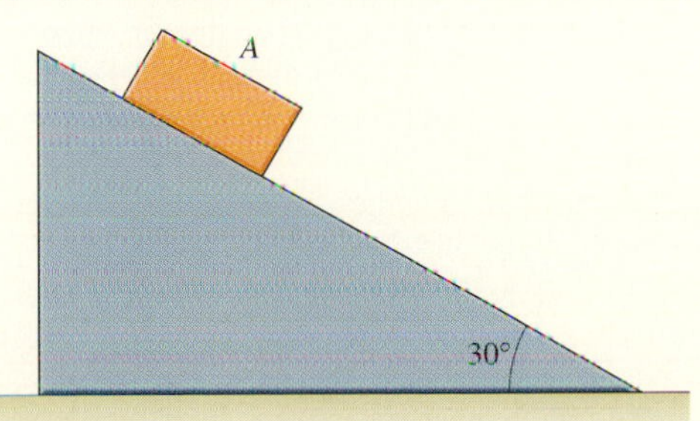

P4.3

4.4 Solve Problem 4.3 if the coefficient of kinetic friction between the box and the inclined surface is $\mu_k = 0.2$.

4.5 The fictional starship *Enterprise* obtains its power by combining matter and antimatter, achieving complete conversion of mass into energy. The energy contained in an amount of matter of mass m is given by Einstein's equation $E = mc^2$, where c is the speed of light (3×10^8 m/s).
(a) The mass of the *Enterprise* is approximately 5×10^9 kg. How much mass must be converted into kinetic energy to accelerate it from rest to one-tenth the speed of light?
(b) How much mass must be converted into kinetic energy to accelerate a 200,000-lb airliner from rest to 600 mi/hr?

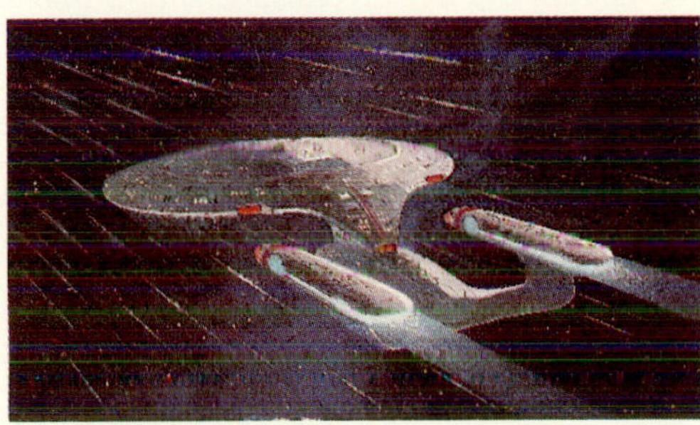

P4.5

4.6 A 2000-lb drag racer can accelerate from rest to 300 mi/hr in a quarter of a mile.
(a) How much work is done on the car?
(b) If you assume as a first approximation that the tangential force exerted on the car is constant, what is the magnitude of the force?

P4.6

4.7 Assume that all the weight of the drag racer in Problem 4.6 acts on its rear (drive) wheels and that the coefficients of friction between the wheels and the road are $\mu_s = \mu_k = 0.9$. Use the principle of work and energy to determine the maximum velocity in miles per hour the car can theoretically reach in a quarter of a mile. What do you think might account for the discrepancy between your answer and the car's actual velocity of 300 mi/hr?

4.8 Assuming as a first approximation that the tangential force exerted on the drag racer in Problem 4.6 is constant, determine (a) the maximum power and (b) the average power transferred to the car as it accelerates from rest to 300 mi/hr.

4.9 A 10,000-kg airplane must reach a velocity of 60 m/s to take off. If the horizontal force exerted by its engine is 60 kN and you neglect other horizontal forces, what length runway is needed?

4.10 Suppose you want to design an auxiliary rocket unit that will allow the airplane in Problem 4.9 to reach its takeoff speed using only 100 m of runway. For your preliminary design calculation, you can assume that the combined mass of the rocket and airplane is constant and equal to 10,500 kg. What horizontal component of thrust must the rocket unit provide?

P4.10

4.11 Determine (a) the maximum power and (b) the average power transferred to the airplane in Problem 4.9 while it takes off.

4.12 (a) Suppose that you hold a 0.15-kg ball 2 m above the ground and release it from rest. Use Eq. (4.8) to determine the work done on the ball by gravity as it falls to the ground. (b) Suppose that you then throw the ball upward at 3 m/s, releasing it at 2 m above the ground. Use Eq. (4.8) to determine the work done on the ball by gravity from the time you release it until it hits the ground. (c) Use the principle of work and energy to determine the magnitude of the ball's velocity just before it hits the ground in cases (a) and (b).

4.13 The force exerted on a car by a prototype crash barrier as the barrier crushes is $F = -(1000 + 10{,}000s)$ lb, where s is the distance in feet from the initial contact. Suppose you want to design the barrier so that it can stop a 5000-lb car traveling at 80 mi/hr. What is the necessary effective length of the barrier? That is, what is the distance required for the barrier to bring the car to a stop?

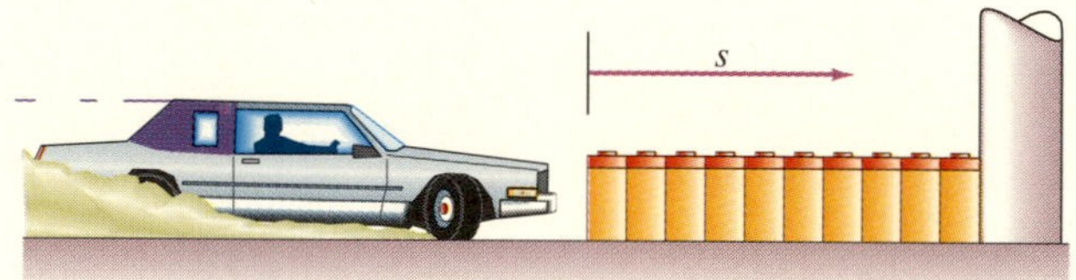

P4.13

4.14 The component of the total external force tangent to a 2-lb object's path is $\Sigma F_t = 4s - s^2$ lb, where s is its position measured along the path in feet. At $s = 0$, the object's velocity is $v = 10$ ft/s.
(a) How much work is done on the object as it moves from $s = 0$ to $s = 4$ ft?
(b) What is its velocity when it reaches $s = 4$ ft?

4.15 The component of the total external force tangent to a 10-kg object's path is $\Sigma F_t = 100 - 20t$ N, where t is in seconds. When $t = 0$, its velocity is $v = 4$ m/s. How much work is done on the object from $t = 2$ to $t = 4$ s?

4.16 A group of engineering students constructs a sun-powered car and tests it on a circular track of 1000-ft radius. The car, which weighs 450 lb with its occupant, starts from rest. The total tangential component of force on the car is $\Sigma F_t = 28 - 0.1s$ lb, where s is the distance the car travels along the track from the point where it starts. (a) Determine the work done on the car when it has gone a distance $s = 100$ ft. (b) Use the principle of work and energy to determine the magnitude of the car's velocity when $s = 100$ ft.

4.17 At the instant shown, the 160-lb vaulter's center of mass is 8.5 ft above the ground and the vertical component of his velocity is 4 ft/s. As his pole straightens, it exerts a vertical force on him of magnitude $180 + 2.8y^2$ lb, where y is the vertical position of his center of mass *relative to its position at the instant shown.* This force is exerted on him from $y = 0$ to $y = 4$ ft, when he releases the pole. What is the maximum height above the ground reached by his center of mass?

P4.17

4.18 The component of the total external force tangent to the path of an object of mass m is $\Sigma F_t = -cv$, where v is the magnitude of the object's velocity and c is a constant. When the position $s = 0$, its velocity is $v = v_0$. How much work is done on the object as it moves from $s = 0$ to a position $s = s_f$?

4.19 The coefficients of friction between the 160-kg crate and the ramp are $\mu_s = 0.3$ and $\mu_k = 0.28$.
(a) What tension T_0 must the winch exert to start the crate moving up the ramp?
(b) If the tension remains at the value T_0 after the crate starts sliding, what total work is done on the crate as it slides a distance $s = 3$ m up the ramp, and what is the resulting velocity of the crate?

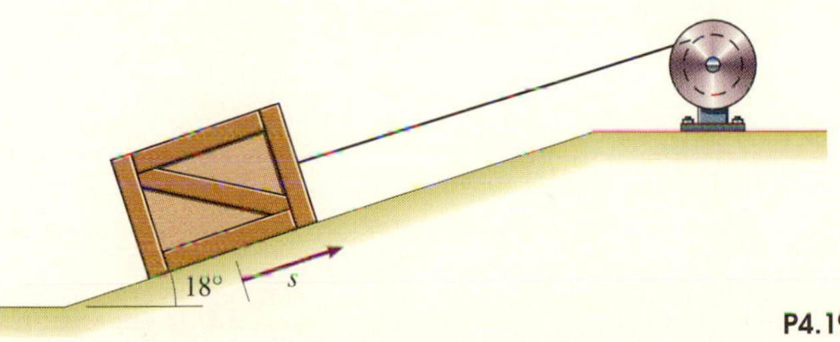

P4.19

4.20 In Problem 4.19, if the winch exerts a tension $T = T_0(1 + 0.1s)$ after the crate starts sliding, what total work is done on the crate as it slides a distance $s = 3$ m up the ramp, and what is the resulting velocity of the crate?

4.21 The 200-mm-diameter gas gun is evacuated on the right of the 8-kg projectile. On the left of the projectile the tube contains gas with pressure $p_0 = 1 \times 10^5$ Pa (N/m^2). The force F is slowly increased, moving the projectile 0.5 m to the left from the position shown. The force is then removed and the projectile accelerates to the right. If you neglect friction and assume that the pressure of the gas is related to its volume by pV = constant, what is the velocity of the projectile when it has returned to its original position?

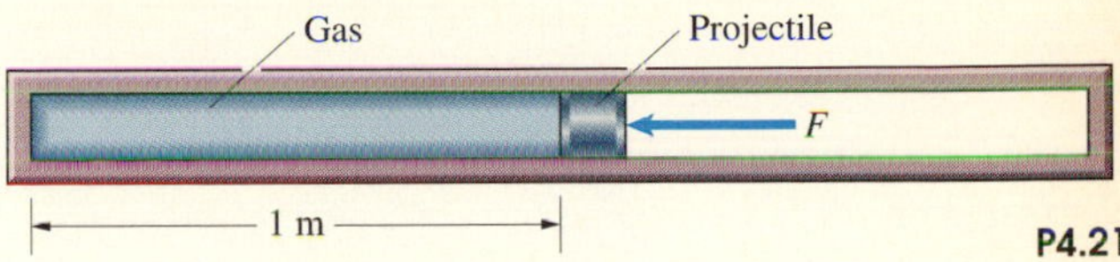

P4.21

4.22 In Problem 4.21, if you assume that the pressure of the gas is related to its volume by pV = constant while it is compressed (an isothermal process) and by $pV^{1.4}$ = constant while it is expanding (an isentropic process), what is the velocity of the projectile when it has returned to its original position?

4.23 The system is released from rest. By applying the principle of work and energy to each weight, determine the magnitude of the velocity of the weights when they have moved 1 ft.

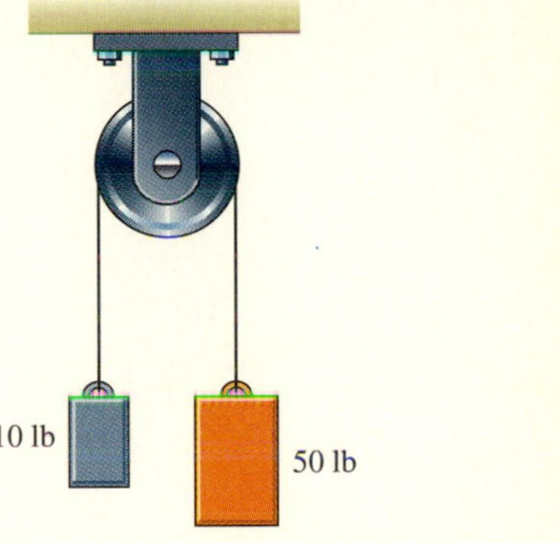

P4.23

4.24 In Problem 4.23, what average power is transmitted to the system during its motion?

4.25 Solve Problem 4.23 by applying the principle of work and energy to the system consisting of the two weights, the cable, and the pulley.

4.26 The mass of each box is 12 kg and the coefficient of kinetic friction between the boxes and the surface is $\mu_k = 0.05$. The system is released from rest. Determine the magnitude of the velocity of the weights when they have moved 1 m.

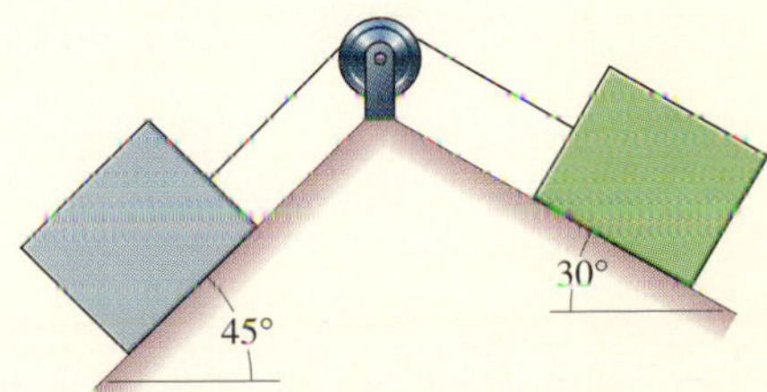

P4.26

4.27 In Problem 4.26, what average power is transmitted to the system during its motion?

4.28 The masses of the three blocks are $m_A = 40$ kg, $m_B = 16$ kg, and $m_C = 12$ kg. Neglect the mass of the bar holding C in place. Friction is negligible. By applying the principle of work and energy to A and B individually, determine the magnitude of their velocity when they have moved 500 mm.

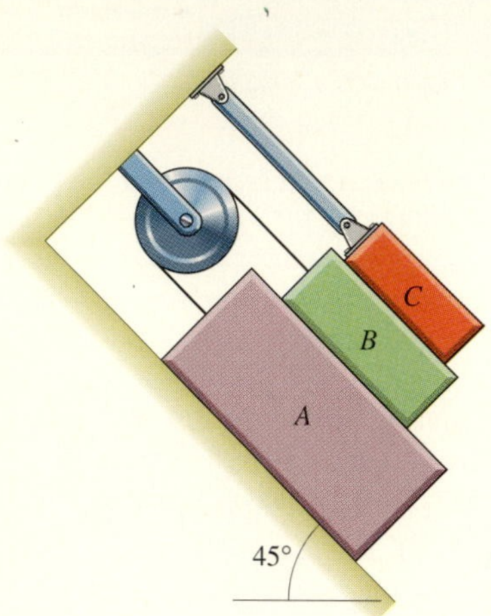

P4.28

4.29 Solve Problem 4.28 by applying the principle of work and energy to the system consisting of A, B, the cable connecting them, and the pulley.

4.30 In Problem 4.28, determine the magnitude of the velocity of A and B when they have moved 500 mm if the coefficient of kinetic friction between all surfaces is $\mu_k = 0.1$.

Strategy: The simplest approach is to apply the principle of work and energy to A and B individually. If you treat them as a single system, you must account for the work done by internal friction forces. See Example 4.3.

4.3 Work Done by Particular Forces

You have seen that if the tangential component of the total external force on an object is known as a function of distance along the object's path, you can use the principle of work and energy to relate a change in position to the change in the object's velocity. For certain types of forces, however, not only can you determine the work without knowing the tangential component of the force as a function of distance along the path, you don't even need to know the path. Two important examples are weight and the force exerted by a spring.

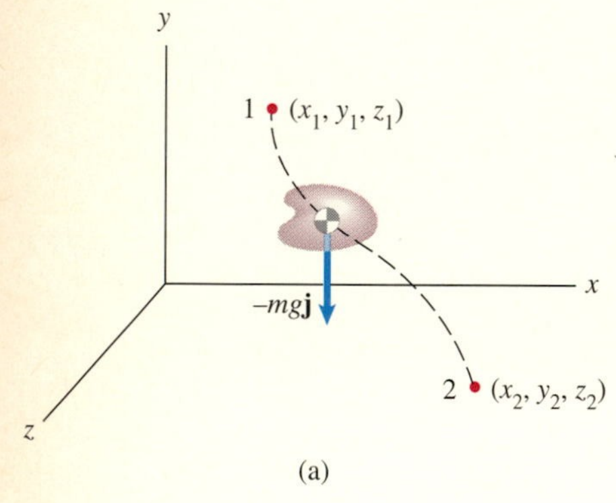

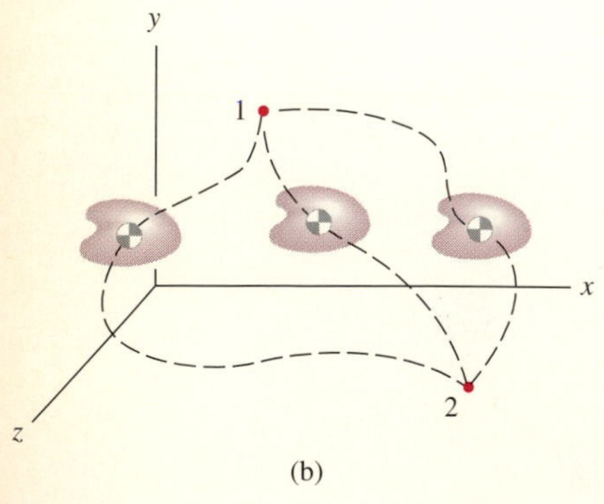

Fig. 4.6
(a) An object moving between two positions.
(b) The work done by the weight is the same for any path.

Weight

To evaluate the work done by an object's weight, we orient a cartesian coordinate system with the y axis upward and suppose that the object moves from position 1 with coordinates (x_1, y_1, z_1) to position 2 with coordinates (x_2, y_2, z_2) (Fig. 4.6a). The force exerted by its weight is $\mathbf{F} = -mg\mathbf{j}$. (Other forces may act on the object, but we are concerned only with the work done by its weight.) Because $\mathbf{v} = d\mathbf{r}/dt$, we can multiply the velocity, expressed in cartesian coordinates, by dt to obtain an expression for the vector $d\mathbf{r}$:

$$d\mathbf{r} = \left(\frac{dx}{dt}\mathbf{i} + \frac{dy}{dt}\mathbf{j} + \frac{dz}{dt}\mathbf{k}\right) dt = dx\mathbf{i} + dy\mathbf{j} + dz\mathbf{k}.$$

Taking the dot product of $\mathbf{F}$ and $d\mathbf{r}$,

$$\mathbf{F} \cdot d\mathbf{r} = (-mg\mathbf{j}) \cdot (dx\mathbf{i} + dy\mathbf{j} + dz\mathbf{k}) = -mg\, dy,$$

the work done as the object moves from position 1 to position 2 reduces to an integral with respect to y:

$$U_{12} = \int_{\mathbf{r}_1}^{\mathbf{r}_2} \mathbf{F} \cdot d\mathbf{r} = \int_{y_1}^{y_2} -mg\, dy.$$

Evaluating the integral, we obtain the work done by the weight of an object as it moves between two positions:

$$\boxed{U_{12} = -mg\,(y_2 - y_1).} \tag{4.15}$$

The work is simply the product of the weight and the change in the object's height. The work done is negative if the height increases and positive if it decreases. Notice that *the work done is the same no matter what path the object follows from position* 1 *to position* 2 (Fig. 4.6b). You don't need to know the path to determine the work done by an object's weight—you only need to know the relative heights of the initial and final positions.

What work is done by an object's weight if we account for its variation with distance from the center of the earth? In terms of polar coordinates, we can write the weight of an object at a distance r from the center of the earth as (Fig. 4.7)

$$\mathbf{F} = -\frac{mgR_E^2}{r^2}\mathbf{e}_r.$$

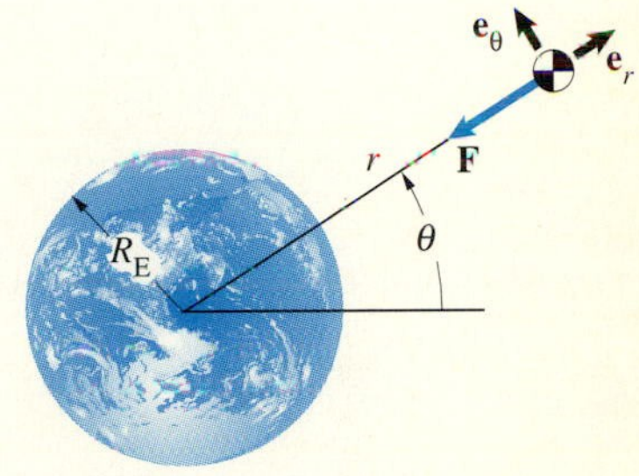

Fig. 4.7
Expressing an object's weight in polar coordinates.

Using the expression for the velocity in polar coordinates, the vector $d\mathbf{r} = \mathbf{v}dt$ is

$$d\mathbf{r} = \left(\frac{dr}{dt}\mathbf{e}_r + r\frac{d\theta}{dt}\mathbf{e}_\theta\right) dt = dr\,\mathbf{e}_r + r\,d\theta\,\mathbf{e}_\theta. \tag{4.16}$$

The dot product of $\mathbf{F}$ and $d\mathbf{r}$ is

$$\mathbf{F} \cdot d\mathbf{r} = \left(-\frac{mgR_E^2}{r^2}\mathbf{e}_r\right) \cdot (dr\,\mathbf{e}_r + r\,d\theta\,\mathbf{e}_\theta) = -\frac{mgR_E^2}{r^2}dr,$$

so the work reduces to an integral with respect to r:

$$U_{12} = \int_{\mathbf{r}_1}^{\mathbf{r}_2} \mathbf{F} \cdot d\mathbf{r} = \int_{r_1}^{r_2} -\frac{mgR_E^2}{r^2}dr.$$

Evaluating the integral, we obtain the work done by an object's weight, accounting for the variation of the weight with height:

$$\boxed{U_{12} = mgR_E^2\left(\frac{1}{r_2} - \frac{1}{r_1}\right).} \tag{4.17}$$

Again, the work is independent of the path from position 1 to position 2. To evaluate it, you only need to know the object's radial distance from the center of the earth at the two positions.

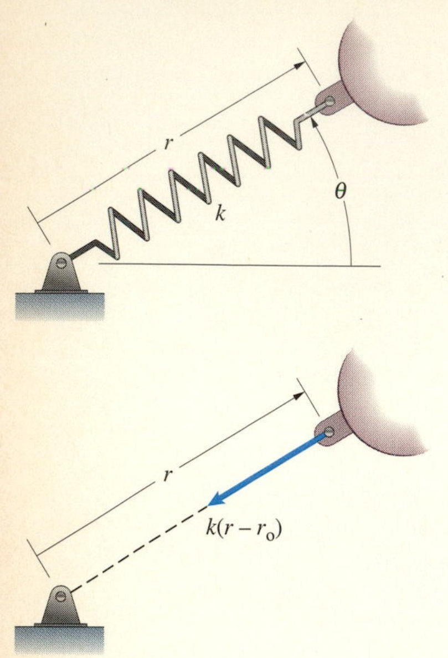

Fig. 4.8
Expressing the force exerted by a linear spring in polar coordinates.

Springs

Suppose that a linear spring connects an object to a fixed support. In terms of polar coordinates (Fig. 4.8), the force exerted on the object is

$$\mathbf{F} = -k(r - r_0)\,\mathbf{e}_r,$$

where k is the spring constant and r_0 is the unstretched length of the spring. Using Eq. (4.16), the dot product of $\mathbf{F}$ and $d\mathbf{r}$ is

$$\mathbf{F}\cdot d\mathbf{r} = [-k(r - r_0)\,\mathbf{e}_r]\cdot(dr\,\mathbf{e}_r + r\,d\theta\,\mathbf{e}_\theta) = -k(r - r_0)\,dr.$$

It is convenient to express the work done by a spring in terms of its **stretch**, defined by $S = r - r_0$. (Although the word *stretch* usually means an increase in length, we use this term more generally to denote the change in length of the spring. A negative stretch is a decrease in length.) In terms of this variable, $\mathbf{F}\cdot d\mathbf{r} = -kS\,dS$, and the work is

$$U_{12} = \int_{\mathbf{r}_1}^{\mathbf{r}_2} \mathbf{F}\cdot d\mathbf{r} = \int_{S_1}^{S_2} -kS\,dS.$$

The work done on an object by a spring attached to a fixed support is

$$\boxed{U_{12} = -\frac{1}{2}k(S_2^2 - S_1^2),} \tag{4.18}$$

where S_1 and S_2 are the values of the stretch at the initial and final positions. You don't need to know the object's path to determine the work done by the spring. You must remember that Eq. (4.18) applies only to a *linear* spring. In Fig. 4.9 we determine the work done in stretching a linear spring by calculating the area defined by the graph of the force as a function of S.

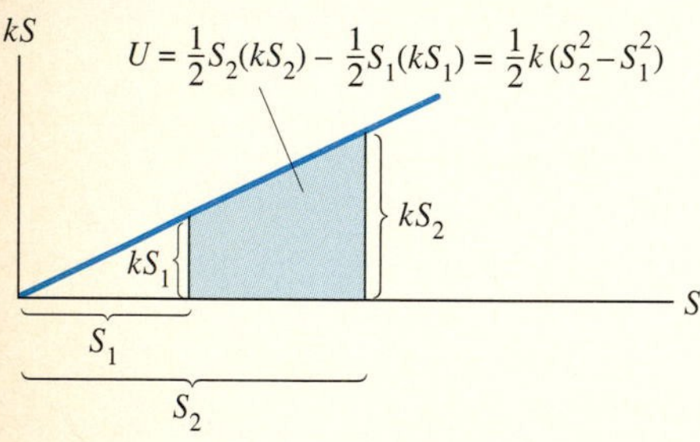

Fig. 4.9
Work done in stretching a linear spring from S_1 to S_2. (If $S_2 > S_1$, the work done *on* the spring is positive, so the work done *by* the spring is negative.)

Example 4.4

At position 1, the skier in Fig. 4.10 is approaching his jump at 15 m/s. When he reaches the horizontal end of the ramp at position 2, 20 m below position 1, he jumps upward, achieving a vertical component of velocity of 3 m/s. (Disregard the small change in the vertical position of his center of mass due to his jumping motion.) Neglect aerodynamic drag and the friction forces on his skis.
(a) What is the magnitude of his velocity as he leaves the ramp at position 2?
(b) At the highest point of his jump, position 3, what are the magnitude of his velocity and the height of his center of mass above position 2?

STRATEGY

(a) If we neglect aerodynamic and friction forces, the only force doing work from position 1 to position 2 is the skier's weight. The normal force exerted on his skis by the ramp does no work because it is perpendicular to his path. We need to know

Fig. 4.10

only the change in his height from position 1 to position 2 to determine the work done by his weight, so we can apply the principle of work and energy to determine his velocity at position 2 before he jumps.

(b) From the time he leaves the ramp at position 2 until he reaches position 3, the only force is his weight, so the horizontal component of his velocity is constant. That means that we know the magnitude of his velocity at position 3, because he is moving horizontally at that point. Therefore we can apply the principle of work and energy to his motion from position 2 to position 3 to determine his height above position 2.

SOLUTION

(a) We will use Eq. (4.15) to evaluate the work done by his weight, measuring the height of his center of mass relative to position 2 (Fig. a). The principle of work and energy from position 1 to position 2 is

$$U_{12} = -mg(y_2 - y_1) = \frac{1}{2}mv_2^2 - \frac{1}{2}mv_1^2:$$

$$-m(9.81)(0 - 20) = \frac{1}{2}mv_2^2 - \frac{1}{2}m(15)^2.$$

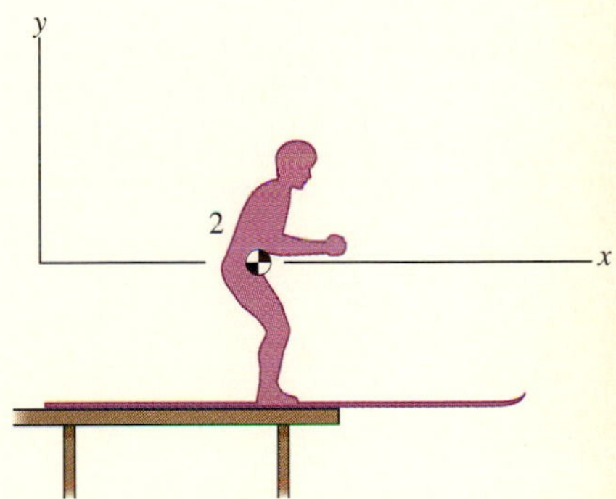

(a) The height of his center of mass is measured relative to position 2.

Solving for v_2, his horizontal velocity at position 2 before he jumps upward is 24.8 m/s. After he jumps upward, the magnitude of his velocity at position 2 is $v'_2 = \sqrt{(24.8)^2 + (3)^2} = 25.0$ m/s.

(b) The magnitude of his velocity at position 3 is equal to the horizontal component of his velocity at position 2: $v_3 = v_2 = 24.8$ m/s. Applying work and energy to his motion from position 2 to position 3,

$$U_{23} = -mg(y_3 - y_2) = \frac{1}{2}mv_3^2 - \frac{1}{2}m(v'_2)^2:$$

$$-m(9.81)(y_3 - 0) = \frac{1}{2}m(24.8)^2 - \frac{1}{2}m(25.0)^2,$$

we obtain $y_3 = 0.459$ m.

DISCUSSION

Although we neglected aerodynamic effects, a ski jumper is actually subjected to substantial aerodynamic forces, both parallel to his path (drag) and perpendicular to it (lift).

Example 4.5

Fig. 4.11

In the forging device shown in Fig. 4.11, the 40-kg hammer is lifted to position 1 and released from rest. It falls and strikes a workpiece when it is in position 2. The spring constant $k = 1500$ N/m, and the tension in each spring is 150 N when the hammer is in position 2. Neglect friction.

(a) What is the velocity of the hammer just before it strikes the workpiece?

(b) Assuming that all the hammer's kinetic energy is transferred to the workpiece, what average power is transferred if the duration of the impact is 0.02 s?

STRATEGY

Work is done on the hammer by its weight and by the forces exerted by the springs (Fig. a). We can apply the principle of work and energy to the motion of the hammer from position 1 to position 2 to determine its velocity at position 2.

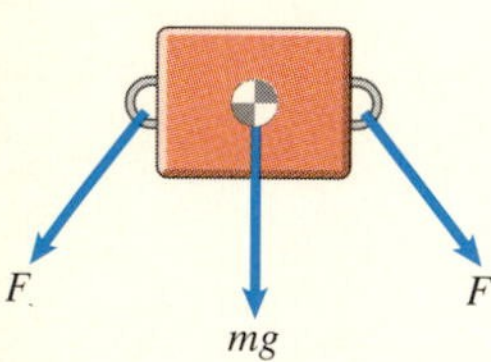

(a) Free-body diagram of the hammer.

SOLUTION

(a) Let r_0 be the unstretched length of one of the springs. In position 2, the tension in the spring is 150 N and its length is 0.3 m. From the relation between the tension in a linear spring and its stretch,

$$150 = k(0.3 - r_0) = (1500)(0.3 - r_0),$$

we obtain $r_0 = 0.2$ m. The values of the stretch of each spring in positions 1 and 2 are $S_1 = \sqrt{(0.4)^2 + (0.3)^2} - 0.2 = 0.3$ m and $S_2 = 0.3 - 0.2 = 0.1$ m. From Eq. (4.18), the total work done on the hammer by the two springs from position 1 to position 2 is

$$U_{\text{springs}} = 2\left[-\frac{1}{2}k(S_2^2 - S_1^2)\right] = -(1500)[(0.1)^2 - (0.3)^2] = 120 \text{ N-m}.$$

The work done by the weight from position 1 to position 2 is positive and equal to the product of the weight and the change in height:

$$U_{\text{weight}} = mg(0.4 \text{ m}) = (40)(9.81)(0.4) = 157 \text{ N-m}.$$

From the principle of work and energy,

$$U_{\text{springs}} + U_{\text{weight}} = \frac{1}{2}mv_2^2 - \frac{1}{2}mv_1^2:$$

$$120 + 157 = \frac{1}{2}(40)v_2^2 - 0,$$

we obtain $v_2 = 3.72$ m/s.

(b) All the hammer's kinetic energy is transferred to the workpiece, so Eq. (4.10) indicates that the average power equals the kinetic energy of the hammer divided by the duration of the impact:

$$P_{\text{av}} = \frac{(1/2)(40 \text{ kg})(3.72 \text{ m/s})^2}{0.02 \text{ s}} = 13.8 \text{ kW (kilowatts)}.$$

Example 4.6

Application to Engineering

Automated Machining

The device in Fig. 4.12a machines a sample of material with the cutting tool at A. The hydraulic actuator attached at B pushes the tool holder to the right on its horizontal guide rail. The spring attached at C returns the tool holder and hydraulic actuator to their initial positions when the cut is completed. The position of the sample of material is changed, and the cycle is repeated. In the free-body diagram of the assembly consisting of the cutting tool and tool holder (Fig. 4.12b), A_x and A_y are the forces exerted on the cutting tool by the machined sample, B is the force exerted by the actuator, F is the force exerted by the spring, and N is the normal force exerted by the guide rail. The mass of the assembly is $m = 22$ kg, the unstretched length of the spring is 200 mm, and the spring constant is $k = 1800$ N/m.

(a) If the assembly starts from rest in the position shown, the horizontal force on the cutting tool is $A_x = 1330$ N, and the force exerted by the actuator is $B = 1700$ N, determine its velocity when it has moved 0.05 m to the right.

(b) Determine the maximum power transferred from the actuator as the assembly moves from $x = 0$ to $x = 0.15$ m.

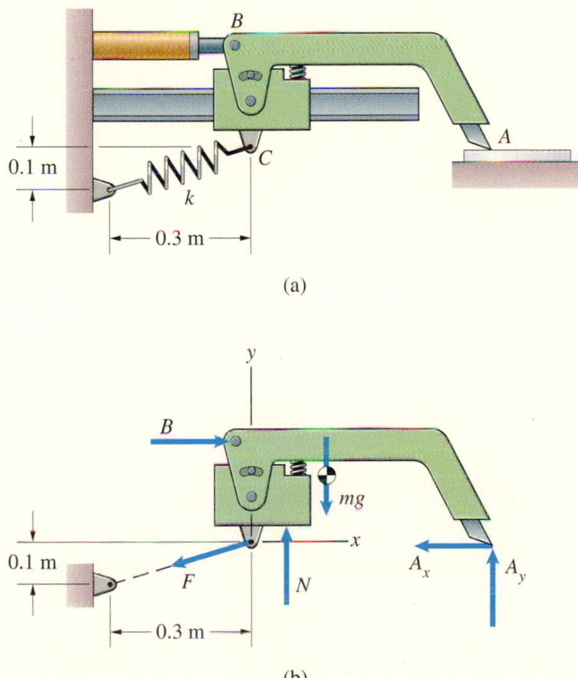

Fig. 4.12

STRATEGY

(a) By determining the total work done by the spring and the horizontal forces A_x and B as a function of x, we can use work and energy to determine the velocity of the assembly as a function of x. (b) The power transferred equals the product of the force exerted by the actuator and the velocity. We will draw a graph of the velocity as a function of x to determine the maximum velocity.

SOLUTION

(a) Let the initial position of the assembly be position 1 and let its position when it has moved a distance x to the right be position 2. The stretch of the spring in position 1 is

$$S_1 = \sqrt{(0.1)^2 + (0.3)^2} - 0.2 \text{ m},$$

and the stretch in position 2 is

$$S_2 = \sqrt{(0.1)^2 + (0.3 + x)^2} - 0.2 \text{ m}.$$

The work done from position 1 to position 2 by the horizontal forces A_x and B and the spring is

$$U_{12} = (B - A_x)x - \frac{1}{2}k(S_2^2 - S_1^2).$$

We apply the principle of work and energy,

$$U_{12} = \frac{1}{2}mv_2^2 - \frac{1}{2}mv_1^2:$$

$$(B - A_x)x - \frac{1}{2}k(S_2^2 - S_1^2) = \frac{1}{2}mv_2^2 - 0.$$

Setting $x = 0.05$ m, $A_x = 1330$ N, $B = 1700$ N, and $m = 22$ kg and solving for the velocity, we obtain $v_2 = 0.766$ m/s.

(b) In Fig. 4.13 we draw a graph of the velocity v_2 as a function of x for $0 \le x \le 0.15$ m. From the graph we estimate that the maximum velocity occurs at $x = 0.1$ m. Solving for the velocity at this position, we obtain $v_2 = 0.884$ m/s. The power transferred from the actuator at this velocity is

$$P = Bv_2 = (1700)(0.884) = 1500 \text{ N-m/s (watts)}.$$

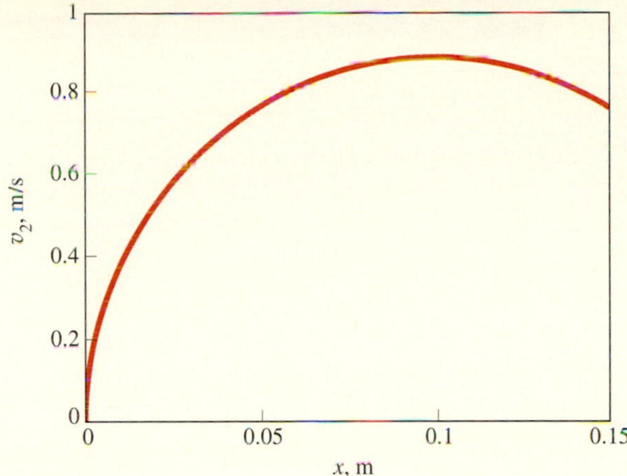

Fig. 4.13
Velocity of the assembly as a function of x.

DESIGN ISSUES

In developing a machine, the engineers must ensure that it will perform the tasks for which it is designed and do so reliably through a design lifetime that may be many years and millions of cycles of use. The machine should be designed for ease of use and maintenance and must be safe for its operators and other persons. It must also be economical. If it is an industrial machine, its original cost and the cost of use and maintenance must be acceptable in comparison to the income derived through its use. The choices implied by these requirements are often mutually contradictory, and the design engineers must make successful compromises.

The use of dynamics to study the motions of machines is essential to designers for analyzing their behavior, predicting their performance, and determining whether their structures will support dynamic loads to which they will be subjected. Using a simple context, we demonstrate in Example 4.6 that the principle of work and energy can be useful in analyzing a machine's motion. It can also provide information on energy requirements of machines. The work done by the hydraulic actuator in Example 4.6 provides an estimate of the energy that must be supplied (in the form of electrical energy to power the hydraulic pump) for each cycle of operation of the machine tool.

Example 4.7

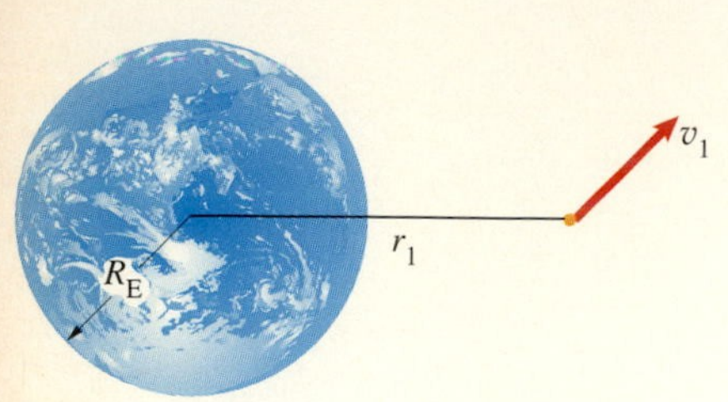

Fig. 4.14

A spacecraft at a distance $r_1 = 2R_E$ from the center of the earth has a velocity of magnitude $v_1 = \sqrt{2gR_E/3}$ relative to a nonrotating reference frame with its origin at the center of the earth (Fig. 4.14). Determine the magnitude of the spacecraft's velocity when it is at a distance $r_2 = 4R_E$ from the center of the earth.

STRATEGY

By applying Eq. (4.17) to determine the work done by the gravitational force on the spacecraft, we can use the principle of work and energy to determine the magnitude of the velocity.

SOLUTION

From Eq. (4.17), the work done by gravity as the spacecraft moves from a distance r_1 from the center of the earth to a distance r_2 is

$$U_{12} = mgR_E^2\left(\frac{1}{r_2} - \frac{1}{r_1}\right).$$

Let v_2 be the magnitude of the spacecraft's velocity when it is at a distance r_2 from the center of the earth. Applying the principle of work and energy,

$$U_{12} = mgR_E^2\left(\frac{1}{r_2} - \frac{1}{r_1}\right) = \frac{1}{2}mv_2^2 - \frac{1}{2}mv_1^2,$$

we can solve for v_2, obtaining

$$\begin{aligned} v_2 &= \sqrt{v_1^2 + 2gR_E^2\left(\frac{1}{r_2} - \frac{1}{r_1}\right)} \\ &= \sqrt{\left(\frac{2gR_E}{3}\right) + 2gR_E^2\left(\frac{1}{4R_E} - \frac{1}{2R_E}\right)} \\ &= \sqrt{\frac{gR_E}{6}}. \end{aligned}$$

The velocity $v_2 = v_1/2$.

DISCUSSION

Notice that we did not need to specify the direction of the spacecraft's initial velocity to determine the magnitude of its velocity at a different distance from the center of the earth. This illustrates the power of the principle of work and energy as well as its limitations. Even if we know the direction of the initial velocity, the principle of work and energy tells us only the magnitude of the velocity at a different distance.

Problems

4.31 (a) Suppose that you hold a 0.15-kg ball 2 m above the ground and release it from rest. Use Eq. (4.15) to determine the work done on the ball by gravity as it falls to the ground. (b) Suppose that you then throw the ball upward at 3 m/s, releasing it at 2 m above the ground. Use Eq. (4.15) to determine the work done on the ball by gravity from the time you release it until it hits the ground. (c) Use the principle of work and energy to determine the magnitude of the ball's velocity just before it hits the ground in cases (a) and (b).

4.32 Suppose that you throw rocks from the top of a 200-m cliff with a velocity of 10 m/s in the three directions shown. Neglecting aerodynamic drag, use the principle of work and energy to determine the magnitude of the velocity of the rock just before it hits the ground in each case.

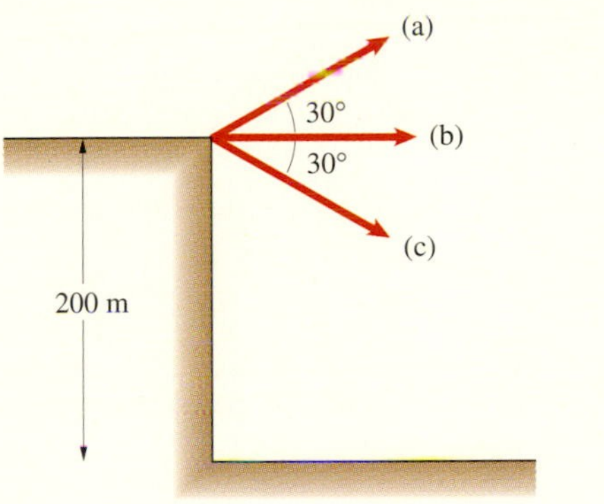

P4.32

4.33 The 30-kg box starts from rest at position 1. Neglect friction. For cases (a) and (b), determine the work done on the box from position 1 to position 2 and the magnitude of the velocity of the box at position 2.

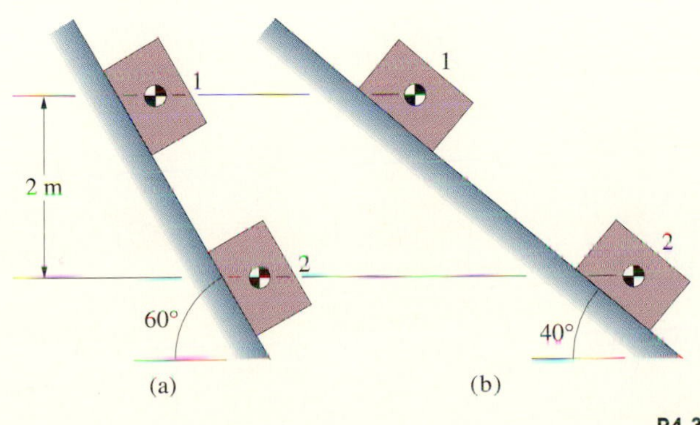

P4.33

4.34 Solve Problem 4.33 if the coefficient of kinetic friction between the box and the inclined surface is $\mu_k = 0.2$.

4.35 In case (a), the 5-oz ball [16 ounces (oz) equal 1 pound] is released from rest at position 1 and falls to position 2. In case (b), the ball is released from rest at position 1 and swings to position 2. For cases (a) and (b), determine the work done on the ball from position 1 to position 2 and the magnitude of the velocity of the ball at position 2.

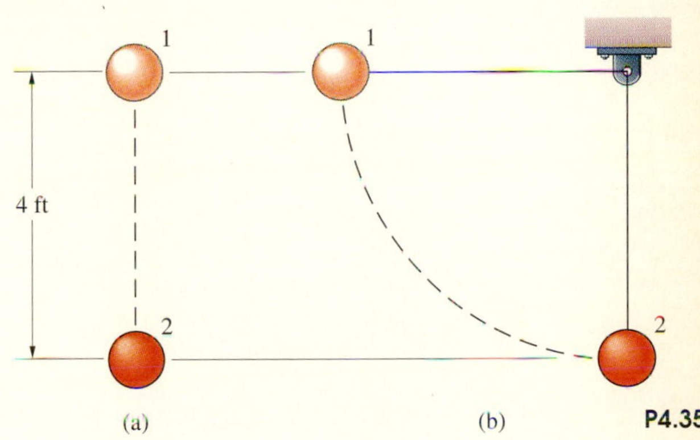

P4.35

4.36 The ball of mass m is released from rest in position 1. Determine the work done on the ball as it swings to position 2 (a) by its weight; (b) by the force exerted on it by the string. (c) What is the magnitude of its velocity at position 2?

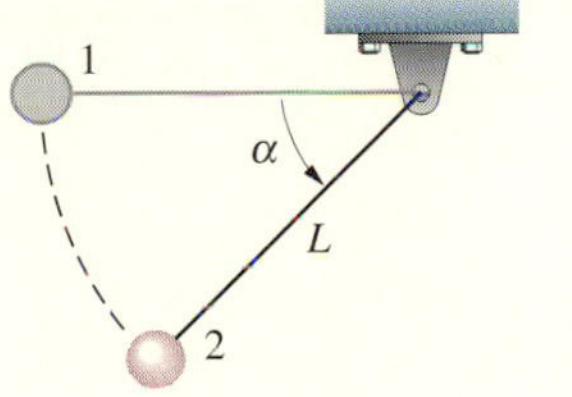

P4.36

4.37 In Problem 4.36, what is the tension in the string in position 2?

4.38 The 200-kg wrecker's ball hangs from a 6-m cable. If it is stationary at position 1, what is the magnitude of its velocity just before it hits the wall at position 2?

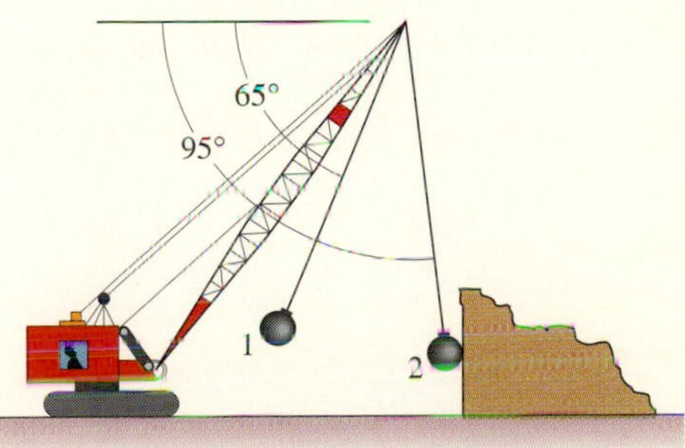

P4.38

4.39 In Problem 4.38, what is the maximum tension in the cable during the motion of the ball from position 1 to position 2?

4.40 A stunt driver wants to drive a car through a circular loop of radius R and hires you as a consultant to tell him the necessary velocity v_0 at which the car must enter the loop so that it can coast through without losing contact with the track.
(a) What is v_0 if you neglect friction and aerodynamic drag for your first rough estimate?
(b) What is the resulting velocity of the car at the top of the loop?

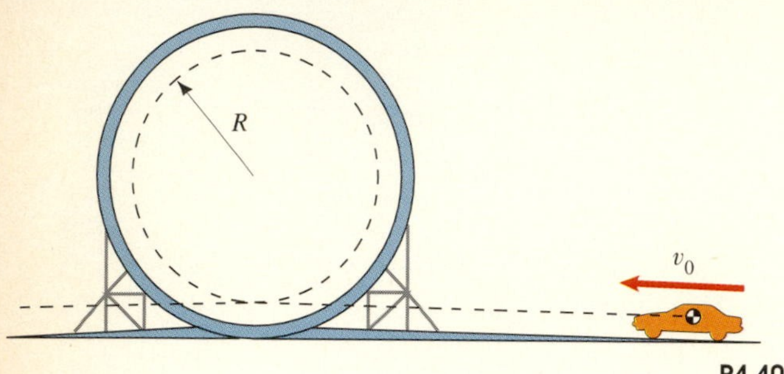

P4.40

4.41 The 2-kg collar starts from rest at position 1 and slides down the smooth rigid wire. The y axis points upward. What is the magnitude of the collar's velocity when it reaches position 2?

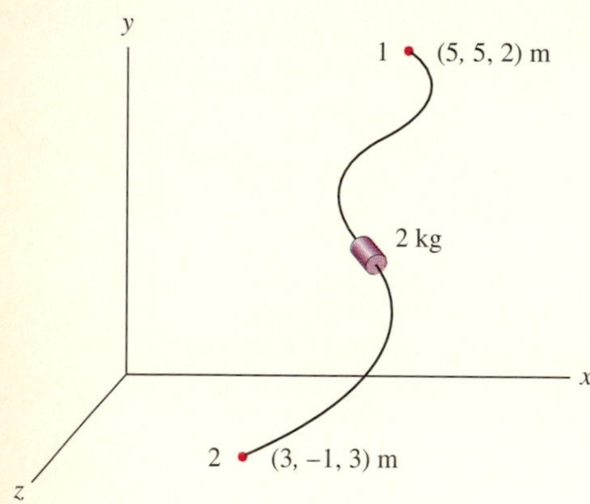

P4.41

4.42 The forces acting on the 24,000-lb airplane are the thrust T and drag D, which are parallel to the airplane's path, the lift L, which is perpendicular to the path, and the weight W. The airplane climbs from a 5000-ft altitude to 10,000 ft. During the climb the magnitude of its velocity decreases from 800 ft/s to 600 ft/s.
(a) What work is done on the airplane by its lift during the climb?
(b) What work is done by the thrust and drag combined?

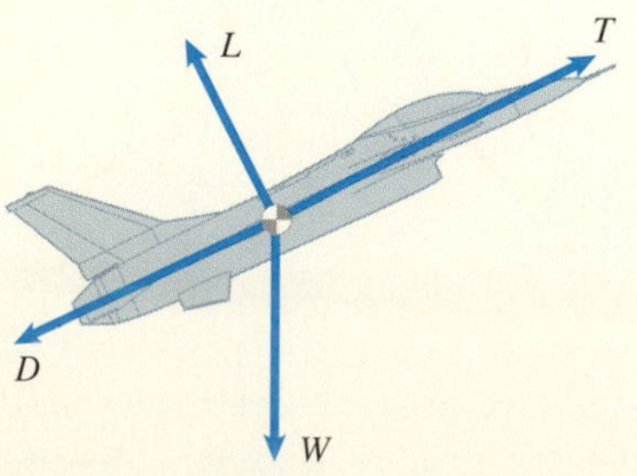

P4.42

4.43 If the airplane in Problem 4.42 is moving at 600 ft/s when it starts its climb at 5000 ft, and the work done by the thrust and drag forces combined as it climbs to 10,000 ft is 1.8×10^8 ft-lb, what is the magnitude of the airplane's velocity when it reaches 10,000 ft?

4.44 The 2400-lb car is traveling 40 mi/hr at position 1. If the combined effect of the aerodynamic drag on the car and the tangential force exerted on its wheels by the road is that they exert no net tangential force on the car, what is the magnitude of its velocity at position 2?

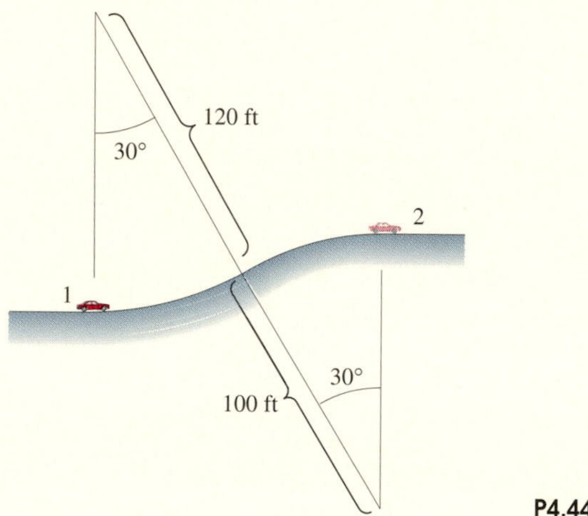

P4.44

4.45 In Problem 4.44, if the combined effect of the aerodynamic drag on the car and the tangential force exerted on its wheels by the road is that they exert a constant 400-lb tangential force on the car in the direction of its motion, what is the magnitude of its velocity at position 2?

4.46 The mass of a rocket is 250 kg, and it has a constant thrust of 6000 N. The total length of the launching ramp is 10 m. Neglecting friction, drag, and the change in mass of the rocket, determine the magnitude of its velocity when it reaches the end of the ramp.

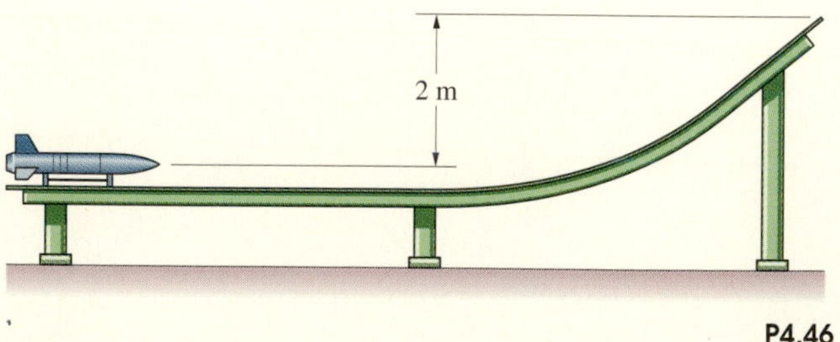

P4.46

4.47 A bioengineer interested in the energy requirements of sports determines from videotape that when the athlete begins his motion to throw the 7.25-kg shot (Fig. a), the shot is stationary and 1.50 m above the ground. At the instant he releases it (Fig. b), the shot is 2.10 m above the ground. The shot reaches a maximum height of 4.60 m above the ground and travels a horizontal distance of 18.66 m from the point where it was released. How much work does the athlete do on the shot from the beginning of his motion to the instant he releases it?

P4.47

4.48 A small pellet of mass m starts from rest at position 1 and slides down the smooth surface of the cylinder.
(a) What work is done on the pellet as it slides from position 1 to position 2?
(b What is the magnitude of the pellet's velocity at position 2?

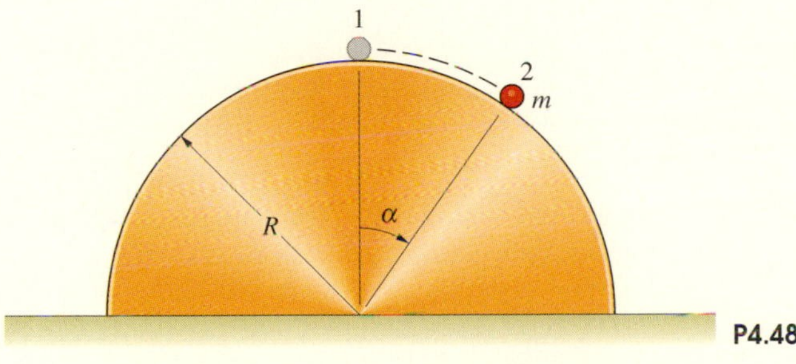

P4.48

4.49 In Problem 4.48, what is the value of the angle α at which the pellet leaves the surface of the cylinder?

4.50 Suppose that you want to design a "bumper" that will bring a 50-lb package moving at 10 ft/s to rest 6 in. from the point of contact. If friction is negligible, what is the necessary spring constant k?

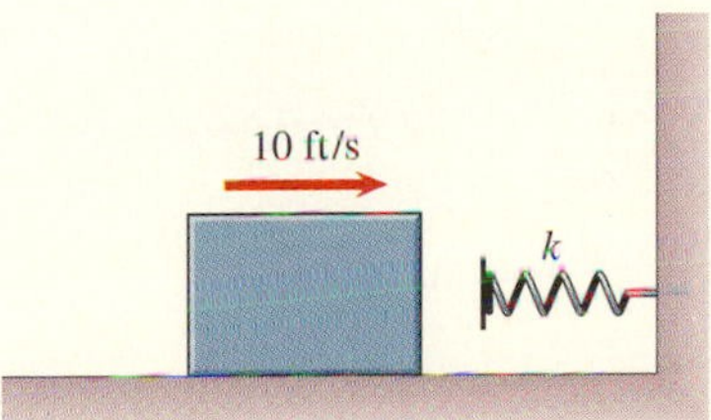

P4.50

4.51 In Problem 4.50, what spring constant is necessary if the coefficient of kinetic friction between the package and the floor is $\mu_k = 0.3$ and the package contacts the spring moving at 10 ft/s?

4.52 An astronaut in an excursion module approaches a space station docking collar. The designer of the collar incorporated a spring to attenuate the shock due to docking. The spring constant is $k = 4800$ N/m. The mass of the astronaut and module is 780 kg. If the module contacts the docking collar moving at 0.1 m/s relative to the collar, what distance is required for the spring to decrease its relative velocity to zero? What is the module's maximum relative deceleration?

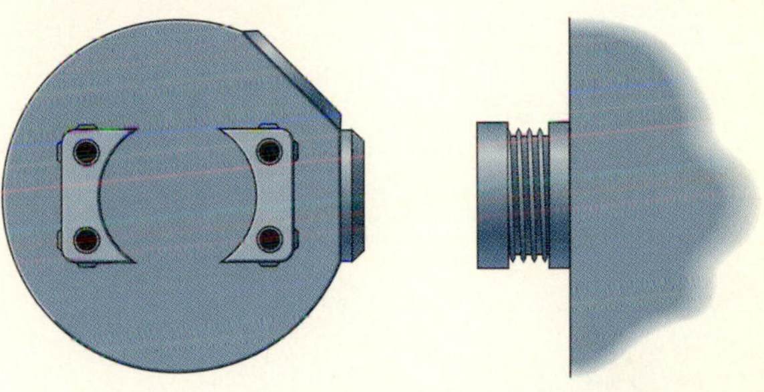

P4.52

4.53 In Problem 4.52, suppose that you design the docking collar so that a 10,000-kg vehicle moving at 0.2 m/s relative to the collar will be brought to rest in a distance of 0.15 m. If the module described in Problem 4.52 contacts the docking collar moving at 0.1 m/s, what is its maximum relative deceleration?

4.54 The system is released from rest with the spring unstretched. If the spring constant is $k = 30$ lb/ft, what maximum velocity do the weights attain?

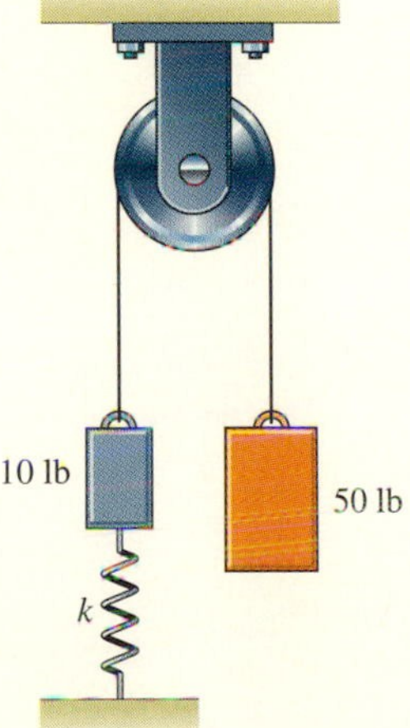

P4.54

4.55 Suppose you don't know the spring constant k of the system in Problem 4.54. If you release the system from rest with the spring unstretched and you observe that the 50-lb weight falls 2 ft before rebounding, what is k?

4.56 In Example 4.5, suppose that the unstretched length of each spring is 200 mm and you want to design the device so that the hammer strikes the workpiece at 5 m/s. Determine the necessary spring constant k.

4.57 The 20-kg crate is released from rest with the spring unstretched. The spring constant is $k = 100$ N/m. Neglect friction.
(a) How far down the inclined surface does the crate slide before it stops?
(b) What maximum velocity does it attain on the way down?

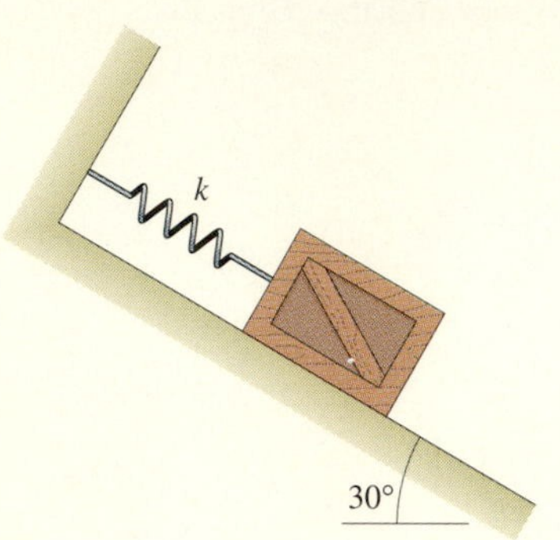

P4.57

4.58 Solve Problem 4.57 if the coefficient of kinetic friction between the crate and the surface is $\mu_k = 0.12$.

4.59 Solve Problem 4.57 if the coefficient of kinetic friction between the crate and the surface is $\mu_k = 0.16$ and the tension in the spring when the crate is released is 20 N.

4.60 The 20-lb collar starts from rest at position 1 with the spring unstretched. The spring constant is $k = 40$ lb/ft. Neglect friction. How far does the collar fall relative to position 1?

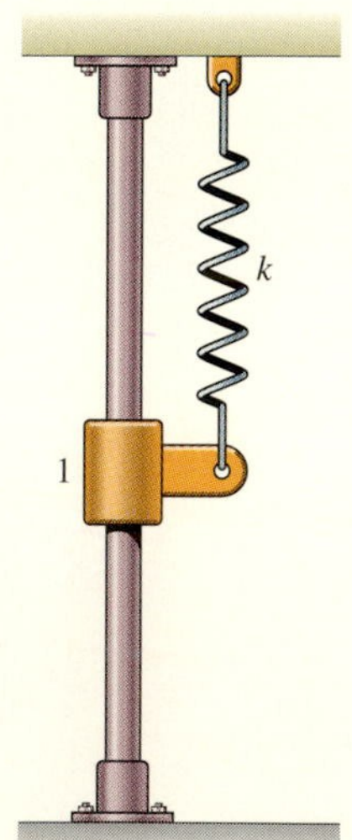

P4.60

4.61 In Problem 4.60, what maximum velocity does the collar attain?

4.62 What is the solution of Problem 4.60 if the tension in the spring in position 1 is 4 lb?

4.63 The 4-kg collar is released from rest at position 1. Neglect friction. If the spring constant is $k = 6$ kN/m and the spring is unstretched in position 2, what is the velocity of the collar when it has fallen to position 2?

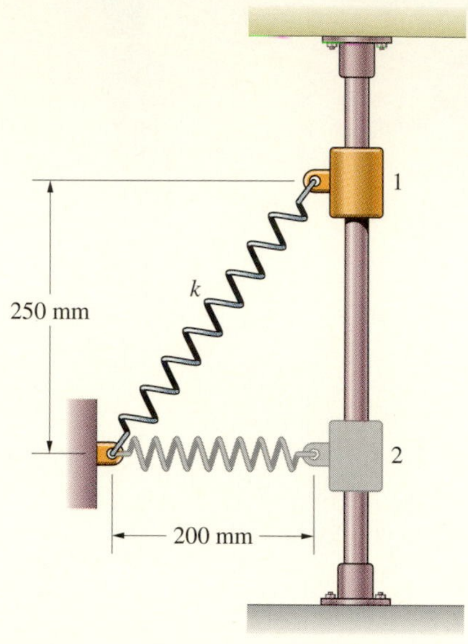

P4.63

4.64 In Problem 4.63, if the spring constant is $k = 4$ kN/m and the tension in the spring in position 2 is 500 N, what is the velocity of the collar when it has fallen to position 2?

4.65 In Problem 4.63, suppose that you don't know the spring constant k. If the spring is unstretched in position 2 and the velocity of the collar when it has fallen to position 2 is 4 m/s, what is k?

4.66 The 10-kg collar starts from rest at position 1 and slides along the smooth bar. The y axis points upward. The spring constant is $k = 100$ N/m and the unstretched length of the spring is 2 m. What is the velocity of the collar when it reaches position 2?

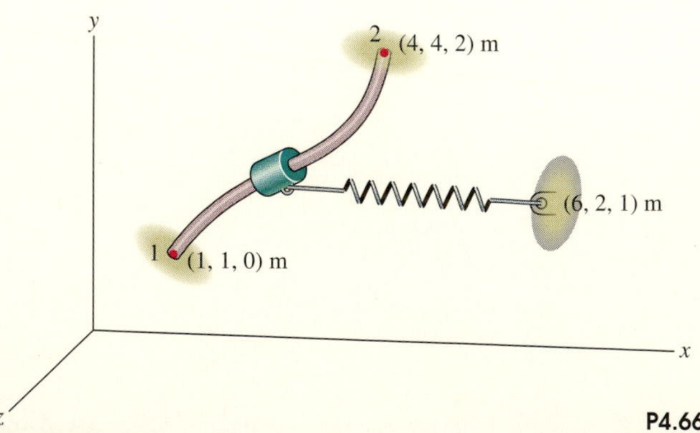

P4.66

4.67 A spring-powered mortar is used to launch 10-lb packages of fireworks into the air. The package starts from rest with the spring compressed to a length of 6 in. The unstretched length of the spring is 30 in. If the spring constant is $k = 1300$ lb/ft, what is the magnitude of the velocity of the package as it leaves the mortar?

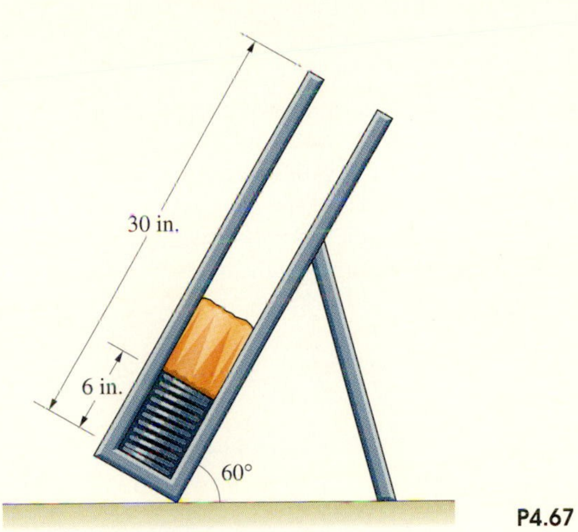

P4.67

4.68 Suppose you want to design the mortar in Problem 4.67 to throw the package to a height of 150 ft above its initial position. Neglecting friction and drag, determine the necessary spring constant.

4.69 Suppose an object has a string or cable with *constant* tension T attached as shown. The force exerted on the object can be expressed in terms of polar coordinates as $\mathbf{F} = -T\,\mathbf{e}_r$. Show that the work done on the object as it moves along an *arbitrary* plane path from a radial position r_1 to a radial position r_2 is $U_{12} = -T(r_2 - r_1)$.

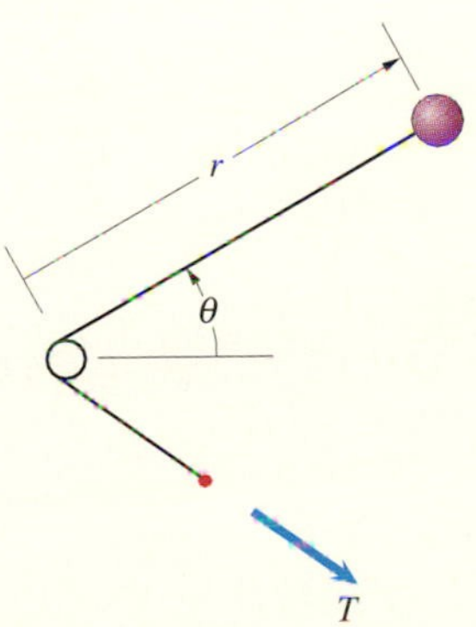

P4.69

4.70 The 2-kg collar is initially at rest at position 1. A constant 100-N force is applied to the string, causing the collar to slide up the smooth vertical bar. What is the velocity of the collar when it reaches position 2?

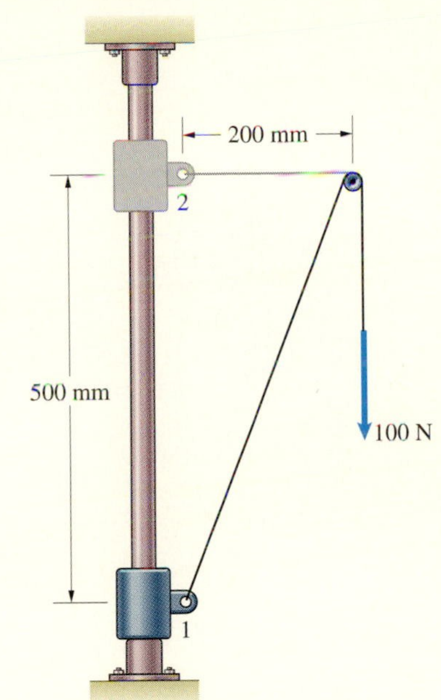

P4.70

4.71 The 10-kg collar starts from rest at position 1. The tension in the string is 200 N, and the y axis points upward. If friction is negligible, what is the magnitude of the collar's velocity when it reaches position 2?

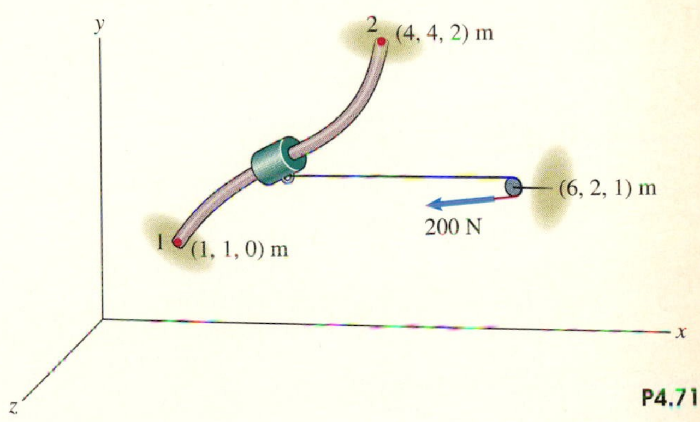

P4.71

4.72 The cable extending from A to B engages the arresting hook of the F/A-18 at C. The arresting mechanism maintains the tension in the cable at a constant value of 880 kN, bringing the 11,800-kg airplane to rest in a distance of 22 m. What was the airplane's initial velocity?

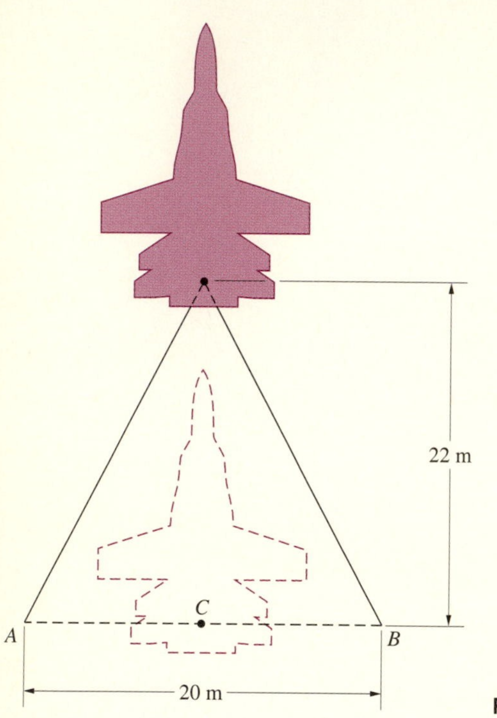

P4.72

4.73 In Problem 4.72, what is the airplane's maximum deceleration? If the 82-kg pilot is subjected to the same deceleration, what maximum force is exerted on him by his restraints?

4.74 A spacecraft 200 mi above the surface of the earth has escape velocity $v_{\text{esc}} = \sqrt{2gR_{\text{E}}^2/r}$, where r is its distance from the center of the earth and $R_{\text{E}} = 3960$ mi is the radius of the earth. What is the magnitude of the spacecraft's velocity when it reaches the moon's orbit 238,000 mi from the center of the earth?

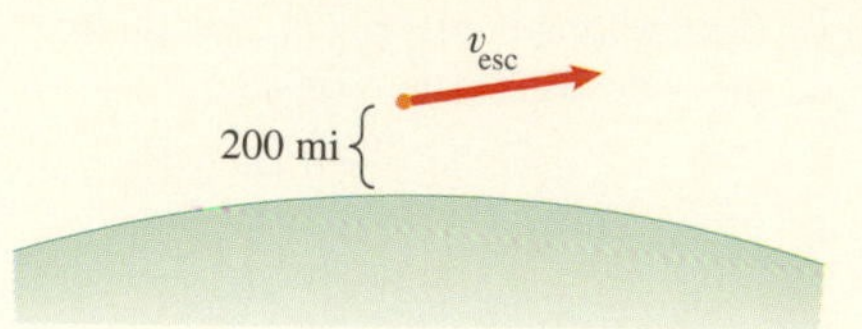

P4.74

4.75 A piece of ejecta thrown up by the impact of a meteor on the moon has a velocity of 200 m/s magnitude, relative to a nonrotating reference frame with its origin at the center of the moon, when it is 1000 km above the moon's surface. What is the magnitude of its velocity just before it strikes the moon's surface? (The acceleration due to gravity at the moon's surface is 1.62 m/s^2, and the moon's radius is 1738 km.)

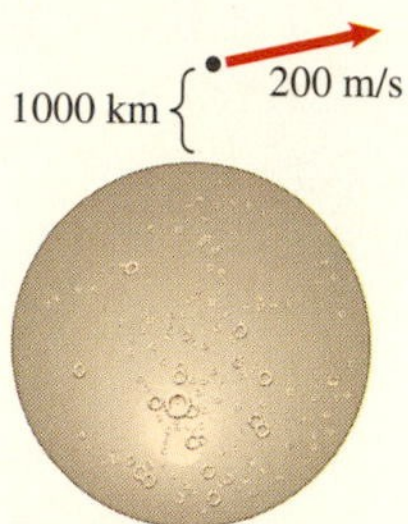

P4.75

4.76 A satellite in a circular orbit of radius r around the earth has velocity $v = \sqrt{gR_{\text{E}}^2/r}$, where $R_E = 6370$ km is the radius of the earth. Suppose you are designing a rocket to transfer a 900-kg communication satellite from a parking orbit with 6700-km radius to a geosynchronous orbit with 42,222-km radius. How much work must the rocket do on the satellite?

4.77 The force exerted on a charged particle by a magnetic field is

$$\mathbf{F} = q\mathbf{v} \times \mathbf{B},$$

where q and $\mathbf{v}$ are the charge and velocity of the particle and $\mathbf{B}$ is the magnetic field vector. If other forces on the particle are negligible, use the principle of work and energy to show that the magnitude of the particle's velocity is constant.

Potential Energy

4.4 Conservation of Energy

The work done on an object by some forces can be expressed as the change of a function of the object's position called the *potential energy.* When all the forces that do work on a system have this property, we can state the principle of work and energy as a conservation law: The sum of the kinetic and potential energies is constant.

When we derived the principle of work and energy by integrating Newton's second law, we were able to evaluate the integral on one side of the equation, obtaining the change in the kinetic energy:

$$U_{12} = \int_{\mathbf{r}_1}^{\mathbf{r}_2} \Sigma \mathbf{F} \cdot d\mathbf{r} = \frac{1}{2}mv_2^2 - \frac{1}{2}mv_1^2. \tag{4.19}$$

Suppose we could determine a scalar function of position V such that

$$\boxed{dV = -\Sigma \mathbf{F} \cdot d\mathbf{r}.} \tag{4.20}$$

Then we could also evaluate the integral defining the work:

$$U_{12} = \int_{\mathbf{r}_1}^{\mathbf{r}_2} \Sigma \mathbf{F} \cdot d\mathbf{r} = \int_{V_1}^{V_2} -dV = -(V_2 - V_1), \tag{4.21}$$

where V_1 and V_2 are the values of V at the positions $\mathbf{r}_1$ and $\mathbf{r}_2$. The principle of work and energy would then have the simple form

$$\frac{1}{2}mv_1^2 + V_1 = \frac{1}{2}mv_2^2 + V_2, \tag{4.22}$$

which means that the sum of the kinetic energy and the function V is constant:

$$\boxed{\frac{1}{2}mv^2 + V = \text{constant.}} \tag{4.23}$$

If the kinetic energy increases, V must decrease, and vice versa, as if V represents a reservoir of "potential" kinetic energy. For this reason, V is called the **potential energy**.

If a potential energy exists for a given force $\mathbf{F}$, which means that a function of position V exists such that $dV = -\mathbf{F} \cdot d\mathbf{r}$, then $\mathbf{F}$ is said to be **conservative**. If all the forces that do work on a system are conservative, the total energy—the sum of the kinetic energy and the potential energies of the forces—is constant, or conserved. In that case, the system is said to be conservative, and you can use conservation of energy instead of the principle of work and energy to relate a change in its position to the change in its kinetic energy. The two approaches are equivalent, and you obtain the same quantitative information. But you gain greater insight by using conservation of energy, because you can interpret the motion of the object or system in terms of transformations between potential and kinetic energies.

4.5 Conservative Forces

You can apply conservation of energy only if the forces doing work on an object or system are conservative and you know (or can determine) their potential energies. In this section, we determine the potential energies of some

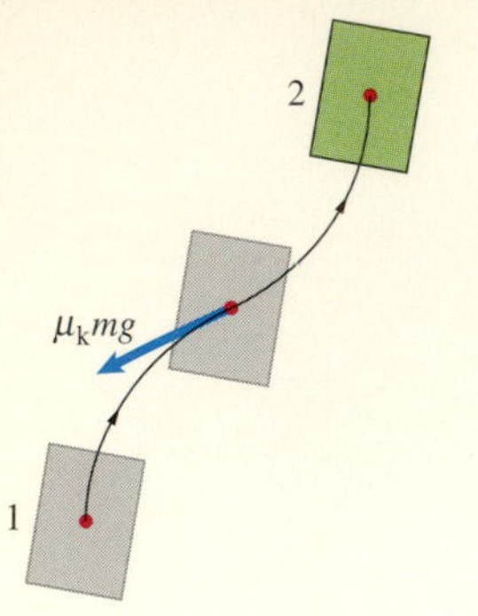

Fig. 4.15

The book's path from position 1 to position 2. The friction force points opposite to the direction of the motion.

conservative forces and use the results to demonstrate applications of conservation of energy. But before discussing forces that are conservative, we demonstrate with a simple example that friction forces are not conservative.

The work done by a conservative force as an object moves from a position 1 to a position 2 is independent of the object's path. This result follows from Eq. (4.21), which states that the work depends only on the values of the potential energy at positions 1 and 2. Equation (4.21) also implies that if the object moves along a closed path, returning to position 1, the work done by a conservative force is zero. Suppose that a book of mass m rests on a table and you push it horizontally so that it slides along a path of length L. The magnitude of the friction force is $\mu_k mg$, and it points opposite to the direction of the book's motion (Fig. 4.15). The work done is

$$U_{12} = \int_0^L -\mu_k mg\, ds = -\mu_k mgL.$$

The work is proportional to the length of the path and therefore is not independent of the object's path. Friction forces are not conservative.

Potential Energies of Particular Forces

The weight of an object and the force exerted by a spring attached to a fixed support are conservative forces. Using them as examples, we demonstrate how you can determine the potential energies of other conservative forces. We also use the potential energies of these forces in examples of the use of conservation of energy to analyze the motions of conservative systems.

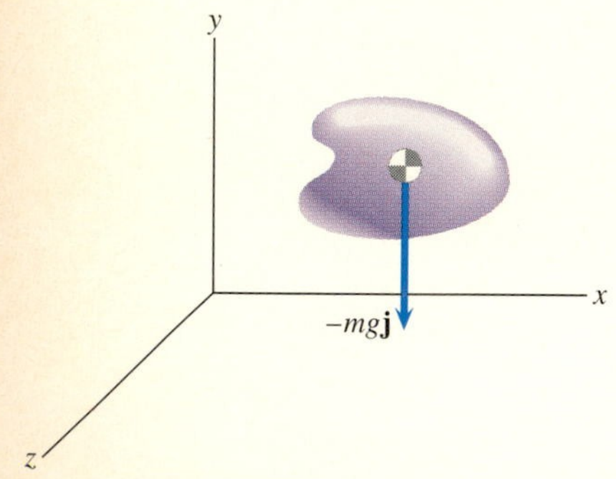

Fig. 4.16

Weight of an object expressed in terms of a coordinate system with the y axis upward.

Weight To determine the potential energy associated with an object's weight, we use a cartesian coordinate system with its y axis upward (Fig. 4.16). The weight is $\mathbf{F} = -mg\mathbf{j}$, and its dot product with the vector $d\mathbf{r}$ is

$$\mathbf{F} \cdot d\mathbf{r} = (-mg\mathbf{j}) \cdot (dx\mathbf{i} + dy\mathbf{j} + dz\mathbf{k}) = -mg\, dy.$$

From Eq. (4.20), the potential energy V must satisfy the relation

$$dV = -\mathbf{F} \cdot d\mathbf{r} = mg\, dy, \tag{4.24}$$

which we can write as

$$\frac{dV}{dy} = mg.$$

Integrating this equation, we obtain

$$V = mgy + C,$$

where C is an integration constant. The constant C is arbitrary, because this expression satisfies Eq. (4.24) for any value of C. Another way of understanding why C is arbitrary is to notice in Eq. (4.22) that it is the *difference* in the potential energy between two positions that determines the change in the kinetic energy. We will let $C = 0$ and write the potential energy of the weight of an object as

$$\boxed{V = mgy.} \tag{4.25}$$

The potential energy is the product of the object's weight and height. The height can be measured from any convenient reference level, or **datum**. Since the difference in potential energy determines the change in the kinetic energy, it is the difference in height that matters, not the level from which the height is measured.

The roller coaster (Fig. 4.17a) is a classic example of conservation of energy. If aerodynamic and friction forces are neglected, the weight is the only force doing work and the system is conservative. The potential energy of the roller coaster is proportional to the height of the track relative to a datum. In Fig. 4.17(b), we assume the roller coaster started from rest at the datum level. The sum of the kinetic and potential energies is constant, so the kinetic energy "mirrors" the potential energy. At points of the track that have equal heights, the magnitudes of the velocities are equal.

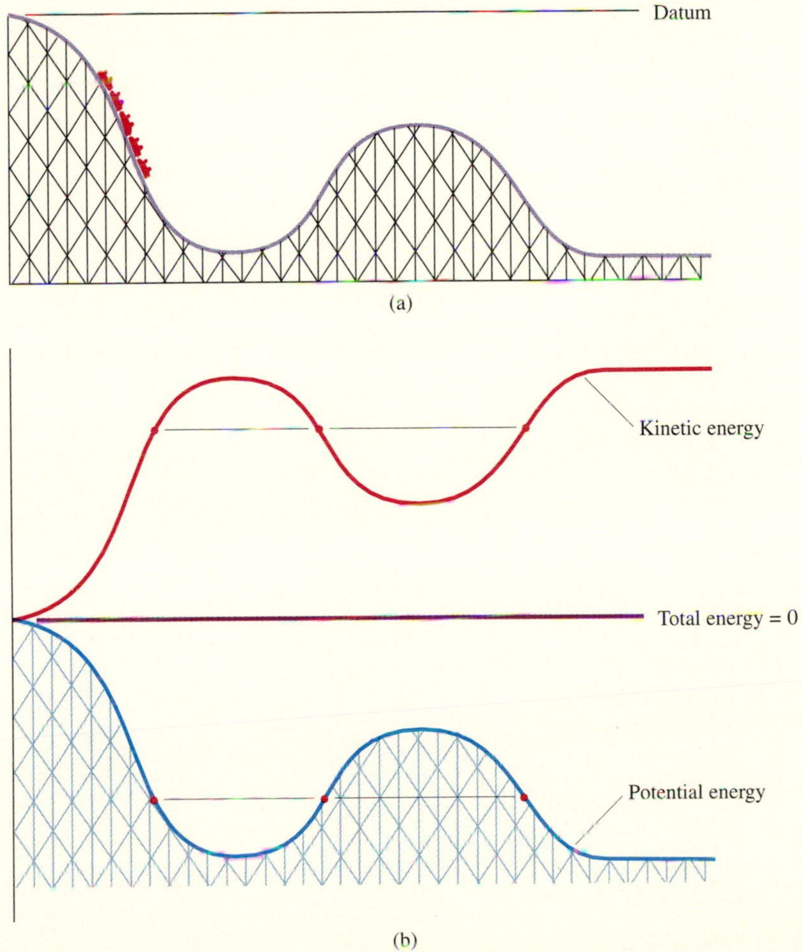

Fig. 4.17

(a) Roller coaster and a reference level, or datum.
(b) The sum of the potential and kinetic energies is constant.

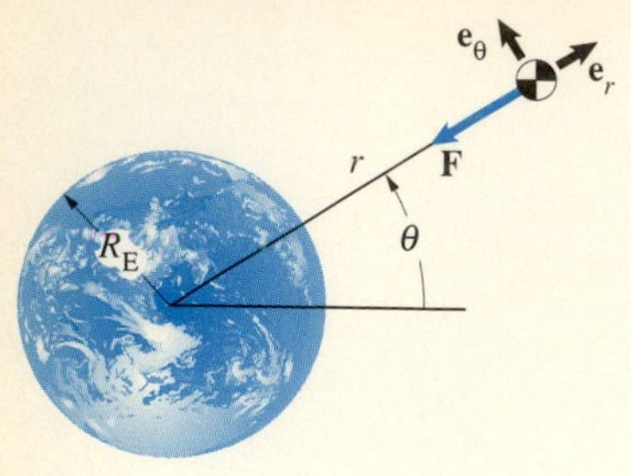

Fig. 4.18
Expressing the weight in terms of polar coordinates.

To account for the variation of the weight with distance from the center of the earth, we can express the weight in polar coordinates as

$$\mathbf{F} = -\frac{mgR_E^2}{r^2}\,\mathbf{e}_r,$$

where r is the distance from the center of the earth (Fig. 4.18). From Eq. (4.16), the vector $d\mathbf{r}$ in terms of polar coordinates is

$$d\mathbf{r} = dr\,\mathbf{e}_r + r\,d\theta\,\mathbf{e}_\theta. \tag{4.26}$$

The potential energy must satisfy

$$dV = -\mathbf{F}\cdot d\mathbf{r} = \frac{mgR_E^2}{r^2}\,dr,$$

or

$$\frac{dV}{dr} = \frac{mgR_E^2}{r^2}.$$

We integrate this equation and let the constant of integration be zero, obtaining the potential energy

$$\boxed{V = -\frac{mgR_E^2}{r}.} \tag{4.27}$$

Springs In terms of polar coordinates, the force exerted on an object by a linear spring is

$$\mathbf{F} = -k(r - r_0)\,\mathbf{e}_r,$$

where r_0 is the unstretched length of the spring (Fig. 4.19). Using Eq. (4.26), the potential energy must satisfy

$$dV = -\mathbf{F}\cdot d\mathbf{r} = k(r - r_0)\,dr.$$

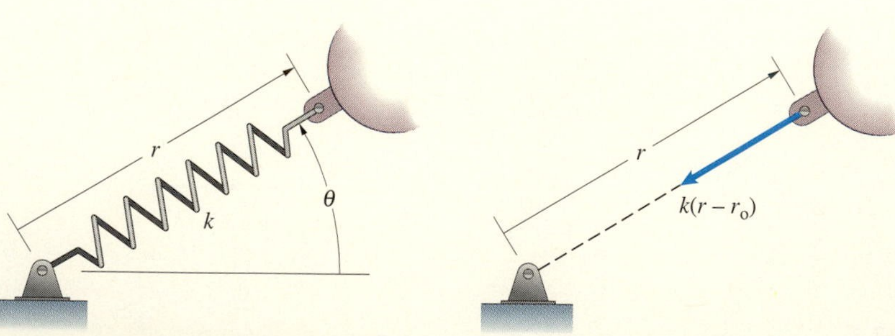

Fig. 4.19
Expressing the force exerted by a linear spring in polar coordinates.

Expressed in terms of the stretch of the spring $S = r - r_0$, this equation is $dV = kS\,dS$, or

$$\frac{dV}{dS} = kS.$$

Integrating this equation, we obtain the potential energy of a linear spring:

$$\boxed{V = \frac{1}{2}kS^2.} \tag{4.28}$$

In the following examples we use conservation of energy to relate changes in the positions of conservative systems to changes in their kinetic energies. This typically involves three steps:

1. Determine whether the system is conservative—*Draw a free-body diagram to identify the forces that do work and confirm that they are conservative.*
2. Determine the potential energy—*Evaluate the potential energies of the forces in terms of the position of the system.*
3. Apply conservation of energy—*Equate the sum of the kinetic and potential energies of the system at two positions to obtain an expression for the change in the kinetic energy.*

Example 4.8

In Example 4.5, the 40-kg hammer is lifted into position 1 and released from rest. Its weight and the two springs ($k = 1500$ N/m) accelerate the hammer downward to position 2, where it strikes a workpiece. Use conservation of energy to determine the hammer's velocity when it reaches position 2.

SOLUTION

Determine Whether the System Is Conservative From the free-body diagram of the hammer (Fig. a), we see that work is done only by its weight and the forces exerted by the springs. The system is conservative.

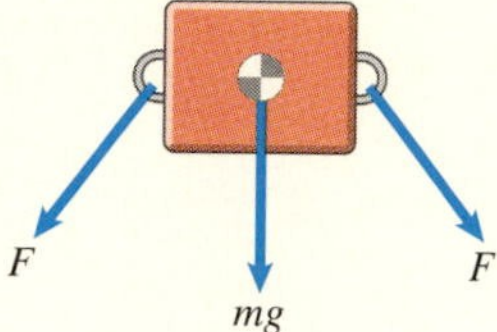

(a) Free-body diagram of the hammer.

Determine the Potential Energy The potential energy of each spring is $\frac{1}{2}kS^2$, where S is the stretch, so the total potential energy of the two springs is

$$V_{\text{springs}} = 2\left(\frac{1}{2}kS^2\right).$$

In Example 4.5 the stretches in positions 1 and 2 were determined to be $S_1 = 0.3$ m, $S_2 = 0.1$ m. The potential energy associated with the weight is

$$V_{\text{weight}} = mgy,$$

where y is the height relative to a convenient datum (Fig. b).

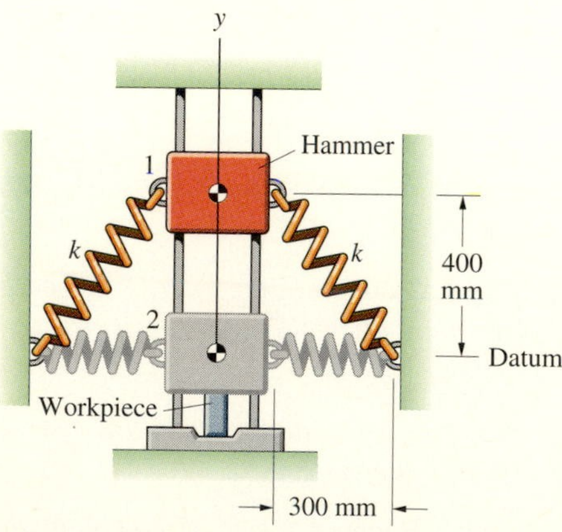

(b) Measuring the height of the hammer relative to position 2.

Apply Conservation of Energy The sums of the potential and kinetic energies at positions 1 and 2 must be equal:

$$2\left(\frac{1}{2}kS_1^2\right) + mgy_1 + \frac{1}{2}mv_1^2 = 2\left(\frac{1}{2}kS_2^2\right) + mgy_2 + \frac{1}{2}mv_2^2:$$

$$(1500)(0.3)^2 + (40)(9.81)(0.4) + 0 = (1500)(0.1)^2 + 0 + \frac{1}{2}(40)v_2^2.$$

Solving this equation, we obtain $v_2 = 3.72$ m/s.

DISCUSSION

From the graphs of the total potential energy associated with the springs and the weight and the kinetic energy of the hammer as functions of y (Fig. 4.20), you can see the transformation of the potential energy into kinetic energy as the hammer falls. Notice that the total energy of the conservative system remains constant.

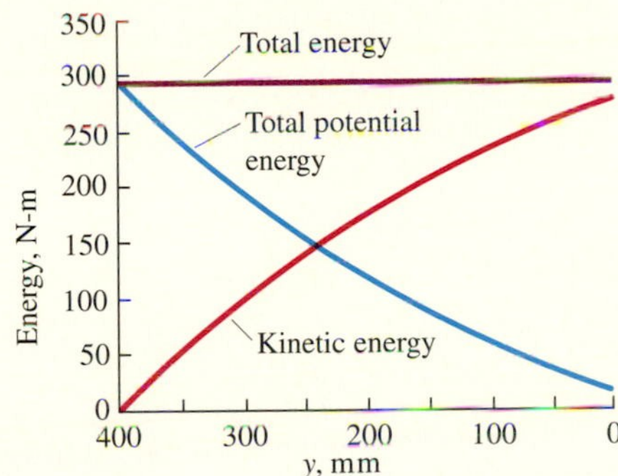

Fig. 4.20

The potential and kinetic energies as functions of the y coordinate of the hammer.

Example 4.9

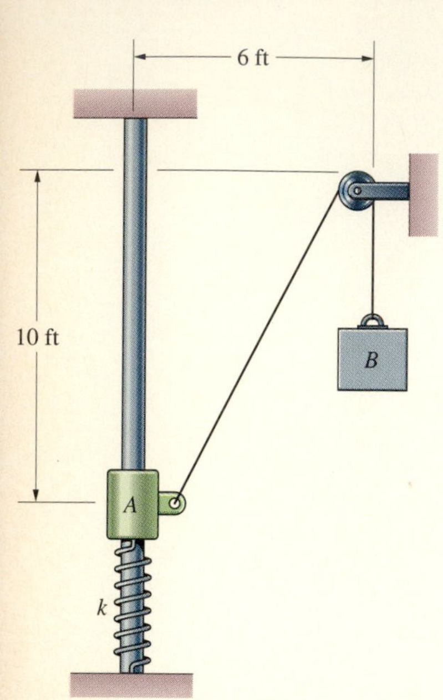

Fig. 4.21

The spring in Fig. 4.21 ($k = 20$ lb/ft) is connected to the floor and to the 200-lb collar A. Collar A is at rest, supported by the spring, when the 300-lb box B is released from rest in the position shown. What are the velocities of the collar and box when the box B has fallen 2 ft?

SOLUTION

Determine Whether the System Is Conservative Let us consider the collar A, the box B, and the pulley as a single system. From the free-body diagram (Fig. a), we see that work is done only by the weights of the collar and box and the spring force F. The system is conservative.

Determine the Potential Energy Using its initial position as the datum, the potential energy associated with the weight of the collar when it has risen a distance x_A (Fig. b) is $V_A = W_A x_A$. Using the initial position of box B as its datum, the potential energy associated with its weight when it has fallen a distance x_B is $V_B = -W_B x_B$. (The minus sign is necessary because we define x_B to be positive downward.)

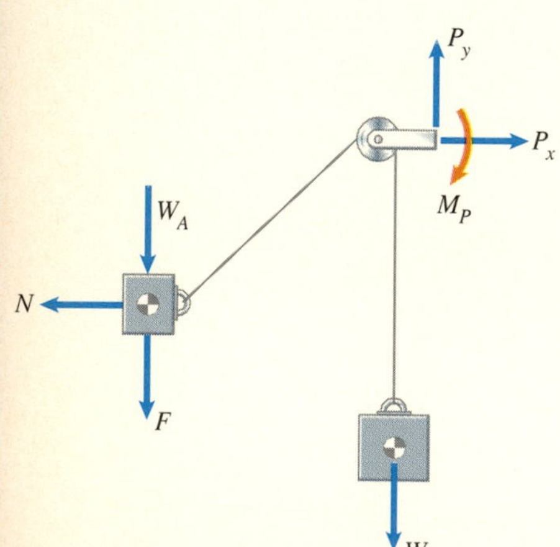

(a) Free-body diagram of the system.

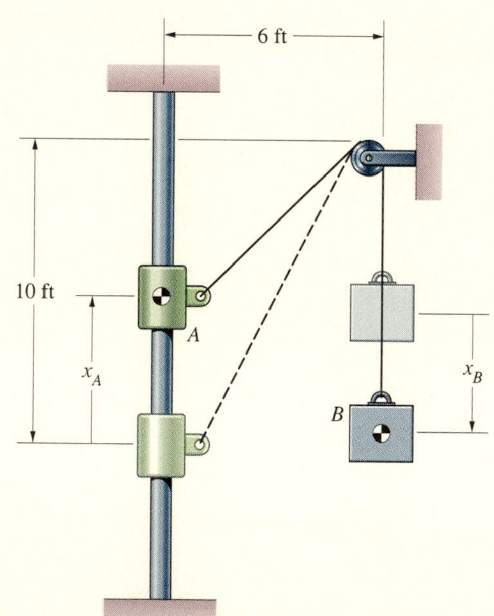

(b) Displacements of the collar and box.

To determine the potential energy associated with the spring force, we need to account for the fact that in position 1 the spring is compressed by the weight of the collar. The spring is initially compressed a distance δ such that $W_A = k\delta$ (Fig. c). When the collar has moved upward a distance x_A, the stretch of the spring is $S = x_A - \delta = x_A - W_A/k$, so its potential energy is

$$V_S = \frac{1}{2}kS^2 = \frac{1}{2}k\left(x_A - \frac{W_A}{k}\right)^2.$$

The potential energy of the system in terms of the displacements of the collar and box is

$$V = V_A + V_B + V_S = W_A x_A - W_B x_B + \frac{1}{2}k\left(x_A - \frac{W_A}{k}\right)^2.$$

Apply Conservation of Energy The sum of the kinetic and potential energies of the system in its initial position and in the position shown in Fig. b must be equal. Denoting the total kinetic energy by T,

$$T_1 + V_1 = T_2 + V_2:$$

$$0 + \frac{1}{2}k\left(-\frac{W_A}{k}\right)^2 = \frac{1}{2}m_A v_A^2 + \frac{1}{2}m_B v_B^2 \qquad (4.29)$$

$$+ W_A x_A - W_B x_B + \frac{1}{2}k\left(x_A - \frac{W_A}{k}\right)^2.$$

We want to determine v_A and v_B when $x_B = 2$ ft, but we have only one equation in terms of x_A, x_B, v_A, and v_B. To complete the solution, we must relate the displacement and velocity of the collar to the displacement and velocity of the box.

From Fig. b, the decrease in the length of the rope from A to the pulley as the collar rises must equal the distance the box falls:

$$\sqrt{(10)^2 + (6)^2} - \sqrt{(10 - x_A)^2 + (6)^2} = x_B.$$

We solve this equation for the value of x_A when $x_B = 2$ ft, obtaining $x_A = 2.43$ ft. By taking the derivative of this equation with respect to time, we also obtain a relation between v_A and v_B:

$$\left[\frac{10 - x_A}{\sqrt{(10 - x_A)^2 + (6)^2}}\right] v_A = v_B.$$

At $x_A = 2.43$ ft, we determine from this equation that

$$v_B = 0.784 v_A.$$

We solve this equation together with Eq. (4.29) for the velocities of the collar and box when $x_B = 2$ ft and $x_A = 2.43$ ft, obtaining $v_A = 3.04$ ft/s and $v_B = 2.38$ ft/s.

DISCUSSION

Why didn't we have to consider the forces exerted on the collar and box by the rope? The reason is that they are internal forces with respect to the system consisting of the collar, box, and pulley.

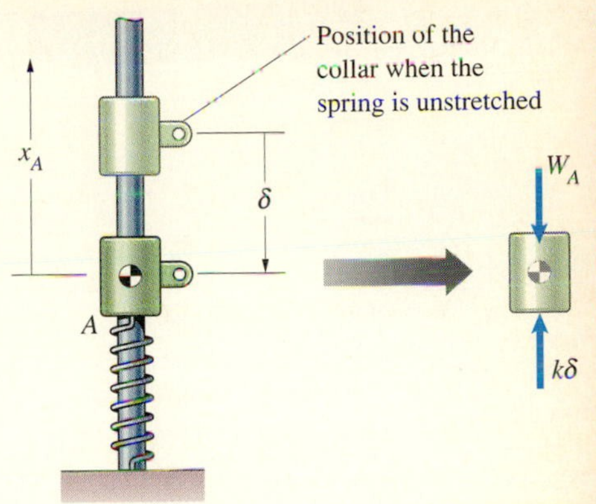

(c) Determining the initial compression of the spring.

Example 4.10

Fig. 4.22

A spacecraft at a distance $r_0 = 2R_E$ from the center of the earth is moving outward with initial velocity $v_0 = \sqrt{2gR_E/3}$ (Fig. 4.22). Determine its velocity as a function of its distance from the center of the earth.

SOLUTION

Determine Whether the System Is Conservative If we assume that work is done on the spacecraft by gravity alone, the system is conservative.

Determine the Potential Energy The potential energy associated with the spacecraft's weight is given in terms of its distance r from the center of the earth by Eq. (4.27):

$$V = -\frac{mgR_E^2}{r}.$$

Apply Conservation of Energy Let v be the magnitude of the spacecraft's velocity at an arbitrary distance r. The sums of the potential and kinetic energies at r_0 and at r must be equal:

$$-\frac{mgR_E^2}{r_0} + \frac{1}{2}mv_0^2 = -\frac{mgR_E^2}{r} + \frac{1}{2}mv^2:$$

$$-\frac{mgR_E^2}{2R_E} + \frac{1}{2}m\left(\frac{2}{3}gR_E\right) = -\frac{mgR_E^2}{r} + \frac{1}{2}mv^2.$$

Solving for v, the spacecraft's velocity as a function of r is

$$v = \sqrt{gR_E\left(\frac{2R_E}{r} - \frac{1}{3}\right)}.$$

DISCUSSION

We show graphs of the kinetic energy, potential energy, and total energy as functions of r/R_E in Fig. 4.23. The kinetic energy decreases and the potential energy increases as the spacecraft moves outward until its velocity decreases to zero at $r = 6R_E$.

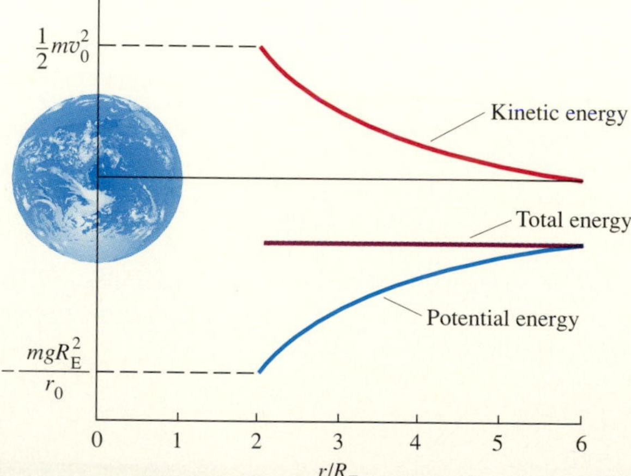

Fig. 4.23
Energies as functions of the radial coordinate.

Example 4.11

The bore of the gas gun shown in Fig. 4.24 has cross-sectional area A. It is evacuated on the right of the projectile of mass m and on the left it contains gas at pressure p. Let the value of the pressure when $s = s_0$ be p_0, and assume that the pressure of the gas is related to its volume V by $pV =$ constant.

(a) Determine the potential energy associated with the force exerted on the projectile in terms of s.

(b) If the projectile starts from rest at $s = s_0$ and friction is negligible, what is its velocity as a function of s?

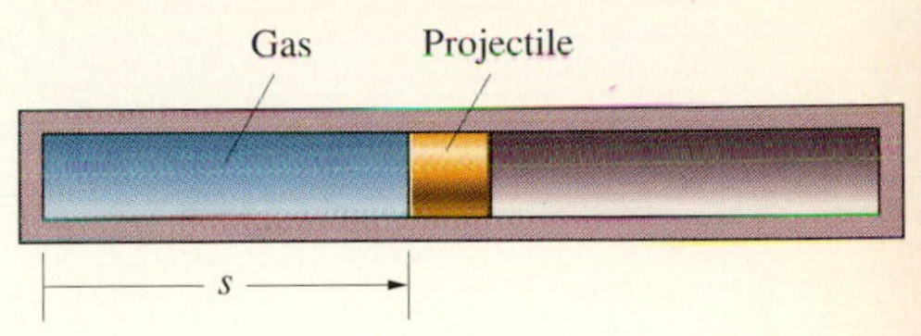

Fig. 4.24

STRATEGY

(a) We can determine the force exerted on the piston by the pressure of the gas as a function of s, then use Eq. (4.20) to determine the potential energy.

(b) Knowing the potential energy, we can use conservation of energy to determine the velocity of the projectile as a function of s.

SOLUTION

(a) The volume of the gas to the left of the projectile is $V = sA$. We know that pV is constant, so psA is constant. The value of the pressure when $s = s_0$ is p_0, so $psA = p_0 s_0 A$. The force F exerted on the projectile is the product of the pressure p and the cross-sectional area A of the projectile. Therefore

$$F = pA = \frac{p_0 s_0 A}{s}.$$

To apply Eq. (4.20), we need to express the total force on the piston and an infinitesimal displacement $d\mathbf{r}$ of the piston as vectors. Let $\mathbf{e}_t$ be a unit vector that points to the right. Neglecting friction, the total force on the piston is

$$\Sigma \mathbf{F} = F\,\mathbf{e}_t = \frac{p_0 s_0 A}{s}\,\mathbf{e}_t,$$

and we can express $d\mathbf{r}$ as

$$d\mathbf{r} = ds\,\mathbf{e}_t.$$

Substituting these expressions into Eq. (4.20), we obtain

$$dV = -\Sigma \mathbf{F} \cdot d\mathbf{r} = -\left(\frac{p_0 s_0 A}{s}\,\mathbf{e}_t\right) \cdot (ds\,\mathbf{e}_t) = -\frac{p_0 s_0 A}{s}\,ds,$$

which we can write as

$$\frac{dV}{ds} = -\frac{p_0 s_0 A}{s}.$$

We integrate this equation, setting the constant of integration equal to zero, to obtain the potential energy:

$$V = -p_0 s_0 A \ln s.$$

(b) If the projectile starts from rest at $s = s_0$, we can use conservation of energy to determine its velocity v at an arbitrary position s:

$$\frac{1}{2}mv_1^2 + V_1 = \frac{1}{2}mv_2^2 + V_2: \quad 0 - p_0 s_0 A \ln s_0 = \frac{1}{2}mv^2 - p_0 s_0 A \ln s.$$

Solving for the velocity, we obtain

$$v = \sqrt{\frac{2p_0 s_0 A}{m} \ln\left(\frac{s}{s_0}\right)}.$$

Relationships Between Force and Potential Energy

Here we consider two questions: (1) Given a potential energy, how can you determine the corresponding force? (2) Given a force, how can you determine whether it is conservative? That is, how can you tell whether an associated potential energy exists?

The potential energy V of a force $\mathbf{F}$ is a function of position that satisfies the relation

$$dV = -\mathbf{F} \cdot d\mathbf{r}. \tag{4.30}$$

If we express V in terms of a cartesian coordinate system,

$$V = V(x, y, z),$$

its differential dV is

$$dV = \frac{\partial V}{\partial x}dx + \frac{\partial V}{\partial y}dy + \frac{\partial V}{\partial z}dz. \tag{4.31}$$

Expressing $\mathbf{F}$ and $d\mathbf{r}$ in terms of their cartesian components, their dot product is

$$\begin{aligned} \mathbf{F} \cdot d\mathbf{r} &= (F_x\mathbf{i} + F_y\mathbf{j} + F_z\mathbf{k}) \cdot (dx\mathbf{i} + dy\mathbf{j} + dz\mathbf{k}) \\ &= F_x\,dx + F_y\,dy + F_z\,dz. \end{aligned}$$

Substituting this expression and Eq. (4.31) into Eq. (4.30), we obtain

$$\frac{\partial V}{\partial x}dx + \frac{\partial V}{\partial y}dy + \frac{\partial V}{\partial z}dz = -(F_x\,dx + F_y\,dy + F_z\,dz),$$

which implies that

$$F_x = -\frac{\partial V}{\partial x}, \qquad F_y = -\frac{\partial V}{\partial y}, \qquad F_z = -\frac{\partial V}{\partial z}. \tag{4.32}$$

Given a potential energy V expressed in cartesian coordinates, you can use these relations to determine the corresponding force. The force $\mathbf{F}$ is

$$\mathbf{F} = -\left(\frac{\partial V}{\partial x}\mathbf{i} + \frac{\partial V}{\partial y}\mathbf{j} + \frac{\partial V}{\partial z}\mathbf{k}\right) = -\nabla V, \tag{4.33}$$

where ∇V is the **gradient** of V. By using expressions for the gradient in terms of other coordinate systems, you can determine the force $\mathbf{F}$ when you know the potential energy in terms of those coordinate systems. For example, in terms of cylindrical coordinates,

$$\mathbf{F} = -\left(\frac{\partial V}{\partial r}\mathbf{e}_r + \frac{1}{r}\frac{\partial V}{\partial \theta}\mathbf{e}_\theta + \frac{\partial V}{\partial z}\mathbf{e}_z\right). \tag{4.34}$$

If a force $\mathbf{F}$ is conservative, its **curl** $\nabla \times \mathbf{F}$ is zero. The expression for the curl of $\mathbf{F}$ in cartesian coordinates is

$$\nabla \times \mathbf{F} = \begin{vmatrix} \mathbf{i} & \mathbf{j} & \mathbf{k} \\ \dfrac{\partial}{\partial x} & \dfrac{\partial}{\partial y} & \dfrac{\partial}{\partial z} \\ F_x & F_y & F_z \end{vmatrix}. \tag{4.35}$$

Substituting Eqs. (4.32) into this expression confirms that $\nabla \times \mathbf{F} = 0$ when $\mathbf{F}$ is conservative. *The converse is also true:* A force $\mathbf{F}$ is conservative if its curl is zero. You can use this condition to determine whether a given force is conservative. In terms of cylindrical coordinates, the curl of $\mathbf{F}$ is

$$\nabla \times \mathbf{F} = \frac{1}{r}\begin{vmatrix} \mathbf{e}_r & r\mathbf{e}_\theta & \mathbf{e}_z \\ \dfrac{\partial}{\partial r} & \dfrac{\partial}{\partial \theta} & \dfrac{\partial}{\partial z} \\ F_r & rF_\theta & F_z \end{vmatrix}. \tag{4.36}$$

Example 4.12

From Eq. (4.27), the potential energy associated with the weight of an object of mass m at a distance r from the center of the earth is (in polar coordinates)

$$V = -\frac{mgR_{\mathrm{E}}^2}{r},$$

where R_E is the radius of the earth. Use this expression to determine the force exerted on the object by its weight.

STRATEGY

The force $\mathbf{F} = -\nabla V$. The potential energy is expressed in terms of polar coordinates, so we can use Eq. (4.34) to determine the force.

SOLUTION

The partial derivatives of V with respect to r, θ, and z are

$$\frac{\partial V}{\partial r} = \frac{mgR_{\mathrm{E}}^2}{r^2}, \qquad \frac{\partial V}{\partial \theta} = 0, \qquad \frac{\partial V}{\partial z} = 0.$$

From Eq. (4.34), the force is

$$\mathbf{F} = -\nabla V = -\frac{mgR_{\mathrm{E}}^2}{r^2}\mathbf{e}_r.$$

DISCUSSION

We already know that the force is conservative, because we know its potential energy, but we can use Eq. (4.36) to confirm that its curl is zero:

$$\nabla \times \mathbf{F} = \frac{1}{r}\begin{vmatrix} \mathbf{e}_r & r\mathbf{e}_\theta & \mathbf{e}_z \\ \dfrac{\partial}{\partial r} & \dfrac{\partial}{\partial \theta} & \dfrac{\partial}{\partial z} \\ -\dfrac{mgR_{\mathrm{E}}^2}{r^2} & 0 & 0 \end{vmatrix} = 0.$$

Although we used cylindrical coordinates in determining $\mathbf{F}$ and in evaluating the cross product, the expression for V and our resulting expression for $\mathbf{F}$ are valid only if the object remains in the plane $z = 0$.

Problems

4.78 Suppose that you throw a ball straight up at 3 m/s, releasing it 2 m above the ground. (a) Use the principle of work and energy to determine the ball's velocity just before it hits the ground. (b) After confirming that the system is conservative, use conservation of energy to determine the ball's velocity just before it hits the ground.

4.79 Suppose that you kick a soccer ball straight up. When it leaves your foot, it is 3 ft above the ground and moving at 40 ft/s. Neglecting drag, use conservation of energy to determine how high above the ground the ball goes and how fast it will be going just before it hits the ground. Obtain the answers by expressing the potential energy in terms of a datum (a) at the level of the ball's initial position; (b) at ground level.

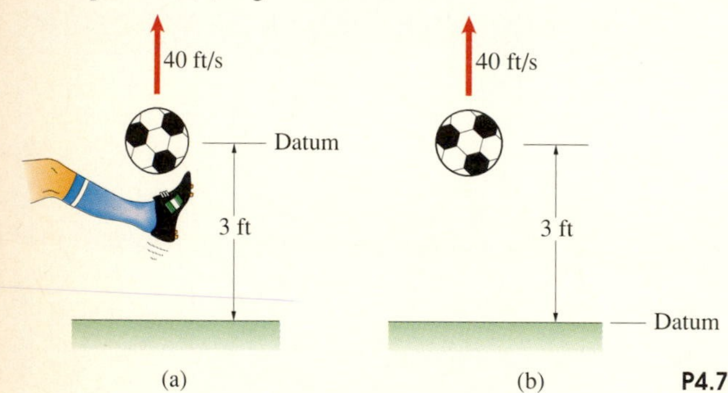

P4.79

4.80 The lunar module could make a safe landing if its vertical velocity at impact is 5 m/s or less. Suppose that you want to determine the greatest height h at which the pilot could shut off the engine if the velocity of the lander relative to the surface was (a) zero; (b) 2 m/s downward; (c) 2 m/s upward. Use conservation of energy to determine h in each case. The acceleration due to gravity at the surface of the moon is 1.62 m/s^2.

P4.80

4.81 The 2-kg collar starts from rest at position 1 and slides down the smooth rigid wire. The y axis points upward. Use conservation of energy to determine the magnitude of the collar's velocity as a function of its y coordinate.

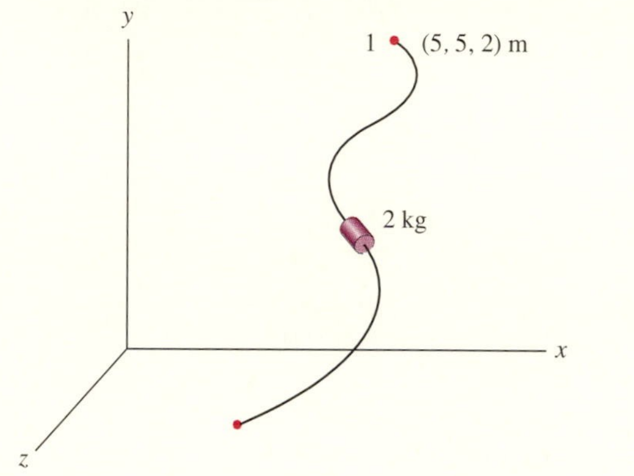

P4.81

4.82 The spring constant $k = 40$ N/m and the mass $m = 12$ kg. The surface is smooth. With the spring unstretched, the mass is given an initial velocity $v_1 = 2$ m/s to the right. The mass moves to the right, stretching the spring, until its velocity decreases to zero and it begins moving back toward its initial position. Determine the sum of the kinetic energy of the mass and the potential energy of the spring: (a) immediately after the mass is given its initial velocity; (b) at the instant the velocity of the mass has decreased to zero; (c) at the instant the mass returns to its original position.

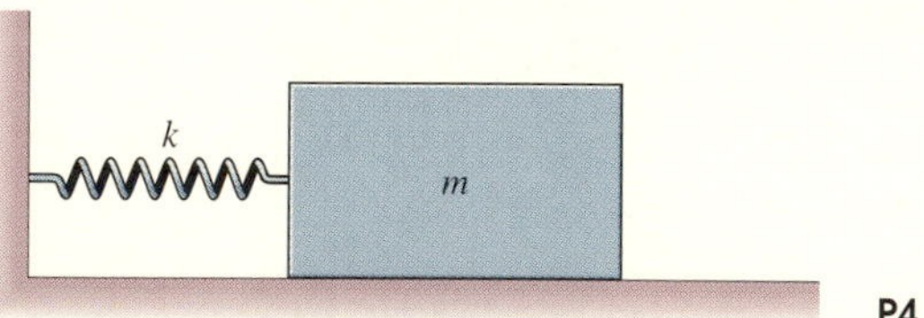

P4.82

4.83 Solve Problem 4.82 if the coefficient of kinetic friction between the mass and the surface is $\mu_k = 0.1$.

4.84 At the instant shown, the 50-lb weight is moving downward at 4 ft/s. Let d be the downward displacement of the 50-lb weight relative to its present position. Use conservation of energy to determine the magnitude of its velocity as a function of d.

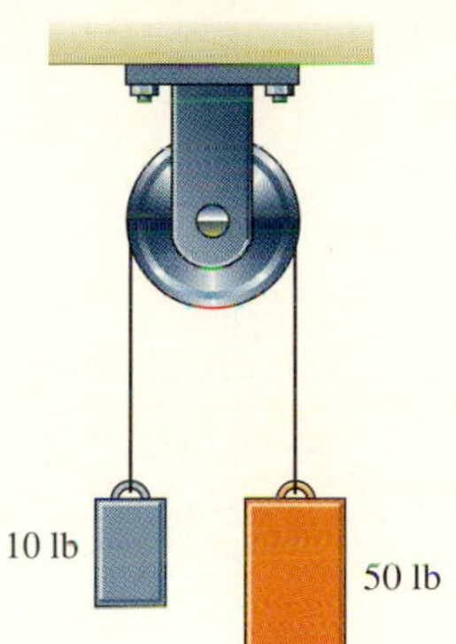

P4.84

4.85 The ball is released from rest in position 1.
(a) Use conservation of energy to determine the magnitude of its velocity at position 2.
(b) Draw graphs of the kinetic energy, the potential energy, and the total energy for values of α from zero to 180°.

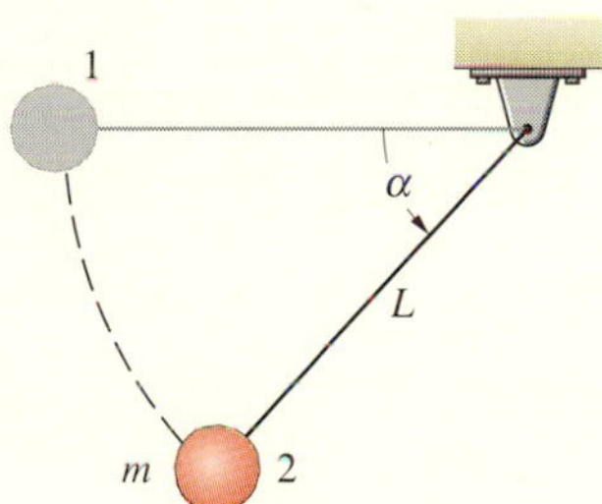

P4.85

4.86 If the ball is released from rest in position 1, use conservation of energy to determine the minimum initial angle α necessary for it to swing to position 2.

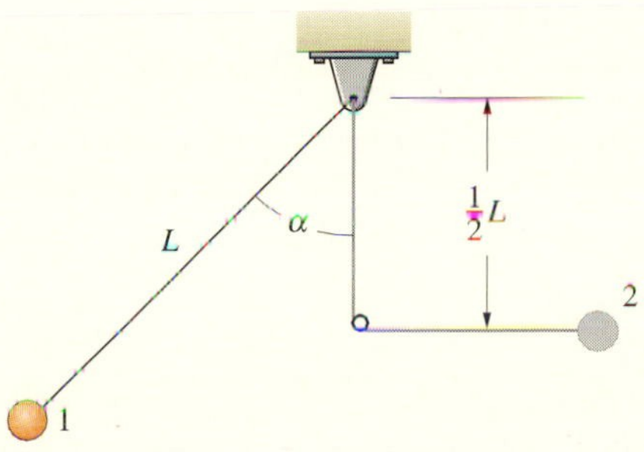

P4.86

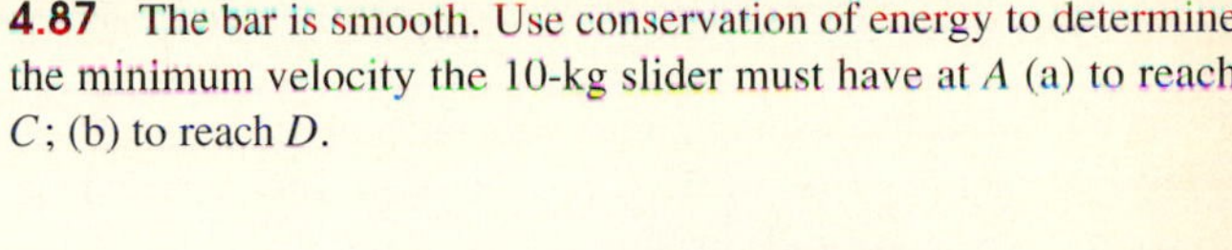

4.87 The bar is smooth. Use conservation of energy to determine the minimum velocity the 10-kg slider must have at A (a) to reach C; (b) to reach D.

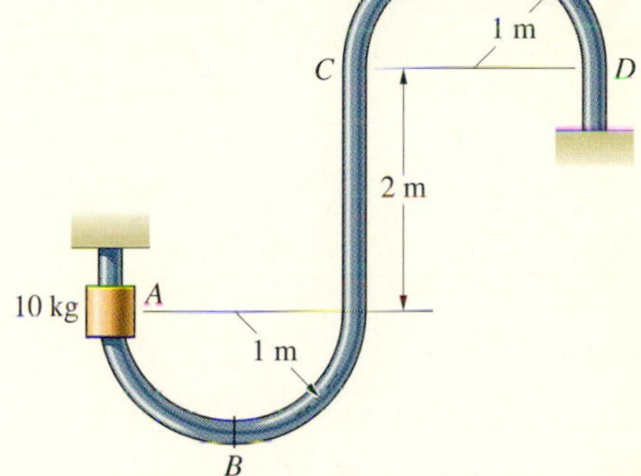

P4.87

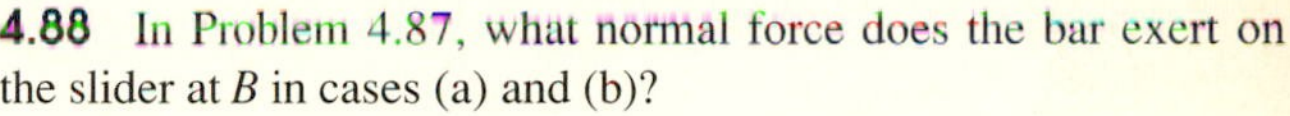

4.88 In Problem 4.87, what normal force does the bar exert on the slider at B in cases (a) and (b)?

4.89 The 10-kg collar starts from rest at position 1 and slides along the bar. The y axis points upward. The spring constant is $k = 100$ N/m, and the unstretched length of the spring is 2 m. Use conservation of energy to determine the magnitude of the collar's velocity when it reaches position 2.

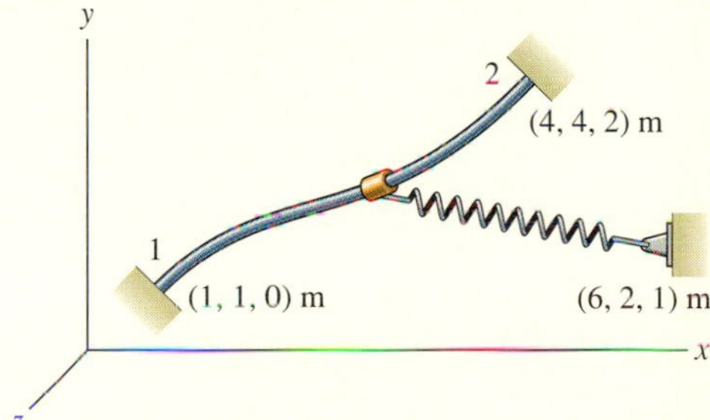

P4.89

4.90 A rock climber of weight W has a rope attached a distance h below him for protection. Suppose that he falls, and assume that the rope behaves like a linear spring with unstretched length h and spring constant $k = C/h$, where C is a constant. Use conservation of energy to determine the maximum force exerted on him by the rope. (Notice that the maximum force is independent of h, which is a reassuring result for climbers—the maximum force resulting from a long fall is the same as that resulting from a very short one.)

P4.90

4.91 The spring constant $k = 700$ N/m, $m_A = 14$ kg, and $m_B = 18$ kg. The collar A slides on the smooth horizontal bar. The system is released from rest with the spring unstretched. Use conservation of energy to determine the velocity of the collar A when it has moved 0.2 m to the right.

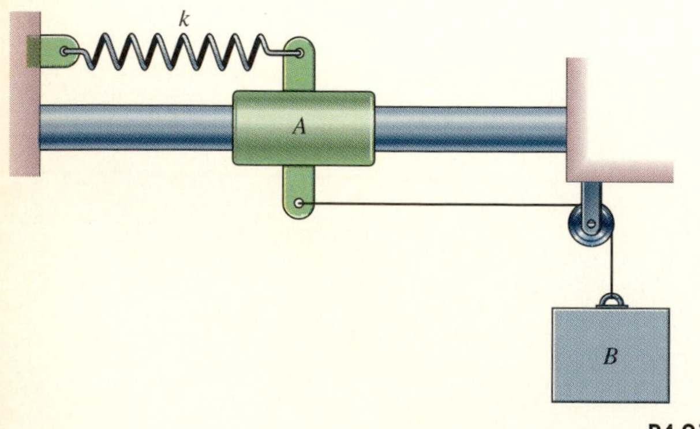

P4.91

4.92 The spring constant $k = 700$ N/m, $m_A = 14$ kg, and $m_B = 18$ kg. The collar A slides on the smooth horizontal bar. The system is released from rest with the spring unstretched. Use conservation of energy to determine the velocity of the collar A when it has moved 0.2 m to the right.

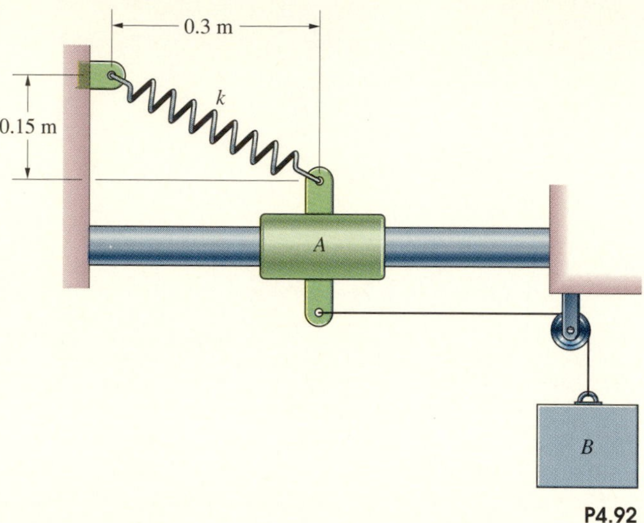

P4.92

4.93 The 5-lb collar starts from rest at A and slides along the semicircular bar. The spring constant is $k = 100$ lb/ft, and the unstretched length of the spring is 1 ft. Use conservation of energy to determine the velocity of the collar at B.

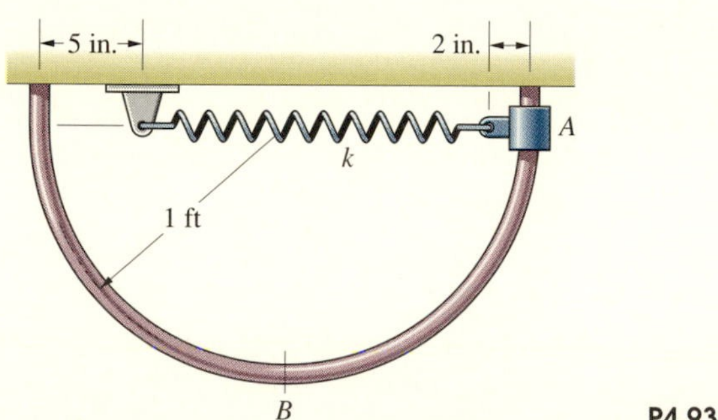

P4.93

4.94 The mass $m = 1$ kg, the spring constant $k = 200$ N/m, and the unstretched length of the spring is 0.1 m. When the system is released from rest in the position shown, the spring contracts, pulling the mass to the right. Use conservation of energy to determine the magnitude of the velocity of the mass when the string and the spring are parallel.

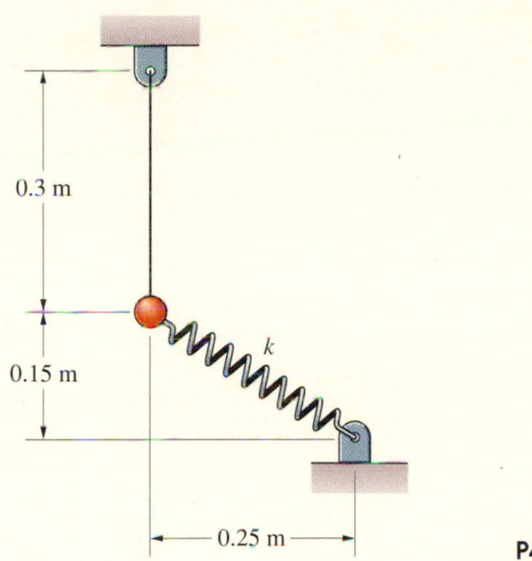

P4.94

4.95 In Problem 4.94, what is the tension in the string when the string and spring are parallel?

4.96 The force exerted on an object by a *nonlinear* spring is

$$\mathbf{F} = -[k(r - r_0) + q(r - r_0)^3]\mathbf{e}_r,$$

where k and q are constants and r_0 is the unstretched length. Determine the potential energy of the spring in terms of its stretch $S = r - r_0$.

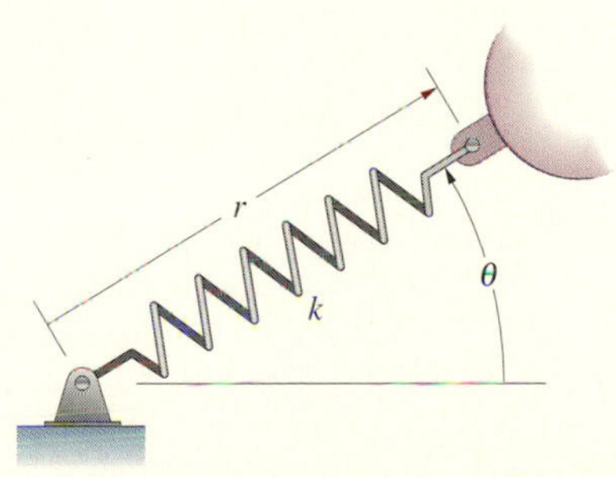

P4.96

4.97 The 20-kg cylinder is released at the position shown and falls onto the linear spring ($k = 3000$ N/m). Use conservation of energy to determine how far down the cylinder moves after contacting the spring.

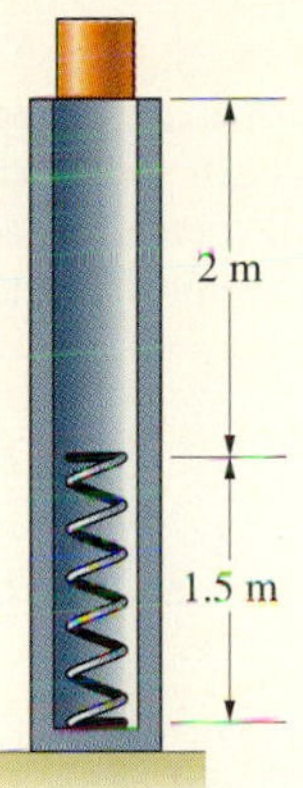

P4.97

4.98 Suppose that the spring in Problem 4.97 is a *nonlinear* spring with potential energy $V = \frac{1}{2}kS^2 + \frac{1}{4}qS^4$, where $k = 3000$ N/m and $q = 4000$ N/m^3. What is the velocity of the cylinder when the spring has been compressed 0.5 m?

4.99 The string exerts a force of constant magnitude T on the object. Determine the potential energy associated with this force in terms of polar coordinates.

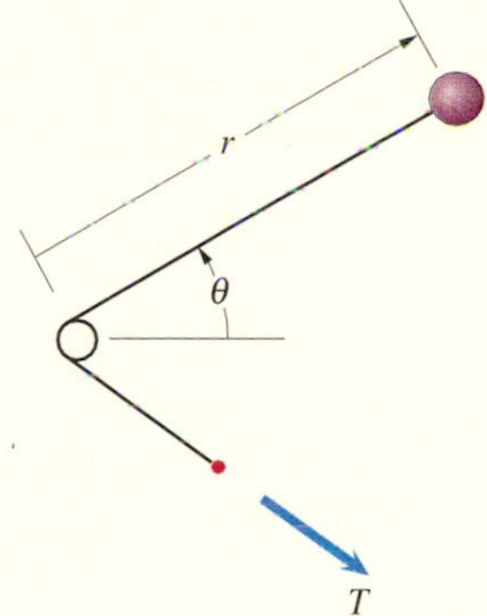

P4.99

4.100 The system is at rest in the position shown, with the 12-lb collar A resting on the spring ($k = 20$ lb/ft), when a constant 30-lb force is applied to the cable. What is the velocity of the collar when it has risen 1 ft?

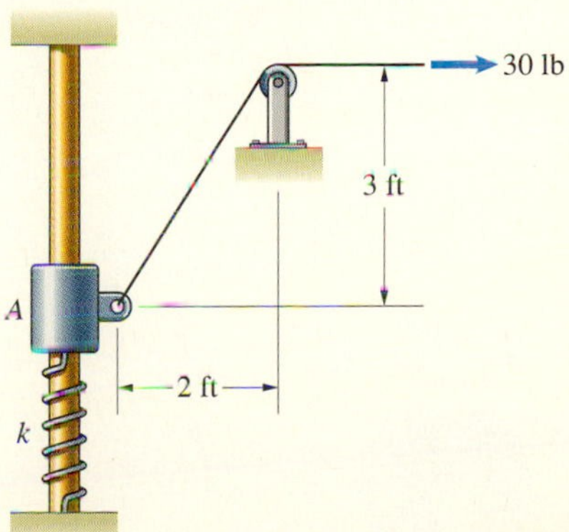

P4.100

4.101 The cable extending from A to B engages the arresting hook of the airplane at C. The arresting mechanism maintains the tension in the cable at a constant value T. The airplane's mass is m. (a) Derive a potential energy V associated with the force exerted on the airplane as a function of its displacement s. (b) The airplane is moving at velocity v_0 when it engages the arresting hook. Use conservation of energy to determine the distance s_0 required to bring it to rest.

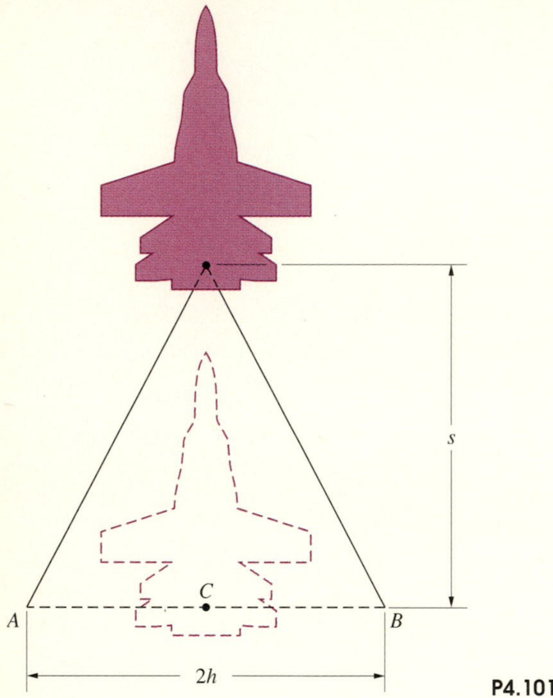

P4.101

4.102 In Example 4.11, assume that the pressure of the gas is related to its volume by $pV^{\gamma} =$ constant, where γ is a constant.
(a) Determine the potential energy associated with the force exerted on the projectile in terms of s.
(b) If the projectile starts from rest at $s = s_0$ and friction is negligible, what is its velocity as a function of s?

4.103 A satellite at a distance r_0 from the center of the earth has a velocity of magnitude v_0. Use conservation of energy to determine the magnitude of its velocity v when it is a distance r from the center of the earth.

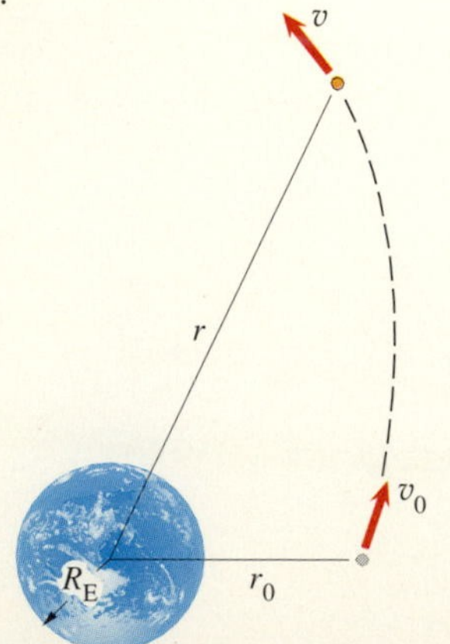

P4.103

4.104 Astronomers detect an asteroid 100,000 km from the earth moving at 2 km/s relative to the center of the earth. If it should strike the earth, use conservation of energy to determine the magnitude of its velocity as it enters the atmosphere. (You can neglect the thickness of the atmosphere in comparison to the earth's 6370-km radius.)

4.105 A satellite is in an elliptic orbit around the earth. Its velocity at the perigee A is 28,280 ft/s. Use conservation of energy to determine its velocity at B. The radius of the earth is 3960 mi.

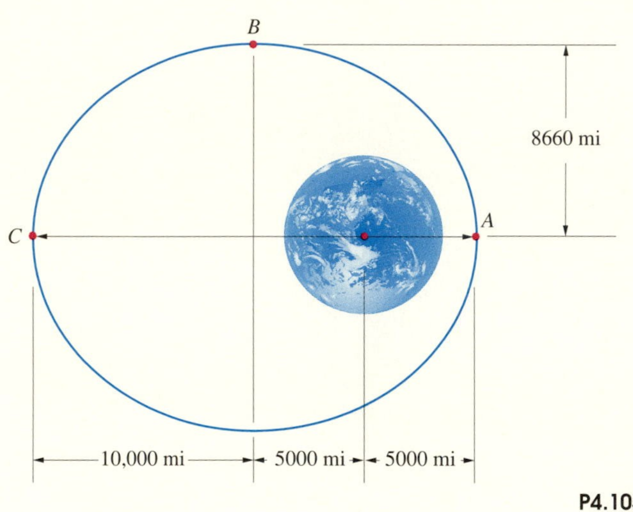

P4.105

4.106 For the satellite orbit in Problem 4.105, use conservation of energy to determine the velocity at the apogee C. Using your result, confirm numerically that the velocities at perigee and apogee satisfy the relation $r_A v_A = r_C v_C$.

4.107 The *Voyager* and *Galileo* spacecraft have observed volcanic plumes, believed to consist of condensed sulfur or sulfur dioxide gas, above the surface of the Jovian satellite Io. The plume observed above a volcano named *Prometheus* was estimated to extend 50 km above the surface. The acceleration due to gravity at the surface is $1.80\ \mathrm{m/s^2}$. Using conservation of energy and neglecting the variation of gravity with height, determine the velocity at which a solid particle would have to be ejected to reach 50 km above the surface.

Computer simulation of a volcanic plume on Io. (Courtesy of Victor Austin and David B. Goldstein.)

P4.107

4.108 Solve Problem 4.107 using conservation of energy and accounting for the variation of gravity with height. The radius of Io is 1815 km.

4.109 The component of the total external force tangent to the path of a 10-kg object moving along the x axis is $\Sigma F_x = 3x^2$ N, where x is in meters. At $x = 2$ m, the object's velocity is $v_x = 4$ m/s.
(a) Use the principle of work and energy to determine the velocity at $x = 6$ m.
(b) Determine the potential energy associated with the force ΣF_x and use conservation of energy to determine the velocity at $x = 6$ m.

4.110 The potential energy associated with a force **F** acting on an object is $V = 2x^2 - y$ N-m, where x and y are in meters.
(a) Determine **F**.
(b) If the object moves from position 1 to position 2 along the paths A and B, determine the work done by **F** along each path.

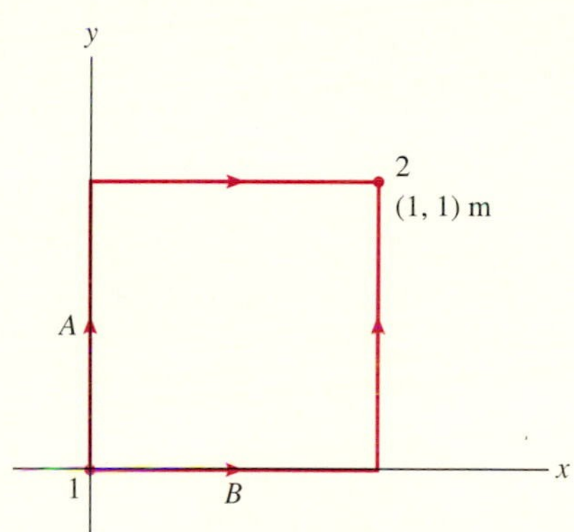

P4.110

4.111 An object is subjected to the force $\mathbf{F} = y\mathbf{i} - x\mathbf{j}$ (N), where x and y are in meters.
(a) Show that **F** is *not* conservative.
(b) If the object moves from point 1 to point 2 along the paths A and B shown in Problem 4.110, determine the work done by **F** along each path.

4.112 In terms of polar coordinates, the potential energy associated with the force **F** exerted on an object by a *nonlinear* spring is

$$V = \frac{1}{2}k(r - r_0)^2 + \frac{1}{4}q(r - r_0)^4,$$

where k and q are constants and r_0 is the unstretched length. Determine **F** in terms of polar coordinates.

4.113 In terms of polar coordinates, the force exerted on an object by a *nonlinear* spring is

$$\mathbf{F} = -[k(r - r_0) + q(r - r_0)^3]\mathbf{e}_r,$$

where k and q are constants and r_0 is the unstretched length. Use Eq. (4.36) to show that **F** is conservative.

4.114 The potential energy associated with a force **F** acting on an object is $V = -r\sin\theta + r^2\cos^2\theta$ ft-lb, where r is in feet.
(a) Determine **F**.
(b) If the object moves from point 1 to point 2 along the circular path, how much work is done by **F**?

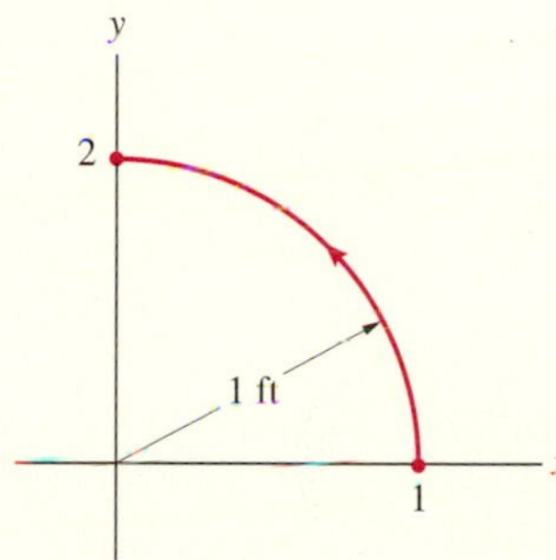

P4.114

4.115 In terms of polar coordinates, the force exerted on an object of mass m by the gravity of a hypothetical two-dimensional planet is $\mathbf{F} = -(mg_T R_T/r)\,\mathbf{e}_r$, where g_T is the acceleration due to gravity at the surface, R_T is the radius of the planet, and r is the distance from the center of the planet.
(a) Determine the potential energy associated with this gravitational force.
(b) If the object is given a velocity v_0 at a distance r_0, what is its velocity v as a function of r?

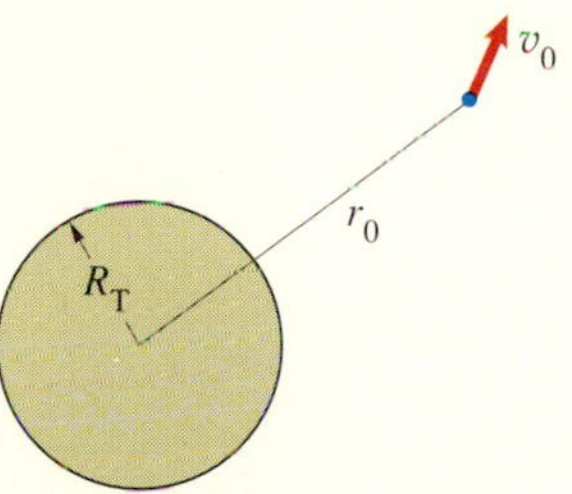

P4.115

4.116 By substituting Eqs. (4.32) into Eq. (4.35), confirm that $\nabla \times \mathbf{F} = 0$ if **F** is conservative.

4.117 Determine which of the following forces are conservative:
(a) $\mathbf{F} = (3x^2 - 2xy)\mathbf{i} - x^2\mathbf{j}$;
(b) $\mathbf{F} = (x - xy^2)\mathbf{i} + x^2y\mathbf{j}$;
(c) $\mathbf{F} = (2xy^2 + y^3)\mathbf{i} + (2x^2y - 3xy^2)\mathbf{j}$.

4.118 Determine which of the following forces are conservative:
(a) $\mathbf{F} = 3r^2\sin^2\theta\,\mathbf{e}_r + 2r^2\sin\theta\cos\theta\,\mathbf{e}_\theta$;
(b) $\mathbf{F} = (2r\sin\theta - \cos\theta)\mathbf{e}_r + (r\cos\theta - \sin\theta)\mathbf{e}_\theta$;
(c) $\mathbf{F} = (\sin\theta + r\cos^2\theta)\mathbf{e}_r + (\cos\theta - r\sin\theta\cos\theta)\mathbf{e}_\theta$.

Computational Mechanics

The following example and problems are designed for the use of a programmable calculator or computer.

Example 4.13

In the mechanical delay switch shown in Fig. 4.25, an electromagnet releases the 1-kg slider at position 1. Under the actions of gravity and the linear spring, the slider moves along the smooth bar from position 1 to position 2, closing the switch. The constant of the spring is $k = 40$ N/m, and its unstretched length is $r_0 = 50$ mm. The dimensions are $R = 200$ mm and $h = 100$ mm. What is the magnitude of the slider's maximum velocity, and where does it occur?

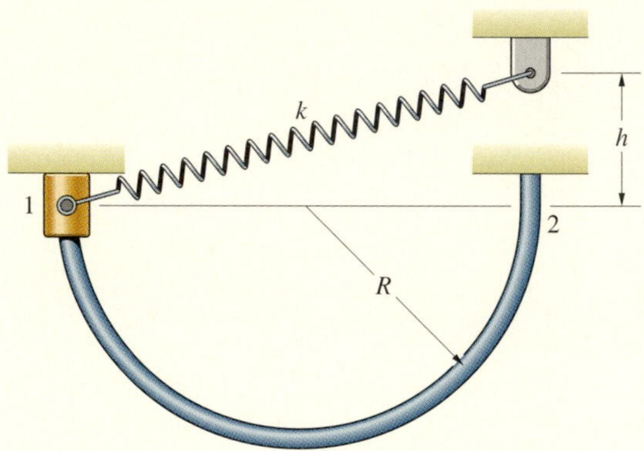

Fig. 4.25

STRATEGY

We can use conservation of energy to obtain an equation relating the magnitude of the slider's velocity to its position. By drawing a graph of the velocity as a function of the position, we can estimate the maximum velocity and the position where it occurs.

SOLUTION

We can specify the slider's position by the angle θ through which it has moved relative to position 1 (Fig. a). In position 1, the stretch of the spring equals its length in position 1 minus its unstretched length:

$$S_1 = \sqrt{(2R)^2 + h^2} - r_0.$$

When the slider has moved through the angle θ, the stretch of the spring is

$$S = \sqrt{(R + R\cos\theta)^2 + (h + R\sin\theta)^2} - r_0.$$

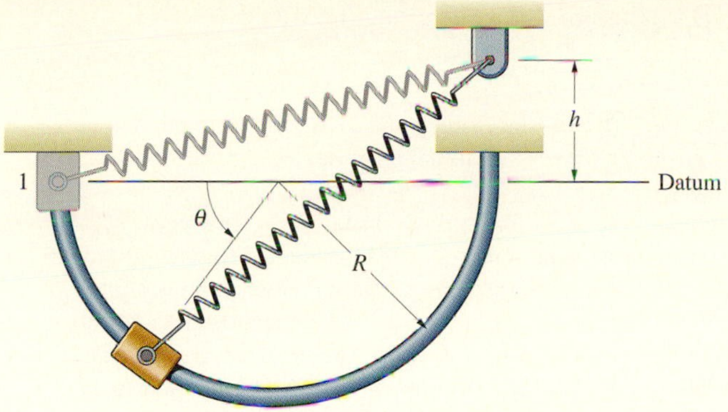

(a) The angle θ specifies the slider's position.

We express the potential energy of the slider's weight using the datum shown in Fig. (a). The sum of the potential and kinetic energies at position 1 must equal the sum of the potential and kinetic energies when the slider has moved through the angle θ:

$$\frac{1}{2}kS_1^2 + mgy_1 + \frac{1}{2}mv_1^2 = \frac{1}{2}kS^2 + mgy + \frac{1}{2}mv^2;$$

$$\frac{1}{2}k\left[\sqrt{(2R)^2 + h^2} - r_0\right]^2 + 0 + 0$$

$$= \frac{1}{2}k\left[\sqrt{(R + R\cos\theta)^2 + (h + R\sin\theta)^2} - r_0\right]^2$$

$$-mgR\sin\theta + \frac{1}{2}mv^2.$$

Solving for v, we obtain

$$v = \left\{(k/m)\left[\sqrt{(2R)^2 + h^2} - r_0\right]^2\right.$$

$$\left. -(k/m)\left[\sqrt{(R + R\cos\theta)^2 + (h + R\sin\theta)^2} - r_0\right]^2 + 2gR\sin\theta\right\}^{1/2}.$$

Computing the values of this expression as a function of θ, we obtain the graph shown in Fig. 4.26. The velocity is a maximum at approximately $\theta = 135°$. By examining the computed results near 135°,

θ	v, m/s
132°	2.5393
133°	2.5397
134°	2.5399
135°	2.5398
136°	2.5394
137°	2.5389
138°	2.5380

we estimate that a maximum velocity of 2.54 m/s occurs at $\theta = 134°$.

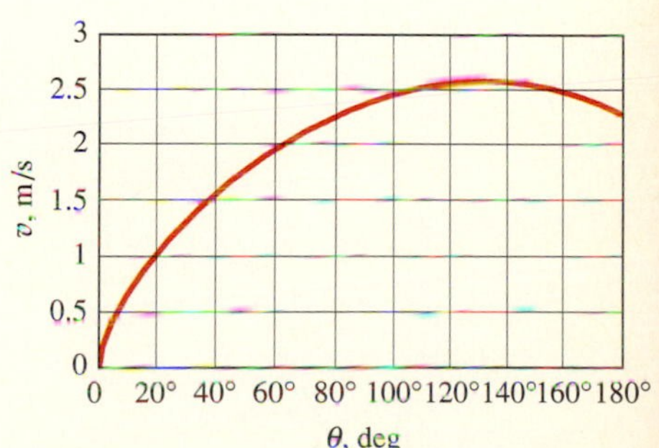

Fig. 4.26
Magnitude of the velocity as a function of θ.

Problems

4.119 The component of the total external force tangent to a 4-kg object's path is $\Sigma F_t = 200 + 2s^2 - 0.2s^3$ N, where s is its position measured along the path in meters. At $s = 0$, the object's velocity is $v = 10$ m/s. What distance along its path has the object traveled when its velocity reaches 30 m/s?

4.120 The 6-kg collar is released from rest in the position shown. If the spring constant is $k = 4$ kN/m and the unstretched length of the spring is 150 mm, how far does the mass fall from its initial position before rebounding?

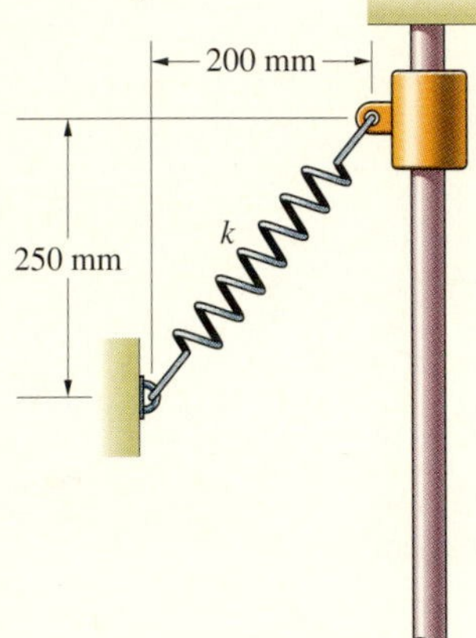

P4.120

4.121 How far below its initial position does the collar in Problem 4.120 reach its maximum velocity, and what is the maximum velocity?

4.122 How far below its initial position does the power being transferred to the collar in Problem 4.120 reach its maximum, and what is the maximum power?

4.123 The system is released from rest in the position shown. The weights are $W_A = 200$ lb and $W_B = 300$ lb. Neglect friction. Determine the maximum velocity attained by A as it rises.

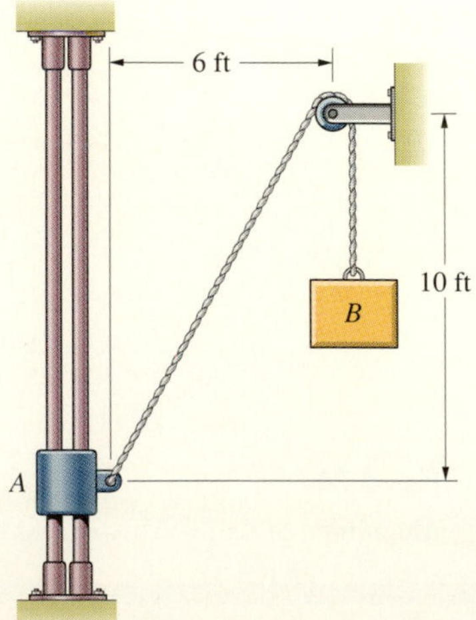

P4.123

4.124 In Problem 4.123, what maximum height is reached by A relative to its initial position?

4.125 The spring constant $k = 2000$ N/m, $m_A = 14$ kg, and $m_B = 18$ kg. The collar A slides on the smooth horizontal bar. The system is released from rest with the spring unstretched. (a) Use conservation of energy to determine the velocity of the collar A when it has moved a distance x to the right. Draw a graph of the velocity for $0 \leq x \leq 0.15$ m. (b) Use your graph to estimate the maximum velocity.

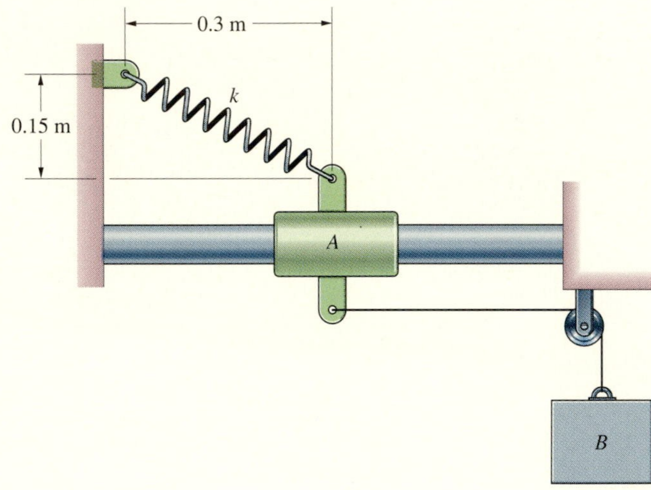

P4.125

4.126 In Problem 4.125, estimate the maximum distance the collar A slides to the right.

4.127 The 16-kg cylinder is released at the position shown and falls onto a nonlinear spring with potential energy $V = \frac{1}{2}kS^2 + \frac{1}{4}qS^4$, where $k = 2400$ N/m and $q = 3000$ N/m^3. Determine how far down the cylinder moves after contacting the spring.

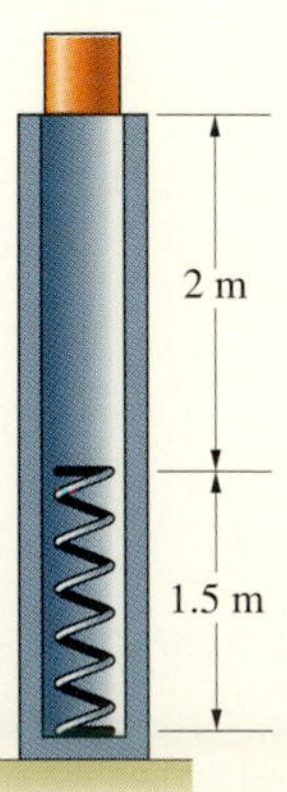

P4.127

4.128 In Problem 4.127, what is the maximum velocity attained by the cylinder?

4.129 The system shown in Fig. (a) is released from rest. The mass $m = 1$ kg, the spring constant $k = 200$ N/m, and the unstretched length of the spring is 0.1 m. (a) Use conservation of energy to determine the magnitude of the velocity of the mass as a function of the angle θ between the string and the vertical (Fig. b) and draw a graph of the velocity for $0 \leq \theta \leq 40°$. (b) Use your graph to estimate the maximum velocity.

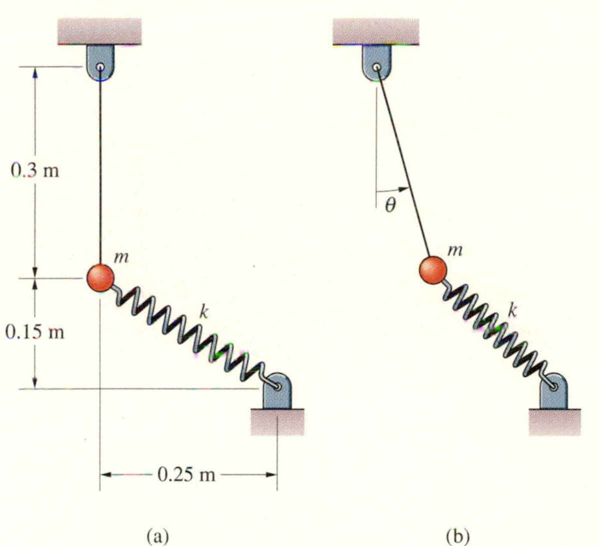

P4.129

4.130 In Problem 4.129, estimate the maximum value of θ reached by the mass.

4.131 A student runs at 15 ft/s, grabs a rope, and swings out over a lake. Determine the angle θ at which he should release the rope to maximize the horizontal distance b. What is the resulting value of b?

P4.131

Chapter Summary

In Chapter 3 we used Newton's second law to determine an object's acceleration when the forces acting on it were known. Once the acceleration was known, it could be integrated to obtain information about the object's velocity and position. In this chapter we have used Newton's second law to derive a new technique, the principle of work and energy, which relates the work done during a change in an object's position to the change in its kinetic energy. This principle is usually applicable only when the forces that do work are known as functions of the object's position, and the only information obtained is the change in the magnitude of the object's velocity. Nevertheless, it can be very useful because the work done by certain forces, including gravitational forces and forces exerted by springs, is easy to determine and is *independent of an object's path*. This property motivated the definitions of the potential energy and conservative forces and resulted in the concept of conservation of energy. In Chapter 5 we will use Newton's second law to derive momentum principles, which are especially useful for applications involving collisions between objects.

Principle of Work and Energy

Defining the **work** done on an object as its center of mass moves from a position $\mathbf{r}_1$ to a position $\mathbf{r}_2$ by

$$U_{12} = \int_{\mathbf{r}_1}^{\mathbf{r}_2} \Sigma \mathbf{F} \cdot d\mathbf{r}, \qquad \textbf{Eq. (4.5)}$$

where $\Sigma \mathbf{F}$ is the sum of the forces acting on the object, the **principle of work and energy** states that the work equals the change in the kinetic energy:

$$U_{12} = \frac{1}{2}mv_2^2 - \frac{1}{2}mv_1^2. \qquad \textbf{Eq. (4.6)}$$

The total work done by external forces on a system of objects equals the change in the total kinetic energy of the system if no net work is done by internal forces.

Power

The **power** is the rate at which work is done. The power transferred to an object by the external forces acting on it is

$$P = \Sigma \mathbf{F} \cdot \mathbf{v}. \qquad \textbf{Eq. (4.9)}$$

The power equals the rate of change of the object's kinetic energy. The average with respect to time of the power during an interval of time from t_1 to t_2 is equal to the change in its kinetic energy, or the work done, divided by the interval of time:

$$P_{\text{av}} = \frac{\frac{1}{2}mv_2^2 - \frac{1}{2}mv_1^2}{t_2 - t_1} = \frac{U_{12}}{t_2 - t_1}. \qquad \textbf{Eq. (4.10)}$$

Evaluating the Work

Let s be the position of an object's center of mass along its path. The work done on the object from a position s_1 to a position s_2 is

$$U_{12} = \int_{s_1}^{s_2} \Sigma F_{\text{t}}\, ds, \qquad \textbf{Eq. (4.7)}$$

where ΣF_{t} is the tangential component of the total external force on the object. *Components of force perpendicular to the path do no work.*

Weight In terms of a coordinate system with the positive y axis upward, the work done by an object's weight as its center of mass moves from position 1 to position 2 is

$$U_{12} = -mg(y_2 - y_1). \qquad \textbf{Eq. (4.15)}$$

The work is the product of the weight and the change in the height of the center of mass. The work is negative if the height increases and positive if it decreases.

When the variation of an object's weight with distance r from the center of the earth is accounted for, the work done by its weight is

$$U_{12} = mgR_E^2\left(\frac{1}{r_2} - \frac{1}{r_1}\right), \qquad \textbf{Eq. (4.17)}$$

where R_E is the radius of the earth.

Springs The work done on an object by a spring attached to a fixed support is

$$U_{12} = -\frac{1}{2}k(S_2^2 - S_1^2), \qquad \textbf{Eq. (4.18)}$$

where S_1 and S_2 are the values of the stretch at the initial and final positions.

Potential Energy

For a given force **F** acting on an object, if a function V of the object's position exists such that

$$dV = -\mathbf{F} \cdot d\mathbf{r}, \qquad \textbf{Eq. (4.20)}$$

then **F** is said to be **conservative** and V is called the **potential energy** associated with **F**. The work done by **F** from a position 1 to a position 2 is

$$U_{12} = -(V_2 - V_1). \qquad \textbf{Eq. (4.21)}$$

If all the forces that do work on an object are conservative, the total energy—the sum of the kinetic energy and the potential energies of the forces—is conserved:

$$\frac{1}{2}mv^2 + V = \text{constant}. \qquad \textbf{Eq. (4.23)}$$

Weight In terms of a cartesian coordinate system with its y axis upward, the potential energy of the weight of an object is

$$V = mgy. \qquad \textbf{Eq. (4.25)}$$

The potential energy is the product of the object's weight and the height of its center of mass measured from any convenient reference level, or **datum**.

When the variation of an object's weight with distance r from the center of the earth is accounted for, the potential energy of its weight is

$$V = -\frac{mgR_E^2}{r}, \qquad \textbf{Eq. (4.27)}$$

where R_E is the radius of the earth.

Springs The potential energy of the force exerted on an object by a linear spring is

$$V = \frac{1}{2}kS^2, \qquad \text{Eq. (4.28)}$$

where S is the stretch of the spring.

Relationships Between Force and Potential Energy A force **F** is related to its associated potential energy by

$$\mathbf{F} = -\left(\frac{\partial V}{\partial x}\mathbf{i} + \frac{\partial V}{\partial y}\mathbf{j} + \frac{\partial V}{\partial z}\mathbf{k}\right) = -\nabla V. \qquad \text{Eq. (4.33)}$$

A force **F** is conservative if and only if its **curl** is zero:

$$\nabla \times \mathbf{F} = \begin{vmatrix} \mathbf{i} & \mathbf{j} & \mathbf{k} \\ \dfrac{\partial}{\partial x} & \dfrac{\partial}{\partial y} & \dfrac{\partial}{\partial z} \\ F_x & F_y & F_z \end{vmatrix} = 0.$$

Review Problems

4.132 The driver of a 3000-lb car moving at 40 mi/hr applies an increasing force on the brake pedal. The magnitude of the resulting friction force exerted on the car by the road is $f = 250 + 6s$ lb, where s is the car's horizontal position (in feet) relative to its position when the brakes were applied. Assuming that the car's tires do not slip, determine the distance required for the car to stop (a) by using Newton's second law; (b) by using the principle of work and energy.

4.133 Suppose that the car in Problem 4.132 is on wet pavement and the coefficients of friction between the tires and the road are $\mu_s = 0.4$, $\mu_k = 0.35$. Determine the distance required for the car to stop.

4.134 An astronaut in a small rocket vehicle (combined mass = 450 kg) is hovering 100 m above the surface of the moon when he discovers he is nearly out of fuel and can exert the thrust necessary to cause the vehicle to hover for only 5 more seconds. He quickly considers two strategies for getting to the surface: (a) Fall 20 m, turn on the thrust for 5 s, then fall the rest of the way; (b) fall 40 m, turn on the thrust for 5 s, then fall the rest of the way. Which strategy gives him the best chance of surviving? How much work is done by the engine's thrust in each case? ($g_{\text{moon}} = 1.62 \text{ m/s}^2$.)

4.135 The coefficients of friction between the 20-kg crate and the inclined surface are $\mu_s = 0.24$ and $\mu_k = 0.22$. If the crate starts from rest and the horizontal force $F = 200$ N, what is the magnitude of its velocity when it has moved 2 m?

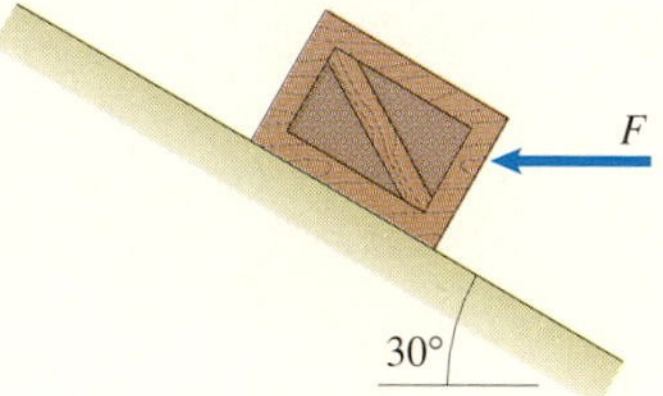

P4.135

4.136 In Problem 4.135, what is the magnitude of the crate's velocity when it has moved 2 m if the horizontal force $F = 40$ N?

4.137 The Union Pacific *Big Boy* locomotive weighs 1.19 million lb, and the tractive effort (tangential force) of its drive wheels is 135,000 lb. If you neglect other tangential forces, what distance is required for it to accelerate from zero to 60 mi/hr?

P4.137

4.138 In Problem 4.137, suppose that the acceleration of the locomotive as it accelerates from zero to 60 mi/hr is $(F_0/m)(1 - v/88)$, where $F_0 = 135{,}000$ lb, m is its mass, and v is its velocity in feet per second.
(a) How much work is done in accelerating it to 60 mi/hr?
(b) Determine its velocity as a function of time.

4.139 If a car traveling 65 mi/hr hits the crash barrier described in Problem 4.13, determine the maximum deceleration to which the passengers are subjected if the car weighs (a) 2500 lb; (b) 5000 lb.

4.140 In a preliminary design for a mail sorting machine, parcels moving at 2 ft/s slide down a smooth ramp and are brought to rest by a linear spring. What should the spring constant be if you don't want a 10-lb parcel to be subjected to a maximum deceleration greater than 10 g's?

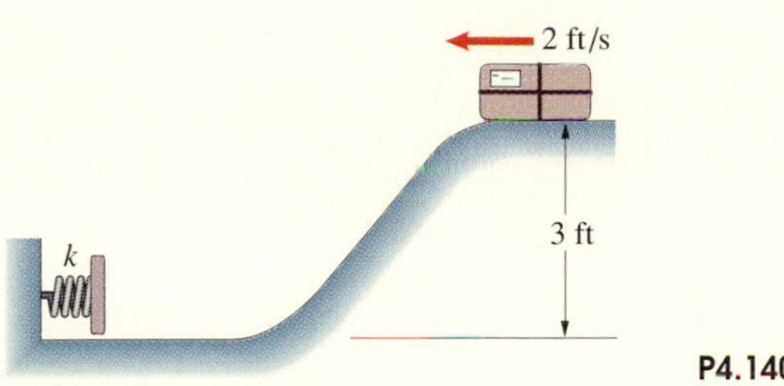

P4.140

4.141 When the 1-kg collar is in position 1, the tension in the spring is 50 N, and the unstretched length of the spring is 260 mm. If the collar is pulled to position 2 and released from rest, what is its velocity when it returns to 1?

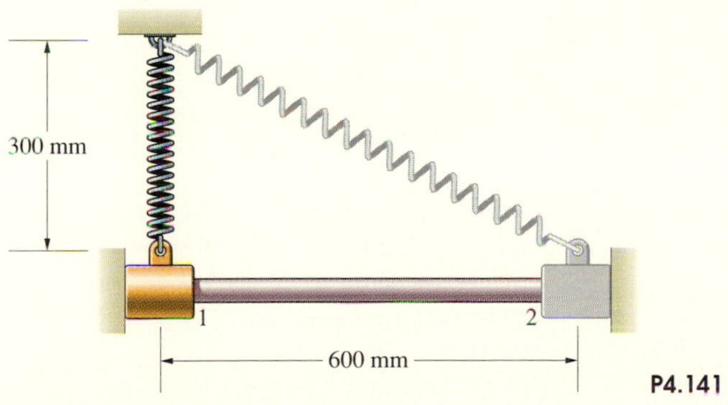

P4.141

4.142 In Problem 4.141, suppose that the tensions in the spring in positions 1 and 2 are 100 N and 400 N, respectively.
(a) What is the spring constant k?
(b) If the collar is given a velocity of 15 m/s at 1, what is its velocity when it reaches 2?

4.143 The 30-lb weight is released from rest with the two springs ($k_A = 30$ lb/ft, $k_B = 15$ lb/ft) unstretched.
(a) How far does the weight fall before rebounding?
(b) What maximum velocity does it attain?

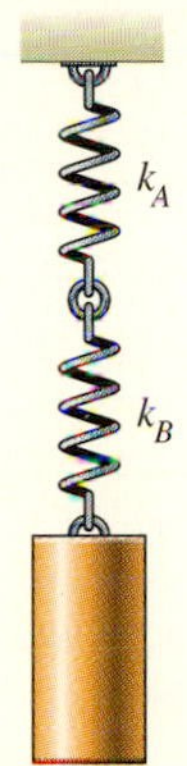

P4.143

4.144 The piston and the load it supports are accelerated upward by the gas in the cylinder. The total weight of the piston and load is 1000 lb. The cylinder wall exerts a constant 50-lb friction force on the piston as it rises. The net force exerted on the piston by pressure is $(p - p_{\text{atm}})A$, where p is the pressure of the gas, $p_{\text{atm}} = 2117$ lb/ft^2 is atmospheric pressure, and $A = 1$ ft^2 is the cross-sectional area of the piston. Assume that the product of p and the volume of the cylinder is constant. When $s = 1$ ft, the piston is stationary and $p = 5000$ lb/ft^2. What is the velocity of the piston when $s = 2$ ft?

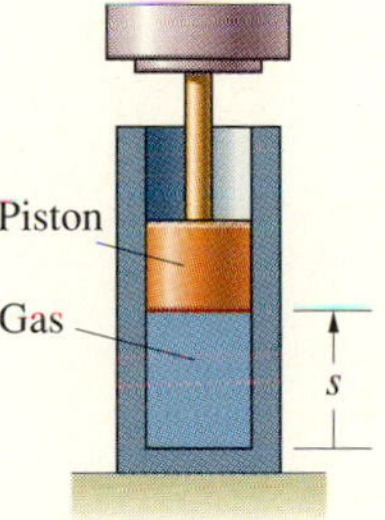

P4.144

4.145 When a 22-Mg rocket's engine burns out at an altitude of 2 km, its velocity is 3 km/s and it is traveling at an angle of 60° relative to the horizontal. Neglect the variation in the gravitational force with altitude.
(a) If you neglect aerodynamic forces, what is the magnitude of the rocket's velocity when it reaches an altitude of 6 km?
(b) If the rocket's actual velocity when it reaches an altitude of 6 km is 2.8 km/s, how much work is done by aerodynamic forces as the rocket moves from 2 km to 6 km altitude?

4.146 The 12-kg collar A is at rest in the position shown at $t = 0$ and is subjected to the tangential force $F = 24 - 12t^2$ N for 1.5 s. Neglecting friction, what maximum height h does it reach?

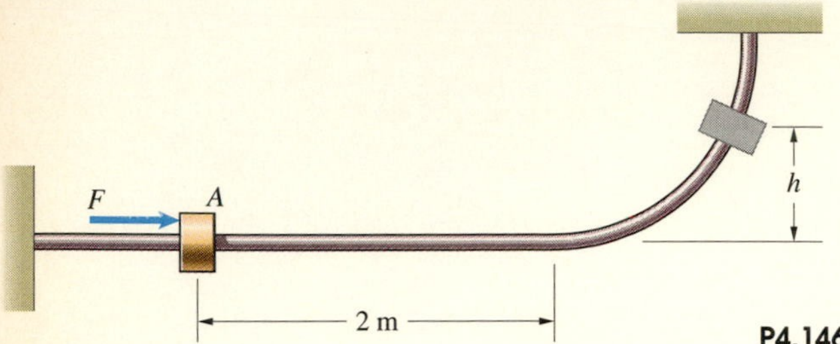

P4.146

4.147 Suppose that in designing a loop for a roller coaster's track, you establish as a safety criterion that at the top of the loop, the normal force exerted on a passenger by the roller coaster should equal 10 percent of the passenger's weight. (That is, the passenger's "effective weight" pressing him down into his seat is 10 percent of his weight.) The roller coaster is moving at 62 ft/s when it enters the loop. What is the necessary instantaneous radius of curvature ρ of the track at the top of the loop?

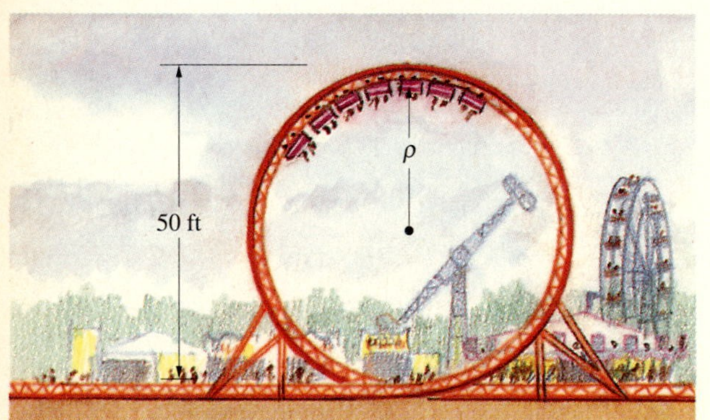

P4.147

4.148 A 180-lb student runs at 15 ft/s, grabs a rope, and swings out over a lake. He releases the rope when his velocity is zero.
(a) What is the angle θ when he releases the rope?
(b) What is the tension in the rope just before he releases it?
(c) What is the maximum tension in the rope?

P4.148

4.149 If the student in Problem 4.148 releases the rope when $\theta = 25°$, what maximum height does he reach relative to his position when he grabs the rope?

4.150 A boy takes a running start and jumps on his sled at 1. He leaves the ground at 2 and lands in deep snow at a distance $b = 25$ ft. How fast was he going at 1?

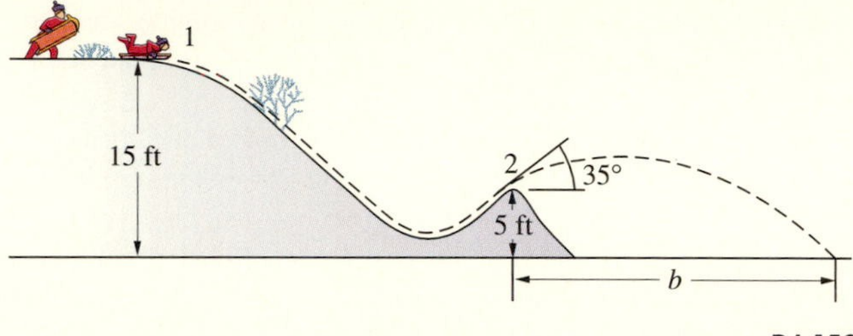

P4.150

4.151 In Problem 4.150, if the boy starts at 1 going 15 ft/s, what distance b does he travel through the air?

4.152 The 1-kg collar A is attached to the linear spring ($k = 500$ N/m) by a string. The collar starts from rest in the position shown, and the initial tension in the string is 100 N. What distance does the collar slide up the smooth bar?

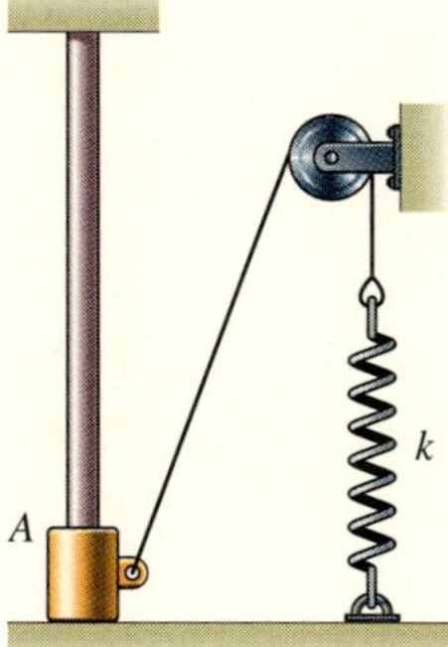

P4.152

4.153 The masses $m_A = 40$ kg and $m_B = 60$ kg. The collar A slides on the smooth horizontal bar. The system is released from rest. Use conservation of energy to determine the velocity of the collar A when it has moved 0.5 m to the right.

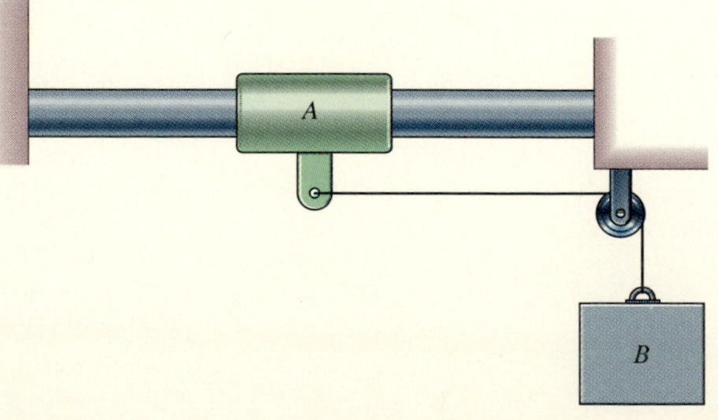

P4.153

4.154 The spring constant is $k = 850$ N/m, $m_A = 40$ kg, and $m_B = 60$ kg. The collar A slides on the smooth horizontal bar. The system is released from rest in the position shown with the spring unstretched. Use conservation of energy to determine the velocity of the collar A when it has moved 0.5 m to the right.

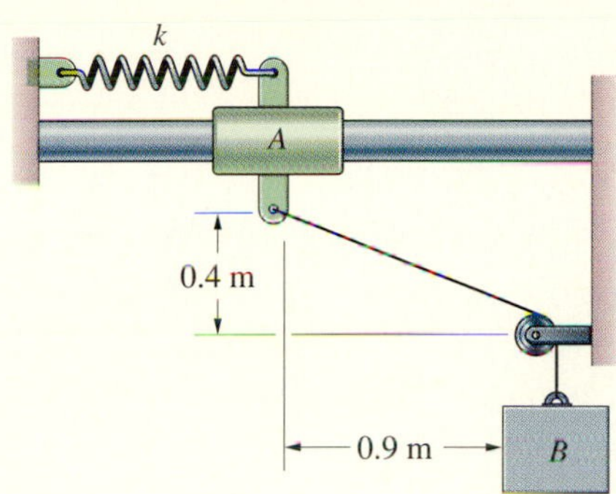

P4.154

4.155 The y axis is vertical and the curved bar is smooth. If the magnitude of the velocity of the 4-lb slider is 6 ft/s at position 1, what is the magnitude of its velocity when it reaches position 2?

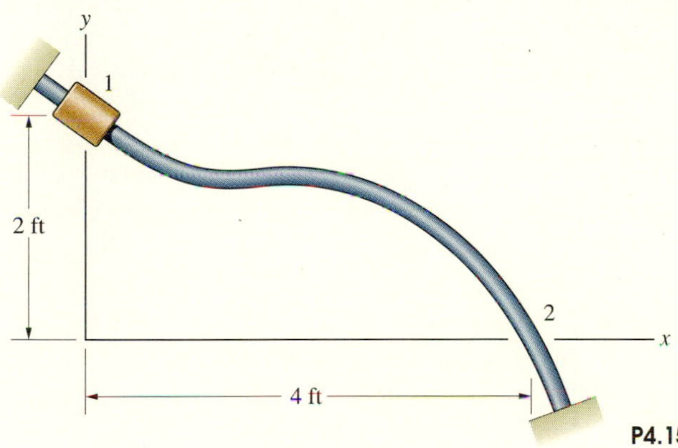

P4.155

4.156 In Problem 4.155, determine the magnitude of the slider's velocity when it reaches position 2 if it is subjected to the additional force $\mathbf{F} = 3x\mathbf{i} - 2\mathbf{j}$ (lb) during its motion.

4.157 Suppose that an object of mass m is beneath the surface of the earth. In terms of a polar coordinate system with its origin at the earth's center, the gravitational force on the object is $-(mgr/R_E)\mathbf{e}_r$, where R_E is the radius of the earth. Show that the potential energy associated with the gravitational force is $V = mgr^2/2R_E$.

4.158 It has been pointed out that if tunnels could be drilled straight through the earth between points on the surface, trains could travel between those points using gravitational force for acceleration and deceleration. (The effects of friction and aerodynamic drag could be minimized by evacuating the tunnels and using magnetically levitated trains.) Suppose that such a train travels from the North Pole to a point on the equator. Determine the magnitude of the train's velocity (a) when it arrives at the equator; (b) when it is halfway from the North Pole to the equator. The radius of the earth is $R_E = 3960$ mi.

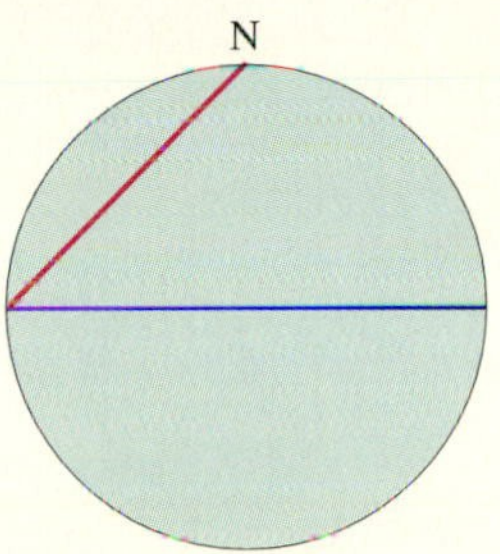

P4.158

4.159 In Problem 4.137, what is the maximum power transferred to the locomotive during its acceleration?

4.160 Just before it lifts off, a 10.5-Mg airplane is traveling at 60 m/s. The total horizontal force exerted by its engines is 189 kN, and the plane is accelerating at 15 m/s^2.
(a) How much power is being transferred to the plane by its engines?
(b) What is the total power being transferred to the plane?

P4.160

4.161 The "Paris Gun" used by Germany in World War I had a range of 120 km, a 37.5-m barrel, a muzzle velocity of 1550 m/s, and fired a 120-kg shell.
(a) If you assume the shell's acceleration to be constant, what maximum power was transferred to it as it traveled along the barrel?
(b) What average power was transferred to the shell?

P4.161

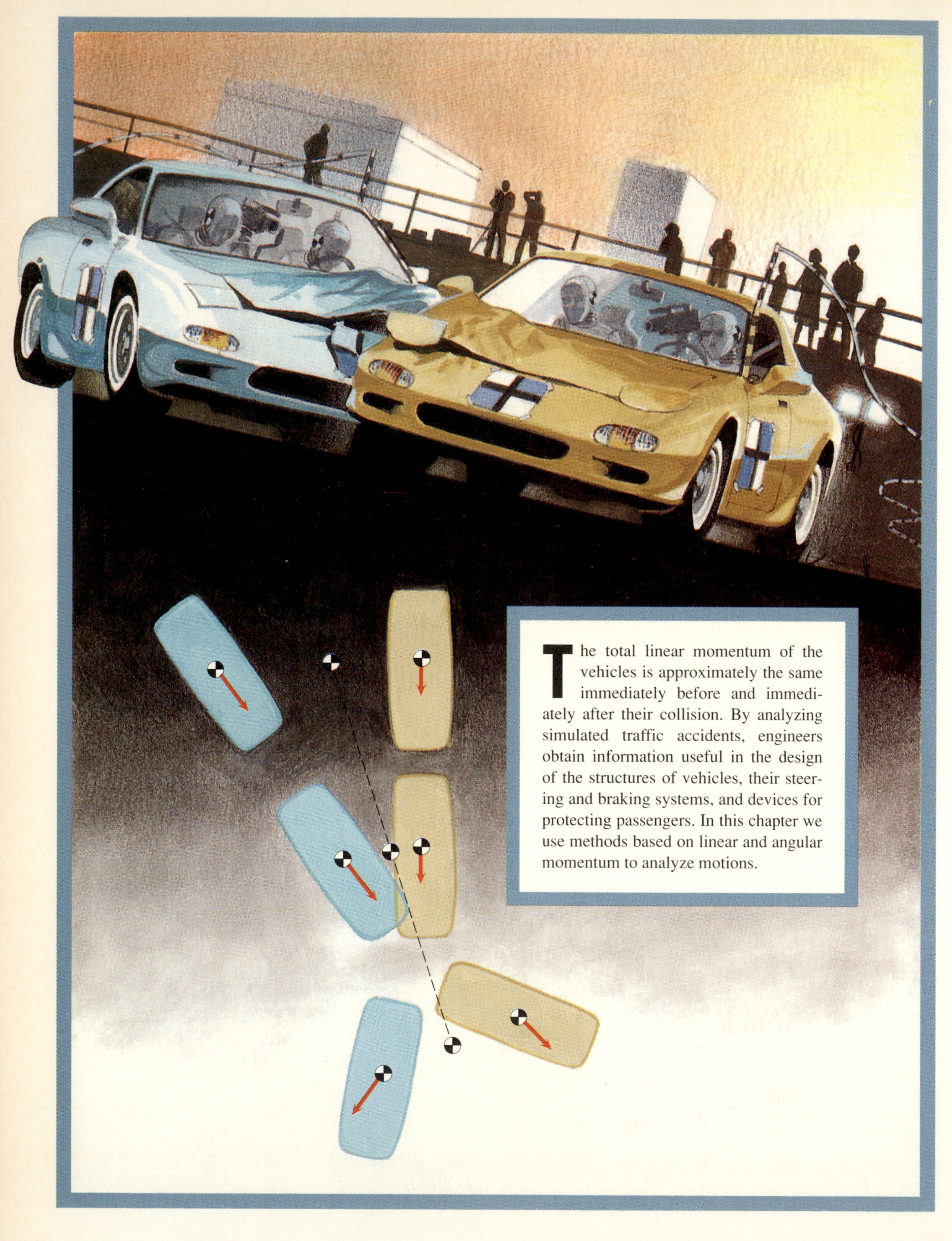

The total linear momentum of the vehicles is approximately the same immediately before and immediately after their collision. By analyzing simulated traffic accidents, engineers obtain information useful in the design of the structures of vehicles, their steering and braking systems, and devices for protecting passengers. In this chapter we use methods based on linear and angular momentum to analyze motions.

Chapter 5

Momentum Methods

IN Chapter 4 we transformed Newton's second law to obtain the principle of work and energy. In this chapter we integrate Newton's second law with respect to time, obtaining a relation between the time integral of the forces acting on an object and the change in the object's linear momentum. With this result, called the principle of impulse and momentum, we can determine the change in an object's velocity when the external forces are known as functions of time.

By applying the principle of impulse and momentum to two or more objects, we obtain the principle of conservation of linear momentum. This conservation law allows us to analyze impacts between objects and evaluate forces exerted by continuous flows of mass, as in jet and rocket engines.

By another transformation of Newton's second law, we obtain a relation between the time integral of the moments exerted on an object and the change in a quantity called angular momentum. We show that in the circumstance called central-force motion, an object's angular momentum is conserved.

5.1 *Principle of Impulse and Momentum*

The principle of work and energy is a very useful tool in mechanics. We can derive another useful tool for the analysis of motion by integrating Newton's second law with respect to time. We express Newton's second law in the form

$$\Sigma \mathbf{F} = m\frac{d\mathbf{v}}{dt}.$$

Then we integrate with respect to time to obtain

$$\int_{t_1}^{t_2} \Sigma \mathbf{F}\, dt = m\mathbf{v}_2 - m\mathbf{v}_1, \tag{5.1}$$

where $\mathbf{v}_1$ and $\mathbf{v}_2$ are the velocities of the center of mass of the object at the times t_1 and t_2. The term on the left is called the **linear impulse**, and $m\mathbf{v}$ is the **linear momentum**. This result is called the **principle of impulse and momentum**: The impulse applied to an object during an interval of time is equal to the change in its linear momentum (Fig. 5.1). The dimensions of the linear impulse and linear momentum are (mass) × (length)/(time).

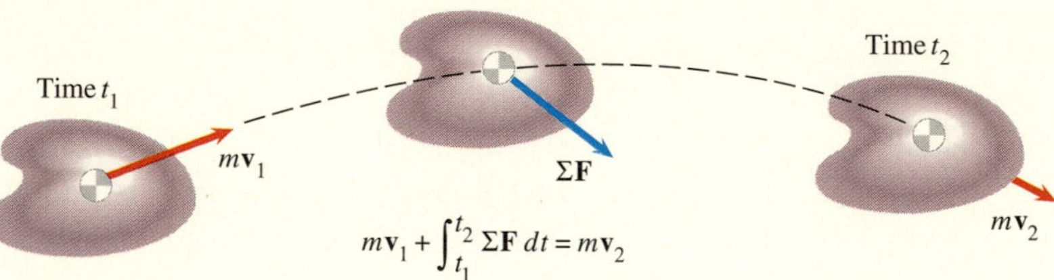

Fig. 5.1
Principle of impulse and momentum.

The average with respect to time of the total force acting on an object from t_1 to t_2 is

$$\Sigma \mathbf{F}_{\text{av}} = \frac{1}{t_2 - t_1}\int_{t_1}^{t_2} \Sigma \mathbf{F}\, dt,$$

so we can write Eq. (5.1) as

$$(t_2 - t_1)\, \Sigma \mathbf{F}_{\text{av}} = m\mathbf{v}_2 - m\mathbf{v}_1. \tag{5.2}$$

With this equation you can determine the average value of the total force acting on an object during a given interval of time if you know the change in its velocity.

A force of relatively large magnitude that acts over a small interval of time is called an **impulsive force** (Fig. 5.2). Determining the actual time history of such a force is often impractical, but with Eq. (5.2) its average value can sometimes be determined. For example, a golf ball struck by a club is subjected to an impulsive force. By making high-speed motion pictures, the duration of the impact and the ball's velocity after the impact can be measured. Knowing the duration of the impact and the ball's change in linear momentum, we can determine the average force exerted by the club. (See Example 5.3.)

We can express Eqs. (5.1) and (5.2) in scalar forms that are often useful. The sum of the forces in the direction tangent to an object's path equals the product of its mass and the rate of change of its velocity along the path (see Eq. 3.7):

$$\Sigma F_t = ma_t = m\frac{dv}{dt}.$$

Integrating this equation with respect to time, we obtain

$$\int_{t_1}^{t_2} \Sigma F_t \, dt = mv_2 - mv_1, \qquad (5.3)$$

where v_1 and v_2 are the velocities along the path at the times t_1 and t_2. The impulse applied to an object by the sum of the forces tangent to its path during an interval of time is equal to the change in its linear momentum along the path. In terms of the average with respect to time of the sum of the forces tangent to the path,

$$\Sigma F_{t\,\mathrm{av}} = \frac{1}{t_2 - t_1}\int_{t_1}^{t_2} \Sigma F_t \, dt,$$

we can write Eq. (5.3) as

$$(t_2 - t_1)\Sigma F_{t\,\mathrm{av}} = mv_2 - mv_1. \qquad (5.4)$$

This equation relates the average of the sum of the forces tangent to the path during an interval of time to the change in the velocity along the path.

Notice that Eq. (5.1) and the principle of work and energy, Eq. (4.6), are quite similar. They both relate an integral of the external forces to the change in an object's velocity. Equation (5.1) is a vector equation that tells you the change in both the magnitude and direction of the velocity, whereas the principle of work and energy, a scalar equation, tells you only the change in the magnitude of the velocity. But there is a greater difference between the two methods: In the case of impulse and momentum, there is no class of forces equivalent to the conservative forces that make work and energy so easy to apply.

When you know the external forces acting on an object as functions of time, the principle of impulse and momentum allows you to determine the change in its velocity during an interval of time. Although this is an important result, it is not new. In Chapter 3, when we used Newton's second law to determine an object's acceleration and then integrated the acceleration with respect to time to determine its velocity, we were effectively applying the principle of impulse and momentum. However, in the rest of this chapter we show that this principle can be extended to new and interesting applications.

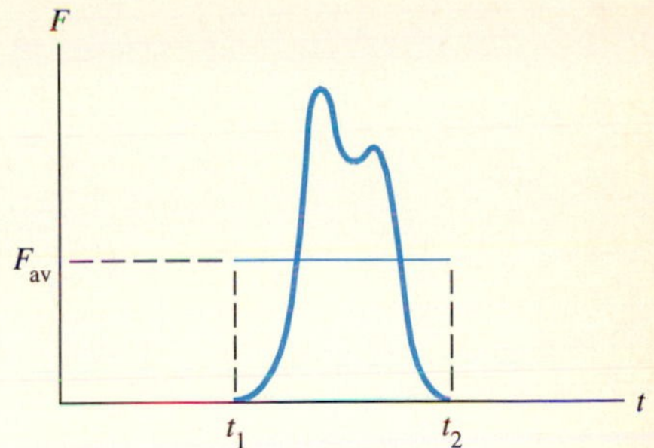

Fig. 5.2
An impulsive force and its average value.

Example 5.1

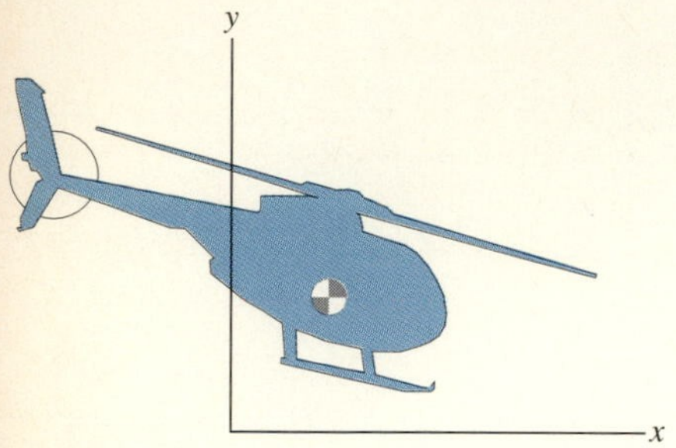

Fig. 5.3

A 1200-kg helicopter starts from rest at $t = 0$ (Fig. 5.3).
(a) The components of the total force on the helicopter from $t = 0$ to $t = 10$ s are given by

$$\Sigma F_x = 720t \text{ N},$$

$$\Sigma F_y = 2160 - 360t \text{ N},$$

$$\Sigma F_z = 0.$$

Determine the helicopter's velocity at $t = 10$ s.
(b) At $t = 20$ s, the helicopter's velocity is $36\mathbf{i} + 8\mathbf{j}$ (m/s). What is the average of the total force acting on it from $t = 10$ s to $t = 20$ s?

STRATEGY

(a) Since we know the helicopter's velocity at $t = 0$ and the components of the total force acting on it as functions of time, we can use the principle of impulse and momentum, Eq. (5.1), to determine its velocity at $t = 10$ s. (b) Knowing the velocity at $t = 10$ s and at $t = 20$ s, we can determine the average of the total force from Eq. (5.2).

SOLUTION

(a) Applying the principle of impulse and momentum from $t = 0$ to $t = 10$ s,

$$\int_{t_1}^{t_2} \Sigma \mathbf{F}\, dt = m\mathbf{v}_2 - m\mathbf{v}_1:$$

$$\int_0^{10} [720t\,\mathbf{i} + (2160 - 360t)\mathbf{j}]\, dt = (1200)\mathbf{v}_2 - (1200)(0),$$

$$36{,}000\mathbf{i} + 3600\mathbf{j} = 1200\mathbf{v}_2,$$

we find that the velocity at $t = 10$ s is $30\mathbf{i} + 3\mathbf{j}$ (m/s).
(b) To determine the average total force from $t = 10$ s to $t = 20$ s, we apply Eq. (5.2) to this interval of time,

$$(t_2 - t_1)\Sigma \mathbf{F}_{av} = m\mathbf{v}_2 - m\mathbf{v}_1:$$

$$(20 - 10)\Sigma \mathbf{F}_{av} = 1200(36\mathbf{i} + 8\mathbf{j}) - 1200(30\mathbf{i} + 3\mathbf{j}),$$

$$\Sigma \mathbf{F}_{av} = 720\mathbf{i} + 600\mathbf{j} \text{ (N)}.$$

DISCUSSION

Although we did not know the total force acting on the helicopter during the interval from $t = 10$ s to $t = 20$ s, we were able to determine the average of the total force because we knew the velocities at the beginning and end of the interval.

Example 5.2

The rocket booster in Fig. 5.4 is traveling straight up when it suddenly starts rotating counterclockwise at one-fourth revolution per second. The range safety officer destroys it 2 s later. The booster's mass is $m = 90$ Mg, its thrust is $T = 1.0$ MN, and it is moving upward at 10 m/s when it starts rotating. If aerodynamic forces are neglected, what is the booster's velocity at the time it is destroyed?

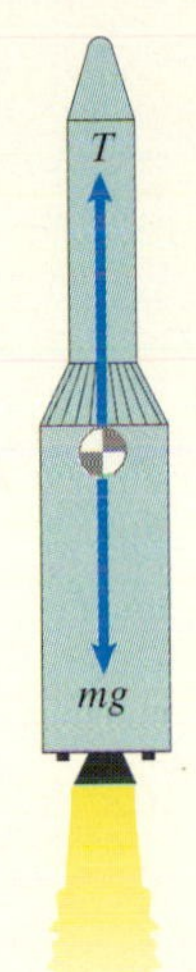

Fig. 5.4

STRATEGY

Because we know the angular velocity, we can determine the direction of the booster's thrust as a function of time and calculate the impulse during the 2-s period.

SOLUTION

The booster's angular velocity is $\pi/2$ rad/s. Letting $t = 0$ be the time at which it starts rotating, the angle between its axis and the vertical is $(\pi/2)t$ (Fig. a). The total force on the booster is

$$\Sigma \mathbf{F} = \left(-T \sin \frac{\pi}{2}t\right)\mathbf{i} + \left(T \cos \frac{\pi}{2}t - mg\right)\mathbf{j},$$

so the impulse from $t = 0$ to $t = 2$ s is

$$\begin{aligned}\int_0^2 \Sigma \mathbf{F}\, dt &= \int_0^2 \left[\left(-T \sin \frac{\pi}{2}t\right)\mathbf{i} + \left(T \cos \frac{\pi}{2}t - mg\right)\mathbf{j}\right] dt \\ &= \left[\left(T\frac{2}{\pi}\cos\frac{\pi}{2}t\right)\mathbf{i} + \left(T\frac{2}{\pi}\sin\frac{\pi}{2}t - mgt\right)\mathbf{j}\right]_0^2 \\ &= -\frac{4}{\pi}T\mathbf{i} - 2mg\mathbf{j}.\end{aligned}$$

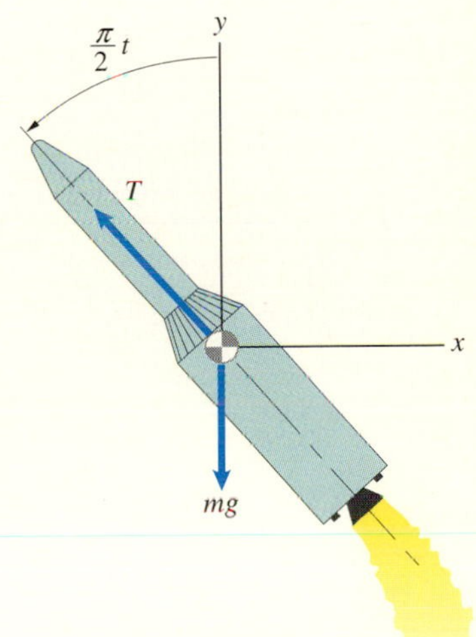

(a) The rotating booster.

From the principle of impulse and momentum,

$$\int_0^2 \Sigma \mathbf{F}\, dt = m\mathbf{v}_2 - m\mathbf{v}_1:$$

$$-\frac{4}{\pi}(1 \times 10^6)\,\mathbf{i} - 2(90 \times 10^3)(9.81)\mathbf{j} = (90 \times 10^3)(\mathbf{v}_2 - 10\mathbf{j}).$$

Solving, we obtain $\mathbf{v}_2 = -14.15\mathbf{i} - 9.62\,\mathbf{j}$ (m/s).

DISCUSSION

Notice that the rocket's thrust has no net effect on its y component of velocity during the 2-s interval. The effect of the positive y component of the thrust during the first one-fourth revolution is canceled by the effect of the negative y component during the second one-fourth revolution. The change in the y component of velocity is caused entirely by the rocket's weight. The thrust has a negative x component during the entire 2-s interval, giving the rocket its negative x component of velocity at the time it is destroyed.

Example 5.3

A golf ball in flight is photographed at intervals of 0.001 s (Fig. 5.5). The 1.62-oz ball is 1.68 in. in diameter. If the club was in contact with the ball for 0.0006 s, estimate the average value of the impulsive force exerted by the club.

Fig. 5.5

STRATEGY

By measuring the distance traveled by the ball in one of the 0.001-s intervals, we can estimate its velocity after being struck, then use Eq. (5.2) to determine the average total force on the ball.

SOLUTION

By comparing the distance moved during one of the 0.001-s intervals with the known diameter of the ball, we estimate that the ball traveled 1.9 in. and that its direction is 21° above the horizontal (Fig. a). The magnitude of the ball's velocity is

$$\frac{(1.9/12)\text{ ft}}{0.001\text{ s}} = 158\text{ ft/s}.$$

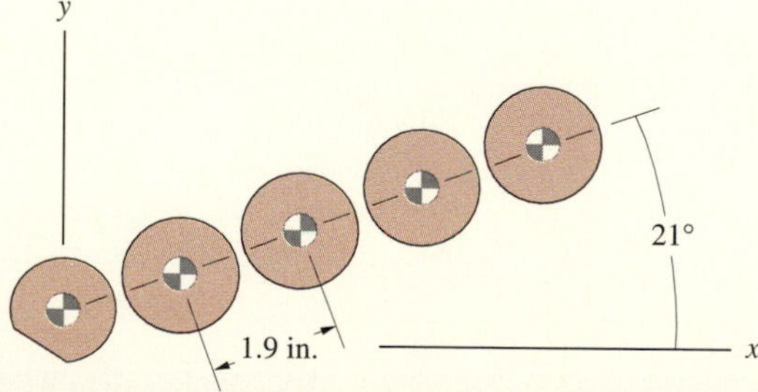

(a) Estimating the distance traveled during one 0.001-s interval.

The weight of the ball is $1.62/16 = 0.101$ lb, so its mass is $0.101/32.2 = 3.14 \times 10^{-3}$ slugs. From Eq. (5.2),

$$(t_2 - t_1)\Sigma \mathbf{F}_{av} = m\mathbf{v}_2 - m\mathbf{v}_1,$$
$$(0.0006)\Sigma \mathbf{F}_{av} = (3.14 \times 10^{-3})(158)(\cos 21°\,\mathbf{i} + \sin 21°\,\mathbf{j}) - 0,$$

we obtain

$$\Sigma \mathbf{F}_{av} = 775\mathbf{i} + 297\mathbf{j} \quad \text{(lb)}.$$

DISCUSSION

The average force during the time the club is in contact with the ball includes both the impulsive force exerted by the club and the ball's weight. In comparison with the large average impulsive force exerted by the club, the weight ($-0.101\mathbf{j}$ lb) is negligible.

Problems

5.1 The aircraft carrier *Nimitz* weighs 91,000 tons. (A ton is 2000 lb.) Suppose that its engines and hydrodynamic drag exert a constant 1,000,000-lb decelerating force on it.
(a) Use the principle of impulse and momentum to determine how long it requires the ship to come to rest from its top speed of approximately 30 knots. (A knot is approximately 6076 ft/hr.)
(b) Use the principle of work and energy to determine the distance the ship travels during the time it takes to come to rest.

P5.1

5.2 The 2000-lb drag racer accelerates from rest to 300 mi/hr in 6 s.
(a) What impulse is applied to the car during the 6 s?
(b) If you assume as a first approximation that the tangential force exerted on the car is constant, what is the magnitude of the force?

P5.2

5.3 The 21,900-kg Gloster Saro Protector, designed for rapid response to airport emergencies, accelerates from rest to 80 km/hr in 35 s.
(a) What impulse is applied to the vehicle during the 35 s?
(b) If you assume as a first approximation that the tangential force exerted on the vehicle is constant, what is the magnitude of the force?
(c) What average power is transferred to the vehicle?

P5.3

5.4 The combined weight of the motorcycle and rider is 300 lb. The coefficient of kinetic friction between the motorcycle's tires and the road is $\mu_k = 0.8$. Suppose that the rider starts from rest and spins the rear (drive) wheel. The normal force between the rear wheel and the road is 250 lb.
(a) What impulse does the friction force on the rear wheel exert in 5 s?
(b) If you neglect other horizontal forces, what velocity is attained in 5 s?

P5.4–5.6

5.5 The combined mass of the motorcycle and rider is 160 kg. The motorcycle starts from rest at $t = 0$. The total horizontal force exerted on it from $t = 0$ to $t = 10$ s is $1200e^{-0.1t}$ N.
(a) Determine the impulse exerted from $t = 0$ to $t = 10$ s.
(b) What is the motorcycle's velocity in km/hr at $t = 10$ s?

5.6 In Problem 5.5, what average total horizontal force acts on the motorcycle from $t = 0$ to $t = 10$ s?

5.7 An astronaut drifts toward a space station at 8 m/s. He carries a maneuvering unit (a small hydrogen peroxide rocket) that can exert a total impulse of 720 N-s. The total mass of the astronaut, his suit, and the maneuvering unit is 120 kg. If he uses all of the impulse to slow himself down, what will be his velocity relative to the station?

P5.7

5.8 The total force on a 20-kg object is $10t^2\mathbf{i} + 60t\mathbf{j}$ (N). At $t = 0$, the object's velocity is $8\mathbf{i} - 4\mathbf{j}$ (m/s).
(a) What impulse is applied to the object from $t = 0$ to $t = 4$ s?
(b) What is the object's velocity at $t = 4$ s?

5.9 In Problem 5.8, what is the average total force on the object from $t = 0$ to $t = 4$ s?

5.10 The 1-lb collar A is initially at rest in the position shown on the smooth horizontal bar. At $t = 0$, a force $\mathbf{F} = \frac{1}{20}t^2\mathbf{i} + \frac{1}{10}t\mathbf{j} - \frac{1}{30}t^3\mathbf{k}$ (lb) is applied to the collar, causing it to slide along the bar. What is the velocity of the collar at $t = 2$ s?

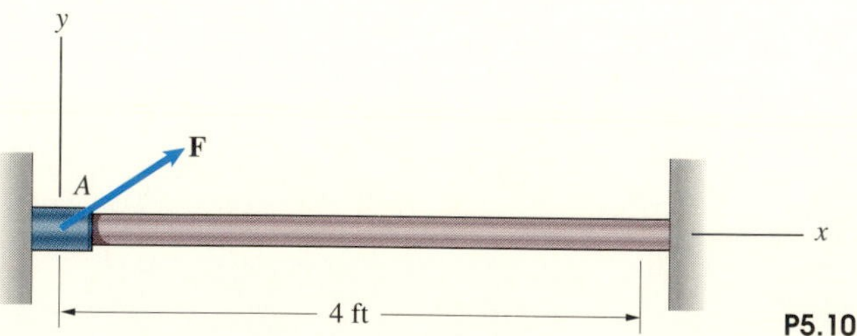

P5.10

5.11 (a) In Problem 5.10, use the principle of impulse and momentum to determine the collar's velocity as a function of time.
(b) Use the result of (a) to determine the time at which the collar reaches the right-hand end of the bar.

5.12 During the first 5 s of a 32,200-lb airplane's takeoff roll, the pilot increases the engine's thrust at a constant rate from 5000 lb to its full thrust of 25,000 lb.

(a) What impulse does the thrust exert on the airplane during the 5 s?
(b) If you neglect other forces, what total time is required for the airplane to reach its takeoff speed of 150 ft/s?

P5.12

5.13 The 100-lb box starts from rest and is subjected to the force shown. If you neglect friction, what is the box's velocity at $t = 8$ s?

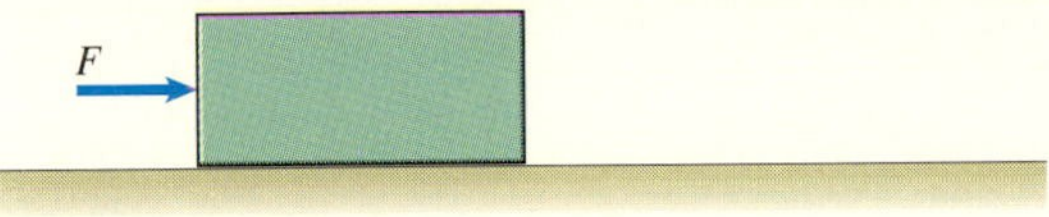

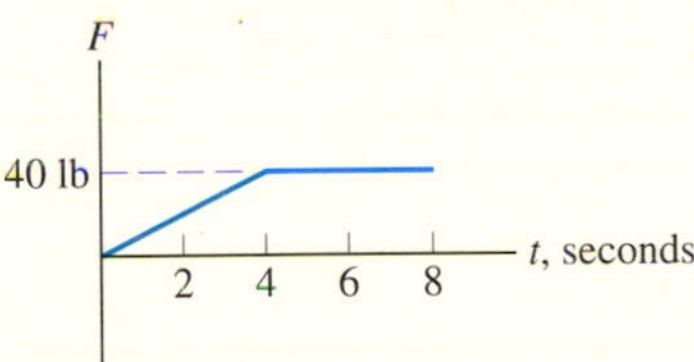

P5.13

5.14 Solve Problem 5.13 if the coefficients of friction between the box and the floor are $\mu_s = \mu_k = 0.2$.

5.15 The crate has a mass of 120 kg, and the coefficients of friction between it and the sloping dock are $\mu_s = 0.6$, $\mu_k = 0.5$. The crate starts from rest, and the winch exerts a tension $T = 1220$ N.
(a) What impulse is applied to the crate during the first second of motion?
(b) What is the crate's velocity after 1 s?

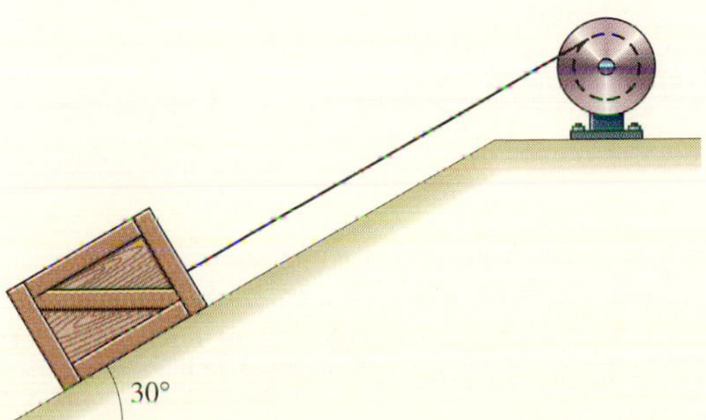

P5.15

5.16 Solve Problem 5.15 if the crate starts from rest at $t = 0$ and the winch exerts a tension $T = 1220 + 200t$ N.

5.17 In an assembly-line process, the 20-kg package A starts from rest and slides down the smooth ramp. Suppose that you want to design the hydraulic device B to exert a constant force of magnitude F on the package and bring it to rest in 0.15 s. What is the required force F?

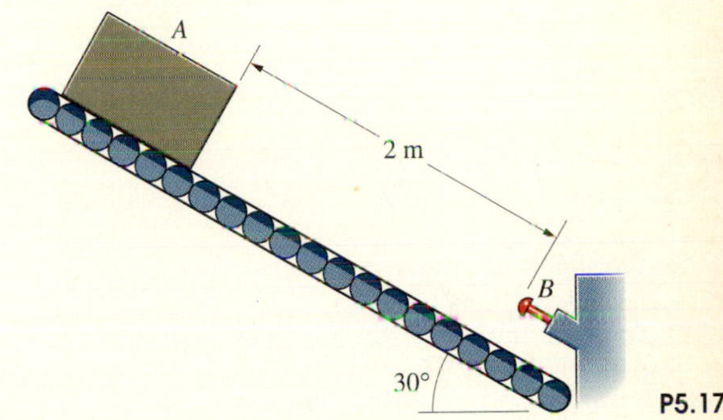

P5.17

5.18 In Problem 5.17, if the hydraulic device B exerts a force of magnitude $F = 540(1 + 0.4t^2)$ N on the package, where t is in seconds measured from the time of first contact, what time is required to bring the package to rest?

5.19 In a cathode-ray tube, an electron (mass $= 9.11 \times 10^{-31}$ kg) is projected at O with velocity $\mathbf{v} = (2.2 \times 10^7)\mathbf{i}$ (m/s). While it is between the charged plates, the electric field generated by the plates subjects it to a force $\mathbf{F} = -eE\mathbf{j}$. The charge of the electron is $e = 1.6 \times 10^{-19}$ C (coulombs), and the electric field strength is $E = 15 \sin(\omega t)$ kN/C, where the frequency $\omega = 2 \times 10^9$ s^{-1}.
(a) What impulse does the electric field exert on the electron while it is between the plates?
(b) What is the velocity of the electron as it leaves the region between the plates?

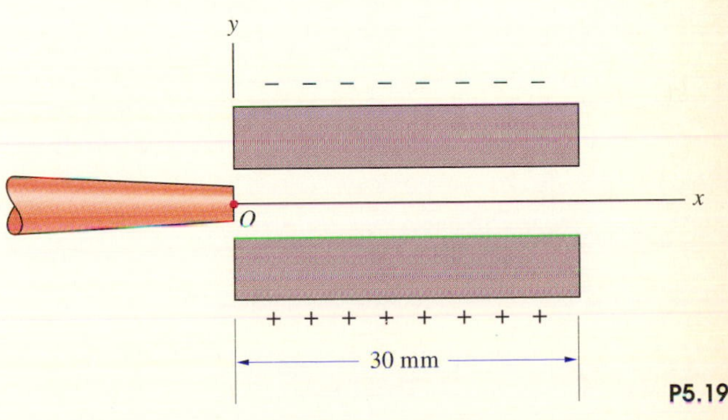

P5.19

5.20 The two weights are released from rest. What is the magnitude of their velocity after one-half second?

Strategy: Apply the principle of impulse and momentum to each weight individually.

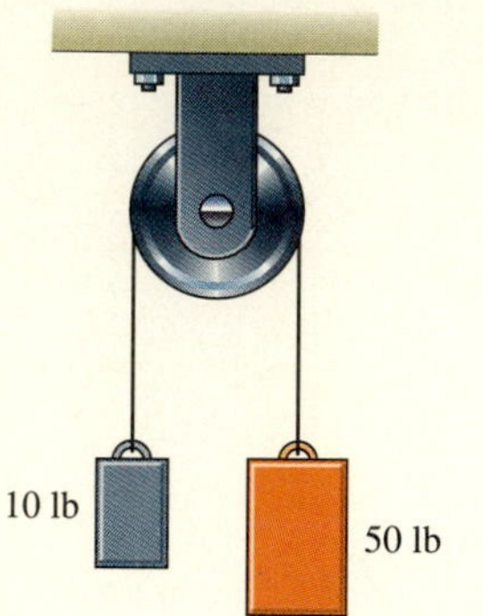

P5.20

5.21 The two crates are released from rest. Their masses are $m_A = 40$ kg and $m_B = 30$ kg, and the coefficient of kinetic friction between crate A and the inclined surface is $\mu_k = 0.15$. What is the magnitude of their velocity after 1 s?

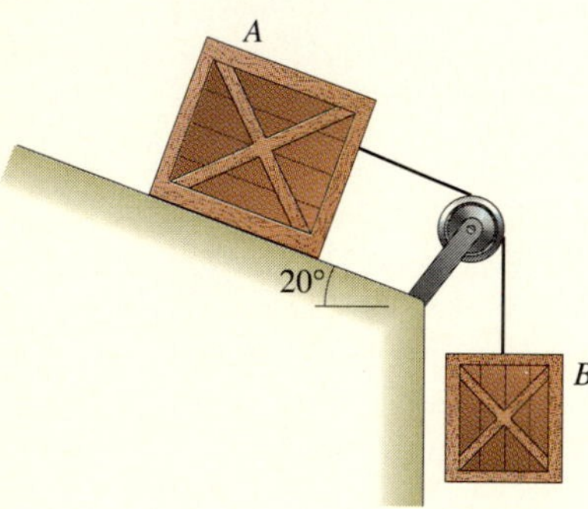

P5.21

5.22 The two crates are released from rest. Their masses are $m_A = 20$ kg and $m_B = 80$ kg, and the surfaces are smooth. The angle $\theta = 20°$. What is the magnitude of the velocity after 1 s?

Strategy: Apply the principle of impulse and momentum to each crate individually.

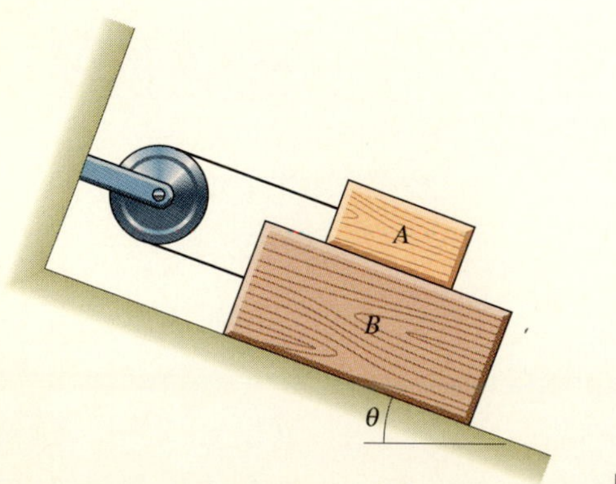

P5.22

5.23 In Problem 5.22, suppose that the coefficient of kinetic friction between the surfaces is $\mu_k = 0.1$. What is the magnitude of the velocity after 1 s?

5.24 In Example 5.2, if the range safety officer destroys the booster 1 s after it starts rotating, what is its velocity at the time it is destroyed?

5.25 An object of mass m slides with constant velocity v_0 on a horizontal table (seen from above in the figure). The object is attached by a string to the fixed point O and is in the position shown, with the string parallel to the x axis, at $t = 0$.

(a) Determine the x and y components of the force exerted on the mass by the string as functions of time.

(b) Use your results from part (a) and the principle of impulse and momentum to determine the velocity vector of the mass when it has traveled one-fourth of a revolution about point O.

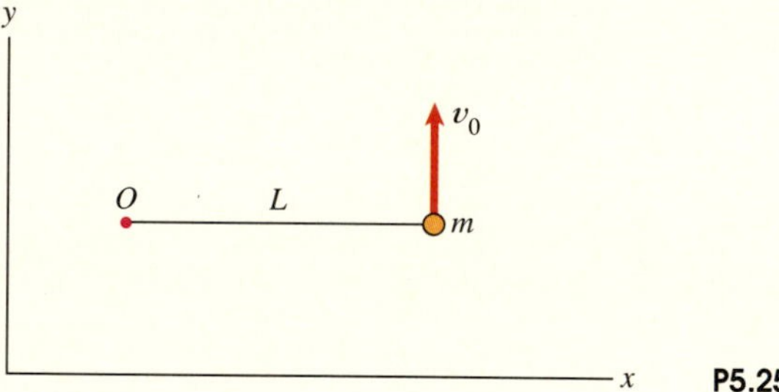

P5.25

5.26 At $t = 0$, a 50-lb projectile is given an initial velocity of 40 ft/s at 60° above the horizontal. Neglect drag.

(a) What impulse is applied to the projectile from $t = 0$ to $t = 2$ s?

(b) What is the projectile's velocity at $t = 2$ s?

5.27 A rail gun, which uses an electromagnetic field to accelerate an object, accelerates a 30-g projectile to 5 km/s in 0.0005 s. What average force is exerted on the projectile?

5.28 The powerboat is going 50 mi/hr when its motor is turned off. In 5 s its velocity decreases to 30 mi/hr. The boat and its passengers weigh 1800 lb. Determine the magnitude of the average force exerted on the boat by hydrodynamic and aerodynamic drag during the 5 s.

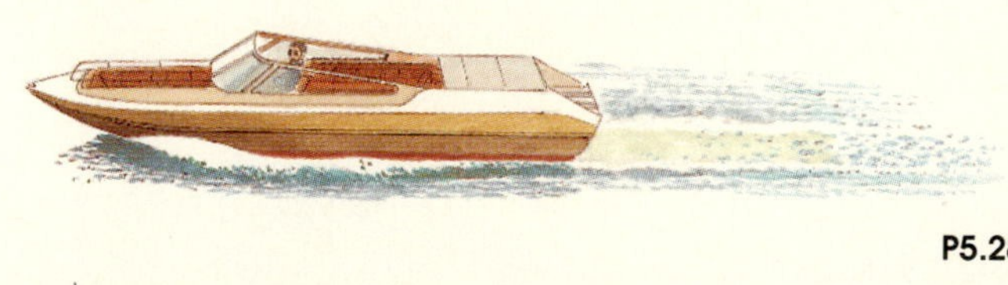

P5.28

5.29 A motorcycle starts from rest at $t = 0$ and travels along a circular track with 400-m radius. The tangential component of the total force on the motorcycle from $t = 0$ to $t = 30$ s is $\Sigma F_t = 100$ (N). The combined mass of the motorcycle and rider is 150 kg. What is the magnitude of the velocity at $t = 30$ s?

Strategy: Use Eq. (5.3).

P5.29

5.30 In Problem 5.29, what is the average of the *normal* component of the total force on the motorcycle from $t = 0$ to $t = 30$ s?

Strategy: Use Eq. (5.3) to determine the magnitude of the velocity as a function of time. With the resulting expression, calculate the average value of the total normal force $\Sigma F_n = mv^2/\rho$.

5.31 A motorcycle starts from rest at $t = 0$ and travels along a circular track with 400-m radius. The tangential component of the total force on the motorcycle from $t = 0$ to $t = 30$ s is $\Sigma F_t = 200 - 6t$ (N). The combined mass of the motorcycle and rider is 150 kg.

(a) What is the magnitude of the velocity at $t = 30$ s?

(b) What is the average of the tangential component of the total force from $t = 0$ to $t = 30$ s?

5.32 In Problem 5.31, what is the average of the *normal* component of the total force on the motorcycle from $t = 0$ to $t = 30$ s?

5.33 The 77-kg skier is traveling at 10 m/s at 1, and he goes from 1 to 2 in 0.7 s.

(a) If you neglect friction and aerodynamic drag, what is the time average of the tangential component of force exerted on him as he moves from 1 to 2?

(b) If his actual velocity is measured at 2 and determined to be 13.1 m/s, what is the time average of the tangential component of force exerted on him as he moves from 1 to 2?

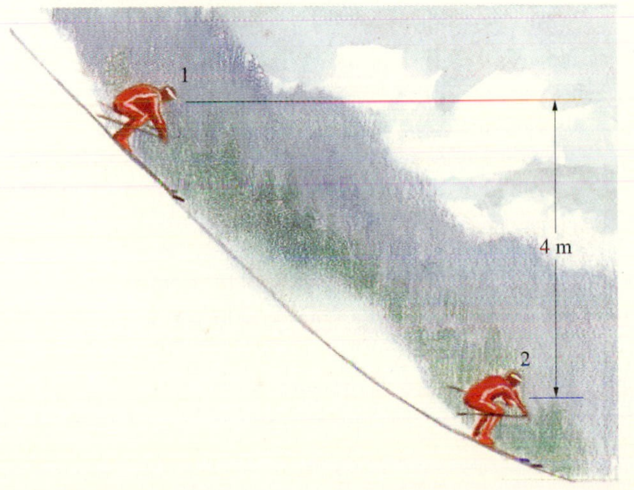

P5.33

5.34 In a test of an energy-absorbing bumper, a 2800-lb car is driven into a barrier at 5 mi/hr. The duration of the impact is 0.4 s, and the car bounces back from the barrier at 1 mi/hr.

(a) What is the magnitude of the average horizontal force exerted on the car during the impact?

(b) What is the average deceleration of the car during the impact?

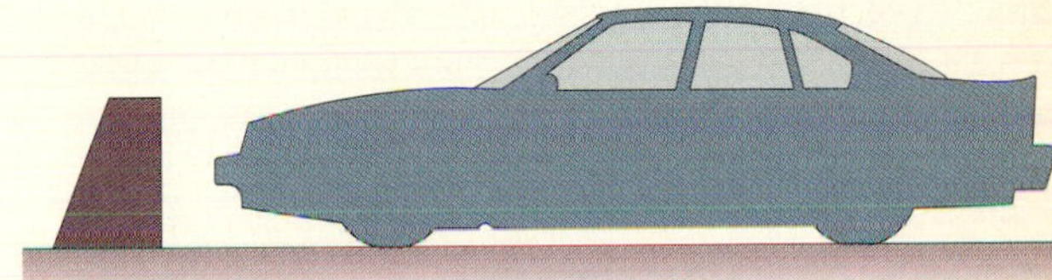

P5.34

5.35 A bioengineer, using an instrumented dummy to test a protective mask for a hockey goalie, launches the 170-g puck so that it strikes the mask moving horizontally at 40 m/s. From photographs of the impact, she estimates its duration to be 0.02 s and observes that the puck rebounds at 5 m/s.

(a) What linear impulse does the puck exert?

(b) What is the average value of the impulsive force exerted on the mask by the puck?

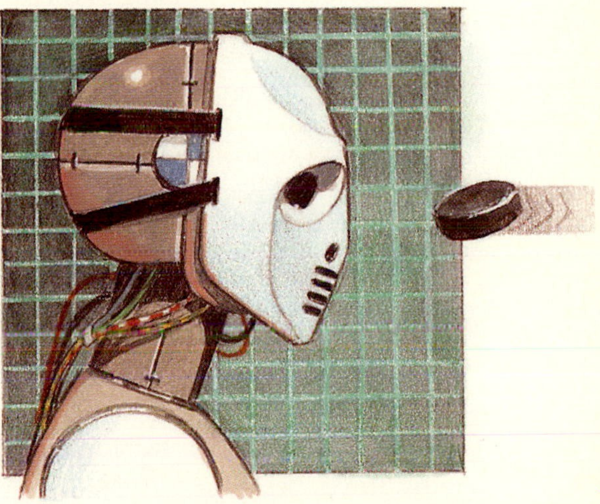

P5.35

5.36 A fragile object dropped onto a hard surface breaks because it is subjected to a large impulsive force. If you drop a 2-oz watch from 4 ft above the floor, the duration of the impact is 0.001 s, and the watch bounces 2 in. above the floor, what is the average value of the impulsive force?

5.37 A 50-lb projectile is subjected to an impulsive force with a duration of 0.01 s that accelerates it from rest to a velocity of 40 ft/s at 60° above the horizontal. What is the average value of the impulsive force?

Strategy: Use Eq. (5.2) to determine the average total force on the projectile. To determine the average value of the impulsive force, you must subtract the projectile's weight.

5.38 An entomologist measures the motion of a 3-g locust during its jump and determines that it accelerates from rest to 3.4 m/s in 25 ms (milliseconds). The angle of takeoff is 55° above the horizontal. What are the horizontal and vertical components of the average impulsive force exerted by the insect's hind legs during the jump?

5.39 A 5-oz baseball is 3 ft above the ground when it is struck by a bat. The horizontal distance to the point where the ball strikes the ground is 180 ft. Photographic studies indicate that the ball was moving approximately horizontally at 100 ft/s before it was struck, the duration of the impact was 0.015 s, and the ball was traveling at 30° above the horizontal after it was struck. What was the magnitude of the average impulsive force exerted on the ball by the bat?

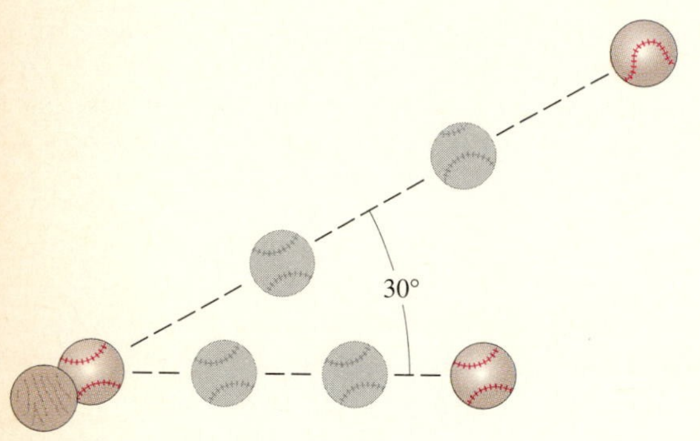

P5.39

5.40 A 1-kg ball is given a horizontal velocity of 1.2 m/s at A. Photographic measurements indicate that $b = 1.2$ m, $h = 1.3$ m, and the duration of the bounce at B is 0.1 s. What are the components of the average impulsive force exerted on the ball by the floor at B?

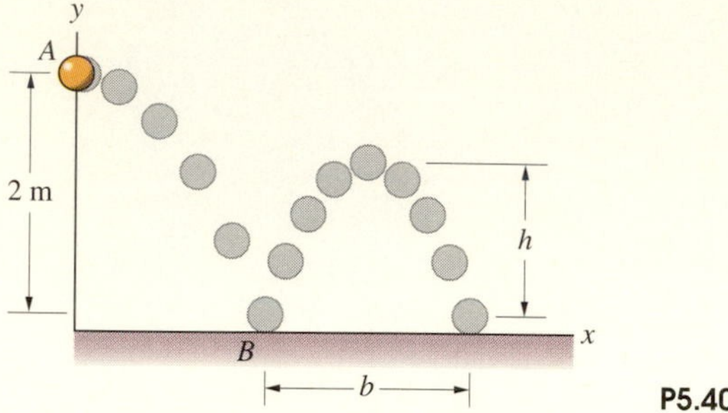

P5.40

5.41 At time $t = 0$, the two masses are released from rest on the smooth surface with the spring stretched. Show that at any later time t, the velocities of the masses are related by

$$m_A \mathbf{v}_A + m_B \mathbf{v}_B = 0.$$

Strategy: Write the principle of impulse and momentum for each mass.

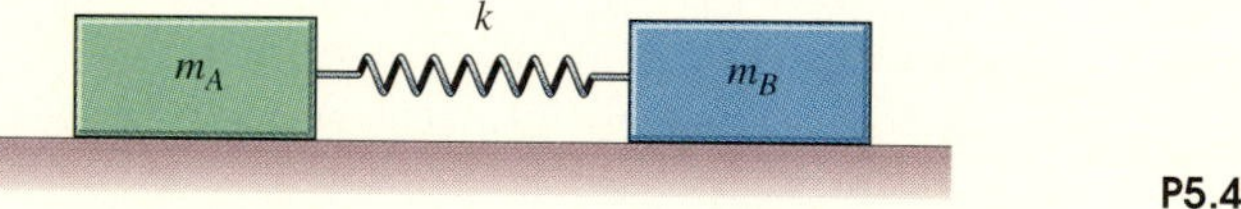

P5.41

5.42 In Problem 5.41, $m_A = 40$ kg, $m_B = 30$ kg, and $k = 400$ N/m. The two masses are released from rest on the smooth surface with the spring stretched 1 m. What are the magnitudes of the velocities of the masses when the spring is unstretched?

5.2 Conservation of Linear Momentum

In this section we consider the motions of several objects and show that if the effects of external forces can be neglected, the total linear momentum of the objects is conserved. (By *external forces* we will mean forces that are not exerted by the objects under consideration.) This result provides you with a powerful tool for analyzing interactions between objects, such as collisions, and also permits you to determine forces exerted on objects as a result of gaining or losing mass.

Consider the objects A and B in Fig. 5.6. $\mathbf{F}_{AB}$ is the force exerted on A by B, and $\mathbf{F}_{BA}$ is the force exerted on B by A. These forces could result from the two objects being in contact, for example, or could be exerted by a spring connecting them. As a consequence of Newton's third law, these forces are equal and opposite:

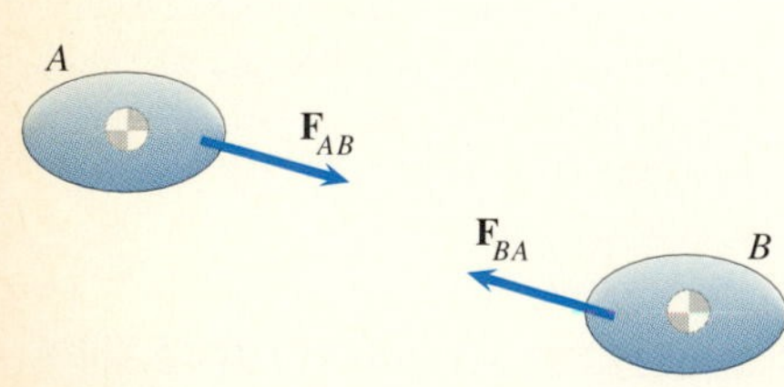

Fig. 5.6
Two objects and the forces they exert on each other.

$$\mathbf{F}_{AB} + \mathbf{F}_{BA} = 0. \tag{5.5}$$

Suppose that no external forces act on A and B, or that external forces are negligible in comparison with the forces A and B exert on each other. We can apply the principle of impulse and momentum to each object for arbitrary times t_1 and t_2:

$$\int_{t_1}^{t_2} \mathbf{F}_{AB}\, dt = m_A \mathbf{v}_{A2} - m_A \mathbf{v}_{A1},$$

$$\int_{t_1}^{t_2} \mathbf{F}_{BA}\, dt = m_B \mathbf{v}_{B2} - m_B \mathbf{v}_{B1}.$$

If we sum these equations, the terms on the left cancel and we obtain

$$m_A \mathbf{v}_{A1} + m_B \mathbf{v}_{B1} = m_A \mathbf{v}_{A2} + m_B \mathbf{v}_{B2},$$

which means that the total linear momentum of A and B is conserved:

$$\boxed{m_A \mathbf{v}_A + m_B \mathbf{v}_B = \text{constant.}} \qquad (5.6)$$

We can show that the velocity of the combined center of mass of the objects A and B (that is, the center of mass of A and B regarded as a single object) is also constant. Let $\mathbf{r}_A$ and $\mathbf{r}_B$ be the position vectors of their individual centers of mass (Fig. 5.7). The position of the combined center of mass is

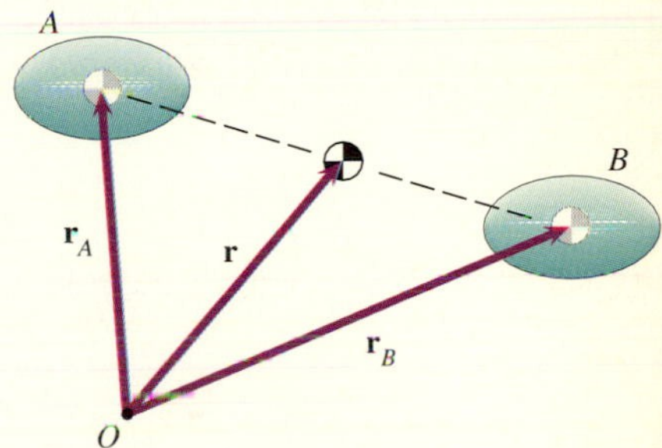

Fig. 5.7

Position vector $\mathbf{r}$ of the center of mass of A and B.

$$\mathbf{r} = \frac{m_A \mathbf{r}_A + m_B \mathbf{r}_B}{m_A + m_B}.$$

By taking the time derivative of this equation and using Eq. (5.6), we obtain

$$(m_A + m_B)\mathbf{v} = m_A \mathbf{v}_A + m_B \mathbf{v}_B = \text{constant}, \qquad (5.7)$$

where $\mathbf{v} = d\mathbf{r}/dt$ is the velocity of the combined center of mass. Although your goal will usually be to determine the individual motions of the objects, knowing that the velocity of the combined center of mass is constant can contribute to your understanding of a problem, and in some instances the motion of the combined center of mass may be the only information you can obtain.

Even when significant external forces act on A and B, if the external forces are negligible in a particular direction, Eqs. (5.6) and (5.7) apply in that direction. These equations also apply to an arbitrary number of objects: If the external forces acting on any collection of objects are negligible, the total linear momentum of the objects is conserved and the velocity of their center of mass is constant.

In the following example we demonstrate the use of Eqs. (5.6) and (5.7) to analyze motions of objects. When you know initial velocities of objects and you can neglect external forces, these equations relate their velocities at any subsequent time.

Example 5.4

A person of mass m_P stands at the center of a stationary barge of mass m_B (Fig. 5.8). Neglect horizontal forces exerted on the barge by the water.
(a) If the person starts running to the right with velocity v_P relative to the water, what is the resulting velocity of the barge relative to the water?
(b) If the person stops when he reaches the right-hand end of the barge, what are his position and the barge's position relative to their original positions?

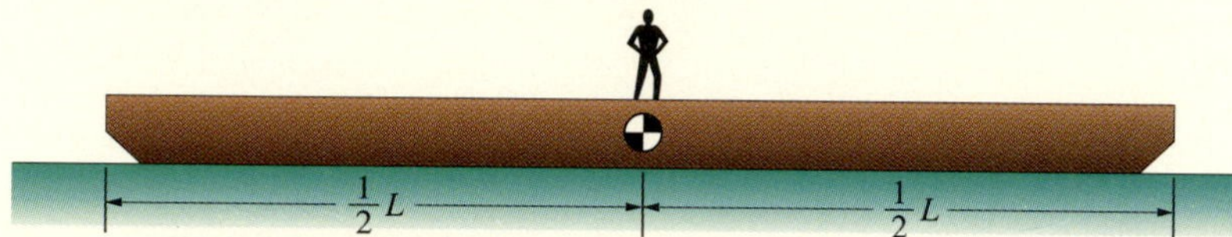

Fig. 5.8

STRATEGY

(a) The only horizontal forces exerted on the person and the barge are the forces they exert on each other. Therefore their total linear momentum *in the horizontal direction* is conserved and we can use Eq. (5.6) to determine the barge's velocity while the person is running.
(b) The combined center of mass of the person and the barge is initially stationary, so it must remain stationary. Knowing the position of the combined center of mass, we can determine the positions of the person and barge when the person is at the right-hand end of the barge.

SOLUTION

(a) Before the person starts running, the total linear momentum of the person and the barge in the horizontal direction is zero, so it must be zero after he starts running. Letting v_B be the value of the barge's velocity *to the left* while the person is running (Fig. a), we obtain

$$m_P v_P + m_B(-v_B) = 0,$$

so the velocity of the barge while he runs is

$$v_B = \left(\frac{m_P}{m_B}\right) v_P.$$

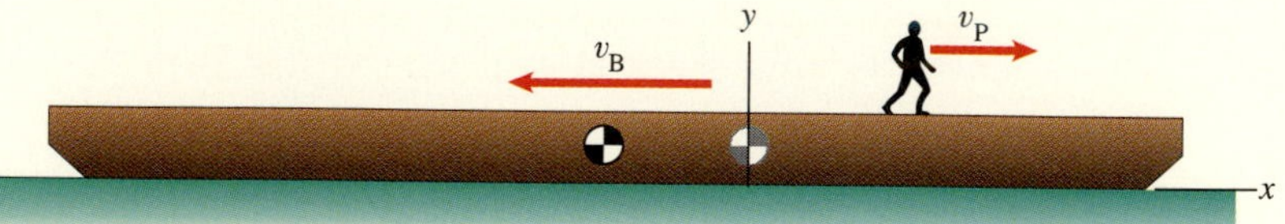

(a) Velocities of the person and barge.

(b) Let the origin of the coordinate system in Fig. (b) be the original horizontal position of the centers of mass of the barge and the person, and let x_B be the position of the barge's center of mass *to the left of the origin*. When the person has stopped at the right-hand end of the barge, the combined center of mass must still be at $x = 0$:

$$\frac{x_P m_P + (-x_B) m_B}{m_P + m_B} = 0.$$

Solving this equation together with the relation $x_P + x_B = L/2$, we obtain

$$x_P = \frac{m_B L}{2(m_P + m_B)}, \qquad x_B = \frac{m_P L}{2(m_P + m_B)}.$$

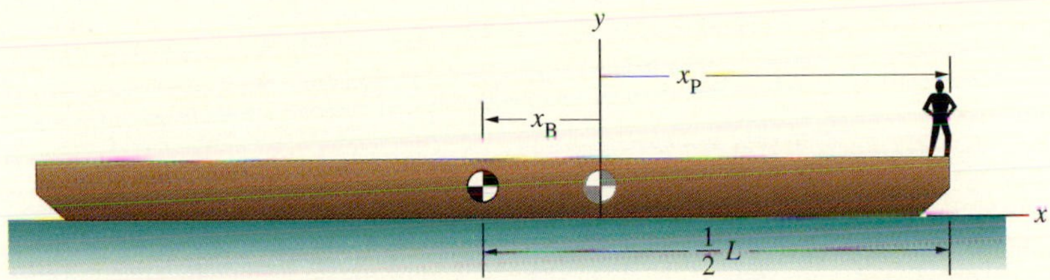

(b) Positions after the person has stopped.

DISCUSSION

This example is a well-known illustration of the power of momentum methods. Notice that we were able to determine the velocity of the barge and the final positions of the person and barge even though we did not know the complicated time dependence of the horizontal forces they exert on each other.

5.3 *Impacts*

In machines that perform stamping or forging operations, dies impact against workpieces. Mechanical printers create images by impacting metal elements against the paper and platen. Vehicles impact each other intentionally, as when railroad cars are rolled against each other to couple them, and unintentionally in accidents. Impacts occur in many situations of concern in engineering. In this section we consider a basic question: If you know the velocities of two objects before they collide, how do you determine their velocities afterward? In other words, what is the effect of the impact on their motions?

If colliding objects are not subjected to external forces, their total linear momentum must be the same before and after the impact. Even when they are subjected to external forces, the force of the impact is often so large, and its duration so brief, that the effect of external forces on their motions during the impact is negligible. Suppose that objects A and B with velocities $\mathbf{v}_A$ and $\mathbf{v}_B$ collide, and let $\mathbf{v}'_A$ and $\mathbf{v}'_B$ be their velocities after the impact (Fig. 5.9a). If the effects of external forces are negligible, their total linear momentum is conserved:

$$m_A \mathbf{v}_A + m_B \mathbf{v}_B = m_A \mathbf{v}'_A + m_B \mathbf{v}'_B. \tag{5.8}$$

Furthermore, the velocity $\mathbf{v}$ of their center of mass is the same before and after the impact. From Eq. (5.7),

$$\mathbf{v} = \frac{m_A \mathbf{v}_A + m_B \mathbf{v}_B}{m_A + m_B}. \tag{5.9}$$

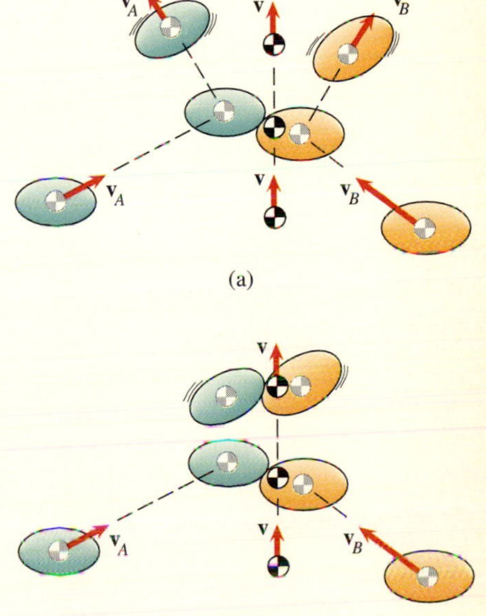

Fig. 5.9

(a) Velocities of A and B before and after the impact, and the velocity $\mathbf{v}$ of their center of mass.
(b) A perfectly plastic impact.

If A and B adhere and remain together after they collide, they are said to undergo a **perfectly plastic impact**. Equation (5.9) gives the velocity of the center of mass of the object they form after the impact (Fig. 5.9b). A remarkable feature of this result is that we determine the velocity following the impact *without considering the physical nature of the impact.*

If A and B do not adhere, linear momentum conservation alone does not provide enough equations to determine their velocities after the impact. We first consider the case in which they travel along the same straight line before and after they collide.

Direct Central Impacts

Suppose that the centers of mass of A and B travel along the same straight line with velocities v_A and v_B before their impact (Fig. 5.10a). Let R be the magnitude of the force they exert on each other during the impact (Fig. 5.10b). We assume that the contacting surfaces are oriented so that R is parallel to the line along which they travel and directed toward their centers of mass. This condition, called **direct central impact**, means that they continue to travel along the same straight line after their impact (Fig. 5.10c). If the effects of external forces during the impact are negligible, their total linear momentum is conserved:

$$m_A v_A + m_B v_B = m_A v'_A + m_B v'_B. \tag{5.10}$$

However, we need another equation to determine the velocities v'_A and v'_B. To obtain it, we must consider the impact in more detail.

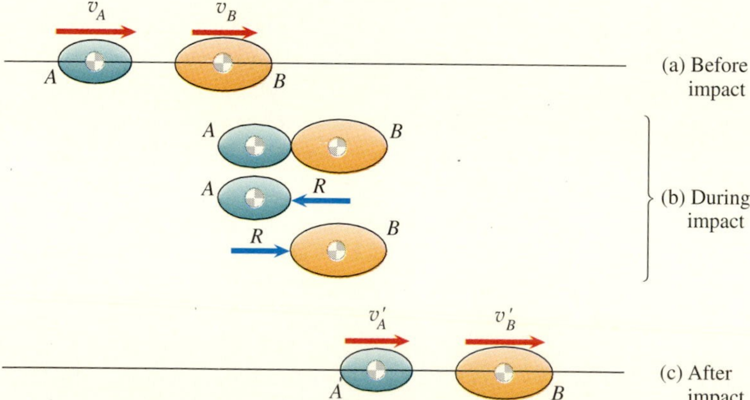

Fig. 5.10
(a) Objects A and B traveling along the same straight line.
(b) During the impact, they exert a force R on each other.
(c) They travel along the same straight line after the central impact.

Let t_1 be the time at which A and B first come into contact (Fig. 5.11a). As a result of the impact, they will deform and their centers of mass will continue to approach each other. At a time t_C, their centers of mass will have reached their nearest proximity (Fig. 5.11b). At this time the relative velocity of the two centers of mass is zero, so they have the same velocity. We denote it by v_C. The objects then begin to move apart and separate at a time t_2 (Fig. 5.11c). We apply the principle of impulse and momentum to A during the intervals of time from t_1 to the time of closest approach t_C and also from t_C to t_2:

$$\int_{t_1}^{t_C} -R\,dt = m_A v_C - m_A v_A, \qquad (5.11)$$

$$\int_{t_C}^{t_2} -R\,dt = m_A v'_A - m_A v_C. \qquad (5.12)$$

Then we apply this principle to B for the same intervals of time:

$$\int_{t_1}^{t_C} R\,dt = m_B v_C - m_B v_B, \qquad (5.13)$$

$$\int_{t_C}^{t_2} R\,dt = m_B v'_B - m_B v_C. \qquad (5.14)$$

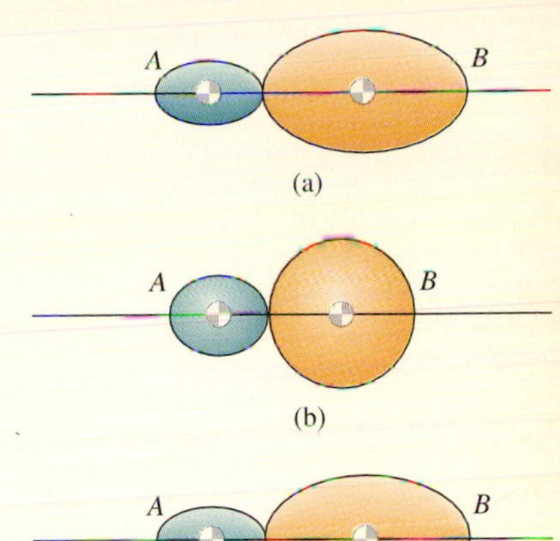

Fig. 5.11
(a) First contact, $t = t_1$.
(b) Closest approach, $t = t_C$.
(c) End of contact, $t = t_2$.

As a result of the impact, part of the objects' kinetic energy can be lost due to a variety of mechanisms, including permanent deformation and generation of heat and sound. As a consequence, the impulse they impart to each other during the "restitution" phase of the impact from t_C to t_2 is in general smaller than the impulse they impart from t_1 to t_C. The ratio of these impulses is called the **coefficient of restitution**:

$$e = \frac{\displaystyle\int_{t_C}^{t_2} R\,dt}{\displaystyle\int_{t_1}^{t_C} R\,dt}. \qquad (5.15)$$

Its value depends on the properties of the objects as well as their velocities and orientations when they collide, and it can be determined only by experiment or by a detailed analysis of the deformations of the objects during the impact.

If we divide Eq. (5.12) by Eq. (5.11) and divide Eq. (5.14) by Eq. (5.13), we can express the resulting equations in the forms

$$(v_C - v_A)e = v'_A - v_C,$$

$$(v_C - v_B)e = v'_B - v_C.$$

Subtracting the first equation from the second one, we obtain

$$\boxed{e = \frac{v'_B - v'_A}{v_A - v_B}.} \qquad (5.16)$$

Thus the coefficient of restitution is related in a simple way to the relative velocities of the objects before and after the impact. If e is known, you can use Eq. (5.16) together with the equation of conservation of linear momentum, Eq. (5.10), to determine v'_A and v'_B.

If $e = 0$, Eq. (5.16) indicates that $v'_B = v'_A$. The objects remain together after the impact, and the impact is perfectly plastic. If $e = 1$, it can be shown that the total kinetic energy is the same before and after the impact:

$$\frac{1}{2}m_A v_A^2 + \frac{1}{2}m_B v_B^2 = \frac{1}{2}m_A(v'_A)^2 + \frac{1}{2}m_B(v'_B)^2 \qquad \text{(when } e = 1\text{)}.$$

An impact in which kinetic energy is conserved is called **perfectly elastic**. Although this is sometimes a useful approximation, energy is lost in any impact in which material objects come into contact. If you can hear a collision, kinetic energy has been converted into sound. Permanent deformations of the colliding objects after the impact also represent losses of kinetic energy.

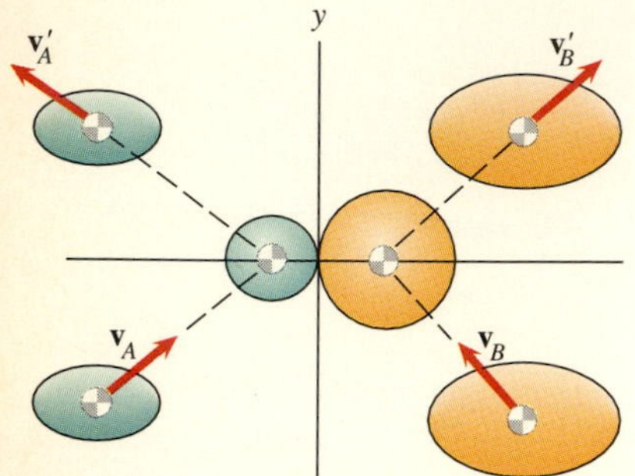

Fig. 5.12
An oblique central impact.

Oblique Central Impacts

We can extend the procedure used to analyze direct central impacts to the case in which the objects approach each other at an oblique angle. Suppose that A and B approach with arbitrary velocities $\mathbf{v}_A$ and $\mathbf{v}_B$ (Fig. 5.12) and that the forces they exert on each other during their impact are parallel to the x axis and point toward their centers of mass. No forces are exerted on them in the y or z directions, so their velocities in those directions are unchanged by the impact:

$$(\mathbf{v}'_A)_y = (\mathbf{v}_A)_y, \qquad (\mathbf{v}'_B)_y = (\mathbf{v}_B)_y,$$
$$(\mathbf{v}'_A)_z = (\mathbf{v}_A)_z, \qquad (\mathbf{v}'_B)_z = (\mathbf{v}_B)_z. \tag{5.17}$$

In the x direction, linear momentum is conserved,

$$m_A(\mathbf{v}_A)_x + m_B(\mathbf{v}_B)_x = m_A(\mathbf{v}'_A)_x + m_B(\mathbf{v}'_B)_x, \tag{5.18}$$

and by the same analysis we used to arrive at Eq. (5.16), the x components of velocity satisfy the relation

$$e = \frac{(\mathbf{v}'_B)_x - (\mathbf{v}'_A)_x}{(\mathbf{v}_A)_x - (\mathbf{v}_B)_x}. \tag{5.19}$$

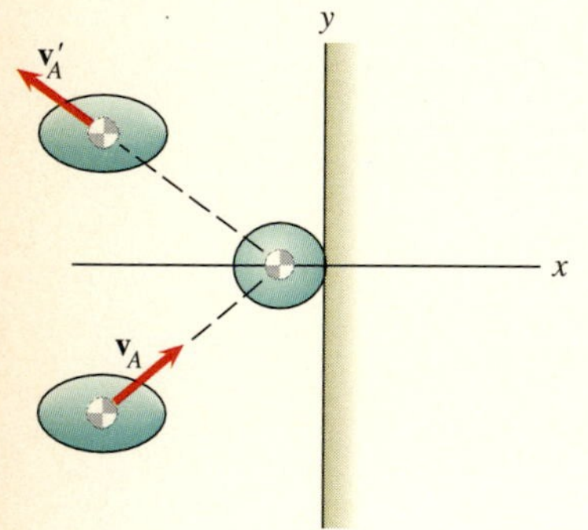

Fig. 5.13
Impact with a stationary object.

We can analyze an oblique central impact in which an object A hits a stationary object B if friction is negligible. Suppose that B is constrained so that it cannot move relative to the inertial reference frame. For example, in Fig. 5.13, A strikes a wall B that is fixed relative to the earth. The y and z components of A's velocity are unchanged, because friction is neglected and the impact exerts no force in those directions. The x component of A's velocity after the impact is given by Eq. (5.19) with B's velocity equal to zero:

$$(\mathbf{v}'_A)_x = -e(\mathbf{v}_A)_x.$$

In the following examples we analyze the impact of two objects. If an impact is perfectly plastic, which means the objects adhere and remain together, you can determine from Eq. (5.9) the velocity of their center of mass after the impact. In an oblique central impact, in terms of the coordinate system shown in Fig. 5.12, the y and z components of the velocities of the objects are unchanged, and you can solve Eqs. (5.18) and (5.19) for the x components of the velocities after the impact.

Example 5.5

The two 10-lb weights in Fig. 5.14 slide on the smooth horizontal bar. Determine their velocities after they collide: (a) if they are coated with Velcro and stick together; (b) if the coefficient of restitution is $e = 0.8$.

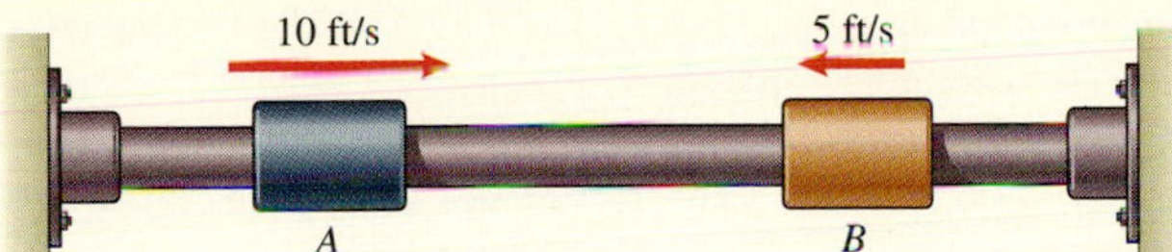

Fig. 5.14

STRATEGY

(a) If the weights stick together, they have the same velocity after their collision. We can determine the velocity from conservation of linear momentum. (b) Knowing the coefficient of restitution, we can determine the velocity of each weight after the collision by using conservation of linear momentum together with the definition of the coefficient of restitution, Eq. (5.16).

SOLUTION

(a) The velocities before the collision are $v_A = 10$ ft/s and $v_B = -5$ ft/s. Let v be their common velocity after the collision. From Eq. (5.10), conservation of linear momentum requires that

$$m_A v_A + m_B v_B = m_A v'_A + m_B v'_B:$$

$$\left(\frac{10}{32.2}\right)(10) + \left(\frac{10}{32.2}\right)(-5) = \left(\frac{10}{32.2} + \frac{10}{32.2}\right) v.$$

Solving, we obtain $v = 2.5$ ft/s. The connected weights move to the right at 2.5 ft/s after the collision.

(b) Conservation of linear momentum requires that

$$m_A v_A + m_B v_B = m_A v'_A + m_B v'_B:$$

$$\left(\frac{10}{32.2}\right)(10) + \left(\frac{10}{32.2}\right)(-5) = \left(\frac{10}{32.2}\right) v'_A + \left(\frac{10}{32.2}\right) v'_B.$$

From Eq. (5.16),

$$e = \frac{v'_B - v'_A}{v_A - v_B}:$$

$$0.8 = \frac{v'_B - v'_A}{10 - (-5)}.$$

We now have two equations in v'_A and v'_B. Solving them, we obtain $v'_A = -3.5$ ft/s and $v'_B = 8.5$ ft/s. A moves to the left at 3.5 ft/s and B moves to the right at 8.5 ft/s after the collision.

Example 5.6

The Apollo CSM (A) attempts to dock with the Soyuz capsule (B), July 15, 1975 (Fig. 5.15). Their masses are $m_A = 18$ Mg and $m_B = 6.6$ Mg. The Soyuz is stationary relative to the reference frame shown, and the CSM approaches with velocity $\mathbf{v}_A = 0.2\mathbf{i} + 0.03\mathbf{j} - 0.02\mathbf{k}$ (m/s).

(a) If the first attempt at docking is successful, what is the velocity of the center of mass of the combined vehicles afterward?

(b) If the first attempt is unsuccessful and the coefficient of restitution of the resulting impact is $e = 0.95$, what are the velocities of the two spacecraft after the impact?

Fig. 5.15

STRATEGY

(a) If the docking is successful, the impact is perfectly plastic and we can use Eq. (5.9) to determine the velocity of the center of mass of the combined object after the impact.

(b) By assuming an oblique central impact with the forces exerted by the docking collars parallel to the x axis, we can use Eqs. (5.18) and (5.19) to determine the velocities of both spacecraft after the impact.

SOLUTION

(a) From Eq. (5.9), the velocity of the center of mass of the combined vehicles is

$$\begin{aligned}\mathbf{v} &= \frac{m_A\mathbf{v}_A + m_B\mathbf{v}_B}{m_A + m_B} \\ &= \frac{(18)(0.2\mathbf{i} + 0.03\mathbf{j} - 0.02\mathbf{k}) + 0}{18 + 6.6} \\ &= 0.146\mathbf{i} + 0.022\mathbf{j} - 0.015\mathbf{k}\ \text{(m/s)}.\end{aligned}$$

(b) The y and z components of the velocities of both spacecraft are unchanged. To determine the x components, we use conservation of linear momentum, Eq. (5.18),

$$m_A(\mathbf{v}_A)_x + m_B(\mathbf{v}_B)_x = m_A(\mathbf{v}'_A)_x + m_B(\mathbf{v}'_B)_x,$$
$$(18)(0.2) + 0 = (18)(\mathbf{v}'_A)_x + (6.6)(\mathbf{v}'_B)_x,$$

and the coefficient of restitution, Eq. (5.19),

$$e = \frac{(\mathbf{v}'_B)_x - (\mathbf{v}'_A)_x}{(\mathbf{v}_A)_x - (\mathbf{v}_B)_x},$$
$$0.95 = \frac{(\mathbf{v}'_B)_x - (\mathbf{v}'_A)_x}{0.2 - 0}.$$

Solving these two equations, we obtain $(\mathbf{v}'_A)_x = 0.095$ (m/s) and $(\mathbf{v}'_B)_x = 0.285$ (m/s), so the velocities of the spacecraft after the impact are

$$\mathbf{v}'_A = 0.095\mathbf{i} + 0.03\mathbf{j} - 0.02\mathbf{k} \text{ (m/s)},$$
$$\mathbf{v}'_B = 0.285\mathbf{i} \text{ (m/s)}.$$

Problems

5.43 A girl weighing 100 lb stands at rest on a barge weighing 500 lb. She starts running at 10 ft/s *relative to the barge* and runs off the end. Neglect the horizontal force exerted on the barge by the water.
(a) Just before she hits the water, what is the horizontal component of her velocity relative to the water?
(b) What is the velocity of the barge relative to the water while she runs?

P5.43

5.44 A 60-kg astronaut aboard the space shuttle kicks off toward the center of mass of the 105-Mg shuttle at 1 m/s relative to the shuttle. He travels 6 m relative to the shuttle before coming to rest at the opposite wall.
(a) What is the magnitude of the change in the velocity of the shuttle while he is in motion?
(b) What is the magnitude of the displacement of the center of mass of the shuttle due to his "flight"?

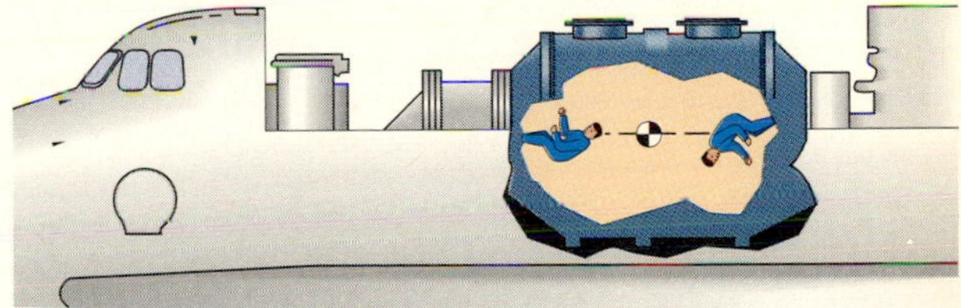

P5.44

5.45 An 80-lb boy sitting in a stationary 20-lb wagon wants to simulate rocket propulsion by throwing bricks out of the wagon. Neglect horizontal forces on the wagon's wheels. If he has three bricks weighing 10 lb each and throws them with a horizontal velocity of 10 ft/s relative to the wagon, determine the velocity he attains (a) if he throws the bricks one at a time; (b) if he throws them all at once.

P5.45

5.46 Two railroad cars ($m_A = 1.7m_B$) collide and become coupled. Car A is full and car B is half-full of carbolic acid. When the cars impact, the acid in B sloshes back and forth violently.
(a) Immediately after the impact, what is the velocity of the common center of mass of the two cars?
(b) A few seconds later, when the sloshing has subsided, what is the velocity of the two cars?

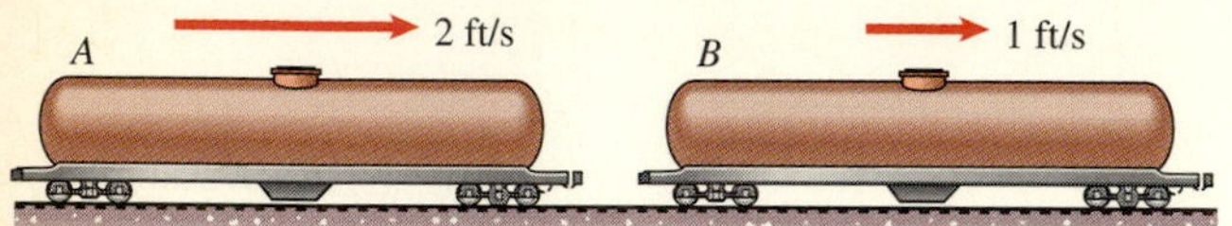

P5.46

5.47 In Problem 5.46, if the track slopes one-half degree upward to the right and the cars are initially 10 ft apart, what is the velocity of their common center of mass immediately after the impact?

5.48 A 400-kg satellite S traveling at 7 km/s is hit by a 1-kg meteor M traveling at 12 km/s. The meteor is embedded in the satellite by the impact. Determine the magnitude of the velocity of their common center of mass after the impact and the angle β between the path of the center of mass and the original path of the satellite.

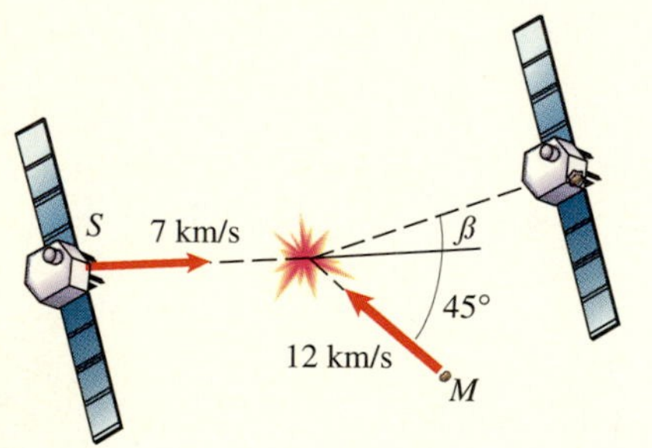

P5.48

5.49 In Problem 5.48, what would the magnitude of the velocity of the 1-kg meteor M need to be to cause the angle between the original path of the satellite and the path of the center of mass of the combined satellite and meteor after the impact to be $\beta = 0.5°$? What is the magnitude of the velocity of the center of mass after the impact?

5.50 A catapult designed to throw a line to ships in distress throws a 2-kg projectile. The mass of the catapult is 36 kg, and it rests on a smooth surface. If the velocity of the projectile *relative to the earth* as it leaves the tube is 50 m/s at $\theta_0 = 30°$ relative to the horizontal, what is the resulting velocity of the catapult toward the left?

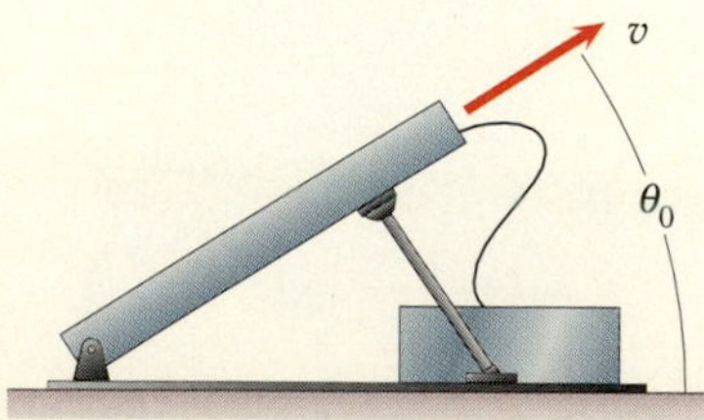

P5.50

5.51 In Problem 5.50, if the velocity of the projectile *relative to the catapult* as it leaves the tube is 50 m/s at $\theta_0 = 30°$ relative to the horizontal, what is the resulting velocity of the catapult toward the left?

5.52 A bullet (mass m) hits a stationary block of wood (mass m_B) and becomes embedded in it. The coefficient of kinetic friction between the block and the floor is μ_k. As a result of the impact, the block slides a distance D before stopping. What was the velocity v of the bullet?

Strategy: First solve the impact problem to determine the velocity of the block and the embedded bullet after the impact in terms of v, then relate the initial velocity of the block and the embedded bullet to the distance D that the block slides.

P5.52

5.53 A 1-oz bullet moving horizontally hits a suspended 100-lb block of wood and becomes embedded in it. If you measure the angle through which the wires supporting the block swing as a result of the impact and determine it to be 7°, what was the bullet's velocity?

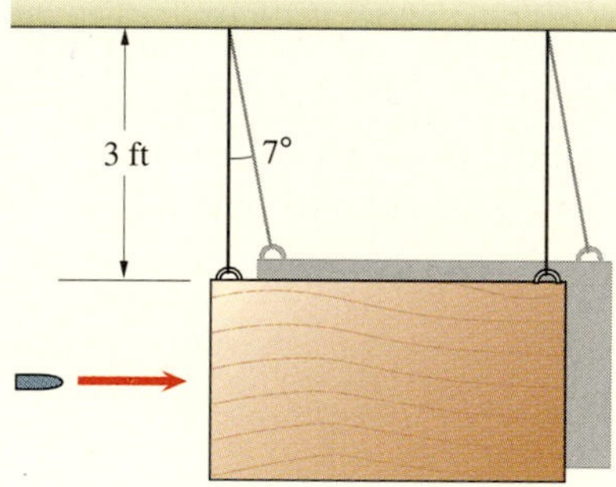

P5.53

5.54 The overhead conveyor drops the 12-kg package A into the 1.6-kg carton B. The package is "tacky" and sticks to the bottom of the carton. If the coefficient of friction between the carton and the horizontal conveyor is $\mu_k = 0.2$, what distance does the carton slide after the impact?

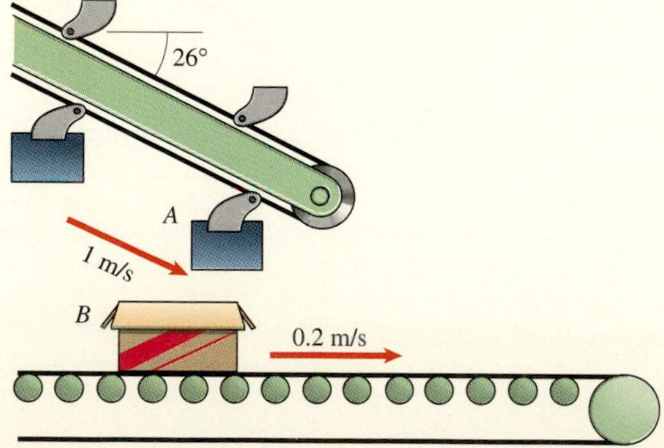

P5.54

5.55 Suppose you investigate an accident in which a 1300-kg car with velocity $\mathbf{v}_C = 36\mathbf{j}$ (km/hr) collided with a 5400-kg bus with velocity $\mathbf{v}_B = 20\mathbf{i}$ (km/hr). The vehicles became entangled and remained together after the collision.
(a) What was the velocity of the common center of mass of the two vehicles after the collision?
(b) If you estimate the coefficient of friction between the sliding vehicles and the road after the collision to be $\mu = 0.4$, what is the approximate final position of their common center of mass relative to its position when the impact occurs?

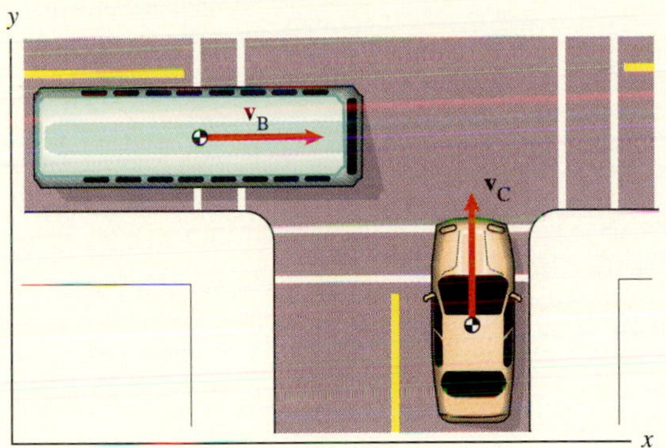

P5.55

5.56 The velocity of the 100-kg astronaut A relative to the space station is $40\mathbf{i} + 30\mathbf{j}$ (mm/s). The velocity of the 200-kg structural member B relative to the station is $-20\mathbf{i} + 30\mathbf{j}$ (mm/s). When they approach each other, the astronaut grasps and clings to the structural member.
(a) Determine the velocity of their common center of mass when they arrive at the station.
(b) Determine the approximate position at which they contact the station.

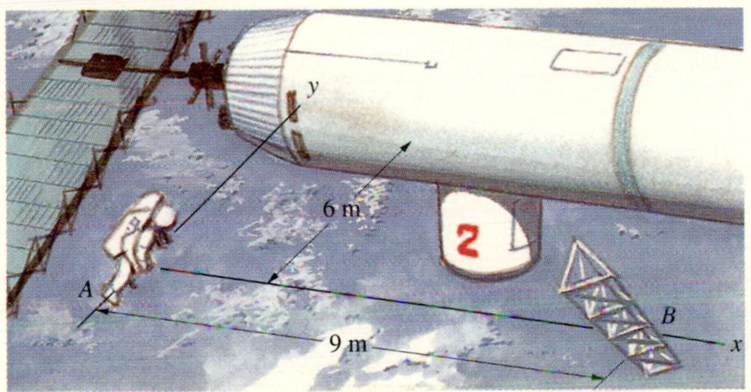

P5.56

5.57 Objects A and B with the same mass m undergo a direct central impact. The velocity of A before the impact is v_A, and B is stationary. Determine the velocities of A and B after the impact if it is (a) perfectly plastic ($e = 0$); (b) perfectly elastic ($e = 1$).

P5.57

5.58 In Problem 5.57, if the velocity of B after the impact is $0.6v_A$, determine the coefficient of restitution e and the velocity of A after the impact.

5.59 Objects A and B with masses m_A and m_B undergo a direct central impact.
(a) If $e = 1$, show that the total kinetic energy after the impact is equal to the total kinetic energy before the impact.
(b) If $e = 0$, how much kinetic energy is lost as a result of the collision?

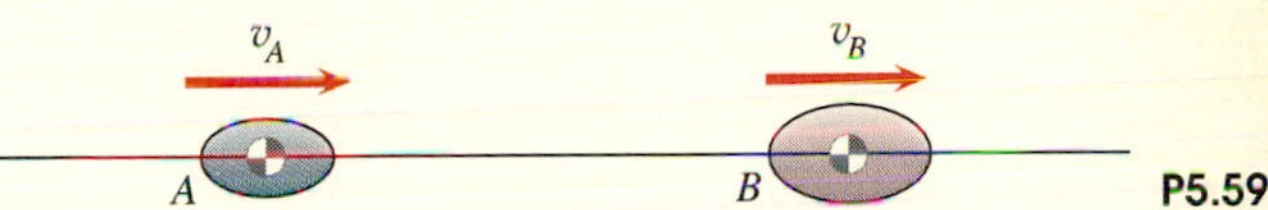

P5.59

5.60 The 20-lb weight A and 30-lb weight B slide on the smooth horizontal bar. Determine their velocities after they collide if the coefficient of restitution is $e = 0.8$.

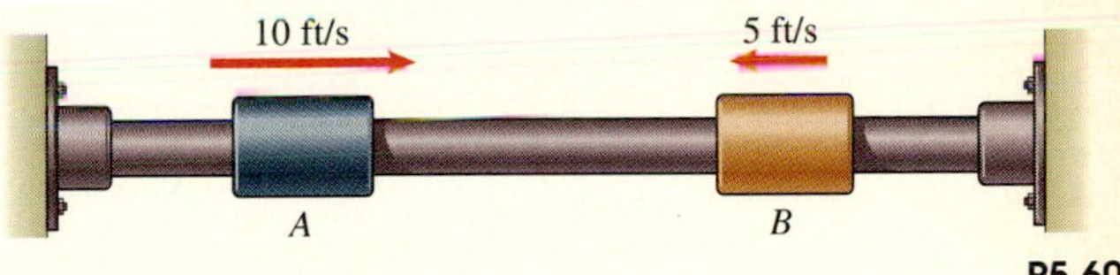

P5.60

5.61 Two cars with energy-absorbing bumpers collide with speeds $v_A = v_B = 5$ mi/hr. Their weights are $W_A = 2800$ lb and $W_B = 4400$ lb. If the coefficient of restitution of the collision is $e = 0.2$, what are the velocities of the cars after the collision?

P5.61

5.62 In Problem 5.61, if the duration of the collision is 0.1 s, what are the magnitudes of the average accelerations to which the occupants of the two cars are subjected?

5.63 The 10-kg mass A is moving at 5 m/s when it is 1 m from the stationary 10-kg mass B. The coefficient of kinetic friction between the floor and the two masses is $\mu_k = 0.6$, and the coefficient of restitution of the impact is $e = 0.5$. Determine how far B moves from its initial position as a result of the impact.

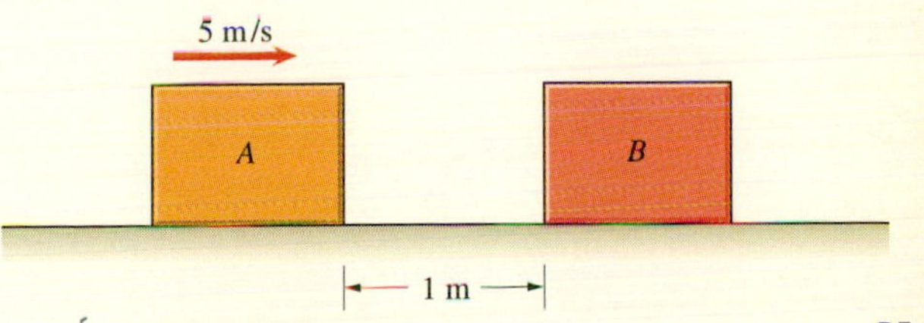

P5.63

5.64 The kinetic coefficients of friction between the 5-kg crates A and B and the inclined surface are 0.1 and 0.4, respectively. The coefficient of restitution between the crates is $e = 0.8$. If the crates are released from rest in the positions shown, what are the magnitudes of their velocities immediately after they collide?

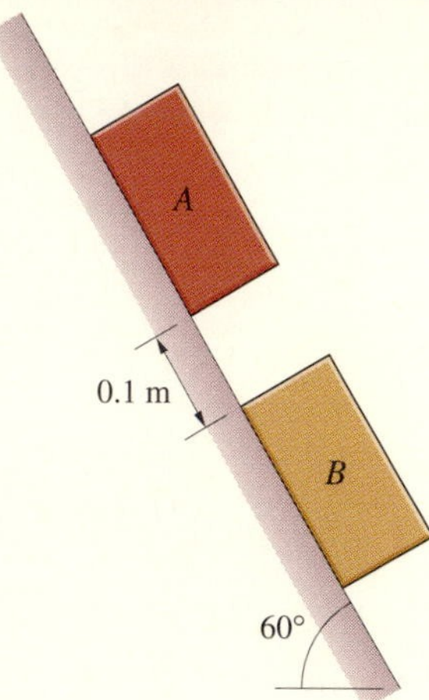

P5.64

5.65 Solve Problem 5.64 if crate A has a velocity of 0.2 m/s down the inclined surface and crate B is at rest when the crates are in the positions shown.

5.66 Suppose you investigate an accident in which a 1300-kg car A struck a parked 1200-kg car B. All four of B's wheels were locked, and skid marks indicate it slid 2 m after the impact. If you estimate the coefficient of friction between B's tires and the road to be $\mu_k = 0.8$ and the coefficient of restitution of the impact to be $e = 0.4$, what was A's velocity just before the impact? (Assume that only one impact occurred.)

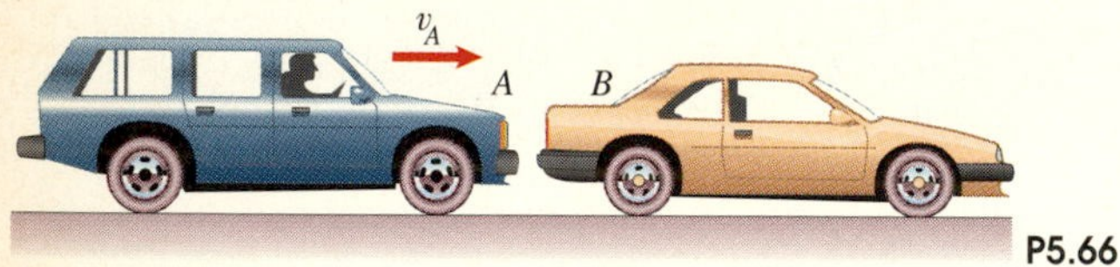

P5.66

5.67 Suppose you drop a basketball 5 ft above the floor and it bounces to a height of 4 ft. If you then throw the ball downward, releasing it 3 ft above the floor moving at 30 ft/s, how high does it bounce?

5.68 The 1-lb soccer ball is 3 ft above the ground when it is kicked upward at 40 ft/s. If the coefficient of restitution between the ball and the ground is $e = 0.6$, how high above the ground does the ball travel on its first bounce?

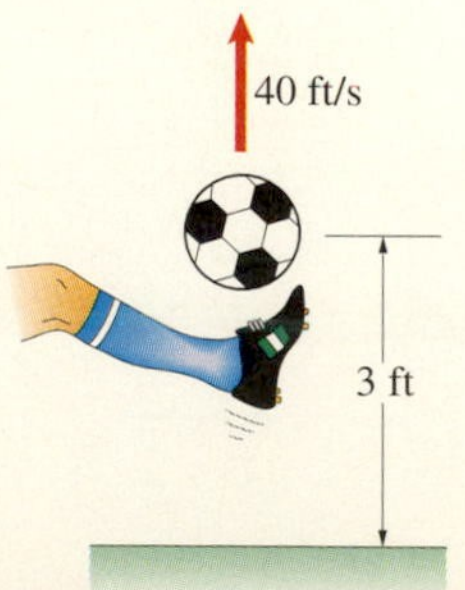

P5.68

5.69 If the soccer ball in Problem 5.68 was stationary just before it was kicked and the impact lasted 0.02 s, what was the average magnitude of the force exerted by the player's foot?

5.70 By making measurements directly from the photograph of the bouncing golf ball, estimate the coefficient of restitution.

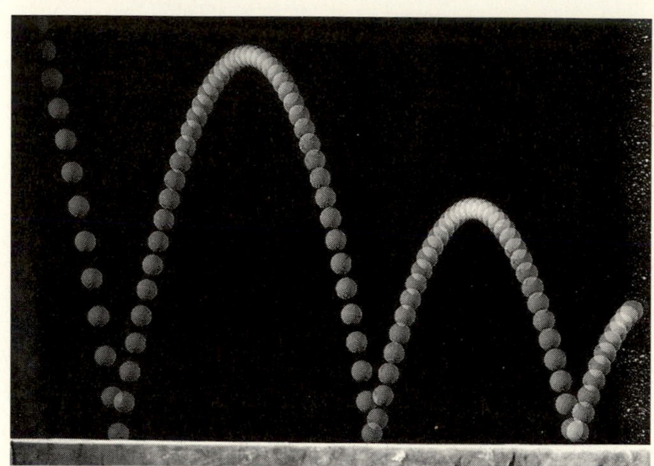

P5.70

5.71 If you throw the golf ball in Problem 5.70 horizontally at 2 ft/s and release it 4 ft above the surface, what is the distance between the first two bounces?

5.72 In a forging operation, the 100-lb weight is lifted into position 1 and released from rest. It falls and strikes a workpiece in position 2. If the weight is moving at 15 ft/s immediately before the impact and the coefficient of restitution is $e = 0.3$, what is its velocity immediately after the impact?

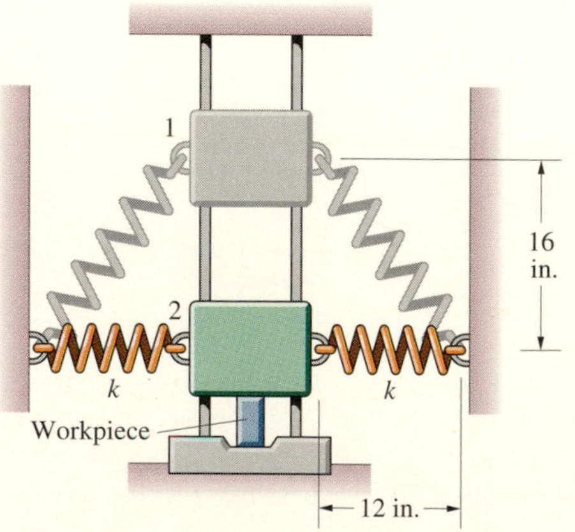

P5.72

5.73 In Problem 5.72, suppose that the spring constant is $k =$ 120 lb/ft, the springs are unstretched in position 2, and the coefficient of restitution is $e = 0.2$. Determine the velocity of the weight immediately after the impact.

5.74 A bioengineer studying helmet design strikes a 2.4-kg helmet containing a 2-kg simulated human head against a rigid surface at 6 m/s. The head, being suspended within the helmet, is not immediately affected by the impact of the helmet with the surface and continues to move to the right at 6 m/s, so it then undergoes an impact with the helmet. If the coefficient of restitution of the helmet's impact with the surface is 0.8 and the coefficient of restitution of the following impact of the head and helmet is 0.2, what are the velocities of the helmet and head after their initial interaction?

P5.74

5.75 (a) In Problem 5.74, if the duration of the impact of the head with the helmet is 0.008 s, to what average force is the head subjected?
(b) Suppose that the simulated head alone strikes the surface at 6 m/s, the coefficient of restitution is 0.3, and the duration of the impact is 0.002 s. To what average force is the head subjected?

5.76 Two small balls, each of mass m, hang from strings of length L. The left ball is released from rest in the position shown. As a result of the first collision, the right ball swings through an angle β. Determine the coefficient of restitution.

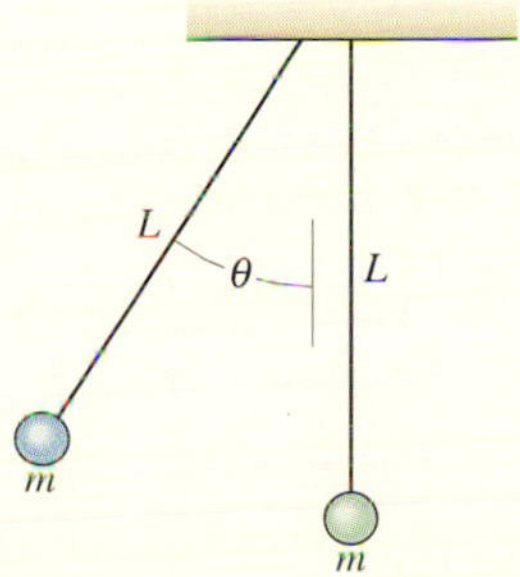

P5.76

5.77 If the duration of the collision in Problem 5.76 is Δt, what is the magnitude of the average force the balls exert on each other?

5.78 In Example 5.6, suppose that the CSM (A) approaches the Soyuz capsule (B) with velocity $\mathbf{v}_A = 0.05\mathbf{i} - 0.002\,\mathbf{j} + 0.007\,\mathbf{k}$ (m/s). The docking is unsuccessful, and a spring in the docking collar of the CSM causes the coefficient of restitution of the impact to be $e = 1$. If you treat the collision as an oblique central impact in which the force is parallel to the x axis, what are the velocities of the centers of mass of the two vehicles afterward?

5.79 A 1-slug object A and a 2-slug object B undergo an oblique central impact. The coefficient of restitution is $e = 0.8$. Before the impact, $\mathbf{v}_B = -10\mathbf{i}$ (ft/s), and after the impact, $\mathbf{v}'_A = -15\mathbf{i} + 4\mathbf{j} + 2\mathbf{k}$ (ft/s). Determine the velocity of A before the impact and the velocity of B after the impact.

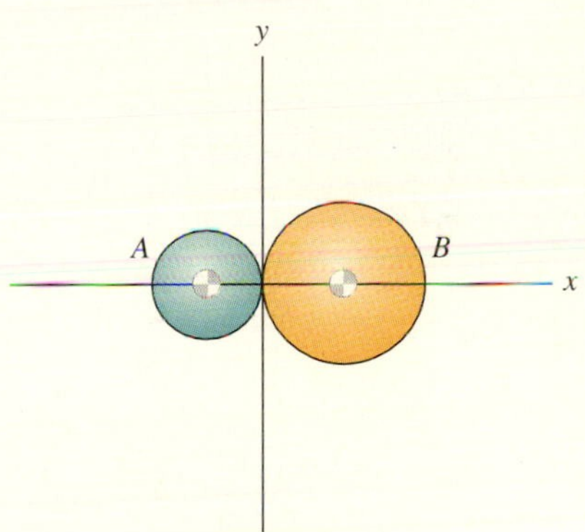

P5.79

5.80 The cue gives the cue ball A a velocity parallel to the y axis. It hits the eight ball B and knocks it straight into the corner pocket. If the magnitude of the velocity of the cue ball just before the impact is 2 m/s and the coefficient of restitution is $e = 1$, what are the velocity vectors of the two balls just after the impact? (The balls are of equal mass.)

P5.80

5.81 In Problem 5.80, what are the velocity vectors of the two balls just after the impact if the coefficient of restitution is $e = 0.9$?

5.82 If the coefficient of restitution is the same for both impacts, show that the cue ball's path after two banks is parallel to its original path.

P5.82

5.83 The velocity of the 170-g hockey puck is $\mathbf{v}_P = 10\mathbf{i} - 4\mathbf{j}$ (m/s). If you neglect the change in the velocity $\mathbf{v}_S = v_S\mathbf{j}$ of the stick resulting from the impact and the coefficient of restitution is $e = 0.6$, what should v_S be to send the puck toward the goal?

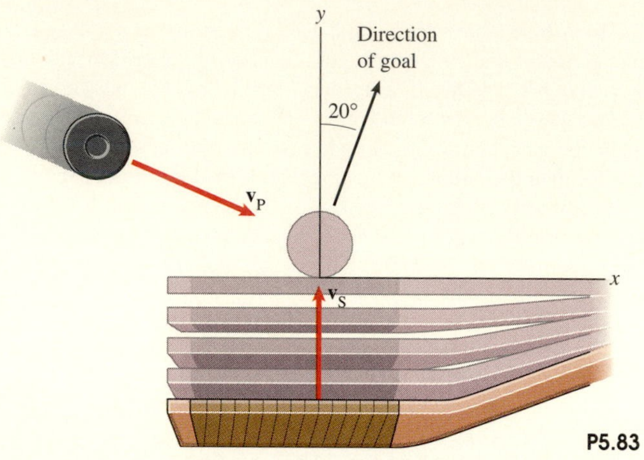

P5.83

5.84 In Problem 5.83, if the stick responds to the impact like an object with the same mass as the puck and the coefficient of restitution is $e = 0.6$, what should v_S be to send the puck toward the goal?

5.4 Angular Momentum

Here we derive a result, analogous to the principle of impulse and momentum, that relates the time integral of a moment to the change in a quantity called the angular momentum.

Principle of Angular Impulse and Momentum

We describe the position of an object relative to an inertial reference frame with origin O by the position vector $\mathbf{r}$ from O to the object's center of mass (Fig. 5.16a). Recall that we obtained the very useful principle of work and energy by taking the dot product of Newton's second law with the velocity. Here we obtain another useful result by taking the cross product of Newton's second law with the position vector. This procedure gives us a relation between the moment of the external forces about O and the object's motion.

We take the cross product of Newton's second law with $\mathbf{r}$:

$$\mathbf{r} \times \Sigma\mathbf{F} = \mathbf{r} \times m\mathbf{a} = \mathbf{r} \times m\frac{d\mathbf{v}}{dt}. \qquad (5.20)$$

Notice that the time derivative of the quantity $\mathbf{r} \times m\mathbf{v}$ is

$$\frac{d}{dt}\left(\mathbf{r} \times m\mathbf{v}\right) = \underbrace{\left(\frac{d\mathbf{r}}{dt} \times m\mathbf{v}\right)}_{=0} + \left(\mathbf{r} \times m\frac{d\mathbf{v}}{dt}\right).$$

(The first term on the right is zero because $d\mathbf{r}/dt = \mathbf{v}$, and the cross product of parallel vectors is zero.) Using this result, we can write Eq. (5.20) as

$$\mathbf{r} \times \Sigma\mathbf{F} = \frac{d\mathbf{H}_O}{dt}, \qquad (5.21)$$

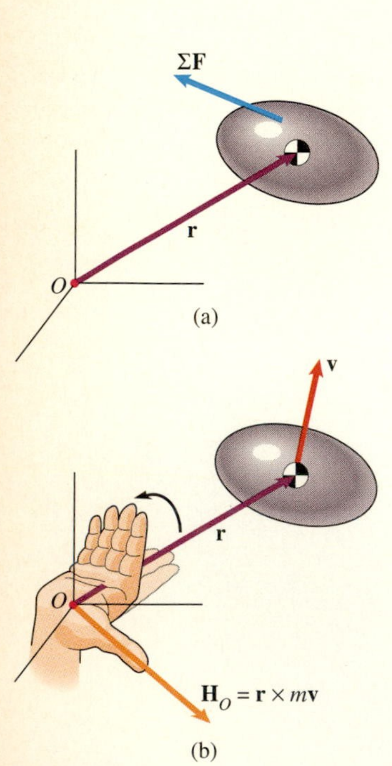

Fig. 5.16
(a) The position vector and the total external force on an object.
(b) The angular momentum vector and the right-hand rule for determining its direction.

where the vector

$$\mathbf{H}_O = \mathbf{r} \times m\mathbf{v} \tag{5.22}$$

is called the **angular momentum** about O (Fig. 5.16b). If we interpret the angular momentum as the moment of the linear momentum of the object about point O, this equation states that the moment $\mathbf{r} \times \Sigma\,\mathbf{F}$ equals the rate of change of the moment of momentum about point O. If the moment is zero during an interval of time, $\mathbf{H}_O$ is constant.

Integrating Eq. (5.21) with respect to time, we obtain

$$\boxed{\int_{t_1}^{t_2} (\mathbf{r} \times \Sigma\,\mathbf{F})\,dt = (\mathbf{H}_O)_2 - (\mathbf{H}_O)_1.} \tag{5.23}$$

The integral on the left is called the **angular impulse**, and this equation is called the **principle of angular impulse and momentum**: The angular impulse applied to an object during an interval of time is equal to the change in its angular momentum. If you know the moment $\mathbf{r} \times \Sigma\,\mathbf{F}$ as a function of time, you can determine the change in the angular momentum. The dimensions of the angular impulse and angular momentum are (mass) $\times$ (length)^2/(time).

Central-Force Motion

If the total force acting on an object remains directed toward a point that is fixed relative to an inertial reference frame, the object is said to be in **central-force motion**. The fixed point is called the **center** of the motion. Orbit problems are the most familiar instances of central-force motion. For example, the gravitational force on an earth satellite remains directed toward the center of the earth.

If we place the reference point O at the center of the motion (Fig. 5.17a), the position vector $\mathbf{r}$ is parallel to the total force, so $\mathbf{r} \times \Sigma\,\mathbf{F}$ equals zero. Therefore, Eq. (5.23) indicates that in central-force motion, an object's angular momentum is conserved:

$$\mathbf{H}_O = \text{constant}. \tag{5.24}$$

In plane central-force motion, we can express $\mathbf{r}$ and $\mathbf{v}$ in cylindrical coordinates (Fig. 5.17b):

$$\mathbf{r} = r\,\mathbf{e}_r, \qquad \mathbf{v} = v_r\,\mathbf{e}_r + v_\theta\,\mathbf{e}_\theta.$$

Substituting these expressions into Eq. (5.22), we obtain the angular momentum:

$$\mathbf{H}_O = (r\,\mathbf{e}_r) \times m(v_r\,\mathbf{e}_r + v_\theta\,\mathbf{e}_\theta) = mrv_\theta\,\mathbf{e}_z.$$

From this expression we see that in plane central-force motion, *the product of the radial distance from the center of the motion and the transverse component of the velocity is constant:*

$$rv_\theta = \text{constant}. \tag{5.25}$$

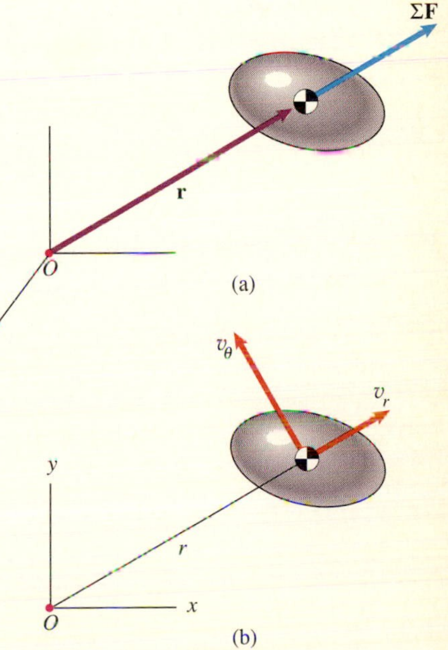

Fig. 5.17
(a) Central-force motion.
(b) Expressing the position and velocity in cylindrical coordinates.

In the following examples we show how you can use the principle of angular impulse and momentum and conservation of angular momentum to analyze motions of objects. If you know the moment* r × Σ F *during an interval of time, you can calculate the angular impulse and determine the change in an object's angular momentum. In central-force motion—the total force acting on an object points toward a fixed point O—you know that the angular momentum about O is conserved.

Example 5.7

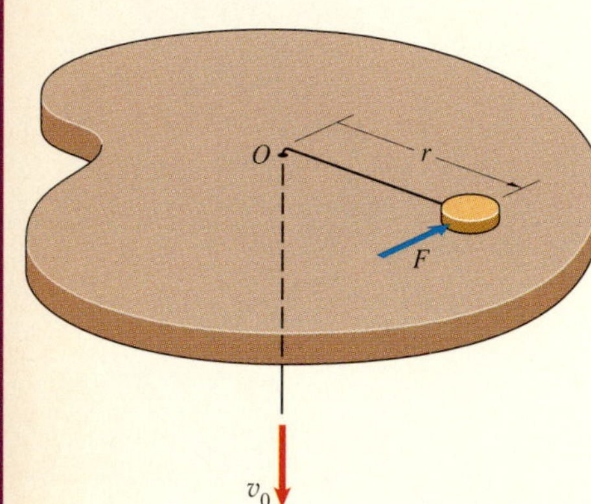

Fig. 5.18

A disk of mass m attached to a string slides on a smooth horizontal table under the action of a constant transverse force F (Fig. 5.18). The string is drawn through a hole in the table at O at constant velocity v_0. At $t = 0$, $r = r_0$ and the transverse velocity of the disk is zero. What is the disk's velocity as a function of time?

STRATEGY

By expressing r as a function of time, we can determine the moment of the forces on the disk about O as a function of time. The disk's angular momentum depends on its velocity, so we can apply the principle of angular impulse and momentum to obtain information about its velocity as a function of time.

SOLUTION

The radial position as a function of time is $r = r_0 - v_0 t$. In terms of polar coordinates (Fig. a), the moment about O of the forces on the disk is

$$\mathbf{r} \times \Sigma \mathbf{F} = r\,\mathbf{e}_r \times (-T\,\mathbf{e}_r + F\,\mathbf{e}_\theta) = F(r_0 - v_0 t)\,\mathbf{e}_z,$$

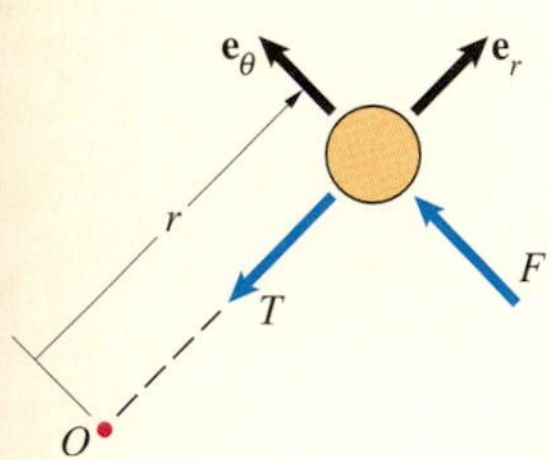

(a) Expressing the moment in terms of polar coordinates.

where T is the tension in the string. The angular momentum at time t is

$$\mathbf{H}_O = \mathbf{r} \times m\mathbf{v} = r\,\mathbf{e}_r \times m(v_r\,\mathbf{e}_r + v_\theta\,\mathbf{e}_\theta)$$
$$= m v_\theta (r_0 - v_0 t)\,\mathbf{e}_z.$$

Substituting these expressions into the principle of angular impulse and momentum, we obtain

$$\int_{t_1}^{t_2} (\mathbf{r} \times \Sigma \mathbf{F})\,dt = (\mathbf{H}_O)_2 - (\mathbf{H}_O)_1:$$

$$\int_0^t F(r_0 - v_0 t)\,\mathbf{e}_z\,dt = m v_\theta (r_0 - v_0 t)\,\mathbf{e}_z - \mathbf{0}.$$

Evaluating the integral, we obtain the transverse component of velocity as a function of time:

$$v_\theta = \frac{[r_0 t - (1/2) v_0 t^2]F}{(r_0 - v_0 t)m}.$$

The disk's velocity as a function of time is

$$\mathbf{v} = -v_0\,\mathbf{e}_r + \frac{[r_0 t - (1/2) v_0 t^2]F}{(r_0 - v_0 t)m}\,\mathbf{e}_\theta.$$

Example 5.8

When an earth satellite is at perigee (the point at which it is nearest to the earth), the magnitude of its velocity is $v_P = 7000$ m/s and its distance from the center of the earth is $r_P = 10{,}000$ km (Fig. 5.19). What are the magnitude of its velocity v_A and its distance r_A from the earth at apogee (the point at which it is farthest from the earth)? The radius of the earth is $R_E = 6370$ km.

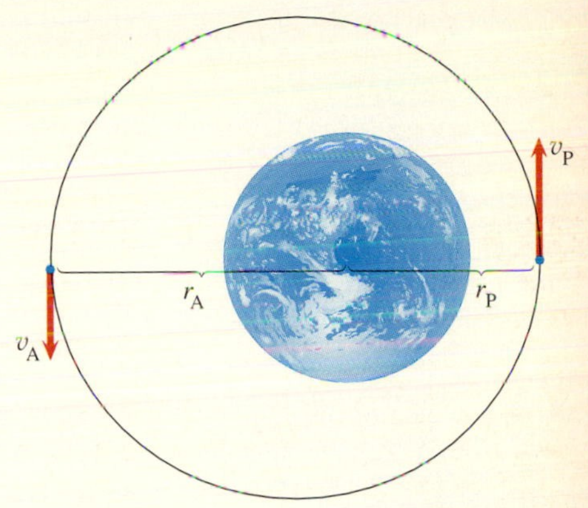

Fig. 5.19

STRATEGY

Because this is central-force motion about the center of the earth, we know that the product of the distance from the center of the earth and the transverse component of the satellite's velocity is constant. This gives us one equation relating v_A and r_A. We can obtain a second equation relating v_A and r_A by using conservation of energy.

SOLUTION

From Eq. (5.25), conservation of angular momentum requires that

$$r_A v_A = r_P v_P.$$

From Eq. (4.27), the satellite's potential energy in terms of distance from the center of the earth is

$$V = -\frac{mgR_E^2}{r}.$$

The sum of the kinetic and potential energies at apogee and perigee must be equal:

$$\frac{1}{2}mv_A^2 - \frac{mgR_E^2}{r_A} = \frac{1}{2}mv_P^2 - \frac{mgR_E^2}{r_P}.$$

Substituting $r_A = r_P v_P / v_A$ into this equation and rearranging, we obtain

$$(v_A - v_P)\left(v_A + v_P - \frac{2gR_E^2}{r_P v_P}\right) = 0.$$

This equation yields the trivial solution $v_A = v_P$ and also the solution for the velocity at apogee:

$$v_A = \frac{2gR_E^2}{r_P v_P} - v_P.$$

Substituting the values of g, R_E, r_P, and v_P, we obtain $v_A = 4370$ m/s and $r_A = 16{,}000$ km.

Problems

5.85 The total external force on a 2-kg object is $\Sigma \mathbf{F} = 2t\mathbf{i} + 4\mathbf{j}$ (N), where t is time in seconds. At time $t_1 = 0$, its position and velocity are $\mathbf{r} = \mathbf{0}$, $\mathbf{v} = \mathbf{0}$.
(a) Use Newton's second law to determine the object's position $\mathbf{r}$ and velocity $\mathbf{v}$ as functions of time.
(b) By integrating $\mathbf{r} \times \Sigma \mathbf{F}$ with respect to time, determine the angular impulse from $t_1 = 0$ to $t_2 = 6$ s.
(c) Use your results from part (a) to determine the change in the object's angular momentum from $t_1 = 0$ to $t_2 = 6$ s.

5.86 A satellite is in an elliptic orbit around the earth. Its velocity at the perigee A is 28,280 ft/s. What is its velocity at the apogee C?

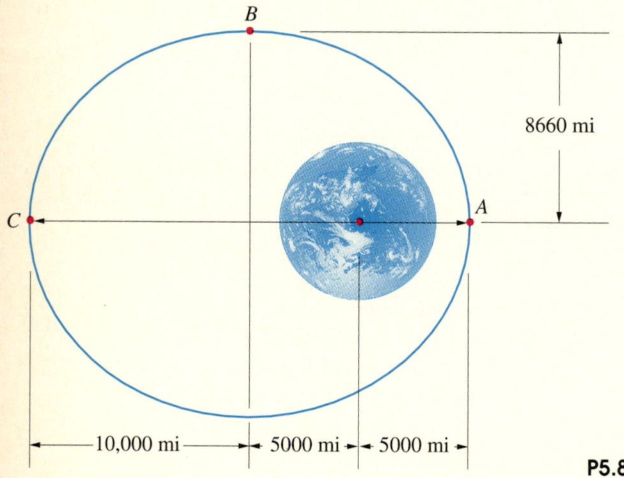

P5.86

5.87 In Problem 5.86, what are the magnitudes of the radial velocity v_r and the transverse velocity v_θ when the satellite is at point B of its elliptic orbit?

5.88 The bar rotates *in the horizontal plane* about a smooth pin at the origin. The 2-kg sleeve C slides on the smooth bar, and the mass of the bar is negligible in comparison to that of the sleeve. The spring constant $k = 40$ N/m, and the unstretched length of the spring is 0.8 m. At $t = 0$, the angular velocity of the bar is $\omega_0 = 6$ rad/s, $r = 0.2$ m, and the radial velocity of the sleeve is $v_r = 0$. What is the angular velocity of the bar when the spring is unstretched?

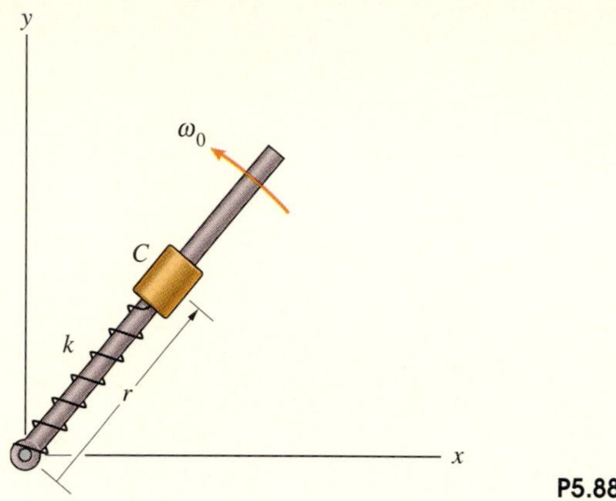

P5.88

5.89 In Problem 5.88, what is the radial velocity of the sleeve when the spring is unstretched?

5.90 In Example 5.7, determine the disk's velocity as a function of time if the force is $F = Ct$, where C is a constant.

5.91 A 2-kg disk slides on a smooth horizontal table and is connected to an elastic cord whose tension is $T = 6r$ N, where r is the radial position of the disk in meters. If the disk is at $r = 1$ m and is given an initial velocity of 4 m/s in the transverse direction, what are the magnitudes of the radial and transverse components of its velocity when $r = 2$ m?

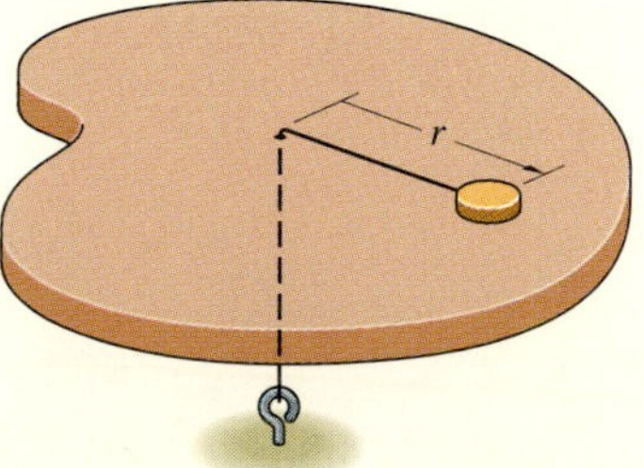

P5.91

5.92 In Problem 5.91, determine the maximum value of r reached by the disk.

5.93 A disk of mass m slides on a smooth horizontal table and is attached to a string that passes through a hole in the table.
(a) If the mass moves in a circular path of radius r_0 with transverse velocity v_0, what is the tension T?
(b) Starting from the initial condition described in (a), the tension is increased in such a way that the string is pulled through the hole at a constant rate until $r = r_0/2$. Determine T as a function of r while this is taking place.

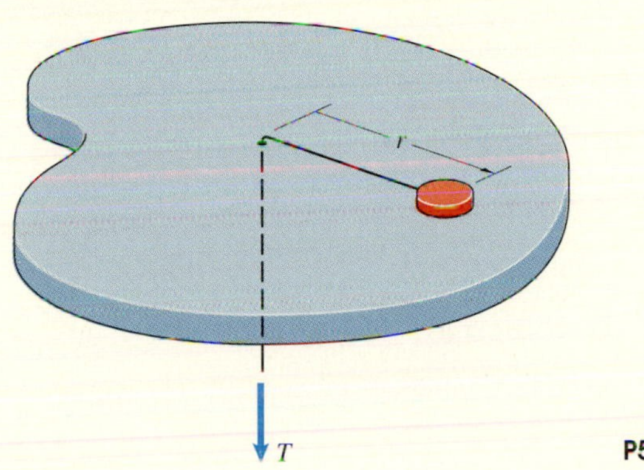

P5.93

5.94 In Problem 5.93, how much work is done on the mass in pulling the string through the hole as described in (b)?

5.95 Two gravity research satellites ($m_A = 250$ kg, $m_B = 50$ kg) are tethered by a cable. The satellites and cable rotate with angular velocity $\omega_0 = 0.25$ revolutions per minute. Ground controllers order satellite A to slowly unreel 6 m of additional cable. What is the angular velocity afterward?

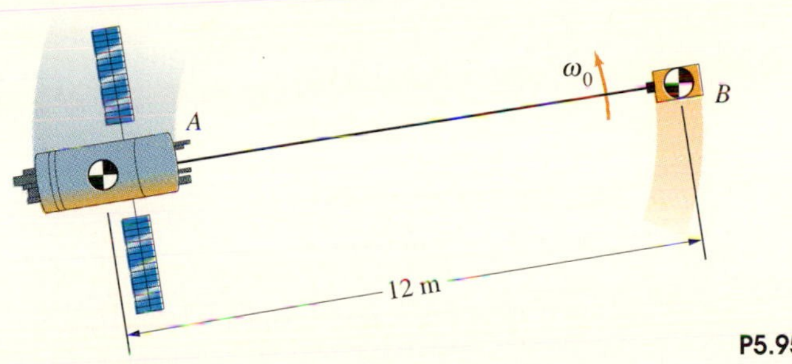

P5.95

5.96 An astronaut moves in the x-y plane at the end of a 10-m tether attached to a large space station at O. The total mass of the astronaut and his equipment is 120 kg.
(a) What is his angular momentum about O before the tether becomes taut?
(b) What is the magnitude of the component of his velocity perpendicular to the tether immediately after the tether becomes taut?

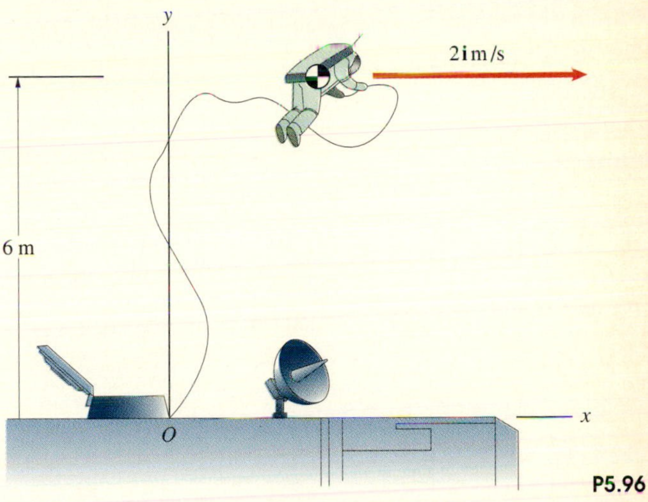

P5.96

5.97 In Problem 5.96, if the coefficient of restitution of the "impact" that occurs when the astronaut reaches the end of the tether is $e = 0.8$, what are the x and y components of his velocity immediately after the tether becomes taut?

5.98 A ball suspended from a string that goes through a hole in the ceiling at O moves with velocity v_A in a horizontal circular path of radius r_A. The string is then drawn through the hole until the ball moves with velocity v_B in a horizontal circular path of radius r_B. Use the principle of angular impulse and momentum to show that $r_A v_A = r_B v_B$.

Strategy: Let $\mathbf{e}$ be a unit vector that is perpendicular to the ceiling. Although this is not a central-force problem—the ball's weight does not point toward O—you can show that $\mathbf{e} \cdot (\mathbf{r} \times \Sigma \mathbf{F}) = 0$, so that $\mathbf{e} \cdot \mathbf{H}_O$ is conserved.

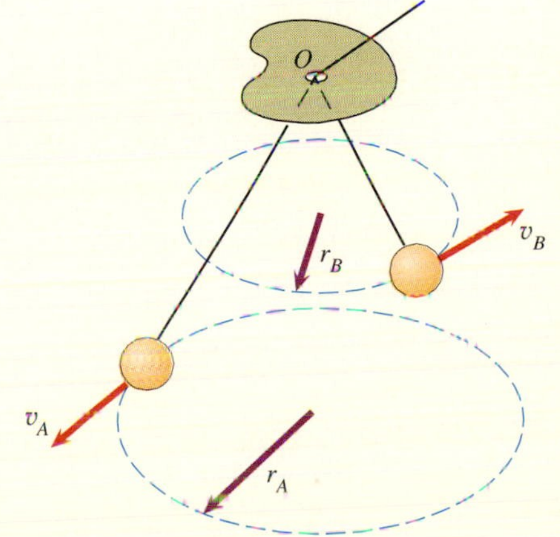

P5.98

5.5 *Mass Flows*

In this section we use conservation of linear momentum to determine the force exerted on an object as a result of emitting or absorbing a continuous flow of mass. The resulting equation applies to a variety of situations including determining the thrust of a rocket and calculating the forces exerted on objects by flows of liquids or granular materials.

Suppose that an object of mass m and velocity $\mathbf{v}$ is subjected to no external forces (Fig. 5.20a) and it emits an element of mass Δm_{f} with velocity $\mathbf{v}_{\text{f}}$ *relative to the object* (Fig. 5.20b). We denote the new velocity of the object by $\mathbf{v} + \Delta\mathbf{v}$. The linear momentum of the object before the element of mass is emitted equals the total linear momentum of the object and the element afterward:

$$m\mathbf{v} = (m - \Delta m_{\text{f}})(\mathbf{v} + \Delta\mathbf{v}) + \Delta m_{\text{f}}(\mathbf{v} + \mathbf{v}_{\text{f}}).$$

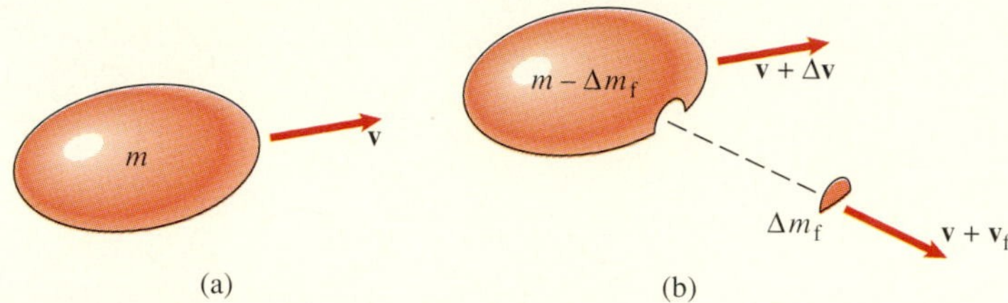

Fig. 5.20
An object's mass and velocity (a) before and (b) after emitting an element of mass.

Evaluating the products and simplifying, we obtain

$$m\,\Delta\mathbf{v} + \Delta m_{\text{f}}\mathbf{v}_{\text{f}} - \Delta m_{\text{f}}\,\Delta\mathbf{v} = 0. \tag{5.26}$$

Now we assume that, instead of a discrete element of mass, the object emits a continuous flow of mass and that Δm_{f} is the amount emitted in an interval of time Δt. We divide Eq. (5.26) by Δt and write the resulting equation as

$$m\frac{\Delta\mathbf{v}}{\Delta t} + \frac{\Delta m_{\text{f}}}{\Delta t}\mathbf{v}_{\text{f}} - \frac{\Delta m_{\text{f}}}{\Delta t}\frac{\Delta\mathbf{v}}{\Delta t}\Delta t = 0.$$

Taking the limit of this equation as $\Delta t \to 0$, we obtain

$$-\frac{dm_{\text{f}}}{dt}\mathbf{v}_{\text{f}} = m\mathbf{a},$$

where $\mathbf{a}$ is the acceleration of the object's center of mass. The term dm_{f}/dt is the **mass flow rate**, the rate at which mass flows from the object. Comparing this equation with Newton's second law, we conclude that a flow of mass *from* an object exerts a force

$$\boxed{\mathbf{F}_{\text{f}} = -\frac{dm_{\text{f}}}{dt}\mathbf{v}_{\text{f}}} \tag{5.27}$$

on the object. The force is proportional to the mass flow rate and to the magnitude of the *relative* velocity of the flow, and its direction is *opposite* to the direction of the relative velocity. Conversely, *a flow of mass to an object exerts a force in the same direction as the relative velocity.*

Example 5.9

The rocket sled in Fig. 5.21 is being slowed by a water brake after its rocket motor has burned out. A tube extends from the sled into a trough of water with its open end pointing forward, so that water enters the tube in the direction parallel to the x axis as the sled moves forward. The other open end of the tube points upward, so that the water flows out in the direction parallel to the y axis. If the sled's velocity is v, the water enters the tube with velocity v relative to the sled and flows out with the same velocity. The mass flow rate of water through the tube is $\rho v A$, where $\rho = 1.94$ slug/ft^3 is the mass density of the water and $A = 0.1$ ft^2 is the cross-sectional area of the tube. At an instant when $v = 1000$ ft/s, what forces are exerted on the sled by the flows of water entering and leaving it?

Fig. 5.21

STRATEGY

We can use Eq. (5.27) to determine the forces exerted by the flows of water entering and leaving the sled.

SOLUTION

Relative to the sled, the velocity vector of the water entering it is $\mathbf{v}_f = -v\mathbf{i}$. Because water is entering the sled, the force exerted is in the direction of the relative velocity:

$$\begin{aligned}\mathbf{F}_f &= -\frac{dm_f}{dt}\mathbf{v}_f \\ &= -\rho v A(-v\mathbf{i}) \\ &= -(1.94)(1000)(0.1)(-1000\mathbf{i}) \\ &= -194{,}000\mathbf{i} \quad \text{(lb)}.\end{aligned}$$

The water leaves the sled with relative velocity $\mathbf{v}_f = v\mathbf{j}$. The force is opposite to the direction of the relative velocity:

$$\begin{aligned}\mathbf{F}_f &= -\frac{dm_f}{dt}\mathbf{v}_f \\ &= -\rho v A(v\mathbf{j}) \\ &= -(1.94)(1000)(0.1)(1000\mathbf{j}) \\ &= -194{,}000\mathbf{j} \quad \text{(lb)}.\end{aligned}$$

Example 5.10

The classic example of a force created by a mass flow is the rocket. The rocket in Fig. 5.22 has a uniform, constant exhaust velocity v_f parallel to the x axis.
(a) What force is exerted on the rocket by the mass flow of its exhaust?
(b) If the force determined in (a) is the only force acting on the rocket, and it starts from rest with an initial mass m_0, determine the rocket's velocity as a function of its mass m.

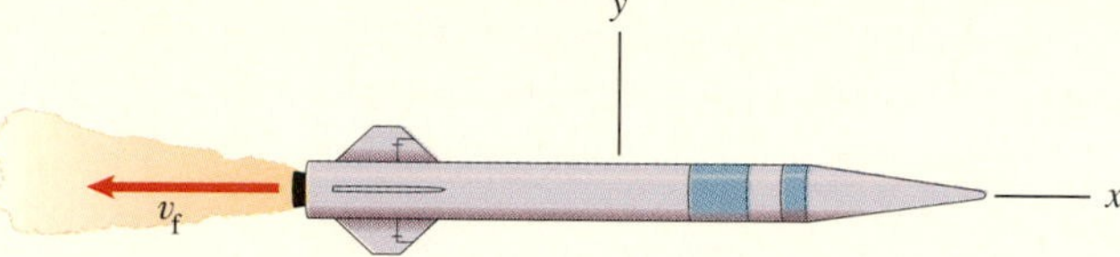

Fig. 5.22

STRATEGY

(a) Equation (5.27) gives the force exerted on the rocket in terms of the exhaust velocity and the mass flow rate of fuel. (b) We can use Newton's second law to obtain an equation for the rocket's velocity as a function of its mass.

SOLUTION

(a) In terms of the coordinate system in Fig. 5.22, the velocity vector of the exhaust is $\mathbf{v}_f = -v_f\mathbf{i}$. From Eq. (5.27), the force exerted on the rocket is

$$\mathbf{F}_f = -\frac{dm_f}{dt}\mathbf{v}_f = \frac{dm_f}{dt}v_f\mathbf{i},$$

where dm_f/dt is the mass flow rate of fuel. The force exerted on the rocket by its exhaust is toward the right, opposite to the direction of the flow of its exhaust.

(b) Newton's second law for the rocket is

$$\Sigma F_x = \frac{dm_f}{dt}v_f = m\frac{dv_x}{dt},$$

where m is the rocket's mass. The mass flow rate of fuel is the rate at which the rocket's mass is being consumed. Therefore the rate of change of the mass of the rocket is

$$\frac{dm}{dt} = -\frac{dm_f}{dt}.$$

Using this expression, we can write Newton's second law as

$$dv_x = -v_f\frac{dm}{m}.$$

Because the exhaust velocity v_f is constant, we can integrate this equation to determine the velocity of the rocket as a function of its mass:

$$\int_0^{v_x} dv_x = -v_f\int_{m_0}^{m}\frac{dm}{m}.$$

The result is

$$v_x = v_f \ln\left(\frac{m_0}{m}\right).$$

DISCUSSION

The velocity attained by the rocket is dependent on the exhaust velocity and the amount of mass expended. Thus a rocket can gain more velocity by expending more of its mass. However, notice that increasing the ratio m_0/m from 10 to 100 increases the velocity attained by only a factor of 2. In contrast, increasing the exhaust velocity results in a proportional increase in the rocket's velocity. Rocket engineers use fuels such as liquid oxygen and liquid hydrogen because they produce a large exhaust velocity. This objective has also led to research on rocket engines that use electromagnetic fields to accelerate charged particles of fuel to large velocities.

Example 5.11

A horizontal stream of water with velocity v_0 and mass flow rate dm_f/dt hits a plate that deflects the water in the horizontal plane through an angle θ (Fig. 5.23). Assume that the magnitude of the velocity of the water when it leaves the plate is approximately equal to v_0. What force is exerted on the plate by the water?

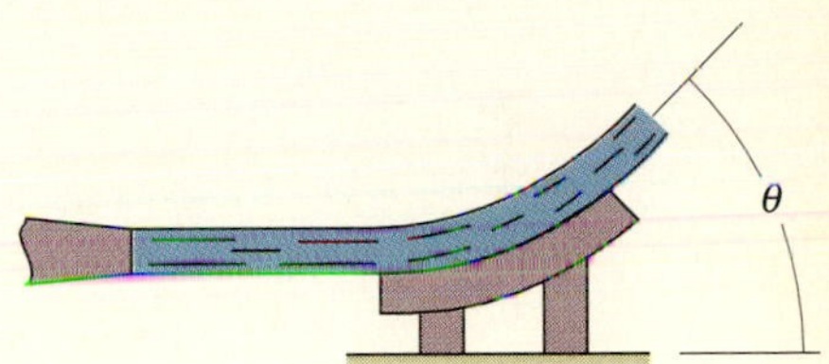

Fig. 5.23

STRATEGY

We can determine the force exerted on the plate by treating the part of the stream in contact with the plate as an "object" with mass flows entering and leaving it.

SOLUTION

In Fig. (a) we draw the free-body diagram of the part of the stream in contact with the plate. Streams of mass with velocity v_0 enter and leave this "object," and $\mathbf{F}_P$ is the force exerted on the stream by the plate. It is the force $-\mathbf{F}_P$ exerted on the plate by the stream that we wish to determine. First we consider the departing stream of water. The mass flow rate of water leaving the free-body diagram must be equal to the mass flow rate entering. In terms of the coordinate system shown, the velocity of the departing stream is

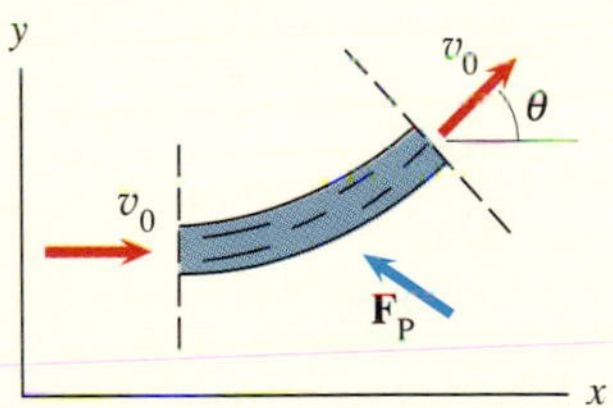

(a) Free-body diagram of the stream.

$$\mathbf{v}_f = v_0 \cos\theta\mathbf{i} + v_0 \sin\theta\mathbf{j}.$$

Let $\mathbf{F}_D$ be the force exerted on the object by the departing stream. From Eq. (5.27),

$$\mathbf{F}_D = -\frac{dm_f}{dt}\mathbf{v}_f = -\frac{dm_f}{dt}(v_0 \cos\theta\mathbf{i} + v_0 \sin\theta\mathbf{j}).$$

The velocity of the entering stream is $\mathbf{v}_f = v_0\mathbf{i}$. Since this flow is entering the object rather than leaving it, the resulting force $\mathbf{F}_E$ is in the same direction as the relative velocity:

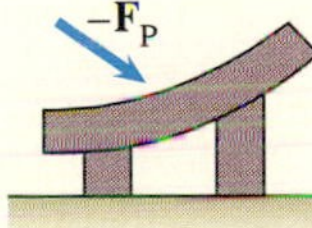

(b) Force exerted on the plate.

$$\mathbf{F}_E = \frac{dm_f}{dt}\mathbf{v}_f = \frac{dm_f}{dt}v_0\mathbf{i}.$$

The sum of the forces on the free-body diagram must equal zero,

$$\mathbf{F}_D + \mathbf{F}_E + \mathbf{F}_P = 0,$$

so the force exerted on the plate by the water is (Fig. b)

$$-\mathbf{F}_P = \mathbf{F}_D + \mathbf{F}_E = \frac{dm_f}{dt}v_0[(1 - \cos\theta)\mathbf{i} - \sin\theta\mathbf{j}].$$

DISCUSSION

This simple example gives you insight into how turbine blades and airplane wings can create forces by deflecting streams of liquid or gas (Fig. c).

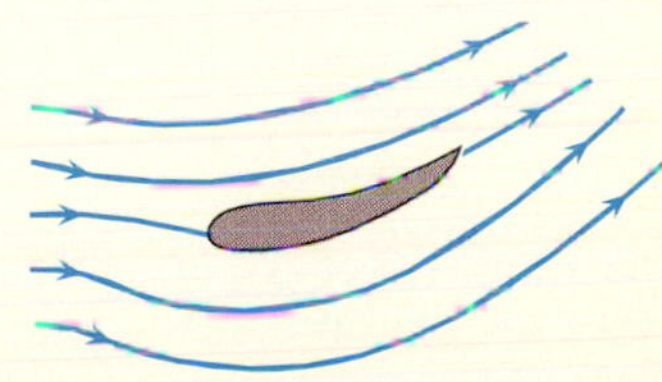

(c) Pattern of moving fluid around an airplane wing.

Example 5.12

Application to Engineering

Jet Engines

In a turbojet engine (Fig. 5.24), a mass flow rate dm_c/dt of inlet air enters the compressor with velocity v_i. The air is mixed with fuel and ignited in the combustion chamber. The mixture then flows through the turbine, which powers the compressor. The exhaust, with a mass flow rate equal to that of the air plus the mass flow rate of the fuel, $dm_c/dt + dm_f/dt$, exits at a high exhaust velocity v_e, exerting a large force on the engine. Suppose that $dm_c/dt = 0.925$ slug/s and $dm_f/dt = 0.009$ slug/s. The inlet air velocity is $v_i = 400$ ft/s, and the exhaust velocity is $v_e = 1605$ ft/s. What is the engine's thrust?

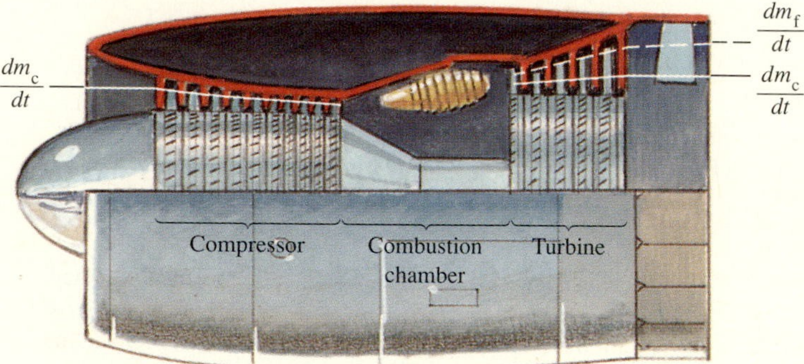

Fig. 5.24

STRATEGY

We can determine the engine's thrust by using Eq. (5.27). We must include both the force exerted by the engine's exhaust and the force exerted by the mass flow of air entering the compressor to determine the net thrust.

SOLUTION

The engine's exhaust exerts a force to the left equal to the product of the mass flow rate of the fuel–air mixture and the exhaust velocity. The inlet air exerts a force to the right equal to the product of the mass flow rate of the inlet air and the inlet velocity. The engine's thrust (the net force to the left) is

$$\begin{aligned} T &= \left(\frac{dm_c}{dt} + \frac{dm_f}{dt}\right) v_e - \frac{dm_c}{dt} v_i \\ &= (0.925 + 0.009)(1605) - (0.925)(400) \\ &= 1130 \text{ lb.} \end{aligned}$$

DESIGN ISSUES

The jet engine was developed in Europe in the years just prior to World War II. Although the turbojet engine in Fig. 5.24 was a very successful design that dominated both military and commercial aviation for many years, it has the drawback of relatively large fuel consumption.

During the last 30 years, the fan-jet engine, shown in Fig. 5.25, has become the most commonly used design, particularly for commercial airplanes. Part of its thrust is provided by air that is accelerated by the fan. The ratio of the mass flow rate of air entering the fan, dm_b/dt, to the mass flow rate of air entering the compressor, dm_c/dt, is called the *bypass ratio*.

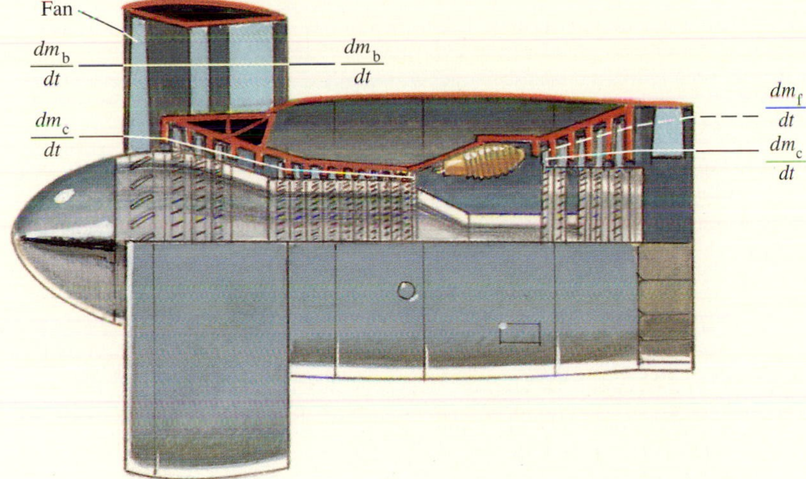

Fig. 5.25
A fan-jet engine. Part of the entering mass flow of air is accelerated by the fan and does not enter the compressor.

The force exerted by a jet engine's exhaust equals the product of the mass flow rate and the exhaust velocity. In the fan-jet engine, the air passing through the fan is not heated by the combustion of fuel and therefore has a higher density than the exhaust of the turbojet engine. As a result, the fan-jet engine can provide a given thrust with a lower average exhaust velocity. Since the work that must be expended to create the thrust depends on the kinetic energy of the exhaust, the fan-jet engine creates thrust more efficiently.

Problems

5.99 The Cheverton fire-fighting and rescue boat can pump 3.8 kg/s of water from each of its two pumps at a velocity of 44 m/s. If both pumps point in the same direction, what total force do they exert on the boat?

P5.99

5.100 The mass flow rate of water through the nozzle is 1.6 slugs/s. Determine the magnitude of the horizontal force exerted on the truck by the flow of water.

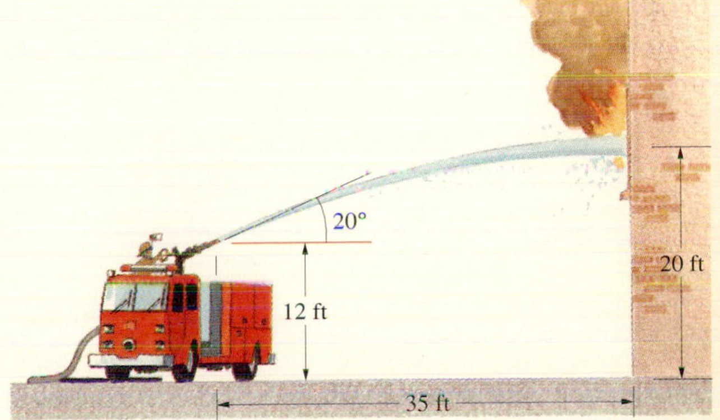

P5.100

5.101 A front-end loader moves at 2 mi/hr and scoops up 66,000 lb of iron ore in 3 s. What horizontal force must its tires exert?

P5.101

5.102 The snowblower moves at 1 m/s and scoops up 750 kg/s of snow. Determine the force exerted by the entering flow of snow.

P5.102

5.103 If you design the snowblower in Problem 5.102 so that it blows snow out at 45° above the horizontal from a port 2 m above the ground and the snow lands 20 m away, what horizontal force is exerted on the blower by the departing flow of snow?

5.104 A nozzle ejects a stream of water horizontally at 40 m/s with a mass flow rate of 30 kg/s, and the stream is deflected in the horizontal plane by a plate. Determine the force exerted on the plate by the stream in cases (a), (b), (c).

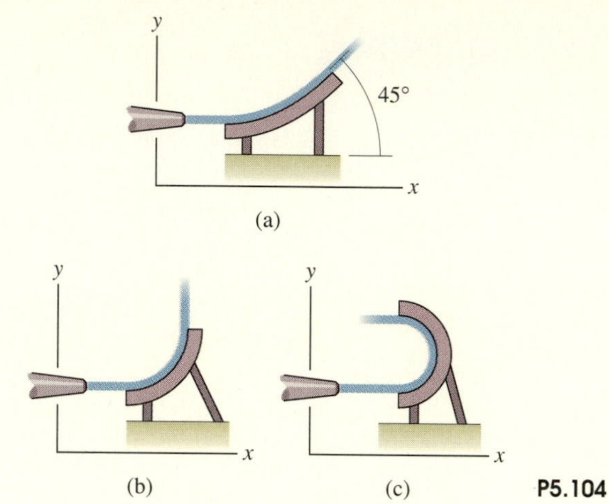

P5.104

5.105 A stream of water with velocity $80\mathbf{i}$ (m/s) and a mass flow rate of 6 kg/s strikes a turbine blade moving with constant velocity $20\mathbf{i}$ (m/s).

(a) What force is exerted on the blade by the water?

(b) What is the magnitude of the velocity of the water as it leaves the blade?

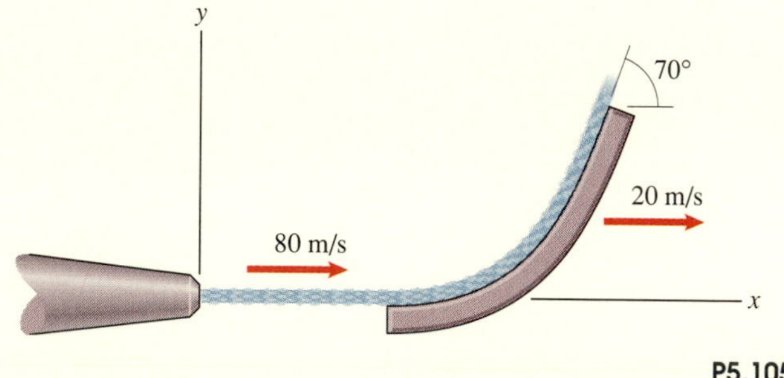

P5.105

5.106 The nozzle A of the lawn sprinkler is located at (7, −0.5, 0.5) in. Water exits each nozzle at 25 ft/s with a weight flow rate of 0.5 lb/s. The direction cosines of the flow direction from A are $(1/\sqrt{3}, -1/\sqrt{3}, 1/\sqrt{3})$. What is the total moment about the z axis exerted on the sprinkler by the flows from all four nozzles?

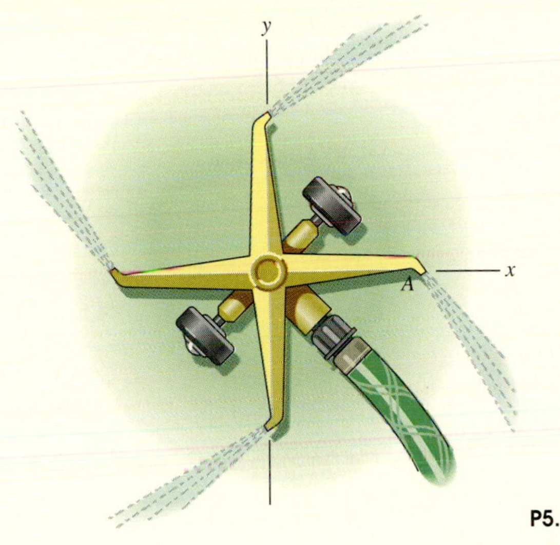

P5.106

5.107 A 45-kg/s flow of gravel exits the chute at 2 m/s and falls onto a conveyer moving at 0.3 m/s. Determine the components of the force exerted on the conveyer by the flow of gravel if $\theta = 0$.

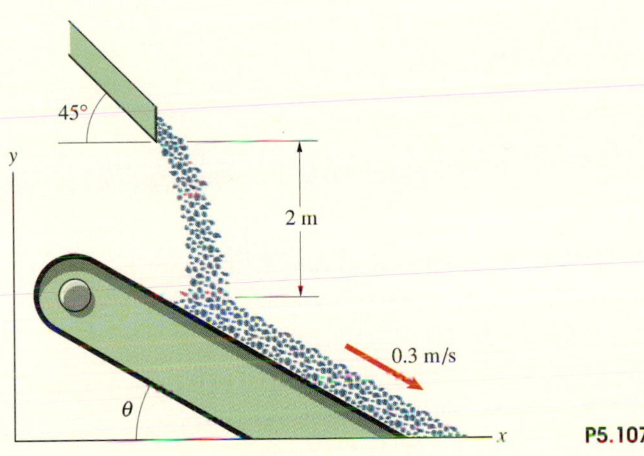

P5.107

5.108 Solve Problem 5.107 if $\theta = 30°$.

5.109 A toy car is propelled by water that squirts from an internal tank at 3 m/s relative to the car. If the mass of the empty car is 1 kg, it holds 2 kg of water, and you neglect other tangential forces, what is its top speed?

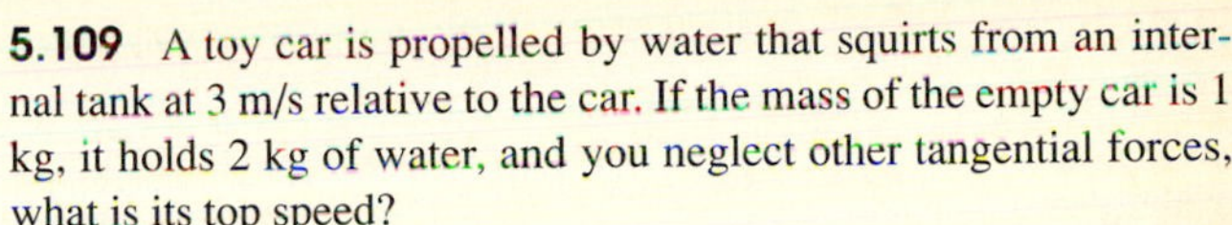

P5.109

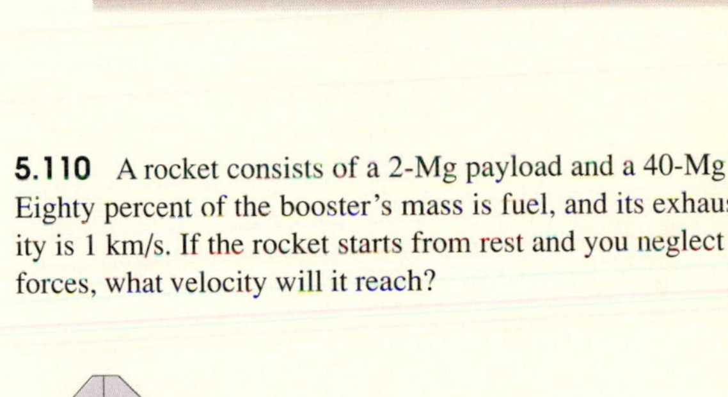

5.110 A rocket consists of a 2-Mg payload and a 40-Mg booster. Eighty percent of the booster's mass is fuel, and its exhaust velocity is 1 km/s. If the rocket starts from rest and you neglect external forces, what velocity will it reach?

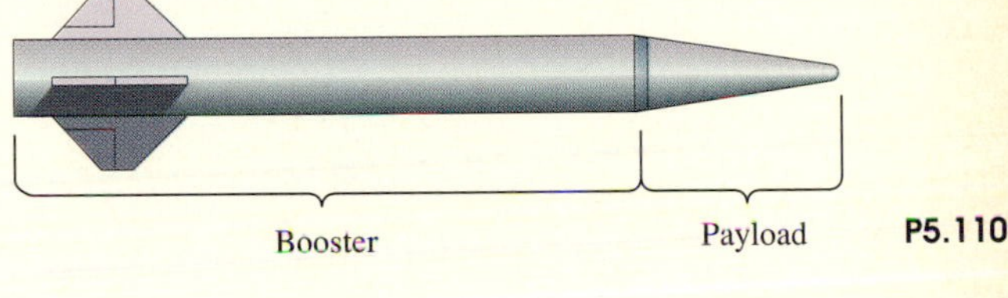

P5.110

5.111 A rocket consists of a 2-Mg payload and a booster. The booster has two stages whose total mass is 40 Mg. Eighty percent of the mass of each stage is fuel. When the fuel of stage 1 is expended, it is discarded and the motor in stage 2 is ignited. The exhaust velocity of both stages is 1 km/s. Assume that the rocket starts from rest and neglect external forces. Determine the velocity reached by the rocket if the two stages are of equal mass, and compare your result to the answer to Problem 5.110.

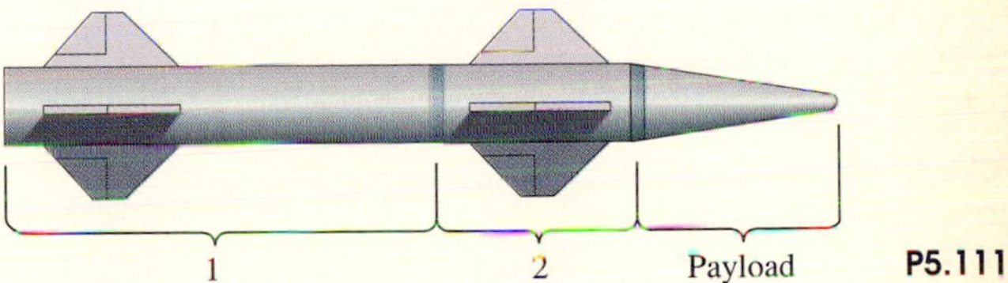

P5.111

5.112 A rocket of initial mass m_0 takes off straight up. Its exhaust velocity v_f and the mass flow rate of its engine $\dot{m}_f = dm_f/dt$ are constant. During the initial part of the flight when aerodynamic drag is negligible, show that the rocket's upward velocity as a function of time is

$$v = v_f \ln\left(\frac{m_0}{m_0 - \dot{m}_f t}\right) - gt.$$

P5.112

5.113 The mass of the rocket sled in Example 5.9 is 30 slugs. The only significant force acting on the sled in the direction of its motion is the force exerted by the flow of water entering it. Determine the time and the distance required for the sled to decelerate from 1000 ft/s to 100 ft/s.

5.114 Suppose that you grasp the end of a chain that weighs 3 lb/ft and lift it straight up off the floor at a constant speed of 2 ft/s.
(a) Determine the upward force F you must exert as a function of the height s.
(b) How much work do you do in lifting the top of the chain to s = 4 ft?

Strategy: Treat the part of the chain you have lifted as an object that is gaining mass.

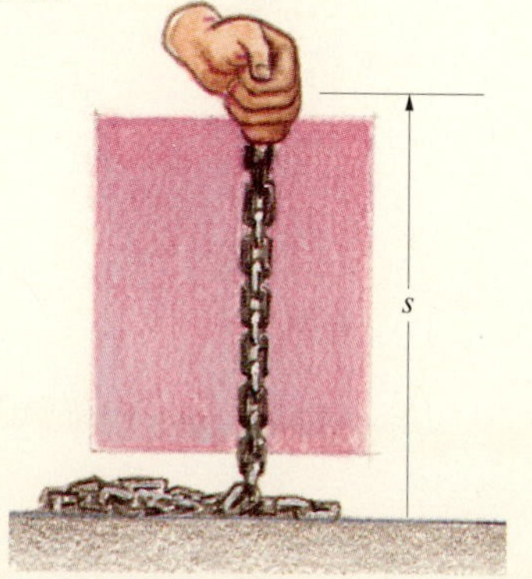

P5.114

5.115 Solve Problem 5.114, assuming that you lift the end of the chain straight up off the floor with a constant acceleration of 2 ft/s^2.

5.116 It has been suggested that a heavy chain could be used to gradually stop an airplane that rolls past the end of the runway. A hook attached to the end of the chain engages the plane's nose wheel, and the plane drags an increasing length of the chain as it rolls. Let m be the airplane's mass and v_0 its initial velocity, and let ρ_L be the mass per unit length of the chain. If you neglect friction and aerodynamic drag, what is the airplane's velocity as a function of s?

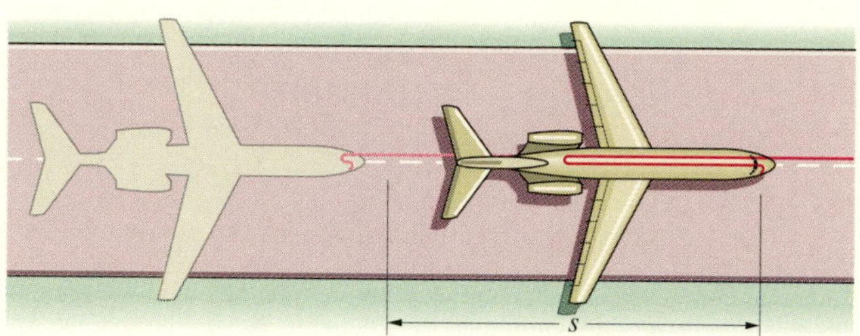

P5.116

5.117 In Problem 5.116, the friction force exerted on the chain by the ground would actually dominate other forces as the distance s increases. If the coefficient of kinetic friction between the chain and the ground is μ_k and you neglect all forces except the friction force, what is the airplane's velocity as a function of s?

Problems 5.118–5.122 are related to Example 5.12.

5.118 The turbojet engine in Fig. 5.24 is being operated on a test stand. The mass flow rate of air entering the compressor is 13.5 kg/s, and the mass flow rate of fuel is 0.13 kg/s. The effective velocity of the air entering the compressor is zero, and the exhaust velocity is 500 m/s. What is the thrust of the engine?

5.119 Suppose that the engine described in Problem 5.118 is in an airplane flying at 400 km/hr. The effective velocity of the air entering the inlet is equal to the airplane's velocity. What is the thrust of the engine?

5.120 A turbojet engine's thrust reverser causes the exhaust to exit the engine at 20° from the engine centerline. The mass flow rate of air entering the compressor is 3 slugs/s, and it enters at 200 ft/s. The mass flow rate of fuel is 0.1 slug/s, and the exhaust velocity is 1200 ft/s. What braking force does the engine exert on the airplane?

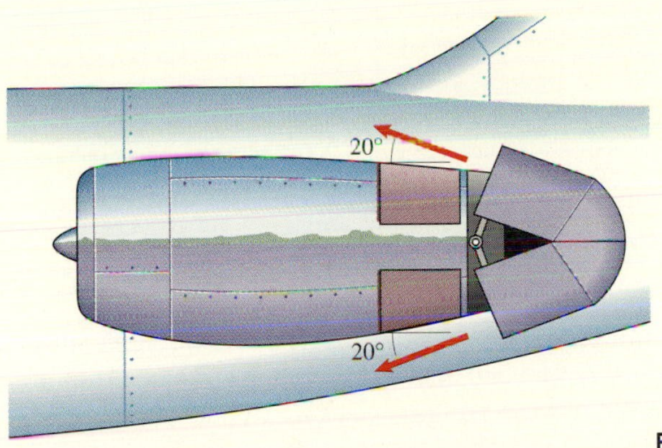

P5.120

5.121 The 13.6-Mg airplane is moving at 400 km/hr. The total mass flow rate of air entering the compressors of its turbojet engines is 280 kg/s, and the total mass flow rate of fuel is 2.6 kg/s. The effective velocity of the air entering the compressors is equal to the airplane's velocity, and the exhaust velocity is 480 m/s. The ratio of the lift force L to the drag force D is 6, and the z component of the airplane's acceleration is zero. What is the x component of its acceleration?

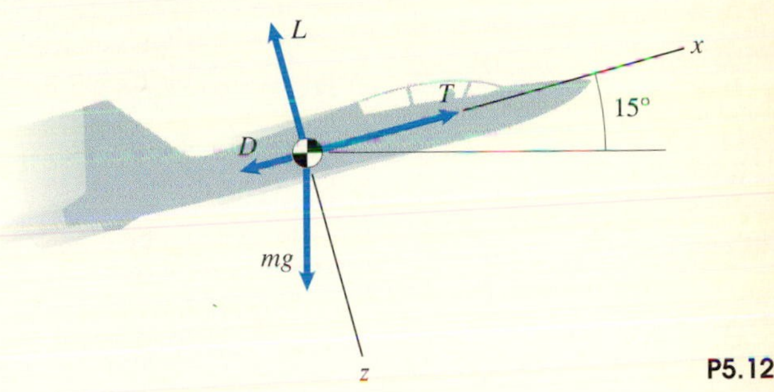

P5.121

5.122 The fan-jet engine in Fig. 5.25 is similar to the Pratt and Whitney JT9D-3A engine used on early models of the Boeing 747. When the airplane begins its takeoff run, the velocity of the air entering the compressor and fan is negligible. A mass flow rate of 38.5 slugs/s enters the fan and is accelerated to 885 ft/s. A mass flow rate of 7.7 slugs/s enters the compressor. The mass flow rate of fuel is 0.23 slugs/s, and the exhaust velocity is 1190 ft/s. (a) What is the bypass ratio? (b) What is the thrust of the engine? (c) If the airplane weighs 500,000 lb, what is its initial acceleration? (It has four engines.)

Chapter Summary

Principle of Impulse and Momentum

The linear impulse applied to an object during an interval of time is equal to the change in its linear momentum:

$$\int_{t_1}^{t_2} \Sigma \mathbf{F}\, dt = m\mathbf{v}_2 - m\mathbf{v}_1. \qquad \textbf{Eq. (5.1)}$$

This result can also be expressed in terms of the average with respect to time of the total force:

$$(t_2 - t_1)\, \Sigma \mathbf{F}_{\text{av}} = m\mathbf{v}_2 - m\mathbf{v}_1. \qquad \textbf{Eq. (5.2)}$$

The principle of impulse and momentum can also be expressed in terms of the tangential component of the total force and the linear momentum along the object's path:

$$\int_{t_1}^{t_2} \Sigma F_t\, dt = mv_2 - mv_1. \qquad \textbf{Eq. (5.3)}$$

The average with respect to time of the tangential component of the total force is related to the change in the linear momentum along the path by

$$(t_2 - t_1)\, \Sigma F_{t\,\text{av}} = mv_2 - mv_1. \qquad \textbf{Eq. (5.4)}$$

Conservation of Linear Momentum

If objects A and B are not subjected to external forces other than the forces they exert on each other (or if the effects of other external forces are negligible), their total linear momentum is conserved,

$$m_A\mathbf{v}_A + m_B\mathbf{v}_B = \text{constant}, \qquad \textbf{Eq. (5.6)}$$

and the velocity of their common center of mass is constant.

Impacts

If colliding objects are not subjected to external forces, their total linear momentum must be the same before and after the impact. Even when they are subjected to external forces, the force of the impact is often so large, and its duration so brief, that the effect of external forces on their motions during the impact is negligible.

If objects A and B adhere and remain together after they collide, they are said to undergo a **perfectly plastic impact**. The velocity of their common center of mass before and after the impact is given by

$$\mathbf{v} = \frac{m_A\mathbf{v}_A + m_B\mathbf{v}_B}{m_A + m_B}. \qquad \textbf{Eq. (5.9)}$$

Central Impacts

In a **direct central impact** (Fig. a), linear momentum is conserved,

$$m_A v_A + m_B v_B = m_A v'_A + m_B v'_B, \qquad \textbf{Eq. (5.10)}$$

and the velocities are related by the **coefficient of restitution**:

$$e = \frac{v'_B - v'_A}{v_A - v_B}. \qquad \textbf{Eq. (5.16)}$$

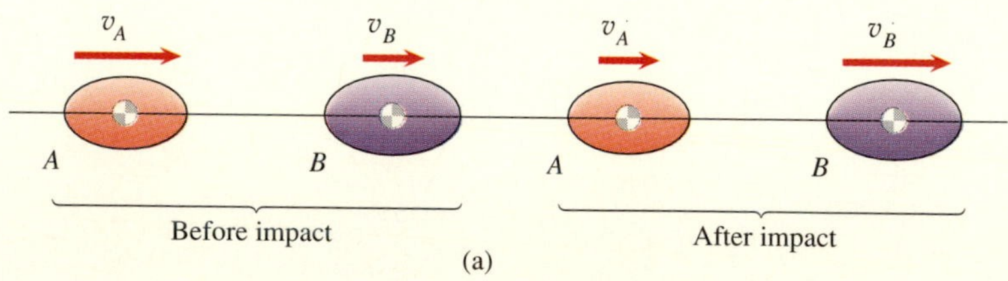

(a)

If $e = 0$, the impact is perfectly plastic. If $e = 1$, the total kinetic energy is conserved and the impact is called **perfectly elastic**.

In an **oblique central impact** (Fig. b), the components of velocity in the y and z directions are unchanged by the impact:

$$(\mathbf{v}'_A)_y = (\mathbf{v}_A)_y, \qquad (\mathbf{v}'_B)_y = (\mathbf{v}_B)_y,$$
$$(\mathbf{v}'_A)_z = (\mathbf{v}_A)_z, \qquad (\mathbf{v}'_B)_z = (\mathbf{v}_B)_z. \qquad \textbf{Eq. (5.17)}$$

In the x direction, linear momentum is conserved,

$$m_A(\mathbf{v}_A)_x + m_B(\mathbf{v}_B)_x = m_A(\mathbf{v}'_A)_x + m_B(\mathbf{v}'_B)_x, \qquad \text{Eq. (5.18)}$$

and the velocity components are related by the coefficient of restitution:

$$e = \frac{(\mathbf{v}'_B)_x - (\mathbf{v}'_A)_x}{(\mathbf{v}_A)_x - (\mathbf{v}_B)_x}. \qquad \text{Eq. (5.19)}$$

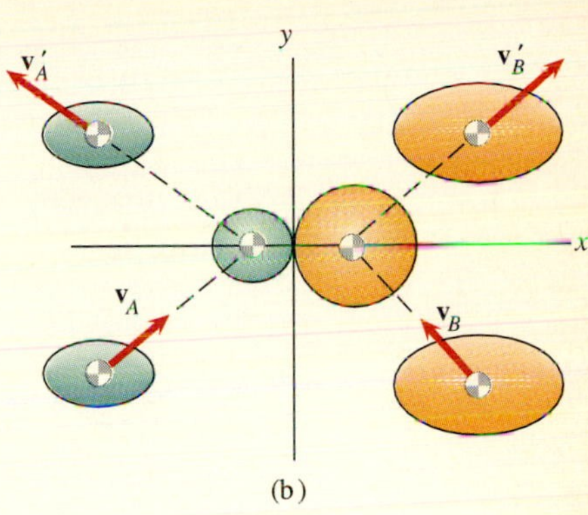

(b)

Principle of Angular Impulse and Momentum

For a fixed point O, the angular impulse applied to an object during an interval of time is equal to the change in its angular momentum about O:

$$\int_{t_1}^{t_2} (\mathbf{r} \times \Sigma\, \mathbf{F})\, dt = (\mathbf{H}_O)_2 - (\mathbf{H}_O)_1, \qquad \text{Eq. (5.23)}$$

where the angular momentum is

$$\mathbf{H}_O = \mathbf{r} \times m\mathbf{v}. \qquad \text{Eq. (5.22)}$$

Central-Force Motion

If the total force acting on an object remains directed toward a fixed point, the object is said to be in **central-force motion**, and its angular momentum about the fixed point is conserved:

$$\mathbf{H}_O = \text{constant}. \qquad \text{Eq. (5.24)}$$

In plane central-force motion, the product of the radial distance and the transverse component of the velocity is constant:

$$r\, v_\theta = \text{constant}. \qquad \text{Eq. (5.25)}$$

Mass Flows

A flow of mass *from* an object with velocity $\mathbf{v}_f$ *relative to the object* exerts a force

$$\mathbf{F}_f = -\frac{dm_f}{dt}\mathbf{v}_f \qquad \text{Eq. (5.27)}$$

on the object, where dm_f/dt is the **mass flow rate**. The direction of the force is opposite to the direction of the relative velocity. A flow of mass *to* an object exerts a force in the same direction as the relative velocity.

Review Problems

5.123 The total external force on a 10-kg object is constant and equal to $90\mathbf{i} - 60\mathbf{j} + 20\mathbf{k}$ (N). At $t = 2$ s, the object's velocity is $-8\mathbf{i} + 6\mathbf{j}$ (m/s).

(a) What impulse is applied to the object from $t = 2$ s to $t = 4$ s?
(b) What is the object's velocity at $t = 4$ s?

5.124 The total external force on an object is $\mathbf{F} = 10t\,\mathbf{i} + 60\,\mathbf{j}$ (lb). At $t = 0$, its velocity is $\mathbf{v} = 20\,\mathbf{j}$ (ft/s). At $t = 12$ s, the x component of its velocity is 48 ft/s.
(a) What impulse is applied to the object from $t = 0$ to $t = 6$ s?
(b) What is its velocity at $t = 6$ s?

5.125 An aircraft arresting system is used to stop airplanes whose braking systems fail. The system stops a 47.5-Mg airplane moving at 80 m/s in 9.15 s.
(a) What impulse is applied to the airplane during the 9.15 s?
(b) What is the average deceleration to which the passengers are subjected?

P5.125

5.126 The 1895 Austrian 150-mm howitzer had a 1.94-m-long barrel, a muzzle velocity of 300 m/s, and fired a 38-kg shell. If the shell took 0.013 s to travel the length of the barrel, what average force was exerted on the shell?

5.127 An athlete throws a shot weighing 16 lb. When he releases it, the shot is 7 ft above the ground and its components of velocity are $v_x = 31$ ft/s, $v_y = 26$ ft/s.
(a) If he accelerates the shot from rest in 0.8 s and you assume as a first approximation that the force **F** he exerts on it is constant, use the principle of impulse and momentum to determine the x and y components of **F**.
(b) What is the horizontal distance from the point where he releases the shot to the point where it strikes the ground?

P5.127

5.128 The 6000-lb pickup truck A moving at 40 ft/s collides with the 4000-lb car B moving at 30 ft/s.
(a) What is the magnitude of the velocity of their common center of mass after the impact?
(b) If you treat the collision as a perfectly plastic impact, how much kinetic energy is lost?

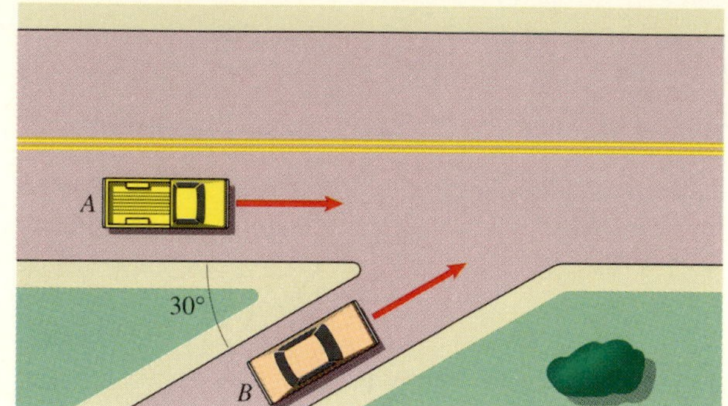

P5.128

5.129 Two hockey players ($m_A = 80$ kg, $m_B = 90$ kg) converging on the puck at $x = 0$, $y = 0$ become entangled and fall. Before the collision, $v_A = 9\mathbf{i} + 4\mathbf{j}$ (m/s) and $v_B = -3\mathbf{i} + 6\mathbf{j}$ (m/s). If the coefficient of kinetic friction between the players and the ice is $\mu_k = 0.1$, what is their approximate position when they stop sliding?

P5.129

5.130 The cannon weighed 400 lb, fired a cannonball weighing 10 lb, and had a muzzle velocity of 200 ft/s. For the 10° elevation angle shown, determine (a) the velocity of the cannon after it was fired; (b) the distance the cannonball traveled. (Neglect drag.)

P5.130

5.131 A 1-kg ball moving horizontally at 12 m/s strikes a 10-kg block. The coefficient of restitution of the impact is $e = 0.6$, and the coefficient of kinetic friction between the block and the inclined surface is $\mu_k = 0.4$. What distance does the block slide before stopping?

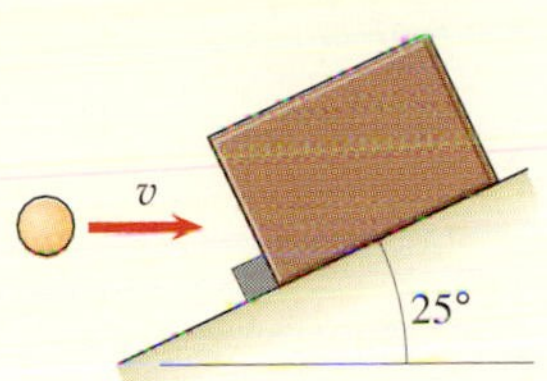

P5.131

5.132 A Peace Corps volunteer designs the simple device shown for drilling water wells in remote areas. A 70-kg "hammer," such as a section of log or a steel drum partially filled with concrete, is hoisted to $h = 1$ m and allowed to drop onto a protective cap on the section of pipe being pushed into the ground. The mass of the cap and section of pipe is 20 kg. Assume the coefficient of restitution is nearly zero.
(a) What is the velocity of the cap and pipe immediately after the impact?
(b) If the pipe moves 30 mm downward when the hammer is dropped, what resistive force was exerted on the pipe by the ground? (Assume the resistive force is constant during the motion of the pipe.)

P5.132

5.133 A tugboat (mass = 40 Mg) and a barge (mass = 160 Mg) are stationary with a slack hawser connecting them. The tugboat accelerates to 2 knots (1 knot = 1852 m/hr) before the hawser becomes taut. Determine the velocities of the tugboat and the barge just after the hawser becomes taut (a) if the "impact" is perfectly plastic ($e = 0$); (b) if the "impact" is perfectly elastic ($e = 1$). Neglect the forces exerted by the water and the tugboat's engines.

P5.133

5.134 In Problem 5.133, determine the magnitude of the impulsive force exerted on the tugboat in the two cases if the duration of the "impact" is 4 s. Neglect the forces exerted by the water and the tugboat's engines during this period.

5.135 The balls are of equal mass m. Balls B and C are connected by an unstretched linear spring and are stationary. Ball A moves toward ball B with velocity v_A. The impact of A with B is perfectly elastic ($e = 1$). Neglect external forces.
(a) What is the velocity of the common center of mass of the balls B and C immediately after the impact?
(b) What is the velocity of the common center of mass of the balls B and C at time t after the impact?

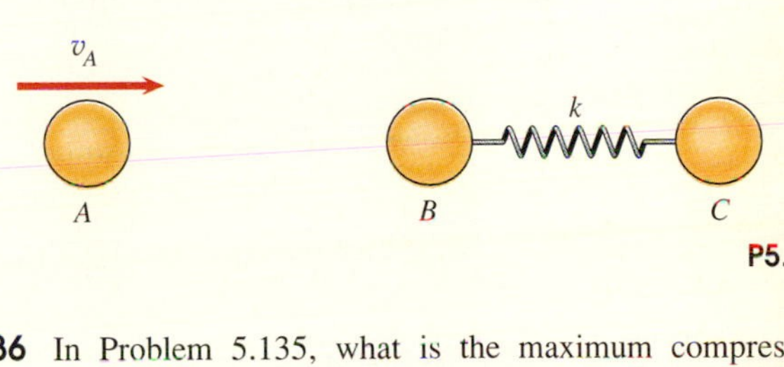

P5.135

5.136 In Problem 5.135, what is the maximum compressive force in the spring as a result of the impact?

5.137 Suppose you interpret Problem 5.135 as an impact between the ball A and an "object" D consisting of the connected balls B and C.
(a) What is the coefficient of restitution of the impact between A and D?
(b) If you consider the total energy after the impact to be the sum of the kinetic energies $\frac{1}{2}m(v'_A)^2 + \frac{1}{2}(2m)(v'_D)^2$, where v'_D is the velocity of the center of mass of D after the impact, how much energy is "lost" as a result of the impact?
(c) How much energy is actually lost as a result of the impact? (This problem is an interesting model for one of the mechanisms of energy loss in impacts between objects. The energy "loss" calculated in part (b) is transformed into "internal energy"—the vibrational motions of B and C relative to their common center of mass.)

5.138 A small object starts from rest at A and slides down the smooth ramp. The coefficient of restitution of its impact with the floor is $e = 0.8$. At what height above the floor does it hit the wall?

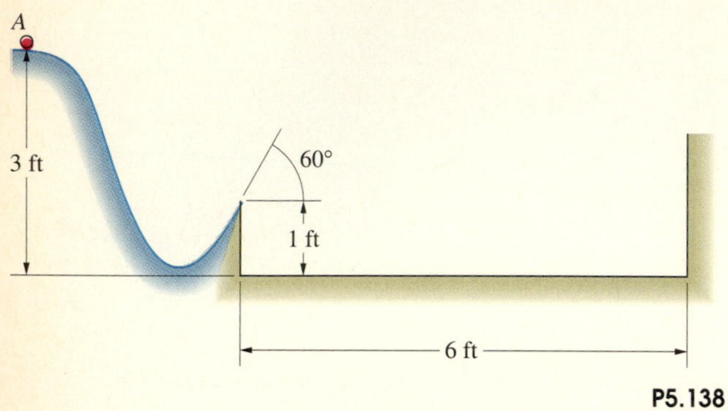

P5.138

5.139 The cue gives the cue ball A a velocity of magnitude 3 m/s. The angle $\beta = 0$ and the coefficient of restitution of the impact of the cue ball and the eight ball B is $e = 1$. If the magnitude of the eight ball's velocity after the impact is 0.9 m/s, what was the coefficient of restitution of the cue ball's impact with the cushion? (The balls are of equal mass.)

P5.139

5.140 What is the solution of Problem 5.139 if the angle $\beta = 10°$?

5.141 What is the solution of Problem 5.139 if the angle $\beta = 15°$ and the coefficient of restitution of the impact between the two balls is $e = 0.9$?

5.142 A ball is given a horizontal velocity of 3 m/s at 2 m above the smooth floor. Determine the distance D between its first and second bounces if the coefficient of restitution is $e = 0.6$.

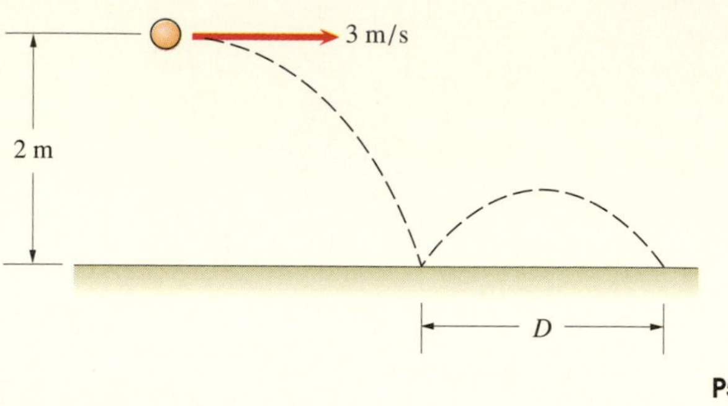

P5.142

5.143 A basketball dropped on the floor from a height of 4 ft rebounds to a height of 3 ft. In the "lay-up" shot shown, the magnitude of the ball's velocity is 5 ft/s and the angles between its velocity vector and the positive coordinate axes are $\theta_x = 42°$, $\theta_y = 68°$, and $\theta_z = 124°$ just before it hits the backboard. What are the magnitude of its velocity and the angles between its velocity vector and the positive coordinate axes just after it hits the backboard?

P5.143

5.144 In Problem 5.143, the basketball's diameter is 9.5 in., the coordinates of the center of the basket rim are $x = 0$, $y = 0$, $z = 12$ in., and the backboard lies in the x-y plane. Determine the x and y coordinates of the point where the ball must hit the backboard so that the center of the ball passes through the center of the basket rim.

5.145 A satellite at $r_0 = 10{,}000$ mi from the center of the earth is given an initial velocity $v_0 = 20{,}000$ ft/s in the direction shown. Determine the magnitude of its transverse component of velocity when $r = 20{,}000$ mi. (The radius of the earth is 3960 mi.)

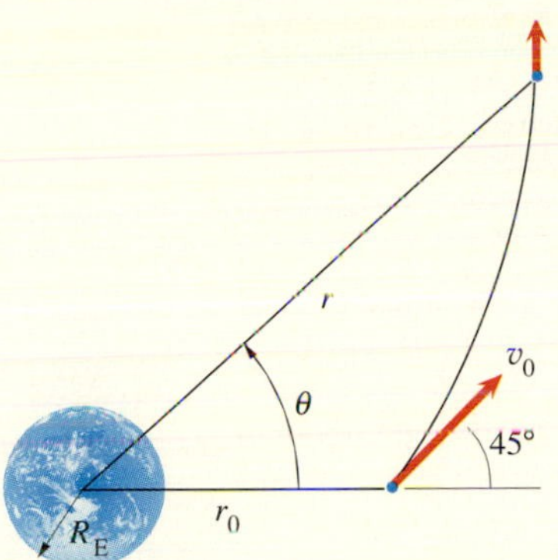

P5.145

5.146 In Problem 5.145, determine the magnitudes of the radial and transverse components of the satellite's velocity when $r = 15{,}000$ mi.

5.147 The snow is 2 ft deep and weighs 20 lb/ft^3, the snowplow is 8 ft wide, and the truck travels at 5 mi/hr. What force does the snow exert on the truck?

P5.147

5.148 An empty 55-lb drum, 3 ft in diameter, stands on a set of scales. Water begins pouring into the drum at 1200 lb/min from 8 ft above the bottom of the drum. The weight density of water is approximately 62.4 lb/ft^3. What do the scales read 40 s after the water starts pouring?

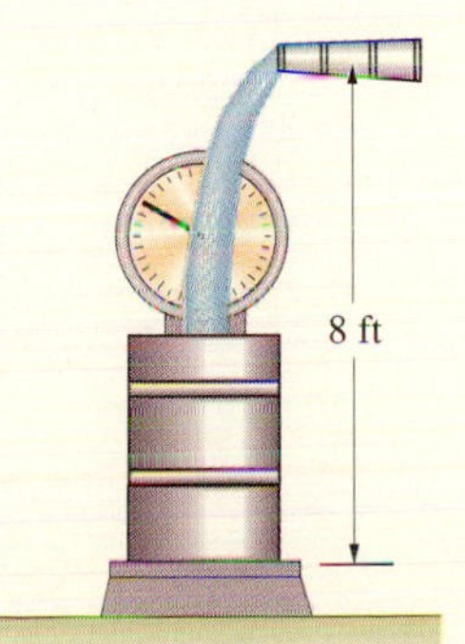

P5.148

5.149 The ski boat's jet propulsive system draws water in at A and expels it at B at 80 ft/s relative to the boat. Assume that the water drawn in enters with no horizontal velocity relative to the surrounding water. The maximum mass flow rate of water through the engine is 2.5 slugs/s. Hydrodynamic drag exerts a force on the boat of magnitude $1.5v$ lb, where v is the boat's velocity in feet per second. If you neglect aerodynamic drag, what is the ski boat's maximum velocity?

P5.149

5.150 The ski boat in Problem 5.149 weighs 2800 lb. The mass flow rate of water through its engine is 2.5 slugs/s, and it starts from rest at $t = 0$. Determine the boat's velocity (a) at $t = 20$ s; (b) at $t = 60$ s.

5.151 A crate of mass m slides across the smooth floor pulling chain from a stationary pile. The mass per unit length of the chain is ρ_L. If the velocity of the crate is v_0 when $s = 0$, what is its velocity as a function of s?

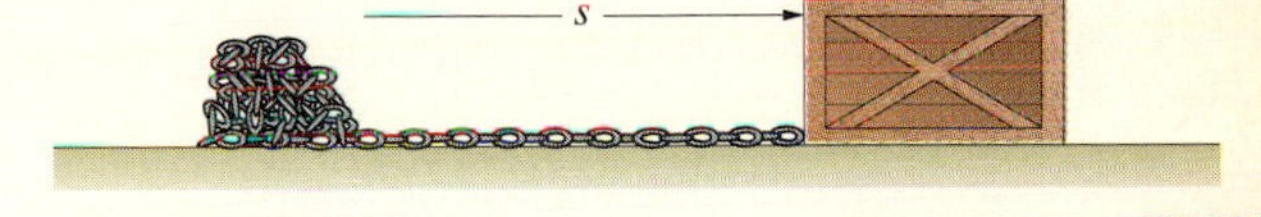

P5.151

Project 5.1 By making measurements, determine the coefficient of restitution of a tennis ball bouncing on a rigid surface. Try to determine whether your result is independent of the velocity with which the ball strikes the surface. Write a brief report describing your procedure and commenting on possible sources of error.

The gear that is engaged determines the ratio of the angular velocity of the pedals and sprocket to that of the bicycle's rear wheel. The ratio of the sprocket's radius to that of the gear equals the ratio of the wheel's angular velocity to that of the sprocket. In this chapter we obtain results of this kind by modeling objects as rigid bodies.

Chapter 6

Planar Kinematics of Rigid Bodies

UNTIL now, we have considered situations in which you could determine the motion of an object's center of mass by using Newton's second law alone. But you must often determine an object's rotational motion as well, even when your only objective is to determine the motion of its center of mass. Moreover, the rotational motion itself can be of interest or even central to the situation you are considering, as in the motions of gears, wheels, generators, turbines, and gyroscopes.

In this chapter we discuss the **kinematics** of objects—the description and analysis of the motions of objects without consideration of the forces and couples causing them. In particular, we show how the motions of individual points of an object are related to the object's angular motion.

6.1 *Rigid Bodies and Types of Motion*

If you throw a brick (Fig. 6.1a), you can determine the motion of its center of mass without having to be concerned about its rotational motion. The only significant force is its weight, and Newton's second law determines the acceleration of its center of mass. But suppose that the brick is standing on the floor, you tip it over (Fig. 6.1b), and you want to determine the motion of its center of mass as it falls. In this case, the brick is subjected to its weight and also a force exerted by the floor. You cannot determine the force exerted by the floor, or the motion of the brick's center of mass, without also analyzing its rotational motion.

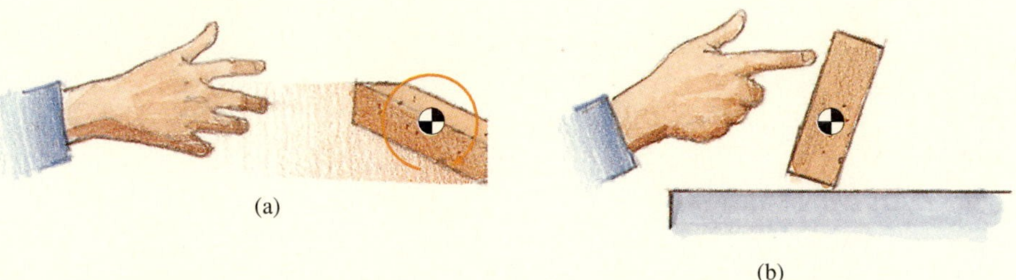

Fig. 6.1
(a) A thrown brick—its rotation doesn't affect the motion of its center of mass.
(b) A tipped brick—its rotation and the motion of its center of mass are interrelated.

Before we can analyze such motions, we must consider how to describe them. A brick is an example of an object whose motion can be described by treating it as a rigid body. A **rigid body** is an idealized model of an object that does not deform, or change shape. The precise definition is that the distance between every pair of points of a rigid body remains constant. Although any object does deform as it moves, if its deformation is small *you can approximate its motion by modeling it as a rigid body.* For example, you can model a twirler's baton in normal use as a rigid body (Fig. 6.2a), but not a fly-casting rod (Fig. 6.2b).

Fig. 6.2
(a) A baton can be modeled as a rigid body.
(b) A fishing rod is too flexible under normal use to model as a rigid body.

Describing the motion of a rigid body requires a reference frame (coordinate system) relative to which the motions of the points of the rigid body and its angular motion are measured. In many situations it is convenient to use a reference frame that is fixed with respect to the earth. For example, we would use such an **earth-fixed** reference frame to describe the motion of the center of mass and the angular motion of the brick in Fig. 6.1. In the following paragraphs we discuss some types of rigid-body motions relative to a given reference frame that occur frequently in applications.

Translation If a rigid body in motion relative to a given reference frame does not rotate, it is said to be in **translation** (Fig. 6.3a). For example, the child's swing in Fig. 6.3b is designed so that the horizontal bar to which the seats are attached is in translation. Although each point of the horizontal bar moves in a circular path, the bar does not rotate. It remains horizontal, making it easier for the child to ride safely. Every point of a rigid body in translation has the same velocity and acceleration, so we completely describe the motion of the rigid body if we describe the motion of a single point.

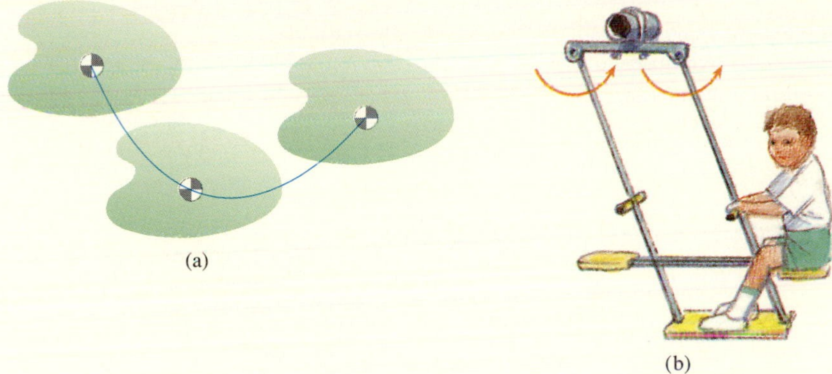

Fig. 6.3
(a) An object in translation does not rotate.
(b) The translating part of the swing on which the child sits remains level.

Rotation About a Fixed Axis After translation, the simplest type of rigid-body motion is rotation about an axis that is fixed relative to a given reference frame (Fig. 6.4a). Each point of the rigid body on the axis is stationary, and each point not on the axis moves in a circular path about the axis as the rigid body rotates. The rotor of an electric motor (Fig. 6.4b) is an example of an object rotating about a fixed axis. The motion of a ship's propeller relative to the ship is rotation about a fixed axis. We discuss this type of motion in more detail in the next section.

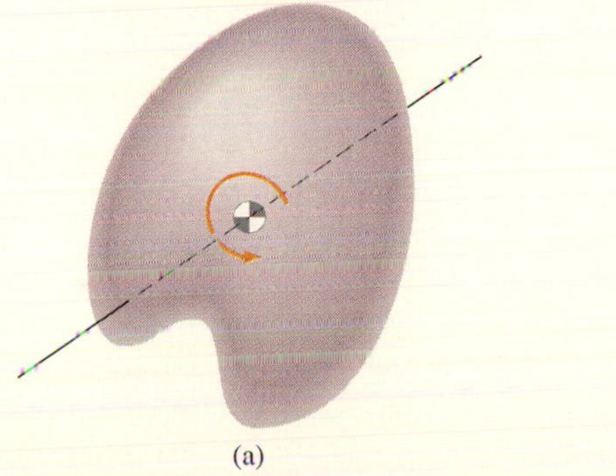

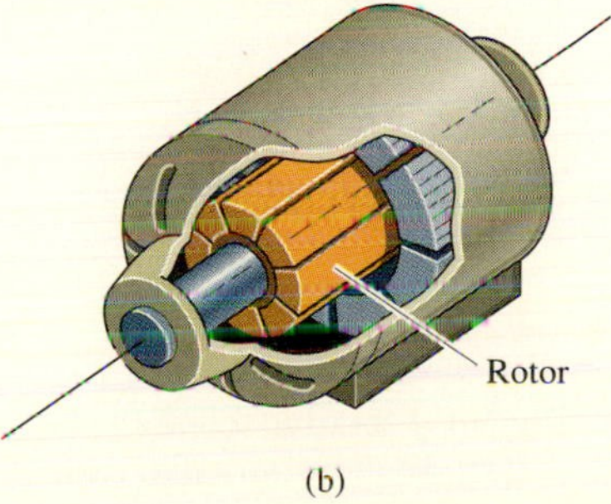

Fig. 6.4
(a) A rigid body rotating about a fixed axis.
(b) Relative to the frame of an electric motor, the rotor rotates about a fixed axis.

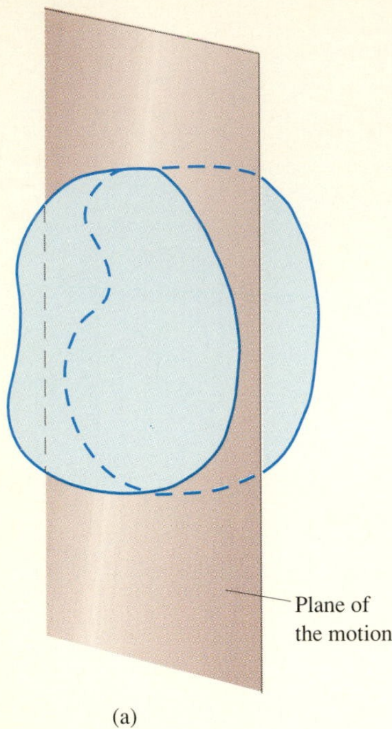

(a)

Planar Motion Consider a plane that is fixed relative to a given reference frame and a rigid body intersected by the plane (Fig. 6.5a). If the rigid body undergoes a motion in which the points intersected by the plane remain in the plane, it is said to be in **two-dimensional**, or **planar**, motion. We refer to the fixed plane as the plane of the motion. Rotation of a rigid body about a fixed axis is a special case of planar motion. As another example, when a car moves in a straight path its wheels are in planar motion (Fig. 6.5b).

The components of an internal combustion engine illustrate these types of motion (Fig. 6.6). Relative to a reference frame that is fixed with respect to the engine, the pistons translate within the cylinders. The connecting rods are in general planar motion, and the crankshaft rotates about a fixed axis.

We begin our analysis of rigid-body motion in the next section with a discussion of rotation about a fixed axis. By using normal and tangential components, we express the velocity and acceleration of a point of the rigid body in terms of the rigid body's angular velocity and angular acceleration. In the following sections we consider general planar motion and obtain expressions relating the relative velocity and acceleration of points of a rigid body to its angular velocity and angular acceleration. With these relations we analyze particular examples of planar motion, such as rolling, and also analyze motions of connected rigid bodies.

(b)

Fig. 6.5

(a) A rigid body intersected by a fixed plane.
(b) A wheel undergoing planar motion.

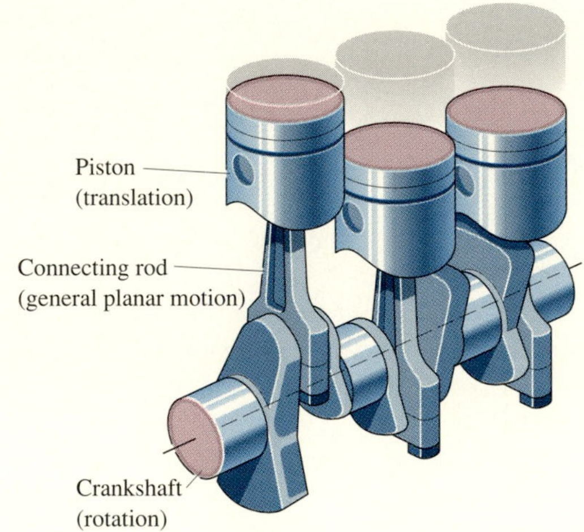

Fig. 6.6

Translation, rotation about a fixed axis, and planar motion in an automobile engine.

6.2 Rotation About a Fixed Axis

By considering rotation of an object about an axis that is fixed relative to a given reference frame, we can introduce some of the concepts of rigid-body motion in a familiar context. In this type of motion, each point of the rigid body moves in a circular path around the fixed axis, so we can analyze motions of points using results developed in Chapter 2.

In Fig. 6.7 we show a rigid body rotating about a fixed axis and introduce two lines perpendicular to the axis. The reference line is fixed, and the body-fixed line rotates with the rigid body. The angle θ between the reference line and the body-fixed line describes the position, or **orientation**, of the rigid body about the fixed axis. The rigid body's **angular velocity**, or rate of rotation, and its **angular acceleration** are

$$\omega = \frac{d\theta}{dt}, \qquad \alpha = \frac{d\omega}{dt} = \frac{d^2\theta}{dt^2}. \tag{6.1}$$

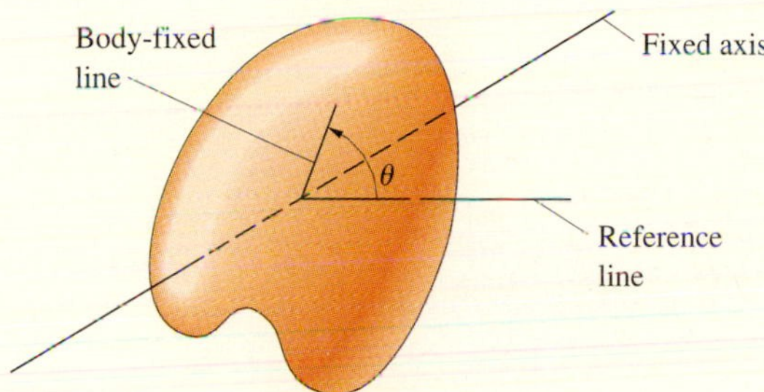

Fig. 6.7
Specifying the orientation of an object rotating about a fixed axis.

Each point of the object not on the fixed axis moves in a circular path about the axis. Using our knowledge of the motion of a point in a circular path, we can relate the velocity and acceleration of a point to the object's angular velocity and angular acceleration. In Fig. 6.8, we view the object in the direction parallel to the fixed axis. The velocity of a point at a distance r from the fixed axis is tangent to its circular path (Fig. 6.8a) and is given in terms of the angular velocity of the object by

$$v = r\omega. \tag{6.2}$$

A point has components of acceleration tangential and normal to its circular path (Fig. 6.8b). In terms of the angular velocity and angular acceleration of the object, the components of acceleration are

$$a_t = r\alpha, \qquad a_n = \frac{v^2}{r} = r\omega^2. \tag{6.3}$$

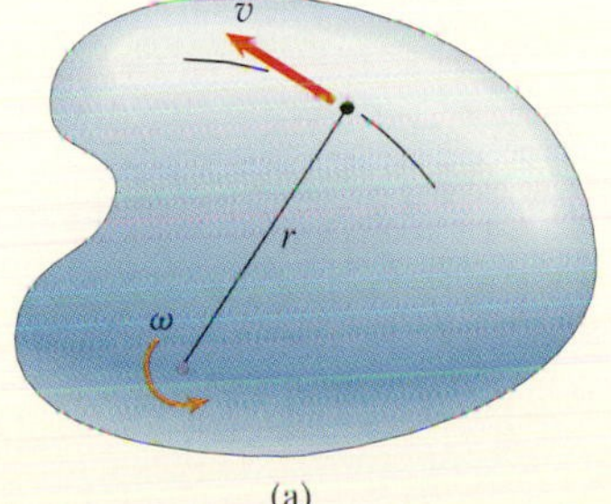

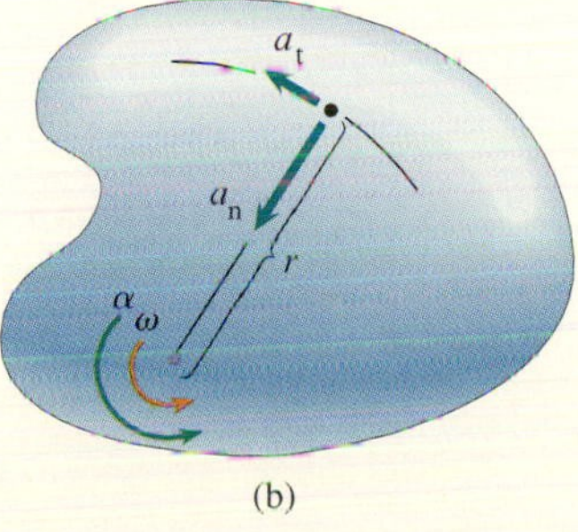

Fig. 6.8
(a) Velocity and (b) acceleration of a point of a rigid body rotating about a fixed axis.

With these relations we can analyze problems involving objects rotating about fixed axes. For example, suppose that we know the angular velocity ω_A and angular acceleration α_A of the gear in Fig. 6.9 relative to a particular reference frame, and we want to determine ω_B and α_B. The velocities of the gears must be equal at P, because there is no relative motion between them in the tangential direction at P. Therefore $r_A\omega_A = r_B\omega_B$, and we find that the angular velocity of gear B is

$$\omega_B = \left(\frac{r_A}{r_B}\right)\omega_A.$$

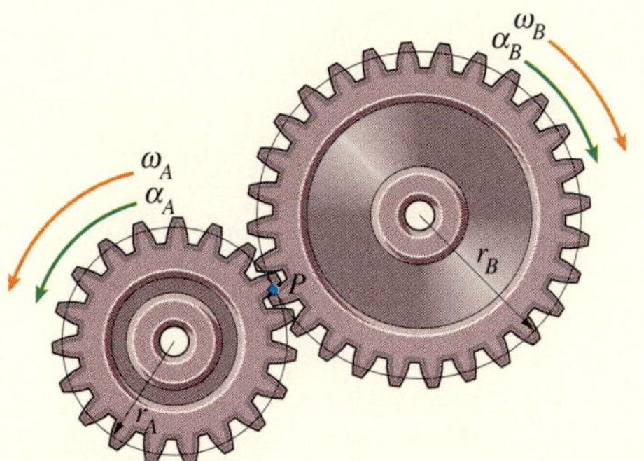

Fig. 6.9
Relating the angular velocities and angular accelerations of meshing gears.

By taking the time derivative of this equation, we determine the angular acceleration of gear B:

$$\alpha_B = \left(\frac{r_A}{r_B}\right)\alpha_A.$$

From this result, we see that the tangential components of the accelerations of the gears at P are equal: $r_A\alpha_A = r_B\alpha_B$. However, the normal components of the accelerations of the gears at P are different in direction and, if the gears have different radii, are different in magnitude as well. The normal component of the acceleration of gear A points toward the center of gear A and its magnitude is $r_A\omega_A^2$. The normal component of the acceleration of gear B points toward the center of gear B and its magnitude is $r_B\omega_B^2 = (r_A/r_B)(r_A\omega_A^2)$.

In the following example we demonstrate the analysis of motions of objects rotating about fixed axes. You can use Eqs. (6.1) to analyze the angular motions and use Eqs. (6.2) and (6.3) to determine the velocities and accelerations of points.

Example 6.1

Gear A of the winch in Fig. 6.10 turns gear B, raising the hook H. If the gear A starts from rest at $t = 0$ and its clockwise angular acceleration is $\alpha_A = 0.2t$ rad/s^2, what vertical distance has the hook H risen and what is its velocity at $t = 10$ s?

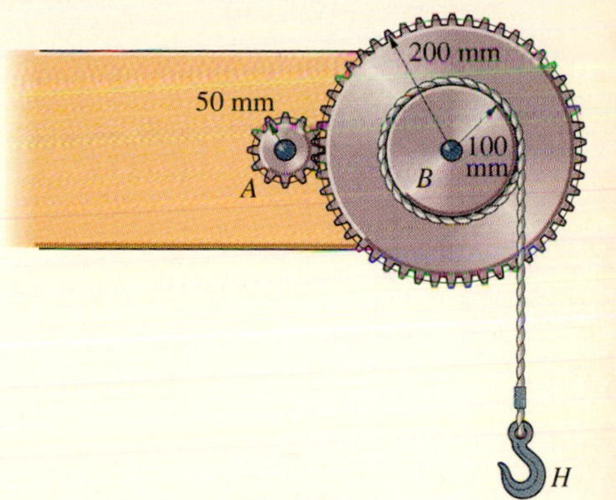

Fig. 6.10

STRATEGY

By equating the tangential components of acceleration of gears A and B at their point of contact, we can determine the angular acceleration of gear B. Then we can integrate to obtain the angular velocity of gear B and the angle through which it has turned at $t = 10$ s.

SOLUTION

The tangential acceleration of the point of contact of the two gears (Fig. a) is

$$a_t = (0.05 \text{ m})(0.2t \text{ rad/s}^2) = (0.2 \text{ m})(\alpha_B).$$

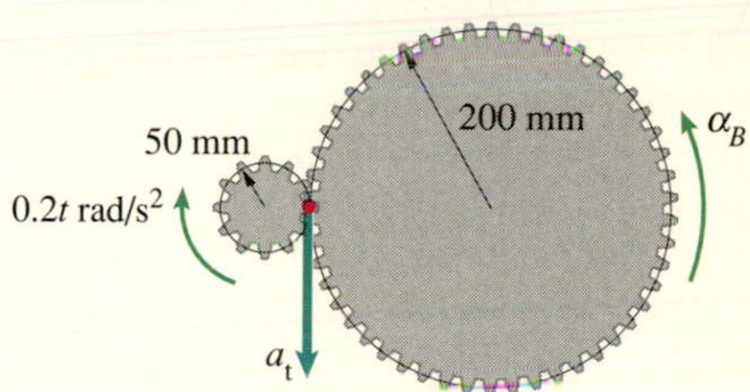

(a) The tangential accelerations of the gears are equal at their point of contact.

Therefore the angular acceleration of gear B is

$$\alpha_B = \frac{d\omega_B}{dt} = \frac{(0.05 \text{ m})(0.2t \text{ rad/s}^2)}{(0.2 \text{ m})} = 0.05t \text{ rad/s}^2.$$

Integrating this equation,

$$\int_0^{\omega_B} d\omega_B = \int_0^t 0.05t \, dt,$$

we obtain the angular velocity of gear B:

$$\omega_B = \frac{d\theta_B}{dt} = 0.025t^2 \text{ rad/s}.$$

Integrating again, we obtain the angle through which gear B has turned:

$$\theta_B = 0.00833t^3 \text{ rad}.$$

At $t = 10$ s, $\theta_B = 8.33$ rad. The amount of cable wound around the drum, which is the distance the hook H has risen, is the product of θ_B and the radius of the drum: $(8.33 \text{ rad})(0.1 \text{ m}) = 0.833$ m.

At $t = 10$ s, $\omega_B = 2.5$ rad/s. The velocity of a point on the rim, which equals the velocity of the hook H (Fig. b), is

$$v_H = (0.1 \text{ m})(2.5 \text{ rad/s}) = 0.25 \text{ m/s}.$$

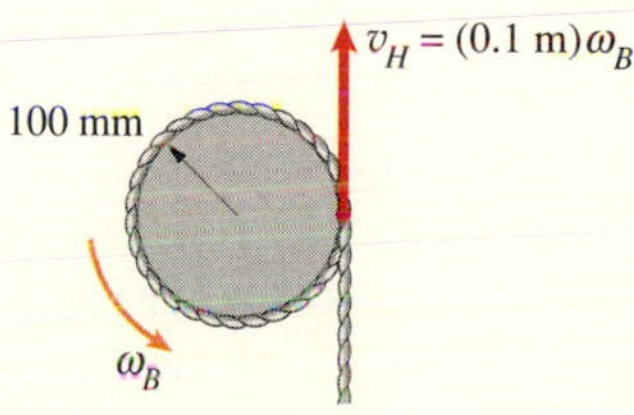

(b) Determining the hook's velocity.

Problems

6.1 The disk rotates relative to the coordinate system about a fixed shaft that is coincident with the z axis. At the instant shown, the disk has a counterclockwise angular velocity of 3 rad/s and a counterclockwise angular acceleration of 4 rad/s^2. What are the x and y components of the velocity and acceleration of point A relative to the coordinate system?

Strategy: Use Eqs. (6.2) and (6.3) and Fig. 6.8 to determine the velocity and acceleration of A.

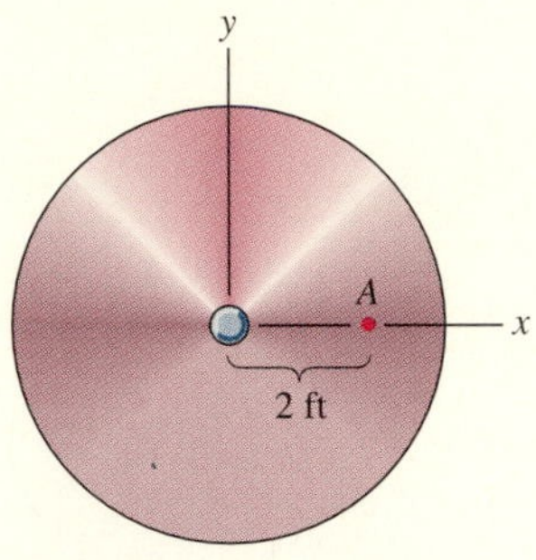

P6.1

6.2 If the angular acceleration of the disk in Problem 6.1 is constant, what are the x and y components of the velocity and acceleration of point A when the disk has rotated 90° relative to the position shown?

6.3 The weight A hangs from a rope that is wrapped around the disk. If the weight is moving downward at 4 m/s, what is the angular velocity of the disk?

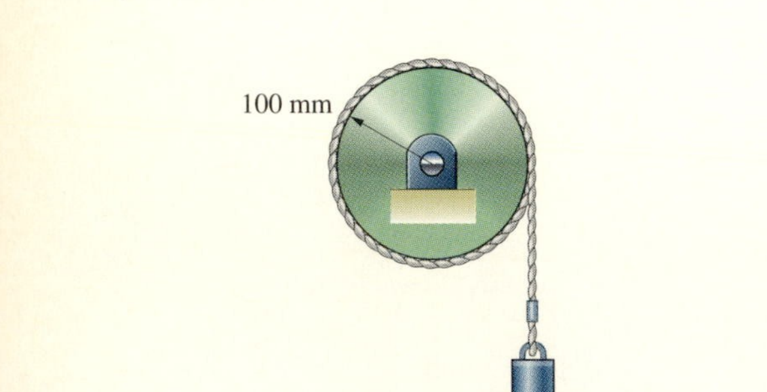

P6.3

6.4 Suppose that the weight A in Problem 6.3 starts from rest at $t = 0$ and falls with a constant acceleration of 2 m/s^2.

(a) What is the angular acceleration of the disk?

(b) How many revolutions has the disk turned at $t = 1$ s?

6.5 Determine ω_B/ω_A and ω_C/ω_A.

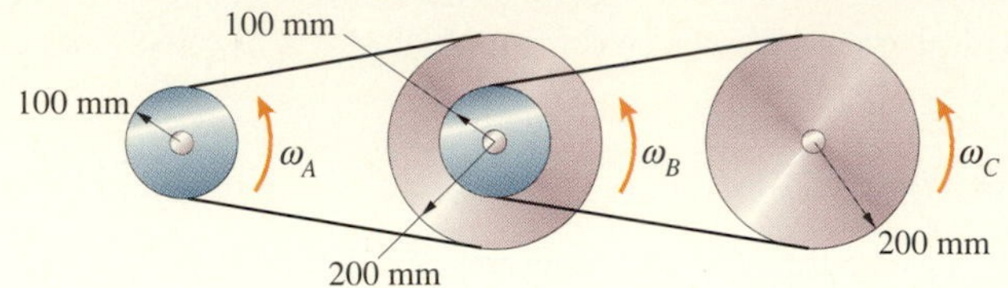

P6.5

6.6 The bicycle's 120-mm sprocket wheel turns at 3 rad/s relative to the frame of the bicycle. What is the angular velocity of the 45-mm gear?

P6.6

6.7 The rear wheel of the bicycle in Problem 6.6 has a 330-mm radius and is rigidly attached to the 45-mm gear. If the rider turns the pedals, which are rigidly attached to the 120-mm sprocket wheel, at one revolution per second, what is the bicycle's velocity?

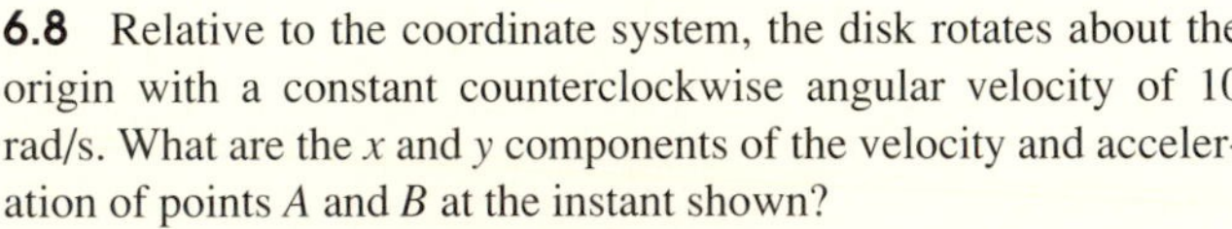

6.8 Relative to the coordinate system, the disk rotates about the origin with a constant counterclockwise angular velocity of 10 rad/s. What are the x and y components of the velocity and acceleration of points A and B at the instant shown?

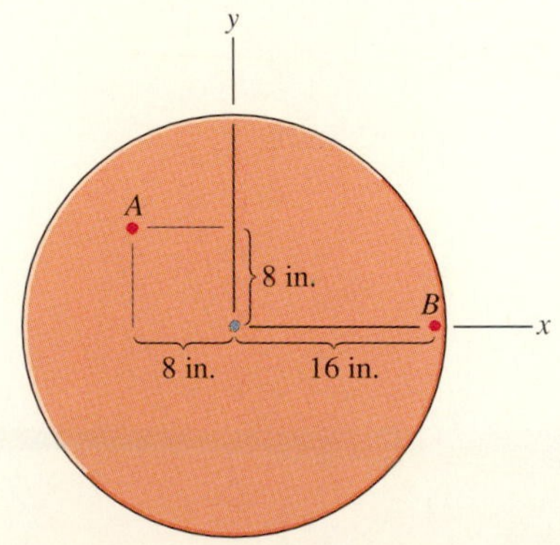

P6.8

6.9 Suppose that the disk in Problem 6.8 starts from rest in the position shown at $t = 0$ and has a constant counterclockwise angular acceleration of 6 rad/s^2 relative to the coordinate system. Determine the x and y components of the velocity and acceleration of point B at $t = 1$ s.

6.10 Suppose that the disk in Problem 6.8 starts from rest in the position shown at $t = 0$ and has a constant counterclockwise angular acceleration of 4 rad/s^2 relative to the coordinate system. Determine the x and y components of the velocity of point A at $t = 2$ s.

6.11 The bracket rotates relative to the coordinate system about a fixed shaft that is coincident with the z axis. If it has a counterclockwise angular velocity of 20 rad/s and a clockwise angular acceleration of 200 rad/s^2, what are the magnitudes of the accelerations of points A and B?

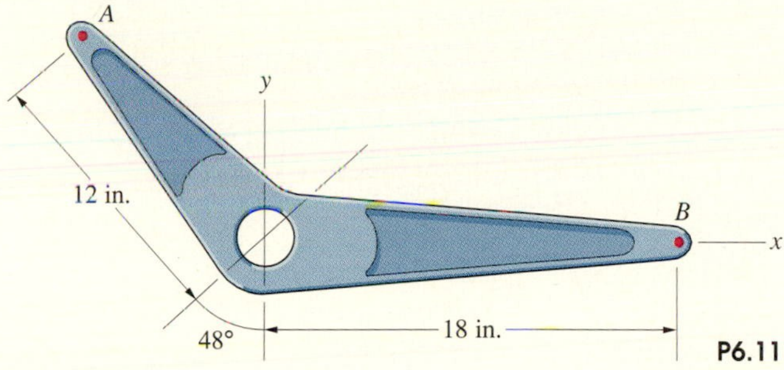

P6.11

6.12 Consider the bracket in Problem 6.11. If $|\mathbf{v}_A| = 10$ ft/s and $|\mathbf{a}_A| = 200$ ft/s^2, what are $|\mathbf{v}_B|$ and $|\mathbf{a}_B|$?

6.13 Consider the bracket in Problem 6.11. If $|\mathbf{v}_A| = 36$ in./s and $|\mathbf{a}_B| = 600$ in./s^2, what are $|\mathbf{v}_B|$ and $|\mathbf{a}_A|$?

6.3 General Motions: Velocities

Each point of a rigid body in translation undergoes the same motion. Each point of a rigid body rotating about a fixed axis undergoes circular motion about the axis. To analyze more complicated motions that combine translation and rotation, we must develop equations that relate the relative motions of points of a rigid body to its angular motion.

Relative Velocities

In Fig. 6.11(a) we view a rigid body perpendicular to the plane of its motion. Points A and B are points of the rigid body contained in that plane, and O is the origin of a given reference frame. The position of A relative to B, $\mathbf{r}_{A/B}$, is related to the positions of A and B relative to O by

$$\mathbf{r}_A = \mathbf{r}_B + \mathbf{r}_{A/B}.$$

Taking the derivative of this equation with respect to time, we obtain

$$\mathbf{v}_A = \mathbf{v}_B + \mathbf{v}_{A/B}, \qquad (6.4)$$

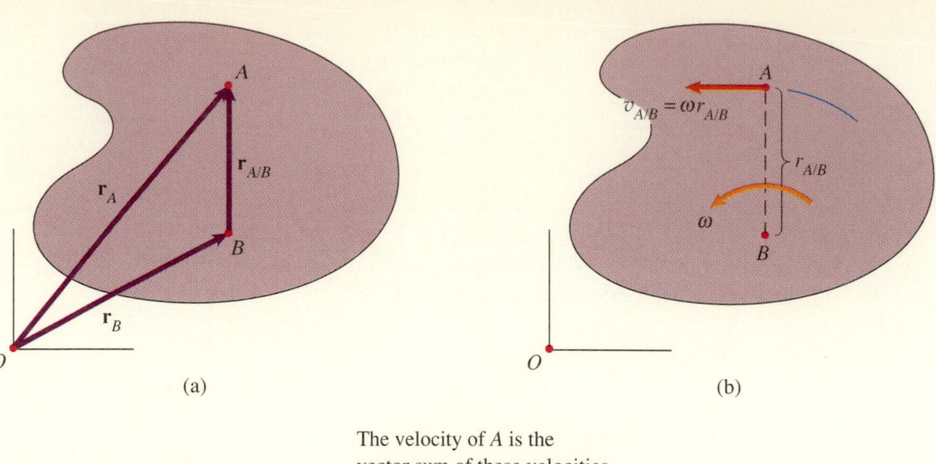

Fig. 6.11

(a) A rigid body in planar motion.
(b) The velocity of A relative to B.
(c) The velocity of A is the sum of its velocity relative to B and the velocity of B.

where $\mathbf{v}_A$ and $\mathbf{v}_B$ are the velocities of A and B relative to the reference frame and $\mathbf{v}_{A/B} = d\mathbf{r}_{A/B}/dt$ is the velocity of A relative to B. (*When we simply speak of the velocity of a point, we will mean its velocity relative to the given reference frame.*) We can show that $\mathbf{v}_{A/B}$ is related in a simple way to the rigid body's angular velocity. Since A and B are points of the rigid body, the distance between them, $r_{A/B} = |\mathbf{r}_{A/B}|$, is constant. That means that relative to B, A moves in a circular path as the rigid body rotates. The velocity of A relative to B is therefore tangent to the circular path and equal to the product of $r_{A/B}$ and the angular velocity ω of the rigid body (Fig. 6.11b). From Eq. (6.4), the velocity of A is the sum of the velocity of B and the velocity of A relative to B (Fig. 6.11c). You can use this result to relate velocities of points of a rigid body in planar motion when you know its angular velocity.

For example, in Fig. 6.12(a) we show a circular disk of radius R rolling on a stationary plane surface with counterclockwise angular velocity ω. Saying that the surface is stationary means we are describing the motion of the disk in terms of a reference frame that is fixed with respect to the surface. By **rolling**, we mean that the velocity of the disk relative to the surface is zero at the point of contact C. The velocity of the center B of the disk relative to C is shown in Fig. 6.12(b). Since $\mathbf{v}_C = 0$, the velocity of B in terms of the fixed coordinate system shown is

$$\mathbf{v}_B = \mathbf{v}_C + \mathbf{v}_{B/C} = -R\omega\,\mathbf{i}.$$

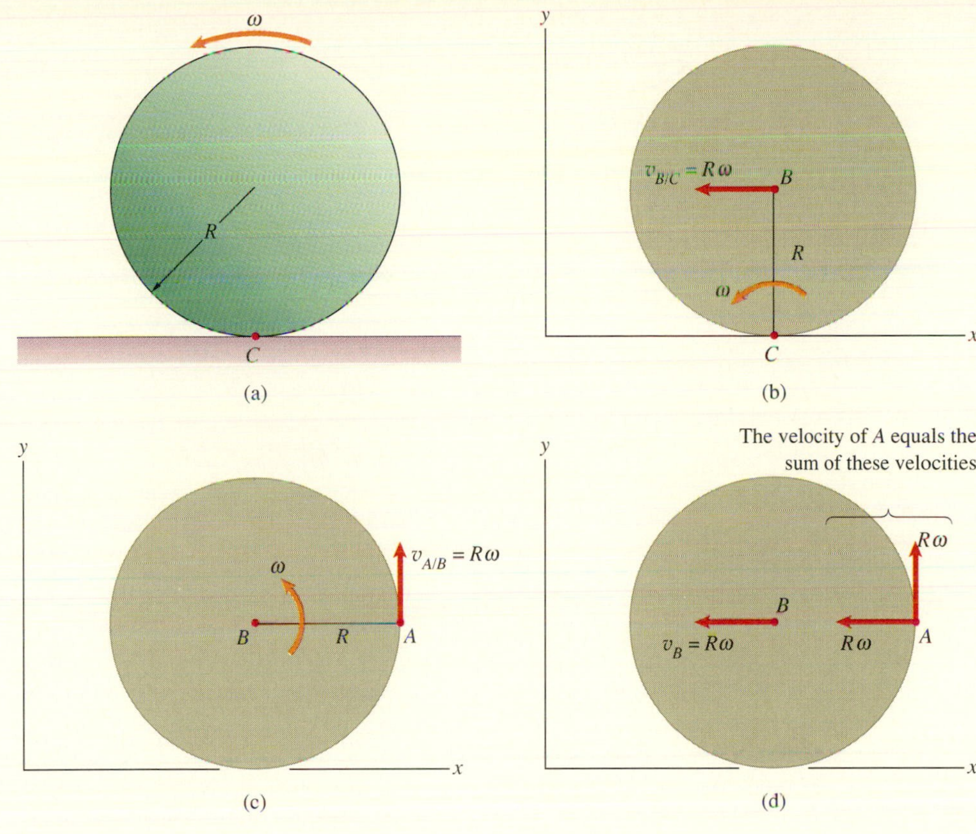

Fig. 6.12

(a) A disk rolling with angular velocity ω.
(b) The velocity of the center B relative to C.
(c) The velocity of A relative to B.
(d) The velocity of A equals the sum of the velocity of B and the velocity of A relative to B.

This result is worth remembering: *The magnitude of the velocity of the center of a round object rolling on a stationary plane surface equals the product of the radius and the magnitude of the angular velocity.*

We can determine the velocity of any other point of the disk in the same way. Figure 6.12(c) shows the velocity of a point A relative to point B. The velocity of A is the sum of the velocity of B and the velocity of A relative to B (Fig. 6.12d):

$$\mathbf{v}_A = \mathbf{v}_B + \mathbf{v}_{A/B} = -R\omega\,\mathbf{i} + R\omega\,\mathbf{j}.$$

The Angular Velocity Vector

We can express the rate of rotation of a rigid body as a vector. **Euler's theorem** states that a rigid body constrained to rotate about a fixed point B can move between any two positions by a single rotation about some axis through B. Suppose that we choose an arbitrary point B of a rigid body that is undergoing an *arbitrary* motion at a time t. Euler's theorem allows us to express the rigid body's change in position relative to B during an interval of time from t to $t + dt$ as a single rotation through an angle $d\theta$ about some axis. At time t, the rigid body's rate of rotation about the axis is its angular velocity $\omega = d\theta/dt$, and the axis about which it rotates is called the **instantaneous axis of rotation**.

The **angular velocity vector**, denoted by $\boldsymbol{\omega}$, specifies both the direction of the instantaneous axis of rotation and the angular velocity. It is defined to be

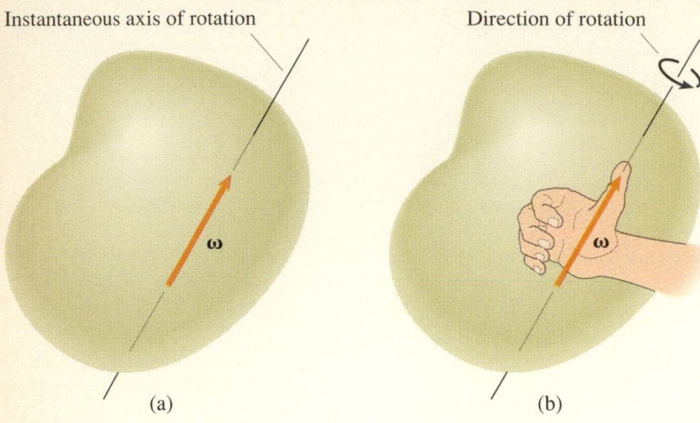

Fig. 6.13
(a) An angular velocity vector.
(b) Right-hand rule for the direction of the vector.

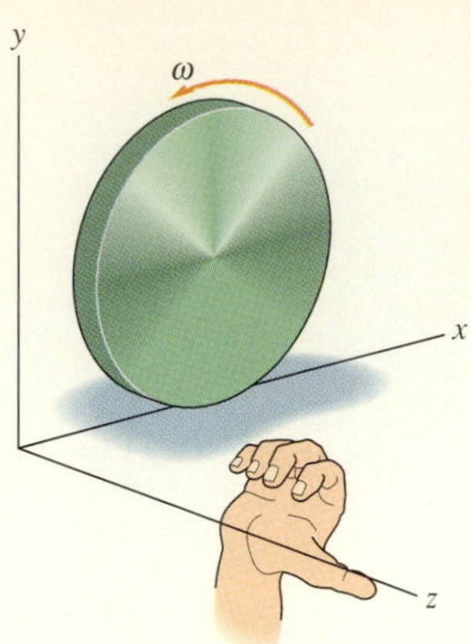

Fig. 6.14
Determining the direction of the angular velocity vector of a rolling disk.

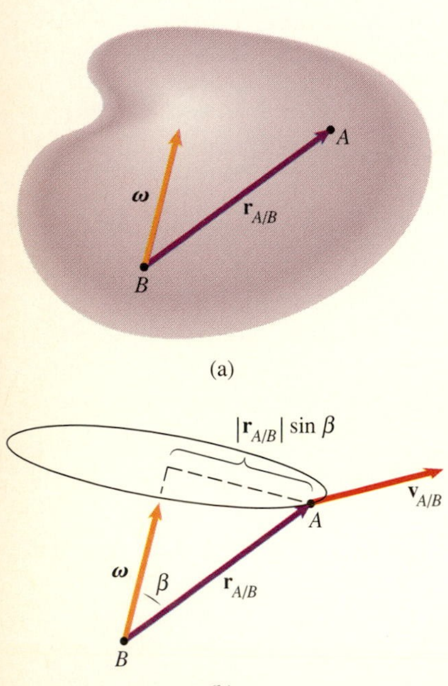

Fig. 6.15
(a) Points A and B of a rotating rigid body.
(b) A is moving in a circular path relative to B.

parallel to the instantaneous axis of rotation (Fig. 6.13a), and its magnitude is the rate of rotation, the absolute value of ω. Its direction is related to the direction of the rigid body's rotation through a right-hand rule: If you point the thumb of your right hand in the direction of $\boldsymbol{\omega}$, the fingers curl around $\boldsymbol{\omega}$ in the direction of the rotation (Fig. 6.13b).

For example, the axis of rotation of the rolling disk in Fig. 6.12 is parallel to the z axis, so its angular velocity vector is parallel to the z axis and its magnitude is ω. If you curl the fingers of your right hand around the z axis in the direction of the rotation, your thumb points in the positive z direction (Fig. 6.14). The angular velocity vector of the disk is $\boldsymbol{\omega} = \omega\mathbf{k}$.

The angular velocity vector allows us to express the results of the previous section in a very convenient form. Let A and B be points of a rigid body with angular velocity $\boldsymbol{\omega}$ (Fig. 6.15a). We can show that the velocity of A relative to B is

$$\mathbf{v}_{A/B} = \frac{d\mathbf{r}_{A/B}}{dt} = \boldsymbol{\omega} \times \mathbf{r}_{A/B}. \qquad (6.5)$$

Relative to B, point A is moving at the present instant in a circular path of radius $|\mathbf{r}_{A/B}| \sin\beta$, where β is the angle between the vectors $\mathbf{r}_{A/B}$ and $\boldsymbol{\omega}$ (Fig. 6.15b). The magnitude of the velocity of A relative to B is equal to the product of the radius of the circular path and the angular velocity of the rigid body, $|\mathbf{v}_{A/B}| = (|\mathbf{r}_{A/B}| \sin\beta)\,|\boldsymbol{\omega}|$, which is the magnitude of the cross product of $\mathbf{r}_{A/B}$ and $\boldsymbol{\omega}$. In addition, $\mathbf{v}_{A/B}$ is perpendicular to $\boldsymbol{\omega}$ and perpendicular to $\mathbf{r}_{A/B}$. But is $\mathbf{v}_{A/B}$ equal to $\boldsymbol{\omega} \times \mathbf{r}_{A/B}$ or $\mathbf{r}_{A/B} \times \boldsymbol{\omega}$? Notice in Fig. 6.15(b) that, pointing the fingers of the right hand in the direction of $\boldsymbol{\omega}$ and closing them toward $\mathbf{r}_{A/B}$, the thumb points in the direction of the velocity of A relative to B, so $\mathbf{v}_{A/B} = \boldsymbol{\omega} \times \mathbf{r}_{A/B}$. Substituting Eq. (6.5) into Eq. (6.4), we obtain an equation for the relation between the velocities of two points of a rigid body in terms of its angular velocity:

$$\boxed{\mathbf{v}_A = \mathbf{v}_B + \underbrace{\boldsymbol{\omega} \times \mathbf{r}_{A/B}}_{\mathbf{v}_{A/B}}.} \qquad (6.6)$$

Let us return to the example of a disk of radius R rolling with angular velocity ω (Fig. 6.16), and use Eq. (6.6) to determine the velocity of point A. The velocity of the center of the disk is given in terms of the angular velocity by $\mathbf{v}_B = -R\omega\,\mathbf{i}$, the disk's angular velocity vector is $\boldsymbol{\omega} = \omega\mathbf{k}$, and the position vector of A relative to the center is $\mathbf{r}_{A/B} = R\mathbf{i}$. The velocity of A is

$$\mathbf{v}_A = \mathbf{v}_B + \boldsymbol{\omega} \times \mathbf{r}_{A/B} = -R\omega\,\mathbf{i} + (\omega\mathbf{k}) \times (R\mathbf{i})$$
$$= -R\omega\,\mathbf{i} + R\omega\,\mathbf{j}.$$

Compare this result with the velocity of A shown in Fig. 6.12(d).

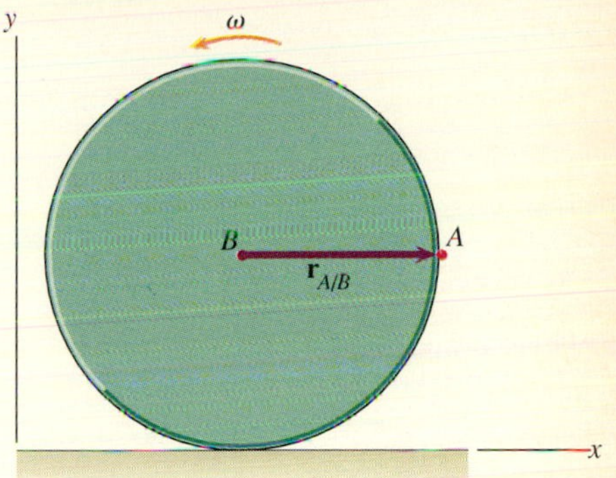

Fig. 6.16
A rolling disk and the position vector of A relative to B.

Beginning with a point of a rigid body whose velocity is known, you can express the velocities of other points of the rigid body in terms of its angular velocity. The following examples demonstrate that repeated application of this procedure allows you to analyze the motions of systems of connected rigid bodies.

Example 6.2

Bar AB in Fig. 6.17 rotates with a clockwise angular velocity of 10 rad/s. Determine the angular velocity of bar BC and the velocity of point C.

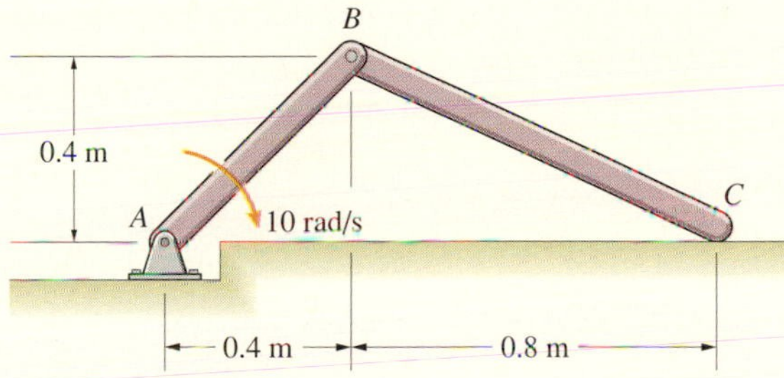

Fig. 6.17

STRATEGY

Bar AB rotates about the fixed point A with a known angular velocity, so we can determine the velocity of B. Then, by expressing the horizontal velocity of C in terms of the velocity of B and the angular velocity of bar BC, we can obtain two equations in terms of the velocity of C and the angular velocity of bar BC. We will first solve the problem using vector diagrams as shown in Fig. 6.11(c), then solve it by applying Eq. (6.6).

SOLUTION

First Method Point B moves in a circular path around the fixed point A. The velocity of B is tangent to the circular path (Fig. a) and its magnitude is

$$v_B = (10 \text{ rad/s})r_{B/A} = 10\sqrt{(0.4)^2 + (0.4)^2} = 5.66 \text{ m/s}.$$

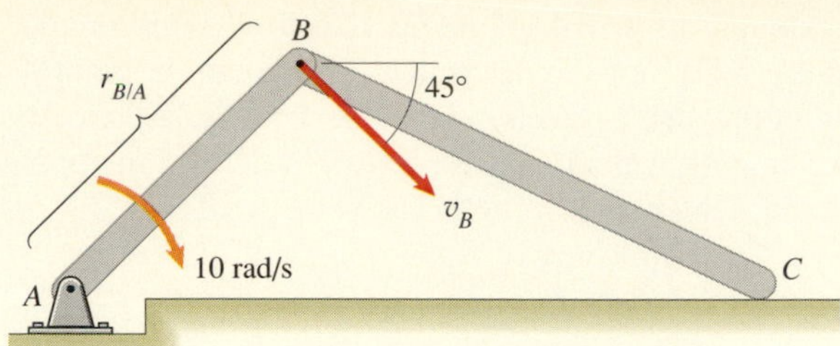

(a) The velocity of B.

Let ω_{BC} be the counterclockwise angular velocity of bar BC. The velocity of C is equal to the velocity of B plus the velocity of C relative to B (Fig. b). The distance $r_{C/B} = \sqrt{(0.4)^2 + (0.8)^2} = 0.894$ m. The velocity component $\omega_{BC} r_{C/B}$ is perpendicular to bar BC, so the angle $\beta = \arctan(0.8/0.4) = 63.4°$. Introducing the coordinate system in Fig. (b), we can express the horizontal velocity of C as $\mathbf{v}_C = v_C\,\mathbf{i}$. From Fig. (b), we see that

$$v_C\,\mathbf{i} = (v_B \cos 45° + \omega_{BC} r_{C/B} \cos\beta)\mathbf{i} + (-v_B \sin 45° + \omega_{BC} r_{C/B} \sin\beta)\mathbf{j}.$$

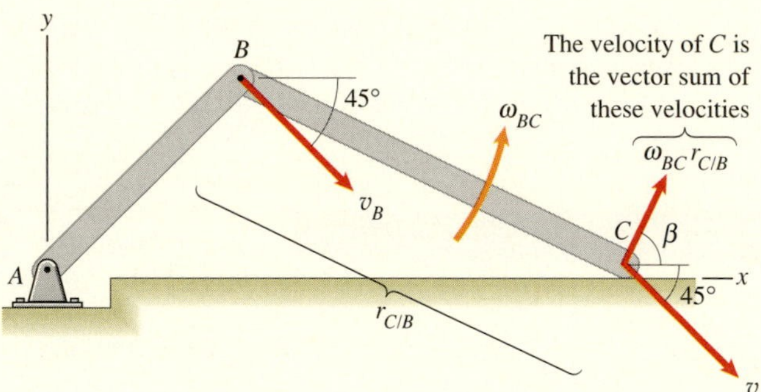

(b) Expressing the velocity of C as the sum of the velocity of B and the velocity of C relative to B.

Equating the **i** and **j** components yields

$$v_C = v_B \cos 45° + \omega_{BC} r_{C/B} \cos\beta,$$
$$0 = -v_B \sin 45° + \omega_{BC} r_{C/B} \sin\beta.$$

Substituting the values of v_B and β, we obtain two equations in terms of v_C and ω_{BC}:

$$v_C = 4 + 0.4\omega_{BC},$$
$$0 = -4 + 0.8\omega_{BC}.$$

Solving, we obtain $v_C = 6$ m/s and $\omega_{BC} = 5$ rad/s. Point C is moving to the right at 6 m/s, and bar BC is rotating at 5 rad/s in the counterclockwise direction.

Second Method We first determine the velocity of B by applying Eq. (6.6) to points A and B. From Fig. (c), the position vector of B relative to A is $\mathbf{r}_{B/A} = 0.4\mathbf{i} + 0.4\mathbf{j}$ (m). The angular velocity vector of bar AB is $\boldsymbol{\omega}_{AB} = -10\mathbf{k}$ (rad/s), so the velocity of B is

$$\mathbf{v}_B = \mathbf{v}_A + \boldsymbol{\omega}_{AB} \times \mathbf{r}_{B/A} = 0 + \begin{vmatrix} \mathbf{i} & \mathbf{j} & \mathbf{k} \\ 0 & 0 & -10 \\ 0.4 & 0.4 & 0 \end{vmatrix}$$
$$= 4\mathbf{i} - 4\mathbf{j}\ \text{(m/s)}.$$

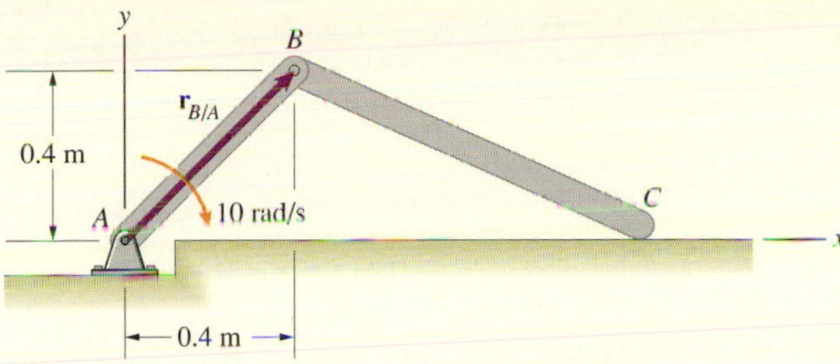

(c) The position vector of B relative to A.

We now apply Eq. (6.6) to points B and C. Let ω_{BC} be the counterclockwise angular velocity of bar BC (Fig. d), so that its angular velocity vector is $\boldsymbol{\omega}_{BC} = \omega_{BC}\,\mathbf{k}$. The position vector of C relative to B is $\mathbf{r}_{C/B} = 0.8\mathbf{i} - 0.4\mathbf{j}$ (m). The velocity of C is

$$v_C\,\mathbf{i} = \mathbf{v}_B + \boldsymbol{\omega}_{BC} \times \mathbf{r}_{B/C}.$$

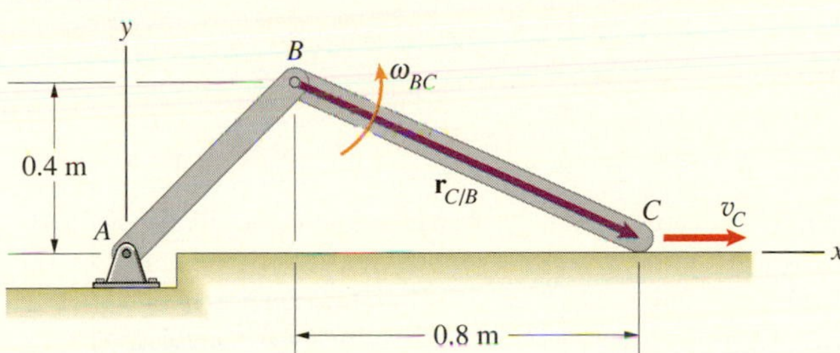

(d) The angular velocity of bar BC, the velocity of C, and the position vector of C relative to B.

We substitute our solution for $\mathbf{v}_B$ and our expressions for $\boldsymbol{\omega}_{BC}$ and $\mathbf{r}_{B/A}$ into this equation, obtaining

$$\begin{aligned} v_C\,\mathbf{i} &= \mathbf{v}_B + \boldsymbol{\omega}_{BC} \times \mathbf{r}_{B/C} \\ &= 4\mathbf{i} - 4\mathbf{j} + \begin{vmatrix} \mathbf{i} & \mathbf{j} & \mathbf{k} \\ 0 & 0 & \omega_{BC} \\ 0.8 & -0.4 & 0 \end{vmatrix} \\ &= 4\mathbf{i} - 4\mathbf{j} + 0.4\omega_{BC}\,\mathbf{i} + 0.8\omega_{BC}\,\mathbf{j}. \end{aligned}$$

Equating the **i** and **j** components yields

$$v_C = 4 + 0.4\omega_{BC},$$

$$0 = -4 + 0.8\omega_{BC}.$$

Solving, we obtain $v_C = 6$ m/s and $\omega_{BC} = 5$ rad/s. The angular velocity of bar BC is $\boldsymbol{\omega}_{BC} = 5\mathbf{k}$ (rad/s), and the velocity of point C is $\mathbf{v}_C = 6\mathbf{i}$ (m/s).

DISCUSSION

You can see that our two methods of solution are equivalent. In the second method, Eq. (6.6) provides the results we obtained from vector diagrams in the first method. Our procedures in this example apply to many problems in which you must determine velocities and angular velocities of connected rigid bodies.

Example 6.3

Bar AB in Fig. 6.18 rotates with a clockwise angular velocity of 10 rad/s. What is the vertical velocity v_R of the rack of the rack and pinion gear?

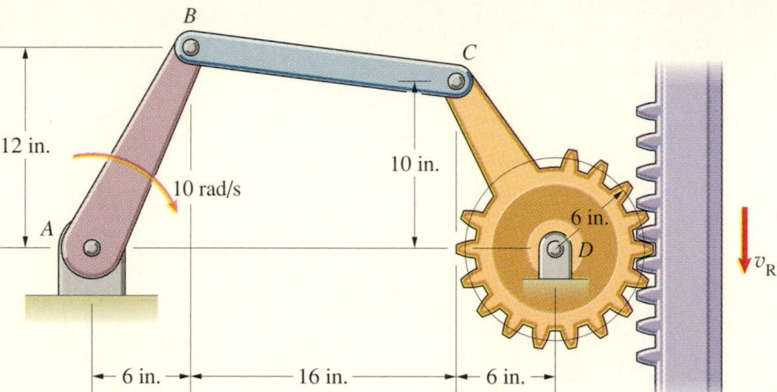

Fig. 6.18

STRATEGY

To determine the velocity of the rack, we must determine the angular velocity of the member CD. Since we know the angular velocity of bar AB, we can apply Eq. (6.6) to points A and B to determine the velocity of point B. Then we can apply Eq. (6.6) to points C and D to obtain an equation for $\mathbf{v}_C$ in terms of the angular velocity of the member CD. We can also apply Eq. (6.6) to points B and C to obtain an equation for $\mathbf{v}_C$ in terms of the angular velocity of bar BC. By equating the two expressions for $\mathbf{v}_C$, we will obtain a vector equation in two unknowns: the angular velocities of bars BC and CD.

SOLUTION

We first apply Eq. (6.6) to points A and B (Fig. a). In terms of the coordinate system shown, the position vector of B relative to A is $\mathbf{r}_{B/A} = 0.5\mathbf{i} + \mathbf{j}$ (ft), and the angular velocity vector of bar AB is $\boldsymbol{\omega}_{AB} = -10\mathbf{k}$ (rad/s). The velocity of B is

$$\mathbf{v}_B = \mathbf{v}_A + \boldsymbol{\omega}_{AB} \times \mathbf{r}_{B/A} = \mathbf{0} + \begin{vmatrix} \mathbf{i} & \mathbf{j} & \mathbf{k} \\ 0 & 0 & -10 \\ 0.5 & 1 & 0 \end{vmatrix}$$

$$= 10\mathbf{i} - 5\mathbf{j} \text{ (ft/s)}.$$

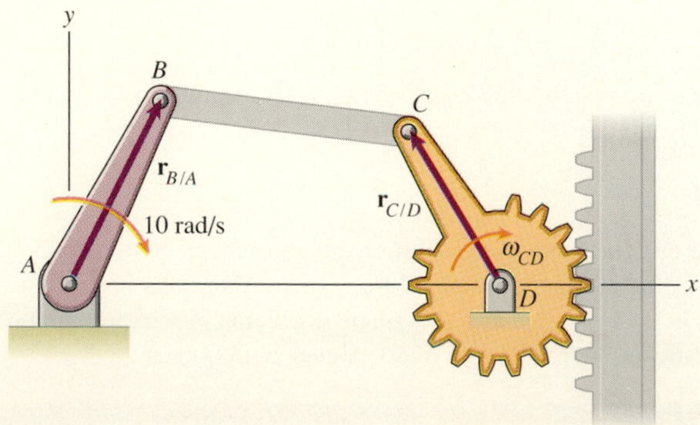

(a) Determining the velocities of points B and C.

We now apply Eq. (6.6) to points C and D. Let ω_{CD} be the unknown angular velocity of member CD (Fig. a). The position vector of C relative to D is $\mathbf{r}_{C/D} = -0.500\mathbf{i} + 0.833\mathbf{j}$ (m), and the angular velocity vector of member CD is $\boldsymbol{\omega}_{CD} = -\omega_{CD}\mathbf{k}$. The velocity of C is

$$\mathbf{v}_C = \mathbf{v}_D + \boldsymbol{\omega}_{CD} \times \mathbf{r}_{C/D} = \mathbf{0} + \begin{vmatrix} \mathbf{i} & \mathbf{j} & \mathbf{k} \\ 0 & 0 & -\omega_{CD} \\ -0.500 & 0.833 & 0 \end{vmatrix}$$

$$= 0.833\omega_{CD}\mathbf{i} + 0.500\omega_{CD}\mathbf{j}.$$

Now we apply Eq. (6.6) to points B and C (Fig. b). We denote the unknown angular velocity of bar BC by ω_{BC}. The position vector of C relative to B is $\mathbf{r}_{C/B} = 1.333\mathbf{i} - 0.167\mathbf{j}$ (ft), and the angular velocity vector of bar BC is $\boldsymbol{\omega}_{BC} = \omega_{BC}\mathbf{k}$. Expressing the velocity of C in terms of the velocity of B, we obtain

$$\mathbf{v}_C = \mathbf{v}_B + \boldsymbol{\omega}_{BC} \times \mathbf{r}_{C/B} = \mathbf{v}_B + \begin{vmatrix} \mathbf{i} & \mathbf{j} & \mathbf{k} \\ 0 & 0 & \omega_{BC} \\ 1.333 & -0.167 & 0 \end{vmatrix}$$

$$= \mathbf{v}_B + 0.167\omega_{BC}\mathbf{i} + 1.333\omega_{BC}\mathbf{j}.$$

Substituting our expressions for $\mathbf{v}_B$ and $\mathbf{v}_C$ into this equation, we obtain

$$0.833\omega_{CD}\mathbf{i} + 0.500\omega_{CD}\mathbf{j} = 10\mathbf{i} - 5\mathbf{j} + 0.167\omega_{BC}\mathbf{i} + 1.333\omega_{BC}\mathbf{j}.$$

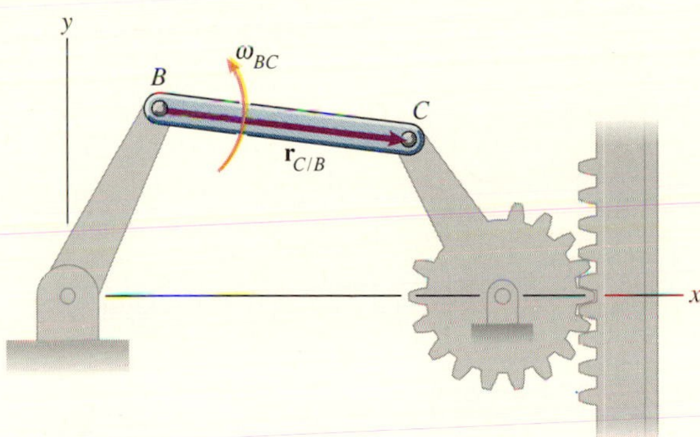

(b) Expressing the velocity of point C in terms of the velocity of point B.

Equating the **i** and **j** components yields two equations in terms of ω_{BC} and ω_{CD}:

$$0.833\omega_{CD} = 10 + 0.167\omega_{BC},$$

$$0.500\omega_{CD} = -5 + 1.333\omega_{BC}.$$

Solving them, we obtain $\omega_{BC} = 8.92$ rad/s and $\omega_{CD} = 13.78$ rad/s.

The vertical velocity of the rack is equal to the velocity of the gear where it contacts the rack:

$$v_R = (0.5 \text{ ft})\omega_{CD} = (0.5)(13.78) = 6.89 \text{ ft/s}.$$

Problems

6.14 The turbine rotates relative to the coordinate system at 30 rad/s about a fixed axis coincident with the x axis. What is its angular velocity vector?

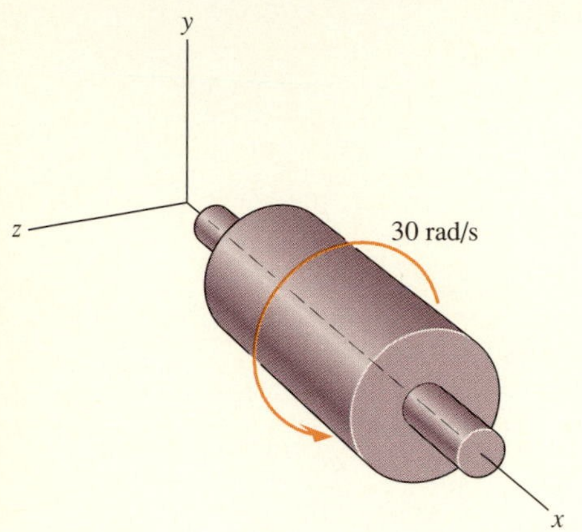

P6.14

6.15 The rectangular plate swings in the x-y plane from arms of equal length. What is the angular velocity vector of (a) the rectangular plate; (b) the bar AB?

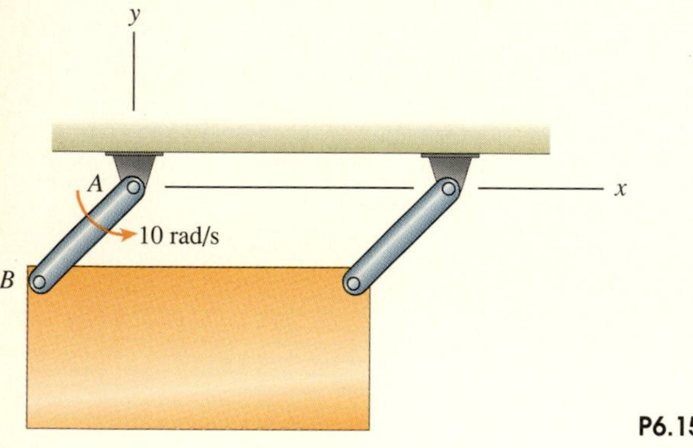

P6.15

6.16 What are the angular velocity vectors of each bar of the linkage?

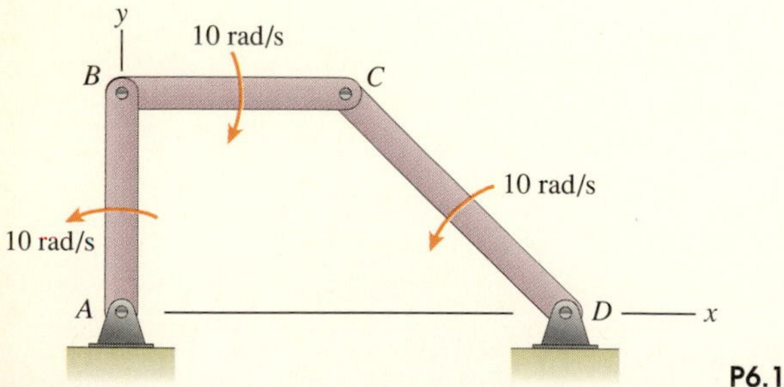

P6.16

6.17 If you model the earth as a rigid body, what is the magnitude of its angular velocity vector $\boldsymbol{\omega}_E$? Does $\boldsymbol{\omega}_E$ point north or south?

6.18 The rigid body rotates with counterclockwise angular velocity ω about a fixed axis through B that is coincident with the z axis. Determine the x and y components of the velocity of A relative to B by representing it as shown in Fig. 6.11(b).

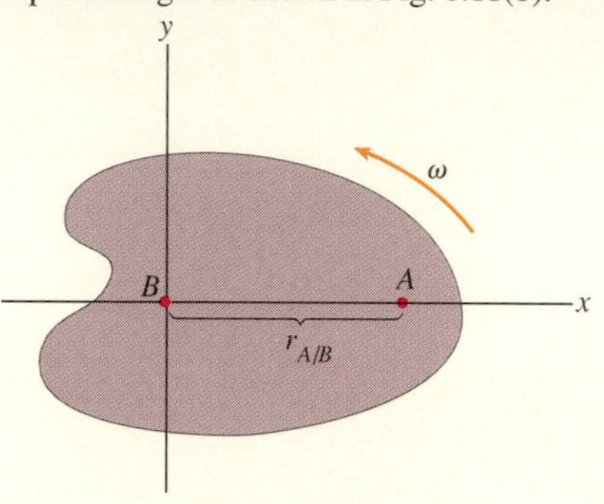

P6.18

6.19 Consider the rotating rigid body in Problem 6.18.
(a) What is its angular velocity vector?
(b) Use Eq. (6.5) to determine the velocity of A relative to B.

6.20 The bar is rotating with a counterclockwise angular velocity of 20 rad/s about the fixed point O. Determine the x and y components of the velocity of A relative to B by representing it as shown in Fig. 6.11(b).

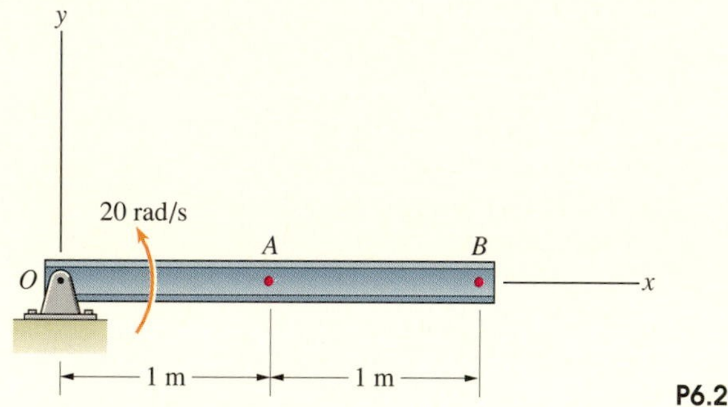

P6.20

6.21 Consider the bar in Problem 6.20.
(a) Use Eq. (6.5) to determine the velocity of A relative to B.
(b) Use the result of (a) and Eq. (6.6) to determine the velocity of A relative to the fixed coordinate system.

6.22 Determine the x and y components of the velocity of point A.

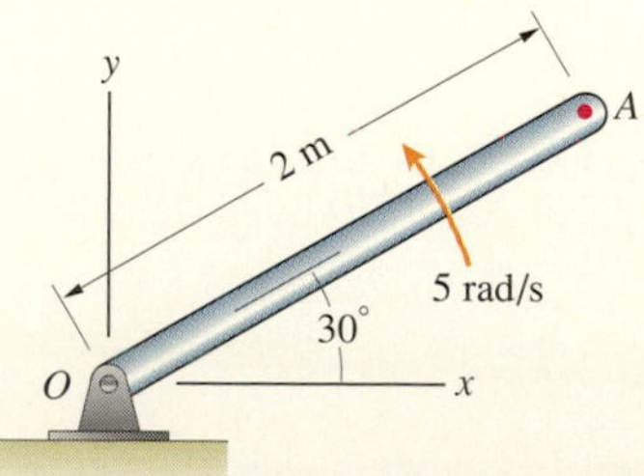

P6.22

6.23 If the angular velocity of the bar in Problem 6.22 is constant, what are the x and y components of the velocity of point A 0.1 s after the instant shown?

6.24 The disk is rotating about the z axis at 50 rad/s in the clockwise direction. Determine the x and y components of the velocities of points A, B, and C.

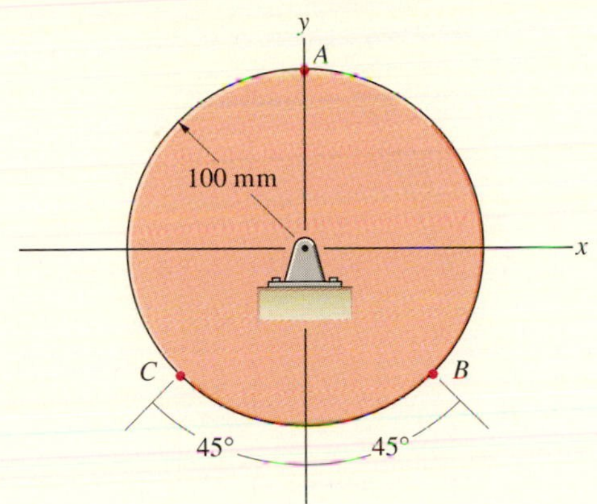

P6.24

6.25 If the angular velocity of the disk in Problem 6.24 is constant, what are the x and y components of the velocity of A 0.02 s after the instant shown?

6.26 The car is moving to the right at 100 km/hr, and its tires are 600 mm in diameter.
(a) What is the angular velocity of its tires?
(b) Which point on the tire shown has the largest velocity relative to a reference frame fixed with respect to the road, and what is the magnitude of the velocity?

P6.26

6.27 The disk rolls on the plane surface. Point A is moving to the right at 6 ft/s.
(a) What is the angular velocity vector of the disk?
(b) Determine the x and y components of the velocities of points B, C, and D.

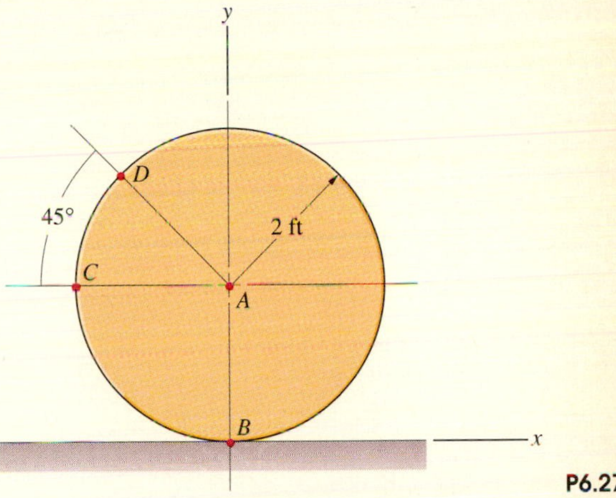

P6.27

6.28 The helicopter is in planar motion in the x-y plane. At the instant shown, the position of its center of mass G is $x = 2$ m, $y = 2.5$ m and its velocity is $\mathbf{v}_G = 12\mathbf{i} + 4\mathbf{j}$ (m/s). The position of point T where the tail rotor is mounted is $x = -3.5$ m, $y = 4.5$ m. The helicopter's angular velocity is 0.2 rad/s clockwise. What is the velocity of point T?

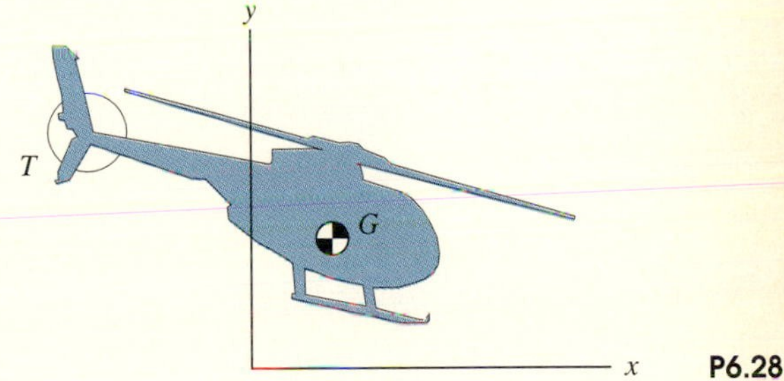

P6.28

6.29 The bar is in two-dimensional motion in the x-y plane. The velocity of point A is $8\mathbf{i}$ (ft/s). The x component of the velocity of point B is 6 ft/s.
(a) What is the angular velocity vector of the bar?
(b) What is the velocity of point B?

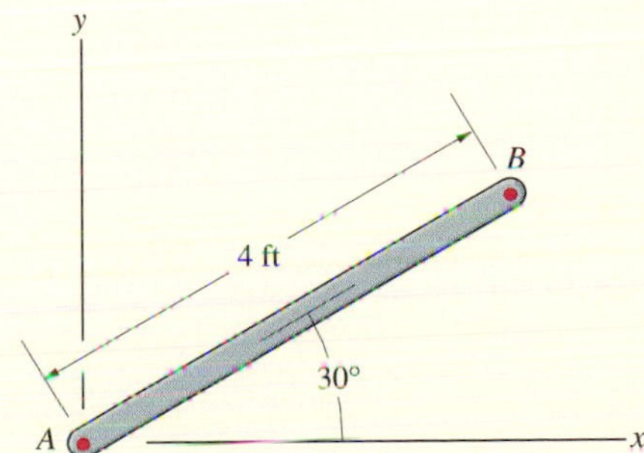

P6.29

6.30 Points A and B of the 1-m bar slide on the plane surfaces. The velocity of point B is $2\mathbf{i}$ (m/s).
(a) What is the angular velocity vector of the bar?
(b) What is the velocity of point A?

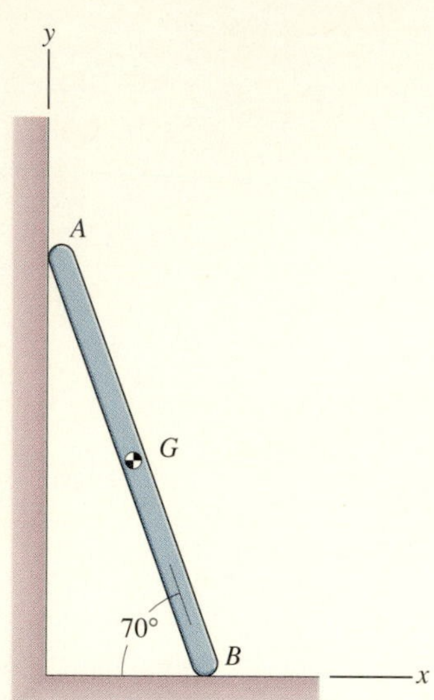

P6.30

6.31 In Problem 6.30, what is the velocity of the midpoint G of the bar?

6.32 If $\theta = 45°$ and the bar OQ is rotating in the counterclockwise direction at 0.2 rad/s, what is the velocity of the sleeve P?

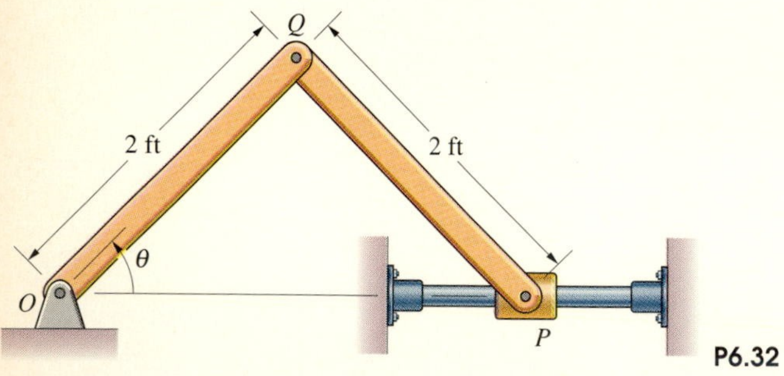

P6.32

6.33 Consider the system shown in Problem 6.32. If $\theta = 40°$ and the sleeve P is moving to the right at 1.0 ft/s, what are the angular velocities of the bars OQ and PQ?

6.34 Bar AB rotates in the counterclockwise direction at 6 rad/s. Determine the angular velocity of bar BCD and the velocity of point D.

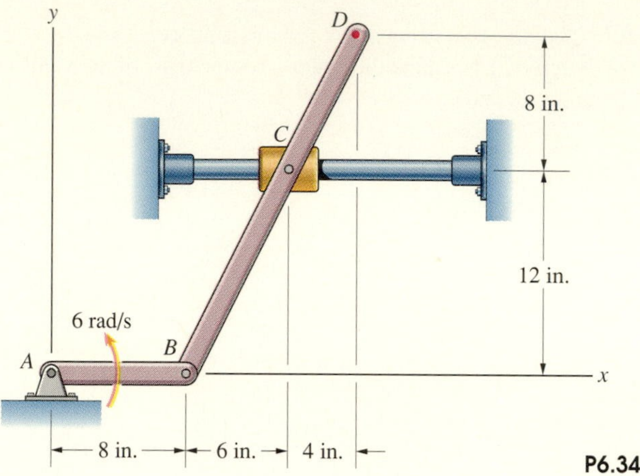

P6.34

6.35 If the crankshaft AB rotates at 6000 rpm in the counterclockwise direction, what is the velocity of the piston at the instant shown?

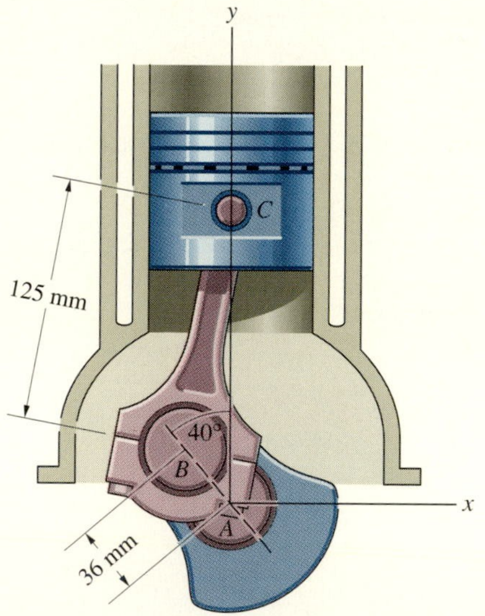

P6.35

6.36 Bar AB rotates at 10 rad/s in the counterclockwise direction. Determine the angular velocity of bar CD.

Strategy: Since you know the angular velocity of the bar AB, you can determine the velocity of B. Then apply Eq. (6.6) to points B and C to obtain an equation for $\mathbf{v}_C$ in terms of the angular velocity of bar BC, and apply it to points C and D to obtain an equation for $\mathbf{v}_C$ in terms of the angular velocity of bar CD. By equating the two expressions, you will obtain a vector equation in two unknowns: the angular velocities of bars BC and CD.

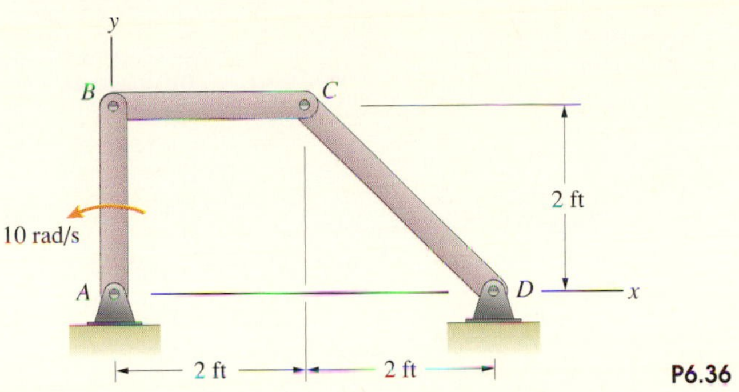

P6.36

6.37 Bar AB rotates at 12 rad/s in the clockwise direction. Determine the angular velocities of bars BC and CD.

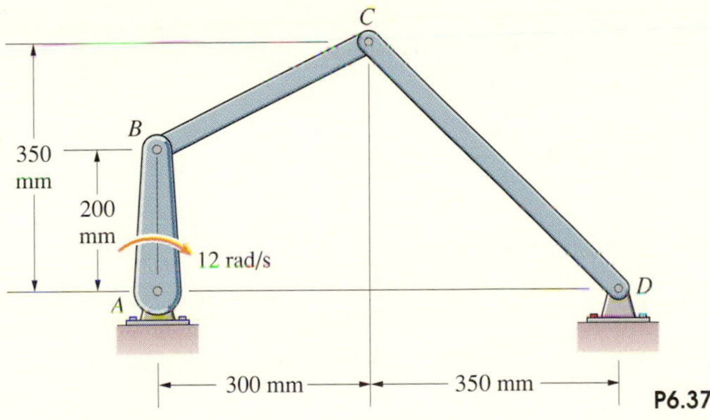

P6.37

6.38 Bar CD rotates at 2 rad/s in the clockwise direction. Determine the angular velocities of bars AB and BC.

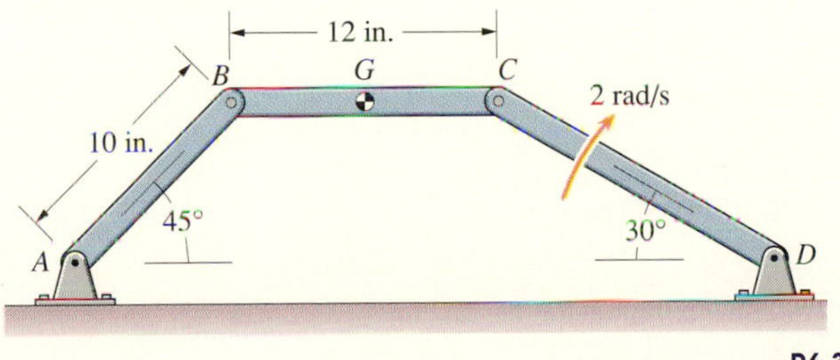

P6.38

6.39 In Problem 6.38, what is the magnitude of the velocity of the midpoint G of bar BC?

6.40 Bar AB rotates at 10 rad/s in the counterclockwise direction. Determine the velocity of point E.

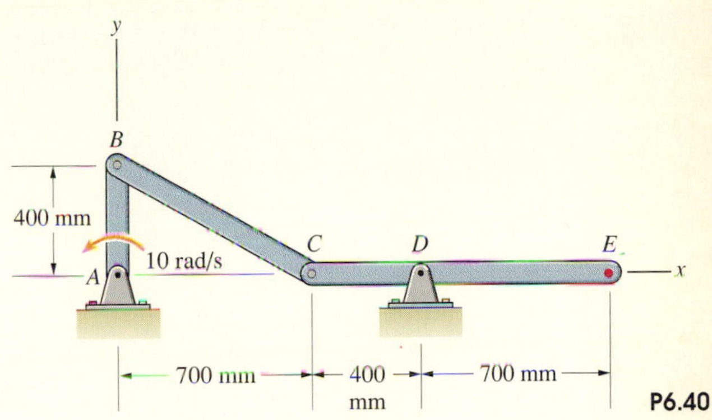

P6.40

6.41 Bar AB rotates at 4 rad/s in the counterclockwise direction. Determine the velocity of point C.

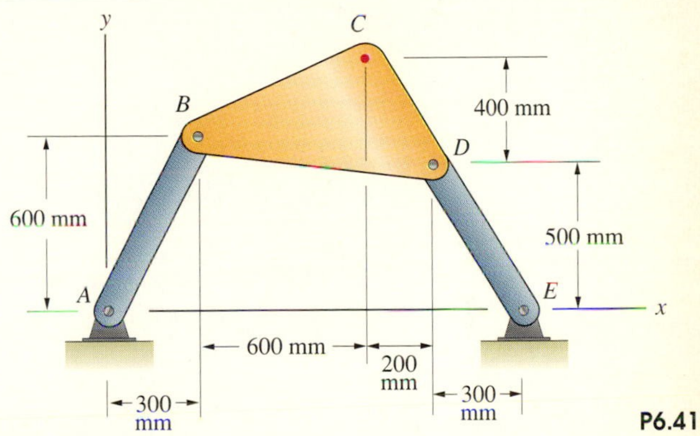

P6.41

6.42 In the system shown in Problem 6.41, if the magnitude of the velocity of point C is $|\mathbf{v}_C| = 2$ m/s, what are the magnitudes of the angular velocities of bars AB and DE?

6.43 The horizontal member ADE supporting the scoop is stationary. If the link BD is rotating in the clockwise direction at 1 rad/s, what is the angular velocity of the scoop?

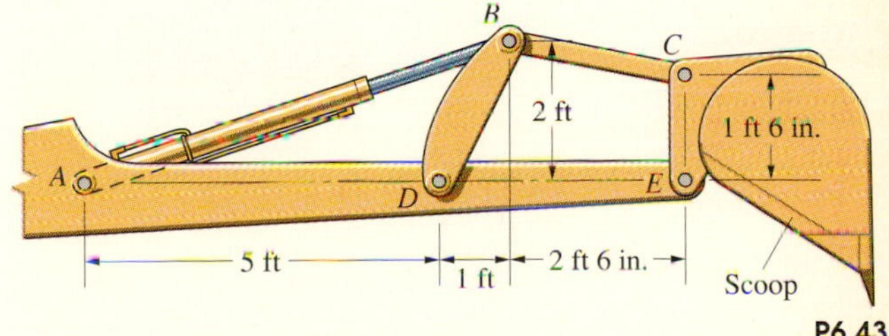

P6.43

6.44 The diameter of the disk is 1 m, and the length of bar AB is 1 m. The disk is rolling, and point B slides on the plane surface. Determine the angular velocity of bar AB and the velocity of point B.

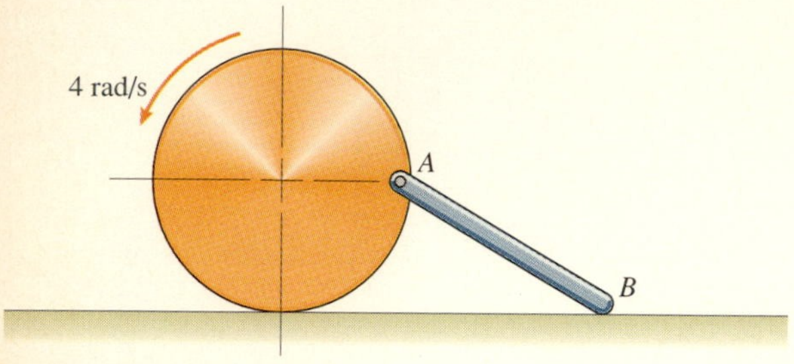

P6.44

6.45 A motor rotates the circular disk mounted at A, moving the saw back and forth. (The saw is supported by a horizontal slot so that point C moves horizontally.) The radius AB is 4 in., and the link BC is 14 in. long. In the position shown, $\theta = 45°$ and the link BC is horizontal. If the angular velocity of the disk is one revolution per second counterclockwise, what is the velocity of the saw?

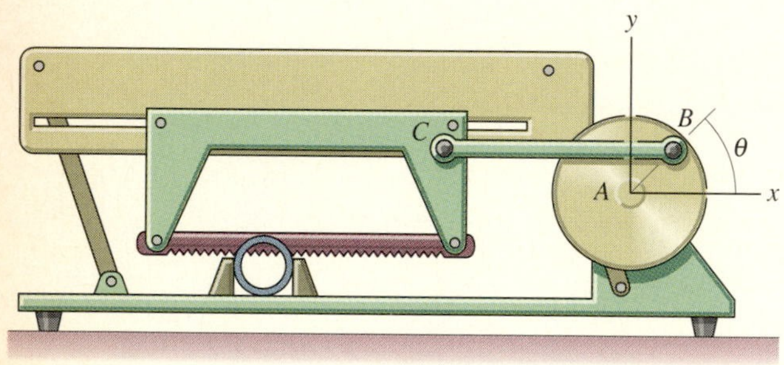

P6.45

6.46 In Problem 6.45, if the angular velocity of the disk is one revolution per second counterclockwise and $\theta = 270°$, what is the velocity of the saw?

6.47 The disks roll on the plane surface. The angular velocity of the left disk is 2 rad/s in the clockwise direction. What is the angular velocity of the right disk?

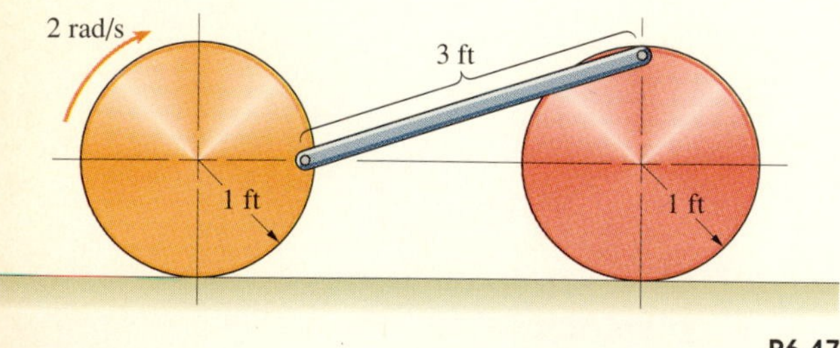

P6.47

6.48 The disk rolls on the curved surface. The bar rotates at 10 rad/s in the counterclockwise direction. Determine the velocity of point A.

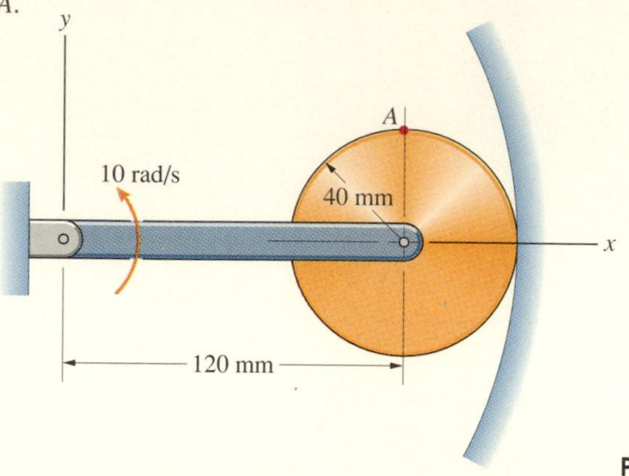

P6.48

6.49 If $\omega_{AB} = 2$ rad/s and $\omega_{BC} = 4$ rad/s, what is the velocity of point C, where the excavator's bucket is attached?

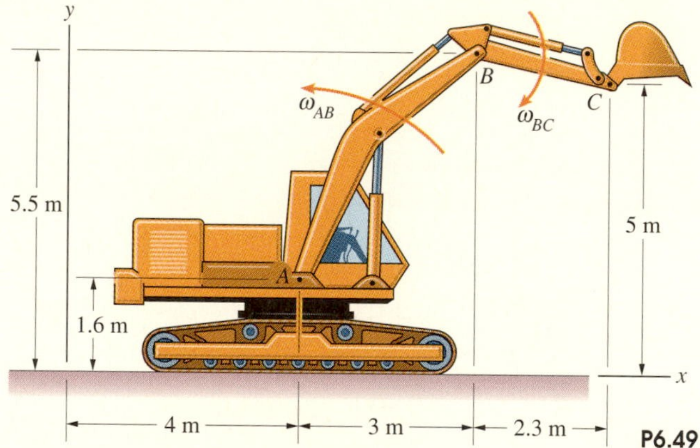

P6.49

6.50 In Problem 6.49, if $\omega_{AB} = 2$ rad/s, what clockwise angular velocity ω_{BC} will cause the vertical component of the velocity of point C to be zero? What is the resulting velocity of point C?

6.51 The motorcycle's rear wheel is rolling on the ground (the velocity of its point of contact with the ground is zero) at 500 rpm, the wheel's radius is 280 mm, and the body of the motorcycle is rotating in the clockwise direction at 6 rad/s. Determine the velocity of the center of mass G.

P6.51

6.52 An athlete exercises his arm by raising the mass m. The shoulder joint A is stationary. The distance AB is 300 mm, and the distance BC is 400 mm. At the instant shown, $\omega_{AB} = 1$ rad/s and $\omega_{BC} = 2$ rad/s. How fast is the mass m rising?

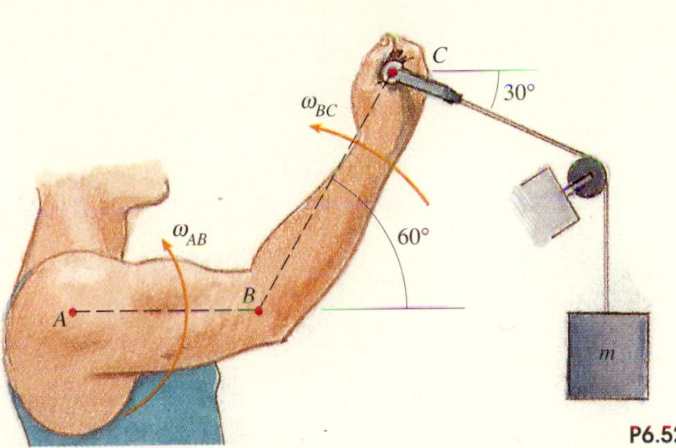

P6.52

6.53 In Problem 6.52, suppose that the distance AB is 12 in., the distance BC is 16 in., $\omega_{AB} = 0.6$ rad/s, and the mass m is rising at 24 in./s. What is the angular velocity ω_{BC}?

6.54 Points B and C are in the x-y plane. The angular velocity vectors of the arms AB and BC are $\boldsymbol{\omega}_{AB} = -0.2\mathbf{k}$ (rad/s) and $\boldsymbol{\omega}_{BC} = 0.4\mathbf{k}$ (rad/s). Determine the velocity of point C.

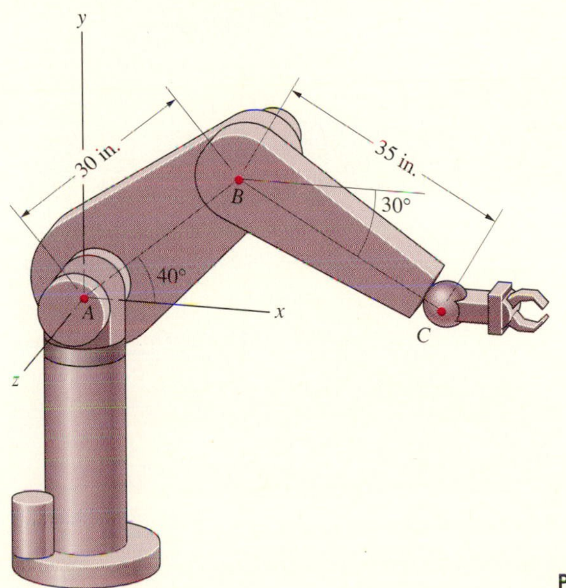

P6.54

6.55 In Problem 6.54, if the velocity of point C is $\mathbf{v}_C = 10\mathbf{j}$ (in./s), what are the angular velocity vectors of the arms AB and BC?

6.56 The link AB of the robot's arm is rotating at 2 rad/s in the counterclockwise direction, the link BC is rotating at 3 rad/s in the clockwise direction, and the link CD is rotating at 4 rad/s in the counterclockwise direction. What is the velocity of point D?

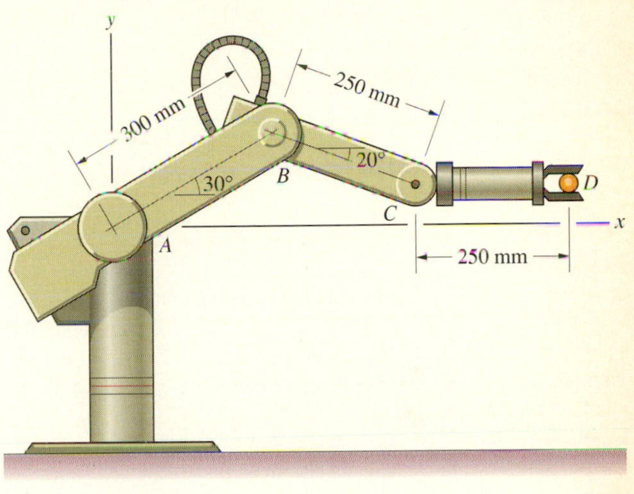

P6.56

6.57 Consider the robot shown in Problem 6.56. Link AB is rotating at 2 rad/s in the counterclockwise direction and link BC is rotating at 3 rad/s in the clockwise direction. If you want the velocity of point D to be parallel to the x axis, what is the necessary angular velocity of link CD? What is the resulting velocity of point D?

6.58 Determine the velocity v_W and the angular velocity of the small pulley.

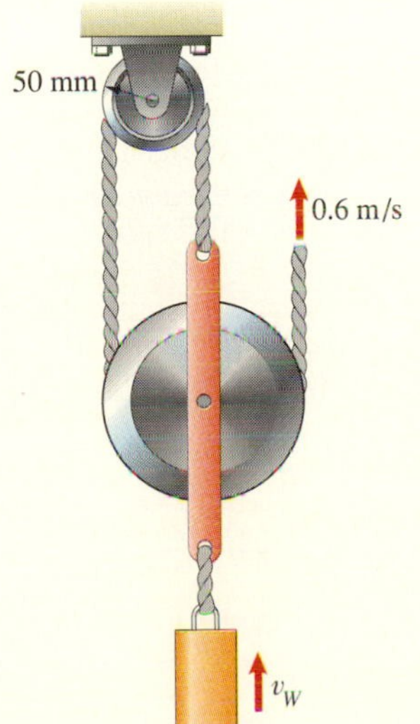

P6.58

6.59 Determine the velocity of the block and the angular velocity of the small pulley.

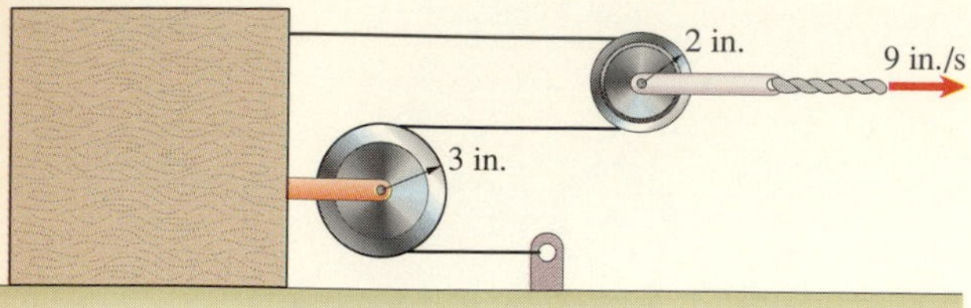

P6.59

6.60 The device shown is used in the semiconductor industry to polish silicon wafers. The wafers are placed on the faces of the carriers. The outer and inner rings are then rotated, causing the wafers to move and rotate against an abrasive surface. If the outer ring rotates in the clockwise direction at 7 rpm and the inner ring rotates in the counterclockwise direction at 12 rpm, what is the angular velocity of the carriers?

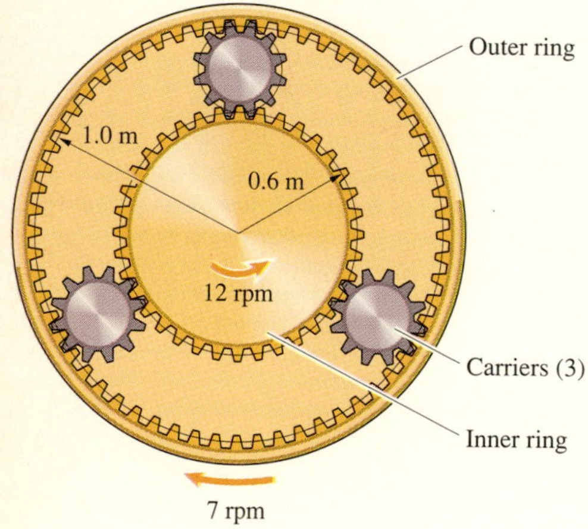

P6.60

6.61 In Problem 6.60, suppose that the outer ring rotates in the clockwise direction at 5 rpm and you want the centerpoints of the carriers to remain stationary during the polishing process. What is the necessary angular velocity of the inner ring?

6.62 The ring gear is fixed and the hub and planet gears are bonded together. The connecting rod rotates in the counterclockwise direction at 60 rpm. Determine the angular velocity of the sun gear and the magnitude of the velocity of point *A*.

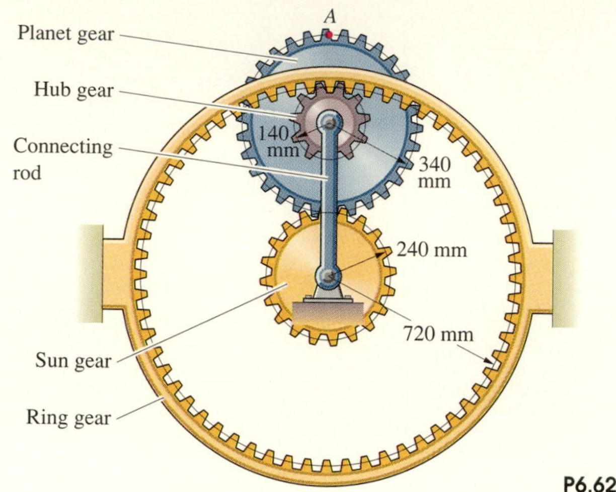

P6.62

6.63 The large gear is fixed. Bar *AB* has a counterclockwise angular velocity of 2 rad/s. What are the angular velocities of bars *CD* and *DE*?

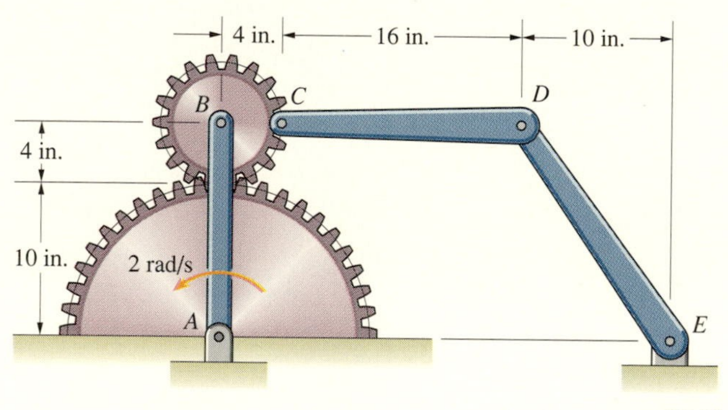

P6.63

Instantaneous Centers

By an **instantaneous center**, we simply mean a point of a rigid body whose velocity is zero at a given instant. "Instantaneous" means it may have zero velocity *only* at the instant under consideration, although we also refer to a fixed point, such as a point of a fixed axis about which a rigid body rotates, as an instantaneous center.

When we know the location of an instantaneous center of a rigid body in two-dimensional motion and we know its angular velocity, the velocities of other points are easy to determine. For example, suppose that point C in Fig. 6.19(a) is the instantaneous center of a rigid body in plane motion with angular velocity ω. Relative to C, a point A moves in a circular path. The velocity of A relative to C is tangent to the circular path and equal to the product of the distance from C to A and the angular velocity. But since C is stationary at this instant, the velocity of A relative to C is the velocity of A. At this instant, every point of the rigid body rotates about C (Fig. 6.19b).

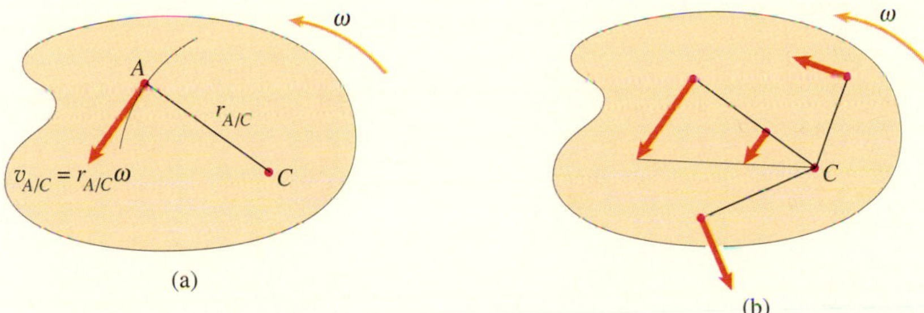

Fig. 6.19
(a) An instantaneous center C and a different point A.
(b) Every point is rotating about the instantaneous center.

You can often locate the instantaneous center of a rigid body in planar motion in a simple way. Suppose that you know the directions of the motions of two points A and B (Fig. 6.20a). If you draw lines through A and B perpendicular to their directions of motion, the point C where the lines intersect is the instantaneous center. To show that this is true, let us express the velocity of C in terms of the velocity of A (Fig. 6.20b):

$$\mathbf{v}_C = \mathbf{v}_A + \boldsymbol{\omega} \times \mathbf{r}_{C/A}.$$

Since the vector $\boldsymbol{\omega} \times \mathbf{r}_{C/A}$ is perpendicular to $\mathbf{r}_{C/A}$, this equation states that the direction of motion of C is parallel to the direction of motion of A. We can also express the velocity of C in terms of the velocity of B:

$$\mathbf{v}_C = \mathbf{v}_B + \boldsymbol{\omega} \times \mathbf{r}_{C/B}.$$

The vector $\boldsymbol{\omega} \times \mathbf{r}_{C/B}$ is perpendicular to $\mathbf{r}_{C/B}$, so this equation states that the direction of motion of C is parallel to the direction of motion of B. But C cannot be moving parallel to A and parallel to B, so these equations are contradictory unless $\mathbf{v}_C = 0$.

The instantaneous center may not be a point of the rigid body (Fig. 6.21a). This simply means that at this instant, the rigid body is rotating about an external point. It's helpful to imagine extending the rigid body so that it includes the instantaneous center (Fig. 6.21b). The velocity of point C of the extended body would be zero at this instant.

Notice in Fig. 6.21(a) that if you change the directions of motion of A and B so that the lines perpendicular to their directions of motion become parallel, C

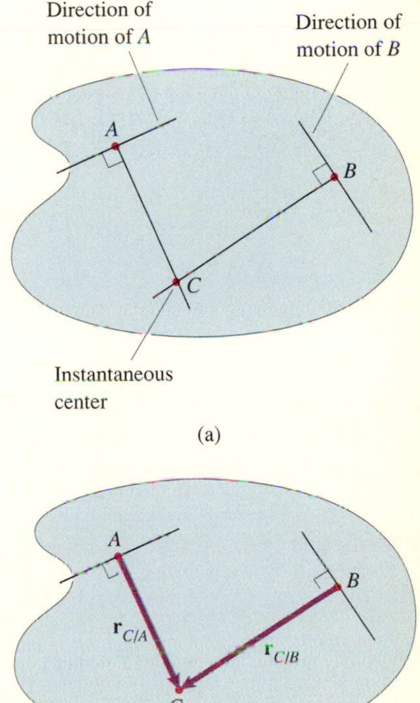

Fig. 6.20
(a) Locating the instantaneous center in planar motion.
(b) Proving that $\mathbf{v}_C = \mathbf{0}$.

moves to infinity. In that case, the rigid body is in translation; its angular velocity is zero.

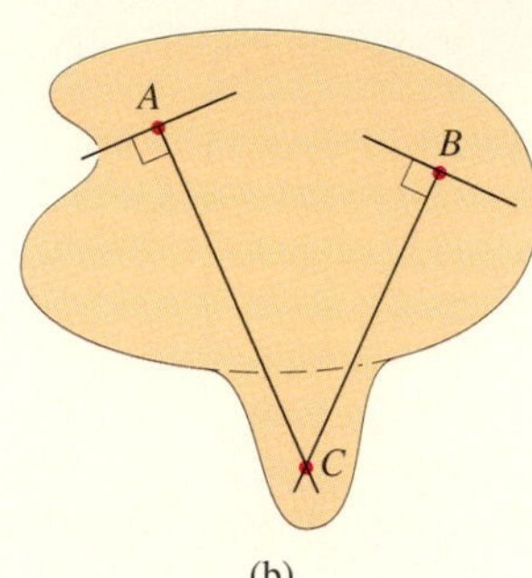

Fig. 6.21
(a) An instantaneous center external to the rigid body.
(b) A hypothetical extended body. Point C would be stationary.

Returning once again to our example of a disk of radius R rolling with angular velocity ω (Fig. 6.22a), the point C in contact with the floor is stationary at that instant—it is the instantaneous center of the disk. Therefore the velocity of any other point is perpendicular to the line from C to the point and its magnitude equals the product of ω and the distance from C to the point. In terms of the coordinate system shown in Fig. 6.22(b), the velocity of point A is

$$\begin{aligned}\mathbf{v}_A &= -\sqrt{2}R\omega\cos 45^\circ\mathbf{i} + \sqrt{2}R\omega\sin 45^\circ\mathbf{j}\\ &= -R\omega\,\mathbf{i} + R\omega\,\mathbf{j}.\end{aligned}$$

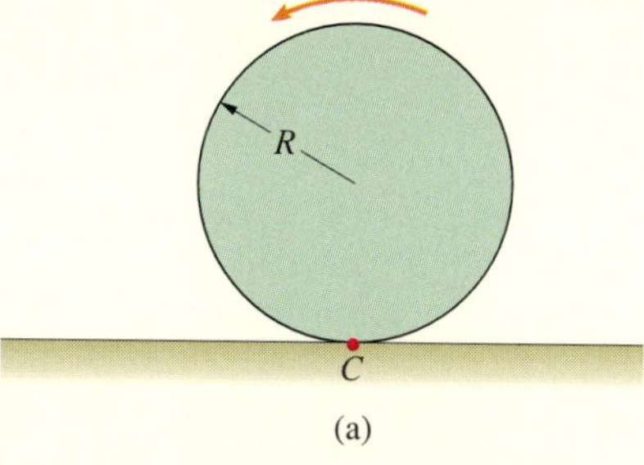

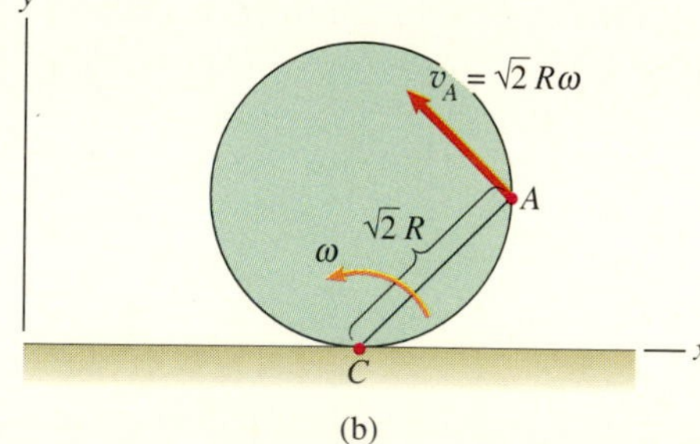

Fig. 6.22
(a) Point C is the instantaneous center of the rolling disk.
(b) Determining the velocity of point A.

In the following example we use instantaneous centers to analyze the motion of a linkage. By identifying the instantaneous center of a rigid body in planar motion, you can express the velocities of its points as products of their distances from the instantaneous center and the angular velocity of the rigid body.

Example 6.4

Bar AB in Fig. 6.23 rotates with a counterclockwise angular velocity of 10 rad/s. What are the angular velocities of bars BC and CD?

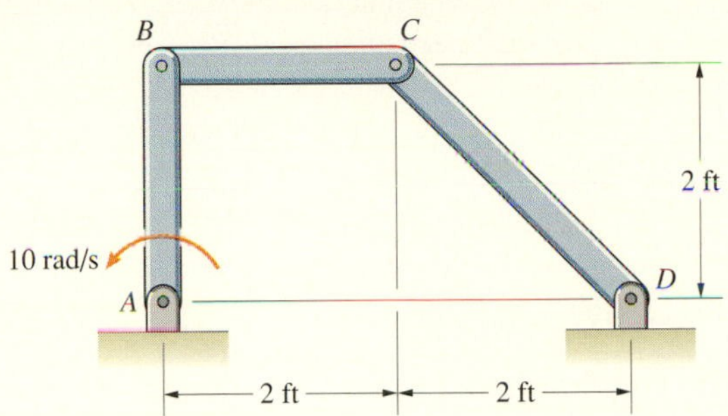

Fig. 6.23

STRATEGY

Because bars AB and CD rotate about fixed axes, we know the directions of motion of points B and C and so can locate the instantaneous center of bar BC. Beginning with bar AB (because we know its angular velocity), we can use the instantaneous centers of the bars to determine both the velocities of the points where they are connected and their angular velocities.

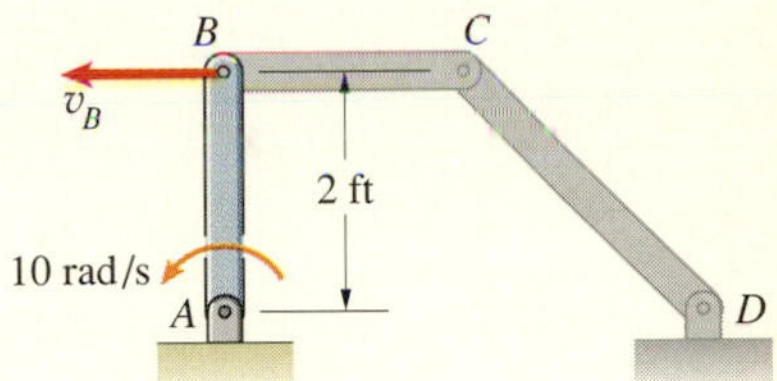

(a) Determining v_B.

SOLUTION

The velocity of B due to the rotation of bar AB about A (Fig. a) is

$$v_B = (2 \text{ ft})(10 \text{ rad/s}) = 20 \text{ ft/s}.$$

Drawing lines perpendicular to the directions of motion of B and C, we locate the instantaneous center of bar BC (Fig. b). The velocity of B is equal to the product of its distance from the instantaneous center of bar BC and the angular velocity ω_{BC},

$$v_B = 20 \text{ ft/s} = (2 \text{ ft})\omega_{BC},$$

so $\omega_{BC} = 10$ rad/s. (Notice that bar BC rotates in the clockwise direction.) Using the instantaneous center of bar BC and its angular velocity ω_{BC}, we can determine the velocity of point C:

$$v_C = (\sqrt{8} \text{ ft})\omega_{BC} = 10\sqrt{8} \text{ ft/s}.$$

Instantaneous center of bar BC

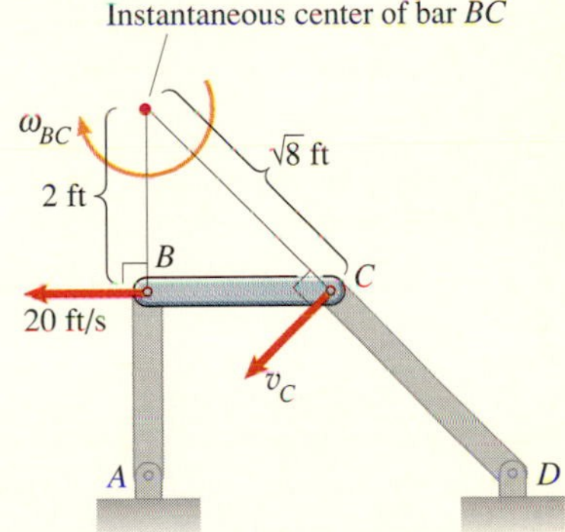

(b) Determining ω_{BC} and v_C.

Our last step is to use the velocity of point C to determine the angular velocity of bar CD about point D (Fig. c),

$$v_C = 10\sqrt{8} \text{ ft/s} = (\sqrt{8} \text{ ft})\omega_{CD},$$

obtaining $\omega_{CD} = 10$ rad/s counterclockwise.

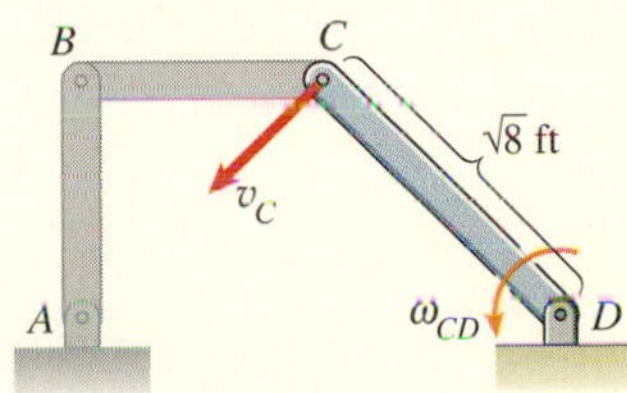

(c) Determining ω_{CD}.

DISCUSSION

In this example, the use of instantaneous centers greatly simplified determining the angular velocities of bars BC and CD in comparison to our previous approach. However, notice that the lengths and positions of the bars made it very easy for us to locate the instantaneous center of bar BC. If the geometry is too complicated, the use of instantaneous centers can be impractical.

Problems

6.64 If the bar has a clockwise angular velocity of 10 rad/s and $v_A = 20$ m/s, what are the coordinates of its instantaneous center and the value of v_B?

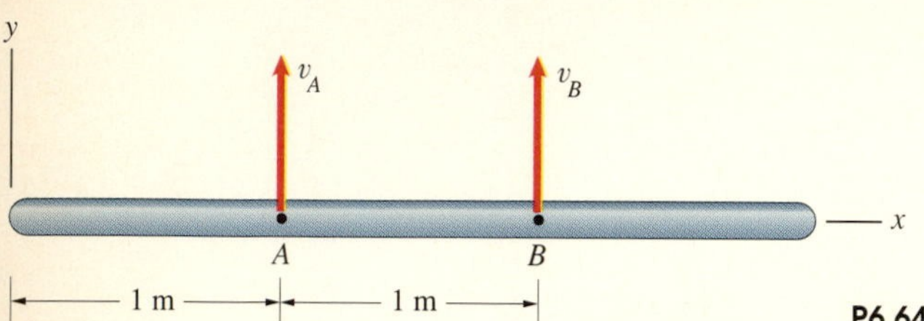

P6.64

6.65 In Problem 6.64, if $v_A = 24$ m/s and $v_B = 36$ m/s, what are the coordinates of the instantaneous center of the bar and its angular velocity?

6.66 The velocity of point O of the bat is $\mathbf{v}_O = -6\mathbf{i} - 1.4\mathbf{j}$ (ft/s), and the bat rotates about the z axis with a counterclockwise angular velocity of 4 rad/s. What are the x and y coordinates of its instantaneous center?

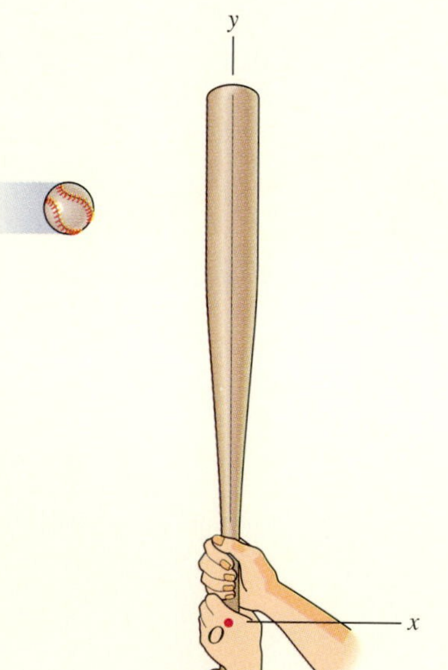

P6.66

6.67 Points A and B of the 1-m bar slide on the plane surfaces. The velocity of B is $\mathbf{v}_B = 2\mathbf{i}$ (m/s).

(a) What are the coordinates of the instantaneous center?

(b) Use the instantaneous center to determine the velocity of A.

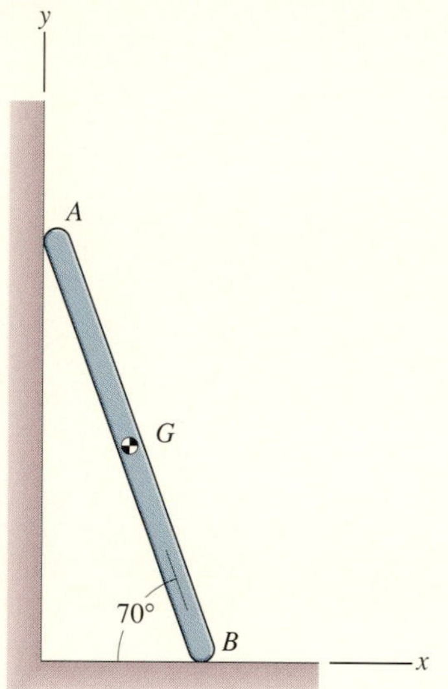

P6.67

6.68 In Problem 6.67, use the instantaneous center to determine the velocity of the bar's midpoint G.

6.69 The bar is in two-dimensional motion in the x-y plane. The velocity of point A is $\mathbf{v}_A = 8\mathbf{i}$ (ft/s), and B is moving in the direction parallel to the bar. Determine the velocity of B (a) by using Eq. (6.6); (b) by using the instantaneous center.

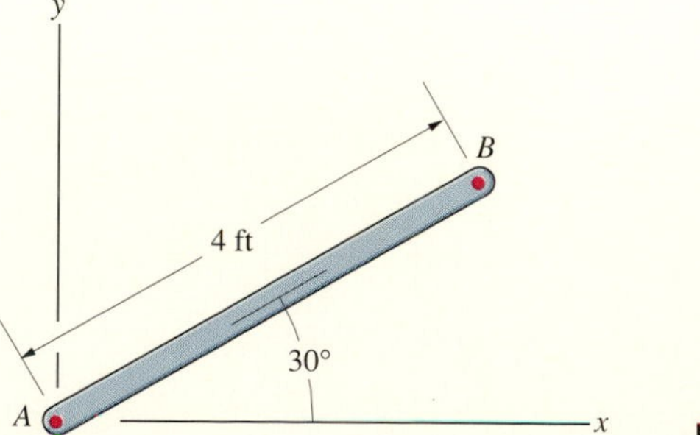

P6.69

6.70 Points A and B of the 4-ft bar slide on the plane surfaces. Point B is sliding down the slanted surface at 2 ft/s.
(a) What are the coordinates of the instantaneous center?
(b) Use the instantaneous center to determine the velocity of A.

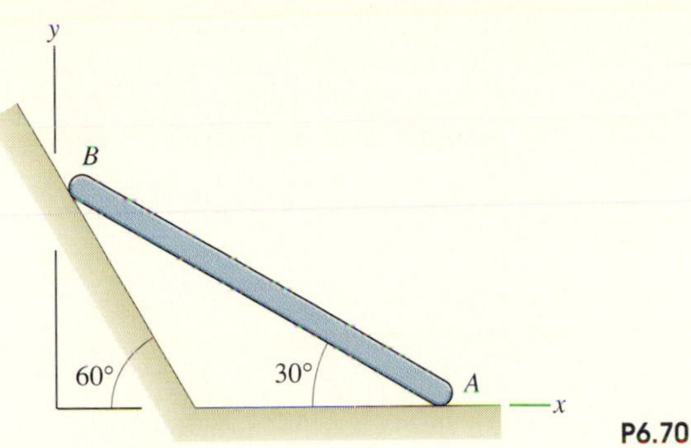

P6.70

6.71 Use instantaneous centers to determine the horizontal velocity of B.

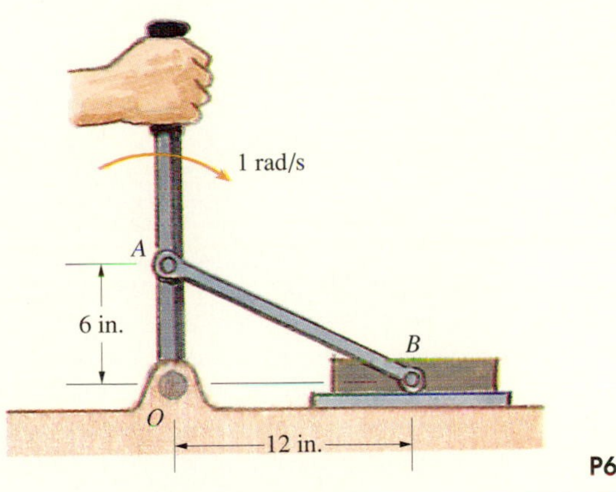

P6.71

6.72 When the mechanism in Problem 6.71 is in this position, use instantaneous centers to determine the horizontal velocity of B.

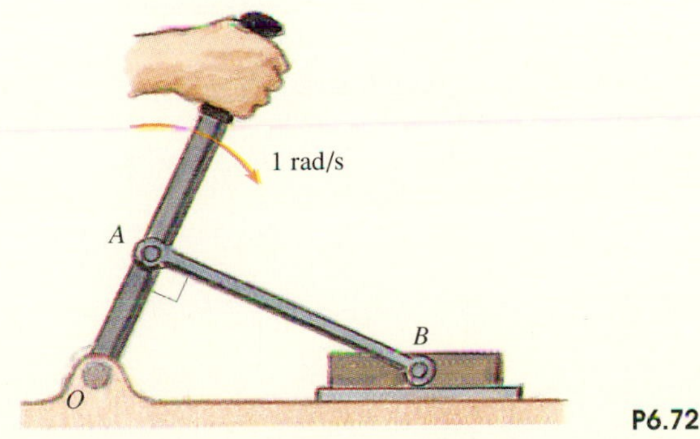

P6.72

6.73 The angle $\theta = 45°$ and bar OQ is rotating in the counterclockwise direction at 0.2 rad/s. Use instantaneous centers to determine the velocity of the sleeve P.

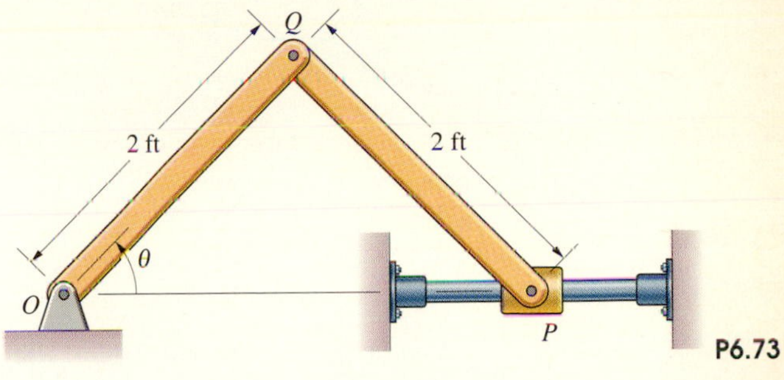

P6.73

6.74 Consider the system shown in Problem 6.73. The angle $\theta = 40°$ and the sleeve P is moving to the right at 1.0 ft/s. Use instantaneous centers to determine the angular velocities of bars OQ and PQ.

6.75 Bar AB rotates at 6 rad/s in the clockwise direction. Use instantaneous centers to determine the angular velocity of bar BC.

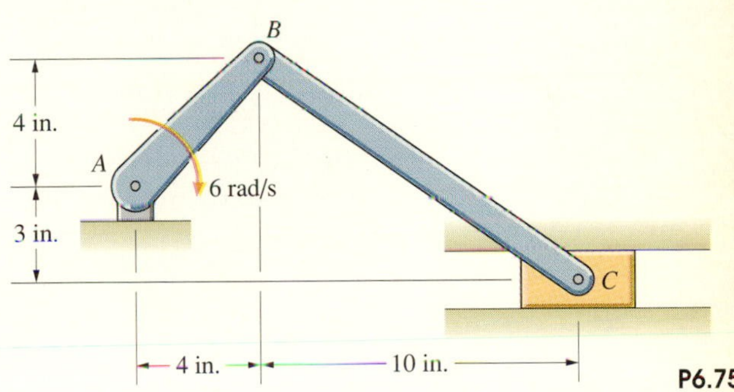

P6.75

6.76 Bar AB rotates at 10 rad/s in the counterclockwise direction. Use instantaneous centers to determine the velocity of point E.

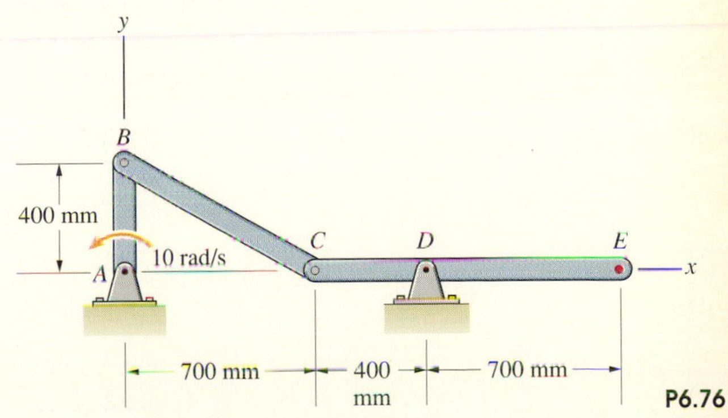

P6.76

6.77 The disks roll on the plane surface. The left disk rotates at 2 rad/s in the clockwise direction. Use instantaneous centers to determine the angular velocities of the bar and the right disk.

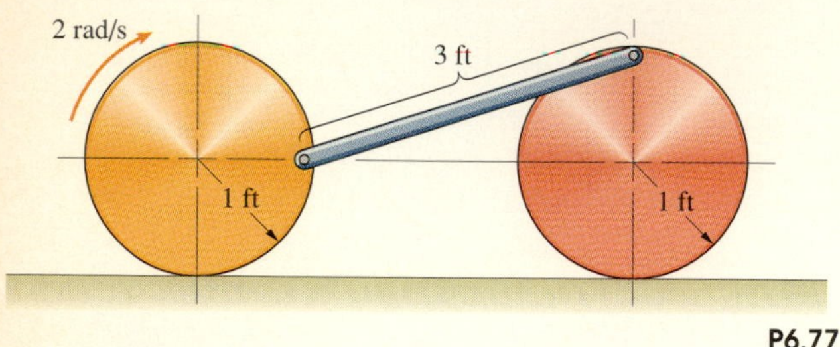

P6.77

6.78 Bar AB rotates at 12 rad/s in the clockwise direction. Use instantaneous centers to determine the angular velocities of bars BC and CD.

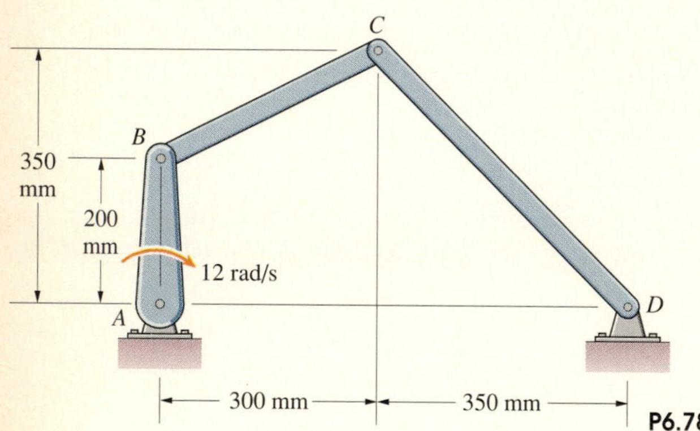

P6.78

6.79 The horizontal member ADE supporting the scoop is stationary. The link BD is rotating in the clockwise direction at 1 rad/s. Use instantaneous centers to determine the angular velocity of the scoop.

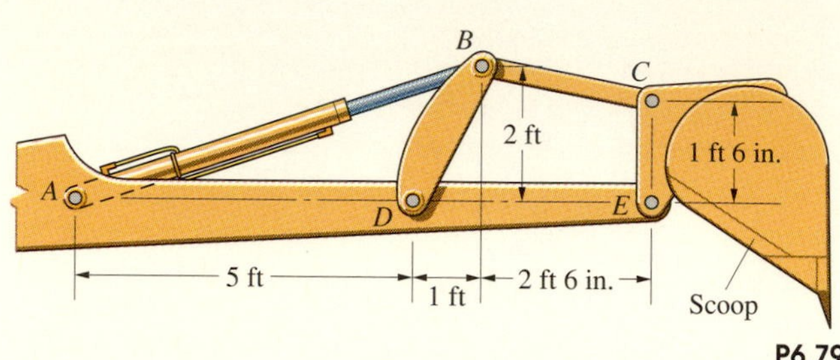

P6.79

6.80 Show that if a rigid body in planar motion has two instantaneous centers, it is stationary at that instant.

6.4 General Motions: Accelerations

In Chapter 7 we will be concerned with determining the motion of a rigid body when we know the external forces and couples acting on it. The governing equations are expressed in terms of the acceleration of the center of mass of the rigid body and its angular acceleration. To solve such problems, you need to understand the relationship between the accelerations of points of a rigid body and its angular acceleration. In this section we extend the methods we have used to analyze velocities of rigid bodies to accelerations.

Consider points A and B of a rigid body in planar motion relative to a given reference frame (Fig. 6.24a). Their velocities are related by

$$\mathbf{v}_A = \mathbf{v}_B + \mathbf{v}_{A/B}.$$

Taking the time derivative of this equation, we obtain

$$\mathbf{a}_A = \mathbf{a}_B + \mathbf{a}_{A/B},$$

where $\mathbf{a}_A$ and $\mathbf{a}_B$ are the accelerations of A and B relative to the reference frame and $\mathbf{a}_{A/B}$ is the acceleration of A relative to B. (*When we simply speak*

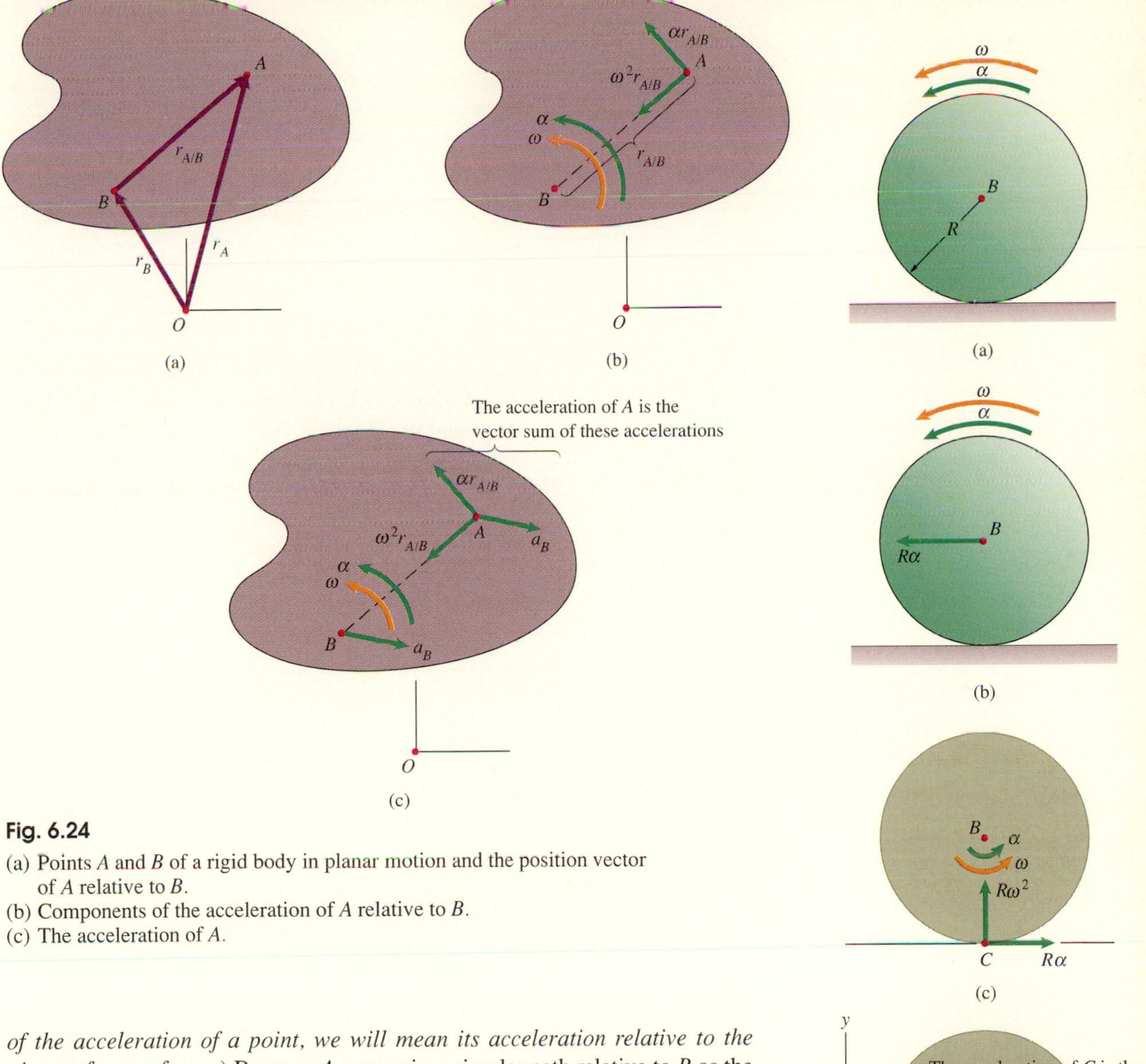

Fig. 6.24

(a) Points A and B of a rigid body in planar motion and the position vector of A relative to B.
(b) Components of the acceleration of A relative to B.
(c) The acceleration of A.

of the acceleration of a point, we will mean its acceleration relative to the given reference frame.) Because A moves in a circular path relative to B as the rigid body rotates, $\mathbf{a}_{A/B}$ has normal and tangential components (Fig. 6.24b). The tangential component equals the product of the distance $r_{A/B} = |\mathbf{r}_{A/B}|$ and the angular acceleration α of the rigid body. The normal component points toward the center of the circular path and its magnitude is $|\mathbf{v}_{A/B}|^2/r_{A/B} = \omega^2 r_{A/B}$. The acceleration of A is the sum of the acceleration of B and the acceleration of A relative to B (Fig. 6.24c).

For example, let us consider a circular disk of radius R rolling on a stationary plane surface with counterclockwise angular velocity ω and counterclockwise angular acceleration α (Fig. 6.25a). The disk's center B is moving in a straight line with velocity $R\omega$, toward the left if ω is positive. Therefore the acceleration of B is $d/dt(R\omega) = R\alpha$, and is toward the left if α is positive (Fig. 6.25b). *The magnitude of the acceleration of the center of a round object rolling on a stationary plane surface is the product of the radius and the angular acceleration.*

Fig. 6.25

(a) A disk rolling with angular velocity ω and angular acceleration α.
(b) Acceleration of the center B.
(c) Components of the acceleration of C relative to B.
(d) The acceleration of C.

Now that we know the acceleration of the disk's center, let us determine the acceleration of the point C in contact with the surface. Relative to B, C moves in a circular path of radius R with angular velocity ω and angular acceleration α. The tangential and normal components of the acceleration of C relative to B are shown in Fig. 6.25(c). The acceleration of C is the sum of the acceleration of B and the acceleration of C relative to B (Fig. 6.25d). In terms of the coordinate system shown,

$$\mathbf{a}_C = \mathbf{a}_B + \mathbf{a}_{C/B} = -R\alpha\,\mathbf{i} + R\alpha\,\mathbf{i} + R\omega^2\mathbf{j}$$
$$= R\omega^2\mathbf{j}.$$

The acceleration of point C parallel to the surface is zero, but it does have an acceleration normal to the surface.

Expressing the acceleration of a point A relative to a point B in terms of A's circular path about B as we have done helps you visualize and understand it. However, just as we did in the case of the relative velocity, we can obtain $\mathbf{a}_{A/B}$ in a form more convenient for applications by using the angular velocity vector $\boldsymbol{\omega}$. The velocity of A relative to B is given in terms of $\boldsymbol{\omega}$ by Eq. (6.5):

$$\mathbf{v}_{A/B} = \boldsymbol{\omega} \times \mathbf{r}_{A/B}.$$

Taking the time derivative of this equation, we obtain

$$\mathbf{a}_{A/B} = \frac{d\boldsymbol{\omega}}{dt} \times \mathbf{r}_{A/B} + \boldsymbol{\omega} \times \mathbf{v}_{A/B}$$

$$= \frac{d\boldsymbol{\omega}}{dt} \times \mathbf{r}_{A/B} + \boldsymbol{\omega} \times (\boldsymbol{\omega} \times \mathbf{r}_{A/B}).$$

Defining the **angular acceleration vector** $\boldsymbol{\alpha}$ to be the rate of change of the angular velocity vector,

$$\boldsymbol{\alpha} = \frac{d\boldsymbol{\omega}}{dt}, \tag{6.7}$$

the acceleration of A relative to B is

$$\mathbf{a}_{A/B} = \boldsymbol{\alpha} \times \mathbf{r}_{A/B} + \boldsymbol{\omega} \times (\boldsymbol{\omega} \times \mathbf{r}_{A/B}).$$

Using this expression, we can write equations relating the velocities and accelerations of two points of a rigid body in terms of its angular velocity and angular acceleration:

$$\mathbf{v}_A = \mathbf{v}_B + \boldsymbol{\omega} \times \mathbf{r}_{A/B}, \tag{6.8}$$

$$\mathbf{a}_A = \mathbf{a}_B + \boldsymbol{\alpha} \times \mathbf{r}_{A/B} + \boldsymbol{\omega} \times (\boldsymbol{\omega} \times \mathbf{r}_{A/B}). \tag{6.9}$$

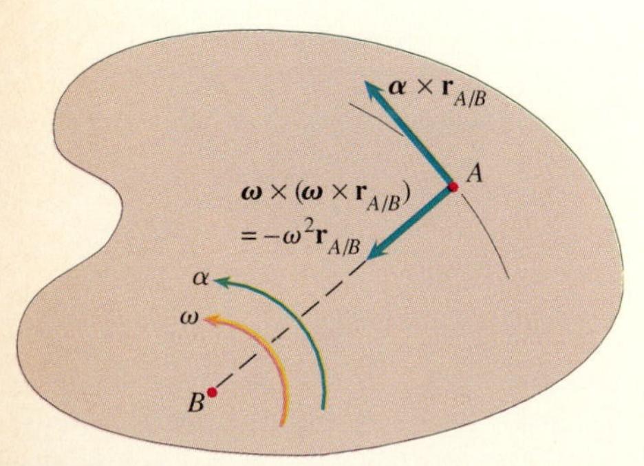

Fig. 6.26
Vector components of the acceleration of A relative to B in planar motion.

In the case of planar motion, the term $\boldsymbol{\alpha} \times \mathbf{r}_{A/B}$ in Eq. (6.9) is the tangential component of the acceleration of A relative to B and $\boldsymbol{\omega} \times (\boldsymbol{\omega} \times \mathbf{r}_{A/B})$ is the normal component (Fig. 6.26). Therefore, for planar motion, we can write Eq. (6.9) in the simpler form

$$\mathbf{a}_A = \mathbf{a}_B + \boldsymbol{\alpha} \times \mathbf{r}_{A/B} - \omega^2 \mathbf{r}_{A/B}. \tag{6.10}$$

In the following examples we use Eqs. (6.8)–(6.10) to analyze motions of rigid bodies. To determine accelerations of points and angular accelerations of rigid bodies, usually you must first determine the velocities of the points and the angular velocities of the rigid bodies, because Eqs. (6.9) and (6.10) contain the angular velocity. When you find a sequence of steps using Eq. (6.8) that determines the velocities and angular velocities, the same sequence of steps using Eq. (6.9) or (6.10) will determine the accelerations and angular accelerations.

Example 6.5

The rolling disk in Fig. 6.27 has counterclockwise angular velocity ω and counterclockwise angular acceleration α. What is the acceleration of point A?

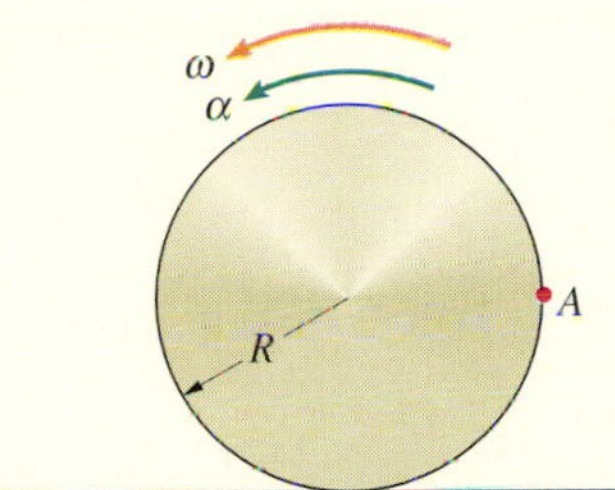

Fig. 6.27

STRATEGY

We know that the magnitude of the acceleration of the center of the disk is the product of the radius and the angular acceleration. Therefore we can express the acceleration of A as the sum of the acceleration of the center and the acceleration of A relative to the center. We will do so first by using vector diagrams as shown in Fig. 6.24(c) and then by using Eq. (6.10).

SOLUTION

First Method In terms of the coordinate system in Fig. (a), the acceleration of the center B is $\mathbf{a}_B = -\alpha R\mathbf{i}$. A's motion in a circular path of radius R relative to B results in the tangential and normal components of relative acceleration shown in Fig. (b):

$$\mathbf{a}_{A/B} = -\omega^2 R\mathbf{i} + \alpha R\mathbf{j}.$$

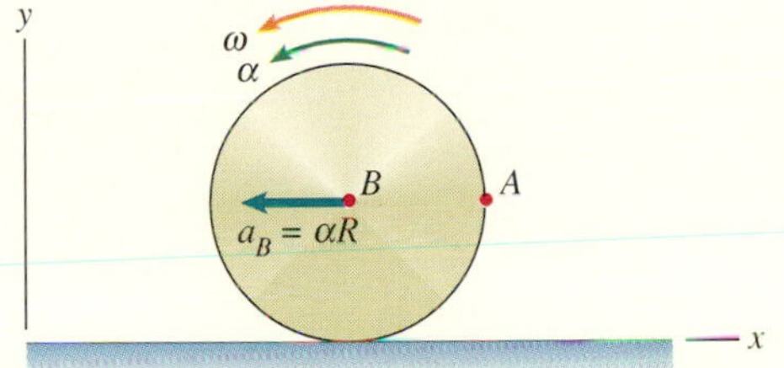

(a) Acceleration of the center of the disk.

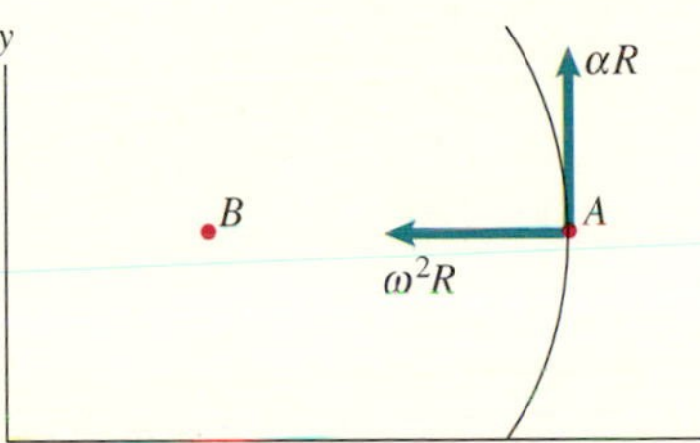

(b) Components of the acceleration of A relative to B.

Therefore the acceleration of A is

$$\mathbf{a}_A = \mathbf{a}_B + \mathbf{a}_{A/B} = -\alpha R\mathbf{i} - \omega^2 R\mathbf{i} + \alpha R\mathbf{j}$$
$$= (-\alpha R - \omega^2 R)\mathbf{i} + \alpha R\mathbf{j}.$$

Second Method The angular acceleration vector of the disk is $\boldsymbol{\alpha} = \alpha\mathbf{k}$, and the position of A relative to B is $\mathbf{r}_{A/B} = R\mathbf{i}$ (Fig. c). From Eq. (6.10), the acceleration of A is

$$\mathbf{a}_A = \mathbf{a}_B + \boldsymbol{\alpha} \times \mathbf{r}_{A/B} - \omega^2\mathbf{r}_{A/B}$$
$$= -\alpha R\mathbf{i} + (\alpha\mathbf{k}) \times (R\mathbf{i}) - \omega^2(R\mathbf{i})$$
$$= (-\alpha R - \omega^2 R)\mathbf{i} + \alpha R\mathbf{j}.$$

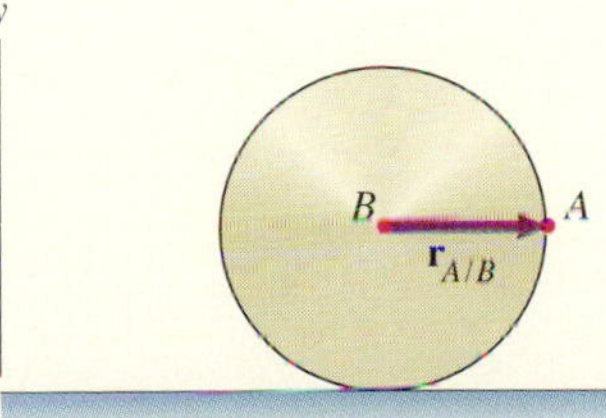

(c) Position of A relative to B.

Example 6.6

Bar AB in Fig. 6.28 has a counterclockwise angular velocity of 10 rad/s and a clockwise angular acceleration of 300 rad/s^2. What are the angular accelerations of bars BC and CD?

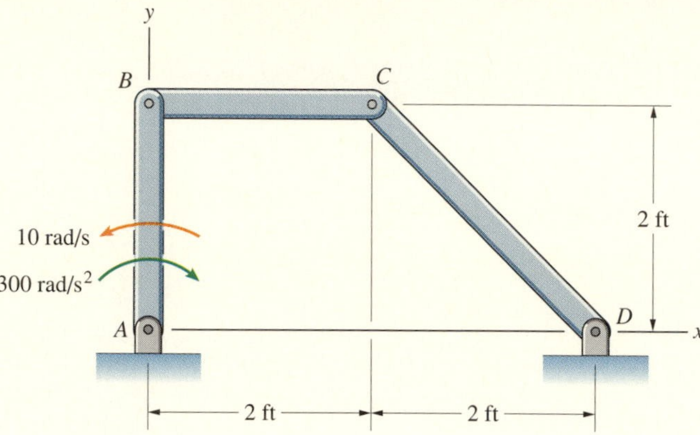

Fig. 6.28

STRATEGY

Since we know the angular velocity of bar AB, we can determine the velocity of point B. Then we can apply Eq. (6.8) to points C and D to obtain an equation for $\mathbf{v}_C$ in terms of the angular velocity of bar CD. We can also apply Eq. (6.8) to points B and C to obtain an equation for $\mathbf{v}_C$ in terms of the angular velocity of bar BC. By equating the two expressions for $\mathbf{v}_C$, we will obtain a vector equation in two unknowns: the angular velocities of bars BC and CD. Then by following the same sequence of steps using Eq. (6.10), we can obtain the angular accelerations of bars BC and CD.

SOLUTION

The velocity of B is (Fig. a)

$$\begin{aligned}\mathbf{v}_B &= \mathbf{v}_A + \boldsymbol{\omega}_{AB} \times \mathbf{r}_{B/A}\\ &= \mathbf{0} + (10\mathbf{k}) \times (2\,\mathbf{j})\\ &= -20\mathbf{i}\ \text{(ft/s)}.\end{aligned}$$

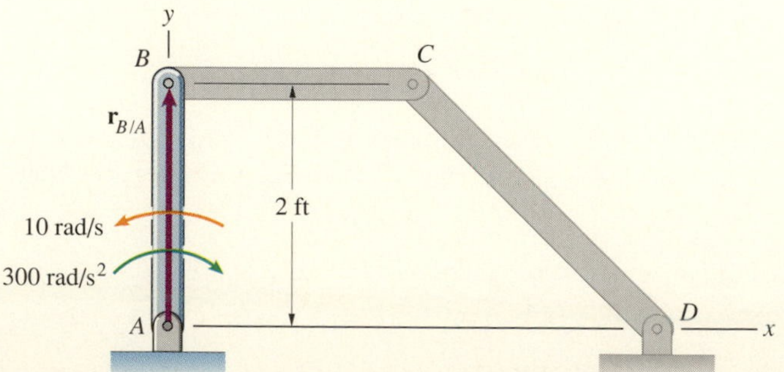

(a) Determining the motion of B.

Let ω_{CD} be the unknown angular velocity of bar CD (Fig. b). The velocity of C in terms of the velocity of D is

$$\mathbf{v}_C = \mathbf{v}_D + \boldsymbol{\omega}_{CD} \times \mathbf{r}_{C/D}$$

$$= \mathbf{0} + \begin{vmatrix} \mathbf{i} & \mathbf{j} & \mathbf{k} \\ 0 & 0 & \omega_{CD} \\ -2 & 2 & 0 \end{vmatrix}$$

$$= -2\omega_{CD}\mathbf{i} - 2\omega_{CD}\mathbf{j}.$$

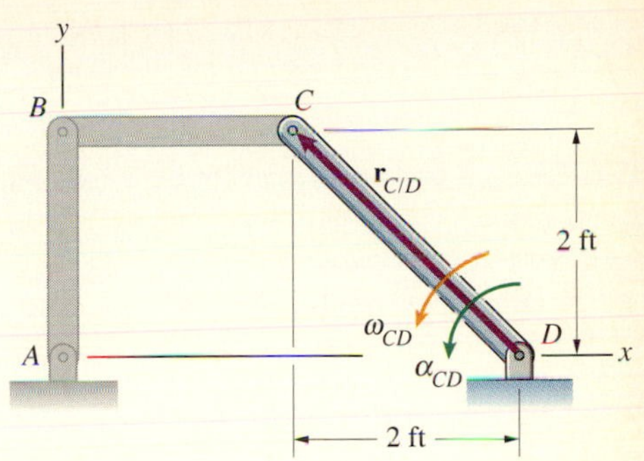

(b) Determining the motion of C in terms of the angular motion of bar CD.

Denoting the angular velocity of bar BC by ω_{BC} (Fig. c), the velocity of C in terms of the velocity of B is

$$\mathbf{v}_C = \mathbf{v}_B + \boldsymbol{\omega}_{BC} \times \mathbf{r}_{C/B}$$

$$= -20\mathbf{i} + (\omega_{BC}\,\mathbf{k}) \times (2\mathbf{i})$$

$$= -20\mathbf{i} + 2\omega_{BC}\,\mathbf{j}.$$

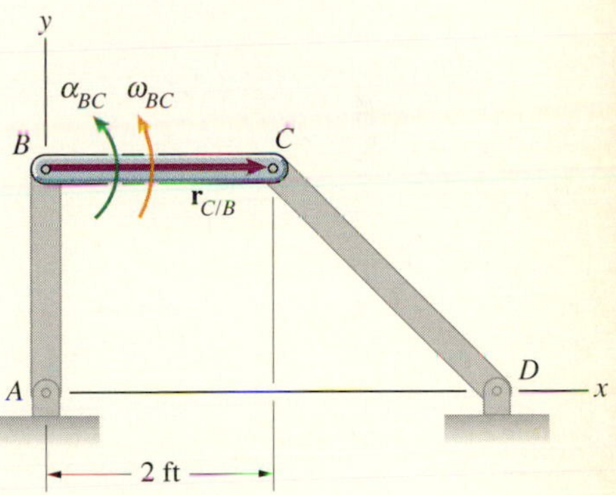

(c) Determining the motion of C in terms of the angular motion of bar BC.

Equating our two expressions for $\mathbf{v}_C$,

$$-2\omega_{CD}\mathbf{i} - 2\omega_{CD}\mathbf{j} = -20\mathbf{i} + 2\omega_{BC}\mathbf{j},$$

and equating the $\mathbf{i}$ and $\mathbf{j}$ components, we obtain $\omega_{CD} = 10$ rad/s and $\omega_{BC} = -10$ rad/s.

We can use the same sequence of steps to determine the angular accelerations. The acceleration of B is (Fig. a)

$$\mathbf{a}_B = \mathbf{a}_A + \boldsymbol{\alpha}_{AB} \times \mathbf{r}_{B/A} - \omega_{AB}^2\mathbf{r}_{B/A}$$

$$= \mathbf{0} + (-300\mathbf{k}) \times (2\,\mathbf{j}) - (10)^2(2\,\mathbf{j})$$

$$= 600\mathbf{i} - 200\mathbf{j}\ (\text{ft/s}^2).$$

The acceleration of C in terms of the acceleration of D is (Fig. b)

$$\mathbf{a}_C = \mathbf{a}_D + \boldsymbol{\alpha}_{CD} \times \mathbf{r}_{C/D} - \omega_{CD}^2\mathbf{r}_{C/D}$$

$$= \mathbf{0} + \begin{vmatrix} \mathbf{i} & \mathbf{j} & \mathbf{k} \\ 0 & 0 & \alpha_{CD} \\ -2 & 2 & 0 \end{vmatrix} - (10)^2(-2\mathbf{i} + 2\,\mathbf{j})$$

$$= (200 - 2\alpha_{CD})\mathbf{i} - (200 + 2\alpha_{CD})\mathbf{j}.$$

The acceleration of C in terms of the acceleration of B is (Fig. c)

$$\mathbf{a}_C = \mathbf{a}_B + \boldsymbol{\alpha}_{BC} \times \mathbf{r}_{C/B} - \omega_{BC}^2\mathbf{r}_{C/B}$$

$$= 600\mathbf{i} - 200\,\mathbf{j} + (\alpha_{BC}\mathbf{k}) \times (2\mathbf{i}) - (-10)^2(2\mathbf{i})$$

$$= 400\mathbf{i} - (200 - 2\alpha_{BC})\mathbf{j}.$$

Equating the expressions for $\mathbf{a}_C$, we obtain

$$(200 - 2\alpha_{CD})\mathbf{i} - (200 + 2\alpha_{CD})\,\mathbf{j} = 400\mathbf{i} - (200 - 2\alpha_{BC})\mathbf{j},$$

and equating $\mathbf{i}$ and $\mathbf{j}$ components, we obtain the angular accelerations $\alpha_{BC} = 100$ rad/s^2 and $\alpha_{CD} = -100$ rad/s^2.

Problems

6.81 The rigid body rotates about the z axis with counterclockwise angular velocity ω and counterclockwise angular acceleration α. Determine the acceleration of point A relative to point B (a) by using Eq. (6.9); (b) by using Eq. (6.10).

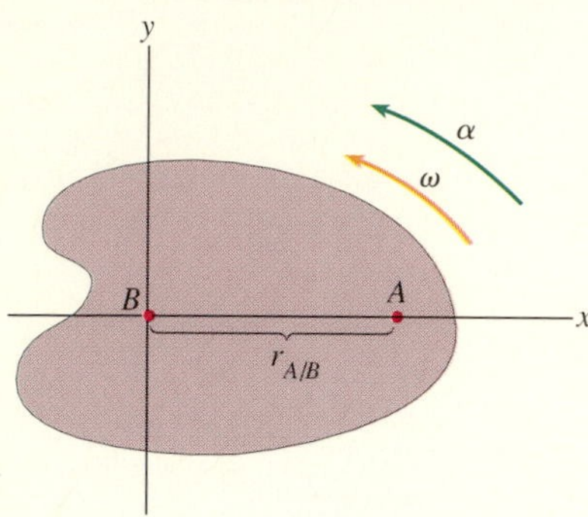

P6.81

6.82 The bar rotates with a counterclockwise angular velocity of 5 rad/s and a counterclockwise angular acceleration of 30 rad/s^2. Determine the acceleration of A (a) by using Eq. (6.9); (b) by using Eq. (6.10).

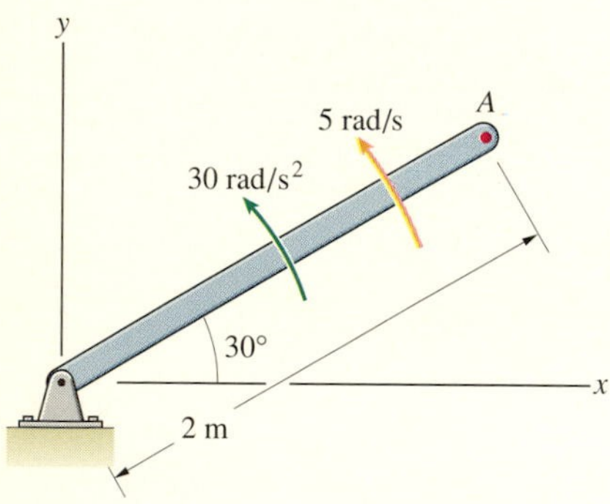

P6.82

6.83 The bar rotates with a constant angular velocity of 20 rad/s in the counterclockwise direction.
(a) Determine the acceleration of point B.
(b) Using your result from (a) and Eq. (6.10), determine the acceleration of point A.

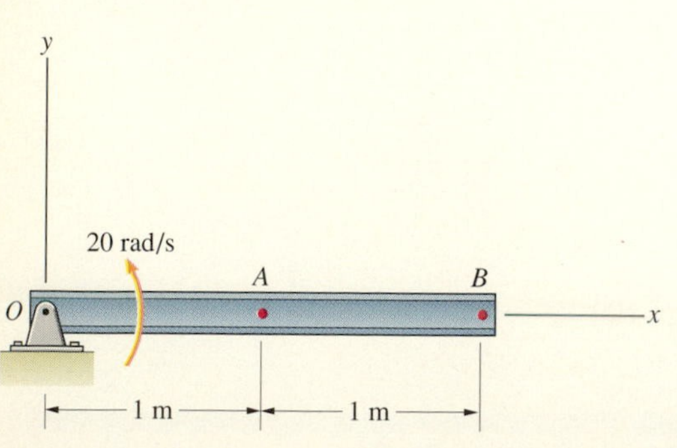

P6.83

6.84 The helicopter is in planar motion in the x-y plane. At the instant shown, the position of its center of mass G is $x = 2$ m, $y = 2.5$ m, its velocity is $\mathbf{v}_G = 12\mathbf{i} + 4\mathbf{j}$ (m/s), and its acceleration is $\mathbf{a}_G = 2\mathbf{i} + 3\mathbf{j}$ (m/s^2). The position of point T where the tail rotor is mounted is $x = -3.5$ m, $y = 4.5$ m. The helicopter's angular velocity is 0.2 rad/s clockwise, and its angular acceleration is 0.1 rad/s^2 counterclockwise. What is the acceleration of point T?

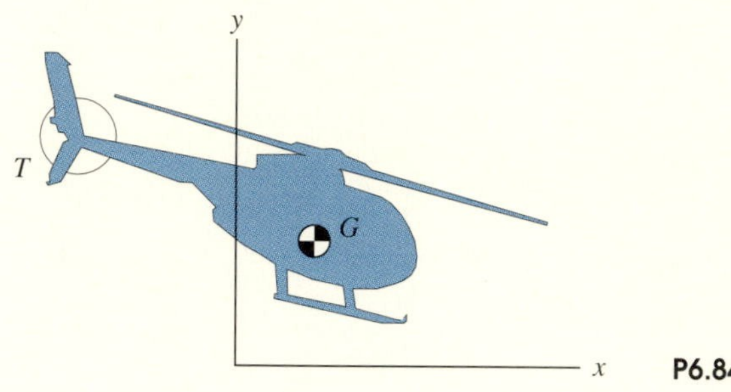

P6.84

6.85 The disk rolls on the plane surface. The velocity of point A is 6 m/s to the right, and its acceleration is 20 m/s^2 to the right.
(a) What is the angular acceleration vector of the disk?
(b) Determine the accelerations of points B, C, and D.

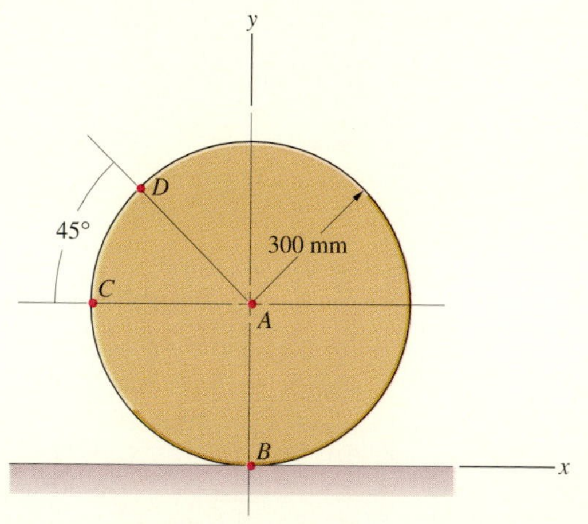

P6.85

6.86 The disk rolls on the circular surface with a constant clockwise angular velocity of 1 rad/s. What are the accelerations of points A and B?

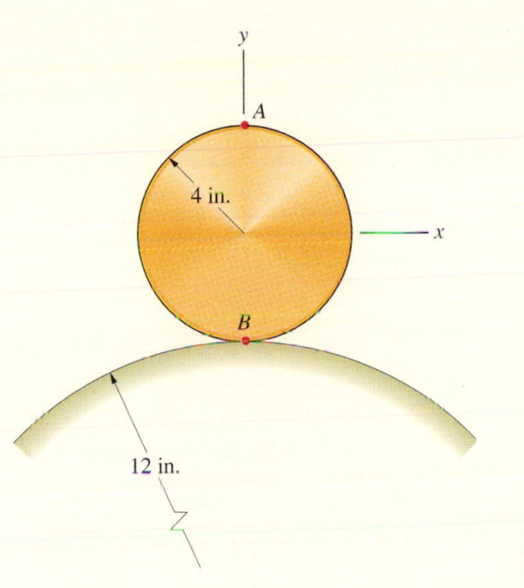

P6.86

6.87 The endpoints of the bar slide on the plane surfaces. Show that the acceleration of the midpoint G is related to the bar's angular velocity and angular acceleration by

$$\mathbf{a}_G = \frac{1}{2}L[(\alpha \cos\theta - \omega^2 \sin\theta)\mathbf{i} - (\alpha \sin\theta + \omega^2 \cos\theta)\mathbf{j}].$$

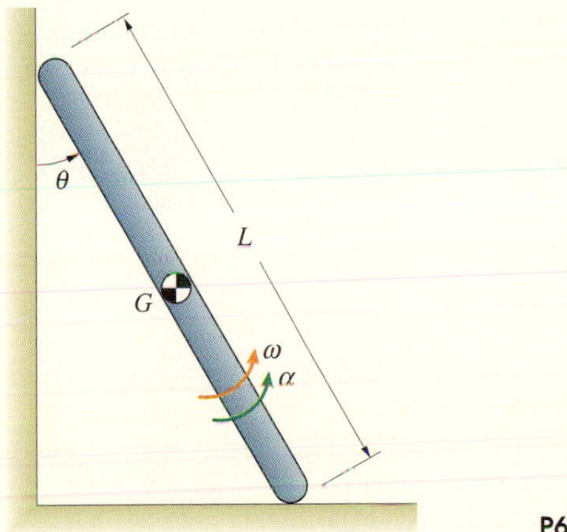

P6.87

6.88 The angular velocity and angular acceleration of bar AB are $\omega_{AB} = 2$ rad/s, $\alpha_{AB} = 10$ rad/s^2. The dimensions of the rectangular plate are 12 in. × 24 in. What are the angular velocity and angular acceleration of the rectangular plate?

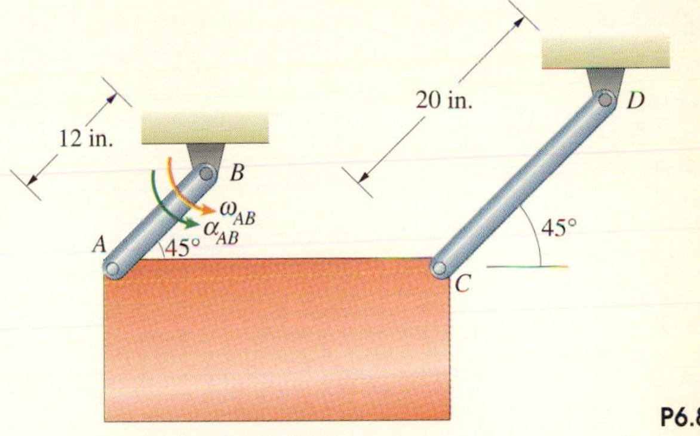

P6.88

6.89 The ring gear is stationary, and the sun gear has an angular acceleration of 10 rad/s^2 in the counterclockwise direction. Determine the angular acceleration of the planet gears.

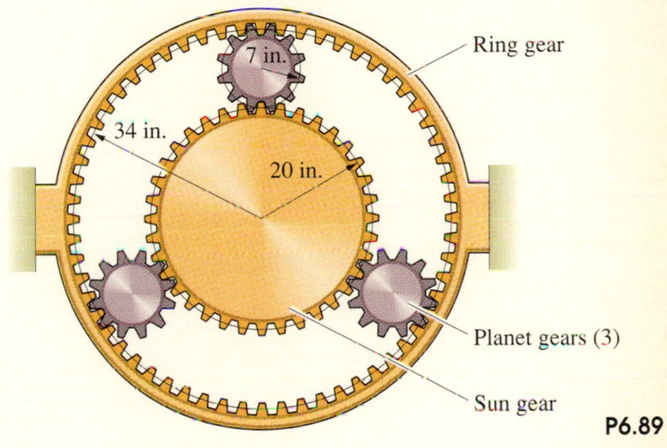

P6.89

6.90 The sun gear in Problem 6.89 has a counterclockwise angular velocity of 4 rad/s and a clockwise angular acceleration of 12 rad/s^2. What is the magnitude of the acceleration of the centerpoints of the planet gears?

6.91 The 1-m-diameter disk rolls and point B of the 1-m-long bar slides on the plane surface. Determine the angular acceleration of the bar and the acceleration of point B.

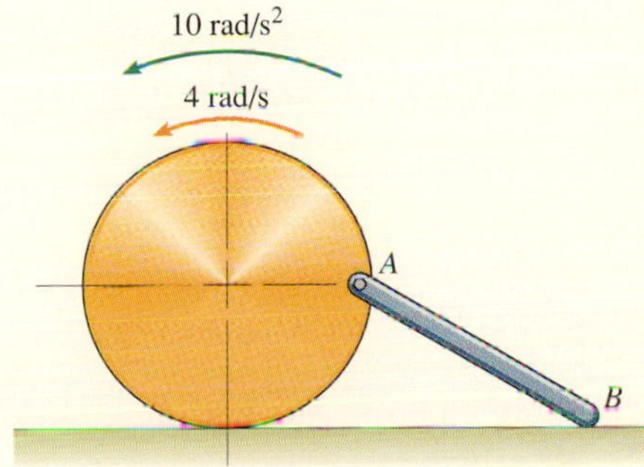

P6.91

6.92 The angle $\theta = 45°$ and bar OQ has a constant counterclockwise angular velocity of 2 rad/s. What is the acceleration of the sleeve P?

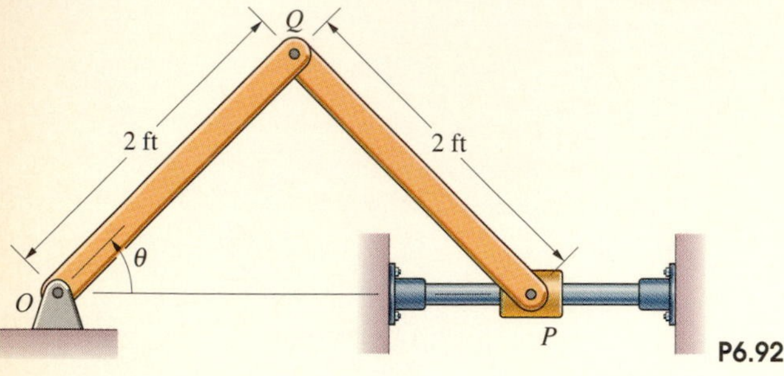

P6.92

6.93 Consider the system shown in Problem 6.92. If $\theta = 40°$ and the sleeve P is moving to the right with a constant velocity of 5 ft/s, what are the angular accelerations of the bars OQ and PQ?

6.94 The angle $\theta = 60°$ and bar OQ has a constant counterclockwise angular velocity of 2 rad/s. What is the angular acceleration of bar PQ?

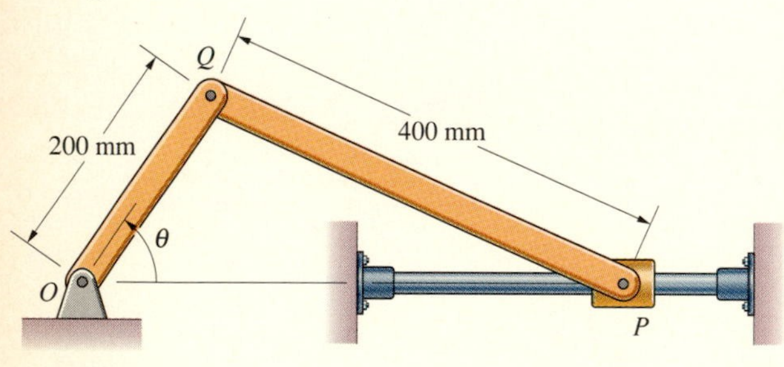

P6.94

6.95 Consider the system shown in Problem 6.94. If $\theta = 55°$ and sleeve P is moving to the right with a constant velocity of 2 m/s, what are the angular accelerations of bars OQ and PQ?

6.96 The angular velocity and acceleration of bar AB are $\omega_{AB} = 2$ rad/s, $\alpha_{AB} = 6$ rad/s^2. What are the angular velocity and angular acceleration of bar BD?

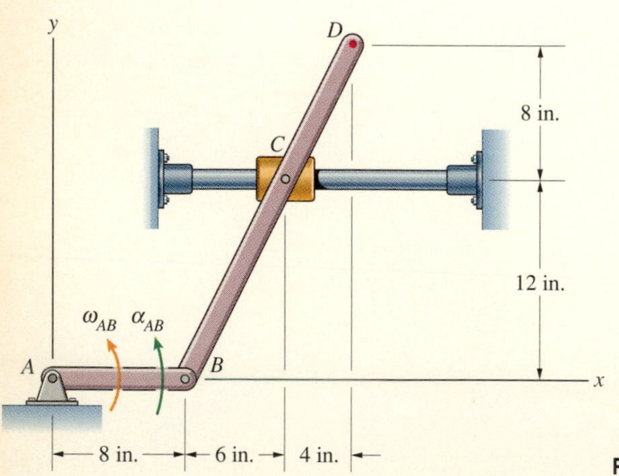

P6.96

6.97 In Problem 6.96, if the angular velocity and acceleration of bar AB are $\omega_{AB} = 2$ rad/s, $\alpha_{AB} = -10$ rad/s^2, what are the velocity and acceleration of point D?

6.98 If $\omega_{AB} = 6$ rad/s and $\alpha_{AB} = 20$ rad/s^2, what are the velocity and acceleration of point C?

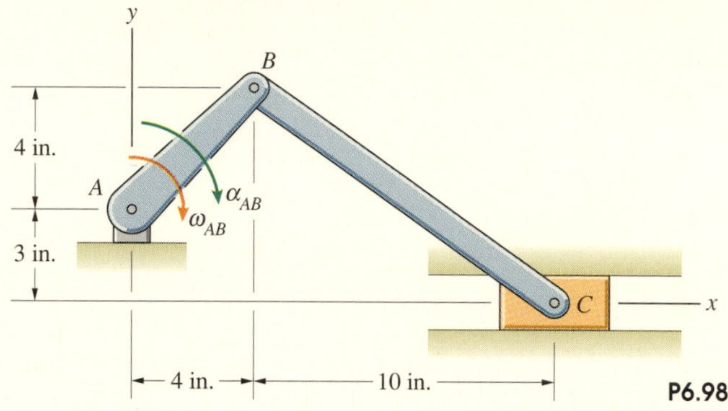

P6.98

6.99 A motor rotates the circular disk mounted at A, moving the saw back and forth. (The saw is supported by a horizontal slot so that point C moves horizontally.) The radius AB is 4 in., and the link BC is 14 in. long. In the position shown, $\theta = 45°$ and the link BC is horizontal. If the disk has a constant angular velocity of one revolution per second counterclockwise, what is the acceleration of the saw?

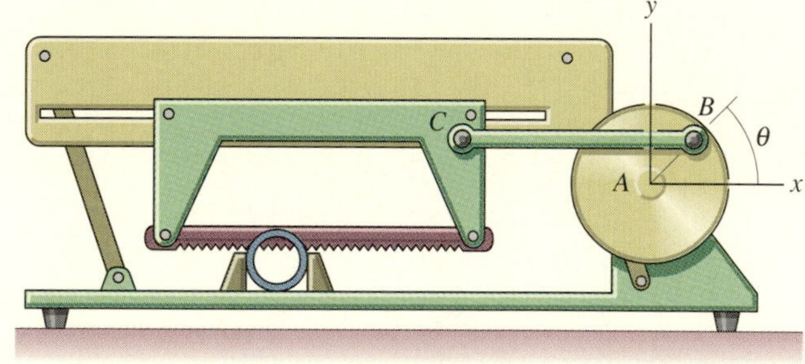

P6.99

6.100 In Problem 6.99, if the disk has a constant angular velocity of one revolution per second counterclockwise and $\theta = 180°$, what is the acceleration of the saw?

6.101 If $\omega_{AB} = 2$ rad/s, $\alpha_{AB} = 2$ rad/s^2, $\omega_{BC} = 1$ rad/s, and $\alpha_{BC} = 4$ rad/s^2, what is the acceleration of point C where the scoop of the excavator is attached?

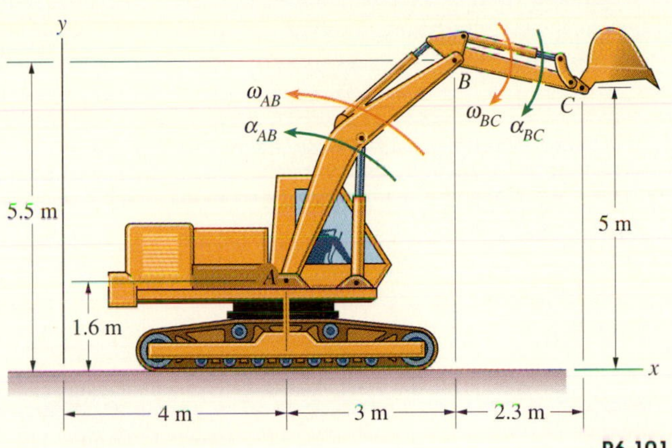

P6.101

6.102 If the velocity of point C of the excavator in Problem 6.101 is $\mathbf{v}_C = 4\mathbf{i}$ (m/s) and is constant, what are ω_{AB}, α_{AB}, ω_{BC}, and α_{BC}?

6.103 Bar AB rotates in the counterclockwise direction with a constant angular velocity of 10 rad/s. What are the angular accelerations of bars BC and CD?

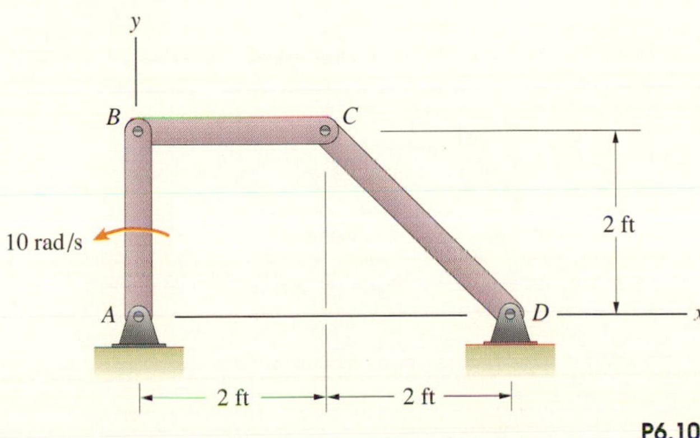

P6.103

6.104 At the instant shown, bar AB has no angular velocity but has a counterclockwise angular acceleration of 10 rad/s^2. Determine the acceleration of point E.

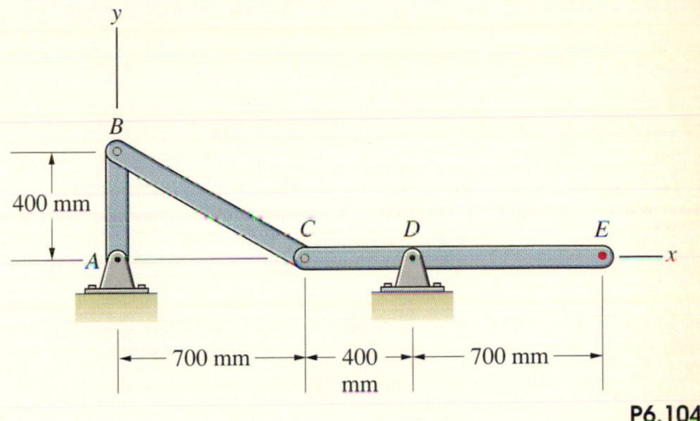

P6.104

6.105 If $\omega_{AB} = 12$ rad/s and $\alpha_{AB} = 100$ rad/s^2, what are the angular accelerations of bars BC and CD?

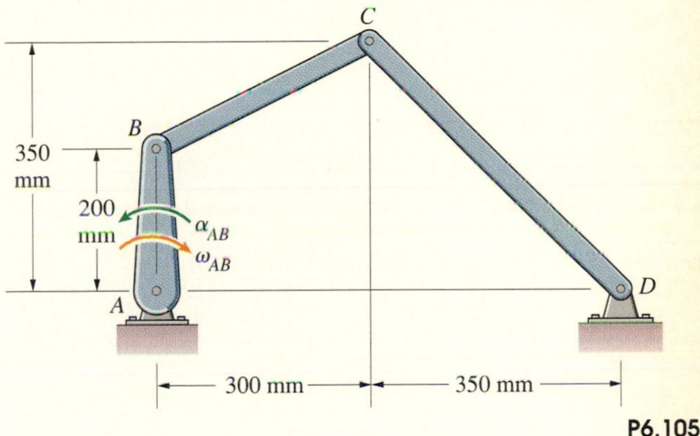

P6.105

6.106 If $\omega_{AB} = 4$ rad/s counterclockwise and $\alpha_{AB} = 12$ rad/s^2 counterclockwise, what is the acceleration of point C?

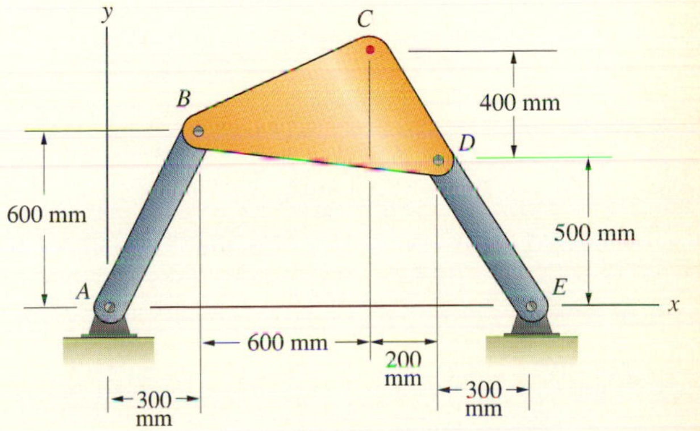

P6.106

6.107 In Problem 6.106, if $\omega_{AB} = 6$ rad/s clockwise and $\alpha_{DE} = 0$, what is the acceleration of point C?

6.108 If arm AB has a constant clockwise angular velocity of 0.8 rad/s, arm BC has a constant clockwise angular velocity of 0.2 rad/s, and arm CD remains vertical, what is the acceleration of the part D?

P6.108

6.109 In Problem 6.108, if arm AB has a constant clockwise angular velocity of 0.8 rad/s and you want D to have zero velocity and acceleration, what are the necessary angular velocities and angular accelerations of arms BC and CD?

6.110 In Problem 6.108, if you want arm CD to remain vertical and you want part D to have velocity $\mathbf{v}_D = 1.0\mathbf{i}$ (m/s) and zero acceleration, what are the necessary angular velocities and angular accelerations of arms AB and BC?

6.111 Link AB of the robot's arm is rotating with a constant counterclockwise angular velocity of 2 rad/s, and link BC is rotating with a constant clockwise angular velocity of 3 rad/s. Link CD is rotating at 4 rad/s in the counterclockwise direction and has a counterclockwise angular acceleration of 6 rad/s^2. What is the acceleration of point D?

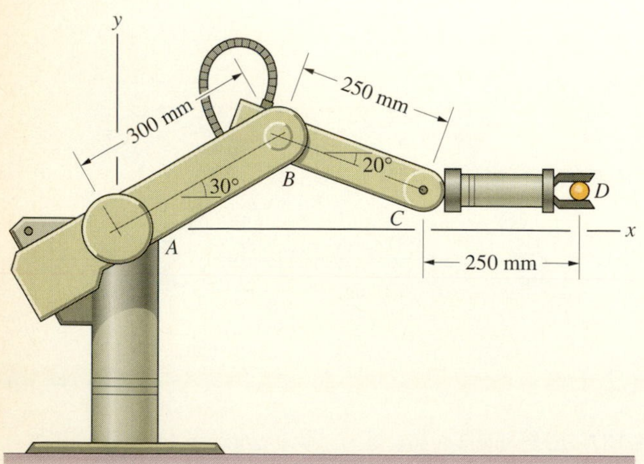

P6.111

6.112 Consider the robot shown in Problem 6.111. Link AB is rotating with a constant counterclockwise angular velocity of 2 rad/s, and link BC is rotating with a constant clockwise angular velocity of 3 rad/s. Link CD is rotating at 4 rad/s. If you want the acceleration of point D to be parallel to the x axis, what is the necessary angular acceleration of link CD? What is the resulting acceleration of point D?

6.113 The horizontal member ADE supporting the scoop is stationary. If link BD has a clockwise angular velocity of 1 rad/s and a counterclockwise angular acceleration of 2 rad/s^2, what is the angular acceleration of the scoop?

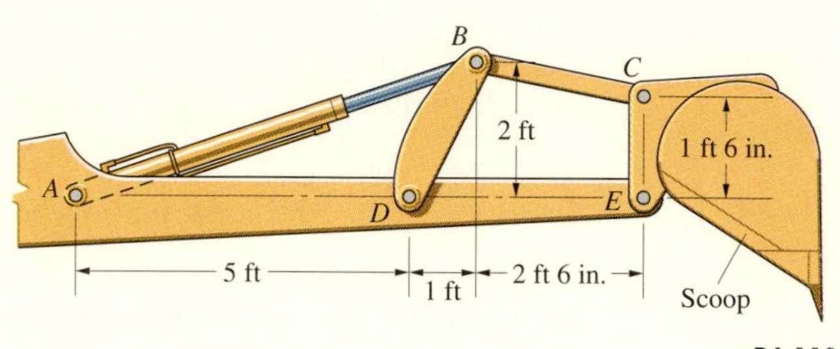

P6.113

6.114 The ring gear is fixed, and the hub and planet gears are bonded together. The connecting rod has a counterclockwise angular acceleration of 10 rad/s^2. Determine the angular accelerations of the planet and sun gears.

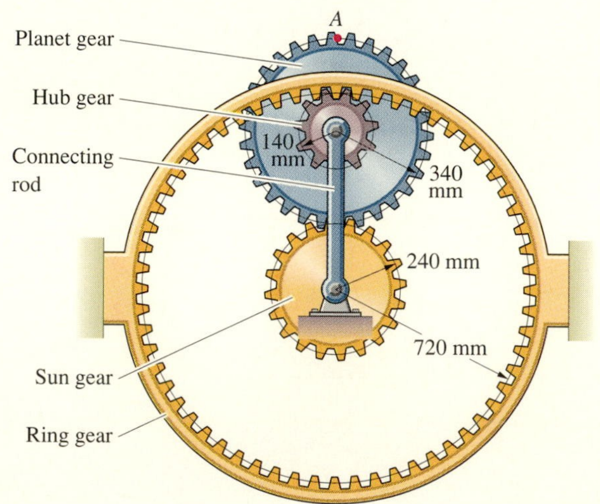

P6.114

6.115 The connecting rod in Problem 6.114 has a counterclockwise angular velocity of 4 rad/s and a clockwise angular acceleration of 12 rad/s^2. Determine the magnitude of the acceleration of point A.

6.116 The large gear is fixed. The angular velocity and angular acceleration of bar AB are $\omega_{AB} = 2$ rad/s, $\alpha_{AB} = 4$ rad/s^2. Determine the angular accelerations of bars CD and DE.

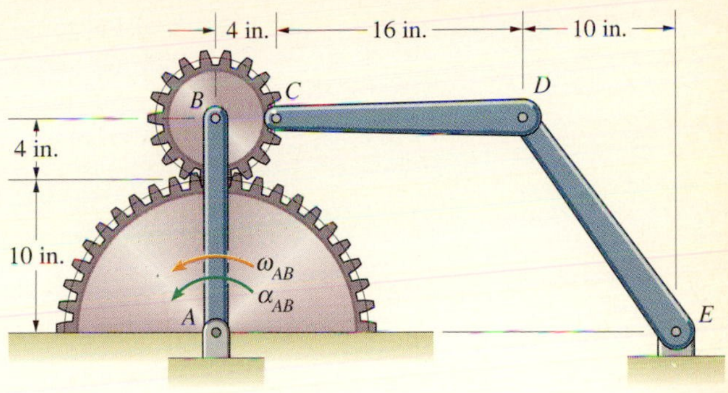

P6.116

6.5 *Sliding Contacts*

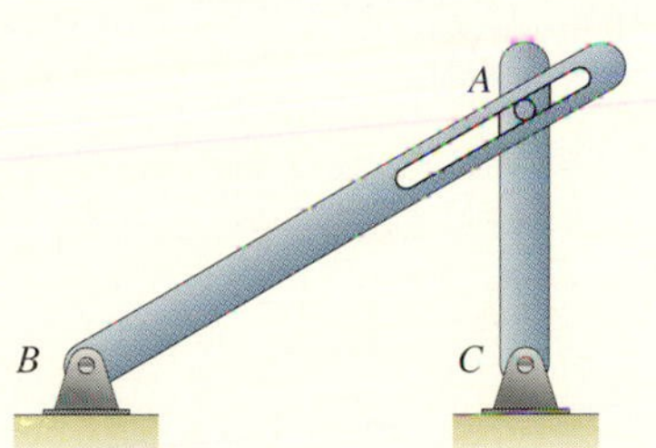

Fig. 6.29

Linkage with a sliding contact.

Here we consider a type of problem that is similar to those we have discussed previously in this chapter, but requires a different method of solution. For example, suppose that we know the angular velocity and angular acceleration of the bar AB in Fig. 6.29, and we want to determine the angular velocity and angular acceleration of bar AC. We cannot use the equation $\mathbf{v}_A = \mathbf{v}_B + \boldsymbol{\omega} \times \mathbf{r}_{A/B}$ to express the velocity of point A in terms of the angular velocity of bar AB, because we derived it under the assumption that points A and B are points of the same rigid body. Point A is not a part of the bar AB, but moves relative to it as the pin slides along the slot. This is an example of a **sliding contact** between rigid bodies. To solve this type of problem, we must rederive Equations (6.8), (6.9), and (6.10) without making the assumption that A is a point of the rigid body.

To describe the motion of a point that moves relative to a given rigid body, it is convenient to use a reference frame that moves with the rigid body. We say that such a reference frame is **body-fixed**. In Fig. 6.30 we introduce a body-fixed reference frame xyz with its origin at a point B of the rigid body in addition to the **primary reference frame** with origin O. We do not assume A to be a point of the rigid body. The position of A relative to O is

$$\mathbf{r}_A = \mathbf{r}_B + \underbrace{x\mathbf{i} + y\mathbf{j} + z\mathbf{k}}_{\mathbf{r}_{A/B}},$$

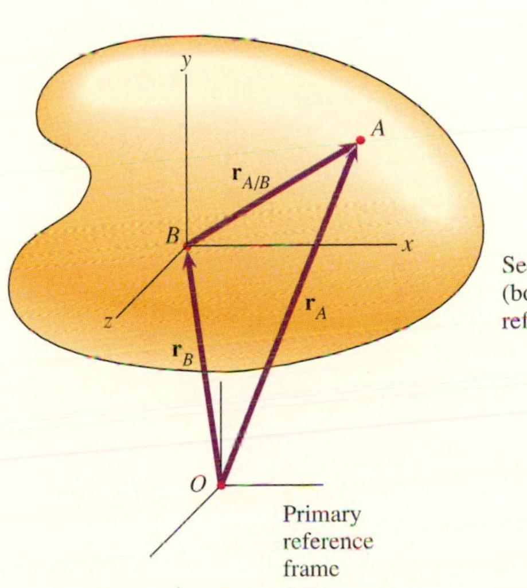

Fig. 6.30

A point B of a rigid body, a body-fixed secondary reference frame, and an arbitrary point A.

where x, y, and z are the coordinates of A in terms of the body-fixed reference frame. Our next step is to take the time derivative of this expression to obtain an equation for the velocity of A. In doing so, we recognize that the unit vectors **i**, **j**, and **k** are not constant, because they rotate with the body-fixed reference frame:

$$\mathbf{v}_A = \mathbf{v}_B + \frac{dx}{dt}\mathbf{i} + x\frac{d\mathbf{i}}{dt} + \frac{dy}{dt}\mathbf{j} + y\frac{d\mathbf{j}}{dt} + \frac{dz}{dt}\mathbf{k} + z\frac{d\mathbf{k}}{dt}.$$

What are the time derivatives of the unit vectors? In Section 6.3 we showed that if $\mathbf{r}_{P/B}$ is the position of a point P of a rigid body relative to another point

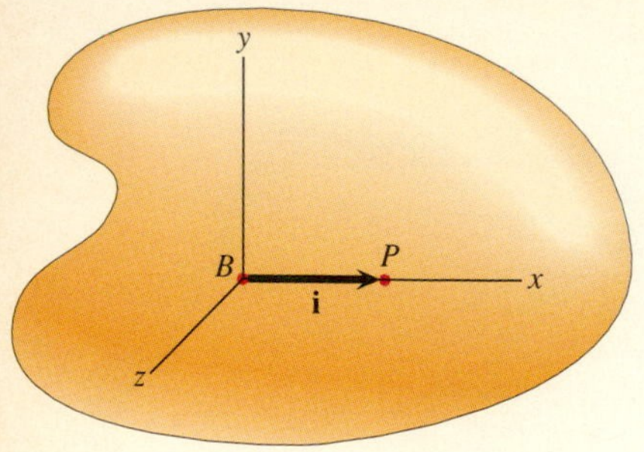

Fig. 6.31
Interpreting **i** as the position vector of a point P relative to B.

B of the same rigid body, $d\mathbf{r}_{P/B}/dt = \mathbf{v}_{P/B} = \boldsymbol{\omega} \times \mathbf{r}_{P/B}$. Since we can regard the unit vector **i** as the position vector of a point P of the rigid body (Fig. 6.31), its time derivative is $d\mathbf{i}/dt = \boldsymbol{\omega} \times \mathbf{i}$. Applying the same argument to the unit vectors **j** and **k**, we obtain

$$\frac{d\mathbf{i}}{dt} = \boldsymbol{\omega} \times \mathbf{i}, \qquad \frac{d\mathbf{j}}{dt} = \boldsymbol{\omega} \times \mathbf{j}, \qquad \frac{d\mathbf{k}}{dt} = \boldsymbol{\omega} \times \mathbf{k}.$$

Using these expressions, we can write the velocity of point A as

$$\boxed{\mathbf{v}_A = \mathbf{v}_B + \underbrace{\mathbf{v}_{A\,\text{rel}} + \boldsymbol{\omega} \times \mathbf{r}_{A/B}}_{\mathbf{v}_{A/B}},} \tag{6.11}$$

where

$$\mathbf{v}_{A\,\text{rel}} = \frac{dx}{dt}\mathbf{i} + \frac{dy}{dt}\mathbf{j} + \frac{dz}{dt}\mathbf{k} \tag{6.12}$$

is the velocity of A relative to the body-fixed reference frame. That is, $\mathbf{v}_A$ is the velocity of A relative to the primary reference frame and $\mathbf{v}_{A\,\text{rel}}$ is the velocity of A relative to the rigid body.

Equation (6.11) expresses the velocity of a point A as the sum of three terms (Fig. 6.32): the velocity of a point B of the rigid body, the velocity $\boldsymbol{\omega} \times \mathbf{r}_{A/B}$ of A relative to B due to the rotation of the rigid body, and the velocity $\mathbf{v}_{A\,\text{rel}}$ of A relative to the rigid body.

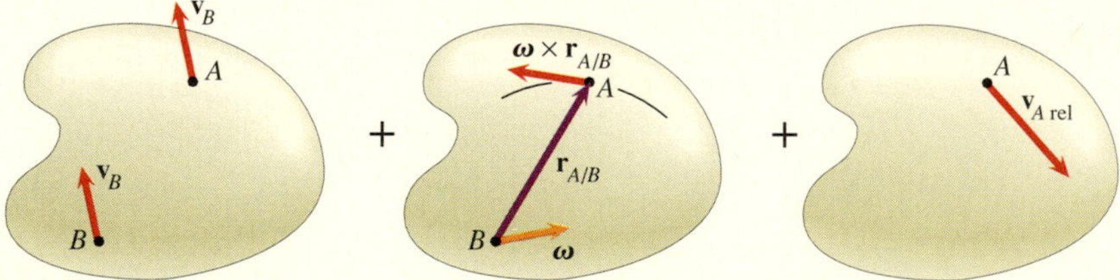

Fig. 6.32
Expressing the velocity of A in terms of the velocity of a point B of the rigid body.

To obtain an equation for the acceleration of point A, we take the time derivative of Eq. (6.11) and use Eq. (6.12). The result is (see Problem 6.142)

$$\boxed{\mathbf{a}_A = \mathbf{a}_B + \underbrace{\mathbf{a}_{A\,\text{rel}} + 2\boldsymbol{\omega} \times \mathbf{v}_{A\,\text{rel}} + \boldsymbol{\alpha} \times \mathbf{r}_{A/B} + \boldsymbol{\omega} \times (\boldsymbol{\omega} \times \mathbf{r}_{A/B})}_{\mathbf{a}_{A/B}},} \tag{6.13}$$

where

$$\mathbf{a}_{A\,\text{rel}} = \frac{d^2x}{dt^2}\mathbf{i} + \frac{d^2y}{dt^2}\mathbf{j} + \frac{d^2z}{dt^2}\mathbf{k} \tag{6.14}$$

is the acceleration of A relative to the body-fixed reference frame. That is, $\mathbf{a}_A$ is the acceleration of A relative to the primary reference frame and $\mathbf{a}_{A\,\text{rel}}$ is the acceleration of A relative to the rigid body.

In the case of planar motion, we can express Eq. (6.13) in the simpler form

$$\boxed{\mathbf{a}_A = \mathbf{a}_B + \underbrace{\mathbf{a}_{A\,\text{rel}} + 2\boldsymbol{\omega} \times \mathbf{v}_{A\,\text{rel}} + \boldsymbol{\alpha} \times \mathbf{r}_{A/B} - \omega^2\mathbf{r}_{A/B}}_{\mathbf{a}_{A/B}}.} \tag{6.15}$$

In summary, $\mathbf{v}_A$ and $\mathbf{a}_A$ are the velocity and acceleration of point A relative to the primary reference frame, the reference frame relative to which the rigid body's motion is being described. The terms $\mathbf{v}_{A\,\text{rel}}$ and $\mathbf{a}_{A\,\text{rel}}$ are the velocity and acceleration of point A relative to the body-fixed reference frame. That is, they are the velocity and acceleration measured by an observer moving with the rigid body (Fig. 6.33). If A is a point of the rigid body, $\mathbf{v}_{A\,\text{rel}}$ and $\mathbf{a}_{A\,\text{rel}}$ are zero, and Eqs. (6.11) and (6.13) are identical to Eqs. (6.8) and (6.9).

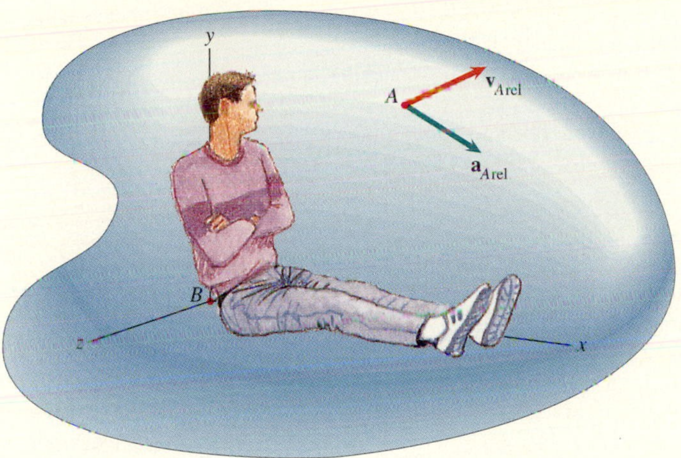

Fig. 6.33
Imagine yourself to be stationary relative to the rigid body.

We can illustrate these concepts with a simple example. Figure 6.34(a) shows a point A moving with velocity v parallel to the axis of a bar. (Imagine that A is a bug walking along the bar.) Suppose that at the same time, the bar is rotating about a fixed point B with a constant angular velocity ω relative to an earth-fixed reference frame (Fig. 6.34b). We will use Eq. (6.11) to determine the velocity of A relative to the earth-fixed reference frame.

Let the coordinate system in Fig. 6.34(c) be fixed with respect to the bar, and let x be the present position of A. In terms of this body-fixed reference frame, the angular velocity vector of the bar (and the reference frame) relative to the primary earth-fixed reference frame is $\boldsymbol{\omega} = \omega\mathbf{k}$. Relative to the body-fixed reference frame, point A moves along the x axis with velocity v, so $\mathbf{v}_{A\,\text{rel}} = v\mathbf{i}$. From Eq. (6.11), the velocity of A relative to the earth-fixed reference frame is

$$\begin{aligned}\mathbf{v}_A &= \mathbf{v}_B + \mathbf{v}_{A\,\text{rel}} + \boldsymbol{\omega} \times \mathbf{r}_{A/B} \\ &= \mathbf{0} + v\mathbf{i} + (\omega\mathbf{k}) \times (x\mathbf{i}) \\ &= v\mathbf{i} + \omega x\mathbf{j}.\end{aligned}$$

Relative to the earth-fixed reference frame, A has a component of velocity parallel to the bar and also a normal component due to the bar's rotation (Fig. 6.34d). Although $\mathbf{v}_A$ is the velocity of A relative to the earth-fixed reference frame, notice that it is expressed in components in terms of the body-fixed reference frame.

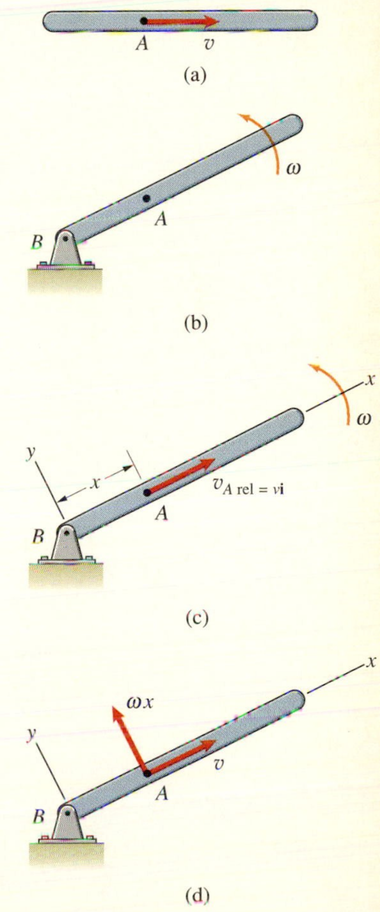

Fig. 6.34
(a) A point moving along a bar.
(b) The bar is rotating.
(c) A body-fixed reference frame.
(d) Components of $\mathbf{v}_A$.

In the following examples we analyze the motions of systems of rigid bodies with sliding contacts. You can use the same general approach you applied to systems of pinned rigid bodies, beginning with points whose velocities and accelerations are known and applying Eqs. (6.11) and (6.15).

Example 6.7

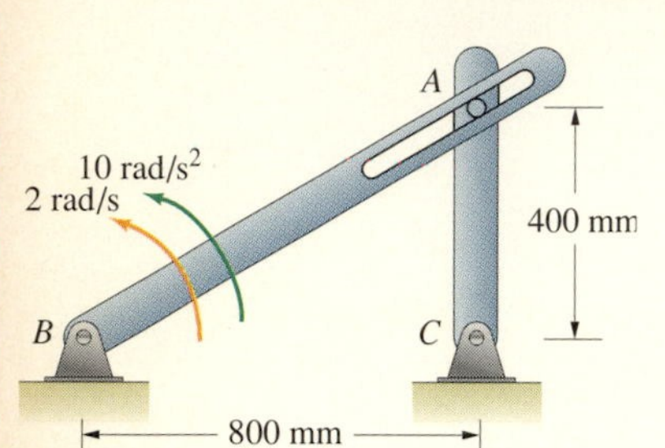

Fig. 6.35

Bar AB in Fig. 6.35 has a counterclockwise angular velocity of 2 rad/s and a counterclockwise angular acceleration of 10 rad/s^2.
(a) Determine the angular velocity of bar AC and the velocity of the pin A relative to the slot in bar AB.
(b) Determine the angular acceleration of bar AC and the acceleration of the pin A relative to the slot in bar AB.

STRATEGY

By using a secondary reference frame that is fixed with respect to the slotted bar, we can use Eq. (6.11) to express the velocity of the pin A in terms of its velocity relative to the slot and the known angular velocity of bar AB. The pins A and C are both points of the bar AC, so we can express $\mathbf{v}_A$ in terms of the angular velocity of the bar AC in the usual way. By equating the resulting expressions for $\mathbf{v}_A$, we will obtain a vector equation in terms of the velocity of A relative to the slot and the angular velocity of bar AC. Then, by following the same sequence of steps but this time using Eq. (6.15), we can obtain the acceleration of A relative to the slot and the angular acceleration of bar AC.

SOLUTION

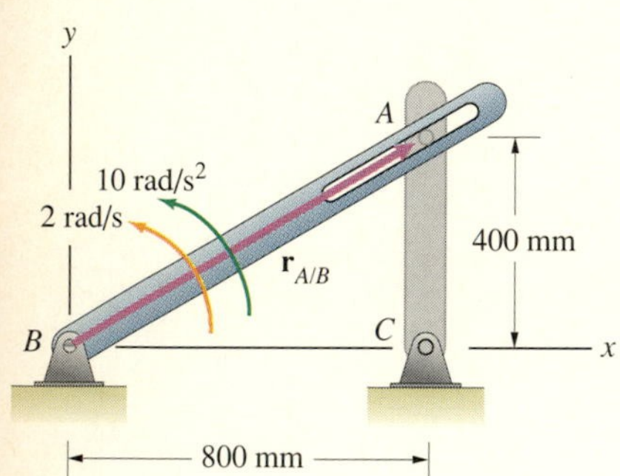

(a) A body-fixed coordinate system and the position vector of the pin A relative to B.

(a) Let the coordinate system in Fig. (a) be body-fixed with respect to the slotted bar. Applying Eq. (6.11) to points A and B, the velocity of A is

$$\mathbf{v}_A = \mathbf{v}_B + \mathbf{v}_{A\,\mathrm{rel}} + \boldsymbol{\omega}_{AB} \times \mathbf{r}_{A/B}$$

$$= 0 + \mathbf{v}_{A\,\mathrm{rel}} + \begin{vmatrix} \mathbf{i} & \mathbf{j} & \mathbf{k} \\ 0 & 0 & 2 \\ 0.8 & 0.4 & 0 \end{vmatrix}.$$

The velocity of the pin A *relative to the body-fixed coordinate system* is parallel to the slot (Fig. b). Therefore we can express it as

$$\mathbf{v}_{A\,\mathrm{rel}} = v_{A\,\mathrm{rel}} \cos\beta\,\mathbf{i} + v_{A\,\mathrm{rel}} \sin\beta\,\mathbf{j},$$

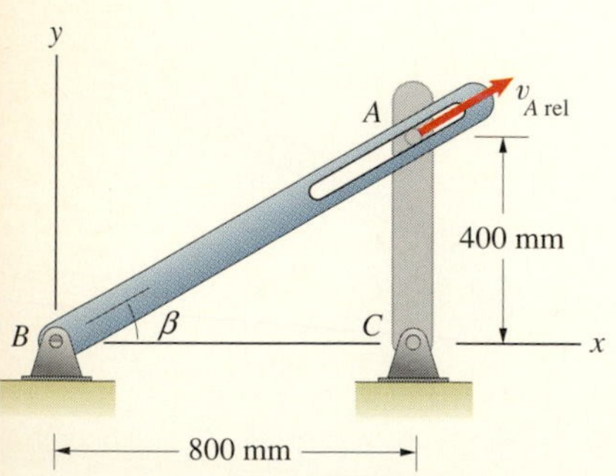

(b) The velocity of the pin A relative to the body-fixed coordinate system.

where $\beta = \arctan(0.4/0.8)$. Substituting this expression into our equation for $\mathbf{v}_A$, we obtain

$$\mathbf{v}_A = (v_{A\,\mathrm{rel}} \cos\beta - 0.8)\mathbf{i} + (v_{A\,\mathrm{rel}} \sin\beta + 1.6)\mathbf{j}.$$

Let ω_{AC} be the angular velocity of bar AC (Fig. c). Expressing the velocity of A in terms of the velocity of C, we obtain

$$\begin{aligned} \mathbf{v}_A &= \mathbf{v}_C + \boldsymbol{\omega}_{AC} \times \mathbf{r}_{A/C} \\ &= \mathbf{0} + (\omega_{AC}\mathbf{k}) \times (0.4\,\mathbf{j}) \\ &= -0.4\omega_{AC}\,\mathbf{i}. \end{aligned}$$

Notice that there is no relative velocity term in this equation, because A is a point of the bar AC. Equating our two expressions for $\mathbf{v}_A$, we obtain

$$(v_{A\,\text{rel}} \cos\beta - 0.8)\mathbf{i} + (v_{A\,\text{rel}} \sin\beta + 1.6)\mathbf{j} = -0.4\omega_{AC}\,\mathbf{i}.$$

Equating $\mathbf{i}$ and $\mathbf{j}$ components yields the two equations

$$v_{A\,\text{rel}} \cos\beta - 0.8 = -0.4\omega_{AC},$$
$$v_{A\,\text{rel}} \sin\beta + 1.6 = 0.$$

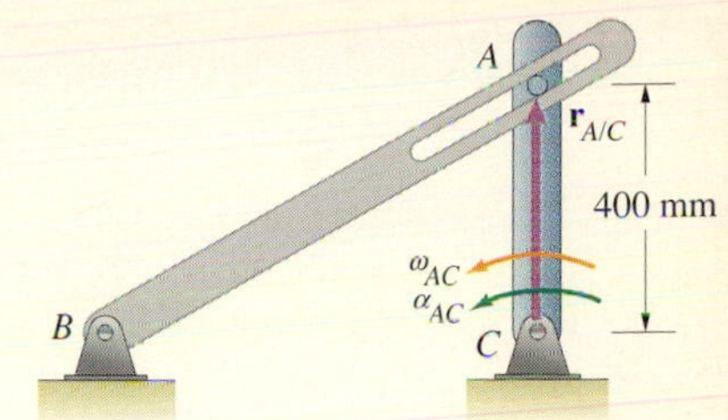

(c) The position vector of A relative to C.

Solving them, we obtain $v_{A\,\text{rel}} = -3.58$ m/s and $\omega_{AC} = 10$ rad/s. At this instant, the pin A is moving relative to the slot at 3.58 m/s toward B. The vector $\mathbf{v}_{A\,\text{rel}}$ is

$$\mathbf{v}_{A\,\text{rel}} = -3.58(\cos\beta\,\mathbf{i} + \sin\beta\,\mathbf{j}) = -3.2\mathbf{i} - 1.6\,\mathbf{j}\ \text{(m/s)}.$$

(b) Applying Eq. (6.15) to bar AB (Fig. b), the acceleration of A is

$$\mathbf{a}_A = \mathbf{a}_B + \mathbf{a}_{A\,\text{rel}} + 2\,\boldsymbol{\omega}_{AB} \times \mathbf{v}_{A\,\text{rel}} + \boldsymbol{\alpha}_{AB} \times \mathbf{r}_{A/B} - \omega_{AB}^2\mathbf{r}_{A/B}$$
$$= \mathbf{0} + \mathbf{a}_{A\,\text{rel}} + 2\begin{vmatrix} \mathbf{i} & \mathbf{j} & \mathbf{k} \\ 0 & 0 & 2 \\ -3.2 & -1.6 & 0 \end{vmatrix} + \begin{vmatrix} \mathbf{i} & \mathbf{j} & \mathbf{k} \\ 0 & 0 & 10 \\ 0.8 & 0.4 & 0 \end{vmatrix}$$
$$-(2)^2(0.8\,\mathbf{i} + 0.4\,\mathbf{j}).$$

The acceleration of A relative to the body-fixed coordinate system is parallel to the slot (Fig. d), so we can write it in the same way we did $\mathbf{v}_{A\,\text{rel}}$:

$$\mathbf{a}_{A\,\text{rel}} = a_{A\,\text{rel}} \cos\beta\mathbf{i} + a_{A\,\text{rel}} \sin\beta\mathbf{j}.$$

Substituting this expression into our equation for $\mathbf{a}_A$ gives

$$\mathbf{a}_A = (a_{A\,\text{rel}} \cos\beta - 0.8)\mathbf{i} + (a_{A\,\text{rel}} \sin\beta - 6.4)\mathbf{j}.$$

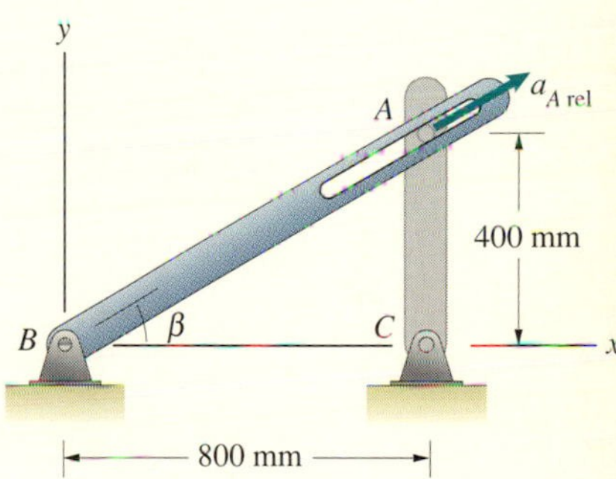

(d) The acceleration of the pin A relative to the body-fixed coordinate system.

Expressing the acceleration of A in terms of the acceleration of C (Fig. c), we obtain

$$\mathbf{a}_A = \mathbf{a}_C + \boldsymbol{\alpha}_{AC} \times \mathbf{r}_{A/C} - \omega_{AC}^2\mathbf{r}_{A/C}$$
$$= \mathbf{0} + (\alpha_{AC}\,\mathbf{k}) \times (0.4\mathbf{j}) - (10)^2(0.4\,\mathbf{j})$$
$$= -0.4\alpha_{AC}\mathbf{i} - 40\mathbf{j}.$$

Equating our expressions for $\mathbf{a}_A$, we obtain

$$(a_{A\,\text{rel}} \cos\beta - 0.8)\mathbf{i} + (a_{A\,\text{rel}} \sin\beta - 6.4)\,\mathbf{j} = -0.4\alpha_{AC}\mathbf{i} - 40\mathbf{j}.$$

Equating $\mathbf{i}$ and $\mathbf{j}$ components yields the two equations

$$a_{A\,\text{rel}} \cos\beta - 0.8 = -0.4\alpha_{AC},$$
$$a_{A\,\text{rel}} \sin\beta - 6.4 = -40.$$

Solving them, we obtain $a_{A\,\text{rel}} = -75.1$ m/s^2 and $\alpha_{AC} = 170$ rad/s^2. At this instant, the pin A is accelerating relative to the slot at 75.1 m/s^2 toward B.

Example 6.8

The collar at B in Fig. 6.36 slides along the circular bar, causing the pin B to move at constant speed v_0 in a circular path of radius R. Bar BC slides in the collar at A. At the instant shown, determine the angular velocity and angular acceleration of bar BC.

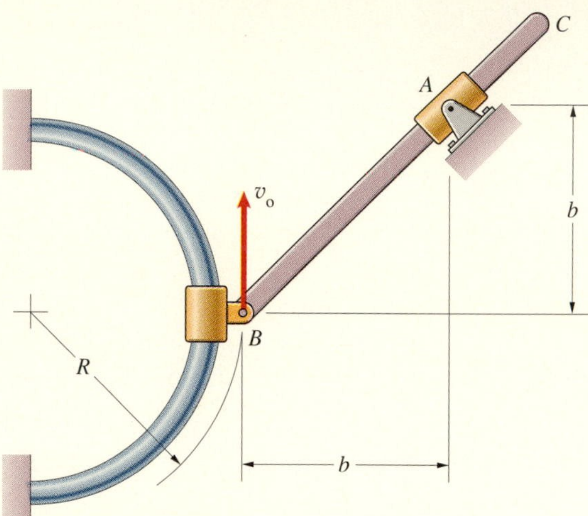

Fig. 6.36

STRATEGY

We will use a secondary reference frame with its origin at B that is body-fixed with respect to bar BC. By using Eqs. (6.11) and (6.15) to express the velocity and acceleration of the stationary pin A in terms of the velocity and acceleration of the pin B, we can determine the angular velocity and angular acceleration of bar BC.

SOLUTION

Angular Velocity Let the angular velocity and angular acceleration of bar BC, which are also the angular velocity and angular acceleration of the body-fixed coordinate system, be ω_{BC} and α_{BC} (Fig. a). The velocity of the stationary pin A is zero. From Eq. (6.11),

$$\mathbf{v}_A = \mathbf{0} = \mathbf{v}_B + \mathbf{v}_{A\,\text{rel}} + \boldsymbol{\omega} \times \mathbf{r}_{A/B}, \tag{6.16}$$

where $\mathbf{v}_{A\,\text{rel}}$ is the velocity of A relative to the body-fixed coordinate system and $\boldsymbol{\omega} = \omega_{BC}\mathbf{k}$ is the angular velocity vector of the coordinate system. The velocity of the pin B is $\mathbf{v}_B = v_0\,\mathbf{j}$. The velocity of the stationary pin A relative to the body-fixed coordinate system is parallel to the bar (Fig. b), so we can express it in the form

$$\mathbf{v}_{A\,\text{rel}} = v_{A\,\text{rel}} \cos 45^\circ\,\mathbf{i} + v_{A\,\text{rel}} \sin 45^\circ\,\mathbf{j},$$

and write Eq. (6.16) as

$$\mathbf{0} = v_0\,\mathbf{j} + v_{A\,\text{rel}} \cos 45^\circ\,\mathbf{i} + v_{A\,\text{rel}} \sin 45^\circ\,\mathbf{j} + \begin{vmatrix} \mathbf{i} & \mathbf{j} & \mathbf{k} \\ 0 & 0 & \omega_{BC} \\ b & b & 0 \end{vmatrix}.$$

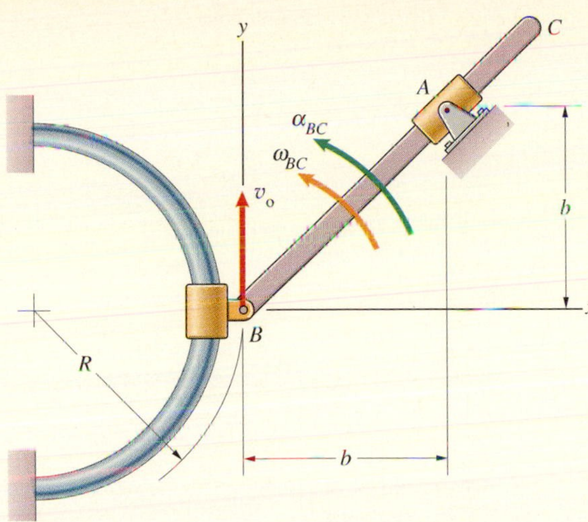

(a) A reference frame fixed with respect to bar BC.

From the **i** and **j** components of this equation, we obtain

$$v_{A\,\text{rel}} \cos 45^\circ - b\omega_{BC} = 0,$$

$$v_0 + v_{A\,\text{rel}} \sin 45^\circ + b\omega_{BC} = 0.$$

Solving these equations, we determine that the velocity of pin A relative to the body-fixed coordinate system is

$$\begin{aligned}\mathbf{v}_{A\,\text{rel}} &= v_{A\,\text{rel}} \cos 45^\circ \mathbf{i} + v_{A\,\text{rel}} \sin 45^\circ \mathbf{j} \\ &= -\frac{v_0}{2}\mathbf{i} - \frac{v_0}{2}\mathbf{j}\end{aligned}$$

and the angular velocity of bar BC is

$$\omega_{BC} = -\frac{v_0}{2b}.$$

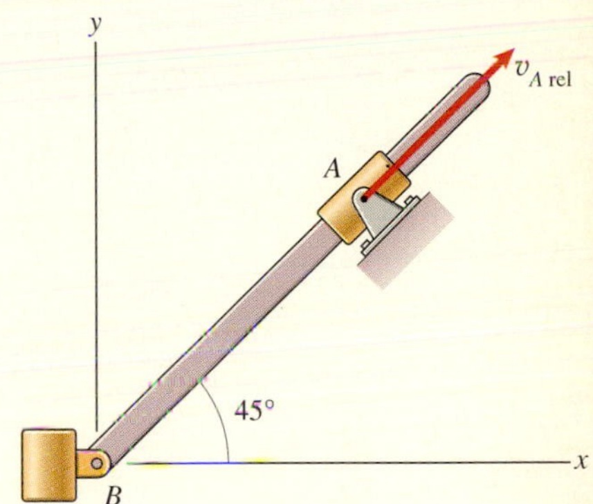

(b) Direction of the velocity of the fixed pin A relative to the body-fixed coordinate system.

Angular Acceleration The acceleration of pin A is zero. From Eq. (6.15),

$$\mathbf{a}_A = \mathbf{0} = \mathbf{a}_B + \mathbf{a}_{A\,\text{rel}} + 2\boldsymbol{\omega} \times \mathbf{v}_{A\,\text{rel}} + \boldsymbol{\alpha} \times \mathbf{r}_{A/B} - \omega^2 \mathbf{r}_{A/B}. \quad (6.17)$$

The acceleration of pin B is $\mathbf{a}_B = -(v_0^2/R)\mathbf{i}$. The acceleration of the pin A relative to the body-fixed coordinate system is parallel to the bar (Fig. c). We can therefore express it as

$$\mathbf{a}_{A\,\text{rel}} = a_{A\,\text{rel}} \cos 45^\circ \mathbf{i} + a_{A\,\text{rel}} \sin 45^\circ \mathbf{j},$$

and write Eq. (6.17) as

$$\begin{aligned}\mathbf{0} = &-\frac{v_0^2}{R}\mathbf{i} + a_{A\,\text{rel}} \cos 45^\circ \mathbf{i} + a_{A\,\text{rel}} \sin 45^\circ \mathbf{j} \\ &+ 2\begin{vmatrix} \mathbf{i} & \mathbf{j} & \mathbf{k} \\ 0 & 0 & \omega_{BC} \\ -v_0/2 & -v_0/2 & 0 \end{vmatrix} + \begin{vmatrix} \mathbf{i} & \mathbf{j} & \mathbf{k} \\ 0 & 0 & \alpha_{BC} \\ b & b & 0 \end{vmatrix} \\ &- \omega_{BC}^2 (b\mathbf{i} + b\mathbf{j}).\end{aligned}$$

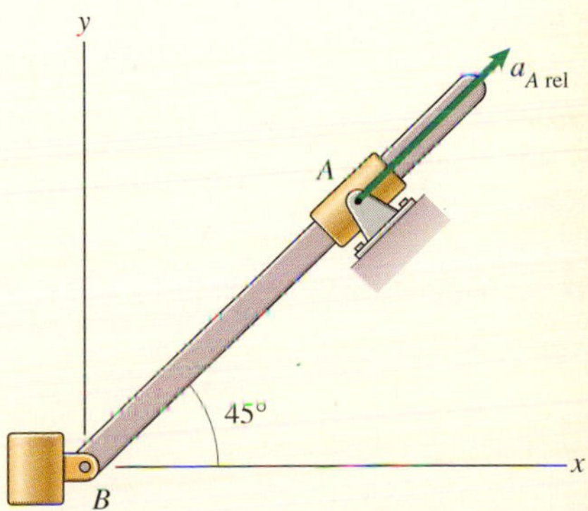

(c) Direction of the acceleration of the fixed pin A relative to the body-fixed coordinate system.

From the **i** and **j** components of this equation, we obtain

$$-\frac{v_0^2}{R} + a_{A\,\text{rel}} \cos 45^\circ + v_0 \omega_{BC} - b\alpha_{BC} - b\omega_{BC}^2 = 0,$$

$$a_{A\,\text{rel}} \sin 45^\circ - v_0 \omega_{BC} + b\alpha_{BC} - b\omega_{BC}^2 = 0.$$

Solving these equations, we determine that the angular acceleration of bar BC is

$$\alpha_{BC} = -\frac{v_0^2}{2b}\left(\frac{1}{R} + \frac{1}{b}\right).$$

Example 6.9

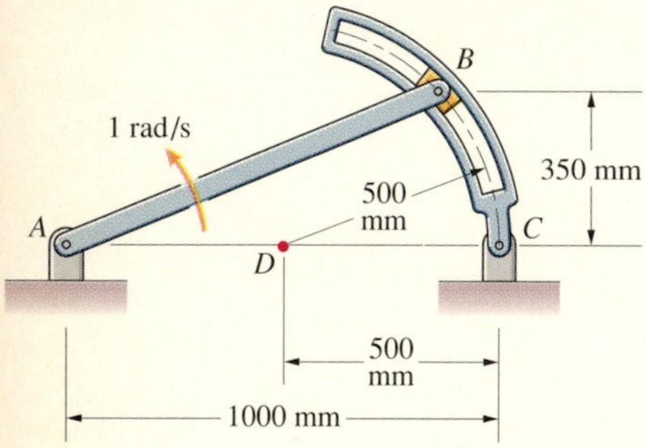

Fig. 6.37

Bar AB in Fig. 6.37 rotates with a constant counterclockwise angular velocity of 1 rad/s. The block B slides in a circular slot in the curved bar BC. At the instant shown, the center of the circular slot is at D. Determine the angular velocity and angular acceleration of bar BC.

STRATEGY

Since we know the angular velocity of bar AB, we can determine the velocity of point B. Because B is not a point of bar BC, we must apply Eq. (6.11) to points B and C. By equating our expressions for $\mathbf{v}_B$, we can solve for the angular velocity of bar BC. Then, by following the same sequence of steps but this time using Eq. (6.15), we can determine the angular acceleration of bar BC.

SOLUTION

To determine the velocity of B, we express it in terms of the velocity of A and the angular velocity of bar AB: $\mathbf{v}_B = \mathbf{v}_A + \boldsymbol{\omega}_{AB} \times \mathbf{r}_{B/A}$. In terms of the coordinate system in Fig. (a), the position vector of B relative to A is

$$\mathbf{r}_{B/A} = (0.500 + 0.500 \cos \beta)\mathbf{i} + 0.350\,\mathbf{j} = 0.857\mathbf{i} + 0.350\mathbf{j} \text{ (m)},$$

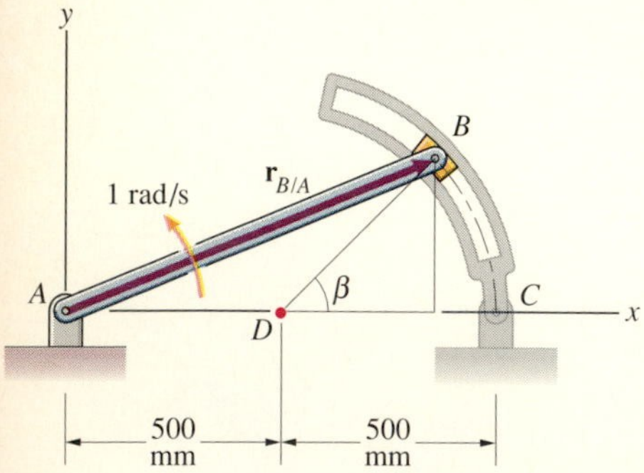

where $\beta = \arcsin(350/500) = 44.4^\circ$. Therefore the velocity of B is

$$\mathbf{v}_B = \mathbf{v}_A + \boldsymbol{\omega}_{AB} \times \mathbf{r}_{B/A} = \mathbf{0} + \begin{vmatrix} \mathbf{i} & \mathbf{j} & \mathbf{k} \\ 0 & 0 & 1 \\ 0.857 & 0.350 & 0 \end{vmatrix} \tag{6.18}$$

$$= -0.350\mathbf{i} + 0.857\mathbf{j} \text{ (m/s)}.$$

(a) Determining the velocity of point B.

To apply Eq. (6.11) to points B and C, we introduce a parallel secondary coordinate system that rotates with the curved bar (Fig. b). The velocity of B is

$$\mathbf{v}_B = \mathbf{v}_C + \mathbf{v}_{B\,\text{rel}} + \boldsymbol{\omega}_{BC} \times \mathbf{r}_{B/C}. \tag{6.19}$$

The position vector of B relative to C is

$$\mathbf{r}_{B/C} = -(0.500 - 0.500 \cos \beta)\mathbf{i} + 0.350\mathbf{j} = -0.143\mathbf{i} + 0.350\mathbf{j} \text{ (m)}.$$

Relative to the body-fixed coordinate system, point B moves in a circular path about point D (Fig. c). In terms of the angle β, the vector $\mathbf{v}_{B\,\text{rel}}$ is

$$\mathbf{v}_{B\,\text{rel}} = -v_{B\,\text{rel}} \sin\beta\mathbf{i} + v_{B\,\text{rel}} \cos\beta\mathbf{j}.$$

We substitute these expressions for $\mathbf{r}_{B/C}$ and $\mathbf{v}_{B\,\text{rel}}$ into Eq. (6.19), obtaining

$$\mathbf{v}_B = -v_{B\,\text{rel}} \sin\beta\mathbf{i} + v_{B\,\text{rel}} \cos\beta\mathbf{j} + \begin{vmatrix} \mathbf{i} & \mathbf{j} & \mathbf{k} \\ 0 & 0 & \omega_{BC} \\ -0.143 & 0.350 & 0 \end{vmatrix}.$$

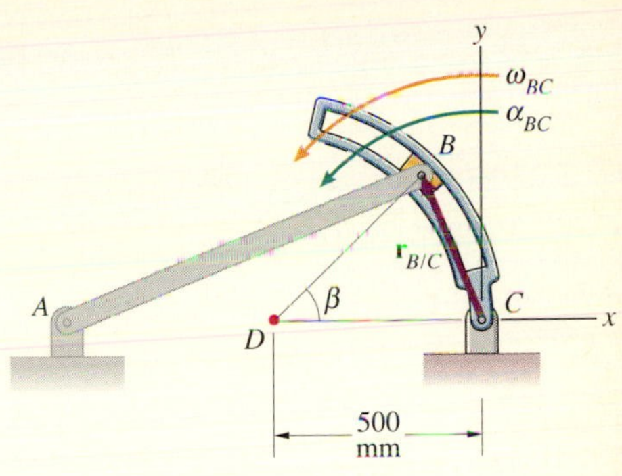

(b) A coordinate system fixed with respect to the curved bar.

Equating this expression for $\mathbf{v}_B$ to its value given in Eq. (6.18) yields the two equations

$$-v_{B\,\text{rel}} \sin\beta - 0.350\omega_{BC} = -0.350,$$

$$v_{B\,\text{rel}} \cos\beta - 0.143\omega_{BC} = 0.857.$$

Solving them, we obtain $v_{B\,\text{rel}} = 1.0$ m/s and $\omega_{BC} = -1.0$ rad/s.

We follow the same sequence of steps to determine the angular acceleration of bar BC. The acceleration of point B is

$$\begin{aligned}\mathbf{a}_B &= \mathbf{a}_A + \boldsymbol{\alpha}_{AB} \times \mathbf{r}_{B/A} - \omega_{AB}^2\mathbf{r}_{B/A} \\ &= \mathbf{0} + \mathbf{0} - (1)^2(0.857\mathbf{i} + 0.350\mathbf{j}) \\ &= -0.857\mathbf{i} - 0.350\mathbf{j}\ (\text{m/s}^2). \end{aligned} \qquad (6.20)$$

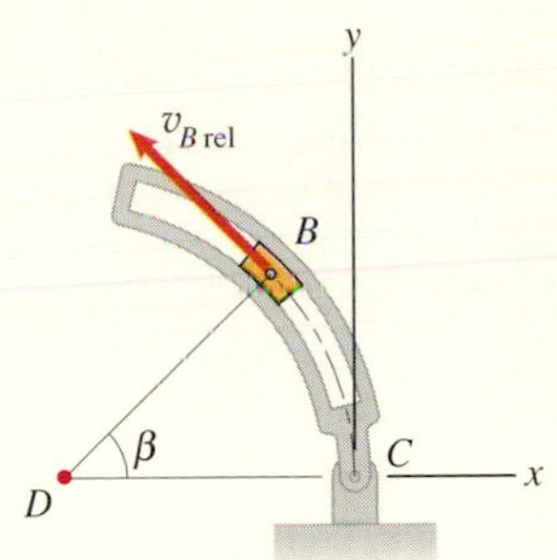

(c) The velocity of B relative to the body-fixed coordinate system.

Because the motion of point B relative to the body-fixed coordinate system is a circular path about point D, there is a tangential component of acceleration, which we denote a_{Bt}, and a normal component of acceleration $v_{B\,\text{rel}}^2/(0.5\text{ m})$. These components are shown in Fig. (d). In terms of the angle β, the vector $\mathbf{a}_{B\,\text{rel}}$ is

$$\begin{aligned}\mathbf{a}_{B\,\text{rel}} &= -a_{Bt} \sin\beta\mathbf{i} + a_{Bt} \cos\beta\mathbf{j} \\ &\quad - (v_{B\,\text{rel}}^2/0.5) \cos\beta\mathbf{i} - (v_{B\,\text{rel}}^2/0.5) \sin\beta\mathbf{j}.\end{aligned}$$

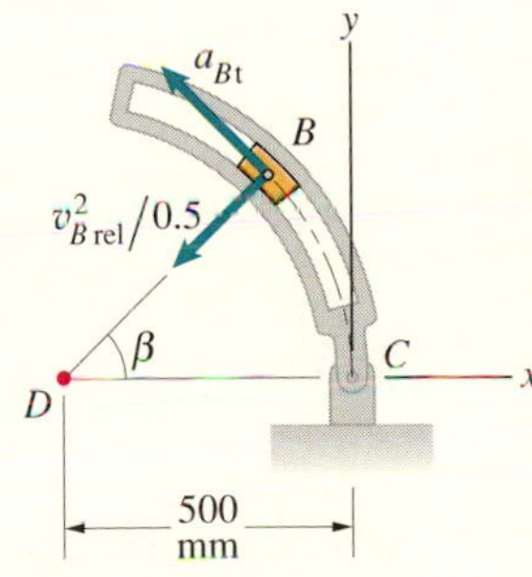

(d) Acceleration of B relative to the body-fixed coordinate system.

Applying Eq. (6.15) to points B and C, the acceleration of B is

$$\begin{aligned}\mathbf{a}_B &= \mathbf{a}_C + \mathbf{a}_{B\,\text{rel}} + 2\boldsymbol{\omega}_{BC} \times \mathbf{v}_{B\,\text{rel}} \\ &\quad + \boldsymbol{\alpha}_{BC} \times \mathbf{r}_{B/C} - \omega_{BC}^2\mathbf{r}_{B/C} \\ &= \mathbf{0} - a_{Bt} \sin\beta\mathbf{i} + a_{Bt} \cos\beta\mathbf{j} \\ &\quad - [(1)^2/0.5] \cos\beta\mathbf{i} - [(1)^2/0.5] \sin\beta\mathbf{j} \\ &\quad + 2\begin{vmatrix} \mathbf{i} & \mathbf{j} & \mathbf{k} \\ 0 & 0 & -1 \\ -(1)\sin\beta & (1)\cos\beta & 0 \end{vmatrix} \\ &\quad + \begin{vmatrix} \mathbf{i} & \mathbf{j} & \mathbf{k} \\ 0 & 0 & \alpha_{BC} \\ -0.143 & 0.350 & 0 \end{vmatrix} - (-1)^2(-0.143\mathbf{i} + 0.350\mathbf{j}).\end{aligned}$$

Equating this expression for $\mathbf{a}_B$ to its value given in Eq. (6.20) yields the two equations

$$-a_{Bt} \sin\beta - 0.350\alpha_{BC} + 0.143 = -0.857,$$

$$a_{Bt} \cos\beta - 0.143\alpha_{BC} - 0.350 = -0.350.$$

Solving them, we obtain $a_{Bt} = 0.408$ m/s^2 and $\alpha_{BC} = 2.040$ rad/s^2.

Problems

6.117 The bar rotates with a constant counterclockwise angular velocity of 10 rad/s, and the sleeve A slides at 4 ft/s relative to the bar. Use Eq. (6.11) and the body-fixed coordinate system shown to determine the velocity of A.

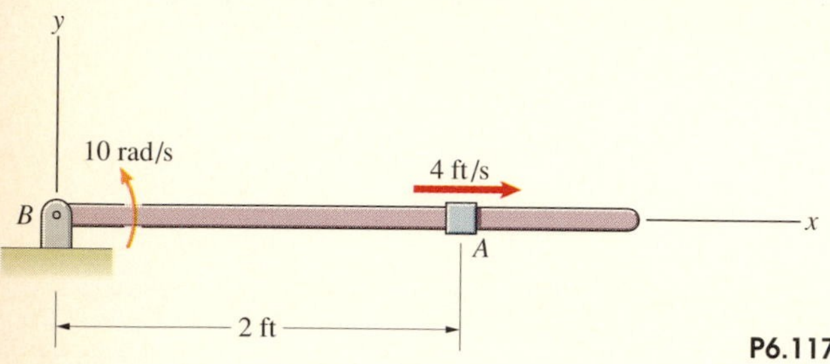

P6.117

6.118 The sleeve A in Problem 6.117 slides relative to the bar at a constant velocity of 4 ft/s. Use Eq. (6.15) to determine the acceleration of A.

6.119 The sleeve C slides at 1 m/s relative to bar BD. Use the body-fixed coordinate system shown to determine the velocity of C.

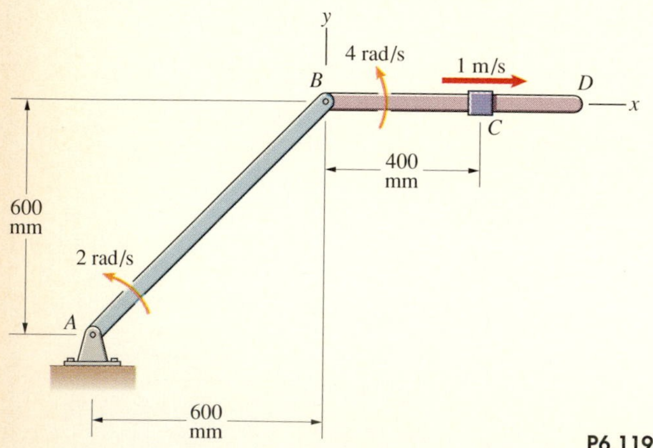

P6.119

6.120 In Problem 6.119, the angular accelerations of the two bars are zero and the sleeve C slides at a constant velocity of 1 m/s relative to bar BD. What is the acceleration of the sleeve C?

6.121 Bar AC has an angular velocity of 2 rad/s in the counterclockwise direction that is decreasing at 4 rad/s^2. The pin at C slides in the slot in bar BD.

(a) Determine the angular velocity of bar BD and the velocity of the pin relative to the slot.

(b) Determine the angular acceleration of bar BD and the acceleration of the pin relative to the slot.

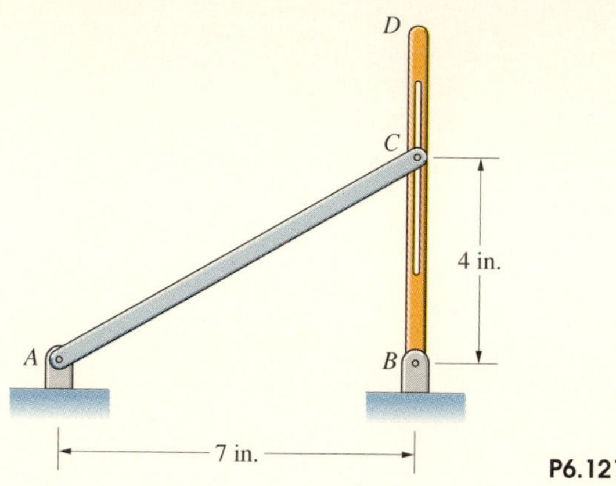

P6.121

6.122 In the system shown in Problem 6.121, the velocity of the pin C relative to the slot is 21 in./s upward and is decreasing at 42 in./s^2. What are the angular velocity and acceleration of bar AC?

6.123 In the system shown in Problem 6.121, what should the angular velocity and acceleration of bar AC be if you want the angular velocity and acceleration of bar BD to be 4 rad/s counterclockwise and 24 rad/s^2 counterclockwise, respectively?

6.124 Bar AB has an angular velocity of 4 rad/s in the clockwise direction. What is the velocity of the pin B relative to the slot?

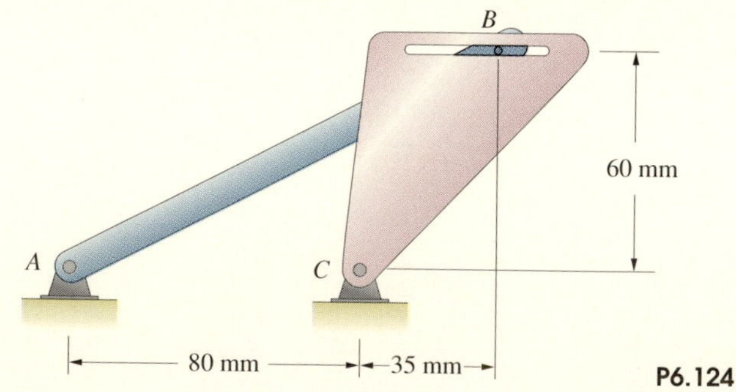

P6.124

6.125 In the system shown in Problem 6.124, bar AB has an angular velocity of 4 rad/s in the clockwise direction and an angular acceleration of 10 rad/s^2 in the counterclockwise direction. What is the acceleration of the pin B relative to the slot?

6.126 Arm AB is rotating at 4 rad/s in the clockwise direction. Determine the angular velocity of arm BC and the velocity of point B relative to the slot in arm BC.

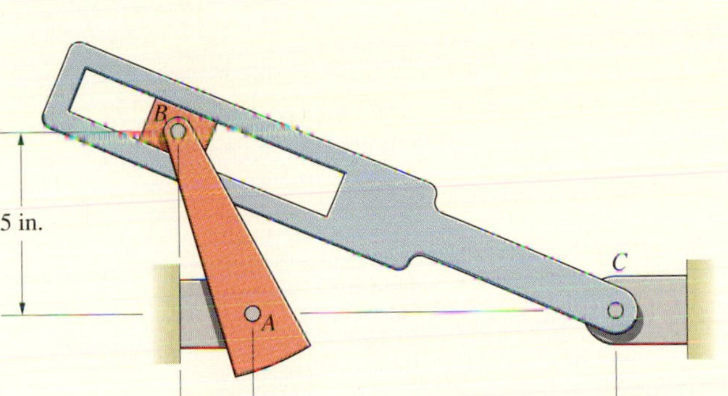

P6.126

6.127 Arm AB in Problem 6.126 is rotating with a constant angular velocity of 4 rad/s in the clockwise direction. Determine the angular acceleration of arm BC and the acceleration of point B relative to the slot in arm BC.

6.128 The angular velocity $\omega_{AC} = 5°$ per second. Determine the angular velocity of the hydraulic actuator BC and the rate at which it is extending.

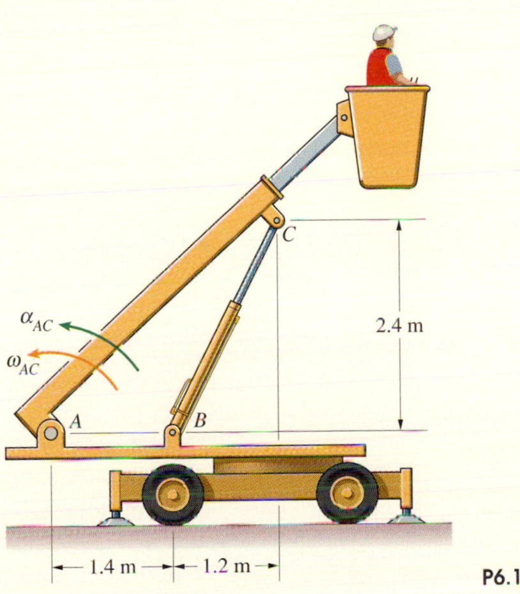

P6.128

6.129 In Problem 6.128, if the angular velocity $\omega_{AC} = 5°$ per second and the angular acceleration $\alpha_{AC} = -2°$ per second squared, determine the angular acceleration of the hydraulic actuator BC and the rate of change of its rate of extension.

6.130 The sleeve at A slides upward at a constant velocity of 10 m/s. Bar AC slides through the sleeve at B. Determine the angular velocity of bar AC and the velocity at which it slides relative to the sleeve at B.

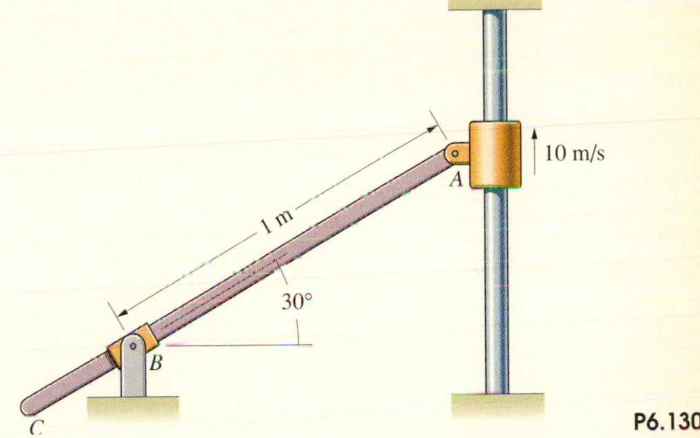

P6.130

6.131 In Problem 6.130, the sleeve at A slides upward at a constant velocity of 10 m/s. Determine the angular acceleration of bar AC and the rate of change of the velocity at which it slides relative to the sleeve at B.

6.132 Block A slides up the inclined surface at 2 ft/s. Determine the angular velocity of bar AC and the velocity of point C.

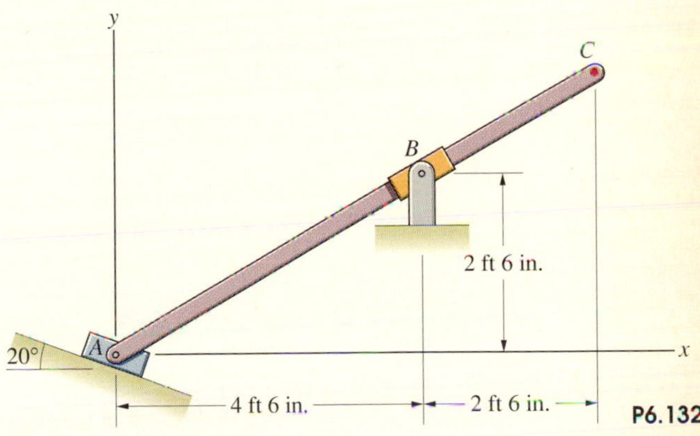

P6.132

6.133 In Problem 6.132, block A slides up the inclined surface at a constant velocity of 2 ft/s. Determine the angular acceleration of bar AC and the acceleration of point C.

6.134 The angular velocity of the scoop is 1.0 rad/s clockwise. Determine the rate at which the hydraulic actuator AB is extending.

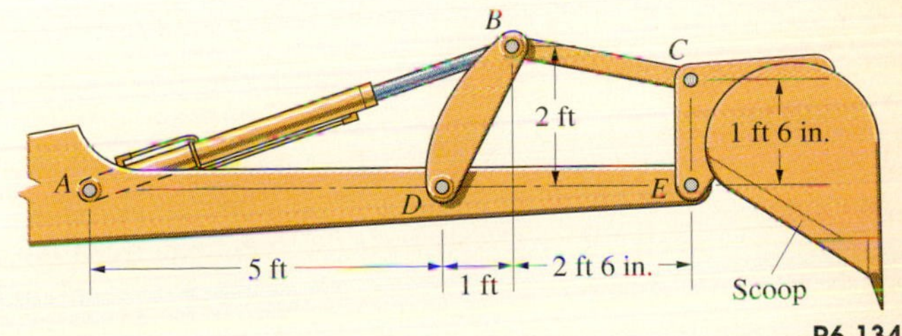

P6.134

6.135 The angular acceleration of the scoop in Problem 6.134 is zero. Determine the rate of change of the rate at which the hydraulic actuator *AB* is extending.

6.136 Suppose that the curved bar in Example 6.9 rotates with a counterclockwise angular velocity of 2 rad/s.
(a) What is the angular velocity of bar *AB*?
(b) What is the velocity of block *B* relative to the slot?

6.137 Suppose that the curved bar in Example 6.9 has a clockwise angular velocity of 4 rad/s and a counterclockwise angular acceleration of 10 rad/s^2. What is the angular acceleration of bar *AB*?

6.138 The disk rolls on the plane surface with a counterclockwise angular velocity of 10 rad/s. Bar *AB* slides on the surface of the disk at *A*. Determine the angular velocity of bar *AB*.

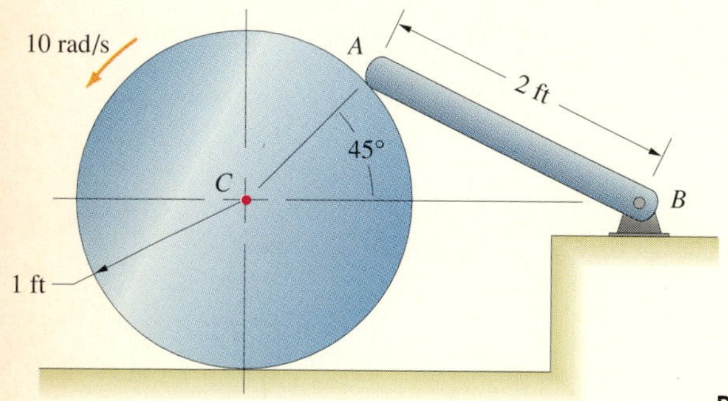

P6.138

6.139 In Problem 6.138, the disk rolls on the plane surface with a constant counterclockwise angular velocity of 10 rad/s. Determine the angular acceleration of bar *AB*.

6.140 Bar *BC* rotates with a counterclockwise angular velocity of 2 rad/s. A pin at *B* slides in a circular slot in the rectangular plate. Determine the angular velocity of the plate and the velocity at which the pin slides relative to the circular slot.

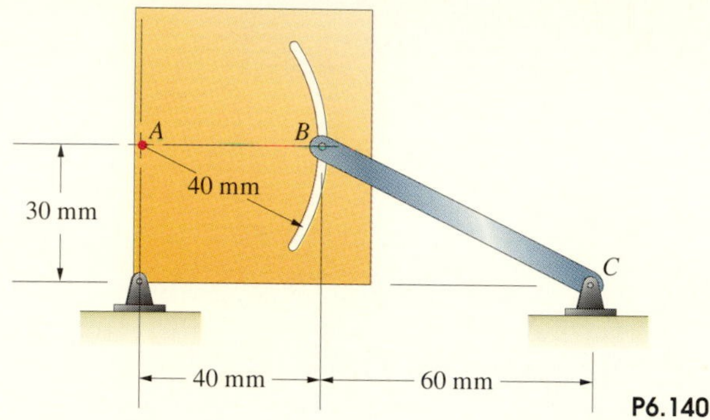

P6.140

6.141 Bar *BC* in Problem 6.140 rotates with a constant counterclockwise angular velocity of 2 rad/s. Determine the angular acceleration of the plate.

6.142 By taking the time derivative of Eq. (6.11) and using Eq. (6.12), derive Eq. (6.13).

6.6 *Moving Reference Frames*

In this section we revisit the subjects of Chapters 2 and 3—the motion of a point and Newton's second law. In many situations it is convenient to describe the motion of a point using a secondary reference frame that moves relative to some primary reference frame. For example, to measure the motion of a point relative to a moving vehicle, you would choose a secondary reference frame that is fixed with respect to the vehicle. Here we show how the velocity and acceleration of a point relative to a primary reference frame are related to their values relative to a moving secondary reference frame. We also discuss how to apply Newton's second law using moving reference frames. In Chapter 3 we mentioned the example of playing tennis on the deck of a cruise ship. If the ship translates with constant velocity, you can use the equation $\Sigma \mathbf{F} = m\mathbf{a}$ expressed in terms of a reference frame fixed with respect to the ship to analyze the ball's motion. You cannot do so if the ship is turning, or changing its speed. However, you *can* apply the second law using reference frames that accelerate and rotate by properly accounting for the acceleration and rotation. Here we explain how this is done.

Motion of a Point Relative to a Moving Reference Frame

Equations (6.11) and (6.13) give the velocity and acceleration of an arbitrary point A relative to a point B of a rigid body in terms of a body-fixed secondary reference frame:

$$\mathbf{v}_A = \mathbf{v}_B + \mathbf{v}_{A\,\text{rel}} + \boldsymbol{\omega} \times \mathbf{r}_{A/B}, \tag{6.21}$$

$$\mathbf{a}_A = \mathbf{a}_B + \mathbf{a}_{A\,\text{rel}} + 2\boldsymbol{\omega} \times \mathbf{v}_{A\,\text{rel}} + \boldsymbol{\alpha} \times \mathbf{r}_{A/B} + \boldsymbol{\omega} \times (\boldsymbol{\omega} \times \mathbf{r}_{A/B}). \tag{6.22}$$

But these results don't require us to assume that the secondary reference frame is connected to some rigid body. They apply to any reference frame having a moving origin B and rotating with angular velocity $\boldsymbol{\omega}$ and angular acceleration $\boldsymbol{\alpha}$ relative to a primary reference frame (Fig. 6.38). The terms $\mathbf{v}_A$ and $\mathbf{a}_A$ are the velocity and acceleration of A relative to the primary reference frame. The terms $\mathbf{v}_{A\,\text{rel}}$ and $\mathbf{a}_{A\,\text{rel}}$ are the velocity and acceleration of A relative to the secondary reference frame. That is, they are the velocity and acceleration measured by an "observer" moving with the secondary reference frame (Fig. 6.39).

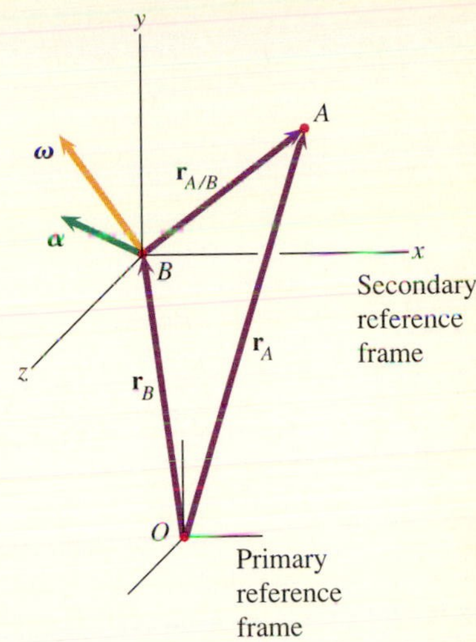

Fig. 6.38
A secondary reference frame with origin B and an arbitrary point A.

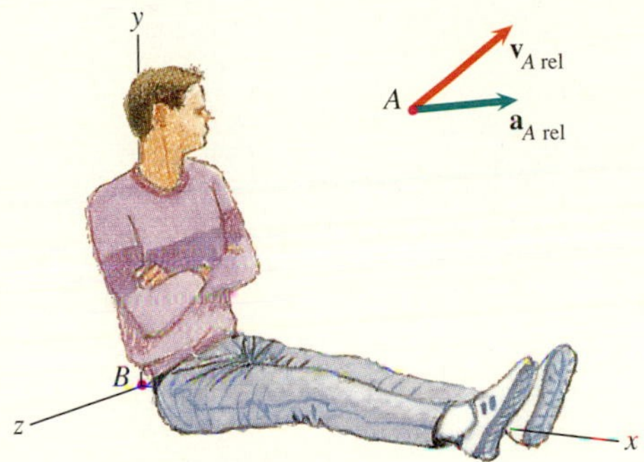

Fig. 6.39
Imagine yourself to be stationary relative to the secondary reference frame.

The following examples demonstrate applications of moving reference frames. If you know the motion of a point A relative to a moving secondary reference frame, you can use Eqs. (6.21) and (6.22) to determine its velocity and acceleration $\mathbf{v}_A$ and $\mathbf{a}_A$ relative to the primary reference frame. In other situations, you will know $\mathbf{v}_A$ and $\mathbf{a}_A$ and will want to use Eqs. (6.21) and (6.22) to determine the velocity and acceleration of A relative to the secondary reference frame.

Example 6.10

The merry-go-round in Fig. 6.40 rotates with constant counterclockwise angular velocity ω. Suppose that you are in the center at B and observe the motion of a second person A, using a coordinate system that rotates with the merry-go-round. Consider two cases.

Case 1 Person A is not on the merry-go-round but stands on the ground next to it. At the instant shown, what are his velocity and acceleration relative to your coordinate system?

Case 2 Person A is on the edge of the merry-go-round and moves with it. What are his velocity and acceleration relative to the earth?

Fig. 6.40

STRATEGY

This simple example clarifies the distinction between the terms $\mathbf{v}_A$, $\mathbf{a}_A$ and the terms $\mathbf{v}_{A\,\mathrm{rel}}$, $\mathbf{a}_{A\,\mathrm{rel}}$ in Eqs. (6.21) and (6.22). We choose the coordinate system that rotates with the merry-go-round as the secondary reference frame, and let the primary reference frame be fixed with respect to the earth. In case 1, A's velocity and acceleration relative to the earth, $\mathbf{v}_A$ and $\mathbf{a}_A$, are known: He is standing still. We can use Eqs. (6.21) and (6.22) to determine $\mathbf{v}_{A\,\mathrm{rel}}$ and $\mathbf{a}_{A\,\mathrm{rel}}$, which are his velocity and acceleration relative to your rotating coordinate system. In case 2, $\mathbf{v}_{A\,\mathrm{rel}}$ and $\mathbf{a}_{A\,\mathrm{rel}}$ are known: A is stationary relative to your coordinate system. We can use Eqs. (6.21) and (6.22) to determine $\mathbf{v}_A$ and $\mathbf{a}_A$.

SOLUTION

Case 1 *A* is standing on the ground, so his velocity relative to the earth is $\mathbf{v}_A = \mathbf{0}$. The angular velocity vector of your coordinate system is $\boldsymbol{\omega} = \omega\mathbf{k}$, and at the instant shown $\mathbf{r}_{A/B} = R\mathbf{i}$. From Eq. (6.21),

$$\mathbf{v}_A = \mathbf{v}_B + \mathbf{v}_{A\,\text{rel}} + \boldsymbol{\omega} \times \mathbf{r}_{A/B}:$$
$$\mathbf{0} = \mathbf{0} + \mathbf{v}_{A\,\text{rel}} + (\omega\mathbf{k}) \times (R\mathbf{i}).$$

We find that $\mathbf{v}_{A\,\text{rel}} = -\omega R\mathbf{j}$. Although *A* is stationary relative to the earth, $\mathbf{v}_{A\,\text{rel}}$ is not zero. What does this term represent? As you sit at the center of the merry-go-round, you see *A* moving around you in a circular path. *Relative to your rotating coordinate system, A* moves in a circular path of radius *R* in the clockwise direction with a velocity of constant magnitude ωR. At the instant shown, *A*'s velocity relative to your coordinate system is $-\omega R\mathbf{j}$.

You know that a point moving in a circular path of radius *R* with velocity v has a normal component of acceleration equal to v^2/R. Relative to your coordinate system, person *A* moves in a circular path of radius *R* with velocity ωR. Therefore, *relative to your coordinate system, A* has a normal component of acceleration $(\omega R)^2/R = \omega^2 R$. At the instant shown, the normal acceleration points in the negative *x* direction. Therefore we conclude that *A*'s acceleration relative to your coordinate system is $\mathbf{a}_{A\,\text{rel}} = -\omega^2 R\mathbf{i}$.

We can confirm this result with Eq. (6.22). *A*'s acceleration relative to the earth is $\mathbf{a}_A = \mathbf{0}$. The angular velocity vector of the coordinate system is constant, so $\boldsymbol{\alpha} = \mathbf{0}$. From Eq. (6.22),

$$\mathbf{a}_A = \mathbf{a}_B + \mathbf{a}_{A\,\text{rel}} + 2\boldsymbol{\omega} \times \mathbf{v}_{A\,\text{rel}} + \boldsymbol{\alpha} \times \mathbf{r}_{A/B} + \boldsymbol{\omega} \times (\boldsymbol{\omega} \times \mathbf{r}_{A/B}):$$
$$\mathbf{0} = \mathbf{0} + \mathbf{a}_{A\,\text{rel}} + 2(\omega\mathbf{k}) \times (-\omega R\mathbf{j}) + \mathbf{0} + (\omega\mathbf{k}) \times [(\omega\mathbf{k}) \times R\mathbf{i})].$$

Solving this equation for $\mathbf{a}_{A\,\text{rel}}$, we obtain $\mathbf{a}_{A\,\text{rel}} = -\omega^2 R\mathbf{i}$. *A*'s velocity and acceleration relative to your coordinate system are shown in Fig. (a).

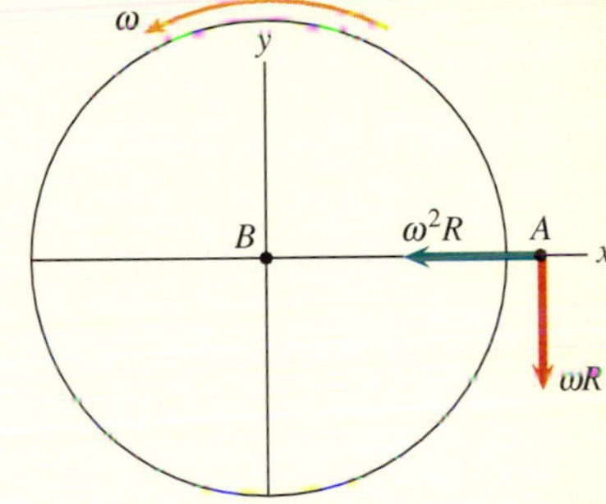

(a) The velocity and acceleration of *A* relative to the rotating coordinate system in case 1.

Case 2 *Relative to your coordinate system, A* is stationary, so $\mathbf{v}_{A\,\text{rel}} = \mathbf{0}$ and $\mathbf{a}_{A\,\text{rel}} = \mathbf{0}$. From Eq. (6.21), *A*'s velocity relative to the earth is

$$\mathbf{v}_A = \mathbf{v}_B + \mathbf{v}_{A\,\text{rel}} + \boldsymbol{\omega} \times \mathbf{r}_{A/B} = \mathbf{0} + \mathbf{0} + (\omega\mathbf{k}) \times (R\mathbf{i})$$
$$= \omega R\mathbf{j}.$$

In this case, *A* is moving in a circular path of radius *R* with a velocity of constant magnitude ωR relative to the earth.

From Eq. (6.22), *A*'s acceleration relative to the earth is

$$\mathbf{a}_A = \mathbf{a}_B + \mathbf{a}_{A\,\text{rel}} + 2\boldsymbol{\omega} \times \mathbf{v}_{A\,\text{rel}} + \boldsymbol{\alpha} \times \mathbf{r}_{A/B} + \boldsymbol{\omega} \times (\boldsymbol{\omega} \times \mathbf{r}_{A/B})$$
$$= \mathbf{0} + \mathbf{0} + \mathbf{0} + \mathbf{0} + (\omega\mathbf{k}) \times [(\omega\mathbf{k}) \times R\mathbf{i})]$$
$$= -\omega^2 R\mathbf{i}.$$

This is *A*'s acceleration relative to the earth due to his circular motion. *A*'s velocity and acceleration relative to the earth are shown in Fig. (b).

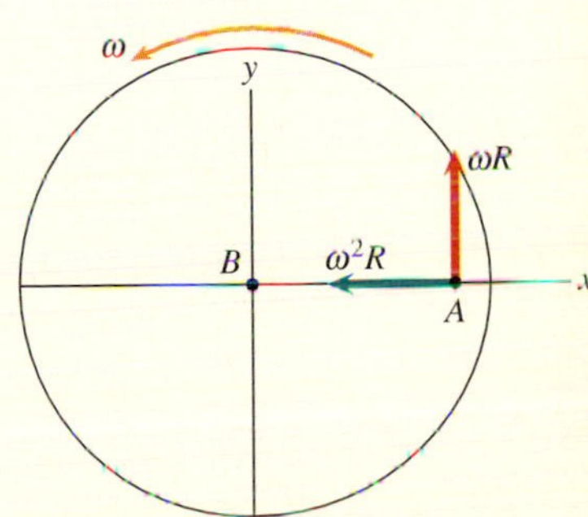

(b) The velocity and acceleration of *A* relative to the earth in case 2.

Example 6.11

At the instant shown, the ship in Fig. 6.41 is moving north at a constant speed of 15.0 m/s relative to the earth and is turning toward the west at a constant rate of 5.0° per second. Relative to the ship's body-fixed coordinate system, its radar indicates that the position, velocity, and acceleration of the helicopter are

$$\mathbf{r}_{A/B} = 420.0\mathbf{i} + 236.2\,\mathbf{j} + 212.0\mathbf{k}\ (\text{m}),$$

$$\mathbf{v}_{A\,\text{rel}} = -53.5\mathbf{i} + 2.0\mathbf{j} + 6.6\mathbf{k}\ (\text{m/s}),$$

$$\mathbf{a}_{A\,\text{rel}} = 0.4\mathbf{i} - 0.2\mathbf{j} - 13.0\mathbf{k}\ (\text{m/s}^2).$$

What are the helicopter's velocity and acceleration relative to the earth?

Fig. 6.41

STRATEGY

We are given the ship's velocity and enough information to determine its acceleration, angular velocity, and angular acceleration relative to the earth. We also know the position, velocity, and acceleration of the helicopter relative to the secondary body-fixed coordinate system. Therefore we can use Eqs. (6.21) and (6.22) to determine the helicopter's velocity and acceleration relative to the earth.

SOLUTION

In terms of the body-fixed coordinate system, the ship's velocity is $\mathbf{v}_B = 15.0\mathbf{i}$ (m/s). The ship's angular velocity due to its rate of turning is $\omega = (5.0/180)\pi = 0.0873$ rad/s. The ship is rotating about the y axis. Pointing the arc of the fingers of the right hand around the y axis in the direction of the ship's rotation, the thumb points in the positive y direction, so the ship's angular velocity vector is $\boldsymbol{\omega} = 0.0873\mathbf{j}$ (rad/s). The helicopter's velocity relative to the earth is

$$\mathbf{v}_A = \mathbf{v}_B + \mathbf{v}_{A\,\text{rel}} + \boldsymbol{\omega} \times \mathbf{r}_{A/B}$$

$$= 15.0\mathbf{i} + (-53.5\mathbf{i} + 2.0\mathbf{j} + 6.6\mathbf{k}) + \begin{vmatrix} \mathbf{i} & \mathbf{j} & \mathbf{k} \\ 0 & 0.0873 & 0 \\ 420.0 & 236.2 & 212.0 \end{vmatrix}$$

$$= -20.0\mathbf{i} + 2.0\mathbf{j} - 30.1\mathbf{k}\ \text{(m/s)}.$$

We can determine the ship's acceleration by expressing it in terms of normal and tangential components in the form given by Eq. (2.37) (Fig. a):

$$\mathbf{a}_B = \frac{dv}{dt}\mathbf{e}_\text{t} + v\frac{d\theta}{dt}\mathbf{e}_\text{n} = \mathbf{0} + (15)(0.0873)\mathbf{e}_\text{n}$$
$$= 1.31\ \mathbf{e}_\text{n}\ (\text{m/s}^2).$$

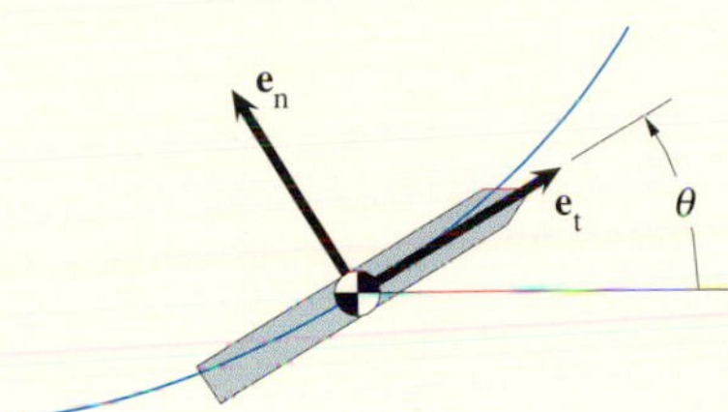

(a) Determining the ship's acceleration.

The z axis is perpendicular to the ship's path and points toward the convex side of the path (Fig. b). Therefore, in terms of the body-fixed coordinate system, the ship's acceleration is $\mathbf{a}_B = -1.31\mathbf{k}$ (m/s^2). The ship's angular velocity vector is constant, so $\boldsymbol{\alpha} = \mathbf{0}$. The helicopter's acceleration relative to the earth is

$$\mathbf{a}_A = \mathbf{a}_B + \mathbf{a}_{A\,\text{rel}} + 2\boldsymbol{\omega} \times \mathbf{v}_{A\,\text{rel}} + \boldsymbol{\alpha} \times \mathbf{r}_{A/B}$$
$$+ \boldsymbol{\omega} \times (\boldsymbol{\omega} \times \mathbf{r}_{A/B})$$

$$= -1.31\mathbf{k} + (0.4\mathbf{i} - 0.2\mathbf{j} - 13.0\mathbf{k}) + 2\begin{vmatrix} \mathbf{i} & \mathbf{j} & \mathbf{k} \\ 0 & 0.0873 & 0 \\ -53.5 & 2.0 & 6.6 \end{vmatrix}$$

$$+ \mathbf{0} + (0.0873\mathbf{j}) \times \begin{vmatrix} \mathbf{i} & \mathbf{j} & \mathbf{k} \\ 0 & 0.0873 & 0 \\ 420.0 & 236.2 & 212.0 \end{vmatrix}$$

$$= -1.65\mathbf{i} - 0.20\mathbf{j} - 6.59\mathbf{k}\ (\text{m/s}^2).$$

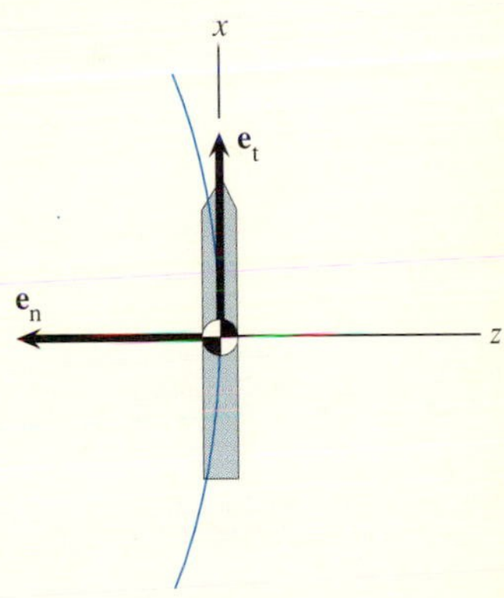

(b) Correspondence between the normal and tangential components and the body-fixed coordinate system.

DISCUSSION

Notice the substantial differences between the helicopter's velocity and acceleration relative to the earth and the values the ship measures using its body-fixed coordinate system.

Example 6.12

The satellite A shown in Fig. 6.42 is in a circular polar orbit (an orbit that intersects the earth's axis of rotation). Relative to a nonrotating primary reference frame with its origin at the center of the earth, the satellite moves in a circular path of radius R with a velocity of constant magnitude v_A. At the present instant, the satellite is above the equator. The secondary earth-fixed reference frame shown is oriented with the y axis in the direction of the north pole and the x axis in the direction of the satellite. What are the satellite's velocity and acceleration relative to the earth-fixed reference frame? Let ω_E be the angular velocity of the earth.

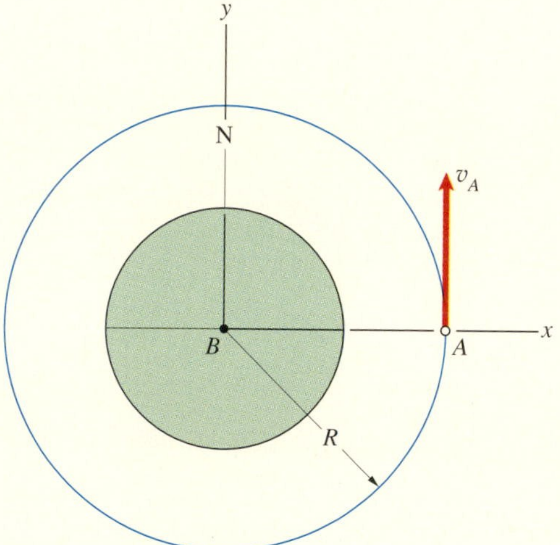

Fig. 6.42

STRATEGY

We are given enough information to determine the satellite's velocity and acceleration $\mathbf{v}_A$ and $\mathbf{a}_A$ relative to the nonrotating primary reference frame and the angular velocity vector $\boldsymbol{\omega}$ of the secondary reference frame. We can therefore use Eqs. (6.21) and (6.22) to determine the satellite's velocity and acceleration $\mathbf{v}_{A\,\mathrm{rel}}$ and $\mathbf{a}_{A\,\mathrm{rel}}$ relative to the earth-fixed reference frame.

SOLUTION

At the present instant, the satellite's velocity and acceleration relative to a nonrotating primary reference frame with its origin at the center of the earth are $\mathbf{v}_A = v_A\mathbf{j}$ and $\mathbf{a}_A = -(v_A^2/R)\mathbf{i}$. The angular velocity vector of the earth points north (confirm this using the right-hand rule), so the angular velocity of the earth-fixed reference frame is $\boldsymbol{\omega} = \omega_E\,\mathbf{j}$. From Eq. (6.21),

$$\mathbf{v}_A = \mathbf{v}_B + \mathbf{v}_{A\,\mathrm{rel}} + \boldsymbol{\omega} \times \mathbf{r}_{A/B}:$$

$$v_A\mathbf{j} = \mathbf{0} + \mathbf{v}_{A\,\mathrm{rel}} + \begin{vmatrix} \mathbf{i} & \mathbf{j} & \mathbf{k} \\ 0 & \omega_E & 0 \\ R & 0 & 0 \end{vmatrix}.$$

Solving for $\mathbf{v}_{A\,\mathrm{rel}}$, we find that the satellite's velocity relative to the earth-fixed reference frame is

$$\mathbf{v}_{A\,\mathrm{rel}} = v_A\mathbf{j} + R\omega_E\mathbf{k}.$$

The second term on the right side of this equation is the satellite's velocity toward the west relative to the rotating earth-fixed reference frame.

From Eq. (6.22),

$$\mathbf{a}_A = \mathbf{a}_B + \mathbf{a}_{A\,\mathrm{rel}} + 2\boldsymbol{\omega} \times \mathbf{v}_{A\,\mathrm{rel}} + \boldsymbol{\alpha} \times \mathbf{r}_{A/B} + \boldsymbol{\omega} \times (\boldsymbol{\omega} \times \mathbf{r}_{A/B}):$$

$$-\frac{v_A^2}{R}\mathbf{i} = \mathbf{0} + \mathbf{a}_{A\,\mathrm{rel}} + 2\begin{vmatrix} \mathbf{i} & \mathbf{j} & \mathbf{k} \\ 0 & \omega_E & 0 \\ 0 & v_A & R\omega_E \end{vmatrix} + \mathbf{0} + \begin{vmatrix} \mathbf{i} & \mathbf{j} & \mathbf{k} \\ 0 & \omega_E & 0 \\ 0 & 0 & -R\omega_E \end{vmatrix}.$$

Solving for $\mathbf{a}_{A\,\mathrm{rel}}$, we find the satellite's acceleration relative to the earth-fixed reference frame to be

$$\mathbf{a}_{A\,\mathrm{rel}} = -\left(\frac{v_A^2}{R} + \omega_E^2 R\right)\mathbf{i}.$$

DISCUSSION

Instruments on earth observing the satellite's motion would measure its velocity and acceleration relative to the earth, $\mathbf{v}_{A\,\mathrm{rel}}$ and $\mathbf{a}_{A\,\mathrm{rel}}$. Equations (6.21) and (6.22) could then be used to calculate $\mathbf{v}_A$ and $\mathbf{a}_A$, the satellite's velocity and acceleration relative to a nonrotating reference frame.

Inertial Reference Frames

We say that a reference frame is inertial if you can use it to apply Newton's second law in the form $\Sigma\,\mathbf{F} = m\mathbf{a}$. Why can you usually assume that an earth-fixed reference frame is inertial, even though it both accelerates and rotates? How can you apply Newton's second law using a reference frame that is fixed with respect to an accelerating, turning ship or airplane? We are now in a position to answer these questions.

Earth-Centered, Nonrotating Reference Frame We begin by showing why a nonrotating reference frame fixed relative to the center of the earth can be assumed to be inertial for the purpose of describing motions of objects near the earth. Figure 6.43(a) shows a hypothetical nonaccelerating, nonrotating reference frame with origin O, and a secondary nonrotating, **earth-centered reference frame**. The earth, and therefore the earth-centered reference frame, accelerates due to the gravitational attractions of the sun, moon, etc. We denote the earth's acceleration by the vector $\mathbf{g}_B$.

Suppose that we want to determine the motion of an object A of mass m (Fig. 6.43b). A is also subject to the gravitational attractions of the sun, moon, etc., and we denote the resulting gravitational acceleration by the vector $\mathbf{g}_A$. The vector $\Sigma\,\mathbf{F}$ is the sum of all *other* external forces acting on A, including the gravitational force exerted on it by the earth. The total external force acting on A is $\Sigma\,\mathbf{F} + m\mathbf{g}_A$. We can apply Newton's second law to A, using our hypothetical inertial reference frame:

$$\Sigma\,\mathbf{F} + m\mathbf{g}_A = m\mathbf{a}_A, \tag{6.23}$$

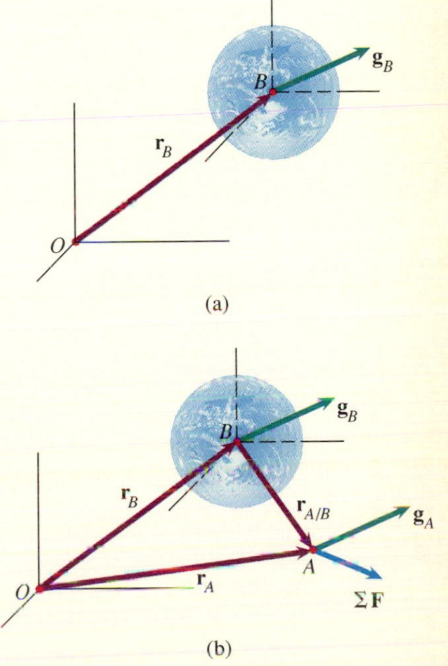

Fig. 6.43
(a) An inertial reference frame and a nonrotating reference frame with its origin at the center of the earth.
(b) Determining the motion of an object A.

where $\mathbf{a}_A$ is the acceleration of A relative to O. Since the earth-centered reference frame does not rotate, we can use Eq. (6.22) to write $\mathbf{a}_A$ as

$$\mathbf{a}_A = \mathbf{a}_B + \mathbf{a}_{A\,\mathrm{rel}},$$

where $\mathbf{a}_{A\,\mathrm{rel}}$ is the acceleration of A relative to the earth-centered reference frame. Using this relation and our definition of the earth's acceleration $\mathbf{a}_B = \mathbf{g}_B$, Eq. (6.23) becomes

$$\Sigma\,\mathbf{F} = m\mathbf{a}_{A\,\mathrm{rel}} + m(\mathbf{g}_B - \mathbf{g}_A). \tag{6.24}$$

If the object A is on or near the earth, its gravitational acceleration $\mathbf{g}_A$ due to the attraction of the sun, etc., is very nearly equal to the earth's gravitational acceleration $\mathbf{g}_B$. If we neglect the difference, Eq. (6.24) becomes

$$\Sigma\,\mathbf{F} = m\mathbf{a}_{A\,\mathrm{rel}}. \tag{6.25}$$

Thus you can apply Newton's second law using a nonrotating, earth-centered reference frame. Even though this reference frame accelerates, *virtually the same gravitational acceleration acts on the object.* Notice that this argument does not hold if the object is not near the earth. If you wanted to analyze the motion of a spacecraft traveling to another planet, for example, you would need to use a nonrotating, sun-centered reference frame.

Earth-Fixed Reference Frame For "down to earth" applications, the most convenient reference frame is a local, **earth-fixed reference frame.** Why can we usually assume that an earth-fixed reference frame is inertial? Figure 6.44 shows a nonrotating reference frame with its origin at the center of the earth O and a secondary earth-fixed reference frame with its origin at a point B. Since we can assume that the earth-centered, nonrotating reference frame is inertial, we can write Newton's second law for an object A of mass m as

$$\Sigma\,\mathbf{F} = m\mathbf{a}_A, \tag{6.26}$$

where $\mathbf{a}_A$ is A's acceleration relative to O. The earth-fixed reference frame rotates with the angular velocity of the earth, which we denote by $\boldsymbol{\omega}_{\mathrm{E}}$. We can use Eq. (6.22) to write Eq. (6.26) in the form

$$\Sigma\,\mathbf{F} = m\mathbf{a}_{A\,\mathrm{rel}} + m[\mathbf{a}_B + 2\boldsymbol{\omega}_{\mathrm{E}} \times \mathbf{v}_{A\,\mathrm{rel}} + \boldsymbol{\omega}_{\mathrm{E}} \times (\boldsymbol{\omega}_{\mathrm{E}} \times \mathbf{r}_{A/B})], \tag{6.27}$$

where $\mathbf{a}_{A\,\mathrm{rel}}$ is A's acceleration relative to the earth-fixed reference frame. If we can neglect the terms in brackets on the right side of Eq. (6.27), the earth-fixed reference frame is inertial. Let's consider each term. (Recall from the definition of the cross product that $|\mathbf{U} \times \mathbf{V}| = |\mathbf{U}|\,|\mathbf{V}|\sin\theta$, where θ is the angle between the two vectors. Therefore the magnitude of the cross product is bounded by the product of the magnitudes of the vectors.)

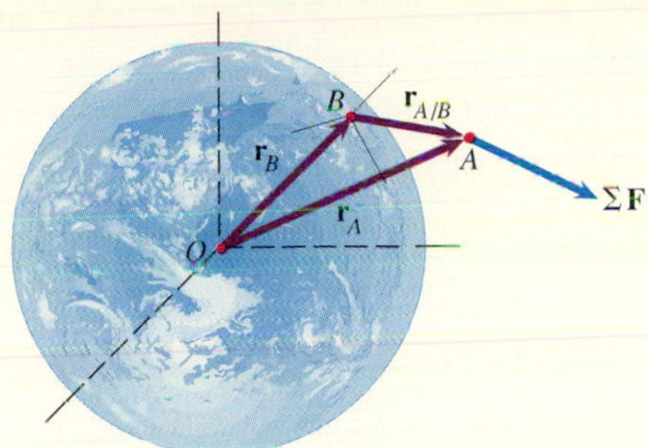

Fig. 6.44

An earth-centered, nonrotating reference frame (origin O), an earth-fixed reference frame (origin B), and an object A.

- *The term* $\boldsymbol{\omega}_E \times (\boldsymbol{\omega}_E \times \mathbf{r}_{A/B})$: The earth's angular velocity ω_E is approximately one revolution per day $= 7.27 \times 10^{-5}$ rad/s. Therefore the magnitude of this term is bounded by $\omega_E^2 \, |\mathbf{r}_{A/B}| = (5.29 \times 10^{-9})|\mathbf{r}_{A/B}|$. For example, if the distance $|\mathbf{r}_{A/B}|$ from the origin of the earth-fixed reference frame to the object A is 10,000 m, this term is no larger than 5.3×10^{-5} m/s^2.
- *The term* $\mathbf{a}_B$: This term is the acceleration of the origin B of the earth-fixed reference frame relative to the center of the earth. B moves in a circular path due to the earth's rotation. If B lies on the earth's surface, this term is bounded by $\omega_E^2 R_E$, where R_E is the radius of the earth. Using the value $R_E = 6370$ km, we find that $\omega_E^2 R_E = 0.0337$ m/s^2. This value is too large to neglect for many purposes. However, under normal circumstances this term is accounted for as a part of the local value of the acceleration due to gravity.
- *The term* $2\boldsymbol{\omega}_E \times \mathbf{v}_{A\,\text{rel}}$: This term is called the **Coriolis acceleration**. Its magnitude is bounded by $2\omega_E \, |\mathbf{v}_{A\,\text{rel}}| = (1.45 \times 10^{-4})|\mathbf{v}_{A\,\text{rel}}|$. For example, if the magnitude of the velocity of A relative to the earth-fixed reference frame is 10 m/s, this term is no larger than 1.45×10^{-3} m/s^2.

We see that in most applications, the terms in brackets in Eq. (6.27) can be neglected. However, in some cases this is not possible. The Coriolis acceleration becomes significant if an object's velocity relative to the earth is large, and even very small accelerations become significant if an object's motion must be predicted over a large period of time. In such cases, you can still use Eq. (6.27) to determine the motion, but you must retain the significant terms. When this is done, the terms in brackets are usually moved to the left side:

$$\Sigma \mathbf{F} - m\mathbf{a}_B - 2m\boldsymbol{\omega}_E \times \mathbf{v}_{A\,\text{rel}} - m\boldsymbol{\omega}_E \times (\boldsymbol{\omega}_E \times \mathbf{r}_{A/B}) = m\mathbf{a}_{A\,\text{rel}}. \qquad (6.28)$$

Written in this way, the equation has the usual form of Newton's second law except that the left side contains additional "forces." We use quotation marks because these terms are not forces, but are artifacts arising from the motion of the earth-fixed reference frame.

Coriolis Effects The term $-2m\boldsymbol{\omega}_{\mathrm{E}} \times \mathbf{v}_{A\ \mathrm{rel}}$ in Eq. (6.28) is called the **Coriolis force**. It explains a number of physical phenomena that exhibit different behaviors in the northern and southern hemispheres, such as the direction a liquid tends to rotate when going down a drain, the direction a vine tends to grow around a vertical shaft, and the direction of rotation of a storm. The earth's angular velocity vector $\boldsymbol{\omega}_{\mathrm{E}}$ points north. When an object in the northern hemisphere that is moving tangent to the earth's surface travels north (Fig. 6.45a), the cross product $\boldsymbol{\omega}_{\mathrm{E}} \times \mathbf{v}_{A\ \mathrm{rel}}$ points west (Fig. 6.45b). Therefore the Coriolis force points east—it causes an object moving north to turn to the right (Fig. 6.45c). If the object is moving south, the direction of $\mathbf{v}_{A\ \mathrm{rel}}$ is reversed and the Coriolis force points west; its effect is to cause the object moving south to turn to the right (Fig. 6.45c). For example, in the northern hemisphere winds converging on a center of low pressure tend to rotate about it in the counterclockwise direction (Fig. 6.46a).

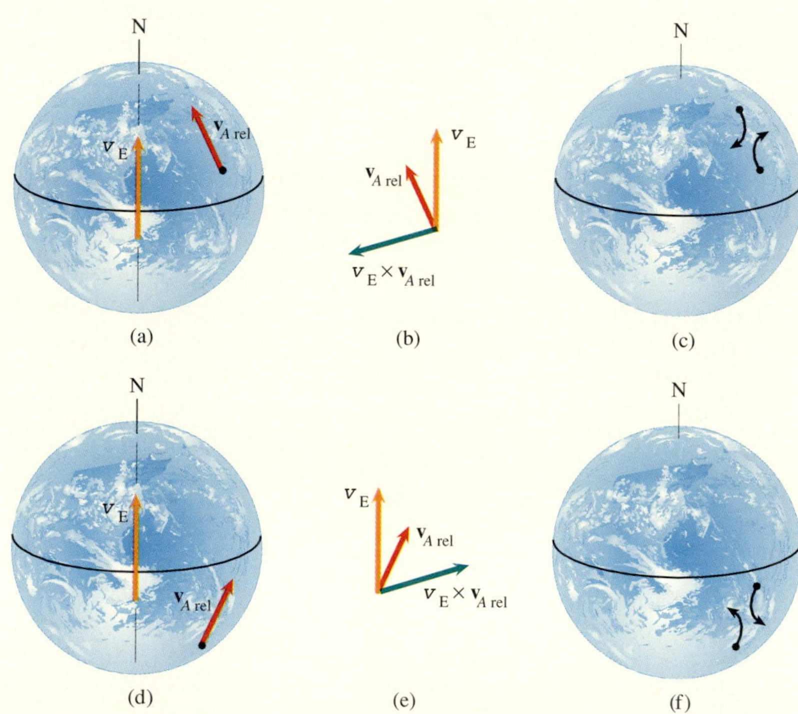

Fig. 6.45

(a) An object in the northern hemisphere moving north.
(b) Cross product of the earth's angular velocity with the object's velocity.
(c) Effects of the Coriolis force in the northern hemisphere.
(d) An object in the southern hemisphere moving north.
(e) Cross product of the earth's angular velocity with the object's velocity.
(f) Effects of the Coriolis force in the southern hemisphere.

When an object in the southern hemisphere travels north (Fig. 6.45d), the cross product $\omega_{\mathrm{E}} \times \mathbf{v}_{A\ \mathrm{rel}}$ points east (Fig. 6.45e). The Coriolis force points west and tends to cause the object to turn to the left (Fig. 6.45f). If the object is moving south, the Coriolis force points east and tends to cause the object to turn to the left (Fig. 6.45f). In the southern hemisphere, winds converging on a center of low pressure tend to rotate about it in the clockwise direction (Fig. 6.46b).

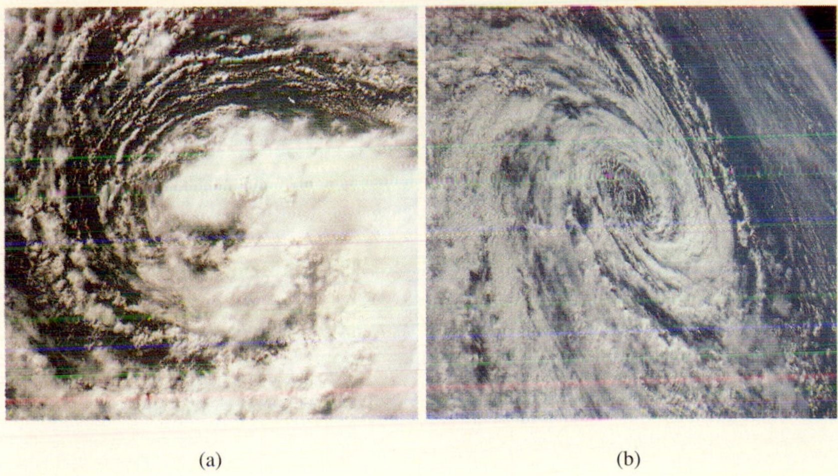

Fig. 6.46
Storms in the (a) northern and (b) southern hemispheres.

Arbitrary Reference Frame How can you analyze an object's motion relative to a reference frame that undergoes an arbitrary motion, such as a reference frame fixed with respect to a moving vehicle? Suppose that the primary reference frame with its origin at O in Fig. 6.47 is inertial and the secondary reference frame with its origin at B undergoes an arbitrary motion with angular velocity $\boldsymbol{\omega}$ and angular acceleration $\boldsymbol{\alpha}$. We can write Newton's second law for an object A of mass m as

$$\Sigma \mathbf{F} = m\mathbf{a}_A, \tag{6.29}$$

where $\mathbf{a}_A$ is A's acceleration relative to O. We use Eq. (6.22) to write Eq. (6.29) in the form

$$\boxed{\begin{aligned}\Sigma \mathbf{F} - m[\mathbf{a}_B + 2\boldsymbol{\omega} \times \mathbf{v}_{A\,\mathrm{rel}} + \boldsymbol{\alpha} \times \mathbf{r}_{A/B} \\ + \boldsymbol{\omega} \times (\boldsymbol{\omega} \times \mathbf{r}_{A/B})] = m\mathbf{a}_{A\,\mathrm{rel}},\end{aligned}} \tag{6.30}$$

where $\mathbf{a}_{A\,\mathrm{rel}}$ is A's acceleration relative to the secondary reference frame. This is Newton's second law expressed in terms of a secondary reference frame undergoing an arbitrary motion relative to an inertial primary reference frame. If you know the forces acting on A and the secondary reference frame's motion, you can use this equation to determine $\mathbf{a}_{A\,\mathrm{rel}}$.

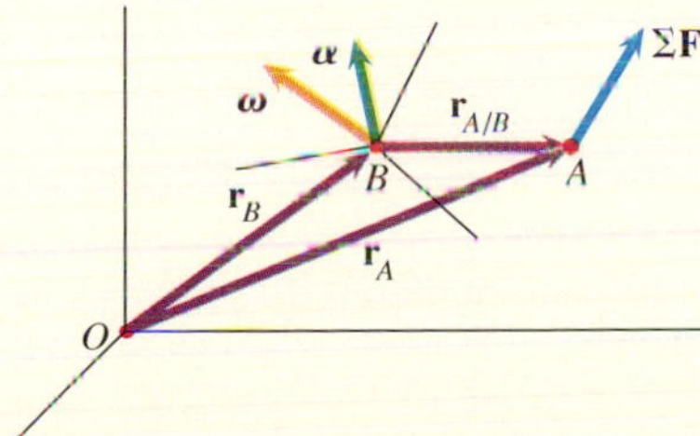

Fig. 6.47
An inertial reference frame (origin O) and a reference frame undergoing an arbitrary motion (origin B).

Example 6.13

Suppose that you and a friend play tennis on the deck of a cruise ship (Fig. 6.48), and you use the ship-fixed coordinate system with origin B to analyze the motion of the ball A. At the instant shown, the ball's position and velocity relative to the ship-fixed coordinate system are $\mathbf{r}_{A/B} = 15\mathbf{i} + 8\mathbf{j} + 36\mathbf{k}$ (ft) and $\mathbf{v}_{A\,\text{rel}} = 2\mathbf{i} - 8\mathbf{j} + 22\mathbf{k}$ (ft/s). The ball weighs 0.125 lb, and the aerodynamic force acting on it at the instant shown is $\mathbf{F} = 0.0250\mathbf{i} + 0.0010\mathbf{j} + 0.0025\mathbf{k}$ (lb). The ship is turning at a constant rate, and as a result the acceleration of point B relative to the earth is $\mathbf{a}_B = -3.0\mathbf{i} + 0.2\mathbf{k}$ (ft/s^2) and the ship's angular velocity is $\boldsymbol{\omega} = 0.1\mathbf{j}$ (rad/s). Determine the ball's acceleration relative to the ship-fixed coordinate system: (a) assuming that the ship-fixed coordinate system in inertial; (b) not assuming that the ship-fixed coordinate system is inertial, but assuming that a local earth-fixed coordinate system is inertial.

Fig. 6.48

STRATEGY

In part (a), we know the ball's mass and the external forces acting on it, so we can simply apply Newton's second law to determine the acceleration. In part (b), we can express Newton's second law in the form given by Eq. (6.30), which applies to a coordinate system undergoing an arbitrary motion relative to an inertial reference frame.

SOLUTION

(a) Assuming that the ship-fixed coordinate system is inertial, Newton's second law is

$$\Sigma \mathbf{F} = m\mathbf{a}_{A\,\text{rel}}:$$

$$-0.125\mathbf{j} + (0.0250\mathbf{i} + 0.0010\mathbf{j} + 0.0025\mathbf{k}) = \left(\frac{0.125}{32.2}\right)\mathbf{a}_{A\,\text{rel}}.$$

Solving this equation, we obtain the ball's acceleration under the assumption that the ship-fixed coordinate system is inertial:

$$\mathbf{a}_{A\,\text{rel}} = 6.44\mathbf{i} - 31.94\mathbf{j} + 0.64\mathbf{k}\ (\text{ft/s}^2).$$

(b) Dividing Eq. (6.30) by m gives

$$\left(\frac{1}{m}\right)\Sigma \mathrm{F} - \mathbf{a}_B - 2\boldsymbol{\omega}\times\mathbf{v}_{A\,\text{rel}} - \boldsymbol{\alpha}\times\mathbf{r}_{A/B} - \boldsymbol{\omega}\times(\boldsymbol{\omega}\times\mathbf{r}_{A/B}) = \mathbf{a}_{A\,\text{rel}}:$$

$$\left[\frac{1}{(0.125/32.2)}\right][-0.125\mathbf{j} + (0.0250\mathbf{i} + 0.0010\mathbf{j} + 0.0025\mathbf{k})]$$

$$- (-3.0\mathbf{i} + 0.2\mathbf{k}) - 2\begin{vmatrix} \mathbf{i} & \mathbf{j} & \mathbf{k} \\ 0 & 0.1 & 0 \\ 2 & -8 & 22 \end{vmatrix} - \mathbf{0}$$

$$- (0.1\mathbf{j}) \times \begin{vmatrix} \mathbf{i} & \mathbf{j} & \mathbf{k} \\ 0 & 0.1 & 0 \\ 15 & 8 & 36 \end{vmatrix} = \mathbf{a}_{A\,\text{rel}}.$$

The ball's acceleration under the assumption that an earth-fixed coordinate system is inertial is

$$\mathbf{a}_{A\,\text{rel}} = 5.19\mathbf{i} - 31.94\mathbf{j} + 1.20\mathbf{k}(\text{ft/s}^2).$$

DISCUSSION

This example illustrates the care you must exercise in applying Newton's second law. The acceleration we predicted by assuming that the ship-fixed coordinate system is inertial does not even approximate the correct value.

Problems

6.143 A merry-go-round rotates at a constant angular velocity of 0.5 rad/s. Person A walks at a constant speed of 1 m/s along a radial line. Determine A's velocity and acceleration *relative to the earth* when she is 2 m from the center of the merry-go-round, using two methods:

(a) Express the velocity and acceleration in terms of polar coordinates.

(b) Use Eqs. (6.21) and (6.22) to express the velocity and acceleration in terms of a body-fixed coordinate system with its x axis aligned with the line along which A walks and its z axis perpendicular to the merry-go-round.

P6.143

6.144 A disk-shaped space station of radius R rotates with constant angular velocity ω about the axis perpendicular to the page. Two persons are stationary relative to the station at A and B, and O is the center of the station. Using Eqs. (6.21) and (6.22) and the body-fixed coordinate system shown, (a) determine A's velocity and acceleration relative to a nonrotating reference frame with its origin at O; (b) determine A's velocity and acceleration relative to a nonrotating reference frame whose origin moves with point B.

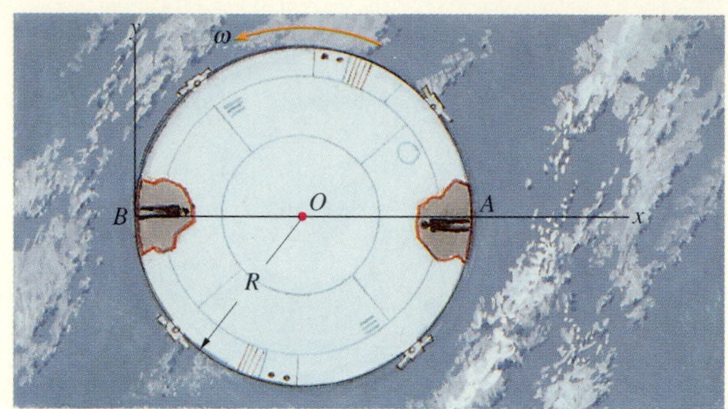

P6.144

6.145 The metal plate is attached to a fixed ball and socket support at O. The pin A slides in a slot in the plate. At the instant shown, $x_A = 1$ m, $dx_A/dt = 2$ m/s, and $d^2x_A/dt^2 = 0$, and the plate's angular velocity and angular acceleration are $\boldsymbol{\omega} = 2\mathbf{k}$ (rad/s) and $\boldsymbol{\alpha} = \mathbf{0}$. What are the x, y, z components of the velocity and acceleration of A relative to a nonrotating reference frame that is stationary with respect to O?

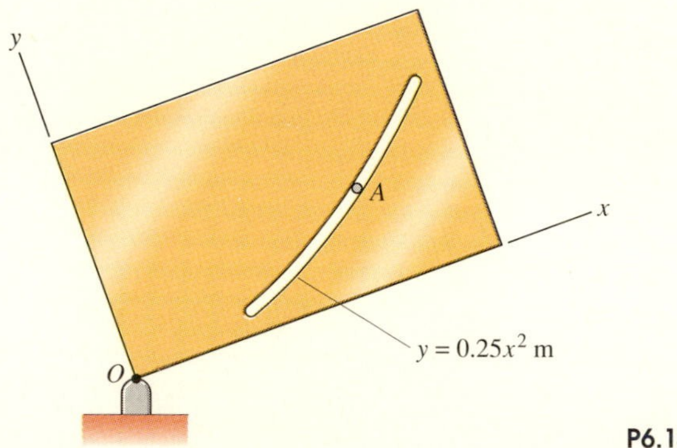

P6.145

6.146 Suppose that at the instant shown in Problem 6.145, $x_A = 1$ m, $dx_A/dt = -3$ m/s, and $d^2x_A/dt^2 = 4$ m/s^2, and the plate's angular velocity and angular acceleration are $\boldsymbol{\omega} = -4\mathbf{j} + 2\mathbf{k}$ (rad/s) and $\boldsymbol{\alpha} = 3\mathbf{i} - 6\mathbf{j}$ (rad/s^2). What are the x, y, z components of the velocity and acceleration of A relative to a nonrotating reference frame that is stationary with respect to O?

6.147 The coordinate system shown is fixed relative to the ship B. At the instant shown, the ship is sailing north at 10 ft/s relative to the earth, and its angular velocity is 0.02 rad/s clockwise. The airplane is flying east at 400 ft/s relative to the earth, and its position relative to the ship is $\mathbf{r}_{A/B} = 2000\mathbf{i} + 2000\mathbf{j} + 1000\mathbf{k}$ (ft). If the ship uses its radar to measure the plane's velocity relative to its body-fixed coordinate system, what is the result?

P6.147

6.148 The space shuttle is attempting to recover a satellite for repair. At the current time, the satellite's position relative to a coordinate system fixed to the shuttle is $50\mathbf{i}$ (m). The rate-gyros on the shuttle indicate that its current angular velocity is $0.05\mathbf{j} + 0.03\mathbf{k}$ (rad/s). The shuttle pilot measures the velocity of the satellite relative to the body-fixed coordinate system and determines it to be $-2\mathbf{i} - 1.5\mathbf{j} + 2.5\mathbf{k}$ (m/s). What are the x, y, z components of the satellite's velocity relative to a nonrotating coordinate system with its origin fixed to the shuttle's center of mass?

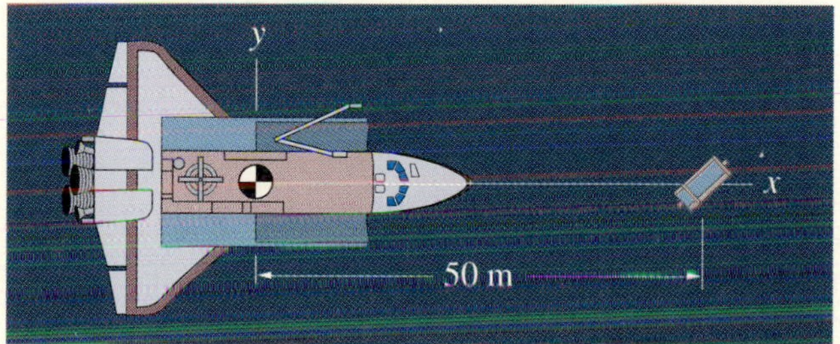

P6.148

6.149 The train on the circular track is traveling at a constant speed of 50 ft/s in the direction shown. The train on the straight track is traveling at 20 ft/s in the direction shown and is increasing its speed at 2 ft/s^2. Determine the velocity of passenger A that passenger B observes relative to the coordinate system shown, which is fixed to the car in which B is riding.

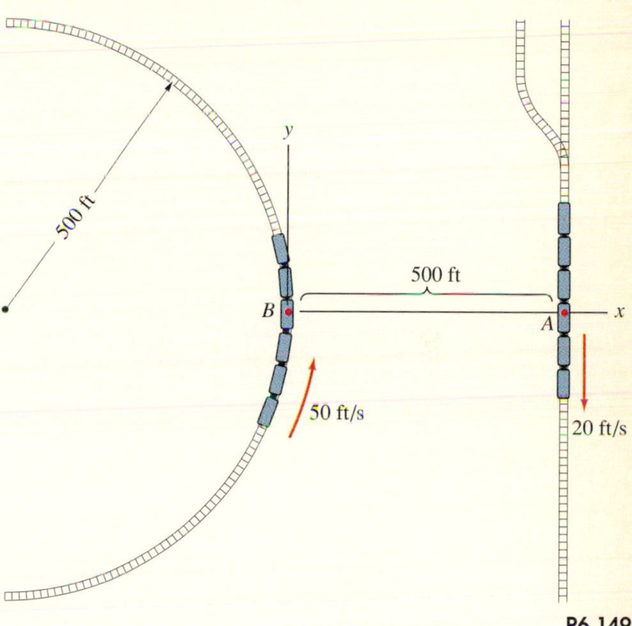

P6.149

6.150 In Problem 6.149, determine the acceleration of passenger A that passenger B observes relative to the coordinate system fixed to the car in which B is riding.

6.151 The satellite A is in a circular polar orbit (a circular orbit that intersects the earth's axis of rotation). The radius of the orbit is R, and the magnitude of the satellite's velocity relative to a nonrotating reference frame with its origin at the center of the earth is v_A. At the instant shown, the satellite is above the equator. An observer B on the earth directly below the satellite measures its motion using the earth-fixed coordinate system shown. What are the velocity and acceleration of the satellite relative to B's earth-fixed coordinate system? The radius of the earth is R_E and its angular velocity is ω_E.

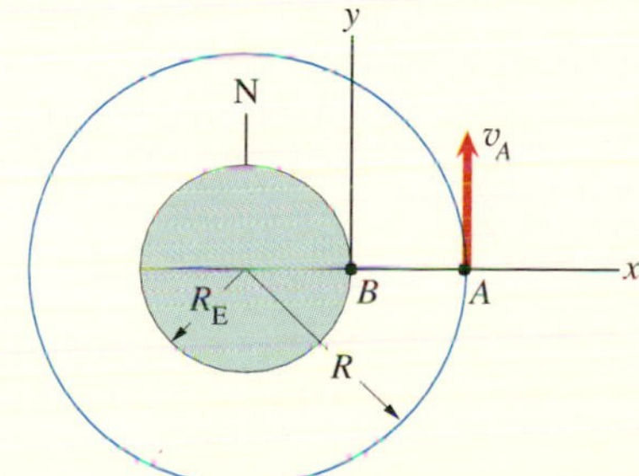

P6.151

6.152 A car A at north latitude L drives north on a north–south highway with constant speed v. The earth's radius is R_E and its angular velocity is ω_E. Determine the x, y, z components of the car's velocity and acceleration relative to the coordinate system shown if (a) the coordinate system is earth fixed; (b) the coordinate system does not rotate.

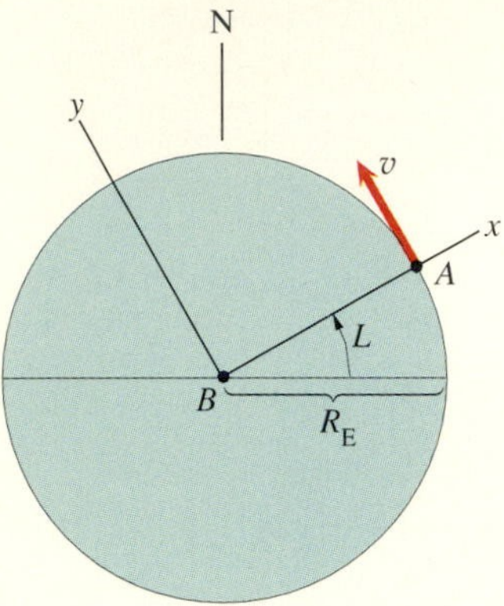

P6.152

6.153 The airplane B conducts flight tests of a missile. At the instant shown, the airplane is traveling at 200 m/s relative to the earth in a circular path of 2000-m radius *in the horizontal plane*. The coordinate system is fixed relative to the airplane. The x axis is tangent to the plane's path and points forward. The y axis points out the plane's right side, and the z axis points out the bottom of the plane. The plane's bank angle (the inclination of the z axis from the vertical) is constant and equal to 20°. *Relative to the airplane's coordinate system,* the pilot measures the missile's position and velocity and determines them to be $\mathbf{r}_{A/B} = 1000\mathbf{i}$ (m) and $\mathbf{v}_{A/B} = 100.0\mathbf{i} + 94.0\mathbf{j} + 34.2\mathbf{k}$ (m/s).

(a) What are the x, y, z components of the airplane's angular velocity vector?

(b) What are the x, y, z components of the missile's velocity relative to the earth?

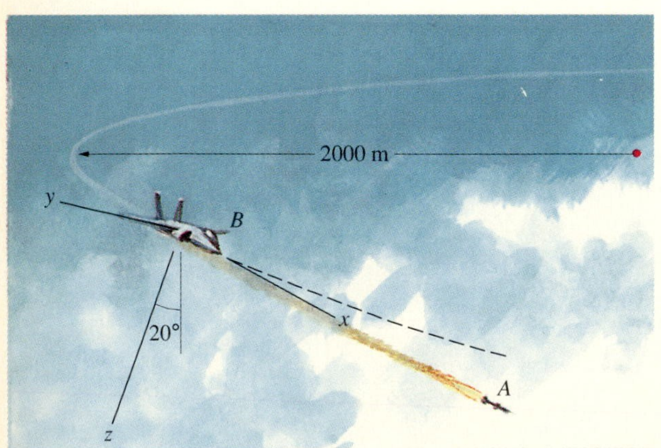

P6.153

6.154 To conduct experiments related to long-term space flight, engineers construct a laboratory on earth that rotates about the vertical axis at B with a constant angular velocity ω of one revolution every 6 s. They establish a laboratory-fixed coordinate system with its origin at B and the z axis upward. An engineer holds an object stationary relative to the laboratory at point A, 3 m from the axis of rotation, and releases it. At the instant he drops the object, determine its acceleration relative to the laboratory-fixed coordinate system (a) assuming that the laboratory-fixed coordinate system is inertial; (b) not assuming that the laboratory-fixed coordinate system is inertial, but assuming that an earth-fixed coordinate system with its origin at B is inertial.

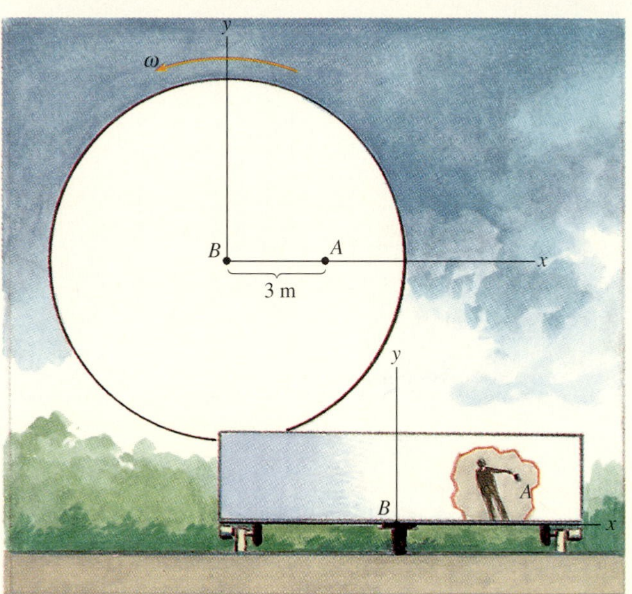

P6.154

6.155 A disk *lying in the horizontal plane* rotates about a fixed shaft at the origin with constant angular velocity ω. The slider A of mass m moves in a smooth slot in the disk. The spring is unstretched when $x = 0$.

(a) By using Eq. (6.30) to express Newton's second law in terms of the body-fixed coordinate system, show that the slider's motion is governed by the equation

$$\frac{d^2x}{dt^2} + \left(\frac{k}{m} - \omega^2\right)x = 0.$$

(b) The slider is given an initial velocity $dx/dt = v_0$ at $x = 0$. Determine its velocity as a function of x.

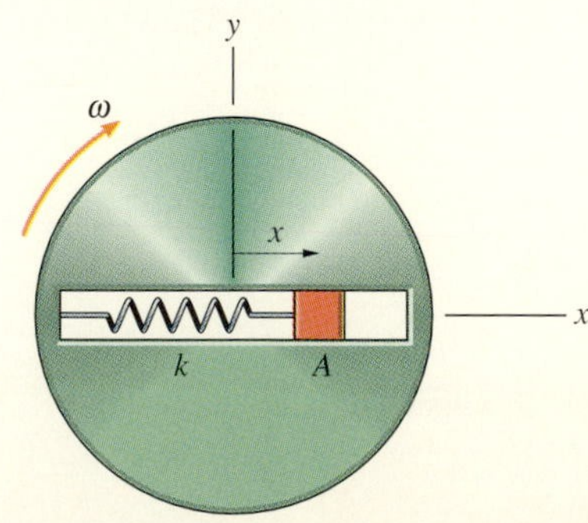

P6.155

6.156 Engineers conduct flight tests of a rocket at 30° north latitude. They measure the rocket's motion using an earth-fixed coordinate system with the x axis upward and the y axis northward. At a particular instant, the mass of the rocket is 4000 kg, its velocity relative to their coordinate system is $2000\mathbf{i} + 2000\mathbf{j}$ (m/s), and the sum of the forces exerted on the rocket by its thrust, weight, and aerodynamic forces is $400\mathbf{i} + 400\mathbf{j}$ (N). Determine the rocket's acceleration relative to their coordinate system (a) assuming that their earth-fixed coordinate system is inertial; (b) not assuming that their earth-fixed coordinate system is inertial.

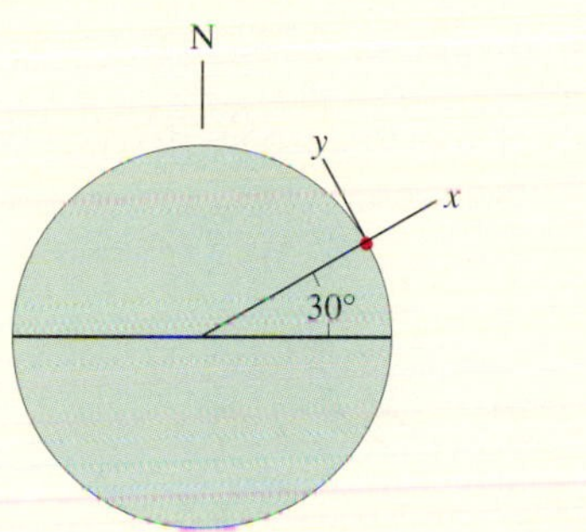

P6.156

6.157 Consider a point A on the surface of the earth at north latitude L. The radius of the earth is R_E and its angular velocity is ω_E. A plumb bob suspended just above the ground at point A will hang at a small angle β relative to the vertical because of the earth's rotation. Show that β is related to the latitude by

$$\tan\beta = \frac{\omega_E^2 R_E \sin L \cos L}{g - \omega_E^2 R_E \cos^2 L}.$$

Strategy: Using the earth-fixed coordinate system shown, express Newton's second law in the form given by Eq. (6.27).

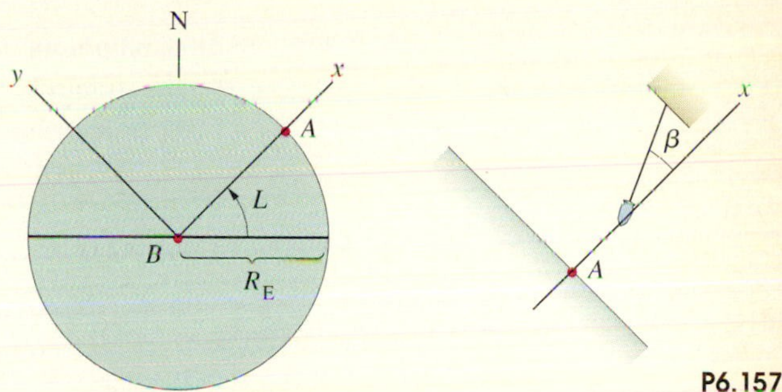

P6.157

6.158 Suppose that a space station is in orbit around the earth and two astronauts on the station toss a ball back and forth. They observe that the ball appears to travel between them in a straight line at constant velocity.
(a) Write Newton's second law for the ball as it travels between them in terms of a nonrotating coordinate system that is stationary relative to the station. What is the term $\Sigma\,\mathbf{F}$? Use the equation to explain the behavior of the ball observed by the astronauts.
(b) Write Newton's second law for the ball as it travels between the astronauts in terms of a nonrotating coordinate system that is stationary relative to the center of the earth. What is the term $\Sigma\,\mathbf{F}$? Explain the difference between this equation and the one you obtained in (a).

Chapter Summary

In this chapter we analyzed the motions of rigid bodies, showing how the velocities and accelerations of their points are related to their angular velocities and angular accelerations. In doing so, we did not consider the forces and couples causing the motions. In Chapter 7 we will use Newton's second law to determine the equations of motion for rigid bodies in planar motion. By drawing a free-body diagram of a rigid body, we will relate the acceleration of its center of mass and its angular acceleration to the forces and couples acting on it.

Types of Motion

If a rigid body in motion does not rotate relative to a given reference frame, it is said to be in **translation**. If the points of a rigid body intersected by a fixed plane remain in that plane, the rigid body is said to undergo **planar** motion. Rotation about a fixed axis is a special case of planar motion.

Relative Velocities and Accelerations

The **angular velocity vector** $\boldsymbol{\omega}$ of a rigid body is parallel to the axis of rotation and its magnitude $|\boldsymbol{\omega}|$ is the rate of rotation. If the thumb of the right hand points in the direction of $\boldsymbol{\omega}$, the fingers curl around $\boldsymbol{\omega}$ in the direction of the rotation. The **angular acceleration vector** $\boldsymbol{\alpha} = d\boldsymbol{\omega}/dt$ is the rate of change of the angular velocity vector.

Consider a point B of a rigid body, a body-fixed reference frame, and an arbitrary point A (Fig. a). The velocities $\mathbf{v}_A$ and $\mathbf{v}_B$ of the points relative to the primary reference frame are related by

$$\mathbf{v}_A = \mathbf{v}_B + \mathbf{v}_{A\,\text{rel}} + \boldsymbol{\omega} \times \mathbf{r}_{A/B}, \qquad \textbf{Eq. (6.11)}$$

where $\mathbf{v}_{A\,\text{rel}}$ is the velocity of A relative to the body-fixed reference frame. If A is a point of the rigid body, $\mathbf{v}_{A\,\text{rel}}$ is zero.

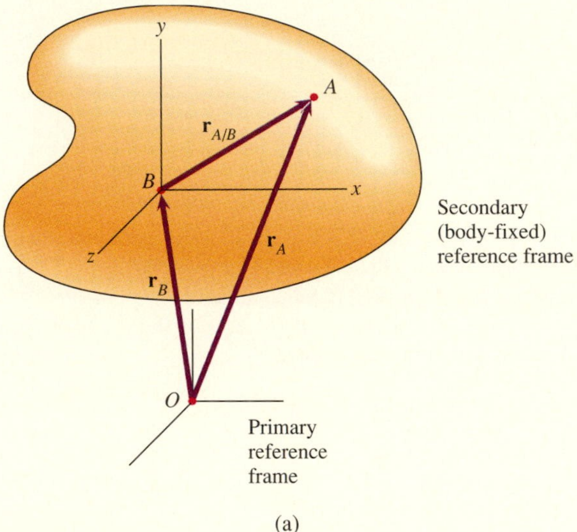

(a)

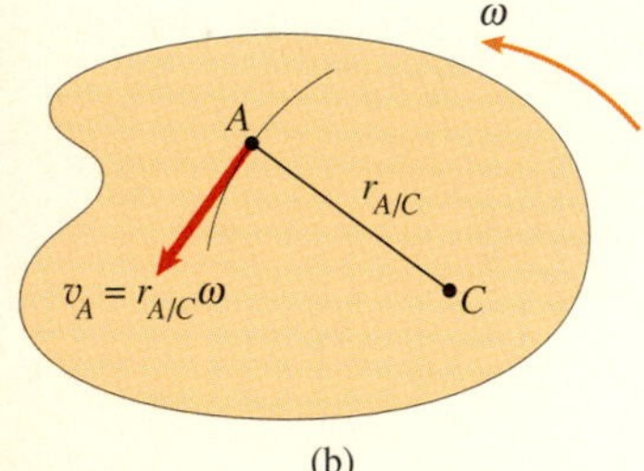

(b)

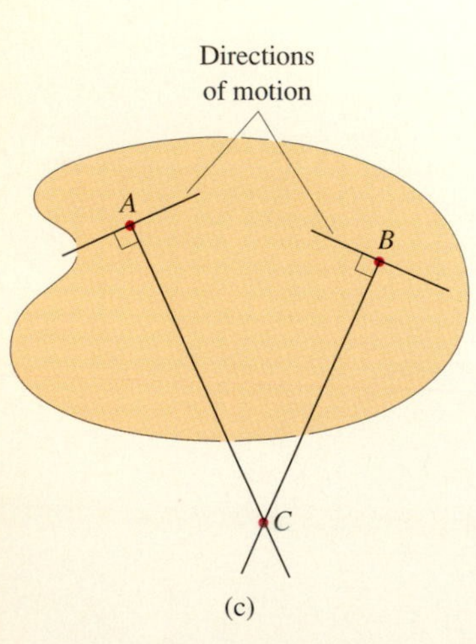

(c)

The accelerations $\mathbf{a}_A$ and $\mathbf{a}_B$ of the points relative to the primary reference frame are related by

$$\mathbf{a}_A = \mathbf{a}_B + \mathbf{a}_{A\,\text{rel}} + 2\boldsymbol{\omega} \times \mathbf{v}_{A\,\text{rel}} + \boldsymbol{\alpha} \times \mathbf{r}_{A/B} + \boldsymbol{\omega} \times (\boldsymbol{\omega} \times \mathbf{r}_{A/B}), \qquad \textbf{Eq. (6.13)}$$

where $\mathbf{a}_{A\,\text{rel}}$ is the acceleration of A relative to the body-fixed reference frame. In planar motion, the term $\boldsymbol{\omega} \times (\boldsymbol{\omega} \times \mathbf{r}_{A/B})$ can be written in the simpler form $-\omega^2 \mathbf{r}_{A/B}$. If A is a point of the rigid body, $\mathbf{a}_{A\,\text{rel}}$ is zero.

Instantaneous Centers

An **instantaneous center** is a point of a rigid body whose velocity is zero at a given instant. Consider a rigid body in planar motion, and suppose that C is an instantaneous center. The velocity of a point A is perpendicular to the line from C to A, and its magnitude is the product of the distance from C to A and the angular velocity (Fig. b).

If you know the directions of the motions of two points A and B of a rigid body in planar motion, lines drawn through A and B perpendicular to their directions of motion intersect at the instantaneous center (Fig. c).

Moving Reference Frames

Consider a point A and a reference frame with origin B that rotates with angular velocity $\boldsymbol{\omega}$ and angular acceleration $\boldsymbol{\alpha}$ relative to a primary reference frame (Fig. d). The velocities of A and B relative to the primary reference frame are related by

$$\mathbf{v}_A = \mathbf{v}_B + \mathbf{v}_{A\,\mathrm{rel}} + \boldsymbol{\omega} \times \mathbf{r}_{A/B}, \qquad \text{Eq. (6.21)}$$

where $\mathbf{v}_{A\,\mathrm{rel}}$ is the velocity of A relative to the moving reference frame. The accelerations of A and B relative to the primary reference frame are related by

$$\mathbf{a}_A = \mathbf{a}_B + \mathbf{a}_{A\,\mathrm{rel}} + 2\boldsymbol{\omega} \times \mathbf{v}_{A\,\mathrm{rel}} + \boldsymbol{\alpha} \times \mathbf{r}_{A/B} + \boldsymbol{\omega} \times (\boldsymbol{\omega} \times \mathbf{r}_{A/B}), \qquad \text{Eq. (6.22)}$$

where $\mathbf{a}_{A\,\mathrm{rel}}$ is the acceleration of A relative to the moving reference frame.

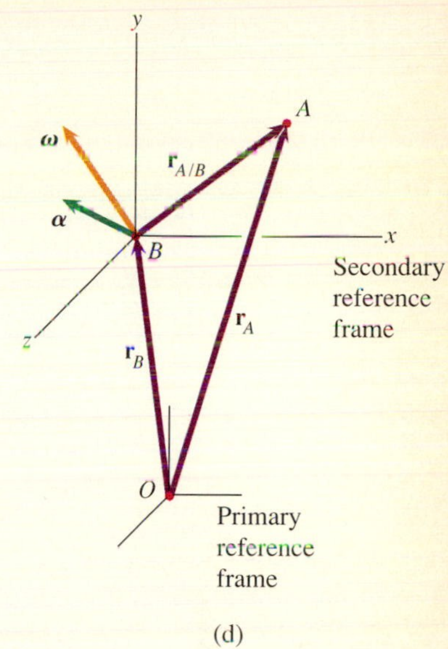

(d)

Review Problems

6.159 If $\theta = 60°$ and bar OQ rotates in the counterclockwise direction at 5 rad/s, what is the angular velocity of bar PQ?

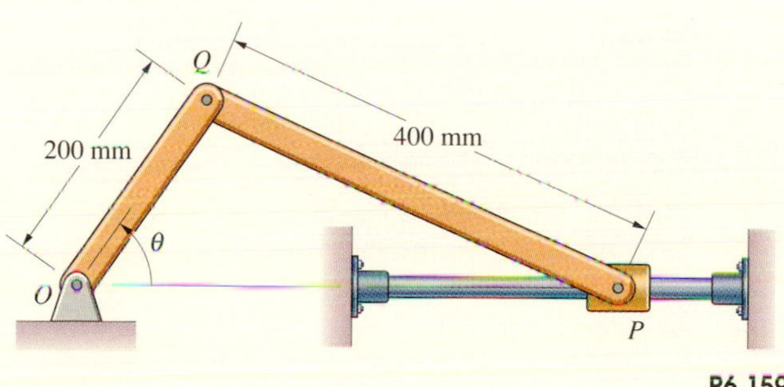

P6.159

6.160 Consider the system shown in Problem 6.159. If $\theta = 55°$ and the sleeve P is moving to the left at 2 m/s, what are the angular velocities of bars OQ and PQ?

6.161 Determine the vertical velocity v_H of the hook and the angular velocity of the small pulley.

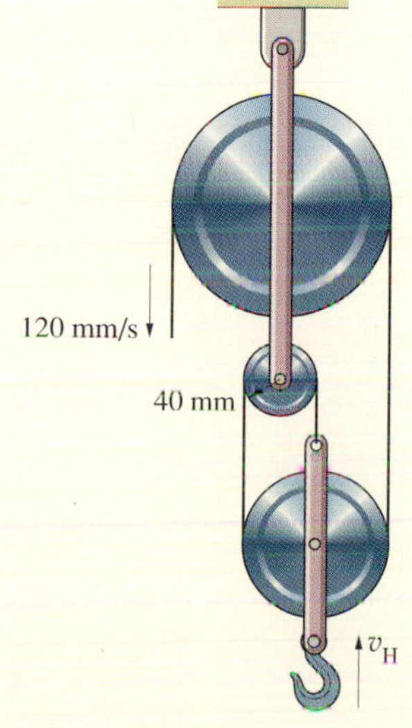

P6.161

6.162 If the crankshaft AB is turning in the counterclockwise direction at 2000 rpm, what is the velocity of the piston?

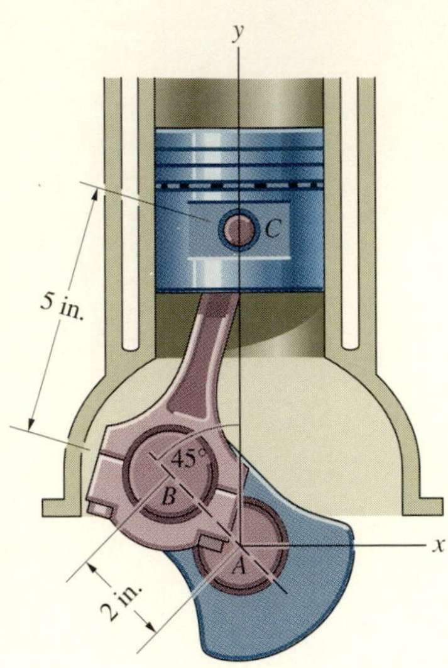

P6.162

6.163 In Problem 6.162, if the piston is moving with velocity $\mathbf{v}_C = 20\mathbf{j}$ (ft/s), what are the angular velocities of the crankshaft AB and the connecting rod BC?

6.164 In Problem 6.162, if the piston is moving with velocity $\mathbf{v}_C = 20\mathbf{j}$ (ft/s) and its acceleration is zero, what are the angular accelerations of the crankshaft AB and the connecting rod BC?

6.165 Bar AB rotates at 6 rad/s in the counterclockwise direction. Use instantaneous centers to determine the angular velocity of bar BCD and the velocity of point D.

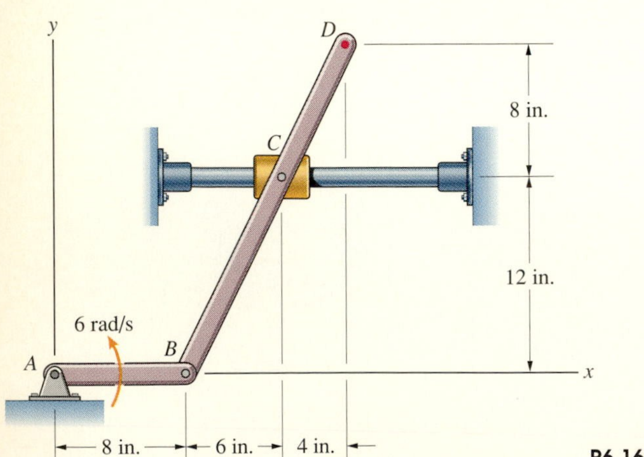

P6.165

6.166 In Problem 6.165, bar AB rotates with a constant angular velocity of 6 rad/s in the counterclockwise direction. Determine the acceleration of point D.

6.167 Point C is moving to the right at 20 in./s. What is the velocity of the midpoint G of bar BC?

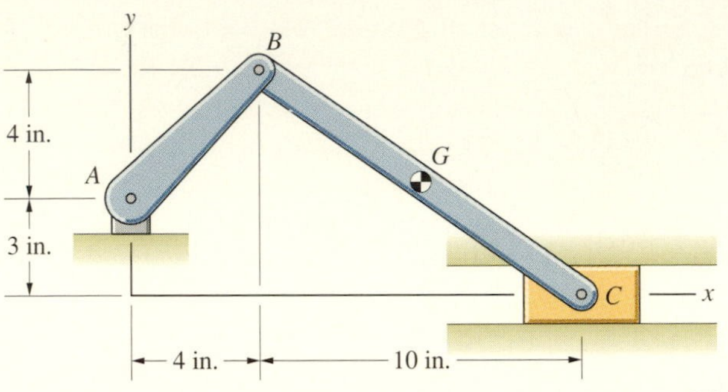

P6.167

6.168 In Problem 6.167, point C is moving to the right with a constant velocity of 20 in./s. What is the acceleration of the midpoint G of bar BC?

6.169 In Problem 6.167, if the velocity of point C is $\mathbf{v}_C = 1.0\mathbf{i}$ (in./s), what are the angular velocity vectors of arms AB and BC?

6.170 Points B and C are in the x-y plane. The angular velocity vectors of arms AB and BC are $\boldsymbol{\omega}_{AB} = -0.5\mathbf{k}$ (rad/s), $\boldsymbol{\omega}_{BC} = 2.0\mathbf{k}$ (rad/s). Determine the velocity of point C.

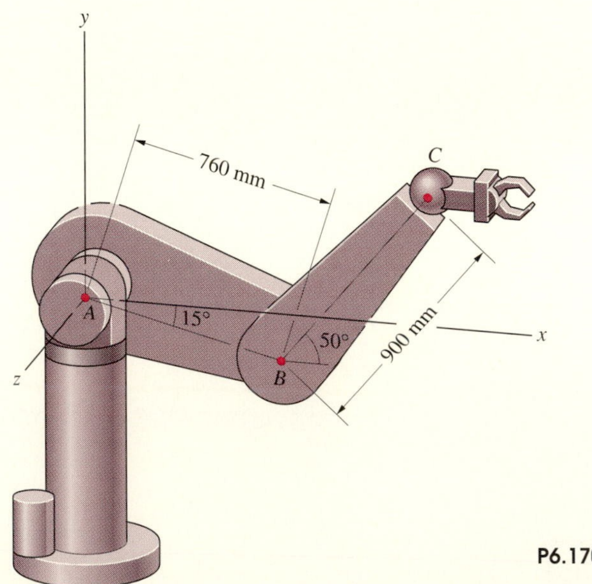

P6.170

6.171 In Problem 6.170, if the velocity of point C is $\mathbf{v}_C = 1.0\mathbf{i}$ (m/s), what are the angular velocity vectors of arms AB and BC?

6.172 In Problem 6.170, if the angular velocity vectors of arms AB and BC are $\boldsymbol{\omega}_{AB} = -0.5\mathbf{k}$ (rad/s), $\boldsymbol{\omega}_{BC} = 2.0\mathbf{k}$ (rad/s), and their angular acceleration vectors are $\boldsymbol{\alpha}_{AB} = 1.0\mathbf{k}$ (rad/s^2), $\boldsymbol{\alpha}_{BC} = 1.0\mathbf{k}$ (rad/s^2), what is the acceleration of point C?

6.173 In Problem 6.170, if the velocity of point C is $\mathbf{v}_C = 1.0\mathbf{i}$ (m/s) and $\mathbf{a}_C = 0$, what are the angular velocity and angular acceleration vectors of arm BC?

6.174 The crank AB has a constant clockwise angular velocity of 200 rpm. What are the velocity and acceleration of the piston P?

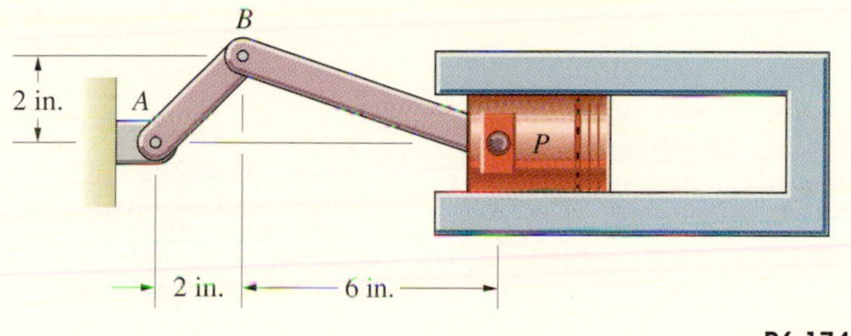

P6.174

6.175 Bar AB has a counterclockwise angular velocity of 10 rad/s and a clockwise angular acceleration of 20 rad/s^2. Determine the angular acceleration of bar BC and the acceleration of point C.

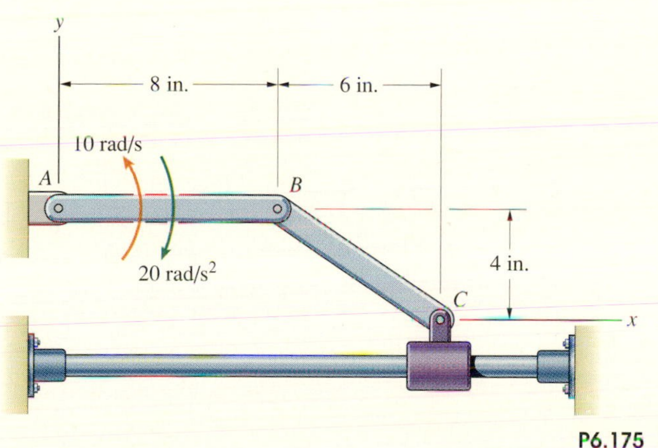

P6.175

6.176 The angular velocity of arm AC is 1 rad/s counterclockwise. What is the angular velocity of the scoop?

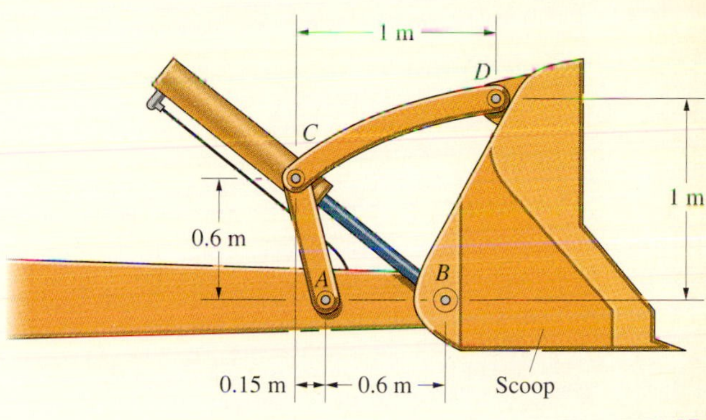

P6.176

6.177 The angular velocity of arm AC in Problem 6.176 is 2 rad/s counterclockwise, and its angular acceleration is 4 rad/s^2 clockwise. What is the angular acceleration of the scoop?

6.178 If you want to program the robot so that, at the instant shown, the velocity of point D is $\mathbf{v}_D = 0.2\mathbf{i} + 0.8\mathbf{j}$ (m/s) and the angular velocity of arm CD is 0.3 rad/s counterclockwise, what are the necessary angular velocities of arms AB and BC?

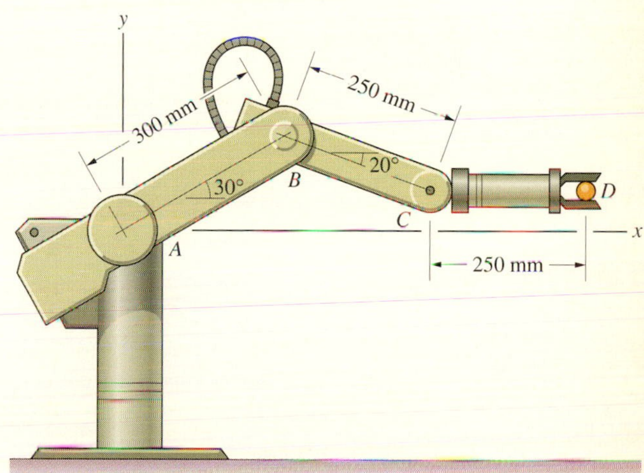

P6.178

6.179 The ring gear is stationary, and the sun gear rotates at 120 rpm in the counterclockwise direction. Determine the angular velocity of the planet gears and the magnitude of the velocity of their centerpoints.

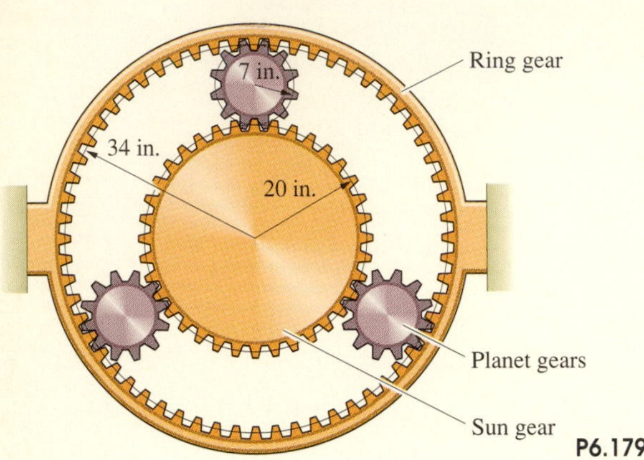

P6.179

6.180 Arm AB is rotating at 10 rad/s in the clockwise direction. Determine the angular velocity of arm BC and the velocity at which it slides relative to the sleeve at C.

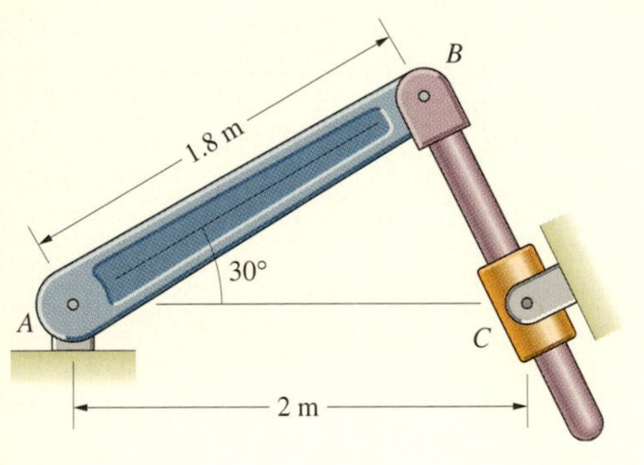

P6.180

6.181 In Problem 6.180, arm AB is rotating with an angular velocity of 10 rad/s and an angular acceleration of 20 rad/s^2, both in the clockwise direction. Determine the angular acceleration of arm BC.

6.182 Arm AB is rotating with a constant counterclockwise angular velocity of 10 rad/s. Determine the vertical velocity and acceleration of the rack R of the rack and pinion gear.

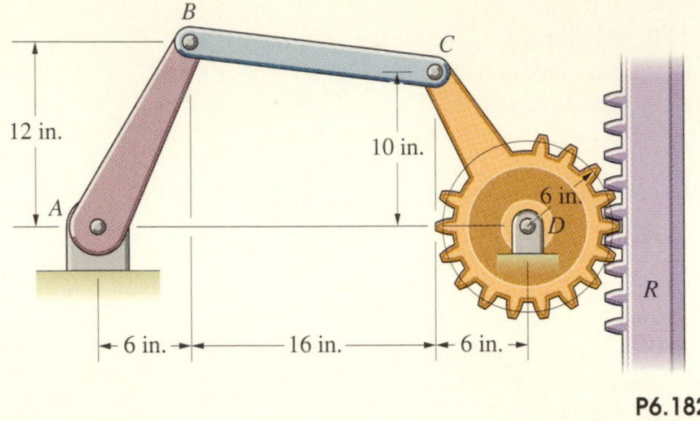

P6.182

6.183 In Problem 6.182, if the rack R of the rack and pinion gear is moving upward with a constant velocity of 10 ft/s, what are the angular velocity and angular acceleration of bar BC?

6.184 Bar AB has a constant counterclockwise angular velocity of 2 rad/s. The 1-kg collar C slides on the smooth horizontal bar. At the instant shown, what is the tension in the cable BC?

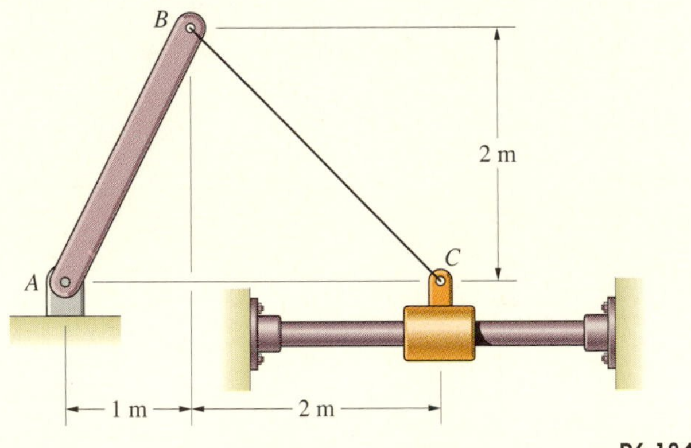

P6.184

6.185 An athlete exercises his arm by raising the 8-kg mass m. The shoulder joint A is stationary. The distance AB is 300 mm, the distance BC is 400 mm, and the distance from C to the pulley is 340 mm. The angular velocities $\omega_{AB} = 1.5$ rad/s and $\omega_{BC} = 2$ rad/s are constant. What is the tension in the cable?

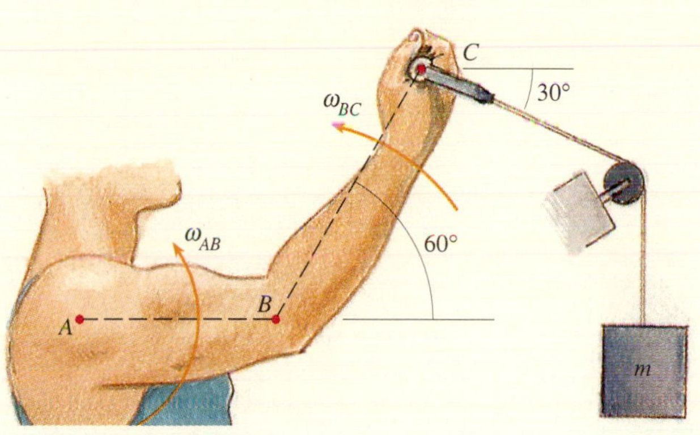

P6.185

6.186 The secondary reference frame shown rotates with a constant angular velocity $\boldsymbol{\omega} = 2\mathbf{k}$ (rad/s) relative to a primary reference frame. The point A moves outward along the x axis at a constant speed of 5 m/s.

(a) What are the velocity and acceleration of A relative to the secondary reference frame?

(b) If the origin B is stationary relative to the primary reference frame, what are the velocity and acceleration of A relative to the primary reference frame when A is at the position $x = 1$ m?

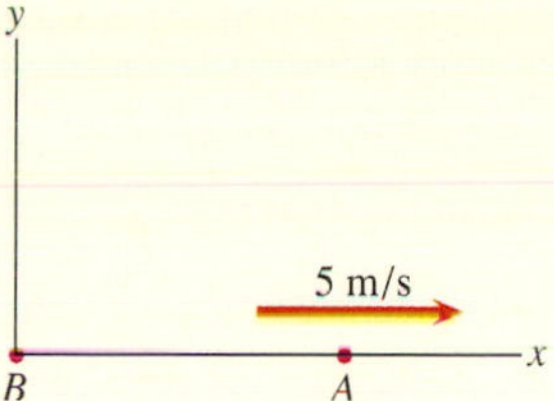

P6.186

6.187 The coordinate system shown is fixed relative to the ship B. The ship uses its radar to measure the position of a stationary buoy A and determines it to be $400\mathbf{i} + 200\mathbf{j}$ (m). The ship also measures the velocity of the buoy relative to its body-fixed coordinate system and determines it to be $2\mathbf{i} - 8\mathbf{j}$ (m/s). What are the ship's velocity and angular velocity relative to the earth? (Assume that the ship's velocity is in the direction of the y axis.)

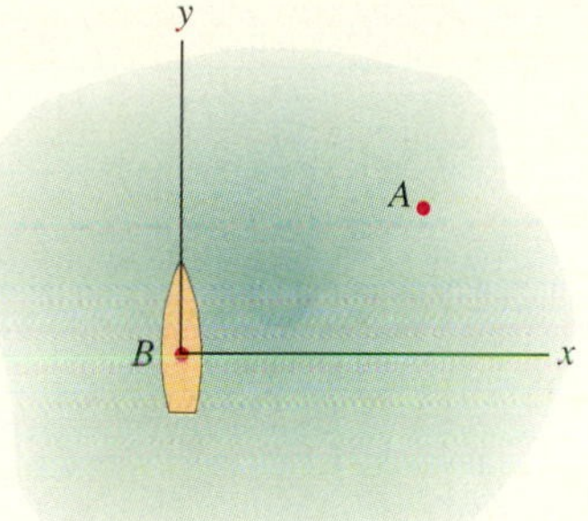

P6.187

The shovel undergoes planar motion as the hydraulic cylinder and supporting members raise it and rotate it in the vertical plane. Newton's second law relates the sum of the forces on the shovel to the acceleration of its center of mass, and an equation of angular motion relates the sum of the moments about the shovel's center of mass to its angular acceleration. In this chapter we use free-body diagrams and the equations of motion for rigid bodies to determine the motions of objects resulting from the forces and couples acting on them.

Chapter 7

Planar Dynamics of Rigid Bodies

IN Chapter 6 we analyzed planar motions of rigid bodies without considering the forces and couples causing them. You have used Newton's second law to determine the motions of the centers of mass of objects, but how can you determine their *rotational* motions? In this chapter we derive planar equations of angular motion for a rigid body. By drawing the free-body diagram of an object such as an excavator's shovel, we can determine both the acceleration of its center of mass and its angular acceleration in terms of the forces and couples to which it is subjected.

7.1 *Preview of the Equations of Motion*

The two-dimensional equations of angular motion for a rigid body are quite simple, but you can easily lose sight of the forest among the trees as we derive them. To help you follow the derivations, we summarize the equations in this section.

The equations of motion of a rigid body include Newton's second law,

$$\Sigma \mathbf{F} = m\mathbf{a},$$

which states that the sum of the external forces acting on the body equals the product of its mass and the acceleration of its center of mass. The equations of motion are completed by an equation of angular motion. If the rigid body rotates about a fixed axis O (Fig. 7.1a), the sum of the moments about the axis due to external forces and couples acting on the body is related to its angular acceleration by

$$\Sigma M_0 = I_0 \alpha,$$

where I_0 is the moment of inertia of the rigid body about O. Just as an object's mass determines the acceleration resulting from the forces acting on it, its moment of inertia I_0 about a fixed axis determines the angular acceleration resulting from the sum of the moments about the axis.

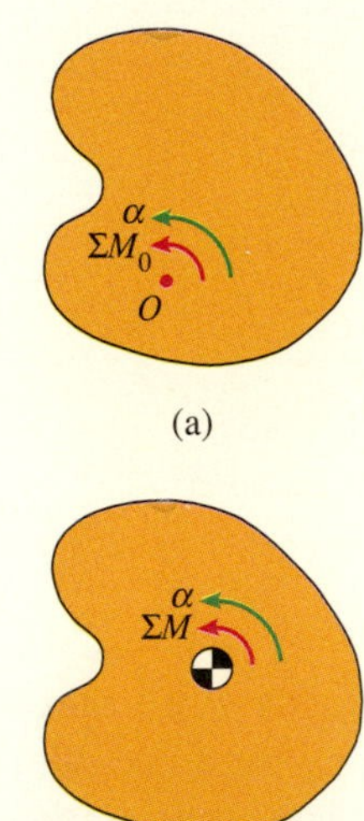

Fig. 7.1
(a) A rigid body rotating about a fixed axis O.
(b) A rigid body in general planar motion.

In the case of general planar motion (Fig. 7.1b), the sum of the moments about the center of mass of a rigid body is related to its angular acceleration by

$$\Sigma M = I\alpha,$$

where I is the moment of inertia of the rigid body about its center of mass. If we know the external forces and couples acting on a rigid body in planar motion, we can use these equations to determine the acceleration of its center of mass and its angular acceleration.

7.2 Momentum Principles for a System of Particles

In this chapter and in our discussion of three-dimensional dynamics of rigid bodies in Chapter 9, our derivations of the equations of motion begin with principles governing the motion of a system of particles. We summarize these general and important principles in this section.

Force–Linear Momentum Principle

We begin by showing that the sum of the external forces on a system of particles equals the rate of change of its total linear momentum. Let us consider a system of N particles. We denote the mass of the ith particle by m_i and denote its position vector relative to the origin O of an inertial reference frame by $\mathbf{r}_i$ (Fig. 7.2). Let $\mathbf{f}_{ij}$ be the force exerted on the ith particle by the jth particle, and let the external force on the ith particle (that is, the total force exerted by objects other than the system of particles we are considering) be $\mathbf{f}_i^E$. Newton's second law states that the total force on the ith particle equals the product of its mass and the rate of change of its linear momentum:

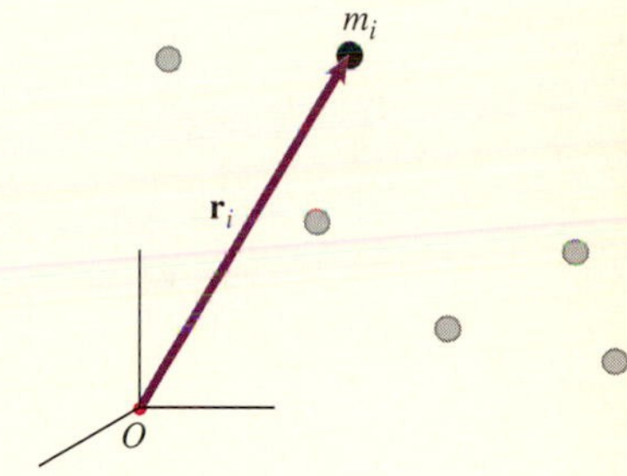

Fig. 7.2
A system of particles. The vector $\mathbf{r}_i$ is the position vector of the ith particle.

$$\sum_j \mathbf{f}_{ij} + \mathbf{f}_i^E = \frac{d}{dt}(m_i \mathbf{v}_i), \tag{7.1}$$

where $\mathbf{v}_i = d\mathbf{r}_i/dt$ is the velocity of the ith particle. Writing this equation for each particle of the system and summing from $i = 1$ to N, we obtain

$$\sum_i \sum_j \mathbf{f}_{ij} + \sum_i \mathbf{f}_i^E = \frac{d}{dt} \sum_i m_i \mathbf{v}_i. \tag{7.2}$$

The first term on the left side of this equation is the sum of the internal forces on the system of particles. As a consequence of Newton's third law ($\mathbf{f}_{ji} + \mathbf{f}_{ij} = 0$), this term equals zero:

$$\sum_i \sum_j \mathbf{f}_{ij} = \mathbf{f}_{12} + \mathbf{f}_{21} + \mathbf{f}_{13} + \mathbf{f}_{31} + \cdots = \mathbf{0}.$$

The second term on the left side of Eq. (7.2) is the sum of the external forces on the system. Denoting it by $\Sigma\,\mathbf{F}$, we conclude that the sum of the external forces on the system equals the rate of change of its total linear momentum:

$$\Sigma\,\mathbf{F} = \frac{d}{dt} \sum_i m_i \mathbf{v}_i. \tag{7.3}$$

Let m be the sum of the masses of the particles:

$$m = \sum_i m_i.$$

The position of the center of mass of the system is

$$\mathbf{r} = \frac{\sum_i m_i \mathbf{r}_i}{m}, \tag{7.4}$$

so the velocity of the center of mass is

$$\mathbf{v} = \frac{d\mathbf{r}}{dt} = \frac{\sum_i m_i \mathbf{v}_i}{m}.$$

By using this expression, we can write Eq. (7.3) as

$$\Sigma \mathbf{F} = \frac{d}{dt}(m\mathbf{v}).$$

The total external force on a system of particles equals the rate of change of the product of its total mass and the velocity of its center of mass. Since any object or collection of objects, including a rigid body, can be regarded as a system of particles, this result is one of the most general and elegant in mechanics. Furthermore, if the total mass m is constant, we obtain

$$\Sigma \mathbf{F} = m\mathbf{a},$$

where $\mathbf{a} = d\mathbf{v}/dt$ is the acceleration of the center of mass. The total external force equals the product of the total mass and the acceleration of the center of mass.

Moment–Angular Momentum Principles

We now obtain relations between the sum of the moments due to the external forces on a system of particles and the rate of change of its total angular momentum. We follow the same procedure used in Section 5.4 to relate the angular impulse to the change in the angular momentum.

The position of the ith particle of the system relative to O is related to its position relative to the center of mass by (Fig. 7.3)

$$\mathbf{r}_i = \mathbf{r} + \mathbf{R}_i. \tag{7.5}$$

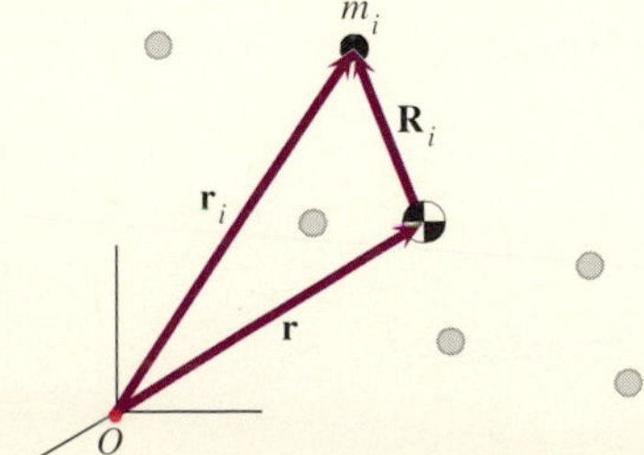

Fig. 7.3
The vector $\mathbf{R}_i$ is the position vector of the ith particle relative to the center of mass.

Multiplying this equation by m_i, summing from 1 to N, and using Eq. (7.4), we find that the positions of the particles relative to the center of mass are related by

$$\sum_i m_i \mathbf{R}_i = \mathbf{0}. \tag{7.6}$$

The total angular momentum of the system about O is the sum of the angular momenta of the particles,

$$\mathbf{H}_O = \sum_i \mathbf{r}_i \times m_i \mathbf{v}_i, \tag{7.7}$$

where $\mathbf{v}_i = d\mathbf{r}_i/dt$. The angular momentum of the system about its center of mass (that is, the angular momentum about the fixed point coincident with the center of mass at the present instant) is

$$\mathbf{H} = \sum_i \mathbf{R}_i \times m_i \mathbf{v}_i. \tag{7.8}$$

By using Eqs. (7.5) and (7.6), it can be shown that

$$\mathbf{H}_O = \mathbf{r} \times m\mathbf{v} + \mathbf{H}. \tag{7.9}$$

This equation expresses the total angular momentum about O as the sum of the angular momentum about O due to the velocity $\mathbf{v}$ of the system's center of mass and the total angular momentum about the center of mass (Fig. 7.4).

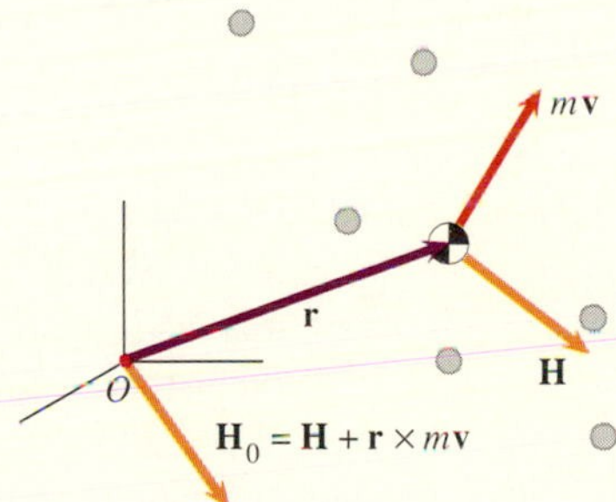

Fig. 7.4
The angular momentum about O equals the sum of the angular momentum about the center of mass and the angular momentum about O due to the velocity of the center of mass.

To obtain relations between the total moment exerted on the system and its total angular momentum, we begin with Newton's second law. We take the cross product of Eq. (7.1) with the position vector $\mathbf{r}_i$ and sum from $i = 1$ to N:

$$\sum_i \sum_j \mathbf{r}_i \times \mathbf{f}_{ij} + \sum_i \mathbf{r}_i \times \mathbf{f}_i^E = \sum_i \mathbf{r}_i \times \frac{d}{dt}(m_i \mathbf{v}_i). \tag{7.10}$$

The term on the right side of this equation is the rate of change of the system's total angular momentum about O:

$$\sum_i \mathbf{r}_i \times \frac{d}{dt}(m_i \mathbf{v}_i) = \sum_i \left[\frac{d}{dt}(\mathbf{r}_i \times m_i \mathbf{v}_i) - \underbrace{\mathbf{v}_i \times m_i \mathbf{v}_i}_{=0} \right] = \frac{d\mathbf{H}_O}{dt}.$$

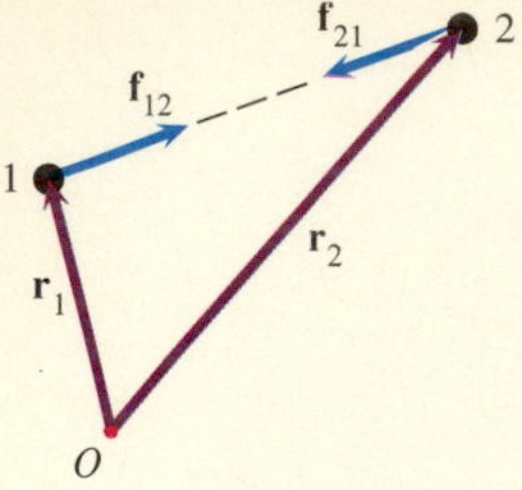

Fig. 7.5
Particles 1 and 2 and the forces they exert on each other. If the forces act along the line between the particles, their total moment about O is zero.

(The second term in brackets vanishes because the cross product of two parallel vectors equals zero.) The first term on the left side of Eq. (7.10) is the sum of the moments about O due to internal forces. This term vanishes if we assume that the internal forces between each pair of particles are not only equal and opposite, but *are directed along the straight line between the two particles*. (This assumption holds except in the case of systems involving electromagnetic forces between charged particles.) For example, consider particles 1 and 2 in Fig. 7.5. If the internal forces are directed along the straight line between the particles, we can write the moment about O due to $\mathbf{f}_{21}$ as $\mathbf{r}_1 \times \mathbf{f}_{21}$, and the total moment about O due to the forces the two particles exert on each other is

$$\mathbf{r}_1 \times \mathbf{f}_{12} + \mathbf{r}_1 \times \mathbf{f}_{21} = \mathbf{r}_1 \times (\mathbf{f}_{12} + \mathbf{f}_{21}) = 0.$$

The second term on the left side of Eq. (7.10) is the sum of the moments about O due to external forces and couples, which we denote by $\Sigma\,\mathbf{M}_O$. Therefore Eq. (7.10) states that the sum of the moments about O due to external forces and couples equals the rate of change of the system's angular momentum about O:

$$\Sigma\,\mathbf{M}_O = \frac{d\mathbf{H}_O}{dt}. \tag{7.11}$$

By using Eq. (7.9), we can also write this result in terms of the total angular momentum about the center of mass,

$$\Sigma\,\mathbf{M}_O = \frac{d}{dt}(\mathbf{r} \times m\mathbf{v} + \mathbf{H}) = \mathbf{r} \times m\mathbf{a} + \frac{d\mathbf{H}}{dt}, \tag{7.12}$$

where $\mathbf{a}$ is the acceleration of the center of mass.

We also need to determine the relation between the sum of the moments about the system's center of mass, which we denote by $\Sigma\,\mathbf{M}$, and the angular momentum about its center of mass. We can obtain this result from Eq. (7.12) by letting the fixed point O be coincident with the center of mass at the present instant. In that case $\Sigma\,\mathbf{M}_O = \Sigma\,\mathbf{M}$ and $\mathbf{r} = \mathbf{0}$, and we see that the sum of the moments about the center of mass equals the rate of change of the angular momentum about the center of mass:

$$\Sigma\,\mathbf{M} = \frac{d\mathbf{H}}{dt}. \tag{7.13}$$

7.3 Derivation of the Equations of Motion

We now derive the equations of motion for a rigid body in planar motion. We have already shown that the total external force on any object equals the product of its mass and the acceleration of its center of mass:

$$\Sigma\,\mathbf{F} = m\mathbf{a}.$$

Therefore this equation, which we refer to as Newton's second law, describes the motion of the center of mass of a rigid body. To derive the equations of angular motion, we first consider rotation about a fixed axis, then general planar motion.

Rotation About a Fixed Axis

Suppose that a rigid body rotates about a fixed axis L_O through a fixed point O. In terms of a coordinate system with the z axis aligned with L_O (Fig. 7.6a), we can express the angular velocity vector as $\boldsymbol{\omega} = \omega\mathbf{k}$, and the velocity of the ith particle is $d\mathbf{r}_i/dt = \boldsymbol{\omega} \times \mathbf{r}_i = \omega\mathbf{k} \times \mathbf{r}_i$. Let $\Sigma M_O = \Sigma \mathbf{M}_O \cdot \mathbf{k}$ be the sum of the moments about L_O. From Eqs. (7.7) and (7.11),

$$\Sigma M_O = \frac{dH_O}{dt}, \tag{7.14}$$

where

$$H_O = \mathbf{H}_O \cdot \mathbf{k} = \sum_i [\mathbf{r}_i \times m_i(\omega\mathbf{k} \times \mathbf{r}_i)] \cdot \mathbf{k} \tag{7.15}$$

is the angular momentum about L_O. Using the identity $\mathbf{U} \cdot (\mathbf{V} \times \mathbf{W}) = (\mathbf{U} \times \mathbf{V}) \cdot \mathbf{W}$, we can write Eq. (7.15) as

$$H_O = \sum_i m_i(\mathbf{k} \times \mathbf{r}_i) \cdot (\mathbf{k} \times \mathbf{r}_i)\omega = \sum_i m_i\,|\mathbf{k} \times \mathbf{r}_i|^2\omega. \tag{7.16}$$

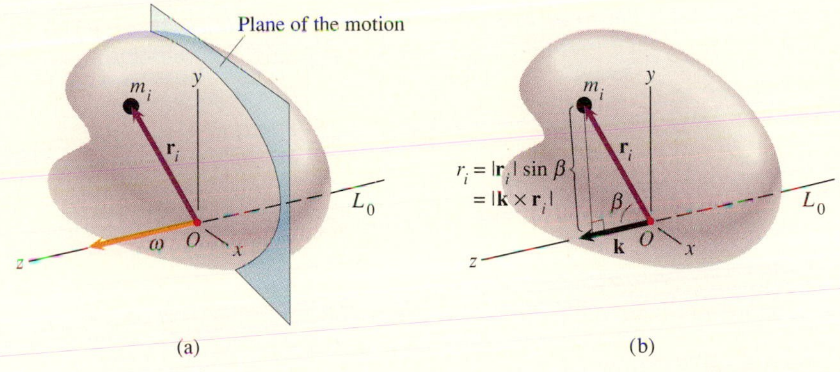

Fig. 7.6
(a) A coordinate system with the z axis aligned with the axis of rotation L_O.
(b) The magnitude of $\mathbf{k} \times \mathbf{r}_i$ is the perpendicular distance from the axis of rotation to m_i.

In Fig. 7.6(b), we show that $|\mathbf{k} \times \mathbf{r}_i|$ is the perpendicular distance from L_O to the ith particle, which we denote by r_i. Using the definition of the moment of inertia of the rigid body about L_O,

$$I_O = \sum_i m_i r_i^2,$$

we can write Eq. (7.16) as

$$H_O = I_O\omega.$$

Substituting this expression into Eq. (7.14), we obtain the equation of angular motion for a rigid body rotating about a fixed axis O:

$$\Sigma M_O = I_O \alpha. \tag{7.17}$$

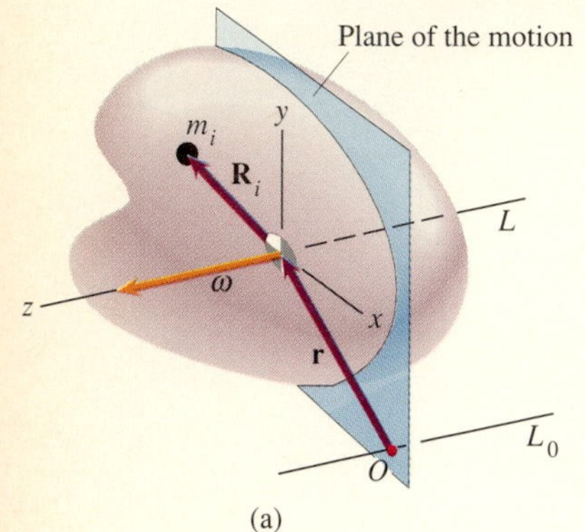

(a)

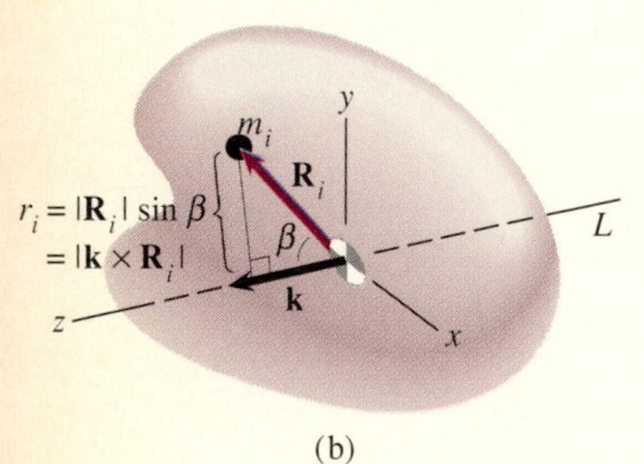

(b)

Fig. 7.7

(a) A coordinate system with the z axis aligned with L.
(b) The magnitude of $\mathbf{k} \times \mathbf{R}_i$ is the perpendicular distance from L to m_i.

General Planar Motion

Let L_O be the axis through a fixed point O that is perpendicular to the plane of the motion of a rigid body, and let L be the parallel axis through the center of mass (Fig. 7.7a). We do *not* assume that the rigid body rotates about L_O. In terms of the coordinate system shown, we can express the angular velocity vector as $\boldsymbol{\omega} = \omega\mathbf{k}$, and the velocity of the ith particle relative to the center of mass is $d\mathbf{R}_i/dt = \omega\mathbf{k} \times \mathbf{R}_i$. From Eqs. (7.8) and (7.12),

$$\Sigma M_O = \frac{d}{dt}[(\mathbf{r} \times m\mathbf{v}) \cdot \mathbf{k} + H], \tag{7.18}$$

where

$$H = \mathbf{H} \cdot \mathbf{k} = \sum_i [\mathbf{R}_i \times m_i(\omega\mathbf{k} \times \mathbf{R}_i)] \cdot \mathbf{k}$$

is the angular momentum about L. Using the same identity we applied to Eq. (7.15), we can write this equation for H as

$$H = \sum_i m_i(\mathbf{k} \times \mathbf{R}_i) \cdot (\mathbf{k} \times \mathbf{R}_i)\omega = \sum_i m_i\,|\mathbf{k} \times \mathbf{R}_i|^2\omega. \tag{7.19}$$

The term $|\mathbf{k} \times \mathbf{R}_i| = r_i$ is the perpendicular distance from L to the ith particle (Fig. 7.7b). In terms of the moment of inertia of the rigid body about L,

$$I = \sum_i m_i r_i^2,$$

Eq. (7.19) states that the rigid body's angular momentum about L is

$$H = I\omega.$$

Substituting this expression into Eq. (7.18), we obtain

$$\Sigma M_O = \frac{d}{dt}[(\mathbf{r} \times m\mathbf{v}) \cdot \mathbf{k} + I\omega] = (\mathbf{r} \times m\mathbf{a}) \cdot \mathbf{k} + I\alpha. \tag{7.20}$$

With this equation we can obtain the relation between the sum of the moments about L, which we denote by ΣM, and the angular acceleration. If we let the fixed axis L_O be coincident with L at the present instant, $\Sigma M_O = \Sigma M$ and $\mathbf{r} = \mathbf{0}$, and from Eq. (7.20) we obtain

$$\Sigma M = I\alpha.$$

The sum of the moments about L equals the product of the moment of inertia about L and the angular acceleration.

7.4 Applications

We have seen that the equations of motion for a rigid body in planar motion include Newton's second law,

$$\boxed{\Sigma \mathbf{F} = m\mathbf{a},} \tag{7.21}$$

where **a** is the acceleration of the center of mass, and an equation relating the moments due to forces and couples to the angular acceleration. If the rigid body rotates about a fixed axis O, the total moment about O equals the product of the moment of inertia about O and the angular acceleration:

$$\boxed{\Sigma M_O = I_O \alpha.} \tag{7.22}$$

In *any* planar motion, the total moment about the center of mass equals the product of the moment of inertia about the center of mass and the angular acceleration:

$$\boxed{\Sigma M = I \alpha.} \tag{7.23}$$

Of course, this equation applies to the case of rotation about a fixed axis, but for that type of motion you will usually find it more convenient to use Eq. (7.22).

When you apply these equations, your objective may be to obtain information about an object's motion, or to determine the values of unknown forces or couples acting on it, or both. This typically involves three steps:

1. **Draw the free-body diagram**—Isolate the object and identify the external forces and couples acting on it.
2. **Apply the equations of motion**—Write equations of motion suitable for the type of motion. You should choose an appropriate coordinate system for applying Newton's second law. For example, if the center of mass moves in a circular path, you may find it advantageous to use normal and tangential components.
3. **Determine kinematic relationships**—If necessary, supplement the equations of motion with relationships between the acceleration of the center of mass and the angular acceleration.

As we show in the following sections, your approach will depend in part on the type of motion involved.

Translation

If a rigid body is in translation (Fig. 7.8), you need only Newton's second law to determine its motion. There is no rotational motion to determine. Nevertheless, you may need to apply the angular equation of motion to determine unknown forces or couples. Since $\alpha = 0$, Eq. (7.23) states that the total moment *about the center of mass* equals zero:

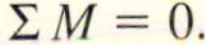

$$\Sigma M = 0.$$

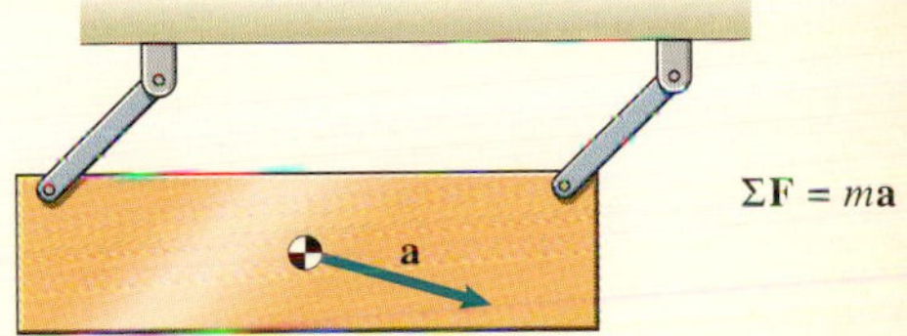

Fig. 7.8
A rigid body in translation. There is no rotational motion to determine.

Example 7.1

The mass of the airplane in Fig. 7.9 is $m = 250$ Mg (megagrams), and the thrust of its engines during its takeoff roll is $T = 700$ kN. Determine the airplane's acceleration and the normal forces exerted on its wheels at A and B. Neglect the horizontal forces exerted on its wheels.

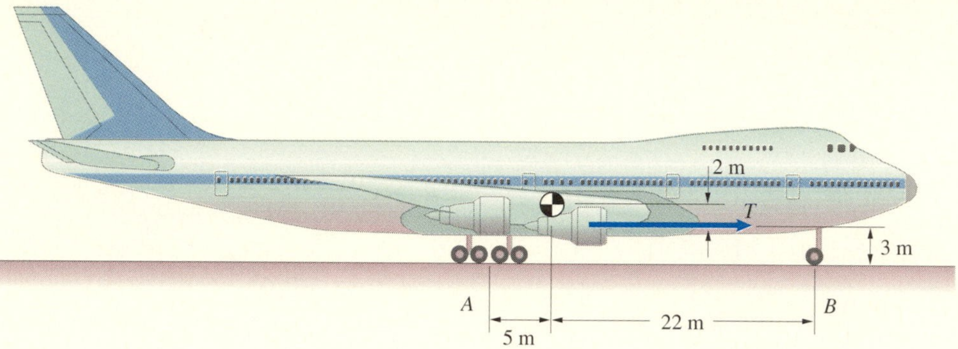

Fig. 7.9

STRATEGY

The airplane is in translation during its takeoff roll, so the sum of the moments about its center of mass equals zero. Using this condition and Newton's second law, we can determine the airplane's acceleration and the normal forces exerted on its wheels.

SOLUTION

Draw the Free-Body Diagram We draw the free-body diagram in Fig. (a), showing the airplane's weight and the normal forces A and B exerted on the wheels.

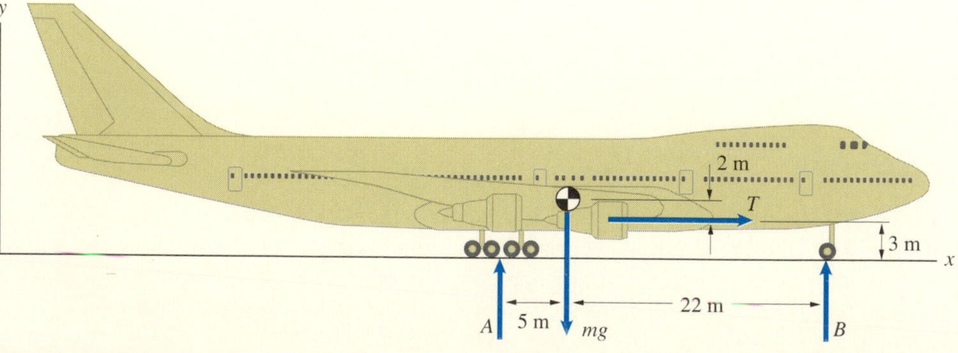

(a) Free-body diagram of the airplane.

Apply the Equations of Motion In terms of the coordinate system in Fig. (a), Newton's second law is

$$\Sigma F_x = T = ma_x,$$

$$\Sigma F_y = A + B - mg = 0.$$

From the first equation, the airplane's acceleration is

$$a_x = \frac{T}{m} = \frac{700{,}000 \text{ N}}{250{,}000 \text{ kg}} = 2.8 \text{ m/s}^2.$$

The angular equation of motion is

$$\Sigma M = (2)T + (22)B - (5)A = 0.$$

Solving this equation together with the second equation we obtained from Newton's second law for A and B, we obtain $A = 2050$ kN, $B = 402$ kN.

DISCUSSION

When an object is in equilibrium, the sum of the moments about any point due to the external forces and couples acting on it is zero. But you must remember that when a translating rigid body is not in equilibrium, you know only that the sum of the moments *about the center of mass* is zero. It would be instructive for you to try reworking this example by assuming that the sum of the moments about A or B is zero. You will not obtain the correct values for the normal forces exerted on the wheels.

Rotation About a Fixed Axis

In the case of rotation about a fixed axis (Fig. 7.10), you need only Eq. (7.22) to determine the rotational motion, although you may also need Newton's second law to determine unknown forces or couples.

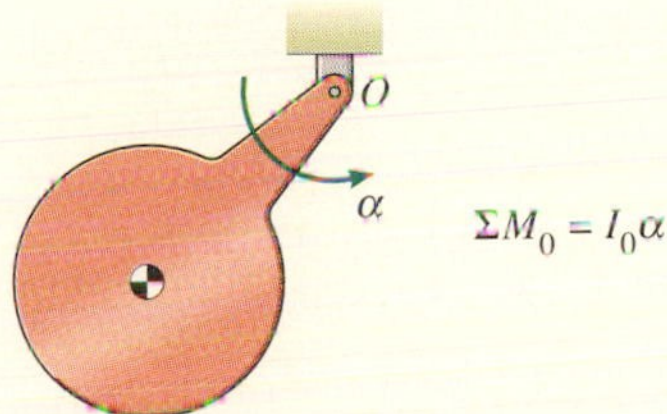

Fig. 7.10

A rigid body rotating about O. You need only the equation of angular motion about O to determine its angular acceleration.

Example 7.2

The 100-lb crate in Fig. 7.11 is pulled up the inclined surface by the winch. The coefficient of kinetic friction between the crate and the surface is $\mu_k = 0.4$. The moment of inertia of the drum on which the cable is wound, including the cable wound on the drum, is $I_A = 3$ slug-ft^2. If the motor exerts a couple $M = 40$ ft-lb on the drum, what is the crate's acceleration?

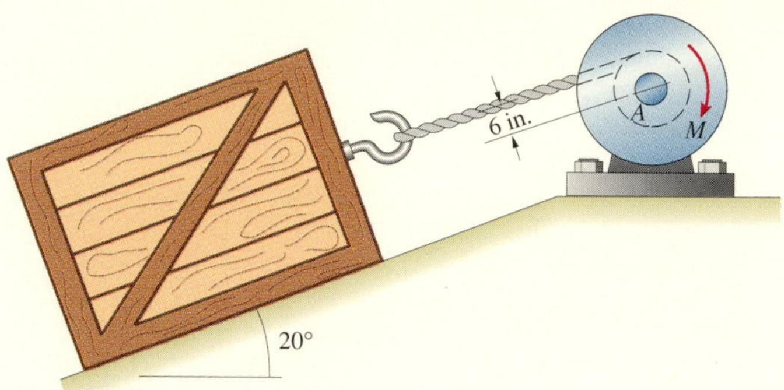

Fig. 7.11

STRATEGY

We will draw separate free-body diagrams of the crate and drum and apply the equations of motion to them individually. The drum rotates about a fixed axis, so we can use the equation of angular motion about the axis to determine its angular acceleration. To complete the solution, we must determine the relationship between the crate's acceleration and the drum's angular acceleration.

SOLUTION

Draw the Free-Body Diagrams We draw the free-body diagrams in Fig. (a), showing the equal forces exerted on the crate and the drum by the cable.

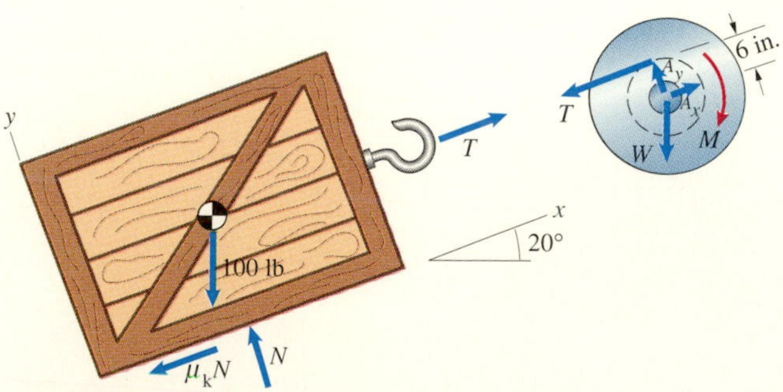

(a) Free-body diagrams of the crate and the drum.

Apply the Equations of Motion We denote the crate's acceleration up the inclined surface by a_x and the *clockwise* angular acceleration of the drum by α (Fig. b). Newton's second law for the crate is

$$\Sigma F_x = T - 100 \sin 20° - \mu_k N = \left(\frac{100}{32.2}\right) a_x,$$

$$\Sigma F_y = N - 100 \cos 20° = 0.$$

(b) The crate's acceleration and the angular acceleration of the drum.

Solving the second equation for N and substituting it into the first one, we obtain

$$T - 100 \sin 20° - (0.4)(100 \cos 20°) = \left(\frac{100}{32.2}\right) a_x.$$

The equation of angular motion for the drum is

$$\Sigma M_A = M - (0.5 \text{ ft})T = I_A \alpha.$$

We eliminate T between these two equations, obtaining

$$2M - 100 \sin 20° - (0.4)(100 \cos 20°) = \left(\frac{100}{32.2}\right) a_x + 2I_A \alpha. \quad (7.24)$$

Our last step is to determine the relation between a_x and α.

Determine Kinematic Relationship The tangential component of acceleration of the drum at the point where the cable begins winding onto it is equal to the crate's acceleration (Fig. b):

$$a_x = (0.5 \text{ ft})\alpha.$$

Using this relation, the solution of Eq. (7.24) for a_x is

$$a_x = \frac{2M - 100 \sin 20° - (0.4)(100 \cos 20°)}{(100/32.2) + 4I_A} = 0.544 \text{ ft/s}^2.$$

DISCUSSION

Notice that, for convenience, we defined the angular acceleration α to be positive in the clockwise direction so that a positive α would correspond to a positive a_x.

Example 7.3

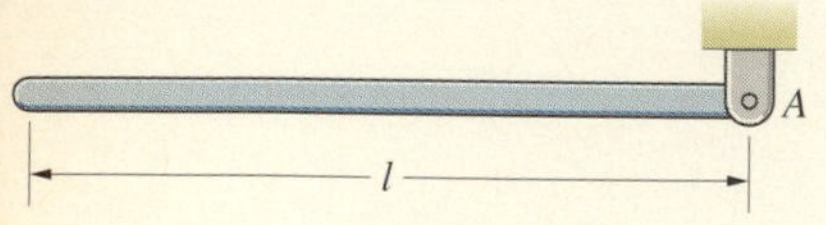

Fig. 7.12

The slender bar of mass m in Fig. 7.12 is released from rest in the horizontal position shown. At that instant, determine the bar's angular acceleration and the force exerted on the bar by the support A.

STRATEGY

Since the bar rotates about a fixed point, we can use Eq. (7.22) to determine its angular acceleration. The advantage of using this equation instead of Eq. (7.23) is that the unknown reactions at A will not appear in the equation of angular motion. Once we know the angular acceleration, we can determine the acceleration of the center of mass and use Newton's second law to obtain the reactions at A.

SOLUTION

Draw the Free-Body Diagram In Fig. (a) we draw the free-body diagram of the bar, showing the reactions at the pin support.

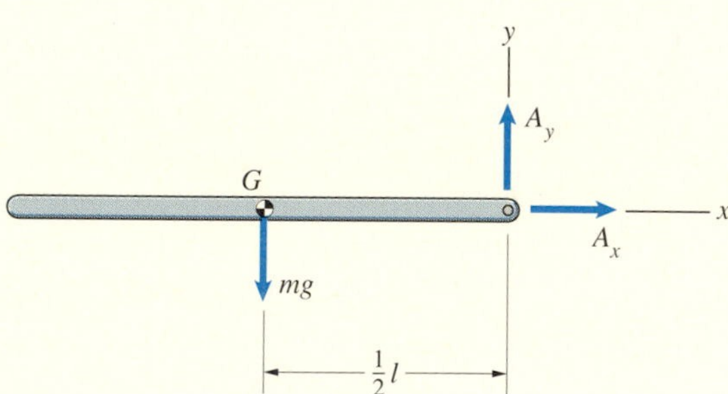

(a) Free-body diagram of the bar.

Apply the Equations of Motion Let the acceleration of the center of mass G of the bar be $\mathbf{a}_G = a_x\mathbf{i} + a_y\mathbf{j}$, and let its counterclockwise angular acceleration be α (Fig. b). Newton's second law for the bar is

$$\Sigma F_x = A_x = ma_x,$$

$$\Sigma F_y = A_y - mg = ma_y.$$

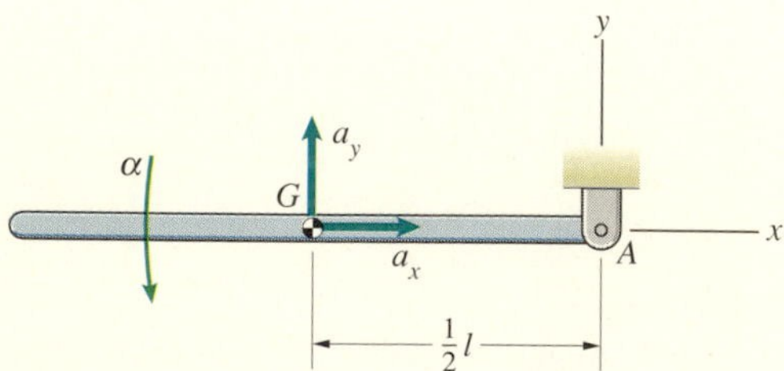

(b) The angular acceleration and components of the acceleration of the center of mass.

The equation of angular motion about the fixed point A is

$$\Sigma M_A = \left(\frac{1}{2}l\right) mg = I_A\alpha. \tag{7.25}$$

The moment of inertia of a slender bar about its center of mass is $I = \frac{1}{12}ml^2$. (See Appendix C.) Using the parallel-axis theorem the moment of inertia of the bar about A is

$$I_A = I + d^2m = \frac{1}{12}ml^2 + \left(\frac{1}{2}l\right)^2 m = \frac{1}{3}ml^2.$$

Substituting this expression into Eq. (7.25), we obtain the angular acceleration:

$$\alpha = \frac{\frac{1}{2}mgl}{\frac{1}{3}ml^2} = \frac{3}{2}\frac{g}{l}.$$

Determine Kinematic Relationships To determine the reactions A_x and A_y, we need to determine the acceleration components a_x and a_y. We can do so by expressing the acceleration of G in terms of the acceleration of A:

$$\mathbf{a}_G = \mathbf{a}_A + \boldsymbol{\alpha} \times \mathbf{r}_{G/A} - \omega^2 \mathbf{r}_{G/A}.$$

At the instant the bar is released, its angular velocity $\omega = 0$. Also, $\mathbf{a}_A = \mathbf{0}$, so we obtain

$$\mathbf{a}_G = a_x\mathbf{i} + a_y\mathbf{j} = (\alpha\,\mathbf{k}) \times \left(-\frac{1}{2}l\mathbf{i}\right) = -\frac{1}{2}l\alpha\,\mathbf{j}.$$

Equating **i** and **j** components, we obtain

$$a_x = 0,$$

$$a_y = -\frac{1}{2}l\alpha = -\frac{3}{4}g.$$

Substituting these acceleration components into Newton's second law, the reactions at A at the instant the bar is released are

$$A_x = 0,$$

$$A_y = mg + m\left(-\frac{3}{4}g\right) = \frac{1}{4}mg.$$

DISCUSSION

We could have determined the acceleration of G in a less formal way. Since G describes a circular path about A, we know the magnitude of the tangential component of acceleration equals the product of the radial distance from A to G and the angular acceleration. Because of the directions in which we define α and a_x to be positive, $a_y = -(\frac{1}{2}l)\alpha$. Also, the normal component of the acceleration of G equals the square of its velocity divided by the radius of its circular path. Since its velocity equals zero at the instant the bar is released, $a_x = 0$.

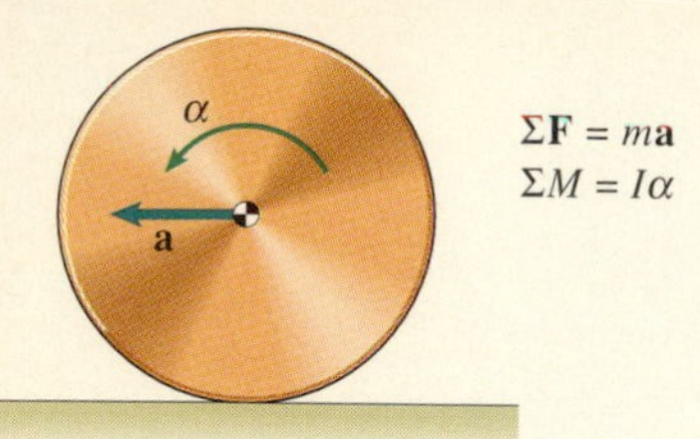

Fig. 7.13

A rigid body in planar motion. You must apply both Newton's second law and the equation of angular motion about the center of mass.

General Planar Motion

If a rigid body undergoes both translation and rotation (Fig. 7.13), you need to use both Newton's second law and the equation of angular motion. If the motion of the center of mass and the rotational motion are not independent—for example, when an object rolls—you will find that there are more unknown quantities than equations of motion. In such cases, you can obtain additional equations by using kinematics to relate the acceleration of the center of mass to the angular acceleration.

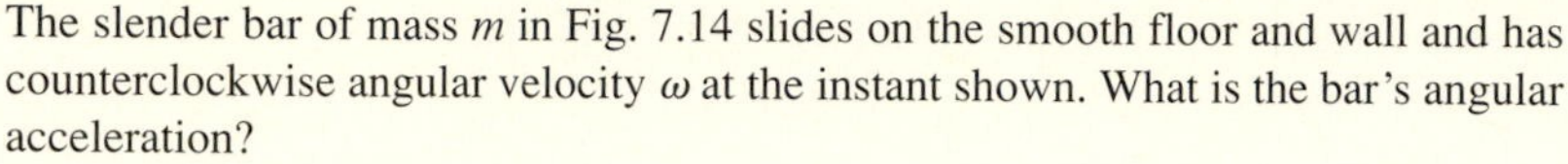

Example 7.4

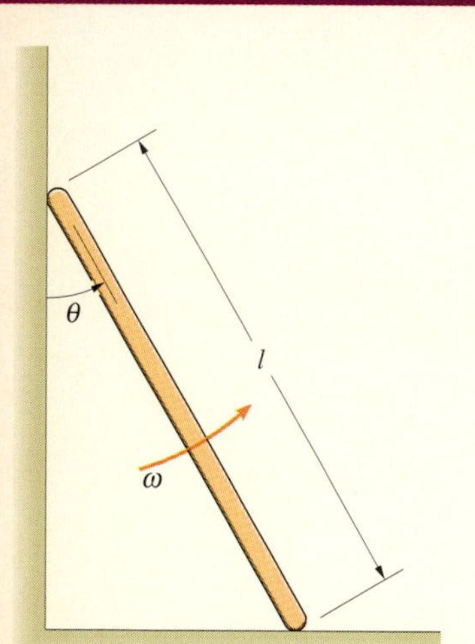

Fig. 7.14

The slender bar of mass m in Fig. 7.14 slides on the smooth floor and wall and has counterclockwise angular velocity ω at the instant shown. What is the bar's angular acceleration?

SOLUTION

Draw the Free-Body Diagram We draw the free-body diagram in Fig. (a), showing the bar's weight and the normal forces exerted by the floor and wall.

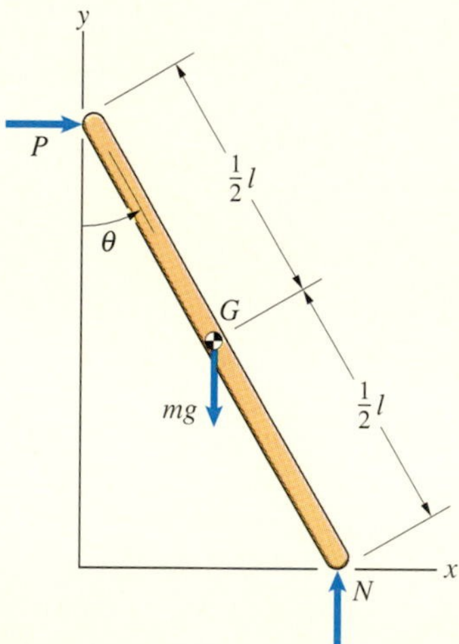

(a) Free-body diagram of the bar.

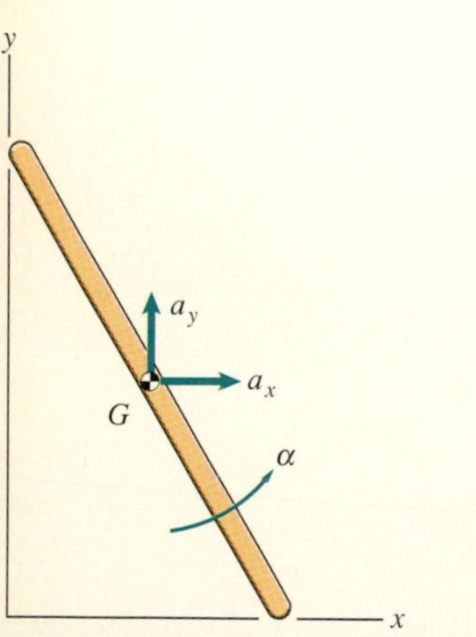

(b) Components of the acceleration of the center of mass and the angular acceleration.

Apply the Equations of Motion Writing the acceleration of the center of mass G as $\mathbf{a}_G = a_x\mathbf{i} + a_y\mathbf{j}$ (Fig. b), Newton's second law is

$$\Sigma F_x = P = ma_x,$$

$$\Sigma F_y = N - mg = ma_y.$$

Let α be the bar's counterclockwise angular acceleration (Fig. b). The equation of angular motion is

$$\Sigma M = N\left(\frac{1}{2}l\sin\theta\right) - P\left(\frac{1}{2}l\cos\theta\right) = I\alpha,$$

where I is the moment of inertia of the bar about its center of mass. We have three equations of motion in terms of five unknowns: P, N, a_x, a_y, and α. To complete the solution, we must relate the acceleration of the center of mass of the bar to its angular acceleration.

Determine Kinematic Relationships Although we don't know the accelerations of the endpoints A and B (Fig. c), we know that A moves horizontally and B moves vertically. We can use this information to obtain the needed relations between the acceleration of the center of mass and the angular acceleration. Expressing the acceleration of A as $\mathbf{a}_A = a_A\mathbf{i}$, we can write the acceleration of the center of mass as (Figs. b and c)

$$\mathbf{a}_G = \mathbf{a}_A + \boldsymbol{\alpha}\times\mathbf{r}_{G/A} - \omega^2\mathbf{r}_{G/A}:$$

$$a_x\mathbf{i} + a_y\mathbf{j} = a_A\mathbf{i} + \begin{vmatrix} \mathbf{i} & \mathbf{j} & \mathbf{k} \\ 0 & 0 & \alpha \\ -\frac{1}{2}l\sin\theta & \frac{1}{2}l\cos\theta & 0 \end{vmatrix} - \omega^2\left(-\frac{1}{2}l\sin\theta\,\mathbf{i} + \frac{1}{2}l\cos\theta\,\mathbf{j}\right).$$

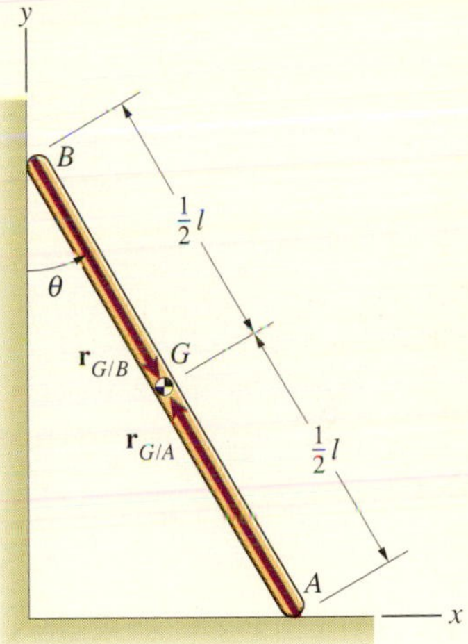

(c) Position vectors of G relative to the endpoints A and B.

Taking advantage of the fact that $\mathbf{a}_A$ has no $\mathbf{j}$ component, we equate the $\mathbf{j}$ components in this equation, obtaining

$$a_y = -\frac{1}{2}l(\alpha\sin\theta + \omega^2\cos\theta).$$

Now we express the acceleration of B as $\mathbf{a}_B = a_B\mathbf{j}$ and write the acceleration of the center of mass as

$$\mathbf{a}_G = \mathbf{a}_B + \boldsymbol{\alpha}\times\mathbf{r}_{G/B} - \omega^2\mathbf{r}_{G/B}:$$

$$a_x\mathbf{i} + a_y\mathbf{j} = a_B\mathbf{j} + \begin{vmatrix} \mathbf{i} & \mathbf{j} & \mathbf{k} \\ 0 & 0 & \alpha \\ \frac{1}{2}l\sin\theta & -\frac{1}{2}l\cos\theta & 0 \end{vmatrix} - \omega^2\left(\frac{1}{2}l\sin\theta\,\mathbf{i} - \frac{1}{2}l\cos\theta\,\mathbf{j}\right).$$

We equate the $\mathbf{i}$ components in this equation, obtaining

$$a_x = \frac{1}{2}l(\alpha\cos\theta - \omega^2\sin\theta).$$

With these two kinematic relationships, we have five equations in five unknowns. Solving them for the angular acceleration and using the relation $I = \frac{1}{12}ml^2$ for the bar's moment of inertia (Appendix C), we obtain

$$\alpha = \frac{3}{2}\frac{g}{l}\sin\theta.$$

DISCUSSION

Notice that by expressing the acceleration of G in terms of the accelerations of the endpoints, we introduced into the solution the constraints imposed on the bar by the floor and wall: We required that point A move horizontally and that point B move vertically.

Example 7.5

The slender bar in Fig. 7.15 has mass m and is pinned at A to a metal block of mass m_B that rests on a smooth level surface. The system is released from rest in the position shown. What is the bar's angular acceleration at the instant of release?

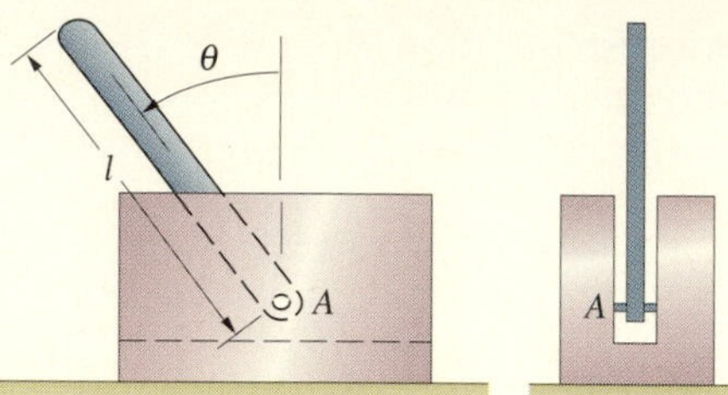

Fig. 7.15

STRATEGY

We must draw free-body diagrams of the bar and the block and apply the equations of motion to them individually. To complete the solution, we must also relate the acceleration of the bar's center of mass and its angular acceleration to the acceleration of the block.

SOLUTION

Draw the Free-Body Diagrams We draw the free-body diagrams of the bar and block in Fig. (a). Notice the opposite forces they exert on each other where they are pinned together.

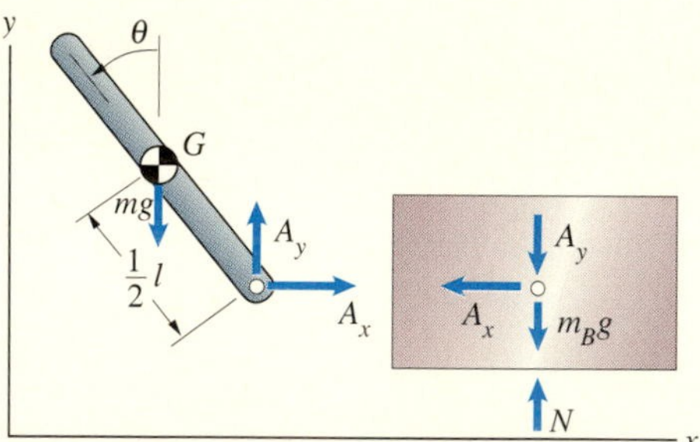

(a) Free-body diagrams of the bar and the block.

Apply the Equations of Motion Writing the acceleration of the center of mass of the bar as $\mathbf{a}_G = a_x\mathbf{i} + a_y\mathbf{j}$ (Fig. b), Newton's second law for the bar is

$$\Sigma F_x = A_x = ma_x,$$

$$\Sigma F_y = A_y - mg = ma_y.$$

Letting α be the bar's counterclockwise angular acceleration (Fig. b), its equation of angular motion is

$$\Sigma M = A_x\left(\frac{1}{2}l\cos\theta\right) + A_y\left(\frac{1}{2}l\sin\theta\right) = I\alpha.$$

(b) Definitions of the accelerations.

We express the block's acceleration as $a_A\mathbf{i}$ (Fig. b) and write Newton's second law for the block:

$$\Sigma F_x = -A_x = m_B a_A,$$

$$\Sigma F_y = N - A_y - m_B g = 0.$$

Determine Kinematic Relationships To relate the bar's motion to that of the block, we express the acceleration of the bar's center of mass in terms of the acceleration of point A (Figs. b and c):

$$\mathbf{a}_G = \mathbf{a}_A + \boldsymbol{\alpha} \times \mathbf{r}_{G/A} - \omega^2 \mathbf{r}_{G/A},$$

$$a_x\mathbf{i} + a_y\mathbf{j} = a_A\mathbf{i} + \begin{vmatrix} \mathbf{i} & \mathbf{j} & \mathbf{k} \\ 0 & 0 & \alpha \\ -\frac{1}{2}l\sin\theta & \frac{1}{2}l\cos\theta & 0 \end{vmatrix} - \mathbf{0}.$$

Equating **i** and **j** components, we obtain

$$a_x = a_A - \frac{1}{2}l\alpha\cos\theta,$$

$$a_y = -\frac{1}{2}l\alpha\sin\theta.$$

We have five equations of motion and two kinematic relations in terms of seven unknowns: A_x, A_y, N, a_x, a_y, α, and a_A. Solving them for the angular acceleration and using the relation $I = \frac{1}{12}ml^2$ for the bar's moment of inertia, we obtain

$$\alpha = \frac{\frac{3}{2}(g/l)\sin\theta}{1 - \frac{3}{4}[m/(m + m_B)]\cos^2\theta}.$$

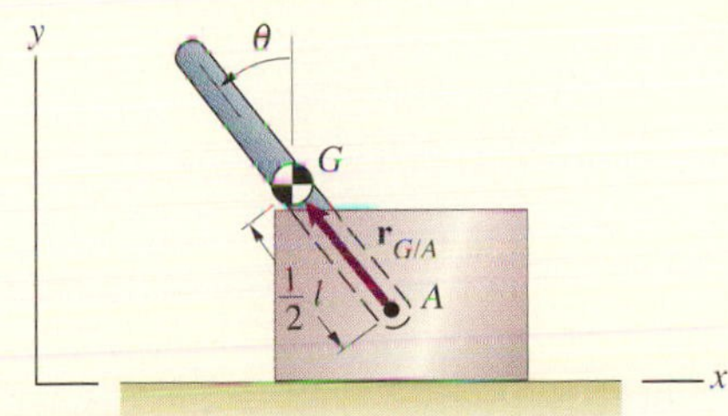

(c) Position vector of G relative to A.

Example 7.6

The drive wheel in Fig. 7.16 rolls on the horizontal track. The wheel is subjected to a downward force F_A by its axle A and a horizontal force F_C by the connecting rod. The mass of the wheel is m and the moment of inertia about its center of mass is I. The center of mass G is offset a distance b from the wheel's center. At the instant shown, the wheel has a counterclockwise angular velocity ω. What is the wheel's angular acceleration?

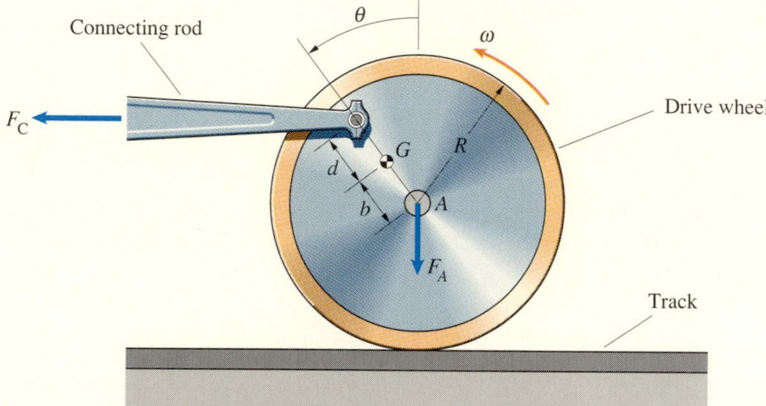

Fig. 7.16

SOLUTION

Draw the Free-Body Diagram We draw the free-body diagram of the drive wheel in Fig. (a), showing its weight and the normal and friction forces exerted by the track.

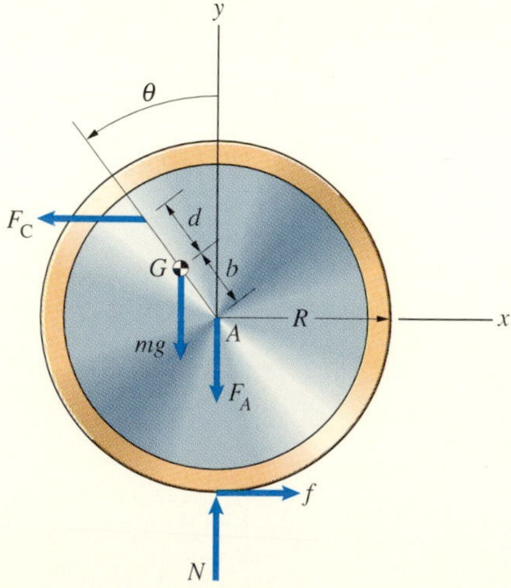

(a) Free-body diagram of the wheel.

Apply the Equations of Motion Writing the acceleration of the center of mass G as $\mathbf{a}_G = a_x\mathbf{i} + a_y\mathbf{j}$ (Fig. b), Newton's second law is

$$\Sigma F_x = f - F_C = ma_x,$$

$$\Sigma F_y = N - F_A - mg = ma_y.$$

Remember that we must express the equation of angular motion in terms of the sum of the moments about the center of mass G, *not the center of the wheel*. The equation of angular motion is (Figs. a and b)

$$\Sigma M = F_C(d\cos\theta) - F_A(b\sin\theta) + N(b\sin\theta) + f(b\cos\theta + R) = I\alpha.$$

We have three equations of motion in terms of five unknowns: N, f, a_x, a_y, and α. To complete the solution, we must relate the acceleration of the wheel's center of mass to its angular acceleration.

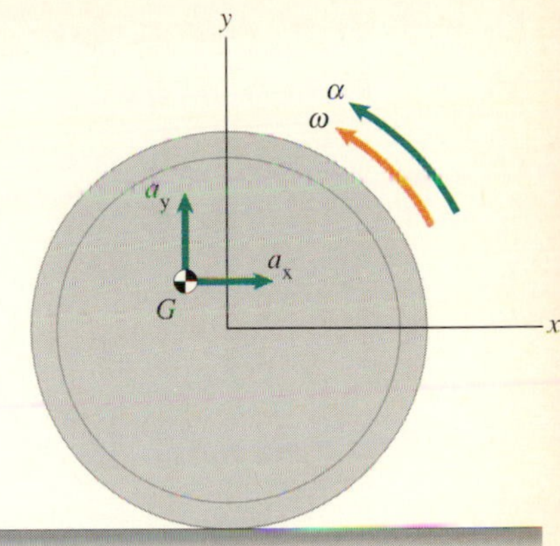

(b) Components of the acceleration of G, the angular velocity, and the angular acceleration.

Determine Kinematic Relationships The acceleration of the center A of the rolling wheel is $\mathbf{a}_A = -R\alpha\mathbf{i}$. By expressing the acceleration of the center of mass, $\mathbf{a}_G$, in terms of $\mathbf{a}_A$ (Figs. b and c), we can obtain relations between the components of $\mathbf{a}_G$ and α:

$$\mathbf{a}_G = \mathbf{a}_A + \boldsymbol{\alpha} \times \mathbf{r}_{G/A} - \omega^2\mathbf{r}_{G/A},$$

$$a_x\mathbf{i} + a_y\mathbf{j} = -R\alpha\mathbf{i} + \begin{vmatrix} \mathbf{i} & \mathbf{j} & \mathbf{k} \\ 0 & 0 & \alpha \\ -b\sin\theta & b\cos\theta & 0 \end{vmatrix} - \omega^2(-b\sin\theta\,\mathbf{i} + b\cos\theta\,\mathbf{j}).$$

We equate the **i** and **j** components in this equation, obtaining

$$a_x = -R\alpha - b\alpha\cos\theta + b\omega^2\sin\theta,$$

$$a_y = -b\alpha\sin\theta - b\omega^2\cos\theta.$$

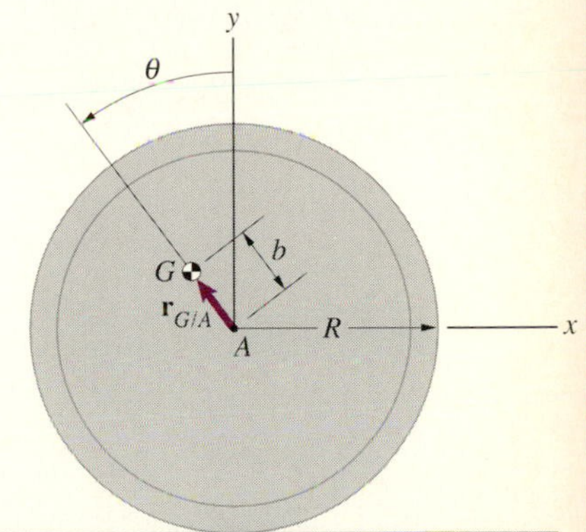

(c) Position vector of G relative to A.

With these two kinematic relationships, we have five equations in five unknowns. Solving them for the angular acceleration, we obtain

$$\alpha = \frac{F_C(R + b\cos\theta + d\cos\theta) + mgb\sin\theta + mbR\omega^2\sin\theta}{m(b^2 + 2bR\cos\theta + R^2) + I}.$$

7.5 D'Alembert's Principle

In this section we describe an alternative method for obtaining the equations of planar motion for a rigid body. By writing Newton's second law as

$$\Sigma \mathbf{F} + (-m\mathbf{a}) = \mathbf{0}, \tag{7.26}$$

we can regard it as an "equilibrium" equation stating that the sum of the forces, including an **inertial force** $-m\mathbf{a}$, equals zero (Fig. 7.17). To state the equation of angular motion in an equivalent way, we use Eq. (7.20), which relates the total moment about a fixed point O to the acceleration of the center of mass and the angular acceleration in general planar motion:

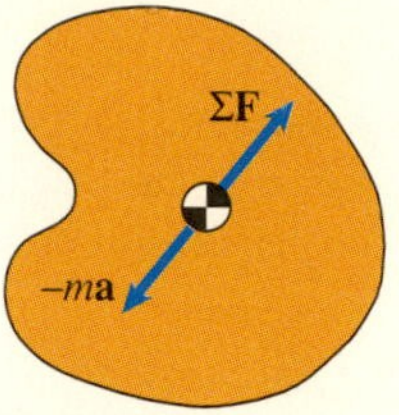

Fig. 7.17

The sum of the external forces and the inertial force is zero.

$$\Sigma M_O = (\mathbf{r} \times m\mathbf{a}) \cdot \mathbf{k} + I\alpha.$$

We write this equation as

$$\Sigma M_O + [\mathbf{r} \times (-m\mathbf{a})] \cdot \mathbf{k} + (-I\alpha) = 0. \tag{7.27}$$

The term $[\mathbf{r} \times (-m\mathbf{a})] \cdot \mathbf{k}$ is the moment about O due to the inertial force $-m\mathbf{a}$. We can therefore regard this equation as an "equilibrium" equation stating that the sum of the moments about any point, including the moment due to the inertial force $-m\mathbf{a}$ acting at the center of mass and an **inertial couple** $-I\alpha$, equals zero.

Stated in this way, the equations of motion for a rigid body are analogous to the equations for static equilibrium: The sum of the forces equals zero and the sum of the moments about any point equals zero when we properly account for inertial forces and couples. This is called **D'Alembert's principle**.

If we define ΣM_0 and α to be positive in the counterclockwise direction, the unit vector $\mathbf{k}$ in Eq. (7.27) points out of the page and the term $[\mathbf{r} \times (-m\mathbf{a})] \cdot \mathbf{k}$ is the counterclockwise moment due to the inertial force. This vector operation determines the moment, or you can evaluate it by using the fact that its magnitude is the product of the magnitude of the inertial force and the perpendicular distance from point O to the line of action of the force (Fig. 7.18a). The moment is positive if it is counterclockwise, as in Fig.

7.18(a), and negative if it is clockwise. Notice that the sense of the inertial couple is opposite to that of the angular acceleration (Fig. 7.18b).

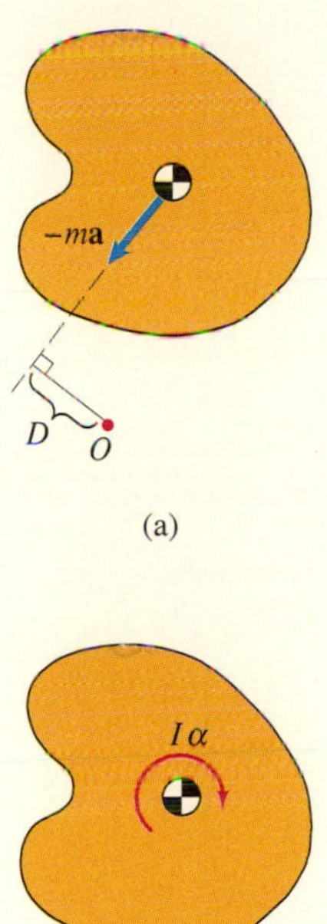

Fig. 7.18

(a) The magnitude of the moment due to the inertial force is $|-m\mathbf{a}|D$.

(b) A clockwise inertial couple results from a counterclockwise angular acceleration.

In the following examples we apply D'Alembert's principle to planar motions of rigid bodies. The sequence of steps—draw the free-body diagram, apply the "equilibrium" equations, and determine kinematic relationships if necessary—is the same as in applying the traditional equations of motion. In using this method you must be particularly careful to assign the correct signs to the terms in your equations.

Example 7.7

The mass of the airplane in Fig. 7.19 is $m = 250$ Mg, and the thrust of its engines during its takeoff roll is $T = 700$ kN. Use D'Alembert's principle to determine the airplane's acceleration and the normal forces exerted on its wheels at A and B. Neglect the horizontal forces exerted on its wheels.

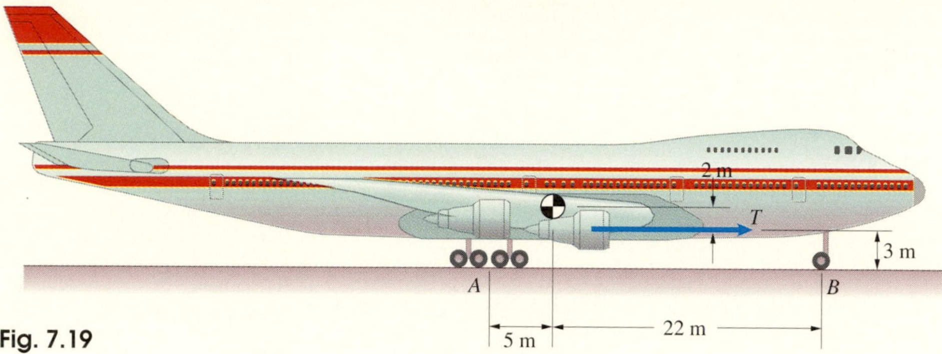

Fig. 7.19

SOLUTION

Draw the Free-Body Diagram In terms of the coordinate system in Fig. (a), we can write the airplane's acceleration as $\mathbf{a} = a_x\mathbf{i}$. On the free-body diagram we show the airplane's weight, the normal forces A and B exerted on the wheels, and the inertial force $-m\mathbf{a} = -ma_x\mathbf{i}$.

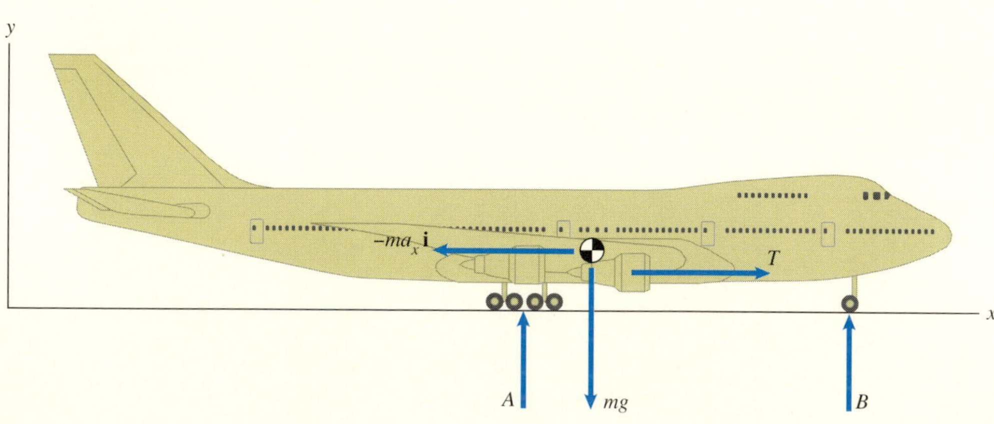

(a) Free-body diagram of the airplane.

Apply the "Equilibrium Equations" Equation (7.26) is

$$\Sigma\mathbf{F} + (-m\mathbf{a}) = \mathbf{0}:$$

$$T\mathbf{i} + (A + B - mg)\mathbf{j} + (-ma_x\mathbf{i}) = \mathbf{0}.$$

Equating **i** and **j** components, we obtain

$$T = ma_x,$$

$$A + B = mg.$$

From the first equation, the airplane's acceleration is

$$a_x = \frac{T}{m} = \frac{700{,}000\ \text{N}}{250{,}000\ \text{kg}} = 2.8\ \text{m/s}^2,$$

and the inertial force is $-ma_x\mathbf{i} = -700\mathbf{i}$ (kN). (See Fig. b.)

In applying Eq. (7.27), we can select any point we wish as the point O. By placing it at A (Fig. b), we will obtain an equation in which the only unknown is the force B. The airplane is translating, so $\alpha = 0$ and there is no inertial couple. The sum of the moments about O, including the moment due to the inertial force, is

$$(5)(700{,}000) - (3)T - (5)mg + (27)B = 0.$$

From this equation we obtain $B = 402$ kN, and then $A = mg - B = 2050$ kN.

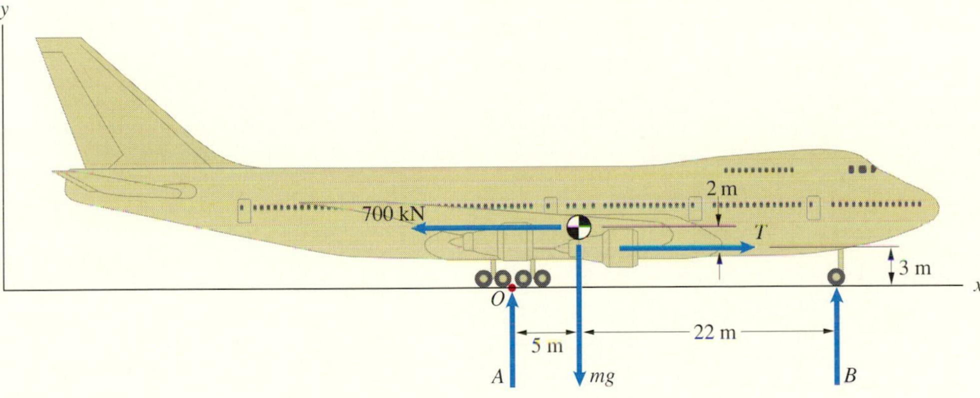

(b) Placing point O at the rear wheels.

DISCUSSION

Notice that we calculated the moment due to the inertial force by multiplying the magnitude of the inertial force by the perpendicular distance to its line of action, (5)(700,000) = 3,500,000 N-m counterclockwise. In this particular example that method is simpler than using the vector expression for the moment,

$$[\mathbf{r} \times (-m\mathbf{a})] \cdot \mathbf{k} = [(5\mathbf{i} + 5\mathbf{j}) \times (-700{,}000\mathbf{i})] \cdot \mathbf{k}$$
$$= 3{,}500{,}000\ \text{N-m counterclockwise},$$

but in some situations you may find that using the vector expression is simpler.

You should compare this application of D'Alembert's principle to our determination of the airplane's acceleration and the normal forces exerted on its wheels in Example 7.1.

Example 7.8

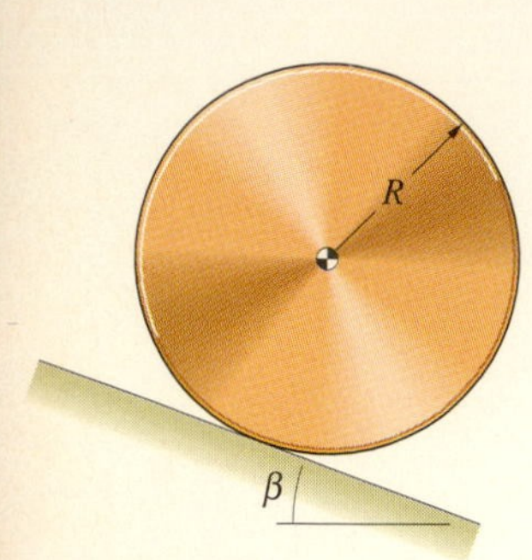

Fig. 7.20

A disk of mass m and moment of inertia I is released from rest on an inclined surface (Fig. 7.20). Assuming that it rolls, use D'Alembert's principle to determine the disk's angular acceleration and the forces exerted on it by the surface.

SOLUTION

Draw the Free-Body Diagram The angular acceleration of the disk and the acceleration of its center are shown in Fig. (a). In Fig. (b) we draw the free-body diagram of the disk showing its weight, the normal and friction forces exerted by the surface, and the inertial force and couple.

Apply the "Equilibrium" Equations Equation (7.26) is

$$\Sigma \mathbf{F} + (-m\mathbf{a}) = 0:$$

$$(mg \sin \beta - f)\mathbf{i} + (N - mg \cos \beta)\mathbf{j} - ma_x \mathbf{i} = 0.$$

From this vector equation we obtain the equations

$$\begin{aligned} mg \sin \beta - f - ma_x &= 0, \\ N - mg \cos \beta &= 0. \end{aligned} \qquad (7.28)$$

We now apply Eq. (7.27), evaluating moments about the point where the disk is in contact with the surface to eliminate f and N from the resulting equation:

$$\begin{aligned} &\Sigma M_0 + [\mathbf{r} \times (-m\mathbf{a})] \cdot \mathbf{k} + (-I\alpha) = 0: \\ &-R(mg \sin \beta) + R(ma_x) - I\alpha = 0. \end{aligned} \qquad (7.29)$$

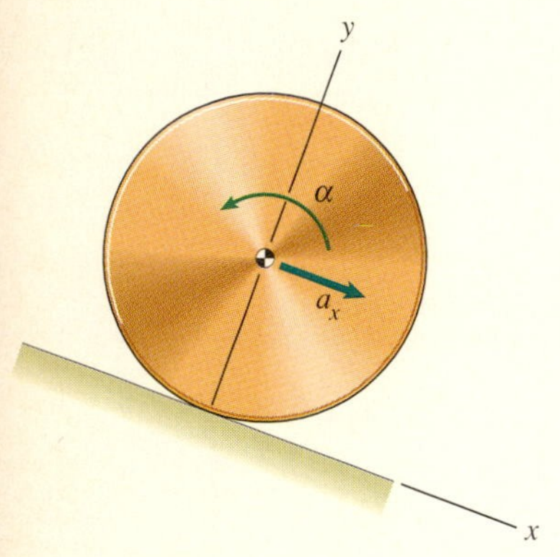

(a) Acceleration of the center of the disk and its angular acceleration.

Determine Kinematic Relationships The acceleration of the center of the rolling disk is related to the counterclockwise angular acceleration by $a_x = -R\alpha$. Substituting this relation into Eq. (7.29) and solving for the angular acceleration, we obtain

$$\alpha = -\frac{mgR \sin \beta}{mR^2 + I}.$$

From this result we also know a_x and can solve Eqs. (7.28) for the normal and friction forces, obtaining

$$N = mg \cos \beta, \qquad f = \frac{mgI \sin \beta}{mR^2 + I}.$$

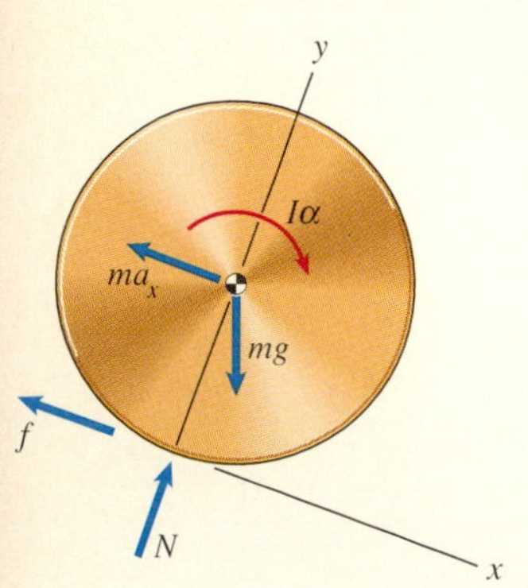

(b) Free-body diagram including the inertial force and couple.

DISCUSSION

In Eq. (7.29) we evaluated the moment due to the inertial force by simply multiplying the magnitude of the force and the perpendicular distance from O to its line of action, but we could have used the vector expression:

$$[\mathbf{r} \times (-m\mathbf{a})] \cdot \mathbf{k} = [(R\mathbf{j}) \times (-ma_x\mathbf{i})] \cdot \mathbf{k} = R(ma_x).$$

This example demonstrates an advantage of D'Alembert's principle. Because the moments in Eq. (7.27) can be evaluated about any point, we chose a point that eliminated f and N from the resulting equation and were able to evaluate α directly.

Problems

7.1 A refrigerator of mass m rests on casters at A and B. Suppose that you push on it with a horizontal force F as shown and that the casters remain on the smooth floor.
(a) What is the refrigerator's acceleration?
(b) What normal forces are exerted on the casters at A and B?

Strategy: Draw the free-body diagram. (a) Use Newton's second law to determine a_x. (b) The sum of the forces in the vertical direction equals zero and, since the refrigerator translates, the sum of the moments *about the center of mass* equals zero. These conditions give you two equations with which to determine A and B.

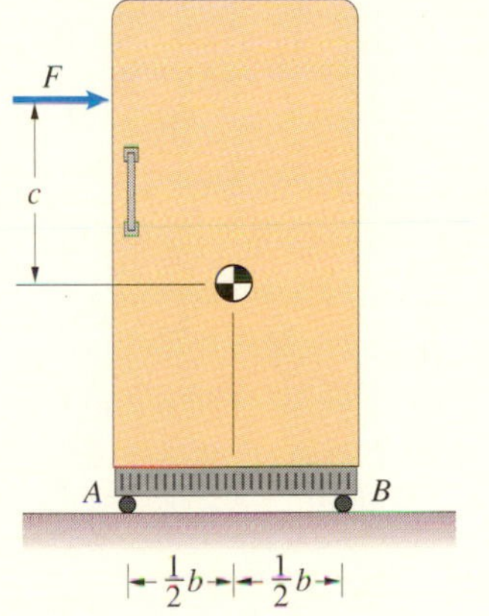

P7.1

7.2 In Problem 7.1, what is the largest force F you can apply if you want the refrigerator to remain on the floor at A and B? (Assume that c is positive.)

7.3 The 14,000-lb airplane's arresting hook exerts the force F and causes the plane to decelerate at six g's. The horizontal forces exerted on the landing gear are negligible. Determine F and the normal forces exerted on the landing gear.

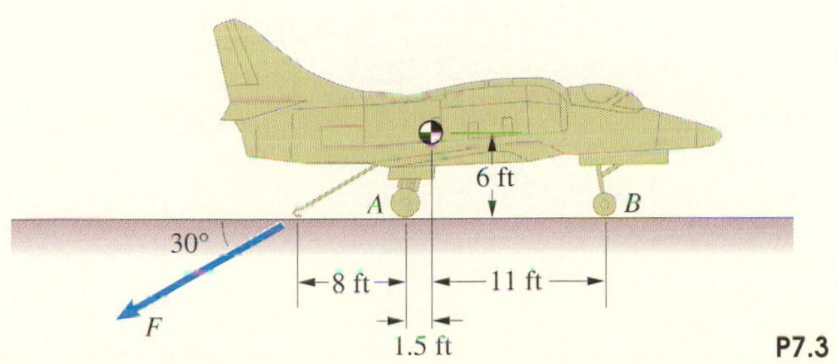

P7.3

7.4 A student catching a ride to his summer job unwisely supports himself in the back of an accelerating truck by exerting a horizontal force F on the truck's cab at A. Determine the horizontal force he must exert in terms of his weight W, the truck's acceleration a, and the dimensions shown.

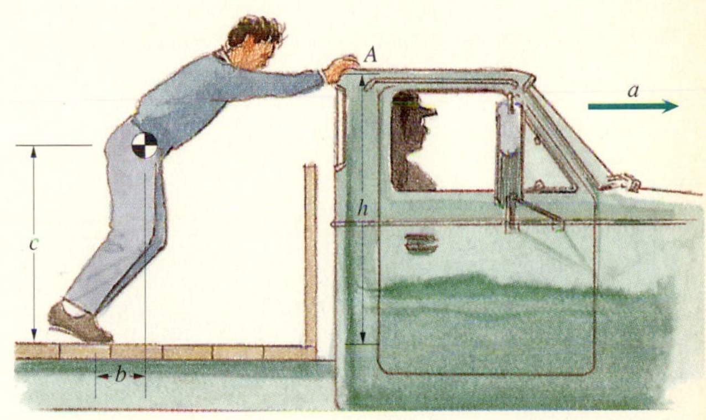

P7.4

7.5 The crane moves to the right with constant acceleration, and the 800-kg load moves without swinging.
(a) What is the acceleration of the crane and load?
(b) What are the tensions in the cables attached at A and B?

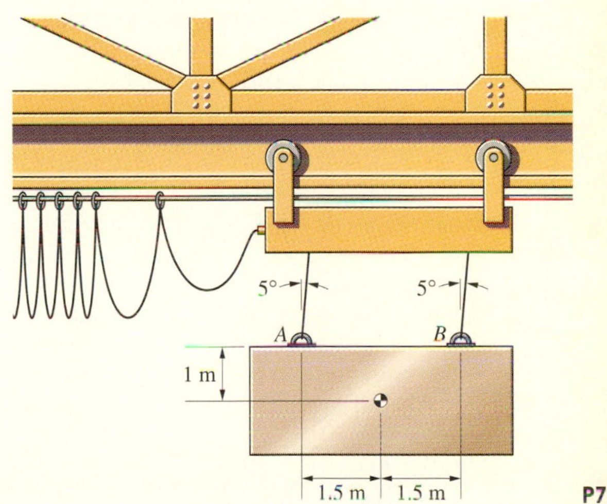

P7.5

7.6 If the acceleration of the crane in Problem 7.5 suddenly decreases to zero, what are the tensions in the cables attached at A and B immediately afterward?

7.7 The combined mass of the person and bicycle is m. The location of their combined center of mass is shown.

(a) If they have acceleration a, what are the normal forces exerted on the wheels by the ground? (Neglect the horizontal force exerted on the ground by the front wheel.)

(b) Based on the results of (a), what is the largest acceleration that can be achieved without causing the front wheel to leave the ground?

P7.7

7.8 In Problem 7.7, $b = 615$ mm, $c = 445$ mm, $h = 985$ mm, and $m = 77$ kg. If the bicycle is traveling at 6 m/s and the person engages the brakes, achieving the largest deceleration for which the rear wheel will not leave the ground, how long does it take the bicycle to stop, and what distance does it travel during that time?

7.9 The combined mass of the motorcycle and rider is 160 kg. The rear wheel exerts a 400-N horizontal force on the road, and you can neglect the horizontal force exerted on the road by the front wheel. Modeling the motorcycle and its wheels as a rigid body, determine (a) the motorcycle's acceleration; (b) the normal forces exerted on the road by the rear and front wheels.

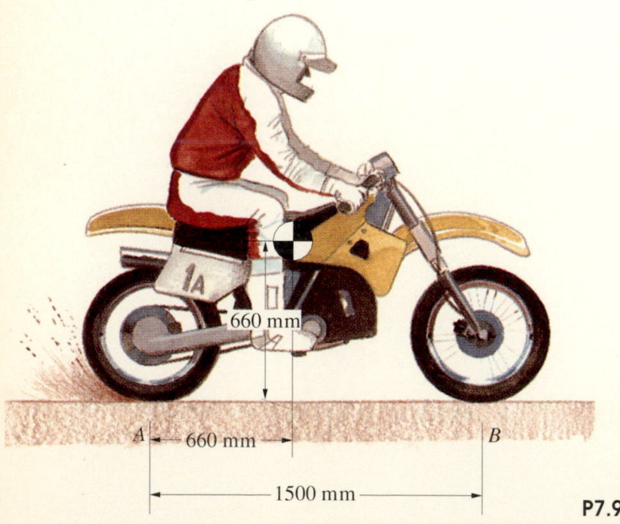

P7.9

7.10 In Problem 7.9, the coefficient of kinetic friction between the motorcycle's rear wheel and the road is $\mu_k = 0.8$. If the rider spins the rear wheel, what is the motorcycle's acceleration and what are the normal forces exerted on the road by the rear and front wheels?

7.11 During extravehicular activity, an astronaut fires a thruster of his maneuvering unit, exerting a force $T = 14.2$ N for 1 s. It requires 60 s from the time the thruster is fired for him to rotate through one revolution. If you model the astronaut and maneuvering unit as a rigid body, what is the moment of inertia about their center of mass?

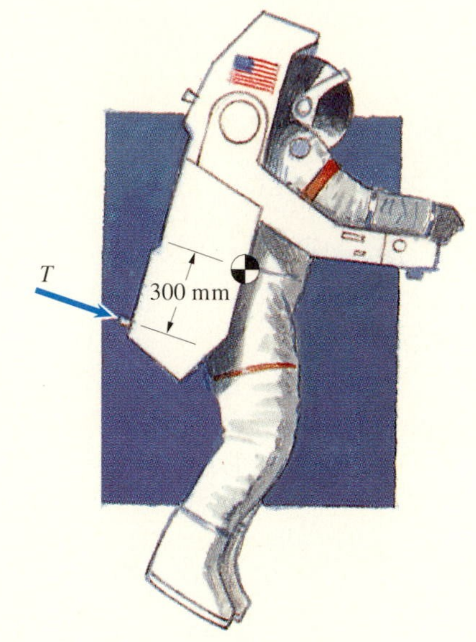

P7.11

7.12 The moment of inertia of the helicopter's rotor is 400 slug-ft^2. If the rotor starts from rest at $t = 0$, the engine exerts a constant torque of 500 ft-lb on the rotor, and aerodynamic drag is neglected, what is the rotor's angular velocity ω at $t = 6$ s?

P7.12

7.13 In Problem 7.12, if aerodynamic drag exerts a torque on the helicopter's rotor of magnitude $20\omega^2$ ft-lb, what is the rotor's angular velocity at $t = 6$ s?

7.14 The moment of inertia of the robotic manipulator arm about the vertical y axis is 8 slug-ft^2. The moment of inertia of the 30-lb casting held by the arm about the y' axis is 0.6 slug-ft^2. What couple about the y axis is necessary to give the manipulator arm an angular acceleration of 2 rad/s^2?

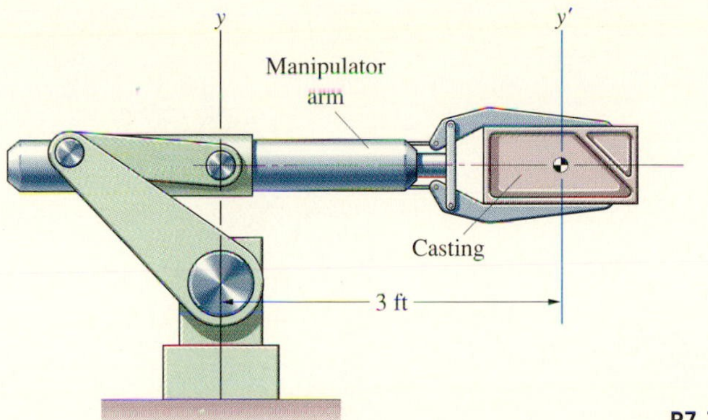

P7.14

7.15 The gears A and B can turn freely on their pin supports. Their moments of inertia are $I_A = 0.002$ kg-m^2 and $I_B = 0.006$ kg-m^2. They are initially stationary, and at $t = 0$ a constant couple $M = 2$ N-m is applied to gear B. How many revolutions has gear A turned at $t = 4$ s?

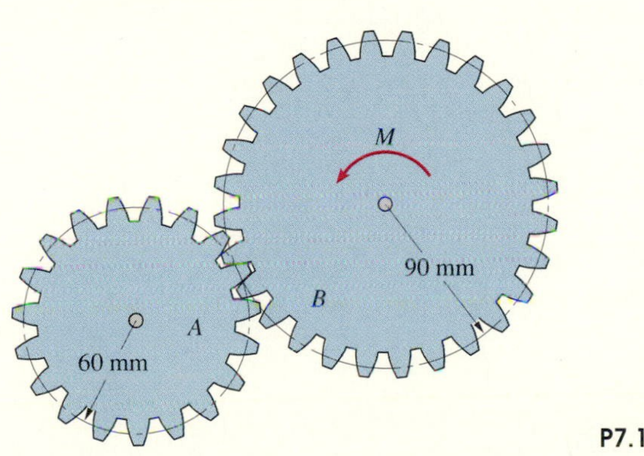

P7.15

7.16 The moment of inertia of the pulley is 0.4 slug-ft^2. Determine the pulley's angular acceleration and the tension in the cable in the two cases.

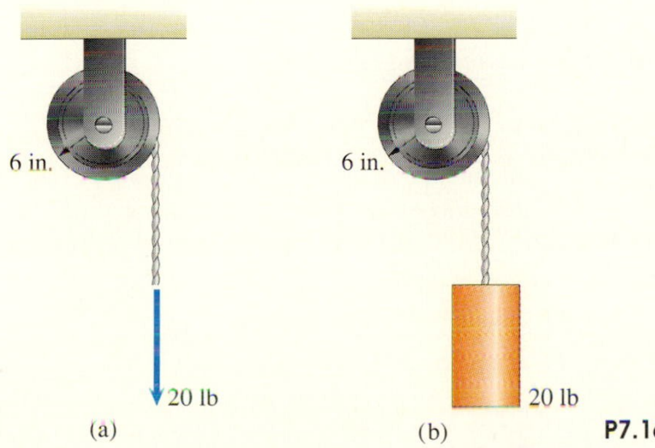

P7.16

7.17 Each box weighs 50 lb, the moment of inertia of the pulley is 0.6 slug-ft^2, and friction can be neglected. If the boxes start from rest at $t = 0$, determine the magnitude of their velocity and the distance they have moved from their initial position at $t = 1$ s.

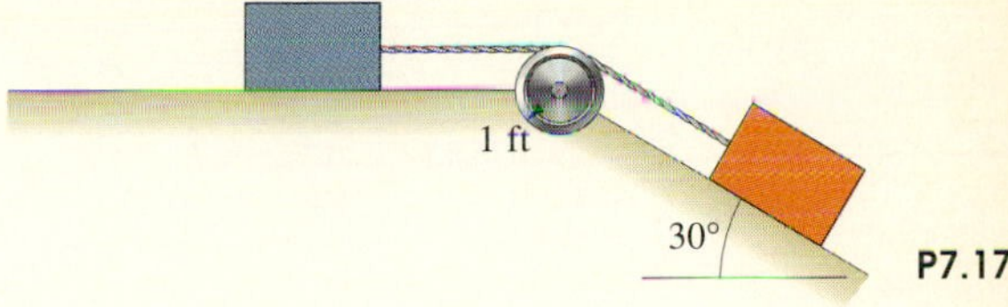

P7.17

7.18 The slender bar weighs 10 lb and the disk weighs 20 lb. The coefficient of kinetic friction between the disk and the horizontal surface is $\mu_k = 0.1$. If the disk has an initial counterclockwise angular velocity of 10 rad/s, how long does it take to stop spinning?

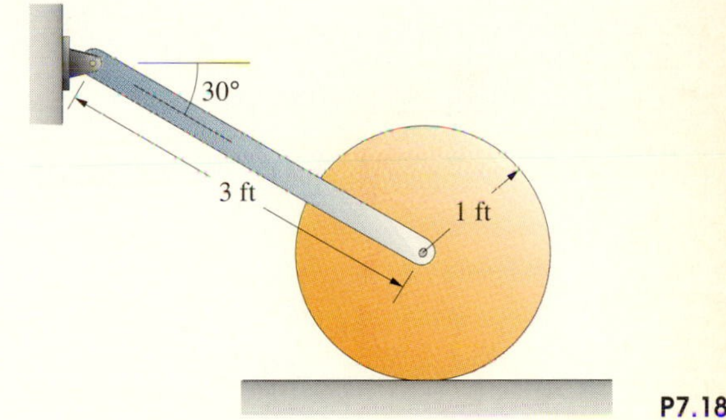

P7.18

7.19 In Problem 7.18, how long does it take the disk to stop spinning if it has an initial clockwise angular velocity of 10 rad/s?

7.20 The objects consist of identical 3-ft, 10-lb bars welded together. If they are released from rest in the positions shown, what are their angular accelerations and what are the components of the reactions at A at that instant? (The y axes are vertical.)

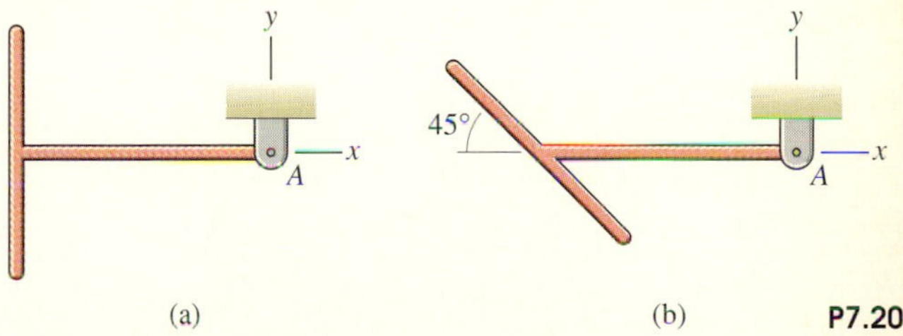

P7.20

7.21 The object consists of identical 1-m, 5-kg bars welded together. If it is released from rest in the position shown, what is its angular acceleration and what are the components of the reaction at A at that instant? (The y axis is vertical.)

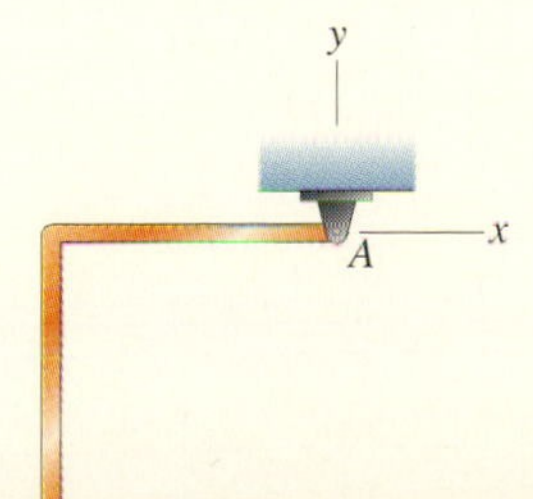

P7.21

7.22 For what value of x is the horizontal bar's angular acceleration a maximum, and what is the maximum angular acceleration?

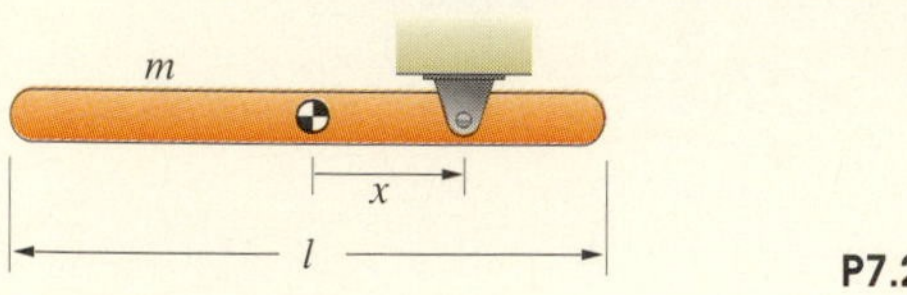

P7.22

7.23 Model the arm ABC as a single rigid body. Its mass is 300 kg, and the moment of inertia about its center of mass is $I = 360$ kg-m^2. If point A is stationary and the angular acceleration of the arm is 0.6 rad/s^2 counterclockwise, what force does the hydraulic cylinder exert on the arm at B? (The arm is actuated by two hydraulic cylinders, one on each side of the vehicle. You are to determine the total force exerted by the two cylinders.)

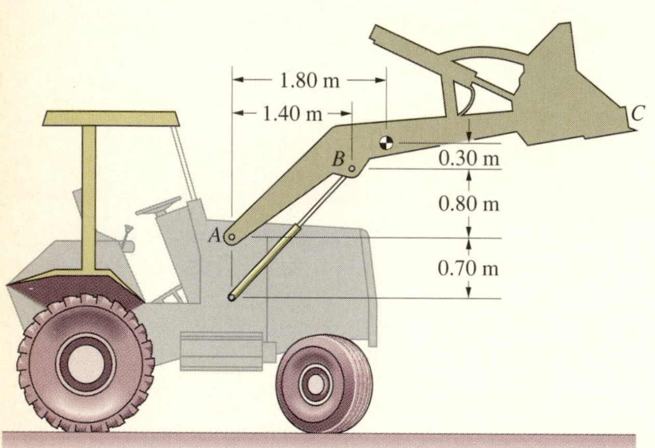

P7.23

7.24 In Problem 7.23, if the angular acceleration of arm ABC is 0.6 rad/s^2 counterclockwise and its angular velocity is 1.4 rad/s clockwise, what are the components of the force exerted on the arm at A? (There are two pin supports, one on each side of the vehicle. You are to determine the components of the total force exerted by the two supports.)

7.25 To lower the drawbridge, the gears that raised it are disengaged and a fraction of a second later a second set of gears that lower it are engaged. At the instant the gears that raised it are disengaged, what are the components of force exerted by the bridge on its support at O? The drawbridge weighs 360 kip, its moment of inertia about O is $I_O = 1.0 \times 10^7$ slug-ft^2, and the coordinates of its center of mass at the instant the gears are disengaged are $\bar{x} = 8$ ft, $\bar{y} = 16$ ft.

P7.25

7.26 Arm BC has a mass of 12 kg and the moment of inertia about its center of mass is 3 kg-m^2. If B is stationary and arm BC has a constant counterclockwise angular velocity of 2 rad/s at the instant shown, determine the couple and the components of force exerted on arm BC at B.

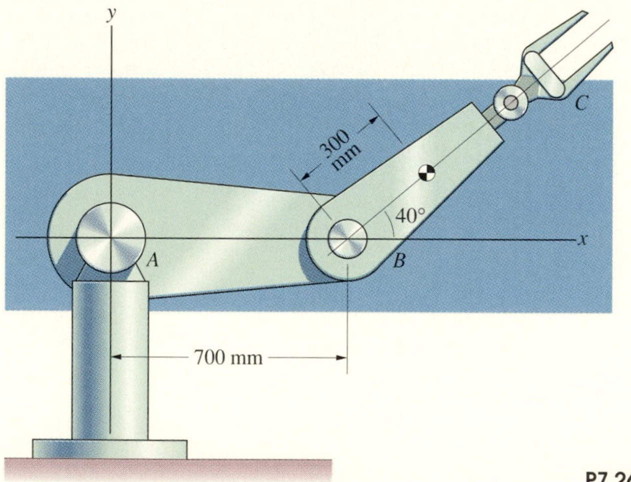

P7.26

7.27 In Problem 7.26, what are the couple and the components of force exerted on arm BC at B if arm AB has a constant clockwise angular velocity of 2 rad/s and arm BC has a counterclockwise angular velocity of 2 rad/s and a clockwise angular acceleration of 4 rad/s^2 at the instant shown?

7.28 The space shuttle's attitude control engines exert two forces $F_f = 8$ kN and $F_r = 2$ kN. The force vectors and the center of mass G lie in the x-y plane of the inertial reference frame. The mass of the shuttle is 54,000 kg, and its moment of inertia about the axis through the center of mass that is parallel to the z axis is 4.5×10^6 kg-m^2. Determine the acceleration of the center of mass and the angular acceleration. (You can ignore the force exerted on the shuttle by its weight.)

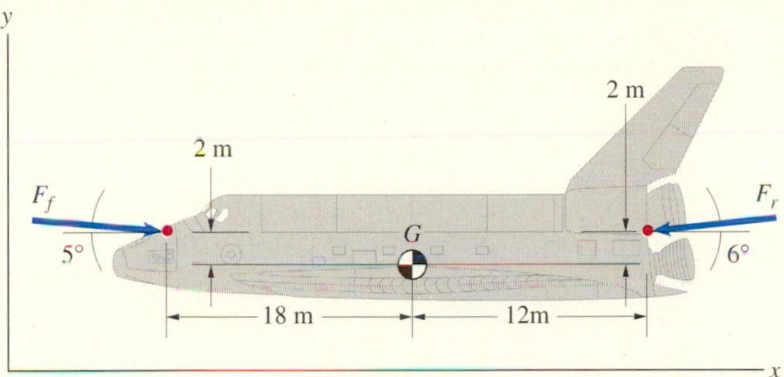

P7.28

7.29 In Problem 7.28, suppose that $F_f = 4$ kN and you want the shuttle's angular acceleration to be zero. Determine the necessary force F_r and the resulting acceleration of the center of mass.

7.30 Points B and C are in the x-y plane. At the instant shown, the angular velocity and angular acceleration vectors of arm AB are $\boldsymbol{\omega}_{AB} = 0.6\mathbf{k}$ (rad/s) and $\boldsymbol{\alpha}_{AB} = -0.3\mathbf{k}$ (rad/s^2). The angular velocity and angular acceleration vectors of arm BC are $\boldsymbol{\omega}_{BC} = 0.4\mathbf{k}$ (rad/s) and $\boldsymbol{\alpha}_{BC} = 2.0\mathbf{k}$ (rad/s^2). The center of mass of the 18-kg arm BC is at the midpoint of the line from B to C, and its moment of inertia about the axis through the center of mass that is parallel to the z axis is 1.5 kg-m^2. The y axis is vertical. Determine the force and couple exerted on arm BC at B.

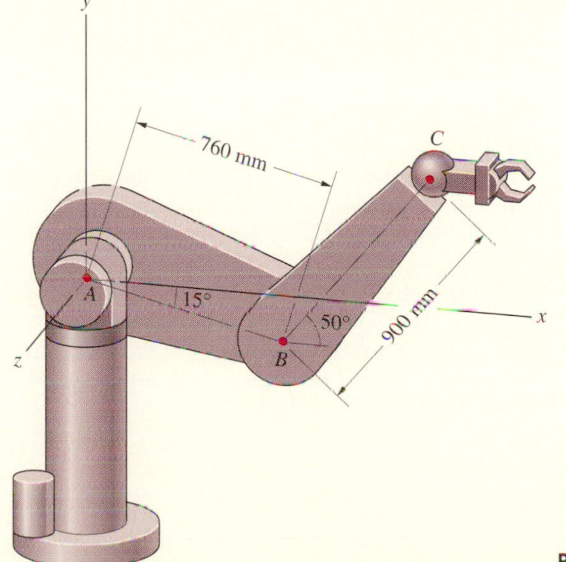

P7.30

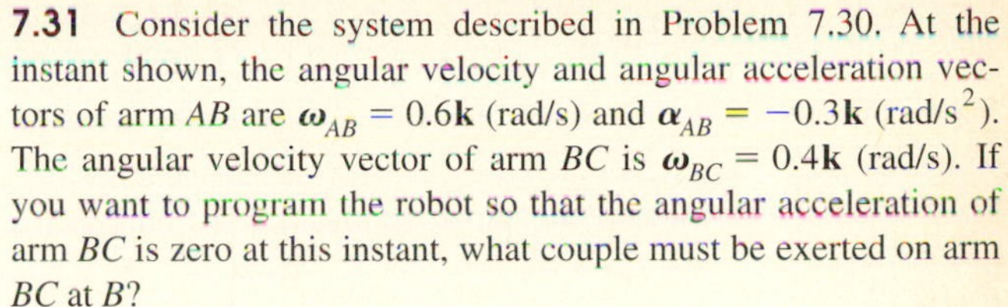

7.31 Consider the system described in Problem 7.30. At the instant shown, the angular velocity and angular acceleration vectors of arm AB are $\boldsymbol{\omega}_{AB} = 0.6\mathbf{k}$ (rad/s) and $\boldsymbol{\alpha}_{AB} = -0.3\mathbf{k}$ (rad/s^2). The angular velocity vector of arm BC is $\boldsymbol{\omega}_{BC} = 0.4\mathbf{k}$ (rad/s). If you want to program the robot so that the angular acceleration of arm BC is zero at this instant, what couple must be exerted on arm BC at B?

7.32 The 9000-kg airplane has just landed. At the instant shown, its angular velocity is zero. Its landing gear are rolling and contact the runway at $x = 10$ m. The friction force on the wheels is negligible. The coordinates of the airplane's center of mass are $x = 10.50$, $y = 3.00$ m. The total aerodynamic force is $-26.8\mathbf{i} + 30.4\mathbf{j}$ (kN), and it effectively acts at the center of pressure located at $x = 10.75$, $y = 3.2$ m. The thrust $T = 4.40$ kN exerts no moment about the center of mass. The moment of inertia of the airplane about its center of mass is 75,000 kg-m^2. Determine the airplane's angular acceleration.

Strategy: Draw a free-body diagram of the airplane including the normal force exerted on the landing gear. To relate the acceleration of the center of mass to the angular acceleration, use the fact that the acceleration of the airplane (treated as a rigid body) is horizontal at the point where the wheels contact the runway.

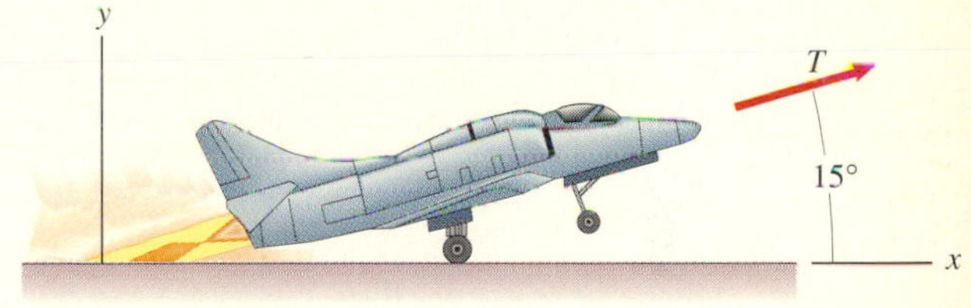

P7.32

7.33 In Problem 7.32, determine the normal force exerted on the airplane by the runway.

7.34 A thin ring and a circular disk, each of mass m and radius R, are released from rest on an inclined surface and allowed to roll a distance D. Determine the ratio of the times required.

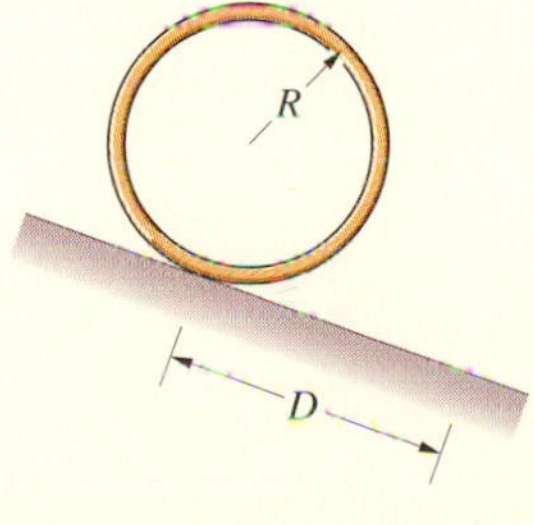

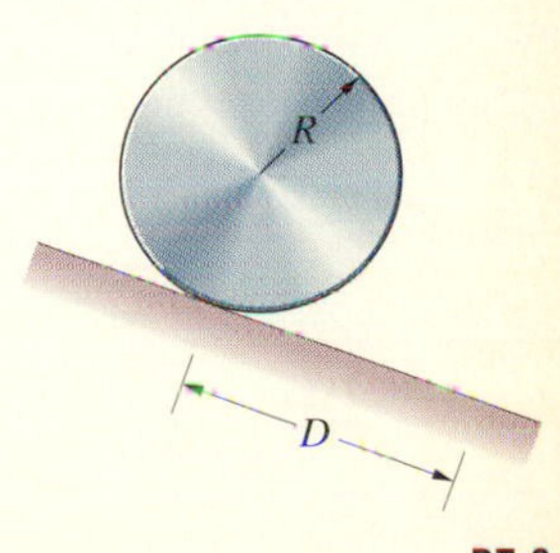

P7.34

7.35 The stepped disk weighs 40 lb and its moment of inertia is $I = 0.2$ slug-ft^2. If it is released from rest, how long does it take the center of the disk to fall 3 ft? (Assume that the string remains vertical.)

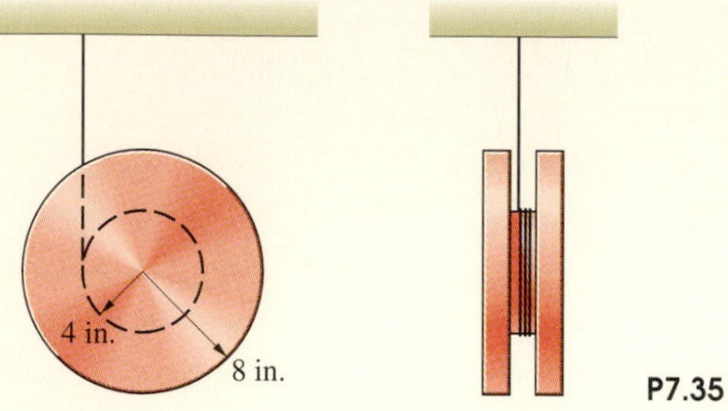

P7.35

7.36 The moment of inertia of the pulley is I. The system is released from rest with the spring unstretched. Determine the velocity of the mass as a function of the distance x it has fallen.

Strategy: By drawing free-body diagrams of the mass and pulley, determine the acceleration of the mass as a function of the distance it has fallen. Then use the chain rule: $a = dv/dt = (dv/dx)(dx/dt) = (dv/dx)v$.

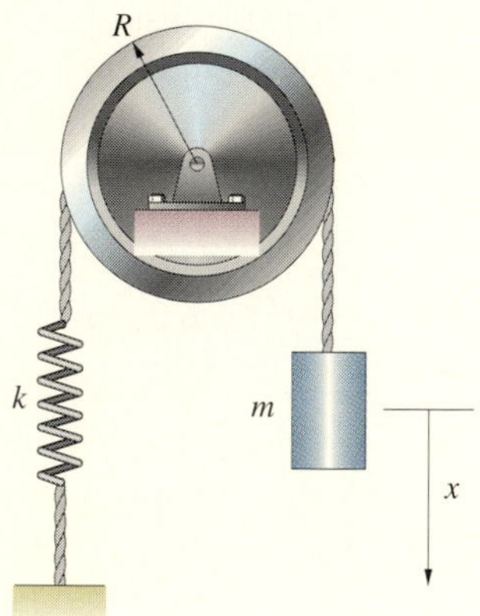

P7.36

7.37 In Problem 7.36, let $R = 100$ mm, $I = 0.1$ kg-m^2, $m = 5$ kg, and $k = 135$ N/m.
(a) What maximum distance x does the mass fall?
(b) At the instant the mass has fallen the maximum distance, what is its acceleration?

7.38 The homogeneous disk weighs 100 lb and its radius is $R = 1$ ft. It rolls on the plane surface. The spring constant is $k = 100$ lb/ft. If the disk is rolled to the left until the spring is compressed 1 ft and released from rest, what is its angular acceleration at the instant it is released?

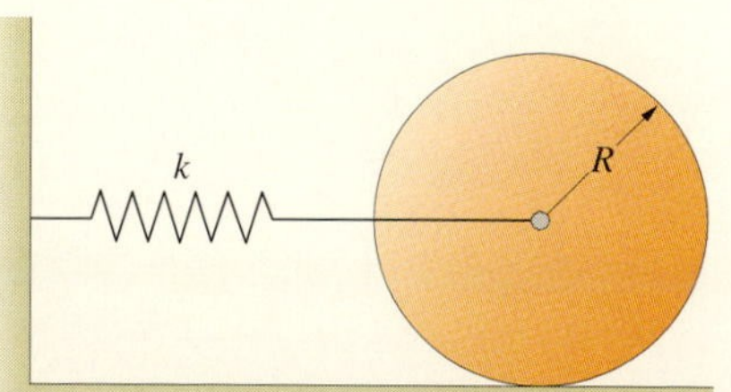

P7.38

7.39 In Problem 7.38, determine the disk's angular velocity when its center has moved 1 ft to the right of the position in which it was released.

7.40 At $t = 0$, a sphere of mass m and radius R ($I = \frac{2}{5}mR^2$) on a flat surface has angular velocity ω_0 and the velocity of its center is zero. The coefficient of kinetic friction between the sphere and the surface is μ_k. What is the maximum velocity the center of the sphere will attain, and how long does it take to reach it?

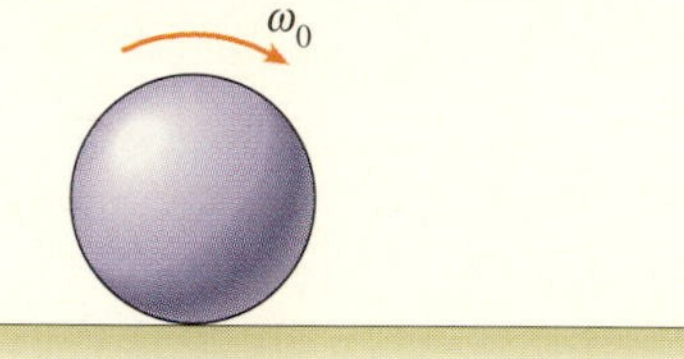

P7.40

7.41 A soccer player kicks the ball to a teammate 20 ft away. The ball leaves his foot moving parallel to the ground at 20 ft/s with no initial angular velocity. The coefficient of kinetic friction between the ball and the grass is $\mu_k = 0.4$. How long does it take the ball to reach his teammate? (The ball is 28 in. in circumference and weighs 14 oz. Estimate its moment of inertia by using the equation for a thin spherical shell: $I = \frac{2}{3}mR^2$.)

P7.41

7.42 The 100-kg cylindrical disk is at rest when the force F is applied to a cord wrapped around it. The static and kinetic coefficients of friction between the disk and the surface are 0.2. Determine the angular acceleration of the disk if (a) $F = 500$ N; (b) $F = 1000$ N.

Strategy: First solve the problem by assuming that the disk does not slip, but rolls on the surface. Determine the friction force and find out whether it exceeds the product of the friction coefficient and the normal force. If it does, you must rework the problem assuming that the disk slips.

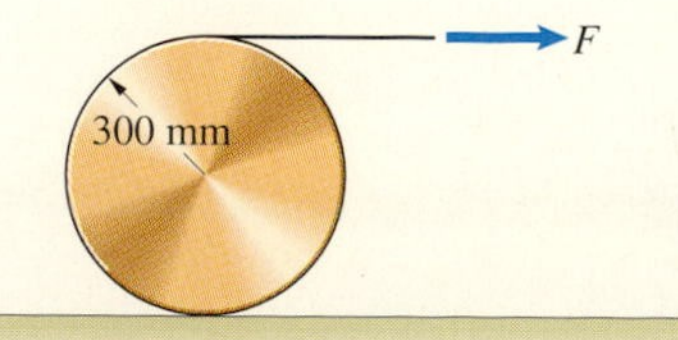

P7.42

7.43 The ring gear is fixed. The mass and moment of inertia of the sun gear are $m_S = 22$ slugs, $I_S = 4400$ slug-ft^2. The mass and moment of inertia of each planet gear are $m_P = 2.7$ slugs, $I_P = 65$ slug-ft^2. If a couple $M = 600$ ft-lb is applied to the sun gear, what is its angular acceleration?

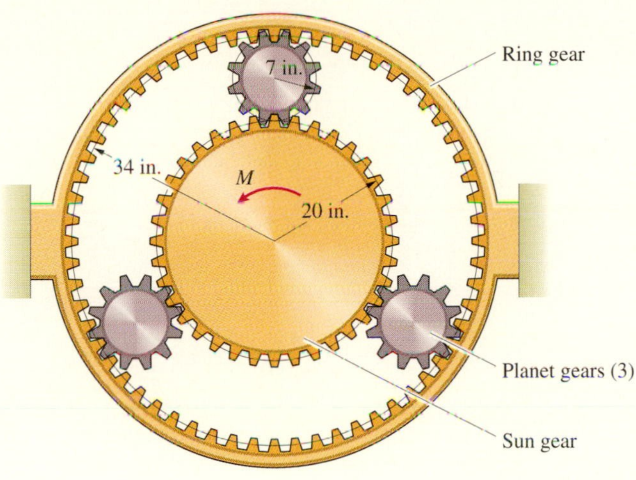

P7.43

7.44 In Problem 7.43, what is the magnitude of the tangential force exerted on the sun gear by each planet gear at their point of contact when the 600 ft-lb couple is applied to the sun gear?

7.45 The 18-kg ladder is released from rest in the position shown. Model it as a slender bar and neglect friction. At the instant of release, determine (a) the angular acceleration; (b) the normal force exerted on the ladder by the floor.

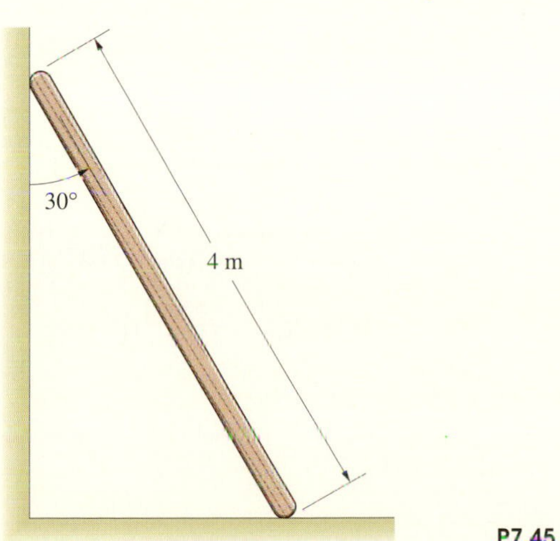

P7.45

7.46 Suppose that the ladder in Problem 7.45 has a counterclockwise angular velocity of 1.0 rad/s in the position shown. Determine (a) the angular acceleration; (b) the normal force exerted on the ladder by the floor.

7.47 Suppose that the ladder in Problem 7.45 has a counterclockwise angular velocity of 1.0 rad/s in the position shown and that the coefficient of kinetic friction at the floor and the wall is $\mu_k = 0.2$. Determine (a) the angular acceleration; (b) the normal force exerted on the ladder by the floor.

7.48 The slender bar weighs 30 lb and the cylindrical disk weighs 20 lb. The system is released from rest with the bar horizontal. Determine the bar's angular acceleration at the instant of release if the bar and disk are welded together at A.

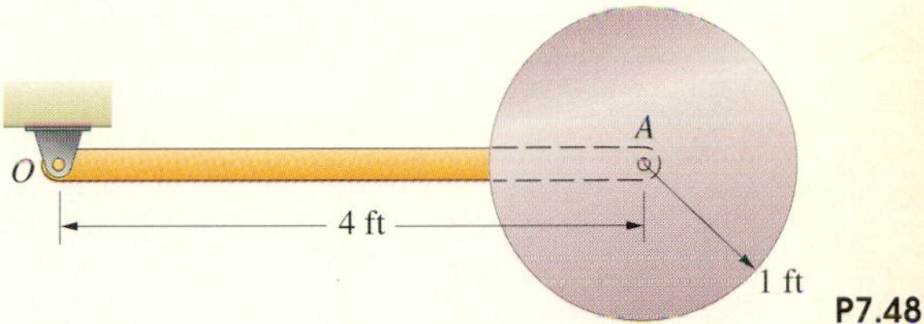

P7.48

7.49 In Problem 7.48, determine the bar's angular acceleration if the bar and disk are connected by a smooth pin at A.

7.50 The 0.1-kg slender bar and 0.2-kg cylindrical disk are released from rest with the bar horizontal. The disk rolls on the curved surface. What is the bar's angular acceleration at that instant?

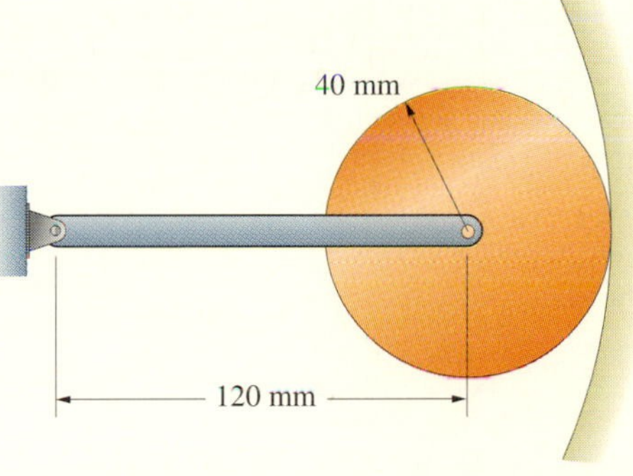

P7.50

7.51 The suspended objects A and B weigh 20 lb and 40 lb, respectively. The left pulley weighs 16 lb, and its moment of inertia is 0.24 slug-ft^2. The right pulley weighs 6 lb, and its moment of inertia is 0.04 slug-ft^2. If the system is released from rest, what is the acceleration of B?

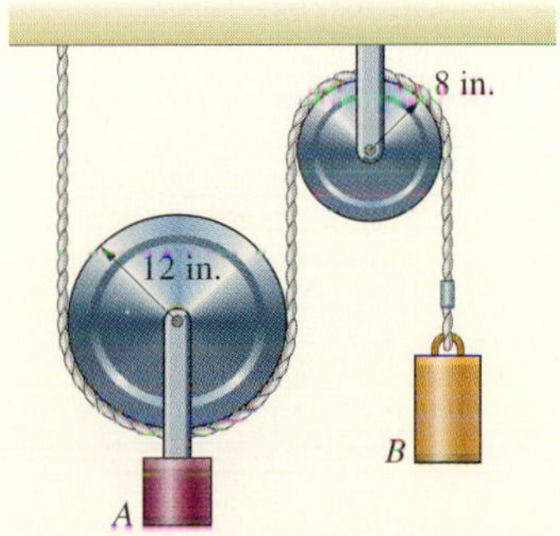

P7.51

7.52 In Problem 7.51, what is the tension in the cable at a point between the two pulleys?

7.53 The 4-lb slender bar and 10-lb block are released from rest in the position shown. If friction is negligible, what is the block's acceleration at that instant?

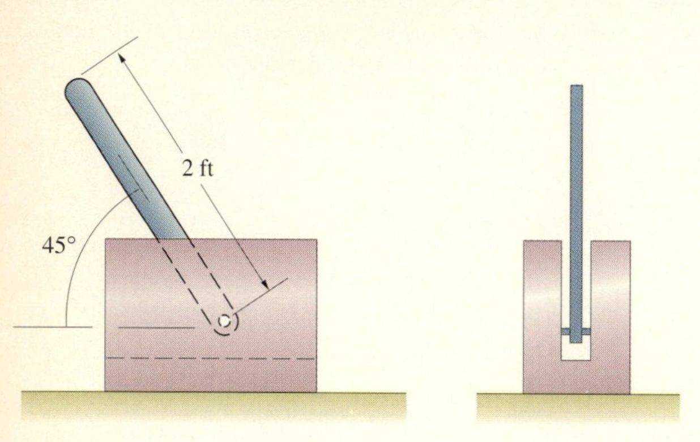

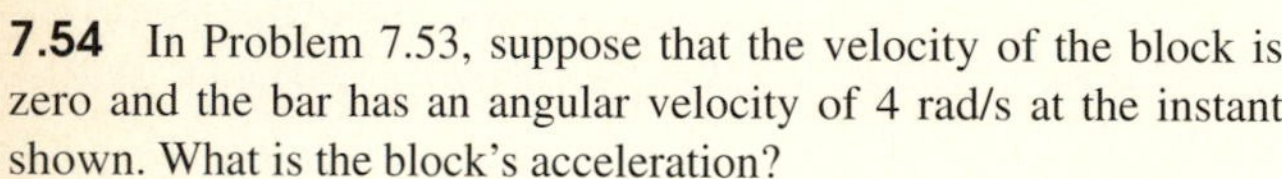

P7.53

7.54 In Problem 7.53, suppose that the velocity of the block is zero and the bar has an angular velocity of 4 rad/s at the instant shown. What is the block's acceleration?

7.55 The 0.4-kg slender bar and 1-kg disk are released from rest in the position shown. If the disk rolls, what is the bar's angular acceleration at that instant?

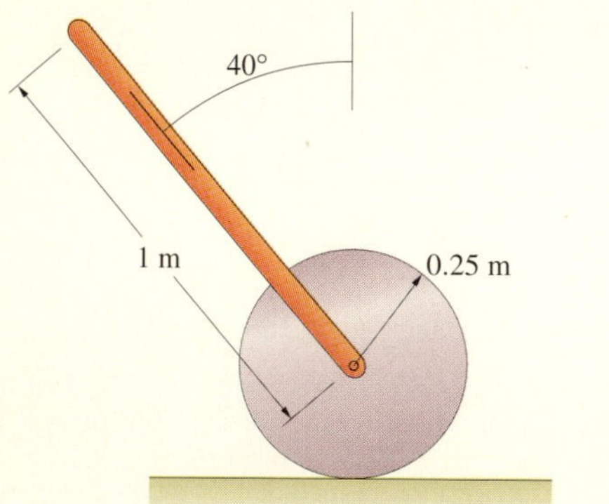

P7.55

7.56 The slender bar weighs 20 lb and the crate weighs 80 lb. The surface the crate rests on is smooth. If the system is stationary at the instant shown, what couple M will cause the crate to accelerate to the left at 4 ft/s^2 at that instant?

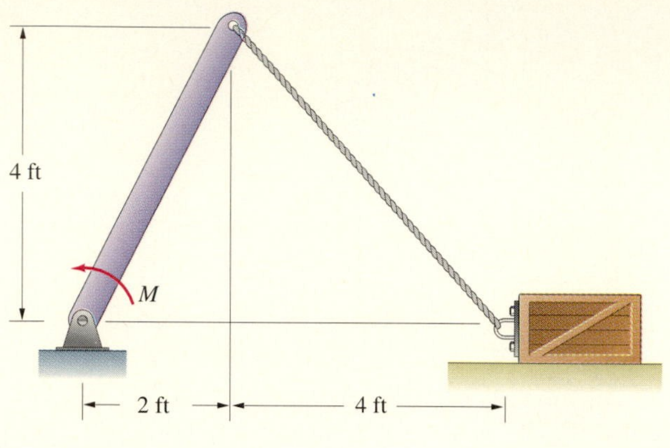

P7.56

7.57 Suppose that the slender bar in Problem 7.56 is rotating in the counterclockwise direction at 2 rad/s at the instant shown and that the coefficient of kinetic friction between the crate and the horizontal surface is $\mu_k = 0.2$. What couple M will cause the crate to accelerate to the left at 4 ft/s^2 at that instant?

7.58 Bar AB rotates with a constant angular velocity of 6 rad/s in the counterclockwise direction. The slender bar BCD weighs 10 lb, and the collar that bar BCD is attached to at C weighs 2 lb. The y axis points upward. Neglecting friction, determine the components of the forces exerted on bar BCD by the pins at B and C at the instant shown.

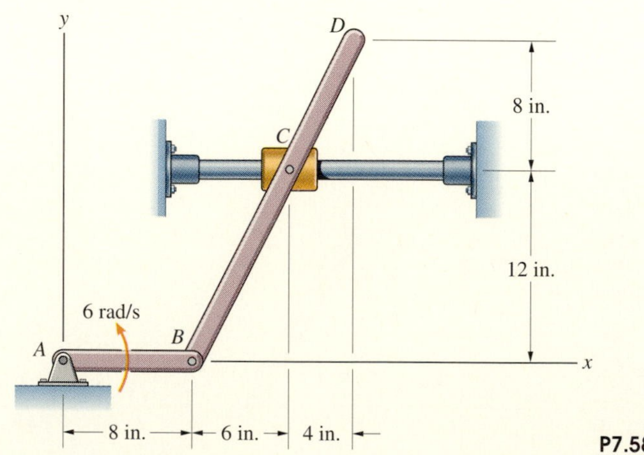

P7.58

7.59 Bar AB weighs 10 lb and bar BC weighs 6 lb. If the system is released from rest in the position shown, what are the angular acceleration of bar AB and the normal force exerted by the floor at C at that instant? Neglect friction.

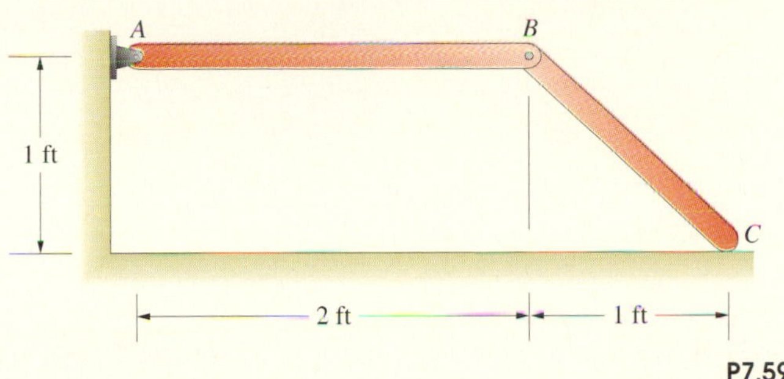

P7.59

7.60 Let the total moment of inertia of the car's two rear wheels and axle be I_R, and let the total moment of inertia of the two front wheels be I_F. The radius of the tires is R, and the total mass of the car including the wheels is m. If the car's engine exerts a torque (couple) T on the rear wheels and the wheels do not slip, show that the car's acceleration is

$$a = \frac{RT}{R^2 m + I_R + I_F}.$$

Strategy: Isolate the wheels and draw three free-body diagrams.

P7.60

7.61 The combined mass of the motorcycle and rider is 160 kg. Each 9-kg wheel has a 330-mm radius and moment of inertia $I = 0.8$ kg-m^2. The engine drives the rear wheel by exerting a couple on it. If the rear wheel exerts a 400-N horizontal force on the road and you do *not* neglect the horizontal force exerted on the road by the front wheel, determine (a) the motorcycle's acceleration; (b) the normal forces exerted on the road by the rear and front wheels. (The location of the center of mass of the motorcycle *not including* its wheels is shown.)

P7.61

7.62 In Problem 7.61, if the front wheel lifts slightly off the road when the rider accelerates, determine (a) the motorcycle's acceleration; (b) the torque exerted by the engine on the rear wheel.

7.63 The moment of inertia of the vertical handle about O is 0.12 slug-ft^2. The object B weighs 15 lb and rests on a smooth surface. The weight of the bar AB is negligible (which means you can treat it as a two-force member). If the person exerts a 0.2-lb horizontal force on the handle 15 in. above O, what is the resulting angular acceleration of the handle?

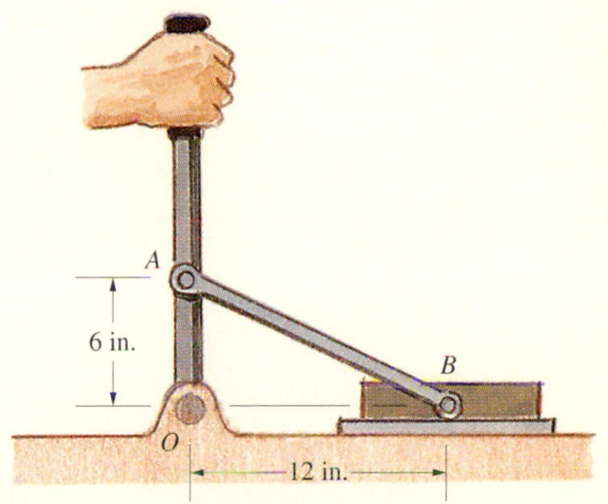

P7.63

7.64 In Problem 7.63, suppose that the kinetic coefficient of friction between the object B and the horizontal surface is 0.05. Immediately after B starts slipping, what horizontal force does the person need to exert on the handle 15 in. above O in order that the angular acceleration of the handle be 1 rad/s^2?

7.65 Bars OQ and PQ each weigh 6 lb. The weight of the collar P and friction between the collar and the horizontal bar are negligible. If the system is released from rest with $\theta = 45°$, what are the angular accelerations of the two bars?

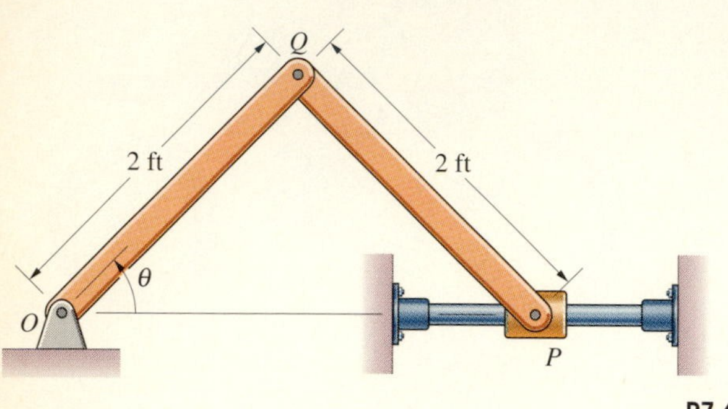

P7.65

7.66 In Problem 7.65, what are the angular accelerations of the two bars if the collar P weighs 2 lb?

7.67 The 4-kg slender bar is pinned to 2-kg sliders at A and B. If friction is negligible and the system is released from rest in the position shown, what is the angular acceleration of the bar at that instant?

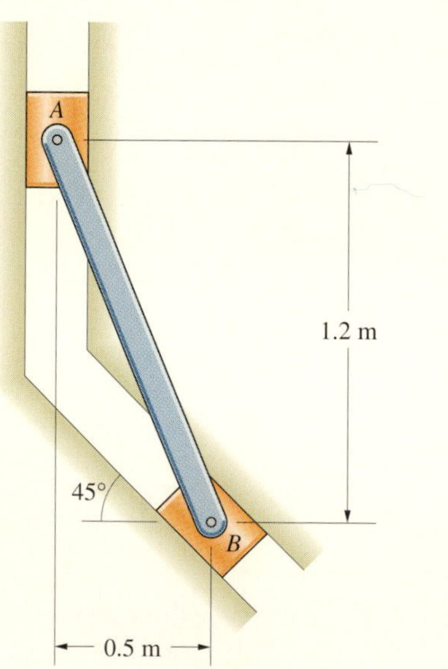

P7.67

7.68 The mass of the slender bar is m and the mass of the homogeneous disk is $4m$. The system is released from rest in the position shown. If the disk rolls and friction between the bar and the horizontal surface is negligible, show that the disk's angular acceleration is $\alpha = 6g/95R$ counterclockwise.

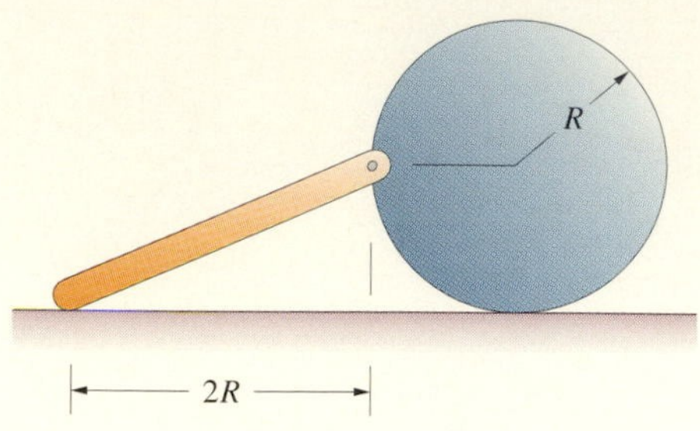

P7.68

7.69 If the disk in Problem 7.68 rolls and the coefficient of kinetic friction between the bar and the horizontal surface is μ_k, what is the disk's angular acceleration at the instant the system is released?

7.70 The 2-kg bar rotates *in the horizontal plane* about the smooth pin. The 6-kg collar A slides on the smooth bar. At the instant shown, $r = 1.2$ m, $\omega = 0.4$ rad/s, and the collar is sliding outward at 0.5 m/s relative to the bar. If you neglect the moment of inertia of the collar (that is, treat the collar as a particle), what is the bar's angular acceleration?

Strategy: Draw individual free-body diagrams of the bar and the collar and write Newton's second law for the collar in terms of polar coordinates.

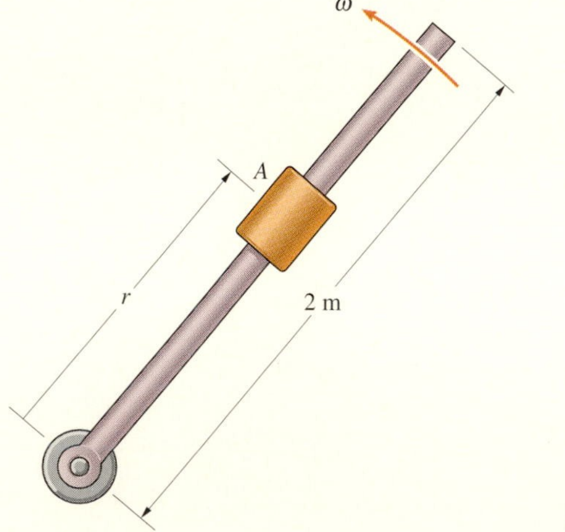

P7.70

7.71 In Problem 7.70, the moment of inertia of the collar about its center of mass is 0.2 kg-m^2. Determine the angular acceleration of the bar and compare your answer with the answer to Problem 7.70.

Example 7.9

Application to Engineering

Internal Forces and Moments in Beams

The slender bar of mass m in Fig. 7.21 starts from rest in the position shown and falls. When it has rotated through an angle θ, what is the maximum bending moment in the bar and where does it occur?

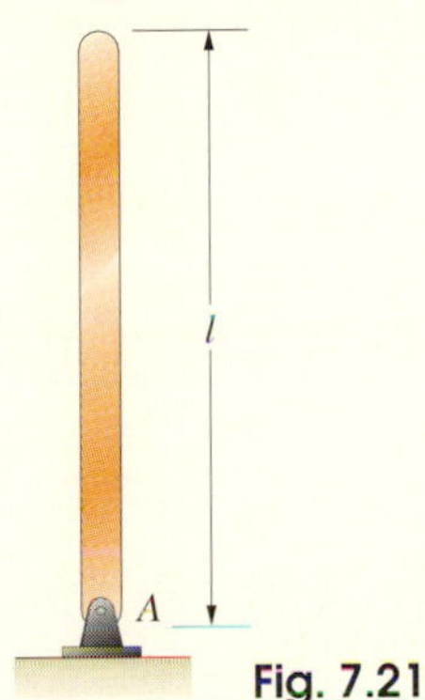

Fig. 7.21

STRATEGY

The internal forces and moments in a beam subjected to two-dimensional loading are the axial force P, shear force V, and bending moment M (Fig. a). We must first use the equation of angular motion to determine the bar's angular acceleration. Then we can cut the bar at an arbitrary distance x from one end and apply *the equations of motion* to determine the bending moment as a function of x.

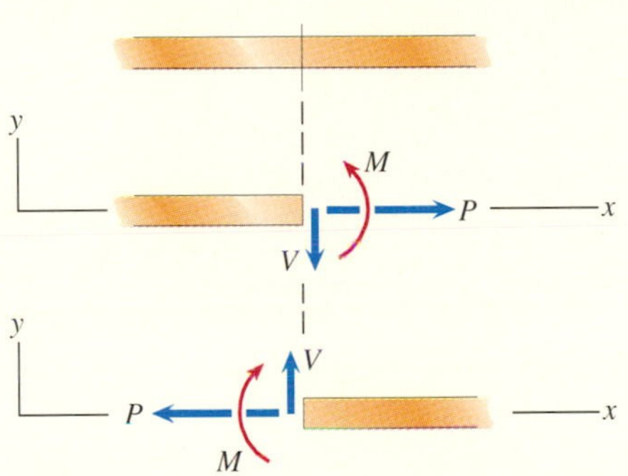

(a) The axial force, shear force, and bending moment in a beam.

SOLUTION

The moment of inertia of the bar about A is

$$I_A = I + d^2 m = \frac{1}{12} ml^2 + \left(\frac{1}{2} l\right)^2 m = \frac{1}{3} ml^2.$$

When the bar has rotated through an angle θ (Fig. b), the total moment about A is $\Sigma M_A = mg(\frac{1}{2} l \sin \theta)$. Point A is fixed, so we can write the equation of angular motion as

$$\Sigma M_A = I_A \alpha:$$

$$\frac{1}{2} mgl \sin \theta = \frac{1}{3} ml^2 \alpha.$$

Solving for the angular acceleration, we obtain

$$\alpha = \frac{3}{2} \frac{g}{l} \sin \theta.$$

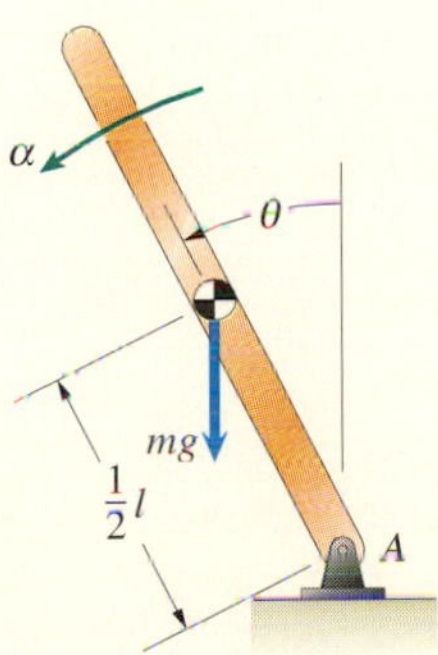

(b) Determining the moment about A.

In Fig. (c) we introduce a coordinate system, cut the bar at a distance x from the top, and draw the free-body diagram of the top part. The center of mass is at the midpoint, and we determine the mass by multiplying the bar's mass by the ratio of the length of the free body to that of the bar. Applying Newton's second law in the y direction, we obtain

$$\Sigma F_y = -V - \frac{x}{l} mg \sin\theta = \frac{x}{l} ma_y.$$

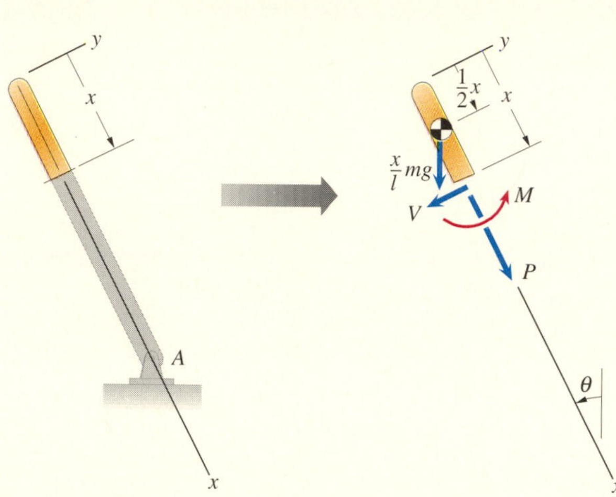

(c) Cutting the bar at an arbitrary distance x.

The moment of inertia of the free body about its center of mass is $\frac{1}{12}[(x/l)m]x^2$, so the equation of angular motion is

$$\Sigma M = I\alpha:$$

$$M - \left(\frac{1}{2}x\right)V = \frac{1}{12}\left(\frac{x}{l}m\right)x^2 \frac{3}{2}\frac{g}{l}\sin\theta.$$

The y component of the acceleration of the center of mass is equal to the product of its radial distance from A and the angular acceleration (Fig. d):

$$a_y = -\left(l - \frac{1}{2}x\right)\alpha = -\left(l - \frac{1}{2}x\right)\frac{3}{2}\frac{g}{l}\sin\theta.$$

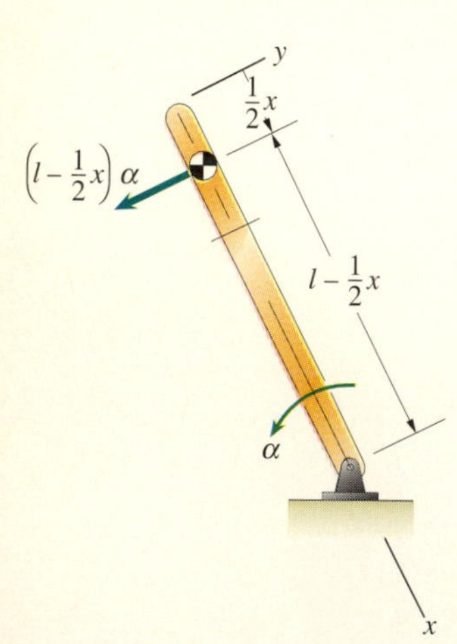

(d) Determining the acceleration of the center of mass of the free body.

Using this expression, we can solve the two equations of motion for V and M in terms of θ. The solution for M is

$$M = \frac{1}{4}mgl\sin\theta\left(\frac{x}{l}\right)^2\left(1 - \frac{x}{l}\right). \tag{7.30}$$

The bending moment equals zero at both ends of the bar. Taking the derivative of this expression with respect to x and equating it to zero to determine where M is a maximum, we obtain $x = \frac{2}{3}l$. Substituting this value of x into Eq. (7.30), we obtain the maximum bending moment:

$$M_{\max} = \frac{1}{27}mgl\sin\theta.$$

The distribution of M is shown in Fig. 7.22.

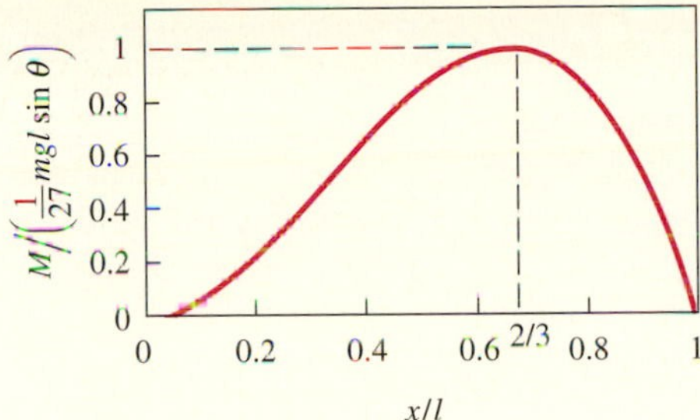

Fig. 7.22
Distribution of the bending moment in a falling bar.

DESIGN ISSUES

To design a member of a structure, engineers must consider both the external and internal forces and moments to which it will be subjected. In the case of a beam, they must determine the distributions of the axial force P, shear force V, and bending moment M as the first step in determining whether the beam will support its design loads without failing. If they know the external loads and reactions and the beam is in equilibrium, they can apply the equilibrium equations to determine the internal forces and moment at a given cross section. But in many situations, a beam will not be in equilibrium. It could be a member of a structure, such as the internal frame of an airplane, that is accelerating, or it could be a connecting rod in an internal combustion engine. In such cases, the maximum internal forces and moments can far exceed the values predicted by a static analysis, and the procedure we describe in this example must be used.

The dynamic bending moment distribution we obtained in Example 7.9 (Fig. 7.22) explains a phenomenon that has been observed during the demolition of masonry chimneys. An explosive charge at the base of the chimney causes it to fall, initially rotating as a rigid body about its base. As the chimney falls, it is observed to fracture near the location of the maximum bending moment (Fig. 7.23).

Fig. 7.23
A falling chimney fractures as it falls due to the bending moment to which it is subjected.

Problems

Problems 7.72-7.78 are related to Example 7.9.

7.72 The 3-Mg rocket is accelerating upward at 2 g's. If you model it as a homogeneous bar, what is the magnitude of the axial force P at the midpoint?

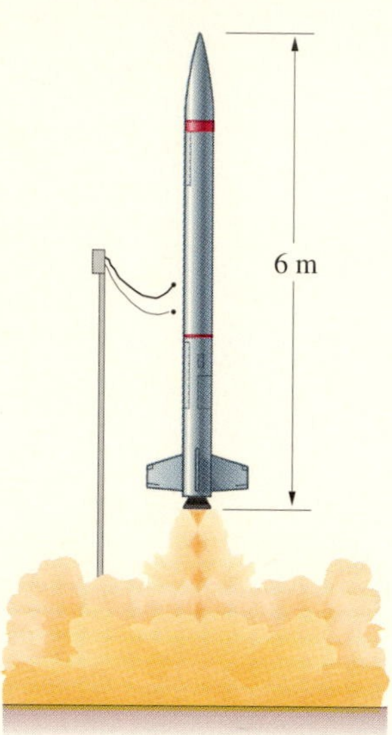

P7.72

7.73 The 20-kg slender bar is attached to a vertical shaft at A and rotates *in the horizontal plane* with a constant angular velocity of 10 rad/s. What is the axial force P at the bar's midpoint?

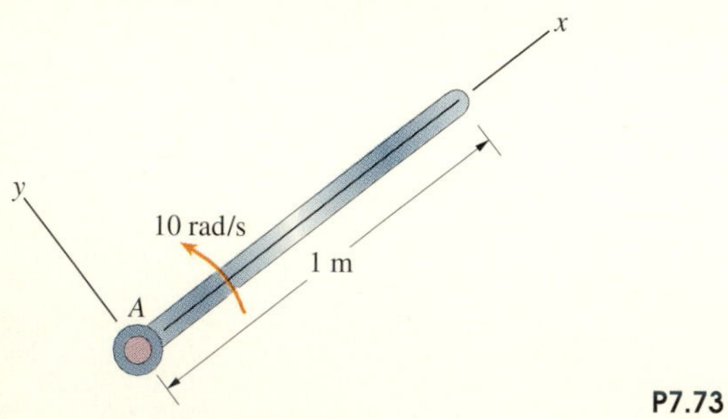

P7.73

7.74 For the rotating bar in Problem 7.73, draw a graph of the axial force as a function of x.

7.75 The 100-lb slender bar AB has a built-in support at A. The y axis points upward. Determine the magnitudes of the shear force and bending moment at the bar's midpoint if (a) the support is stationary; (b) the support is accelerating upward at 10 ft/s^2.

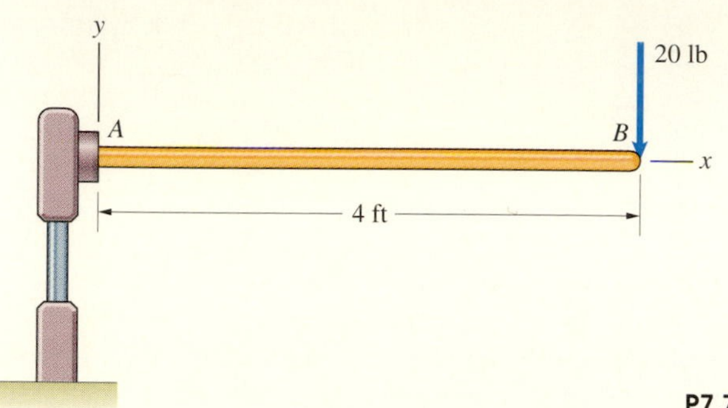

P7.75

7.76 For the bar in Problem 7.75, draw the shear force and bending moment diagrams for the two cases.

7.77 The 18-kg ladder is held in equilibrium in the position shown by the force F. Model the ladder as a slender bar and neglect friction.
(a) What are the axial force, shear force, and bending moment at the ladder's midpoint?
(b) If the force F is suddenly removed, what are the axial force, shear force, and bending moment at the ladder's midpoint at that instant?

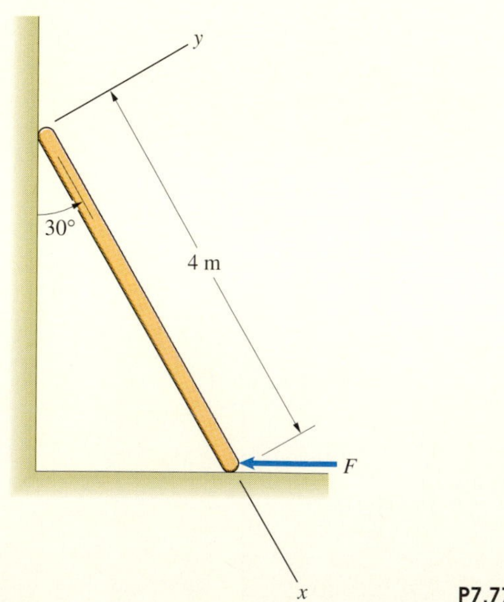

P7.77

7.78 For the ladder in Problem 7.77, draw the shear force and bending moment diagrams for the two cases.

Computational Mechanics

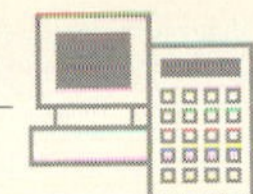

The material in this section is designed for the use of a programmable calculator or computer.

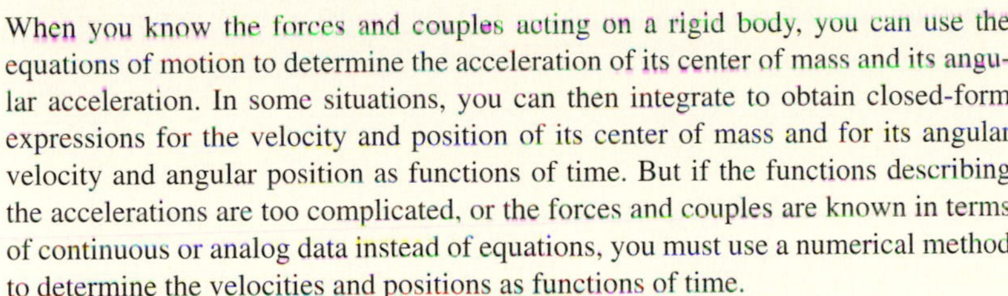

When you know the forces and couples acting on a rigid body, you can use the equations of motion to determine the acceleration of its center of mass and its angular acceleration. In some situations, you can then integrate to obtain closed-form expressions for the velocity and position of its center of mass and for its angular velocity and angular position as functions of time. But if the functions describing the accelerations are too complicated, or the forces and couples are known in terms of continuous or analog data instead of equations, you must use a numerical method to determine the velocities and positions as functions of time.

In Chapter 3 we described a simple finite-difference method for determining the position and velocity of the center of mass of an object as functions of time. You can determine the angular position and angular velocity in the same way. Let the angular acceleration of a rigid body be a function of time, its angular position, and its angular velocity:

$$\alpha = \alpha(t, \theta, \omega).$$

Suppose that at a particular time t_0, we know the angle $\theta(t_0)$ and angular velocity $\omega(t_0)$. The angular acceleration at t_0 is

$$\frac{d\omega}{dt}(t_0) = \alpha(t_0, \theta(t_0), \omega(t_0)). \tag{7.31}$$

To determine the angular velocity at a time $t + \Delta t$, we express it as a Taylor series:

$$\omega(t_0 + \Delta t) = \omega(t_0) + \frac{d\omega}{dt}(t_0)\,\Delta t + \frac{1}{2}\frac{d^2\omega}{dt^2}(\Delta t)^2 + \cdots .$$

By choosing a sufficiently small value of Δt, we can neglect terms in this equation of second and higher order in Δt and substitute Eq. (7.31) to obtain an approximation for the angular velocity at $t_0 + \Delta t$:

$$\omega(t_0 + \Delta t) = \omega(t_0) + \alpha(t_0, \theta(t_0), \omega(t_0))\,\Delta t. \tag{7.32}$$

We approximate the angle at $t_0 + \Delta t$ in the same way. Expressing it as a Taylor series,

$$\theta(t_0 + \Delta t) = \theta(t_0) + \frac{d\theta}{dt}(t_0)\,\Delta t + \frac{1}{2}\frac{d^2\theta}{dt^2}(\Delta t)^2 + \cdots ,$$

and neglecting higher-order terms in Δt, we obtain

$$\theta(t_0 + \Delta t) = \theta(t_0) + \omega(t_0)\,\Delta t. \tag{7.33}$$

With Eqs. (7.32) and (7.33), we can determine the approximate values of the angular velocity and position at $t_0 + \Delta t$. Using these values as initial conditions, we can repeat the procedure to determine the angular velocity and position at $t_0 + 2\,\Delta t$, and so forth.

Example 7.10

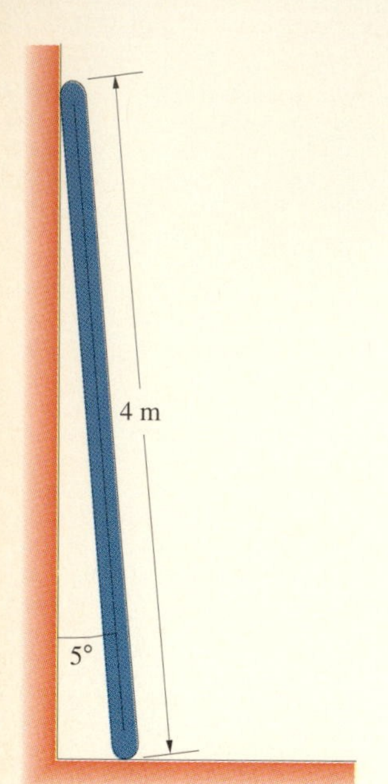

Fig. 7.24

The 18-kg ladder in Fig. 7.24 is released from rest in the position shown at $t = 0$. Neglecting friction, determine its angular position and angular velocity as functions of time. Use time increments Δt of 0.1 s, 0.01 s, and 0.001 s.

STRATEGY

The initial steps—drawing the free-body diagram of the ladder, applying the equations of motion, and determining the angular acceleration—are presented in Example 7.4. The ladder's angular acceleration is

$$\alpha = \frac{3g}{2l} \sin \theta,$$

where θ is the angle between the ladder and the wall and l is its length. With this expression, we can use Eqs. (7.32) and (7.33) to approximate the ladder's angular position and angular velocity as functions of time.

SOLUTION

The angular acceleration is

$$\alpha = \frac{(3)(9.81)}{(2)(4)} \sin \theta = 3.68 \sin \theta \text{ rad/s}^2.$$

Let $\Delta t = 0.1$ s. At the initial time $t_0 = 0$, $\theta(t_0) = 5° = 0.0873$ rad and $\omega(t_0) = 0$. We can use Eqs. (7.32) and (7.33) to determine the angular velocity and position at time $t_0 + \Delta t = 0.1$ s. The angular position is

$$\theta(t_0 + \Delta t) = \theta(t_0) + \omega(t_0)\,\Delta t:$$

$$\theta(0.1) = \theta(0) + \omega(0)\,\Delta t$$

$$= 0.0873 + (0)(0.1) = 0.0873 \text{ rad}.$$

The angular velocity is

$$\omega(t_0 + \Delta t) = \omega(t_0) + \alpha(t_0)\,\Delta t:$$

$$\omega(0.1) = 0 + [3.68 \sin(0.0873)](0.1) = 0.0321 \text{ rad/s}.$$

Using these values as the initial conditions for the next time step, the angular position at $t = 0.2$ s is

$$\theta(0.2) = \theta(0.1) + \omega(0.1)\,\Delta t$$

$$= 0.0873 + (0.0321)(0.1) = 0.0905 \text{ rad},$$

and the angular velocity is

$$\omega(0.2) = \omega(0.1) + \alpha(0.1)\,\Delta t$$

$$= 0.0321 + [3.68 \sin(0.0873)](0.1) = 0.0641 \text{ rad/s}.$$

Continuing in this way, we obtain the following values for the first five time steps:

Time, s	θ, rad	ω, rad/s
0.0	0.0873	0.0000
0.1	0.0873	0.0321
0.2	0.0905	0.0641
0.3	0.0969	0.0974
0.4	0.1066	0.1329
0.5	0.1199	0.1721

Figures 7.25 and 7.26 show the numerical solutions for the angular position and angular velocity obtained using $\Delta t = 0.1$ s, $\Delta t = 0.01$ s, and $\Delta t = 0.001$ s. Trials with smaller time intervals indicate that $\Delta t = 0.001$ s closely approximates the exact solution. We show the positions of the falling ladder at 0.2-s intervals in Fig. 7.27.

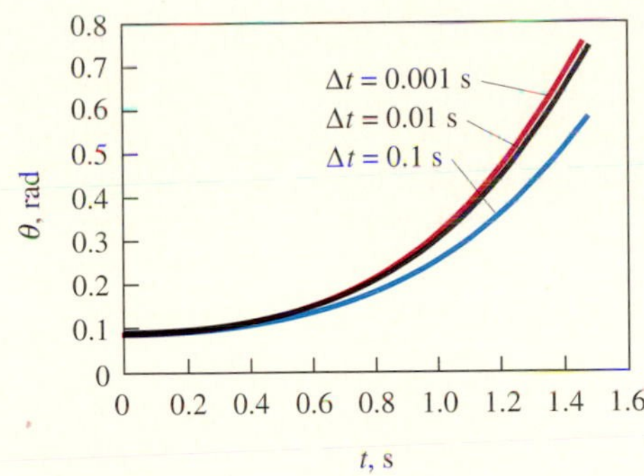

Fig. 7.25
Numerical solutions for the ladder's angular position.

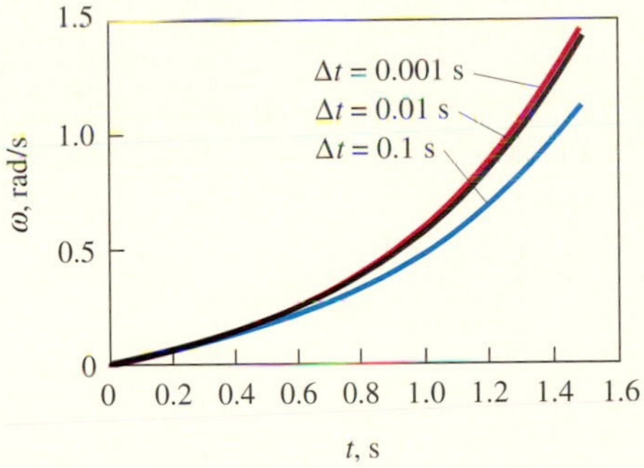

Fig. 7.26
Numerical solutions for the ladder's angular velocity.

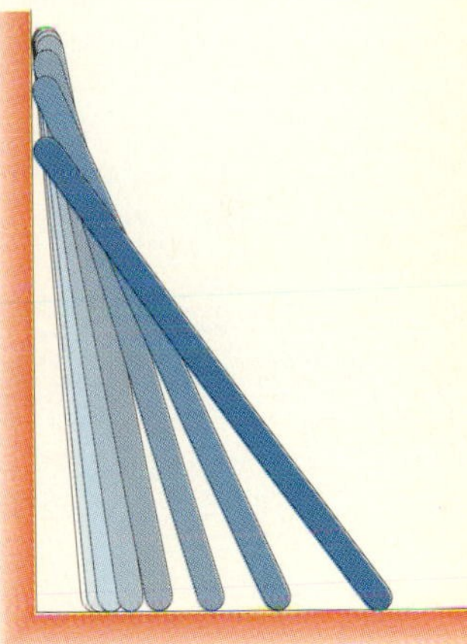

Fig. 7.27
Position of the falling ladder at 0.2-s intervals from $t = 0$ to $t = 1.4$ s.

DISCUSSION

By using the chain rule, we can write the ladder's angular acceleration as

$$\alpha = \frac{d\omega}{dt} = \frac{d\omega}{d\theta}\omega = \frac{3g}{2l}\sin\theta.$$

Separating variables, we can integrate to determine the angular velocity as a function of the angular position:

$$\int_0^{\omega} \omega \, d\omega = \int_{5^\circ}^{\theta} \frac{3g}{2l}\sin\theta \, d\theta.$$

We obtain

$$\omega = \sqrt{\left(\frac{3g}{l}\right)(\cos 5^\circ - \cos\theta)}.$$

This closed-form result is compared with the graph of our numerical solution (using $\Delta t = 0.001$ s) in Fig. 7.28. The curves are indistinguishable.

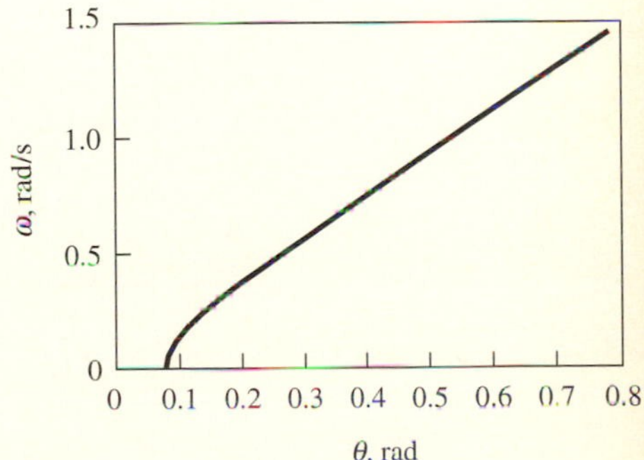

Fig. 7.28
Analytical and numerical solutions for the ladder's angular velocity as a function of its angular position.

Problems

7.79 Continue the calculations presented in Example 7.10, using $\Delta t = 0.1$ s, and determine the ladder's angular position and angular velocity at $t = 0.6$ s and $t = 0.7$ s.

7.80 The moment of inertia of the helicopter's rotor is 400 slug-ft^2. It starts from rest at $t = 0$, the engine exerts a constant torque of 500 ft-lb, and aerodynamic drag exerts a torque of magnitude $20\omega^2$ ft-lb, where ω is the rotor's angular velocity in radians per second. Using $\Delta t = 0.2$ s, determine the rotor's angular position and angular velocity for the first five time steps. Compare your results for the angular velocity with the closed-form solution.

P7.80

7.81 In Problem 7.80, draw a graph of the rotor's angular velocity as a function of time from $t = 0$ to $t = 10$ s, comparing the closed-form solution, the numerical solution using $\Delta t = 1.0$ s, and the numerical solution using $\Delta t = 0.2$ s.

7.82 The slender 10-kg bar is released from rest in the horizontal position shown. Using $\Delta t = 0.1$ s, determine the bar's angular position and angular velocity for the first five time steps.

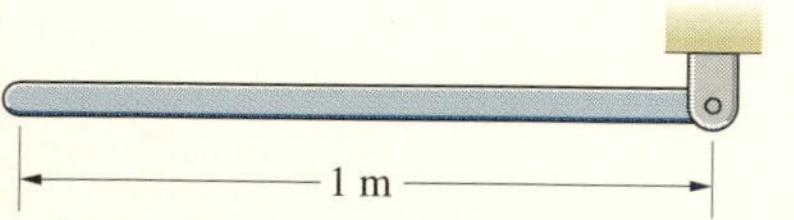

P7.82

7.83 In Problem 7.82, determine the bar's angular position and angular velocity as functions of time from $t = 0$ to $t = 0.8$ s using $\Delta t = 0.1$ s, $\Delta t = 0.01$ s, and $\Delta t = 0.001$ s. Draw the graphs of the angular velocity as a function of the angular position for these three cases and compare them with the graph of the closed-form solution for the angular velocity as a function of the angular position.

7.84 In Problem 7.82, suppose that the bar's pin support contains a damping device that exerts a resisting couple on the bar of magnitude $c\omega$ (N-m), where ω is the angular velocity in radians per second. Using $\Delta t = 0.001$ s, draw graphs of the bar's angular velocity as a function of time from $t = 0$ to $t = 0.8$ s for the cases $c = 0$, $c = 2$, $c = 4$, and $c = 8$.

7.85 The falling ladder in Example 7.10 will lose contact with the wall before it hits the floor. Using $\Delta t = 0.001$ s, estimate the time and the value of the angle between the wall and the ladder when this occurs.

7.86 A torsional spring at A exerts a counterclockwise couple $k\theta$ on the bar, where $k = 20$ N-m and θ is in radians. The 2-kg bar is 1 m long. At $t = 0$, the bar is released from rest in the horizontal position ($\theta = 0$). Using $\Delta t = 0.01$ s, determine the bar's angular position and angular velocity for the first five time steps.

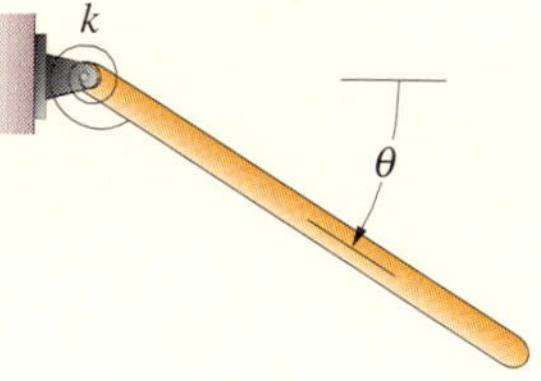

P7.86

7.87 Using a numerical solution with $\Delta t = 0.001$ s, estimate the maximum angle θ reached by the bar in Problem 7.86 when it is released from rest in the horizontal position. At what time after release does the maximum angle occur?

Appendix: Moments of Inertia

When a rigid body is subjected to forces and couples, the rotational motion that results depends not only on its mass, but also on how its mass is *distributed*. Although the two objects in Fig. 7.29 have the same mass, the angular accelerations caused by the couple M are different. This difference is reflected in the equation of angular motion $M = I\alpha$ through the moment of inertia I. The object in Fig. 7.29(a) has a smaller moment of inertia about the axis L, so its angular acceleration is greater.

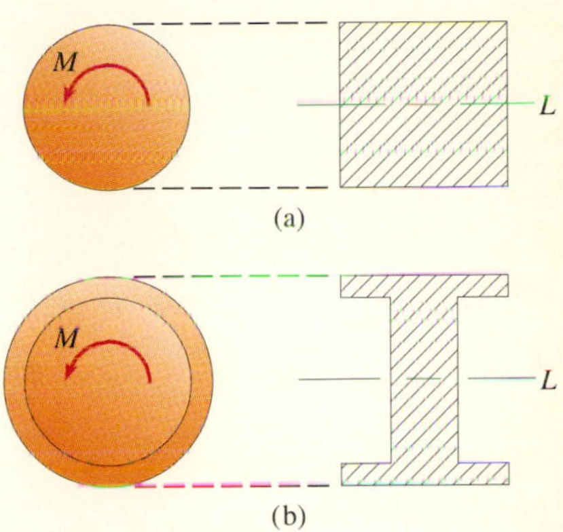

Fig. 7.29

Objects of equal mass that have different moments of inertia about L.

In deriving the equations of motion of a rigid body in Sections 7.2 and 7.3, we regarded it as a finite number of particles and expressed its moment of inertia about an axis L_O as

$$I_O = \sum_i m_i r_i^2,$$

where m_i is the mass of the ith particle and r_i is the perpendicular distance from L_O to the ith particle (Fig. 7.30a). To calculate moments of inertia of objects, it is often more convenient to model them as continuous distributions of mass and express the moment of inertia about L_O as

$$I_O = \int_m r^2\, dm, \tag{7.34}$$

where r is the perpendicular distance from L_O to the differential element of mass dm (Fig. 7.30b). When the axis passes through the center of mass of the object, we denote the axis by L and the moment of inertia about L by I.

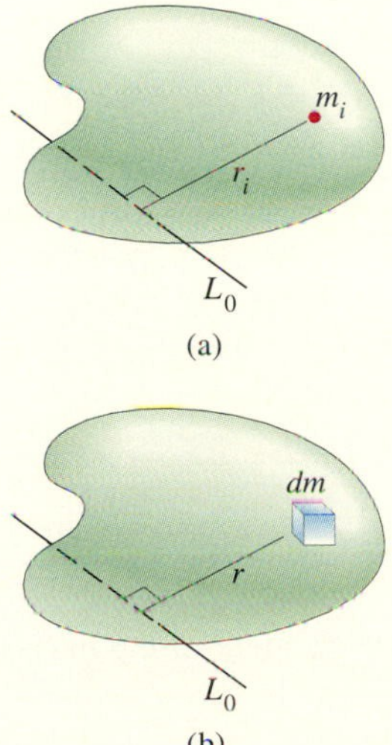

Fig. 7.30

Determining the moment of inertia by modeling an object as (a) a finite number of particles and (b) a continuous distribution of mass.

Simple Objects

We begin by determining moments of inertia of some simple objects. Then in the next section we describe the parallel-axis theorem, which simplifies the task of determining moments of inertia of objects composed of combinations of simple parts.

Slender Bars We will determine the moment of inertia of a straight slender bar about a perpendicular axis L through the center of mass of the bar (Fig. 7.31a). "Slender" means we assume that the bar's length is much greater than its width. Let the bar have length l, cross-sectional area A, and mass m. We assume that A is uniform along the length of the bar and that the material is homogeneous.

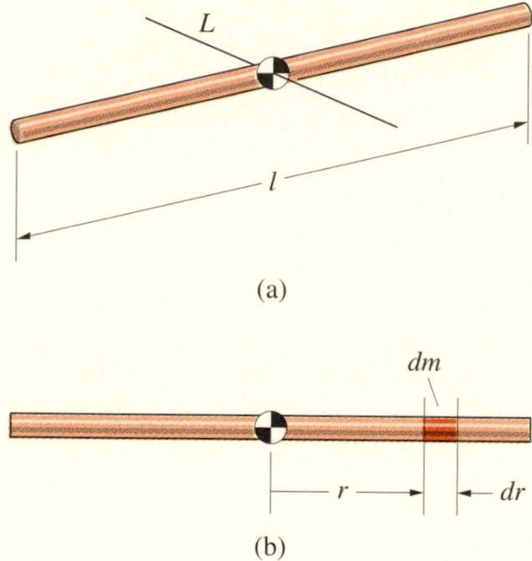

Fig. 7.31
(a) A slender bar.
(b) A differential element of length dr.

Consider a differential element of the bar of length dr at a distance r from the center of mass (Fig. 7.31b). The element's mass is equal to the product of its volume and the mass density: $dm = \rho A\ dr$. Substituting this expression into Eq. (7.34), we obtain the moment of inertia of the bar about a perpendicular axis through its center of mass:

$$I = \int_m r^2\, dm = \int_{-l/2}^{l/2} \rho A r^2\, dr = \frac{1}{12}\rho A l^3.$$

The mass of the bar equals the product of the mass density and the volume of the bar, $m = \rho A l$, so we can express the moment of inertia as

$$I = \frac{1}{12} m l^2. \qquad (7.35)$$

We have neglected the lateral dimensions of the bar in obtaining this result. That is, we treated the differential element of mass dm as if it were concentrated on the axis of the bar. As a consequence, Eq. (7.35) is an approximation for the moment of inertia of a bar. Later in this section, we will determine the moments of inertia for a bar of finite lateral dimension and show that Eq. (7.35) is a good approximation when the width of the bar is small in comparison to its length.

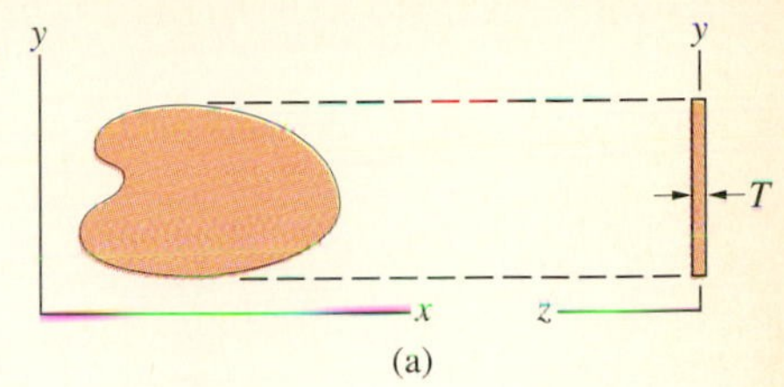

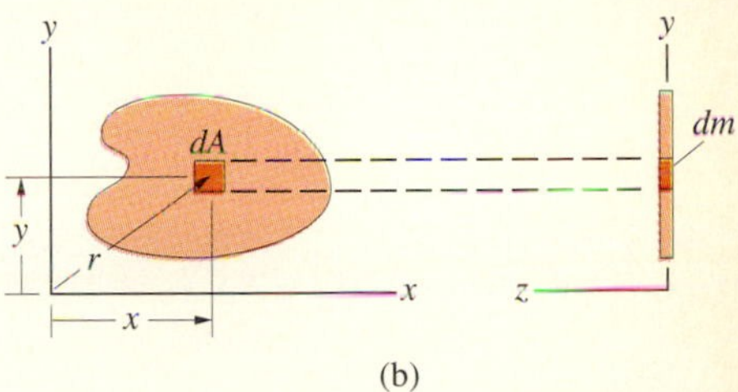

Fig. 7.32

(a) A plate of arbitrary shape and uniform thickness T.

(b) An element of volume obtained by projecting an element of area dA through the plate.

Thin Plates Consider a homogeneous flat plate that has mass m and uniform thickness T. We will leave the shape of the cross-sectional area of the plate unspecified. Let a cartesian coordinate system be oriented so that the plate lies in the x-y plane (Fig. 7.32a). Our objective is to determine the moments of inertia of the plate about the x, y, and z axes.

We can obtain a differential element of volume of the plate by projecting an element of area dA through the thickness T of the plate (Fig. 7.32b). The resulting volume is $T\,dA$. The mass of this element of volume is equal to the product of the mass density and the volume: $dm = \rho T\,dA$. Substituting this expression into Eq. (7.34), we obtain the moment of inertia of the plate about the z axis in the form

$$I_{(z\text{ axis})} = \int_m r^2\,dm = \rho T \int_A r^2\,dA,$$

where r is the distance from the z axis to dA. Since the mass of the plate is $m = \rho TA$, where A is the cross-sectional area of the plate, the product $\rho T = m/A$. The integral on the right is the polar moment of inertia J_0 of the cross-sectional area of the plate. Therefore we can write the moment of inertia of the plate about the z axis as

$$I_{(z\text{ axis})} = \frac{m}{A} J_0. \tag{7.36}$$

From Fig. 7.32(b), we see that the perpendicular distance from the x axis to the element of area dA is the y coordinate of dA. Therefore the moment of inertia of the plate about the x axis is

$$I_{(x\text{ axis})} = \int_m y^2\,dm = \rho T \int_A y^2\,dA = \frac{m}{A} I_x, \tag{7.37}$$

where I_x is the moment of inertia of the cross-sectional area of the plate about the x axis. The moment of inertia of the plate about the y axis is

$$I_{(y\text{ axis})} = \int_m x^2\,dm = \rho T \int_A x^2\,dA = \frac{m}{A} I_y, \tag{7.38}$$

where I_y is the moment of inertia of the cross-sectional area of the plate about the y axis.

Thus we have expressed the moments of inertia of a thin homogeneous plate of uniform thickness in terms of the moments of inertia of the cross-sectional area of the plate. In fact, these results explain why the area integrals I_x, I_y, and J_0 are called moments of inertia.

Since the sum of the area moments of inertia I_x and I_y is equal to the polar moment of inertia J_0, the moment of inertia of the thin plate about the z axis is equal to the sum of its moments of inertia about the x and y axes:

$$I_{(z\text{ axis})} = I_{(x\text{ axis})} + I_{(y\text{ axis})}. \tag{7.39}$$

In the following example we use integration to determine the moment of inertia of an object consisting of two slender bars welded together. We then present an example that demonstrates the use of Eqs. (7.36)–(7.38) to determine the moments of inertia of a thin, homogeneous plate with a specific cross-sectional area.

Example 7.11

Two homogeneous, slender bars, each of length l, mass m, and cross-sectional area A, are welded together to form an L-shaped object (Fig. 7.33). Determine the moment of inertia of the object about the axis L_O through point O. (The axis L_O is perpendicular to the two bars.)

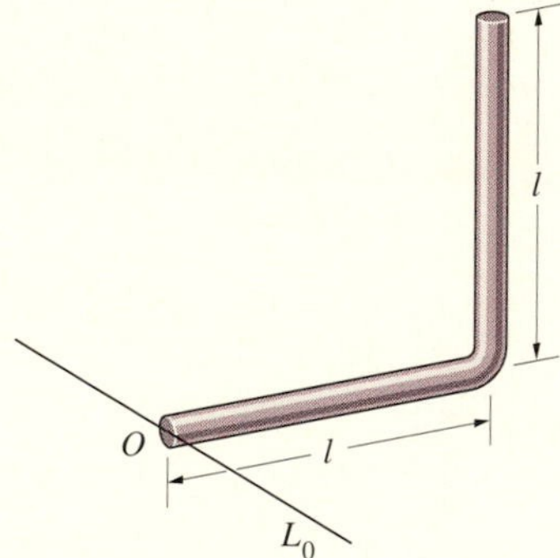

Fig. 7.33

STRATEGY

Using the same integration procedure we used for a single bar, we can determine the moment of inertia of each bar about L_O and sum the results.

SOLUTION

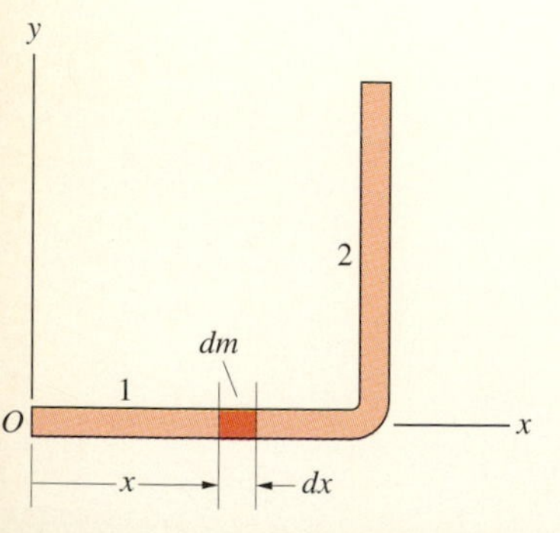

(a) Differential element of bar 1.

We orient a coordinate system with the z axis along L_O and the x axis colinear with bar 1 (Fig. a). The mass of the differential element of bar 1 of length dx is $dm = \rho A\, dx$. The moment of inertia of bar 1 about L_O is

$$(I_O)_1 = \int_m r^2\, dm = \int_0^l \rho A x^2\, dx = \frac{1}{3}\rho A l^3.$$

In terms of the mass of the bar, $m = \rho A l$, we can write this result as

$$(I_O)_1 = \frac{1}{3} m l^2.$$

The mass of the element of bar 2 of length dy shown in Fig. (b) is $dm = \rho A\, dy$. From the figure we see that the perpendicular distance from L_O to the element is $r = \sqrt{l^2 + y^2}$. Therefore the moment of inertia of bar 2 about L_O is

$$(I_O)_2 = \int_m r^2\, dm = \int_0^l \rho A(l^2 + y^2)\, dy = \frac{4}{3}\rho A l^3.$$

In terms of the mass of the bar, we obtain

$$(I_O)_2 = \frac{4}{3} m l^2.$$

The moment of inertia of the L-shaped object about L_O is

$$I_O = (I_O)_1 + (I_O)_2 = \frac{1}{3} m l^2 + \frac{4}{3} m l^2 = \frac{5}{3} m l^2.$$

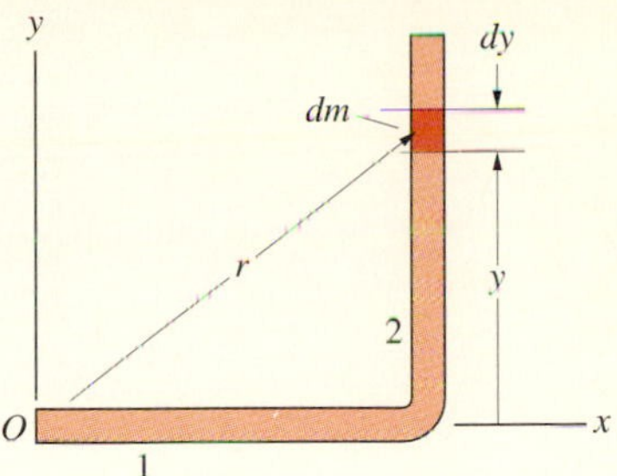

(b) Differential element of bar 2.

Example 7.12

The thin, homogeneous plate in Fig. 7.34 is of uniform thickness and mass m. Determine its moments of inertia about the x, y, and z axes.

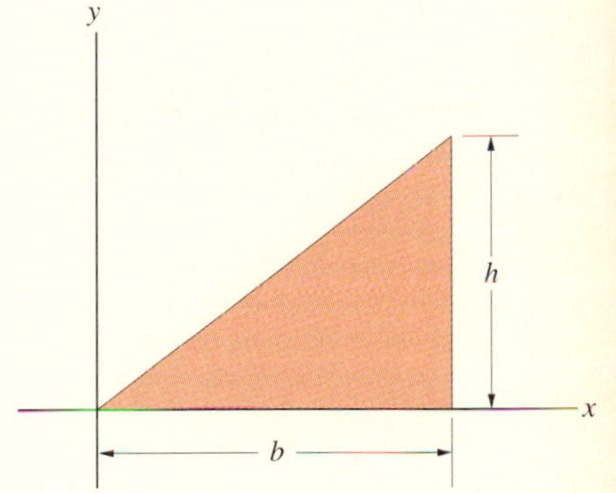

Fig. 7.34

STRATEGY

The moments of inertia about the x and y axes are given by Eqs. (7.37) and (7.38) in terms of the moments of inertia of the cross-sectional area of the plate. We can determine the moment of inertia of the plate about the z axis from Eq. (7.39).

SOLUTION

From Appendix B, the moments of inertia of the triangular area about the x and y axes are $I_x = \frac{1}{12}bh^3$ and $I_y = \frac{1}{4}hb^3$. Therefore the moments of inertia of the plate about the x and y axes are

$$I_{(x\text{ axis})} = \frac{m}{A} I_x = \frac{m}{\frac{1}{2}bh}\left(\frac{1}{12}bh^3\right) = \frac{1}{6}mh^2,$$

$$I_{(y\text{ axis})} = \frac{m}{A} I_y = \frac{m}{\frac{1}{2}bh}\left(\frac{1}{4}hb^3\right) = \frac{1}{2}mb^2.$$

The moment of inertia about the z axis is

$$I_{(z\text{ axis})} = I_{(x\text{ axis})} + I_{(y\text{ axis})} = m\left(\frac{1}{6}h^2 + \frac{1}{2}b^2\right).$$

Parallel-Axis Theorem

This theorem allows us to determine the moment of inertia of a composite object when we know the moments of inertia of its parts. Suppose that we know the moment of inertia I about an axis L through the center of mass of an object, and we wish to determine its moment of inertia I_O about a parallel axis L_O (Fig. 7.35a). To determine I_O, we introduce parallel coordinate systems xyz and $x'y'z'$ with the z axis along L_O and the z' axis along L, as shown in Fig. 7.35(b). (In this figure the axes L_O and L are perpendicular to the page.) The origin O of the xyz coordinate system is contained in the x'-y' plane. The terms d_x and d_y are the coordinates of the center of mass relative to the xyz coordinate system.

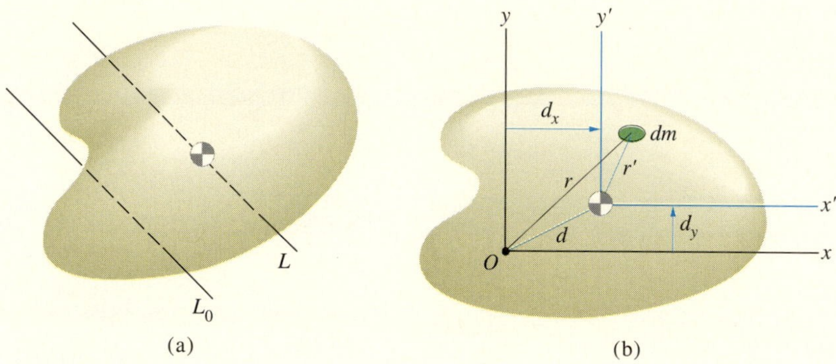

Fig. 7.35

(a) An axis L through the center of mass of an object and a parallel axis L_0.
(b) The xyz and $x'y'z'$ coordinate systems.

The moment of inertia of the object about L_O is

$$I_O = \int_m r^2\,dm = \int_m (x^2 + y^2)\,dm, \tag{7.40}$$

where r is the perpendicular distance from L_O to the differential element of mass dm, and x, y are the coordinates of dm in the x-y plane. The x-y coordinates of dm are related to its x'-y' coordinates by

$$x = x' + d_x, \qquad y = y' + d_y.$$

By substituting these expressions into Eq. (7.40), we can write it as

$$I_O = \int_m [(x')^2 + (y')^2]\,dm + 2d_x \int_m x'\,dm + 2d_y \int_m y'\,dm + \int_m (d_x^2 + d_y^2)\,dm. \tag{7.41}$$

Since $(x')^2 + (y')^2 = (r')^2$, where r' is the perpendicular distance from L to dm, the first integral on the right side of this equation is the moment of inertia I of the object about L. Recall that the x' and y' coordinates of the center of mass of the object relative to the $x'y'z'$ coordinate system are defined by

$$\bar{x}' = \frac{\int_m x'\,dm}{\int_m dm}, \qquad \bar{y}' = \frac{\int_m y'\,dm}{\int_m dm}.$$

Because the center of mass of the object is at the origin of the $x'y'z'$ system, $\bar{x}' = 0$ and $\bar{y}' = 0$. Therefore the integrals in the second and third terms on the right side of Eq. (7.41) are equal to zero. From Fig. 7.35(b), we see that $d_x^2 + d_y^2 = d^2$, where d is the perpendicular distance between the axes L and L_O. Therefore we obtain

$$I_O = I + d^2 m, \tag{7.42}$$

where m is the mass of the object. This is the **parallel-axis theorem**. If you know the moment of inertia of an object about a given axis, you can use this theorem to determine its moment of inertia about any parallel axis.

In the next two examples we use the parallel-axis theorem to determine moments of inertia of composite objects. Determining the moment of inertia about a given axis L_O typically requires three steps:

1. Choose the parts—*Try to divide the object into parts whose moments of inertia you know or can easily determine.*

2. Determine the moments of inertia of the parts—*You must first determine the moment of inertia of each part about the axis through its center of mass parallel to L_O. Then you can use the parallel-axis theorem to determine its moment of inertia about L_O.*

3. Sum the results—*Sum the moments of inertia of the parts (or subtract in the case of a hole or cutout) to obtain the moment of inertia of the composite object.*

Example 7.13

Two homogeneous, slender bars, each of length l and mass m, are welded together to form an L-shaped object (Fig. 7.36). Determine the moment of inertia of the object about the axis L_O through point O. (The axis L_O is perpendicular to the two bars.)

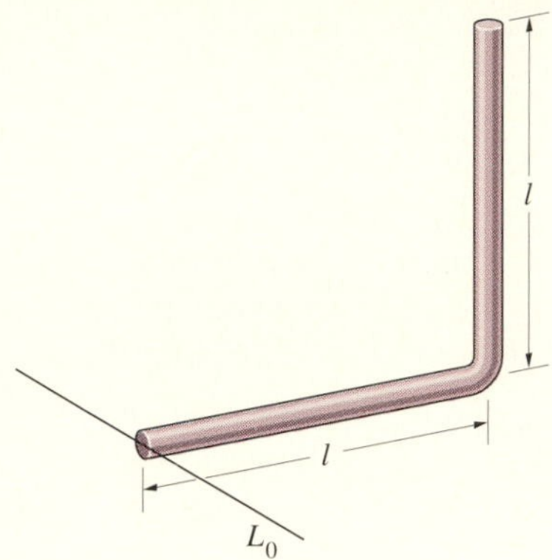

Fig. 7.36

SOLUTION

Choose the Parts The parts are the two bars, which we call bar 1 and bar 2 (Fig. a).

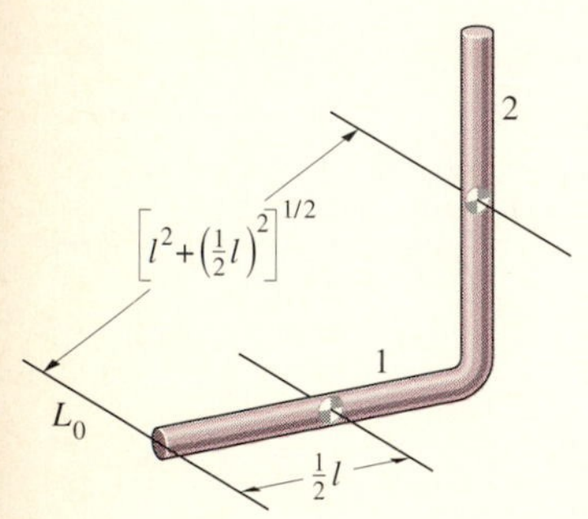

(a) The distances from L_O to parallel axes through the centers of mass of bars 1 and 2.

Determine the Moments of Inertia of the Parts From Appendix C, the moment of inertia of each bar about a perpendicular axis through its center of mass is $I = \frac{1}{12}ml^2$. The distance from L_O to the parallel axis through the center of mass of bar 1 is $\frac{1}{2}l$ (Fig. a). Therefore the moment of inertia of bar 1 about L_O is

$$(I_O)_1 = I + d^2m = \frac{1}{12}ml^2 + \left(\frac{1}{2}l\right)^2 m = \frac{1}{3}ml^2.$$

The distance from L_0 to the parallel axis through the center of mass of bar 2 is $[l^2 + (\frac{1}{12}l)^2]^{1/2}$. The moment of inertia of bar 2 about L_O is

$$(I_O)_2 = I + d^2m = \frac{1}{12}ml^2 + \left[l^2 + \left(\frac{1}{2}l\right)^2\right] m = \frac{4}{3}ml^2.$$

Sum the Results The moment of inertia of the L-shaped object about L_O is

$$I_O = (I_O)_1 + (I_O)_2 = \frac{1}{3}ml^2 + \frac{4}{3}ml^2 = \frac{5}{3}ml^2.$$

DISCUSSION

Compare this solution to Example 7.11, in which we used integration to determine the moment of inertia of this object about L_O. We obtained the result much more easily with the parallel-axis theorem, but of course we needed to know the moments of inertia of the bars about the axes through their centers of mass.

Example 7.14

The object in Fig. 7.37 consists of a slender, 3-kg bar welded to a thin, circular, 2-kg disk. Determine its moment of inertia about the axis L through its center of mass. (The axis L is perpendicular to the bar and disk.)

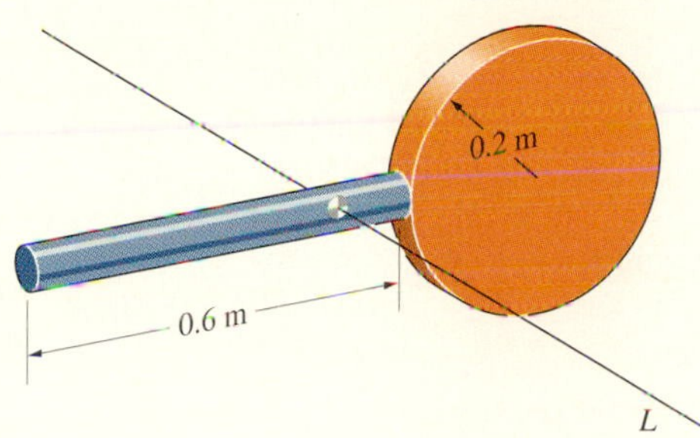

Fig. 7.37

STRATEGY

We must locate the center of mass of the composite object, then apply the parallel-axis theorem. We can obtain the moments of inertia of the bar and disk from Appendix C.

SOLUTION

Choose the Parts The parts are the bar and the disk. Introducing the coordinate system in Fig. (a), the x coordinate of the center of mass of the composite object is

$$\bar{x} = \frac{\bar{x}_{(\text{bar})} m_{(\text{bar})} + \bar{x}_{(\text{disk})} m_{(\text{disk})}}{m_{(\text{bar})} + m_{(\text{disk})}} = \frac{(0.3)(3) + (0.6 + 0.2)(2)}{3 + 2} = 0.5 \text{ m}.$$

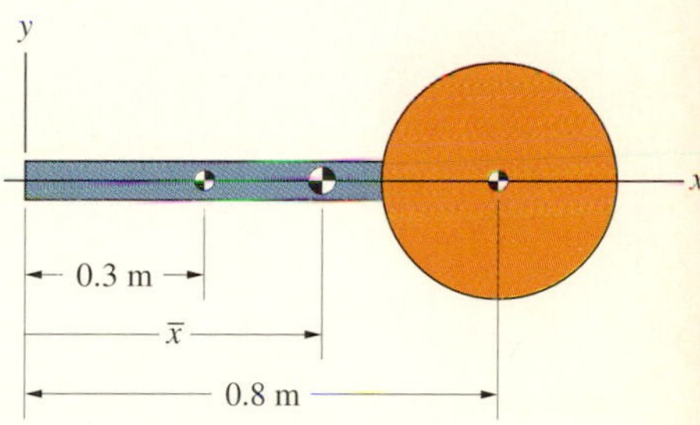

(a) The coordinate $\bar{x}$ of the center of mass of the object.

Determine the Moments of Inertia of the Parts The distance from the center of mass of the bar to the center of mass of the composite object is 0.2 m (Fig. b). Therefore the moment of inertia of the bar about L is

$$I_{(\text{bar})} = \frac{1}{12}(3)(0.6)^2 + (0.2)^2(3) = 0.210 \text{ kg-m}^2.$$

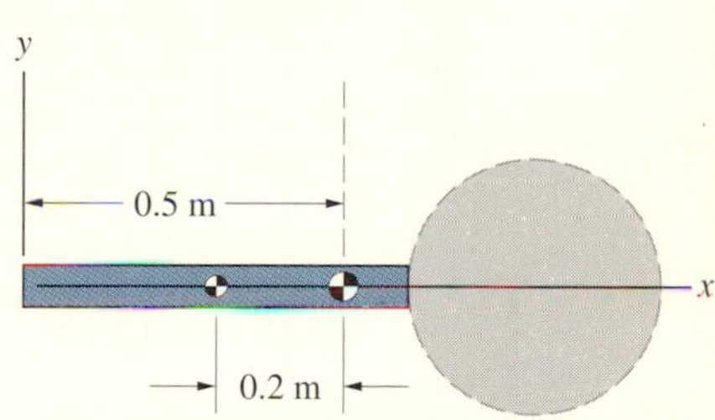

(b) Distance from L to the center of mass of the bar.

The distance from the center of mass of the disk to the center of mass of the composite object is 0.3 m (Fig. c). The moment of inertia of the disk about L is

$$I_{(\text{disk})} = \frac{1}{2}(2)(0.2)^2 + (0.3)^2(2) = 0.220 \text{ kg-m}^2.$$

Sum the Results The moment of inertia of the composite object about L is

$$I = I_{(\text{bar})} + I_{(\text{disk})} = 0.430 \text{ kg-m}^2.$$

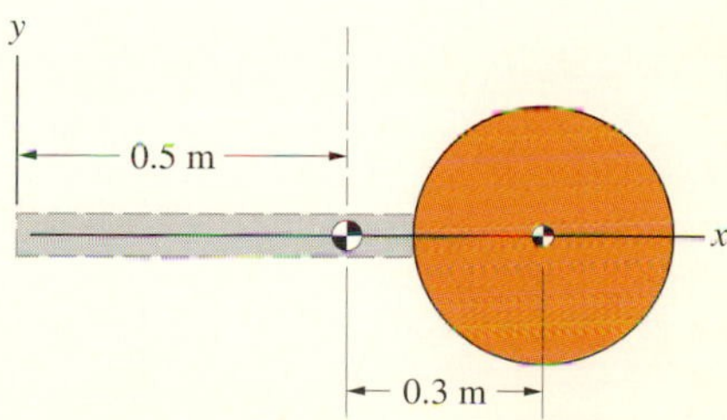

(c) Distance from L to the center of mass of the disk.

Example 7.15

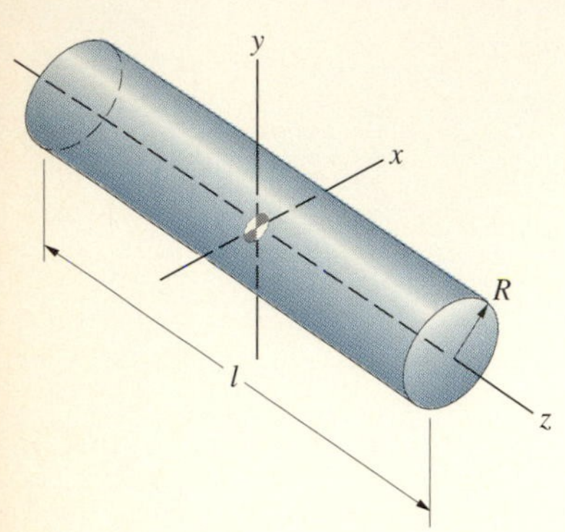

Fig. 7.38

The homogeneous cylinder in Fig. 7.38 has mass m, length l, and radius R. Determine its moments of inertia about the x, y, and z axes.

STRATEGY

We can determine the moments of inertia of the cylinder by an interesting application of the parallel-axis theorem. We use it to determine the moments of inertia about the x, y, and z axes of an infinitesimal element of the cylinder consisting of a disk of thickness dz. Then we integrate the results with respect to z to obtain the moments of inertia of the cylinder.

SOLUTION

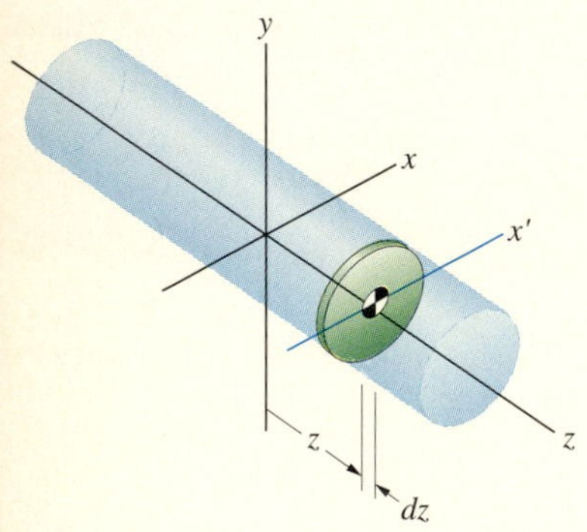

(a) A differential element of the cylinder in the form of a disk.

Consider an element of the cylinder of thickness dz at a distance z from the center of the cylinder (Fig. a). (You can imagine obtaining this element by "slicing" the cylinder perpendicular to its axis.) The mass of the element is equal to the product of the mass density and the volume of the element, $dm = \rho(\pi R^2\, dz)$. We obtain the moments of inertia of the element by using the values for a thin circular plate given in Appendix C. The moment of inertia about the z axis is

$$dI_{(z\text{ axis})} = \frac{1}{2}\, dm\, R^2 = \frac{1}{2}(\rho\pi R^2\, dz)R^2.$$

We integrate this result with respect to z from $-l/2$ to $l/2$, thereby summing the moments of inertia of the infinitesimal disk elements that make up the cylinder. The result is the moment of inertia of the cylinder about the z axis:

$$I_{(z\text{ axis})} = \int_{-l/2}^{l/2} \frac{1}{2}\rho\pi R^4\, dz = \frac{1}{2}\rho\pi R^4 l.$$

We can write this result in terms of the mass of the cylinder, $m = \rho(\pi R^2 l)$, as

$$I_{(z\text{ axis})} = \frac{1}{2}mR^2.$$

The moment of inertia of the disk element about the x' axis is

$$dI_{(x'\text{axis})} = \frac{1}{4}dm\, R^2 = \frac{1}{4}(\rho\pi R^2\, dz)R^2.$$

We use this result and the parallel-axis theorem to determine the moment of inertia of the element about the x axis:

$$dI_{(x\text{ axis})} = dI_{(x'\text{axis})} + z^2\, dm = \frac{1}{4}(\rho\pi R^2\, dz)R^2 + z^2(\rho\pi R^2\, dz).$$

Integrating this expression with respect to z from $-l/2$ to $l/2$, we obtain the moment of inertia of the cylinder about the x axis:

$$I_{(x\text{ axis})} = \int_{-l/2}^{l/2}\left(\frac{1}{4}\rho\pi R^4 + \rho\pi R^2 z^2\right) dz = \frac{1}{4}\rho\pi R^4 l + \frac{1}{12}\rho\pi R^2 l^3.$$

In terms of the mass of the cylinder,

$$I_{(x \text{ axis})} = \frac{1}{4}mR^2 + \frac{1}{12}ml^2.$$

Due to the symmetry of the cylinder,

$$I_{(y \text{ axis})} = I_{(x \text{ axis})}.$$

DISCUSSION

When the cylinder is very long in comparison to its width, $l \gg R$, the first term in the equation for $I_{(x \text{ axis})}$ can be neglected and we obtain the moment of inertia of a slender bar about a perpendicular axis, Eq. (7.35). On the other hand, when the radius of the cylinder is much greater than its length, $R \gg l$, the second term in the equation for $I_{(x \text{ axis})}$ can be neglected and we obtain the moment of inertia for a thin circular disk about an axis parallel to the disk. This indicates the sizes of the terms you neglect when you use the approximate expressions for the moments of inertia of a "slender" bar and a "thin" disk.

Problems

7.88 The homogeneous, slender bar has mass m and length l. Use integration to determine its moment of inertia about the perpendicular axis L_O.

Strategy: Use the same approach we used to obtain Eq. (7.35). You need to change only the limits of integration.

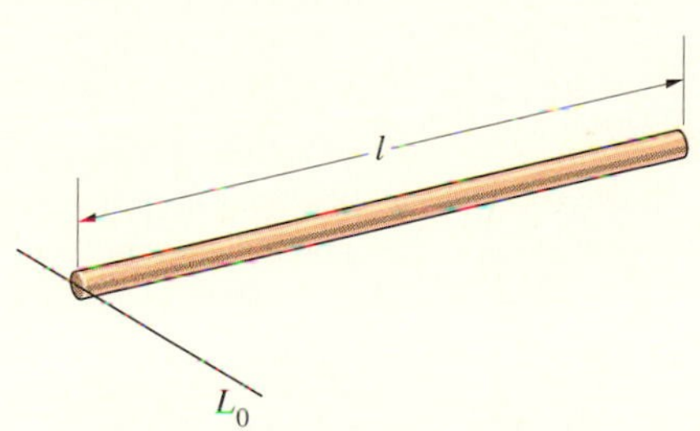

P7.88

7.89 Two homogeneous, slender bars, each of mass m and length l, are welded together to form the T-shaped object. Use integration to determine the moment of inertia of the object about the axis through point O that is perpendicular to the bars.

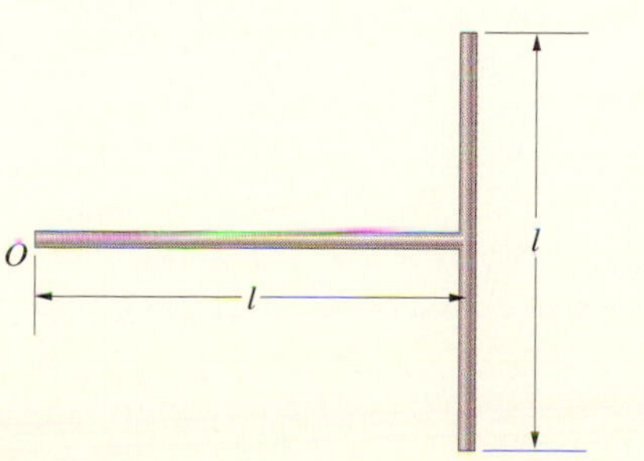

P7.89

7.90 The homogeneous, slender bar has mass m and length l. Use integration to determine the moment of inertia of the bar about the axis L.

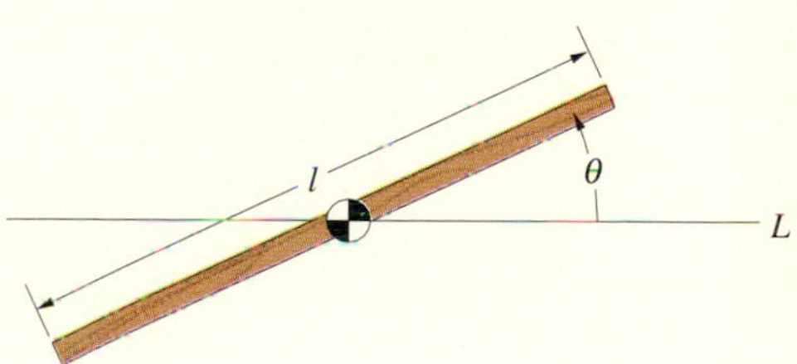

P7.90

7.91 A homogeneous, slender bar is bent into a circular ring of mass m and radius R. Determine the moment of inertia of the ring (a) about the axis through its center of mass that is perpendicular to the ring; (b) about the axis L.

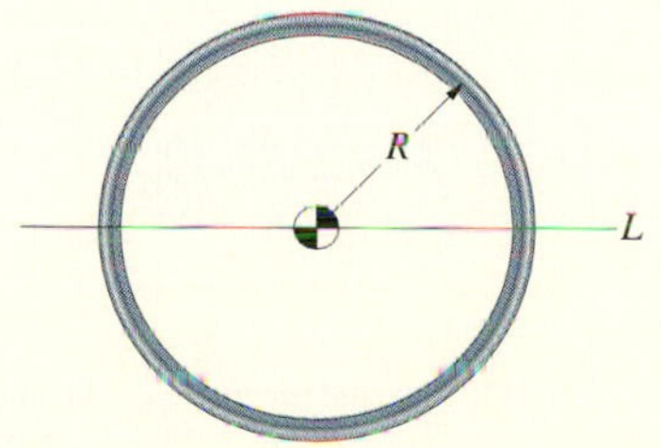

P7.91

7.92 The homogeneous, thin plate is of uniform thickness and mass m. Determine its moments of inertia about the x, y, and z axes.

Strategy: The moments of inertia of a thin plate of arbitrary shape are given by Eqs. (7.37)–(7.39) in terms of the moments of inertia of the cross-sectional area of the plate. You can obtain the moments of inertia of the rectangular area from Appendix B.

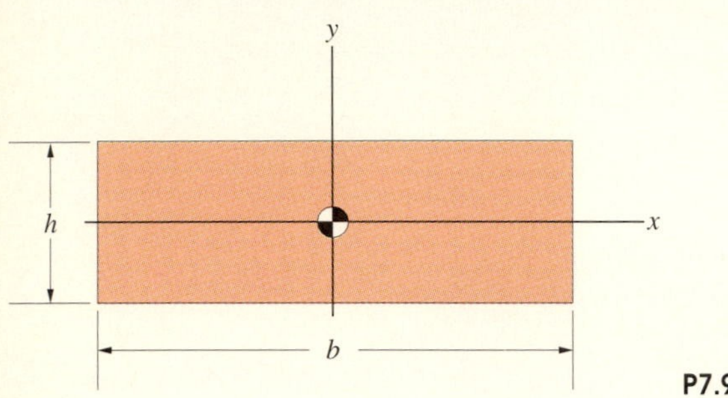

P7.92

7.93 The brass washer is of uniform thickness and mass m.
(a) Determine its moments of inertia about the x and z axes.
(b) Let $R_i = 0$ and compare your results with the values given in Appendix C for a thin circular plate.
(c) Let $R_i \to R_o$, and compare your results with the solutions of Problem 7.91.

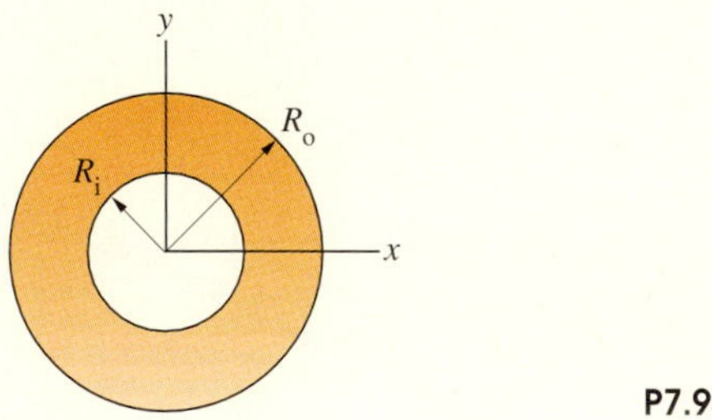

P7.93

7.94 The homogeneous, thin plate is of uniform thickness and weighs 20 lb. Determine its moment of inertia about the y axis.

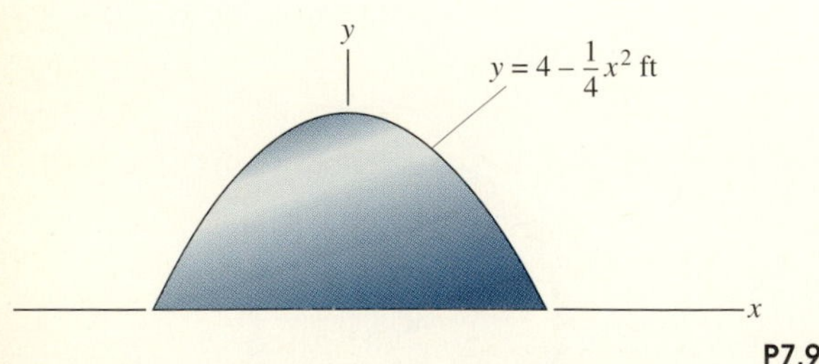

P7.94

7.95 Determine the moment of inertia of the plate in Problem 7.94 about the x axis.

7.96 The mass of the object is 10 kg. Its moment of inertia about L_1 is 10 kg-m^2. What is its moment of inertia about L_2? (The three axes lie in the same plane.)

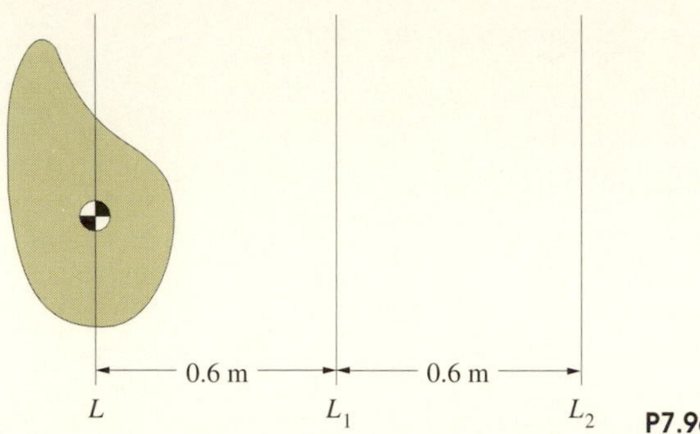

P7.96

7.97 An engineer gathering data for the design of a maneuvering unit determines that the astronaut's center of mass is at $x = 1.01$ m, $y = 0.16$ m and that his moment of inertia about the z axis is 105.6 kg-m^2. His mass is 81.6 kg. What is his moment of inertia about the z' axis through his center of mass?

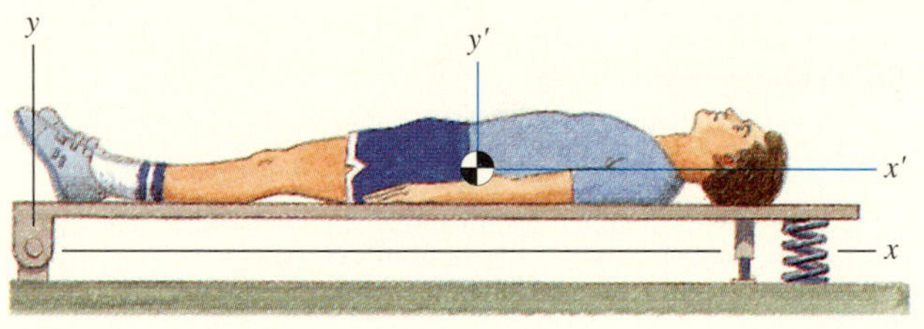

P7.97

7.98 Two homogeneous, slender bars, each of mass m and length l, are welded together to form the T-shaped object. Use the parallel-axis theorem to determine the moment of inertia of the object about the axis through point O that is perpendicular to the bars.

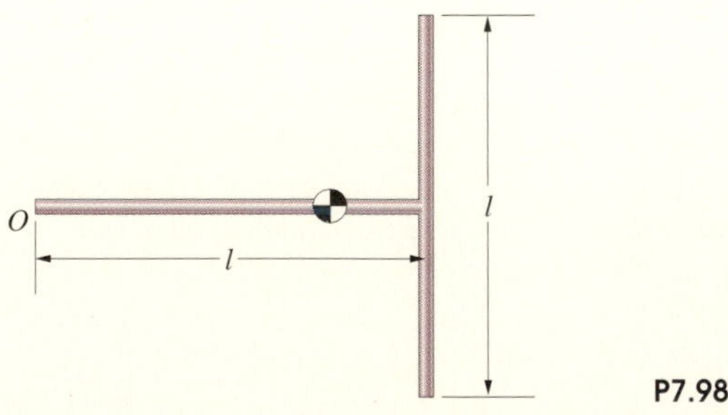

P7.98

7.99 Use the parallel-axis theorem to determine the moment of inertia of the T-shaped object in Problem 7.98 about the axis through the center of mass of the object that is perpendicular to the two bars.

7.100 The mass of the homogeneous, slender bar is 20 kg. Determine its moment of inertia about the z axis.

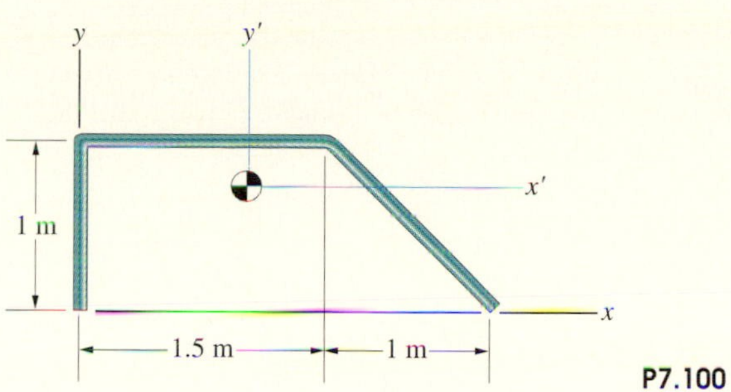

P7.100

7.101 Determine the moment of inertia of the bar in Problem 7.100 about the z' axis through its center of mass.

7.102 The homogeneous, slender bar weighs 5 lb. Determine its moment of inertia about the z axis.

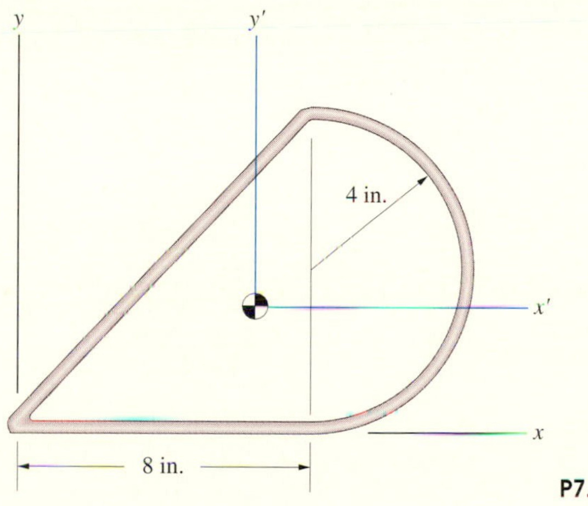

P7.102

7.103 Determine the moment of inertia of the bar in Problem 7.102 about the z' axis through its center of mass.

7.104 The rocket is used for atmospheric research. Its weight and its moment of inertia about the z axis through its center of mass (including its fuel) are 10,000 lb and 10,200 slug-ft^2, respectively. The rocket's fuel weighs 6000 lb, its center of mass is located at $x = -3$ ft, $y = 0$, $z = 0$, and the moment of inertia of the fuel about the axis through the fuel's center of mass parallel to the z axis is 2200 slug-ft^2. When the fuel is exhausted, what is the rocket's moment of inertia about the axis through its new center of mass parallel to the z axis?

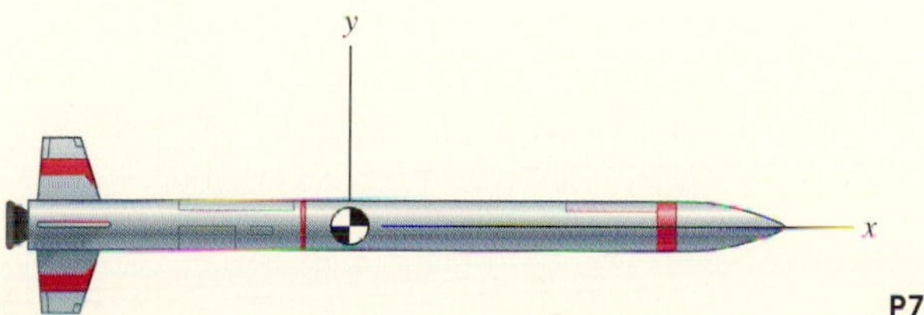

P7.104

7.105 The mass of the homogeneous, thin plate is 36 kg. Determine its moment of inertia about the x axis.

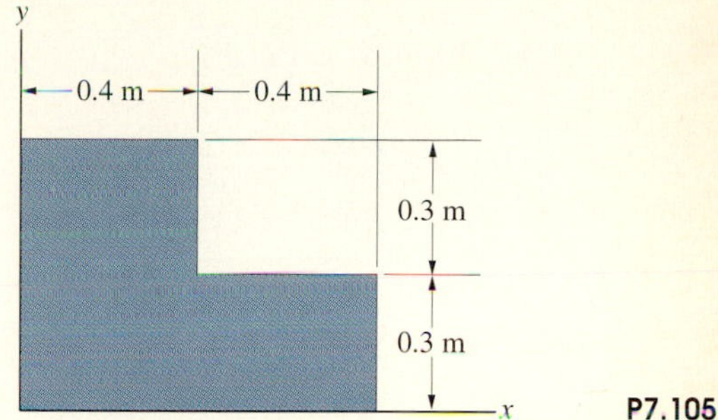

P7.105

7.106 Determine the moment of inertia of the plate in Problem 7.105 about the z axis.

7.107 The homogeneous, thin plate weighs 10 lb. Determine its moment of inertia about the x axis.

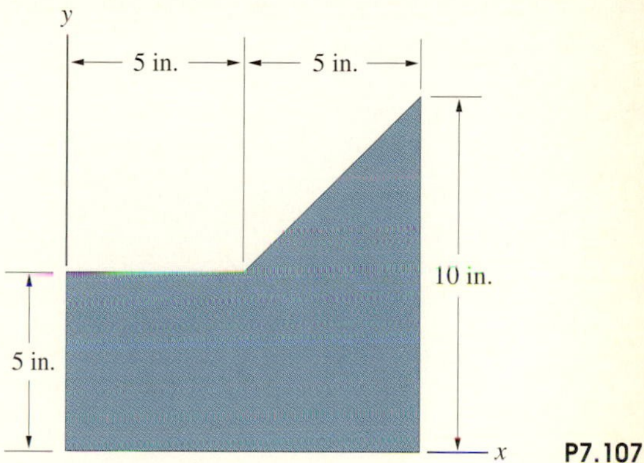

P7.107

7.108 Determine the moment of inertia of the plate in Problem 7.107 about the y axis.

7.109 The thermal radiator (used to eliminate excess heat from a satellite) can be modeled as a homogeneous, thin, rectangular plate. Its mass is 5 slugs. Determine its moments of inertia about the x, y, and z axes.

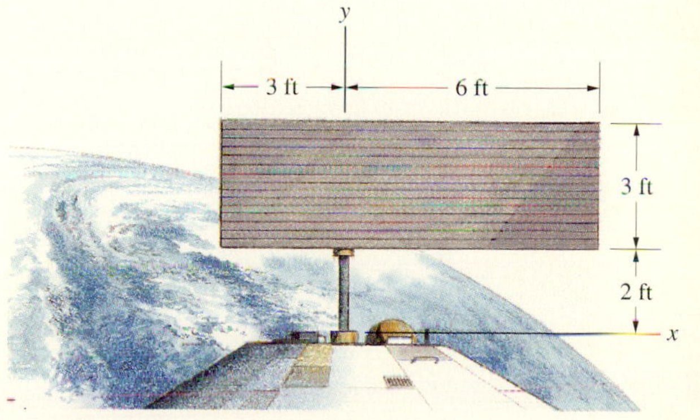

P7.109

7.110 The mass of the homogeneous, thin plate is 2 kg. Determine its moment of inertia about the axis through point O that is perpendicular to the plate.

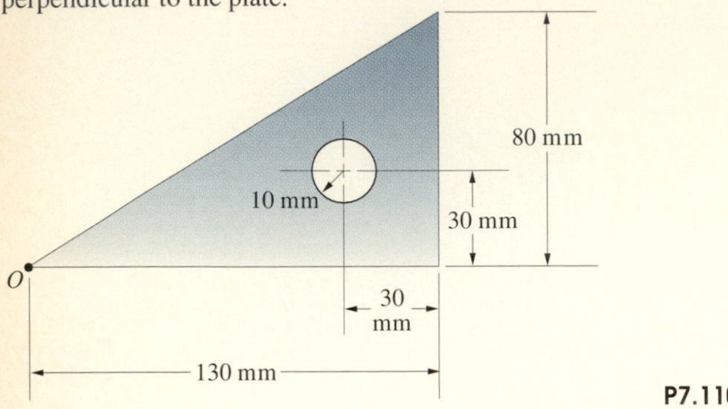

P7.110

7.111 The homogeneous cone is of mass m. Determine its moment of inertia about the z axis and compare your result with the value given in Appendix C.

Strategy: Use the same approach we used in Example 7.15 to obtain the moments of inertia of a homogeneous cylinder.

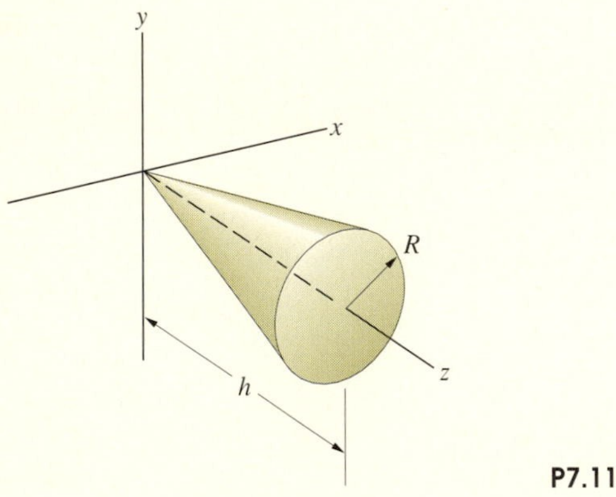

P7.111

7.112 Determine the moments of inertia of the homogeneous cone in Problem 7.111 about the x and y axes and compare your results with the values given in Appendix C.

7.113 The homogeneous object has the shape of a truncated cone and consists of bronze with mass density $\rho = 8200$ kg/m^3. Determine its mass moment of inertia about the z axis.

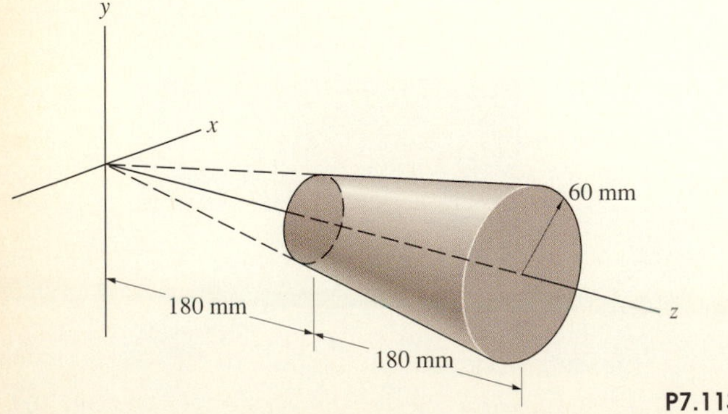

P7.113

7.114 Determine the mass moment of inertia of the object in Problem 7.113 about the x axis.

7.115 The homogeneous, rectangular parallelepiped is of mass m. Determine its moments of inertia about the x, y, and z axes and compare your results with the values given in Appendix C.

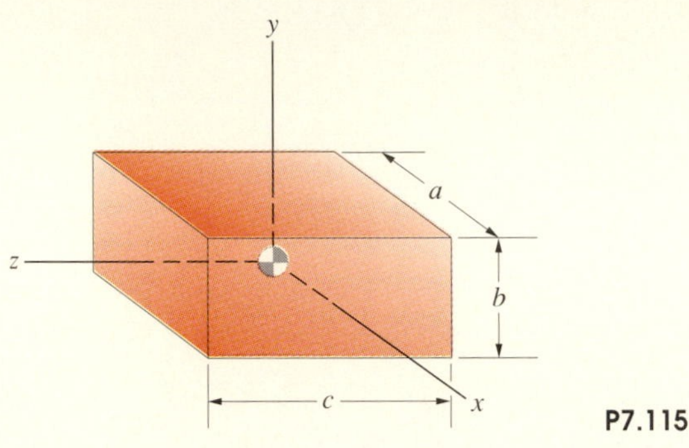

P7.115

7.116 The L-shaped machine part is composed of two homogeneous bars. Bar 1 is tungsten alloy with mass density 14,000 kg/m^3 and bar 2 is steel with mass density 7800 kg/m^3. Determine its moment of inertia about the x axis.

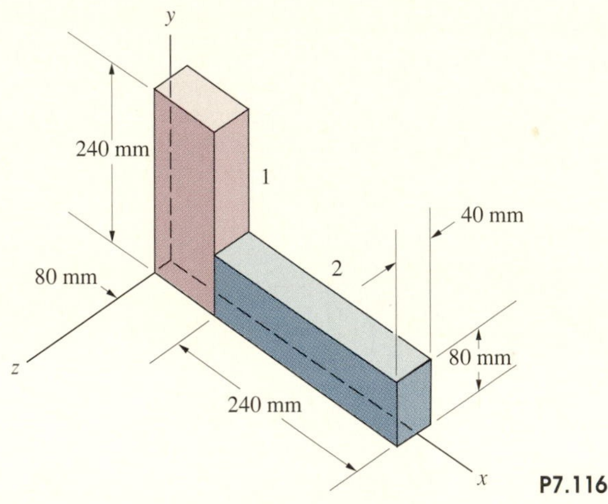

P7.116

7.117 Determine the moment of inertia of the L-shaped machine part in Problem 7.116 about the z axis.

7.118 The homogeneous ring consists of steel of density $\rho = 15$ slug/ft^3. Determine its moment of inertia about the axis L through its center of mass.

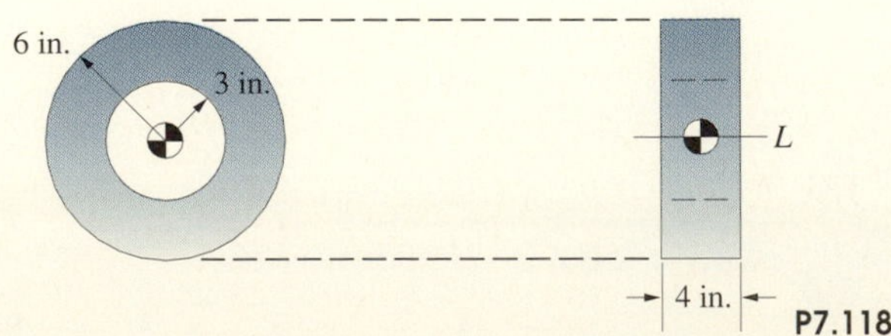

P7.118

7.119 The homogeneous half-cylinder is of mass m. Determine its moment of inertia about the axis L through its center of mass.

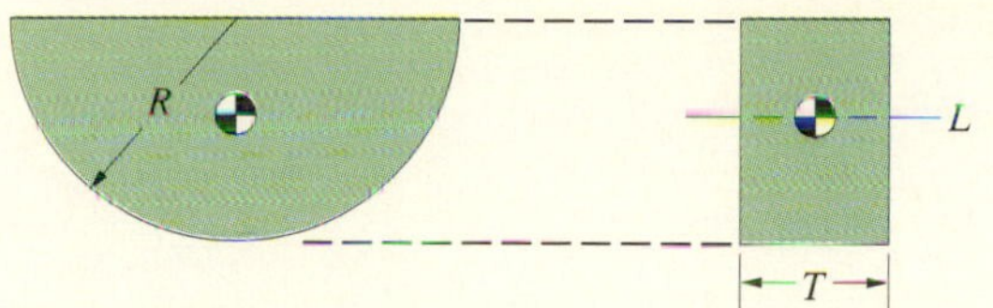

P7.119

7.120 The homogeneous machine part is made of aluminum alloy with mass density $\rho = 2800\ \text{kg/m}^3$. Determine its moment of inertia about the z axis.

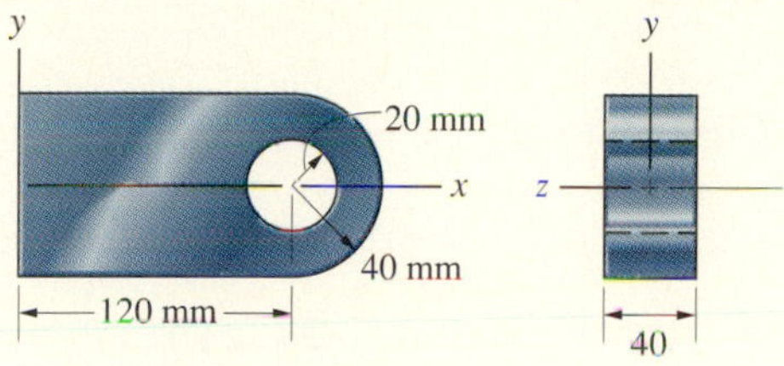

P7.120

7.121 Determine the moment of inertia of the machine part in Problem 7.120 about the x axis.

7.122 The object shown consists of steel of density $\rho = 7800\ \text{kg/m}^3$. Determine its moment of inertia about the axis L_O through point O.

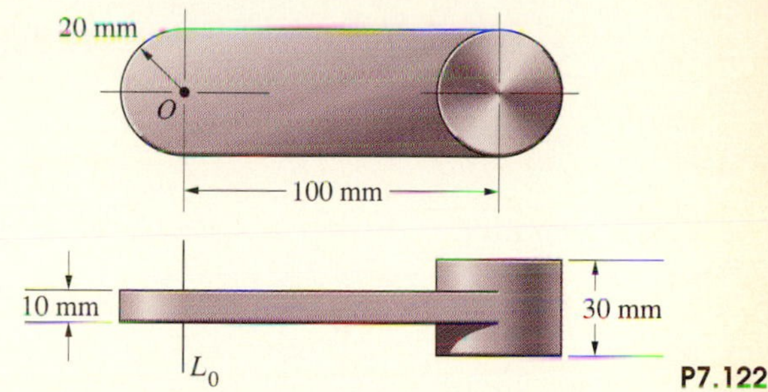

P7.122

7.123 Determine the moment of inertia of the object in Problem 7.122 about the axis through the center of mass of the object parallel to L_O.

7.124 The thick plate consists of steel of density $\rho = 15\ \text{slug/ft}^3$. Determine its moment of inertia about the z axis.

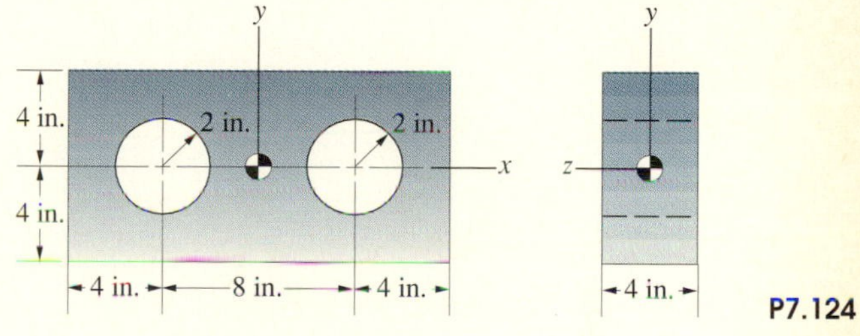

P7.124

7.125 Determine the moment of inertia of the object in Problem 7.124 about the x axis.

Chapter Summary

In this chapter we derived the equations of planar motion for a rigid body. Together with the kinematics relationships developed in Chapter 6, we used them to determine motions of rigid bodies resulting from the forces and couples acting on them. In Chapter 8 we will apply energy and momentum methods to the motions of rigid bodies and show that they can greatly simplify the solution of particular types of problems.

Moment-Angular Momentum Relations

Let $\mathbf{r}_i$ be the position of the ith particle of a system of particles relative to the origin O of an inertial reference frame, and let $\boldsymbol{R}_i$ be its position relative to the center of mass. The angular momentum of the system about O is the sum of the angular momenta of the particles,

$$\mathbf{H}_O = \sum_i \mathbf{r}_i \times m_i \mathbf{v}_i, \qquad \text{Eq. (7.7)}$$

where $\mathbf{v}_i = d\mathbf{r}_i/dt$, and the angular momentum about the center of mass is

$$\mathbf{H} = \sum_i \mathbf{R}_i \times m_i \mathbf{v}_i. \qquad \textbf{Eq. (7.8)}$$

These angular momenta are related by

$$\mathbf{H}_O = \mathbf{r} \times m\mathbf{v} + \mathbf{H}, \qquad \textbf{Eq. (7.9)}$$

where $\mathbf{v} = d\mathbf{r}/dt$ is the velocity of the center of mass.

The total moment about O equals the rate of change of the angular momentum about O:

$$\Sigma\,\mathbf{M}_O = \frac{d\mathbf{H}_O}{dt}. \qquad \textbf{Eq. (7.11)}$$

This result can also be expressed in terms of the angular momentum about the center of mass:

$$\Sigma\,\mathbf{M}_O = \frac{d}{dt}(\mathbf{r} \times m\mathbf{v} + \mathbf{H}) = \mathbf{r} \times m\mathbf{a} + \frac{d\mathbf{H}}{dt}, \qquad \textbf{Eq. (7.12)}$$

where $\mathbf{a}$ is the acceleration of the center of mass.

The total moment about the center of mass equals the rate of change of the angular momentum about the center of mass:

$$\Sigma\,\mathbf{M} = \frac{d\mathbf{H}}{dt}. \qquad \textbf{Eq. (7.13)}$$

Equations of Planar Motion

The equations of motion for a rigid body in planar motion include Newton's second law,

$$\Sigma\,\mathbf{F} = m\mathbf{a}, \qquad \textbf{Eq. (7.21)}$$

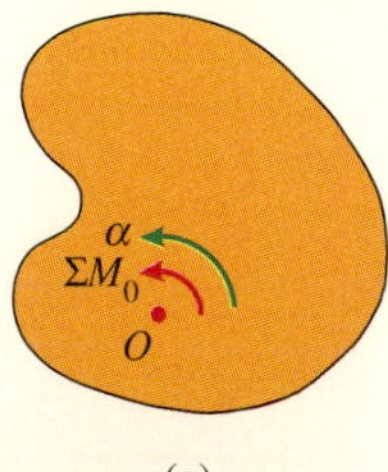

(a)

where $\mathbf{a}$ is the acceleration of the center of mass. If the rigid body rotates about a fixed point O, the total moment about O equals the product of the moment of inertia about O and the angular acceleration (Fig. a):

$$\Sigma\,M_O = I_O\alpha. \qquad \textbf{Eq. (7.22)}$$

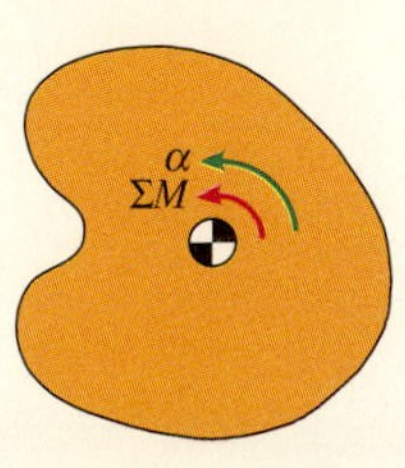

(b)

In any planar motion, the total moment about the center of mass equals the product of the moment of inertia about the center of mass and the angular acceleration (Fig. b):

$$\Sigma\,M = I\alpha. \qquad \textbf{Eq. (7.23)}$$

If a rigid body is in translation, Newton's second law is sufficient to determine its motion. Nevertheless, the angular equation of motion may be needed to determine unknown forces or couples. Since $\alpha = 0$, the total moment *about the center of mass* equals zero. In the case of rotation about a fixed axis, Eq. (7.22) is sufficient to determine the rotational motion, although Newton's second law may be needed to determine unknown forces or couples. If a rigid body undergoes general planar motion, both Newton's second law and the equation of angular motion are needed.

D'Alembert's Principle

If Newton's second law is written as

$$\Sigma \mathbf{F} + (-m\mathbf{a}) = \mathbf{0}, \qquad \text{Eq. (7.26)}$$

it can be regarded as an "equilibrium" equation stating that the sum of the external forces, including an **inertial force** $-m\mathbf{a}$, equals zero. The equation of angular motion can be written as

$$\Sigma M_O + [\mathbf{r} \times (-m\mathbf{a})] \cdot \mathbf{k} + (-I\alpha) = 0, \qquad \text{Eq. (7.27)}$$

stating that the sum of the moments about *any point O* due to external forces and couples, including the moment due to the inertial force $-m\mathbf{a}$ acting at the center of mass and an **inertial couple** $-I\alpha$, equals zero. Stated in this way, the equations of motion of a rigid body are analogous to the equations for static equilibrium.

Moments of Inertia

The moment of inertia of an object about an axis L_O is

$$I_O = \int_m r^2\, dm, \qquad \text{Eq. (7.34)}$$

where r is the perpendicular distance from L_O to the differential element of mass dm (Fig. c).

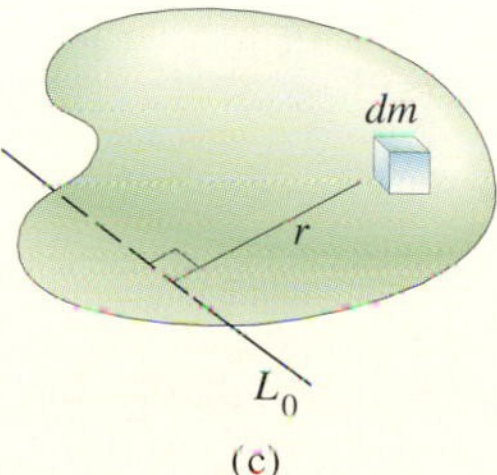

(c)

Let L be an axis through the center of mass of an object, and let L_O be a parallel axis. The moment of inertia I_O about L_O is given in terms of the moment of inertia I about L by the **parallel-axis theorem**

$$I_O = I + d^2 m, \qquad \text{Eq. (7.42)}$$

where m is the mass of the object and d is the distance between L and L_O.

Review Problems

7.126 The airplane is at the beginning of its takeoff run. Its weight is 1000 lb, and the initial thrust T exerted by its engine is 300 lb. Assume that the thrust is horizontal, and neglect the tangential forces exerted on its wheels.
(a) If the acceleration of the airplane remains constant, how long will it take to reach its takeoff speed of 80 mi/hr?
(b) Determine the normal force exerted on the forward landing gear at the beginning of the takeoff run.

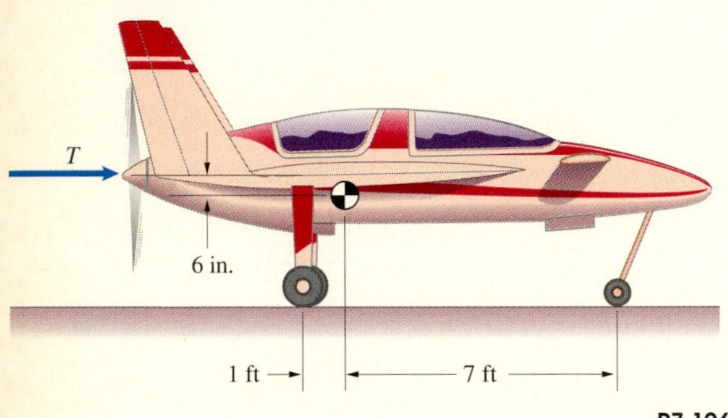

P7.126

7.127 The pulleys can turn freely on their pin supports. Their moments of inertia are $I_A = 0.002$ kg-m^2, $I_B = 0.036$ kg-m^2, and $I_C = 0.032$ kg-m^2. They are initially stationary, and at $t = 0$ a constant couple $M = 2$ N-m is applied to pulley A. What is the angular velocity of pulley C and how many revolutions has it turned at $t = 2$ s?

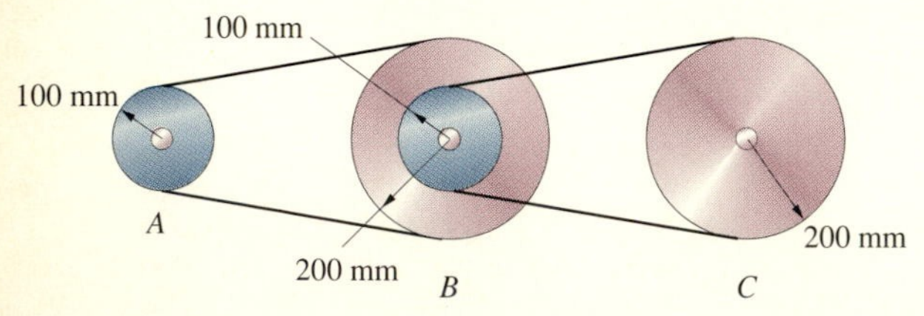

P7.127

7.128 A 2-kg box is subjected to a 40-N horizontal force. Neglect friction.
(a) If the box remains on the floor, what is its acceleration?
(b) Determine the range of values of c for which the box will remain on the floor when the force is applied.

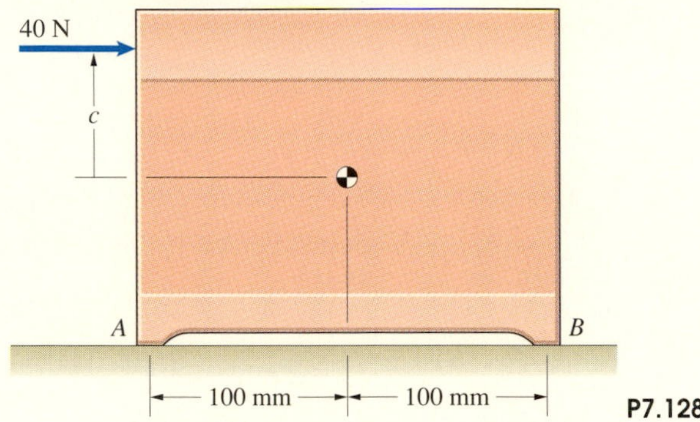

P7.128

7.129 The slender, 2-slug bar AB is 3 ft long. It is pinned to the cart at A and leans against it at B.
(a) If the acceleration of the cart is $a = 20$ ft/s^2, what normal force is exerted on the bar by the cart at B?
(b) What is the largest acceleration a for which the bar will remain in contact with the surface at B?

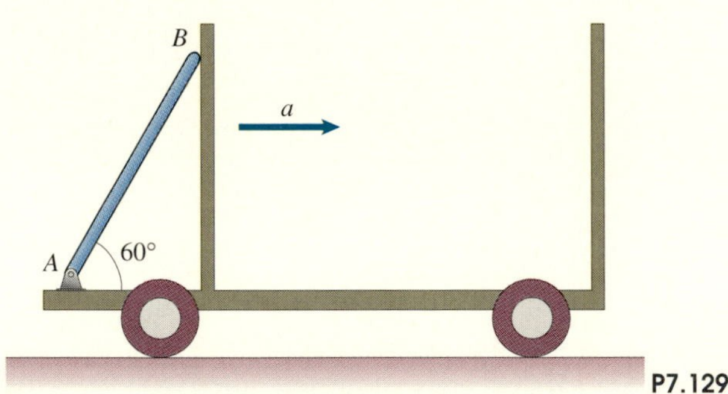

P7.129

7.130 To determine a 4.5-kg tire's moment of inertia, an engineer lets it roll down an inclined surface. If it takes 3.5 s to start from rest and roll 3 m down the surface, what is the tire's moment of inertia about its center of mass?

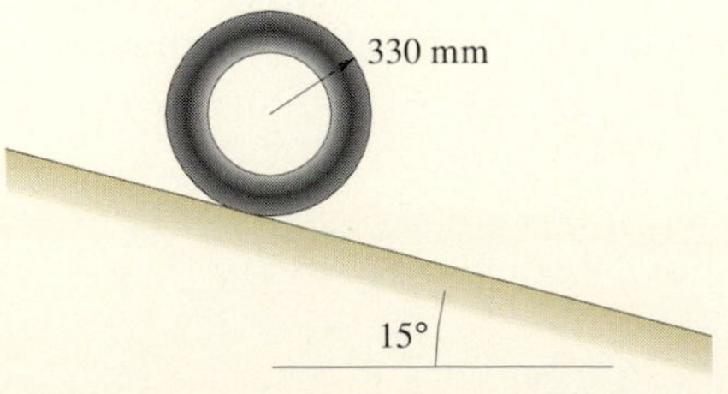

P7.130

7.131 Pulley A weighs 4 lb, $I_A = 0.060$ slug-ft^2, and $I_B = 0.014$ slug-ft^2. If the system is released from rest, what distance does the 16-lb weight fall in 0.5 s?

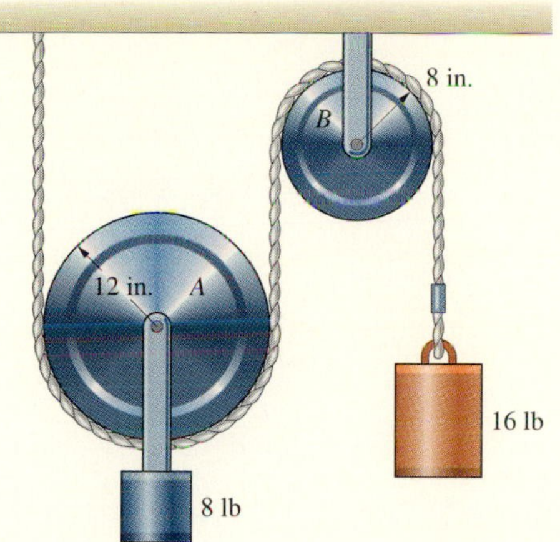

P7.131

7.132 Model the excavator's arm ABC as a single rigid body. Its mass is 1200 kg, and the moment of inertia *about its center of mass* is $I = 3600$ kg-m^2. If point A is stationary, the angular velocity of the arm is zero, and its angular acceleration is 1.0 rad/s^2 counterclockwise, what force does the vertical hydraulic cylinder exert on the arm at B?

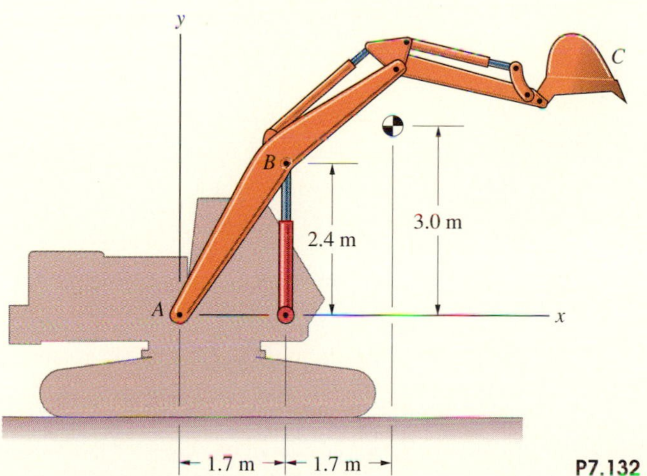

P7.132

7.133 In Problem 7.132, if the angular acceleration of arm ABC is 1.0 rad/s^2 counterclockwise and its angular velocity is 2.0 rad/s counterclockwise, what are the components of the force exerted on the arm at A?

7.134 To decrease the angle of elevation of the stationary 200-kg ladder, the gears that raised it are disengaged and a fraction of a second later a second set of gears that lower it are engaged. At the instant the gears that raised it are disengaged, what is the ladder's angular acceleration and what are the components of force exerted on the ladder by its support at O? The moment of inertia of the ladder about O is $I_O = 14{,}000$ kg-m^2, and the coordinates of its center of mass at the instant the gears are disengaged are $\bar{x} = 3$ m, $\bar{y} = 4$ m.

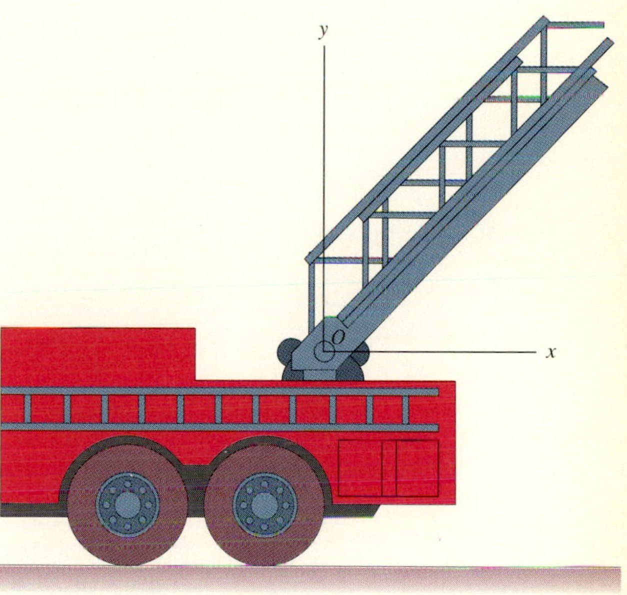

P7.134

7.135 The slender bars each weigh 4 lb and are 10 in. long. The homogeneous plate weighs 10 lb. If the system is released from rest in the position shown, what is the angular acceleration of the bars at that instant?

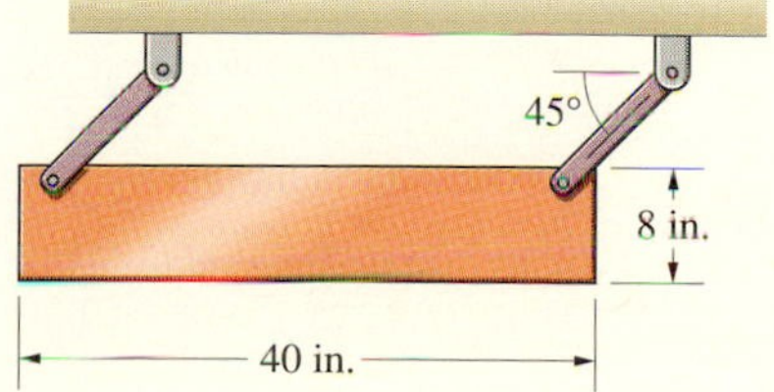

P7.135

7.136 A slender bar of mass m is released from rest in the position shown. The static and kinetic coefficients of friction at the floor and wall have the same value μ. If the bar slips, what is its angular acceleration at the instant of release?

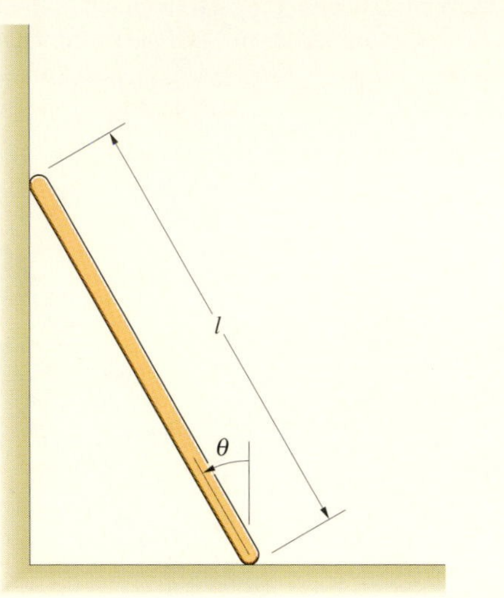

P7.136

7.137 Each of the go-cart's front wheels weighs 5 lb and has a moment of inertia of 0.01 slug-ft^2. The two rear wheels and rear axle form a single rigid body weighing 40 lb and having a moment of inertia of 0.1 slug-ft^2. The total weight of the go-cart and driver is 240 lb. (The location of the center of mass of the go-cart and driver *not including* the front wheels or the rear wheels and rear axle is shown.) If the engine exerts a torque of 12 ft-lb on the rear axle, what is the go-cart's acceleration?

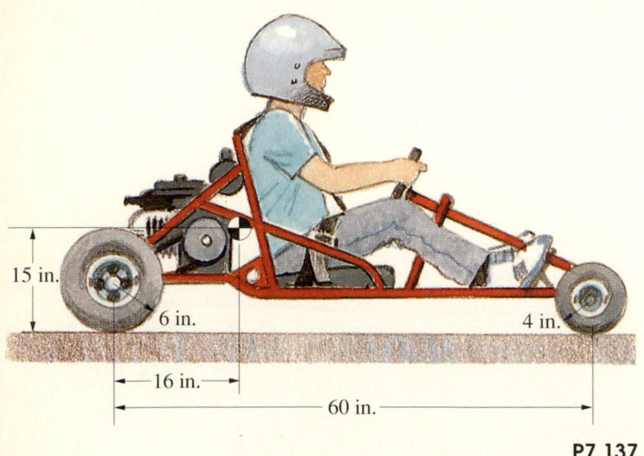

P7.137

7.138 Bar AB rotates with a constant angular velocity of 10 rad/s in the counterclockwise direction. The masses of the slender bars BC and CDE are 2 kg and 3.6 kg, respectively. The y axis points upward. Determine the components of the forces exerted on bar BC by the pins at B and C at the instant shown.

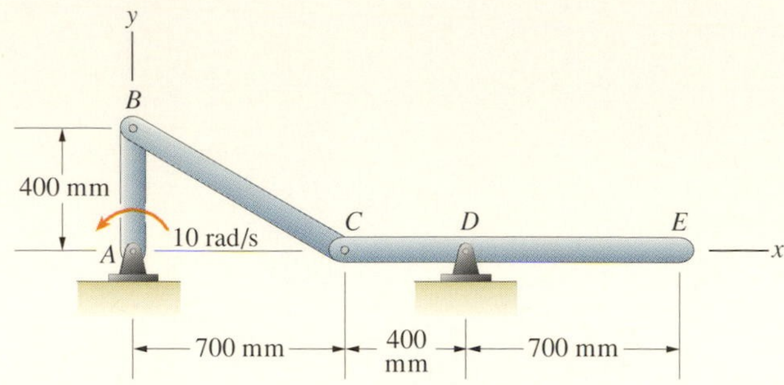

P7.138

7.139 At the instant shown, the arms of the robotic manipulator have constant counterclockwise angular velocities $\omega_{AB} = -0.5$ rad/s, $\omega_{BC} = 2$ rad/s, and $\omega_{CD} = 4$ rad/s. The mass of arm CD is 10 kg, and its center of mass is at its midpoint. At this instant, what force and couple are exerted on arm CD at C?

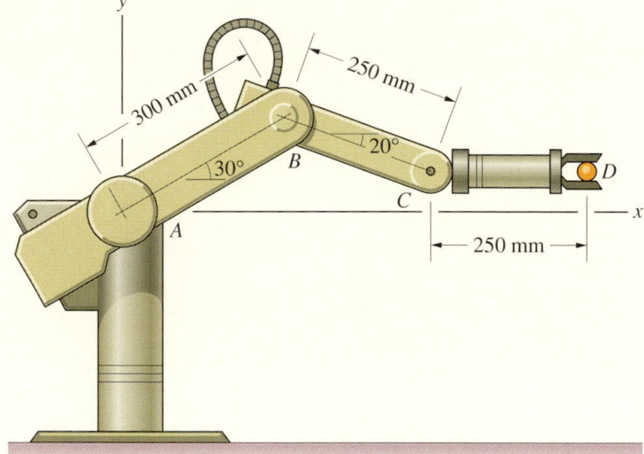

P7.139

7.140 Each bar is 1 m in length and has a mass of 4 kg. The inclined surface is smooth. If the system is released from rest in the position shown, what are the angular accelerations of the bars at that instant?

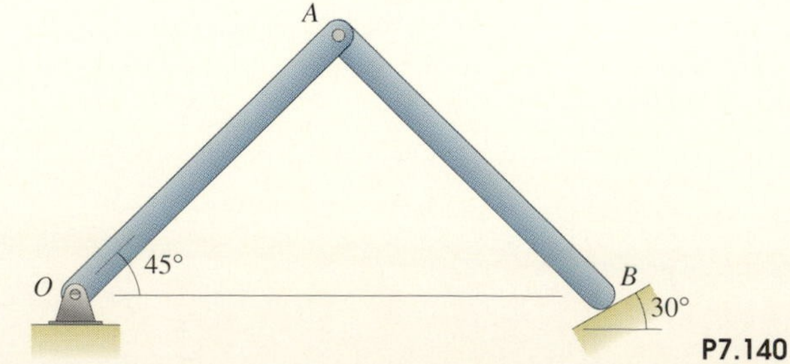

P7.140

7.141 At the instant the system in Problem 7.140 is released, what is the magnitude of the force exerted on bar OA by the support at O?

7.142 The fixed ring gear lies *in the horizontal plane*. The hub and planet gears are bonded together. The mass and moment of inertia of the combined hub and planet gears are $m_{HP} = 130$ kg and $I_{HP} = 130$ kg-m^2. The moment of inertia of the sun gear is $I_S = 60$ kg-m^2. The mass of the connecting rod is 5 kg, and it can be modeled as a slender bar. If a 1 kN-m counterclockwise couple is applied to the sun gear, what is the resulting angular acceleration of the bonded hub and planet gears?

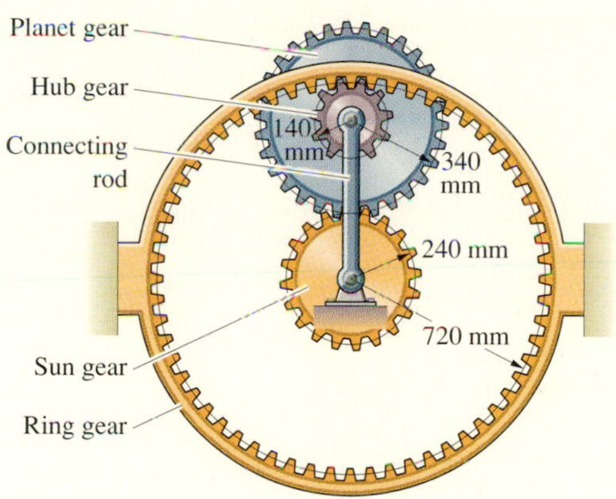

P7.142

7.143 The system is stationary at the instant shown. The net force exerted on the piston by the exploding fuel–air mixture and friction is 5 kN to the left. A clockwise couple $M = 200$ N-m acts on the crank AB. The moment of inertia of the crank about A is 0.0003 kg-m^2. The mass of the connecting rod BC is 0.36 kg, and its center of mass is 40 mm from B on the line from B to C. The connecting rod's moment of inertia about its center of mass is 0.0004 kg-m^2. The mass of the piston is 4.6 kg. What is the piston's acceleration at this instant? (Neglect the gravitational forces on the crank and connecting rod.)

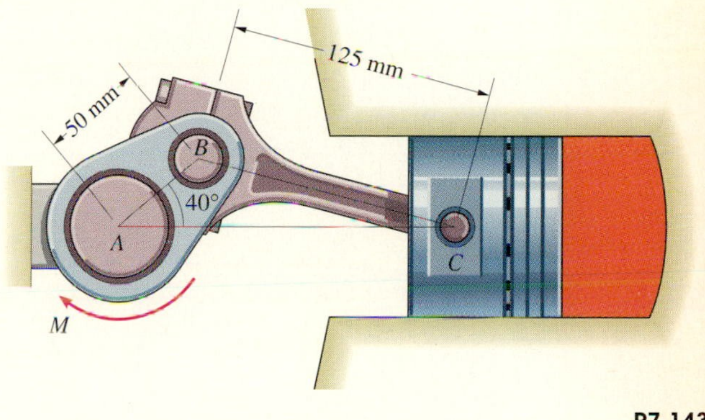

P7.143

7.144 If the crank AB in Problem 7.143 has a counterclockwise angular velocity of 2000 rpm at the instant shown, what is the piston's acceleration?

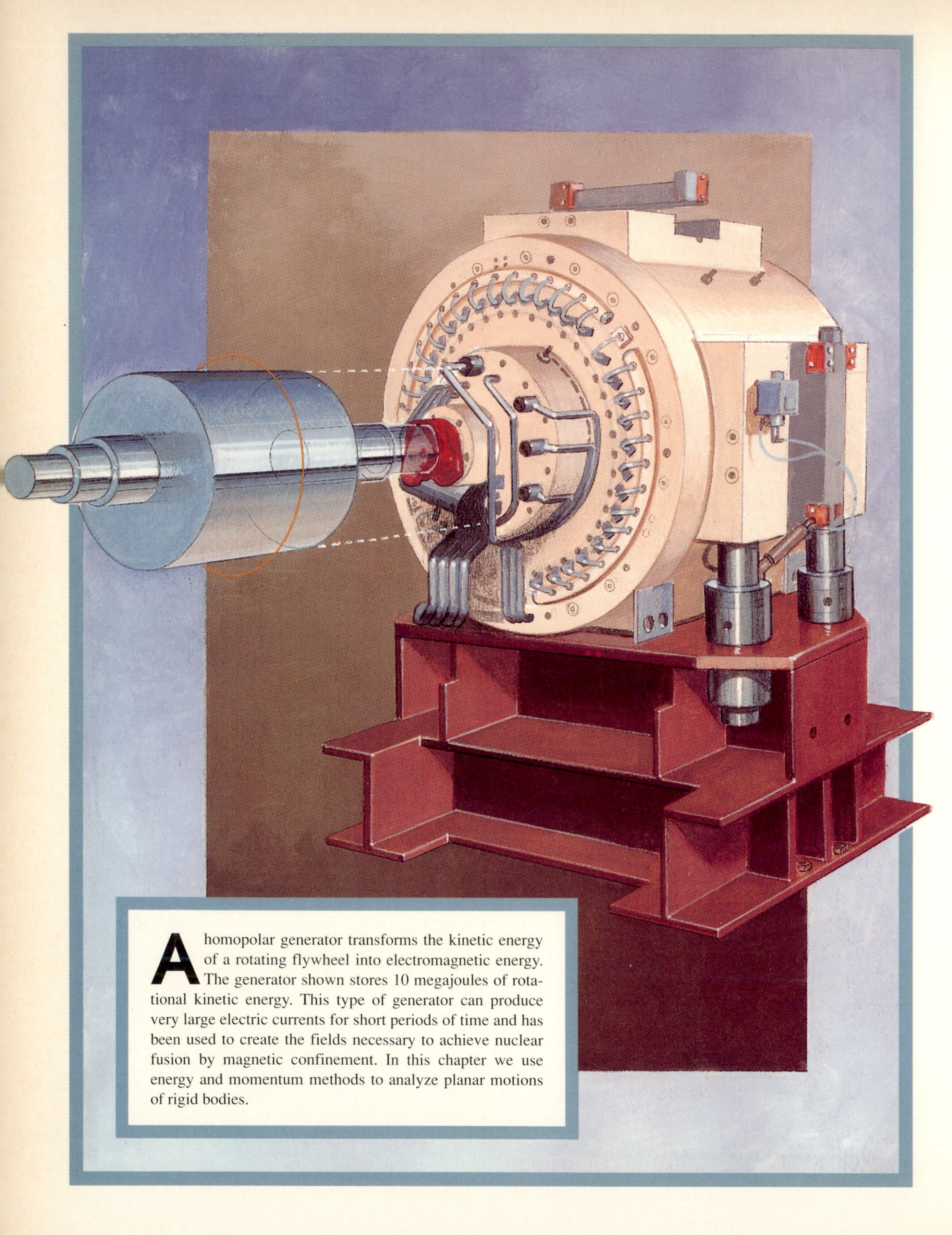

A homopolar generator transforms the kinetic energy of a rotating flywheel into electromagnetic energy. The generator shown stores 10 megajoules of rotational kinetic energy. This type of generator can produce very large electric currents for short periods of time and has been used to create the fields necessary to achieve nuclear fusion by magnetic confinement. In this chapter we use energy and momentum methods to analyze planar motions of rigid bodies.

Chapter 8

Energy and Momentum in Rigid-Body Dynamics

YOU have seen in Chapters 4 and 5 that energy and momentum methods are very useful for particular types of problems in dynamics. If the forces on an object are known functions of position, you can use the principle of work and energy to relate the change in the magnitude of the object's velocity to the change in its position. If the forces are known functions of time, you can use the principle of impulse and momentum to determine the change in the object's velocity during an interval of time. In this chapter we extend these methods to situations in which you must consider both the translational and rotational motions of objects.

8.1 *Principle of Work and Energy*

The principle of work and energy for a rigid body in planar motion is a simple statement and involves simple equations, although its derivation is rather involved. To help you follow our derivation, we begin by summarizing the principle. Let T be the kinetic energy of a rigid body. The principle of work and energy states that the work U_{12} done by external forces and couples as the rigid body moves between two positions 1 and 2 equals the change in its kinetic energy:

$$U_{12} = T_2 - T_1.$$

In general planar motion, the kinetic energy is

$$T = \frac{1}{2}mv^2 + \frac{1}{2}I\omega^2,$$

where v is the magnitude of the velocity of the center of mass and I is the moment of inertia about the center of mass. In the case of rotation about a fixed axis O, the kinetic energy can also be expressed as

$$T = \frac{1}{2}I_O\omega^2.$$

To derive these results, we adopt the same approach used in Chapter 7 to derive the equations of motion for a rigid body. We obtain the principle of work and energy for a system of particles and use it to obtain the principle for a rigid body.

System of Particles

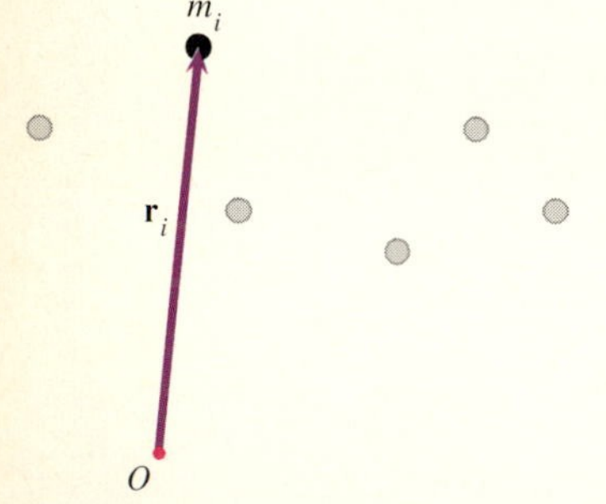

Fig. 8.1

A system of particles. The vector $\mathbf{r}_i$ is the position vector of the ith particle.

Let m_i be the mass of the ith particle of a system of N particles, and let $\mathbf{r}_i$ be its position relative to a point O that is fixed relative to an inertial reference frame (Fig. 8.1). We denote the sum of the kinetic energies of the particles by T,

$$T = \sum_i \frac{1}{2}m_i\mathbf{v}_i \cdot \mathbf{v}_i, \tag{8.1}$$

where $\mathbf{v}_i = d\mathbf{r}_i/dt$ is the velocity of the ith particle. Our objective is to relate the work done on the system of particles to the change in T. We begin with Newton's second law for the ith particle,

$$\sum_j \mathbf{f}_{ij} + \mathbf{f}_i^{\mathrm{E}} = \frac{d}{dt}(m_i\mathbf{v}_i), \tag{8.2}$$

where $\mathbf{f}_{ij}$ is the force exerted on the ith particle by the jth particle and $\mathbf{f}_i^{\mathrm{E}}$ is the external force on the ith particle. We take the dot product of this equation with $\mathbf{v}_i$ and sum from $i = 1$ to N:

$$\sum_i \sum_j \mathbf{f}_{ij} \cdot \mathbf{v}_i + \sum_i \mathbf{f}_i^{\mathrm{E}} \cdot \mathbf{v}_i = \sum_i \mathbf{v}_i \cdot \frac{d}{dt}(m_i\mathbf{v}_i). \tag{8.3}$$

We can express the term on the right side of this equation as the rate of change of the total kinetic energy:

$$\sum_i \mathbf{v}_i \cdot \frac{d}{dt}(m_i\mathbf{v}_i) = \frac{d}{dt}\sum_i \frac{1}{2}m_i\mathbf{v}_i \cdot \mathbf{v}_i = \frac{dT}{dt}.$$

Therefore multiplying Eq. (8.3) by dt yields

$$\sum_i \sum_j \mathbf{f}_{ij} \cdot d\mathbf{r}_i + \sum_i \mathbf{f}_i^{\mathrm{E}} \cdot d\mathbf{r}_i = dT.$$

We integrate this equation, obtaining

$$\sum_i \sum_j \int_{(\mathbf{r}_i)_1}^{(\mathbf{r}_i)_2} \mathbf{f}_{ij} \cdot d\mathbf{r}_i + \sum_i \int_{(\mathbf{r}_i)_1}^{(\mathbf{r}_i)_2} \mathbf{f}_i^{\mathrm{E}} \cdot d\mathbf{r}_i = T_2 - T_1.$$

The terms on the left side are the work done on the system by internal and external forces as the particles move from positions $(\mathbf{r}_i)_1$ to positions $(\mathbf{r}_i)_2$. Denoting the work by U_{12}, we obtain the **principle of work and energy for a system of particles**: The work done by internal and external forces equals the change in the total kinetic energy:

$$\boxed{U_{12} = T_2 - T_1.} \qquad (8.4)$$

This result applies to any object or collection of objects, including a rigid body.

Rigid Body in Planar Motion

We have shown that the work done on a rigid body by internal and external forces as it moves between two positions equals the change in its kinetic energy. If we assume that the internal forces between each pair of particles are directed along the straight line between the two particles, *the work done on a rigid body by internal forces is zero.* To show that this is true, we consider two particles of a rigid body designated 1 and 2 (Fig. 8.2). The sum of the forces the two particles exert on each other is zero, $\mathbf{f}_{12} + \mathbf{f}_{21} = \mathbf{0}$, so the rate at which the forces do work (the power) is

$$\mathbf{f}_{12} \cdot \mathbf{v}_1 + \mathbf{f}_{21} \cdot \mathbf{v}_2 = \mathbf{f}_{21} \cdot (\mathbf{v}_2 - \mathbf{v}_1).$$

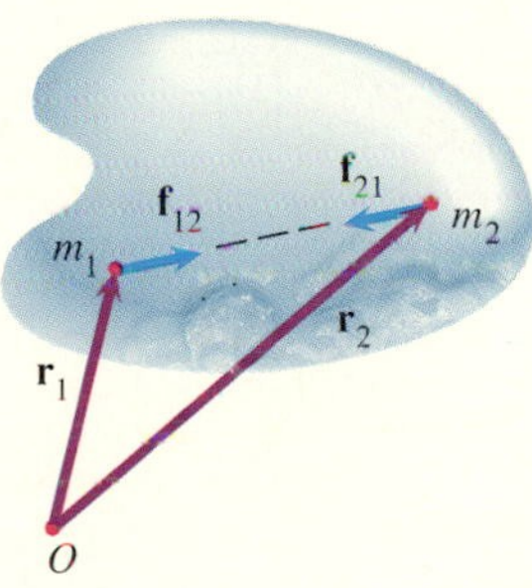

Fig. 8.2
Particles 1 and 2 and the forces they exert on each other.

We can show that $\mathbf{f}_{21}$ is perpendicular to $\mathbf{v}_2 - \mathbf{v}_1$, and therefore the rate at which work is done by the internal forces between these two particles is zero. Because the particles are points of a rigid body, we can express their relative velocity in terms of the rigid body's angular velocity $\boldsymbol{\omega}$ as

$$\mathbf{v}_2 - \mathbf{v}_1 = \boldsymbol{\omega} \times (\mathbf{r}_2 - \mathbf{r}_1). \qquad (8.5)$$

This equation shows that the relative velocity $\mathbf{v}_2 - \mathbf{v}_1$ is perpendicular to $\mathbf{r}_2 - \mathbf{r}_1$, which is the position vector from particle 1 to particle 2. Since the force $\mathbf{f}_{21}$ is parallel to $\mathbf{r}_2 - \mathbf{r}_1$, it is perpendicular to $\mathbf{v}_2 - \mathbf{v}_1$. We can repeat this argument for each pair of particles of the rigid body, so the total rate at which work is done by internal forces is zero. This implies that the work done by internal forces as the rigid body moves between two positions is zero.

The system of external forces on a rigid body may be represented as forces and couples, so we obtain the **principle of work and energy for a rigid body**: The work done by external forces and couples as a rigid body moves between two positions equals the change in its kinetic energy. We can also state this principle for a system of rigid bodies: The work done by external *and internal* forces and couples as a system of rigid bodies moves between two positions equals the change in their total kinetic energy.

To complete our derivation of the principle of work and energy for a rigid body in planar motion, we must express the kinetic energy in terms of the velocity of the center of mass of the rigid body and its angular velocity. We first consider general planar motion, then rotation about a fixed axis.

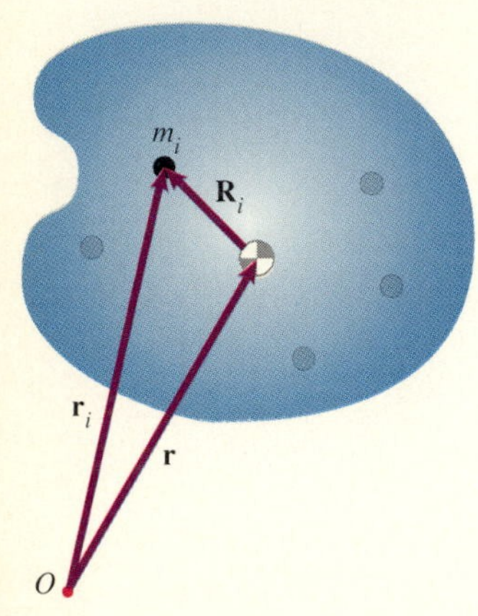

Fig. 8.3

Representing a rigid body as a system of particles.

Kinetic Energy in General Planar Motion Let us represent a rigid body as a system of particles, and let $\mathbf{R}_i$ be the position vector of the ith particle relative to the center of mass (Fig. 8.3). The position of the center of mass is

$$\mathbf{r} = \frac{\sum_i m_i \mathbf{r}_i}{m}, \tag{8.6}$$

where m is the mass of the rigid body. The position of the ith particle relative to O is related to its position relative to the center of mass by

$$\mathbf{r}_i = \mathbf{r} + \mathbf{R}_i, \tag{8.7}$$

and the vectors $\mathbf{R}_i$ satisfy the relation

$$\sum_i m_i \mathbf{R}_i = \mathbf{0}. \tag{8.8}$$

The kinetic energy of the rigid body is the sum of the kinetic energies of its particles, given by Eq. (8.1):

$$T = \sum_i \frac{1}{2} m_i \mathbf{v}_i \cdot \mathbf{v}_i. \tag{8.9}$$

By taking the time derivative of Eq. (8.7), we obtain

$$\mathbf{v}_i = \mathbf{v} + \frac{d\mathbf{R}_i}{dt},$$

where $\mathbf{v}$ is the velocity of the center of mass. Substituting this expression into Eq. (8.9) and using Eq. (8.8), we obtain the kinetic energy in the form

$$T = \frac{1}{2} m v^2 + \sum_i \frac{1}{2} m_i \frac{d\mathbf{R}_i}{dt} \cdot \frac{d\mathbf{R}_i}{dt}, \tag{8.10}$$

where v is the magnitude of the velocity of the center of mass.

Let L_0 be the axis through a fixed point O that is perpendicular to the plane of the motion, and let L be the parallel axis through the center of mass (Fig. 8.4a). In terms of the coordinate system shown, we can express the angular velocity vector as $\boldsymbol{\omega} = \omega \mathbf{k}$. The velocity of the ith particle relative to the center of mass is $d\mathbf{R}_i/dt = \omega \mathbf{k} \times \mathbf{R}_i$, so we can write Eq. (8.10) as

$$T = \frac{1}{2} m v^2 + \frac{1}{2}\left[\sum_i m_i (\mathbf{k} \times \mathbf{R}_i) \cdot (\mathbf{k} \times \mathbf{R}_i)\right]\omega^2. \tag{8.11}$$

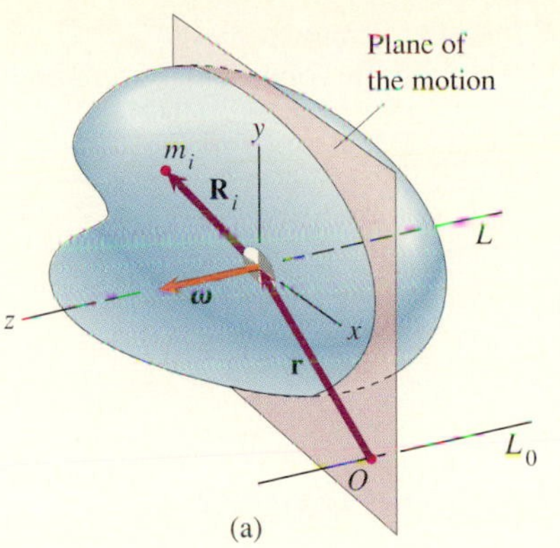

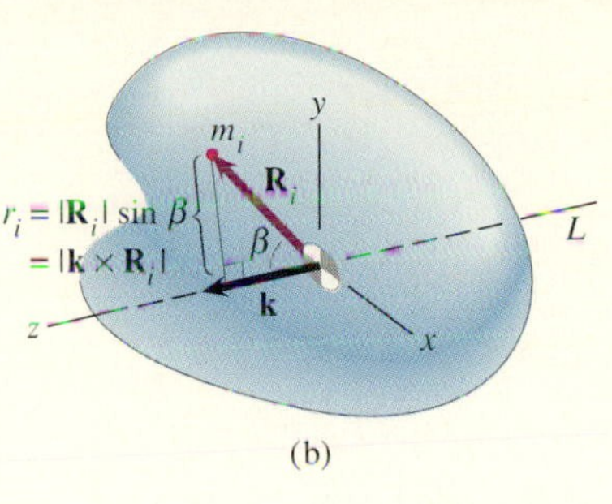

Fig. 8.4

(a) A coordinate system with the z axis aligned with L.
(b) The magnitude of $\mathbf{k} \times \mathbf{R}_i$ is the perpendicular distance from L to m_i.

The magnitude of the vector $\mathbf{k} \times \mathbf{R}_i$ is the perpendicular distance r_i from L to the ith particle (Fig. 8.4b), so the term in brackets in Eq. (8.11) is the moment of inertia about L:

$$\sum_i m_i(\mathbf{k} \times \mathbf{R}_i) \cdot (\mathbf{k} \times \mathbf{R}_i) = \sum_i m_i |\mathbf{k} \times \mathbf{R}_i|^2 = \sum_i m_i r_i^2 = I.$$

Thus we obtain the kinetic energy of a rigid body in general planar motion in the form

$$\boxed{T = \frac{1}{2}mv^2 + \frac{1}{2}I\omega^2.} \qquad (8.12)$$

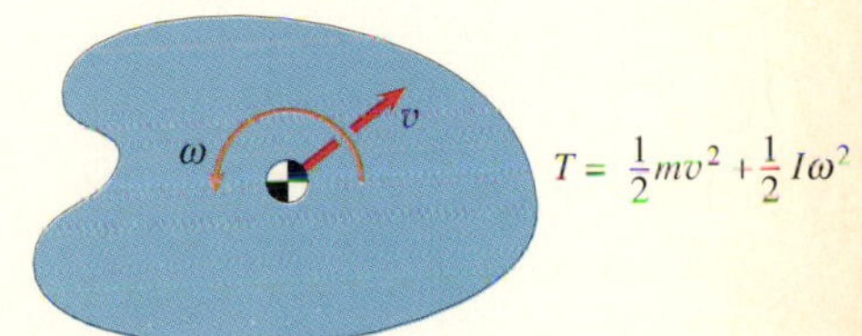

Fig. 8.5

Kinetic energy in general planar motion.

The kinetic energy consists of two terms: the **translational kinetic energy**, expressed in terms of the velocity of the center of mass, and the **rotational kinetic energy** (Fig. 8.5).

Kinetic Energy in Fixed Axis Rotation An object rotating about a fixed axis is in general planar motion, and its kinetic energy is given by Eq. (8.12). But there is another expression for the kinetic energy that you will often find convenient. Suppose that a rigid body rotates with angular velocity ω about a fixed axis O. In terms of the distance d from O to the center of mass, the velocity of the center of mass is $v = \omega d$ (Fig. 8.6a). From Eq. (8.12), the kinetic energy is

$$T = \frac{1}{2}m(\omega d)^2 + \frac{1}{2}I\omega^2 = \frac{1}{2}(I + d^2m)\omega^2.$$

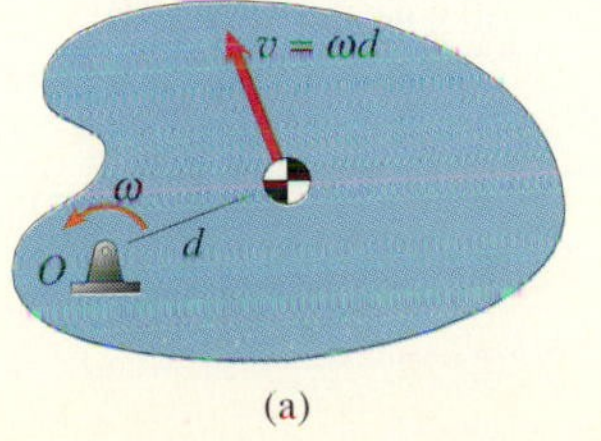

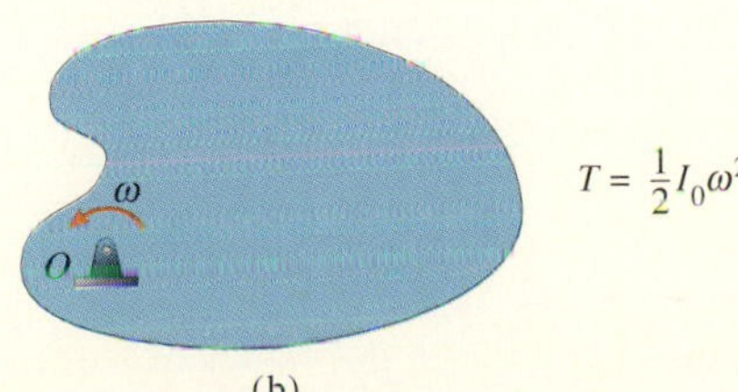

Fig. 8.6

(a) Velocity of the center of mass.
(b) Kinetic energy of a rigid body rotating about a fixed axis.

According to the parallel-axis theorem, the moment of inertia about O is $I_O = I + d^2 m$, so we obtain the kinetic energy of a rigid body rotating about a fixed axis O in the form (Fig. 8.6b)

$$\boxed{T = \frac{1}{2} I_O \omega^2.} \tag{8.13}$$

8.2 Work and Potential Energy

The procedures for determining the work done by different types of forces and the expressions you learned in Chapter 4 for the potential energies of forces provide you with the essential tools for applying the principle of work and energy to a rigid body. The work done on a rigid body by a force $\mathbf{F}$ is given by

$$\boxed{U_{12} = \int_{(\mathbf{r}_p)_1}^{(\mathbf{r}_p)_2} \mathbf{F} \cdot d\mathbf{r}_p,} \tag{8.14}$$

where $\mathbf{r}_p$ is the position of the *point of application* of $\mathbf{F}$ (Fig. 8.7). If the point of application is stationary, or if its direction of motion is perpendicular to $\mathbf{F}$, no work is done.

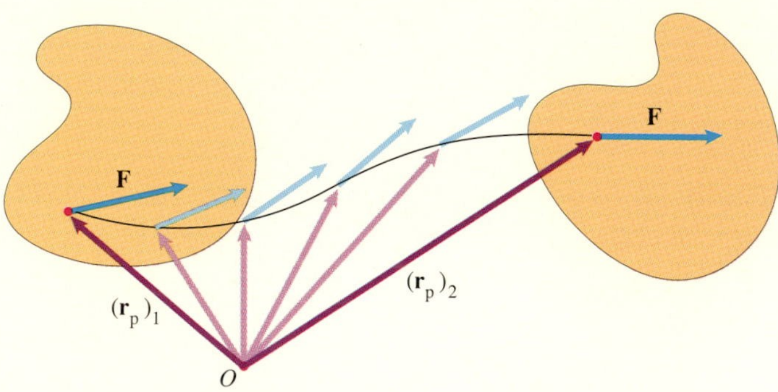

Fig. 8.7
The work done by a force on a rigid body is determined by the path of the point of application.

A force $\mathbf{F}$ is conservative if a potential energy V exists such that

$$\mathbf{F} \cdot d\mathbf{r}_p = -dV. \tag{8.15}$$

In terms of its potential energy, the work done by a conservative force $\mathbf{F}$ is

$$U_{12} = \int_{(\mathbf{r}_p)_1}^{(\mathbf{r}_p)_2} \mathbf{F} \cdot d\mathbf{r}_p = \int_{V_1}^{V_2} -dV = -(V_2 - V_1),$$

where V_1 and V_2 are the values of V at $(\mathbf{r}_p)_1$ and $(\mathbf{r}_p)_2$.

If a rigid body is subjected to a couple M (Fig. 8.8a), what work is done as it moves between two positions? We can evaluate the work by representing the couple by forces (Fig. 8.8b) and determining the work done by the forces. If the rigid body rotates through an angle $d\theta$ in the direction of the couple (Fig. 8.8c), the work done by each force is $(\frac{1}{2}D\,d\theta)F$, so the total work is $DF\,d\theta = M\,d\theta$. Integrating this expression, we obtain the work done by a couple M as the rigid body rotates from θ_1 to θ_2 in the direction of M:

$$U_{12} = \int_{\theta_1}^{\theta_2} M\,d\theta. \qquad (8.16)$$

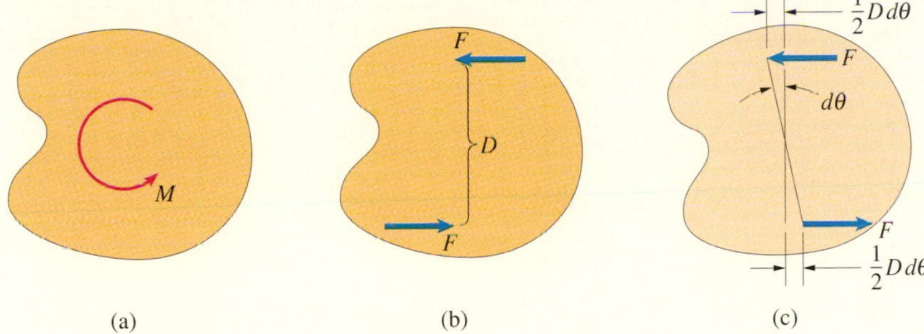

Fig. 8.8

(a) A rigid body subjected to a couple.
(b) An equivalent couple consisting of two forces: $DF = M$.
(c) Determining the work done by the forces.

If M is constant between θ_1 and θ_2, the work is simply the product of the couple and the angular displacement:

$$U_{12} = M(\theta_2 - \theta_1). \qquad \textbf{Constant couple}$$

A couple M is conservative if a potential energy V exists such that

$$M\,d\theta = -dV. \qquad (8.17)$$

We can express the work done by a conservative couple in terms of its potential energy:

$$U_{12} = \int_{\theta_1}^{\theta_2} M\,d\theta = \int_{V_1}^{V_2} -dV = -(V_2 - V_1).$$

For example, in Fig. 8.9 a torsional spring exerts a couple on a bar that is proportional to the bar's angle of rotation: $M = -k\theta$. From the relation

$$M\,d\theta = -k\theta\,d\theta = -dV,$$

we see that the potential energy must satisfy the relation

$$\frac{dV}{d\theta} = k\theta.$$

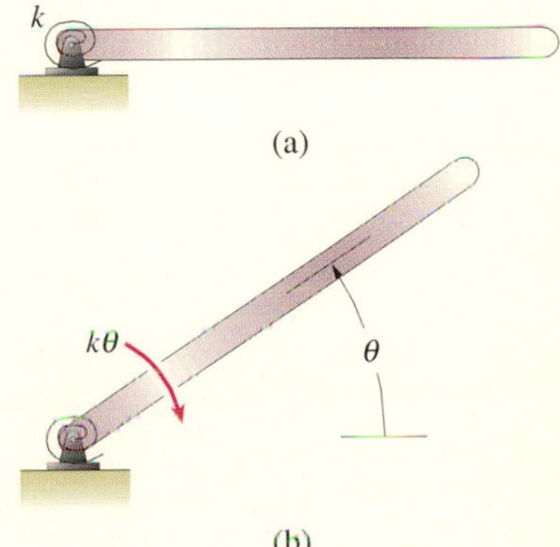

Fig. 8.9

(a) A linear torsional spring connected to a bar.
(b) The spring exerts a couple of magnitude $k\theta$ in the direction opposite to the bar's rotation.

Integrating this equation, we find that the potential energy of the torsional spring is

$$V = \frac{1}{2}k\theta^2. \tag{8.18}$$

If *all* the forces and couples that do work on a rigid body are conservative, we can express the total work done as it moves between two positions 1 and 2 in terms of the total potential energy of the forces and couples:

$$U_{12} = V_1 - V_2.$$

Combining this relation with the principle of work and energy, we conclude that the sum of the kinetic energy and the total potential energy is constant—energy is conserved:

$$\boxed{T + V = \text{constant.}} \tag{8.19}$$

8.3 *Power*

The work done on a rigid body by a force $\mathbf{F}$ during an infinitesimal displacement $d\mathbf{r}_\text{p}$ of its point of application is

$$\mathbf{F} \cdot d\mathbf{r}_\text{p}.$$

We obtain the power P transmitted to the rigid body, the rate at which work is done on it, by dividing this expression by the interval of time dt during which the displacement takes place:

$$P = \mathbf{F} \cdot \mathbf{v}_\text{p}, \tag{8.20}$$

where $\mathbf{v}_\text{p}$ is the velocity of the point of application of $\mathbf{F}$.

Similarly, the work done on a rigid body in planar motion by a couple M during an infinitesimal rotation $d\theta$ in the direction of M is

$$M\, d\theta.$$

Dividing this expression by dt, the power transmitted to the rigid body is the product of the couple and the angular velocity:

$$P = M\omega. \tag{8.21}$$

The total work done on a rigid body during an interval of time equals the change in its kinetic energy, so the total power transmitted equals the rate of change of its kinetic energy:

$$P = \frac{dT}{dt}.$$

The average with respect to time of the power during an interval of time from t_1 to t_2 is

$$P_{av} = \frac{1}{t_2 - t_1}\int_{t_1}^{t_2} P\,dt = \frac{1}{t_2 - t_1}\int_{T_1}^{T_2} dT = \frac{T_2 - T_1}{t_2 - t_1}.$$

This expression shows that we can determine the average power transferred to or from a rigid body during an interval of time by dividing the change in its kinetic energy, or the total work done, by the interval of time:

$$P_{av} = \frac{T_2 - T_1}{t_2 - t_1} = \frac{U_{12}}{t_2 - t_1}. \tag{8.22}$$

In the following examples we use energy methods to analyze motions of rigid bodies and systems of rigid bodies. You should consider using energy methods when you want to relate changes in the translational and angular velocities of an object to a change in its position. This typically involves three steps:

1. Identify the forces and couples that do work—*You must use free-body diagrams to determine which external forces and couples do work.*
2. Apply work and energy or conservation of energy—*Either equate the total work done during a change in position to the change in the kinetic energy or equate the sum of the kinetic and potential energies at two positions.*
3. Determine kinematic relationships—*To complete your solution, you often will need to relate the velocity of the center of mass of a rigid body to its angular velocity.*

Example 8.1

A disk of mass m and moment of inertia I is released from rest on an inclined surface (Fig. 8.10). Assuming that it rolls, what is the velocity of the disk's center when it has moved a distance b?

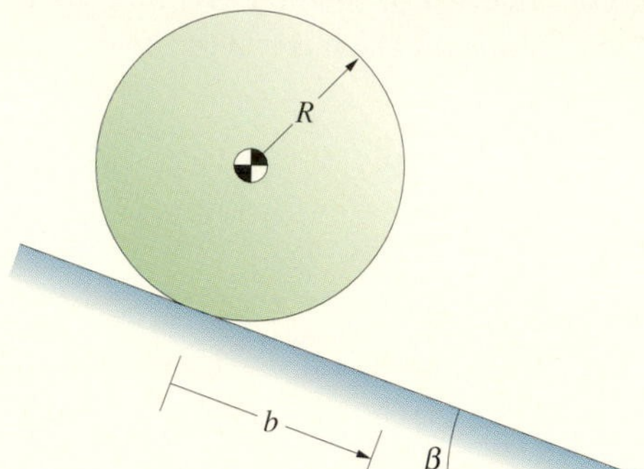

Fig. 8.10

STRATEGY

We can determine the velocity by equating the total work done as the disk rolls a distance b to the change in its kinetic energy.

SOLUTION

Identify the Forces and Couples That Do Work We draw the free-body diagram of the disk in Fig. (a). The disk's weight does work as it rolls, but the normal force N and the friction force f do not. To help you understand why the friction force does no work, we can write the work done by a force $\mathbf{F}$ as

$$\int_{(\mathbf{r}_\mathrm{p})_1}^{(\mathbf{r}_\mathrm{p})_2} \mathbf{F} \cdot d\mathbf{r}_\mathrm{p} = \int_{t_1}^{t_2} \mathbf{F} \cdot \frac{d\mathbf{r}_\mathrm{p}}{dt}\, dt = \int_{t_1}^{t_2} \mathbf{F} \cdot \mathbf{v}_\mathrm{p}\, dt,$$

where $\mathbf{v}_\mathrm{p}$ is the velocity of the point of application of $\mathbf{F}$. Since the velocity of the point where f acts is zero as the disk rolls, the work done by f is zero.

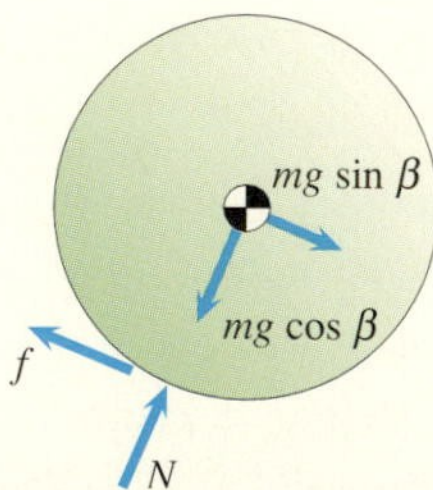

(a) Free-body diagram of the disk.

Apply Work and Energy We can determine the work done by the weight by multiplying the component in the direction of the motion of the center of the disk by the distance b:

$$U_{12} = (mg \sin \beta)b.$$

Letting v and ω be the velocity of the center and the angular velocity of the disk when it has moved a distance b (Fig. b), we equate the work to the change in the disk's kinetic energy:

$$mgb \sin \beta = \frac{1}{2}mv^2 + \frac{1}{2}I\omega^2 - 0. \tag{8.23}$$

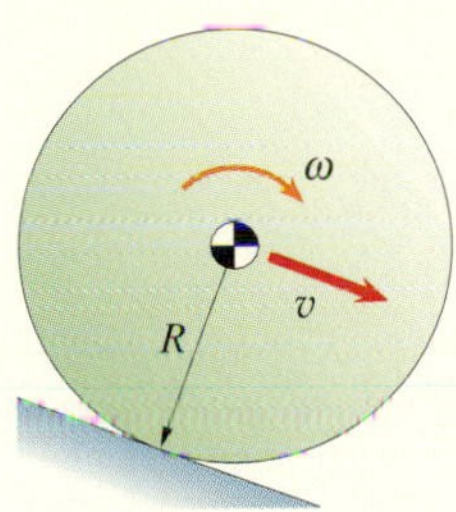

(b) Velocity of the center and the angular velocity when the disk has moved a distance b.

Determine Kinematic Relationship The angular velocity ω of the rolling disk is related to the velocity v by $\omega = v/R$. Substituting this relation into Eq. (8.23) and solving for v, we obtain

$$v = \sqrt{\frac{2gb \sin \beta}{1 + I/mR^2}}.$$

DISCUSSION

Suppose that the surface is smooth, so that the disk slides instead of rolling. In this case, the disk has no angular velocity, so Eq. (8.23) becomes

$$mgb \sin \beta = \frac{1}{2}mv^2 - 0,$$

and the velocity of the center of the disk is

$$v = \sqrt{2gb \sin \beta}.$$

The velocity is greater when the disk slides. You can see why by comparing the two expressions for the principle of work and energy. The work done by the disk's weight is the same in each case. When the disk rolls, part of the work increases the disk's translational kinetic energy and part increases its rotational kinetic energy. When the disk slides, all of the work increases its translational kinetic energy.

Example 8.2

Each wheel of the motorcycle in Fig. 8.11 has mass $m_W = 9$ kg, radius $R = 330$ mm, and moment of inertia $I = 0.8$ kg-m^2. The combined mass of the rider and the motorcycle, not including the wheels, is $m_C = 142$ kg. The motorcycle starts from rest, and its engine exerts a constant couple $M = 140$ N-m on the rear wheel. Assume that the wheels do not slip.

(a) What horizontal distance b must the motorcycle travel to reach a velocity of 25 m/s?

(b) What is the maximum power transmitted to the motorcycle by its engine during the motion described in (a)?

Fig. 8.11

STRATEGY

(a) We can apply the principle of work and energy to the system consisting of the rider and the motorcycle, including its wheels, to determine the distance b.

(b) The power transmitted by the couple exerted on the rear wheel is given by Eq. (8.21). To determine the maximum power, we must determine the wheel's maximum angular velocity.

SOLUTION

(a) Determining the distance b requires three steps.

Identify the Forces and Couples That Do Work We draw the free-body diagram of the system in Fig. (a). The weights do no work because the motion is horizontal, and the forces exerted on the wheels by the road do no work because the velocity of their point of application is zero. (See Example 8.1.) No work is done by external forces and couples! However, work is done by the couple M exerted on the rear wheel by the engine (Fig. b). Although this is an internal couple for the system we are considering—the wheel exerts an opposite couple on the body of the motorcycle—net work is done because the wheel rotates whereas the body does not.

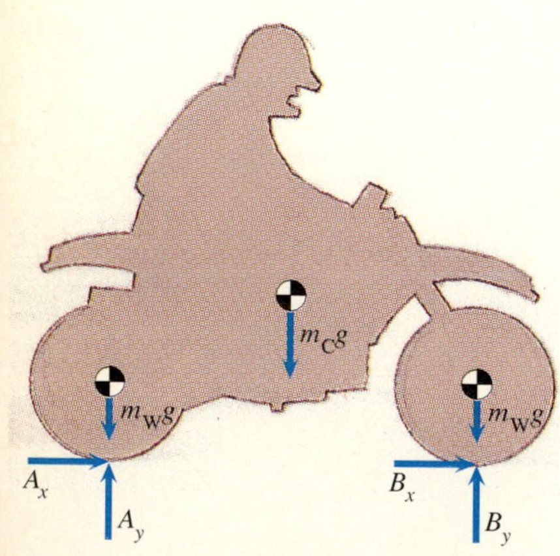

(a) Free-body diagram of the system.

Apply Work and Energy If the motorcycle moves a horizontal distance b, the wheels turn through an angle b/R rad and the work done by the constant couple M is

$$U_{12} = M(\theta_2 - \theta_1) = M\left(\frac{b}{R}\right).$$

Let v be the motorcycle's velocity and ω the angular velocity of the wheels when the motorcycle has moved a distance b. The work equals the change in the total kinetic energy:

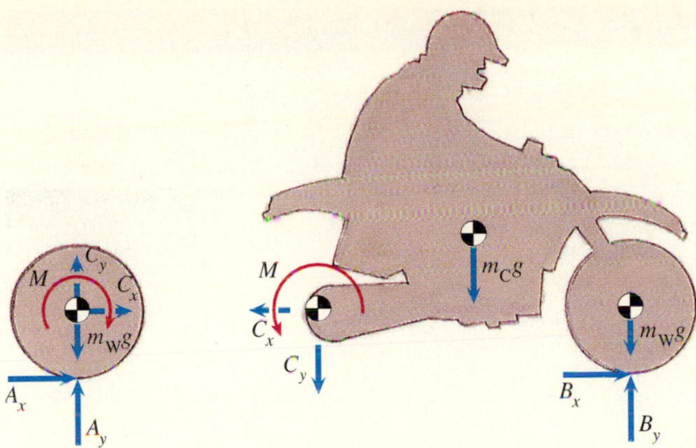

(b) Isolating the rear wheel.

$$M\left(\frac{b}{R}\right) = \frac{1}{2}m_C v^2 + 2\left[\frac{1}{2}m_W v^2 + \frac{1}{2}I\omega^2\right] - 0. \quad (8.24)$$

Determine Kinematic Relationship The angular velocity of the rolling wheels is related to the velocity v by $\omega = v/R$. Substituting this relation into Eq. (8.24) and solving for b, we obtain

$$b = \left(\frac{1}{2}m_C + m_W + \frac{I}{R^2}\right)\frac{Rv^2}{M}$$

$$= \left[\frac{1}{2}(142) + (9) + \frac{(0.8)}{(0.33)^2}\right]\frac{(0.33)(25)^2}{(140)}$$

$$= 129 \text{ m}.$$

(b) The angular velocity of the wheels when the motorcycle reaches its maximum velocity is

$$\omega = \frac{v}{R} = \frac{25}{0.33} = 75.8 \text{ rad/s}.$$

From Eq. (8.21), the maximum power is

$$P = M\omega = (140)(75.8) = 10{,}600 \text{ W}.$$

DISCUSSION

Although we drew separate free-body diagrams of the motorcycle and its rear wheel to clarify the work done by the couple exerted by the engine, notice that we treated the motorcycle, including its wheels, as a single system in applying the principle of work and energy. By doing so, we did not need to consider the work done by the internal forces between the motorcycle's body and its wheels. When applying the principle of work and energy to a system of rigid bodies, you will usually find it simplest to express the principle for the system as a whole. This is in contrast to determining the motion of a system of rigid bodies by using the equations of motion, which usually requires that you draw free-body diagrams of each rigid body and apply the equations to them individually.

Example 8.3

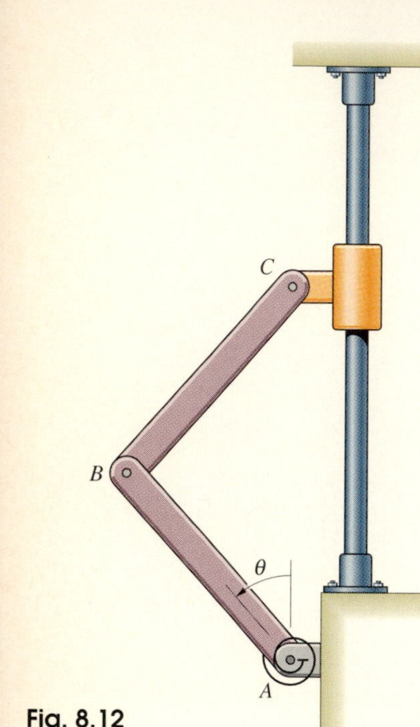

Fig. 8.12

The slender bars AB and BC of the linkage in Fig. 8.12 have mass m and length l, and the collar C has mass m_C. A torsional spring at A exerts a clockwise couple $k\theta$ on bar AB. The system is released from rest in the position $\theta = 0$ and allowed to fall. Neglecting friction, determine the angular velocity $\omega = d\theta/dt$ of bar AB as a function of θ.

SOLUTION

Identify the Forces and Couples That Do Work We draw the free-body diagram of the system in Fig. (a). The forces and couples that do work—the weights of the bars and collar and the couple exerted by the torsional spring—are conservative. We can use conservation of energy and the kinematic relationships between the angular velocities of the bars and the velocity of the collar to determine ω as a function of θ.

Apply Conservation of Energy We denote the center of mass of bar BC by G and the angular velocity of bar BC by ω_{BC} (Fig. b). The moment of inertia of each bar about its center of mass is $I = \frac{1}{12}ml^2$. Since bar AB rotates about the fixed point A, we can write its kinetic energy as

$$T_{\text{bar }AB} = \frac{1}{2}I_A\omega^2 = \frac{1}{2}\left[I + \left(\frac{1}{2}l\right)^2 m\right]\omega^2 = \frac{1}{6}ml^2\omega^2.$$

The kinetic energy of bar BC is

$$T_{\text{bar }BC} = \frac{1}{2}mv_G^2 + \frac{1}{2}I\omega_{BC}^2 = \frac{1}{2}mv_G^2 + \frac{1}{24}ml^2\omega_{BC}^2.$$

The kinetic energy of the collar C is

$$T_{\text{collar}} = \frac{1}{2}m_C v_C^2.$$

Using the datum in Fig. (a), we obtain the potential energies of the weights:

$$V_{\text{bar }AB} + V_{\text{bar }BC} + V_{\text{collar}} = mg\left(\frac{1}{2}l\cos\theta\right) + mg\left(\frac{3}{2}l\cos\theta\right) + m_C g(2l\cos\theta).$$

The potential energy of the torsional spring is given by Eq. (8.18):

$$V_{\text{spring}} = \frac{1}{2}k\theta^2.$$

We now have all the ingredients to apply conservation of energy. We equate the sum of the kinetic and potential energies at the position $\theta = 0$ to the sum of the kinetic and potential energies at an arbitrary value of θ:

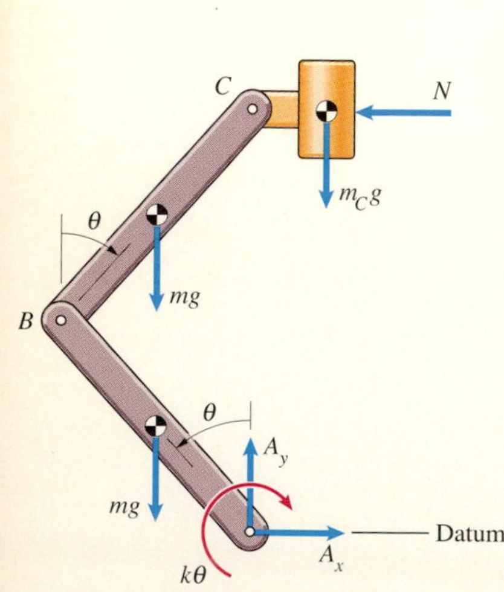

(a) Free-body diagram of the system.

$$T_1 + V_1 = T_2 + V_2:$$

$$0 + 2mgl + 2m_Cgl = \frac{1}{6}ml^2\omega^2 + \frac{1}{2}mv_G^2 + \frac{1}{24}ml^2\omega_{BC}^2 + \frac{1}{2}m_Cv_C^2$$

$$+ 2mgl\cos\theta + 2m_Cgl\cos\theta + \frac{1}{2}k\theta^2.$$

To determine ω from this equation, we must express the velocities v_G, v_C, and ω_{BC} in terms of ω.

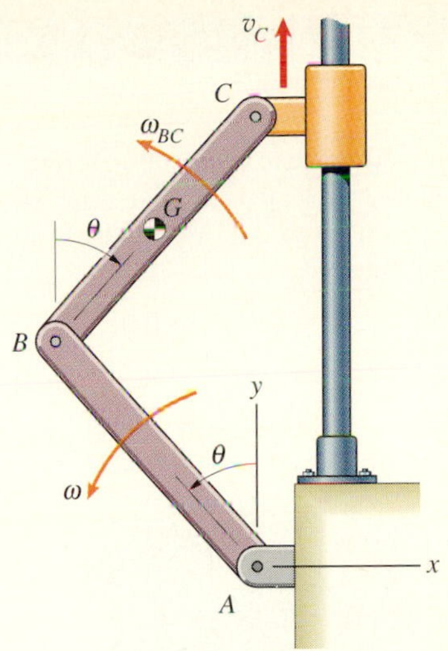

(b) Angular velocities of the bars and the velocity of the collar.

Determine Kinematic Relationships We can determine the velocity of point B in terms of ω and then express the velocity of point C in terms of the velocity of point B and the angular velocity ω_{BC}.

The velocity of B is

$$\mathbf{v}_B = \mathbf{v}_A + \boldsymbol{\omega}_{AB} \times \mathbf{r}_{B/A}$$

$$= \mathbf{0} + \begin{vmatrix} \mathbf{i} & \mathbf{j} & \mathbf{k} \\ 0 & 0 & \omega \\ -l\sin\theta & l\cos\theta & 0 \end{vmatrix}$$

$$= -l\omega\cos\theta\,\mathbf{i} - l\omega\sin\theta\,\mathbf{j}.$$

The velocity of C expressed in terms of the velocity of B is

$$v_C\mathbf{j} = \mathbf{v}_B + \boldsymbol{\omega}_{BC} \times \mathbf{r}_{C/B}$$

$$= -l\omega\cos\theta\,\mathbf{i} - l\omega\sin\theta\,\mathbf{j} + \begin{vmatrix} \mathbf{i} & \mathbf{j} & \mathbf{k} \\ 0 & 0 & \omega_{BC} \\ l\sin\theta & l\cos\theta & 0 \end{vmatrix}.$$

Equating $\mathbf{i}$ and $\mathbf{j}$ components, we obtain

$$\omega_{BC} = -\omega, \qquad v_C = -2l\omega\sin\theta.$$

(The minus signs indicate that the directions of the velocities are opposite to the directions we assumed in Fig. b.) Now that we know the angular velocity of bar BC in terms of ω, we can determine the velocity of its center of mass in terms of ω by expressing it in terms of $\mathbf{v}_B$:

$$\mathbf{v}_G = \mathbf{v}_B + \boldsymbol{\omega}_{BC} \times \mathbf{r}_{G/B}$$

$$= -l\omega\cos\theta\,\mathbf{i} - l\omega\sin\theta\,\mathbf{j} + \begin{vmatrix} \mathbf{i} & \mathbf{j} & \mathbf{k} \\ 0 & 0 & -\omega \\ \frac{1}{2}l\sin\theta & \frac{1}{2}l\cos\theta & 0 \end{vmatrix}$$

$$= -\frac{1}{2}l\omega\cos\theta\,\mathbf{i} - \frac{3}{2}l\omega\sin\theta\,\mathbf{j}.$$

Substituting these expressions for ω_{BC}, v_C, and $\mathbf{v}_G$ into our equation of conservation of energy and solving for ω, we obtain

$$\omega = \left[\frac{2gl(m + m_C)(1 - \cos\theta) - \frac{1}{2}k\theta^2}{\frac{1}{3}ml^2 + (m + 2m_C)l^2\sin^2\theta}\right]^{1/2}.$$

Problems

8.1 A main landing gear wheel of a Boeing 747 weighs 240 lb, has a moment of inertia of 17 slug-ft^2, and has a radius of 2 ft. If the airplane is moving at 245 ft/s and the wheel rolls, what is the wheel's kinetic energy?

P8.1

8.2 The 8-kg slender bar is released from rest in the horizontal position. When it has fallen to the position shown, its angular velocity is 3.226 rad/s.
(a) Use the given value of the angular velocity to determine the bar's kinetic energy.
(b) How much work is done by the bar's weight as it falls to the position shown?

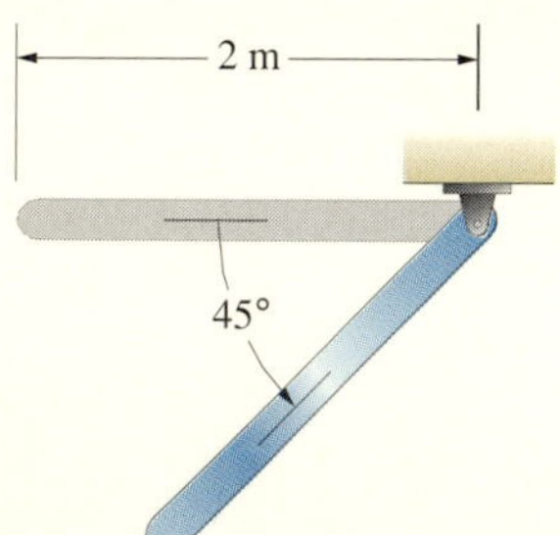

P8.2

8.3 The angular velocity of the space station is 1 rpm. Use work and energy to determine the constant couple the station's reaction control system would have to exert to reduce its angular velocity to zero in 100 revolutions. The moment of inertia of the station is $I = 1.5 \times 10^{10}$ kg-m^2.

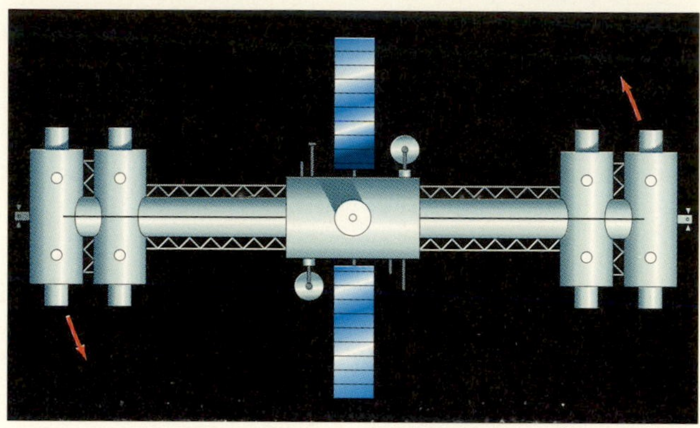

P8.3

8.4 Determine the average power transferred from the space station in Problem 8.3 as the control system reduces its angular velocity to zero.

8.5 The moment of inertia of the helicopter's rotor is 400 slug-ft^2. If the rotor starts from rest, the engine exerts a constant torque of 500 ft-lb on it, and aerodynamic drag is neglected, use the principle of work and energy to determine how many revolutions the rotor must turn to reach an angular velocity of 2 revolutions per second.

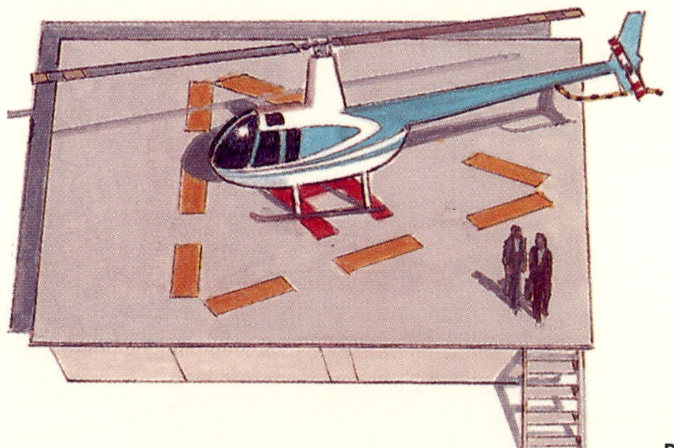

P8.5

8.6 What average power is transmitted to the rotor in Problem 8.5 in accelerating it from rest to 2 revolutions per second?

8.7 During extravehicular activity, an astronaut activates two thrusters of her maneuvering unit, exerting constant equal and opposite forces $T = 2$ N. The moment of inertia of the astronaut and her equipment about the axis through their center of mass perpendicular to the page is 45 kg-m^2. Using the principle of work and energy, determine her rate of rotation in revolutions per second when she has rotated one-fourth of a revolution from her initial orientation.

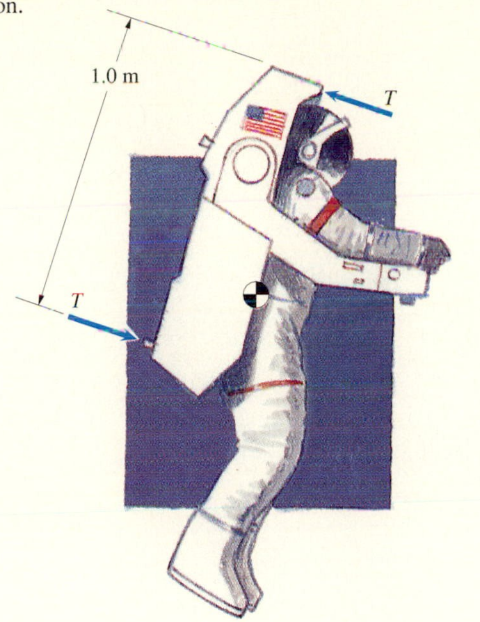

P8.7

8.8 What average power is transmitted to the astronaut in Problem 8.7 as she rotates one-fourth of a revolution from her initial orientation?

8.9 A slender bar of mass m is released from rest in the horizontal position shown. Determine its angular velocity when it is vertical (a) by using the principle of work and energy; (b) by using conservation of energy.

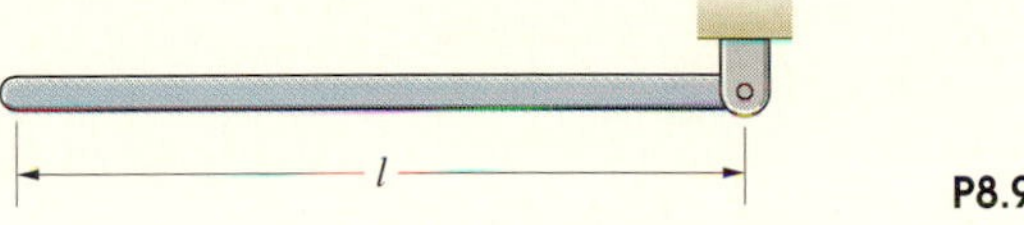

P8.9

8.10 The moment of inertia of the pulley is 0.4 slug-ft^2. The pulley starts from rest. For both cases, use the principle of work and energy to determine the pulley's angular velocity when it has turned 1 revolution.

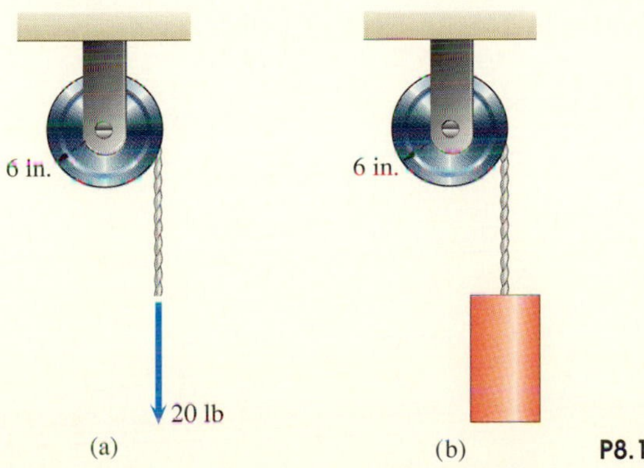

P8.10

8.11 The object consists of identical 1-m, 5-kg bars welded together. If it is released from rest in the position shown, what is its angular velocity when the bar attached at A is vertical?

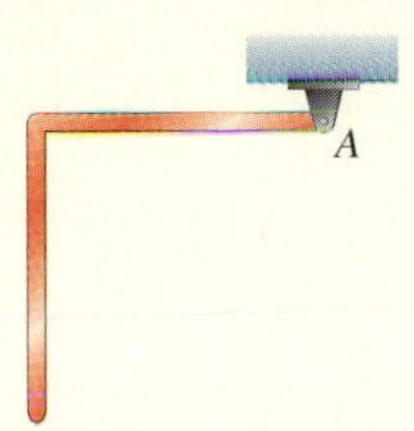

P8.11

8.12 The objects consist of identical 3-ft, 10-lb bars welded together. If they are released from rest in the positions shown, what are their angular velocities when the bars attached at A are vertical?

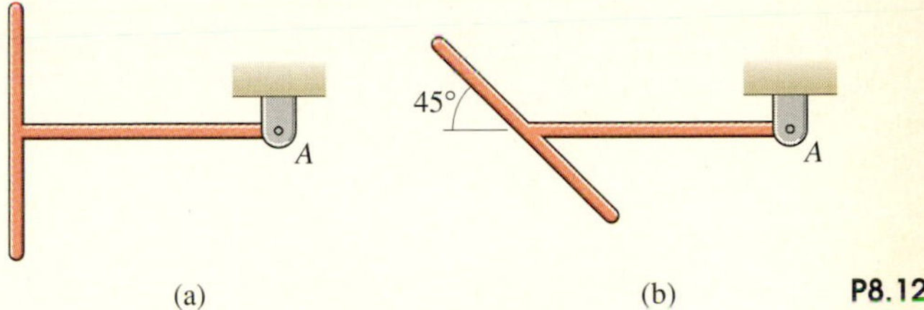

P8.12

8.13 The 8-kg slender bar is released from rest in the horizontal position. When it has fallen to the position shown, what are the x and y components of force exerted on the bar by the pin support A?

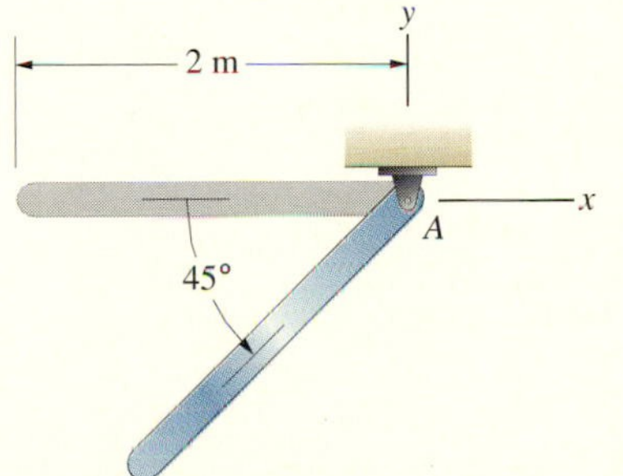

P8.13

8.14 The slender bar is released from rest in the position shown.
(a) Use conservation of energy to determine the angular velocity when the bar is vertical.
(b) For what value of x is the angular velocity determined in (a) a maximum?

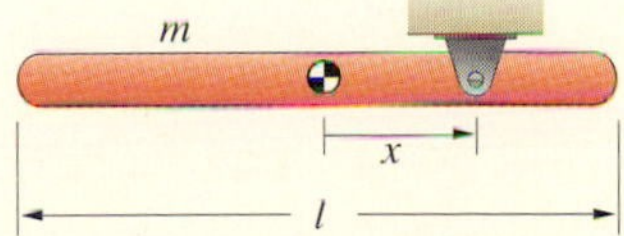

P8.14

8.15 The gears can turn freely on their pin supports. Their moments of inertia are $I_A = 0.002$ kg-m^2 and $I_B = 0.006$ kg-m^2. They are at rest when a constant couple $M = 2$ N-m is applied to gear B. Neglecting friction, use the principle of work and energy to determine the angular velocities of the gears when gear A has turned 100 revolutions.

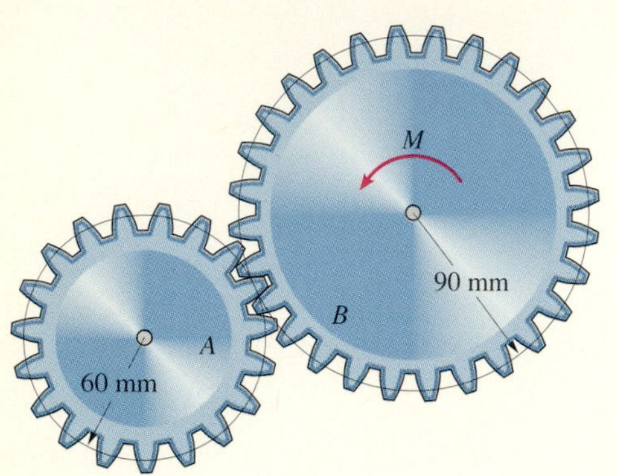

P8.15

8.16 The moments of inertia of gears A and B are $I_A = 0.02$ kg-m^2 and $I_B = 0.09$ kg-m^2. Gear A is connected to a torsional spring with constant $k = 12$ N-m/rad. If gear B is given an initial counterclockwise angular velocity of 10 rad/s with the torsional spring unstretched, through what maximum counterclockwise angle does gear B rotate?

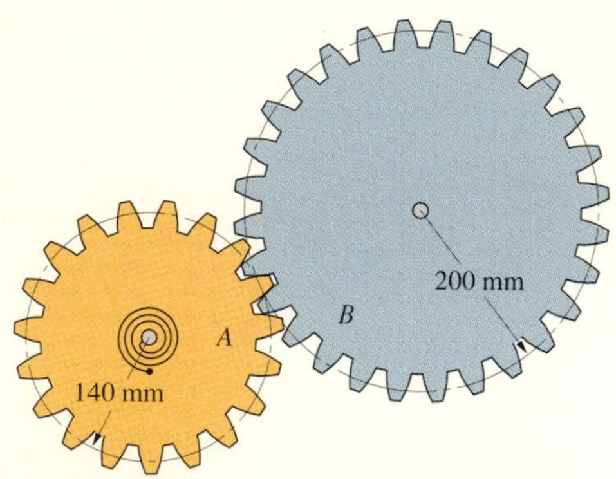

P8.16

8.17 The pulleys can turn freely on their pin supports. Their moments of inertia are $I_A = 0.002$ kg-m^2, $I_B = 0.036$ kg-m^2, and $I_C = 0.032$ kg-m^2. They are stationary when a constant couple $M = 2$ N-m is applied to pulley A. What is the angular velocity of pulley A when it has turned 10 revolutions?

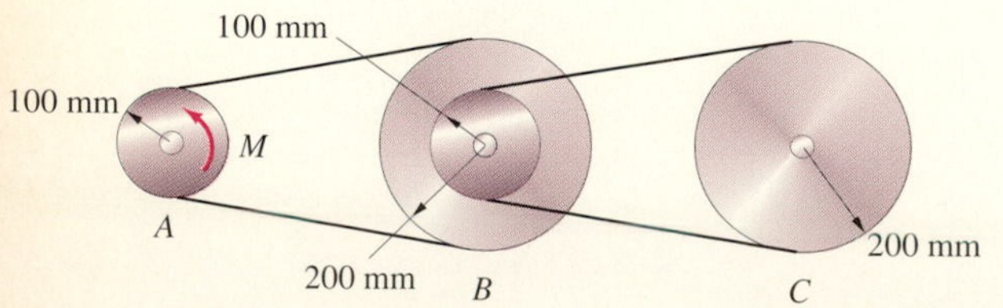

P8.17

8.18 Model the arm ABC as a single rigid body. Its mass is 300 kg, and the moment of inertia about its center of mass is $I = 360$ kg-m^2. Starting from rest with its center of mass 2 m above the ground (position 1), the hydraulic cylinders push arm ABC upward. When it is in the position shown (position 2), its counterclockwise angular velocity is 1.4 rad/s. How much work do the hydraulic cylinders do on the arm in moving it from position 1 to position 2?

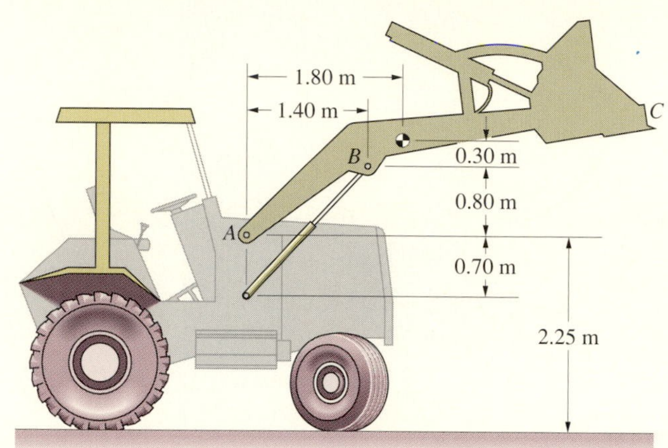

P8.18

8.19 The mass of the homogeneous cylindrical disk is m, and its radius is R. The disk is stationary when a constant clockwise couple M is applied to it. Use work and energy to determine the disk's angular velocity when it has rolled a distance b.

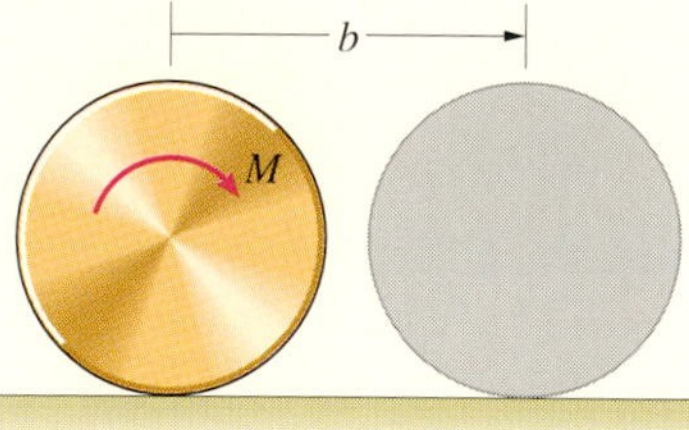

P8.19

8.20 A disk of mass m and moment of inertia I starts from rest on an inclined surface and is subjected to a constant clockwise couple M. Assuming that it rolls, what is the angular velocity of the disk when it has moved a distance b?

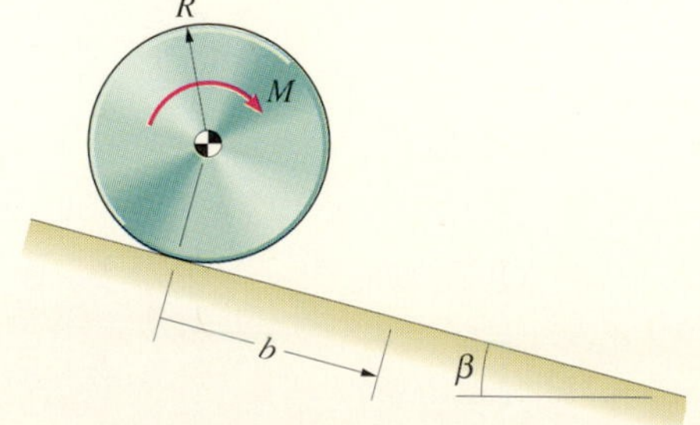

P8.20

8.21 The stepped disk weighs 40 lb, and its moment of inertia is $I = 0.2$ slug-ft^2. If it is released from rest, what is its angular velocity when the center of the disk has fallen 3 ft?

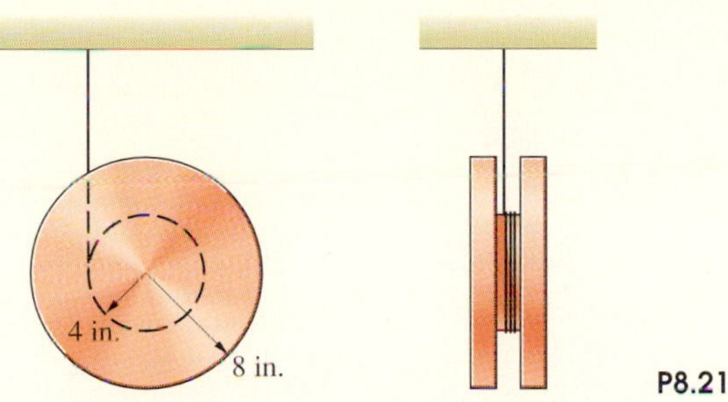

P8.21

8.22 The 100-kg homogeneous cylindrical disk is at rest when the force $F = 500$ N is applied to a cord wrapped around it, causing the disk to roll. Use the principle of work and energy to determine the disk's angular velocity when it has turned 1 revolution.

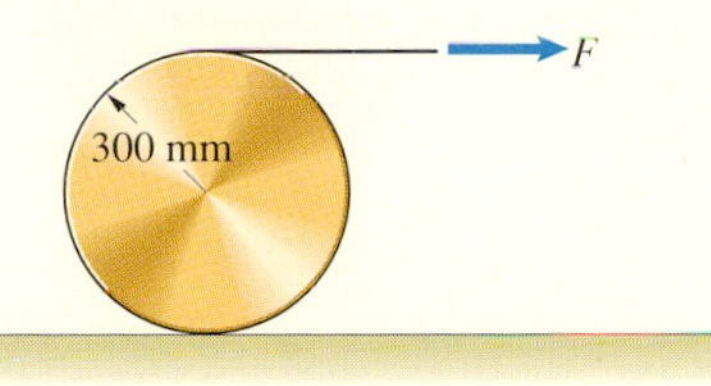

P8.22

8.23 The 1-slug homogeneous cylindrical disk is given a clockwise angular velocity of 2 rad/s with the spring unstretched. The spring constant is $k = 3$ lb/ft. If the disk rolls, how far will its center move to the right?

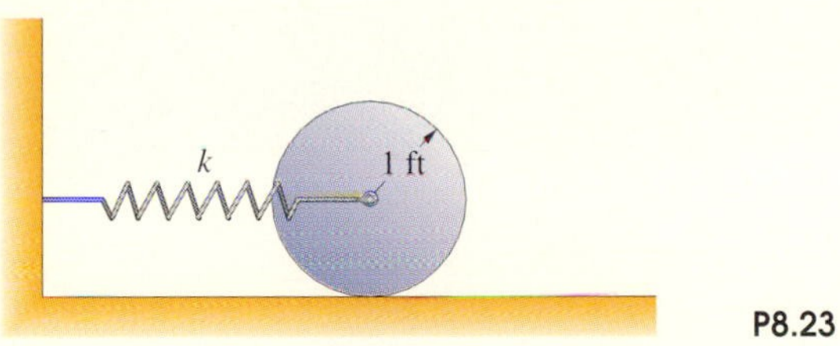

P8.23

8.24 The 22-kg platen P rests on four roller bearings. The roller bearings can be modeled as 1-kg homogeneous cylinders with 30-mm radii. The platen is stationary and the spring ($k = 900$ N/m) is unstretched when a constant horizontal force $F = 100$ N is applied as shown. What is the platen's velocity when it has moved 200 mm to the right?

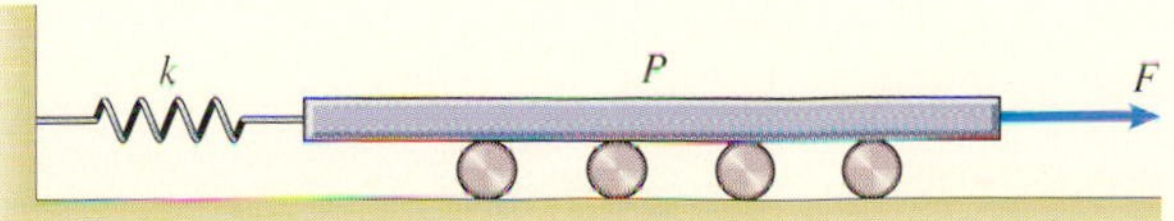

P8.24

8.25 Consider the system described in Problem 8.24.
(a) What maximum distance does the platen move to the right when the force F is applied?
(b) What maximum velocity does the platen achieve, and how far has the platen moved to the right when it occurs?

8.26 The rules of a soapbox derby specify the required combined weight of the car and driver and the radius of the wheels. A young contestant designing her car ponders two possibilities: (a) use heavy wheels; (b) use light wheels, making up the weight by adding ballast. Analyze this problem using the principle of work and energy, and explain the advice you would give her.

P8.26

8.27 The total moment of inertia of the car's two rear wheels and axle is I_R, and the total moment of inertia of the two front wheels is I_F. The radius of the tires is R, and the total mass of the car including the wheels is m. The car is moving at velocity v_0 when the driver applies the brakes. If the car's brakes exert a constant retarding couple M on each wheel and the tires do not slip, determine the car's velocity as a function of the distance s from the point where the brakes are applied.

P8.27

8.28 In Problem 8.27, $I_R = 0.24$ kg-m^2, $I_F = 0.20$ kg-m^2, $R = 0.30$ m, and $m = 1480$ kg. If the car's brakes exert a constant retarding couple $M = 650$ N-m on each wheel and the tires do not slip, determine the distance s required for the car to come to a stop if it is moving at 100 km/hr.

8.29 Each box weighs 50 lb, the moment of inertia of the pulley is 0.6 slug-ft^2, and friction can be neglected. If the boxes start from rest, determine the magnitude of their velocity when they have moved 4 ft from their initial positions.

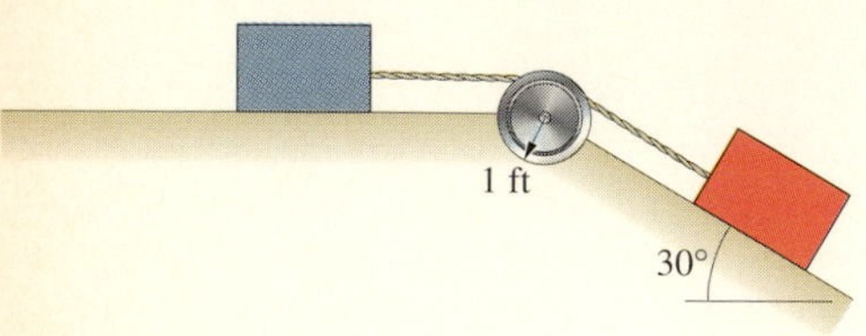

P8.29

8.30 The slender bar weighs 30 lb, and the cylindrical disk weighs 20 lb. The system is released from rest with the bar horizontal. Determine the magnitude of the bar's angular velocity when it is vertical if the bar and disk are welded together at A.

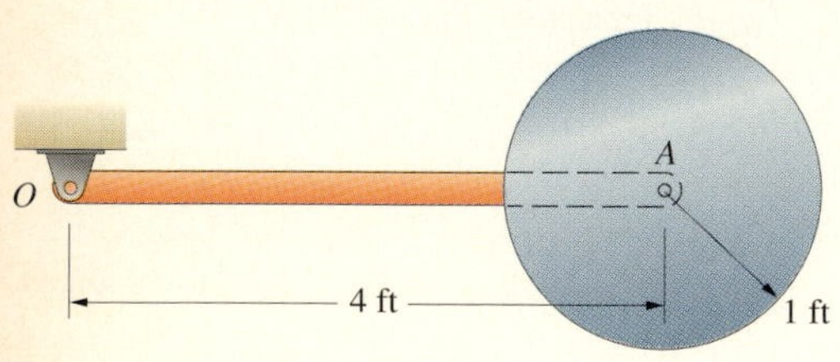

P8.30

8.31 In Problem 8.30, determine the magnitude of the bar's angular velocity when it has reached the vertical position if the bar and disk are connected by a smooth pin at A.

8.32 The 100-lb crate is pulled up the inclined surface by the winch. The coefficient of kinetic friction between the crate and the surface is $\mu_k = 0.4$. The moment of inertia of the drum on which the cable is wound, including the cable wound on the drum, is $I_A = 3$ slug-ft^2. The motor exerts a constant couple $M = 40$ ft-lb on the drum. If the crate starts from rest, use the principle of work and energy to determine its velocity when it has moved 2 ft.

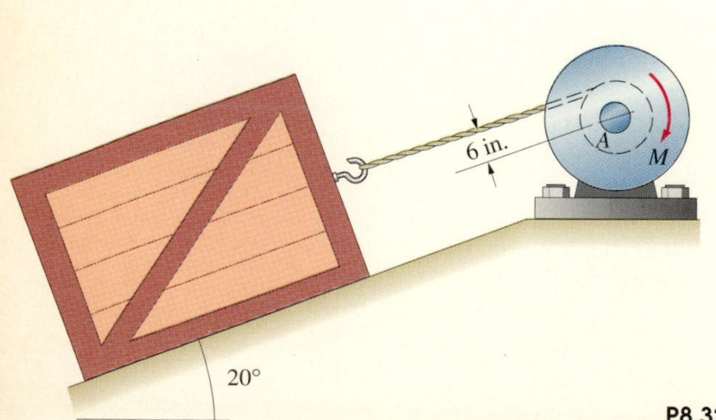

P8.32

8.33 The 2-ft slender bars each weigh 4 lb, and the rectangular plate weighs 20 lb. If the system is released from rest in the position shown, what is the velocity of the plate when the bars are vertical?

P8.33

8.34 The slender bar has mass m and length l. A torsional spring with constant k is attached to the bar at the pin support. The spring is unstretched when the bar is vertical. If the bar is released from rest in the position shown, what is its angular velocity as a function of the angle θ between the bar's axis and the vertical?

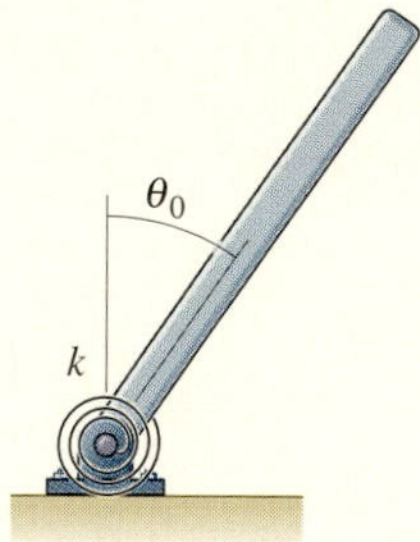

P8.34

8.35 The unstretched length of the spring is 1.5 m, and its constant is $k = 50$ N/m. When the 15-kg slender bar is horizontal, its angular velocity is 0.1 rad/s. What is its angular velocity when it is in the position shown?

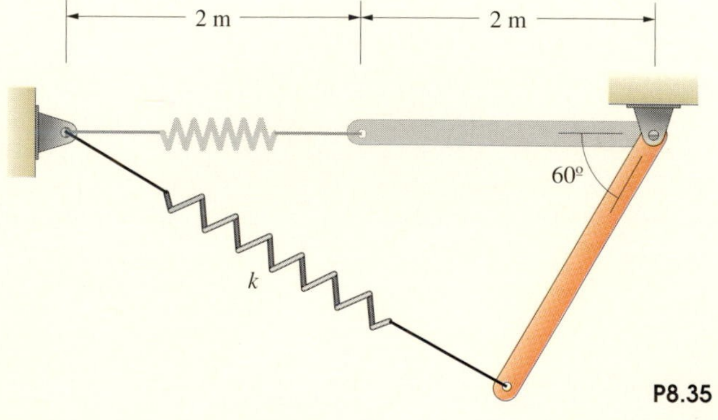

P8.35

8.36 Pulley A weighs 4 lb, $I_A = 0.060$ slug-ft^2, and $I_B = 0.014$ slug-ft^2. If the system is released from rest, what is the velocity of the 16-lb weight when it has fallen 2 ft?

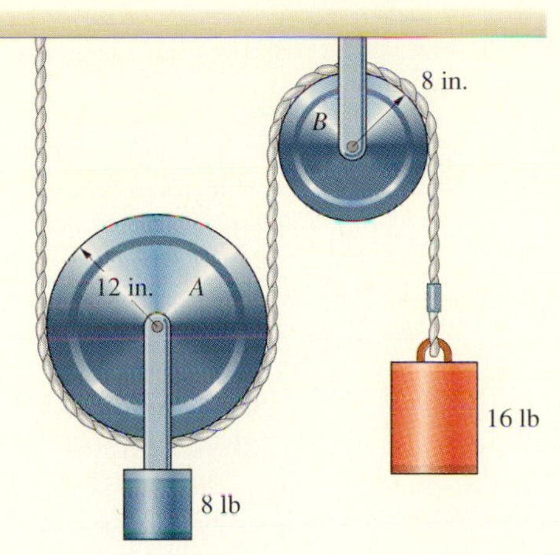

P8.36

8.37 The 18-kg ladder is released from rest with $\theta = 10°$. The wall and floor are smooth. Modeling the ladder as a slender bar, use conservation of energy to determine its angular velocity when $\theta = 40°$.

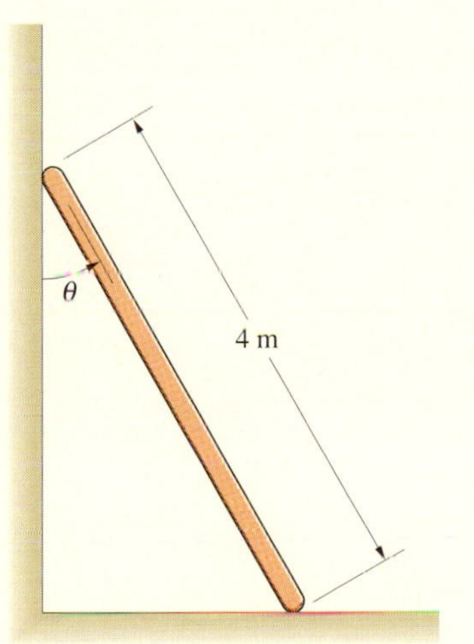

P8.37

8.38 The spring attached to the slender bar of mass m is unstretched when $\theta = 0$. If the bar falls from rest in the vertical position, what is its angular velocity as a function of θ? Friction is negligible.

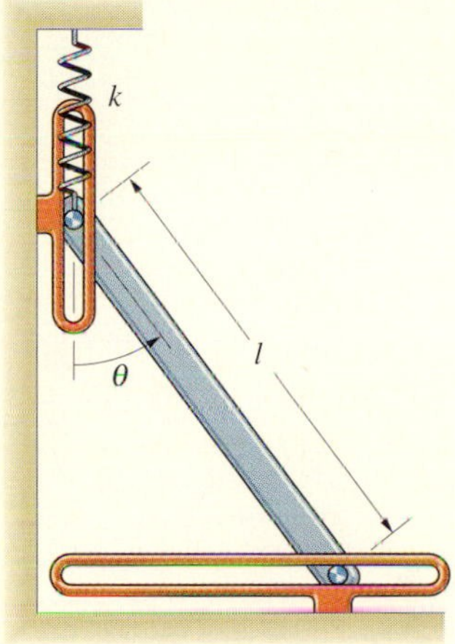

P8.38

8.39 Consider the system shown in Problem 8.38. The mass and length of the bar are $m = 4$ kg and $l = 1.2$ m. The spring constant is $k = 180$ N/m. If the bar is released from rest in the position $\theta = 10°$, what is its angular velocity when it has fallen to $\theta = 20°$?

8.40 The 4-kg slender bar is pinned to a 2-kg slider at A and to a 4-kg homogeneous cylindrical disk at B. Neglect the friction force on the slider and assume that the disk rolls. If the system is released from rest with $\theta = 60°$, what is the bar's angular velocity when $\theta = 0$?

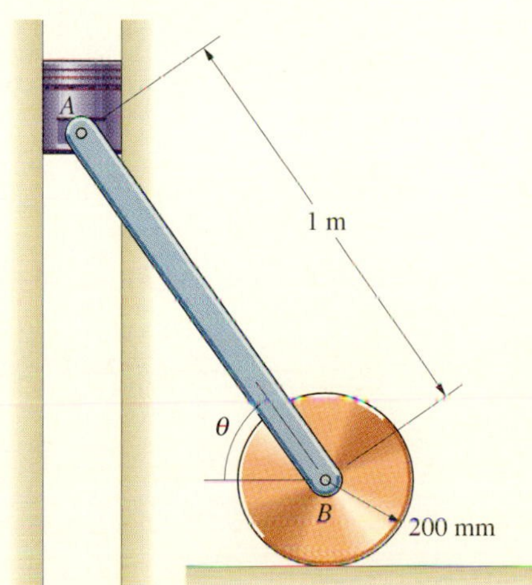

P8.40

8.41 If the system in Problem 8.40 is released from rest with $\theta = 80°$, what is the bar's angular velocity when $\theta = 20°$?

8.42 The system is in equilibrium in the position shown. The mass of the slender bar ABC is 6 kg, the mass of the slender bar BD is 3 kg, and the mass of the slider at C is 1 kg. The spring constant is $k = 200$ N/m. If a constant 100-N downward force is applied at A, what is the angular velocity of bar ABC when it has rotated 20° from its initial position?

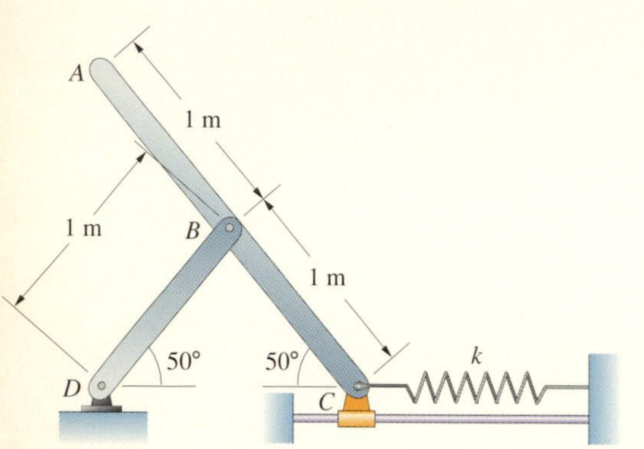

P8.42

8.43 Bar AB weighs 10 lb and bar BC weighs 6 lb. If the system is released from rest in the position shown, what are the angular velocities of the bars at the instant just before the joint B hits the smooth floor?

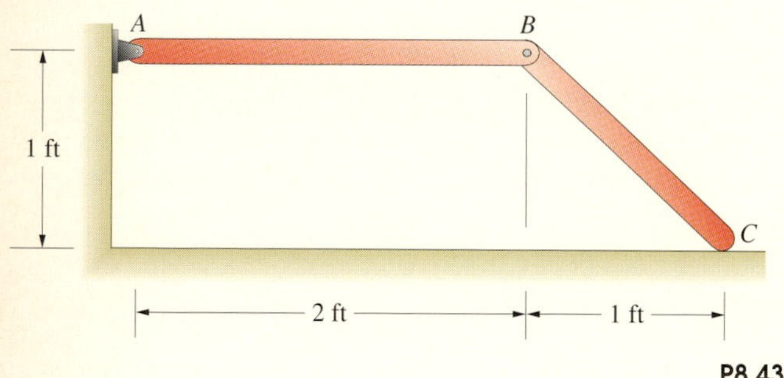

P8.43

8.44 If bar AB in Problem 8.43 is rotating at 1 rad/s in the clockwise direction at the instant shown, what are the angular velocities of the bars at the instant just before the joint B hits the smooth floor?

8.45 Each bar is of mass m and length l. The spring is unstretched when $\theta = 0$. If the system is released from rest with the bars vertical, determine the angular velocity of the lower bar as a function of θ.

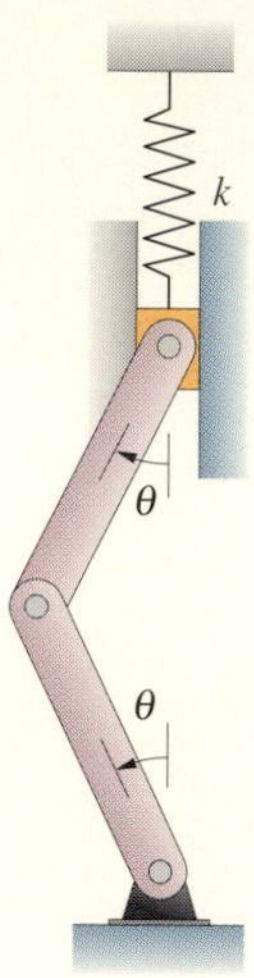

P8.45

8.46 The system starts from rest with the crank AB vertical. A constant couple M exerted on the crank causes it to rotate in the clockwise direction, compressing the gas in the cylinder. Let s be the displacement (in meters) of the piston to the right relative to its initial position. The net force toward the left exerted on the piston by atmospheric pressure and the gas in the cylinder is $350/(1 - 10s)$ N. The moment of inertia of the crank about A is 0.0003 kg-m^2. The mass of the connecting rod BC is 0.36 kg, and its center of mass is at its midpoint. The connecting rod's moment of inertia about its center of mass is 0.0004 kg-m^2. The mass of the piston is 4.6 kg. If the clockwise angular velocity of the crank AB is 200 rad/s when it has rotated 90° from its initial position, what is M? (Neglect the work done by the weights of the crank and connecting rod.)

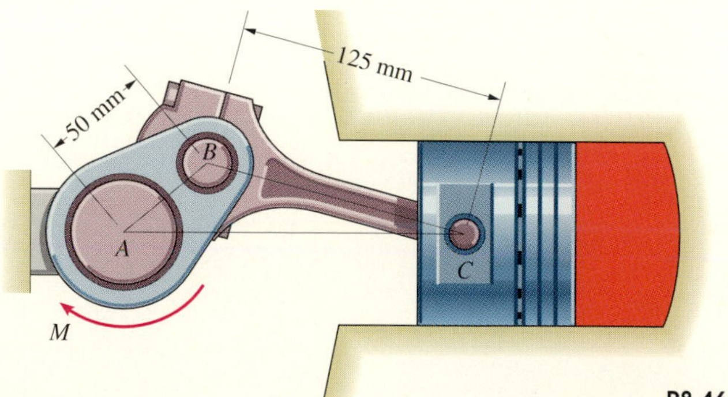

P8.46

8.47 In Problem 8.46, if the system starts from rest with the crank AB vertical and the couple $M = 40$ N-m, what is the clockwise angular velocity of the crank AB when it has rotated 45° from its initial position?

8.4 Principles of Impulse and Momentum

In this section we review our discussion of the principle of linear impulse and momentum from Chapter 5 and then derive the principle of angular impulse and momentum for a rigid body. These principles relate time integrals of the forces and couples acting on a rigid body to changes in the velocity of its center of mass and its angular velocity. You can use these principles to determine the effects of impulsive forces and couples on the motion of a rigid body. They also allow you to determine both the velocities of the centers of mass and the angular velocities of objects after they undergo collisions.

Linear Momentum

Integrating Newton's second law with respect to time yields the principle of linear impulse and momentum for a rigid body:

$$\int_{t_1}^{t_2} \Sigma \mathbf{F}\, dt = m\mathbf{v}_2 - m\mathbf{v}_1, \tag{8.25}$$

where $\mathbf{v}_1$ and $\mathbf{v}_2$ are the velocities of the center of mass at the times t_1 and t_2 (Fig. 8.13). If you know the external forces acting on a rigid body as functions of time, this principle allows you to determine the change in the velocity of its center of mass during an interval of time. In terms of the average with respect to time of the total force from t_1 to t_2,

$$\Sigma \mathbf{F}_{\mathrm{av}} = \frac{1}{t_2 - t_1}\int_{t_1}^{t_2} \Sigma\ \mathbf{F}\, dt,$$

we can write Eq. (8.25) as

$$(t_2 - t_1)\Sigma \mathbf{F}_{\mathrm{av}} = m\mathbf{v}_2 - m\mathbf{v}_1. \tag{8.26}$$

This form of the principle of linear impulse and momentum is often useful when an object is subjected to impulsive forces.

If the only forces acting on two rigid bodies A and B are the forces they exert on each other, or if other forces are negligible, their total linear momentum is conserved:

$$m_A\mathbf{v}_A + m_B\mathbf{v}_B = \text{constant}. \tag{8.27}$$

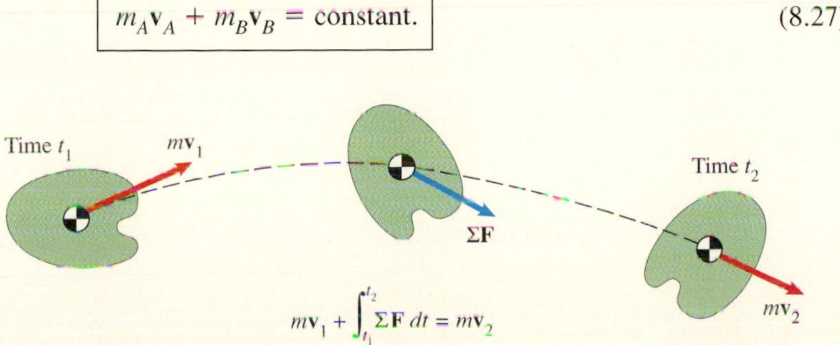

Fig. 8.13
Principle of linear impulse and momentum.

Angular Momentum

When you apply momentum principles to rigid bodies, you will often be interested in determining both the velocities of their centers of mass and their angular velocities. For this task, linear momentum principles alone are not sufficient. In this section we derive angular momentum principles for a rigid body in planar motion.

Principles of Angular Impulse and Momentum The total moment about the center of mass of a rigid body in planar motion equals the product of the moment of inertia about the center of mass and the angular acceleration:

$$\Sigma M = I\alpha.$$

We can write this equation in the form

$$\Sigma M = \frac{dH}{dt}, \tag{8.28}$$

where

$$\boxed{H = I\omega} \tag{8.29}$$

is the rigid body's angular momentum about its center of mass. Integrating Eq. (8.28) with respect to time, we obtain one form of the principle of angular impulse and momentum:

$$\boxed{\int_{t_1}^{t_2} \Sigma M \, dt = H_2 - H_1,} \tag{8.30}$$

where H_1 and H_2 are the values of the angular momentum at the times t_1 and t_2. The angular impulse about the center of mass during the interval of time from t_1 to t_2 is equal to the change in the rigid body's angular momentum about its center of mass. If the total moment about the center of mass is known as a function of time, you can use this equation to determine the change in the angular velocity from t_1 to t_2.

We can derive another useful form of this principle. Let $\mathbf{r}$ be the position vector of the center of mass of the rigid body relative to a fixed point O (Fig. 8.14). In Chapter 7, we derived a relationship between the total moment about O due to external forces and couples and the rate of change of the rigid body's angular momentum about O (Eq. 7.20):

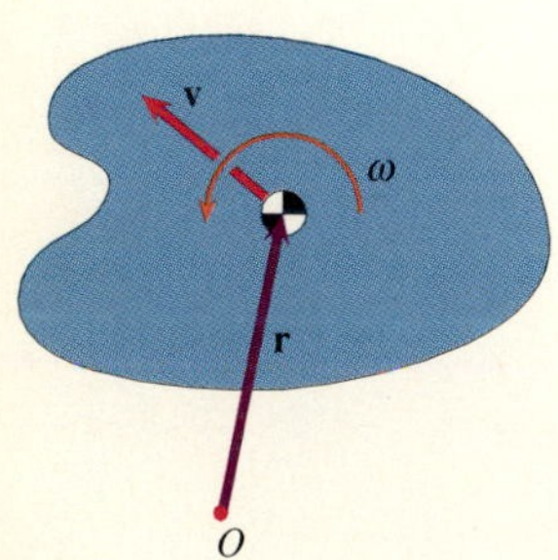

Fig. 8.14
A rigid body in planar motion with velocity **v** and angular velocity ω.

$$\Sigma M_O = \frac{dH_O}{dt}, \tag{8.31}$$

where

$$\boxed{H_O = (\mathbf{r} \times m\mathbf{v}) \cdot \mathbf{k} + I\omega.} \tag{8.32}$$

Integrating Eq. (8.31) with respect to time, we obtain a second form of the principle of angular impulse and momentum:

$$\int_{t_1}^{t_2} \Sigma M_O \, dt = H_{O2} - H_{O1}. \tag{8.33}$$

The angular impulse about a fixed point O during the interval of time from t_1 to t_2 is equal to the change in the rigid body's angular momentum about O (Fig. 8.15).

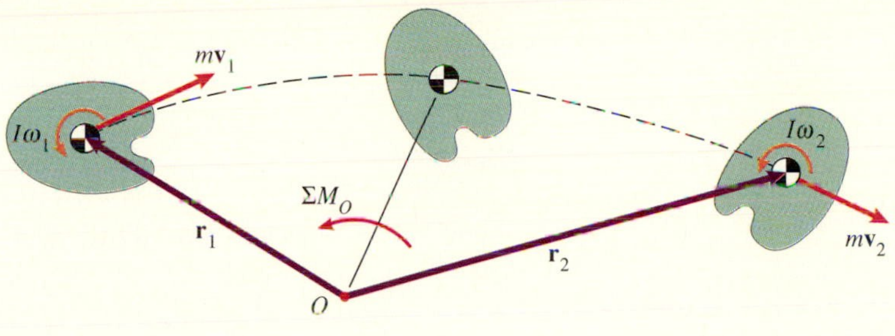

Fig. 8.15
The impulse about O equals the change in the angular momentum about O.

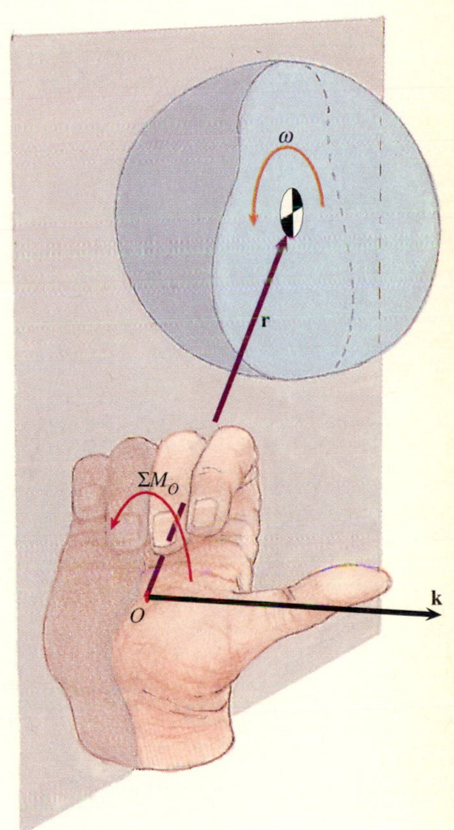

Fig. 8.16
The direction of **k**.

The term $(\mathbf{r} \times m\mathbf{v}) \cdot \mathbf{k}$ in Eq. (8.32) is the rigid body's angular momentum about O due to the velocity of its center of mass. This term has the same form as the moment of a force, with the linear momentum $m\mathbf{v}$ in place of the force. If we define ΣM_O and ω to be positive in the counterclockwise direction, the unit vector **k** points out of the page (Fig. 8.16) and $(\mathbf{r} \times m\mathbf{v}) \cdot \mathbf{k}$ is the counterclockwise "moment" of the linear momentum. You can use the vector expression to calculate it, but it is often easier to use the fact that its magnitude is the product of the magnitude of the linear momentum and the perpendicular distance from point O to the line of action of the velocity. The "moment" is positive if it is counterclockwise (Fig. 8.17a) and negative if it is clockwise (Fig. 8.17b).

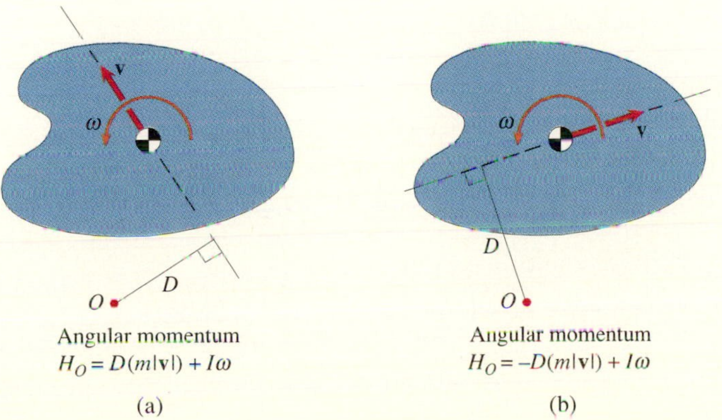

Fig. 8.17
Determining the angular momentum about O by calculating the "moment" of the linear momentum.

Impulsive Forces and Couples The average with respect to time of the moment about the center of mass from t_1 to t_2 is

$$\Sigma M_{\mathrm{av}} = \frac{1}{t_2 - t_1} \int_{t_1}^{t_2} \Sigma M \, dt.$$

Using this equation, we can write Eq. (8.30) as

$$\boxed{(t_2 - t_1)\Sigma M_{\mathrm{av}} = H_2 - H_1.} \tag{8.34}$$

In the same way, we can express Eq. (8.33) in terms of the average moment about point O:

$$\boxed{(t_2 - t_1)(\Sigma M_O)_{\mathrm{av}} = H_{O2} - H_{O1}.} \tag{8.35}$$

When the average value of the moment and its duration are known, you can use Eq. (8.34) or (8.35) to determine the change in the angular momentum. These equations are often useful when a rigid body is subjected to impulsive forces and couples.

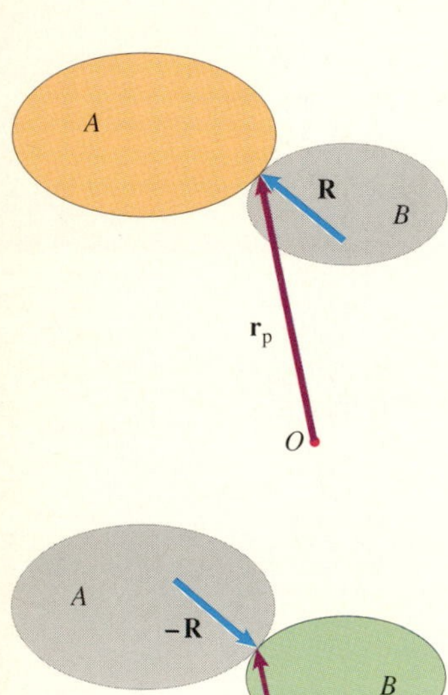

Fig. 8.18
Rigid bodies A and B exerting forces on each other by contact.

Conservation of Angular Momentum We can use Eq. (8.33) to obtain an equation of conservation of total angular momentum for two rigid bodies. Let A and B be rigid bodies in two-dimensional motion in the same plane, and suppose that they are subjected only to the forces and couples they exert on each other, or that other forces and couples are negligible. Let M_{OA} be the moment about a fixed point O due to the forces and couples acting on A, and let M_{OB} be the moment about O due to the forces and couples acting on B. Under the same assumption we made in deriving the equations of motion—the forces between each pair of particles are directed along the line between the particles—the moment $M_{OB} = -M_{OA}$. For example, in Fig. 8.18, A and B exert forces on each other by contact. The resulting moments about O are $M_{OA} = (\mathbf{r}_\mathrm{p} \times \mathbf{R}) \cdot \mathbf{k}$ and $M_{OB} = [\mathbf{r}_\mathrm{p} \times (-\mathbf{R})] \cdot \mathbf{k} = -M_{OA}$.

We apply Eq. (8.33) to A and B for arbitrary times t_1 and t_2, obtaining

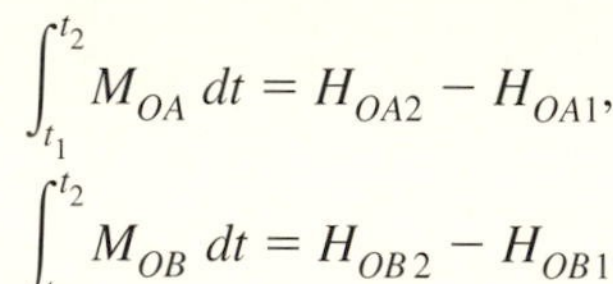

$$\int_{t_1}^{t_2} M_{OA} \, dt = H_{OA2} - H_{OA1},$$

$$\int_{t_1}^{t_2} M_{OB} \, dt = H_{OB2} - H_{OB1}.$$

Summing these equations, the terms on the left cancel and we obtain

$$H_{OA1} + H_{OB1} = H_{OA2} + H_{OB2}.$$

We see that the total angular momentum of A and B about O is conserved:

$$\boxed{H_{OA} + H_{OB} = \text{constant.}} \tag{8.36}$$

Notice that this result holds even when A and B are subjected to significant external forces and couples *if the total moment about O due to the external forces and couples is zero*. You can sometimes choose the point O so that this condition is satisfied. This result also applies to an arbitrary number of rigid bodies: Their total angular momentum about O is conserved if the total moment about O due to external forces and couples is zero.

In the following examples we demonstrate the use of the principles of linear and angular impulse and momentum to analyze motions of rigid bodies.

Example 8.4

Disk A in Fig. 8.19 initially has a counterclockwise angular velocity ω_0, and disk B is stationary. At $t = 0$, the disks are moved into contact. As a result of friction at the point of contact, the angular velocity of A decreases and the angular velocity of B increases until there is no slip between them. What are their final angular velocities ω_A and ω_B? The disks are supported at their centers of mass and their moments of inertia are I_A, I_B.

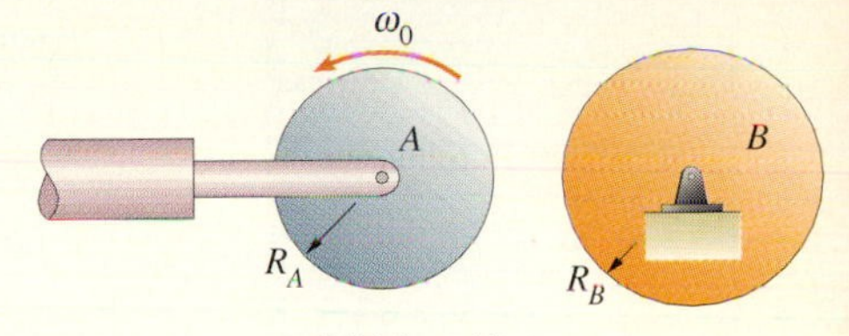

Initial position

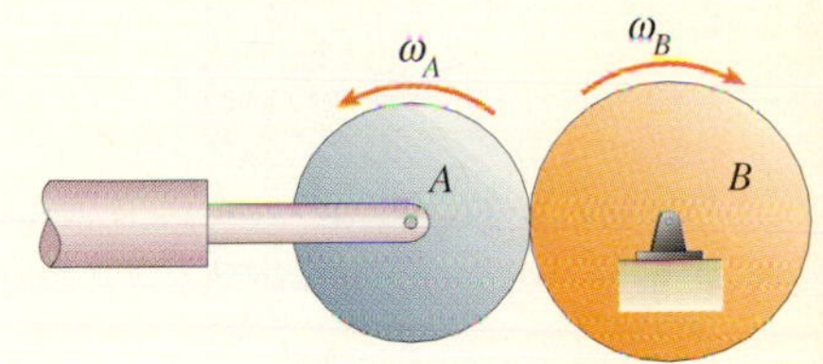

Time $t = 0$

Fig. 8.19

STRATEGY

Since the disks rotate about fixed axes through their centers of mass while they are in contact, we can apply the principle of angular impulse and momentum in the form given by Eq. (8.30) to each disk. When there is no longer any slip between the disks, their velocities are equal at their point of contact. With this kinematic relationship and the equations we obtain with the principle of angular impulse and momentum, we can determine the final angular velocities.

SOLUTION

We draw the free-body diagrams of the disks while slip occurs in Fig. (a), showing the normal and friction forces they exert on each other. Let t_f be the time at which slip ceases. We apply Eq. (8.30) to disk A for the interval of time from $t = 0$ to $t = t_f$:

$$\int_{t_1}^{t_2} \Sigma M\, dt = H_2 - H_1 = I\omega_2 - I\omega_1:$$

$$\int_0^{t_f} -R_A f\, dt = I_A\omega_A - I_A\omega_0.$$

We also apply Eq. (8.30) to disk B:

$$\int_{t_1}^{t_2} \Sigma M\, dt = H_2 - H_1 = I\omega_2 - I\omega_1:$$

$$\int_0^{t_f} -R_B f\, dt = -I_B\omega_B - 0.$$

(Notice that because ω_B is clockwise, $\omega_2 = -\omega_B$.) We divide the first equation by the second one and write the resulting equation as

$$\omega_A + \frac{R_A I_B}{R_B I_A}\omega_B = \omega_0.$$

When there is no slip, the velocities of the disks are equal at their point of contact:

$$R_A\omega_A = R_B\omega_B.$$

Solving these two equations, we obtain

$$\omega_A = \omega_0\left[\frac{1}{1 + \dfrac{R_A^2 I_B}{R_B^2 I_A}}\right], \qquad \omega_B = \omega_0\left[\frac{R_A/R_B}{1 + \dfrac{R_A^2 I_B}{R_B^2 I_A}}\right].$$

Notice that if the disks have the same radius and moment of inertia, $\omega_A = \frac{1}{2}\omega_0$ and $\omega_B = \frac{1}{2}\omega_0$.

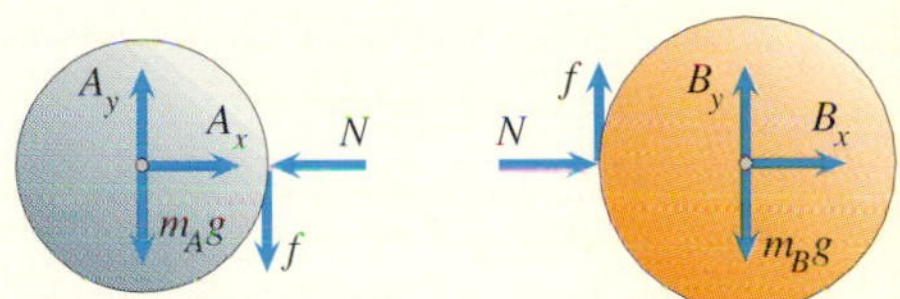

(a) Free-body diagrams of the disks.

Example 8.5

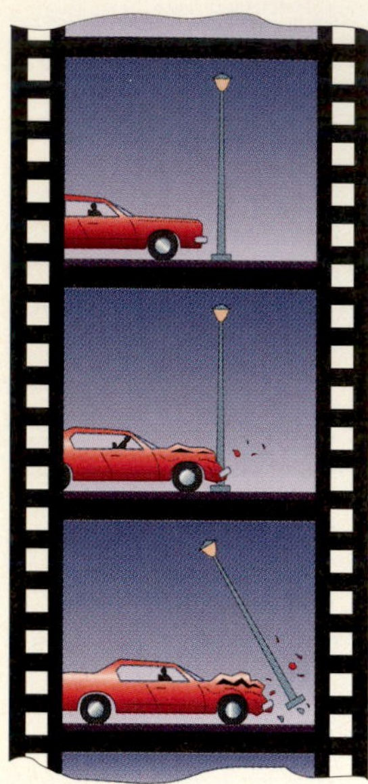

Fig. 8.20

Engineers design a streetlight to shear off at ground level when struck by a vehicle, to help prevent injuries to passengers (Fig. 8.20). From videotape of a test impact, they estimate the angular velocity of the pole to be 0.74 rad/s and the horizontal velocity of its centerpoint to be 22 ft/s after the impact, and they estimate the duration of the impact to be $\Delta t = 0.01$ s. If the pole can be modeled as a 20-ft, 140-lb slender bar, the car strikes it 2 ft above the ground, and the couple exerted on the pole by its support can be neglected, what average force was required to shear off the bolts supporting the pole?

STRATEGY

We will determine the average force by applying the principles of linear and angular impulse and momentum expressed in terms of the average forces and moments exerted on the pole. We can apply the principle of angular momentum by using either Eq. (8.34) or Eq. (8.35). We will use Eq. (8.35) to demonstrate its use.

SOLUTION

We draw the free-body diagram of the pole in Fig. (a), where F is the average force exerted by the car and S is the average shearing force exerted on the pole by the bolts. Let m be the mass of the pole, and let v and ω be the velocity of its center of mass and its angular velocity at the end of the impact (Fig. b). From Eq. (8.26), the principle of linear impulse and momentum expressed in terms of the average horizontal force is

$$(t_2 - t_1)(\Sigma F_x)_{\text{av}} = (mv_x)_2 - (mv_x)_1:$$
$$\Delta t(F - S) = mv - 0. \tag{8.37}$$

To apply the principle of angular impulse and momentum, we use Eq. (8.35), placing the fixed point O at the bottom of the pole (Figs. a and b). The pole's angular momentum about O at the end of the impact is

$$H_{O2} = [(\mathbf{r} \times m\mathbf{v}) \cdot \mathbf{k} + I\omega]_2 = [(10\mathbf{j}) \times m(v\mathbf{i})] \cdot \mathbf{k} + I\omega = -10mv + I\omega.$$

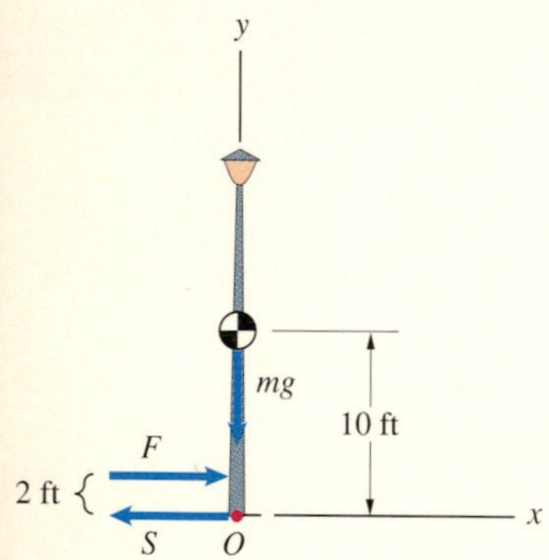

(a) Free-body diagram of the pole.

(We can also obtain this result by calculating the "moment" of the linear momentum about O and adding the term $I\omega$. The magnitude of the "moment" is the product of the magnitude of the linear momentum (mv) and the perpendicular distance from O to the line of action of the linear momentum (10 ft), and it is negative because the "moment" is clockwise. See Fig. 8.17. From Eq. (8.35), we obtain

$$(t_2 - t_1)(\Sigma M_0)_{\text{av}} = H_{O2} - H_{O1}:$$
$$\Delta t(-2F) = -10mv + I\omega - 0.$$

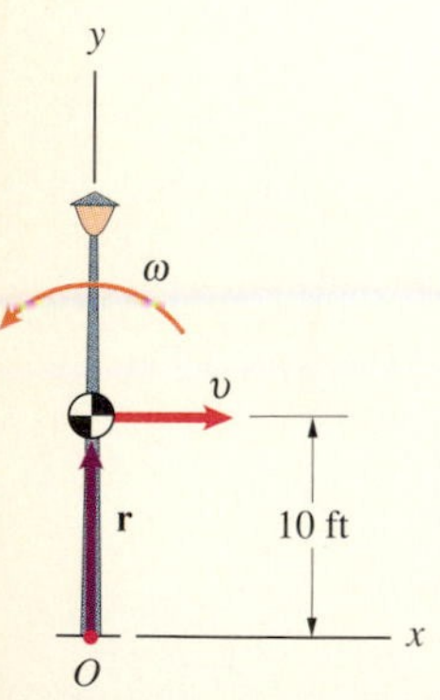

(b) Velocity and angular velocity at the end of the impact.

Solving this equation together with Eq. (8.37) for the average shear force S, we obtain

$$S = \frac{8mv - I\omega}{2\,\Delta t} = \frac{8(140/32.2)(22) - \frac{1}{12}(140/32.2)(20)^2(0.74)}{2(0.01)}$$
$$= 32{,}900 \text{ lb}.$$

Example 8.6

In a well-known demonstration of conservation of angular momentum, a person stands on a rotating platform holding a mass m in each hand (Fig. 8.21). The moment of inertia of the person and platform is $I_P = 0.4$ kg-m^2, the mass $m = 4$ kg, and the moment of inertia of each mass about the vertical axis through its center of mass is $I_M = 0.001$ kg-m^2. If the person's angular velocity with her arms extended to $r_1 = 0.6$ m is $\omega_1 = 1$ revolution per second, what is her angular velocity ω_2 when she pulls the masses inward to $r_2 = 0.2$ m? Neglect the change in her moment of inertia due to the change in the position of her arms. (You have observed skaters using this phenomenon to control their angular velocity in a spin by altering the positions of their arms.)

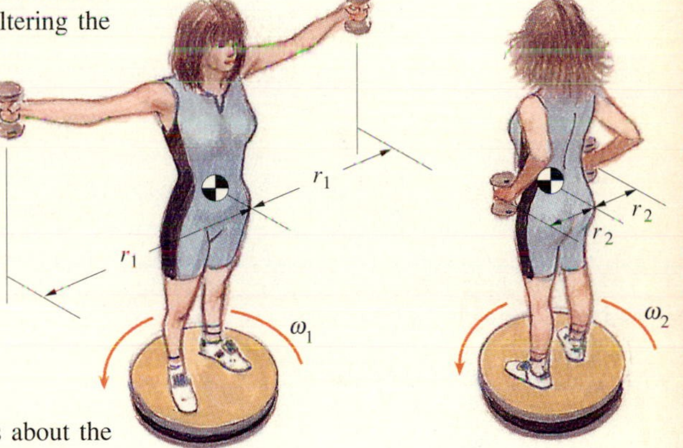

Fig. 8.21

STRATEGY

If we neglect friction in the rotating platform, the total angular momentum of the person, platform, and masses about the vertical axis of rotation is conserved. We can use this condition to determine ω_2.

SOLUTION

We begin by determining the angular momentum of one of the masses about the axis of rotation when her arms are extended. The magnitude of its velocity about the axis of rotation is $r_1\omega_1$, so the "moment" of its linear momentum about the axis of rotation is $r_1(mr_1\omega_1) = mr_1^2\omega_1$. The total angular momentum of the mass about the axis of rotation when her arms are extended is

$$(\mathbf{r} \times m\mathbf{v}) \cdot \mathbf{k} + I\omega = mr_1^2\omega_1 + I_M\omega_1.$$

Since the center of mass of the person and platform lie on the axis of rotation, their angular momentum about the axis of rotation is $I_P\omega_1$. Therefore the combined angular momentum of the person, platform, and masses when her arms are extended is

$$H_{O1} = I_P\omega_1 + 2(mr_1^2\omega_1 + I_M\omega_1) = [I_P + 2(I_M + mr_1^2)]\omega_1.$$

By replacing ω_1 by ω_2 and r_1 by r_2 in this expression, we obtain the combined angular momentum when she has pulled the masses inward. Angular momentum is conserved, so

$$H_{O1} = H_{O2}:$$

$$[I_P + 2(I_M + mr_1^2)]\omega_1 = [I_P + 2(I_M + mr_2^2)]\omega_2,$$

$$\{0.4 + 2[0.001 + 4(0.6)^2]\}\omega_1 = \{0.4 + 2[0.001 + 4(0.2)^2]\}\omega_2.$$

Solving, we find that $\omega_2 = 4.55\omega_1 = 4.55$ revolutions per second.

DISCUSSION

If the person holds the masses at a distance r from the axis of rotation, the parallel-axis theorem states that the moment of inertia of each mass *about the axis of rotation* is $I_M + mr^2$. The term $I_P + 2(I_M + mr^2)$ that appears in our equation of conservation of angular momentum is simply the total moment of inertia of the person and the two masses about the axis of rotation.

Problems

8.48 The 8-kg slender bar is released from rest in the horizontal position. When it has fallen to the position shown, its angular velocity is 3.226 rad/s. Determine the bar's angular momentum (a) about its center of mass; (b) about point A.

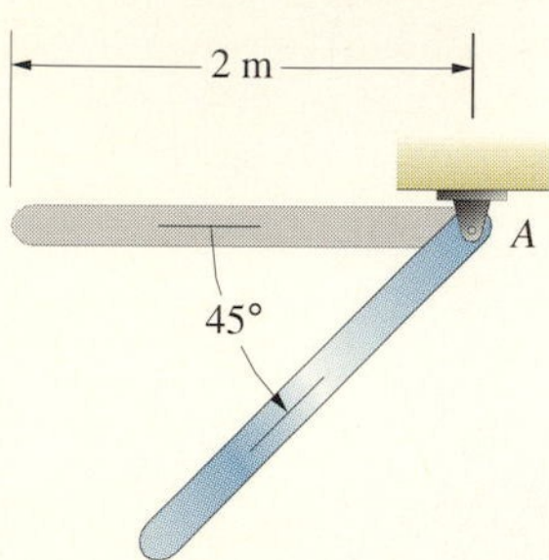

P8.48

8.49 The moment of inertia of the pulley is 0.4 slug-ft^2. The pulley starts from rest at $t = 0$. For both cases, use momentum principles to determine the pulley's angular velocity at $t = 1$ s.

Strategy: In case (a), apply the principle of angular impulse and momentum to the pulley. In case (b), draw separate free-body diagrams and apply the principle of angular impulse and momentum to the pulley and the principle of linear impulse and momentum to the weight.

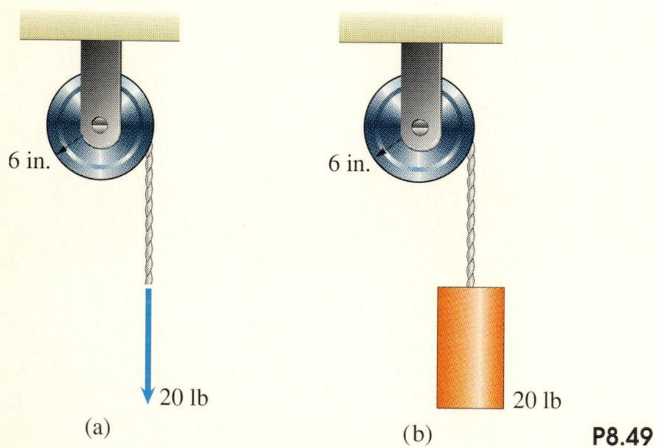

P8.49

8.50 An astronaut fires a thruster of his maneuvering unit, exerting a force $T = 2(1 + t)$ N, where t is in seconds. The combined mass of the astronaut and his equipment is 122 kg, and the moment of inertia about their center of mass is 45 kg-m^2. Modeling the astronaut and his equipment as a rigid body, use the principle of angular impulse and momentum to determine how long it takes for his angular velocity to reach 0.1 rad/s.

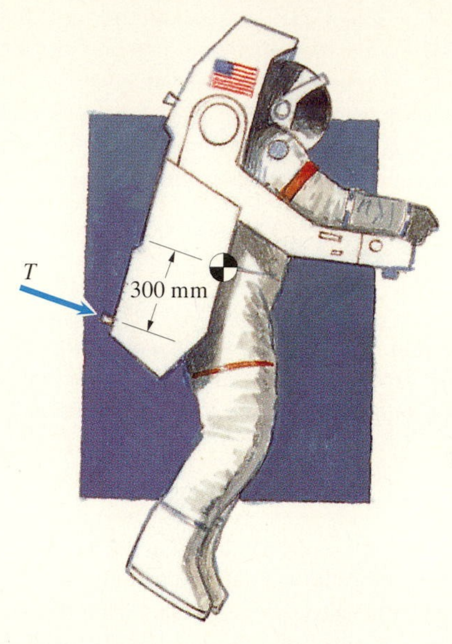

P8.50

8.51 The maneuvering unit in Problem 8.50 exerts an impulsive force T of 0.2-s duration, giving the astronaut a counterclockwise angular velocity of 1 rpm.

(a) What is the average value of the impulsive force?

(b) What is the magnitude of the change in the velocity of his center of mass?

8.52 A flywheel attached to an electric motor is initially at rest. At $t = 0$, the motor exerts a couple $M = 200e^{-0.1t}$ N-m on the flywheel. The moment of inertia of the flywheel is 10 kg-m^2.

(a) What is the flywheel's angular velocity at $t = 10$ s?

(b) What maximum angular velocity will the flywheel attain?

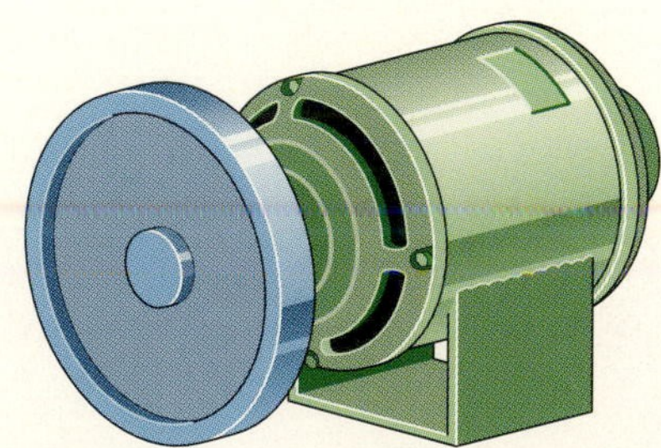

P8.52

8.53 A main landing gear wheel of a Boeing 747 has a moment of inertia of 17 slug-ft^2 and a 2-ft radius. The airplane is moving at 245 ft/s when it touches down. Suppose that you measure the skid marks where the plane touches down and find that they are 30 ft long. Assuming that the airplane's velocity and the normal force on the wheel are constant while the wheel skids, use the principle of angular impulse and momentum to estimate the friction force exerted on the wheel while it skids.

8.54 The force a club exerts on a 1.62-oz golf ball is shown. The ball is 1.68 in. in diameter and can be modeled as a homogeneous sphere. The club is in contact with the ball for 0.0006 s, and the magnitude of the velocity of the ball's center of mass after it is struck is 160 ft/s. What is the ball's angular velocity after it is struck?

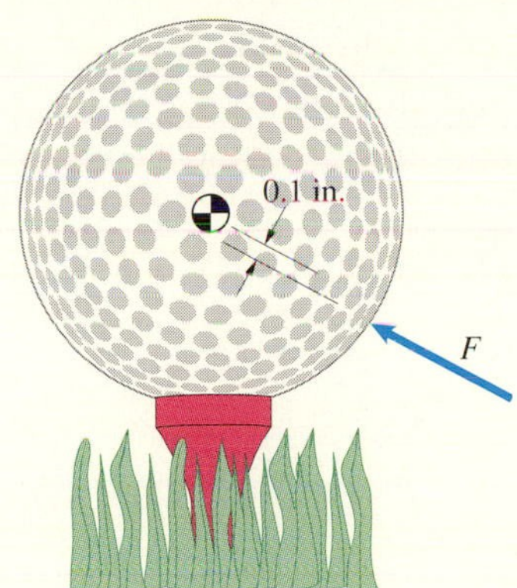

P8.54

8.55 The suspended 8-kg slender bar is subjected to a horizontal impulsive force at B. The average value of the force is 1000 N, and its duration is 0.03 s. If the force causes the bar to swing to the horizontal position before coming to a stop, what is the distance h?

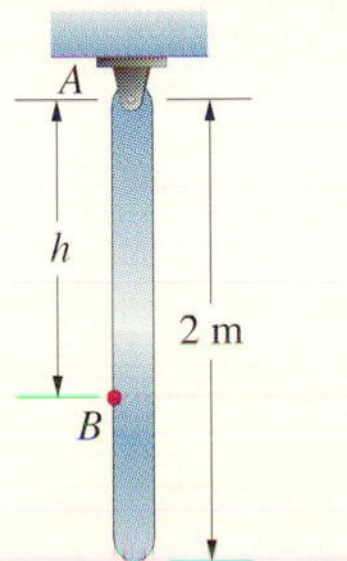

P8.55

8.56 For what value of the distance h in Problem 8.55 will no average horizontal force be exerted on the bar by the support A when the horizontal impulsive force is applied at B? What is the angular velocity of the bar just after the impulsive force is applied?

8.57 The force exerted on the cue ball by the cue is horizontal. Determine the value of h for which the ball rolls without slipping. (Assume that the friction force exerted on the ball by the table is negligible.)

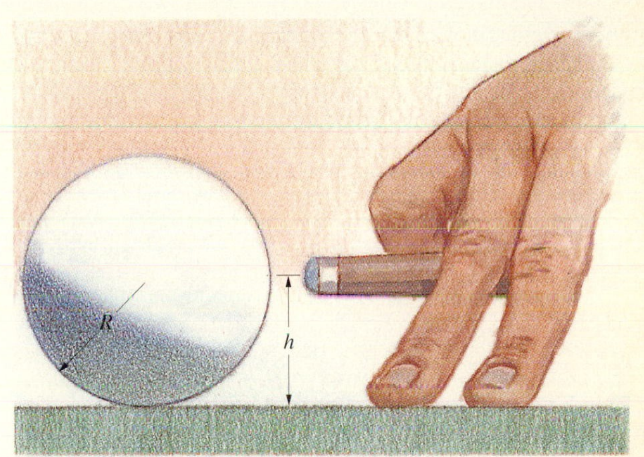

P8.57

8.58 Two gravity research satellites ($m_A = 250$ kg, $I_A = 350$ kg-m^2; $m_B = 50$ kg, $I_B = 16$ kg-m^2) are tethered by a cable. The satellites and cable rotate with angular velocity $\omega_0 = 0.25$ rpm. Ground controllers order satellite A to slowly unreel 6 m of additional cable. What is the angular velocity afterward?

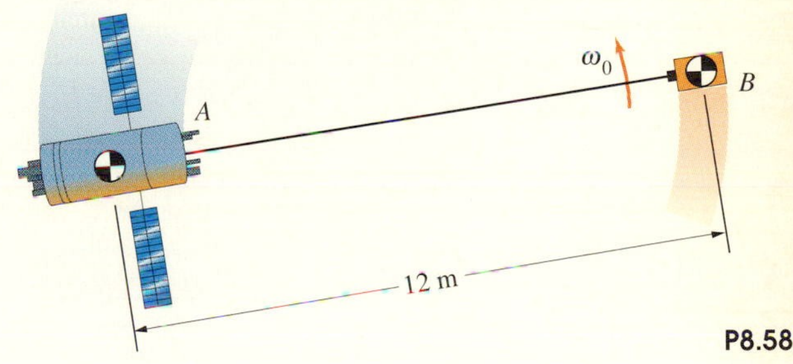

P8.58

8.59 Solve Problem 8.58 by treating the satellites as particles (that is, neglect their moments of inertia I_A and I_B) and compare your answer to that of Problem 8.58.

8.60 The 2-kg bar is 1 m in length. It rotates *in the horizontal plane* about the smooth pin. The 6-kg collar A slides on the smooth bar. The unstretched length of the spring is 0.2 m, and its spring constant is $k = 10$ N/m. At the instant shown, the angular velocity of the bar is $\omega_0 = 2$ rad/s, the distance from the pin to the collar is $r = 0.6$ m, and the radial velocity of the collar is zero. Use conservation of angular momentum to determine the bar's angular velocity when the distance from the pin to the collar is $r = 0.8$ m. Neglect the moment of inertia of the collar about its center of mass; that is, treat the collar as a particle.

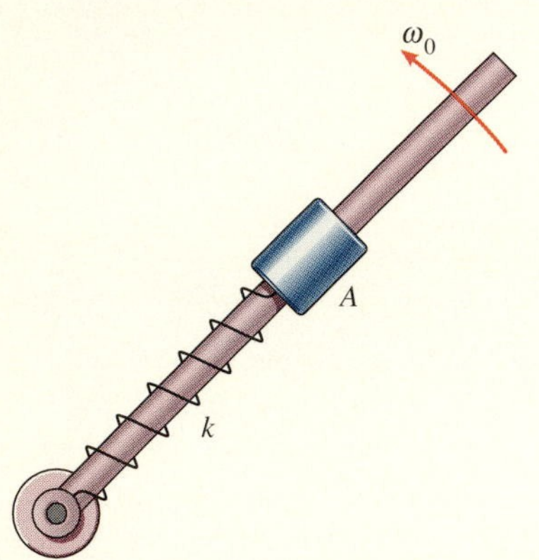

P8.60

8.61 In Problem 8.60, what is the collar's radial velocity when the distance from the pin to the collar is $r = 0.8$ m?

8.62 In Problem 8.60, suppose that the moment of inertia of the collar about its center of mass is 0.2 kg-m^2. Determine the bar's angular velocity when the distance from the pin to the collar is $r = 0.8$ m and compare your answer with that of Problem 8.60.

8.63 The circular bar is welded to the vertical shafts, which can rotate freely in bearings at A and B. Let I be the moment of inertia of the circular bar and shafts about the vertical axis. The circular bar has an initial angular velocity ω_0 and the mass m is released in the position shown with no velocity relative to the bar. Determine the angular velocity of the circular bar as a function of the angle β between the vertical and the position of the mass. Neglect the moment of inertia of the mass about its center of mass; that is, treat the mass as a particle.

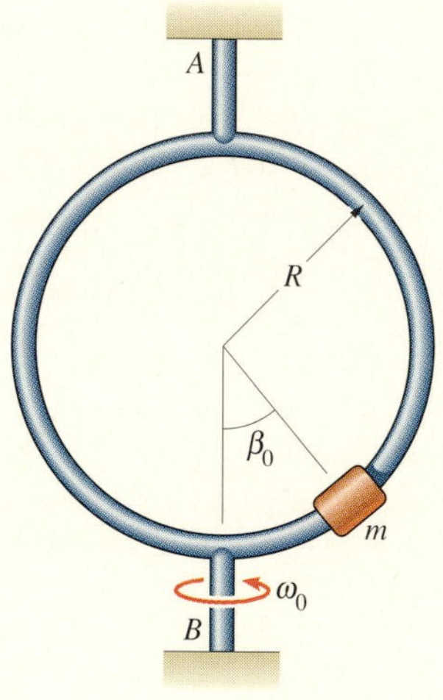

P8.63

8.5 *Impacts*

In Chapter 5, we analyzed impacts between objects with the objective of determining their velocities—the velocities of their centers of mass—after the collision. We now discuss how you can determine the velocities of the centers of mass *and the angular velocities* of rigid bodies after they collide.

Fig. 8.22
Rigid bodies A and B colliding. Because of the pin support, their total linear momentum is *not* conserved, but their total angular momentum about O is conserved.

Conservation of Momentum

Suppose that two rigid bodies A and B, in two-dimensional motion in the same plane, collide. What do the principles of linear and angular momentum tell us about their motions after the collision?

Linear Momentum If other forces are negligible in comparison to the impact forces A and B exert on each other, their total linear momentum is the same before and after the impact. But you must use care in applying this result. For example, if one of the rigid bodies has a pin support (Fig. 8.22), the reactions exerted by the support cannot be neglected and linear momentum is not conserved.

Angular Momentum If other forces and couples are negligible in comparison to the impact forces and couples A and B exert on each other, their total angular momentum about *any* fixed point O is the same before and after the impact. (See Eq. 8.36.) If, in addition, A and B exert only forces on each other at their point of impact P, and no couples, the angular momentum about P of *each* rigid body is the same before and after the impact (Fig. 8.23). This result follows from the principle of angular impulse and momentum, Eq. (8.33), because the impact forces on A and B exert no moment about P. If one of the rigid bodies has a pin support at a point O, as in Fig. 8.22, their total angular momentum about O is the same before and after the impact.

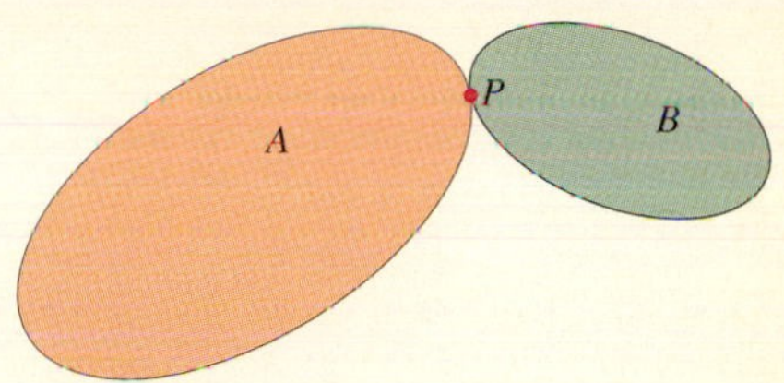

Fig. 8.23
Rigid bodies A and B colliding at P. If forces are exerted only at P, the angular momentum of A about P and the angular momentum of B about P are each conserved.

Coefficient of Restitution

If two rigid bodies adhere and move as a single rigid body after colliding, you can determine their velocities and angular velocity using momentum conservation and kinematic relationships alone. These relationships are not sufficient if the objects do not adhere. But you can analyze some impacts of the latter type by also using the concept of the coefficient of restitution.

Let P be the point of contact of rigid bodies A and B during an impact (Fig. 8.24), and let their velocities at P be $\mathbf{v}_{AP}$ and $\mathbf{v}_{BP}$ just before the impact and $\mathbf{v}'_{AP}$ and $\mathbf{v}'_{BP}$ just afterward. The x axis is perpendicular to the contacting surfaces at P. If the friction forces resulting from the impact are negligible, we can show that the components of the velocities normal to the surfaces at P are related to the coefficient of restitution e by

$$e = \frac{(\mathbf{v}'_{BP})_x - (\mathbf{v}'_{AP})_x}{(\mathbf{v}_{AP})_x - (\mathbf{v}_{BP})_x}.$$

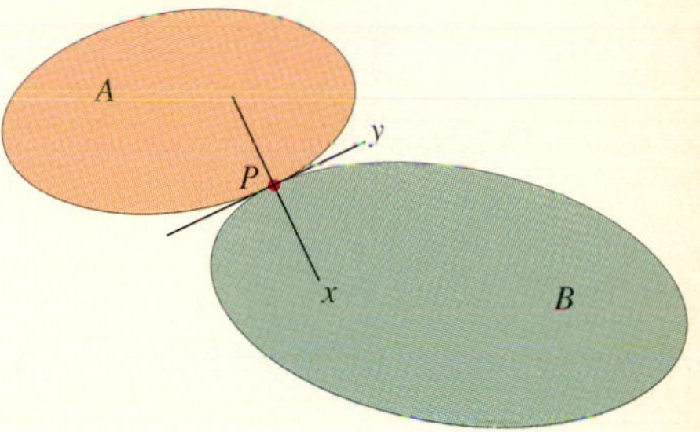

Fig. 8.24
Rigid bodies A and B colliding at P. The x axis is perpendicular to the contacting surfaces.

To derive this result, we must consider the effects of the impact on the individual objects. Let t_1 be the time at which they first come into contact. The objects are not actually rigid, but will deform as a result of the collision. At a time t_C, the maximum deformation will occur and the objects will begin a "recovery" phase in which they tend to resume their original shapes. Let t_2 be the time at which they separate.

Our first step is to apply the principle of linear impulse and momentum to A and B for the intervals of time from t_1 to t_C and from t_C to t_2. Let R be the magnitude of the normal force exerted during the impact (Fig. 8.25). We denote the velocity of the center of mass of A at the times t_1, t_C, and t_2 by $\mathbf{v}_A$, $\mathbf{v}_{AC}$, and $\mathbf{v}'_A$, and denote the corresponding velocities of the center of mass of B by $\mathbf{v}_B$, $\mathbf{v}_{BC}$, and $\mathbf{v}'_B$. For A, we have

$$\int_{t_1}^{t_C} -R\,dt = m_A(\mathbf{v}_{AC})_x - m_A(\mathbf{v}_A)_x, \tag{8.38}$$

$$\int_{t_C}^{t_2} -R\,dt = m_A(\mathbf{v}'_A)_x - m_A(\mathbf{v}_{AC})_x, \tag{8.39}$$

and for B,

$$\int_{t_1}^{t_C} R\,dt = m_B(\mathbf{v}_{BC})_x - m_B(\mathbf{v}_B)_x, \tag{8.40}$$

$$\int_{t_C}^{t_2} R\,dt = m_B(\mathbf{v}'_B)_x - m_B(\mathbf{v}_{BC})_x. \tag{8.41}$$

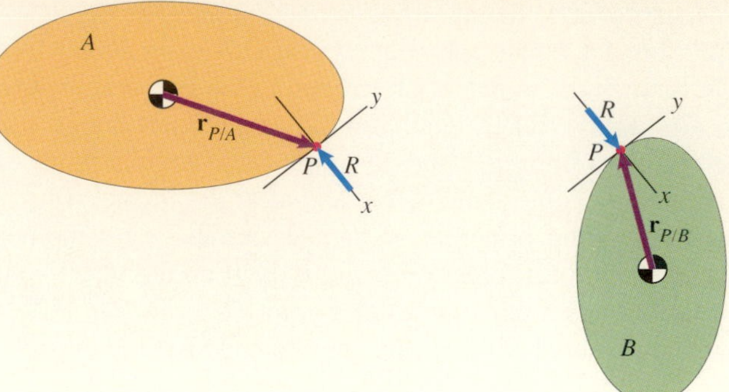

Fig. 8.25
The normal force R resulting from the impact.

The coefficient of restitution is the ratio of the linear impulse during the recovery phase to the linear impulse during the deformation phase:

$$e = \frac{\displaystyle\int_{t_C}^{t_2} R\,dt}{\displaystyle\int_{t_1}^{t_C} R\,dt}.$$

If we divide Eq. (8.39) by Eq. (8.38) and divide Eq. (8.41) by Eq. (8.40), we can write the resulting equations as

$$\begin{aligned}(\mathbf{v}'_A)_x &= -(\mathbf{v}_A)_x e + (\mathbf{v}_{AC})_x(1 + e),\\ (\mathbf{v}'_B)_x &= -(\mathbf{v}_B)_x e + (\mathbf{v}_{BC})_x(1 + e).\end{aligned} \tag{8.42}$$

We now apply the principle of angular impulse and momentum to A and B for the intervals of time from t_1 to t_C and from t_C to t_2. We denote the counterclockwise angular velocity of A at the times t_1, t_C, and t_2 by ω_A, ω_{AC}, and ω'_A, and denote the corresponding angular velocities of B by ω_B, ω_{BC}, and ω'_B. We write the position vectors of P relative to the centers of mass of A and B as (Fig. 8.25)

$$\begin{aligned}\mathbf{r}_{P/A} &= x_A\mathbf{i} + y_A\mathbf{j},\\ \mathbf{r}_{P/B} &= x_B\mathbf{i} + y_B\mathbf{j}.\end{aligned}$$

The moment about the center of mass of A due to the force exerted on A by the impact is $\mathbf{r}_{P/A} \times (-R\mathbf{i}) = y_A R\mathbf{k}$. From Eq. (8.30), we obtain

$$\int_{t_1}^{t_C} y_A R\,dt = I_A\omega_{AC} - I_A\omega_A, \tag{8.43}$$

$$\int_{t_C}^{t_2} y_A R\,dt = I_A\omega'_A - I_A\omega_{AC}. \tag{8.44}$$

The corresponding equations for B are

$$\int_{t_1}^{t_C} -y_B R\,dt = I_B\omega_{BC} - I_B\omega_B, \tag{8.45}$$

$$\int_{t_C}^{t_2} -y_B R\,dt = I_B\omega'_B - I_B\omega_{BC}. \tag{8.46}$$

Dividing Eq. (8.44) by Eq. (8.43) and dividing Eq. (8.46) by Eq. (8.45), we can write the resulting equations as

$$\begin{aligned} \omega'_A &= -\omega_A e + \omega_{AC}(1 + e), \\ \omega'_B &= -\omega_B e + \omega_{BC}(1 + e). \end{aligned} \tag{8.47}$$

By expressing the velocity of the point of A at P in terms of the velocity of the center of mass of A and the angular velocity of A, and expressing the velocity of the point of B at P in terms of the velocity of the center of mass of B and the angular velocity of B, we obtain

$$\begin{aligned} (\mathbf{v}_{AP})_x &= (\mathbf{v}_A)_x - \omega_A y_A, \\ (\mathbf{v}'_{AP})_x &= (\mathbf{v}'_A)_x - \omega'_A y_A, \\ (\mathbf{v}_{BP})_x &= (\mathbf{v}_B)_x - \omega_B y_B, \\ (\mathbf{v}'_{BP})_x &= (\mathbf{v}'_B)_x - \omega'_B y_B. \end{aligned} \tag{8.48}$$

At time t_C, the x components of the velocities of the two objects are equal at P, which yields the relation

$$(\mathbf{v}_{AC})_x - \omega_{AC} y_A = (\mathbf{v}_{BC})_x - \omega_{BC} y_B. \tag{8.49}$$

From Eqs. (8.48),

$$\frac{(\mathbf{v}'_{BP})_x - (\mathbf{v}'_{AP})_x}{(\mathbf{v}_{AP})_x - (\mathbf{v}_{BP})_x} = \frac{(\mathbf{v}'_B)_x - \omega'_B y_B - (\mathbf{v}'_A)_x + \omega'_A y_A}{(\mathbf{v}_A)_x - \omega_A y_A - (\mathbf{v}_B)_x + \omega_B y_B}.$$

Substituting Eqs. (8.42) and (8.47) into this equation and collecting terms, we obtain

$$\frac{(\mathbf{v}'_{BP})_x - (\mathbf{v}'_{AP})_x}{(\mathbf{v}_{AP})_x - (\mathbf{v}_{BP})_x} = e - \left[\frac{(\mathbf{v}_{AC})_x - \omega_{AC} y_A - (\mathbf{v}_{BC})_x + \omega_{BC} y_B}{(\mathbf{v}_A)_x - \omega_A y_A - (\mathbf{v}_B)_x + \omega_B y_B}\right](e + 1).$$

The term in brackets vanishes due to Eq. (8.49), and we obtain the equation relating the normal components of the velocities at the point of contact to the coefficient of restitution:

$$\boxed{e = \frac{(\mathbf{v}'_{BP})_x - (\mathbf{v}'_{AP})_x}{(\mathbf{v}_{AP})_x - (\mathbf{v}_{BP})_x}.} \tag{8.50}$$

In obtaining this equation, we assumed that the contacting surfaces are smooth, so *the collision exerts no force on A or B in the direction tangential to their contacting surfaces.*

Although we derived Eq. (8.50) under the assumption that the motions of A and B are unconstrained, it also holds if they are not—for example, if one of them is connected to a pin support.

In the following examples we analyze collisions of rigid bodies in planar motion. Your approach will depend on the type of collision. If other forces are negligible in comparison to the impact forces, total linear momentum is conserved. If other forces and couples are negligible in comparison to the impact forces and couples, total angular momentum about any fixed point is conserved. If, in addition, only forces are exerted at the point of impact P, the angular momentum about P of each rigid body is conserved. If one of the rigid bodies has a pin support at a point O, the total angular momentum about O is conserved. If the impact is assumed to exert no forces on the colliding objects in the direction tangential to their surface of contact, the coefficient of restitution e relates the normal components of the velocities at the point of contact through Eq. (8.50).

Example 8.7

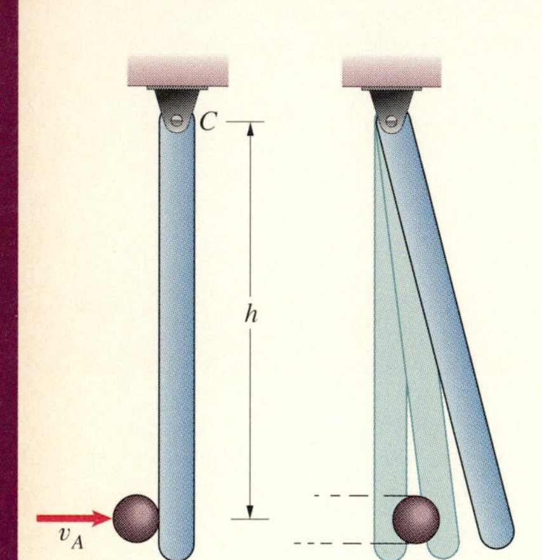

Fig. 8.26

The homogeneous sphere in Fig. 8.26 is moving horizontally with velocity v_A and no angular velocity when it strikes the stationary slender bar. The sphere has mass m_A, and the bar has mass m_B and length l. The coefficient of restitution of the impact is e.

(a) What is the angular velocity of the bar after the impact?

(b) If the duration of the impact is Δt, what average horizontal force is exerted on the bar by the pin support C as a result of the impact?

STRATEGY

(a) From the definition of the coefficient of restitution, we can obtain an equation relating the horizontal velocity of the sphere and the velocity of the bar at the point of impact after the collision occurs. In addition, the total angular momentum of the sphere and bar about the pin C is conserved. With these two equations and kinematic relationships, we can determine the velocity of the sphere and the angular velocity of the bar. (b) We can determine the average force exerted on the bar by the support by applying the principle of angular impulse and momentum to the bar.

SOLUTION

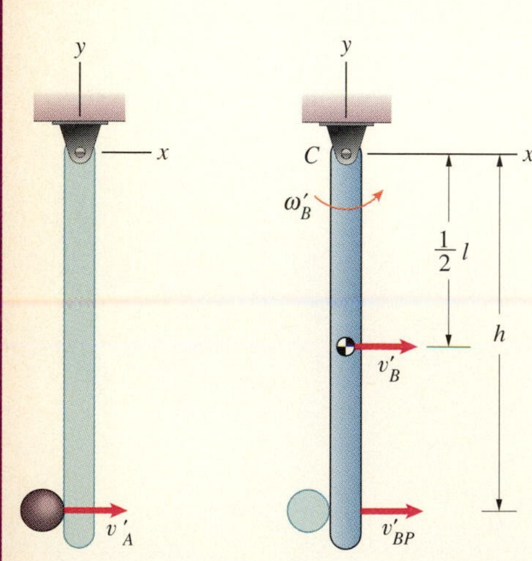

(a) Velocities of the sphere and bar after the impact.

(a) In Fig. (a) we show the velocities just after the impact, where v'_{BP} is the bar's velocity at the point of impact. From the definition of the coefficient of restitution Eq. (8.50), we obtain

$$e = \frac{v'_{BP} - v'_A}{v_A - 0}.$$

The equation of conservation of total angular momentum about C is

$$H_{CA} + H_{CB} = H'_{CA} + H'_{CB}:$$

$$(\mathbf{r}_A \times m_A\mathbf{v}_A) \cdot \mathbf{k} + I_A\omega_A + (\mathbf{r}_B \times m_B\mathbf{v}_B) \cdot \mathbf{k} + I_B\omega_B$$
$$= (\mathbf{r}_A \times m_A\mathbf{v}'_A) \cdot \mathbf{k} + I_A\omega'_A + (\mathbf{r}_B \times m_B\mathbf{v}'_B) \cdot \mathbf{k} + I_B\omega'_B,$$

$$(-h\mathbf{j} \times m_A v_A\mathbf{i}) \cdot \mathbf{k} = (-h\mathbf{j} \times m_A v'_A\mathbf{i}) \cdot \mathbf{k} + \left(-\frac{1}{2}l\mathbf{j} \times m_B v'_B\mathbf{i}\right) \cdot \mathbf{k} + I_B\omega'_B.$$

Carrying out the vector operations, we obtain

$$hm_A v_A = hm_A v'_A + \frac{1}{2} l m_B v'_B + I_B \omega'_B.$$

Notice in Fig. (a) that the velocities v'_B and v'_{BP} are related to the angular velocity of the bar ω'_B by

$$v'_B = \frac{1}{2} l \omega'_B, \qquad v'_{BP} = h\omega'_B.$$

We now have four equations in the four unknowns v'_A, v'_B, v'_{BP}, and ω'_B. Solving them for the angular velocity of the bar and using the expression $I_B = \frac{1}{12} m_B l^2$, we obtain

$$\omega'_B = \frac{(1+e)hm_A v_A}{h^2 m_A + \frac{1}{3} m_B l^2}.$$

(b) Let the forces on the free-body diagram of the bar in Fig. (b) represent the average forces exerted during the impact. We apply the principle of angular impulse and momentum, in the form given by Eq. (8.35), about the point of impact:

$$(t_2 - t_1)(\Sigma M_P)_{\text{av}} = H'_B - H_B:$$

$$(t_2 - t_1)(\Sigma M_P)_{\text{av}} = [(\mathbf{r}_B \times m_B \mathbf{v}'_B) \cdot \mathbf{k} + I_B \omega'_B] - [(\mathbf{r}_B \times m_B \mathbf{v}_B) \cdot \mathbf{k} + I_B \omega_B],$$

$$\Delta t(-hC_x) = \left[\left(h - \frac{1}{2} l \right) \mathbf{j} \times m_B v'_B \mathbf{i} \right] \cdot \mathbf{k} + I_B \omega'_B - 0.$$

Solving for C_x, we obtain

$$C_x = \frac{(h - \frac{1}{2} l) m_B v'_B - I_B \omega'_B}{h\,\Delta t}.$$

Using our solution for ω'_B from part (a) and the relation $v'_B = \frac{1}{2} l \omega'_B$, we obtain the average horizontal force exerted by the support:

$$C_x = \frac{(1+e)(\frac{1}{2}h - \frac{1}{3}l) l m_A m_B v_A}{(h^2 m_A + \frac{1}{3} m_B l^2)\,\Delta t}.$$

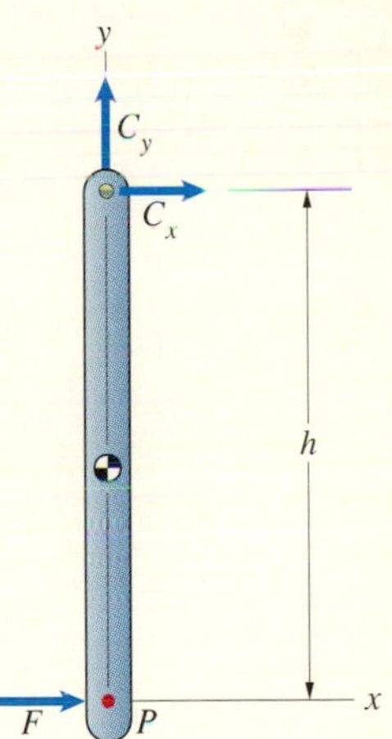

(b) Average forces exerted on the bar during the impact.

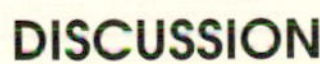

DISCUSSION

Notice that the average horizontal force exerted on the bar by the support can be in either direction or can be zero, depending on where the impact occurs. The force is zero if $h = \frac{2}{3} l$.

Example 8.8

The combined mass of the motorcycle and rider in Fig. 8.27 is $m = 170$ kg, and their combined moment of inertia about their center of mass is 22 kg-m^2. Following a jump over an obstacle, the motorcycle and rider are in the position shown just before the rear wheel contacts the ground. The velocity of their center of mass is of magnitude $|\mathbf{v}_G| = 8.8$ m/s and their angular velocity is $\omega = 0.2$ rad/s. If the motorcycle and rider are modeled as a single rigid body and the coefficient of restitution of the impact is $e = 0.8$, what are the angular velocity ω' and velocity $\mathbf{v}'_G$ after the impact? Neglect the tangential component of force exerted on the motorcycle's wheel during the impact.

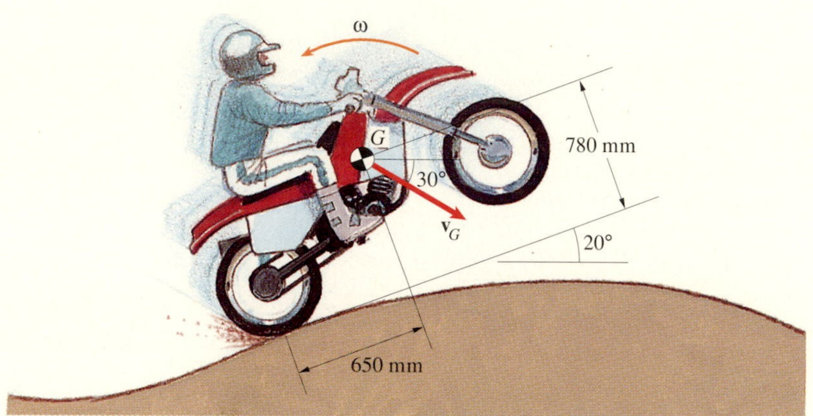

Fig. 8.27

STRATEGY

Since the tangential component of force on the motorcycle's wheel during the impact is neglected, the component of the velocity of the center of mass parallel to the ground is unchanged by the impact. The coefficient of restitution relates the motorcycle's velocity normal to the ground *at the point of impact* before the impact to its value after the impact. Also, the force of the impact exerts no moment about the point of impact, so the motorcycle's angular momentum about that point is conserved. (We assume the impact to be so brief that the angular impulse due to the weight is negligible.) With these three relations we can determine the two components of the velocity of the center of mass and the angular velocity after the impact.

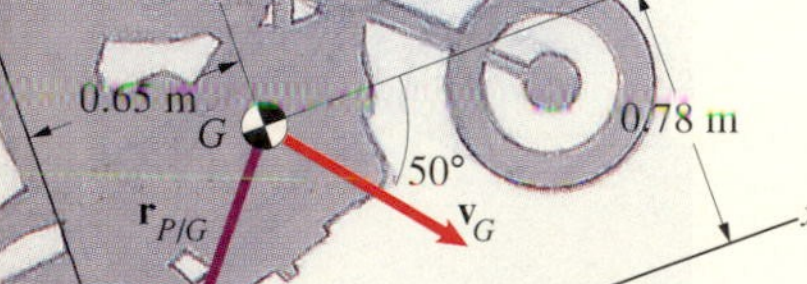

(a) Aligning the x axis of the coordinate system tangent to the ground at P.

SOLUTION

In Fig. (a) we align a coordinate system parallel and perpendicular to the ground at the point P where the impact occurs. Let the components of the velocity of the center of mass before and after the impact be $\mathbf{v}_G = v_x\mathbf{i} + v_y\mathbf{j}$ and $\mathbf{v}'_G = v'_x\mathbf{i} + v'_y\mathbf{j}$, respectively. The components v_x and v_y are

$$v_x = 8.8 \cos 50° = 5.66 \text{ m/s},$$

$$v_y = -8.8 \sin 50° = -6.74 \text{ m/s}.$$

Because the component of the impact force tangential to the ground is neglected, the x component of the velocity of the center of mass is unchanged:

$$v'_x = v_x = 5.66 \text{ m/s}.$$

We can express the y component of the wheel's velocity at P before the impact in terms of the velocity of the center of mass and the angular velocity (Fig. a):

$$\begin{aligned}\mathbf{j}\cdot\mathbf{v}_P &= \mathbf{j}\cdot(\mathbf{v}_G + \boldsymbol{\omega}\times\mathbf{r}_{P/G})\\ &= \mathbf{j}\cdot\left\{v_x\mathbf{i} + v_y\mathbf{j} + \begin{vmatrix}\mathbf{i} & \mathbf{j} & \mathbf{k}\\ 0 & 0 & \omega\\ -0.65 & -0.78 & 0\end{vmatrix}\right\}\\ &= v_y - 0.65\omega.\end{aligned}$$

(Notice that this expression gives the y component of the velocity at P even though the wheel is spinning.) The y component of the wheel's velocity at P after the impact is

$$\begin{aligned}\mathbf{j}\cdot\mathbf{v}'_P &= \mathbf{j}\cdot(\mathbf{v}'_G + \boldsymbol{\omega}'\times\mathbf{r}_{P/G})\\ &= v'_y - 0.65\omega'.\end{aligned}$$

The coefficient of restitution relates the y components of the wheel's velocity at P before and after the impact:

$$e = \frac{-(\mathbf{j}\cdot\mathbf{v}'_P)}{(\mathbf{j}\cdot\mathbf{v}_P)} = \frac{-(v'_y - 0.65\omega')}{(v_y - 0.65\omega)}. \qquad (8.51)$$

The force of the impact exerts no moment about P, so angular momentum about P is conserved:

$$H_P = H'_P:$$

$$[(\mathbf{r}_{G/P}\times m\mathbf{v}_G)\cdot\mathbf{k} + I\omega] = [(\mathbf{r}_{G/P}\times m\mathbf{v}'_G)\cdot\mathbf{k} + I\omega'],$$

$$\begin{vmatrix}\mathbf{i} & \mathbf{j} & \mathbf{k}\\ 0.65 & 0.78 & 0\\ mv_x & mv_y & 0\end{vmatrix}\cdot\mathbf{k} + I\omega = \begin{vmatrix}\mathbf{i} & \mathbf{j} & \mathbf{k}\\ 0.65 & 0.78 & 0\\ mv'_x & mv'_y & 0\end{vmatrix}\cdot\mathbf{k} + I\omega'.$$

Expanding the determinants and evaluating the dot products, we obtain

$$0.65mv_y - 0.78mv_x + I\omega = 0.65mv'_y - 0.78mv'_x + I\omega'. \qquad (8.52)$$

Since we have already determined v'_x, we can solve Eqs. (8.51) and (8.52) for v'_y and ω'. The results are

$$v'_y = -3.84 \text{ m/s}, \qquad \omega' = -14.4 \text{ rad/s}.$$

The velocity of the center of mass after the impact is $\mathbf{v}'_G = 5.66\mathbf{i} - 3.84\mathbf{j}$ m/s, and the angular velocity is 14.4 rad/s in the clockwise direction.

Example 8.9

An engineer simulates a collision between two 1600-kg cars by modeling them as rigid bodies (Fig. 8.28). The moment of inertia of each car about its center of mass is 960 kg-m^2. He assumes the contacting surfaces at P to be smooth and parallel to the x axis and assumes the coefficient of restitution to be $e = 0.2$. What are the angular velocities of the cars and the velocities of their centers of mass after the collision?

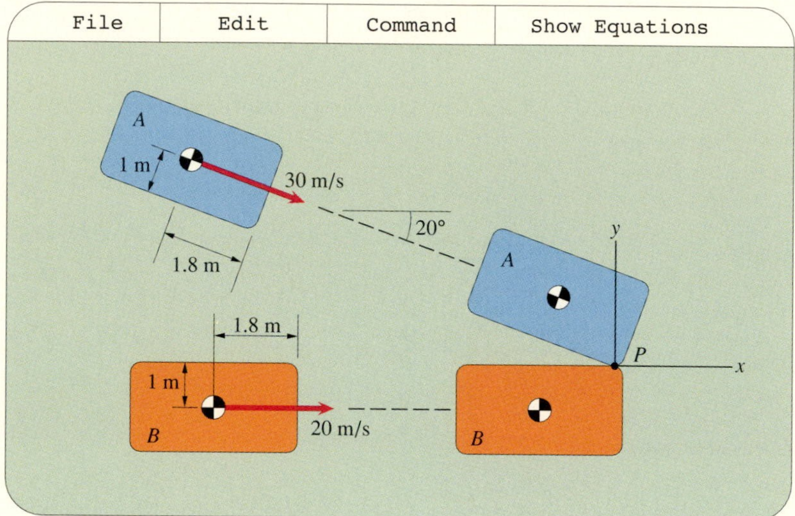

Fig. 8.28

STRATEGY

Since the contacting surfaces are smooth, the x components of the velocities of the centers of mass are unchanged by the collision. The y components of the velocities must satisfy conservation of linear momentum, and the y components of the velocities *at the point of impact* before and after the impact are related by the coefficient of restitution. The force of the impact exerts no moment about P on either car, so the angular momentum *of each car* about P is conserved. From these conditions and kinematic relations between the velocities of the centers of mass and the velocities at P, we can determine the angular velocities and velocities of the centers of mass after the impact.

SOLUTION

The components of the velocities of the centers of mass before the impact are

$$\mathbf{v}_A = 30 \cos 20^\circ \, \mathbf{i} - 30 \sin 20^\circ \, \mathbf{j}$$
$$= 28.2\mathbf{i} - 10.3\mathbf{j} \text{ (m/s)}$$

and

$$\mathbf{v}_B = 20\mathbf{i} \text{ (m/s)}.$$

The x components of the velocities are unchanged by the impact:

$$v'_{Ax} = v_{Ax} = 28.2 \text{ m/s}, \qquad v'_{Bx} = v_{Bx} = 20 \text{ m/s}.$$

The y components of the velocities must satisfy conservation of linear momentum:

$$m_A v_{Ay} + m_B v_{By} = m_A v'_{Ay} + m_B v'_{By}. \qquad (8.53)$$

Let the velocities of the two cars at P before the collision be $\mathbf{v}_{AP}$ and $\mathbf{v}_{BP}$. The coefficient of restitution $e = 0.2$ relates the y components of the velocities at P:

$$0.2 = \frac{v'_{BPy} - v'_{APy}}{v_{APy} - v_{BPy}}. \qquad (8.54)$$

We can express the velocities at P after the impact in terms of the velocities of the centers of mass and the angular velocities after the impact (Fig. a). The position of P relative to the center of mass of car A is

$$\mathbf{r}_{P/A} = [(1.8)\cos 20° - (1)\sin 20°]\mathbf{i} - [(1.8)\sin 20° + (1)\cos 20°]\mathbf{j}$$
$$= 1.35\mathbf{i} - 1.56\mathbf{j} \text{ (m)}.$$

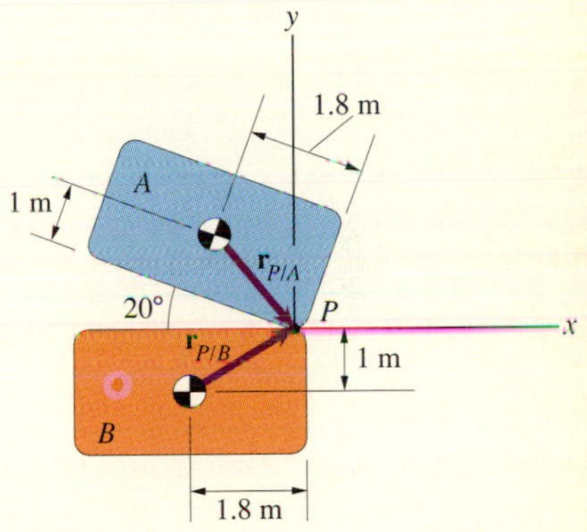

(a) Position vectors of P relative to the centers of mass.

Therefore we can express the velocity of point P of car A after the impact as

$$\mathbf{v}'_{AP} = \mathbf{v}'_A + \boldsymbol{\omega}'_A \times \mathbf{r}_{P/A}:$$

$$v'_{APx}\mathbf{i} + v'_{APy}\mathbf{j} = v'_{Ax}\mathbf{i} + v'_{Ay}\mathbf{j} + \begin{vmatrix} \mathbf{i} & \mathbf{j} & \mathbf{k} \\ 0 & 0 & \omega'_A \\ 1.35 & -1.56 & 0 \end{vmatrix}.$$

Equating $\mathbf{i}$ and $\mathbf{j}$ components in this equation, we obtain

$$v'_{APx} = v'_{Ax} + 1.56\omega'_A, \qquad v'_{APy} = v'_{Ay} + 1.35\omega'_A. \qquad (8.55)$$

The position of P relative to the center of mass of car B is

$$\mathbf{r}_{P/B} = 1.8\mathbf{i} + \mathbf{j} \text{ (m)}.$$

We can express the velocity of point P of car B after the impact as

$$\mathbf{v}'_{BP} = \mathbf{v}'_B + \boldsymbol{\omega}'_B \times \mathbf{r}_{P/B}:$$

$$v'_{BPx}\mathbf{i} + v'_{BPy}\mathbf{j} = v'_{Bx}\mathbf{i} + v'_{By}\mathbf{j} + \begin{vmatrix} \mathbf{i} & \mathbf{j} & \mathbf{k} \\ 0 & 0 & \omega'_B \\ 1.8 & 1 & 0 \end{vmatrix}.$$

Equating $\mathbf{i}$ and $\mathbf{j}$ components, we obtain

$$v'_{BPx} = v'_{Bx} - \omega'_B, \qquad v'_{BPy} = v'_{By} + 1.8\omega'_B. \qquad (8.56)$$

The angular momentum of car A about P is conserved:

$$H_{PA} = H'_{PA}:$$

$$[(\mathbf{r}_{A/P} \times m_A\mathbf{v}_A)\cdot\mathbf{k} + I_A\omega_A] = [(\mathbf{r}_{A/P} \times m_A\mathbf{v}'_A)\cdot\mathbf{k} + I_A\omega'_A],$$

$$\begin{vmatrix} \mathbf{i} & \mathbf{j} & \mathbf{k} \\ -1.35 & 1.56 & 0 \\ m_A v_{Ax} & m_A v_{Ay} & 0 \end{vmatrix} \cdot \mathbf{k} + 0 = \begin{vmatrix} \mathbf{i} & \mathbf{j} & \mathbf{k} \\ -1.35 & 1.56 & 0 \\ m_A v'_{Ax} & m_A v'_{Ay} & 0 \end{vmatrix} \cdot \mathbf{k} + I_A\omega'_A.$$

Expanding the determinants and evaluating the dot products, we obtain

$$\begin{aligned} -1.35 m_A v_{Ay} - 1.56 m_A v_{Ax} \\ = -1.35 m_A v'_{Ay} - 1.56 m_A v'_{Ax} + I_A \omega'_A. \end{aligned} \tag{8.57}$$

The angular momentum of car B about P is also conserved,

$$H_{PB} = H'_{PB}:$$

$$[(\mathbf{r}_{B/P} \times m_B \mathbf{v}_B) \cdot \mathbf{k} + I_B \omega_B] = [(\mathbf{r}_{B/P} \times m_B \mathbf{v}'_B) \cdot \mathbf{k} + I_B \omega'_B],$$

$$\begin{vmatrix} \mathbf{i} & \mathbf{j} & \mathbf{k} \\ -1.8 & -1 & 0 \\ m_B v_{Bx} & 0 & 0 \end{vmatrix} \cdot \mathbf{k} + 0 = \begin{vmatrix} \mathbf{i} & \mathbf{j} & \mathbf{k} \\ -1.8 & -1 & 0 \\ m_B v'_{Bx} & m_B v'_{By} & 0 \end{vmatrix} \cdot \mathbf{k} + I_B \omega'_B.$$

From this equation we obtain

$$m_B v_{Bx} = -1.8 m_B v'_{By} + m_B v'_{Bx} + I_B \omega'_B. \tag{8.58}$$

We can solve Eqs. (8.53)–(8.58) for $\mathbf{v}'_A$, $\mathbf{v}'_{AP}$, ω'_A, $\mathbf{v}'_B$, $\mathbf{v}'_{BP}$, and ω'_B. The results for the velocities of the centers of mass of the cars and their angular velocities are

$$\mathbf{v}'_A = 28.2\mathbf{i} - 9.08\mathbf{j} \text{ (m/s)}, \qquad \omega'_A = 2.65 \text{ rad/s},$$

$$\mathbf{v}'_B = 20.0\mathbf{i} - 1.18\mathbf{j} \text{ (m/s)}, \qquad \omega'_B = -3.54 \text{ rad/s}.$$

Problems

8.64 The 2-kg slender bar starts from rest in the vertical position and falls, striking the smooth surface at P. The coefficient of restitution of the impact is $e = 0.5$. When the bar rebounds, through what angle relative to the horizontal will it rotate?

Strategy: Use the coefficient of restitution to relate the bar's velocity at P just after the impact to its value just before the impact.

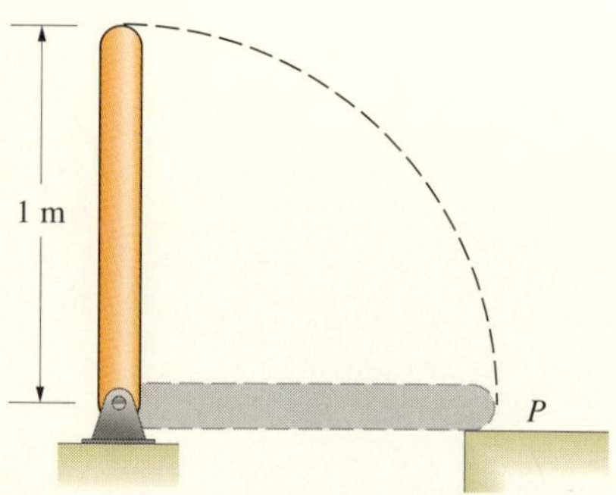

P8.64

8.65 The slender bar of mass m falls from rest in the position shown and hits the smooth projection at A. The coefficient of restitution is e. Show that the velocity of the center of mass of the bar is zero immediately after the impact if $b^2 = el^2/12$.

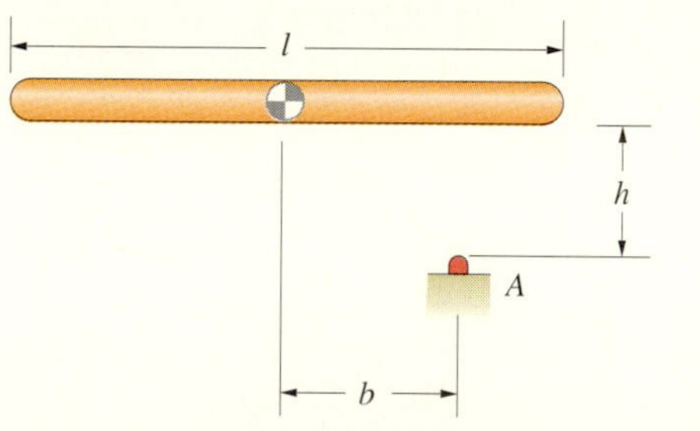

P8.65

8.66 In Problem 8.65, if $m = 2$ kg, $l = 1$ m, $b = 350$ mm, $h = 200$ mm, and the coefficient of restitution of the impact is $e = 0.4$, determine the bar's angular velocity after the impact.

8.67 If the duration of the impact described in Problem 8.66 is 0.02 s, what average force is exerted on the bar by the projection at A during the impact?

8.68 Wind causes the 600-ton ship to drift sideways at 1 ft/s and strike the stationary quay at P. The ship's moment of inertia about its center of mass is 3×10^8 slug-ft^2, and the coefficient of restitution of the impact is $e = 0.2$. What is the ship's angular velocity after the impact?

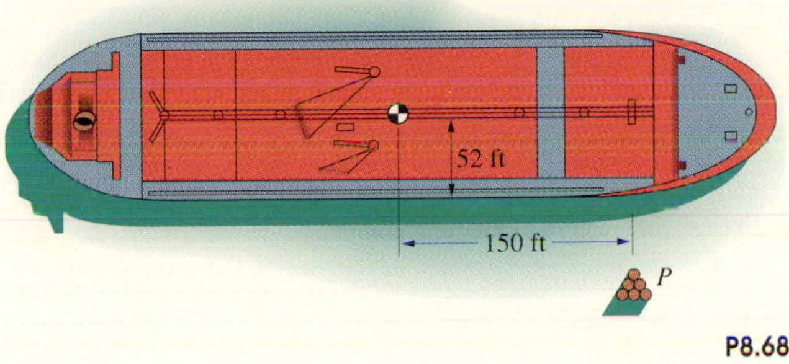

P8.68

8.69 In Problem 8.68, if the duration of the ship's impact with the quay is 10 s, what is the average value of the force exerted on the ship by the impact?

8.70 A 1-lb sphere A translating at 20 ft/s strikes the end of a stationary 10-lb slender bar B. The bar is pinned to a fixed support at O. What is the angular velocity of the bar after the impact if the sphere adheres to the bar?

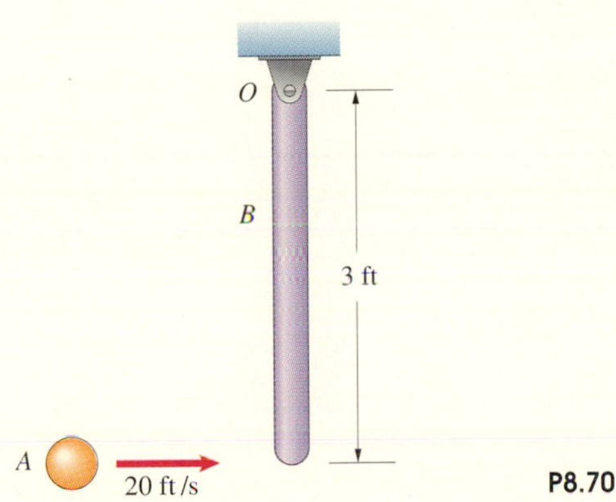

P8.70

8.71 In Problem 8.70, determine the velocity of the smooth sphere and the angular velocity of the bar after the impact if the coefficient of restitution is $e = 0.8$.

8.72 The 1-kg sphere A is moving at 10 m/s when it strikes the end of the 4-kg stationary slender bar B. If the sphere adheres to the bar, what is the bar's angular velocity after the impact?

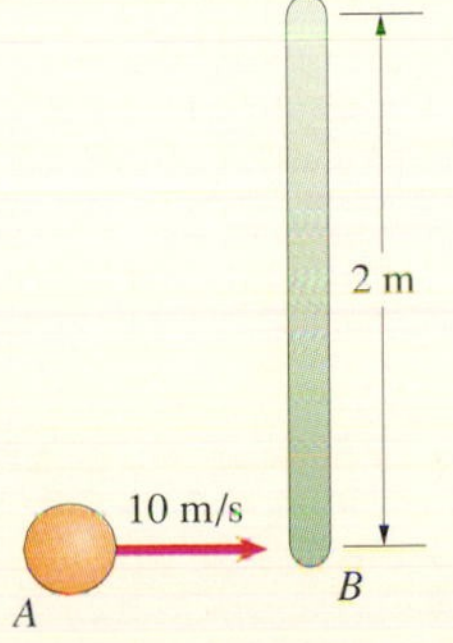

P8.72

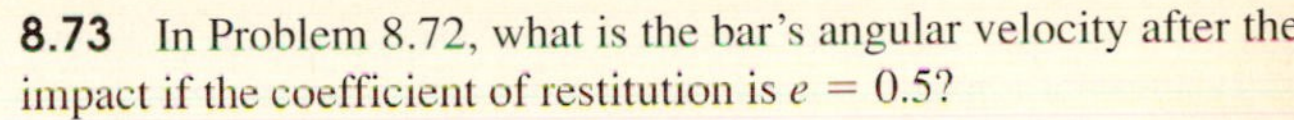

8.73 In Problem 8.72, what is the bar's angular velocity after the impact if the coefficient of restitution is $e = 0.5$?

8.74 In Problem 8.72, determine the total kinetic energy of the sphere and bar before and after the impact if (a) $e = 0.5$; (b) $e = 1$.

8.75 The 5-oz ball is translating with velocity $v_A = 80$ ft/s perpendicular to the bat just before impact. The player is swinging the 31-oz bat with angular velocity $\omega = 6\pi$ rad/s before the impact. Point C is the bat's instantaneous center both before and after the impact. The distances $b = 14$ in. and $\bar{y} = 26$ in. The bat's moment of inertia about its center of mass is $I_B = 0.033$ slug-ft^2. The coefficient of restitution is $e = 0.6$, and the duration of the impact is 0.008 s. Determine the magnitude of the velocity of the ball after the impact and the average force A_x exerted on the bat by the player during the impact if (a) $d = 0$; (b) $d = 3$ in.; (c) $d = 8$ in.

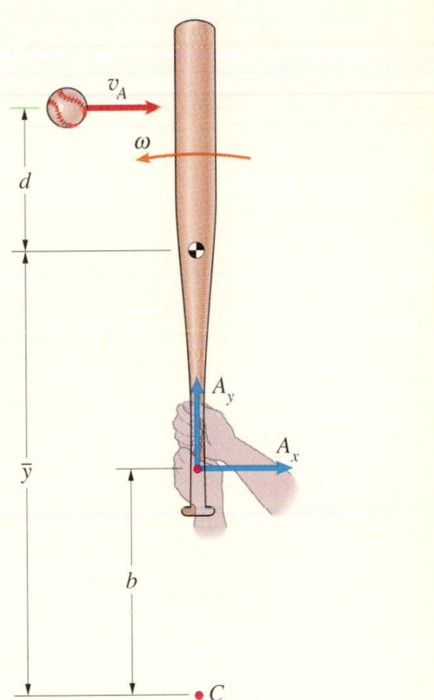

P8.75

8.76 In Problem 8.75, show that the force A_x is zero if $d = I_B/(m_B\bar{y})$, where m_B is the mass of the bat.

8.77 A slender bar of mass m is released from rest in the horizontal position at a height h above a peg (Fig. a). A small hook at the end of the bar engages the peg, and the bar swings from the peg (Fig. b). What minimum height h is necessary for the bar to swing 270° from its position when it engages the peg?

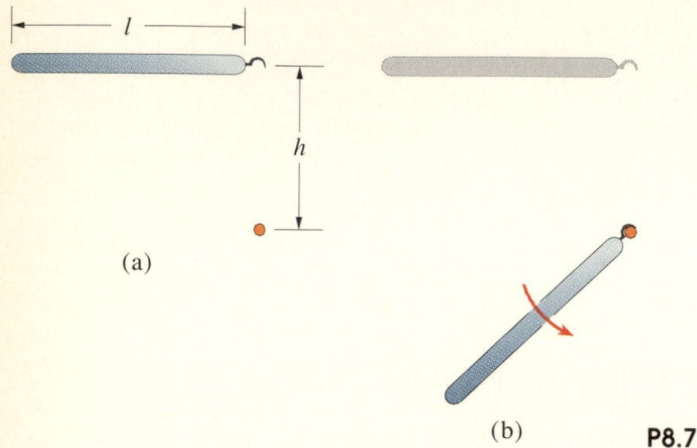

P8.77

8.78 Is energy conserved in Problem 8.77? If not, how much energy is lost?

8.79 A wheel that can be modeled as a 1-slug homogeneous cylindrical disk rolls at 10 ft/s on a horizontal surface toward a 6-in. step. If the wheel remains in contact with the step and does not slip while rolling up onto it, what is the wheel's velocity once it is on the step?

P8.79

8.80 In Problem 8.79, what is the minimum velocity the wheel must have rolling toward the step in order to climb up onto it?

8.81 The slender bar is shown just before it hits the smooth floor. The length of the bar is 1 m and its mass is 2 kg. The bar's angular velocity is zero, and it is moving downward at 4 m/s. If the coefficient of restitution of the impact is $e = 0.2$, what is the bar's angular velocity after the impact?

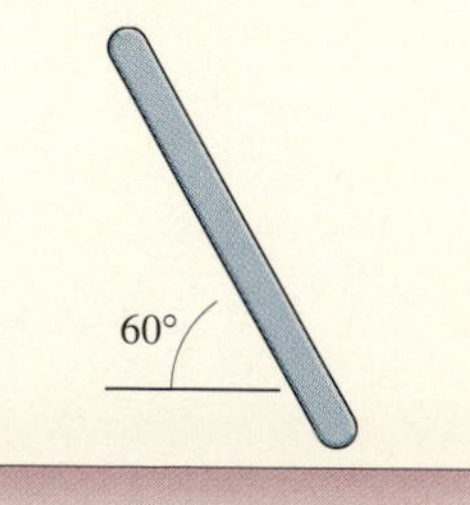

P8.81

8.82 If the duration of the impact in Problem 8.81 is 0.03 s, what average force is exerted on the bar by the floor?

8.83 In Problem 8.81, suppose that the center of mass of the bar is moving downward at 4 m/s and the bar has a clockwise angular velocity ω just before it hits the floor. What value of ω would cause the bar to have no angular velocity after the impact?

8.84 During her parallel-bars routine, the velocity of the 90-lb gymnast's center of mass is $4\mathbf{i} - 10\mathbf{j}$ (ft/s) and her angular velocity is zero just before she grasps the bar at A. In the position shown, her moment of inertia about her center of mass is 1.8 slug-ft^2. If she stiffens her shoulders and legs so that she can be modeled as a rigid body, what is the velocity of her center of mass and her angular velocity just after she grasps the bar?

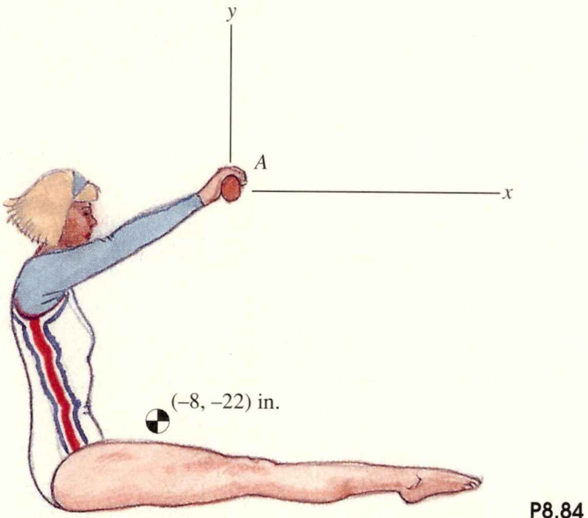

P8.84

8.85 The 20-kg homogeneous rectangular plate is released from rest (Fig. a) and falls 200 mm before coming to the end of the string attached at the corner A (Fig. b). Assuming that the vertical component of the velocity of A is zero just after the plate reaches the end of the string, determine the angular velocity of the plate and the magnitude of the velocity of the corner B at that instant.

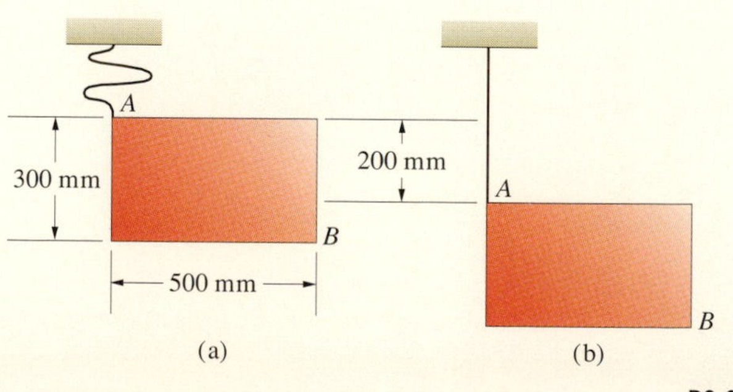

P8.85

8.86 Two bars A and B are each 2 m in length and each has a mass of 4 kg. In Fig. (a), bar A has no angular velocity and is moving to the right at 1 m/s, and bar B is stationary. If the bars bond together on impact (Fig. b), what is their angular velocity ω' after the impact?

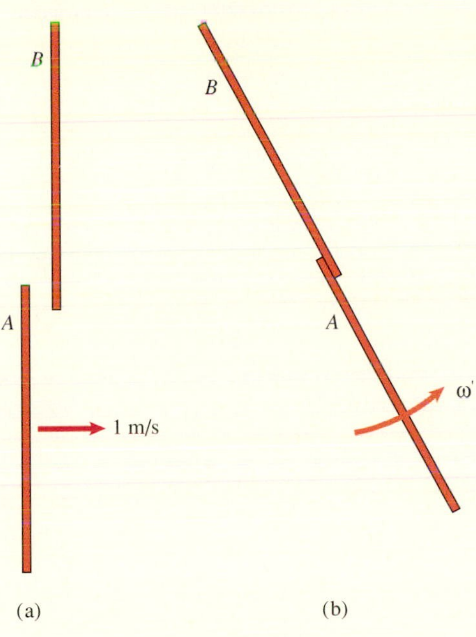

P8.86

8.87 In Problem 8.86, what is the velocity of the center of mass of bar A just after the impact?

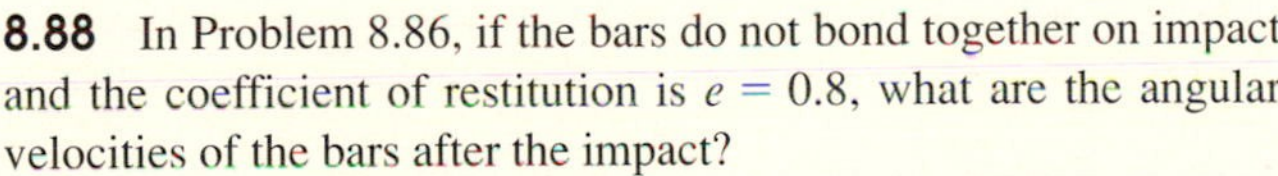

8.88 In Problem 8.86, if the bars do not bond together on impact and the coefficient of restitution is $e = 0.8$, what are the angular velocities of the bars after the impact?

8.89 The horizontal velocity of the landing airplane is 50 m/s, its vertical velocity (rate of descent) is 2 m/s, and its angular velocity is zero. The mass of the airplane is 12 Mg and the moment of inertia about its center of mass is 1×10^5 kg-m^2. When the rear wheels touch the runway, they remain in contact with it. Neglecting the horizontal force exerted on the wheels by the runway, determine the airplane's angular velocity just after it touches down.

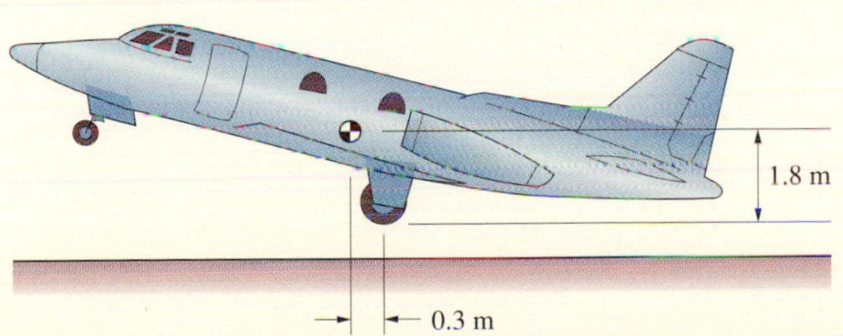

P8.89

8.90 Determine the angular velocity of the airplane in Problem 8.89 just after it touches down if its wheels don't stay in contact with the runway and the coefficient of restitution of the impact is $e = 0.4$.

8.91 While attempting to drive on an icy street for the first time, a student skids his 1260-kg car (A) into the university president's unoccupied 2700-kg Rolls-Royce Corniche (B). The point of impact is P. Assume that the impacting surfaces are smooth and parallel to the y axis, and that the coefficient of restitution of the impact is $e = 0.5$. The moments of inertia of the cars about their centers of mass are $I_A = 2400$ kg-m^2 and $I_B = 7600$ kg-m^2. Determine the angular velocities of the cars and the velocities of their centers of mass after the collision.

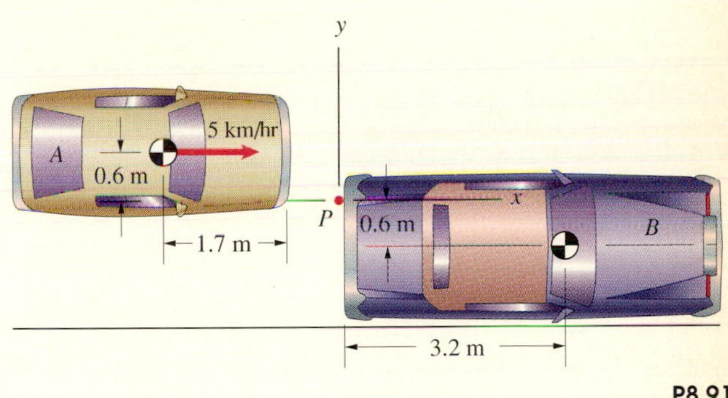

P8.91

8.92 The student in Problem 8.91 claimed he was moving at 5 km/hr prior to the collision, but investigating police estimate that the center of mass of the Rolls-Royce was moving at 1.7 m/s after the collision. What was the student's actual speed?

8.93 Each slender bar is 48 in. long and weighs 20 lb. Bar A is released in the horizontal position shown. The bars are smooth and the coefficient of restitution of their impact is $e = 0.8$. Determine the angle through which B swings afterward.

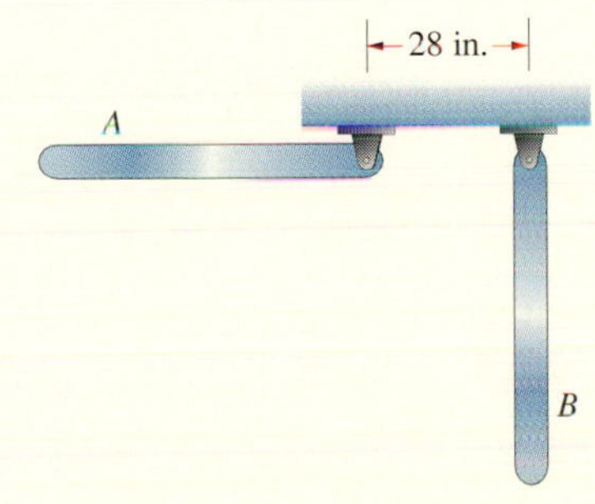

P8.93

8.94 The *Apollo* CSM (A) approaches the Soyuz Space Station (B). The mass of the *Apollo* is $m_A = 18$ Mg, and the moment of inertia about the axis through its center of mass parallel to the z axis is $I_A = 114$ Mg-m^2. The mass of the Soyuz is $m_B = 6.6$ Mg, and the moment of inertia about the axis through its center of mass parallel to the z axis is $I_B = 70$ Mg-m^2. The Soyuz is stationary relative to the reference frame shown and the CSM approaches with velocity $\mathbf{v}_A = 0.2\mathbf{i} + 0.05\mathbf{j}$ (m/s) and no angular velocity. What is their angular velocity after docking?

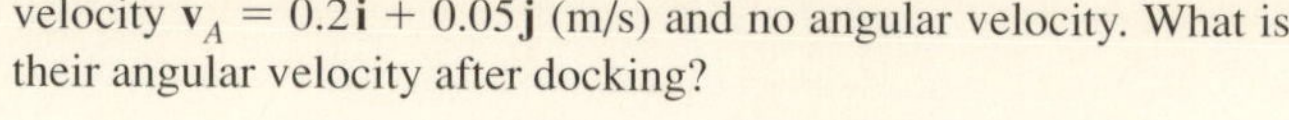

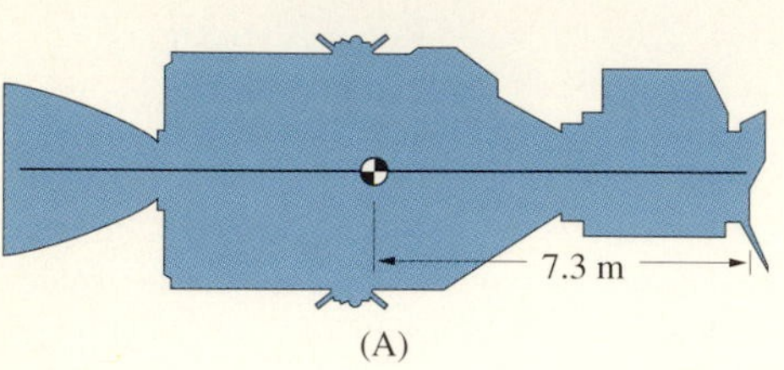

(A)

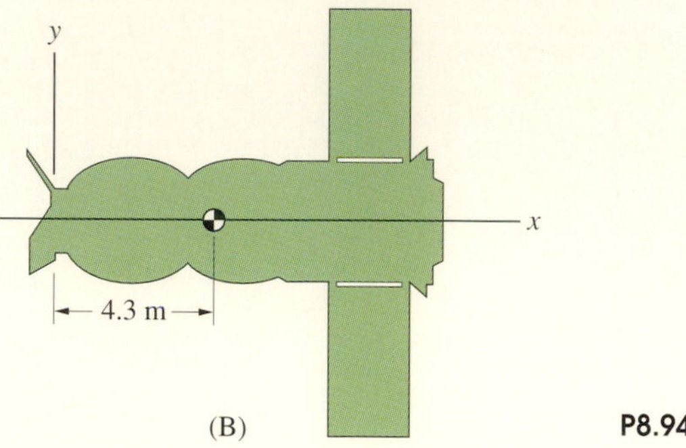

(B)

P8.94

Chapter Summary

Work and Energy

The work done by external forces and couples as a rigid body moves between two positions is equal to the change in its kinetic energy:

$$U_{12} = T_2 - T_1. \qquad \textbf{Eq. (8.4)}$$

The work done on a system of rigid bodies by external and internal forces and couples equals the change in the total kinetic energy.

The kinetic energy of a rigid body in general planar motion is

$$T = \frac{1}{2}mv^2 + \frac{1}{2}I\omega^2, \qquad \textbf{Eq. (8.12)}$$

where v is the magnitude of the velocity of the center of mass and I is the moment of inertia about the center of mass. If a rigid body rotates about a fixed axis O, its kinetic energy can also be expressed as

$$T = \frac{1}{2}I_0\omega^2. \qquad \textbf{Eq. (8.13)}$$

The work done on a rigid body by a force $\mathbf{F}$ is

$$U_{12} = \int_{(\mathbf{r}_p)_1}^{(\mathbf{r}_p)_2} \mathbf{F} \cdot d\mathbf{r}_p, \qquad \textbf{Eq. (8.14)}$$

where $\mathbf{r}_p$ is the position of the *point of application* of $\mathbf{F}$. If the point of application is stationary, or if its direction of motion is perpendicular to $\mathbf{F}$, no work is done.

The work done by a couple M on a rigid body in planar motion as it rotates from θ_1 to θ_2 in the direction of M is

$$U_{12} = \int_{\theta_1}^{\theta_2} M\, d\theta. \qquad \textbf{Eq. (8.16)}$$

A couple M is conservative if a potential energy V exists such that

$$M\,d\theta = -dV. \qquad \textbf{Eq. (8.17)}$$

The potential energy of a linear torsional spring that exerts a couple $k\theta$ in the direction opposite to its angular displacement θ (Fig. a) is $\frac{1}{2}k\theta^2$.

If all the forces and couples that do work on a rigid body are conservative, the sum of the kinetic energy and the total potential energy is constant:

$$T + V = \text{constant}. \qquad \textbf{Eq. (8.19)}$$

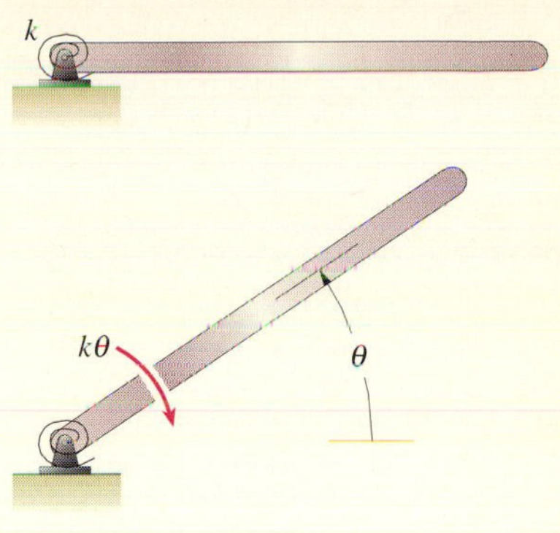

(a)

Power

The power transmitted to a rigid body by a force $\mathbf{F}$ is

$$P = \mathbf{F} \cdot \mathbf{v}_{\mathrm{P}}, \qquad \textbf{Eq. (8.20)}$$

where $\mathbf{v}_{\mathrm{P}}$ is the velocity of the point of application of $\mathbf{F}$. The power transmitted to a rigid body in planar motion by a couple M is

$$P = M\omega. \qquad \textbf{Eq. (8.21)}$$

The average power transferred to a rigid body during an interval of time is equal to the change in its kinetic energy, or the total work done, divided by the interval of time:

$$P_{\mathrm{av}} = \frac{T_2 - T_1}{t_2 - t_1} = \frac{U_{12}}{t_2 - t_1}. \qquad \textbf{Eq. (8.22)}$$

Impulse and Momentum

The principle of linear impulse and momentum states that the linear impulse applied to a rigid body during an interval of time is equal to the change in its linear momentum:

$$\int_{t_1}^{t_2} \Sigma \mathbf{F}\, dt = m\mathbf{v}_2 - m\mathbf{v}_1. \qquad \textbf{Eq. (8.25)}$$

This result can also be expressed in terms of the average with respect to time of the total force:

$$(t_2 - t_1)\Sigma \mathbf{F}_{\mathrm{av}} = m\mathbf{v}_2 - m\mathbf{v}_1. \qquad \textbf{Eq. (8.26)}$$

If the only forces acting on two rigid bodies A and B are the forces they exert on each other, or if other forces are negligible, their total linear momentum is conserved:

$$m_A\mathbf{v}_A + m_B\mathbf{v}_B = \text{constant}. \qquad \textbf{Eq. (8.27)}$$

The angular momentum about the center of mass of a rigid body in planar motion is

$$H = I\omega, \qquad \textbf{Eq. (8.29)}$$

where I is the moment of inertia about the center of mass. One form of the principle of angular impulse and momentum states that the angular impulse about the center of mass during the interval of time from t_1 to t_2 is equal to the change in the rigid body's angular momentum about its center of mass:

$$\int_{t_1}^{t_2} \Sigma M \, dt = H_2 - H_1. \qquad \textbf{Eq. (8.30)}$$

The angular momentum of a rigid body about a fixed point O is

$$H_O = (\mathbf{r} \times m\mathbf{v}) \cdot \mathbf{k} + I\omega, \qquad \textbf{Eq. (8.32)}$$

where $\mathbf{r}$ is the position of the center of mass relative to O and $\mathbf{v}$ is the velocity of the center of mass. A second form of the principle of angular impulse and momentum states that the angular impulse about O during the interval of time from t_1 to t_2 is equal to the change in the rigid body's angular momentum about O:

$$\int_{t_1}^{t_2} \Sigma M_O \, dt = H_{O2} - H_{O1}. \qquad \textbf{Eq. (8.33)}$$

In terms of the averages of the total moments with respect to time, Eqs. (8.30) and (8.32) are

$$(t_2 - t_1)\Sigma M_{\text{av}} = H_2 - H_1 \qquad \textbf{Eq. (8.34)}$$

and

$$(t_2 - t_1)(\Sigma M_O)_{\text{av}} = H_{O2} - H_{O1}. \qquad \textbf{Eq. (8.35)}$$

If two rigid bodies A and B in planar motion are subjected to only internal forces and couples, or if the total moment due to external forces and couples about a fixed point O is zero, the total angular momentum of A and B about O is conserved:

$$H_{OA} + H_{OB} = \text{constant}. \qquad \textbf{Eq. (8.36)}$$

Impacts

Suppose that two rigid bodies A and B, in two-dimensional motion in the same plane, collide. If other forces and couples are negligible in comparison to the impact forces and couples A and B exert on each other, their total linear momentum and their total angular momentum about any fixed point O is conserved. If, in addition, A and B exert only forces on each other at their point of impact P, the angular momentum about P of *each* rigid body is conserved. If one of the rigid bodies has a pin support at a point O, their total angular momentum about that point is conserved.

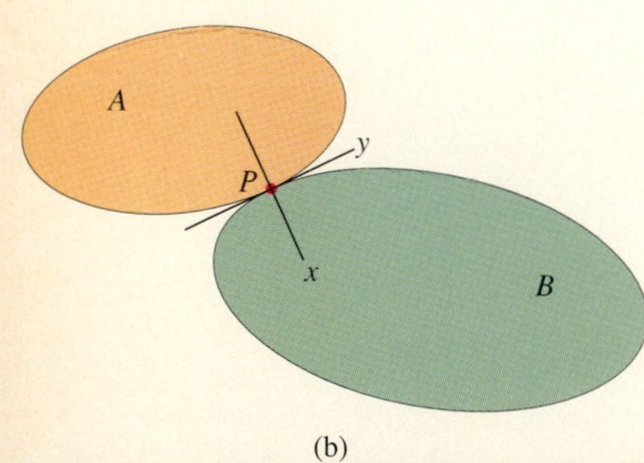

(b)

Let P be the point of impact (Fig. b). The normal components of the velocities at P are related to the coefficient of restitution e by

$$e = \frac{(\mathbf{v}'_{BP})_x - (\mathbf{v}'_{AP})_x}{(\mathbf{v}_{AP})_x - (\mathbf{v}_{BP})_x}. \qquad \textbf{Eq. (8.50)}$$

Review Problems

8.95 The moment of inertia of the pulley is 0.2 kg-m^2. The system is released from rest. Use the principle of work and energy to determine the velocity of the 10-kg cylinder when it has fallen 1 m.

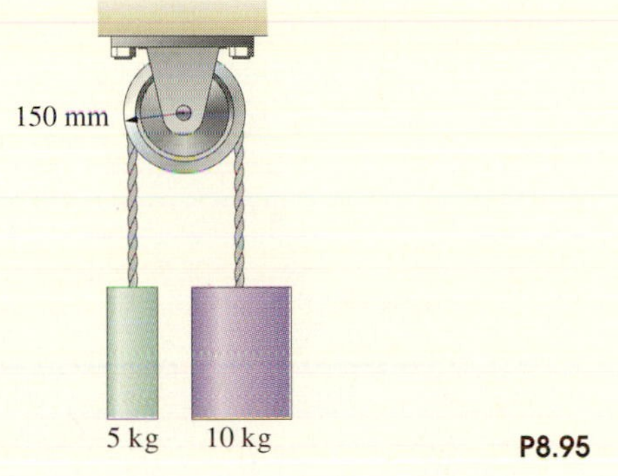

P8.95

8.96 Use momentum principles to determine the velocity of the 10-kg cylinder in Problem 8.95 1 s after the system is released from rest.

8.97 Arm BC has a mass of 12 kg, and the moment of inertia about its center of mass is 3 kg-m^2. Point B is stationary. Arm BC is initially aligned with the (horizontal) x axis with zero angular velocity, and a constant couple M applied at B causes it to rotate upward. When it is in the position shown, its counterclockwise angular velocity is 2 rad/s. Determine M.

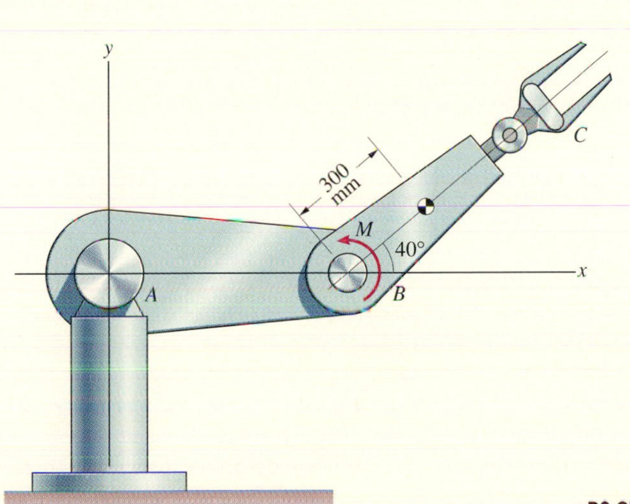

P8.97

8.98 The cart is stationary when a constant force F is applied to it. What will its velocity be when it has rolled a distance b? The mass of the body of the cart is m_c and each of the four wheels has mass m, radius R, and moment of inertia I.

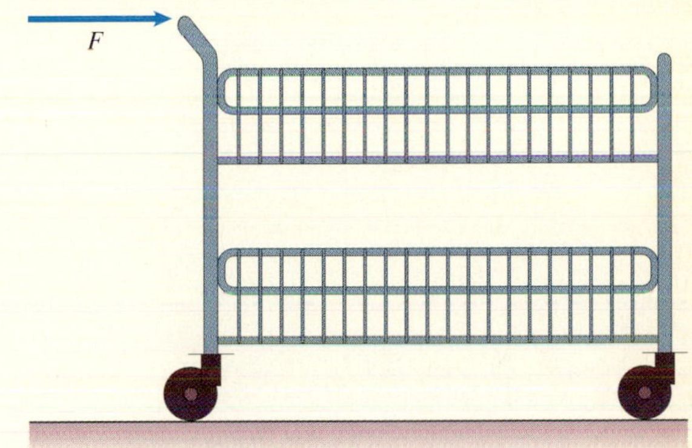

P8.98

8.99 Each pulley has moment of inertia $I = 0.003$ kg-m^2, and the mass of the belt is 0.2 kg. If a constant couple $M = 4$ N-m is applied to the bottom pulley, what will its angular velocity be when it has turned 10 revolutions?

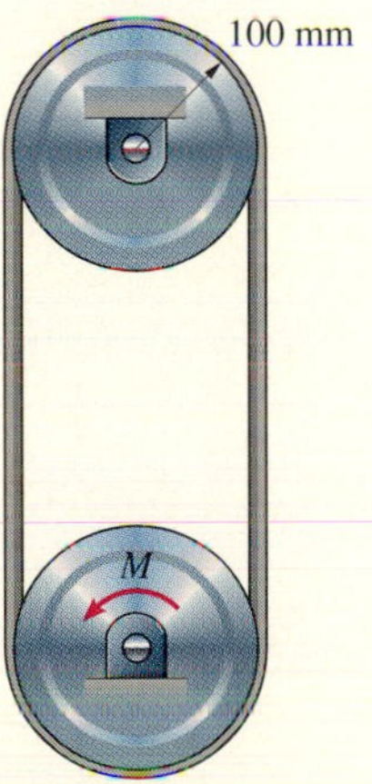

P8.99

8.100 The ring gear is fixed. The mass and moment of inertia of the sun gear are $m_S = 22$ slugs, $I_S = 4400$ slug-ft^2. The mass and moment of inertia of each planet gear are $m_P = 2.7$ slugs, $I_P = 65$ slug-ft^2. A couple $M = 600$ ft-lb is applied to the sun gear. Use work and energy to determine the sun gear's angular velocity after it has turned 100 revolutions.

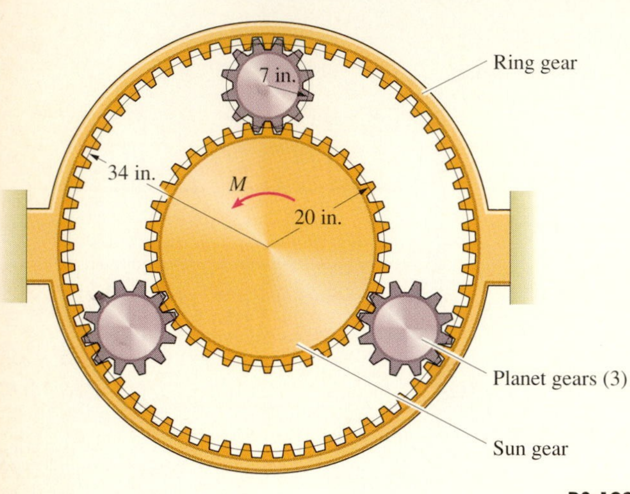

P8.100

8.101 The moments of inertia of gears A and B are $I_A = 0.014$ slug-ft^2 and $I_B = 0.100$ slug-ft^2. Gear A is connected to a torsional spring with constant $k = 0.2$ ft-lb/rad. If the spring is unstretched, and the surface supporting the 5-lb weight is removed, what is the weight's velocity when it has fallen 3 in.?

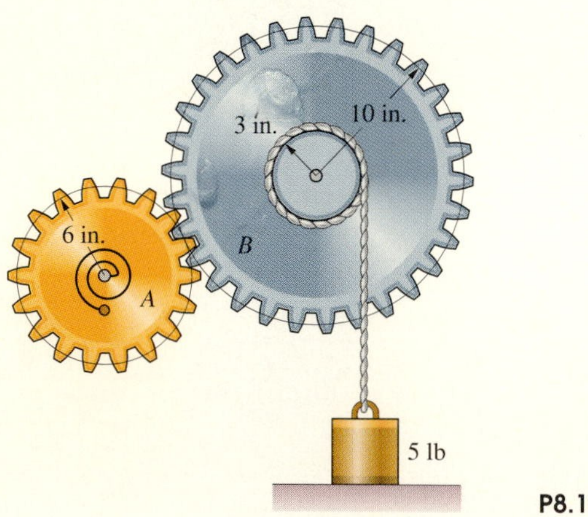

P8.101

8.102 Consider the system in Problem 8.101.
(a) What maximum distance does the 5-lb weight fall when the supporting surface is removed?
(b) What maximum velocity does the weight achieve?

8.103 Each of the go-cart's front wheels weighs 5 lb and has a moment of inertia of 0.01 slug-ft^2. The two rear wheels and rear axle form a single rigid body weighing 40 lb and having a moment of inertia of 0.1 slug-ft^2. The total weight of the rider and go-cart, including its wheels, is 240 lb. The go-cart starts from rest, its engine exerts a constant torque of 15 ft-lb on the rear axle, and its wheels do not slip. If you neglect friction and aerodynamic drag, how fast is it moving when it has traveled 50 ft?

P8.103

8.104 Determine the maximum power and the average power transmitted to the go-cart in Problem 8.103 by its engine.

8.105 The system starts from rest with the 4-kg slender bar horizontal. The mass of the suspended cylinder is 10 kg. What is the bar's angular velocity when it is in the position shown?

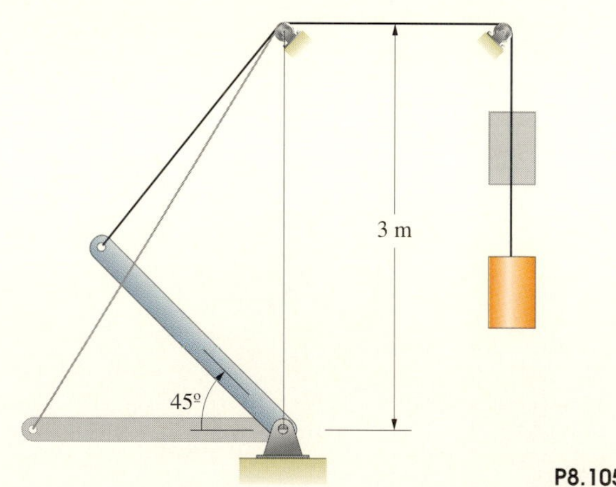

P8.105

8.106 The 0.1-kg slender bar and 0.2-kg cylindrical disk are released from rest with the bar horizontal. The disk rolls on the curved surface. What is the bar's angular velocity when it is vertical?

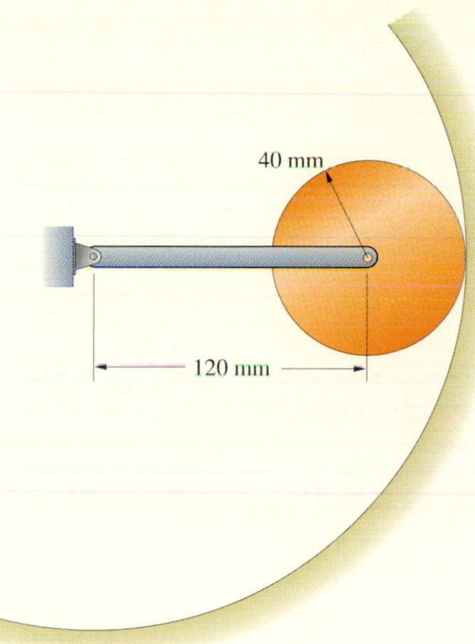

P8.106

8.107 A slender bar of mass m is released from rest in the vertical position and allowed to fall. Neglecting friction and assuming that it remains in contact with the floor and wall, determine its angular velocity as a function of θ.

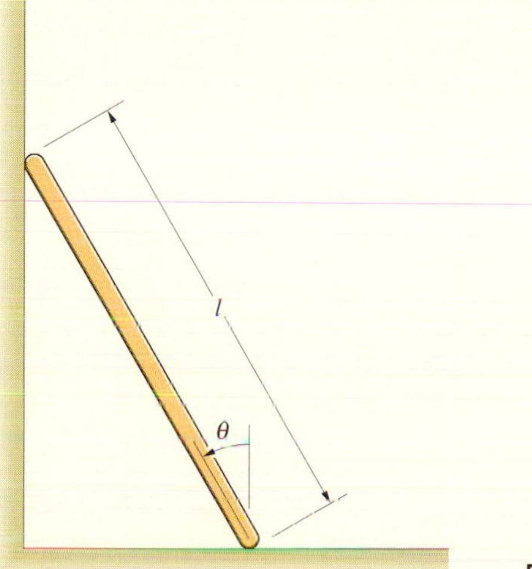

P8.107

8.108 The 4-kg slender bar is pinned to 2-kg sliders at A and B. If friction is negligible and the system starts from rest in the position shown, what is the bar's angular velocity when the slider at A has fallen 0.5 m?

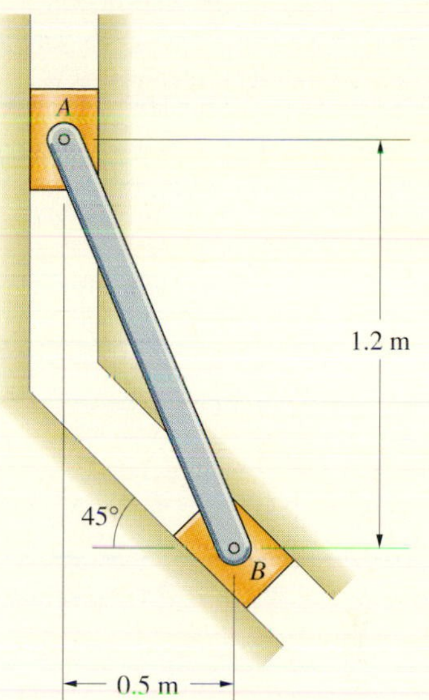

P8.108

8.109 A homogeneous hemisphere of mass m is released from rest in the position shown. If it rolls on the horizontal surface, what is its angular velocity when its flat surface is horizontal?

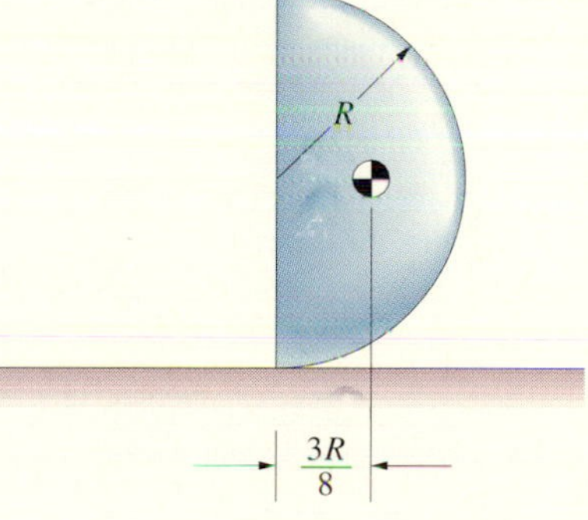

P8.109

8.110 What normal force is exerted on the hemisphere in Problem 8.109 by the horizontal surface at the instant its flat surface is horizontal?

8.111 The slender bar rotates freely *in the horizontal plane* about a vertical shaft at O. The bar weighs 20 lb and its length is 6 ft. The slider A weighs 2 lb. If the bar's angular velocity is $\omega = 10$ rad/s and the radial component of the velocity of A is zero when $r = 1$ ft, what is the angular velocity of the bar when $r = 4$ ft? (The moment of inertia of A about its center of mass is negligible; that is, treat A as a particle.)

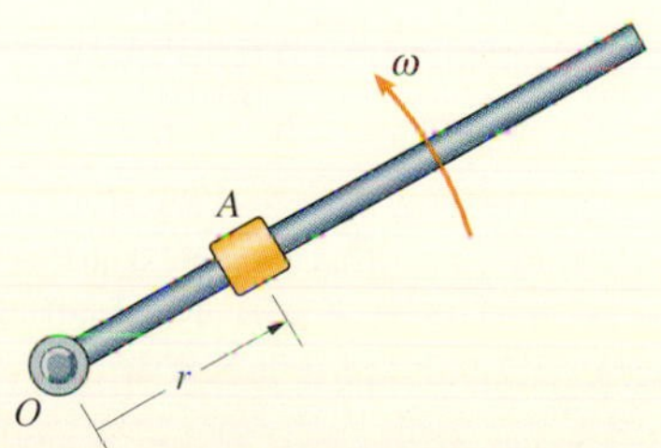

P8.111

8.112 A satellite is deployed with angular velocity $\omega = 1$ rad/s (Fig. a). Two internally stored antennas that span the diameter of the satellite are then extended, and the satellite's angular velocity decreases to ω' (Fig. b). By modeling the satellite as a 500-kg sphere of 1.2-m radius and each antenna as a 10-kg slender bar, determine ω'.

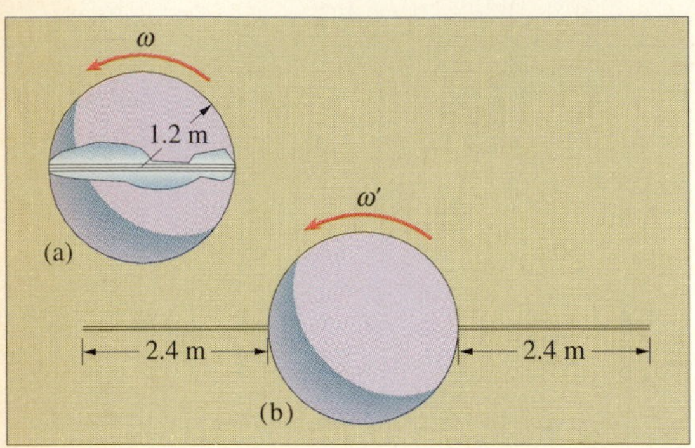

P8.112

8.113 An engineer decides to control the angular velocity of a satellite by deploying small masses attached to cables. If the angular velocity of the satellite in configuration (a) is 4 rpm, determine the distance d in configuration (b) that will cause the angular velocity to be 1 rpm. The moment of inertia of the satellite is $I = 500$ kg-m^2 and each mass is 2 kg. (Assume that the cables and masses rotate with the same angular velocity as the satellite. Neglect the masses of the cables and the moments of inertia of the masses about their centers of mass.)

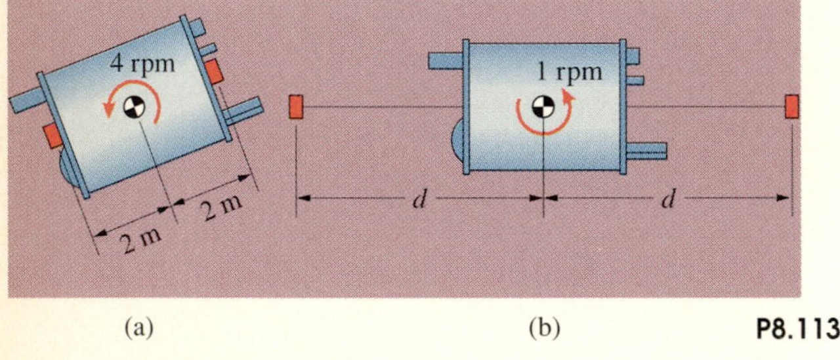

(a) (b) P8.113

8.114 A homogeneous cylindrical disk of mass m rolls on the horizontal surface with angular velocity ω. If it does not slip or leave the slanted surface when it comes into contact with it, what is the angular velocity ω' of the disk immediately afterward?

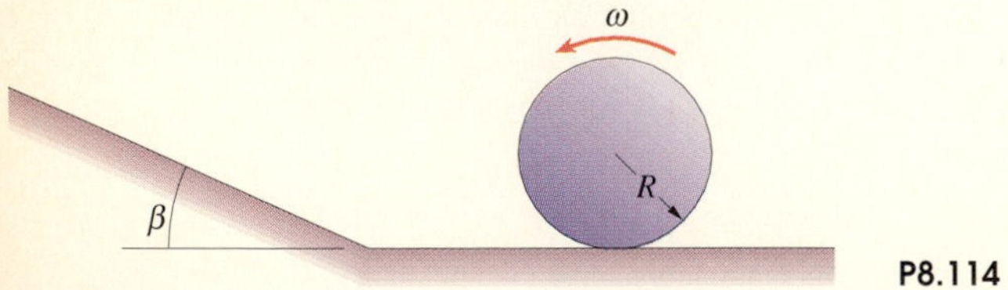

P8.114

8.115 The 10-lb slender bar falls from rest in the vertical position and hits the smooth projection at B. The coefficient of restitution of the impact is $e = 0.6$, the duration of the impact is 0.1 s, and $b = 1$ ft. Determine the average force exerted on the bar at B as a result of the impact.

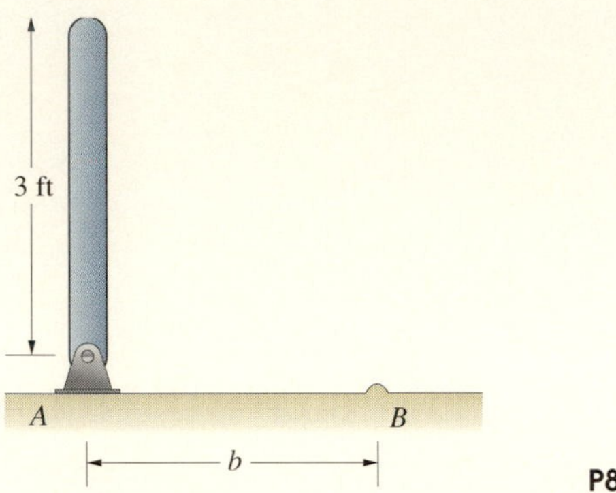

P8.115

8.116 In Problem 8.115, determine the value of b for which the average force exerted on the bar at A as a result of the impact is zero.

8.117 The 1-kg sphere A is moving at 2 m/s when it strikes the end of the 2-kg stationary slender bar B. If the velocity of the sphere after the impact is 0.8 m/s to the right, what is the coefficient of restitution?

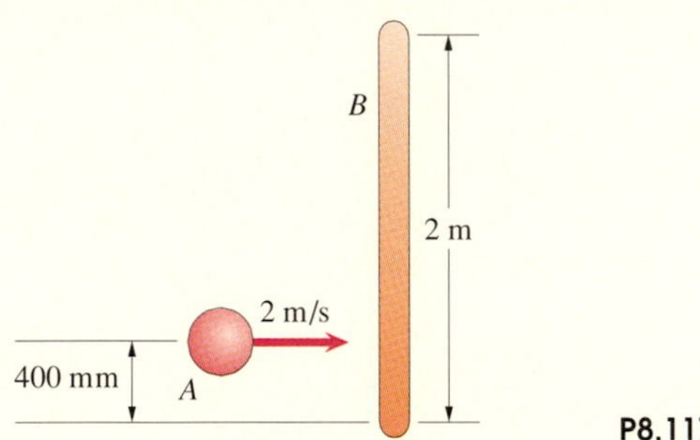

P8.117

8.118 The slender bar is released from rest in the position shown in Fig. (a) and falls a distance $h = 1$ ft. When the bar hits the floor, its tip is supported by a depression and remains on the floor (Fig. b). The length of the bar is 1 ft and its weight is 4 oz. What is its angular velocity ω just after it hits the floor?

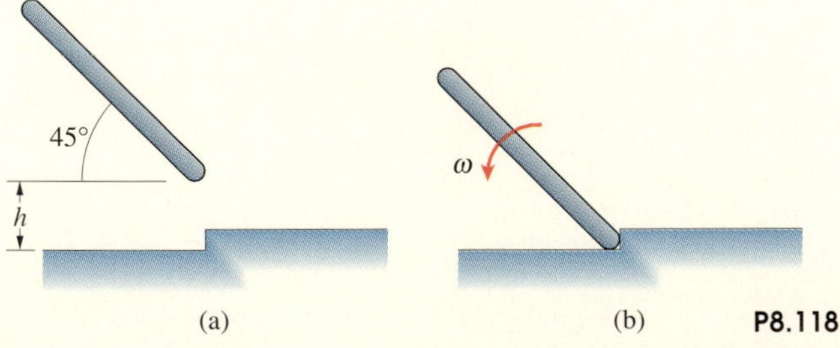

(a) (b) P8.118

8.119 The slender bar is released from rest with $\theta = 45°$ and falls a distance $h = 1$ m onto the smooth floor. The length of the bar is 1 m and its mass is 2 kg. If the coefficient of restitution of the impact is $e = 0.4$, what is the bar's angular velocity just after it hits the floor?

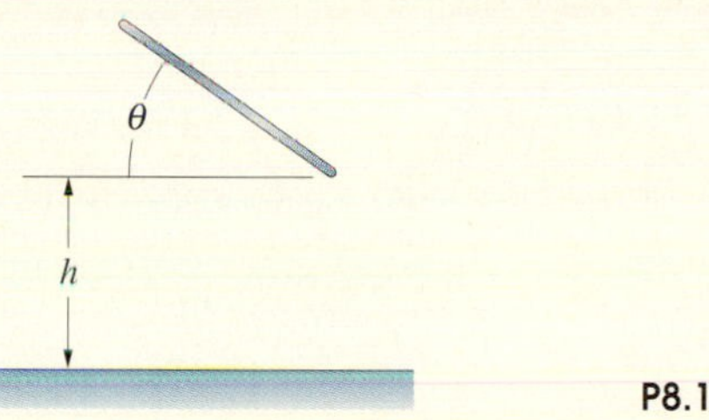

P8.119

8.120 In Problem 8.119, determine the angle θ for which the angular velocity of the bar just after it hits the floor is a maximum. What is the maximum angular velocity?

8.121 A nonrotating slender bar A moving with velocity v_0 strikes a stationary slender bar B. Each bar has mass m and length l. If the bars adhere when they collide, what is their angular velocity after the impact?

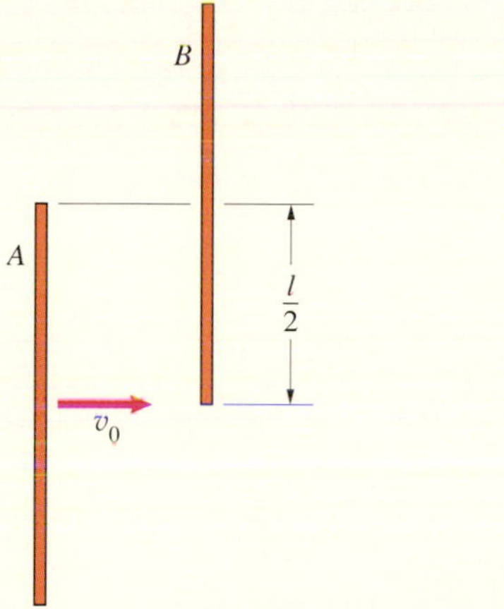

P8.121

8.122 An astronaut translates toward a nonrotating satellite at $1.0\mathbf{i}$ (m/s) relative to the satellite. Her mass is 136 kg, and the moment of inertia about the axis through her center of mass parallel to the z axis is 45 kg-m^2. The mass of the satellite is 450 kg and its moment of inertia about the z axis is 675 kg-m^2. At the instant she attaches to the satellite and begins moving with it, the position of her center of mass is $(-1.8, -0.9, 0)$ m. The axis of rotation of the satellite after she attaches is parallel to the z axis. What is their angular velocity?

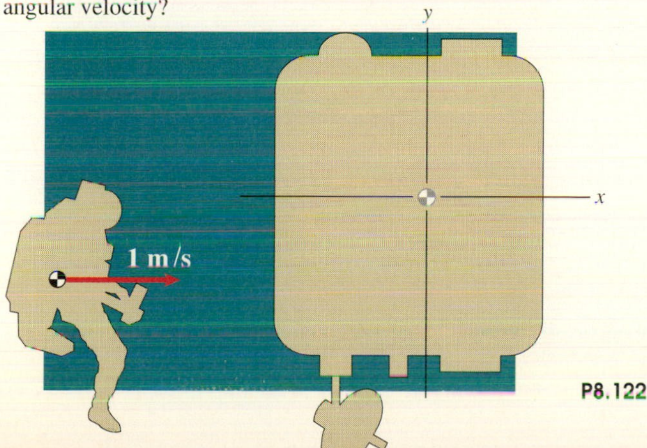

P8.122

8.123 In Problem 8.122, suppose that the design parameters of the satellite's control system require that its angular velocity not exceed 0.02 rad/s. If the astronaut is moving parallel to the x axis and the position of her center of mass when she attaches is $(-1.8, -0.9, 0)$ m, what is the maximum relative velocity at which she should approach the satellite?

8.124 A 2800-lb car skidding on ice strikes a concrete abutment at 3 mi/hr. The car's moment of inertia about its center of mass is 1800 slug-ft^2. Assume that the impacting surfaces are smooth and parallel to the y axis and that the coefficient of restitution of the impact is $e = 0.8$. What are the car's angular velocity and the velocity of its center of mass after the impact?

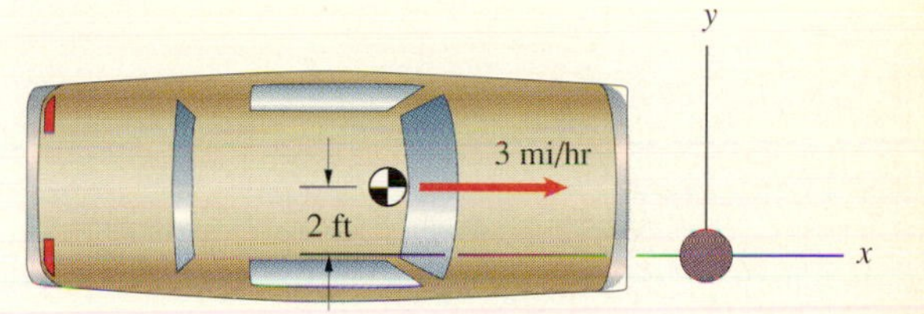

P8.124

8.125 A 170-lb wide receiver jumps vertically to receive a pass and is stationary at the instant he catches the ball. At the same instant, he is hit at P by a 180-lb linebacker moving horizontally at 15 ft/s. The wide receiver's moment of inertia about his center of mass is 7 slug-ft^2. If you model the players as rigid bodies and assume that the coefficient of restitution is $e = 0$, what is the wide receiver's angular velocity immediately after the impact?

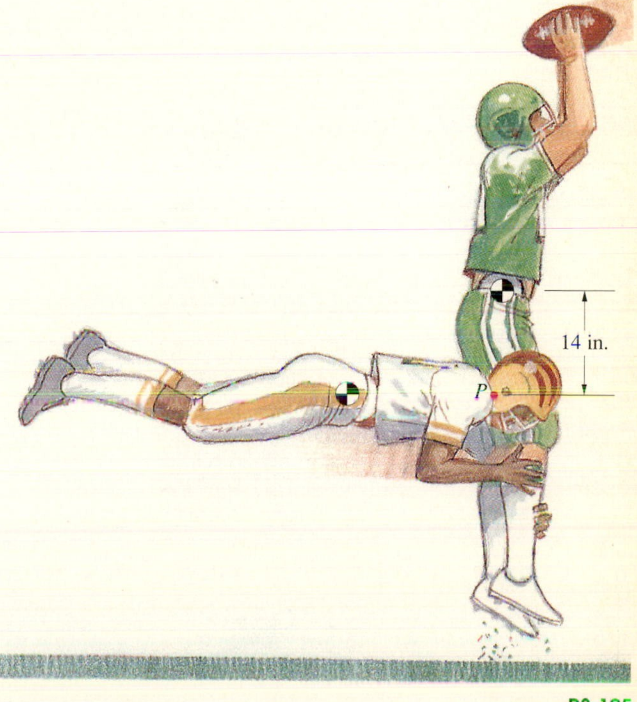

P8.125

The orientation of the biplane, or any rigid body, can be described by three angles specifying rotations of a body-fixed coordinate system relative to a primary reference frame. The change in the biplane's orientation as a function of time is governed by equations of three-dimensional angular motion that relate the forces and couples acting on the biplane to its angular acceleration. In this chapter we analyze three-dimensional motions of rigid bodies.
Z
y
x
z
θ
Y
φ
ψ
X

Chapter 9

Three-Dimensional Kinematics and Dynamics of Rigid Bodies

UNTIL now, our discussion of the dynamics of rigid bodies has dealt only with planar motion. But for many engineering applications of dynamics, such as the design of airplanes and other vehicles, we must consider three-dimensional motion. Our first step is to explain how three-dimensional motion of a rigid body is described. We then derive the equations of motion and use them to analyze simple three-dimensional motions. Finally, we introduce the Eulerian angles used to specify the orientation of a rigid body in three dimensions and express the equations of angular motion in terms of them.

9.1 *Kinematics*

If you ride a bicycle in a straight path, the wheels undergo planar motions; but if you are turning, their motions are three-dimensional (Fig. 9.1a). An airplane can remain in planar motion while in level flight, descending, climbing, or performing loops. But if it banks and turns, it is in three-dimensional motion (Fig. 9.1b). If you spin a top, it may remain in planar motion for a brief period, rotating about a fixed vertical axis; but eventually the top's axis begins to tilt and rotate. The top is then in three-dimensional motion and exhibits interesting, apparently gravity-defying behavior (Fig. 9.1c). In this section we begin the analysis of such motions by discussing kinematics of rigid bodies in three-dimensional motion.

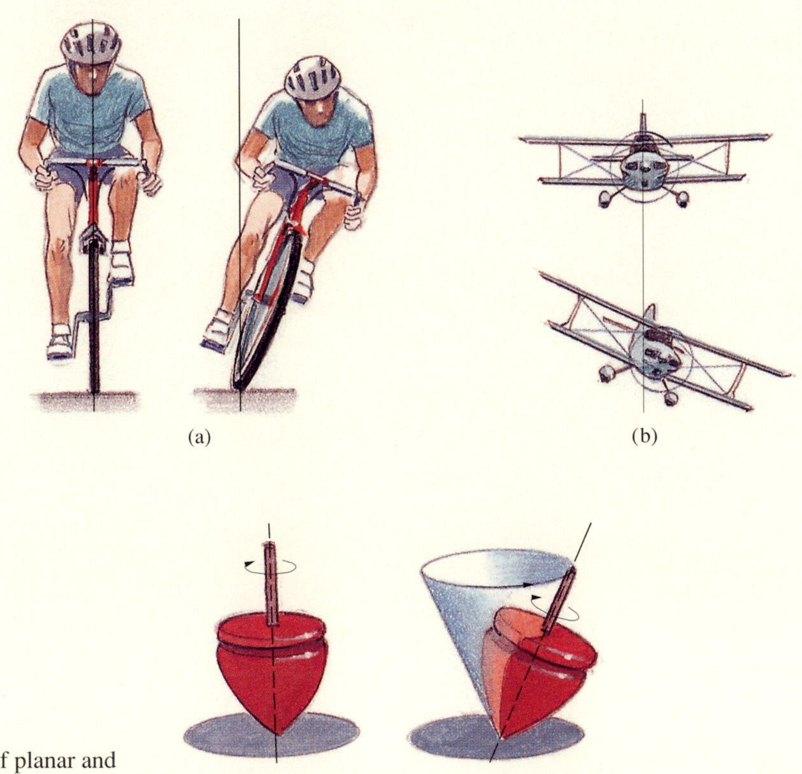

Fig. 9.1
Examples of planar and three-dimensional motions.

Velocities and Accelerations

We have already discussed some of the concepts needed to describe three-dimensional motion of a rigid body relative to a given reference frame. In Chapter 6, we showed that Euler's theorem implies that a rigid body undergoing any motion other than translation has an instantaneous axis of rotation. The direction of this axis at a particular instant and the rate at which the rigid body rotates about the axis are specified by the angular velocity vector $\boldsymbol{\omega}$.

Furthermore, we have shown that a rigid body's velocity is completely specified by its angular velocity and the velocity of a single point of the rigid body. For the rigid body and reference frame in Fig. 9.2, suppose that we know the angular velocity $\boldsymbol{\omega}$ and the velocity $\mathbf{v}_B$ of a point B. Then the velocity of any other point A of the rigid body is given by Eq. (6.8):

$$\mathbf{v}_A = \mathbf{v}_B + \boldsymbol{\omega} \times \mathbf{r}_{A/B}. \tag{9.1}$$

A rigid body's acceleration is completely specified by its angular acceleration vector $\boldsymbol{\alpha} = d\boldsymbol{\omega}/dt$, its angular velocity vector, and the acceleration of a single point. If we know $\boldsymbol{\alpha}$, $\boldsymbol{\omega}$, and the acceleration $\mathbf{a}_B$ of the point B in Fig. 9.2, the acceleration of any other point A of the rigid body is given by Eq. (6.9):

$$\mathbf{a}_A = \mathbf{a}_B + \boldsymbol{\alpha} \times \mathbf{r}_{A/B} + \boldsymbol{\omega} \times (\boldsymbol{\omega} \times \mathbf{r}_{A/B}). \tag{9.2}$$

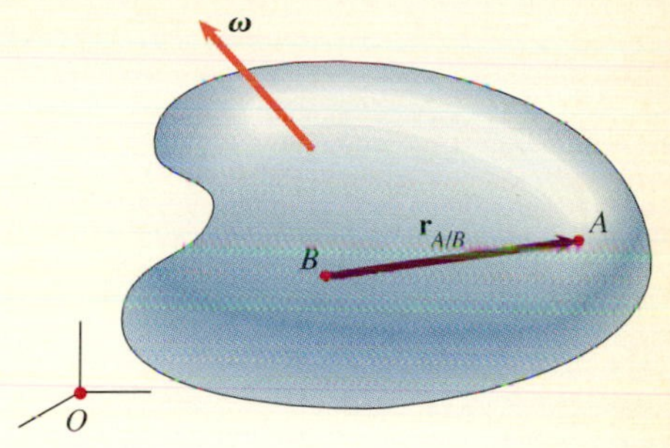

Fig. 9.2
Points A and B of a rigid body. The velocity of A can be determined if the velocity of B and the rigid body's angular velocity $\boldsymbol{\omega}$ are known. The acceleration of A can be determined if the acceleration of B, the angular velocity, and the angular acceleration are known.

Moving Reference Frames

The velocities and accelerations in Eqs. (9.1) and (9.2) are measured relative to the reference frame shown in Fig. 9.2, which we refer to as the primary reference frame. In this chapter we also use secondary reference frames that move relative to the primary reference frame. In some situations, the secondary reference frame will be body-fixed. In other situations, we will find it convenient to use a secondary reference frame that moves relative to the primary reference frame but is not body-fixed. The secondary reference frame and its motion will be chosen for convenience in describing the motion of a particular rigid body.

Let us consider a secondary reference frame xyz whose rate of rotation relative to the primary reference frame is described by an angular velocity vector $\boldsymbol{\Omega}$, and a rigid body whose angular velocity vector relative to the primary reference frame is $\boldsymbol{\omega}$ (Fig. 9.3). If the secondary reference frame is body-fixed, $\boldsymbol{\Omega} = \boldsymbol{\omega}$. If we express $\boldsymbol{\omega}$ in terms of its components in the secondary reference frame,

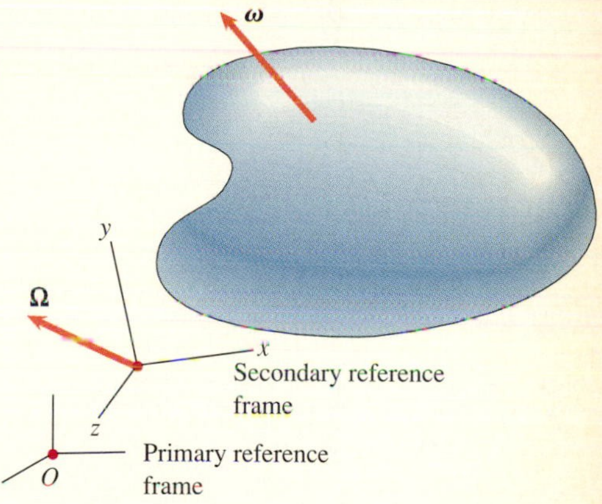

Fig. 9.3
The primary and secondary reference frames. The vector $\boldsymbol{\Omega}$ is the angular velocity of the secondary reference frame relative to the primary reference frame. The vector $\boldsymbol{\omega}$ is the angular velocity of the rigid body relative to the primary reference frame.

$$\boldsymbol{\omega} = \omega_x\mathbf{i} + \omega_y\mathbf{j} + \omega_z\mathbf{k},$$

the rigid body's angular acceleration relative to the primary reference frame is

$$\boldsymbol{\alpha} = \frac{d\boldsymbol{\omega}}{dt} = \frac{d\omega_x}{dt}\mathbf{i} + \omega_x\frac{d\mathbf{i}}{dt} + \frac{d\omega_y}{dt}\mathbf{j} + \omega_y\frac{d\mathbf{j}}{dt} + \frac{d\omega_z}{dt}\mathbf{k} + \omega_z\frac{d\mathbf{k}}{dt}. \tag{9.3}$$

In terms of the angular velocity $\boldsymbol{\Omega}$ of the secondary reference frame, the time derivatives of the unit vectors $\mathbf{i}$, $\mathbf{j}$, and $\mathbf{k}$ of the secondary reference frame are (see Section 6.5)

$$\frac{d\mathbf{i}}{dt} = \boldsymbol{\Omega} \times \mathbf{i}, \qquad \frac{d\mathbf{j}}{dt} = \boldsymbol{\Omega} \times \mathbf{j}, \qquad \frac{d\mathbf{k}}{dt} = \boldsymbol{\Omega} \times \mathbf{k}.$$

Substituting these expressions into Eq. (9.3), we obtain the angular acceleration of the rigid body relative to the primary reference frame in the form

$$\boldsymbol{\alpha} = \frac{d\omega_x}{dt}\mathbf{i} + \frac{d\omega_y}{dt}\mathbf{j} + \frac{d\omega_z}{dt}\mathbf{k} + \boldsymbol{\Omega} \times \boldsymbol{\omega}. \tag{9.4}$$

When the secondary reference frame is not body-fixed, it is often convenient to express the rigid body's angular velocity $\boldsymbol{\omega}$ as the sum of the angular velocity $\boldsymbol{\Omega}$ of the secondary reference frame and the angular velocity $\boldsymbol{\omega}_{\text{rel}}$ of the rigid body relative to the secondary reference frame:

$$\boldsymbol{\omega} = \boldsymbol{\Omega} + \boldsymbol{\omega}_{\text{rel}}. \tag{9.5}$$

The following examples demonstrate the use of Eqs. (9.1), (9.2), (9.4), and (9.5) to analyze three-dimensional motions of rigid bodies. You will often find that the simplest way to determine the angular velocity $\boldsymbol{\omega}$ and angular acceleration $\boldsymbol{\alpha}$ of a rotating rigid body is to first determine its angular velocity $\boldsymbol{\omega}_{\text{rel}}$ relative to a secondary reference frame and then use Eqs. (9.5) and (9.4).

Example 9.1

At the instant shown in Fig. 9.4, the velocity of the biplane's center of mass G expressed in terms of the body-fixed coordinate system is $20\mathbf{i}$ (m/s) and its acceleration is $-6\mathbf{i} + 4\mathbf{k}$ (m/s^2). Onboard rate gyros indicate that the biplane's angular velocity and angular acceleration are $\boldsymbol{\omega} = -6\mathbf{i} + 2\mathbf{j}$ (rad/s) and $\boldsymbol{\alpha} = 20\mathbf{i} - 12\mathbf{j}$ (rad/s^2). Determine the velocity and acceleration of point A of the airplane with coordinates $x_A = -1.4$ m, $y_A = 0.1$ m, and $z_A = 0.2$ m.

Fig. 9.4

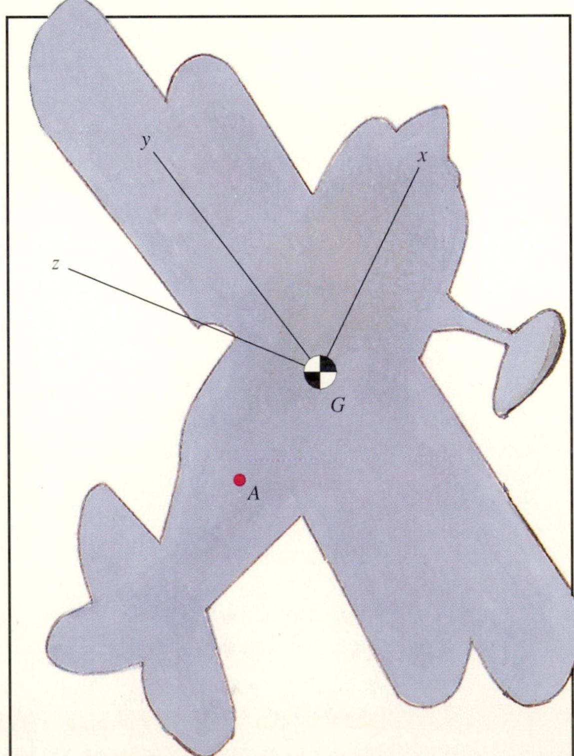

STRATEGY

In this example the primary reference frame (the reference frame relative to which the biplane's motion is specified) is fixed relative to the earth, and we want to determine the velocity and acceleration of point A relative to the earth. We know the velocity and acceleration of point G and the biplane's angular velocity and angular acceleration, so we can use Eqs. (9.1) and (9.2) to determine the velocity and acceleration of point A.

SOLUTION

Velocity With Eq. (9.1) we can express the velocity of point A in terms of the velocity of the center of mass G and the biplane's angular velocity:

$$\mathbf{v}_A = \mathbf{v}_G + \boldsymbol{\omega} \times r_{A/G}.$$

The position of point A relative to point G is

$$\begin{aligned}\mathbf{r}_{A/G} &= x_A\mathbf{i} + y_A\mathbf{j} + z_A\mathbf{k}\\ &= -1.4\mathbf{i} + 0.1\mathbf{j} + 0.2\mathbf{k}\ \text{(m)}.\end{aligned}$$

Evaluating the term $\boldsymbol{\omega} \times \mathbf{r}_{A/G}$,

$$\boldsymbol{\omega} \times \mathbf{r}_{A/G} = \begin{vmatrix} \mathbf{i} & \mathbf{j} & \mathbf{k} \\ -6 & 2 & 0 \\ -1.4 & 0.1 & 0.2 \end{vmatrix} = 0.4\mathbf{i} + 1.2\mathbf{j} + 2.2\mathbf{k}\ \text{(m/s)},$$

we obtain the velocity of point A:

$$\begin{aligned}\mathbf{v}_A &= \mathbf{v}_G + \boldsymbol{\omega} \times \mathbf{r}_{A/G}\\ &= 20\mathbf{i} + (0.4\mathbf{i} + 1.2\mathbf{j} + 2.2\mathbf{k})\\ &= 20.4\mathbf{i} + 1.2\mathbf{j} + 2.2\mathbf{k}\ \text{(m/s)}.\end{aligned}$$

Acceleration From Eq. (9.2), the acceleration of point A is

$$\begin{aligned}\mathbf{a}_A &= \mathbf{a}_G + \boldsymbol{\alpha} \times \mathbf{r}_{A/G} + \boldsymbol{\omega} \times (\boldsymbol{\omega} \times \mathbf{r}_{A/G})\\ &= -6\mathbf{i} + 4\mathbf{k} + \begin{vmatrix} \mathbf{i} & \mathbf{j} & \mathbf{k} \\ 20 & -12 & 0 \\ -1.4 & 0.1 & 0.2 \end{vmatrix} + \begin{vmatrix} \mathbf{i} & \mathbf{j} & \mathbf{k} \\ -6 & 2 & 0 \\ 0.4 & 1.2 & 2.2 \end{vmatrix}\\ &= -4\mathbf{i} + 9.2\mathbf{j} - 18.8\mathbf{k}\ (\text{m/s}^2).\end{aligned}$$

DISCUSSION

Although $\mathbf{v}_A$ and $\mathbf{a}_A$ are the velocity and acceleration of point A relative to the primary earth-fixed reference frame, notice that they are expressed in components in terms of the body-fixed reference frame.

Example 9.2

The tire in Fig. 9.5 is rolling on the level surface. As the car turns, the midpoint B of the tire moves at 5 m/s in a circular path about the fixed point P (see the top view), and the tire remains perpendicular to the line from B to P.
(a) What is the tire's angular velocity $\boldsymbol{\omega}$ relative to an earth-fixed reference frame?
(b) Determine the velocity of point A, the rearmost point of the tire at the instant shown.

Fig. 9.5

STRATEGY

(a) In Fig. (a) we introduce a secondary reference frame with its origin at B and its y axis along the line from B to P. We assume that the x axis *remains horizontal*. The motion of this reference frame is simple: It rotates about its z axis as the car turns. The motion of the tire relative to this secondary reference frame is also simple: It rotates about the y axis. By determining the angular velocity $\boldsymbol{\Omega}$ of the secondary reference frame and the angular velocity $\boldsymbol{\omega}_{\text{rel}}$ of the tire relative to the secondary reference frame, we can use Eq. (9.5) to determine the tire's angular velocity relative to an earth-fixed reference frame.
(b) Knowing the velocity of point B and the tire's angular velocity $\boldsymbol{\omega}$, we can use Eq. (9.1) to determine the velocity of point A.

SOLUTION

(a) The magnitude of the angular velocity of the line PB about P is (5 m/s)/(10 m) = 0.5 rad/s. The secondary coordinate system rotates about its z axis in the clockwise direction as viewed in Fig. (a), so its angular velocity is

$$\boldsymbol{\Omega} = -0.5\mathbf{k} \text{ (rad/s)}.$$

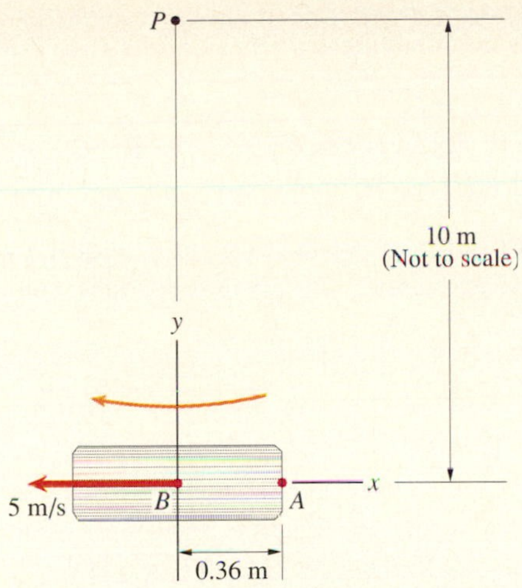

(a) A rotating secondary coordinate system. The y axis remains aligned with the line BP, and the x axis remains horizontal.

The center of the tire moves at 5 m/s, so the magnitude of the tire's angular velocity about the y axis is (5 m/s)/(0.36 m) = 13.9 rad/s. The tire's angular velocity relative to the secondary coordinate system is

$$\boldsymbol{\omega}_{\text{rel}} = -13.9\mathbf{j} \text{ (rad/s)}.$$

Therefore the tire's angular velocity relative to an earth-fixed reference frame is

$$\boldsymbol{\omega} = \boldsymbol{\Omega} + \boldsymbol{\omega}_{\text{rel}} = -13.9\mathbf{j} - 0.5\mathbf{k} \text{ (rad/s)}.$$

(b) The position vector of point A relative to point B is $\mathbf{r}_{A/B} = 0.36\mathbf{i}$ (m). From Eq. (9.1), the velocity of point A is

$$\begin{aligned} \mathbf{v}_A &= \mathbf{v}_B + \boldsymbol{\omega} \times r_{A/B} \\ &= -5\mathbf{i} + \begin{vmatrix} \mathbf{i} & \mathbf{j} & \mathbf{k} \\ 0 & -13.9 & -0.5 \\ 0.36 & 0 & 0 \end{vmatrix} \\ &= -5\mathbf{i} - 0.18\mathbf{j} + 5\mathbf{k} \text{ (m/s)}. \end{aligned}$$

DISCUSSION

Notice how using a secondary coordinate system rotating with the car simplified the determination of the tire's angular velocity. Although the tire's motion in space is quite complicated, its motion relative to the secondary coordinate system is simple.

Example 9.3

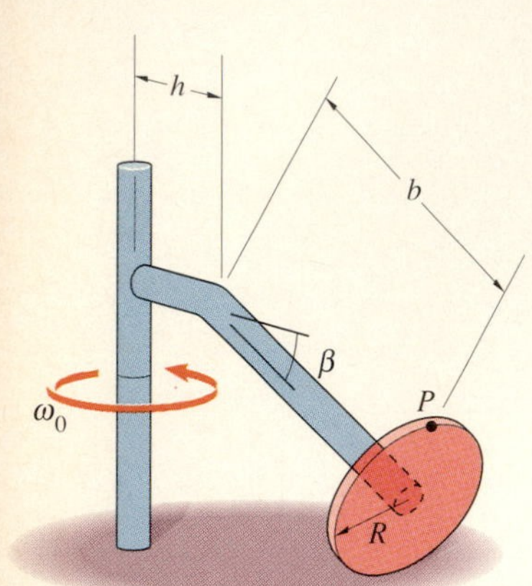

Fig. 9.6

The bent bar in Fig. 9.6 is rigidly attached to the vertical shaft, which rotates with constant angular velocity ω_0. The circular disk is pinned to the bent bar and rolls on the horizontal surface.

(a) Determine the disk's angular velocity $\boldsymbol{\omega}_{\text{disk}}$ and angular acceleration $\boldsymbol{\alpha}_{\text{disk}}$.

(b) Determine the velocity of point P, which is the uppermost point of the circular disk at the present instant.

STRATEGY

(a) In this example the primary reference frame is one that is fixed with respect to the surface on which the disk rolls. To simplify our analysis of the disk's angular motion, we will use a secondary coordinate system that is body-fixed with respect to the bent bar. By applying Eq. (9.1) to the bent bar, we will determine the velocity of the center of the disk. We will determine the disk's angular velocity by recognizing that the velocity of the point of the disk in contact with the horizontal surface is zero. We can then use Eq. (9.4) to determine the disk's angular acceleration. (b) Knowing the velocity of the center of the disk and the disk's angular velocity, we can apply Eq. (9.1) to the disk to determine the velocity of point P.

SOLUTION

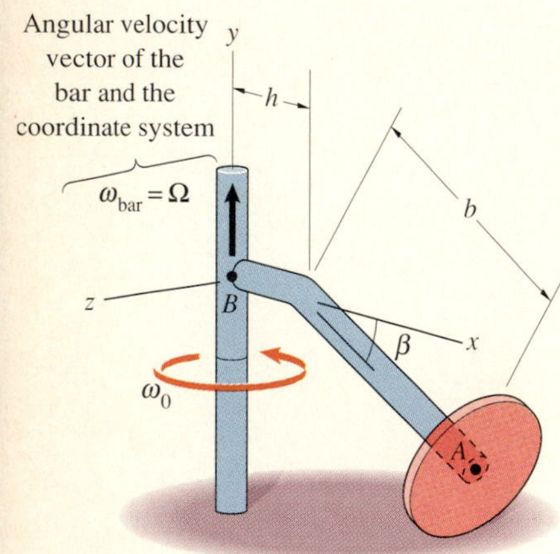

(a) A secondary coordinate system fixed with respect to the bent bar.

(a) Let the coordinate system in Fig. (a) be body-fixed with respect to the bent bar. The x axis coincides with the horizontal part of the bar, and the y axis coincides with the vertical shaft. The angular velocity $\boldsymbol{\omega}_{\text{bar}}$ of the bar and the angular velocity $\boldsymbol{\Omega}$ of the coordinate system are equal:

$$\boldsymbol{\omega}_{\text{bar}} = \boldsymbol{\Omega} = \omega_0 \mathbf{j}.$$

Let point B be the stationary origin of the coordinate system and let point A be the center of the disk (Fig. a). The position vector of A relative to B is

$$\mathbf{r}_{A/B} = (h + b\cos\beta)\mathbf{i} - b\sin\beta\mathbf{j}.$$

From Eq. (9.1), the velocity of point A is

$$\begin{aligned}\mathbf{v}_A &= \mathbf{v}_B + \boldsymbol{\omega}_{\text{bar}} \times \mathbf{r}_{A/B} \\ &= \mathbf{0} + \begin{vmatrix} \mathbf{i} & \mathbf{j} & \mathbf{k} \\ 0 & \omega_0 & 0 \\ h + b\cos\beta & -b\sin\beta & 0 \end{vmatrix} \\ &= -\omega_0(h + b\cos\beta)\mathbf{k}.\end{aligned}$$

Because the coordinate system is body-fixed with respect to the bent bar, we can write the angular velocity of the disk relative to the coordinate system as (Fig. b)

$$\boldsymbol{\omega}_{\text{rel}} = \omega_{\text{rel}}\cos\beta\mathbf{i} - \omega_{\text{rel}}\sin\beta\mathbf{j}.$$

Let point C in Fig. (b) be the point of the disk in contact with the surface. To determine ω_{rel}, we use the condition that $\mathbf{v}_C = \mathbf{0}$. The position of C relative to A is

$$\mathbf{r}_{C/A} = -R\sin\beta\,\mathbf{i} - R\cos\beta\,\mathbf{j}.$$

Therefore

$$\mathbf{v}_C = \mathbf{v}_A + \boldsymbol{\omega}_{\text{disk}} \times \mathbf{r}_{C/A}$$

$$= -\omega_0(h + b\cos\beta)\mathbf{k} + \begin{vmatrix} \mathbf{i} & \mathbf{j} & \mathbf{k} \\ \omega_{\text{rel}}\cos\beta & \omega_0 - \omega_{\text{rel}}\sin\beta & 0 \\ -R\sin\beta & -R\cos\beta & 0 \end{vmatrix} = \mathbf{0}.$$

Solving this equation for ω_{rel}, we obtain

$$\omega_{\text{rel}} = -\omega_0\left(\frac{h}{R} + \frac{b}{R}\cos\beta - \sin\beta\right).$$

The angular velocity of the coordinate system is $\boldsymbol{\Omega} = \omega_0\,\mathbf{j}$, so the disk's angular velocity $\boldsymbol{\omega}_{\text{disk}}$ is

$$\boldsymbol{\omega}_{\text{disk}} = \boldsymbol{\Omega} + \boldsymbol{\omega}_{\text{rel}} = \omega_{\text{rel}}\cos\beta\,\mathbf{i} + (\omega_0 - \omega_{\text{rel}}\sin\beta)\mathbf{j}.$$

Even though the components of $\boldsymbol{\omega}_{\text{disk}}$ are constants, we find from Eq. (9.4) that the disk's angular acceleration $\boldsymbol{\alpha}_{\text{disk}}$ is not zero:

$$\boldsymbol{\alpha}_{\text{disk}} = \boldsymbol{\Omega} \times \boldsymbol{\omega}_{\text{disk}} = \begin{vmatrix} \mathbf{i} & \mathbf{j} & \mathbf{k} \\ 0 & \omega_0 & 0 \\ \omega_{\text{rel}}\cos\beta & \omega_0 - \omega_{\text{rel}}\sin\beta & 0 \end{vmatrix}$$

$$= -\omega_0\omega_{\text{rel}}\cos\beta\,\mathbf{k}.$$

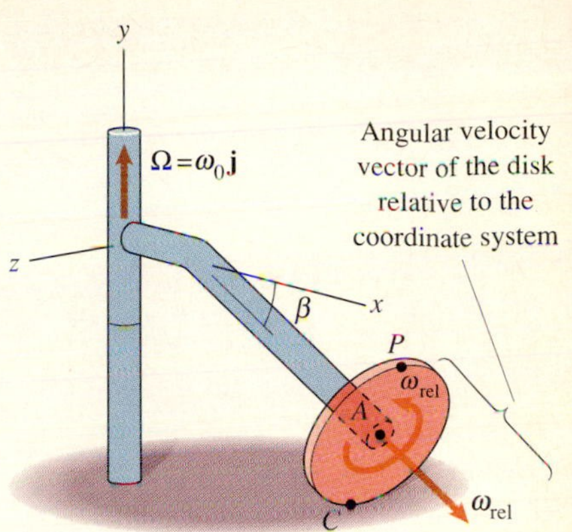

(b) Analyzing the motion of the disk.

(b) The position vector of point P relative to the center of the disk is

$$\mathbf{r}_{P/A} = R\sin\beta\,\mathbf{i} + R\cos\beta\,\mathbf{j}.$$

Using Eq. (9.1) and our result for the velocity $\mathbf{v}_A$ of the center of the disk, we determine the velocity of point P:

$$\mathbf{v}_P = \mathbf{v}_A + \boldsymbol{\omega}_{\text{disk}} \times \mathbf{r}_{P/A}$$

$$= -\omega_0(h + b\cos\beta)\mathbf{k} + \begin{vmatrix} \mathbf{i} & \mathbf{j} & \mathbf{k} \\ \omega_{\text{rel}}\cos\beta & \omega_0 - \omega_{\text{rel}}\sin\beta & 0 \\ R\sin\beta & R\cos\beta & 0 \end{vmatrix}$$

$$= [-\omega_0(h + b\cos\beta + R\sin\beta) + \omega_{\text{rel}}R]\mathbf{k}$$

$$= -2\omega_0(h + b\cos\beta)\mathbf{k}.$$

DISCUSSION

The use of a secondary coordinate system rotating with the bent bar simplified our determination of the disk's angular velocity $\boldsymbol{\omega}_{\text{disk}}$ and angular acceleration $\boldsymbol{\alpha}_{\text{disk}}$. Although the disk's angular motion is complicated, its angular motion relative to the bar is simple.

Example 9.4

The bar AB in Fig. 9.7 is connected by ball and socket joints to collars sliding on the fixed bars CD and EF. At the present instant the collar at B has constant velocity $\mathbf{v}_B = 20\mathbf{i}$ (m/s). The angular velocity and angular acceleration of bar AB about its axis are zero. Determine the velocity and acceleration of the collar at A and the angular velocity and angular acceleration of bar AB.

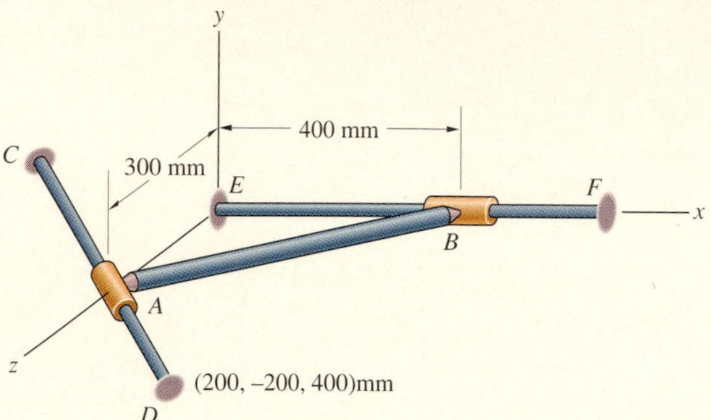

Fig. 9.7

STRATEGY

Since we know the velocity of the collar at B, if we apply Eq. (9.1) to points A and B we will obtain three scalar equations in four unknowns: the magnitude of the velocity of point A and the three components of the angular velocity $\boldsymbol{\omega}$ of the bar AB. Because the angular velocity of bar AB about its axis is zero, $\mathbf{r}_{A/B} \cdot \boldsymbol{\omega} = 0$, which supplies the required fourth equation. We can then determine the acceleration of the collar at A and the angular acceleration of bar AB using the same procedure, applying Eq. (9.2) to points A and B.

SOLUTION

Velocity By dividing the position vector of point D relative to point A by its magnitude, we obtain a unit vector that points from A toward D:

$$\mathbf{e}_{AD} = \frac{0.2\mathbf{i} - 0.2\mathbf{j} + 0.1\mathbf{k}}{\sqrt{(0.2)^2 + (-0.2)^2 + (0.1)^2}} = 0.667\mathbf{i} - 0.667\mathbf{j} + 0.333\mathbf{k}.$$

We can write the velocity of the collar at A as

$$\mathbf{v}_A = v_A \mathbf{e}_{AD} = v_A(0.667\mathbf{i} - 0.667\mathbf{j} + 0.333\mathbf{k}).$$

We apply Eq. (9.1) to the bar AB:

$$\mathbf{v}_A = \mathbf{v}_B + \boldsymbol{\omega} \times \mathbf{r}_{A/B}:$$

$$v_A(0.667\mathbf{i} - 0.667\mathbf{j} + 0.333\mathbf{k}) = 20\mathbf{i} + \begin{vmatrix} \mathbf{i} & \mathbf{j} & \mathbf{k} \\ \omega_x & \omega_y & \omega_z \\ -0.4 & 0 & 0.3 \end{vmatrix}.$$

Equating **i**, **j**, and **k** components, we obtain

$$0.667v_A = 20 + 0.3\omega_y,$$
$$-0.667v_A = -0.4\omega_z - 0.3\omega_x,$$
$$0.333v_A = 0.4\omega_y.$$

Since the angular velocity of bar AB about its axis is zero, we know that

$$\mathbf{r}_{A/B}\cdot\boldsymbol{\omega} = -0.4\omega_x + 0.3\omega_z = 0.$$

We therefore have four equations in terms of v_A, ω_x, ω_y, and ω_z. Solving, we obtain

$$v_A = 48 \text{ m/s}, \qquad \boldsymbol{\omega} = 38.4\mathbf{i} + 40\mathbf{j} + 51.2\mathbf{k} \text{ (rad/s)}.$$

Acceleration We can write the acceleration of the collar at A as

$$\mathbf{a}_A = a_A\mathbf{e}_{AD} = a_A(0.667\mathbf{i} - 0.667\mathbf{j} + 0.333\mathbf{k}).$$

We apply Eq. (9.2) to the bar AB:

$$\mathbf{a}_A = \mathbf{a}_B + \boldsymbol{\alpha}\times\mathbf{r}_{A/B} + \boldsymbol{\omega}\times(\boldsymbol{\omega}\times\mathbf{r}_{A/B}):$$

$$a_A(0.667\mathbf{i} - 0.667\mathbf{j} + 0.333\mathbf{k}) = \mathbf{0} + \begin{vmatrix} \mathbf{i} & \mathbf{j} & \mathbf{k} \\ \alpha_x & \alpha_y & \alpha_z \\ -0.4 & 0 & 0.3 \end{vmatrix}$$

$$+ \begin{vmatrix} \mathbf{i} & \mathbf{j} & \mathbf{k} \\ \omega_x & \omega_y & \omega_z \\ 0.3\omega_y & -0.4\omega_z - 0.3\omega_x & 0.4\omega_y \end{vmatrix}.$$

Equating the **i**, **j**, and **k** components yields

$$0.667a_A = 0.3\alpha_y + 0.4\omega_y^2 + 0.4\omega_z^2 + 0.3\omega_x\omega_z,$$
$$-0.667a_A = -0.4\alpha_z - 0.3\alpha_x + 0.3\omega_y\omega_z - 0.4\omega_x\omega_y,$$
$$0.333a_A = 0.4\alpha_y - 0.4\omega_x\omega_z - 0.3\omega_x^2 - 0.3\omega_y^2.$$

Because the angular acceleration of bar AB about its axis equals zero,

$$\mathbf{r}_{A/B}\cdot\boldsymbol{\alpha} = -0.4\alpha_x + 0.3\alpha_z = 0.$$

We know the components of the angular velocity, so we have four equations in terms of a_A, α_x, α_y, and α_z. Solving, we obtain

$$a_A = 8540 \text{ m/s}^2, \qquad \boldsymbol{\alpha} = 6840\mathbf{i} + 11{,}390\mathbf{j} + 9110\mathbf{k} \text{ (rad/s}^2\text{)}.$$

Problems

9.1 Relative to a primary reference frame, a rigid body's angular velocity is $\boldsymbol{\omega} = 200\mathbf{i} + 900\mathbf{j} - 600\mathbf{k}$ (rad/s). The position of its center of mass relative to the origin O of the reference frame is $\mathbf{r}_G = 6\mathbf{i} + 6\mathbf{j} + 2\mathbf{k}$ (m), and the velocity of its center of mass is $\mathbf{v}_G = 100\mathbf{i} + 80\mathbf{j} - 60\mathbf{k}$ (m/s). What is the velocity of a point A of the rigid body whose position relative to O is $\mathbf{r}_A = 5.8\mathbf{i} + 6.4\mathbf{j} + 1.6\mathbf{k}$ (m)?

Strategy: Use Eq. (9.1) to express $\mathbf{v}_A$ in terms of $\mathbf{v}_G$, $\boldsymbol{\omega}$, and $\mathbf{r}_{A/G}$.

9.2 The angular acceleration of the rigid body in Problem 9.1 is $\boldsymbol{\alpha} = 8000\mathbf{i} - 8000\mathbf{j} - 4000\mathbf{k}$ (rad/s^2), and the acceleration of its center of mass is zero. What is the acceleration of point A?

9.3 The airplane's rate gyros indicate that its angular velocity is $\boldsymbol{\omega} = 4.0\mathbf{i} + 6.4\mathbf{j} + 0.2\mathbf{k}$ (rad/s). What is the velocity relative to the center of mass of the point A with coordinates (8, 2, 2) ft?

P9.3

9.4 The rate gyros of the airplane in Problem 9.3 indicate that its angular acceleration is $\boldsymbol{\alpha} = -4\mathbf{i} + 12\mathbf{j} + 2\mathbf{k}$ (rad/s^2). What is the acceleration of point A relative to the center of mass?

9.5 Relative to the reference frame shown, the rectangular parallelepiped is rotating about a fixed axis through points A and B. Its direction of rotation is clockwise when the axis of rotation is viewed from point A toward point B.

(a) What is its angular velocity vector $\boldsymbol{\omega}$?

(b) What are the velocities of points C and D?

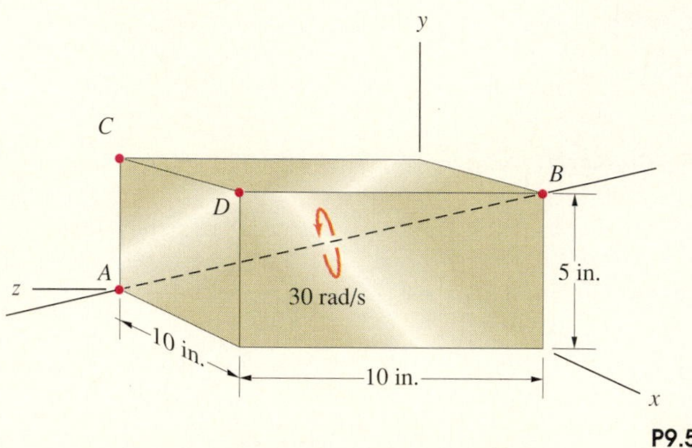

P9.5

9.6 If the angular velocity of the rectangular parallelepiped in Problem 9.5 is constant, what is the acceleration of point C?

9.7 Relative to the reference frame shown, the turbine is rotating about a fixed axis coincident with the line OA.

(a) What is its angular velocity vector?

(b) What is the velocity of the point of the turbine with coordinates (3, 2, 2) m?

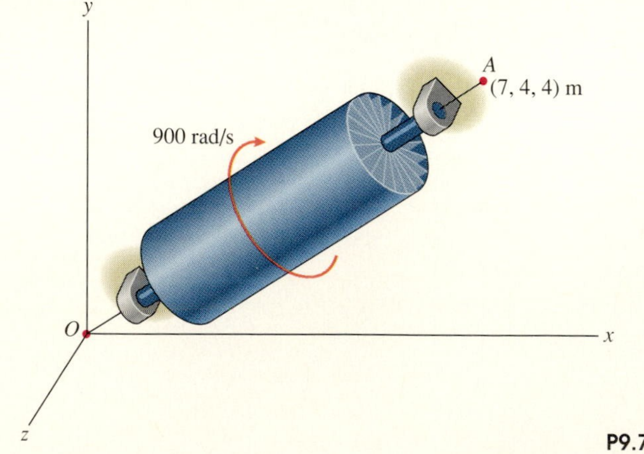

P9.7

9.8 The 900-rad/s angular velocity of the turbine in Problem 9.7 is decreasing at 100 rad/s^2.

(a) What is the turbine's angular acceleration vector?

(b) What is the acceleration of the point of the turbine with coordinates (3, 2, 2) m?

9.9 The disk of radius R is supported by the vertical shaft. Relative to an earth-fixed reference frame, the shaft rotates with constant angular velocity ω_0. The disk rotates with constant angular velocity ω_d relative to the shaft. Determine:
(a) the disk's angular velocity $\boldsymbol{\omega}$ relative to the earth-fixed reference frame;
(b) the velocity of point A of the disk relative to the earth-fixed reference frame.

Strategy: (a) Assume that the y axis remains vertical and the x axis remains perpendicular to the disk. Determine $\boldsymbol{\omega}$ by expressing it as the sum of the angular velocity $\boldsymbol{\Omega}$ of the coordinate system and the angular velocity $\boldsymbol{\omega}_{rel}$ of the disk relative to the secondary coordinate system. (b) Use Eq. (9.1) to express $\mathbf{v}_A$ in terms of the stationary center of the disk.

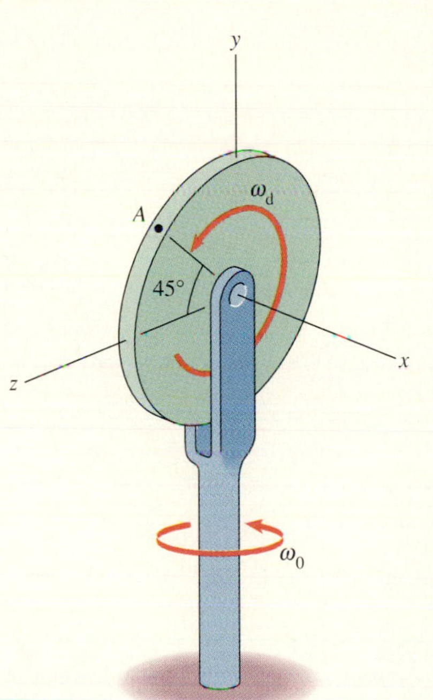

P9.9

9.10 For the disk in Problem 9.9, determine: (a) the disk's angular acceleration $\boldsymbol{\alpha}$ relative to the earth-fixed reference frame; (b) the acceleration of point A of the disk.

9.11 The base of the dish antenna is rotating at 1 rad/s. The angle $\theta = 30°$ and is increasing at 20°/s.
(a) What are the components of the antenna's angular velocity vector $\boldsymbol{\omega}$ in terms of the body-fixed coordinate system shown?
(b) What is the velocity of the point of the antenna with coordinates (2, 2, −2) m?

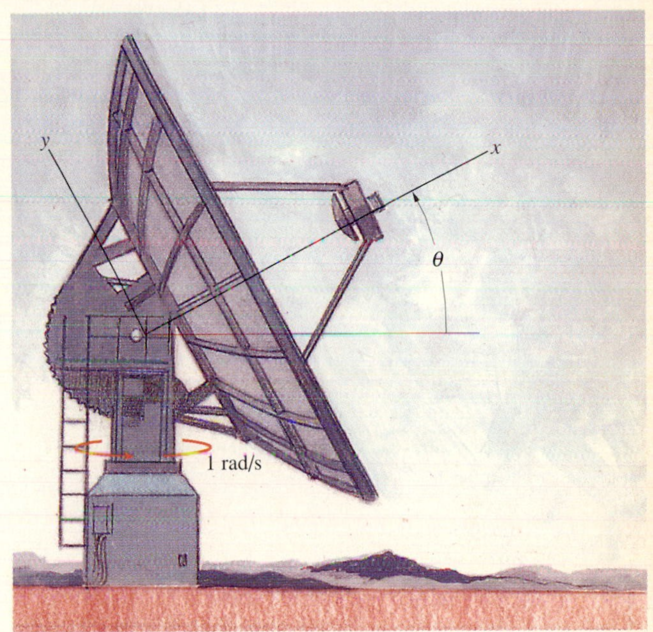

P9.11

9.12 The circular disk rotates with angular velocity ω_d relative to the horizontal bar, the horizontal bar rotates with angular velocity ω_b about the z axis, and the vertical shaft rotates with angular velocity ω_0. In terms of components in the coordinate system shown, determine: (a) the angular velocity $\boldsymbol{\omega}_{disk}$ of the disk; (b) the velocity of point P, which is the uppermost point of the disk at the present instant.

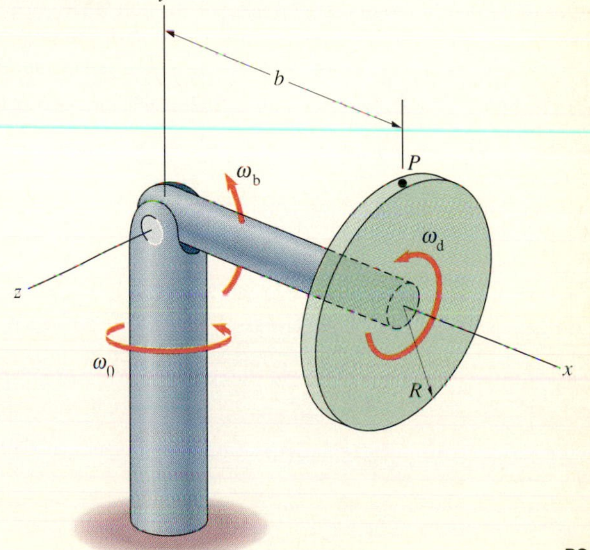

P9.12

9.13 The angular velocities given in Problem 9.12 are constant. Determine: (a) the angular acceleration $\boldsymbol{\alpha}_{disk}$ of the disk; (b) the acceleration of point P.

9.14 The bent bar is rigidly attached to the vertical shaft, which rotates with constant angular velocity ω_0. The circular disk is pinned to the bent bar and rotates with constant angular velocity ω_d relative to the bar.
(a) Determine the disk's angular velocity $\boldsymbol{\omega}_{disk}$.
(b) Determine the velocity of point P, which is the uppermost point of the circular disk at the present instant.

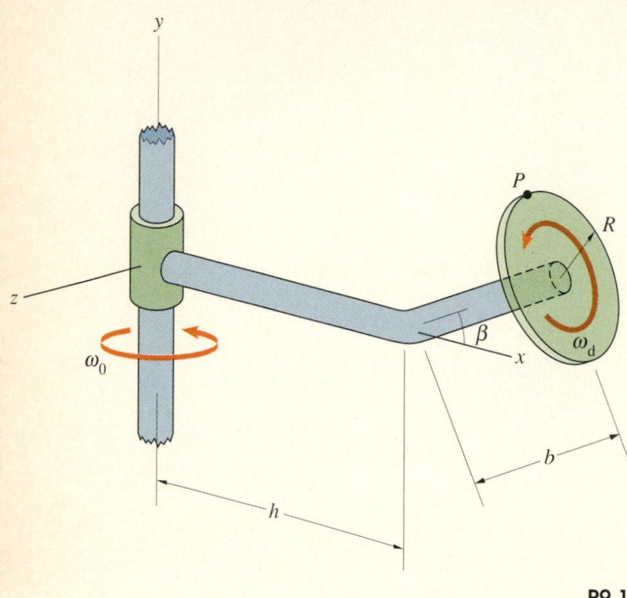

P9.14

9.15 In Problem 9.14, determine: (a) the disk's angular acceleration $\boldsymbol{\alpha}_{disk}$; (b) the acceleration of point P.

9.16 Relative to a primary reference frame, the gyroscope's circular frame rotates about the vertical axis at 2 rad/s in the counterclockwise direction when viewed from above. The 2.4-in.-diameter wheel rotates relative to the frame at 10 rad/s. Determine the velocities of points A and B relative to the primary reference frame.

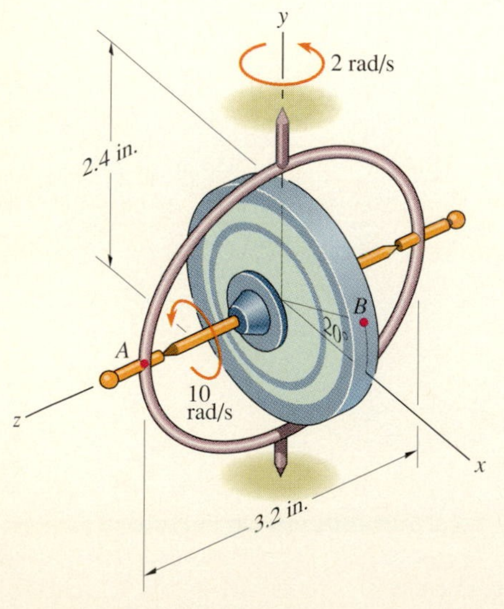

P9.16

9.17 If the angular velocities of the frame and wheel of the gyroscope in Problem 9.16 are constant, what are the accelerations of points A and B relative to the primary reference frame?

9.18 Relative to an earth-fixed reference frame, the manipulator rotates about the vertical axis with angular velocity $\omega_y = 0.1$ rad/s. The y axis of the secondary coordinate system remains vertical, and the x axis rotates with the manipulator so that points A, B, and C remain in the x-y plane. The angular velocities of the arms AB and BC *relative to the secondary coordinate system* are $-0.2\mathbf{k}$ (rad/s) and $0.4\mathbf{k}$ (rad/s), respectively.
(a) What is the angular velocity $\boldsymbol{\omega}_{BC}$ of arm BC relative to the earth-fixed reference frame?
(b) What is the velocity of point C relative to the earth-fixed reference frame?

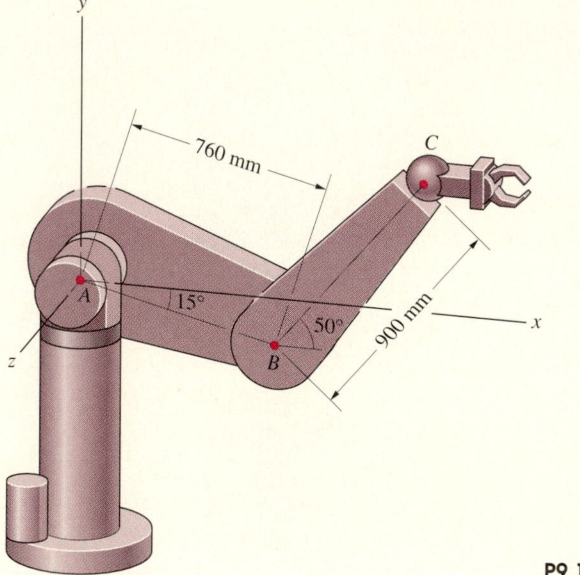

P9.18

9.19 The angular velocity of the manipulator in Problem 9.18 about the vertical axis is constant. The angular accelerations of the arms AB and BC relative to the secondary coordinate system are zero. What is the acceleration of point C relative to the earth-fixed reference frame?

9.20 The cone's curved surface rolls on the horizontal surface. The x axis of the secondary coordinate system remains coincident with the cone's axis and the z axis remains horizontal. The z axis has a constant angular velocity ω_0 in the horizontal plane. In terms of components in the secondary coordinate system, determine:
(a) the angular velocity $\boldsymbol{\Omega}$ of the secondary coordinate system;
(b) the angular velocity $\boldsymbol{\omega}_{rel}$ of the cone relative to the secondary

coordinate system; (c) the cone's angular velocity $\boldsymbol{\omega}$ relative to the primary reference frame.

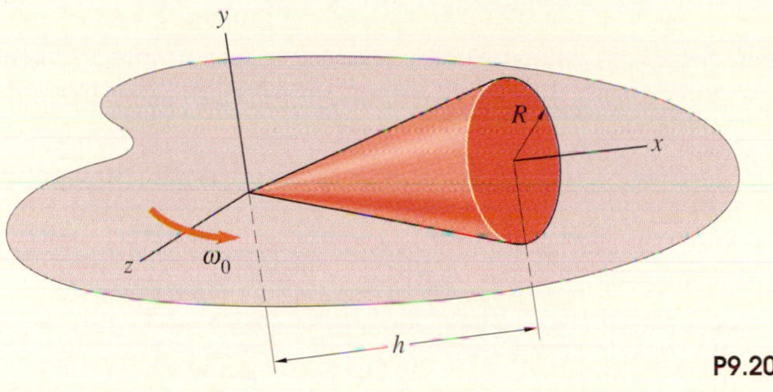

P9.20

9.21 In Problem 9.20, determine the velocity relative to the primary reference frame of the point on the base of the cone with coordinates $x = h$, $y = 0$, $z = R$.

9.22 In Problem 9.20, determine: (a) the cone's angular acceleration $\boldsymbol{\alpha}$ relative to the primary reference frame; (b) the acceleration relative to the primary reference frame of the point on the base of the cone with coordinates $x = h$, $y = 0$, $z = R$.

9.23 A tilted cylinder of length l and radius R undergoes a steady motion in which one end rolls on the plane surface while the center of the cylinder remains stationary relative to the surface. The z axis of the secondary coordinate system remains coincident with the cylinder's axis and the y axis remains horizontal. The angular velocity of the y axis in the horizontal plane is ω_0. In terms of components in the secondary coordinate system, determine: (a) the angular velocity $\boldsymbol{\Omega}$ of the secondary coordinate system; (b) the angular velocity $\boldsymbol{\omega}_{\text{rel}}$ of the cylinder relative to the secondary coordinate system; (c) the cone's angular velocity $\boldsymbol{\omega}$ relative to the primary reference frame.

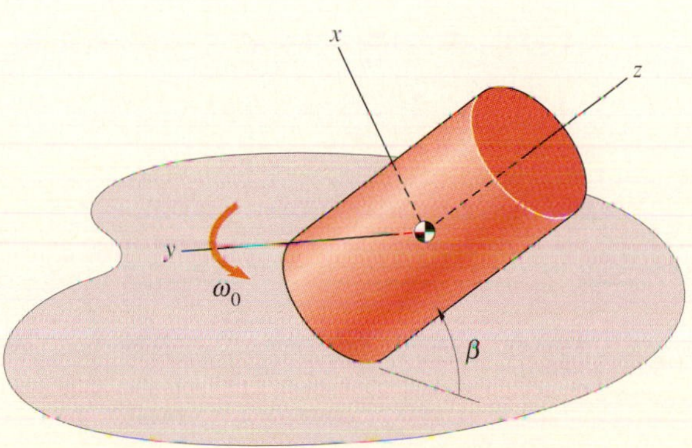

P9.23

9.24 In Problem 9.23, determine the velocity relative to the primary reference frame of the point on the upper base of the cylinder with coordinates $x = R$, $y = 0$, $z = \frac{1}{2}l$.

9.25 In Problem 9.23, determine: (a) the cylinder's angular acceleration $\boldsymbol{\alpha}$ relative to the primary reference frame; (b) the acceleration relative to the primary reference frame of the point on the upper base of the cylinder with coordinates $x = R$, $y = 0$, $z = \frac{1}{2}l$.

9.26 The bar AB is connected by ball and socket joints to the edge of the horizontal circular disk at A and to a collar that slides on the vertical bar at B. The disk rotates with constant angular velocity $\omega_{\text{d}} = 4$ rad/s and the angular velocity of bar AB about its axis is zero. Determine the velocity of the collar at B and the angular velocity of bar AB.

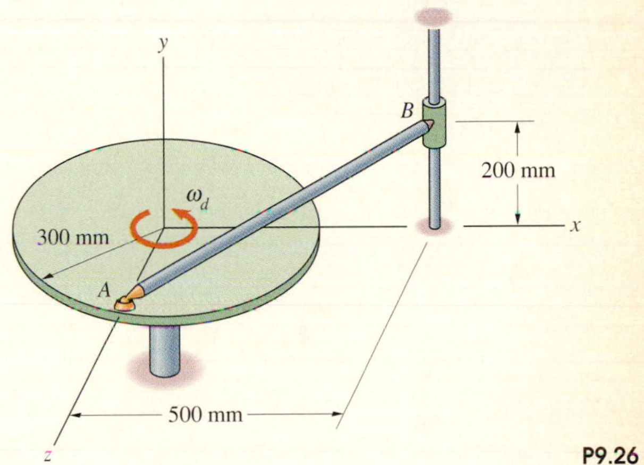

P9.26

9.27 In Problem 9.26, the angular acceleration of bar AB about its axis is zero. Determine the acceleration of the collar at B and the angular acceleration of bar AB.

9.2 Angular Momentum

Just as in the case of planar motion, the equations governing three-dimensional motion of a rigid body consist of Newton's second law and equations of angular motion. In comparison to the simple equations governing angular motion in two dimensions, the equations of angular motion in three dimensions are more complicated. Three equations relate the components of the total moment about each coordinate axis to the components of the rigid body's angular acceleration and angular velocity. In this section we begin deriving the equations of angular motion by obtaining expressions for the angular momentum of a rigid body in three-dimensional motion. We first consider rotation about a fixed point, then general three-dimensional motion.

Rotation About a Fixed Point

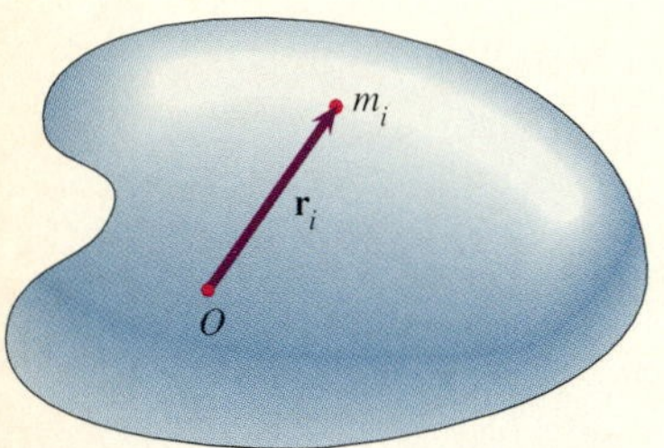

Fig. 9.8
Mass and position of the ith particle of a rigid body.

Let m_i be the mass of the ith particle of a rigid body, and let $\mathbf{r}_i$ be its position relative to a point O that is fixed with respect to an inertial primary reference frame (Fig. 9.8). The angular momentum of the rigid body about O is the sum of the angular momenta of its particles,

$$\mathbf{H}_O = \sum_i \mathbf{r}_i \times m_i \mathbf{v}_i,$$

where $\mathbf{v}_i = d\mathbf{r}_i/dt$. Let us assume that the rigid body rotates about the fixed point O with angular velocity $\boldsymbol{\omega}$. Then we can express the velocity of the ith particle as $\mathbf{v}_i = \boldsymbol{\omega} \times \mathbf{r}_i$, and the angular momentum is

$$\mathbf{H}_O = \sum_i \mathbf{r}_i \times m_i(\boldsymbol{\omega} \times \mathbf{r}_i). \tag{9.6}$$

In terms of a coordinate system with its origin at O (Fig. 9.9), we can express the vectors $\boldsymbol{\omega}$ and $\mathbf{r}_i$ in terms of their components as

$$\boldsymbol{\omega} = \omega_x \mathbf{i} + \omega_y \mathbf{j} + \omega_z \mathbf{k},$$
$$\mathbf{r}_i = x_i \mathbf{i} + y_i \mathbf{j} + z_i \mathbf{k},$$

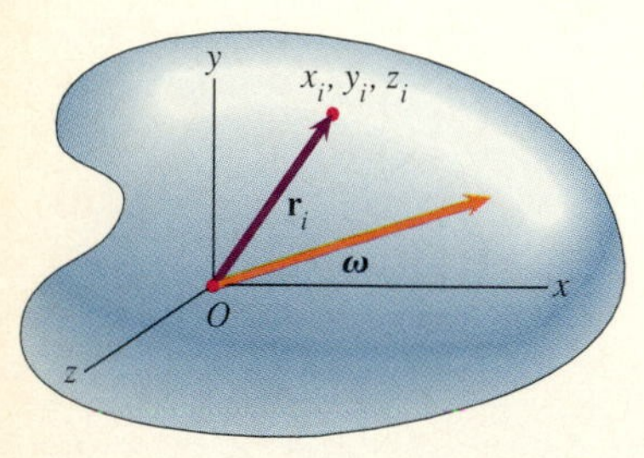

Fig. 9.9
Introducing a secondary coordinate system with its origin at O.

where (x_i, y_i, z_i) are the coordinates of the ith particle. Substituting these expressions into Eq. (9.6) and evaluating the cross products, we can write the resulting components of the angular momentum vector in the forms

$$\begin{aligned} H_{Ox} &= I_{xx}\omega_x - I_{xy}\omega_y - I_{xz}\omega_z, \\ H_{Oy} &= -I_{yx}\omega_x + I_{yy}\omega_y - I_{yz}\omega_z, \\ H_{Oz} &= -I_{zx}\omega_x - I_{zy}\omega_y + I_{zz}\omega_z. \end{aligned} \tag{9.7}$$

The coefficients

$$I_{xx} = \sum_i m_i(y_i^2 + z_i^2), \qquad I_{yy} = \sum_i m_i(x_i^2 + z_i^2), \qquad I_{zz} = \sum_i m_i(x_i^2 + y_i^2) \tag{9.8}$$

are called the **moments of inertia** about the x, y, and z axes. The coefficients

$$I_{xy} = I_{yx} = \sum_i m_i x_i y_i, \qquad I_{yz} = I_{zy} = \sum_i m_i y_i z_i, \\ I_{xz} = I_{zx} = \sum_i m_i x_i z_i \tag{9.9}$$

are called the **products of inertia**. We can write Eqs. (9.7) as the matrix equation

$$\begin{bmatrix} H_{Ox} \\ H_{Oy} \\ H_{Oz} \end{bmatrix} = \begin{bmatrix} I_{xx} & -I_{xy} & -I_{xz} \\ -I_{yx} & I_{yy} & -I_{yz} \\ -I_{zx} & -I_{zy} & I_{zz} \end{bmatrix} \begin{bmatrix} \omega_x \\ \omega_y \\ \omega_z \end{bmatrix}, \tag{9.10}$$

where

$$\begin{bmatrix} I_{xx} & -I_{xy} & -I_{xz} \\ -I_{yx} & I_{yy} & -I_{yz} \\ -I_{zx} & -I_{zy} & I_{zz} \end{bmatrix} = [I]$$

is called the **inertia matrix** of the rigid body.

Although Eqs. (9.7) appear complicated in comparison to the simple equation $H_O = I_O\omega$ for the angular momentum of a rigid body in planar motion about a fixed axis, we can point out simple correspondences between them. Suppose that the rigid body rotates about a fixed axis L_O coinciding with the x axis (Fig. 9.10). From Eqs. (9.7), the angular momentum about the x axis is

$$H_{Ox} = I_{xx}\omega_x.$$

The term $y_i^2 + z_i^2$ appearing in the definition of I_{xx} is the square of the perpendicular distance from the x axis to the ith particle, so $I_{xx} = I_O$. Therefore this equation relating H_{Ox} to ω_x is equivalent to the planar equation. Notice from Eqs. (9.7), however, that if the axis of rotation does not coincide with one of the coordinate axes, the component of the angular momentum about a coordinate axis depends in general not only on the component of the angular velocity about that axis, but also on the components of the angular velocity about the other axes through the products of inertia I_{xy}, I_{yz}, and I_{zx}.

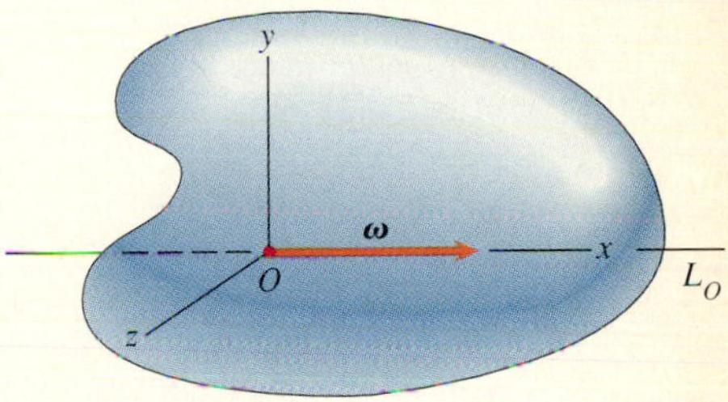

Fig. 9.10
The axis of rotation L_O coincident with the x axis.

General Motion

Here we obtain the angular momentum for a rigid body undergoing general three-dimensional motion. The derivation and the resulting equations are very similar to those for rotation about a fixed point.

Let $\mathbf{R}_i$ be the position of the ith particle of a rigid body relative to the center of mass (Fig. 9.11). The rigid body's angular momentum about its center of mass is

$$\mathbf{H} = \sum_i \mathbf{R}_i \times m_i \frac{d\mathbf{R}_i}{dt} = \sum_i \mathbf{R}_i \times m_i(\boldsymbol{\omega} \times \mathbf{R}_i).$$

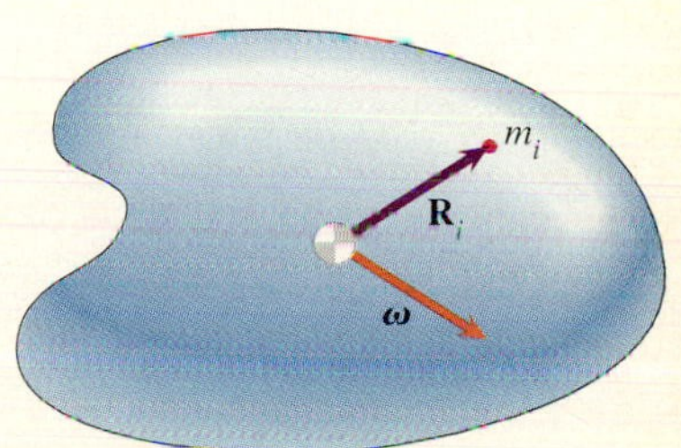

Fig. 9.11
Position of the ith particle of a rigid body relative to the center of mass.

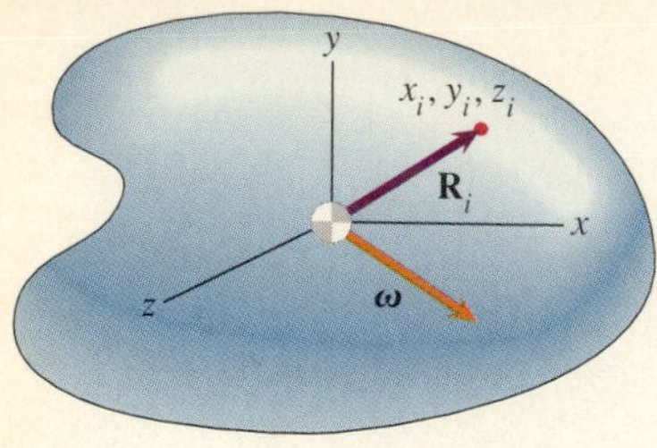

Fig. 9.12

Introducing a coordinate system with its origin at the center of mass.

Introducing a coordinate system with its origin at the center of mass (Fig. 9.12), we express $\boldsymbol{\omega}$ and $\mathbf{R}_i$ in terms of their components as

$$\boldsymbol{\omega} = \omega_x \mathbf{i} + \omega_y \mathbf{j} + \omega_z \mathbf{k},$$

$$\mathbf{R}_i = x_i \mathbf{i} + y_i \mathbf{j} + z_i \mathbf{k},$$

where (x_i, y_i, z_i) are the coordinates of the ith particle relative to the center of mass. The resulting components of the angular momentum vector are

$$\begin{aligned} H_x &= I_{xx}\omega_x - I_{xy}\omega_y - I_{xz}\omega_z, \\ H_y &= -I_{yx}\omega_x + I_{yy}\omega_y - I_{yz}\omega_z, \\ H_z &= -I_{zx}\omega_x - I_{zy}\omega_y + I_{zz}\omega_z, \end{aligned} \qquad (9.11)$$

or

$$\begin{bmatrix} H_x \\ H_y \\ H_z \end{bmatrix} = \begin{bmatrix} I_{xx} & -I_{xy} & -I_{xz} \\ -I_{yx} & I_{yy} & -I_{yz} \\ -I_{zx} & -I_{zy} & I_{zz} \end{bmatrix} \begin{bmatrix} \omega_x \\ \omega_y \\ \omega_z \end{bmatrix}. \qquad (9.12)$$

The moments and products of inertia are defined by Eqs. (9.8) and (9.9).

These equations for the angular momentum in general motion are identical in form to those we obtained for rotation about a fixed point. The expressions for the moments and products of inertia are the same. However, in the case of rotation about a fixed point, the moments and products of inertia are expressed in terms of a coordinate system with its origin at the fixed point, whereas in the case of general motion they are expressed in terms of a coordinate system with its origin at the center of mass.

9.3 *Moments and Products of Inertia*

To determine the angular momentum of a given rigid body in three-dimensional motion, you must know its moments and products of inertia. From your experience with planar rigid-body dynamics, you are familiar with evaluating an object's moment of inertia about a given axis. The same techniques apply in three-dimensional problems. We demonstrate this by evaluating the moments and products of inertia for a slender bar and a thin plate. We then extend the parallel-axis theorem to three dimensions to simplify the evaluation of the moments and products of inertia of composite objects.

Simple Objects

If we model an object as a continuous distribution of mass, we can express the inertia matrix as

$$[I] = \begin{bmatrix} I_{xx} & -I_{xy} & -I_{xz} \\ -I_{yx} & I_{yy} & -I_{yz} \\ -I_{zx} & -I_{zy} & I_{zz} \end{bmatrix}$$

$$= \begin{bmatrix} \int_m (y^2 + z^2)\,dm & -\int_m xy\,dm & -\int_m xz\,dm \\ -\int_m yx\,dm & \int_m (x^2 + z^2)\,dm & -\int_m yz\,dm \\ -\int_m zx\,dm & -\int_m zy\,dm & \int_m (x^2 + y^2)\,dm \end{bmatrix}, \quad (9.13)$$

where x, y, and z are the coordinates of the differential element of mass dm (Fig. 9.13).

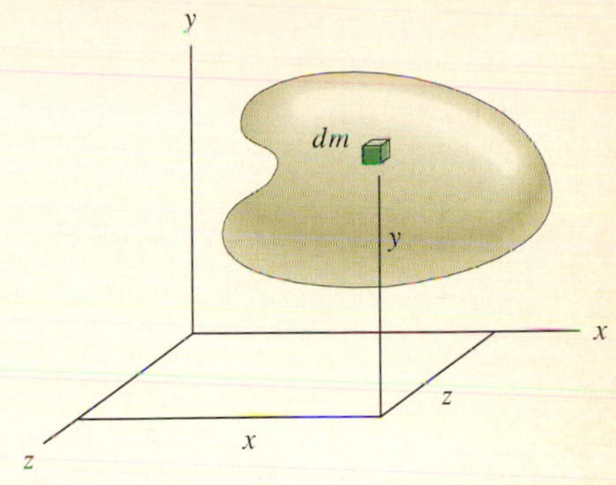

Fig. 9.13

Determining the moments and products of inertia by modeling an object as a continuous distribution of mass.

Slender Bars Let the origin of the coordinate system be at a slender bar's center of mass with the x axis along the bar (Fig. 9.14a). The bar has length l, cross-sectional area A, and mass m. We assume that A is uniform along the length of the bar and that the material is homogeneous.

Consider a differential element of the bar of length dx at a distance x from the center of mass (Fig. 9.14b). The mass of the element is $dm = \rho A\,dx$, where ρ is the mass density. We neglect the lateral dimensions of the bar, assuming the coordinates of the differential element dm to be $(x, 0, 0)$. As a consequence of this approximation, the moment of inertia of the bar about the x axis is zero:

$$I_{xx} = \int_m (y^2 + z^2)\,dm = 0.$$

The moment of inertia about the y axis is

$$I_{yy} = \int_m (x^2 + z^2)\,dm = \int_{-l/2}^{l/2} \rho A x^2\,dx = \frac{1}{12}\rho A l^3.$$

Expressing this result in terms of the mass of the bar $m = \rho A l$, we obtain

$$I_{yy} = \frac{1}{12} m l^2.$$

The moment of inertia about the z axis is equal to the moment of inertia about the y axis:

$$I_{zz} = \int_m (x^2 + y^2)\,dm = \frac{1}{12} m l^2.$$

Because the y and z coordinates of dm are zero, the products of inertia are zero, so the inertia matrix for the slender bar is

$$[I] = \begin{bmatrix} 0 & 0 & 0 \\ 0 & \frac{1}{12} m l^2 & 0 \\ 0 & 0 & \frac{1}{12} m l^2 \end{bmatrix}. \quad (9.14)$$

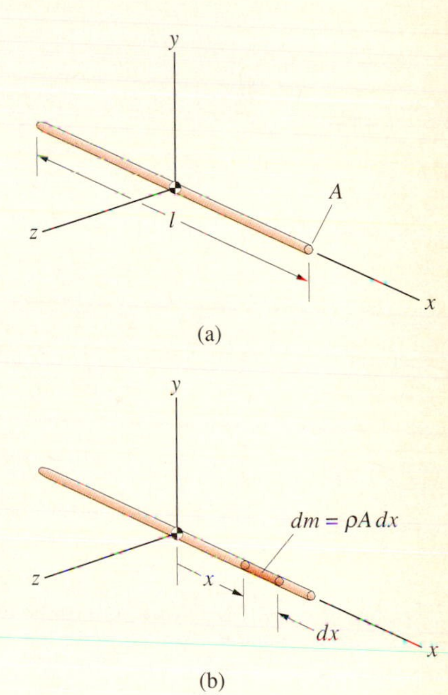

Fig. 9.14

(a) A slender bar and a coordinate system with the x axis aligned with the bar.
(b) A differential element of mass of length dx.

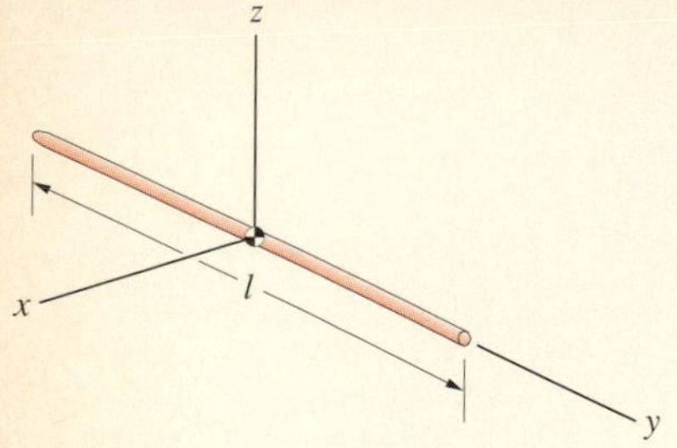

Fig. 9.15
Aligning the y axis with the bar.

You must remember that the moments and products of inertia depend on the orientation of the coordinate system relative to the object. In terms of the alternative coordinate system shown in Fig. 9.15, the bar's inertia matrix is

$$[I] = \begin{bmatrix} \frac{1}{12}ml^2 & 0 & 0 \\ 0 & 0 & 0 \\ 0 & 0 & \frac{1}{12}ml^2 \end{bmatrix}.$$

Thin Plates Suppose that a homogeneous plate of uniform thickness T, area A, and unspecified shape lies in the x-y plane (Fig. 9.16a). We can express its moments of inertia in terms of the moments of inertia of its cross-sectional area.

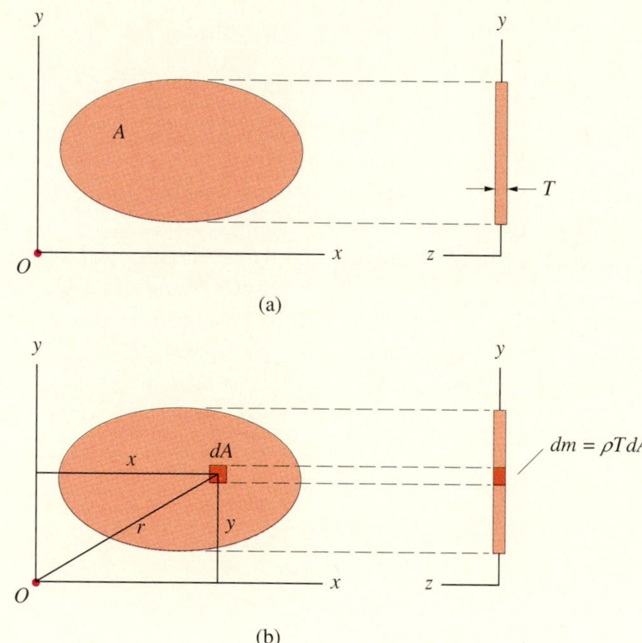

Fig. 9.16
(a) A thin plate lying in the x-y plane.
(b) Obtaining a differential element of mass by projecting an element of area dA through the plate.

By projecting an element of area dA through the thickness T of the plate (Fig. 9.16b), we obtain a differential element of mass $dm = \rho T\, dA$. We neglect the thickness of the plate in calculating the moments of inertia, so the coordinates of the element dm are $(x, y, 0)$. The plate's moment of inertia about the x axis is

$$I_{xx} = \int_m (y^2 + z^2)\, dm = \rho T \int_A y^2\, dA = \rho T I_x,$$

where I_x is the moment of inertia of the plate's cross-sectional area about the x axis. Since the mass of the plate is $m = \rho T A$, the product $\rho T = m/A$, and we obtain the moment of inertia in the form

$$I_{xx} = \frac{m}{A} I_x.$$

The moment of inertia about the y axis is

$$I_{yy} = \int_m (x^2 + z^2)\, dm = \rho T \int_A x^2\, dA = \frac{m}{A} I_y,$$

where I_y is the moment of inertia of the cross-sectional area about the y axis. The moment of inertia about the z axis is

$$I_{zz} = \int_m (x^2 + y^2)\, dm = \frac{m}{A} J_O,$$

where $J_O = I_x + I_y$ is the polar moment of inertia of the cross-sectional area. The product of inertia I_{xy} is

$$I_{xy} = \int_m xy\, dm = \frac{m}{A} I_{xy}^A,$$

where

$$I_{xy}^A = \int_A xy\, dA$$

is the product of inertia of the cross-sectional area. (We use a superscript A to distinguish the product of inertia of the plate's cross-sectional area from the product of inertia of its mass.) If the cross-sectional area A is symmetric about either the x axis or the y axis, $I_{xy}^A = 0$.

Because the z coordinate of dm is zero, the products of inertia I_{xz} and I_{yz} are zero. The inertia matrix for the thin plate is

$$[I] = \begin{bmatrix} \frac{m}{A} I_x & -\frac{m}{A} I_{xy}^A & 0 \\ -\frac{m}{A} I_{xy}^A & \frac{m}{A} I_y & 0 \\ 0 & 0 & \frac{m}{A} J_O \end{bmatrix}. \qquad (9.15)$$

If you know, or can determine, the moments of inertia and product of inertia of the plate's cross-sectional area, you can use these expressions to obtain the moments and products of inertia of its mass.

Parallel-Axis Theorems

Suppose that we know an object's inertia matrix $[I']$ in terms of a coordinate system $x'y'z'$ with its origin at the center of mass, and we want to determine the inertia matrix $[I]$ in terms of a parallel coordinate system xyz (Fig. 9.17). Let (d_x, d_y, d_z) be the coordinates of the center of mass in the xyz coordinate system. The coordinates of a differential element of mass dm in the xyz system are given in terms of its coordinates in the $x'y'z'$ system by

$$x = x' + d_x, \qquad y = y' + d_y, \qquad z = z' + d_z. \qquad (9.16)$$

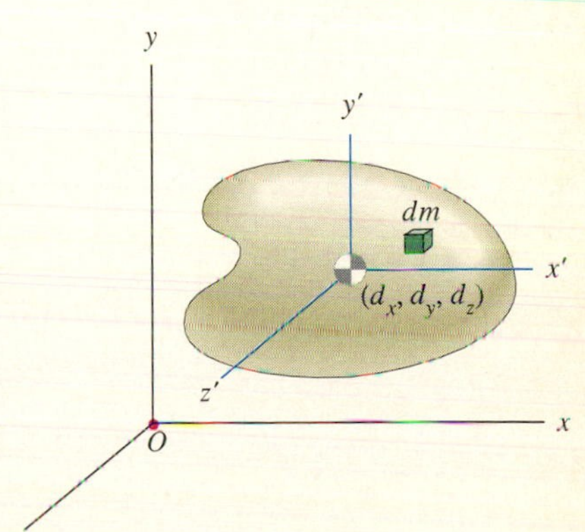

Fig. 9.17
A coordinate system $x'y'z'$ with its origin at the center of mass and a parallel coordinate system xyz.

Substituting these expressions into the definition of I_{xx}, we obtain

$$I_{xx} = \int_m [(y')^2 + (z')^2]\, dm + 2d_y \int_m y'\, dm + 2d_z \int_m z'\, dm + (d_y^2 + d_z^2) \int_m dm. \quad (9.17)$$

The first integral on the right is the object's moment of inertia about the x' axis. We can show that the second and third integrals are zero by using the definitions of the object's center of mass expressed in terms of the $x'y'z'$ coordinate system:

$$\bar{x}' = \frac{\int_m x'\, dm}{\int_m dm}, \qquad \bar{y}' = \frac{\int_m y'\, dm}{\int_m dm}, \qquad \bar{z}' = \frac{\int_m z'\, dm}{\int_m dm}.$$

The object's center of mass is at the origin of the $x'y'z'$ system, so $\bar{x}' = \bar{y}' = \bar{z}' = 0$. Therefore the second and third integrals on the right of Eq. (9.17) are zero, and we obtain

$$I_{xx} = I_{x'x'} + (d_y^2 + d_z^2)m,$$

where m is the mass of the object. Substituting Eqs. (9.16) into the definition of I_{xy}, we obtain

$$\begin{aligned} I_{xy} &= \int_m x'y'\, dm + d_x \int_m y'\, dm + d_y \int_m x'\, dm + d_x d_y \int_m dm \\ &= I_{x'y'} + d_x d_y m. \end{aligned}$$

Proceeding in this way for each of the moments and products of inertia, we obtain the **parallel-axis theorems**:

$$\begin{aligned} I_{xx} &= I_{x'x'} + (d_y^2 + d_z^2)m, \\ I_{yy} &= I_{y'y'} + (d_x^2 + d_z^2)m, \\ I_{zz} &= I_{z'z'} + (d_x^2 + d_y^2)m, \\ I_{xy} &= I_{x'y'} + d_x d_y m, \\ I_{yz} &= I_{y'z'} + d_y d_z m, \\ I_{zx} &= I_{z'x'} + d_z d_x m. \end{aligned} \quad (9.18)$$

If you know an object's inertia matrix in terms of a particular coordinate system, you can use these theorems to determine its inertia matrix in terms of any parallel coordinate system. You can also use them to determine the inertia matrices of composite objects.

Moment of Inertia About an Arbitrary Axis

If we know a rigid body's inertia matrix in terms of a given coordinate system with origin O, we can determine its moment of inertia about an arbitrary axis through O. Suppose that the rigid body rotates with angular velocity $\boldsymbol{\omega}$ about an arbitrary fixed axis L_O through O, and let $\mathbf{e}$ be a unit vector with the same direction as $\boldsymbol{\omega}$ (Fig. 9.18). In terms of the moment of inertia I_O about L_O, the rigid body's angular momentum about L_O is

$$H_O = I_O|\boldsymbol{\omega}|.$$

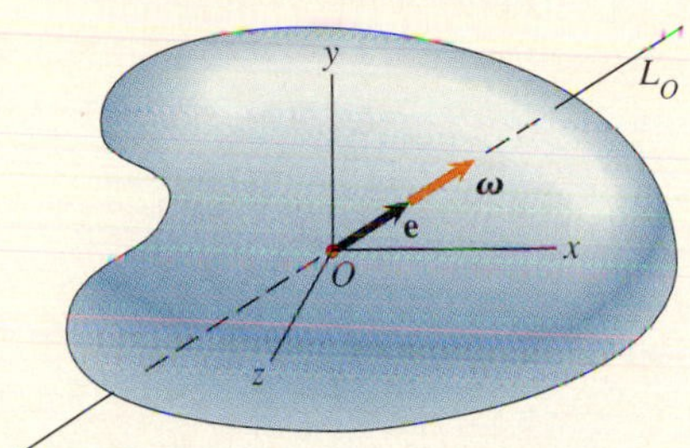

Fig. 9.18
Rigid body rotating about L_O.

We can express the angular velocity vector as

$$\boldsymbol{\omega} = |\boldsymbol{\omega}|(e_x\mathbf{i} + e_y\mathbf{j} + e_z\mathbf{k}),$$

so that $\omega_x = |\boldsymbol{\omega}|e_x$, $\omega_y = |\boldsymbol{\omega}|e_y$, and $\omega_z = |\boldsymbol{\omega}|e_z$. Using these expressions and Eqs. (9.7), the angular momentum about L_O is

$$\begin{aligned} H_O = \mathbf{H}_O \cdot \mathbf{e} = {} & (I_{xx}|\boldsymbol{\omega}|e_x - I_{xy}|\boldsymbol{\omega}|e_y - I_{xz}|\boldsymbol{\omega}|e_z)e_x \\ & + (-I_{yx}|\boldsymbol{\omega}|e_x + I_{yy}|\boldsymbol{\omega}|e_y - I_{yz}|\boldsymbol{\omega}|e_z)e_y \\ & + (-I_{zx}|\boldsymbol{\omega}|e_x - I_{zy}|\boldsymbol{\omega}|e_y + I_{zz}|\boldsymbol{\omega}|e_z)e_z. \end{aligned}$$

Equating our two expressions for H_O, we obtain

$$\begin{aligned} I_O = {} & I_{xx}e_x^2 + I_{yy}e_y^2 + I_{zz}e_z^2 - 2I_{xy}e_xe_y - 2I_{yz}e_ye_z \\ & - 2I_{zx}e_ze_x. \end{aligned} \tag{9.19}$$

Notice that the moment of inertia about an arbitrary axis depends on the products of inertia, in addition to the moments of inertia about the coordinate axes. If you know an object's inertia matrix, you can use Eq. (9.19) to determine its moment of inertia about an axis through O whose direction is specified by the unit vector $\mathbf{e}$.

Principal Axes

For *any* object and origin O, at least one coordinate system exists for which the products of inertia are zero:

$$[I] = \begin{bmatrix} I_{xx} & 0 & 0 \\ 0 & I_{yy} & 0 \\ 0 & 0 & I_{zz} \end{bmatrix}. \tag{9.20}$$

These coordinate axes are called **principal axes**, and the moments of inertia are called the **principal moments of inertia**.

If you know the inertia matrix of a rigid body in terms of a coordinate system $x'y'z'$ and the products of inertia are zero, $x'y'z'$ is a set of principal axes. Suppose that the products of inertia are not zero, and you want to find a set of principal axes xyz and the corresponding principal moments of inertia (Fig. 9.19). It can be shown that the principal moments of inertia are roots of the cubic equation

$$\begin{aligned} & I^3 - (I_{x'x'} + I_{y'y'} + I_{z'z'})I^2 \\ & + (I_{x'x'}I_{y'y'} + I_{y'y'}I_{z'z'} + I_{z'z'}I_{x'x'} - I_{x'y'}^2 - I_{y'z'}^2 - I_{z'x'}^2)I \\ & - (I_{x'x'}I_{y'y'}I_{z'z'} - I_{x'x'}I_{y'z'}^2 - I_{y'y'}I_{x'z'}^2 - I_{z'z'}I_{x'y'}^2 - 2I_{x'y'}I_{y'z'}I_{z'x'}) = 0. \end{aligned} \qquad (9.21)$$

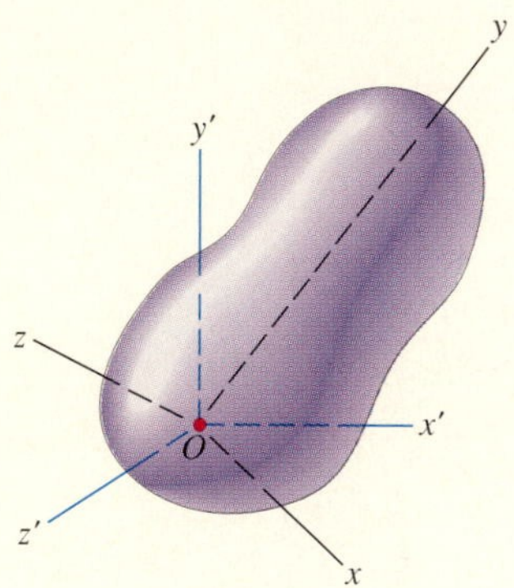

Fig. 9.19

The $x'y'z'$ system with its origin at O and a set of principal axes xyz.

For each principal moment of inertia I, the vector $\mathbf{V}$ with components

$$\begin{aligned} V_{x'} &= (I_{y'y'} - I)(I_{z'z'} - I) - I_{y'z'}^2, \\ V_{y'} &= I_{x'y'}(I_{z'z'} - I) + I_{x'z'}I_{y'z'}, \\ V_{z'} &= I_{x'z'}(I_{y'y'} - I) + I_{x'y'}I_{y'z'} \end{aligned} \qquad (9.22)$$

is parallel to the corresponding principal axis.

To determine the principal moments of inertia, you must obtain the roots of Eq. (9.21). Then substitute one of the principal moments of inertia into Eqs. (9.22) to obtain the components of a vector parallel to the corresponding principal axis. By repeating this step for each principal moment of inertia, you can determine the three principal axes. If you do not obtain a solution from Eqs. (9.22), try one of the other principal moments of inertia. You can choose the axes you identify as x, y, and z arbitrarily, although you must make sure your coordinate system is right-handed. See Example 9.7.

Axes through O about which an object's moment of inertia is a minimum or maximum are principal axes. If the three principal moments of inertia are equal, any coordinate system with its origin at O is a set of principal axes, and the moment of inertia has the same value about any axis through O. This is the case, for example, if the object is a homogeneous sphere and the origin is at its center (Fig. 9.20a). If two of the principal moments of inertia are equal, you can determine a unique principal axis from the third one, and any axes perpendicular to the unique principal axis are principal axes. This is the case when an object has an axis of rotational symmetry and the origin is a point on the axis (Fig. 9.20b). The axis of symmetry is the unique principal axis.

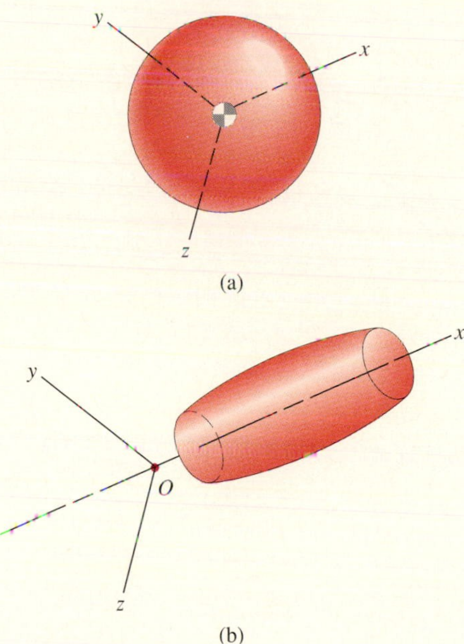

Fig. 9.20

(a) A homogeneous sphere. Any coordinate system with its origin at the center is a set of principal axes.
(b) A rotationally symmetric object. The axis of symmetry is a principal axis, and any perpendicular axes are principal axes.

In the following examples we determine moments and products of inertia of simple objects, apply the parallel-axis theorems, and evaluate the angular momenta of rigid bodies.

Example 9.5

The boom AB of the crane in Fig. 9.21 has a mass of 4800 kg, and the boom BC has a mass of 1600 kg and is perpendicular to boom AB. Modeling each boom as a slender bar and treating them as a single object, determine the moments and products of inertia of the object in terms of the coordinate system shown.

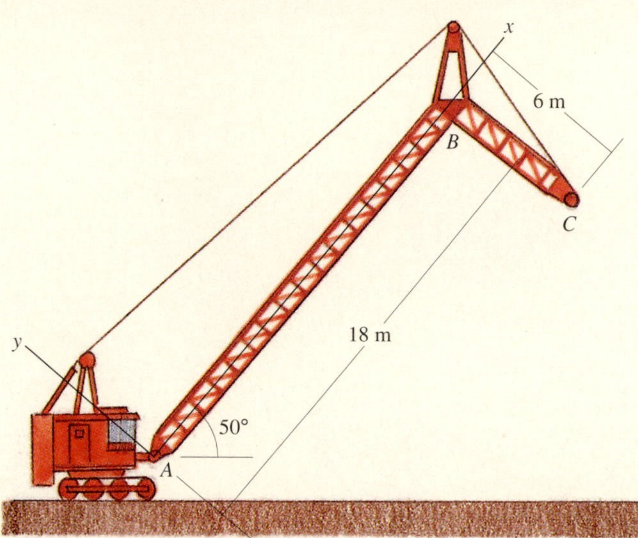

Fig. 9.21

STRATEGY

We can apply the parallel-axis theorems to each boom to determine its moments and products of inertia in terms of the given coordinate system. The moments and products of inertia of the combined object are the sums of those for the two booms.

SOLUTION

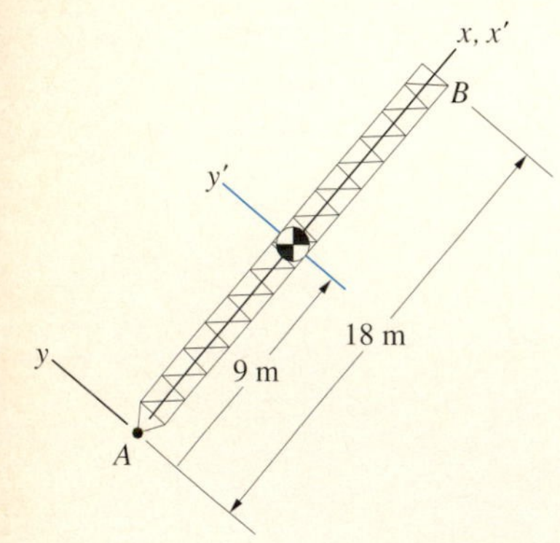

(a) Applying the parallel-axis theorems to boom AB.

Boom *AB* In Fig. (a) we introduce a parallel coordinate system $x'y'z'$ with its origin at the center of mass of boom AB. In terms of the $x'y'z'$ system, the inertia matrix of boom AB is

$$[I'] = \begin{bmatrix} 0 & 0 & 0 \\ 0 & \frac{1}{12}ml^2 & 0 \\ 0 & 0 & \frac{1}{12}ml^2 \end{bmatrix}$$

$$= \begin{bmatrix} 0 & 0 & 0 \\ 0 & \frac{1}{12}(4800)(18)^2 & 0 \\ 0 & 0 & \frac{1}{12}(4800)(18)^2 \end{bmatrix} \text{kg-m}^2.$$

The coordinates of the origin of the $x'y'z'$ system relative to the xyz system are $d_x = 9$ m, $d_y = 0$, $d_z = 0$. Applying the parallel-axis theorems, we obtain

$$I_{xx} = I_{x'x'} + (d_y^2 + d_z^2)m = 0,$$

$$I_{yy} = I_{y'y'} + (d_x^2 + d_z^2)m = \frac{1}{12}(4800)(18)^2 + (9)^2(4800)$$

$$= 518{,}400 \text{ kg-m}^2,$$

$$I_{zz} = I_{z'z'} + (d_x^2 + d_y^2)m = \frac{1}{12}(4800)(18)^2 + (9)^2(4800)$$

$$= 518{,}400 \text{ kg-m}^2,$$

$$I_{xy} = I_{x'y'} + d_x d_y m = 0,$$

$$I_{yz} = I_{y'z'} + d_y d_z m = 0,$$

$$I_{zx} = I_{z'x'} + d_z d_x m = 0.$$

Boom *BC* In Fig. (b) we introduce a parallel coordinate system $x'y'z'$ with its origin at the center of mass of boom BC. In terms of the $x'y'z'$ system, the inertia matrix of boom BC is

$$[I'] = \begin{bmatrix} \frac{1}{12}ml^2 & 0 & 0 \\ 0 & 0 & 0 \\ 0 & 0 & \frac{1}{12}ml^2 \end{bmatrix}$$

$$= \begin{bmatrix} \frac{1}{12}(1600)(6)^2 & 0 & 0 \\ 0 & 0 & 0 \\ 0 & 0 & \frac{1}{12}(1600)(6)^2 \end{bmatrix} \text{kg-m}^2.$$

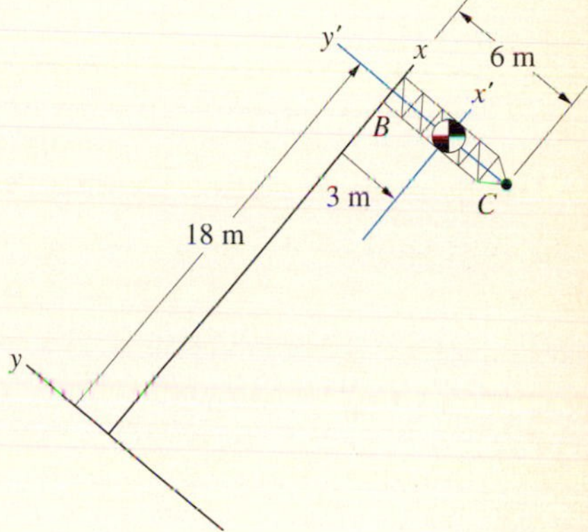

(b) Applying the parallel-axis theorems to boom BC.

The coordinates of the origin of the $x'y'z'$ system relative to the xyz system are $d_x =$ 18 m, $d_y = -3$ m, $d_z = 0$. Applying the parallel-axis theorems, we obtain

$$I_{xx} = I_{x'x'} + (d_y^2 + d_z^2)m = \frac{1}{12}(1600)(6)^2 + (-3)^2(1600)$$

$$= 19{,}200 \text{ kg-m}^2,$$

$$I_{yy} = I_{y'y'} + (d_x^2 + d_z^2)m = 0 + (18)^2(1600) = 518{,}400 \text{ kg-m}^2,$$

$$I_{zz} = I_{z'z'} + (d_x^2 + d_y^2)m = \frac{1}{12}(1600)(6)^2 + [(18)^2 + (-3)^2](1600)$$

$$= 537{,}600 \text{ kg-m}^2,$$

$$I_{xy} = I_{x'y'} + d_x d_y m = 0 + (18)(-3)(1600) = -86{,}400 \text{ kg-m}^2,$$

$$I_{yz} = I_{y'z'} + d_y d_z m = 0,$$

$$I_{zx} = I_{z'x'} + d_z d_x m = 0.$$

Summing the results for the two booms, we obtain the inertia matrix for the single object:

$$[I] = \begin{bmatrix} 19{,}200 & -(-86{,}400) & 0 \\ -(-86{,}400) & 518{,}400 + 518{,}400 & 0 \\ 0 & 0 & 518{,}400 + 537{,}600 \end{bmatrix}$$

$$= \begin{bmatrix} 19{,}200 & 86{,}400 & 0 \\ 86{,}400 & 1{,}036{,}800 & 0 \\ 0 & 0 & 1{,}056{,}000 \end{bmatrix} \text{kg-m}^2.$$

Example 9.6

The 4-kg rectangular plate in Fig. 9.22 lies in the x-y plane of the body-fixed coordinate system.

(a) Determine the plate's moments and products of inertia.

(b) Determine the plate's moment of inertia about the diagonal axis L_O.

(c) If the plate is rotating about the fixed point O with angular velocity $\boldsymbol{\omega} = 4\mathbf{i} - 2\mathbf{j}$ (rad/s), what is the plate's angular momentum about O?

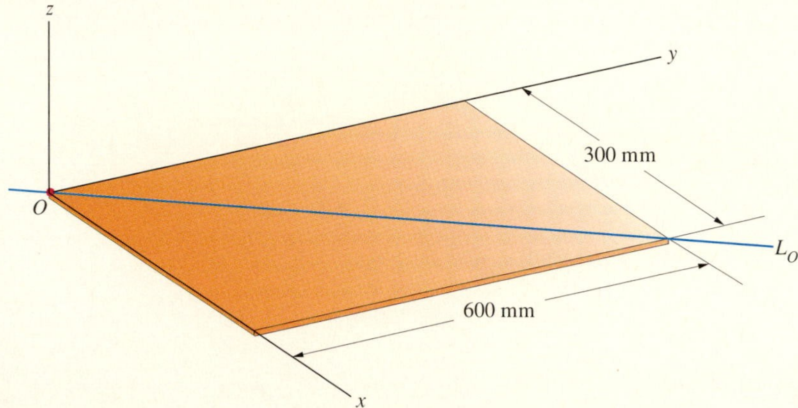

Fig. 9.22

STRATEGY

(a) We can obtain the moments and products of inertia of the plate's rectangular area from Appendix B and use Eqs. (9.15) to obtain the moments and products of inertia of its mass.

(b) Once we know the moments and products of inertia, we can use Eq. (9.19) to determine the moment of inertia about L_O.

(c) The angular momentum about O is given by Eq. (9.10).

SOLUTION

(a) From Appendix B, the moments of inertia of the plate's cross-sectional area are (Fig. a):

$$I_x = \frac{1}{3}bh^3, \qquad I_y = \frac{1}{3}hb^3,$$

$$I_{xy}^A = \frac{1}{4}b^2h^2, \qquad J_O = \frac{1}{3}(bh^3 + hb^3).$$

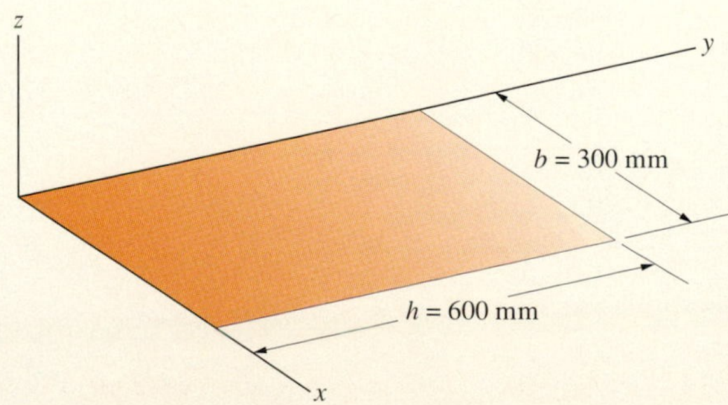

(a) Determining the moments of inertia of the plate's area.

Therefore the moments and products of inertia of the plate are

$$I_{xx} = \frac{m}{A} I_x = \frac{(4)}{(0.3)(0.6)} \left(\frac{1}{3}\right)(0.3)(0.6)^3 = 0.48 \text{ kg-m}^2,$$

$$I_{yy} = \frac{m}{A} I_y = \frac{(4)}{(0.3)(0.6)} \left(\frac{1}{3}\right)(0.6)(0.3)^3 = 0.12 \text{ kg-m}^2,$$

$$I_{xy} = \frac{m}{A} I_{xy}^A = \frac{(4)}{(0.3)(0.6)} \left(\frac{1}{4}\right)(0.3)^2(0.6)^2 = 0.18 \text{ kg-m}^2,$$

$$I_{zz} = \frac{m}{A} J_O = \frac{(4)}{(0.3)(0.6)} \left(\frac{1}{3}\right)[(0.3)(0.6)^3 + (0.6)(0.3)^3] = 0.60 \text{ kg-m}^2,$$

$$I_{xz} = I_{yz} = 0.$$

(b) To apply Eq. (9.19), we must determine the components of a unit vector parallel to L_O:

$$\mathbf{e} = \frac{300\mathbf{i} + 600\mathbf{j}}{|300\mathbf{i} + 600\mathbf{j}|} = 0.447\mathbf{i} + 0.894\mathbf{j}.$$

The moment of inertia about L_O is

$$\begin{aligned} I_O &= I_{xx}e_x^2 + I_{yy}e_y^2 + I_{zz}e_z^2 - 2I_{xy}e_xe_y - 2I_{yz}e_ye_z - 2I_{zx}e_ze_x \\ &= (0.48)(0.447)^2 + (0.12)(0.894)^2 - 2(0.18)(0.447)(0.894) \\ &= 0.048 \text{ kg-m}^2. \end{aligned}$$

(c) The plate's angular momentum about O is

$$\begin{bmatrix} H_{Ox} \\ H_{Oy} \\ H_{Oz} \end{bmatrix} = \begin{bmatrix} I_{xx} & -I_{xy} & -I_{xz} \\ -I_{yx} & I_{yy} & -I_{yz} \\ -I_{zx} & -I_{zy} & I_{zz} \end{bmatrix} \begin{bmatrix} \omega_x \\ \omega_y \\ \omega_z \end{bmatrix}$$

$$= \begin{bmatrix} 0.48 & -0.18 & 0 \\ -0.18 & 0.12 & 0 \\ 0 & 0 & 0.6 \end{bmatrix} \begin{bmatrix} 4 \\ -2 \\ 0 \end{bmatrix}$$

$$= \begin{bmatrix} 2.28 \\ -0.96 \\ 0 \end{bmatrix} \text{kg-m}^2\text{/s}.$$

Example 9.7

In terms of a coordinate system $x'y'z'$ with its origin at the center of mass, the inertia matrix of a rigid body is

$$[I'] = \begin{bmatrix} 4 & -2 & 1 \\ -2 & 2 & -1 \\ 1 & -1 & 3 \end{bmatrix} \text{kg-m}^2.$$

Determine the principal moments of inertia and the directions of a set of principal axes relative to the $x'y'z'$ system.

SOLUTION

Substituting the moments and products of inertia into Eq. (9.21), we obtain the equation

$$I^3 - 9I^2 + 20I - 10 = 0.$$

We show the value of the left side of this equation as a function of I in Fig. 9.23. The three roots, which are the values of the principal moments of inertia in kg-m^2, are $I_1 = 0.708$, $I_2 = 2.397$, and $I_3 = 5.895$.

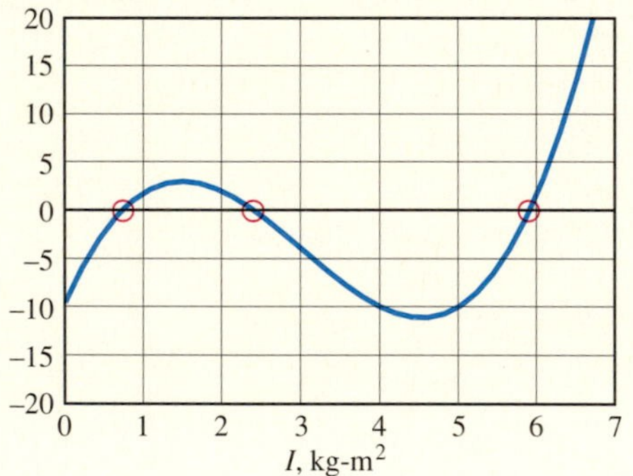

Fig. 9.23
Graph of $I^3 - 9I^2 + 20I - 10$.

Substituting the principal moment of inertia $I_1 = 0.708$ kg-m^2 into Eqs. (9.22) and dividing the resulting vector $\mathbf{V}$ by its magnitude, we obtain a unit vector parallel to the corresponding principal axis:

$$\mathbf{e}_1 = 0.473\mathbf{i} + 0.864\mathbf{j} + 0.171\mathbf{k}.$$

Substituting $I_2 = 2.397$ kg-m^2 into Eqs. (9.22), we obtain the unit vector

$$\mathbf{e}_2 = -0.458\mathbf{i} + 0.076\mathbf{j} + 0.886\mathbf{k},$$

and substituting $I_3 = 5.895$ kg-m^2 into Eqs. (9.22), we obtain the unit vector

$$\mathbf{e}_3 = 0.753\mathbf{i} - 0.497\mathbf{j} + 0.432\mathbf{k}.$$

We have determined the principal moments of inertia and the components of unit vectors parallel to the corresponding principal axes. In Fig. 9.24, we show the principal axes, arbitrarily designating them so that $I_{xx} = 5.895$ kg-m^2, $I_{yy} = 0.708$ kg-m^2, and $I_{zz} = 2.397$ kg-m^2.

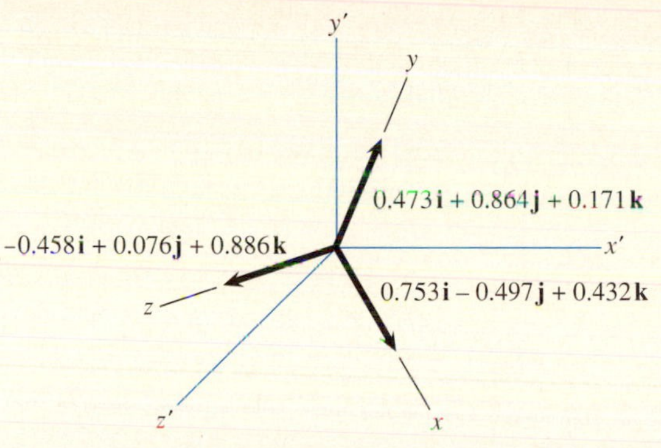

Fig. 9.24

The principal axes. Our choice of which to call x, y, and z is arbitrary.

Problems

9.28 The inertia matrix of a rigid body in terms of a body-fixed coordinate system with its origin at the center of mass is

$$[I] = \begin{bmatrix} 8.8 & 4.0 & 0 \\ 4.0 & 6.6 & 0 \\ 0 & 0 & 15.4 \end{bmatrix} \text{kg-m}^2.$$

If the rigid body's angular velocity is $\boldsymbol{\omega} = -6\mathbf{i} + 2\mathbf{j} - 2\mathbf{k}$ (rad/s), what is its angular momentum about its center of mass?

9.29 What is the moment of inertia of the rigid body in Problem 9.28 about the axis that passes through the origin and the point (1, 2, −2) m?

Strategy: Determine the components of a unit vector parallel to the axis and use Eq. (9.19).

9.30 A rigid body rotates about a fixed point O. Its inertia matrix in terms of a body-fixed coordinate system with its origin at O is

$$[I] = \begin{bmatrix} 1 & -1 & 0 \\ -1 & 5 & 1 \\ 0 & 1 & 7 \end{bmatrix} \text{slug-ft}^2.$$

If the rigid body's angular velocity is $\boldsymbol{\omega} = 6\mathbf{i} + 6\mathbf{j} - 4\mathbf{k}$ (rad/s), what is its angular momentum about O?

9.31 What is the moment of inertia of the rigid body in Problem 9.30 about the axis that passes through the origin and the point (−1, 5, 2) ft?

9.32 The mass of the homogeneous slender bar is 6 kg. Determine its moments and products of inertia in terms of the coordinate system shown.

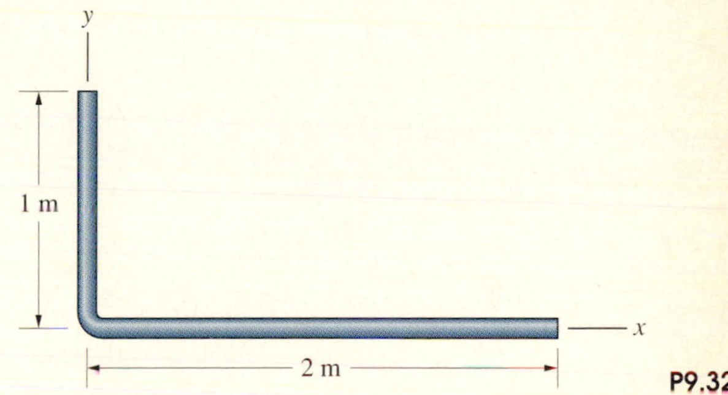

P9.32

9.33 Consider the slender bar in Problem 9.32.
(a) Determine its moments and products of inertia in terms of a parallel coordinate system $x'y'z'$ with its origin at the bar's center of mass.
(b) If the bar is rotating with angular velocity $\boldsymbol{\omega} = 4\mathbf{i}$ (rad/s), what is its angular momentum about its center of mass?

9.34 The 4-kg thin rectangular plate lies in the x-y plane. Determine its moments and products of inertia in terms of the coordinate system shown.

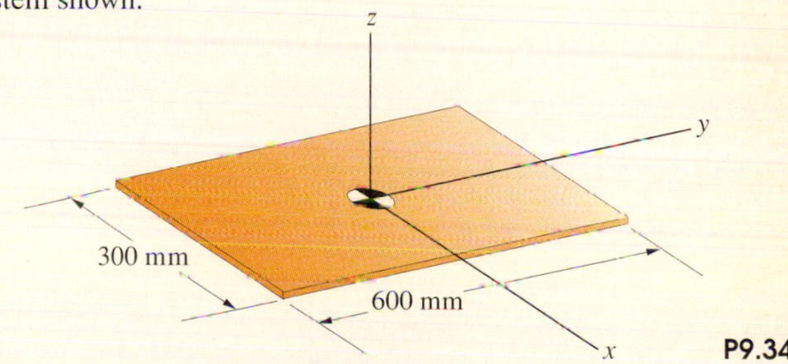

P9.34

9.35 If the plate in Problem 9.34 is rotating with angular velocity $\boldsymbol{\omega} = 6\mathbf{i} + 4\mathbf{j} - 2\mathbf{k}$ (rad/s), what is its angular momentum about its center of mass?

9.36 The 30-lb thin triangular plate lies in the x-y plane. Determine its moments and products of inertia in terms of the coordinate system shown.

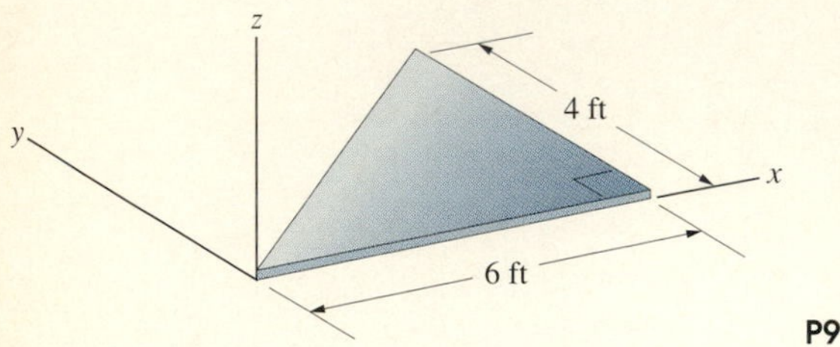

P9.36

9.37 Consider the triangular plate in Problem 9.36.
(a) Determine its moments and products of inertia in terms of a parallel coordinate system $x'y'z'$ with its origin at the plate's center of mass.
(b) If the plate is rotating with angular velocity $\boldsymbol{\omega} = 20\mathbf{i} - 12\mathbf{j} + 16\mathbf{k}$ (rad/s), what is its angular momentum about its center of mass?

9.38 Determine the inertia matrix of the 2.4-kg steel plate in terms of the coordinate system shown.

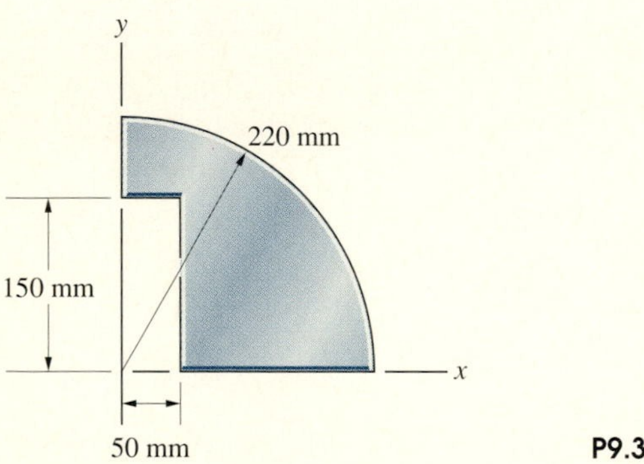

P9.38

9.39 Consider the steel plate in Problem 9.38.
(a) Determine its moments and products of inertia in terms of a parallel coordinate system $x'y'z'$ with its origin at the plate's center of mass.
(b) If the plate is rotating with angular velocity $\boldsymbol{\omega} = 20\mathbf{i} + 10\mathbf{j} - 10\mathbf{k}$ (rad/s), what is its angular momentum about its center of mass?

9.40 The slender bar of mass m rotates about the fixed point O with angular velocity $\boldsymbol{\omega} = \omega_y\,\mathbf{j} + \omega_z\mathbf{k}$. Determine its angular momentum (a) about its center of mass; (b) about O.

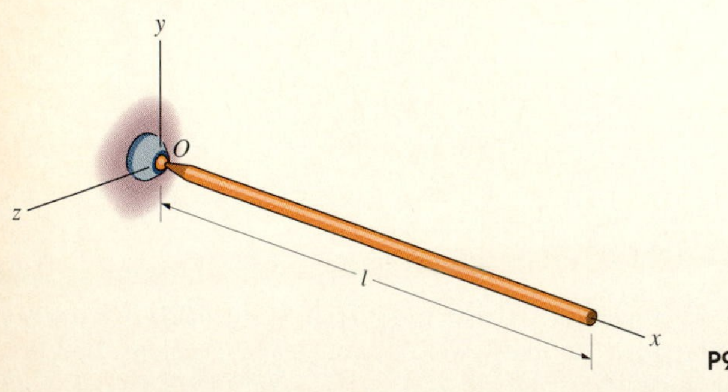

P9.40

9.41 The slender bar of mass m is parallel to the x axis. If the coordinate system is body-fixed and its angular velocity about the fixed point O is $\boldsymbol{\omega} = \omega_y\,\mathbf{j}$, what is the bar's angular momentum about O?

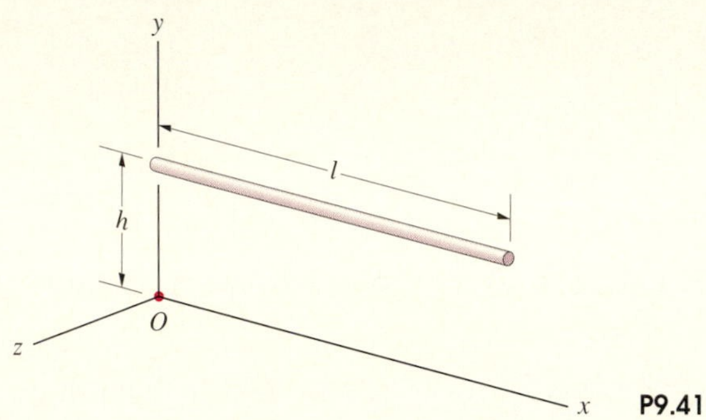

P9.41

9.42 In Example 9.5, the moments and products of inertia of the object consisting of the booms AB and BC were determined in terms of the coordinate system shown in Fig. 9.21. Determine the moments and products of inertia of the object in terms of a parallel coordinate system $x'y'z'$ with its origin at the center of mass of the object.

9.43 Suppose that the crane described in Example 9.5 undergoes a rigid-body rotation about the vertical axis at 0.1 rad/s in the counterclockwise direction when viewed from above.
(a) What is its angular velocity vector $\boldsymbol{\omega}$ in terms of the body-fixed coordinate system shown in Fig. 9.21?
(b) What is the angular momentum of the object consisting of the booms AB and BC *about its center of mass*?

9.44 A 3-kg slender bar is rigidly attached to a 2-kg thin circular disk. In terms of the body-fixed coordinate system shown, the angular velocity of the composite object is $\boldsymbol{\omega} = 100\mathbf{i} - 4\mathbf{j} + 6\mathbf{k}$ (rad/s). What is the object's angular momentum about its center of mass?

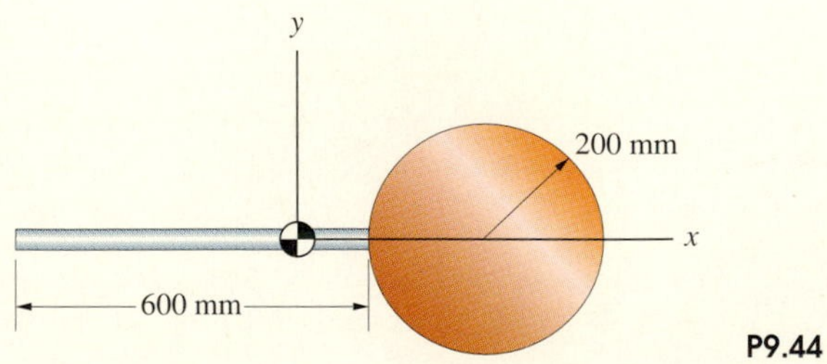

P9.44

9.45 The mass of the homogeneous slender bar is m. If the bar rotates with angular velocity $\boldsymbol{\omega} = \omega_0(24\mathbf{i} + 12\mathbf{j} - 6\mathbf{k})$, what is its angular momentum about its center of mass?

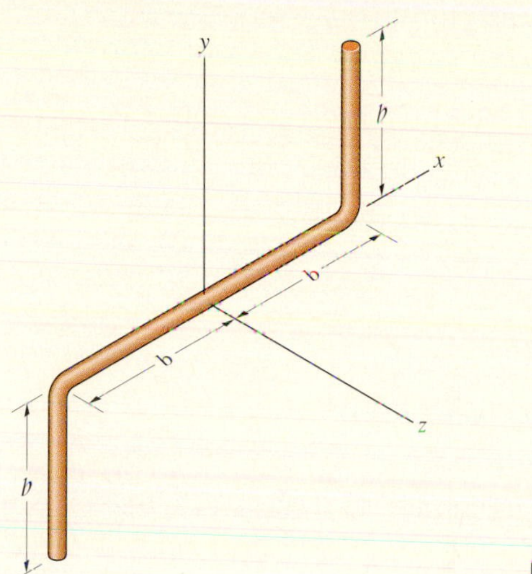

P9.45

9.46 The 8-kg homogeneous slender bar has ball and socket supports at A and B.
(a) What is the bar's moment of inertia about the axis AB?
(b) If the bar rotates about the axis AB at 4 rad/s, what is the magnitude of its angular momentum about its axis of rotation?

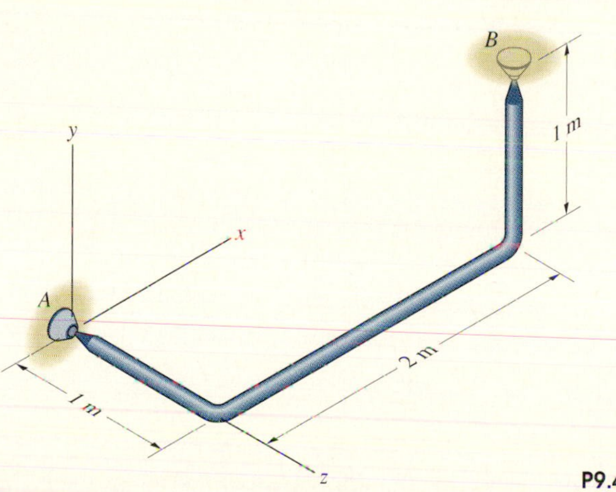

P9.46

9.47 The 8-kg homogeneous slender bar in Problem 9.46 is released from rest in the position shown. (The x-z plane is horizontal.) At that instant, what is the magnitude of the bar's angular acceleration about the axis AB?

9.48 In terms of a coordinate system $x'y'z'$ with its origin at the center of mass, the inertia matrix of a rigid body is

$$[I'] = \begin{bmatrix} 20 & 10 & -10 \\ 10 & 60 & 0 \\ -10 & 0 & 80 \end{bmatrix} \text{kg-m}^2.$$

Determine the principal moments of inertia and unit vectors parallel to the corresponding principal axes.

9.49 For the steel plate and coordinate system shown in Problem 9.38, determine the principal moments of inertia and unit vectors parallel to the corresponding principal axes. Draw a sketch of the plate showing the principal axes.

9.50 The 1-kg, 1-m-long slender bar lies in the x-y plane. Its moment of inertia matrix is

$$[I'] = \begin{bmatrix} \frac{1}{12}\sin^2\beta & -\frac{1}{12}\sin\beta\cos\beta & 0 \\ -\frac{1}{12}\sin\beta\cos\beta & \frac{1}{12}\cos^2\beta & 0 \\ 0 & 0 & \frac{1}{12} \end{bmatrix}.$$

Use Eqs. (9.21) and (9.22) to determine the principal moments of inertia and unit vectors parallel to the corresponding principal axes.

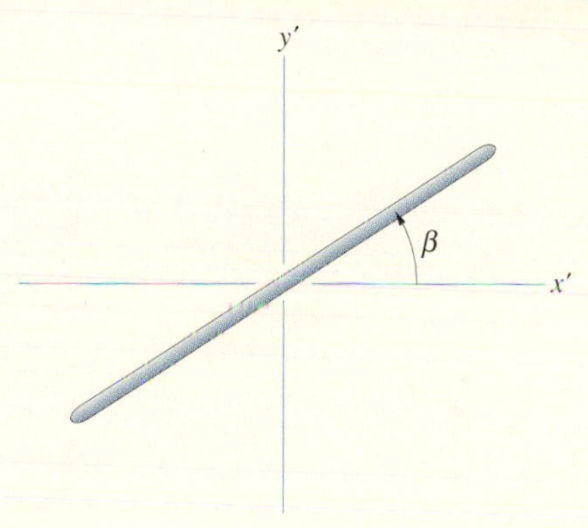

P9.50

9.51 The mass of the homogeneous thin plate is 3 slugs. For a coordinate system with its origin at O, determine the principal moments of inertia and unit vectors parallel to the corresponding principal axes.

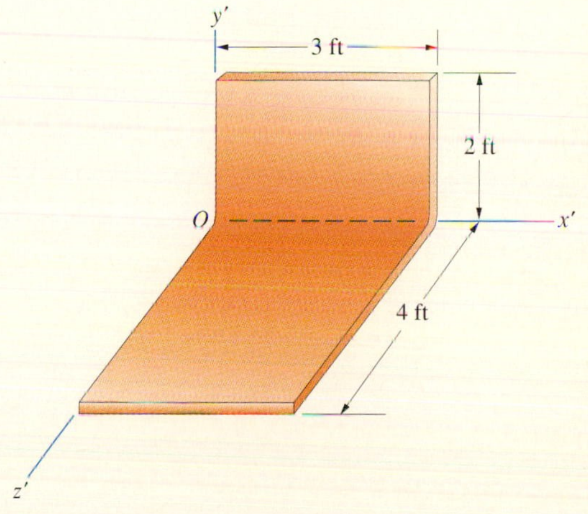

P9.51

9.4 Euler's Equations

The equations governing three-dimensional motion of a rigid body, which are known as Euler's equations, consist of Newton's second law

$$\Sigma \mathbf{F} = m\mathbf{a}$$

and equations of angular motion. In the following sections we derive the equations of angular motion, beginning with momentum principles for a system of particles developed in Chapter 7 and using our expressions for the angular momentum of a rigid body in three dimensions.

Rotation About a Fixed Point

If a rigid body rotates about a fixed point O, the sum of the moments about O due to external forces and couples equals the rate of change of the angular momentum about O (Eq. 7.11):

$$\Sigma \mathbf{M}_O = \frac{d\mathbf{H}_O}{dt}. \tag{9.23}$$

To obtain the equations of angular motion, we must substitute the components of the angular momentum given by Eqs. (9.7) into this equation. The secondary coordinate system used to express these components is usually body-fixed and so rotates with the angular velocity $\boldsymbol{\omega}$ of the rigid body. In some situations, it is convenient to use a secondary coordinate system that rotates but is not body-fixed. We denote the coordinate system's angular velocity by $\boldsymbol{\Omega}$, where $\boldsymbol{\Omega} = \boldsymbol{\omega}$ if the coordinate system is body-fixed. Expressing $\mathbf{H}_O$ in terms of its components,

$$\mathbf{H}_O = H_{Ox}\mathbf{i} + H_{Oy}\mathbf{j} + H_{Oz}\mathbf{k},$$

the rate of change of the angular momentum is

$$\frac{d\mathbf{H}_O}{dt} = \frac{dH_{Ox}}{dt}\mathbf{i} + H_{Ox}\frac{d\mathbf{i}}{dt} + \frac{dH_{Oy}}{dt}\mathbf{j} + H_{Oy}\frac{d\mathbf{j}}{dt} + \frac{dH_{Oz}}{dt}\mathbf{k} + H_{Oz}\frac{d\mathbf{k}}{dt}.$$

By expressing the time derivatives of the unit vectors in terms of the coordinate system's angular velocity $\boldsymbol{\Omega}$,

$$\frac{d\mathbf{i}}{dt} = \boldsymbol{\Omega} \times \mathbf{i}, \qquad \frac{d\mathbf{j}}{dt} = \boldsymbol{\Omega} \times \mathbf{j}, \qquad \frac{d\mathbf{k}}{dt} = \boldsymbol{\Omega} \times \mathbf{k},$$

we can write Eq. (9.23) as

$$\Sigma \mathbf{M}_O = \frac{dH_{Ox}}{dt}\mathbf{i} + \frac{dH_{Oy}}{dt}\mathbf{j} + \frac{dH_{Oz}}{dt}\mathbf{k} + \boldsymbol{\Omega} \times \mathbf{H}_O. \tag{9.24}$$

Substituting the components of $\mathbf{H}_O$ from Eq. (9.7) into this equation, we obtain the equations of angular motion (see Problem 9.84):

$$\begin{aligned}
\Sigma M_{Ox} &= I_{xx}\frac{d\omega_x}{dt} - I_{xy}\frac{d\omega_y}{dt} - I_{xz}\frac{d\omega_z}{dt} \\
&\quad - \Omega_z(-I_{yx}\omega_x + I_{yy}\omega_y - I_{yz}\omega_z) \\
&\quad + \Omega_y(-I_{zx}\omega_x - I_{zy}\omega_y + I_{zz}\omega_z), \\
\Sigma M_{Oy} &= -I_{yx}\frac{d\omega_x}{dt} + I_{yy}\frac{d\omega_y}{dt} - I_{yz}\frac{d\omega_z}{dt} \\
&\quad + \Omega_z(I_{xx}\omega_x - I_{xy}\omega_y - I_{xz}\omega_z) \\
&\quad - \Omega_x(-I_{zx}\omega_x - I_{zy}\omega_y + I_{zz}\omega_z), \\
\Sigma M_{Oz} &= -I_{zx}\frac{d\omega_x}{dt} - I_{zy}\frac{d\omega_y}{dt} + I_{zz}\frac{d\omega_z}{dt} \\
&\quad - \Omega_y(I_{xx}\omega_x - I_{xy}\omega_y - I_{xz}\omega_z) \\
&\quad + \Omega_x(-I_{yx}\omega_x + I_{yy}\omega_y - I_{yz}\omega_z).
\end{aligned} \tag{9.25}$$

We can write these equations as the matrix equation

$$\begin{bmatrix} \Sigma M_{Ox} \\ \Sigma M_{Oy} \\ \Sigma M_{Oz} \end{bmatrix} = \begin{bmatrix} I_{xx} & -I_{xy} & -I_{xz} \\ -I_{yx} & I_{yy} & -I_{yz} \\ -I_{zx} & -I_{zy} & I_{zz} \end{bmatrix} \begin{bmatrix} d\omega_x/dt \\ d\omega_y/dt \\ d\omega_z/dt \end{bmatrix} + \begin{bmatrix} 0 & -\Omega_z & \Omega_y \\ \Omega_z & 0 & -\Omega_x \\ -\Omega_y & \Omega_x & 0 \end{bmatrix} \begin{bmatrix} I_{xx} & -I_{xy} & -I_{xz} \\ -I_{yx} & I_{yy} & -I_{yz} \\ -I_{zx} & -I_{zy} & I_{zz} \end{bmatrix} \begin{bmatrix} \omega_x \\ \omega_y \\ \omega_z \end{bmatrix}. \tag{9.26}$$

General Motion

The sum of the moments about the center of mass of a rigid body due to external forces and couples equals the rate of change of the angular momentum about the center of mass (Eq. 7.13):

$$\Sigma \mathbf{M} = \frac{d\mathbf{H}}{dt}. \tag{9.27}$$

By following the same steps that led from Eq. (9.23) to Eq. (9.24), we can write Eq. (9.27) as

$$\Sigma \mathbf{M} = \frac{dH_x}{dt}\mathbf{i} + \frac{dH_y}{dt}\mathbf{j} + \frac{dH_z}{dt}\mathbf{k} + \mathbf{\Omega} \times \mathbf{H}, \tag{9.28}$$

where $\boldsymbol{\Omega}$ is the angular velocity of the coordinate system. Substituting Eqs. (9.11) into this equation, we obtain the equations of angular motion

$$
\begin{aligned}
\Sigma M_x &= I_{xx}\frac{d\omega_x}{dt} - I_{xy}\frac{d\omega_y}{dt} - I_{xz}\frac{d\omega_z}{dt} \\
&\quad - \Omega_z(-I_{yx}\omega_x + I_{yy}\omega_y - I_{yz}\omega_z) \\
&\quad + \Omega_y(-I_{zx}\omega_x - I_{zy}\omega_y + I_{zz}\omega_z), \\
\Sigma M_y &= -I_{yx}\frac{d\omega_x}{dt} + I_{yy}\frac{d\omega_y}{dt} - I_{yz}\frac{d\omega_z}{dt} \\
&\quad + \Omega_z(I_{xx}\omega_x - I_{xy}\omega_y - I_{xz}\omega_z) \\
&\quad - \Omega_x(-I_{zx}\omega_x - I_{zy}\omega_y + I_{zz}\omega_z), \\
\Sigma M_z &= -I_{zx}\frac{d\omega_x}{dt} - I_{zy}\frac{d\omega_y}{dt} + I_{zz}\frac{d\omega_z}{dt} \\
&\quad - \Omega_y(I_{xx}\omega_x - I_{xy}\omega_y - I_{xz}\omega_z) \\
&\quad + \Omega_x(-I_{yx}\omega_x + I_{yy}\omega_y - I_{yz}\omega_z),
\end{aligned}
\tag{9.29}
$$

or

$$
\begin{aligned}
\begin{bmatrix} \Sigma M_x \\ \Sigma M_y \\ \Sigma M_z \end{bmatrix} &= \begin{bmatrix} I_{xx} & -I_{xy} & -I_{xz} \\ -I_{yx} & I_{yy} & -I_{yz} \\ -I_{zx} & -I_{zy} & I_{zz} \end{bmatrix} \begin{bmatrix} d\omega_x/dt \\ d\omega_y/dt \\ d\omega_z/dt \end{bmatrix} \\
&\quad + \begin{bmatrix} 0 & -\Omega_z & \Omega_y \\ \Omega_z & 0 & -\Omega_x \\ -\Omega_y & \Omega_x & 0 \end{bmatrix} \begin{bmatrix} I_{xx} & -I_{xy} & -I_{xz} \\ -I_{yx} & I_{yy} & -I_{yz} \\ -I_{zx} & -I_{zy} & I_{zz} \end{bmatrix} \begin{bmatrix} \omega_x \\ \omega_y \\ \omega_z \end{bmatrix}.
\end{aligned}
\tag{9.30}
$$

The equations of angular motion for general motion, and the expressions for the moments and products of inertia, are identical in form to those we obtained for rotation about a fixed point. However, in the case of general motion the equations of angular motion are expressed in terms of the components of the moment about the center of mass, and the moments and products of inertia are expressed in terms of a secondary coordinate system with its origin at the center of mass.

Since the components of the angular velocity in Eqs. (9.25) and (9.29) are expressed in terms of a secondary coordinate system rotating with angular velocity $\boldsymbol{\Omega}$, the rigid body's angular acceleration is (Eq. 9.4)

$$
\boldsymbol{\alpha} = \frac{d\boldsymbol{\omega}}{dt} = \frac{d\omega_x}{dt}\mathbf{i} + \frac{d\omega_y}{dt}\mathbf{j} + \frac{d\omega_z}{dt}\mathbf{k} + \boldsymbol{\Omega} \times \boldsymbol{\omega}. \tag{9.31}
$$

If the secondary coordinate system does not rotate or is body-fixed, the terms $d\omega_x/dt$, $d\omega_y/dt$, and $d\omega_z/dt$ are the components of the rigid body's angular acceleration. Otherwise, you must use Eq. (9.31) to determine the angular acceleration.

In the following examples we use the Euler equations to analyze three-dimensional motions of rigid bodies. This typically involves three steps:

1. Choose a coordinate system—*If an object rotates about a fixed point O, it usually simplifies the equations of angular motion if you express them in terms of a coordinate system with its origin at O. Otherwise, you must use a coordinate system with its origin at the center of mass. In either case, be sure to choose the coordinate system's orientation to simplify your determination of the moments and products of inertia.*

2. Draw the free-body diagram—*Isolate the object and identify the external forces and couples acting on it.*

3. Apply the equations of motion—*Use Newton's second law and the equations of angular motion to relate the forces and couples acting on the object to the acceleration of its center of mass and its angular acceleration.*

Example 9.8

During an assembly process, the 4-kg rectangular plate in Fig. 9.25 is held at O by a robotic manipulator. Point O is stationary. At the instant shown, the plate is horizontal, its angular velocity is $\boldsymbol{\omega} = 4\mathbf{i} - 2\mathbf{j}$ (rad/s), and its angular acceleration is $\boldsymbol{\alpha} = -10\mathbf{i} + 6\mathbf{j}$ (rad/s^2). Determine the couple exerted on the plate by the manipulator.

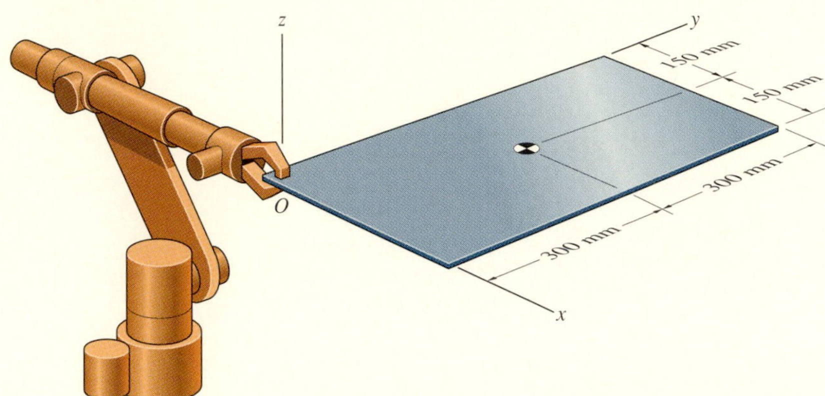

Fig. 9.25

STRATEGY

The plate rotates about the fixed point O, so we can use Eq. (9.26) to determine the total moment exerted on the plate about O.

SOLUTION

Draw the Free-Body Diagram We denote the force and couple exerted on the plate by the manipulator by **F** and **C** (Fig. a).

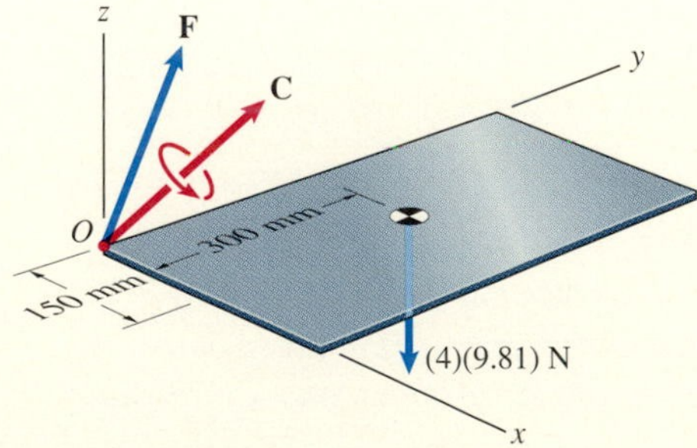

(a) Free-body diagram of the plate.

Apply the Equations of Motion The total moment about O is the sum of the couple exerted by the manipulator and the moment about O due to the plate's weight:

$$\begin{aligned}\Sigma \mathbf{M}_O &= \mathbf{C} + (0.15\mathbf{i} + 0.30\mathbf{j}) \times [-(4)(9.81)\mathbf{k}] \\ &= \mathbf{C} - 11.77\mathbf{i} + 5.89\mathbf{j} \text{ (N-m).}\end{aligned} \tag{9.32}$$

To obtain the unknown couple $\mathbf{C}$, we can determine the total moment about O from Eq. (9.26).

We let the secondary coordinate system be body-fixed, so its angular velocity $\mathbf{\Omega}$ equals the plate's angular velocity $\boldsymbol{\omega}$. We determined the plate's inertia matrix in Example 9.6, obtaining

$$[I] = \begin{bmatrix} 0.48 & -0.18 & 0 \\ -0.18 & 0.12 & 0 \\ 0 & 0 & 0.6 \end{bmatrix} \text{kg-m}^2.$$

Therefore the total moment about O exerted on the plate is

$$\begin{aligned}\begin{bmatrix} \Sigma M_{Ox} \\ \Sigma M_{Oy} \\ \Sigma M_{Oz} \end{bmatrix} &= \begin{bmatrix} I_{xx} & -I_{xy} & -I_{xz} \\ -I_{yx} & I_{yy} & -I_{yz} \\ -I_{zx} & -I_{zy} & I_{zz} \end{bmatrix} \begin{bmatrix} d\omega_x/dt \\ d\omega_y/dt \\ d\omega_z/dt \end{bmatrix} \\ &\quad + \begin{bmatrix} 0 & -\omega_z & \omega_y \\ \omega_z & 0 & -\omega_x \\ -\omega_y & \omega_x & 0 \end{bmatrix} \begin{bmatrix} I_{xx} & -I_{xy} & -I_{xz} \\ -I_{yx} & I_{yy} & -I_{yz} \\ -I_{zx} & -I_{zy} & I_{zz} \end{bmatrix} \begin{bmatrix} \omega_x \\ \omega_y \\ \omega_z \end{bmatrix} \\ &= \begin{bmatrix} 0.48 & -0.18 & 0 \\ -0.18 & 0.12 & 0 \\ 0 & 0 & 0.6 \end{bmatrix} \begin{bmatrix} -10 \\ 6 \\ 0 \end{bmatrix} \\ &\quad + \begin{bmatrix} 0 & 0 & -2 \\ 0 & 0 & -4 \\ 2 & 4 & 0 \end{bmatrix} \begin{bmatrix} 0.48 & -0.18 & 0 \\ -0.18 & 0.12 & 0 \\ 0 & 0 & 0.6 \end{bmatrix} \begin{bmatrix} 4 \\ -2 \\ 0 \end{bmatrix} \\ &= \begin{bmatrix} -5.88 \\ 2.52 \\ 0.72 \end{bmatrix} \text{N-m.}\end{aligned}$$

We substitute this result into Eq. (9.32),

$$\Sigma \mathbf{M}_O = \mathbf{C} - 11.77\mathbf{i} + 5.89\mathbf{j} = -5.88\mathbf{i} + 2.52\mathbf{j} + 0.72\mathbf{k},$$

and solve for the couple $\mathbf{C}$:

$$\mathbf{C} = 5.89\mathbf{i} - 3.37\mathbf{j} + 0.72\mathbf{k} \text{ (N-m).}$$

Example 9.9

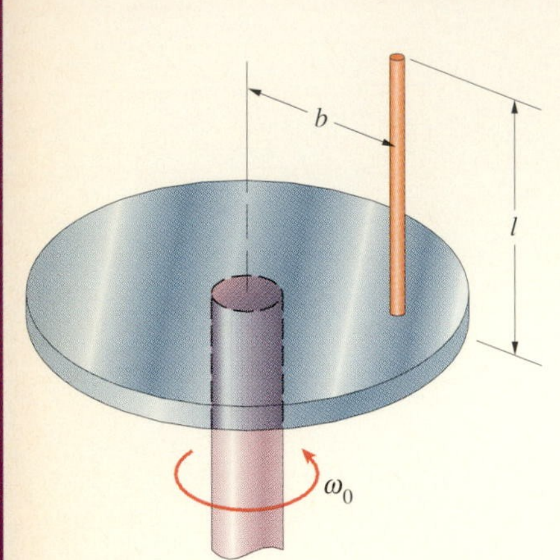

Fig. 9.26

A slender vertical bar of mass m is rigidly attached to a horizontal disk rotating with constant angular velocity ω_0 (Fig. 9.26). What force and couple are exerted on the bar by the disk?

STRATEGY

The external forces and couples on the bar are its weight and the force and couple exerted on it by the disk. The angular velocity and acceleration of the bar are given and we can determine the acceleration of its center of mass, so we can use the Euler equations to determine the total force and couple exerted on the bar.

SOLUTION

Choose a Coordinate System In Fig. (a) we place the origin of a body-fixed coordinate system at the center of mass with the y axis vertical and the x axis in the radial direction. With this orientation we will obtain simple expressions for the bar's angular velocity and the acceleration of its center of mass.

Draw the Free-Body Diagram We draw the free-body diagram of the bar in Fig. (a), showing the force $\mathbf{F}$ and couple $\mathbf{C}$ exerted by the disk.

Apply the Equations of Motion The acceleration of the center of mass of the bar due to its motion along its circular path is $\mathbf{a} = -\omega_0^2 b\mathbf{i}$. From Newton's second law,

$$\Sigma \mathbf{F} = \mathbf{F} - mg\mathbf{j} = m(-\omega_0^2 b\mathbf{i}),$$

we obtain the force exerted on the bar by the disk:

$$\mathbf{F} = -m\omega_0^2 b\mathbf{i} + mg\mathbf{j}.$$

The total moment about the center of mass is the sum of the couple $\mathbf{C}$ and the moment due to $\mathbf{F}$:

$$\Sigma \mathbf{M} = \mathbf{C} + \left(-\frac{1}{2}l\mathbf{j}\right) \times (-m\omega_0^2 b\mathbf{i} + mg\mathbf{j})$$

$$= C_x\mathbf{i} + C_y\mathbf{j} + \left(C_z - \frac{1}{2}mlb\omega_0^2\right)\mathbf{k}.$$

The bar's inertia matrix in terms of the coordinate system in Fig. (a) is

$$[I] = \begin{bmatrix} \frac{1}{12}ml^2 & 0 & 0 \\ 0 & 0 & 0 \\ 0 & 0 & \frac{1}{12}ml^2 \end{bmatrix},$$

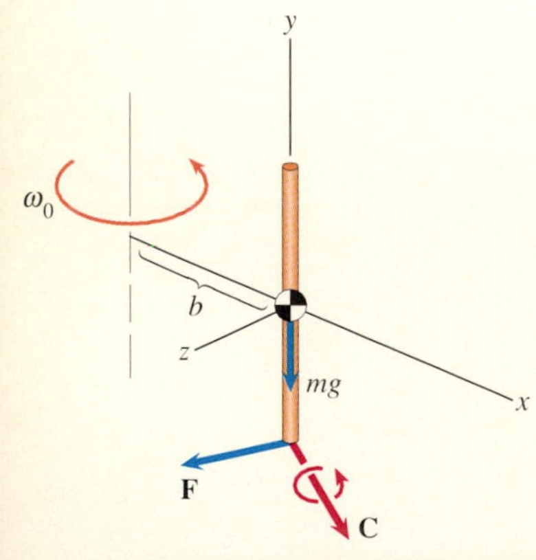

(a) Free-body diagram of the bar.

and its angular velocity, $\boldsymbol{\omega} = \omega_0 \mathbf{j}$, is constant. The equation of angular motion, Eq. (9.30), is

$$\begin{bmatrix} C_x \\ C_y \\ C_z - \frac{1}{2}mlb\omega_0^2 \end{bmatrix} = \begin{bmatrix} 0 & 0 & \omega_0 \\ 0 & 0 & 0 \\ -\omega_0 & 0 & 0 \end{bmatrix} \begin{bmatrix} \frac{1}{12}ml^2 & 0 & 0 \\ 0 & 0 & 0 \\ 0 & 0 & \frac{1}{12}ml^2 \end{bmatrix} \begin{bmatrix} 0 \\ \omega_0 \\ 0 \end{bmatrix}.$$

The right side of this equation equals zero, so the components of the couple exerted on the bar by the disk are $C_x = 0$, $C_y = 0$, and $C_z = \frac{1}{2}mlb\omega_0^2$.

Alternative Solution The bar rotates about a fixed axis, so we can also determine the couple **C** by using Eq. (9.26). Let the fixed point O be the center of the disk (Fig. b), and let the body-fixed coordinate system be oriented with the x axis through the bottom of the bar. The total moment about O is

$$\begin{aligned} \Sigma \mathbf{M}_O &= \mathbf{C} + (b\mathbf{i}) \times (-m\omega_0^2 b\mathbf{i} + mg\mathbf{j}) + \left(b\mathbf{i} + \frac{1}{2}l\mathbf{j}\right) \times (-mg\mathbf{j}) \\ &= \mathbf{C}. \end{aligned}$$

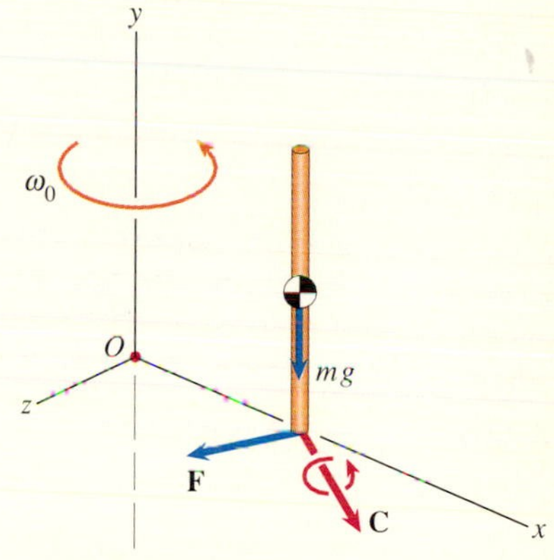

(b) Expressing the equation of angular motion in terms of the fixed point O.

Thus the only moment about O is the couple exerted by the disk. Applying the parallel-axis theorems, the bar's moments and products of inertia are (Fig. c)

$$\begin{aligned} I_{xx} &= I_{x'x'} + (d_y^2 + d_z^2)m = \frac{1}{12}ml^2 + \left(\frac{1}{2}l\right)^2 m = \frac{1}{3}ml^2, \\ I_{yy} &= I_{y'y'} + (d_x^2 + d_z^2)m = mb^2, \\ I_{zz} &= I_{z'z'} + (d_x^2 + d_y^2)m = \frac{1}{12}ml^2 + \left[b^2 + \left(\frac{1}{2}l\right)^2\right]m = \frac{1}{3}ml^2 + mb^2, \\ I_{xy} &= I_{x'y'} + d_x d_y m = 0 + (b)\left(\frac{1}{2}l\right)m = \frac{1}{2}mbl, \\ I_{yz} &= I_{y'z'} + d_y d_z m = 0, \\ I_{zx} &= I_{z'x'} + d_z d_x m = 0. \end{aligned}$$

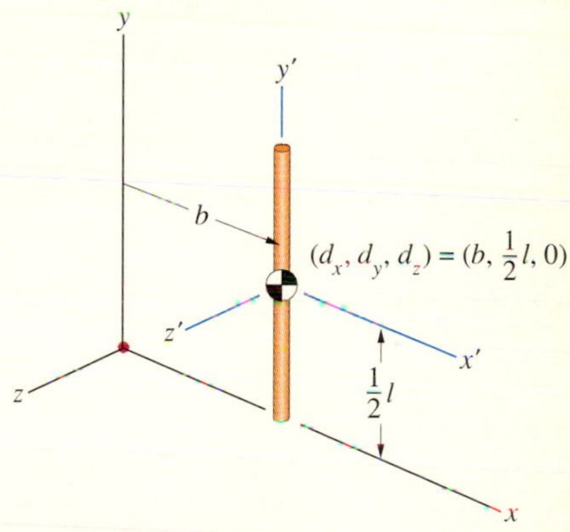

(c) Applying the parallel-axis theorem.

Substituting these results into Eq. (9.26), we obtain

$$\begin{aligned} \begin{bmatrix} C_x \\ C_y \\ C_z \end{bmatrix} &= \begin{bmatrix} 0 & 0 & \omega_0 \\ 0 & 0 & 0 \\ -\omega_0 & 0 & 0 \end{bmatrix} \begin{bmatrix} \frac{1}{3}ml^2 & -\frac{1}{2}mbl & 0 \\ -\frac{1}{2}mbl & mb^2 & 0 \\ 0 & 0 & \frac{1}{3}ml^2 + mb^2 \end{bmatrix} \begin{bmatrix} 0 \\ \omega_0 \\ 0 \end{bmatrix} \\ &= \begin{bmatrix} 0 \\ 0 \\ \frac{1}{2}mlb\omega_0^2 \end{bmatrix}. \end{aligned}$$

DISCUSSION

If the bar were attached to the disk by a ball and socket support instead of a built-in support, you can see that the bar would rotate outward due to the disk's rotation. We have determined the couple the built-in support exerts on the bar that prevents it from rotating outward.

Example 9.10

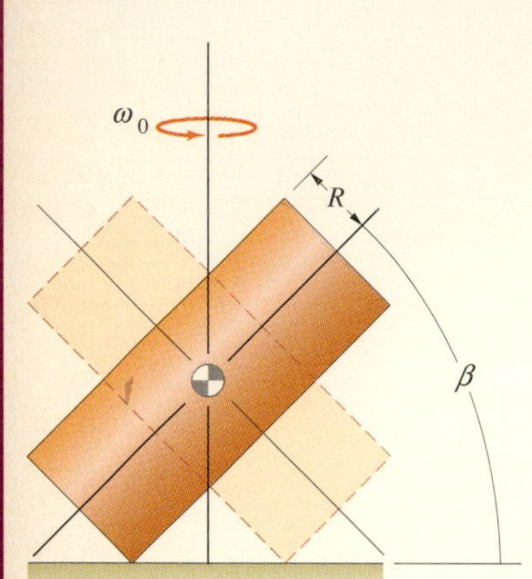

Fig. 9.27

The tilted homogeneous cylinder in Fig. 9.27 undergoes a steady motion in which one end rolls on the floor while its center of mass remains stationary. The angle β between the cylinder axis and the horizontal remains constant, and the cylinder axis rotates about the vertical axis with constant angular velocity ω_0. The cylinder has mass m, radius R, and length l. What is ω_0?

STRATEGY

By expressing the equations of angular motion in terms of ω_0, we can determine the value of ω_0 necessary for the equations to be satisfied. Therefore our first task is to determine the cylinder's angular velocity $\boldsymbol{\omega}$ in terms of ω_0. We can simplify this task by using a secondary coordinate system that is not body-fixed.

SOLUTION

Choose a Coordinate System We use a secondary coordinate system in which the z axis remains aligned with the cylinder axis and the y axis remains horizontal (Fig. a). The reason for this choice is that the angular velocity of the coordinate system is easy to describe—the coordinate system rotates about the vertical axis with the angular velocity ω_0—and the rotation of the cylinder relative to the coordinate system is also easy to describe. The angular velocity vector of the coordinate system is

$$\boldsymbol{\Omega} = \omega_0 \cos\beta\, \mathbf{i} + \omega_0 \sin\beta\, \mathbf{k}.$$

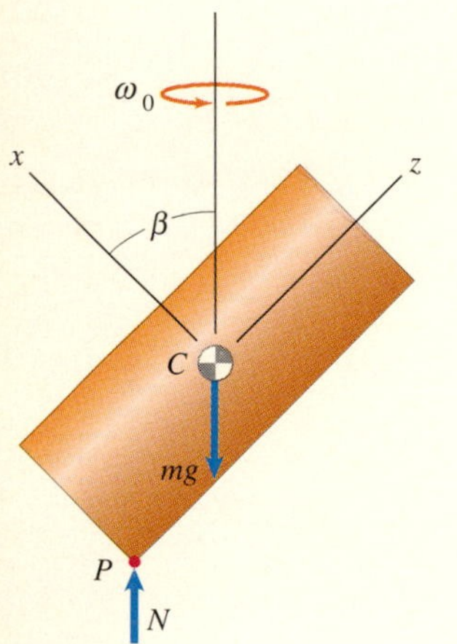

(a) Coordinate system with the z axis aligned with the cylinder axis and the y axis horizontal.

Relative to the coordinate system, the cylinder rotates about the z axis. Writing its angular velocity relative to the coordinate system as $\omega_{\text{rel}}\mathbf{k}$, the angular velocity vector of the cylinder is

$$\boldsymbol{\omega} = \boldsymbol{\Omega} + \omega_{\text{rel}}\,\mathbf{k} = \omega_0 \cos\beta\mathbf{i} + (\omega_0 \sin\beta + \omega_{\text{rel}})\mathbf{k}.$$

We can determine ω_{rel} from the condition that the velocity of the point P in contact with the floor is zero. Expressing the velocity of P in terms of the velocity of the center of mass C, we obtain

$$\mathbf{v}_P = \mathbf{v}_C + \boldsymbol{\omega} \times \mathbf{r}_{P/C}:$$

$$\mathbf{0} = \mathbf{0} + [\omega_0 \cos\beta\mathbf{i} + (\omega_0 \sin\beta + \omega_{\text{rel}})\mathbf{k}] \times \left[-R\mathbf{i} - \frac{1}{2}l\mathbf{k}\right]$$

$$= \left[\frac{1}{2}l\omega_0 \cos\beta - R(\omega_0 \sin\beta + \omega_{\text{rel}})\right]\mathbf{j}.$$

Solving for ω_{rel}, we obtain

$$\omega_{\text{rel}} = \left[\frac{1}{2}\left(\frac{l}{R}\right)\cos\beta - \sin\beta\right]\omega_0.$$

Therefore the cylinder's angular velocity vector is

$$\boldsymbol{\omega} = \omega_0 \cos\beta\,\mathbf{i} + \frac{1}{2}\left(\frac{l}{R}\right)\omega_0 \cos\beta\,\mathbf{k}.$$

Draw the Free-Body Diagram We draw the free-body diagram of the cylinder in Fig. (a), showing its weight and the normal force exerted by the floor. Because the center of mass is stationary, we know that the floor exerts no horizontal force on the cylinder and the normal force is $N = mg$.

Apply the Equations of Motion The moment about the center of mass due to the normal force is

$$\Sigma\,\mathbf{M} = \left(mg\,R\sin\beta - \frac{1}{2}mgl\cos\beta\right)\mathbf{j}.$$

From Appendix C, the inertia matrix is

$$\begin{bmatrix} \frac{1}{4}mR^2 + \frac{1}{12}ml^2 & 0 & 0 \\ 0 & \frac{1}{4}mR^2 + \frac{1}{12}ml^2 & 0 \\ 0 & 0 & \frac{1}{2}mR^2 \end{bmatrix}.$$

Substituting our expressions for $\boldsymbol{\Omega}$, $\boldsymbol{\omega}$, $\Sigma\,\mathbf{M}$, and the moments and products of inertia into the equation of angular motion, Eq. (9.30), and evaluating the matrix products, we obtain the equation

$$mg\left(R\sin\beta - \frac{1}{2}l\cos\beta\right) = \left(\frac{1}{4}mR^2 + \frac{1}{12}ml^2\right)\omega_0^2\sin\beta\cos\beta - \frac{1}{2}\left(\frac{1}{2}mR^2\right)\omega_0^2\left(\frac{l}{R}\right)\cos^2\beta.$$

We solve this equation for ω_0^2:

$$\omega_0^2 = \frac{g(R\sin\beta - \frac{1}{2}l\cos\beta)}{(\frac{1}{4}R^2 + \frac{1}{12}l^2)\sin\beta\cos\beta - \frac{1}{4}lR\cos^2\beta}. \tag{9.33}$$

DISCUSSION

If our solution yields a negative value for ω_0^2 for a given value of β, the assumed steady motion of the cylinder is not possible. For example, if the cylinder's diameter is equal to its length, $2R = l$, we can write Eq. (9.33) as

$$\frac{R\omega_0^2}{g} = \frac{\sin\beta - \cos\beta}{\frac{7}{12}\sin\beta\cos\beta - \frac{1}{2}\cos^2\beta}.$$

In Fig. 9.28 we show the graph of this equation as a function of β. For values of β from approximately 40° to 45°, there is no real solution for ω_0. Notice that at $\beta =$ 45°, $\omega_0 = 0$, which means that the cylinder is stationary and balanced with the center of mass directly above point P.

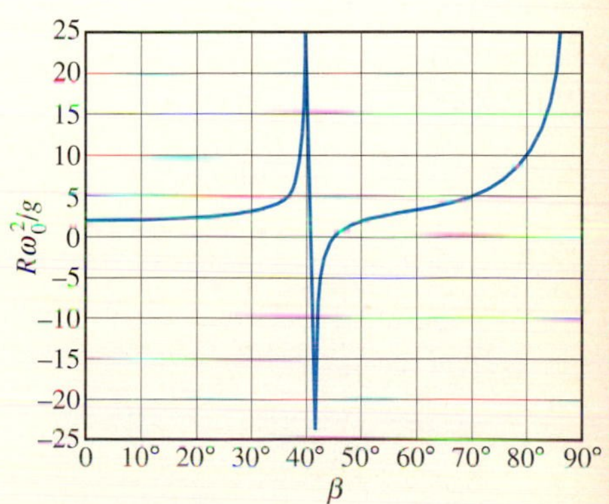

Fig. 9.28
Graph of $R\omega_0^2/g$ as a function of β.

Problems

9.52 The inertia matrix of a rigid body in terms of a body-fixed coordinate system with its origin at the center of mass is

$$[I] = \begin{bmatrix} 8.8 & 4.0 & 0 \\ 4.0 & 6.6 & 0 \\ 0 & 0 & 15.4 \end{bmatrix} \text{kg-m}^2.$$

If the rigid body's angular velocity is $\boldsymbol{\omega} = -6\mathbf{i} + 2\mathbf{j} - 2\mathbf{k}$ (rad/s) and its angular acceleration is zero, what are the components of the total moment about its center of mass?

9.53 If the total moment about the center of mass of the rigid body in Problem 9.52 is zero, what are the components of its angular acceleration?

9.54 A rigid body rotates about a fixed point O. Its inertia matrix in terms of a body-fixed coordinate system with its origin at O is

$$[I] = \begin{bmatrix} 1 & -1 & 0 \\ -1 & 5 & 1 \\ 0 & 1 & 7 \end{bmatrix} \text{slug-ft}^2.$$

If the rigid body's angular velocity is $\boldsymbol{\omega} = 6\mathbf{i} + 6\mathbf{j} - 4\mathbf{k}$ (rad/s) and its angular acceleration is zero, what are the components of the total moment about O?

9.55 If the total moment about O due to the forces and couples acting on the rigid body in Problem 9.54 is zero, what are the components of its angular acceleration?

9.56 At $t = 0$, the stationary rectangular plate of mass m is subjected to the force F perpendicular to the plate. No other external forces or couples act on the plate. What is the magnitude of the acceleration of point A at $t = 0$?

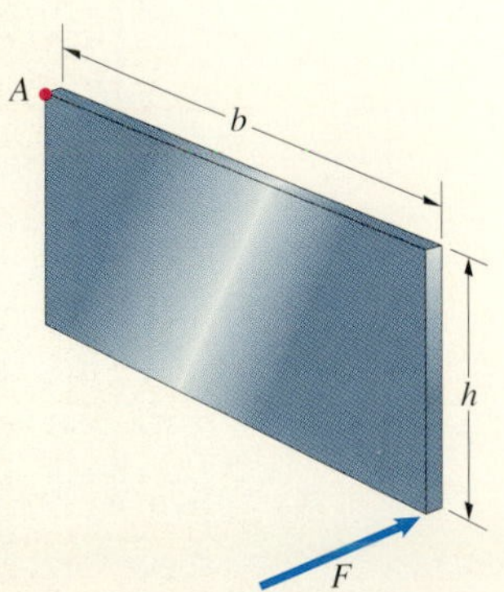

P9.56

9.57 The mass of the homogeneous slender bar is 6 kg. At $t = 0$, the stationary bar is subjected to the force $\mathbf{F} = 12\,\mathbf{k}$ (N) at the point $x = 2$ m, $y = 0$. No other external forces or couples act on the bar.

(a) What is the bar's angular acceleration at $t = 0$?

(b) What is the acceleration of the point $x = 2$ m, $y = 0$ at $t = 0$?

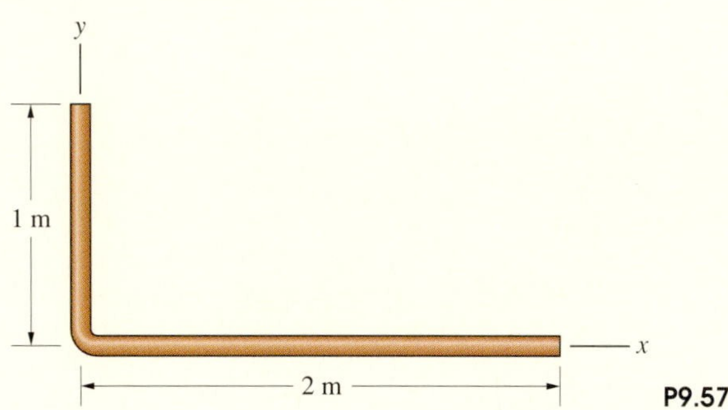

P9.57

9.58 The mass of the homogeneous slender bar is 1.2 slug. At $t = 0$, the stationary bar is subjected to the force $\mathbf{F} = 2\mathbf{i} + 4\mathbf{k}$ (lb) at the point $x = 1$ ft, $y = 1$ ft. No other external forces or couples act on the bar.

(a) What is the bar's angular acceleration at $t = 0$?

(b) What is the acceleration of the point $x = -1$ ft, $y = -1$ ft at $t = 0$?

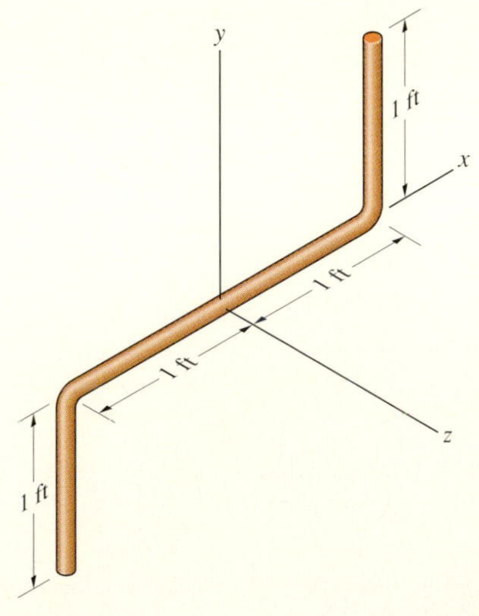

P9.58

9.59 In terms of the coordinate system shown, the inertia matrix of the 2.4-kg steel plate is

$$[I] = \begin{bmatrix} 0.0318 & -0.0219 & 0 \\ -0.0219 & 0.0357 & 0 \\ 0 & 0 & 0.0674 \end{bmatrix} \text{kg-m}^2.$$

The angular velocity of the plate is $\boldsymbol{\omega} = 6.4\mathbf{i} + 8.2\mathbf{j} + 14.0\mathbf{k}$ (rad/s), and its angular acceleration is $\boldsymbol{\alpha} = 60\mathbf{i} + 40\mathbf{j} - 120\mathbf{k}$ (rad/s^2). What are the components of the total moment exerted on the plate about its center of mass?

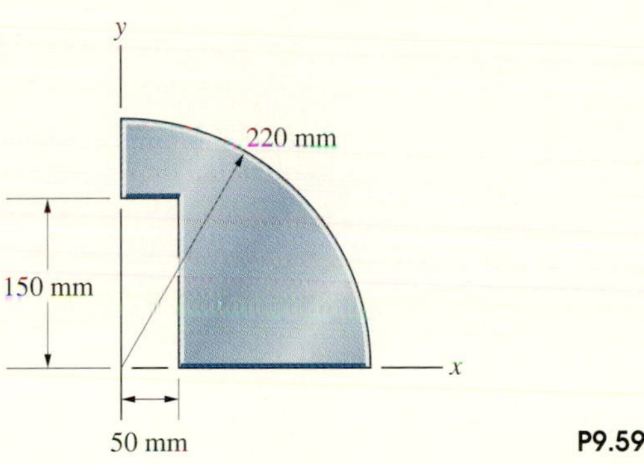

P9.59

9.60 At $t = 0$, the plate in Problem 9.59 is stationary and is subjected to a force $\mathbf{F} = -10\mathbf{k}$ (N) at the point (220, 0, 0) mm. No other forces or couples act on the plate. At that instant, determine (a) the acceleration of the plate's center of mass; (b) the plate's angular acceleration.

9.61 A 3-kg slender bar is rigidly attached to a 2-kg thin circular disk. In terms of the body-fixed coordinate system shown, the angular velocity of the composite object is $\boldsymbol{\omega} = 100\mathbf{i} - 4\mathbf{j} + 6\mathbf{k}$ (rad/s) and its angular acceleration is zero. What are the components of the total moment exerted on the object about its center of mass?

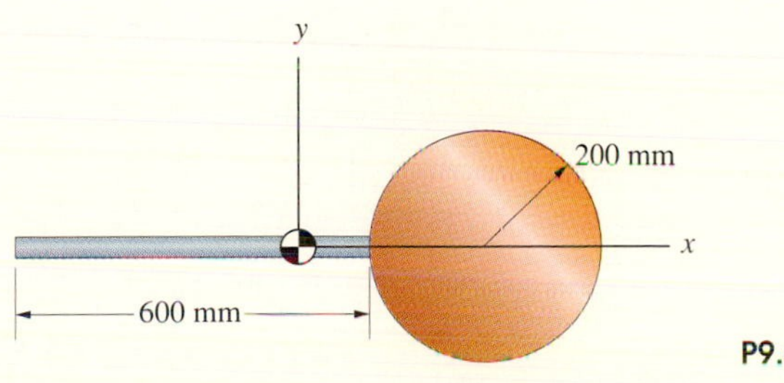

P9.61

9.62 At $t = 0$, the composite object in Problem 9.61 is stationary and is subjected to the moment $\Sigma\,\mathbf{M} = -10\mathbf{i} + 10\mathbf{j}$ (N-m) about its center of mass. No other forces or couples act on the object. What are the components of its angular acceleration at $t = 0$?

9.63 The base of the dish antenna is rotating with a constant angular velocity of 1 rad/s. The angle $\theta = 30°$, $d\theta/dt = 20°/\text{s}$, and $d^2\theta/dt^2 = -40°/\text{s}^2$. The mass of the antenna is 280 kg, and its moments and products of inertia in kg-m^2 are $I_{xx} = 140$, $I_{yy} = I_{zz} = 220$, $I_{xy} = I_{yz} = I_{zx} = 0$. Determine the couple exerted on the antenna by its support A at this instant.

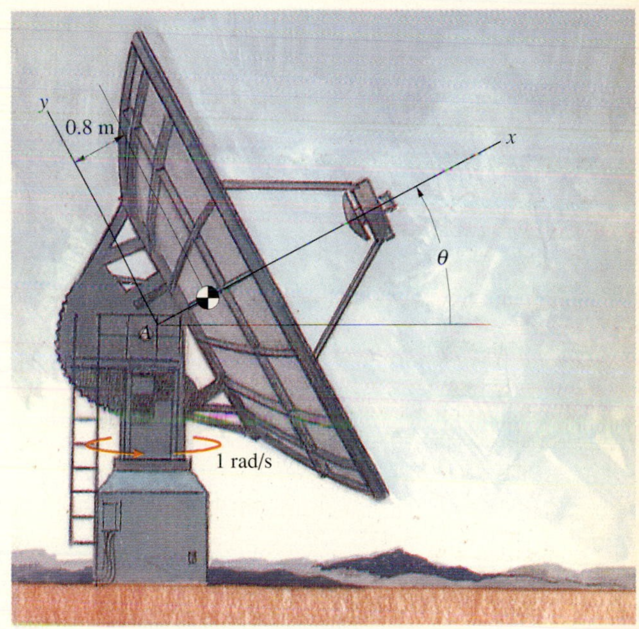

P9.63

9.64 A thin triangular plate of mass m is supported by a ball and socket at O. If it is held in the horizontal position and released from rest, what are the components of its angular acceleration at that instant?

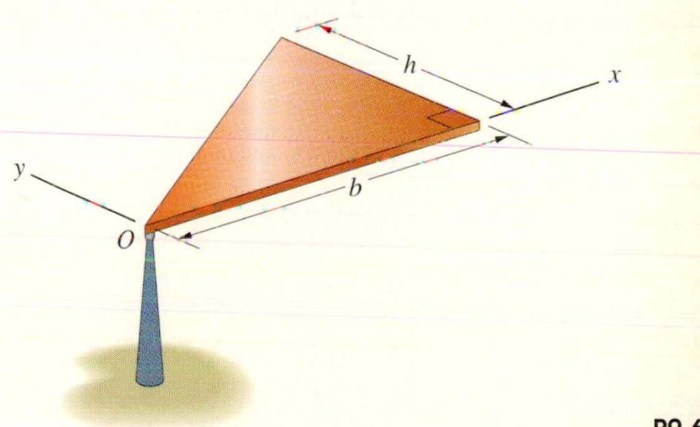

P9.64

9.65 Determine the force exerted on the triangular plate in Problem 9.64 by the ball and socket support at the instant of release.

9.66 In Problem 9.64, the mass of the plate is 5 kg, $b = 900$ mm, and $h = 600$ mm. If the plate is released in the horizontal position with angular velocity $\boldsymbol{\omega} = 4\mathbf{i}$ (rad/s), what are the components of its angular acceleration at that instant?

9.67 A subassembly of a space station can be modeled as two rigidly connected slender bars, each of mass 5 Mg. The subassembly is not rotating at $t = 0$ when a reaction control motor exerts a force $\mathbf{F} = 400\mathbf{k}$ (N) at B. What is the acceleration of point A at that instant?

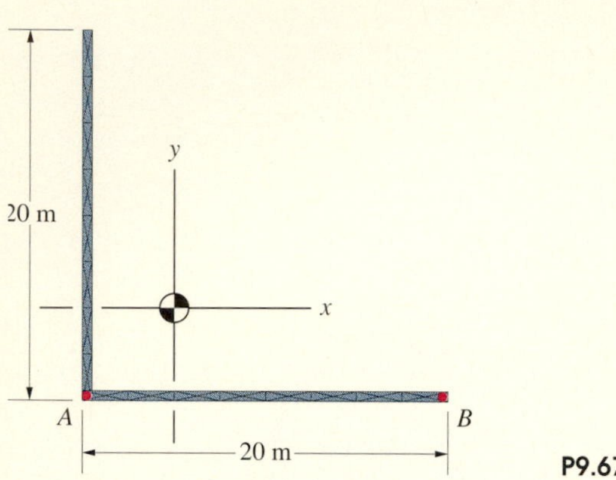

P9.67

9.68 If the subassembly described in Problem 9.67 rotates about the x axis at a constant rate of 1 revolution every 10 minutes, what is the magnitude of the couple its reaction control system must exert on it?

9.69 The thin circular disk of radius R and mass m is attached rigidly to the vertical shaft. The disk is slanted at an angle β relative to the horizontal plane. The shaft rotates with constant angular velocity ω_0. What is the magnitude of the couple exerted on the disk by the shaft?

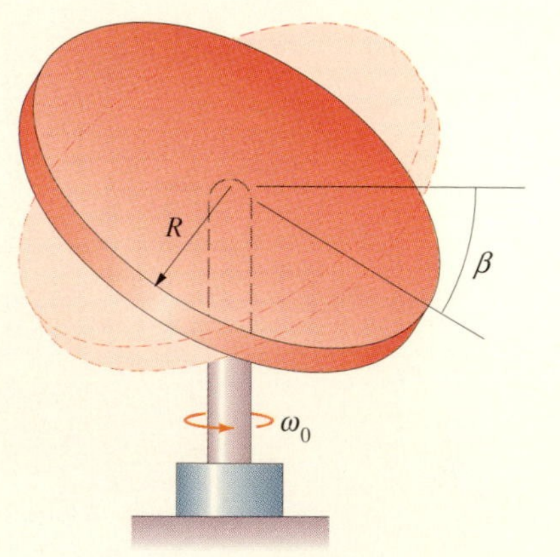

P9.69

9.70 A slender bar of mass m and length l is welded to a horizontal shaft that rotates with constant angular velocity ω_0. Determine the magnitudes of the force $\mathbf{F}$ and couple $\mathbf{C}$ exerted on the bar by the shaft. (Write the equations of angular motion in terms of the body-fixed coordinate system shown.)

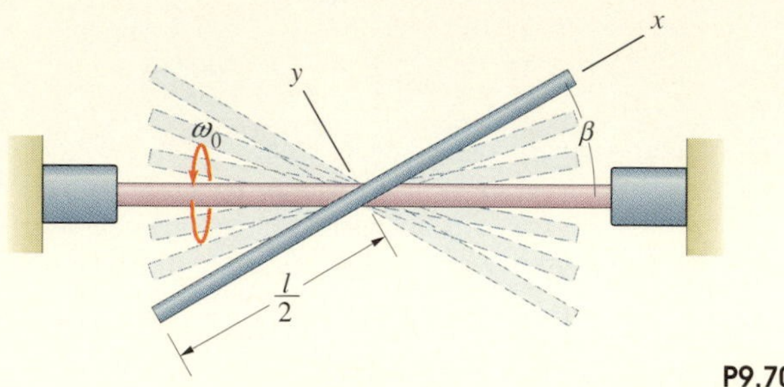

P9.70

9.71 A slender bar of mass m and length l is welded to a horizontal shaft that rotates with constant angular velocity ω_0. Determine the magnitude of the couple $\mathbf{C}$ exerted on the bar by the shaft. (Write the equation of angular motion in terms of the body-fixed coordinate system shown.)

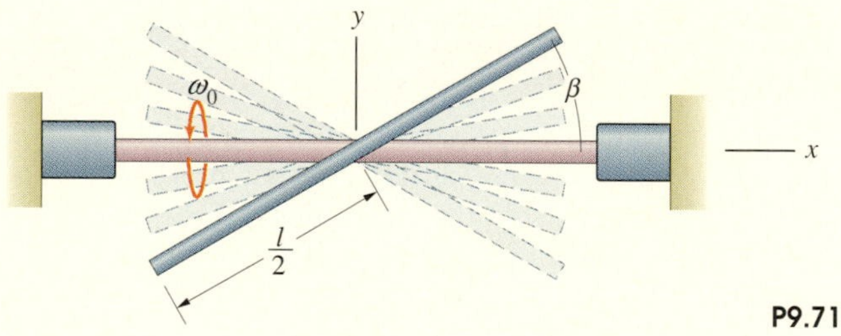

P9.71

9.72 The slender bar of length l and mass m is pinned to the vertical shaft at O. The vertical shaft rotates with a constant angular velocity ω_0. Show that the value of ω_0 necessary for the bar to remain at a constant angle β relative to the vertical is

$$\omega_0 = \sqrt{3g/(2l\cos\beta)}.$$

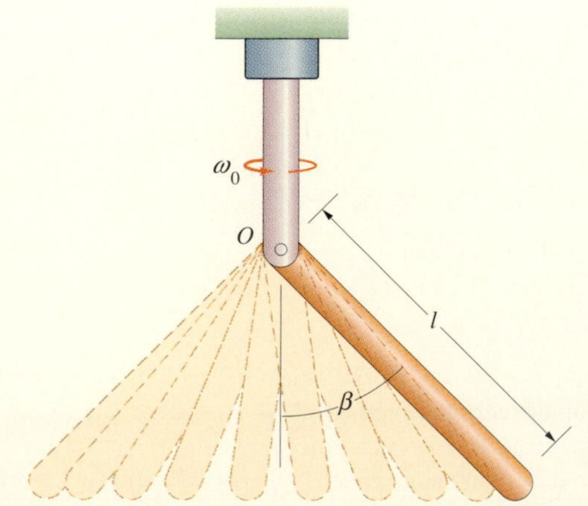

P9.72

9.73 The vertical shaft rotates with constant angular velocity ω_0. The 35° angle between the edge of the 10-lb thin rectangular plate pinned to the shaft and the shaft remains constant. Determine ω_0.

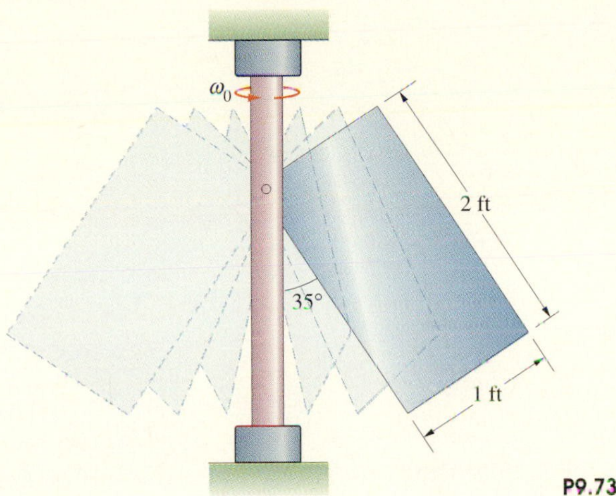

P9.73

9.74 A thin circular disk of mass m mounted on a horizontal shaft rotates relative to the shaft with constant angular velocity ω_d. The horizontal shaft is rigidly attached to a vertical shaft rotating with constant angular velocity ω_0. Determine the magnitude of the couple exerted on the disk by the horizontal shaft.

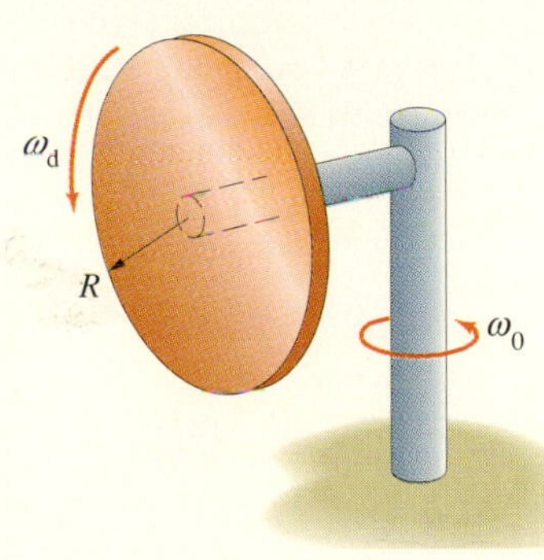

P9.74

9.75 The thin triangular plate has ball and socket supports at A and B. The y axis is vertical. If the plate rotates with constant angular velocity ω_0, what are the horizontal components of the reactions on the plate at A and B?

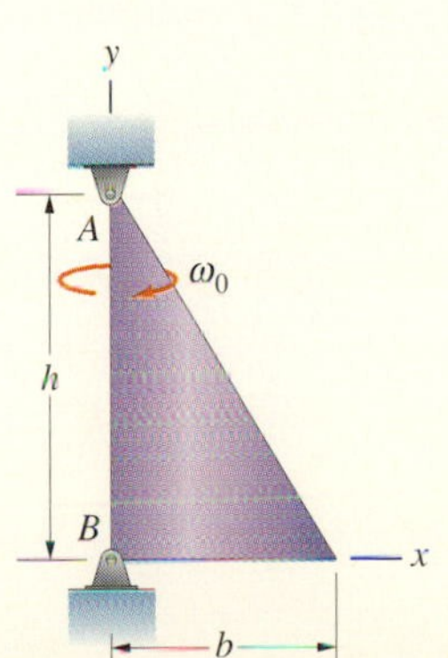

P9.75

9.76 The 10-lb thin circular disk is rigidly attached to the 12-lb slender horizontal shaft. The disk and horizontal shaft rotate about the axis of the shaft with constant angular velocity $\omega_d = 20$ rad/s. The entire assembly rotates about the vertical axis with constant angular velocity $\omega_0 = 4$ rad/s. Determine the components of the force and couple exerted on the horizontal shaft by the disk.

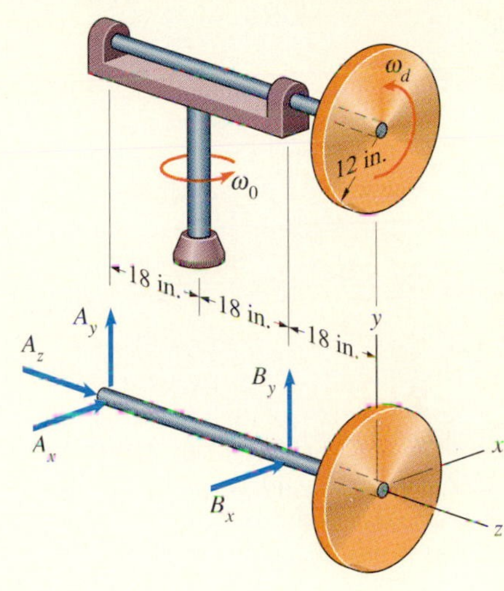

P9.76

9.77 In Problem 9.76, determine the reactions exerted on the horizontal shaft by the two bearings.

9.78 The thin rectangular plate is attached to the rectangular frame by pins. The frame rotates with constant angular velocity ω_0. Show that

$$\frac{d^2\beta}{dt^2} = -\omega_0^2 \sin\beta\cos\beta.$$

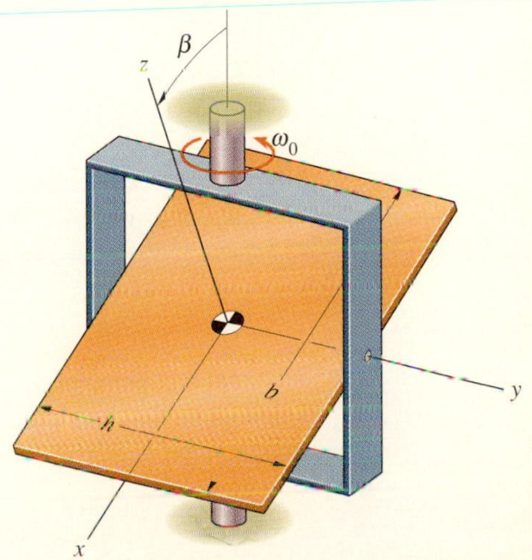

P9.78

9.79 The axis of the right circular cone of mass m, height h, and radius R spins about the vertical axis with constant angular velocity ω_0. Its center of mass is stationary and its base rolls on the floor. Show that the angular velocity ω_0 necessary for this motion is $\omega_0 = \sqrt{10g/3R}$.

Strategy: Let the z axis remain aligned with the axis of the cone and the x axis remain vertical.

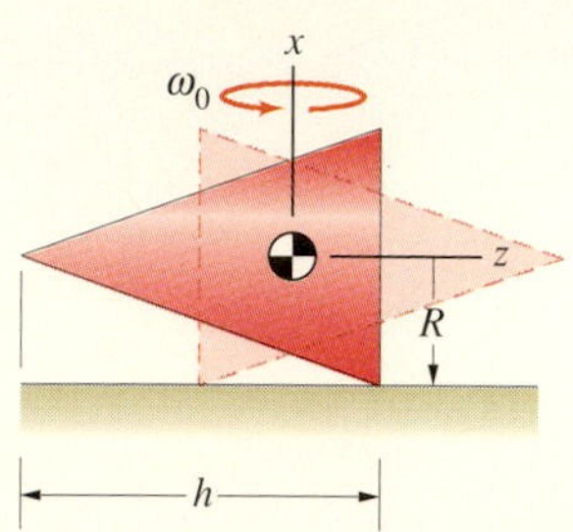

P9.79

9.80 A thin circular disk of radius R and mass m rolls along a circular path of radius r. The magnitude v of the velocity of the center of the disk and the angle θ between the disk's axis and the vertical are constants. Show that v satisfies the equation

$$v^2 = \frac{\frac{2}{3}g \cot\theta (r - R\cos\theta)^2}{r - \frac{5}{6}R\cos\theta}.$$

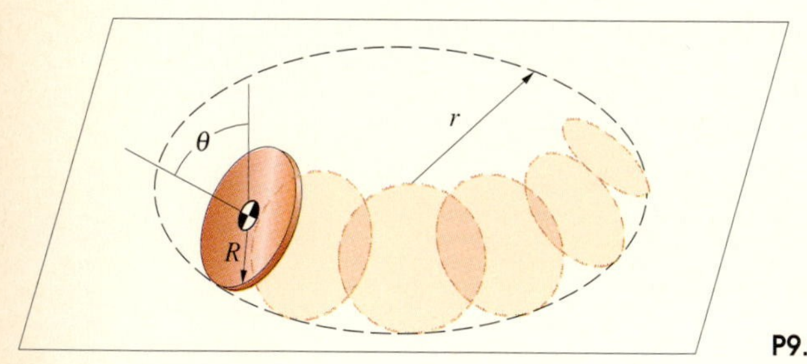

P9.80

9.81 The vertical shaft rotates with constant angular velocity ω_0, causing the grinding mill to roll on the horizontal surface. Assume that point P of the mill is stationary at the instant shown, and that the force N exerted on the mill by the surface is perpendicular to the surface and acts at P. The mass of the mill is m, and its moments and products of inertia in terms of the coordinate system shown are I_{zz}, $I_{xx} = I_{yy}$, and $I_{xy} = I_{yz} = I_{zx} = 0$. Determine N.

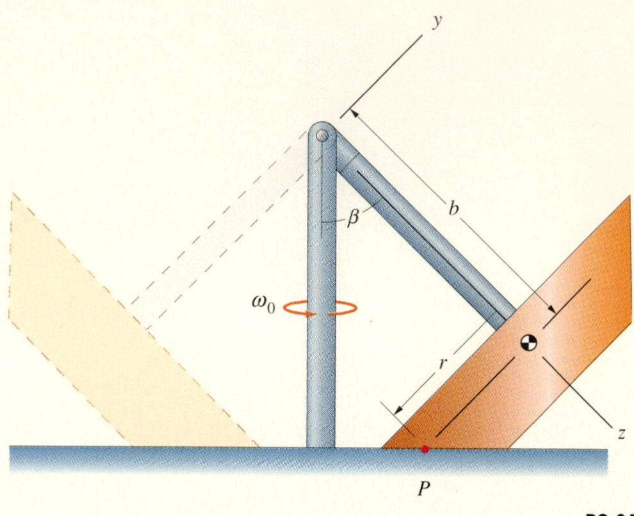

P9.81

9.82 The view of an airplane's landing gear looking from behind the airplane is shown in Fig. (a). The radius of the wheel is 300 mm, and its moment of inertia is 2 kg-m^2. The airplane takes off at 30 m/s. After takeoff, the landing gear retracts by rotating toward the right side of the airplane as shown in Fig. (b). Determine the magnitude of the couple exerted by the wheel on its support. (Neglect the airplane's angular motion.)

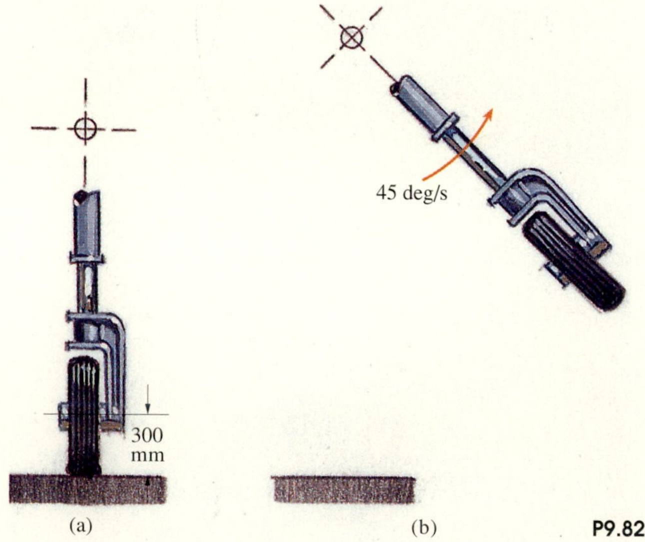

P9.82

9.83 If the rider turns to his left, will the couple exerted on the motorcycle by its wheels tend to cause the motorcycle to lean toward the rider's left side or his right side?

P9.83

9.84 By substituting the components of $\mathbf{H}_O$ from Eqs. (9.7) into Eq. (9.24), derive Eqs. (9.25).

9.5 *Eulerian Angles*

The equations of angular motion relate the total moment acting on a rigid body to its angular velocity and acceleration. If we know the total moment and the angular velocity, we can determine the angular acceleration. But how can we use the angular acceleration to determine the rigid body's angular position, or orientation, as a function of time? To explain how this is done, we must first show how to specify the orientation of a rigid body in three dimensions.

You have seen that describing the orientation of a rigid body in planar motion requires only the angle θ that specifies the body's rotation relative to some reference orientation. In three-dimensional motion, three angles are required. To understand why, consider a particular axis that is fixed relative to the rigid body. Two angles are necessary to specify the direction of the axis, and a third angle is needed to specify the rigid body's orientation about the axis. Although several systems of angles for describing the orientation of a rigid body are commonly used, the best-known system is the one called the Eulerian angles. In this section we define these angles and express the equations of angular motion in terms of them.

Objects with an Axis of Symmetry

We first explain how the Eulerian angles are used to describe the orientation of an object with an axis of rotational symmetry, because this case results in simpler equations of angular motion.

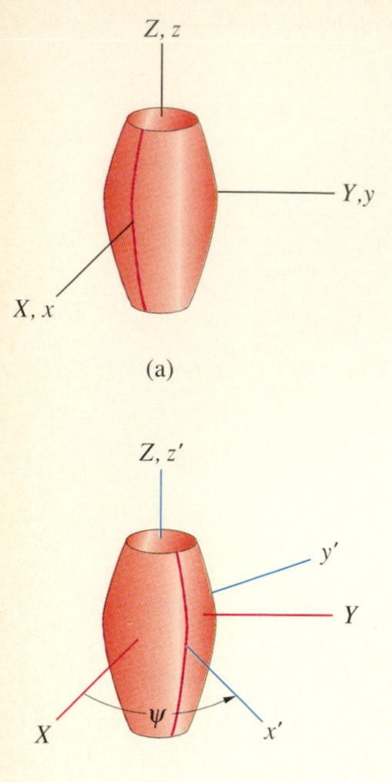

Definitions We assume that an object has an axis of rotational symmetry, and introduce two reference frames: a secondary coordinate system xyz, with its z axis coincident with the object's axis of symmetry, and an inertial primary coordinate system XYZ. We begin with the object in a reference position in which xyz and XYZ are superimposed (Fig. 9.29a).

Our first step is to rotate the object and the xyz system together through an angle ψ about the Z axis (Fig. 9.29b). In this intermediate orientation, we denote the secondary coordinate system by $x'y'z'$. Next, we rotate the object and the xyz system together through an angle θ about the x' axis (Fig. 9.29c). Finally, we rotate the object relative to the xyz system through an angle ϕ about its axis of symmetry (Fig. 9.29d). Notice that the x axis remains in the XY plane.

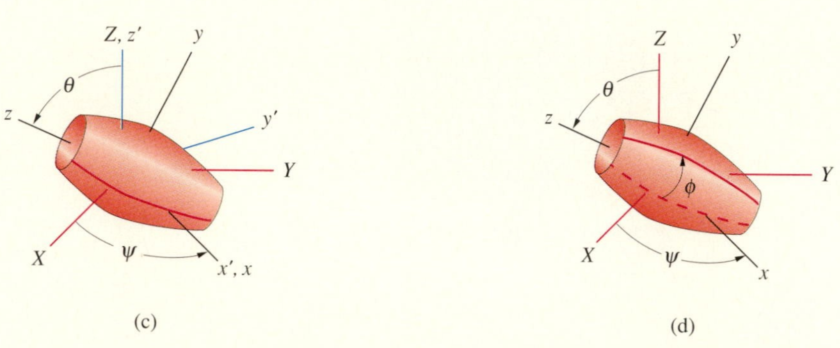

Fig. 9.29
(a) The reference position.
(b) The rotation ψ about the Z axis.
(c) The rotation θ about the x' axis.
(d) The rotation ϕ of the object relative to the xyz system.

The angles ψ and θ specify the orientation of the secondary xyz system relative to the primary XYZ system. ψ is called the **precession angle**, and θ is called the **nutation angle**. The angle ϕ specifying the rotation of the rigid body relative to the xyz system is called the **spin angle**. These three angles specify the orientation of the rigid body relative to the primary coordinate system and are called the **Eulerian angles**. We can obtain any orientation of the object relative to the primary coordinate system by appropriate choices of these angles: We choose ψ and θ to obtain the desired direction of the axis of symmetry, then choose ϕ to obtain the desired rotational position of the object about its axis of symmetry.

Equations of Angular Motion To analyze an object's motion in terms of the Eulerian angles, we must express the equations of angular motion in terms of them. Figure 9.30(a) shows the rotation ψ from the reference orientation of the xyz system to its intermediate orientation $x'y'z'$. We represent the angular velocity of the coordinate system due to the rate of change of ψ by the angular velocity vector $\dot{\psi}$ pointing in the z' direction. (We use a dot to denote the derivative with respect to time.) Figure 9.30(b) shows the second rotation θ. We represent the angular velocity due to the rate of change of θ by the vector $\dot{\theta}$ pointing in the x direction. We also resolve the angular velocity vector $\dot{\psi}$ into components in the y and z directions. The components of the angular velocity of the xyz system relative to the primary coordinate system are

$$\Omega_x = \dot{\theta},$$
$$\Omega_y = \dot{\psi} \sin \theta, \quad (9.34)$$
$$\Omega_z = \dot{\psi} \cos \theta.$$

In Fig. 9.30(c), we represent the angular velocity of the rigid body relative to the *xyz* system by the vector $\dot{\phi}$. Adding this angular velocity to the angular velocity of the *xyz* system, we obtain the components of the angular velocity of the rigid body relative to the *XYZ* system:

$$\omega_x = \dot{\theta},$$
$$\omega_y = \dot{\psi} \sin \theta, \quad (9.35)$$
$$\omega_z = \dot{\phi} + \dot{\psi} \cos \theta.$$

Taking the time derivatives of these equations, we obtain

$$\frac{d\omega_x}{dt} = \ddot{\theta},$$
$$\frac{d\omega_y}{dt} = \ddot{\psi} \sin \theta + \dot{\psi}\dot{\theta} \cos \theta, \quad (9.36)$$
$$\frac{d\omega_z}{dt} = \ddot{\phi} + \ddot{\psi} \cos \theta - \dot{\psi}\dot{\theta} \sin \theta.$$

As a consequence of the object's rotational symmetry, the products of inertia I_{xy}, I_{xz}, and I_{yz} are zero and $I_{xx} = I_{yy}$. The inertia matrix is of the form

$$[I] = \begin{bmatrix} I_{xx} & 0 & 0 \\ 0 & I_{xx} & 0 \\ 0 & 0 & I_{zz} \end{bmatrix}. \quad (9.37)$$

Substituting Eqs. (9.34)–(9.37) into Eqs. (9.29), we obtain the equations of angular motion in terms of the Eulerian angles:

$$\Sigma M_x = I_{xx}\ddot{\theta} + (I_{zz} - I_{xx})\dot{\psi}^2 \sin \theta \cos \theta + I_{zz}\dot{\phi}\dot{\psi} \sin \theta, \quad (9.38)$$

$$\Sigma M_y = I_{xx}(\ddot{\psi} \sin \theta + 2\dot{\psi}\dot{\theta} \cos \theta) - I_{zz}(\dot{\phi}\dot{\theta} + \dot{\psi}\dot{\theta} \cos \theta), \quad (9.39)$$

$$\Sigma M_z = I_{zz}(\ddot{\phi} + \ddot{\psi} \cos \theta - \dot{\psi}\dot{\theta} \sin \theta). \quad (9.40)$$

To determine the Eulerian angles as functions of time when the total moment is known, these equations usually must be solved by numerical integration. However, we can obtain an important class of closed-form solutions by assuming a specific type of motion.

Steady Precession The motion called steady precession is commonly observed in tops and gyroscopes. The object's rate of spin $\dot{\phi}$ relative to the *xyz* coordinate system is assumed to be constant (Fig. 9.31). The nutation angle θ,

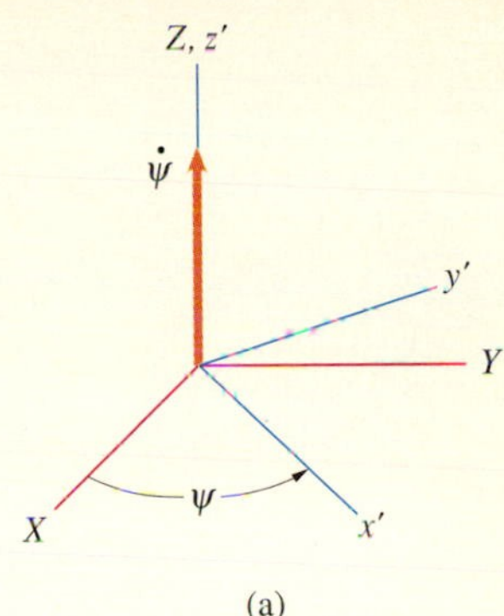

(a)

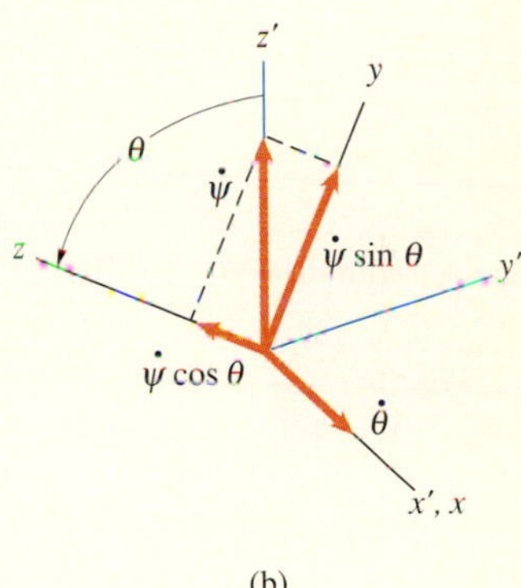

(b)

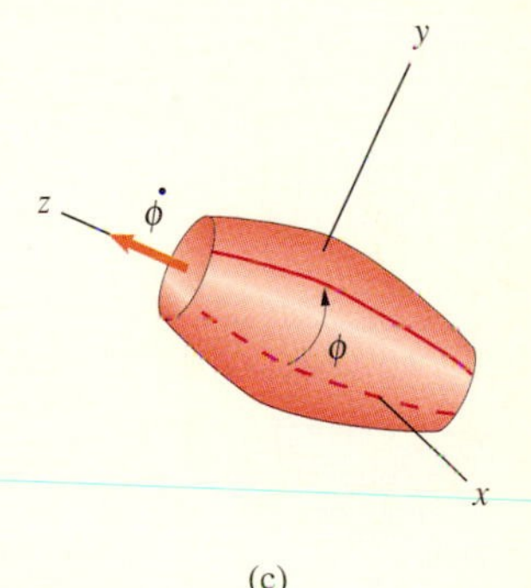

(c)

Fig 9.30
(a) The rotation ψ and the angular velocity $\dot{\psi}$.
(b) The rotation θ, the angular velocity $\dot{\theta}$, and the components of the angular velocity $\dot{\psi}$.
(c) The rotation ϕ and the angular velocity $\dot{\phi}$.

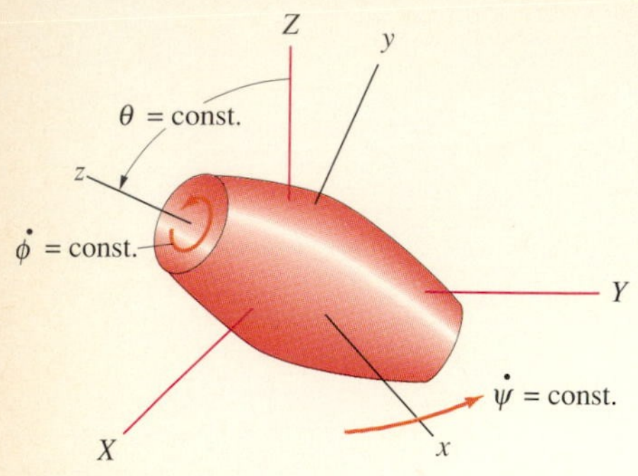

Fig. 9.31
Steady precession.

the inclination of the **spin axis** z relative to the Z axis, is assumed to be constant, and the **precession rate** $\dot{\psi}$, the rate at which the xyz system rotates about the Z axis, is assumed to be constant. The last assumption explains the name given to this motion.

With these assumptions, Eqs. (9.38)–(9.40) reduce to

$$\Sigma M_x = (I_{zz} - I_{xx})\dot{\psi}^2 \sin\theta\cos\theta + I_{zz}\dot{\phi}\dot{\psi}\sin\theta, \tag{9.41}$$

$$\Sigma M_y = 0, \tag{9.42}$$

$$\Sigma M_z = 0. \tag{9.43}$$

We discuss two examples: the steady precession of a spinning top and the steady precession of an axially symmetric object that is free of external moments.

Precession of a Top The peculiar behavior of a top (Fig. 9.32a) inspired some of the first analytical studies of three-dimensional motions of rigid bodies. When a top is set into motion, its spin axis may initially remain vertical, a motion called **sleeping**. As friction reduces the spin rate, the spin axis begins to lean over and rotate about the vertical axis. This phase of the top's motion approximates steady precession. (The top's spin rate continuously decreases due to friction, whereas in steady precession we assume the spin rate to be constant.)

To analyze the motion, we place the primary coordinate system XYZ with its origin at the point of the top and the Z axis upward. Then we align the z axis of the xyz system with the spin axis (Fig. 9.32b). We assume that the top's point rests in a small depression so that it remains at a fixed point on the floor. The precession angle ψ and nutation angle θ specify the orientation of the spin axis, and the spin rate of the top relative to the xyz system is $\dot{\phi}$.

The top's weight exerts a moment $\Sigma M_x = mgh\sin\theta$ about the origin, and the moments $\Sigma M_y = 0$ and $\Sigma M_z = 0$. Substituting $\Sigma M_x = mgh\sin\theta$ into Eq. (9.41), we obtain

$$mgh = (I_{zz} - I_{xx})\dot{\psi}^2\cos\theta + I_{zz}\dot{\phi}\dot{\psi}, \tag{9.44}$$

and Eqs. (9.42) and (9.43) are identically satisfied. Equation (9.44) relates the spin rate, nutation angle, and rate of precession. For example, if we know the spin rate $\dot{\phi}$ and nutation angle θ, we can solve for the top's precession rate $\dot{\psi}$.

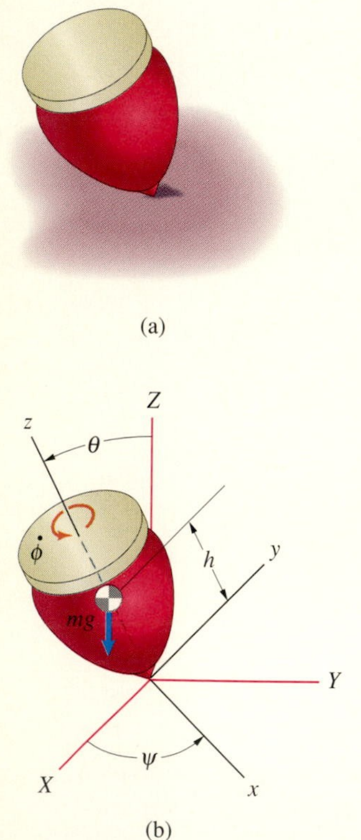

Fig. 9.32
(a) A spinning top seems to defy gravity.
(b) The precession angle ψ and nutation angle θ specify the orientation of the spin axis.

Moment-Free Steady Precession A spinning axisymmetric object that is free of external moments, such as an axisymmetric satellite in orbit, can exhibit a motion similar to the steady precessional motion of a top. We observe this motion when an American football is thrown in a "wobbly" spiral. To analyze it, we place the origin of the xyz system at the object's center of mass (Fig. 9.33a). Equation (9.41) becomes

$$(I_{zz} - I_{xx})\dot{\psi}\cos\theta + I_{zz}\dot{\phi} = 0, \tag{9.45}$$

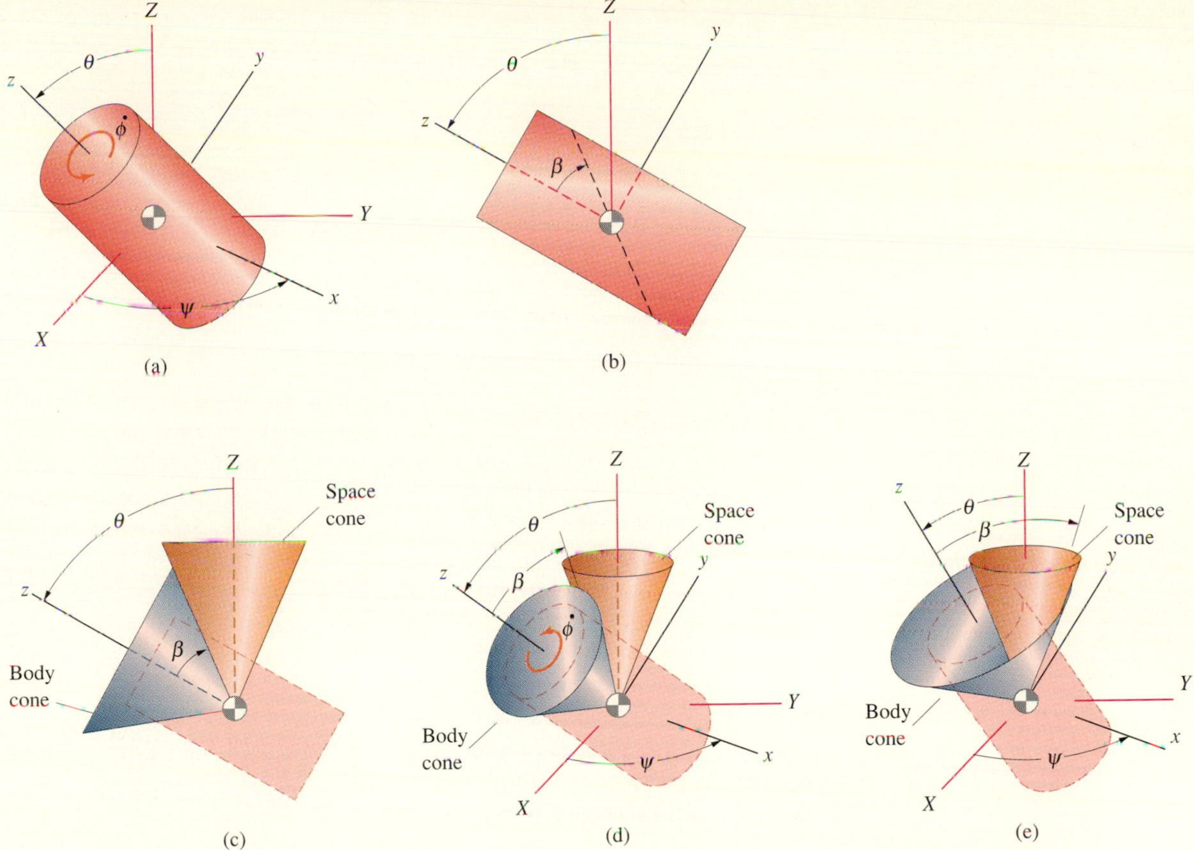

Fig. 9.33

(a) An axisymmetric object.
(b) Points on the straight line at an angle β from the z axis are stationary relative to the XYZ coordinate system.
(c), (d) The body and space cones. The body cone rolls on the stationary space cone.
(e) When $\beta > \theta$, the interior surface of the body cone rolls on the stationary space cone.

and Eqs. (9.42) and (9.43) are identically satisfied. For a given value of the nutation angle, Eq. (9.45) relates the object's rates of precession and spin.

We can interpret Eq. (9.45) in a way that enables you to visualize the motion. We look for a point in the y-z plane at which the object's velocity relative to the center of mass is zero at the current instant. We want to find a point with coordinates $(0, y, z)$ such that

$$\begin{aligned}\boldsymbol{\omega} \times (y\mathbf{j} + z\mathbf{k}) &= [(\dot{\psi} \sin\theta)\mathbf{j} + (\dot{\phi} + \dot{\psi}\cos\theta)\mathbf{k}] \times [y\mathbf{j} + z\mathbf{k}] \\ &= [z\dot{\psi}\sin\theta - y(\dot{\phi} + \dot{\psi}\cos\theta)]\mathbf{i} = 0.\end{aligned}$$

This equation is satisfied at points in the y-z plane such that

$$\frac{y}{z} = \frac{\dot{\psi}\sin\theta}{\dot{\phi} + \dot{\psi}\cos\theta}.$$

This relation is satisfied by points on the straight line at an angle β relative to the z axis in Fig. 9.33(b), where

$$\tan\beta = \frac{y}{z} = \frac{\dot{\psi}\sin\theta}{\dot{\phi} + \dot{\psi}\cos\theta}.$$

Solving Eq. (9.45) for $\dot{\phi}$ and substituting the result into this equation, we obtain

$$\tan\beta = \left(\frac{I_{zz}}{I_{xx}}\right)\tan\theta.$$

If $I_{xx} > I_{zz}$, the angle $\beta < \theta$. In Fig. 9.33(c), we show an imaginary cone of half-angle β, called the **body cone**, whose axis is coincident with the z axis. The body cone is in contact with a fixed cone, called the **space cone**, whose axis is coincident with the Z axis. If the body cone rolls on the curved surface of the space cone as the z axis precesses about the Z axis (Fig. 9.33d), the points of the body cone lying on the straight line in Fig. 9.33(b) have zero velocity relative to the XYZ system. That means that *the motion of the body cone is identical to the motion of the object.* You can visualize the object's motion by visualizing the motion of the body cone as it rolls around the outer surface of the space cone. This motion is called **direct precession.**

If $I_{xx} < I_{zz}$, the angle $\beta > \theta$. In this case you must visualize the *interior* surface of the body cone rolling on the fixed space cone (Fig. 9.33e). This motion is called **retrograde precession.**

Arbitrary Objects

In our analysis of axially symmetric objects, the object moves relative to the secondary xyz coordinate system, rotating about the z axis. As a consequence, only two angles, the precession angle ψ and nutation angle θ, are needed to specify the orientation of the xyz coordinate system, and this simplifies the equations of angular motion. The object must be axially symmetric about the z axis so that the moments and products of inertia will not vary as it rotates. In the case of an arbitrary object, the moments and products of inertia will be constants only if the xyz coordinate system is body-fixed. This means that three angles are needed to specify the orientation of the coordinate system, and the resulting equations of angular motion are more complicated.

Definitions We begin with a reference position in which the body-fixed xyz and primary XYZ coordinate systems are superimposed (Fig. 9.34a). First, we rotate the xyz system through the precession angle ψ about the Z axis (Fig. 9.34b) and denote it by $x'y'z'$ in this intermediate orientation. Then we rotate the xyz system through the nutation angle θ about the x' axis (Fig. 9.34c), denoting it by $x''y''z''$. We obtain the final orientation of the xyz system by rotating it through the angle ϕ about the z'' axis (Fig. 9.34d). Notice that we have used one more rotation of the xyz system than in the case of an axially symmetric object.

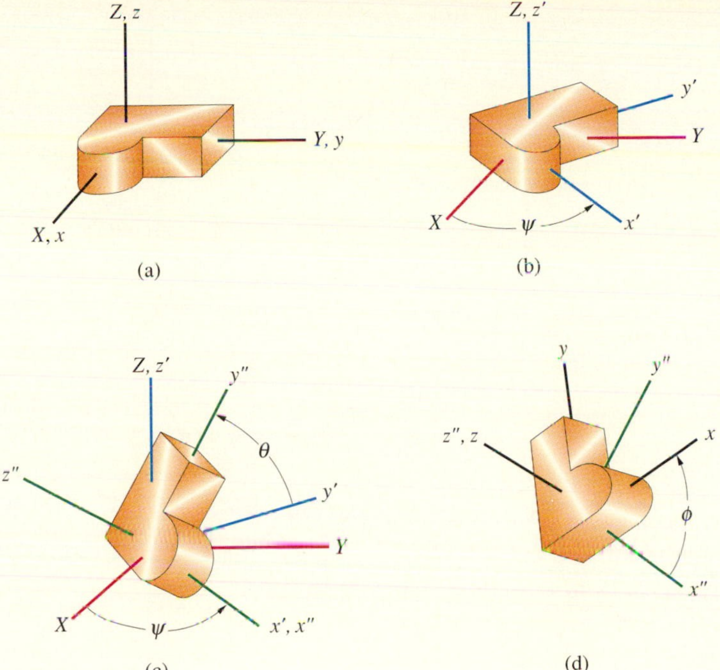

Fig. 9.34
(a) The reference position.
(b) The rotation ψ about the Z axis.
(c) The rotation θ about the x' axis.
(d) The rotation ϕ about the z'' axis.

We can obtain any orientation of the body-fixed coordinate system relative to the reference coordinate system by these three rotations. We choose ψ and θ to obtain the desired direction of the z axis, then choose ϕ to obtain the desired orientation of the x and y axes.

Just as in the case of an object with rotational symmetry, we must express the components of the rigid body's angular velocity in terms of the Eulerian angles to obtain the equations of angular motion. Figure 9.35(a) shows the rotation ψ from the reference orientation of the xyz system to the intermediate orientation $x'y'z'$. We represent the angular velocity of the body-fixed coordinate system due to the rate of change of ψ by the vector $\dot{\psi}$ pointing in the z' direction. Figure 9.35(b) shows the next rotation θ that takes the body-fixed coordinate system to the intermediate orientation $x''y''z''$. We represent the angular velocity due to the rate of change of θ by the vector $\dot{\theta}$ pointing in the x'' direction. In this figure we also show the components of the angular velocity vector $\dot{\psi}$ in the y'' and z'' directions. Figure 9.35(c) shows the third rotation ϕ that takes the body-fixed coordinate system to its final orientation defined by the three Eulerian angles. We represent the angular velocity due to the rate of change of ϕ by the vector $\dot{\phi}$ pointing in the z direction.

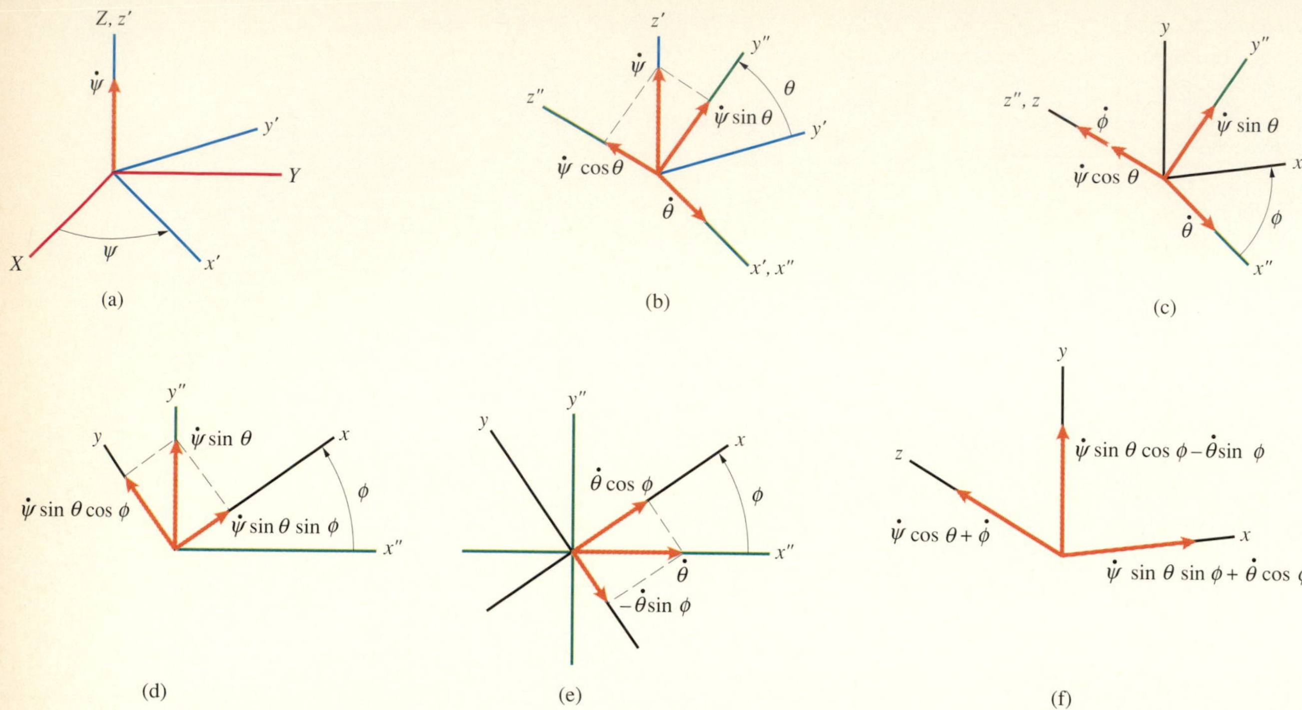

Fig. 9.35

(a) The rotation ψ and the angular velocity $\dot{\psi}$.
(b) The rotation θ, the angular velocity $\dot{\theta}$, and the components of $\dot{\psi}$ in the $x''y''z''$ system.
(c) The rotation ϕ and the angular velocity $\dot{\phi}$.
(d), (e) The components of the angular velocities $\dot{\psi}\sin\theta$ and $\dot{\theta}$ in the xyz system.
(f) The angular velocities ω_x, ω_y, ω_z.

To determine ω_x, ω_y, and ω_z in terms of the Eulerian angles, we need to determine the components of the angular velocities shown in Fig. 9.35(c) in the x, y, and z axis directions. The vectors $\dot{\phi}$ and $\dot{\psi}\cos\theta$ point in the z axis direction. In Figs. 9.35(d) and (e), which are drawn with the z axis pointing out of the page, we determine the components of the vectors $\dot{\psi}\sin\theta$ and $\dot{\theta}$ in the x and y axis directions.

By summing the components of the angular velocities in the three coordinate directions (Fig. 9.35f), we obtain

$$\begin{aligned}
\omega_x &= \dot{\psi}\sin\theta\sin\phi + \dot{\theta}\cos\phi,\\
\omega_y &= \dot{\psi}\sin\theta\cos\phi - \dot{\theta}\sin\phi,\\
\omega_z &= \dot{\psi}\cos\theta + \dot{\phi}.
\end{aligned} \tag{9.46}$$

The time derivatives of these equations are

$$\begin{aligned}
\frac{d\omega_x}{dt} &= \ddot{\psi}\sin\theta\sin\phi + \dot{\psi}\dot{\theta}\cos\theta\sin\phi + \dot{\psi}\dot{\phi}\sin\theta\cos\phi\\
&\quad + \ddot{\theta}\cos\phi - \dot{\theta}\dot{\phi}\sin\phi,\\
\frac{d\omega_y}{dt} &= \ddot{\psi}\sin\theta\cos\phi + \dot{\psi}\dot{\theta}\cos\theta\cos\phi - \dot{\psi}\dot{\phi}\sin\theta\sin\phi\\
&\quad - \ddot{\theta}\sin\phi - \dot{\theta}\dot{\phi}\cos\phi,\\
\frac{d\omega_z}{dt} &= \ddot{\psi}\cos\theta - \dot{\psi}\dot{\theta}\sin\theta + \ddot{\phi}.
\end{aligned} \tag{9.47}$$

Equations of Angular Motion With Eqs. (9.46) and (9.47), we can express the equations of angular motion in terms of the three Eulerian angles. To simplify the equations, *we assume that the body-fixed coordinate system xyz is a set of principal axes.* Then the equations of angular motion, Eqs. (9.29), are

$$\Sigma M_x = I_{xx}\frac{d\omega_x}{dt} - (I_{yy} - I_{zz})\omega_y\omega_z,$$

$$\Sigma M_y = I_{yy}\frac{d\omega_y}{dt} - (I_{zz} - I_{xx})\omega_z\omega_x,$$

$$\Sigma M_z = I_{zz}\frac{d\omega_z}{dt} - (I_{xx} - I_{yy})\omega_x\omega_y.$$

Substituting Eqs. (9.46) and (9.47) into these equations, we obtain the equations of angular motion in terms of Eulerian angles:

$$\begin{aligned}
\Sigma M_x &= I_{xx}\ddot{\psi}\sin\theta\sin\phi + I_{xx}\ddot{\theta}\cos\phi \\
&\quad + I_{xx}(\dot{\psi}\dot{\theta}\cos\theta\sin\phi + \dot{\psi}\dot{\phi}\sin\theta\cos\phi - \dot{\theta}\dot{\phi}\sin\phi) \\
&\quad - (I_{yy} - I_{zz})(\dot{\psi}\sin\theta\cos\phi - \dot{\theta}\sin\phi)(\dot{\psi}\cos\theta + \dot{\phi}), \\
\Sigma M_y &= I_{yy}\ddot{\psi}\sin\theta\cos\phi - I_{yy}\ddot{\theta}\sin\phi \\
&\quad + I_{yy}(\dot{\psi}\dot{\theta}\cos\theta\cos\phi - \dot{\psi}\dot{\phi}\sin\theta\sin\phi - \dot{\theta}\dot{\phi}\cos\phi) \\
&\quad - (I_{zz} - I_{xx})(\dot{\psi}\cos\theta + \dot{\phi})(\dot{\psi}\sin\theta\sin\phi + \dot{\theta}\cos\phi), \\
\Sigma M_z &= I_{zz}\ddot{\psi}\cos\theta + I_{zz}\ddot{\phi} - I_{zz}\dot{\psi}\dot{\theta}\sin\theta \\
&\quad - (I_{xx} - I_{yy})(\dot{\psi}\sin\theta\sin\phi + \dot{\theta}\cos\phi)(\dot{\psi}\sin\theta\cos\phi - \dot{\theta}\sin\phi).
\end{aligned} \tag{9.48}$$

If you know the Eulerian angles and their first and second time derivatives, you can solve these equations for the components of the total moment. Or, if you know the total moment, the Eulerian angles, and the first time derivatives of the Eulerian angles, you can determine the second time derivatives of the Eulerian angles. You can use these equations to determine the Eulerian angles as functions of time when you know the total moment, but numerical integration is usually necessary.

In the following example we analyze the motion of an object in steady precession. By aligning the coordinate system as shown in Fig. 9.31, you can use Eq. (9.41) to relate the total moment about the x axis to the nutation angle θ, precession rate $\dot{\psi}$, and spin rate $\dot{\phi}$.

Example 9.11

The thin circular disk of radius R and mass m in Fig. 9.36 rolls along a horizontal circular path of radius r. The angle θ between the disk's axis and the vertical remains constant. Determine the magnitude v of the velocity of the center of the disk as a function of the angle θ.

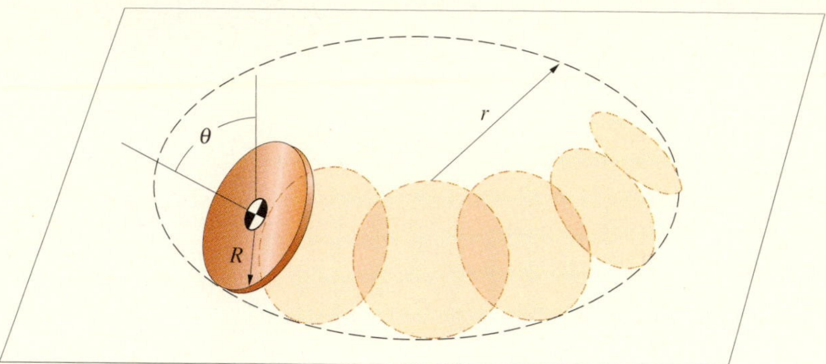

Fig. 9.36

STRATEGY

We can obtain the velocity of the center of the disk by assuming the disk is in steady precession and determining the conditions necessary for the equations of motion to be satisfied.

SOLUTION

In Fig. (a) we align the z axis of the secondary coordinate system with the disk's spin axis and assume that the x axis remains parallel to the surface on which the disk rolls. The angle θ is the nutation angle. The center of mass moves in a circular path of radius $r_G = r - R\cos\theta$. Therefore the precession rate, the rate at which the x axis rotates in the horizontal plane, is

$$\dot{\psi} = \frac{v}{r_G}.$$

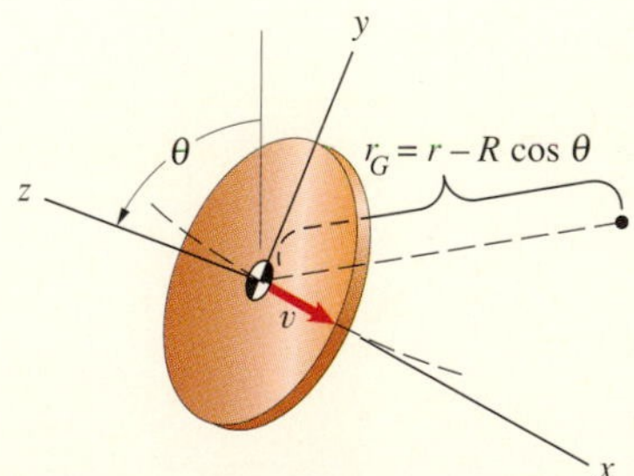

(a) Aligning the z axis with the spin axis. The x axis is horizontal.

From Eqs. (9.35), the components of the disk's angular velocity are

$$\omega_x = \dot{\theta} = 0,$$

$$\omega_y = \dot{\psi}\sin\theta = \frac{v}{r_G}\sin\theta,$$

$$\omega_z = \dot{\phi} + \dot{\psi}\cos\theta = \dot{\phi} + \frac{v}{r_G}\cos\theta,$$

where $\dot{\phi}$ is the spin rate. To determine $\dot{\phi}$, we use the condition that the velocity of the point of the disk in contact with the surface is zero. In terms of the velocity of the center, the velocity of the point of contact is

$$\mathbf{0} = v\mathbf{i} + \boldsymbol{\omega} \times (-R\mathbf{j}) = v\mathbf{i} + \begin{vmatrix} \mathbf{i} & \mathbf{j} & \mathbf{k} \\ 0 & \dfrac{v}{r_G}\sin\theta & \dot{\phi} + \dfrac{v}{r_G}\cos\theta \\ 0 & -R & 0 \end{vmatrix}.$$

Expanding the determinant and solving for $\dot{\phi}$, we obtain

$$\dot{\phi} = -\frac{v}{R} - \frac{v}{r_G}\cos\theta.$$

We draw the free-body diagram of the disk in Fig. (b). Because the center of mass moves in the horizontal plane, its acceleration in the vertical direction is zero. Therefore the normal force $N = mg$. The acceleration of the center of mass in the direction perpendicular to its circular path is $a_n = v^2/r_G$. Newton's second law in the direction perpendicular to the circular path is

$$T = m\frac{v^2}{r_G}.$$

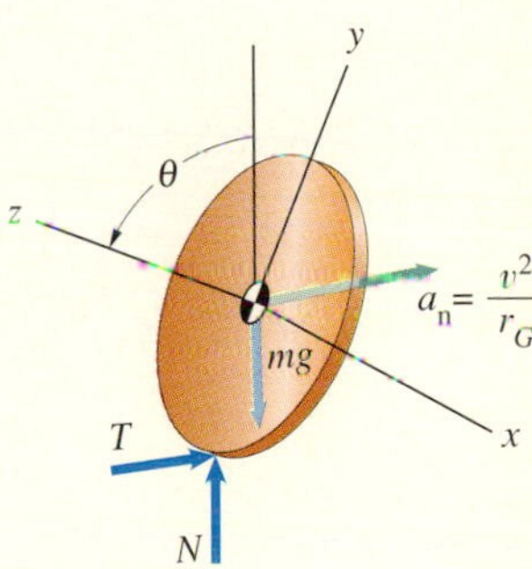

(b) Free-body diagram of the disk showing the normal acceleration of the center of mass.

Therefore the components of the total moment about the center of mass are

$$\Sigma M_x = TR\sin\theta - NR\cos\theta = m\frac{v^2}{r_G}R\sin\theta - mgR\cos\theta,$$

$$\Sigma M_y = 0,$$

$$\Sigma M_z = 0.$$

Substituting our expressions for $\dot{\psi}$, $\dot{\phi}$, and ΣM_x into the equation of angular motion for steady precession, Eq. (9.41), and solving for v, we obtain

$$v = \sqrt{\frac{\frac{2}{3}g\cot\theta(r - R\cos\theta)^2}{r - \frac{5}{6}R\cos\theta}}.$$

Problems

9.85 A ship has a turbine engine. The spin axis of the axisymmetric turbine is horizontal and aligned with the ship's longitudinal axis. The turbine rotates at 10,000 rpm. Its moment of inertia about its spin axis is 1000 kg-m^2. If the ship turns at a constant rate of 20 degrees per minute, what is the magnitude of the moment exerted on the ship by the turbine?

Strategy: Treat the turbine's motion as steady precession with nutation angle $\theta = 90°$.

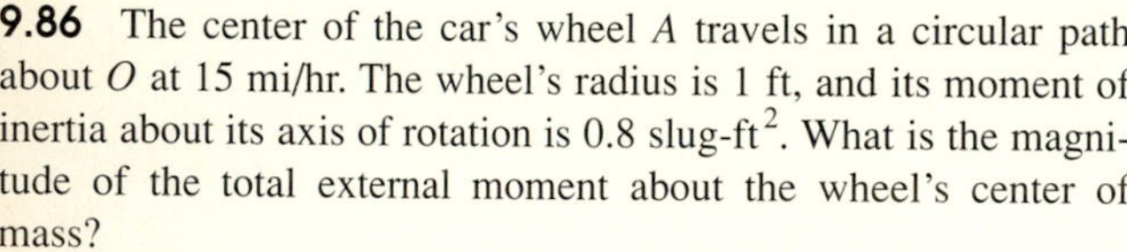

P9.85

9.86 The center of the car's wheel A travels in a circular path about O at 15 mi/hr. The wheel's radius is 1 ft, and its moment of inertia about its axis of rotation is 0.8 slug-ft^2. What is the magnitude of the total external moment about the wheel's center of mass?

Strategy: Treat the wheel's motion as steady precession with nutation angle $\theta = 90°$.

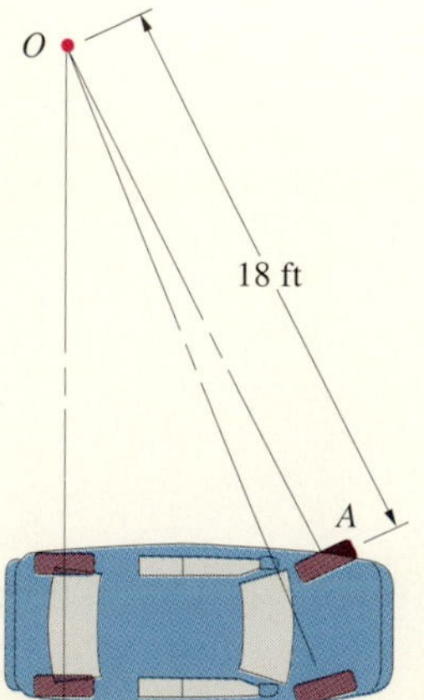

P9.86

9.87 Solve Problem 9.74 by treating the motion as steady precession.

9.88 Solve Problem 9.79 by treating the motion as steady precession.

9.89 The bent bar is rigidly attached to the vertical shaft, which rotates with constant angular velocity ω_0. The thin circular disk of mass m and radius R is pinned to the bent bar and rotates with constant angular velocity ω_d relative to the bar. Determine the magnitudes of the force and couple exerted on the disk by the bar.

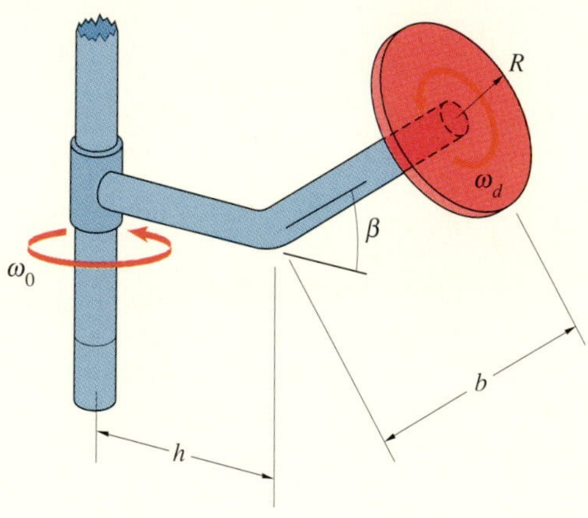

P9.89

9.90 In Problem 9.89, determine the value of the angular velocity ω_d which causes no couple to be exerted on the disk by the bar.

9.91 A thin circular disk undergoes moment-free steady precession. The z axis is perpendicular to the disk. Show that the disk's precession rate is $\dot{\psi} = -2\dot{\phi}/\cos\theta$. (Notice that when the nutation angle is small, the precession rate is approximately two times the spin rate.)

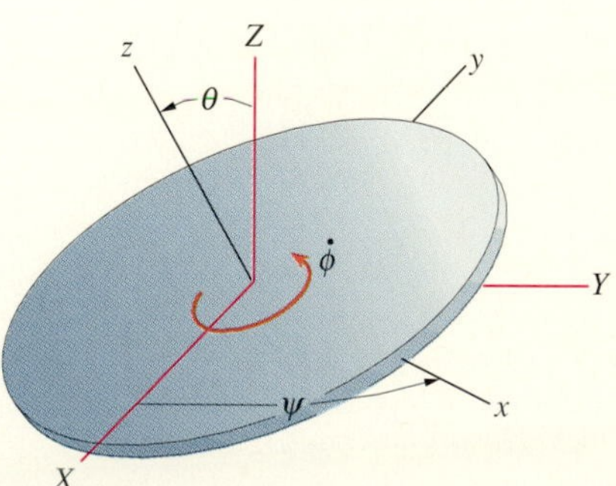

P9.91

9.92 The rocket is in moment-free steady precession with nutation angle $\theta = 40°$ and spin rate $\dot{\phi} = 4$ revolutions per second. Its moments of inertia are $I_{xx} = 10{,}000$ kg-m^2 and $I_{zz} = 2000$ kg-m^2. What is the rocket's precession rate $\dot{\psi}$ in revolutions per second?

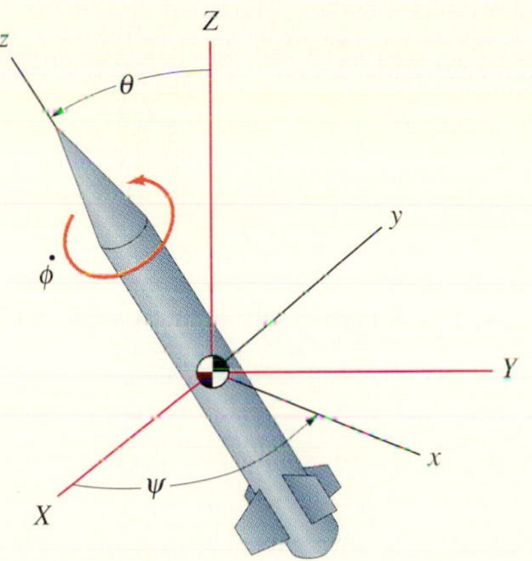

P9.92

9.93 Sketch the body and space cones for the motion of the rocket in Problem 9.92.

9.94 The top is in steady precession with nutation angle $\theta = 15°$ and precession rate $\dot{\psi} = 1$ revolution per second. The mass of the top is 8×10^{-4} slugs, its center of mass is 1 in. from the point, and its moments of inertia are $I_{xx} = 6 \times 10^{-6}$ slug-ft^2 and $I_{zz} = 2 \times 10^{-6}$ slug-ft^2. What is the spin rate $\dot{\phi}$ of the top in revolutions per second?

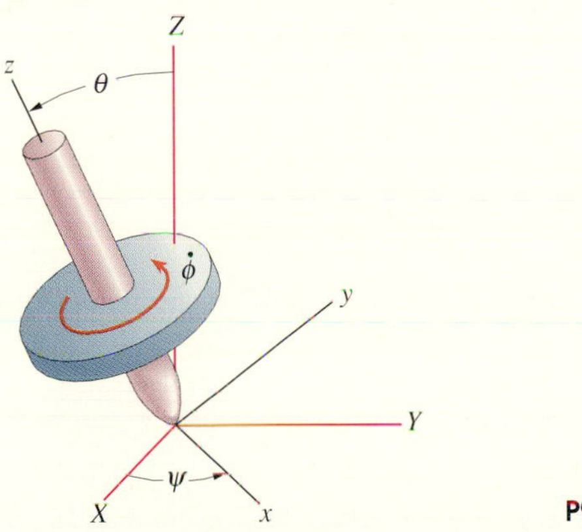

P9.94

9.95 The top described in Problem 9.94 has a spin rate $\dot{\phi} = 15$ revolutions per second. Draw a graph of the precession rate (in revolutions per second) as a function of the nutation angle θ for values of θ from zero to 45°.

9.96 The rotor of a tumbling gyroscope can be modeled as being in moment-free steady precession. Its moments of inertia are $I_{xx} = I_{yy} = 0.04$ kg-m^2, $I_{zz} = 0.18$ kg-m^2. Its spin rate is $\dot{\phi} = 1500$ rpm and its nutation angle is $\theta = 20°$.
(a) What is its precession rate in rpm?
(b) Sketch the body and space cones.

9.97 A satellite can be modeled as an 800-kg cylinder 4 m in length and 2 m in diameter. If the nutation angle is $\theta = 20°$ and the spin rate $\dot{\phi}$ is one revolution per second, what is the satellite's precession rate $\dot{\psi}$ in revolutions per second?

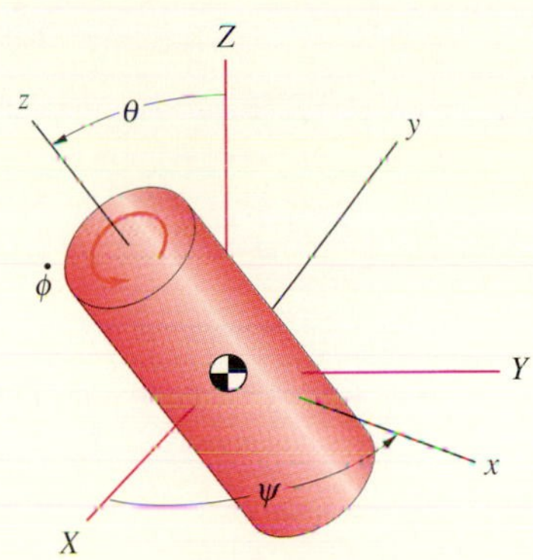

P9.97

9.98 Solve Problem 9.81 by treating the motion as steady precession.

9.99 Solve Problem 9.82 by treating the motion as steady precession.

9.100 Solve Problem 9.83 by treating the motion as steady precession.

9.101 Suppose that you are testing a car and use accelerometers and gyroscopes to measure its Eulerian angles and their derivatives relative to a reference coordinate system. At a particular instant, $\psi = 15°$, $\theta = 4°$, $\phi = 15°$, the rates of change of the Eulerian angles are zero, and their second time derivatives are $\ddot{\psi} = 0$, $\ddot{\theta} = 1$ rad/s^2, and $\ddot{\phi} = -0.5$ rad/s^2. The car's principal moments of inertia in kg-m^2 are $I_{xx} = 2200$, $I_{yy} = 480$, and $I_{zz} = 2600$. What are the components of the total moment about the car's center of mass?

P9.101

9.102 If the Eulerian angles and their second derivatives of the car described in Problem 9.101 have the given values but their rates of change are $\dot{\psi} = 0.2$ rad/s, $\dot{\theta} = -2$ rad/s, and $\dot{\phi} = 0$, what are the components of the total moment about the car's center of mass?

9.103 Suppose that the Eulerian angles of the car described in Problem 9.101 are $\psi = 40°$, $\theta = 20°$, and $\phi = 5°$, their rates of change are zero, and the components of the total moment about the car's center of mass are $\Sigma M_x = -400$ N-m, $\Sigma M_y = 200$ N-m, and $\Sigma M_z = 0$. What are the x, y, and z components of the car's angular acceleration?

Chapter Summary

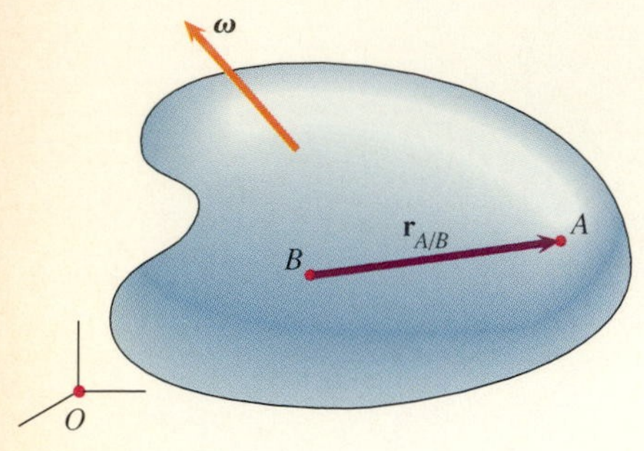

(a)

Kinematics

The velocity of a point A of a rigid body relative to a given reference frame is given in terms of the velocity of a point B of the rigid body and the rigid body's angular velocity by (Fig. a)

$$\mathbf{v}_A = \mathbf{v}_B + \boldsymbol{\omega} \times \mathbf{r}_{A/B}. \qquad \text{Eq. (9.1)}$$

The acceleration of point A is given in terms of the acceleration of point B, the rigid body's angular acceleration, and its angular velocity by

$$\mathbf{a}_A = \mathbf{a}_B + \boldsymbol{\alpha} \times \mathbf{r}_{A/B} + \boldsymbol{\omega} \times (\boldsymbol{\omega} \times \mathbf{r}_{A/B}). \qquad \text{Eq. (9.2)}$$

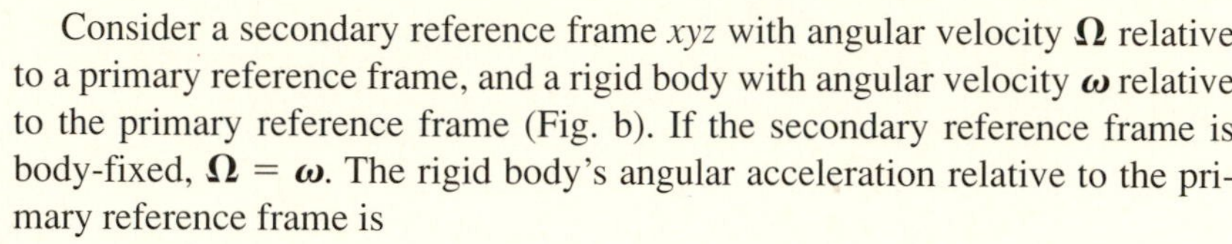

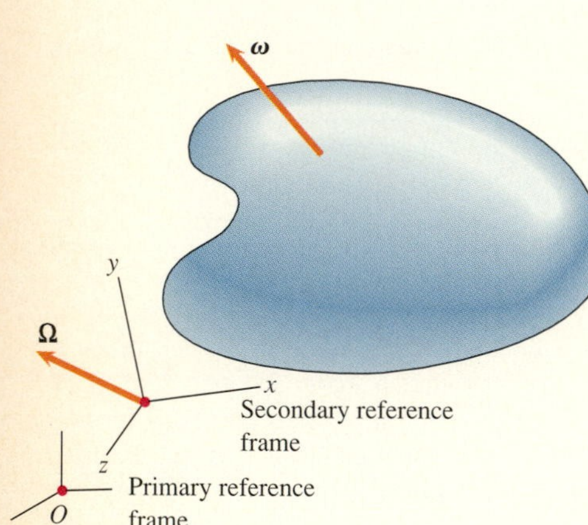

(b)

Consider a secondary reference frame xyz with angular velocity $\boldsymbol{\Omega}$ relative to a primary reference frame, and a rigid body with angular velocity $\boldsymbol{\omega}$ relative to the primary reference frame (Fig. b). If the secondary reference frame is body-fixed, $\boldsymbol{\Omega} = \boldsymbol{\omega}$. The rigid body's angular acceleration relative to the primary reference frame is

$$\boldsymbol{\alpha} = \frac{d\omega_x}{dt}\mathbf{i} + \frac{d\omega_y}{dt}\mathbf{j} + \frac{d\omega_z}{dt}\mathbf{k} + \boldsymbol{\Omega} \times \boldsymbol{\omega}. \qquad \text{Eq. (9.4)}$$

When the secondary reference frame is not body-fixed, it is often convenient to express $\boldsymbol{\omega}$ as the sum of $\boldsymbol{\Omega}$ and the angular velocity $\boldsymbol{\omega}_{\text{rel}}$ of the rigid body relative to the secondary reference frame:

$$\boldsymbol{\omega} = \boldsymbol{\Omega} + \boldsymbol{\omega}_{\text{rel}}. \qquad \text{Eq. (9.5)}$$

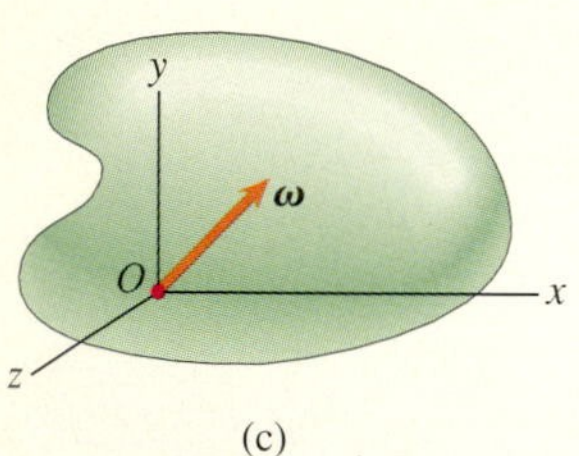

(c)

Angular Momentum

If a rigid body rotates about a fixed point O with angular velocity $\boldsymbol{\omega}$ (Fig. c), the components of its angular momentum about O are given by

$$\begin{bmatrix} H_{Ox} \\ H_{Oy} \\ H_{Oz} \end{bmatrix} = \begin{bmatrix} I_{xx} & -I_{xy} & -I_{xz} \\ -I_{yx} & I_{yy} & -I_{yz} \\ -I_{zx} & -I_{zy} & I_{zz} \end{bmatrix} \begin{bmatrix} \omega_x \\ \omega_y \\ \omega_z \end{bmatrix}. \qquad \text{Eq. (9.10)}$$

(d)

This equation also gives the components of the rigid body's angular momentum about the center of mass in general three-dimensional motion (Fig. d). In that case the moments of inertia are evaluated in terms of a coordinate system with its origin at the center of mass.

Euler's Equations

The equations governing three-dimensional motion of a rigid body include Newton's second law and equations of angular motion. For a rigid body rotating about a fixed point O (Fig. c), the equations of angular motion are expressed in terms of the components of the total moment about O [Eq. (9.26)]:

$$\begin{bmatrix} \Sigma M_{Ox} \\ \Sigma M_{Oy} \\ \Sigma M_{Oz} \end{bmatrix} = \begin{bmatrix} I_{xx} & -I_{xy} & -I_{xz} \\ -I_{yx} & I_{yy} & -I_{yz} \\ -I_{zx} & -I_{zy} & I_{zz} \end{bmatrix} \begin{bmatrix} d\omega_x/dt \\ d\omega_y/dt \\ d\omega_z/dt \end{bmatrix} + \begin{bmatrix} 0 & -\Omega_z & \Omega_y \\ \Omega_z & 0 & -\Omega_x \\ -\Omega_y & \Omega_x & 0 \end{bmatrix} \begin{bmatrix} I_{xx} & -I_{xy} & -I_{xz} \\ -I_{yx} & I_{yy} & -I_{yz} \\ -I_{zx} & -I_{zy} & I_{zz} \end{bmatrix} \begin{bmatrix} \omega_x \\ \omega_y \\ \omega_z \end{bmatrix},$$

where $\mathbf{\Omega}$ is the angular velocity of the coordinate system. If the coordinate system is body-fixed, $\mathbf{\Omega} = \boldsymbol{\omega}$. In the case of general three-dimensional motion (Fig. d), the equations of angular motion are identical except that they are expressed in terms of the components of the total moment about the center of mass.

Moments and Products of Inertia

In terms of a given coordinate system xyz, the **inertia matrix** of an object is defined by [Eq. (9.13)]

$$[I] = \begin{bmatrix} I_{xx} & -I_{xy} & -I_{xz} \\ -I_{yx} & I_{yy} & -I_{yz} \\ -I_{zx} & -I_{zy} & I_{zz} \end{bmatrix} = \begin{bmatrix} \int_m (y^2 + z^2)\, dm & -\int_m xy\, dm & -\int_m xz\, dm \\ -\int_m yx\, dm & \int_m (x^2 + z^2)\, dm & -\int_m yz\, dm \\ -\int_m zx\, dm & -\int_m zy\, dm & \int_m (x^2 + y^2)\, dm \end{bmatrix},$$

where x, y, and z are the coordinates of the differential element of mass dm. The terms I_{xx}, I_{yy}, and I_{zz} are the moments of inertia about the x, y, and z axes, and I_{xy}, I_{yz}, and I_{zx} are the products of inertia.

If $x'y'z'$ is a coordinate system with its origin at the center of mass of an object and xyz is a parallel system (Fig. e), the **parallel-axis theorems** state that

$$\begin{aligned} I_{xx} &= I_{x'x'} + (d_y^2 + d_z^2)m, \\ I_{yy} &= I_{y'y'} + (d_x^2 + d_z^2)m, \\ I_{zz} &= I_{z'z'} + (d_x^2 + d_y^2)m, \\ I_{xy} &= I_{x'y'} + d_x d_y m, \\ I_{yz} &= I_{y'z'} + d_y d_z m, \\ I_{zx} &= I_{z'x'} + d_z d_x m, \end{aligned} \qquad \textbf{Eq. (9.18)}$$

where (d_x, d_y, d_z) are the coordinates of the center of mass in the xyz coordinate system.

The moment of inertia about an axis through the origin parallel to a unit vector $\mathbf{e}$ is given by

$$I_O = I_{xx}e_x^2 + I_{yy}e_y^2 + I_{zz}e_z^2 - 2I_{xy}e_xe_y - 2I_{yz}e_ye_z - 2I_{zx}e_ze_x. \qquad \textbf{Eq. (9.19)}$$

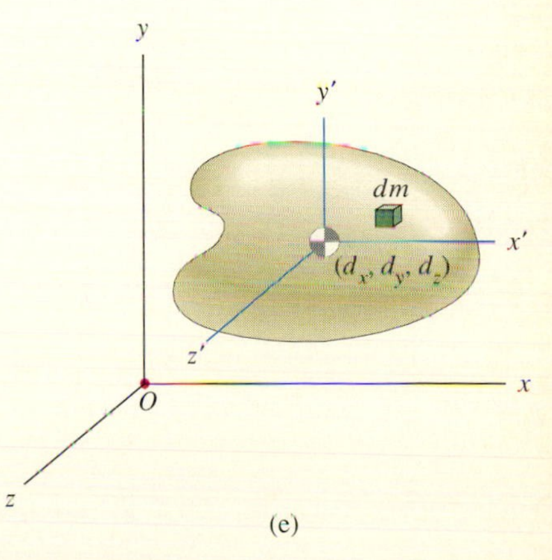

(e)

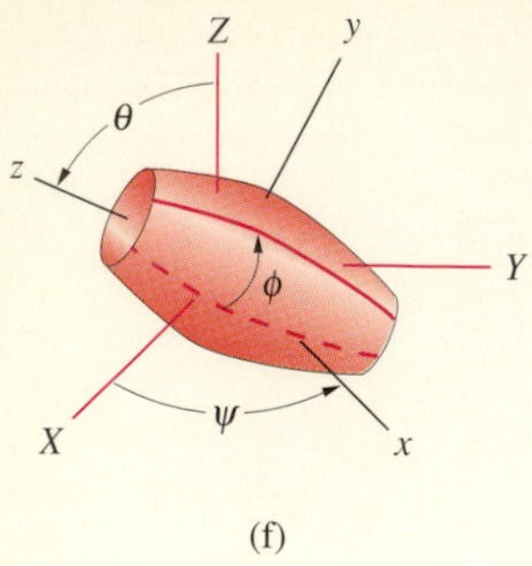

(f)

Eulerian Angles: Axisymmetric Objects

In the case of an object with an axis of rotational symmetry, the orientation of the xyz system relative to the reference XYZ system is specified by the **precession angle** ψ and the **nutation angle** θ (Fig. f). The rotation of the object relative to the xyz system is specified by the **spin angle** ϕ.

The components of the rigid body's angular velocity relative to the XYZ system are given by

$$\omega_x = \dot{\theta},$$
$$\omega_y = \dot{\psi}\sin\theta, \qquad \textbf{Eq. (9.35)}$$
$$\omega_z = \dot{\phi} + \dot{\psi}\cos\theta.$$

The equations of angular motion expressed in terms of the **Eulerian angles** are [Eqs. (9.38)–(9.40)]:

$$\Sigma M_x = I_{xx}\ddot{\theta} + (I_{zz} - I_{xx})\dot{\psi}^2 \sin\theta\cos\theta + I_{zz}\dot{\phi}\dot{\psi}\sin\theta,$$
$$\Sigma M_y = I_{xx}(\ddot{\psi}\sin\theta + 2\dot{\psi}\dot{\theta}\cos\theta) - I_{zz}(\dot{\phi}\dot{\theta} + \dot{\psi}\dot{\theta}\cos\theta),$$
$$\Sigma M_z = I_{zz}(\ddot{\phi} + \ddot{\psi}\cos\theta - \dot{\psi}\dot{\theta}\sin\theta).$$

In **steady precession** of an axisymmetric spinning object, the **spin rate** $\dot{\phi}$, the **nutation angle** θ, and the **precession rate** $\dot{\psi}$ are assumed to be constant. With these assumptions, the equations of angular motion reduce to [Eqs. (9.41)–(9.43)]

$$\Sigma M_x = (I_{zz} - I_{xx})\dot{\psi}^2 \sin\theta\cos\theta + I_{zz}\dot{\phi}\dot{\psi}\sin\theta,$$
$$\Sigma M_y = 0,$$
$$\Sigma M_z = 0.$$

Review Problems

9.104 The circular disk remains perpendicular to the horizontal shaft and rotates relative to it with angular velocity ω_d. The horizontal shaft is rigidly attached to a vertical shaft rotating with angular velocity ω_0.

(a) What is the disk's angular velocity vector $\boldsymbol{\omega}$?

(b) What is the velocity of point A of the disk?

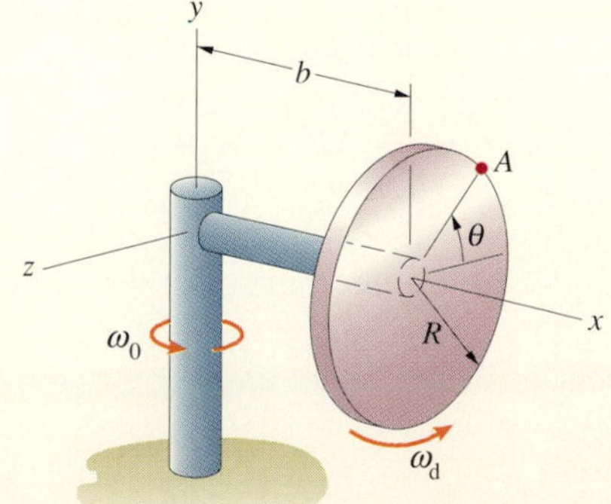

P9.104

9.105 If the angular velocities ω_d and ω_0 in Problem 9.104 are constant, what is the acceleration of point A of the disk?

9.106 The cone is connected by a ball and socket joint at its vertex to a 100-mm post. The radius of its base is 100 mm, and the base rolls on the floor. The velocity of the center of the base is $\mathbf{v}_C = 2\mathbf{k}$ (m/s).
(a) What is the cone's angular velocity vector $\boldsymbol{\omega}$?
(b) What is the velocity of point A?

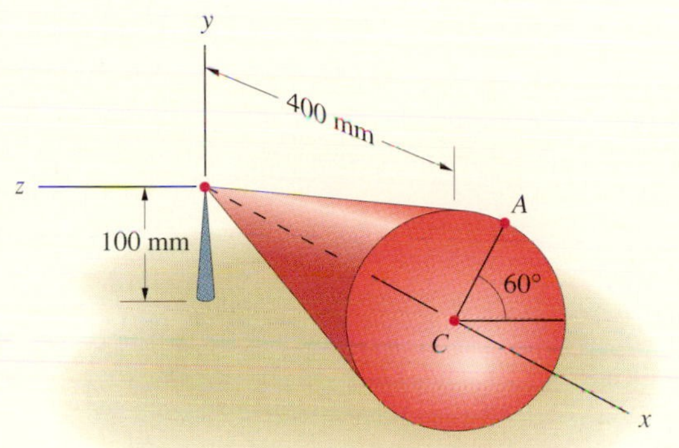

P9.106

9.107 The bar AB is connected by ball and socket joints to collars sliding on the fixed bars CD and EF. The bar EF is parallel to the y axis. At the present instant the collar at A has velocity $\mathbf{v}_A = 20\mathbf{k}$ (ft/s) and the angular velocity of bar AB about its axis is zero. Determine the velocity of the collar at B and the angular velocity of bar AB.

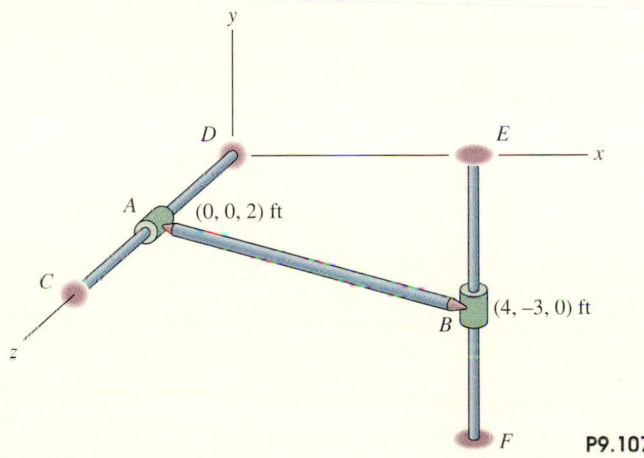

P9.107

9.108 The collar at A in Problem 9.107 has acceleration $\mathbf{a}_A = -40\mathbf{k}$ (ft/s^2) and the angular acceleration of bar AB about its axis is zero. Determine the acceleration of the collar at B and the angular acceleration of bar AB.

9.109 The mechanism shown is a type of universal joint called a yoke and spider. The axis L lies in the x-z plane. Determine the angular velocity ω_L and the angular velocity vector $\boldsymbol{\omega}_S$ of the cross-shaped "spider" in terms of the angular velocity ω_R at the instant shown.

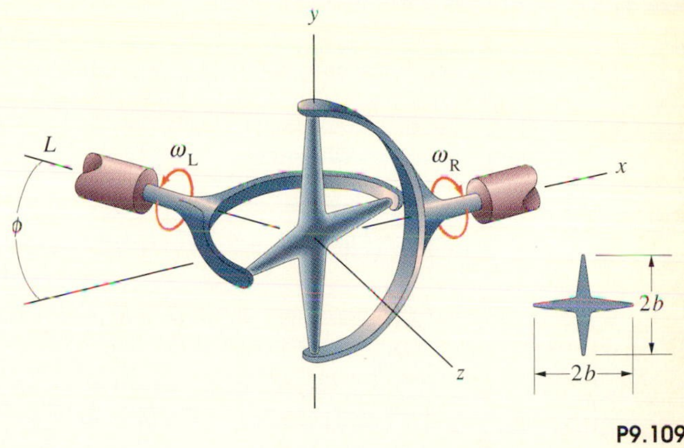

P9.109

9.110 The inertia matrix of a rigid body in terms of a body-fixed coordinate system with its origin at the center of mass is

$$[I] = \begin{bmatrix} 4 & 1 & -1 \\ 1 & 2 & 0 \\ -1 & 0 & 6 \end{bmatrix} \text{kg-m}^2.$$

If the rigid body's angular velocity is $\boldsymbol{\omega} = 10\mathbf{i} - 5\mathbf{j} + 10\mathbf{k}$ (rad/s), what is its angular momentum about its center of mass?

9.111 What is the moment of inertia of the rigid body in Problem 9.110 about the axis that passes through the origin and the point (4, −4, 7) m?

Strategy: Determine the components of a unit vector parallel to the axis and use Eq. (9.19).

9.112 Determine the inertia matrix of the 0.6-slug thin plate in terms of the coordinate system shown.

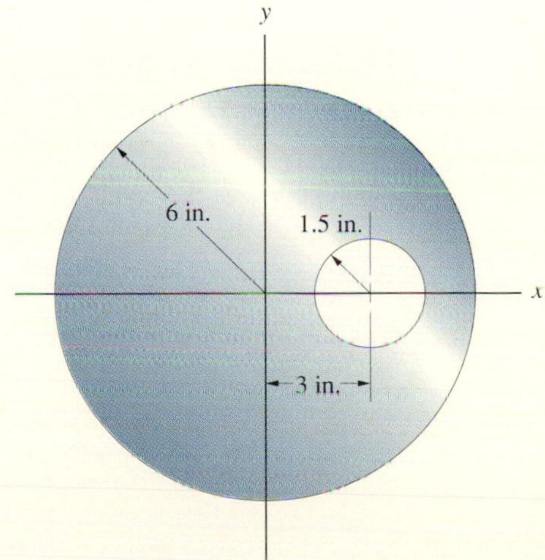

P9.112

9.113 At $t = 0$, the plate in Problem 9.112 has angular velocity $\boldsymbol{\omega} = 10\mathbf{i} + 10\mathbf{j}$ (rad/s) and is subjected to the force $\mathbf{F} = -10\mathbf{k}$ (lb) acting at the point (0, 6, 0) in. No other forces or couples act on the plate. What are the components of its angular acceleration at that instant?

9.114 The inertia matrix of a rigid body in terms of a body-fixed coordinate system with its origin at the center of mass is

$$[I] = \begin{bmatrix} 4 & 1 & -1 \\ 1 & 2 & 0 \\ -1 & 0 & 6 \end{bmatrix} \text{kg-m}^2.$$

If the rigid body's angular velocity is $\boldsymbol{\omega} = 10\mathbf{i} - 5\mathbf{j} + 10\mathbf{k}$ (rad/s) and its angular acceleration is zero, what are the components of the total moment about its center of mass?

9.115 If the total moment about the center of mass of the rigid body in Problem 9.114 is zero, what are the components of its angular acceleration?

9.116 The slender bar of length l and mass m is pinned to the L-shaped bar at O. The L-shaped bar rotates about the vertical axis with a constant angular velocity ω_0. Determine the value of ω_0 necessary for the bar to remain at a constant angle β relative to the vertical.

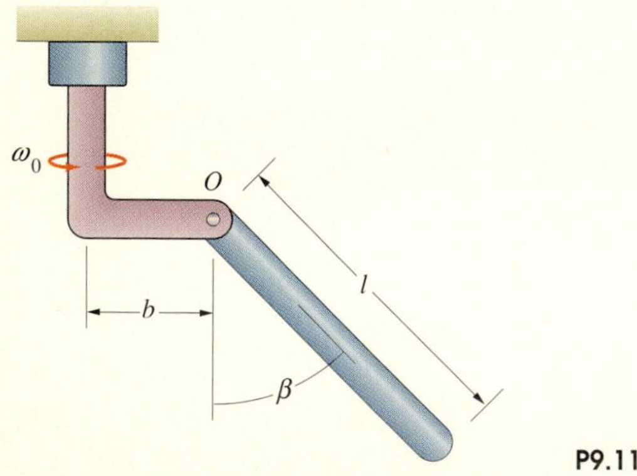

P9.116

9.117 A slender bar of length l and mass m is rigidly attached to the center of a thin circular disk of radius R and mass m. The composite object undergoes a motion in which the bar rotates in the horizontal plane with constant angular velocity ω_0 about the center of mass of the composite object and the disk rolls on the floor. Show that $\omega_0 = 2\sqrt{g/R}$.

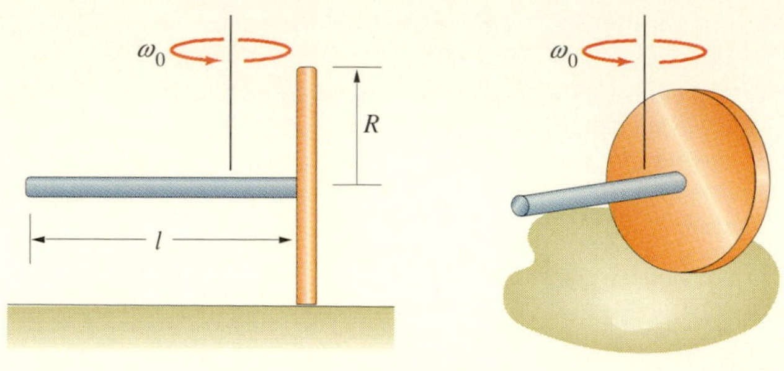

P9.117

9.118 The thin plate of mass m spins about a vertical axis with the plane of the plate perpendicular to the floor. The corner of the plate at O rests in an indentation so that it remains at the same point on the floor. The plate rotates with constant angular velocity ω_0 and the angle β is constant.

(a) Show that the angular velocity ω_0 is related to the angle β by

$$\frac{h\omega_0^2}{g} = \frac{2\cos\beta - \sin\beta}{\sin^2\beta - 2\sin\beta\cos\beta - \cos^2\beta}.$$

(b) The equation you obtained in (a) indicates that $\omega_0 = 0$ when $2\cos\beta - \sin\beta = 0$. What is the interpretation of this result?

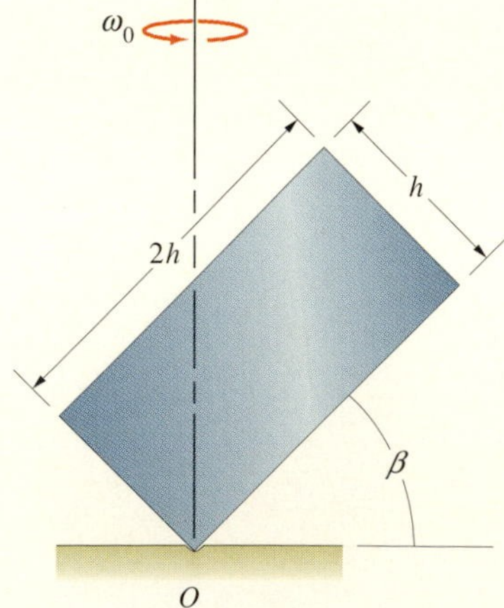

P9.118

9.119 In Problem 9.118, determine the range of values of the angle β for which the plate will remain in the steady motion described.

9.120 Arm BC has a mass of 12 kg, and its moments and products of inertia in terms of the coordinate system shown are $I_{xx} = 0.03$ kg-m^2, $I_{yy} = I_{zz} = 4$ kg-m^2, $I_{xy} = I_{yz} = I_{xz} = 0$. At the instant shown, arm AB is rotating in the horizontal plane with a constant angular velocity of 1 rad/s in the counterclockwise direction viewed from above. Relative to arm AB, arm BC is rotating about the z axis with a constant angular velocity of 2 rad/s. Determine the force and couple exerted on arm BC at B.

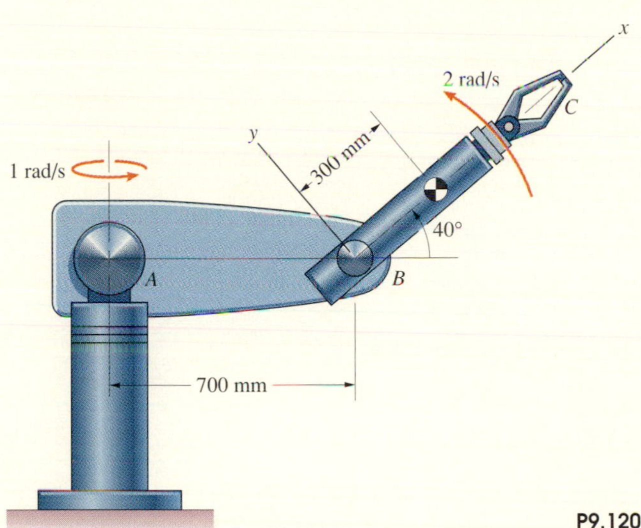

P9.120

9.121 Suppose that you throw a football in a wobbly spiral with a nutation angle of 25°. The football's moments of inertia are $I_{xx} = I_{yy} = 0.003$ slug-ft^2 and $I_{zz} = 0.001$ slug-ft^2. If the spin rate is $\dot{\phi} =$ 4 revolutions per second, what is the magnitude of the precession rate (the rate at which it wobbles)?

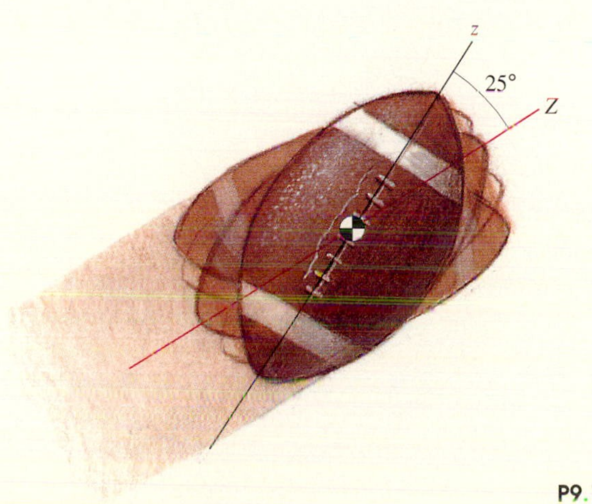

P9.121

9.122 Sketch the body and space cones for the motion of the football in Problem 9.121.

9.123 The mass of the homogeneous thin plate is 1 kg. For a coordinate system with its origin at O, determine the principal moments of inertia and the directions of unit vectors parallel to the corresponding principal axes.

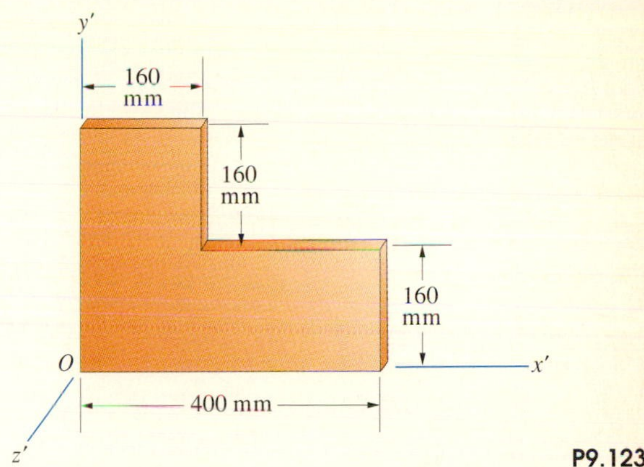

P9.123

9.124 The airplane's principal moments of inertia in slug-ft^2 are $I_{xx} = 8000$, $I_{yy} = 48{,}000$, and $I_{zz} = 50{,}000$.
(a) The airplane begins in the reference position shown and maneuvers into the orientation $\psi = \theta = \phi = 45°$. Draw a sketch showing its orientation relative to the XYZ system.
(b) If the airplane is in the orientation described in (a), the rates of change of the Eulerian angles are $\dot{\psi} = 0$, $\dot{\theta} = 0.2$ rad/s, and $\dot{\phi} = 0.2$ rad/s, and their second time derivatives are zero, what are the components of the total moment about the airplane's center of mass?

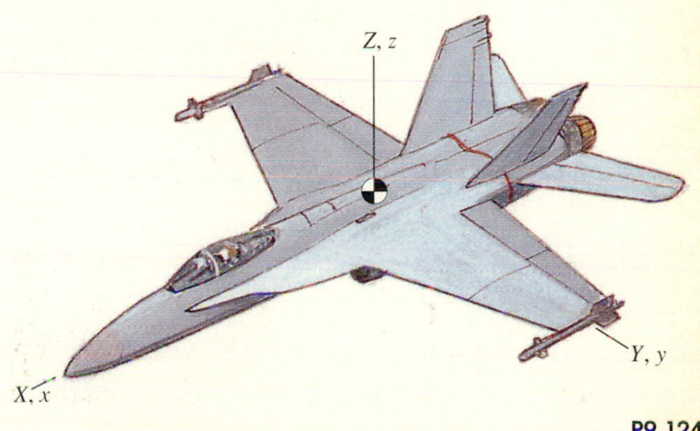

P9.124

9.125 What are the x, y, and z components of the angular acceleration of the airplane described in Problem 9.124?

9.126 If the orientation of the airplane in Problem 9.124 is $\psi = 45°$, $\theta = 60°$, $\phi = 45°$, the rates of change of the Eulerian angles are $\dot{\psi} = 0$, $\dot{\theta} = 0.2$ rad/s, and $\dot{\phi} = 0.1$ rad/s, and the components of the total moment about the center of mass are $\Sigma M_x = 400$ ft-lb, $\Sigma M_y = 1200$ ft-lb, and $\Sigma M_z = 0$, what are the x, y, and z components of the airplane's angular acceleration?

Engineers use "shake tables" to simulate the vibrations of buildings and other structures during earthquakes and investigate methods for minimizing structural damage. The tables can be programmed to simulate the magnitudes and time histories of the ground vibrations measured during actual earthquakes. In this chapter we analyze the vibrations of simple mechanical systems.

Chapter 10

Vibrations

VIBRATIONS have been of concern in engineering at least since the beginning of the industrial revolution. The oscillatory motions of rotating and reciprocating engines subject their parts to large loads that must be considered in their design. Operators and passengers of vehicles powered by these engines must be isolated from their vibrations. Beginning with the development of electromechanical devices capable of creating and measuring mechanical vibrations, engineering applications of vibrations have included the various areas of acoustics, from architectural acoustics to earthquake detection and analysis.

In this chapter we consider vibrating systems with one degree of freedom; that is, the position, or configuration, of each system can be specified by a single variable. Many actual vibrating systems either have only one degree of freedom or their motions can be modeled by a one-degree-of-freedom system in particular circumstances. We introduce fundamental concepts, including amplitude, frequency, period, damping, and resonance, that are also used in the analysis of systems with multiple degrees of freedom.

10.1 Conservative Systems

We begin by presenting different examples of one-degree-of-freedom systems subjected to conservative forces, demonstrating that their motions are described by the same differential equation. We then examine solutions of this equation and use them to describe the vibrations of one-degree-of-freedom conservative systems.

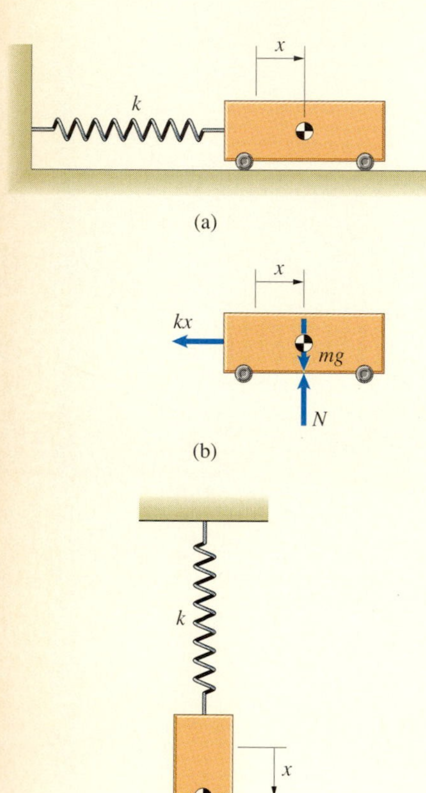

Fig. 10.1

(a) The spring-mass oscillator has one degree of freedom.
(b) Free-body diagram of the mass.
(c) Suspending the mass.

Examples

The spring-mass oscillator (Fig. 10.1a) is the simplest example of a one-degree-of-freedom vibrating system. A single coordinate x measuring the displacement of the mass relative to a reference point is sufficient to specify the position of the system. We draw the free-body diagram of the mass in Fig. 10.1(b), neglecting friction and assuming that the spring is unstretched when $x = 0$. Applying Newton's second law, we can write the equation describing the horizontal motion of the mass as

$$\frac{d^2x}{dt^2} + \frac{k}{m}x = 0. \tag{10.1}$$

We can also obtain this equation by using a different method that you will find very useful. The only force that does work on the mass, the force exerted by the spring, is conservative, which means that the sum of the kinetic and potential energies is constant:

$$\frac{1}{2}m\left(\frac{dx}{dt}\right)^2 + \frac{1}{2}kx^2 = \text{constant}.$$

Taking the time derivative of this equation, we can write the resulting equation as

$$\left(\frac{dx}{dt}\right)\left(\frac{d^2x}{dt^2} + \frac{k}{m}x\right) = 0,$$

again obtaining Eq. (10.1).

Suppose that the mass is suspended from the spring, as shown in Fig. 10.1(c), and it undergoes vertical motion. If the spring is unstretched when $x = 0$, you can easily confirm that the equation of motion is

$$\frac{d^2x}{dt^2} + \frac{k}{m}x = g.$$

If the suspended mass is stationary, the magnitude of the force exerted by the spring must equal the weight, $kx = mg$, so the equilibrium position is $x = mg/k$. (Notice that we can also determine the equilibrium position by setting the acceleration equal to zero in the equation of motion.) Let us introduce a new variable $\tilde{x}$ that measures the position of the mass relative to its equilibrium position: $\tilde{x} = x - mg/k$. Writing the equation of motion in terms of this variable, we obtain

$$\frac{d^2\tilde{x}}{dt^2} + \frac{k}{m}\tilde{x} = 0,$$

which is identical to Eq. (10.1). The vertical motion of the mass in Fig. 10.1(c) *relative to its equilibrium position* is described by the same equation that describes the horizontal motion of the mass in Fig. 10.1(a) relative to its equilibrium position.

Let's consider a different one-degree-of-freedom system. If we rotate the slender bar in Fig. 10.2(a) through some angle and release it, it will oscillate back and forth. (An object swinging from a fixed point is called a **pendulum**.) There is only one degree of freedom, since θ specifies the bar's position.

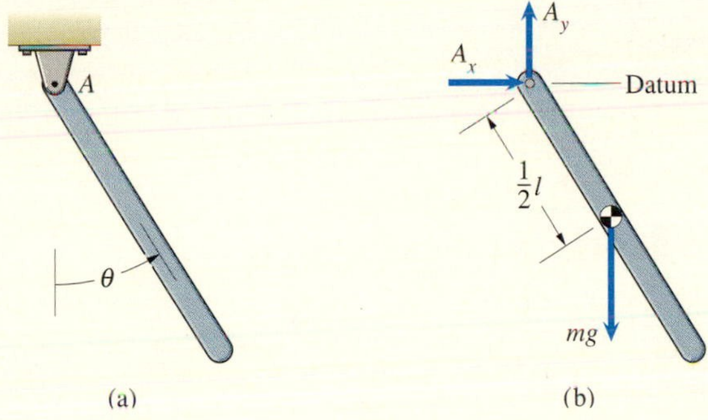

Fig. 10.2
(a) A pendulum consisting of a slender bar.
(b) Free-body diagram of the bar.

Drawing the free-body diagram of the bar (Fig. 10.2b) and writing the equation of angular motion about A, we obtain

$$\frac{d^2\theta}{dt^2} + \frac{3g}{2l}\sin\theta = 0. \tag{10.2}$$

We can also obtain this equation by using conservation of energy. The bar's kinetic energy is $T = \frac{1}{2}I_A(d\theta/dt)^2$. If we place the datum at the level of point A (Fig. 10.2b), the potential energy associated with the bar's weight is $V = -mg(\frac{1}{2}l\cos\theta)$, so

$$T + V = \frac{1}{2}\left(\frac{1}{3}ml^2\right)\left(\frac{d\theta}{dt}\right)^2 - \frac{1}{2}mgl\cos\theta = \text{constant}.$$

Taking the time derivative of this equation and writing the result in the form

$$\left(\frac{d\theta}{dt}\right)\left(\frac{d^2\theta}{dt^2} + \frac{3g}{2l}\sin\theta\right) = 0,$$

we obtain Eq. (10.2).

Equation (10.2) does not have the same form as Eq. (10.1). However, if we express $\sin\theta$ in terms of its Taylor series,

$$\sin\theta = \theta - \frac{1}{6}\theta^3 + \frac{1}{120}\theta^5 + \cdots,$$

and assume that θ remains small enough to approximate $\sin\theta$ by θ, then Eq. (10.2) becomes identical in form to Eq. (10.1):

$$\frac{d^2\theta}{dt^2} + \frac{3g}{2l}\theta = 0. \tag{10.3}$$

Our analyses of the spring-mass oscillator and pendulum resulted in equations of motion that are identical in form. To accomplish this in the case of the suspended spring-mass oscillator, we had to express the equation of motion in terms of displacement relative to the equilibrium position. In the case of the pendulum, we needed to assume that the motions are small. But within those restrictions, you will see that the equation we obtained describes the motions of many one-degree-of-freedom conservative systems.

Solutions

Let us consider the differential equation

$$\boxed{\frac{d^2x}{dt^2} + \omega^2 x = 0,} \tag{10.4}$$

where ω is a constant. We have seen that with $\omega^2 = k/m$, this equation describes the motion of a spring-mass oscillator, and with $\omega^2 = 3g/2l$, it describes small motions of a suspended slender bar. Equation (10.4) is an **ordinary differential equation**, because it is expressed in terms of ordinary (not partial) derivatives of the dependent variable x with respect to the independent variable t. It is **linear**, meaning there are no nonlinear terms in x or its derivatives, and it is **homogeneous**, meaning that each term contains x or one of its derivatives. The general solution of this differential equation is

$$\boxed{x = A\sin\omega t + B\cos\omega t,} \tag{10.5}$$

where A and B are arbitrary constants.

Although in practical problems you will usually find Eq. (10.5) to be the most convenient form of the solution of Eq. (10.4), we can describe the properties of the solution more easily by expressing it in the alternative form

$$x = E\sin(\omega t - \phi), \tag{10.6}$$

where E and ϕ are constants. To show that these two solutions are equivalent, we can use the identity

$$\begin{aligned} E\sin(\omega t - \phi) &= E(\sin\omega t\cos\phi - \cos\omega t\sin\phi) \\ &= (E\cos\phi)\sin\omega t + (-E\sin\phi)\cos\omega t. \end{aligned}$$

This expression is identical to Eq. (10.5) if the constants A and B are related to E and ϕ by

$$A = E\cos\phi, \qquad B = -E\sin\phi. \tag{10.7}$$

Equation (10.6) clearly demonstrates the oscillatory nature of the solution of Eq. (10.4). Called **simple harmonic motion**, it describes a sinusoidal function of ωt. The constant ϕ determines the horizontal placement of the sinusoidal function relative to the origin $\omega t = 0$, and is called the **phase**. We define ϕ to be the distance to the right of $\omega t = 0$ at which the solution first crosses the horizontal axis with positive slope (Fig. 10.3). The positive constant E is called the **amplitude** of the vibration. By squaring Eqs. (10.7) and adding them, we obtain a relation between the amplitude and the constants A and B:

$$E = \sqrt{A^2 + B^2}. \tag{10.8}$$

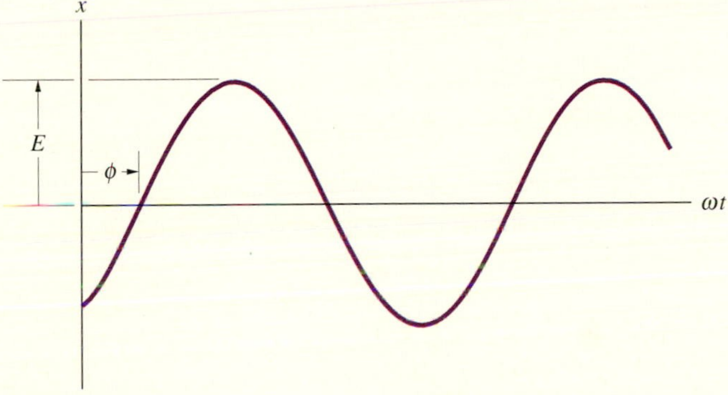

Fig. 10.3
Graph of x as a function of ωt.

We can interpret Eq. (10.6) in terms of the uniform motion of a point along a circular path. We draw a circle whose radius equals the amplitude (Fig. 10.4) and assume that the line from O to P rotates in the counterclockwise direction with constant angular velocity ω. If we choose the position of P at $t = 0$ as shown, the projection of the line OP onto the vertical axis is $E\sin(\omega t - \phi)$.

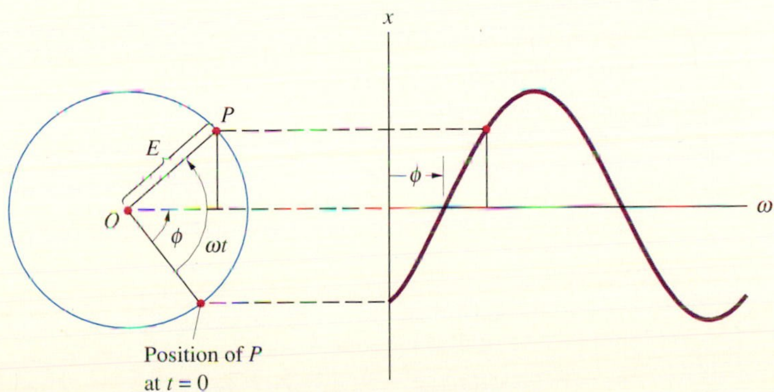

Fig. 10.4
Correspondence of simple harmonic motion with circular motion of a point.

Thus there is a one-to-one correspondence between the circular motion of P and Eq. (10.6). Point P makes one complete revolution, or **cycle**, during the time required for the angle ωt to increase by 2π radians. The time $\tau = 2\pi/\omega$ required for one cycle is called the **period** of the vibration. Since τ is the time required for one cycle, its inverse $f = 1/\tau$ is the number of cycles per unit time, or **natural frequency** of the vibration. The frequency is usually expressed in cycles per second, or Hertz (Hz). We illustrate the effect of changing the period and frequency in Fig. 10.5.

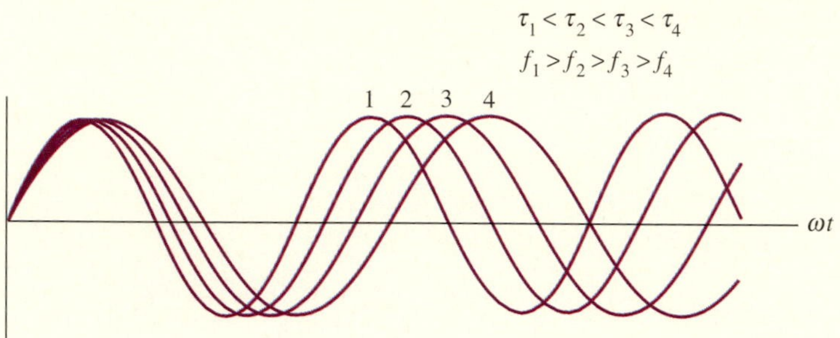

Fig. 10.5

Effect of increasing the period (decreasing the frequency) of simple harmonic motion.

In summary, the period and natural frequency are

$$\tau = \frac{2\pi}{\omega}, \tag{10.9}$$

$$f = \frac{\omega}{2\pi}. \tag{10.10}$$

A system's period and natural frequency are determined by its physical properties and do not depend on the functional form in which its motion is expressed.

The natural frequency f is the number of revolutions the point P moves around the circular path in Fig. 10.4 per unit time, so $\omega = 2\pi f$ is the number of radians per unit time. Therefore ω is also a measure of the frequency and is expressed in rad/s. To distinguish it from f, ω is called the **circular natural frequency**. The term ωt, and the variable ϕ that specifies the phase, can be specified in either radians or degrees.

Suppose that Eq. (10.6) describes the displacement of the spring-mass oscillator in Fig. 10.1(a), so that $\omega^2 = k/m$. The kinetic energy of the mass is

$$T = \frac{1}{2}m\left(\frac{dx}{dt}\right)^2 = \frac{1}{2}mE^2\omega^2 \cos^2(\omega t - \phi),$$

and the potential energy of the spring is

$$V = \frac{1}{2}kx^2 = \frac{1}{2}mE^2\omega^2 \sin^2(\omega t - \phi).$$

The sum of the kinetic and potential energies, $T + V = \frac{1}{2}mE^2\omega^2$, is constant (Fig. 10.6). As the system vibrates, its total energy oscillates between kinetic and potential energy. Notice that the total energy is proportional to the square of the amplitude and the square of the natural frequency.

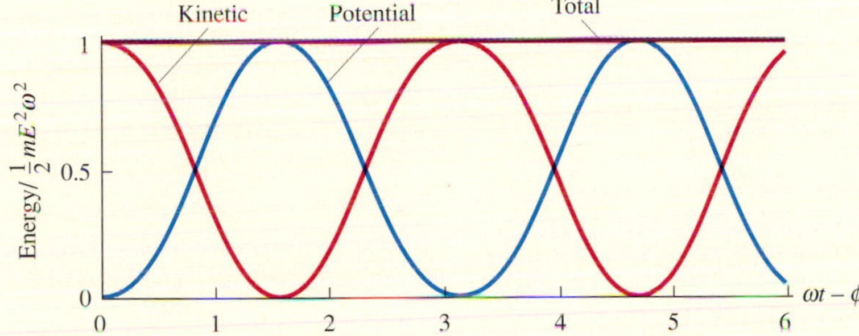

Fig. 10.6
Kinetic, potential, and total energies of a spring-mass oscillator.

In the following examples, we analyze one-degree-of-freedom vibrating systems. Your first objective will usually be to determine the equation of motion of the system and express it in the form of Eq. (10.4):

$$\frac{d^2x}{dt^2} + \omega^2 x = 0.$$

To do so, you must write the equation of motion in terms of the displacement of the system relative to its equilibrium position. You may also need to linearize the equation by assuming that the displacement is small, as we did in obtaining Eq. (10.3) from Eq. (10.2). Once you have the equation of motion in this form, you know the value of ω for the system and can use it to obtain the period and natural frequency from Eqs. (10.9) and (10.10). You can also determine the motion of the system from Eq. (10.5) or Eq. (10.6) if you are given sufficient information to evaluate the arbitrary constants.

Example 10.1

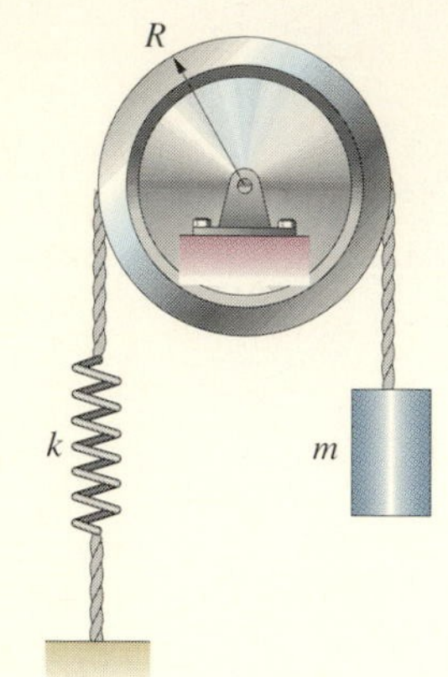

Fig. 10.7

The pulley in Fig. 10.7 has radius R and moment of inertia I, and the cable does not slip relative to the pulley. The mass m is displaced downward a distance h from its equilibrium position and released from rest at $t = 0$.

(a) What is the natural frequency of the resulting vibrations?

(b) Determine the position of the mass relative to its equilibrium position as a function of time.

STRATEGY

A single coordinate specifying the vertical displacement of the mass specifies the positions of both the mass and pulley, so there is one degree of freedom. We can obtain the equation of motion of the system either by writing the individual equations of motion of the mass and pulley or by using conservation of energy.

SOLUTION

Let x be the downward displacement of the mass relative to its position when the spring is unstretched. We draw the free-body diagrams of the pulley and mass in Fig. (a), where T_C is the tension in the cable and α is the angular acceleration of the pulley. Applying Newton's second law to the mass, we obtain

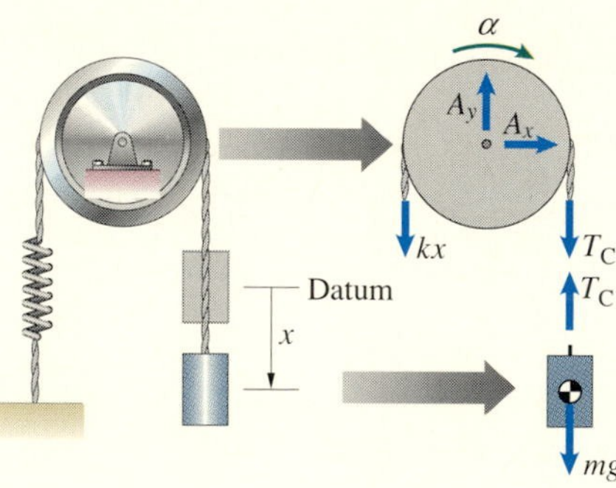

(a) Free-body diagrams of the pulley and mass.

$$mg - T_C = m\frac{d^2x}{dt^2}. \tag{10.11}$$

The equation of angular motion for the pulley is

$$T_C R - (kx)R = I\alpha. \tag{10.12}$$

The angular acceleration of the pulley and the acceleration of the mass are related by $\alpha = (d^2x/dt^2)/R$, so we can write Eq. (10.12) as

$$T_C - kx = \left(\frac{I}{R^2}\right)\frac{d^2x}{dt^2}.$$

Summing this equation and Eq. (10.11), we obtain the equation of motion

$$\left(m + \frac{I}{R^2}\right)\frac{d^2x}{dt^2} + kx = mg. \tag{10.13}$$

Alternative Solution In terms of the velocity of the mass, the angular velocity of the pulley is $(dx/dt)/R$. Therefore we can write the total kinetic energy of the mass and pulley as

$$T = \frac{1}{2}m\left(\frac{dx}{dt}\right)^2 + \frac{1}{2}I\left[\frac{1}{R}\left(\frac{dx}{dt}\right)\right]^2.$$

Placing the datum for the potential energy associated with the weight of the mass at $x = 0$, the total potential energy is

$$V = -mgx + \frac{1}{2}kx^2.$$

The sum of the kinetic and potential energies is constant:

$$T + V = \frac{1}{2}\left(m + \frac{I}{R^2}\right)\left(\frac{dx}{dt}\right)^2 - mgx + \frac{1}{2}kx^2 = \text{constant}.$$

Taking the time derivative of this equation, we again obtain Eq. (10.13).

By setting $d^2x/dt^2 = 0$ in Eq. (10.13), we see that the equilibrium position is $x = mg/k$. By expressing Eq. (10.13) in terms of a new variable $\tilde{x} = x - mg/k$ that measures the position of the mass relative to its equilibrium position, we obtain the equation

$$\frac{d^2\tilde{x}}{dt^2} + \omega^2\tilde{x} = 0,$$

where

$$\omega^2 = \frac{k}{m + I/R^2}.$$

(a) The natural frequency of vibration of the system is

$$f = \frac{\omega}{2\pi} = \frac{1}{2\pi}\sqrt{\frac{k}{m + I/R^2}}.$$

(b) From Eq. (10.5), we can write the general solution for $\tilde{x}$ in the form

$$\tilde{x} = A \sin \omega t + B \cos \omega t.$$

When $t = 0$, $\tilde{x} = h$ and $d\tilde{x}/dt = 0$. The derivative of the general solution is

$$\frac{d\tilde{x}}{dt} = A\omega \cos \omega t - B\omega \sin \omega t.$$

The initial conditions yield the equations

$$h = B, \qquad 0 = A\omega,$$

so the position of the mass relative to its equilibrium position is

$$\tilde{x} = h \cos \omega t.$$

Example 10.2

Fig. 10.8

The spring attached to the slender bar of mass m in Fig. 10.8 is unstretched when $\theta = 0$. Neglecting friction, determine the natural frequency of small vibrations of the bar relative to its equilibrium position.

STRATEGY

The angle θ specifies the bar's position, so there is one degree of freedom. We can express the kinetic and potential energies in terms of θ and its time derivative and take the time derivative of the total energy to obtain the equation of motion.

SOLUTION

The kinetic energy of the bar is

$$T = \frac{1}{2}mv^2 + \frac{1}{2}I\left(\frac{d\theta}{dt}\right)^2,$$

where v is the velocity of the center of mass and $I = \frac{1}{12}ml^2$. The distance from the bar's instantaneous center to its center of mass is $\frac{1}{2}l$ (Fig. a), so $v = (\frac{1}{2}l)(d\theta/dt)$, and the kinetic energy is

$$T = \frac{1}{2}m\left[\frac{1}{2}l\left(\frac{d\theta}{dt}\right)\right]^2 + \frac{1}{2}\left(\frac{1}{12}ml^2\right)\left(\frac{d\theta}{dt}\right)^2 = \frac{1}{6}ml^2\left(\frac{d\theta}{dt}\right)^2.$$

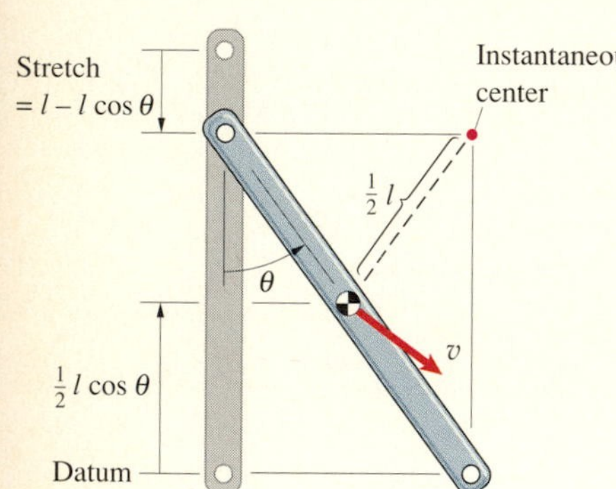

(a) Determining the velocity of the center of mass, the stretch of the spring, and the height of the center of mass above the datum.

In terms of θ, the stretch of the spring is $l - l\cos\theta$. We place the datum for the potential energy associated with the weight at the bottom of the bar (Fig. a), so the total potential energy is

$$V = mg\left(\frac{1}{2}l\cos\theta\right) + \frac{1}{2}k(l - l\cos\theta)^2.$$

The sum of the kinetic and potential energies is constant:

$$T + V = \frac{1}{6}ml^2\left(\frac{d\theta}{dt}\right)^2 + \frac{1}{2}mgl\cos\theta + \frac{1}{2}kl^2(1 - \cos\theta)^2 = \text{constant}.$$

Taking the time derivative of this equation, we obtain the equation of motion:

$$\frac{1}{3}ml^2\frac{d^2\theta}{dt^2} - \frac{1}{2}mgl\sin\theta + kl^2(1 - \cos\theta)\sin\theta = 0. \qquad (10.14)$$

To express this equation in the form of Eq. (10.4), we need to write it in terms of small vibrations relative to the equilibrium position. Let θ_e be the value of θ when the bar is in equilibrium. By setting $d^2\theta/dt^2 = 0$ in Eq. (10.14), we find that θ_e must satisfy the relation

$$\cos\theta_e = 1 - \frac{mg}{2kl}. \qquad (10.15)$$

We define $\tilde{\theta} = \theta - \theta_e$, and expand $\sin\theta$ and $\cos\theta$ in Taylor series in terms of $\tilde{\theta}$:

$$\sin\theta = \sin(\theta_e + \tilde{\theta}) = \sin\theta_e + \cos\theta_e\tilde{\theta} + \cdots,$$

$$\cos\theta = \cos(\theta_e + \tilde{\theta}) = \cos\theta_e - \sin\theta_e\tilde{\theta} + \cdots.$$

Substituting these expressions into Eq. (10.14), neglecting terms in $\tilde{\theta}$ of second and higher orders, and using Eq. (10.15), we obtain

$$\frac{d^2\tilde{\theta}}{dt^2} + \omega^2\tilde{\theta} = 0,$$

where

$$\omega^2 = \frac{3g}{l}\left(1 - \frac{mg}{4kl}\right).$$

From Eq. (10.10), the natural frequency of small vibrations of the bar is

$$f = \frac{\omega}{2\pi} = \frac{1}{2\pi}\sqrt{\frac{3g}{l}\left(1 - \frac{mg}{4kl}\right)}.$$

Problems

10.1 Confirm that $x = A\sin\omega t + B\cos\omega t$, where A and B are arbitrary constants, satisfies Eq. (10.4).

10.2 Confirm that $x = E\sin(\omega t - \phi)$, where E and ϕ are arbitrary constants, satisfies Eq. (10.4).

10.3 (a) Show that $x = G\cos(\omega t - \psi)$, where G and ψ are arbitrary constants, satisfies Eq. (10.4). (b) Determine the constants A and B in the form of the solution given by Eq. (10.5) in terms of the constants G and ψ.

10.4 The position of a vibrating system is

$$x = (1/\sqrt{2})\sin\omega t - (1/\sqrt{2})\cos\omega t \text{ ft}.$$

(a) Determine the amplitude of the vibration.
(b) Draw a sketch of x for values of ωt from zero to 4π radians.

10.5 The position of a vibrating system is

$$x = -\sqrt{2}\sin\omega t + \sqrt{2}\cos\omega t \text{ m}.$$

(a) Determine the amplitude of the vibration.
(b) Draw a sketch of x for values of ωt from zero to 4π radians.

10.6 The mass $m = 10$ kg and $k = 90$ N/m. The coordinate x measures the displacement of the mass relative to its equilibrium position. At $t = 0$, the mass is released from rest in the position $x = 0.1$ m.
(a) Determine the period and natural frequency of the resulting vibrations.
(b) Determine x as a function of time.

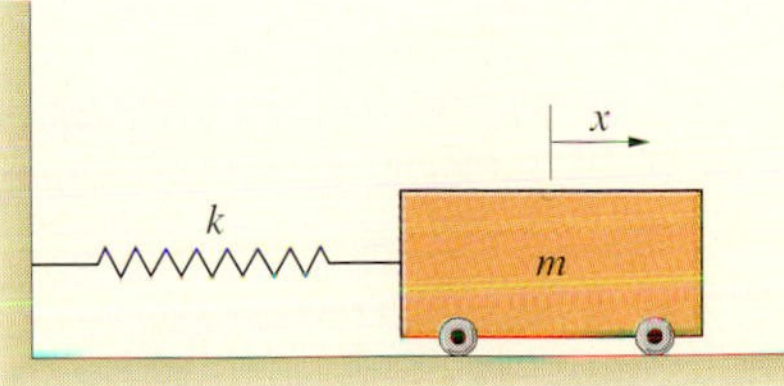

P10.6

10.7 The suspended object weighs 30 lb and $k = 20$ lb/ft. At $t = 0$, the displacement of the object *relative to its equilibrium position* is $\tilde{x} = 0.25$ ft and it is moving downward at 1.5 ft/s.
(a) Determine the period and natural frequency of the resulting vibrations.

(b) Determine $\tilde{x}$ as a function of time.
(c) Draw a graph of $\tilde{x}$ as a function of time from $t = 0$ to $t = 3$ s.

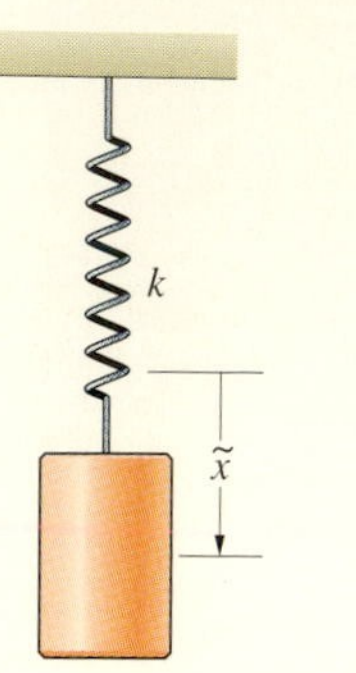

P10.7

10.8 Determine the natural frequency of vibration of the mass relative to its equilibrium position.

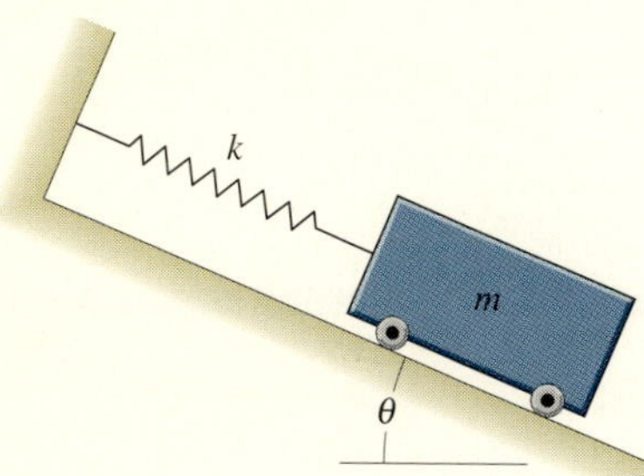

P10.8

10.9 The thin rectangular plate is attached to the rectangular frame by pins. The frame rotates with constant angular velocity ω_0. The angle β between the z axis of the body-fixed coordinate system and the vertical is governed by the equation

$$\frac{d^2\beta}{dt^2} = -\omega_0^2 \sin\beta \cos\beta.$$

Determine the natural frequency of small vibrations of the plate relative to its horizontal position.

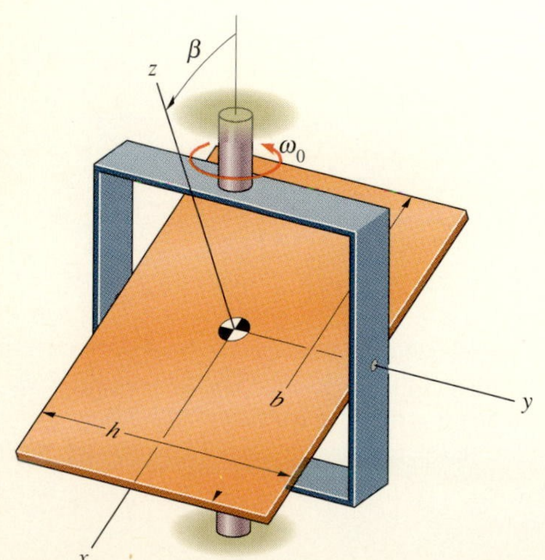

P10.9

10.10 The rectangular frame in Problem 10.9 rotates with a constant angular velocity $\omega_0 = 6$ rad/s. At $t = 0$, the angle $\beta = 0.01$ rad and $d\beta/dt = 0$. Determine β as a function of time.

10.11 A 200-lb "bungee jumper" jumps from a bridge above a river. The bungee cord has an unstretched length of 60 ft, and it stretches an additional 40 ft before he rebounds. If you model the cord as a linear spring, what is the period of his vertical oscillations?

P10.11

10.12 The total mass of the piston and the load it supports is 90 kg. In terms of the distance s in meters, the net upward force exerted on the piston by the gas in the cylinder and atmospheric pressure is $(780/s) - 3000$ N. Determine the frequency of small vibrations of the piston and load relative to their equilibrium position.

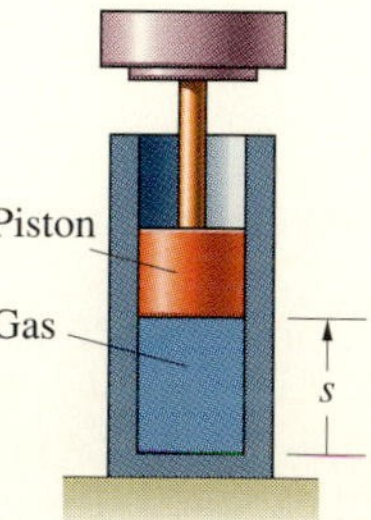

P10.12

10.13 The piston and load in Problem 10.12 are displaced downward 2 mm from their equilibrium position and released from rest. Determine their position relative to the equilibrium position as a function of time.

10.14 The pendulum consists of a homogeneous 1-kg disk attached to a 0.2-kg slender bar. What is the natural frequency of small vibrations of the pendulum?

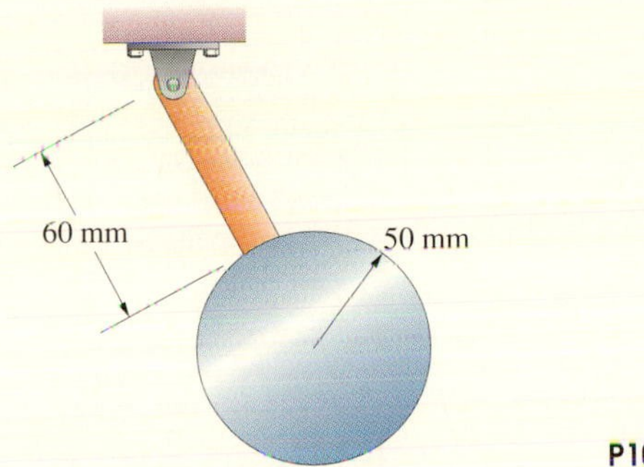

P10.14

10.15 The homogeneous disk weighs 100 lb and its radius is $R = 1$ ft. It rolls on the plane surface. The spring constant is $k = 100$ lb/ft.

(a) Determine the natural frequency of vibrations of the disk relative to its equilibrium position.

(b) At $t = 0$, the spring is unstretched and the disk has a clockwise angular velocity of 2 rad/s. What is the amplitude of the resulting vibrations of the disk and what is the angular velocity of the disk when $t = 3$ s?

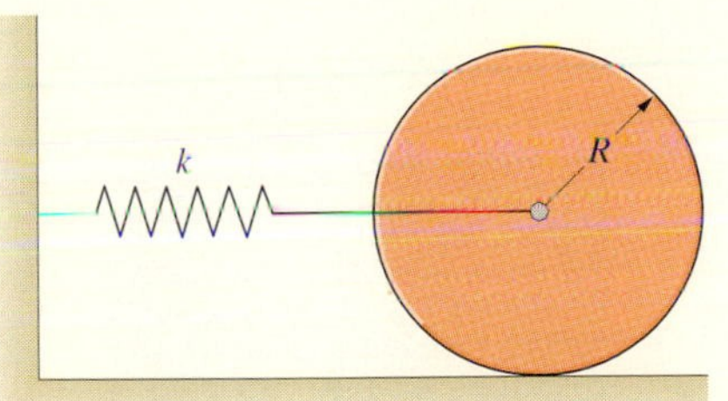

P10.15

10.16 The radius of the disk is $R = 100$ mm, and its moment of inertia is $I = 0.1$ kg-m^2. The mass $m = 5$ kg, and the spring constant is $k = 135$ N/m. The cable does not slip relative to the disk. The coordinate x measures the displacement of the mass relative to the position in which the spring is unstretched.

(a) What are the period and natural frequency of vertical vibrations of the mass relative to its equilibrium position?

(b) Determine x as a function of time if the system is released from rest with $x = 0$.

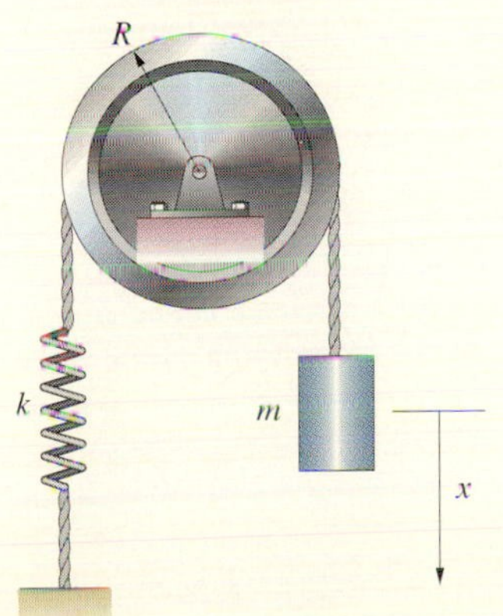

P10.16

10.17 If the spring constant is $k = 30$ lb/ft and the moment of inertia of the pulley is negligible, what is the frequency of vertical vibrations of the weights relative to their equilibrium positions?

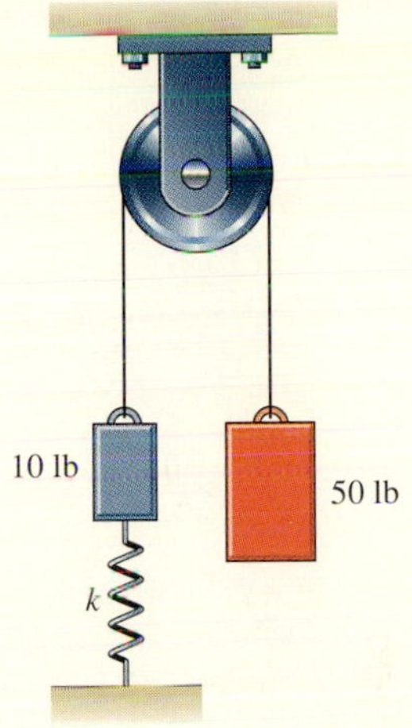

P10.17

10.18 In Problem 10.17, the spring constant is $k = 30$ lb/ft, the radius of the pulley is 0.5 ft, and the moment of inertia of the pulley is 0.25 slug-ft^2. At $t = 0$, the weights are released from rest with the spring unstretched. Determine the position of the 10-lb weight relative to its position at $t = 0$ as a function of time.

10.19 A homogeneous disk of mass m and radius r rolls on a curved surface of radius R. Show that the natural frequency of small vibrations of the disk relative to its equilibrium position is

$$f = \frac{1}{\pi}\sqrt{\frac{g}{6(R - r)}}.$$

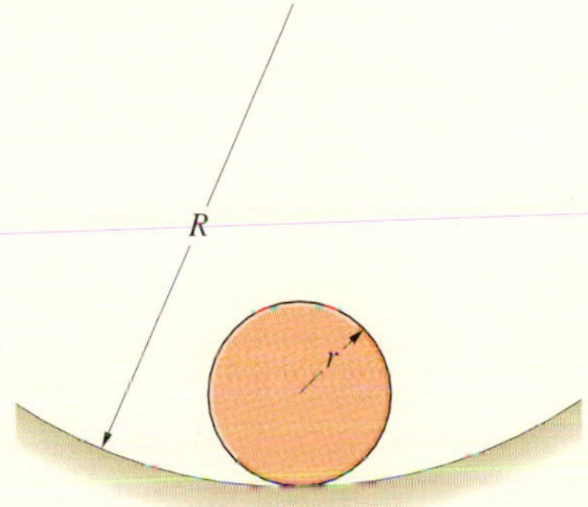

P10.19

10.20 The slender bar has roller supports at its ends and is at rest in a circular depression with an 8-ft radius. What is the frequency of small vibrations of the bar relative to its equilibrium position?

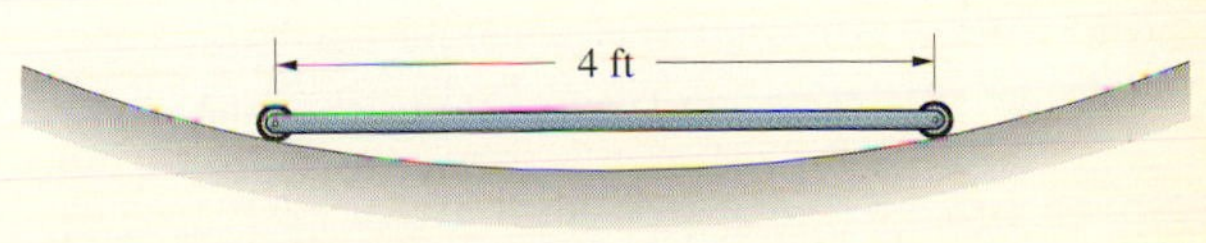

P10.20

10.21 A slender bar of mass m and length l is pinned to a fixed support as shown. A torsional spring of constant k attached to the bar at the support is unstretched when the bar is vertical. Show that the equation governing small vibrations of the bar from its vertical equilibrium position is

$$\frac{d^2\theta}{dt^2} + \omega^2\theta = 0, \qquad \text{where } \omega^2 = \frac{(k - \frac{1}{2}mgl)}{\frac{1}{3}ml^2}.$$

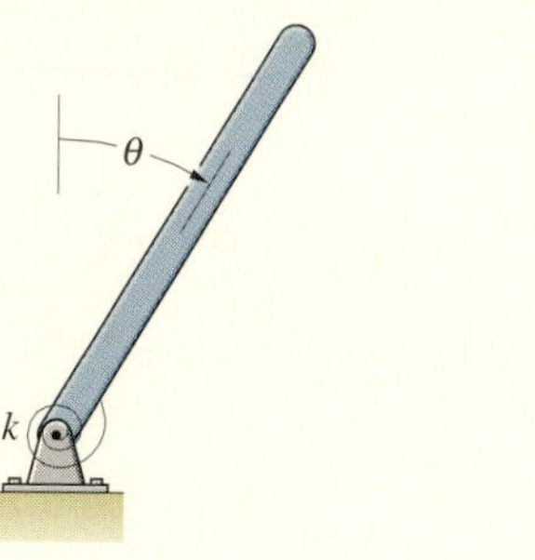

P10.21

10.22 The initial conditions of the slender bar in Problem 10.21 are

$$t = 0 \begin{cases} \theta = 0 \\ \dfrac{d\theta}{dt} = \dot{\theta}_0. \end{cases}$$

(a) If $k > \frac{1}{2}mgl$, show that θ is given as a function of time by

$$\theta = \frac{\dot{\theta}_0}{\omega}\sin\omega t, \qquad \text{where } \omega^2 = \frac{(k - \frac{1}{2}mgl)}{\frac{1}{3}ml^2}.$$

(b) If $k < \frac{1}{2}mgl$, show that θ is given as a function of time by

$$\theta = \frac{\dot{\theta}_0}{2h}(e^{ht} - e^{-ht}), \qquad \text{where } h^2 = \frac{(\frac{1}{2}mgl - k)}{\frac{1}{3}ml^2}.$$

Strategy: To do (b), seek a solution of the equation of motion of the form $x = Ce^{\lambda t}$, where C and λ are constants.

10.23 Engineers use the device shown to measure an astronaut's moment of inertia. The horizontal board is pinned at O and supported by the linear spring with constant $k = 12$ kN/m. When the astronaut is not present, the frequency of small vibrations of the board about O is measured and determined to be 6.0 Hz. When the astronaut is lying on the board as shown, the frequency of small vibrations of the board about O is 2.8 Hz. What is the astronaut's moment of inertia about the z axis?

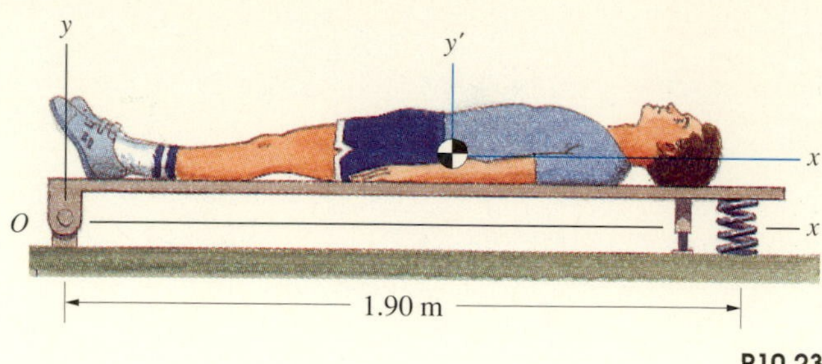

P10.23

10.24 In Problem 10.23, the astronaut's center of mass is at $x = 1.01$ m, $y = 0.16$ m and his mass is 81.6 kg. What is his moment of inertia about the z' axis through his center of mass?

10.25 A floating sonobuoy (sound measuring device) is in equilibrium in the vertical position shown. (Its center of mass is low enough that it is stable in this position.) It is a 10-kg cylinder 1 m in length and 125 mm in diameter. The water density is 1025 kg/m^3, and the buoyancy force supporting the buoy equals the weight of the water that would occupy the volume of the part of the cylinder below the surface. If you push the sonobuoy slightly deeper and release it, what is the natural frequency of the resulting vertical vibrations?

P10.25

10.26 A disk rotates about a fixed *vertical* axis with constant angular velocity Ω. (The plane of the disk is horizontal.) A mass m slides in a smooth slot in the disk and is attached to a spring with constant k. The distance from the center of the disk to the mass when the spring is unstretched is r_0. Show that if $k/m > \Omega^2$, the natural frequency of vibration of the mass is $f = (1/2\pi)\sqrt{k/m - \Omega^2}$.

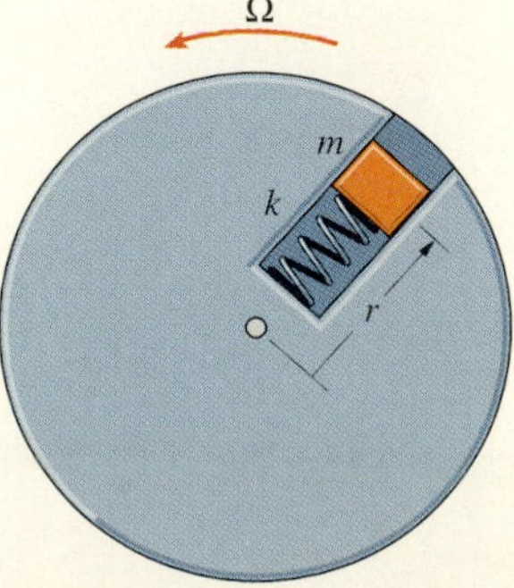

P10.26

10.27 Suppose that at $t = 0$, the mass described in Problem 10.26 is located at $r = r_0$ and its radial velocity is $dr/dt = 0$. Determine the position r of the mass as a function of time.

10.28 A homogeneous 100-lb disk with radius $R = 1$ ft is attached to two identical cylindrical steel bars of length $L = 1$ ft. The relation between the moment M exerted on the disk by one of the bars and the angle of rotation θ of the disk is

$$M = \frac{GJ}{L}\theta,$$

where J is the polar moment of inertia of the cross section of the bar and $G = 1.7 \times 10^9$ lb/ft^2 is the shear modulus of the steel. Determine the required radius of the bars if the natural frequency of rotational vibrations of the disk is to be 10 Hz.

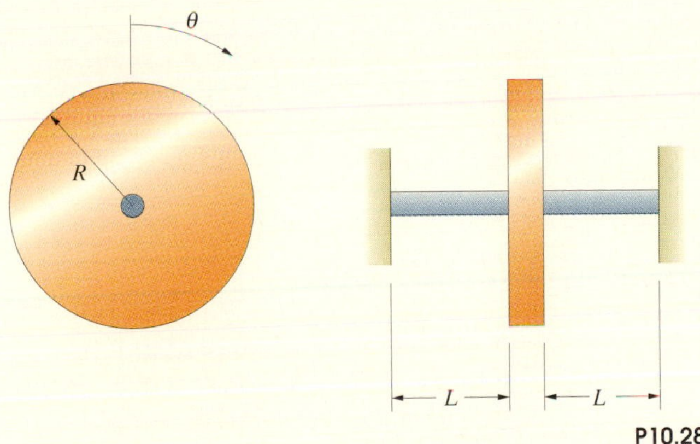

P10.28

10.29 The moments of inertia of gears A and B are $I_A = 0.025$ kg-m^2 and $I_B = 0.100$ kg-m^2. Gear A is connected to a torsional spring with constant $k = 10$ N-m/rad. What is the natural frequency of small angular vibrations of the gears?

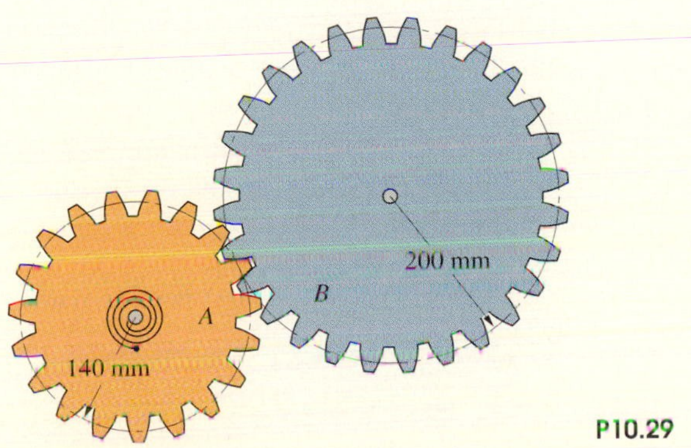

P10.29

10.30 At $t = 0$, the torsional spring in Problem 10.29 is unstretched and gear B has a counterclockwise angular velocity of 2 rad/s. Determine the counterclockwise angular position of gear B relative to its equilibrium position as a function of time.

10.31 Each slender bar is of mass m and length l. Determine the natural frequency of small vibrations of the system.

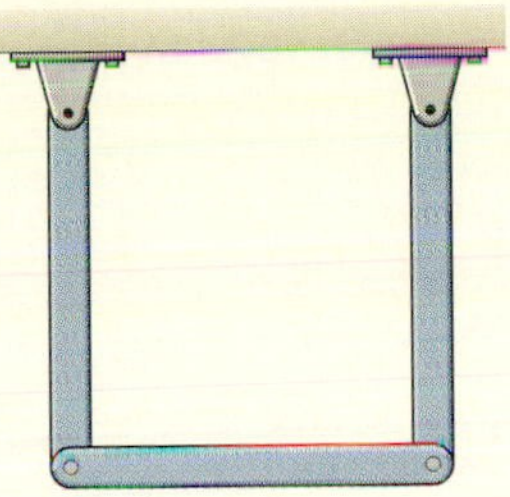

P10.31

10.32 The masses of the slender bar and the homogeneous disk are m and m_d, respectively. The spring is unstretched when $\theta = 0$. Assume that the disk rolls on the horizontal surface.

(a) Show that the motion is governed by the equation

$$\left(\frac{1}{3} + \frac{3m_d}{2m}\cos^2\theta\right)\frac{d^2\theta}{dt^2} - \frac{3m_d}{2m}\sin\theta\cos\theta\left(\frac{d\theta}{dt}\right)^2 - \frac{g}{2l}\sin\theta + \frac{k}{m}(1 - \cos\theta)\sin\theta = 0.$$

(b) If the system is in equilibrium at the angle $\theta = \theta_e$ and $\tilde{\theta} = \theta - \theta_e$, show that the equation governing small vibrations relative to the equilibrium position is

$$\left(\frac{1}{3} + \frac{3m_d}{2m}\cos^2\theta_e\right)\frac{d^2\tilde{\theta}}{dt^2} + \left[\frac{k}{m}\left(\cos\theta_e - \cos^2\theta_e + \sin^2\theta_e\right) - \frac{g}{2l}\cos\theta_e\right]\tilde{\theta} = 0.$$

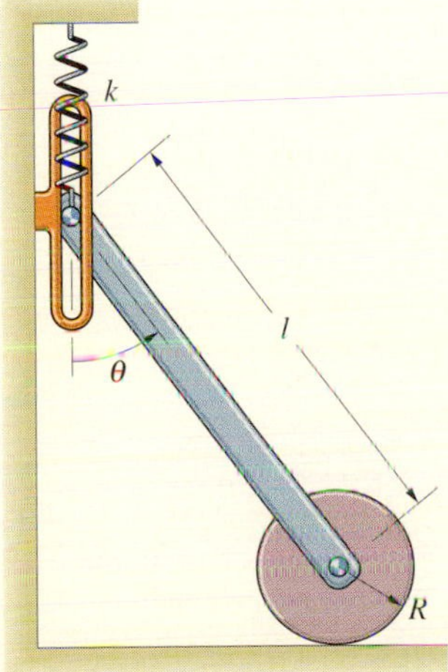

P10.32

10.33 The masses of the bar and disk in Problem 10.34 are $m = 2$ kg and $m_d = 4$ kg, respectively. The dimensions $l = 1$ m and $R = 0.28$ m, and the spring constant is $k = 70$ N/m.
(a) Determine the angle θ_e at which the system is in equilibrium.
(b) The system is at rest in the equilibrium position, and the disk is given a clockwise angular velocity of 0.1 rad/s. Determine θ as a function of time.

10.34 The mass of each slender bar is 1 kg. If the natural frequency of small vibrations of the system is 0.935 Hz, what is the mass of the object A?

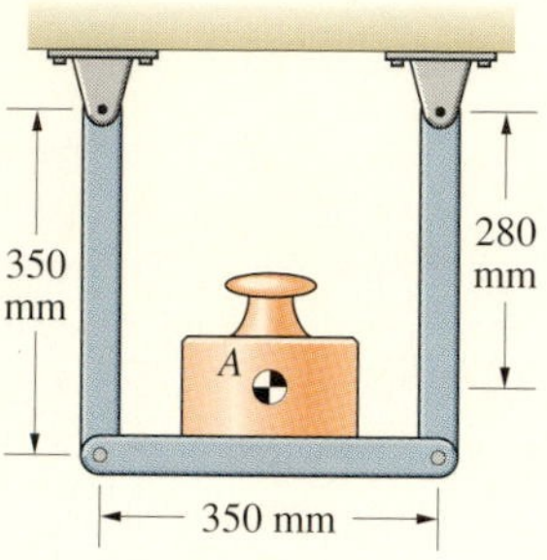

P10.34

10.35 The slender bar of mass m and length l is held in equilibrium in the position shown by a torsional spring with constant k. The spring is unstretched when the bar is vertical. Determine the natural frequency of small vibrations relative to the equilibrium position shown.

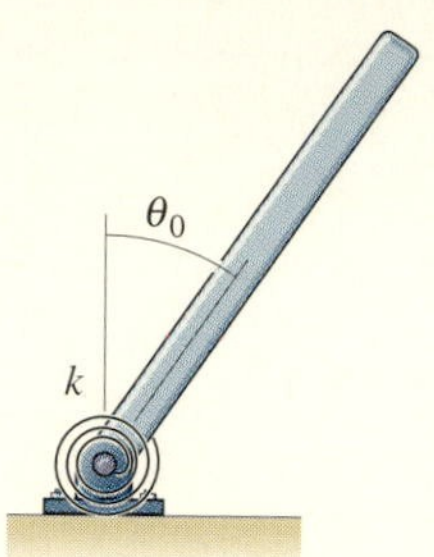

P10.35

10.2 *Damped Vibrations*

If you displace the mass of a spring-mass oscillator and release it, you know that it won't continue to vibrate indefinitely. It will slow down and eventually stop as a result of frictional forces, or **damping mechanisms**, acting on the system. Damping mechanisms damp out, or **attenuate**, the vibration. In some cases, engineers intentionally include damping mechanisms in vibrating systems. For example, the shock absorbers in a car are designed to damp out vibrations of the suspension relative to the frame. In the previous section we neglected damping, so the solutions we obtained describe only motions of systems over periods of time brief enough that the effects of damping can be neglected. We now discuss a simple method for modeling damping in vibrating systems.

The spring-mass oscillator in Fig. 10.9(a) has a **damping element**. The schematic diagram for the damping element represents a piston moving in a cylinder of viscous fluid, which is called a **dashpot**. The force required to lengthen or shorten a damping element is defined to be the product of a constant c, the **damping constant**, and the rate of change of its length (Fig. 10.9b). Therefore the equation of motion of the mass is

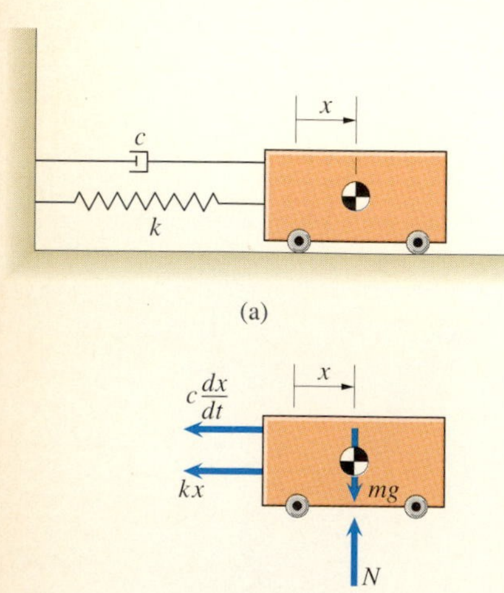

Fig. 10.9
(a) Damped spring-mass oscillator.
(b) Free-body diagram of the mass.

$$-c\frac{dx}{dt} - kx = m\frac{d^2x}{dt^2}.$$

By defining $\omega = \sqrt{k/m}$ and $d = c/2m$, we can write this equation in the form

$$\boxed{\frac{d^2x}{dt^2} + 2d\frac{dx}{dt} + \omega^2 x = 0.} \qquad (10.16)$$

This equation describes the vibrations of many damped, one-degree-of-freedom systems. The form of its solution, and consequently the character of the predicted behavior of a system, depends on whether d is less than, equal to, or greater than ω. We discuss these cases in the following sections.

Subcritical Damping

If $d < \omega$, a system is said to be **subcritically damped**. By assuming a solution of the form

$$x = Ce^{\lambda t}, \tag{10.17}$$

where C and λ are constants, and substituting it into Eq. (10.16), we obtain

$$\lambda^2 + 2d\lambda + \omega^2 = 0.$$

This quadratic equation yields two roots for the constant λ that we can write as

$$\lambda = -d \pm i\omega_d,$$

where $i = \sqrt{-1}$ and

$$\omega_d = \sqrt{\omega^2 - d^2}. \tag{10.18}$$

Because we are assuming that $d < \omega$, the constant ω_d is a real number. The two roots for λ give us two solutions of the form of Eq. (10.17). The resulting general solution of Eq. (10.16) is

$$x = e^{-dt}(Ce^{i\omega_d t} + De^{-i\omega_d t}),$$

where C and D are constants. By using the identity $e^{i\theta} = \cos\theta + i\sin\theta$, we can express the general solution in the form

$$\boxed{x = e^{-dt}(A\sin\omega_d t + B\cos\omega_d t),} \tag{10.19}$$

where A and B are constants. Equation (10.19) is the product of an exponentially decaying function of time and an expression identical in form to the solution we obtained for an undamped system. The exponential function causes the expected effect of damping: The amplitude of the vibration attenuates with time. The coefficient d determines the rate at which the amplitude decreases.

Damping has an important effect in addition to causing attenuation. Because the oscillatory part of the solution is identical in form to Eq. (10.5) except that the circular natural frequency ω is replaced by ω_d, it follows from Eqs. (10.9) and (10.10) that the period and natural frequency of the damped system are

$$\boxed{\tau_d = \frac{2\pi}{\omega_d}, \qquad f_d = \frac{1}{\tau_d} = \frac{\omega_d}{2\pi}.} \tag{10.20}$$

From Eq. (10.18) we see that $\omega_d < \omega$, so *the period of the vibration is increased and its natural frequency is decreased as a result of subcritical damping*.

The rate of damping is often expressed in terms of the **logarithmic decrement** δ, which is the natural logarithm of the ratio of the amplitude at a time t to the amplitude at time $t + \tau_d$. Since the amplitude is proportional to e^{-dt}, we can obtain a simple relation between the logarithmic decrement, the coefficient d, and the period:

$$\boxed{\delta = \ln\left[\frac{e^{-dt}}{e^{-d(t + \tau_d)}}\right] = d\tau_d.}$$

Critical and Supercritical Damping

When $d \geq \omega$, the character of the solution of Eq. (10.16) is very different from the case of subcritical damping. Suppose that $d > \omega$. When this is the case, a system is said to be **supercritically damped**. We again substitute a solution of the form

$$x = Ce^{\lambda t} \tag{10.21}$$

into Eq. (10.16), obtaining

$$\lambda^2 + 2d\lambda + \omega^2 = 0. \tag{10.22}$$

We can write the roots of this equation as

$$\lambda = -d \pm h,$$

where

$$h = \sqrt{d^2 - \omega^2}. \tag{10.23}$$

The resulting general solution of Eq. (10.16) is

$$\boxed{x = Ce^{-(d - h)t} + De^{-(d + h)t},} \tag{10.24}$$

where C and D are constants.

When $d = \omega$, a system is said to be **critically damped**. The constant $h = 0$, so Eq. (10.22) has a repeated root, $\lambda = -d$, and we obtain only one solution of the form (10.21). In this case it can be shown that the general solution of Eq. (10.16) is

$$\boxed{x = Ce^{-dt} + Dte^{-dt},} \tag{10.25}$$

where C and D are constants.

Equations (10.24) and (10.25) indicate that the motion of a system is not oscillatory when $d \geq \omega$. They are expressed in terms of exponential functions and do not contain sines and cosines. The condition $d = \omega$ defines the minimum amount of damping necessary to avoid oscillatory behavior, which is why it is referred to as the critically damped case. Figure 10.10 shows the

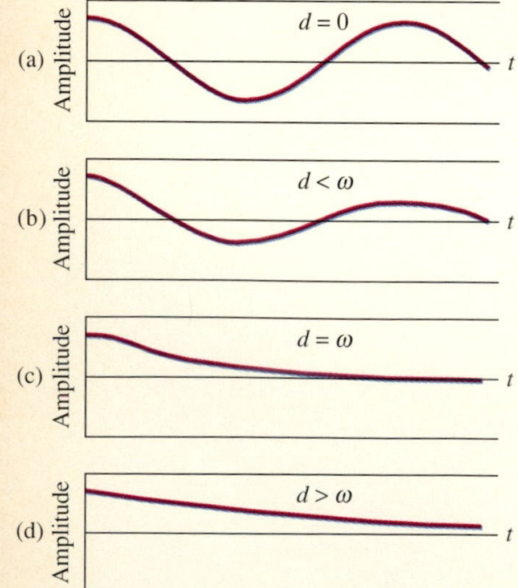

Fig. 10.10
Amplitude history of a vibrating system that is (a) undamped; (b) subcritically damped; (c) critically damped; (d) supercritically damped.

effect of increasing amounts of damping on the behavior of a vibrating system.

The concept of critical damping has important implications in the design of many systems. For example, it is desirable to introduce enough damping into a car's suspension so that its motion is not oscillatory, but too much damping would cause the suspension to be too "stiff."

In the following examples we analyze damped one-degree-of-freedom systems. By expressing the equation of motion of the system in the form of Eq. (10.16), you can determine d and ω. Their values tell you whether the damping is subcritical, critical, or supercritical, which indicates the form of solution you should use:

	Type of Damping	Solution
$d < \omega$:	Subcritical	Eq. (10.19)
$d = \omega$:	Critical	Eq. (10.25)
$d > \omega$:	Supercritical	Eq. (10.24)

Example 10.3

The damped spring-mass oscillator in Fig. 10.11 has mass $m = 2$ kg, spring constant $k = 8$ N/m, and damping constant $c = 1$ N-s/m. At $t = 0$, the mass is released from rest in the position $x = 0.1$ m. Determine its position as a function of time.

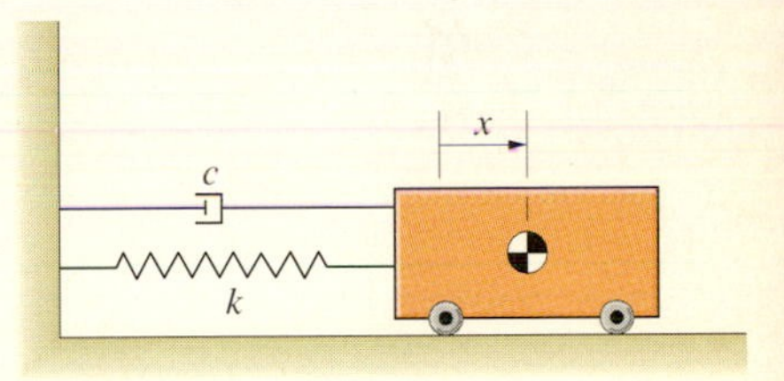

Fig. 10.11

SOLUTION

The constants $\omega = \sqrt{k/m} = 2$ rad/s and $d = c/2m = 0.25$ rad/s, so the damping is subcritical and the motion is described by Eq. (10.19). From Eq. (10.18),

$$\omega_d = \sqrt{\omega^2 - d^2} = 1.98 \text{ rad/s}.$$

From Eq. (10.19),

$$x = e^{-0.25t}(A \sin 1.98t + B \cos 1.98t),$$

and the velocity of the mass is

$$\frac{dx}{dt} = -0.25e^{-0.25t}(A \sin 1.98t + B \cos 1.98t) + e^{-0.25t}(1.98A \cos 1.98t - 1.98B \sin 1.98t).$$

From the conditions $x = 0.1$ m and $dx/dt = 0$ at $t = 0$, we obtain $A = 0.0126$ m and $B = 0.1$ m, so the position of the mass is

$$x = e^{-0.25t}(0.0126 \sin 1.98t + 0.1 \cos 1.98t) \text{ m}.$$

The graph of x for the first 10 s of motion in Fig. 10.12 clearly exhibits the attenuation of the amplitude.

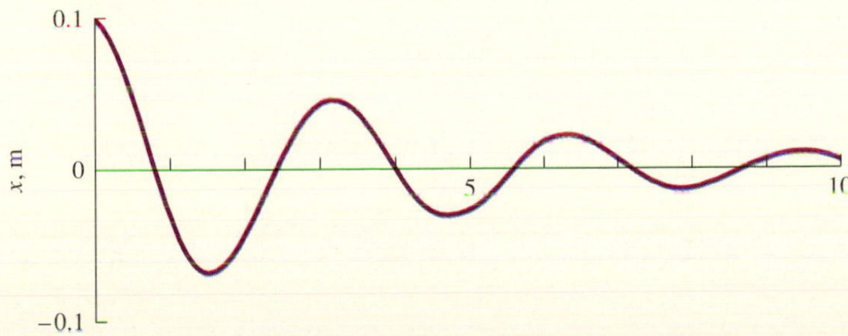

Fig. 10.12
Position of the mass as a function of time.

Example 10.4

Fig. 10.13

The 40-lb stepped disk in Fig. 10.13 is released from rest with the spring unstretched. Determine the position of the center of the disk as a function of time if $R = 1$ ft, $k = 10$ lb/ft, $c = 4$ lb-s/ft, and the moment of inertia expressed in terms of the mass m of the disk is $I = 3mR^2$.

SOLUTION

Let x be the downward displacement of the center of the disk relative to its position when the spring is unstretched. From the position of the disk's instantaneous center (Fig. a), we can see that the rate at which the spring is stretched is $2(dx/dt)$ and the rate at which the damping element is lengthened is $3(dx/dt)$. When the center of the disk is displaced a distance x, the stretch of the spring is $2x$.

We draw the free-body diagram of the disk in Fig. (b), showing the forces exerted by the spring, the damping element, and the tension in the cable. Newton's second law is

$$mg - T - 2kx - 3c\frac{dx}{dt} = m\frac{d^2x}{dt^2},$$

and the equation of angular motion is

$$RT - R(2kx) - 2R\left(3c\frac{dx}{dt}\right) = (3mR^2)\alpha.$$

The angular acceleration is related to the acceleration of the center of the disk by $\alpha = (d^2x/dt^2)/R$. Adding the equation of angular motion to Newton's second law, we obtain the equation of motion

$$4m\frac{d^2x}{dt^2} + 9c\frac{dx}{dt} + 4kx = mg.$$

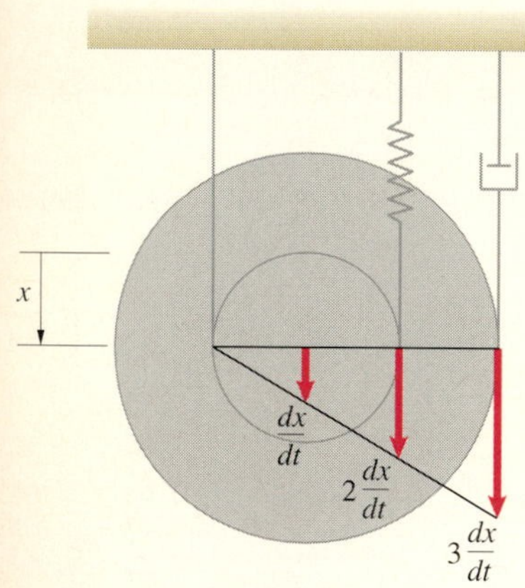

(a) Using the instantaneous center to determine the relationships between the velocities.

By setting d^2x/dt^2 and dx/dt equal to zero in this equation, we determine that the equilibrium position of the disk is $x = mg/4k$. Rewriting the equation of motion in terms of the variable $\tilde{x} = x - mg/4k$, we obtain

$$\frac{d^2\tilde{x}}{dt^2} + \left(\frac{9c}{4m}\right)\frac{d\tilde{x}}{dt} + \frac{k}{m}\tilde{x} = 0.$$

This equation is identical in form to Eq. (10.16), where the constants d and ω are

$$d = \frac{9c}{8m} = \frac{(9)(4)}{(8)(40/32.2)} = 3.62 \text{ rad/s},$$

$$\omega = \sqrt{\frac{k}{m}} = \sqrt{\frac{10}{(40/32.2)}} = 2.84 \text{ rad/s}.$$

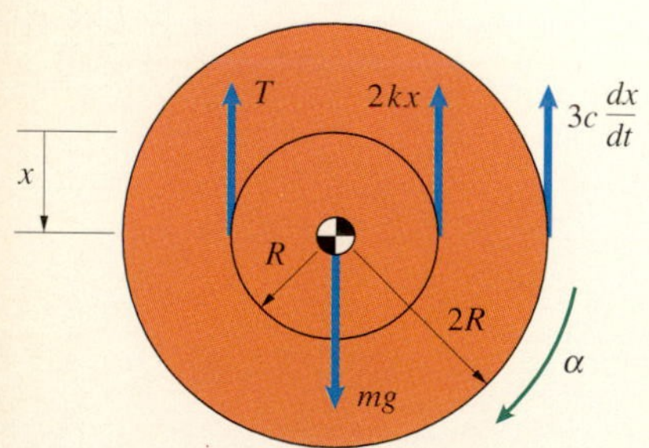

(b) Free-body diagram of the disk.

The damping is supercritical, so the motion is described by Eq. (10.24) with $h = \sqrt{d^2 - \omega^2} = 2.25$ rad/s:

$$\tilde{x} = Ce^{-(d-h)t} + De^{-(d+h)t} = Ce^{-1.37t} + De^{-5.87t}.$$

The velocity is

$$\frac{d\tilde{x}}{dt} = -1.37Ce^{-1.37t} - 5.87De^{-5.87t}.$$

At $t = 0$, $\tilde{x} = -mg/4k = -1$ ft and $d\tilde{x}/dt = 0$. From these conditions, we obtain $C = -1.304$ ft and $D = 0.304$ ft, so the position of the center of the disk relative to its equilibrium position is

$$\tilde{x} = -1.304e^{-1.37t} + 0.304e^{-5.87t} \text{ ft.}$$

We show the graph of the position for the first 4 s of motion in Fig. 10.14.

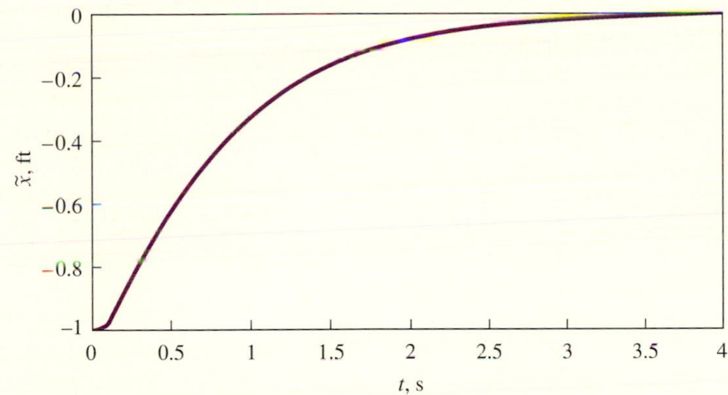

Fig. 10.14
Position of the center of the disk as a function of time.

Problems

10.36 (a) What are the natural frequency and period of the spring-mass oscillator described in Example 10.3? (b) What are the natural frequency and period if the damping element is removed?

10.37 (a) What value of c is necessary for the stepped disk in Example 10.4 to be critically damped? (b) If c equals the value you determined in (a) and the disk is released from rest with the spring unstretched, determine the position of the center of the disk relative to its equilibrium position as a function of time.

10.38 The damping constant of the damped spring-mass oscillator is $c = 20$ N-s/m. What are the period and natural frequency of the system? Compare them to the period and natural frequency when the system is undamped.

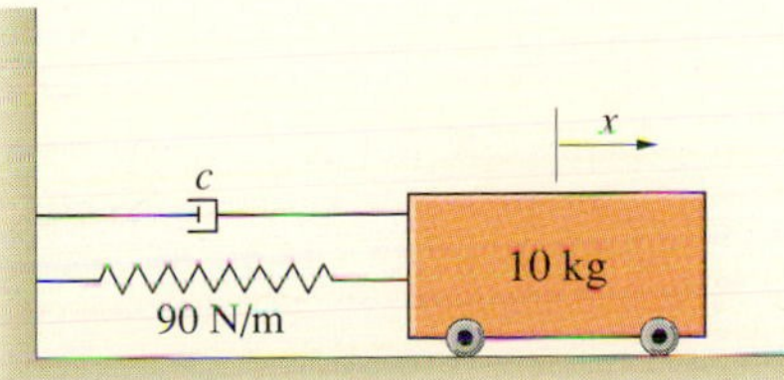

P10.38

10.39 At $t = 0$, the position of the mass in Problem 10.38 relative to its equilibrium position is $x = 0$ and its velocity is 1 m/s to the right. Determine x as a function of time.

10.40 In Problem 10.38, what value of the damping constant c will cause the amplitude of vibration of the system to decrease to one-half of its initial value in 10 s?

10.41 At $t = 0$, the position of the mass in Problem 10.38 is $x = 0$ and it has a velocity of 1 m/s to the right. Determine x as a function of time if c has twice the value necessary for the system to be critically damped.

10.42 For small vertical displacements of the tire and wheel, the motion of the car's suspension can be modeled by the damped spring-mass oscillator in Fig. 10.9 with $m = 36$ kg and $k = 22$ kN/m. Determine the value of the damping constant c that must be provided by the suspension's shock absorber to achieve critical damping.

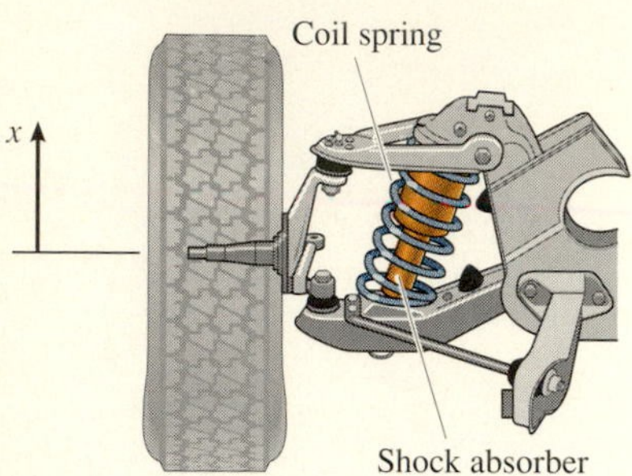

P10.42

10.43 The motion of the car's suspension shown in Problem 10.42 can be modeled by the damped spring-mass oscillator in Fig. 10.9 with $m = 36$ kg, $k = 22$ kN/m, and $c = 2.2$ kN-s/m. Assume that no external forces act on the tire and wheel. At $t = 0$, the spring is unstretched and the tire and wheel are given a velocity $dx/dt = 10$ m/s. Determine the position x as a function of time.

10.44 The homogeneous slender bar is 4 ft long and weighs 10 lb. Aerodynamic drag and friction at the support exert a resisting moment on the bar of magnitude $0.5(d\theta/dt)$ ft-lb, where $d\theta/dt$ is the angular velocity of the bar in rad/s.
(a) What are the period and natural frequency of small vibrations of the bar?
(b) How long does it take for the amplitude of vibration to decrease to one-half of its initial value?

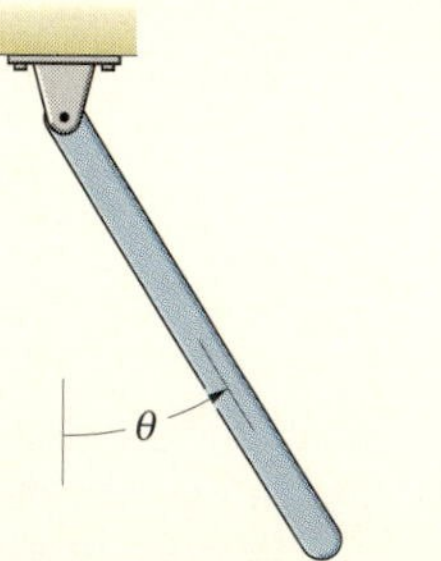

P10.44

10.45 If the bar in Problem 10.44 is displaced a small angle θ_0 and released from rest at $t = 0$, what is θ as a function of time?

10.46 The radius of the pulley is $R = 100$ mm and its moment of inertia is $I = 0.1$ kg-m^2. The mass $m = 5$ kg, and the spring constant is $k = 135$ N/m. The cable does not slip relative to the pulley. The coordinate x measures the displacement of the mass relative to the position in which the spring is unstretched. Determine x as a function of time if $c = 60$ N-s/m and the system is released from rest with $x = 0$.

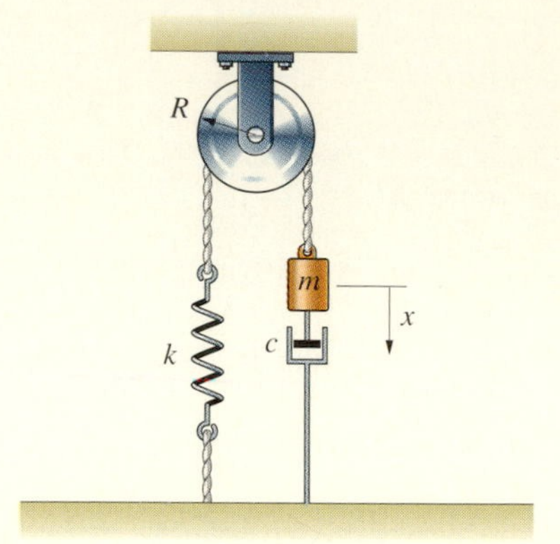

P10.46

10.47 For the system described in Problem 10.46, determine x as a function of time if $c = 120$ N-s/m and the system is released from rest with $x = 0$.

10.48 For the system described in Problem 10.46, choose the value of c so that the system is critically damped and determine x as a function of time if the system is released from rest with $x = 0$.

10.49 The spring constant is $k = 30$ lb/ft and the damping constant is $c = 3.5$ lb-s/ft. The radius of the pulley is 0.5 ft, and its moment of inertia is 0.25 slug-ft^2. What is the frequency of small vibrations of the system relative to its equilibrium position?

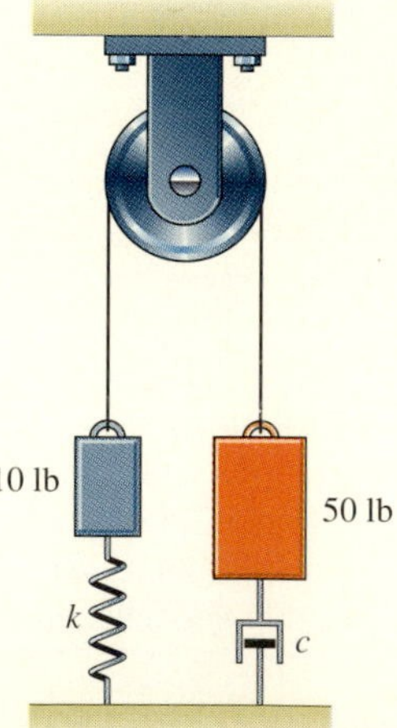

P10.49

10.50 The system described in Problem 10.49 is released from rest with the spring unstretched. Determine the position of the 10-lb weight relative to its position at $t = 0$ as a function of time.

10.51 The homogeneous disk weighs 100 lb and its radius is $R = 1$ ft. It rolls on the plane surface. The spring constant is $k = 100$ lb/ft and the damping constant is $c = 3$ lb-s/ft. Determine the natural frequency of small vibrations of the disk relative to its equilibrium position.

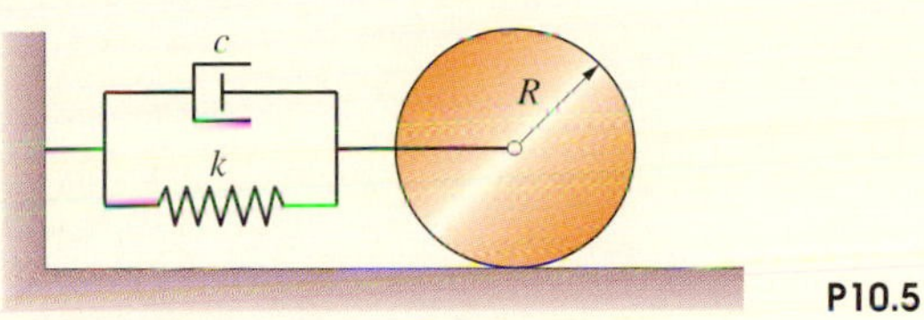

P10.51

10.52 In Problem 10.51, the spring is unstretched at $t = 0$ and the disk has a clockwise angular velocity of 2 rad/s. What is the angular velocity of the disk when $t = 3$ s?

10.53 The moment of inertia of the stepped disk is I. Let θ be the angular displacement of the disk relative to its position when the spring is unstretched. Show that the equation governing θ is identical in form to Eq. (10.16), where

$$d = \frac{R^2 c}{2I} \quad \text{and} \quad \omega^2 = \frac{4R^2 k}{I}.$$

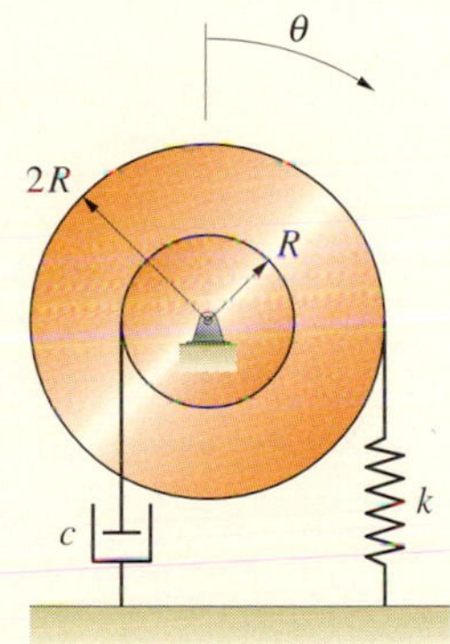

P10.53

10.54 In Problem 10.53, the radius $R = 250$ mm, $k = 150$ N/m, and the moment of inertia of the disk is $I = 2$ kg-m^2.
(a) What value of c will cause the system to be critically damped?
(b) At $t = 0$, the spring is unstretched and the clockwise angular velocity of the disk is 10 rad/s. Determine θ as a function of time if the system is critically damped.
(c) Using the result of (b), determine the maximum resulting angular displacement of the disk and the time at which it occurs.

10.55 The moments of inertia of gears A and B are $I_A = 0.025$ kg-m^2 and $I_B = 0.100$ kg-m^2. Gear A is connected to a torsional spring with constant $k = 10$ N-m/rad. The bearing supporting gear B incorporates a damping element that exerts a resisting moment on gear B of magnitude $2(d\theta_B/dt)$ N-m, where $d\theta_B/dt$ is the angular velocity of gear B in rad/s. What is the frequency of small angular vibrations of the gears?

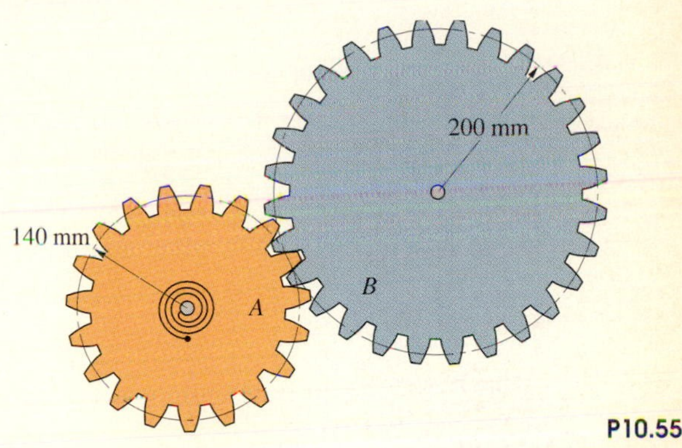

P10.55

10.56 At $t = 0$, the torsional spring in Problem 10.55 is unstretched and gear B has a counterclockwise angular velocity of 2 rad/s. Determine the counterclockwise angular position of gear B relative to its equilibrium position as a function of time.

10.57 For the case of critically damped motion, confirm that the expression

$$x = Ce^{-dt} + Dte^{-dt}$$

is a solution of Eq. (10.16).

10.3 *Forced Vibrations*

The term **forced vibrations** means that external forces affect the vibrations of a system. Until now, we have discussed **free vibrations** of systems, vibrations unaffected by external forces. For example, during an earthquake, a building undergoes forced vibrations induced by oscillatory forces exerted on its foundations. After the earthquake subsides, the building vibrates freely until its motion damps out.

The damped spring-mass oscillator in Fig. 10.15(a) is subjected to a horizontal time-dependent force $F(t)$. From the free-body diagram of the mass (Fig. 10.15b), its equation of motion is

$$F(t) - kx - c\frac{dx}{dt} = m\frac{d^2x}{dt^2}.$$

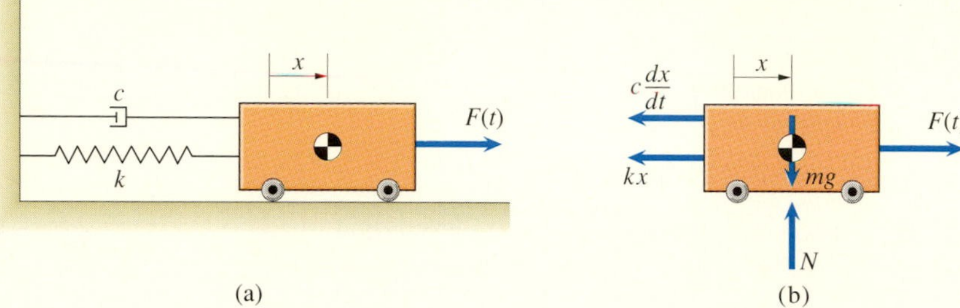

Fig. 10.15
(a) A damped spring-mass oscillator subjected to a time-dependent force.
(b) Free-body diagram of the mass.

Defining $d = c/2m$, $\omega^2 = k/m$, and $a(t) = F(t)/m$, we can write this equation in the form

$$\boxed{\frac{d^2x}{dt^2} + 2d\frac{dx}{dt} + \omega^2 x = a(t).} \tag{10.26}$$

We call $a(t)$ the **forcing function**. Equation (10.26) describes the forced vibrations of many damped, one-degree-of-freedom systems. It is nonhomogeneous, because the forcing function does not contain x or one of its derivatives. Its general solution consists of two parts, the homogeneous and particular solutions:

$$x = x_h + x_p.$$

The **homogeneous solution** x_h is the general solution of Eq. (10.26) with the right side set equal to zero. Therefore the homogeneous solution is the general solution for free vibrations, which we described in Section 10.2. The **particular solution** x_p is a solution that satisfies Eq. (10.26). In the following sections we discuss the particular solutions for two types of forcing functions that occur frequently in applications.

Oscillatory Forcing Function

Unbalanced wheels and shafts exert forces that oscillate at their frequency of rotation. When your car's wheels are out of balance, they exert oscillatory forces that cause vibrations you can feel. Engineers design electromechanical devices that transform oscillating currents into oscillating forces for use in testing vibrating systems. But the principal reason we are interested in this type of forcing function is that nearly any forcing function can be represented as a sum of oscillatory forcing functions with several different frequencies or with a continuous spectrum of frequencies.

By studying the motion of a vibrating system subjected to an oscillatory forcing function, we can determine its response as a function of the frequency of the force. Suppose that the forcing function is an oscillatory function of the form

$$a(t) = a_0 \sin \omega_0 t + b_0 \cos \omega_0 t, \tag{10.27}$$

where a_0, b_0, and the circular frequency of the forcing function ω_0 are given constants. We can obtain the particular solution to Eq. (10.26) by seeking a solution of the form

$$x_{\text{p}} = A_{\text{p}} \sin \omega_0 t + B_{\text{p}} \cos \omega_0 t, \tag{10.28}$$

where A_{p} and B_{p} are constants we must determine. Substituting this expression and Eq. (10.27) into Eq. (10.26), we can write the resulting equation as

$$(-\omega_0^2 A_{\text{p}} - 2d\omega_0 B_{\text{p}} + \omega^2 A_{\text{p}} - a_0) \sin \omega_0 t + (-\omega_0^2 B_{\text{p}} + 2d\omega_0 A_{\text{p}} + \omega^2 B_{\text{p}} - b_0) \cos \omega_0 t = 0.$$

Equating the coefficients of $\sin \omega_0 t$ and $\cos \omega_0 t$ to zero and solving for A_{p} and B_{p}, we obtain

$$\begin{aligned} A_{\text{p}} &= \frac{(\omega^2 - \omega_0^2)a_0 + 2d\omega_0 b_0}{(\omega^2 - \omega_0^2)^2 + 4d^2\omega_0^2}, \\ B_{\text{p}} &= \frac{-2d\omega_0 a_0 + (\omega^2 - \omega_0^2)b_0}{(\omega^2 - \omega_0^2)^2 + 4d^2\omega_0^2}. \end{aligned} \tag{10.29}$$

Substituting these results into Eq. (10.28), the particular solution is

$$\begin{aligned} x_{\text{p}} = &\left[\frac{(\omega^2 - \omega_0^2)a_0 + 2d\omega_0 b_0}{(\omega^2 - \omega_0^2)^2 + 4d^2\omega_0^2}\right] \sin \omega_0 t \\ &+ \left[\frac{-2d\omega_0 a_0 + (\omega^2 - \omega_0^2)b_0}{(\omega^2 - \omega_0^2)^2 + 4d^2\omega_0^2}\right] \cos \omega_0 t. \end{aligned} \tag{10.30}$$

The amplitude of the particular solution is

$$E_{\text{p}} = \sqrt{A_{\text{p}}^2 + B_{\text{p}}^2} = \frac{\sqrt{a_0^2 + b_0^2}}{\sqrt{(\omega^2 - \omega_0^2)^2 + 4d^2\omega_0^2}}. \tag{10.31}$$

We showed in Section 10.2 that the solution of the equation describing free vibration of a damped system attenuates with time. For this reason, the particular solution for the motion of a damped vibrating system subjected to an oscillatory external force is also called the **steady-state solution**. The motion approaches the steady-state solution with increasing time. (See Example 10.5.)

We illustrate the effects of damping and the frequency of the forcing function on the amplitude of the particular solution in Fig. 10.16. We plot the nondimensional expression $\omega^2 E_p / \sqrt{a_0^2 + b_0^2}$ as a function of ω_0/ω for several values of the parameter d/ω. When there is no damping ($d = 0$), the amplitude of the particular solution approaches infinity as the circular frequency ω_0 of the forcing function approaches the circular natural frequency ω. When the damping is small, the amplitude of the particular solution approaches a finite maximum value at a value of ω_0 that is smaller than ω. The frequency at which the amplitude of the particular solution is a maximum is called the **resonant frequency**.

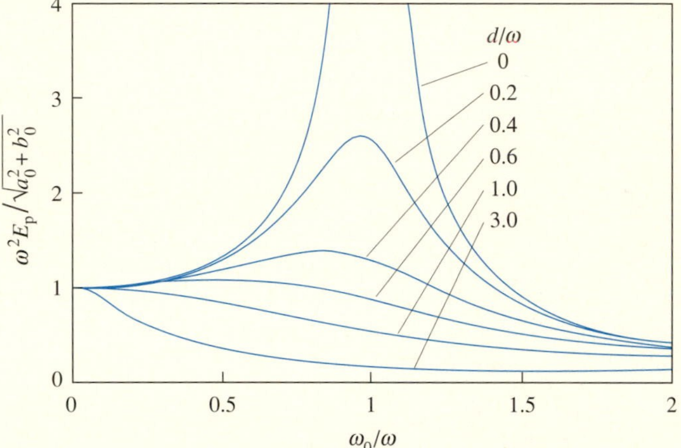

Fig. 10.16
Amplitude of the particular (steady-state) solution as a function of the frequency of the forcing function.

The phenomenon of resonance is a familiar one in our everyday experience. For example, when a wheel of your car is out of balance, you notice the resulting vibrations when the car is moving at a certain speed. At that speed, the wheel rotates at the resonant frequency of your car's suspension. Resonance is of practical importance in many applications, because relatively small oscillatory forces can result in large vibration amplitudes that may cause damage or interfere with the functioning of a system. The classic example is that of soldiers marching across a bridge. If their steps in unison coincide with one of the bridge's resonant frequencies, they may damage it even though the bridge can safely support their weight.

Polynomial Forcing Function

Suppose that the forcing function $a(t)$ in Eq. (10.26) is a polynomial function of time:

$$a(t) = a_0 + a_1 t + a_2 t^2 + \cdots + a_N t^N,$$

where $a_1, a_2, \ldots, a_N$ are given constants. This forcing function is important in applications because you can approximate many smooth functions by polynomials over a given interval of time. In this case, we can obtain the particular solution to Eq. (10.26) by seeking a solution of the same form:

$$x_p = A_0 + A_1 t + A_2 t^2 + \cdots + A_N t^N, \tag{10.32}$$

where $A_0, A_1, A_2, \ldots, A_N$ are constants we must determine.

For example, if $a(t) = a_0 + a_1 t$, Eq. (10.26) is

$$\frac{d^2x}{dt^2} + 2d\frac{dx}{dt} + \omega^2 x = a_0 + a_1 t, \tag{10.33}$$

and we seek a particular solution of the form $x_p = A_0 + A_1 t$. Substituting this solution into Eq. (10.33), we can write the resulting equation as

$$(2d A_1 + \omega^2 A_0 - a_0) + (\omega^2 A_1 - a_1)t = 0.$$

This equation can be satisfied over an interval of time only if

$$2d A_1 + \omega^2 A_0 - a_0 = 0$$

and

$$\omega^2 A_1 - a_1 = 0.$$

Solving these two equations for A_0 and A_1, we obtain the particular solution:

$$x_p = \frac{a_0 - 2da_1/\omega^2 + a_1 t}{\omega^2}.$$

You should confirm that this is a particular solution by substituting it into Eq. (10.33).

In the following examples we analyze forced vibrations of one-degree-of-freedom systems. After expressing the equation of motion of the system in the form

$$\frac{d^2x}{dt^2} + 2d\frac{dx}{dt} + \omega^2 x = a(t),$$

you must usually determine the homogeneous and particular solutions. The forms of the homogeneous solution are given in Section 10.2.

Example 10.5

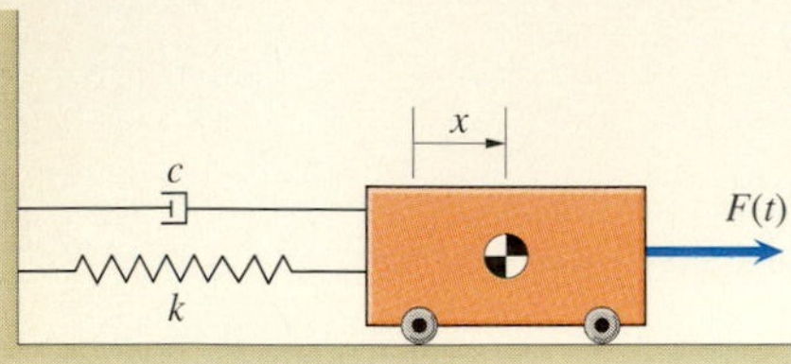

Fig. 10.17

An engineer designing a vibration isolation system for an instrument console models the console and isolation system by the damped spring-mass oscillator in Fig. 10.17 with mass $m = 2$ kg, spring constant $k = 8$ N/m, and damping constant $c = 1$ N-s/m. To determine the system's response to external vibration, she assumes that the mass is initially stationary with the spring unstretched, and at $t = 0$ a force

$$F(t) = 20 \sin 4t \text{ N}$$

is applied to the mass.
(a) What is the amplitude of the particular (steady-state) solution?
(b) What is the position of the mass as a function of time?

STRATEGY

The forcing function is $a(t) = F(t)/m = 10 \sin 4t$ m/s^2, which is an oscillatory function of the form of Eq. (10.27) with $a_0 = 10$ m/s^2, $b_0 = 0$, and $\omega_0 = 4$ rad/s. The amplitude of the particular solution is given by Eq. (10.31), and the particular solution is given by Eq. (10.30). We must also determine whether the damping is subcritical, critical, or supercritical and choose the appropriate form of the homogeneous solution.

SOLUTION

(a) The circular natural frequency of the undamped system is $\omega = \sqrt{k/m} = 2$ rad/s and the constant $d = c/2m = 0.25$ rad/s. Therefore the amplitude of the particular solution is

$$E_p = \frac{a_0}{\sqrt{(\omega^2 - \omega_0^2)^2 + 4d^2\omega_0^2}} = \frac{10}{\sqrt{[(2)^2 - (4)^2]^2 + 4(0.25)^2(4)^2}}$$
$$= 0.822 \text{ m}.$$

(b) Since $d < \omega$, the system is subcritically damped and the homogeneous solution is given by Eq. (10.19). The circular frequency of the damped system is $\omega_d = \sqrt{\omega^2 - d^2} = 1.98$ rad/s, so the homogeneous solution is

$$x_h = e^{-0.25t}(A \sin 1.98t + B \cos 1.98t).$$

From Eq. (10.30), the particular solution is

$$x_p = -0.811 \sin 4t - 0.135 \cos 4t,$$

and the complete solution is

$$x = x_h + x_p$$
$$= e^{-0.25t}(A \sin 1.98t + B \cos 1.98t) - 0.811 \sin 4t - 0.135 \cos 4t.$$

At $t = 0$, $x = 0$ and $dx/dt = 0$. Using these conditions to determine the constants A and B, we obtain $A = 1.651$ m and $B = 0.135$ m. The position of the mass as a function of time is

$$x = e^{-0.25t}(1.651 \sin 1.98t + 0.135 \cos 1.98t) - 0.811 \sin 4t - 0.135 \cos 4t \text{ m.}$$

Figure 10.18 shows the homogeneous, particular, and complete solutions for the first 25 s of motion. The complete solution has an initial "transient" phase due to the homogeneous part of the solution. As the homogeneous solution attenuates, the complete solution approaches the particular, or steady-state, solution.

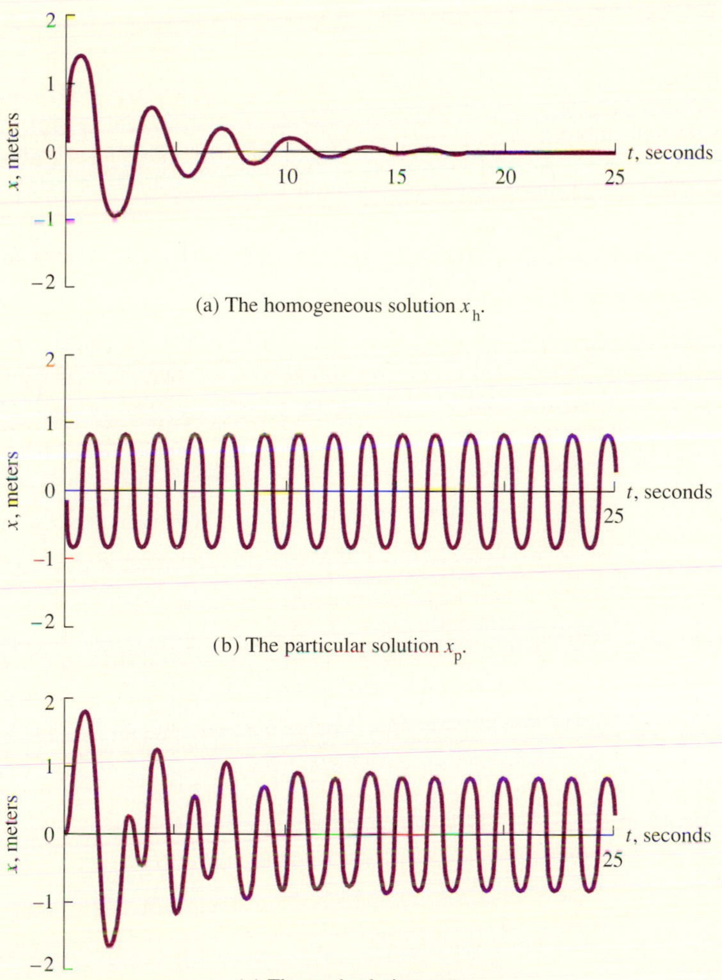

Fig. 10.18

The homogeneous, particular, and complete solutions.

Example 10.6

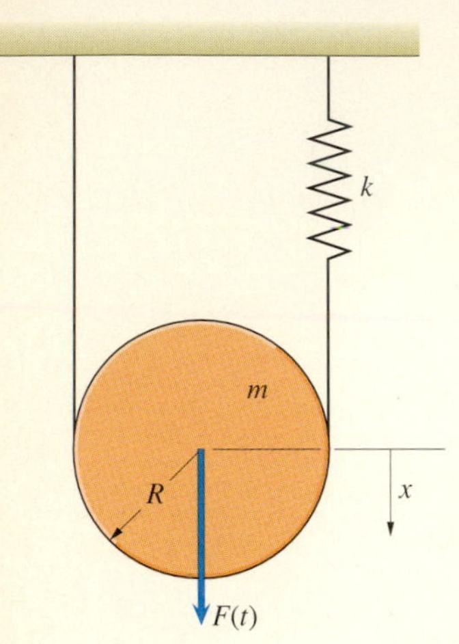

Fig. 10.19

The homogeneous disk in Fig. 10.19 has radius $R = 2$ ft and mass $m = 4$ slugs. The spring constant is $k = 30$ lb/ft. The disk is initially stationary in its equilibrium position, and at $t = 0$ a downward force $F(t) = 12 + 12t - 0.6t^2$ lb is applied to the center of the disk. Determine the position of the center of the disk as a function of time.

STRATEGY

The force $F(t)$ is a polynomial, so we can seek a particular solution of the form of Eq. (10.32).

SOLUTION

Let x be the displacement of the center of the disk relative to its position when the spring is unstretched. We draw the free-body diagram of the disk in Fig. (a), where T is the tension in the cable on the left side of the disk. From Newton's second law,

$$F(t) + mg - 2kx - T = m\frac{d^2x}{dt^2}. \tag{10.34}$$

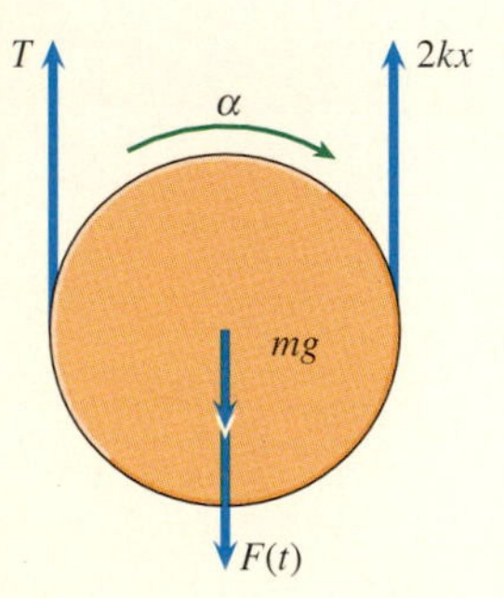

(a) Free-body diagram of the disk.

The angular acceleration of the disk in the clockwise direction is related to the acceleration of the center of the disk by $\alpha = (d^2x/dt^2)/R$. Using this expression, we can write the equation of angular motion of the disk as

$$\Sigma M = I\alpha:$$

$$TR - 2kxR = \left(\frac{1}{2}mR^2\right)\left(\frac{1}{R}\frac{d^2x}{dt^2}\right).$$

Solving this equation for T and substituting the result into Eq. (10.34), we obtain the equation of motion

$$\frac{3}{2}m\frac{d^2x}{dt^2} + 4kx = F(t) + mg. \tag{10.35}$$

Setting $d^2x/dt^2 = 0$ and $F(t) = 0$ in this equation, we find that the equilibrium position of the disk is $x = mg/4k$. In terms of the position of the center of the disk relative to its equilibrium position, $\tilde{x} = x - mg/4k$, the equation of motion is

$$\frac{d^2\tilde{x}}{dt^2} + \frac{8k}{3m}\tilde{x} = \frac{2F(t)}{3m}.$$

This equation is identical in form to Eq. (10.26). Substituting the values of k and m and the polynomial function $F(t)$, we obtain

$$\frac{d^2\tilde{x}}{dt^2} + 20\tilde{x} = 2 + 2t - 0.1t^2. \tag{10.36}$$

Comparing this equation with Eq. (10.26), we see that $d = 0$ (there is no damping) and $\omega^2 = 20\ (\text{rad/s})^2$. From Eq. (10.19), the homogeneous solution is

$$\tilde{x}_{\text{h}} = A \sin 4.472t + B \cos 4.472t.$$

To obtain the particular solution, we seek a solution in the form of a polynomial of the same order as $F(t)$:

$$\tilde{x}_{\text{p}} = A_0 + A_1 t + A_2 t^2,$$

where A_0, A_1, and A_2 are constants we must determine. We substitute this expression into Eq. (10.36) and collect terms of equal powers in t:

$$(2A_2 + 20A_0 - 2) + (20A_1 - 2)t + (20A_2 + 0.1)t^2 = 0.$$

This equation is satisfied if the coefficients multiplying each power of t equal zero:

$$\begin{aligned} 2A_2 + 20A_0 &= 2, \\ 20A_1 &= 2, \\ 20A_2 &= -0.1. \end{aligned}$$

Solving these three equations for A_0, A_1, and A_2, we obtain the particular solution:

$$\tilde{x}_{\text{p}} = 0.101 + 0.100t - 0.005t^2.$$

The complete solution is

$$\begin{aligned} \tilde{x} &= \tilde{x}_{\text{h}} + \tilde{x}_{\text{p}} \\ &= A \sin 4.472t + B \cos 4.472t + 0.101 + 0.100t - 0.005t^2. \end{aligned}$$

At $t = 0$, $\tilde{x} = 0$ and $d\tilde{x}/dt = 0$. Using these conditions to determine A and B, we obtain the position of the center of the disk as a function of time:

$$\tilde{x} = -0.022 \sin 4.472t - 0.101 \cos 4.472t + 0.101 + 0.100t - 0.005t^2.$$

The position is shown for the first 30 seconds of motion in Fig. 10.20. You can see the undamped, oscillatory homogeneous solution superimposed on the slowly varying particular solution.

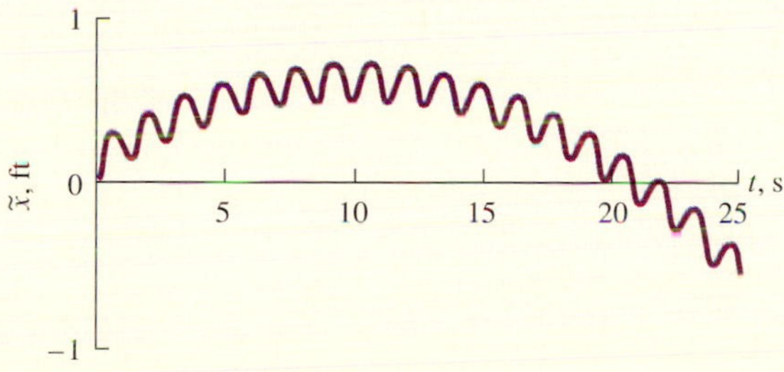

Fig. 10.20
Position of the center of the disk as a function of time.

Example 10.7

Application to Engineering

Displacement Transducers

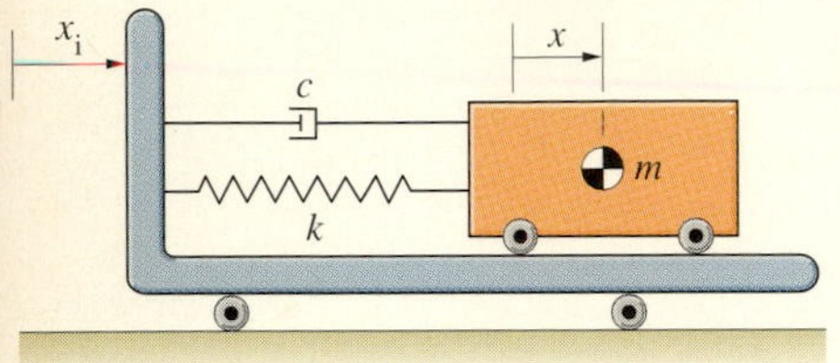

Fig. 10.21

A damped spring-mass oscillator, *or a device that can be modeled as a damped spring-mass oscillator*, can be used to measure an object's displacement. Suppose that the base of the spring-mass oscillator in Fig. 10.21 is attached to an object, and the coordinate x_i is a displacement to be measured relative to an inertial reference frame. The coordinate x measures the displacement of the mass *relative to the base*. When $x = 0$, the spring is unstretched. Suppose that the system is initially stationary, and at $t = 0$ the base undergoes the oscillatory motion

$$x_i = a_i \sin \omega_i t + b_i \cos \omega_i t. \tag{10.37}$$

If $m = 2$ kg, $k = 8$ N/m, $c = 4$ N-s/m, $a_i = 0.1$ m, $b_i = 0.1$ m, and $\omega_i = 10$ rad/s, what is the resulting steady-state amplitude of the displacement of the mass relative to the base?

SOLUTION

The acceleration of the mass relative to the base is d^2x/dt^2, so its acceleration relative to the inertial reference frame is $(d^2x/dt^2) + (d^2x_i/dt^2)$. Newton's second law for the mass is

$$-c\frac{dx}{dt} - kx = m\left(\frac{d^2x}{dt^2} + \frac{d^2x_i}{dt^2}\right).$$

We can write this equation as

$$\frac{d^2x}{dt^2} + 2d\frac{dx}{dt} + \omega^2 x = a(t),$$

where $d = c/2m = 1$ rad/s, $\omega = \sqrt{k/m} = 2$ rad/s, and the function $a(t)$ is

$$a(t) = -\frac{d^2x_i}{dt^2} = a_i\omega_i^2 \sin \omega_i t + b_i\omega_i^2 \cos \omega_i t. \tag{10.38}$$

Thus we obtain an equation of motion identical in form to that for a spring-mass oscillator subjected to an oscillatory force. Comparing Eq. (10.38) to Eq. (10.27), we can obtain the amplitude of the particular (steady-state) solution from Eq. (10.31) by setting $a_0 = a_i\omega_i^2$, $b_0 = b_i\omega_i^2$, and $\omega_0 = \omega_i$:

$$E_p = \frac{\omega_i^2\sqrt{a_i^2 + b_i^2}}{\sqrt{(\omega^2 - \omega_i^2)^2 + 4d^2\omega_i^2}}. \tag{10.39}$$

Therefore the steady-state amplitude of the displacement of the mass relative to its base is

$$E_{\mathrm{p}} = \frac{(10)^2\sqrt{(0.1)^2 + (0.1)^2}}{\sqrt{[(2)^2 - (10)^2]^2 + 4(1)^2(10)^2}} = 0.144 \text{ m}.$$

DESIGN ISSUES

A microphone transforms sound waves into a varying voltage that can be recorded or transformed back into sound waves by a speaker. A device that transforms a mechanical input into an electromagnetic output, or an electromagnetic input into a mechanical output, is called a **transducer**. Transducers can be used to measure displacements, velocities, and accelerations by transforming them into measurable voltages or currents.

In Fig. 10.21, the coordinate x_{i} is the displacement to be measured, the input. To use the spring-mass oscillator as a tranducer, it would be designed to produce a voltage or current proportional to the displacement x, the output. If the relationship between the input and output is known, the displacement x_{i} can be determined. Some seismographs (Fig. 10.22) measure motions of the earth in this way.

Fig. 10.22
A seismograph that measures the local displacement of the earth.

If the input is an oscillatory displacement given by Eq. (10.37), the amplitude of the output is given by Eq. (10.39). We can write the latter equation as

$$\frac{E_{\mathrm{p}}}{E_{\mathrm{i}}} = \frac{(\omega_{\mathrm{i}}/\omega)^2}{\sqrt{[1 - (\omega_{\mathrm{i}}/\omega)^2]^2 + 4(d/\omega)^2(\omega_{\mathrm{i}}/\omega)^2}},$$

where $E_{\mathrm{i}} = \sqrt{a_{\mathrm{i}}^2 + b_{\mathrm{i}}^2}$ is the amplitude of the input. In Fig. 10.23 we show the ratio $E_{\mathrm{p}}/E_{\mathrm{i}}$ as a function of the ratio of the input frequency to the natural frequency of the undamped system, $\omega_{\mathrm{i}}/\omega$, for several values of d/ω. If the parameters of the spring-mass oscillator are known, you can use a graph of this type to determine the amplitude of the input by measuring the amplitude of the output.

In practice, the input displacement does not usually have a single frequency, but consists of a combination of different frequencies or even a continuous *spectrum* of frequencies. For example, the displacements resulting from earthquakes have a spectrum of frequencies. In that case, it is desirable for the ratio of the output amplitude to the input amplitude to be approximately equal to one over the range of the input frequencies. The response of the instrument is said to be "flat." In Fig. 10.23 you can see that the response is approximately flat for frequencies ω_{i} greater than about 2ω if the damping of the system is chosen so that d/ω is in the range 0.6–0.7. Also, notice that making the natural frequency ω small increases the range of input frequencies over which the response of the instrument is flat. For that reason, seismographs are often designed with large masses and relatively weak springs.

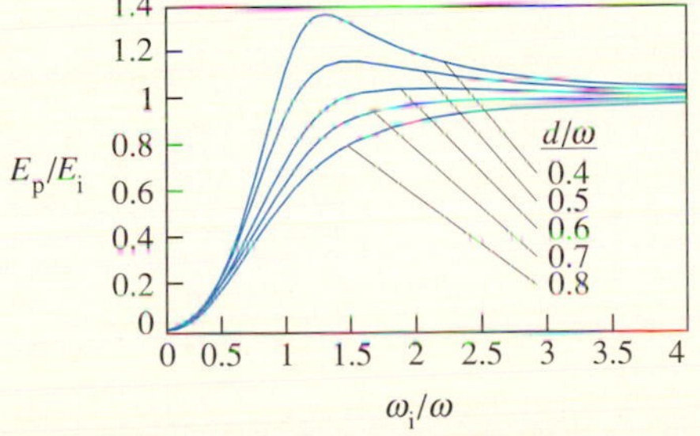

Fig. 10.23
Ratio of the output amplitude to the input amplitude.

Problems

10.58 The mass $m = 2$ slugs and $k = 200$ lb/ft. Let x be the position of the mass relative to its position when the spring is unstretched. The force $F(t) = 36 \sin 8t$ lb.
(a) Determine the particular solution.
(b) At $t = 0$, $x = 1$ ft and the velocity of the mass is zero. Determine x as a function of time.

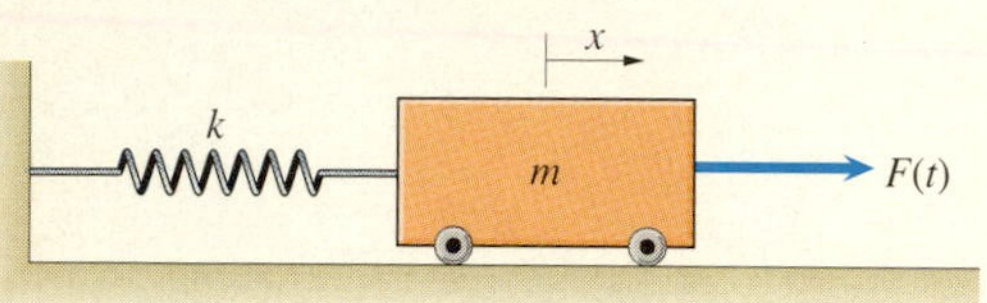

P10.58

10.59 Consider the spring-mass oscillator in Problem 10.58. The mass is $m = 2$ slugs and $k = 200$ lb/ft. The mass is initially stationary with the spring unstretched. At $t = 0$, a force $F(t) = 200 - 80t^2$ lb is applied to the mass.
(a) What is the steady-state (particular) solution?
(b) Determine the position of the mass as a function of time.

10.60 The damped spring-mass oscillator is initially stationary with the spring unstretched. At $t = 0$, a constant force $F(t) = 6$ N is applied to the mass.
(a) What is the steady-state (particular) solution?
(b) Determine the position of the mass as a function of time.

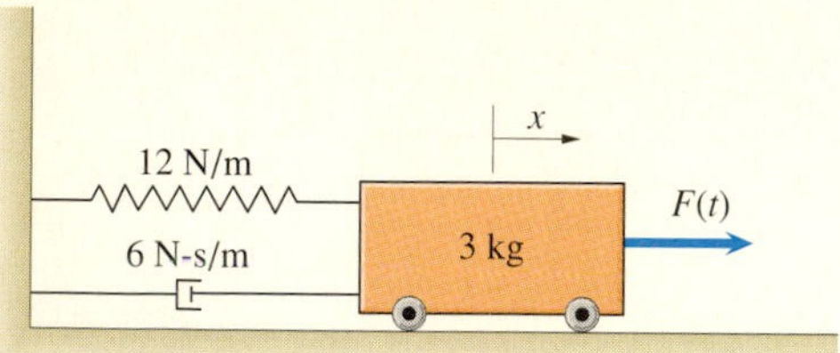

P10.60

10.61 The damped spring-mass oscillator shown in Problem 10.60 is initially stationary with the spring unstretched. At $t = 0$, a force $F(t) = 6 \cos 1.6t$ N is applied to the mass.
(a) What is the steady-state (particular) solution?
(b) Determine the position of the mass as a function of time.

10.62 The disk has moment of inertia $I = 3$ kg-m^2. It rotates about a fixed shaft and is attached to a torsional spring with constant $k = 20$ N-m/rad. At $t = 0$, the angle $\theta = 0$, the angular velocity is $d\theta/dt = 4$ rad/s, and the disk is subjected to a couple $M(t) = 10 \sin 2t$ N-m. Determine θ as a function of time.

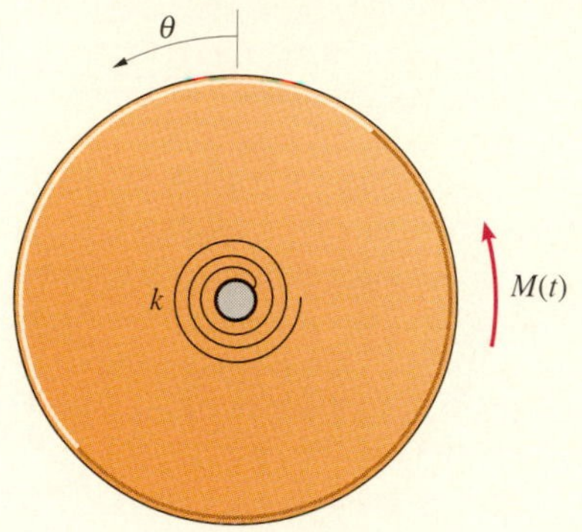

P10.62

10.63 The stepped disk weighs 20 lb and its moment of inertia is $I = 0.6$ slug-ft^2. It rolls on the horizontal surface. The disk is initially stationary with the spring unstretched, and at $t = 0$ a constant force $F = 10$ lb is applied as shown. Determine the position of the center of the disk as a function of time.

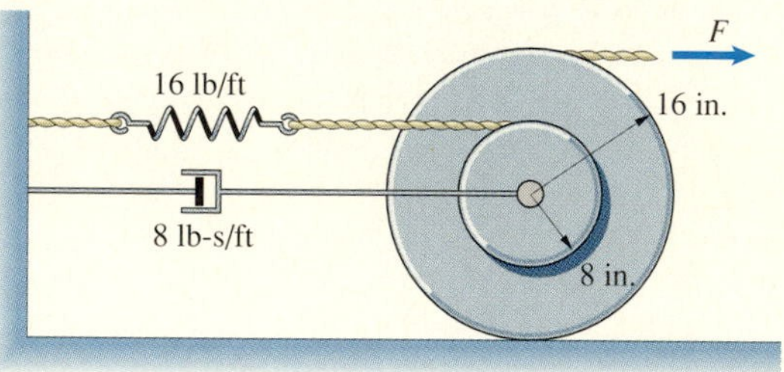

P10.63

10.64 An electric motor is bolted to a metal table. When the motor is on, it causes the tabletop to vibrate horizontally. Assume that the legs of the table behave like linear springs and neglect damping. The total weight of the motor and the tabletop is 150 lb. When the motor is not turned on, the frequency of horizontal vibration of the tabletop and motor is 5 Hz. When the motor is running at 600 rpm, the amplitude of the horizontal vibration is 0.01

in. What is the magnitude of the oscillatory force exerted on the table by the motor at this speed?

P10.64

10.65 The moments of inertia of gears A and B are $I_A = 0.014$ slug-ft^2 and $I_B = 0.100$ slug-ft^2. Gear A is connected to a torsional spring with constant $k = 2$ ft-lb/rad. The system is in equilibrium at $t = 0$ when it is subjected to an oscillatory force $F(t) = 4 \sin 3t$ lb. What is the downward displacement of the 5-lb weight as a function of time?

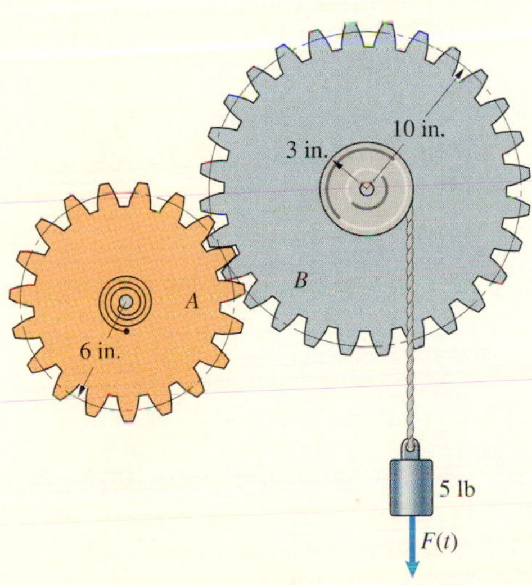

P10.65

10.66 A 1.5-kg cylinder is mounted on a "sting" in a wind tunnel with the cylinder axis transverse to the flow direction. When there is no flow, a 10-N vertical force applied to the cylinder causes it to deflect 0.15 mm. When air flows in the wind tunnel, vortices subject the cylinder to alternating lateral forces. The velocity of the air is 5 m/s, the distance between vortices is 80 mm, and the magnitude of the lateral forces is 1 N. If you model the lateral forces by

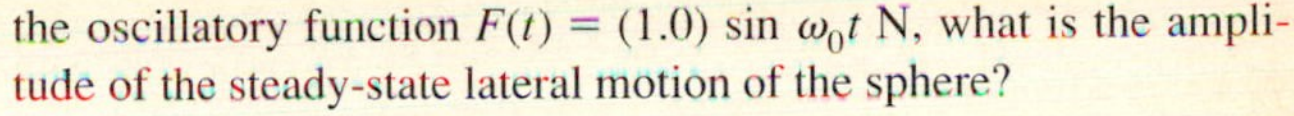

the oscillatory function $F(t) = (1.0) \sin \omega_0 t$ N, what is the amplitude of the steady-state lateral motion of the sphere?

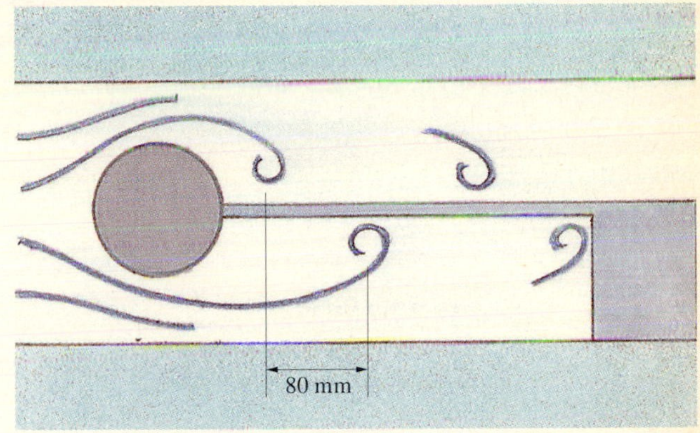

P10.66

10.67 Show that the amplitude of the particular solution given by Eq. (10.31) is a maximum when the frequency of the oscillatory forcing function is $\omega_0 = \sqrt{\omega^2 - 2d^2}$.

10.68 A sonobuoy (sound-measuring device) floats in a standing wave tank. It is a cylinder of mass m and cross-sectional area A. The water density is ρ, and the buoyancy force supporting the buoy equals the weight of the water that would occupy the volume of the part of the cylinder below the surface. When the water in the tank is stationary, the buoy is in equilibrium in the vertical position shown in Fig. (a). Waves are then generated in the tank, causing the depth of the water at the sonobuoy's position *relative to its original depth* to be $d = d_0 \sin \omega_0 t$. Let y be the sonobuoy's vertical position relative to its original position. Show that the sonobuoy's vertical position is governed by the equation

$$\frac{d^2y}{dt^2} + \left(\frac{A\rho g}{m}\right) y = \left(\frac{A\rho g}{m}\right) d_0 \sin \omega_0 t.$$

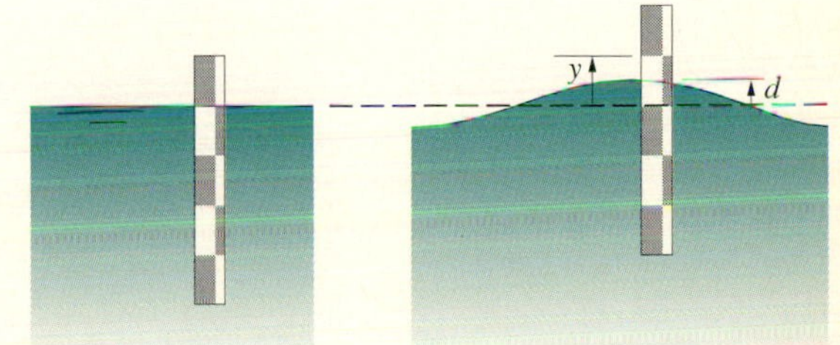

P10.68

10.69 Suppose that the mass of the sonobuoy in Problem 10.68 is $m = 10$ kg, its diameter is 125 mm, and the water density is $\rho = 1025$ kg/m^3. If $d = 0.1 \sin 2t$ m, what is the magnitude of the steady-state vertical vibrations of the sonobuoy?

Problems 10.70–10.73 are related to Example 10.7.

10.70 The mass in Fig. 10.21 weighs 50 lb. The spring constant is $k = 200$ lb/ft, and $c = 10$ lb-s/ft. If the base is subjected to an oscillatory displacement x_i of amplitude 10 in. and circular frequency $\omega_i = 15$ rad/s, what is the resulting steady-state amplitude of the displacement of the mass relative to the base?

10.71 The mass in Fig. 10.21 is 100 kg. The spring constant is $k = 4$ N/m, and $c = 24$ N-s/m. The base is subjected to an oscillatory displacement of circular frequency $\omega_i = 0.2$ rad/s. The steady-state amplitude of the displacement of the mass relative to the base is measured and determined to be 200 mm. What is the amplitude of the displacement of the base?

10.72 A team of engineering students builds the simple seismograph shown. The coordinate x_i measures the local horizontal ground motion. The coordinate x measures the position of the mass relative to the frame of the seismograph. The spring is unstretched when $x = 0$. The mass $m = 1$ kg, $k = 10$ N/m, and $c = 2$ N-s/m. Suppose that the seismograph is initially stationary and at $t = 0$ it is subjected to an oscillatory ground motion $x_i = 10 \sin 2t$ mm. What is the amplitude of the steady-state response of the mass?

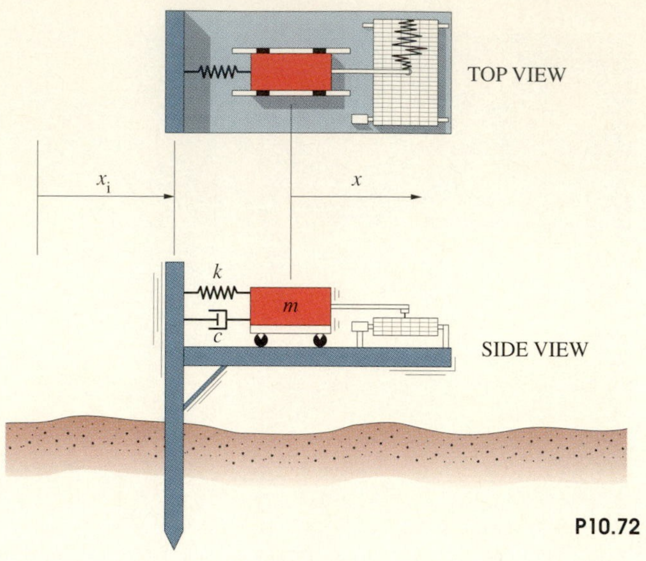

P10.72

10.73 In Problem 10.72, determine the position x of the mass relative to the base as a function of time.

Chapter Summary

Conservative Systems

Small vibrations of many one-degree-of-freedom conservative systems relative to an equilibrium position are governed by the equation

$$\frac{d^2x}{dt^2} + \omega^2 x = 0, \qquad \textbf{Eq. (10.4)}$$

where ω is a constant determined by the properties of the system. Its general solution is

$$x = A \sin \omega t + B \cos \omega t, \qquad \textbf{Eq. (10.5)}$$

where A and B are constants. Its general solution can also be expressed in the form

$$x = E \sin (\omega t - \phi), \qquad \textbf{Eq. (10.6)}$$

where the constants E and ϕ are related to A and B by

$$A = E \cos \phi, \qquad B = -E \sin \phi. \qquad \textbf{Eq. (10.7)}$$

The **amplitude** of the vibration is

$$E = \sqrt{A^2 + B^2}. \qquad \textbf{Eq. (10.8)}$$

The **period** τ of the vibration is the time required for one complete oscillation, or **cycle**. The **natural frequency** f is the number of cycles per unit time. The period and natural frequency are related to ω by

$$\tau = \frac{2\pi}{\omega}, \qquad \textbf{Eq. (10.9)}$$

$$f = \frac{\omega}{2\pi}. \qquad \textbf{Eq. (10.10)}$$

The term $\omega = 2\pi f$ is called the **circular natural frequency**.

Damped Vibrations

Small vibrations of many damped one-degree-of-freedom systems relative to an equilibrium position are governed by the equation

$$\frac{d^2x}{dt^2} + 2d\frac{dx}{dt} + \omega^2 x = 0. \qquad \textbf{Eq. (10.16)}$$

Subcritical Damping If $d < \omega$, the system is said to be subcritically damped. In this case, the general solution of Eq. (10.16) is

$$x = e^{-dt}(A \sin \omega_d t + B \cos \omega_d t), \qquad \textbf{Eq. (10.19)}$$

where A and B are constants and ω_d is defined by

$$\omega_d = \sqrt{\omega^2 - d^2}. \qquad \textbf{Eq. (10.18)}$$

The period and frequency of the damped vibrations are

$$\tau_d = \frac{2\pi}{\omega_d}, \qquad f_d = \frac{\omega_d}{2\pi}. \qquad \textbf{Eq. (10.20)}$$

Critical and Supercritical Damping If $d > \omega$, the system is said to be supercritically damped. The general solution is

$$x = Ce^{-(d-h)t} + De^{-(d+h)t}, \qquad \textbf{Eq. (10.24)}$$

where C and D are constants and h is defined by

$$h = \sqrt{d^2 - \omega^2}. \qquad \textbf{Eq. (10.23)}$$

If $d = \omega$, the system is said to be critically damped. The general solution is

$$x = Ce^{-dt} + Dte^{-dt}, \qquad \textbf{Eq. (10.25)}$$

where C and D are constants.

Forced Vibrations

The forced vibrations of many damped, one-degree-of-freedom systems are governed by the equation

$$\frac{d^2x}{dt^2} + 2d\frac{dx}{dt} + \omega^2 x = a(t), \qquad \textbf{Eq. (10.26)}$$

where $a(t)$ is the **forcing function**. The general solution of Eq. (10.26) consists of the homogeneous and particular solutions:

$$x = x_\text{h} + x_\text{p}.$$

The **homogeneous solution** x_h is the general solution of Eq. (10.26) with the right side set equal to zero, and the **particular solution** x_p is a solution that satisfies Eq. (10.26).

Oscillatory Forcing Function If $a(t)$ is an oscillatory function of the form

$$a(t) = a_0 \sin \omega_0 t + b_0 \cos \omega_0 t,$$

where a_0, b_0, and ω_0 are constants, the particular solution is (Eq. 10.30)

$$x_\text{p} = \left[\frac{(\omega^2 - \omega_0^2)a_0 + 2d\omega_0 b_0}{(\omega^2 - \omega_0^2)^2 + 4d^2\omega_0^2}\right] \sin \omega_0 t + \left[\frac{-2d\omega_0 a_0 + (\omega^2 - \omega_0^2)b_0}{(\omega^2 - \omega_0^2)^2 + 4d^2\omega_0^2}\right] \cos \omega_0 t,$$

and its amplitude is (Eq. 10.31)

$$E_\text{p} = \frac{\sqrt{a_0^2 + b_0^2}}{\sqrt{(\omega^2 - \omega_0^2)^2 + 4d^2\omega_0^2}}.$$

The particular solution for the motion of a damped vibrating system subjected to an oscillatory external force is also called the **steady-state solution**. The motion approaches the steady-state solution with increasing time.

Polynomial Forcing Function If $a(t)$ is a polynomial of the form

$$a(t) = a_0 + a_1 t + a_2 t^2 + \cdots + a_N t^N,$$

where $a_1, a_2, \ldots, a_N$ are constants, the particular solution can be obtained by seeking a solution of the same form:

$$x_\text{p} = A_0 + A_1 t + A_2 t^2 + \cdots + A_N t^N, \qquad \textbf{Eq. (10.32)}$$

where $A_0, A_1, A_2, \ldots, A_N$ are constants that must be determined.

Review Problems

10.74 The coordinate x measures the displacement of the mass relative to the position in which the spring is unstretched. The mass is given the initial conditions

$$t = 0 \begin{cases} x = 0.1 \text{ m}, \\ \dfrac{dx}{dt} = 0. \end{cases}$$

(a) Determine the position of the mass as a function of time.
(b) Draw graphs of the position and velocity of the mass as functions of time for the first 5 s of motion.

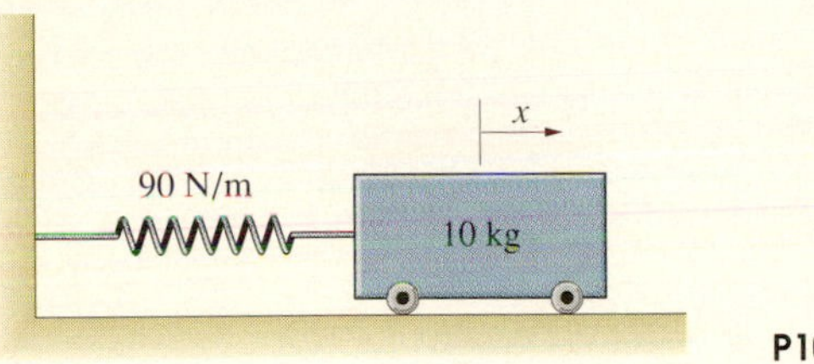

P10.74

10.75 When $t = 0$, the mass in Problem 10.74 is in the position in which the spring is unstretched and has a velocity of 0.3 m/s to the right. Determine the position of the mass as a function of time and the amplitude of the vibration: (a) by expressing the solution in the form given by Eq. (10.5); (b) by expressing the solution in the form given by Eq. (10.6).

10.76 A homogeneous disk of mass m and radius R rotates about a fixed shaft and is attached to a torsional spring with constant k. (The torsional spring exerts a restoring moment of magnitude $k\theta$, where θ is the angle of rotation of the disk relative to its position in which the spring is unstretched.) Show that the period of rotational vibrations of the disk is

$$\tau = \pi R\sqrt{2m/k}.$$

P10.76

10.77 Assigned to determine the moments of inertia of astronaut candidates, an engineer attaches a horizontal platform to a vertical steel bar. The moment of inertia of the platform about L is 7.5 kg-m^2, and the natural frequency of torsional oscillations of the unloaded platform is 1 Hz. With an astronaut candidate in the position shown, the natural frequency of torsional oscillations is 0.520 Hz. What is the candidate's moment of inertia about L?

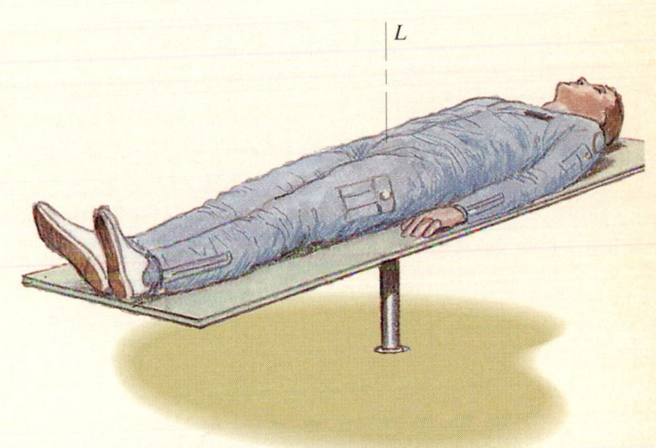

P10.77

10.78 The 22-kg platen P rests on four roller bearings. The roller bearings can be modeled as 1-kg homogeneous cylinders with 30-mm radii, and the spring constant is $k = 900$ N/m. What is the natural frequency of horizontal vibrations of the platen relative to its equilibrium position?

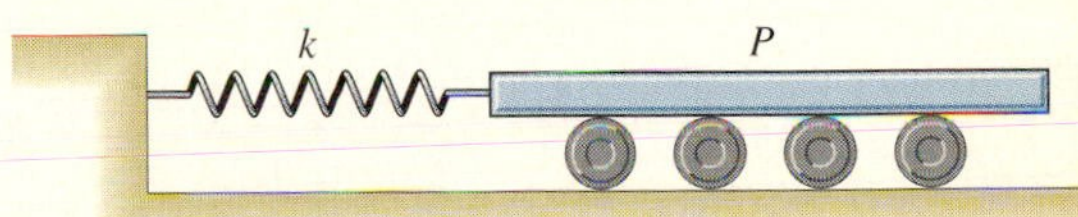

P10.78

10.79 At $t = 0$, the platen described in Problem 10.78 is 0.1 m to the left of its equilibrium position and is moving to the right at 2 m/s. What are the platen's position and velocity at $t = 4$ s?

10.80 The moments of inertia of gears A and B are $I_A = 0.014$ slug-ft^2 and $I_B = 0.100$ slug-ft^2. Gear A is connected to a torsional spring with constant $k = 2$ ft-lb/rad. What is the natural frequency of small angular vibrations of the gears relative to their equilibrium position?

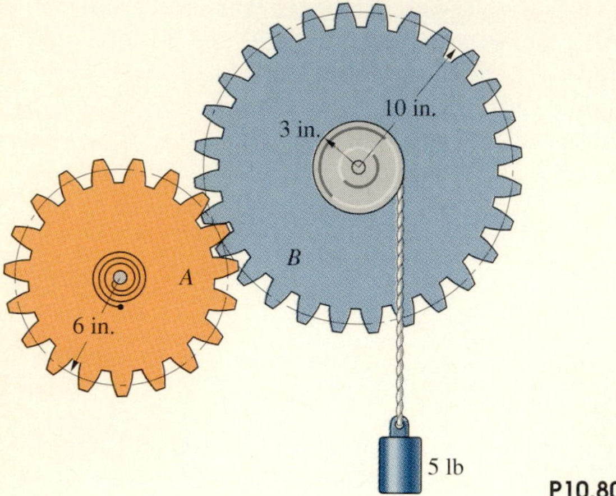

P10.80

10.81 The 5-lb weight in Problem 10.80 is raised 0.5 in. from its equilibrium position and released from rest at $t = 0$. Determine the counterclockwise angular position of gear B relative to its equilibrium position as a function of time.

10.82 The mass of the slender bar is m. The spring is unstretched when the bar is vertical. The light collar C slides on the smooth vertical bar so that the spring remains horizontal. Determine the natural frequency of small vibrations of the bar.

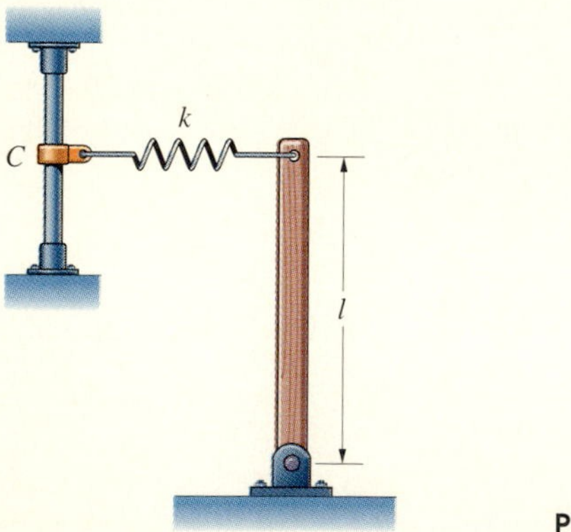

P10.82

10.83 A homogeneous hemisphere of radius R and mass m rests on a level surface. If you rotate the hemisphere slightly from its equilibrium position and release it, what is the natural frequency of its vibrations?

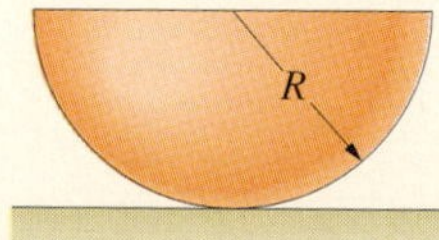

P10.83

10.84 The frequency of the spring-mass oscillator is measured and determined to be 4.00 Hz. The spring-mass oscillator is then placed in a barrel of oil, and its frequency is determined to be 3.80 Hz. What is the logarithmic decrement of vibrations of the mass when the oscillator is immersed in oil?

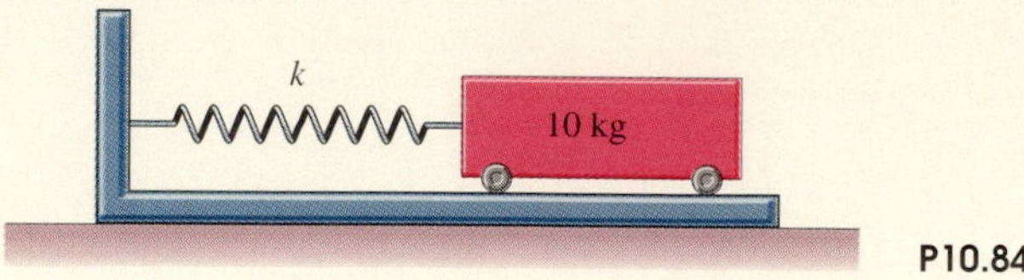

P10.84

10.85 Consider the oscillator immersed in oil described in Problem 10.84. If the mass is displaced 0.1 m to the right of its equilibrium position and released from rest, what is its position relative to the equilibrium position as a function of time?

10.86 The stepped disk weighs 20 lb, and its moment of inertia is $I = 0.6$ slug-ft^2. It rolls on the horizontal surface. If $c = 8$ lb-s/ft, what is the frequency of small vibrations of the disk?

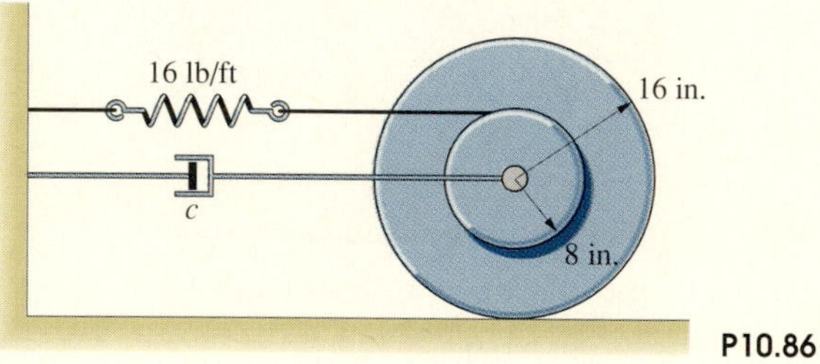

P10.86

10.87 The stepped disk described in Problem 10.86 is initially in equilibrium, and at $t = 0$ it is given a clockwise angular velocity of 1 rad/s. Determine the position of the center of the disk relative to its equilibrium position as a function of time.

10.88 The stepped disk described in Problem 10.86 is initially in equilibrium, and at $t = 0$ it is given a clockwise angular velocity of 1 rad/s. Determine the position of the center of the disk relative to its equilibrium position as a function of time if $c = 16$ lb-s/ft.

10.89 The 22-kg platen P rests on four roller bearings. The roller bearings can be modeled as 1-kg homogeneous cylinders with 30-mm radii, and the spring constant is $k = 900$ N/m. The platen is subjected to a force $F(t) = 100 \sin 3t$ N. What is the magnitude of the platen's steady-state horizontal vibration?

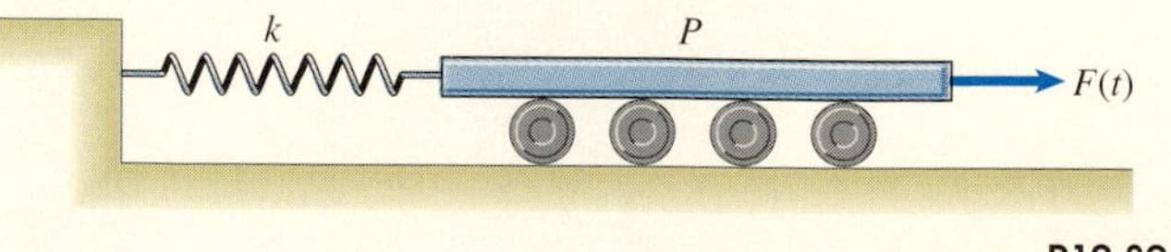

P10.89

10.90 At $t = 0$, the platen described in Problem 10.91 is 0.1 m to the right of its equilibrium position and is moving to the right at 2 m/s. Determine the platen's position relative to its equilibrium position as a function of time.

10.91 The moments of inertia of gears A and B are $I_A = 0.014$ slug-ft^2 and $I_B = 0.100$ slug-ft^2. Gear A is connected to a torsional spring with constant $k = 2$ ft-lb/rad. The bearing supporting gear B incorporates a damping element that exerts a resisting moment on gear B of magnitude $1.5(d\theta_B/dt)$ ft-lb, where $d\theta_B/dt$ is the angular velocity of gear B in rad/s. What is the frequency of small angular vibrations of the gears?

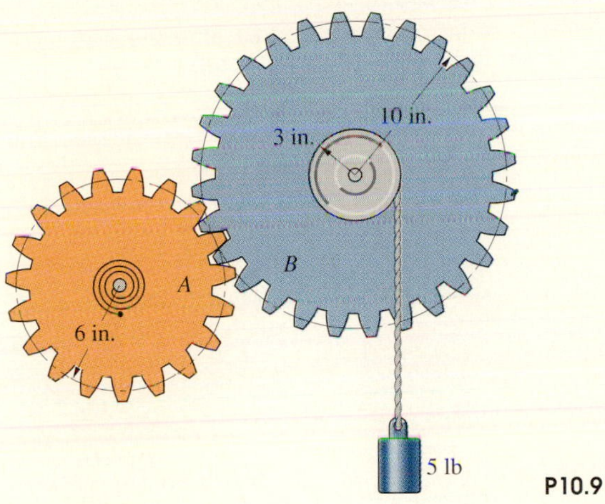

P10.91

10.92 The 5-lb weight in Problem 10.89 is raised 0.5 in. from its equilibrium position and released from rest at $t = 0$. Determine the counterclockwise angular position of gear B relative to its equilibrium position as a function of time.

10.93 The base and mass m are initially stationary. The base is then subjected to a vertical displacement $h \sin \omega_i t$ relative to its original position. What is the magnitude of the resulting steady-state vibration of the mass m *relative to the base*?

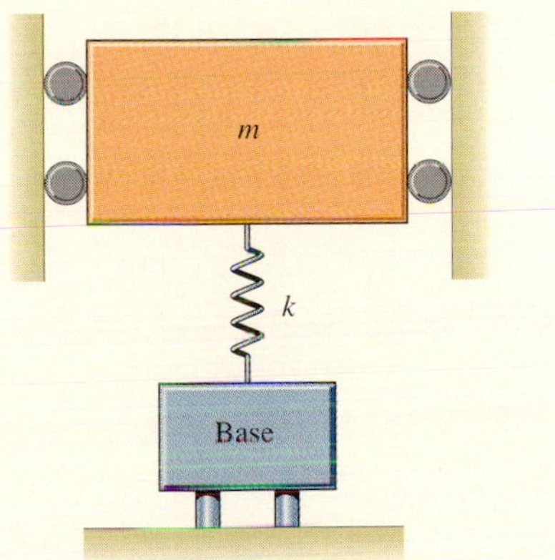

P10.93

10.94 The mass of the trailer, not including its wheels and axle, is m, and the spring constant of its suspension is k. To analyze the suspension's behavior, an engineer assumes that the height of the road surface relative to its mean height is $h \sin(2\pi x/\lambda)$. Assume that the trailer's wheels remain on the road and its horizontal component of velocity is v. Neglect the damping due to the suspension's shock absorbers.

(a) Determine the magnitude of the trailer's vertical steady-state vibration *relative to the road surface*.
(b) At what velocity v does resonance occur?

P10.94

10.95 The trailer in Problem 10.94, not including its wheels and axle, weighs 1000 lb. The spring constant of its suspension is $k = 2400$ lb/ft, and the damping coefficient due to its shock absorbers is $c = 200$ lb-s/ft. The road surface parameters are $h = 2$ in. and $\lambda = 8$ ft. The trailer's horizontal velocity is $v = 6$ mi/hr. Determine the magnitude of the trailer's vertical steady-state vibration relative to the road surface: (a) neglecting the damping due to the shock absorbers; (b) not neglecting the damping.

10.96 A disk with moment of inertia I rotates about a fixed shaft and is attached to a torsional spring with constant k. The angle θ measures the angular position of the disk relative to its position when the spring is unstretched. The disk is initially stationary with the spring unstretched. At $t = 0$, a time-dependent moment $M(t) = M_0(1 - e^{-t})$ is applied to the disk, where M_0 is a constant. Show that the angular position of the disk as a function of time is

$$\theta = \frac{M_0}{I}\left[-\frac{1}{\omega(1+\omega^2)}\sin \omega t - \frac{1}{\omega^2(1+\omega^2)}\cos \omega t + \frac{1}{\omega^2} - \frac{1}{(1+\omega^2)}e^{-t}\right].$$

Strategy: To determine the particular solution, seek a solution of the form

$$\theta_p = A_p + B_p e^{-t},$$

where A_p and B_p are constants that you must determine.

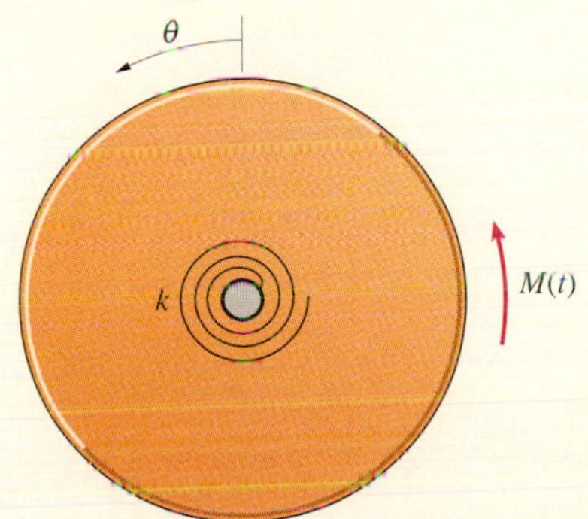

P10.96

Appendix A

Review of Mathematics

A.1 Algebra

Quadratic Equations

The solutions of the quadratic equation

$$ax^2 + bx + c = 0$$

are

$$x = \frac{-b \pm \sqrt{b^2 - 4ac}}{2a}.$$

Natural Logarithms

The natural logarithm of a positive real number x is denoted by $\ln x$. It is defined to be the number such that

$$e^{\ln x} = x,$$

where $e = 2.7182\ldots$ is the base of natural logarithms.

Logarithms have the following properties:

$$\ln (xy) = \ln x + \ln y,$$

$$\ln (x/y) = \ln x - \ln y,$$

$$\ln y^x = x \ln y.$$

A.2 Trigonometry

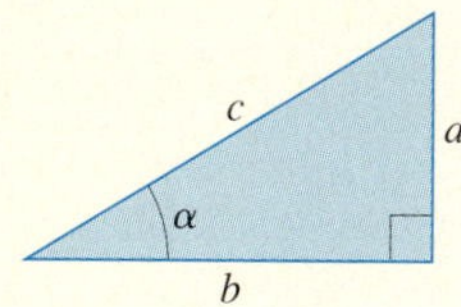

The trigonometric functions for a right triangle are

$$\sin \alpha = \frac{1}{\csc \alpha} = \frac{a}{c}, \qquad \cos \alpha = \frac{1}{\sec \alpha} = \frac{b}{c}, \qquad \tan \alpha = \frac{1}{\cot \alpha} = \frac{a}{b}.$$

The sine and cosine satisfy the relation

$$\sin^2 \alpha + \cos^2 \alpha = 1,$$

and the sine and cosine of the sum and difference of two angles satisfy

$$\sin (\alpha + \beta) = \sin \alpha \cos \beta + \cos \alpha \sin \beta,$$

$$\sin (\alpha - \beta) = \sin \alpha \cos \beta - \cos \alpha \sin \beta,$$

$$\cos (\alpha + \beta) = \cos \alpha \cos \beta - \sin \alpha \sin \beta,$$

$$\cos (\alpha - \beta) = \cos \alpha \cos \beta + \sin \alpha \sin \beta.$$

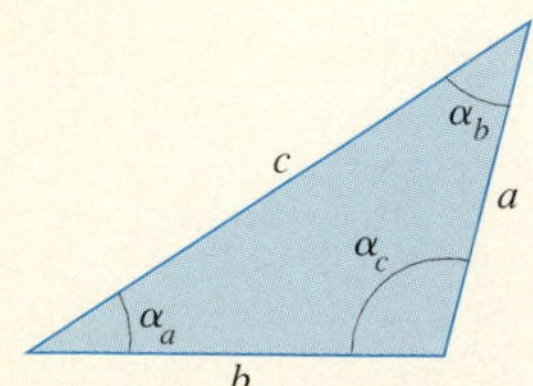

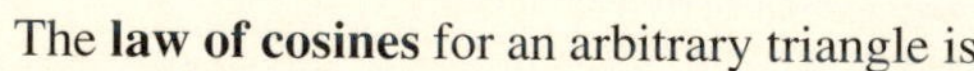

The **law of cosines** for an arbitrary triangle is

$$c^2 = a^2 + b^2 - 2ab \cos \alpha_c,$$

and the **law of sines** is

$$\frac{\sin \alpha_a}{a} = \frac{\sin \alpha_b}{b} = \frac{\sin \alpha_c}{c}.$$

A.3 *Derivatives*

$$\frac{d}{dx}x^n = nx^{n-1} \qquad \frac{d}{dx}\sin x = \cos x \qquad \frac{d}{dx}\sinh x = \cosh x$$

$$\frac{d}{dx}e^x = e^x \qquad \frac{d}{dx}\cos x = -\sin x \qquad \frac{d}{dx}\cosh x = \sinh x$$

$$\frac{d}{dx}\ln x = \frac{1}{x} \qquad \frac{d}{dx}\tan x = \frac{1}{\cos^2 x} \qquad \frac{d}{dx}\tanh x = \frac{1}{\cosh^2 x}$$

A.4 *Integrals*

$$\int x^n\,dx = \frac{x^{n+1}}{n+1} \qquad (n \neq -1)$$

$$\int x^{-1}\,dx = \ln x$$

$$\int (a+bx)^{1/2}\,dx = \frac{2}{3b}(a+bx)^{3/2}$$

$$\int x(a+bx)^{1/2}\,dx = -\frac{2(2a-3bx)(a+bx)^{3/2}}{15b^2}$$

$$\int (1+a^2x^2)^{1/2}\,dx = \frac{1}{2}\left\{x(1+a^2x^2)^{1/2} + \frac{1}{a}\ln\left[x + \left(\frac{1}{a^2}+x^2\right)^{1/2}\right]\right\}$$

$$\int x(1+a^2x^2)^{1/2}\,dx = \frac{a}{3}\left(\frac{1}{a^2}+x^2\right)^{3/2}$$

$$\int x^2(1+a^2x^2)^{1/2}\,dx = \frac{1}{4}ax\left(\frac{1}{a^2}+x^2\right)^{3/2} - \frac{1}{8a^2}x(1+a^2x^2)^{1/2} - \frac{1}{8a^3}\ln\left[x + \left(\frac{1}{a^2}+x^2\right)^{1/2}\right]$$

$$\int (1-a^2x^2)^{1/2}\,dx = \frac{1}{2}\left[x(1-a^2x^2)^{1/2} + \frac{1}{a}\arcsin ax\right]$$

$$\int x(1-a^2x^2)^{1/2}\,dx = -\frac{a}{3}\left(\frac{1}{a^2}-x^2\right)^{3/2}$$

$$\int x^2(a^2-x^2)^{1/2}\,dx = -\frac{1}{4}x(a^2-x^2)^{3/2} + \frac{1}{8}a^2\left[x(a^2-x^2)^{1/2} + a^2\arcsin\frac{x}{a}\right]$$

$$\int \frac{dx}{(1+a^2x^2)^{1/2}} = \frac{1}{a}\ln\left[x + \left(\frac{1}{a^2}+x^2\right)^{1/2}\right]$$

$$\int \frac{dx}{(1-a^2x^2)^{1/2}} = \frac{1}{a}\arcsin ax, \text{ or } -\frac{1}{a}\arccos ax$$

$$\int \sin x\,dx = -\cos x$$

$$\int \cos x\,dx = \sin x$$

$$\int \sin^2 x\,dx = -\frac{1}{2}\sin x\cos x + \frac{1}{2}x$$

$$\int \cos^2 x\,dx = \frac{1}{2}\sin x\cos x + \frac{1}{2}x$$

$$\int \sin^3 x\,dx = -\frac{1}{3}\cos x(\sin^2 x + 2)$$

$$\int \cos^3 x\,dx = \frac{1}{3}\sin x(\cos^2 x + 2)$$

$$\int \cos^4 x\,dx = \frac{3}{8}x + \frac{1}{4}\sin 2x + \frac{1}{32}\sin 4x$$

$$\int \sin^n x\cos x\,dx = \frac{(\sin x)^{n+1}}{n+1} \qquad (n \neq -1)$$

$$\int \sinh x\,dx = \cosh x$$

$$\int \cosh x\,dx = \sinh x$$

$$\int \tanh x\,dx = \ln\cosh x$$

$$\int e^{ax}\,dx = \frac{e^{ax}}{a}$$

$$\int x\,e^{ax}\,dx = \frac{e^{ax}}{a^2}(ax-1)$$

A.5 *Taylor Series*

The Taylor series of a function $f(x)$ is

$$f(a+x) = f(a) + f'(a)x + \frac{1}{2!}f''(a)x^2 + \frac{1}{3!}f'''(a)x^3 + \cdots,$$

where the primes indicate derivatives.

Some useful Taylor series are

$$e^x = 1 + x + \frac{x^2}{2!} + \frac{x^3}{3!} + \cdots,$$

$$\sin(a+x) = \sin a + (\cos a)x - \frac{1}{2}(\sin a)x^2 - \frac{1}{6}(\cos a)x^3 + \cdots,$$

$$\cos(a+x) = \cos a - (\sin a)x - \frac{1}{2}(\cos a)x^2 + \frac{1}{6}(\sin a)x^3 + \cdots,$$

$$\tan(a+x) = \tan a + \left(\frac{1}{\cos^2 a}\right)x + \left(\frac{\sin a}{\cos^3 a}\right)x^2 + \left(\frac{\sin^2 a}{\cos^4 a} + \frac{1}{3\cos^2 a}\right)x^3 + \cdots.$$

A.6 *Vector Analysis*

Cartesian Coordinates

The gradient of a scalar field ψ is

$$\nabla\psi = \frac{\partial\psi}{\partial x}\mathbf{i} + \frac{\partial\psi}{\partial y}\mathbf{j} + \frac{\partial\psi}{\partial z}\mathbf{k}.$$

The divergence and curl of a vector field $\mathbf{v} = v_x\mathbf{i} + v_y\mathbf{j} + v_z\mathbf{k}$ are

$$\nabla\cdot\mathbf{v} = \frac{\partial v_x}{\partial x} + \frac{\partial v_y}{\partial y} + \frac{\partial v_z}{\partial z},$$

$$\nabla\times\mathbf{v} = \begin{vmatrix} \mathbf{i} & \mathbf{j} & \mathbf{k} \\ \dfrac{\partial}{\partial x} & \dfrac{\partial}{\partial y} & \dfrac{\partial}{\partial z} \\ v_x & v_y & v_z \end{vmatrix}.$$

Cylindrical Coordinates

The gradient of a scalar field ψ is

$$\nabla\psi = \frac{\partial\psi}{\partial r}\mathbf{e}_r + \frac{1}{r}\frac{\partial\psi}{\partial\theta}\mathbf{e}_\theta + \frac{\partial\psi}{\partial z}\mathbf{e}_z.$$

The divergence and curl of a vector field $\mathbf{v} = v_r\mathbf{e}_r + v_\theta\mathbf{e}_\theta + v_z\mathbf{e}_z$ are

$$\nabla\cdot\mathbf{v} = \frac{\partial v_r}{\partial r} + \frac{v_r}{r} + \frac{1}{r}\frac{\partial v_\theta}{\partial\theta} + \frac{\partial v_z}{\partial z},$$

$$\nabla\times\mathbf{v} = \frac{1}{r}\begin{vmatrix} \mathbf{e}_r & r\mathbf{e}_\theta & \mathbf{e}_z \\ \dfrac{\partial}{\partial r} & \dfrac{\partial}{\partial\theta} & \dfrac{\partial}{\partial z} \\ v_r & rv_\theta & v_z \end{vmatrix}.$$

Appendix B Properties of Areas and Lines

B.1 Areas

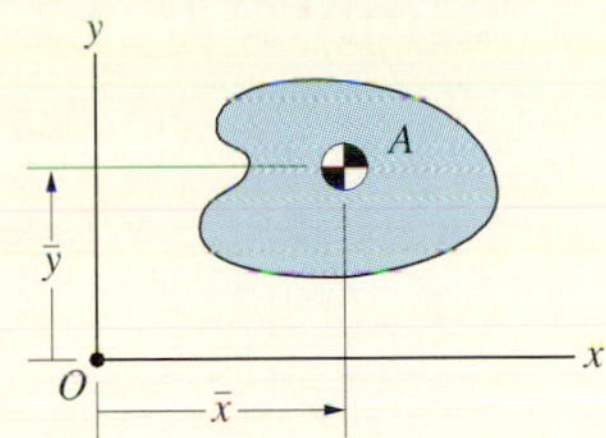

The coordinates of the centroid of the area A are

$$\bar{x} = \frac{\int_A x\,dA}{\int_A dA}, \qquad \bar{y} = \frac{\int_A y\,dA}{\int_A dA}.$$

The moment of inertia about the x axis I_x, the moment of inertia about the y axis I_y, and the product of inertia I_{xy} are

$$I_x = \int_A y^2\,dA, \qquad I_y = \int_A x^2\,dA, \qquad I_{xy} = \int_A xy\,dA.$$

The polar moment of inertia about O is

$$J_O = \int_A r^2\,dA = \int_A (x^2 + y^2)\,dA = I_x + I_y.$$

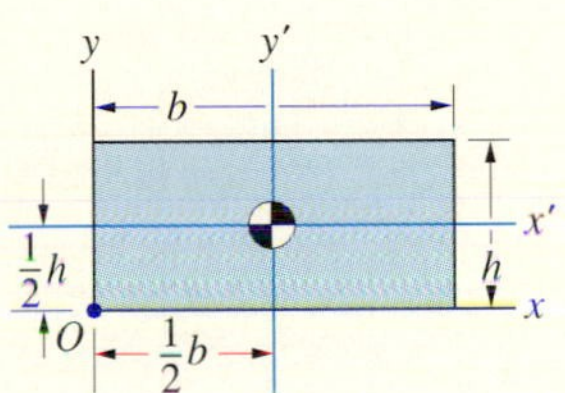

Rectangular Area

Area $= bh$

$$I_x = \frac{1}{3}bh^3, \qquad I_y = \frac{1}{3}hb^3, \qquad I_{xy} = \frac{1}{4}b^2h^2$$

$$I_{x'} = \frac{1}{12}bh^3, \qquad I_{y'} = \frac{1}{12}hb^3, \qquad I_{x'y'} = 0$$

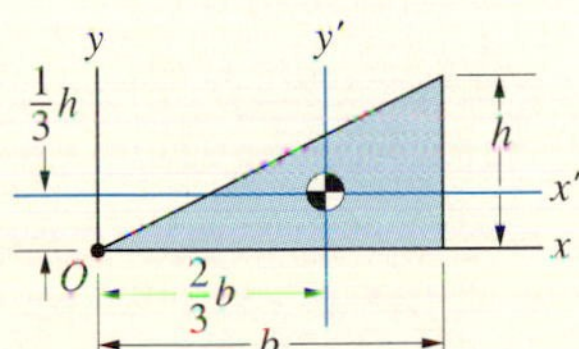

Triangular Area

Area $= \frac{1}{2}bh$

$$I_x = \frac{1}{12}bh^3, \qquad I_y = \frac{1}{4}hb^3, \qquad I_{xy} = \frac{1}{8}b^2h^2$$

$$I_{x'} = \frac{1}{36}bh^3, \qquad I_{y'} = \frac{1}{36}hb^3, \qquad I_{x'y'} = \frac{1}{72}b^2h^2$$

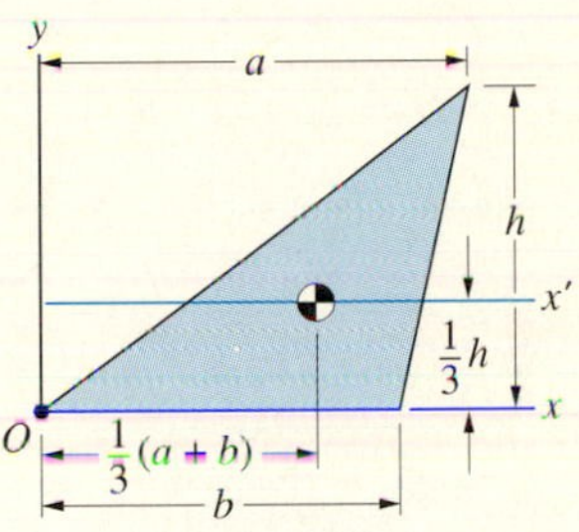

Triangular Area

Area $= \frac{1}{2}bh$

$$I_x = \frac{1}{12}bh^3, \qquad I_{x'} = \frac{1}{36}bh^3$$

Circular Area

Area $= \pi R^2$ $\qquad I_{x'} = I_{y'} = \frac{1}{4}\pi R^4, \qquad I_{x'y'} = 0$

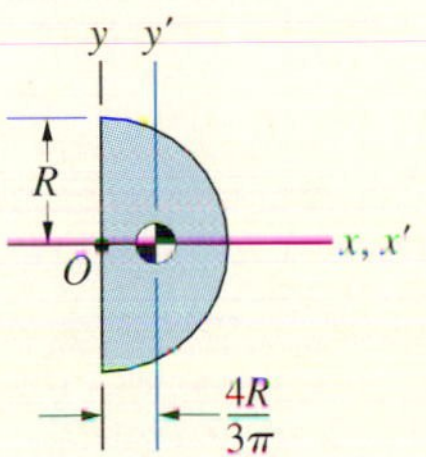

Semicircular Area

Area $= \frac{1}{2}\pi R^2$

$$I_x = I_y = \frac{1}{8}\pi R^4, \qquad I_{xy} = 0$$

$$I_{x'} = \frac{1}{8}\pi R^4, \qquad I_{y'} = \left(\frac{\pi}{8} - \frac{8}{9\pi}\right)R^4, \qquad I_{x'y'} = 0$$

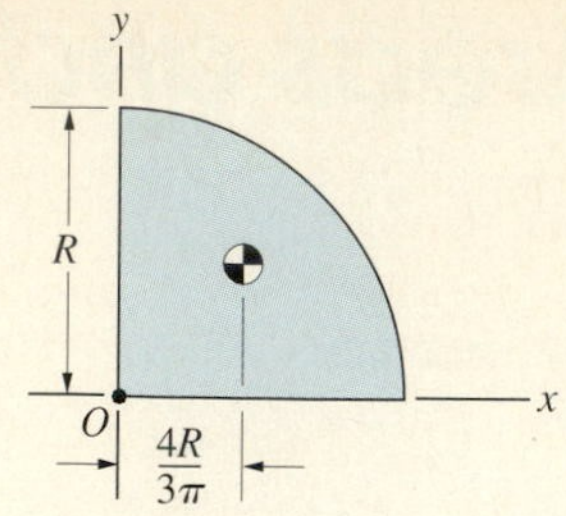

Quarter-Circular Area

$$\text{Area} = \frac{1}{4}\pi R^2$$

$$I_x = I_y = \frac{1}{16}\pi R^4, \qquad I_{xy} = \frac{1}{8}R^4$$

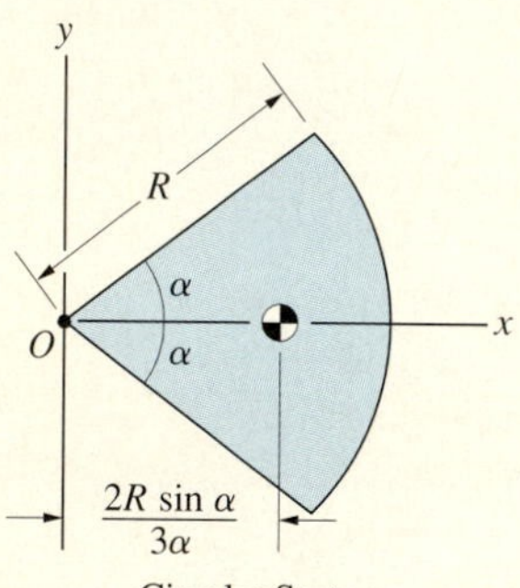

Circular Sector

$$\text{Area} = \alpha R^2$$

$$I_x = \frac{1}{4}R^4\left(\alpha - \frac{1}{2}\sin 2\alpha\right), \qquad I_y = \frac{1}{4}R^4\left(\alpha + \frac{1}{2}\sin 2\alpha\right),$$

$$I_{xy} = 0$$

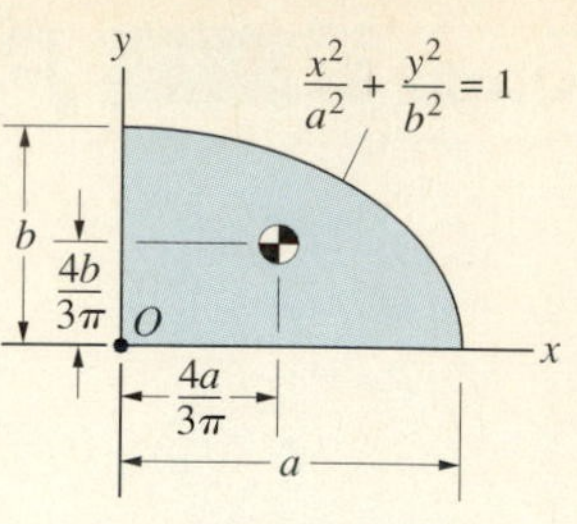

Quarter-Elliptical Area

$$\text{Area} = \frac{1}{4}\pi ab$$

$$I_x = \frac{1}{16}\pi ab^3, \qquad I_y = \frac{1}{16}\pi a^3 b, \qquad I_{xy} = \frac{1}{8}a^2 b^2$$

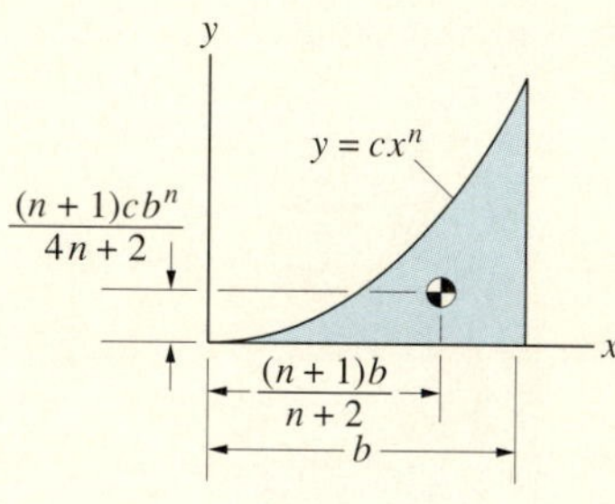

Spandrel

$$\text{Area} = \frac{cb^{n+1}}{n+1}$$

$$I_x = \frac{c^3 b^{3n+1}}{9n+3}, \qquad I_y = \frac{cb^{n+3}}{n+3}, \qquad I_{xy} = \frac{c^2 b^{2n+2}}{4n+4}$$

B.2 *Lines*

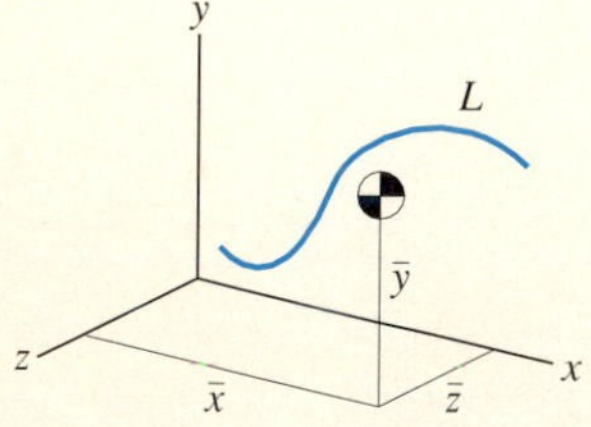

The coordinates of the centroid of the line L are

$$\bar{x} = \frac{\int_L x\,dL}{\int_L dL}, \qquad \bar{y} = \frac{\int_L y\,dL}{\int_L dL}, \qquad \bar{z} = \frac{\int_L z\,dL}{\int_L dL}.$$

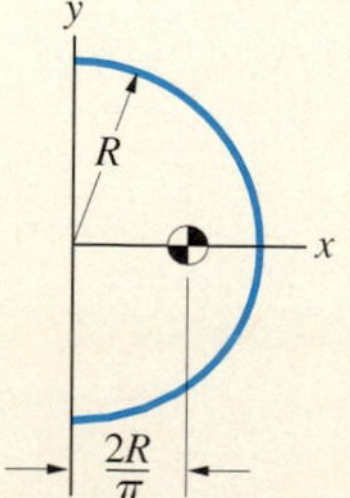

Semicircular Arc

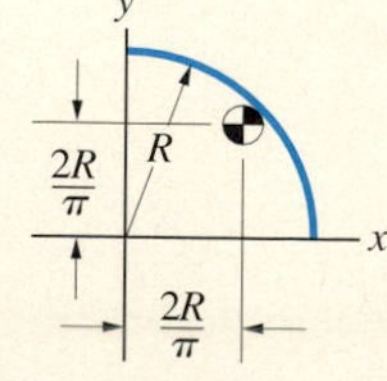

Quarter-Circular Arc

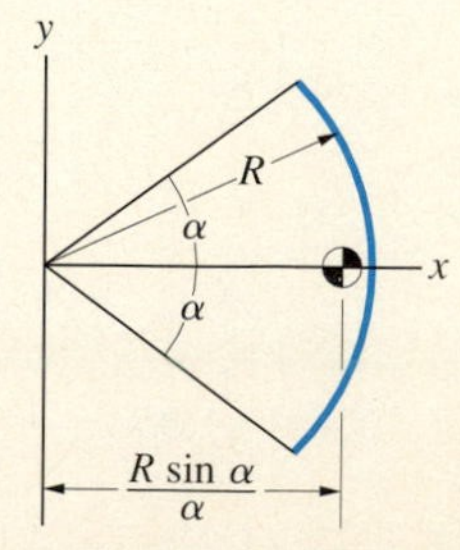

Circular Arc

Appendix C Properties of Volumes and Homogeneous Objects

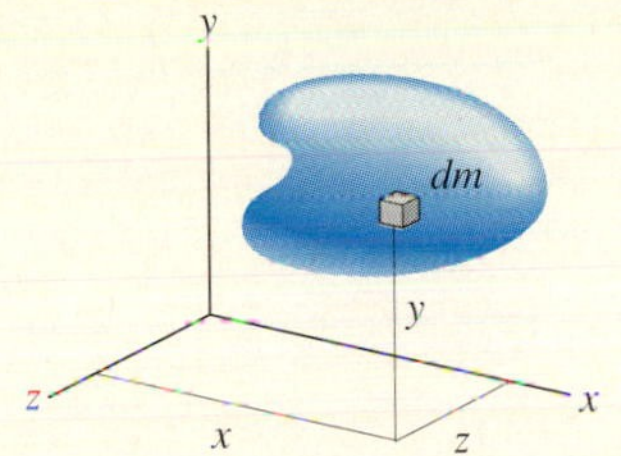

The moments and products of inertia of the object in terms of the xyz coordinate system are

$$I_{(x\text{ axis})} = I_{xx} = \int_m (y^2 + z^2)\, dm,$$

$$I_{(y\text{ axis})} = I_{yy} = \int_m (x^2 + z^2)\, dm,$$

$$I_{(z\text{ axis})} = I_{zz} = \int_m (x^2 + y^2)\, dm,$$

$$I_{xy} = \int_m xy\, dm, \qquad I_{yz} = \int_m yz\, dm,$$

$$I_{zx} = \int_m zx\, dm.$$

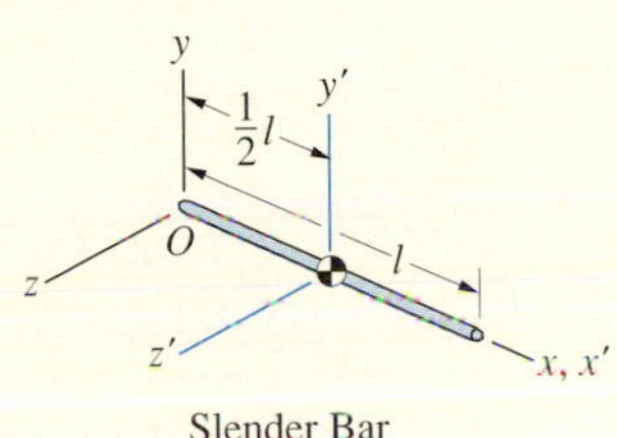

Slender Bar

$$I_{(x\text{ axis})} = 0, \qquad I_{(y\text{ axis})} = I_{(z\text{ axis})} = \frac{1}{3}ml^2,$$

$$I_{xy} = I_{yz} = I_{zx} = 0.$$

$$I_{(x'\text{ axis})} = 0, \qquad I_{(y'\text{ axis})} = I_{(z'\text{ axis})} = \frac{1}{12}ml^2,$$

$$I_{x'y'} = I_{y'z'} = I_{z'x'} = 0.$$

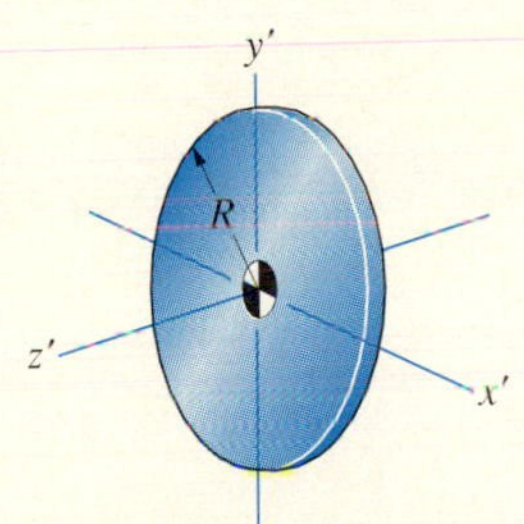

Thin Circular Plate

$$I_{(x'\text{ axis})} = I_{(y'\text{ axis})} = \frac{1}{4}mR^2, \qquad I_{(z'\text{ axis})} = \frac{1}{2}mR^2,$$

$$I_{xy} = I_{yz} = I_{zx} = 0.$$

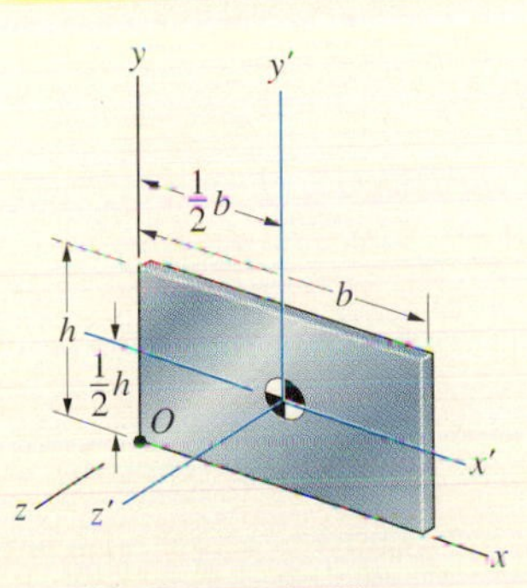

Thin Rectangular Plate

$$I_{(x\text{ axis})} = \frac{1}{3}mh^2, \qquad I_{(y\text{ axis})} = \frac{1}{3}mb^2, \qquad I_{(z\text{ axis})} = \frac{1}{3}m(b^2 + h^2),$$

$$I_{xy} = \frac{1}{4}mbh, \qquad I_{yz} = I_{zx} = 0.$$

$$I_{(x'\text{ axis})} = \frac{1}{12}mh^2, \qquad I_{(y'\text{ axis})} = \frac{1}{12}mb^2, \qquad I_{(z'\text{ axis})} = \frac{1}{12}m(b^2 + h^2),$$

$$I_{x'y'} = I_{y'z'} = I_{z'x'} = 0.$$

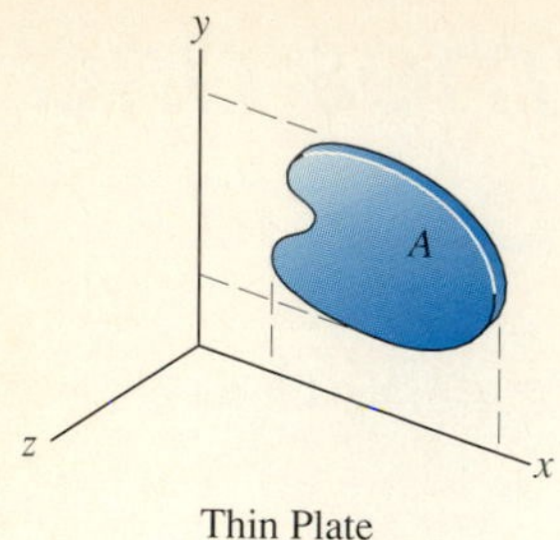

Thin Plate

$$I_{(x \text{ axis})} = \frac{m}{A} I_x^A, \qquad I_{(y \text{ axis})} = \frac{m}{A} I_y^A, \qquad I_{(z \text{ axis})} = I_{(x \text{ axis})} + I_{(y \text{ axis})},$$

$$I_{xy} = \frac{m}{A} I_{xy}^A, \qquad I_{yz} = I_{zx} = 0.$$

(The superscripts A denote moments of inertia of the plate's cross-sectional area A).

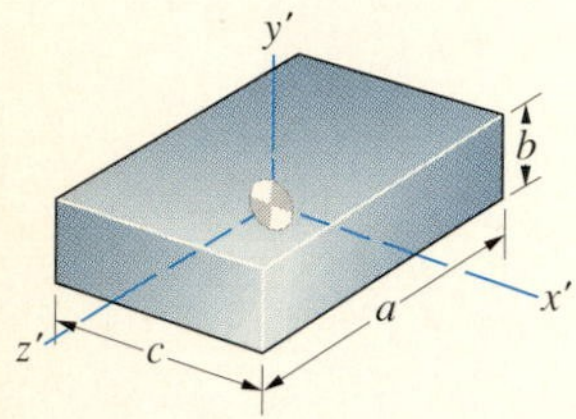

Rectangular Prism

$$\text{Volume} = abc$$

$$I_{(x' \text{ axis})} = \frac{1}{12} m(a^2 + b^2), \qquad I_{(y' \text{ axis})} = \frac{1}{12} m(a^2 + c^2),$$

$$I_{(z' \text{ axis})} = \frac{1}{12} m(b^2 + c^2), \qquad I_{x'y'} = I_{y'z'} = I_{z'x'} = 0.$$

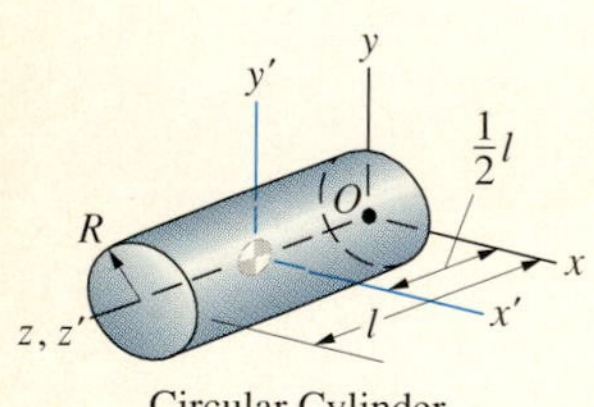

Circular Cylinder

$$\text{Volume} = \pi R^2 l.$$

$$I_{(x \text{ axis})} = I_{(y \text{ axis})} = m\left(\frac{1}{3} l^2 + \frac{1}{4} R^2\right), \qquad I_{(z \text{ axis})} = \frac{1}{2} mR^2,$$

$$I_{xy} = I_{yz} = I_{zx} = 0.$$

$$I_{(x' \text{ axis})} = I_{(y' \text{ axis})} = m\left(\frac{1}{12} l^2 + \frac{1}{4} R^2\right), \qquad I_{(z' \text{ axis})} = \frac{1}{2} mR^2,$$

$$I_{x'y'} = I_{y'z'} = I_{z'x'} = 0.$$

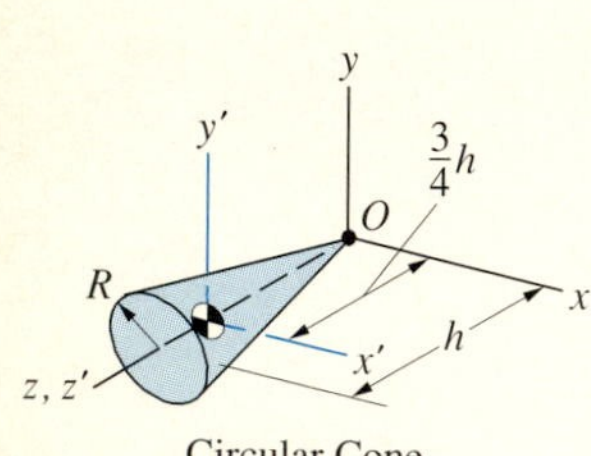

Circular Cone

$$\text{Volume} = \frac{1}{3} \pi R^2 h.$$

$$I_{(x \text{ axis})} = I_{(y \text{ axis})} = m\left(\frac{3}{5} h^2 + \frac{3}{20} R^2\right), \qquad I_{(z \text{ axis})} = \frac{3}{10} mR^2,$$

$$I_{xy} = I_{yz} = I_{zx} = 0.$$

$$I_{(x' \text{ axis})} = I_{(y' \text{ axis})} = m\left(\frac{3}{80} h^2 + \frac{3}{20} R^2\right), \qquad I_{(z' \text{ axis})} = \frac{3}{10} mR^2.$$

$$I_{x'y'} = I_{y'z'} = I_{z'x'} = 0.$$

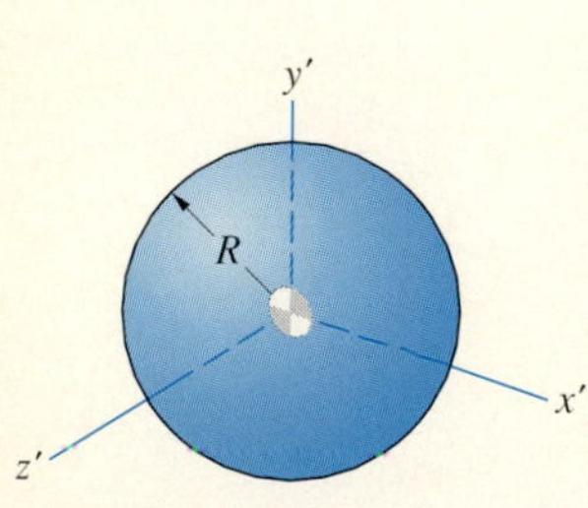

Sphere

$$\text{Volume} = \frac{4}{3} \pi R^3.$$

$$I_{(x' \text{ axis})} = I_{(y' \text{ axis})} = I_{(z' \text{ axis})} = \frac{2}{5} mR^2,$$

$$I_{x'y'} = I_{y'z'} = I_{z'x'} = 0.$$

Appendix D Spherical Coordinates

This appendix summarizes the equations of kinematics and vector calculus in spherical coordinates.

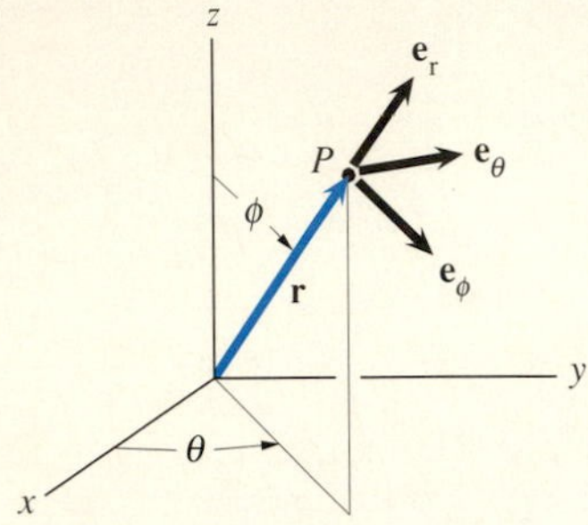

The position vector, velocity, and acceleration are

$$\mathbf{r} = r\mathbf{e}_r,$$

$$\mathbf{v} = \frac{dr}{dt}\mathbf{e}_r + r\frac{d\phi}{dt}\mathbf{e}_\phi + r\frac{d\theta}{dt}\sin\phi\,\mathbf{e}_\theta,$$

$$\mathbf{a} = \left[\frac{d^2r}{dt^2} - r\left(\frac{d\phi}{dt}\right)^2 - r\left(\frac{d\theta}{dt}\right)^2\sin^2\phi\right]\mathbf{e}_r$$

$$+ \left[r\frac{d^2\phi}{dt^2} + 2\frac{dr}{dt}\frac{d\phi}{dt} - r\left(\frac{d\theta}{dt}\right)^2\sin\phi\cos\phi\right]\mathbf{e}_\phi$$

$$+ \left[r\frac{d^2\theta}{dt^2}\sin\phi + 2\frac{dr}{dt}\frac{d\theta}{dt}\sin\phi + 2r\frac{d\phi}{dt}\frac{d\theta}{dt}\cos\phi\right]\mathbf{e}_\theta.$$

The gradient of a scalar field ψ is

$$\nabla\psi = \frac{\partial\psi}{\partial r}\mathbf{e}_r + \frac{1}{r}\frac{\partial\psi}{\partial\phi}\mathbf{e}_\phi + \frac{1}{r\sin\phi}\frac{\partial\psi}{\partial\theta}\mathbf{e}_\theta.$$

The divergence and curl of a vector field $\mathbf{v} = v_r\,\mathbf{e}_r + v_\theta\mathbf{e}_\theta + v_\phi\mathbf{e}_\phi$ are

$$\nabla\cdot\mathbf{v} = \frac{1}{r^2}\frac{\partial}{\partial r}(r^2v_r) + \frac{1}{r\sin\phi}\frac{\partial}{\partial\phi}(v_\phi\sin\phi) + \frac{1}{r\sin\phi}\frac{\partial v_\theta}{\partial\theta},$$

$$\nabla\times\mathbf{v} = \frac{1}{r^2\sin\phi}\begin{vmatrix} \mathbf{e}_r & r\mathbf{e}_\phi & r\sin\phi\,\mathbf{e}_\theta \\ \dfrac{\partial}{\partial r} & \dfrac{\partial}{\partial\phi} & \dfrac{\partial}{\partial\theta} \\ v_r & rv_\phi & r\sin\phi\,v_\theta \end{vmatrix}.$$

Answers to Even-Numbered Problems

Chapter 1

1.2 2.7183.

1.4 7.32 m wide, 2.44 m high.

1.6 The 1-in. wrench fits the 25-mm nut.

1.8 149 mi/hr.

1.10 (a) 5000 m/s; (b) 3.11 mi/s.

1.12 $g = 32.2$ ft/s^2.

1.14 0.310 m^2.

1.16 2.07×10^6 Pa.

1.18 $G = 3.44 \times 10^{-8}$ lb-ft^2/slug^2.

1.20 (a) The SI units of T are kg-m^2/s^2; (b) $T = 73.8$ slug-ft^2/s^2.

1.22 (a) Fully fueled, 8490 slugs; fuel expended, 1950 slugs; (b) fully fueled, 1.22 MN; fuel expended, 0.279 MN.

1.24 (a) 491 N; (b) 81.0 N.

1.26 163 lb.

1.28 32.1 km.

1.30 345,000 km.

Chapter 2

2.2 (a) $s = -0.3t + 0.15$ m. (b) $v = -0.3$ m/s.

2.4 (a) $\Delta s = 32$ ft. (b) $v = 0$, $a = 4$ ft/s^2. (c) $v = 16$ ft/s, $a = 4$ ft/s^2.

2.6 (a) $\Delta s = 132$ m. (b) $v = 28$ m/s at $t = 4$ s. (c) $a = 0$.

2.8 $v = 0.508$ m/s, $a = 1.16$ m/s^2.

2.10 (a) 0.628 m/s. (b) 3.95 m/s^2.

2.12 (a) $v = 126$ ft/s at $t = 11.8$ s. (b) $a = 14.2$ ft/s^2 at $t = 4.2$ s.

2.14 $s = 100$ m, $v = 80$ m/s.

2.16 $a = 8.07$ ft/s^2, distance = 403 ft.

2.18 0.0273 s, $\Delta s = 1.20$ ft.

2.20 2.73 m/s.

2.22 $s = 505$ ft.

2.24 $s = 1070$ m, $a = -24$ m/s^2.

2.26 6.77 s.

2.28 $s = 833$ m.

2.30 Yes, the car travels 94.7 m in 5 s.

2.32 51.9 solar years.

2.34 (a) 45.5 s. (b) 3390 m.

2.36 10 s.

2.38 $v = 6.73$ ft/s, $a = -8.64$ ft/s^2.

2.40 $v = 2e^{-2t}$ m/s.

2.42 $v = 4$ ft/s.

2.44 $v = \dfrac{0.9g}{c}(1 - e^{-ct})$.

2.46 6.98 mi.

2.48 (a) $v^2 = \dfrac{g}{c}(1 - e^{-2cs})$. (b) $v^2 = 2gs$.

2.50 3190 m.

2.54 $v = 3$ ft/s.

2.56 $v = 60.0$ m/s.

2.58 $v = -2(1 - s^2)^{1/2}$ m/s.

2.60 (a) 1.29 ft. (b) 4.55 ft/s.

2.62 $v_0 = 35{,}900$ ft/s = 24,500 mi/hr.

2.64 2370 m/s.

2.66 (a) $v = \pm\sqrt{v_0^2 - 2g_T R_T \ln(s/s_0)}$.

2.68 $\mathbf{r} = 3\mathbf{i} + 10\mathbf{j}$ (ft).

2.70 $x = 29.3$ m, $y = 63.3$ m, $z = -58.0$ m.

2.72 179 ft/s.

2.74 $\theta_0 = 45°$, $R_{max} = v_0^2/g$.

2.76 $31.2 < v_0 < 34.2$ ft/s.

2.78 (a) Yes. (b) No.

2.80 82.5 m.

2.82 (b) $v_0 = 64.8$ ft/s, $\theta_0 = 47.4°$.

2.84 $R = 18.6$ m.

2.86 $\mathbf{v} = 0.602\mathbf{i} - 4.66\mathbf{j}$ (m/s).

2.88 $v_x = 1.77$ m/s, $v_y = 2.21$ m/s, $v_z = 0.89$ m/s.

2.90 $\mathbf{a} = -0.099\mathbf{i} + 0.414\mathbf{j}$ (m/s^2).

2.94 (a) Positive. (b) Approximately 2π radians in 24 hr.

2.96 $\omega = 4.54$ rad/s, $\alpha = 0.98$ rad/s^2.

2.98 (a) 27 rad/s. (b) 149 rad.

2.100 $\omega = \pm 2$ rad/s.

2.102 $d\mathbf{e}/dt = -8.8\mathbf{i} + 13.4\mathbf{j}$.

2.104 (a) $a_t = 0$, $a_n = 200$ m/s^2. (b) 50 rad/s.

2.106 (a) 1730 rpm. (b) 3.01 rad/s^2.

2.108 $a_t = 2.27$ m/s^2, $a_n = 1754.25$ m/s^2.

2.110 (a) $|\mathbf{v}| = 16$ ft/s, $|\mathbf{a}| = 66.0$ ft/s^2. (b) 8 ft.

2.112 $|\mathbf{v}| = 1520$ ft/s = 1040 mi/hr.

2.114 $|\mathbf{v}| = 1320$ ft/s, $|\mathbf{a}| = 0.0958$ ft/s^2.

2.116 54.2 m/s^2.

2.118 $v = 17.9$ ft/s, $|\mathbf{a}| = 1.24$ ft/s^2.

2.120 (a) $|\mathbf{a}| = 45.0$ ft/s^2. (b) $|\mathbf{a}| = 59.9$ ft/s^2.

2.122 (a) $\mathbf{a} = -4.75\mathbf{e}_t + 8.23\mathbf{e}_n$ (m/s^2).

2.124 (a) $\mathbf{v} = 17.7\mathbf{e}_t$ (ft/s), $\mathbf{a} = -6.5\mathbf{e}_t + 31.5\mathbf{e}_n$ (ft/ (b) $\rho = 9.91$ ft.

2.126 218 m.

2.128 $\rho = 81.0$ m.

2.130 $a_n = g/\sqrt{1 + (gt/v_0)^2}$.

2.132 $dy/dt = 0.260$ m/s, $d^2y/dt^2 = -0.150$ m/s^2.

2.134 (a) $\rho = 16.7$ ft. (b) $a_n = 77.3$ ft/s^2.

2.138 (a) $|\mathbf{a}| = 22.4$ ft/s^2. (b) $\rho = 6.03$ ft.

2.140 $\mathbf{a} = 9\mathbf{e}_r - 8\mathbf{e}_\theta$ (m/s^2).

2.142 $\mathbf{v} = 5.30\mathbf{e}_r + 8.48\mathbf{e}_\theta$ (ft/s), $\mathbf{a} = -17.1\mathbf{e}_r - 27.3\mathbf{e}_\theta$ (ft/s^2).

2.144 (a) $\mathbf{v} = 0.32\mathbf{e}_r + 2.03\mathbf{e}_\theta$ (m/s).
(b) $\mathbf{v} = 0.32\mathbf{i} + 2.03\mathbf{j}$ (m/s).

2.148 (a) $\mathbf{a} = 5.97\mathbf{e}_r - 4.03\mathbf{e}_\theta$ (m/s^2).
(b) $\mathbf{a} = 7.17\mathbf{i} - 0.70\mathbf{j}$ (m/s^2).

2.152 $\mathbf{v} = \sqrt{v_0^2 + (\omega_0^2 - K)(r^2 - r_0^2)}\,\mathbf{e}_r + r\omega_0\mathbf{e}_\theta$.

2.154 $a_r = 0$, $a_\theta = 0$, $\alpha = -0.0065$ rad/s^2.

2.156 (a) $\mathbf{a} = -3.52\mathbf{e}_r + 4.06\mathbf{e}_\theta$ (m/s^2).
(b) $\mathbf{a} = -0.38\mathbf{i} - 5.36\mathbf{j}$ (m/s^2).

2.158 (a) $\mathbf{a} = -225\mathbf{e}_r - 173\mathbf{e}_\theta$ (ft/s^2).
(b) $\mathbf{a} = -108\mathbf{i} - 263\mathbf{j}$ (ft/s^2).

2.160 $\mathbf{v} = 1047\mathbf{e}_\theta + 587\mathbf{e}_z$ (ft/s), $\mathbf{a} = -219{,}000\mathbf{e}_r$ (ft/s^2).

2.162 $\mathbf{v}_{A/B} = -\mathbf{v}_{B/A} = -20\mathbf{i} - 10\mathbf{j}$ (m/s).

2.166 (a) $\mathbf{a}_{A/B} = 0$. (b) $\mathbf{v}_{A/B} = -36.6\mathbf{i} + 36.6\mathbf{j}$ (ft/s).
(c) $\mathbf{r}_{A/B} = -56.8\mathbf{i} + 56.8\mathbf{j}$ (ft).

2.168 $|\mathbf{a}_{A/B}| = 283$ ft/s^2.

2.170 $|\mathbf{a}_{A/B}| = 0.625$ m/s^2.

2.172 $\mathbf{a}_{B/A} = -5\mathbf{i} - 8.66\mathbf{j}$ (m/s^2).

2.174 $\mathbf{a}_B = -41.6\mathbf{i} - 24.8\mathbf{j}$ (m/s^2).

2.176 $|\mathbf{a}_{A/B}| = 191$ m/s^2.

2.178 $\mathbf{a}_{A/B} = -5\mathbf{i} - 2\mathbf{j}$ (ft/s^2).

2.180 Velocity = $44.4\mathbf{i} - 7.6\mathbf{j}$ (ft/s),
acceleration = $1.32\mathbf{i} + 0.76\mathbf{j}$ (ft/s^2).

2.182 $x = -4.99$ ft.

2.184 53.1° west of north, 424 mi/hr.

2.186 2.34 m/s.

2.188 $\mathbf{v}_{A/B} = v_A\mathbf{j} + R_E\omega\mathbf{k}$.

2.190 (b) $v = 3.72$ ft/s at $t = 2.31$ s.

2.192 $\theta_0 = 33.4°$.

2.194 60.9° and 74.1°.

2.196 $c = 1.31$ s^{-1}.

2.198 (b) $|\mathbf{a}|_{min} = 2.07$ (m/s^2) at $t = 0.310$ s.

2.200 $|\mathbf{a}|_{max} = 22.6$ m/s^2 at $r = 1.49$ m, $\theta = 20.5°$.

2.202 (a) 7.10 m. (b) 2.22 s. (c) 11.8 m/s.

2.204 $v = 42.3$ m/s.

2.206 13.1 s.

2.208 68.7 ft/s.

2.210 $|\mathbf{v}| = 2.19$ m/s, $|\mathbf{a}| = 5.58$ m/s^2.

2.212 $\mathbf{a} = -2.75\mathbf{e}_r - 4.86\mathbf{e}_\theta$ (m/s^2).

2.214 (a) $\mathbf{v} = -2.13\mathbf{e}_r + 6.64\mathbf{e}_\theta$ (m/s).
(b) $\mathbf{v} = -5.90\mathbf{i} + 3.71\mathbf{j}$ (m/s).

2.216 $|\mathbf{v}_M| = 7.65$ km/s.

2.218 $|\mathbf{v}_{C/A}| = 37.0$ ft/s.

2.220 $\omega = \pm 29.6$ rad/s. No.

Chapter 3

3.2 (a) 1.02 m/s. (b) 0.51 m.

3.4 (a) 2.06 ft/s. (b) 1.03 ft.

3.6 The elevator's acceleration is $a = -g$.

3.8 $|\mathbf{a}| = 11$ m/s^2.

3.10 100 slugs.

3.12 21.5 ft.

3.14 27.8 ft/s^2.

3.16 $x = F_0t_0^2/m$.

3.18 $-373\mathbf{i} + 276\mathbf{j}$ (lb).

3.20 $W = 88.3$ kN, $T = 61.7$ kN, $L = 66.9$ kN.

3.22 (a) $dv/dt = 4.03$ m/s^2. (b) $\rho = 4350$ m.

3.24 (66, 138, −58) m.

3.26 $0.023\mathbf{i} + 10.017\mathbf{j}$ (lb).

3.28 2.68 ft.

3.30 Velocity = 8.05 ft/s, distance = 4.02 ft.

3.32 $F = 331$ N, $a = 3.92$ m/s^2.

3.34 $F = 125$ N.

3.36 4.43 ft up the surface.

3.38 (a) 1200 N. (b) 1.84 m/s.

3.40 (a) 2.1 ft. (b) 6.9 ft.

3.42 (a) $a = -5$ m/s^2. (b) $v = -2.24$ m/s.

3.44 $t = 48.3$ s.

3.46 (a) 11.9 ft. (b) 813 lb.

3.48 $y = -18.8$ mm.

3.50 (a) 50 s. (b) 40.8 N. (c) $4.8\mathbf{i} + \mathbf{j}$ (m/s).

3.52 2.06 m/s^2 up the bar.

3.54 $F = 19.4$ lb.

3.56 $t = 0.600$ s.

3.58 $F_x = -0.544$ N.

3.60 $\Sigma F_x = 813$ lb, $\Sigma F_y = 604$ lb.

3.62 0.284 s.

3.64 26.4 ft.

3.66 (a) $\Sigma F_t = 0$. (b) $\Sigma F_n = 489$ lb. (c) Zero.

3.68 35.1 m/s.

3.70 6.48 kN.

3.72 8.08 lb.

3.76 70.9 lb.

3.78 $2.62 \le v \le 3.74$ m/s.

3.80 (a) 207,000 lb. (b) 41,700 ft.

3.82 (a) 401 ft/s. (b) 450 lb downward.

3.86 $\theta = L_0/R - \sqrt{(L_0/R)^2 - (2v_0/R)t}$.

3.88 $\rho = 697$ m, $\mathbf{e} = 0.916\mathbf{i} - 0.308\mathbf{j} + 0.256\mathbf{k}$.

3.90 400 ft.

3.92 11.4 m/s^2.

3.94 $\Sigma F_r = -138$ lb, $\Sigma F_\theta = 138$ lb.

3.96 $9.46\mathbf{e}_r + 3.44\mathbf{e}_\theta$ (N).

3.98 $|\mathbf{v}| = 16.6$ ft/s.

3.100 $|\mathbf{v}| = \sqrt{2gL \sin\theta}$, $T = 3mg \sin\theta$.

3.102 $N = 1.02$ lb.

3.104 $\mu_s = 0.406$. The mass slips toward O.

3.106 $-1.48\mathbf{e}_r - 0.20\mathbf{e}_\theta$ (lb).

3.108 $k = 2m\omega_0^2$.

3.110 $11.5\mathbf{e}_r + 44.2\mathbf{e}_\theta + 10\mathbf{e}_z$ (N).

3.112 $|\Sigma\mathbf{F}| = 8.36$ N at $t = 4.39$ s.

3.114 2.36×10^6 s, or 27.3 days.

3.116

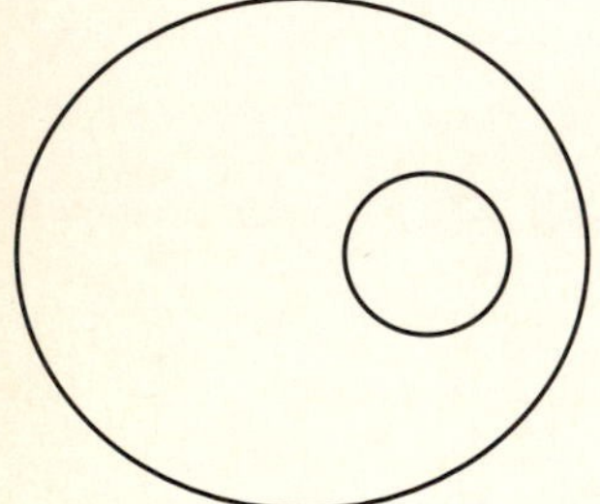

3.118 $v_0 = 10.8$ km/s.

3.120

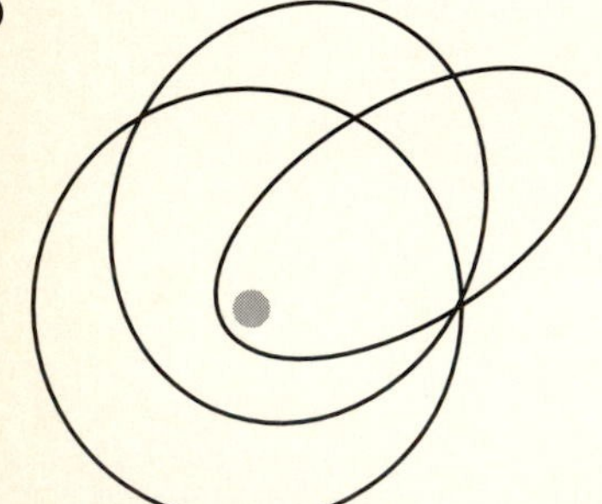

3.122

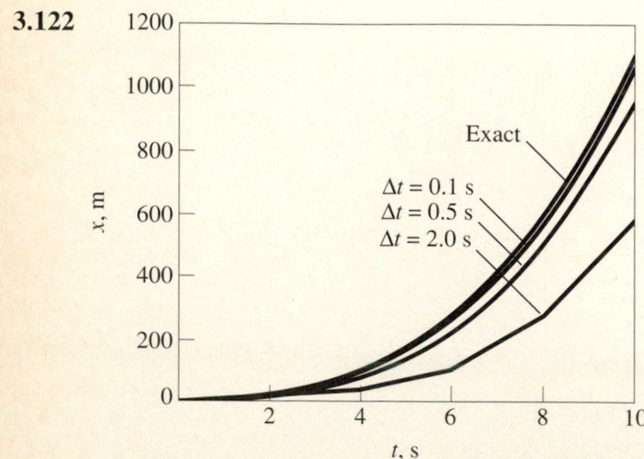

3.124

Closed - form
$\Delta t = 0.05$ s
$\Delta t = 0.5$ s

x, m (0–80); t, s (0–4)

3.126

time, s	position, m	velocity, m/s
0.00	1.0000	0.0000
0.01	1.0000	−0.0100
0.02	0.9999	−0.0200
0.03	0.9997	−0.0300
0.04	0.9994	−0.0400
0.05	0.9990	−0.0500

3.128 $x = 22.5$ ft, $v_x = 23.1$ ft/s.

3.130

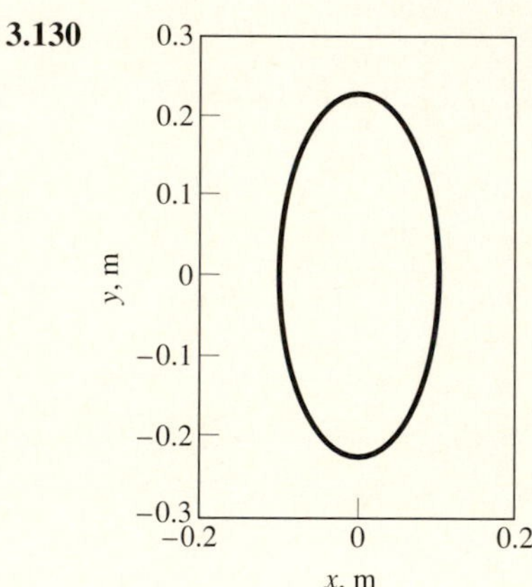

3.132 $C = 0.0217$.

3.134 $|\mathbf{a}| = 136.0$ ft/s^2 at $t = 0.084$ s.

3.136 (a) 34.6 ft/s^2, 3490 lb. (b) $\mu_s = 1.07$.

3.138 (a) $F_1 = 63$ kN, $F_2 = 126$ kN, $F_3 = 189$ kN.
(b) $F_1 = 75$ kN, $F_2 = 150$ kN, $F_3 = 225$ kN.

3.140 Deceleration is 1.49 m/s^2, compared with 8.83 m/s^2 on a level road.

3.142 14,300 ft (2.71 mi).

3.144 10.5 lb.

3.146 $a_A = 4.02$ ft/s^2, $T = 17.5$ lb.

3.148 29.7 ft.

3.150 9.30 N.

3.152 $\tan\alpha = v^2/(\rho g)$.

3.154 $\Sigma\mathbf{F} = -10.7\mathbf{e}_r + 2.55\mathbf{e}_\theta$ (N).

3.156 5.46 ft/s^2.

Chapter 4

4.2 (a) 8.79 ft/s. (b) 24 ft-lb. (c) 8.79 ft/s.

4.4 (a), (b) 3.58 m/s.

4.6 (a) 6.01×10^6 ft-lb. (b) 4560 lb.

4.8 (a) 2.00×10^6 ft-lb/s, or 3640 hp.
(b) 1.00×10^6 ft-lb/s, or 1820 hp.

4.10 129 kN.

4.12 (a), (b) 2.94 N-m. (c) 6.26 m/s, 6.95 m/s.

4.14 (a) 10.7 ft-lb. (b) 21.1 ft/s.

4.16 (a) 2300 ft-lb. (b) $|\mathbf{v}| = 18.1$ ft/s.

4.18 $U_{12} = \frac{1}{2}m[(v_0 - cs_f/m)^2 - v_0^2]$.

4.20 Work = 509 N-m, $v = 2.52$ m/s.

4.22 $v = 21.8$ m/s.

4.24 131 ft-lb/s.

4.26 1.12 m/s.

4.28 $v = 1.72$ m/s.

4.30 $v = 1.14$ m/s.

4.32 (a), (b), (c) 63.4 m/s.

4.34 (a) Work = 521 N-m, $v = 5.89$ m/s.
(b) Work = 448 N-m, $v = 5.47$ m/s.

4.36 (a) $mgL \sin \alpha$. (b) Zero. (c) $v = \sqrt{2gL \sin \alpha}$.

4.38 $v = 3.25$ m/s.

4.40 (a) $v_0 = \sqrt{5gR}$. (b) $\sqrt{gR}$.

4.42 (a) Zero. (b) 1.57×10^7 ft-lb.

4.44 39.3 ft/s, or 26.8 mi/hr.

4.46 21.0 m/s.

4.48 (a) $U_{12} = mgR(1 - \cos \alpha)$.
(b) $v = \sqrt{2gR(1 - \cos \alpha)}$.

4.50 621 lb/ft.

4.52 Distance is 40.3 mm, deceleration is 0.248 m/s^2.

4.54 5.35 ft/s.

4.56 $k = 4290$ N/m.

4.58 (a) 1.55 m. (b) 1.74 m/s.

4.60 1 ft.

4.62 0.8 ft.

4.64 $v_2 = 7.03$ m/s.

4.66 5.77 m/s.

4.68 $k = 997$ lb/ft.

4.70 4.90 m/s.

4.72 $v = 65.0$ m/s.

4.74 4730 ft/s.

4.76 2.25×10^{10} N-m.

4.78 6.95 m/s.

4.80 (a) 7.72 m. (b), (c) 6.48 m.

4.82 (a), (b), (c) 24 N-m.

4.84 $|\mathbf{v}| = \sqrt{16 + 42.9d}$.

4.86 60°.

4.88 (a) 687 N. (b) 883 N.

4.90 $W\left[1 + \sqrt{1 + 4C/W}\right]$.

4.92 1.20 m/s.

4.94 1.99 m/s.

4.96 $V = \frac{1}{2}kS^2 + \frac{1}{4}qS^4$.

4.98 $v = 2.30$ m/s.

4.100 $v = 8.45$ ft/s.

4.102 (a) $V = p_0 s_0^\gamma A s^{1-\gamma}/(\gamma - 1)$.
(b) $v = \sqrt{2p_0 s_0^\gamma A(s_0^{1-\gamma} - s^{1-\gamma})/[m(\gamma - 1)]}$.

4.104 $v = 11.0$ km/s.

4.106 $v_C = 9420$ ft/s.

4.108 $v = 419$ m/s.

4.110 (a) $\mathbf{F} = -4x\mathbf{i} + \mathbf{j}$ (N). (b) The work is -1 N-m for any path from position 1 to position 2.

4.112 $\mathbf{F} = -[k(r - r_0) + q(r - r_0)^3]\mathbf{e}_r$.

4.114 (a) $\mathbf{F} = (\sin \theta - 2r \cos^2 \theta)\, \mathbf{e}_r + (\cos \theta + 2r \sin \theta \cos \theta)\, \mathbf{e}_\theta$.
(b) The work is 2 ft-lb for any path from 1 to 2.

4.118 (a) and (c).

4.120 553 mm.

4.122 58.7 mm, 1280 N-m/s (watts).

4.124 8.01 ft.

4.126 Maximum distance is 0.203 m.

4.128 $v = 6.32$ m/s.

4.130 $\theta_{max} = 51°$.

4.132 (a), (b) 193 ft.

4.134 He should choose (b). Impact velocity is 11.8 m/s, work is -251 kN-m. In (a), impact velocity is 13.9 m/s, work is -119 kN-m.

4.136 $v = 2.08$ m/s.

4.138 (a) 14.3×10^7 ft-lb. (b) $v = 88(1 - e^{-(F_0/88m)t})$.

4.140 $k = 163$ lb/ft.

4.142 (a) $k = 809$ N/m. (b) $v_2 = 6.29$ m/s.

4.144 4.39 ft/s.

4.146 $h = 0.179$ m.

4.148 (a) $\theta = 27.9°$. (b) 159 lb. (c) 222 lb.

4.150 $v_1 = 4.73$ ft/s.

4.152 1.02 m.

4.154 2.00 m/s.

4.156 24.8 ft/s.

4.158 (a) $v = 0$. (b) $v = \sqrt{gR_E/2} = 18{,}300$ ft/s, or 12,500 mi/hr.

4.160 (a) 11.3 MW (megawatts). (b) 9.45 MW.

Chapter 5

5.2 (a) 27.3 kip-s. (b) 4560 lb.

5.4 (a) 1000 lb-s. (b) 107 ft/s, or 73.2 mi/hr.

5.6 759 kN.

5.8 (a) 213**i** + 480**j** (N-s). (b) 18.7i + 20**j** (m/s).

5.10 **v** = 4.29**i** (ft/s).

5.12 (a) 75 kip-s. (b) 8 s.

5.14 32.2 ft/s.

5.16 (a) 222 N-s. (b) 1.85 m/s.

5.18 0.199 s.

5.20 10.7 ft/s.

5.22 2.01 m/s.

5.24 −7.07**i** + 7.26**j** (m/s).

5.26 (a) −100**j** (lb-s). (b) **v** = 20**i** − 29.8**j** (ft/s).

5.28 328 lb.

5.30 $\Sigma F_{n\,\mathrm{av}} = 50$ N.

5.32 $\Sigma F_{n\,\mathrm{av}} = 89.3$ N.

5.34 (a) 1910 lb. (b) 22 ft/s^2.

5.36 75.1 lb (approximately 600 times the watch's weight).

5.38 Horizontal force is 0.234 N, vertical force is 0.364 N.

5.40 −0.35**i** + 122.96**j** (N).

5.42 $|\mathbf{v}_A| = 2.07$ m/s, $|\mathbf{v}_B| = 2.76$ m/s.

5.44 (a) 0.00057 m/s. (b) 0.00343 m.

5.46 (a), (b) 1.63 ft/s.

5.48 Velocity is 6.96 km/s, $\beta = 0.174°$.

5.50 2.41 m/s.

5.52 $v = \sqrt{2\mu_k g D}\,(1 + m_B/m)$.

5.54 48.6 mm.

5.56 (a) 30**j** (mm/s). (b) 6**i** + 6**j** (m).

5.58 $e = 0.2$, $v'_A = 0.4v_A$.

5.60 *A*: 6.2 ft/s toward the left. *B*: 5.8 ft/s toward the right.

5.62 *A*: 107.6 ft/s^2. *B*: 68.4 ft/s^2.

5.64 $|\mathbf{v}_A| = 2.46$ m/s, $|\mathbf{v}_B| = 2.90$ m/s.

5.66 $v_A = 7.70$ m/s.

5.68 10.0 ft.

5.70 $e = 0.77$.

5.72 4.5 ft/s upward.

5.74 Helmet: 1.09 m/s to the right; head: 1.07 m/s to the left.

5.76 $e = 2\sqrt{(1 - \cos\beta)/(1 - \cos\theta)} - 1$.

5.78 $\mathbf{v}_A = 0.0232\mathbf{i} - 0.002\mathbf{j} + 0.007\mathbf{k}$ (m/s), $\mathbf{v}_B = 0.0732\mathbf{i}$ (m/s).

5.80 $\mathbf{v}'_A = \mathbf{i} + \mathbf{j}$ (m/s), $\mathbf{v}'_B = -\mathbf{i} + \mathbf{j}$ (m/s).

5.84 $v_S = 35.3$ m/s.

5.86 9430 ft/s.

5.88 $\omega = 0.375$ rad/s.

5.90 $\mathbf{v} = -v_0\mathbf{e}_r + \dfrac{(\frac{1}{2}r_0t^2 - \frac{1}{3}v_0t^3)C}{(r_0 - v_0t)m}\mathbf{e}_\theta$.

5.92 2.31 m.

5.94 $3mv_0^2/2$.

5.96 (a) −1440**k** (kg-m^2/s). (b) 1.2 m/s.

5.100 103 lb.

5.102 750 N.

5.104 (a) 351**i** − 849**j** (N). (b) 1200**i** − 1200**j** (N). (c) 2400**i** (N).

5.106 0.486**k** (ft-lb).

5.108 51.9**i** − 282.2**j** (N).

5.110 1.44 km/s.

5.112 $v = v_f \ln\left(\dfrac{m_0}{m_0 - \dot{m}_f t}\right) - gt$.

5.114 (a) $F = 3s + 12/32.2$ lb. (b) 25.5 ft-lb.

5.116 $v = v_0/[1 + (\rho_L/2m)s]$.

5.118 6820 N.

5.120 4100 lb.

5.122 (a) 5. (b) 43,500 lb. (c) 11.2 ft/s^2.

5.124 (a) 180**i** + 360**j** (lb-s). (b) 12**i** + 44**j** (ft/s).

5.126 877 kN.

5.128 (a) 34.9 ft/s. (b) 15,700 ft-lb.

5.130 (a) 4.80 ft/s. (b) 415 ft.

5.132 (a) 3.45 m/s. (b) 18.7 kN.

5.134 (a) 8.23 kN. (b) 16.5 kN.

5.136 $v_A\sqrt{mk/2}$.

5.138 1.57 ft.

5.140 $e = 0.304$.

5.142 2.30 m.

5.144 $x = -11.13$ in., $y = 6.42$ in.

5.146 $|v_r| = 11{,}550$ ft/s, $|v_\theta| = 9430$ ft/s.

5.148 867 lb (including the weight of the drum).

5.150 (a) 30.1 ft/s. (b) 46.8 ft/s.

Chapter 6

6.2 $\mathbf{v}_A = -9.29\mathbf{i}$ (ft/s), $\mathbf{a}_A = -8\mathbf{i} - 43.1\mathbf{j}$ (ft/s^2).

6.4 (a) 20 rad/s^2 clockwise. (b) 1.59 revolutions.

6.6 8 rad/s.

6.8 $\mathbf{v}_A = -6.67\mathbf{i} - 6.67\mathbf{j}$ (ft/s), $\mathbf{a}_A = 66.7\mathbf{i} - 66.7\mathbf{j}$ (ft/s^2), $\mathbf{v}_B = 13.3\mathbf{j}$ (ft/s), $\mathbf{a}_B = -133\mathbf{i}$ (ft/s^2).

6.10 $\mathbf{v}_A = 6.05\mathbf{i} - 4.50\mathbf{j}$ (ft/s), $\mathbf{a}_A = 39.0\mathbf{i} + 46.2\mathbf{j}$ (ft/s^2).

6.12 $|\mathbf{v}_B| = 15$ ft/s, $|\mathbf{a}_B| = 300$ ft/s^2.

6.14 $\boldsymbol{\omega} = 30\mathbf{i}$ (rad/s).

6.16 $\boldsymbol{\omega}_{AB} = 10\mathbf{k}$ (rad/s), $\boldsymbol{\omega}_{BC} = -10\mathbf{k}$ (rad/s), $\boldsymbol{\omega}_{CD} = 10\mathbf{k}$ (rad/s).

6.18 $\mathbf{v}_{A/B} = r_{A/B}\omega\mathbf{j}$.

6.20 $\mathbf{v}_{A/B} = -20\mathbf{j}$ (m/s).

6.22 $\mathbf{v}_A = -5\mathbf{i} + 8.66\mathbf{j}$ (m/s).

6.24 $\mathbf{v}_A = 5\mathbf{i}$ (m/s), $\mathbf{v}_B = -3.54\mathbf{i} - 3.54\mathbf{j}$ (m/s), $\mathbf{v}_C = -3.54\mathbf{i} + 3.54\mathbf{j}$ (m/s).

6.26 (a) 92.6 rad/s clockwise.
(b) The top; 200 km/hr, or 55.6 m/s.

6.28 $\mathbf{v}_T = 12.4\mathbf{i} + 5.1\mathbf{j}$ (m/s).

6.30 (a) $\boldsymbol{\omega} = 2.13\mathbf{k}$ (rad/s). (b) $\mathbf{v}_A = -0.73\mathbf{j}$ (m/s).

6.32 0.566 ft/s toward the left.

6.34 $\omega_{BCD} = 8$ rad/s clockwise, $\mathbf{v}_D = 160\mathbf{i} - 32\mathbf{j}$ (in./s).

6.36 $\omega_{CD} = 10$ rad/s counterclockwise.

6.38 $\omega_{AB} = 2$ rad/s clockwise,
$\omega_{BC} = 3.22$ rad/s counterclockwise.

6.40 $\mathbf{v}_E = -12.3\mathbf{j}$ (m/s).

6.42 $|\boldsymbol{\omega}_{AB}| = 4.91$ rad/s, $|\boldsymbol{\omega}_{DE}| = 6.77$ rad/s.

6.44 $\omega_{AB} = 2.31$ rad/s clockwise, $v_B = 3.15$ m/s to the left.

6.46 $\mathbf{v}_C = 25.1\mathbf{i}$ in./s.

6.48 $\mathbf{v}_A = 1.2\mathbf{i} + 1.2\mathbf{j}$ (m/s).

6.50 $\omega_{BC} = 2.61$ rad/s, $\mathbf{v}_C = -9.1\mathbf{i}$ (m/s).

6.52 0.95 m/s.

6.54 $\mathbf{v}_C = 12.86\mathbf{i} + 7.53\mathbf{j}$ (in./s).

6.56 $\mathbf{v}_D = -0.557\mathbf{i} + 0.815\mathbf{j}$ (m/s).

6.58 $v_W = 0.2$ m/s, 4 rad/s counterclockwise.

6.60 35.5 rpm clockwise.

6.62 Angular velocity = 52.1 rad/s, or 497 rpm,
$|\mathbf{v}_A| = 5.21$ m/s.

6.64 $x_C = 3$ m, $y_C = 0$, $v_B = 10$ m/s.

6.66 $x = 0.35$ ft, $y = -1.5$ ft.

6.68 $\mathbf{v}_G = 1.00\mathbf{i} - 0.36\mathbf{j}$ (m/s).

6.70 (a) (3.46, 4) ft. (b) $\mathbf{v}_A = 2\mathbf{i}$ (ft/s).

6.72 6.57 in./s.

6.74 *OQ*: 0.389 rad/s clockwise;
PQ: 0.389 rad/s counterclockwise.

6.76 $\mathbf{v}_E = -12.25\mathbf{j}$ (m/s).

6.78 $\omega_{BC} = 5.33$ rad/s counterclockwise,
$\omega_{CD} = 4.57$ rad/s clockwise.

6.82 (a), (b) $\mathbf{a}_A = -73.3\mathbf{i} + 27.0\mathbf{j}$ (m/s^2).

6.84 $\mathbf{a}_T = 2.02\mathbf{i} + 2.37\mathbf{j}$ (m/s^2).

6.86 $\mathbf{a}_A = -5\mathbf{j}$ (in./s^2), $\mathbf{a}_B = 3\mathbf{j}$ (in./s^2).

6.88 $\omega_{AC} = 0$, $\alpha_{AC} = 1.13$ rad/s^2 clockwise.

6.90 Acceleration = 134 in./s^2.

6.92 11.3 ft/s^2 toward the left.

6.94 1.77 rad/s^2 counterclockwise.

6.96 $\omega_{BD} = 2.67$ rad/s clockwise,
$\alpha_{BD} = 3.85$ rad/s counterclockwise.

6.98 $\mathbf{v}_C = 40.8\mathbf{i}$ (in./s), $\mathbf{a}_C = 6.98\mathbf{i}$ (in./s^2).

6.100 $a_C = 206\mathbf{i}$ (in./s^2).

6.102 $\omega_{AB} = -0.879$ rad/s, $\alpha_{AB} = -1.06$ rad/s^2,
$\omega_{BC} = -1.15$ rad/s, $\alpha_{BC} = -2.41$ rad/s^2.

6.104 $\mathbf{a}_E = -12.3\mathbf{j}$ (m/s^2).

6.106 $\mathbf{a}_C = -7.78\mathbf{i} - 33.54\mathbf{j}$ (m/s^2).

6.108 $\mathbf{a}_D = -0.135\mathbf{i} - 0.144\mathbf{j}$ (m/s^2).

6.110 $\omega_{AB} = 3.55$ rad/s clockwise,
$\alpha_{AB} = 12.1$ rad/s^2 clockwise,
$\omega_{BC} = 2.36$ rad/s counterclockwise,
$\alpha_{BC} = 16.5$ rad/s^2 counterclockwise.

6.112 0.678 rad/s^2 clockwise. $\mathbf{a}_D = -7.15\mathbf{i}$ (m/s^2).

6.114 $\alpha_{\text{planet}} = 41.4$ rad/s^2 clockwise,
$\alpha_{\text{sun}} = 82.9$ rad/s^2 counterclockwise.

6.116 $\alpha_{CD} = 26.5$ rad/s^2 clockwise,
$\alpha_{DE} = 31.1$ rad/s^2 counterclockwise.

6.118 $\mathbf{a}_A = -200\mathbf{i} + 80\mathbf{j}$ (ft/s^2).

6.120 $\mathbf{a}_C = -8.80\mathbf{i} + 5.60\mathbf{j}$ (m/s^2).

6.122 $\omega_{AC} = 3$ rad/s counterclockwise,
$\alpha_{AC} = 6$ rad/s counterclockwise.

6.124 0.549 m/s toward the left.

6.126 1.16 rad/s clockwise, 15.4 in./s toward *C*.

6.128 $\omega_{BC} = 6.17°$ per second counterclockwise, rate of extension is 0.109 m/s.

6.130 $\omega_{AC} = 8.66$ rad/s counterclockwise, and bar *AC* slides through the sleeve at 5 m/s toward *A*.

6.132 $\omega_{AC} = 0.293$ rad/s clockwise,
$\mathbf{v}_C = -0.738\mathbf{i} - 1.370\mathbf{j}$ (ft/s).

6.134 0.801 ft/s.

6.136 $\omega_{AB} = 2$ rad/s clockwise, $v_{B\,\text{rel}} = 2$ m/s toward *C*.

6.138 $\omega_{AB} = 5.18$ rad/s counterclockwise.

6.140 $\omega_{\text{plate}} = 2$ rad/s counterclockwise, and the velocity at which the pin slides relative to the slot is 0.2 m/s downward.

6.144 (a) $\mathbf{v}_A = R\omega\mathbf{j}$, $\mathbf{a}_A = -R\omega^2\mathbf{i}$.
(b) $\mathbf{v}_A = 2R\omega\mathbf{j}$, $\mathbf{a}_A = -2R\omega^2\mathbf{i}$.

6.146 $\mathbf{v}_A = -3.5\mathbf{i} + 0.5\mathbf{j} + 4\mathbf{k}$ (m/s),
$\mathbf{a}_A = -10\mathbf{i} - 6.5\mathbf{j} - 19.25\mathbf{k}$ (m/s^2).

6.148 $-2\mathbf{i}$ (m/s).

6.150 $\mathbf{a}_{A\,\text{rel}} = -14\mathbf{i} - 2\mathbf{j}$ (ft/s^2).

6.152 (a) $\mathbf{v}_{A\,\text{rel}} = v\mathbf{j}$, $\mathbf{a}_{A\,\text{rel}} = -(v^2/R_{\text{E}})\mathbf{i}$.
(b) $\mathbf{v}_A = v\mathbf{j} - \omega_{\text{E}} R_{\text{E}} \cos L\mathbf{k}$,
$\mathbf{a}_A = (-v^2/R_{\text{E}} - \omega_{\text{E}}^2 R_{\text{E}} \cos^2 L)\mathbf{i} + \omega_{\text{E}}^2 R_{\text{E}} \sin L \cos L\,\mathbf{j} + 2\omega_{\text{E}} v \sin L\mathbf{k}$.

6.154 (a) $-9.81\mathbf{k}$ (m/s^2). (b) $3.29\mathbf{i} - 9.81\mathbf{k}$ (m/s^2).

6.156 (a) $0.1\mathbf{i} + 0.1\mathbf{j}$ (m/s^2).
(b) $0.125\mathbf{i} + 0.085\mathbf{j} + 0.106\mathbf{k}$ (m/s^2).

6.160 OQ: 9.29 rad/s counterclockwise;
PQ: 2.92 rad/s clockwise.

6.162 $\mathbf{v}_C = -32.0\mathbf{j}$ (ft/s).

6.164 $\alpha_{AB} = 13.60 \times 10^3$ rad/s^2 clockwise,
$\alpha_{BC} = 8.64 \times 10^3$ rad/s^2 counterclockwise.

6.166 $\mathbf{a}_D = -3490\mathbf{i}$ (in./s^2).

6.168 $\mathbf{a}_G = -8.18\mathbf{i} - 26.4\mathbf{j}$ (in./s^2).

6.170 $\mathbf{v}_C = -1.48\mathbf{i} + 0.79\mathbf{j}$ (m/s).

6.172 $\mathbf{a}_C = -2.99\mathbf{i} - 1.40\mathbf{j}$ (m/s^2).

6.174 Velocity is 55.9 in./s to the right;
acceleration is 390 in./s^2 to the left.

6.176 $\omega_{BD} = 0.733$ rad/s counterclockwise.

6.178 $\omega_{AB} = 0.261$ rad/s counterclockwise,
$\omega_{BC} = 2.80$ rad/s counterclockwise.

6.180 $\omega_{BC} = 1.22$ rad/s clockwise, 18.0 m/s from B toward C.

6.182 Velocity = 6.89 ft/s upward; acceleration = 169 ft/s^2 upward.

6.184 5.66 N.

6.186 (a) $\mathbf{v}_{A\,\mathrm{rel}} = 5\mathbf{i}$ m/s, $\mathbf{a}_{A\,\mathrm{rel}} = 0$.
(b) $\mathbf{v}_{A/B} = 5\mathbf{i} + 2\mathbf{j}$ (m/s),
$\mathbf{a}_{A/B} = -4\mathbf{i} + 20\mathbf{j}$ (m/s^2).

Chapter 7

7.2 $F = (b/2c)mg$.

7.4 $F = W(b - ca/g)/h$.

7.6 $T_A = 3.68$ kN, $T_B = 4.14$ kN.

7.8 Time = 0.980 s, distance = 2.94 m.

7.10 6.78 m/s^2, $N_A = 1356$ N, $N_B = 213$ N.

7.12 $\omega = 7.5$ rad/s.

7.14 34.0 ft-lb.

7.16 (a) $\alpha = 25$ rad/s^2 clockwise, $T = 20$ lb.
(b) $\alpha = 18.0$ rad/s^2 clockwise, $T = 14.4$ lb.

7.18 Time = 1.17 s.

7.20 (a), (b) $\alpha = 11.4$ rad/s^2 counterclockwise, $A_x = 0$,
$A_y = 4.12$ lb.

7.22 $x = l/\sqrt{12}$, $\alpha_{\max} = \sqrt{3g/l}$.

7.24 $A_x = -10.28$ kN, $A_y = -7.04$ kN.

7.26 $M_B = 27.1$ N-m counterclockwise, $B_x = -11.0$ N,
$B_y = 108.5$ N.

7.28 $\mathbf{a}_G = 0.1108\mathbf{i} - 0.0168\mathbf{j}$ (m/s), $\alpha = -0.000427$ rad/s^2.

7.30 $\mathbf{F}_B = -19.1\mathbf{i} + 183.3\mathbf{j}$ (N), $\mathbf{M}_B = 62.6\mathbf{k}$ (N-m).

7.32 $\alpha = 0.200$ rad/s^2 clockwise.

7.34 $t_{\mathrm{ring}}/t_{\mathrm{disk}} = \sqrt{4/3}$.

7.36 $v = \sqrt{(2mgx - kx^2)/(m + I/R^2)}$.

7.38 21.5 rad/s^2 clockwise.

7.40 $v_{\max} = \frac{2}{7}R\omega_0$, time $= \frac{2}{7}R\omega_0/(\mu_k g)$.

7.42 (a) It doesn't slip, $\alpha = 22.2$ rad/s^2 clockwise.
(b) It does slip, $\alpha = 53.6$ rad/s^2 clockwise.

7.44 10.1 lb.

7.46 (a) $\alpha = 1.84$ rad/s^2 counterclockwise. (b) 112 N.

7.48 9.20 rad/s^2 clockwise.

7.50 61.3 rad/s^2 clockwise.

7.52 22.5 lb.

7.54 (a) 0.244 ft/s^2 to the right.

7.56 $M = 82.4$ ft-lb.

7.58 $B_x = -60.3$ lb, $B_y = -125.6$ lb, $C_x = 11.4$ lb,
$C_y = 135.6$ lb.

7.62 (a) 9.34 m/s^2. (b) 516 N-m.

7.64 0.499 lb.

7.66 $\alpha_{OQ} = 4.88$ rad/s^2 clockwise,
$\alpha_{PQ} = 4.88$ rad/s^2 counterclockwise.

7.70 $\alpha = 0.255$ rad/s^2 clockwise.

7.72 $|P| = 44.1$ kN.

7.74

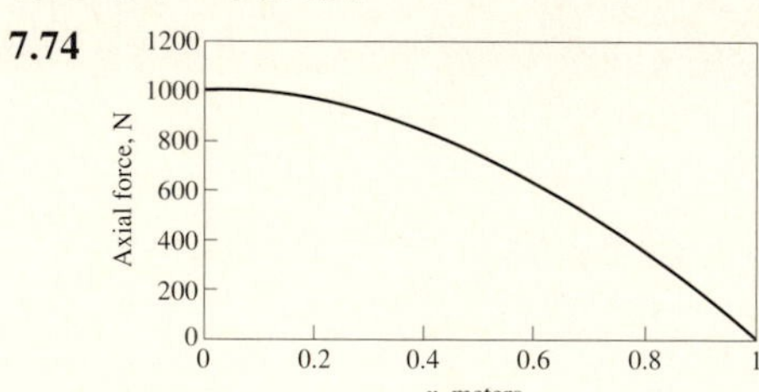

7.76

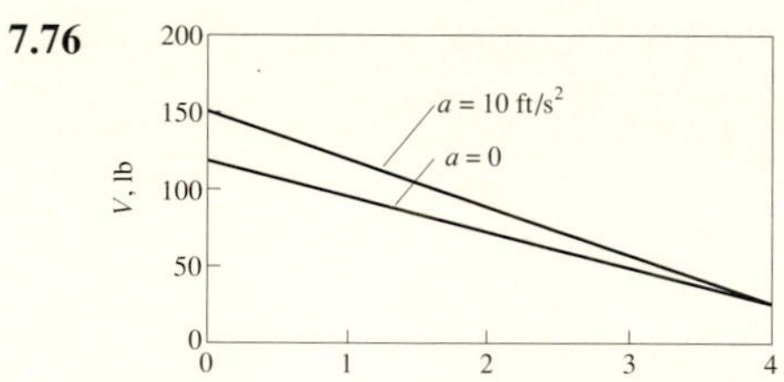

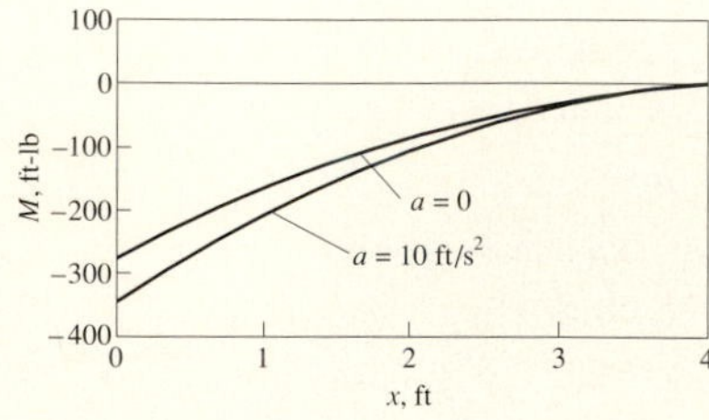

7.78

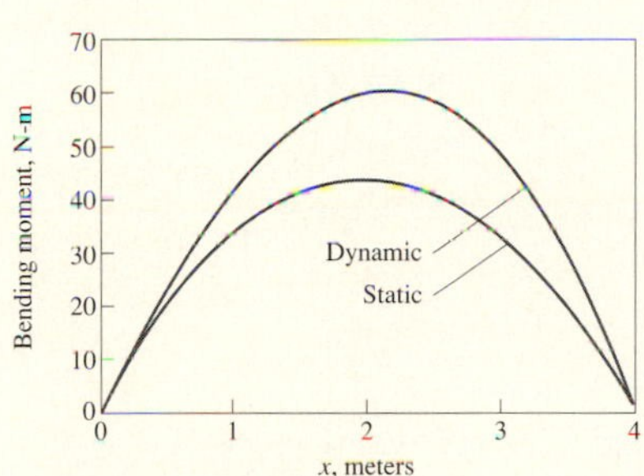

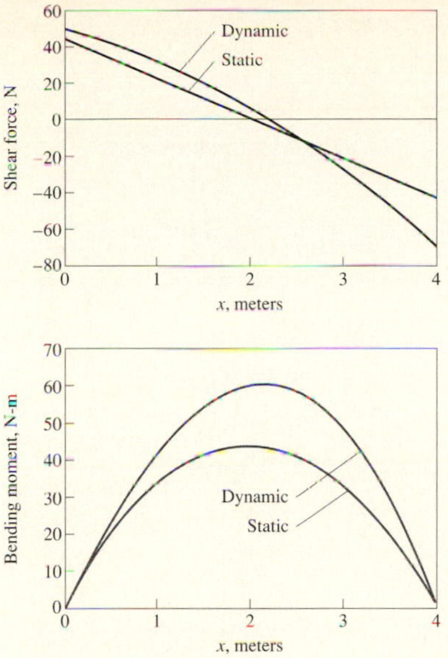

7.80

time, s	θ, rad	ω, rad/s	closed-form ω, rad/s
0.0	0.000	0.000	0.000
0.2	0.000	0.250	0.250
0.4	0.050	0.499	0.498
0.6	0.150	0.747	0.744
0.8	0.299	0.991	0.987
1.0	0.498	1.232	1.225

7.82

time, s	θ, rad	ω, rad/s
0.0	0.000	0.000
0.1	0.000	1.472
0.2	0.147	2.943
0.3	0.441	4.399
0.4	0.881	5.729
0.5	1.454	6.665

7.84

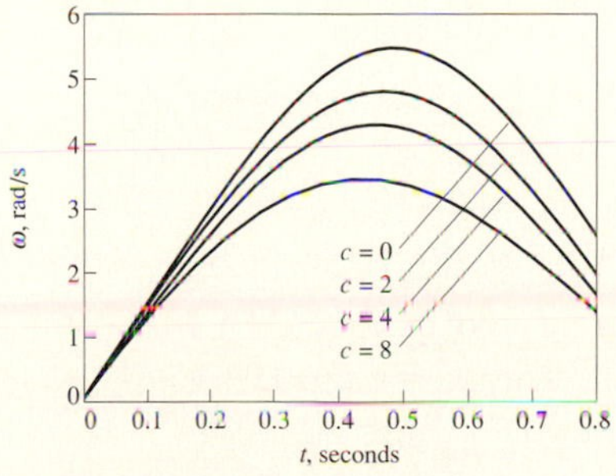

7.86

time, s	θ, rad	ω, rad/s
0.00	0.0000	0.0000
0.01	0.0000	0.1472
0.02	0.0015	0.2943
0.03	0.0044	0.4410
0.04	0.0088	0.5868
0.05	0.0147	0.7313

7.88 $I_0 = \frac{1}{3}ml^2$.

7.90 $I = \frac{1}{12}ml^2 \sin^2 \theta$.

7.92 $I_{(x\text{ axis})} = \frac{1}{12}mh^2$, $I_{(y\text{ axis})} = \frac{1}{12}mb^2$, $I_{(z\text{ axis})} = \frac{1}{12}m(b^2 + h^2)$.

7.94 $I_{(y\text{ axis})} = 1.99$ slug-ft^2.

7.96 20.8 kg-m^2.

7.98 $I_0 = \frac{17}{12}ml^2$.

7.100 $I_{(z\text{ axis})} = 47.0$ kg-m^2.

7.102 $I_{(z\text{ axis})} = 0.0803$ slug-ft^2.

7.104 3810 slug-ft^2.

7.106 $I_{(z\text{ axis})} = 9.00$ kg-m^2.

7.108 $I_{(y\text{ axis})} = 0.0881$ slug-ft^2.

7.110 $I_0 = 0.0188$ kg-m^2.

7.112 $I_{(x\text{ axis})} = I_{(y\text{ axis})} = m(\frac{3}{20}R^2 + \frac{3}{5}h^2)$.

7.114 $I_{(x\text{ axis})} = 0.844$ kg-m^2.

7.116 $I_{(x\text{ axis})} = 0.221$ kg-m^2.

7.118 $I = 0.460$ slug-ft^2.

7.120 $I_{(z\text{ axis})} = 0.00911$ kg-m^2.

7.122 $I_0 = 0.00367$ kg-m^2.

7.124 $I_{(z\text{ axis})} = 0.714$ slug-ft^2.

7.126 (a) 12.1 s. (b) 144 lb.

7.128 (a) 20 m/s^2. (b) $c \leq 49.1$ mm.

7.130 $I = 2.05$ kg-m^2.

7.132 40.2 kN.

7.134 $\alpha = -0.420$ rad/s^2, $F_x = 336$ N, $F_y = 1710$ N.

7.136 $\alpha = (g/l)[3(1 - \mu^2)\sin\theta - 6\mu\cos\theta]/(2 - \mu^2)$ counterclockwise.

7.138 $B_x = -1959$ N, $B_y = 1238$ N, $C_x = 2081$ N, $C_y = -922$ N.

7.140 $\alpha_{OA} = 0.425$ rad/s^2 counterclockwise, $\alpha_{AB} = 1.586$ rad/s^2 clockwise.

7.142 $\alpha_{HP} = 5.37$ rad/s^2 clockwise.

7.144 208 m/s^2 to the left.

Chapter 8

8.2 (a), (b) 55.5 N-m.

8.4 6850 W (watts).

8.6 3140 ft-lb/s (5.71 hp).

8.8 $P_{av} = 0.374$ W.

8.10 (a) 17.7 rad/s. (b) 15.0 rad/s.

8.12 (a), (b) 4.77 rad/s.

8.14 (a) $\omega = \sqrt{2gx/(\frac{1}{12}l^2 + x^2)}$. (b) $x = l/\sqrt{12}$.

8.16 0.731 rad = 41.9°.

8.18 $U_{12} = 5630$ N-m.

8.20 $\omega = \sqrt{2b(mg\sin\beta + M/R)/(I + R^2m)}$.

8.22 16.7 rad/s clockwise.

8.24 $v = 0.413$ m/s.

8.28 $s = 66.1$ m.

8.30 $|\omega| = 4.29$ rad/s.

8.32 1.47 ft/s.

8.34 $\omega = \pm\sqrt{(3/l)[g(\cos\theta_0 - \cos\theta) + (k/ml)(\theta_0^2 - \theta^2)]}$.

8.36 7.93 ft/s downward.

8.38 $\omega = \pm\sqrt{3[(g/l)(1 - \cos\theta) - (k/m)(1 - \cos\theta)^2]}$.

8.40 4.52 rad/s counterclockwise.

8.42 2.80 rad/s counterclockwise.

8.44 $\omega_{AB} = 4.64$ rad/s clockwise,
$\omega_{BC} = 5.68$ rad/s counterclockwise.

8.46 $M = 28.2$ N-m.

8.48 (a) 8.60 kg-m^2/s. (b) 34.4 kg-m^2/s.

8.50 3 s.

8.52 (a) 126 rad/s. (b) 200 rad/s.

8.54 $\omega = 680$ rad/s.

8.56 $h = 1.33$ m, $\omega = 3.75$ rad/s.

8.58 0.115 rpm.

8.60 1.25 rad/s.

8.62 1.29 rad/s.

8.64 14.5°.

8.66 4.72 rad/s counterclockwise.

8.68 0.00589 rad/s counterclockwise.

8.70 1.54 rad/s counterclockwise.

8.72 3.75 rad/s counterclockwise.

8.74 (a) 50 N-m before, 31.3 N-m after.
(b) 50 N-m before, 50 N-m after.

8.78 Energy lost is $\frac{1}{6}mgl$.

8.80 5.96 ft/s.

8.82 183 N upward.

8.84 Velocity $= 5.77\mathbf{i} - 2.10\mathbf{j}$ (ft/s),
angular velocity = 3.15 rad/s counterclockwise.

8.86 $\omega' = 0.375$ rad/s.

8.88 $\omega'_A = \omega'_B = 0.675$ rad/s counterclockwise.

8.90 $\omega = 0.0997$ rad/s = 5.71 deg/s counterclockwise.

8.92 15.0 km/hr.

8.94 0.00336 rad/s clockwise.

8.96 $v = 2.05$ m/s.

8.98 $v = \sqrt{Fb/[\frac{1}{2}m_c + 2(m + I/R^2)]}$.

8.100 $\omega_S = 12.5$ rad/s.

8.102 (a) 1.13 ft. (b) 1.09 ft/s.

8.104 $P_{max} = 580$ ft-lb/s (1.05 hp), $P_{av} = 290$ ft-lb/s (0.53 hp).

8.106 11.1 rad/s.

8.108 1.77 rad/s counterclockwise.

8.110 $N = (373/283)mg$.

8.112 $\omega' = 0.721$ rad/s.

8.114 $\omega' = (\frac{1}{3} + \frac{2}{3}\cos\beta)\omega$.

8.116 $b = 2$ ft.

8.118 $\omega = 8.51$ rad/s.

8.120 $\theta = 54.7°$, $\omega = 10.7$ rad/s counterclockwise.

8.122 $0.0822\mathbf{k}$ (rad/s).

8.124 0.641 rad/s clockwise, $\mathbf{v}' = -2.24\mathbf{i}$ (ft/s).

Chapter 9

9.2 $\mathbf{a}_A = 358.8\mathbf{i} + 24.0\mathbf{j} + 149.6\mathbf{k}$ (km/s^2).

9.4 $\mathbf{a}_A = -255\mathbf{i} + 199\mathbf{j} - 209\mathbf{k}$ (ft/s^2).

9.6 $\mathbf{a}_C = 1000\mathbf{i} - 4000\mathbf{j} - 1000\mathbf{k}$ (in./s^2).

9.8 (a) $\boldsymbol{\alpha} = 77.8\mathbf{i} + 44.4\mathbf{j} + 44.4\mathbf{k}$ (rad/s^2).
(b) $160\mathbf{i} - 140\mathbf{j} - 140\mathbf{k}$ (km/s^2).

9.10 (a) $\boldsymbol{\alpha} = -\omega_0\omega_d\,\mathbf{k}$.
(b) $\mathbf{a}_A = R\sin 45°[2\omega_0\omega_d\mathbf{i} - \omega_d^2\,\mathbf{j} - (\omega_0^2 + \omega_d^2)\mathbf{k}]$.

9.12 $\boldsymbol{\omega}_{disk} = \omega_d\mathbf{i} + \omega_0\,\mathbf{j} + \omega_b\mathbf{k}$,
$\mathbf{v}_P = -\omega_b R\mathbf{i} + \omega_b b\mathbf{j} + (\omega_d R - \omega_0 b)\mathbf{k}$.

9.14 (a) $\boldsymbol{\omega}_{disk} = \omega_d\cos\beta\mathbf{i} + (\omega_0 + \omega_d\sin\beta)\mathbf{j}$.
(b) $\mathbf{v}_P = [\omega_d R - \omega_0(h + b\cos\beta - R\sin\beta)]\mathbf{k}$.

9.16 $\mathbf{v}_A = 3.20\mathbf{i}$ (in./s),
$\mathbf{v}_B = -4.10\mathbf{i} + 11.28\mathbf{j} - 2.26\mathbf{k}$ (in./s).

9.18 (a) $\boldsymbol{\omega}_{BC} = 0.1\mathbf{j} + 0.4\mathbf{k}$ (rad/s).
(b) $\mathbf{v}_C = -0.315\mathbf{i} + 0.085\mathbf{j} - 0.131\mathbf{k}$ (m/s).

9.20 (a) $\boldsymbol{\Omega} = (\omega_0/\sqrt{h^2 + R^2})(R\mathbf{i} + h\mathbf{j})$.
(b) $\boldsymbol{\omega}_{rel} = -(\omega_0\sqrt{h^2 + R^2}/R)\mathbf{i}$.
(c) $\boldsymbol{\omega} = (\omega_0/\sqrt{h^2 + R^2})[-(h^2/R)\mathbf{i} + h\mathbf{j}]$.

9.22 (a) $\boldsymbol{\alpha} = (\omega_0^2\,h/R)\mathbf{k}$.
(b) $\mathbf{a} = -[\omega_0^2 h^2/(h^2 + R^2)][h\mathbf{i} - R\mathbf{j} + (R + h^2/R)\mathbf{k}]$.

9.24 Zero.

9.26 $\mathbf{v}_B = 3\mathbf{j}$ (m/s), $\boldsymbol{\omega} = 2.37\mathbf{i} + 0.95\mathbf{j} + 4.58\mathbf{k}$ (rad/s).

9.28 $\mathbf{H} = -44.8\mathbf{i} - 10.8\mathbf{j} - 30.8\mathbf{k}$ (kg-m^2/s).

9.30 $\mathbf{H}_O = 20\mathbf{j} - 22\mathbf{k}$ (slug-ft^2/s).

9.32 $I_{xx} = 0.67$ kg-m^2, $I_{yy} = 5.33$ kg-m^2,
$I_{zz} = 6$ kg-m^2, $I_{xy} = I_{yz} = I_{zx} = 0$.

9.34 $I_{xx} = 0.12$ kg-m^2, $I_{yy} = 0.03$ kg-m^2,
$I_{zz} = 0.15$ kg-m^2, $I_{xy} = I_{yz} = I_{zx} = 0$.

9.36 $I_{xx} = 2.48$ slug-ft^2, $I_{yy} = 16.77$ slug-ft^2,
$I_{zz} = 19.25$ slug-ft^2, $I_{xy} = 5.59$ slug-ft^2, $I_{yz} = I_{zx} = 0$.

9.38 $I_{xx} = 0.0318$ kg-m^2, $I_{yy} = 0.0357$ kg-m^2,
$I_{zz} = 0.0674$ kg-m^2, $I_{xy} = 0.0219$ kg-m^2, $I_{yz} = I_{zx} = 0$.

9.40 (a) $\mathbf{H} = \frac{1}{12}ml^2(\omega_y\mathbf{j} + \omega_z\mathbf{k})$.
(b) $\mathbf{H}_O = \frac{1}{3}ml^2(\omega_y\mathbf{j} + \omega_z\mathbf{k})$.

9.42 $I_{x'x'} = 15{,}600$ kg-m^2, $I_{y'y'} = 226{,}800$ kg-m^2,
$I_{z'z'} = 242{,}400$ kg-m^2, $I_{x'y'} = -32{,}400$ kg-m^2,
$I_{y'z'} = I_{z'x'} = 0$.

9.44 $\mathbf{H} = 2.00\mathbf{i} - 1.64\mathbf{j} + 2.58\mathbf{k}$ (kg-m^2/s).

9.46 (a) $I = 3.56$ kg-m^2. (b) 14.22 kg-m^2/s.

9.48 $I_1 = 16.15, I_2 = 62.10, I_3 = 81.75$ kg-m^2,
$\mathbf{e}_1 = 0.964\mathbf{i} - 0.220\mathbf{j} + 0.151\mathbf{k}$,
$\mathbf{e}_2 = -0.204\mathbf{i} - 0.972\mathbf{j} - 0.114\mathbf{k}$,
$\mathbf{e}_3 = 0.172\mathbf{i} + 0.079\mathbf{j} - 0.982\mathbf{k}$.

9.50 $I_1 = 0, I_2 = 1/12, I_3 = 1/12$ kg-m^2,
$\mathbf{e}_1 = \cos\beta\mathbf{i} + \sin\beta\mathbf{j}, \mathbf{e}_2 = -\sin\beta\mathbf{i} + \cos\beta\mathbf{j}$,
$\mathbf{e}_3 = \mathbf{k}$.

9.52 $\Sigma\mathbf{M} = -83.2\mathbf{i} - 95.2\mathbf{j} + 154.4\mathbf{k}$ (N-m).

9.54 $\Sigma\mathbf{M}_O = -52\mathbf{i} + 132\mathbf{j} + 120\mathbf{k}$ (ft-lb).

9.56 $|\mathbf{a}_A| = 5F/m$.

9.58 (a) $\boldsymbol{\alpha} = 28.57\mathbf{i} + 5.71\mathbf{j} - 2.00\mathbf{k}$ (rad/s^2).
(b) $-0.33\mathbf{i} + 2.00\mathbf{j} - 19.54\mathbf{k}$ (ft/s^2).

9.60 (a) $-4.17\mathbf{k}$ m/s^2. (b) $\boldsymbol{\alpha} = 49.5\mathbf{i} + 137.7\mathbf{j}$ (rad/s^2).

9.62 $\boldsymbol{\alpha} = -500.0\mathbf{i} + 24.4\mathbf{j}$ (rad/s^2).

9.64 $\boldsymbol{\alpha} = \frac{4}{3}(g/b)\mathbf{j}$.

9.66 $\boldsymbol{\alpha} = 14.53\mathbf{j} + 4.65\mathbf{k}$ (rad/s^2).

9.68 27.4 N-m.

9.70 $|\mathbf{F}| = mg, |\mathbf{C}| = \frac{1}{12}ml^2\omega_0^2|\sin\beta\cos\beta|$.

9.74 $\frac{1}{2}mR^2\omega_0\omega_d$.

9.76 $-10\mathbf{j} + 14.91\mathbf{k}$ (lb), $-12.4\mathbf{i}$ (ft-lb).

9.82 157 N-m.

9.86 21.5 ft-lb.

9.90 $\omega_d = -\frac{1}{2}\omega_0 \sin\beta$.

9.92 $\dot{\psi} = 1.31$ rev/s.

9.94 29.1 rev/s.

9.96 (a) $\dot{\psi} = -2050$ rpm.
(b)

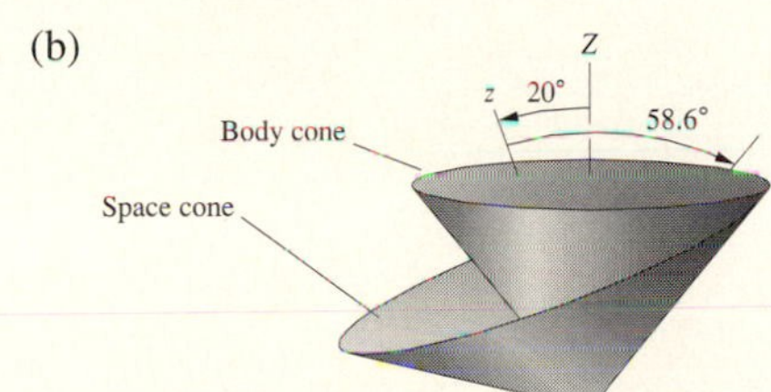

9.98 $N = \omega_0^2[(b/r)I_{zz}\sin^2\beta - I_{xx}\sin\beta\cos\beta]/$
$(b\sin\beta - r\cos\beta)$.

9.100 His right side.

9.102 $\Sigma\mathbf{M} = 2123\mathbf{i} - 155\mathbf{j} - 534\mathbf{k}$ (N-m).

9.104 (a) $\boldsymbol{\omega} = \omega_d\mathbf{i} + \omega_0\mathbf{j}$.
(b) $\mathbf{v}_A = -R\omega_0\cos\theta\mathbf{i} + R\omega_d\cos\theta\mathbf{j} + (R\omega_d\sin\theta - b\omega_0)\mathbf{k}$.

9.106 (a) $\boldsymbol{\omega} = 20\mathbf{i} - 5\mathbf{j}$ (rad/s).
(b) $\mathbf{v}_A = 0.25\mathbf{i} + 1.00\mathbf{j} + 3.73\mathbf{k}$ (m/s).

9.108 $\mathbf{a}_B = 166\mathbf{j}$ (ft/s^2),
$\boldsymbol{\alpha} = 7.31\mathbf{i} - 5.52\mathbf{j} + 22.89\mathbf{k}$ (rad/s^2).

9.110 $\mathbf{H} = 25\mathbf{i} + 50\mathbf{k}$ (kg-m^2/s).

9.112 $I_{xx} = 0.0398$ slug-ft^2, $I_{yy} = 0.0373$ slug-ft^2,
$I_{zz} = 0.0772$ slug-ft^2, $I_{xy} = I_{yz} = I_{zx} = 0$.

9.114 $\Sigma\mathbf{M} = -250\mathbf{i} - 250\mathbf{j} + 125\mathbf{k}$ (N-m).

9.116 $\omega_0 = \sqrt{g\sin\beta/(\frac{2}{3}l\sin\beta\cos\beta + b\cos\beta)}$.

9.118 (b) If $\omega_0 = 0$, the plate is stationary. The solution of the equation $2\cos\beta - \sin\beta = 0$ is the value of β for which the center of mass of the plate is directly above point O; the plate is balanced on one corner.

9.120 $\mathbf{F}_B = 52.72\mathbf{i} + 97.35\mathbf{j} + 9.26\mathbf{k}$ (N),
$\mathbf{M}_B = 0.05\mathbf{i} - 10.25\mathbf{j} + 30.63\mathbf{k}$ (N-m).

9.122

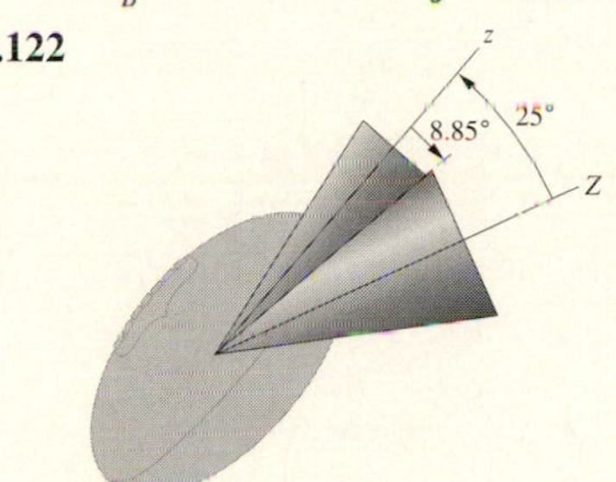

9.124 $\Sigma\mathbf{M} = -283\mathbf{i} - 2546\mathbf{j} - 800\mathbf{k}$ (ft-lb).

9.126 $\boldsymbol{\alpha} = 0.0535\mathbf{i} + 0.0374\mathbf{j} + 0.0160$ (rad/s^2).

Chapter 10

10.4 (a) $E = 1, \phi = 45°$.
(b)

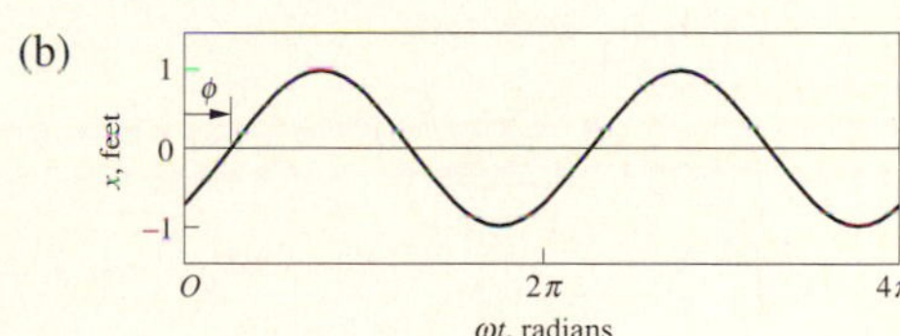

10.6 (a) $\tau = 2.09$ s, $f = 0.477$ Hz.
(b) $x = (0.1)\cos 3t$ m.

10.8 $f = (1/2\pi)\sqrt{k/m}$.

10.10 $\beta = 0.01\cos 6t$ rad.

10.12 $f = 2.33$ Hz.

10.14 1.46 Hz.

10.16 (a) $\tau = 2.09$ s, $f = 0.48$ Hz.
(b) $x = 0.36(1 - \cos 3t)$ m.

10.18 $x = 1.33(1 - \cos 3.24t)$ ft.

10.20 $f = 0.321$ Hz.

10.24 $I_{(z' \text{ axis})} = 24.2$ kg-m^2.

10.28 0.39 in.

10.30 $\theta_B = 0.172\sin 11.6t$ rad.

10.32 $m_A = 4.38$ kg.

10.36 (a) $f_d = 0.316$ Hz, $\tau_d = 3.17$ s.
(b) $f = 0.318$ Hz, $\tau = 3.14$ s.

10.38 $\tau_d = 2.22$ s, $f_d = 0.450$ Hz. (The undamped values are $\tau = 2.09$ s, $f = 0.477$ Hz.)

10.40 $c = 1.39$ N-s/m.

10.42 $c = 1780$ N-s/m.

10.44 (a) $\tau_d = 1.81$ s, $f_d = 0.55$ Hz. (b) 4.59 s.

10.46 $x = e^{-2t}(-0.325 \sin 2.24t - 0.363 \cos 2.24t) + 0.363$ m.

10.48 $x = -(0.363 + 1.090t)e^{-3t} + 0.363$ m.

10.50 $x = 1.333 + e^{-0.611t}(-0.256 \sin 3.18t - 1.333 \cos 3.18t)$ ft.

10.52 0.153 rad/s clockwise.

10.54 (a) 277 N-s/m. (b) $\theta = 10te^{-4.33t}$ rad. (c) $\theta_{max} = 0.850$ rad at $t = 0.231$ s.

10.56 $\theta_B = 0.209e^{-6.62t} \sin 9.55t$ rad.

10.58 (a) $x_p = 0.5 \sin 8t$ ft. (b) $x = -0.4 \sin 10t + \cos 10t + 0.5 \sin 8t$ ft.

10.60 (a) $x_p = 0.5$ m. (b) $x = e^{-t}(-0.289 \sin 1.73t - 0.5 \cos 1.73t) + 0.5$ m.

10.62 $\theta = 0.581 \sin 2.58t + 1.250 \sin 2t$ rad.

10.64 11.5 lb.

10.66 0.113 mm.

10.70 16.5 in.

10.72 5.5 mm.

10.74 (a) $x = (0.1) \cos 3t$ m.

(b)

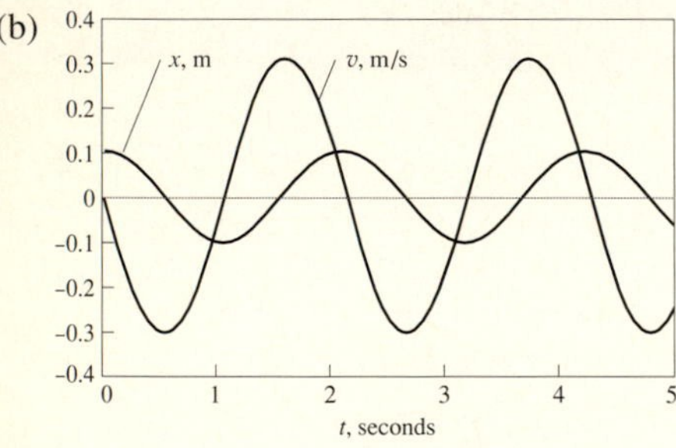

10.78 $f = 0.985$ Hz.

10.80 1.03 Hz.

10.82 $f = (1/2\pi)\sqrt{3[(k/m) - (g/2l)]}$.

10.84 $\delta = 2.07$.

10.86 $f_d = 0.714$ Hz.

10.88 $x = 0.118e^{-2.68t} - 0.118e^{-14.01t}$ ft.

10.90 $\theta_B = e^{-5.05t}(0.244 \sin 3.45t + 0.167 \cos 3.45t)$ rad.

10.92 $x = 0.253 \sin 6.19t + 0.100 \cos 6.19t + 0.145 \sin 3.00t$ m.

10.94 (a) $E_p = (2\pi v/\lambda)^2 h/[(k/m) - (2\pi v/\lambda)^2]$. (b) $v = \lambda\sqrt{k/m}/2\pi$.

Index

A

Acceleration, 4, 20
  angular, 54, 265, 292
  cartesian coordinates, 44
  centripetal, 75
  constant, 24
  Coriolis, 75, 321, 322
  cylindrical coordinates, 77
  normal and tangential components, 61
  polar coordinates, 75
  spherical coordinates, 574
  straight-line motion, 22
    as a function of position, 36
    as a function of time, 23
    as a function of velocity, 35
Acceleration due to gravity, 6
Amplitude, 529
Angular acceleration, 54, 265
Angular acceleration vector, 292
  of a rotating coordinate system, 313
Angular impulse, 239, 426
Angular impulse and momentum, principle of, 239, 426
Angular momentum, 239, 340, 426, 428, 472
Angular motion, 53, 490, 505
  equation of, three-dimensional, 490
  equation of, two-dimensional, 338, 342, 345
Angular units, 9
Angular velocity, 53, 265, 271
Angular velocity vector, 271
  of a rotating coordinate system, 313, 459
Apogee, 145
Attenuation, 540
Average force, 214, 425
Axial force, 373

B

Base units:
  International System, 8
  U.S. Customary, 8
Beams, internal forces and moments in, 373
Bending moment, 373
Body cone, 510
Body-fixed coordinate system, 301, 459, 490

C

Cartesian coordinates, 43
  acceleration, 44
  Newton's second law, 110
  position vector, 43
  velocity, 44
Celestial mechanics, see Orbital mechanics
Central-force motion, 239
Centripetal acceleration, 75
Centroids, table of, 570
Chain rule, 25, 35, 36, 37
Circular motion, 62, 76
Circular natural frequency, 530
Circular orbit, 66, 144
Collisions, see Impacts
Conic section, 144
Conservation of angular momentum, 239, 428
Conservation of energy, 182, 526
Conservation of linear momentum, 224, 227, 425, 434
Conservative force, 183, 194
  force exerted by a linear spring, 186
  weight of an object, 184
Conversion of units, 9
Coordinate system,
  body-fixed, 301, 459, 490
  cartesian, 43
  cylindrical, 76
  moving, 312
  polar, 73
  spherical, 574
Coriolis acceleration, 75, 321, 322
Coriolis force, 322
Couple,
  inertial, 358
  work done by, 409
Critical damping, 542
Curl, 194, 569, 574
Curvilinear motion, 43
Cylindrical coordinates, 76, 136
  acceleration, 77
  position vector, 77
  velocity, 77

D

D'Alembert's principle, 358
Damped vibrations, 540
  critical damping, 542
  logarithmic decrement, 542
  subcritical damping, 541
  supercritical damping, 542
Damping element, 540
Damping mechanism, 540
Dashpot, 540
Datum, 185
Degree, 9
Degree of freedom, 525
Derivative of a vector, 19
Derivatives, table of, 568
Derived units, 8
Differential equation, 528, 540
Direct central impacts, 228
Directrix, 144
Displacement, 19
  in straight-line motion, 21
Displacement transducer, 556
Dynamics, 2
Dynamics of a system of particles, 339
  principle of work and energy, 404

E

Earth-centered reference frame, 319
Earth-fixed reference frame, 320
Eccentricity, 144
Energy, 158, 404
  conservation of, 182, 410, 526
  kinetic, 158, 404
  potential, 183, 408
  of a system of particles, 404
  use of in vibrations, 526
Escape velocity, 40, 144

Euler, Leonhard, 147
Eulerian angles, 505
Euler's equations, 490
Euler's theorem, 271, 458

Finite-difference method, 146, 377
Fixed-axis rotation, 263
Flat response, 557
Foot, 4, 8
Force, 4
  average, 214
  conservative, 183, 194
  external, 107, 339
  gravitational, 6
  impulsive, 214
  inertial, 358
  internal, 107, 339, 405
  work done by, 158, 408
Forced vibrations, 547
Forcing function, 548
  oscillatory, 548
  polynomial, 550
Forward differencing, 147
Free vibrations, 528, 540
Frequency,
  circular, 530
  damped, 541
  natural, 530
  spectrum, 557

Gradient, 194, 569, 574
Gravity, 6
  of the earth, 6
  force between two particles, 6
  at sea level, 7
Gyroscope, 470

Homogeneous differential equation, 528
Homogeneous solution, 548

Impacts, 227, 434
  direct central, 228
  oblique central, 230
  perfectly elastic, 229
  perfectly plastic, 228
  of rigid bodies, 434
Impulse,
  angular, 239, 426
  linear, 214, 425
Impulse and momentum, principle of,
  angular, 239, 426
  linear, 214, 425
Impulsive force, 214
Inertial couple, 358
Inertial force, 358
Inertial reference frame, 108, 319
Inertia matrix, 473
Instantaneous axis of rotation, 271, 458
Instantaneous center, 284
Instantaneous radius of curvature, 61
Integrals, table of, 568
Internal forces and moments, 373
International System of units, 4, 8

Jet engine, 248

Kepler, Johannes, 141
Kilogram, 6, 8
Kinematics, 261, 458
Kinetic energy, 158, 404, 407

Linear differential equation, 528
Linear impulse, 214, 425
Linear impulse and momentum, principle of, 214, 425
Linear momentum, 106, 214, 224
  conservation of, 224, 227
Logarithmic decrement, 542

Mass, 5, 8
Mass flow, 244
Mass flow rate, 244
Mass moment of inertia, see Moments of inertia
Mechanics, 2
Meter, 4, 8
Moments of inertia,
  area, 383, 476
    table of, 570
  mass, 338, 381, 472
    about an arbitrary axis, 478
    cylinder, 390
    inertia matrix, 473
    parallel-axis theorems, 386, 477
    principal axes and moments of inertia, 479
    products of inertia, 473
    slender bar, 382, 475
    table of, 572
    thin plate, 383, 476
Momentum,
  angular, 239, 340, 472
  linear, 106, 214, 224
  of a system of particles, 339
Motion of a point, 17
  along a curvilinear path, 18, 43
  along a straight line, 21
  in cartesian coordinates, 43
  in cylindrical coordinates, 76
  in normal and tangential components, 58
  in polar coordinates, 73
Motion of a rigid body, see Rigid body
Motion of the center of mass, 106, 339, 490
Motor vehicles, dynamics of, 130

Natural frequency, 530
Newton, Isaac, 4, 6, 141
Newtonian gravitation, 6
Newton's laws, 4
Newton's second law, 126
  in cartesian coordinates, 110
  for the center of mass of an object, 106
  in normal and tangential components, 126
  in polar coordinates, 136
  in straight-line motion, 110
  in terms of a moving reference frame, 323
Newton's third law, 339
Normal and tangential components, 58
  acceleration, 59, 61

Newton's second law, 126
three-dimensional motion, 62
velocity, 59, 61
Numerical solutions, 146, 377
Nutation angle, 506, 510

O

Oblique central impacts, 230
Orbit, 141
circular, 66, 144
elliptic, 144
hyperbolic, 145
parabolic, 144
Orbital mechanics, 141
Orientation of a rigid body, 505
Oscillatory motion, see Vibrations
Osculating plane, 63
Overdamped systems, see Supercritical damping

P

Parallel-axis theorems, 386, 477
Particle kinematics, see Motion of a point
Particles, system of, 339, 404
Particular solution, 548
Path angle, 60
Pendulum, 527
Perigee, 145
Period, 530
damped, 541
Phase, 529
Planar motion, 264, also see Rigid body, dynamics and kinematics
Polar coordinates, 73
acceleration, 75
Newton's second law, 136
position vector, 74
velocity, 75
Position vector, 18
cartesian coordinates, 43
cylindrical coordinates, 77
polar coordinates, 73
spherical coordinates, 574
Potential energy, 182, 194
of a couple, 409
of a linear spring, 186
of an object's weight, 184
of a torsional spring, 409
Pound, 6, 9
Pound mass, 9
Power, 160, 410
Precession,
moment-free, 508
steady, 507
Precession angle, 506, 510
Principal axes, 479
Principal moments of inertia, 480
Products of inertia, 473
Projectile, 44

R

Radian, 9
Rectilinear motion, see Straight-line motion
Reference frame, 18
Relative acceleration,
in terms of a moving reference frame, 313
of two arbitrary points, 85
of two points of a rigid body, 292, 459
Relative motion of two points, 85
acceleration, 85, 313, 459
position, 85
velocity, 85, 313, 459
Relative velocity,
in terms of a moving reference frame, 313
of two arbitrary points, 85
of two points of a rigid body, 272, 459
Restitution, coefficient of, 229, 435
Right-hand rule, 272
Rigid body, 262
dynamics,
conservation of energy, 410
D'Alembert's principle, 358
equations of angular motion in three dimensions, 490
equations of angular motion in two dimensions, 338, 342
impacts, 434
numerical solutions, 146, 377
planar, 337
principle of angular impulse and momentum, 426
principle of linear impulse and momentum, 425
principle of work and energy, 404
three-dimensional, 490
two-dimensional, 337
kinematics,
angular velocity vector, 271
Eulerian angles, 505
Euler's theorem, 271, 458
relative acceleration of two points, 292, 459
relative velocity of two points, 85, 272, 459
rolling, 270
rotation about a fixed axis, 263, 264
sliding contact, 301
three-dimensional, 458
translation, 263
two-dimensional, 264
Rocket, 246
Rolling, 270
Rotation,
about a fixed axis, 263, 264
about a fixed point, 472, 490

S

Seismograph, 557
Separating variables, 35, 36, 37
Shear force, 373
Significant digits, 7
Simple harmonic motion, 529
SI units, see International System of units
Sliding contact, 301
Slug, 6, 9
Space, 4
Space cone, 510
Spectrum, 557
Spherical coordinates, 574
Spin angle, 506
Spring-mass oscillator, 526
Statics, 2
Steady-state solution, 549
Straight-line motion, 21
Strategy, 3
Subcritical damping, 541
Supercritical damping, 542

T

Tables:
centroids of areas, 570
centroids of lines, 571

centroids of volumes, 572
derivatives, 568
integrals, 568
moments of inertia of areas, 570
moments of inertia of homogeneous objects, 572
Taylor series, 569
Time, 4
Torsional springs, 380
potential energy, 409
Transducer, 557
Translation, 263
Trigonometry, 567
Two-dimensional equations of rigid-body motion, 338, 342
Two-dimensional motion, see Planar motion

U

Underdamped systems, see Subcritical damping
Units,
conversion of, 9
International System, 4, 8
U.S. Customary, 4, 8
Unit vectors,
in cartesian coordinates, 44
in cylindrical coordinates, 76
in normal and tangential components, 59, 61, 64
in polar coordinates, 73
rotating, 54
Universal gravitational constant, 6
U.S. Customary units, 4, 8

V

Velocity, 4
cartesian coordinates, 44
cylindrical coordinates, 77
determined using instantaneous center, 284
escape, 40, 144
normal and tangential components, 59, 61
polar coordinates, 75
spherical coordinates, 574
straight-line motion, 21
Vibrations, 525
amplitude, 529
conservative systems, 526
damped and unforced, 540
forced, 547
frequency, 530, 541
period, 530, 541
phase, 529
steady-state, 549
undamped and unforced, 528

W

Weight, 6
Work, 158, 168, 408
done by a couple, 409
done by a linear spring, 170
done by an object's weight, 168
relation to potential energy, 183
Work and energy, principle of, 158, 404

Appendix C

Properties of Volumes and Homogeneous Objects

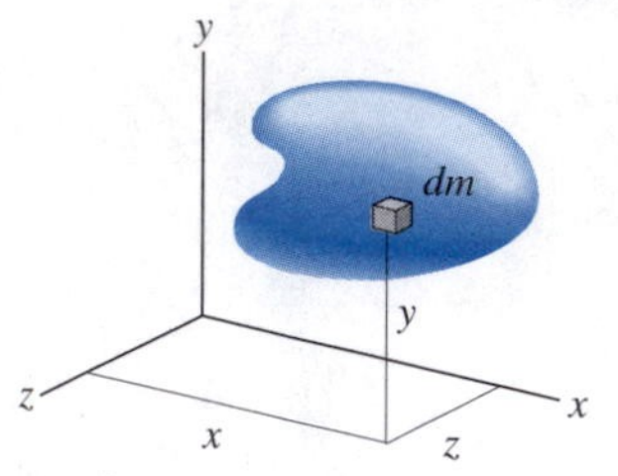

The moments and products of inertia of the object in terms of the xyz coordinate system are

$$I_{(x\text{ axis})} = I_{xx} = \int_m (y^2 + z^2)\, dm,$$

$$I_{(y\text{ axis})} = I_{yy} = \int_m (x^2 + z^2)\, dm,$$

$$I_{(z\text{ axis})} = I_{zz} = \int_m (x^2 + y^2)\, dm,$$

$$I_{xy} = \int_m xy\, dm, \qquad I_{yz} = \int_m yz\, dm,$$

$$I_{zx} = \int_m zx\, dm.$$

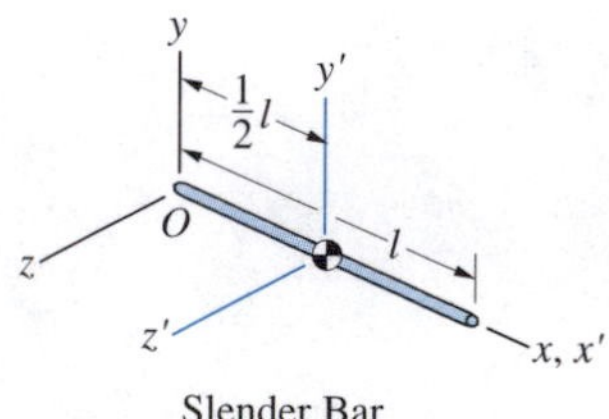

Slender Bar

$$I_{(x\text{ axis})} = 0, \qquad I_{(y\text{ axis})} = I_{(z\text{ axis})} = \frac{1}{3}ml^2,$$

$$I_{xy} = I_{yz} = I_{zx} = 0.$$

$$I_{(x'\text{ axis})} = 0, \qquad I_{(y'\text{ axis})} = I_{(z'\text{ axis})} = \frac{1}{12}ml^2,$$

$$I_{x'y'} = I_{y'z'} = I_{z'x'} = 0.$$

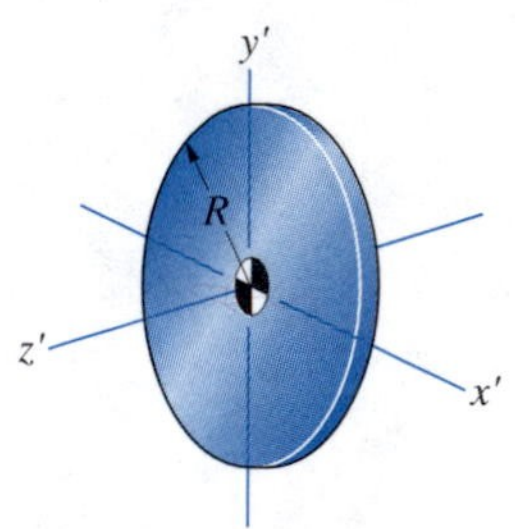

Thin Circular Plate

$$I_{(x'\text{ axis})} = I_{(y'\text{ axis})} = \frac{1}{4}mR^2, \qquad I_{(z'\text{ axis})} = \frac{1}{2}mR^2,$$

$$I_{xy} = I_{yz} = I_{zx} = 0.$$

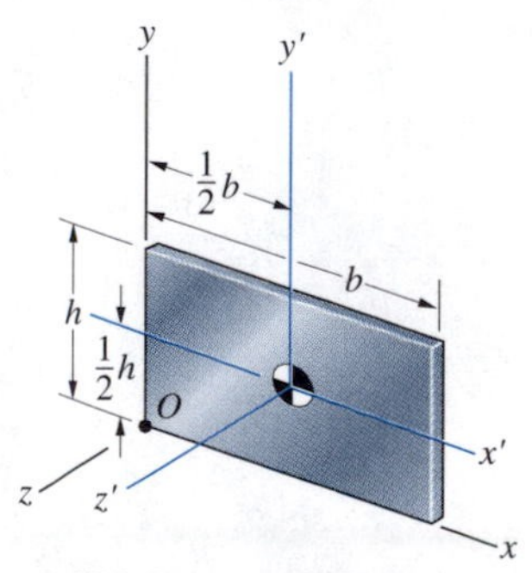

Thin Rectangular Plate

$$I_{(x\text{ axis})} = \frac{1}{3}mh^2, \qquad I_{(y\text{ axis})} = \frac{1}{3}mb^2, \qquad I_{(z\text{ axis})} = \frac{1}{3}m(b^2 + h^2),$$

$$I_{xy} = \frac{1}{4}mbh, \qquad I_{yz} = I_{zx} = 0.$$

$$I_{(x'\text{ axis})} = \frac{1}{12}mh^2, \qquad I_{(y'\text{ axis})} = \frac{1}{12}mb^2, \qquad I_{(z'\text{ axis})} = \frac{1}{12}m(b^2 + h^2),$$

$$I_{x'y'} = I_{y'z'} = I_{z'x'} = 0.$$

D0146403
Follett's Bookstore
FOR OFFICIAL USED TEXTBOOKS
Top Cash for Books Year 'Round

Social security numbers, number possible, 740
Sociological survey, 754
Softball diamond, size, 527, 528
Speaker placement, 528
Television screen, size, 528
Track records, 79, 94, 106, 444, 456
Turkey consumption, 621, 662
Typing speed, 621
VCR counter reading, 357, 591
Walking speed, 667
Work problems, 307, 313, 314, 316, 318, 324, 326, 327, 482, 564, 567, 609, 610, 669, 767
Work rate, 371, 374, 482
Zip codes, number possible, 739, 740

GEOMETRIC APPLICATIONS

Area
- of a circle, 78, 160, 167, 189, 224, 286, 561, 562, 596, 707
- of a parallelogram, 7, 189
- of a rectangle, 3, 105, 106, 143, 164, 167, 174, 175, 182, 183, 189, 195, 209, 224, 259, 260, 261, 262, 286, 293, 316, 462, 479, 482, 508, 561, 594, 609, 702, 703, 705, 707, 708, 767
- of a sector of a circle, 79
- of a square, 107, 144, 154, 165, 189, 190, 254, 259, 260, 261, 262, 286, 327, 562, 705, 707, 708
- of a triangle, 3, 95, 107, 175, 189, 224, 255, 259, 356, 523, 561, 597
- of a trapezoid, 317, 321
- *See also* Surface area.

Circumference, 76, 160, 375, 707
Complementary angles, 382, 397
Diagonals of a polygon, 529, 541, 557, 560, 562, 596, 606
Perimeter
- of a general polygon, 167, 209, 611, 767
- of an octagon, 375
- of a rectangle, 78, 89, 93, 94, 106, 109, 110, 111, 143, 209, 293, 300, 388, 401, 406, 470, 479, 482, 508, 589, 462, 669, 682, 702, 704, 705, 707, 708, 767
- of a square, 144, 470, 605, 611, 707
- of a triangle, 605

Right triangle problems, 258, 260, 523–530, 541, 542, 559, 560, 561, 562, 598, 703, 704, 705
Similar triangles, 312, 315, 316, 325
Supplementary angles, 388
Surface area
- of a capsule, 596
- of a cone, 596
- of a cube, 596, 598
- of a rectangular solid, 174, 260, 596
- of a right cylindrical solid, 184, 320, 596, 597, 598
- of a silo, 188, 190
- of a sphere, 356, 596

Volume
- of a cube, 154, 157, 174
- of a rectangular solid, 174, 183, 211, 261, 595, 705
- of a right circular cylinder, 375
- of a sphere, 321

PHYSICAL SCIENCE

Aperture, relative, 374
Atmospheric pressure, 664
Aviation, 141
Bridge expansion, 529
Carbon dating, 661, 663, 666, 668
Centripetal force, 767
Distance
- to the horizon, 523
- over water, 528
- traveled, 7, 140, 374, 402
- *See also* General Interest: Motion problems.

Earthquake magnitude, 656, 666, 668
Energy-mass relationship, 596
Falling object
- distance fallen, 160, 320, 375, 593, 596, 597
- time of fall, 558, 560, 596, 597, 609

Gas, volume and pressure, 374, 375
Grade, 329, 336, 337, 338, 378
Gravitational force, 317, 596
Guy wire, 524, 527, 561
Hang time, 593, 597
Hooke's law, 373
Horizon, distance to, 523
Horsepower, 682
Intensity of a signal, 375
Melting snow, 368
Ohm's law, 373
Olympic arrow, path of, 161
Orbiting time of a satellite, 7
Pendulum, period, 508, 592, 596
pH of substances, 655, 663, 666, 668
Pitch of a musical tone, 374
Power consumption, 375
Pressure at sea depth, 455
Projectile, height, 161, 260, 606
Pumping rate, 374
Relativity, 596
Resistance, 319, 321, 374, 375
Skidding car, speed, 503
Skydiving, 160
Stopping distance, 375
Television signal, intensity, 375
Temperature
- conversion, 79, 106, 107, 321, 322, 356, 447, 455
- of liquids, 455
- wind chill, 503

Thunderstorm, distance from, 75
Water from melting snow, 368
Weight
- of an astronaut, 375
- on Mars, 316, 374
- on the moon, 315

Whispering gallery, 687
Wind chill temperature, 503

Elementary and Intermediate Algebra

Concepts and Applications: A Combined Approach

Elementary and Intermediate Algebra

Concepts and Applications: A Combined Approach

MARVIN L. BITTINGER
Indiana University–Purdue University at Indianapolis

DAVID J. ELLENBOGEN
St. Michael's College

BARBARA JOHNSON
Indiana University–Purdue University at Indianapolis

ADDISON-WESLEY PUBLISHING COMPANY
Reading, Massachusetts • Menlo Park, California • New York
Don Mills, Ontario • Wokingham, England • Amsterdam
Bonn • Sydney • Singapore • Tokyo • Madrid
San Juan • Milan • Paris

Sponsoring Editor Jason Jordan
Managing Editor Karen Guardino
Production Supervisors Jennifer Bagdigian and Peggy McMahon
Design, Editorial, and Production Services Quadrata, Inc.
Illustrator Scientific Illustrators and Leo Harrington
Manufacturing Supervisor Roy Logan
Cover Designer Leslie Haimes
Cover Photograph Larry Dale Gordon, The Image Bank

Photo Credits

p. 1, Barrie Rokeach/The Image Bank **p. 61,** Dick Morton **p. 113,** David F. Wisse **p. 147,** Steve Brown/Leo de Wys **p. 213,** Grant V. Faint/The Image Bank **p. 265,** Lou Jones/The Image Bank **p. 307,** ©Todd Phillips/Third Coast Stock Source Inc. **p. 329,** ©David Ball, The Stock Market **p. 330,** AP/Wide World Photos **p. 381,** ©Fundamental Photographs, NYC, Richard Megna **p. 439,** Keith Wood/Tony Stone Images **p. 444,** AP/Wide World Photos **p. 471,** Comstock, Inc./Russ Kinne **p. 483,** ©Mark and Audrey Gibson, The Stock Market **p. 545,** Bettmann Archive **p. 560,** Bettmann Archive **p. 597,** AP/Wide World Photos **p. 613,** Sotheby's Inc. **p. 664,** AP/Wide World **p. 671,** The Finest Image Photography Studio **p. 709,** Grant V. Faint/The Image Bank

Library of Congress Cataloging-in-Publication Data

Bittinger, Marvin L.
Elementary and intermediate algebra : concepts and applications : a combined approach / Marvin L. Bittinger, David Ellenbogen, Barbara Johnson. — 1st ed.
p. cm
Includes index.
ISBN 0-201-76559-4
1. Algebra. I. Ellenbogen, David. II. Johnson, Barbara L. (Barbara Loreen), 1962– . III. Title.
QA152.2.B5797 1995
512.9—dc20 95-14467
CIP

Copyright © 1996 by Addison-Wesley Publishing Company, Inc. All rights reserved. No part of this publication may be reproduced, stored in a retrieval system, or transmitted, in any form or by any means, electronic, mechanical, photocopying, recording, or otherwise, without the prior written permission of the publisher. Printed in the United States of America. Published simultaneously in Canada.

4 5 6 7 8 9 10—DOW—9897

For our spouses: Elaine, Peggy and Jeff

M.L.B.
D.J.E.
B.L.J.

PREFACE

Our goal in preparing *Elementary and Intermediate Algebra Concepts and Applications: A Combined Approach* was to address the major challenges for teachers of developmental mathematics courses that we have seen emerging during the early 1990s. The first challenge is to prepare students of developmental mathematics to make the transition from ''skills-oriented'' elementary and intermediate algebra courses to the more ''concept-oriented'' presentation of college algebra or other college-level mathematics courses. The second is to teach these same students critical-thinking skills: to reason mathematically, to communicate mathematically, and to solve mathematical problems. The third challenge is to eliminate the content overlap between elementary and intermediate algebra texts.

Appropriate for a course combining the study of elementary and intermediate algebra, this text covers both elementary and intermediate algebra topics without the repetition of instruction necessary in two separate texts. Topics introduced early are reinforced later without overlapping content. This text is designed to prepare students for any mathematics course at the college algebra level.

APPROACH

Our approach, which has been developed over many years, is designed to help today's students both learn and retain mathematical concepts. The following are some aspects of the approach that we have used in the text to help meet the challenges we all face teaching developmental mathematics.

PROBLEM SOLVING AND APPLICATIONS

One distinguishing feature of our approach is our treatment of and emphasis on problem solving. We use problem solving and applications to motivate the material wherever possible, and we include problem-solving techniques and a variety of real-life applications throughout the text. We feel that problem solving encourages students to think about how mathematics can be used as a tool throughout their lives. It also challenges students and helps to prepare them for more difficult material in later courses.

- In Chapter 2, we introduce the five-step process for solving problems: (1) Familiarize, (2) Translate, (3) Carry out, (4) Check, and (5) State the Answer. These steps are used throughout the text whenever we encounter a problem-solving situation. Repeated use of this algorithm gives students a sense that they have a starting point for any type of problem they encounter, and frees them to focus on the mathematics necessary to successfully translate the problem situation. (See pages 84–86, 312, 313.)
- Because so many students remain convinced that they ''cannot do word problems,'' we make use of guessing as a means of familiarizing oneself with a problem-solving situation. By checking to see if a guess is correct, students can more easily discover an algebraic translation of the problem. (See pages 88–91.)
- A list of real-world applications, which appeal to both faculty and students, appears on the inside front cover of the text. Applications come from a wide variety of disciplines and fields, such as natural and social sciences, business and economics, and health.

GRAPHING AND FUNCTIONS

Chapter 3 contains an intuitive introduction to graphing, a topic which is integrated throughout the text. This helps students visualize the mathematics of many concepts while at the same time preparing them for the more formal discussion of graphing in Chapter 7. (See pages 116, 124.)

Functions are also introduced in Chapter 7. Throughout the rest of the text, we present functions and graphs to help students develop an intuitive understanding of different types of equations and their solutions. (See pages 576, 614.)

PEDAGOGY

Skill Maintenance Exercises and *Cumulative Reviews.* Retention of skills is critical to the future success of students. In nearly all exercise sets, we include carefully chosen exercises that review skills and concepts from preceding chapters of the text. Each chapter test includes Skill Maintenance Exercises selected to test material from the three or four text sections that are identified at the beginning of each chapter. After every three chapters, and at the end of the text, we have also included a Cumulative Review, which reviews skills and concepts from all preceding chapters of the text. (See pages 164, 326, 481, 669, and 776.)

Synthesis Exercises. Each exercise set ends with a set of Synthesis Exercises. These problems offer opportunities for students to synthesize skills and concepts from earlier sections with the present material, or can provide students with deeper insights into the current topic. Synthesis Exercises are generally more challenging than those in the main body of the exercise set. (See pages 21, 79, 389 and 390.)

Verbalization Skills. Wherever appropriate throughout the text, we have discussed how mathematical terms are used in common language. In addition, thinking and writing exercises are included in the Synthesis Exercises. These encourage students to verbalize mathematical concepts, leading to a better understanding of them. The Summary and Review sections of each chapter also emphasize key terms and important properties and formulas. (See pages 413 and 538.)

FEATURES

Design

- The design is open and uncluttered, so students can easily read the text. Pedagogical use of color makes it easier to see where exercises, explanations, and examples begin and end.
- The art has been carefully chosen and designed. We have ensured the accuracy of the graphical art through the use of computer-generated graphs. Color is used pedagogically and precisely to help the student visualize the mathematics. (See pages 524 and 619.)

Technology Connections

- Technology Connections features occur throughout the book; they integrate graphing technology, increase the understanding of concepts through visualization, encourage exploration, and motivate discovery learning.
- Many section exercise sets contain problems which are designed to be solved using a grapher and are marked with an icon {}.
- Some section exercise sets also include exercises which are designed to be solved using a scientific calculator and are marked with an icon {}.

Writing Exercises

- Nearly every set of Synthesis Exercises begins with two writing exercises. These exercises require written answers that aid in student comprehension, critical thinking, and conceptualization. Answers to writing exercises are provided for the chapter reviews only because some instructors may collect answers to writing exercises and more than one answer may be correct.

SUPPLEMENTS FOR THE INSTRUCTOR

INSTRUCTOR'S SOLUTIONS MANUAL
by Judith A. Penna

This supplement contains worked-out solutions to all exercises in the text.

PRINTED TEST BANK/INSTRUCTOR'S RESOURCE GUIDE
By Donna DeSpain

This supplement contains the following:

- Six alternative test forms for each chapter and six final examinations
- Extra practice problems for challenging topics in the text
- Black-line masters of grids and number lines for transparency masters or test preparation
- Videotape index and section cross references to the tutorial software packages available with this text

COMPUTERIZED TESTING

Omnitest^3 (for IBM and Macintosh). This computerized test bank allows you to create up to 99 versions of a customized test with just a few keystrokes, and allows the option of choosing items by chapter, section, or objective. It contains over 400 multiple-choice and open-ended algorithms. You may enter your own test items, edit existing items, and define the level of difficulty of problems.

SUPPLEMENTS FOR THE STUDENT

STUDENT'S SOLUTIONS MANUAL
by Judith A. Penna

This manual contains completely worked-out solutions with step-by-step annotations for all the odd-numbered exercises in the text, and answers for all even-numbered exercises in the text.

VIDEOTAPES

Developed especially for this text, these videotapes feature an engaging team of lecturers presenting material from each section of the text in an interactive format that includes a group of students. The lecturers' presentation also incorporates slides, computer-generated graphics, and a white board to support an approach that emphasizes visualization and problem solving.

INTERACT MATH TUTORIAL SOFTWARE

InterAct Math Tutorial Software has been developed and designed by professional software engineers working closely with a team of experienced developmental math teachers.

InterAct Math Tutorial Software includes exercises that are linked with every section in the textbook and require the same computational and problem-solving skills as their companion exercises in the text. Each exercise has an example and an interactive guided solution that are designed to involve students in the solution process and to help them identify precisely where they are having trouble. In addition, the software recognizes common student errors and provides students with appropriate customized feedback.

With its sophisticated answer recognition capabilities, InterAct Math Tutorial Software recognizes appropriate forms of the same answer for any kind of input. It also tracks student activity and scores for each section, which can then be printed out.

Available for DOS-based, Windows, and Macintosh computers, the software is free to qualifying adopters.

ACKNOWLEDGMENTS

We wish to express our appreciation to the many people who helped with the development of this book. Laurie A. Hurley and Bettie Bedan deserve special thanks for their many fine suggestions. Their proofreading of the text contributed immeasurably to the accuracy and readability of the text. Judy Penna also merits special thanks for her preparation of the *Student's Solutions Manual,* the *Instructor's Solution Manual,* and the indexes. Judy's work is always performed with a thoroughness that amounts to another proofreading of the book and for that we are grateful. We are also indebted to Stuart Ball for his expert guidance in preparing the Technology Connections and the associated artwork.

This book's sponsoring editor, Jason Jordan, did an excellent job coordinating the project; the production supervisors, Jenny Bagdigian and Peggy McMahon, performed admirably directing the production, with careful attention to detail; George and Brian Morris of Scientific Illustrators generated a remarkable set of graphs and illustrations that are both precise and easily understood; and Leo Harrington drew the many fine sketches that enhance our exercises and examples.

In addition, we thank the following professors for their thoughtful reviews and insightful comments.

Diane Bender, *Highline Community College*
John Coburn, *St. Louis Community College*
Jane Hammontree, *Tulsa Junior College NE*
Joanne Thomasson, *Pellissippi State Technical Community College*
Karl Zilm, *Lewis and Clark Community College*

Finally, a special thank you to all those who so generously agreed to discuss their professional uses for mathematics in our chapter openers. These dedicated people all share a desire to make math more meaningful to students. We cannot imagine a finer set of role models.

M.L.B.
D.J.E.
B.L.J.

CONTENTS

1 Introduction to Algebra and Algebraic Expressions

1.1 Introduction to Algebra 2
1.2 The Commutative, Associative, and Distributive Laws 8
1.3 Fractional Notation 14
1.4 Positive and Negative Real Numbers 22
1.5 Addition of Real Numbers 30
1.6 Subtraction of Real Numbers 36
1.7 Multiplication and Division of Real Numbers 42
1.8 Exponential Notation and Order of Operations 49
Summary and Review 56
Test 58

2 Equations, Inequalities, and Problem Solving

2.1 Solving Equations 62
2.2 Using the Principles Together 68
2.3 Formulas 75
2.4 Applications with Percent 80
2.5 Problem Solving 84
2.6 Solving Inequalities 95
2.7 Problem Solving Using Inequalities 103
Summary and Review 107
Test 110

3 Introduction to Graphing

3.1 Ordered Pairs and Graphs 114
3.2 Graphing Linear Equations 122
3.3 More on Graphing Linear Equations 129
3.4 Graphs and Problem Solving 135
Summary and Review 142
Test 144

Cumulative Review: Chapters 1–3 145

4
Polynomials

4.1 Exponents and Their Properties 148
4.2 Poynomials 154
4.3 Addition and Subtraction of Polynomials 162
4.4 Multiplication of Polynomials 169
4.5 Special Products 175
4.6 Polynomials in Several Variables 183
4.7 Division of Polynomials 190
4.8 Synthetic Division 195
4.9 Negative Exponents and Scientific Notation 199
Summary and Review 207
Test 210

5
Polynomials and Factoring

5.1 Introduction to Factoring 214
5.2 Factoring Trinomials of the Type $x^2 + bx + c$ 218
5.3 Factoring Trinomials of the Type $ax^2 + bx + c$ 224
5.4 Factoring Trinomial Squares and Differences of Squares 232
5.5 Factoring Sums and Differences of Cubes 239
5.6 Factoring: A General Strategy 242
5.7 Solving Quadratic Equations by Factoring 247
5.8 Problem Solving 254
Summary and Review 261
Test 263

6
Rational Expressions and Equations

6.1 Rational Expressions 266
6.2 Multiplication and Division 271
6.3 Addition and Subtraction 276
6.4 Least Common Multiples and Denominators 281
6.5 Addition and Subtraction with Unlike Denominators 287
6.6 Complex Rational Expressions 293
6.7 Solving Rational Equations 300
6.8 Problem Solving: Rational Equations and Proportions 306
6.9 Formulas and More Problem Solving 317
Summary and Review 322
Test 325

Cumulative Review: Chapters 1–6 326

7 Graphs and Functions

7.1 Slope 330
7.2 Slope-Intercept Form 338
7.3 Point-Slope Form 344
7.4 Functions 350
7.5 The Algebra of Functions 359
7.6 Variation and Problem Solving 367
Summary and Review 376
Test 379

8 Systems of Equations and Problem Solving

8.1 Systems of Equations in Two Variables 382
8.2 Solving by Substitution or Elimination 391
8.3 Problem Solving Using Systems of Two Equations in Two Variables 397
8.4 Systems of Equations in Three Variables 407
8.5 Problem Solving Using Systems of Three Equations 414
8.6 Elimination Using Matrices 419
8.7 Determinants and Cramer's Rule 423
8.8 Business and Economic Applications 428
Summary and Review 433
Test 436

9 Inequalities and Linear Programming

9.1 Interval Notation and Problem Solving 440
9.2 Compound Inequalities 449
9.3 Absolute-Value Equations and Inequalities 456
9.4 Inequalities in Two Variables 463
9.5 Problem Solving Using Linear Programming 471
Summary and Review 477
Test 480

Cumulative Review: Chapters 1–9 481

10 Rational Exponents and Radicals

10.1 Radical Expressions 484
10.2 Rational Numbers as Exponents 491
10.3 Multiplying and Simplifying with Radical Expressions 496

10.4 Dividing and Simplifying Radical Expressions 504
10.5 Addition, Subtraction, and More Multiplication 509
10.6 Rationalizing Numerators and Denominators 513
10.7 Solving Radical Equations 518
10.8 Geometric Applications 523
10.9 The Complex Numbers 530
Summary and Review 538
Test 542

11 Quadratic Equations and Functions

11.1 Solving Quadratic Equations: The Principle of Square Roots 546
11.2 Solving Quadratic Equations: Completing the Square 549
11.3 The Quadratic Formula and Problem Solving 554
11.4 Rational Equations and Problem Solving 562
11.5 The Discriminant and Solutions to Quadratic Equations 568
11.6 Equations Reducible to Quadratic 571
11.7 Quadratic Functions and Their Graphs 575
11.8 More About Graphing Quadratic Functions 583
11.9 Problem Solving and Quadratic Functions 588
11.10 Polynomial and Rational Inequalities 598
Summary and Review 606
Test 610

12 Exponential and Logarithmic Functions

12.1 Exponential Functions 614
12.2 Composite and Inverse Functions 622
12.3 Logarithmic Functions 633
12.4 Properties of Logarithmic Functions 638
12.5 Common and Natural Logarithms 644
12.6 Solving Exponential and Logarithmic Equations 651
12.7 Applications of Exponential and Logarithmic Functions 655
Summary and Review 664
Test 667

Cumulative Review: Chapters 1–12 669

13 Conic Sections

13.1 The Distance and Midpoint Formulas 672
13.2 Conic Sections: Parabolas and Circles 676
13.3 Conic Sections: Ellipses and Hyperbolas 684
13.4 Nonlinear Systems of Equations 697
Summary and Review 706
Test 708

14 Sequences, Series, and Combinatorics

14.1 Sequences and Series 710
14.2 Arithmetic Sequences and Series 716
14.3 Geometric Sequences and Series 724
14.4 Combinatorics: Permutations 732
14.5 Combinatorics: Combinations 741
14.6 The Binomial Theorem 747
14.7 Probability 753
Summary and Review 762
Test 765

Cumulative Review: Chapters 1–14 766

Tables

Table 1 Powers, Roots, and Reciprocals 769
Table 2 Common Logarithms 770
Table 3 Geometric Formulas 772
Table 4 Fractional and Decimal Equivalents 773

Answers A-1

Index I-1

Elementary and Intermediate Algebra

Concepts and Applications: A Combined Approach

CHAPTER 1

Introduction to Algebra and Algebraic Expressions

AN APPLICATION

In the course of one four-month period, the water level of Lake Champlain went down 2 ft, up 1 ft, down 5 ft, and up 3 ft. How much had the lake level changed at the end of the four months?

This problem appears as Example 7 in Section 1.5.

Sharon Meyer
TELEVISION WEATHER FORECASTER

"Whether it's in calculating the number of heating degree days, the probability of precipitation, or changes in lake level, a forecaster's career relies heavily on a solid understanding of mathematics. I use math every single day as part of my job."

Problem solving is the focus of this text. In this chapter we discuss some preliminaries that are needed for the problem-solving approach that we develop and begin to use in Chapter 2. We also review some arithmetic, discuss real numbers and their properties, and examine how real numbers are added, subtracted, multiplied, divided, and raised to powers.

1.1 Introduction to Algebra

Algebraic Expressions • Translating to Algebraic Expressions • Translating to Equations

This section introduces some basic concepts of algebra. Since equation solving is central to the study of algebra, we concentrate on the expressions that appear in equations and some important words for translating English to mathematics.

Algebraic Expressions

Probably the greatest difference between arithmetic and algebra is the extensive use of *variables* in algebra. When a letter is used to stand for any number chosen from a variety of numbers, we call the number a **variable.** For example, if n represents the number of students registered for a college's 8 AM section of Elementary Algebra, the number n will vary from semester to semester, if not from day to day. If each student in the 8:00 AM Elementary Algebra section paid $500 to take the class, the college would collect a total of $500 \cdot n$ dollars. Since the cost per student, $500, is the same regardless of how many students are registered, the number 500 is called a **constant.** The number of registered students n can vary, so n is a variable.

Cost Per Student (in Dollars)	Number of Students Registered	Total Collected (in Dollars)
500	n	$500 \cdot n$

The expression $500 \cdot n$ is called a **variable expression** because its value varies with the choice of n. Of course, the total amount collected, $500 \cdot n$, will grow as the number of students registered grows. We now replace n with a variety of values and compute the total amount collected. In doing so, we say that we are **evaluating** the expression $500 \cdot n$.

Cost Per Student (in Dollars) 500	Number of Students Registered n	Total Collected (in Dollars) $500 \cdot n$
500	20	10,000
500	25	12,500
500	30	15,000

Variable expressions are examples of *algebraic expressions*. An **algebraic expression** consists of variables, numerals, operation signs, and/or grouping symbols. Examples of algebraic expressions are

$$x + 35, \quad 7 \cdot t, \quad 13a - b, \quad 15 \div z, \quad \tfrac{9}{7}, \quad \text{and} \quad 3x(a + b).$$

Note that a fraction bar is a division symbol: $\frac{9}{7}$ means $9 \div 7$. Similarly, multiplication can be written in several ways. For example, "7 times t" can be written as $7 \cdot t$, $7 \times t$, $7(t)$, or simply $7t$.

To evaluate an algebraic expression, we **substitute** a number for each variable in the expression.

EXAMPLE 1

Evaluate the expression for the given values: **(a)** $x + y$ when $x = 37$ and $y = 28$; **(b)** $5ab$ when $a = 2$ and $b = 3$.

Solution

a) We substitute 37 for x and 28 for y and carry out the addition:

$$x + y = 37 + 28 = 65.$$

The number 65 is called the **value** of the expression.

b) We substitute 2 for a and 3 for b and multiply:

$$5ab = 5 \cdot 2 \cdot 3 = 10 \cdot 3 = 30.$$ ❑

EXAMPLE 2

The area A of a rectangle of length l and width w is given by the formula $A = lw$. Find the area when l is 24.5 in. and w is 10 in.

Solution We evaluate, using 24.5 in. for l and 10 in. for w and carry out the multiplication:

$$\begin{aligned} A = lw &= (24.5 \text{ in.})(10 \text{ in.}) \\ &= (24.5)(10)(\text{in.})(\text{in.}) \\ &= 245 \text{ in}^2, \quad \text{or } 245 \text{ square inches.} \end{aligned}$$

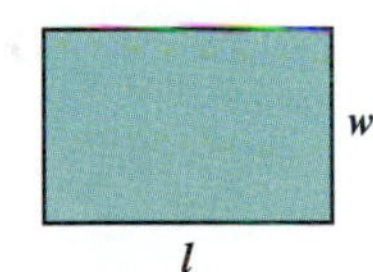

Note that $(\text{in.})(\text{in.}) = \text{in}^2$. Exponents are discussed in detail in Section 1.8. ❑

EXAMPLE 3

The area of a triangle with a base of length b and a height of length h is given by the formula $A = \frac{1}{2}bh$. Find the area when b is 8 m and h is 6.4 m.

Solution We substitute 8 m for b and 6.4 m for h and carry out the multiplication:

$$\begin{aligned} A = \tfrac{1}{2}bh &= \tfrac{1}{2}(8 \text{ m})(6.4 \text{ m}) \\ &= \tfrac{1}{2}(8)(6.4)(\text{m})(\text{m}) \\ &= 4(6.4) \text{ m}^2 \\ &= 25.6 \text{ m}^2, \quad \text{or } 25.6 \text{ square meters.} \end{aligned}$$

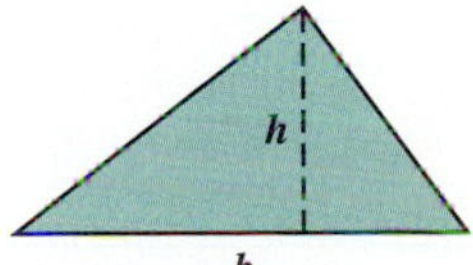

❑

Translating to Algebraic Expressions

Before attempting to translate problems to equations, we need to be able to translate certain phrases to algebraic expressions.

KEY WORDS

Addition (+)	Subtraction (−)	Multiplication (·)	Division (÷)
add sum plus more than increased by greater than	subtract difference minus less than decreased by take from	multiply product times twice of	divide quotient divided by ratio of

EXAMPLE 4

Translate each phrase to an algebraic expression.

a) Twice (or two times) some number
b) Seven less than some number
c) Eighteen more than a number
d) A number divided by five

Solution

a) Think of some number, say 8. What number is twice 8? It is 16. To get 16, you multiplied by 2. Now consider a variable. Use y to represent "some number" and multiply by 2. The expression

$$y \times 2, \quad 2 \times y, \quad 2 \cdot y, \quad \text{or} \quad 2y$$

is the translation of "Twice (or two times) some number."

b) We let x represent "some number." Now if the number were 23, then the translation would be $23 - 7$. If the number were 345, the translation would be $345 - 7$. If the number is x, the translation of "Seven less than some number" is

$$x - 7.$$

c) If we knew the number to be 10, the translation would be $10 + 18$, or $18 + 10$. We let t represent "a number," so the translation of "Eighteen more than a number" is

$$t + 18, \quad \text{or} \quad 18 + t.$$

d) We let m represent "a number." If the number were 9, the translation would be $9 \div 5$, or $\frac{9}{5}$. Thus our translation of "A number divided by five" is

$$m \div 5, \quad \text{or} \quad \frac{m}{5}.$$

❑

CAUTION! Because the order in which we subtract and divide affects the answer, answering $7 - x$ or $5 \div m$ in Examples 4(b) and 4(d) is incorrect.

EXAMPLE 5

Translate each of the following.

a) Some number, increased by 5
b) Half of a number
c) Five more than three times some number
d) Six less than the product of two numbers
e) Seventy-six percent of some number

Solution

	Phrase	*Algebraic Expression*
a)	Some number, increased by 5	$n + 5$, or $5 + n$
b)	Half of a number	$\frac{1}{2}t$, or $\frac{t}{2}$
c)	Five more than three times some number	$3p + 5$, or $5 + 3p$
d)	Six less than the product of two numbers	$mn - 6$
e)	Seventy-six percent of some number	$76\%z$, or $0.76z$

❑

Translating to Equations

The symbol = (''equals'') is used to indicate that the algebraic expressions on either side of the equals sign represent, or name, the same number. An **equation** is a number sentence with the verb = . Equations may be true, false, or neither true nor false.

EXAMPLE 6

Determine whether the equation is true, false, or neither: **(a)** $3 + 2 = 5$; **(b)** $7 - 2 = 4$; **(c)** $x + 6 = 13$.

Solution

a) $3 + 2 = 5$ The equation is *true*.
b) $7 - 2 = 4$ The equation is *false*.
c) $x + 6 = 13$ The equation is *neither* true nor false, because we do not know what number x represents. ❑

Solution

A replacement or substitution that makes an equation true is called a *solution*. Some equations may have more than one solution, and some may have no solution. When we have found all the solutions, we say that we have *solved* the equation.

One way to determine whether a number is a solution of an equation is to use that number to evaluate the expressions on each side of the equation by substitution. If the values are the same, then the number is a solution.

EXAMPLE 7

Determine whether 7 is a solution of $x + 6 = 13$.

Solution

$x + 6 = 13$ Writing the equation

$7 + 6 \ ? \ 13$ Substituting 7 for x

$13 \mid 13$ $13 = 13$ is TRUE.

Since the left-hand and the right-hand sides are the same, we have a solution. No other number makes the equation true, so the only solution is the number 7. ❑

Although we do not study solving equations until Chapter 2, we can translate certain problem situations to equations now.

EXAMPLE 8

Translate the following problem to an equation.

What number plus 478 is 1019?

Solution We let y represent the unknown number. In this example, the translation comes almost directly from the English sentence.

What number plus 478 is 1019?

$$y \quad + \quad 478 \quad = \quad 1019$$

Note that "is" translates to "=" and "plus" translates to "+." ❑

Sometimes it helps to reword a problem before translating.

EXAMPLE 9

Translate the following problem to an equation.

The elevation of Denver, 5280 ft, is 88 times the elevation of Houston. What is the elevation of Houston?

Solution We let h represent the elevation of Houston. The rewording and translation follow:

Rewording: 88 times the elevation of Houston is 5280.

Translating: $88 \cdot h = 5280$

Note that "times" translates to "·." ❑

EXERCISE SET 1.1

Evaluate.

1. $6x$, when $x = 7$
2. $7y$, when $y = 7$
3. $9 + a$, when $a = 7$
4. $15 - a$, when $a = 11$
5. $\dfrac{3p}{q}$, when $p = 2$ and $q = 6$
6. $\dfrac{5y}{z}$, when $y = 15$ and $z = 25$
7. $\dfrac{x + y}{5}$, when $x = 10$ and $y = 20$
8. $\dfrac{p + q}{2}$, when $p = 2$ and $q = 16$
9. $\dfrac{x - y}{8}$, when $x = 20$ and $y = 4$
10. $\dfrac{m - n}{5}$, when $m = 16$ and $n = 6$
11. $\dfrac{x}{y}$, when $x = 3$ and $y = 6$
12. $\dfrac{p}{q}$, when $p = 4$ and $q = 16$
13. $\dfrac{5z}{y}$, when $z = 8$ and $y = 2$
14. $\dfrac{9m}{q}$, when $m = 4$ and $q = 18$

Substitute to find the value of each expression.

15. A driver who drives at a speed of r miles per hour for t hours will travel a distance of rt miles. How far will a driver travel at a speed of 55 mph for 4 hr?

16. A baseball player's batting average is h/a, where h is the number of hits and a is the number of "at bats." Find the batting average of a batter who had 13 hits in 25 at bats.

17. The area of a parallelogram with base b and height h is bh. Find the area of the parallelogram when the height is 15.4 cm (centimeters) and the base is 6.5 cm.

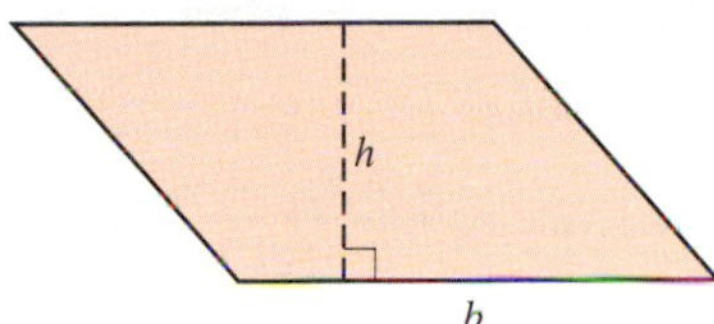

18. A satellite orbiting 300 mi above the earth's surface travels about 27,000 mi in one orbit. The time, in hours, that it takes to orbit the earth one time is given by

$$\frac{27{,}000}{v},$$

where v is the velocity of the satellite in miles per hour. Find the orbiting time of the satellite when the velocity v is 10,000 mph.

19. Enrico takes five times as long to do a job as Rosa does. Suppose t represents the time it takes Rosa to do the job. Then $5t$ represents the time it takes Enrico. How long does it take Enrico if Rosa takes 30 sec? 90 sec? 2 min?

20. Theresa is six years younger than her husband Frank. Suppose the variable x represents Frank's age. Then $x - 6$ represents Theresa's age. How old is Theresa when Frank is 29? 34? 47?

Translate to an algebraic expression.

21. 6 more than b

22. 8 more than t

23. 9 less than c

24. 4 less than d

25. 6 increased by q

26. 11 increased by z

27. b more than a

28. c more than d

29. x less than y

30. c less than h

31. x divided by w

32. s divided by t

33. m subtracted from n

34. p subtracted from q

35. The sum of r and s

36. The sum of d and f

37. Twice x

38. Three times p

39. One third of t

40. One quarter of d

41. 97% of some number

42. 43% of some number

43. Mandy had d dollars before going to the bookstore. She bought a book for \$29.95. How much did Mandy have after the purchase?

44. Dan drove at a speed of 65 mph for t hours. How far did Dan travel?

Determine whether the given number is a solution of the given equation.

45. 15; $x + 17 = 32$

46. 75; $y + 28 = 93$

47. 21; $x - 7 = 12$

48. 27; $y - 8 = 19$

49. 7; $6x = 54$

50. 9; $8y = 72$

51. 30; $\frac{x}{6} = 5$

52. 49; $\frac{y}{8} = 6$

53. 19; $5x + 7 = 107$

54. 9; $9x + 5 = 86$

Translate each problem to an equation. Do not solve.

55. What number added to 60 is 112?

56. Seven times what number is 2233?

57. When 42 is multiplied by a number, the result is 2352. Find the number.

58. When 345 is added to a number, the result is 987. Find the number.

59. A game board has 64 squares. If you win 35 squares and your opponent wins the rest, how many does your opponent get?

60. A consultant charges \$80 an hour. How many hours did the consultant work in order to make \$53,400?

61. In a recent year, the cost of four 12-oz boxes of Post® Raisin Bran was \$7.96. How much did one box cost?

62. The total amount spent on women's blouses in a recent year was \$6.5 billion. This was \$0.2 billion more than was spent on women's dresses. How much was spent on women's dresses?

Synthesis

To the student and the instructor: Synthesis exercises are designed to challenge students to extend the concepts or skills studied in that section. Some synthesis exercises will require the assimilation of skills and concepts from several sections.

Writing exercises, denoted by ◈, should be answered using one or more complete English sentences. In virtually every section, two writing exercises appear as the first synthesis exercises. These two exercises are not as challenging as those exercises appearing later in the exercise set and can be assigned to students who might not otherwise attempt synthesis exercises. Because many writing exercises are open-ended, their solutions are not listed in the answer section.

63. ◈ How does a variable expression differ from a variable, and how does it differ from an equation?

64. ◈ Write a problem, in words, that translates to the equation $35x = 840$.

Translate to an algebraic expression.

65. A number y plus two times x

66. A number a plus 2 plus b

67. A number that is 3 less than twice x

68. Your age in 5 years, if you are a years old now

69. Your age two years ago, if you are b years old now

70. Some number x increased by itself

71. The perimeter of a square with side s (perimeter means distance around)

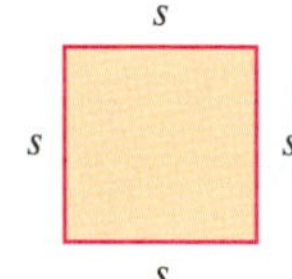

72. The perimeter of a rectangle with length l and width w

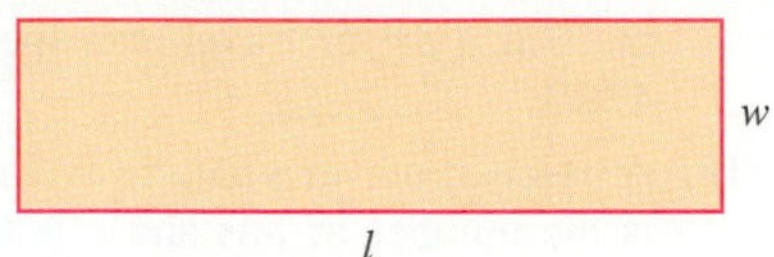

73. Evaluate $\frac{x + y}{4}$ when $y = 8$ and x is twice y.

74. Evaluate $\frac{x - y}{7}$ when $y = 35$ and x is twice y.

75. Evaluate $\frac{y - x}{3}$ when $x = 9$ and y is three times x.

76. Evaluate $\frac{y + x}{2} + \frac{3 \cdot y}{x}$ for $x = 2$ and $y = 4$.

Answer each question with an algebraic expression.

77. If $w + 3$ is a whole number, what is the next whole number after it?

78. If $d + 2$ is an odd number, what is the preceding odd number?

79. The difference between two numbers is 3. One number is t. What are two possible values for the other number?

80. You invest n dollars at 10% interest per year. Write an expression for the number of dollars in the bank a year from now.

1.2 The Commutative, Associative, and Distributive Laws

Equivalent Expressions • The Commutative Laws • The Associative Laws • Using the Laws Together • The Distributive Law • The Distributive Law and Factoring

In order to solve equations, it is important to be able to manipulate algebraic expressions. The commutative, associative, and distributive laws discussed in this section enable us to write *equivalent expressions* that can streamline our work.

Equivalent Expressions

In arithmetic we learned that expressions like $4 + 4 + 4$, $3 \cdot 4$, and $4 \cdot 3$ all represent the same number, 12. Expressions that represent the same number are said to be **equivalent.** The expressions $t + 18$ and $18 + t$ were used in Section 1.1 when we translated "eighteen more than some number." To illustrate that these expressions are equivalent, we can make some choices for t:

When $t = 3$, $t + 18 = 3 + 18 = 21$ and $18 + t = 18 + 3 = 21$.

When $t = 40$, $t + 18 = 40 + 18 = 58$ and $18 + t = 18 + 40 = 58$.

The Commutative Laws

We have seen that changing the order in addition or multiplication does not change the result. Equations like $3 + 18 = 18 + 3$ and $3 \cdot 4 = 4 \cdot 3$ illustrate this idea, and show that addition and multiplication are **commutative.**

The Commutative Laws

For Addition. For any numbers a and b,

$$a + b = b + a.$$

(We can change the order when adding without affecting the answer.)

For Multiplication. For any numbers a and b,

$$ab = ba.$$

(We can change the order when multiplying without affecting the answer.)

EXAMPLE 1

Use the commutative laws to write an expression equivalent to each of the following: **(a)** $y + 5$; **(b)** $9x$; **(c)** $7 + ab$.

Solution

a) An expression equivalent to $y + 5$ is $5 + y$ by the commutative law of addition.

b) An expression equivalent to $9x$ is $x9$ by the commutative law of multiplication.

c) An expression equivalent to $7 + ab$ is $ab + 7$ by the commutative law of *addition*.

Another expression equivalent to $7 + ab$ is $7 + ba$ by the commutative law of *multiplication*.

Also equivalent to $7 + ab$ is $ba + 7$ by both commutative laws. ❑

The Associative Laws

Parentheses are used to indicate groupings. We normally simplify within the parentheses first. For example,

$$3 + (8 + 4) = 3 + 12 = 15$$

and

$$(3 + 8) + 4 = 11 + 4 = 15.$$

Similarly,

$$4(2 \cdot 3) = 4(6) = 24$$

and

$$(4 \cdot 2)3 = (8)3 = 24.$$

Note that, so long as only addition or only multiplication appears in an expression, changing the grouping does not change the result. Equations such as $3 + (8 + 4) = (3 + 8) + 4$ and $4(2 \cdot 3) = (4 \cdot 2)3$ illustrate that addition and multiplication are **associative.**

The Associative Laws

For Addition. For any numbers a, b, and c,

$$a + (b + c) = (a + b) + c.$$

(Numbers can be grouped in any manner for addition.)

For Multiplication. For any numbers a, b, and c,

$$a \cdot (b \cdot c) = (a \cdot b) \cdot c.$$

(Numbers can be grouped in any manner for multiplication.)

EXAMPLE 2

Use an associative law to write an expression equivalent to each of the following: **(a)** $y + (z + 3)$; **(b)** $(8x)y$.

Solution

a) An expression equivalent to $y + (z + 3)$ is $(y + z) + 3$ by the associative law of addition.

b) An expression equivalent to $(8x)y$ is $8(xy)$ by the associative law of multiplication. ❑

Using the Laws Together

When only additions or only multiplications are involved, parentheses can be placed any way we please. For that reason, we often omit them. For example,

$x + (y + 7)$ means $x + y + 7$, and $l(wh)$ means lwh.

A sum such as $(5 + 1) + (3 + 5) + 9$ can be simplified by looking for pairs of numbers that add to 10:

$$\begin{aligned}(5 + 1) + (3 + 5) + 9 &= 5 + 5 + 9 + 1 + 3 \\ &= 10 + 10 + 3 = 23.\end{aligned}$$

EXAMPLE 3

Use the commutative and/or associative laws of addition to write at least two expressions equivalent to $(x + 5) + y$.

Solution

a) $(x + 5) + y = x + (5 + y)$ Using the associative law

$= x + (y + 5)$ Using the commutative law

b) $(x + 5) + y = y + (x + 5)$ Using the commutative law

$= y + (5 + x)$ Using the commutative law again ❑

EXAMPLE 4

Use the commutative and/or associative laws of multiplication to rewrite $2(x3)$ as $6x$. Show and give reasons for each step.

Solution

$2(x3) = 2(3x)$ Using the commutative law

$= (2 \cdot 3)x$ Using the associative law

$= 6x$ Simplifying ❑

The Distributive Law

The *distributive law* is probably the single most important law for manipulating algebraic expressions. Unlike the commutative and associative laws, the distributive law uses multiplication together with addition.

You have already used the distributive law although you probably didn't realize it. To see this, try to multiply $3 \cdot 21$ mentally. Most people find the product, 63, by thinking of 21 as $20 + 1$ and then multiplying 20 by 3 and 1 by 3. The sum of the two products, $60 + 3$, is 63. Note that if the 3 is not used as a multiplier twice, the result will not be correct.

EXAMPLE 5

Compute in two ways: $4(3 + 2)$.

Solution

a) As in the multiplication of $3(20 + 1)$ above, we can multiply by 4 twice and add the results:

$4(3 + 2) = 4 \cdot 3 + 4 \cdot 2$ Multiplying both 3 and 2 by 4

$= 12 + 8 = 20.$ Adding

b) By first adding inside the parentheses, we get the same result in a different way:

$4(3 + 2) = 4(5)$ Adding; $3 + 2 = 5$

$= 20.$ Multiplying ❑

The Distributive Law

For any numbers a, b, and c,

$$a(b + c) = ab + ac.$$

EXAMPLE 6

Multiply: $3(x + 2)$.

Solution Since $x + 2$ cannot be simplified unless a value for x is given, we use the distributive law:

$3(x + 2) = 3x + 3 \cdot 2$ Using the distributive law

$= 3x + 6.$ ❑

In the expression $x + 2$, the parts separated by the plus sign are called **terms.*** The distributive law can also be used when more than two terms are being multiplied.

EXAMPLE 7

Multiply: $6(s + 2 + 5w)$.

Solution

$$6(s + 2 + 5w) = 6s + 6 \cdot 2 + 6 \cdot 5w \quad \text{Using the distributive law}$$
$$= 6s + 12 + (6 \cdot 5)w \quad \text{Using the associative law for multiplication}$$
$$= 6s + 12 + 30w$$ ❑

Because of the commutative law of multiplication, the distributive law can be used on the ''right'': $(b + c)a = ba + ca$.

EXAMPLE 8

Multiply: $(c + 4)5$.

Solution

$$(c + 4)5 = c \cdot 5 + 4 \cdot 5 \quad \text{Using the distributive law on the right}$$
$$= 5c + 20$$ ❑

> CAUTION! The distributive law provides a useful way of removing parentheses. However, do not forget to multiply each number inside the parentheses by the number outside:
>
> $a(b + c) \neq ab + c.$

The Distributive Law and Factoring

If we reverse the statement of the distributive law, we have the basis of a process called **factoring:** $ab + ac = a(b + c)$. To **factor** an expression means to write an equivalent expression that is a product. The parts of the product are then called **factors.**

EXAMPLE 9

Use the distributive law to factor: **(a)** $3x + 3y$; **(b)** $7x + 21y + 7$.

Solution

a) By the distributive law,

$$3x + 3y = 3(x + y). \quad \text{The } \textit{common factor} \text{ is 3.}$$

b) $$7x + 21y + 7 = 7 \cdot x + 7 \cdot 3y + 7 \cdot 1 \quad \text{The common factor is 7.}$$
$$= 7(x + 3y + 1) \quad \text{Using the distributive law}$$

Be sure not to omit the 1 or the common factor, 7. ❑

*Terms are discussed in greater detail in Sections 1.5–1.8.

To check our factoring, we can multiply and see if the original expression is obtained. For example, since

$$7(x + 3y + 1) = 7x + 7 \cdot 3y + 7 \cdot 1$$
$$= 7x + 21y + 7,$$

the factoring of Example 9(b) is correct.

EXERCISE SET 1.2

Use the commutative law of addition to write an equivalent expression.

1. $y + 5$
2. $x + 6$
3. $5 + ab$
4. $x + 3y$
5. $9x + 3y$
6. $3a + 7b$
7. $2(a + 3)$
8. $9(x + 5)$

Use the commutative law of multiplication to write an equivalent expression.

9. rt
10. mn
11. $5a$
12. $7b$
13. $5 + ab$
14. $x + 3y$
15. $2(a + 3)$
16. $9(x + 5)$

Use the associative law of addition to write an equivalent expression.

17. $x + (y + 2)$
18. $(a + 3) + b$
19. $(9 + m) + 2$
20. $x + (2 + y)$
21. $(ab + c) + d$
22. $(m + np) + r$

Use the associative law of multiplication to write an equivalent expression.

23. $5(ab)$
24. $(7x)y$
25. $(6m)n$
26. $9(rp)$
27. $3[2(a + b)]$
28. $5[x(2 + y)]$

Use the commutative and/or associative laws to write two equivalent expressions.

29. $(a + b) + 2$
30. $5 + (v + w)$
31. $7(ab)$
32. $(xy)3$

Use the commutative and/or associative laws to rewrite each of the following. Label each step with a reason, as in Example 4.

33. $(3a)4$ as $12a$
34. $(2 + m) + 3$ as $m + 5$
35. $5 + (2 + x)$ as $x + 7$
36. $(a3)5$ as $15a$

Multiply.

37. $2(b + 5)$
38. $4(x + 3)$
39. $7(1 + t)$
40. $6(v + 4)$
41. $3(x + 1)$
42. $7(x + 8)$
43. $4(1 + y)$
44. $9(s + 1)$
45. $6(5x + 2)$
46. $9(6m + 7)$
47. $7(x + 4 + 6y)$
48. $4(5x + 8 + 3p)$
49. $(a + b)2$
50. $(x + 2)7$
51. $(x + y + 2)5$
52. $(2 + a + b)6$

Use the distributive law to factor each of the following. Check by multiplying.

53. $2x + 2y$
54. $5y + 5z$
55. $5 + 5y$
56. $13 + 13x$
57. $3x + 12y$
58. $5x + 20y$
59. $5x + 10 + 15y$
60. $3 + 27b + 6c$
61. $9x + 9$
62. $6x + 6$
63. $9x + 3y$
64. $15x + 5y$
65. $2a + 16b + 64$
66. $5 + 20x + 35y$
67. $11x + 44y + 121$
68. $7 + 14b + 56w$

Skill Maintenance

To the student and the instructor: Skill maintenance exercises review skills studied in earlier sections of the text. These exercises appear in almost every exercise set.

Translate to an algebraic expression.

69. 9 less than t
70. Half of m

Synthesis

71. ◈ Are subtraction and division commutative? Why or why not?

72. ◈ Are subtraction and division associative? Why or why not?

Tell whether the following expressions are equivalent. Also, explain why.

73. $5m + 6$ and $6 + 5m$

74. $3(2 + x + y)$ and $6 + 3(x + y)$

75. $axy + ax$ and $xa(1 + y)$

76. $3a(b + c)$ and $(ca + ba)3$

77. ◈ Factor $17x + 34$. Then evaluate $17x + 34$ and its factorization when $x = 10$. Do your results indicate that $17x + 34$ and its factorization are equivalent? Why or why not?

78. When you put money in the bank and draw simple interest, the amount in your account later on is given by the expression $P + Prt$, where P is the principal, r is the rate of interest, and t is the time. Factor the expression.

79. ◈ Evaluate the expressions $3(2 + x)$ and $6 + x$ when $x = 0$. Do your results indicate that $3(2 + x)$ and $6 + x$ are equivalent? Why or why not?

1.3 Fractional Notation

Factors and Prime Factorizations • Fractional Notation • Multiplication and Simplification • Canceling • Addition, Subtraction, and Division

This section reviews multiplication, addition, subtraction, and division with fractional notation. Although much of this may be review, note that fractional expressions that contain variables are also introduced.

Factors and Prime Factorizations

In order to be able to study addition and subtraction using fractional notation, we first review how *natural numbers* are factored. **Natural numbers** can be thought of as the counting numbers:

$$1, 2, 3, 4, 5, \ldots .$$

The dots indicate that the pattern of the preceding numbers continues without ending.*

In Section 1.2, we factored expressions by writing an equivalent product. For example, $3x + 3y$ was factored as $3(x + y)$. Natural numbers can also be factored by writing an equivalent product. For instance, 30 can be factored as $3 \cdot 10$. To **factor** a number means to write it as a product; each number appearing in a product is called **a factor.** Note that the word factor can be used as both a verb and a noun.

EXAMPLE 1

Factor the number 8. List the factors.

Solution The number 8 can be factored in several ways:

$$2 \cdot 4, \quad 1 \cdot 8, \quad 2 \cdot 2 \cdot 2.$$

The factors of 8 are 1, 2, 4, and 8. ❑

*A less frequently used set of numbers, the **whole numbers,** includes 0: 0, 1, 2, 3, ...

A collection of symbols expressing a number as a product is called a **factorization** of the number.

EXAMPLE 2

Write several factorizations of the number 12.

Solution

$1 \cdot 12, \quad 2 \cdot 6, \quad 3 \cdot 4, \quad 2 \cdot 2 \cdot 3$ □

Some numbers have only two factors, the number itself and 1. Such numbers are called **prime.**

Prime Number

A *prime number* is a natural number that has exactly two different factors.

EXAMPLE 3

Which of these numbers are prime? 7, 4, 1

Solution

7 is prime. It has exactly two different factors, 7 and 1.

4 is not prime. It has three different factors, 1, 2, and 4.

1 is not prime. It does not have two *different* factors. □

If a natural number, other than 1, is not prime, we call it **composite.** Every composite number can be factored into a product of prime numbers. Such a factorization is called a **prime factorization.**

EXAMPLE 4

Find the prime factorization of 36.

Solution We begin by factoring 36 in any way that we can. One way is like this:

$$36 = 4 \cdot 9.$$

The factors 4 and 9 are not prime, so we factor them:

$$36 = 4 \cdot 9 = 2 \cdot 2 \cdot 3 \cdot 3. \qquad \text{2 and 3 are both prime.}$$

The prime factorization of 36 is $2 \cdot 2 \cdot 3 \cdot 3$. □

Fractional Notation

An example of **fractional notation** for a number is:

Numerator

$$\frac{2}{3}.$$

Denominator

The top number is called the **numerator,** and the bottom number is called the **denominator.** When the numerator and the denominator are the same nonzero number, we have fractional notation for the number 1.

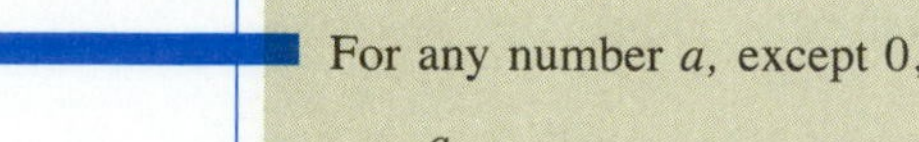

For any number a, except 0,

$$\frac{a}{a} = 1.$$

(Any nonzero number divided by itself is 1.)

Multiplication and Simplification

Recall from arithmetic that fractions are multiplied according to the following rule.

Multiplication of Fractions

For any two fractions a/b and c/d,

$$\frac{a}{b} \cdot \frac{c}{d} = \frac{ac}{bd}.$$

(The numerator of the product is the product of the individual numerators. The denominator of the product is the product of the individual denominators.)

EXAMPLE 5

Multiply: **(a)** $\frac{2}{3} \cdot \frac{7}{5}$; **(b)** $\frac{4}{x} \cdot \frac{8}{y}$.

Solution

a) We multiply numerators as well as denominators:

$$\frac{2}{3} \cdot \frac{7}{5} = \frac{2 \cdot 7}{3 \cdot 5} = \frac{14}{15}.$$

b) $\frac{4}{x} \cdot \frac{8}{y} = \frac{4 \cdot 8}{x \cdot y} = \frac{32}{xy}$ ❑

When one of the fractions being multiplied is 1, multiplying yields an equivalent expression because of the *identity property of* 1.

The Identity Property of 1

For any number a,

$$a \cdot 1 = a.$$

(Multiplying a number by 1 gives that same number.)

EXAMPLE 6

Multiply: $\frac{4}{5} \cdot \frac{6}{6}$.

Solution Since $\frac{6}{6} = 1$, the expression $\frac{4}{5} \cdot \frac{6}{6}$ is equivalent to $\frac{4}{5} \cdot 1$, or simply $\frac{4}{5}$. We have

$$\frac{4}{5} \cdot \frac{6}{6} = \frac{4 \cdot 6}{5 \cdot 6} = \frac{24}{30}.$$

Note that $\frac{24}{30}$ is equivalent to $\frac{4}{5}$. □

The steps of Example 6 can be reversed by "removing a factor of 1"—in this case, $\frac{6}{6}$. By removing a factor of 1, we can *simplify* an expression like $\frac{24}{30}$ to an equivalent expression like $\frac{4}{5}$.

To simplify, we factor the numerator and the denominator, looking for the largest factor common to both. This is sometimes made easier by writing the prime factorizations. Once common factors have been identified, the fraction can be expressed as a product of two fractions, one of which is in the form a/a.

EXAMPLE 7

Simplify: **(a)** $\frac{15}{40}$; **(b)** $\frac{36}{24}$.

Solution

a) Observe that 5 is a factor of both 15 and 40:

$$\frac{15}{40} = \frac{3 \cdot 5}{8 \cdot 5}$$ Factoring the numerator and the denominator, using the common factor, 5

$$= \frac{3}{8} \cdot \frac{5}{5}$$ Rewriting as a product of two fractions

$$= \frac{3}{8} \cdot 1$$ $\frac{5}{5} = 1$

$$= \frac{3}{8}.$$ Using the identity property of 1 (removing a factor of 1)

b) $$\frac{36}{24} = \frac{2 \cdot 2 \cdot 3 \cdot 3}{2 \cdot 2 \cdot 2 \cdot 3}$$ Writing the prime factorizations and identifying common factors

$$= \frac{3}{2} \cdot \frac{2 \cdot 2 \cdot 3}{2 \cdot 2 \cdot 3}$$ Rewriting as a product of two fractions

$$= \frac{3}{2} \cdot 1$$ $\frac{2 \cdot 2 \cdot 3}{2 \cdot 2 \cdot 3} = 1$

$$= \frac{3}{2}$$ Using the identity property of 1 □

It is always wise to check your result to see if any common factors of the numerator and the denominator remain. (This will never happen if prime factorizations are used correctly.) If common factors remain, repeat the process and simplify again.

Canceling

Canceling is a shortcut that you may have used for removing a factor of 1 when working with fractional notation. With *great* concern, we mention it as a possible way to speed up your work. You should use canceling only when removing common factors in numerators and denominators. Canceling *may not* be done in sums or when adding expressions together. Our concern is that "canceling" be done with care and understanding. Example 7(b) might have been done faster as follows:

$$\frac{36}{24} = \frac{\not{2} \cdot \not{2} \cdot 3 \cdot \not{3}}{\not{2} \cdot \not{2} \cdot 2 \cdot \not{3}} = \frac{3}{2}, \quad \text{or} \quad \frac{36}{24} = \frac{3 \cdot \not{12}}{2 \cdot \not{12}} = \frac{3}{2}, \quad \text{or} \quad \frac{\overset{\overset{3}{\not{18}}}{\not{36}}}{\underset{\underset{2}{\not{12}}}{\not{24}}} = \frac{3}{2}$$

> CAUTION! The difficulty with canceling is that it is often applied incorrectly:
>
> $$\frac{\not{2} + 3}{\not{2}} = 3, \qquad \frac{\not{4} + 1}{\not{4} + 2} = \frac{1}{2}, \qquad \frac{1\not{5}}{\not{5}4} = \frac{1}{4}.$$
>
> ↓ Wrong! ↓ Wrong! ↓ Wrong!
>
> $$\frac{2 + 3}{2} = \frac{5}{2} \qquad \frac{4 + 1}{4 + 2} = \frac{5}{6} \qquad \frac{15}{54} = \frac{5 \cdot 3}{18 \cdot 3} = \frac{5}{18}$$
>
> In each of these situations, the expressions canceled out were *not* factors. Factors are parts of products. For example, in $2 \cdot 3$, 2 and 3 are factors, but in $2 + 3$, 2 and 3 are *not* factors. **If you can't factor, you can't cancel! If in doubt, don't cancel!**

The number of factors in the numerator and the denominator may not always be the same. If not, the identity property of 1 allows us to insert the number 1 as a factor.

EXAMPLE 8

Simplify: $\frac{9}{72}$.

Solution

$$\frac{9}{72} = \frac{1 \cdot 9}{8 \cdot 9}$$ Factoring and using the identity property of 1 to write 9 as $1 \cdot 9$

$$= \frac{1 \cdot \not{9}}{8 \cdot \not{9}}$$ Removing a factor of 1: $\frac{9}{9} = 1$

$$= \frac{1}{8}$$ Simplifying ❑

Addition, Subtraction, and Division

When denominators are the same, fractions are added or subtracted by adding or subtracting numerators and keeping the same denominator.

EXAMPLE 9

Add and simplify: $\frac{4}{8} + \frac{5}{8}$.

Solution The common denominator is 8. We add the numerators and keep the common denominator:

$$\frac{4}{8} + \frac{5}{8} = \frac{4 + 5}{8} = \frac{9}{8}.$$ ❑

In arithmetic, you usually write $1\frac{1}{8}$ rather than the *improper* fraction $\frac{9}{8}$. In algebra, symbols such as $\frac{9}{8}$ are more useful and are quite "proper" for our purposes.

When denominators are different, we use the property of 1 and multiply to find a common denominator.

EXAMPLE 10

Add or subtract as indicated: **(a)** $\frac{7}{8} + \frac{5}{12}$; **(b)** $\frac{9}{8} - \frac{4}{5}$.

Solution

a) The number 24 is divisible by both 8 and 12. We multiply both $\frac{7}{8}$ and $\frac{5}{12}$ by suitable factors of 1 to obtain two fractions with denominators of 24:

$$\frac{7}{8} + \frac{5}{12} = \frac{7}{8}\cdot\frac{3}{3} + \frac{5}{12}\cdot\frac{2}{2}$$

Multiplying by 1. Since $3\cdot 8 = 24$, we multiply the first number by $\frac{3}{3}$. Since $2\cdot 12 = 24$, we multiply the second number by $\frac{2}{2}$.

$$= \frac{21}{24} + \frac{10}{24}$$

Performing the multiplications

$$= \frac{31}{24}.$$

b)

$$\frac{9}{8} - \frac{4}{5} = \frac{9}{8}\cdot\frac{5}{5} - \frac{4}{5}\cdot\frac{8}{8}$$

Using 40 as a common denominator

$$= \frac{45}{40} - \frac{32}{40}$$

$$= \frac{13}{40}$$

❑

Two numbers whose product is 1 are called **reciprocals,** or **multiplicative inverses,** of each other. All numbers, except zero, have reciprocals. For example,

the reciprocal of $\frac{2}{3}$ is $\frac{3}{2}$ because $\frac{2}{3}\cdot\frac{3}{2} = \frac{6}{6} = 1$,
the reciprocal of 9 is $\frac{1}{9}$ because $9\cdot\frac{1}{9} = \frac{9}{9} = 1$, and
the reciprocal of $\frac{1}{4}$ is 4 because $\frac{1}{4}\cdot 4 = 1$.

Any division problem can be rewritten as multiplication.

Division of Fractions

To divide, multiply by the reciprocal of the divisor:

$$\frac{a}{b} \div \frac{c}{d} = \frac{a}{b}\cdot\frac{d}{c}.$$

EXAMPLE 11

Divide: $\frac{1}{2} \div \frac{3}{5}$.

Solution

$$\frac{1}{2} \div \frac{3}{5} = \frac{1}{2}\cdot\frac{5}{3}$$

$\frac{5}{3}$ is the reciprocal of $\frac{3}{5}$

$$= \frac{5}{6}$$

❑

After we have performed an operation of multiplication, addition, subtraction, or division, the answer may need to be simplified.

EXAMPLE 12

Perform the indicated operation and simplify: **(a)** $\frac{7}{10} - \frac{1}{5}$; **(b)** $\frac{5}{6} \cdot \frac{9}{25}$; **(c)** $\frac{2}{3} \div \frac{4}{9}$.

Solution

a) $\frac{7}{10} - \frac{1}{5} = \frac{7}{10} - \frac{1}{5} \cdot \frac{2}{2}$ Using 10 as the common denominator

$= \frac{7}{10} - \frac{2}{10}$

$= \frac{5}{10} = \frac{1 \cdot \cancel{5}}{2 \cdot \cancel{5}} = \frac{1}{2}$ Removing a factor of 1: $\frac{5}{5} = 1$

b) $\frac{5}{6} \cdot \frac{9}{25} = \frac{5 \cdot 9}{6 \cdot 25}$ Multiplying numerators and denominators

$= \frac{5 \cdot 3 \cdot 3}{2 \cdot 3 \cdot 5 \cdot 5}$ Factoring

$= \frac{3 \cdot \cancel{3} \cdot \cancel{5}}{2 \cdot 5 \cdot \cancel{3} \cdot \cancel{5}}$ Removing a factor of 1: $\frac{3 \cdot 5}{3 \cdot 5} = 1$

$= \frac{3}{10}$ Simplifying

c) $\frac{2}{3} \div \frac{4}{9} = \frac{2}{3} \cdot \frac{9}{4}$ Multiplying by the reciprocal of the divisor

$= \frac{\cancel{2} \cdot \cancel{3} \cdot 3}{\cancel{3} \cdot \cancel{2} \cdot 2}$ Factoring and removing a factor of 1: $\frac{2 \cdot 3}{2 \cdot 3} = 1$

$= \frac{3}{2}$ ❑

EXERCISE SET 1.3

Write at least one factorization of each number. There can be more than one correct answer.

1. 56 **2.** 102
3. 93 **4.** 144

Find the prime factorization of each number. If the number is prime, state so.

5. 14 **6.** 15 **7.** 33
8. 55 **9.** 9 **10.** 25
11. 49 **12.** 121 **13.** 18
14. 24 **15.** 40 **16.** 56
17. 90 **18.** 120 **19.** 210
20. 330 **21.** 79 **22.** 143
23. 119 **24.** 221

Simplify.

25. $\frac{18}{45}$ **26.** $\frac{16}{56}$ **27.** $\frac{49}{14}$
28. $\frac{72}{27}$ **29.** $\frac{6}{42}$ **30.** $\frac{13}{104}$
31. $\frac{56}{7}$ **32.** $\frac{132}{11}$ **33.** $\frac{19}{76}$
34. $\frac{17}{51}$ **35.** $\frac{100}{20}$ **36.** $\frac{150}{25}$

37. $\frac{425}{525}$
38. $\frac{625}{325}$
39. $\frac{2600}{1400}$
40. $\frac{4800}{1600}$
41. $\frac{8 \cdot x}{6 \cdot x}$
42. $\frac{13 \cdot v}{39 \cdot v}$

Perform the indicated operation and simplify.

43. $\frac{1}{4} \cdot \frac{1}{2}$
44. $\frac{11}{10} \cdot \frac{8}{5}$
45. $\frac{17}{2} \cdot \frac{3}{4}$
46. $\frac{11}{12} \cdot \frac{12}{11}$
47. $\frac{1}{2} + \frac{1}{2}$
48. $\frac{1}{2} + \frac{1}{4}$
49. $\frac{4}{9} + \frac{13}{18}$
50. $\frac{4}{5} + \frac{8}{15}$
51. $\frac{3}{a} \cdot \frac{b}{7}$
52. $\frac{x}{5} \cdot \frac{y}{z}$
53. $\frac{3}{x} + \frac{2}{x}$
54. $\frac{7}{a} - \frac{5}{a}$
55. $\frac{3}{10} + \frac{8}{15}$
56. $\frac{9}{8} + \frac{7}{12}$
57. $\frac{5}{4} - \frac{3}{4}$
58. $\frac{12}{5} - \frac{2}{5}$
59. $\frac{13}{18} - \frac{4}{9}$
60. $\frac{13}{15} - \frac{8}{45}$
61. $\frac{11}{12} - \frac{2}{5}$
62. $\frac{15}{16} - \frac{2}{3}$
63. $\frac{7}{6} \div \frac{3}{5}$
64. $\frac{7}{5} \div \frac{3}{4}$
65. $\frac{8}{9} \div \frac{4}{15}$
66. $\frac{3}{4} \div \frac{3}{7}$
67. $\frac{1}{4} \div \frac{1}{2}$
68. $\frac{1}{10} \div \frac{1}{5}$
69. $\frac{\frac{13}{12}}{\frac{39}{5}}$
70. $\frac{\frac{17}{6}}{\frac{3}{8}}$
71. $100 \div \frac{1}{5}$
72. $78 \div \frac{1}{6}$
73. $\frac{3}{4} \div 10$
74. $\frac{5}{6} \div 15$
75. $\frac{5}{3} \div \frac{a}{b}$
76. $\frac{x}{7} \div \frac{4}{y}$
77. $\frac{x}{6} - \frac{1}{3}$
78. $\frac{9}{10} + \frac{x}{2}$

Skill Maintenance

Use a commutative law to write an equivalent expression. There can be more than one correct answer.

79. $5(x + 3)$
80. $7 + (a + b)$

Synthesis

81. ◈ Is multiplication of fractions commutative? Why or why not?

82. ◈ Use the word factor in two sentences—once as a noun and once as a verb.

Simplify.

83. $\frac{128}{192}$
84. $\frac{pqrs}{qrst}$
85. $\frac{33sba}{2(11a)}$
86. $\frac{4 \cdot 9 \cdot 16}{2 \cdot 8 \cdot 15}$
87. $\frac{36 \cdot (2rh)}{8 \cdot (9hg)}$
88. $\frac{3 \cdot (4xy) \cdot (5)}{2 \cdot (3x) \cdot (4y)}$

89. A candy company uses two sizes of boxes, 6 in. and 8 in. long. These are packed in bigger cartons to be shipped. What is the shortest-length carton that will accommodate boxes of either size without any room left over? (Each carton can contain only boxes of one size; no mixing is allowed.)

90. In the following table, the top number has been factored in such a way that the sum of the factors is the bottom number. For example, in the first column, 56 has been factored as $7 \cdot 8$, and $7 + 8 = 15$, the bottom number. Find the missing numbers in the table.

Product	56	63	36	72	140	96
Factor	7					
Factor	8					
Sum	15	16	20	38	24	20

Product		168	110			
Factor				9	24	3
Factor	8	8		10	18	
Sum	14		21			24

Find the area of each figure.

91.

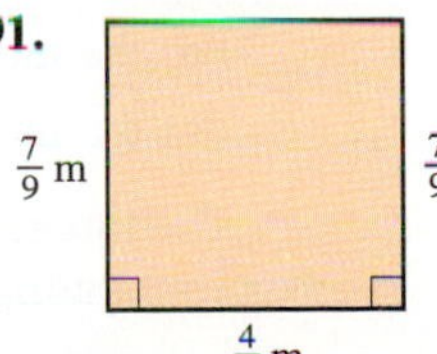

92.

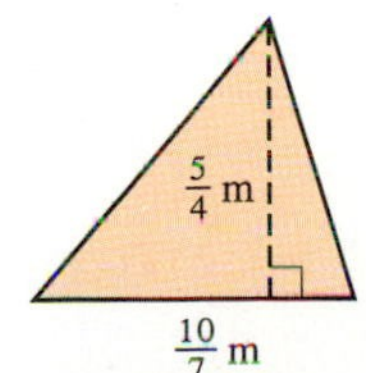

93. Find the perimeter of a square with sides of length $\frac{5}{9}$ m.

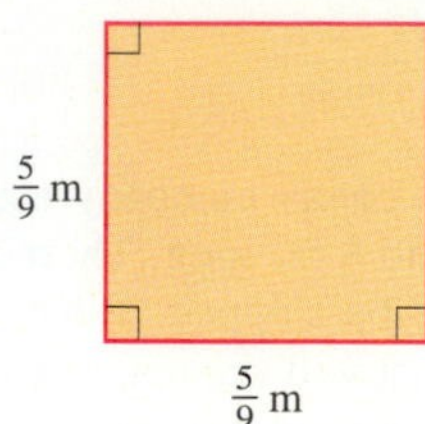

94. Find the perimeter of the rectangle in Exercise 91.

95. ◈ Make use of the properties and laws discussed in Sections 1.2 and 1.3 to explain why $x + y$ is equivalent to $(2y + 2x)/2$.

1.4 Positive and Negative Real Numbers

The Integers • Integers and the Real World • The Rational Numbers • The Real Numbers and Order • Translating to Inequalities • Absolute Value

A **set** is a collection of objects. The set containing the numbers 1, 3, and 7 is generally written $\{1, 3, 7\}$ and is said to be a **subset** if it is part of some other set. In this section, we examine the set of *real numbers* and some of its important subsets. More on sets can be found in Appendix A.

The Integers

Two sets of numbers have already been discussed. We represent these sets using dots on a number line.

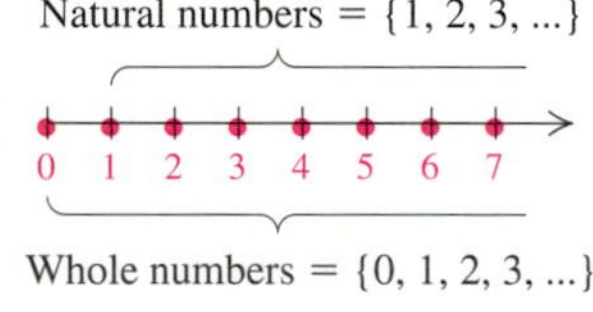

To create a new set, called the *integers,* we start with the whole numbers, 0, 1, 2, 3, and so on. For each natural number 1, 2, 3, and so on, we include in the set a new number the same distance to the left of 0 on the number line:

For the number 1, include the *opposite* number -1 (negative 1).

For the number 2, include the *opposite* number -2 (negative 2).

For the number 3, include the *opposite* number -3 (negative 3), and so on.

The **integers** consist of the whole numbers and these new numbers. We picture them on a number line as follows.

Integers

Negative integers — Positive integers

−6 −5 −4 −3 −2 −1 0 1 2 3 4 5 6

Opposites

The new numbers to the left of 0 are called *negative integers*. The natural numbers are called *positive integers*. Zero is neither positive nor negative, so the whole numbers are sometimes referred to as *nonnegative integers*. Numbers like −1 and 1 or 3 and −3 are said to be *opposites* of each other, with 0 acting as its own opposite.

Set of Integers

The set of integers = {. . . , −5, −4, −3, −2, −1, 0, 1, 2, 3, 4, 5, . . .}.

Integers and the Real World

Integers are associated with many real-world problems and situations.

EXAMPLE 1

State the integer that corresponds to each situation: **(a)** The temperature is 3 degrees below zero; **(b)** Losing 21 points in a card game; **(c)** Death Valley is 280 ft below sea level; **(d)** A business made \$145 on Monday, but lost \$68 on Tuesday.

Solution

a) Since 3° below zero is −3°, the corresponding integer is −3.

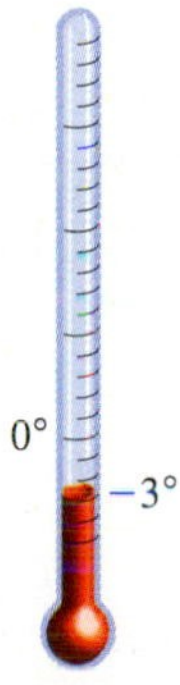

b) Losing 21 points in a card game gives you −21 points.

c) The integer -280 corresponds to the situation (see the figure at right). The elevation is -280 ft.

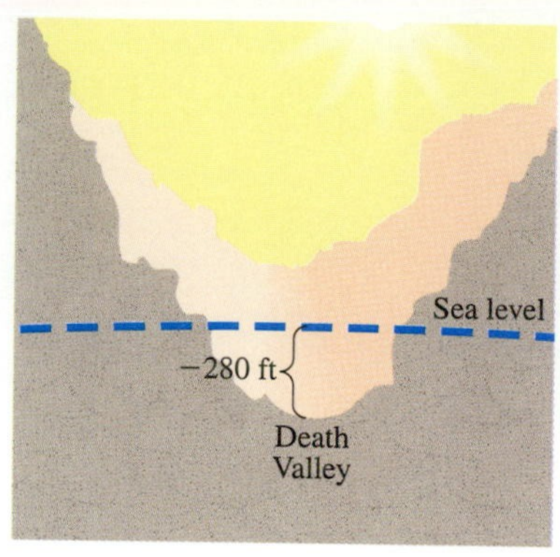

d) The integer 145 corresponds to the profit on Monday and -68 corresponds to the loss on Tuesday. ❑

The Rational Numbers

We now examine a set of numbers that contains the set of integers as a subset. This set, the **rational numbers,** contains fractions and decimals in addition to the integers. The following are rational numbers:

$$\frac{2}{3},\quad -\frac{2}{3},\quad \frac{7}{1},\quad 4,\quad -3,\quad 0,\quad \frac{23}{-8},\quad 2.4,\quad -0.17.$$

The number $-\frac{2}{3}$ (read "negative two-thirds") can also be named $\frac{2}{-3}$ or $\frac{-2}{3}$. The number 2.4 can be named $\frac{24}{10}$ or $\frac{12}{5}$, and -0.17 can be named $-\frac{17}{100}$.

Note that the set of rational numbers contains the whole numbers, the integers, and all fractions and decimals commonly seen in arithmetic. We cannot list all rational numbers, but we can describe the set of rational numbers as follows.

Set of Rational Numbers

The set of rational numbers $= \left\{ \frac{a}{b} \,\middle|\, a \text{ and } b \text{ are integers and } b \neq 0 \right\}$.

This is read "the set of all numbers $\frac{a}{b}$, where a and b are integers and $b \neq 0$."

Every rational number can be **graphed** by marking its location on a number line.

EXAMPLE 2

Graph each of the following rational numbers: **(a)** $\frac{5}{2}$; **(b)** -3.2; **(c)** $\frac{11}{8}$.

Solution

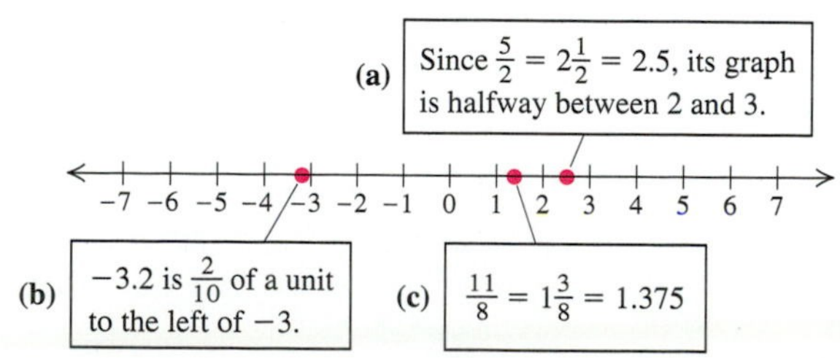

❑

Every rational number can be written using fractional or decimal notation.

EXAMPLE 3 Convert to decimal notation: $-\frac{5}{8}$.

Solution We first find decimal notation for $\frac{5}{8}$. Since $\frac{5}{8}$ means $5 \div 8$, we divide.

$$\begin{array}{rl} & 0.625 \\ 8 & \overline{)5.000} \\ & \underline{4\,8} \\ & 20 \\ & \underline{16} \\ & 40 \\ & \underline{40} \\ & 0 \end{array}$$

Thus, $\frac{5}{8} = 0.625$, so $-\frac{5}{8} = -0.625$. ❑

Decimal notation for $-\frac{5}{8}$ is -0.625. We consider -0.625 to be a **terminating decimal** because we reached a remainder of 0. Decimal notation for some numbers, however, repeats.

EXAMPLE 4 Convert to decimal notation: $\frac{7}{11}$.

Solution We divide:

$$\begin{array}{rl} & 0.6363\ldots \\ 11 & \overline{)7.0000} \\ & \underline{6\,6} \\ & 40 \\ & \underline{33} \\ & 70 \\ & \underline{66} \\ & 40 \\ & \underline{33} \\ & 7 \end{array}$$

We abbreviate repeating decimals by writing a bar over the repeating part, in this case, $0.\overline{63}$. ❑

The Real Numbers and Order

Every rational number has a point on the number line. However, not every point on the number line corresponds to a rational number. Some points correspond to what are called **irrational numbers.**

What kinds of numbers are irrational numbers? One example is the number π, which is used to find the area and circumference of a circle: $A = \pi r^2$ and $C = 2\pi r$.

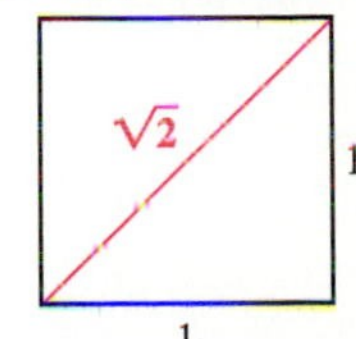

Another irrational number, $\sqrt{2}$ (read "the square root of 2"), is the length of the diagonal of a square with sides of length 1. It is also the number that, when multiplied by itself, gives 2. No rational number can be multiplied by itself to get 2, although the following approximations come close:

1.4 is an *approximation* of $\sqrt{2}$ because $(1.4)^2 = 1.96$;

1.41 is a better approximation because $(1.41)^2 = 1.9881$;

1.4142 is an even better approximation because $(1.4142)^2 = 1.99996164$.

To approximate $\sqrt{2}$ on most calculators, simply press [2] and then the [√] key. On other calculators, press [√], [2], and [ENTER], or consult an owner's manual.

Decimal notation for rational numbers *either* terminates *or* repeats. Decimal notation for irrational numbers *neither* terminates *nor* repeats. Examples of irrational numbers are π, $\sqrt{3}$, $-\sqrt{8}$, $\sqrt{11}$, and 0.121221222122221 . . . .

The rational numbers and the irrational numbers together correspond to all the points on a number line and make up what is called the **real-number system.**

Set of Real Numbers

The set of real numbers = The set of all numbers corresponding to points on the number line.

The following figure shows the relationships among various kinds of numbers.

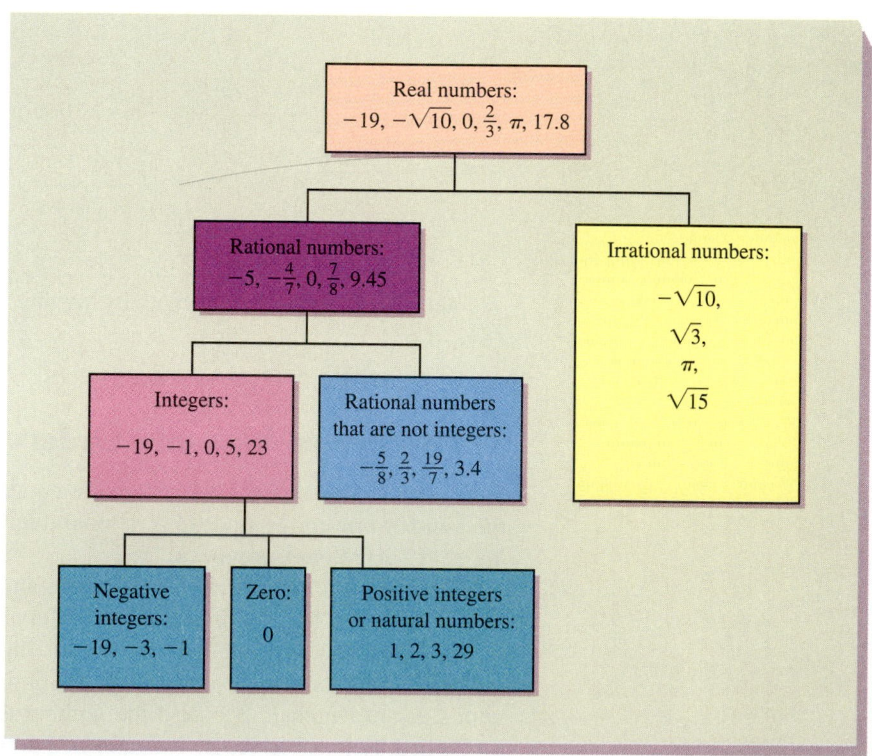

EXAMPLE 5

Graph the real number $\sqrt{3}$ on a number line.

Solution We use a calculator or Table 1 at the back of the book and approximate: $\sqrt{3} \approx 1.732$ ("$\approx$" means "approximately equals"). Then we locate this number on a number line.

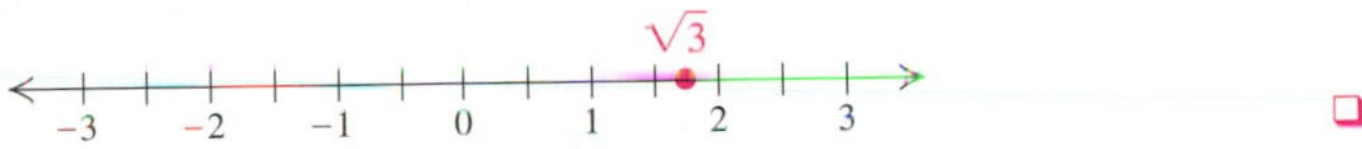

Real numbers are named in order on the number line, with larger numbers further to the right. For any two numbers on the line, the one to the left is less than the one to the right. We use the symbol < to mean "**is less than.**" The sentence $-8 < 6$ means "-8 is less than 6." The symbol > means "**is greater than.**" The sentence $-3 > -7$ means "-3 is greater than -7."

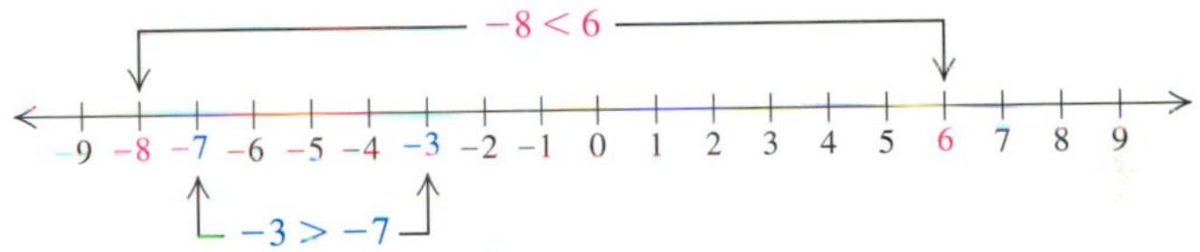

EXAMPLE 6

Use either < or > for ■ to write a true sentence: **(a)** 2 ■ 9; **(b)** -3.45 ■ 1.32; **(c)** 6 ■ -12; **(d)** -18 ■ -5; **(e)** $\frac{7}{11}$ ■ $\frac{5}{8}$.

Solution

a) Since 2 is to the left of 9, 2 is less than 9, so $2 < 9$.

b) Since -3.45 is to the left of 1.32, we have $-3.45 < 1.32$.

c) Since 6 is to the right of -12, we have $6 > -12$.

d) Since -18 is to the left of -5, we have $-18 < -5$.

e) We convert to decimal notation: $\frac{7}{11} = 0.\overline{63}\ldots$ and $\frac{5}{8} = 0.625$. Thus, $\frac{7}{11} > \frac{5}{8}$. We also could have used a common denominator: $\frac{7}{11} = \frac{56}{88} > \frac{55}{88} = \frac{5}{8}$. ❑

Sentences like "$a < -5$" and "$-3 > -8$" are called **inequalities.** It is useful to remember that every inequality can be written two ways. For instance,

$$-3 > -8 \quad \text{has the same meaning as} \quad -8 < -3.$$

It may be helpful to think of an inequality sign as an "arrow" with the smaller side pointing to the smaller number.

Note that all positive real numbers are greater than zero and all negative real numbers are less than zero.

> If x is a positive real number, then $x > 0$.
> If x is a negative real number, then $x < 0$.

Expressions like $a \leq b$ and $b \geq a$ are also inequalities. We read $a \leq b$ as "**a is less than or equal to b.**" We read $a \geq b$ as "**a is greater than or equal to b.**"

EXAMPLE 7

Write true or false for each inequality: **(a)** $-3 \leqslant 5$; **(b)** $-3 \leqslant -3$; **(c)** $-5 \geqslant 4$.

Solution

a) $-3 \leqslant 5$ is *true* because $-3 < 5$ is true.

b) $-3 \leqslant -3$ is *true* because $-3 = -3$ is true.

c) $-5 \geqslant 4$ is *false* since neither $-5 > 4$ nor $-5 = 4$ is true. ❑

Translating to Inequalities

In the following example, we see some ways in which English sentences can be translated to inequalities.

EXAMPLE 8

Translate each sentence to mathematical language: **(a)** A number is less than -3; **(b)** The temperature is at least 75°; **(c)** A debt of \$150 is worse than a debt of \$100.

Solution

a) *English:* A number is less than -3.

Translation: $x < -3$.

b) *English:* The temperature is at least 75°.

Translation: $t \geqslant 75$.

c) *English:* A debt of \$150 is worse than a debt of \$100.

Translation: $-150 < -100$. ❑

Absolute Value

In Section 1.5, we will need terminology for a number's distance from zero. We call a number's distance from zero on a number line its **absolute value.** Thus the absolute value of 4 is 4 and the absolute value of -4 is also 4.

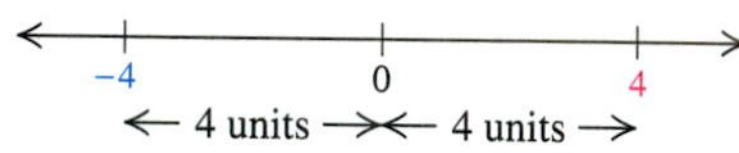

Absolute Value

We write $|a|$, read "the absolute value of a," to represent the number of units that a is from zero.

EXAMPLE 9

Find the absolute value: **(a)** $|-3|$; **(b)** $|7.2|$; **(c)** $|0|$.

Solution

a) $|-3| = 3$ since -3 is 3 units from 0.

b) $|7.2| = 7.2$ since 7.2 is 7.2 units from 0.

c) $|0| = 0$ since 0 is 0 units from itself. ❑

Distance is never negative, so numbers that are opposites have the same absolute value. If a number is nonnegative, its absolute value is the number itself. If a number is negative, its absolute value is its opposite.

EXERCISE SET | 1.4

Tell which real numbers correspond to each situation.

1. In a game, Mo won 5 points. In the next game, he lost 12 points.
2. The temperature on Wednesday was 18° above zero. On Thursday, it was 2° below zero.
3. A family owes \$170. The same family has \$950 in its bank account.
4. A printer earned \$1200 one week and lost \$560 the next.
5. The Dead Sea is 1286 feet below sea level, whereas Mt. Everest is 29,028 feet above sea level.
6. In bowling, the Jets are 34 pins behind the Strikers after the first game. Describe the situation in two ways.

7. Janice deposited \$750 in a savings account. Two weeks later, she withdrew \$125.
8. During a certain time period, the United States had a deficit of \$3 million in foreign trade.
9. During a video game, Cindy intercepted a missile worth 20 points, lost a starship worth 150 points, and captured a base worth 300 points.
10. Ignition occurs 10 seconds before liftoff. A spent fuel tank is detached 235 seconds after liftoff.

Graph the rational number on a number line.

11. $\frac{10}{3}$ **12.** $-\frac{17}{5}$ **13.** -4.3

14. 3.87 **15.** -2 **16.** 5

Find decimal notation.

17. $-\frac{3}{8}$ **18.** $-\frac{1}{8}$ **19.** $\frac{5}{3}$

20. $\frac{5}{6}$ **21.** $\frac{7}{6}$ **22.** $\frac{5}{12}$

23. $\frac{2}{3}$ **24.** $\frac{1}{4}$ **25.** $-\frac{1}{2}$

26. $\frac{5}{8}$ **27.** $\frac{1}{10}$ **28.** $-\frac{7}{20}$

Write a true sentence using either $<$ or $>$.

29. $5 \blacksquare 0$ **30.** $9 \blacksquare 0$ **31.** $-9 \blacksquare 5$

32. $8 \blacksquare -8$ **33.** $-6 \blacksquare 6$ **34.** $0 \blacksquare -7$

35. $-8 \blacksquare -5$ **36.** $-4 \blacksquare -3$

37. $-5 \blacksquare -11$ **38.** $-3 \blacksquare -4$

39. $-12.5 \blacksquare -9.4$ **40.** $-10.3 \blacksquare -14.5$

41. $2.14 \blacksquare 1.24$ **42.** $-3.3 \blacksquare -2.2$

43. $\frac{5}{12} \blacksquare \frac{11}{25}$ **44.** $-\frac{14}{17} \blacksquare -\frac{27}{35}$

Write an inequality with the same meaning as each of the following.

45. $-6 > x$ **46.** $x < 8$

47. $-10 \leqslant y$ **48.** $12 \geqslant t$

Write true or false.

49. $-3 \geqslant -11$ **50.** $5 \leqslant -5$

51. $0 \geqslant 8$ **52.** $-5 \leqslant 7$

53. $-8 \leqslant -8$ **54.** $8 \geqslant 8$

Translate to mathematical language.

55. -5 is greater than some number.
56. Some number is less than -1.
57. In cards, a score of 120 is better than one of -20.
58. A deposit of \$20 in a savings account is better than a withdrawal of \$25.
59. In trade, a deficit of \$500,000 is worse than an excess of \$1,000,000.

60. In bowling, it is better to be 60 pins ahead than to be 20 pins ahead.

61. Alice's test score was at most 95.

62. Some number never exceeds -9.

63. Fran's Franks considers profits to be poor if they don't exceed \$15,000.

64. A number is at most zero.

Find the absolute value.

65. $|-3|$ **66.** $|-7|$ **67.** $|10|$

68. $|11|$ **69.** $|0|$ **70.** $|-4|$

71. $|-24|$ **72.** $|325|$ **73.** $\left|-\frac{2}{3}\right|$

74. $\left|-\frac{10}{7}\right|$ **75.** $|43.9|$ **76.** $|14.8|$

77. $|x|$ when $x = 5$ **78.** $|b|$ when $b = -\frac{7}{8}$

79. List ten examples of rational numbers.

80. List ten examples of rational numbers that are *not* integers.

81. List three examples of irrational numbers.

82. List three examples of negative integers.

Skill Maintenance

83. Multiply and simplify: $\frac{21}{5} \cdot \frac{1}{7}$.

84. Evaluate $3xy$ when $x = 2$ and $y = 7$.

85. Use a commutative law to write an expression equivalent to $ab + 5$.

86. Factor: $3x + 9 + 12y$.

Synthesis

87. ◈ Is every nonnegative integer a whole number? Why or why not?

88. ◈ Why is it impossible for the absolute value of a number to be negative?

List in order from least to greatest.

89. $13, -12, 5, -17$ **90.** $-23, 4, 0, -17$

91. $\frac{4}{5}, \frac{4}{3}, \frac{4}{8}, \frac{4}{6}, \frac{4}{9}, \frac{4}{2}, -\frac{4}{3}$

92. $-\frac{2}{3}, \frac{1}{2}, -\frac{3}{4}, -\frac{5}{6}, \frac{3}{8}, \frac{1}{6}$

Write a true sentence using either $<$, $>$, or $=$.

93. $|-5| \quad |-2|$ **94.** $|4| \quad |-7|$

95. $|-8| \quad |8|$ **96.** $|23| \quad |-23|$

97. $|-3| \quad |5|$ **98.** $|-19| \quad |-27|$

Solve.

99. $|x| = 7$

100. $|x| < 2$ (Consider only integer replacements.)

101. We know that 0.3333. . . is $\frac{1}{3}$ and 0.6666. . . is $\frac{2}{3}$. What rational number is named by each of the following?

a) 0.5555. . . **b)** 0.1111. . .

c) 0.2222. . . **d)** 0.9999. . .

1.5 Addition of Real Numbers

Adding with a Number Line • Adding without a Number Line • Problem Solving • Collecting Like Terms

We now consider addition of real numbers. To gain understanding, we will use a number line first. After we observe the principles involved, we will develop rules that will enable us to work more quickly.

Adding with a Number Line

To perform the addition $a + b$ on a number line, we start at a and then move according to b.

a) If b is positive, we move to the right (the positive direction).
b) If b is negative, we move to the left (the negative direction).
c) If b is 0, we stay at a.

EXAMPLE 1

Add: $-4 + 9$.

Solution To add on a number line, we locate the first number, -4, and then move 9 units to the right. Note that it requires 4 units to reach 0. The difference between 9 and 4 is where we finish.

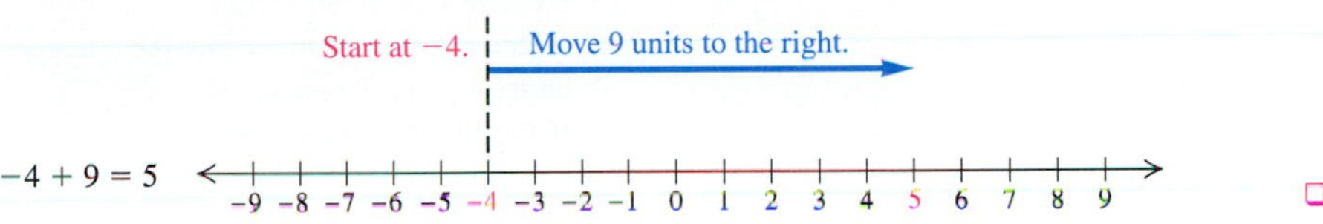

EXAMPLE 2

Add: $3 + (-5)$.

Solution We locate the first number, 3, and then move 5 units to the left. Note that it requires 3 units to reach 0. The difference between 5 and 3 is 2, so we finish 2 units to the left of 0.

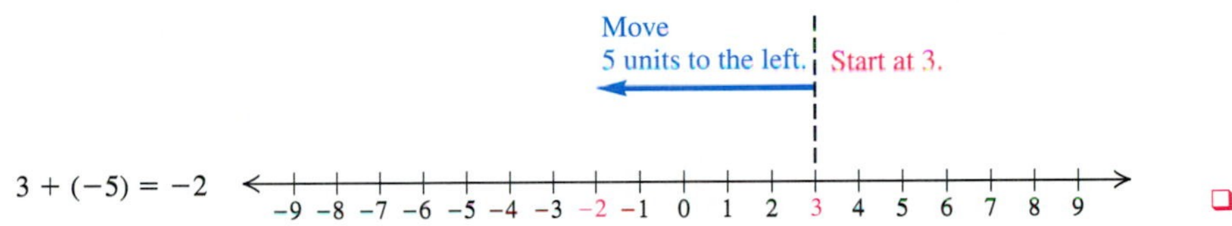

EXAMPLE 3

Add: $-4 + (-3)$.

Solution After locating -4, we move 3 units to the left. We finish a total of 7 units to the left of 0.

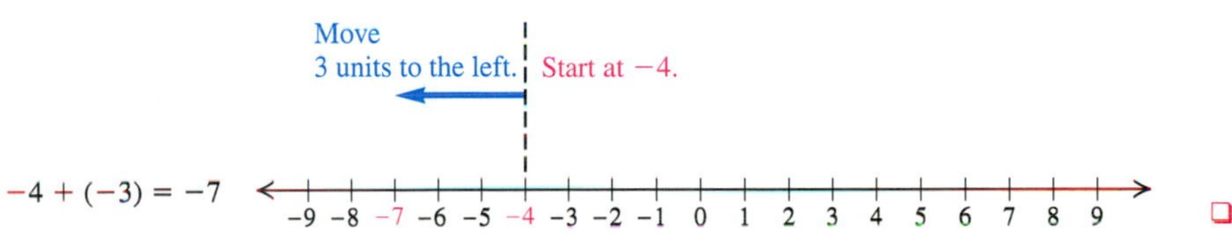

EXAMPLE 4

Add: $-5.2 + 0$.

Solution We locate -5.2 and move 0 units. Thus, we finish where we started.

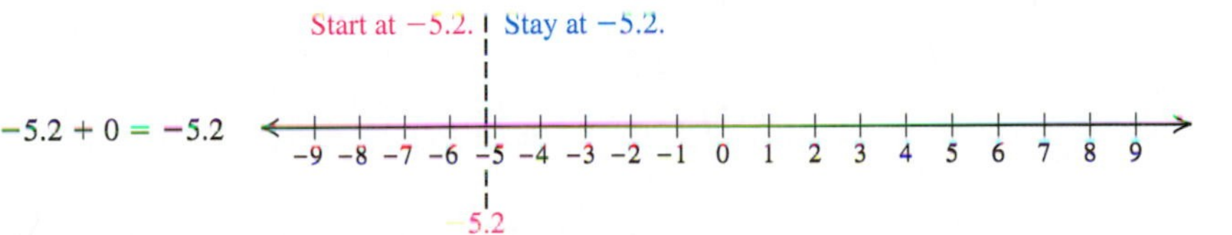

From Examples 1–4, we develop the following rules.

Rules for Addition of Real Numbers

1. *Positive numbers:* Add as usual. The answer is positive.
2. *Negative numbers:* Add absolute values and make the answer negative (see Example 3).
3. *A positive and a negative number:* Subtract absolute values. Then:
   a) If the positive number has the greater absolute value, the answer is positive (see Example 1).
   b) If the negative number has the greater absolute value, the answer is negative (see Example 2).
   c) If the numbers have the same absolute value, the answer is 0.
4. *One number is zero:* The sum is the other number (see Example 4).

Rule 4 is known as the **Identity Property of 0.** It says that for any real number a, $a + 0 = a$.

Adding without a Number Line

The rules for addition that are listed above can be used without drawing a number line.

EXAMPLE 5

Add without using a number line.

a) $-12 + (-7)$ **b)** $-1.4 + 8.5$ **c)** $-36 + 21$
d) $1.5 + (-1.5)$ **e)** $-\frac{7}{8} + 0$ **f)** $\frac{2}{3} + \left(-\frac{5}{8}\right)$

Solution

a) $-12 + (-7) = -19$ — Two negatives. *Think:* Add the absolute values, 12 and 7, to get 19. Make the answer *negative,* -19.

b) $-1.4 + 8.5 = 7.1$ — A negative and a positive. *Think:* The difference of absolute values is $8.5 - 1.4$, or 7.1. The positive number has the larger absolute value, so the answer is *positive,* 7.1.

c) $-36 + 21 = -15$ — A negative and a positive. *Think:* The difference of absolute values is $36 - 21$, or 15. The negative number has the larger absolute value, so the answer is *negative,* -15.

d) $1.5 + (-1.5) = 0$ — A negative and a positive. *Think:* Since the numbers are opposites, they have the same absolute value and the answer is 0.

e) $-\frac{7}{8} + 0 = -\frac{7}{8}$ — One number is zero. The sum is $-\frac{7}{8}$.

f) $\frac{2}{3} + \left(-\frac{5}{8}\right) = \frac{16}{24} + \left(-\frac{15}{24}\right) = \frac{1}{24}$ — A negative and a positive ❑

To add several numbers, some positive and some negative, the commutative and associative laws can be used. We add all the positives, then all the negatives, and then add the results. Of course, we can also add from left to right, if we prefer.

EXAMPLE 6

Add: $15 + (-2) + 7 + 14 + (-5) + (-12)$.

Solution

$$\begin{aligned} &15 + (-2) + 7 + 14 + (-5) + (-12) \\ &\quad = 15 + 7 + 14 + (-2) + (-5) + (-12) && \text{Using the commutative law of addition} \\ &\quad = (15 + 7 + 14) + [(-2) + (-5) + (-12)] && \text{Using the associative law of addition} \\ &\quad = 36 + (-19) && \text{Adding the positives; adding the negatives} \\ &\quad = 17 && \text{Adding a positive and a negative} \end{aligned}$$

❑

Problem Solving

The addition of real numbers is used in many real-world applications.

EXAMPLE 7

In the course of one four-month period, the water level of Lake Champlain went down 2 ft, up 1 ft, down 5 ft, and up 3 ft. How much had the lake level changed at the end of the four months?

Solution The problem translates to a sum:

Rewording:	The 1st change	plus	the 2nd change	plus	the 3rd change	plus	the 4th change	is	the total change.
	↓	↓	↓	↓	↓	↓	↓	↓	↓
Translating:	-2	$+$	1	$+$	(-5)	$+$	3	$=$	Total change.

Since

$$\begin{aligned} -2 + 1 + (-5) + 3 &= -1 + (-5) + 3 \\ &= -6 + 3 \\ &= -3, \end{aligned}$$

the lake level has dropped 3 ft at the end of the four months. ❑

Collecting Like Terms

The rules for addition apply to variable expressions as well as to numbers. When two terms have variable factors that are exactly the same, like $5ab$ and $7ab$, the terms are called **like, or similar, terms.** The distributive law enables us to **collect,** or **combine, like terms.**

EXAMPLE 8

Collect like terms.

a) $-7x + 9x$

b) $2a + (-3b) + (-5a) + 9b$

c) $7 + y + (-3.5y) + 2$

Solution

a)
$$\begin{aligned} -7x + 9x &= (-7 + 9)x && \text{Using the distributive law} \\ &= 2x && \text{Adding} \end{aligned}$$

b) $2a + (-3b) + (-5a) + 9b = 2a + (-5a) + (-3b) + 9b$ — Using the commutative law of addition

$= (2 + (-5))a + (-3 + 9)b$ — Using the distributive law

$= -3a + 6b$ — Adding

c) $7 + y + (-3.5y) + 2 = y + (-3.5y) + 7 + 2$ — Using the commutative law of addition

$= (1 + (-3.5))y + 7 + 2$ — Using the distributive law

$= -2.5y + 9$ — Adding ❑

With practice we can leave out some steps, collecting like terms mentally. Numbers like 7 and 2 in the expression $7 + y + (-3.5y) + 2$ are constants and are also considered to be like terms.

EXERCISE SET 1.5

Add using a number line.

1. $-9 + 2$
2. $2 + (-5)$
3. $-10 + 6$
4. $8 + (-3)$
5. $-8 + 8$
6. $6 + (-6)$
7. $-3 + (-5)$
8. $-4 + (-6)$

Add. Do not use a number line except as a check.

9. $-7 + 0$
10. $-13 + 0$
11. $0 + (-27)$
12. $0 + (-35)$
13. $17 + (-17)$
14. $-15 + 15$
15. $-17 + (-25)$
16. $-24 + (-17)$
17. $-18 + 18$
18. $11 + (-11)$
19. $8 + (-5)$
20. $-7 + 8$
21. $-4 + (-5)$
22. $10 + (-12)$
23. $13 + (-6)$
24. $-3 + 14$
25. $11 + (-9)$
26. $-14 + (-19)$
27. $-20 + (-6)$
28. $19 + (-19)$
29. $-15 + (-7)$
30. $23 + (-5)$
31. $40 + (-8)$
32. $-23 + (-9)$
33. $-25 + 25$
34. $40 + (-40)$
35. $63 + (-18)$
36. $85 + (-65)$
37. $-6.5 + 4.7$
38. $-3.6 + 1.9$
39. $-2.8 + (-5.3)$
40. $-7.9 + (-6.5)$
41. $-\frac{3}{5} + \frac{2}{5}$
42. $-\frac{4}{3} + \frac{2}{3}$
43. $-\frac{3}{7} + \left(-\frac{5}{7}\right)$
44. $-\frac{4}{9} + \left(-\frac{6}{9}\right)$
45. $-\frac{5}{8} + \frac{1}{4}$
46. $-\frac{5}{6} + \frac{2}{3}$
47. $-\frac{3}{7} + \left(-\frac{2}{5}\right)$
48. $-\frac{5}{8} + \left(-\frac{1}{3}\right)$
49. $75 + (-14) + (-17) + (-5)$
50. $28 + (-44) + 17 + 31 + (-94)$
51. $-44 + \left(-\frac{3}{8}\right) + 95 + \left(-\frac{5}{8}\right)$
52. $24 + 3.1 + (-44) + (-8.2) + 63$
53. $98 + (-54) + 113 + (-998) + 44 + (-612) + (-18) + 334$
54. $-455 + (-123) + 1026 + (-919) + 213 + 111 + (-874)$

Problem Solving

55. In a college football game, the quarterback attempted passes with the following results.

First try	13-yd gain
Second try	incomplete
Third try	12-yd loss
Fourth try	21-yd gain
Fifth try	14-yd loss

Find the total gain (or loss).

56. The following table shows the profits and losses of a small business over a five-year period. Find the profit or loss after this period of time.

Year	*Profit or loss*
1989	+\$32,056
1990	−\$2,925
1991	+\$81,429
1992	−\$19,365
1993	−\$13,875

57. The barometric pressure at Omaha dropped 6 millibars (mb); then it rose 3 mb. After that, it dropped 14 mb and then rose 4 mb. What was the total change in pressure?

58. Monday the value of a share of IBM stock dropped $\$\frac{1}{4}$. Tuesday it rose in value $\$\frac{5}{8}$ and on Wednesday it lost $\$\frac{3}{8}$. How much did the stock's value rise or fall at the end of the three-day period?

59. Kyle's credit card bill is \$470. She sends a check to the credit card company for \$45, charges another \$160 in merchandise, and then pays off another \$500 of her bill. How much does either Kyle owe the company or the company owe Kyle?

60. Tony has \$460 in a checking account. He writes a check for \$530, makes a deposit of \$75, and then writes a check for \$90. What is the balance in the account?

Collect like terms.

61. $3a + 8a$

62. $4x + 8x$

63. $-2x + 15x$

64. $2m + (-7m)$

65. $4x + 7x$

66. $5a + 9a$

67. $7m + (-9m)$

68. $-4x + 9x$

69. $-6a + 10a$

70. $10n + (-17n)$

71. $-3 + 8x + 4 + (-10x)$

72. $8a + 5 + (-a) + (-3)$

Find the perimeter of the figure.

73.

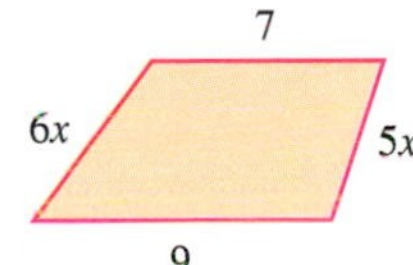

74.

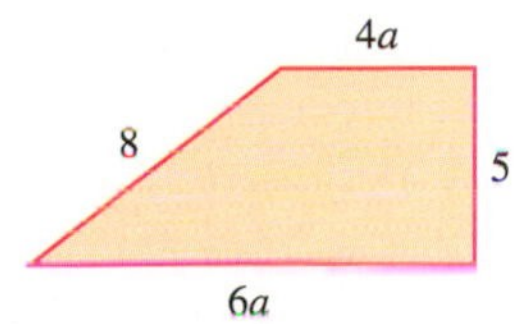

75.

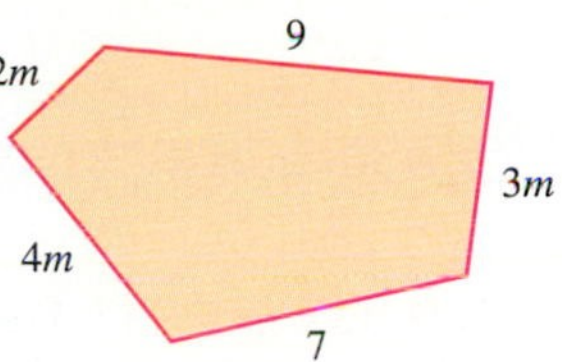

76.

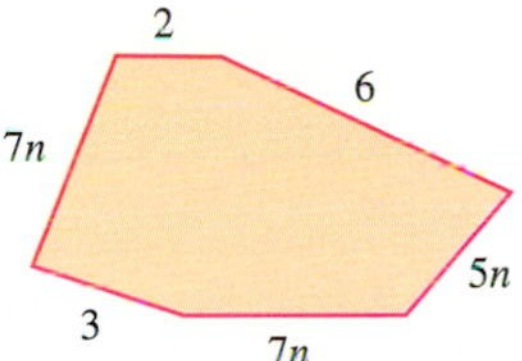

Skill Maintenance

77. Multiply: $7(3z + y + 2)$.

78. Divide and simplify: $\frac{7}{2} \div \frac{3}{8}$.

Synthesis

79. ◈ Without performing the actual addition, explain why the sum of all integers from -50 to 50 is 0.

80. ◈ Write a problem for a classmate to solve. Devise the problem so that it translates to a sum of negative and positive integers.

81. A stock's value rose $\$2\frac{3}{8}$ and then dropped $\$3\frac{1}{4}$ before finishing at $\$64\frac{3}{8}$. What was the stock's original value?

82. A sports card's value dropped \$12 and then rose \$3.70 before settling at \$32.50. What was its original value?

Find the missing term.

83. $7x + \_\_\_ + (-9x) + (-2y) = -2x - 7y$

84. $-3a + 9b + \_\_\_ + 5a = 2a - 2b$

85. $3m + 2n + \_\_\_ + (-2m) = 2n + (-6m)$

86. $\_\_\_ + 9x + (-4y) + x = 10x - 7y$

87. The perimeter of a rectangle is $6x + 10$. If the rectangle's length is 5, determine its width.

88. After five rounds of golf, a golf pro was 3 under par twice, 2 over par once, 2 under par once, and 1 over par once. On average, how far above or below par was the golfer?

1.6 Subtraction of Real Numbers

Opposites and Additive Inverses • Subtraction • Problem Solving

In arithmetic, when a number b is subtracted from another number a, the difference, $a - b$, is the number that when added to b gives a. For example, $45 - 17 = 28$ because $28 + 17 = 45$.

In this section, we use the above approach to subtraction to develop a more efficient way of finding the difference $a - b$ for any real numbers a and b. Before doing so, however, we must develop some terminology.

Opposites and Additive Inverses

Numbers such as 6 and -6 are known as **opposites,** or **additive inverses,** of each other. Whenever opposites are added, the result is 0; and whenever two numbers add to 0, the numbers are opposites.

EXAMPLE 1

Find the opposite of each number: **(a)** 34; **(b)** -8.3; **(c)** 0; **(d)** $-\frac{7}{8}$.

Solution

a) The opposite of 34 is -34: $34 + (-34) = 0$.
b) The opposite of -8.3 is 8.3: $-8.3 + 8.3 = 0$.
c) The opposite of 0 is 0: $0 + 0 = 0$.
d) The opposite of $-\frac{7}{8}$ is $\frac{7}{8}$: $-\frac{7}{8} + \frac{7}{8} = 0$. ❑

To name the opposite, we use the symbol $-$, as follows.

Opposite

The *opposite,* or *additive inverse,* of a number a can be named $-a$ (read "the opposite of a" or "the additive inverse of a").

Note that if we take a number, say 8, and find its opposite, -8, and then find the opposite of the result, we will have the original number, 8, again. Thus, for any number a,

$$-(-a) = a.$$

EXAMPLE 2

Find $-x$ and $-(-x)$ when $x = 16$.

Solution

a) If $x = 16$, then $-x = -16$. The opposite of 16 is -16.
b) If $x = 16$, then $-(-x) = -(-16) = 16$. The opposite of the opposite of 16 is 16. ❑

EXAMPLE 3

Find $-x$ and $-(-x)$ when $x = -3$.

Solution

a) If $x = -3$, then $-x = -(-3) = 3$.
b) If $x = -3$, then $-(-x) = -(-(-3)) = -(\ \ 3\ \ ) = -3$. ❑

Note in Example 3 that an extra set of parentheses is used to show that we are substituting the negative number -3 for x. Symbolism like $- \ -x$ is not used.

A symbol such as -8 is usually read "negative 8." It could be read "the additive inverse of 8," because the additive inverse of 8 is negative 8. It could also be read "the opposite of 8," because the opposite of 8 is -8. Thus a symbol like -8 can be read in more than one way. A symbol like $-x$, which has a variable, should be read "the opposite of x" or "the additive inverse of x" and *not* "negative x," since to do so suggests that $-x$ represents a negative number. As we saw in Example 3, $-x$ can represent a positive number. This notation can be used to restate Rule 3(c) from Section 1.5 as the *law of opposites:*

The Law of Opposites

For any two numbers a and $-a$,

$$a + (-a) = 0.$$

(When opposites are added, their sum is 0.)

Signs of Numbers

A negative number is said to have a "negative *sign*." A positive number is said to have a "positive *sign*." When we replace a number by its opposite, or additive inverse, we can say that we have "changed or reversed its sign."

EXAMPLE 4

Change the sign (find the opposite) of each number: **(a)** -3; **(b)** -10; **(c)** 14.

Solution

a) When we change the sign of -3, we obtain 3.
b) When we change the sign of -10, we obtain 10.
c) When we change the sign of 14, we obtain -14. ❑

Subtraction

Opposites are helpful when subtraction involves negative numbers. To see how, look for a pattern in the following examples:

Subtracting		*Adding an Opposite*
$5 - 8 = -3$	since $-3 + 8 = 5$	$5 + (-8) = -3$
$-6 - 4 = -10$	since $-10 + 4 = -6$	$-6 + (-4) = -10$
$-7 - (-10) = 3$	since $3 + (-10) = -7$	$-7 + 10 = 3$
$-7 - (-2) = -5$	since $-5 + (-2) = -7$	$-7 + 2 = -5$

Perhaps you noticed that we can subtract by adding the opposite of the number being subtracted. This can always be done and often provides the quickest method for subtraction of real numbers.

For any real numbers a and b,

$$a - b = a + (-b).$$

(To subtract, add the opposite, or additive inverse, of the number being subtracted.)

EXAMPLE 5

Subtract the following and then check by addition: **(a)** $2 - 6$; **(b)** $4 - (-9)$; **(c)** $-4.2 - (-3.6)$.

Solution

a) $2 - 6 = 2 + (-6) = -4$ — The opposite of 6 is -6. We change the subtraction to addition and add the opposite. *Check:* $-4 + 6 = 2$.

b) $4 - (-9) = 4 + 9 = 13$ — The opposite of -9 is 9. We change the subtraction to addition and add the opposite. *Check:* $13 + (-9) = 4$.

c) $-4.2 - (-3.6) = -4.2 + 3.6 = -0.6$ — Adding the opposite. *Check:* $-0.6 + (-3.6) = -4.2$. ❑

EXAMPLE 6

Subtract $-\frac{3}{5}$ from $-\frac{2}{5}$.

Solution A common denominator already exists so we subtract as follows:

$$-\frac{2}{5} - \left(-\frac{3}{5}\right) = -\frac{2}{5} + \frac{3}{5} \quad \text{Adding the opposite}$$

$$= \frac{-2 + 3}{5} = \frac{1}{5}.$$

Check: $\frac{1}{5} + \left(-\frac{3}{5}\right) = \frac{1 + (-3)}{5} = \frac{-2}{5}.$ ❑

The symbol "$-$" is read differently depending on where it appears. For instance, the expression $-5 - (-x)$ is read "negative five minus the opposite of x."

EXAMPLE 7

Read each of the following and then subtract: **(a)** $3 - 5$; **(b)** $-4.6 - (-9.8)$; **(c)** $-\frac{3}{4} - \frac{7}{5}$.

Solution

a) $3 - 5$; Read "three minus five"

$3 - 5 = 3 + (-5) = -2$ — Adding the opposite

b) $-4.6 - (-9.8)$; Read "negative four point six minus negative nine point eight"

$-4.6 - (-9.8) = -4.6 + 9.8 = 5.2$ — Adding the opposite

c) $-\frac{3}{4} - \frac{7}{5}$; Read "negative three-fourths minus seven-fifths"

$-\frac{3}{4} - \frac{7}{5} = -\frac{15}{20} + \left(-\frac{28}{20}\right) = -\frac{43}{20}$ — Finding a common denominator and adding the opposite ❑

When several additions and subtractions occur together, we can make them all additions.

EXAMPLE 8

Simplify: $8 - (-4) - 2 - (-5) + 3$.

Solution

$$8 - (-4) - 2 - (-5) + 3 = 8 + 4 + (-2) + 5 + 3$$ To subtract, we add the opposite.

$$= 18$$

❑

The **terms** of an algebraic expression are separated by plus signs. For instance, the terms of the expression $5x - 7y - 9$ are $5x$, $-7y$, and -9 since $5x - 7y - 9 = 5x + (-7y) + (-9)$.

EXAMPLE 9

Identify the terms of the expression $4 - 2ab + 7a - 9$.

Solution We have

$$4 - 2ab + 7a - 9 = 4 + (-2ab) + 7a + (-9),$$ Rewriting as addition

so the terms are 4, $-2ab$, $7a$, and -9. ❑

EXAMPLE 10

Collect like terms.

a) $1 + 3x - 7x$

b) $-5a - 7b - 4a + 10b$

c) $9 - 3m - 14 + 7m$

Solution

a) $1 + 3x - 7x = 1 + 3x + (-7x)$ Adding the opposite

$= 1 + (3 + (-7))x$ Using the distributive law

$= 1 + (-4x)$

$= 1 - 4x$ Rewriting as subtraction to be more concise

b) $-5a - 7b - 4a + 10b = -5a + (-7b) + (-4a) + 10b$ Rewriting as addition

$= -5a + (-4a) + (-7b) + 10b$ Using the commutative law of addition

$= -9a + 3b$ Adding like terms mentally

c) $9 - 3m - 14 + 7m = 9 + (-3m) + (-14) + 7m$ Rewriting

$= 9 + (-14) + (-3m) + 7m$ Using a commutative law

$= -5 + 4m$ ❑

Problem Solving

Subtraction is used to solve problems involving differences.

EXAMPLE 11

The lowest point in Asia is the Dead Sea, which is 400 m below sea level. The lowest point in the United States is Death Valley, which is 86 m below sea level. What is the difference in elevation between the Dead Sea and Death Valley?

Solution It is helpful to draw a picture of the situation.

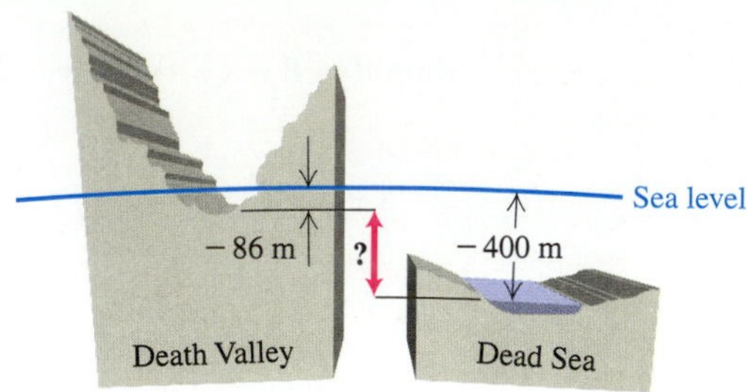

To find the difference in elevation, we always subtract the lower elevation, -400 m, from the higher elevation, -86 m:

$$-86 - (-400) = -86 + 400 = 314.$$

Death Valley is 314 m higher than the Dead Sea. ❑

EXERCISE SET 1.6

Find the opposite, or additive inverse.

1. 24 **2.** -64 **3.** -9

4. $\frac{7}{2}$ **5.** -26.9 **6.** 48.2

Find $-x$ when x is each of the following.

7. 9 **8.** -26 **9.** $-\frac{14}{3}$

10. $\frac{1}{328}$ **11.** 0.101 **12.** 0

Find $-(-x)$ when x is each of the following.

13. -65 **14.** 29

15. $\frac{5}{3}$ **16.** -9.1

Change the sign. (Find the opposite.)

17. -1 **18.** -7

19. 7 **20.** 10

Subtract.

21. $3 - 7$ **22.** $4 - 9$

23. $0 - 7$ **24.** $0 - 10$

25. $-8 - (-2)$ **26.** $-6 - (-8)$

27. $-10 - (-10)$ **28.** $-8 - (-8)$

29. $12 - 16$ **30.** $14 - 19$

31. $20 - 27$ **32.** $30 - 4$

33. $-9 - (-3)$ **34.** $-7 - (-9)$

35. $-40 - (-40)$ **36.** $-9 - (-9)$

37. $7 - 7$ **38.** $9 - 9$

39. $7 - (-7)$ **40.** $4 - (-4)$

41. $8 - (-3)$ **42.** $-7 - 4$

43. $-6 - 8$ **44.** $6 - (-10)$

45. $-4 - (-9)$ **46.** $-14 - 2$

47. $-6 - (-5)$ **48.** $-4 - (-3)$

49. $8 - (-10)$ **50.** $5 - (-6)$

51. $0 - 5$ **52.** $0 - 6$

53. $-5 - (-2)$ **54.** $-3 - (-1)$

55. $-7 - 14$ **56.** $-9 - 16$

57. $0 - (-5)$ **58.** $0 - (-1)$

59. $-8 - 0$ **60.** $-9 - 0$

61. $7 - (-5)$ **62.** $20 - (-15)$

63. $2 - 25$ **64.** $18 - 63$

65. $-42 - 26$ **66.** $-18 - 63$

67. $-71 - 2$ **68.** $-49 - 3$

69. $24 - (-92)$ **70.** $48 - (-73)$

71. $-50 - (-50)$ **72.** $-70 - (-70)$

73. $\frac{3}{8} - \frac{5}{8}$ **74.** $\frac{3}{9} - \frac{9}{9}$

75. $\frac{3}{4} - \frac{2}{3}$ **76.** $\frac{5}{8} - \frac{3}{4}$

77. $-\frac{3}{4} - \frac{2}{3}$
78. $-\frac{5}{8} - \frac{3}{4}$
79. $-2.8 - 0$
80. $6.04 - 1.1$
81. $0.99 - 1$
82. $0.87 - 1$
83. $\frac{1}{6} - \frac{2}{3}$
84. $-\frac{3}{8} - \left(-\frac{1}{2}\right)$
85. $-\frac{4}{7} - \left(-\frac{10}{7}\right)$
86. $\frac{12}{5} - \frac{12}{5}$

Translate the phrase to mathematical language and simplify. See Example 11.

87. The difference between 1.5 and -3.5
88. The difference between -2.1 and -5.9
89. The difference between -79 and 114
90. The difference between 23 and -17
91. Subtract 41 from -13.
92. Subtract 19 from -7.
93. Subtract -25 from 9.
94. Subtract -31 from -5.

Write words for each of the following and then perform the subtraction.

95. $-3.2 - 5.8$
96. $-2.7 - 5.9$
97. $-230 - (-500)$
98. $-350 - (-1000)$

Simplify.

99. $18 - (-15) - 3 - (-5) + 2$
100. $22 - (-18) + 7 + (-42) - 27$
101. $-31 + (-28) - (-14) - 17$
102. $-43 - (-19) - (-21) + 25$
103. $-34 - 28 + (-33) - 44$
104. $39 + (-88) - 29 - (-83)$
105. $-93 - (-84) - 41 - (-56)$
106. $84 + (-99) + 44 - (-18) - 43$

Identify the terms in each expression.

107. $3x - 2y$
108. $7a - 9b$
109. $-5 + 3m - 6mn$
110. $-9 - 4t + 10rt$
111. $5 - a - 6b + 2$
112. $-2 + 3x - y - 8$

Collect like terms.

113. $7a - 12a$
114. $3x - 15x$
115. $-3m - 5 + m$
116. $-7 + 9n - 8$
117. $3x + 5 - 9x$
118. $2 + 3a - 7$
119. $2 - 6t - 9 - 2t$
120. $-5 + 3b - 7 - 5b$
121. $-5 - (-3x) + 3x + 4x - (-12)$
122. $14 - (-5x) + 2x - (-32)$
123. $13x - (-2x) + 45 - (-21)$
124. $8x - (-2x) - 14 - (-5x) + 53$

Problem Solving

125. Your total assets are \$619.46. You borrow \$950 for the purchase of a stereo system. What are your total assets now?

126. You owe a friend \$420. The friend decides to cancel \$156 of the debt. How much do you owe now?

127. In Churchill, Manitoba, Canada, the average daily low temperature in January is -31°C. The average daily low temperature in Key West, Florida, is 19°C. What is the difference in the average daily low temperature of the two cities?

128. On a winter night, the temperature dropped from 5°C to -12°C. How many degrees did it drop?

129. The lowest point in Africa is Lake Assal, which is 156 m below sea level. The lowest point in South America is the Valdes Peninsula, which is 40 m below sea level. How much lower is Lake Assal than the Valdes Peninsula?

130. The deepest point in the Pacific Ocean is the Marianas Trench, with a depth of 10,415 m. The deepest point in the Atlantic Ocean is the Puerto Rico Trench, with a depth of 8648 m. What is the difference in elevation of the two trenches?

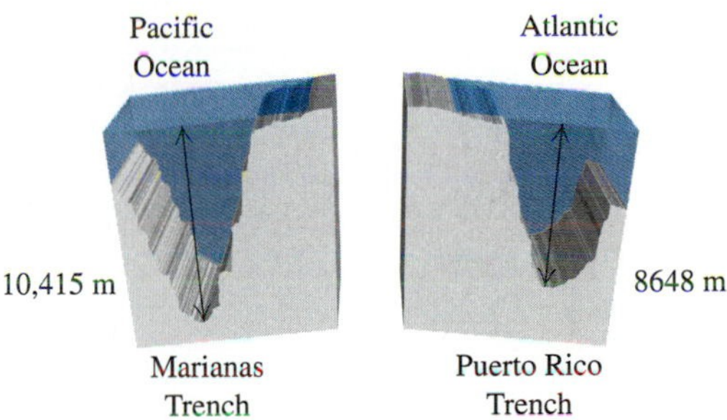

Skill Maintenance

131. Find the area of a rectangle when the length is 36 ft and the width is 12 ft.

132. Find the prime factorization of 864.

Synthesis

133. ◈ Explain why $-a + b$ is the opposite of $a + (-b)$.

134. ◈ A student claims to be able to add real numbers but unable to subtract them. What advice would you offer this student?

Tell whether each statement is true or false for all real numbers m and n. Use various replacements for m and n to support your answer.

135. If $m > n$, then $m - n > 0$.

136. If $m > n$, then $m + n > 0$.

137. If m and n are opposites, then $m - n = 0$.

138. If $m = -n$, then $m + n = 0$.

139. ◈ A gambler loses a wager and then loses "double or nothing" (meaning the gambler owes twice as much) twice more. After the three losses, the gambler's assets are $-\$20$. Explain how much the gambler originally bet and how the \$20 debt occurred.

140. ◈ If n is positive and m is negative, what is the sign of $n + (-m)$? Why?

1.7 Multiplication and Division of Real Numbers

Multiplication • Division

We now develop rules for multiplication and division of real numbers. Since multiplication and division are closely related, it should come as no surprise that the rules are quite similar.

Multiplication

We are already familiar with how to multiply two nonnegative numbers. To see how to multiply a positive number and a negative number, consider the following pattern in which multiplication is regarded as repeated addition:

This number decreases by 1 each time. →

$$4(-5) = (-5) + (-5) + (-5) + (-5) = -20$$
$$3(-5) = (-5) + (-5) + (-5) = -15$$
$$2(-5) = (-5) + (-5) = -10$$
$$1(-5) = (-5) = -5$$
$$0(-5) = 0 = 0$$

← This number increases by 5 each time.

This pattern illustrates that the product of a negative number and a positive number is negative.

> To multiply a positive number and a negative number, multiply their absolute values. The answer is negative.

EXAMPLE 1

Multiply: **(a)** $8(-5)$; **(b)** $-\frac{1}{3} \cdot \frac{5}{7}$.

Solution

a) $8(-5) = -40$

b) $-\frac{1}{3} \cdot \frac{5}{7} = -\frac{5}{21}$ ❑

The pattern developed above includes not just products of positive and negative numbers, but a product involving zero as well.

The Multiplicative Property of Zero

For any real number a,

$$0 \cdot a = a \cdot 0 = 0.$$

(The product of 0 and any real number is 0).

EXAMPLE 2

Multiply: $173(-452)0$.

Solution

$$173(-452)0 = 173[(-452)0]$$ Using the associative law of multiplication

$$= 173[0]$$ Using the multiplicative property of zero

$$= 0$$ Using the multiplicative property of zero again

Note that whenever 0 appears as a factor, the product will be 0. ❑

We can extend the above pattern still further to examine the product of two negative numbers.

This number decreases by 1 each time. →

$$2(-5) = (-5) + (-5) = -10$$
$$1(-5) = (-5) = -5$$
$$0(-5) = 0 = 0$$
$$-1(-5) = -(-5) = 5$$
$$-2(-5) = -(-5) - (-5) = 10$$

← This number increases by 5 each time.

According to the pattern, the product of two negative numbers is positive.

To multiply two negative numbers, multiply their absolute values. The answer is positive.

EXAMPLE 3

Multiply: **(a)** $(-5)(-7)$; **(b)** $(-1.2)(-3)$.

Solution

a) The absolute value of -5 is 5 and the absolute value of -7 is 7. Thus,

$$(-5)(-7) = 5 \cdot 7$$ Multiplying absolute values

$$= 35.$$

b) $(-1.2)(-3) = (1.2)(3)$ Multiplying absolute values

$$= 3.6$$ Try to go directly to this step. ❑

When three or more numbers are multiplied, we can order and group the numbers as we please, because of the commutative and associative laws.

EXAMPLE 4

Multiply: **(a)** $-3(-2)(-5)$; **(b)** $-4(-6)(-1)(-2)$.

Solution

a) $-3(-2)(-5) = 6(-5)$ — Multiplying the first two numbers. The product of two negatives is positive.

$= -30$ — The product of a positive and a negative is negative.

b) $-4(-6)(-1)(-2) = 24 \cdot 2$ — Multiplying the first two numbers and the last two numbers

$= 48$ ❑

We can see the following pattern in the results of Example 4.

The product of an even number of negative numbers is positive.
The product of an odd number of negative numbers is negative.

Division

Because the definition of division makes use of multiplication, the rules for multiplication are used to develop rules for division.

The quotient $\frac{a}{b}$ (or $a \div b$) is the number, if there is one, that when multiplied by b gives a. ($a \div b = c$ if $c \cdot b = a$.)

EXAMPLE 5

Divide, if possible, and check your answer: **(a)** $14 \div (-7)$; **(b)** $\frac{-32}{-4}$; **(c)** $\frac{-10}{7}$; **(d)** $\frac{-17}{0}$.

Solution

a) $14 \div (-7) = -2$ — We look for a number that when multiplied by -7 gives 14. That number is -2. *Check:* $(-2)(-7) = 14$.

b) $\dfrac{-32}{-4} = 8$ — We look for a number that when multiplied by -4 gives -32. That number is 8. *Check:* $8(-4) = -32$.

c) $\dfrac{-10}{7} = -\dfrac{10}{7}$ — We look for a number that when multiplied by 7 gives -10. That number is $-\frac{10}{7}$. *Check:* $-\frac{10}{7} \cdot 7 = -10$.

d) $\dfrac{-17}{0}$ is **undefined.** — We look for a number that when multiplied by 0 gives -17. There is no such number because the product of 0 and *any* number is 0, not -17. ❑

The rules for division are the same as those for multiplication. We state them together.

Rules for Multiplication and Division

To multiply or divide two real numbers:

1. Using the absolute values, multiply or divide, as indicated.
2. If the signs are the same, the answer is positive.
3. If the signs are different, the answer is negative.

Had Example 5(a) been written as $-14 \div 7$ instead of $14 \div (-7)$, the result would have still been -2. Similarly, had Example 5(c) been written as $10/-7$ instead of $-10/7$, the result, $-\frac{10}{7}$, would not have changed. In short, our rules for division give us the following:

$$\frac{-a}{b} = \frac{a}{-b} = -\frac{a}{b} \quad \text{and} \quad \frac{-a}{-b} = \frac{a}{b}.$$

EXAMPLE 6

Rewrite each of the following in two equivalent forms: **(a)** $\frac{5}{-2}$; **(b)** $-\frac{3}{10}$.

Solution We use the property listed above.

a) $\frac{5}{-2} = \frac{-5}{2}$ and $\frac{5}{-2} = -\frac{5}{2}$

b) $-\frac{3}{10} = \frac{-3}{10}$ and $-\frac{3}{10} = \frac{3}{-10}$ ❑

In some situations, it may help to rewrite a fraction that has a negative sign in an equivalent form.

EXAMPLE 7

Perform the indicated operation: **(a)** $\left(-\frac{4}{5}\right)\left(\frac{-7}{3}\right)$; **(b)** $-\frac{2}{7} + \frac{9}{-7}$.

Solution

a) $\left(-\frac{4}{5}\right)\left(\frac{-7}{3}\right) = \left(\frac{-4}{5}\right)\left(\frac{-7}{3}\right)$ Rewriting $-\frac{4}{5}$ as $\frac{-4}{5}$

$= \frac{28}{15}$ Try to go directly to this step.

b) Given a choice, we generally choose a positive denominator, although this is not a "must":

$-\frac{2}{7} + \frac{9}{-7} = \frac{-2}{7} + \frac{-9}{7}$ Rewriting both fractions with a common denominator of 7

$= \frac{-11}{7}$, or $-\frac{11}{7}$. ❑

EXAMPLE 8

Find the reciprocal: **(a)** -27; **(b)** $\frac{-3}{4}$; **(c)** $-\frac{1}{5}$.

Solution

a) The reciprocal of -27 is $\frac{1}{-27}$. More often, this number is written as $-\frac{1}{27}$.

b) The reciprocal of $\frac{-3}{4}$ is $\frac{4}{-3}$, or, equivalently, $-\frac{4}{3}$.

c) The reciprocal of $-\frac{1}{5}$ is -5. ❑

Keep in mind that the opposite, or additive inverse, of a number is what we add to the number to get 0, whereas a reciprocal is what we multiply the number by to get 1. Compare the following.

Number	Opposite (Change the sign.)	Reciprocal (Invert but do not change the sign.)
$-\frac{3}{8}$	$\frac{3}{8}$	$-\frac{8}{3}$
19	-19	$\frac{1}{19}$
0	0	Undefined

$\left(-\frac{3}{8}\right)\left(-\frac{8}{3}\right) = 1$

$-\frac{3}{8} + \frac{3}{8} = 0$

When dividing with fractional notation, it is usually easier to multiply by a reciprocal. With decimal notation, it is usually easier to carry out long division.

EXAMPLE 9

Divide: **(a)** $-\frac{2}{3} \div \left(-\frac{5}{4}\right)$; **(b)** $-\frac{3}{4} \div \frac{3}{10}$; **(c)** $27.9 \div (-3)$.

Solution

a) $-\frac{2}{3} \div \left(-\frac{5}{4}\right) = -\frac{2}{3} \cdot \left(-\frac{4}{5}\right) = \frac{8}{15}$ Multiplying by the reciprocal

Be careful not to change the sign when taking a reciprocal!

b) $-\frac{3}{4} \div \frac{3}{10} = -\frac{3}{4} \cdot \left(\frac{10}{3}\right) = -\frac{30}{12} = -\frac{5}{2} \cdot \frac{6}{6} = -\frac{5}{2}$ Removing a factor of 1: $\frac{6}{6} = 1$

c) $27.9 \div (-3) = \frac{27.9}{-3} = -9.3$ Do the long division $3\overline{)27.9}$ giving 9.3. The answer is negative. ❑

In Example 5(d), we explained why we cannot divide -17 by 0. This also explains why *no* nonzero number b can be divided by 0: Consider $b \div 0$. Is there a number that when multiplied by 0 gives b? No, because the product of 0 and any number is 0, not b. We say that $b \div 0$ is **undefined** for $b \neq 0$.

On the other hand, if we divide 0 by 0, we look for a number r such that $0 \div 0 = r$ and $r \cdot 0 = 0$. But, $r \cdot 0 = 0$ for *any* number r. Thus it appears that $0 \div 0$ could be any number we choose. Getting any answer we want when we divide 0 by 0 would lead to contradictions. Thus we say that $0 \div 0$ is **indeterminate.**

Finally, note that $0 \div 7 = 0$ since $0 \cdot 7 = 0$. This can be written $0/7 = 0$.

EXAMPLE 10

Divide, if possible: **(a)** $\frac{0}{-2}$; **(b)** $\frac{5}{0}$; **(c)** $\frac{0}{0}$.

Solution

a) $\frac{0}{-2} = 0$ *Check:* $0(-2) = 0$.

b) $\frac{5}{0}$ is undefined.

c) $\frac{0}{0}$ is indeterminate. ❑

Division Involving Zero

For any nonzero real number a,

$$\frac{0}{a} = 0 \quad \text{and} \quad \frac{a}{0} \text{ is undefined.}$$

The expression $\frac{0}{0}$ is indeterminate.

EXERCISE SET 1.7

Multiply.

1. $-8 \cdot 2$
2. $-2 \cdot 5$
3. $-7 \cdot 6$
4. $-9 \cdot 2$
5. $8 \cdot (-3)$
6. $9 \cdot (-5)$
7. $-9 \cdot 8$
8. $-10 \cdot 3$
9. $-8 \cdot (-2)$
10. $-2 \cdot (-5)$
11. $-7 \cdot (-6)$
12. $-9 \cdot (-2)$
13. $15 \cdot (-8)$
14. $-12 \cdot (-10)$
15. $-14 \cdot 17$
16. $-13 \cdot (-15)$
17. $-25 \cdot (-48)$
18. $39 \cdot (-43)$
19. $-3.5 \cdot (-28)$
20. $97 \cdot (-2.1)$
21. $9 \cdot (-8)$
22. $7 \cdot (-9)$
23. $-7 \cdot (-3.1)$
24. $-4 \cdot (-3.2)$
25. $\frac{2}{3} \cdot \left(-\frac{3}{5}\right)$
26. $\frac{5}{7} \cdot \left(-\frac{2}{3}\right)$
27. $-\frac{3}{8} \cdot \left(-\frac{2}{9}\right)$
28. $-\frac{5}{8} \cdot \left(-\frac{2}{5}\right)$
29. -6.3×2.7
30. -4.1×9.5
31. $-\frac{5}{9} \cdot \frac{3}{4}$
32. $-\frac{8}{3} \cdot \frac{9}{4}$
33. $7 \cdot (-4) \cdot (-3) \cdot 5$
34. $9 \cdot (-2) \cdot (-6) \cdot 7$
35. $-\frac{2}{3} \cdot \frac{1}{2} \cdot \left(-\frac{6}{7}\right)$
36. $-\frac{1}{8} \cdot \left(-\frac{1}{4}\right) \cdot \left(-\frac{3}{5}\right)$
37. $-3 \cdot (-4) \cdot (-5)$
38. $-2 \cdot (-5) \cdot (-7)$
39. $-2 \cdot (-5) \cdot (-3) \cdot (-5)$
40. $-3 \cdot (-5) \cdot (-2) \cdot (-1)$
41. $(-14) \cdot (-27) \cdot 0$
42. $7 \cdot (-6) \cdot 5 \cdot (-4) \cdot 3 \cdot (-2) \cdot 1 \cdot 0$
43. $(-8)(-9)(-10)$
44. $(-7)(-8)(-9)(-10)$
45. $(-6)(-7)(-8)(-9)(-10)$
46. $(-5)(-6)(-7)(-8)(-9)(-10)$

Divide, if possible, and check. If a quotient is undefined or indeterminate, state so.

47. $36 \div (-6)$
48. $\frac{28}{-7}$
49. $\frac{26}{-2}$
50. $26 \div (-13)$
51. $\frac{-16}{8}$
52. $-22 \div (-2)$
53. $\frac{-48}{-12}$
54. $-63 \div (-9)$
55. $\frac{-72}{9}$
56. $\frac{-50}{25}$
57. $-100 \div (-50)$
58. $\frac{-200}{8}$
59. $-108 \div 9$
60. $\frac{-64}{-7}$
61. $\frac{200}{-25}$
62. $-300 \div (-13)$
63. $\frac{75}{0}$
64. $\frac{0}{-5}$
65. $\frac{88}{-9}$
66. $\frac{0}{0}$
67. $\frac{0}{-9}$
68. $\frac{-35}{0}$
69. $0 \div 0$
70. $0 \div (-47)$

Write the number in two equivalent forms, as in Example 6.

71. $\frac{9}{-5}$ **72.** $\frac{-12}{7}$

73. $\frac{-36}{11}$ **74.** $\frac{9}{-14}$

75. $-\frac{7}{3}$ **76.** $-\frac{4}{15}$

77. $\frac{-x}{2}$ **78.** $\frac{9}{-a}$

Find the reciprocal.

79. $\frac{-3}{7}$ **80.** $\frac{2}{-9}$

81. $-\frac{47}{13}$ **82.** $-\frac{31}{12}$

83. -10 **84.** 13

85. 4.3 **86.** -8.5

87. $\frac{5}{-3}$ **88.** $\frac{-6}{11}$

89. -1 **90.** $\frac{1}{1/2}$

Perform the indicated operation and simplify, if possible. If a quotient is undefined or indeterminate, state so.

91. $\left(-\frac{3}{7}\right)\left(\frac{2}{-5}\right)$ **92.** $\left(\frac{-4}{9}\right)\left(-\frac{2}{3}\right)$

93. $\left(\frac{7}{-2}\right)\left(\frac{-5}{6}\right)$ **94.** $\left(\frac{-6}{5}\right)\left(\frac{2}{-11}\right)$

95. $\frac{-4}{5}+\frac{7}{-5}$ **96.** $\frac{3}{-8}+\frac{-5}{8}$

97. $\left(-\frac{2}{7}\right)\left(\frac{5}{-8}\right)$ **98.** $\left(\frac{-9}{5}\right)\left(-\frac{10}{7}\right)$

99. $\frac{-9}{7}+\left(-\frac{4}{7}\right)$ **100.** $\left(-\frac{3}{11}\right)+\frac{5}{-11}$

101. $\frac{3}{4}\div\left(-\frac{2}{3}\right)$ **102.** $\frac{7}{8}\div\left(-\frac{1}{2}\right)$

103. $\frac{-5}{12}\cdot\frac{7}{15}$ **104.** $\frac{9}{5}\cdot\frac{-20}{3}$

105. $\left(-\frac{12}{5}\right)+\left(-\frac{3}{5}\right)$ **106.** $\left(-\frac{18}{7}\right)+\left(-\frac{3}{7}\right)$

107. $-\frac{5}{4}\div\left(-\frac{3}{4}\right)$ **108.** $-\frac{5}{9}\div\left(-\frac{5}{6}\right)$

109. $-6.6\div 3.3$ **110.** $-44.1\div(-6.3)$

111. $\frac{-3}{7}-\frac{2}{7}$ **112.** $\frac{-5}{9}-\frac{2}{9}$

113. $\frac{-5}{9}+\frac{2}{-3}$ **114.** $\frac{-3}{10}+\frac{2}{-5}$

115. $\left(\frac{-3}{5}\right)\div\frac{6}{15}$ **116.** $\frac{7}{10}\div\left(\frac{-3}{5}\right)$

117. $\frac{4}{9}-\frac{1}{-9}$ **118.** $\frac{5}{7}-\frac{1}{-7}$

119. $\frac{3}{-10}+\frac{-1}{5}$ **120.** $\frac{-4}{15}+\frac{2}{-3}$

121. $\frac{-2}{3}-\frac{1}{-6}$ **122.** $\frac{-7}{10}-\frac{1}{-5}$

Skill Maintenance

123. Simplify: $\frac{264}{468}$.

124. Collect like terms: $x+12y+11x-14y-9$.

Synthesis

125. ◈ Most calculators have a key, often appearing as [1/x], for finding reciprocals. To use this key, enter a number and then press [1/x] to find its reciprocal. What should happen if you enter a number on a calculator and press the reciprocal key twice? Why?

126. ◈ What advice would you offer a student who claims to be able to multiply, but not divide, any two real numbers?

127. Determine those real numbers a for which the opposite of a is the same as the reciprocal of a.

128. Determine those real numbers that are their own reciprocals.

Tell whether the expression represents a positive number or a negative number when m and n are negative.

129. $\frac{-n}{m}$ **130.** $\frac{-n}{-m}$

131. $-\left(\frac{-n}{m}\right)$ **132.** $-\left(\frac{n}{-m}\right)$

133. $-\left(\frac{-n}{-m}\right)$

134. What must be true of m and n if $-mn$ is to be **(a)** positive? **(b)** zero? **(c)** negative?

135. The following is a proof that a positive number times a negative number is negative. Explain the reason for each step. Assume that $a>0$ and $b>0$.

$$\begin{aligned} a(-b)+ab &= a[-b+b] \\ &= a(0) \\ &= 0 \end{aligned}$$

Therefore, $a(-b)=-ab$.

136. ◈ Is it true that for any numbers a and b, if a is larger than b, then the reciprocal of a is smaller than the reciprocal of b? Why or why not?

1.8 Exponential Notation and Order of Operations

Exponential Notation • Order of Operations • Simplifying and the Distributive Law • The Opposite of a Sum

Algebraic expressions often contain *exponential notation*. In this section, we learn how to use exponential notation as well as rules for the *order of operations* in performing certain algebraic manipulations.

Exponential Notation

A product like $3 \cdot 3 \cdot 3 \cdot 3$, in which the factors are the same, is called a **power.** Powers occur often enough that a simpler notation called **exponential notation** is often used. For

$$\underbrace{3 \cdot 3 \cdot 3 \cdot 3}_{4 \text{ factors}}, \quad \text{we write} \quad 3^4.$$

This is read "three to the fourth power," or, simply, "three to the fourth." The number 4 is called an **exponent** and the number 3 a **base.**

Expressions like s^2 and s^3 are usually read "*s* squared" and "*s* cubed," respectively. This comes from the fact that a square of side *s* has an area *A* given by $A = s^2$ and a cube of side *s* has a volume *V* given by $V = s^3$.

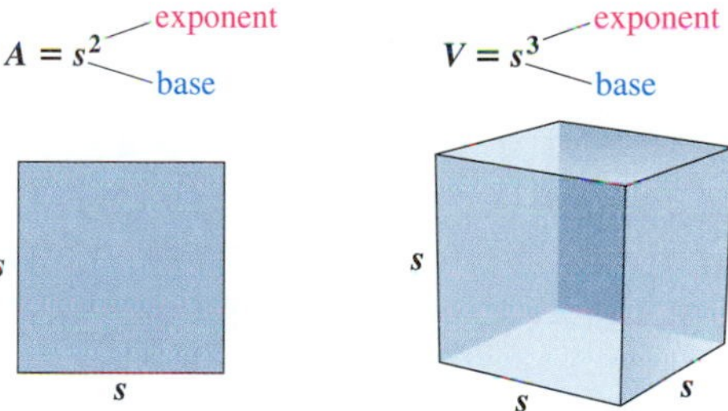

EXAMPLE 1

Write exponential notation for $10 \cdot 10 \cdot 10 \cdot 10 \cdot 10$.

Solution

Exponential notation is 10^5. 5 is the *exponent*. 10 is the *base*. □

EXAMPLE 2

Evaluate: **(a)** 5^2; **(b)** $(-5)^3$; **(c)** $(2n)^3$.

Solution

a) $5^2 = 5 \cdot 5 = 25$ The second power indicates two factors of 5.

b) $(-5)^3 = (-5)(-5)(-5)$
$= 25(-5)$ Using the associative law of multiplication
$= -125$

c) $(2n)^3 = (2n)(2n)(2n)$
$= 2 \cdot 2 \cdot 2 \cdot n \cdot n \cdot n$ Using the associative and commutative laws of multiplication
$= 8n^3$ ❑

Look for a pattern in the following:

$8 \cdot 8 \cdot 8 \cdot 8 = 8^4$
$8 \cdot 8 \cdot 8 = 8^3$ We divide by 8 each time.
$8 \cdot 8 = 8^2$
$8 = 8^?$.

The exponents decrease by 1 each time. To continue the pattern, we say that

$8 = 8^1$.

Exponential Notation

$b^1 = b$, for any number b.

For any natural number n greater than or equal to 2,

$$b^n \text{ means } \overbrace{b \cdot b \cdot b \cdot b \cdots b}^{n \text{ factors}}.$$

Order of Operations

What does $5 \times 2 + 4$ mean? If we multiply 5 by 2 and add 4, we get 14. If we add 2 and 4 and multiply by 5, we get 30. Since our results are different, we see that the order in which we carry out operations is important. To tell which operation to do first, we use grouping symbols such as parentheses ( ), brackets [ ], braces { }, absolute value bars | |, or fraction bars. For example,

$(3 \times 5) + 6 = 15 + 6 = 21$,

but

$3 \times (5 + 6) = 3 \times 11 = 33$.

Besides grouping symbols, there are rules for the order in which operations should be done.

Rules for Order of Operations

1. Perform all calculations within grouping symbols.
2. Evaluate all exponential expressions.
3. Perform all multiplications and divisions in order from left to right.
4. Perform all additions and subtractions in order from left to right.

EXAMPLE 3

Simplify: $15 - 2 \times 5 + 3$.

Solution When no groupings or exponents appear, we always multiply or divide before adding or subtracting:

$$15 - 2 \times 5 + 3 = 15 - 10 + 3 \quad \text{Multiplying}$$
$$= 5 + 3$$
$$= 8. \quad \text{Subtracting and adding from left to right}$$

❑

Always calculate within parentheses first. When there are exponents and no parentheses, simplify powers before multiplying or dividing.

EXAMPLE 4

Simplify: **(a)** $(3 \times 4)^2$; **(b)** 3×4^2.

Solution

a) $(3 \times 4)^2 = (12)^2$ Working within parentheses first
$= 144$

b) $3 \times 4^2 = 3 \times 16$ Simplifying the power
$= 48$ Multiplying

Note that $(3 \times 4)^2 \neq 3 \times 4^2$. ❑

> CAUTION! Example 4 illustrates that, in general, $(ab)^2 \neq ab^2$.

EXAMPLE 5

Evaluate for $x = 5$: **(a)** $(-x)^2$; **(b)** $-x^2$.

Solution

a) $(-x)^2 = (-5)^2 = (-5)(-5) = 25$ Substitute 5 for x, take the opposite, and then evaluate the power.

b) $-x^2 = -(5)^2 = -25$ Substitute 5 for x. Evaluate the power. Then find the opposite. ❑

> CAUTION! Example 5 illustrates that, in general, $(-x)^2 \neq -x^2$.

EXAMPLE 6

Evaluate $-15 \div 3(6 - a)^3$ for $a = 4$.

Solution

$$-15 \div 3(6 - a)^3 = -15 \div 3(6 - 4)^3 \quad \text{Substituting 4 for } a$$
$$= -15 \div 3(2)^3 \quad \text{Working within parentheses first}$$
$$= -15 \div 3 \cdot 8 \quad \text{Simplifying the exponential expression}$$
$$= -5 \cdot 8$$
$$= -40 \quad \text{Dividing and multiplying from left to right}$$

❑

When combinations of grouping symbols are used, the rules still apply. We begin with the innermost grouping symbols and work to the outside.

EXAMPLE 7

Simplify: $16 \div (-2) + 3[10 + 2(3 - 5)^3]$.

Solution

$$16 \div (-2) + 3[10 + 2(3 - 5)^3]$$

$= 16 \div (-2) + 3[10 + 2(-2)^3]$ Doing the calculations in the innermost parentheses first

$= 16 \div (-2) + 3[10 + 2(-8)]$ $(-2)^3 = (-2)(-2)(-2) = -8$

$= 16 \div (-2) + 3[10 + (-16)]$

$= 16 \div (-2) + 3[-6]$ Completing the calculations within the brackets

$= -8 + (-18)$ Multiplying and dividing from left to right

$= -26$ ❑

EXAMPLE 8

Calculate: $\dfrac{12(9 - 7) + 4 \cdot 5}{3^4 + 2^3}$.

Solution An equivalent expression with brackets as grouping symbols is

$$[12(9 - 7) + 4 \cdot 5] \div [3^4 + 2^3].$$

What this shows, in effect, is that we do the calculations in the numerator and then in the denominator, and divide the results:

$$\frac{12(9 - 7) + 4 \cdot 5}{3^4 + 2^3} = \frac{12(2) + 4 \cdot 5}{81 + 8} = \frac{24 + 20}{89} = \frac{44}{89}.$$

❑

Simplifying and the Distributive Law

Sometimes we cannot simplify within parentheses. When a sum or difference is within the parentheses, we can use the distributive law to help simplify.

EXAMPLE 9

Simplify: $5x - 9 + 2(4x + 5)$.

Solution

$5x - 9 + 2(4x + 5) = 5x - 9 + 8x + 10$ Using the distributive law

$= 13x + 1$ Collecting like terms ❑

Now that exponents have been introduced, we can make our definition of *like* or *similar terms* more precise. **Like,** or **similar, terms** are either constant terms or terms containing the same variable(s) raised to the same power(s). Thus, 5 and -7, $19xy$ and $-xy$, and $4a^3b$ and $7a^3b$ are all pairs of like terms.

EXAMPLE 10

Simplify: $7x^2 + 3(x^2 + 2x) - 5x$.

Solution

$$7x^2 + 3(x^2 + 2x) - 5x = 7x^2 + 3x^2 + 6x - 5x \quad \text{Using the distributive law}$$
$$= 10x^2 + x \quad \text{Collecting like terms}$$

The Opposite of a Sum

Multiplication by -1 changes a number's sign. For example, $-1(7) = -7$ and $-1(-5) = 5$. Thus, when a number is multiplied by -1, we get the opposite of that number.

The Property of -1

For any real number a,

$$-1 \cdot a = -a.$$

(Negative one times a is the opposite of a.)

When grouping symbols are preceded by a "$-$" symbol, we can multiply the grouping by -1 and use the distributive law. In this manner, we can find the *opposite*, or *additive inverse, of a sum.*

EXAMPLE 11

Write an expression equivalent to $-(3x + 2y + 4)$ without using parentheses.

Solution

$$-(3x + 2y + 4) = -1(3x + 2y + 4) \quad \text{Using the property of } -1$$
$$= -1(3x) + (-1)(2y) + (-1)4 \quad \text{Using the distributive law}$$
$$= -3x - 2y - 4 \quad \text{Using the property of } -1$$

Example 11 illustrates an important property of real numbers.

The Opposite of a Sum

For any real numbers a and b,

$$-(a + b) = -a + (-b).$$

(The opposite of a sum is the sum of the opposites.)

To remove parentheses from an expression like $-(x - 7y + 5)$, we can first rewrite the subtraction as addition:

$$-(x - 7y + 5) = -(x + (-7y) + 5) \quad \text{Rewriting as addition}$$
$$= -x + 7y - 5. \quad \text{Taking the opposite of a sum}$$

This procedure is normally streamlined to one step in which we find the opposite by "removing parentheses and changing the sign of every term":

$$-(x - 7y + 5) = -x + 7y - 5.$$

EXAMPLE 12

Simplify: $3x - (4x + 2)$.

Solution

$3x - (4x + 2) = 3x + [-(4x + 2)]$ Adding the opposite of $(4x + 2)$

$= 3x + [-4x - 2]$ Changing the sign of each term inside the parentheses

$= 3x - 4x - 2$

$= -x - 2$ Collecting like terms ❑

In practice, the first two steps of Example 12 are often skipped.

EXAMPLE 13

Simplify: **(a)** $5y - (3y + 4)$; **(b)** $3y - 2 - (2y - 4)$.

Solution

a) $5y - (3y + 4) = 5y - 3y - 4$ Removing parentheses by changing the sign of every term inside the parentheses

$= 2y - 4$ Collecting like terms

b) $3y - 2 - (2y - 4) = 3y - 2 - 2y + 4$

$= y + 2$ ❑

Expressions such as $7 - 3(x + 2)$ can be simplified as follows:

$7 - 3(x + 2) = 7 + [-3(x + 2)]$ Adding the opposite of $3(x + 2)$

$= 7 + [-3x - 6]$ Multiplying $x + 2$ by -3

$= 7 - 3x - 6$ Try to go directly to this step.

$= 1 - 3x.$ Collecting like terms

EXAMPLE 14

Simplify: **(a)** $3y - 2(4y - 5)$; **(b)** $2a + 3b - 7 - 4(-5a - 6b + 12)$.

Solution

a) $3y - 2(4y - 5) = 3y - 8y + 10$ Multiplying each term in the parentheses by -2

$= -5y + 10$

b) $2a + 3b - 7 - 4(-5a - 6b + 12) = 2a + 3b - 7 + 20a + 24b - 48$

$= 22a + 27b - 55$ ❑

EXERCISE SET 1.8

Write exponential notation.

1. $10 \times 10 \times 10$

2. $6 \times 6 \times 6 \times 6$

3. $x \cdot x \cdot x \cdot x \cdot x \cdot x \cdot x$

4. $y \cdot y \cdot y \cdot y \cdot y \cdot y$

5. $3y \cdot 3y \cdot 3y \cdot 3y$

6. $5m \cdot 5m \cdot 5m \cdot 5m \cdot 5m$

Simplify.

7. 2^4

8. 5^3

9. $(-3)^2$

10. $(-7)^2$

11. 1^5

12. $(-1)^5$

13. 4^3 **14.** 9^1 **15.** $(-4)^3$

16. 5^4 **17.** 7^1 **18.** 1^7

19. $(4a)^2$ **20.** $(3x)^2$ **21.** $(-7x)^3$

22. $(-5x)^4$ **23.** $7 + 2 \times 6$

24. $11 + 4 \times 4$ **25.** $8 \times 7 + 6 \times 5$

26. $10 \times 5 + 1 \times 1$ **27.** $19 - 5 \times 3 + 3$

28. $14 - 2 \times 6 + 7$ **29.** $9 \div 3 + 16 \div 8$

30. $32 - 8 \div 4 - 2$ **31.** $7 + 10 - 10 \div 2$

32. $(2 - 5)^2$ **33.** $(3 - 5)^3$

34. $3 \cdot 2^3$ **35.** $8 - 2 \cdot 3 - 9$

36. $8 - (2 \cdot 3 - 9)$ **37.** $(8 - 2 \cdot 3) - 9$

38. $(8 - 2)(3 - 9)$

39. $(-24) \div (-3) \cdot \left(-\frac{1}{2}\right)$

40. $32 \div (-2) \cdot (-2)$

41. $16 \cdot (-24) + 50$

42. $10 \cdot 20 - 15 \cdot 24$

43. $2^4 + 2^3 - 10$

44. $40 - 3^2 - 2^3$

45. $5^3 + 26 \cdot 71 - (16 + 25 \cdot 3)$

46. $4^3 + 10 \cdot 20 + 8^2 - 23$

47. $[2 \cdot (5 - 3)]^2$

48. $5^3 - 7^2$ **49.** $\dfrac{7 + 2}{5^2 - 4^2}$

50. $\dfrac{5^2 - 3^2}{2 \cdot 6 - 4}$ **51.** $8(-7) + |6(-5)|$

52. $|10(-5)| + 1(-1)$ **53.** $19 - 5(-3) + 3$

54. $14 - 2(-6) + 7$ **55.** $9 \div (-3) \cdot 16 \div 8$

56. $-32 - 8 \div 4 \cdot (-2)$ **57.** $20 + 4^3 \div (-8) \cdot 2$

58. $2 \times 10^3 - 5000$

59. $8|(6 - 13) - 11|$

60. $6|9 - (3 - 4)|$

61. $256 \div (-32) \div (-4)$

62. $-1000 \div (-100) \div 10$

63. $\dfrac{5^2 - 4^3 - 3}{9^2 - 2^2 - 1^5}$

64. $\dfrac{3(6 - 7) - 5 \cdot 4}{6 \cdot 7 - 8(4 - 1)}$

65. $\dfrac{20(8 - 3) - 4(10 - 3)}{10(2 - 6) - 2(5 + 2)}$

66. $\dfrac{2^3 - 3^2 + 12 \cdot 5}{-32 \div (-16) \div (-4)}$

Evaluate.

67. $7 - 3x$, when $x = 5$

68. $9 - x^2$, when $x = 4$

69. $a \div 6 \cdot 2$, when $a = 12$

70. $10 \div a \cdot 5$, when $a = 2$

71. $-20 \div t^2 - 3(t - 1)$, when $t = -4$

72. $-30 \div t(t + 4)^2$, when $t = -6$

73. $-x^2 - 5x$, when $x = -3$

74. $(-x)^2 - 5x$, when $x = -3$

Rename the expression without using parentheses.

75. $-(2x + 7)$ **76.** $-(3x + 5)$

77. $-(5x - 8)$ **78.** $-(6x - 7)$

79. $-(4a - 3b + 7c)$ **80.** $-(5x - 2y - 3z)$

81. $-(3x^2 + 5x - 1)$ **82.** $-(8x^3 - 6x + 5)$

Remove parentheses and simplify.

83. $9x - (4x + 3)$ **84.** $7y - (2y + 9)$

85. $2a - (5a - 9)$ **86.** $11n - (3n - 7)$

87. $2x + 7x - (4x + 6)$

88. $3a + 2a - (4a + 7)$

89. $2x - 4y - 3(7x - 2y)$

90. $3a - 7b - 1(4a - 3b)$

91. $15x - y - 5(3x - 2y + 5z)$

92. $4a - b - 4(5a - 7b + 8c)$

93. $3x^2 + 7 - (2x^2 + 5)$

94. $7x^4 + 9x - (5x^4 + 3x)$

95. $9x^3 + x - 2(x^3 + 3x)$

96. $-7x^2 + 5x - 3(x^2 - 4x)$

97. $12a^2 - 3ab + 5b^2 - 5(-5a^2 + 4ab - 6b^2)$

98. $-8a^2 + 5ab - 12b^2 - 6(2a^2 - 4ab - 10b^2)$

99. $-7t^3 - t^2 - 3(5t^3 - 3t)$

100. $9t^4 + 7t - 5(9t^3 - 2t)$

101. $[10(x + 3) - 4] + [2(x - 1) + 6]$

102. $[9(x + 5) - 7] + [4(x - 12) + 9]$

103. $[7(x^2 + 5) - 19] - [4(x^2 - 6) + 10]$

104. $[6(x^3 + 4) - 12] - [5(x^3 - 8) + 11]$

105. $3\{[7(x - 2) + 4] - [2(2x - 5) + 6]\}$

106. $4\{[8(x - 3) + 9] - [4(3x - 7) + 2]\}$

107. $4\{[5(x^3 - 3) + 2] - 3[2(x^3 + 5) - 9]\}$

108. $3\{[6(x^2 - 4) + 5] - 2[5(x^2 + 8) - 10]\}$

Skill Maintenance

Translate to an algebraic expression.

109. Nine more than twice a number

110. Half of the sum of two numbers

Synthesis

111. ◈ Write the sentence $(-x)^2 \neq -x^2$ in words. Explain why $(-x)^2$ and $-x^2$ are not equivalent.

112. ◈ Write the sentence $-|x| \neq -x$ in words. Explain why $-|x|$ and $-x$ are not equivalent.

Simplify.

113. $z - \{2z - [3z - (4z - 5z) - 6z] - 7z\} - 8z$

114. $\{x - [f - (f - x)] + [x - f]\} - 3x$

115. $x - \{x - 1 - [x - 2 - (x - 3 - \{x - 4 - [x - 5 - (x - 6)]\})]\}$

116. ◈ Determine whether it is true that, for any real numbers a and b, $ab = (-a)(-b)$. Explain why or why not.

117. ◈ Determine whether it is true that, for any real numbers a and b, $-(ab) = (-a)b = a(-b)$. Explain why or why not.

If $n > 0$, $m > 0$, and $n \neq m$, determine whether each of the following is true.

118. $-n + m = n - m$

119. $-n + m = -(n + m)$

120. $-n - m = -(n + m)$

121. $-n - m = -(n - m)$

122. $n(-n - m) = -n^2 + nm$

123. $-m(n - m) = -(mn + m^2)$

124. $-m(-n + m) = m(n - m)$

125. $-n(-n - m) = n(n + m)$

SUMMARY AND REVIEW 1

KEY TERMS

Variable, p. 2
Constant, p. 2
Variable expression, p. 2
Evaluate an expression, p. 2
Algebraic expression, p. 3
Substitute, p. 3
Value of an expression, p. 3
Equation, p. 5
Equivalent expressions, p. 9
Term, p. 12
Factors, p. 12
Natural number, p. 14
Whole number, p. 14
Factorization, p. 15
Prime number, p. 15
Composite number, p. 15
Prime factorization, p. 15
Fractional notation, p. 15
Numerator, p. 15
Denominator, p. 15
Simplifying, p. 17
Reciprocal, p. 19
Set, p. 22
Subset, p. 22
Integer, p. 22
Negative integer, p. 23
Positive integer, p. 23
Rational number, p. 24
Terminating decimal, p. 25
Irrational number, p. 25
Real-number system, p. 26
Less than, p. 27
Greater than, p. 27
Inequality, p. 27
Absolute value, p. 28
Collect like terms, p. 33
Opposite, p. 36
Undefined, p. 46
Indeterminate, p. 46
Power, p. 49
Exponential notation, p. 49
Exponent, p. 49
Base, p. 49
Like terms, p. 52

IMPORTANT PROPERTIES AND FORMULAS

Area of a rectangle:	$A = lw$
Area of a triangle:	$A = \frac{1}{2}bh$
Area of a parallelogram:	$A = bh$
Commutative laws:	$a + b = b + a, \quad ab = ba$

Associative laws: $a + (b + c) = (a + b) + c$, $a(bc) = (ab)c$

Distributive law: $a(b + c) = ab + ac$

Identity property of 1: $1 \cdot a = a \cdot 1 = a$

Law of opposites: $a + (-a) = 0$

Multiplicative property of 0: $0 \cdot a = a \cdot 0 = 0$

$$\frac{-a}{b} = \frac{a}{-b} = -\frac{a}{b}, \quad \frac{-a}{-b} = \frac{a}{b}$$

Property of -1: $-1 \cdot a = -a$

Opposite of a sum: $-(a + b) = -a + (-b)$

Rules for Order of Operations

1. Perform all calculations within grouping symbols.
2. Evaluate all exponential expressions.
3. Perform all multiplications and divisions in order from left to right.
4. Perform all additions and subtractions in order from left to right.

REVIEW EXERCISES

Evaluate.

1. $3a$, when $a = 5$
2. $\frac{x}{y}$, when $x = 12$ and $y = 2$
3. $\frac{2 \cdot p}{q}$, when $p = 20$ and $q = 8$
4. $\frac{x - y}{3}$, when $x = 17$ and $y = 5$
5. $10 - y^2$, when $y = 5$
6. $-10 + a^2 \div (b + 1)$, when $a = 5$ and $b = 4$

Translate to an algebraic expression.

7. 8 less than z
8. Three times x
9. One-third of y

10. Determine whether 35 is a solution of $x/5 = 8$.

11. Translate to an equation: Six times what number is 6768?

12. Use the commutative law of addition to write an expression equivalent to $2x + y$.

13. Use the commutative law of multiplication to write an expression equivalent to $2x + y$.

14. Use the associative law of addition to write an expression equivalent to $(2x + y) + z$.

15. Use the commutative and associative laws to write three expressions equivalent to $4(xy)$.

Multiply.

16. $6(3x + 5y)$
17. $8(5x + 3y + 2)$

Factor.

18. $21x + 7y$
19. $35x + 14 + 7y$

Simplify.

20. $\frac{20}{48}$
21. $\frac{180}{18}$

Perform the indicated operation and simplify.

22. $\frac{4}{9} + \frac{5}{12}$
23. $\frac{3}{4} \div 3$
24. $\frac{2}{3} - \frac{1}{15}$
25. $\frac{9}{10} \cdot \frac{16}{5}$

26. Tell which integers correspond to this situation: Renir has a debt of \$45 and Raoul has \$72 in his savings account.

27. Translate to mathematical language: A bowling score is at most 300.

28. Graph on a number line: $\frac{-1}{3}$.

29. Write an inequality with the same meaning as $-3 < x$.

30. Write true or false: $0 \leq -1$.

31. Find decimal notation: $-\frac{7}{8}$.

32. Find the absolute value: $|-1|$.

33. Find $-(-x)$ when x is -5.

Simplify.

34. $4 + (-7)$

35. $-\frac{2}{3} + \frac{1}{12}$

36. $6 + (-9) + (-8) + 7$

37. $-3.8 + 5.1 + (-12) + (-4.3) + 10$

38. $-3 - (-7)$

39. $-\frac{9}{10} - \frac{1}{2}$

40. $-3.8 - 4.1$

41. $-9 \cdot (-6)$

42. $-2.7(3.4)$

43. $\frac{2}{3} \cdot \left(-\frac{3}{7}\right)$

44. $3 \cdot (-7) \cdot (-2) \cdot (-5)$

45. $35 \div (-5)$

46. $-5.1 \div 1.7$

47. $-\frac{3}{5} \div \left(-\frac{4}{5}\right)$

48. $|-3 \cdot 4 - 12 \cdot 2| - 8(-7)$

49. $|-12(-3) - 2^3 - (-9)(-10)|$

50. $120 - 6^2 \div 4 \cdot 8$

51. $(120 - 6^2) \div 4 \cdot 8$

52. $(120 - 6^2) \div (4 \cdot 8)$

53. $\dfrac{4(18 - 8) + 7 \cdot 9}{9^2 - 8^2}$

Collect like terms.

54. $11a + 2b + (-4a) + (-5b)$

55. $7x - 3y - 9x + 8y$

56. Find the opposite of -7.

57. Find the reciprocal of -7.

58. Write exponential notation for $2x \cdot 2x \cdot 2x \cdot 2x$.

59. Simplify: $(-3y)^3$.

Remove parentheses and simplify.

60. $2a - (5a - 9)$

61. $3(b + 7) - 5b$

62. $3[11x - 3(4x - 1)]$

63. $2[6(y - 4) + 7]$

64. $[8(x + 4) - 10] - [3(x - 2) + 4]$

65. $5\{[6(x - 1) + 7] - [3(3x - 4) + 8]\}$

Synthesis

66. ◈ Explain at least three uses of the distributive law considered in this chapter.

67. ◈ Devise a rule for determining the sign of a negative quantity raised to a power.

68. Evaluate $a^{50} - 20a^{25}b^4 + 100b^8$ when $a = 1$ and $b = 2$.

69. If $0.090909\ldots = \frac{1}{11}$ and $0.181818\ldots = \frac{2}{11}$, what rational number is named by each of the following?

a) $0.272727\ldots$ **b)** $0.909090\ldots$

Simplify.

70. $-\left|\frac{7}{8} - \left(-\frac{1}{2}\right) - \frac{3}{4}\right|$

71. $(|2.7 - 3| + 3^2 - |-3|) \div (-3)$

CHAPTER TEST 1

1. Evaluate $\dfrac{3x}{y}$ when $x = 10$ and $y = 5$.

2. Write an algebraic expression: Nine less than some number.

3. Find the area of a triangle when the height h is 30 ft and the base b is 16 ft.

4. Use the commutative law of addition to write an expression equivalent to $3p + q$.

5. Use the associative law of multiplication to write an expression equivalent to $x \cdot (4 \cdot y)$.

Multiply.

6. $3(6 - x)$

7. $-5(y - 1)$

Factor.

8. $11 - 44x$

9. $7x + 21 + 14y$

10. Find the prime factorization of 300.

Write a true sentence using either $<$ or $>$.

11. $-4 \;\blacksquare\; 0$

12. $-3 \;\blacksquare\; -8$

13. $-0.78 \;\blacksquare\; -0.87$

14. $-\frac{1}{8} \;\blacksquare\; \frac{1}{2}$

Find the absolute value.

15. $|-7|$

16. $\left|\frac{9}{4}\right|$

17. $|-2.7|$

18. Find the opposite of $\frac{2}{3}$.

19. Find the reciprocal of $-\frac{4}{7}$.

20. Find $-x$ when x is -8.

21. Write an inequality with the same meaning as $x \leqslant -2$.

Compute and simplify.

22. $3.1 - (-4.7)$

23. $-8 + 4 + (-7) + 3$

24. $-\frac{1}{5} + \frac{3}{8}$

25. $2 - (-8)$

26. $3.2 - 5.7$

27. $\frac{1}{8} - \left(-\frac{3}{4}\right)$

28. $4 \cdot (-12)$

29. $-\frac{1}{2} \cdot \left(-\frac{3}{8}\right)$

30. $-45 \div 5$

31. $-\frac{3}{5} \div \left(-\frac{4}{5}\right)$

32. $4.864 \div (-0.5)$

33. $-2(16) - |2(-8) - 5^3|$

34. $6 + 7 - 4 - (-3)$

35. $256 \div (-16) \div 4$

36. $2^3 - 10[4 - (-2 + 18)3]$

37. Collect like terms: $18y + 30a - 9a + 4y$.

38. Simplify: $(-2x)^4$.

Remove parentheses and simplify.

39. $5x - (3x - 7)$

40. $4(2a - 3b) + a - 7$

41. $4\{3[5(y - 3) + 9] + 2(y + 8)\}$

Synthesis

42. Evaluate $\dfrac{5y - x}{4}$ when $x = 20$ and y is 4 less than x.

Simplify.

43. $\dfrac{13{,}800}{42{,}000}$

44. $|-27 - 3(4)| - |-36| + |-12|$

45. $a - \{3a - [4a - (2a - 4a)]\}$

CHAPTER 2

Equations, Inequalities, and Problem Solving

AN APPLICATION

A rectangular community vegetable garden is to be enclosed with 92 m of wooden fencing. In order to allow for compost storage, the garden must be 4 m longer than it is wide. Determine the dimensions of the garden.

This problem appears as Example 5 in Section 2.5.

Robert E. Romero
ZONING ENFORCEMENT MANAGER

"My primary responsibility is to enforce city ordinances relating to the development and use of private property. I use mathematics every day—for example, when determining lot size for irregularly shaped lots, solar access angles for tall buildings, parking layouts, and area available for signage."

Solving equations and inequalities is a recurring theme in much of mathematics. In this chapter, we will study some of the principles used to solve equations and inequalities. Then we will use equations and inequalities to solve applied problems.

In addition to material from this chapter, the review and test for Chapter 2 include material from Sections 1.1, 1.2, 1.7, and 1.8.

2.1 Solving Equations

Equations and Solutions • The Addition Principle • The Multiplication Principle

Solving equations is essential for problem solving in algebra. In this section, we study two of the most important principles used to solve equations.

Equations and Solutions

We have already seen that an equation is a number sentence stating that the expressions on either side of the equals sign represent the same number. Some equations, like $3 + 2 = 5$ or $2x + 6 = 2(x + 3)$, are *always* true and some, like $3 + 2 = 6$ or $x + 2 = x + 3$, are *never* true. In this text, we will concentrate on equations like $x + 6 = 13$ or $7x = 141$ that are either true or false, depending on the replacement value.

Solution of an Equation

Any replacement for the variable that makes an equation true is called a *solution* of the equation. To solve an equation means to find *all* of its solutions.

One way to determine whether a number is a solution of an equation is to evaluate the algebraic expression on each side of the equation by substitution. If the values are the same, then the number is a solution.

EXAMPLE 1

Determine whether 7 is a solution of $x + 6 = 13$.

Solution We have

$x + 6 = 13$ Writing the equation

$7 + 6 \;?\; 13$ Substituting 7 for x

$13 \mid 13$ TRUE

Since the left-hand and the right-hand sides are the same, 7 is a solution. ❑

EXAMPLE 2

Determine whether 19 is a solution of $7x = 141$.

Solution We have

$7x = 141$	Writing the equation
$7(19) \; ? \; 141$	Substituting 19 for x
$133 \mid 141$ FALSE	

Since the left-hand and the right-hand sides are not the same, 19 is not a solution. ❑

The Addition Principle

Consider the equation

$x = 7.$

We can easily see that the solution of this equation is 7. If we replace x by 7, we get

$7 = 7,$ which is true.

Now consider the equation of Example 1:

$x + 6 = 13.$

There we discovered that the solution of $x + 6 = 13$ is also 7, but the solution of $x = 7$ seems more obvious. We now begin to consider principles that allow us to start with one equation and end up with an equation like $x = 7$, in which the variable is alone on one side and for which the solution is easy to see. The equations $x + 6 = 13$ and $x = 7$ are **equivalent.**

Equivalent Equations

Equations with the same solutions are called *equivalent equations*.

One of the principles that we use in solving equations concerns adding. An equation $a = b$ says that a and b stand for the same number. Suppose this is true, and we add a number c to the number a. We get the same answer if we add c to b, because a and b are the same number.

The Addition Principle

For any real numbers a, b, and c,

if $a = b$, then $a + c = b + c$.

To visualize the addition principle, consider a balance similar to one a jeweler might use. When the two sides of the balance hold

quantities of equal weight, the balance is level. If weight is added or removed, equally, on both sides, the balance will remain level.

When we use the addition principle, we sometimes say that we "add the same number on both sides of an equation." This is also true for subtraction, since every subtraction can be regarded as the addition of an opposite.

EXAMPLE 3

Solve: $x + 5 = -7$.

Solution

$$x + 5 = -7$$

$$x + 5 - 5 = -7 - 5 \quad \text{Using the addition principle: adding } -5 \text{ on both sides or subtracting 5 on both sides}$$

$$x + 0 = -12 \quad \text{Simplifying}$$

$$x = -12 \quad \text{Identity property of 0}$$

It is obvious that the solution of $x = -12$ is the number -12. To check the answer in the original equation, we substitute.

Check:

$$\begin{array}{r|l} x + 5 & -7 \\ \hline -12 + 5 \ ? & -7 \\ -7 & -7 \quad \text{TRUE} \end{array}$$

The solution of the original equation is -12. ❑

In Example 3, to get x alone, we used the addition principle and subtracted 5 on both sides. This eliminated the 5 on the left and produced a simpler, but equivalent, equation, $x = -12$.

Next we use the addition principle to solve a subtraction problem.

EXAMPLE 4

Solve: $-6.5 = y - 8.4$.

Solution

$$-6.5 = y - 8.4$$

$$-6.5 + 8.4 = y - 8.4 + 8.4 \quad \text{Using the addition principle: adding 8.4 on both sides eliminates } -8.4 \text{ on the right}$$

$$1.9 = y$$

Check:

$$\begin{array}{r|l} -6.5 & y - 8.4 \\ \hline -6.5 \ ? & 1.9 - 8.4 \\ -6.5 & -6.5 \quad \text{TRUE} \end{array}$$

The solution is 1.9. ❑

Note that the equations $a = b$ and $b = a$ have the same meaning. Thus to solve $-6.5 = y - 8.4$, we can reverse it and solve $y - 8.4 = -6.5$ if we wish.

EXAMPLE 5

Solve: $-\frac{2}{3} + x = \frac{5}{2}$.

Solution

$$-\tfrac{2}{3} + x = \tfrac{5}{2}$$

$$-\tfrac{2}{3} + x + \tfrac{2}{3} = \tfrac{5}{2} + \tfrac{2}{3} \qquad \text{Adding } \tfrac{2}{3}$$

$$x = \tfrac{5}{2} + \tfrac{2}{3}$$

$$x = \tfrac{5}{2}\cdot\tfrac{3}{3} + \tfrac{2}{3}\cdot\tfrac{2}{2} \qquad \text{Multiplying by 1 to obtain a common denominator}$$

$$x = \tfrac{15}{6} + \tfrac{4}{6}$$

$$x = \tfrac{19}{6}$$

The check is left to the student. The solution is $\frac{19}{6}$. ❑

The Multiplication Principle

An equation like $\frac{5}{4}x = \frac{3}{8}$ says that $\frac{5}{4}x$ and $\frac{3}{8}$ represent the same number. Because of this, if $\frac{5}{4}x$ and $\frac{3}{8}$ are both multiplied by some number c, the products $c \cdot \frac{5}{4}x$ and $c \cdot \frac{3}{8}$ will represent the same number.

The Multiplication Principle

For any real numbers a, b, and c,

if $a = b$, then $c \cdot a = c \cdot b$.

EXAMPLE 6

Solve: $\frac{5}{4}x = \frac{3}{8}$.

Solution

$$\tfrac{5}{4}x = \tfrac{3}{8} \qquad \text{Note that the reciprocal of } \tfrac{5}{4} \text{ is } \tfrac{4}{5} \text{ and that } \tfrac{4}{5}\cdot\tfrac{5}{4} = 1.$$

$$\tfrac{4}{5}\cdot\tfrac{5}{4}x = \tfrac{4}{5}\cdot\tfrac{3}{8} \qquad \text{Using the multiplication principle: multiplying on both sides by } \tfrac{4}{5} \text{ eliminates } \tfrac{5}{4} \text{ on the left}$$

$$1 \cdot x = \tfrac{3}{10} \qquad \text{Simplifying}$$

$$x = \tfrac{3}{10} \qquad \text{Using the identity property of 1: } 1 \cdot x = x$$

Check:

$$\tfrac{5}{4}x = \tfrac{3}{8}$$

$$\tfrac{5}{4}\left(\tfrac{3}{10}\right) \;?\; \tfrac{3}{8}$$

$$\tfrac{3}{8} \;\big|\; \tfrac{3}{8} \quad \text{TRUE}$$

The solution is $\frac{3}{10}$. ❑

In Example 6, to get x alone, we multiplied by the *multiplicative inverse,* or *reciprocal,* of $\frac{5}{4}$. When we multiplied, we got the *multiplicative identity,* 1, times x, which simplified to x. This enabled us to eliminate the $\frac{5}{4}$ on the left.

Because division is the same as multiplying by a reciprocal, the multiplication principle also tells us that we can "divide on both sides by the same nonzero number." That is,

$$\text{if } a = b, \text{ then } \frac{1}{c}\cdot a = \frac{1}{c}\cdot b \quad \text{and} \quad \frac{a}{c} = \frac{b}{c} \quad (\text{provided } c \neq 0).$$

In an expression like $3x$, the number 3 is called the **coefficient.** In practice, it is usually more convenient to divide on both sides of an equation if the coefficient of the variable is in decimal notation or is an integer. When the coefficient is in fractional notation, it is more convenient to multiply by a reciprocal.

EXAMPLE 7

Solve: **(a)** $-4x = 92$; **(b)** $12.6 = 3x$; **(c)** $-x = 9$.

Solution

a) $-4x = 92$

$\frac{-4x}{-4} = \frac{92}{-4}$ Using the multiplication principle. Dividing on both sides by -4 is the same as multiplying by $-\frac{1}{4}$.

$1 \cdot x = -23$ Simplifying

$x = -23$ Using the identity property of 1

Check:

$-4x = 92$	
$-4(-23)$	? 92
92	92 TRUE

The solution is -23.

b) $12.6 = 3x$

$\frac{12.6}{3} = \frac{3x}{3}$ Dividing on both sides by 3 or multiplying by $\frac{1}{3}$

$4.2 = 1x$

$4.2 = x$ Simplifying

Check:

$12.6 = 3x$	
12.6	? $3 \cdot 4.2$
12.6	12.6 TRUE

The solution is 4.2.

c) To solve an equation like $-x = 9$, remember that when an expression is multiplied or divided by -1, its sign is changed. Here we multiply on both sides by -1:

$-x = 9$

$(-1)(-x) = (-1) \cdot 9$ Multiplying by -1 on both sides

$x = -9.$ Note that $(-1)(-x)$ is the same as $(-1)(-1)x$.

Check:

$-x = 9$	
$-(-9)$	? 9
9	9 TRUE

The solution is -9. □

Consider an equation like $y/-9 = 14$. The left side can be rewritten as $(1/-9) \cdot y$. Using the multiplication principle, we can multiply on both sides by -9 to solve, as in the following example.

EXAMPLE 8 Solve: $\frac{y}{-9} = 14$.

Solution

$$\frac{y}{-9} = 14$$

$$\frac{1}{-9} \cdot y = 14 \quad \text{Rewriting division as multiplication}$$

$$-9\left(\frac{1}{-9}\right)y = -9 \cdot 14 \quad \text{Multiplying by } -9 \text{ on both sides}$$

$$y = -126 \quad \text{Simplifying}$$

Check:

$\frac{y}{-9} = 14$	
$\frac{-126}{-9}$	? 14
14	14 TRUE

The solution is -126. ❑

EXERCISE SET 2.1

Solve using the addition principle. Don't forget to check!

1. $x + 2 = 6$
2. $x + 5 = 8$
3. $x + 15 = -5$
4. $y + 9 = 43$
5. $x + 6 = -8$
6. $t + 9 = -12$
7. $-2 = x + 16$
8. $-6 = y + 25$
9. $x - 9 = 6$
10. $x - 8 = 5$
11. $x - 7 = -21$
12. $x - 3 = -14$
13. $5 + t = 7$
14. $8 + y = 12$
15. $13 = -7 + y$
16. $15 = -9 + z$
17. $-3 + t = -9$
18. $-6 + y = -21$
19. $r + \frac{1}{3} = \frac{8}{3}$
20. $t + \frac{3}{8} = \frac{5}{8}$
21. $m + \frac{5}{6} = -\frac{11}{12}$
22. $x + \frac{2}{3} = -\frac{5}{6}$
23. $x - \frac{5}{6} = \frac{7}{8}$
24. $y - \frac{3}{4} = \frac{5}{6}$
25. $-\frac{1}{5} + z = -\frac{1}{4}$
26. $-\frac{1}{8} + y = -\frac{3}{4}$
27. $x + 2.3 = 7.4$
28. $y + 4.6 = 9.3$
29. $-9.7 = -4.7 + y$
30. $-7.8 = 2.8 + x$

Solve using the multiplication principle. Don't forget to check!

31. $6x = 36$
32. $3x = 39$
33. $5x = 45$
34. $9x = 72$
35. $84 = 7x$
36. $56 = 8x$
37. $-x = 40$
38. $100 = -x$
39. $-x = -1$
40. $-68 = -r$
41. $7x = -49$
42. $9x = -36$
43. $-12x = 72$
44. $-15x = 105$
45. $-21x = -126$
46. $-13x = -104$
47. $\frac{t}{7} = -9$
48. $\frac{y}{-8} = 11$
49. $\frac{3}{4}x = 27$
50. $\frac{4}{5}x = 16$
51. $\frac{-t}{3} = 7$
52. $\frac{-x}{6} = 9$
53. $\frac{1}{5} = -\frac{m}{3}$
54. $\frac{1}{9} = -\frac{z}{7}$

55. $-\frac{3}{5}r = -\frac{9}{10}$

56. $-\frac{2}{5}y = -\frac{4}{15}$

57. $\frac{-3r}{2} = -\frac{27}{4}$

58. $\frac{5x}{7} = -\frac{10}{14}$

59. $6.3x = 44.1$

60. $2.7y = 54$

Solve.

61. $3.7 + t = 8.2$

62. $\frac{3}{4}x = 18$

63. $18 = -\frac{2}{3}x$

64. $t - 7.4 = -12.9$

65. $17 = y + 29$

66. $96 = -\frac{3}{4}t$

67. $y - \frac{2}{3} = -\frac{1}{6}$

68. $-\frac{x}{7} = \frac{2}{9}$

69. $-24 = \frac{8x}{5}$

70. $\frac{1}{5} + y = -\frac{3}{10}$

71. $-4.1t = 10.25$

72. $\frac{19}{23} = -x$

Skill Maintenance

Collect like terms.

73. $3x + 4x$

74. $6x + 5 - 7x$

Remove parentheses and simplify.

75. $3x - (4 + 2x)$

76. $2 - 5(x + 5)$

Synthesis

77. ◈ Why are equivalent equations useful when solving equations?

78. ◈ Explain why it is not necessary to prove a subtraction principle: If $a = b$, then $a - c = b - c$.

Solve. The icon ▦ indicates an exercise designed to give practice in the use of a calculator.

79. ▦ $-356.788 = -699.034 + t$

80. ▦ $-0.2344m = 2028.732$

81. $0 \cdot x = 9$

82. $x + 3 = 3 + x$

83. $4|x| = 48$

84. $2|x| = -12$

85. $0 \cdot x = 0$

86. $x + x = x$

87. $x + 4 = 5 + x$

88. $|3x| = 6$

Solve for x.

89. $ax = 5a$

90. $x - 4 = a$

91. $3x = \frac{b}{a}$

92. $cx = a^2 + 1$

93. $1 - c = a + x$

94. $|x| + 6 = 19$

95. If $x - 4720 = 1634$, find $x + 4720$.

96. ▦ A student makes a calculation and gets an answer of 22.5. On the last step, the student multiplies by 0.3 when a division by 0.3 should have been done. What should the correct answer be?

97. ◈ Are the equations $x = 5$ and $x^2 = 25$ equivalent? Why or why not?

2.2 Using the Principles Together

Applying Both Principles • Collecting Like Terms • Clearing Fractions and Decimals • Equations Containing Parentheses

The equations in Section 2.1 required use of *either* the addition principle *or* the multiplication principle. Now we will consider equations in which *both* principles are used.

Applying Both Principles

Consider the equation $3x + 5 = 17$. To solve such an equation, we first isolate the variable term, $3x$, using the addition principle. We then use the multiplication principle to get the variable by itself.

EXAMPLE 1 Solve: $3x + 5 = 17$.

Solution

First isolate the x-term.

$$3x + 5 = 17$$

$$3x + 5 - 5 = 17 - 5$$ Using the addition principle: subtracting 5 on both sides (adding -5)

$$3x = 12$$ Simplifying

$$\frac{3x}{3} = \frac{12}{3}$$ Using the multiplication principle: dividing by 3 on both sides (multiplying by $\frac{1}{3}$)

Then isolate x.

$$x = 4$$ Simplifying

Check:

$3x + 5$	$= 17$
$3 \cdot 4 + 5$	? 17
$12 + 5$	
17	17 TRUE

We use the rules for order of operations: Find the product, $3 \cdot 4$, and then add.

The solution is 4. ❑

EXAMPLE 2 Solve: $-5x - 6 = 16$.

Solution

$$-5x - 6 = 16$$

$$-5x - 6 + 6 = 16 + 6$$ Adding 6 on both sides

$$-5x = 22$$

$$\frac{-5x}{-5} = \frac{22}{-5}$$ Dividing on both sides by -5

$$x = -\frac{22}{5}, \quad \text{or } -4\frac{2}{5}$$ Simplifying

Check:

$-5x - 6$	$= 16$
$-5(-\frac{22}{5}) - 6$	? 16
$22 - 6$	
16	16 TRUE

The solution is $-\frac{22}{5}$. ❑

EXAMPLE 3 Solve: $45 - t = 13$.

Solution

$$45 - t = 13$$

$$45 - t - 45 = 13 - 45$$ Subtracting 45 on both sides

$$-t = -32$$

$$(-1)(-t) = (-1)(-32)$$ Multiplying on both sides by -1 (Dividing on both sides by -1 would also change the sign on both sides.)

$$t = 32$$

We leave the check to the student. The solution is 32. ❑

As we improve our equation-solving skills, we begin to shorten some of our writing. Thus we may not always write a number being added, subtracted, multiplied, or divided on both sides. We simply write it on the opposite side, as in the following example.

EXAMPLE 4

Solve: $16.3 - 7.2y = -8.18$.

Solution

$$16.3 - 7.2y = -8.18$$

$$-7.2y = -8.18 - 16.3$$ Subtracting 16.3 on both sides. We write the subtraction of 16.3 on the right side and remove 16.3 on the left side.

$$-7.2y = -24.48$$

$$y = \frac{-24.48}{-7.2}$$ Dividing by -7.2 on both sides. We write the division by -7.2 on the right side and remove the -7.2 on the left side.

$$y = 3.4$$

Check:

$16.3 - 7.2y = -8.18$	
$16.3 - 7.2(3.4)$?	-8.18
$16.3 - 24.48$	
-8.18	-8.18 TRUE

The solution is 3.4. ❑

Collecting Like Terms

If like terms appear in an equation, we first collect them and then solve. When the like terms are not on the same side of an equation, we can use the addition principle to write all variable terms on one side.

EXAMPLE 5

Solve.

a) $3x + 4x = -14$
b) $2x - 4 = -3x + 1$
c) $6x + 5 - 7x = 10 - 4x + 3$

Solution

a) $3x + 4x = -14$

$$7x = -14$$ Collecting like terms

$$x = \frac{-14}{7}$$ Dividing by 7 on both sides

$$x = -2$$

The check is left to the student. The solution is -2.

b)

Isolate variable terms on one side and constant terms on the other side.

$$2x - 4 = -3x + 1$$

$$2x - 4 + 4 = -3x + 1 + 4$$ Adding 4

$$2x = -3x + 5$$ Simplifying

$$2x + 3x = -3x + 3x + 5$$ Adding $3x$

$$5x = 5$$ Collecting like terms and simplifying

and

$$\frac{5x}{5} = \frac{5}{5} \quad \text{Dividing by 5}$$

$$x = 1 \quad \text{Simplifying}$$

Check:

$$\begin{array}{r|l} 2x - 4 & -3x + 1 \\ \hline 2 \cdot 1 - 4 \ ? & -3 \cdot 1 + 1 \\ 2 - 4 & -3 + 1 \\ -2 & -2 \end{array} \quad \text{TRUE}$$

The solution is 1.

c)

$$6x + 5 - 7x = 10 - 4x + 3$$

$$-x + 5 = 13 - 4x \quad \text{Collecting like terms}$$

$$-x + 5 + 4x = 13 - 4x + 4x \quad \text{Adding } 4x$$

$$3x + 5 = 13 \quad \text{Simplifying}$$

$$3x + 5 - 5 = 13 - 5 \quad \text{Subtracting 5}$$

$$3x = 8 \quad \text{Simplifying}$$

$$\frac{3x}{3} = \frac{8}{3} \quad \text{Dividing by 3}$$

$$x = \frac{8}{3} \quad \text{Simplifying}$$

The student can confirm that $\frac{8}{3}$ checks and is the solution. ❑

Clearing Fractions and Decimals

Equations are usually easier to solve when they do not contain fractions or decimals. Consider, for example,

$$\tfrac{1}{2}x + 5 = \tfrac{3}{4} \quad \text{and} \quad 2.3x + 7 = 5.4.$$

If we multiply on both sides of the first equation by 4 and on both sides of the second equation by 10, we have

$$4\left(\tfrac{1}{2}x + 5\right) = 4 \cdot \tfrac{3}{4} \quad \text{and} \quad 10(2.3x + 7) = 10 \cdot 5.4,$$

or

$$2x + 20 = 3 \qquad \text{and} \qquad 23x + 70 = 54.$$

The first equation has been "cleared of fractions" and the second equation has been "cleared of decimals."

The easiest way to clear an equation of fractions is to multiply *every term on both sides* of the equation by the smallest, or *least,* common denominator.

EXAMPLE 6

Solve: $\frac{2}{3}x - \frac{1}{6} = 2x$.

Solution The number 6 is the least common denominator, so we multiply by 6 on both sides.

$$6\left(\frac{2}{3}x - \frac{1}{6}\right) = 6 \cdot 2x \quad \text{Multiplying by 6 on both sides}$$

and

$$6 \cdot \frac{2}{3}x - 6 \cdot \frac{1}{6} = 6 \cdot 2x$$ Using the distributive law. (*Caution*! Be sure to multiply *all* the terms by 6.)

$$4x - 1 = 12x$$ Simplifying. Note that the fractions are cleared.

$$-1 = 8x$$ Subtracting $4x$ on both sides

$$-\frac{1}{8} = x$$ Dividing by 8

The number $-\frac{1}{8}$ checks and is the solution. ❑

To clear an equation of decimals, we count the greatest number of decimal places in any one number. If the greatest number of decimal places is 1, we multiply both sides by 10^1, or 10; if it is 2, we multiply by 10^2, or 100; and so on.

EXAMPLE 7

Solve: $16.3 - 7.2y = -8.18$.

Solution The greatest number of decimal places in any one number is *two*. Multiplying by 100, which has *two* 0's, will clear *all* decimals.

$$100(16.3 - 7.2y) = 100(-8.18)$$ Multiplying by 100 on both sides

$$100(16.3) - 100(7.2y) = 100(-8.18)$$ Using the distributive law

$$1630 - 720y = -818$$ Simplifying. Note that the decimals are cleared.

$$-720y = -818 - 1630$$ Subtracting 1630 on both sides

$$-720y = -2448$$ Collecting like terms

$$y = \frac{-2448}{-720}$$ Dividing by -720 on both sides

$$y = 3.4$$

In Example 4, the same solution was found without clearing decimals. Finding the same answer in two different ways is a good check. The solution is 3.4. ❑

Equations Containing Parentheses

To solve equations that contain parentheses, we can use the distributive law to first remove the parentheses. Then we proceed as before.

EXAMPLE 8

Solve: **(a)** $3x = 2(5x - 7)$; **(b)** $2 - 5(x + 5) = 3(x - 2) - 1$.

Solution

a)

$$3x = 2(5x - 7)$$

$$3x = 10x - 14$$ Using the distributive law

$$3x - 10x = -14$$ Subtracting $10x$ to get all x-terms on one side

$$-7x = -14$$ Collecting like terms

$$\frac{-7x}{-7} = \frac{-14}{-7}$$ Dividing by -7

$$x = 2$$

Check:

$3x = 2(5x - 7)$	
$3 \cdot 2$?	$2(5 \cdot 2 - 7)$
6	$2(10 - 7)$
	$2 \cdot 3$
6	6 TRUE

The solution is 2.

b) $2 - 5(x + 5) = 3(x - 2) - 1$

$2 - 5x - 25 = 3x - 6 - 1$ Using the distributive law to multiply and remove parentheses

$-5x - 23 = 3x - 7$ Simplifying

$-23 + 7 = 3x + 5x$ Adding $5x$ and 7 to get all x-terms on one side and all constant terms on the other side

$-16 = 8x$ Simplifying

$-2 = x$ Dividing by 8

Check:

$2 - 5(x + 5) = 3(x - 2) - 1$	
$2 - 5(-2 + 5)$?	$3(-2 - 2) - 1$
$2 - 5(3)$	$3(-4) - 1$
$2 - 15$	$-12 - 1$
-13	-13 TRUE

The solution is -2. ❑

Here is a procedure for solving the types of equations discussed in this section.

An Equation-Solving Procedure

1. If necessary, use the distributive law to remove parentheses. Then collect like terms on each side.
2. Multiply on both sides to clear fractions or decimals. (This is optional, but it can ease computations.)
3. Get all terms with variables on one side and all constant terms on the other side, using the addition principle.
4. Collect like terms again, if necessary.
5. Multiply or divide to solve for the variable, using the multiplication principle.
6. Check all possible solutions in the original equation.

EXERCISE SET 2.2

Solve and check.

1. $5x + 6 = 31$
2. $3x + 6 = 30$
3. $8x + 4 = 68$
4. $7z + 9 = 72$
5. $4x - 6 = 34$
6. $6x - 3 = 15$
7. $3x - 9 = 33$
8. $5x - 7 = 48$
9. $7x + 2 = -54$
10. $5x + 4 = -41$

11. $-45 = 3 + 6y$

12. $-91 = 9t + 8$

13. $-4x + 7 = 35$

14. $-5x - 7 = 108$

15. $-7x - 24 = -129$

16. $-6z - 18 = -132$

17. $5x + 7x = 72$

18. $4x + 5x = 45$

19. $8x + 7x = 60$

20. $3x + 9x = 96$

21. $4x + 3x = 42$

22. $6x + 19x = 100$

23. $-6y - 3y = 27$

24. $-4y - 8y = 48$

25. $-7y - 8y = -15$

26. $-10y - 3y = -39$

27. $10.2y - 7.3y = -58$

28. $6.8y - 2.4y = -88$

29. $x + \frac{1}{3}x = 8$

30. $x + \frac{1}{4}x = 10$

31. $8y - 35 = 3y$

32. $4x - 6 = 6x$

33. $8x - 1 = 23 - 4x$

34. $5y - 2 = 28 - y$

35. $2x - 1 = 4 + x$

36. $5x - 2 = 6 + x$

37. $6x + 3 = 2x + 11$

38. $5y + 3 = 2y + 15$

39. $5 - 2x = 3x - 7x + 25$

40. $10 - 3x = 2x - 8x + 40$

41. $4 + 3x - 6 = 3x + 2 - x$

42. $5 + 4x - 7 = 4x - 2 - x$

43. $4y - 4 + y + 24 = 6y + 20 - 4y$

44. $5y - 7 + y = 7y + 21 - 5y$

Solve and check. Clear fractions or decimals first.

45. $\frac{7}{2}x + \frac{1}{2}x = 3x + \frac{3}{2} + \frac{5}{2}x$

46. $\frac{7}{8}x - \frac{1}{4} + \frac{3}{4}x = \frac{1}{16} + x$

47. $\frac{2}{3} + \frac{1}{4}t = 6$

48. $-\frac{3}{2} + x = -\frac{5}{6} - \frac{4}{3}$

49. $\frac{2}{3} + 3y = 5y - \frac{2}{15}$

50. $\frac{1}{2} + 4m = 3m - \frac{5}{2}$

51. $\frac{5}{3} + \frac{2}{3}x = \frac{25}{12} + \frac{5}{4}x + \frac{3}{4}$

52. $1 - \frac{2}{3}y = \frac{9}{5} - \frac{1}{5}y + \frac{3}{5}$

53. $2.1x + 45.2 = 3.2 - 8.4x$

54. $0.96y - 0.79 = 0.21y + 0.46$

55. $1.03 - 0.6x = 0.71 - 0.2x$

56. $1.7t + 8 - 1.62t = 0.4t - 0.32 + 8$

57. $\frac{2}{7}x - \frac{1}{2}x = \frac{3}{4}x + 1$

58. $\frac{5}{16}y + \frac{3}{8}y = 2 + \frac{1}{4}y$

Solve and check.

59. $3(2y - 3) = 27$

60. $4(2y - 3) = 28$

61. $40 = 5(3x + 2)$

62. $9 = 3(5x - 2)$

63. $2(3 + 4m) - 9 = 45$

64. $3(5 + 3m) - 8 = 88$

65. $5r - (2r + 8) = 16$

66. $6b - (3b + 8) = 16$

67. $6 - 2(3x - 1) = 2$

68. $10 - 3(2x - 1) = 1$

69. $5(d + 4) = 7(d - 2)$

70. $3(t - 2) = 9(t + 2)$

71. $8(2t + 1) = 4(7t + 7)$

72. $7(5x - 2) = 6(6x - 1)$

73. $3(r - 6) + 2 = 4(r + 2) - 21$

74. $5(t + 3) + 9 = 3(t - 2) + 6$

75. $19 - (2x + 3) = 2(x + 3) + x$

76. $13 - (2c + 2) = 2(c + 2) + 3c$

77. $\frac{1}{3}(6x + 24) - 20 = -\frac{1}{4}(12x - 72)$

78. $\frac{1}{4}(8y + 4) - 17 = -\frac{1}{2}(4y - 8)$

79. $2[4 - 2(3 - x)] - 1 = 4[2(4x - 3) + 7] - 25$

80. $5[3(7 - t) - 4(8 + 2t)] - 20 = -6[2(6 + 3t) - 4]$

81. $\frac{2}{3}(2x - 1) = 10$

82. $\frac{4}{5}(3x + 4) = 20$

83. $\frac{3}{4}\left(3x - \frac{1}{2}\right) - \frac{2}{3} = \frac{1}{3}$

84. $\frac{2}{3}\left(\frac{7}{8} - 4x\right) - \frac{5}{8} = \frac{3}{8}$

85. $0.7(3x + 6) = 1.1 - (x + 2)$

86. $0.9(2x + 8) = 20 - (x + 5)$

87. $a + (a - 3) = (a + 2) - (a + 1)$

88. $0.8 - 4(b - 1) = 0.2 + 3(4 - b)$

Skill Maintenance

89. Divide: $-22.1 \div 3.4$.

90. Factor: $7x - 21 - 14y$.

91. Use $<$ or $>$ for ■ to write a true sentence: -15 ■ -13.

92. Find $-(-x)$ when $x = -14$.

Synthesis

93. What procedure would you follow for solving an equation like $0.23x + \frac{17}{3} = -0.8 + \frac{3}{4}x$? Could your procedure be streamlined? If so, how?

94. Dave is determined to solve the equation $3x + 4 = -11$ by first using the multiplication principle to "eliminate" the 3. How should Dave proceed and why?

Solve.

95. $0.008 + 9.62x - 42.8 = 0.944x + 0.0083 - x$

96. $-2[3(x - 2) + 4] = 4(1 - x) + 8$

97. $0 = y - (-14) - (-3y)$

98. $3(x + 4) = 3(4 + x)$

99. $475(54x + 7856) + 9762 = 402(83x + 975)$

100. $30{,}000 + 20{,}000x = 55{,}000(1 + 12{,}500x)$

101. $x(x - 4) = 3x(x + 1) - 2(x^2 + x - 5)$

102. $0.05y - 1.82 = 0.708y - 0.504$

103. $-2y + 5y = 6y$

104. $3x = 4x$

105. $\dfrac{5 + 2y}{3} = \dfrac{25}{12} + \dfrac{5y + 3}{4}$

106. $\dfrac{4 - 3x}{7} = \dfrac{2 + 5x}{49} - \dfrac{x}{14}$

2.3 Formulas

Evaluating Formulas • Solving for a Letter

Many applications of mathematics involve relationships between two or more quantities. An equation that represents such a relationship will use two or more letters and is known as a **formula.**

EXAMPLE 1

The formula $M = \frac{1}{5}n$ can be used to determine how far you are from a thunderstorm. Here n is the number of seconds that it takes the sound of thunder to reach you once lightning appears and M is the distance, in miles, that you are from the lightning. If it takes 10 sec for the sound of thunder to reach you after you have seen lightning, how far away is the storm?

Solution We substitute 10 for n and calculate M: $M = \frac{1}{5}n = \frac{1}{5}(10) = 2$. The storm is 2 mi away. ❑

Suppose that we are told how far away a storm is and we want to predict how long it will take the sound of thunder to reach us. We could substitute the distance, say 2, for M, and then solve for n:

$2 = \frac{1}{5}n$ Replacing M with 2

$10 = n.$ Multiplying by 5

Were we to do this for a variety of distances, it might be easier to first solve for n and then substitute values for M.

EXAMPLE 2

Solve for n: $M = \frac{1}{5}n$.

Solution We have

$$M = \frac{1}{5}n \qquad \text{We want this letter alone.}$$

$$5 \cdot M = 5 \cdot \frac{1}{5}n \qquad \text{Multiplying on both sides by 5}$$

$$5M = n.$$

The equation $5M = n$ gives a quick, easy way to find the number of seconds it takes thunder to reach us when a storm is M miles away. ❑

To see how the addition and multiplication principles apply to formulas, compare the following. In (A), we solve as we did before; in (B), we do not carry out the calculations; and in (C), we cannot carry out calculations since the numbers are unknown.

A.
$$5x + 2 = 12$$
$$5x = 12 - 2$$
$$5x = 10$$
$$x = \frac{10}{5} = 2$$

B.
$$5x + 2 = 12$$
$$5x = 12 - 2$$
$$x = \frac{12 - 2}{5}$$

C.
$$ax + b = c$$
$$ax = c - b$$
$$x = \frac{c - b}{a}$$

EXAMPLE 3

Solve for r: $C = 2\pi r$.

Solution Although π is a constant (approximately 3.14), to keep our solution precise we do not replace it with an approximation.

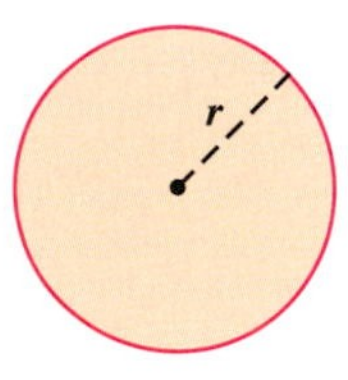

Given a radius, r, we can use this equation to find a circle's circumference, C.

$$C = 2\pi r \qquad \text{We want this letter alone.}$$

$$\frac{C}{2\pi} = \frac{2\pi r}{2\pi} \qquad \text{Dividing by } 2\pi$$

$$\frac{C}{2\pi} = r$$

Given a circle's circumference, C, we can use this equation to find the radius, r. ❑

EXAMPLE 4

Solve for a: $T = \dfrac{a + b + c}{3}$.

Solution This is a formula for the average, T, of three numbers a, b, and c.

$$T = \frac{a + b + c}{3} \qquad \text{We want the letter } a \text{ alone.}$$

$$3T = a + b + c \qquad \text{Multiplying by 3 to clear the fraction}$$

$$3T - b - c = a \qquad \text{Using the addition principle to get } a \text{ by itself (subtracting } b \text{ and subtracting } c \text{ on both sides)}$$

This formula can be used when two test grades are known and we wish to know what grade we need on a third test in order to have an average of T after three tests. ❑

EXAMPLE 5

Solve for C: $Q = \frac{100M}{C}$.

Solution This is a formula used in psychology for intelligence quotient Q, where M is mental age and C is chronological, or actual, age.

$$Q = \frac{100M}{C} \quad \text{We want the letter } C \text{ alone.}$$

$$CQ = 100M \quad \text{Multiplying by } C \text{ to clear the fraction}$$

$$C = \frac{100M}{Q} \quad \text{Dividing by } Q$$

This formula can be used to determine a person's chronological, or actual, age once the person's mental age and intelligence quotient are known. ❑

With the formulas in this section, we can use a procedure like that described in Section 2.2 to solve for a given letter.

A Formula-Solving Procedure

To solve a formula for a given letter:

1. Identify the letter being solved for and multiply on both sides to clear fractions or decimals, if necessary.
2. Collect like terms on each side, if necessary. This may require factoring.
3. Get all terms with the letter to be solved for on one side of the equation and all other terms on the other side.
4. Collect like terms again, if necessary. This may require factoring.
5. Multiply or divide to solve for the variable in question.

EXAMPLE 6

Solve for x: $y = ax + bx - 4$.

Solution We solve as follows:

$$y = ax + bx - 4 \quad \text{We want this letter alone.}$$

$$y + 4 = ax + bx \quad \text{Adding 4}$$

$$y + 4 = (a + b)x \quad \text{Collecting like terms by factoring out } x$$

$$\frac{y + 4}{a + b} = x. \quad \text{Multiplying by } \frac{1}{a + b}$$

We can also write this as

$$x = \frac{y + 4}{a + b}.$$ ❑

CAUTION! Had we performed the following steps in Example 6, we would *not* have solved for x:

$$y = ax + bx - 4$$

$$y - ax + 4 = bx \quad \text{Subtracting } ax \text{ and adding 4}$$

Two occurrences of x

$$\frac{y - ax + 4}{b} = x \quad \text{Dividing by } b$$

The mathematics of each step is correct, but note that x occurs on both sides of the formula. Thus *we have not solved the formula for* x. Remember that the variable being solved for should be alone on one side of the equation, with *no* occurrence of that variable on the other side!

EXERCISE SET 2.3

Solve the formula for the indicated letter.

1. $A = bh$, for b
(Area of a parallelogram with base b and height h)

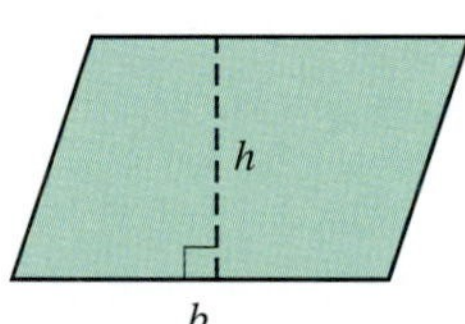

2. $A = bh$, for h
3. $d = rt$, for r
(A distance formula, where d is distance, r is speed, and t is time)
4. $d = rt$, for t
5. $I = Prt$, for P
(Simple-interest formula, where I is interest, P is principal, r is interest rate, and t is time)
6. $I = Prt$, for t
7. $F = ma$, for a
(A physics formula, where F is force, m is mass, and a is acceleration)
8. $F = ma$, for m
9. $P = 2l + 2w$, for w
(Perimeter of a rectangle of length l and width w)

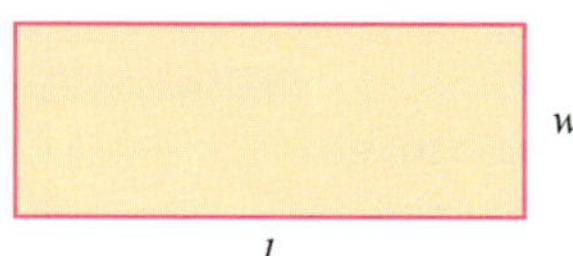

10. $P = 2l + 2w$, for l
11. $A = \pi r^2$, for r^2
(Area of a circle with radius r)

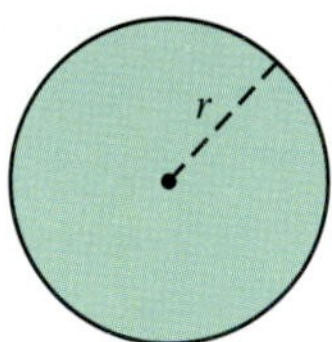

12. $A = \pi r^2$, for π
13. $A = \frac{1}{2}bh$, for b
(Area of a triangle with base b and height h)

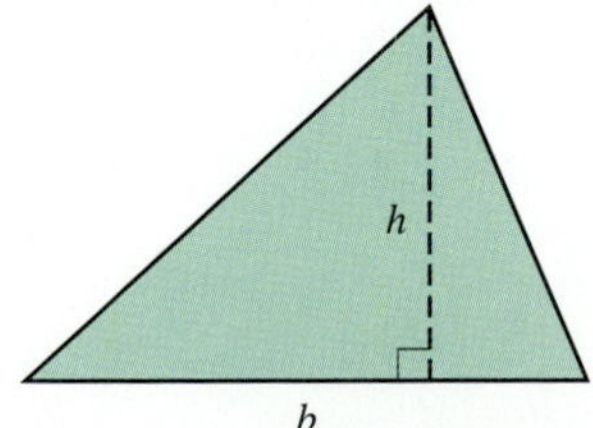

14. $A = \frac{1}{2}bh$, for h

15. $E = mc^2$, for m
(A relativity formula)

16. $E = mc^2$, for c^2

17. $Q = \dfrac{c + d}{2}$, for d

18. $Q = \dfrac{p - q}{2}$, for p

19. $A = \dfrac{a + b + c}{3}$, for b

20. $A = \dfrac{a + b + c}{3}$, for c

21. $v = \dfrac{3k}{t}$, for t

22. $P = \dfrac{ab}{c}$, for c

23. $Ax + By = C$, for y

24. $Ax + By = C$, for x

25. $A = \frac{1}{2}ah + \frac{1}{2}bh$, for b

26. $A = \frac{1}{2}ah - \frac{1}{2}bh$, for a; for h

27. $Q = 3a + 5ca$, for a

28. $P = 4m + 7mn$, for m

29. The formula
$$A = P + Prt$$
is used to find the amount A in an account when simple interest is added to an investment of P dollars (see Exercise 5). Solve for P.

30. The formula
$$S = P - P(0.01r)$$
can be used to find the sale price S of an item when the regular price P is reduced r percent. Solve for P.

31. The area of a sector of a circle is given by
$$A = \frac{\pi r^2 S}{360},$$
where r is the radius and S is the angle measure, in degrees, of the sector. Solve for S.

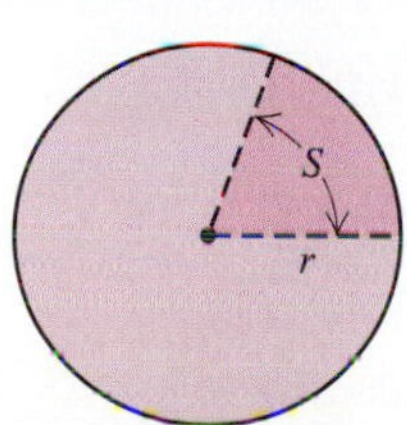

32. Solve for r^2: $A = \dfrac{\pi r^2 S}{360}$.

33. The formula $R = -0.0075t + 3.85$ can be used to estimate the world record in the 1500-m run t years after 1930. Solve for t.

34. The formula $F = \frac{9}{5}C + 32$ can be used to convert from Celsius, or Centigrade, temperature C to Fahrenheit temperature F. Solve for C.

Skill Maintenance

Multiply.

35. $7(-3)2$

36. $-\frac{2}{3} \cdot \frac{9}{10}$

Simplify.

37. $10 \div (-2) \cdot 5 - 4$

38. $3|7 - (2 - 5)|$

Synthesis

39. ◈ Devise an application in which it would be useful to solve the equation $d = rt$ for r (see Exercise 3).

40. ◈ A meteorologist claims to be able to convert Celsius temperatures to Fahrenheit temperatures but not Fahrenheit to Celsius. What advice would you offer the meteorologist?

41. Solve for y:
$$\frac{\left(\frac{y}{z}\right)}{\left(\frac{z}{t}\right)} = 1.$$

42. Solve for F: $\dfrac{1}{E + F} = G$.

43. Solve for t: $q = r(s + t)$.

44. Solve for c: $ac = bc + d$.

45. Solve for x: $a = c(x + y) + bx$.

46. Solve for a: $3a = c - a(b + d)$.

47. ◈ In the formula $A = l \cdot w$, suppose that l and w are both doubled. What is the effect on A? Why?

48. ◈ The equations
$$P = 2l + 2w \quad \text{and} \quad w = \frac{P}{2} - l$$
are equivalent formulas involving the perimeter P, length l, and width w of a rectangle. Devise a problem for which the second of the two formulas would be more useful.

2.4 Applications with Percent

Converting Between Percent Notation and Decimal Notation • Solving Percent Problems

Formulas like those in Section 2.3 appear in a wide variety of applications. In this section, we examine applications involving percent.

Suppose that Village Stationers installs a new cash register and the sales clerks inadvertently print out "totals" on each receipt without separating each transaction into "merchandise" and the five-percent "sales tax." In order to process any refunds, the shop needs a formula for separating each total into the amount spent on merchandise and the amount spent on tax.

Before such a formula can be developed, we must review the basics of percent problems.

Converting Between Percent Notation and Decimal Notation

The average family spends 26% of its income for food. What does this mean? It means that out of every \$100 earned, \$26 is spent for food. Thus 26% is a ratio of 26 to 100.

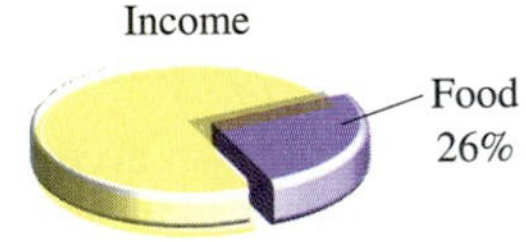

The percent symbol % means "per hundred." We can regard the percent symbol as part of a name for a number. For example,

$$26\% \quad \text{is defined to mean} \quad 26 \times 0.01, \quad \text{or} \quad 26 \times \frac{1}{100}, \quad \text{or} \quad \frac{26}{100}.$$

In general,

Percent Notation

$$n\% \quad \text{means} \quad n \times 0.01, \quad \text{or} \quad n \times \frac{1}{100}, \quad \text{or} \quad \frac{n}{100}.$$

EXAMPLE 1 Convert to decimal notation: **(a)** 78%; **(b)** 1.3%.

Solution

a) $78\% = 78 \times 0.01$ Replacing % by $\times\, 0.01$

$= 0.78$

b) $1.3\% = 1.3 \times 0.01$ Replacing % by × 0.01
$= 0.013$ ❑

> To convert from percent notation to decimal notation, move the decimal point two places to the left and drop the percent symbol.

EXAMPLE 2

Convert 43.67% to decimal notation.

Solution

43.67% 0.43.67 43.67% = 0.4367

Move the decimal point two places to the left. ❑

The procedure used in Example 2 can be reversed. Consider 0.38:

$0.38 = \frac{38}{100}$ Converting to fractional notation
$= 38\%.$ $\frac{n}{100}$ means $n\%$.

> To convert from decimal notation to percent notation, move the decimal point two places to the right and write a percent symbol.

EXAMPLE 3

Convert to percent notation: **(a)** 1.27; **(b)** $\frac{1}{4}$; **(c)** 0.3.

Solution

a) We first move the decimal point two places to the right: 1.27.
and then write a % symbol: 127%

b) Note that $\frac{1}{4} = 0.25$. We move the decimal point two places to the right: 0.25.
and then write a % symbol: 25%

c) We first move the decimal point two places to the right (recall that 0.3 = 0.30): 0.30.
and then write a % symbol: 30% ❑

There is a table of fractional, decimal, and percent equivalents at the back of the text. If you do not already know these facts, it might be helpful to memorize some or all of them.

Solving Percent Problems

To solve problems involving percents, we translate to mathematical language and then solve an equation.

EXAMPLE 4

What is 11% of 49?

Solution

Translate: What is 11% of 49?

$$a = 11\% \cdot 49$$

$$a = 0.11 \cdot 49 \qquad 11\% = 0.11$$

Since a is already by itself, we just multiply:

$$\begin{array}{r} 49 \\ \times 0.11 \\ \hline 49 \\ 490 \\ \hline a = 5.39 \end{array}$$

> A way of checking answers is by estimating as follows:
>
> $$11\% \times 49 \approx 10\% \times 50$$
> $$= 0.10 \times 50 = 5.$$
>
> Since 5 is close to 5.39, our answer is reasonable.

Thus, 5.39 is 11% of 49. The answer is 5.39. ❑

EXAMPLE 5

3 is 16 percent of what?

Solution Translate: 3 is 16 percent of what?

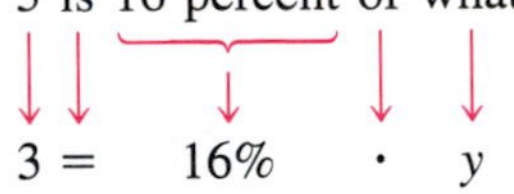

$$3 = 16\% \cdot y$$

We solve the equation:

$$3 = 0.16y$$

$$\frac{3}{0.16} = y \qquad \text{Dividing on both sides by 0.16}$$

$$18.75 = y.$$

Thus, 3 is 16 percent of 18.75. The answer is 18.75. ❑

EXAMPLE 6

What percent of \$50 is \$16?

Solution Translate: What percent of \$50 is \$16?

$$n \cdot 50 = 16$$

We solve the equation and then convert to percent notation:

$$n \cdot 50 = 16$$

$$n = \frac{16}{50} \qquad \text{Dividing on both sides by 50}$$

$$n = 0.32 = 32\%.$$

Thus, \$16 is 32% of \$50. The answer is 32%. ❑

Examples 4–6 represent the three basic types of percent problems.

EXAMPLE 7

A receipt from Village Stationers indicates the total paid (including tax), but not the price of the merchandise. If the sales tax is 5%, find the following:

a) The cost of the merchandise when the total is \$31.50.

b) A formula for the cost of the merchandise c when the total is T dollars.

Solution

a) When tax is added to the cost of an item, the customer actually pays more than 100% of the item's price. When sales tax is 5%, the total paid is 105% of the price of the merchandise. Thus if c = the cost of the merchandise, we have

$$\underbrace{\$31.50}\ \text{is}\ 105\%\ \text{of}\ c$$

$$\downarrow \quad \downarrow \quad \downarrow \quad \downarrow \quad \downarrow$$

$$31.50 = 1.05 \cdot c$$

$$\frac{31.50}{1.05} = c \qquad \text{Dividing by 1.05}$$

$$30 = c. \qquad \text{Simplifying}$$

The merchandise cost \$30 before tax.

b) When the total is T dollars, we modify the approach used in part (a):

$$T = 1.05c$$

$$\frac{T}{1.05} = c. \qquad \text{Dividing by 1.05}$$

To check this result, note that the answer in part (a) shows that when T is replaced with \$31.50, c = \$30. The sales tax of 5% would add $0.05 \cdot \$30$, or \$1.50, to the bill, for a total of \$30 + \$1.50 = \$31.50. Thus our answer in part (a) checks and the formula $T/1.05 = c$ appears to yield correct values for c.

The formula $c = T/1.05$ can be used to find the cost of the merchandise when the total T is known and the sales tax is 5%. ❑

EXERCISE SET 2.4

Find decimal notation.

1. 76% **2.** 54% **3.** 54.7%

4. 96.2% **5.** 100% **6.** 1%

7. 0.61% **8.** 125% **9.** 240%

10. 0.73% **11.** 3.25% **12.** 2.3%

Find percent notation.

13. 4.54 **14.** 1 **15.** 0.998

16. 0.73 **17.** 2 **18.** 0.0057

19. 0.072 **20.** 1.34 **21.** 9.2

22. 0.013 **23.** 0.0068 **24.** 0.675

25. $\frac{1}{8}$ **26.** $\frac{1}{3}$ **27.** $\frac{17}{25}$

28. $\frac{11}{20}$ **29.** $\frac{3}{4}$ **30.** $\frac{2}{5}$

31. $\frac{7}{10}$ **32.** $\frac{8}{10}$ **33.** $\frac{3}{5}$

34. $\frac{17}{50}$ **35.** $\frac{2}{3}$ **36.** $\frac{3}{8}$

Solve.

37. What percent of 68 is 17?

38. What percent of 75 is 36?

39. What percent of 125 is 30?

40. What percent of 300 is 57?

41. 45 is 30% of what number?

42. 20.4 is 24% of what number?

43. 0.3 is 12% of what number?

44. 7 is 175% of what number?

45. What number is 65% of 840?

46. What number is 1% of a million?

47. What percent of 80 is 100?

48. What percent of 10 is 205?

49. What is 2% of 40?

50. What is 40% of 2?

51. 2 is what percent of 40?

52. 40 is 2% of what number?

53. The FBI annually receives 16,000 applications for agents. It accepts 600 of these applicants. What percent does it accept?

54. The U.S. Postal Service reports that we open and read 78% of the junk mail that we receive. A business sends out 9500 advertising brochures. How many of them can it expect to be opened and read?

55. It has been determined by sociologists that 17% of the population is left-handed. Each week 160 bowlers enter a tournament conducted by the Professional Bowlers Association. How many would you expect to be left-handed? Round to the nearest one.

56. In a medical study, it was determined that if 800 people kiss someone else who has a cold, only 56 will actually catch the cold. What percent is this?

57. On a test of 88 items, a student got 76 correct. What percent were correct?

58. A baseball player had 13 hits in 25 times at bat. What percent were hits?

59. A bill at Officeland totaled $37.80. How much did the merchandise cost if the sales tax is 5%?

60. Doreen's checkbook shows that she wrote a check for $987 for building materials. What was the price of the materials if the sales tax was 5%?

Skill Maintenance

61. Convert to decimal notation: $\frac{23}{25}$.

62. Add: $-23 + (-67)$.

63. Subtract: $-45.8 - (-32.6)$.

64. Remove parentheses and simplify:

$$4a - 8b - 5(5a - 4b).$$

Synthesis

65. ◈ Mary Alice bought a dress for $120 after it had been reduced 20%. When Mary Alice returned the dress to the shop, the salesclerk incorrectly said that the original price must have been $144. What mistake did the salesclerk make?

66. ◈ Would it be better to receive a 5% raise and then an 8% raise or the other way around? Why?

67. Rollie's Music charges $11.99 for a compact disc. Sound Warp charges $13.99, but you have a coupon for $2 off. In both cases, a 7% sales tax is charged on the *regular* price. How much does the disc cost at each store?

68. The weather report is "a 60% chance of showers during the day, 30% tonight, and 5% tomorrow morning." What are the chances that it won't rain during the day? tonight? tomorrow morning?

69. If x is 160% of y, y is what percent of x?

70. The new price of a car is 25% higher than the old price of $10,400. The old price is what percent lower than the new price?

71. Generalize the result of Example 7. That is, find a formula for the cost of the merchandise c when the total is T dollars and the tax rate is r.

2.5 Problem Solving

Five Steps for Problem Solving • Applying the Five Steps

One of the most important uses of algebra is as a tool for problem solving. In this section, we develop a problem-solving approach that will be used throughout the remainder of the text.

Five Steps for Problem Solving

In Section 2.4, we solved a problem in which Village Stationers needed a formula. Our solution of the problem required us to *familiarize* ourselves with percent notation so that we could then *translate* a percent problem into an equation. We then *solved* the equation, *checked* the solution, and *stated* the answer to the problem at the end of the section.

Five Steps for Problem Solving in Algebra

1. *Familiarize* yourself with the problem situation.
2. *Translate* to mathematical language. (This often means write an equation.)
3. *Carry out* some mathematical manipulation. (This often means *solve* an equation.)
4. *Check* your possible answer in the original problem.
5. *State* the answer clearly.

Of the five steps, the most important is probably the first one: becoming familiar with the problem situation. Here are some hints for familiarization.

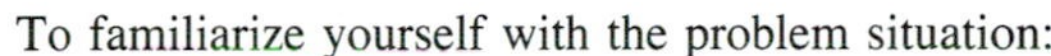

To familiarize yourself with the problem situation:

1. Read the problem carefully.
2. Reread the problem, perhaps aloud. Try to visualize the problem.
3. List the information given and the questions to be answered. Choose a variable (or variables) to represent the unknown and clearly state what the variable represents. For example, let L = length in centimeters, d = distance in miles, and so on.
4. Gather further information. Look up a formula in this book or in a reference book. Talk to a reference librarian or an expert in the field.
5. Make a table of the given information and the information you have collected. Look for patterns that may help in the translation to an equation.
6. Make a drawing and label it with known and unknown information, using specific units if given.
7. Guess the answer and check the guess. Observe the manner in which the guess is checked.

EXAMPLE 1

A 72-in. board is cut into two pieces. One piece is twice as long as the other. How long are the pieces?

Solution

1. **FAMILIARIZE.** We first draw a picture. We let

x = the length of the shorter piece.

Then

$2x$ = the length of the longer piece.

(We can also let y = the length of the longer piece. Then $\frac{1}{2}y$ = the length of the shorter piece. This, however, introduces fractions and will make the solution somewhat more difficult.)

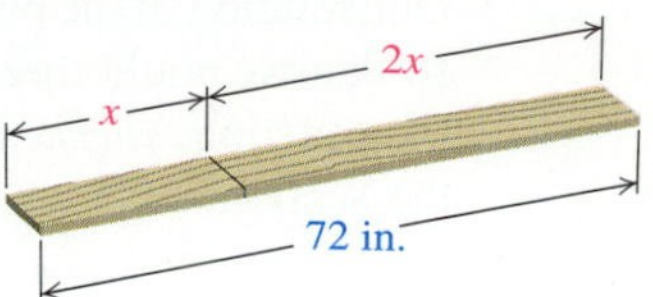

We can further familiarize ourselves with the problem by making a guess. Suppose $x = 31$ in. Then $2x = 62$ in. and $x + 2x = 93$ in. This is not correct but making the guess does help us become familiar with the problem.

2. **TRANSLATE.** From the figure, we can see that the lengths of the two pieces must add up to 72 in. This gives us our translation.

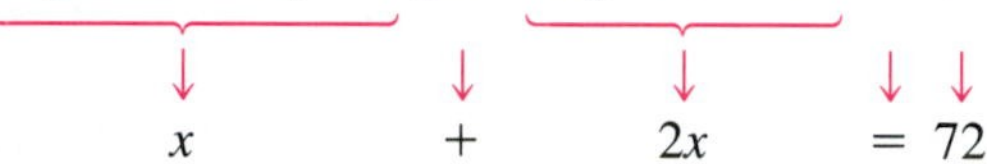

$$x \quad + \quad 2x \quad = 72$$

3. **CARRY OUT.** We solve the equation:

$$\begin{aligned} x + 2x &= 72 \\ 3x &= 72 \qquad \text{Collecting like terms} \\ x &= 24. \qquad \text{Dividing by 3} \end{aligned}$$

4. **CHECK.** If one piece is 24 in. long, the other, to be twice as long, must be 48 in. long. The lengths of the pieces add up to 72 in. This checks.

5. **STATE.** One piece is 24 in. long, and the other is 48 in. long. ❑

EXAMPLE 2

Five more than twice a number is nineteen. What is the number?

Solution

1. **FAMILIARIZE.** Let $x =$ the number. Then ''twice a number'' translates to $2x$ and ''five more than twice a number'' translates to $2x + 5$.

2. **TRANSLATE.** The familiarization leads us to the following translation:

$$2x + 5 \quad = \quad 19.$$

3. **CARRY OUT.** We solve the equation:

$$\begin{aligned} 2x + 5 &= 19 \\ 2x &= 14 \qquad \text{Subtracting 5} \\ x &= 7. \qquad \text{Dividing by 2} \end{aligned}$$

4. **CHECK.** Twice, or two times, 7 is 14. Adding 5 to 14, we get 19. This checks.

5. **STATE.** The number is 7. ❑

The following are examples of **consecutive integers:** 16, 17, 18, 19; and

$-31, -30, -29$. Unknown consecutive integers can be represented in the form x, $x + 1$, $x + 2$, and so on.

The following are examples of **consecutive even integers:** 16, 18, 20, 22; and $-52, -50, -48$. Unknown consecutive even integers can be represented in the form x, $x + 2$, $x + 4$, and so on.

The following are examples of **consecutive odd integers:** 21, 23, 25, 27; and $-71, -69, -67$. Unknown consecutive odd integers can be represented in the form x, $x + 2$, $x + 4$, and so on.

EXAMPLE 3

A book is opened. The sum of the page numbers on the facing pages is 233. Find the page numbers.

Solution

1. FAMILIARIZE. Page numbers on facing pages are consecutive integers. Thus if we let x = the smaller number, then $x + 1$ = the larger number. (Another way such numbers could be represented is to let y = the larger number and $y - 1$ = the smaller.) Had this problem appeared and you did not understand the word "sum," you would have needed to learn its meaning. This is a general problem-solving tip: You may need to look up a definition or a formula or some other piece of information in order to solve a problem.

   To become more familiar with the problem, we can make a table. How do we get the entries in the table? First, we just guess a value for x. Then we find $x + 1$. Finally, we add the two numbers to find their sum.

x	$x+1$	Sum of x and $x+1$	
14	15	29	⟵ This guess was much too small.
102	103	205	⟵ This guess was a bit too small.

   Our second guess leads us to suspect that the value of x is not much greater than 102. The problem could actually be solved with further guessing, but we need to practice using algebra.

2. TRANSLATE. We reword the problem and translate as follows.

   Rewording: First integer + second integer = 233

   Translating: $x + (x + 1) = 233$

3. CARRY OUT. We solve the equation:

$$x + (x + 1) = 233$$
$$2x + 1 = 233 \quad \text{Using an associative law and collecting like terms}$$
$$2x = 232 \quad \text{Subtracting 1}$$
$$x = 116. \quad \text{Dividing by 2}$$

   If x is 116, then $x + 1$ is 117.

4. CHECK. Our possible answers are 116 and 117. These are consecutive integers and their sum is 233, so the answers check in the original problem.

5. STATE. The page numbers are 116 and 117. ❑

EXAMPLE 4

Truck-Rite Rentals rents trucks at a daily rate of \$49.95 plus 39¢ per mile. Concert Productions has budgeted \$100 for renting a truck to haul equipment to an upcoming concert. How many miles can a rental truck be driven on a \$100 budget?

Solution

1. FAMILIARIZE. Suppose Concert Productions drives 75 mi. Then the cost is

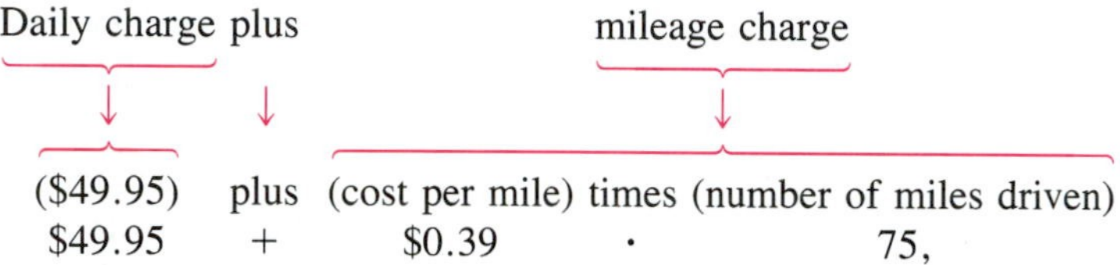

Daily charge plus mileage charge

(\$49.95) plus (cost per mile) times (number of miles driven)

$$\$49.95 + \$0.39 \cdot 75,$$

which is \$49.95 + \$29.25, or \$79.20. This familiarizes us with the way in which a calculation is made. Note that we convert 39 cents to \$0.39 so that we are using the same units, dollars, throughout the problem.

Let m = the number of miles that can be driven for \$100.

2. TRANSLATE. We reword the problem and translate as follows.

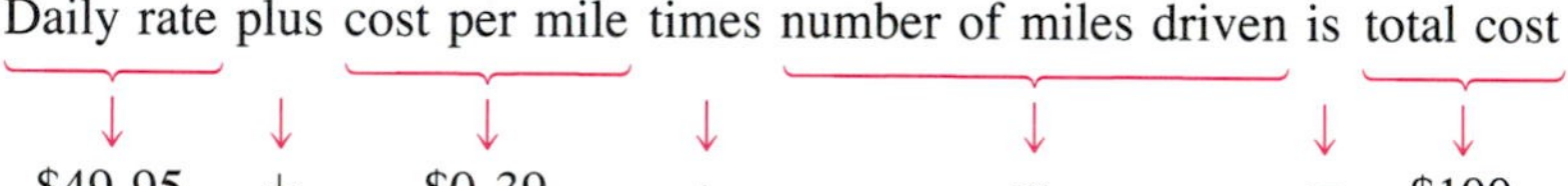

Daily rate plus cost per mile times number of miles driven is total cost

$$\$49.95 + \$0.39 \cdot m = \$100$$

3. CARRY OUT. We solve the equation:

$$49.95 + 0.39m = 100$$

$$100(49.95 + 0.39m) = 100(100) \quad \text{Multiplying by 100 on both sides to clear decimals}$$

$$100(49.95) + 100(0.39m) = 10{,}000 \quad \text{Using the distributive law}$$

$$4995 + 39m = 10{,}000$$

$$39m = 5005 \quad \text{Subtracting 4995}$$

$$m = \frac{5005}{39} \quad \text{Dividing by 39}$$

$$m \approx 128.3 \quad \text{Rounding to the nearest tenth}$$

4. CHECK. We check in the original problem. We multiply 128.3 by \$0.39, getting \$50.037. Then we add \$50.037 to \$49.95 and get \$99.987, which is just about the \$100 allotted.

5. STATE. The truck can be driven about 128.3 mi on the rental allotment of \$100. ❑

EXAMPLE 5

A rectangular community garden is to be enclosed with 92 m of fencing. In order to allow for compost storage, the garden must be 4 m longer than it is wide. Determine the dimensions of the garden.

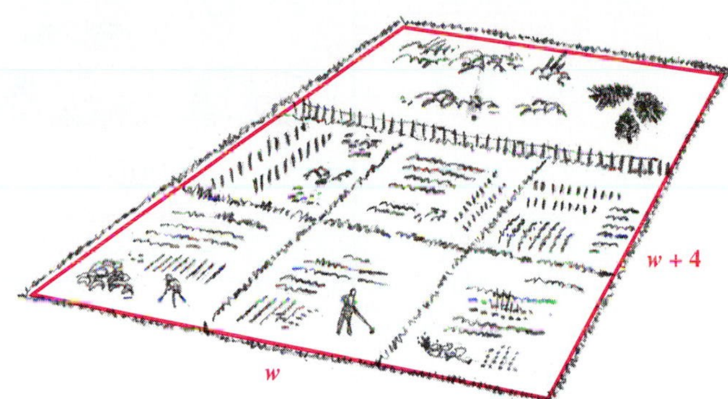

Solution

1. **FAMILIARIZE.** Suppose the garden were 30 m wide. The length would then be 30 + 4 m, or 34 m, and the perimeter would be $2 \cdot 30 + 2 \cdot 34$ m, or 128 m. Our guess was too big, but we have at least gained familiarity with the problem.

   We let w = the width of the garden. Since the garden is "4 m longer than it is wide," we have $w + 4$ = the length. Recall that the perimeter P of a rectangle is the distance around it and is given by the formula $2l + 2w = P$, where l = the length and w = the width.

2. **TRANSLATE.** To translate the problem, we substitute $w + 4$ for l and 92 for P, as follows:

$$2l + 2w = P$$
$$2(w + 4) + 2w = 92.$$

3. **CARRY OUT.** We solve the equation:

$$
\begin{aligned}
2(w + 4) + 2w &= 92 \\
2w + 8 + 2w &= 92 \qquad \text{Using the distributive law} \\
4w + 8 &= 92 \qquad \text{Collecting like terms} \\
4w &= 84 \\
w &= 21.
\end{aligned}
$$

   The dimensions appear to be $w = 21$ m and l, or $w + 4$, = 25 m.

4. **CHECK.** If the width is 21 m and the length 25 m, the perimeter is 2(25 m) + 2(21 m), or 92 m. Since 92 m of fencing is available, the dimensions 21 m and 25 m check.

5. **STATE.** The garden should be 21 m wide and 25 m long. □

EXAMPLE 6

The price of a car was reduced to a sale price of $13,559. This was a 9% reduction. What was the original price?

Solution

1. FAMILIARIZE. Suppose the original price were $16,000. A 9% reduction can be found by taking 9% of $16,000, that is,

   9% of $16,000 = 0.09($16,000) = $1440.

   The sale price is then found by subtracting the amount of reduction:

   $16,000 − $1440 = $14,560.

   In becoming familiar with the problem, we find that our guess of $16,000 was too high. We let x = the original price of the car. Because it is reduced by 9%, the sale price $= x - 9\%x$.

2. TRANSLATE. We reword and then translate.

   Rewording: Original price − reduction is sale price

   Translating: $x \quad - \quad 9\%x \quad = \quad \$13{,}559$

3. CARRY OUT. We solve the equation:

$$x - 9\%x = 13{,}559$$

$$x - 0.09x = 13{,}559 \qquad \text{Converting to decimal notation}$$

$$1x - 0.09x = 13{,}559$$

$$(1 - 0.09)x = 13{,}559 \qquad \text{Factoring out the } x$$

$$0.91x = 13{,}559$$

Collecting like terms. Had we noted that the sale price is 91% of the original price, we could have begun with this equation.

$$x = \frac{13{,}559}{0.91} \qquad \text{Dividing by 0.91}$$

$$x = 14{,}900.$$

4. CHECK. To check, we find 9% of $14,900 and subtract:

   9% × $14,900 = 0.09 × $14,900 = $1341

   $14,900 − $1341 = $13,559.

   Since we get the sale price, $13,559, the $14,900 checks.

5. STATE. The original price was \$14,900. ☐

> CAUTION! A common error in a problem like this is to take 9% of the sale price and subtract or add. Note that 9% of the original price is not equal to 9% of the sale price!

EXAMPLE 7

The second angle of a triangle is 20° greater than the first. The third angle is twice as large as the first. How large are the angles?

Solution

1. FAMILIARIZE. We draw a picture. Here the measure of the first angle = x, the measure of the second angle = $x + 20$, and the measure of the third angle = $2x$.

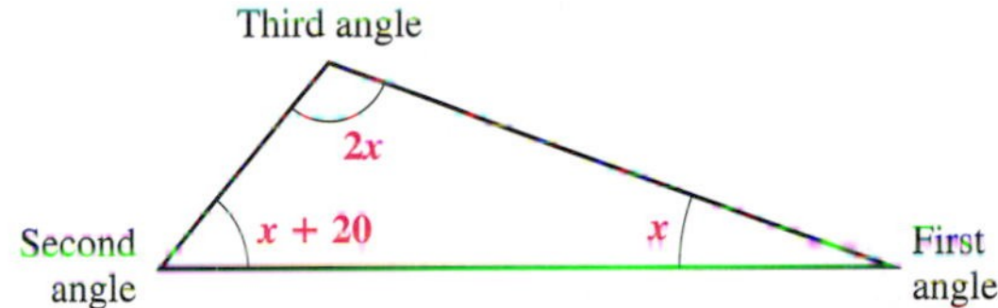

2. TRANSLATE. To translate, we need to recall a geometric fact (you might, as part of step 1, look it up in a geometry book or in the list of formulas at the back of this book). The measures of the angles of any triangle add up to 180°.

$$\underbrace{\text{Measure of first angle}} + \underbrace{\text{measure of second angle}} + \underbrace{\text{measure of third angle}} = 180°$$

$$x + (x + 20) + 2x = 180.$$

3. CARRY OUT. We solve:

$$\begin{aligned} x + (x + 20) + 2x &= 180 \\ 4x + 20 &= 180 \\ 4x &= 160 \\ x &= 40. \end{aligned}$$

The measures for the angles appear to be:

First angle: $x = 40°$,

Second angle: $x + 20 = 40 + 20 = 60°$,

Third angle: $2x = 2(40) = 80°$.

4. CHECK. Consider 40°, 60°, and 80°. The second is 20° greater than the first, the third is twice the first, and the sum is 180°. These numbers check.

5. STATE. The measures of the angles are 40°, 60°, and 80°. ☐

We close this section with some tips to aid you in problem solving.

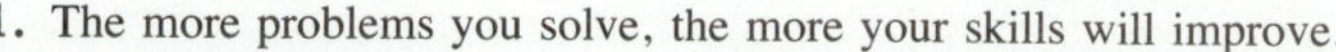

Problem-Solving Tips

1. The more problems you solve, the more your skills will improve.
2. Look for patterns when solving problems. Each time you study an example in a text, you may observe a pattern for problems that you will encounter later in the exercise sets or some other practical situation.
3. When translating to mathematics, consider the dimensions of the variables and constants in the equation. The variables that represent length should all be in the same unit, those that represent money should all be in dollars or all in cents, and so on.
4. Make sure that units appear in the answer whenever appropriate.

EXERCISE SET 2.5

Translate to an algebraic expression. Do not solve.

1. Three less than twice a number
2. Five less than a number divided by 8
3. One half of the product of a number and 7
4. Two fewer than ten times a number
5. Five times the sum of 3 and some number
6. The sum of two numbers times 6
7. An 8-ft board is cut into two pieces. One piece is 2 ft longer than the other. Let $L =$ the length of the shorter piece. Write an expression for the longer piece.

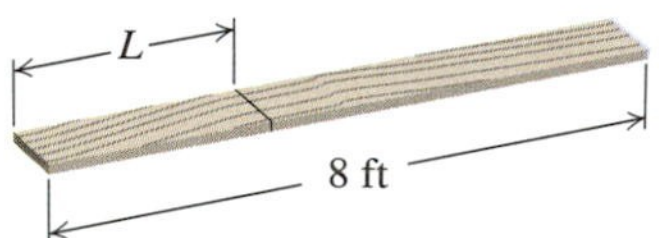

8. A 240-in. pipe is cut into two pieces. One piece is three times the length of the other. Let $x =$ the length of the longer piece. Write an expression for the shorter piece.

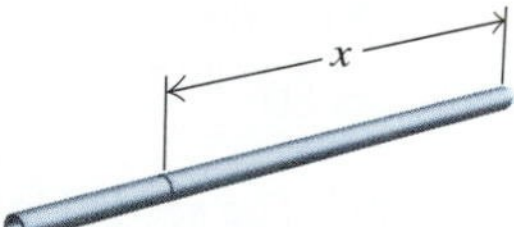

9. The price of a book is decreased by 30% during a sale. Let $b =$ the price of the book before the reduction. Write an expression for the sale price.
10. On sale, the price of a compact disc player was reduced by 20%. Let $p =$ the price before the reduction. Write an expression for the sale price.
11. Write an expression for the sum of three consecutive even integers.
12. Write an expression for the sum of three consecutive integers.
13. Jetz rents compact cars at a rate of \$34.95 plus 27¢ per mile. Let $m =$ the total number of miles driven. Write an expression for the cost of driving m miles.
14. Stacy scored two more points than half the number scored by the entire team. Let $t =$ the number of points scored by the entire team. Write an expression for the number of points Stacy scored.
15. The length of a rectangle is twice the width.
    a) Let $w =$ the width. Write an expression for the length.
    b) Let $l =$ the length. Write an expression for the width.
16. The base of a triangle is three times five more than the height. Let $h =$ the height. Write an expression for the base.
17. The second angle of a triangle is three times as large as the first. The third angle measures 30° more than the first. Let $x =$ the measure of the first angle. Write expressions for the other angles.
18. The second angle of a triangle is four times as large as the first. The third angle is 45° less than the sum of the other two angles. Let $x =$ the measure of the first angle. Write expressions for the other angles.

Solve these problems. Even though you might find the answer quickly in some other way, practice using the five-step problem-solving process.

19. Three less than twice a number is 25. What is the number?

20. Two fewer than ten times a number is 118. What is the number?

21. Five times the sum of 3 and some number is 70. What is the number?

22. Twice the sum of 4 and some number is 34. What is the number?

23. When 18 is subtracted from six times a certain number, the result is 96. What is the number?

24. When 28 is subtracted from five times a certain number, the result is 232. What is the number?

25. If you double a number and then add 16, you get $\frac{2}{5}$ of the original number. What is the original number?

26. If you double a number and then add 85, you get $\frac{3}{4}$ of the original number. What is the original number?

27. If you add two fifths of a number to the number itself, you get 56. What is the number?

28. If you add one third of a number to the number itself, you get 48. What is the number?

29. A 180-m rope is cut into three pieces. The second piece is twice as long as the first. The third piece is three times as long as the second. How long is each piece of rope?

30. A 480-m wire is cut into three pieces. The second piece is three times as long as the first. The third piece is four times as long as the second. How long is each piece?

31. The sum of the page numbers on the facing pages of a book is 273. What are the page numbers?

32. The sum of the page numbers on the facing pages of a book is 281. What are the page numbers?

33. The sum of two consecutive even integers is 114. What are the integers?

34. The sum of two consecutive even integers is 106. What are the integers?

35. The sum of three consecutive integers is 108. What are the integers?

36. The sum of three consecutive integers is 126. What are the integers?

37. The sum of three consecutive odd integers is 189. What are the integers?

38. The sum of three consecutive even integers is 396. What are the integers?

39. The top of the John Hancock Building in Chicago is a rectangle whose length is 60 ft more than the width. The perimeter is 520 ft. Find the width and the length of the rectangle. Find the area of the rectangle.

40. The ground floor of the John Hancock Building is a rectangle whose length is 100 ft more than the width. The perimeter is 860 ft. Find the width and the length of the rectangle. Find the area of the rectangle.

41. The perimeter of a standard-size piece of typewriter paper is 99 cm. The width is 6.3 cm less than the length. Find the length and the width.

42. The perimeter of the state of Wyoming is 1280 mi. The width is 90 mi less than the length. Find the width and the length.

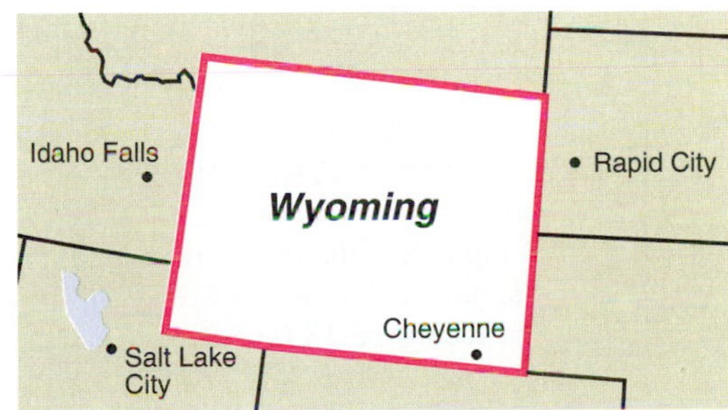

43. The second angle of a triangle is four times as large as the first. The third angle is 45° less than the sum of the other two angles. Find the measure of the first angle.

44. The second angle of a triangle is three times as large as the first. The third angle is 25° less than the sum of the other two angles. Find the measure of the first angle.

45. After a 40% price reduction, a shirt is on sale at \$9.60. What was the original price (that is, the price before reduction)?

46. After a 34% price reduction, a blouse is on sale at \$9.24. What was the original price?

47. Money is invested in a savings account at a rate of 6% simple interest. After one year, there is \$4664 in the account. How much was originally invested?

48. Money is borrowed at a rate of 10% simple interest. After one year, \$7194 pays off the loan. How much was originally borrowed?

49. Badger Rent-A-Car rents a compact car at a daily rate of \$34.95 plus 10¢ per mile. A businessperson is allotted \$80 for car rental. How many miles can the businessperson travel on the \$80 budget?

50. Badger rents midsized cars at a rate of \$43.95 plus 10¢ per mile. A tourist has a car-rental budget of \$90. How many miles can the tourist travel on the \$90?

51. The second angle of a triangle is three times as large as the first. The measure of the third angle is 40° greater than that of the first. How large are the angles?

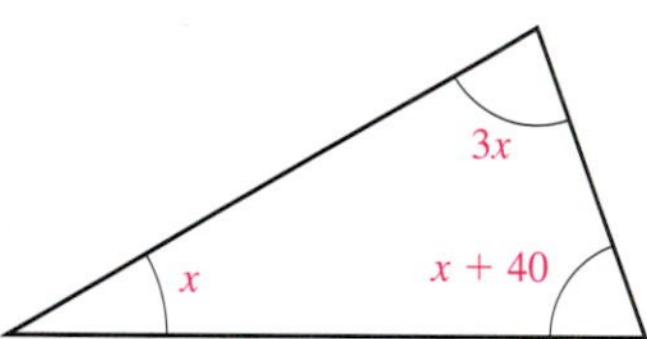

52. One angle of a triangle is 32 times as large as another. The measure of the third angle is 10° greater than that of the smallest angle. How large are the angles?

53. The equation $R = -0.028t + 20.8$ can be used to predict the world record in the 200-m dash, where R represents the record in seconds and t represents the number of years since 1920. In what year will the record be 18.0 sec?

54. The equation $F = \frac{1}{4}N + 40$ can be used to determine the temperature given the number of times a cricket chirps per minute, where F represents the temperature in degrees and N represents the number of chirps per minute. Determine the number of chirps per minute for a temperature of 80°.

Skill Maintenance

55. Factor: $3x - 12y + 60$.

56. Simplify: $5x - 3(9 - 2x) + 14$.

Synthesis

57. ◈ A student claims to be able to solve most of the problems of this section by guessing. Is there anything wrong with this approach? Why or why not?

58. ◈ Write a problem for a classmate to solve. Devise it so that the problem can be translated to the equation

$$x + (x + 2) + (x + 4) = 375.$$

59. Abraham Lincoln's 1863 Gettysburg Address refers to the year 1776 as "four *score* and seven years ago." Write an equation and find what a score is.

60. One number is 25% of another. The larger number is 12 more than the smaller. What are the numbers?

61. If the daily rental for a car is \$18.90 plus a certain price per mile and a person must drive 190 mi and stay within a \$55.00 budget, what is the highest price per mile that the person can afford?

62. Hal scored 78 on a test that had 4 fill-ins worth 7 points each and 24 multiple-choice questions worth 3 points each. He had one fill-in wrong. How many multiple-choice questions did Hal get right?

63. The width of a rectangle is three fourths of the length. The perimeter of the rectangle becomes 50 cm when the length and the width are each increased by 2 cm. Find the length and the width.

64. Apples are collected in a basket for six people. One third, one fourth, one eighth, and one fifth of the apples are given to four people, respectively. The fifth person gets ten apples, and one apple remains for the sixth person. Find the original number of apples in the basket.

65. In a basketball league, the Falcons won 15 of their first 20 games. In order to win 60% of the total number of games, how many more games will they have to play, assuming they win only half the remaining games?

66. Luke spends 12% of his weekly salary on dining out. Of the money spent dining out, 55%, or \$39.60, is spent on fast-food. What is Luke's weekly salary?

67. The area of a triangle is 2.9047 in^2. Find the height of the triangle if the base is 8 in.

68. Ella has an average score of 82 on three tests. Her average score on the first two tests is 85. What was the score on the third test?

69. A school purchases a piano and must choose between paying $2000 at the time of purchase or $2150 at the end of one year. Which option should the school select and why?

2.6 Solving Inequalities

Solutions of Inequalities • Graphs of Inequalities • Solving Inequalities Using the Addition Principle • Solving Inequalities Using the Multiplication Principle • Using the Principles Together

Many real-world situations can be translated to *inequalities*. For instance, a student might be interested in what test scores will assure *at least* a 90 average; an elevator might be designed to hold *at most* 2000 pounds; a tax credit might be allowable for families with incomes *less than* $25,000; and so on. Before we can solve applications of this type, we must adapt our equation-solving principles to the solving of inequalities.

Solutions of Inequalities

In Section 1.4, we learned that an inequality is a number sentence with > (is greater than), < (is less than), ≥ (is greater than or equal to), or ≤ (is less than or equal to) as its verb. Inequalities like

$$-7 > x, \quad t < 5, \quad 5x - 2 \geq 9, \quad \text{and} \quad -3y + 8 \leq -7$$

are true for some replacements of the variable and false for others.

EXAMPLE 1

Determine whether the given number is a solution of $x < 2$: **(a)** -2; **(b)** 2.

Solution

a) Since $-2 < 2$ is true, -2 is a solution.
b) Since $2 < 2$ is false, 2 is not a solution. □

EXAMPLE 2

Determine whether the given number is a solution of $y \geq 6$: **(a)** 6; **(b)** -4.

a) Since $6 \geq 6$ is true, 6 is a solution.
b) Since $-4 \geq 6$ is false, -4 is not a solution. □

Graphs of Inequalities

Because the solutions of inequalities like $x < 2$ are too numerous to list, it is helpful to make a drawing that represents all the solutions. A **graph** of an inequality is just such a drawing. Graphs of inequalities in one variable can be drawn on a number line by shading all points that are solutions. Open dots are used to indicate endpoints that are *not* solutions and closed dots indicate endpoints that *are* solutions.

EXAMPLE 3

Graph each inequality: **(a)** $x < 2$; **(b)** $y \geq -3$; **(c)** $-2 < x \leq 3$.

Solution

a) The solutions of $x < 2$ are those numbers less than 2. They are shown on the graph by shading all points to the left of 2. The open dot at 2 indicates that 2 *is not* part of the graph, but numbers like 1.2 and 1.99 are.

b) The solutions of $y \geq -3$ are shown on the number line by shading the point for -3 and all points to the right of -3. The closed dot at -3 indicates that -3 *is* part of the graph.

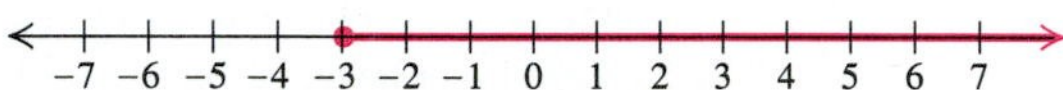

c) The inequality $-2 < x \leq 3$ is read "-2 is less than x *and* x is less than or equal to 3," or "x is greater than -2 *and* less than or equal to 3." To be a solution of this inequality, a number must be a solution of both $-2 < x$ *and* $x \leq 3$. The number 1 is a solution, as are -0.5, 2, 2.5, and 3. The solution set is graphed as follows:

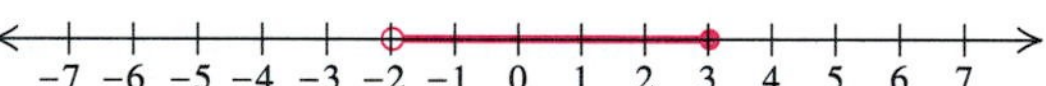

The open dot at -2 means that -2 is not part of the graph. The closed dot at 3 means that 3 is part of the graph. The other solutions are shaded. ❑

Solving Inequalities Using the Addition Principle

Consider a balance similar to one that appears in Section 2.1. Should one side of the balance weigh more than the other, the balance will tip in that direction. If equal amounts of weight are then added or subtracted on each side of the balance, the balance will remain tipped in the same direction.

The balance illustrates the idea that when a number, such as 2, is added (or subtracted) on both sides of an inequality, such as $3 < 7$, we get another true inequality:

$$3 + 2 < 7 + 2, \quad \text{or} \quad 5 < 9.$$

Similarly, if we add -3 on both sides, we get another true inequality:

$$3 + (-3) < 7 + (-3), \quad \text{or} \quad 0 < 4.$$

The Addition Principle for Inequalities

For any real numbers a, b, and c,

if $a < b$, then $a + c < b + c$; if $a > b$, then $a + c > b + c$;

if $a \leq b$, then $a + c \leq b + c$; if $a \geq b$, then $a + c \geq b + c$.

As with equation solving, when solving inequalities, our objective is to isolate a variable on one side.

EXAMPLE 4

Solve $x + 2 > 8$ and then graph the solution.

Solution We use the addition principle, subtracting 2 on both sides:

$$x + 2 - 2 > 8 - 2 \qquad \text{Subtracting 2 or adding } -2 \text{ on both sides}$$
$$x > 6.$$

Using the addition principle, we found an inequality, $x > 6$, from which we can determine the solutions easily.

Any number greater than 6 makes $x > 6$ true and is a solution of that inequality as well as the inequality $x + 2 > 8$. The graph is as follows:

Because there are infinitely many solutions of most inequalities, we cannot possibly check them all. A partial check can be made using one of the possible solutions. In this case, we can check by substituting any number greater than 6, say 6.1, into the original inequality:

$x + 2 > 8$	
$6.1 + 2$?	8
8.1	8 TRUE

Since $8.1 > 8$ is true, 6.1 is a solution. Any number greater than 6 is a solution. ❑

Although the inequality $x > 6$ is easy to solve (we merely replace x with numbers greater than 6), it is important to note that $x > 6$ is an *inequality,* not a *solution.* In fact, the solutions of $x > 6$ are numbers. To describe the set of all solutions, we will use **set-builder notation** to write the *solution set* of Example 4 as

$$\{x \mid x > 6\}.$$

This notation is read

"The set of all x such that x is greater than 6."

Thus a number is in $\{x| \ x > 6\}$ if that number is greater than 6. From now on, solutions of inequalities will be written using set-builder notation.

EXAMPLE 5

Solve $3x - 1 \leq 2x - 5$ and then graph the solution.

Solution

$$3x - 1 \leq 2x - 5$$

$$3x - 1 + 1 \leq 2x - 5 + 1 \quad \text{Adding 1}$$

$$3x \leq 2x - 4 \quad \text{Simplifying}$$

$$3x - 2x \leq 2x - 4 - 2x \quad \text{Subtracting } 2x$$

$$x \leq -4 \quad \text{Simplifying}$$

The graph is as follows:

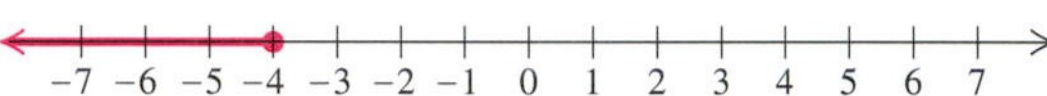

Any number less than or equal to -4 is a solution so the solution set is $\{x| \ x \leq -4\}$. □

Solving Inequalities Using the Multiplication Principle

There is a multiplication principle for inequalities similar to that for equations, but it must be modified when multiplying on both sides by a negative number. Consider the inequality

$$3 < 7.$$

If we multiply on both sides by a *positive* number like 2, we get another true inequality:

$$3 \cdot 2 < 7 \cdot 2, \quad \text{or} \quad 6 < 14. \qquad \text{TRUE}$$

If we multiply on both sides by a negative number like -2, we get a *false* inequality:

$$3 \cdot (-2) < 7 \cdot (-2), \quad \text{or} \quad -6 < -14. \qquad \text{FALSE}$$

The fact that $6 < 14$ is true, but $-6 < -14$ is false, stems from the fact that the negative numbers, in a sense, mirror the positive numbers. That is, whereas 14 is to the *right* of 6, the number -14 is to the *left* of -6. Thus if we reverse the inequality symbol in $-6 < -14$, we get a true inequality:

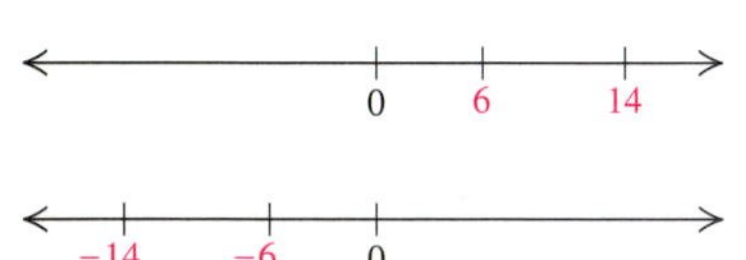

$$-6 > -14. \qquad \text{TRUE}$$

The Multiplication Principle for Inequalities

For any real numbers a and b, and for any *positive* number c,

if $a < b$, then $ac < bc$ and if $a > b$, then $ac > bc$.

For any real numbers a and b, and for any *negative* number c,

if $a < b$, then $ac > bc$ and if $a > b$, then $ac < bc$.

Similar statements hold for $\leq$ and $\geq$.

EXAMPLE 6

Solve and graph each inequality: **(a)** $\frac{1}{4}x < 7$; **(b)** $-2y < 18$.

Solution

a) $\frac{1}{4}x < 7$

$4 \cdot \frac{1}{4}x < 4 \cdot 7$ Multiplying by 4, the reciprocal of $\frac{1}{4}$

The symbol stays the same.

$x < 28$ Simplifying

The solution set is $\{x \mid x < 28\}$. The graph is as follows:

b) $-2y < 18$

$\frac{-2y}{-2} > \frac{18}{-2}$ Multiplying by $-\frac{1}{2}$ or dividing by -2

The symbol must be reversed!

$y > -9$ Simplifying

As a partial check we substitute a number greater than -9, say -8, into the original inequality:

$$\begin{array}{c|l} -2y < 18 & \\ \hline -2(-8) \ ? & 18 \\ 16 & 18 \quad \text{TRUE} \end{array}$$

The solution set is $\{y \mid y > -9\}$. The graph is as follows:

❑

Using the Principles Together

We use the addition and multiplication principles together to solve inequalities much as we did when solving equations. We generally use the addition principle first.

EXAMPLE 7

Solve: **(a)** $6 - 5y > 7$; **(b)** $2x - 9 \leq 7x + 1$.

Solution

a) $6 - 5y > 7$

$-6 + 6 - 5y > -6 + 7$ Adding -6

$-5y > 1$ Simplifying

$-\frac{1}{5} \cdot (-5y) < -\frac{1}{5} \cdot 1$ Multiplying by $-\frac{1}{5}$

The symbol must be reversed.

$y < -\frac{1}{5}$ Simplifying

As a check, we substitute a number smaller than $-\frac{1}{5}$, say -1, into the original inequality:

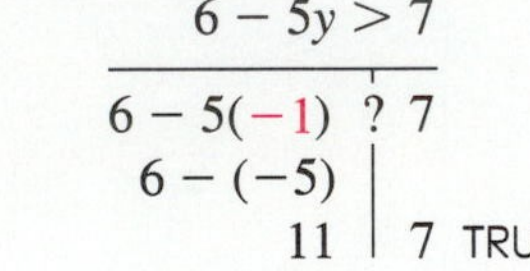

$$6 - 5y > 7$$

$6 - 5(-1)$	? 7
$6 - (-5)$	
11	7 TRUE

The solution set is $\{y \mid y < -\frac{1}{5}\}$.

b)
$$\begin{aligned}
2x - 9 &\leqslant 7x + 1 \\
2x - 9 - 1 &\leqslant 7x + 1 - 1 && \text{Subtracting 1} \\
2x - 10 &\leqslant 7x && \text{Simplifying} \\
2x - 10 - 2x &\leqslant 7x - 2x && \text{Subtracting } 2x \\
-10 &\leqslant 5x && \text{Simplifying} \\
\frac{-10}{5} &\leqslant \frac{5x}{5} && \text{Dividing by 5} \\
-2 &\leqslant x && \text{Simplifying}
\end{aligned}$$

The solution set is $\{x \mid -2 \leqslant x\}$, or $\{x \mid x \geqslant -2\}$. ❑

All of the equation-solving techniques used in Sections 2.1 and 2.2 can be used with inequalities provided we remember to reverse the inequality symbol when multiplying or dividing on both sides by a negative number.

EXAMPLE 8

Solve: **(a)** $16.3 - 7.2p \leqslant -8.18$; **(b)** $3(x - 2) - 1 < 2 - 5(x + 6)$.

Solution

a) The greatest number of decimal places in any one number is *two*. Multiplying by 100, which has two 0's, will clear decimals. Then we proceed as before.

$$\begin{aligned}
16.3 - 7.2p &\leqslant -8.18 \\
100(16.3 - 7.2p) &\leqslant 100(-8.18) && \text{Multiplying by 100 on both sides} \\
100(16.3) - 100(7.2p) &\leqslant 100(-8.18) && \text{Using the distributive law} \\
1630 - 720p &\leqslant -818 && \text{Simplifying} \\
-720p &\leqslant -818 - 1630 && \text{Subtracting 1630 on both sides} \\
-720p &\leqslant -2448 && \text{Simplifying} \\
p &\geqslant \frac{-2448}{-720} && \text{Multiplying by } -\frac{1}{720}
\end{aligned}$$

The symbol must be reversed.

$$p \geqslant 3.4$$

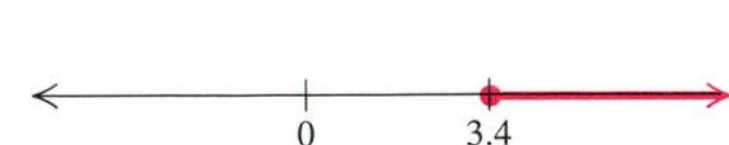

The solution set is $\{p \mid p \geqslant 3.4\}$.

b)
$$\begin{aligned}
3(x - 2) - 1 &< 2 - 5(x + 6) \\
3x - 6 - 1 &< 2 - 5x - 30 && \text{Using the distributive law to multiply and remove parentheses} \\
3x - 7 &< -5x - 28 && \text{Simplifying} \\
3x + 5x &< -28 + 7 && \text{Adding } 5x \text{ and also 7, to get all } x\text{-terms on one side and all other terms on the other side}
\end{aligned}$$

and

$$8x < -21 \quad \text{Simplifying}$$

$$x < -\frac{21}{8} \quad \text{Multiplying by } \frac{1}{8}$$

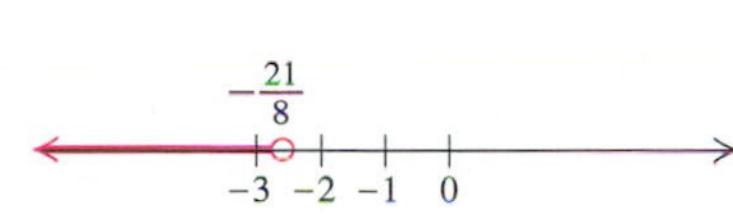

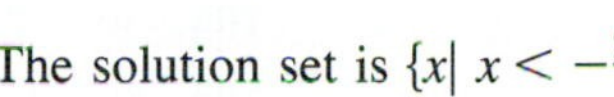

The solution set is $\{x \mid x < -\frac{21}{8}\}$. □

EXERCISE SET 2.6

Determine whether each number is a solution of the given inequality.

1. $x > -4$
  a) 4 **b)** 0 **c)** −4.1
  d) −3.9 **e)** 5.6

2. $y < 5$
  a) 0 **b)** 5 **c)** 4.99
  d) −13 **e)** $7\frac{1}{4}$

3. $x \geq 6$
  a) −6 **b)** 0 **c)** 6
  d) 6.01 **e)** $-3\frac{1}{2}$

4. $x \leq 10$
  a) 4 **b)** −10 **c)** 0
  d) 10.2 **e)** −4.7

Graph on a number line.

5. $x > 4$ **6.** $y < 0$ **7.** $t < -3$

8. $y > 5$ **9.** $m \geq -1$ **10.** $p \leq 3$

11. $-3 < x \leq 4$ **12.** $-5 \leq x < 2$

13. $0 < x < 3$ **14.** $-5 \leq x \leq 0$

Describe the graph using set-builder notation.

15.

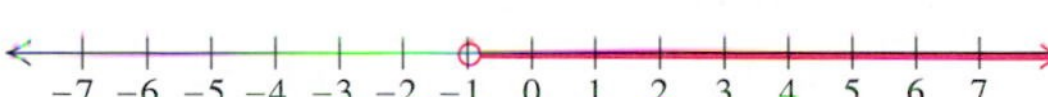

16.

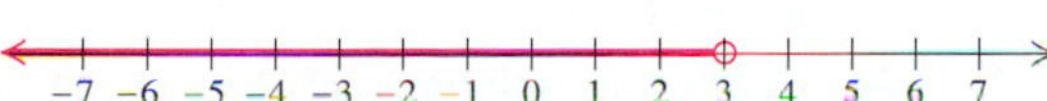

17.

18.

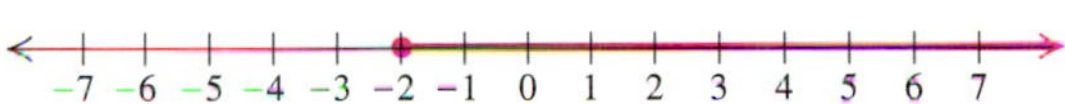

19.

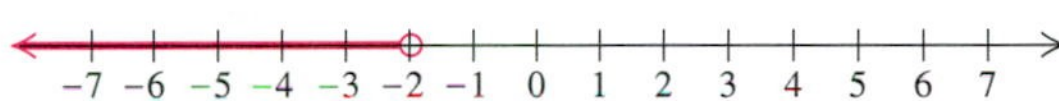

20.

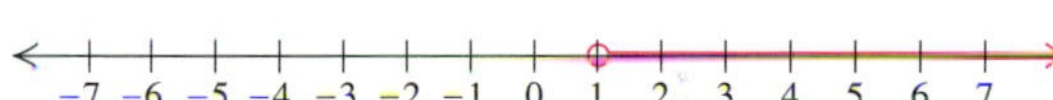

21.

22.

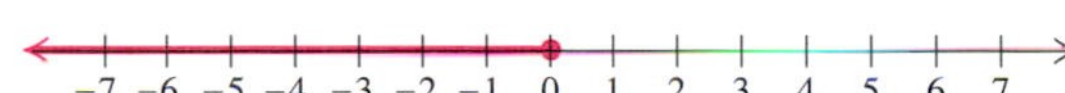

Solve using the addition principle. Graph and write set-builder notation for the answers.

23. $y + 5 > 8$ **24.** $y + 7 > 9$

25. $x + 8 \leq -10$ **26.** $x + 9 \leq -12$

27. $x - 7 < 9$ **28.** $x - 3 < 14$

29. $x - 6 \geq 2$ **30.** $x - 9 \geq 4$

31. $y - 7 > -12$ **32.** $y - 10 > -16$

33. $2x + 3 \leq x + 5$ **34.** $2x + 4 \leq x + 7$

Solve using the addition principle. Write the answers in set-builder notation.

35. $3x - 6 \geq 2x + 7$ **36.** $3x - 9 \geq 2x + 11$

37. $5x - 6 < 4x - 2$

38. $6x - 8 < 5x - 9$

39. $7 + c > 7$

40. $-9 + c > 9$

41. $y + \frac{1}{4} \leqslant \frac{1}{2}$

42. $y + \frac{1}{3} \leqslant \frac{5}{6}$

43. $x - \frac{1}{3} > \frac{1}{4}$

44. $x - \frac{1}{8} > \frac{1}{2}$

45. $-14x + 21 > 21 - 15x$

46. $-10x + 15 > 18 - 11x$

Solve using the multiplication principle. Graph and write set-builder notation for the answers.

47. $5x < 35$

48. $8x \geqslant 32$

49. $9y \leqslant 81$

50. $10x > 240$

51. $7x < 13$

52. $8y < 17$

53. $12x > -36$

54. $16x < -64$

Solve using the multiplication principle. Write the answers in set-builder notation.

55. $5y \geqslant -2$

56. $7x > -4$

57. $-2x \leqslant 12$

58. $-3y \leqslant 15$

59. $-4y \geqslant -16$

60. $-7x < -21$

61. $-3x < -17$

62. $-5y > -23$

63. $-2y > \frac{1}{7}$

64. $-4x \leqslant \frac{1}{9}$

65. $-\frac{6}{5} \leqslant -4x$

66. $-\frac{7}{8} > -56t$

Solve using the addition and multiplication principles.

67. $4 + 3x < 28$

68. $5 + 4y < 37$

69. $6 + 5y \geqslant 36$

70. $7 + 8x \geqslant 71$

71. $3x - 5 \leqslant 13$

72. $5y - 9 \leqslant 21$

73. $13x - 7 < -46$

74. $8y - 4 < -52$

75. $5x + 3 \geqslant -7$

76. $7y + 4 \geqslant -10$

77. $13 < 4 - 3y$

78. $22 < 6 - 8x$

79. $30 > 3 - 9x$

80. $40 > 5 - 7y$

81. $3 - 6y > 23$

82. $8 - 2y > 14$

83. $-3 < 8x + 7 - 7x$

84. $-5 < 9x + 8 - 8x$

85. $6 - 4y > 4 - 3y$

86. $7 - 8y > 5 - 7y$

87. $5 - 9y \leqslant 2 - 8y$

88. $6 - 13y \leqslant 4 - 12y$

89. $21 - 8y < 6y + 49$

90. $33 - 12x < 4x + 97$

91. $27 - 11x > 14x - 18$

92. $42 - 13y > 15y - 19$

93. $2.1x + 45.2 > 3.2 - 8.4x$

94. $0.96y - 0.79 \leqslant 0.21y + 0.46$

95. $0.7n - 15 + n \geqslant 2n - 8 - 0.4n$

96. $1.7t + 8 - 1.62t < 0.4t - 0.32 + 8$

97. $\frac{x}{3} - 2 \leqslant 1$

98. $\frac{2}{3} - \frac{x}{5} < \frac{4}{15}$

99. $\frac{y}{5} + 1 \leqslant \frac{2}{5}$

100. $\frac{3x}{5} \geqslant -15$

101. $3(2y - 3) < 27$

102. $4(2y - 3) > 28$

103. $5(d + 4) \leqslant 7(d - 2)$

104. $3(t - 2) \geqslant 9(t + 2)$

105. $8(2t + 1) > 4(7t + 7)$

106. $7(5x - 2) < 6(6x - 1)$

107. $3(r - 6) + 2 < 4(r + 2) - 21$

108. $5(t + 3) + 9 > 3(t - 2) + 6$

109. $\frac{2}{3}(2x - 1) \geqslant 10$

110. $\frac{4}{5}(3x + 4) \leqslant 20$

111. $\frac{3}{4}\left(3x - \frac{1}{2}\right) - \frac{2}{3} < \frac{1}{3}$

112. $\frac{2}{3}\left(\frac{7}{8} - 4x\right) - \frac{5}{8} < \frac{3}{8}$

Skill Maintenance

Simplify.

113. $10 \div 2 \cdot 5 - 3^2 + (-4)^2$

114. $7 - 3^2 + (8 - 3)^2 \cdot 4$

Synthesis

115. ◈ Are all solutions of $x > -5$ solutions of $-x < 5$? Why or why not?

116. ◈ Explain in your own words why it is necessary to reverse the inequality symbol when multiplying both sides of an inequality by a negative number.

Solve.

117. $2[4 - 2(3 - x)] - 1 \geqslant 4[2(4x - 3) + 7] - 25$

118. $5[3(7 - t) - 4(8 + 2t)] - 20 < -6[2(6 + 3t) - 4]$

Solve for x.

119. $-(x + 5) \geqslant 4a - 5$

120. $\frac{1}{2}(2x + 2b) > \frac{1}{3}(21 + 3b)$

121. $y < ax + b$ (Assume $a > 0$.)

122. $y < ax + b$ (Assume $a < 0$.)

123. Determine whether each number is a solution of the inequality $|x| < 3$.

a) 0 **b)** -2
c) -3 **d)** 4
e) 3 **f)** 1.7
g) -2.8

124. Graph the solutions of $|x| < 3$ on a number line.

2.7 Problem Solving Using Inequalities

Translating to Inequalities • Solving Problems

The five steps for problem solving can be used for problems involving inequalities.

Translating to Inequalities

Before solving problems that involve inequalities, we list some important phrases to look for. Sample translations are listed as well.

Important Words	Sample Sentence	Translation
is at least	Bill is at least 21 years old.	$b \geq 21$
is at most	At most 5 students dropped the course.	$n \leq 5$
cannot exceed	To qualify, earnings cannot exceed $12,000.	$r \leq 12{,}000$
must exceed	The speed must exceed 15 mph.	$s > 15$
is less than	Spot's weight is less than 50 lb.	$w < 50$
is more than	Boston is more than 200 miles away.	$d > 200$
is between	The film was between 90 and 100 minutes long.	$90 < t < 100$

Solving Problems

EXAMPLE 1

Martha is taking an introductory algebra course in which four tests are to be given. To get an A, a student must average at least 90 on the four tests. Martha got scores of 96, 82, and 91 on the first three tests. Determine (in terms of an inequality) what scores on the last test will earn her an A.

Solution

1. FAMILIARIZE. Many students in Martha's situation might make a guess, like 87, and then compute the average of the four scores. The average of the four scores is their sum divided by the number of tests, 4:

$$\frac{96 + 82 + 91 + 87}{4} = 89.$$

Our work shows that an 87 will not earn Martha her A, since the average is not *at least* 90. Rather than make a second guess, let's translate to an inequality and solve. We let x = Martha's score on the last test.

2. TRANSLATE. Having familiarized ourselves with the meaning of average,

we can reword and translate, using x as the fourth score:

Rewording: The average of the four scores must be at least 90

Translating: $\dfrac{96 + 82 + 91 + x}{4} \geq 90.$

3. **CARRY OUT.** To solve the inequality, we first multiply by 4 to clear fractions:

$$\frac{96 + 82 + 91 + x}{4} \geq 90$$

$$4\left(\frac{96 + 82 + 91 + x}{4}\right) \geq 4 \cdot 90 \qquad \text{Multiplying by 4}$$

$$96 + 82 + 91 + x \geq 360$$

$$269 + x \geq 360 \qquad \text{Simplifying}$$

$$x \geq 91.$$

The solution set is $\{x | x \geq 91\}$.

4. **CHECK.** We can obtain a partial check by substituting a number greater than or equal to 91. We leave it to the student to try 92 in a manner similar to what was done in the familiarization step.

5. **STATE.** A score of 91 or better on the last test will give Martha an A in the course. ❑

EXAMPLE 2

The women's volleyball team can spend at most $400 for its awards banquet at a local restaurant. If the restaurant charges a $45 set-up fee plus $12.50 per person, at most how many can attend?

Solution

1. **FAMILIARIZE.** Suppose 20 people were to attend the dinner. The cost would then be $45 + $12.50 · 20, or $295. This shows that more than 20 people could attend without exceeding $400. We could next make another guess, or subtract $295 from $400 and ''use up'' the difference. Instead, we translate to an inequality and solve. We let n = the number of people in attendance.

2. **TRANSLATE.** The cost of the banquet will be $45 for the set-up fee plus $12.50 times the number of people attending. We can reword as follows:

Rewording: The set-up fee plus the cost of the dinners cannot exceed $400.

Translating: $45 + 12.50 \cdot n \leq 400.$

3. **CARRY OUT.** We solve the inequality:

$$45 + 12.50n \leq 400$$

$$12.5n \leq 355 \qquad \text{Subtracting 45}$$

$$n \leq \frac{355}{12.5} \qquad \text{Dividing by 12.5}$$

$$n \leq 28.4 \qquad \text{Simplifying}$$

The solution set of the inequality is $\{n \mid n \leq 28.4\}$.

4. **CHECK.** Although the solution set of the inequality is all numbers less than or equal to 28.4, since n represents the number of people in attendance, we round down to 28. If 28 people attend, the cost will be $\$45 + \$12.50 \cdot 28$, or \$395, and if 29 attend, the cost will exceed \$400.

5. **STATE.** At most 28 people can attend the banquet. ❑

In the check to Example 2, an important point is made: Solutions of equations or inequalities do not always solve the problem from which the equation or inequality originates. In some cases, answers must be nonnegative, and in other cases, answers must be integers. Thus it is important to always check that the original problem has been solved.

EXERCISE SET 2.7

Translate to an inequality.

1. A number is greater than 4.
2. A number is less than 7.
3. A number is less than or equal to -6.
4. A number is greater than or equal to 13.
5. The temperature is at most 80°.
6. The bag weighs at least 2 pounds.
7. Between 75 and 100 people attended the concert.
8. The average speed was between 90 and 110 mph.
9. The number of people is at least 1200.
10. The cost is at most \$3457.95.
11. The amount of acid is not to exceed 500 liters.
12. The cost of gasoline is no less than 99 cents per gallon.
13. Two more than 3 times a number is less than 13.
14. Five less than one half of a number is greater than 17.

Solve. Use the five steps for problem solving.

15. Your quiz grades are 73, 75, 89, and 91. What scores on a fifth quiz will make your average quiz grade at least 85? Use an inequality.
16. Alvin is taking a literature course in which four tests are to be given. To get a B, a student must average at least 80 on the four tests. Alvin got scores of 82, 76, and 78 on the first three tests. What scores on the last test will earn him at least a B? Use an inequality.
17. Ridem rents trucks at a daily rate of \$42.95 plus \$0.46 per mile. A family wants a one-day truck rental, but must stay within a budget of \$200. What mileages will allow the family to stay within budget? Use an inequality and round to the nearest tenth of a mile.
18. Atlas rents a cargo van at a daily rate of \$44.95 plus \$0.39 per mile. A business has budgeted \$250 for a one-day van rental. What mileages will allow the business to stay within budget? Use an inequality and round to the nearest tenth of a mile.
19. The width of a rectangle is fixed at 4 cm. For what lengths will the area be less than 86 cm^2? Use an inequality.

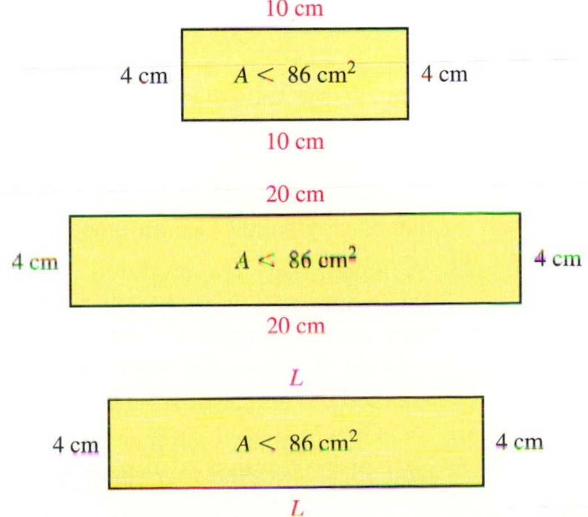

20. The width of a rectangle is fixed at 16 yd. For what lengths will the area be at least 264 yd^2? Use an inequality.

21. Laura is certain that every time she parks in the municipal garage it costs her at least \$2.20. If the garage charges 45¢ plus 25¢ for each half hour, how long is Laura's car parked? Use an inequality.

22. Simon claims that it costs him at least \$3.00 every time he calls a customer from a pay phone. If a typical call costs 75¢ plus 45¢ for each minute, how long do his calls last? Use an inequality.

23. The formula $R = -0.075t + 3.85$ can be used to predict the world record in the 1500-m run t years after 1930. Determine (in terms of an inequality) those years for which the world record will be less than 3.5 min.

24. The formula $R = -0.028t + 20.8$ can be used to predict the world record in the 200-m dash t years after 1920. Determine (in terms of an inequality) those years for which the world record will be less than 19.0 sec.

25. A 9-lb puppy is gaining weight at a rate of $\frac{3}{4}$ lb per week. When will the puppy's weight exceed $22\frac{1}{2}$ lb? Use an inequality.

26. On July 1, Garrett's Pond was 25 ft deep. Since that date, the water level has dropped $\frac{2}{3}$ ft per week. For what dates will the water level not exceed 21 ft? Use an inequality.

27. Butter stays solid at Fahrenheit temperatures below 88°. The formula $F = \frac{9}{5}C + 32$ can be used to convert Celsius temperatures C to Fahrenheit temperatures F. For what Celsius temperatures does butter stay solid? Use an inequality.

28. A human body is considered to be feverish when its temperature is higher than 98.6°F. Using the formula in Exercise 27, determine (in terms of an inequality) those Celsius temperatures for which the body is feverish.

29. Find all numbers such that the sum of the number and 15 is less than four times the number.

30. Find all numbers such that three times the number minus ten times the number is greater than or equal to eight times the number.

31. The length of a rectangle is fixed at 26 cm. What widths will make the perimeter greater than 80 cm?

32. The width of a rectangle is fixed at 8 ft. What lengths will make the perimeter at least 200 ft? at most 200 ft?

33. One side of a triangle is 2 cm shorter than the base. The other side is 3 cm longer than the base. What lengths of the base will allow the perimeter to be greater than 19 cm?

34. The perimeter of a rectangular swimming pool is not to exceed 70 ft. The length is to be twice the width. What widths will meet these conditions?

35. Dot's Electric made 17 customer calls last week and 22 calls this week. How many calls must be made next week in order to maintain an average of at least 20 for the three-week period?

36. George and Joan do volunteer work at a hospital. Joan worked 3 more hours than George, and together they worked more than 27 hours. What possible numbers of hours did each work?

37. Angelo is shopping for a new pair of jeans and two sweaters of the same kind. He is determined to spend no more than \$120.00 for the clothes. He buys jeans for \$21.95. What is the most that Angelo can spend for each sweater?

38. The medium-sized box of dog food weighs 1 lb more than the small size. The large size weighs 2 lb more than the small size. The total weight of the three boxes is at most 30 lb. What are the possible weights of the small box?

39. The width of a rectangle is 32 km. What lengths will make the area at least 2048 km^2?

40. The height of a triangle is 20 cm. What lengths of the base will make the area at most 40 cm^2?

41. The average price of a movie ticket can be estimated by the equation $P = 0.1522Y - 298.592$, where Y represents the year and P the average price, in dollars. The price is lower than what might be expected due to senior-citizen discounts, children's prices, and special volume discounts. For what years will the average price of a movie ticket be at least $6?

42. The equation $y = 0.027x + 0.19$ can be used to determine the approximate cost y, in dollars, of driving x miles on the Indiana toll road. For what mileages x will the cost be at most $6?

Skill Maintenance

Simplify.

43. $-3 + 2(-5)^2(-3) - 7$

44. $7 - a^2 - 9 + 5a^2$

45. $9x - 5 + 4x^2 - 2 - 13x$

46. $3x + 2[4 - 5(2x - 1)]$

Synthesis

47. ◈ The symbols $\nless$, $\ngtr$, $\nleqslant$, and $\ngeqslant$ have not been discussed. Do you feel that this was an oversight? Why or why not?

48. ◈ Suppose that t = Todd's age and f = Frances's age. Write a sentence that would translate to the inequality

$$t \geqslant f + 10.$$

49. The area of a square can be no more than 64 cm^2. What lengths of a side will allow this?

50. The sum of two consecutive odd integers is less than 100. What is the largest possible pair of such integers?

51. Mack's Parking Garage charges $4.00 for the first hour and $2.50 for each additional hour. For how long has a car been parked when the charge exceeds $16.50?

52. A salesperson can choose to be paid in one of two ways.

Plan A: A salary of $600 per month, plus a commission of 4% of gross sales

Plan B: A salary of $800 per month, plus a commission of 6% of gross sales over $10,000

For what gross sales is plan A better than plan B, assuming that gross sales are always more than $10,000?

53. When asked how much the parking charge is for a certain car (see Exercise 51), Mack replies "between 14 and 24 dollars." For how long has the car been parked?

54. Green ski wax works best at Fahrenheit temperatures between 5° and 15°. Determine those Celsius temperatures for which green ski wax works best. (See Exercise 27.)

55. ◈ Chassman and Bem Booksellers offers a preferred customer card for $25. The card entitles a customer to a 10% discount on all purchases for a period of one year. Under what circumstances would an individual save money by purchasing a card?

56. ◈ After 9 quizzes, Jackie's average is 84. Is it possible for Jackie to improve her average two points with the next quiz? Why or why not?

SUMMARY AND REVIEW 2

KEY TERMS

Equivalent equations, p. 63
Coefficient, p. 66
Clearing fractions and decimals, p. 71
Formula, p. 75
Consecutive integers, p. 86
Graph of an inequality, p. 95
Solution set, p. 97
Set-builder notation, p. 97

IMPORTANT PROPERTIES AND FORMULAS

Solving Equations

The Addition Principle: For any real numbers a, b, and c,
if $a = b$, then $a + c = b + c$.

The Multiplication Principle: For any real numbers a, b, and c,
if $a = b$, then $ac = bc$.

Solving Inequalities

The Addition Principle: For any real numbers a, b, and c,
if $a < b$, then $a + c < b + c$, and
if $a > b$, then $a + c > b + c$.

The Multiplication Principle: For any real numbers a and b and any *positive* number c,
if $a < b$, then $ac < bc$, and
if $a > b$, then $ac > bc$.

For any real numbers a and b and any *negative* number c,
if $a < b$, then $ac > bc$, and
if $a > b$, then $ac < bc$.

An Equation-Solving Procedure

1. If necessary, use the distributive law to remove parentheses. Then collect like terms on each side.
2. Multiply on both sides to clear fractions or decimals. (This is optional, but it can ease computations.)
3. Get all terms with variables on one side and all constant terms on the other side, using the addition principle.
4. Collect like terms again, if necessary.
5. Multiply or divide to solve for the variable, using the multiplication principle.
6. Check all possible solutions in the original equation.

Percent Notation

$n\%$ means $n \times 0.01$, or $n \times \frac{1}{100}$, or $\frac{n}{100}$.

Five Steps for Problem Solving in Algebra

1. *Familiarize* yourself with the problem situation.
2. *Translate* to mathematical language. (This often means write an equation.)
3. *Carry out* some mathematical manipulation. (This often means *solve* an equation.)
4. *Check* your possible answer in the original problem.
5. *State* the answer clearly.

REVIEW EXERCISES

This chapter's review and test include Skill Maintenance exercises from Sections 1.1, 1.2, 1.7, and 1.8.

Solve.

1. $x + 5 = -17$
2. $-8x = -56$
3. $-\frac{x}{4} = 48$
4. $n - 7 = -6$
5. $15x = -35$
6. $x - 11 = 14$
7. $-\frac{2}{3} + x = -\frac{1}{6}$
8. $\frac{4}{5}y = -\frac{3}{16}$
9. $y - 0.9 = 9.09$
10. $5 - x = 13$
11. $5t + 9 = 3t - 1$
12. $7x - 6 = 25x$
13. $\frac{1}{4}x - \frac{5}{8} = \frac{3}{8}$
14. $14y = 23y - 17 - 10$
15. $0.22y - 0.6 = 0.12y + 3 - 0.8y$
16. $\frac{1}{4}x - \frac{1}{8}x = 3 - \frac{1}{16}x$
17. $4(x + 3) = 36$
18. $3(5x - 7) = -66$
19. $8(x - 2) = 5(x + 4)$
20. $-5x + 3(x + 8) = 16$
21. $C = \pi d$, for d
22. $V = \frac{1}{3}Bh$, for B
23. $A = \frac{a + b}{2}$, for a
24. Find decimal notation: 0.7%.
25. Find percent notation: $\frac{11}{25}$.
26. What percent of 60 is 12?
27. 198 is 55% of what number?

Determine whether the given number is a solution of the inequality $x \leq 4$.

28. -3
29. 7
30. 4

Graph on a number line.

31. $4x - 6 < x + 3$
32. $-2 < x \leq 5$
33. $y > 0$

Solve. Write the answers in set-builder notation.

34. $y + \frac{2}{3} \geq \frac{1}{6}$
35. $9x \geq 63$
36. $2 + 6y > 14$
37. $7 - 3y \geq 27 + 2y$
38. $3x + 5 < 2x - 6$
39. $-4y < 28$
40. $3 - 4x < 27$
41. $4 - 8x < 13 + 3x$
42. $-3y \geq -21$
43. $-4x \leq \frac{1}{3}$
44. A color television sold for $629 in May. This was $38 more than the cost in January. Find the cost in January.
45. Selma gets a $4 commission for each appliance that she sells. One week she received $108 in commissions. How many appliances did she sell?
46. An 8-m board is cut into two pieces. One piece is 2 m longer than the other. How long are the pieces?
47. If 14 is added to three times a certain number, the result is 41. Find the number.
48. The sum of two consecutive odd integers is 116. Find the integers.
49. The perimeter of a rectangle is 56 cm. The width is 6 cm less than the length. Find the width and the length.

50. After a 30% reduction, an item is on sale for \$154. What was the marked price (the price before reducing)?

51. A businessperson's salary is \$30,000, which is a 15% increase over the previous year's salary. What was the previous salary (to the nearest dollar)?

52. The measure of the second angle of a triangle is 50° more than that of the first. The measure of the third angle is 10° less than twice the first. Find the measures of the angles.

53. Steve's quiz grades are 71, 75, 82, and 86. What is the lowest grade that he can get on the next quiz and still have an average of at least 80?

54. The length of a rectangle is 43 cm. What widths will make the perimeter greater than 120 cm?

Skill Maintenance

55. Evaluate $\frac{x-y}{5}$ when $x = 27$ and $y = 2$.

56. Multiply: $4(3t + 2 + s)$.

57. Divide: $12.42 \div (-5.4)$.

58. Remove parentheses and simplify:

$$5x - 8(6x - y).$$

Synthesis

59. ◈ What is the difference between using the multiplication principle for solving equations and for solving inequalities?

60. ◈ Explain how checking the solutions of an equation differs from checking the solutions of an inequality.

61. The total length of the Nile and Amazon Rivers is 13,108 km. If the Amazon were 234 km longer, it would be as long as the Nile. Find the length of each river.

62. Consumer experts advise us never to pay the sticker price for a car. A rule of thumb is to pay the sticker price minus 20% of the sticker price, plus \$200. A car is purchased for \$11,520 using the rule. What was the sticker price?

Solve.

63. $2|n| + 4 = 50$

64. $|3n| = 60$

65. $y = 2a - ab + 3$, for a

CHAPTER TEST 2

Solve.

1. $x + 7 = 15$

2. $t - 9 = 17$

3. $3x = -18$

4. $-\frac{4}{7}x = -28$

5. $3t + 7 = 2t - 5$

6. $\frac{1}{2}x - \frac{3}{5} = \frac{2}{5}$

7. $8 - y = 16$

8. $-\frac{2}{5} + x = -\frac{3}{4}$

9. $3(x + 2) = 27$

10. $-3x + 6(x + 4) = 9$

11. $0.4p + 0.2 = 4.2p - 7.8 - 0.6p$

Solve. Write the answers in set-builder notation.

12. $x + 6 \leqslant 2$

13. $14x + 9 > 13x - 4$

14. $12x \leqslant 60$

15. $-2y \geqslant 26$

16. $-4y \leqslant -32$

17. $-5x \geqslant \frac{1}{4}$

18. $4 - 6x > 40$

19. $5 - 9x \geqslant 19 + 5x$

Solve the formula for the given letter.

20. $A = 2\pi rh$, for r

21. $w = \frac{P - 2l}{2}$, for l

22. Find decimal notation: 200%.

23. Find percent notation: 0.054.

24. What number is 42% of 50?

25. What percent of 75 is 33?

Graph on a number line.

26. $y < 9$

27. $-2 \leqslant x \leqslant 2$

Solve.

28. The perimeter of a rectangle is 36 cm. The length is 4 cm greater than the width. Find the width and the length.

29. If you triple a number and then subtract 14, you get two thirds of the original number. What is the original number?

30. The sum of three consecutive odd integers is 249. Find the integers.

31. Money is invested in a savings account at 6% simple interest. After 1 year, there is \$2650 in the account. How much was originally invested?

32. Find all numbers such that six times the number is greater than the number plus 30.

33. The width of a rectangle is 96 yd. Find all possible lengths so that the perimeter of the rectangle will be at least 540 yd.

Skill Maintenance

34. Translate to an algebraic expression: 10 less than x.

35. Factor: $3a + 24b + 12$.

36. Multiply: $-\frac{3}{8} \cdot \left(-\frac{4}{5}\right)$.

37. Simplify: $8 + (-10) \div 5 \cdot 2 + 1$.

Synthesis

38. Solve $c = \dfrac{1}{a - d}$, for d.

39. Solve: $3|w| - 8 = 37$.

40. A movie theater had a certain number of tickets to give away. Five people got the tickets. The first got one third of the tickets, the second got one fourth of the tickets, and the third got one fifth of the tickets. The fourth person got eight tickets, and there were five tickets left for the fifth person. Find the total number of tickets given away.

CHAPTER 3

Introduction to Graphing

AN APPLICATION

A smoker is 15 times more likely to die from lung cancer than a nonsmoker. An exsmoker who stopped smoking t years ago is w times more likely to die from lung cancer than a nonsmoker, where $t + w = 15$.

a) Sandy gave up smoking $2\frac{1}{2}$ years ago. How much more likely is she to die from lung cancer than Polly, who never smoked?

b) Graph the equation, using t as the first coordinate.

This problem appears as Exercise 21 in Section 3.4.

Don Kissack
DIRECTOR OF BENEFITS

"Corporations are increasingly looking at how factors like diet, exercise, smoking, alcohol, and seat belt use affect the welfare of employees and the health-care costs for a business.

"Algebra and graphing are important tools in human resources, especially in the areas of compensation and benefits."

We now begin our study of graphing. First we will examine graphs as they commonly appear in newspapers or magazines and develop some terminology. Following that, we will practice graphing equations whose graphs are lines. Finally we will practice using graphs as a problem-solving tool for certain applications.

In addition to material from this chapter, the review and test for Chapter 3 include material from Sections 1.3, 1.4, 2.2, and 2.3.

3.1 Ordered Pairs and Graphs

Problem Solving with Graphs • Points and Ordered Pairs • Quadrants • Finding Coordinates

Today's print and electronic media make almost constant use of graphs. This is due in part to the ease with which some graphs are prepared by computer, and in part to the large quantity of information that a graph can display. We first consider problem solving with bar graphs, line graphs, and circle graphs. Then we examine graphs that use a coordinate system.

Problem Solving with Graphs

Bar Graphs

A *bar graph* is convenient for showing comparisons. The bars can be either vertical or horizontal. Typically, certain units, such as body weight in the graph of Example 1, are shown horizontally. With each horizontal number, there is associated a vertical number, or unit. In Example 1, the vertical unit is the number of drinks required to raise the blood alcohol level to a point at which driving is illegal in all 50 states, 0.10%.

EXAMPLE 1

Although some states use a cut-off of 0.08%, in *all* states, a blood-alcohol level of 0.10% or higher indicates that an individual has consumed too much alcohol to drive. The following bar graph* shows the number of drinks that a person of a certain

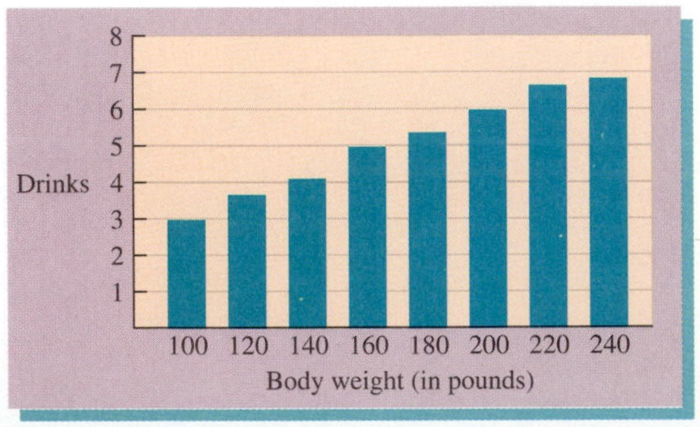

*Adapted from *Neighborhood Digest*, Vol. 7, No. 12.

weight would consume to achieve a blood-alcohol level of 0.10%. Note that a 12-ounce beer, a 5-ounce glass of wine, or a cocktail containing $1\frac{1}{2}$ ounces of distilled liquor all count as one drink.

a) Approximately how many drinks would a 200-pound person have consumed in order to have a blood-alcohol level of 0.10%?

b) At least how much would an individual have to weigh in order to consume 4 drinks without reaching a blood-alcohol level of 0.10%?

Solution

a) We go to the top of the bar that is above the body weight 200 lb. Then we move horizontally from the top of the bar to the vertical scale listing numbers of drinks. It appears that approximately 6 drinks will give a 200-lb person a blood-alcohol level of 0.10%.

b) By moving up the vertical scale to the number 4, and then moving horizontally, we see that the first bar to reach a height of 4 corresponds to a weight of 140 lb. An individual should weigh at least 140 lb if he or she wishes to consume 4 drinks without exceeding a blood-alcohol level of 0.10%. ❑

EXAMPLE 2

These reasons for dropping out of high school were given in a recent National Assessment of Educational Progress survey.

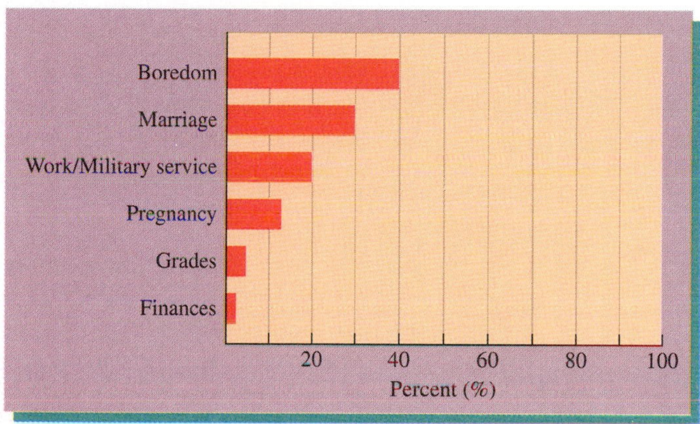

a) Approximately what percent of those surveyed dropped out because of pregnancy?

b) What reason for dropping out was given least often?

c) What reason for dropping out was given by about 30% of those surveyed?

Solution

a) We go to the right of the bar representing pregnancy and then look down to the percent scale. We find that approximately 12% dropped out because of pregnancy. Note that this is only an estimate.

b) The reason given least often is finances, since that is represented by the shortest bar.

c) We locate the 30% mark on the percent scale and then look up for a bar ending at approximately 30%.We then go across to the left and read the reason. The reason given by about 30% was marriage. ❑

Line Graphs

Line graphs are often used to show change over time. Certain points are drawn to represent given information. When segments are drawn to connect the points, a line graph is formed.

Sometimes it is impractical to begin the listing of horizontal or vertical values with zero. When this occurs, as in Example 3, the symbol ⌇ is used to indicate a break in the listing of values.

EXAMPLE 3

The following line graph shows the relationship between a person's resting pulse rate and months of regular exercise.*

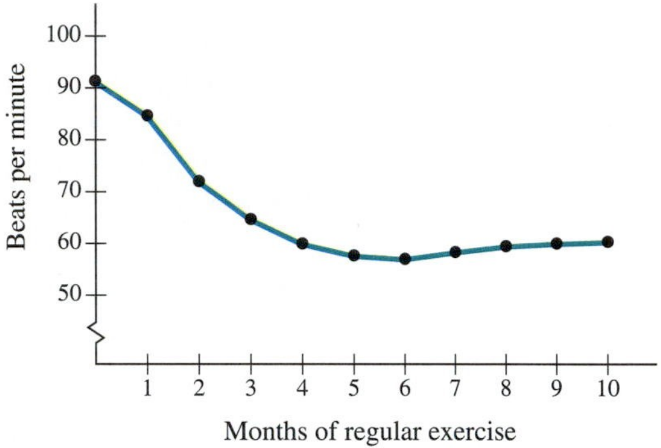

a) How many months of regular exercise are required to lower the pulse rate as much as possible?

b) How many months of regular exercise are needed to achieve a pulse rate of 65 beats per minute?

Solution

a) The lowest point on the graph occurs above the number 6. After 6 months of regular exercise, the pulse rate is lowered as much as possible.

b) We locate 65 on the vertical scale and then move right until the line is reached. At that point, we move down to the horizontal scale and read the information we are seeking.

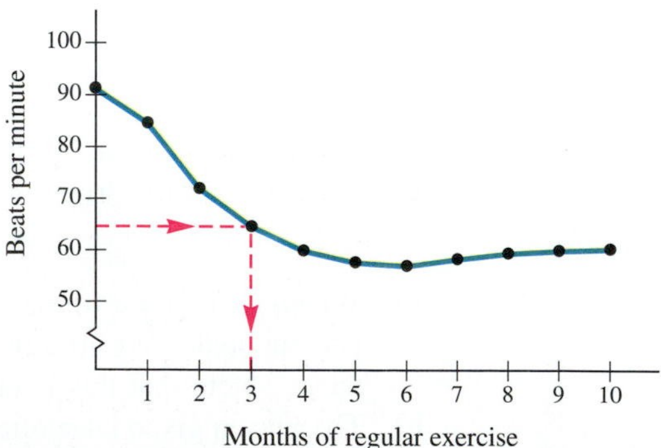

The pulse rate is 65 beats per minute after 3 months of regular exercise. ❑

*Data from *Body Clock* by Dr. Martin Hughes, p. 60. New York: Facts on File, Inc.

Circle Graphs

Circle graphs, or *pie charts,* are often used to show what percent of a whole each particular item in a group represents.

EXAMPLE 4

This pie chart shows expenses as a percent of income for a family of four, according to the Bureau of Labor Statistics.

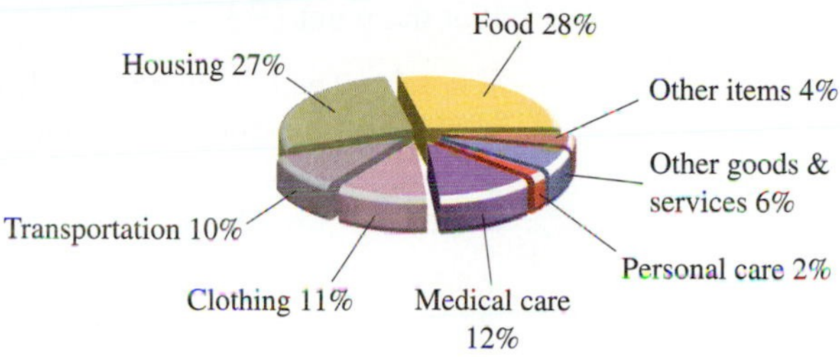

A family with a monthly income of \$2500 would typically spend how much on housing per month?

Solution

1. **FAMILIARIZE.** The chart tells us that housing is 27% of income. We let $y =$ the amount spent on housing.
2. **TRANSLATE.** We reword and translate the problem as follows:

   Reword: What is 27% of income?

   Translate: $y = 27\% \cdot \$2500$
3. **CARRY OUT.** We solve by carrying out the computation:

   $y = 0.27 \cdot \$2500 = \$675.$
4. **CHECK.** We leave the check to the student.
5. **STATE.** The family would spend \$675 on housing. ❑

Points and Ordered Pairs

We have already graphed numbers on a number line. In order to graph an equation containing two variables, we must learn to graph pairs of numbers on a plane.

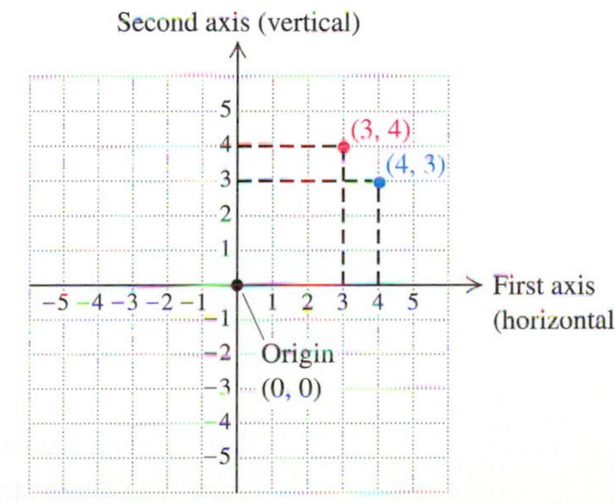

To graph pairs of numbers on a plane, we use two perpendicular number lines called **axes** (singular, **axis**). The axes cross at a point called the **origin.** Arrows on the axes show the positive directions.

Consider the pair (3, 4). The numbers in such a pair are called **coordinates.** In (3, 4), the **first coordinate** is 3 and the **second coordinate** is 4. To plot (3, 4), we start at the origin and move horizontally to the 3. Then we move up vertically 4 units and make a "dot." Note that (3, 4) is located above 3 on the first axis and to the right of 4 on the second axis.

The point (4, 3) is also plotted in the figure on the preceding page. Note that (3, 4) and (4, 3) give different points. They are called **ordered pairs** because the order in which the numbers appear is important.

EXAMPLE 5

Plot the point $(-3, 4)$.

Solution The first number, -3, is negative. Starting at the origin, we move 3 units in the negative horizontal direction (3 units to the left). The second number, 4, is positive, so we move 4 units in the positive vertical direction (up). The point $(-3, 4)$ is above -3 on the first axis and to the left of 4 on the second axis.

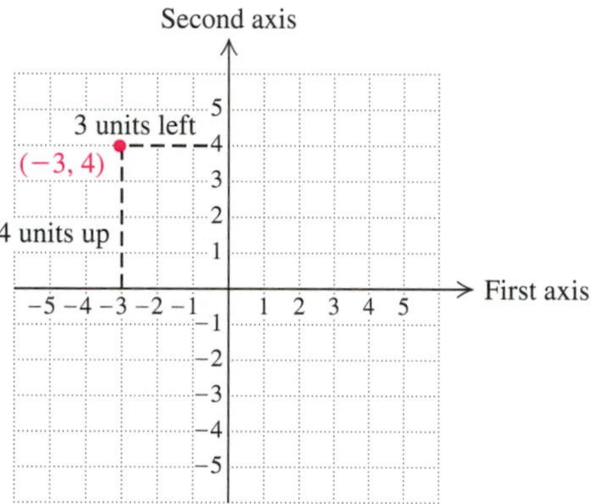

Quadrants

The following figure shows some points and their coordinates. In region I (the *first quadrant*), both coordinates of any point are positive. In region II (the *second quadrant*), the first coordinate is negative and the second is positive. In region III (the *third quadrant*), both coordinates are negative. In region IV (the *fourth quadrant*), the first coordinate is positive and the second is negative.

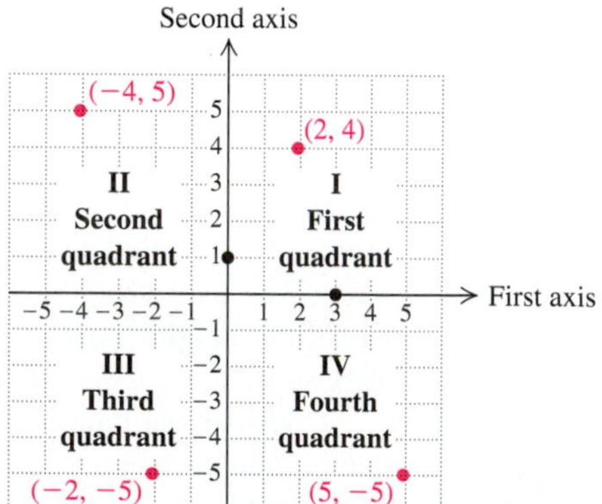

Note that the point $(-4, 5)$ is in the second quadrant and the point $(5, -5)$ is in the fourth quadrant. The points (3, 0) and (0, 1) are on the axes and are not considered to be in any quadrant.

Finding Coordinates

To find the coordinates of a point, we see how far to the right or left of the vertical axis it is and how far above or below the horizontal axis it is.

EXAMPLE 6 Find the coordinates of points A, B, C, D, E, F, and G.

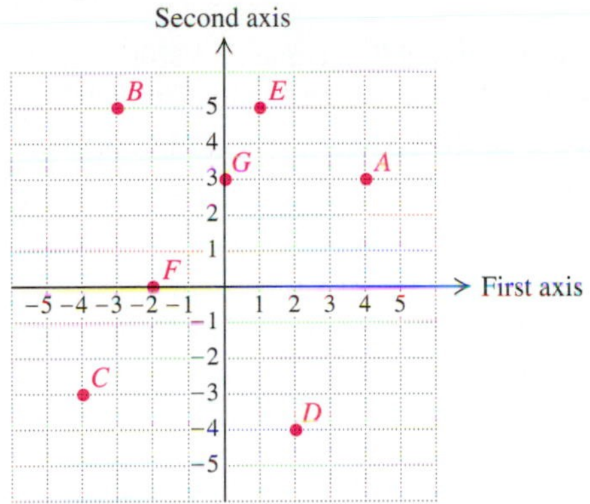

Solution Point A is 4 units to the right (horizontal direction) and 3 units up (vertical direction). Its coordinates are (4, 3). The coordinates of the other points are as follows:

B: $(-3, 5)$; C: $(-4, -3)$; D: $(2, -4)$;
E: $(1, 5)$; F: $(-2, 0)$; G: $(0, 3)$. □

EXERCISE SET 3.1

Use the bar graph in Example 1 to answer Exercises 1–4.

1. Approximately how many drinks would it take for a 100-lb person to reach a blood-alcohol level of 0.10%?
2. Approximately how many drinks would it take for a 160-lb person to reach a blood-alcohol level of 0.10%?
3. At least how much does an individual weigh if she has consumed $3\frac{1}{2}$ drinks without reaching a blood-alcohol level of 0.10%?
4. At least how much does an individual weigh if he has consumed $4\frac{1}{2}$ drinks without reaching a blood-alcohol level of 0.10%?

Use the bar graph in Example 2 to answer Exercises 5–8.

5. What reason was given most often for dropping out?
6. What reason was given by about 20% for dropping out?
7. Approximately what percent in the survey dropped out because of grades?
8. Approximately what percent in the survey dropped out because of boredom?

Use the line graph in Example 3 to answer Exercises 9–12.

9. Approximately how many months of regular exercise are needed to achieve a pulse rate of 85 beats per minute?
10. Approximately how many months of regular exercise are needed to achieve a pulse rate of 56 beats per minute?
11. What month caused the greatest drop in pulse rate?
12. During what month did the pulse rate first drop below 60?

Use the pie chart in Example 4 to answer Exercises 13–16.

13. What percent of income is spent on medical expenses?

14. A family with a monthly income of $2000 would typically spend how much on clothing per month?

15. A family with a monthly income of $2400 would typically spend how much on food per month?

16. What percent of the income is spent on food and housing combined?

Use the following line graphs to answer Exercises 17–26. The graphs show the percentages of the Gross National Product (GNP) spent on health, education, and defense over a period of 40 years.*

"Health vs. Education vs. Defense." Copyright 1992 by Consumers Union of U.S., Inc., Yonkers, NY 10703–1057. Adapted with permission from CONSUMER REPORTS, July 1992. Although this material originally appeared in CONSUMER REPORTS, the selective adaptation and resulting conclusions presented are those of the author(s) and are not sanctioned or endorsed in any way by Consumers Union, the publisher of CONSUMER REPORTS.

Percentage of GNP 1950-1990

Elementary Algebra

Defense expenditures

All health care expenditures

Public education expenditures

14%, 12%, 10%, 8%, 6%, 4%, 2%, 0%

1950 '55 '60 '65 '70 '75 '80 '85 '90

Sources: U.S. National Center for Education Statistics, Health Care Financing Administration, U.S. Office of Management and Budget

17. Approximately what percent of the GNP was spent on public education in 1965?

18. Approximately what percent of the GNP was spent on defense in 1970?

19. In what year did health care costs represent about 10% of the GNP?

20. In what year did health care costs exceed 12% of the GNP?

*From *Consumer Reports*, July 1992, p. 439. Reprinted with permission.

21. In what three years were defense expenditures approximately 8% of the GNP?

22. In what year did public education expenditures (as a percent of GNP) peak?

23. In what year did defense expenditures (as a percent of GNP) peak?

24. In what two years were public education expenditures about $4\frac{1}{2}\%$?

25. Approximately how much did health care expenditures (as a percent of GNP) grow over the years 1970–1990?

26. Approximately how much did defense expenditures (as a percent of GNP) grow over the years 1950–1990?

Use the following circle graph to answer Exercises 27–32.

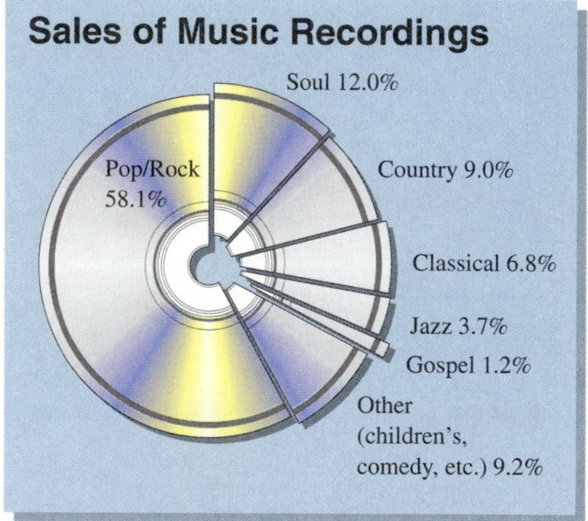

Source: National Association of Recording Merchandisers

27. What percent of all recordings sold are jazz?

28. What percent of all recordings sold are country?

29. Together, what percent of all recordings sold are either soul or pop/rock?

30. Together, what percent of all recordings sold are either classical or jazz?

31. A music store sells 3000 recordings a month. How many would you expect to be pop/rock? soul? country?

32. A music store sells 2500 recordings a month. How many would you expect to be pop/rock? classical? gospel?

33. Plot these points.

$(2, 5), (-1, 3), (3, -2), (-2, -4),$
$(0, 4), (0, -5), (5, 0), (-5, 0)$

34. Plot these points.

(4, 4), (−2, 4), (5, −3), (−5, −5), (0, 4), (0, −4), (3, 0), (−4, 0)

In which quadrant is the point located?

35. (−5, 3)
36. (−12, 1)
37. (100, −1)
38. (35.6, −2.5)
39. (−6, −29)
40. (−3.6, −105.9)
41. (3.8, 9.2)
42. (1895, 1492)

43. In quadrant III, first coordinates are always ________ and second coordinates are always ________.

44. In quadrant II, ________ coordinates are always positive and ________ coordinates are always negative.

45. Find the coordinates of points A, B, C, D, and E.

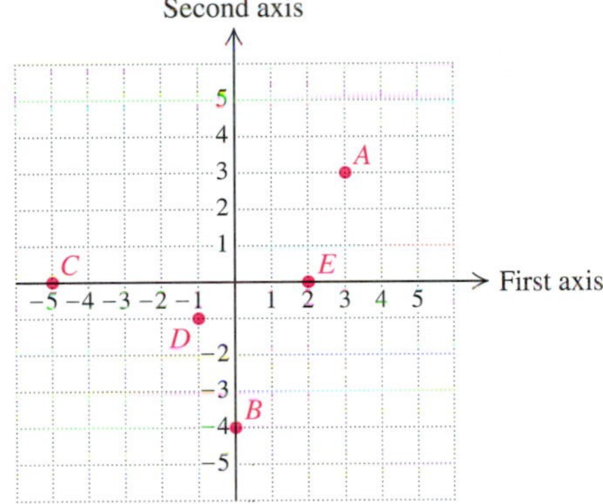

46. Find the coordinates of points A, B, C, D, and E.

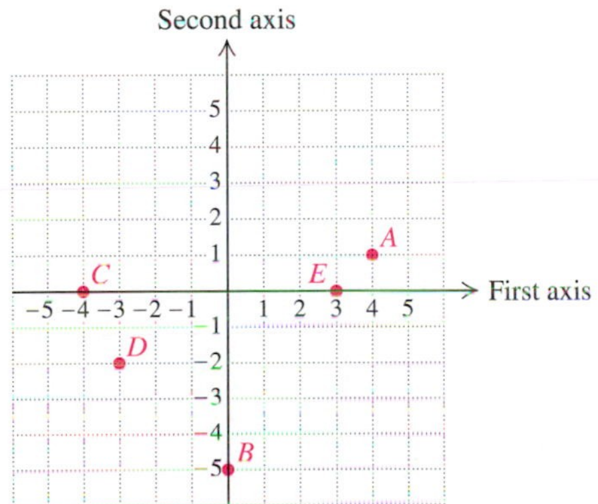

Skill Maintenance

Perform the indicated operation and simplify.

47. $\frac{3}{5} \cdot \frac{10}{9}$
48. $\frac{2}{3} + \frac{1}{5}$
49. $\frac{3}{7} - \frac{4}{5}$
50. $-\frac{2}{3} \div 5$

Synthesis

51. ◈ The graph accompanying Example 3 flattens out. Explain why the graph does not continue to slope downward.

52. ◈ Loreena and Phil's yearly income totals $29,000. They are considering renting an apartment for $700 per month. In light of Example 4, would you recommend that they rent the apartment? Why or why not?

In Exercises 53–56, tell in which quadrant(s) the given point could be located.

53. The first coordinate is positive.

54. The second coordinate is negative.

55. The first and second coordinates are equal.

56. The first coordinate is the additive inverse of the second coordinate.

57. The points (−1, 1), (4, 1), and (4, −5) are three vertices of a rectangle. Find the coordinates of the fourth vertex.

58. Three parallelograms share the vertices (−2, −3), (−1, 2), and (4, −3). Find the fourth vertex of each parallelogram.

59. Graph eight points such that the sum of the coordinates in each pair is 6.

60. Graph eight points such that the first coordinate minus the second coordinate is 1.

61. Find the perimeter of a rectangle whose vertices have coordinates (5, 3), (5, −2), (−3, −2), and (−3, 3).

62. Find the area of a triangle whose vertices have coordinates (0, 9), (0, −4), and (5, −4).

Coordinates on the globe. Coordinates can also be used to describe the location of three-dimensional objects: 0° latitude is the equator and 0° longitude is a line from the North Pole to the South Pole through France and Algeria. In the figure below, the hurricane Clara is at a point about 260 mi northwest of Bermuda near latitude 36.0° North, longitude 69.0° West.

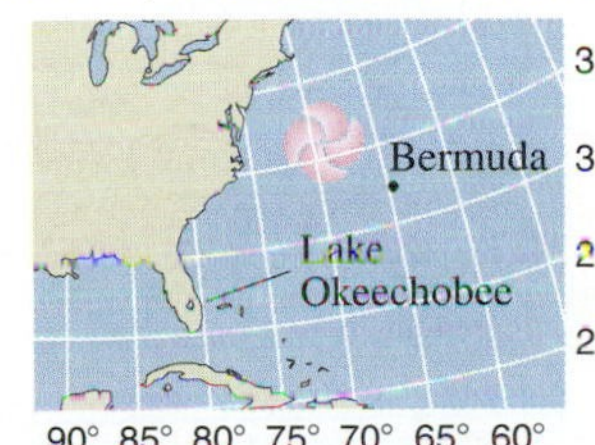

63. Approximate the latitude and the longitude of Bermuda.

64. Approximate the latitude and the longitude of Lake Okeechobee.

3.2 Graphing Linear Equations

Solutions of Equations • Graphing Equations of the Type $y = mx$ and $y = mx + b$

We have seen how bar, line, and circle graphs are used and how points are plotted using a coordinate system. Now we begin to learn how graphs can be used to represent solutions of equations.

Equations like $5x + 2y = 7$, $3y = 4 - 2x$, and $y = \frac{2}{3}x + 1$ are called **linear equations.** In general, any equation that can be written in the form $Ax + By = C$, where A, B, and C are real numbers with A and B not both zero, is a linear equation. We will find that when the solutions of a linear equation are graphed, the result is a straight line.

Solutions of Equations

When an equation contains two variables, solutions must be ordered pairs in which each number in the pair replaces a letter in the equation. Unless directed otherwise, the first number in the pair generally replaces the variable that occurs first alphabetically.

EXAMPLE 1

Determine whether each of the following pairs is a solution of $4q - 3p = 22$: **(a)** (2, 7); **(b)** (1, 6).

Solution

a) We substitute 2 for p and 7 for q (alphabetical order of variables):

$$\begin{array}{r|l} 4q - 3p & = 22 \\ \hline 4 \cdot 7 - 3 \cdot 2 & ?\ 22 \\ 28 - 6 & \\ 22 & 22 \quad \text{TRUE} \end{array}$$

Since $22 = 22$ is *true,* the pair (2, 7) *is* a solution.

b) In this case, we replace p by 1 and q by 6:

$$\begin{array}{r|l} 4q - 3p & = 22 \\ \hline 4 \cdot 6 - 3 \cdot 1 & ?\ 22 \\ 24 - 3 & \\ 21 & 22 \quad \text{FALSE} \end{array}$$

Since $21 = 22$ is *false,* the pair (1, 6) is *not* a solution. ❑

EXAMPLE 2

Show that the pairs (3, 7), (0, 1), and (−3, −5) are solutions of $y = 2x + 1$.

Solution We substitute, replacing x with the first coordinate and y with the second coordinate of each pair:

$y = 2x + 1$			$y = 2x + 1$			$y = 2x + 1$		
7	? $2 \cdot 3 + 1$		1	? $2 \cdot 0 + 1$		-5	? $2(-3) + 1$	
	$6 + 1$			$0 + 1$			$-6 + 1$	
7	7	TRUE	1	1	TRUE	-5	-5	TRUE

In each of the three cases, the substitution results in a true equation. Thus the pairs (3, 7), (0, 1), and (−3, −5) are all solutions. ❑

In Example 2, three pairs were shown to solve the equation $y = 2x + 1$. The pairs (5, 11), (10, 21), and (−5, −9) are also solutions. In fact, there are infinitely many pairs that are solutions of $y = 2x + 1$. Similarly, there are infinitely many solutions of the equation in Example 1, $4q - 3p = 22$.

For any linear equation, there are infinitely many solutions.

Graphing Equations of the Type $y = mx$ and $y = mx + b$

Because every linear equation has infinitely many solutions, we cannot possibly list them all. Instead, we can make a drawing that represents the solutions. Since all solutions are ordered pairs, each pair can be represented as a point on a plane. The drawing that represents all such points is called the **graph** of the equation.

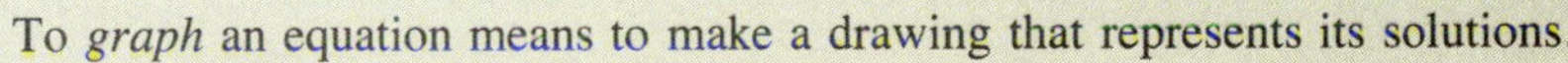

To *graph* an equation means to make a drawing that represents its solutions.

In the following examples, equations of the form $y = mx$ or $y = mx + b$ are graphed. As with any type of linear equation, once a few points have been plotted, we will discover that the graph is a straight line.

EXAMPLE 3

Graph: $y = 2x$.

Solution Before a graph of the equation can be drawn, we need to find some ordered pairs that are solutions. Rather than attempt this by trial and error, we will *choose* a number for x, the first coordinate, and then find the associated value for y by substitution. For example,

if $x = 3$,	then $y = 2 \cdot 3 = 6$;	We get a solution: the ordered pair (3, 6).
if $x = 1$,	then $y = 2 \cdot 1 = 2$;	We get another solution: the ordered pair (1, 2).
if $x = 0$,	then $y = 2 \cdot 0 = 0$;	We get another solution: the ordered pair (0, 0).
if $x = -2$,	then $y = 2(-2) = -4$	We get another solution: the ordered pair (−2, −4).

These results are often listed in a table, as shown below. The points corresponding to each pair are then plotted and we look for a pattern.

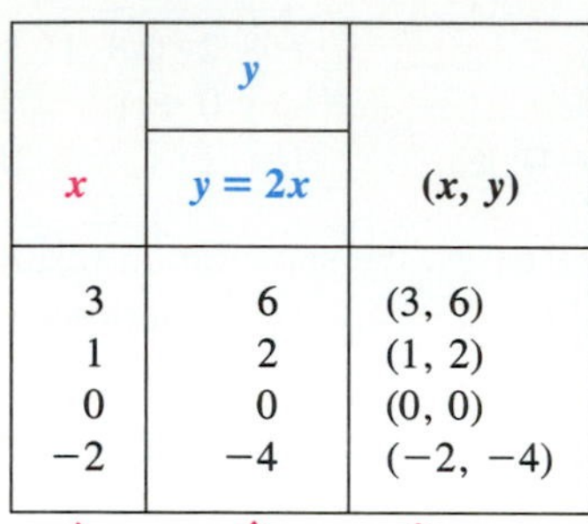

x	y $y = 2x$	(x, y)
3	6	(3, 6)
1	2	(1, 2)
0	0	(0, 0)
−2	−4	(−2, −4)

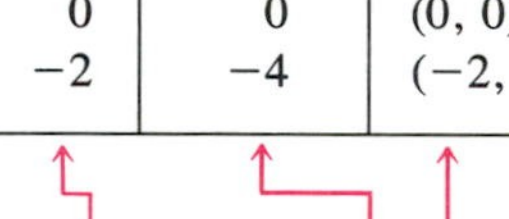

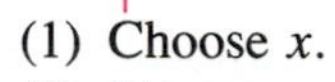

(1) Choose x.
(2) Compute y.
(3) Form the pair (x, y).
(4) Plot the points.

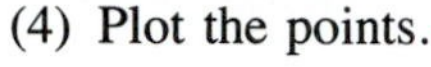

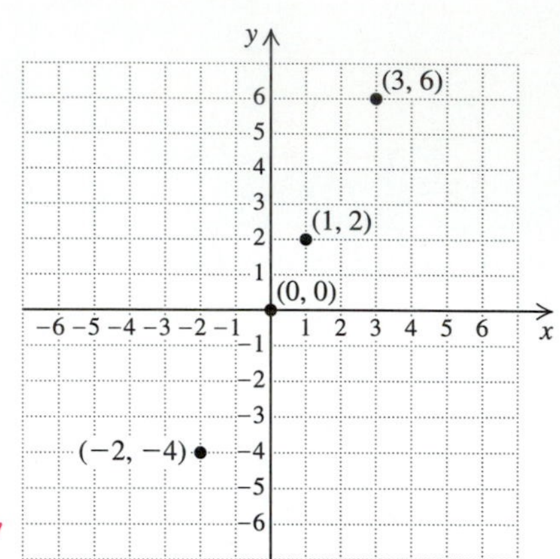

Note that the points resemble a straight line. We draw the line using a ruler or some other straightedge. Every point on the line (for instance, (1.5, 3)) represents a solution of $y = 2x$.

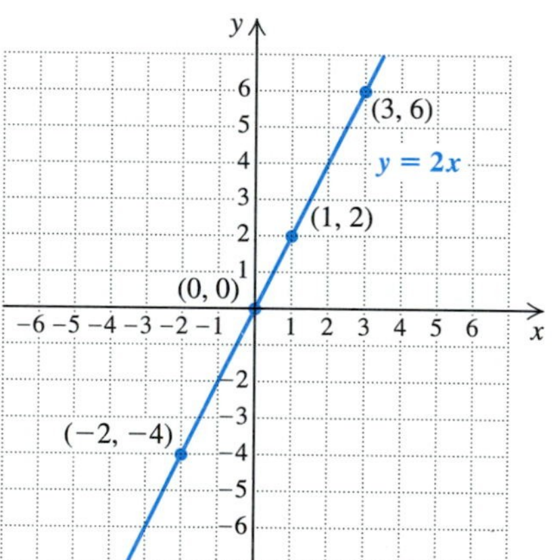

❑

EXAMPLE 4

Graph: $y = -3x$.

Solution Since two points determine a line, we need find only two ordered pairs that are solutions and plot the corresponding points. As a check against making a mistake, however, we generally plot *three* points. If all three points do not line up, we know that we have made a mistake.

We choose a number for x, the first coordinate, and calculate the associated y-value by substitution.

If $x = 1$, then $y = -3 \cdot 1 = -3$. We get the ordered pair $(1, -3)$.

If $x = -1$, then $y = -3(-1) = 3$. We get the ordered pair $(-1, 3)$.

If $x = 2$, then $y = -3 \cdot 2 = -6$. We get the ordered pair $(2, -6)$.

The following table lists these solutions. Next, we plot the points and see that they form a line. Finally, we draw and label the line.

x	y $y = -3x$	(x, y)
1	−3	(1, −3)
−1	3	(−1, 3)
2	−6	(2, −6)

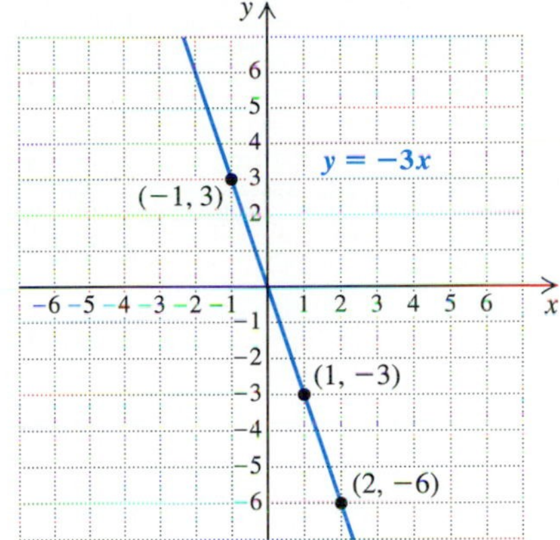

When choosing values for the x-coordinate, we try to avoid numbers that take us off the graph paper. If a fraction appears as the coefficient of x, we can choose x-coordinates that are multiples of the denominator and thereby avoid y-coordinates that are fractions.

EXAMPLE 5

Graph: $y = \frac{2}{3}x$.

Solution We make a table of solutions.

When $x = 3$, $y = \frac{2}{3} \cdot 3 = 2$.
When $x = -3$, $y = \frac{2}{3}(-3) = -2$.
When $x = 6$, $y = \frac{2}{3} \cdot 6 = 4$.
When $x = 1$, $y = \frac{2}{3} \cdot 1 = \frac{2}{3}$.

Note that when multiples of 3 are substituted for x, the y-coordinates are not fractions.

Next, we plot the points and complete the graph by drawing a line through them.

x	y $y = \frac{2}{3}x$	(x, y)
3	2	(3, 2)
−3	−2	(−3, −2)
6	4	(6, 4)
1	$\frac{2}{3}$	$\left(1, \frac{2}{3}\right)$

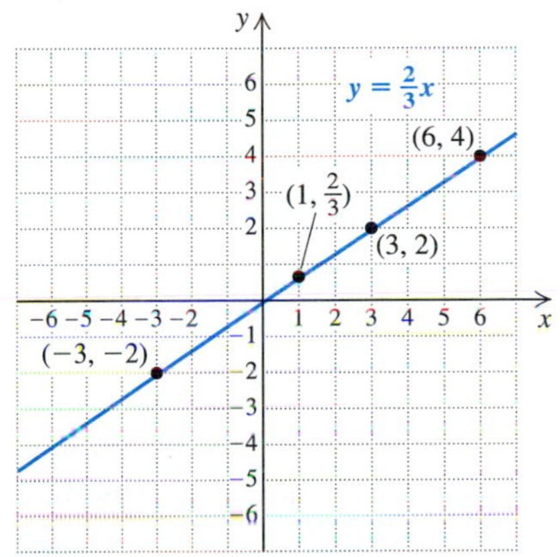

The lines in Examples 3–5 all contain the origin, (0, 0), because all three equations are of the form $y = mx$. What will happen if we add a number b on the right side to get an equation of the form $y = mx + b$?

EXAMPLE 6

Graph $y = 2x + 3$ and compare the graph with that of $y = 2x$.

Solution The equation $y = 2x$ was graphed on page 124. For the purpose of comparison, we graph $y = 2x + 3$ using the same choices for x. Note that:

if $x = 3$, then $y = 2 \cdot 3 + 3 = 9$; We get a solution: the ordered pair (3, 9).

if $x = 1$, then $y = 2 \cdot 1 + 3 = 5$; We get another solution: the ordered pair (1, 5).

if $x = 0$, then $y = 2 \cdot 0 + 3 = 3$; We get another solution: the ordered pair (0, 3).

if $x = -2$, then $y = 2(-2) + 3 = -1$. We get another solution: the ordered pair $(-2, -1)$.

The table of values and the graph of $y = 2x + 3$ are shown below.

x	y $y = 2x + 3$	(x, y)
3	9	(3, 9)
1	5	(1, 5)
0	3	(0, 3)
−2	−1	(−2, −1)

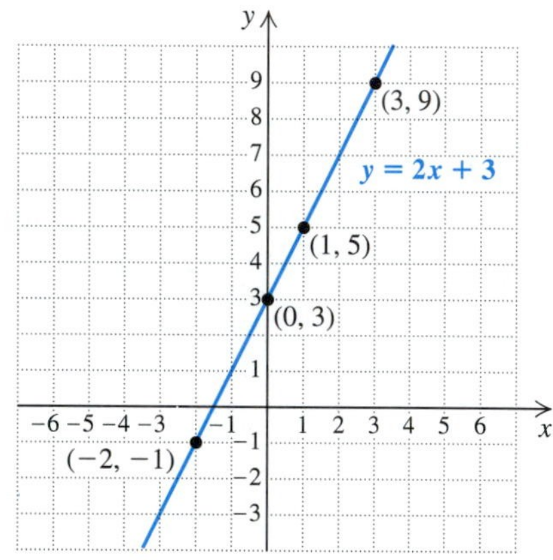

Note that the graph of $y = 2x + 3$ looks just like the graph of $y = 2x$ but shifted 3 units up. In particular, instead of crossing the y-axis at (0, 0), the graph now crosses the y-axis at (0, 3). ❑

In Example 6, when we replaced x with 0, the corresponding y-value was $2 \cdot 0 + 3$, or 3. In general, if $y = mx + b$ and x is replaced with 0, the corresponding y-value is $m \cdot 0 + b$, or b. Thus all graphs of equations of the form $y = mx + b$ will contain the point $(0, b)$.

EXAMPLE 7

Graph: $y = \frac{2}{5}x + 4$.

Solution Note that when $x = 0$, we have $y = \frac{2}{5} \cdot 0 + 4 = 4$ so the graph will contain (0, 4) as expected. We find two other pairs, using multiples of 5 to avoid fractions.

When $x = 5$, $y = \frac{2}{5} \cdot 5 + 4 = 2 + 4 = 6$.

When $x = -5$, $y = \frac{2}{5}(-5) + 4 = -2 + 4 = 2$.

x	y
0	4
5	6
−5	2

Now we can draw the graph of $y = \frac{2}{5}x + 4$.

We have seen that any equation of the form $y = mx + b$ contains the point $(0, b)$. That point is called the **y-intercept** of the graph since it is the point at which the graph crosses the y-axis. Sometimes, for convenience, we may refer to b as the y-intercept rather than list both coordinates.

y-intercept

The graph of any equation $y = mx + b$ passes through the y-intercept $(0, b)$.

EXAMPLE 8

Graph: $x + 3y = -6$.

Solution To make use of what we've studied, we first solve for y to write an equivalent equation in the form $y = mx + b$:

$$x + 3y = -6$$

$$3y = -x - 6 \quad \text{Adding } -x \text{ on both sides}$$

$$y = \tfrac{1}{3}(-x - 6) \quad \text{Multiplying by } \tfrac{1}{3} \text{ on both sides}$$

$$y = -\tfrac{1}{3}x - 2. \quad \text{Using the distributive law}$$

TECHNOLOGY CONNECTION

Beginning in this chapter, we will include in some sections and exercise sets activities that utilize graphing calculators or computer graphing software. Such calculators and software will be referred to simply as *graphers*. Most activities will use only basic features common to virtually all graphers. All will be presented in a generic form—check with a user's manual or ask your instructor for more exact procedures.

All graphers have a *window*, the rectangular portion of the screen in which a graph appears. We will describe a window with four numbers, [L, R, B, T], that represent the *L*eft and *R*ight endpoints of the x-axis and the *B*ottom and *T*op endpoints of the y-axis. A *Range* feature is sometimes used to set these dimensions.

The primary use for graphers is to graph equations. For example, let's graph the equation $y = -\frac{4}{5}x + \frac{13}{5}$. Selecting the window $[-10, 10, -10, 10]$ results in the graph shown at the top of the next column.

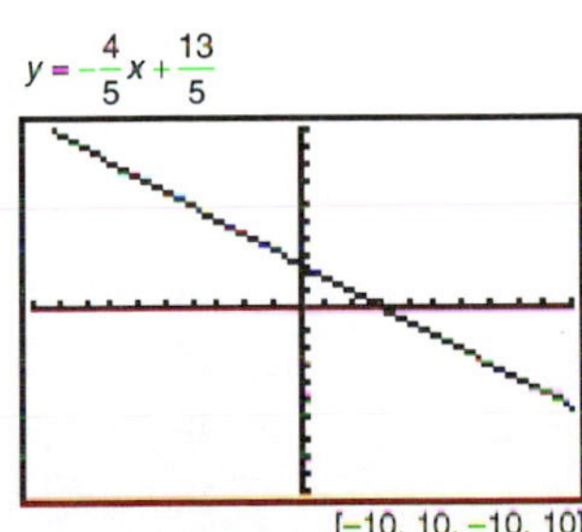

Use a grapher to draw each of the following lines. Select the window $[-10, 10, -10, 10]$ for each graph.

TC1. $y = -5x + 6.5$

TC2. $y = 3x - 4.5$

TC3. $y = \frac{4}{7}x - \frac{22}{7}$

TC4. $y = -\frac{11}{5}x - 4$

Thus, $x + 3y = -6$ is equivalent to $y = -\frac{1}{3}x - 2$, or $y = -\frac{1}{3}x + (-2)$. The y-intercept is therefore $(0, -2)$. We find two other pairs using multiples of 3 for x to avoid fractions.

When $x = 6$, $\quad y = -\frac{1}{3} \cdot 6 - 2 = -2 - 2 = -4$.
When $x = -6$, $\quad y = -\frac{1}{3}(-6) - 2 = 2 - 2 = 0$.

x	y
0	−2
6	−4
−6	0

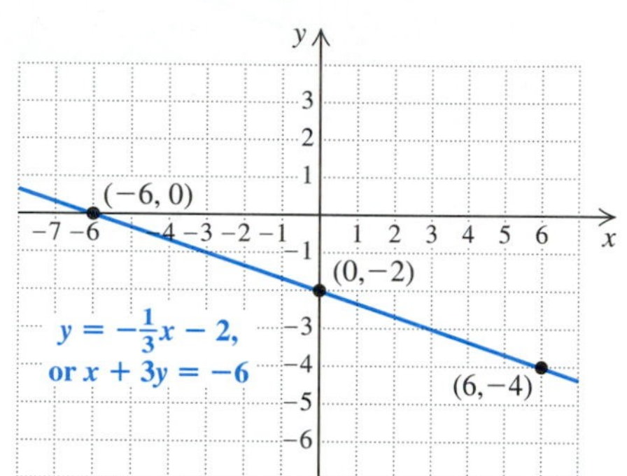

❑

EXERCISE SET 3.2

Determine whether the given point is a solution of the equation.

1. $(2, 5)$; $y = 3x - 1$
2. $(1, 7)$; $y = 2x + 5$
3. $(2, -3)$; $3x - y = 4$
4. $(-1, 4)$; $2x + y = 6$
5. $(-2, -1)$; $2c + 2d = -7$
6. $(0, -4)$; $4p + 2q = -9$

Graph.

7. $y = x$
8. $y = -x$
9. $y = -2x$
10. $y = -4x$
11. $y = \frac{1}{3}x$
12. $y = \frac{1}{2}x$
13. $y = -\frac{3}{2}x$
14. $y = -\frac{5}{4}x$
15. $y = x + 1$
16. $y = x + 3$
17. $y = 2x + 2$
18. $y = 3x - 2$
19. $y = \frac{1}{3}x - 1$
20. $y = \frac{1}{2}x + 1$
21. $y + x = -3$
22. $y + x = -2$
23. $y = \frac{5}{2}x + 3$
24. $y = \frac{5}{3}x - 2$
25. $y = -\frac{5}{2}x - 2$
26. $y = -\frac{5}{3}x - 2$
27. $y = \frac{1}{2}x - 5$
28. $y = \frac{3}{2}x - 6$
29. $2x + y = 3$
30. $5x + y = 7$
31. $y = x - \frac{1}{2}$
32. $y = x + \frac{2}{3}$
33. $x + 2y = -4$
34. $x + 2y = 8$
35. $6x - 3y = 9$
36. $8x - 4y = 12$
37. $6y + 2x = 8$
38. $8y + 2x = -4$

Skill Maintenance

Solve and check.

39. $3x - 7 = -34$
40. $2(x - 9) + 4 = 2 - 3x$
41. Solve $Ax + By = C$ for y.
42. Solve $A = \dfrac{T + Q}{2}$ for Q.

Synthesis

43. ◈ An equation is graphed by plotting first two points and then a third point as a check. If the three points form a line, is it still possible that the graph is incorrect? Why or why not?
44. ◈ Do all graphs of linear equations have y-intercepts? Why or why not?

45. Complete the following table for $y = x^2 + 1$. Plot the points on graph paper and draw the graph.

x	0	−1	1	−2	2	−3	3
y							

46. Find all whole-number solutions of

$$x + y = 6.$$

47. Find all whole-number solutions of

$$x + 3y = 15.$$

48. Translate to an equation: n nickels and d dimes total $1.95. Find three solutions.

49. Translate to an equation: n nickels and q quarters total $2.35. Find three solutions.

50. Find three solutions of $y = |x|$.

For Exercises 51–54, use a grapher to graph the equation. Use a $[-10, 10, -10, 10]$ window.

51. $y = -2.8x + 3.5$

52. $y = 4.5x + 2.1$

53. $y = \frac{2}{7}x - \frac{24}{5}$

54. $y = -\frac{33}{8}x - \frac{45}{7}$

55. Are the equations

$$y = -\tfrac{2}{5}x + 3 \quad \text{and} \quad 2x + 5y = 15$$

equivalent? Why or why not?

3.3 More on Graphing Linear Equations

Graphing Using Intercepts • Graphing Horizontal or Vertical Lines

Although equations like $4x + 3y = 12$ can be rewritten in the form $y = mx + b$ and then graphed as in Section 3.2, there is a faster way to graph equations of the form $Ax + By = C$.

Graphing Using Intercepts

We have seen that the y-intercept of a graph occurs where the graph crosses the y-axis and that the first coordinate of the y-intercept is always 0.

The point at which a graph crosses the x-axis is called the **x-intercept.** Since x-intercepts are always on the x-axis, the second coordinate of the x-intercept will always be 0.

Intercepts are found as follows.

Intercepts

The y-intercept is $(0, b)$. To find b, let $x = 0$ and solve the original equation for y.

The x-intercept is $(a, 0)$. To find a, let $y = 0$ and solve the original equation for x.

EXAMPLE 1

Find the x- and y-intercepts of the graph of $4x + 3y = 12$.

Solution

To find the y-intercept, let $x = 0$. Then solve for y:

$$4 \cdot 0 + 3y = 12 \quad \text{Replacing } x \text{ with } 0$$

$$3y = 12$$

$$y = 4.$$

Thus, $(0, 4)$ is the y-intercept.

To find the x-intercept, let $y = 0$. Then solve for x:

$$4x + 3 \cdot 0 = 12 \quad \text{Replacing } y \text{ with } 0$$

$$4x = 12$$

$$x = 3.$$

Thus, $(3, 0)$ is the x-intercept. ❑

By plotting both intercepts and a third point as a check, we can quickly graph any linear equation of the form $Ax + By = C$.

EXAMPLE 2

Graph $4x + 3y = 12$ using intercepts.

Solution In Example 1, we found that the y-intercept is $(0, 4)$ and the x-intercept is $(3, 0)$. Before drawing a line, we plot a third point as a check. We substitute any convenient value for x and solve for y.

If we let $x = 1$, then

$$4 \cdot 1 + 3y = 12 \quad \text{Substituting 1 for } x$$

$$4 + 3y = 12$$

$$3y = 12 - 4 = 8$$

$$y = \tfrac{8}{3}, \quad \text{or } 2\tfrac{2}{3}. \quad \text{Solving for } y$$

Since the point $(1, 2\frac{2}{3})$ appears to line up with the intercepts, we draw and label the graph.

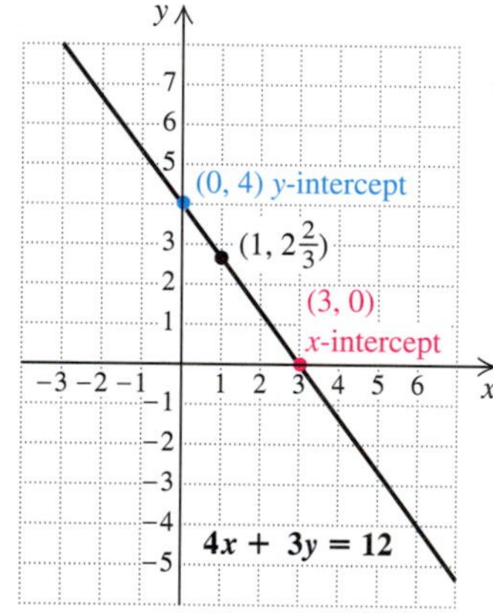

❑

In Example 1, the equation $4x + 3y = 12$ simplified to $3y = 12$ when we solved for the y-intercept. Thus, to find the y-intercept, we can simply "cover up" the x-term and solve the remaining equation.

In a similar manner, $4x + 3y = 12$ simplified to $4x = 12$ when we solved for the x-intercept. Thus, to find the x-intercept, we can simply "cover up" the y-term and solve the remaining equation.

EXAMPLE 3

Graph $3x - 2y = 6$ using intercepts.

Solution To find the y-intercept, let $x = 0$. This amounts to covering up the x-term and then solving:

$$-2y = 6 \quad \text{When } x \text{ is 0, } 3x - 2y \text{ is simply } -2y.$$

$$y = -3.$$

The y-intercept is $(0, -3)$.

To find the x-intercept, let $y = 0$. This is the same as covering up the y-term and then solving:

$3x = 6$ When y is 0, $3x - 2y$ is simply $3x$.

$x = 2$.

The x-intercept is (2, 0).

To find a third point, we replace x with 4 and solve for y:

$3 \cdot 4 - 2y = 6$ Other numbers besides 4 can be used for x.

$12 - 2y = 6$

$-2y = -6$

$y = 3$.

The point (4, 3) appears to line up with the intercepts so we draw the graph.

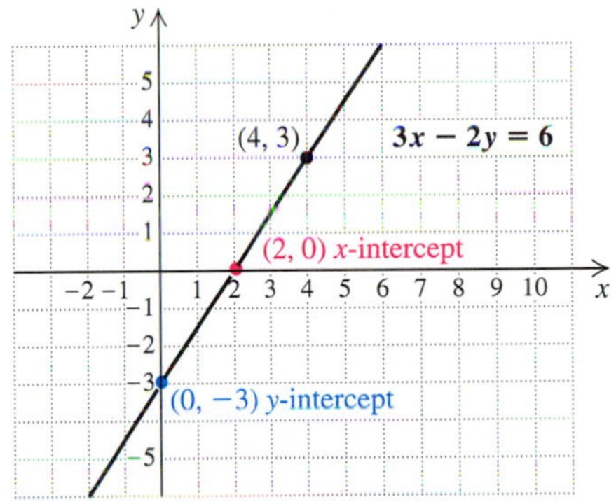

Graphing Horizontal or Vertical Lines

Equations like $y = 3$ or $x = -4$ are in the form $Ax + By = C$, but with $A = 0$ or $B = 0$. The graph of such an equation will be either a horizontal or a vertical line. There will be a y-intercept or an x-intercept but never both.

EXAMPLE 4

Graph: $y = 3$.

Solution We can regard the equation $y = 3$ as $0 \cdot x + y = 3$. No matter what number we choose for x, we find that y must be 3 if the equation is to be solved. Consider the following table.

Choose any number for x. →

x	y $y = 3$	(x, y)
−2	3	(−2, 3)
0	3	(0, 3)
4	3	(4, 3)

All pairs will have 3 as the y-coordinate.

y must be 3.

When we plot the ordered pairs $(-2, 3)$, $(0, 3)$, and $(4, 3)$ and connect the points, we obtain a horizontal line. Any ordered pair $(x, 3)$ is a solution. So the line is parallel to the x-axis with y-intercept $(0, 3)$.

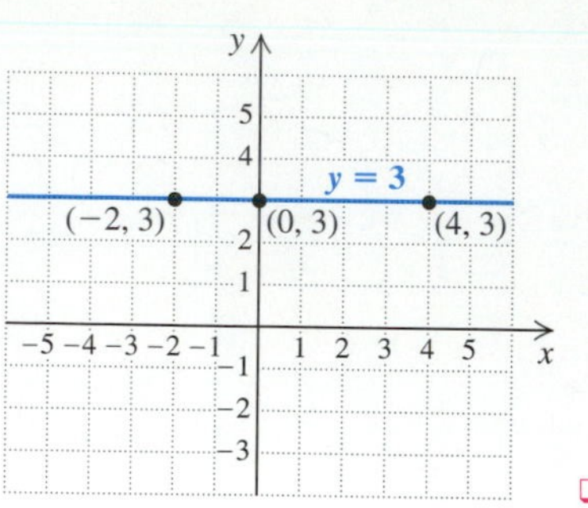

EXAMPLE 5

Graph: $x = -4$.

Solution We can regard the equation $x = -4$ as $x + 0 \cdot y = -4$. We make up a table with all -4's in the x-column.

x must be −4. →

x		
$x = -4$	y	(x, y)
-4	-5	$(-4, -5)$
-4	1	$(-4, 1)$
-4	3	$(-4, 3)$

All pairs will have −4 as the x-coordinate.

Choose any number for y.

When we plot the ordered pairs $(-4, -5)$, $(-4, 1)$, and $(-4, 3)$ and connect them, we obtain a vertical line. Any ordered pair $(-4, y)$ is a solution. So the line is parallel to the y-axis with x-intercept $(-4, 0)$.

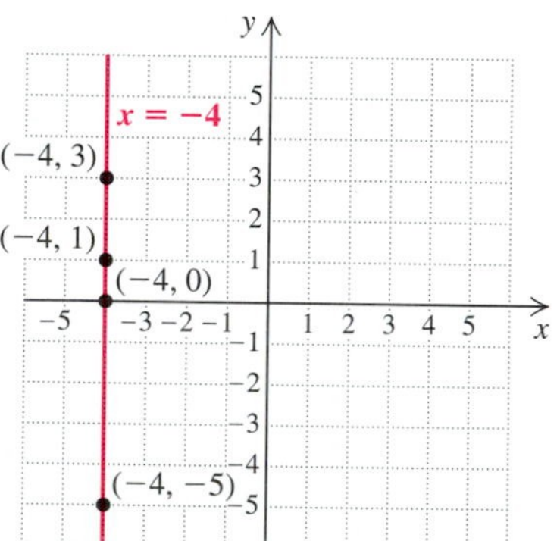

The graph of $y = b$ is a horizontal line, with y-intercept $(0, b)$.
The graph of $x = a$ is a vertical line, with x-intercept $(a, 0)$.

The following is a general procedure for graphing linear equations.

To Graph Linear Equations

1. If the equation is of the type $x = a$ or $y = b$, the graph will be a line parallel to an axis.

   Examples.

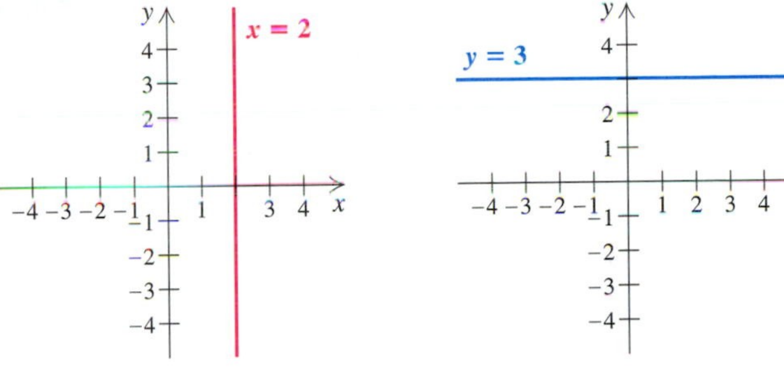

2. If the equation is of the type $y = mx$, both intercepts are the origin, (0, 0). Plot (0, 0) and one other point. A third point can be calculated as a check.

   Example.

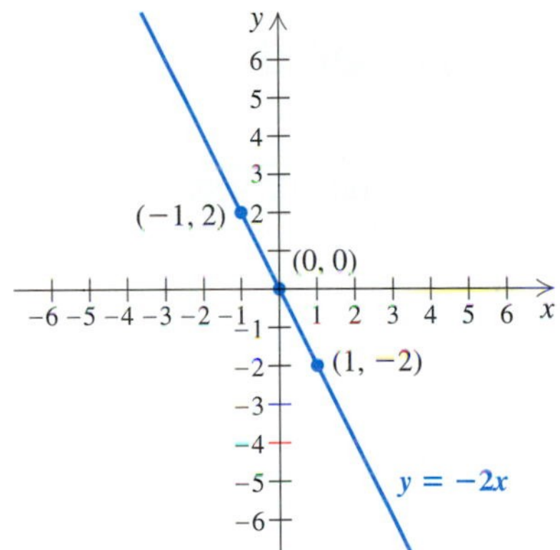

3. If the equation is of the type $y = mx + b$, the y-intercept is $(0, b)$. Plot $(0, b)$ and one other point. A third point can be calculated as a check.

   Example.

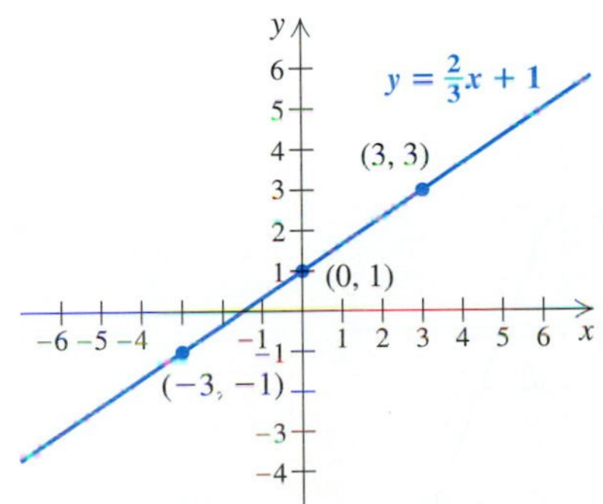

4. If the equation is of the type $Ax + By = C$, but not of the type $x = a$ or $y = b$, graph using intercepts. A third point can be used as a check.
Example.

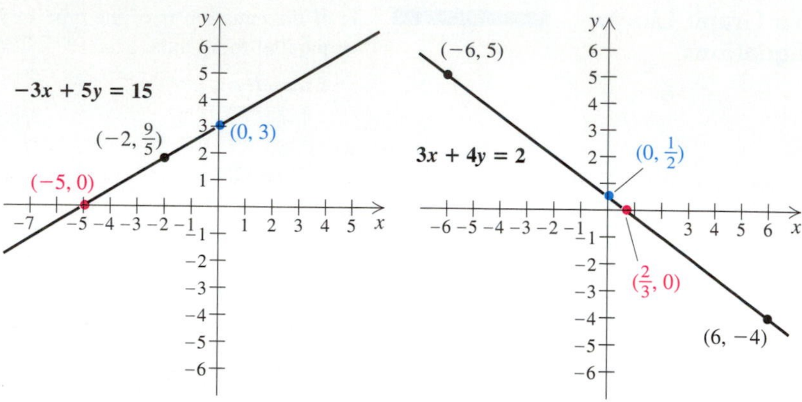

EXERCISE SET 3.3

For Exercises 1–4, find **(a)** the coordinates of the y-intercept and **(b)** the coordinates of the x-intercept.

1.

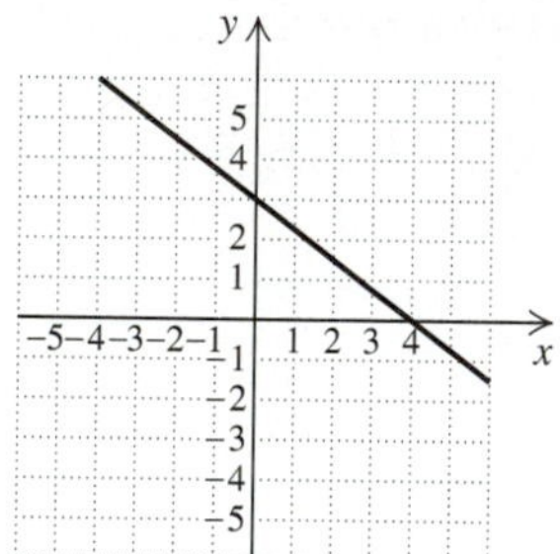

2.

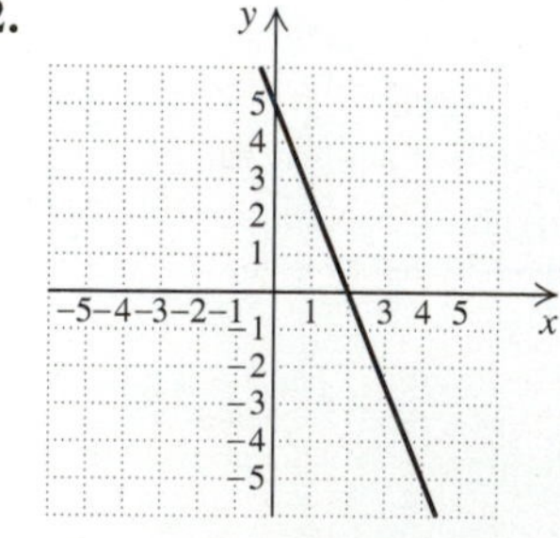

3.

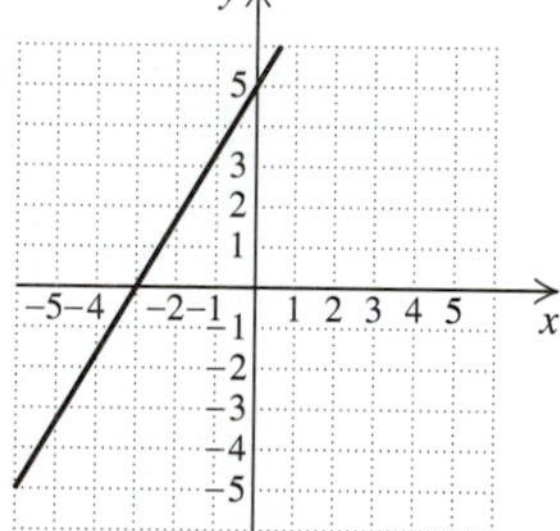

4.

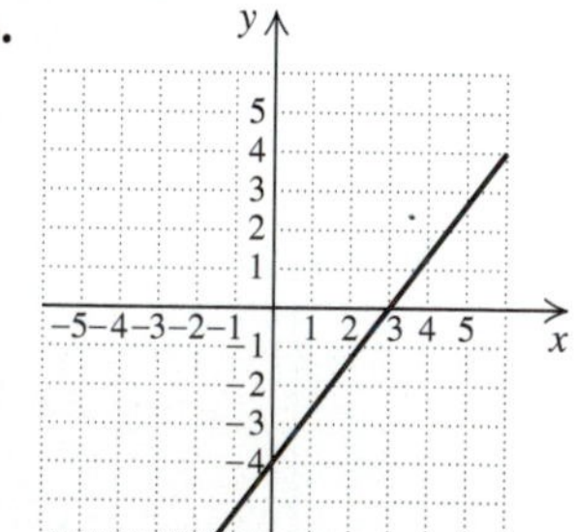

For Exercises 5–12, find **(a)** the coordinates of the y-intercept and **(b)** the coordinates of the x-intercept. Do not graph.

5. $2x + 5y = 20$

6. $5x + 3y = 15$

7. $4x - 3y = 24$
8. $2x - 7y = 28$
9. $-6x + y = 8$
10. $-8x + y = 10$
11. $2y - 4 = 6x$
12. $3y + 6 = 9x$

Find the intercepts. Then graph.

13. $3x + 2y = 12$
14. $2x + 4y = 16$
15. $x + 3y = 6$
16. $x + 2y = 8$
17. $-x + 2y = 4$
18. $-x + 3y = 9$
19. $3x + y = 9$
20. $2x + y = 6$
21. $2y - 2 = 6x$
22. $3y - 6 = 9x$
23. $3x - 9 = 3y$
24. $5x - 10 = 5y$
25. $2x - 3y = 6$
26. $2x - 5y = 10$
27. $4x + 5y = 20$
28. $2x + 6y = 12$
29. $2x + 3y = 8$
30. $x - 1 = y$
31. $x - 3 = y$
32. $2x - 1 = y$
33. $3x - 2 = y$
34. $4x - 3y = 12$
35. $6x - 2y = 18$
36. $7x + 2y = 6$
37. $3x + 4y = 5$
38. $y = -4 - 4x$
39. $y = -3 - 3x$
40. $-3x = 6y - 2$
41. $-4x = 8y - 5$
42. $3 = 2x - 5y$
43. $y - 3x = 0$
44. $x + 2y = 0$

Graph.

45. $x = -2$
46. $x = -1$
47. $y = 2$
48. $y = 4$
49. $x = 7$
50. $x = 3$
51. $y = 0$
52. $y = -1$
53. $x = \frac{3}{2}$
54. $x = -\frac{5}{2}$
55. $3y = -5$
56. $12y = 45$
57. $4x + 3 = 0$
58. $-3x + 12 = 0$
59. $18 - 3y = 0$
60. $63 + 7y = 0$

Skill Maintenance

Write the prime factorization of the number.

61. 98
62. 240

Simplify.

63. $\frac{36}{90}$
64. $\frac{12x}{84x}$

Synthesis

65. ◈ Explain in your own words why equations of the form $y = C$ have graphs that are horizontal lines.
66. ◈ Can any equation of the type $Ax + By = C$ be written in the form $y = mx + b$? Why or why not?
67. Write an equation for the y-axis.
68. Write an equation for the x-axis.
69. Find the coordinates of the point of intersection of the graphs of the equations $x = -3$ and $y = 6$.
70. Write an equation of a line parallel to the x-axis and 5 units below it.
71. Write an equation of a line parallel to the y-axis and 13 units to the right of it.
72. Write an equation of a line parallel to the x-axis and intersecting the y-axis at $(0, 2.8)$.
73. Find the value of m in the equation $y = mx + 3$ so that the x-intercept of its graph will be $(2, 0)$.
74. Find the value of b in the equation $2y = -5x + 3b$ so that the y-intercept of its graph will be $(0, -12)$.
75. ◈ For A and B nonzero, the graphs of $Ax + D = C$ and $By + D = C$ will be parallel to an axis. Explain why.

3.4 Graphs and Problem Solving

Translating to Equations • Graphing • Problem Solving

Suppose we are asked to find a pair of numbers whose sum is 5. There are infinitely many such pairs, so a graph would provide a convenient way of showing all solutions.

Translating to Equations

When a problem-solving situation involves pairs of numbers and many solutions, we can often translate it to an equation with two variables.

EXAMPLE 1

Translate to an equation.

a) The sum of two numbers is 5.
b) Harry is two years older than Jane.
c) The express train is twice as fast as the local.
d) Sondra earns $100 less than twice Jim's salary.

Solution

a) Let x and y represent the two numbers whose sum is 5. The translation follows immediately: $x + y = 5$.

b) Let h = Harry's age and j = Jane's age. We reword and then translate.

Reword: Harry's age is two more than Jane's age.

Translate: $h = j + 2$

c) Let s = the speed of the express train and l = the speed of the local train. Then we have

Reword: The speed of the express train is twice the speed of the local.

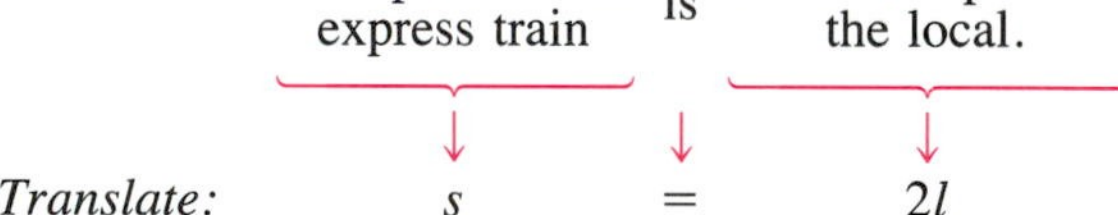

Translate: $s = 2l$

d) Let s = Sondra's salary and j = Jim's salary. Then we have

Reword: Sondra's salary is $100 less than twice Jim's salary.

Translate: $s = 2j - 100$ ❑

Graphing

When a problem-solving situation translates to a linear equation, we can label axes and draw a graph. The graph provides a visual representation of the situation.

EXAMPLE 2

The total number of calories C expended by a 150-lb person who is walking briskly for t hours can be estimated by the equation $C = 300t$.

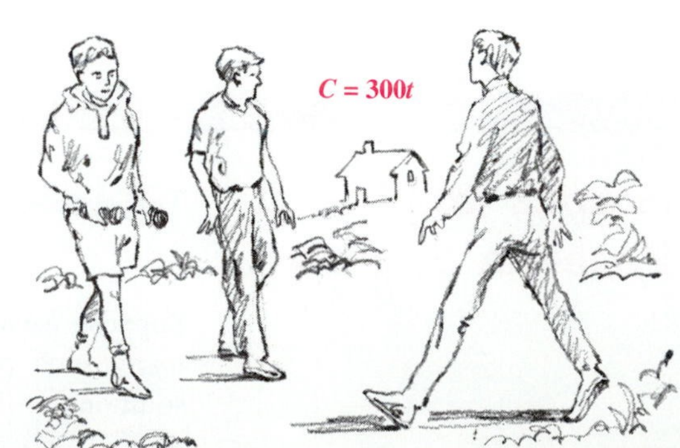

a) Use the equation to estimate the energy expended in walking for 2 hr, 3.5 hr, and 5 hr.

b) Graph the equation. Assume that t is the first coordinate and C is the second.

Solution

a) To find the number of calories used to walk 2 hr, 3.5 hr, and 5 hr, we substitute and calculate as follows:

$C = 300(2) \quad = \ 600$ calories; Walking 2 hr burns 600 Cal.

$C = 300(3.5) = 1050$ calories; Walking 3.5 hr burns 1050 Cal.

$C = 300(5) \quad = 1500$ calories. Walking 5 hr burns 1500 Cal.

b) We use the values computed in part (a) and any others that we may calculate to make a table. Each pair of values represents a point, which we can plot using the horizontal axis for time and the vertical axis for the total number of calories burned.

Often, in problems such as this, it is impractical to count by 1's. In this case, we count by 100's on the vertical axis. By altering the *scale* in this way, we can plot all the pairs listed in the table.

Once the points are graphed, we draw a straight line through them.

Time (in Hours)	Calories Burned
2	600
3.5	1050
5	1500

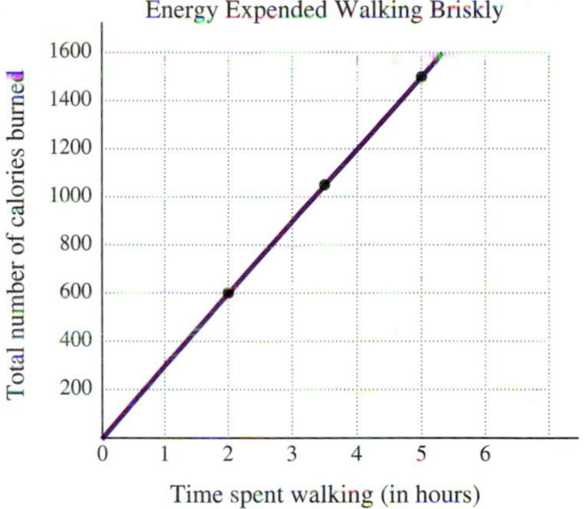

Problem Solving

When a problem is translated and graphed, the graph often provides a quick and useful way of approximating values that could be calculated more precisely (but also more slowly) from a formula.

EXAMPLE 3

Ridem Trucks charges $49.95 per day plus 35¢ per mile for the rental of an 18-ft truck. To help customers predict the cost of a rental, the firm wishes to draw a graph in which mileage is measured on the horizontal axis and cost on the vertical axis. Using such a graph, predict the cost of renting an 18-ft truck for one day and driving 125 miles.

Solution

1. FAMILIARIZE. In Section 2.5, another truck rental problem was solved, so we already have some familiarity. Since the axes will represent mileage and cost, we let m = mileage and c = cost.

2. TRANSLATE. Since the cost of a rental is \$49.95 plus 35¢ for each mile, and since m miles are to be driven, we have the translation

$$c = 49.95 + 0.35m.$$

3. CARRY OUT. We make a table of values using some convenient choices for m.

When $m = 50$, $c = 49.95 + 0.35(50) = 67.45$.
When $m = 150$, $c = 49.95 + 0.35(150) = 102.45$.
When $m = 300$, $c = 49.95 + 0.35(300) = 154.95$.

TECHNOLOGY CONNECTION

The *Trace* feature found on most graphers allows you to find the coordinates of a point on a line. When the Trace feature is activated, a cursor (often blinking) appears on the graph and its x- and y-coordinates are displayed. These coordinates change as you move the cursor along the graph.

Let's consider the problem described in Example 3. Since most graphers use x- and y-terms, we rewrite the equation as $y = 49.95 + 0.35x$. Selecting the window [0, 400, 0, 250] gives the following graph.

$y = 49.95 + 0.35x$

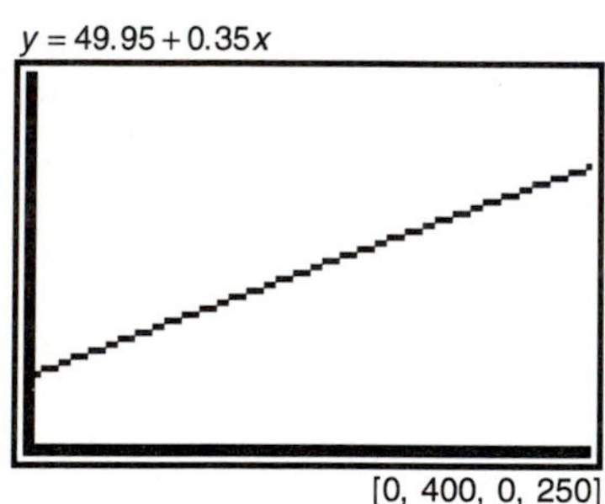

[0, 400, 0, 250]

Note that the axes appear heavyset because the *scales* remain at 1. If this is bothersome, the *Range* feature can be used to adjust the scales for x and y. Using a scale of 50 on both axes results in this graph:

$y = 49.95 + 0.35x$ Xscl = 50 Yscl = 50

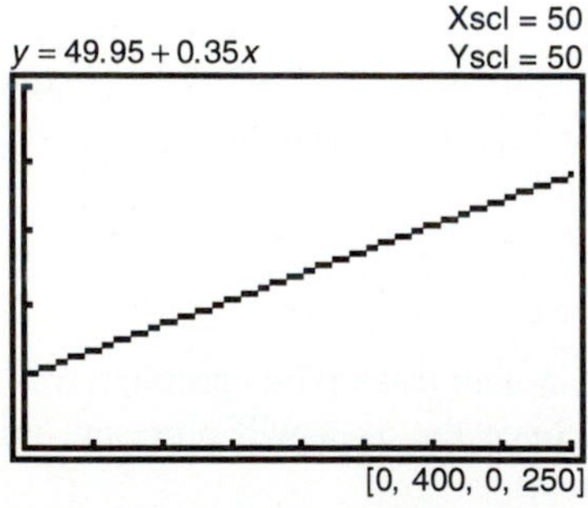

[0, 400, 0, 250]

Let's use the graph to determine the cost of driving 210 mi. To do so, we activate the Trace feature and move the cursor until its x-coordinate is approximately 210. The displayed coordinates indicate a cost of approximately \$124.

$y = 49.95 + 0.35x$

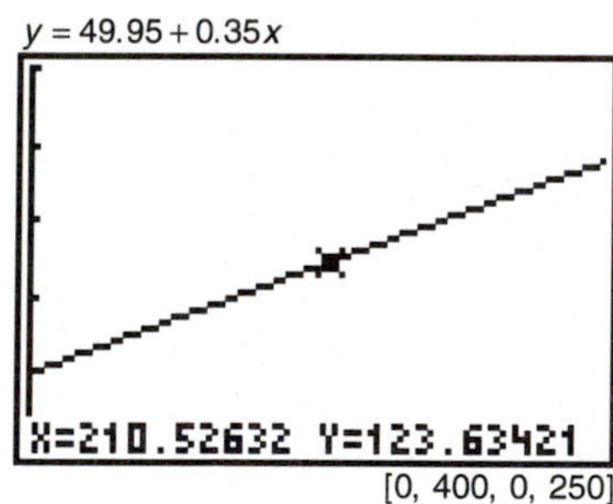

[0, 400, 0, 250]

For more precision, we can zoom in on the point in question. To do so, we can shrink the window's dimensions or use a *Zoom* feature to magnify a portion of the graph. (Consult a manual or your instructor on how to best utilize the Zoom feature.) By zooming in or shrinking the window (say, to [209, 211, 122, 124]), we can use the Trace feature again for a better approximation: \$123.45.

Solve each of the following using a grapher.

TC1. Use the Trace feature and reset the range or zoom in to approximate to the nearest cent the cost of driving the rental truck 190 mi.

TC2. Starting at an altitude of 2000 ft, a pilot climbs at the rate of 3500 ft per minute. What is the plane's altitude after 4.5 min? (*Hint:* Use the window [0, 10, 2000, 40,000] with the y-scale 4000.)

TC3. Sally's Plumbing Service charges \$35 for each visit plus \$45 per hour. If a job lasts 6.5 hr, how much will it cost?

Mileage	Cost
50	\$67.45
150	\$102.45
300	\$154.95

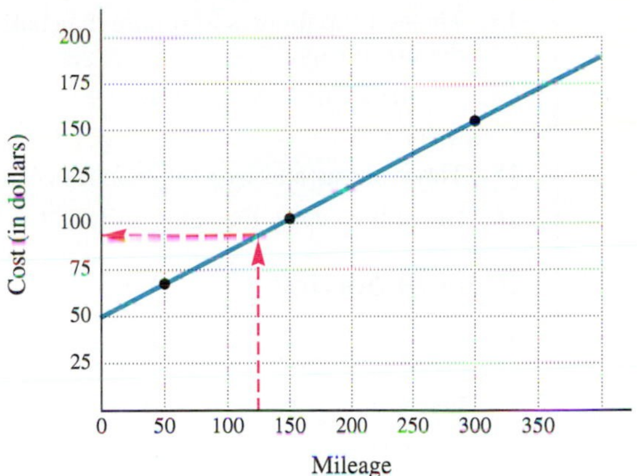

Being careful to label the axes correctly, we draw the graph by plotting the points listed in the table and then drawing a line through them.

To predict the cost of driving 125 mi, we locate 125 on the horizontal axis. From there we trace a path *up* to the line and then *left* from the line to the vertical axis. Since the predicted cost is closer to \$100 than to \$75, we estimate the cost of the rental at \$90.

4. CHECK. The three points that we graphed form a line, so the graph is probably correct. To check the predicted cost of \$90, we could calculate the cost precisely. Doing so may make the graph seem unnecessary, but the rental firm might still want the graph for other quick cost estimates:

$$\text{Cost of rental} = 49.95 + 0.35 \cdot (\text{miles driven}),$$
$$\text{Cost of driving } 125 \text{ mi} = 49.95 + 0.35(125) = \$93.70.$$

Our prediction, \$90, was close enough to serve as a good approximation.

5. STATE. Using the graph, we could predict that the cost of renting an 18-ft truck and driving it 125 mi is about \$90. ❑

EXERCISE SET 3.4

Translate each sentence to an equation containing two variables. Be sure to state what each variable represents.

1. The sum of two numbers is 27.

2. The sum of two numbers is 53.

3. A number plus twice another number is 65.

4. The sum of a number and twice another number is 93.

5. One number is three times as great as another.

6. One number is half as great as another.

7. One number is 5 more than another.

8. One number is 7 less than another.

9. Hank's age plus 7 is twice Nanette's age.

10. Lisa's age is 5 less than twice Lou's.

11. Lois earns \$170 more than three times Roberta's salary.

12. Evelyn's salary plus \$200 is four times Eric's salary.

13. The nonstop flight is $\frac{3}{4}$ hr more than half the time of the direct flight.
14. The delay took 5 min more than half the flight time.
15. Three pizzas and two sandwiches cost $37.
16. One entree costs the same as two desserts.

Problem Solving

17. *Car traveling at a fixed speed.* A car travels at a speed of 55 mph for time t, in hours. The distance that it travels is given by

$$d = 55t.$$

a) How far does the car travel in 1 hr? in 2 hr? in 5 hr? in 10 hr?
b) Graph the equation. Assume that t is the first coordinate and d is the second.

18. *Hair growth.* Hair will normally grow at a rate of about $\frac{1}{2}$ in. per month. In t months, it will grow N in., where N is given by

$$N = \tfrac{1}{2}t.$$

a) How long will hair grow in 2 months? in 5 months? in 8 months? in 12 months?
b) Graph the equation. Assume that t is the first coordinate and N is the second.

19. *Women's shoe sizes in the United States and Britain.* Shoe sizes vary from country to country. The equation

$$y = x - 2$$

can be used to convert a women's shoe size x in the United States to its corresponding size y in Britain.

a) What women's shoe size in Britain corresponds to the U.S. size of 4? of 5? of 6? of 7? of 8?
b) Graph the equation.

20. *Toll road charges.* The equation

$$y = 0.027x + 0.19$$

can be used to determine the approximate cost y, in dollars, of driving x miles on the Indiana toll road.

a) Use the equation to predict the cost of driving 10 miles, 100 miles, and 150 miles.
b) Graph the equation.

21. *Increasing life expectancy.** A regular smoker is 15 times more likely to die from lung cancer than a nonsmoker. An exsmoker who stopped smoking t years ago is w times more likely to die from lung cancer than a nonsmoker, where

$$t + w = 15.$$

a) Sandy gave up smoking $2\frac{1}{2}$ years ago. How much more likely is she to die from lung cancer than Polly, who never smoked?
b) Graph the equation, using t as the first coordinate.

22. *Allocating resources.* Servemaster food services has $240 to spend on turkey at $4.00 per pound and/or roast beef at $6.00 per pound. If t pounds of turkey and r pounds of roast beef are bought, the equation

$$4t + 6r = 240$$

must be satisfied.

a) If 10 lb of turkey are bought, how much roast beef will be purchased?
b) Graph the equation, using r as the first coordinate.

23. *Truck rentals.* Iron Mike's Trucks charges $39.95 plus 55¢ per mile for the rental of a 20-ft truck. Draw a graph in which mileage is measured on the horizontal axis and cost on the vertical axis. Then use the graph to predict the cost of renting a 20-ft truck for one day and driving 180 mi.

24. *Van rentals.* Rent King charges $59.95 plus 45¢ per mile for the rental of its 20-ft van. Draw a graph in which mileage is measured on the horizontal axis and cost on the vertical axis. Then use the graph to predict the cost of renting the 20-ft van for one day and driving 340 mi.

25. *Wages and commissions.* Each salesperson at the Shoe Box is paid a weekly salary of $150 plus 4% of that person's sales. Draw a graph in which sales are measured on the horizontal axis and wages on the vertical axis. Then use the graph to estimate the wages paid when a salesperson sells $4500 in merchandise in one week.

26. *Wages and commissions.* Each salesperson at Grand Buy Appliances is paid a weekly salary of $200 plus 5% of that person's sales. Draw a graph in which sales are measured on the horizontal axis and wages on the vertical axis. Then use the graph to estimate the wages paid when a salesperson sells $3700 in appliances in one week.

27. *Cost of a road call.* Dave's Foreign Auto Village charges $35 for a road call plus $10 for each 15-min unit of time. Draw a graph that can be used to predict the cost of a service call. Use the horizontal axis for time and the vertical axis for cost. Then

---

**Source:* Data from *Body Clock* by Dr. Martin Hughes, p. 60. New York: Facts on File, Inc.

use the graph to estimate the cost of a $1\frac{1}{2}$-hr road call.

28. *Parking fees.* Karla's Parking charges \$3.00 to park plus 50¢ for each 15-min unit of time. Draw a graph that can be used to predict the cost of parking at Karla's. Use the horizontal axis for time and the vertical axis for cost. Then use the graph to estimate how much it will cost to park for $3\frac{1}{2}$ hr.

29. *Food preparation.* Harriet's Catering believes that parties for more than 10 should include a 3-lb wheel of cheese and an additional $\frac{2}{9}$ lb for each person in excess of 10. Draw a graph that can be used to predict how much cheese should be purchased for parties of 10 or more. Use the horizontal axis for the number of people and the vertical axis for the number of pounds. Then use the graph to estimate the amount of cheese needed for a party of 21.

30. *Real-estate depreciation.* Because of wear and tear, rental property can be depreciated each year that it is in service. The depreciated value for some real estate is found by subtracting $\frac{1}{18}$ of the original value for each year that the property is rented. Draw a graph that can be used to find the depreciated value of a house that was valued at \$150,000 when it was first rented. Use the horizontal axis for time and estimate the depreciated value after 8 yr of renting.

31. *Aviation.* Captain Hsu is landing a 747 from its cruising altitude of 32,000 ft. The jet descends at a rate of 3000 ft/min. Draw a graph in which the plane's altitude is measured on the vertical axis and use the graph to predict the altitude 8 min into the descent.

32. *Aviation.* Helga is landing a single-engine Tandem Taildragger from a cruising altitude of 6000 ft. Her plane is descending at a rate of 50 ft/min. Draw a graph in which the plane's altitude is measured on the vertical axis and use the graph to predict the altitude 17 min into the descent.

Skill Maintenance

33. Solve $s = vt + d$ for t.

34. Solve: $3(x - 4) + 7 = -2x + 6(x - 5)$.

Synthesis

35. ◈ Write a problem for a classmate to solve. Devise the problem so that a graph can be drawn to represent the situation and then predict a value (see Exercises 23–32).

36. ◈ Exercises 27 and 28 refer to units of time and Exercise 29 refers to the number of people at a party. Is it misleading to use solid lines rather than points in these graphs? Why or why not?

37. A Boeing 737 climbs from sea level to a cruising altitude of 34,000 ft at a rate of 6500 ft/min. After cruising for 3 min, the jet is forced to land, descending at a rate of 3500 ft/min. Draw a graph in which the plane's altitude is measured on the vertical axis and time on the horizontal axis.

38. Each salesperson at Mike's Bikes is paid \$140 a week plus 13% of all sales up to \$2000, and then 20% on any sales in excess of \$2000. Draw a graph in which sales are measured on the horizontal axis and wages on the vertical axis. Then use the graph to estimate the wages paid when a salesperson sells \$2700 in merchandise in one week.

39. Peggy earns \$150 less than twice Paul's weekly salary. Paul's salary is \$70 more than half of Jenna's. Draw a graph in which Jenna's salary is listed on the horizontal axis and Peggy's salary on the vertical axis. Then write an equation that relates Peggy's salary, p, to Jenna's salary, j.

40. Fast Eddie claims his truck rentals are always 5% less expensive than Rental Rick's. Rental Rick charges \$40 plus 50¢ per mile for a 20-ft rental truck. Draw a graph that Rental Rick could use to calculate prices. Measure mileage on the horizontal axis and cost on the vertical axis. Then write an equation that relates Fast Eddie's cost, c, to the mileage driven, m.

Solve using a grapher.

41. Weekly pay at Bikes for Hikes is \$148 plus a 3.5% sales commission. If a salesperson sells \$3775 worth of merchandise, what is the weekly pay?

42. It costs Bert's Shirts \$38 plus \$2.35 a shirt to print tee-shirts for a day camp. Find Bert's cost for producing 144 shirts.

SUMMARY AND REVIEW 3

KEY TERMS

Bar graph, p. 114
Line graph, p. 116
Circle graph (or pie chart), p. 117
Axes (singular, axis), p. 117
Origin, p. 117
Coordinate, p. 117
Ordered pair, p. 118
Quadrant, p. 118
Linear equation, p. 122
Graph, p. 123
y-intercept, p. 127
x-intercept, p. 129

IMPORTANT PROPERTIES AND FORMULAS

To Graph Linear Equations

1. If the equation is of the type $x = a$ or $y = b$, the graph will be a line parallel to an axis.
2. If the equation is of the type $y = mx$, both intercepts are the origin, $(0, 0)$. Plot $(0, 0)$ and one other point. A third point can be calculated as a check.
3. If the equation is of the type $y = mx + b$, the y-intercept is $(0, b)$. Plot $(0, b)$ and one other point. A third point can be calculated as a check.
4. If the equation is of the type $Ax + By = C$, graph using intercepts. A third point can be used as a check.

REVIEW EXERCISES

This chapter's review and test include Skill Maintenance exercises from Sections 1.3, 1.4, 2.2, and 2.3.

This line graph shows the prime rate (the interest rate charged by banks to their best customers) in June for several years.

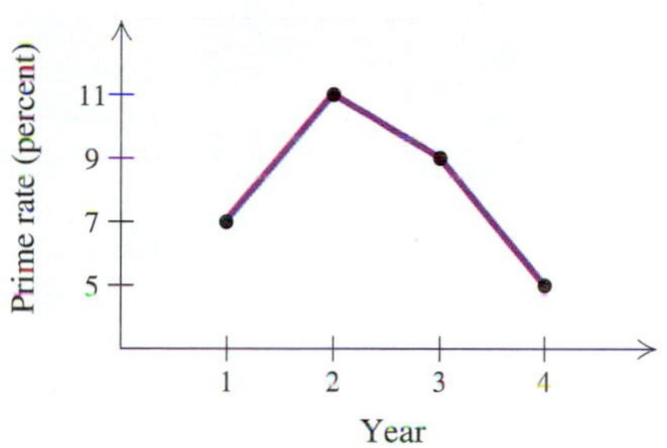

1. What was the highest prime rate?
2. Between what two consecutive years did the prime rate decrease the most?

Plot the point.

3. (2, 5)
4. (0, −3)
5. (−4, −2)

In which quadrant is the point located?

6. (3, −8)
7. (−20, −14)
8. (4.9, 1.3)

Find the coordinates of each point in the figure.

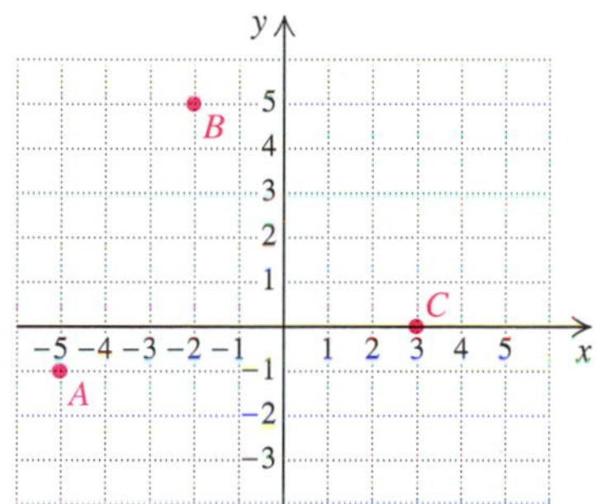

9. *A*
10. *B*
11. *C*

Determine whether the point is a solution of the equation $2y - x = 10$.

12. (2, −6)
13. (0, 5)

Graph.

14. $y = 2x - 5$
15. $y = -\frac{3}{4}x$
16. $y = -x + 4$
17. $y = 3 - 4x$
18. $5x - 2y = 10$
19. $4x + 3 = 0$

Find the x- and y-intercepts. Do not graph.

20. $x - 2y = 6$
21. $2x + 3y = 27$
22. Translate to an equation containing two variables. State what each variable represents.

    One number is 3 less than twice another number.

23. *Impulses in nerve fibers.* Impulses in nerve fibers travel at a speed of 293 ft/sec. The distance d traveled in t seconds is given by

    $$d = 293t.$$

    a) How far will a nerve impulse travel in 0.01 sec? in 0.4 sec? in 1 sec?
    b) Graph the equation using decimal values for t from 0 to 1.

24. *Water rates.* Bay City Water Company charges each customer a \$6.40 monthly service charge, plus \$1.03 for each 100 cubic feet of water used. Draw a graph in which water usage is measured on the horizontal axis and the monthly bill on the vertical axis. Then use the graph to estimate the water bill for a month in which a customer used 800 cubic feet of water.

Skill Maintenance

25. Add and simplify: $\frac{3}{8} + \frac{5}{12}$.
26. Find decimal notation: $-\frac{7}{8}$.
27. Solve: $2(x - 3) = 10$.
28. Solve $A = \dfrac{m + n}{2}$ for m.

Synthesis

29. ◈ Describe two ways in which a small business might make use of graphs.
30. ◈ Explain why the first coordinate of the y-intercept is always 0.
31. Find the value of m in $y = mx + 3$ such that $(-2, 5)$ is on the graph.
32. Find the value of b in $y = -5x + b$ such that (3, 4) is on the graph.
33. Find the area and the perimeter of a rectangle for which $(-2, 2)$, $(7, 2)$, and $(7, -3)$ are three of the vertices.
34. Find three solutions of $y = 4 - |x|$.

CHAPTER TEST 3

Consider the bar graph at right for Exercises 1–4.

1. What kind of degree was awarded most?
2. How many more bachelor's degrees than associate degrees were awarded?
3. How many more master's degrees than doctoral degrees were awarded?
4. In all, how many graduate degrees were awarded; that is, how many master's, doctoral, and professional degrees were awarded?

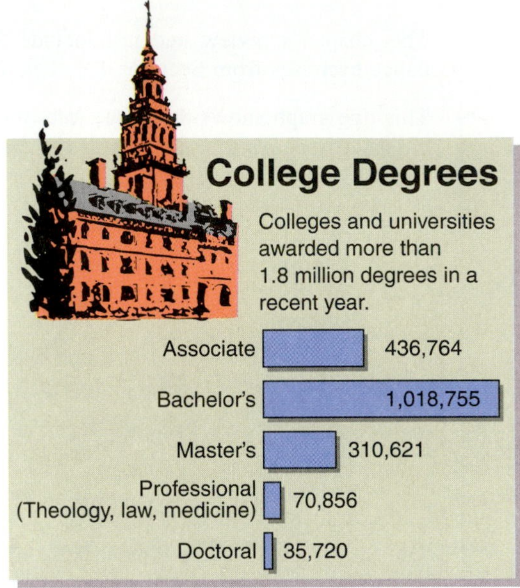

Source: National Center for Education Statistics, Digest of Education Statistics, 1992.

In which quadrant is the point located?

5. $\left(-\frac{1}{2}, 7\right)$

6. $(-5, -6)$

Find the coordinates of each point in the figure.

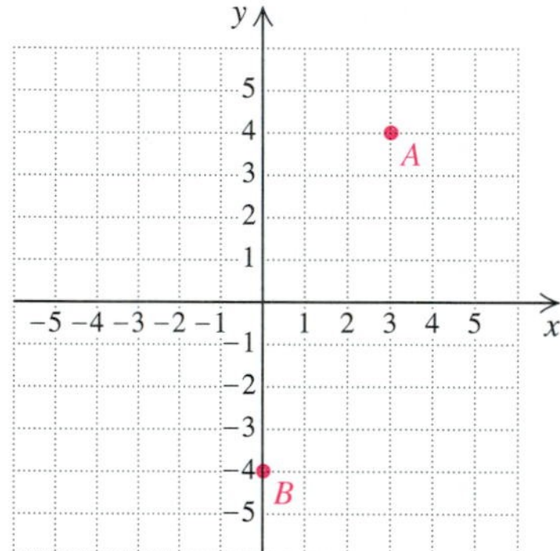

7. A

8. B

9. Determine whether the ordered pair $(2, -4)$ is a solution of the equation $y - 3x = -10$.

Graph.

10. $y = 2x - 1$

11. $2x - 4y = -8$

12. $y = 5$

13. $y = -\frac{3}{2}x$

14. $2x + 8 = 0$

Find the coordinates of the x- and y-intercepts. Do not graph.

15. $5x - 3y = 45$

16. $x = 10 - 4y$

17. Translate to an equation containing two variables. State what each variable represents.

    Greta earns \$50 more than twice Alice's salary.

18. Bright Electrical Service charges \$25 for a service call plus \$45 per hour. Draw a graph that could be used to predict the cost of a service call. Use the horizontal axis for time and the vertical axis for cost. Then use the graph to estimate the cost of a $\frac{1}{2}$-hr service call.

Skill Maintenance

19. Divide and simplify: $\frac{3}{5} \div \frac{3}{11}$.
20. Write a true sentence using either $<$ or $>$: $-9.7 \;\blacksquare\; -11.3$.
21. Solve: $\frac{1}{3} + 2p = 4p - \frac{1}{5}$.
22. Solve $mx = b - nx$ for x.

Synthesis

23. A diagonal of a square connects the points $(-3, -1)$ and $(2, 4)$. Find the area and the perimeter of the square.
24. Write an equation of a line parallel to the x-axis and 3 units above it.

CUMULATIVE REVIEW 1–3

1. Evaluate $\frac{x}{2y}$ when $x = 10$ and $y = 2$.

2. Multiply: $3(4x - 5y + 7)$.

3. Factor: $3x + 9y + 15$.

4. Find the prime factorization of 42.

5. Find decimal notation: $\frac{9}{20}$.

6. Find the absolute value: $|-4|$.

7. Find the opposite of 5.

8. Find the reciprocal of 5.

9. Collect like terms: $2x - 5y + (-3x) + 4y$.

10. Find decimal notation: 78.5%.

Simplify.

11. $\frac{3}{5} - \frac{5}{12}$

12. $3.4 + (-0.8)$

13. $(-2)(-1.4)(2.6)$

14. $\frac{3}{8} \div \left(-\frac{9}{10}\right)$

15. $2 - [32 \div (4 + 2^2)]$

16. $-5 + 16 \div 2 \cdot 4$

17. $y - (3y + 7)$

18. $3(x - 1) - 2[x - (2x + 7)]$

Solve.

19. $1.5 = 2.7 + x$

20. $\frac{2}{7}x = -6$

21. $5x - 9 = 36$

22. $\frac{2}{3} = \frac{-m}{10}$

23. $5.4 - 1.9x = 0.8x$

24. $x - \frac{7}{8} = \frac{3}{4}$

25. $2(2 - 3x) = 3(5x + 7)$

26. $\frac{1}{4}x - \frac{2}{3} = \frac{3}{4} + \frac{1}{3}x$

27. $y + 5 - 3y = 5y - 9$

28. $x - 28 < 20 - 2x$

29. $2(x + 2) \geq 5(2x + 3)$

30. Solve $A = 2\pi rh + \pi r^2$, for h.

31. In which quadrant is the point $(3, -1)$ located?

32. Graph on a number line: $-1 < x \leq 2$.

Graph.

33. $y = -2$

34. $2x + 5y = 10$

35. $y = -2x + 1$

36. $y = \frac{2}{3}x$

Find the coordinates of the x- and y-intercepts. Do not graph.

37. $2x - 7y = 21$

38. $y = 4x + 5$

Solve.

39. A bill for a tie at Great Necks totaled \$14.28. How much did the tie cost if the sales tax is 5%?

40. If 25 is subtracted from a certain number, the result is 129. Find the number.

41. Sara and Becky purchased two dresses for a total of \$107. Sara paid \$17 more for her dress than Becky did. What did Becky pay?

42. A 143-m wire is cut into three pieces. The second is 3 m longer than the first. The third is four fifths as long as the first. How long is each piece?

43. Money is invested in a savings account at 4% simple interest. After 1 year, there is \$1560 in the account. How much was originally invested?

44. Sven's test grades are 75, 82, 86, and 79. Determine, in terms of an inequality, what scores on the last test will assure Sven an average test score of at least 80.

45. Translate to an equation containing two variables. Be sure to state what each variable represents.

> The steak cost \$5 more than three times the cost of the chicken.

46. Sparkle's Cleaning charges \$20 a visit plus \$15 an hour for commercial cleaning jobs. Draw a graph that can be used to predict the cost of a cleaning job. Use the horizontal axis for time and the vertical axis for cost. Then use the graph to estimate the cost of a $2\frac{1}{2}$-hr job.

Synthesis

47. Paula's salary at the end of a year is \$26,780. This reflects a 4% salary increase that preceded a 3% cost-of-living adjustment during the year. What was her salary at the beginning of the year?

Solve.

48. $4|x| - 13 = 3$

49. $4(x + 2) = 4(x - 2) + 16$

50. $0(x + 3) + 4 = 0$

51. $\frac{2 + 5x}{4} = \frac{11}{28} + \frac{8x + 3}{7}$

52. $5(7 + x) = (x + 7)5$

53. Solve $p = \frac{2}{m + Q}$, for Q.

CHAPTER 4

Polynomials

AN APPLICATION

During the first 13 seconds of a jump, the number of feet that a skydiver falls in t seconds is approximated by the polynomial

$$11.12t^2.$$

Approximately how far has a skydiver fallen 10 seconds after jumping from a plane?

This problem appears as Exercise 71 in Section 4.2.

Fran Strimenos
SKYDIVING INSTRUCTOR

"We use math every day in running our parachute company and school. Every jump requires calculations of wind velocity, the plane's speed, and the fall rate, and must take into consideration parachute design and the jumper's weight."

Algebraic expressions like $16t^2$, $5a^2 - 45$, and $3x^2 - 7x + 5$ are called polynomials. Polynomials occur frequently in applications and appear in most branches of mathematics. Thus learning to add, subtract, multiply, and divide polynomials is an important part of most courses in elementary algebra and is the focus of this chapter.

In addition to the material from this chapter, the review and test for Chapter 4 include material from Sections 1.5, 2.5, 2.6, and 3.1.

4.1 Exponents and Their Properties

Multiplying Powers with Like Bases • Dividing Powers with Like Bases • Zero as an Exponent • Raising a Power to a Power • Raising a Product or a Quotient to a Power

Because most of the polynomials with which we will work contain exponents, it is important that we develop some rules for simplifying exponential expressions.

Multiplying Powers with Like Bases

We know that an expression like a^3 means $a \cdot a \cdot a$ and that a^1 means a. Now consider multiplying powers with like bases:

$$\begin{aligned} a^3 \cdot a^2 &= (a \cdot a \cdot a)(a \cdot a) && \text{There are three factors in } a^3\text{; two factors in } a^2. \\ &= a \cdot a \cdot a \cdot a \cdot a && \text{Using an associative law} \\ &= a^5. \end{aligned}$$

Note that the exponent in a^5 is the sum of the exponents in $a^3 \cdot a^2$. That is, $3 + 2 = 5$. Similarly,

$$b^4 \cdot b^3 = (b \cdot b \cdot b \cdot b)(b \cdot b \cdot b) = b^7, \quad \text{where } 4 + 3 = 7.$$

Adding the exponents gives the correct result.

The Product Rule

For any number a and any positive integers m and n,

$$a^m \cdot a^n = a^{m+n}.$$

(When multiplying with exponential notation, if the bases are the same, keep the base and add the exponents.)

EXAMPLE 1

Multiply and simplify each of the following. (Here "simplify" means express the product as one base to a power whenever possible.)

a) $x^2 \cdot x^9$ b) $8^4 \cdot 8^3$ c) $x \cdot x^8$

d) $m^5 m^{10} m^3$ e) $(a^3b^2)(a^3b^5)$

Solution

a) $x^2 \cdot x^9 = x^{2+9}$ Adding exponents: $a^m \cdot a^n = a^{m+n}$
$= x^{11}$

b) $8^4 \cdot 8^3 = 8^{4+3}$
$= 8^7$

c) $x \cdot x^8 = x^1 \cdot x^8 = x^{1+8}$
$= x^9$

d) $m^5 m^{10} m^3 = m^{5+10+3}$
$= m^{18}$

e) $(a^3b^2)(a^3b^5) = a^3b^2a^3b^5$ Using an associative law
$= a^3a^3b^2b^5$ Using a commutative law
$= a^6b^7$ Adding exponents ❑

Dividing Powers with Like Bases

The following suggests a rule for dividing powers with like bases, such as a^5/a^2:

$$\frac{a^5}{a^2} = \frac{a \cdot a \cdot a \cdot a \cdot a}{a \cdot a} = \frac{a \cdot a \cdot a \cdot a \cdot a}{1 \cdot a \cdot a} = \frac{a \cdot a \cdot a}{1} \cdot \frac{a \cdot a}{a \cdot a} = \frac{a \cdot a \cdot a}{1} \cdot 1$$
$$= a \cdot a \cdot a = a^3.$$

Note that the exponent in a^3 is the difference of the exponents in a^5/a^2. Similarly,

$$\frac{x^4}{x^3} = \frac{x \cdot x \cdot x \cdot x}{x \cdot x \cdot x} = \frac{x}{1} \cdot \frac{x \cdot x \cdot x}{x \cdot x \cdot x} = \frac{x}{1} \cdot 1 = x^1, \quad \text{or } x.$$

Subtracting exponents gives the correct result.

The Quotient Rule

For any nonzero number a and any positive integers m and n for which $m > n$,

$$\frac{a^m}{a^n} = a^{m-n}.$$

(To divide with exponential notation, if the bases are the same, keep the base and subtract the exponent of the denominator from the exponent of the numerator.)

EXAMPLE 2

Divide and simplify. (Here "simplify" means express the quotient as one base to a power whenever possible.)

a) $\dfrac{x^8}{x^2}$ b) $\dfrac{6^5}{6^3}$ c) $\dfrac{t^{12}}{t}$ d) $\dfrac{p^5q^7}{p^2q^5}$

Solution

a) $\dfrac{x^8}{x^2} = x^{8-2}$ Subtracting exponents: $\dfrac{a^m}{a^n} = a^{m-n}$
$= x^6$

b) $\dfrac{6^5}{6^3} = 6^{5-3}$
$= 6^2$

c) $\dfrac{t^{12}}{t} = t^{12-1} = t^{11}$ d) $\dfrac{p^5q^7}{p^2q^5} = p^{5-2}q^{7-5} = p^3q^2$ ❑

Zero as an Exponent

The quotient rule can be used to help determine what 0 should mean when it appears as an exponent. Consider a^4/a^4, where a is nonzero. Since the numerator and the denominator are the same,

$$\frac{a^4}{a^4} = 1.$$

On the other hand, using the quotient rule gives us

$$\frac{a^4}{a^4} = a^{4-4} = a^0. \quad \text{Subtracting exponents}$$

Since $a^0 = a^4/a^4 = 1$, it follows that $a^0 = 1$ for any nonzero value of a.

The Exponent Zero

For any real number a, $a \neq 0$,

$$a^0 = 1.$$

(Any nonzero number raised to the 0 power is 1.)

EXAMPLE 3

Simplify: **(a)** 1957^0; **(b)** $(-7)^0$; **(c)** $(-1)7^0$; **(d)** $(3x)^0$.

Solution

a) $1957^0 = 1$ Any nonzero number raised to the 0 power is 1.
b) $(-7)^0 = 1$ Any nonzero number raised to the 0 power is 1.
c) We have

$$(-1)7^0 = (-1)1 = -1.$$

Since multiplying by -1 is the same as finding the opposite, the expression $(-1)7^0$ could have been written as -7^0. Note that while $-7^0 = -1$, part (b) shows that $(-7)^0 = 1$.

d) The parentheses indicate that the base is $3x$. Thus,

$$(3x)^0 = 1 \quad \text{for any } x \neq 0.$$ ❑

To see why 0^0 is not defined, note that $0^0 = 0^{1-1} = 0^1/0^1 = 0/0$. As we saw in Section 1.7, 0/0 has not been defined. Thus, 0^0 is not defined either. Henceforth in this text we will assume that expressions like a^m do not represent 0^0.

Raising a Power to a Power

Consider an expression like $(5^2)^4$.

$$\begin{aligned}(5^2)^4 &= (5^2)(5^2)(5^2)(5^2) && \text{There are four factors of } 5^2.\\ &= (5\cdot5)(5\cdot5)(5\cdot5)(5\cdot5)\\ &= 5\cdot5\cdot5\cdot5\cdot5\cdot5\cdot5\cdot5 && \text{Using an associative law}\\ &= 5^8.\end{aligned}$$

Note that in this case we could have multiplied the exponents:

$$(5^2)^4 = 5^{2 \cdot 4} = 5^8.$$

Likewise, $(y^7)^3 = (y^7)(y^7)(y^7) = y^{21}$. Once again, we get the same result if we multiply the exponents:

$$(y^7)^3 = y^{7 \cdot 3} = y^{21}.$$

The Power Rule

For any number a and any whole numbers m and n,

$$(a^m)^n = a^{mn}.$$

(To raise a power to a power, multiply the exponents and leave the base unchanged.)

EXAMPLE 4

Simplify: **(a)** $(m^2)^5$; **(b)** $(3^5)^4$.

Solution

a) $(m^2)^5 = m^{2 \cdot 5}$
$= m^{10}$

Multiplying exponents: $(a^m)^n = a^{mn}$

b) $(3^5)^4 = 3^{5 \cdot 4}$
$= 3^{20}$ ❑

Raising a Product or a Quotient to a Power

When an expression inside parentheses is raised to a power, the inside expression is the base. Let us compare $2a^3$ and $(2a)^3$:

$2a^3 = 2 \cdot a \cdot a \cdot a$; The base is a.

$(2a)^3 = (2a)(2a)(2a)$ The base is $2a$.

$= (2 \cdot 2 \cdot 2)(a \cdot a \cdot a)$ Using an associative and a commutative law

$= 2^3a^3$

$= 8a^3$.

We see that $2a^3$ and $(2a)^3$ are *not* equivalent. Note too that $(2a)^3$ can be simplified by raising each factor to the power 3. This leads us to the following rule for raising a product to a power.

Raising a Product to a Power

For any numbers a and b and any whole number n,

$$(ab)^n = a^nb^n.$$

(To raise a product to the nth power, raise each factor to the nth power.)

EXAMPLE 5

Simplify: **(a)** $(4a)^3$; **(b)** $(5x^4)^2$; **(c)** $(-4a^5b^3)^3$.

Solution

a) $(4a)^3 = 4^3a^3 = 64a^3$ Raising each factor to the third power and simplifying

b) $(5x^4)^2 = 5^2(x^4)^2$ — Raising each factor to the second power

$= 25x^8$ — Simplifying 5^2; using the power rule

c) $(-4a^5b^3)^3 = (-4)^3(a^5)^3(b^3)^3$ — Cubing each factor

$= -64a^{15}b^9$ — A negative number raised to an odd power is negative; using the power rule. □

There is a similar rule for raising a quotient to a power.

Raising a Quotient to a Power

For any numbers a and b, $b \neq 0$, and any whole number n,

$$\left(\frac{a}{b}\right)^n = \frac{a^n}{b^n}.$$

(To raise a quotient to a power, raise the numerator to the power and divide by the denominator to the power.)

EXAMPLE 6

Simplify: **(a)** $\left(\frac{x}{5}\right)^2$; **(b)** $\left(\frac{5}{a^4}\right)^3$; **(c)** $\left(\frac{3a^4}{b^3}\right)^2$.

Solution

a) $\left(\frac{x}{5}\right)^2 = \frac{x^2}{5^2} = \frac{x^2}{25}$ — Squaring the numerator and the denominator

b) $\left(\frac{5}{a^4}\right)^3 = \frac{5^3}{(a^4)^3} = \frac{125}{a^{4\cdot 3}} = \frac{125}{a^{12}}$

c) $\left(\frac{3a^4}{b^3}\right)^2 = \frac{(3a^4)^2}{(b^3)^2} = \frac{3^2(a^4)^2}{b^{3\cdot 2}} = \frac{9a^8}{b^6}$ □

In the following summary of definitions and rules, we assume that no denominators are 0 and that 0^0 is not considered.

For any whole numbers m and n,

1 as an exponent:	$a^1 = a$
0 as an exponent:	$a^0 = 1$
The Product Rule:	$a^m \cdot a^n = a^{m+n}$
The Quotient Rule:	$\frac{a^m}{a^n} = a^{m-n}$
The Power Rule:	$(a^m)^n = a^{mn}$
Raising a product to a power:	$(ab)^n = a^nb^n$
Raising a quotient to a power:	$\left(\frac{a}{b}\right)^n = \frac{a^n}{b^n}$

EXERCISE SET 4.1

Multiply and simplify.

1. $2^4 \cdot 2^3$
2. $3^5 \cdot 3^2$
3. $8^5 \cdot 8^9$
4. $n^3 \cdot n^{20}$
5. $x^4 \cdot x^3$
6. $y^7 \cdot y^9$
7. $9^{17} \cdot 9^{21}$
8. $t^0 \cdot t^{16}$
9. $(3y)^4(3y)^8$
10. $(2t)^8(2t)^{17}$
11. $(7y)^1(7y)^{16}$
12. $(8x)^0(8x)^1$
13. $(a^2b^7)(a^3b^2)$
14. $(m^5n^4)(m^6n^2)$
15. $(xy^9)(x^3y^5)$
16. $(a^8b^3)(a^4b)$
17. $r^3 \cdot r^7 \cdot r^2$
18. $s^4 \cdot s^5 \cdot s^2$
19. $x^3(xy^4)(xy)$
20. $a^4(a^3b)(ab)$

Divide and simplify.

21. $\frac{7^5}{7^2}$
22. $\frac{4^7}{4^3}$
23. $\frac{8^{12}}{8^6}$
24. $\frac{9^{14}}{9^2}$
25. $\frac{y^9}{y^5}$
26. $\frac{x^{12}}{x^{11}}$
27. $\frac{(5a)^7}{(5a)^6}$
28. $\frac{(3m)^9}{(3m)^8}$
29. $\frac{6^5x^8}{6^2x^3}$
30. $\frac{3^9a^7}{3^4a^5}$
31. $\frac{18m^5}{6m^2}$
32. $\frac{30n^7}{6n^3}$
33. $\frac{a^9b^7}{a^2b}$
34. $\frac{r^{10}s^7}{r^3s^0}$
35. $\frac{m^9n^8}{m^0n^4}$
36. $\frac{a^{10}b^{12}}{a^2b^3}$

Simplify.

37. x^0 when $x = -12$
38. y^0 when $y = 23$
39. $5x^0$ when $x = -4$
40. $7m^0$ when $m = 1.7$
41. n^0, $n \neq 0$
42. t^0, $t \neq 0$
43. 10^0
44. 9^0
45. $5^1 - 5^0$
46. $8^0 - 8^1$

Simplify. Answers should not contain parentheses.

47. $(x^3)^4$
48. $(a^4)^6$
49. $(2^3)^8$
50. $(5^7)^3$
51. $(m^7)^5$
52. $(n^9)^2$
53. $(a^{25})^3$
54. $(a^3)^{25}$
55. $(3x)^2$
56. $(5a)^2$
57. $(-2a)^3$
58. $(-3x)^3$
59. $(4m^3)^2$
60. $(5n^4)^2$
61. $(3a^2b)^3$
62. $(5xy^2)^3$
63. $(a^3b^2)^5$
64. $(m^4n^5)^6$
65. $(-5x^4y^5)^2$
66. $(-3a^5b^7)^4$
67. $\left(\frac{a}{4}\right)^3$
68. $\left(\frac{3}{x}\right)^4$
69. $\left(\frac{7}{5a}\right)^2$
70. $\left(\frac{4x}{3}\right)^3$
71. $\left(\frac{a^2}{b^3}\right)^4$
72. $\left(\frac{x^3}{y^4}\right)^5$
73. $\left(\frac{y^3}{2}\right)^2$
74. $\left(\frac{a^5}{3}\right)^3$
75. $\left(\frac{5x^2}{y^3}\right)^3$
76. $\left(\frac{7y^5}{x^4}\right)^2$
77. $\left(\frac{a^3}{-2b^5}\right)^4$
78. $\left(\frac{x^5}{-3y^3}\right)^4$
79. $\left(\frac{2a^2}{3b^4}\right)^3$
80. $\left(\frac{3x^5}{4y^3}\right)^2$
81. $\left(\frac{4x^3y^5}{3z^7}\right)^2$
82. $\left(\frac{5a^7}{2b^5c}\right)^3$

Skill Maintenance

83. Factor: $3s + 3t + 24$.
84. Factor: $-7x - 14$.
85. Collect like terms: $9x + 2y - 4x - 2y$.
86. 24 is what percent of 64?

Synthesis

87. ◈ Under what conditions does a^n represent a negative number? Why?
88. ◈ Using the quotient rule, explain why 9^0 is 1.
89. Solve for x: $\frac{w^{50}}{w^x} = w^x$.

Simplify.

90. $(y^{2x})(y^{3x})$
91. $a^{5k} \div a^{3k}$
92. $\frac{a^{6t}(a^{7t})}{a^{9t}}$
93. $\frac{\left(\frac{1}{2}\right)^4}{\left(\frac{1}{2}\right)^5}$
94. $\frac{(0.4)^5}{(0.4)^3(0.4)^2}$

Use $>$, $<$, or $=$ for ■ to write a true sentence.

95. 3^5 ■ 3^4
96. 4^2 ■ 4^3
97. 4^3 ■ 5^3
98. 4^3 ■ 3^4

Find a value of the variable that shows that the two expressions are *not* equivalent.

99. $3x^2$; $(3x)^2$
100. $(a + 3)^2$; $a^2 + 3^2$

101. $\frac{x+2}{2}$; x

102. $\frac{y^6}{y^3}$; y^2

Interest compounded annually. If a principal P is invested at interest rate r, compounded annually, in t years it will grow to an amount A given by

$$A = P(1 + r)^t.$$

103. Suppose \$10,400 is invested at 8.5% compounded annually. How much is in the account at the end of 5 years?

104. Suppose \$20,800 is invested at 4.5%, compounded annually. How much is in the account at the end of 6 years?

105. Find the area of a square if each side has length $5x$.

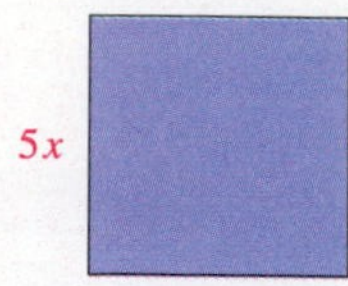

106. Find the volume of a cube if each side has length $7a$.

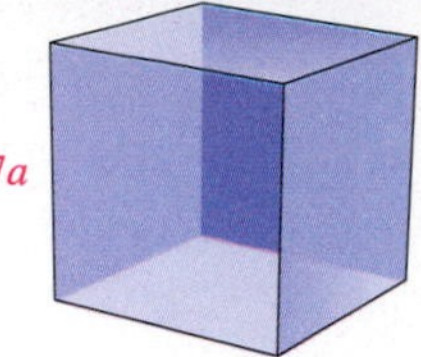

4.2 Polynomials

Terms • Collecting Like Terms • Evaluating Polynomials and Applications

We now begin our study of an important algebraic expression known as a *polynomial*. Certain polynomials have appeared earlier in this text so you already have some experience working with them.

One type of polynomial that we have already seen is a *monomial*. A **monomial** is a constant or the product of some constant and variable(s) raised to a whole-number power. Examples of monomials are

$$5, \quad x, \quad -7a, \quad 3m^2, \quad \text{and} \quad \tfrac{3}{4}x^2y^5.$$

Algebraic expressions like the following are **polynomials:**

$$\tfrac{3}{4}y^5, \quad -2, \quad 5y+3, \quad 3x^2+2x-5, \quad -7a^3+4ab, \quad 6x, \quad 37p^4, \quad x, \quad 0.$$

Polynomial

A *polynomial* is a monomial or a combination of sums and/or differences of monomials.

The following algebraic expressions are *not* polynomials:

$$\textbf{(1)}\ \frac{x+3}{x-4}, \qquad \textbf{(2)}\ 5x^3 - 2x^2 + \frac{1}{x}, \qquad \textbf{(3)}\ \frac{1}{x^3-2}.$$

Expressions (1) and (3) are not polynomials because they represent quotients, not sums or differences. Expression (2) is not a polynomial because $1/x$ is not a monomial.

A sum or difference of two monomials is called a **binomial.** When three monomials are added and/or subtracted, we have what is called a **trinomial.** Polynomials composed of four or more monomials have no special name.

Monomials	Binomials	Trinomials	No Special Name
$4x^2$	$2x + 4$	$3t^3 + 4t + 7$	$4x^3 - 5x^2 + x - 8$
9	$3a^5 + 6bc$	$6x^7 - 7x^2 + 4$	$z^5 + 2z^4 - z^3 + 7z + 3$
$-7a^{19}$	$-9x^7 - 6$	$4x^2 - 6x - \frac{1}{2}$	$4x^6 - 3x^5 + x^4 - x^3 + 2x - 1$

Terms

Although every monomial is a term, some terms (like $5/x^2$) are not monomials. To identify the terms of a polynomial, we look at the polynomial as a sum of monomials.

EXAMPLE 1

Identify the terms of the polynomial $3t^4 - 5t^6 - 4t + 2$.

Solution The terms are $3t^4$, $-5t^6$, $-4t$, and 2. If this is difficult to see, try rewriting all subtractions as additions of opposites:

$$3t^4 - 5t^6 - 4t + 2 = \underset{\uparrow}{3t^4} + \underset{\uparrow}{(-5t^6)} + \underset{\uparrow}{(-4t)} + \underset{\uparrow}{2}.$$

These are the terms of the polynomial. ❑

The **coefficient** of a term is the number multiplying the variable(s) in the term. Thus the coefficient of the term $5x^3$ is 5.

EXAMPLE 2

Identify the coefficient of each term in the polynomial

$$4x^3 - 7x^2y + x - 8.$$

Solution

The coefficient of the first term is 4.

The coefficient of the second term is -7.

The coefficient of the third term is 1, since $x = 1x$.

The coefficient of the fourth term is -8. ❑

The **degree of a term** is the exponent of the variable.* The degree of the term $5x^3$ is 3.

EXAMPLE 3

Identify the degree of each term of $8x^4 + 3x + 7$.

Solution

The degree of $8x^4$ is 4.

The degree of $3x$ is 1. Recall that $x = x^1$.

The degree of 7 is 0. Think of 7 as $7x^0$. Recall that $x^0 = 1$. ❑

---

*A more detailed definition for terms with several variables is given in Section 4.6.

The **degree of a polynomial** is the largest of the degrees of its terms, unless it is the polynomial 0. We agree that 0 has *no* degree either as a term or as a polynomial. This is because we can express 0 as $0 = 0x^5 = 0x^7$, and so on, using any exponent we wish.

EXAMPLE 4

Identify the degree of the polynomial $5x^3 - 6x^4 + 7$.

Solution We have

$$5x^3 - 6x^4 + 7. \quad \text{The largest exponent is 4.}$$

The degree of the polynomial is 4. ❑

Let us summarize the terminology we have learned for the polynomial

$$3x^4 - 8x^3 + 5x^2 + 7x - 6.$$

Term	Coefficient	Degree of the Term	Degree of the Polynomial
$3x^4$	3	4	4
$-8x^3$	-8	3	
$5x^2$	5	2	
$7x$	7	1	
-6	-6	0	

Collecting Like Terms

As we saw in Section 1.8, when a polynomial contains *like,* or *similar, terms*—that is, terms with the same variable(s) raised to the same power(s)—those terms can be *combined* or *collected.*

EXAMPLE 5

Identify the like terms in $4x^3 + 5x - 7x^2 + 2x^3 + x^2$.

Solution

Like terms: $4x^3$ and $2x^3$ Same variable and exponent

Like terms: $-7x^2$ and x^2 Same variable and exponent ❑

EXAMPLE 6

Collect like terms:

a) $2x^3 - 6x^3$

b) $5x^2 + 7 + 2x^4 + 4x^2 - 11 - 2x^4$

c) $7a^3 - 5a^2 + 9a^3 + a^2$

d) $\frac{2}{3}x^4 - x^3 - \frac{1}{6}x^4 + \frac{2}{5}x^3 - \frac{3}{10}x^3$

Solution

a) $2x^3 - 6x^3 = (2 - 6)x^3$ Using the distributive law

$= -4x^3$

b) $5x^2 + 7 + 2x^4 + 4x^2 - 11 - 2x^4$

$= (5 + 4)x^2 + (2 - 2)x^4 + (7 - 11)$

$= 9x^2 + 0x^4 + (-4)$

These steps are often done mentally.

$= 9x^2 - 4$

c) $7a^3 - 5a^2 + 9a^3 + a^2 = 7a^3 - 5a^2 + 9a^3 + 1a^2$

When a variable to a power appears without a coefficient, we can write in 1.

$$= 16a^3 - 4a^2$$

d) $\frac{2}{3}x^4 - x^3 - \frac{1}{6}x^4 + \frac{2}{5}x^3 - \frac{3}{10}x^3 = \left(\frac{2}{3} - \frac{1}{6}\right)x^4 + \left(-1 + \frac{2}{5} - \frac{3}{10}\right)x^3$

$$= \left(\frac{4}{6} - \frac{1}{6}\right)x^4 + \left(-\frac{10}{10} + \frac{4}{10} - \frac{3}{10}\right)x^3$$
$$= \frac{3}{6}x^4 - \frac{9}{10}x^3$$
$$= \frac{1}{2}x^4 - \frac{9}{10}x^3$$

❑

Note in Examples 6(b), (c), and (d) that the solutions are written so that the term of highest degree appears first, followed by the term of next highest degree, and so on. This is known as *descending order* and is the form in which answers will normally appear.

Evaluating Polynomials and Applications

Recall from Chapter 1 that when we replace the variable in a polynomial with a number, the polynomial then represents a number, or *value*, that can be calculated using the rules for the order of operations.

EXAMPLE 7

Evaluate $-x^2 + 3x + 9$ when $x = -2$.

Solution When $x = -2$, we have

$$-x^2 + 3x + 9 = -(-2)^2 + 3(-2) + 9$$
$$= -4 + (-6) + 9$$
$$= -10 + 9 = -1.$$

❑

Polynomials occur in many real-world situations and are used in problem solving. The following examples are three such applications. Because the examples involve only the evaluation of a polynomial, we do not apply all five problem-solving steps.

EXAMPLE 8

Volume of a cube. The volume of a cube with sides of length x is given by the polynomial

$$x^3.$$

Find the volume of a cube with sides of length 5 cm.

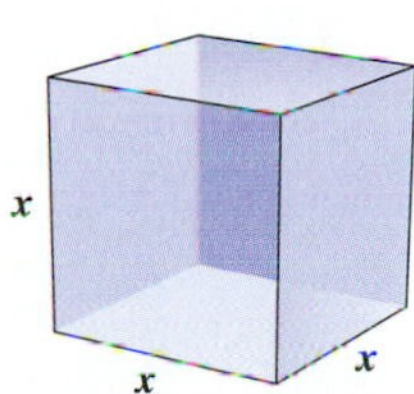

Solution We evaluate the polynomial for $x = 5$:

$$x^3 = 5^3 = 125.$$

The volume is 125 cm^3 (cubic centimeters).

❑

EXAMPLE 9

Games in a sports league. In a sports league of n teams in which each team plays every other team twice, the total number of games to be played is given by the polynomial

$$n^2 - n.$$

A women's softball league has 10 teams. What is the total number of games to be played?

Solution We evaluate the polynomial for $n = 10$:

$$\begin{aligned} n^2 - n &= 10^2 - 10 \\ &= 100 - 10 \\ &= 90. \end{aligned}$$

The league plays 90 games. ❑

EXAMPLE 10

Medical dosage. The concentration, in parts per million, of a certain medication in the bloodstream after t hours is given by the polynomial

$$-0.05t^2 + 2t + 2.$$

Find the concentration after 2 hr.

Solution To find the concentration after 2 hr, we evaluate the polynomial for $t = 2$:

$$\begin{aligned} -0.05t^2 + 2t + 2 &= -0.05(2)^2 + 2(2) + 2 \\ &= -0.05(4) + 2(2) + 2 \\ &= -0.2 + 4 + 2 \\ &= 5.8. \end{aligned}$$

The concentration after 2 hr is 5.8 parts per million. ❑

The polynomial in Example 10 can be graphed if we evaluate it for several values of t. Note that the concentration peaks at the 20-hr mark and after a bit more than 40 hr, the concentration is 0. Since neither time nor concentration can be negative, our graph uses only the first quadrant.

t	$-0.05t^2 + 2t + 2$
0	2
2	5.8
10	17
20	22
30	17

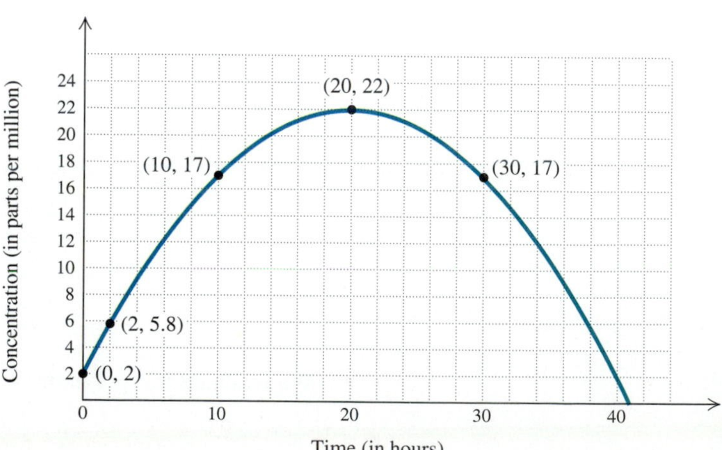

EXERCISE SET 4.2

Classify the polynomial as a monomial, binomial, trinomial, or none of these.

1. $x^2 - 10x + 25$

2. $-6x^4$

3. $x^3 - 7x^2 + 2x - 4$

4. $x^2 - 9$

5. $4x^2 - 25$

6. $2x^4 - 7x^3 + x^2 + x - 6$

7. $40x$

8. $4x^2 + 12x + 9$

Identify the terms of the polynomial.

9. $2 - 3x + x^2$

10. $2x^2 + 3x - 4$

Identify the like terms in the polynomial.

11. $5x^3 + 6x^2 - 3x^2$

12. $3x^2 + 4x^3 - 2x^2$

13. $2x^4 + 5x - 7x - 3x^4$

14. $-3t + t^3 - 2t - 5t^3$

Identify the coefficient of each term of the polynomial.

15. $-3x + 6$

16. $2x - 4$

17. $5x^2 + 3x + 3$

18. $3x^2 - 5x + 2$

19. $-7x^3 + 6x^2 + 3x + 7$

20. $5x^4 + x^2 - x + 2$

21. $-5x^4 + 6x^3 - 3x^2 + 8x - 2$

22. $7x^3 - 4x^2 - 4x + 5$

Determine the degree of each term of the polynomial and the degree of the polynomial.

23. $2x - 4$

24. $6 - 3x$

25. $3x^2 - 5x + 2$

26. $5x^3 - 2x^2 + 3$

27. $-7x^3 + 6x^2 + 3x + 7$

28. $5x^4 + x^2 - x + 2$

29. $x^2 - 3x + x^6 - 9x^4$

30. $8x - 3x^2 + 9 - 8x^3$

31. For the polynomial $-7x^4 + 6x^3 - 3x^2 + 8x - 2$, complete the following table.

Term	Coefficient	Degree of the Term	Degree of the Polynomial
$6x^3$	6		
		2	
$8x$		1	
	-2		

32. For the polynomial $3x^2 + 8x^5 - 46x^3 + 6x - 2.4 - \frac{1}{2}x^4$, complete the following table.

Term	Coefficient	Degree of the Term	Degree of the Polynomial
		5	
$-\frac{1}{2}x^4$		4	
	-46		
$3x^2$		2	
	6		
-2.4			

Collect like terms. Write all answers in descending order.

33. $2x - 5x$

34. $2x^2 + 8x^2$

35. $x - 9x$

36. $x - 5x$

37. $5x^3 + 6x^3 + 4$

38. $6x^4 - 2x^4 + 5$

39. $5x^3 + 6x - 4x^3 - 7x$

40. $3a^4 - 2a + 2a + a^4$

41. $6b^5 + 3b^2 - 2b^5 - 3b^2$

42. $2x^2 - 6x + 3x + 4x^2$

43. $\frac{1}{4}x^5 - 5 + \frac{1}{2}x^5 - 2x - 37$

44. $\frac{1}{3}x^3 + 2x - \frac{1}{6}x^3 + 4 - 16$

45. $6x^2 + 2x^4 - 2x^2 - x^4 - 4x^2$

46. $8x^2 + 2x^3 - 3x^3 - 4x^2 - 4x^2$

47. $\frac{1}{4}x^3 - x^2 - \frac{1}{6}x^2 + \frac{3}{8}x^3 + \frac{5}{16}x^3$

48. $\frac{1}{5}x^4 + \frac{1}{5} - 2x^2 + \frac{1}{10} - \frac{3}{15}x^4 + 2x^2 - \frac{3}{10}$

49. $3x^4 - 5x^6 - 2x^4 + 6x^6$

50. $-1 + 5x^3 - 3 - 7x^3 + x^4 + 5$

51. $-2x + 4x^3 - 7x + 9x^3 + 8$

52. $-6x^2 + x - 5x + 7x^2 + 1$

53. $3x + 3x + 3x - x^2 - 4x^2$

54. $-2x - 2x - 2x + x^3 - 5x^3$

55. $-x + \frac{3}{4} + 15x^4 - x - \frac{1}{2} - 3x^4$

56. $2x - \frac{5}{6} + 4x^3 + x + \frac{1}{3} - 2x$

Evaluate the polynomial for $x = 4$.

57. $-5x + 2$

58. $-3x + 1$

59. $2x^2 - 5x + 7$

60. $3x^2 + x + 7$

61. $x^3 - 5x^2 + x$

62. $7 - x + 3x^2$

Evaluate the polynomial for $x = -1$.

63. $3x + 5$

64. $6 - 2x$

65. $x^2 - 2x + 1$

66. $5x - 6 + x^2$

67. $-3x^3 + 7x^2 - 3x - 2$

68. $-2x^3 - 5x^2 + 4x + 3$

Daily accidents. The average number of accidents per day involving drivers of age r can be approximated by the polynomial

$0.4r^2 - 40r + 1039$.

69. Evaluate the polynomial for $r = 18$ to find the daily number of accidents involving 18-year-old drivers.

70. Evaluate the polynomial for $r = 20$ to find the daily number of accidents involving 20-year-old drivers.

71. *Skydiving.* During the first 13 sec of a jump, the number of feet that a skydiver falls in t seconds is approximated by the polynomial

$11.12t^2$.

Approximately how far has a skydiver fallen 10 sec after jumping from a plane?

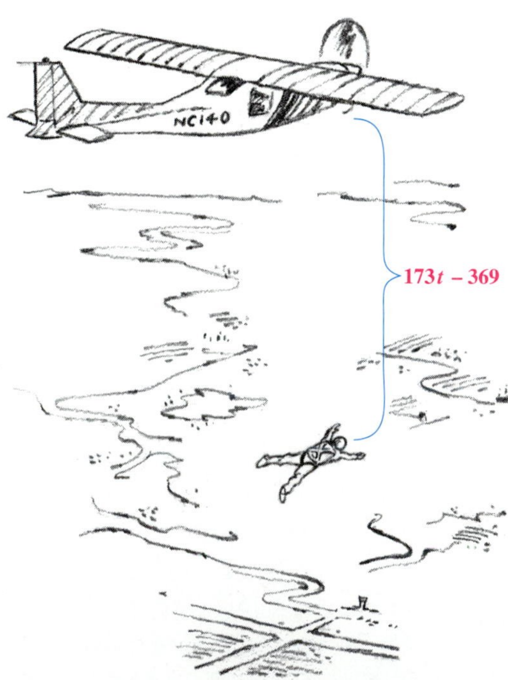

72. *Skydiving.* For jumps that exceed 13 sec, the polynomial $173t - 369$ can be used to approximate the distance, in feet, that a skydiver has fallen in t seconds. Approximately how far has a skydiver fallen 20 sec after jumping from a plane?

Total revenue. An electronics firm is marketing a new kind of stereo. Total revenue is the total amount of money taken in. The firm determines that when it sells x stereos, it will take in

$280x - 0.4x^2$ dollars.

73. What is the total revenue from the sale of 75 stereos?

74. What is the total revenue from the sale of 100 stereos?

Total cost. The electronics firm determines that the total cost of producing x stereos is given by

$5000 + 0.6x^2$ dollars.

75. What is the total cost of producing 500 stereos?

76. What is the total cost of producing 650 stereos?

Circumference. The circumference of a circle of radius r is given by the polynomial $2\pi r$,

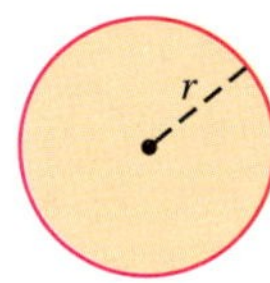

where π is an irrational number. For an approximation of π, use 3.14 unless indicated otherwise.

77. Find the circumference of a circle with radius 10 cm.

78. Find the circumference of a circle with radius 5 ft.

Area of a circle. The area of a circle of radius r is given by the polynomial πr^2.

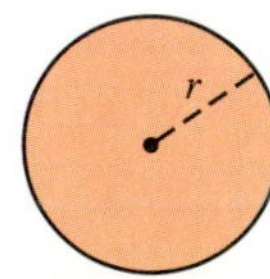

79. Find the area of a circle with radius 5 m. Use 3.14 for π.

80. Find the area of a circle with radius 10 in. Use 3.14 for π.

In Exercises 81 and 82, complete the table for the given choices of t. Then plot the points and connect them with a smooth curve representing the graph of the polynomial.

81.

t	$-t^2 + 6t - 4$
1	
2	
3	
4	
5	

82.

t	$-t^2 + 10t - 18$
3	
4	
5	
6	
7	

Skill Maintenance

83. The sum of the page numbers on the facing pages of a book is 549. What are the page numbers?

84. A family spent \$2011 to drive a car one year, during which the car was driven 7400 mi. The family spent \$972 for insurance and \$114 for a license registration fee. The only other cost was for gasoline. How much did gasoline cost per mile?

Synthesis

85. ◈ Explain why an understanding of the rules for order of operations is essential when evaluating polynomials.

86. ◈ Is it better to evaluate a polynomial before or after like terms have been combined? Why?

Simplify.

87. $\frac{9}{2}x^8 + \frac{1}{9}x^2 + \frac{1}{2}x^9 + \frac{9}{2}x + \frac{9}{2}x^9 + \frac{8}{9}x^2 + \frac{1}{2}x - \frac{1}{2}x^8$

88. $(3x^2)^3 + 4x^2 \cdot 4x^4 - x^4(2x)^2 + ((2x)^2)^3 - 100x^2(x^2)^2$

89. Evaluate both $s^2 - 50s + 675$ and $-s^2 + 50s - 675$ when $s = 18$, $s = 25$, and $s = 32$.

90. *Daily accidents.* The average number of accidents per day involving drivers of age r can be approximated by the polynomial

$$0.4r^2 - 40r + 1039.$$

For what age is the number of daily accidents smallest?

91. Construct a polynomial in x (meaning that x is the variable) of degree 5 with four terms and coefficients that are integers.

92. Construct a trinomial in y of degree 4 with coefficients that are rational numbers.

93. What is the degree of $(5m^5)^2$?

94. Construct three like terms of degree 4.

95. *Path of the Olympic arrow.* The Olympic flame at the 1992 Summer Olympics was lit by a flaming arrow. As the arrow moved d meters horizontally from the archer, its height, in meters, was approximated by the polynomial

$$-0.0064d^2 + 0.8d + 2.$$

Complete the table for the choices of d given. Then plot the points and draw a graph representing the path of the arrow.

d	$-0.0064d^2 + 0.8d + 2$
0	
30	
60	
90	
120	

96. A polynomial in x has degree 3. The coefficient of x^2 is 3 less than the coefficient of x^3. The coefficient of x is 3 times the coefficient of x^2. The remaining constant is 2 more than the coefficient of x^3. The sum of the coefficients is -4. Find the polynomial.

4.3 Addition and Subtraction of Polynomials

Addition of Polynomials • Opposites of Polynomials • Subtraction of Polynomials • Problem Solving

Addition of Polynomials

To add two polynomials, we write a plus sign between them and then collect like terms.

EXAMPLE 1

Add.

a) $(-3x^3 + 2x - 4) + (4x^3 + 3x^2 + 2)$
b) $\left(\frac{2}{3}x^4 + 3x^2 - 2x + \frac{1}{2}\right) + \left(-\frac{1}{3}x^4 + 5x^3 - 3x^2 + 3x - \frac{1}{2}\right)$

Solution

a) $(-3x^3 + 2x - 4) + (4x^3 + 3x^2 + 2)$

$= (-3 + 4)x^3 + 3x^2 + 2x + (-4 + 2)$ Collecting like terms; using the distributive law

$= x^3 + 3x^2 + 2x - 2$

b) $\left(\frac{2}{3}x^4 + 3x^2 - 2x + \frac{1}{2}\right) + \left(-\frac{1}{3}x^4 + 5x^3 - 3x^2 + 3x - \frac{1}{2}\right)$

$= \left(\frac{2}{3} - \frac{1}{3}\right)x^4 + 5x^3 + (3 - 3)x^2 + (-2 + 3)x + \left(\frac{1}{2} - \frac{1}{2}\right)$ Collecting like terms

$= \frac{1}{3}x^4 + 5x^3 + x$ ❑

After some practice, polynomial addition is often performed mentally.

EXAMPLE 2

Add: $(3x^2 - 2x + 2) + (5x^3 - 2x^2 + 3x - 4)$.

Solution

$(3x^2 - 2x + 2) + (5x^3 - 2x^2 + 3x - 4)$

$= 5x^3 + (3 - 2)x^2 + (-2 + 3)x + (2 - 4)$ You might do this step mentally.

$= 5x^3 + x^2 + x - 2$ Then you would write only this. ❑

We can also add polynomials by writing like terms in columns. Sometimes this makes like terms easier to see.

EXAMPLE 3

Add: $9x^5 - 2x^3 + 6x^2 + 3$ and $5x^4 - 7x^2 + 6$ and $3x^6 - 5x^5 + x^2 + 5$.

Solution We arrange the polynomials with like terms in columns.

$$\begin{array}{rrrrrrl}
 & 9x^5 & & -2x^3 & +6x^2 & +3 & \\
 & & 5x^4 & & -7x^2 & +6 & \text{We leave spaces for missing terms.} \\
3x^6 & -5x^5 & & & +1x^2 & +5 & \text{Writing } x^2 \text{ as } 1x^2 \\
\hline
3x^6 & +4x^5 & +5x^4 & -2x^3 & & +14 & \text{Adding}
\end{array}$$

We write the answer as $3x^6 + 4x^5 + 5x^4 - 2x^3 + 14$ without the extra space. □

Opposites of Polynomials

In Section 1.8, we used the property of negative one to determine that the opposite of a sum is the sum of the opposites. This idea can be extended to polynomials with any number of terms.

> To find an equivalent polynomial for the *opposite,* or *additive inverse,* of a polynomial, replace each term by its opposite—that is, *change the sign of every term.*

EXAMPLE 4

Find two equivalent expressions for the opposite of $4x^5 - 7x^3 - 8x + \frac{5}{6}$.

Solution

i) $-(4x^5 - 7x^3 - 8x + \frac{5}{6})$

ii) $-4x^5 + 7x^3 + 8x - \frac{5}{6}$ Changing the sign of every term

Thus, $-(4x^5 - 7x^3 - 8x + \frac{5}{6})$ is equivalent to $-4x^5 + 7x^3 + 8x - \frac{5}{6}$, and each is the opposite of the polynomial $4x^5 - 7x^3 - 8x + \frac{5}{6}$. □

EXAMPLE 5

Simplify: $-(-7x^4 - \frac{5}{9}x^3 + 8x^2 - x + 67)$.

Solution

$$-(-7x^4 - \tfrac{5}{9}x^3 + 8x^2 - x + 67) = 7x^4 + \tfrac{5}{9}x^3 - 8x^2 + x - 67$$ □

Subtraction of Polynomials

We can now subtract one polynomial from another by adding the opposite of the polynomial being subtracted.

EXAMPLE 6

Subtract.

a) $(9x^5 + x^3 - 2x^2 + 4) - (-2x^5 + x^4 - 4x^3 - 3x^2)$

b) $(7x^5 + x^3 - 9x) - (3x^5 - 4x^3 + 5)$

Solution

a) $(9x^5 + x^3 - 2x^2 + 4) - (-2x^5 + x^4 - 4x^3 - 3x^2)$

$= 9x^5 + x^3 - 2x^2 + 4 + 2x^5 - x^4 + 4x^3 + 3x^2$ Adding the opposite

$= 11x^5 - x^4 + 5x^3 + x^2 + 4$ Collecting like terms

b) $(7x^5 + x^3 - 9x) - (3x^5 - 4x^3 + 5)$

$= 7x^5 + x^3 - 9x + (-3x^5) + 4x^3 - 5$ Adding the opposite

$= 7x^5 + x^3 - 9x - 3x^5 + 4x^3 - 5$ Try to go directly to this step

$= 4x^5 + 5x^3 - 9x - 5$ Collecting like terms □

To use columns to subtract, we replace coefficients by their opposites and then add.

EXAMPLE 7

Write in columns and subtract: $(5x^2 - 3x + 6) - (9x^2 - 5x - 3)$.

Solution

i) $\begin{array}{r} 5x^2 - 3x + 6 \\ -(9x^2 - 5x - 3) \\ \hline \end{array}$ Writing similar terms in columns

ii) $\begin{array}{r} 5x^2 - 3x + 6 \\ -9x^2 + 5x + 3 \\ \hline \end{array}$ Changing signs and removing parentheses

iii) $\begin{array}{r} 5x^2 - 3x + 6 \\ -9x^2 + 5x + 3 \\ \hline -4x^2 + 2x + 9 \end{array}$ Adding ❑

If you can do so without error, you can arrange the polynomials in columns and write just the answer.

EXAMPLE 8

Write in columns and subtract: $(x^3 + x^2 + 2x - 12) - (-2x^3 + x^2 - 3x)$.

Solution We have

$$\begin{array}{r} x^3 + x^2 + 2x - 12 \\ -(-2x^3 + x^2 - 3x \qquad) \\ \hline 3x^3 \qquad + 5x - 12. \end{array}$$

Leaving space for the missing term ❑

Problem Solving

EXAMPLE 9

Find a polynomial for the sum of the areas of these rectangles.

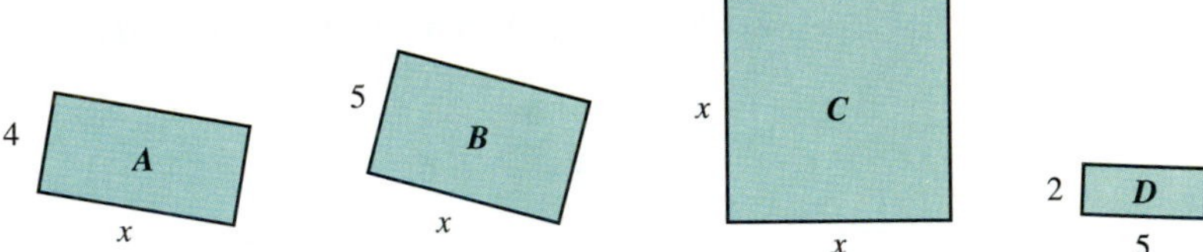

Solution

1. **FAMILIARIZE.** Recall that the area of a rectangle is the product of the length and the width.
2. **TRANSLATE.** We translate the problem to mathematical language. The sum of the areas is a sum of products. We find each product and then add:

$$\underbrace{\text{Area of } A}_{4x} \text{ plus } \underbrace{\text{area of } B}_{5x} \text{ plus } \underbrace{\text{area of } C}_{x \cdot x} \text{ plus } \underbrace{\text{area of } D}_{2 \cdot 5}$$

$$4x + 5x + x \cdot x + 2 \cdot 5.$$

3. **CARRY OUT.** We collect like terms:

$$4x + 5x + x^2 + 10 = x^2 + 9x + 10.$$

4. **CHECK.** We can check by going over our calculations. Another way to check is to replace x with a number, say 3. Then we find each of the areas and add the results:

$$4 \cdot 3 + 5 \cdot 3 + 3 \cdot 3 + 2 \cdot 5 = 12 + 15 + 9 + 10 = 46.$$

When we substitute 3 for x in the polynomial $x^2 + 9x + 10$, we should also get 46:

$$x^2 + 9x + 10 = 3^2 + 9(3) + 10 = 46.$$

Our check is only a partial check, since it is also possible for an incorrect answer to equal 46 when evaluated for $x = 3$. This would be very unlikely, especially if a second choice of x, say $x = 5$, also checks. We leave that check to the student.

5. **STATE.** A polynomial for the sum of the areas is $x^2 + 9x + 10$. ❑

EXAMPLE 10

A 4-ft by 4-ft sandbox is placed on a square lawn x ft on a side. Find a polynomial for the remaining area.

Solution

1. **FAMILIARIZE.** We draw a picture of the situation as follows.

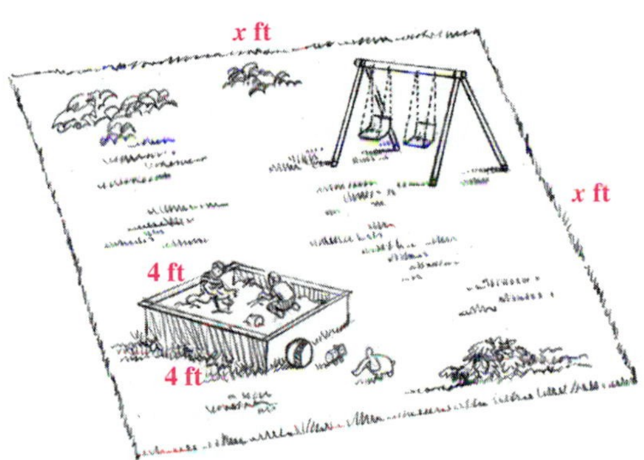

2. **TRANSLATE.** We reword the problem and translate as follows.

Reword:	Area of lawn	−	area of sandbox	=	area left over
	↓	↓	↓	↓	↓
Translate:	$x \cdot x$	−	$4 \cdot 4$	=	Area left over

3. **CARRY OUT.** We carry out the manipulation by multiplying the numbers:

$$x^2 - 16 = \text{Area left over}.$$

4. **CHECK.** We can perform a partial check by assigning some value to x—say, 10—and carrying out the computation of the area in two ways:

Area of lawn $= 10 \cdot 10 = 100$ ft^2;

Area of sandbox $= 4 \cdot 4 = 16$ ft^2;

Area left over $= 100 - 16 = 84$ ft^2.

This is the same as substituting 10 for x in $x^2 - 16$:

$$10^2 - 16 = 100 - 16 = 84.$$

Thus our solution is probably correct.

5. **STATE.** The remaining area is $x^2 - 16$ ft^2. ❑

EXERCISE SET 4.3

Add.

1. $(3x + 2) + (-4x + 3)$
2. $(6x + 1) + (-7x + 2)$
3. $(-6x + 2) + (x^2 + x - 3)$
4. $(x^2 - 5x + 4) + (8x - 9)$
5. $(x^2 - 9) + (x^2 + 9)$
6. $(x^3 + x^2) + (2x^3 - 5x^2)$
7. $(3x^2 - 5x + 10) + (2x^2 + 8x - 40)$
8. $(6x^4 + 3x^3 - 1) + (4x^2 - 3x + 3)$
9. $(1.2x^3 + 4.5x^2 - 3.8x) + (-3.4x^3 - 4.7x^2 + 23)$
10. $(0.5x^4 - 0.6x^2 + 0.7) + (2.3x^4 + 1.8x - 3.9)$
11. $(1 + 4x + 6x^2 + 7x^3) + (5 - 4x + 6x^2 - 7x^3)$
12. $(3x^4 - 6x - 5x^2 + 5) + (6x^2 - 4x^3 - 1 + 7x)$
13. $(9x^8 - 7x^4 + 2x^2 + 5) + (8x^7 + 4x^4 - 2x)$
14. $(4x^5 - 6x^3 - 9x + 1) + (6x^3 + 9x^2 + 9x)$
15. $\left(\frac{1}{4}x^4 + \frac{2}{3}x^3 + \frac{5}{8}x^2 + 7\right) + \left(-\frac{3}{4}x^4 + \frac{3}{8}x^2 - 7\right)$
16. $\left(\frac{1}{3}x^9 + \frac{1}{5}x^5 - \frac{1}{2}x^2 + 7\right) + \left(-\frac{1}{5}x^9 + \frac{1}{4}x^4 - \frac{3}{5}x^5 + \frac{3}{4}x^2 + \frac{1}{2}\right)$
17. $(0.02x^5 - 0.2x^3 + x + 0.08) + (-0.01x^5 + x^4 - 0.8x - 0.02)$
18. $(0.03x^6 + 0.05x^3 + 0.22x + 0.05) + \left(\frac{7}{100}x^6 - \frac{3}{100}x^3 + 0.5\right)$
19. $$\begin{array}{r} -3x^4 + 6x^2 + 2x - 1 \\ -\ 3x^2 + 2x + 1 \\ \hline \end{array}$$
20. $$\begin{array}{r} -4x^3 + 8x^2 + 3x - 2 \\ -\ 4x^2 + 3x + 2 \\ \hline \end{array}$$
21. $$\begin{array}{lllll} 0.15x^4 & + 0.10x^3 & - 0.9x^2 & & \\ & - 0.01x^3 & + 0.01x^2 & + x & \\ 1.25x^4 & & + 0.11x^2 & & + 0.01 \\ & 0.27x^3 & & & + 0.99 \\ -0.35x^4 & & + 15x^2 & & - 0.03 \\ \hline \end{array}$$
22. $$\begin{array}{lllll} 0.05x^4 & + 0.12x^3 & - 0.5x^2 & & \\ & - 0.02x^3 & + 0.02x^2 & + 2x & \\ 1.5x^4 & & + 0.01x^2 & & + 0.15 \\ & 0.25x^3 & & & + 0.85 \\ -0.25x^4 & & + 10x^2 & & - 0.04 \\ \hline \end{array}$$

Find two equivalent expressions for the opposite of the polynomial.

23. $-5x$
24. $x^2 - 3x$
25. $-x^2 + 10x - 2$
26. $-4x^3 - x^2 - x$
27. $12x^4 - 3x^3 + 3$
28. $4x^3 - 6x^2 - 8x + 1$

Simplify.

29. $-(3x - 7)$
30. $-(-2x + 4)$
31. $-(4x^2 - 3x + 2)$
32. $-(-6a^3 + 2a^2 - 9a + 1)$
33. $-(-4x^4 + 6x^2 + \frac{3}{4}x - 8)$
34. $-(-5x^4 + 4x^3 - x^2 + 0.9)$

Subtract.

35. $(3x + 2) - (-4x + 3)$
36. $(6x + 1) - (-7x + 2)$
37. $(-6x + 2) - (x^2 + x - 3)$
38. $(x^2 - 5x + 4) - (8x - 9)$
39. $(x^2 - 9) - (x^2 + 9)$
40. $(x^3 + x^2) - (2x^3 - 5x^2)$
41. $(6x^4 + 3x^3 - 1) - (4x^2 - 3x + 3)$
42. $(-4x^2 + 2x) - (3x^3 - 5x^2 + 3)$
43. $(1.2x^3 + 4.5x^2 - 3.8x) - (-3.4x^3 - 4.7x^2 + 23)$
44. $(0.5x^4 - 0.6x^2 + 0.7) - (2.3x^4 + 1.8x - 3.9)$
45. $(5x^2 + 6) - (3x^2 - 8)$
46. $(7x^3 - 2x^2 + 6) - (7x^2 + 2x - 4)$
47. $(6x^5 - 3x^4 + x + 1) - (8x^5 + 3x^4 - 1)$

48. $\left(\frac{1}{2}x^2 - \frac{3}{2}x + 2\right) - \left(\frac{3}{2}x^2 + \frac{1}{2}x - 2\right)$

49. $(6x^2 + 2x) - (-3x^2 - 7x + 8)$

50. $7x^3 - (-3x^2 - 2x + 1)$

51. $\left(\frac{5}{8}x^3 - \frac{1}{4}x - \frac{1}{3}\right) - \left(-\frac{1}{8}x^3 + \frac{1}{4}x - \frac{1}{3}\right)$

52. $\left(\frac{1}{5}x^3 + 2x^2 - 0.1\right) - \left(-\frac{2}{5}x^3 + 2x^2 + 0.01\right)$

53. $(0.08x^3 - 0.02x^2 + 0.01x) - (0.02x^3 + 0.03x^2 - 1)$

54. $(0.8x^4 + 0.2x - 1) - \left(\frac{7}{10}x^4 + \frac{1}{5}x - 0.1\right)$

55. $\begin{array}{r} x^2 + 5x + 6 \\ -(x^2 + 2x \quad\ ) \\ \hline \end{array}$

56. $\begin{array}{r} x^3 \qquad + 1 \\ -(x^3 + x^2 \quad\ ) \\ \hline \end{array}$

57. $\begin{array}{r} 5x^4 + 6x^3 - 9x^2 \qquad\qquad \\ -(-6x^4 - 6x^3 \qquad + 8x + 9) \\ \hline \end{array}$

58. $\begin{array}{r} 5x^4 \qquad + 6x^2 - 3x + 6 \\ -(\quad 6x^3 + 7x^2 - 8x - 9) \\ \hline \end{array}$

59. $\begin{array}{r} 3x^4 + 6x^2 + 8x - 1 \\ -(4x^5 - 6x^4 \qquad - 8x - 7) \\ \hline \end{array}$

60. $\begin{array}{r} 6x^5 \qquad + 3x^2 - 7x + 2 \\ -(10x^5 + 6x^3 - 5x^2 - 2x + 4) \\ \hline \end{array}$

61. $\begin{array}{r} x^5 \qquad\qquad\qquad\qquad - 1 \\ -(x^5 - x^4 + x^3 - x^2 + x - 1) \\ \hline \end{array}$

62. $\begin{array}{r} x^5 + x^4 - x^3 + x^2 - x + 2 \\ -(x^5 - x^4 + x^3 - x^2 - x + 2) \\ \hline \end{array}$

Problem Solving

63. Solve.

a) Find a polynomial for the sum of the areas of these rectangles.

b) Find the sum of the areas when $x = 3$ and $x = 8$.

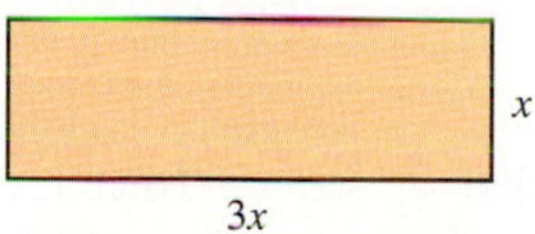

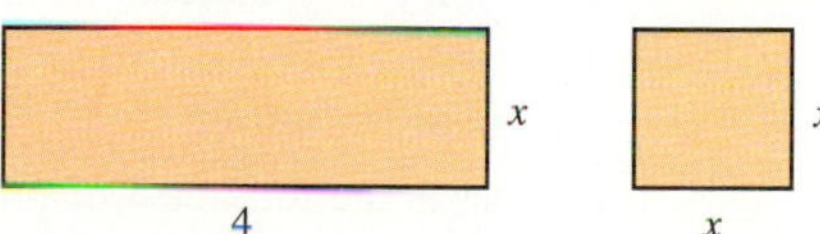

64. Solve.

a) Find a polynomial for the sum of the areas of these circles.

b) Find the sum of the areas when $r = 5$ and $r = 11.3$.

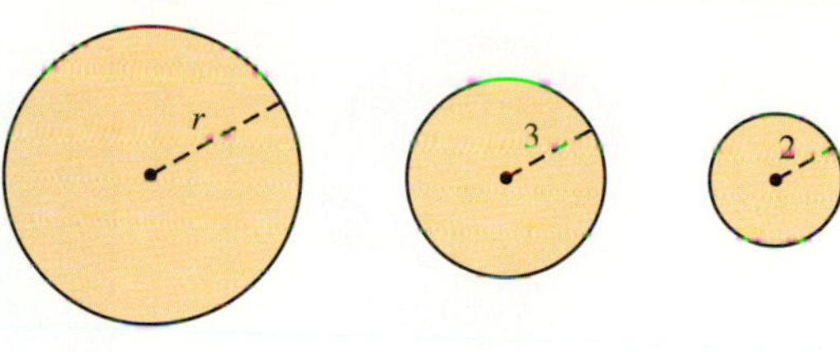

Find a polynomial for the perimeter of the figure.

65.

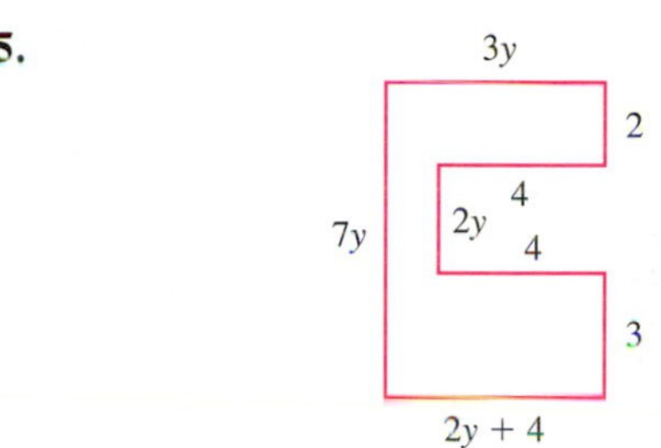

66.

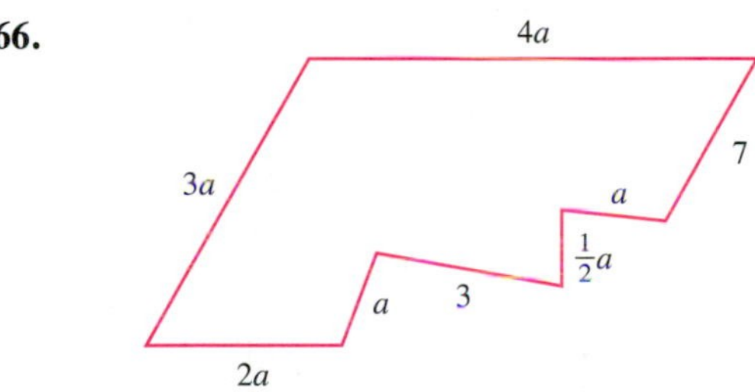

Find two algebraic expressions for the area of the figure. For one expression, view the figure as one large rectangle, and for the other, view the figure as a sum of four smaller rectangles.

67.

68.

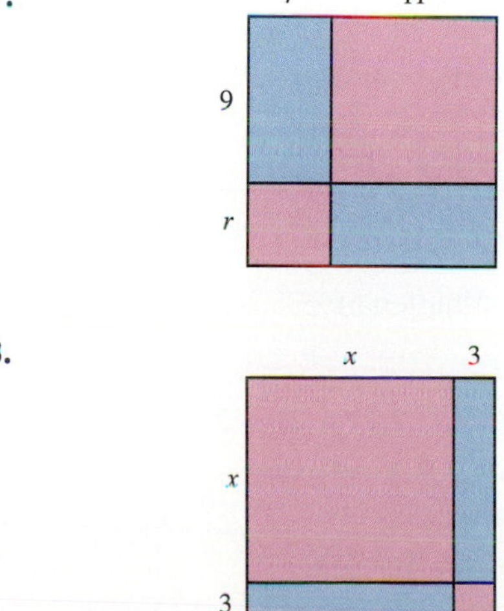

Find a polynomial for the shaded area of the figure.

69.

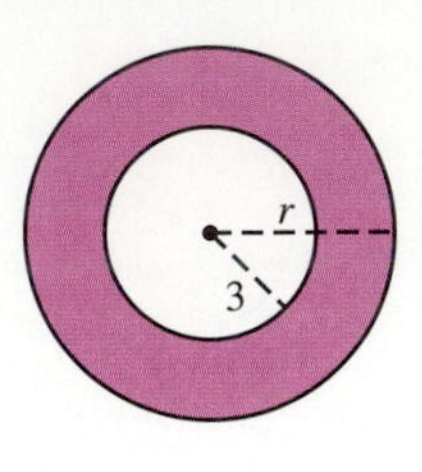

70.

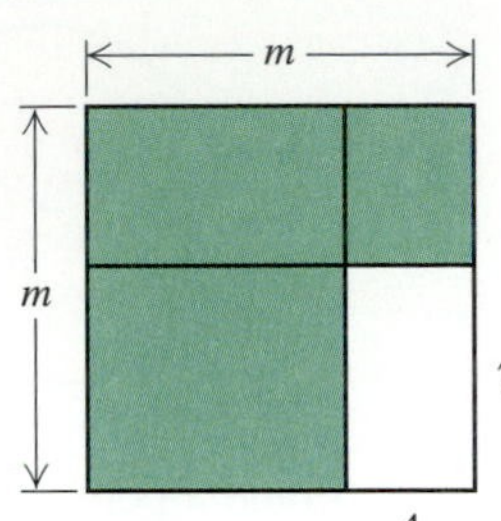

71.

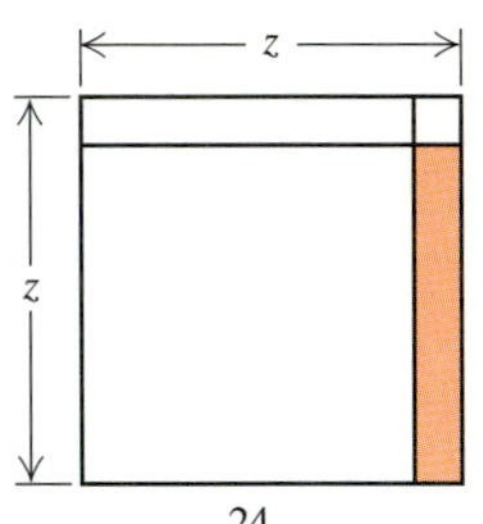

72.

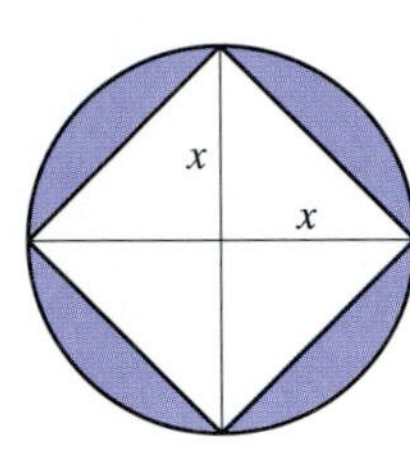

73.

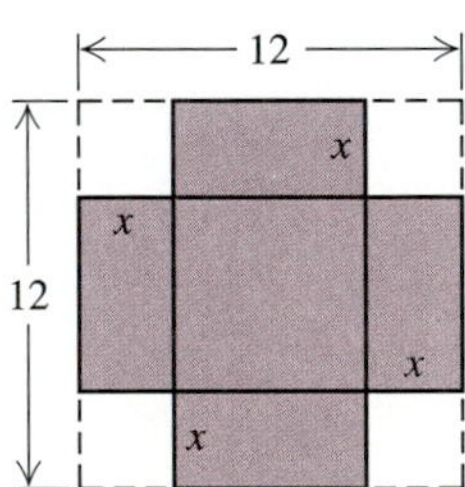

74. Find $(y - 2)^2$ using the four parts of this square.

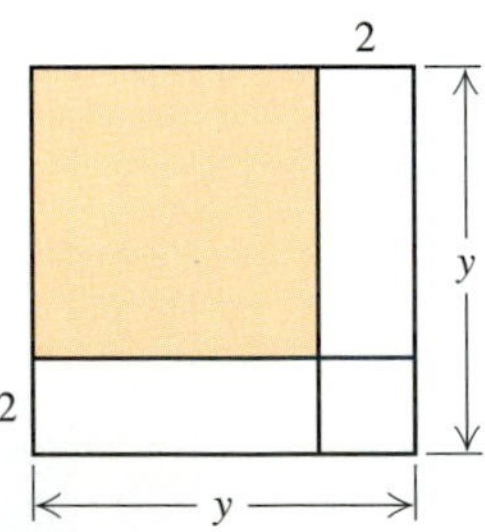

Skill Maintenance

Solve.

75. $1.5x - 2.7x = 23 - 5.6x$

76. $3x - 3 = -4x + 4$

77. $8(x - 2) = 16$

78. $4(x - 5) = 7(x + 8)$

Synthesis

79. ◈ Which, if any, of the commutative, associative, and distributive laws are needed for adding polynomials? Why?

80. ◈ Is the sum of two binomials ever a trinomial? Why or why not?

Simplify.

81. $(4a^2 - 3a) + (7a^2 - 9a - 13) - (6a - 9)$

82. $(3x^2 - 4x + 6) - (-2x^2 + 4) + (-5x - 3)$

83. $(-8y^2 - 4) - (3y + 6) - (2y^2 - y)$

84. $(5x^3 - 4x^2 + 6) - (2x^3 + x^2 - x) + (x^3 - x)$

85. $(-y^4 - 7y^3 + y^2) + (-2y^4 + 5y - 2) - (-6y^3 + y^2)$

86. $(-4 + x^2 + 2x^3) - (-6 - x + 3x^3) - (-x^2 - 5x^3)$

87. ▦ $(345.099x^3 - 6.178x) - (-224.508x^3 + 8.99x)$

Find a polynomial for the surface area of the right rectangular solid.

88.

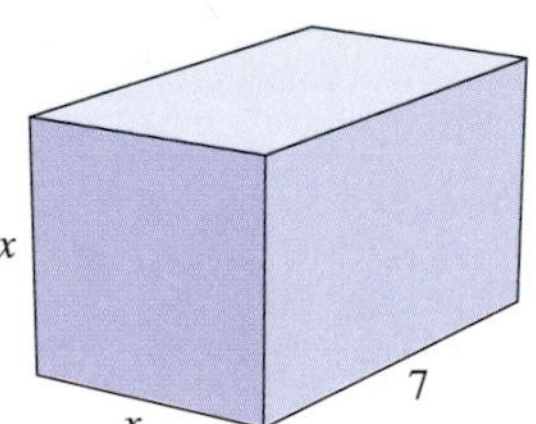

89.

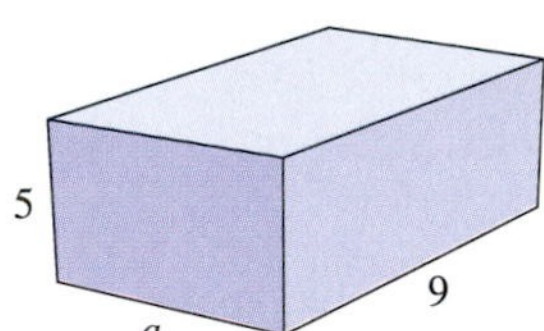

90. ◈ Does replacing each occurrence of the variable x in $5x^3 - 3x^2 + 2x$ with its opposite result in the opposite of the polynomial? Why or why not?

91. *Total profit.* An electronics firm is marketing a new kind of stereo. Total revenue is the total amount of money taken in. The firm determines that when it sells x stereos, its total revenue is given by

$$R = 280x - 0.4x^2.$$

Total cost is the total cost of producing x stereos. The electronics firm determines that the total cost of producing x stereos is given by

$$C = 5000 + 0.6x^2.$$

The total profit is

(Total Revenue) − (Total Cost) = $R - C$.

a) Find a polynomial for total profit.

b) What is the total profit on the production and sale of 75 stereos?

c) What is the total profit on the production and sale of 100 stereos?

4.4 Multiplication of Polynomials

Multiplying Monomials • Multiplying a Monomial and a Polynomial • Multiplying Two Binomials • Multiplying Any Polynomials

We now multiply polynomials using techniques based largely on the distributive, associative, and commutative laws and the rules for exponents.

Multiplying Monomials

Consider $(3x)(4x)$. We multiply as follows:

$$\begin{aligned}(3x)(4x) &= 3 \cdot x \cdot 4 \cdot x && \text{By an associative law}\\ &= 3 \cdot 4 \cdot x \cdot x && \text{By a commutative law}\\ &= (3 \cdot 4) \cdot x \cdot x && \text{By an associative law}\\ &= 12x^2.\end{aligned}$$

To find an equivalent expression for the product of two monomials, multiply the coefficients and then multiply the variables using the product rule for exponents.

EXAMPLE 1

Multiply: **(a)** $(5x)(6x)$; **(b)** $(3a)(-a)$; **(c)** $(-7x^5)(4x^3)$.

Solution

a) $(5x)(6x) = (5 \cdot 6)(x \cdot x)$ Multiplying the coefficients; multiplying the variables

$= 30x^2$ Simplifying

b) $(3a)(-a) = (3a)(-1a)$ Writing $-a$ as $-1a$ can ease calculations.

$= (3)(-1)(a \cdot a)$ Using an associative and a commutative law

$= -3a^2$

c) $(-7x^5)(4x^3) = (-7 \cdot 4)(x^5 \cdot x^3)$

$= -28x^{5+3}$

$= -28x^8$ Using the product rule for exponents ❑

After some practice, you can try writing only the answer.

Multiplying a Monomial and a Polynomial

To find an equivalent expression for the product of a monomial, such as $2x$, and a polynomial, such as $5x + 3$, we use the distributive law.

EXAMPLE 2

Multiply: **(a)** x and $x + 3$; **(b)** $5x(2x^2 - 3x + 4)$.

Solution

a) $x(x + 3) = x \cdot x + x \cdot 3$ Using the distributive law

$= x^2 + 3x$

b) $5x(2x^2 - 3x + 4) = (5x)(2x^2) - (5x)(3x) + (5x)(4)$ Using the distributive law

$= 10x^3 - 15x^2 + 20x$ Multiplying the three pairs of terms ❑

The product in Example 2(a) can be visualized as the area of a rectangle with width x and length $x + 3$.

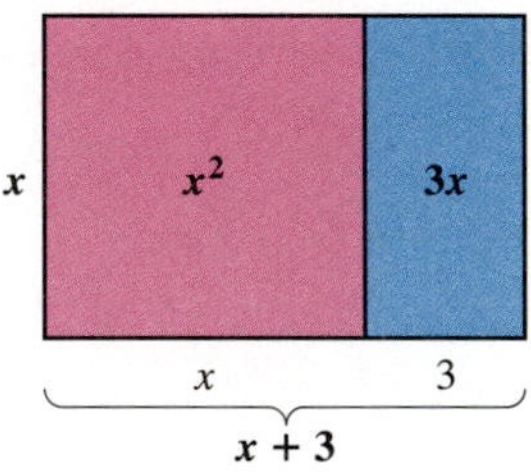

Note that the total area can be expressed as $x(x + 3)$ or, by adding the two smaller areas, $x^2 + 3x$.

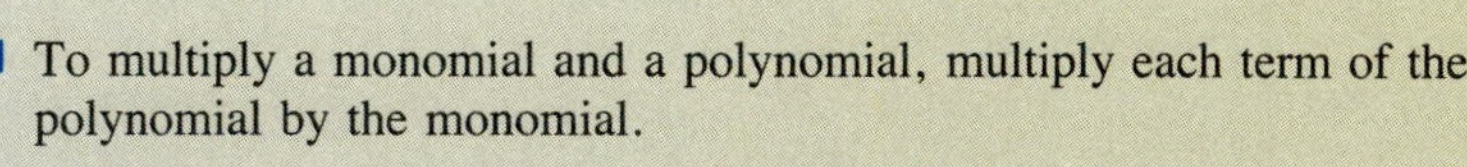
To multiply a monomial and a polynomial, multiply each term of the polynomial by the monomial.

Try to do this mentally, if possible.

EXAMPLE 3

Multiply: $2x^2(x^3 - 7x^2 + 10x - 4)$.

Solution

Think: $2x^2 \cdot x^3 - 2x^2 \cdot 7x^2 + 2x^2 \cdot 10x - 2x^2 \cdot 4$

$2x^2(x^3 - 7x^2 + 10x - 4) = 2x^5 - 14x^4 + 20x^3 - 8x^2$ ❑

Multiplying Two Binomials

To find an equivalent expression for the product of two binomials, we again begin by using the distributive law. This time, however, it is a *binomial* rather than a monomial that is being distributed.

EXAMPLE 4

Multiply: **(a)** $x + 5$ and $x + 4$; **(b)** $4x - 3$ and $x - 2$.

Solution

a) $(x + 5)(x + 4) = (x + 5)x + (x + 5)4$ — Using the distributive law

$= x(x + 5) + 4(x + 5)$ — Using the commutative law for multiplication

$= x \cdot x + x \cdot 5 + 4 \cdot x + 4 \cdot 5$ — Using the distributive law (twice)

$= x^2 + 5x + 4x + 20$ — Multiplying the monomials

$= x^2 + 9x + 20$ — Collecting like terms

b) $(4x - 3)(x - 2) = (4x - 3)x - (4x - 3)2$ — Using the distributive law

$= x(4x - 3) - 2(4x - 3)$ — Using the commutative law for multiplication. This step is often omitted.

$= x \cdot 4x - x \cdot 3 - 2 \cdot 4x - 2(-3)$ — Using the distributive law (twice)

$= 4x^2 - 3x - 8x + 6$ — Multiplying the monomials

$= 4x^2 - 11x + 6$ — Collecting like terms ❑

To visualize the product in Example 4(a), consider a rectangle of length $x + 5$ and width $x + 4$.

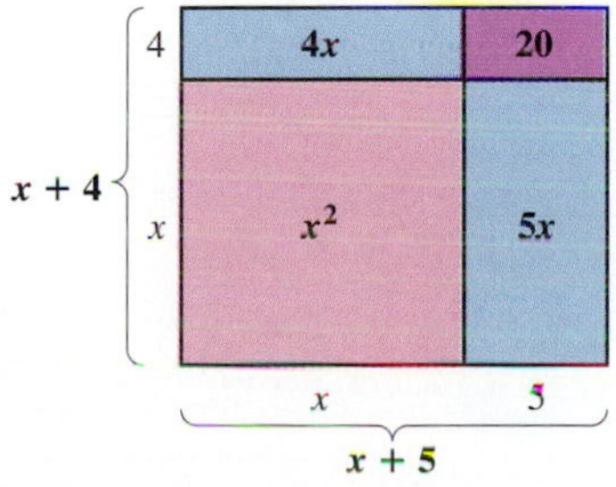

The total area can be expressed as $(x + 5)(x + 4)$ or, by adding the four smaller areas, $x^2 + 5x + 4x + 20$.

Multiplying Any Polynomials

Let us consider the product of a binomial and a trinomial. Again we make repeated use of the distributive law.

EXAMPLE 5

Multiply: $(x^2 + 2x - 3)(x + 4)$.

Solution

$(x^2 + 2x - 3)(x + 4)$

$= (x^2 + 2x - 3)x + (x^2 + 2x - 3)4$	Using the distributive law
$= x(x^2 + 2x - 3) + 4(x^2 + 2x - 3)$	Using a commutative law
$= x \cdot x^2 + x \cdot 2x - x \cdot 3 + 4 \cdot x^2 + 4 \cdot 2x - 4 \cdot 3$	Using the distributive law (twice)
$= x^3 + 2x^2 - 3x + 4x^2 + 8x - 12$	Multiplying the monomials
$= x^3 + 6x^2 + 5x - 12$	Collecting like terms

❑

Perhaps you have discovered the following in the preceding examples.

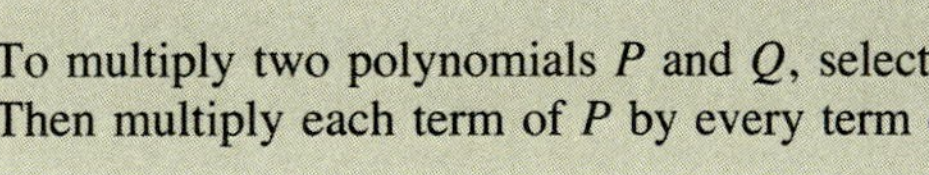

To multiply two polynomials P and Q, select one of the polynomials, say P. Then multiply each term of P by every term of Q and collect like terms.

To use columns for long multiplications, multiply each term at the top by every term at the bottom. We write like terms in columns, and then add the results. Such multiplication is like multiplying with whole numbers:

$$\begin{array}{rrrr} & 3 & 2 & 1 \\ \times & & 1 & 2 \\ \hline & 6 & 4 & 2 \\ 3 & 2 & 1 & \\ \hline 3 & 8 & 5 & 2 \end{array} \qquad \begin{array}{rrrr} & 300 + & 20 + & 1 \\ \times & & 10 + & 2 \\ \hline & 600 + & 40 + & 2 \\ 3000 + & 200 + & 10 & \\ \hline 3000 + & 800 + & 50 + & 2 \end{array}$$

EXAMPLE 6

Multiply: $(4x^3 - 2x^2 + 3x)(x^2 + 2x)$.

Solution

$$\begin{array}{rrrrl} & 4x^3 & - 2x^2 & + 3x & \\ & & x^2 & + 2x & \\ \hline & 8x^4 & - 4x^3 & + 6x^2 & \text{Multiplying the top row by } 2x \\ 4x^5 - 2x^4 & + 3x^3 & & & \text{Multiplying the top row by } x^2 \\ \hline 4x^5 + 6x^4 & - x^3 & + 6x^2 & & \text{Collecting like terms} \end{array}$$

Line up like terms in columns.

❑

When terms are missing, it helps to leave space for them and align like terms as we multiply.

EXAMPLE 7

Multiply: $(-2x^2 - 3)(5x^3 - 3x + 4)$.

Solution

$$\begin{array}{rrrrrl} & 5x^3 & & -3x & +4 & \\ & & -2x^2 & & -3 & \\ \hline & -15x^3 & & +9x & -12 & \text{Multiplying by } -3 \\ -10x^5 & +6x^3 & -8x^2 & & & \text{Multiplying by } -2x^2 \\ \hline -10x^5 & -9x^3 & -8x^2 & +9x & -12 & \text{Collecting like terms} \end{array}$$

❑

With practice some steps can be skipped. Sometimes we can multiply horizontally, while still aligning like terms.

EXAMPLE 8

Multiply: $(2x^3 + 3x^2 - 4x + 6)(3x + 5)$.

Solution

$$\begin{aligned}(3x + 5)(2x^3 + 3x^2 - 4x + 6) &= 6x^4 + 9x^3 - 12x^2 + 18x && \text{Multiplying by } 3x\\ &\quad + 10x^3 + 15x^2 - 20x + 30 && \text{Multiplying by } 5\\ &= 6x^4 + 19x^3 + 3x^2 - 2x + 30\end{aligned}$$

❑

EXERCISE SET 4.4

Multiply.

1. $(6x^2)(7)$
2. $(5x^2)(-2)$
3. $(-x^3)(-x)$
4. $(-x^4)(x^2)$
5. $(-x^5)(x^3)$
6. $(-x^6)(-x^2)$
7. $(3x^4)(2x^2)$
8. $(5x^3)(4x^5)$
9. $(7t^5)(4t^3)$
10. $(10a^2)(3a^2)$
11. $(-0.1x^6)(0.2x^4)$
12. $(0.3x^3)(-0.4x^6)$
13. $\left(-\frac{1}{5}x^3\right)\left(-\frac{1}{3}x\right)$
14. $\left(-\frac{1}{4}x^4\right)\left(\frac{1}{5}x^8\right)$
15. $(-4x^2)(0)$
16. $(-4m^5)(-1)$
17. $(3x^2)(-4x^3)(2x^6)$
18. $(-2y^5)(10y^4)(-3y^3)$
19. $3x(-x + 5)$
20. $2x(4x - 6)$
21. $4x(x + 1)$
22. $3x(x + 2)$
23. $(x + 7)5x$
24. $(x - 6)3x$
25. $x^2(x^3 + 1)$
26. $-2x^3(x^2 - 1)$
27. $3x(2x^2 - 6x + 1)$
28. $-4x(2x^3 - 6x^2 - 5x + 1)$
29. $4x^2(3x + 6)$
30. $5x^2(-2x + 1)$
31. $-6x^2(x^2 + x)$
32. $-4x^2(x^2 - x)$
33. $3y^2(6y^4 + 8y^3)$
34. $4y^4(y^3 - 6y^2)$
35. $3x^4(14x^{50} + 20x^{11} + 6x^{57} + 60x^{15})$
36. $5x^6(4x^{32} - 10x^{19} + 5x^8)$
37. $(x + 6)(x + 3)$
38. $(x + 5)(x + 2)$
39. $(x + 5)(x - 2)$
40. $(x + 6)(x - 2)$
41. $(x - 4)(x - 3)$
42. $(x - 7)(x - 3)$
43. $(x + 3)(x - 3)$
44. $(x + 6)(x - 6)$
45. $(5 - x)(5 - 2x)$
46. $(3 + x)(6 + 2x)$
47. $(2x + 5)(2x + 5)$

48. $(3x - 4)(3x - 4)$

49. $(3y - 4)(3y + 4)$

50. $(2y + 1)(2y - 1)$

51. $\left(x - \frac{5}{2}\right)\left(x + \frac{2}{5}\right)$

52. $\left(x + \frac{4}{3}\right)\left(x + \frac{3}{2}\right)$

53. $(x^2 + x + 1)(x - 1)$

54. $(x^2 - x + 2)(x + 2)$

55. $(2x + 1)(2x^2 + 6x + 1)$

56. $(3x - 1)(4x^2 - 2x - 1)$

57. $(y^2 - 3)(3y^2 - 6y + 2)$

58. $(3y^2 - 3)(y^2 + 6y + 1)$

59. $(x^3 + x^2)(x^3 + x^2 - x)$

60. $(x^3 - x^2)(x^3 - x^2 + x)$

61. $(-5x^3 - 7x^2 + 1)(2x^2 - x)$

62. $(-4x^3 + 5x^2 - 2)(5x^2 + 1)$

63. $(1 + x + x^2)(-1 - x + x^2)$

64. $(1 - x + x^2)(1 - x + x^2)$

65. $(2x^2 + 3x - 4)(2x^2 + x - 2)$

66. $(2x^2 - x - 3)(2x^2 - 5x - 2)$

67. $(x + 1)(x^3 + 7x^2 + 5x + 4)$

68. $(x + 2)(x^3 + 5x^2 + 9x + 3)$

69. $(2x^2 + x - 2)(-2x^2 + 4x - 5)$

70. $(3x^2 - 8x + 1)(-2x^2 - 4x + 2)$

71. $(2x + 1)(x^3 - 4x^2 + 3x - 2)$

72. $(4x + 3)(x^3 - 2x^2 + 5x - 1)$

73. $(x^3 + x^2 + x + 1)(x - 1)$

74. $(x^3 - x^2 + x - 2)(x - 2)$

75. $(x^3 + x^2 - x - 3)(x - 3)$

76. $(x^3 - x^2 - x + 4)(x + 4)$

Skill Maintenance

77. Subtract: $-\frac{1}{4} - \frac{1}{2}$.

78. Factor: $16x - 24y + 36$.

Synthesis

79. ◈ The polynomials

$$(a + b + c + d) \quad \text{and} \quad (r + s + m + p)$$

are multiplied. Without performing the multiplication, determine how many terms the product will contain. Provide a justification for your answer.

80. ◈ Is it possible to understand polynomial multiplication without first understanding the distributive law? Why or why not?

Find a polynomial for the shaded area of the figure.

81.

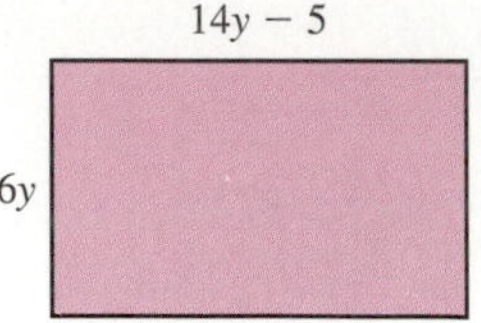

82.

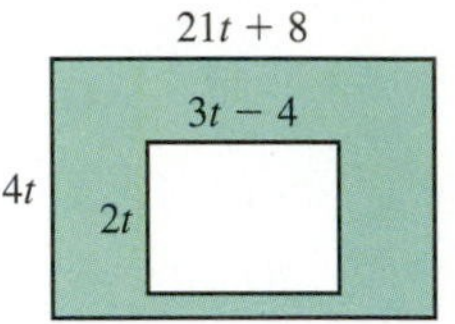

83. A box with a square bottom is to be made from a 12-in.-square piece of cardboard. Squares with side x are cut out of the corners and the sides are folded up. Find polynomials for the volume and the outside surface area of the box.

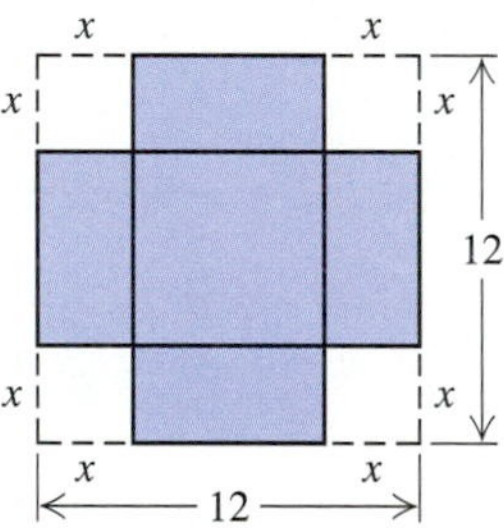

84. An open wooden box is a cube with side x cm. The wood from which the box is made is 1 cm thick. Find a polynomial for the interior volume of the cube.

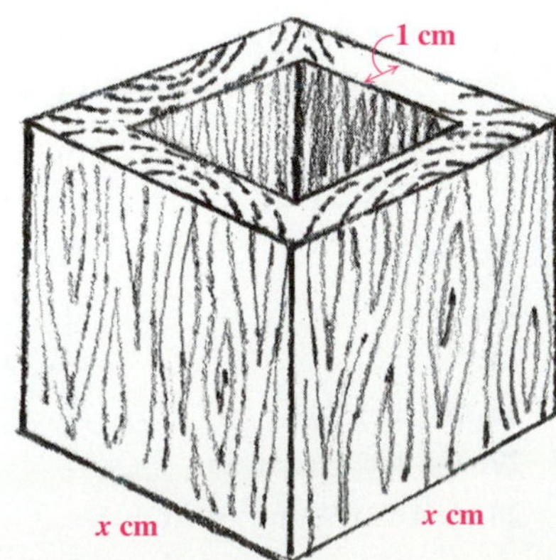

85. The height of a triangle is 4 ft longer than its base. Find a polynomial for the area.

86. A rectangular garden is twice as long as it is wide (see the figure at right). It is surrounded by a sidewalk that is 4 ft wide. The area of the garden and the sidewalk together is 256 ft^2 more than the area of the garden alone. Find the dimensions of the garden.

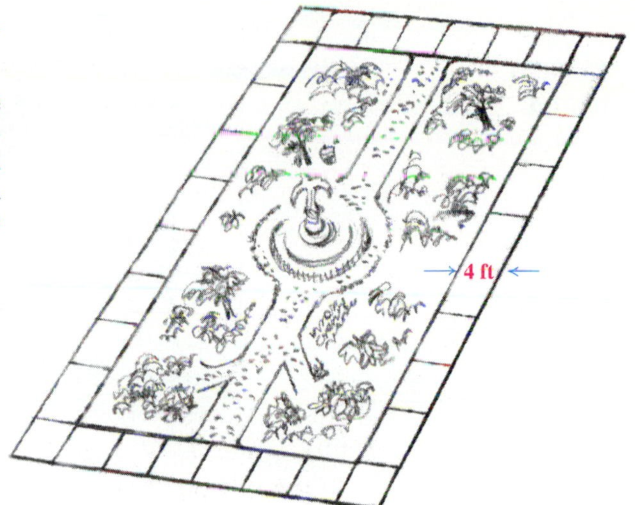

Compute and simplify.

87. $(x + 3)(x + 6) + (x + 3)(x + 6)$

88. $(x - 2)(x - 7) + (x - 2)(x - 7)$

89. $(x + 5)^2 - (x - 3)^2$

90. $(x - 6)^2 + (4 - x)^2$

4.5 Special Products

Products of Two Binomials • Multiplying Sums and Differences of Two Terms • Squaring Binomials • Multiplications of Various Types

Certain products of two binomials occur so often that it is helpful to be able to compute them quickly. In this section, we develop methods for computing "special" products more quickly than we were able to in Section 4.4.

Products of Two Binomials

To multiply two binomials, we can select one binomial and multiply each term of that binomial by every term of the other. Then we collect like terms. Consider the product $(x + 5)(x + 4)$:

$$\begin{aligned}(x + 5)(x + 4) &= x \cdot x + x \cdot 4 + 5 \cdot x + 5 \cdot 4 \\ &= x^2 + 4x + 5x + 20 \\ &= x^2 + 9x + 20.\end{aligned}$$

Note that the product $x \cdot x$ is found by multiplying the *First* terms of each binomial, $x \cdot 4$ is found by multiplying the *Outside* terms of the two binomials, $5 \cdot x$ is the product of the *Inside* terms of the two binomials, and $5 \cdot 4$ is the product of the *Last* terms of each binomial:

$$(x + 5)(x + 4) = \underbrace{x \cdot x}_{\text{First terms}} + \underbrace{4 \cdot x}_{\text{Outside terms}} + \underbrace{5 \cdot x}_{\text{Inside terms}} + \underbrace{5 \cdot 4}_{\text{Last terms}}.$$

To remember this method of multiplying, use the initials **FOIL.**

The FOIL Method

To multiply two binomials, $A + B$ and $C + D$, multiply the First terms AC, the Outside terms AD, the Inside terms BC, and then the Last terms BD. Then collect like terms, if possible.

$$(A + B)(C + D) = AC + AD + BC + BD$$

1. Multiply First terms: AC.
2. Multiply Outside terms: AD.
3. Multiply Inside terms: BC.
4. Multiply Last terms: BD.

↓
FOIL

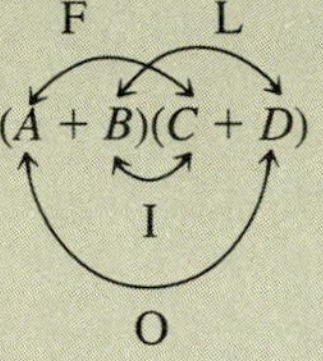

EXAMPLE 1

Multiply: $(x + 8)(x^2 + 5)$.

Solution

F L
F O I L

$$(x + 8)(x^2 + 5) = x^3 + 5x + 8x^2 + 40$$
$$= x^3 + 8x^2 + 5x + 40$$

I
O
❑

After multiplying, any like terms should be collected.

EXAMPLE 2

Multiply.

a) $(x + 7)(x + 4)$

b) $(y + 3)(y - 2)$

c) $(4t^3 + 5t)(3t^2 - 2)$

d) $(3 - 4x)(7 - 5x^3)$

Solution

a) $(x + 7)(x + 4) = x^2 + 4x + 7x + 28$ Using FOIL
$= x^2 + 11x + 28$ Collecting like terms

b) $(y + 3)(y - 2) = y^2 - 2y + 3y - 6$
$= y^2 + y - 6$

c) $(4t^3 + 5t)(3t^2 - 2) = 12t^5 - 8t^3 + 15t^3 - 10t$ Remember to add exponents when multiplying terms with the same base.
$= 12t^5 + 7t^3 - 10t$

d) $(3 - 4x)(7 - 5x^3) = 21 - 15x^3 - 28x + 20x^4$
$= 21 - 28x - 15x^3 + 20x^4$ Because the original binomials are in *ascending* order, we write the answer that way. ❑

Multiplying Sums and Differences of Two Terms

Consider the product of the sum and difference of the same two terms, such as

$$(x + 2)(x - 2).$$

Since this is the product of two binomials, we can use FOIL. This product occurs so often, however, that it will be even quicker to use another method. To find a faster way to compute such a product, look for a pattern in the following:

a) $(x + 2)(x - 2) = x^2 - 2x + 2x - 4$
$= x^2 - 4;$

b) $(3a - 5)(3a + 5) = 9a^2 + 15a - 15a - 25$
$= 9a^2 - 25;$

c) $\left(x^3 + \frac{2}{7}\right)\left(x^3 - \frac{2}{7}\right) = x^6 - \frac{2}{7}x^3 + \frac{2}{7}x^3 - \frac{4}{49}$ $\quad x^3 \cdot x^3 = (x^3)^2 = x^6$
$= x^6 - \frac{4}{49}.$

Perhaps you discovered in each case that when we multiply the two binomials, the "outer" and "inner" products add to 0 and "drop out."

The Product of a Sum and Difference

The product of the sum and difference of the same two terms is the square of the first term minus the square of the second term:

$$(A + B)(A - B) = A^2 - B^2.$$

This is called a *difference of squares*.

EXAMPLE 3

Multiply.

a) $(x + 4)(x - 4)$ **b)** $(5 + 2w)(5 - 2w)$

c) $(3a^4 - 5)(3a^4 + 5)$ **d)** $(-4x - 10)(-4x + 10)$

Solution

$(A + B)(A - B) = A^2 - B^2$ Saying the words can help.

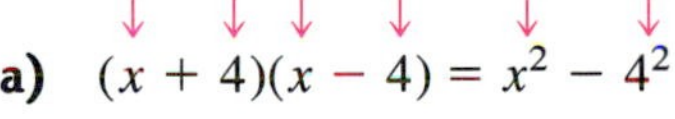

a) $(x + 4)(x - 4) = x^2 - 4^2$ "The square of the first term, x^2, minus the square of the second, 4^2."
$= x^2 - 16$ Simplifying

b) $(5 + 2w)(5 - 2w) = 5^2 - (2w)^2$
$= 25 - 4w^2$ Squaring both 2 and w

c) $(3a^4 - 5)(3a^4 + 5) = (3a^4)^2 - 5^2$
$= 9a^8 - 25$ Using the rules for exponents. Remember to multiply exponents when raising a power to a power.

d) $(-4x - 10)(-4x + 10) = (-4x)^2 - 10^2$
$= 16x^2 - 100$ Squaring both -4 and x ❑

Squaring Binomials

Consider the square of a binomial, such as $(x + 3)^2$. This can be expressed as $(x + 3)(x + 3)$. Since this is the product of two binomials, we can use FOIL. But again, this product occurs so often that it will speed up our work to be able to use an even faster method. Look for a pattern in the following:

a) $(x + 3)^2 = (x + 3)(x + 3)$
$= x^2 + 3x + 3x + 9$
$= x^2 + 6x + 9;$

b) $(5 + 3p)^2 = (5 + 3p)(5 + 3p)$
$= 25 + 15p + 15p + 9p^2$
$= 25 + 30p + 9p^2;$

c) $(a^3 - 1)^2 = (a^3 - 1)(a^3 - 1)$
$= a^6 - a^3 - a^3 + 1$
$= a^6 - 2a^3 + 1;$

d) $(7x - 2)^2 = (7x - 2)(7x - 2)$
$= 49x^2 - 14x - 14x + 4$
$= 49x^2 - 28x + 4.$

Perhaps you noticed that in each product the "outer" and "inner" products are identical. The other two terms, the "first" and "last" products, are squares.

The Square of a Binomial

The square of a binomial is the square of the first term, plus twice the product of the two terms, plus the square of the last term:

$(A + B)^2 = A^2 + 2AB + B^2;$
$(A - B)^2 = A^2 - 2AB + B^2.$

EXAMPLE 4

Multiply: **(a)** $(x + 7)^2$; **(b)** $(t - 5)^2$; **(c)** $(3a + 0.4)^2$; **(d)** $(5x - 3x^4)^2$.

Solution

$(A + B)^2 = A^2 + 2 \cdot A \cdot B + B^2$ — Saying the words can help.

a) $(x + 7)^2 = x^2 + 2 \cdot x \cdot 7 + 7^2$ — "The square of the first term, x^2, plus twice the product of the terms, $2 \cdot 7x$, plus the square of the second term, 7^2."

$= x^2 + 14x + 49$

b) $(t - 5)^2 = t^2 - 2 \cdot t \cdot 5 + 5^2$
$= t^2 - 10t + 25$

c) $(3a + 0.4)^2 = (3a)^2 + 2 \cdot 3a \cdot 0.4 + 0.4^2$
$= 9a^2 + 2.4a + 0.16$

d) $(5x - 3x^4)^2 = (5x)^2 - 2 \cdot 5x \cdot 3x^4 + (3x^4)^2$
$= 25x^2 - 30x^5 + 9x^8$ Using the rules for exponents ❑

CAUTION! Note carefully in Example 4 that the square of a sum is *not* the sum of the squares:

The middle term $2AB$ is missing.

$$(A + B)^2 \neq A^2 + B^2.$$

To confirm this inequality, note that

$$(20 + 5)^2 = 25^2 = 625,$$

whereas

$$20^2 + 5^2 = 400 + 25 = 425, \quad \text{and } 425 \neq 625.$$

Geometrically, $(A + B)^2$ can be viewed as follows.

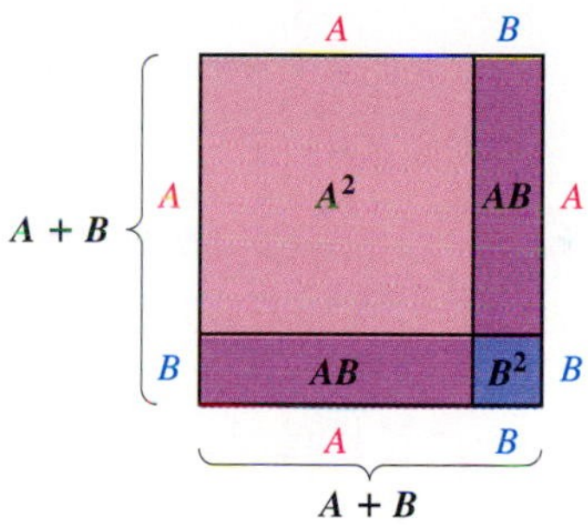

The area of the large square is

$$(A + B)(A + B) = (A + B)^2.$$

This is equal to the sum of the areas of the smaller rectangles:

$$A^2 + AB + AB + B^2 = A^2 + 2AB + B^2.$$

Thus,

$$(A + B)^2 = A^2 + 2AB + B^2.$$

Multiplications of Various Types

Let us now try a variety of multiplications mixed together so that we can learn to sort them out. First try to determine what kind of product you have. Then use the best method. The formulas you should know and the questions you should ask yourself are as follows.

Multiplying Two Polynomials

1. Is the product the square of a binomial? If so, use the following:

$$(A + B)(A + B) = (A + B)^2 = A^2 + 2AB + B^2,$$

or

$$(A - B)(A - B) = (A - B)^2 = A^2 - 2AB + B^2.$$

The square of a binomial is the square of the first term, plus *twice* the product of the two terms, plus the square of the last term.

[The answer has 3 terms.]

2. Is it the product of the sum and difference of the *same* two terms? If so, use the following:

$$(A + B)(A - B) = A^2 - B^2.$$

The product of the sum and difference of the same two terms is a difference of squares.

[The answer has 2 terms.]

3. Is it the product of two binomials other than those above? If so, use FOIL.

[The answer will have 3 or 4 terms.]

4. Is it the product of a monomial and a polynomial? If so, multiply each term of the polynomial by the monomial.
5. Is it the product of two polynomials other than those above? If so, multiply each term of one by every term of the other. Use columns if you wish.

Note that FOIL will work instead of either of the first two methods, but not as quickly.

EXAMPLE 5

Multiply.

a) $(x + 3)(x - 3)$ **b)** $(t + 7)(t - 5)$ **c)** $(x + 7)(x + 7)$

d) $2x^3(9x^2 + x - 7)$ **e)** $(5x^3 - 7x)^2$ **f)** $(p + 3)(p^2 + 2p - 1)$

g) $\left(3x + \frac{1}{4}\right)^2$

Solution

a) $(x + 3)(x - 3) = x^2 - 9$ — The product of the sum and difference of the same two terms

b) $(t + 7)(t - 5) = t^2 - 5t + 7t - 35$ — Using FOIL

$= t^2 + 2t - 35$ — Try to go directly to this step.

c) $(x + 7)(x + 7) = x^2 + 14x + 49$ — The product is the square of a binomial.

d) $2x^3(9x^2 + x - 7) = 18x^5 + 2x^4 - 14x^3$ — Multiplying each term of the trinomial by the monomial

e) $(5x^3 - 7x)^2 = 25x^6 - 2(5x^3)(7x) + 49x^2$ — Squaring a binomial; remember the rules for powers.

$= 25x^6 - 70x^4 + 49x^2$

f)
$$\begin{array}{r} p^2 + 2p - 1 \\ p + 3 \\ \hline 3p^2 + 6p - 3 \\ p^3 + 2p^2 - p \quad \\ \hline p^3 + 5p^2 + 5p - 3 \end{array}$$

Using columns to multiply a binomial and a trinomial

Multiplying by 3

Multiplying by p

g) $\left(3x + \frac{1}{4}\right)^2 = 9x^2 + 2(3x)\left(\frac{1}{4}\right) + \frac{1}{16}$ Squaring a binomial

$= 9x^2 + \frac{3}{2}x + \frac{1}{16}$ ❑

EXERCISE SET 4.5

Multiply. Try to write only the answer. If you need more steps, by all means use them.

1. $(x + 1)(x^2 + 3)$ **2.** $(x^2 - 3)(x - 1)$

3. $(x^3 + 2)(x + 1)$ **4.** $(x^4 + 2)(x + 12)$

5. $(y + 2)(y - 3)$ **6.** $(a + 2)(a + 2)$

7. $(3x + 2)(3x + 3)$ **8.** $(4x + 1)(2x + 2)$

9. $(5x - 6)(x + 2)$ **10.** $(x - 8)(x + 8)$

11. $(3t - 1)(3t + 1)$ **12.** $(2m + 3)(2m + 3)$

13. $(4x - 2)(x - 1)$ **14.** $(2x - 1)(3x + 1)$

15. $\left(p - \frac{1}{4}\right)\left(p + \frac{1}{4}\right)$ **16.** $\left(q + \frac{3}{4}\right)\left(q + \frac{3}{4}\right)$

17. $(x - 0.1)(x + 0.1)$ **18.** $(x + 0.3)(x - 0.4)$

19. $(2x^2 + 6)(x + 1)$ **20.** $(2x^2 + 3)(2x - 1)$

21. $(-2x + 1)(x + 6)$ **22.** $(3x + 4)(2x - 4)$

23. $(a + 7)(a + 7)$ **24.** $(2y + 5)(2y + 5)$

25. $(1 + 2x)(1 - 3x)$ **26.** $(-3x - 2)(x + 1)$

27. $(x^2 + 3)(x^3 - 1)$ **28.** $(x^4 - 3)(2x + 1)$

29. $(3x^2 - 2)(x^4 - 2)$ **30.** $(x^{10} + 3)(x^{10} - 3)$

31. $(3x^5 + 2)(2x^2 + 6)$ **32.** $(1 - 2x)(1 + 3x^2)$

33. $(8x^3 + 1)(x^3 + 8)$ **34.** $(4 - 2x)(5 - 2x^2)$

35. $(4x^2 + 3)(x - 3)$ **36.** $(7x - 2)(2x - 7)$

37. $(4y^4 + y^2)(y^2 + y)$

38. $(5y^6 + 3y^3)(2y^6 + 2y^3)$

Multiply mentally, writing only the answer if possible. If you need extra steps, by all means use them.

39. $(x + 4)(x - 4)$ **40.** $(x + 1)(x - 1)$

41. $(2x + 1)(2x - 1)$ **42.** $(x^2 + 1)(x^2 - 1)$

43. $(5m - 2)(5m + 2)$ **44.** $(3x^4 + 2)(3x^4 - 2)$

45. $(2x^2 + 3)(2x^2 - 3)$ **46.** $(6x^5 - 5)(6x^5 + 5)$

47. $(3x^4 - 4)(3x^4 + 4)$ **48.** $(t^2 - 0.2)(t^2 + 0.2)$

49. $(x^6 - x^2)(x^6 + x^2)$

50. $(2x^3 - 0.3)(2x^3 + 0.3)$

51. $(x^4 + 3x)(x^4 - 3x)$ **52.** $\left(\frac{3}{4} + 2x^3\right)\left(\frac{3}{4} - 2x^3\right)$

53. $(x^{12} - 3)(x^{12} + 3)$

54. $(12 - 3x^2)(12 + 3x^2)$

55. $(2y^8 + 3)(2y^8 - 3)$ **56.** $\left(m - \frac{2}{3}\right)\left(m + \frac{2}{3}\right)$

57. $(x + 2)^2$ **58.** $(2x - 1)^2$

59. $(3x^2 + 1)^2$ **60.** $\left(3x + \frac{3}{4}\right)^2$

61. $\left(a - \frac{1}{2}\right)^2$ **62.** $\left(2a - \frac{1}{5}\right)^2$

63. $(3 + x)^2$ **64.** $(x^3 - 1)^2$

65. $(x^2 + 1)^2$ **66.** $(8x - x^2)^2$

67. $(2 - 3x^4)^2$ **68.** $(6x^3 - 2)^2$

69. $(5 + 6t^2)^2$ **70.** $(3p^2 - p)^2$

71. $(7x - 0.3)^2$ **72.** $(4a - 0.6)^2$

73. $5a^3(2a^2 - 1)$

74. $(a - 3)(a^2 + 2a - 4)$

75. $(x^2 - 5)(x^2 + x - 1)$

76. $9x^4(3x^2 - x)$ **77.** $(3 - 2x^3)^2$

78. $(x - 4x^3)^2$ **79.** $4x(x^2 + 6x - 3)$

80. $8x(-x^5 + 6x^2 + 9)$ **81.** $\left(2x^2 - \frac{1}{2}\right)\left(2x^2 - \frac{1}{2}\right)$

82. $(-x^2 + 1)^2$ **83.** $(-1 + 3p)(1 + 3p)$

84. $(-3q + 2)(3q + 2)$ **85.** $3t^2(5t^3 - t^2 + t)$

86. $-6x^2(x^3 + 8x - 9)$ **87.** $(6x^4 + 4)^2$

88. $(8a + 5)^2$ **89.** $(3x + 2)(4x^2 + 5)$

90. $(2x^2 - 7)(3x^2 + 9)$ **91.** $(8 - 6x^4)^2$

92. $\left(\frac{1}{5}x^2 + 9\right)\left(\frac{3}{5}x^2 - 7\right)$

93. $(t - 1)(t^2 + t + 1)$

94. $(y + 5)(y^2 - 5y + 25)$

Compute and compare.

95. $3^2 + 4^2$; $(3 + 4)^2$

96. $6^2 + 7^2$; $(6 + 7)^2$

97. $9^2 - 5^2$; $(9 - 5)^2$

98. $11^2 - 4^2$; $(11 - 4)^2$

Find the total shaded area.

99.

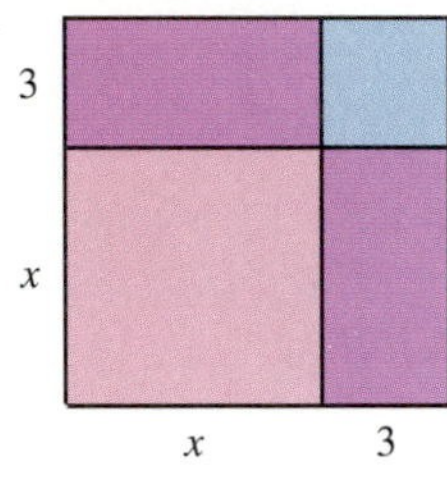

100.

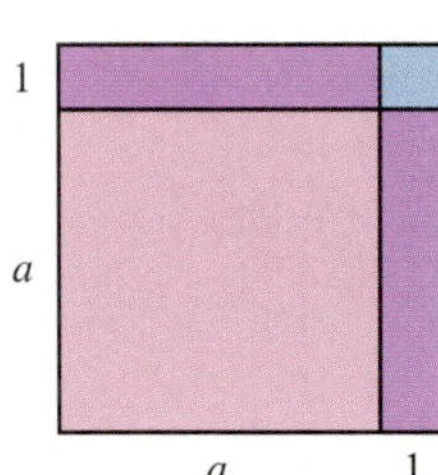

101.

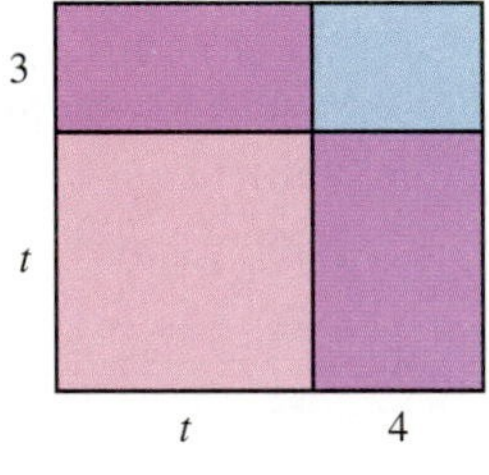

102.

103.

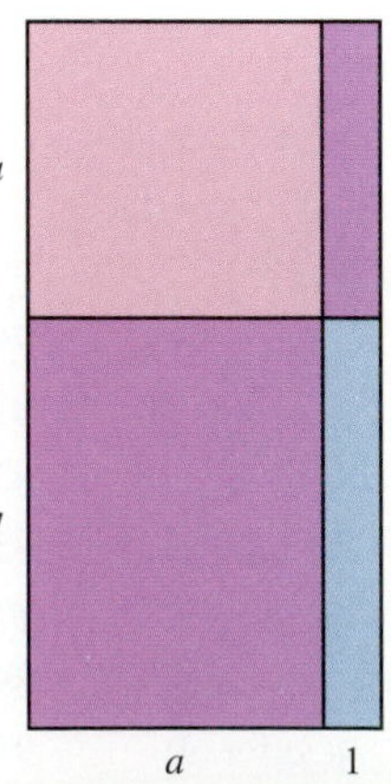

104.

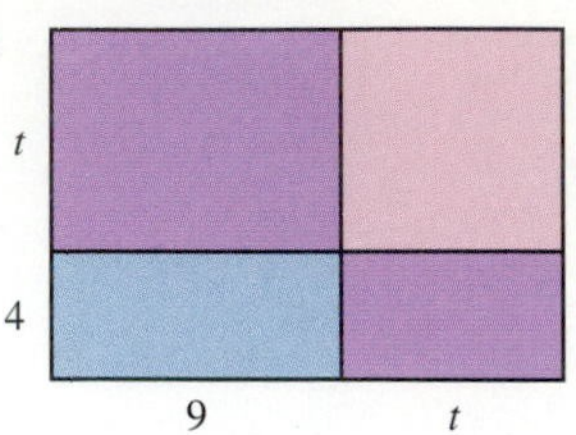

Skill Maintenance

105. In an apartment, lamps, an air conditioner, and a television set are all operating at the same time. The lamps take 10 times as many watts as the television set, and the air conditioner takes 40 times as many watts as the television set. The total wattage used in the apartment is 2550 watts. How many watts are used by each appliance?

106. Solve: $3x - 8x = 4(7 - 8x)$.

Synthesis

107. ◈ The product $(A + B)^2$ can be regarded as the sum of the areas of four regions (as shown following Example 4). How might one visually represent $(A + B)^3$? Why?

108. ◈ Anais claims that by writing $19 \cdot 21$ as $(20 - 1)(20 + 1)$ she can find the product mentally. How is this possible?

Multiply.

109. $4y(y + 5)(2y + 8)$

110. $8x(2x - 3)(5x + 9)$

111. $[(3x - 2)(3x + 2)](9x^2 + 4)$

112. $[(2x - 1)(2x + 1)](4x^2 + 1)$

113. $(5t^3 - 3)^2(5t^3 + 3)^2$
[*Hint:* Regroup as two pairs of binomials.]

114. $[3a - (2a - 3)][3a + (2a - 3)]$
[*Hint:* Do not collect like terms before multiplying.]

115. ▦ $(67.58x + 3.225)^2$

116. $[(x + 3) - 7]^2$

Calculate as the difference of squares.

117. 18×22 [*Hint:* $(20 - 2)(20 + 2)$.]

118. 93×107

Solve.

119. $(x + 2)(x - 5) = (x + 1)(x - 3)$

120. $(2x + 5)(x - 4) = (x + 5)(2x - 4)$

The height of a box is one more than its length l, and the length is one more than its width w. Find a polynomial for the volume V in terms of the following.

121. The width w

122. The length l

123. The height h

Find two expressions for the total shaded area.

124.

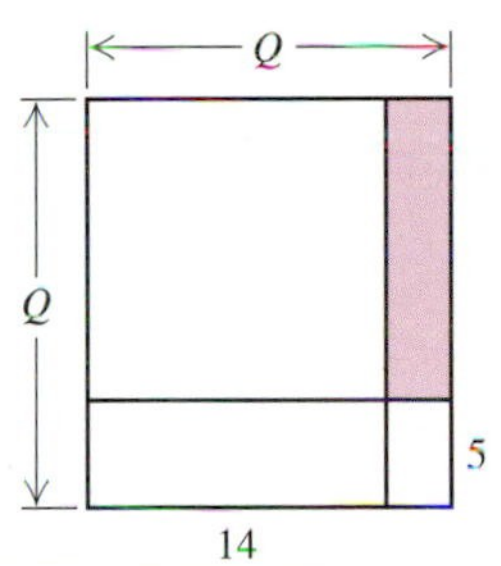

125.

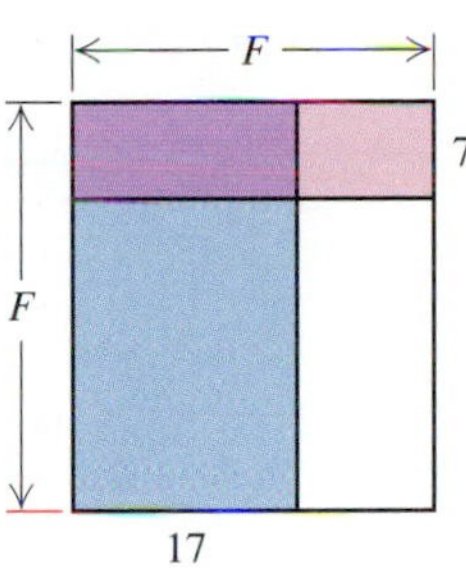

126.

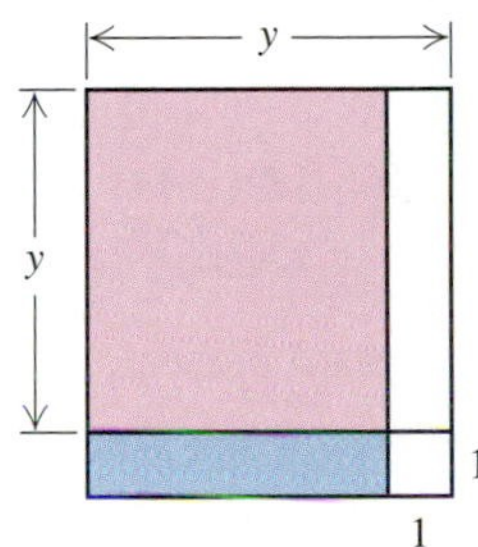

127. A polynomial for the shaded area in this rectangle is $(A + B)(A - B)$.

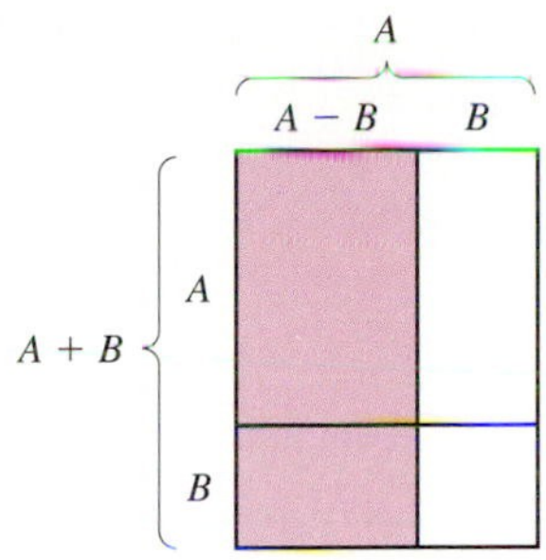

a) Find a polynomial for the area of the entire rectangle.

b) Find a polynomial for the sum of the areas of the two small unshaded rectangles.

c) Find a polynomial for the area in part (a) minus the area in part (b).

d) Find a polynomial for the area of the shaded region and compare this with the polynomial found in part (c).

128. Find three consecutive integers for which the sum of the squares is 65 more than three times the square of the smallest integer.

129. Find $(10x + 5)^2$. Use your result to show how to mentally square any two-digit number ending in 5.

4.6 Polynomials in Several Variables

Evaluating Polynomials • Coefficients and Degrees • Collecting Like Terms • Addition and Subtraction • Multiplication

The polynomials that we have been studying have only one variable. When a polynomial contains two or more variables, it is referred to as a **polynomial in several variables.** Here are some examples:

$$3a + ab^2 + 5b + 4, \qquad 8xy^2z - 2x^3z - 13x^4y^2 + 15, \qquad 9m^2 - 4n^2.$$

In this section, we will find that polynomials in several variables can be evaluated, added, subtracted, and multiplied much like polynomials in one variable.

Evaluating Polynomials

To evaluate a polynomial in two or more variables, we substitute numbers for the variables. Then we compute, using the rules for order of operations.

EXAMPLE 1

Evaluate the polynomial $4 + 3x + xy^2 + 8x^3y^3$ when $x = -2$ and $y = 5$.

Solution We replace x by -2 and y by 5:

$$4 + 3x + xy^2 + 8x^3y^3 = 4 + 3(-2) + (-2)\cdot 5^2 + 8(-2)^3\cdot 5^3$$
$$= 4 - 6 - 50 - 8000 = -8052.$$

❑

EXAMPLE 2

Surface area of a right circular cylinder. The surface area of a right circular cylinder is given by the polynomial

$$2\pi rh + 2\pi r^2,$$

where h is the height and r is the radius of the base. A 12-oz beverage can has a height of 4.7 in. and a radius of 1.2 in. Find its surface area, using 3.14 as an approximation for π.*

Solution We evaluate the polynomial for $h = 4.7$, $r = 1.2$, and $\pi \approx 3.14$:

$$2\pi rh + 2\pi r^2 \approx 2(3.14)(1.2)(4.7) + 2(3.14)(1.2)^2$$
$$= 2(3.14)(1.2)(4.7) + 2(3.14)(1.44)$$
$$= 35.4192 + 9.0432 = 44.4624.$$

The surface area is about 44.4624 in^2 (square inches). ❑

Coefficients and Degrees

The **degree of a term** is the sum of the exponents of the variables. The **degree of a polynomial** is the degree of the term of highest degree.

EXAMPLE 3

Identify the coefficient and the degree of each term and the degree of the polynomial

$$9x^2y^3 - 14xy^2z^3 + xy + 4y + 5x^2 + 7.$$

Solution

Term	Coefficient	Degree	Degree of the Polynomial
$9x^2y^3$	9	5	6
$-14xy^2z^3$	-14	6	
xy	1	2	
$4y$	4	1	
$5x^2$	5	2	
7	7	0	

❑

*Many pocket calculators have a key that gives a better approximation of π. Often the label, π, is printed above the key itself, meaning that another key, often labeled [SHIFT] or [2ND FCN], must be pressed before the key labeled π. Pressing [SHIFT] [π], we have $\pi \approx 3.141592654$.

Collecting Like Terms

Like terms (or **similar terms**) have exactly the same variables with exactly the same exponents. For example,

$3x^2y^3$ and $-7x^2y^3$ are like terms;

and

$9a^4b^7$ and $12a^4b^7$ are like terms.

On the other hand,

$13xy^5$ and $-2x^2y^5$ are *not* like terms, because the x-factors have different exponents;

and

$3abc^2$ and $4ab$ are *not* like terms, because the factor c^2 does not appear in the second expression.

As always, collecting like terms is based on the distributive law.

EXAMPLE 4

Collect like terms.

a) $9x^2y + 3xy^2 - 5x^2y - xy^2$
b) $7ab - 5ab^2 + 3ab^2 + 6a^3 + 9ab - 11a^3 + b - 1$

Solution

a) $9x^2y + 3xy^2 - 5x^2y - xy^2 = (9 - 5)x^2y + (3 - 1)xy^2$

$= 4x^2y + 2xy^2$ Try to go directly to this step.

b) $7ab - 5ab^2 + 3ab^2 + 6a^3 + 9ab - 11a^3 + b - 1$

$= -2ab^2 + 16ab - 5a^3 + b - 1$ ❑

Addition and Subtraction

The procedure used for adding polynomials in one variable is used to add polynomials in several variables.

EXAMPLE 5

Add.

a) $(-5x^3 + 3y - 5y^2) + (8x^3 + 4x^2 + 7y^2)$
b) $(5ab^2 - 4a^2b + 5a^3 + 2) + (3ab^2 - 2a^2b + 3a^3b - 5)$

Solution

a) $(-5x^3 + 3y - 5y^2) + (8x^3 + 4x^2 + 7y^2)$

$= (-5 + 8)x^3 + 4x^2 + 3y + (-5 + 7)y^2$ Try to do this step mentally.

$= 3x^3 + 4x^2 + 3y + 2y^2$

b) $(5ab^2 - 4a^2b + 5a^3 + 2) + (3ab^2 - 2a^2b + 3a^3b - 5)$

$= 8ab^2 - 6a^2b + 5a^3 + 3a^3b - 3$ ❑

When subtracting a polynomial, remember to find the opposite of each term in that polynomial and then add.

EXAMPLE 6

Subtract: $(4x^2y + x^3y^2 + 3x^2y^3 + 6y) - (4x^2y - 6x^3y^2 + x^2y^2 - 5y)$.

Solution

$$(4x^2y + x^3y^2 + 3x^2y^3 + 6y) - (4x^2y - 6x^3y^2 + x^2y^2 - 5y)$$

$$= 4x^2y + x^3y^2 + 3x^2y^3 + 6y - 4x^2y + 6x^3y^2 - x^2y^2 + 5y \quad \text{Adding the opposite}$$

$$= 7x^3y^2 + 3x^2y^3 - x^2y^2 + 11y \quad \text{Collecting like terms}$$

❑

Multiplication

To multiply polynomials in several variables, multiply each term of one polynomial by every term of the other, much as we did in Sections 4.4 and 4.5.

EXAMPLE 7

Multiply: $(3x^2y - 2xy + 3y)(xy + 2y)$.

Solution

$$\begin{array}{r} 3x^2y - 2xy + 3y \\ xy + 2y \\ \hline 6x^2y^2 - 4xy^2 + 6y^2 \\ 3x^3y^2 - 2x^2y^2 + 3xy^2 \qquad\qquad \\ \hline 3x^3y^2 + 4x^2y^2 - xy^2 + 6y^2 \end{array}$$

Multiplying by $2y$

Multiplying by xy

Adding

❑

The special products that we have studied can be used to speed up our multiplication of polynomials in several variables.

EXAMPLE 8

Multiply.

a) $(p + 5q)(2p - 3q)$
b) $(3x + 2y)^2$
c) $(a^3 - 7a^2b)^2$
d) $(3x^2y + 2y)(3x^2y - 2y)$
e) $(-2x^3y^2 + 5t)(2x^3y^2 + 5t)$
f) $(2x + 3 - 2y)(2x + 3 + 2y)$

Solution

F O I L

a) $(p + 5q)(2p - 3q) = 2p^2 - 3pq + 10pq - 15q^2$

$= 2p^2 + 7pq - 15q^2$ Collecting like terms

$(A + B)^2 = A^2 + 2 \cdot A \cdot B + B^2$

↓ ↓ ↓ ↓ ↓ ↓

b) $(3x + 2y)^2 = (3x)^2 + 2(3x)(2y) + (2y)^2$ Squaring a binomial

$= 9x^2 + 12xy + 4y^2$

$(A - B)^2 = A^2 - 2 \cdot A \cdot B + B^2$

↓ ↓ ↓ ↓ ↓ ↓ ↓

c) $(a^3 - 7a^2b)^2 = (a^3)^2 - 2(a^3)(7a^2b) + (7a^2b)^2$ Squaring a binomial

$= a^6 - 14a^5b + 49a^4b^2$ Using the rules for exponents

$$\begin{array}{cccccc} (A & + B) & (A & - B) = & A^2 & - B^2 \\ \downarrow & \downarrow & \downarrow & \downarrow & \downarrow & \downarrow \end{array}$$

d) $(3x^2y + 2y)(3x^2y - 2y) = (3x^2y)^2 - (2y)^2$ Multiplying the sum and the difference of the same two terms

$= 9x^4y^2 - 4y^2$

e) $(-2x^3y^2 + 5t)(2x^3y^2 + 5t) = (5t - 2x^3y^2)(5t + 2x^3y^2)$ Using the commutative law for addition twice

$= (5t)^2 - (2x^3y^2)^2$ Multiplying the sum and the difference of the same two terms

$= 25t^2 - 4x^6y^4$

$$\begin{array}{cccccc} (A & - B) & (A & + B) = & A^2 & - B^2 \\ \downarrow & \downarrow & \downarrow & \downarrow & \downarrow & \downarrow \end{array}$$

f) $(2x + 3 - 2y)(2x + 3 + 2y) = (2x + 3)^2 - (2y)^2$ Multiplying a sum and a difference

$= 4x^2 + 12x + 9 - 4y^2$ Squaring a binomial

❑

Note that in Example 8 we recognized patterns that might have evaded some students, particularly in parts (e) and (f). In part (e), we could have used FOIL, and in part (f), we could have used long multiplication, but doing so would have been slower. By carefully inspecting a problem before "jumping in," we can often save ourselves considerable work.

EXERCISE SET 4.6

Evaluate the polynomial when $x = 3$ and $y = -2$.

1. $x^2 - y^2 + xy$

2. $x^2 + y^2 - xy$

Evaluate the polynomial when $x = 2$, $y = -3$, and $z = -1$.

3. $xyz^2 + z$

4. $xy - xz + yz$

Interest compounded annually for two years. An amount of money P is invested at interest rate r. In 2 years, it will grow to an amount given by the polynomial

$A = P(1 + r)^2$.

5. Evaluate the polynomial when $P = 10{,}000$ and $r = 0.08$ to find the amount to which \$10,000 will grow at 8% interest for 2 years.

6. Evaluate the polynomial when $P = 10{,}000$ and $r = 0.07$ to find the amount to which \$10,000 will grow at 7% interest for 2 years.

Interest compounded annually for three years. An amount of money P is invested at interest rate r. In 3 years, it will grow to an amount given by the polynomial

$A = P(1 + r)^3$.

7. Evaluate the polynomial when $P = 10{,}000$ and $r = 0.08$ to find the amount to which \$10,000 will grow at 8% interest for 3 years.

8. Evaluate the polynomial when $P = 10{,}000$ and $r = 0.07$ to find the amount to which \$10,000 will grow at 7% interest for 3 years.

Surface area of a silo. A silo is a structure that is shaped like a right circular cylinder with a half sphere on top. The surface area of a silo of height h and radius r (including the area of the base) is given by the polynomial

$$2\pi rh + \pi r^2.$$

9. A $1\frac{1}{2}$-oz bottle of roll-on deodorant has a height of 4 in. and a radius of $\frac{3}{4}$ in. Find the surface area of the bottle if the bottle is shaped like a silo. Use 3.14 for π.

10. A container of tennis balls is silo-shaped, with a height of $7\frac{1}{2}$ in. and a radius of $1\frac{1}{4}$ in. Find the surface area of the container. Use 3.14 for π.

Identify the coefficient and the degree of each term of the polynomial. Then find the degree of the polynomial.

11. $x^3y - 2xy + 3x^2 - 5$

12. $5y^3 - y^2 + 15y + 1$

13. $17x^2y^3 - 3x^3yz - 7$

14. $6 - xy + 8x^2y^2 - y^5$

Collect like terms.

15. $a + b - 2a - 3b$

16. $y^2 - 1 + y - 6 - y^2$

17. $3x^2y - 2xy^2 + x^2$

18. $m^3 + 2m^2n - 3m^2 + 3mn^2$

19. $2u^2v - 3uv^2 + 6u^2v - 2uv^2$

20. $3x^2 + 6xy + 3y^2 - 5x^2 - 10xy - 5y^2$

21. $6au + 3av + 14au + 7av$

22. $3x^2y - 2z^2y + 3xy^2 + 5z^2y$

Add or subtract, as indicated.

23. $(2x^2 - xy + y^2) + (-x^2 - 3xy + 2y^2)$

24. $(2z - z^2 + 5) + (z^2 - 3z + 1)$

25. $(r^3 + 3rs - 5s^2) - (5r^3 + rs + 4s^2)$

26. $(7a^4 - 5ab + 6ab^2) - (9a^4 + 3ab - ab^2)$

27. $(r - 2s + 3) + (2r + 3s - 7)$

28. $(b^3a^2 - 2b^2a^3 + 3ba + 4) + (b^2a^3 - 4b^3a^2 + 2ba - 1)$

29. $(2x^2 - 3xy + y^2) + (-4x^2 - 6xy - y^2) + (x^2 + xy - y^2)$

30. $(x^3 - y^3) - (-2x^3 + x^2y - xy^2 + 2y^3)$

31. $(xy - ab) - (xy - 3ab)$

32. $(3y^4x^2 + 2y^3x - 3y) - (2y^4x^2 + 2y^3x - 4y - 2x)$

33. $(-2a + 7b - c) + (-3b + 4c - 8d)$

34. $(5a^2b + 7ab) + (9a^2b - 5ab) + (a^2b - 6ab)$

35. Subtract $7x + 3y$ from the sum of $4x + 5y$ and $-5x + 6y$.

36. Subtract $5a + 2b$ from the sum of $2a + b$ and $3a - 4b$.

Multiply.

37. $(3z - u)(2z + 3u)$

38. $(a - b)(a^2 + b^2 + 2ab)$

39. $(a^2b - 2)(a^2b - 5)$

40. $(xy + 7)(xy - 4)$

41. $(a^3 + bc)(a^3 - bc)$

42. $(m^2 + n^2 - mn)(m^2 + mn + n^2)$

43. $(y^4x + y^2 + 1)(y^2 + 1)$

44. $(a - b)(a^2 + ab + b^2)$

45. $(3xy - 1)(4xy + 2)$

46. $(m^3n + 8)(m^3n - 6)$

47. $(3 - c^2d^2)(4 + c^2d^2)$

48. $(6x - 2y)(5x - 3y)$

49. $(m^2 - n^2)(m + n)$

50. $(pq + 0.2)(0.4pq - 0.1)$

51. $(xy + x^5y^5)(x^4y^4 - xy)$

52. $(x - y^3)(2y^3 + x)$

53. $(x + h)^2$

54. $(3a + 2b)^2$

55. $(r^3t^2 - 4)^2$

56. $(3a^2b - b^2)^2$

57. $(p^4 + m^2n^2)^2$

58. $(ab + cd)^2$

59. $(2a - b)(2a + b)$

60. $(x - y)(x + y)$

61. $(c^2 - d)(c^2 + d)$

62. $(p^3 - 5q)(p^3 + 5q)$

63. $(ab + cd^2)(ab - cd^2)$

64. $(xy + pq)(xy - pq)$

65. $(x + y - 3)(x + y + 3)$

66. $(p + q + 4)(p + q - 4)$

67. $[x + y + z][x - (y + z)]$

68. $[a + b + c][a - (b + c)]$

69. $(a + b + c)(a - b - c)$

70. $(3x + 2 - 5y)(3x + 2 + 5y)$

Find a polynomial for the shaded area.

71.

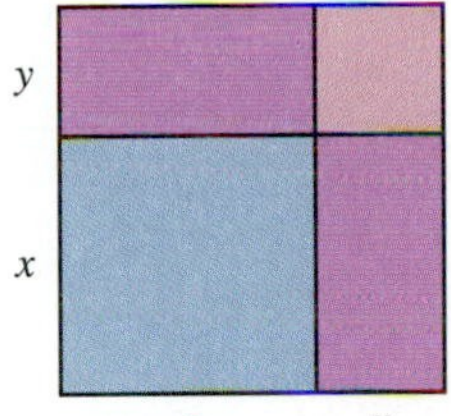

72.

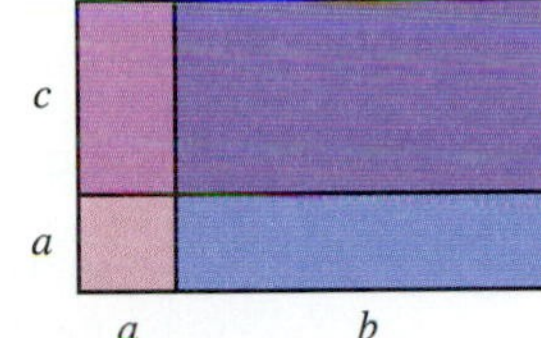

73.

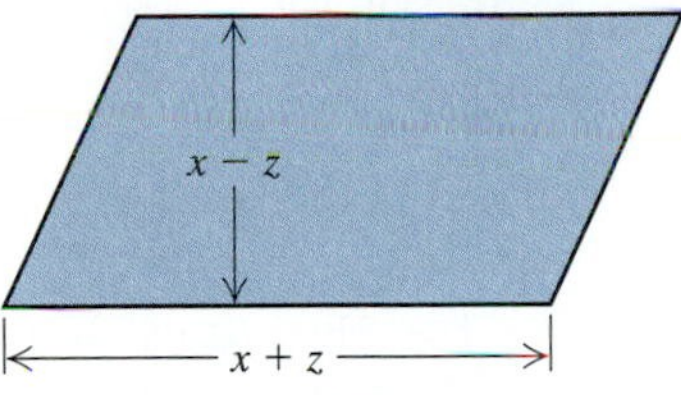

74.

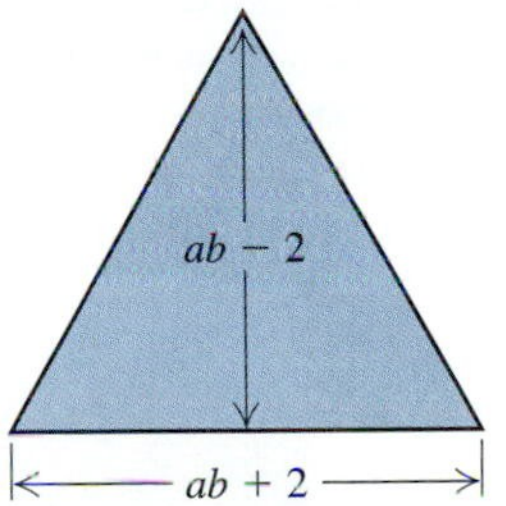

Skill Maintenance

The graph at right* shows the prices paid for a ton of white office paper and for a ton of newsprint by recyclers in the Boston market.

75. How much was being paid for white office paper in December 1989?

76. At what date did recyclers switch from paying for newsprint to charging for collecting it?

77. When did the value of newsprint peak?

78. When did the price that was paid for white office paper first drop to $20 per ton?

*From Burlington Free Press, 7/13/92, Burlington, VT. Reprinted with permission.

Synthesis

79. ◈ Is it possible for a polynomial in 4 variables to have a degree less than 4? Why or why not?

80. ◈ Explain how the formulas for the surface area of a right circular cylinder (Example 2) and a silo (Exercise 9) can be used to find a formula for the surface area of a sphere.

Find a polynomial for the shaded area. (Leave results in terms of π where appropriate.)

81.

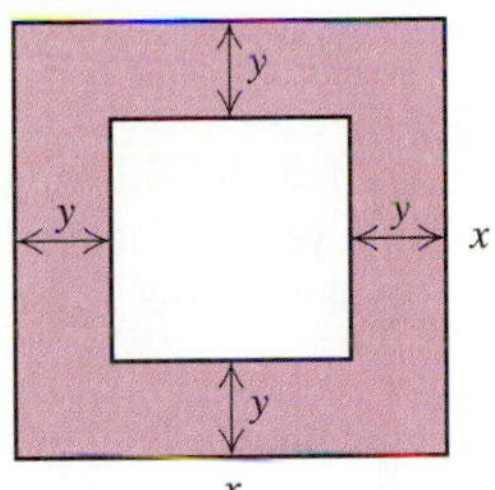

82.

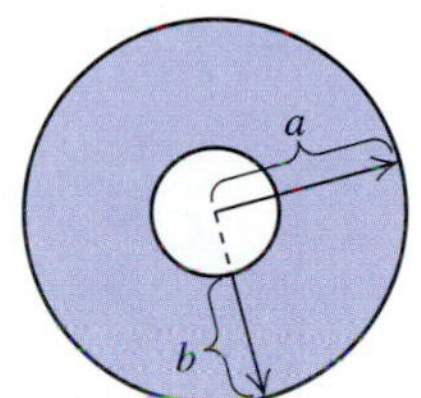

83.

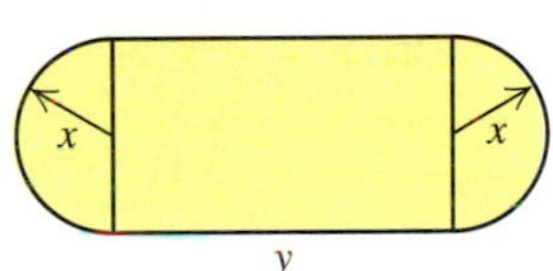

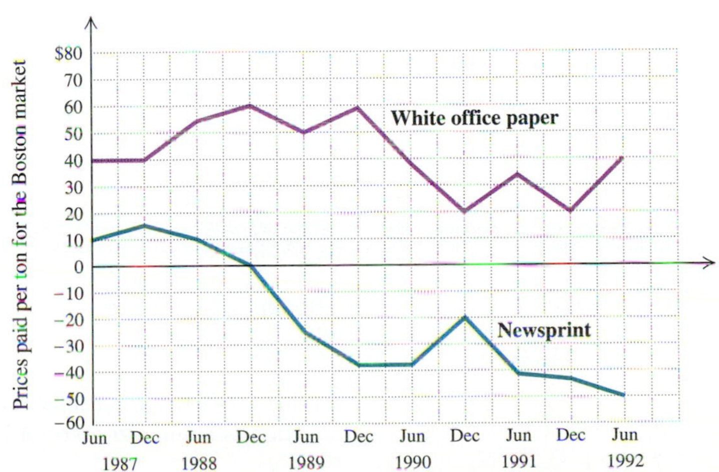

84. 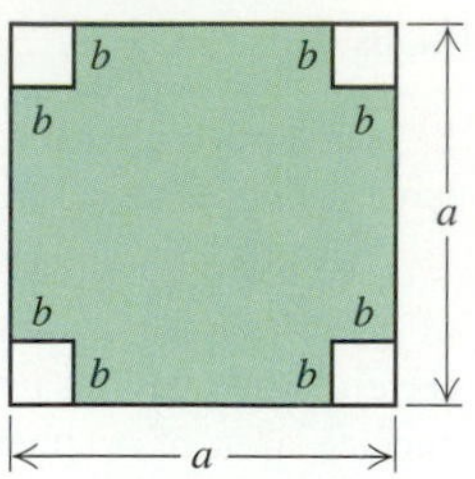

85. *Lung capacity.* The polynomial

$$0.041h - 0.018A - 2.69$$

can be used to estimate the lung capacity, in liters, of a female with height h, in centimeters, and age A, in years. Find the lung capacity of a 30-year-old woman who is 165 cm tall.

86. *The magic number.* The Boston Red Sox are leading the New York Yankees for the Eastern Division championship of the American League. The magic number is 8. This means that any combination of Red Sox wins and Yankee losses that totals 8 will ensure the championship for the Red Sox. The magic number is given by the polynomial

$$G - P - L + 1,$$

where G is the number of games in the season, P is the number of games that the leading team has played, and L is the number of games by which the leading team is ahead of the second-place team in the loss column.

Given the situation shown in the table and assuming a 162-game season, what is the magic number for the Philadelphia Phillies?

	W	L
Philadelphia	77	40
Pittsburgh	65	53
New York	61	60
Chicago	55	67
St. Louis	51	65
Florida	46	68
Montreal	41	73

87. The observatory at Danville University is shaped like a silo that is 40 ft high and 30 ft wide (see Exercise 9). Rick and Annie are to paint the exterior of the observatory using paint that covers 250 ft^2 per gallon. How many gallons should they purchase? Explain your reasoning.

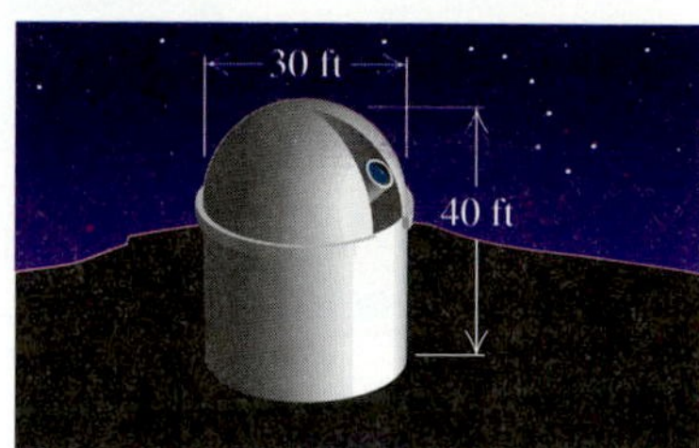

4.7 Division of Polynomials

Divisor a Monomial • Divisor a Binomial

In this section, we consider division of polynomials. You will see that polynomial division is similar to division in arithmetic.

Divisor a Monomial

We first consider division by a monomial. When dividing a monomial by a monomial, we use the quotient rule of Section 4.1 to subtract exponents when bases are the same. For example,

$$\frac{15x^{10}}{3x^4} = 5x^{10-4} = 5x^6;$$

and

$$\frac{42a^2b^5}{-3ab^2} = \frac{42}{-3}a^{2-1}b^{5-2} = -14ab^3.$$

When we are dividing a monomial into a polynomial, we use the rule for addition using fractional notation. That is, since

$$\frac{A}{C} + \frac{B}{C} = \frac{A+B}{C},$$

we know that

$$\frac{A+B}{C} = \frac{A}{C} + \frac{B}{C}.$$

You have actually used this rule for divisions. Consider $86 \div 2$: Although we might simply write

$$\frac{86}{2} = 43,$$

we are really saying

$$\frac{80+6}{2} = \frac{80}{2} + \frac{6}{2} = 40 + 3.$$

EXAMPLE 1

Divide $9x^8 + 12x^6$ by $3x^2$.

Solution This is equivalent to

$$\frac{9x^8}{3x^2} + \frac{12x^6}{3x^2}.$$ To see this, add and get the original expression.

We can now perform the separate divisions. *Caution*! The coefficients are *divided*, but the exponents are *subtracted*.

$$\frac{9x^8}{3x^2} + \frac{12x^6}{3x^2} = \frac{9}{3}x^{8-2} + \frac{12}{3}x^{6-2} = 3x^6 + 4x^4.$$

To check, we multiply the quotient by $3x^2$:

$$(3x^6 + 4x^4)3x^2 = 9x^8 + 12x^6.$$ The answer checks. ❑

EXAMPLE 2

Divide and check.

a) $(x^3 + 10x^2 - 8x) \div 2x$

b) $(10a^5b^4 - 2a^3b^2 + 6a^2b) \div 2a^2b$

Solution

a) $$\frac{x^3 + 10x^2 - 8x}{2x} = \frac{x^3}{2x} + \frac{10x^2}{2x} - \frac{8x}{2x}$$

$$= \frac{1}{2}x^{3-1} + \frac{10}{2}x^{2-1} - \frac{8}{2}x^{1-1}$$ Dividing coefficients and subtracting exponents

$$= \frac{1}{2}x^2 + 5x - 4$$

Check: We check by multiplying the quotient by $2x$:

$$\begin{array}{r} \frac{1}{2}x^2 + 5x - 4 \\ 2x \\ \hline x^3 + 10x^2 - 8x \end{array}$$

Multiplying

The answer checks.

b) $$\frac{10a^5b^4 - 2a^3b^2 + 6a^2b}{2a^2b} = \frac{10a^5b^4}{2a^2b} - \frac{2a^3b^2}{2a^2b} + \frac{6a^2b}{2a^2b}$$

$$= \frac{10}{2}a^{5-2}b^{4-1} - \frac{2}{2}a^{3-2}b^{2-1} + \frac{6}{2}$$

$$= 5a^3b^3 - ab + 3$$

Check:

$$\begin{array}{r} 5a^3b^3 - ab + 3 \\ 2a^2b \\ \hline 10a^5b^4 - 2a^3b^2 + 6a^2b \end{array}$$

Multiplying

The answer checks. □

To divide a polynomial by a monomial, divide each term by the monomial.

Divisor a Binomial

When the divisor has more than one term, we use long division very much as we do in arithmetic. We write polynomials in descending order and write in any missing terms.

EXAMPLE 3

Divide $x^2 + 5x + 6$ by $x + 2$.

Solution We have

$$\begin{array}{r} x \\ x + 2\overline{)x^2 + 5x + 6} \\ x^2 + 2x \\ \hline 3x \end{array}$$

← Divide the first term, x^2, by the first term in the divisor: $x^2/x = x$. Ignore the term 2.

← Multiply x above by the divisor, $x + 2$.

← Subtract: $(x^2 + 5x) - (x^2 + 2x) = x^2 + 5x - x^2 - 2x = 3x$.

Now we "bring down" the next term—in this case, 6—and repeat the procedure:

$$\begin{array}{r} x + 3 \\ x+2\overline{)x^2 + 5x + 6} \\ \underline{x^2 + 2x} \\ 3x + 6 \\ \underline{3x + 6} \\ 0 \end{array}$$

Consider the "remainder" $3x + 6$. Divide its first term by the first term of the divisor: $3x/x = 3$.

The 6 has been "brought down."

Multiply 3 by the divisor, $x + 2$.

Subtract: $(3x + 6) - (3x + 6) = 0$.

The quotient is $x + 3$. The remainder is 0, usually expressed as R0. A remainder of 0 is generally not listed in an answer.

To check, we multiply the quotient by the divisor and add the remainder, if any, to see if we get the dividend:

$$\underbrace{(x+2)}_{\text{Divisor}}\ \underbrace{(x+3)}_{\text{Quotient}} + \underbrace{0}_{\text{Remainder}} = \underbrace{x^2 + 5x + 6}_{\text{Dividend}}.$$

The division checks. ❑

EXAMPLE 4

Divide and check: $(x^2 + 2x - 12) \div (x - 3)$.

Solution We have

$$\begin{array}{r} x \\ x-3\overline{)x^2 + 2x - 12} \\ \underline{x^2 - 3x} \\ 5x \end{array}$$

Divide the first term by the first term: $x^2/x = x$.

Multiply x above by the divisor, $x - 3$.

Subtract: $(x^2 + 2x) - (x^2 - 3x) = x^2 + 2x - x^2 + 3x = 5x$.

Next we bring down the next term of the dividend, -12.

$$\begin{array}{r} x + 5 \\ x-3\overline{)x^2 + 2x - 12} \\ \underline{x^2 - 3x} \\ 5x - 12 \\ \underline{5x - 15} \\ 3 \end{array}$$

Divide the first term of $5x - 12$ by the first term of the divisor: $5x/x = 5$.

The -12 has been "brought down."

Multiply 5 by the divisor, $x - 3$.

Subtract: $(5x - 12) - (5x - 15) = 5x - 12 - 5x + 15 = 3$.

The answer is $x + 5$ with R3, or

$$\underbrace{x + 5}_{\text{Quotient}} + \frac{3 \rightarrow \text{Remainder}}{x - 3 \rightarrow \text{Divisor}}$$

(This is the way answers will be given at the back of the book.)

Check: In arithmetic, to check that $9 \div 4 = 2\frac{1}{4}$, we can multiply and add: $4 \cdot 2 + 1$. A similar procedure, used to check that

$$(x^2 + 2x - 12) \div (x - 3) = x + 5 + \frac{3}{x - 3},$$

is as follows:

$$(x - 3)(x + 5) + 3 = x^2 + 2x - 15 + 3 = x^2 + 2x - 12.$$ ❑

Our division procedure ends when the degree of the remainder is less than the degree of the divisor. Check this for Example 4.

EXAMPLE 5

Divide: **(a)** $(x^3 + 1) \div (x + 1)$; **(b)** $(x^4 - 3x^2 + 1) \div (x - 4)$.

Solution

a)

$$\begin{array}{r l}
 & x^2 - x + 1 \\
x + 1 \,\big) & x^3 + 0x^2 + 0x + 1 \quad \longleftarrow \text{Fill in the missing terms.} \\
 & x^3 + x^2 \\
 & \quad - x^2 + 0x \quad \longleftarrow \text{Subtracting } x^3 + x^2 \text{ from } x^3 + 0x^2 \text{ and bringing down the } 0x \\
 & \quad - x^2 - x \\
 & \qquad\quad x + 1 \quad \longleftarrow \text{Subtracting } -x^2 - x \text{ from } -x^2 + 0x \text{ and bringing down the } 1 \\
 & \qquad\quad x + 1 \\
 & \qquad\qquad 0
\end{array}$$

The answer is $x^2 - x + 1$.

Check: $(x + 1)(x^2 - x + 1) + 0 = x^3 - x^2 + x + x^2 - x + 1 + 0 = x^3 + 1.$

b)

$$\begin{array}{r l}
 & x^3 + 4x^2 + 13x + 52 \\
x - 4 \,\big) & x^4 + 0x^3 - 3x^2 + 0x + 1 \quad \longleftarrow \text{Fill in the missing terms.} \\
 & x^4 - 4x^3 \\
 & \quad 4x^3 - 3x^2 \qquad\qquad x^4 + 0x^3 - (x^4 - 4x^3) = 4x^3 \\
 & \quad 4x^3 - 16x^2 \\
 & \qquad 13x^2 + 0x \qquad\qquad (4x^3 - 3x^2) - (4x^3 - 16x^2) = 13x^2 \\
 & \qquad 13x^2 - 52x \\
 & \qquad\qquad 52x + 1 \\
 & \qquad\qquad 52x - 208 \\
 & \qquad\qquad\qquad 209
\end{array}$$

The answer is $x^3 + 4x^2 + 13x + 52$, with R209, or

$$x^3 + 4x^2 + 13x + 52 + \frac{209}{x - 4}.$$

Check:

$$\begin{aligned}
&(x - 4)(x^3 + 4x^2 + 13x + 52) + 209 \\
&= x^4 + 4x^3 + 13x^2 + 52x - 4x^3 - 16x^2 - 52x - 208 + 209 \\
&= x^4 - 3x^2 + 1
\end{aligned}$$

❑

EXERCISE SET 4.7

Divide and check.

1. $\dfrac{24x^4 - 4x^3}{8}$

2. $\dfrac{12a^4 - 3a^2}{6}$

3. $\dfrac{u - 2u^2 - u^5}{u}$

4. $\dfrac{50x^5 - 7x^4 + x^2}{x}$

5. $(15t^3 + 24t^2 - 6t) \div 3t$

6. $(25t^3 + 15t^2 - 30t) \div 5t$

7. $(20x^6 - 20x^4 - 5x^2) \div (-5x^2)$

8. $(24x^6 + 32x^5 - 8x^2) \div (-8x^2)$

9. $(24x^5 - 40x^4 + 6x^3) \div (4x^3)$

10. $(18x^6 - 27x^5 - 3x^3) \div (9x^3)$

11. $\dfrac{8x^2 - 3x + 1}{2}$

12. $\dfrac{6x^2 + 3x - 2}{3}$

13. $\dfrac{2x^3 + 6x^2 + 4x}{2x}$

14. $\dfrac{2x^4 - 3x^3 + 5x^2}{x^2}$

15. $\dfrac{9r^2s^2 + 3r^2s - 6rs^2}{-3rs}$

16. $\dfrac{4x^4y - 8x^6y^2 + 12x^8y^6}{4x^4y}$

17. $(x^2 + 4x + 4) \div (x + 2)$

18. $(x^2 - 6x + 9) \div (x - 3)$

19. $(x^2 - 10x - 25) \div (x - 5)$

20. $(x^2 + 8x - 16) \div (x + 4)$

21. $(x^2 + 4x - 14) \div (x + 6)$

22. $(x^2 + 5x - 9) \div (x - 2)$

23. $\dfrac{x^2 - 9}{x + 3}$

24. $\dfrac{x^2 - 25}{x + 5}$

25. $\dfrac{x^5 + 1}{x + 1}$

26. $\dfrac{x^5 - 1}{x - 1}$

27. $\dfrac{8x^3 - 22x^2 - 5x + 12}{4x + 3}$

28. $\dfrac{2x^3 - 9x^2 + 11x - 3}{2x - 3}$

29. $(x^6 - 13x^3 + 42) \div (x^3 - 7)$

30. $(x^6 + 5x^3 - 24) \div (x^3 - 3)$

31. $(x^4 - 16) \div (x - 2)$

32. $(x^4 - 81) \div (x - 3)$

33. $(t^3 - t^2 + t - 1) \div (t - 1)$

34. $(t^3 - t^2 + t - 1) \div (t + 1)$

Skill Maintenance

35. The perimeter of a rectangle is 640 ft. The length is 15 ft greater than the width. Find the area of the rectangle.

36. Solve: $2x > 12 + 7x$.

37. Plot the points $(4, -1)$, $(0, 5)$, $(-2, 3)$, and $(-3, 0)$.

38. In which quadrant are both coordinates negative?

Synthesis

39. On an assignment, a student writes

$$\frac{12x^3 - 6x}{3x} = 4x^2 - 6x.$$

What mistake is the student making and how might you convince the person that a mistake has been made?

40. Explain how you might divide $x^2 - 49$ by $x - 7$ without doing long division.

Divide.

41. $(x^4 + 9x^2 + 20) \div (x^2 + 4)$

42. $(y^4 + a^2) \div (y + a)$

43. $(5a^3 + 8a^2 - 23a - 1) \div (5a^2 - 7a - 2)$

44. $(15y^3 - 30y + 7 - 19y^2) \div (3y^2 - 2 - 5y)$

45. Divide the sum of $4x^5 - 14x^3 - x^2 + 3$ and $2x^5 + 3x^4 + x^3 - 3x^2 + 5x$ by $3x^3 - 2x - 1$.

46. Divide $5x^7 - 3x^4 + 2x^2 - 10x + 2$ by the sum of $(x - 3)^2$ and $5x - 8$.

47. Divide $6a^{3h} + 13a^{2h} - 4a^h - 15$ by $2a^h + 3$.

If the remainder is 0 when one polynomial is divided by another, the divisor is a *factor* of the dividend. Find the value(s) of c for which $x - 1$ is a factor of the polynomial.

48. $x^2 + 4x + c$

49. $2x^2 + 3cx - 8$

50. $c^2x^2 - 2cx + 1$

4.8 Synthetic Division

Streamlining Long Division • Using Synthetic Division

To divide a polynomial by a binomial of the type $x - a$, we can streamline the usual procedure to develop a process called *synthetic division*.

Compare the following. In each stage, we attempt to write a bit less than in the previous stage, while retaining enough essentials to solve the problem. At the end, we will return to the usual polynomial notation.

Stage 1

When a polynomial is written in descending order, the coefficients provide the essential information:

$$\begin{array}{r} 4x^2+5x+11 \\ x-2\overline{)4x^3-3x^2+x+7} \\ 4x^3-8x^2 \\ \hline 5x^2+x \\ 5x^2-10x \\ \hline 11x+7 \\ 11x-22 \\ \hline 29 \end{array} \qquad \begin{array}{r} 4+5+11 \\ 1-2\overline{)4-3+1+7} \\ 4-8 \\ \hline 5+1 \\ 5-10 \\ \hline 11+7 \\ 11-22 \\ \hline 29 \end{array}$$

Because the coefficient of x is 1 in the divisor, each time we multiply the divisor by a term in the answer, the leading coefficient of that product duplicates a coefficient in the answer. In the next stage, we don't bother to duplicate these numbers. We also show where -2 is used in the problem and stop writing 1 in the divisor.

Stage 2

$$\begin{array}{r} 4x^2+5x+11 \\ x-2\overline{)4x^3-3x^2+x+7} \\ 4x^3-8x^2 \\ \hline 5x^2+x \\ 5x^2-10x \\ \hline 11x+7 \\ 11x-22 \\ \hline 29 \end{array} \qquad \begin{array}{rl} 4+5+11 & \\ -2\overline{)4-3+1+7} & \\ (-2)4 & \leftarrow \text{Multiply: } 4(-2)=-8. \\ \hline & \leftarrow \text{Subtract: } -3-(-8)=5. \\ 5+1 & \\ (-2)5 & \leftarrow \text{Multiply: } 5(-2)=-10. \\ \hline & \leftarrow \text{Subtract: } 1-(-10)=11. \\ 11+7 & \\ (-2)11 & \leftarrow \text{Multiply: } 11(-2)=-22. \\ \hline 29 & \leftarrow \text{Subtract: } 7-(-22)=29. \end{array}$$

To simplify further, we now change the sign of the -2 in the divisor and, in exchange, *add* at each step in the long division.

Stage 3

$$\begin{array}{r} 4x^2+5x+11 \\ x-2\overline{)4x^3-3x^2+x+7} \\ 4x^3-8x^2 \\ \hline 5x^2+x \\ 5x^2-10x \\ \hline 11x+7 \\ 11x-22 \\ \hline 29 \end{array} \qquad \begin{array}{rl} 4+5+11 & \quad \text{Replace the } -2 \text{ with } 2. \\ 2\overline{)4-3+1+7} & \\ 8 & \leftarrow \text{Multiply: } 4\cdot 2=8. \\ \hline & \leftarrow \text{Add: } -3+8=5. \\ 5+1 & \\ 10 & \leftarrow \text{Multiply: } 5\cdot 2=10. \\ \hline & \leftarrow \text{Add: } 1+10=11. \\ 11+7 & \\ 22 & \leftarrow \text{Multiply: } 11\cdot 2=22. \\ \hline 29 & \leftarrow \text{Add: } 7+22=29. \end{array}$$

As you can see from the blue numbers, there is still some duplication that we can eliminate.

Stage 4

$$
\begin{array}{r}
4x^2 + 5x + 11 \\
x - 2\overline{)4x^3 - 3x^2 + x + 7} \\
\underline{4x^3 - 8x^2} \\
5x^2 + x \\
\underline{5x^2 - 10x} \\
11x + 7 \\
\underline{11x - 22} \\
29
\end{array}
\qquad
\begin{array}{r}
4 \quad 5 \quad 11 \\
2\overline{)4 \quad -3 \quad 1 \quad 7} \\
\underline{8} \quad \underline{10} \quad \underline{22} \\
5 \quad 11 \quad 29
\end{array}
$$

Don't lose sight of how the products 8, 10, and 22 are found. Also, keep in mind that the 5 and 11 preceding the remainder 29 coincide with the 5 and 11 following the 4 on the top line. If we write a 4 to the left of 5 on the bottom line, we can dispense with the top line and read our answer from the bottom line. This final stage is commonly called **synthetic division.**

Stage 5

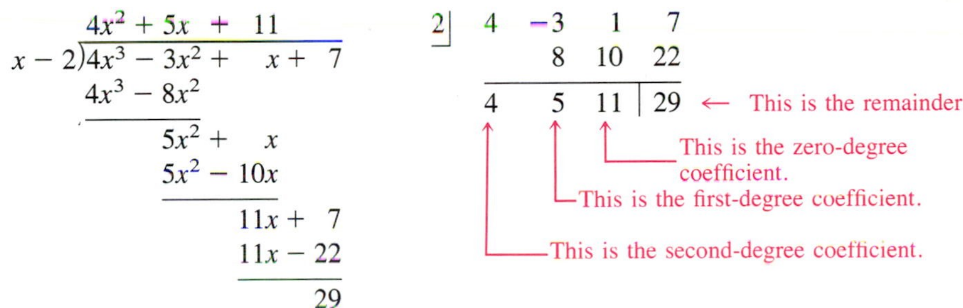

The quotient is $4x^2 + 5x + 11$. The remainder is 29.

> Remember that in order for this method to work, the divisor must be of the form $x - a$, that is, a variable minus a constant. The coefficient of the variable must be 1.

EXAMPLE 1

Use synthetic division to divide:

$$(x^3 + 6x^2 - x - 30) \div (x - 2).$$

Solution

A.
$$
\begin{array}{r|rrrr}
2 & 1 & 6 & -1 & -30 \\
 & & & & \\
\hline
 & 1 & & &
\end{array}
$$

Write the 2 of $x - 2$ and the coefficients of the dividend.

Bring down the first coefficient.

B.
$$
\begin{array}{r|rrrr}
2 & 1 & 6 & -1 & -30 \\
 & & 2 & & \\
\hline
 & 1 & 8 & &
\end{array}
$$

Multiply 1 by 2 to get 2.

Add 6 and 2.

C.
$$
\begin{array}{r|rrrr}
2 & 1 & 6 & -1 & -30 \\
 & & 2 & 16 & \\
\hline
 & 1 & 8 & 15 &
\end{array}
$$

Multiply 8 by 2.

Add -1 and 16.

D.
$$\begin{array}{r|rrrr} 2 & 1 & 6 & -1 & -30 \\ & & 2 & 16 & 30 \\ \hline & 1 & 8 & 15 & 0 \end{array}$$

Multiply 15 by 2 and add.

The answer is $x^2 + 8x + 15$ with R0, or just $x^2 + 8x + 15$. ❑

EXAMPLE 2

Use synthetic division to divide.

a) $(2x^3 + 7x^2 - 5) \div (x + 3)$

b) $(x^3 + 4x^2 - x - 4) \div (x + 4)$

c) $(8x^5 - 6x^3 + x - 8) \div (x - 2)$

Solution

a) $(2x^3 + 7x^2 - 5) \div (x + 3)$

The dividend has no x-term, so we must write a 0 for its coefficient of x. Note that $x + 3 = x - (-3)$.

$$\begin{array}{r|rrrr} -3 & 2 & 7 & 0 & -5 \\ & & -6 & -3 & 9 \\ \hline & 2 & 1 & -3 & 4 \end{array}$$

The answer is $2x^2 + x - 3$, with R4, or $2x^2 + x - 3 + \dfrac{4}{x + 3}$.

b) $(x^3 + 4x^2 - x - 4) \div (x + 4)$

$$\begin{array}{r|rrrr} -4 & 1 & 4 & -1 & -4 \\ & & -4 & 0 & 4 \\ \hline & 1 & 0 & -1 & 0 \end{array}$$

The answer is $x^2 - 1$.

c) $(8x^5 - 6x^3 + x - 8) \div (x - 2)$

$$\begin{array}{r|rrrrrr} 2 & 8 & 0 & -6 & 0 & 1 & -8 \\ & & 16 & 32 & 52 & 104 & 210 \\ \hline & 8 & 16 & 26 & 52 & 105 & 202 \end{array}$$

The answer is $8x^4 + 16x^3 + 26x^2 + 52x + 105$ with R202, or

$$8x^4 + 16x^3 + 26x^2 + 52x + 105 + \frac{202}{x - 2}.$$ ❑

EXERCISE SET 4.8

Use synthetic division to divide.

1. $(x^3 - 2x^2 + 2x - 5) \div (x - 1)$
2. $(x^3 - 2x^2 + 2x - 5) \div (x + 1)$
3. $(a^2 + 11a - 19) \div (a + 4)$
4. $(a^2 + 11a - 19) \div (a - 4)$
5. $(x^3 - 7x^2 - 13x + 3) \div (x - 2)$
6. $(x^3 - 7x^2 - 13x + 3) \div (x + 2)$
7. $(3x^3 + 7x^2 - 4x + 3) \div (x + 3)$
8. $(3x^3 + 7x^2 - 4x + 3) \div (x - 3)$
9. $(y^3 - 3y + 10) \div (y - 2)$

10. $(x^3 - 2x^2 + 8) \div (x + 2)$
11. $(3x^4 - 25x^2 - 18) \div (x - 3)$
12. $(6y^4 + 15y^3 + 28y + 6) \div (y + 3)$
13. $(x^3 - 27) \div (x - 3)$
14. $(y^3 + 27) \div (y + 3)$
15. $(y^5 - 1) \div (y - 1)$
16. $(x^5 - 32) \div (x - 2)$
17. $(3x^4 + 8x^3 + 2x^2 - 7x - 4) \div (x + 2)$
18. $(2x^4 - x^3 - 5x^2 + x + 7) \div (x + 1)$
19. $(3x^3 + 7x^2 - x + 1) \div \left(x + \frac{1}{3}\right)$
20. $(8x^3 - 6x^2 + 7x - 1) \div \left(x - \frac{1}{2}\right)$

Skill Maintenance

Collect like terms.

21. $3y + (-10y)$
22. $-4x + 5x$
23. $10m + 5 + m + 1$
24. $-2 + 8x + (-4) + 5x$

Synthesis

25. ◈ Describe how synthetic division can be used when the divisor is a linear polynomial $ax + b$, with $a > 1$.
26. ◈ For the polynomial
$P = 8x^5 - 3x^4 + 7x - 4$:
    a) By synthetic division, find the remainder when P is divided by $x - 2$.
    b) Find P when $x = 2$.
    c) Compare the answers from parts (a) and (b).
27. Let $Q = x^3 - 5x^2 + 5x - 4$.
    a) Use synthetic division to show that $x - 4$ is a factor of $x^3 - 5x^2 + 5x - 4$.
    b) Why does the result in part (a) indicate that $Q = 0$ when $x = 4$?
    c) Check by substituting 4 into the given polynomial.
28. Let $P = 2x^4 + 7x^3 - 4x^2 - 27x - 18$.
    a) Use synthetic division to show that $x - 2$ is a factor of P.
    b) Let Q_1 be the quotient found in part (a). Write P as a product of $x - 2$ and Q_1.
    c) Use synthetic division to show that $x + 3$ is a factor of Q_1.
    d) Let Q_2 be the quotient found in part (c). Write P as a product of $x - 2$, $x + 3$, and Q_2.
    e) Use synthetic division to show that $x + 1$ is a factor of Q_2.
    f) Let Q_3 be the quotient found in part (e). Write P as a product of $x - 2$, $x + 3$, $x + 1$, and Q_3. This product is the *factorization* of P.

4.9 Negative Exponents and Scientific Notation

Negative Integers as Exponents • Scientific Notation • Multiplying and Dividing Using Scientific Notation • Problem Solving with Scientific Notation

We now develop a definition of negative exponents. Once we are able to consider both positive and negative powers, we will study a method of writing numbers known as *scientific notation*.

Negative Integers as Exponents

No meaning has yet been attached to negative exponents. By defining negative exponents in a certain way, however, all the rules that apply to whole-number exponents will hold for integer exponents as well.

Consider $5^3/5^7$ and first simplify by removing a factor of 1:

$$\frac{5^3}{5^7} = \frac{5 \cdot 5 \cdot 5}{5 \cdot 5 \cdot 5 \cdot 5 \cdot 5 \cdot 5 \cdot 5}$$

$$= \frac{5 \cdot 5 \cdot 5}{5 \cdot 5 \cdot 5} \cdot \frac{1}{5 \cdot 5 \cdot 5 \cdot 5} = \frac{1}{5^4}.$$

Next, simplify the same expression using the quotient rule:

$$\frac{5^3}{5^7} = 5^{3-7} = 5^{-4}.$$

From these two expressions for $5^3/5^7$, it follows that

$$5^{-4} = \frac{1}{5^4}.$$

This leads to our definition of negative exponents.

Negative Exponents

For any real number a that is nonzero and any integer n,

$$a^{-n} = \frac{1}{a^n}.$$

(The numbers a^{-n} and a^n are reciprocals.)

EXAMPLE 1

Express using positive exponents, and then simplify: **(a)** m^{-3}; **(b)** 4^{-2}; **(c)** $(-3)^{-2}$; **(d)** ab^{-1}; **(e)** $3c^{-5}$; **(f)** $1/x^{-3}$.

Solution

a) $m^{-3} = \frac{1}{m^3}$ — m^{-3} is the reciprocal of m^3.

b) $4^{-2} = \frac{1}{4^2} = \frac{1}{16}$ — 4^{-2} is the reciprocal of 4^2. Note that $4^{-2} \neq 4(-2)$.

c) $(-3)^{-2} = \frac{1}{(-3)^2} = \frac{1}{(-3)(-3)} = \frac{1}{9}$ — $(-3)^{-2}$ is the reciprocal of $(-3)^2$. Note that $(-3)^{-2} \neq -\frac{1}{3^2}$.

d) $ab^{-1} = a\left(\frac{1}{b^1}\right) = a\left(\frac{1}{b}\right) = \frac{a}{b}$ — b^{-1} is the reciprocal of b^1.

e) $3c^{-5} = 3\left(\frac{1}{c^5}\right) = \frac{3}{c^5}$

f) $\frac{1}{x^{-3}} = x^{-(-3)} = x^3$ — The reciprocal of x^{-3} is $x^{-(-3)}$, or x^3. ❑

Notice in Examples 1(b) and 1(c) that a negative exponent does not, in itself, indicate that an expression represents a negative number.

The following is another way to understand why negative exponents are defined as they are:

On this side, we divide by 5 at each step. ↓		On this side, the exponents decrease by 1. ↓
	$5 \cdot 5 \cdot 5 = 5^3$	
	$5 \cdot 5 = 5^2$	
	$5 = 5^1$	
	$1 = 5^0$	
	$\frac{1}{5} = 5^?$	
	$\frac{1}{25} = 5^?$	

To continue the pattern, it should follow that

$$\frac{1}{5} = \frac{1}{5^1} = 5^{-1} \quad \text{and} \quad \frac{1}{25} = \frac{1}{5^2} = 5^{-2}.$$

The rules for powers still hold when exponents are negative.

EXAMPLE 2

Simplify: **(a)** $a^5 \cdot a^{-2}$; **(b)** x/x^7; **(c)** b^{-4}/b^{-5}; **(d)** $(y^{-5})^{-7}$; **(e)** $(5x^2y^{-3})^4$.

Solution

a) $a^5 \cdot a^{-2} = a^{5+(-2)}$ Adding exponents

$= a^3$

b) $\frac{x}{x^7} = x^{1-7}$ Subtracting exponents

$= x^{-6}$, or $\frac{1}{x^6}$

c) $\frac{b^{-4}}{b^{-5}} = b^{-4-(-5)} = b^1 = b$ We subtract exponents even if the exponent in the denominator is negative.

d) $(y^{-5})^{-7} = y^{(-5)(-7)}$ Multiplying exponents

$= y^{35}$

e) $(5x^2y^{-3})^4 = 5^4(x^2)^4(y^{-3})^4$ Raising each factor to the fourth power

$= 625x^8y^{-12}$, or $\frac{625x^8}{y^{12}}$ □

Some manipulations with negative exponents can be performed quickly when certain patterns are discovered. For example, since $m^{-5} = 1/m^5$ and $1/x^{-3} = x^3$, we have

$$\frac{m^{-5}}{x^{-3}} = m^{-5} \cdot \frac{1}{x^{-3}} = \frac{1}{m^5} \cdot x^3 = \frac{x^3}{m^5}.$$

Note how the signs of the exponents change.

EXAMPLE 3 Simplify: $\left(\frac{y^3}{5}\right)^{-2}$.

Solution

$$\left(\frac{y^3}{5}\right)^{-2} = \frac{(y^3)^{-2}}{5^{-2}} \quad \text{Raising a quotient to a power}$$

$$= \frac{y^{-6}}{5^{-2}} \quad \text{Using the power rule}$$

$$= \frac{5^2}{y^6} \quad \text{Rewriting with positive exponents}$$

$$= \frac{25}{y^6}$$

❑

The following summary of definitions and rules assumes that no denominators are 0 and that 0^0 is not considered.

Definitions and Rules for Exponents

For any integers m and n,

1 as an exponent: $a^1 = a$

0 as an exponent: $a^0 = 1$

Negative exponents: $a^{-n} = \frac{1}{a^n}$

The Product Rule: $a^m \cdot a^n = a^{m+n}$

The Quotient Rule: $\frac{a^m}{a^n} = a^{m-n}$

The Power Rule: $(a^m)^n = a^{mn}$

Raising a product to a power: $(ab)^n = a^n b^n$

Raising a quotient to a power: $\left(\frac{a}{b}\right)^n = \frac{a^n}{b^n}$

Scientific Notation

When working with the very large or very small numbers that frequently arise in science, **scientific notation** is an especially useful way of writing numbers. The following are examples of scientific notation.

The distance from the earth to the sun:

9.3×10^7 mi = 93,000,000 mi

The mass of a hydrogen atom:

1.7×10^{-24} gm = 0.0000000000000000000000017 gm

Scientific Notation

Scientific notation for a number is an expression of the type

$N \times 10^n$,

where N is at least 1 but less than 10 ($1 \leq N < 10$), N is expressed in decimal notation, and n is an integer.

Converting from scientific to decimal notation involves multiplying by a power of 10. Consider the following.

Scientific Notation $N \times 10^n$	*Multiplication*	*Decimal Notation*
4.52×10^2	4.52×100	452.
4.52×10^1	4.52×10	45.2
4.52×10^0	4.52×1	4.52
4.52×10^{-1}	4.52×0.1	0.452
4.52×10^{-2}	4.52×0.01	0.0452

Note that when n, the power of 10, is positive, the decimal point moves right n places in decimal notation. When n is negative, the decimal point moves left n places. We generally try to perform this multiplication mentally.

EXAMPLE 4

Convert to decimal notation: **(a)** 7.893×10^5; **(b)** 4.7×10^{-8}.

Solution

a) Since the exponent is positive, the decimal point moves to the right:

7.89300. 5 places $\quad 7.893 \times 10^5 = 789{,}300$ The decimal point moves 5 places to the right.

b) Since the exponent is negative, the decimal point moves to the left:

0.00000004.7 8 places $\quad 4.7 \times 10^{-8} = 0.000000047$ The decimal point moves 8 places to the left. ❑

To convert from decimal to scientific notation, we reverse the above procedure.

EXAMPLE 5

Write in scientific notation: **(a)** 7800; **(b)** 0.0549.

Solution

a) We must have $7800 = N \times 10^n$, where $1 \leq N < 10$. Because multiplication by 10^n moves only the decimal point, we must have $N = 7.8$:

$$7800 = 7.8 \times 10^n.$$

Multiplying 7.8 by 10^3 moves the decimal point 3 places to the right. Thus, n is 3 and

$$7800 = 7.8 \times 10^3.$$

b) In scientific notation, 0.0549 is written as 5.49×10^n. Multiplying 5.49 by 10^{-2} moves the decimal point 2 places to the left. Thus, n is -2 and

$$0.0549 = 5.49 \times 10^{-2}.$$ ❑

You should try to make conversions to scientific notation mentally as much as possible. Remember that positive exponents are used when representing large numbers and negative exponents are used when representing numbers between 0 and 1.

Multiplying and Dividing Using Scientific Notation

Products and quotients of numbers written in scientific notation are found using the rules for exponents.

EXAMPLE 6

Simplify: **(a)** $(1.8 \times 10^6) \cdot (2.3 \times 10^{-4})$; **(b)** $(3.41 \times 10^5) \div (1.1 \times 10^{-3})$.

Solution

a)
$$\begin{aligned}(1.8 \times 10^6) \cdot (2.3 \times 10^{-4}) &= (1.8 \cdot 2.3) \times (10^6 \cdot 10^{-4}) \\ &= 4.14 \times 10^{6+(-4)} \qquad \text{Adding exponents} \\ &= 4.14 \times 10^2\end{aligned}$$

b)
$$\begin{aligned}(3.41 \times 10^5) \div (1.1 \times 10^{-3}) &= \frac{3.41 \times 10^5}{1.1 \times 10^{-3}} \\ &= \frac{3.41}{1.1} \times \frac{10^5}{10^{-3}} \\ &= 3.1 \times 10^{5-(-3)} \qquad \text{Subtracting exponents} \\ &= 3.1 \times 10^8\end{aligned}$$ ❑

When a problem is stated using scientific notation, it is customary to use scientific notation for the answer.

EXAMPLE 7

Simplify: **(a)** $(3.1 \times 10^5) \cdot (4.5 \times 10^{-3})$; **(b)** $(7.2 \times 10^{-7}) \div (8.0 \times 10^6)$.

Solution

a) We have

$$\begin{aligned}(3.1 \times 10^5) \cdot (4.5 \times 10^{-3}) &= (3.1 \times 4.5)(10^5 \cdot 10^{-3}) \\ &= 13.95 \times 10^2.\end{aligned}$$

Our answer is not yet in scientific notation because 13.95 is not between 1 and 10. We convert to scientific notation as follows:

$$\begin{aligned}13.95 \times 10^2 &= 1.395 \times 10^1 \times 10^2 \qquad \text{Substituting } 1.395 \times 10^1 \text{ for } 13.95 \\ &= 1.395 \times 10^3 \qquad \text{Adding exponents}\end{aligned}$$

b)
$$\begin{aligned}(7.2 \times 10^{-7}) \div (8.0 \times 10^6) &= \frac{7.2 \times 10^{-7}}{8.0 \times 10^6} = \frac{7.2}{8.0} \times \frac{10^{-7}}{10^6} \\ &= 0.9 \times 10^{-13} \\ &= 9.0 \times 10^{-1} \times 10^{-13} \qquad \text{Substituting } 9.0 \times 10^{-1} \text{ for } 0.9 \\ &= 9.0 \times 10^{-14} \qquad \text{Adding exponents}\end{aligned}$$ ❑

Problem Solving with Scientific Notation

EXAMPLE 8

Light traveling at a rate of 300,000 kilometers per second (km/s) takes 499 seconds to reach the earth from the sun. Find the distance, expressed in scientific notation, from the sun to the earth.

Solution

1. FAMILIARIZE. The time t that it takes for light to reach the earth from the sun is 4.99×10^2 sec (s). The speed r is 3.0×10^5 km/s. Recall that distance can be expressed in terms of speed and time:

$$d = rt$$

Distance = Speed × Time.

(If you did not know this formula, you might look it up in a reference book.)

2. TRANSLATE. We translate the problem to mathematical language by substituting 3.0×10^5 for r and 4.99×10^2 for t:

$$\begin{aligned} d &= rt \\ &= (3.0 \times 10^5)(4.99 \times 10^2). \end{aligned}$$

3. CARRY OUT. We carry out the computation and express the results using scientific notation for the answer:

$$\begin{aligned} d &= (3.0 \times 10^5)(4.99 \times 10^2) \\ &= 14.97 \times 10^7 \\ &= 1.497 \times 10^8 \text{ km.} \end{aligned}$$

Converting to scientific notation

4. CHECK. We can check by reviewing our computations. Note too that our answer seems reasonable since it far exceeds the time or the rate alone.

5. STATE. The distance from the earth to the sun is 1.497×10^8 km. □

EXERCISE SET 4.9

Express using positive exponents. Then simplify.

1. 3^{-2}
2. 2^{-3}
3. 10^{-4}
4. 5^{-6}
5. 7^{-3}
6. 5^{-2}
7. a^{-3}
8. x^{-2}
9. $\frac{1}{y^{-4}}$
10. $\frac{1}{t^{-7}}$
11. $\frac{1}{z^{-n}}$
12. $\frac{1}{h^{-m}}$
13. 2^{-1}
14. $\left(\frac{2}{3}\right)^{-1}$
15. $\left(\frac{1}{4}\right)^{-2}$
16. $\left(\frac{4}{5}\right)^{-2}$

Express using negative exponents.

17. $\frac{1}{4^3}$
18. $\frac{1}{5^2}$
19. $\frac{1}{x^3}$
20. $\frac{1}{y^2}$
21. $\frac{1}{a^4}$
22. $\frac{1}{t^5}$
23. $\frac{1}{p^n}$
24. $\frac{1}{m^n}$
25. $\frac{1}{5}$

26. $\frac{1}{8}$ 27. $\frac{1}{t}$ 28. $\frac{1}{m}$

Simplify.

29. $3^{-5} \cdot 3^{8}$
30. $5^{-8} \cdot 5^{9}$
31. $x^{-2} \cdot x$
32. $x \cdot x^{-1}$
33. $x^{-7} \cdot x^{-6}$
34. $y^{-5} \cdot y^{-8}$
35. $\frac{m^6}{m^{12}}$
36. $\frac{p^4}{p^5}$
37. $\frac{(8x)^6}{(8x)^{10}}$
38. $\frac{(9t)^4}{(9t)^{11}}$
39. $\frac{18^9}{18^9}$
40. $\frac{(6y)^7}{(6y)^7}$
41. $(a^{-3}b^{-5})(a^{-4}b^{-6})$
42. $(x^{-2}y^{-7})(x^{-3}y^{-2})$
43. $\frac{x^7}{x^{-2}}$
44. $\frac{t^8}{t^{-3}}$
45. $\frac{z^{-6}}{z^{-2}}$
46. $\frac{y^{-7}}{y^{-3}}$
47. $\frac{x^{-5}}{x^{-8}}$
48. $\frac{y^{-4}}{y^{-9}}$
49. $\frac{x}{x^{-1}}$
50. $\frac{x^6}{x}$
51. $(a^{-3})^5$
52. $(x^{-5})^6$
53. $(5^2)^{-3}$
54. $(9^3)^{-4}$
55. $(x^{-3})^{-4}$
56. $(a^{-5})^{-6}$
57. $(m^{-3})^7$
58. $(n^{-2})^8$
59. $(ab)^{-3}$
60. $(mn)^{-5}$
61. $(5ab)^{-2}$
62. $(4xy)^{-2}$
63. $(6x^{-5})^2$
64. $(3a^{-4})^4$
65. $(x^4y^5)^{-3}$
66. $(t^5x^3)^{-4}$
67. $(x^{-6}y^{-2})^{-4}$
68. $(x^{-2}y^{-7})^{-5}$
69. $(3x^3y^{-8}z^{-3})^2$
70. $(2a^2y^{-4}z^{-5})^3$
71. $(x^3y^{-4}z^{-5})(x^{-4}y^{-2}z^9)$
72. $(a^{-5}b^7c^{-2})(a^{-3}b^{-2}c^6)$
73. $(m^{-4}n^7p^3)(m^9n^{-2}p^{-10})$
74. $(t^{-9}p^{10}m^8)(t^{-5}p^{-7}m^{-2})$
75. $\left(\frac{y^2}{2}\right)^{-3}$
76. $\left(\frac{a^4}{3}\right)^{-2}$
77. $\left(\frac{3}{a^2}\right)^3$
78. $\left(\frac{7}{x^7}\right)^2$
79. $\left(\frac{x^2y}{z}\right)^3$
80. $\left(\frac{m}{n^4p}\right)^3$
81. $\left(\frac{a^2b}{cd^3}\right)^{-2}$
82. $\left(\frac{2a^2}{3b^4}\right)^{-3}$

Convert to decimal notation.

83. 2.14×10^3
84. 8.92×10^2
85. 6.92×10^{-3}
86. 7.26×10^{-4}
87. 7.84×10^8
88. 1.35×10^7
89. 8.764×10^{-10}
90. 9.043×10^{-3}
91. 10^8
92. 10^4
93. 10^{-4}
94. 10^{-7}

Convert to scientific notation.

95. 25,000
96. 71,500
97. 0.00371
98. 0.0814
99. 78,000,000,000
100. 3,700,000,000,000
101. 907,000,000,000,000,000
102. 168,000,000,000,000
103. 0.00000374
104. 0.000000000275
105. 0.000000018
106. 0.00000000002
107. 10,000,000
108. 100,000,000,000
109. 0.000000001
110. 0.0000001

Multiply or divide, and write scientific notation for the result.

111. $(3 \times 10^4)(2 \times 10^5)$
112. $(1.9 \times 10^8)(3.4 \times 10^{-3})$
113. $(5.2 \times 10^5)(6.5 \times 10^{-2})$
114. $(7.1 \times 10^{-7})(8.6 \times 10^{-5})$
115. $(9.9 \times 10^{-6})(8.23 \times 10^{-8})$
116. $(1.123 \times 10^4) \times 10^{-9}$
117. $\frac{8.5 \times 10^8}{3.4 \times 10^{-5}}$
118. $\frac{5.6 \times 10^{-2}}{2.5 \times 10^5}$
119. $(3.0 \times 10^6) \div (6.0 \times 10^9)$
120. $(1.5 \times 10^{-3}) \div (1.6 \times 10^{-6})$
121. $\frac{7.5 \times 10^{-9}}{2.5 \times 10^{12}}$
122. $\frac{4.0 \times 10^{-3}}{8.0 \times 10^{20}}$

Problem Solving

Write scientific notation for each answer.

123. There are 3064 members of the Professional Bowlers Association. There are 249 million people in the United States. What fractional part of the population are members of the Professional Bowlers Association?

124. There are 300,000 words in the English language. The average person knows about 10,000 of them. What fractional part of the total number of words does the average person know?

125. Americans eat 6.5 million gal of popcorn each day. How much popcorn do they eat in one year?

126. Americans drink 3 million gal of orange juice each day. How much orange juice do Americans consume in one year?

127. The average discharge at the mouth of the Amazon River is 4,200,000 cubic feet per second. How much water is discharged from the Amazon River in one hour? in one year?

128. There are 300,000 words in the English language. The exceptional person knows about 20,000 of them. What fractional part of the total number of words does the exceptional person know?

Skill Maintenance

Collect like terms.

129. $-9a + 17a$

130. $-12x + (-5x)$

131. Plot the points $(-4, 1)$, $(-3, -2)$, $(5, 2)$, and $(-1, 4)$.

132. In which two quadrants is the first coordinate positive?

Synthesis

Carry out the indicated operations. Write scientific notation for the result.

133. $\dfrac{(5.2 \times 10^6)(6.1 \times 10^{-11})}{1.28 \times 10^{-3}}$

134. $\dfrac{3.9 \times 10^{15}}{(8.0 \times 10^{-12})(3.2 \times 10^{19})}$

135. Perform the indicated operations. Express the result in scientific notation.

$$\{2.1 \times 10^6[(2.5 \times 10^{-3}) \div (5.0 \times 10^{-5})]\} \div (3.0 \times 10^{17})$$

136. Find the reciprocal. Express in scientific notation.

a) 6.25×10^{-3}

b) 4.0×10^{10}

137. Write $4^{-3} \cdot 8 \cdot 16$ as a power of 2.

138. Write $2^8 \cdot 16^{-3} \cdot 64$ as a power of 4.

Simplify.

139. $(5^{-12})^2 \cdot 5^{25}$

140. $49^{18} \cdot 7^{-35}$

141. $\left(\dfrac{1}{a}\right)^{-n}$

142. $\dfrac{(0.4)^5}{[(0.4)^3]^2}$

Determine whether each of the following is true for any pairs of integers m and n and any positive numbers x and y.

143. $x^m \cdot y^n = (xy)^{mn}$

144. $x^m \cdot y^m = (xy)^{2m}$

145. $(x - y)^m = x^m - y^m$

SUMMARY AND REVIEW 4

KEY TERMS

Polynomial, p. 154
Monomial, p. 154
Binomial, p. 155
Trinomial, p. 155
Coefficient, p. 155
Degree, pp. 155, 184
Descending order, p. 157
Opposite of a polynomial, p. 163
FOIL, p. 176
Difference of squares, p. 177
Polynomial in several variables, p. 183
Like terms, p. 185
Synthetic division, p. 197
Scientific notation, p. 202

IMPORTANT PROPERTIES AND FORMULAS

Definitions and Rules for Exponents

Assuming that no denominator is 0 and that 0^0 is not considered, for any integers m and n,

1 as an exponent:	$a^1 = a$
0 as an exponent:	$a^0 = 1$
Negative exponents:	$a^{-n} = \frac{1}{a^n}$
The Product Rule:	$a^m \cdot a^n = a^{m+n}$
The Quotient Rule:	$\frac{a^m}{a^n} = a^{m-n}$
The Power Rule:	$(a^m)^n = a^{mn}$
Raising a product to a power:	$(ab)^n = a^n b^n$
Raising a quotient to a power:	$\left(\frac{a}{b}\right)^n = \frac{a^n}{b^n}$

Special Products of Polynomials

$$(A + B)(A - B) = A^2 - B^2$$
$$(A + B)(A + B) = A^2 + 2AB + B^2$$
$$(A - B)(A - B) = A^2 - 2AB + B^2$$

Scientific Notation: $N \times 10^n$, where $1 \leq N < 10$

REVIEW EXERCISES

This chapter's review and test include Skill Maintenance exercises from Sections 1.5, 2.5, 2.6, and 3.1.

Simplify.

1. $y^7 \cdot y^3 \cdot y$
2. $(3x)^5 \cdot (3x)^9$
3. $t^8 \cdot t^0$
4. $\frac{4^5}{4^2}$
5. $\frac{(7x)^4}{(7x)^4}$
6. $\left(\frac{3t^4}{2s^3}\right)^2$
7. $(-2xy^2)^3$
8. $(2x^3)^2(-3x)^2$
9. $2(x^3)^2(-3x)^2$

Identify the terms of the polynomial.

10. $3x^2 + 6x + \frac{1}{2}$
11. $-4y^5 + 7y^2 - 3y - 2$

Identify the coefficient of each term of the polynomial.

12. $6x^2 + 17$
13. $4x^3 + 6x^2 - 5x + \frac{5}{3}$

Determine the degree of each term and the degree of the polynomial.

14. $x^3 + 4x - 6$
15. $3 - 2x^4 + 3x^9 + x^6 - \frac{3}{4}x^3$

Classify the polynomial as a monomial, a binomial, a trinomial, or none of these.

16. $4x^3 - 1$
17. $4 - 9t^3 - 7t^4 + 10t^2$
18. $7y^2$

Collect like terms and write in descending order.

19. $5x - x^2 + 4x$
20. $\frac{3}{4}x^3 + 4x^2 - x^3 + 7$
21. $-2x^4 + 16 + 2x^4 + 9 - 3x^5$
22. $3x^2 - 2x + 3 - 5x^2 - 1 - x$
23. $-x + \frac{1}{2} + 14x^4 - 7x^2 - 1 - 4x^4$

Evaluate the polynomial when $x = -1$.

24. $7x - 10$
25. $x^2 - 3x + 6$

Add or subtract.

26. $(3x^4 - x^3 + x - 4) + (x^5 + 7x^3 - 3x^2 - 5) + (-5x^4 + 6x^2 - x)$
27. $(3x^5 - 4x^4 + x^3 - 3) + (3x^4 - 5x^3 + 3x^2) + (4x^5 + 4x^3) + (-5x^5 - 5x^2) + (-5x^4 + 2x^3 + 5)$
28. $(5x^2 - 4x + 1) - (3x^2 + 7)$
29. $(3x^5 - 4x^4 + 2x^2 + 3) - (2x^5 - 4x^4 + 3x^3 + 4x^2 - 5)$
30.
$$\begin{array}{rrrrr} -\frac{3}{4}x^4 & +\frac{1}{2}x^3 & & & +\frac{7}{8} \\ & -\frac{1}{4}x^3 & -x^2 & -\frac{7}{4}x & \\ +\frac{3}{2}x^4 & & +\frac{2}{3}x^2 & & -\frac{1}{2} \\ \hline \end{array}$$
31.
$$\begin{array}{l} \quad 2x^5 \qquad\quad - x^3 \qquad\qquad + x + 3 \\ -(3x^5 - x^4 + 4x^3 + 2x^2 - x + 3) \\ \hline \end{array}$$

32. The length of a rectangle is 4 m greater than its width.

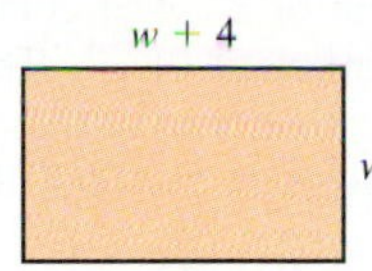

a) Find a polynomial for the perimeter.
b) Find a polynomial for the area.

Multiply.

33. $3x(-4x^2)$
34. $(7x + 1)^2$
35. $\left(x + \frac{2}{3}\right)\left(x + \frac{1}{2}\right)$
36. $(1.5x - 6.5)(0.2x + 1.3)$
37. $(4x^2 - 5x + 1)(3x - 2)$
38. $(x - 9)^2$
39. $5x^4(3x^3 - 8x^2 + 10x + 2)$
40. $(x + 4)(x - 7)$
41. $(x - 0.3)(x - 0.75)$
42. $(x^4 - 2x + 3)(x^3 + x - 1)$
43. $(3y^2 - 2y)^2$
44. $(2t^2 + 3)(t^2 - 7)$
45. $(x^3 - 2x + 3)(4x^2 - 5x)$
46. $(3x^2 + 4)(3x^2 - 4)$
47. $(2 - x)(2 + x)$
48. $(13x - 3)(x - 13)$
49. Evaluate $2 - 5xy + y^2 - 4xy^3 + x^6$ when $x = -1$ and $y = 2$.

Identify the coefficient and the degree of each term of the polynomial. Then find the degree of the polynomial.

50. $x^5y - 7xy + 9x^2 - 8$
51. $x^2y^5z^9 - y^{40} + x^{13}z^{10}$

Collect like terms.

52. $y + w - 2y + 8w - 5$
53. $m^6 - 2m^2n + m^2n^2 + n^2m - 6m^3 + m^2n^2 + 7n^2m$

Add or subtract.

54. $(5x^2 - 7xy + y^2) + (-6x^2 - 3xy - y^2) + (x^2 + xy - 2y^2)$
55. $(6x^3y^2 - 4x^2y - 6x) - (-5x^3y^2 + 4x^2y + 6x^2 - 6)$

Multiply.

56. $(p - q)(p^2 + pq + q^2)$
57. $(3a^4 - \frac{1}{3}b^3)^2$

Divide.

58. $(10x^3 - x^2 + 6x) \div 2x$
59. $(6x^3 - 5x^2 - 13x + 13) \div (2x + 3)$
60. $\dfrac{t^4 + t^3 + 2t^2 - t - 3}{t + 1}$
61. $\dfrac{2x^5 + x^3 - x^2 + 1}{x^3 - 1}$

Divide using synthetic division.

62. $(x^3 + 3x^2 + 2x - 6) \div (x - 3)$
63. $(4x^3 + 6x^2 - 5) \div (x + 3)$
64. Express using a positive exponent: y^{-4}.
65. Express using a negative exponent: $\dfrac{1}{t^5}$.

Simplify.

66. $7^2 \cdot 7^{-4}$
67. $\dfrac{a^{-5}b}{a^8b^8}$
68. $(x^3)^{-4}$
69. $(2x^{-3}y)^{-2}$
70. $\left(\dfrac{2x}{y}\right)^{-3}$
71. Convert to decimal notation: 8.3×10^6.
72. Convert to scientific notation: 0.0000328.

Multiply or divide and write scientific notation for the result.

73. $(3.8 \times 10^4)(5.5 \times 10^{-1})$
74. $\dfrac{1.28 \times 10^{-8}}{2.5 \times 10^{-4}}$
75. Each day Americans eat 170 million eggs. How many eggs do Americans eat in one year? Write scientific notation for the answer.

Skill Maintenance

76. Find the perimeter of the figure.

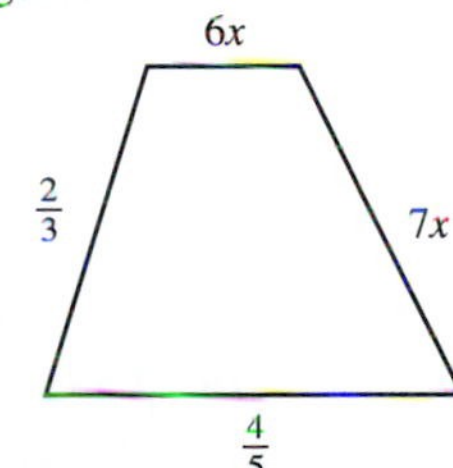

77. The perimeter of a rectangle is 540 m. The width is 19 m less than the length. Find the width and the length.
78. Solve: $3 - 2x \leq 7$.
79. In which quadrant is the point $(3, -1)$ located?

Synthesis

80. ◈ Explain why $5x^3$ and $(5x)^3$ are not equivalent expressions.

81. ◈ If two polynomials of degree n are added, is the sum also of degree n? Why or why not?

82. If a and b are positive, how many terms are there in each of the following?

a) $(x - a)(x - b) + (x - a)(x - b)$
b) $(x + a)(x - b) + (x - a)(x + b)$

83. Collect like terms:

$$-3x^5 \cdot 3x^3 - x^6(2x)^2 + (3x^4)^2 + (2x^2)^4 - 40x^2(x^3)^2.$$

84. A polynomial has degree 4. The x^2-term is missing. The coefficient of x^4 is two times the coefficient of x^3. The coefficient of x is 3 less than the coefficient of x^4. The remaining coefficient is 7 less than the coefficient of x. The sum of the coefficients is 15. Find the polynomial.

85. Multiply: $[(x - 4) - x^3][(x + 4) + 4x^3]$.

86. Solve: $(x - 7)(x + 10) = (x - 4)(x - 6)$.

CHAPTER TEST 4

Simplify.

1. $x^6 \cdot x^2 \cdot x$ **2.** $(4a)^3 \cdot (4a)^8$ **3.** $\dfrac{3^5}{3^2}$

4. $\dfrac{(2x)^5}{(2x)^5}$ **5.** $(x^3)^2$ **6.** $(-3y^2)^3$

7. $(3x^2)^3(-2x^5)^3$ **8.** $3(x^2)^3(-2x^5)^3$

9. Classify the polynomial as a monomial, a binomial, a trinomial, or none of these:

$$7 - x.$$

10. Identify the coefficient of each term of the polynomial:

$$\tfrac{1}{3}x^5 - x + 7.$$

11. Determine the degree of each term and the degree of the polynomial:

$$2x^3 - 4 + 5x + 3x^6.$$

12. Evaluate the polynomial $x^2 + 5x - 1$ when $x = -2$.

Collect like terms and write in descending order.

13. $4a^2 - 6 + a^2$ **14.** $y^2 - 3y - y + \frac{3}{4}y^2$

15. $3 - x^2 + 2x^3 + 5x^2 - 6x - 2x + x^5$

Add or subtract.

16. $(3x^5 + 5x^3 - 5x^2 - 3) + (x^5 + x^4 - 3x^3 - 3x^2 + 2x - 4)$

17. $\left(x^4 + \frac{2}{3}x + 5\right) + \left(4x^4 + 5x^2 + \frac{1}{3}x\right)$

18. $(2x^4 + x^3 - 8x^2 - 6x - 3) - (6x^4 - 8x^2 + 2x)$

19. $(x^3 - 0.4x^2 - 12) - (x^5 + 0.3x^3 + 0.4x^2 + 9)$

Multiply.

20. $-3x^2(4x^2 - 3x - 5)$ **21.** $\left(x - \frac{1}{3}\right)^2$

22. $(3x + 10)(3x - 10)$ **23.** $(3b + 5)(b - 3)$

24. $(x^6 - 4)(x^8 + 4)$ **25.** $(8 - y)(6 + 5y)$

26. $(2x + 1)(3x^2 - 5x - 3)$

27. $(5t + 2)^2$

28. Collect like terms:

$$x^3y - y^3 + xy^3 + 8 - 6x^3y - x^2y^2 + 11.$$

29. Subtract:

$$(8a^2b^2 - ab + b^3) - (-6ab^2 - 7ab - ab^3 + 5b^3).$$

30. Multiply: $(3x^5 - 4y^5)(3x^5 + 4y^5)$.

Divide.

31. $(12x^4 + 9x^3 - 15x^2) \div 3x^2$

32. $(6x^3 - 8x^2 - 14x + 13) \div (3x + 2)$

33. Divide using synthetic division:

$(x^3 + 5x^2 + 4x - 7) \div (x - 4)$.

34. Express using a positive exponent: 5^{-3}.

35. Express using a negative exponent: $\dfrac{1}{y^8}$.

Simplify.

36. $6^{-2} \cdot 6^{-3}$ **37.** $\dfrac{x^3y^2}{x^8y^{-3}}$

38. $(2a^3b^{-1})^{-4}$ **39.** $\left(\dfrac{ab}{c}\right)^{-3}$

40. Convert to scientific notation: 3,900,000,000.

41. Convert to decimal notation: 5×10^{-8}.

Multiply or divide and write scientific notation for the result.

42. $\dfrac{5.6 \times 10^6}{3.2 \times 10^{-11}}$

43. $(2.4 \times 10^5)(5.4 \times 10^{16})$

44. Each day Americans eat 170 million eggs. There are 249 million people in this country. How many eggs does the average person eat in one year? Write scientific notation for the answer.

Skill Maintenance

45. Solve: $7x - 4x - 2 > 37$.

46. Plot the point $(-1, 5)$.

47. Add: $\frac{2}{5} + \left(-\frac{3}{4}\right)$.

48. The first angle of a triangle is four times as large as the second. The measure of the third angle is 30° greater than that of the second. How large are the angles?

Synthesis

49. The height of a box is 1 less than its length, and the length is 2 more than its width. Find the volume in terms of the length.

50. Solve: $x^2 + (x - 7)(x + 4) = 2(x - 6)^2$.

CHAPTER 5

Polynomials and Factoring

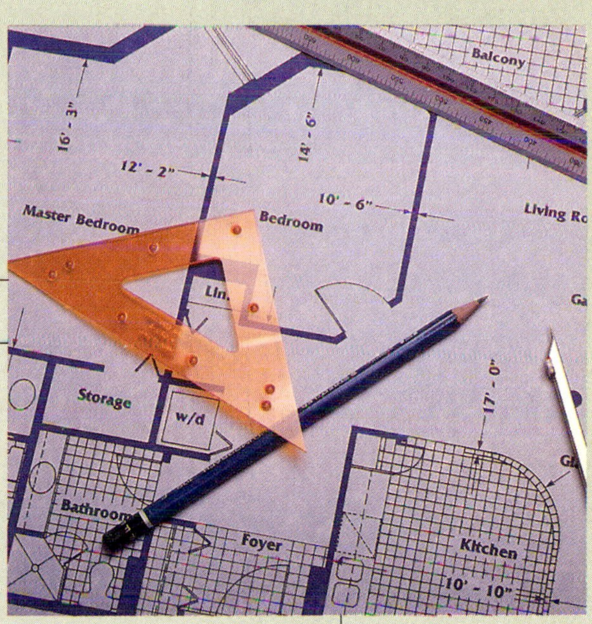

AN APPLICATION

An architect has allocated a rectangular space of 264 ft^2 for a square dining room and a 10-ft wide kitchen. Determine the dimensions of each room.

This problem appears as Exercise 31 in Section 5.8.

Paula Haynes
ARCHITECT

"In general, people think of architecture as art only. But art is only one portion of the picture. Architecture integrates both art and science. A solid background in mathematics is essential for anyone pursuing a future in architecture."

In Chapter 1, we learned that *factoring* is multiplying reversed. To *factor* a polynomial is to find an equivalent expression that is a product. Factoring polynomials requires a solid command of the multiplication methods studied in Chapter 4.

One important reason for studying factoring is that it can be used to solve certain "new" equations that occur in problem-solving situations. Factoring is also very important elsewhere in algebra, as we will see in the remaining chapters of this text.

In addition to the material from this chapter, the review and test for Chapter 5 include material from Sections 1.6, 2.7, 3.2, and 4.1.

5.1 Introduction to Factoring

Factoring Monomials • Factoring When Terms Have a Common Factor • Factoring by Grouping

Just as a number, like 15, can be factored as $3 \cdot 5$, a polynomial, like $x^2 + 7x$, can be factored as $x(x + 7)$. In both cases, we ask ourselves, "What was multiplied to obtain the given result?" The situation is much like a popular television game show in which an "answer" is given and participants must find a "question" to which the answer corresponds.

Factoring

To *factor* a polynomial is to find an equivalent expression that is a product.

Factoring Monomials

To factor a monomial, we find two monomials whose product is equivalent to the original monomial. For example, $20x^2$ can be factored as $2 \cdot 10x^2$, $4x \cdot 5x$, or $10x \cdot 2x$ in addition to other ways that we haven't listed.

EXAMPLE 1

Find three factorizations of $15x^3$.

Solution

a) $15x^3 = (3 \cdot 5)(x \cdot x^2)$

$= (3x)(5x^2)$ The factors here are $3x$ and $5x^2$.

b) $15x^3 = (3 \cdot 5)(x^2 \cdot x)$

$= (3x^2)(5x)$ The factors here are $3x^2$ and $5x$.

c) $15x^3 = ((-5)(-3))x^3$

$= (-5)(-3x^3)$ The factors here are -5 and $-3x^3$. ❑

Note that each part of a factorization is referred to as a *factor*, so the word "factor" operates as a verb or a noun, depending on the context in which it appears.

Factoring When Terms Have a Common Factor

To multiply a polynomial of two or more terms by a monomial, we multiply each term by the monomial, using the distributive law $a(b + c) = ab + ac$. To factor, we do the reverse. We express a polynomial as a product, using the same law, read from right to left: $ab + ac = a(b + c)$. Consider the following:

Multiply	*Factor*
$3x(x^2 + 2x - 4)$	$3x^3 + 6x^2 - 12x$
$= 3x \cdot x^2 + 3x \cdot 2x - 3x \cdot 4$	$= 3x \cdot x^2 + 3x \cdot 2x - 3x \cdot 4$
$= 3x^3 + 6x^2 - 12x;$	$= 3x(x^2 + 2x - 4).$

In the factorization on the right, note that since $3x$ appears as a factor of $3x^3$, $6x^2$, and $-12x$, it is a *common factor* for all the terms of the trinomial $3x^3 + 6x^2 - 12x$.

To factor a polynomial with two or more terms, always try to first find a factor common to all terms. In some cases, there may not be a common factor (other than 1). If a common factor *does* exist, we generally use the common factor with the largest possible coefficient and the largest possible exponent. Such a factor is called the *largest common factor*.

EXAMPLE 2

Factor: $5x^2 + 15$.

Solution We have

$$5x^2 + 15 = 5 \cdot x^2 + 5 \cdot 3 \qquad \text{Factoring each term}$$
$$= 5(x^2 + 3). \qquad \text{Factoring out the common factor, 5}$$

To check, we multiply: $5(x^2 + 3) = 5 \cdot x^2 + 5 \cdot 3 = 5x^2 + 15$. ❑

CAUTION! $5 \cdot x^2 + 5 \cdot 3$ is a factorization of the *terms* of $5x^2 + 15$, but not the polynomial itself. The factorization of $5x^2 + 15$ is $5(x^2 + 3)$.

When all terms in a polynomial contain the same letter raised to various powers, we factor out the largest power possible.

EXAMPLE 3

Factor: $24x^3 + 30x^2$.

Solution The largest factor common to 24 and 30 is 6. The largest power of x common to x^3 and x^2 is x^2. (To see this, write x^3 as $x^2 \cdot x$.) Thus the largest common factor of $24x^3$ and $30x^2$ is $6x^2$. We factor as follows:

$$24x^3 + 30x^2 = 6x^2 \cdot 4x + 6x^2 \cdot 5 \qquad \text{Factoring each term}$$
$$= 6x^2(4x + 5). \qquad \text{Factoring out } 6x^2$$

❑

Suppose in Example 3 that you had not recognized the *largest* common factor and removed only part of it, as follows:

$$24x^3 + 30x^2 = 2x^2 \cdot 12x + 2x^2 \cdot 15$$
$$= 2x^2(12x + 15). \qquad 12x + 15 \text{ still has a common factor.}$$

Note that $12x + 15$ still has a common factor, 3. To find the largest common factor, continue factoring out common factors, as follows, until no more exist:

$$= 2x^2[3(4x + 5)]$$
$$= 6x^2(4x + 5). \quad \text{Using an associative law}$$

EXAMPLE 4

Factor: $15x^5 - 12x^4 + 27x^3 - 3x^2$.

Solution

$$15x^5 - 12x^4 + 27x^3 - 3x^2$$
$$= 3x^2 \cdot 5x^3 - 3x^2 \cdot 4x^2 + 3x^2 \cdot 9x - 3x^2 \cdot 1 \quad \text{Try to do this mentally.}$$
$$= 3x^2(5x^3 - 4x^2 + 9x - 1) \quad \text{Factoring out } 3x^2$$

CAUTION! Don't forget the term -1.

❑

If you can spot the largest common factor without writing out a factorization of each term, you can write the answer in one step.

EXAMPLE 5

Factor: **(a)** $8m^3 - 16m$; **(b)** $14p^2y^3 - 8py^2 + 2py$; **(c)** $\frac{4}{5}x^2 + \frac{1}{5}x + \frac{2}{5}$.

Solution

a) $8m^3 - 16m = 8m(m^2 - 2)$

b) $14p^2y^3 - 8py^2 + 2py = 2py(7py^2 - 4y + 1)$

c) $\frac{4}{5}x^2 + \frac{1}{5}x + \frac{2}{5} = \frac{1}{5}(4x^2 + x + 2)$

Determine the largest common factor by inspection; then carefully fill in the parentheses.

❑

Below are two of the most important points to keep in mind as we study this chapter.

1. Before doing any other kind of factoring, first try to factor out the largest common factor.
2. You can always check your factoring by multiplying.

Factoring by Grouping

Sometimes algebraic expressions contain a common factor that is a polynomial with two or more terms.

EXAMPLE 6

Factor: $x^2(x + 1) + 2(x + 1)$.

Solution The binomial $x + 1$ is a factor of both $x^2(x + 1)$ and $2(x + 1)$. Thus, $x + 1$ is a common factor:

$$x^2(x + 1) + 2(x + 1) = (x + 1)x^2 + (x + 1)2 \quad \text{Using a commutative law twice}$$
$$= (x + 1)(x^2 + 2). \quad \text{Factoring out the common factor, } x + 1$$

The factorization is $(x + 1)(x^2 + 2)$. ❑

Next consider a four-term polynomial like $x^3 + x^2 + 2x + 2$. Although there is no factor, other than 1, common to all terms, we can factor $x^3 + x^2$ and $2x + 2$ separately:

$$x^3 + x^2 = x^2(x + 1) \quad \text{and} \quad 2x + 2 = 2(x + 1).$$

Note that $x^3 + x^2$ and $2x + 2$ share a common factor of $x + 1$.

When a polynomial can be split into two groups of terms that share a common factor, we can factor out that common factor. This gives a factorization of the original polynomial:

$$\begin{aligned} x^3 + x^2 + 2x + 2 &= (x^3 + x^2) + (2x + 2) && \text{Using an associative law} \\ &= x^2(x + 1) + 2(x + 1) && \text{Factoring each binomial} \\ &= (x + 1)(x^2 + 2). && \text{Using Example 6} \end{aligned}$$

This method is called **factoring by grouping.** Factoring by grouping can be tried on any polynomial with four terms.

EXAMPLE 7 Factor by grouping.

a) $6x^3 - 9x^2 + 4x - 6$

b) $12x^5 + 20x^2 - 21x^3 - 35$

Solution

a) $$\begin{aligned} 6x^3 - 9x^2 + 4x - 6 &= (6x^3 - 9x^2) + (4x - 6) \\ &= 3x^2(2x - 3) + 2(2x - 3) && \text{Factoring each binomial} \\ &= (2x - 3)(3x^2 + 2) && \text{Factoring out the common factor, } 2x - 3 \end{aligned}$$

b) $$\begin{aligned} 12x^5 + 20x^2 - 21x^3 - 35 &= 4x^2(3x^3 + 5) - 7(3x^3 + 5) && \text{Factoring two binomials. Using } -7 \text{ gives a common binomial factor.} \\ &= (3x^3 + 5)(4x^2 - 7) \end{aligned}$$ ❑

Although factoring by grouping is a very useful skill, many polynomials, like $x^3 + x^2 + 2x - 2$, cannot be factored by grouping:

$$x^3 + x^2 + 2x - 2 = x^2(x + 1) + 2(x - 1). \qquad \text{There is no common factor.}$$

EXERCISE SET 5.1

Find three factorizations for the monomial.

1. $6x^3$
2. $9x^4$
3. $-9x^5$
4. $-12x^6$
5. $24x^4$
6. $15x^5$

Factor. Remember to use the largest common factor and to check by multiplying.

7. $x^2 - 4x$
8. $x^2 + 8x$
9. $2x^2 + 6x$
10. $3x^2 - 3x$
11. $x^3 + 6x^2$
12. $4x^4 + x^2$
13. $8x^4 - 24x^2$
14. $5x^5 + 10x^3$
15. $2x^2 + 2x - 8$
16. $6x^2 + 3x - 15$
17. $17x^5y^3 + 34x^3y^2 + 51xy$
18. $16x^6y^4 - 32x^5y^3 - 48xy^2$

19. $6x^4 - 10x^3 + 3x^2$

20. $5x^5 + 10x^2 - 8x$

21. $x^5y^5 + x^4y^3 + x^3y^3 - x^2y^2$

22. $x^9y^6 - x^7y^5 + x^4y^4 + x^3y^3$

23. $2x^7 - 2x^6 - 64x^5 + 4x^3$

24. $10x^3 + 25x^2 + 15x - 20$

25. $1.6x^4 - 2.4x^3 + 3.2x^2 + 6.4x$

26. $2.5x^6 - 0.5x^4 + 5x^3 + 10x^2$

27. $\frac{5}{3}x^6 + \frac{4}{3}x^5 + \frac{1}{3}x^4 + \frac{1}{3}x^3$

28. $\frac{5}{7}x^7 + \frac{3}{7}x^5 - \frac{6}{7}x^3 - \frac{1}{7}x$

Factor.

29. $y(y + 3) + 4(y + 3)$

30. $b(b - 5) - 3(b - 5)$

31. $x^2(x + 3) + 2(x + 3)$

32. $3z^2(2z + 1) + (2z + 1)$

33. $y^2(y + 8) + (y + 8)$

34. $x^2(x - 7) - 3(x - 7)$

Factor by grouping, if possible.

35. $x^3 + 3x^2 + 2x + 6$

36. $6z^3 + 3z^2 + 2z + 1$

37. $2x^3 + 6x^2 + x + 3$

38. $3x^3 + 2x^2 + 3x + 2$

39. $8x^3 - 12x^2 + 6x - 9$

40. $10x^3 - 25x^2 + 4x - 10$

41. $12x^3 - 16x^2 + 3x - 4$

42. $18x^3 - 21x^2 + 30x - 35$

43. $x^3 + 8x^2 - 3x - 24$

44. $2x^3 + 12x^2 - 5x - 30$

45. $w^3 - 7w^2 + 4w - 28$

46. $y^3 + 8y^2 - 2y - 16$

47. $x^3 - x^2 - 2x + 5$

48. $p^3 + p^2 - 3p + 10$

49. $2x^3 - 8x^2 - 9x + 36$

50. $20g^3 - 4g^2 - 25g + 5$

Skill Maintenance

51. Graph: $y = x - 6$.

52. Solve: $4x - 8x + 16 \geqslant 6(x - 2)$.

53. Subtract: $-13 - (-25)$.

54. Solve $A = \dfrac{p + q}{2}$ for p.

Multiply.

55. $(y + 5)(y + 7)$

56. $(y + 7)^2$

57. $(y + 7)(y - 7)$

58. $(y - 7)^2$

Synthesis

59. ◈ Write a two-sentence paragraph in which the word "factor" is used at least once as a noun and once as a verb.

60. ◈ In answering a factoring problem, Taylor says the largest common factor is $-5x^2$ and Natasha says the largest common factor is $5x^2$. Can they both be correct? Why or why not?

Factor, if possible.

61. $4x^5 + 6x^3 + 6x^2 + 9$

62. $x^6 + x^4 + x^2 + 1$

63. $x^{12} + x^7 + x^5 + 1$

64. $x^3 + x^2 - 2x + 2$

65. $p^3 - p^2 + 3p + 3$

66. $ax^2 + 2ax + 3a + x^2 + 2x + 3$

67. ◈ Explain how to construct a four-term polynomial that can be factored by grouping.

68. ◈ Explain how to construct a polynomial of degree 9 for which $5x^3y^2$ is the largest common factor.

5.2 Factoring Trinomials of the Type $x^2 + bx + c$

Constant Term Positive • Constant Term Negative

We now learn how to factor trinomials like

$$x^2 + 5x + 4 \quad \text{or} \quad x^2 + 3x - 10,$$

for which no common factor exists. We will limit our attention to trinomials of the type $ax^2 + bx + c$, where $a = 1$. The coefficient a is often called the **leading coefficient.**

Constant Term Positive

Recall the FOIL method of multiplying two binomials:

$$\begin{aligned}(x + 2)(x + 5) &= \overset{F}{x^2} + \underbrace{\overset{O}{5x} + \overset{I}{2x}} + \overset{L}{10} \\ &= x^2 + 7x + 10.\end{aligned}$$

To factor $x^2 + 7x + 10$, we think of FOIL in reverse. We multiplied x times x to get the first term of the trinomial, so we know that the first term of each binomial factor is x. Next we look for numbers p and q such that

$$x^2 + 7x + 10 = (x + p)(x + q).$$

To get the middle term and the last term of the trinomial, we look for two numbers p and q whose product is 10 and whose sum is 7. Those numbers are 2 and 5. Thus the factorization is

$$(x + 2)(x + 5).$$

EXAMPLE 1 Factor: $x^2 + 5x + 6$.

Solution Think of FOIL in reverse. The first term of each factor is x:

$$(x + p)(x + q).$$

We then look for two numbers p and q whose product is 6 and whose sum is 5.

Pairs of Factors of 6	Sums of Factors
1, 6	7
2, 3	5 ← The numbers we seek are 2 and 3.
−1, −6	−7
−2, −3	−5

Since $2 \cdot 3 = 6$ and $2 + 3 = 5$, the factorization of $x^2 + 5x + 6$ is $(x + 2)(x + 3)$. To check, we simply multiply the two binomials to see whether we get the original trinomial.

Check: $(x + 2)(x + 3) = x^2 + 3x + 2x + 6 = x^2 + 5x + 6$.

Note that since 5 and 6 are both positive, when factoring $x^2 + 5x + 6$ we need not consider negative factors of 6. Note too that changing the signs of the factors changes the sign of the sum. ❑

We began this section examining the product of two sums: $(x + 2)(x + 5)$. Note that a product of two differences, like $(x - 2)(x - 5)$, also contains a positive constant term:

$$\begin{aligned}(x - 2)(x - 5) &= \overset{F}{x^2} - \underbrace{\overset{O}{5x} - \overset{I}{2x}} + \overset{L}{10} \\ &= x^2 - 7x + 10.\end{aligned}$$

When the constant term of a trinomial is positive, we look for two numbers with the same sign. The sign is that of the middle term:

$$(x^2 - 7x + 10) = (x - 2)(x - 5).$$

EXAMPLE 2

Factor: $y^2 - 8y + 12$.

Solution Since the constant term is positive and the coefficient of the middle term is negative, we look for a factorization of 12 in which both factors are negative. Their sum must be -8.

Pairs of Factors of 12	Sums of Factors
$-1, -12$	-13
$-2, -6$	-8 ← The numbers we need are -2 and -6.
$-3, -4$	-7

The factorization is $(y - 2)(y - 6)$. The check is left for the student. ❑

Constant Term Negative

Sometimes when we use FOIL, the product has a negative constant term. Consider these multiplications:

a) $$(x - 5)(x + 2) = \overset{F}{x^2} + \overset{O}{2x} - \overset{I}{5x} - \overset{L}{10}$$
$$= x^2 - 3x - 10;$$

b) $$(x + 5)(x - 2) = \overset{F}{x^2} - \overset{O}{2x} + \overset{I}{5x} - \overset{L}{10}$$
$$= x^2 + 3x - 10.$$

Reversing the signs of -5 and 2 changes only the sign of the middle term.

When the constant term of a trinomial is negative, we look for two numbers whose product is negative. One of them must be positive and the other negative. Their sum must be the coefficient of the middle term.

EXAMPLE 3

Factor: $x^2 - 8x - 20$.

Solution The factorization of the constant term, -20, must have one factor positive and one factor negative. The sum must be -8, so the negative factor must have the larger absolute value. Thus we consider only pairs of factors in which the negative factor has the larger absolute value.

Pairs of Factors of −20	Sums of Factors
1, −20	−19
2, −10	−8
4, −5	−1
5, −4	1
10, −2	8
20, −1	19

The numbers we need are 2 and −10.

Since the positive factor in each of these pairs has the larger absolute value, the sums are all positive. For this problem, we can disregard these pairs. Note that changing the signs of the factors changes the sign of the sum.

The numbers we need are 2 and −10. Thus the factorization is $(x + 2)(x - 10)$.

Check: $(x + 2)(x - 10) = x^2 - 10x + 2x - 20 = x^2 - 8x - 20.$ ❑

EXAMPLE 4

Factor: $t^2 - 24 + 5t$.

Solution It helps to first write the trinomial in descending order: $t^2 + 5t - 24$. The factorization of the constant term, −24, must have one factor positive and one factor negative. The sum must be 5, so the positive factor must have the larger absolute value. Thus we consider only pairs of factors in which the positive factor has the larger absolute value.

Pairs of Factors of −24	Sums of Factors
−1, 24	23
−2, 12	10
−3, 8	5
−4, 6	2

The numbers we need are −3 and 8.

The factorization is $(t - 3)(t + 8)$. The check is left for the student. ❑

Polynomials in two or more variables, such as $a^2 + 4ab - 21b^2$, can be factored in a similar manner.

EXAMPLE 5

Factor: $a^2 + 4ab - 21b^2$.

Solution It may help to write the trinomial in the equivalent form

$$a^2 + 4ba - 21b^2.$$

This way we think of $-21b^2$ as the "constant" term and $4b$ as the "coefficient" of the middle term. Then we try to express $-21b^2$ as a product of two factors whose sum is $4b$. Those factors are $-3b$ and $7b$. Thus the factorization is

$$(a - 3b)(a + 7b).$$

Check: $(a - 3b)(a + 7b) = a^2 + 7ab - 3ba - 21b^2 = a^2 + 4ab - 21b^2.$ ❑

EXAMPLE 6

Factor: $x^2 - x + 5$.

Solution Since 5 has very few factors, we can easily check all possibilities.

Pairs of Factors of 5	Sums of Factors
5, 1	6
−5, −1	−6

Since there are no factors whose sum is -1, the polynomial is *not* factorable into binomials. ❑

A polynomial like $x^2 - x + 5$ that cannot be factored further is said to be **prime.**

Often factoring requires two or more steps. In general, when told to factor, we should *factor completely*. This means that the final factorization should not contain any factors that can be factored further.

EXAMPLE 7

Factor: $2x^3 - 20x^2 + 50x$.

Solution *Always* look first for a common factor. This time there is one, $2x$, which we factor out first:

$$2x^3 - 20x^2 + 50x = 2x(x^2 - 10x + 25).$$

Now consider $x^2 - 10x + 25$. Since the constant term is positive and the coefficient of the middle term is negative, we look for a factorization of 25 in which both factors are negative. Their sum must be -10.

Pairs of Factors of 25	Sums of Factors
−25, −1	−26
−5, −5	−10 ← The numbers we need are −5 and −5.

The factorization of

$$x^2 - 10x + 25$$

is

$$(x - 5)(x - 5), \quad \text{or} \quad (x - 5)^2,$$

but we must not forget the common factor, $2x$. The factorization of

$$2x^3 - 20x^2 + 50x$$

is

$$2x(x - 5)(x - 5), \quad \text{or} \quad 2x(x - 5)^2.$$

Check: $2x(x - 5)(x - 5) = 2x[x^2 - 10x + 25]$ Multiplying binomials

$= 2x^3 - 20x^2 + 50x.$ Using the distributive law ❑

Once any common factors are factored out, the following summary can be used to factor $x^2 + bx + c$.

To factor $x^2 + bx + c$:

1. First arrange in descending order. Use a trial-and-error process to express c as a product of two factors whose sum is b.
   a) If c is positive, the signs of the factors are the same as the sign of b.
   b) If c is negative, one factor is positive and the other is negative. Select the factors so that the factor with the larger absolute value is the factor with the same sign as b.
2. Check by multiplying.

EXERCISE SET 5.2

Factor completely. Remember that you can check by multiplying.

1. $x^2 + 8x + 15$
2. $x^2 + 5x + 6$
3. $x^2 + 7x + 12$
4. $x^2 + 9x + 8$
5. $x^2 - 6x + 9$
6. $y^2 + 11y + 28$
7. $x^2 + 9x + 14$
8. $a^2 + 11a + 30$
9. $b^2 + 5b + 4$
10. $x^2 - \frac{2}{5}x + \frac{1}{25}$
11. $x^2 + \frac{2}{3}x + \frac{1}{9}$
12. $z^2 - 8z + 7$
13. $d^2 - 7d + 10$
14. $x^2 - 8x + 15$
15. $y^2 - 11y + 10$
16. $x^2 - 2x - 15$
17. $x^2 + x - 42$
18. $x^2 + 2x - 15$
19. $2x^2 - 14x - 36$
20. $3y^2 - 9y - 84$
21. $x^3 - 6x^2 - 16x$
22. $x^3 - x^2 - 42x$
23. $y^2 - 4y - 45$
24. $x^2 - 7x - 60$
25. $-2x - 99 + x^2$
26. $x^2 - 72 + 6x$
27. $c^4 + c^3 - 56c^2$
28. $5b^2 + 25b - 120$
29. $2a^2 + 4a - 70$
30. $x^5 + x^4 - 2x^3$
31. $x^2 + x + 1$
32. $x^2 + 2x + 3$
33. $7 - 2p + p^2$
34. $11 - 3w + w^2$
35. $x^2 + 20x + 100$
36. $x^2 + 20x + 99$
37. $3x^3 - 63x^2 - 300x$
38. $2x^3 - 40x^2 + 192x$
39. $x^2 - 21x - 72$
40. $4x^2 + 40x + 100$
41. $x^2 - 25x + 144$
42. $y^2 - 21y + 108$
43. $a^4 + a^3 - 132a^2$
44. $a^6 + 9a^5 - 90a^4$
45. $120 - 23x + x^2$
46. $96 + 22d + d^2$
47. $108 - 3x - x^2$
48. $112 + 9y - y^2$
49. $y^2 - 0.2y - 0.08$
50. $t^2 - 0.3t - 0.10$
51. $p^2 + 3pq - 10q^2$
52. $a^2 - 2ab - 3b^2$
53. $m^2 + 5mn + 5n^2$
54. $x^2 - 11xy + 24y^2$
55. $s^2 - 2st - 15t^2$
56. $b^2 + 8bc - 20c^2$
57. $2x^3 - 10x^2 + 12x$
58. $3a^6 - 24a^5 + 36a^4$
59. $7a^9 - 28a^8 - 35a^7$
60. $6x^{10} - 30x^9 - 84x^8$

Skill Maintenance

Multiply.

61. $(x + 6)(3x + 4)$
62. $(7w + 6)^2$
63. In a recent year, 29,090 people were arrested for counterfeiting. This figure was down 1.2% from the year before. How many people were arrested the year before?
64. The first angle of a triangle is four times as large as the second. The measure of the third angle is 30° greater than that of the second. How large are the angles?

Synthesis

65. ◈ Without multiplying $(x - 17)(x - 18)$, explain why it cannot possibly be a factorization of $x^2 + 35x + 306$.
66. ◈ A student factors $x^3 - 8x^2 + 15x$ as $(x^2 - 5x)(x - 3)$. Is the student wrong? Why or why not? What advice would you offer the student?
67. Find all integers m for which $y^2 + my + 50$ can be factored.
68. Find all integers b for which $a^2 + ba - 50$ can be factored.

Factor completely.

69. $x^2 - \frac{1}{2}x - \frac{3}{16}$

70. $x^2 - \frac{1}{4}x - \frac{1}{8}$

71. $x^2 + \frac{30}{7}x - \frac{25}{7}$

72. $\frac{1}{3}x^3 + \frac{1}{3}x^2 - 2x$

73. $b^{2n} + 7b^n + 10$

74. $a^{2m} - 11a^m + 28$

75. $(x + 1)a^2 + (x + 1)3a + (x + 1)2$

76. $ax^2 - 5x^2 + 8ax - 40x - (a - 5)9$
(*Hint:* See Exercise 75.)

Find a polynomial in factored form for the shaded area in the figure. (Leave answers in terms of π.)

77.

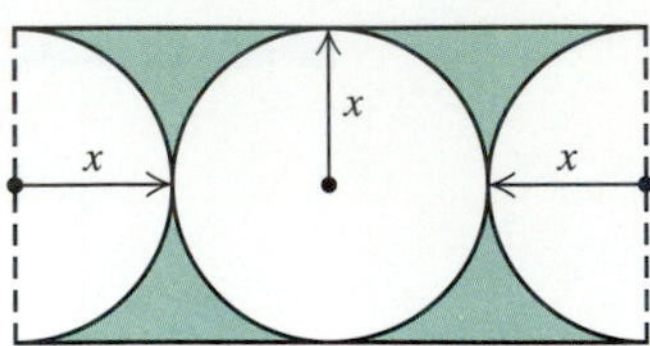

78.

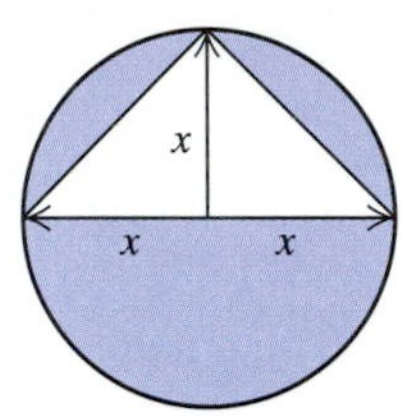

5.3 Factoring Trinomials of the Type $ax^2 + bx + c$, $a \neq 1$

Factoring with FOIL • The Grouping Method

In Section 5.2, we learned a trial-and-error method for factoring trinomials of the type $x^2 + bx + c$. Now we learn to factor trinomials in which the leading, or x^2, coefficient is not 1. First we will study the standard trial-and-error method and then we will consider an alternative method that involves factoring by grouping.

Factoring with FOIL

We want to factor trinomials of the type $ax^2 + bx + c$. Consider the following multiplication:

$$\begin{aligned} & \quad\;\; \text{F} \quad\;\; \text{O} \quad\;\; \text{I} \quad\; \text{L} \\ (2x + 5)(3x + 4) &= 6x^2 + 8x + 15x + 20 \\ &= 6x^2 + \quad 23x \quad + 20 \end{aligned}$$

To factor $6x^2 + 23x + 20$, we reverse the above multiplication and look for two binomials whose product is this trinomial. The product of the First terms must be $6x^2$. The product of the Outside terms plus the product of the Inside terms must be $23x$. The product of the Last terms must be 20. We know from the preceding discussion that the answer is

$$(2x + 5)(3x + 4).$$

Generally, however, finding such an answer is a trial-and-error process. It turns out that $(-2x - 5)(-3x - 4)$ is also a correct answer, but we usually choose an answer in which the first coefficients are positive.

We will use the following trial-and-error method.

To factor $ax^2 + bx + c$, $a \neq 1$, using the FOIL method:

1. Factor out the largest common factor, if one exists.
2. Find two First terms whose product is ax^2:

$$(\blacksquare x + \quad)(\blacksquare x + \quad) = ax^2 + bx + c.$$

FOIL

3. Find two Last terms whose product is c:

$$(\quad x + \blacksquare)(\quad x + \blacksquare) = ax^2 + bx + c.$$

FOIL

4. Redo steps (2) and (3) until a combination is found for which the sum of the Outer and Inner products is bx:

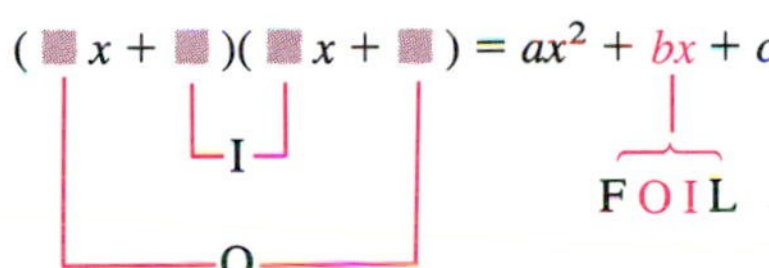

EXAMPLE 1

Factor: $3x^2 - 10x - 8$.

Solution

1) First, factor out a common factor, if any. There is none (other than 1 or -1).

2) Find two **F**irst terms whose product is $3x^2$.

The only possibilities for the **F**irst terms are $3x$ and x, so any factorization must be of the form

$$(3x + \quad)(x + \quad).$$

3) Find two **L**ast terms whose product is -8.

Possible factorizations of -8 are

$$(-8)\cdot 1, \quad 8\cdot(-1), \quad (-2)\cdot 4, \quad \text{and} \quad 2\cdot(-4).$$

Since the First terms are not identical, we must also consider

$$1\cdot(-8), \quad (-1)\cdot 8, \quad 4\cdot(-2), \quad \text{and} \quad (-4)\cdot 2.$$

4) Inspect the **O**uter and **I**nner products resulting from steps (2) and (3). Look for a combination in which the sum of the products is the middle term, $-10x$:

Trial	*Product*	
$(3x - 8)(x + 1)$	$3x^2 + 3x - 8x - 8$	
	$= 3x^2 - 5x - 8$	← Wrong middle term
$(3x + 8)(x - 1)$	$3x^2 - 3x + 8x - 8$	
	$= 3x^2 + 5x - 8$	← Wrong middle term
$(3x - 2)(x + 4)$	$3x^2 + 12x - 2x - 8$	
	$= 3x^2 + 10x - 8$	← Wrong middle term

(continued)

$(3x + 2)(x - 4)$	$3x^2 - 12x + 2x - 8$	
	$= 3x^2 - 10x - 8$	← Correct middle term!
$(3x + 1)(x - 8)$	$3x^2 - 24x + x - 8$	
	$= 3x^2 - 23x - 8$	← Wrong middle term
$(3x - 1)(x + 8)$	$3x^2 + 24x - x - 8$	
	$= 3x^2 + 23x - 8$	← Wrong middle term
$(3x + 4)(x - 2)$	$3x^2 - 6x + 4x - 8$	
	$= 3x^2 - 2x - 8$	← Wrong middle term
$(3x - 4)(x + 2)$	$3x^2 + 6x - 4x - 8$	
	$= 3x^2 + 2x - 8$	← Wrong middle term

The correct factorization is $(3x + 2)(x - 4)$. ❑

Two observations can be made from Example 1. First, we listed all possible trials even though we could have stopped after finding the correct factorization. We did this to show that each trial differs only in the middle term of the product. Second, note that as in Section 5.2, only the sign of the middle term changes when the signs in the binomials are reversed.

If a trial produces a middle term that is the opposite of the term we want, we need only change the signs in the two binomials.

EXAMPLE 2

Factor: $24x^2 - 76x + 40$.

Solution

1) First we factor out the largest common factor, 4:

$$4(6x^2 - 19x + 10).$$

Now we factor the trinomial $6x^2 - 19x + 10$.

2) Because $6x^2$ can be factored as $3x \cdot 2x$ or $6x \cdot x$, we have these possibilities for factorizations:

$$(3x + \quad)(2x + \quad) \quad \text{or} \quad (6x + \quad)(x + \quad).$$

3) There are four pairs of factors of 10 and they each can be listed in two ways:

$$10, 1 \qquad -10, -1 \qquad 5, 2 \qquad -5, -2$$

and

$$1, 10 \qquad -1, -10 \qquad 2, 5 \qquad -2, -5.$$

4) The two possibilities from step (2) and the eight possibilities from step (3) give $2 \cdot 8$, or 16 possibilities for factorizations. We look for **O**uter and **I**nner products resulting from steps (2) and (3) for which the sum is the middle term, $-19x$. Since the sign of the middle term is negative, but the sign of the last term, 10, is positive, the signs of both factors of the last term, 10, must be negative. This means only four pairings from step (3) need be considered. We first try these factors with $(3x + \quad)(2x + \quad)$. If none gives the correct factorization, we will consider $(6x + \quad)(x + \quad)$.

Trial	*Product*	
$(3x - 10)(2x - 1)$	$6x^2 - 3x - 20x + 10$	
	$= 6x^2 - 23x + 10$	← Wrong middle term
$(3x - 1)(2x - 10)$	$6x^2 - 30x - 2x + 10$	
	$= 6x^2 - 32x + 10$	← Wrong middle term
$(3x - 5)(2x - 2)$	$6x^2 - 6x - 10x + 10$	
	$= 6x^2 - 16x + 10$	← Wrong middle term
$(3x - 2)(2x - 5)$	$6x^2 - 15x - 4x + 10$	
	$= 6x^2 - 19x + 10$	← Correct middle term!

Since we have a correct factorization, we need not consider

$$(6x + \quad)(x + \quad).$$

Look again at the possibility $(3x - 5)(2x - 2)$. Without multiplying, we can reject such a possibility. To see why, consider the following:

$$(3x - 5)(2x - 2) = 2(3x - 5)(x - 1).$$

The expression $2x - 2$ has a common factor, 2. But we removed the *largest* common factor before we began. If this expression were a factorization, then 2 would have to be a common factor in addition to the original 4. Thus, $(2x - 2)$ cannot be part of the factorization of the original trinomial.

> Given that we factored out the largest common factor at the outset, we can eliminate factorizations that have a common factor.

The factorization of $6x^2 - 19x + 10$ is $(3x - 2)(2x - 5)$, but *do not forget the common factor*! We must include it in order to factor the original trinomial:

$$\begin{aligned} 24x^2 - 76x + 40 &= 4(6x^2 - 19x + 10) \\ &= 4(3x - 2)(2x - 5). \end{aligned}$$

❑

EXAMPLE 3

Factor: $10x^2 + 37x + 7$.

Solution

1) There is no common factor (other than 1 or -1).

2) Because $10x^2$ factors as $10x \cdot x$ or $5x \cdot 2x$, we have these possibilities for factorizations:

$$(10x + \quad)(x + \quad)$$

and

$$(5x + \quad)(2x + \quad).$$

3) There are two pairs of factors of 7 and they each can be listed in two ways:

$1, 7 \qquad -1, -7$

and

$7, 1 \qquad -7, -1.$

4) From steps (2) and (3), we see that there are 8 possibilities for factorizations. Look for **O**uter and **I**nner products for which the sum is the middle term. Because all coefficients in $10x^2 + 37x + 7$ are positive, we need consider only positive factors of 7. The possibilities are

$$(10x + 1)(x + 7) = 10x^2 + 71x + 7,$$
$$(10x + 7)(x + 1) = 10x^2 + 17x + 7,$$
$$(5x + 7)(2x + 1) = 10x^2 + 19x + 7,$$
$$(5x + 1)(2x + 7) = 10x^2 + 37x + 7.$$

The factorization is $(5x + 1)(2x + 7)$. ❑

Tips for factoring $ax^2 + bx + c$, $a \neq 1$:

1. Reversing the signs in the binomials reverses the sign of the middle term.
2. If the largest common factor has been factored out of the original trinomial, then no binomial factor can have a common factor (other than 1 or -1).
3. If c is positive, then the signs in both binomial factors must match the sign of b.
4. Be systematic about your trials. Keep track of those you have tried and those you have not.

Keep in mind that this method of factoring trinomials of the type $ax^2 + bx + c$ involves trial and error. As you practice, you will find that you can make fewer trials. Don't forget: When factoring any polynomial, always look first for a common factor. Failure to do so is such a common error that this caution bears repeating.

EXAMPLE 4

Factor: $6p^2 - 13pq - 28q^2$.

Solution Since no common factor exists, we examine the first term, $6p^2$. We have these possibilities:

$$(2p + \quad)(3p + \quad) \quad \text{and} \quad (6p + \quad)(p + \quad).$$

The last term, $-28q^2$, has the following pairs of factors:

$$28q, -q \qquad 14q, -2q \qquad 7q, -4q$$

and

$$-28q, q \qquad -14q, 2q \qquad -7q, 4q,$$

as well as each of the pairings reversed.

Some trials, like $(2p + 28q)(3p - q)$ and $(2p + 14q)(3p - 2q)$, cannot be correct because they contain a common factor, 2. We try $(2p + 7q)(3p - 4q)$:

$$\begin{aligned}(2p + 7q)(3p - 4q) &= 6p^2 - 8pq + 21pq - 28q^2\\ &= 6p^2 + 13pq - 28q^2.\end{aligned}$$

Our trial is incorrect, but only because of the sign of the middle term. The correct factorization is found by changing the signs in the binomials:

$$(2p - 7q)(3p + 4q).$$

Check: $(2p - 7q)(3p + 4q) = 6p^2 - 13pq - 28q^2$. ❑

The Grouping Method

Another method of factoring trinomials of the type $ax^2 + bx + c$ is known as the *grouping method*. The grouping method relies on finding two numbers, p and q, for which $p + q = b$ and $ax^2 + px + qx + c$ can be factored by grouping. The method is outlined as follows.

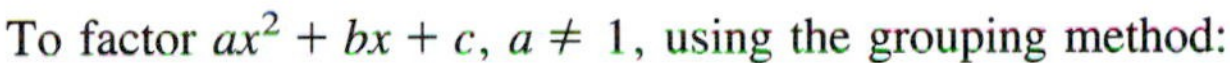

To factor $ax^2 + bx + c$, $a \neq 1$, using the grouping method:

1. Factor out the largest common factor, if one exists.
2. Multiply the leading coefficient a and the constant c.
3. List pairs of factors of ac until you find two numbers whose product is ac and whose sum is b.
4. Rewrite the middle term. That is, write it as a sum or difference using the factors found in step (3).
5. Then factor by grouping.

EXAMPLE 5

Factor: $3x^2 - 10x - 8$.

Solution

1) First note that there is no common factor (other than 1 or -1).
2) We multiply the leading coefficient, 3, and the constant, -8:

$$3(-8) = -24.$$

3) We then look for a factorization of -24 in which the sum of the factors is the coefficient of the middle term, -10.

Pairs of Factors of -24	Sums of Factors
1, -24	-23
-1, 24	23
2, -12	-10 ← $2 + (-12) = -10$
-2, 12	10
3, -8	-5
-3, 8	5
4, -6	-2
-4, 6	2

We generally stop listing pairs of factors once we have found the one we are after. (rows -2, 12 through -4, 6)

4) Next, we express the middle term as a sum or difference using the factors found in step (3):

$$-10x = -12x + 2x.$$

5) We now factor by grouping as follows:

$$3x^2 - 10x - 8 = 3x^2 - 12x + 2x - 8 \quad \text{Substituting } -12x + 2x \text{ for } -10x$$

$$= 3x(x - 4) + 2(x - 4) \quad \text{Factoring by grouping; see Section 5.1}$$

$$= (x - 4)(3x + 2). \quad \text{Factoring out the common factor, } x - 4$$

Check: $(x - 4)(3x + 2) = 3x^2 - 10x - 8$. ❑

EXAMPLE 6

Factor: $8x^3 + 22x^2 - 6x$.

Solution

1) We factor out the largest common factor, $2x$:

$$8x^3 + 22x^2 - 6x = 2x(4x^2 + 11x - 3).$$

2) To factor $4x^2 + 11x - 3$ by grouping, we multiply the leading coefficient, 4, and the constant term, -3:

$$4(-3) = -12.$$

3) We next look for factors of -12 that add to 11.

Pairs of Factors of -12	Sums of Factors
1, -12	-11
-1, 12	11
⋮	⋮

Since $-1 + 12 = 11$, there is no need to list other pairs of factors.

4) We then rewrite the middle term, $11x$, as follows:

$$11x = 12x - 1x.$$

5) Next, we factor by grouping:

$$\begin{aligned} 4x^2 + 11x - 3 &= 4x^2 + 12x - 1x - 3 && \text{Rewriting the middle term} \\ &= 4x(x + 3) - 1(x + 3) && \text{Factoring by grouping. Removing } -1 \text{ gives a common factor of } x + 3. \\ &= (x + 3)(4x - 1). && \text{Factoring out the common factor} \end{aligned}$$

The factorization of $4x^2 + 11x - 3$ is $(x + 3)(4x - 1)$. But don't forget the common factor, $2x$, when giving the factorization of the original trinomial:

$$8x^3 + 22x^2 - 6x = 2x(x + 3)(4x - 1).$$ ❑

EXERCISE SET 5.3

Factor completely. If a polynomial is prime, state so.

1. $2x^2 - 7x - 4$
2. $3x^2 - x - 4$
3. $5x^2 + x - 18$
4. $3x^2 - 4x - 15$
5. $6x^2 + 23x + 7$
6. $6x^2 + 13x + 6$
7. $3x^2 + 4x + 1$
8. $7x^2 + 15x + 2$
9. $4x^2 + 4x - 15$
10. $9x^2 + 6x - 8$
11. $2x^2 - x - 1$
12. $15x^2 - 19x - 10$
13. $9x^2 + 18x - 16$
14. $2x^2 + 5x + 2$
15. $3x^2 - 5x - 2$
16. $18x^2 - 3x - 10$
17. $12x^2 + 31x + 20$
18. $15x^2 + 19x - 10$
19. $14x^2 + 19x - 3$
20. $35x^2 + 34x + 8$
21. $9x^2 + 18x + 8$
22. $4 - 13x + 6x^2$
23. $49 - 42x + 9x^2$
24. $25x^2 + 40x + 16$
25. $24x^2 + 47x - 2$
26. $16a^2 + 78a + 27$
27. $35x^2 - 57x - 44$
28. $9a^2 + 12a - 5$
29. $2x^2 - 6x - 19$
30. $2x^2 - x - 15$
31. $12x^2 + 28x - 24$
32. $6x^2 + 33x + 15$
33. $30x^2 - 24x - 54$
34. $20x^2 - 25x + 5$

35. $4x + 6x^2 - 10$
36. $-9 + 18x^2 - 21x$
37. $3x^2 - 4x + 1$
38. $6x^2 - 13x + 6$
39. $12x^2 - 28x - 24$
40. $6x^2 - 33x + 15$
41. $-1 + 2x^2 - x$
42. $-19x + 15x^2 + 6$
43. $9x^2 - 18x - 16$
44. $14x^2 + 35x + 14$
45. $15x^2 - 25x - 10$
46. $18x^2 + 3x - 10$
47. $12x^3 + 31x^2 + 20x$
48. $15x^3 + 19x^2 - 10x$
49. $14x^4 + 19x^3 - 3x^2$
50. $70x^4 + 68x^3 + 16x^2$
51. $168x^3 - 45x^2 + 3x$
52. $144x^5 + 168x^4 + 48x^3$
53. $15x^2 - 19x + 6$
54. $9x^2 + 18x + 8$
55. $25t^2 + 80t + 64$
56. $9x^2 - 42x + 49$
57. $6x^3 + 4x^2 - 10x$
58. $18x^3 - 21x^2 - 9x$
59. $25x^2 + 89x + 64$
60. $9y^2 - 42y + 47$
61. $x^2 + 3x - 7$
62. $x^2 + 13x - 12$
63. $12m^2 + mn - 20n^2$
64. $12a^2 + 17ab + 6b^2$
65. $6a^2 - ab - 15b^2$
66. $3p^2 - 16pq - 12q^2$
67. $9a^2 + 18ab + 8b^2$
68. $10s^2 + 4st - 6t^2$
69. $35p^2 + 34pq + 8q^2$
70. $30a^2 + 87ab + 30b^2$
71. $18x^2 - 6xy - 24y^2$
72. $15a^2 - 5ab - 20b^2$

Factor. Use factoring by grouping even though it would seem reasonable to first collect like terms.

73. $y^2 + 4y + y + 4$
74. $x^2 + 5x + 2x + 10$
75. $x^2 - 4x - x + 4$
76. $a^2 + 5a - 2a - 10$
77. $6x^2 + 4x + 9x + 6$
78. $3x^2 - 2x + 3x - 2$
79. $3x^2 - 4x - 12x + 16$
80. $24 - 18y - 20y + 15y^2$
81. $35x^2 - 40x + 21x - 24$
82. $8x^2 - 6x - 28x + 21$
83. $4x^2 + 6x - 6x - 9$
84. $2x^4 - 6x^2 - 5x^2 + 15$

Factor by grouping. If a polynomial is prime, state so.

85. $2x^2 - 7x - 4$
86. $3x^2 - x - 4$
87. $5x^2 + x - 18$
88. $3x^2 - 4x - 15$
89. $6x^2 + 23x + 7$
90. $6x^2 + 13x + 6$
91. $3x^2 + 4x + 1$
92. $7x^2 + 15x + 2$
93. $4x^2 + 4x - 15$
94. $9x^2 + 6x - 8$
95. $2x^2 - x - 1$
96. $15x^2 - 19x - 10$
97. $9x^2 + 18x - 16$
98. $2x^2 + 5x + 2$
99. $3x^2 - 5x - 2$
100. $18x^2 - 3x - 10$
101. $12x^2 + 31x + 20$
102. $15x^2 + 19x - 10$
103. $14x^2 + 19x - 3$
104. $35x^2 + 34x + 8$
105. $9x^2 + 18x + 8$
106. $6 - 13x + 6x^2$
107. $49 - 42x + 9x^2$
108. $25x^2 + 40x + 16$

Skill Maintenance

109. The earth is a sphere (or ball) that is about 40,000 km in circumference. Find the radius of the earth, in kilometers and in miles. Use 3.14 for π. (*Hint:* 1 km $\approx$ 0.62 mi.)

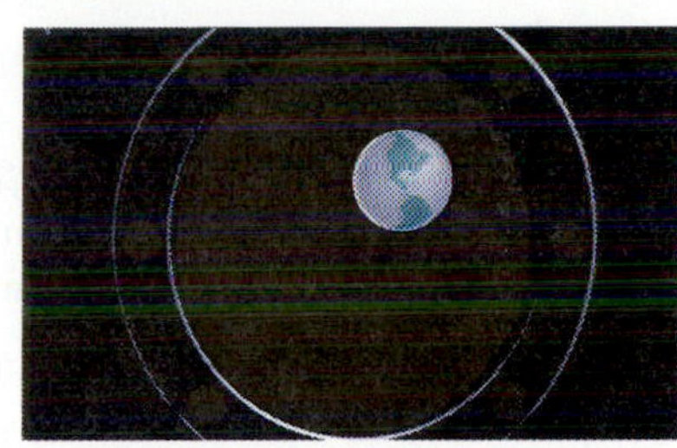

110. The second angle of a triangle is 10° less than twice the first. The third angle is 15° more than four times the first. Find the measure of the second angle.

111. Graph: $y = \frac{2}{5}x - 1$.

112. Divide: $\frac{y^{12}}{y^4}$.

Synthesis

113. ◈ A student presents the following work:

$$4x^2 + 28x + 48 = (2x + 6)(2x + 8) = 2(x + 3)(x + 4).$$

Is this correct? Explain.

114. ◈ If a trinomial's leading coefficient and constant term are both prime numbers, at most how many trials can be made when factoring the trinomial? Why?

Factor.

115. $9x^{10} - 12x^5 + 4$
116. $16x^{10} + 8x^5 + 1$
117. $20x^{2n} + 16x^n + 3$
118. $-15x^{2m} + 26x^m - 8$
119. $3x^{6a} - 2x^{3a} - 1$
120. $x^{2n+1} - 2x^{n+1} + x$
121. $3(a + 1)^{n+1}(a + 3)^2 - 5(a + 1)^n(a + 3)^3$

5.4

Factoring Trinomial Squares and Differences of Squares

Recognizing Trinomial Squares • Factoring Trinomial Squares • Recognizing Differences of Squares • Factoring Differences of Squares • Factoring Completely • More Factoring by Grouping

In Chapter 4, we studied some shortcuts for finding certain products of binomials. We now reverse these procedures to discover shortcuts for factoring certain polynomials.

Recognizing Trinomial Squares

Some trinomials are squares of binomials. For example, the trinomial $x^2 + 10x + 25$ is the square of the binomial $x + 5$. To see this, we can calculate $(x + 5)^2$. It is $x^2 + 2 \cdot x \cdot 5 + 5^2$, or $x^2 + 10x + 25$. A trinomial that is the square of a binomial is called a **trinomial square.**

In Chapter 4, we considered squaring binomials as a special-product rule:

$$(A + B)^2 = A^2 + 2AB + B^2;$$
$$(A - B)^2 = A^2 - 2AB + B^2.$$

Written from right to left, these equations can be used to factor trinomial squares. Note that in order for a trinomial to be the square of a binomial, we must have the following:

A. Two terms, A^2 and B^2, must be squares, such as

$4, \quad x^2, \quad 81m^2, \quad 16t^2.$

B. There must be no minus sign before A^2 or B^2.

C. If we multiply A and B (the square roots of A^2 and B^2) and double the result, we get the remaining term, $2 \cdot A \cdot B$, or its opposite, $-2 \cdot A \cdot B$.

EXAMPLE 1

Determine whether $x^2 + 6x + 9$ is a trinomial square.

Solution

A. We know that x^2 and 9 are squares.

B. There is no minus sign before x^2 or 9.

C. If we multiply the square roots, x and 3, and double the product, we get the remaining term: $2 \cdot x \cdot 3 = 6x$.

Thus, $x^2 + 6x + 9$ is a trinomial square. ❑

EXAMPLE 2

Determine whether $x^2 + 6x + 11$ is a trinomial square.

Solution The answer is no, because only one term is a square. ❑

EXAMPLE 3

Determine whether $16x^2 + 49 - 56x$ is a trinomial square.

Solution It helps to first write the trinomial in descending order:

$16x^2 - 56x + 49.$

A. We know that $16x^2$ and 49 are squares.
B. There is no minus sign before $16x^2$ or 49.
C. If we multiply the square roots, $4x$ and 7, and double the product, we get

$$2 \cdot 4x \cdot 7, \quad \text{or} \quad 56x;$$

$56x$ is the opposite of the remaining term.

Thus, $16x^2 + 49 - 56x$ is a trinomial square. ❑

Factoring Trinomial Squares

We can use the trial-and-error or grouping methods from Sections 5.2 and 5.3 to factor trinomial squares, but a faster method uses the following equations:

$$A^2 + 2AB + B^2 = (A + B)^2;$$
$$A^2 - 2AB + B^2 = (A - B)^2.$$

The factorization uses the square roots of the squared terms and the sign of the remaining term.

EXAMPLE 4

Factor: **(a)** $x^2 + 6x + 9$; **(b)** $x^2 + 49 - 14x$; **(c)** $16x^2 - 40x + 25$.

Solution

a) $x^2 + 6x + 9 = x^2 + 2 \cdot x \cdot 3 + 3^2 = (x + 3)^2$ — The sign of the middle term is positive.

$$A^2 + 2\ A\ B + B^2 = (A + B)^2$$

b) $x^2 + 49 - 14x = x^2 - 14x + 49$ — Using a commutative law to write descending order

$= x^2 - 2 \cdot x \cdot 7 + 7^2$ — The sign of the middle term is negative.

$= (x - 7)^2$ — Factoring the trinomial square

c) $16x^2 - 40x + 25 = (4x)^2 - 2 \cdot 4x \cdot 5 + 5^2 = (4x - 5)^2$

$$A^2\ -2\ A\ B + B^2 = (A - B)^2$$ ❑

With practice, you will be able to spot trinomial squares whenever they occur and factor them quickly.

EXAMPLE 5

Factor: **(a)** $4p^2 - 12pq + 9q^2$; **(b)** $75m^3 + 60m^2 + 12m$.

Solution

a) $4p^2 - 12pq + 9q^2 = (2p)^2 - 2(2p)(3q) + (3q)^2$ — Recognizing the trinomial square

$= (2p - 3q)^2$ — The sign of the middle term is negative.

Check: $(2p - 3q)(2p - 3q) = 4p^2 - 12pq + 9q^2$.

b) *Always* look first for a common factor. This time there is one, $3m$:

$$75m^3 + 60m^2 + 12m = 3m[25m^2 + 20m + 4]$$ Factoring out the largest common factor

$$= 3m[(5m)^2 + 2(5m)(2) + 2^2]$$ Recognizing the trinomial square. Try to do this mentally.

$$= 3m(5m + 2)^2.$$

Check: $3m(5m + 2)^2 = 3m(5m + 2)(5m + 2)$
$= 3m(25m^2 + 20m + 4)$
$= 75m^3 + 60m^2 + 12m.$ ❑

Recognizing Differences of Squares

Some binomials represent the difference of two squares. For example, the binomial $16x^2 - 9$ is a difference of two expressions, $16x^2$ and 9, that are squares. To see this, note that $16x^2 = (4x)^2$ and $9 = 3^2$.

Any expression, like $16x^2 - 9$, that can be written in the form $A^2 - B^2$ is called a **difference of squares.** Note that in order for a binomial to be a difference of squares, we must have the following:

A. There must be two expressions, both squares, such as

$$4x^2, \quad 9, \quad 4x^2y^2, \quad 1, \quad x^6, \quad 49y^8.$$

B. The terms in the binomial must have different signs.

Note that in order for a term to be a square, its coefficient must be a perfect square and the power(s) of the variable(s) must be even.

EXAMPLE 6

Is $9x^2 - 64$ a difference of squares?

Solution

A. The first expression is a square: $9x^2 = (3x)^2$.
The second expression is a square: $64 = 8^2$.

B. The terms have different signs.

Thus we have a difference of squares, $(3x)^2 - 8^2$. ❑

EXAMPLE 7

Is $25 - t^3$ a difference of squares?

Solution

A. The expression t^3 is not a square.

Thus, $25 - t^3$ is not a difference of squares. ❑

EXAMPLE 8

Is $-4x^2 + 16$ a difference of squares?

Solution

A. The expressions $4x^2$ and 16 are squares: $4x^2 = (2x)^2$ and $16 = 4^2$.

B. The terms have different signs.

Thus we have a difference of squares. We can also see this by rewriting in the equivalent form: $16 - 4x^2$. □

Factoring Differences of Squares

To factor a difference of squares, we use an equation that first appeared, written from right to left, in Chapter 4:

$$A^2 - B^2 = (A + B)(A - B).$$

To factor a difference of squares $A^2 - B^2$, we first find the square roots A and B. Then we use A and B to form two factors. One factor is $A + B$, and the other is $A - B$.

EXAMPLE 9

Factor: **(a)** $x^2 - 4$; **(b)** $m^2 - 9p^2$.

Solution

a) $x^2 - 4 = x^2 - 2^2 = (x + 2)(x - 2)$

$$\uparrow \quad \uparrow \quad \uparrow \quad \uparrow \quad \uparrow \quad \uparrow$$
$$A^2 - B^2 = (A + B)(A - B)$$

b) $m^2 - 9p^2 = m^2 - (3p)^2 = (m + 3p)(m - 3p)$

$$\uparrow \quad \uparrow \quad \uparrow \quad \uparrow \quad \uparrow \quad \uparrow$$
$$A^2 - B^2 = (A + B)(A - B)$$
□

When powers larger than 2 arise, it is important to remember that $(a^r)^s = a^{r \cdot s}$ and $a^n \cdot a^m = a^{n+m}$.

EXAMPLE 10

Factor: **(a)** $9 - 16t^{10}$; **(b)** $18x^2 - 50x^6$; **(c)** $49x^4 - 9x^6$.

Solution

a) $9 - 16t^{10} = 3^2 - (4t^5)^2$ Using the rules for powers

$$\uparrow \quad \uparrow$$
$$A^2 - B^2$$

$= (3 + 4t^5)(3 - 4t^5)$ Try to go directly to this step.

$$\uparrow \quad \uparrow \quad \uparrow \quad \uparrow$$
$$(A + B)(A - B)$$

b) *Always* look first for a common factor. This time there is one, $2x^2$:

$18x^2 - 50x^6 = 2x^2(9 - 25x^4)$ Factoring out the largest common factor

$= 2x^2[3^2 - (5x^2)^2]$ Recognizing $A^2 - B^2$. Try to do this mentally.

$= 2x^2(3 + 5x^2)(3 - 5x^2)$ Factoring the difference of squares

Check: $2x^2(3 + 5x^2)(3 - 5x^2) = 2x^2(9 - 25x^4) = 18x^2 - 50x^6$.

c) $49x^4 - 9x^6 = x^4(49 - 9x^2)$ Factoring out the largest common factor
$= x^4(7 + 3x)(7 - 3x)$ Factoring the difference of squares

Check: $x^4(7 + 3x)(7 - 3x) = x^4(49 - 9x^2) = 49x^4 - 9x^6$. ❑

CAUTION! Note carefully in these examples that a difference of squares is *not* the square of the difference; that is,

$$A^2 - B^2 \neq (A - B)^2.$$

For example,

$$8^2 - 3^2 = 64 - 9 = 55,$$

but

$$(8 - 3)^2 = 5^2 = 25.$$

Factoring Completely

If a factor with more than one term can still be factored, you should do so. When no factor can be factored further, you have factored completely. Always factor completely whenever you are asked to factor.

EXAMPLE 11

Factor: $p^4 - 16$.

Solution

$p^4 - 16 = (p^2)^2 - 4^2$
$= (p^2 + 4)(p^2 - 4)$ Factoring a difference of squares
$= (p^2 + 4)(p + 2)(p - 2)$ Factoring further. The factor $p^2 - 4$ is itself a difference of squares.

Check: $(p^2 + 4)(p + 2)(p - 2) = (p^2 + 4)(p^2 - 4) = p^4 - 16$. ❑

Observe in Example 11 that the factor $p^2 + 4$ is a *sum* of squares that cannot be factored further.

CAUTION! If the largest common factor has been removed, then you cannot factor a sum of squares further. In particular,

$$A^2 + B^2 \neq (A + B)^2.$$

Consider $25x^2 + 100$. This is a case in which we have a sum of squares with a common factor, 25. Factoring, we get $25(x^2 + 4)$, where $x^2 + 4$ is prime.

More Factoring by Grouping

Sometimes when factoring a polynomial with four terms, we may be able to factor further.

EXAMPLE 12

Factor: $x^3 + 3x^2 - 4x - 12$.

Solution

$$\begin{aligned} x^3 + 3x^2 - 4x - 12 &= x^2(x+3) - 4(x+3) \\ &= (x+3)(x^2-4) \\ &= (x+3)(x+2)(x-2) \end{aligned}$$ □

A difference of squares can have more than two terms. For example, one of the squares may be a trinomial. We can factor by a different type of grouping.

EXAMPLE 13

Factor: $x^2 + 6x + 9 - y^2$.

Solution

$$x^2 + 6x + 9 - y^2 = (x^2 + 6x + 9) - y^2$$ Grouping as a trinomial minus y^2 to show a difference of squares

$$= (x+3)^2 - y^2$$

$$= (x + 3 + y)(x + 3 - y)$$ □

EXAMPLE 14

Factor: $a^2 - b^2 + 8b - 16$.

Solution Grouping into two groups of two terms each does not yield a common binomial factor, so we look for a trinomial square. In this case, the trinomial square is being subtracted from a^2:

$$a^2 - b^2 + 8b - 16 = a^2 - (b^2 - 8b + 16)$$ Factoring out -1 and rewriting as subtraction

$$= a^2 - (b-4)^2$$ Factoring the trinomial square

$$= (a + (b-4))(a - (b-4))$$ Factoring a difference of squares

$$= (a + b - 4)(a - b + 4)$$ Removing parentheses □

As you proceed through the exercises, these suggestions may prove helpful.

1. Always look first for a common factor! If there is one, factor it out.
2. Be alert for trinomial squares and differences of squares. Once recognized, they can be factored without trial and error.
3. Always factor completely.
4. Check by multiplying.

EXERCISE SET 5.4

Determine whether each of the following is a trinomial square.

1. $x^2 - 14x + 49$
2. $x^2 - 16x + 64$
3. $x^2 + 16x - 64$
4. $x^2 - 14x - 49$
5. $x^2 - 3x + 9$
6. $x^2 + 2x + 4$
7. $9x^2 - 36x + 24$
8. $36x^2 - 24x + 16$

Factor completely. Remember to look first for a common factor and to check by multiplying.

9. $x^2 - 14x + 49$
10. $x^2 - 16x + 64$
11. $x^2 + 16x + 64$
12. $x^2 + 14x + 49$
13. $x^2 - 2x + 1$
14. $x^2 + 2x + 1$
15. $4 + 4x + x^2$
16. $4 + x^2 - 4x$
17. $9x^2 + 6x + 1$
18. $25x^2 - 10x + 1$
19. $49 - 56y + 16y^2$
20. $120m + 75 + 48m^2$
21. $2x^2 - 4x + 2$
22. $2x^2 - 40x + 200$
23. $x^3 - 18x^2 + 81x$
24. $x^3 + 24x^2 + 144x$
25. $20x^2 + 100x + 125$
26. $12x^2 + 36x + 27$
27. $49 - 42x + 9x^2$
28. $64 - 112x + 49x^2$
29. $5y^2 + 10y + 5$
30. $2a^2 + 28a + 98$
31. $2 + 20x + 50x^2$
32. $7 - 14a + 7a^2$
33. $4p^2 + 12pq + 9q^2$
34. $25m^2 + 20mn + 4n^2$
35. $a^2 - 14ab + 49b^2$
36. $x^2 - 6xy + 9y^2$
37. $64m^2 + 16mn + n^2$
38. $81p^2 - 18pq + q^2$
39. $16s^2 - 40st + 25t^2$
40. $36a^2 + 96ab + 64b^2$

Determine whether each of the following is a difference of squares.

41. $x^2 - 4$
42. $x^2 - 36$
43. $x^2 + 36$
44. $x^2 + 4$
45. $x^2 - 35$
46. $x^2 - 50y^2$
47. $16x^2 - 25y^2$
48. $-1 + 36x^2$

Factor completely. Remember to look first for a common factor.

49. $y^2 - 4$
50. $x^2 - 36$
51. $p^2 - 9$
52. $q^2 - 1$
53. $-49 + t^2$
54. $-64 + m^2$
55. $a^2 - b^2$
56. $p^2 - q^2$
57. $25t^2 - m^2$
58. $w^2 - 49z^2$
59. $100 - k^2$
60. $81 - w^2$
61. $16a^2 - 9$
62. $25x^2 - 4$
63. $4x^2 - 25y^2$
64. $9a^2 - 16b^2$
65. $8x^2 - 98$
66. $24x^2 - 54$
67. $36x - 49x^3$
68. $16x - 81x^3$
69. $49a^4 - 81$
70. $25a^4 - 9$
71. $x^4 - 1$
72. $x^4 - 16$
73. $4x^4 - 64$
74. $5x^4 - 80$
75. $1 - y^8$
76. $x^8 - 1$
77. $3x^3 - 24x^2 + 48x$
78. $2a^4 - 36a^3 + 162a^2$
79. $x^{12} - 16$
80. $x^8 - 81$
81. $y^2 - \frac{1}{16}$
82. $x^2 - \frac{1}{25}$
83. $a^8 - 2a^7 + a^6$
84. $x^8 - 8x^7 + 16x^6$
85. $25 - \frac{1}{49}x^2$
86. $4 - \frac{1}{9}y^2$
87. $16m^4 - t^4$
88. $1 - a^4b^4$
89. $m^3 - 7m^2 - 4m + 28$
90. $x^3 + 8x^2 - x - 8$
91. $a^3 - ab^2 - 2a^2 + 2b^2$
92. $p^2q - 25q + 3p^2 - 75$
93. $(a + b)^2 - 100$
94. $(p - 7)^2 - 144$
95. $a^2 + 2ab + b^2 - 9$
96. $x^2 - 2xy + y^2 - 25$
97. $r^2 - 2r + 1 - 4s^2$
98. $c^2 + 4cd + 4d^2 - 9p^2$
99. $50a^2 - 2m^2 - 4mn - 2n^2$
100. $3x^2 - 12y^2 - 12y - 3$
101. $9 - a^2 + 2ab - b^2$
102. $16 - x^2 + 2xy - y^2$

Skill Maintenance

103. Bonnie is taking an astronomy course. To get an A, a student must average at least 90 after four hour exams. Bonnie scored 96, 98, and 89 on the first three tests. Determine (in terms of an inequality) what scores on the last test will earn her an A.

104. About 5 L of oxygen can be dissolved in 100 L of water at 0°C. This is 1.6 times the amount that can be dissolved in the same volume of water at 20°C. How much oxygen can be dissolved at 20°C?

Simplify.

105. $(x^3y^5)(x^9y^7)$
106. $(5a^2b^3)^2$

Synthesis

107. A student concludes that since $x^2 - 9 = (x-3)(x+3)$, it must follow that $x^2 + 9 = (x+3)(x-3)$. What mistake(s) is the student making?

108. Explain in your own words how to determine if a polynomial is a trinomial square.

Factor completely. If a polynomial is prime, state so.

109. $49x^2 - 216$
110. $x^2 - 5x + 25$
111. $18x^3 + 12x^2 + 2x$
112. $162x^2 - 82$
113. $x^8 - 2^8$
114. $4x^4 - 4x^2$
115. $3x^5 - 12x^3$
116. $3x^2 - \frac{1}{3}$
117. $18x^3 - \frac{8}{25}x$
118. $x^2 - 2.25$
119. $0.49p - p^3$
120. $0.64x^2 - 1.21$
121. $(x+3)^2 - 9$
122. $(y-5)^2 - 36q^2$
123. $x^2 - \left(\frac{1}{x}\right)^2$
124. $a^{2n} - 49b^{2n}$
125. $81 - b^{4k}$
126. $x^4 - 8x^2 - 9$
127. $9b^{2n} + 12b^n + 4$
128. $16x^4 - 96x^2 + 144$
129. $(y+3)^2 + 2(y+3) + 1$
130. $49(x+1)^2 - 42(x+1) + 9$
131. $27x^3 - 63x^2 - 147x + 343$
132. Subtract $(x^2+1)^2$ from $x^2(x+1)^2$ and factor the result.

Find c so that the polynomial will be the square of a binomial.

133. $cy^2 + 6y + 1$
134. $cy^2 - 24y + 9$
135. Show that the difference of the squares of two consecutive integers is the sum of the integers. (*Hint:* Use x for the smaller number.)
136. Find the value of a if $x^2 + a^2x + a^2$ factors into $(x+a)^2$.

5.5 Factoring Sums or Differences of Cubes

Formulas for Factoring Sums or Differences of Cubes • Using the Formulas

Although a sum of two squares cannot be factored using real-number coefficients (unless it has a common factor), a sum of two cubes can. In this section, we develop a method for factoring sums or differences of two cubes.

Consider the following products:

$$\begin{aligned}(A+B)(A^2 - AB + B^2) &= A(A^2 - AB + B^2) + B(A^2 - AB + B^2)\\ &= A^3 - A^2B + AB^2 + A^2B - AB^2 + B^3\\ &= A^3 + B^3 \quad \text{Combining like terms}\end{aligned}$$

and

$$\begin{aligned}(A-B)(A^2 + AB + B^2) &= A(A^2 + AB + B^2) - B(A^2 + AB + B^2)\\ &= A^3 + A^2B + AB^2 - A^2B - AB^2 - B^3\\ &= A^3 - B^3. \quad \text{Combining like terms}\end{aligned}$$

These equations (reversed) allow us to factor a sum or a difference of two cubes.

$$A^3 + B^3 = (A+B)(A^2 - AB + B^2)$$
$$A^3 - B^3 = (A-B)(A^2 + AB + B^2)$$

The table of cubes below can help in the following examples.

N	0.2	0.1	0	1	2	3	4	5	6	7	8	9	10
N^3	0.008	0.001	0	1	8	27	64	125	216	343	512	729	1000

EXAMPLE 1

Factor: $x^3 - 27$.

Solution We have

$$x^3 - 27 = x^3 - 3^3.$$

In one set of parentheses, we write the first quantity that was cubed, x, then a minus sign, and then the second quantity that was cubed, 3. This gives us the expression $x - 3$:

$$(x - 3)(\qquad).$$

To get the other factor, we think of $x - 3$ and do the following:

Square the first term: x^2.
Multiply the terms and then change the sign: $3x$.
Square the second term: $(-3)^2$, or 9.

$$(x - 3)(x^2 + 3x + 9).$$

Note that we cannot factor $x^2 + 3x + 9$. (It is not a trinomial square nor can it be factored by trial and error.) ❑

EXAMPLE 2

Factor: **(a)** $125x^3 + y^3$; **(b)** $128y^7 - 250x^6y$; **(c)** $a^6 - b^6$.

Solution

a) We have

$$125x^3 + y^3 = (5x)^3 + y^3.$$

In one set of parentheses, we write the first quantity that is cubed, $5x$, then a plus sign, and then the second quantity that is cubed, y:

$$(5x + y)(\qquad).$$

To get the other factor, we think of $5x + y$ and do the following:

Square the first term: $(5x)^2$, or $25x^2$.
Multiply the terms and then change the sign: $-5xy$.
Square the second term: y^2.

$$(5x + y)(25x^2 - 5xy + y^2).$$

b) We have

$$128y^7 - 250x^6y = 2y(64y^6 - 125x^6)$$ Remember: *Always* look for a common factor.

$$= 2y[(4y^2)^3 - (5x^2)^3]$$ Rewriting as quantities cubed

$$= 2y(4y^2 - 5x^2)(16y^4 + 20x^2y^2 + 25x^4).$$

c) We have

$$a^6 - b^6 = (a^3)^2 - (b^3)^2.$$

We factor as follows:

$$(a^3 + b^3)(a^3 - b^3).$$

One factor is a sum of two cubes, and the other factor is a difference of two cubes. We factor them:

$$(a + b)(a^2 - ab + b^2)(a - b)(a^2 + ab + b^2).$$ □

In Example 2(c), had we thought of factoring first as a difference of two cubes, we would have had

$$\begin{aligned}(a^2)^3 - (b^2)^3 &= (a^2 - b^2)(a^4 + a^2b^2 + b^4)\\ &= (a + b)(a - b)(a^4 + a^2b^2 + b^4).\end{aligned}$$

In this case, we might have missed some factors; $a^4 + a^2b^2 + b^4$ can be factored as $(a^2 - ab + b^2)(a^2 + ab + b^2)$, but we probably would never have suspected that such a factorization exists.

EXAMPLE 3

Factor: $64a^6 - 729b^6$.

Solution We have

$$\begin{aligned}64a^6 - 729b^6 &= (8a^3 - 27b^3)(8a^3 + 27b^3) && \text{Factoring a difference of squares}\\ &= [(2a)^3 - (3b)^3][(2a)^3 + (3b)^3].\end{aligned}$$

Each factor is a sum or a difference of cubes. We factor each:

$$= (2a - 3b)(4a^2 + 6ab + 9b^2)(2a + 3b)(4a^2 - 6ab + 9b^2).$$ □

Remember the following:

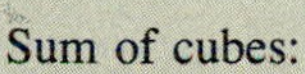

Sum of cubes:	$A^3 + B^3 = (A + B)(A^2 - AB + B^2)$;
Difference of cubes:	$A^3 - B^3 = (A - B)(A^2 + AB + B^2)$;
Difference of squares:	$A^2 - B^2 = (A + B)(A - B)$;
Sum of squares:	$A^2 + B^2$ cannot be factored using real numbers if the largest common factor has been removed.

EXERCISE SET 5.5

Factor.

1. $x^3 + 8$
2. $c^3 + 27$
3. $y^3 - 64$
4. $z^3 - 1$
5. $w^3 + 1$
6. $x^3 + 125$
7. $8a^3 + 1$
8. $27x^3 + 1$
9. $y^3 - 8$
10. $p^3 - 27$
11. $8 - 27b^3$
12. $64 - 125x^3$
13. $64y^3 + 1$
14. $125x^3 + 1$
15. $8x^3 + 27$
16. $27y^3 + 64$
17. $a^3 - b^3$
18. $x^3 - y^3$

19. $a^3 + \frac{1}{8}$
20. $b^3 + \frac{1}{27}$
21. $2y^3 - 128$
22. $3z^3 - 3$
23. $24a^3 + 3$
24. $54x^3 + 2$
25. $rs^3 + 64r$
26. $ab^3 + 125a$
27. $5x^3 - 40z^3$
28. $2y^3 - 54z^3$
29. $x^3 + 0.001$
30. $y^3 + 0.125$
31. $64x^6 - 8t^6$
32. $125c^6 - 8d^6$
33. $2y^4 - 128y$
34. $3z^5 - 3z^2$
35. $z^6 - 1$
36. $t^6 + 1$
37. $t^6 + 64y^6$
38. $p^6 - q^6$

Skill Maintenance

39. Your quiz grades are 78, 76, 82, and 93. What scores on a fifth quiz will make your average quiz grade at least 80? Use an inequality.

Simplify.

40. $38 + (-47) - 16 - (-81)$
41. $-\frac{2}{3} - \frac{3}{4} + \frac{1}{2}$
42. Collect like terms: $3y - 6 - 5y + 4$.

Synthesis

43. ◈ Explain how the formula for factoring a *difference* of two cubes can be used to factor $x^3 + 8$.
44. ◈ How could you use factoring to convince someone that $x^3 + y^3 \neq (x + y)^3$?

Factor. Assume that variables in exponents represent natural numbers.

45. $x^{6a} + y^{3b}$
46. $a^3x^3 - b^3y^3$
47. $3x^{3a} + 24y^{3b}$
48. $\frac{8}{27}x^3 + \frac{1}{64}y^3$
49. $\frac{1}{24}x^3y^3 + \frac{1}{3}z^3$
50. $\frac{1}{16}x^{3a} + \frac{1}{2}y^{6a}z^{9b}$
51. $7x^3 + \frac{7}{8}$
52. $[(c - d)^3 - d^3]^2$
53. $(x + y)^3 - x^3$
54. $(1 - x)^3 + (x - 1)^6$
55. $(a + 2)^3 - (a - 2)^3$
56. $y^4 - 8y^3 - y + 8$

5.6 Factoring: A General Strategy

Choosing the Right Method • Checking by Evaluating

We now combine all of our factoring techniques and consider a general strategy for factoring polynomials. Here we will encounter polynomials of all the types we have considered, in random order, so you will have to determine which method to use.

Choosing the Right Method

A. Always factor out the largest common factor. Be sure to include it in your final answer.

B. Look at the number of terms.

Two terms: Try factoring as a difference of squares first. Next, try factoring as a sum or a difference of cubes. Do *not* try to factor a *sum* of squares.

Three terms: Try factoring as a trinomial square. Next, try trial and error, using the FOIL method or the grouping method.

Four or more terms: Try factoring by grouping and factoring out a common binomial factor. Next, try grouping into a difference of squares, one of which is a trinomial.

C. Always *factor completely*. If a factor with more than one term can itself be factored further, do so.

D. Check by multiplying.

EXAMPLE 1

Factor: $5t^4 - 80$.

Solution

A. We look for a common factor:

$$5t^4 - 80 = 5(t^4 - 16).$$

B. The factor $t^4 - 16$ has only two terms. It is a difference of squares: $(t^2)^2 - 4^2$. We factor it, being careful to rewrite the common factor:

$$5t^4 - 80 = 5(t^2 + 4)(t^2 - 4).$$

C. We see that one of the factors is again a difference of squares. We factor it:

$$5t^4 - 80 = 5(t^2 + 4)(t - 2)(t + 2).$$

↑ This is a sum of squares. It cannot be factored!

We have factored completely because no factor with more than one term can be factored further.

D. *Check:* $5(t^2 + 4)(t - 2)(t + 2) = 5(t^2 + 4)(t^2 - 4)$
$= 5(t^4 - 16) = 5t^4 - 80.$ □

EXAMPLE 2

Factor: $2x^3 + 10x^2 + x + 5$.

Solution

A. We look for a common factor. There isn't one.

B. There are four terms. We try factoring by grouping:

$2x^3 + 10x^2 + x + 5$

$= (2x^3 + 10x^2) + (x + 5)$ Separating into two binomials

$= 2x^2(x + 5) + 1(x + 5)$ Factoring out the largest common factor from each binomial

$= (x + 5)(2x^2 + 1)$ Factoring out the common factor, $x + 5$

C. No factor with more than one term can be factored further, so we have factored completely.

D. *Check:* $(x + 5)(2x^2 + 1) = 2x^3 + x + 10x^2 + 5 = 2x^3 + 10x^2 + x + 5.$ □

EXAMPLE 3

Factor: $x^5 - 2x^4 - 35x^3$.

Solution

A. We look first for a common factor. This time there is one, x^3:

$$x^5 - 2x^4 - 35x^3 = x^3(x^2 - 2x - 35).$$

B. The factor $x^2 - 2x - 35$ has three terms, but it is not a trinomial square. We factor it using trial and error:

$x^5 - 2x^4 - 35x^3 = x^3(x^2 - 2x - 35)$

$= x^3(x - 7)(x + 5).$ Don't forget to rewrite the common factor.

C. No factor with more than one term can be factored further, so we have factored completely.

D. *Check:* $x^3(x - 7)(x + 5) = x^3(x^2 - 2x - 35) = x^5 - 2x^4 - 35x^3.$ □

EXAMPLE 4 Factor: $x^2 - 20x + 100$.

Solution

A. We look first for a common factor. There isn't one.

B. There are three terms. This polynomial is a trinomial square, so we factor it accordingly:

$$x^2 - 20x + 100 = x^2 - 2 \cdot x \cdot 10 + 10^2 \quad \text{Try to do this step mentally.}$$
$$= (x - 10)^2.$$

C. No factor with more than one term can be factored further, so we have factored completely.

D. *Check:* $(x - 10)(x - 10) = x^2 - 20x + 100$. ❑

EXAMPLE 5 Factor: $6x^2y^4 - 21x^3y^5 + 3x^2y^6$.

Solution

A. We look first for a common factor:

$$6x^2y^4 - 21x^3y^5 + 3x^2y^6 = 3x^2y^4(2 - 7xy + y^2).$$

B. There are three terms in $2 - 7xy + y^2$. Since only y^2 is a square, we do not have a trinomial square. Can the trinomial be factored by trial and error? A key to the answer is that x is only in the term $-7xy$. If the polynomial factored into a form like $(1 - y)(2 - y)$, there would be no x in the middle term. Thus, $2 - 7xy + y^2$ cannot be factored.

C. Have we factored completely? Yes, because no factor with more than one term can be factored further.

D. *Check:* $3x^2y^4(2 - 7xy + y^2) = 6x^2y^4 - 21x^3y^5 + 3x^2y^6$. ❑

EXAMPLE 6 Factor: $(p + q)(x + 2) + (p + q)(x + y)$.

Solution

A. We look for a common factor:

$$(p + q)(x + 2) + (p + q)(x + y) = (p + q)[(x + 2) + (x + y)]$$
$$= (p + q)(2x + y + 2). \quad \text{Collecting like terms}$$

B. There are three terms in $2x + y + 2$, but this trinomial cannot be factored further.

C. No factor with more than one term can be factored further, so we have factored completely.

D. To check, the student can reverse the steps in part (A). ❑

EXAMPLE 7 Factor: $a^6 - 64$.

Solution

A. Look for a common factor. There is none (other than 1 or -1).

B. There are two terms, a difference of squares: $(a^3)^2 - (8)^2$. We factor it:

$$(a^3 + 8)(a^3 - 8).$$

C. One factor is a sum of two cubes, and the other factor is a difference of two cubes. We factor them:

$$(a + 2)(a^2 - 2a + 4)(a - 2)(a^2 + 2a + 4).$$

The factorization is complete because no factor can be factored further.

D. *Check:* $(a + 2)(a^2 - 2a + 4)(a - 2)(a^2 + 2a + 4)$

$= [(a + 2)(a^2 - 2a + 4)][(a - 2)(a^2 + 2a + 4)]$ Using an associative law

$= [a^3 + 2a^2 - 2a^2 - 4a + 4a + 8][a^3 - 2a^2 + 2a^2 - 4a + 4a - 8]$

$= [a^3 + 8][a^3 - 8]$

$= a^6 - 64.$ ❑

EXAMPLE 8

Factor: $6x^2 - 20x - 16$.

Solution

A. Factor out the largest common factor: $2(3x^2 - 10x - 8)$.

B. In the parentheses, there are three terms. The trinomial is not a square. We factor by trial: $2(x - 4)(3x + 2)$.

C. We cannot factor further.

D. *Check:* $2(x - 4)(3x + 2) = 2(3x^2 - 10x - 8) = 6x^2 - 20x - 16.$ ❑

EXAMPLE 9

Factor: $y^2 - 9a^2 + 12y + 36$.

Solution

A. There is no common factor (other than 1 or -1).

B. There are four terms. We try grouping to remove a common binomial factor, but find none. Next, we try grouping as a difference of squares:

$(y^2 + 12y + 36) - 9a^2$ Grouping

$= (y + 6)^2 - (3a)^2$ Factoring the trinomial square

$= (y + 6 + 3a)(y + 6 - 3a).$ Factoring the difference of squares

C. No factor with more than one term can be factored further.

D. *Check:* $(y + 6 + 3a)(y + 6 - 3a) = [(y + 6) + 3a][(y + 6) - 3a]$

$= y^2 + 12y + 36 - 9a^2$

$= y^2 - 9a^2 + 12y + 36.$ ❑

EXAMPLE 10

Factor: $p^2q^2 + 7pq + 12$.

Solution

A. We look first for a common factor. There isn't one.

B. There are three terms. Since the first term is a square, but neither of the other terms is, we do not have a trinomial square. We use trial and error, thinking of the product pq as a single variable. The binomials will then have the following form:

$(pq + \quad)(pq + \quad).$

We factor the last term, 12. All the signs are positive, so we consider only positive factors. Possibilities are 1, 12 and 2, 6 and 3, 4. The pair 3, 4 gives a sum of 7 for the coefficient of the middle term. Thus,

$p^2q^2 + 7pq + 12 = (pq + 3)(pq + 4).$

C. No factor with more than one term can be factored further, so we have factored completely.

D. The check is left to the student. ❑

Checking by Evaluating

Multiplication is but one way of checking a factorization. Another method, which serves as a partial check, is to evaluate the original polynomial and the proposed factorization using the same replacement(s). If the factorization is correct, the polynomial and the factorization will have the same value for any replacement(s) of the variable(s).

EXAMPLE 11 Check the factorization of Example 3, $x^5 - 2x^4 - 35x^3 = x^3(x - 7)(x + 5)$, by evaluating.

Solution We choose a convenient value for x, say 1, and evaluate both expressions:

$$x^5 - 2x^4 - 35x^3 = 1^5 - 2 \cdot 1^4 - 35 \cdot 1^3 = 1 - 2 - 35 = -36;$$

$$x^3(x - 7)(x + 5) = 1^3(1 - 7)(1 + 5) = 1(-6)(6) = -36.$$

Since the value of both expressions is -36, the factorization is probably correct. ❑

Evaluating both a polynomial and its factorization is a quick way to check without multiplying out the factorization. Because checking by evaluating is not foolproof, however, it is a good idea to use this method as only a partial check.

EXERCISE SET 5.6

Factor completely. If a polynomial is prime, state so.

1. $x^2 - 144$

2. $y^2 - 81$

3. $p^2 + 16p + 64$

4. $y^2 - 10y + 25$

5. $2x^2 - 11x + 12$

6. $8y^2 - 18y - 5$

7. $x^3 + 24x^2 + 144x$

8. $x^3 - 18x^2 + 81x$

9. $x^3 + 3x^2 - 4x - 12$

10. $x^3 - 5x^2 - 25x + 125$

11. $9x^2 - 25y^2$

12. $8x^2 - 98y^2$

13. $20x^3 - 4x^2 - 72x$

14. $9x^3 + 12x^2 - 45x$

15. $a^3b - 8b$

16. $5 + 5y^3$

17. $x^4 + 7x^2 - 3x^3 - 21x$

18. $m^4 + 8m^3 + 8m^2 + 64m$

19. $x^5 - 14x^4 + 49x^3$

20. $2x^6 + 8x^5 + 8x^4$

21. $m^6 - 1$

22. $64t^6 - 1$

23. $x^2 + 3x + 1$

24. $x^2 + 5x + 2$

25. $4x^4 - 64$

26. $5x^5 - 80x$

27. $t^2 + 25$

28. $x^2 + 4$

29. $x^5 - 4x^4 + 3x^3$

30. $x^6 - 2x^5 + 7x^4$

31. $x^2 + 6x + 9 - 16y^2$

32. $t^2 + 10t + 25 - p^2$

33. $12n^2 + 24n^3$

34. $ax^2 + ay^2$

35. $9x^2y^2 - 36xy$

36. $x^2y - xy^2$

37. $2\pi rh + 2\pi r^2$

38. $10p^4q^4 + 35p^3q^3 + 10p^2q^2$

39. $(a + b)(x - 3) + (a + b)(x + 4)$

40. $5c(a^3 + b) - (a^3 + b)$

41. $16x^3 + 54y^3$

42. $250a^3 - 54b^3$

43. $n^2 + 2n + np + 2p$

44. $(x - 2)(x + 5) + (x - 2)(x + 8)$

45. $ac + cd - ab - bd$

46. $6y^2 - 3y + 2py - p$

47. $x^2 + y^2 - 2xy$

48. $4b^2 + a^2 - 4ab$

49. $9c^2 + 6cd + d^2$

50. $16x^2 + 24xy + 9y^2$

51. $7p^4 - 7q^4$

52. $4x^2y^2 + 12xyz + 9z^2$

53. $25z^2 + 10zy + y^2$

54. $a^4b^4 - 16$

55. $a^5 + 4a^4b - 5a^3b^2$

56. $4p^2q + pq^2 + 4p^3$

57. $a^2 - ab - 2b^2$

58. $3b^2 - 17ab - 6a^2$

59. $2mn - 360n^2 + m^2$

60. $15 + x^2y^2 + 8xy$

61. $m^2n^2 - 4mn - 32$

62. $p^2q^2 + 7pq + 6$

63. $a^5b^2 + 3a^4b - 10a^3$

64. $m^2n^6 + 4mn^5 - 32n^4$

65. $49m^2 - 112mn + 64n^2$

66. $2s^6t^2 + 10s^3t^3 + 12t^4$

67. $x^6 + x^5y - 2x^4y^2$

68. $a^2 + 2a^2bc + a^2b^2c^2$

69. $36a^2 - 15a + \frac{25}{16}$

70. $\frac{1}{81}x^2 - \frac{8}{27}x + \frac{16}{9}$

71. $\frac{1}{4}a^2 + \frac{1}{3}ab + \frac{1}{9}b^2$

72. $0.01x^2 - 0.1xy + 0.25y^2$

73. $81a^4 - b^4$

74. $1 - n^8$

75. $w^3 - 7w^2 - 4w + 28$

76. $y^3 + 8y^2 - y - 8$

Skill Maintenance

77. Show that the pairs $(-1, 11)$, $(0, 7)$, and $(3, -5)$ are solutions of $y = -4x + 7$.

78. Graph: $y = -\frac{1}{2}x + 4$.

79. Solve $A = aX + bX - 7$ for X.

80. Solve: $4(x - 9) - 2(x + 7) < 14$.

Synthesis

81. ◈ In your own words, describe a strategy that can be used to factor polynomials.

82. ◈ Kelly factored $16 - 8x + x^2$ as $(x - 4)^2$, while Tony factored it as $(4 - x)^2$. Evaluate each expression for several values of x. Then explain why $(x - 4)^2$ and $(4 - x)^2$ are equivalent.

Check that the factorization is most likely correct by evaluating with the values given.

83. $6x^2 - xy - 15y^2 = (2x + 3y)(3x - 5y)$;
$x = 1, y = 1$

84. $6x^2y^2 - 23xy + 20 = (2xy - 5)(3xy - 4)$;
$x = 1, y = -1$

Factor.

85. $18 + y^3 - 9y - 2y^2$

86. $-(x^4 - 7x^2 - 18)$

87. $a^3 + 4a^2 + a + 4$

88. $x^3 + x^2 - (4x + 4)$

89. $x^4 - 7x^2 - 18$

90. $3x^4 - 15x^2 + 12$

91. $x^3 - x^2 - 4x + 4$

92. $y^2(y + 1) - 4y(y + 1) - 21(y + 1)$

93. $y^2(y - 1) - 2y(y - 1) + (y - 1)$

94. $6(x - 1)^2 + 7y(x - 1) - 3y^2$

95. $(y + 4)^2 + 2x(y + 4) + x^2$

96. $2(a + 3)^2 - (a + 3)(b - 2) - (b - 2)^2$

97. Factor $x^{2k} - 2^{2k}$ when $k = 4$.

98. ◈ At most how many factors can a seventh-degree polynomial in x have? Why?

5.7 Solving Quadratic Equations by Factoring

The Principle of Zero Products • Factoring to Solve Equations • Graphing and Quadratic Equations

In this section, we will use the factoring skills we have just learned to solve equations like $x^2 - 8x = -16$ and $x^2 + x - 156 = 0$. Second-degree equations of this type are said to be **quadratic.**

Quadratic Equation

A *quadratic equation* is an equation equivalent to one of the form

$$ax^2 + bx + c = 0, \quad \text{where } a \neq 0.$$

The Principle of Zero Products

The product of two numbers is 0 if one or both of the numbers is 0. Furthermore, *if any product is* 0, *then at least one of the factors must be* 0. For example, if $7x = 0$, then we can conclude that x must be 0. If $x(2x - 9) = 0$, we can conclude that $x = 0$ and/or $2x - 9 = 0$. If $(x + 3)(x - 2) = 0$, we can conclude that $x + 3 = 0$ and/or $x - 2 = 0$. In a product like $ab = 24$, we cannot conclude that either factor must be a specific value.

EXAMPLE 1

Solve: $(x + 3)(x - 2) = 0$.

Solution We have a product of 0. This equation will be true when either factor is 0. Hence it is true when

$$x + 3 = 0 \quad or \quad x - 2 = 0.$$

Here we have two simple equations that we know how to solve:

$$x = -3 \quad or \quad x = 2.$$

Each of the numbers -3 and 2 is a solution of the original equation, as we can see in the following checks.

Check: For -3:

$(x + 3)(x - 2)$	$= 0$
$(-3 + 3)(-3 - 2)$	? 0
$0(-5)$	
0	0 TRUE

For 2:

$(x + 3)(x - 2)$	$= 0$
$(2 + 3)(2 - 2)$	? 0
$5(0)$	
0	0 TRUE

We now have a principle to help in solving quadratic equations.

The Principle of Zero Products

An equation $ab = 0$ is true if and only if $a = 0$ or $b = 0$, or both. (A product is 0 if and only if at least one factor is 0.)

EXAMPLE 2

Solve: $(5x + 1)(x - 7) = 0$.

Solution

$$(5x + 1)(x - 7) = 0$$

$5x + 1 = 0 \quad or \quad x - 7 = 0$ — Using the principle of zero products

$5x = -1 \quad or \quad x = 7$ — Solving the two equations separately

$x = -\frac{1}{5} \quad or \quad x = 7$

Check: For $-\frac{1}{5}$:

$$\begin{array}{r|l} (5x+1)(x-7) & = 0 \\ \hline (5(-\frac{1}{5})+1)(-\frac{1}{5}-7) & ?\ 0 \\ (-1+1)(-7\frac{1}{5}) & \\ 0(-7\frac{1}{5}) & \\ 0 & 0 \quad \text{TRUE} \end{array}$$

For 7:

$$\begin{array}{r|l} (5x+1)(x-7) & = 0 \\ \hline (5(7)+1)(7-7) & ?\ 0 \\ (35+1)0 & \\ 36 \cdot 0 & \\ 0 & 0 \quad \text{TRUE} \end{array}$$

The solutions are $-\frac{1}{5}$ and 7. ❑

The principle of zero products can be used any time a product equals 0—even if a factor has only one term.

EXAMPLE 3

Solve: $x(2x - 9) = 0$.

Solution

$$x(2x-9) = 0$$

$$x = 0 \quad or \quad 2x - 9 = 0 \qquad \text{Using the principle of zero products}$$

$$x = 0 \quad or \quad 2x = 9$$

$$x = 0 \quad or \quad x = \frac{9}{2}$$

The solutions are 0 and $\frac{9}{2}$. The check is left to the student. ❑

Factoring to Solve Equations

By factoring and using the principle of zero products, we can now solve quadratic equations.

EXAMPLE 4

Solve: $x^2 + 5x + 6 = 0$.

Solution This equation differs from those that we solved in Chapter 2. There are no like terms to collect, and we have a squared term. We first factor the polynomial. Then we use the principle of zero products:

$$x^2 + 5x + 6 = 0$$

$$(x+2)(x+3) = 0 \qquad \text{Factoring}$$

$$x + 2 = 0 \quad or \quad x + 3 = 0 \qquad \text{Using the principle of zero products}$$

$$x = -2 \quad or \quad x = -3.$$

Check: For -2:

$$\begin{array}{r|l} x^2+5x+6 & = 0 \\ \hline (-2)^2+5(-2)+6 & ?\ 0 \\ 4-10+6 & \\ -6+6 & \\ 0 & 0 \quad \text{TRUE} \end{array}$$

For -3:

$$\begin{array}{r|l} x^2+5x+6 & = 0 \\ \hline (-3)^2+5(-3)+6 & ?\ 0 \\ 9-15+6 & \\ -6+6 & \\ 0 & 0 \quad \text{TRUE} \end{array}$$

The solutions are -2 and -3. ❑

> CAUTION! We *must* have 0 on one side before using the principle of zero products. Get all nonzero terms on one side and 0 on the other.

EXAMPLE 5

Solve: **(a)** $x^2 - 8x = -16$; **(b)** $x^2 + 5x = 0$; **(c)** $4x^2 = 25$.

Solution

a) We first add 16 to get 0 on one side:

$$x^2 - 8x = -16$$

$$x^2 - 8x + 16 = 0 \qquad \text{Adding 16 on both sides}$$

$$(x - 4)(x - 4) = 0 \qquad \text{Factoring}$$

$$x - 4 = 0 \quad or \quad x - 4 = 0 \qquad \text{Using the principle of zero products}$$

$$x = 4 \quad or \quad x = 4.$$

There is only one solution, 4. The check is left to the student.

b) $x^2 + 5x = 0$

$$x(x + 5) = 0 \qquad \text{Factoring out a common factor}$$

$$x = 0 \quad or \quad x + 5 = 0 \qquad \text{Using the principle of zero products}$$

$$x = 0 \quad or \quad x = -5$$

The solutions are 0 and -5. The check is left to the student.

c)

$$4x^2 = 25$$

$$4x^2 - 25 = 0 \qquad \text{Subtracting 25 on both sides to get 0 on one side}$$

$$(2x - 5)(2x + 5) = 0 \qquad \text{Factoring a difference of squares}$$

$$2x - 5 = 0 \quad or \quad 2x + 5 = 0$$

$$2x = 5 \quad or \quad 2x = -5$$

$$x = \tfrac{5}{2} \quad or \quad x = -\tfrac{5}{2}$$

The solutions are $\frac{5}{2}$ and $-\frac{5}{2}$. The check is left to the student. ❑

EXAMPLE 6

Solve: $(x + 3)(2x - 1) = 9$.

Solution Be careful with an equation like this! Remember that since we must have 0 on one side, we multiply out the product on the left and then subtract 9 from both sides.

$$(x + 3)(2x - 1) = 9$$

$$2x^2 + 5x - 3 = 9 \qquad \text{Multiplying on the left}$$

$$2x^2 + 5x - 3 - 9 = 9 - 9 \qquad \text{Subtracting 9 on both sides}$$

$$2x^2 + 5x - 12 = 0$$

$$(2x - 3)(x + 4) = 0 \qquad \text{Factoring}$$

$$2x - 3 = 0 \quad or \quad x + 4 = 0 \qquad \text{Using the principle of zero products}$$

$$2x = 3 \quad or \quad x = -4$$

$$x = \tfrac{3}{2} \quad or \quad x = -4$$

Check: For $\frac{3}{2}$:

$(x + 3)(2x - 1)$	$= 9$
$(\frac{3}{2} + 3)(2 \cdot \frac{3}{2} - 1)$	? 9
$(\frac{9}{2})(2)$	
9	9 TRUE

For -4:

$(x + 3)(2x - 1)$	$= 9$
$(-4 + 3)(2(-4) - 1)$	? 9
$(-1)(-9)$	
9	9 TRUE

The solutions are $\frac{3}{2}$ and -4. ❑

Graphing and Quadratic Equations

In Chapter 3, we graphed linear equations of the form $Ax + By = C$ and $y = mx + b$. Recall that to find the x-intercept, we replaced y with 0 and solved for x. This same procedure can be used to find the x-intercepts when an equation of the form $y = ax^2 + bx + c$ $(a \neq 0)$ is graphed. Equations like this are graphed in Chapter 11. Their graphs are shaped like the following curves:

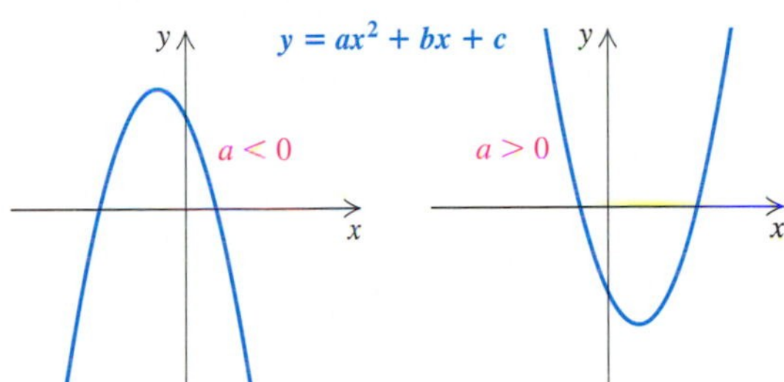

A grapher can help us solve quadratic equations by zooming in on any x-intercepts that may exist. This technique works whether the equation is factorable or not. As an example, consider the quadratic equation $y = x^2 - 3x - 5$. Graphing this equation in a $[-10, 10, -10, 10]$ window gives the following graph:

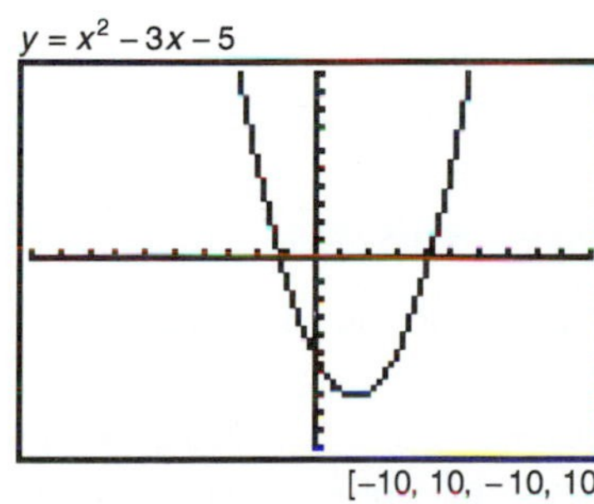

There appears to be an x-intercept between -2 and -1 and another one between 4 and 5. Let's first examine the negative intercept using a $[-2, -1, -1, 1]$ window or the Zoom feature:

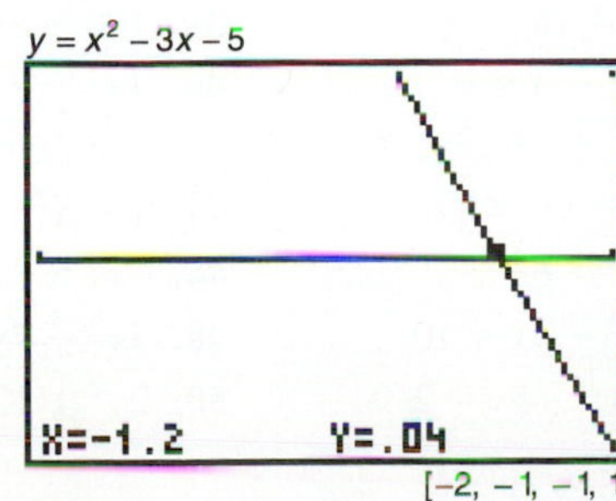

Using the Trace feature, we can see that the intercept is close to -1.2. If we change the window dimensions to $[-1.22, -1.18, -0.1, 0.1]$ or zoom in again, it becomes clear that the intercept is very close to -1.19:

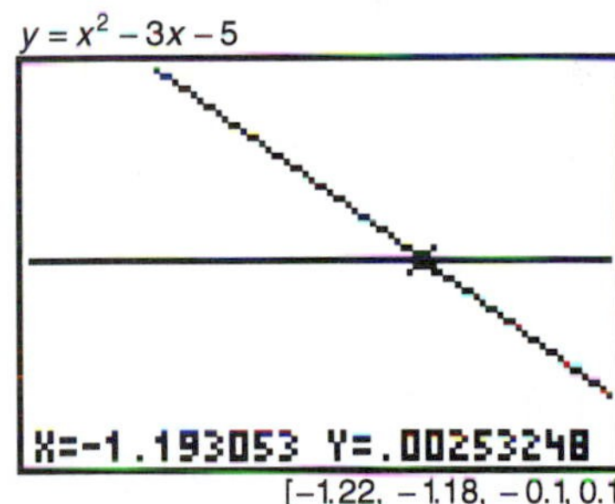

By similarly zooming in on the positive intercept, we can conclude that the other solution of $x^2 - 3x - 5 = 0$ is close to 4.19.

Use a grapher to find the solutions, if they exist, accurate to two decimal places.

TC1. $x^2 + 4x - 3 = 0$

TC2. $x^2 - 5x - 2 = 0$

TC3. $x^2 + 13.54x + 40.95 = 0$

TC4. $x^2 - 4.43x + 6.32 = 0$

EXAMPLE 7 Find the x-intercepts for the graph of the equation shown.

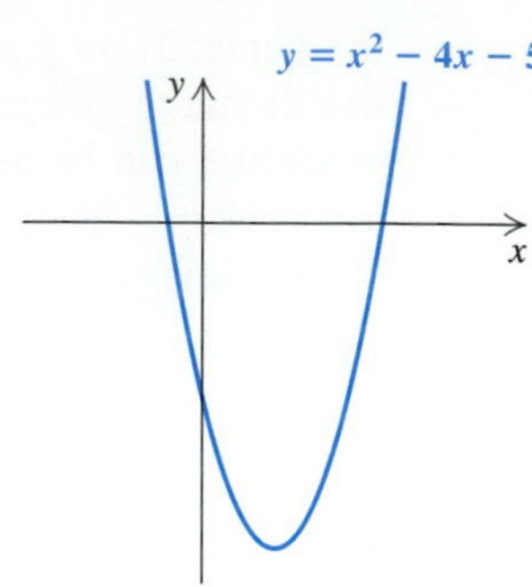

Solution To find the x-intercepts, we let $y = 0$ and solve for x:

$$\begin{aligned} 0 &= x^2 - 4x - 5 && \text{Substituting 0 for } y \\ 0 &= (x-5)(x+1) && \text{Factoring} \\ x - 5 = 0 \quad &or \quad x + 1 = 0 && \text{Using the principle of zero products} \\ x = 5 \quad &or \quad x = -1. && \text{Solving for } x \end{aligned}$$

The x-intercepts are $(5, 0)$ and $(-1, 0)$. ❑

EXERCISE SET 5.7

Solve using the principle of zero products.

1. $(x+8)(x+6) = 0$
2. $(x+3)(x+2) = 0$
3. $(x-3)(x+5) = 0$
4. $(x+9)(x-3) = 0$
5. $(x+12)(x-11) = 0$
6. $(x-13)(x+53) = 0$
7. $x(x+5) = 0$
8. $y(y+7) = 0$
9. $0 = y(y+10)$
10. $0 = x(x-21)$
11. $(2x+5)(x+4) = 0$
12. $(2x+9)(x+8) = 0$
13. $(5x+1)(4x-12) = 0$
14. $(4x+9)(14x-7) = 0$
15. $(7x-28)(28x-7) = 0$
16. $(12x-11)(8x-5) = 0$
17. $2x(3x-2) = 0$
18. $75x(8x-9) = 0$
19. $\frac{1}{2}x(\frac{2}{3}x - 12) = 0$
20. $\frac{5}{7}x(\frac{3}{4}x - 6) = 0$
21. $(\frac{1}{5} + 2x)(\frac{1}{9} - 3x) = 0$
22. $(\frac{7}{4}x - \frac{1}{12})(\frac{2}{3}x - \frac{12}{11}) = 0$
23. $(0.3x - 0.1)(0.05x - 1) = 0$
24. $(0.1x - 0.3)(0.4x - 20) = 0$
25. $9x(3x-2)(2x-1) = 0$
26. $(x-5)(x+55)(5x-1) = 0$

Solve by factoring and using the principle of zero products.

27. $x^2 + 6x + 5 = 0$
28. $x^2 + 7x + 6 = 0$
29. $x^2 + 7x - 18 = 0$
30. $x^2 + 4x - 21 = 0$
31. $x^2 - 8x + 15 = 0$
32. $x^2 - 9x + 14 = 0$
33. $x^2 - 8x = 0$
34. $x^2 - 3x = 0$
35. $x^2 + 19x = 0$
36. $x^2 + 12x = 0$
37. $x^2 = 16$
38. $100 = x^2$
39. $9x^2 - 4 = 0$
40. $4x^2 - 9 = 0$
41. $0 = 6x + x^2 + 9$
42. $0 = 25 + x^2 + 10x$
43. $x^2 + 16 = 8x$
44. $1 + x^2 = 2x$
45. $5x^2 = 6x$
46. $7x^2 = 8x$
47. $6x^2 - 4x = 10$
48. $3x^2 - 7x = 20$
49. $12y^2 - 5y = 2$
50. $2y^2 + 12y = -10$

51. $x(x - 5) = 14$

52. $t(3t + 1) = 2$

53. $64m^2 - 25 = 56$

54. $100t^2 - 9 = 40$

55. $3x^2 + 8x = 9 + 2x$

56. $x^2 - 5x = 18 + 2x$

57. $(3x + 5)(x + 3) = 7$

58. $(5x + 4)(x - 1) = 2$

Find the x-intercepts for the graph of each equation.

59. $y = x^2 - x - 6$

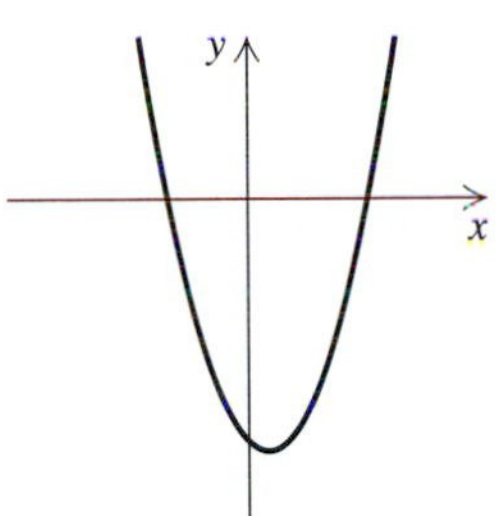

60. $y = x^2 + 3x - 4$

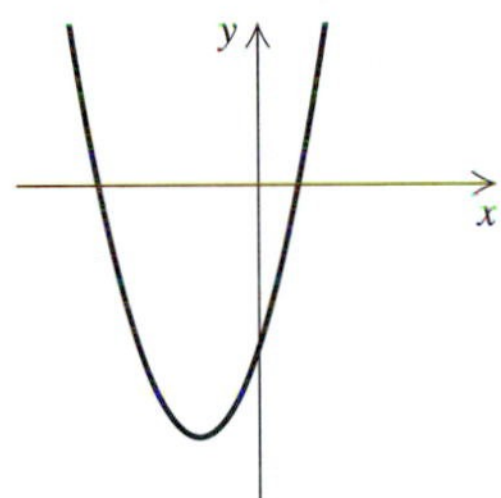

61. $y = x^2 + 2x - 8$

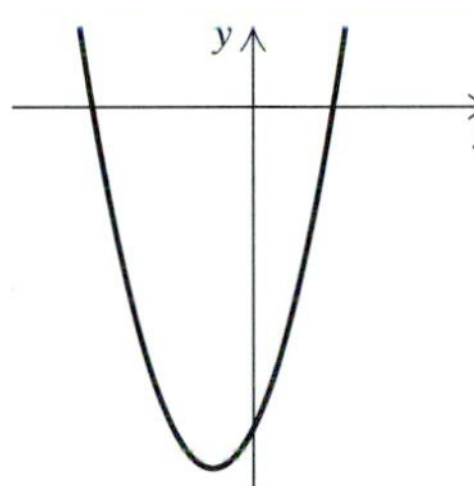

62. $y = x^2 - 2x - 15$

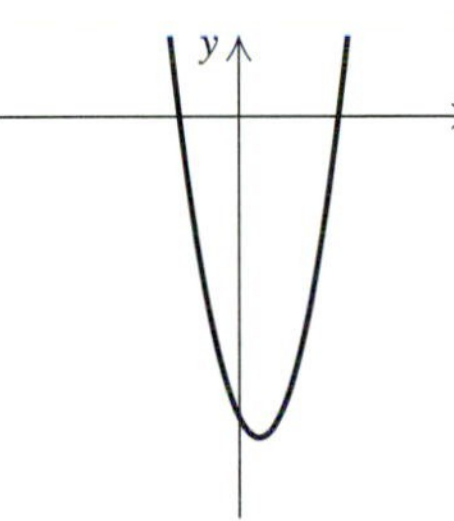

63. $y = 2x^2 + 3x - 9$

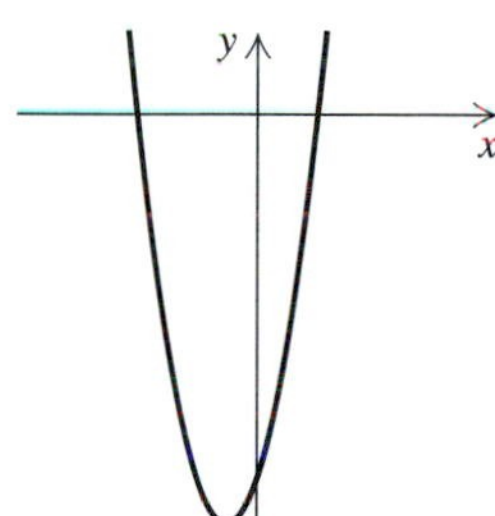

64. $y = 2x^2 + x - 10$

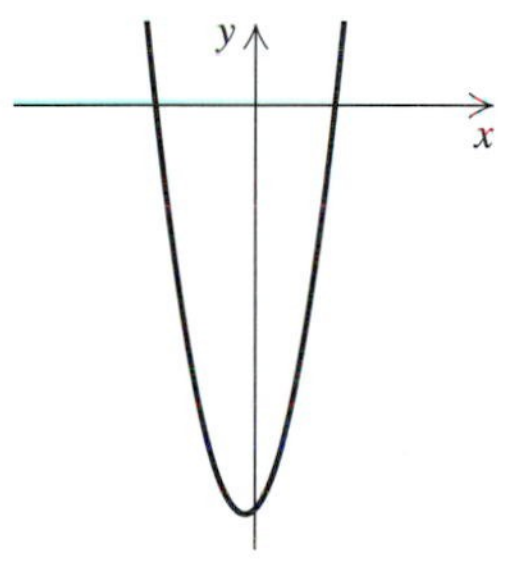

Skill Maintenance

Translate to an algebraic expression.

65. The square of the sum of a and b

66. The sum of the squares of a and b

Translate to an inequality.

67. 5 more than twice a number is less than 19.

68. 7 less than half of a number exceeds 24.

Synthesis

69. ◈ What is the difference between a trinomial and a quadratic equation?

70. ◈ The equation $x^2 + 1$ has no real-number solutions. What implications does this have for the graph of $y = x^2 + 1$?

71. ◈ What is wrong with solving $x^2 = 3x$ by dividing both sides of the equation by x?

72. ◈ When the principle of zero products is used to solve an equation, will there always be two solutions? Why or why not?

73. Find an equation with integer coefficients that has the given numbers as solutions. For example, 3 and -2 are solutions to $x^2 - x - 6 = 0$.

a) $-3, 4$ **b)** $-3, -4$ **c)** $\frac{1}{2}, \frac{1}{2}$
d) $5, -5$ **e)** $0, 0.1, \frac{1}{4}$

Solve.

74. $b(b + 9) = 4(5 + 2b)$

75. $y(y + 8) = 16(y - 1)$

76. $(t - 5)^2 = 2(5 - t)$

77. $x^2 - \frac{1}{64} = 0$

78. $x^2 - \frac{25}{36} = 0$

79. $\frac{5}{16}x^2 = 5$

80. $\frac{27}{25}x^2 = \frac{1}{3}$

81. For each equation on the left, find an equivalent equation on the right.

a) $3x^2 - 4x + 8 = 0$ — $4x^2 + 8x + 36 = 0$
b) $(x - 6)(x + 3) = 0$ — $(2x + 8)(2x - 5) = 0$
c) $x^2 + 2x + 9 = 0$ — $9x^2 - 12x + 24 = 0$
d) $(2x - 5)(x + 4) = 0$ — $(x + 1)(5x - 5) = 0$
e) $5x^2 - 5 = 0$ — $x^2 - 3x - 18 = 0$
f) $x^2 + 10x - 2 = 0$ — $2x^2 + 20x - 4 = 0$

82. ◈ Explain how to construct an equation that has seven solutions.

83. ◈ Explain how the graph in Exercise 59 can be used to visualize the solutions of

$$x^2 - x - 6 = -4.$$

Use a grapher to find the solutions of the equation, accurate to two decimal places.

84. $x^2 - 1.80x - 5.69 = 0$

85. $x^2 + 9.10x + 15.77 = 0$

86. $-x^2 + 0.63x + 0.22 = 0$

87. $x^2 + 13.74x + 42.00 = 0$

88. $6.4x^2 - 8.45x - 94.06 = 0$

89. $-0.25x^2 - 2.50x - 5.48 = 0$

5.8 Problem Solving

Applications • The Pythagorean Theorem

We can use our five-step problem-solving process and our new methods for solving quadratic equations to solve problems.

EXAMPLE 1

One more than a number times one less than the number is 8. Find all such numbers.

Solution

1. **FAMILIARIZE.** Let's make some guesses. Try 5. One more than 5 is 6. One less than the number is 4. The product of one more than the number and one less than the number is $6 \cdot 4$, or 24, which is too large. Let's try 3. One more than 3 is 4. One less than 3 is 2. The product of these numbers is $4 \cdot 2$, or 8. This checks and we have guessed one of the desired numbers. Are there more? We use our algebra skills to find out. Let $x =$ a number that satisfies the requirements of the problem.

2. **TRANSLATE.** From the familiarization, we can translate as follows:

$$\underbrace{\text{One more than a number}}_{(x+1)} \quad \underbrace{\text{times}}_{\cdot} \quad \underbrace{\text{one less than that number}}_{(x-1)} \quad \underbrace{\text{is}}_{=} \quad \underbrace{8.}_{8}$$

3. **CARRY OUT.** We solve the equation as follows:

$$(x+1)(x-1) = 8$$
$$x^2 - 1 = 8 \qquad \text{Multiplying}$$
$$x^2 - 1 - 8 = 0 \qquad \text{Subtracting 8 to get 0 on one side}$$
$$x^2 - 9 = 0$$
$$(x-3)(x+3) = 0 \qquad \text{Factoring}$$
$$x - 3 = 0 \quad or \quad x + 3 = 0 \qquad \text{Using the principle of zero products}$$
$$x = 3 \quad or \quad x = -3.$$

4. **CHECK.** We already guessed and checked one of the solutions, 3, in the *Familiarize* step. To check -3, note that one more than -3 is -2 and one less than -3 is -4. The product of -2 and -4 is 8. Thus, -3 also checks.

5. **STATE.** There are two such numbers, 3 and -3. ❑

EXAMPLE 2

A vacant square-shaped lot is being turned into a community garden. Because a path 2 m wide is needed at one end, only 48 m^2 of the lot will be garden space. Find the dimensions of the lot.

Solution

1. **FAMILIARIZE.** We first make a drawing. Recall that the area of any rectangle, including a square, is length · width. We let $x =$ the length (or width) of the square lot. Note that the path is a rectangle 2 m wide and x m long.

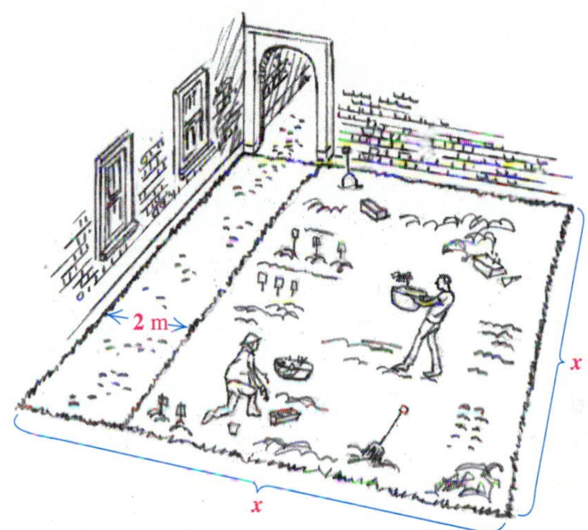

2. **TRANSLATE.** It helps to reword this problem before translating:

Reword:	The area of the lot	minus	the area of the path	is	48 m^2.
	↓	↓	↓	↓	↓
Translate:	x^2	$-$	$2 \cdot x$	$=$	48

3. **CARRY OUT.** We solve the equation as follows:

$$x^2 - 2x = 48$$

$$x^2 - 2x - 48 = 0 \qquad \text{Subtracting 48 to get 0 on one side}$$

$$(x - 8)(x + 6) = 0 \qquad \text{Factoring}$$

$$x - 8 = 0 \quad or \quad x + 6 = 0 \qquad \text{Using the principle of zero products}$$

$$x = 8 \quad or \quad x = -6.$$

4. **CHECK.** Since measurements cannot be negative, we must disregard -6 as a solution of the original problem. If $x = 8$, the area of the lot is $8 \cdot 8 = 64$ m^2 and the area of the path is $2 \cdot 8 = 16$ m^2, leaving $64 - 16 = 48$ m^2 for the garden. Thus, 8 checks. Another approach to this problem is to express the garden's area as $x(x - 2)$ and set this equal to 48. This provides a second check, since $8(8 - 2) = 8 \cdot 6 = 48$.

5. **STATE.** The lot is 8 m long and 8 m wide. ❑

EXAMPLE 3

The height of a triangular sail is 7 ft more than the base. The area of the triangle is 30 ft^2. Find the height and the base.

Solution

1. **FAMILIARIZE.** We first make a drawing. The formula for the area of a triangle is Area $= \frac{1}{2} \cdot (\text{base}) \cdot (\text{height})$. We let $b =$ the length, in feet, of the triangle's base.

2. **TRANSLATE.** We reword the problem and translate:

Reword:	$\frac{1}{2}$	times	the base	times	the base plus 7	is	30.
	↓	↓	↓	↓	↓	↓	↓
Translate:	$\frac{1}{2}$	$\cdot$	b	$\cdot$	$(b + 7)$	$=$	30

3. **CARRY OUT.** We solve the equation as follows:

$$\frac{1}{2} \cdot b \cdot (b + 7) = 30$$

$$\frac{1}{2}(b^2 + 7b) = 30 \qquad \text{Multiplying}$$

$$b^2 + 7b = 60 \qquad \text{Multiplying by 2 to clear fractions}$$

$$b^2 + 7b - 60 = 0 \qquad \text{Subtracting 60 to get 0 on one side}$$

$$(b + 12)(b - 5) = 0 \qquad \text{Factoring}$$

$$b + 12 = 0 \quad \text{or} \quad b - 5 = 0 \qquad \text{Using the principle of zero products}$$

$$b = -12 \quad \text{or} \quad b = 5.$$

4. **CHECK.** The base of a triangle cannot have a negative length, so -12 cannot be a solution. Suppose the base is 5 ft. Then the height is 7 ft more than the base, so the height is 12 ft and the area is $\frac{1}{2}(5)(12)$, or 30 ft^2. These numbers check in the original problem.

5. **STATE.** The height is 12 ft and the base is 5 ft. ❑

EXAMPLE 4

In a sports league of n teams in which each team plays every other team twice, the total number N of games to be played is given by

$$n^2 - n = N.$$

If a basketball league plays a total of 240 games, how many teams are in the league?

Solution

1. FAMILIARIZE. To familiarize yourself with this equation, reread Example 9 in Section 4.2, where we first considered it.

2. TRANSLATE. We are trying to find the number of teams n in a league in which 240 games are played. We substitute 240 for N in order to solve for n:

$$n^2 - n = 240. \quad \text{Substituting 240 for } N$$

3. CARRY OUT. We solve the equation as follows:

$$\begin{aligned} n^2 - n &= 240 \\ n^2 - n - 240 &= 0 && \text{Subtracting 240 to get 0 on one side} \\ (n - 16)(n + 15) &= 0 && \text{Factoring} \\ n - 16 = 0 \quad &or \quad n + 15 = 0 && \text{Using the principle of zero products} \\ n = 16 \quad &or \quad n = -15. \end{aligned}$$

4. CHECK. The solutions of the equation are 16 and -15. Since the number of teams cannot be negative, -15 cannot be a solution. However, 16 checks, since $16^2 - 16 = 256 - 16 = 240$.

5. STATE. There are 16 teams in the league. ❑

EXAMPLE 5

The product of the page numbers on two consecutive pages of a book is 156. Find the page numbers.

Solution

1. FAMILIARIZE. Recall that consecutive page numbers are one apart, like 49 and 50. Let x = the first page number; then $x + 1$ = the next page number.

2. TRANSLATE. We reword the problem before translating:

Reword:	The first page number	times	the next page number	is	156.
	↓	↓	↓	↓	↓
Translate:	x	$\cdot$	$(x + 1)$	$=$	156

3. CARRY OUT. We solve the equation as follows:

$$\begin{aligned} x(x + 1) &= 156 \\ x^2 + x &= 156 && \text{Multiplying} \\ x^2 + x - 156 &= 0 && \text{Subtracting 156 to get 0 on one side} \\ (x - 12)(x + 13) &= 0 && \text{Factoring} \\ x - 12 = 0 \quad &or \quad x + 13 = 0 && \text{Using the principle of zero products} \\ x = 12 \quad &or \quad x = -13. \end{aligned}$$

4. CHECK. The solutions of the equation are 12 and -13. Since page numbers cannot be negative, -13 can't be a solution. On the other hand, if x is 12, then $x + 1$ is 13 and $12 \cdot 13 = 156$. Thus, 12 checks.

5. STATE. The pair of page numbers is 12 and 13. ❑

The following problem involves the Pythagorean theorem, which relates the lengths of the sides of a right triangle. A **right triangle** has a 90° angle. The side opposite the 90° angle is called the **hypotenuse.** The other sides are called **legs.**

The Pythagorean Theorem

The sum of the squares of the legs of a right triangle is equal to the square of the hypotenuse:

$$a^2 + b^2 = c^2.$$

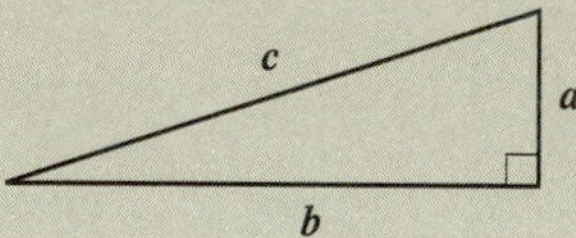

EXAMPLE 6

The length of one leg of a right triangle is 7 ft longer than the other. The length of the hypotenuse is 13 ft. Find the lengths of the legs.

Solution

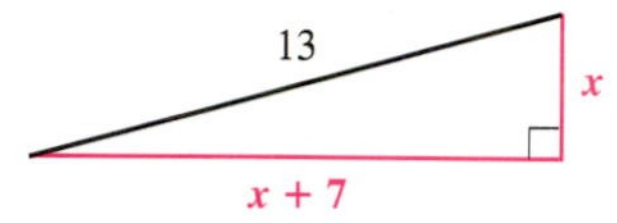

1. FAMILIARIZE. We make a drawing and let x = the length of one leg. Since the other leg is 7 ft longer, we know that $x + 7$ = the length of the other leg. The hypotenuse has length 13 ft.

2. TRANSLATE. Applying the Pythagorean theorem, we obtain the following translation:

$$a^2 + b^2 = c^2$$
$$x^2 + (x + 7)^2 = 13^2. \quad \text{Substituting}$$

3. CARRY OUT. We solve the equation as follows:

$$x^2 + (x^2 + 14x + 49) = 169 \quad \text{Squaring the binomial and 13}$$
$$2x^2 + 14x + 49 = 169 \quad \text{Collecting like terms}$$
$$2x^2 + 14x - 120 = 0 \quad \text{Subtracting 169 to get 0 on side}$$
$$2(x^2 + 7x - 60) = 0 \quad \text{Factoring out a common factor}$$
$$2(x + 12)(x - 5) = 0 \quad \text{Factoring}$$
$$x + 12 = 0 \quad or \quad x - 5 = 0 \quad \text{Using the principle of zero products}$$
$$x = -12 \quad or \quad x = 5.$$

4. CHECK. The integer -12 cannot be a length of a side because it is negative. When $x = 5$, $x + 7 = 12$, and $5^2 + 12^2 = 13^2$. So 5 checks.

5. STATE. The lengths of the legs are 5 ft and 12 ft. ❑

EXERCISE SET 5.8

Solve.

1. If you subtract a number from four times its square, the result is 3. Find all such numbers.
2. If 7 is added to the square of a number, the result is 32. Find all such numbers.
3. Eight more than the square of a number is six times the number. Find all such numbers.
4. Fifteen more than the square of a number is eight times the number. Find all such numbers.

5. The product of the page numbers on two facing pages of a book is 210. Find the page numbers.

6. The product of the page numbers on two facing pages of a book is 110. Find the page numbers.

7. The product of two consecutive even integers is 168. Find the integers.

8. The product of two consecutive even integers is 224. Find the integers.

9. The product of two consecutive odd integers is 255. Find the integers.

10. The product of two consecutive odd integers is 143. Find the integers.

11. The length of a rectangular garden is 4 m greater than the width. The area of the rectangle is 96 m^2. Find the length and the width.

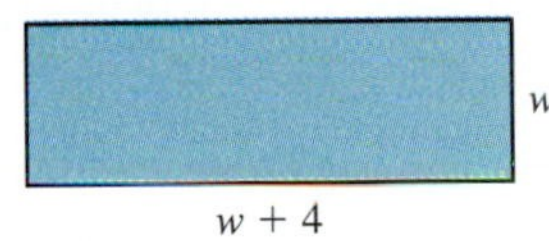

12. The length of a rectangular calculator is 5 cm greater than the width. The area of the rectangle is 84 cm^2. Find the length and the width.

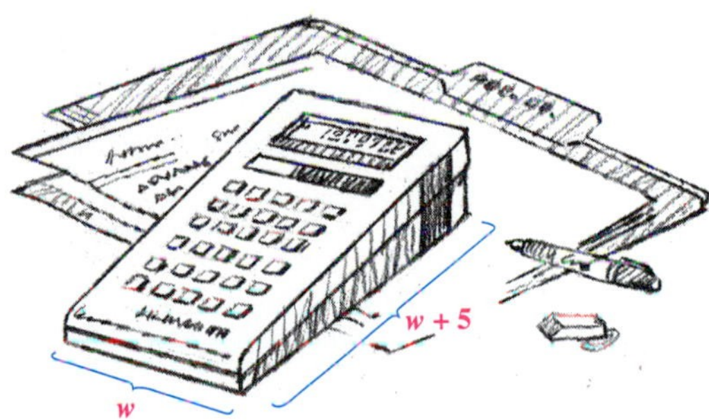

13. The area of a square bookcase is 5 more than the perimeter. Find the length of a side.

14. The perimeter of a square porch is 3 more than the area. Find the length of a side.

15. The base of a triangle is 10 cm greater than the height. The area is 28 cm^2. Find the height and the base.

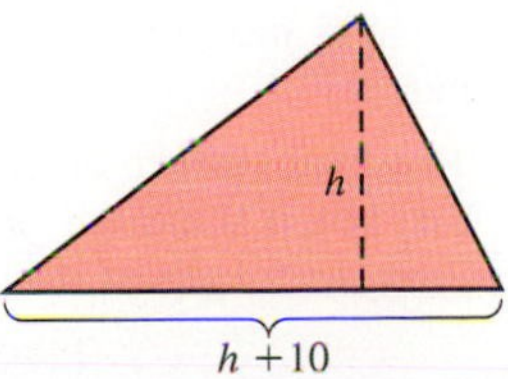

16. The height of a triangle is 8 m less than the base. The area is 10 m^2. Find the height and the base.

17. If the sides of a square are lengthened by 3 m, the area becomes 81 m^2. Find the length of a side of the original square.

18. If the sides of a square are lengthened by 7 km, the area becomes 121 km^2. Find the length of a side of the original square.

19. The sum of the squares of two consecutive odd whole numbers is 74. Find the numbers.

20. The sum of the squares of two consecutive odd whole numbers is 130. Find the numbers.

Use the formula from Example 4, $n^2 - n = N$, for Exercises 21–24.

21. A women's volleyball league has 23 teams. What is the total number of games to be played?

22. A chess league has 14 teams. What is the total number of games to be played?

23. A women's softball league plays a total of 132 games. How many teams are in the league?

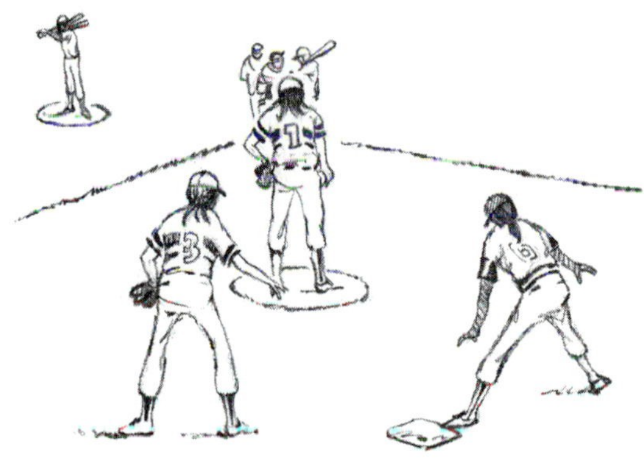

24. A basketball league plays a total of 90 games. How many teams are in the league?

The number of possible handshakes within a group of n people is given by $N = \frac{1}{2}(n^2 - n)$.

25. At a meeting, there are 40 people. How many handshakes are possible?

26. At a party, there are 100 people. How many handshakes are possible?

27. During a toast at a party, there were 190 "clicks" of glasses. How many people took part in the toast?

28. After winning the championship, all members of a team exchanged "high fives." Altogether, there were 300 high fives. How many people were on the team?

29. The length of one leg of a right triangle is 8 ft. The length of the hypotenuse is 2 ft longer than the other leg. Find the lengths of the hypotenuse and the other leg.

30. The length of one leg of a right triangle is 24 ft. The length of the other leg is 16 ft shorter than the hypotenuse. Find the lengths of the hypotenuse and the other leg.

31. An architect has allocated a rectangular space of 264 ft^2 for a square dining room and a 10-ft wide kitchen. Find the dimensions of each room.

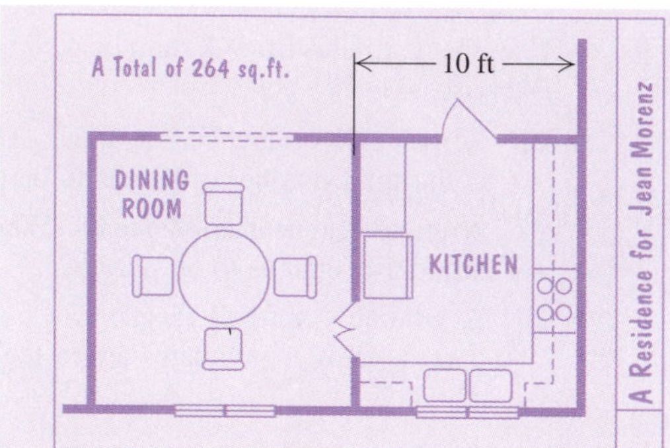

32. A television antenna's guy wire is 1 m longer than the height of the antenna. If the guy wire is anchored 3 m from the foot of the antenna, how tall is the antenna?

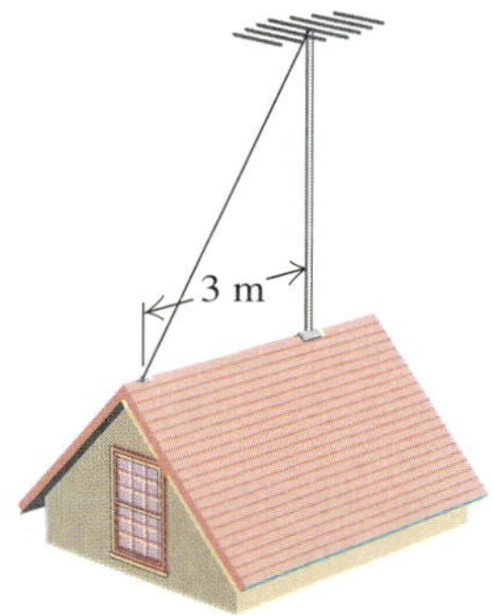

Skill Maintenance

Graph.

33. $y = -\frac{2}{3}x + 1$

34. $y = \frac{3}{5}x - 1$

Simplify.

35. $7x^0$ when $x = -4$

36. $\frac{m^7n^9}{mn^3}$

Synthesis

37. ◈ Write a problem in which a quadratic equation must be solved in order to solve the problem.

38. ◈ Write a problem for a classmate to solve so that only one of two solutions of a quadratic equation can be used as an answer.

39. A cement walk of uniform width is built around a 20-ft by 40-ft rectangular pool. The total area of the pool and the walk is 1500 ft^2. Find the width of the walk.

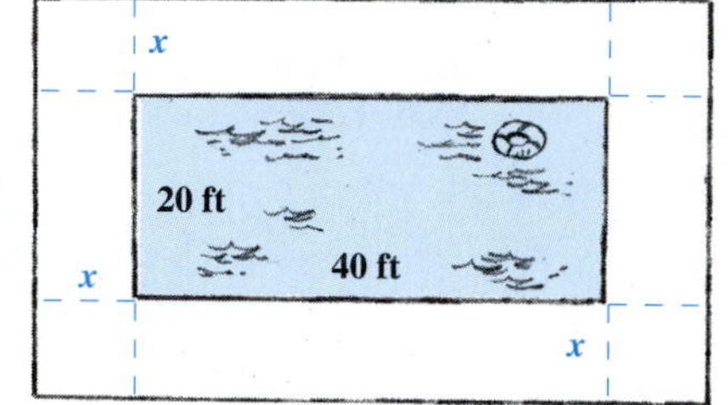

40. A model rocket is launched with an initial velocity of 180 ft/sec. Its height h, in feet, after t seconds is given by the formula $h = 180t - 16t^2$.

a) After how many seconds will the rocket first reach a height of 464 ft?

b) How many seconds after first reaching a height of 464 ft will it be at that height again?

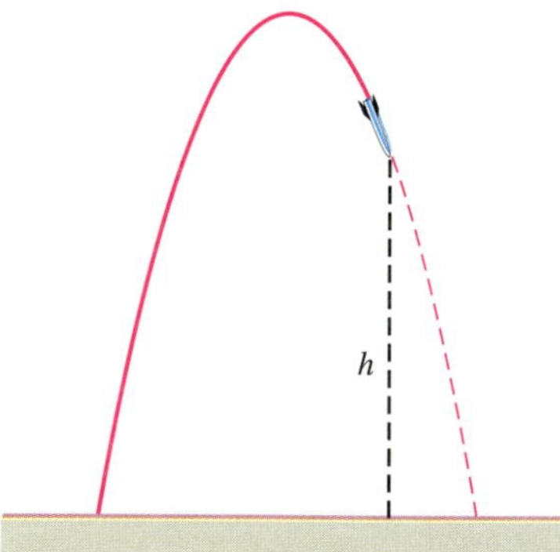

41. The one's digit of a number less than 100 is four greater than the ten's digit. The sum of the number and the product of the digits is 58. Find the number.

42. The total surface area of a closed box is 350 m^2. The box is 9 m high and has a square base and lid. Find the length of the side of the base.

43. A rectangular piece of cardboard is twice as long as it is wide. A 4-cm square is cut out of each corner, and the sides are turned up to make a box with an open top. The volume of the box is 616 cm^3. Find the original dimensions of the cardboard.

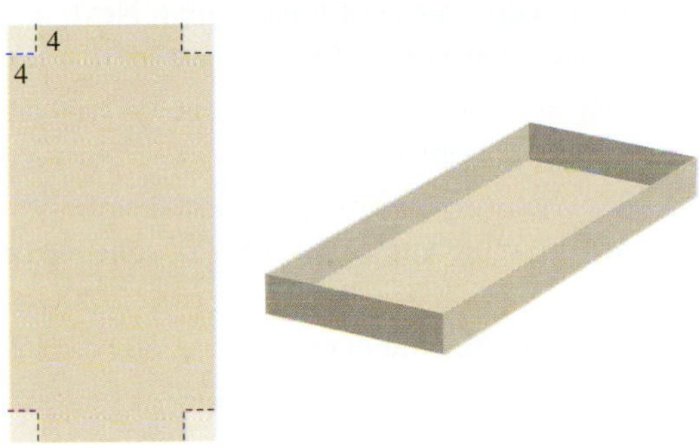

44. An open rectangular gutter is made by turning up the sides of a piece of metal 20 in. wide. The area of the cross-section of the gutter is 50 in^2. Find the depth of the gutter.

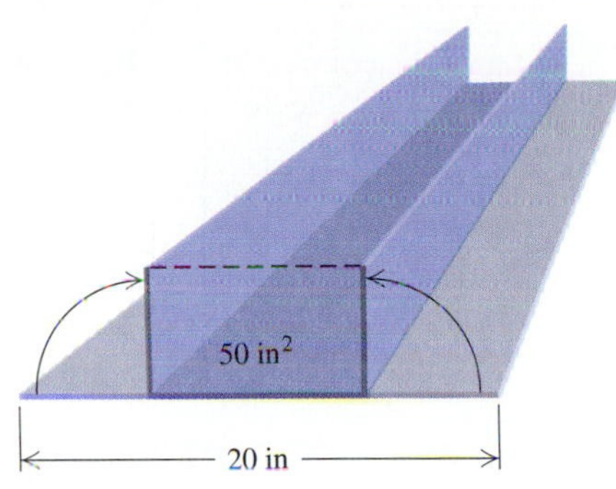

45. The length of each side of a square is increased by 5 cm to form a new square. The area of the new square is $2\frac{1}{4}$ times the area of the original square. Find the area of each square.

SUMMARY AND REVIEW 5

KEY TERMS

Factor, p. 214
Common factor, p. 215
Factoring by grouping, p. 217
Leading coefficient, p. 218
Prime polynomial, p. 222
Factor completely, p. 222
Trinomial square, p. 232
Difference of squares, p. 234
Sum or difference of cubes, p. 239
Quadratic equation, p. 247
Right triangle, p. 257
Hypotenuse, p. 257
Leg, p. 257

IMPORTANT PROPERTIES AND FORMULAS

Factoring Formulas

$$A^2 + 2AB + B^2 = (A + B)^2,$$
$$A^2 - 2AB + B^2 = (A - B)^2,$$
$$A^2 - B^2 = (A + B)(A - B),$$
$$A^3 + B^3 = (A + B)(A^2 - AB + B^2),$$
$$A^3 - B^3 = (A - B)(A^2 + AB + B^2)$$

To factor a polynomial:

A. Always factor out the largest common factor. Be sure to include it in your final answer.

B. Look at the number of terms.

Two terms: Try factoring as a difference of squares first. Next, try factoring as a sum or a difference of cubes. Do *not* try to factor a *sum* of squares.

Three terms: Try factoring as a trinomial square. Next, try trial and error, using the FOIL method or the grouping method.

Four or more terms: Try factoring by grouping and factoring out a common binomial factor. Next, try grouping into a difference of squares, one of which is a trinomial.

C. Always *factor completely*. If a factor with more than one term can itself be factored further, do so.

D. Check by multiplying.

The Principle of Zero Products: $ab = 0$ is true if and only if $a = 0$ or $b = 0$, or both.

The Pythagorean theorem: $a^2 + b^2 = c^2$

REVIEW EXERCISES

This chapter's review and test include Skill Maintenance exercises from Sections 1.6, 2.7, 3.2, and 4.1.

Find three factorizations of the monomial.

1. $-10x^2$

2. $36x^5$

Factor completely. If a polynomial is prime, state so.

3. $x^2 - 5x$

4. $12y^4 - 3y^2$

5. $9x^2 - 4$

6. $x^2 + 4x - 12$

7. $x^2 + 14x + 49$

8. $x^3 + 2x^2 - 9x - 18$

9. $x^3 + x^2 + 3x + 3$

10. $6x^2 - 5x + 1$

11. $x^4 - 81$

12. $9x^3 + 12x^2 - 45x$

13. $2x^2 - 50$

14. $x^4 + 4x^3 - 2x - 8$

15. $16x^4 - 1$

16. $8x^6 - 32x^5 + 4x^4$

17. $75 + 12x^2 + 60x$

18. $x^2 + 9$

19. $x^3 - x^2 - 30x$

20. $1 + a^3$

21. $9x^2 + 25 - 30x$

22. $6x^2 - 28x - 48$

23. $x^2 - 6x + 9$

24. $2x^2 - 7x - 4$

25. $18x^2 - 12x + 2$

26. $27x^3 - 8$

27. $54x^6y - 2y$

28. $25x^2 - 20x + 4$

29. $x^2y^2 + xy - 12$

30. $12a^2 + 84ab + 147b^2$

31. $a^2 - 2ab + b^2 - 4t^2$

32. $32x^4 - 128y^4z^4$

Solve.

33. $(x - 1)(x + 3) = 0$

34. $x^2 + 2x - 35 = 0$

35. $x^2 + x - 12 = 0$

36. $6b^2 + 6 = 13b$

37. $8y^2 + 5 = 14y$

38. $16 = x(x - 6)$

39. The square of a number is 6 more than the number. Find the number.

40. Find the x-intercepts for the graph of $y = 2x^2 - 3x - 5$.

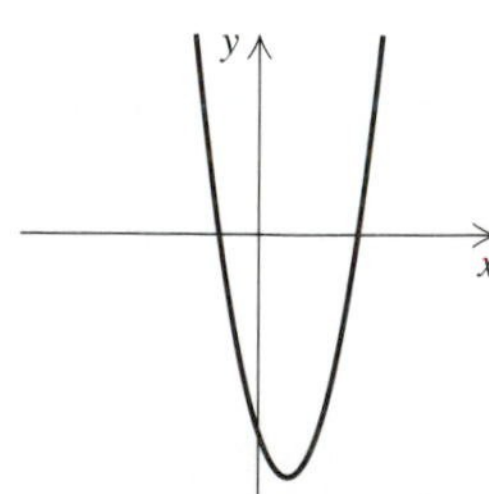

41. The product of two consecutive odd integers is 323. Find the integers.

42. The area of a square is 5 more than 4 times the length of a side. What is the length of a side of the square?

Skill Maintenance

43. Subtract: $\frac{2}{5} - \left(-\frac{1}{10}\right)$.

44. Graph: $y = -\frac{3}{4}x$.

45. Divide and simplify: $\dfrac{m^3n^{10}}{mn^5}$.

46. Translate to an inequality: 2 less than half a number is no more than 10.

Synthesis

47. ◈ Compare the two methods of checking factorizations that were discussed.

48. ◈ How do the equations solved in this chapter differ from those solved in previous chapters?

Solve.

49. The pages of a book measure 15 cm by 20 cm. Margins of equal width surround the printing on each page and constitute one half of the area of the page. Find the width of the margins.

15 cm

20 cm

50. The cube of a number is the same as twice the square of the number. Find the number.

51. The length of a rectangle is two times its width. When the length is increased by 20 and the width is decreased by 1, the area is 160. Find the original length and width.

Solve.

52. $x^2 + 25 = 0$

53. $(x - 2)(x + 3)(2x - 5) = 0$

CHAPTER TEST 5

1. Find three factorizations of $4x^3$.

Factor completely.

2. $x^2 - 7x + 10$

3. $x^2 + 25 - 10x$

4. $6y^2 - 8y^3 + 4y^4$

5. $x^3 + x^2 + 2x + 2$

6. $x^2 - 5x$

7. $x^3 + 2x^2 - 3x$

8. $28x - 48 + 10x^2$

9. $4x^2 - 25$

10. $x^2 - x - 12$

11. $24m^3 + 40m^2 + 3m$

12. $3w^2 - 75$

13. $16a^7b + 54ab^7$

14. $3x^4 - 48$

15. $49x^2 - 84x + 36$

16. $5x^2 - 26x + 5$

17. $y^2 + 8y + 16 - 100t^2$

18. $3r^3 - 3$

19. $x^3 + 5x^2 - 4x - 20$

20. $6t^3 + 9t^2 - 15t$

21. $3m^2 - 9mn - 30n^2$

Solve.

22. $x^2 - x - 20 = 0$

23. $2x^2 + 7x = 15$

24. $x(x - 3) = 28$

25. The square of a number is 24 more than five times the number. Find the number.

26. A photograph is 3 cm longer than it is wide. Its area is 40 cm^2. Find its length and its width.

Skill Maintenance

27. Simplify: $-2.8 - 3.5 + 4.2 - (-1.7)$.

28. The width of a rectangle must be 8 cm. For what lengths will the area be less than 104 cm^2? Use an inequality.

29. Graph: $y = \frac{3}{4}x + 1$.

30. Simplify: $(-3xy^5)^3$.

Synthesis

31. The length of a rectangle is five times its width. When the length is decreased by 3 and the width is increased by 2, the area of the new rectangle is 60. Find the original length and width.

32. Factor: $(a + 3)^2 - 2(a + 3) - 35$.

CHAPTER 6

Rational Expressions and Equations

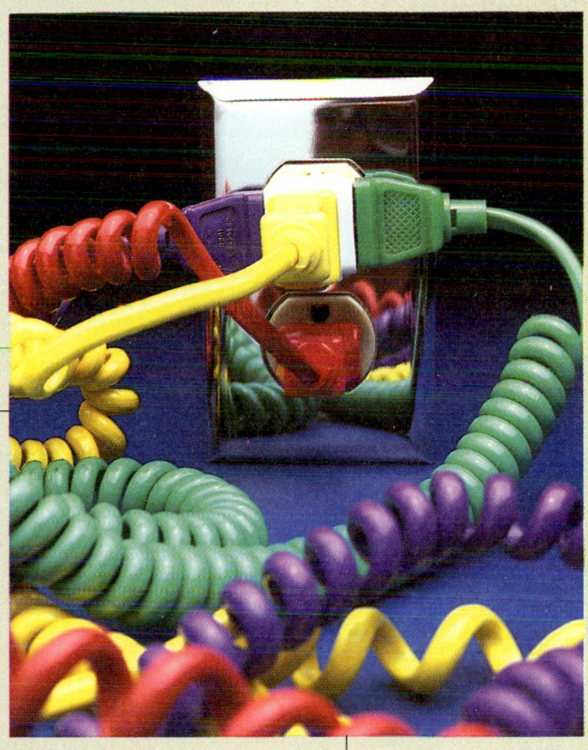

AN APPLICATION

The formula

$$\frac{1}{R}=\frac{1}{r_1}+\frac{1}{r_2}$$

gives the resistance R of two resistors r_1 and r_2 connected in parallel. Solve for r_2.

This problem appears in Example 5 of Section 6.9.

Michael Holzhausen
ELECTRICIAN

"All electrician training programs are rich in math, and provide an important background for electrical work. Electricians regularly use math when laying out work, bending conduit, sizing wire, and more."

Just as fractional notation helps us to solve arithmetic problems, rational expressions similar to those in the following pages will help us solve algebra problems. We now learn how to simplify, as well as add, subtract, multiply, and divide, rational expressions. These skills are important for solving problems like the one on the preceding page.

In addition to the material from this chapter, the review and test for Chapter 6 include material from Sections 2.1, 3.3, 4.3, and 5.2.

6.1 Rational Expressions

Simplifying Rational Expressions • Factors That Are Opposites

Whereas a rational number is a quotient of two integers, a **rational expression** is a quotient of two polynomials. The following are rational expressions:

$$\frac{7}{3}, \quad \frac{5}{x+6}, \quad \frac{t^2-5t+6}{4t^2-7}.$$

Rational expressions are examples of what are sometimes called *algebraic fractions* or *fractional expressions*.

Because rational expressions indicate division, we must be careful to avoid denominators that are 0. When a variable is replaced by a number that produces a denominator of 0, the rational expression is undefined. For example, in the expression

$$\frac{x+3}{x-7},$$

when x is replaced with 7, the denominator is 0, and the expression is undefined:

$$\frac{x+3}{x-7} = \frac{7+3}{7-7} = \frac{10}{0} \leftarrow \text{Undefined.}$$

When x is replaced with a number other than 7, like 6, the expression *is* defined because the denominator is nonzero:

$$\frac{x+3}{x-7} = \frac{6+3}{6-7} = \frac{9}{-1} = -9.$$

EXAMPLE 1

Find all numbers for which the rational expression $\dfrac{x+4}{x^2-3x-10}$ is undefined.

Solution To determine which numbers make the rational expression undefined, we set the denominator equal to 0 and solve:

$$\begin{aligned} x^2 - 3x - 10 &= 0 \\ (x-5)(x+2) &= 0 && \text{Factoring} \\ x-5 = 0 \quad &or \quad x+2 = 0 && \text{Using the principle of zero products} \\ x = 5 \quad &or \quad x = -2. \end{aligned}$$

Check:

For $x = 5$:

$$\frac{x+4}{x^2-3x-10} = \frac{5+4}{5^2-3\cdot5-10} = \frac{9}{25-15-10} = \frac{9}{0},$$

which is undefined.

For $x = -2$:

$$\frac{x+4}{x^2-3x-10} = \frac{-2+4}{(-2)^2-3(-2)-10} = \frac{2}{4+6-10} = \frac{2}{0},$$

which is undefined.

Thus, $\dfrac{x+4}{x^2-3x-10}$ is undefined for $x = 5$ and $x = -2$. □

Simplifying Rational Expressions

Simplifying rational expressions is similar to simplifying the fractional expressions studied in Section 1.3. We saw, for example, that an expression like $\frac{15}{40}$ could be simplified as follows:

$\dfrac{15}{40} = \dfrac{3\cdot5}{8\cdot5}$ Factoring the numerator and the denominator. Note the common factor of 5.

$= \dfrac{3}{8}\cdot\dfrac{5}{5}$ Rewriting as a product of two fractions

$= \dfrac{3}{8}\cdot 1$ $\dfrac{5}{5} = 1$

$= \dfrac{3}{8}.$ Using the identity property of 1. We call this "removing a factor of 1."

The same steps are followed when simplifying rational expressions: We factor and then remove a factor of 1.

EXAMPLE 2

Simplify: $\dfrac{8x^2}{24x}$.

Solution

$\dfrac{8x^2}{24x} = \dfrac{8\cdot x\cdot x}{3\cdot8\cdot x}$ Factoring the numerator and the denominator. Note the common factor of $8\cdot x$.

$= \dfrac{x}{3}\cdot\dfrac{8x}{8x}$ Rewriting as a product of two rational expressions

$= \dfrac{x}{3}\cdot 1$ $\dfrac{8x}{8x} = 1$

$= \dfrac{x}{3}$ Removing a factor of 1 □

When two or more terms appear in a numerator or a denominator, we factor as we did in Chapter 5. Then we try to remove a factor of 1.

EXAMPLE 3

Simplify: **(a)** $\dfrac{5a + 15}{10}$; **(b)** $\dfrac{6x + 12}{7x + 14}$; **(c)** $\dfrac{6a^2 + 4a}{2a^2 + 2a}$; **(d)** $\dfrac{x^2 + 3x + 2}{x^2 - 1}$.

Solution

a) $\dfrac{5a + 15}{10} = \dfrac{5(a + 3)}{5 \cdot 2}$ Factoring the numerator and the denominator

$= \dfrac{5}{5} \cdot \dfrac{a + 3}{2}$ Rewriting as a product of two rational expressions

$= 1 \cdot \dfrac{a + 3}{2}$

$= \dfrac{a + 3}{2}$ Removing a factor of 1: $\dfrac{5}{5} = 1$

b) $\dfrac{6x + 12}{7x + 14} = \dfrac{6(x + 2)}{7(x + 2)}$ Factoring the numerator and the denominator

$= \dfrac{6}{7} \cdot \dfrac{x + 2}{x + 2}$ Rewriting as a product of two rational expressions

$= \dfrac{6}{7} \cdot 1$

$= \dfrac{6}{7}$ Removing a factor of 1: $\dfrac{x + 2}{x + 2} = 1$

c) $\dfrac{6a^2 + 4a}{2a^2 + 2a} = \dfrac{2a(3a + 2)}{2a(a + 1)}$ Factoring the numerator and the denominator

$= \dfrac{2a}{2a} \cdot \dfrac{3a + 2}{a + 1}$ Rewriting as a product of two rational expressions

$= 1 \cdot \dfrac{3a + 2}{a + 1}$ $\dfrac{2a}{2a} = 1$

$= \dfrac{3a + 2}{a + 1}$ Removing a factor of 1. Note in this step that you *cannot* remove the a's because they are not factors of the entire numerator and the entire denominator.

d) $\dfrac{x^2 + 3x + 2}{x^2 - 1} = \dfrac{(x + 2)(x + 1)}{(x + 1)(x - 1)}$ Factoring

$= \dfrac{x + 1}{x + 1} \cdot \dfrac{x + 2}{x - 1}$ Rewriting as a product of two rational expressions

$= 1 \cdot \dfrac{x + 2}{x - 1}$

$= \dfrac{x + 2}{x - 1}$ ❑

Canceling

Canceling is a shortcut that can be used—and easily *misused*—when working with rational expressions. As we stated in Section 1.3, canceling must be done with care and understanding. Essentially, canceling streamlines the steps in which a rational expression is simplified by removing a factor of 1. Example 3(d) could have been done faster as follows:

$$\frac{x^2+3x+2}{x^2-1} = \frac{(x+2)(x+1)}{(x+1)(x-1)}$$

When a factor of 1 is noted, it is "canceled": $\frac{x+1}{x+1} = 1$.

$$= \frac{x+2}{x-1}.$$

Simplifying

CAUTION! Canceling is often used incorrectly. The following cancellations are *incorrect*:

$$\frac{x+2}{x+3}; \quad \frac{a^2-5}{5}; \quad \frac{6x^2+5x+1}{4x^2-3x}.$$

Wrong! Wrong! Wrong!

None of the above cancellations represents removing a factor of 1. Factors are parts of products. For example, in $x \cdot 2$, x and 2 are factors, but in $x + 2$, x and 2 are *not* factors. If you can't factor, you can't cancel! When in doubt, don't cancel!

EXAMPLE 4

Simplify: $\frac{3x^2-2x-1}{x^2-3x+2}$.

Solution We factor the numerator and the denominator and look for common factors:

$$\frac{3x^2-2x-1}{x^2-3x+2} = \frac{(3x+1)(x-1)}{(x-2)(x-1)}$$

Try to visualize this as $\frac{3x+1}{x-2} \cdot \frac{x-1}{x-1}$.

$$= \frac{3x+1}{x-2}.$$

Removing a factor of 1: $\frac{x-1}{x-1} = 1$ ❑

Checking

When a rational expression is simplified, the result is an equivalent expression. Example 3(a) says that

$$\frac{5a+15}{10} \text{ is equivalent to } \frac{a+3}{2}.$$

This result can be partially checked using a value of a for which both expressions are defined. For instance, if $a = 2$, then

$$\frac{5a+15}{10} = \frac{5\cdot 2+15}{10} = \frac{25}{10} = \frac{5}{2} \quad \text{and} \quad \frac{a+3}{2} = \frac{2+3}{2} = \frac{5}{2}.$$

Had we evaluated both expressions and obtained differing results, we would have known that a mistake had been made.

For instance, had $(5a + 15)/10$ been incorrectly simplified as $(a + 15)/2$ and we had evaluated using $a = 2$, we would have found that

$$\frac{5a+15}{10} = \frac{5\cdot 2+15}{10} = \frac{5}{2}, \quad \text{whereas} \quad \frac{a+15}{2} = \frac{2+15}{2} = \frac{17}{2}.$$

Factors That Are Opposites

Consider

$$\frac{x-4}{4-x}.$$

At first glance the numerator and the denominator do not appear to have any common factors other than 1. But $x - 4$ and $4 - x$ are opposites, or additive inverses, of each other. Thus we can find a common factor by factoring out -1 in one expression.

EXAMPLE 5

Simplify: $\dfrac{x-4}{4-x}$.

Solution

$$\frac{x-4}{4-x} = \frac{x-4}{-1(-4+x)} \qquad \text{Factoring out } -1 \text{ in the denominator}$$

$$= \frac{x-4}{-1(x-4)} \qquad -4 + x = x + (-4) = x - 4$$

$$= \frac{1}{-1} \cdot \frac{x-4}{x-4} \qquad \text{Rewriting as a product. It helps to write the 1 in the numerator.}$$

$$= -1$$

As a partial check, note that for any choice of x other than 4, the value of the rational expression is -1. For instance, if $x = 6$, then

$$\frac{x-4}{4-x} = \frac{6-4}{4-6} = \frac{2}{-2}$$

$$= -1.$$

EXERCISE SET 6.1

List all numbers for which the rational expression is undefined.

1. $\dfrac{-5}{2x}$
2. $\dfrac{14}{-5y}$
3. $\dfrac{a+7}{a-8}$
4. $\dfrac{a-8}{a+7}$
5. $\dfrac{3}{2y+5}$
6. $\dfrac{x^2-9}{4x-12}$
7. $\dfrac{x^2+11}{x^2-3x-28}$
8. $\dfrac{p^2-9}{p^2-7p+10}$
9. $\dfrac{m^3-2m}{m^2-25}$
10. $\dfrac{7-3x+x^2}{49-x^2}$

Simplify by removing a factor of 1. Show all steps.

11. $\dfrac{10a^3b}{30ab^2}$
12. $\dfrac{45x^3y^2}{9x^5y}$
13. $\dfrac{35x^2y}{14x^3y^5}$
14. $\dfrac{12a^5b^6}{18a^3b}$
15. $\dfrac{9x+15}{6x+10}$
16. $\dfrac{14x-7}{10x-5}$
17. $\dfrac{a^2-25}{a^2+6a+5}$
18. $\dfrac{a^2+5a+6}{a^2-9}$

Simplify, if possible. Then check by evaluating.

19. $\dfrac{48x^4}{18x^6}$
20. $\dfrac{76a^5}{24a^3}$

21. $\dfrac{4x - 12}{4x}$

22. $\dfrac{-2y + 6}{-4y}$

23. $\dfrac{3m^2 + 3m}{6m^2 + 9m}$

24. $\dfrac{4y^2 - 2y}{5y^2 - 5y}$

25. $\dfrac{a^2 - 9}{a^2 + 5a + 6}$

26. $\dfrac{t^2 - 25}{t^2 + t - 20}$

27. $\dfrac{2t^2 + 6t + 4}{4t^2 - 12t - 16}$

28. $\dfrac{3a^2 - 9a - 12}{6a^2 + 30a + 24}$

29. $\dfrac{x^2 - 25}{x^2 - 10x + 25}$

30. $\dfrac{x^2 + 8x + 16}{x^2 - 16}$

31. $\dfrac{a^2 - 1}{a - 1}$

32. $\dfrac{t^2 - 1}{t + 1}$

33. $\dfrac{x^2 + 1}{x + 1}$

34. $\dfrac{y^2 + 4}{y + 2}$

35. $\dfrac{6x^2 - 54}{4x^2 - 36}$

36. $\dfrac{8x^2 - 32}{4x^2 - 16}$

37. $\dfrac{6t + 12}{t^2 - t - 6}$

38. $\dfrac{5y + 5}{y^2 + 7y + 6}$

39. $\dfrac{a^2 - 10a + 21}{a^2 - 11a + 28}$

40. $\dfrac{y^2 - 3y - 18}{y^2 - 2y - 15}$

41. $\dfrac{t^2 - 4}{(t + 2)^2}$

42. $\dfrac{(a - 3)^2}{a^2 - 9}$

43. $\dfrac{6 - x}{x - 6}$

44. $\dfrac{x - 8}{8 - x}$

45. $\dfrac{a - b}{b - a}$

46. $\dfrac{q - p}{-p + q}$

47. $\dfrac{6t - 12}{2 - t}$

48. $\dfrac{5a - 15}{3 - a}$

49. $\dfrac{a^2 - 1}{1 - a}$

50. $\dfrac{a^2 - b^2}{b^2 - a^2}$

Skill Maintenance

Factor.

51. $x^2 + 8x + 7$

52. $x^2 - 9x + 14$

Find the intercepts. Then graph.

53. $5x + 2y = 20$

54. $2x - 4y = 8$

Synthesis

55. ◈ How is canceling related to the identity property of 1?

56. ◈ Explain why evaluating is not a foolproof check when simplifying rational expressions.

Simplify.

57. $\dfrac{x^4 - 16y^4}{(x^2 + 4y^2)(x - 2y)}$

58. $\dfrac{(a - b)^2}{b^2 - a^2}$

59. $\dfrac{(t^4 - 1)(t^2 - 9)(t - 9)^2}{(t^4 - 81)(t^2 + 1)(t + 1)^2}$

60. $\dfrac{(t + 2)^3(t^2 + 2t + 1)(t + 1)}{(t + 1)^3(t^2 + 4t + 4)(t + 2)}$

61. $\dfrac{(x^2 - y^2)(x^2 - 2xy + y^2)}{(x - y)^2(x^2 - 4xy - 5y^2)}$

62. $\dfrac{(x - 1)(x^4 - 1)(x^2 - 1)}{(x^2 + 1)(x - 1)^2(x^4 - 2x^2 + 1)}$

63. ◈ Select any number x, multiply by 2, add 5, multiply by 5, subtract 25, and divide by 10. What do you get? Explain how this procedure can be used for a number trick.

6.2 Multiplication and Division

Multiplication • **Division**

Multiplication and division of rational expressions is similar to multiplication and division with fractional notation.

Multiplication

Recall that to multiply fractions, we simply multiply their numerators and multiply their denominators. Rational expressions are multiplied the same way.

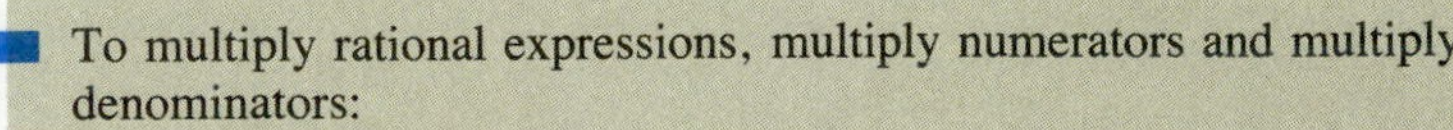

To multiply rational expressions, multiply numerators and multiply denominators:

$$\frac{A}{B} \cdot \frac{C}{D} = \frac{AC}{BD}.$$

Then factor and simplify the result if possible.

For example,

$$\frac{3}{5} \cdot \frac{8}{11} = \frac{24}{55}$$

and

$$\frac{x}{3} \cdot \frac{x+2}{y} = \frac{x(x+2)}{3y}.$$

Fraction bars are grouping symbols, so parentheses are needed when writing some products. Because we normally simplify, it is best to leave parentheses in the product. There is no need to multiply further.

EXAMPLE 1 Multiply and simplify.

a) $\dfrac{5a^3}{4} \cdot \dfrac{2}{5a}$ **b)** $\dfrac{x^2+6x+9}{x^2-4} \cdot \dfrac{x-2}{x+3}$

c) $\dfrac{x^2+x-2}{15} \cdot \dfrac{5}{2x^2-3x+1}$

Solution

a) $\dfrac{5a^3}{4} \cdot \dfrac{2}{5a} = \dfrac{5a^3(2)}{4(5a)}$ Multiplying the numerators and the denominators

$= \dfrac{2 \cdot 5 \cdot a \cdot a \cdot a}{2 \cdot 2 \cdot 5 \cdot a}$ Factoring the numerator and the denominator

$= \dfrac{\cancel{2} \cdot \cancel{5} \cdot \cancel{a} \cdot a \cdot a}{\cancel{2} \cdot 2 \cdot \cancel{5} \cdot \cancel{a}}$

$= \dfrac{a^2}{2}$ Removing a factor of 1: $\dfrac{2 \cdot 5 \cdot a}{2 \cdot 5 \cdot a} = 1$

b) $\dfrac{x^2+6x+9}{x^2-4} \cdot \dfrac{x-2}{x+3} = \dfrac{(x^2+6x+9)(x-2)}{(x^2-4)(x+3)}$ Multiplying the numerators and the denominators

$= \dfrac{(x+3)(x+3)(x-2)}{(x+2)(x-2)(x+3)}$ Factoring the numerator and the denominator

$= \dfrac{\cancel{(x+3)}(x+3)\cancel{(x-2)}}{(x+2)\cancel{(x-2)}\cancel{(x+3)}}$

$= \dfrac{x+3}{x+2}$ Removing a factor of 1: $\dfrac{(x+3)(x-2)}{(x+3)(x-2)} = 1$

c) $\frac{x^2+x-2}{15} \cdot \frac{5}{2x^2-3x+1} = \frac{(x^2+x-2)5}{15(2x^2-3x+1)}$ Multiplying the numerators and the denominators

$= \frac{(x+2)(x-1)5}{5(3)(x-1)(2x-1)}$ Factoring the numerator and the denominator. Try to go directly to this step.

$= \frac{(x+2)\cancel{(x-1)}\cancel{5}}{\cancel{5}(3)\cancel{(x-1)}(2x-1)}$

$= \frac{x+2}{3(2x-1)}$

Removing a factor of 1: $\frac{5(x-1)}{5(x-1)} = 1$

You need not carry out this multiplication.

Division

As with fractions, reciprocals of rational expressions are found by interchanging the numerator and the denominator. For example,

the reciprocal of $\frac{2}{7}$ is $\frac{7}{2}$, and the reciprocal of $\frac{3x}{x+5}$ is $\frac{x+5}{3x}$.

To divide by a rational expression, multiply by its reciprocal:

$$\frac{A}{B} \div \frac{C}{D} = \frac{A}{B} \cdot \frac{D}{C} = \frac{AD}{BC}.$$

Then factor and simplify if possible.

EXAMPLE 2

Divide: **(a)** $\frac{x}{5} \div \frac{7}{y}$; **(b)** $(x+1) \div \frac{x-1}{x+3}$.

Solution

a) $\frac{x}{5} \div \frac{7}{y} = \frac{x}{5} \cdot \frac{y}{7}$ Multiplying by the reciprocal of the divisor

$= \frac{xy}{35}$ Multiplying rational expressions

b) $(x+1) \div \frac{x-1}{x+3} = \frac{x+1}{1} \cdot \frac{x+3}{x-1}$ Multiplying by the reciprocal of the divisor. Writing $x+1$ as $\frac{x+1}{1}$ can be helpful.

$= \frac{(x+1)(x+3)}{x-1}$

As usual, we should simplify when possible.

EXAMPLE 3 Divide and simplify.

a) $\dfrac{x+1}{x^2-1} \div \dfrac{x+1}{x^2-2x+1}$

b) $\dfrac{a^2+3a+2}{a^2+4} \div (5a^2+10a)$

c) $\dfrac{x^2-2x-3}{x^2-4} \div \dfrac{x+1}{x+5}$

Solution

a)
$$\frac{x+1}{x^2-1} \div \frac{x+1}{x^2-2x+1} = \frac{x+1}{x^2-1} \cdot \frac{x^2-2x+1}{x+1}$$
Multiplying by the reciprocal

$$= \frac{(x+1)(x-1)(x-1)}{(x+1)(x-1)(x+1)}$$
Multiplying rational expressions and factoring numerators and denominators

$$= \frac{\cancel{(x+1)}\cancel{(x-1)}(x-1)}{\cancel{(x+1)}\cancel{(x-1)}(x+1)}$$
$$= \frac{x-1}{x+1}$$
Removing a factor of 1: $\dfrac{(x+1)(x-1)}{(x+1)(x-1)} = 1$

b)
$$\frac{a^2+3a+2}{a^2+4} \div (5a^2+10a) = \frac{a^2+3a+2}{a^2+4} \cdot \frac{1}{5a^2+10a}$$
Multiplying by the reciprocal

$$= \frac{(a+2)(a+1)}{(a^2+4)5a(a+2)}$$
Multiplying rational expressions and factoring

$$= \frac{\cancel{(a+2)}(a+1)}{(a^2+4)5a\cancel{(a+2)}}$$
$$= \frac{a+1}{(a^2+4)5a}$$
Removing a factor of 1: $\dfrac{a+2}{a+2} = 1$

c)
$$\frac{x^2-2x-3}{x^2-4} \div \frac{x+1}{x+5} = \frac{x^2-2x-3}{x^2-4} \cdot \frac{x+5}{x+1}$$
Multiplying by the reciprocal

$$= \frac{(x-3)(x+1)(x+5)}{(x-2)(x+2)(x+1)}$$
Multiplying rational expressions and factoring

$$= \frac{(x-3)\cancel{(x+1)}(x+5)}{(x-2)(x+2)\cancel{(x+1)}}$$
$$= \frac{(x-3)(x+5)}{(x-2)(x+2)}$$
Removing a factor of 1: $\dfrac{x+1}{x+1} = 1$

❑

EXERCISE SET 6.2

Multiply. Leave parentheses in the product.

1. $\dfrac{3x}{2} \cdot \dfrac{x+4}{x-1}$

2. $\dfrac{4x}{5} \cdot \dfrac{x-3}{x+2}$

3. $\dfrac{x-1}{x+2} \cdot \dfrac{x+1}{x+2}$

4. $\dfrac{x-2}{x-5} \cdot \dfrac{x-2}{x+5}$

5. $\frac{2x+3}{4}\cdot\frac{x+1}{x-5}$

6. $\frac{-5}{3x-4}\cdot\frac{-6}{5x+6}$

7. $\frac{a-5}{a^2+1}\cdot\frac{a+2}{a^2-1}$

8. $\frac{t+3}{t^2-2}\cdot\frac{t+3}{t^2-2}$

9. $\frac{x+1}{2+x}\cdot\frac{x-1}{x+1}$

10. $\frac{m^2+5}{m+8}\cdot\frac{m^2-4}{m^2-4}$

Multiply and simplify, if possible.

11. $\frac{4x^3}{3x}\cdot\frac{14}{x}$

12. $\frac{32}{b^4}\cdot\frac{3b^2}{8}$

13. $\frac{3c}{d^2}\cdot\frac{4d}{6c^3}$

14. $\frac{3x^2y}{2}\cdot\frac{4}{xy^3}$

15. $\frac{x^2-3x-10}{(x-2)^2}\cdot\frac{x-2}{x-5}$

16. $\frac{t^2}{t^2-4}\cdot\frac{t^2-5t+6}{t^2-3t}$

17. $\frac{a^2-25}{a^2-4a+3}\cdot\frac{2a-5}{2a+5}$

18. $\frac{x+3}{x^2+9}\cdot\frac{x^2+5x+4}{x+9}$

19. $\frac{a^2-9}{a^2}\cdot\frac{a^2-3a}{a^2+a-12}$

20. $\frac{x^2+10x-11}{x^2-1}\cdot\frac{x+1}{x+11}$

21. $\frac{4a^2}{3a^2-12a+12}\cdot\frac{3a-6}{2a}$

22. $\frac{5v+5}{v-2}\cdot\frac{v^2-4v+4}{v^2-1}$

23. $\frac{t^2+2t-3}{t^2+4t-5}\cdot\frac{t^2-3t-10}{t^2+5t+6}$

24. $\frac{x^2+5x+4}{x^2-6x+8}\cdot\frac{x^2+5x-14}{x^2+8x+7}$

25. $\frac{5a^2-180}{10a^2-10}\cdot\frac{20a+20}{2a-12}$

26. $\frac{2t^2-98}{4t^2-4}\cdot\frac{8t+8}{16t-112}$

27. $\frac{x^2-1}{x^2-9}\cdot\frac{(x-3)^4}{(x+1)^2}$

28. $\frac{(x+2)^5}{(x-1)^3}\cdot\frac{x^2-1}{x^2+5x+6}$

29. $\frac{a^2-4}{a^2+2a+1}\cdot\frac{a-1}{a^4+1}$

30. $\frac{a^2+4}{a^2-6a+9}\cdot\frac{a^2+6a+9}{a^4+16}$

31. $\frac{(t-2)^3}{(t-1)^3}\cdot\frac{t^2-2t+1}{t^2-4t+4}$

32. $\frac{(y+4)^3}{(y+2)^3}\cdot\frac{y^2+4y+4}{y^2+8y+16}$

Find the reciprocal.

33. $\frac{4}{x}$

34. $\frac{a+3}{a-1}$

35. x^2-y^2

36. $\frac{1}{a+b}$

37. $\frac{x^2+2x-5}{x^2-4x+7}$

38. $\frac{x^2-3xy+y^2}{x^2+7xy-y^2}$

Divide and simplify, if possible.

39. $\frac{2}{5}\div\frac{4}{3}$

40. $\frac{5}{6}\div\frac{2}{3}$

41. $\frac{2}{x}\div\frac{8}{x}$

42. $\frac{x}{2}\div\frac{3}{x}$

43. $\frac{x^2}{y}\div\frac{x^3}{y^3}$

44. $\frac{a}{b^2}\div\frac{a^2}{b^3}$

45. $\frac{a+2}{a-3}\div\frac{a-1}{a+3}$

46. $\frac{y+2}{4}\div\frac{y}{2}$

47. $\frac{x^2-1}{x}\div\frac{x+1}{x-1}$

48. $\frac{4y-8}{y+2}\div\frac{y-2}{y^2-4}$

49. $\frac{x+1}{6}\div\frac{x+1}{3}$

50. $\frac{a}{a-b}\div\frac{b}{a-b}$

51. $(y^2-9)\div\frac{y^2-2y-3}{y^2+1}$

52. $(x^2-5x-6)\div\frac{x^2-1}{x+6}$

53. $\frac{5x-5}{16}\div\frac{x-1}{6}$

54. $\frac{-4+2x}{8}\div\frac{x-2}{2}$

55. $\frac{-6+3x}{5}\div\frac{4x-8}{25}$

56. $\frac{-12+4x}{4}\div\frac{-6+2x}{6}$

57. $\frac{a+2}{a-1}\div\frac{3a+6}{a-5}$

58. $\frac{t-3}{t+2}\div\frac{4t-12}{t+1}$

59. $(x-5)\div\frac{2x^2-11x+5}{4x^2-1}$

60. $(a+7)\div\frac{3a^2+14a-49}{a^2+8a+7}$

61. $\frac{x^2-4}{x}\div\frac{x-2}{x+2}$

62. $\frac{x+y}{x-y}\div\frac{x^2+y}{x^2-y^2}$

63. $\dfrac{x^2-9}{4x+12} \div \dfrac{x-3}{6}$

64. $\dfrac{x-b}{2x} \div \dfrac{x^2-b^2}{5x^2}$

65. $\dfrac{c^2+3c}{c^2+2c-3} \div \dfrac{c}{c+1}$

66. $\dfrac{x-5}{2x} \div \dfrac{x^2-25}{4x^2}$

67. $\dfrac{2y^2-7y+3}{2y^2+3y-2} \div \dfrac{6y^2-5y+1}{3y^2+5y-2}$

68. $\dfrac{x^2-x-20}{x^2+7x+12} \div \dfrac{x^2-10x+25}{x^2+6x+9}$

69. $\dfrac{c^2+10c+21}{c^2-2c-15} \div (c^2+2c-35)$

70. $\dfrac{1-z}{1+2z-z^2} \div (1-z)$

71. $\dfrac{(t+5)^3}{(t-5)^3} \div \dfrac{(t+5)^2}{(t-5)^2}$

72. $\dfrac{(y-3)^3}{(y+3)^3} \div \dfrac{(y-3)^2}{(y+3)^2}$

Skill Maintenance

73. Sixteen more than the square of a number is eight times the number. Find the number.

Subtract.

74. $(6x^2+7)-(4x^2-9)$

75. $(8x^3-3x^2+7)-(8x^2+3x-5)$

76. $(0.08y^3-0.04y^2+0.01y)-(0.02y^3+0.05y^2+1)$

Synthesis

77. ◈ Explain why the quotient

$$\frac{x+3}{x-5} \div \frac{x-7}{x+1}$$

is undefined for $x=5$, $x=-1$, and $x=7$.

78. ◈ A student claims to be able to divide, but not multiply, rational expressions. Why is this claim difficult to believe?

Simplify.

79. $\dfrac{2a^2-5ab}{c-3d} \div (4a^2-25b^2)$

80. $(x-2a) \div \dfrac{a^2x^2-4a^4}{a^2x+2a^3}$

81. $\dfrac{3a^2-5ab-12b^2}{3ab+4b^2} \div (3b^2-ab)$

82. $\dfrac{3x^2-2xy-y^2}{x^2-y^2} \div (3x^2+4xy+y^2)$

83. $xy \cdot \dfrac{y^2-4xy}{y-x} \div \dfrac{16x^2y^2-y^4}{4x^2-3xy-y^2}$

84. $\dfrac{z^2-8z+16}{z^2+8z+16} \div \dfrac{(z-4)^5}{(z+4)^5}$

85. $\dfrac{x^2-x+xy-y}{x^2+6x-7} \div \dfrac{x^2+2xy+y^2}{4x+4y}$

86. $\dfrac{3x+3y+3}{9x} \div \dfrac{x^2+2xy+y^2-1}{x^4+x^2}$

87. $\dfrac{t^4-1}{t^4-81} \cdot \dfrac{t^2-9}{t^2+1} \div \dfrac{(t+1)^2}{(t-9)^2}$

88. $\dfrac{(t+2)^3}{(t+1)^3} \div \dfrac{t^2+4t+4}{t^2+2t+1} \cdot \dfrac{t+1}{t+2}$

89. $\left(\dfrac{y^2+5y+6}{y^2} \cdot \dfrac{3y^3+6y^2}{y^2-y-12}\right) \div \dfrac{y^2-y}{y^2-2y-8}$

90. $\dfrac{a^4-81b^4}{a^2c-6abc+9b^2c} \cdot \dfrac{a+3b}{a^2+9b^2} \div \dfrac{a^2+6ab+9b^2}{(a-3b)^2}$

6.3 Addition and Subtraction

Addition When Denominators Are the Same • Subtraction When Denominators Are the Same • Addition and Subtraction When Denominators Are Opposites

Addition When Denominators Are the Same

Recall that to add fractions having the same denominator, like $\frac{2}{7}$ and $\frac{3}{7}$, we add the numerators: $\frac{2}{7}+\frac{3}{7}=\frac{5}{7}$. The same procedure is used to add rational expressions.

To add when the denominators are the same, add the numerators and keep the same denominator:

$$\frac{A}{B} + \frac{C}{B} = \frac{A + C}{B}.$$

Whenever possible, we will simplify the final result.

EXAMPLE 1

Add.

a) $\frac{4}{a} + \frac{3 + a}{a}$

b) $\frac{3x}{x - 5} + \frac{2x + 1}{x - 5}$

c) $\frac{2x^2 + 3x - 7}{2x + 1} + \frac{x^2 + x - 8}{2x + 1}$

d) $\frac{x - 5}{x^2 - 9} + \frac{2}{x^2 - 9}$

Solution

a) $\frac{4}{a} + \frac{3 + a}{a} = \frac{7 + a}{a}$ When the denominators are alike, add the numerators.

b) $\frac{3x}{x - 5} + \frac{2x + 1}{x - 5} = \frac{5x + 1}{x - 5}$ Adding the numerators

c) $\frac{2x^2 + 3x - 7}{2x + 1} + \frac{x^2 + x - 8}{2x + 1} = \frac{(2x^2 + 3x - 7) + (x^2 + x - 8)}{2x + 1}$

$= \frac{3x^2 + 4x - 15}{2x + 1}$ Collecting like terms in the numerator

d) $\frac{x - 5}{x^2 - 9} + \frac{2}{x^2 - 9} = \frac{(x - 5) + 2}{x^2 - 9}$

$= \frac{x - 3}{x^2 - 9}$ Collecting like terms in the numerator

$= \frac{x - 3}{(x - 3)(x + 3)}$ Factoring

$= \frac{\cancel{x - 3}}{(\cancel{x - 3})(x + 3)}$

$= \frac{1}{x + 3}$ Removing a factor of 1: $\frac{x - 3}{x - 3} = 1$

Subtraction When Denominators Are the Same

Recall that to subtract fractions having the same denominator, we subtract the numerators—for example, $\frac{5}{7} - \frac{2}{7} = \frac{3}{7}$. The same procedure is used to subtract rational expressions.

To subtract when the denominators are the same, subtract the numerators and keep the same denominator:

$$\frac{A}{B} - \frac{C}{B} = \frac{A - C}{B}.$$

CAUTION! Keep in mind that a fraction bar is a grouping symbol. When a numerator is being subtracted, remember to subtract *every* term in that expression.

EXAMPLE 2

Subtract: **(a)** $\frac{3x}{x+2} - \frac{x-5}{x+2}$; **(b)** $\frac{x^2}{x-4} - \frac{x+12}{x-4}$.

Solution

a) $\frac{3x}{x+2} - \frac{x-5}{x+2} = \frac{3x - (x-5)}{x+2}$ The parentheses are needed to make sure that we subtract both terms.

$= \frac{3x - x + 5}{x+2}$ Removing the parentheses and changing signs

$= \frac{2x+5}{x+2}$ Collecting like terms

b) $\frac{x^2}{x-4} - \frac{x+12}{x-4} = \frac{x^2 - (x+12)}{x-4}$ Remember the parentheses!

$= \frac{x^2 - x - 12}{x-4}$ Removing parentheses

$= \frac{(x-4)(x+3)}{x-4}$

$= \frac{\cancel{(x-4)}(x+3)}{\cancel{x-4}}$

$= x + 3$

Simplifying by factoring and removing a factor of 1: $\frac{x-4}{x-4} = 1$

❑

Addition and Subtraction When Denominators Are Opposites

Recall from Chapter 1 that

$$\frac{a}{-b} = \frac{-a}{b} = -\frac{a}{b} \quad \text{and} \quad m - (-n) = m + n.$$

These results can be used to add or subtract when denominators are opposites.

EXAMPLE 3

Perform the indicated operation: **(a)** $\frac{x}{2} + \frac{3}{-2}$; **(b)** $\frac{x}{5} - \frac{3x-4}{-5}$.

Solution

a) We rewrite $\frac{3}{-2}$ as $\frac{-3}{2}$ to obtain a common denominator of 2. Then we add and, if possible, simplify:

$$\frac{x}{2}+\frac{3}{-2}=\frac{x}{2}+\frac{-3}{2} \qquad \text{Since } \frac{a}{-b}=\frac{-a}{b}$$

$$=\frac{x+(-3)}{2}=\frac{x-3}{2}.$$

b)
$$\frac{x}{5}-\frac{3x-4}{-5}=\frac{x}{5}-\left(-\frac{3x-4}{5}\right) \qquad \text{Since } \frac{a}{-b}=-\frac{a}{b}$$

$$=\frac{x}{5}+\frac{3x-4}{5} \qquad \text{Since } m-(-n)=m+n$$

$$=\frac{x+(3x-4)}{5}$$

$$=\frac{4x-4}{5}$$

❑

Sometimes denominators are of the form $x - a$ and $a - x$. As we saw in Section 6.1, expressions of this form are opposites of each other.

EXAMPLE 4

Perform the indicated operation: **(a)** $\frac{5}{x-2}-\frac{3}{2-x}$; **(b)** $\frac{3y}{y-5}+\frac{y+1}{5-y}$.

Solution

a) Note that $x - 2$ and $2 - x$ are opposites. Thus,

$$\frac{5}{x-2}-\frac{3}{2-x}=\frac{5}{x-2}-\frac{3}{-(x-2)} \qquad \text{Since } 2-x \text{ is the opposite of } x-2$$

$$=\frac{5}{x-2}-\left(-\frac{3}{x-2}\right) \qquad \frac{a}{-b}=-\frac{a}{b}$$

$$=\frac{5}{x-2}+\frac{3}{x-2} \qquad m-(-n)=m+n$$

$$=\frac{8}{x-2}.$$

b)
$$\frac{3y}{y-5}+\frac{y+1}{5-y}=\frac{3y}{y-5}+\frac{y+1}{-(y-5)} \qquad \text{Since } 5-y \text{ is the opposite of } y-5$$

$$=\frac{3y}{y-5}+\frac{-(y+1)}{y-5} \qquad \frac{a}{-b}=\frac{-a}{b}$$

$$=\frac{3y}{y-5}+\frac{-y-1}{y-5} \qquad \text{The opposite of } y+1 \text{ is } -y-1.$$

$$=\frac{3y+(-y-1)}{y-5} \qquad \text{Adding rational expressions}$$

$$=\frac{2y-1}{y-5}$$

❑

Note that the second step of Example 4(b) could have been written as

$$\frac{3y}{y-5}+\left(-\frac{y+1}{y-5}\right) \quad \text{or} \quad \frac{3y}{y-5}-\frac{y+1}{y-5}.$$

Either expression would have led to the same result.

EXERCISE SET 6.3

Perform the indicated operation. Simplify, if possible.

1. $\frac{3}{x}+\frac{5}{x}$
2. $\frac{4}{a^2}+\frac{9}{a^2}$
3. $\frac{x}{15}+\frac{2x+1}{15}$
4. $\frac{a}{7}+\frac{3a-4}{7}$
5. $\frac{2}{a+3}+\frac{4}{a+3}$
6. $\frac{5}{x+2}+\frac{8}{x+2}$
7. $\frac{9}{a+6}-\frac{5}{a+6}$
8. $\frac{8}{x+7}-\frac{2}{x+7}$
9. $\frac{3y+9}{2y}-\frac{y+1}{2y}$
10. $\frac{5+3t}{4t}-\frac{2t+1}{4t}$
11. $\frac{9x+5}{x+1}+\frac{2x+3}{x+1}$
12. $\frac{3a+2}{a+4}+\frac{2a+7}{a+4}$
13. $\frac{9x+5}{x+1}-\frac{2x+3}{x+1}$
14. $\frac{3a+2}{a+4}-\frac{2a+7}{a+4}$
15. $\frac{a^2}{a-4}+\frac{a-20}{a-4}$
16. $\frac{x^2}{x+5}+\frac{7x+10}{x+5}$
17. $\frac{x^2}{x-2}-\frac{6x-8}{x-2}$
18. $\frac{a^2}{a+3}-\frac{2a+15}{a+3}$
19. $\frac{t^2+4t}{t-1}+\frac{2t-7}{t-1}$
20. $\frac{y^2+6y}{y+2}+\frac{2y+12}{y+2}$
21. $\frac{x+1}{x^2+5x+6}+\frac{2}{x^2+5x+6}$
22. $\frac{-7}{x^2-4x+3}+\frac{x+4}{x^2-4x+3}$
23. $\frac{a^2+3}{a^2+5a-6}-\frac{4}{a^2+5a-6}$
24. $\frac{a^2-3}{a^2-7a+12}-\frac{6}{a^2-7a+12}$
25. $\frac{t^2-3t}{t^2+6t+9}+\frac{2t-12}{t^2+6t+9}$
26. $\frac{y^2-7y}{y^2+8y+16}+\frac{6y-20}{y^2+8y+16}$
27. $\frac{2x^2+3}{x^2-6x+5}-\frac{x^2-5x+9}{x^2-6x+5}$
28. $\frac{2x^2+x}{x^2-8x+12}-\frac{x^2-2x+10}{x^2-8x+12}$
29. $\frac{3x}{8}+\frac{x}{-8}$
30. $\frac{5a}{6}+\frac{a}{-6}$
31. $\frac{3}{t}+\frac{4}{-t}$
32. $\frac{5}{-a}+\frac{8}{a}$
33. $\frac{2x+7}{x-6}+\frac{3x}{6-x}$
34. $\frac{3x-2}{4x-3}+\frac{2x-5}{3-4x}$
35. $\frac{a}{6}-\frac{7a}{-6}$
36. $\frac{x}{8}-\frac{5x}{-8}$
37. $\frac{5}{a}-\frac{8}{-a}$
38. $\frac{3}{t}-\frac{4}{-t}$
39. $\frac{x}{4}-\frac{3x-5}{-4}$
40. $\frac{2}{x-1}-\frac{2}{1-x}$
41. $\frac{y^2}{y-3}+\frac{9}{3-y}$
42. $\frac{t^2}{t-2}+\frac{4}{2-t}$
43. $\frac{b-7}{b^2-16}+\frac{7-b}{16-b^2}$
44. $\frac{a-3}{a^2-25}+\frac{a-3}{25-a^2}$
45. $\frac{3-2t}{t^2-5t+4}+\frac{2-3t}{t^2-5t+4}$
46. $\frac{7-2x}{x^2-6x+8}+\frac{3-3x}{x^2-6x+8}$
47. $\frac{3-x}{x-7}-\frac{2x-5}{7-x}$
48. $\frac{t^2}{t-2}-\frac{4}{2-t}$
49. $\frac{x-8}{x^2-16}-\frac{x-8}{16-x^2}$
50. $\frac{x-2}{x^2-25}-\frac{6-x}{25-x^2}$
51. $\frac{4-x}{x-9}-\frac{3x-8}{9-x}$
52. $\frac{3-x}{x-7}-\frac{2x-5}{7-x}$
53. $\frac{5-3x}{x^2-2x+1}-\frac{x+1}{x^2-2x+1}$

54. $\dfrac{x-7}{x^2+3x-4}-\dfrac{2x-3}{x^2+3x-4}$

Skill Maintenance

Graph.

55. $y=-1$

56. $x=4$

57. $y=x-1$

58. $y=\frac{1}{2}x-1$

Synthesis

59. ◈ Are parentheses as important for adding rational expressions as they are for subtracting rational expressions? Why or why not?

60. ◈ Explain in your own words why the expressions

$$\frac{1}{3-x} \quad \text{and} \quad \frac{1}{x-3}$$

are opposites of each other.

Perform the indicated operations and simplify.

61. $\dfrac{3(2x+5)}{x-1}-\dfrac{3(2x-3)}{1-x}+\dfrac{6x-1}{x-1}$

62. $\dfrac{2x-y}{x-y}+\dfrac{x-2y}{y-x}-\dfrac{3x-3y}{x-y}$

63. $\dfrac{x-y}{x^2-y^2}+\dfrac{x+y}{x^2-y^2}-\dfrac{2x}{x^2-y^2}$

64. $\dfrac{x+y}{2(x-y)}-\dfrac{2x-2y}{2(x-y)}+\dfrac{x-3y}{2(y-x)}$

65. $\dfrac{10}{2y-1}-\dfrac{6}{1-2y}+\dfrac{y}{2y-1}+\dfrac{y-4}{1-2y}$

66. $\dfrac{(x+3)(2x-1)}{(2x-3)(x-3)}-\dfrac{(x-3)(x+1)}{(3-x)(3-2x)}+\dfrac{(2x+1)(x+3)}{(3-2x)(x-3)}$

67. $\dfrac{x}{(x-y)(y-z)}-\dfrac{x}{(y-x)(z-y)}$

68. $\dfrac{x}{x-y}+\dfrac{y}{y-x}+\dfrac{x+y}{x-y}+\dfrac{x-y}{y-x}$

69. $\dfrac{x^2}{3x^2-5x-2}-\dfrac{2x}{3x+1}\cdot\dfrac{1}{x-2}$

70. $\dfrac{3}{x+4}\cdot\dfrac{2x+11}{x-3}-\dfrac{-1}{4+x}\cdot\dfrac{6x+3}{3-x}$

71. ◈ Explain how evaluating can be used to perform a partial check on the result of Example 1(d):

$$\frac{x-5}{x^2-9}+\frac{2}{x^2-9}=\frac{1}{x+3}.$$

6.4 Least Common Multiples and Denominators

Least Common Multiples • Equivalent Expressions and Least Common Denominators

Like fractions, rational expressions must have common denominators before they can be added or subtracted. Our work will be easier if we use the smallest, or *least*, common denominator.

Least Common Multiples

To add fractions like $\frac{5}{12}$ and $\frac{7}{30}$, we first look for the smallest number that contains both 12 and 30 as factors. Such a number, the **least common multiple**, or **LCM**, of 12 and 30, will then be used as the **least common denominator**, or **LCD**.

Let's find the LCM of 12 and 30 using a method that can also be used with polynomials. We begin by writing the prime factorization of 12:

$$12 = 2 \cdot 2 \cdot 3.$$

Two factors of 2 and a factor of 3 must appear in the LCM if 12 is to be a factor of the LCM.

Next we write the prime factorization of 30:

$30 = 2 \cdot 3 \cdot 5.$

The factors 2, 3, and 5 must appear in the LCM if 30 is to be a factor of the LCM.

Note that the prime factorization of 12 already includes a 2 and a 3, but lacks a 5. The smallest product that contains both 12 and 30 as factors, 60, is found by multiplying $2 \cdot 2 \cdot 3$ by 5:

$\text{LCM} = 2 \cdot 2 \cdot 3 \cdot 5$

12 is a factor of the LCM. (the factors $2 \cdot 2 \cdot 3$)

30 is a factor of the LCM. (the factors $2 \cdot 3 \cdot 5$)

The factors common to the two denominators—in this case, 2 and 3—are used the greatest number of times that they appear in either of the individual factorizations.

To find the least common denominator (LCD):

1. Write the prime factorization of each denominator.
2. Select one of the factorizations and inspect it to see if it contains the other.
   a) If it does, it represents the LCD.
   b) If it does not, multiply that factorization by any factors of the other denominator that it lacks. The final product is the LCD.
3. The LCD should include each factor the greatest number of times that it occurs in any one factorization.

Let's finish adding $\frac{5}{12}$ and $\frac{7}{30}$:

$$\frac{5}{12} + \frac{7}{30} = \frac{5}{2 \cdot 2 \cdot 3} + \frac{7}{2 \cdot 3 \cdot 5}.$$

The least common denominator, or LCD, is $2 \cdot 2 \cdot 3 \cdot 5$. To get the LCD in the first denominator, we need a 5. To get the LCD in the second denominator, we need another 2. We get these numbers by multiplying by 1:

$$\frac{5}{12} + \frac{7}{30} = \frac{5}{2 \cdot 2 \cdot 3} \cdot \frac{5}{5} + \frac{7}{2 \cdot 3 \cdot 5} \cdot \frac{2}{2} \qquad \frac{5}{5} = 1 \text{ and } \frac{2}{2} = 1$$

$$= \frac{25}{2 \cdot 2 \cdot 3 \cdot 5} + \frac{14}{2 \cdot 3 \cdot 5 \cdot 2} \qquad \text{The denominators are now the LCD.}$$

$$= \frac{39}{2 \cdot 2 \cdot 3 \cdot 5} \qquad \text{Adding the numerators and keeping the LCD}$$

$$= \frac{\cancel{3} \cdot 13}{2 \cdot 2 \cdot \cancel{3} \cdot 5}$$

$$= \frac{13}{20}. \qquad \text{Simplifying by removing a factor of 1: } \frac{3}{3} = 1$$

Rational expressions, like $7/(24x)$ and $5/(36x^2)$, can be added in much the same way that we added $\frac{5}{12}$ and $\frac{7}{30}$.

EXAMPLE 1

Find the LCM of $24x$ and $36x^2$.

Solution

1. We begin by writing the prime factorizations of $24x$ and $36x^2$:

$$24x = 2 \cdot 2 \cdot 2 \cdot 3 \cdot x,$$
$$36x^2 = 2 \cdot 2 \cdot 3 \cdot 3 \cdot x \cdot x.$$

2. Note that the factorization of $36x^2$ contains the entire factorization of $24x$ except for a third factor of 2. To find the smallest product that contains both $24x$ and $36x^2$ as factors, we need to add a third factor of 2 to $2 \cdot 2 \cdot 3 \cdot 3 \cdot x \cdot x$:

$$\text{LCM} = 2 \cdot 2 \cdot 3 \cdot 3 \cdot x \cdot x \cdot 2.$$

$36x^2$ is a factor.

$24x$ is a factor.

The LCM is thus $2^3 \cdot 3^2 \cdot x^2$, or $72x^2$. □

Now let's finish adding $\dfrac{7}{24x}$ and $\dfrac{5}{36x^2}$:

$$\frac{7}{24x} + \frac{5}{36x^2} = \frac{7}{2 \cdot 2 \cdot 2 \cdot 3 \cdot x} + \frac{5}{2 \cdot 2 \cdot 3 \cdot 3 \cdot x \cdot x}.$$

In Example 1, we found that the LCD is $2 \cdot 2 \cdot 2 \cdot 3 \cdot 3 \cdot x \cdot x$. To obtain equivalent expressions with this LCD, we multiply each expression by 1, using the missing factors of the LCD to write 1:

$$\frac{7}{24x} + \frac{5}{36x^2} = \frac{7}{2 \cdot 2 \cdot 2 \cdot 3 \cdot x} \cdot \frac{3 \cdot x}{3 \cdot x} + \frac{5}{2 \cdot 2 \cdot 3 \cdot 3 \cdot x \cdot x} \cdot \frac{2}{2}$$

The LCD requires additional factors of 3 and x.

The LCD requires another factor of 2.

$$= \frac{21x}{2 \cdot 2 \cdot 2 \cdot 3 \cdot x \cdot 3 \cdot x} + \frac{10}{2 \cdot 2 \cdot 3 \cdot 3 \cdot x \cdot x \cdot 2}$$

Both denominators are now the LCD.

$$= \frac{21x + 10}{72x^2}.$$

You now have the "big" picture of how LCMs are used with rational expressions. For the remainder of this section, we will practice finding LCMs and rewriting rational expressions so that they have the LCD as the denominator. In Section 6.5, we will carry out the actual addition and subtraction of rational expressions.

EXAMPLE 2

For each pair of polynomials, find the least common multiple.

a) $15a$ and $35b$

b) $21x^3y^6$ and $7x^5y^2$

c) $x^2 + 5x - 6$ and $x^2 - 1$

Solution

a) We write the prime factorizations and then construct the LCM:

$15a = 3 \cdot 5 \cdot a$
$35b = 5 \cdot 7 \cdot b$

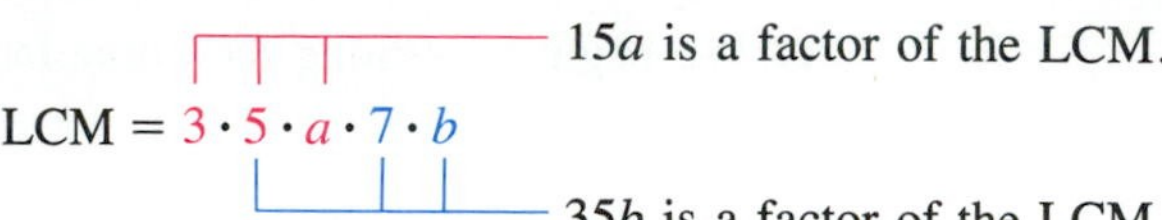

The LCM is $3 \cdot 5 \cdot a \cdot 7 \cdot b$, or $105ab$.

b) $21x^3y^6 = 3 \cdot 7 \cdot x \cdot x \cdot x \cdot y \cdot y \cdot y \cdot y \cdot y \cdot y$
$7x^5y^2 = 7 \cdot x \cdot x \cdot x \cdot x \cdot x \cdot y \cdot y$

Try to visualize the factors of x and y mentally.

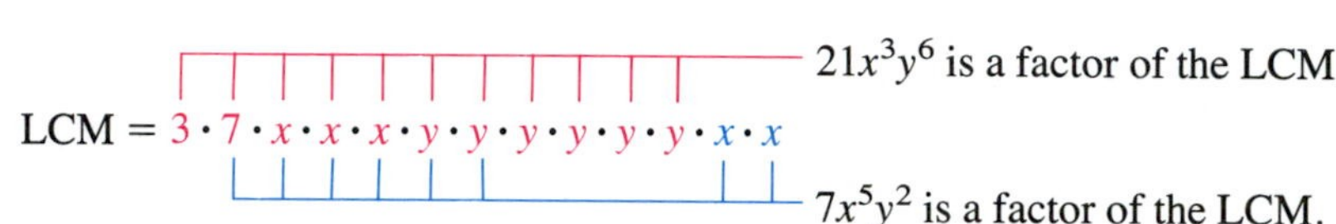

Note that we used the highest power of each factor in $3 \cdot 7 \cdot x^3y^6$ and $7x^5y^2$. The LCM is $21x^5y^6$.

c) $x^2 + 5x - 6 = (x - 1)(x + 6)$
$x^2 - 1 = (x - 1)(x + 1)$

$x^2 + 5x - 6$ is a factor of the LCM.

$\text{LCM} = (x - 1)(x + 6)(x + 1)$

$x^2 - 1$ is a factor of the LCM.

The LCM is $(x - 1)(x + 6)(x + 1)$. There is no need to multiply this out. ❑

The above procedure can be used to find the LCM of three polynomials as well. We factor each polynomial and then construct the LCM using each factor the greatest number of times that it appears in any one factorization.

EXAMPLE 3

For each group of polynomials, find the LCM.

a) $12x$, $16y$, and $8xyz$

b) $x^2 + 4$, $x + 1$, and 5

Solution

a) $12x = 2 \cdot 2 \cdot 3 \cdot x$
$16y = 2 \cdot 2 \cdot 2 \cdot 2 \cdot y$
$8xyz = 2 \cdot 2 \cdot 2 \cdot x \cdot y \cdot z$

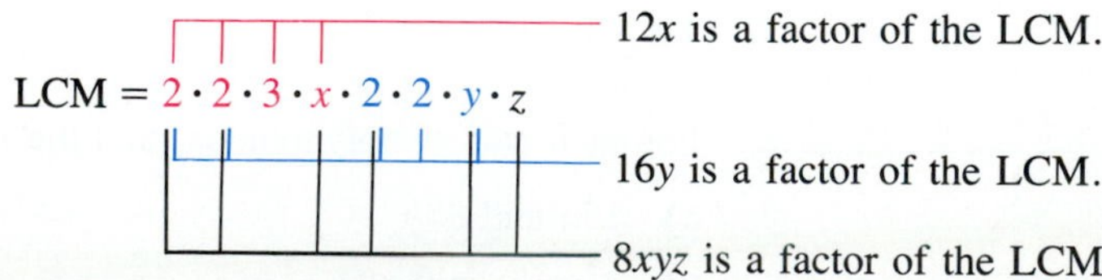

The LCM is $2^4 \cdot 3 \cdot xyz$, or $48xyz$.

b) Since $x^2 + 4$, $x + 1$, and 5 are not factorable, the LCM is their product: $5(x^2 + 4)(x + 1)$. ❑

When two or more rational expressions have different denominators, it is important to be able to write equivalent expressions that have the LCD.

EXAMPLE 4

Find equivalent expressions that have the LCD:

$$\frac{x+3}{x^2+5x-6}, \qquad \frac{x+7}{x^2-1}.$$

Solution Look at Example 2(c). Note that the LCD is $(x + 6)(x - 1)(x + 1)$. Since $x^2 + 5x - 6 = (x + 6)(x - 1)$, the factor of the LCD that is missing from the first denominator is $x + 1$. We multiply by 1 using $(x + 1)/(x + 1)$:

$$\frac{x+3}{x^2+5x-6} = \frac{x+3}{(x+6)(x-1)} \cdot \frac{x+1}{x+1} = \frac{(x+3)(x+1)}{(x+6)(x-1)(x+1)}.$$

For the second expression, we have $x^2 - 1 = (x + 1)(x - 1)$. The factor of the LCD that is missing is $x + 6$. We multiply by 1 using $(x + 6)/(x + 6)$:

$$\frac{x+7}{x^2-1} = \frac{x+7}{(x+1)(x-1)} \cdot \frac{x+6}{x+6} = \frac{(x+7)(x+6)}{(x+1)(x-1)(x+6)}.$$

We leave the answers in factored form. ❑

EXERCISE SET 6.4

Find the LCM.

1. 12, 27
2. 10, 15
3. 8, 9
4. 12, 15
5. 6, 9, 21
6. 8, 36, 40
7. 24, 36, 40
8. 3, 4, 5
9. 28, 42, 60
10. 10, 100, 500

Add, first finding the LCM of the denominators. Simplify, if possible.

11. $\frac{7}{24} + \frac{11}{18}$
12. $\frac{7}{60} + \frac{6}{75}$
13. $\frac{1}{6} + \frac{3}{40} + \frac{2}{75}$
14. $\frac{5}{24} + \frac{3}{20} + \frac{7}{30}$
15. $\frac{2}{15} + \frac{5}{9} + \frac{3}{20}$
16. $\frac{1}{20} + \frac{1}{30} + \frac{2}{45}$

Find the LCM.

17. $6x^2$, $12x^3$
18. $2a^2b$, $8ab^2$
19. $2x^2$, $6xy$, $18y^2$
20. c^2d, cd^2, c^3d
21. $2(y - 3)$, $6(y - 3)$
22. $4(x - 1)$, $8(x - 1)$
23. t, $t + 2$, $t - 2$
24. x, $x + 3$, $x - 3$
25. $x^2 - 4$, $x^2 + 5x + 6$
26. $x^2 + 3x + 2$, $x^2 - 4$
27. $t^3 + 4t^2 + 4t$, $t^2 - 4t$
28. $y^3 - y^2$, $y^4 - y^2$
29. $9a^5b^2$, $6ab^6$
30. $10a^4b^7$, $15ab^8$
31. $10x^2y$, $6y^2z$, $5xz^3$
32. $8x^3z$, $12xy^2$, $4y^5z^2$
33. $a + 1$, $(a - 1)^2$, $a^2 - 1$
34. $x^2 - y^2$, $2x + 2y$, $x^2 + 2xy + y^2$
35. $m^2 - 5m + 6$, $m^2 - 4m + 4$
36. $2x^2 + 5x + 2$, $2x^2 - x - 1$
37. $2 + 3x$, $4 - 9x^2$, $2 - 3x$
38. $3 - 2x$, $9 - 4x^2$, $3 + 2x$

39. $10v^2 + 30v,\ 5v^2 + 35v + 60$

40. $12a^2 + 24a,\ 4a^2 + 20a + 24$

41. $9x^3 - 9x^2 - 18x,\ 6x^5 - 24x^4 + 24x^3$

42. $x^5 - 4x^3,\ x^3 + 4x^2 + 4x$

43. $x^5 + 4x^4 + 4x^3,\ 3x^2 - 12,\ 2x + 4$

44. $x^5 + 2x^4 + x^3,\ 2x^3 - 2x,\ 5x - 5$

Find equivalent expressions that have the LCD.

45. $\frac{7}{6x^5}, \frac{y}{12x^3}$

46. $\frac{3}{10a^3}, \frac{b}{5a^6}$

47. $\frac{3}{2a^2b}, \frac{5}{8ab^2}$

48. $\frac{7}{3x^4y^2}, \frac{4}{9xy^3}$

49. $\frac{x+1}{x^2-4}, \frac{x-2}{x^2+5x+6}$

50. $\frac{x-4}{x^2-9}, \frac{x+2}{x^2+11x+24}$

51. $\frac{3}{t}, \frac{4}{t+2}, \frac{t}{t-2}$

52. $\frac{1}{x}, \frac{-2}{x+3}, \frac{x^2}{x-3}$

53. $\frac{x+1}{2x-3}, \frac{x-2}{4x^2-9}, \frac{x+1}{2x+3}$

54. $\frac{x}{x^5+4x^4+4x^3}, \frac{3}{3x^2-12}$
(*Hint:* Simplify first.)

Skill Maintenance

Factor.

55. $x^2 - 19x + 60$

56. $x^2 + 9x - 36$

Find a polynomial that can represent the shaded area of the figure.

57.

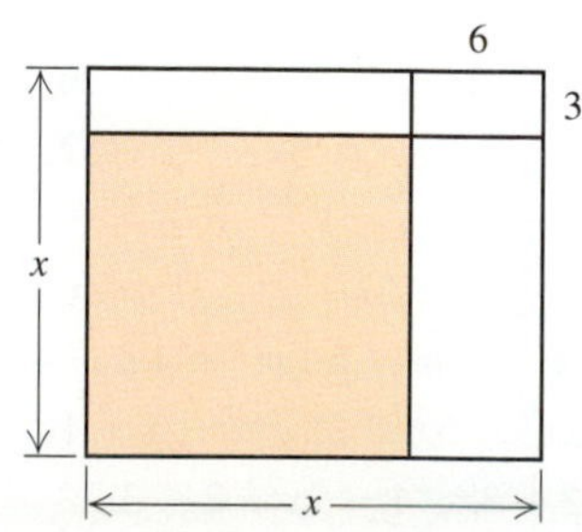

58.

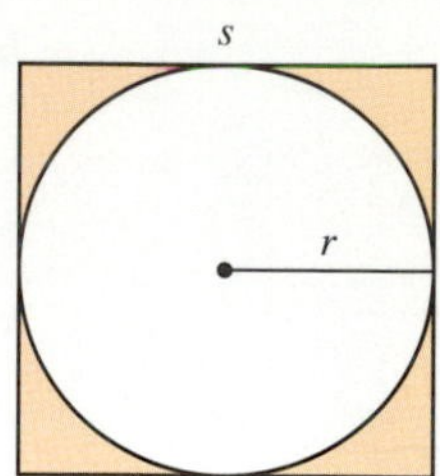

Synthesis

59. ◈ Explain why the product of two numbers is not always their LCM.

60. ◈ Is every LCD an LCM? Why or why not?

Find the LCM.

61. 72, 90, 96

62. $8x^2 - 8,\ 6x^2 - 12x + 6,\ 10 - 10x$

63. Two joggers leave the starting point of a fitness loop at the same time. One jogger completes a lap in 6 min and the second jogger in 8 min. Assuming they continue to run at the same pace, when will they next meet at the starting place?

64. ◈ If the LCM of two expressions is the same as one of the expressions, what relationship exists between the two expressions?

6.5 Addition and Subtraction with Unlike Denominators

Using LCDs for Addition and Subtraction • Combined Additions and Subtractions

We now know how to rewrite any two rational expressions in an equivalent form that uses the LCD. Once rational expressions share a common denominator, they can be added or subtracted just as they were in Section 6.3. The procedure is as follows.

To add or subtract rational expressions that have different denominators:

1. Find the LCD.
2. Multiply each rational expression by an expression for 1 made up of the factors of the LCD missing from that expression's denominator.
3. Add or subtract the numerators, as indicated. Write the sum or difference over the LCD.
4. Simplify, if possible.

EXAMPLE 1

Add: $\dfrac{5x^2}{8} + \dfrac{7x}{12}$.

Solution First, we find the LCD:

$$\left.\begin{aligned} 8 &= 2 \cdot 2 \cdot 2 \\ 12 &= 2 \cdot 2 \cdot 3 \end{aligned}\right\} \quad \text{LCD} = 2 \cdot 2 \cdot 2 \cdot 3, \text{ or } 24.$$

The denominator 8 needs to be multiplied by 3 to obtain the LCD. The denominator 12 needs to be multiplied by 2 to obtain the LCD. Thus we multiply by $\frac{3}{3}$ and $\frac{2}{2}$ to get the LCD. Then we add and, if possible, simplify.

$$\frac{5x^2}{8} + \frac{7x}{12} = \frac{5x^2}{2 \cdot 2 \cdot 2} + \frac{7x}{2 \cdot 2 \cdot 3}$$

$$= \frac{5x^2}{2 \cdot 2 \cdot 2} \cdot \frac{3}{3} + \frac{7x}{2 \cdot 2 \cdot 3} \cdot \frac{2}{2}$$ Multiplying twice by an expression for 1 to get the LCD

$$= \frac{15x^2}{24} + \frac{14x}{24}$$

$$= \frac{15x^2 + 14x}{24}$$ ❑

Subtraction is performed in much the same way.

EXAMPLE 2

Subtract: $\dfrac{7}{8x} - \dfrac{5}{12x^2}$.

Solution First, we find the LCD:

$$\left.\begin{aligned} 8x &= 2\cdot 2\cdot 2\cdot x \\ 12x^2 &= 2\cdot 2\cdot 3\cdot x\cdot x \end{aligned}\right\} \quad \text{LCD} = 2\cdot 2\cdot 3\cdot x\cdot x\cdot 2, \text{ or } 24x^2.$$

The denominator $8x$ must be multiplied by $3x$ to obtain the LCD. The denominator $12x^2$ must be multiplied by 2 to obtain the LCD. Thus we multiply by $3x/3x$ and $2/2$ to get the LCD. Then we subtract and, if possible, simplify.

$$\frac{7}{8x} - \frac{5}{12x^2} = \frac{7}{8x}\cdot\frac{3\cdot x}{3\cdot x} - \frac{5}{12x^2}\cdot\frac{2}{2}$$

$$= \frac{21x}{24x^2} - \frac{10}{24x^2}$$

← CAUTION! Do not simplify *these* rational expressions or you will lose the LCD.

$$= \frac{21x - 10}{24x^2}$$

□

When denominators contain polynomials with two or more terms, the same steps are used.

EXAMPLE 3

Perform the indicated operations.

a) $\dfrac{2a}{a^2 - 1} + \dfrac{1}{a^2 + a}$

b) $\dfrac{x + 4}{x - 2} - \dfrac{x - 7}{x + 5}$

c) $\dfrac{x}{x^2 + 11x + 30} + \dfrac{-5}{x^2 + 9x + 20}$

d) $\dfrac{x}{x^2 + 5x + 6} - \dfrac{2}{x^2 + 3x + 2}$

Solution

a) First, we find the LCD:

$$\left.\begin{aligned} a^2 - 1 &= (a - 1)(a + 1) \\ a^2 + a &= a(a + 1) \end{aligned}\right\} \quad \text{LCD} = (a - 1)(a + 1)a.$$

We multiply by 1 to get the LCD in each expression. Then we add and, if possible, simplify.

$$\frac{2a}{a^2 - 1} + \frac{1}{a^2 + a} = \frac{2a}{(a - 1)(a + 1)}\cdot\frac{a}{a} + \frac{1}{a(a + 1)}\cdot\frac{a - 1}{a - 1}$$

Writing 1 as $\dfrac{a}{a}$ and as $\dfrac{a - 1}{a - 1}$ to get the LCD

$$= \frac{2a^2}{(a - 1)(a + 1)a} + \frac{a - 1}{(a - 1)(a + 1)a}$$

Then

$$\frac{2a}{a^2 - 1} + \frac{1}{a^2 + a} = \frac{2a^2 + a - 1}{(a - 1)(a + 1)a}$$

$$= \frac{(2a - 1)(a + 1)}{(a - 1)(a + 1)a}$$

$$= \frac{2a - 1}{(a - 1)a}$$

Simplifying by factoring and removing a factor of 1: $\frac{a+1}{a+1} = 1$

b) First, we find the LCD. It is just the product of the denominators:

$$\text{LCD} = (x - 2)(x + 5).$$

We multiply by 1 to get the LCD in each expression. Then we subtract and, if possible, simplify.

$$\frac{x + 4}{x - 2} - \frac{x - 7}{x + 5} = \frac{x + 4}{x - 2} \cdot \frac{x + 5}{x + 5} - \frac{x - 7}{x + 5} \cdot \frac{x - 2}{x - 2}$$

$$= \frac{x^2 + 9x + 20}{(x - 2)(x + 5)} - \frac{x^2 - 9x + 14}{(x - 2)(x + 5)}$$

Multiplying out numerators (but not denominators)

$$= \frac{x^2 + 9x + 20 - (x^2 - 9x + 14)}{(x - 2)(x + 5)}$$

When we are subtracting a numerator with more than one term, parentheses are important.

$$= \frac{x^2 + 9x + 20 - x^2 + 9x - 14}{(x - 2)(x + 5)}$$

Removing parentheses in the numerator

$$= \frac{18x + 6}{(x - 2)(x + 5)}$$

Although $18x + 6$ can be factored as $6(3x + 1)$, doing so will not enable us to simplify our result.

c) $\frac{x}{x^2 + 11x + 30} + \frac{-5}{x^2 + 9x + 20}$

$$= \frac{x}{(x + 5)(x + 6)} + \frac{-5}{(x + 5)(x + 4)}$$

Factoring the denominators in order to find the LCD. The LCD is $(x + 5)(x + 6)(x + 4)$.

$$= \frac{x}{(x + 5)(x + 6)} \cdot \frac{x + 4}{x + 4} + \frac{-5}{(x + 5)(x + 4)} \cdot \frac{x + 6}{x + 6}$$

Multiplying to get the LCD

$$= \frac{x^2 + 4x}{(x + 5)(x + 6)(x + 4)} + \frac{-5x - 30}{(x + 5)(x + 6)(x + 4)}$$

Multiplying in each numerator

$$= \frac{x^2 + 4x - 5x - 30}{(x + 5)(x + 6)(x + 4)}$$

Adding numerators

$$= \frac{x^2 - x - 30}{(x + 5)(x + 6)(x + 4)}$$

$$= \frac{(x + 5)(x - 6)}{(x + 5)(x + 6)(x + 4)}$$

$$= \frac{x - 6}{(x + 6)(x + 4)}$$

Always simplify the result, if possible, by removing a factor of 1. (Here $(x + 5)/(x + 5) = 1$.)

d) $\dfrac{x}{x^2+5x+6} - \dfrac{2}{x^2+3x+2}$

$= \dfrac{x}{(x+2)(x+3)} - \dfrac{2}{(x+2)(x+1)}$ Factoring denominators. The LCD is $(x+2)(x+3)(x+1)$.

$= \dfrac{x}{(x+2)(x+3)} \cdot \dfrac{x+1}{x+1} - \dfrac{2}{(x+2)(x+1)} \cdot \dfrac{x+3}{x+3}$

$= \dfrac{x^2+x}{(x+2)(x+3)(x+1)} - \dfrac{2x+6}{(x+2)(x+3)(x+1)}$

$= \dfrac{x^2+x-(2x+6)}{(x+2)(x+3)(x+1)}$ Don't forget the parentheses!

$= \dfrac{x^2+x-2x-6}{(x+2)(x+3)(x+1)}$

$= \dfrac{x^2-x-6}{(x+2)(x+3)(x+1)}$

$= \dfrac{(x+2)(x-3)}{(x+2)(x+3)(x+1)}$

$= \dfrac{x-3}{(x+3)(x+1)}$ Simplifying; $\dfrac{x+2}{x+2} = 1$ ❑

Suppose that after factoring to find the LCD, we find a factor in one denominator that is the opposite of a factor in the other denominator. When this happens, we reuse a "trick" from Section 6.3: $b-a$ can be rewritten as $-(a-b)$ since $b-a$ is the opposite of $a-b$.

EXAMPLE 4

Add: $\dfrac{x}{x^2-25} + \dfrac{3}{5-x}$.

Solution

$\dfrac{x}{x^2-25} + \dfrac{3}{5-x} = \dfrac{x}{(x-5)(x+5)} + \dfrac{3}{-(x-5)}$ Since $b-a=-(a-b)$

$= \dfrac{x}{(x-5)(x+5)} + \dfrac{-3}{(x-5)}$ Since $\dfrac{a}{-b} = \dfrac{-a}{b}$

$= \dfrac{x}{(x-5)(x+5)} + \dfrac{-3}{(x-5)} \cdot \dfrac{x+5}{x+5}$ The LCD is $(x-5)(x+5)$.

$= \dfrac{x}{(x-5)(x+5)} + \dfrac{-3x-15}{(x-5)(x+5)}$

$= \dfrac{-2x-15}{(x-5)(x+5)}$ ❑

Combined Additions and Subtractions

EXAMPLE 5

Perform the indicated operations and simplify.

a) $\dfrac{x+9}{x^2-4} + \dfrac{5-x}{4-x^2} - \dfrac{2+x}{x^2-4}$ **b)** $\dfrac{1}{x} - \dfrac{1}{x^2} + \dfrac{2}{x+1}$

Solution

a) We have

$$\frac{x+9}{x^2-4}+\frac{5-x}{4-x^2}-\frac{2+x}{x^2-4}$$

$$=\frac{x+9}{x^2-4}+\frac{5-x}{-(x^2-4)}-\frac{2+x}{x^2-4}$$ Recognizing the middle denominator as the opposite of the others

$$=\frac{x+9}{x^2-4}+\frac{-(5-x)}{x^2-4}-\frac{2+x}{x^2-4}$$ Using $\frac{a}{-b}=\frac{-a}{b}$

$$=\frac{x+9-5+x-(2+x)}{x^2-4}$$

$$=\frac{x+9-5+x-2-x}{x^2-4}$$ Be careful with parentheses.

$$=\frac{x+2}{x^2-4}$$

$$=\frac{(x+2)\cdot 1}{(x+2)(x-2)}$$ Simplifying

$$=\frac{1}{x-2}.$$

b) The LCD $= x\cdot x\cdot(x+1)$, or $x^2(x+1)$.

$$\frac{1}{x}-\frac{1}{x^2}+\frac{2}{x+1}=\frac{1}{x}\cdot\frac{x(x+1)}{x(x+1)}-\frac{1}{x^2}\cdot\frac{(x+1)}{(x+1)}+\frac{2}{x+1}\cdot\frac{x^2}{x^2}$$

$$=\frac{x(x+1)}{x^2(x+1)}-\frac{x+1}{x^2(x+1)}+\frac{2x^2}{x^2(x+1)}$$

$$=\frac{x^2+x-(x+1)+2x^2}{x^2(x+1)}$$ Subtract this numerator. Don't forget the parentheses.

$$=\frac{x^2+x-x-1+2x^2}{x^2(x+1)}$$

$$=\frac{3x^2-1}{x^2(x+1)}$$ ❑

EXERCISE SET 6.5

Perform the indicated operation. Simplify, if possible.

1. $\frac{2}{x}+\frac{5}{x^2}$
2. $\frac{4}{x}+\frac{8}{x^2}$
3. $\frac{5}{6r}-\frac{7}{8r}$
4. $\frac{2}{9t}-\frac{11}{6t}$
5. $\frac{4}{xy^2}+\frac{6}{x^2y}$
6. $\frac{2}{c^2d}+\frac{7}{cd^3}$
7. $\frac{2}{9t^3}-\frac{1}{6t^2}$
8. $\frac{-2}{3xy^2}-\frac{6}{x^2y^3}$
9. $\frac{x+5}{8}+\frac{x-3}{12}$
10. $\frac{x-4}{9}+\frac{x+5}{6}$

11. $\frac{x-2}{6} - \frac{x+1}{3}$

12. $\frac{a+2}{2} - \frac{a-4}{4}$

13. $\frac{a+4}{16a} + \frac{3a+4}{4a^2}$

14. $\frac{2a-1}{3a^2} + \frac{5a+1}{9a}$

15. $\frac{4z-9}{3z} - \frac{3z-8}{4z}$

16. $\frac{x-1}{4x} - \frac{2x+3}{x}$

17. $\frac{x+y}{xy^2} + \frac{3x+y}{x^2y}$

18. $\frac{2c-d}{c^2d} + \frac{c+d}{cd^2}$

19. $\frac{4x+2t}{3xt^2} - \frac{5x-3t}{x^2t}$

20. $\frac{5x+3y}{2x^2y} - \frac{3x+4y}{xy^2}$

21. $\frac{3}{x-2} + \frac{3}{x+2}$

22. $\frac{2}{x-1} + \frac{2}{x+1}$

23. $\frac{5}{x+5} - \frac{3}{x-5}$

24. $\frac{2z}{z-1} - \frac{3z}{z+1}$

25. $\frac{3}{x+1} + \frac{2}{3x}$

26. $\frac{2}{x+5} + \frac{3}{4x}$

27. $\frac{3}{2t^2-2t} - \frac{5}{2t-2}$

28. $\frac{8}{x^2-4} - \frac{3}{x+2}$

29. $\frac{2x}{x^2-16} + \frac{x}{x-4}$

30. $\frac{4x}{x^2-25} + \frac{x}{x+5}$

31. $\frac{6}{z+4} - \frac{2}{3z+12}$

32. $\frac{t}{t-3} - \frac{5}{4t-12}$

33. $\frac{3}{x-1} + \frac{2}{(x-1)^2}$

34. $\frac{2}{x+3} + \frac{4}{(x+3)^2}$

35. $\frac{2t}{t^2-9} - \frac{3}{t-3}$

36. $\frac{2}{5x^2+5x} - \frac{4}{3x+3}$

37. $\frac{4a}{5a-10} + \frac{3a}{10a-20}$

38. $\frac{3a}{4a-20} + \frac{9a}{6a-30}$

39. $\frac{a}{x+a} - \frac{a}{x-a}$

40. $\frac{t}{y-t} - \frac{y}{y+t}$

41. $\frac{x+4}{x} + \frac{x}{x+4}$

42. $\frac{x}{x-5} + \frac{x-5}{x}$

43. $\frac{x}{x^2+5x+6} - \frac{2}{x^2+3x+2}$

44. $\frac{x}{x^2+11x+30} - \frac{5}{x^2+9x+20}$

45. $\frac{x}{x^2+2x+1} + \frac{1}{x^2+5x+4}$

46. $\frac{7}{a^2+a-2} + \frac{5}{a^2-4a+3}$

47. $\frac{x}{x^2+15x+56} - \frac{6}{x^2+13x+42}$

48. $\frac{-5}{x^2+17x+16} - \frac{3}{x^2+9x+8}$

49. $\frac{10}{x^2+x-6} + \frac{3x}{x^2-4x+4}$

50. $\frac{2}{z^2-z-6} + \frac{3}{z^2-9}$

51. $\frac{y+2}{y-7} + \frac{3-y}{49-y^2}$

52. $\frac{4-p}{25-p^2} + \frac{p+1}{p-5}$

53. $\frac{8x}{16-x^2} - \frac{5}{x-4}$

54. $\frac{5x}{x^2-9} - \frac{4}{3-x}$

55. $\frac{a}{a^2-1} + \frac{2a}{a-a^2}$

56. $\frac{3x+2}{3x+6} + \frac{x}{4-x^2}$

57. $\frac{4x}{x^2-y^2} - \frac{6}{y-x}$

58. $\frac{4-a^2}{a^2-9} - \frac{a-2}{3-a}$

Simplify.

59. $\frac{4y}{y^2-1} - \frac{2}{y} - \frac{2}{y+1}$

60. $\frac{x+6}{4-x^2} - \frac{x+3}{x+2} + \frac{x-3}{2-x}$

61. $\frac{2z}{1-2z} + \frac{3z}{2z+1} - \frac{3}{4z^2-1}$

62. $\frac{1}{x+y} + \frac{1}{x-y} - \frac{2x}{x^2-y^2}$

63. $\frac{5}{3-2x} + \frac{3}{2x-3} - \frac{x-3}{2x^2-x-3}$

64. $\frac{2r}{r^2-s^2} + \frac{1}{r+s} - \frac{1}{r-s}$

65. $\frac{3}{2c-1} - \frac{1}{c+2} - \frac{5}{2c^2+3c-2}$

66. $\frac{3y-1}{2y^2+y-3} - \frac{2-y}{y-1}$

67. $\frac{1}{x+y} - \frac{1}{x-y} + \frac{2x}{x^2-y^2}$

68. $\frac{1}{a-b} - \frac{1}{a+b} + \frac{2b}{a^2-b^2}$

Skill Maintenance

Graph.

69. $y = \frac{1}{2}x - 5$

70. $y = -\frac{1}{2}x - 5$

71. $y = 3$

72. $x = -5$

Synthesis

73. ◈ In your own words, describe the procedure for adding any two rational expressions.

74. Under what circumstances is the product of the denominators the LCD?

Write expressions for the perimeter and the area of the rectangle.

75.

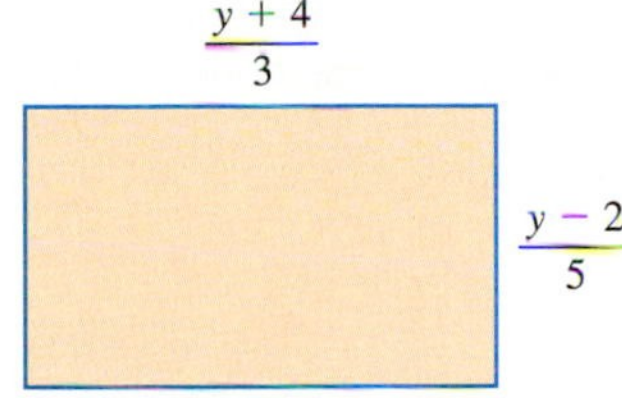

76.

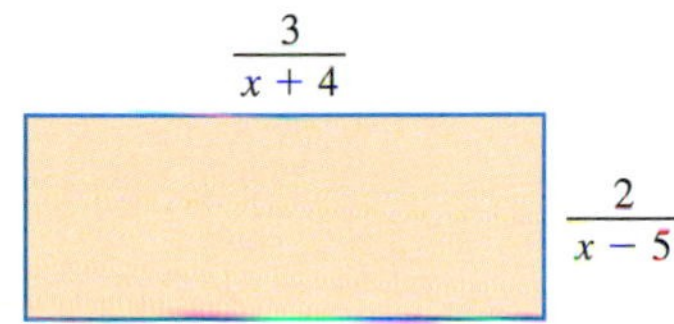

Add or subtract as indicated. Simplify, if possible.

77. $\frac{5}{z+2} + \frac{4z}{z^2-4} + 2$

78. $\frac{3z^2}{z^4-4} + \frac{5z^2-3}{2z^4+z^2-6}$

79. $\frac{1}{2xy-6x+ay-3a} - \frac{ay+xy}{(a^2-4x^2)(y^2-6y+9)}$

80. $\frac{x}{x^4-y^4} - \frac{1}{x^2+2xy+y^2}$

81. Express

$$\frac{a-3b}{a-b}$$

as a sum of two rational expressions with denominators that are opposites of each other. Answers may vary.

6.6 Complex Rational Expressions

Multiplying by the LCD • Simplifying by Adding or Subtracting

A **complex rational expression,** or **complex fractional expression,** is a rational expression that has one or more rational expressions within its numerator or denominator. Here are some examples:

$$\frac{1+\frac{2}{x}}{3}, \quad \frac{\frac{x+y}{2}}{\frac{2x}{x+1}}, \quad \frac{\frac{1}{3}+\frac{1}{5}}{\frac{2}{x}-\frac{x}{y}}.$$

These are rational expressions within the complex rational expression.

There are two methods to simplify complex rational expressions. We will consider them both. Use the one that works best for you or the one that your instructor directs you to use.

Multiplying by the LCD (Method 1)

Our first method of simplifying complex rational expressions relies on multiplying by an expression for 1.

To simplify a complex rational expression by using the LCD:

1. Find the LCD of all expressions *within* the complex rational expression.
2. Multiply the complex rational expression by 1, using the LCD to construct the 1.
3. Distribute and simplify. No rational expressions should remain within the complex rational expression.
4. Factor and, if possible, simplify.

EXAMPLE 1

Simplify: $\dfrac{\frac{1}{2}+\frac{3}{4}}{\frac{5}{6}-\frac{3}{8}}$.

Solution The denominators *within* the complex rational expression are 2, 4, 6, and 8. Their LCD is 24, so we multiply by $\frac{24}{24}$:

$$\frac{\frac{1}{2}+\frac{3}{4}}{\frac{5}{6}-\frac{3}{8}} = \frac{\frac{1}{2}+\frac{3}{4}}{\frac{5}{6}-\frac{3}{8}} \cdot \frac{24}{24}$$

Multiplying by 1, using the LCD: $\frac{24}{24} = 1$

$$= \frac{\left(\frac{1}{2}+\frac{3}{4}\right)24}{\left(\frac{5}{6}-\frac{3}{8}\right)24}.$$

Multiplying the numerator by 24

Don't forget the parentheses!

Multiplying the denominator by 24

Using the distributive law, we carry out the multiplications:

$$= \frac{\frac{1}{2}(24)+\frac{3}{4}(24)}{\frac{5}{6}(24)-\frac{3}{8}(24)}$$

$$= \frac{12+18}{20-9} \quad \text{Simplifying}$$

$$= \frac{30}{11}.$$

□

Multiplying in this manner has the effect of clearing fractions in both the top and bottom of the complex rational expression.

EXAMPLE 2

Simplify: **(a)** $\dfrac{\frac{3}{x}+\frac{1}{2x}}{\frac{1}{3x}-\frac{3}{4x}}$; **(b)** $\dfrac{1-\frac{1}{x}}{1-\frac{1}{x^2}}$; **(c)** $\dfrac{\frac{3}{2x-2}-\frac{1}{x+1}}{\frac{1}{x-1}+\frac{x}{x^2-1}}$.

Solution

a) The denominators within the complex expression are x, $2x$, $3x$, and $4x$, so the LCD is $12x$. We multiply by 1 using $(12x)/(12x)$:

$$\frac{\frac{3}{x}+\frac{1}{2x}}{\frac{1}{3x}-\frac{3}{4x}} = \frac{\frac{3}{x}+\frac{1}{2x}}{\frac{1}{3x}-\frac{3}{4x}} \cdot \frac{12x}{12x}$$

$$= \frac{\frac{3}{x}(12x)+\frac{1}{2x}(12x)}{\frac{1}{3x}(12x)-\frac{3}{4x}(12x)}$$ Using the distributive law

$$= \frac{36+6}{4-9}$$ Simplifying. All fractions have been cleared in both the numerator and the denominator.

$$= -\frac{42}{5}.$$

b) $$\frac{1-\frac{1}{x}}{1-\frac{1}{x^2}} = \frac{1-\frac{1}{x}}{1-\frac{1}{x^2}} \cdot \frac{x^2}{x^2}$$ The LCD is x^2 so we multiply by 1 using x^2/x^2.

$$= \frac{1 \cdot x^2 - \frac{1}{x} \cdot x^2}{1 \cdot x^2 - \frac{1}{x^2} \cdot x^2}$$ Using the distributive law

$$= \frac{x^2-x}{x^2-1}$$ All fractions have been cleared within the complex rational expression.

$$= \frac{x(x-1)}{(x+1)(x-1)}$$

$$= \frac{x}{x+1}$$ Factoring and simplifying; $\frac{x-1}{x-1} = 1$

c) Note that to find the LCD, we may have to factor first:

$$\frac{\frac{3}{2x-2}-\frac{1}{x+1}}{\frac{1}{x-1}+\frac{x}{x^2-1}}=\frac{\frac{3}{2(x-1)}-\frac{1}{x+1}}{\frac{1}{x-1}+\frac{x}{(x-1)(x+1)}}$$

The LCD is $2(x-1)(x+1)$.

$$=\frac{\frac{3}{2(x-1)}-\frac{1}{x+1}}{\frac{1}{x-1}+\frac{x}{(x-1)(x+1)}}\cdot\frac{2(x-1)(x+1)}{2(x-1)(x+1)}$$

Multiplying by 1

$$=\frac{\frac{3}{2(x-1)}\cdot 2(x-1)(x+1)-\frac{1}{x+1}\cdot 2(x-1)(x+1)}{\frac{1}{x-1}\cdot 2(x-1)(x+1)+\frac{x}{(x-1)(x+1)}\cdot 2(x-1)(x+1)}$$

Using the distributive law

$$=\frac{\frac{\cancel{2(x-1)}}{\cancel{2(x-1)}}\cdot 3(x+1)-\frac{\cancel{x+1}}{\cancel{x+1}}\cdot 2(x-1)}{\frac{\cancel{x-1}}{\cancel{x-1}}\cdot 2(x+1)+\frac{\cancel{(x-1)(x+1)}}{\cancel{(x-1)(x+1)}}\cdot 2x}$$

Removing factors of 1

$$=\frac{3(x+1)-2(x-1)}{2(x+1)+2x}$$

Simplifying

$$=\frac{3x+3-2x+2}{2x+2+2x}$$

Using the distributive law

$$=\frac{x+5}{4x+2}.$$

❑

Simplifying by Adding or Subtracting (Method 2)

A second method of simplifying complex rational expressions involves rewriting the expression as a quotient of two rational expressions.

To simplify a complex rational expression by first adding or subtracting:

1. Add or subtract, as indicated, to get a single rational expression in the numerator.
2. Add or subtract, as indicated, to get a single rational expression in the denominator.
3. Divide the numerator by the denominator (invert and multiply).
4. If possible, simplify by removing a factor of 1.

We will redo Examples 1 and 2 using this method.

EXAMPLE 3

Simplify.

a) $\dfrac{\frac{1}{2}+\frac{3}{4}}{\frac{5}{6}-\frac{3}{8}}$ b) $\dfrac{\frac{3}{x}+\frac{1}{2x}}{\frac{1}{3x}-\frac{3}{4x}}$ c) $\dfrac{1-\frac{1}{x}}{1-\frac{1}{x^2}}$ d) $\dfrac{\frac{3}{2x-2}-\frac{1}{x+1}}{\frac{1}{x-1}+\frac{x}{x^2-1}}$

Solution

a) $\dfrac{\frac{1}{2}+\frac{3}{4}}{\frac{5}{6}-\frac{3}{8}}=\dfrac{\frac{1}{2}\cdot\frac{2}{2}+\frac{3}{4}}{\frac{5}{6}\cdot\frac{4}{4}-\frac{3}{8}\cdot\frac{3}{3}}$ ← Multiplying by 1 to get the LCD, 4, in the numerator; ← Multiplying by 1 to get the LCD, 24, in the denominator

$=\dfrac{\frac{2}{4}+\frac{3}{4}}{\frac{20}{24}-\frac{9}{24}}=\dfrac{\frac{5}{4}}{\frac{11}{24}}$ Adding in the numerator; subtracting in the denominator

$=\dfrac{5}{4}\cdot\dfrac{24}{11}$ Multiplying by the reciprocal of the divisor

$=\dfrac{5}{2\cdot2}\cdot\dfrac{3\cdot2\cdot2\cdot2}{11}$ Factoring

$=\dfrac{5}{\cancel{2}\cdot\cancel{2}}\cdot\dfrac{3\cdot2\cdot\cancel{2}\cdot\cancel{2}}{11}$ Removing a factor of 1: $\dfrac{2\cdot2}{2\cdot2}=1$

$=\dfrac{30}{11}$

b) $\dfrac{\frac{3}{x}+\frac{1}{2x}}{\frac{1}{3x}-\frac{3}{4x}}=\dfrac{\frac{3}{x}\cdot\frac{2}{2}+\frac{1}{2x}}{\frac{1}{3x}\cdot\frac{4}{4}-\frac{3}{4x}\cdot\frac{3}{3}}$ ← Multiplying by 1 to get the LCD, $2x$, in the numerator; ← Multiplying by 1 to get the LCD, $12x$, in the denominator

$=\dfrac{\frac{6}{2x}+\frac{1}{2x}}{\frac{4}{12x}-\frac{9}{12x}}=\dfrac{\frac{7}{2x}}{\frac{-5}{12x}}$ Adding in the numerator; subtracting in the denominator

$=\dfrac{7}{2x}\cdot\dfrac{12x}{-5}$ Multiplying by the reciprocal of the divisor

$=\dfrac{7}{2x}\cdot\dfrac{6(2x)}{-5}$ Factoring

$=\dfrac{7}{\cancel{2x}}\cdot\dfrac{6(\cancel{2x})}{-5}$ Removing a factor of 1: $\dfrac{2x}{2x}=1$

$=\dfrac{42}{-5}$

$=-\dfrac{42}{5}$

c)

$$\frac{1-\frac{1}{x}}{1-\frac{1}{x^2}} = \frac{\frac{x}{x}-\frac{1}{x}}{\frac{x^2}{x^2}-\frac{1}{x^2}}$$

← Rewriting the numerator using the LCD

← Rewriting the denominator using the LCD

$$= \frac{\frac{x-1}{x}}{\frac{x^2-1}{x^2}}$$

Subtracting in the numerator; subtracting in the denominator

$$= \frac{x-1}{x}\cdot\frac{x^2}{x^2-1}$$

Multiplying by the reciprocal of the divisor

$$= \frac{(x-1)x\cdot x}{x(x-1)(x+1)}$$

Factoring

$$= \frac{\cancel{(x-1)}\cancel{x}\cdot x}{\cancel{x}\cancel{(x-1)}(x+1)}$$

Removing a factor of 1: $\frac{x(x-1)}{x(x-1)} = 1$

$$= \frac{x}{x+1}$$

d)

$$\frac{\frac{3}{2x-2}-\frac{1}{x+1}}{\frac{1}{x-1}+\frac{x}{x^2-1}} = \frac{\frac{3}{2(x-1)}\cdot\frac{x+1}{x+1}-\frac{1}{x+1}\cdot\frac{2(x-1)}{2(x-1)}}{\frac{1}{x-1}\cdot\frac{x+1}{x+1}+\frac{x}{(x+1)(x-1)}}$$

Multiplying by 1 to get the LCD, $2(x-1)(x+1)$, in the numerator

Multiplying by 1 to get the LCD, $(x+1)(x-1)$, in the denominator

$$= \frac{\frac{3x+3}{2(x-1)(x+1)}-\frac{2x-2}{2(x-1)(x+1)}}{\frac{x+1}{(x-1)(x+1)}+\frac{x}{(x+1)(x-1)}}$$

Multiplying rational expressions

$$= \frac{\frac{3x+3-(2x-2)}{2(x-1)(x+1)}}{\frac{x+1+x}{(x-1)(x+1)}} = \frac{\frac{x+5}{2(x-1)(x+1)}}{\frac{2x+1}{(x-1)(x+1)}}$$

Subtracting in the numerator; adding in the denominator

$$= \frac{x+5}{2(x-1)(x+1)}\cdot\frac{(x-1)(x+1)}{2x+1}$$

Multiplying by the reciprocal of the divisor

$$= \frac{x+5}{2\cancel{(x-1)}\cancel{(x+1)}}\cdot\frac{\cancel{(x-1)}\cancel{(x+1)}}{2x+1}$$

Removing a factor of 1: $\frac{(x-1)(x+1)}{(x-1)(x+1)} = 1$

$$= \frac{x+5}{2(2x+1)}$$

$$= \frac{x+5}{4x+2}$$

❑

EXERCISE SET 6.6

Simplify.

1. $\dfrac{1+\frac{9}{16}}{1-\frac{3}{4}}$

2. $\dfrac{9-\frac{1}{4}}{3+\frac{1}{2}}$

3. $\dfrac{1-\frac{3}{5}}{1+\frac{1}{5}}$

4. $\dfrac{\frac{5}{27}-5}{\frac{1}{3}+1}$

5. $\dfrac{\frac{1}{x}+4}{\frac{1}{x}-3}$

6. $\dfrac{\frac{3}{s}+s}{\frac{s}{3}+s}$

7. $\dfrac{\frac{1}{2}+\frac{3}{4}}{\frac{5}{8}-\frac{5}{6}}$

8. $\dfrac{\frac{2}{3}-\frac{5}{6}}{\frac{3}{4}+\frac{7}{8}}$

9. $\dfrac{\frac{2}{y}+\frac{1}{2y}}{y+\frac{y}{2}}$

10. $\dfrac{4-\frac{1}{x^2}}{2-\frac{1}{x}}$

11. $\dfrac{8+\frac{8}{d}}{1+\frac{1}{d}}$

12. $\dfrac{2-\frac{3}{b}}{2-\frac{b}{3}}$

13. $\dfrac{\frac{x}{8}-\frac{8}{x}}{\frac{1}{8}+\frac{1}{x}}$

14. $\dfrac{\frac{2}{m}+\frac{m}{2}}{\frac{m}{2}-\frac{2}{m}}$

15. $\dfrac{1+\frac{1}{y}}{1-\frac{1}{y^2}}$

16. $\dfrac{\frac{1}{q^2}-1}{\frac{1}{q}+1}$

17. $\dfrac{\frac{25-a^2}{5a}}{\frac{a+5}{5}}$

18. $\dfrac{\frac{x^2-y^2}{xy}}{\frac{x-y}{y}}$

19. $\dfrac{\frac{x}{x-y}}{\frac{x^2}{x^2-y^2}}$

20. $\dfrac{\frac{x}{y}-\frac{y}{x}}{\frac{1}{y}+\frac{1}{x}}$

21. $\dfrac{\frac{3}{m}+\frac{2}{m^3}}{\frac{4}{m^2}-\frac{3}{m}}$

22. $\dfrac{\frac{a}{a^2-b^2}}{\frac{a^2}{a+b}}$

23. $\dfrac{\frac{5}{4x^3}-\frac{3}{8x}}{\frac{3}{2x}+\frac{3}{4x^3}}$

24. $\dfrac{\frac{2}{7a^4}-\frac{1}{14a}}{\frac{3}{5a^2}+\frac{2}{15a}}$

25. $\dfrac{\frac{a}{6b^3}+\frac{4}{9b^2}}{\frac{5}{6b}-\frac{1}{9b^3}}$

26. $\dfrac{\frac{x}{5y^3}-\frac{3}{10y}}{\frac{x}{10y}+\frac{3}{y^4}}$

27. $\dfrac{\frac{2}{x^2y}+\frac{3}{xy^2}}{\frac{2}{xy^3}+\frac{1}{x^2y}}$

28. $\dfrac{\frac{5}{ab^4}+\frac{2}{a^3b}}{\frac{5}{a^3b}-\frac{3}{ab}}$

29. $\dfrac{3-\frac{2}{a^4}}{2+\frac{3}{a^3}}$

30. $\dfrac{2-\frac{3}{x^2}}{2+\frac{3}{x^4}}$

31. $\dfrac{\frac{1}{x+h}-\frac{1}{x}}{h}$

32. $\dfrac{\frac{1}{a-h}-\frac{1}{a}}{h}$

33. $\dfrac{\dfrac{y^2-y-6}{y^2-5y-14}}{\dfrac{y^2+6y+5}{y^2-6y-7}}$

34. $\dfrac{\dfrac{x^2-x-12}{x^2-2x-15}}{\dfrac{x^2+8x+12}{x^2-5x-14}}$

35. $\dfrac{\dfrac{x+5}{x^2}}{\dfrac{2}{x}-\dfrac{3}{x^2}}$

36. $\dfrac{\dfrac{a-7}{a^3}}{\dfrac{3}{a^2}+\dfrac{2}{a}}$

37. $\dfrac{x-3+\dfrac{2}{x}}{x-4+\dfrac{3}{x}}$

38. $\dfrac{x-2+\dfrac{x}{3}}{x+7-\dfrac{4}{5x}}$

39. $\dfrac{\dfrac{1}{x-2}+\dfrac{3}{x-1}}{\dfrac{2}{x-1}+\dfrac{5}{x-2}}$

40. $\dfrac{\dfrac{2}{y-3}+\dfrac{1}{y+1}}{\dfrac{3}{y+1}+\dfrac{4}{y-3}}$

41. $\dfrac{\dfrac{3}{a^2-9}+\dfrac{2}{a+3}}{\dfrac{4}{a^2-9}+\dfrac{1}{a+3}}$

42. $\dfrac{\dfrac{2}{a^2-1}+\dfrac{1}{a+1}}{\dfrac{3}{a^2-1}+\dfrac{2}{a-1}}$

43. $\dfrac{\dfrac{1}{x^2-1}+\dfrac{1}{x^2+4x+3}}{\dfrac{1}{x^2-1}+\dfrac{1}{x^2-3x+2}}$

44. $\dfrac{\dfrac{1}{x^2-4}+\dfrac{1}{x^2+3x+2}}{\dfrac{1}{x^2-4}+\dfrac{1}{x^2-4x+4}}$

Skill Maintenance

45. Subtract:

$$(5x^4-6x^3+23x^2-79x+24)$$
$$-(-18x^4-56x^3+84x-17).$$

46. The length of a rectangle is 3 yd greater than the width. The area of the rectangle is 10 yd^2. Find the perimeter.

Synthesis

47. ◈ Which of the two methods presented would you use to simplify Exercise 22? Why?

48. ◈ Which of the two methods presented would you use to simplify Exercise 28? Why?

49. Find simplified form for the reciprocal of

$$\frac{2}{x-1}-\frac{1}{3x-2}.$$

Simplify.

50. $\dfrac{\dfrac{a}{b}+\dfrac{c}{d}}{\dfrac{b}{a}+\dfrac{d}{c}}$

51. $\dfrac{\dfrac{a}{b}-\dfrac{c}{d}}{\dfrac{b}{a}-\dfrac{d}{c}}$

52. $\left[\dfrac{\dfrac{x+1}{x-1}+1}{\dfrac{x+1}{x-1}-1}\right]^5$

53. $1+\dfrac{1}{1+\dfrac{1}{1+\dfrac{1}{x}}}$

54. $\dfrac{\dfrac{z}{1-\dfrac{z}{2+2z}}-2z}{\dfrac{2z}{5z-2}-3}$

55. $\dfrac{a^{-1}+b^{-1}}{\dfrac{a^2-b^2}{ab}}$

6.7 Solving Rational Equations

A New Type of Equation • A Visual Interpretation

In Chapters 1 and 2, we first distinguished between *equivalent expressions*, like $x+x$ and $2x$, and *equivalent equations*, like $3(x+2)=18$ and $3x+6=18$. Recall that equivalent equations have the same solution sets. In Sections 6.1–6.6, we saw how to write equivalent expressions but nowhere did we solve an equation. We now begin to solve equations and, in so doing, we will use the equation-solving principles discussed in Sections 2.1 and 5.7.

A **rational,** or **fractional, equation** is an equation containing one or more rational expressions. Here are some examples:

$$\frac{2}{3}+\frac{5}{6}=\frac{x}{9}, \qquad x+\frac{6}{x}=-5, \qquad \frac{x^2}{x-1}=\frac{1}{x-1}.$$

To solve a rational equation:

1. Clear the equation of fractions by multiplying on both sides by the LCD of all rational expressions in the equation.
2. Solve the resulting equation using the addition principle, the multiplication principle, and the principle of zero products.
3. Check the possible solution(s) in the original equation.

As we proceed, you will see why the third step, checking, is so important.

EXAMPLE 1

Solve: $\frac{x}{6}-\frac{x}{8}=\frac{1}{12}$.

Solution The LCD is 24. We multiply on both sides by 24:

$$24\left(\frac{x}{6}-\frac{x}{8}\right)=24\cdot\frac{1}{12}$$ Using the multiplication principle to multiply both sides by the LCD. Parentheses are important!

$$24\cdot\frac{x}{6}-24\cdot\frac{x}{8}=24\cdot\frac{1}{12}$$ Using the distributive law

Be sure to multiply *each* term by the LCD.

$$4x-3x=2$$ Simplifying

$$x=2.$$

Check:

$$\begin{array}{c|c} \frac{x}{6}-\frac{x}{8} & \frac{1}{12} \\ \hline \frac{2}{6}-\frac{2}{8} \; ? & \frac{1}{12} \\ \frac{1}{3}-\frac{1}{4} & \\ \frac{4}{12}-\frac{3}{12} & \\ \frac{1}{12} & \frac{1}{12} \quad \text{TRUE} \end{array}$$

This checks, so the solution is 2. ❑

Up to now, the multiplication principle has been used only to multiply both sides of an equation by a constant. Because rational equations often contain variable expressions in a denominator, we will now have occasion to multiply both sides of an equation by a variable expression. Since variable expressions are sometimes equal to 0, multiplying both sides of an equation by a variable expression does not always produce an equivalent equation. This is why checking in the original equation is so important.

EXAMPLE 2 Solve.

a) $\dfrac{2}{3x} + \dfrac{1}{x} = 10$

b) $x + \dfrac{6}{x} = -5$

c) $1 + \dfrac{3x}{x+2} = \dfrac{-6}{x+2}$

d) $\dfrac{3}{x-5} + \dfrac{1}{x+5} = \dfrac{2}{x^2-25}$

Solution

a) The LCD is $3x$. We multiply on both sides by $3x$:

$$\frac{2}{3x} + \frac{1}{x} = 10$$

$$3x\left(\frac{2}{3x} + \frac{1}{x}\right) = 3x \cdot 10$$ Using the multiplication principle to multiply both sides by the LCD. Don't forget the parentheses!

$$3x \cdot \frac{2}{3x} + 3x \cdot \frac{1}{x} = 3x \cdot 10$$ Multiplying to remove parentheses

$$2 + 3 = 30x$$ Simplifying

$$5 = 30x$$

$$\frac{5}{30} = x, \quad \text{so } x = \frac{1}{6}.$$

Check:

$$\frac{2}{3x} + \frac{1}{x} = 10$$

$$\frac{2}{3 \cdot \frac{1}{6}} + \frac{1}{\frac{1}{6}} \; ? \; 10$$

$$\frac{2}{\frac{1}{2}} + \frac{1}{\frac{1}{6}}$$

$$4 + 6$$

$$10 \quad | \quad 10 \quad \text{TRUE}$$

The solution is $\frac{1}{6}$.

b)

$$x + \frac{6}{x} = -5$$ The LCD is x.

$$x\left(x + \frac{6}{x}\right) = -5x$$ Multiplying on both sides by x

$$x \cdot x + x \cdot \frac{6}{x} = -5x$$ Note that each term on the left is now multiplied by x.

$$x^2 + 6 = -5x$$ Simplifying. Note that we have a quadratic equation.

$$x^2 + 5x + 6 = 0$$ Using the addition principle to add $5x$ on both sides

$$(x+3)(x+2) = 0$$ Factoring

$$x + 3 = 0 \quad or \quad x + 2 = 0$$ Using the principle of zero products

$$x = -3 \quad or \quad x = -2$$

Check: For -3: For -2:

$$\begin{array}{c|c} x + \frac{6}{x} = -5 & \\ \hline -3 + \frac{6}{-3} \;?\; & -5 \\ -3 - 2 & \\ -5 & -5 \text{ TRUE} \end{array} \qquad \begin{array}{c|c} x + \frac{6}{x} = -5 & \\ \hline -2 + \frac{6}{-2} \;?\; & -5 \\ -2 - 3 & \\ -5 & -5 \text{ TRUE} \end{array}$$

Both of these check, so there are two solutions, -3 and -2.

c)

$$1 + \frac{3x}{x+2} = \frac{-6}{x+2}$$ The LCD is $x + 2$.

$$(x+2)\left(1 + \frac{3x}{x+2}\right) = (x+2)\frac{-6}{x+2}$$ Multiplying on both sides by $x + 2$

$$(x+2)\cdot 1 + (x+2)\frac{3x}{x+2} = (x+2)\frac{-6}{x+2}$$ Using the distributive law

$$x + 2 + 3x = -6$$ Simplifying

$$4x + 2 = -6$$

$$4x = -8$$

$$x = -2$$

Check:

$$\begin{array}{c|c} 1 + \frac{3x}{x+2} = \frac{-6}{x+2} & \\ \hline 1 + \frac{3(-2)}{-2+2} \;?\; & \frac{-6}{-2+2} \\ 1 + \frac{-6}{0} & \frac{-6}{0} \end{array}$$

$\frac{-6}{0}$ is undefined.

We see that -2 is *not* a solution of the original equation because it results in division by 0. The equation has no solution.

d)

$$\frac{3}{x-5} + \frac{1}{x+5} = \frac{2}{x^2-25}$$ Note that $x^2 - 25 = (x-5)(x+5)$.

$$(x-5)(x+5)\left(\frac{3}{x-5} + \frac{1}{x+5}\right) = \frac{2}{(x-5)(x+5)}(x-5)(x+5)$$ Multiplying both sides by the LCD

$$\frac{\cancel{(x-5)}(x+5)\cdot 3}{\cancel{x-5}} + \frac{(x-5)\cancel{(x+5)}}{\cancel{x+5}} = \frac{2\cancel{(x-5)}\cancel{(x+5)}}{\cancel{(x-5)}\cancel{(x+5)}}$$ Using the distributive law

$$(x+5)3 + (x-5) = 2$$ Removing factors of 1: $\frac{x-5}{x-5} = 1$, $\frac{x+5}{x+5} = 1$, and $\frac{(x-5)(x+5)}{(x-5)(x+5)} = 1$

$$3x + 15 + x - 5 = 2$$ Using the distributive law

$$4x + 10 = 2$$

$$4x = -8$$

$$x = -2$$

We leave it to the student to check that the number -2 is the solution. ❑

A Visual Interpretation

It is possible to solve a rational equation by graphing. The procedure consists of graphing each side of the equation and then determining the first coordinate(s) of any point(s) of intersection. (With the advent of the graphing calculator, producing such graphs requires little work.) For example, the equation

$$\frac{x}{4} + \frac{x}{2} = 6$$

can be solved by graphing the equations

$$y = \frac{x}{4} + \frac{x}{2} \quad \text{and} \quad y = 6$$

on the same set of axes.

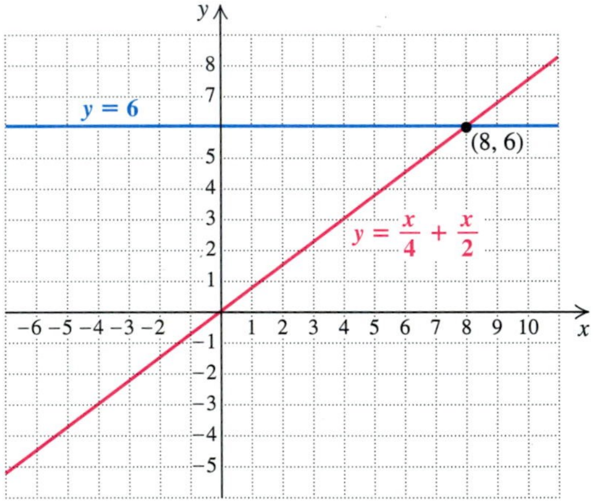

As you can see in the graph above, when $x = 8$, the value of $x/4 + x/2$ is 6. Thus, 8 is the solution of $x/4 + x/2 = 6$. This solution can be checked by substitution:

$$\frac{x}{4} + \frac{x}{2} = \frac{8}{4} + \frac{8}{2} = 2 + 4 = 6.$$

A grapher can be used to check that Example 2(c),

$$1 + \frac{3x}{x+2} = \frac{-6}{x+2},$$

has no solution. To do so, graph

$$y = 1 + \frac{3x}{x+2}$$

and

$$y = \frac{-6}{x+2}$$

on the same set of axes. Then use the Zoom feature to convince yourself that no points of intersection exist.

EXERCISE SET 6.7

Solve.

1. $\frac{3}{8}+\frac{4}{5}=\frac{x}{20}$

2. $\frac{3}{5}+\frac{2}{3}=\frac{x}{9}$

3. $\frac{2}{3}-\frac{5}{6}=\frac{1}{x}$

4. $\frac{1}{8}-\frac{3}{5}=\frac{1}{x}$

5. $\frac{1}{6}+\frac{1}{8}=\frac{1}{t}$

6. $\frac{1}{8}+\frac{1}{10}=\frac{1}{t}$

7. $x+\frac{4}{x}=-5$

8. $x+\frac{3}{x}=-4$

9. $\frac{x}{4}-\frac{4}{x}=0$

10. $\frac{x}{5}-\frac{5}{x}=0$

11. $\frac{5}{x}=\frac{6}{x}-\frac{1}{3}$

12. $\frac{4}{x}=\frac{5}{x}-\frac{1}{2}$

13. $\frac{5}{3x}+\frac{3}{x}=1$

14. $\frac{3}{4x}+\frac{5}{x}=1$

15. $\frac{x-7}{x+2}=\frac{1}{4}$

16. $\frac{a-2}{a+3}=\frac{3}{8}$

17. $\frac{2}{x+1}=\frac{1}{x-2}$

18. $\frac{5}{x-1}=\frac{3}{x+2}$

19. $\frac{x}{6}-\frac{x}{10}=\frac{1}{6}$

20. $\frac{x}{8}-\frac{x}{12}=\frac{1}{8}$

21. $\frac{x+1}{3}-1=\frac{x-1}{2}$

22. $\frac{x+2}{5}-1=\frac{x-2}{4}$

23. $\frac{a-3}{3a+2}=\frac{1}{5}$

24. $\frac{x-1}{2x+5}=\frac{1}{4}$

25. $\frac{x-1}{x-5}=\frac{4}{x-5}$

26. $\frac{x-7}{x-9}=\frac{2}{x-9}$

27. $\frac{2}{x+3}=\frac{5}{x}$

28. $\frac{3}{x+4}=\frac{4}{x}$

29. $\frac{x-2}{x-3}=\frac{x-1}{x+1}$

30. $\frac{2b-3}{3b+2}=\frac{2b+1}{3b-2}$

31. $\frac{1}{x+3}+\frac{1}{x-3}=\frac{1}{x^2-9}$

32. $\frac{4}{x-3}+\frac{2x}{x^2-9}=\frac{1}{x+3}$

33. $\frac{x}{x+4}-\frac{4}{x-4}=\frac{x^2+16}{x^2-16}$

34. $\frac{5}{y-3}-\frac{30}{y^2-9}=1$

35. $\frac{-3}{y-7}=\frac{-10-y}{7-y}$

36. $\frac{4-m}{8-m}=\frac{4}{m-8}$

Skill Maintenance

Simplify.

37. $(a^2b^5)^{-3}$

38. $(x^{-2}y^{-3})^{-4}$

39. $\left(\frac{2x}{t^2}\right)^4$

40. $\left(\frac{y^3}{w^2}\right)^{-2}$

Synthesis

41. ◈ Explain the difference between adding rational expressions and solving rational equations.

42. ◈ Without multiplying by the LCD and solving, explain why the rational equation

$$\frac{x}{x+2}=\frac{-2}{x+2}$$

cannot have a solution. (*Hint:* Examine both numerators and denominators carefully.)

Solve.

43. $\frac{4}{y-2}-\frac{2y-3}{y^2-4}=\frac{5}{y+2}$

44. $\frac{x}{x^2+3x-4}+\frac{x+1}{x^2+6x+8}=\frac{2x}{x^2+x-2}$

45. $\frac{12-6x}{x^2-4}=\frac{3x}{x+2}-\frac{2x-3}{x-2}$

46. $\frac{x^2}{x^2-4}=\frac{x}{x+2}-\frac{2x}{2-x}$

47. $4a-3=\frac{a+13}{a+1}$

48. $\frac{3x-9}{x-3}=\frac{5x-4}{2}$

49. $\frac{y^2-4}{y+3}=2-\frac{y-2}{y+3}$

50. $\frac{3a-5}{a^2+4a+3}+\frac{2a+2}{a+3}=\frac{a-3}{a+1}$

51. Use a grapher to check the solutions to Examples 1–2(b). Be sure to use the Zoom and Trace features carefully.

52. Use a grapher to confirm the answers to Exercises 9, 25, and 43.

6.8 Problem Solving: Rational Equations and Proportions

Problem Solving • Problems Involving Work • Problems Involving Motion • Problems Involving Proportions

In many areas of study, applications involving rates, proportions, or reciprocals translate to rational equations. By using the five steps for problem solving and the lessons of Section 6.7, we can now solve such problems.

Problem Solving

EXAMPLE 1

A number plus three times its reciprocal is -4. Find the number.

Solution

1. FAMILIARIZE. Let's try to guess the number. Try 2: $2 + 3\cdot\frac{1}{2} = \frac{7}{2}$. Although $\frac{7}{2} \neq -4$, the guess helps us to better understand how the problem can be translated. We can also see that a positive number cannot be a solution of the problem. Let x = the number.

2. TRANSLATE. From the familiarization step, we can translate directly:

A number plus three times its reciprocal is -4.

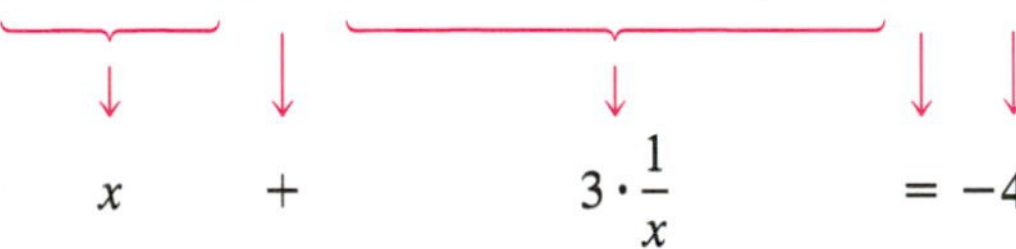

$$x \quad + \quad 3\cdot\frac{1}{x} \quad = -4$$

3. CARRY OUT. We solve the equation:

$$x + 3\cdot\frac{1}{x} = -4$$

$$x\left(x + \frac{3}{x}\right) = x(-4) \quad \text{Multiplying by the LCD, } x\text{, on both sides}$$

$$x\cdot x + x\cdot\frac{3}{x} = -4x \quad \text{Using the distributive law}$$

$$x^2 + 3 = -4x \quad \text{Simplifying}$$

$$\left.\begin{array}{l} x^2 + 4x + 3 = 0 \\ (x+3)(x+1) = 0 \\ x + 3 = 0 \quad or \quad x + 1 = 0 \end{array}\right\} \text{Using the principle of zero products}$$

$$x = -3 \quad or \quad x = -1.$$

4. CHECK. Three times the reciprocal of -3 is $3\cdot\frac{1}{-3}$, or -1. Since $-3 + (-1) = -4$, the number -3 is a solution.

   Three times the reciprocal of -1 is $3\cdot\frac{1}{-1}$, or -3. Since $-1 + (-3) = -4$, the number -1 is a solution.

5. STATE. The solutions are -3 and -1. ❑

Problems Involving Work

EXAMPLE 2

Cecilia and Aaron work as volunteers at a town's recycling depot. Cecilia can sort a day's accumulation of recyclables in 4 hr, while Aaron requires 6 hr to do the same job. How long would it take them, working together, to sort the recyclables?

Solution

1. FAMILIARIZE. We familiarize ourselves with the problem by considering two *incorrect* ways of translating the problem to mathematical language.

   a) A common incorrect way to translate the problem is to just add the two times:

   $$4 \text{ hr} + 6 \text{ hr} = 10 \text{ hr}.$$

   Let's think about this. If Cecilia can do the sorting alone in 4 hr, then Cecilia and Aaron together should take *less* than 4 hr. Thus we reject 10 hr as a solution and reason that the answer must be less than 4 hr.

   b) Another incorrect way to translate the problem is to assume that Cecilia does half the sorting and Aaron does the other half. Then

   Cecilia sorts $\frac{1}{2}$ of the accumulation in $\frac{1}{2}(4 \text{ hr})$, or 2 hr, and
   Aaron sorts $\frac{1}{2}$ of the accumulation in $\frac{1}{2}(6 \text{ hr})$, or 3 hr.

   This would waste time since Cecilia would finish 1 hr earlier than Aaron. If Cecilia helps Aaron after completing her half, the entire job should take between 2 hr and 3 hr.

   We proceed to a translation by considering how much of the sorting is finished in 1 hr, 2 hr, 3 hr, and so on. It takes Cecilia 4 hr to sort the recyclables alone. Then, in 1 hr, she can do $\frac{1}{4}$ of the job. It takes Aaron 6 hr to do the sorting alone. Then, in 1 hr, he can do $\frac{1}{6}$ of the job. Working together, they can complete

   $$\frac{1}{4} + \frac{1}{6}, \quad \text{or} \quad \frac{5}{12} \text{ of the sorting in 1 hr.}$$

In 2 hr, Cecilia can do $2(\frac{1}{4})$ of the sorting and Aaron can do $2(\frac{1}{6})$ of the sorting. Working together, they can complete

$$2\left(\frac{1}{4}\right) + 2\left(\frac{1}{6}\right), \quad \text{or } \frac{5}{6} \text{ of the sorting in 2 hr.}$$

Continuing this reasoning, we can form a table like the following one.

Time	Fraction of the Sorting Completed		
	Cecilia	Aaron	Together
1 hr	$\frac{1}{4}$	$\frac{1}{6}$	$\frac{1}{4} + \frac{1}{6}$, or $\frac{5}{12}$
2 hr	$2\left(\frac{1}{4}\right)$	$2\left(\frac{1}{6}\right)$	$2\left(\frac{1}{4}\right) + 2\left(\frac{1}{6}\right)$, or $\frac{5}{6}$
3 hr	$3\left(\frac{1}{4}\right)$	$3\left(\frac{1}{6}\right)$	$3\left(\frac{1}{4}\right) + 3\left(\frac{1}{6}\right)$, or $1\frac{1}{4}$
t hr	$t\left(\frac{1}{4}\right)$	$t\left(\frac{1}{6}\right)$	$t\left(\frac{1}{4}\right) + t\left(\frac{1}{6}\right)$

From the table, we see that if they work 3 hr, the fraction of the sorting that they complete is $1\frac{1}{4}$, which is more of the job than needs to be done. We also see that the answer is somewhere between 2 hr and 3 hr. What we want is a number t such that the fraction of the sorting that is completed in t hours is 1; that is, the job is just completed—not more and not less.

2. TRANSLATE. From the table, we see that the time we want is some number t for which

$$t\left(\frac{1}{4}\right) + t\left(\frac{1}{6}\right) = 1,$$

Portion of work done by Cecilia in t hr; Portion of work done by Aaron in t hr

or

$$\frac{t}{4} + \frac{t}{6} = 1.$$

3. CARRY OUT. We solve the equation:

$$\frac{t}{4} + \frac{t}{6} = 1$$

$$12\left(\frac{t}{4} + \frac{t}{6}\right) = 12 \cdot 1 \quad \text{The LCD is } 2 \cdot 2 \cdot 3, \text{ or } 12.$$

$$12 \cdot \frac{t}{4} + 12 \cdot \frac{t}{6} = 12$$

$$3t + 2t = 12$$

$$5t = 12$$

$$t = \frac{12}{5}, \quad \text{or } 2\frac{2}{5} \text{ hr.}$$

4. CHECK. The check can be done following the pattern used in the table of the *Familiarize* step above:

$$\frac{12}{5}\left(\frac{1}{4}\right) + \frac{12}{5}\left(\frac{1}{6}\right) = \frac{3}{5} + \frac{2}{5} = \frac{5}{5} = 1.$$

We also have a partial check in that we expected the answer to be between 2 hr and 3 hr.

5. STATE. It takes $2\frac{2}{5}$ hr for them to complete the sorting working together. ❑

The Work Principle

Suppose that a = the time it takes A to complete a task, b = the time it takes B to complete the same task, and t = the time it takes them to complete the task working together. Then

$$t\left(\frac{1}{a}\right) + t\left(\frac{1}{b}\right) = 1, \quad \text{or} \quad \frac{t}{a} + \frac{t}{b} = 1.$$

Problems Involving Motion

Problems that deal with distance, speed (or rate), and time are called **motion problems.** Translation of these problems involves the distance formula, $d = r \cdot t$, and/or the equivalent formulas $r = d/t$ and $t = d/r$.

EXAMPLE 3

One car travels 20 km/h faster than another. In the same time that one car travels 240 km, the other travels 160 km. Find their speeds.

Solution

1. FAMILIARIZE. Let's guess that the slow car is moving 30 km/h. The fast car would then be traveling 30 + 20, or 50 km/h. Thus, if r represents the slow car's speed in kilometers per hour, then the fast car's speed can be represented by $r + 20$.

If the fast car were traveling 50 km/h, it would travel 240 km in 240/50, or $4\frac{4}{5}$ hr. At 30 km/h, the slow car would travel 160 km in 160/30, or $5\frac{1}{3}$ hr. Because we are told that both cars spend the same amount of time traveling, and because

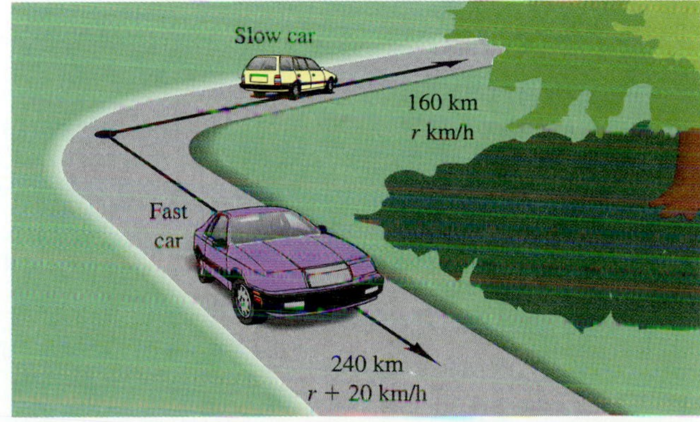

$4\frac{4}{5}$ hr $\neq$ $5\frac{1}{3}$ hr, we see that our guess of 30 km/h is incorrect. We let t = the time, in hours, that the cars spend traveling and organize the given information in a table.

	d = Distance	r · Speed	t Time
Slow Car	160	r	t
Fast Car	240	$r + 20$	t

2. **TRANSLATE.** Examine how we checked our guess. We found, and then compared, the times of the two cars. To find the times, we divided the distances, 240 km and 160 km, by the rates, 50 km/h and 30 km/h, respectively. Thus the t's in the above table can be replaced, using the formula $t = d/r$.

	Distance	Speed	Time
Slow Car	160	r	$160/r$
Fast Car	240	$r + 20$	$240/(r + 20)$

The times are the same.

Since the times are the same for both cars, we have the equation

$$\frac{160}{r} = \frac{240}{r + 20}.$$

3. **CARRY OUT.** To solve the equation, we first multiply on both sides by the LCD, $r(r + 20)$:

$$r(r + 20) \cdot \frac{160}{r} = r(r + 20) \cdot \frac{240}{r + 20} \quad \text{Multiplying on both sides by the LCD, } r(r + 20)$$

$$160(r + 20) = 240r \quad \text{Simplifying}$$

$$160r + 3200 = 240r \quad \text{Removing parentheses}$$

$$3200 = 80r \quad \text{Subtracting } 160r$$

$$\frac{3200}{80} = r \quad \text{Dividing by 80}$$

$$40 = r.$$

We now have a possible solution. The speed of the slow car is 40 km/h, and the speed of the fast car is 40 + 20, or 60 km/h.

4. **CHECK.** We first reread the problem to see what we were to find. We check the speeds of 40 km/h for the slow car and 60 km/h for the fast car. The fast car does travel 20 km/h faster than the slow car. If the fast car goes 240 km at 60 km/h, it has traveled for 240/60, or 4 hr. If the slow car goes 160 km at 40 km/h, it has traveled for 160/40, or 4 hr. Since the times are the same, the speeds check.

5. **STATE.** The slow car has a speed of 40 km/h, and the fast car has a speed of 60 km/h. ❑

Problems Involving Proportions

A **ratio** of two quantities is their quotient. For example, 37% is the ratio of 37 to 100, $\frac{37}{100}$. A **proportion** is an equation stating that two ratios are equal.

Proportion

An equality of ratios, $A/B = C/D$, is called a *proportion*. The numbers named in a proportion are said to be *proportional* to each other.

Proportions can be used to solve applied problems by expressing a single ratio in two ways.

EXAMPLE 4

A car travels 135 mi on 6 gal of gas. Find the amount of gas required for a 360-mi trip.

Solution By assuming that the car always burns gas at the same rate, we can form a proportion in which the ratio of miles to gallons is expressed in two ways:

$$\begin{matrix}\text{Miles} \rightarrow \\ \text{Gas} \rightarrow\end{matrix} \frac{135}{6} = \frac{360}{x}. \begin{matrix}\leftarrow \text{Miles} \\ \leftarrow \text{Gas}\end{matrix}$$

To solve for x, we multiply both sides of the equation by the LCD, $6x$:

$$6x \cdot \frac{135}{6} = 6x \cdot \frac{360}{x}$$

$$\cancel{6} \cdot \frac{135x}{\cancel{6}} = \cancel{x} \cdot \frac{6 \cdot 360}{\cancel{x}}. \quad \text{Removing factors of 1: } \frac{6}{6} = 1 \text{ and } \frac{x}{x} = 1$$

Note that the equation $135x = 6 \cdot 360$ could have been obtained from the original equation by *cross-multiplying:*

$$\frac{135}{6} = \frac{360}{x}. \quad 135x \text{ and } 6 \cdot 360 \text{ are called } \textit{cross-products}.$$

We complete the problem as follows:

$$135x = 6 \cdot 360$$

$$x = \frac{6 \cdot 360}{135} \quad \text{Dividing by 135 on both sides}$$

$$x = 16. \quad \text{Simplifying}$$

The trip will require 16 gal of gas. ❑

Proportions occur in geometry when we are studying *similar triangles*. Two triangles are said to be **similar** when corresponding angles have the same measure and corresponding sides are proportional. To illustrate, if triangle ABC is similar to triangle RST, the measure of angle A = the measure of angle R, the measure of angle B = the measure of angle S, the measure of angle C = the measure of angle T, and

$$\frac{a}{r} = \frac{b}{s} = \frac{c}{t}.$$

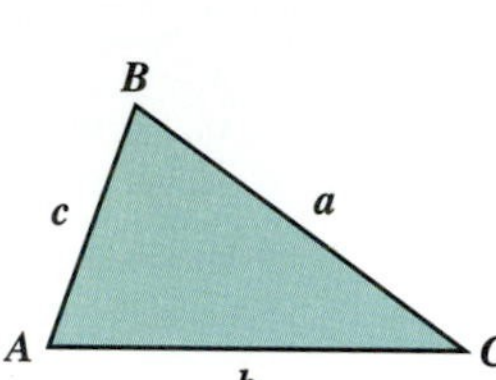

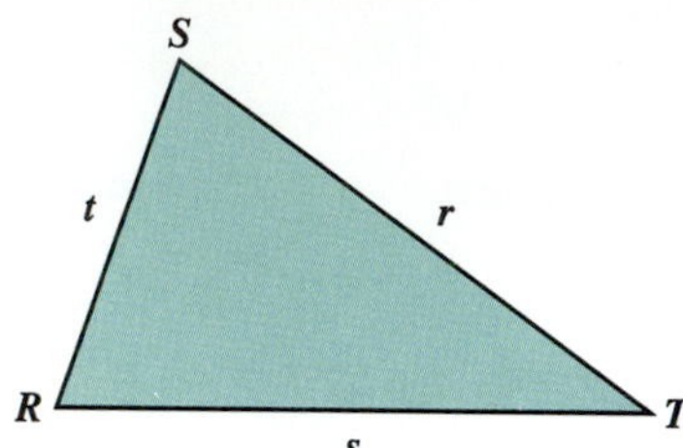

EXAMPLE 5

Triangles ABC and XYZ are similar. Solve for z if $x = 10$, $a = 8$, and $c = 5$.

Solution We make a sketch, write a proportion, and then solve. Note that side a is always opposite angle A, side x is always opposite angle X, and so on.

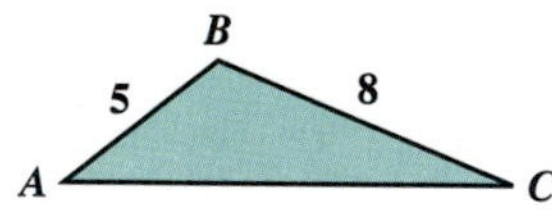

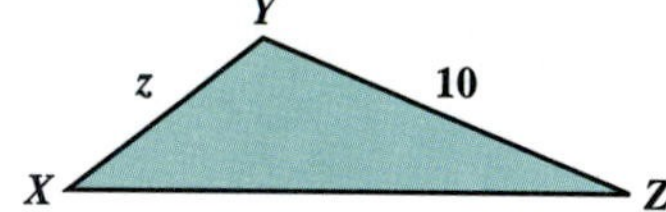

We have

$$\frac{z}{5} = \frac{10}{8}$$ The proportions $\frac{5}{z} = \frac{8}{10}$, $\frac{5}{8} = \frac{z}{10}$, or $\frac{8}{5} = \frac{10}{z}$ could also be used.

$$z = \frac{10}{8} \cdot 5$$ Multiplying by 5 on both sides

$$z = \frac{50}{8}, \quad \text{or } 6.25.$$ ❑

EXAMPLE 6

To determine the number of fish in a lake, a park ranger catches 225 fish, tags them, and throws them back into the lake. Later, 108 fish are caught, and 15 are found to be tagged. Estimate how many fish are in the lake.

Solution

1. FAMILIARIZE. If we knew that the 225 tagged fish constituted, say, 10% of the fish population, we could easily calculate the total fish population from the proportion

$$\frac{225}{F} = \frac{10}{100},$$

where F is the fish population. Unfortunately, we are *not* told the percentage of fish that were tagged. We must reread the problem, looking for numbers that could be used to approximate the percentage of the total fish population that was tagged.

2. TRANSLATE. Since 15 of 108 fish that were later caught had tags, we can use the ratio 15/108 to estimate the percentage of fish that had tags. Then we can

translate to a proportion.

$$\text{Fish tagged originally} \rightarrow \frac{225}{F} = \frac{15}{108} \leftarrow \text{Tagged fish caught later}$$
$$\text{Fish in lake} \rightarrow F \qquad 108 \leftarrow \text{Fish caught later}$$

3. **CARRY OUT.** To solve the proportion, we multiply by the LCD, $108F$:

$$108F \cdot \frac{225}{F} = 108F \cdot \frac{15}{108} \qquad \text{Multiplying by } 108F$$
$$108 \cdot 225 = F \cdot 15$$
$$\frac{108 \cdot 225}{15} = F \quad \text{or} \quad F = 1620.$$

4. **CHECK.** We leave the check to the student.
5. **STATE.** There are about 1620 fish in the lake. ❑

EXERCISE SET 6.8

Solve.

1. A number minus twice its reciprocal is 1. Find the number.
2. A number minus four times its reciprocal is 3. Find the number.
3. The sum of a number and its reciprocal is 2. Find the number.
4. The sum of a number and five times its reciprocal is 6. Find the number.
5. It takes David 4 hr to paint a certain area of a house. It takes Sierra 5 hr to do the same job. How long would it take them, working together, to complete the painting job?

6. By checking work records, a carpenter finds that Juanita can build a certain type of garage in 12 hr. Antoine can do the same job in 16 hr. How long would it take if they worked together?
7. Vern can shovel his driveway in 45 min after a snowfall. Nina can do the same job in 60 min. How long would it take Nina and Vern to shovel the driveway if they worked together?
8. Zoë can rake her yard in 4 hr. Steffi does the same job in 3 hr. How long would it take the two of them, working together, to rake the yard?
9. By checking work records, a plumber finds that Rory can do a certain job in 12 hr. Mira can do the same job in 9 hr. How long would it take if they worked together?
10. A tank can be filled in 18 hr by pipe A alone and in 24 hr by pipe B alone. How long would it take to fill the tank if both pipes were working?
11. By checking work records, a contractor finds that it takes Red Bryck 6 hr to construct a wall of a certain size. It takes Lotta Mudd 8 hr to construct the same wall. How long would it take if they worked together?

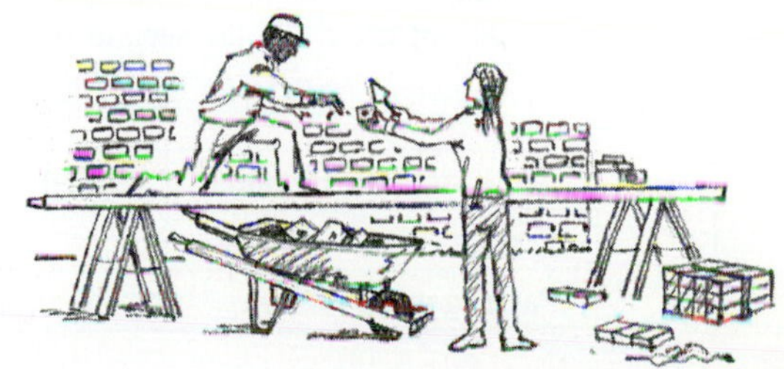

12. Bobbi can pick a quart of raspberries in 20 min. Blanche can pick a quart in 25 min. How long would it take if Bobbi and Blanche worked together?

13. One car travels 40 km/h faster than another. In the same time that one travels 150 km, the other goes 350 km. Find their speeds.

Complete the tables as part of the familiarization. Do not use t's in the second table.

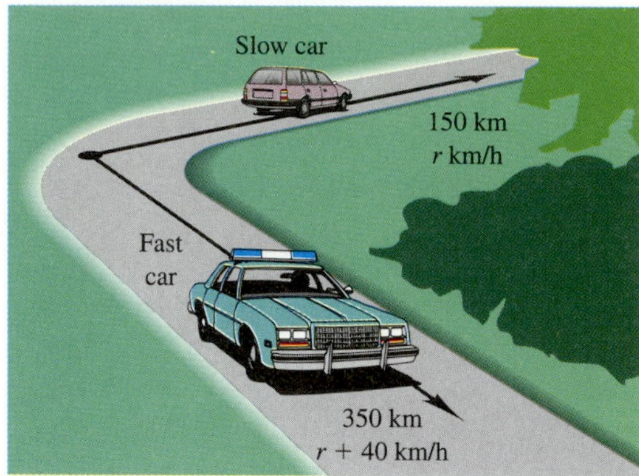

$$d \quad = \quad r \quad \cdot \quad t$$

	Distance	Speed	Time
Slow car	150	r	
Fast car	350		t

	Distance	Speed	Time
Slow car	150	r	$\frac{150}{r}$
Fast car	350		

14. One car travels 30 km/h faster than another. In the same time that one goes 250 km, the other goes 400 km. Find their speeds.

15. The speed of a freight train is 14 km/h slower than the speed of a passenger train. The freight train travels 330 km in the same time that it takes the passenger train to travel 400 km. Find the speed of each train.

Complete the tables as part of the familiarization. Do not use t's in the second table.

$$d \quad = \quad r \quad \cdot \quad t$$

	Distance	Speed	Time
Freight	330		t
Passenger	400	r	

	Distance	Speed	Time
Freight	330		
Passenger	400	r	

16. The speed of a freight train is 15 km/h slower than the speed of a passenger train. The freight train travels 390 km in the same time that it takes the passenger train to travel 480 km. Find the speed of each train.

17. Gail and Dexter bicycle at the same rate. After 16 mi, Dexter stopped for a repair. Gail continued for another 3 hr, biking a total of 50 mi that day. How many hours did Dexter ride before stopping?

18. Tucker and Jasmine's snowmobiles travel at the same rate. After 50 km, Tucker stops to fish, while Jasmine continues for another 4 hr. If Jasmine rides a total of 150 km, how many hours did Tucker travel before stopping?

19. Manley's tractor is just as fast as Caledonia's. It takes Manley 1 hr more than it takes Caledonia to drive to town. If Manley is 20 mi from town and Caledonia is 15 mi from town, how long does it take Caledonia to drive to town?

20. Tory and Emilio's motorboats both travel at the same speed. Tory pilots her boat 40 km before docking. Emilio continues for another 2 hr, traveling a total of 100 km before docking. How long did it take Tory to navigate the 40 km?

Find the ratio of the following. Simplify, if possible.

21. 54 days, 6 days

22. 800 mi, 50 gal

Solve.

23. A black racer snake travels 4.6 km in 2 hr. What is its speed in kilometers per hour?

24. Light travels 558,000 mi in 3 sec. What is its speed in miles per second?

25. The coffee beans from 14 trees are needed to produce 7.7 kg of coffee. (This is the average that each person in the United States consumes each year.) How many trees are needed to produce 320 kg of coffee?

26. Last season a minor-league baseball player got 240 hits in 600 times at bat. This season, his ratio of hits to number of times at bat is the same. He batted 500 times. How many hits has he had?

27. Wanda walked 234 km in 14 days. At this rate, how far would she walk in 42 days?

28. In a potato bread recipe, the ratio of milk to flour is $\frac{3}{13}$. If 5 cups of milk are used, how many cups of flour are used?

29. A normal 10-cc specimen of human blood contains 1.2 g of hemoglobin. How many grams would 16 cc of the same blood contain?

30. The winner of an election for class president won by a vote of 3 to 2, with 324 votes. How many votes did the loser get?

For each pair of similar triangles, find the value of the indicated letter.

31. b

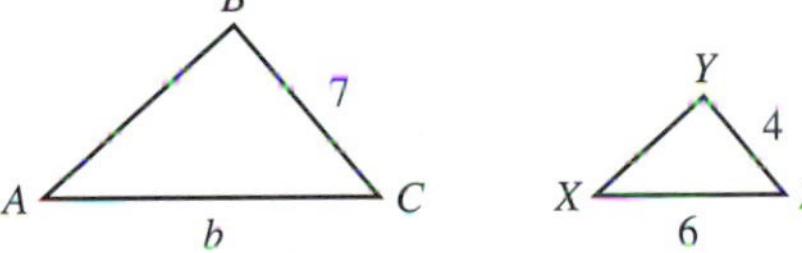

32. a

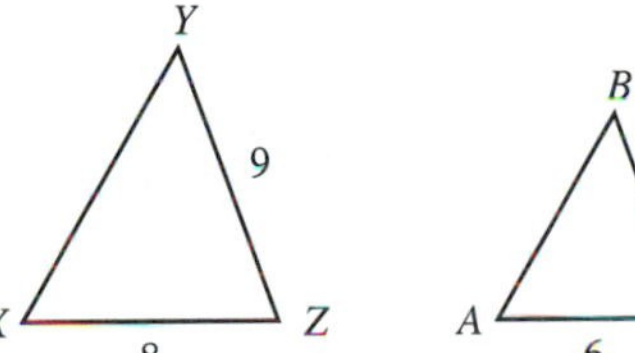

33. f

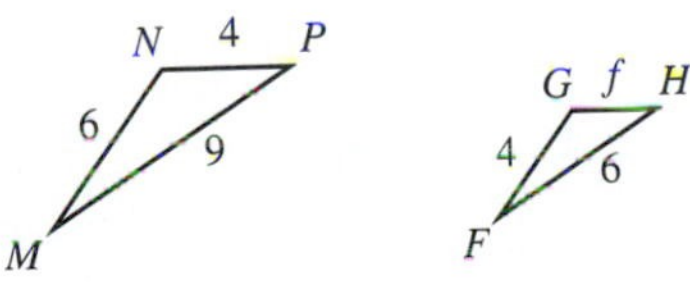

34. r

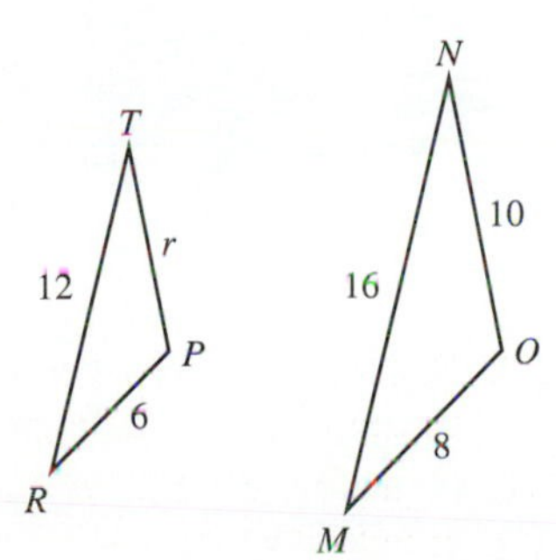

35. n

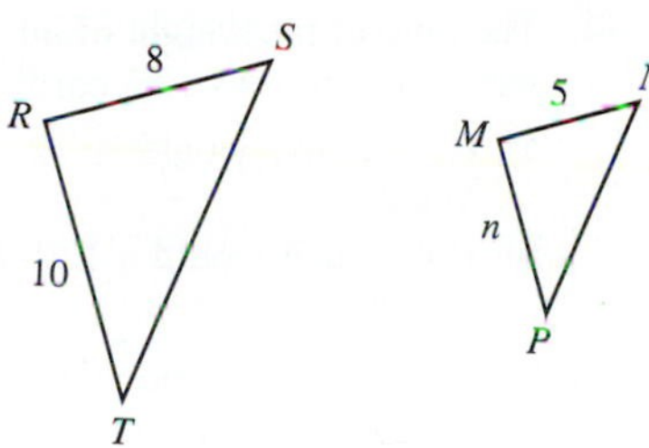

36. h

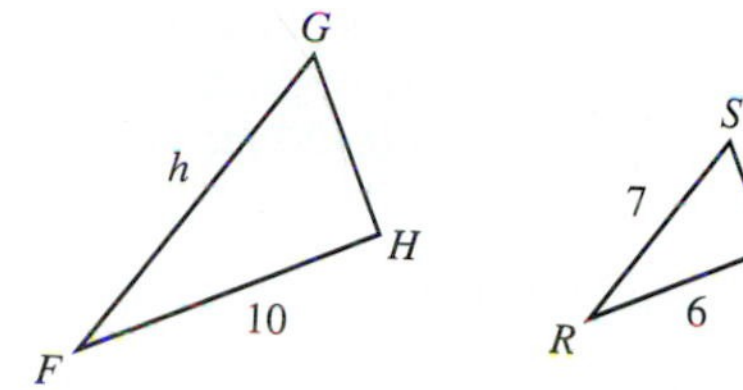

Solve.

37. To determine the number of trout in a lake, a naturalist catches 112 trout, tags them, and throws them back into the lake. Later, 82 trout are caught; 32 of them have tags. Estimate the number of trout in the lake.

38. To determine the number of deer in a game preserve, a game warden catches 318 deer, tags them, and lets them loose. Later, 168 deer are caught; 56 of them have tags. Estimate the number of deer in the preserve.

39. To determine the number of deer in a forest, a game warden catches 612 deer, tags them, and lets them loose. Later, 244 deer are caught and 72 of them have tags. Estimate how many deer are in the forest.

40. A sample of 184 light bulbs contained 6 defective bulbs. How many defective bulbs would you expect in a sample of 1288 bulbs?

41. A sample of 144 firecrackers contained 9 "duds." How many duds would you expect in a sample of 320 firecrackers?

42. To determine the number of moose in a park, a naturalist catches, tags, and then releases 25 moose. Later, 36 moose are caught; 4 of them have tags. Estimate the moose population of the park.

43. The ratio of the weight of an object on the moon to the weight of an object on earth is 0.16 to 1.

a) How much would a 12-ton rocket weigh on the moon?

b) How much would a 180-lb astronaut weigh on the moon?

44. The ratio of the weight of an object on Mars to the weight of an object on earth is 0.4 to 1.

a) How much would a 12-ton rocket weigh on Mars?

b) How much would a 120-lb astronaut weigh on Mars?

45. Simplest fractional notation for a rational number is $\frac{9}{17}$. Find an equal ratio where the sum of the numerator and the denominator is 104.

46. A baseball team has 12 more games to play. They have won 25 out of the 36 games they have played. How many more games must they win in order to finish with a 0.750 record?

Skill Maintenance

Subtract.

47. $(x + 2) - (x + 1)$

48. $(x^2 + x) - (x + 1)$

49. $(4y^3 - 5y^2 + 7y - 24) - (-9y^3 + 9y^2 - 5y + 49)$

50. The perimeter of a rectangle is 642 ft. The length is 15 ft greater than the width. Find the area of the rectangle.

Synthesis

51. ◈ Write a problem similar to Example 2 for a classmate to solve. Design the problem so that the translation step is

$$\frac{t}{7} + \frac{t}{5} = 1.$$

52. ◈ Write a problem similar to Example 3 for a classmate to solve. Design the problem so that the translation step is

$$\frac{30}{r + 4} = \frac{18}{r}.$$

53. The denominator of a fraction is 1 more than the numerator. If 2 is subtracted from both the numerator and the denominator, the resulting fraction is $\frac{1}{2}$. Find the original fraction.

54. Ann and Betty work together and complete a job in 4 hr. Working alone, it would take Betty 6 hr longer to complete the job than it would Ann. How long would it take each of them to complete the job working alone?

55. The speed of a boat in still water is 10 mph. It travels 24 mi upstream and 24 mi downstream in a total time of 5 hr. What is the speed of the current?

56. Express 100 as the sum of two numbers for which the ratio of one number, increased by 5, to the other number, decreased by 5, is 4.

57. Given that

$$\frac{A}{B} = \frac{C}{D},$$

write three different proportions using A, B, C, and D.

58. How soon after 5 o'clock will the hands on a clock first be together?

59. Rosina, Ng, and Oscar can write a computer program in 3 days. Rosina can write the program in 8 days and Ng can do it in 10 days. How many days will it take Oscar to write the program?

60. Together, Michelle, Sal, and Kristen can wax a car in 1 hr and 20 min. To complete the job alone, Michelle needs twice the time that Sal needs and 2 hr more than Kristen. How long would it take each to wax the car working alone?

61. To reach an appointment 50 mi away, Dr. Wright allowed 1 hr. After driving 30 mi, she realized that her speed would have to be increased 15 mph for the remainder of the trip. What was her speed for the first 30 mi?

62. ◈ Are the equations

$$\frac{A + B}{B} = \frac{C + D}{D} \quad \text{and} \quad \frac{A}{B} = \frac{C}{D}$$

equivalent? Why or why not?

63. The shadow from a 40-ft cliff just reaches across a water-filled quarry at the same time that a 6-ft tall diver casts a 10-ft shadow. How wide is the pond?

6.9 Formulas and More Problem Solving

Solving for a Letter • Solving and Evaluating

Solving for a Letter

Formulas arise frequently in the natural and social sciences, business, and engineering. When a formula takes the form of a rational equation, we can use our equation-solving techniques to solve for any specified letter.

EXAMPLE 1

Gravitational force. The gravitational force f between objects of mass M and m, at a distance d from each other, is given by

$$f = \frac{kMm}{d^2},$$

where k represents a fixed number constant. Solve for m.

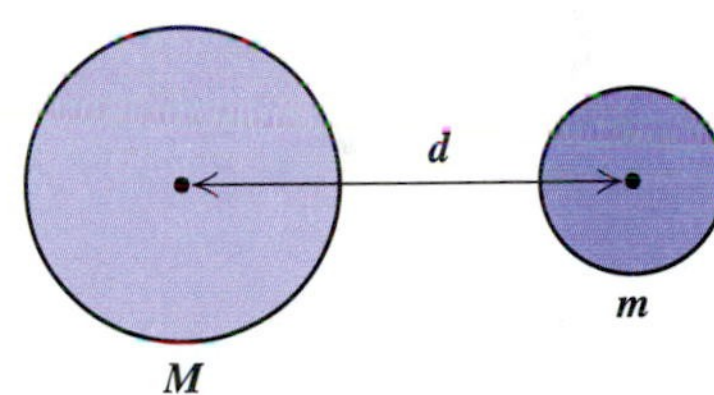

Solution

$$f = \frac{kMm}{d^2}$$

$$fd^2 = kMm \qquad \text{Multiplying by the LCD, } d^2$$

$$\frac{fd^2}{kM} = m \qquad \text{Dividing by } kM$$

❑

The procedure we follow is similar to the one used in Section 2.3.

To solve a formula for a given letter:

1. Identify the letter being solved for and multiply on both sides to clear fractions or decimals, if necessary.
2. Multiply, if necessary, to remove parentheses.
3. Get all terms with the letter to be solved for on one side of the equation and all other terms on the other side, using the addition principle.
4. Factor out the letter being solved for if it appears in more than one term.
5. Multiply or divide to solve for the letter in question.

EXAMPLE 2

The area of a trapezoid. The area A of a trapezoid is half the product of the height h and the sum of the lengths a and b of the parallel sides:

$$A = \tfrac{1}{2}(a + b)h.$$

Solve for b.

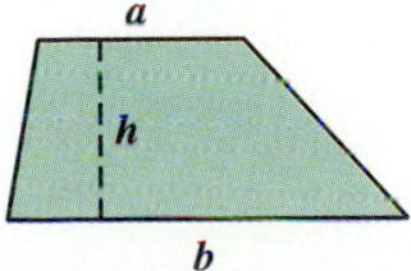

Solution

$$A = \frac{1}{2}(a + b)h$$

$2A = (a + b)h$ Multiplying by 2 to clear fractions

$2A = ah + bh$ Using the distributive law to remove parentheses

$2A - ah = bh$ Subtracting ah to get the b-term alone

$\dfrac{2A - ah}{h} = b$ Dividing by h ❑

EXAMPLE 3

A work formula. The formula $t/a + t/b = 1$ was used in Section 6.8. Solve it for t.

Solution

$$\frac{t}{a} + \frac{t}{b} = 1$$

$ab\left(\dfrac{t}{a} + \dfrac{t}{b}\right) = ab \cdot 1$ Multiplying by the LCD, ab, to clear fractions

$\dfrac{\not{a}bt}{\not{a}} + \dfrac{a\not{b}t}{\not{b}} = ab$ Using the distributive law to remove parentheses

$bt + at = ab$ Removing factors of 1: $\dfrac{a}{a} = 1$ and $\dfrac{b}{b} = 1$

$(b + a)t = ab$ Factoring out t, the letter for which we are solving

$t = \dfrac{ab}{b + a}$ Dividing by $b + a$ ❑

The answer to Example 3 can be used to find solutions to problems such as Example 2 in Section 6.8:

$$t = \frac{4 \cdot 6}{6 + 4} = \frac{24}{10} = 2\frac{2}{5}.$$

In Examples 1 and 2, the letter for which we solved was on the right side of the equation. In Example 3, the letter was on the left. The location of the letter is unimportant, since all equations are reversible.

Recall from Section 2.3 that the variable to be solved for should be alone on one side of the equation, with *no* occurrence of that variable on the other side.

EXAMPLE 4

Solve $y = \dfrac{x + y}{a}$ for y.

Solution

$$y = \frac{x + y}{a}$$

$a \cdot y = a \cdot \dfrac{x + y}{a}$ Multiplying by a to clear fractions

$ay = x + y$ Simplifying

CAUTION! If we next divide by a, we will not isolate y since y would still appear on both sides of the equation.

$$ay - y = x \quad \text{Subtracting } y \text{ to get all terms involving } y \text{ on one side}$$

$$y(a - 1) = x \quad \text{Factoring out } y$$

$$y = \frac{x}{a - 1} \quad \text{Dividing by } a - 1$$

Solving and Evaluating

We often solve a formula for a letter when we know the values of the other variables in the formula for a specific situation. Then finding the unknown value for that situation is a matter of simply evaluating the new expression.

EXAMPLE 5

Resistance. The formula

$$\frac{1}{R} = \frac{1}{r_1} + \frac{1}{r_2}$$

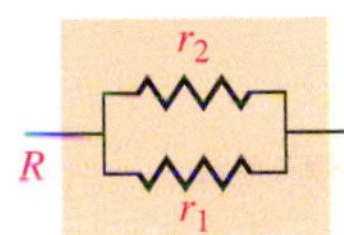

gives the resistance R of two resistors r_1 and r_2 connected in parallel.

a) Solve for r_2.
b) Find r_2 when $R = 3.75$ ohms and $r_1 = 6$ ohms.

Solution

a) To solve for r_2, we multiply by the LCD, Rr_1r_2:

$$\frac{1}{R} = \frac{1}{r_1} + \frac{1}{r_2}$$

$$Rr_1r_2 \cdot \frac{1}{R} = Rr_1r_2 \cdot \left[\frac{1}{r_1} + \frac{1}{r_2}\right] \quad \text{Multiplying by the LCD}$$

$$Rr_1r_2 \cdot \frac{1}{R} = Rr_1r_2 \cdot \frac{1}{r_1} + Rr_1r_2 \cdot \frac{1}{r_2} \quad \text{Using the distributive law}$$

$$\left.\begin{aligned} \frac{Rr_1r_2}{R} &= \frac{Rr_1r_2}{r_1} + \frac{Rr_1r_2}{r_2} \\ r_1r_2 &= Rr_2 + Rr_1. \end{aligned}\right\} \quad \text{Simplifying by removing factors of 1: } \frac{R}{R} = 1;\ \frac{r_1}{r_1} = 1;\ \frac{r_2}{r_2} = 1$$

You might be tempted at this point to multiply by $1/r_1$ to get r_2 alone on the left, but note that there is an r_2 on the right. We must get all the terms involving r_2 on the *same side* of the equation:

$$r_1r_2 - Rr_2 = Rr_1 \quad \text{Adding } -Rr_2$$

$$r_2(r_1 - R) = Rr_1. \quad \text{Factoring out } r_2$$

Multiplying by $1/(r_1 - R)$ on both sides, we obtain the equation

$$r_2 = \frac{Rr_1}{r_1 - R}.$$

b) We use the equation for r_2, replacing R with 3.75 and r_1 with 6:

$$r_2 = \frac{3.75 \cdot 6}{6 - 3.75}$$

$$r_2 = 10.$$

The second resistor has a resistance of 10 ohms. ❑

EXERCISE SET 6.9

Solve.

1. $S = 2\pi rh$, for r
2. $A = P(1 + rt)$, for t (An interest formula)
3. $A = \frac{1}{2}bh$, for b (The area of a triangle)

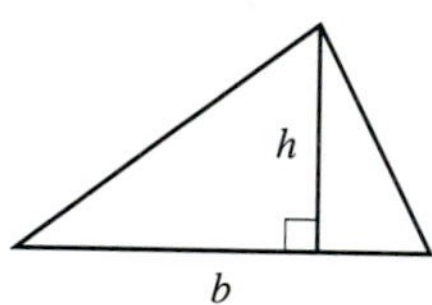

4. $s = \frac{1}{2}gt^2$, for g (A physics formula for distance)

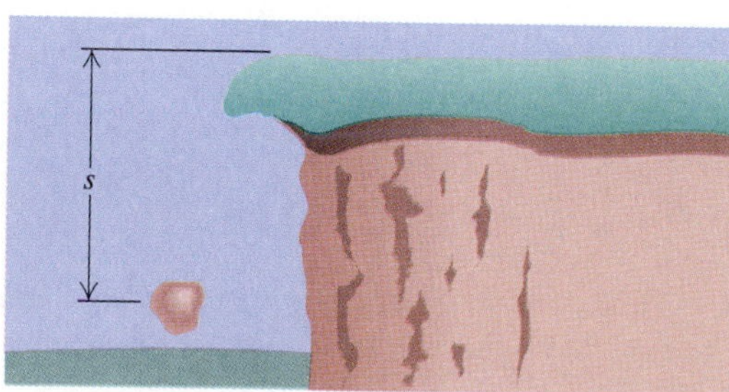

5. $\frac{1}{180} = \frac{n-2}{s}$, for n
6. $S = \frac{n}{2}(a + l)$, for a
7. $V = \frac{1}{3}k(B + b + 4M)$, for b
8. $A = P + Prt$, for P (*Hint:* Factor the right-hand side.)
9. $rl - rS = L$, for r
10. $T = mg - mf$, for m (*Hint:* Factor the right-hand side.)
11. $A = \frac{1}{2}h(b_1 + b_2)$, for h
12. $S = 2\pi r(r + h)$, for h (The surface area of a right circular cylinder)

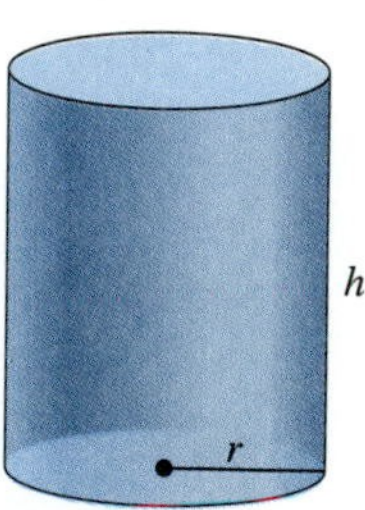

13. $ab = ac + d$, for a
14. $mn + p = np$, for n
15. $\frac{r}{p} = q$, for p
16. $n = \frac{m}{v}$, for v

17. $a + b = \frac{c}{d}$, for d

18. $\frac{m}{n} = p - q$, for n

19. $\frac{x - y}{z} = p + q$, for z

20. $\frac{M - g}{t} = r + s$, for t

21. $\frac{1}{p} + \frac{1}{q} = \frac{1}{f}$, for f
(An optics formula)

22. $\frac{1}{R} = \frac{1}{r_1} + \frac{1}{r_2}$, for R
(An electricity formula)

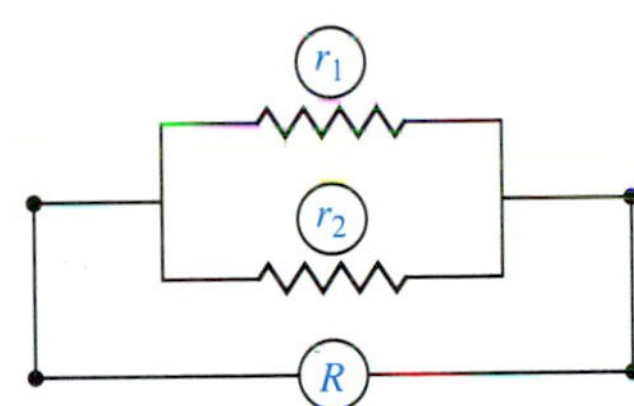

23. $r = \frac{v^2pL}{a}$, for p

24. $P = 2(l + w)$, for l

25. $\frac{a}{c} = n + bn$, for n

26. $ab - ac = \frac{Q}{M}$, for a

27. $S = \frac{a + 2b}{3b}$, for b

28. $C = \frac{Ka - b}{a}$, for a

29. $C = \frac{5}{9}(F - 32)$, for F
(A temperature conversion formula)

30. $V = \frac{4}{3}\pi r^3$, for r^3
(The volume of a sphere)

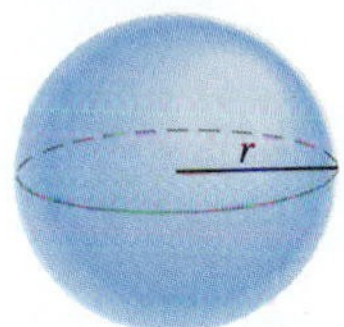

31. $\frac{W_1}{W_2} = \frac{d_1}{d_2}$, for d_2

32. $\frac{W_1}{W_2} = \frac{d_1}{d_2}$, for W_2

33. $S = \frac{a - ar^n}{1 - r}$, for a

34. $S = \frac{a}{1 - r}$, for r

35. $f = \frac{gm - t}{m}$, for m

36. $S = \frac{rl - a}{r - l}$, for r

Problem Solving

37. The formula $A = 9R/I$ gives a pitcher's earned run average, where A is the earned run average, R is the number of earned runs, and I is the number of innings pitched. How many earned runs were given up if a pitcher's earned run average is 2.4 after 45 innings?

38. Two resistors are connected in parallel. Their resistances are, respectively, 8 ohms and 15 ohms. What is the resistance of the combination?

39. A resistor has a resistance of 50 ohms. What size resistor should be put with it, in parallel, in order to obtain a resistance of 5 ohms?

40. The area of a certain trapezoid is 25cm^2. Its height is 5 cm and the length of one base is 4 cm. Find the length of the other base.

41. The formula
$$P = \frac{A}{1 + r}$$
can be used to determine the principal P that must be invested for one year at $(100 \cdot r)\%$ simple interest in order to have A dollars after a year. Solve for r.

42. At what yearly interest rate should \$1600 be invested in order to have a total of \$1712 after one year? (See Exercise 41.)

Skill Maintenance

43. Solve: $-\frac{3}{5}x = \frac{9}{20}$.

44. Find the intercepts and then graph $3x + 4y = 24$.

45. Factor: $x^2 - 13x - 30$.

46. Subtract: $(5x^3 - 7x^2 + 9) - (8x^3 - 2x^2 + 4)$.

Synthesis

47. Is it easier to solve

$$\frac{1}{25} + \frac{1}{23} = \frac{1}{x} \quad \text{for } x,$$

or to solve

$$\frac{1}{p} + \frac{1}{q} = \frac{1}{f} \quad \text{for } f?$$

Explain why.

48. Explain why someone might want to solve

$$A = \tfrac{1}{2}bh \quad \text{for } b.$$

(See Exercise 3.)

Solve.

49. $u = -F\left(E - \dfrac{P}{T}\right)$, for T

50. $\dfrac{n_1}{p_1} + \dfrac{n_2}{p_2} = \dfrac{n_2 - n_1}{R}$, for n_2

51. The formula

$$C = \tfrac{5}{9}(F - 32)$$

is used to convert Fahrenheit temperatures to Celsius temperatures. At what temperature are the Fahrenheit and Celsius readings the same?

52. In

$$N = \frac{a}{c},$$

what is the effect on N when c increases? when c decreases? Assume that a, c, and N are positive.

53. The formula

$$I_t = \frac{I_f}{1 - T}$$

gives the *taxable interest rate* I_t equivalent to the *tax-free interest rate* I_f for a person in the $(100 \cdot T)\%$ tax bracket. Solve for T.

54. Zoë invested the identical amounts in tax-free bonds at 3% and taxable certificates of deposit at $4\frac{1}{6}\%$. For her income, these rates are equivalent. What is her tax bracket? (See Exercise 53.)

SUMMARY AND REVIEW 6

KEY TERMS

Rational expression, p. 266
Simplifying, p. 267
Least Common Multiple, LCM, p. 281
Least Common Denominator, LCD, p. 281
Complex rational expression, p. 293
Rational equation, p. 301
Motion problem, p. 309
Ratio, p. 311
Proportion, p. 311
Cross-multiply, p. 311
Similar triangles, p. 311

IMPORTANT PROPERTIES AND FORMULAS

To add, subtract, multiply, and divide rational expressions:

$$\frac{A}{B} \cdot \frac{C}{D} = \frac{AC}{BD}; \qquad \frac{A}{B} \div \frac{C}{D} = \frac{A}{B} \cdot \frac{D}{C} = \frac{AD}{BC};$$

$$\frac{A}{B} + \frac{C}{B} = \frac{A + C}{B}; \qquad \frac{A}{B} - \frac{C}{B} = \frac{A - C}{B}.$$

The Work Principle: $t\left(\frac{1}{a}\right) + t\left(\frac{1}{b}\right) = 1$, or $\frac{t}{a} + \frac{t}{b} = 1$

To find the least common denominator (LCD):

1. Write the prime factorization of each denominator.
2. Select one of the factorizations and inspect it to see if it contains the other.
   a) If it does, it represents the LCD.
   b) If it does not, multiply that factorization by any factors of the other denominator that it lacks. The final product is the LCD.
3. The LCD should include each factor the greatest number of times that it occurs in any one factorization.

To add or subtract rational expressions that have different denominators:

1. Find the LCD.
2. Multiply each rational expression by an expression for 1 made up of the factors of the LCD missing from that expression's denominator.
3. Add or subtract the numerators, as indicated. Write the sum or difference over the LCD.
4. Simplify, if possible.

To simplify a complex rational expression by using the LCD:

1. Find the LCD of all expressions *within* the complex rational expression.
2. Multiply the complex rational expression by 1, using the LCD to construct the 1.
3. Distribute and simplify. No rational expressions should remain within the complex rational expression.
4. Factor and, if possible, simplify.

To simplify a complex rational expression by first adding or subtracting:

1. Add or subtract, as indicated, to get a single rational expression in the numerator.
2. Add or subtract, as indicated, to get a single rational expression in the denominator.
3. Divide the numerator by the denominator (invert and multiply).
4. If possible, simplify by removing a factor of 1.

To solve a rational equation:

1. Clear the equation of fractions by multiplying on both sides by the LCD of all rational expressions in the equation.
2. Solve the resulting equation using the addition principle, the multiplication principle, and the principle of zero products.
3. Check the possible solution(s) in the original equation.

REVIEW EXERCISES

This chapter's review and test include Skill Maintenance exercises from Sections 2.1, 3.3, 4.3, and 5.2.

List all numbers for which the expression is undefined.

1. $\frac{3}{x}$

2. $\frac{4}{x-6}$

3. $\frac{x+5}{x^2-36}$

4. $\frac{x^2-3x+2}{x^2+x-30}$

5. $\frac{-4}{(x+2)^2}$

6. $\frac{x-5}{x^3-8x^2+15x}$

Simplify.

7. $\frac{4x^2-8x}{4x^2+4x}$

8. $\frac{14x^2-x-3}{2x^2-7x+3}$

9. $\frac{(y-5)^2}{y^2-25}$

Multiply or divide and simplify, if possible.

10. $\frac{a^2-36}{10a} \cdot \frac{2a}{a+6}$

11. $\frac{6t-6}{2t^2+t-1} \cdot \frac{t^2-1}{t^2-2t+1}$

12. $\frac{10-5t}{3} \div \frac{t-2}{12t}$

13. $\frac{4x^4}{x^2-1} \div \frac{2x^3}{x^2-2x+1}$

14. $\frac{x^2+1}{x-2} \cdot \frac{2x+1}{x+1}$

15. $(t^2+3t-4) \div \frac{t^2-1}{t+4}$

Find the LCM.

16. $3x^2$, $10xy$, $15y^2$

17. x^2-x, x^5-x^3, x^4

18. y^2-y-2, y^2-4

Add or subtract and simplify, if possible.

19. $\frac{x+8}{x+7} + \frac{10-4x}{x+7}$

20. $\frac{3}{3x-9} + \frac{x-2}{3-x}$

21. $\frac{6x-3}{x^2-x-12} - \frac{2x-15}{x^2-x-12}$

22. $\frac{3x-1}{2x} - \frac{x-3}{x}$

23. $\frac{x+3}{x-2} - \frac{x}{2-x}$

24. $\frac{2a}{a+1} - \frac{4a}{1-a^2}$

25. $\frac{d^2}{d-c} + \frac{c^2}{c-d}$

26. $\frac{1}{x^2-25} - \frac{x-5}{x^2-4x-5}$

27. $\frac{3x}{x+2} - \frac{x}{x-2} + \frac{8}{x^2-4}$

28. $\frac{3}{2x} + \frac{1}{2x+1}$

Simplify.

29. $\dfrac{\frac{1}{z}+1}{\frac{1}{z^2}-1}$

30. $\dfrac{2+\frac{1}{xy^2}}{\frac{1+x}{x^4y}}$

31. $\dfrac{\frac{c}{d}-\frac{d}{c}}{\frac{1}{c}+\frac{1}{d}}$

Solve.

32. $\frac{3}{y} - \frac{1}{4} = \frac{1}{y}$

33. $\frac{5}{x+3} = \frac{3}{x+2}$

34. $\frac{15}{x} - \frac{15}{x+2} = 2$

Problem Solving

35. In checking records, a contractor finds that Sean can build a deck in 9 hr. Shane can do the same job in 12 hr. How long would it take if they worked together?

36. A lab is testing two high-speed trains. One train travels 40 km/h faster than the other. In the same time that one train travels 70 km, the other travels 60 km. Find the speed of each train.

37. The reciprocal of 1 more than a number is twice the reciprocal of the number itself. What is the number?

38. A sample of 250 batteries contained 8 defective batteries. How many defective batteries would you expect among 5000 batteries?

39. Triangles ABC and XYZ are similar. Find the value of x.

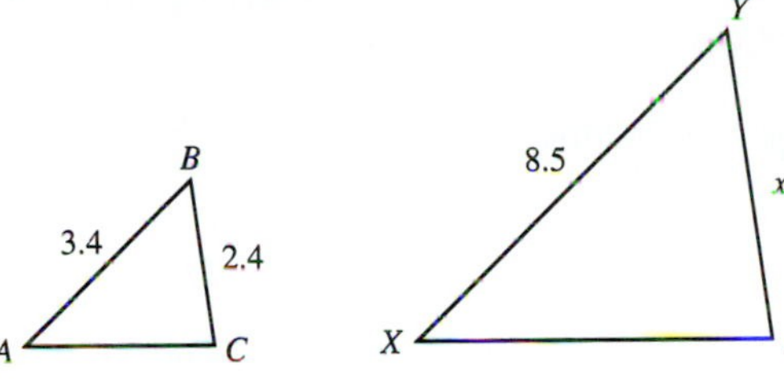

Solve.

40. $\frac{1}{r} + \frac{1}{s} = \frac{1}{t}$, for s

41. $F = \frac{9C + 160}{5}$, for C

Skill Maintenance

42. Solve: $-3 + x = 8$.

43. Find the intercepts and graph: $2x - y = 6$.

44. Factor: $x^2 + 8x - 48$.

45. Subtract:
$$(5x^3 - 4x^2 + 3x - 4) - (7x^3 - 7x^2 - 9x + 14).$$

Synthesis

46. ◈ Why is factoring an important skill to master before beginning a study of rational equations?

47. ◈ A student insists on finding a common denominator by always multiplying the denominators of the expressions being added. How could the student's approach be improved?

Simplify.

48. $\frac{2a^2 + 5a - 3}{a^2} \cdot \frac{5a^3 + 30a^2}{2a^2 + 7a - 4} \div \frac{a^2 + 6a}{a^2 + 7a + 12}$

49. $\frac{12a}{(a - b)(b - c)} - \frac{2a}{(b - a)(c - b)}$

CHAPTER TEST 6

List all numbers for which the expression is undefined.

1. $\frac{8}{2x}$

2. $\frac{5}{x + 8}$

3. $\frac{x - 7}{x^2 - 49}$

4. $\frac{x^2 + x - 30}{x^2 - 3x + 2}$

5. $\frac{11}{(x - 1)^2}$

6. $\frac{x + 2}{x^3 + 8x^2 + 15x}$

7. Simplify: $\frac{6x^2 + 17x + 7}{2x^2 + 7x + 3}$.

8. Multiply and simplify: $\frac{a^2 - 25}{6a} \cdot \frac{3a}{a - 5}$.

9. Divide and simplify:
$$\frac{25x^2 - 1}{9x^2 - 6x} \div \frac{5x^2 + 9x - 2}{3x^2 + x - 2}.$$

10. Find the LCM:
$$y^2 - 9,\ y^2 + 10y + 21,\ y^2 + 4y - 21.$$

Add or subtract. Simplify, if possible.

11. $\frac{16 + x}{x^3} + \frac{7 - 4x}{x^3}$

12. $\frac{5 - t}{t^2 + 1} - \frac{t - 3}{t^2 + 1}$

13. $\frac{x - 4}{x - 3} + \frac{x - 1}{3 - x}$

14. $\frac{x - 4}{x - 3} - \frac{x - 1}{3 - x}$

15. $\frac{5}{t - 1} + \frac{3}{t}$

16. $\frac{1}{x^2 - 16} - \frac{x + 4}{x^2 - 3x - 4}$

17. $\frac{1}{x - 1} + \frac{4}{x^2 - 1} - \frac{2}{x^2 - 2x + 1}$

Simplify.

18. $\frac{9 - \frac{1}{y^2}}{3 - \frac{1}{y}}$

19. $\frac{\frac{3}{a^2b} - \frac{2}{ab^3}}{\frac{1}{ab} + \frac{2}{a^4b}}$

Solve.

20. $\frac{7}{y} - \frac{1}{3} = \frac{1}{4}$

21. $\frac{15}{x} - \frac{15}{x - 2} = -2$

Problem Solving

22. The reciprocal of 3 less than a number is four times the reciprocal of the number itself. What is the number?
23. A sample of 125 spark plugs contained 4 defective spark plugs. How many defective spark plugs would you expect among 500 spark plugs?
24. One car travels 20 km/h faster than another. In the same time that one goes 225 km, the other goes 325 km. Find the speed of each car.
25. Solve $d = rt + wt$ for t.

Skill Maintenance

26. Solve: $-3y = \frac{9}{7}$.
27. Find the intercepts and graph: $2x + 5y = 20$.
28. Factor: $x^2 - 4x - 45$.
29. Subtract:

$$(5x^2 - 19x + 34) - (-8x^2 + 10x - 42).$$

Synthesis

30. Reggie and Rema work together to mulch the flower beds around an office complex in $2\frac{6}{7}$ hr. Working alone, it would take Reggie 6 hr more than it would take Rema. How long would it take each of them to complete the landscaping working alone?
31. Simplify: $1 + \cfrac{1}{1 + \cfrac{1}{1 + \cfrac{1}{a}}}$.

CUMULATIVE REVIEW 1–6

1. Use the commutative law of addition to write an expression equivalent to $a + 2b$.
2. Write a true sentence using either $<$ or $>$:

   $-3.1 \blacksquare -3.15$.
3. Evaluate $(y - 1)^2$ when $y = -6$.
4. Remove parentheses and simplify:

   $-4[2(x - 3) - 1]$.

Simplify.

5. $-\frac{1}{2} + \frac{3}{8} + (-6) + \frac{3}{4}$
6. $-\frac{72}{108} \div \left(-\frac{2}{3}\right)$
7. $-6.262 \div 1.01$
8. $4 \div (-2) \cdot 2 + 3 \cdot 4$

Solve.

9. $3(x - 2) = 24$
10. $6y + 3 = -15$
11. $-4x = -18$
12. $5x + 7 = -3x - 9$
13. $4(y - 5) = -2(y + 2)$
14. $-6x - 2(x - 4) = 10$
15. $\frac{1}{3}x - \frac{2}{9} = \frac{2}{3} + \frac{4}{9}x$
16. $-\frac{5}{6} = x - \frac{1}{3}$
17. $3 - y \geqslant 2y + 5$
18. $(3x - 4)(2x + 5) = 0$
19. $2x^2 + 7x = 4$
20. $16 = x^2$
21. $\frac{x^2}{x + 2} = \frac{4}{x + 2}$
22. $x + \frac{1}{x} = 2$
23. $\frac{2}{x^2 - 9} + \frac{5}{x - 3} = \frac{3}{x + 3}$

Solve the formula.

24. $A = \frac{4b}{t}$, for t
25. $\frac{1}{t} = \frac{1}{m} - \frac{1}{n}$, for n
26. $r = \frac{a - b}{c}$, for c

Collect like terms.

27. $x + 2y - 2z + \frac{1}{2}x - z$
28. $2x^3 - 7 + \frac{3}{7}x^2 - 6x^3 - \frac{4}{7}x^2 + 5$

Graph.

29. $y = 1 - \frac{1}{2}x$
30. $x = -3$
31. $x - 6y = 6$

Simplify.

32. $x^8 \cdot x^2$
33. $\frac{z^4}{z^{-7}}$
34. $-(3x^2y)^3$
35. Subtract:

$$(-8y^2 - y + 2) - (y^3 - 6y^2 + y - 5).$$

Multiply.

36. $4(3x + 4y + z)$

37. $(2.5a + 7.5)(0.4a - 1.2)$

38. $(2x^2 - 1)(x^3 + x - 3)$

39. $(6x - 5y)^2$

40. $(2x^5 + 3)(3x^2 - 6)$

41. $(2x^3 + 1)(2x^3 - 1)$

Factor.

42. $6x - 2x^2 - 24x^4$

43. $16x^2 - 81$

44. $x^2 - 10x + 24$

45. $8x^2 + 10x + 3$

46. $6x^2 - 28x + 16$

47. $2x^2 - 18$

48. $16x^2 + 40x + 25$

49. $3x^2 + 10x - 8$

50. $x^4 + 2x^3 - 3x - 6$

Simplify.

51. $\frac{y^2 - 36}{2y + 8} \cdot \frac{y + 4}{y + 6}$

52. $\frac{x^2 - 1}{x^2 - x - 2} \div \frac{x - 1}{x - 2}$

53. $\frac{5ab}{a^2 - b^2} + \frac{a + b}{a - b}$

54. $\frac{x + 2}{4 - x} - \frac{x + 3}{x - 4}$

55. $\frac{1 + \frac{2}{x}}{1 - \frac{4}{x^2}}$

56. $\frac{\frac{1}{t} + 2t}{t - \frac{2}{t^2}}$

Divide.

57. $\frac{15x^4 - 12x^3 + 6x^2 + 2x + 18}{3x^2}$

58. $(15x^4 - 12x^3 + 6x^2 + 2x + 18) \div (x + 3)$

Problem Solving

59. The sum of two consecutive even integers is -554. Find the integers.

60. What number is 96% of 567?

61. If you double a number and then add 20, you get $\frac{2}{3}$ of the original number. What is the original number?

62. Linnae has \$36 budgeted for stationery. Engraved stationery costs \$20 for the first 25 sheets and \$0.08 for each additional sheet. Use an inequality to express the amounts of stationery that will enable Linnae to stay within her budget.

63. If the sides of a square are increased by 2 ft, the sum of the areas of the two squares is 452. Find the length of a side of the original square.

64. One car travels 10 km/h faster than another. In the same time it takes one car to go 120 km, the other car goes 150 km. Find the speed of each car.

65. By checking work records, a contractor finds that it takes Rita 6 hr to construct a wall of a certain size. It takes Lotta 8 hr to construct the same wall. How long would it take if they worked together?

Synthesis

66. Simplify:

$$(x + 7)(x - 4) - (x + 8)(x - 5).$$

67. Solve: $\frac{1}{3}|n| + 8 = 56$.

68. Multiply:

$$[4y^3 - (y^2 - 3)][4y^3 + (y^2 - 3)].$$

69. Factor: $2a^{32} - 13{,}122b^{40}$.

70. Solve: $(x - 4)(x + 7)(x - 12) = 0$.

71. Simplify: $-|0.875 - \left(-\frac{1}{8}\right) - 8|$.

CHAPTER 7

Graphs and Functions

AN APPLICATION

The grade of a road tells us how steep a road up a hill is. For example, a 3% grade means that the road rises 3 ft for every 100 ft that it runs horizontally. If a road rises 50 ft vertically over a horizontal distance of 1250 ft, find the grade of the road.

This problem appears as Example 5 in Section 7.1.

Cheryl Adams
ROADWAY DESIGNER

"I use math daily in evaluating designs that are submitted by other engineers in the Department of Transportation. We must use math for everything from setting the grade to identifying the quantities of material and the costs of a project."

The basics of graphing were first introduced in Chapter 3. There we learned how to graph equations by using tables or intercepts. The treatment of graphing in this chapter focuses on a topic that is of great importance in mathematics, science, and business: slope. Once slope is understood, we can develop faster and easier ways of graphing certain equations.

In addition to the material from this chapter, the review and test for Chapter 7 include material from Sections 4.4, 4.5, 5.1, and 6.1.

7.1 Slope

Rate and Slope • Horizontal and Vertical Lines • Applications

Rate and Slope

Suppose that a car manufacturer is operating two plants: one in Michigan and one in Pennsylvania. If we know that the Michigan plant produces 3 cars every 2 hours and the Pennsylvania plant produces 5 cars every 4 hours, we can set up tables listing the number of cars produced after various amounts of time.

Michigan Plant	
Hours Elapsed	**Cars Produced**
0	0
2	3
4	6
6	9
8	12

Pennsylvania Plant	
Hours Elapsed	**Cars Produced**
0	0
4	5
8	10
12	15
16	20

By comparing the number of cars produced at each plant over a specified period of time, we can compare the **rates** at which the plants produce cars. For example, the Michigan plant produces 3 cars every 2 hours, so its *rate* is $3 \div 2 = 1\frac{1}{2}$, or $\frac{3}{2}$ cars per hour. Since the Pennsylvania plant produces 5 cars every 4 hours, its rate is $5 \div 4 = 1\frac{1}{4}$, or $\frac{5}{4}$ cars per hour.

Let's now graph the pairs of numbers listed in the tables, using the horizontal axis for time and the vertical axis for the number of cars produced.

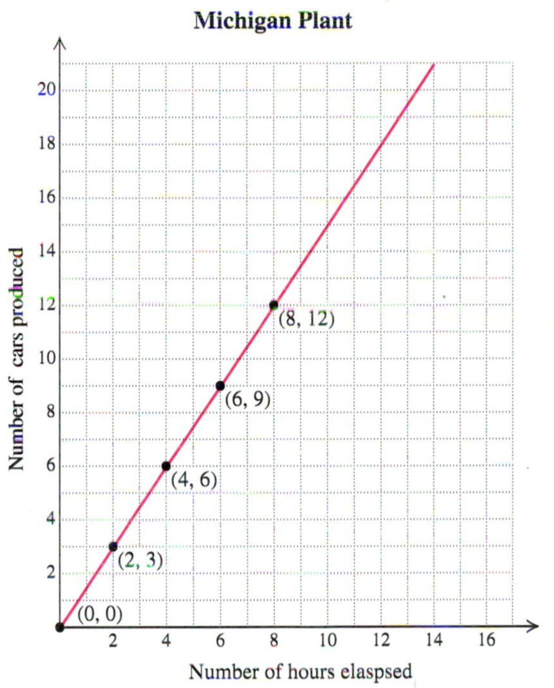

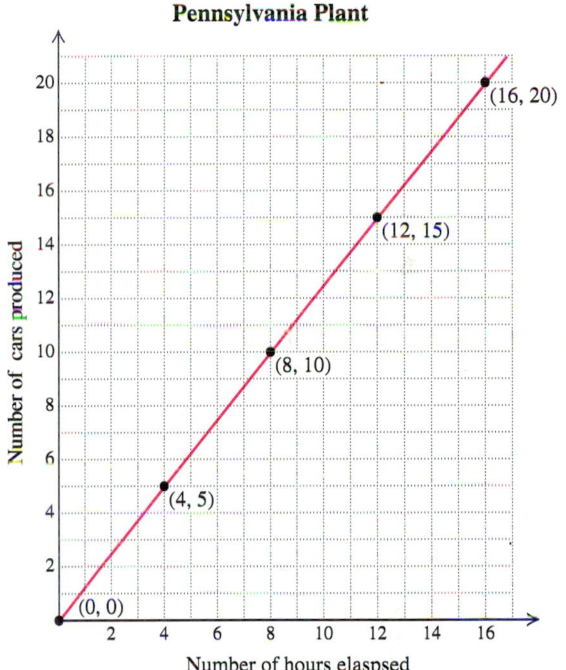

The rates of $\frac{3}{2}$ and $\frac{5}{4}$ can also be found using the coordinates of any two points that are on the line. For example, we can use the points (6, 9) and (8, 12) to find the production rate for the Michigan plant. To do so, remember that these coordinates tell us that after 6 hr, 9 cars have been produced, and after 8 hr, 12 cars have been produced. In the 2 hr between the 6-hr and 8-hr points, $12 - 9$, or 3 cars were produced. Thus,

$$\begin{aligned} \text{Michigan rate} &= \frac{\text{change in number of cars produced}}{\text{corresponding change in time}} \\ &= \frac{12 - 9 \text{ cars}}{8 - 6 \text{ hr}} \\ &= \frac{3 \text{ cars}}{2 \text{ hr}} = \frac{3}{2} \text{ cars per hour.} \end{aligned}$$

The same rate can be found using other points on that line, such as (0, 0) and (4, 6):

$$\text{Michigan rate} = \frac{6 - 0 \text{ cars}}{4 - 0 \text{ hr}} = \frac{6 \text{ cars}}{4 \text{ hr}} = \frac{3}{2} \text{ cars per hour.}$$

Note that the rate is always the vertical change divided by the associated horizontal change.

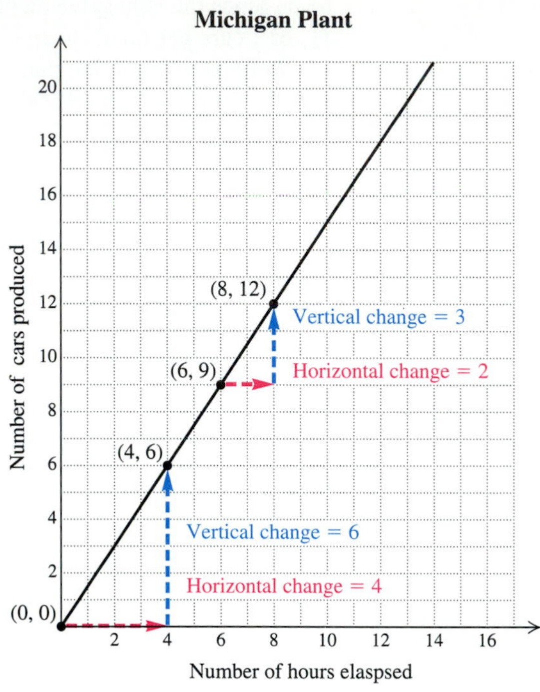

EXAMPLE 1

Use the graph of car production at the Pennsylvania plant to find the rate of production.

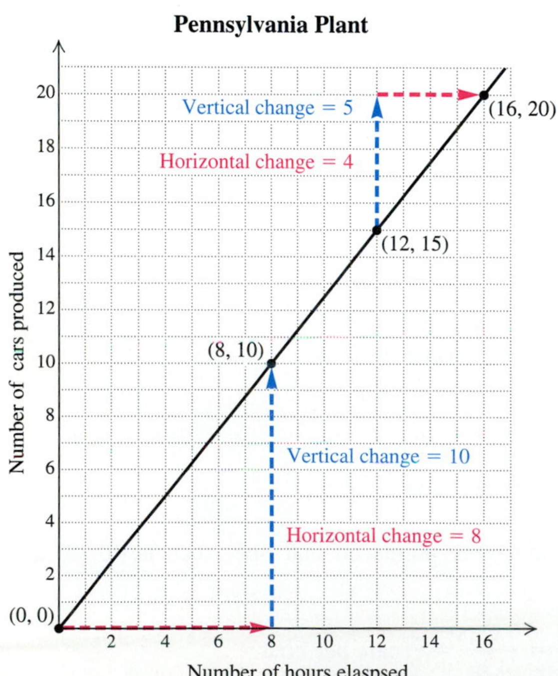

Solution We can use any two points on the line, such as (12, 15) and (16, 20):

$$\text{Pennsylvania rate} = \frac{\text{change in number of cars produced}}{\text{corresponding change in time}}$$

$$= \frac{20 - 15 \text{ cars}}{16 - 12 \text{ hr}}$$

$$= \frac{5 \text{ cars}}{4 \text{ hr}}$$

$$= \frac{5}{4} \text{ cars per hour.}$$

As a check, we can use another pair of points, like (0, 0) and (8, 10):

$$\text{Pennsylvania rate} = \frac{10 - 0 \text{ cars}}{8 - 0 \text{ hr}}$$

$$= \frac{10 \text{ cars}}{8 \text{ hr}}$$

$$= \frac{5}{4} \text{ cars per hour.}$$

❑

Even when a graph's axes are labeled simply x and y, it is useful to form the ratio of vertical change to horizontal change. This ratio gives a measure of a line's slant, or *slope*.

Consider a line passing through the points (2, 3) and (6, 5), as shown. We find the ratio of vertical change, or *rise*, to horizontal change, or *run*, as follows:

$$\text{Ratio of vertical change to horizontal change} = \frac{\text{change in } y}{\text{change in } x} = \frac{\text{rise}}{\text{run}}$$

$$= \frac{5-3}{6-2}$$

$$= \frac{2}{4}, \quad \text{or} \quad \frac{1}{2}.$$

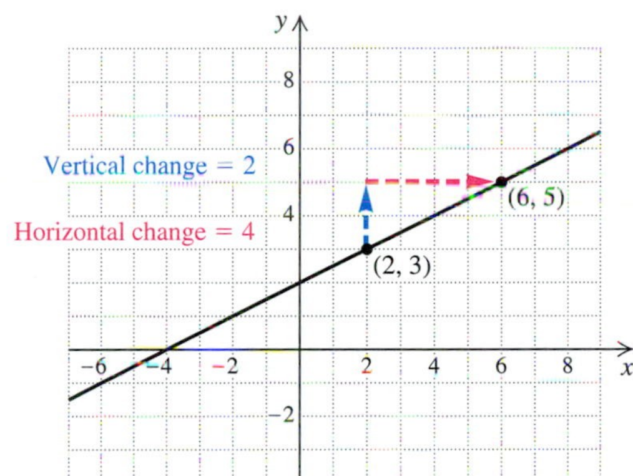

Thus the y-coordinate of a point on the line increases 2 units for every 4-unit increase in x, 1 unit for every 2-unit increase in x, and $\frac{1}{2}$ unit for every 1-unit increase in x. The slope of the line is $\frac{1}{2}$.

Slope

The *slope* of a line containing points (x_1, y_1) and (x_2, y_2) is given by

$$m = \frac{\text{change in } y}{\text{change in } x} = \frac{\text{rise}}{\text{run}} = \frac{y_2 - y_1}{x_2 - x_1}.$$

EXAMPLE 2

Graph the line containing the points $(-4, 3)$ and $(2, -6)$ and find the slope.

Solution The graph is on the next page. From $(-4, 3)$ to $(2, -6)$, the change in y, or rise, is $-6 - 3$, or -9. The change in x, or run, is $2 - (-4)$, or 6.

Thus,

$$\text{Slope} = \frac{\text{change in } y}{\text{change in } x} = \frac{\text{rise}}{\text{run}}$$
$$= \frac{-6 - 3}{2 - (-4)}$$
$$= \frac{-9}{6}$$
$$= -\frac{9}{6}, \quad \text{or } -\frac{3}{2}.$$

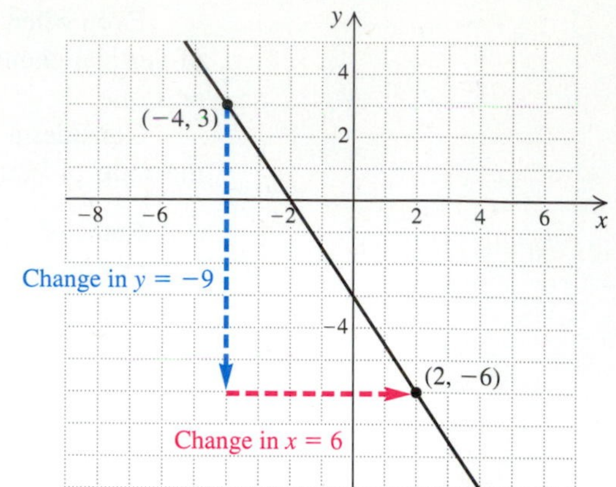

CAUTION! When we use the formula

$$m = \frac{y_2 - y_1}{x_2 - x_1},$$

it makes no difference which of the two points is considered (x_1, y_1) so long as we subtract the y-coordinates in the same order that we subtract the x-coordinates.

To illustrate, we can reverse the subtractions in Example 2 and still obtain a slope of $-\frac{3}{2}$:

$$\text{Slope} = \frac{\text{change in } y}{\text{change in } x} = \frac{3 - (-6)}{-4 - 2} = \frac{9}{-6} = -\frac{3}{2}.$$

If a line has a positive slope, it slants up from left to right. The larger the slope, the steeper the slant. A line with negative slope slants downward from left to right.

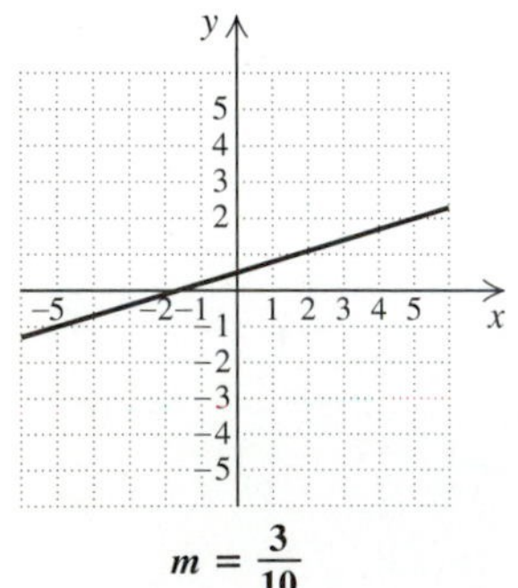

$m = \frac{3}{10}$

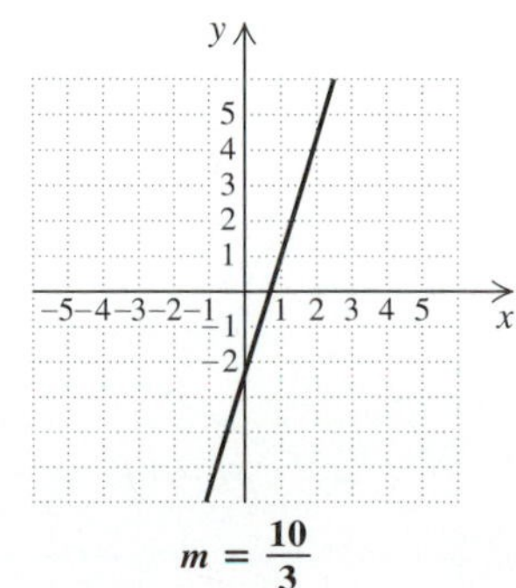

$m = \frac{10}{3}$

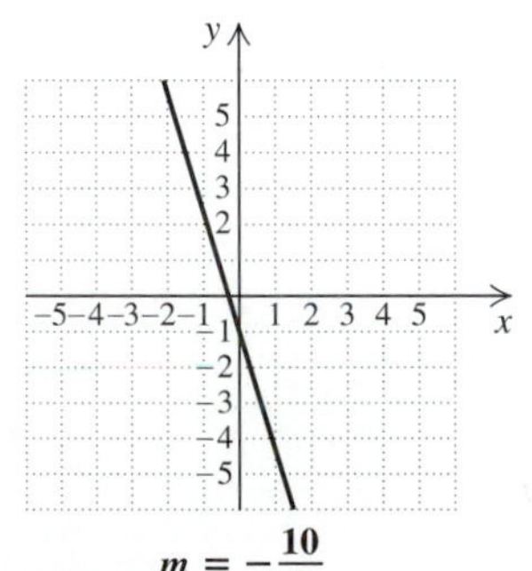

$m = -\frac{10}{3}$

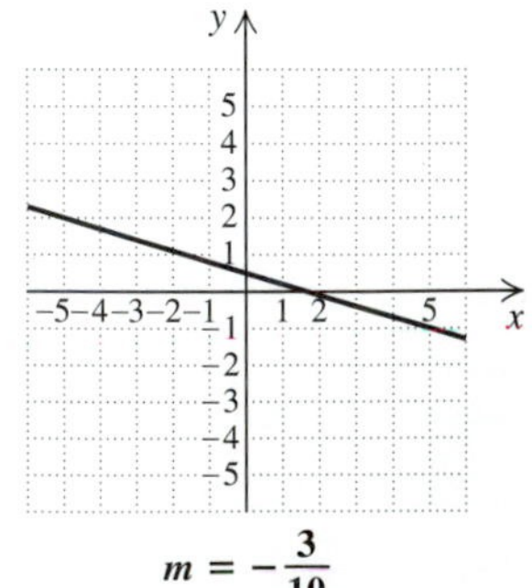

$m = -\frac{3}{10}$

Horizontal and Vertical Lines

What about the slope of a horizontal or a vertical line?

EXAMPLE 3

Find the slope of the line $y = 4$.

Solution Consider the points $(-3, 4)$ and $(2, 4)$, which are on the line.

The change in $y = 4 - 4$, or 0.
The change in $x = -3 - 2$, or -5.

$$m = \frac{4-4}{-3-2} = \frac{0}{-5} = 0$$

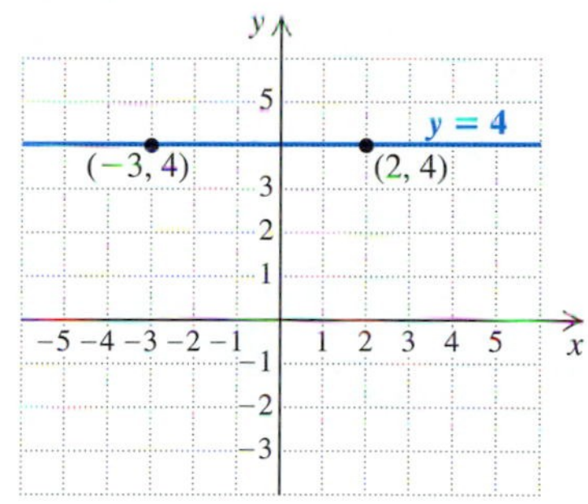

Any two points on a horizontal line have the same y-coordinate. Thus the change in y is 0, so the slope is 0. ❑

A horizontal line has slope 0.

EXAMPLE 4

Find the slope of the line $x = -3$.

Solution Consider the points $(-3, 4)$ and $(-3, -2)$, which are on the line.

The change in $y = 4 - (-2)$, or 6.
The change in $x = -3 - (-3)$, or 0.

$$m = \frac{4-(-2)}{-3-(-3)} = \frac{6}{0} \quad \text{(undefined)}$$

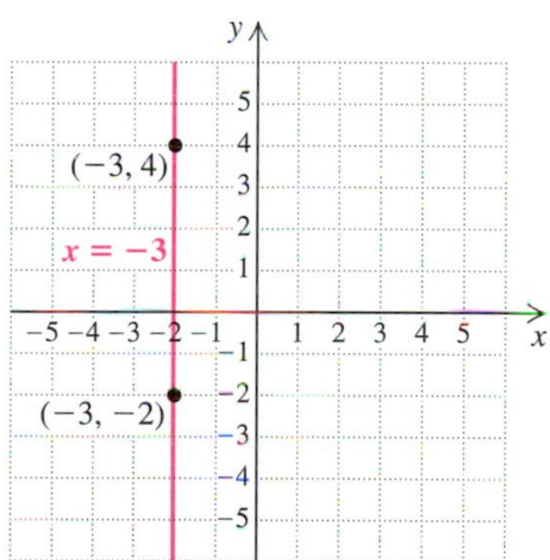

Since division by 0 is not defined, the slope of this line is not defined. The answer to this example is "The slope of this line is undefined." ❑

The slope of a vertical line is undefined.

Applications

Slope has many real-world applications. For example, numbers like 2%, 3%, and 6% are often used to represent the **grade** of a road, a measure of how steep a road on a hill or mountain is. For example, a 3% grade means that for every horizontal distance of 100 ft, the road rises or drops 3 ft. The concept of grade also occurs in fitness training when a person runs on a treadmill. A trainer may change the slope or grade of a treadmill to measure its effect on heartbeat. Another application of slope occurs when engineering a dam. A river's force or strength depends on how much the river drops over a specified horizontal distance.

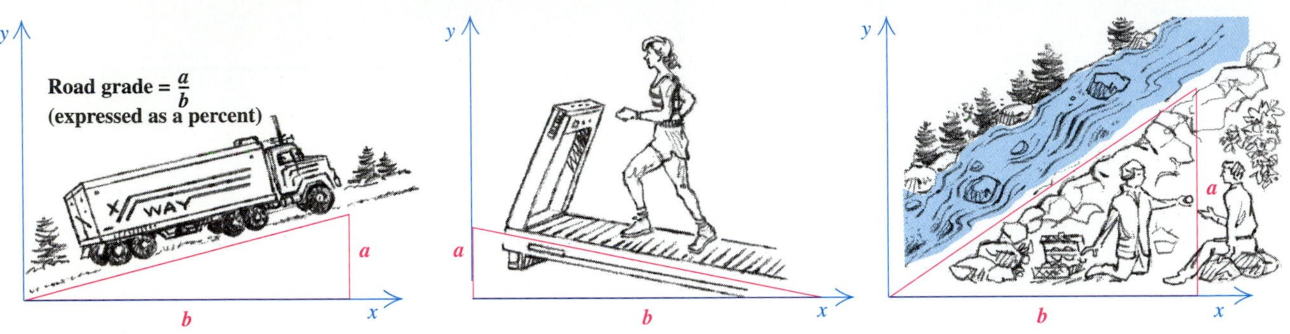

EXAMPLE 5

A road rises 50 ft vertically over a horizontal distance of 1250 ft. Find the grade of the road.

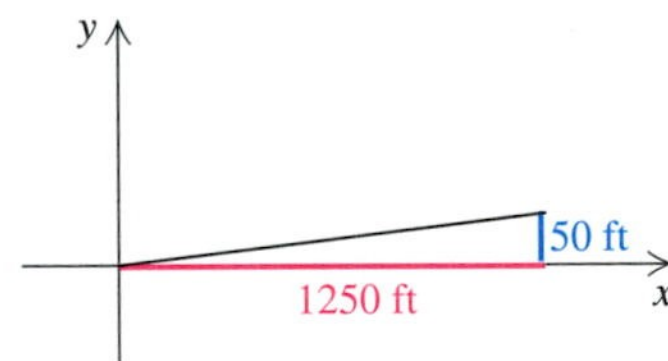

Solution The grade of the road is the slope of the line:

$$m = \frac{50}{1250} = 0.04 = 4\%. \quad \square$$

EXERCISE SET 7.1

Find the slope, if it is defined, of the line.

1.

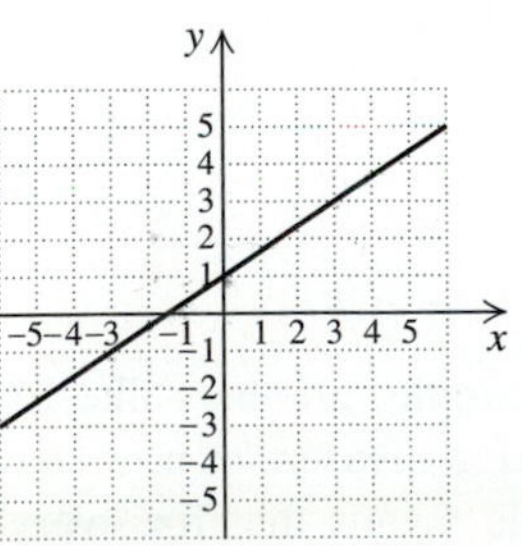

2.

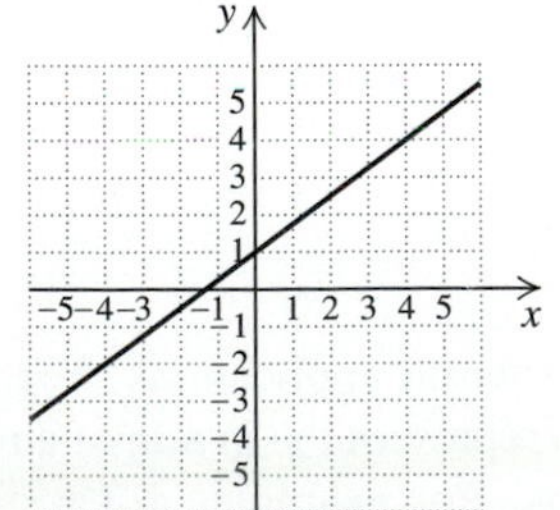

3.

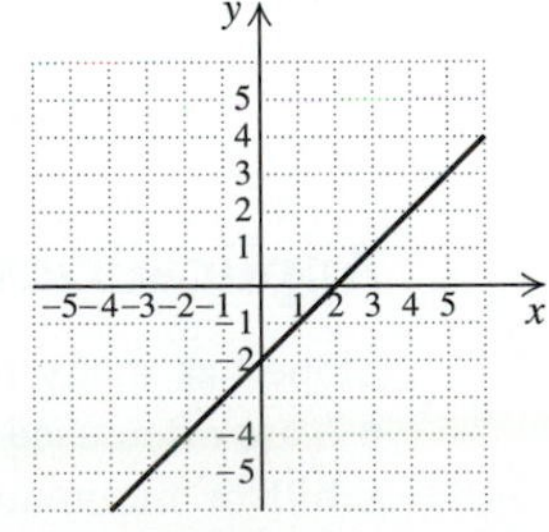

4.

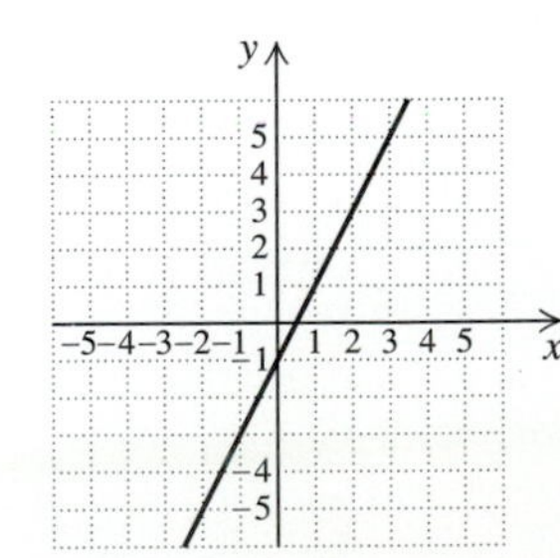

5.

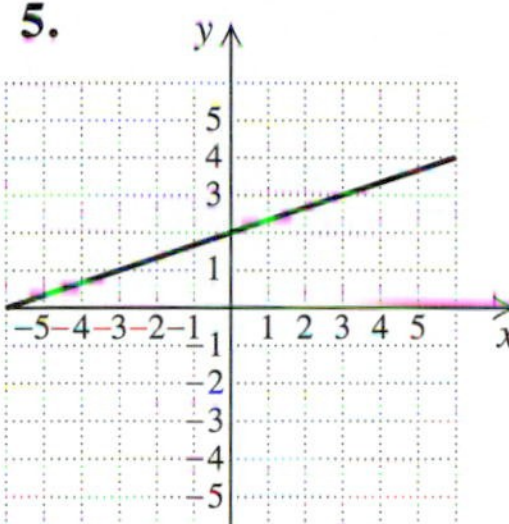

6.

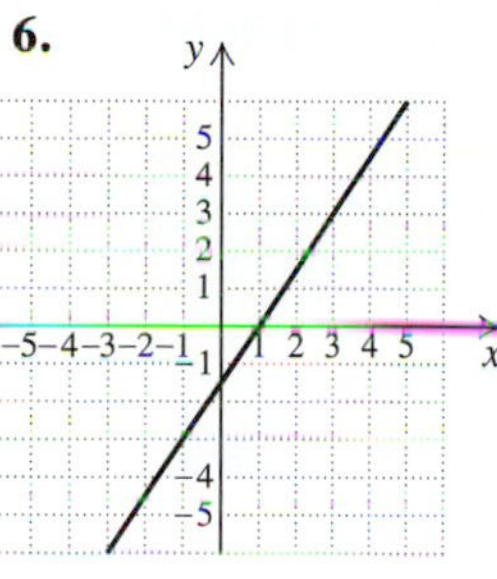

7.

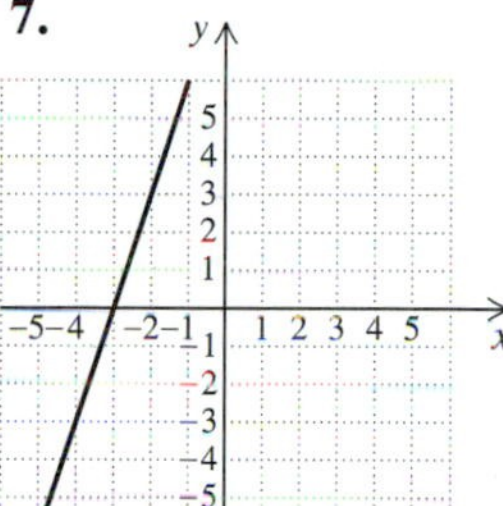

8.

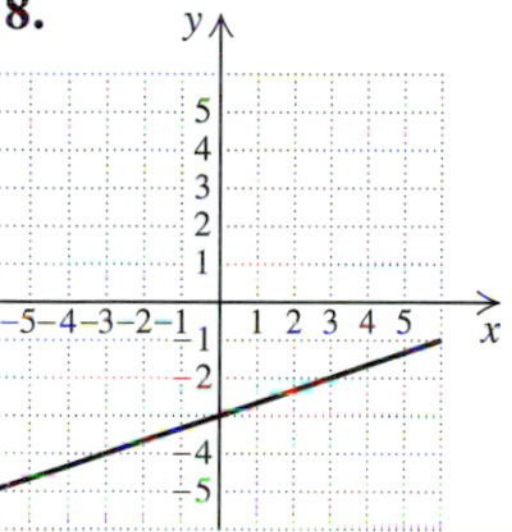

9.

10.

11.

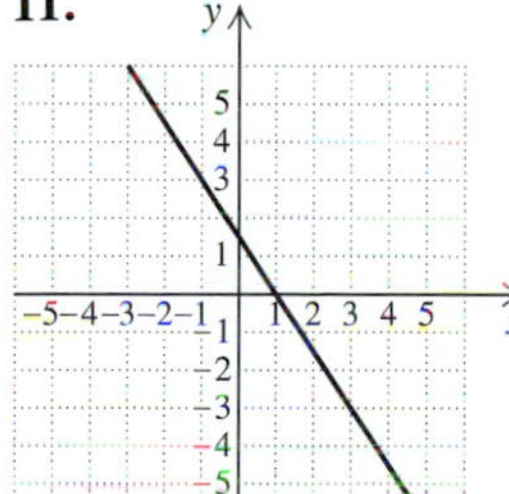

12.

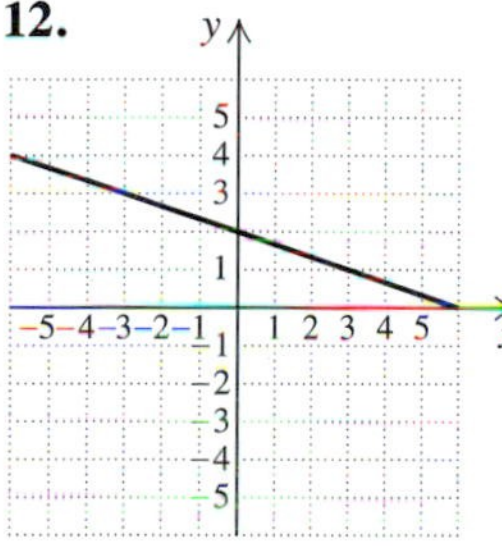

13.

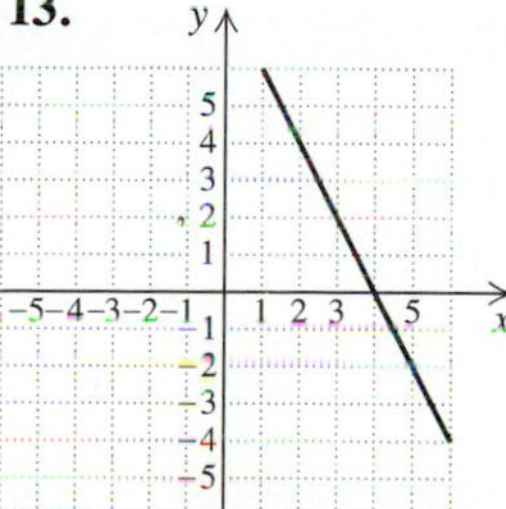

14.

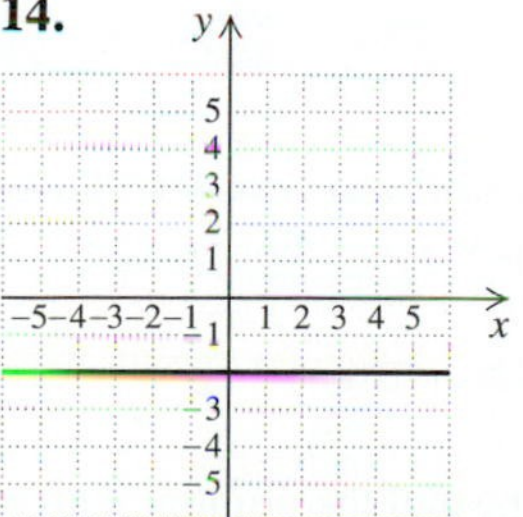

15.

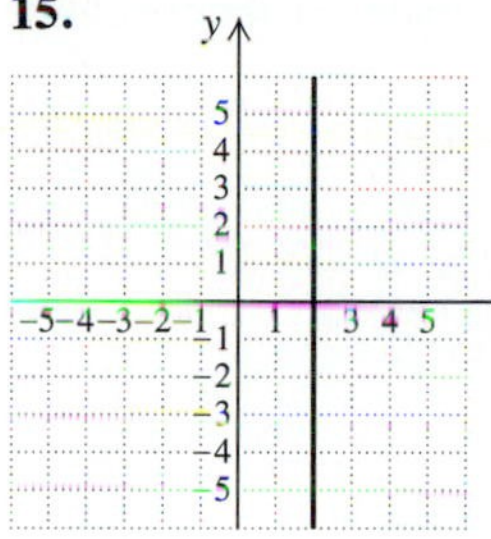

16.

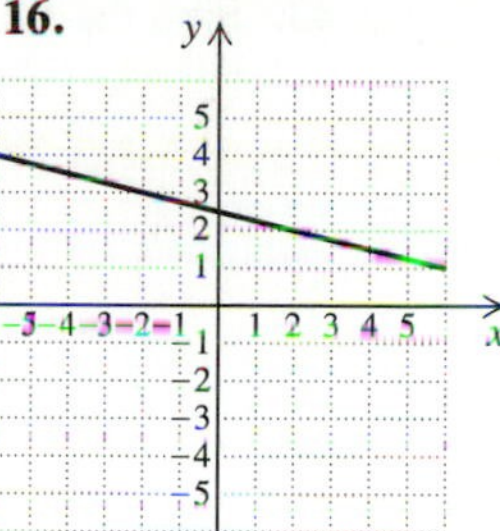

Find the slope of the line containing the given pair of points.

17. (3, 2) and (−1, 5)

18. (4, 1) and (−2, −3)

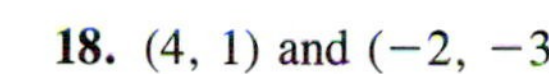

19. (−2, 4) and (3, 0)

20. (−4, 2) and (2, −3)

21. (4, 0) and (5, 7)

22. (3, 0) and (6, 2)

23. (0, 8) and (−3, 10)

24. (0, 9) and (4, 7)

25. (3, −2) and (5, −6)

26. (−2, 4) and (6, −7)

27. $\left(-2, \frac{1}{2}\right)$ and $\left(-5, \frac{1}{2}\right)$

28. (8, −3) and (10, −3)

29. (9, −4) and (9, −7)

30. (−10, 3) and (−10, 4)

31. (−1, 5) and (4, 5)

32. (−4, −2) and (−4, 7)

Find the slope of the line.

33. $x = -8$

34. $x = -4$

35. $y = 2$

36. $y = 17$

37. $x = 9$

38. $x = 6$

39. $y = -9$

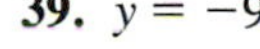

40. $y = -4$

41. Find the grade of the road.

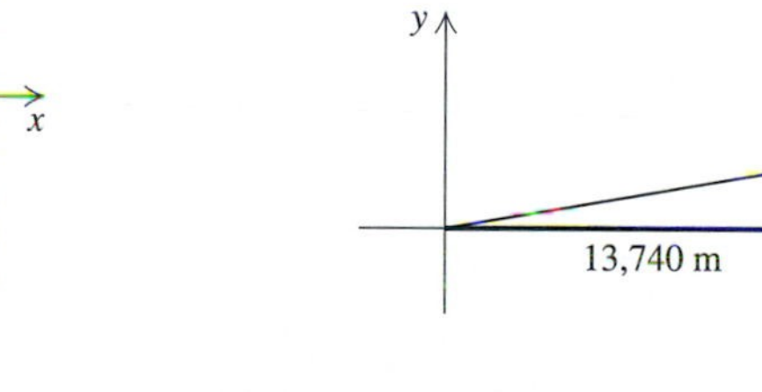

42. Find the slope (or pitch) of the roof.

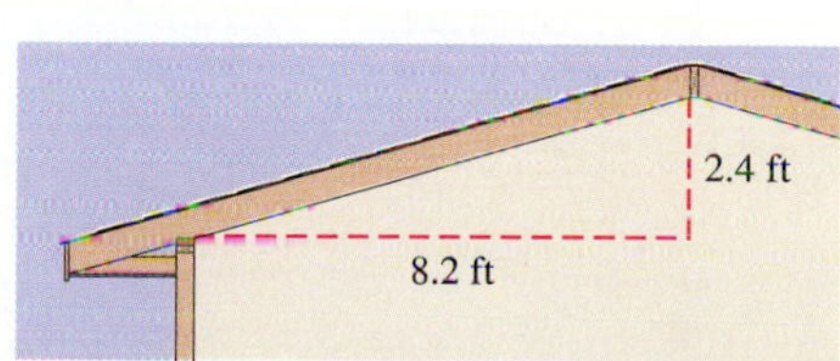

43. Find the slope (or grade) of the treadmill.

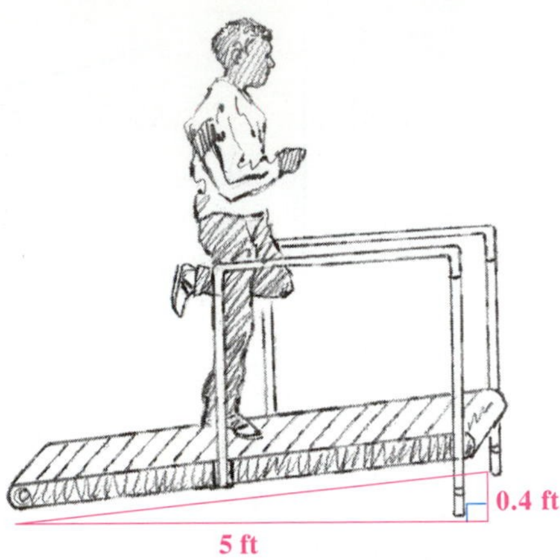

44. Find the slope of the river.

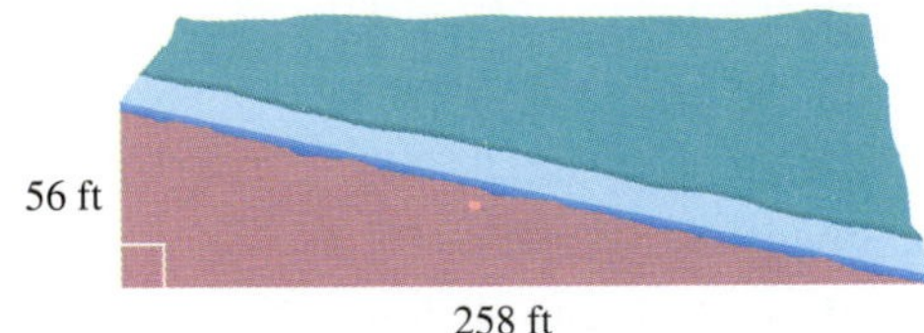

45. A road drops 158.4 ft vertically over a horizontal distance of 5280 ft. What is the grade of the road?

46. A river drops 55.71 ft vertically over a horizontal distance of 1238 ft. What is the slope of the river?

47. A treadmill that casts a shadow 5 ft long is set at a 12% grade when a heart arrhythmia occurs. How far off the floor is the end of the treadmill if the light source is directly overhead?

48. A river flows at a slope of 0.12. How many feet does it fall vertically over a horizontal distance of 250 ft?

Skill Maintenance

Multiply.

49. $5x(9x - 3)$

50. $(x - 2)(x^2 + 3x - 5)$

51. $(x - 7)(x + 7)$

52. $(5x + 2)^2$

Synthesis

53. ◈ If one line has a slope of -3 and another has a slope of 2, which line is steeper? Why?

54. ◈ Explain why the order in which coordinates are subtracted to find slope does not matter so long as y-coordinates and x-coordinates are subtracted in the same order.

55. A nonvertical line passes through (3, 4). What numbers could the line have for its slope if the line never enters the second quadrant?

56. A nonvertical line passes through $(5, -6)$. What numbers could the line have for its slope if the line never enters the first quadrant?

57. By 3:00, Catanya and Chad had already made 46 candles. Forty minutes later, the total reached 64 candles. Find the rate at which Catanya and Chad made candles. Give your answer in number of candles per hour.

58. Marcy picks apples twice as fast as Ryan. By 4:30, Ryan had already picked 4 bushels of apples; 50 minutes later, his total reached $5\frac{1}{2}$ bushels. Find Marcy's picking rate. Give your answer in number of bushels per hour.

59. ◈ The points $(-4, -3)$, $(1, 4)$, $(4, 2)$, and $(-1, -5)$ are vertices of a quadrilateral. Use slopes to explain why the quadrilateral is a parallelogram.

60. ◈ Can the points $(-4, 0)$, $(-1, 5)$, $(6, 2)$, and $(2, -3)$ be vertices of a parallelogram? Why or why not?

7.2 Slope–Intercept Form

Using the *y*-intercept and the Slope to Graph a Line • Equations in Slope–Intercept Form • Graphing and Slope–Intercept Form

If we know a line's slope and the point at which the y-axis is crossed, it is possible to draw a graph of the line. In this section, we will learn how to find a line's slope and

y-intercept from its equation. We will then be able to graph certain equations quite easily.

Using the y-intercept and the Slope to Graph a Line

Let's return to the car production situation described at the beginning of Section 7.1. Now suppose that as a new workshift begins, 4 cars have already been produced. At the Michigan plant, 3 cars were being produced every 2 hours, a rate of $\frac{3}{2}$ cars per hour. If this rate remains the same regardless of how many cars have already been produced, the following table and graph can be made.

Michigan Plant	
Hours Elapsed	**Cars Produced**
0	4
2	7
4	10
6	13
8	16

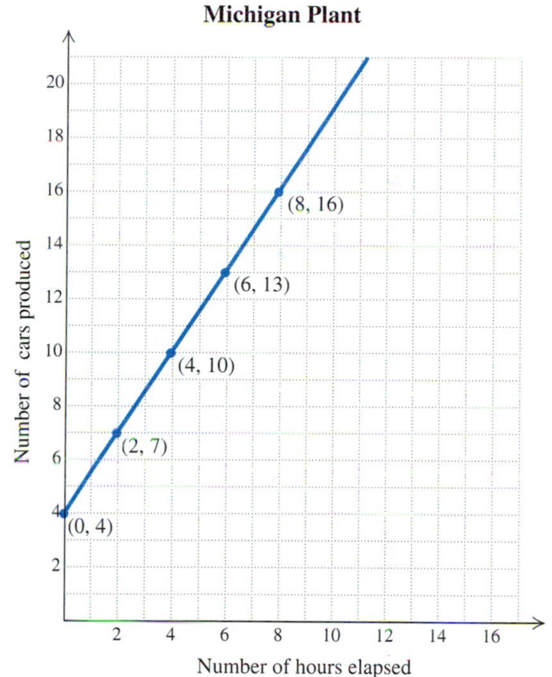

To confirm that the production rate is still $\frac{3}{2}$, we compute the slope using any pair of points on the graph. We choose (0, 4) and (2, 7):

$$\text{Slope} = \frac{\text{change in } y}{\text{change in } x}$$
$$= \frac{7-4}{2-0} = \frac{3}{2}.$$

Note that had we simply plotted the point (0, 4) and from there moved *up* 3 units and *to the right* 2 units, we could have located the point (2, 7) without consulting the table. Using (0, 4) and (2, 7), we then could have drawn the line. This is the method used in the following example.

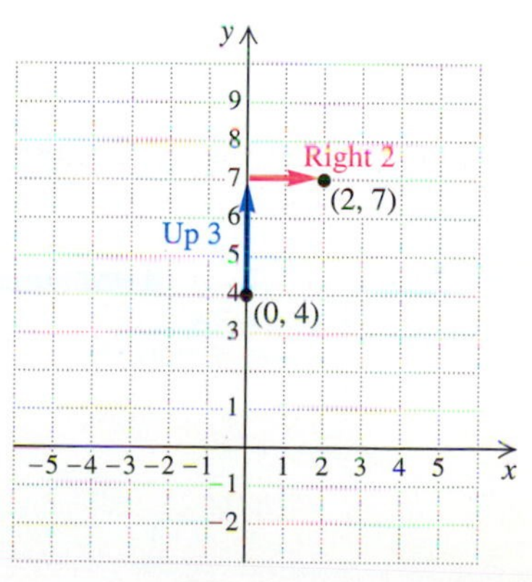

EXAMPLE 1

Draw a line that has slope $\frac{1}{4}$ and y-intercept (0, 2).

Solution We plot (0, 2) and from there move *up* 1 unit and *to the right* 4 units. This locates the point (4, 3). We plot (4, 3) and draw a line passing through (0, 2) and (4, 3), as shown on the right below.

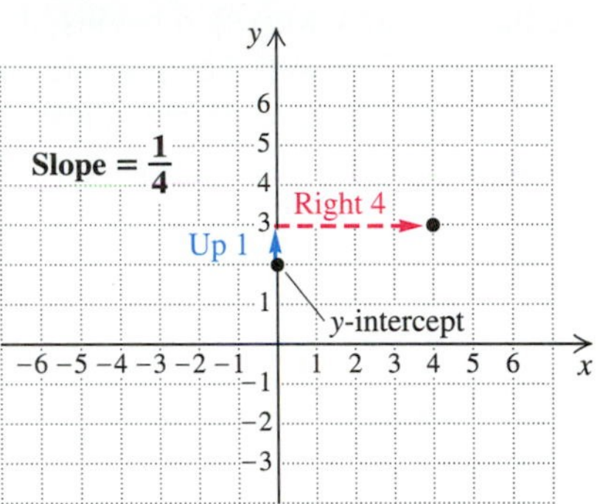

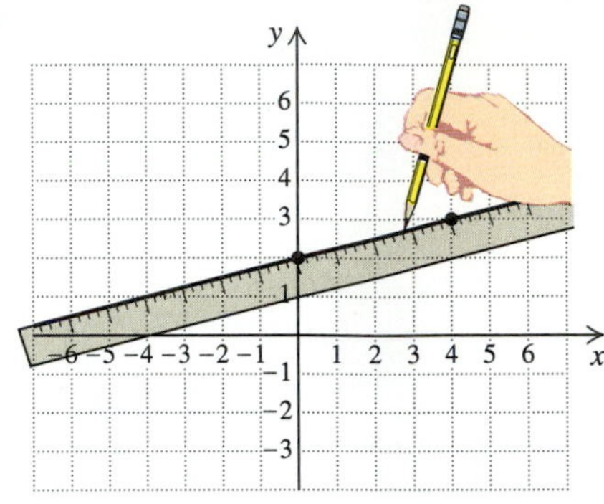

Equations in Slope–Intercept Form

It is not difficult to find a line's slope and y-intercept from its equation. Recall from Chapter 3 that to find the y-intercept of an equation's graph, we replace x with 0 and solve the resulting equation for y. For example, to find the y-intercept of the graph of $y = 2x + 3$, we replace x with 0 and solve as follows:

$$y = 2 \cdot 0 + 3 = 0 + 3 = 3.$$

The y-intercept of the graph of $y = 2x + 3$ is (0, 3). In a similar manner, it can be shown that the y-intercept of the graph of any equation $y = mx + b$ is $(0, b)$.

To calculate the slope of the graph of $y = 2x + 3$, we need two points. The y-intercept (0, 3) is one point and a second point, (1, 5), can be found easily by substituting 1 for x. We then have

$$\text{Slope} = \frac{\text{change in } y}{\text{change in } x} = \frac{5 - 3}{1 - 0} = \frac{2}{1} = 2.$$

Note that the slope, 2, is also the coefficient of x in $y = 2x + 3$. In a similar manner, it can be shown that the slope of the graph of any equation $y = mx + b$ is m (see Exercise 63).

The Slope–Intercept Equation

The equation $y = mx + b$ is called the *slope–intercept equation.* The equation represents a line of slope m with y-intercept $(0, b)$.

EXAMPLE 2

Find the slope and the y-intercept of the line: **(a)** $y = \frac{4}{5}x - 8$; **(b)** $2x + y = 5$; **(c)** $3x + 4y = 7$.

Solution

a) We rewrite $y = \frac{4}{5}x - 8$ as $y = \frac{4}{5}x + (-8)$. Now we simply read the slope and the y-intercept from the equation:

$$y = \frac{4}{5}x + (-8).$$

The slope is $\frac{4}{5}$. The y-intercept is $(0, -8)$.

b) We first solve for y to rewrite the equation in the form $y = mx + b$:

$$2x + y = 5$$

$$y = -2x + 5. \quad \text{Adding } -2x \text{ on both sides}$$

The slope is -2. The y-intercept is $(0, 5)$.

c) We rewrite the equation in the form $y = mx + b$:

$$3x + 4y = 7$$

$$4y = -3x + 7 \quad \text{Adding } -3x \text{ on both sides}$$

$$y = \frac{1}{4}(-3x + 7) \quad \text{Multiplying by } \tfrac{1}{4} \text{ on both sides}$$

$$y = -\frac{3}{4}x + \frac{7}{4} \quad \text{Using the distributive law}$$

The slope is $-\frac{3}{4}$. The y-intercept is $\left(0, \frac{7}{4}\right)$. □

EXAMPLE 3

A line has slope $-\frac{12}{5}$ and y-intercept $(0, 11)$. Find an equation of the line.

Solution We use the slope–intercept equation, substituting $-\frac{12}{5}$ for m and 11 for b:

$$y = mx + b$$

$$y = -\frac{12}{5}x + 11. \quad □$$

Graphing and Slope–Intercept Form

Our work in Examples 1 and 2 can be easily combined.

EXAMPLE 4

Graph: **(a)** $y = \frac{2}{5}x + 4$; **(b)** $2x + 3y = 3$.

Solution

a) First we plot the y-intercept, $(0, 4)$. We then consider the slope, $\frac{2}{5}$. Starting at the y-intercept and using the slope, we plot a second point by moving *up* 2 units (since the numerator is *positive* and corresponds to the change in y) and *to the* *right* 5 units (since the denominator is *positive* and corresponds to the change in x). We reach a new point, (5, 6).

We can also rewrite the slope as $\frac{-2}{-5}$. We again start at the y-intercept, (0, 4), but move *down* 2 units (since the numerator is *negative* and corresponds to the change in y) and *to the left* 5 units (since the denominator is *negative* and corresponds to the change in x). We reach another new point, $(-5, 2)$. Once two or three points have been plotted, the line can be drawn.

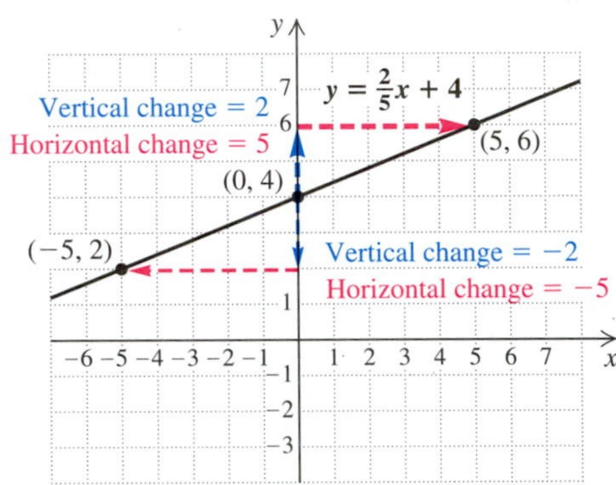

b) We graph the equation $2x + 3y = 3$ by first rewriting it in slope–intercept form:

$$2x + 3y = 3$$

$$3y = -2x + 3 \qquad \text{Adding } -2x \text{ on both sides}$$

$$y = \tfrac{1}{3}(-2x + 3) \qquad \text{Multiplying by } \tfrac{1}{3} \text{ on both sides}$$

$$y = -\tfrac{2}{3}x + 1. \qquad \text{Using the distributive law}$$

TECHNOLOGY CONNECTION

Using a standard $[-10, 10, -10, 10]$ window, graph the equations $y_1 = \frac{2}{3}x + 1$, $y_2 = \frac{3}{8}x + 1$, $y_3 = \frac{2}{3}x + 5$, and $y_4 = \frac{3}{8}x + 5$. If you can, use your grapher in the *Mode* that will graph equations *simultaneously*. Once the four lines have been drawn, try to decide which equation corresponds to each line. After matching equations with lines, you can check your matches by graphing the same equations in a *sequence* mode, if your grapher has one. In the sequence mode, equation y_1 is drawn first, y_2 is drawn next, and so on.

TC1. Graph the equations $y_1 = -\frac{3}{4}x - 2$, $y_2 = -\frac{1}{5}x - 2$, $y_3 = -\frac{3}{4}x - 5$, and $y_4 = -\frac{1}{5}x - 5$ using a grapher. If possible, use the simultaneous mode. Then match each line with the corresponding equation. Check using the sequence mode.

TC2. Write four different slope–intercept equations, two of which have the same slope but different y-intercepts, and two of which have the same y-intercepts but different slopes. Then use a grapher to draw all four lines and ask a classmate to match each equation with the appropriate line.

To graph $y = -\frac{2}{3}x + 1$, we first plot the y-intercept, (0, 1). We can think of the slope as $\frac{-2}{3}$. Starting at the y-intercept and using the slope, we find a second point by moving *down* 2 units (since the numerator is *negative*) and *to the right* 3 units (since the denominator is *positive*). We plot the new point, (3, −1). In a similar manner, we can move from the point (3, −1) to locate a third point, (6, −3). The line can then be drawn.

If the slope is thought of as $\frac{2}{-3}$, we can start at (0, 1), but this time move *up* 2 units (since the numerator is *positive*) and *to the left* 3 units (since the denominator is *negative*). We get another point on the graph, (−3, 3).

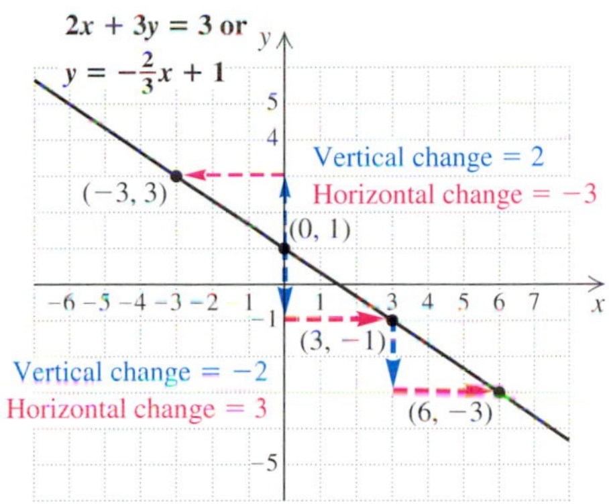

❑

EXERCISE SET 7.2

Draw a line that has the given slope and y-intercept.

1. Slope $\frac{2}{5}$; y-intercept (0, 1)
2. Slope $\frac{5}{3}$; y-intercept (0, −1)
3. Slope $\frac{5}{2}$; y-intercept (0, −3)
4. Slope $\frac{3}{5}$; y-intercept (0, 2)
5. Slope $-\frac{3}{4}$; y-intercept (0, 5)
6. Slope $-\frac{4}{5}$; y-intercept (0, 6)
7. Slope 2; y-intercept (0, −4)
8. Slope −2; y-intercept (0, −3)
9. Slope −3; y-intercept (0, 2)
10. Slope 3; y-intercept (0, 4)

Find the slope and the y-intercept of the line.

11. $y = 2x + 1$
12. $y = -3x + 7$
13. $y = -\frac{5}{6}x + 2$
14. $y = \frac{7}{2}x + 4$
15. $y = \frac{9}{4}x - 7$
16. $y = \frac{2}{9}x - 1$
17. $y = -\frac{2}{5}x$
18. $y = \frac{4}{3}x$
19. $-2x + y = 4$
20. $-5x + y = 5$
21. $4x - 3y = -12$
22. $x - 2y = 9$
23. $x - 3y = -2$
24. $x + y = 7$
25. $-2x + 4y = 8$
26. $-5x + 7y = 2$
27. $y = 5$
28. $y + 2 = 6$

Find the slope–intercept equation for the line with the indicated slope and y-intercept.

29. Slope 3; y-intercept (0, 1)
30. Slope −5; y-intercept (0, 3)

31. Slope $\frac{3}{5}$; y-intercept $(0, -5)$
32. Slope $-\frac{5}{2}$; y-intercept $(0, -1)$
33. Slope $-\frac{5}{3}$; y-intercept $(0, -8)$
34. Slope $\frac{3}{4}$; y-intercept $(0, 23)$
35. Slope -2; y-intercept $(0, 3)$
36. Slope 7; y-intercept $(0, -6)$
37. Slope 1; y-intercept $(0, -2)$
38. Slope -1; y-intercept $(0, 1)$

Graph.

39. $y = \frac{3}{5}x + 2$
40. $y = -\frac{3}{5}x - 1$
41. $y = -\frac{3}{5}x + 4$
42. $y = \frac{3}{5}x - 2$
43. $y = \frac{5}{3}x + 3$
44. $y = \frac{5}{3}x - 2$
45. $y = -\frac{3}{2}x - 2$
46. $y = -\frac{4}{3}x + 3$
47. $2x + y = 1$
48. $3x + y = 2$
49. $3x - y = 4$
50. $2x - y = 5$
51. $2x + 3y = 9$
52. $4x + 5y = 15$
53. $x - 4y = 12$
54. $x + 5y = 20$
55. $5x - 6y = 24$
56. $6x - 7y = 56$

Skill Maintenance

57. Solve: $3x^2 - 9x = 0$.
58. Factor: $y^3 - y^2 - 30y$.
59. The product of two consecutive odd integers is 195. Find the integers.
60. Eleven less than the square of a number is ten times the number. Find the number.

Synthesis

61. ◈ Under what circumstances might you draw an incorrect graph of a line even though three points were plotted and lined up with each other before the line was drawn?
62. ◈ Can an equation of a horizontal line be written in slope–intercept form? Why or why not?
63. Show that the slope of the line given by $y = mx + b$ is m. (*Hint:* Substitute both 0 and 1 for x to find two pairs of coordinates. Then use the formula Slope = change in y/change in x.)
64. Find an equation that can be used to predict the total number of cars N produced after t hours by the Michigan plant discussed in this section.
65. Find an equation of the line with the same slope as the line $3x - 2y = 8$ and the same y-intercept as the line $2y + 3x = -4$.

7.3 Point–Slope Form

Writing Equations in Point–Slope Form • Graphing and Point–Slope Form • Parallel and Perpendicular Lines

We now learn how to write an equation of a line using the line's slope and any one point through which the line passes.

Writing Equations in Point–Slope Form

Consider a line with slope 2 passing through the point (4, 1), as shown in the figure. In order for a point (x, y) to be on the line, the coordinates x and y must solve the slope equation

$$\frac{y-1}{x-4} = 2.$$

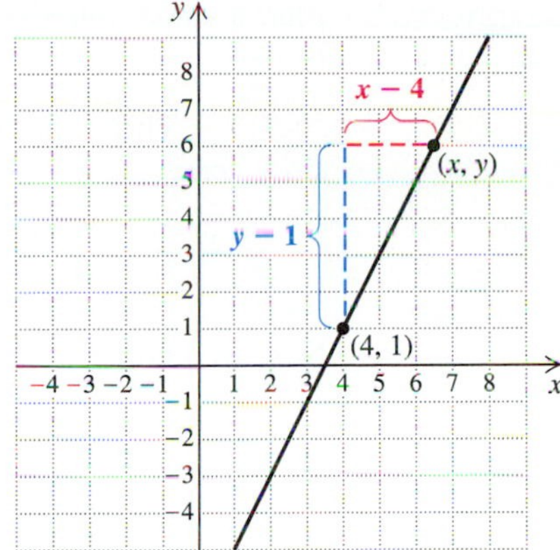

Multiplying on both sides by $x - 4$, we have

$$y - 1 = 2(x - 4).$$

This is considered **point–slope form** for the line shown.

To generalize, a line with slope m passing through the point (x_1, y_1) will include a point (x, y) if the coordinates x and y solve the slope equation

$$\frac{y - y_1}{x - x_1} = m.$$

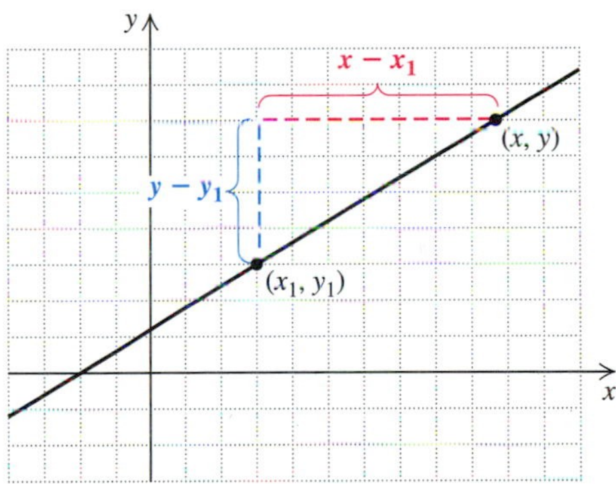

Multiplying on both sides by $x - x_1$ gives us the *point–slope equation*

$$y - y_1 = m(x - x_1).$$

The Point–Slope Equation

The equation $y - y_1 = m(x - x_1)$ is called the *point–slope equation* for the line with slope m that contains the point (x_1, y_1).

EXAMPLE 1

Find a point–slope equation for the line with slope $\frac{1}{5}$ that contains the point $(-2, -3)$.

Solution We substitute $\frac{1}{5}$ for m, -2 for x_1, and -3 for y_1:

$y - y_1 = m(x - x_1)$ Using the point–slope equation

$y - (-3) = \frac{1}{5}(x - (-2)).$ Substituting ❑

EXAMPLE 2

Find a slope–intercept equation for the line with slope 3 that contains the point (1, −5).

Solution There are two parts to this solution. First we write an equation in point–slope form:

$$y - y_1 = m(x - x_1)$$

$$y - (-5) = 3(x - 1). \quad \text{Substituting}$$

Next we find an equivalent equation of the form $y = mx + b$:

$$y - (-5) = 3(x - 1)$$

$$y + 5 = 3x - 3 \quad \text{Simplifying the subtraction and using the distributive law}$$

$$y = 3x - 8. \quad \text{This is in slope–intercept form.}$$

❑

EXAMPLE 3

A line passes through the points (3, −5) and (−4, 9). Find an equation for the line **(a)** in point–slope form and **(b)** in slope–intercept form.

Solution

a) To find a point–slope equation, we first compute the slope:

$$m = \frac{9 - (-5)}{-4 - 3} = \frac{14}{-7} = -2.$$

Next we use the point–slope equation and substitute, using −2 for m and either (3, −5) or (−4, 9) as (x_1, y_1):

$$y - y_1 = m(x - x_1)$$

$$y - (-5) = -2(x - 3). \quad \text{Using } (3, -5) \text{ for } (x_1, y_1)$$

b) To write an equation in slope–intercept form, we use the result of part (a) above:

$$y - (-5) = -2(x - 3)$$

$$y + 5 = -2x + 6 \quad \text{Simplifying the subtraction and using the distributive law}$$

$$y = -2x + 1. \quad \text{This is in slope–intercept form.}$$

Had we used (−4, 9) as (x_1, y_1) in part (a), we would have obtained the same slope–intercept equation:

$$y - 9 = -2(x - (-4)) \quad \text{Using } (-4, 9) \text{ for } (x_1, y_1)$$

$$y - 9 = -2(x + 4)$$

$$y - 9 = -2x - 8$$

$$y = -2x + 1.$$

❑

Graphing and Point–Slope Form

Equations written in point–slope form are easily graphed.

EXAMPLE 4

Graph: $y - 2 = 3(x - 4)$.

Solution Since $y - 2 = 3(x - 4)$ is in point–slope form, we know that the line has slope 3, or $\frac{3}{1}$, and passes through the point (4, 2). We plot (4, 2) and then find a second point by moving *up* 3 units and *to the right* 1 unit. The line can then be drawn, as shown below.

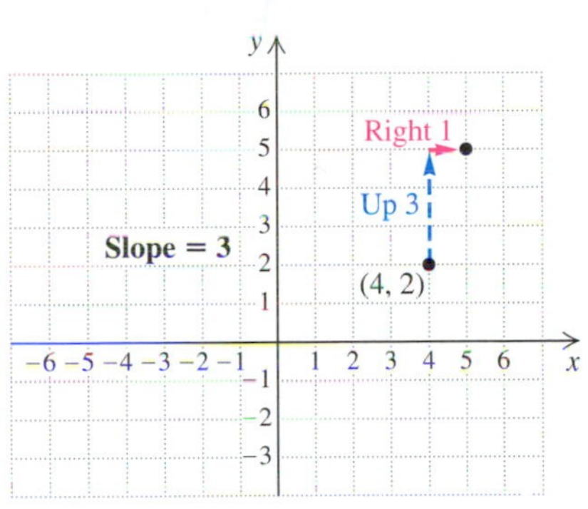

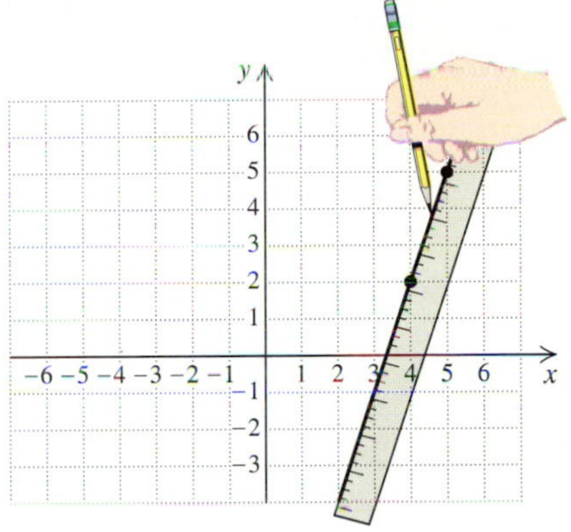

Parallel and Perpendicular Lines

When we graph a pair of linear equations, there are three possibilities.

1. The graphs are the same.
2. The graphs intersect at exactly one point.
3. The graphs are parallel (they do not intersect).

If two lines are vertical, they are parallel. How can we tell whether nonvertical lines are parallel? The answer is simple: We look at their slopes.

> Two nonvertical lines are parallel if they have the same slope.

EXAMPLE 5

Determine whether the line passing through the points (1, 7) and (4, −2) is parallel to the line $y = -3x + 4$.

Solution The slope of the line passing through (1, 7) and (4, −2) is given by

$$m = \frac{7 - (-2)}{1 - 4} = \frac{9}{-3} = -3.$$

Because the line given by $y = -3x + 4$ also has a slope of −3, we conclude that the lines are parallel. ❑

If one line is vertical and another is horizontal, they are perpendicular. There are other instances in which two lines are perpendicular.

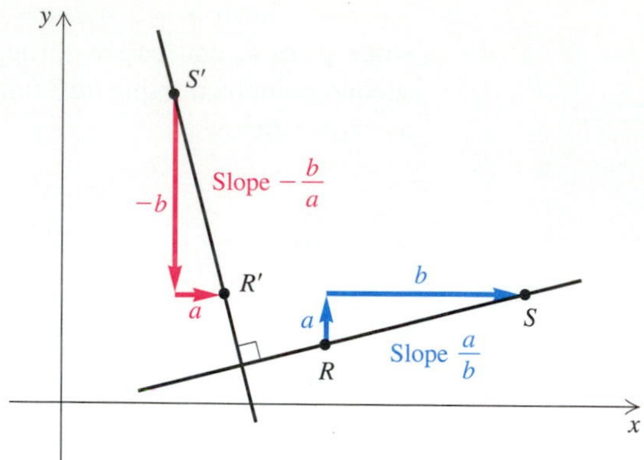

Consider a line $\overleftrightarrow{RS}$, as shown in the graph, with slope a/b. Then think of rotating the figure 90° to get a line $\overleftrightarrow{R'S'}$ perpendicular to $\overleftrightarrow{RS}$. For the new line, the rise and the run are interchanged, but the run is now negative. Thus the slope of the new line is $-b/a$. Let us multiply the slopes:

$$\frac{a}{b}\left(-\frac{b}{a}\right) = -1.$$

This is the condition under which lines are perpendicular.

Slope and Perpendicular Lines

Two lines are perpendicular if the product of their slopes is -1. (If one line has slope m, the slope of a line perpendicular to it is $-1/m$. That is, we take the reciprocal and change the sign.) Lines are also perpendicular if one of them is vertical and the other is horizontal.

EXAMPLE 6

Determine whether the graphs of $y = -3x + 1$ and $x - 3y = 10$ are perpendicular.

Solution The slope of the line $y = -3x + 1$ is -3. To find the slope of the second line, we rewrite $x - 3y = 10$ in slope-intercept form:

$$x - 3y = 10$$

$$-3y = -x + 10 \quad \text{Adding } -x \text{ on both sides}$$

$$y = \tfrac{1}{3}x - \tfrac{10}{3}. \quad \text{Dividing by } -3 \text{ on both sides}$$

To determine whether the lines are perpendicular, multiply the slopes:

$$-3\left(\tfrac{1}{3}\right) = -1.$$

Since the product of the slopes is -1, the lines are perpendicular. ☐

EXAMPLE 7

Consider the line given by the equation $8y = 7x - 24$.

a) Find an equation for a parallel line passing through $(-1, 2)$.
b) Find an equation for a perpendicular line passing through $(-1, 2)$.

Solution To find the slope of the line $8y = 7x - 24$, we solve for y to obtain slope–intercept form:

$$8y = 7x - 24$$
$$y = \tfrac{7}{8}x - 3. \quad \text{Multiplying by } \tfrac{1}{8} \text{ on both sides}$$

↑ The slope is $\frac{7}{8}$.

a) Using the point–slope equation, we have

$$y - 2 = \tfrac{7}{8}[x - (-1)] \quad \text{Substituting } \tfrac{7}{8} \text{ for the slope and } (-1, 2) \text{ for the point}$$
$$y = \tfrac{7}{8}x + \tfrac{23}{8}.$$

b) The slope of a perpendicular line is given by the opposite of the reciprocal of $\frac{7}{8}$, $-\frac{8}{7}$. The point–slope equation yields

$$y - 2 = \tfrac{8}{7}[x - (-1)] \quad \text{Substituting } -\tfrac{8}{7} \text{ for the slope and } (-1, 2) \text{ for the point}$$
$$y = -\tfrac{8}{7}x + \tfrac{6}{7}.$$

❑

EXERCISE SET 7.3

Find a point–slope equation for the line containing the given point and having the given slope.

1. $(2, 5)$, $m = 5$

2. $(-3, 0)$, $m = -2$

3. $(2, 4)$, $m = \frac{3}{4}$

4. $\left(\frac{1}{2}, 2\right)$, $m = -1$

5. $(2, -6)$, $m = 1$

6. $(4, -2)$, $m = 6$

7. $(-7, 0)$, $m = -3$

8. $(0, 3)$, $m = -3$

9. $(5, 6)$, $m = \frac{2}{3}$

10. $(2, 7)$, $m = \frac{5}{6}$

Find the slope–intercept equation for the line containing the given point and having the given slope.

11. $(3, 7)$, $m = 2$

12. $(1, 5)$, $m = 4$

13. $(4, 5)$, $m = -1$

14. $(2, -3)$, $m = 1$

15. $(-2, 3)$, $m = \frac{1}{2}$

16. $(6, -4)$, $m = -\frac{1}{2}$

17. $(-6, -5)$, $m = -\frac{1}{3}$

18. $(-5, 0)$, $m = \frac{1}{5}$

19. $(4, 0)$, $m = \frac{5}{4}$

20. $(-3, 8)$, $m = \frac{4}{3}$

Find the slope–intercept equation for the line containing the given pair of points. (*Hint:* First use point–slope form.)

21. $(-6, 1)$ and $(2, 3)$

22. $(12, 16)$ and $(1, 5)$

23. $(0, 4)$ and $(4, 2)$

24. $(0, 0)$ and $(4, 2)$

25. $(3, 2)$ and $(1, 5)$

26. $(-4, 1)$ and $(-1, 4)$

27. $(5, 0)$ and $(0, -2)$

28. $(-2, -2)$ and $(1, 3)$

29. $(-2, -4)$ and $(2, -1)$

30. $(-3, 5)$ and $(-1, -3)$

Graph.

31. $y - 5 = \frac{1}{2}(x - 3)$

32. $y - 2 = \frac{1}{3}(x - 5)$

33. $y - 3 = -\frac{1}{2}(x - 5)$

34. $y - 1 = -\frac{1}{4}(x - 3)$

35. $y + 5 = \frac{1}{2}(x - 3)$

36. $y - 2 = \frac{1}{3}(x + 5)$

37. $y + 2 = 3(x + 1)$

38. $y + 4 = 2(x + 1)$

39. $y - 4 = -2(x + 1)$

40. $y + 3 = -1(x - 4)$

41. $y + 3 = -(x + 2)$

42. $y + 4 = 2(x + 2)$

Determine whether the graphs of each pair of equations are parallel.

43. $y = \frac{2}{3}x + 1$,
$y = \frac{2}{3}x - 1$

44. $y = -\frac{5}{8}x + 3$,
$y = \frac{5}{8}x + 4$

45. $x + 6 = y$,
$y + x = -2$

46. $2x - 7 = y$,
$y - 2x = 8$

47. $3x + 4y = 8$,
$7 - 12y = 9x$

48. $3x = 5y - 2$,
$10y = 4 - 6x$

Find the slope–intercept equation of the line containing the specified point and parallel to the indicated line.

49. $(3, 7)$, $x + 2y = 6$

50. $(0, 3)$, $3x - y = 7$

51. $(2, -1)$, $5x - 7y = 8$

52. $(-4, -5)$, $2x + y = -3$

53. $(-6, 2)$, $3x - 9y = 2$

54. $(-7, 0)$, $5x + 2y = 6$

Determine whether the graphs of each pair of equations are perpendicular.

55. $y = 4x - 5$,
$y = -\frac{1}{4}x + 8$

56. $2x - 5y = -3$,
$2x + 5y = 4$

57. $3x + 5y = 10$,
$15x + 9y = 18$

58. $y = -x + 7$,
$y = x + 3$

59. $x = 5,$
$y = \frac{1}{2}$

60. $y = -2,$
$x = 2$

Find the slope–intercept equation of the line containing the specified point and perpendicular to the indicated line.

61. $(2, 5)$, $2x + y = -3$

62. $(4, 0)$, $x - 3y = 0$

63. $(3, -2)$, $3x + 4y = 5$

64. $(-3, -5)$, $5x - 2y = 4$

65. $(0, 9)$, $2x + 5y = 7$

66. $(-3, -4)$, $-3x + 6y = 2$

Skill Maintenance

Factor.

67. $7x^3y^2 + 35x^2y^6$

68. $5x^2(3x - 1) + (3x - 1)$

Simplify.

69. $\dfrac{5x^2 + 5x}{10x^3 - 10x^2}$

70. $\dfrac{x^2 - 7x + 10}{x^2 - 4}$

Synthesis

71. In your own words, describe a procedure that can be used to write a slope–intercept equation for any line passing through two given points.

72. In your own words, describe a procedure that can be used for graphing any equation of the form $y - k = m(x - h)$.

73. Find an equation of the line that contains the point $(2, -3)$ and that has the same slope as the line $3x - y + 4 = 0$.

74. Find an equation of the line that has the same y-intercept as the line $x - 3y = 6$ and contains the point $(5, -1)$.

75. Find an equation of the line that has x-intercept $(-2, 0)$ and is parallel to $4x - 8y = 12$.

76. Why is slope–intercept form more useful than point–slope form when using a grapher? How can point–slope form be modified so that it better accommodates graphers?

7.4 Functions

Notation for Functions • Functions and Graphs • The Vertical-Line Test

We now develop the idea of a *function*—one of the most important concepts in mathematics. In much the same way that ordered pairs form correspondences between first coordinates and second coordinates, a function is a correspondence from one set to another. For example:

To each person in a class	there corresponds	his or her mother.
To each item in a store	there corresponds	its price.
To each real number	there corresponds	the cube of that number.

In each example, the first set is called the **domain.** The second set is called the **range.** Given a member of the domain, there is *just one* member of the range to which it corresponds. This kind of correspondence is called a **function.**

Domain — Correspondence → Range

EXAMPLE 1

Determine whether the correspondence is a function.

a) $a \rightarrow 4$, $b \rightarrow 0$, $c \rightarrow 0$

b) San Francisco → Giants; New York → Giants; New York → Mets; Cleveland → Browns

Solution

a) The correspondence *is* a function because each member of the domain corresponds to just one member of the range.

b) The correspondence *is not* a function because a member of the domain (New York) corresponds to more than one member of the range. ❑

Function

A *function* is a correspondence between a first set, called the *domain,* and a second set, called the *range,* such that each member of the domain corresponds to *exactly one* member of the range.

EXAMPLE 2

Determine whether the correspondence is a function.

	Domain	*Correspondence*	*Range*
a)	A family	Each person's weight	A set of positive numbers
b)	The natural numbers	Each number's square	A set of natural numbers
c)	The set of all states	Each state's members of the U.S. Senate	A set of U.S. senators

Solution

a) The correspondence *is* a function, because each person has *only one* weight.

b) The correspondence *is* a function, because each natural number has *only one* square.

c) The correspondence *is not* a function, because each state has two U.S. senators. ❑

When a correspondence between two sets is not a function, it is still an example of a **relation.**

Relation

A *relation* is a correspondence between a first set, called the *domain,* and a second set, called the *range,* such that each member of the domain corresponds to *at least one* member of the range.

Thus, although the correspondences of Examples 1 and 2 are not all functions, they *are* all relations. A function is a special type of relation—one in which each member of the domain is paired with *exactly one* member of the range.

Notation for Functions

To understand function notation, it helps to imagine a "function machine." Think of putting a member of the domain (an *input*) into the machine. The machine knows the correspondence and gives you a member of the range (the *output*).

The function has been named f. We call the input x, and its output $f(x)$. This is read "f of x," or "f at x" or "the value of f at x." Note that $f(x)$ does *not* mean "f times x."

Most functions are described by equations. For example, $f(x) = 2x + 3$ describes the function that takes an input x, multiplies it by 2, and then adds 3.

$$f(x) = \underbrace{2}_{\text{Double}}\overset{\text{Input}}{x} \underbrace{+\,3}_{\text{Add 3}}$$

To find the output $f(4)$, we take the input 4, double it, and add 3 to get 11. That is, we substitute 4 into the formula for $f(x)$:

$$f(4) = 2 \cdot 4 + 3$$
$$= 11.$$

Sometimes, instead of writing $f(x) = 2x + 3$, we might write $y = 2x + 3$, where it is understood that the value of y, the *dependent variable*, is calculated after first choosing a value for x, the *independent variable*. To understand why $f(x)$ notation is so useful, consider two equivalent statements:

a) If $f(x) = 2x + 3$, then $f(4) = 11$.
b) If $y = 2x + 3$, then the value of y is 11 when x is 4.

The notation used in part (a) is far more concise.

EXAMPLE 3

Find the indicated function value.

a) $f(5)$, for $f(x) = 3x + 2$
b) $g(3)$, for $g(z) = 5z^2 - 4$
c) $A(-2)$, for $A(r) = 3r^2 + 2r$
d) $F(a + 1)$, for $F(x) = 3x + 2$

Solution

a) $f(5) = 3 \cdot 5 + 2 = 17$
b) $g(3) = 5(3)^2 - 4 = 41$
c) $A(-2) = 3(-2)^2 + 2(-2) = 8$
d) $F(a + 1) = 3(a + 1) + 2 = 3a + 3 + 2 = 3a + 5$ ❑

When all of a function's inputs share one output value, we say that we have a *constant function*. Thus, for the constant function $h(x) = 7$, we have $h(3) = 7$ and $h(5) = 7$. When the graph of a function is a line, we say that we have a *linear function*. The functions $f(x) = 2x - 5$ and $g(x) = -\frac{1}{2}x + 1$ are linear functions.

Note that whether we write $f(x) = 3x + 2$, or $f(t) = 3t + 2$, or $f(\square) = 3\square + 2$, we still have $f(5) = 17$. Thus the independent variable can be thought of as a *dummy variable*. The letter chosen for the dummy variable is not as important as the algebraic manipulations to which it is subjected.

Although you have probably already used functions that are described by formulas, you may not have seen function notation in those applications. For example, the formula for finding the area A of a circle with radius r is

$$A = \pi r^2.$$

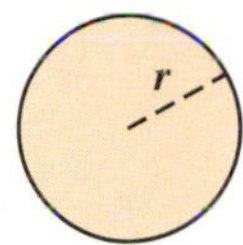

We say that the area A is *a function of* r. To emphasize that fact, we often write

$$A(r) = \pi r^2.$$

Thus, to find the area of a circle with a radius of 12 cm, we might write

$$\begin{aligned} A(12) &= \pi(12)^2 && \text{Substituting for } r \\ &= \pi \times 144 \\ &\approx 452 \text{ cm}^2. && \text{Using 3.14 for } \pi \end{aligned}$$

Functions and Graphs

Functions are often described by graphs. To use a graph in problem solving, we note that each point on the graph represents a pair of values—one from the horizontal axis (the domain) and one from the vertical axis (the range). In the following example, we first draw a graph and then estimate a function value from it.

EXAMPLE 4

According to the Federal Center for Disease Control, there were 309 newly reported cases of AIDS in the United States in 1981, 4436 cases in 1984, 21,114 cases in 1987, 43,339 cases in 1990, and 47,095 cases in 1992. Estimate the number of newly reported cases for the years 1983 and 1988.

Solution

1. and **2.** FAMILIARIZE and TRANSLATE. The given information enables us to plot five points on a graph. We let the horizontal axis represent the year and the vertical axis the number of reported cases.

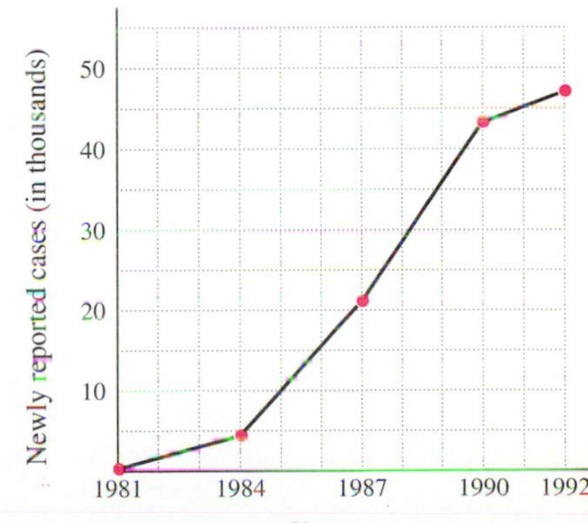

3. **CARRY OUT.** To estimate the reported number of cases in 1983, we locate the point that is directly above the year 1983. After doing so, we estimate its second coordinate by moving horizontally from the point to the vertical axis.

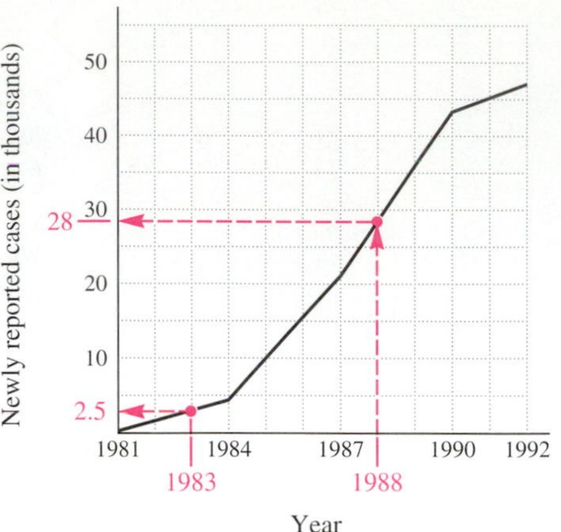

We can estimate from the graph that the function value is 2.5 thousand. Following a similar procedure, we estimate that 1988 is paired with 28 thousand.

4. **CHECK.** Although a precise check is impossible, note that 2.5 thousand is between 309 and 4436. Similarly, 28 thousand is between 21,114 and 43,339. Thus both answers are at least plausible.

5. **STATE.** There were about 2500 newly reported cases of AIDS in 1983 and about 28,000 newly reported cases in 1988. ❑

The Vertical-Line Test

When we are graphing functions, the domain consists of all values on the horizontal axis that serve as a first coordinate for some point on the graph. The range consists of all values on the vertical axis that serve as a second coordinate for some point on the graph. If any value on the horizontal axis is the first coordinate of more than one point on the graph, the graph cannot represent a function (otherwise one member of the domain would correspond to more than one member of the range). This observation is the basis of the *vertical-line test*.

The Vertical-Line Test

A graph is that of a function provided it is not possible to draw a vertical line that intersects the graph more than once.

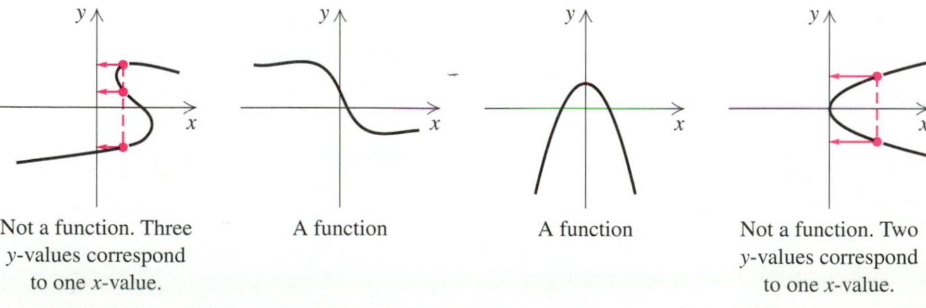

Not a function. Three y-values correspond to one x-value.

A function

A function

Not a function. Two y-values correspond to one x-value.

EXERCISE SET 7.4

Determine whether the correspondence is a function.

1. 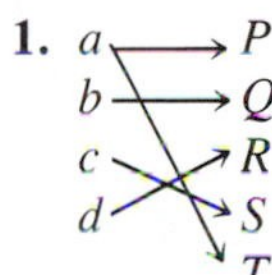

2. 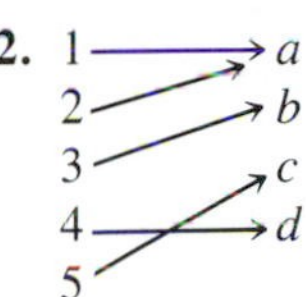

3.

Firm	*Number of Partners*
Brown & Jones	→850
Smith & Hawkens	→850
Hernandez & Rowle	→1900
Ciani & Ross	→1270

4.

Firm	*Number of Female Partners*
Brown & Jones	→38
Smith & Hawkens	→44
Hernandez & Rowle	→38
Ciani & Ross	→27
Morong & Davis	→54

5. 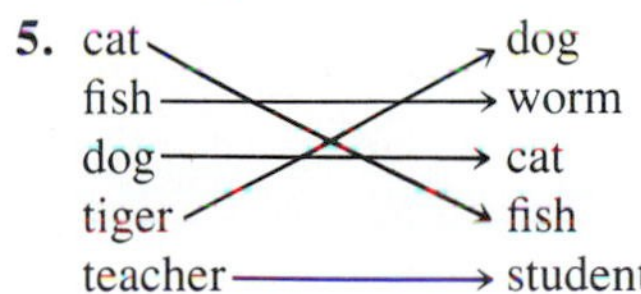

6. 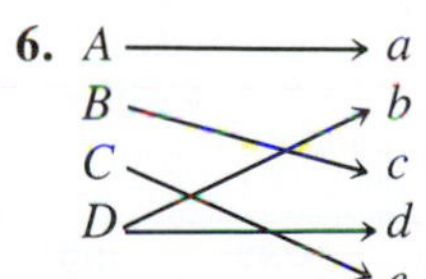

7. 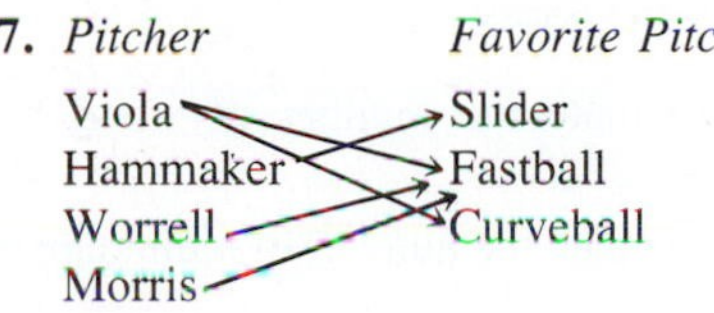

8. 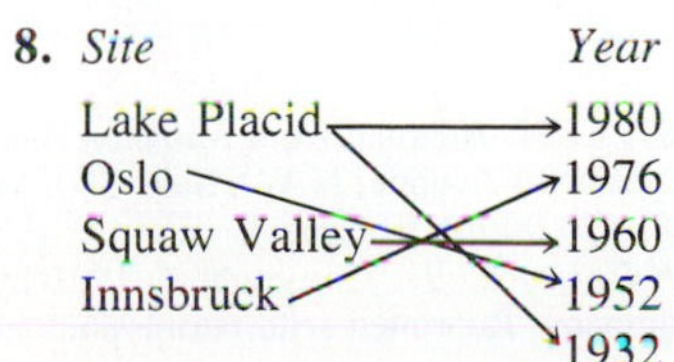

9. 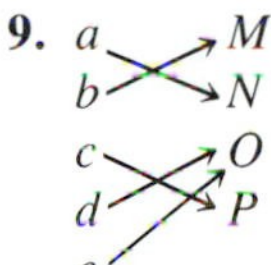

10. 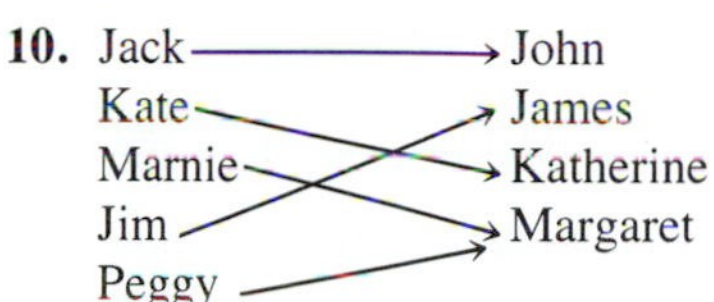

Determine whether each of the following is a function. Identify any relations that are not functions.

	Domain	*Correspondence*	*Range*
11.	A math class	Each person's seat number	A set of numbers
12.	A set of numbers	Square each number and then add 4.	A set of numbers
13.	A set of shapes	Find the area of each shape.	A set of numbers
14.	A family	Each person's eye color	A set of colors
15.	The people in a town	Each person's aunt	A set of females
16.	A set of avenues	Find an intersecting road.	A set of cross streets

Find the function values.

17. $g(x) = x + 1$

a) $g(0)$ **b)** $g(-4)$ **c)** $g(-7)$
d) $g(8)$ **e)** $g(a + 2)$

18. $h(x) = x - 4$

a) $h(4)$ **b)** $h(8)$ **c)** $h(-3)$
d) $h(-4)$ **e)** $h(a - 1)$

19. $f(n) = 5n^2 + 4$

a) $f(0)$ **b)** $f(-1)$ **c)** $f(3)$
d) $f(t)$ **e)** $f(2a)$

20. $g(n) = 3n^2 - 2$

a) $g(0)$ **b)** $g(-1)$ **c)** $g(3)$
d) $g(t)$ **e)** $g(2a)$

21. $g(r) = 3r^2 + 2r - 1$

a) $g(2)$ **b)** $g(3)$ **c)** $g(-3)$
d) $g(1)$ **e)** $g(3r)$

22. $h(r) = 4r^2 - r + 2$

a) $h(3)$ **b)** $h(0)$ **c)** $h(-1)$
d) $h(-2)$ **e)** $h(3r)$

23. $f(x) = \dfrac{x-3}{2x-5}$

a) $f(0)$ **b)** $f(4)$ **c)** $f(-1)$
d) $f(3)$ **e)** $f(x+2)$

24. $s(x) = \dfrac{3x-4}{2x+5}$

a) $s(10)$ **b)** $s(2)$ **c)** $s\left(-\frac{5}{2}\right)$
d) $s(-1)$ **e)** $s(x+3)$

The function A described by $A(s) = s^2\dfrac{\sqrt{3}}{4}$ gives the area of an equilateral triangle with side s.

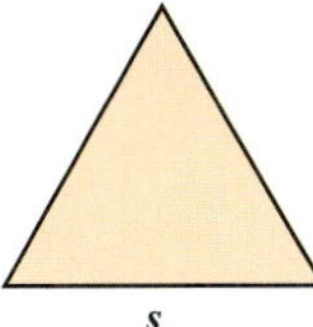

25. Find the area when a side measures 4 cm.

26. Find the area when a side measures 6 in.

The function V described by $V(r) = 4\pi r^2$ gives the surface area of a sphere with radius r.

27. Find the area when the radius is 3 in.

28. Find the area when the radius is 5 cm.

The function F described by $F(C) = \frac{9}{5}C + 32$ gives the Fahrenheit temperature corresponding to the Celsius temperature C.

29. Find the Fahrenheit temperature equivalent to $-10°$C.

30. Find the Fahrenheit temperature equivalent to $5°$C.

The function H described by $H(x) = 2.75x + 71.48$ can be used to predict the height, in centimeters, of a woman whose *humerus* (the bone from the elbow to the shoulder) is x cm long. Predict the height of a woman whose humerus is the length given.

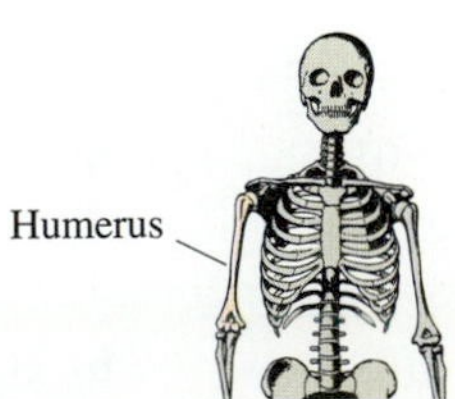

31. 32 cm **32.** 35 cm

For Exercises 33 and 34, use the following graph, which shows the annual heart attack rate per 10,000 men as a function of blood cholesterol level.*

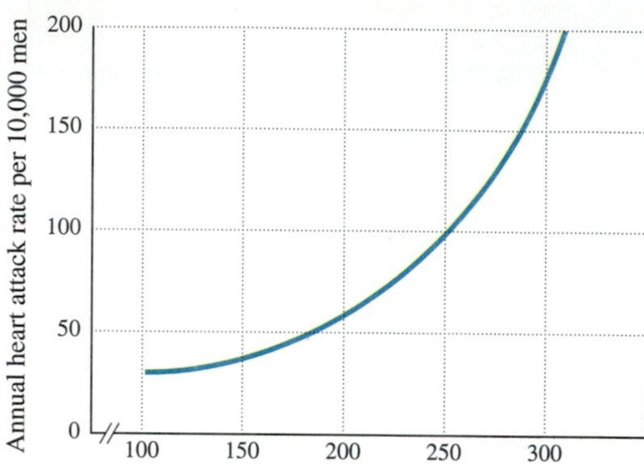

33. Approximate the annual heart attack rate per 10,000 men for those whose blood cholesterol level is 225 mg/dl.

34. Approximate the annual heart attack rate per 10,000 men for those whose blood cholesterol level is 275 mg/dl.

For Exercises 35 and 36, use the following graph, which shows the number of baseball bats sold as a function of time.†

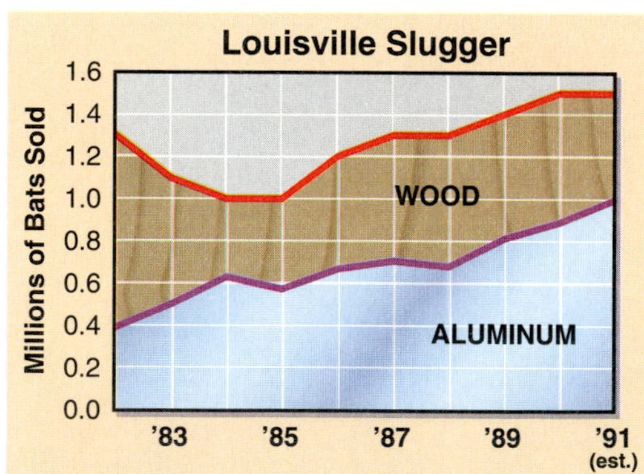

35. Approximate the number of wood bats sold in 1989.

36. Approximate the number of aluminum bats sold in 1987.

*Copyright 1989, CSPI. Adapted from *Nutrition Action Healthletter* (1875 Connecticut Avenue, N.W., Suite 300, Washington, DC 20009-5728. $20.00 for 10 issues).

†*The New York Times,* 7/7/91. Copyright © 1991 by The New York Times Company. Reprinted with permission.

The following table can be used to predict the number of drinks required for a person of a specified weight to be legally intoxicated (blood alcohol level of 0.08 or above) in Vermont. One 12-oz glass of beer, a 5-oz glass of wine, or a cocktail containing 1 oz of a distilled liquor all count as one drink. Assume that all drinks are consumed within one hour.

Input, Body Weight (in Pounds)	Output, Number of Drinks
100	2.5
160	4
180	4.5
200	5

37. Use the table above to draw a graph and to estimate the number of drinks that a 140-lb person would have to drink to be considered intoxicated.

38. Use the graph from Exercise 37 to estimate the number of drinks a 120-lb person would have to drink to be considered intoxicated.

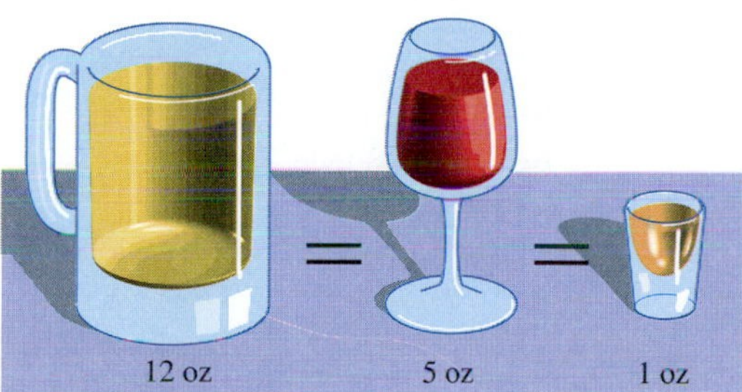

Jamaal buys a video cassette recorder on which there is a revolution counter. There is also a booklet with a table that relates the counter reading and the time for which the tape has run.

Counter Reading	Time of Tape (in Hours)
000	0
300	1
500	2
675	3
800	4

39. Use the data in the table to draw a graph of the time that a tape has run as a function of the counter reading and then estimate the time elapsed when the counter has reached 600.

40. Use the graph from Exercise 39 to estimate the time elapsed when the counter has reached 200.

A city experiencing rapid growth recorded the following dates and populations.

Input, Year	Output, Population (in Tens of Thousands)
1985	5.8
1987	6
1989	7
1991	10

41. Use the data in the table to draw a graph of the population as a function of time. Then estimate what the population was in 1988.

42. Use the graph in Exercise 41 to predict the city's population in the year 1993.

43. A gift shop experiencing constant growth totalled $250,000 in sales in 1988 and $285,000 in 1993. Use a graph that displays the store's total sales as a function of time to predict total sales for the year 1997.

44. Use the graph in Exercise 43 to estimate what the total sales were in 1991.

Determine whether each of the following is the graph of a function.

45.

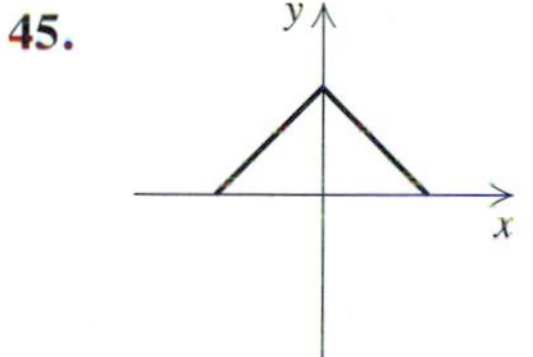

46.

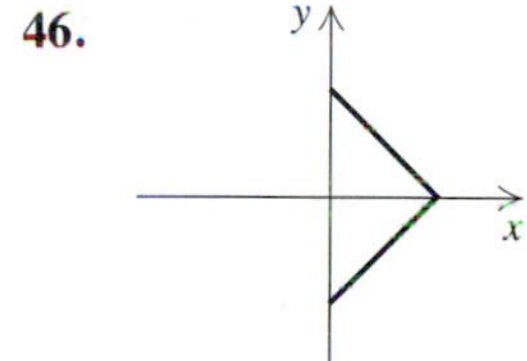

47.

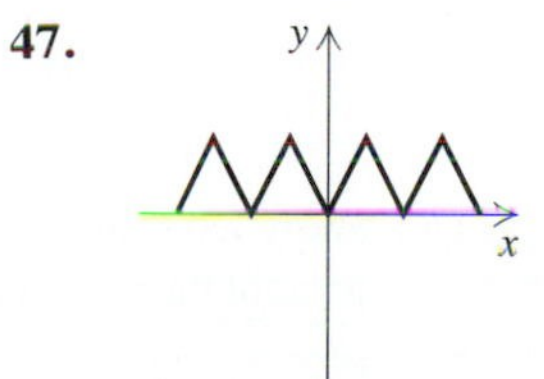

48.

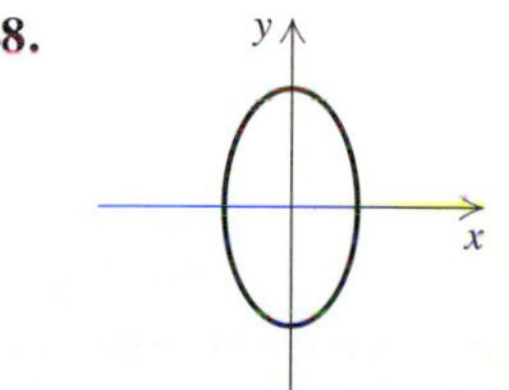

49.

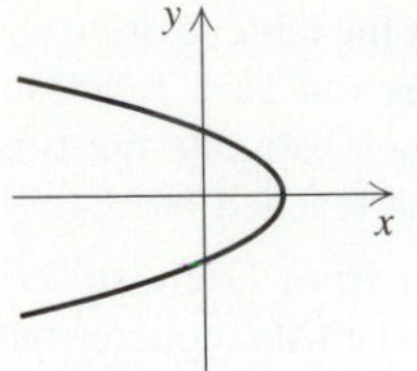

50.

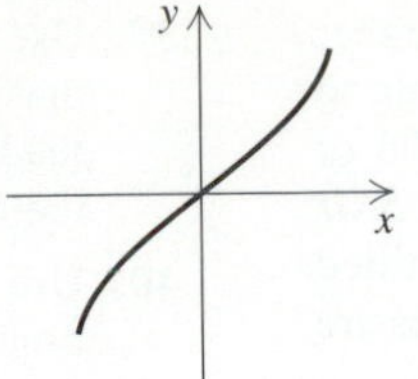

51.

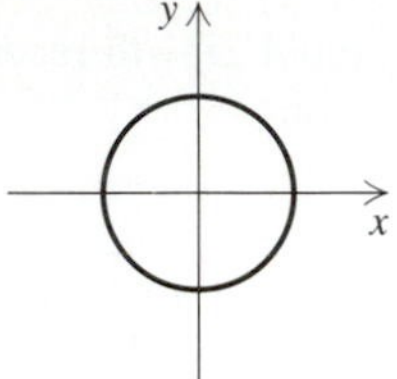

52.

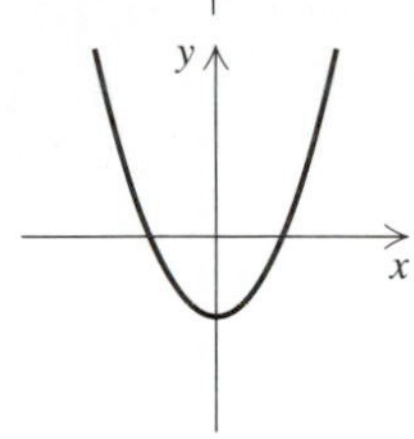

Skill Maintenance

Factor.

53. $3x^4 - 9x^3$

54. $7y + 21z - 14$

55. $x^3 - 7x^2 + 3x - 21$

56. $y^3 - y^2 - 2y + 2$

Synthesis

57. ◈ Explain why every function is a relation, but not every relation is a function.

58. ◈ Example 4 contains a graph showing the number of newly reported cases of AIDS in the United States. Can a linear function be used to express the number of cases as a function of the number of years since 1981? Why or why not?

For each function, find the indicated function values.

59. $f(x) = 4.3x^2 - 1.4x$

a) $f(1.034)$ **b)** $f(-3.441)$
c) $f(27.35)$ **d)** $f(-16.31)$

60. $g(x) = 2.2x^3 + 3.5$

a) $g(17.3)$ **b)** $g(-64.2)$
c) $g(0.095)$ **d)** $g(-6.33)$

61. Suppose that a function g is such that $g(-1) = -7$ and $g(3) = 8$. Find a formula for g if $g(x)$ is of the form $g(x) = mx + b$, where m and b are constants.

62. Suppose that for some function f, $f(x - 1) = 5x$. What is $f(6)$?

Researchers at Yale University have suggested that the following graphs* may represent three different aspects of love.

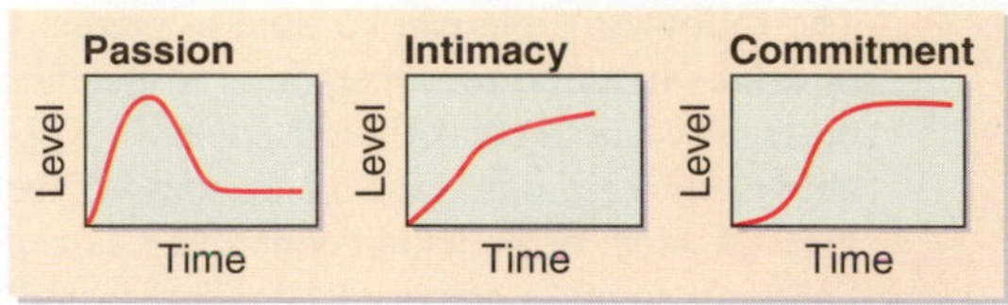

63. ◈ In what unit would you measure time if the horizontal length of each graph were ten units? Why?

64. ◈ Do you agree with the researchers that these graphs should be shaped as they are? Why or why not?

65. ◈ Does the following chart constitute a function? Why or why not?

APPROXIMATE ENERGY EXPENDITURE BY A 150-POUND PERSON IN VARIOUS ACTIVITIES

Activity	Calories per Hour
Lying down or sleeping	80
Sitting	100
Driving an automobile	120
Standing	140
Domestic work	180
Walking, $2\frac{1}{2}$ mph	210
Bicycling, $5\frac{1}{2}$ mph	210
Gardening	220
Golf; lawn mowing, power mower	250
Bowling	270
Walking, $3\frac{3}{4}$ mph	300
Swimming, $\frac{1}{4}$ mph	300
Square dancing, volleyball, roller skating	350
Wood chopping or sawing	400
Tennis	420
Skiing, 10 mph	600
Squash and handball	600
Bicycling, 13 mph	660
Running, 10 mph	900

Source: Based on material prepared by Robert E. Johnson, M.D., Ph.D., and colleagues, University of Illinois.

*From "A Triangular Theory of Love," by R. J. Sternberg, 1986, *Psychological Review*, **93**(2), 119–135. Copyright 1986 by the American Psychological Association, Inc. Reprinted by permission.

For Exercises 66–69, use the following graph of a woman's "stress test." This graph shows the size of a pregnant woman's contractions as a function of time.

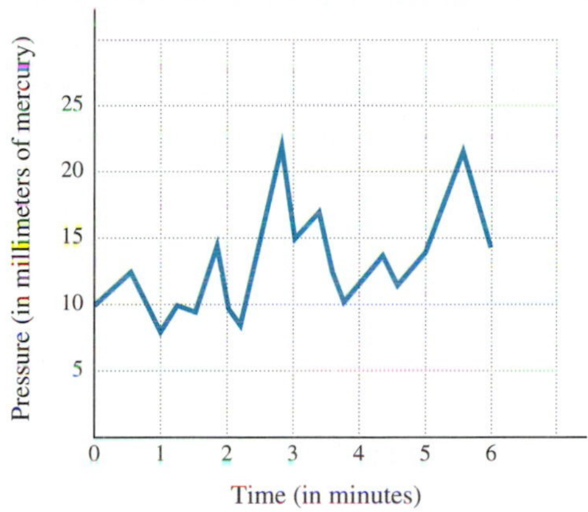

66. How large is the largest contraction that occurred during the test?

67. At what time during the test did the largest contraction occur?

68. On the basis of the information provided, how large a contraction would you expect 60 seconds from the end of the test? Why?

69. What is the frequency of the woman's largest contractions?

70. The *greatest integer function* $f(x) = [\![x]\!]$ is defined as follows: $[\![x]\!]$ is the greatest integer that is less than or equal to x. For example, if $x = 3.74$, then $[\![x]\!] = 3$; and if $x = -0.98$, then $[\![x]\!] = -1$. Graph the greatest integer function for values of x such that $-5 \leq x \leq 5$. (The notation $f(x) = \text{INT}[x]$ is used in many computer programs for the greatest integer function.)

7.5 The Algebra of Functions

The Domain of a Sum, Difference, Product, or Quotient of Two Functions • Domains and Graphs

Let's now return to the idea of a function as a machine. Suppose that a is in the domain of two functions, f and g. The input a is paired with $f(a)$ by f and with $g(a)$ by g. The outputs can then be added to get $f(a) + g(a)$.

EXAMPLE 1 Let $f(x) = x + 5$ and $g(x) = x^2$. Find $f(2) + g(2)$.

Solution We visualize two function machines. Because 2 is in the domain of each function, we can compute $f(2)$ and $g(2)$.

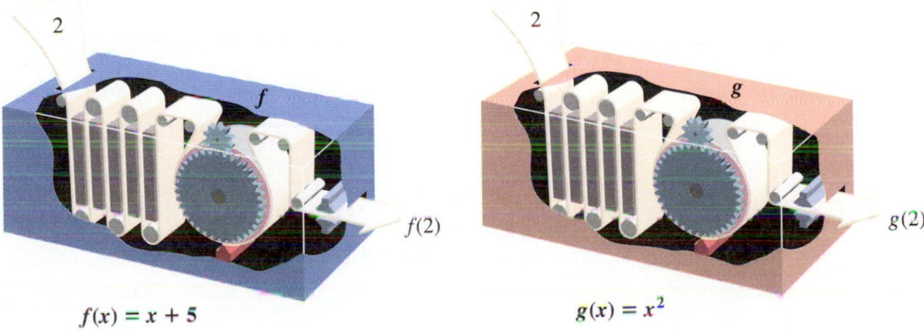

Since

$$f(2) = 2 + 5 = 7 \quad \text{and} \quad g(2) = 2^2 = 4,$$

we have

$$f(2) + g(2) = 7 + 4 = 11.$$

□

In Example 1, suppose that we were to write $f(x) + g(x)$ as $(x + 5) + x^2$ so that $f(x) + g(x) = x^2 + x + 5$. This could then be regarded as a "new" function. The notation $(f + g)(x)$ is generally used to denote a function formed in this manner. Similar notations exist for subtraction, multiplication, and division of functions.

The Algebra of Functions

If f and g are functions and x is in the domain of both functions, then:

1. $(f + g)(x) = f(x) + g(x)$;
2. $(f - g)(x) = f(x) - g(x)$;
3. $(f \cdot g)(x) = f(x) \cdot g(x)$;
4. $(f/g)(x) = f(x)/g(x)$, provided $g(x) \neq 0$.

EXAMPLE 2

For $f(x) = x^2 - 1$ and $g(x) = x + 2$, find the following.

a) $(f + g)(3)$
b) $(f - g)(x)$ and $(f - g)(-1)$
c) $(f/g)(x)$ and $(f/g)(-4)$
d) $(f \cdot g)(3)$

Solution

a) Since $f(3) = 3^2 - 1 = 8$ and $g(3) = 3 + 2 = 5$, we have

$$\begin{aligned}(f + g)(3) &= f(3) + g(3)\\ &= 8 + 5 \qquad \text{Substituting}\\ &= 13.\end{aligned}$$

Alternatively, we could first find $(f + g)(x)$:

$$\begin{aligned}(f + g)(x) &= f(x) + g(x)\\ &= x^2 - 1 + x + 2\\ &= x^2 + x + 1. \qquad \text{Combining like terms}\end{aligned}$$

Thus,

$$(f + g)(3) = 3^2 + 3 + 1 = 13.$$

b) We have

$$\begin{aligned}(f - g)(x) &= f(x) - g(x)\\ &= x^2 - 1 - (x + 2) \qquad \text{Substituting}\\ &= x^2 - x - 3. \qquad \text{Removing parentheses and combining like terms}\end{aligned}$$

Thus

$$\begin{aligned}(f - g)(-1) &= (-1)^2 - (-1) - 3\\ &= -1. \qquad \text{Simplifying}\end{aligned}$$

c) We have

$$\begin{aligned}(f/g)(x) &= f(x)/g(x)\\ &= \frac{x^2 - 1}{x + 2}.\end{aligned}$$

Thus,

$$(f/g)(-4) = \frac{(-4)^2 - 1}{-4 + 2} \quad \text{Substituting}$$
$$= \frac{15}{-2}$$
$$= -7.5.$$

d) Using our work in part (a), we have

$$(f \cdot g)(3) = f(3) \cdot g(3)$$
$$= 8 \cdot 5$$
$$= 40.$$

It is also possible to compute $(f \cdot g)(3)$ by first multiplying $x^2 - 1$ and $x + 2$. ❑

Although it is usually difficult to visualize the product or quotient of two or more functions, sums and differences *can* be visualized. In the following graph,* the total number of airline passengers, $F(t)$, in the New York area is regarded as a function of time. The number of passengers using Kennedy Airport is denoted by $k(t)$, the number of passengers using LaGuardia Airport is $l(t)$, and the number of passengers using Newark Airport is $n(t)$. Although separate graphs for k, l, and n have not been drawn, we can see that

$$F(t) = k(t) + l(t) + n(t).$$

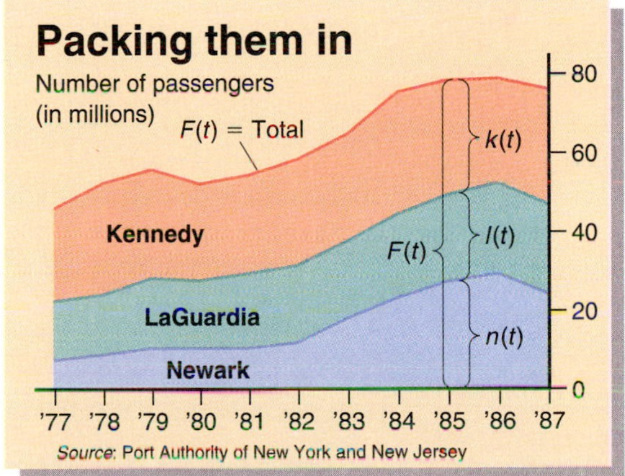

In the next graph, the functions *are* graphed separately before being added. Here all braces extend to the horizontal axis.

*Copyright ©1988 by the New York Times Company. Reprinted with permission.

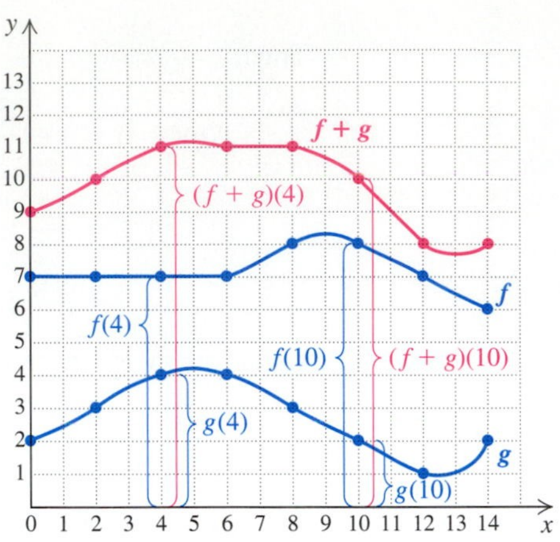

The Domain of a Sum, Difference, Product, or Quotient of Two Functions

It makes sense that in order to find $(f+g)(a)$, $(f-g)(a)$, $(f \cdot g)(a)$, or $(f/g)(a)$, we must first be able to find $f(a)$ and $g(a)$. Thus we need to determine whether a is in the domains of f and g.

When the domain of a function is not specified, we will assume the domain to be all numbers that could work as inputs. For a function like $f(x) = 2x + 3$, any real number can be an input. Thus the domain of f is the set of all real numbers. For some other functions, certain numbers must be excluded from the domain.

EXAMPLE 3

Find the domain of the function:

a) $g(x) = \dfrac{x-2}{x-4}$; **b)** $h(x) = \dfrac{x}{x^2-9}$; **c)** $r(x) = \dfrac{x^2-2x}{x^2-4x}$.

Solution Recall from Chapter 6 that a rational expression is undefined when the denominator is 0.

a) Since g is described by a rational expression, any value of x that would cause division by 0 is not in the domain. To determine which x-value causes the denominator to be 0, set $x - 4$ equal to 0 and solve.

$x - 4 = 0$

$x = 4$ Adding 4 on both sides

Therefore 4 is not in the domain of g. We write

Domain of $g = \{x \mid x \text{ is a real number and } x \neq 4\}$.

b) Again, set the denominator equal to 0.

$x^2 - 9 = 0$

$(x+3)(x-3) = 0$ $A^2 - B^2 = (A+B)(A-B)$

$x + 3 = 0$ *or* $x - 3 = 0$ Using the principle of zero products

$x = -3$ *or* $x = 3$

Thus the domain of $h = \{x \mid x \text{ is a real number and } x \neq -3 \text{ and } x \neq 3\}$.

c) Set the denominator equal to 0.

$$x^2 - 4x = 0$$
$$x(x-4) = 0$$
$$x = 0 \quad or \quad x - 4 = 0 \qquad \text{Using the principle of zero products}$$
$$x = 0 \quad or \quad x = 4$$

We write

Domain of $r = \{x| \ x \text{ is a real number and } x \neq 0 \text{ and } x \neq 4\}$. □

Although the rational expression in Example 3(c) can be simplified, we cannot determine the domain of r from the simplified expression.

The expressions

$$\frac{x^2 - 2x}{x^2 - 4x} \quad \text{and} \quad \frac{x-2}{x-4}$$

are equivalent for all replacements for x *except* $x = 0$. So the functions $g(x)$ and $r(x)$ in Example 3 all have different domains.

EXAMPLE 4

Let

$$f(x) = \frac{5}{x} \quad \text{and} \quad g(x) = \frac{2x-6}{x+1}.$$

Find the domain of $f + g$, the domain of $f - g$, and the domain of $f \cdot g$.

Solution Note that because division by 0 is undefined, we have

Domain of $f = \{x| \ x \text{ is a real number and } x \neq 0\}$

and

Domain of $g = \{x| \ x \text{ is a real number and } x \neq -1\}$.

In order to find $f(a) + g(a)$, $f(a) - g(a)$, or $f(a) \cdot g(a)$, we must know that a is in *both* of the above domains. Thus,

Domain of $f + g$ = Domain of $f - g$ = Domain of $f \cdot g$
$= \{x| \ x \text{ is a real number and } x \neq 0 \text{ and } x \neq -1\}$. □

Suppose that in Example 2(c) we needed to find $(f/g)(-2)$. Finding $f(-2)$ and $g(-2)$ poses no problem:

$$f(-2) = (-2)^2 - 1 = 3 \quad \text{and} \quad g(-2) = -2 + 2 = 0;$$

but then

$$(f/g)(-2) = f(-2)/g(-2)$$
$$= 3/0.$$

Thus, $(f/g)(-2)$ is undefined. Although -2 is in the domain of both f and g, it is not in the domain of f/g.

EXAMPLE 5

Let $F(x) = x^3$ and $G(x) = 4x - 3$. Find the domain of F/G.

Solution The domain of F and G is all real numbers. However, the domain of

$$(F/G)(x) = \frac{x^3}{4x - 3}$$

is the set of all real numbers such that $4x - 3 \neq 0$. Because $4x - 3 = 0$ when x is $\frac{3}{4}$, we conclude that the domain of F/G is the set of all real numbers except $\frac{3}{4}$. In set-builder notation,

$$\text{Domain of } F/G = \{x \mid x \text{ is a real number and } x \neq \tfrac{3}{4}\}.$$ ❑

EXAMPLE 6

Find the domain of f/g, if

$$f(x) = \frac{3}{x - 4} \quad \text{and} \quad g(x) = \frac{6}{x + 2}.$$

Solution Since the domain of $f = \{x \mid x \text{ is a real number and } x \neq 4\}$ and the domain of $g = \{x \mid x \text{ is a real number and } x \neq -2\}$, we conclude that the domain of f/g is the set of all real numbers except -2, 4, and any x-values for which $g(x) = 0$. Because $6/(x + 2)$ is never 0, there is no x-value such that $g(x) = 0$. Thus,

$$\text{Domain of } f/g = \{x \mid x \text{ is a real number and } x \neq 4 \text{ and } x \neq -2\}.$$ ❑

Comment: It is tempting to write

$$(f/g)(x) = \frac{\frac{3}{x-4}}{\frac{6}{x+2}} = \frac{3}{x - 4} \div \frac{6}{x + 2}$$

$$= \frac{3}{x - 4} \cdot \frac{x + 2}{6}$$

$$= \frac{x + 2}{2(x - 4)},$$

in which case the domain of f/g would exclude only 4. Because of the fact that $(f/g)(x)$ is defined as $f(x)/g(x)$, we can use only x-values that are in the domains of *both* functions. Thus we stipulate

$$(f/g)(x) = \frac{x + 2}{2(x - 4)}, \qquad \textit{provided } x \neq -2.$$

EXAMPLE 7

Find the domain of p/q, if

$$p(x) = \frac{5}{x} \quad \text{and} \quad q(x) = \frac{2x - 6}{x + 1}.$$

Solution We have

$$\text{Domain of } p = \{x \mid x \text{ is a real number and } x \neq 0\},$$
$$\text{Domain of } q = \{x \mid x \text{ is a real number and } x \neq -1\}.$$

Since $q(x) = 0$ when $2x - 6 = 0$, we have $q(x) = 0$ when x is 3. We conclude that

$$\text{Domain of } p/q = \{x|\ x \text{ is a real number and } x \neq 0,\ x \neq -1, \text{ and } x \neq 3\}.$$ ❑

Determining the Domain

To find the domain of a sum, difference, product, or quotient of two functions:

1. Determine the domain of each function individually.
2. The domain of the sum, difference, or product is the set of all values common to both domains.
3. The domain of the quotient is the set of all values common to both domains, excluding any value that would lead to division by 0.

Domains and Graphs

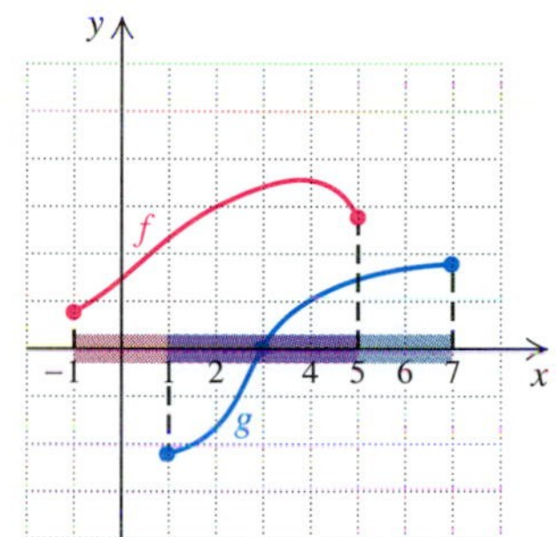

There is an interesting visual interpretation to the algebra of functions. Consider f and g as sketched in the figure.

Note that

$$\text{Domain of } f = \{x|\ -1 \leq x \leq 5\}$$

and

$$\text{Domain of } g = \{x|\ 1 \leq x \leq 7\}$$

can be regarded as the *projections*, or "shadows," of f and g on the x-axis. Thus for $f + g$, $f - g$, or $f \cdot g$, the domain is $\{x|\ 1 \leq x \leq 5\}$, the values common to the domains of f and g. Since $g(3) = 0$,

$$\text{Domain of } f/g = \{x|\ 1 \leq x \leq 5 \text{ and } x \neq 3\}.$$

EXERCISE SET 7.5

Let $f(x) = -3x + 1$ and $g(x) = x^2 + 2$. Find the following.

1. $f(1) + g(1)$ **2.** $f(2) + g(2)$

3. $f(-1) + g(-1)$ **4.** $f(-2) + g(-2)$

5. $f(-7) - g(-7)$ **6.** $f(-5) - g(-5)$

7. $f(5) - g(5)$ **8.** $f(4) - g(4)$

9. $f(2) \cdot g(2)$ **10.** $f(3) \cdot g(3)$

11. $f(-3) \cdot g(-3)$ **12.** $f(-4) \cdot g(-4)$

13. $f(0)/g(0)$ **14.** $f(1)/g(1)$

15. $f(-3)/g(-3)$ **16.** $g(-3)/f(-3)$

Let $F(x) = x^2 - 3$ and $G(x) = 4 - x$. Find the following.

17. $(F + G)(-3)$ **18.** $(F + G)(-2)$

19. $(F + G)(x)$ **20.** $(F + G)(a)$

21. $(F - G)(-4)$ **22.** $(F - G)(-5)$

23. $(F \cdot G)(2)$ **24.** $(F \cdot G)(3)$

25. $(F \cdot G)(-3)$ **26.** $(F \cdot G)(-4)$

27. $(F/G)(0)$ **28.** $(F/G)(1)$

29. $(F/G)(-2)$ **30.** $(G/F)(-2)$

Find the domain of the function.

31. $f(x) = \dfrac{7}{3 - x}$

32. $g(x) = \dfrac{3}{2x + 1}$

33. $v(x) = \dfrac{x}{x^2 + 4x}$

34. $h(x) = \dfrac{x + 1}{x^2 - 25}$

35. $f(x) = x^2 + x - 30$

36. $r(x) = x^3 - 1$

37. $g(x) = \dfrac{x^2 - 4}{x^2 - 8x + 12}$

38. $h(x) = \dfrac{9 - x^2}{x^2 - 6x + 8}$

For each pair of functions f and g, determine the domain of the sum, difference, and product of the two functions.

39. $f(x) = x^2$, $g(x) = 3x - 4$

40. $f(x) = 5x - 1$, $g(x) = 2x^3$

41. $f(x) = \dfrac{1}{x - 2}$, $g(x) = 4x^3$

42. $f(x) = 3x^2$, $g(x) = \dfrac{1}{x + 4}$

43. $f(x) = 4x + \dfrac{2}{x - 1}$, $g(x) = 3x^3$

44. $f(x) = 9 - x^2$, $g(x) = \dfrac{3}{x - 5} + 2x$

45. $f(x) = \dfrac{3}{x - 2}$, $g(x) = \dfrac{5}{4 - x}$

46. $f(x) = \dfrac{7}{x + 3}$, $g(x) = \dfrac{1}{x - 2}$

47. $f(x) = \dfrac{3}{x + 2}$, $g(x) = \dfrac{x}{3x - 4}$

48. $f(x) = \dfrac{2x}{3 - x}$, $g(x) = \dfrac{4}{2x - 5}$

For each pair of functions f and g, determine the domain of f/g.

49. $f(x) = x^4$, $g(x) = x - 3$

50. $f(x) = 2x^3$, $g(x) = 5 - x$

51. $f(x) = 3x - 2$, $g(x) = 2x - 8$

52. $f(x) = 5 + x$, $g(x) = 6 - 2x$

53. $f(x) = \dfrac{3}{x - 4}$, $g(x) = 5 - x$

54. $f(x) = \dfrac{1}{2 - x}$, $g(x) = 7 - x$

55. $f(x) = \dfrac{x - 1}{x - 4}$, $g(x) = 3x^2$

56. $f(x) = \dfrac{x + 2}{x + 3}$, $g(x) = 4x^3$

57. $f(x) = 3x^2$, $g(x) = \dfrac{x - 1}{3x - 4}$

58. $f(x) = 4x^3$, $g(x) = \dfrac{x + 2}{x + 3}$

59. $f(x) = \dfrac{x - 1}{x - 2}$, $g(x) = \dfrac{x - 3}{x - 4}$

60. $f(x) = \dfrac{x + 4}{x + 3}$, $g(x) = \dfrac{x + 2}{2x + 1}$

For Exercises 61–64, consider the functions F and G as shown below.

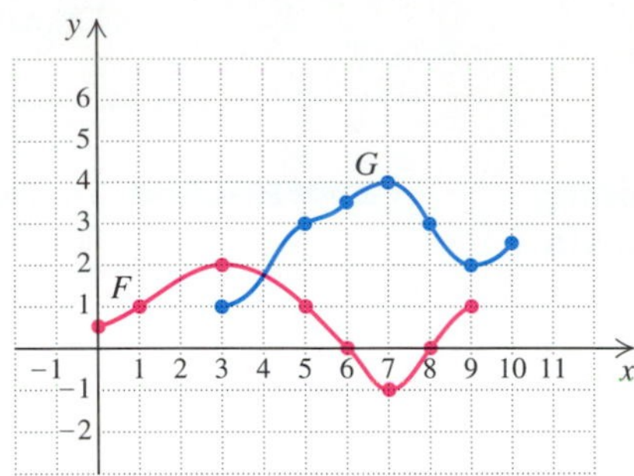

61. Find the domain of F, the domain of G, the domain of $F + G$, and the domain of F/G.

62. Find the domain of $F - G$, the domain of $F \cdot G$, and the domain of G/F.

63. Graph $F + G$.

64. Graph $G - F$.

Skill Maintenance

Multiply.

65. $(2x^2 - x - 1)(x + 3)$

66. $(3x - 5)(3x + 5)$

67. Factor: $3a^3 + 18a^2 - 4a - 24$.

68. Simplify: $\dfrac{x^2 - 9x + 14}{x^2 + 3x - 10}$.

Synthesis

In the graph* that follows, $W(t)$ represents the number of gallons of whole milk, $L(t)$ the number of gallons of lowfat milk, and $S(t)$ the number of gallons of skim milk consumed by the average American in a year.

*Copyright 1990, CSPI. Adapted from Nutrition Action Healthletter (1875 Connecticut Avenue, N.W., Suite 300, Washington, DC 20009-5728. $20.00 for 10 issues); 1991 data U.S.D.A.

69. ◈ Explain in words what $(W - L)(t)$ represents and what it would mean to have $(W - L)(t) < 0$.

70. ◈ Consider $(W + L + S)(t)$ and explain why you feel that total milk consumption per person has or has not changed over the years 1971–1991.

71. Find the domain of p/q, if

$$p(x) = \begin{cases} 2x, & \text{if } x > 1, \\ x^2, & \text{if } x < 1 \end{cases}$$

and

$$q(x) = \frac{x - 3}{x - 2}.$$

72. Find the domain of m/n, if

$$m(x) = 3x \text{ for } -1 < x < 5$$

and

$$n(x) = 2x - 3.$$

73. Find the domains of $f + g$, $f - g$, $f \cdot g$, and f/g, if

$$f = \{(-2, 1), (-1, 2), (0, 3), (1, 4), (2, 5)\}$$

and

$$g = \{(-4, 4), (-3, 3), (-2, 4), (-1, 0), (0, 5), (1, 6)\}.$$

74. For f and g as defined in Exercise 73, find $(f + g)(-2)$, $(f \cdot g)(0)$, and $(f/g)(1)$.

75. Find the domain of F/G, if

$$F(x) = \frac{1}{x^2 - 1} \quad \text{and} \quad G(x) = \frac{x^2 - 4}{x - 3}.$$

76. Find the domain of f/g, if

$$f(x) = \frac{3x}{2x + 5} \quad \text{and} \quad g(x) = \frac{x^4 - 1}{3x + 9}.$$

77. Write equations for two functions f and g such that the domain of $f + g$ is

$$\{x \mid x \text{ is a real number and } x \neq -2 \text{ and } x \neq 5\}.$$

78. Sketch the graph of two functions f and g such that the domain of f/g is

$$\{x \mid -2 \leq x \leq 3 \text{ and } x \neq 1\}.$$

79. Use the graph of the number of airline passengers that appears after Example 2 to estimate each of the following.

a) $(k + l + n)(1983)$
b) $(l + n)(1986)$
c) $(l + k)(1986)$

7.6 Variation and Problem Solving

Direct Variation • Inverse Variation • Combined Variation

We extend our study of formulas and functions by examining three situations that frequently arise in problem solving: direct variation, inverse variation, and combined variation.

Direct Variation

Let's say that a worker earns \$18 per hour. In 1 hr \$18 is earned. In 2 hr \$36 is earned. In 3 hr \$54 is earned, and so on. This gives rise to a set of ordered pairs of numbers:

$$(1, 18), (2, 36), (3, 54), (4, 72), \quad \text{and so on.}$$

The ratio of earnings to time is $\frac{18}{1}$ in every case.

Whenever a situation gives rise to pairs of numbers in which the ratio is constant, we say that there is **direct variation.** Here the earnings *vary directly* as the time:

$$E = 18t \quad \text{or, using function notation,} \quad E(t) = 18t.$$

Direct Variation

Whenever a situation gives rise to a linear function $f(x) = kx$, or $y = kx$, where k is a nonzero constant, we say that there is *direct variation,* that y *varies directly as* x, or that y is *proportional to* x. The number k is called the *variation constant,* or *constant of proportionality.*

The graph of $y = kx$, $k > 0$, always goes through the origin and rises from left to right. Note that as x increases, y increases. The constant k is also the slope of the line.

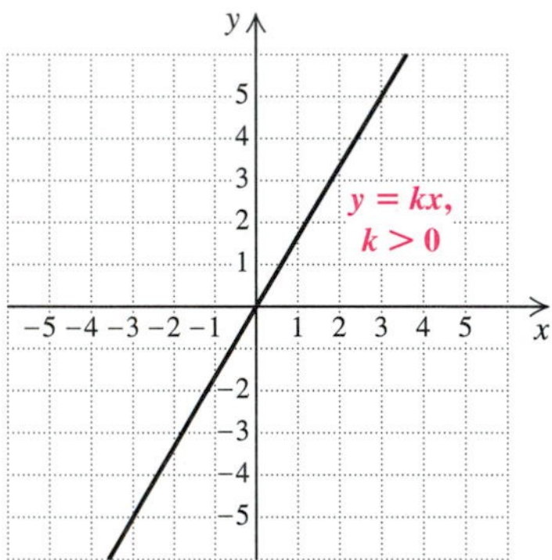

EXAMPLE 1

Find the variation constant and an equation of variation in which y varies directly as x, and $y = 32$ when $x = 2$.

Solution We know that (2, 32) is a solution of $y = kx$. Therefore,

$32 = k \cdot 2$ Substituting

$\frac{32}{2} = k$, or $k = 16$. Solving for k

The variation constant is 16. The equation of variation is $y = 16x$. The notation $y(x) = 16x$ or $f(x) = 16x$ is also used. ❑

EXAMPLE 2

Water from melting snow. The number of centimeters W of water produced from melting snow varies directly as the number of centimeters S of snow. Meteorologists have found that under certain conditions, 150 cm of snow will melt to 16.8 cm of water. To how many centimeters of water will 200 cm of snow melt under these conditions?

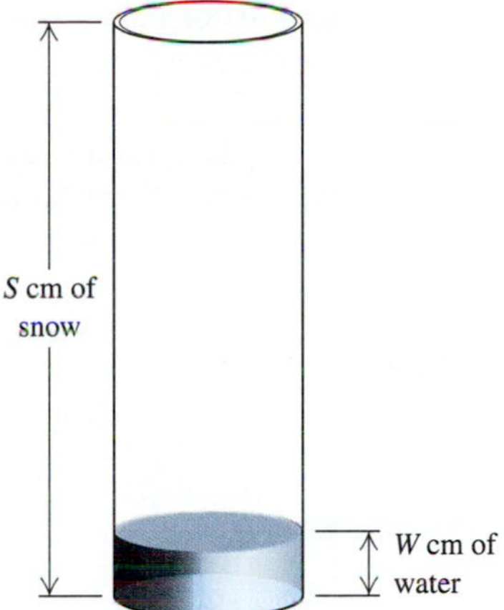

Solution

1. FAMILIARIZE. Because of the phrase "W . . . varies directly as . . . S," we decide to express the amount of water as a function of the amount of snow. Thus, $W(S) = kS$, where k is the variation constant. From the information provided, we know that $W(150) = 16.8$. That is, 150 cm of snow melts to 16.8 cm of water.

2. TRANSLATE. We find the variation constant using the data and then find the equation of variation:

$$W(S) = kS$$

$$W(150) = k \cdot 150 \quad \text{Replacing } S \text{ with } 150$$

$$16.8 = k \cdot 150 \quad \text{Substituting}$$

$$\frac{16.8}{150} = k \quad \text{Solving for } k$$

$$0.112 = k. \quad \text{This is the variation constant.}$$

The equation of variation is $W(S) = 0.112S$. This is the translation.

3. CARRY OUT. To find how much water 200 cm of snow will melt to, we compute $W(200)$:

$$W(S) = 0.112S$$

$$W(200) = 0.112(200) \quad \text{Replacing } S \text{ with } 200$$

$$= 22.4.$$

4. CHECK. To check, we could reexamine all our calculations. Note that our answer seems reasonable since 200/22.4 and 150/16.8 are equal.

5. STATE. 200 cm of snow will melt into 22.4 cm of water. □

Inverse Variation

To see what we mean by inverse variation, consider the following situation.

A bus is traveling a distance of 20 mi. At a speed of 20 mph, the trip will take 1 hr. At 40 mph, it will take $\frac{1}{2}$ hr. At 60 mph, it will take $\frac{1}{3}$ hr, and so on. This gives rise to a set of pairs of numbers, all having the same product:

$(20, 1), (40, \frac{1}{2}), (60, \frac{1}{3}), (80, \frac{1}{4}),$ and so on.

Whenever a situation gives rise to pairs of numbers whose product is constant, we say that there is **inverse variation.** The time t required for the bus to travel 20 mi at rate r is given by

$$t = \frac{20}{r} \quad \text{or, using function notation,} \quad t(r) = \frac{20}{r}.$$

Inverse Variation

> Whenever a situation gives rise to a function $f(x) = k/x$, or $y = k/x$, where k is a nonzero constant, we say that there is *inverse variation*, that *y varies inversely as x*, or that y is *inversely proportional to x*. The number k is called the *variation constant*, or *constant of proportionality*.

Although we will not study such graphs until Chapter 13, it is helpful to look at the graph of $y = k/x$, for $k > 0$ and $x > 0$. The graph is like the one shown below. Note that as x increases, y decreases.

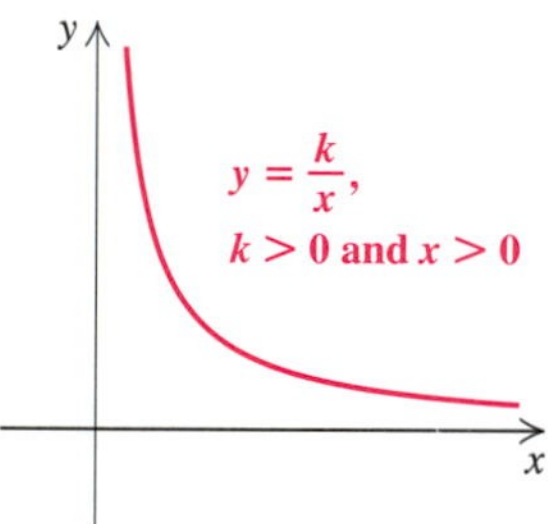

EXAMPLE 3 Find the variation constant and an equation of variation in which y varies inversely as x, and $y = 32$ when $x = 0.2$.

Solution We know that $(0.2, 32)$ is a solution of

$$y = \frac{k}{x}.$$

Therefore,

$$32 = \frac{k}{0.2} \quad \text{Substituting}$$

$$(0.2)32 = k$$

$$6.4 = k. \quad \text{Solving for } k$$

The variation constant is 6.4. The equation of variation is

$$y = \frac{6.4}{x}.$$

□

There are many problems that translate to an equation of inverse variation.

EXAMPLE 4

The time t required to do a certain job varies inversely as the number of people P who work on the job (assuming that all do the same amount of work). It takes 4 hr for 12 people to build a woodshed. How long would it take 3 people to do the same job?

Solution

1. FAMILIARIZE. Because of the phrase "t . . . varies inversely as . . . P," we decide to express the amount of time required, in hours, as a function of the number of people working. Thus we have $t(P) = k/P$. From the information provided, we know that $t(12) = 4$. That is, it takes 4 hr for 12 people to do the job.

2. TRANSLATE. We find the variation constant using the data and then find the equation of variation:

$$t(P) = \frac{k}{P} \quad \text{Using function notation}$$

$$t(12) = \frac{k}{12} \quad \text{Replacing } P \text{ with } 12$$

$$4 = \frac{k}{12} \quad \text{Substituting 4 for } t(12)$$

$$48 = k. \quad \text{Solving for } k \text{, the variation constant}$$

The equation of variation is $t(P) = 48/P$. This is the translation.

3. CARRY OUT. To find how long it would take 3 people to do the job, we compute $t(3)$:

$$t(P) = \frac{48}{P}$$

$$t(3) = \frac{48}{3} \quad \text{Replacing } P \text{ with } 3$$

$$t = 16. \quad t = 16 \text{ when } P = 3$$

4. CHECK. We could now recheck each step. Note that, as expected, as the number of people working goes *down*, the time required for the job goes *up*.

5. STATE. It will take 3 people 16 hr to build a woodshed. ❑

Combined Variation

Often one variable varies directly or inversely with more than one other variable. For example, in the formula for the volume of a right circular cylinder, $V = \pi r^2 h$, we say that V varies *jointly* as h and the square of r.

Joint Variation

y varies *jointly* as x and z if there is some nonzero constant k such that $y = kxz$.

EXAMPLE 5

Find an equation of variation in which y varies jointly as x and z, and $y = 42$ when $x = 2$ and $z = 3$.

Solution We have

$$y = kxz, \quad \text{so } 42 = k \cdot 2 \cdot 3 \quad \text{and} \quad k = 7.$$

Thus, $y = 7xz$. ❑

EXAMPLE 6

Find an equation of variation in which y varies jointly as x and z and inversely as the square of w, and $y = 105$ when $x = 3$, $z = 20$, and $w = 2$.

Solution The equation of variation is of the form

$$y = k \cdot \frac{xz}{w^2},$$

so

$$105 = k \cdot \frac{3 \cdot 20}{2^2}, \quad \text{or} \quad 105 = k \cdot 15, \quad \text{and} \quad k = 7.$$

Thus, $y = 7 \cdot \dfrac{xz}{w^2}$. ❑

EXAMPLE 7

The volume of a tree trunk. The volume of wood V in a tree trunk varies jointly as the height h and the square of the girth g (girth is distance around). If the volume is 35 ft^3 when the height is 20 ft and the girth is 5 ft, what is the girth when the volume is 85.75 ft^3 and the height is 25 ft?

Solution

1. FAMILIARIZE. We'll make a table, including the data from the problem and the data we need to find.

	Volume of Wood	Height of Tree	Girth of Tree
Smaller Tree	35 ft^3	20 ft	5 ft
Larger Tree	85.75 ft^3	25 ft	g

Let h, g, and V represent the height, girth, and volume of a tree, respectively. We wish to determine g when V is 85.75 ft^3 and h is 25 ft.

We know from the statement of the problem that in this situation the volume varies jointly as the height and the square of the girth.

2. TRANSLATE. First we find k using the first set of data. Then we solve for g using the second set of data:

$$V = khg^2$$
$$35 = k \cdot 20 \cdot 5^2$$
$$0.07 = k. \quad \text{This is the variation constant.}$$

The equation of variation is $V = 0.07hg^2$.

3. CARRY OUT. The translation is $V = 0.07hg^2$. We substitute and solve for g:

$$85.75 = 0.07 \cdot 25 \cdot g^2$$
$$85.75 = 1.75g^2$$
$$49 = g^2$$
$$0 = g^2 - 49 \quad \text{Subtracting 49 to get 0 on one side}$$
$$0 = (g + 7)(g - 7)$$
$$g + 7 = 0 \quad or \quad g - 7 = 0 \quad \text{Using the principle of zero products}$$
$$g = -7 \quad or \quad g = 7$$

We could have first solved the formula for g and then substituted; either approach is valid.

4. CHECK. We should now recheck all our calculations and perhaps make an estimate to see whether our answer is reasonable. We leave this for the student to do. Since a tree's girth must be positive, we accept only 7 as a solution. This seems to be a reasonable figure.

5. STATE. The answer is that the girth of the tree is 7 ft. ❑

EXERCISE SET 7.6

Find the variation constant and an equation of variation in which y varies directly as x and the following conditions exist.

1. $y = 24$ when $x = 3$
2. $y = 5$ when $x = 12$
3. $y = 3.6$ when $x = 1$
4. $y = 2$ when $x = 5$
5. $y = 30$ when $x = 8$
6. $y = 1$ when $x = \frac{1}{3}$
7. $y = 0.8$ when $x = 0.5$
8. $y = 0.6$ when $x = 0.4$

Solve.

9. *Ohm's law.* The electric current I, in amperes, in a circuit varies directly as the voltage V. When 12 volts are applied, the current is 4 amperes. What is the current when 18 volts are applied?

10. *Hooke's law.* Hooke's law states that the distance d that a spring is stretched by a hanging object varies directly as the mass m of the object. If the distance is 40 cm when the mass is 3 kg, what is the distance when the mass is 5 kg?

11. *Weekly allowance.* According to Fidelity Investments *Investment Vision Magazine,* the average weekly allowance A of children varies directly as their grade level, G. It is known that the average allowance of a 9th-grade student is \$9.66 per week. What then is the average weekly allowance of a 4th-grade student?

12. *Nutrition.* The maximum number of grams of fat N that should be in a diet varies directly as a person's weight W. A person weighing 120 lb should have no more than 60 g of fat a day. What is the maximum daily fat intake for a person weighing 145 lb?

13. *Mass of water in body.* The number of kilograms W of water in a human body varies directly as the mass of the body. A 96-kg person contains 64 kg of water. How many kilograms of water are in a 75-kg person?

14. *Weight on Mars.* The weight M of an object on Mars varies directly as its weight E on earth. A person who weighs 95 lb on earth weighs 38 lb on Mars. How much would a 100-lb person weigh on Mars?

15. *Relative aperture.* The relative aperture, or f-stop, of a 23.5-mm lens is directly proportional to the focal length F of the lens. If a 150-mm focal length has an f-stop of 6.3, find the f-stop of a 23.5-mm lens with a focal length of 80 mm.

16. *Computer processor speed.* The number of computer instructions per second N varies directly as the processor speed S. A 25-megahertz processor performs 2,000,000 instructions per second. How many instructions will the same processor perform if it is running at 40 megahertz?

Find the variation constant and an equation of variation in which y varies inversely as x, and the following conditions exist.

17. $y = 6$ when $x = 10$

18. $y = 16$ when $x = 4$

19. $y = 4$ when $x = 3$

20. $y = 4$ when $x = 9$

21. $y = 12$ when $x = 3$

22. $y = 9$ when $x = 5$

23. $y = 27$ when $x = \frac{1}{3}$

24. $y = 81$ when $x = \frac{1}{9}$

Solve.

25. *Current and resistance.* The current I in an electrical conductor varies inversely as the resistance R of the conductor. If the current is 2 amperes when the resistance is 960 ohms, what is the current when the resistance is 540 ohms?

26. *Pumping rate.* The time t required to empty a tank varies inversely as the rate r of pumping. If a pump can empty a tank in 45 min at the rate of 600 kL/min, how long will it take the pump to empty the same tank at the rate of 1000 kL/min?

27. *Volume and pressure.* The volume V of a gas varies inversely as the pressure P upon it. The volume of a gas is 200 cm^3 under a pressure of 32 kg/cm^2. What will be its volume under a pressure of 40 kg/cm^2?

28. *Work rate.* The time T required to do a job varies inversely as the number of people P working. It takes 5 hr for 7 bricklayers to complete a certain job. How long will it take 10 bricklayers to complete the job?

29. *Rate of travel.* The time t required to drive a fixed distance varies inversely as the speed r. It takes 5 hr at a speed of 80 km/h to drive a fixed distance. How long will it take to drive the fixed distance at a speed of 60 km/h?

30. *Pitch.* The pitch P of a musical tone varies inversely as its wavelength W. One tone has a pitch of 660 vibrations per second and a wavelength of 1.6 ft. Find the wavelength of another tone that has a pitch of 440 vibrations per second.

Find an equation of variation in which:

31. y varies directly as the square of x, and $y = 0.15$ when $x = 0.1$.

32. y varies directly as the square of x, and $y = 6$ when $x = 3$.

33. y varies inversely as the square of x, and $y = 0.15$ when $x = 0.1$.

34. y varies inversely as the square of x, and $y = 6$ when $x = 3$.

35. y varies jointly as x and z, and $y = 56$ when $x = 7$ and $z = 8$.

36. y varies directly as x and inversely as z, and $y = 4$ when $x = 12$ and $z = 15$.

37. y varies jointly as x and the square of z, and $y = 105$ when $x = 14$ and $z = 5$.

38. y varies jointly as x and z and inversely as w, and $y = \frac{3}{2}$ when $x = 2$, $z = 3$, and $w = 4$.

39. y varies jointly as w and the square of x and inversely as z, and $y = 49$ when $w = 3$, $x = 7$, and $z = 12$.

40. y varies directly as x and inversely as w and the square of z, and $y = 4.5$ when $x = 15$, $w = 5$, and $z = 2$.

41. y varies jointly as x and z and inversely as the product of w and p, and $y = \frac{3}{28}$ when $x = 3$, $z = 10$, $w = 7$, and $p = 8$.

42. y varies jointly as x and z and inversely as the square of w, and $y = \frac{12}{5}$ when $x = 16$, $z = 3$, and $w = 5$.

Solve.

43. *Stopping distance of a car.* The stopping distance d of a car after the brakes have been applied varies directly as the square of the speed r. If a car traveling 60 mph can stop in 200 ft, how fast can a car go and still stop in 72 ft?

44. *Volume of a gas.* The volume V of a given mass of a gas varies directly as the temperature T and inversely as the pressure P. If $V = 231$ cm^3 when $T = 42°$ and $P = 20$ kg/cm^2, what is the volume when $T = 30°$ and $P = 15$ kg/cm^2?

45. *Intensity of a signal.* The intensity I of a television signal varies inversely as the square of the distance d from the transmitter. If the intensity is 25 watts per square meter (W/m^2) at a distance of 2 km, how far from the transmitter are you when the intensity is 2.56 W/m^2?

46. *Distance of a fall.* The distance d that an object falls varies directly as the square of the amount of time t that it is falling. If an object falls 64 ft in 2 sec, how long will it take to fall 400 ft?

47. *Weight of an astronaut.* The weight W of an object varies inversely as the square of the distance d from the center of the earth. At sea level (6400 km from the center of the earth), an astronaut weighs 100 lb. How far *above the earth* must the astronaut be in order to weigh 64 lb?

48. *Volume of a can.* The volume V of a can varies jointly as its height h and the square of its radius r. If a 12-fluid-ounce soda comes in a can that is 12 cm high with a 3.2-cm radius, what is the radius of a 9-fluid-ounce can that is 4 cm high?

49. *Electrical resistance.* At a fixed temperature, the resistance R of a wire varies directly as the length l and inversely as the square of its diameter d. If the resistance is 0.1 ohm when the diameter is 1 mm and the length is 50 cm, what is the diameter when the resistance is 1 ohm and the length is 2000 cm?

50. *Power consumption.* The number of kilowatt-hours per year N that an appliance uses varies jointly as the number of watts w the appliance consumes and the number of hours h per day it is used. A hair dryer that consumes 1200 watts and is used $\frac{1}{4}$ of an hour each day uses 109.5 kilowatt-hours per year. How many kilowatt-hours does a 100 watt lightbulb use each year if it is turned on 2 hours a day?

Skill Maintenance

51. Factor: $ac + bc - a - b$.

52. Simplify: $\dfrac{x^3 - x}{x^2 + 5x + 4}$.

Multiply.

53. $(2x + 0.1)^2$

54. $(x + 1)(x^2 - x + 1)$

Synthesis

State whether the situation represents direct variation, inverse variation, or neither. Give reasons for your answers.

55. ◈ The cost of mailing a letter in the United States and the distance it travels

56. ◈ A runner's speed in a race and the time it takes to run the race

57. ◈ The weight of a turkey and the cooking time

58. ◈ The number of plays it takes to go 80 yd for a touchdown and the average gain per play

Write an equation of direct variation to describe the situation. If possible, give a value for k and graph the equation.

59. The perimeter P of an equilateral octagon varies directly as the length S of a side.

60. The circumference C of a circle varies directly as the radius r.

61. The number of bags of peanuts B sold at a baseball game varies directly as the number of people N in attendance.

62. The cost C of building a new house varies directly as the area A of the floor space of the house.

63. ▦ *The gravity model.* It has been determined that the average number of telephone calls in a day N, between two cities, is directly proportional to the populations P_1 and P_2 of the cities and inversely proportional to the square of the distance between the cities. This model is called the *gravity model* because the equation of variation resembles the equation that applies to Newton's law of gravity.

a) In 1986, the population of Indianapolis was 744,624 and the population of Cincinnati was 452,524. The average number of daily phone calls between the two cities was 11,153. Find the value k and write the equation of variation given that the cities are 174 km apart.

b) In 1986, the average number of daily phone calls between Indianapolis and New York was 4270, and the population of New York was 7,895,563. Estimate the distance between Indianapolis and New York.

64. *Golf distance finder.* A device used in golf to estimate the distance d to a hole measures the size s that the 7-ft pin *appears* to be in a viewfinder. The viewfinder uses the principle, diagrammed here, that s gets bigger when d gets smaller. If $s = 0.56$ in. when $d = 50$ yd, find an equation of variation that expresses d as a function of s. What is d when $s = 0.40$ in.?

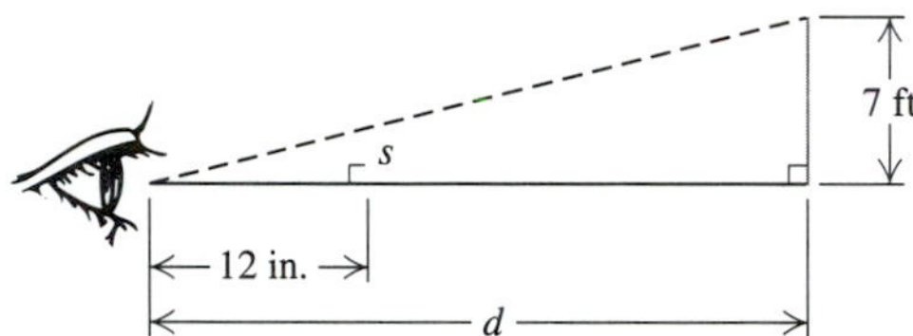

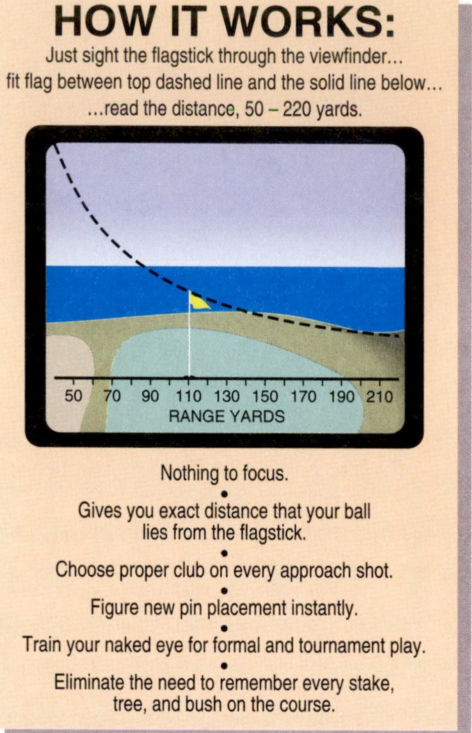

Describe, in words, the variation given by the equation.

65. $Q = \dfrac{kp^2}{q^3}$

66. $W = \dfrac{km_1M_1}{d^2}$

67. Show that if p varies directly as q, then q varies directly as p.

68. If a varies directly as b and b varies directly as c, does it follow that a varies directly as c? Why or why not?

SUMMARY AND REVIEW 7

KEY TERMS

Rate, p. 331
Slope, p. 333
Rise, p. 333
Run, p. 333
Grade, p. 335
Slope–intercept equation, p. 340
Point–slope equation, p. 345
Parallel lines, p. 347
Perpendicular lines, p. 348
Domain, p. 350
Range, p. 350
Function, p. 350
Relation, p. 351
Input, p. 352
Output, p. 352
Dependent variable, p. 352
Independent variable, p. 352
Constant function, p. 353
Linear function, p. 353
Dummy variable, p. 353
Direct variation, p. 368
Variation constant, p. 368
Constant of proportionality, p. 368
Inverse variation, p. 370
Joint variation, p. 371

IMPORTANT PROPERTIES AND FORMULAS

$$\text{Slope} = m = \frac{y_2 - y_1}{x_2 - x_1} = \frac{\text{change in } y}{\text{change in } x} = \frac{\text{rise}}{\text{run}}$$

Horizontal line:	Slope 0
Vertical line:	Slope undefined
Slope–intercept equation:	$y = mx + b$
Point–slope equation:	$y - y_1 = m(x - x_1)$
Parallel lines:	Slopes equal, y-intercepts different
Perpendicular lines:	Product of slopes $= -1$

The Vertical-Line Test

A graph is that of a function provided it is not possible to draw a vertical line that intersects the graph more than once.

The Algebra of Functions

1. $(f + g)(x) = f(x) + g(x)$
2. $(f - g)(x) = f(x) - g(x)$
3. $(f \cdot g)(x) = f(x) \cdot g(x)$
4. $(f/g)(x) = f(x)/g(x)$, provided $g(x) \neq 0$

To find the domain of a sum, difference, product, or quotient of two functions:

1. Determine the domain of each function.
2. The domain of the sum, difference, or product is the set of all values common to both domains.
3. The domain of the quotient is the set of all values common to both domains, excluding any value that would lead to division by 0.

Variation

y varies directly as x if there is some nonzero constant k such that $y = kx$.

y varies inversely as x if there is some nonzero constant k such that $y = k/x$.

y varies jointly as x and z if there is some nonzero constant k such that $y = kxz$.

REVIEW EXERCISES

This chapter's review and test include Skill Maintenance exercises from Sections 4.4, 4.5, 5.1, and 6.1.

Find the slope of the line.

1.

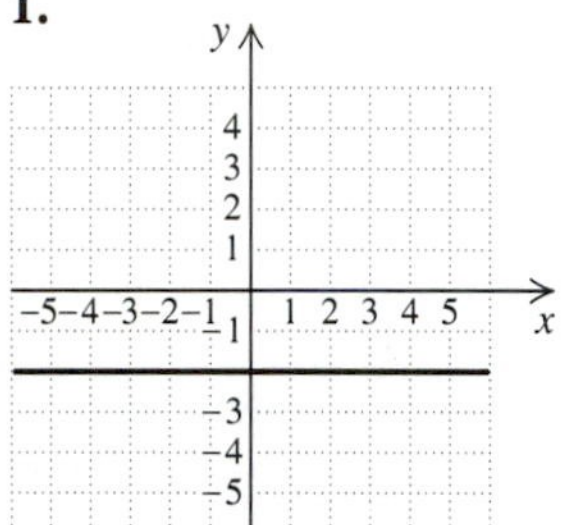

2.

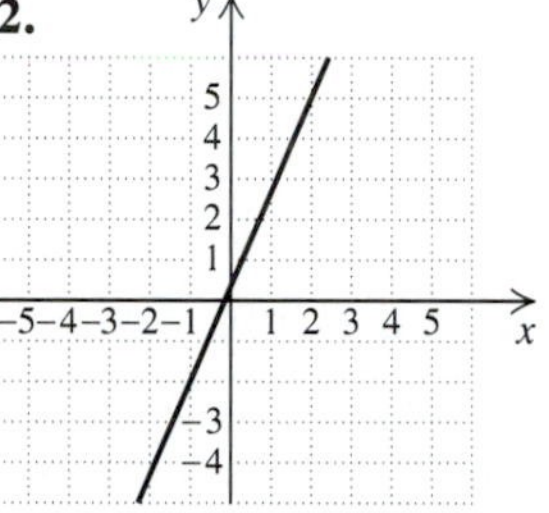

3.

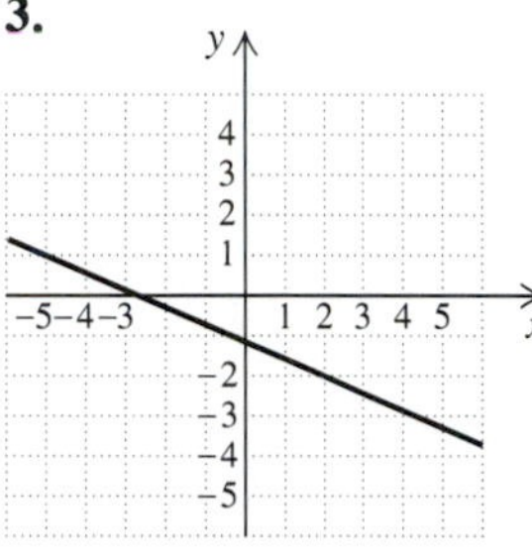

Find the slope of the line containing the given pair of points.

4. $(6, 8)$ and $(-2, -4)$
5. $(5, 1)$ and $(-1, 1)$
6. $(-3, 0)$ and $(-3, 5)$
7. $(-8.3, 4.6)$ and $(-9.9, 1.4)$
8. A road drops 369.6 ft vertically over a horizontal distance of 5280 ft. What is the grade of the road?

Find the slope of the line.

9. $y = -6$
10. $3x - 5y = 4$
11. $2x + y = 6$
12. $x = 90$

Find the slope and the y-intercept of the line.

13. $y = -9x + 46$
14. $x + y = 9$
15. $2x - 6y = 4$

Find the slope–intercept equation for the line with the indicated slope and y-intercept.

16. Slope -2; y-intercept $(0, -4)$
17. Slope 1.5; y-intercept $(0, 1)$

Graph.

18. $y = -\frac{3}{4}x - 2$
19. $y + \frac{1}{2}x = 2$
20. $y - 2 = 3(x - 6)$

Find a point–slope equation for the line containing the given point and having the given slope.

21. $(1, 2)$, $m = 3$
22. $(-2, -5)$, $m = \frac{2}{3}$

Find the slope–intercept equation for the line containing the given pair of points.

23. $(5, 7)$ and $(-1, 1)$
24. $(2, 0)$ and $(-4, -3)$

Determine whether the pair of equations represents parallel lines.

25. $4x + y = 6$,
$4x = 8 - y$
26. $3x - y = 6$,
$3x + y = 8$

Determine whether the pair of equations represents perpendicular lines.

27. $x - y = 4$,
$y = x + 1$
28. $2x + y = 5$,
$x - 2y = 1$
29. Find the slope–intercept equation of the line containing $(1, -1)$ and parallel to the line $y = 2x - 5$.
30. Find the slope–intercept equation of the line containing $(1, -1)$ and perpendicular to the line $y = 2x - 5$.

Let $g(x) = 2x - 5$ and $h(x) = 3x + 7$. Find the following.

31. $g(0)$
32. $h(-5)$
33. $(g \cdot h)(4)$
34. $(g - h)(-2)$
35. $(g/h)(-1)$
36. $g(a + b)$
37. The domain of $g + h$ and $g \cdot h$
38. The domain of h/g
39. Determine whether the graph on the following page is that of a function.

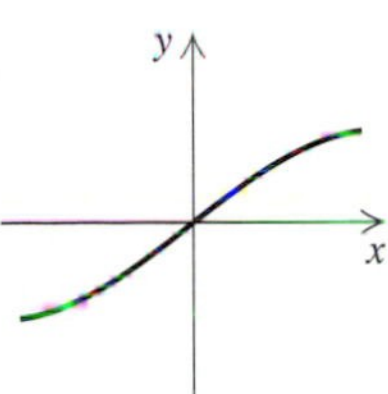

40. The number of sandwiches S that can be made at a buffet varies directly as the number of pounds of cold cuts C in the buffet. From 6 lb of cold cuts, 25 sandwiches can be made. How many pounds of cold cuts are needed for 40 sandwiches?

41. The apparent size s of an object varies inversely as the distance d of the object from the eye. A framed photograph 30 cm from an observer appears to be 27.5 cm tall. How tall will the object appear to be if it is 100 cm from the eye?

Skill Maintenance

Multiply.

42. $(x - 2)(3x^2 - 2x + 1)$ 43. $\left(\frac{1}{2}y + \frac{1}{4}\right)^2$

44. Factor: $x^3 - x^2 + 2x - 2$.

45. Simplify: $\dfrac{a^2 - 4}{2a^2 - 3a - 2}$.

Synthesis

46. ◈ Describe a situation in which point–slope form would be more useful than slope–intercept form.

47. ◈ Explain why the slope of a vertical line is undefined whereas the slope of a horizontal line is 0.

48. Find an equation of the line having the same y-intercept as the line $2x - y = 3$ and the same slope as the line $2x + y = 3$.

49. Determine the value of a so that the lines $3x - 4y = 12$ and $ax + 6y = -9$ are parallel.

50. Find an equation of the line for which the second coordinate is the opposite of the first coordinate.

51. Find the slope and the intercepts of a line whose equation is

$$\frac{x}{a} + \frac{y}{b} = 1, \quad a \neq 0 \text{ and } b \neq 0.$$

CHAPTER TEST 7

Find the slope of the line containing the pair of points.

1. (4, 7) and (4, −1)
2. (9, 2) and (−3, −5)

Find the slope of the line.

3. $2x + y = \frac{1}{3}$ 4. $y = -7$
5. $x = 6$

Find the slope and the y-intercept of the line.

6. $y = 2x - \frac{1}{4}$ 7. $-4x + 3y = -6$

Find the slope–intercept equation for the line with the indicated slope and y-intercept.

8. Slope $\frac{1}{2}$; y-intercept (0, −7)
9. Slope −4; y-intercept (0, 3)

Find a point–slope equation for the line containing the given point and having the given slope.

10. (3, 5), $m = 1$ 11. (−2, 0), $m = -3$

Find the slope–intercept equation for the line containing the given pair of points.

12. (1, 1) and (2, −2)
13. (4, −1) and (−4, −3)

Graph.

14. $2x + y = 5$ 15. $y - 5 = \frac{2}{3}(x - 6)$

16. Determine whether the following pair of equations represents parallel lines:

$$2x + y = 8,$$
$$2x + y = 4.$$

17. Determine whether the following pair of equations represents perpendicular lines:

$$4y + 2 = 3x,$$
$$3x + 4y = -12.$$

18. Find the following function values, given that $g(x) = -3x - 4$ and $h(x) = x^2 + 1$.

a) $g(0)$ **b)** $h(-2)$
c) $(g/h)(2)$ **d)** $(g - h)(-3)$

19. Let $f(x) = \frac{1}{x}$ and $g(x) = x + 5$.

Find the following.

a) The domain of f
b) The domain of g
c) The domain of $f + g$
d) The domain of f/g.

20. If y varies inversely as x and $y = 81$ when $x = 3$, find the equation of variation.

21. Find the equation of variation if A varies directly as the square of r and $A = 3.4$ when $r = 5$.

Skill Maintenance

Multiply.

22. $(-3y^4)(y^3 - 3y + 7)$ **23.** $(x + 0.1)(x - 0.1)$

24. Factor: $6x^3 + 3x^2 - 3x$.

25. Simplify: $\dfrac{-2x + 8}{2x}$.

Synthesis

26. A line contains the points $(-100, 4)$ and $(0, 0)$. List four more points of the line.

27. The function $f(t) = 5 + 15t$ can be used to determine a bicycle racer's position, in miles from the starting line, measured in hours since passing the 5-mile mark.

a) How far from the start will the racer be 1 hour and 40 minutes after passing the 5-mile mark?

b) Assuming a constant rate, how fast is the racer traveling?

CHAPTER 8

Systems of Equations and Problem Solving

AN APPLICATION

"Arctic Antifreeze" is 18% alcohol. "Frost-No-More" is 10% alcohol. How many liters of each should be mixed together in order to get 20 L of a mixture that is 15% alcohol?

This problem can be solved using a *system of equations*. It appears as Exercise 19 in Section 8.3.

James C. Letton
CHEMIST

"I learned very early that studies in science require a solid math background. Today, I often use systems of equations to determine the rate of change in the concentration of reactants, intermediates, and products during a chemical reaction."

The most difficult part of solving a problem in algebra is almost always translating the problem situation to mathematical language. Once an equation is translated, the rest is usually straightforward. In this chapter, we study *systems of equations* and how to solve them using graphing, substitution, and elimination. One of the great advantages of using a system of equations is that many problem situations then become easier to translate to mathematical language.

Systems of equations have extensive application to many fields, such as psychology, sociology, business, education, engineering, and science. This chapter includes a study of *matrices*, which can also be used to solve systems of equations.

In addition to the material in this chapter, the review and test for Chapter 8 include material from Sections 5.2, 5.3, 6.5, and 7.5.

8.1 Systems of Equations in Two Variables

Translating • Systems of Equations and Solutions • Solving Systems Graphically

Translating

Let us see how more than one equation can be used when translating a problem to mathematical language. Many problems involving more than one unknown can be more easily solved by translating to two or more equations than by translating to a single equation.

EXAMPLE 1

Translate the following problem situation to mathematical language, using two equations.

> Two angles are complementary. One angle is 12° less than three times the other. Find the measures of the angles.

Solution

1. FAMILIARIZE. Recall that we may sometimes have to look up some kind of information when problem solving. If you did not know the definition of complementary angles, you might look it up in a geometry book. It turns out that two angles are complementary if the sum of their measures is 90°. Suppose we make a guess about the answers. The measures 30° and 60° have a sum that is 90°. One angle is supposed to be 12° less than three times the other. Thus, $3 \cdot 30° - 12° = 90° - 12° = 78°$, which is not the same as the second part of our guess, 60°. Although our guess was wrong, the guessing has familiarized us with the problem. We decide to let x = the measure of one angle and y = the measure of the other angle.

Complementary angles

2. **TRANSLATE.** There are two statements in the problem. The fact that the angles are complementary can be reworded and translated as follows:

Two angles are complementary.

Rewording: The sum of the measures is 90°.

Translating: $x + y \qquad = 90$

The second statement of the problem can be translated directly, again using x and y:

Wording: One angle is 12° less than three times the other.

Translating: $y \quad = \quad 3x - 12$

The second statement could have been translated as $x = 3y - 12$, which would also have been correct. The problem has been translated to a pair, or **system of equations:**

$$x + y = 90,$$
$$y = 3x - 12.$$

❑

EXAMPLE 2

In one day, Glovers, Inc., sold 20 pairs of gloves. Cloth gloves sold for $24.95 per pair and pigskin gloves sold for $37.50 per pair. The company took in $687.25. Write a system of equations that could be used to find how many of each kind were sold.

Solution

1. **FAMILIARIZE.** To familiarize ourselves with this problem, let's make a guess:

Glovers sold 12 pairs of cloth gloves and
11 pairs of pigskin gloves.

Does the guess check? Does it make sense?

The total number of pairs of gloves sold was supposed to be 20, so our guess cannot be right. Let's try another guess:

12 pairs of cloth gloves and
8 pairs of pigskin gloves.

Now the total number sold is 20, so our guess is right in that respect.

How much money would then have been taken in? Since cloth gloves sold for $24.95, Glovers would have taken in

12($24.95)

from the cloth gloves. They would have taken in

8($37.50)

from the pigskin gloves. This makes the total received

12($24.95) + 8($37.50) = $299.40 + $300.00 = $599.40.

Our guess is not the answer to the problem, because the total, according to the problem, was \$687.25. Since \$599.40 is smaller than \$687.25, it seems reasonable that more of the expensive gloves were sold than we had guessed. We could now adjust our guess accordingly. Instead, let's work toward an algebraic approach that avoids guessing.

2. **TRANSLATE.** We let c = the number of pairs of cloth gloves sold and p = the number of pairs of pigskin gloves sold. The information can be organized in a table, which will help with the translating.

Kind of Glove	Cloth	Pigskin	Total	
Number Sold	c	p	20	$\longrightarrow c + p = 20$
Price	\$24.95	\$37.50		
Amount Taken In	$24.95c$	$37.50p$	687.25	$\longrightarrow 24.95c + 37.50p = 687.25$

The first row of the table and the first sentence of the problem indicate that a total of 20 pairs of gloves were sold:

$$c + p = 20.$$

Since each pair of cloth gloves cost \$24.95 and c pairs were sold, $24.95c$ represents the amount taken in from the sale of cloth gloves. Similarly, $37.50p$ represents the amount taken in from the sale of p pairs of pigskin gloves. From the third row of the table and the third sentence of the problem, we get the second equation:

$$24.95c + 37.50p = 687.25.$$

Multiplying both sides by 100, we clear the decimals. Thus we have the translation, a system of equations:

$$c + p = 20,$$
$$2495c + 3750p = 68{,}725.$$

❑

Systems of Equations and Solutions

A *solution* of a system of equations in two variables is an ordered pair of numbers that makes *both* equations true.

EXAMPLE 3

Determine whether $(-4, 7)$ is a solution of the system

$$x + y = 3,$$
$$5x - y = -27.$$

Solution We use alphabetical order of the variables. Thus we replace x by -4 and y by 7:

$$\begin{array}{c|l} \multicolumn{2}{c}{x + y = 3} \\ \hline -4 + 7 \ ? & 3 \\ 3 & 3 \quad \text{TRUE} \end{array} \qquad \begin{array}{r|l} \multicolumn{2}{c}{5x - y = -27} \\ \hline 5(-4) - (7) \ ? & -27 \\ -20 - 7 & \\ -27 & -27 \quad \text{TRUE} \end{array}$$

The pair $(-4, 7)$ makes both equations true, so it is a solution of the system. We sometimes describe such a solution by saying, in this case, that $x = -4$ and $y = 7$. Set notation can also be used to list the solution as $\{(-4, 7)\}$. ❑

Solving Systems Graphically

Recall that the graph of an equation is a drawing that represents its solution set. If the graph of an equation is a line, then every point on that line corresponds to an ordered pair that is a solution of the equation. If we graph a *system* of two linear equations, any point at which the lines intersect is a solution of *both* equations. Such a point is a solution of the system.

EXAMPLE 4

Solve the system graphically.

a) $y - x = 1,$
$y + x = 3$

b) $y = -3x + 5,$
$y = -3x - 2$

c) $3y - 2x = 6,$
$-12y + 8x = -24$

Solution

a) We draw the graphs of each equation using any method studied in Chapters 3 or 7. All ordered pairs from line L_1 are solutions of the first equation. All ordered pairs from line L_2 are solutions of the second equation. The point of intersection has coordinates that make *both* equations true. The solution seems to be the point (1, 2). Graphing is not perfectly accurate, so solving by graphing may get only approximate answers. Since the pair (1, 2) does check, it is the solution.

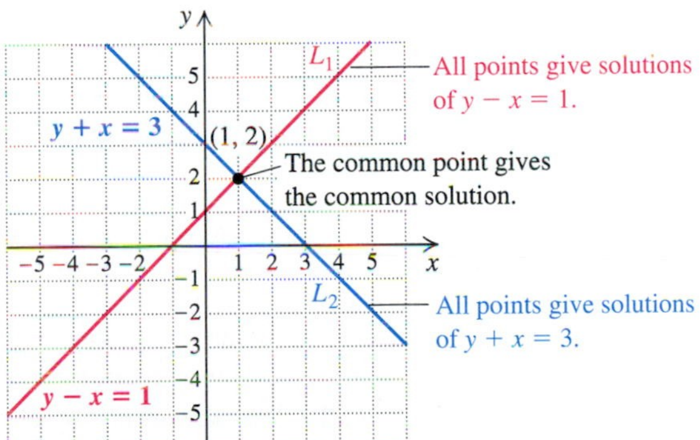

Check:

$$\begin{array}{c|c} y - x & 1 \\ \hline 2 - 1 & 1 \\ 1 & 1 \end{array} \text{ TRUE} \qquad \begin{array}{c|c} y + x & 3 \\ \hline 2 + 1 & 3 \\ 3 & 3 \end{array} \text{ TRUE}$$

b) We graph the equations. The lines have the same slope, −3, and different y-intercepts, so they are parallel. There is no point at which they cross, so the system has no solution.

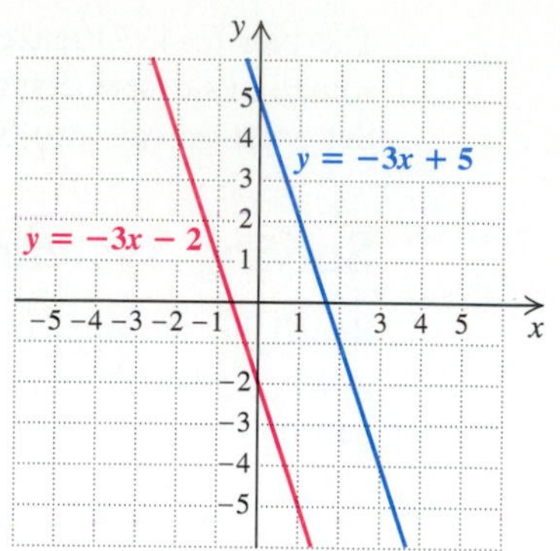

No matter what point we try, it will *not* check in *both* equations. There is no solution. The solution set is thus the empty set, denoted $\varnothing$ or $\{\ \}$.

c) We graph the equations and see that the graphs are the same. Thus any solution of one of the equations is a solution of the other. Each equation has an infinite number of solutions, some of which are listed on the graph. Each of these is also a solution of the other equation. We check one solution, (0, 2), which is the y-intercept of each equation.

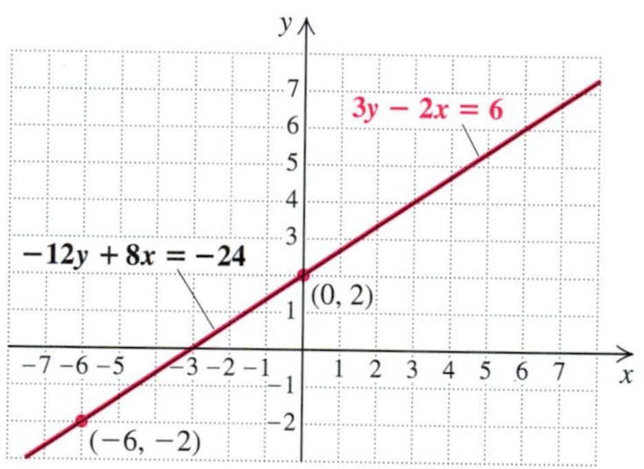

$3y - 2x = 6$	
$3(2) - 2(0)$	? 6
$6 - 0$	
6	6 TRUE

$-12y + 8x = -24$	
$-12(2) + 8(0)$	? -24
$-24 + 0$	
-24	-24 TRUE

You can check that $(-6, -2)$ is another solution of both equations. In fact, any pair that is a solution of one equation is a solution of the other equation as well. Thus the solution set is $\{(x, y) \mid 3y - 2x = 6\}$ or, in words, "the set of all pairs (x, y) for which $3y - 2x = 6$." In place of $3y - 2x = 6$, we could have used $-12y + 8x = -24$ since the two equations are equivalent. ❑

Example 4 illustrates that when we graph a system of two linear equations in two variables, one of the following three outcomes will occur.

1. The lines have one point in common, and that point is the only solution of the system (see part (a)).

2. The lines are parallel, with no point in common, and the system has no solution (see part (b)). This system is called **inconsistent.**
3. The lines coincide, sharing the same graph. Since every solution of one equation is a solution of the other, the system has an infinite number of solutions (see part (c)). The equations are said to be **dependent.**

Any system of equations that has at least one solution is said to be **consistent.** The systems of Examples 4(a) and 4(c) are both consistent.

Graphing is helpful when solving systems because it allows us to "see" the solution, and in many practical situations, an estimate made graphically is quite satisfactory. However, graphing has the disadvantage of not yielding exact answers when fractional or decimal solutions are involved. In Section 8.2, we will develop two algebraic methods of solving systems. Both methods produce exact answers.

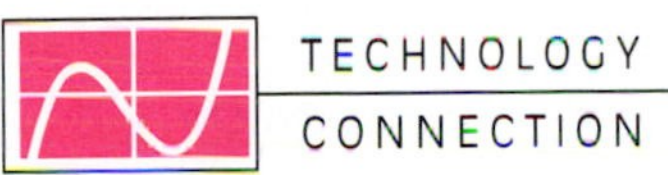
TECHNOLOGY CONNECTION

A grapher can be used to solve systems of equations, especially if it has a Zoom feature that allows a small portion of the graph to be magnified. Since most graphers require equations to have y alone on one side, the first step is to solve for y in each equation. After graphing both equations, use the Zoom feature to magnify the intersection (shrink the window's dimensions if your grapher lacks the Zoom feature). By doing this repeatedly and using the Trace feature, you can determine the coordinates of the point of intersection to the desired accuracy.

A grapher is especially useful when the equations in a system include fractions or decimals or when the coordinates of the intersection are not integers. To illustrate, let's solve the following system:

$$2.34x + 5.71y = -12.45,$$
$$-9.08x + 14.45y = 9.45.$$

After solving each equation for y, enter them into the grapher. The dimensions of the "standard" window for most graphers are $[-10, 10, -10, 10]$. In such a window, these two lines should appear as shown here:

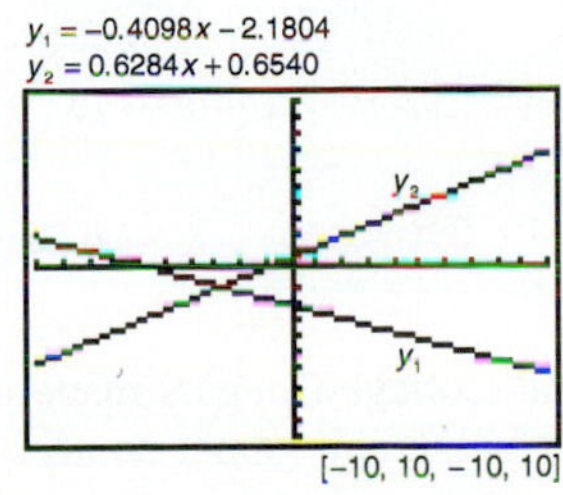

Now use the Zoom and Trace features to magnify the region in which the intersection appears. (Changing the window dimensions to $[-3, -2, -2, 0]$ gives a similar result.) You will now see the following:

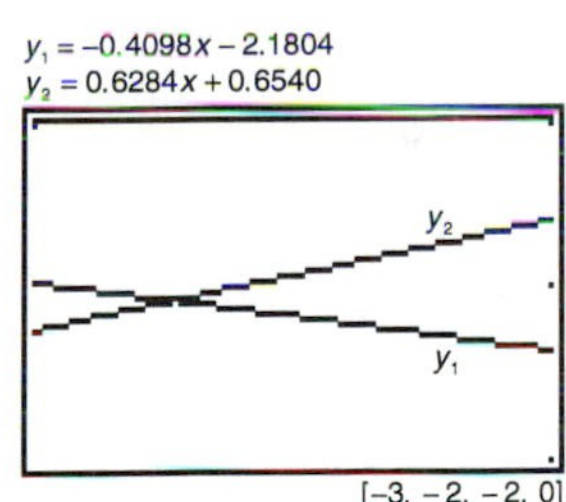

By "zooming" and "tracing" over and over, we can determine that, to the nearest hundredth, the coordinates of the intersection are $(-2.73, -1.06)$.

Use a grapher to find the solution to the system. Round the x- and y-coordinates to the nearest hundredth.

TC1. $y = -4.56x + 12.95,$
$y = 7.88x - 6.77$

TC2. $y = 123.52x + 89.32,$
$y = -89.22x + 33.76$

TC3. $1.76x + 8.21y = 12.22,$
$6.02x - 3.22y = 9.18$

TC4. $-9.25x - 12.94y = -3.88,$
$21.83x + 16.33y = 13.69$

EXERCISE SET 8.1

Translate the problem situation to a system of equations. Do not attempt to solve, but save for later use.

1. The sum of two numbers is -42. The first number minus the second number is 52. What are the numbers?

2. The difference between two numbers is 11. Twice the smaller plus three times the larger is 123. What are the numbers?

3. Windy City Mufflers sold 40 scarves. White ones sold for \$4.95 and printed ones sold for \$7.95. In all, \$282 worth of scarves were sold. How many of each kind were sold?

4. Twin City Pens sold 45 pens, one kind at \$8.50 and another kind at \$9.75. In all, \$398.75 was taken in. How many of each kind were sold?

5. Two angles are complementary. The sum of the measures of the first angle and half the second angle is 64°. Find the measures of the angles.

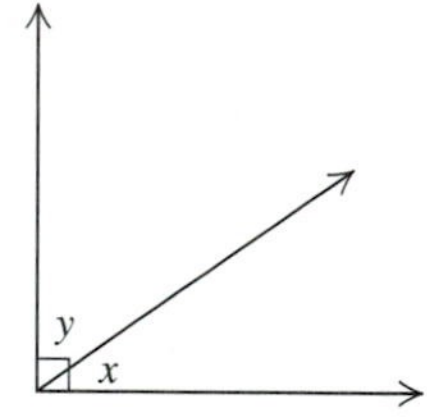

Complementary angles

6. Two angles are supplementary. One angle is 3° less than twice the other. Find the measures of the angles.

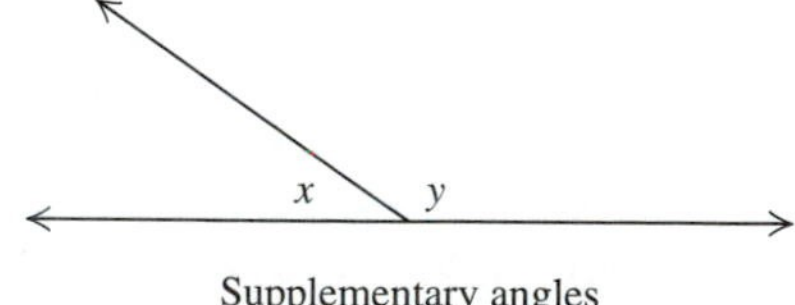

Supplementary angles

7. At a barbecue, there were 250 dinners served. A child's plate cost \$3.50 and an adult's plate cost \$7.00. If the total amount of money collected for dinners was \$1347.50, how many of each type of plate was served?

8. Mary Monroe scored 18 times during one basketball game. She scored a total of 30 points, two for each field goal and one for each free throw. How many field goals did she make? How many free throws?

9. The perimeter of a standard tennis court when used for doubles play is 228 ft. The width is 42 ft less than the length. Find the dimensions.

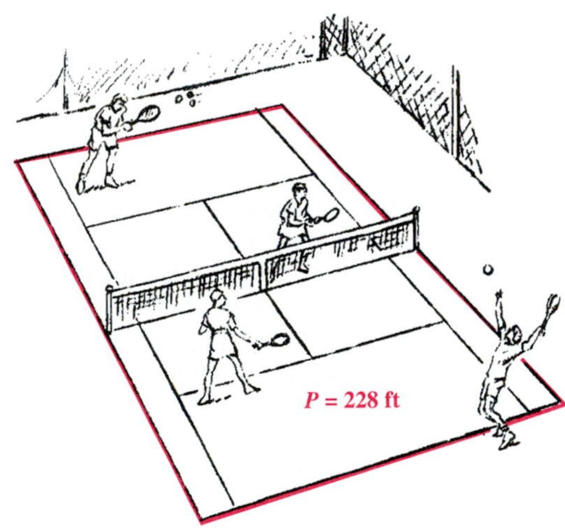

10. The perimeter of a standard basketball court is 288 ft. The length is 44 ft longer than the width. Find the dimensions.

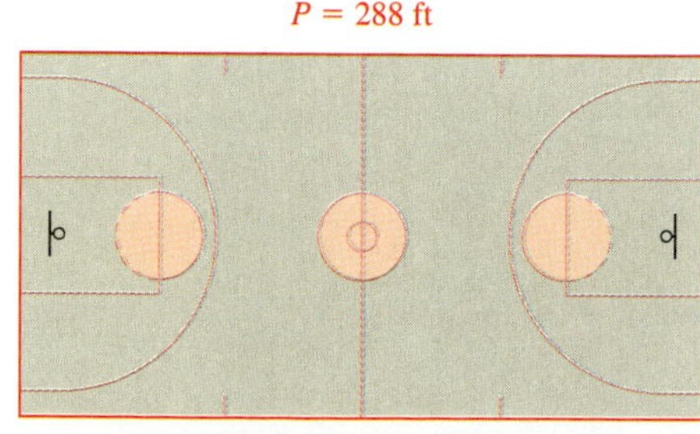

11. The Central College Cougars made 40 field goals in a recent basketball game, some 2-pointers and

some 3-pointers. Altogether the 40 baskets counted for 89 points. How many of each type of field goal was made?

12. Acme Video rents general interest films for \$3.00 and children's films for \$1.50. In one day, a total of \$213 was taken in from the rental of 77 videos. How many of each type of video was rented?

13. A lumber company can convert logs into either lumber or plywood. In a given day, the mill turns out a total of 400 units of lumber and plywood. It makes a profit of \$20 on a unit of lumber and \$30 on a unit of plywood. How many of each unit must be produced and sold in order to make a profit of \$11,000?

14. Hockey teams receive 2 points when they win and 1 point when they tie. One season, the Wildcats won a championship with 60 points. They won 9 more games than they tied. How many wins and how many ties did the Wildcats have?

15. A disc jockey must play 12 commercial spots during one hour of a radio show. Each commercial is either 30 sec or 60 sec long. If the total commercial time during that hour is 10 min, how many 30-sec commercials were played that hour? How many 60-sec commercials?

16. An airplane has a total of 152 seats. The number of coach-class seats is five more than six times the number of first-class seats. How many of each type of seat are there on the plane?

Determine whether the ordered pair is a solution of the given system of equations. Remember to use alphabetical order of variables.

17. (1, 2); $4x - y = 2,$ $10x - 3y = 4$

18. (−1, −2); $2x + y = -4,$ $x - y = 1$

19. (2, 5); $y = 3x - 1,$ $2x + y = 4$

20. (−1, −2); $x + 3y = -7,$ $3x - 2y = 12$

21. (1, 5); $x + y = 6,$ $y = 2x + 3$

22. (5, 2); $a + b = 7,$ $2a - 8 = b$

23. (2, −7); $3a + b = -1,$ $2a - 3b = -8$

24. (2, 1); $3p + 2q = 5,$ $4p + 5q = 2$

25. (3, 1); $3x + 4y = 13,$ $5x - 4y = 11$

26. (4, −2); $-3x - 2y = -8,$ $y = 2x - 5$

Solve the system graphically. Be sure to check.

27. $x + y = 4,$ $x - y = 2$

28. $x - y = 3,$ $x + y = 5$

29. $2x - y = 4,$ $5x - y = 13$

30. $3x + y = 5,$ $x - 2y = 4$

31. $4x - y = 9,$ $x - 3y = 16$

32. $4y = x + 8,$ $3x - 2y = 6$

33. $a = 1 + b,$ $b = -2a + 5$

34. $x = y - 1,$ $2x = 3y$

35. $2u + v = 3,$ $2u = v + 7$

36. $2b + a = 11,$ $a - b = 5$

37. $y = -\frac{1}{3}x - 1,$ $4x - 3y = 18$

38. $y = -\frac{1}{4}x + 1,$ $2y = x - 4$

39. $6x - 2y = 2,$ $9x - 3y = 1$

40. $y - x = 5,$ $2x - 2y = 10$

41. $x = 4,$ $y = -5$

42. $x = -3,$ $y = 2$

43. $y = -x - 1,$ $4x - 3y = 24$

44. $2a + b = 4,$ $b = 4a + 1$

45. $2x - 3y = 6,$ $3y - 2x = -6$

46. $y = 3 - x,$ $2x + 2y = 6$

47. For the systems in the odd-numbered exercises 27–45, which are consistent?

48. For the systems in the even-numbered exercises 28–46, which are consistent?

49. For the systems in the odd-numbered exercises 27–45, which are dependent?

50. For the systems in the even-numbered exercises 28–46, which are dependent?

Skill Maintenance

Solve.

51. $3x + 4 = x - 2$

52. $\frac{3}{5}x + 2 = \frac{2}{5}x - 5$

53. $4x - 5x = 8x - 9 + 11x$

54. Solve $Q = \frac{1}{4}(a - b)$ for b.

Synthesis

55. ◈ Write a problem for a classmate to solve that can be translated into a system of two equations. Devise the problem so that the solution is "Lucy sold 5 apple pies and 7 cherry pies."

56. Write a problem for a classmate to solve that requires writing a system of two equations. Devise the problem so that the solution is "The Lakers made 6 three-point baskets and 31 two-point baskets."

For Exercises 57–59, consider the following graph.*

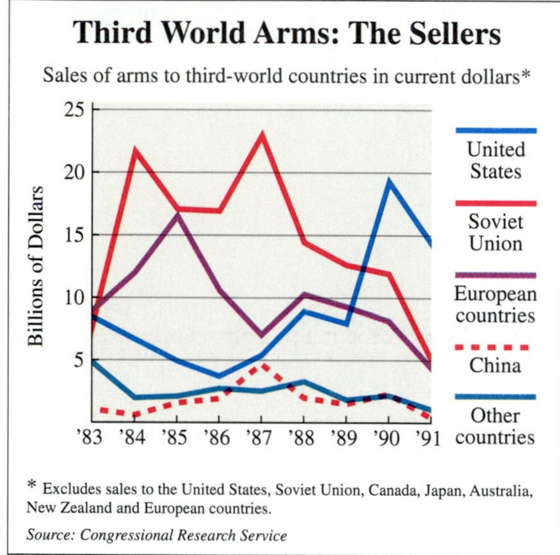

57. In what years did U.S. arms sales to third-world countries exceed Soviet arms sales to third-world countries?

58. What was the last year in which European arms sales to third-world countries exceeded U.S. arms sales to third-world countries?

59. Determine the most recent year in which Soviet arms sales to third-world countries exceeded the combined sales of the United States, Europe, and China.

60. The solution of the following system is $(4, -5)$. Find A and B.

$$Ax - 6y = 13,$$
$$x - By = -8.$$

61. Write a system of equations for which:

a) $(5, 1)$ is a solution,
b) there is no solution, and
c) there are infinitely many solutions.

---

62. A system of linear equations has $(1, -1)$ and $(-2, 3)$ as solutions. Determine:

a) a third point that is a solution, and
b) how many solutions there are.

Translate to a system of equations. Do not solve.

63. Burl is twice as old as his son. Ten years ago, Burl was three times as old as his son. How old are they now?

64. Lou and Juanita have been mathematics professors at a state university. Together, they have 46 years of service. Two years ago, Lou had taught 2.5 times as many years as Juanita. How long has each taught at the university?

65. A piece of posterboard has a perimeter of 156 in. If you cut 6 in. off the width, the length becomes four times the width. What are the dimensions of the original piece of posterboard?

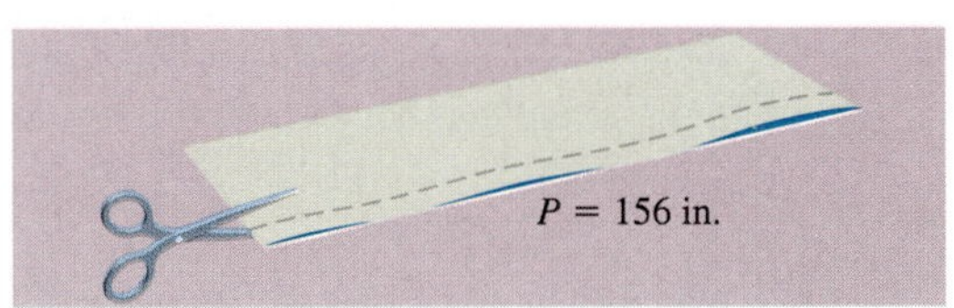

66. The Mudhens have played 104 baseball games. Before winning their last eight games in a row, they had won three times as many games as they had lost. What is their present won–loss record?

Solve graphically.

67. $y = |x|,$
$x + 4y = 15$

68. $x - y = 0,$
$y = x^2$

In Exercises 69–72, use a grapher to solve the system of linear equations for x and y. Round all coordinates to the nearest hundredth.

69. $y = 8.23x + 2.11,$
$y = -9.11x - 4.66$

70. $y = -3.44x - 7.72,$
$y = 4.19x - 8.22$

71. $14.12x + 7.32y = 2.98,$
$21.88x - 6.45y = -7.22$

72. $5.22x - 8.21y = -10.21,$
$-12.67x + 10.34y = 12.84$

*_The New York Times_, 8/11/91. Copyright © 1991 by the New York Times Company. Reprinted with permission.

8.2

Solving by Substitution or Elimination

The Substitution Method • The Elimination Method • Comparing Methods

The Substitution Method

One nongraphical method for solving systems of equations, the *substitution method,* relies on having one variable isolated.

EXAMPLE 1

Solve the system

$$x + y = 4, \qquad (1)$$
$$x = y + 1. \qquad (2)$$

Solution Equation (2) says that x and $y + 1$ name the same number. Thus we can substitute $y + 1$ for x in Equation (1):

$$x + y = 4 \qquad \text{Equation (1)}$$
$$(y + 1) + y = 4. \qquad \text{Substituting } y + 1 \text{ for } x$$

We solve this last equation, using methods learned earlier:

$$(y + 1) + y = 4$$
$$2y + 1 = 4 \qquad \text{Removing parentheses and collecting like terms}$$
$$2y = 3 \qquad \text{Subtracting 1 on both sides}$$
$$y = \tfrac{3}{2}. \qquad \text{Dividing by 2}$$

We now return to the original pair of equations and substitute $\frac{3}{2}$ for y in either equation so that we can solve for x. Calculations will be easier if we choose Equation (2):

$$x = y + 1 \qquad \text{Equation (2)}$$
$$x = \tfrac{3}{2} + 1 \qquad \text{Substituting } \tfrac{3}{2} \text{ for } y$$
$$x = \tfrac{3}{2} + \tfrac{2}{2} = \tfrac{5}{2}.$$

We obtain the ordered pair $(\frac{5}{2}, \frac{3}{2})$. We check to be sure that it is a solution.

Check:

$x + y = 4$	
$\frac{5}{2} + \frac{3}{2}$	? 4
$\frac{8}{2}$	
4	4 TRUE

$x = y + 1$	
$\frac{5}{2}$	? $\frac{3}{2} + 1$
	$\frac{3}{2} + \frac{2}{2}$
$\frac{5}{2}$	$\frac{5}{2}$ TRUE

Since $(\frac{5}{2}, \frac{3}{2})$ checks, it is the solution. ❑

The solution to Example 1 would have been difficult to find graphically because it involves fractions.

If neither equation in a system has a variable alone on one side, we first isolate a variable in one equation and then substitute.

EXAMPLE 2

Solve the system

$$2x + y = 6, \qquad (1)$$
$$3x + 4y = 4. \qquad (2)$$

Solution First we solve one equation for one variable. To isolate y, we can add $-2x$ to both sides of Equation (1) to get

$$y = 6 - 2x. \qquad (3)$$

Then we substitute $6 - 2x$ for y in Equation (2) and solve for x:

$$3x + 4(6 - 2x) = 4 \quad \text{Substituting } 6 - 2x \text{ for } y. \text{ Use parentheses!}$$
$$3x + 24 - 8x = 4 \quad \text{Multiplying to remove parentheses}$$
$$3x - 8x = 4 - 24$$
$$-5x = -20$$
$$x = 4.$$

Now we can substitute in either Equation (1), (2), or (3). It is easiest to use Equation (3) since it is already solved for y:

$$y = 6 - 2x = 6 - 2(4) = 6 - 8 = -2.$$

The pair $(4, -2)$ appears to be the solution.

Check:

$2x + y = 6$	
$2(4) + (-2)$	? 6
$8 - 2$	
6	6 TRUE

$3x + 4y = 4$	
$3(4) + 4(-2)$	? 4
$12 - 8$	
4	4 TRUE

Since $(4, -2)$ checks, it is the solution. ❑

Some systems have no solution, as we saw graphically in Section 8.1. How do we recognize such systems if we are solving using an algebraic method?

EXAMPLE 3

Solve the system

$$y = -3x + 5,$$
$$y = -3x - 2.$$

Solution We solved this problem graphically in Example 4(b) of Section 8.1, and found that the lines are parallel and the system has no solution. Let us now try to solve the system using the substitution method. We substitute $-3x - 2$ for y in the first equation:

$$-3x - 2 = -3x + 5 \quad \text{Substituting } -3x - 2 \text{ for } y$$
$$-2 = 5. \quad \text{Adding } 3x$$

When we add $3x$ to get the x-terms on one side, the x-terms drop out and we end up with a *false* equation. When solving algebraically yields a false equation, the system has no solution. ❑

The Elimination Method

The *elimination method* for solving systems of equations makes use of the *addition principle:* If $a = b$, then $a + c = b + c$. Consider the following system:

$$\begin{aligned} 2x - 3y &= 0, && (1) \\ -4x + 3y &= -1. && (2) \end{aligned}$$

The key to the advantage of the elimination method for solving this system involves the $-3y$ in one equation and the $3y$ in the other. These terms are opposites. If we add the terms on the left side of the equations, these terms will add to 0, and in effect, the variable y will be "eliminated."

To use the addition principle for equations, note that according to Equation (2), $-4x + 3y$ and -1 are the same number. Thus we can use a vertical form and add $-4x + 3y$ to the left side of Equation (1) and -1 to the right side:

$$\begin{aligned} 2x - 3y &= 0 && (1) \\ -4x + 3y &= -1 && (2) \\ \hline -2x + 0y &= -1. && \text{Adding} \end{aligned}$$

We have eliminated the variable y, which is why we call this the *elimination method.* We now have an equation with just one variable, x, which we solve for:

$$\begin{aligned} -2x &= -1 \\ x &= \tfrac{1}{2}. \end{aligned}$$

Next we substitute $\frac{1}{2}$ for x in Equation (1) and solve for y:

$$\begin{aligned} 2 \cdot \tfrac{1}{2} - 3y &= 0 && \text{Substituting} \\ 1 - 3y &= 0 && \\ y &= \tfrac{1}{3}. && \text{Solving for } y \end{aligned}$$

Check:

$$\begin{array}{c|c} 2x - 3y = 0 & \\ \hline 2(\frac{1}{2}) - 3(\frac{1}{3}) \; ? \; 0 & \\ 1 - 1 & \\ 0 & 0 \quad \text{TRUE} \end{array} \qquad \begin{array}{c|c} -4x + 3y = -1 & \\ \hline -4(\frac{1}{2}) + 3(\frac{1}{3}) \; ? \; -1 & \\ -2 + 1 & \\ -1 & -1 \quad \text{TRUE} \end{array}$$

Since $(\frac{1}{2}, \frac{1}{3})$ checks, it is the solution.

In order to eliminate a variable, we sometimes must multiply before adding.

EXAMPLE 4

Solve the system

$$\begin{aligned} 3x + 3y &= 15, && (1) \\ 2x + 6y &= 22. && (2) \end{aligned}$$

Solution If we add, we will not eliminate a variable. However, if the $3y$ in Equation (1) were $-6y$, we would. So we multiply by -2 on both sides of the first equation:

$$\begin{aligned} -6x - 6y &= -30 && \text{Multiplying by } -2 \text{ on both sides of Equation (1)} \\ 2x + 6y &= 22 && \\ \hline -4x + 0 &= -8 && \text{Adding} \\ x &= 2. && \text{Solving for } x \end{aligned}$$

Then

$$2 \cdot 2 + 6y = 22 \quad \text{Substituting 2 for } x \text{ in Equation (2)}$$
$$4 + 6y = 22$$
$$y = 3. \quad \text{Solving for } y$$

We obtain (2, 3), or $x = 2$, $y = 3$. This checks, so it is the solution. ❑

Sometimes we must multiply twice in order to make two terms become opposites.

EXAMPLE 5

Solve the system

$$2x + 3y = 17, \quad (1)$$
$$5x + 7y = 29. \quad (2)$$

Solution We have

$$2x + 3y = 17, \longrightarrow \text{Multiplying Equation (1) by 5} \longrightarrow \quad 10x + 15y = 85$$
$$5x + 7y = 29 \longrightarrow \text{Multiplying Equation (2) by } -2 \longrightarrow \quad -10x - 14y = -58$$
$$0 + y = 27 \quad \text{Adding}$$
$$y = 27.$$

We then find that

$$2x + 3 \cdot 27 = 17 \quad \text{Substituting 27 for } y \text{ in Equation (1)}$$
$$2x + 81 = 17$$
$$2x = -64$$
$$x = -32. \quad \text{Solving for } x$$

Check:

$2x + 3y = 17$	
$2(-32) + 3(27)$	? 17
$-64 + 81$	
17	17 TRUE

$5x + 7y = 29$	
$5(-32) + 7(27)$	? 29
$-160 + 189$	
29	29 TRUE

We obtain $(-32, 27)$, or $x = -32$, $y = 27$, as the solution. ❑

EXAMPLE 6

Solve the system

$$3y - 2x = 6, \quad (1)$$
$$-12y + 8x = -24. \quad (2)$$

Solution We graphed this system in Example 4(c) of Section 8.1, and found that the lines coincided and the system has an infinite number of solutions. Suppose we try to solve this system using the elimination method:

$$12y - 8x = 24 \quad \text{Multiplying Equation (1) by 4}$$
$$-12y + 8x = -24$$
$$0 = 0. \quad \text{Adding, we obtain a true equation.}$$

Note that we have eliminated both variables and what remains is a true equation. If a pair solves Equation (1), then it will also solve Equation (2). The equations are dependent and the solution set contains infinitely many pairs: $\{(x, y)|3y - 2x = 6\}$. ❑

When solving a system of two linear equations in two variables:

1. If a false equation is obtained, such as $0 = 7$, then the system has no solution. The system is inconsistent and the equations are independent.*
2. If a true equation is obtained, such as $0 = 0$, then the system has an infinite number of solutions. The system is consistent and the equations are dependent.

Should decimals or fractions appear, it often helps to *clear* before solving.

EXAMPLE 7

Solve the system

$$0.2x + 0.3y = 1.7,$$
$$\tfrac{1}{7}x + \tfrac{1}{5}y = \tfrac{29}{35}.$$

Solution We have

$$0.2x + 0.3y = 1.7, \quad \longrightarrow \text{Multiplying by 10} \longrightarrow \quad 2x + 3y = 17$$
$$\tfrac{1}{7}x + \tfrac{1}{5}y = \tfrac{29}{35} \quad \longrightarrow \text{Multiplying by 35} \longrightarrow \quad 5x + 7y = 29.$$

We multiplied by 10 to clear the decimals. Multiplication by 35, the least common multiple of the denominators 7, 5, and 35, clears the fractions. The problem is now identical to the one solved in Example 5. The solution is $(-32, 27)$, or $x = -32$, $y = 27$. ❑

Comparing Methods

The following table is a summary that compares the graphical, substitution, and elimination methods for solving systems of equations.

Method	Strengths	Weaknesses
Graphical	Can "see" solutions.	Inexact when solutions involve numbers that are not integers.
Substitution	Yields exact solutions. Easy to use when a variable is alone on one side.	Introduces extensive computations with fractions when solving more complicated systems. Cannot "see" solutions quickly.
Elimination	Yields exact solutions. Easy to use when fractions or decimals appear in the system.	Cannot "see" solutions quickly.

*Consistency and dependency are discussed in detail in Section 8.4.

When deciding which method to use, consider this table and directions from your instructor. The situation is analogous to having a piece of wood to cut and three saws with which to cut it. The right saw for the job depends on the wood, the result you want, and how you want to go about it.

EXERCISE SET 8.2

Solve using the substitution method.

1. $3x + 5y = 3,$ $x = 8 - 4y$

2. $2x - 3y = 13,$ $y = 5 - 4x$

3. $9x - 2y = 3,$ $3x - 6 = y$

4. $x = 3y - 3,$ $x + 2y = 9$

5. $5m + n = 8,$ $3m - 4n = 14$

6. $4x + y = 1,$ $x - 2y = 16$

7. $4x + 12y = 4,$ $-5x + y = 11$

8. $-3b + a = 7,$ $5a + 6b = 14$

9. $3x - y = 1,$ $2x + 2y = 2$

10. $5p + 7q = 1,$ $4p - 2q = 16$

11. $3x - y = 7,$ $2x + 2y = 5$

12. $5x + 3y = 4,$ $x - 4y = 3$

13. $x + 2y = 6,$ $x = 4 - 2y$

14. $y - 2x = 1,$ $2x - 3 = y$

15. $x - 3 = y,$ $2x - 2y = 6$

16. $3y = x - 2,$ $x = 2 + 3y$

Solve using the elimination method.

17. $x + 3y = 7,$ $-x + 4y = 7$

18. $x + y = 9,$ $2x - y = -3$

19. $2x + y = 6,$ $x - y = 3$

20. $x - 2y = 6,$ $-x + 3y = -4$

21. $9x + 3y = -3,$ $2x - 3y = -8$

22. $6x - 3y = 18,$ $6x + 3y = -12$

23. $5x + 3y = 19,$ $2x - 5y = 11$

24. $3x + 2y = 3,$ $9x - 8y = -2$

25. $5r - 3s = 24,$ $3r + 5s = 28$

26. $5x - 7y = -16,$ $2x + 8y = 26$

27. $0.3x - 0.2y = 4,$ $0.2x + 0.3y = 1$

28. $0.7x - 0.3y = 0.5,$ $-0.4x + 0.7y = 1.3$

29. $\frac{1}{2}x + \frac{1}{3}y = 4,$ $\frac{1}{4}x + \frac{1}{3}y = 3$

30. $\frac{2}{3}x + \frac{1}{7}y = -11,$ $\frac{1}{7}x - \frac{1}{3}y = -10$

31. $\frac{2}{5}x + \frac{1}{2}y = 2,$ $\frac{1}{2}x - \frac{1}{6}y = 3$

32. $\frac{1}{3}x + \frac{1}{5}y = 7,$ $\frac{1}{6}x - \frac{2}{5}y = -4$

33. $2x + 3y = 1,$ $4x + 6y = 2$

34. $3x - 2y = 1,$ $-6x + 4y = -2$

35. $2x - 4y = 5,$ $2x - 4y = 6$

36. $3x - 5y = -2,$ $5y - 3x = 7$

Solve.

37. $5x - 9y = 7,$ $7y - 3x = -5$

38. $a - 2b = 16,$ $b + 3 = 3a$

39. $3(a - b) = 15,$ $4a = b + 1$

40. $10x + y = 306,$ $10y + x = 90$

41. $x - \frac{1}{10}y = 100,$ $y - \frac{1}{10}x = -100$

42. $\frac{1}{8}x + \frac{3}{5}y = \frac{19}{2},$ $-\frac{3}{10}x - \frac{7}{20}y = -1$

43. $0.05x + 0.25y = 22,$ $0.15x + 0.05y = 24$

44. $1.3x - 0.2y = 12,$ $0.4x + 17y = 89$

Skill Maintenance

Find the LCM.

45. $9a^5b^2$, $6ab^6$

46. t, $t + 2$, $t - 2$

Synthesis

47. ◈ Write a system of linear equations that would be most easily solved using the substitution method. Explain why substitution would be easier to use than the elimination method.

48. ◈ Write a system of linear equations that would be most easily solved using the elimination method. Explain why elimination would be easier to use than the substitution method.

Solve.

49. 🖩 $3.5x - 2.1y = 106.2,$ $4.1x + 16.7y = -106.28$

50. $\dfrac{x+y}{2} - \dfrac{x-y}{5} = 1,$ $\dfrac{x-y}{2} + \dfrac{x+y}{6} = -2$

51. Solve for x and y in terms of a and b:

$$5x + 2y = a,$$
$$x - y = b.$$

52. Determine a and b for which $(-4, -3)$ will be a solution of the system

$$ax + by = -26,$$
$$bx - ay = 7.$$

53. The points $(0, -3)$ and $\left(-\frac{3}{2}, 6\right)$ are two of the solutions of the equation $px - qy = -1$. Find p and q.

54. For $f(x) = mx + b$, two solutions are $(1, 2)$ and $(-3, 4)$. Find m and b.

Each of the following is a system of equations that is *not* linear. But each is *reducible to linear*, because an appropriate substitution (say, u for $1/x$ and v for $1/y$) yields a linear system. Solve for the new variable and then solve for the original variable.

55. $\frac{1}{x} - \frac{3}{y} = 2,$

$\frac{6}{x} + \frac{5}{y} = -34$

56. $\frac{2}{x} + \frac{1}{y} = 0,$

$\frac{5}{x} + \frac{2}{y} = -5$

8.3

Problem Solving Using Systems of Two Equations in Two Variables

Problems Involving Money • Problems Involving Mixtures • Problems Involving Motion

You are in a much better position to solve problems now that systems of equations can be used. Using systems often makes the translating step easier.

EXAMPLE 1

Two angles are complementary. One angle is 12° less than three times the other. Find the measures of the angles.

Solution The **Familiarize** and **Translate** steps have been done in Example 1 of Section 8.1. The resulting system of equations is

$$x + y = 90,$$
$$y = 3x - 12,$$

where $x =$ the measure of the first angle and $y =$ the measure of the other angle.

3. CARRY OUT. We solve the system of equations. What method should we use?

Since we have a variable alone on one side, let's use the substitution method:

$$x + y = 90$$
$$x + (3x - 12) = 90 \quad \text{Substituting } 3x - 12 \text{ for } y$$
$$4x - 12 = 90 \quad \text{Collecting like terms}$$
$$4x = 102$$
$$x = 25.5.$$

We return to the second equation and substitute 25.5 for x and compute y:

$$y = 3x - 12 = 3(25.5) - 12 = 64.5.$$

The angles seem to be 25.5° and 64.5°.

4. **CHECK.** The sum of the angles is 90°, so they are complementary. Also, three times the 25.5° angle minus 12°, is 64.5°, the measure of the second angle. Thus the measures of the angles check.

5. **STATE.** The measure of one angle is 25.5°, and the measure of the other is 64.5°. ❑

EXAMPLE 2

In one day, Glovers, Inc., sold 20 pairs of gloves. Cloth gloves sold for \$24.95 per pair and pigskin gloves sold for \$37.50 per pair. The company took in \$687.25. How many of each kind were sold?

Solution The **Familiarize** and **Translate** steps were done in Example 2 of Section 8.1.

3. **CARRY OUT.** We are to solve the system of equations

$$\begin{aligned} c + \quad p &= 20, & (1)\\ 2495c + 3750p &= 68{,}725, & (2) \end{aligned}$$

where $c =$ the number of pairs of cloth gloves sold and $p =$ the number of pairs of pigskin gloves sold. What method should we use? We could try graphing, but the large numbers would make that cumbersome. Since no variable appears alone and the equations are in the form $Ax + By = C$, let's try the elimination method. We eliminate c by multiplying Equation (1) by -2495 and adding it to Equation (2):

$$\begin{aligned} -2495c - 2495p &= -49{,}900 && \text{Multiplying Equation (1) by } -2495\\ 2495c + 3750p &= 68{,}725 \\ \hline 1255p &= 18{,}825 && \text{Adding}\\ p &= 15. && \text{Solving for } p \end{aligned}$$

To find c, we substitute 15 for p in Equation (1) and solve for c:

$$\begin{aligned} c + p &= 20 && \text{Equation (1)}\\ c + 15 &= 20 && \text{Substituting 15 for } p\\ c &= 5. && \text{Solving for } c \end{aligned}$$

We obtain (5, 15), or $c = 5$, $p = 15$.

4. **CHECK.** We check in the original problem. Remember that c is the number of cloth gloves and p is the number of pigskin gloves. Thus:

Number of gloves: $c + p = 5 + 15 = 20$

Money from cloth gloves: $\$24.95c = 24.95 \times 5 = \$124.75$

Money from pigskin gloves: $\$37.50p = 37.50 \times 15 = \$562.50$

Total = \$687.25

The numbers check.

5. **STATE.** Glovers sold 5 pairs of cloth gloves and 15 pairs of pigskin gloves. ❑

EXAMPLE 3

Yardbirds Gardening, Inc., carries two brands of solutions containing weedkiller and water. "Gently Green" is 5% weedkiller and "Sun Saver" is 15% weedkiller. Yardbirds Gardening needs to combine the two types of solutions to make 100 L of a solution that is 12% weedkiller. How much of each brand should be used?

Solution

1. **FAMILIARIZE.** Suppose that 40 L of Gently Green and 60 L of Sun Saver are mixed. The resulting mixture will be the right size, 100 L, but will it be the right strength? To find out, note that 40 L of Gently Green would contribute $0.05(40) = 2$ L of weedkiller to the mixture. Since the Sun Saver is 15% weedkiller, 60 L would contribute $0.15(60) = 9$ L of weedkiller to the mixture. Altogether, 40 L of Gently Green and 60 L of Sun Saver would make 100 L of a mixture that has $2 + 9 = 11$ L of weedkiller. Since this would mean that the final mixture is 11% weedkiller, our guess of 40 L and 60 L is incorrect. Still, the process of checking our guess has familiarized us with the problem.
2. **TRANSLATE.** Let g = the number of liters of Gently Green and s = the number of liters of Sun Saver. The information can be organized in a table.

	Gently Green	Sun Saver	Mixture	
Number of Liters	g	s	100	$\longrightarrow g + s = 100$
Percent of Weedkiller	5%	15%	12%	
Amount of Weedkiller	$0.05g$	$0.15s$	0.12×100, or 12 liters	$\longrightarrow 0.05g + 0.15s = 12$

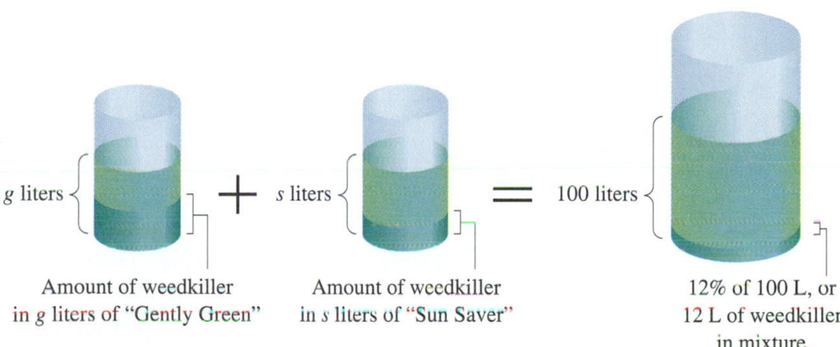

If we add g and s in the first row, we get one equation. It represents the total amount of mixture: $g + s = 100$.

If we add the amounts of weedkiller listed in the third row, we get a second equation. This equation represents the amount of weedkiller in the mixture: $0.05g + 0.15s = 12$.

After clearing decimals, we have the problem translated to the system

$$g + s = 100, \quad (1)$$
$$5g + 15s = 1200. \quad (2)$$

3. **CARRY OUT.** We use the elimination method to solve the system:

$$\begin{aligned} -5g - 5s &= -500 && \text{Multiplying Equation (1) by } -5 \\ 5g + 15s &= 1200 \\ \hline 10s &= 700 && \text{Adding} \\ s &= 70; && \text{Solving for } s \end{aligned}$$

$$\begin{aligned} g + 70 &= 100 && \text{Substituting into Equation (1)} \\ g &= 30. && \text{Solving for } g \end{aligned}$$

4. **CHECK.** Remember, g is the number of liters of Gently Green and s is the number of liters of Sun Saver.

Total number of liters of mixture: $g + s = 30 + 70 = 100$

Total amount of weedkiller: $5\% \times 30 + 15\% \times 70 = 1.5 + 10.5 = 12$

Percentage of weedkiller in mixture: $\dfrac{\text{Total amount of weedkiller}}{\text{Total number of liters of mixture}} = \dfrac{12}{100} = 12\%$

The numbers do check in the original problem.

5. **STATE.** Yardbird Gardening should mix 30 L of Gently Green with 70 L of Sun Saver. ❑

EXAMPLE 4

Two investments are made totaling \$4800. Part of the money is invested at 8% and the rest at 9%. In the first year, the total yield is \$412 in simple interest. Find the amount invested at each rate of interest.

Solution

1. **FAMILIARIZE.** As in Example 3, we can begin with a guess. If \$1000 was invested at 8% and \$3800 was invested at 9%, the two investments would total \$4800. The interest would then be 8%(\$1000), or \$80, and 9%(\$3800), or \$342, for a total of \$422 in interest. Our guess was wrong, but the manner in which we checked the guess has familiarized us with the problem.

2. **TRANSLATE.** Let x = the amount of money invested at 8% and y = the amount of money invested at 9%. We can then organize the information in a table. Each column in the table comes from the formula for simple interest: *Interest = Principal · Rate · Time*.

	First Investment	Second Investment	Total	
Principal	x	y	\$4800	→ $x + y = \$4800$
Rate of Interest	8%	9%		
Time	1 yr	1 yr		
Interest	$0.08x$	$0.09y$	\$412	→ $0.08x + 0.09y = \$412$

The total of the amounts invested is found in the first row. This gives us one equation:

$$x + y = 4800.$$

Look at the last row. The interest, or *yield,* totals \$412. This gives us a second equation:

$$8\%x + 9\%y = 412, \quad \text{or} \quad 0.08x + 0.09y = 412.$$

After we multiply on both sides to clear the decimals, we have

$$8x + 9y = 41{,}200.$$

3. **CARRY OUT.** The following system can be solved by elimination or substitution:

$$x + y = 4800,$$
$$8x + 9y = 41{,}200.$$

We find that $x = 2000$ and $y = 2800$.

4. **CHECK.** The sum is \$2000 + \$2800, or \$4800. The interest from \$2000 at 8% for 1 yr is 8%(\$2000), or \$160. The interest from \$2800 at 9% for 1 yr is 9%(\$2800), or \$252. The total interest is \$160 + \$252, or \$412, so the numbers check.

5. **STATE.** An amount of \$2000 was invested at 8% and \$2800 at 9%. ❑

EXAMPLE 5

The ground floor of the John Hancock building in Chicago is a rectangle whose perimeter is 860 ft. The length is 100 ft more than the width. Find the length and the width.

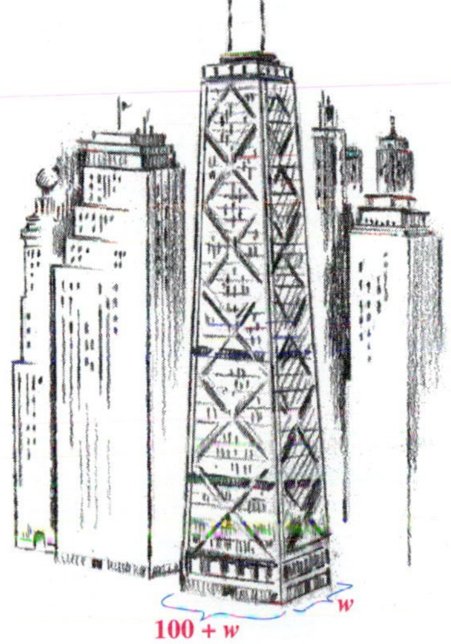

Solution

1. **FAMILIARIZE.** We make a drawing and label it. We recall, or look up, the definition of *perimeter:* $P = 2l + 2w$. We let l = the length of the ground floor of the building and w = the width. Guesses could be made now, but this time we proceed to the next step.

2. **TRANSLATE.** We translate as follows:

The perimeter is 860.

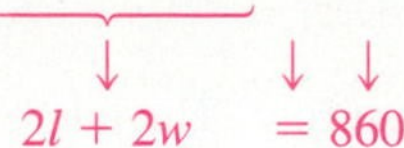

$2l + 2w \quad = 860$

We translate the second statement:

The length is 100 ft more than the width.

$l \quad = \quad w + 100$

This gives us the system of equations

$$2l + 2w = 860, \qquad (1)$$
$$l = w + 100. \qquad (2)$$

3. **CARRY OUT.** In this case, it is probably easier to use the substitution method:

$$2(w + 100) + 2w = 860 \qquad \text{Substituting } w + 100 \text{ for } l \text{ in Equation (1)}$$
$$2w + 200 + 2w = 860 \qquad \text{Multiplying to remove parentheses}$$
$$4w + 200 = 860 \qquad \text{Collecting like terms}$$
$$4w = 660 \qquad \text{Adding } -200$$
$$w = 165. \qquad \text{Multiplying by } \tfrac{1}{4}$$

Then we substitute 165 for w in Equation (2):

$$l = 165 + 100 = 265.$$

4. **CHECK.** The length is 100 ft more than the width, and the perimeter is 2(265 ft) + 2(165 ft), or 860 ft. The numbers check in the original problem.

5. **STATE.** The length is 265 ft and the width is 165 ft. ❑

Problems Involving Motion

When a problem deals with distance, speed (rate), and time, we need to recall the following.

> If r represents rate, t represents time, and d represents distance, then:
>
> $$d = rt, \qquad r = \frac{d}{t}, \quad \text{and} \quad t = \frac{d}{r}.$$

You should remember at least one of these equations and obtain the others by using algebraic manipulations as needed in a problem situation.

EXAMPLE 6

A freight train leaves Ames traveling east at a speed of 60 km/h. Two hours later, a passenger train leaves Ames traveling in the same direction on a parallel track at 90 km/h. At what point will the passenger train catch up to the freight train?

Solution

1. FAMILIARIZE. To familiarize ourselves with the problem, we make a guess—say, 180 km—and check to see if it is correct. A freight train, traveling 60 km/h, would reach a point 180 km from Ames in $\frac{180}{60} = 3$ hr. The passenger train, traveling 90 km/h, would cover 180 km in $\frac{180}{90} = 2$ hr. Since the problem states that the freight train has been running for two hours before the passenger train departs, and since $\frac{180}{60} = 3$ hr is *not* two hours more than $\frac{180}{90} = 2$ hr, we see that our guess of 180 km is incorrect. Although our guess was wrong, we now see that the time that the trains are running is an unknown as well as the point at which they meet up. Let $t =$ the time the freight train is running, $t - 2 =$ the time the passenger train is running, and $d =$ the distance at which the trains meet. Then make a sketch.

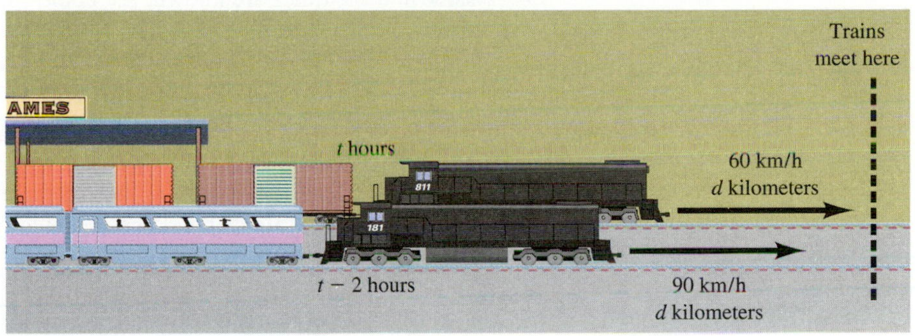

2. TRANSLATE. We can organize the information in a chart. Each row is determined by the formula $d = rt$.

	Distance (d)	= Rate (r)	· Time (t)	
Freight Train	d	60	t	→ $d = 60t$
Passenger Train	d	90	$t - 2$	→ $d = 90(t - 2)$

Using $d = rt$ in each row of the table, we get an equation. Thus we have a system of equations:

$$d = 60t, \qquad (1)$$
$$d = 90(t - 2). \qquad (2)$$

3. CARRY OUT. We solve the system:

$$60t = 90(t - 2) \qquad \text{Substituting } 60t \text{ for } d \text{ in Equation (2)}$$
$$60t = 90t - 180$$
$$-30t = -180$$
$$t = 6.$$

The time for the freight train is 6 hr, which means that the time for the passenger train is 6 − 2, or 4 hr. For $t = 6$, we have $d = 60 \cdot 6 = 360$ km.

4. **CHECK.** At 60 km/h, the freight train will travel $60 \cdot 6$, or 360 km, in 6 hr. At 90 km/h, the passenger train will travel $90 \cdot (6 - 2) = 360$ km in 4 hr. Remember that it is distance, not time, that the problem asked for. The numbers check.

5. **STATE.** The trains will meet at a point 360 km east of Ames. ❑

EXAMPLE 7

A motorboat took 4 hr to make a trip downstream with a 6-mph current. The return trip against the same current took 5 hr. Find the speed of the boat in still water.

Solution

1. **FAMILIARIZE.** We visualize the problem mentally and then make a sketch. Observe that the current *slows* the boat traveling upstream and *speeds up* the boat traveling downstream. Since the distances traveled each way are the same and the times are known, we could make a guess of, say, 40 mph for the speed of the boat in still water. The boat would then go $40 - 6 = 34$ mph upstream and would travel $34 \cdot 5 = 170$ mi against the current. The boat would go $40 + 6 = 46$ mph downstream, and would travel $46 \cdot 4 = 184$ mi downstream. Since $170 \neq 184$, our guess of 40 mph is incorrect. Rather than guess again, we decide to let r = the speed, in miles per hour, of the boat in still water, $r - 6$ = the boat's speed going upstream, and $r + 6$ = the boat's speed going downstream. We also let d = the distance traveled, in miles.

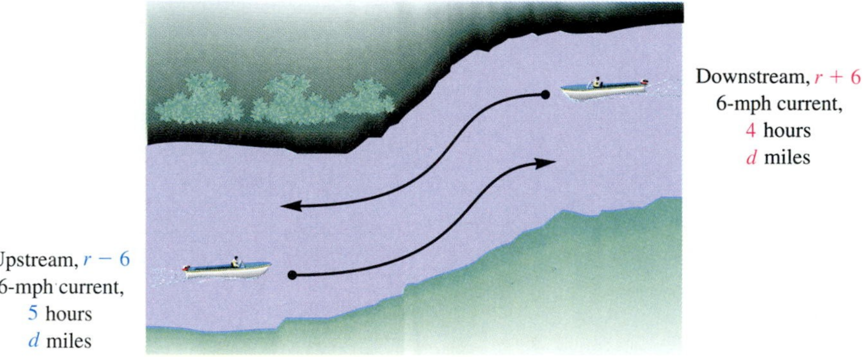

2. **TRANSLATE.** The information can be organized in a chart. The distances traveled are the same, so we use the formula *Distance* = *Rate* (or *Speed*) $\cdot$ *Time*. Each row of the chart gives an equation.

	Distance	Rate	Time	
Downstream	d	$r + 6$	4	⟶ $d = (r + 6)4$
Upstream	d	$r - 6$	5	⟶ $d = (r - 6)5$

The two equations constitute a system:

$$d = (r + 6)4, \quad (1)$$
$$d = (r - 6)5. \quad (2)$$

3. CARRY OUT. We solve the system:

$(r - 6)5 = (r + 6)4$ Substituting $(r - 6)5$ for d in Equation (1)

$5r - 30 = 4r + 24$ Using the distributive law

$r = 54.$ Solving for r

4. CHECK. When $r = 54$, $r + 6 = 60$, and $60 \cdot 4 = 240$, the distance. When $r = 54$, $r - 6 = 48$, and $48 \cdot 5 = 240$. In both cases, we get the same distance.

5. STATE. The speed of the boat in still water is 54 mph. ☐

Tips for Solving Motion Problems

1. Draw a diagram using an arrow or arrows to represent distance and the direction of each object in motion.
2. Organize the information in a chart.
3. Look for as many things as you can that are the same, so you can write equations.
4. Translating to a system of equations eases the solution of many motion problems.
5. When checking, be sure that you have solved for what the problem asked for.

EXERCISE SET 8.3

1.–16. For Exercises 1–16, solve Exercises 1–16 from Exercise Set 8.1.

17. Soybean meal is 16% protein and corn meal is 9% protein. How many pounds of each should be mixed together in order to get a 350-lb mixture that is 12% protein?

18. Lucinda has one solution that is 25% acid and a second that is 50% acid. How many liters of each should be mixed together in order to get 10 L of a solution that is 40% acid?

19. "Arctic Antifreeze" is 18% alcohol and "Frost-No-More" is 10% alcohol. How many liters of each should be mixed together in order to get 20 L of a mixture that is 15% alcohol?

20. "Orange-Thirst" is 15% orange juice and "Quencho" is 5% orange juice. How many liters of each should be mixed together in order to get 10 L of a mixture that is 10% orange juice?

21. Two investments are made totaling $15,000. For a certain year, these investments yield $1432 in simple interest. Part of the $15,000 is invested at 9% and part at 10%. Find the amount invested at each rate.

22. Two investments are made totaling $8800. For a certain year, these investments yield $1326 in simple interest. Part of the $8800 is invested at 14% and part at 16%. Find the amount invested at each rate.

23. $27,000 is invested, part of it at 10% and part of it at 12%. The total yield at simple interest for one year is $2990. How much was invested at each rate?

24. $1150 is invested, part of it at 12% and part of it at 11%. The total yield at simple interest for one year is $133.75. How much was invested at each rate?

25. At a club play, 117 tickets were sold. Adults' tickets cost $1.25 each and children's tickets cost $0.75 each. In all, $129.75 was taken in. How many of each kind of ticket were sold?

26. I.Q. Bean's sold 30 sweatshirts. White ones cost $9.95 and yellow ones cost $10.50. In all, $310.60

worth of sweatshirts were sold. How many of each color were sold?

27. Paula is 12 years older than her brother Bob. Four years from now, Bob will be $\frac{2}{3}$ as old as Paula. How old are they now?

28. Carlos is 8 years older than his sister Maria. Four years ago, Maria was $\frac{2}{3}$ as old as Carlos. How old are they now?

29. The perimeter of a lot is 190 m. The width is one fourth of the length. Find the dimensions.

30. The perimeter of a rectangular field is 194 yd. The length is two more than four times the width. Find the dimensions.

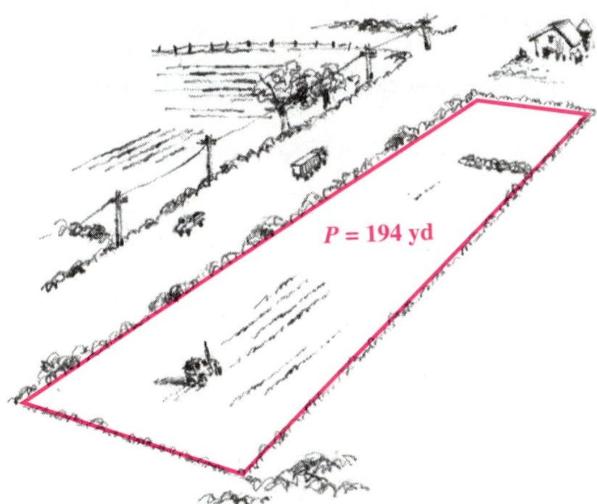

31. A customer goes to a bank and gets change for a \$50 bill consisting of all \$5 bills and \$1 bills. There are 22 bills in all. How many of each kind are there?

32. Cecilia makes a \$9.25 purchase at the bookstore with a \$20 bill. The store has no bills and gives her the change in quarters and fifty-cent pieces. There are 30 coins in all. How many of each kind are there?

33. A train leaves Danville Junction and travels north at a speed of 75 km/h. Two hours later, a second train leaves on a parallel track and travels north at 125 km/h. How far from the station will they meet?

34. Two cars leave Denver traveling in opposite directions. One car travels at a speed of 80 km/h and the other at 96 km/h. In how many hours will they be 528 km apart?

35. Two motorcycles travel toward each other from Chicago and Indianapolis, which are about 350 km apart, at rates of 110 km/h and 90 km/h. They started at the same time. In how many hours will they meet?

36. Two planes travel toward each other from cities that are 780 km apart at rates of 190 km/h and 200 km/h. They started at the same time. In how many hours will they meet?

37. A motorboat took 3 hr to make a trip downstream with a 6-mph current. The return trip against the same current took 5 hr. Find the speed of the boat in still water.

38. A canoeist paddled for 4 hr with a 6-km/h current to reach a campsite. The return trip against the same current took 10 hr. Find the speed of the canoe in still water.

Skill Maintenance

Factor.

39. $x^2 + 6x + 5$

40. $2x^2 - 9x - 5$

41. Write an equation of a line containing $(2, -5)$ with slope $-\frac{3}{4}$.

42. Graph: $y = -\frac{1}{2}x + 5$.

Synthesis

43.–46. For Exercises 43–46, solve Exercises 63–66 from Exercise Set 8.1.

47. The radiator in Michelle's car contains 16 L of antifreeze and water. This mixture is 30% antifreeze. How much of this mixture should she drain and replace with pure antifreeze so that there will be a mixture of 50% antifreeze?

48. Natalie jogs and walks to school each day. She averages 4 km/h walking and 8 km/h jogging. The distance from home to school is 6 km and she makes the trip in 1 hr. How far does she jog in a trip?

49. The ten's digit of a two-digit positive integer is two more than three times the unit's digit. If the digits are interchanged, the new number is thirteen less than half the given number. Find the given integer. (*Hint:* Let $x =$ the ten's-place digit and $y =$ the unit's-place digit; then $10x + y$ is the number.)

50. A limited edition of a book published by a historical society was offered for sale to its membership. The cost was one book for \$12 or two books for \$20. The society sold 880 books, and the total amount of money taken in was \$9840. How many members ordered two books?

51. A train leaves Union Station for Central Station, 216 km away, at 9 A.M. One hour later, a train leaves Central Station for Union Station. They meet at noon. If the second train had started at 9 A.M. and the first train at 10:30 A.M., they would still have met at noon. Find the speed of each train.

52. Dianne Osborne's station wagon gets 18 miles per gallon (mpg) in city driving and 24 mpg in highway driving. The car is driven 465 mi on 23 gal of gasoline. How many miles were driven in the city and how many were driven on the highway?

53. Phil and Phyllis are siblings. Phyllis has twice as many brothers as she has sisters. Phil has the same number of brothers as sisters. How many girls and how many boys are in the family?

54. ◈ Write three or four study tips of your own for use by someone beginning this exercise set.

8.4 Systems of Equations in Three Variables

Identifying Solutions • Solving Systems in Three Variables • Dependency, Inconsistency, and Geometric Considerations

Some problem situations easily translate to two equations. Others more naturally call for a translation to three or more equations. In this section, we consider how to solve systems of three linear equations. Later, we will use such systems in problem-solving situations.

Identifying Solutions

A **linear equation in three variables** is an equation equivalent to one in the form $Ax + By + Cz = D$, where A, B, C, and D are real numbers. We will refer to the form $Ax + By + Cz = D$ as *standard form* for a linear equation in three variables.

A solution of a system of three equations in three variables is an ordered triple (p, q, r) that makes *all three* equations true.

EXAMPLE 1

Determine whether $\left(\frac{3}{2}, -4, 3\right)$ is a solution of the system

$$\begin{aligned} 4x - 2y - 3z &= 5, \\ -8x - y + z &= -5, \\ 2x + y + 2z &= 5. \end{aligned}$$

Solution We substitute $\left(\frac{3}{2}, -4, 3\right)$ into the three equations, using alphabetical order.

$$\begin{array}{r|l} \multicolumn{2}{c}{4x - 2y - 3z = 5} \\ \hline 4\cdot\frac{3}{2} - 2(-4) - 3\cdot 3 & ?\ 5 \\ 6 + 8 - 9 & \\ 5 & 5 \quad \text{TRUE} \end{array}$$

$$\begin{array}{r|l} \multicolumn{2}{c}{-8x - y + z = -5} \\ \hline -8\cdot\frac{3}{2} - (-4) + 3 & ?\ -5 \\ -12 + 4 + 3 & \\ -5 & -5 \quad \text{TRUE} \end{array}$$

$$\begin{array}{r|l} \multicolumn{2}{c}{2x + y + 2z = 5} \\ \hline 2\cdot\frac{3}{2} + (-4) + 2\cdot 3 & ?\ 5 \\ 3 - 4 + 6 & \\ 5 & 5 \quad \text{TRUE} \end{array}$$

The triple makes all three equations true, so it is a solution. ❑

Solving Systems in Three Variables

Graphical methods for solving linear equations in three variables are unsatisfactory, because a three-dimensional coordinate system is required and the graph of a linear equation in three variables is a plane. The substitution method can be used in any situation, but it is not as useful unless one or more of the equations has only two variables. Fortunately, we can use the elimination method to eliminate a variable and obtain a system of two equations in two variables.

EXAMPLE 2

Solve the following system of equations:

$$\begin{aligned} x + y + z &= 4, & (1)\\ x - 2y - z &= 1, & (2)\\ 2x - y - 2z &= -1. & (3) \end{aligned}$$

Solution We first use *any* two of the three equations to get an equation in two variables. Let's use Equations (1) and (2) and add to eliminate z:

$$\begin{aligned} x + y + z &= 4 & (1)\\ x - 2y - z &= 1 & (2)\\ \hline 2x - y \quad &= 5. & (4) \quad \text{Adding} \end{aligned}$$

Next we use a different pair of equations and eliminate the *same variable* we did above. Let's use Equations (1) and (3) to again eliminate z. Be careful here! A common error is to eliminate a different variable in this step.

$x + y + z = 4$, ⟶ Multiplying Equation (1) by 2 ⟶ $2x + 2y + 2z = 8$

$2x - y - 2z = -1$ $\qquad\qquad\qquad\qquad$ $2x - y - 2z = -1$

$\qquad\qquad\qquad\qquad\qquad\qquad\qquad\qquad$ $4x + y \quad = 7$ (5)

Now we solve the resulting system of Equations (4) and (5). That solution will give us two of the numbers.

$$\begin{aligned} 2x - y &= 5 & (4)\\ 4x + y &= 7 & (5)\\ \hline 6x \quad &= 12 & \text{Adding}\\ x &= 2 \end{aligned}$$

Note that we now have two equations in two variables. Had we eliminated different variables above, this would not be the case.

We can use either Equation (4) or (5) to find y. We choose Equation (5):

$$\begin{aligned} 4x + y &= 7 & (5)\\ 4 \cdot 2 + y &= 7 & \text{Substituting 2 for } x \text{ in Equation (5)}\\ 8 + y &= 7\\ y &= -1. \end{aligned}$$

We have $x = 2$ and $y = -1$. To find the value for z, we use any of the original three equations and substitute to find the third number, z. Let's use Equation (1) and substitute our two numbers in it:

$$\begin{aligned} x + y + z &= 4 & (1)\\ 2 + (-1) + z &= 4 & \text{Substituting 2 for } x \text{ and } -1 \text{ for } y\\ 1 + z &= 4\\ z &= 3. \end{aligned}$$

We have obtained the triple (2, −1, 3). We now check in *all three* equations:

$$\begin{array}{c|l} x + y + z = 4 & \\ \hline 2 + (-1) + 3 \ ? & 4 \\ 4 & 4 \quad \text{TRUE} \end{array} \qquad \begin{array}{c|l} x - 2y - z = 1 & \\ \hline 2 - 2(-1) - 3 \ ? & 1 \\ 1 & 1 \quad \text{TRUE} \end{array}$$

$$\begin{array}{c|l} 2x - y - 2z = -1 & \\ \hline 2 \cdot 2 - (-1) - 2 \cdot 3 \ ? & -1 \\ -1 & -1 \quad \text{TRUE} \end{array}$$

The solution is (2, −1, 3). ☐

To use the elimination method to solve systems of three equations:

1. Write all equations in the standard form $Ax + By + Cz = D$.
2. Clear any decimals or fractions.
3. Choose a variable to eliminate. Then use any two of the three equations to get an equation in two variables.
4. Next use a different pair of equations and get another equation *in the same two variables*. That is, eliminate the same variable that you did in step (3).
5. Solve the resulting system (pair) of equations. That will give two of the numbers.
6. Then use any of the original three equations to find the third number.

EXAMPLE 3

Solve the system

$$\begin{aligned} 4x - 2y - 3z &= 5, && (1) \\ -8x - y + z &= -5, && (2) \\ 2x + y + 2z &= 5. && (3) \end{aligned}$$

Solution

1., 2. The equations are already in standard form and there are no fractions or decimals.

3. We must choose a variable to eliminate. We decide on y because the y-terms are opposites of each other in Equations (2) and (3). We add:

$$\begin{array}{rl} -8x - y + z = -5 & (2) \\ 2x + y + 2z = 5 & (3) \\ \hline -6x \qquad + 3z = 0. & (4) \quad \text{Adding} \end{array}$$

4. We use another pair of equations to get an equation in the same two variables, x and z. That is, we eliminate the same variable, y, that we did in step (3). We use Equations (1) and (3) and eliminate y:

$$\begin{array}{lcrl} 4x - 2y - 3z = 5, & & 4x - 2y - 3z = 5 & \\ 2x + y + 2z = 5 & \longrightarrow \text{Multiplying Equation (3) by 2} \longrightarrow & 4x + 2y + 4z = 10 & \\ \hline & & 8x \qquad + z = 15. & (5) \end{array}$$

5. Now we solve the resulting system of Equations (4) and (5). That will give us two of the numbers.

$$\begin{aligned} -6x + 3z &= 0, \\ 8x + z &= 15 \end{aligned} \longrightarrow \text{Multiplying Equation (5) by } -3 \longrightarrow \begin{aligned} -6x + 3z &= 0 \\ -24x - 3z &= -45 \\ \hline -30x \quad &= -45 \\ x &= \tfrac{-45}{-30} = \tfrac{3}{2} \end{aligned}$$

We use Equation (5) to find z:

$$\begin{aligned} 8x + z &= 15 \\ 8 \cdot \tfrac{3}{2} + z &= 15 \qquad \text{Substituting } \tfrac{3}{2} \text{ for } x \\ 12 + z &= 15 \\ z &= 3. \end{aligned}$$

6. Then we use any of the original equations and substitute to find the third number, y. We choose Equation (3):

$$\begin{aligned} 2x + y + 2z &= 5 \\ 2 \cdot \tfrac{3}{2} + y + 2 \cdot 3 &= 5 \qquad \text{Substituting } \tfrac{3}{2} \text{ for } x \text{ and 3 for } z \\ 3 + y + 6 &= 5 \\ y + 9 &= 5 \\ y &= -4. \end{aligned}$$

The solution is $(\frac{3}{2}, -4, 3)$. The check is in Example 1. ❑

Sometimes, certain variables are missing at the outset.

EXAMPLE 4

Solve the system

$$\begin{aligned} x + y + z &= 180, \qquad (1) \\ x \qquad - z &= -70, \qquad (2) \\ 2y - z &= 0. \qquad (3) \end{aligned}$$

Solution

1., 2. The equations are already in standard form with no fractions or decimals.

3., 4. Observe that there is no y in Equation (2). Thus, at the outset, we already have y eliminated from one equation. We need another equation with y eliminated. Let's use Equations (1) and (3):

$$\begin{aligned} x + y + z &= 180, \\ 2y - z &= 0 \end{aligned} \rightarrow \text{Multiplying Equation (1) by } -2 \rightarrow \begin{aligned} -2x - 2y - 2z &= -360 \\ 2y - z &= 0 \\ \hline -2x \qquad - 3z &= -360. \qquad (4) \end{aligned}$$

5., 6. Now we solve the resulting system of Equations (2) and (4):

$$\begin{aligned} x - z &= -70, \\ -2x - 3z &= -360 \end{aligned} \rightarrow \text{Multiplying Equation (2) by 2} \rightarrow \begin{aligned} 2x - 2z &= -140 \\ -2x - 3z &= -360 \\ \hline -5z &= -500 \\ z &= 100. \end{aligned}$$

Continuing as in previous examples, we get the solution (30, 50, 100). ❑

Dependency, Inconsistency, and Geometric Considerations

Each equation in Examples 2, 3, and 4 has a graph that is a plane in three dimensions. The solution in each case is a point common to all the planes of the system. Since it is possible for three planes to have infinitely many points in common or no points at all in common, we need to extend the notions of dependent equations and inconsistent systems to systems of equations in three variables.

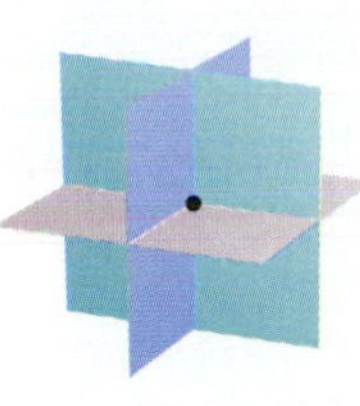

One solution: planes intersecting in exactly one point.

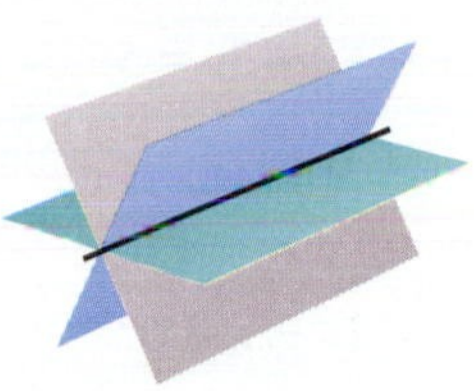

The planes intersect along a common line. There are infinitely many points common to the three planes.

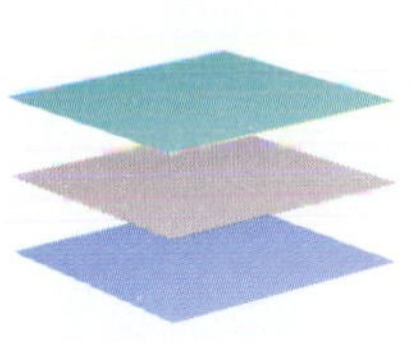

Three parallel planes. There is no common point of intersection.

Planes intersect two at a time, but there is no point common to all three.

A system of equations that has no solution is called **inconsistent.**
A system of equations that has at least one solution is called **consistent.**

EXAMPLE 5

Solve:

$$\begin{aligned} y + 3z &= 4, &\quad (1)\\ -x - y + 2z &= 0, &\quad (2)\\ x + 2y + z &= 1. &\quad (3) \end{aligned}$$

Solution The variable x is missing in Equation (1). Thus we can use Equations (2) and (3) and add to eliminate x. We get

$$\begin{aligned} -x - y + 2z &= 0 &\quad (2)\\ x + 2y + z &= 1 &\quad (3)\\ \hline y + 3z &= 1. &\quad (4) \quad \text{Adding} \end{aligned}$$

Equations (1) and (4) are both in y and z. We multiply Equation (1) by -1 and add:

$$\begin{aligned} y + 3z = 4, &\rightarrow \text{Multiplying Equation (1) by } -1 \rightarrow & -y - 3z &= -4\\ y + 3z = 1 & & y + 3z &= 1\\ & & \hline 0 &= -3. \quad \text{Adding} \end{aligned}$$

Since we end up with a *false* equation, we know that the system has no solution. It is *inconsistent*. The solution set is $\varnothing$. ❑

We now extend our definition of the word dependent from its meaning in Section 8.1.

> If a system of n linear equations is equivalent to a system of fewer than n of them, we say that the equations are *dependent*. If such is not the case, we call the equations *independent*.

EXAMPLE 6

Solve:

$$\begin{aligned} 2x + y + z &= 3, \quad (1)\\ x - 2y - z &= 1, \quad (2)\\ 3x + 4y + 3z &= 5. \quad (3) \end{aligned}$$

Solution We use Equations (1) and (2) and add to eliminate z:

$$\begin{aligned} 2x + y + z &= 3\\ x - 2y - z &= 1\\ \hline 3x - y \quad &= 4. \quad (4) \end{aligned}$$

Next we use Equations (2) and (3) to eliminate z again:

$$\begin{aligned} x - 2y - z &= 1, \rightarrow \text{Multiplying Equation (2) by 3} \rightarrow & 3x - 6y - 3z &= 3\\ 3x + 4y + 3z &= 5 & 3x + 4y + 3z &= 5\\ & & \hline 6x - 2y \quad &= 8. \quad (5) \end{aligned}$$

We now try to solve the resulting system of Equations (4) and (5):

$$\begin{aligned} 3x - y &= 4, \rightarrow \text{Multiplying Equation (4) by } -2 \rightarrow & -6x + 2y &= -8\\ 6x - 2y &= 8 & 6x - 2y &= 8\\ & & \hline 0 &= 0. \quad (6) \end{aligned}$$

Equation (6) indicates that Equations (1), (2), and (3) are *dependent*. To see that the original system of three equations is equivalent to a system of two equations, note that two times Equation (1), minus Equation (2), is Equation (3). Thus, removing Equation (3) from the system does not affect the solution of the system. In writing an answer to this problem, we simply state that the equations are dependent. □

The observant student might have noticed that when dependent equations appeared in Section 8.1, the solution sets were always infinite in size and were written in set-builder notation. That is, in Section 8.1, all systems of dependent equations were *consistent*. This is not always the case for systems of three equations. The following figure illustrates some possibilities geometrically.

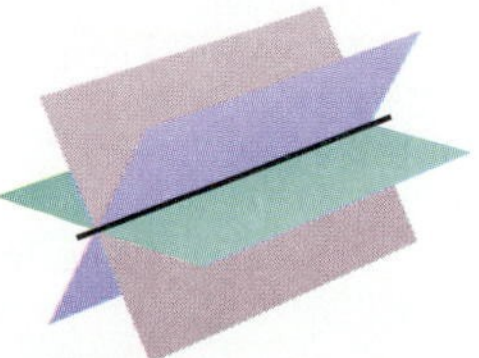

The planes intersect along a common line. The equations are dependent and the system is consistent. There is an infinite number of solutions.

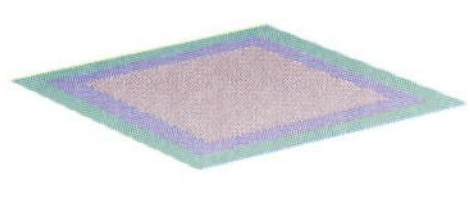

The planes coincide. The equations are dependent and the system is consistent. There is an infinite number of solutions.

Two planes coincide. The third plane is parallel. The equations are dependent and the system is inconsistent. There is no solution.

EXERCISE SET 8.4

1. Determine whether $(1, -2, 3)$ is a solution of the system
$$x + y + z = 2,\quad x - 2y - z = 2,\quad 3x + 2y + z = 2.$$

2. Determine whether $(2, -1, -2)$ is a solution of the system
$$x + y - 2z = 5,\quad 2x - y - z = 7,\quad -x - 2y + 3z = 6.$$

Solve.

3. $x + y + z = 6,$ $2x - y + 3z = 9,$ $-x + 2y + 2z = 9$

4. $2x - y + z = 10,$ $4x + 2y - 3z = 10,$ $x - 3y + 2z = 8$

5. $2x - y - 3z = -1,$ $2x - y + z = -9,$ $x + 2y - 4z = 17$

6. $x - y + z = 6,$ $2x + 3y + 2z = 2,$ $3x + 5y + 4z = 4$

7. $2x - 3y + z = 5,$ $x + 3y + 8z = 22,$ $3x - y + 2z = 12$

8. $6x - 4y + 5z = 31,$ $5x + 2y + 2z = 13,$ $x + y + z = 2$

9. $3a - 2b + 7c = 13,$ $a + 8b - 6c = -47,$ $7a - 9b - 9c = -3$

10. $x + y + z = 0,$ $2x + 3y + 2z = -3,$ $-x + 2y - 3z = -1$

11. $2x + 3y + z = 17,$ $x - 3y + 2z = -8,$ $5x - 2y + 3z = 5$

12. $2x + y - 3z = -4,$ $4x - 2y + z = 9,$ $3x + 5y - 2z = 5$

13. $2x + y + z = -2,$ $2x - y + 3z = 6,$ $3x - 5y + 4z = 7$

14. $2x + y + 2z = 11,$ $3x + 2y + 2z = 8,$ $x + 4y + 3z = 0$

15. $x - y + z = 4,$ $5x + 2y - 3z = 2,$ $4x + 3y - 4z = -2$

16. $-2x + 8y + 2z = 4,$ $x + 6y + 3z = 4,$ $3x - 2y + z = 0$

17. $4x - y - z = 4,$ $2x + y + z = -1,$ $6x - 3y - 2z = 3$

18. $a + 2b + c = 1,$ $7a + 3b - c = -2,$ $a + 5b + 3c = 2$

19. $2r + 3s + 12t = 4,$ $4r - 6s + 6t = 1,$ $r + s + t = 1$

20. $10x + 6y + z = 7,$ $5x - 9y - 2z = 3,$ $15x - 12y + 2z = -5$

21. $4a + 9b = 8,$ $8a + 6c = -1,$ $6b + 6c = -1$

22. $3p + 2r = 11,$ $q - 7r = 4,$ $p - 6q = 1$

23. $x + y + z = 57,$ $-2x + y = 3,$ $x - z = 6$

24. $x + y + z = 105,$ $10y - z = 11,$ $2x - 3y = 7$

25. $2a - 3b = 2,$ $7a + 4c = \frac{3}{4},$ $2c - 3b = 1$

26. $a - 3c = 6,$ $b + 2c = 2,$ $7a - 3b - 5c = 14$

27. $x + y + z = 180,$ $y = 2 + 3x,$ $z = 80 + x$

28. $l + m = 7,$ $3m + 2n = 9,$ $4l + n = 5$

29. $x + z = 0,$ $x + y + 2z = 3,$ $y + z = 2$

30. $x + y = 0,$ $x + z = 1,$ $2x + y + z = 2$

31. $x + y + z = 1,$ $-x + 2y + z = 2,$ $2x - y = -1$

32. $y + z = 1,$ $x + y + z = 1,$ $x + 2y + 2z = 2$

Skill Maintenance

Let $g(x) = 2x - 7$. Find the following.

33. $g(-1)$

34. $g(a + h)$

Synthesis

35. ◈ Suggest a procedure that could be used to solve a system of four equations in four variables.

36. ◈ Explain, in your own words, what it means for the equations of a system to be dependent.

Solve.

37. $\dfrac{x+2}{3} - \dfrac{y+4}{2} + \dfrac{z+1}{6} = 0,$ $\dfrac{x-4}{3} + \dfrac{y+1}{4} - \dfrac{z-2}{2} = -1,$ $\dfrac{x+1}{2} + \dfrac{y}{2} + \dfrac{z-1}{4} = \dfrac{3}{4}$

38. $0.2x + 0.3y + 1.1z = 1.6,$ $0.5x - 0.2y + 0.4z = 0.7,$ $-1.2x + y - 0.7z = -0.9$

39. $w + x + y + z = 2,$ $w + 2x + 2y + 4z = 1,$ $w - x + y + z = 6,$ $w - 3x - y + z = 2$

40. $w + x - y + z = 0,$ $w - 2x - 2y - z = -5,$ $w - 3x - y + z = 4,$ $2w - x - y + 3z = 7$

For Exercises 41 and 42, let u represent $1/x$, v represent $1/y$, and w represent $1/z$. Solve for u, v, and w, and then solve for x, y, and z.

41. $\frac{2}{x} - \frac{1}{y} - \frac{3}{z} = -1,$
$\frac{2}{x} - \frac{1}{y} + \frac{1}{z} = -9,$
$\frac{1}{x} + \frac{2}{y} - \frac{4}{z} = 17$

42. $\frac{2}{x} + \frac{2}{y} - \frac{3}{z} = 3,$
$\frac{1}{x} - \frac{2}{y} - \frac{3}{z} = 9,$
$\frac{7}{x} - \frac{2}{y} + \frac{9}{z} = -39$

Determine k so that the system is dependent.

43. $x - 3y + 2z = 1,$
$2x + y - z = 3,$
$9x - 6y + 3z = k$

44. $5x - 6y + kz = -5,$
$x + 3y - 2z = 2,$
$2x - y + 4z = -1$

In each case, three solutions of an equation are given. Find the equation.

45. $Ax + By + Cz = 12;$
$(1, \frac{3}{4}, 3)$, $(\frac{4}{3}, 1, 2)$, and $(2, 1, 1)$

46. $z = b - mx - ny;$
$(1, 1, 2)$, $(3, 2, -6)$, and $(\frac{3}{2}, 1, 1)$

8.5 Problem Solving Using Systems of Three Equations

Applications of Three Equations in Three Unknowns

Solving systems of three or more equations is important in many applications. Systems of equations occur very often in such fields as social science, business, natural science, and engineering.

EXAMPLE 1

The sum of three numbers is 4. The first number minus twice the second, minus the third is 1. Twice the first number minus the second, minus twice the third is -1. Find the numbers.

Solution

1. **FAMILIARIZE.** In this case, there are three statements in the problem. The translation looks as though it can be made directly from these statements once we decide what letters to assign to the unknown numbers.
2. **TRANSLATE.** Let's call the three numbers x, y, and z. Then we can translate directly, from the words of the problem, as follows.

The sum of the three numbers is 4.

$$x + y + z = 4$$

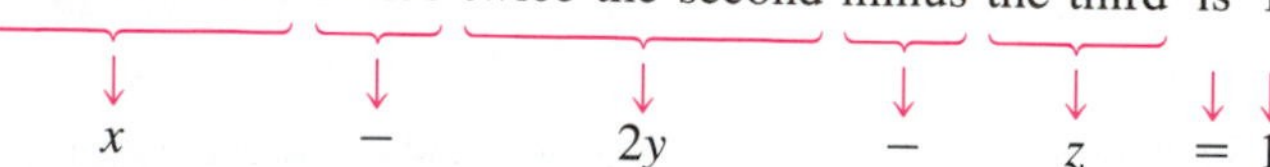

Twice the first number minus the second minus twice the third is -1.

$$2x - y - 2z = -1$$

We now have a system of three equations:

$$\begin{aligned} x + y + z &= 4, \\ x - 2y - z &= 1, \\ 2x - y - 2z &= -1. \end{aligned}$$

3. **CARRY OUT.** We need to solve the system of equations. Note that we have already solved the system as Example 2 in Section 8.4. We obtained $(2, -1, 3)$ for the solution.

4. **CHECK.** We go to the original problem. The first statement says that the sum of the three numbers is 4. That checks. The second statement says that the first number minus twice the second, minus the third is 1. We calculate: $2 - 2(-1) - 3 = 1$. That checks. We leave the check of the third statement to the student.

5. **STATE.** The three numbers are 2, -1, and 3. ❑

EXAMPLE 2

In a triangle, the largest angle is 70° greater than the smallest angle. The largest angle is also twice as large as the remaining angle. Find the measure of each angle.

Solution

1. **FAMILIARIZE.** The first thing we do is make a drawing, or a sketch.

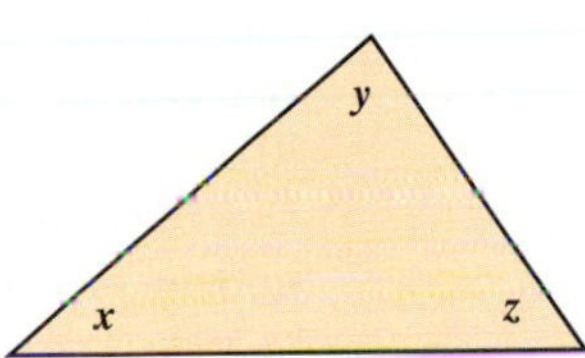

Since we don't know the size of any angle, we use x, y, and z for the measures of the angles. Recall that the measures of the angles of a triangle add up to 180°.

2. **TRANSLATE.** This geometric fact about triangles gives us one equation:

$$x + y + z = 180.$$

There are two statements in the problem that we can translate almost directly.

The largest angle is 70° greater than the smallest angle.

The largest angle is twice as large as the remaining angle.

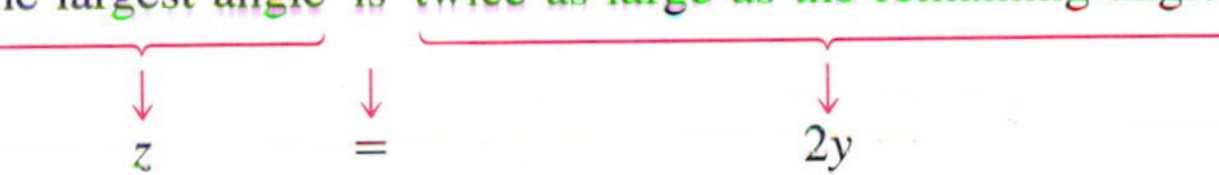

We now have a system of three equations:

$$\begin{aligned} x + y + z &= 180, \\ x + 70 &= z, \\ 2y &= z; \end{aligned} \qquad \text{or} \qquad \begin{aligned} x + y + z &= 180, \\ x \qquad - z &= -70, \\ 2y - z &= 0. \end{aligned}$$

Converting to standard form

3. **CARRY OUT.** The system was solved in Example 4 of Section 8.4. The solution is (30, 50, 100).

4. **CHECK.** The sum of the numbers is 180, so that checks.

   The largest angle measures 100° and the smallest measures 30°. The largest angle is 70° greater than the smallest.

   The remaining angle measures 50°. The largest angle measures 100°, so it is twice as large. We do have an answer to the problem.

5. **STATE.** The measures of the angles of the triangle are 30°, 50°, and 100°. ❑

EXAMPLE 3

Americans are becoming very conscious of their cholesterol levels. Recent studies indicate that a child should ingest no more than 300 mg of cholesterol per day. By eating 1 egg, 1 cupcake, and 1 slice of pizza, a child would ingest 302 mg of cholesterol. If the child eats 2 cupcakes and 3 slices of pizza, he or she ingests 65 mg of cholesterol. By eating 2 eggs and 1 cupcake, the child consumes 567 mg of cholesterol. How much cholesterol is in each item?

Solution

1. **FAMILIARIZE.** After we have read the problem a few times, it becomes clear that an egg contains considerably more cholesterol than the other foods. Let's guess that one egg contains 200 mg of cholesterol and one cupcake contains 50 mg. Because of the third sentence in the problem, it would then follow that a slice of pizza contains 52 mg of cholesterol since $200 + 50 + 52 = 302$.

   To see if our guess satisfies the other statements in the problem, we find the amount of cholesterol that 2 cupcakes and 3 slices of pizza would contain: $2 \cdot 50 + 3 \cdot 52 = 256$. Since this does not match the 65 mg listed in the fourth sentence of the problem, we know that our guess is incorrect. Rather than guess again, we examine how we checked our guess. We decide to let e, c, and s = the number of milligrams of cholesterol in an egg, a cupcake, and a slice of pizza, respectively.

2. **TRANSLATE.** By rewording some of the sentences in the problem, we can translate it into three equations.

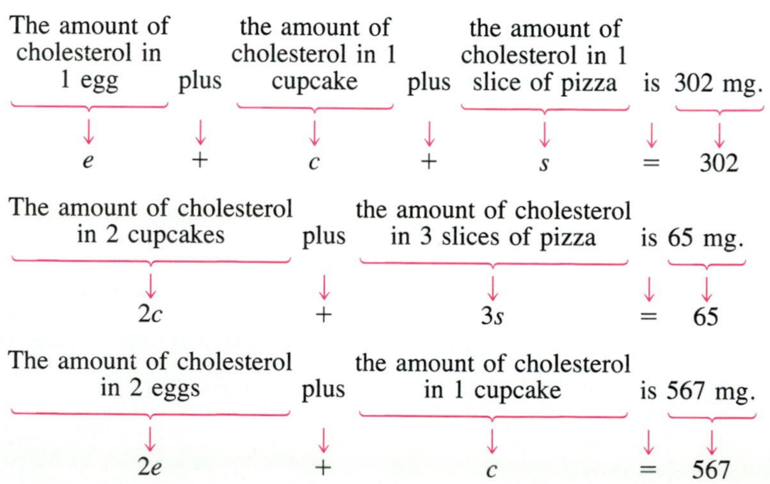

We now have a system of three equations:

$$\begin{aligned} e + c + s &= 302, \\ 2c + 3s &= 65, \\ 2e + c &= 567. \end{aligned}$$

3. **CARRY OUT.** We solve and get $e = 274$, $c = 19$, $s = 9$, or $(274, 19, 9)$.

4. **CHECK.** The sum of 274, 19, and 9 is 302 so the total cholesterol in 1 egg, 1 cupcake, and 1 slice of pizza checks. Two cupcakes and three slices of pizza would contain $2 \cdot 19 + 3 \cdot 9 = 65$ mg, so that checks. Finally, two eggs and one cupcake would contain $2 \cdot 274 + 19 = 567$ mg of cholesterol. The answer checks.

5. **STATE.** An egg contains 274 mg of cholesterol, a cupcake contains 19 mg of cholesterol, and a slice of pizza contains 9 mg of cholesterol. ❑

EXERCISE SET 8.5

Solve.

1. The sum of three numbers is 105. The third is 11 less than 10 times the second. Twice the first is 7 more than 3 times the second. Find the numbers.

2. The sum of three numbers is 57. The second is 3 more than the first. The third is 6 more than the first. Find the numbers.

3. The sum of three numbers is 5. The first number minus the second plus the third is 1. The first minus the third is 3 more than the second. Find the numbers.

4. The sum of three numbers is 26. Twice the first minus the second is 2 less than the third. The third is the second minus three times the first. Find the numbers.

5. In triangle *ABC*, the measure of angle *B* is 2° more than three times the measure of angle *A*. The measure of angle *C* is 8° more than the measure of angle *A*. Find the angle measures.

6. In triangle *ABC*, the measure of angle *B* is three times the measure of angle *A*. The measure of angle *C* is 30° greater than the measure of angle *A*. Find the angle measures.

7. In triangle *ABC*, the measure of angle *B* is twice the measure of angle *A*. The measure of angle *C* is 80° more than that of angle *A*. Find the angle measures.

8. In triangle *ABC*, the measure of angle *B* is three times that of angle *A*. The measure of angle *C* is 20° more than that of angle *A*. Find the angle measures.

9. In a recent year, companies spent a total of $84.8 billion on newspaper, television, and radio ads. The total amount spent on television and radio ads was only $2.6 billion more than the amount spent on newspaper ads alone. The amount spent on newspaper ads was $5.1 billion more than what was spent on television ads. How much was spent on each form of advertising? (*Hint:* Let the variables represent numbers of billions of dollars.)

10. A recent basic model of a particular automobile had a cost of $12,685. The basic model with the added features of automatic transmission and power door locks was $14,070. The basic model with air conditioning (AC) and power door locks was $13,580. The basic model with AC and automatic transmission was $13,925. What was the individual cost of each of the three options?

11. A dietician in a hospital prepares meals under the guidance of a physician. Suppose that for a particular patient a physician prescribes a meal to have 800 Calories, 55 g of protein, and 220 mg of vitamin C. The dietician prepares the meal using steak (each 3-oz serving contains 300 Cal, 20 g of protein, and no vitamin C), baked potatoes (one baked

potato contains 100 Cal, 5 g of protein, and 20 mg of vitamin C), and broccoli (one 156-g serving contains 50 Cal, 5 g of protein, and 100 mg of vitamin C). How many servings of each food are needed in order to satisfy the physician's requirements? (*Hint:* Let s = the number of servings of steak, p = the number of baked potatoes, and b = the number of servings of broccoli. Find an equation for the total number of calories, the total amount of protein, and the total amount of vitamin C.)

12. Repeat Exercise 11 but replace the broccoli with asparagus, for which one 180-g serving contains 50 calories, 5 g of protein, and 44 mg of vitamin C. Which meal would you prefer eating?

13. In the United States, the highest incidence of fraternal twin births occurs among Orientals, then African-Americans, and then Caucasians. Out of every 15,400 births, the total number of fraternal twin births for all three is 739, where there are 185 more for Orientals than African-Americans and 231 more for Orientals than Caucasians. How many births of fraternal twins are there for each race out of every 15,400 births?

14. The sum of the average number of times a man, a woman, and a one-year-old child cry each month is 71.7. A one-year-old cries 46.4 more times than a man. The average number of times a one-year-old cries per month is 28.3 more than the average number of times combined that a man and a woman cry. What is the average number of times per month that each cries?

15. In a factory there are three polishing machines, A, B, and C. When all three of them are working, 5700 lenses can be polished in one week. When only A and B are working, 3400 lenses can be polished in one week. When only B and C are working, 4200 lenses can be polished in one week. How many lenses can be polished in a week by each machine?

16. Sawmills A, B, and C can produce 7400 board-feet of lumber per day. Mills A and B together can produce 4700 board-feet per day, while mills B and C together can produce 5200 board-feet per day. How many board-feet can each mill produce by itself?

17. When three pumps, A, B, and C, are running together, they can pump 3700 gal/hr. When only A and B are running, 2200 gal/hr can be pumped. When only A and C are running, 2400 gal/hr can be pumped. What is the pumping capacity of each pump?

18. Pat, Chris, and Jean can weld 37 linear feet per hour when working together. Pat and Chris together can weld 22 linear feet per hour, while Pat and Jean can weld 25 linear feet per hour. How many linear feet per hour can each weld alone?

19. One year an investment of $80,000 was made by a business club. The investment was split into three parts and lasted for one year. The first part of the investment earned 8% interest, the second 6%, and the third 9%. Total interest from the investments was $6300. The interest from the first investment was four times the interest from the second. Find the amounts of the three parts of the investment.

20. Find the year in which the first U.S. transcontinental railroad was completed. The following are some facts about the number. The sum of the digits in the year is 24. The one's digit is one more than the hundred's digit. Both the ten's and the one's digits are multiples of three.

Skill Maintenance

Add and simplify.

21. $\dfrac{2}{t+1}+\dfrac{t}{t+1}$

22. $\dfrac{t-2}{t-1}+\dfrac{1}{t-1}$

23. $\dfrac{2}{t}+\dfrac{t}{t+1}$

24. $\dfrac{1}{t+1}+\dfrac{2}{t-1}$

Synthesis

25. Tammy's age is the sum of the ages of Carmen and Dennis. Carmen's age is two more than the sum of the ages of Dennis and Mark. Dennis's age is four times Mark's age. The sum of all four ages is 42. How old is Tammy?

26. Find a three-digit positive integer such that the sum of all three digits is 14, the ten's digit is two more than the one's digit, and if the digits are reversed, the number is unchanged.

27. Find the sum of the angle measures at the tips of the star in this figure.

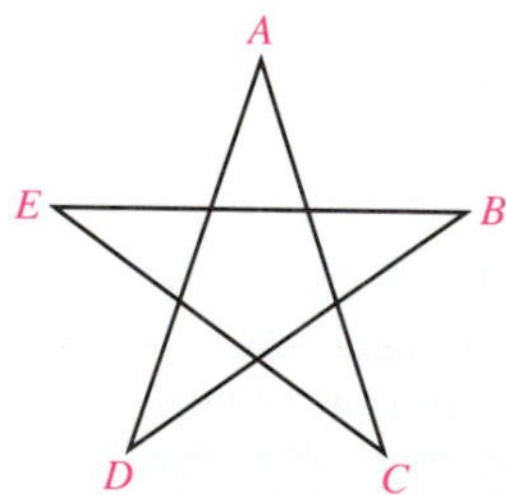

28. A theater audience of 100 people consists of adults, students, and children. The ticket prices are \$10 for adults, \$3 for students, and 50¢ for children. The total amount of money taken in is \$100. How many adults, students, and children are in attendance? Does there seem to be some information missing? Do some more careful reasoning.

29. Hal gives Tom as many raffle tickets as Tom has and Gary as many as Gary has. In like manner, Tom then gives Hal and Gary as many tickets as each then has. Similarly, Gary gives Hal and Tom as many tickets as each then has. If each finally has 40 tickets, with how many tickets does Tom begin?

8.6 Elimination Using Matrices

Definition of a Matrix • Row-Equivalent Operations

In solving systems of equations, we perform computations with the constants. The variables play no important role until the end. Thus we can simplify writing a system by omitting the variables. For example, the system

$$\begin{aligned} 3x + 4y &= 5, \\ x - 2y &= 1 \end{aligned} \quad \text{simplifies to} \quad \begin{matrix} 3 & 4 & 5 \\ 1 & -2 & 1 \end{matrix}$$

if we leave off the variables, omit the operation of addition, and omit the equals signs.

In this example, we have written a rectangular array of numbers. Such an array is called a **matrix** (plural, **matrices**). We ordinarily write brackets around matrices. The following are matrices:

$$\begin{bmatrix} 4 & 1 & 3 & 5 \\ 1 & 0 & 1 & 2 \\ 6 & 3 & -2 & 0 \end{bmatrix}, \quad \begin{bmatrix} 6 & 2 & 1 & 4 & 7 \\ 1 & 2 & 1 & 3 & 1 \\ 4 & 0 & -2 & 0 & -3 \end{bmatrix}, \quad \begin{bmatrix} 1 & 2 \\ 145 & 0 \\ -7 & 9 \\ 8 & 1 \\ 0 & 0 \end{bmatrix}.$$

The **rows** of a matrix are horizontal, and the **columns** are vertical.

$$\begin{bmatrix} 5 & -2 & 2 \\ 1 & 0 & 1 \\ 0 & 1 & 2 \end{bmatrix} \begin{matrix} \leftarrow \text{row 1} \\ \leftarrow \text{row 2} \\ \leftarrow \text{row 3} \end{matrix}$$

column 1 column 2 column 3

Let us now use matrices to solve systems of linear equations.

EXAMPLE 1

Solve the system

$$\begin{aligned} 5x - 4y &= -1, \\ -2x + 3y &= 2. \end{aligned}$$

Solution We write a matrix using only the constants, keeping in mind that x corresponds to the first column and y to the second. A dashed line separates the coefficients from the constants at the end of each equation:

$$\left[\begin{array}{rr|r} 5 & -4 & -1 \\ -2 & 3 & 2 \end{array}\right].$$

The individual numbers are called *elements* or *entries*.

Our goal is to transform this matrix into one of the form

$$\left[\begin{array}{cc|c} a & b & c \\ 0 & d & e \end{array}\right].$$

The variables can then be reinserted to form equations from which we can complete the solution.

We do calculations that are similar to those that we would do if we wrote the entire equations. The first step, if possible, is to multiply and/or interchange the rows so that each number in the first column below the first number is a multiple of that number. In this case, we do so by multiplying Row 2 by 5. This corresponds to multiplying the second equation by 5.

$$\left[\begin{array}{rr|r} 5 & -4 & -1 \\ -10 & 15 & 10 \end{array}\right] \qquad \text{New Row 2} = 5(\text{Row 2})$$

Next, we multiply the first row by 2 and add the result to the second row. This corresponds to multiplying the first equation by 2 and adding the result to the second equation. You should write out these computations if necessary—we perform them mentally.

$$\left[\begin{array}{rr|r} 5 & -4 & -1 \\ 0 & 7 & 8 \end{array}\right] \qquad \text{New Row 2} = 2(\text{Row 1}) + (\text{Row 2})$$

If we now reinsert the variables, we have

$$5x - 4y = -1, \qquad (1)$$
$$7y = 8. \qquad (2)$$

We can now proceed as before, solving Equation (2) for y:

$$7y = 8 \qquad (2)$$
$$y = \tfrac{8}{7}.$$

Next we substitute $\frac{8}{7}$ for y back in Equation (1). This procedure is called *back-substitution*.

$$5x - 4y = -1 \qquad (1)$$
$$5x - 4 \cdot \tfrac{8}{7} = -1 \qquad \text{Substituting } \tfrac{8}{7} \text{ for } y \text{ in Equation (1)}$$
$$x = \tfrac{5}{7} \qquad \text{Solving for } x$$

The solution is $\left(\frac{5}{7}, \frac{8}{7}\right)$. ❑

EXAMPLE 2

Solve the system

$$2x - y + 4z = -3,$$
$$x \qquad - 4z = 5,$$
$$6x - y + 2z = 10.$$

Solution We first write a matrix, using only the constants. Where there are missing terms, we must write 0's:

$$\begin{bmatrix} 2 & -1 & 4 & | & -3 \\ 1 & 0 & -4 & | & 5 \\ 6 & -1 & 2 & | & 10 \end{bmatrix} \begin{matrix} \text{(P1)} \\ \text{(P2)} \\ \text{(P3)} \end{matrix}$$

(P1), (P2), and (P3) designate the equations that are in the first, second, and third position, respectively.

Our goal is to find an equivalent matrix of the form

$$\begin{bmatrix} a & b & c & | & d \\ 0 & e & f & | & g \\ 0 & 0 & h & | & i \end{bmatrix}.$$

A matrix of this form can be rewritten as a system of equations from which a solution can be found easily.

The first step, if possible, is to interchange the rows so that each number in the first column below the first number is a multiple of that number. In this case, we do so by interchanging Rows 1 and 2:

$$\begin{bmatrix} 1 & 0 & -4 & | & 5 \\ 2 & -1 & 4 & | & -3 \\ 6 & -1 & 2 & | & 10 \end{bmatrix}$$

This corresponds to interchanging the first two equations.

Next, we multiply the first row by -2 and add it to the second row:

$$\begin{bmatrix} 1 & 0 & -4 & | & 5 \\ 0 & -1 & 12 & | & -13 \\ 6 & -1 & 2 & | & 10 \end{bmatrix}.$$

This corresponds to multiplying new equation (P1) by -2 and adding it to new equation (P2). We perform the calculations mentally.

Now we multiply the first row by -6 and add it to the third row:

$$\begin{bmatrix} 1 & 0 & -4 & | & 5 \\ 0 & -1 & 12 & | & -13 \\ 0 & -1 & 26 & | & -20 \end{bmatrix}.$$

This corresponds to multiplying equation (P1) by -6 and adding it to equation (P3).

Next, we multiply Row 2 by -1 and add it to the third row:

$$\begin{bmatrix} 1 & 0 & -4 & | & 5 \\ 0 & -1 & 12 & | & -13 \\ 0 & 0 & 14 & | & -7 \end{bmatrix}.$$

This corresponds to multiplying equation (P2) by -1 and adding it to equation (P3).

Reinserting the variables gives us

$$\begin{aligned} x \qquad - 4z &= 5, && \text{(P1)} \\ -y + 12z &= -13, && \text{(P2)} \\ 14z &= -7. && \text{(P3)} \end{aligned}$$

We now solve (P3) for z and get $z = -\frac{1}{2}$. Next we back-substitute $-\frac{1}{2}$ for z in (P2) and solve for y: $-y + 12\left(-\frac{1}{2}\right) = -13$, so $y = 7$. Since there is no y-term in (P1), we need only substitute $-\frac{1}{2}$ for z in (P1) and solve for x: $x - 4\left(-\frac{1}{2}\right) = 5$, so $x = 3$. The solution is $\left(3, 7, -\frac{1}{2}\right)$. ❑

All the operations used in the preceding example correspond to operations with the equations and produce equivalent systems of equations. We call the matrices **row-equivalent** and the operations that produce them **row-equivalent operations.**

Row-Equivalent Operations

Each of the following row-equivalent operations produces an equivalent matrix:

a) Interchanging any two rows.
b) Multiplying each element of a row by the same nonzero number.
c) Multiplying each element of a row by a nonzero number and adding the result to another row.

The best overall method for solving systems of equations is by row-equivalent matrices; even computers are programmed to use them. Matrices are part of a branch of mathematics known as linear algebra. They are also studied in more detail in many courses in finite mathematics.

EXERCISE SET 8.6

Solve using matrices.

1. $4x + 2y = 11,$
$3x - y = 2$

2. $3x - 3y = 11,$
$9x - 2y = 5$

3. $x + 4y = 8,$
$3x + 5y = 3$

4. $x + 4y = 5,$
$-3x + 2y = 13$

5. $5x - 3y = -2,$
$4x + 2y = 5$

6. $3x + 4y = 7,$
$-5x + 2y = 10$

7. $4x - y - 3z = 1,$
$8x + y - z = 5,$
$2x + y + 2z = 5$

8. $3x + 2y + 2z = 3,$
$x + 2y - z = 5,$
$2x - 4y + z = 0$

9. $p + q + r = 1,$
$p - 2q - 3r = 3,$
$4p + 5q + 6r = 4$

10. $x + 2y - 3z = 9,$
$2x - y + 2z = -8,$
$3x - y - 4z = 3$

11. $3p + 2r = 11,$
$q - 7r = 4,$
$p - 6q = 1$

12. $4a + 9b = 8,$
$8a + 6c = -1,$
$6b + 6c = -1$

13. $2x + 2y - 2z - 2w = -10,$
$w + y + z + x = -5,$
$x - y + 4z + 3w = -2,$
$w - 2y + 2z + 3x = -6$

14. $-w - 3y + z + 2x = -8,$
$x + y - z - w = -4,$
$w + y + z + x = 22,$
$x - y - z - w = -14$

Problem Solving

15. A collection of 34 coins consists of dimes and nickels. The total value is $1.90. How many dimes and how many nickels are there?

16. A collection of 43 coins consists of dimes and quarters. The total value is $7.60. How many dimes and how many quarters are there?

17. A grocer has two kinds of granola. One is worth $4.05 per pound and the other is worth $2.70 per pound. The grocer wants to blend the two granolas to get a 15-lb mixture worth $3.15 per pound. How much of each kind of granola should be used?

18. A grocer mixes candy worth $0.80 per pound with nuts worth $0.70 per pound to get a 20-lb mixture worth $0.77 per pound. How many pounds of candy and how many pounds of nuts should be used?

19. Elena receives $212 per year in simple interest from three investments totaling $2500. Part is invested at 7%, part at 8%, and part at 9%. There is $1100 more invested at 9% than at 8%. Find the amount invested at each rate.

20. Miguel receives $306 per year in simple interest from three investments totaling $3200. Part is invested at 8%, part at 9%, and part at 10%. There is $1900 more invested at 10% than at 9%. Find the amount invested at each rate.

Skill Maintenance

Factor.

21. $x^2 + x - 6$

22. $4x^2 - 4x - 3$

23. $x^2 + \frac{3}{4}x + \frac{1}{8}$

24. $12x^2 + 20x + 3$

Synthesis

25. ◈ Explain how you can recognize a dependent system when solving with matrices.

26. ◈ Explain how you can recognize an inconsistent system when solving with matrices.

27. The sum of the digits in a four-digit number is 10. Twice the sum of the thousand's digit and the ten's digit is one less than the sum of the other two digits. The ten's digit is twice the thousand's digit. The one's digit equals the sum of the thousand's digit and the hundred's digit. Find the four-digit number.

28. Solve for x and y:

$$ax + by = c,$$
$$dx + ey = f.$$

8.7 Determinants and Cramer's Rule

Determinants of 2 × 2 Matrices • Cramer's Rule: 2 × 2 Systems • Cramer's Rule: 3 × 3 Systems

Determinants of 2 × 2 Matrices

When a matrix has m rows and n columns, it is called an "m by n" matrix. Thus its *dimensions* are denoted by $m \times n$. If a matrix has the same number of rows and columns, it is called a **square matrix.** With every square matrix is associated a number called its **determinant,** defined as follows for 2 × 2 matrices.

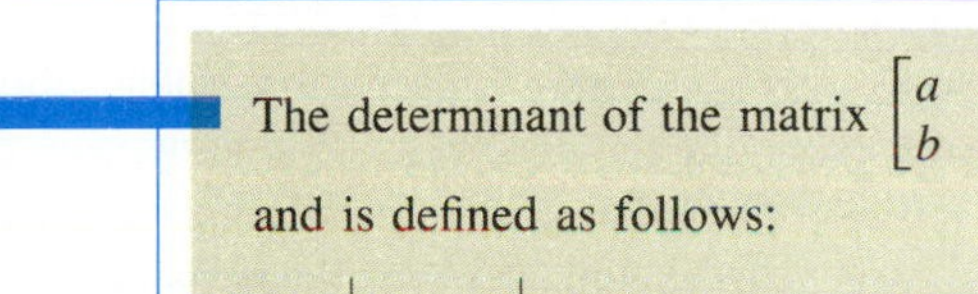

The determinant of the matrix $\begin{bmatrix} a & c \\ b & d \end{bmatrix}$ is denoted $\begin{vmatrix} a & c \\ b & d \end{vmatrix}$ and is defined as follows:

$$\begin{vmatrix} a & c \\ b & d \end{vmatrix} = ad - bc.$$

EXAMPLE 1

Evaluate: $\begin{vmatrix} 2 & -5 \\ 6 & 7 \end{vmatrix}$.

Solution We multiply and subtract as follows:

$$\begin{vmatrix} 2 & -5 \\ 6 & 7 \end{vmatrix} = 2 \cdot 7 - 6 \cdot (-5) = 14 + 30 = 44.$$ ❑

Cramer's Rule: 2 × 2 Systems

One of the many uses for determinants is in solving systems of linear equations in which the number of variables is the same as the number of equations and the constants are not all 0. Let us consider a system of two equations:

$$a_1x + b_1y = c_1,$$
$$a_2x + b_2y = c_2.$$

Using the elimination method, we can solve to obtain

$$x = \frac{c_1b_2 - c_2b_1}{a_1b_2 - a_2b_1}, \qquad y = \frac{a_1c_2 - a_2c_1}{a_1b_2 - a_2b_1}.$$

The numerators and the denominators of the expressions for x and y can be written as determinants.

Cramer's Rule: 2 × 2 Systems

The solution of the system

$$a_1x + b_1y = c_1,$$
$$a_2x + b_2y = c_2,$$

if it is unique, is given by

$$x = \frac{\begin{vmatrix} c_1 & b_1 \\ c_2 & b_2 \end{vmatrix}}{\begin{vmatrix} a_1 & b_1 \\ a_2 & b_2 \end{vmatrix}}, \qquad y = \frac{\begin{vmatrix} a_1 & c_1 \\ a_2 & c_2 \end{vmatrix}}{\begin{vmatrix} a_1 & b_1 \\ a_2 & b_2 \end{vmatrix}}.$$

The equations above make sense only if the determinant in the denominator is not 0. If the denominator *is* 0, then one of two things happens.

1. If the denominator is 0 and the other two determinants in the numerators are also 0, then the equations in the system are dependent.
2. If the denominator is 0 and at least one of the other determinants in the numerators is not 0, then the system is inconsistent.

To use Cramer's rule, we find the three determinants and compute x and y as shown above. Note that the denominator in both cases contains a_1, a_2, b_1, and b_2 in the same position as in the original equations. For x, the numerator is obtained by replacing a_1 and a_2 by c_1 and c_2. For y, the numerator is obtained by replacing b_1 and b_2 by c_1 and c_2.

EXAMPLE 2

Solve using Cramer's rule:

$$2x + 5y = 7,$$
$$5x - 2y = -3.$$

Solution We have

$$x = \frac{\begin{vmatrix} 7 & 5 \\ -3 & -2 \end{vmatrix}}{\begin{vmatrix} 2 & 5 \\ 5 & -2 \end{vmatrix}} \qquad \text{Using Cramer's rule}$$

$$= \frac{7(-2) - (-3)5}{2(-2) - 5 \cdot 5} = -\frac{1}{29},$$

and

$$y = \frac{\begin{vmatrix} 2 & 7 \\ 5 & -3 \end{vmatrix}}{\begin{vmatrix} 2 & 5 \\ 5 & -2 \end{vmatrix}} \qquad \text{Using Cramer's rule}$$

$$= \frac{2(-3) - 5 \cdot 7}{-29} = \frac{41}{29}.$$

The solution is $\left(-\frac{1}{29}, \frac{41}{29}\right)$. □

Cramer's Rule: 3 × 3 Systems

A similar method has been developed for solving systems of three equations in three unknowns. Before stating the rule, though, we must develop some terminology.

The *determinant* of a three-by-three matrix is defined as follows:

$$\begin{vmatrix} a_1 & b_1 & c_1 \\ a_2 & b_2 & c_2 \\ a_3 & b_3 & c_3 \end{vmatrix} = a_1 \begin{vmatrix} b_2 & c_2 \\ b_3 & c_3 \end{vmatrix} - a_2 \begin{vmatrix} b_1 & c_1 \\ b_3 & c_3 \end{vmatrix} + a_3 \begin{vmatrix} b_1 & c_1 \\ b_2 & c_2 \end{vmatrix}$$

Subtract. Add.

Note that the a's come from the first column.

Note too that the second-order determinants above can be obtained by crossing out the row and column in which the a occurs.

For a_1: $\begin{vmatrix} a_1 & b_1 & c_1 \\ a_2 & b_2 & c_2 \\ a_3 & b_3 & c_3 \end{vmatrix}$ For a_2: $\begin{vmatrix} a_1 & b_1 & c_1 \\ a_2 & b_2 & c_2 \\ a_3 & b_3 & c_3 \end{vmatrix}$ For a_3: $\begin{vmatrix} a_1 & b_1 & c_1 \\ a_2 & b_2 & c_2 \\ a_3 & b_3 & c_3 \end{vmatrix}$

EXAMPLE 3

Evaluate:

$$\begin{vmatrix} -1 & 0 & 1 \\ -5 & 1 & -1 \\ 4 & 8 & 1 \end{vmatrix}.$$

Solution We have

$$\begin{vmatrix} -1 & 0 & 1 \\ -5 & 1 & -1 \\ 4 & 8 & 1 \end{vmatrix} = -1 \begin{vmatrix} 1 & -1 \\ 8 & 1 \end{vmatrix} - (-5) \begin{vmatrix} 0 & 1 \\ 8 & 1 \end{vmatrix} + 4 \begin{vmatrix} 0 & 1 \\ 1 & -1 \end{vmatrix}$$

Subtract. Add.

$$= -1(1 + 8) + 5(0 - 8) + 4(0 - 1) \qquad \text{Evaluating the three determinants}$$

$$= -9 - 40 - 4 = -53.$$ □

Cramer's Rule: 3 × 3 Systems

The solution of the system

$$a_1x + b_1y + c_1z = d_1,$$
$$a_2x + b_2y + c_2z = d_2,$$
$$a_3x + b_3y + c_3z = d_3$$

is found by considering the following determinants:

$$D = \begin{vmatrix} a_1 & b_1 & c_1 \\ a_2 & b_2 & c_2 \\ a_3 & b_3 & c_3 \end{vmatrix}, \qquad D_x = \begin{vmatrix} d_1 & b_1 & c_1 \\ d_2 & b_2 & c_2 \\ d_3 & b_3 & c_3 \end{vmatrix},$$

$$D_y = \begin{vmatrix} a_1 & d_1 & c_1 \\ a_2 & d_2 & c_2 \\ a_3 & d_3 & c_3 \end{vmatrix}, \qquad D_z = \begin{vmatrix} a_1 & b_1 & d_1 \\ a_2 & b_2 & d_2 \\ a_3 & b_3 & d_3 \end{vmatrix}.$$

The solution, if it is unique, is given by

$$x = \frac{D_x}{D}, \quad y = \frac{D_y}{D}, \quad z = \frac{D_z}{D}.$$

EXAMPLE 4

Solve using Cramer's rule:

$$x - 3y + 7z = 13,$$
$$x + y + z = 1,$$
$$x - 2y + 3z = 4.$$

Solution We compute D, D_x, D_y and D_z:

$$D = \begin{vmatrix} 1 & -3 & 7 \\ 1 & 1 & 1 \\ 1 & -2 & 3 \end{vmatrix} = -10; \qquad D_x = \begin{vmatrix} 13 & -3 & 7 \\ 1 & 1 & 1 \\ 4 & -2 & 3 \end{vmatrix} = 20;$$

$$D_y = \begin{vmatrix} 1 & 13 & 7 \\ 1 & 1 & 1 \\ 1 & 4 & 3 \end{vmatrix} = -6; \qquad D_z = \begin{vmatrix} 1 & -3 & 13 \\ 1 & 1 & 1 \\ 1 & -2 & 4 \end{vmatrix} = -24.$$

Then

$$x = \frac{D_x}{D} = \frac{20}{-10} = -2;$$

$$y = \frac{D_y}{D} = \frac{-6}{-10} = \frac{3}{5};$$

$$z = \frac{D_z}{D} = \frac{-24}{-10} = \frac{12}{5}.$$

The solution is $\left(-2, \frac{3}{5}, \frac{12}{5}\right)$. ❑

In Example 4, we would not have needed to evaluate D_z. Once we found x and y, we could have substituted them into one of the equations to find z. In practice, it is faster to use determinants to find only two of the numbers; then we find the third by substitution into an equation.

In using Cramer's rule, we divide by D. If D should be 0, however, we could not do so. If $D = 0$ and at least one of the other determinants is not 0, then the system is inconsistent. If $D = 0$ and all the other determinants are also 0, then the equations in the system are dependent.

EXERCISE SET 8.7

Evaluate.

1. $\begin{vmatrix} 2 & 7 \\ 1 & 5 \end{vmatrix}$

2. $\begin{vmatrix} 3 & 2 \\ 2 & -3 \end{vmatrix}$

3. $\begin{vmatrix} 6 & -9 \\ 2 & 3 \end{vmatrix}$

4. $\begin{vmatrix} 3 & 2 \\ -7 & 5 \end{vmatrix}$

5. $\begin{vmatrix} 0 & 2 & 0 \\ 3 & -1 & 1 \\ 1 & -2 & 2 \end{vmatrix}$

6. $\begin{vmatrix} 3 & 0 & -2 \\ 5 & 1 & 2 \\ 2 & 0 & -1 \end{vmatrix}$

7. $\begin{vmatrix} -1 & -2 & -3 \\ 3 & 4 & 2 \\ 0 & 1 & 2 \end{vmatrix}$

8. $\begin{vmatrix} 1 & 2 & 2 \\ 2 & 1 & 0 \\ 3 & 3 & 1 \end{vmatrix}$

9. $\begin{vmatrix} 3 & 2 & -2 \\ -2 & 1 & 4 \\ -4 & -3 & 3 \end{vmatrix}$

10. $\begin{vmatrix} 2 & -1 & 1 \\ 1 & 2 & -1 \\ 3 & 4 & -3 \end{vmatrix}$

Solve using Cramer's rule.

11. $3x - 4y = 6,$
$5x + 9y = 10$

12. $5x + 8y = 1,$
$3x + 7y = 5$

13. $-2x + 4y = 3,$
$3x - 7y = 1$

14. $5x - 4y = -3,$
$7x + 2y = 6$

15. $3x + 2y - z = 4,$
$3x - 2y + z = 5,$
$4x - 5y - z = -1$

16. $3x - y + 2z = 1,$
$x - y + 2z = 3,$
$-2x + 3y + z = 1$

17. $2x - 3y + 5z = 27,$
$x + 2y - z = -4,$
$5x - y + 4z = 27$

18. $x - y + 2z = -3,$
$x + 2y + 3z = 4,$
$2x + y + z = -3$

19. $r - 2s + 3t = 6,$
$2r - s - t = -3,$
$r + s + t = 6$

20. $a - 3c = 6,$
$b + 2c = 2,$
$7a - 3b - 5c = 14$

Skill Maintenance

Let $f(x) = 2x + 5$ and $g(x) = x^2$. Find the following.

21. $f(x) + g(x)$

22. $(f + g)(x)$

23. The domain of $(g/f)(x)$

Synthesis

Solve.

24. $\begin{vmatrix} y & -2 \\ 4 & 3 \end{vmatrix} = 44$

25. $\begin{vmatrix} 2 & x & -1 \\ -1 & 3 & 2 \\ -2 & 1 & 1 \end{vmatrix} = -12$

26. $\begin{vmatrix} m+1 & -2 \\ m-2 & 1 \end{vmatrix} = 27$

27. Show that an equation of the line through (x_1, y_1) and (x_2, y_2) can be written

$$\begin{vmatrix} x & y & 1 \\ x_1 & y_1 & 1 \\ x_2 & y_2 & 1 \end{vmatrix} = 0.$$

28. ◈ Cramer's rule states that whenever the equations $a_1x + b_1y = c_1$ and $a_2x + b_2y = c_2$ are dependent, we have

$$\begin{vmatrix} a_1 & b_1 \\ a_2 & b_2 \end{vmatrix} = 0.$$

Explain why this occurs.

8.8 Business and Economic Applications

Break-Even Analysis • Supply and Demand

Break-Even Analysis

When a company manufactures x units of a product, it invests money. This is **total cost** and can be thought of as a function C, where $C(x)$ is the total cost of producing x units. When the company sells x units of the product, it takes in money. This is **total revenue** and can be thought of as a function R, where $R(x)$ is the total revenue from the sale of x units. **Total profit** is the money taken in less the money spent, or total revenue minus total cost. Total profit from the production and sale of x units is a function P given by

Profit = Revenue − Cost, or $P(x) = R(x) - C(x)$.

If $R(x)$ is greater than $C(x)$, then the company makes money. If $C(x)$ is greater than $R(x)$, then the company has a loss. If $R(x) = C(x)$, then the company breaks even.

There are two kinds of costs. First, there are costs like rent, insurance, machinery, and so on. These costs, which must be paid whether a product is produced or not, are called *fixed costs*. When a product is being produced, there are costs for labor, materials, marketing, and so on. These are called *variable costs*. They vary according to the amount being produced. Adding these together, we find the *total cost* of producing a product.

EXAMPLE 1

Ergs, Inc., is planning to make a new kind of radio. During the first year, fixed costs will be $90,000, and it will cost $15 to produce each radio (variable costs). Each radio will sell for $26.

a) Find the total cost $C(x)$ of producing x radios.
b) Find the total revenue $R(x)$ from the sale of x radios.
c) Find the total profit $P(x)$ from the production and sale of x radios.
d) What profit or loss will the company realize from the production and sale of 3000 radios? of 14,000 radios?
e) Graph the total-cost, total-revenue, and total-profit functions using the same set of axes. Determine the break-even point.

Solution

a) Total cost is given by

$$C(x) = \text{(Fixed costs) plus (Variable costs)},$$
$$\text{or}\quad C(x) = 90{,}000 \quad + \quad 15x,$$

where x is the number of radios produced.

b) Total revenue is given by

$$R(x) = 26x.$$ $26 times the number of radios sold. We are assuming that all radios produced are sold.

c) Total profit is given by

$$\begin{aligned} P(x) &= R(x) - C(x) \\ &= 26x - (90{,}000 + 15x) \\ &= 11x - 90{,}000. \end{aligned}$$

d) Total profit is found by

$$P(3000) = 11 \cdot 3000 - 90{,}000 = -\$57{,}000$$

when 3000 radios are produced and sold, and

$$P(14{,}000) = 11 \cdot 14{,}000 - 90{,}000 = \$64{,}000$$

when 14,000 radios are produced and sold. Thus the company loses money from the production and sale of 3000 radios, but makes money from the production and sale of 14,000 radios.

e) The graphs of each of the three functions are shown below:

$$R(x) = 26x, \qquad (1)$$

$$C(x) = 90{,}000 + 15x, \qquad (2)$$

$$P(x) = 11x - 90{,}000. \qquad (3)$$

$R(x)$, $C(x)$, and $P(x)$ are all in dollars.

Equation (1) has a graph that goes through the origin and has a slope of 26. Equation (2) has an intercept on the \$-axis of 90,000 and has a slope of 15. Equation (3) has an intercept on the \$-axis of −90,000 and has a slope of 11. It is shown by the dashed line. The color dashed line shows a ''negative'' profit, which is a loss. (That is what is known as ''being in the red.'') The black dashed line shows a ''positive'' profit, or gain. (That is what is known as ''being in the black.'')

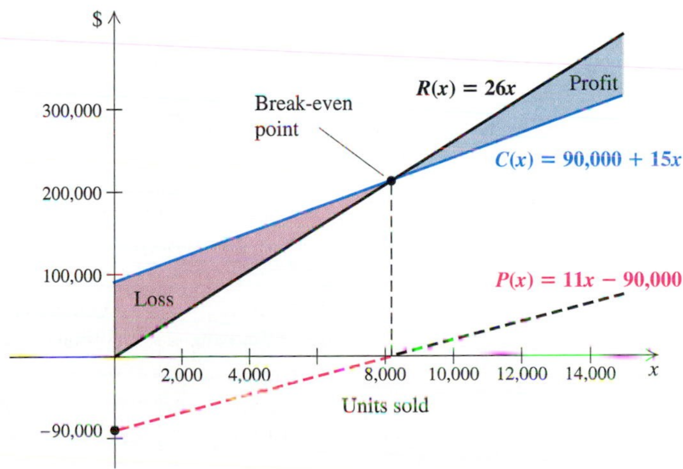

Profits occur where the revenue is greater than the cost. Losses occur where the revenue is less than the cost. The **break-even point** occurs where the graphs of R and C cross. Thus to find the break-even point, we solve a system:

$$R(x) = 26x,$$

$$C(x) = 90{,}000 + 15x.$$

Since both revenue and cost are in *dollars* and they are equal at the break-even point, the system can be rewritten as

$$d = 26x, \quad (1)$$
$$d = 90{,}000 + 15x \quad (2)$$

and solved using substitution:

$$26x = 90{,}000 + 15x \qquad \text{Substituting } 26x \text{ for } d \text{ in Equation (2)}$$
$$11x = 90{,}000$$
$$x \approx 8181.8.$$

The firm will break even if it produces and sells about 8182 radios (8181 will yield a tiny loss and 8182 a tiny gain), and takes in $R(8182) = 26 \cdot 8182 = \$212{,}732$ in revenue. Note that the x-coordinate of the break-even point can also be found by solving $P(x) = 0$. ❑

Supply and Demand

Demand Varies with Price. As the price of coffee varied over a period of years, the amount sold varied. The table and graph both show that the amount *demanded* by consumers goes down as the price goes up. As price goes down, demand goes up.

DEMAND FUNCTION, D

Price, p, per Kilogram	Quantity, $D(p)$, in Millions of Kilograms
\$ 8.00	25
9.00	20
10.00	15
11.00	10
12.00	5

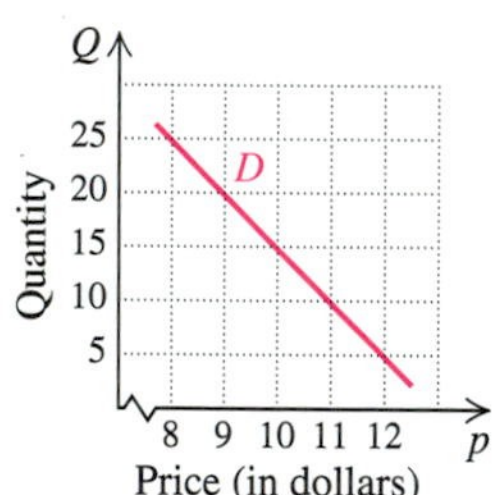

Supply Varies with Price. As the price of coffee varied, we see that the amount available varied. The table and graph show that the seller *supplies* less as the price goes down, but is willing to supply more as the price goes up.

SUPPLY FUNCTION, S

Price, p, per Kilogram	Quantity, $S(p)$, in Millions of Kilograms
\$ 9.00	5
9.50	10
10.00	15
10.50	20
11.00	25

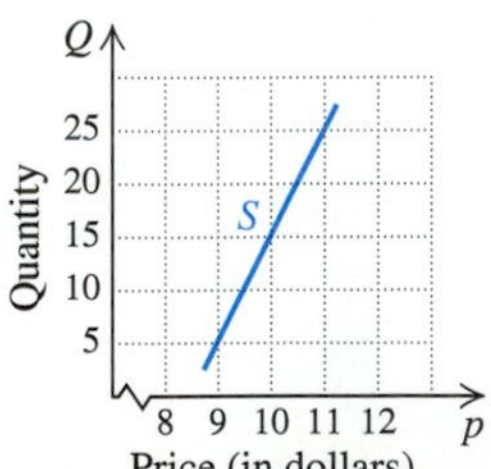

The Equilibrium Point. Let us look at the above graphs together. We see that as price increases, demand decreases. As price increases, supply increases. The point of intersection is called the *equilibrium point*. At that price, the amount that the seller will supply is the same amount that the consumer will buy. The situation is analogous to a buyer and a seller negotiating the price of an item. The equilibrium point, or selling price, is what they finally agree on.

Any ordered pair of coordinates from the graph is (price, quantity), because the horizontal axis is the price axis and the vertical axis is the quantity axis. If D is the demand function for a product and S is the supply function, then the equilibrium point is where demand equals supply:

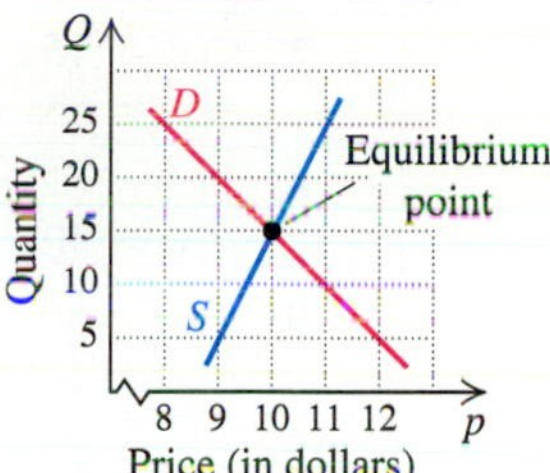

$$D(p) = S(p).$$

EXAMPLE 2

Find the equilibrium point for the demand and supply functions given:

$$D(p) = 1000 - 60p, \qquad (1)$$
$$S(p) = 200 + 4p. \qquad (2)$$

Solution Since both demand and supply are *quantities* and they are equal at the equilibrium point, the system can be rewritten as

$$q = 1000 - 60p, \qquad (1)$$
$$q = 200 + 4p. \qquad (2)$$

We substitute $200 + 4p$ for q in Equation (1) and solve:

$$200 + 4p = 1000 - 60p$$
$$200 + 64p = 1000 \qquad \text{Adding } 60p$$
$$64p = 800 \qquad \text{Adding } -200$$
$$p = \tfrac{800}{64} = 12.5.$$

Thus the equilibrium price is \$12.50 per unit.

To find the equilibrium quantity, we substitute \$12.50 into either $D(p)$ or $S(p)$. We use $S(p)$:

$$S(12.5) = 200 + 4(12.5) = 200 + 50 = 250.$$

Thus the equilibrium quantity is 250 units, and the equilibrium point is (\$12.50, 250). ❑

EXERCISE SET 8.8

For each of the following pairs of total-cost and total-revenue functions, find (a) the total-profit function and (b) the break-even point.

1. $C(x) = 45x + 600{,}000$; $R(x) = 65x$
2. $C(x) = 25x + 360{,}000$; $R(x) = 70x$
3. $C(x) = 10x + 120{,}000$; $R(x) = 60x$
4. $C(x) = 30x + 49{,}500$; $R(x) = 85x$
5. $C(x) = 20x + 10{,}000$; $R(x) = 100x$

6. $C(x) = 40x + 22{,}500$;
   $R(x) = 85x$

7. $C(x) = 15x + 75{,}000$;
   $R(x) = 55x$

8. $C(x) = 22x + 16{,}000$;
   $R(x) = 40x$

9. $C(x) = 50x + 195{,}000$;
   $R(x) = 125x$

10. $C(x) = 34x + 928{,}000$;
    $R(x) = 128x$

Find the equilibrium point for each of the following pairs of demand and supply functions.

11. $D(p) = 2000 - 60p$,
    $S(p) = 460 + 94p$

12. $D(p) = 1000 - 10p$,
    $S(p) = 250 + 5p$

13. $D(p) = 760 - 13p$,
    $S(p) = 430 + 2p$

14. $D(p) = 800 - 43p$,
    $S(p) = 210 + 16p$

15. $D(p) = 7500 - 25p$,
    $S(p) = 6000 + 5p$

16. $D(p) = 8800 - 30p$,
    $S(p) = 7000 + 15p$

17. $D(p) = 1600 - 53p$,
    $S(p) = 320 + 75p$

18. $D(p) = 5500 - 40p$,
    $S(p) = 1000 + 85p$

Solve.

19. Sky View Electronics is planning to introduce a new line of computers. For the first year, the fixed costs for setting up the production line are \$125,000. The variable costs for producing each computer are \$750. The revenue from each computer is \$1050.
    a) Find the total cost $C(x)$ of producing x computers.
    b) Find the total revenue $R(x)$ from the sale of x computers.
    c) Find the total profit $P(x)$ from the production and sale of x computers.
    d) What profit or loss will the company realize from the production and sale of 400 computers? of 700 computers?
    e) Find the break-even point.

20. City Lights, Inc., is planning a new type of lamp. For the first year, the fixed costs for setting up the production line are \$22,500. The variable costs for producing each lamp are estimated to be \$40. The revenue from each lamp is to be \$85.
    a) Find the total cost $C(x)$ of producing x lamps.
    b) Find the total revenue $R(x)$ from the sale of x lamps.
    c) Find the total profit $P(x)$ from the production and sale of x lamps.
    d) What profit or loss will the company realize from the production and sale of 3000 lamps? of 400 lamps?
    e) Find the break-even point.

21. Sarducci's is planning a new line of sport coats. For the first year, the fixed costs for setting up the production line are \$10,000. The variable costs for producing each coat are \$20. The revenue from each coat is to be \$100.
    a) Find the total cost $C(x)$ of producing x coats.
    b) Find the total revenue $R(x)$ from the sale of x coats.
    c) Find the total profit $P(x)$ from the production and sale of x coats.
    d) What profit or loss will the company realize from the production and sale of 2000 coats? of 50 coats?
    e) Find the break-even point.

22. Martina's Custom Printing is planning on adding painter's caps to its product line. For the first year, the fixed costs for setting up the production line are \$16,400. The variable costs for producing a dozen caps are \$6.00. The revenue on each dozen caps will be \$18.00.
    a) Find the total cost $C(x)$ of producing x dozen caps.
    b) Find the total revenue $R(x)$ from the sale of x dozen caps.
    c) Find the total profit $P(x)$ from the production and sale of x dozen caps.
    d) What profit or loss will the company realize from the production and sale of 3000 dozen caps? of 1000 dozen caps?
    e) Find the break-even point.

Skill Maintenance

Factor.

23. $x^2 + 7x + 12$

24. $6x^2 - x - 2$

25. Add: $\frac{2}{x} + \frac{x+1}{x^2}$.

26. Let $f(x) = x^2 + 3$ and $g(x) = x + 1$. Find $(f \cdot g)(x)$.

Synthesis

27. Bing Boing Hobbies is willing to produce 100 yo-yo's at \$2.00 each and 500 yo-yo's at \$8.00 each. Research indicates that the public will buy 500 yo-yo's at \$1.00 each and 100 yo-yo's at \$9.00 each. Find the equilibrium point.

28. Fidelity Speakers, Inc., has fixed costs of \$15,400 and variable costs of \$100 for each pair of speakers produced. If the speakers sell for \$250 a pair, how many pairs of speakers need to be produced (and sold) in order to have enough profit to cover the fixed costs of two new facilities? Assume that all fixed costs are identical.

29. ◈ Variable costs and fixed costs are often compared to the slope and the y-intercept, respectively, of an equation for a line. Explain why you feel this analogy is or is not valid.

30. ◈ In this section, we examined supply and demand functions for coffee. Does it seem realistic to you for the graph of D to have a constant slope? Why or why not?

In Exercises 31 and 32, use a grapher to solve each of the problems.

31. **a)** The Number Cruncher Computer Corporation is planning a new line of computers, each of which will sell for \$970. The fixed costs in setting up the production line are \$1,235,580 and the variable costs for each computer are \$697. What is the break-even point? (Round to the nearest whole number.)

b) The marketing department at Number Cruncher is not sure that the \$970 price is the best price. They have determined that the demand function for the new computers will be $D(p) = -304.5p + 374{,}580$ and the supply function will be $S(p) = 788.7p - 576{,}504$. To the nearest dollar, what price p would result in equilibrium between supply and demand?

32. **a)** Puppy Love, Inc., will soon begin producing a new line of puppy food. The marketing department predicts that the demand function will be $D(p) = -14.97p + 987.35$ and the supply function will be $S(p) = 98.55p - 5.13$. To the nearest cent, what price per unit should be charged in order to have equilibrium between supply and demand?

b) The production of the puppy food involves \$87,985 in fixed costs and \$5.15 per unit in variable costs. If the price per unit is the value you found in part (a), how many units must be sold in order to break even?

SUMMARY AND REVIEW 8

KEY TERMS

System of equations, p. 383
Solution of a system, p. 384
Inconsistent, p. 387
Dependent, p. 387
Consistent, p. 387
Substitution method, p. 391
Elimination method, p. 393
Independent, p. 395
Matrix (matrices), p. 419
Row, p. 419
Column, p. 419
Element, p. 420
Back-substitution, p. 420
Row-equivalent, p. 421
Determinant, p. 423
Cramer's rule, p. 424
Total cost, p. 428
Total revenue, p. 428
Total profit, p. 428
Break-even point, p. 429
Demand function, p. 430
Supply function, p. 430
Equilibrium point, p. 430

IMPORTANT PROPERTIES AND FORMULAS

To use the elimination method to solve systems of three equations:

1. Write all equations in the standard form $Ax + By + Cz = D$.
2. Clear any decimals or fractions.
3. Choose a variable to eliminate. Then use any two of the three equations to get an equation in two variables.
4. Next use a different pair of equations and get another equation *in the same two variables*. That is, eliminate the same variable that you did in step (3).
5. Solve the resulting system (pair) of equations. That will give two of the numbers.
6. Then use any of the original three equations to find the third number.

Row-Equivalent Operations

Each of the following row-equivalent operations produces an equivalent matrix:

a) Interchanging any two rows.
b) Multiplying each element of a row by the same nonzero number.
c) Multiplying each element of a row by a nonzero number and adding the result to another row.

Determinant of a 2 × 2 Matrix

$$\begin{vmatrix} a & c \\ b & d \end{vmatrix} = ad - bc$$

Determinant of a 3 × 3 Matrix

$$\begin{vmatrix} a_1 & b_1 & c_1 \\ a_2 & b_2 & c_2 \\ a_3 & b_3 & c_3 \end{vmatrix} = a_1 \begin{vmatrix} b_2 & c_2 \\ b_3 & c_3 \end{vmatrix} - a_2 \begin{vmatrix} b_1 & c_1 \\ b_3 & c_3 \end{vmatrix} + a_3 \begin{vmatrix} b_1 & c_1 \\ b_2 & c_2 \end{vmatrix}$$

Cramer's Rule: 2 × 2 Systems

The solution of the system

$$a_1x + b_1y = c_1,$$
$$a_2x + b_2y = c_2,$$

if it is unique, is given by

$$x = \frac{\begin{vmatrix} c_1 & b_1 \\ c_2 & b_2 \end{vmatrix}}{\begin{vmatrix} a_1 & b_1 \\ a_2 & b_2 \end{vmatrix}}, \qquad y = \frac{\begin{vmatrix} a_1 & c_1 \\ a_2 & c_2 \end{vmatrix}}{\begin{vmatrix} a_1 & b_1 \\ a_2 & b_2 \end{vmatrix}}.$$

Cramer's Rule: 3 × 3 Systems

The solution of the system

$$a_1x + b_1y + c_1z = d_1,$$
$$a_2x + b_2y + c_2z = d_2,$$
$$a_3x + b_3y + c_3z = d_3$$

is found by considering the following determinants:

$$D = \begin{vmatrix} a_1 & b_1 & c_1 \\ a_2 & b_2 & c_2 \\ a_3 & b_3 & c_3 \end{vmatrix}, \qquad D_x = \begin{vmatrix} d_1 & b_1 & c_1 \\ d_2 & b_2 & c_2 \\ d_3 & b_3 & c_3 \end{vmatrix},$$

$$D_y = \begin{vmatrix} a_1 & d_1 & c_1 \\ a_2 & d_2 & c_2 \\ a_3 & d_3 & c_3 \end{vmatrix}, \qquad D_z = \begin{vmatrix} a_1 & b_1 & d_1 \\ a_2 & b_2 & d_2 \\ a_3 & b_3 & d_3 \end{vmatrix}.$$

The solution, if it is unique, is given by

$$x = \frac{D_x}{D}, \qquad y = \frac{D_y}{D}, \qquad z = \frac{D_z}{D}.$$

REVIEW EXERCISES

This chapter's review and test include Skill Maintenance exercises from Sections 5.2, 5.3, 6.5, and 7.5.

Solve graphically.

1. $4x - y = 10,$
   $2x + 3y = 12$

2. $y = 3x + 7,$
   $3x + 2y = -4$

Solve using the substitution method.

3. $7x - 4y = 6,$
   $y - 3x = -2$

4. $y = x + 2,$
   $y - x = 8$

5. $9x - 6y = 2,$
   $x = 4y + 5$

Solve using the elimination method.

6. $8x - 2y = 10,$
   $-4y - 3x = -17$

7. $4x - 7y = 18,$
   $9x + 14y = 40$

8. $3x - 5y = -4,$
   $5x - 3y = 4$

Solve.

9. Sean has $37 to spend. He can spend all the money for two compact discs and a cassette, or he can buy one CD and two cassettes and have $5.00 left over. What is the price of a CD? What is the price of a cassette?

10. A train leaves Watsonville at noon traveling north at a speed of 44 mph. One hour later, another train, going 55 mph, travels north on a parallel track. How many hours will the second train travel before it overtakes the first train?

11. Cleanse-O is 30% alcohol. Tingle is 50% alcohol. How much of each should be mixed in order to obtain 40 L of a solution that is 45% alcohol?

Solve.

12. $x + 2y + z = 10,$
    $2x - y + z = 8,$
    $3x + y + 4z = 2$

13. $3x + 2y + z = 3,$
    $6x - 4y - 2z = -34,$
    $-x + 3y - 3z = 14$

14. $2x - 5y - 2z = -4,$
    $7x + 2y - 5z = -6,$
    $-2x + 3y + 2z = 4$

15. $-5x + 5y = -6,$
    $2x - 2y = 4$

16. $3x + y = 2,$
    $x + 3y + z = 0,$
    $x + z = 2$

17. $3x + 4y = 6,$
    $1.5x - 3 = -2y$

18. $x + y + 2z = 1,$
    $x - y + z = 1,$
    $x + 2y + z = 2$

Solve.

19. In triangle ABC, the measure of angle A is four times the measure of angle C, and the measure of angle B is 45° more than the measure of angle C. What are the measures of the angles of the triangle?

20. Find the three-digit number in which the sum of the digits is 11, the ten's digit is three less than the sum of the hundred's and one's digits, and the one's digit is five less than the hundred's digit.

21. Lynn has $194 in her purse, consisting of $20, $5, and $1 bills. The number of $1 bills is one less than the total number of $20 and $5 bills. If she has 39 bills in her purse, how many of each denomination does she have?

Solve using matrices. Show your work.

22. $3x + 4y = -13,$
    $5x + 6y = 8$

23. $3x - y + z = -1,$
    $2x + 3y + z = 4,$
    $5x + 4y + 2z = 5$

Evaluate.

24. $\begin{vmatrix} -2 & 4 \\ -3 & 5 \end{vmatrix}$

25. $\begin{vmatrix} 2 & 3 & 0 \\ 1 & 4 & -2 \\ 2 & -1 & 5 \end{vmatrix}$

Solve using Cramer's rule. Show your work.

26. $2x + 3y = 6,$
    $x - 4y = 14$

27. $2x + y + z = -2,$
    $2x - y + 3z = 6,$
    $3x - 5y + 4z = 7$

28. Find the equilibrium point for the demand and supply functions

$$D(p) = 60 + 7p \quad \text{and} \quad S(p) = 120 - 13p.$$

29. Kregel Furniture is planning to produce a new type of bed. For the first year, the fixed costs for setting up the production line are $35,000. The variable costs for producing each bed are $175. The revenue from each bed is $225.

    a) Find the total cost $C(x)$ of producing x beds.
    b) Find the total revenue $R(x)$ from the sale of x beds.
    c) Find the total profit $P(x)$ from the production and sale of x beds.
    d) What profit or loss will the company realize

from the production and sale of 1200 beds? of 500 beds?

e) Find the break-even point.

Skill Maintenance

Factor.

30. $x^2 + 12x + 11$

31. $11x^2 + 122x + 11$

32. Subtract and simplify, if possible: $\dfrac{x}{x+1} - \dfrac{1}{x-1}$.

33. Let $f(x) = x + 5$ and $g(x) = x^2 + 1$. Find $(f - g)(x)$.

Synthesis

34. ◈ Explain why the solution of a system of two linear equations is the point of intersection of the graphs of the lines.

35. ◈ Explain how a system of equations can be both dependent and inconsistent.

36. Solve using the substitution method:

$$\begin{aligned} x - y + 2z &= -3, \\ 2x + y - 3z &= 11, \\ z &= -2. \end{aligned}$$

37. The graph of $f(x) = ax^2 + bx + c$ contains the points $(-2, 3)$, $(1, 1)$, and $(0, 3)$. Find a, b, and c and give a formula for the function.

CHAPTER TEST 8

Solve using the substitution method.

1. $x + 3y = -8,$
$4x - 3y = 23$

2. $2x + 4y = -6,$
$y = 3x - 9$

Solve using the elimination method.

3. $4x - 6y = 3,$
$5x + 3y = 2$

4. $4y + 2x = 18,$
$3x + 6y = 26$

5. The perimeter of a rectangle is 96. The length of the rectangle is 6 less than twice the width. Find the dimensions of the rectangle.

6. Tyler bought 18 sheets of plywood for \$750. Oak plywood cost \$45 a sheet and pine cost \$25. How many sheets of each kind of plywood did he buy?

Solve.

7.
$$\begin{aligned} -3x + y - 2z &= 8, \\ -x + 2y - z &= 5, \\ 2x + y + z &= -3 \end{aligned}$$

8.
$$\begin{aligned} 6x + 2y - 4z &= 15, \\ -3x - 4y + 2z &= -6, \\ 4x - 6y + 3z &= 8 \end{aligned}$$

9.
$$\begin{aligned} 2x + 2y &= 0, \\ 4x + 4z &= 4, \\ 2x + y + z &= 2 \end{aligned}$$

10.
$$\begin{aligned} 3x + 3z &= 0, \\ 2x + 2y &= 2, \\ 3y + 3z &= 3 \end{aligned}$$

Solve using matrices.

11. $7x - 8y = 10,$
$9x + 5y = -2$

12.
$$\begin{aligned} x + 3y - 3z &= 12, \\ 3x - y + 4z &= 0, \\ -x + 2y - z &= 1 \end{aligned}$$

Evaluate.

13. $\begin{vmatrix} 4 & -2 \\ 3 & 7 \end{vmatrix}$

14. $\begin{vmatrix} 3 & 4 & 2 \\ 2 & -5 & 4 \\ 1 & 5 & -3 \end{vmatrix}$

15. Solve using Cramer's rule:

$$\begin{aligned} 8x - 3y &= 5, \\ 2x + 6y &= 3. \end{aligned}$$

16. An electrician, a carpenter, and a plumber are hired to work on a house. The electrician earns \$21 per hour, the carpenter earns \$19.50 per hour, and the plumber earns \$24 per hour. The first day on the job, they worked a total of 21.5 hr and earned a total of \$469.50. If the plumber worked two more hours than the carpenter did, how many hours did the electrician work?

17. Find the equilibrium point for the demand and supply functions

$$D(p) = 79 - 8p \quad \text{and} \quad S(p) = 37 + 6p.$$

18. A sporting goods manufacturer is planning a new type of tennis racket. For the first year, the fixed costs for setting up production lines are $40,000. The variable costs for producing each racket are $45. The sales department predicts that 1500 rackets can be sold during the first year. The revenue from each racket is $80.

a) Find the total cost $C(x)$ of producing x tennis rackets.

b) Find the total revenue $R(x)$ from the sale of x tennis rackets.

c) Find the total profit $P(x)$.

d) What profit or loss will the company realize if the expected sales of 1500 rackets occur?

e) Find the break-even point.

Skill Maintenance

Factor.

19. $6x^2 - 5x - 25$

20. $x^2 - 14x + 13$

21. Add and simplify, if possible: $\frac{2}{t^2 - 4} + \frac{3}{t - 2}$.

22. Let $f(x) = x + 1$ and $g(x) = x - 1$. Find the domain of $(f/g)(x)$.

Synthesis

23. The graph of the function $f(x) = mx + b$ contains the points $(-1, 3)$ and $(-2, -4)$. Find m and b.

24. At a county fair, an adult's ticket sold for $5.50, a senior citizen's ticket sold for $4.00, and a child's ticket sold for $1.50. On the opening day, the number of tickets sold to children and senior citizens was 30 more than the number of tickets sold to adults. The number of tickets sold to senior citizens was 6 more than four times the number of tickets sold to children. Total receipts from the ticket sales were $14,967. How many of each type of ticket were sold?

CHAPTER 9

Inequalities and Linear Programming

AN APPLICATION

An insurance company offers two types of medical coverage. With plan A, the employee pays the first $100 of medical bills and the insurance company pays 80% of the rest. With plan B, the employee pays the first $250 of medical bills and the insurer pays 90% of the rest. For what amount of medical bills would plan B save an employee money?

This problem appears as Exercise 43 in Section 9.1.

Mike Mezo
PRESIDENT,
USWA LOCAL 1010

"Successful contract negotiations often rely on a solid understanding of math. A group insurance package is typically an important part of our contracts, and failure to procure the most beneficial plan could cost our members thousands of dollars."

Inequalities are mathematical sentences containing symbols such as < (is less than). In this chapter, we use the principles for solving inequalities developed in Chapter 2 to solve compound inequalities. We also combine our knowledge of inequalities and systems of equations to solve systems of inequalities.

In addition to material from this chapter, the review and test for Chapter 9 include material from Sections 2.2, 3.2, 8.2, and 8.3.

9.1 Interval Notation and Problem Solving

Solving Inequalities • Interval Notation • Problem Solving

Solving Inequalities

Recall from Chapter 1 that an **inequality** is any sentence having one of the verbs $<$, $>$, $\leqslant$, or $\geqslant$ (see Section 1.4)—for example,

$$-2 < a, \quad x > 4, \quad x + 3 \leqslant 6, \quad \text{and} \quad 16 - 7y \geqslant 10y - 4.$$

Some replacements for the variable in an inequality make it true, and some make it false. A replacement that makes it true is called a **solution.** The set of all solutions is called the **solution set.** When we have found the set of all solutions of an inequality, we say that we have **solved** the inequality.

We can use two principles, developed in Chapter 2, to solve inequalities.

The Addition Principle for Inequalities

Adding the same number on both sides of an inequality forms an equivalent inequality.

The Multiplication Principle for Inequalities

Multiplying on both sides of a true inequality by a *positive* number produces another true inequality. If we multiply by a *negative* number on both sides, the inequality symbol must be reversed to produce another true inequality.

When we solve an inequality using the multiplication principle, we can multiply on both sides by any number except zero.

The solution set of an inequality can be written using *set-builder notation* (see Chapter 2).

EXAMPLE 1

Solve: **(a)** $x + 5 > 3$; **(b)** $-5x \geqslant -80$; **(c)** $3x + 10 < 5$; **(d)** $16 - 7y \geqslant 10y - 4$.

Solution

a)

$$x + 5 > 3$$

$$x + 5 + (-5) > 3 + (-5) \quad \text{Using the addition principle, add } -5.$$

$$x > -2$$

The solution set is $\{x \mid x > -2\}$.

b)

$$-5x \geqslant -80$$

The symbol must be reversed.

$$-\tfrac{1}{5} \cdot (-5x) \leqslant -\tfrac{1}{5} \cdot (-80) \quad \text{Multiplying by } -\tfrac{1}{5}$$

$$x \leqslant 16$$

The solution set is $\{x \mid x \leqslant 16\}$.

c)

$$3x + 10 < 5$$

$$3x + 10 + (-10) < 5 + (-10) \quad \text{Adding } -10$$

$$3x < -5$$

The symbol stays the same.

$$\tfrac{1}{3} \cdot 3x < \tfrac{1}{3} \cdot (-5) \quad \text{Multiplying by } \tfrac{1}{3}$$

$$x < -\tfrac{5}{3}$$

The solution set is $\{x \mid x < -\tfrac{5}{3}\}$.

d)

$$16 - 7y \geqslant 10y - 4$$

$$-16 + 16 - 7y \geqslant -16 + 10y - 4 \quad \text{Adding } -16$$

$$-7y \geqslant 10y - 20$$

$$-10y + (-7y) \geqslant -10y + 10y - 20 \quad \text{Adding } -10y$$

$$-17y \geqslant -20$$

The symbol must be reversed.

$$-\tfrac{1}{17} \cdot (-17y) \leqslant -\tfrac{1}{17} \cdot (-20) \quad \text{Multiplying by } -\tfrac{1}{17}$$

$$y \leqslant \tfrac{20}{17}$$

The solution set is $\{y \mid y \leqslant \tfrac{20}{17}\}$. ❑

A *graph* of an inequality is a drawing that represents its solutions. An inequality in one variable can be graphed on a number line. Inequalities in two variables can be graphed on a coordinate plane, and will be considered later in this chapter.

EXAMPLE 2

Graph $x < 4$ on a number line.

Solution The solutions are all real numbers less than 4, so we shade all numbers less than 4. Since 4 is not a solution, we use an open circle at 4.

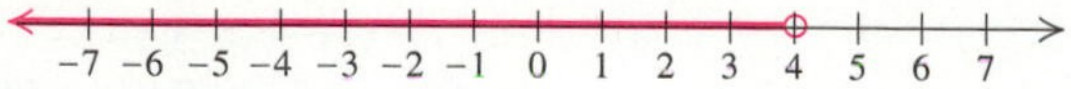

❑

Interval Notation

Another way to list the solutions of an inequality in one variable is to use **interval notation.** Pay special attention to the manner in which parentheses, ( ), and brackets, [ ], are used.

If a and b are real numbers such that $a < b$, we define the **open interval (a, b)** as the set of all numbers x for which $a < x < b$. Thus,

$$(a, b) = \{x \mid a < x < b\}.$$

Its graph excludes the endpoints:

(a, b)
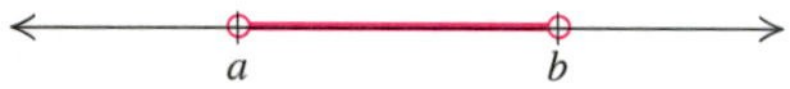

Be careful not to confuse the *interval* (a, b) with the *ordered pair* (a, b). The context of the discussion usually makes the meaning clear.

The **closed interval $[a, b]$** is defined as the set of all numbers x for which $a \leq x \leq b$. Thus,

$$[a, b] = \{x \mid a \leq x \leq b\}.$$

Its graph includes the endpoints:

[a, b]
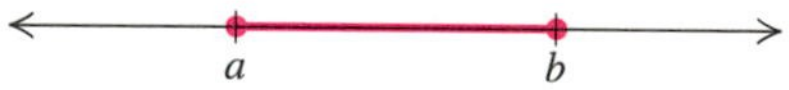

There are two kinds of **half-open intervals** defined as follows:

1. $(a, b] = \{x \mid a < x \leq b\}$. This is open on the left. Its graph is as follows:

(a, b]
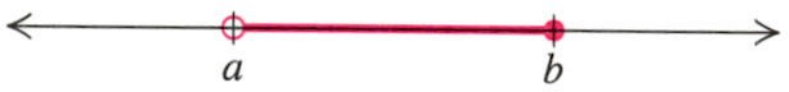

2. $[a, b) = \{x \mid a \leq x < b\}$. This is open on the right. Its graph is as follows:

[a, b)
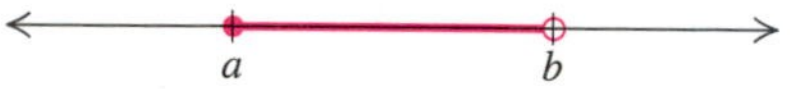

We use the symbols ∞ and $-\infty$ to represent positive and negative infinity, respectively. Thus the notation (a, ∞) represents the set of all real numbers greater than a, and $(-\infty, a)$ represents the set of all real numbers less than a.

(a, ∞)
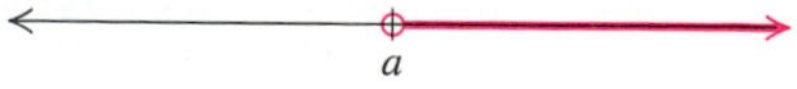

(−∞, a)
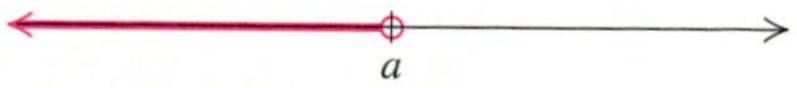

The notations $[a, \infty)$ and $(-\infty, a]$ are used when we want to include the endpoint a.

EXAMPLE 3

Graph $y \geq -2$ on a number line and write the solution set using both set-builder and interval notations.

Solution Using set-builder notation, we write the solution set as $\{y \mid y \geq -2\}$.

Using interval notation, we write the solution set as $[-2, \infty)$.

To graph the solution, we shade all numbers to the right of -2 and use a solid endpoint to indicate that -2 is also a solution.

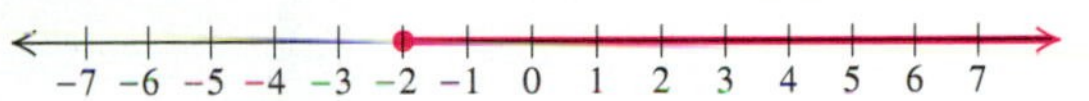

□

EXAMPLE 4

Solve and graph: **(a)** $4x - 1 \geq 5x - 2$; **(b)** $-3(x + 8) - 5x > 4x - 9$.

Solution

a)

$4x - 1 \geq 5x - 2$	
$4x - 1 + 2 \geq 5x - 2 + 2$	Adding 2
$4x + 1 \geq 5x$	Simplifying
$4x + 1 - 4x \geq 5x - 4x$	Adding $-4x$
$1 \geq x$	Simplifying

We know that $1 \geq x$ has the same meaning as $x \leq 1$. Thus any number less than or equal to 1 is a solution. We can express the solution set as $\{x \mid 1 \geq x\}$ or as $\{x \mid x \leq 1\}$. The latter is probably used most often. Using interval notation, we write the solution set as $(-\infty, 1]$. The graph is as follows:

b)

$-3(x + 8) - 5x > 4x - 9$	
$-3x - 24 - 5x > 4x - 9$	Using the distributive law
$-24 - 8x > 4x - 9$	
$-24 - 8x + 8x > 4x - 9 + 8x$	Adding $8x$ on both sides
$-24 > 12x - 9$	
$-24 + 9 > 12x - 9 + 9$	Adding 9
$-15 > 12x$	The symbol stays the same.
$-\frac{5}{4} > x$	Multiplying by $\frac{1}{12}$ and simplifying

The solution set is $\{x \mid -\frac{5}{4} > x\}$, or $\{x \mid x < -\frac{5}{4}\}$, or $(-\infty, -\frac{5}{4})$. The graph is as follows:

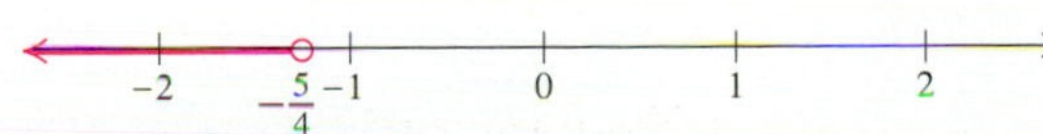

□

Problem Solving

EXAMPLE 5

Records in the women's 100-m dash. Florence Griffith Joyner set a world record of 10.49 sec in the women's 100-m dash in the 1988 Olympics. The formula

$$R = -0.0433t + 10.49$$

can be used to predict the world record in the women's 100-m dash t years after 1988. For example, to find the record in 2008, we would subtract 1988, to get $t = 20$. Determine (in terms of an inequality) those years for which the world record will be less than 10.0 sec.

Solution

1. **FAMILIARIZE.** We already have a formula. To become more familiar with it, we might make a substitution for t. Suppose we want to know the record after 30 years, in the year 2018. We substitute 30 for t:

$$\begin{aligned} R &= -0.0433t + 10.49 \\ &= -0.0433(30) + 10.49 = 9.191 \text{ sec.} \end{aligned}$$

Since $9.191 < 10.0$, the year 2018 is a solution. It is not, however, the only year for which the record will be less than 10.0 sec.

We continue by translating to an inequality.

2. **TRANSLATE.** The record R is to be *less than* 10.0 sec. Thus we have the inequality

$$R < 10.0.$$

We substitute $-0.0433t + 10.49$ for R to find the times t that satisfy the inequality:

$$-0.0433t + 10.49 < 10.0.$$

3. **CARRY OUT.** We solve the inequality:

$$-0.0433t + 10.49 < 10.0$$

$$-0.0433t < -0.49 \quad \text{Adding } -10.49$$

$$t > 11.32. \quad \text{Dividing by } -0.0433 \text{ or multiplying by } -\tfrac{1}{0.0433} \text{ and rounding. You might use a calculator.}$$

4. **CHECK.** We can check by substituting a value for t greater than 11.32. We did that in the familiarization step.

5. **STATE.** The record will be less than 10.0 for those races occurring more than 11.32 years after 1988, which is approximately $\{x | x \geq 1999\}$. ❑

EXAMPLE 6

On your new job, you can be paid in one of two ways:

Plan A: A salary of $600 per month, plus a commission of 4% of sales;

Plan B: A salary of $800 per month, plus a commission of 6% of sales in excess of $10,000.

For what amount of sales is plan A better than plan B, if we assume that sales are always more than $10,000?

Solution

1. **FAMILIARIZE.** Listing the given information in a table will be helpful.

Plan A: Monthly Income	Plan B: Monthly Income
$600 salary 4% of sales *Total:* $600 + 4% of sales	$800 salary 6% of sales over $10,000 *Total:* $800 + 6% of sales over $10,000

Next, suppose that you were to sell a certain amount—say, $12,000—in one month. Which plan would be better? Under plan A, you would earn $600 plus 4% of $12,000, or

$$600 + 0.04(12{,}000) = \$1080.$$

Since with plan B commissions are paid only on sales in excess of $10,000, you would earn $800 plus 6% of (12,000 − 10,000), or

$$800 + 0.06(12{,}000 - 10{,}000) = \$920.$$

This shows that for monthly sales of $12,000, plan A is better. Similar calculations will show that for sales of $30,000 a month, plan B is better. To determine *all* values for which plan A earns more money, we must solve an inequality that is based on the above calculations.

2. **TRANSLATE.** We let S represent the amount of monthly sales. Examining the calculations in the *Familiarize* step, we see that monthly income from plan A

is $600 + 0.04S$ and from plan B is $800 + 0.06(S - 10{,}000)$. We want to find all values of S for which

Income from plan A	is greater than	income from plan B
↓	↓	↓
$600 + 0.04S$	$>$	$800 + 0.06(S - 10{,}000)$.

3. CARRY OUT. We solve the inequality:

$$600 + 0.04S > 800 + 0.06(S - 10{,}000)$$
$$600 + 0.04S > 800 + 0.06S - 600 \qquad \text{Using the distributive law}$$
$$600 + 0.04S > 200 + 0.06S \qquad \text{Collecting like terms}$$
$$400 > 0.02S \qquad \text{Subtracting both 200 and } 0.04S$$
$$20{,}000 > S. \qquad \text{Dividing by 0.02}$$

4. CHECK. For $S = 20{,}000$, the income from plan A is

$$600 + 4\% \cdot 20{,}000, \quad \text{or} \quad \$1400.$$

The income from plan B is

$$800 + 6\% \cdot 10{,}000, \quad \text{or} \quad \$1400.$$

In the *Familiarize* step, we saw that for sales of \$12,000, plan A pays more. Since $20{,}000 > 12{,}000$, we have a partial check. We cannot check all possible values of S so we will stop here.

5. STATE. For monthly sales of less than \$20,000, plan A is better. ☐

EXERCISE SET 9.1

Graph the inequality, and write the solution set using both set-builder and interval notation.

1. $x > 4$
2. $y < 5$
3. $t \leqslant 6$
4. $x \geqslant -4$
5. $y < -3$
6. $t > -2$
7. $x \geqslant -6$
8. $x \leqslant -5$

Solve. Then graph.

9. $x + 8 > 3$
10. $y + 4 < 10$
11. $a + 9 \leqslant -12$
12. $a + 7 \leqslant -13$
13. $8x \geqslant 24$
14. $9t < -81$
15. $-9x \geqslant -8.1$
16. $-8y \leqslant 3.2$
17. $-\frac{3}{4}x \geqslant -\frac{5}{8}$
18. $-\frac{5}{6}y \leqslant -\frac{3}{4}$
19. $2x + 7 < 19$
20. $5y + 13 > 28$
21. $5y + 2y \leqslant -21$
22. $-9x + 3x \geqslant -24$
23. $3x - \frac{1}{8} \leqslant \frac{3}{8} + 2x$
24. $2x - 3 < \frac{13}{4}x + 10 - 4.25x$

Solve.

25. $3(2x + 1) \geqslant 4(3x - 2)$
26. $4(1 - 3y) + 2y < 3(5 - y)$
27. $5[3m - (m + 4)] > -2(m - 4)$
28. $[8x - 3(3x + 2)] - 5 \geqslant 3(x + 4) - 2x$
29. $2[4 - 2(3 - x)] - 1 \geqslant 4[2(4x - 3) + 7] - 25$
30. $5[3(7 - t) - 4(8 + 2t)] - 20 \leqslant -6[2(6 + 3t) - 4]$

Solve.

31. Ridem rents trucks at a daily rate of \$42.95 plus \$0.46 per mile. Atlas rents similar trucks for \$75 per day with unlimited mileage. For what daily mileages would it be less expensive to rent a Ridem truck?

32. A cargo van can be rented for \$35 per day with unlimited mileage, or for \$28 per day plus 19¢ per mile. For what daily mileages would the unlimited mileage plan save you money?

33. A long-distance telephone call using Down East Calling costs 20 cents for the first minute and 16 cents for each additional minute. The same call, placed on Long Call Systems, costs 19 cents for the first minute and 18 cents for each additional minute. For what length phone calls is Down East Calling less expensive?

34. Quick Move charges \$90 plus \$45 an hour to move households across town. Lug-a-lot Movers charges \$60 an hour for cross-town moves. For what lengths of times is Lug-a-lot the more expensive mover?

35. On your new job, you can be paid in one of two ways:

Plan A: A salary of \$500 per month, plus a commission of 4% of gross sales;
Plan B: A salary of \$750 per month plus a commission of 5% of gross sales over \$8000.

For what gross sales is plan B better than plan A, assuming that gross sales are always more than \$8000?

36. On your new job, you can be paid in one of two ways:

Plan A: A salary of \$25,000 per year;
Plan B: A salary of \$1500 per month plus a commission of 6% on gross sales.

For what gross sales is plan A better than plan B?

37. A painter can be paid in one of two ways:

Plan A: \$500 plus \$6.00 per hour;
Plan B: Straight \$11.00 per hour.

Suppose that the job takes n hours. For what values of n is plan A better for the painter?

38. A mason can be paid in one of two ways:

Plan A: \$300 plus \$9.00 per hour;
Plan B: Straight \$12.50 per hour.

Suppose that the job takes n hours. For what values of n is plan B better for the mason?

39. You are going to invest \$25,000, part at 7% and part at 8%. What is the most that can be invested at 7% in order to make at least \$1800 interest per year?

40. You are going to invest \$20,000, part at 6% and part at 8%. What is the most that can be invested at 6% in order to make at least \$1500 interest per year?

41. In planning for a college dance, you find that one band will play for \$250 plus 50% of the total ticket sales. Another band will play for a flat fee of \$550. In order for the first band to produce more profit for the school than the other band, what is the highest price you can charge per ticket, assuming that 300 people will attend?

42. It costs \$3.00 to cross a toll bridge connecting an island in the Gulf of Mexico with the mainland. Visitors can purchase a 6-month pass for \$15 and pay \$0.50 each time they cross. How many crossings would it take for the pass to save you money?

43. An insurance company offers two plans. With plan A, the employee pays the first \$100 of medical bills and the insurance company pays 80% of the rest. With plan B, the employee pays the first \$250 of medical bills and the insurance company pays 90% of the rest. For what amount of medical bills will plan B save the employee money?

44. You can spend \$3.50 at the laundromat washing your clothes, or you can have them do the laundry for 40¢ per pound. For what weights of clothes will it save you money to wash your clothes yourself?

45. The formula

$$C = \tfrac{5}{9}(F - 32)$$

can be used to convert Fahrenheit temperatures F to Celsius temperatures C.

a) Gold is solid at Celsius temperatures less than 1063° C. Find the Fahrenheit temperatures for which gold is solid.

b) Silver is solid at Celsius temperatures less than 960.8° C. Find the Fahrenheit temperatures for which silver is solid.

46. The percentage of the total active military duty force that is women has been steadily increasing. The number N of women in the active duty force t years since 1971 is approximated by

$$N = 12{,}197.8t + 44{,}000.$$

a) How many women were in the military in 1971 ($t = 0$)? in 1981 ($t = 10$)? in 1990 ($t = 19$)?

b) For what years will the number of women be at least 250,000?

47. Ergs, Inc., is planning to make a new kind of radio. Fixed costs will be $90,000, and variable costs will be $15 for the production of each radio. The total-cost function is

$$C(x) = 90{,}000 + 15x.$$

The company makes $26 in revenue for each radio sold. The total-revenue function is

$$R(x) = 26x.$$

(See Section 8.8.)

a) When $R(x) < C(x)$, the company loses money. Find the values of x for which the company loses money.

b) When $R(x) > C(x)$, the company makes a profit. Find the values of x for which the company makes a profit.

48. The demand and supply functions for a certain product are given by

$$D(p) = 2000 - 60p \quad \text{and}$$
$$S(p) = 460 + 94p.$$

(See Section 8.8.)

a) Find those values of p for which demand exceeds supply.

b) Find those values of p for which demand is less than supply.

Skill Maintenance

49. Graph: $5y - 10 = 2x$.

50. Solve: $-2x - 3 = 5$.

51. Simplify: $|-7|$.

52. Simplify: $-|-7|$.

Synthesis

53. ◈ Explain in your own words why the inequality symbol must be reversed when both sides of an inequality are multiplied by a negative number.

54. ◈ Presto photocopiers cost $510 and Exact Image photocopiers cost $590. Write a problem that involves the cost of the copiers, the cost per page of photocopies, and the number of copies for which the Presto machine is the more expensive machine to own.

Solve. Assume that a, b, c, d, and m are positive constants.

55. $3ax + 2x \geqslant 5ax - 4$; assume $a > 1$

56. $6by - 4y \leqslant 7by + 10$

57. $a(by - 2) \geqslant b(2y + 5)$; assume $a > 2$

58. $c(6x - 4) < d(3 + 2x)$; assume $3c > d$

59. $c(2 - 5x) + dx > m(4 + 2x)$; assume $5c + 2m < d$

60. $a(3 - 4x) + cx < d(5x + 2)$; assume $c > 4a + 5d$

Determine whether the statement is true or false. If false, give an example that shows this.

61. For any real numbers a, b, c, and d, if $a < b$ and $c < d$, then $a - c < b - d$.

62. For all real numbers x and y, if $x < y$, then $x^2 < y^2$.

63. ◈ Determine whether the inequalities

$$x < 3 \quad \text{and} \quad x + \frac{1}{x} < 3 + \frac{1}{x}$$

are equivalent. Give reasons to support your answer.

64. ◈ Determine whether the inequalities

$$x < 3 \quad \text{and} \quad 0 \cdot x < 0 \cdot 3$$

are equivalent. Give reasons to support your answer.

Solve.

65. $x + 5 \leqslant 5 + x$

66. $x + 8 < 3 + x$

67. $x^2 > 0$

68. $x^2 + 1 > 0$

9.2 Compound Inequalities

Intersections of Sets and Conjunctions of Sentences • Unions of Sets and Disjunctions of Sentences

We now consider **compound inequalities,** that is, sentences formed by two or more inequalities, joined with the word *and* or the word *or*.

Intersections of Sets and Conjunctions of Sentences

The **intersection** of two sets A and B is the set of all members that are common to both A and B. We denote the intersection of sets A and B as

$$A \cap B.$$

The intersection of two sets is often pictured as follows:

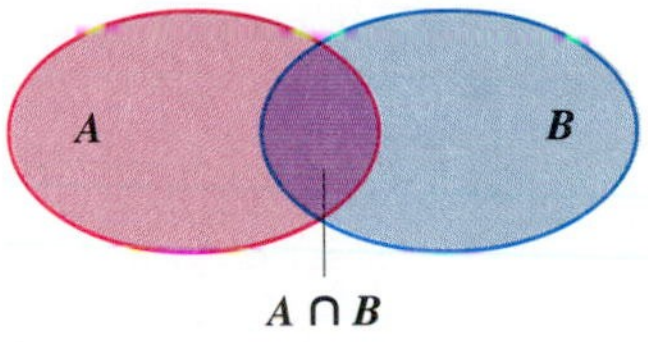

EXAMPLE 1

Find the intersection: $\{1, 2, 3, 4, 5\} \cap \{-2, -1, 0, 1, 2, 3\}$.

Solution The numbers 1, 2, and 3 are common to both sets, so the intersection is $\{1, 2, 3\}$. ❑

When two or more sentences are joined by the word *and* to make a compound sentence, the new sentence is called a **conjunction** of the sentences. The following is a conjunction of inequalities:

$$-2 < x \quad \textit{and} \quad x < 1.$$

For a conjunction to be true, all the individual sentences must be true. The solution set of a conjunction is the intersection of the solution sets of the individual sentences. Let us consider the conjunction

$$-2 < x \quad \textit{and} \quad x < 1.$$

The graphs of each separate sentence are shown below, and the intersection is the last graph. We use both set-builder and interval notations.

$\{x \mid -2 < x\}$

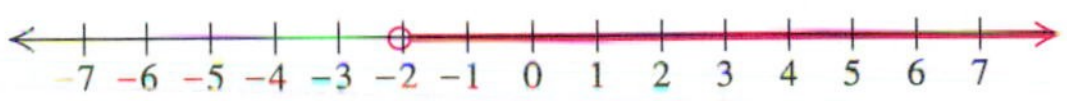

$(-2, \infty)$

$\{x \mid x < 1\}$

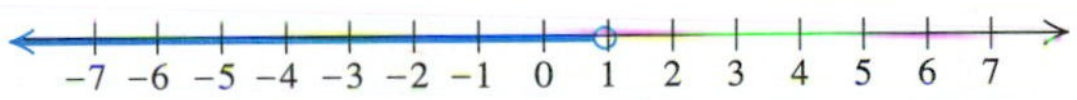

$(-\infty, 1)$

$\{x \mid -2 < x\} \cap \{x \mid x < 1\}$
$= \{x \mid -2 < x \textit{ and } x < 1\}$

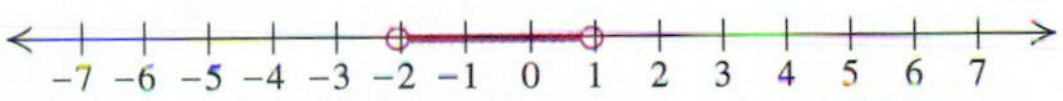

$(-2, 1)$

The conjunction $-2 < x$ *and* $x < 1$ can be abbreviated by $-2 < x < 1$. Thus the interval $(-2, 1)$ can be represented as $\{x| -2 < x < 1\}$, the set of all numbers that are simultaneously greater than -2 *and* less than 1.

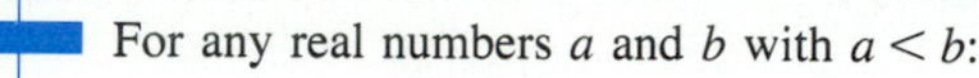

For any real numbers a and b with $a < b$:

The conjunction $a < x$ *and* $x < b$ can be abbreviated as $a < x < b$.

The conjunction $b > x$ *and* $x > a$ can be abbreviated as $b > x > a$.

EXAMPLE 2

Solve and graph: $-1 \leq 2x + 5 < 13$.

Solution This inequality is an abbreviation for the following conjunction:

$$-1 \leq 2x + 5 \quad \textit{and} \quad 2x + 5 < 13.$$

The word *and* corresponds to set *intersection*. The solution set is thus the intersection of the solution set of $-1 \leq 2x + 5$ and the solution set of $2x + 5 < 13$:

$$\{x| -1 \leq 2x + 5\} \cap \{x| 2x + 5 < 13\}.$$

Method 1. We write the conjunction with the word *and:*

$$\begin{aligned} -1 &\leq 2x + 5 && \textit{and} && 2x + 5 < 13 \\ -6 &\leq 2x && \textit{and} && 2x < 8 \\ -3 &\leq x && \textit{and} && x < 4. \end{aligned}$$

We now abbreviate the answer:

$$-3 \leq x < 4.$$

The solution set is $\{x| -3 \leq x < 4\}$, or, in interval notation, $[-3, 4)$.

Method 2. Using Method 1, we did the same thing to each inequality. We can shorten the writing as follows:

$$\begin{aligned} -1 &\leq 2x + 5 < 13 \\ -1 - 5 &\leq 2x + 5 - 5 < 13 - 5 && \text{Subtracting 5} \\ -6 &\leq 2x < 8 \\ -3 &\leq x < 4. && \text{Dividing by 2} \end{aligned}$$

The solution set is $\{x| -3 \leq x < 4\}$, or $[-3, 4)$.

The graph is the intersection of the individual graphs.

$\{x| -3 \leq x\}$

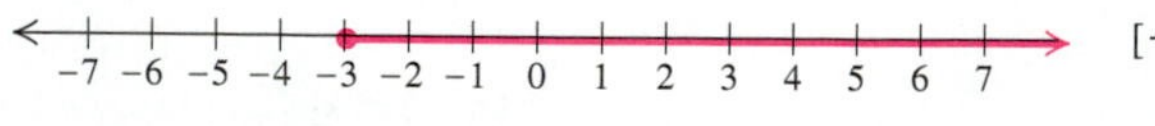

$[-3, \infty)$

$\{x| x < 4\}$

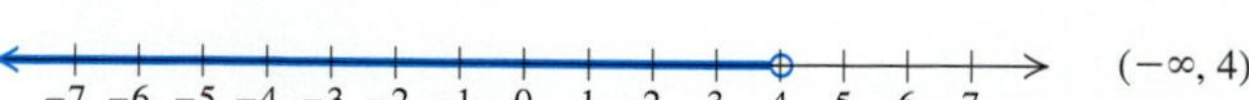

$(-\infty, 4)$

$\{x| -3 \leq x\} \cap \{x| x < 4\}$
$= \{x| -3 \leq x < 4\}$

-7 -6 -5 -4 -3 -2 -1 0 1 2 3 4 5 6 7

$[-3, 4)$

❑

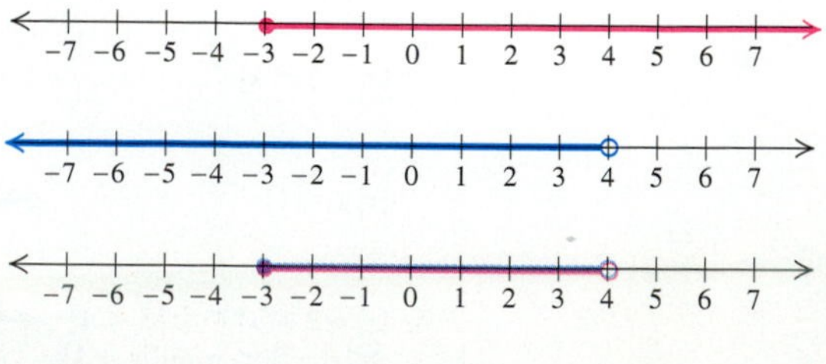

EXAMPLE 3

Solve and graph: $2x - 5 \geq -3$ *and* $5x + 2 \geq 17$.

Solution We solve each inequality separately:

$$\begin{aligned} 2x - 5 &\geq -3 \quad \textit{and} \quad 5x + 2 \geq 17 \\ 2x &\geq 2 \quad \textit{and} \quad 5x \geq 15 \\ x &\geq 1 \quad \textit{and} \quad x \geq 3. \end{aligned}$$

The solution set is the intersection of the solution sets of the individual inequalities.

$\{x|\, x \geq 1\}$

$[1, \infty)$

$\{x|\, x \geq 3\}$

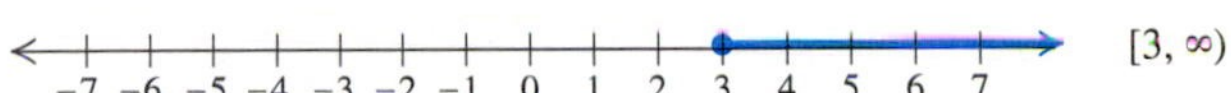

$[3, \infty)$

$\{x|\, x \geq 1\} \cap \{x|\, x \geq 3\}$
$= \{x|\, x \geq 3\}$

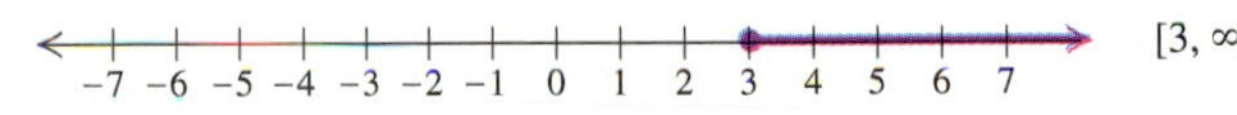

$[3, \infty)$

The numbers common to both sets are those that are greater than or equal to 3. Thus the solution set is $\{x|\, x \geq 3\}$, or, in interval notation, $[3, \infty)$. □

Intersection

The word "and" corresponds to "intersection" and to the symbol "$\cap$". For a number to be a solution of the conjunction, it must be in *both* solution sets.

If sets have no common members, we say that their intersection is the empty set, $\varnothing$. The following two sets have an empty intersection:

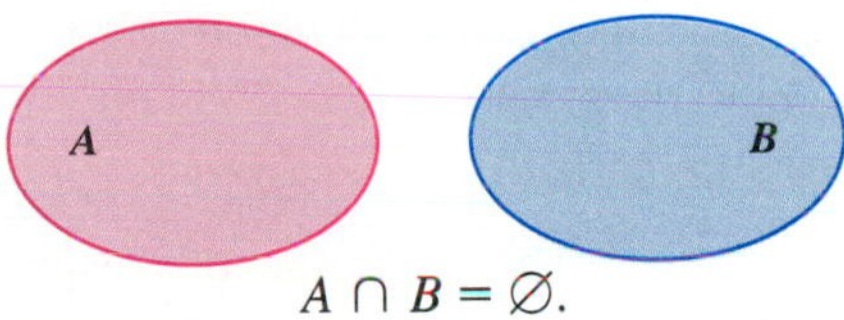

$A \cap B = \varnothing$.

EXAMPLE 4

Solve and graph: $2x - 3 > 1$ *and* $3x - 1 < 2$.

Solution We solve each inequality separately:

$$\begin{aligned} 2x - 3 &> 1 \quad \textit{and} \quad 3x - 1 < 2 \\ 2x &> 4 \quad \textit{and} \quad 3x < 3 \\ x &> 2 \quad \textit{and} \quad x < 1. \end{aligned}$$

The solution set is the intersection of the individual inequalities.

$\{x|\, x > 2\}$

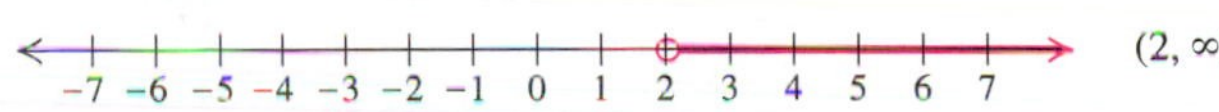

$(2, \infty)$

$\{x|\, x < 1\}$

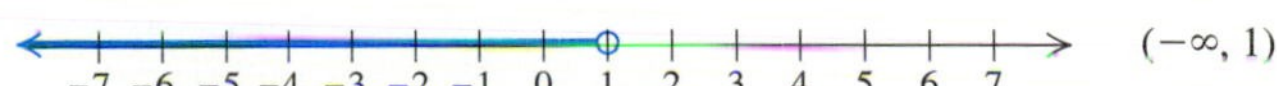

$(-\infty, 1)$

$\{x|\, x > 2\} \cap \{x|\, x < 1\}$
$= \{x|\, x > 2 \textit{ and } x < 1\} = \varnothing$

−7 −6 −5 −4 −3 −2 −1 0 1 2 3 4 5 6 7

$\varnothing$

Since no number is both greater than 2 and less than 1, the solution set is the empty set, $\varnothing$. ❑

Unions of Sets and Disjunctions of Sentences

The **union** of two sets A and B is formed by putting the sets together. We denote the union of sets A and B as

$$A \cup B.$$

The union of two sets is often pictured as shown below.

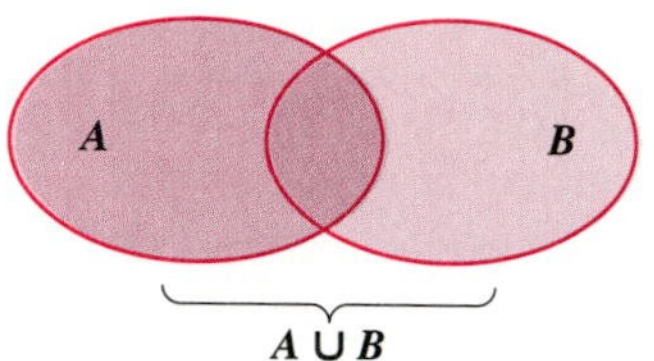

EXAMPLE 5

Find the union: $\{2, 3, 4\} \cup \{3, 5, 7\}$.

Solution The numbers in either or both sets are 2, 3, 4, 5, and 7, so the union is $\{2, 3, 4, 5, 7\}$. ❑

When two or more sentences are joined by the word *or* to make a compound sentence, the new sentence is called a **disjunction** of the sentences. Here are three examples:

$x < -3$ *or* $x > 3$;

y is an odd number *or* y is a prime number;

$x < 0$ *or* $x = 0$ *or* $x > 0$.

For a disjunction to be true, at least one of the individual sentences must be true. The solution set of a disjunction is the union of the individual solution sets. Consider the disjunction

$x < -3$ *or* $x > 3$.

The graphs of each separate sentence are shown below, and the union is the last graph. Again, we use both set-builder and interval notations.

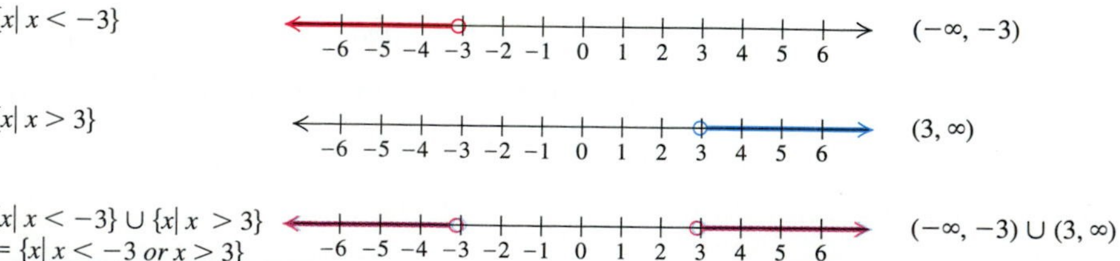

The interval notation $(-\infty, -3) \cup (3, \infty)$ cannot be simplified. The solution set of $x < -3$ *or* $x > 3$ is most commonly written $\{x| \, x < -3 \text{ or } x > 3\}$

Union

The word "or" corresponds to "union" and to the symbol "$\cup$". In order for a number to be a solution of the disjunction, it must be in *at least one* of the solution sets.

EXAMPLE 6

Solve and graph: $7 + 2x < 1$ *or* $13 - 5x \leqslant 3$.

Solution We solve each inequality separately:

$$7 + 2x < 1 \quad or \quad 13 - 5x \leqslant 3$$
$$2x < -6 \quad or \quad -5x \leqslant -10$$
$$x < -3 \quad or \quad x \geqslant 2.$$

To find the solution set of the disjunction, we consider the individual graphs. We graph $x < -3$. We also graph $x \geqslant 2$. Then we take the union of the two graphs:

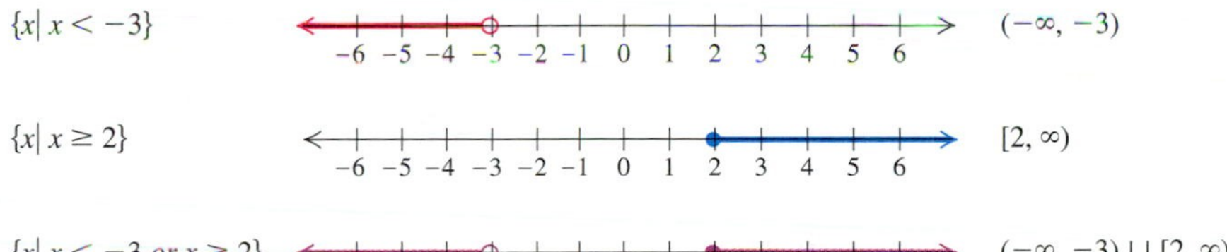

The solution set is $\{x \mid x < -3$ *or* $x \geqslant 2\}$, or, in interval notation, $(-\infty, -3) \cup [2, \infty)$. ❑

CAUTION! A compound inequality like

$$x < -3 \quad or \quad x \geqslant 2,$$

as in Example 6, cannot be abbreviated to one like $2 \leqslant x < -3$ because to do so would be to say that x is *simultaneously* less than -3 and greater than or equal to 2. When the word *or* appears, you must keep that word.

EXAMPLE 7

Solve: $-2x - 5 < -2$ *or* $x - 3 < -10$.

Solution We solve the individual inequalities separately, retaining the word *or:*

$$-2x - 5 < -2 \quad or \quad x - 3 < -10$$
$$-2x < 3 \quad or \quad x < -7$$
$$x > -\tfrac{3}{2} \quad or \quad x < -7.$$

Reverse the symbol. (beside $-2x < 3$ → $x > -\tfrac{3}{2}$)

Keep the word "or."

The solution set consists of all numbers that are less than -7 or greater than $-\frac{3}{2}$. We write the solution set as $\left\{x \mid x < -7 \text{ or } x > -\frac{3}{2}\right\}$, or, in interval notation, $(-\infty, -7) \cup \left(-\frac{3}{2}, \infty\right)$. ❑

EXAMPLE 8

Solve: $3x - 11 < 4$ *or* $4x + 9 \geq 1$.

Solution We solve the individual inequalities separately, retaining the word *or*.

$$3x - 11 < 4 \quad or \quad 4x + 9 \geq 1$$
$$3x < 15 \quad or \quad 4x \geq -8$$
$$x < 5 \quad or \quad x \geq -2$$

Keep the word "or."

To find the solution set, we look at the individual graphs.

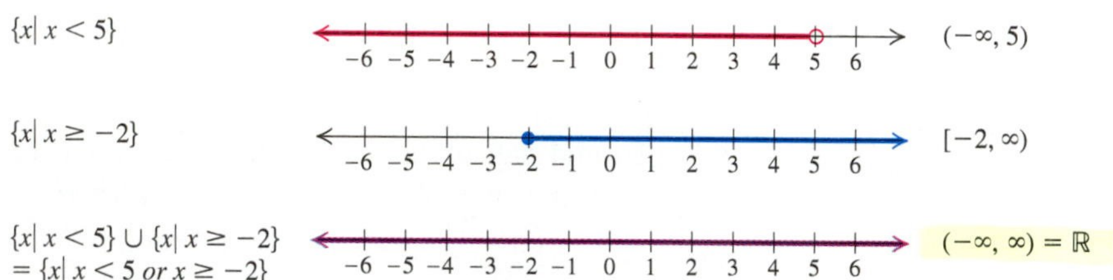

Since *all* numbers are less than 5 or greater than or equal to −2, the two sets fill up the entire number line. Thus the solution set is ℝ, the set of all real numbers. ❑

EXERCISE SET 9.2

Find the intersection.

1. $\{5, 6, 7, 8\} \cap \{4, 6, 8, 10\}$
2. $\{9, 10, 27\} \cap \{8, 10, 38\}$
3. $\{2, 4, 6, 8\} \cap \{1, 3, 5\}$
4. $\{-4, -2, 0\} \cap \{-1, 0, 1\}$
5. $\{1, 2, 3, 4\} \cap \{1, 2, 3, 4\}$
6. $\{8, 9, 10\} \cap \varnothing$

Graph and write interval notation.

7. $1 < x < 6$
8. $0 \leq y \leq 3$
9. $-7 \leq y \leq -3$
10. $-9 \leq x < -5$
11. $-4 \leq -x < 3$
12. $x > -8$ *and* $x < -3$
13. $6 > -x \geq -2$
14. $x > -4$ *and* $x < 2$
15. $5 > x \geq -2$
16. $3 > x \geq 0$
17. $x < 5$ *and* $x \geq 1$
18. $x \geq -2$ *and* $x < 2$

Solve.

19. $-2 < x + 2 < 8$
20. $-1 < x + 1 \leq 6$
21. $1 < 2y + 5 \leq 9$
22. $3 \leq 5x + 3 \leq 8$
23. $-10 \leq 3x - 5 \leq -1$
24. $-18 \leq -2x - 7 < 0$
25. $2 < x + 3 \leq 9$
26. $-6 \leq x + 1 < 9$
27. $-6 \leq 2x - 3 < 6$
28. $4 > -3m - 7 \geq 2$
29. $-\frac{1}{2} < \frac{1}{4}x - 3 \leq \frac{1}{2}$
30. $-\frac{2}{3} \leq 4 - \frac{1}{4}x < \frac{2}{3}$

Find the union.

31. $\{4, 5, 6, 7, 8\} \cup \{1, 4, 6, 11\}$
32. $\{8, 9, 27\} \cup \{2, 8, 27\}$
33. $\{2, 4, 6, 8\} \cup \{1, 3, 5\}$
34. $\{8, 9, 10\} \cup \varnothing$
35. $\{4, 8, 11\} \cup \varnothing$
36. $\varnothing \cup \varnothing$

Graph.

37. $x < -1$ *or* $x > 2$
38. $x < -2$ *or* $x > 0$
39. $x \leq -3$ *or* $x > 1$
40. $x \leq -1$ *or* $x > 3$

Solve.

41. $x + 7 < -2$ *or* $x + 7 > 2$

42. $x + 9 < -4$ *or* $x + 9 > 4$

43. $2x - 8 \leq -3$ *or* $x - 8 \geq 3$

44. $x + 7 \leq -2$ *or* $3x - 7 \geq 2$

45. $7x + 4 \geq -17$ *or* $6x + 5 \geq -7$

46. $4x - 4 < -8$ *or* $4x - 4 < 12$

47. $7 > -4x + 5$ *or* $10 \leq -4x + 5$

48. $6 > 2x - 1$ *or* $-4 \leq 2x - 1$

49. $3x - 7 > -10$ *or* $5x + 2 \leq 22$

50. $3x + 2 < 2$ *or* $4 - 2x < 14$

51. $-2x - 2 < -6$ *or* $-2x - 2 > 6$

52. $-3m - 7 < -5$ *or* $-3m - 7 > 5$

53. $\frac{2}{3}x - 14 < -\frac{5}{6}$ *or* $\frac{2}{3}x - 14 > \frac{5}{6}$

54. $\frac{1}{4} - 3x \leq -3.7$ *or* $\frac{1}{4} - 5x \geq 4.8$

55. $\frac{2x - 5}{6} \leq -3$ *or* $\frac{2x - 5}{6} \geq 4$

56. $\frac{7 - 3x}{5} < -4$ *or* $\frac{7 - 3x}{5} > 4$

57. $5x - 7 \leq 13$ *or* $2x - 1 \geq -7$

58. $5x + 4 \leq 14$ *or* $7 - 2x \geq 9$

Skill Maintenance

Solve.

59. $2x - 3y = 7,$
$3x + 2y = -10$

60. $3x - 9(x + 4) = 20(3x + 7)$

61. $5(2x + 3) = 3(x - 4)$

62. Graph: $3x - 4y = -12$.

Synthesis

63. ◈ Explain how the use of the word *or* in a compound inequality differs from the use of the word *or* in everyday English. (*Hint:* Consider the expression and/or.)

64. ◈ Explain why the conjunction $3 < x$ *and* $x < 5$ can be rewritten as $3 < x < 5$, but the disjunction $3 < x$ *or* $x < 5$ cannot be rewritten as $3 < x < 5$.

65. *Temperatures of liquids.* The formula

$$C = \frac{5}{9}(F - 32)$$

can be used to convert Fahrenheit temperatures F to Celsius temperatures C.

a) Gold is liquid for Celsius temperatures C such that $1063° \leq C < 2660°$. Find a similar such inequality for the corresponding Fahrenheit temperatures.

b) Silver is liquid for Celsius temperatures C such that $960.8° \leq C < 2180°$. Find a similar such inequality for the corresponding Fahrenheit temperatures.

66. We say that x is *between* a and b if $a < x < b$. Find all the numbers on a number line from which you can subtract 3, and still be between -8 and 8.

67. *Converting dress sizes.* The function

$$f(x) = 2(x + 10)$$

can be used to convert dress sizes x in the United States to dress sizes $f(x)$ in Italy. For what dress sizes in the United States will dress sizes in Italy be between 32 and 46?

68. *Pressure at sea depth.* The function

$$P(d) = 1 + \frac{d}{33}$$

gives the pressure, in atmospheres (atm), at a depth of d feet in the sea. For what depths d is the pressure at least 1 atm and at most 7 atm?

69. *Women in the military ranks.* The percentage of the total active military duty force that is women has been steadily increasing. The number N of women in the active duty force t years since 1971 can be predicted by

$$N = 12{,}197.8t + 44{,}000.$$

For what years will the number of women always be at least 50,000 and at most 250,000? (Measure from the end of 1971 and make sure that $N \nless 50{,}000$ and $N \ngtr 250{,}000$ at any point in the time span.)

70. *Records in the women's 100-m dash.* Florence Griffith Joyner set a world record of 10.49 sec in the women's 100-m dash in 1988. The formula

$$R = -0.0433t + 10.49$$

can be used to predict the world record in the women's 100-m dash t years after 1988. Predict (in terms of an inequality) those years for which the world record was between 11.5 and 10.8 sec. (Measure from the end of 1988.)

Solve and graph.

71. $4a - 2 \leq a + 1 \leq 3a + 4$

72. $4m - 8 > 6m + 5$ *or* $5m - 8 < -2$

73. $x - 10 < 5x + 6 \leq x + 10$

74. $2[5(3 - y) - 2(y - 2)] > y + 4$

75. $3x < 4 - 5x < 5 + 3x$

76. $(x + 6)(x - 4) > (x + 1)(x - 3)$

Determine whether the sentence is true or false for all real numbers a, b, and c.

77. If $b > c$, then $b \nleq c$.

78. If $-b < -a$, then $a < b$.

79. If $c \neq a$, then $a < c$.

80. If $a < c$ and $c < b$, then $b \ngeq a$.

81. If $a < c$ and $b < c$, then $a < b$.

82. If $-a < c$ and $-c > b$, then $a < b$.

Solve.

83. $[4x - 2 < 8$ *or* $3(x - 1) < -2]$ *and* $-2 \leq 5x \leq 10$

84. $-2 \leq 4m + 3 < 7$ *and* $[m - 5 \geq 4$ *or* $3 - m > 12]$

9.3 Absolute-Value Equations and Inequalities

Equations with Absolute Value • Inequalities with Absolute Value

Equations with Absolute Value

Recall from Section 1.4 the definition of absolute value.

Absolute Value

The absolute value of x, denoted $|x|$, is defined as

$$|x| = \begin{cases} x, & \text{if } x \geq 0, \\ -x, & \text{if } x < 0. \end{cases}$$

In words, the definition states that the absolute value of a nonnegative number is the number itself, and the absolute value of a negative number is the opposite of the number.

Since distance is always nonnegative, we can think of a number's absolute value as its distance from zero on a number line.

EXAMPLE 1

Find the solution set: **(a)** $|x| = 4$; **(b)** $|x| = 0$; **(c)** $|x| = -7$.

Solution

a) We interpret $|x| = 4$ to mean that the number x is 4 units from zero on a number line. There are two such numbers, 4 and -4. Thus the solution set is $\{-4, 4\}$.

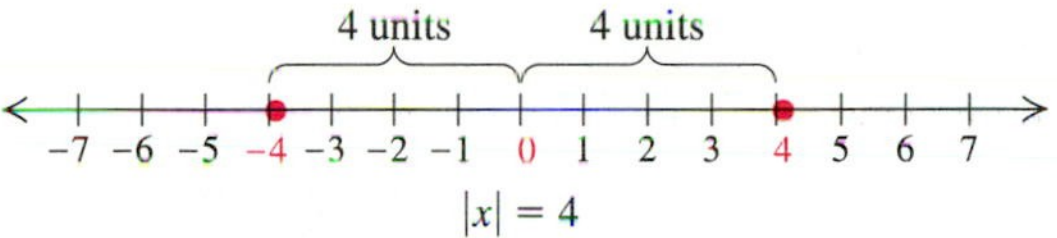

$|x| = 4$

b) We interpret $|x| = 0$ to mean that x is 0 units from zero on a number line. The only number that satisfies this is zero itself. Thus the solution set is $\{0\}$.

c) Since distance is always nonnegative, it doesn't make sense to talk about a number that is -7 units from zero. Remember that the absolute value of a number is never negative. Thus there is no solution; the solution set is $\varnothing$. ❑

Example 1 leads us to the following principle for solving equations with absolute value.

The Absolute-Value Principle for Equations

For any positive number p and any algebraic expression X:

The solutions of $|X| = p$ are those numbers that satisfy $X = -p$ *or* $X = p$.

If $|X| = 0$, then $X = 0$. If p is negative, then $|X| = p$ has no solution.

EXAMPLE 2

Find the solution set: **(a)** $|2x + 5| = 13$; **(b)** $|4 - 7x| = -8$.

Solution

a) We use the absolute-value principle, replacing X by $2x + 5$ and p by 13:

$$|X| = p$$

$$|2x + 5| = 13$$

$$2x + 5 = -13 \quad or \quad 2x + 5 = 13 \qquad \text{Using the absolute-value principle}$$

$$2x = -18 \quad or \quad 2x = 8$$

$$x = -9 \quad or \quad x = 4.$$

Check: For -9:

$\lvert 2x + 5\rvert$	$= 13$
$\lvert 2(-9) + 5\rvert$	? 13
$\lvert -18 + 5\rvert$	
$\lvert -13\rvert$	
13	13 TRUE

For 4:

$\lvert 2x + 5\rvert$	$= 13$
$\lvert 2 \cdot 4 + 5\rvert$	? 13
$\lvert 8 + 5\rvert$	
$\lvert 13\rvert$	
13	13 TRUE

The number $2x + 5$ is 13 units from 0 if x is replaced by -9 or 4. The solution set is $\{-9, 4\}$.

b) The absolute-value principle reminds us that absolute value is always nonnegative. The equation $|4 - 7x| = -8$ has no solution. The solution set is $\varnothing$. ❑

The absolute-value principle can be used together with the addition and multiplication principles to solve many types of equations with absolute value.

EXAMPLE 3

Solve: $2|x + 3| + 1 = 15$.

Solution We first isolate $|x + 3|$. Then we use the absolute-value principle:

$$2|x + 3| + 1 = 15$$

$$2|x + 3| = 14 \qquad \text{Adding } -1 \text{ on both sides}$$

$$|x + 3| = 7 \qquad \text{Multiplying by } \tfrac{1}{2} \text{ on both sides}$$

$$x + 3 = -7 \quad \text{or} \quad x + 3 = 7 \qquad \text{Replacing } X \text{ by } x + 3 \text{ and } p \text{ by 7 in the absolute-value principle}$$

$$x = -10 \quad \text{or} \quad x = 4.$$

We leave the check to the student. The solutions are -10 and 4. The solution set is $\{-10, 4\}$. ❑

EXAMPLE 4

Solve: $|x - 2| = 3$.

Solution This equation can be solved in two different ways.

Method 1. This approach is helpful in calculus. The expressions $|a - b|$ and $|b - a|$ can be used to represent the *distance between a and b on the number line*. For example, the distance between 7 and 8 is given by $|8 - 7|$ or $|7 - 8|$. From this viewpoint, the equation $|x - 2| = 3$ states that the distance between x and 2 is 3 units. We draw a number line and locate those numbers that are 3 units from 2.

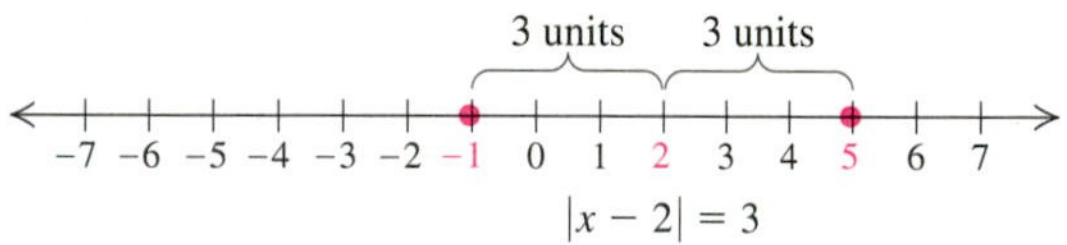

The solutions of $|x - 2| = 3$ are -1 and 5.

Method 2. Using the absolute-value principle, we can interpret the equation as stating that the number $x - 2$ is 3 units from zero. Thus we replace X by $x - 2$ and p by 3:

$$|X| = p$$

$$|x - 2| = 3$$

$$x - 2 = -3 \quad \text{or} \quad x - 2 = 3 \qquad \text{Using the absolute-value principle}$$

$$x = -1 \quad \text{or} \quad x = 5.$$

The check consists of observing that both methods gave the same solutions. The solution set is $\{-1, 5\}$. ❑

Sometimes an equation has two absolute-value expressions. Consider $|a| = |b|$. This means that a and b are the same distance from zero.

If a and b are the same distance from zero, then either they are the same number or they are opposites.

EXAMPLE 5

Solve: $|2x - 3| = |x + 5|$.

Solution Either $2x - 3 = x + 5$ or $2x - 3 = -(x + 5)$. We solve each equation separately:

$$\begin{aligned} 2x - 3 &= x + 5 \quad &\text{or} \quad 2x - 3 &= -(x + 5) \\ x - 3 &= 5 \quad &\text{or} \quad 2x - 3 &= -x - 5 \\ x &= 8 \quad &\text{or} \quad 3x - 3 &= -5 \\ & & 3x &= -2 \\ & & x &= -\tfrac{2}{3}. \end{aligned}$$

We leave the check to the student. The solutions are 8 and $-\frac{2}{3}$. The solution set is $\{8, -\frac{2}{3}\}$. ❑

Inequalities with Absolute Value

Our methods for solving equations with absolute value can be extended for solving inequalities.

EXAMPLE 6

Solve $|x| < 4$. Then graph.

Solution The solutions of $|x| < 4$ are those numbers whose *distance from zero is less than* 4. By substituting or by looking at the number line, we can see that numbers like $-3, -2, -1, -\frac{1}{2}, -\frac{1}{4}, 0, \frac{1}{4}, \frac{1}{2}, 1, 2,$ and 3 are all solutions. In fact, the solutions are all the numbers between -4 and 4. The solution set is $\{x \mid -4 < x < 4\}$ or, in interval notation, $(-4, 4)$. The graph is as follows:

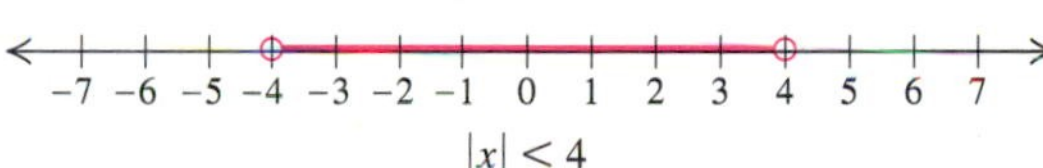

$|x| < 4$ ❑

EXAMPLE 7

Solve $|x| \geq 4$. Then graph.

Solution The solutions of $|x| \geq 4$ are those numbers whose *distance from zero is greater than or equal to* 4—in other words, those numbers x such that $x \leq -4$ or $4 \leq x$. The solution set is $\{x \mid x \leq -4 \text{ or } x \geq 4\}$, or, in interval notation, $(-\infty, -4] \cup [4, \infty)$. We check with numbers like -4.1, -5, 4.1, and 5. Note also that -3.9 and 3.9 are *not* solutions. The graph is as follows:

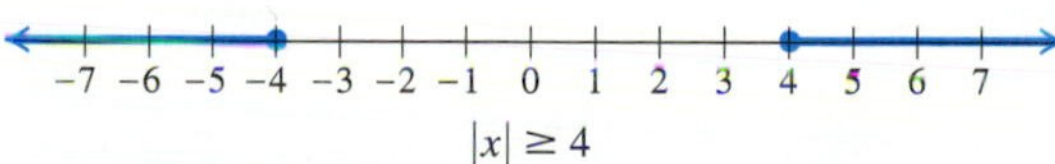

$|x| \geq 4$ ❑

Examples 1, 6, and 7 illustrate three types of problems in which absolute-value signs appear. The following is a general principle for solving such problems.

Principles for Solving Absolute-Value Problems

For any positive number p and any expression X:

a) The solutions of $|X| = p$ are those numbers that satisfy $X = -p$ *or* $X = p$.

b) The solutions of $|X| < p$ are those numbers that satisfy $-p < X < p$.

c) The solutions of $|X| > p$ are those numbers that satisfy $X < -p$ *or* $p < X$.

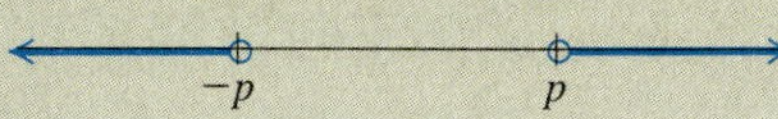

Of course, if p is negative, any value of X will satisfy the inequality $|X| > p$ since absolute value is never negative. By the same reasoning, $|X| < p$ has no solution when p is not positive. Thus the inequality $|2x - 7| > -3$ is true for any real number x, and the inequality $|2x - 7| < -3$ has no solution.

Note that an inequality of the form $|X| < p$ corresponds to a *con*junction, whereas an inequality of the form $|X| > p$ corresponds to a *dis*junction.

EXAMPLE 8 Solve $|3x - 2| < 4$. Then graph.

Solution We use part (b) of the principles listed above. In this case, X is $3x - 2$ and p is 4:

$$|X| < p$$

$|3x - 2| < 4$ Replacing X by $3x - 2$ and p by 4

$-4 < 3x - 2 < 4$ The number $3x - 2$ must be within 4 units of zero.

$-2 < 3x < 6$ Adding 2

$-\frac{2}{3} < x < 2.$ Multiplying by $\frac{1}{3}$

The solution set is $\{x \mid -\frac{2}{3} < x < 2\}$, or, in interval notation, $(-\frac{2}{3}, 2)$. The graph is as follows:

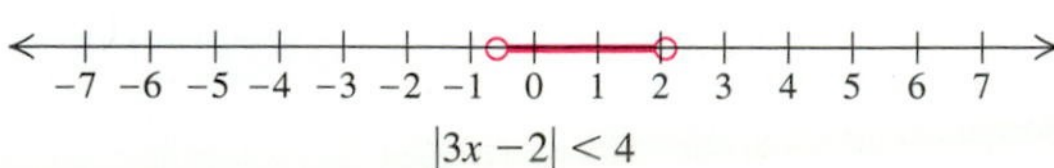

$|3x - 2| < 4$

EXAMPLE 9

Solve $|4x + 2| \geq 6$. Then graph.

Solution We use part (c) of the principles listed above. In this case, X is $4x + 2$ and p is 6:

$\lvert X\rvert \geq p$			
$\lvert 4x + 2\rvert \geq 6$			Replacing X by $4x + 2$ and p by 6
$4x + 2 \leq -6$	*or*	$6 \leq 4x + 2$	The number $4x + 2$ must be at least 6 units from zero.
$4x \leq -8$	*or*	$4 \leq 4x$	Adding -2
$x \leq -2$	*or*	$1 \leq x.$	Multiplying by $\frac{1}{4}$

The solution set is $\{x \mid x \leq -2 \text{ or } x \geq 1\}$, or, in interval notation, $(-\infty, -2] \cup [1, \infty)$. The graph is as follows:

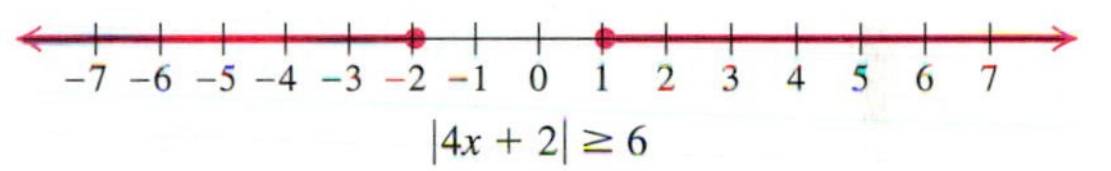

$|4x + 2| \geq 6$

To solve an inequality like $|4x + 2| \geq 6$ with a grapher, simply graph the equation $y_1 = |4x + 2|$ using the ABS key or by typing ABS for the absolute-value function. On the same set of axes, graph $y_2 = 6$. Using the Trace and Zoom features, locate those points on the graph of $y_1 = |4x + 2|$ that are *on or above* the line $y = 6$. The x-values of those points solve the inequality. How can the same graph be used to solve the inequality $|4x + 2| < 6$ or the equation $|4x + 2| = 6$? Try using this procedure to solve Example 8 on your grapher.

EXERCISE SET 9.3

Solve.

1. $|x| = 3$
2. $|x| = 5$
3. $|x| = -3$
4. $|x| = -5$
5. $|p| = 0$
6. $|y| = 8.6$
7. $|t| = 5.5$
8. $|m| = 0$
9. $|x - 3| = 12$
10. $|3x - 2| = 6$
11. $|2x - 3| = 4$
12. $|5x + 2| = 3$
13. $|2y - 7| = 10$
14. $|3y - 4| = 8$
15. $|3x - 10| = -8$
16. $|7x - 2| = -9$
17. $|x| + 7 = 18$
18. $|x| - 2 = 6.3$
19. $|5x| - 3 = 37$
20. $|2y| - 5 = 13$
21. $5|q| - 2 = 9$
22. $7|z| + 2 = 16$
23. $\left|\frac{2x - 1}{3}\right| = 5$
24. $\left|\frac{4 - 5x}{6}\right| = 7$
25. $|m + 5| + 9 = 16$
26. $|t - 7| + 3 = 4$
27. $\left|\frac{1 - 2x}{3}\right| = 1$
28. $\left|\frac{3x - 2}{5}\right| = 2$
29. $5 - 2|3x - 4| = -5$
30. $3|2x - 5| - 7 = -1$

Each pair of numbers represents two points on a number line. Find the distance between the points.

31. 13, 17
32. 9, 15
33. 25, 14
34. 32, 17
35. −9, 24
36. −18, −37
37. −8, −42
38. −9, −36

Solve.

39. $|3x + 4| = |x - 7|$
40. $|2x - 8| = |x + 3|$
41. $|x + 5| = |x - 2|$
42. $|x - 7| = |x + 8|$
43. $|2a + 4| = |3a - 1|$
44. $|5p + 7| = |4p + 3|$
45. $|y - 3| = |3 - y|$
46. $|m - 7| = |7 - m|$
47. $|5 - p| = |p + 8|$
48. $|8 - q| = |q + 19|$
49. $\left|\frac{2x - 3}{6}\right| = \left|\frac{4 - 5x}{8}\right|$
50. $\left|\frac{6 - 8x}{5}\right| = \left|\frac{7 + 3x}{2}\right|$
51. $|\frac{1}{2}x - 5| = |\frac{1}{4}x + 3|$
52. $|2 - \frac{2}{3}x| = |4 + \frac{7}{8}x|$

Solve and graph.

53. $|x| < 3$

54. $|x| \leqslant 5$

55. $|x| \geqslant 2$

56. $|y| > 8$

57. $|t| \geqslant 5.5$

58. $|m| > 0$

59. $|x - 3| < 1$

60. $|x - 2| < 6$

61. $|x + 2| \leqslant 5$

62. $|x + 4| \leqslant 1$

63. $|x - 3| > 1$

64. $|x - 2| > 6$

65. $|2x - 3| \leqslant 4$

66. $|5x + 2| \leqslant 3$

67. $|2y - 7| > -1$

68. $|3y - 4| > 8$

69. $|4x - 9| \geqslant 14$

70. $|9y - 1| \geqslant -3$

71. $|y - 3| < 12$

72. $|p - 2| < 3$

73. $|2x + 3| \leqslant 4$

74. $|5x - 2| \leqslant 3$

75. $|4 - 3y| > 8$

76. $|7 - 2y| < -6$

77. $|9 - 4x| \leqslant 14$

78. $|2 - 9p| \geqslant 17$

79. $|3 - 4x| < -5$

80. $|-5 - 7x| \leqslant 30$

81. $7 + |2x - 1| > 16$

82. $5 + |3x + 2| > 19$

83. $\left|\dfrac{x - 7}{3}\right| < 4$

84. $\left|\dfrac{x + 5}{4}\right| \leqslant 2$

85. $\left|\dfrac{2 - 5x}{4}\right| \geqslant \dfrac{2}{3}$

86. $\left|\dfrac{1 + 3x}{5}\right| > \dfrac{7}{8}$

87. $|m + 5| + 9 \leqslant 16$

88. $|t - 7| + 3 \geqslant 4$

89. $|g + 7| + 13 > 9$

90. $2|2x - 7| + 11 > 10$

91. $\left|\dfrac{2x - 1}{3}\right| \leqslant 1$

92. $\left|\dfrac{3x - 2}{5}\right| \leqslant 2$

Skill Maintenance

93. The perimeter of a rectangular field is 628 m. The length of the field is 6 m greater than the width. Find the area of the field.

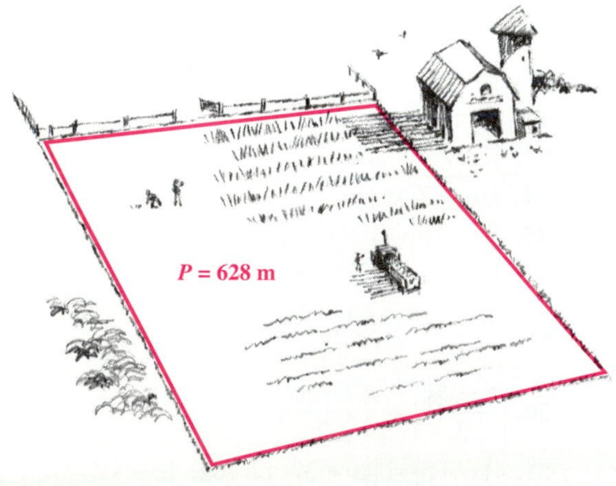

94. At a barbecue, there were 250 dinners served. The cost of a dinner was \$1.50 each for children and \$4.00 each for adults. The total amount of money collected was \$705. How many of each type of plate was served?

Synthesis

95. ◈ Explain why the inequality $|x + 7| < 1$ can be interpreted as "the distance between x and -7 is less than 1."

96. ◈ Explain why the inequality $|x + 5| \geqslant 2$ can be interpreted as "the number x is at least 2 units from -5."

97. From the definition of absolute value, $|x| = x$ only when $x \geqslant 0$. Thus, $|x + 3| = x + 3$ only when $x + 3 \geqslant 0$, which means that $x \geqslant -3$. Solve $|2x - 5| = 2x - 5$ using this same argument.

Solve.

98. $1 - |\frac{1}{4}x + 8| = \frac{3}{4}$

99. $|x + 5| = x + 5$

100. $|x - 1| = x - 1$

101. $|7x - 2| = x + 4$

102. $|3.7x - \frac{4}{9}| > -2$

103. $|5.2x - \frac{6}{7}| \leqslant -8$

104. $|\frac{5}{9} + 3x| < -\frac{1}{6}$

105. $|x + 5| > x$

106. $2 \leqslant |x - 1| \leqslant 5$

Find an equivalent inequality with absolute value.

107. $-3 < x < 3$

108. $-5 \leqslant y \leqslant 5$

109. $x \leqslant -6$ *or* $6 \leqslant x$

110. $x < -4$ *or* $4 < x$

111. $x < -8$ *or* $2 < x$

112. $-5 < x < 1$

113. Pipe is being constructed so that it has a length of 5 ft with a tolerance of $\frac{1}{8}$ in. This means that the length of the pipe can be at most 5 ft plus $\frac{1}{8}$ in. and at least 5 ft minus $\frac{1}{8}$ in. Suppose that p = the length of such a pipe. Find an inequality with absolute value whose solutions are all the possible lengths p.

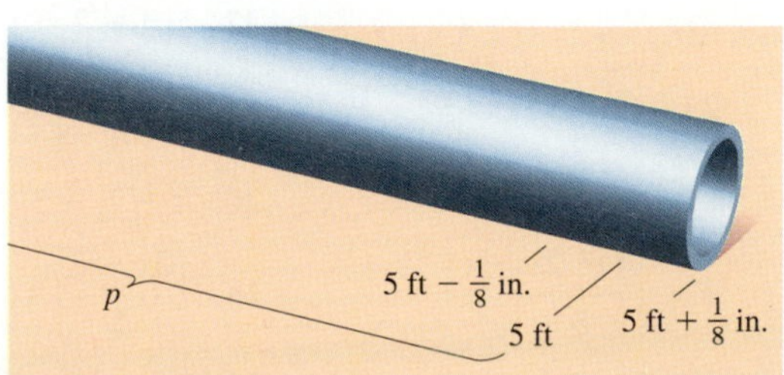

114. A weighted spring is bouncing up and down so that its distance d above the ground satisfies the inequality $|d - 6\text{ ft}| \leq \frac{1}{2}\text{ ft}$ (see the figure at right). Find all possible distances d.

115. Use a grapher to check your solutions to Exercises 1, 9, 13, 41, 53, 63, 87, 99, and 105.

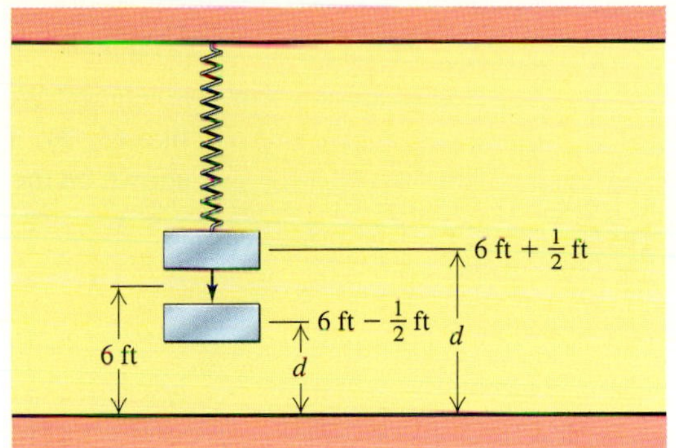

9.4 Inequalities in Two Variables

Solutions of Inequalities in Two Variables • Systems of Linear Inequalities

In Section 9.1, we graphed inequalities in one variable on a number line. Now we graph inequalities in two variables on a plane.

Solutions of Inequalities in Two Variables

The solutions of inequalities in two variables are ordered pairs.

EXAMPLE 1

Determine whether $(-3, 2)$ and $(6, -7)$ are solutions of the inequality $5x - 4y > 13$.

Solution Below, on the left, we replace x by -3 and y by 2. On the right, we replace x by 6 and y by -7.

$5x - 4y$	> 13		$5x - 4y$	> 13	
$5(-3) - 4\cdot 2$	? 13		$5(6) - 4(-7)$	? 13	
$-15 - 8$			$30 + 28$		
-23	13	FALSE	58	13	TRUE

Since $-23 > 13$ is false, $(-3, 2)$ is not a solution.

Since $58 > 13$ is true, $(6, -7)$ is a solution. ❑

We now consider graphs of inequalities in two variables.

EXAMPLE 2

Graph: $y < x$.

Solution We first graph the line $y = x$. Every solution of $y = x$ is an ordered pair like (3, 3). The first and second coordinates are the same. The graph of $y = x$ is shown on the left below. We draw it dashed because these points are *not* solutions of $y < x$.

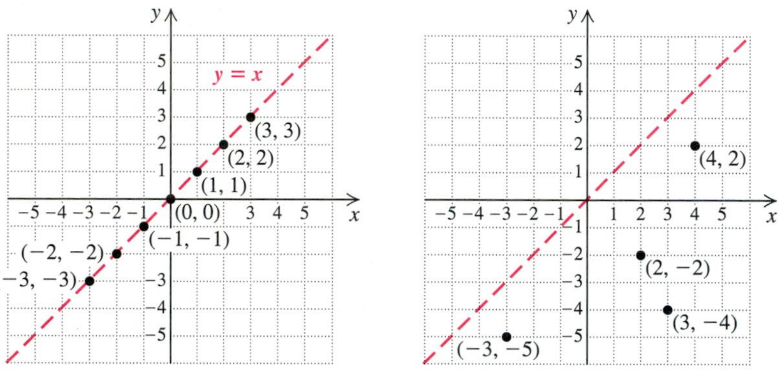

Notice that in the graph on the right each of the ordered pairs on the half-plane below $y = x$ contains a y-coordinate that is less than the x-coordinate. Thus all the pairs shown represent solutions of $y < x$. We can check a pair, (4, 2), as follows:

$$\begin{array}{c|c} y & < x \\ \hline 2 & 4 \end{array} \quad \text{TRUE}$$

It turns out that *any* point on the same side of $y = x$ as (4, 2) is also a solution. Thus, if one point in a half-plane is a solution, then all points in that half-plane are solutions. In this text, we will usually indicate this by color shading. We shade the half-plane below $y = x$.

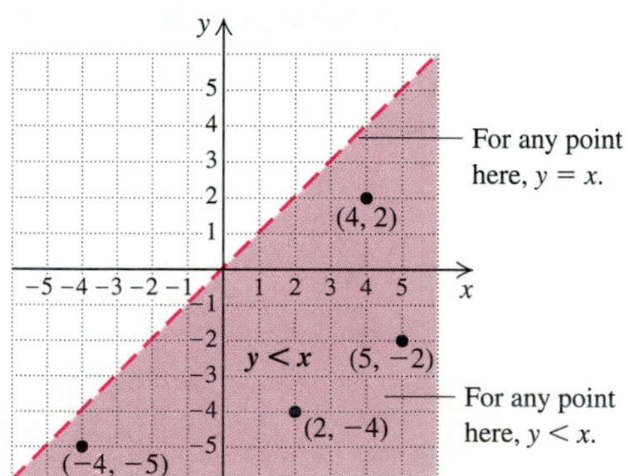

❑

EXAMPLE 3

Graph: $8x + 3y \geqslant 24$.

Solution First we sketch the line $8x + 3y = 24$. Since the inequality sign is $\geqslant$, points on the line $8x + 3y = 24$ are also in the graph of $8x + 3y \geqslant 24$, so we draw a solid line. This indicates that all points on the line are solutions. The rest of the solutions are either in the half-plane above the line or the half-plane below the line. To determine which, we select a point that is not on the line and determine whether it is a solution of $8x + 3y \geqslant 24$. We try $(-3, 4)$ as a test point:

$$\begin{array}{r|l} \multicolumn{2}{c}{8x + 3y \geqslant 24} \\ \hline 8(-3) + 3 \cdot 4 & ?\ 24 \\ -24 + 12 & \\ -12 & 24 \quad \text{FALSE} \end{array}$$

We see that $-12 \geqslant 24$ is *false*. Since $(-3, 4)$ is not a solution, no point in the half-plane containing $(-3, 4)$ is a solution. Thus the points in the other half-plane are solutions. We shade that half-plane and obtain the graph:

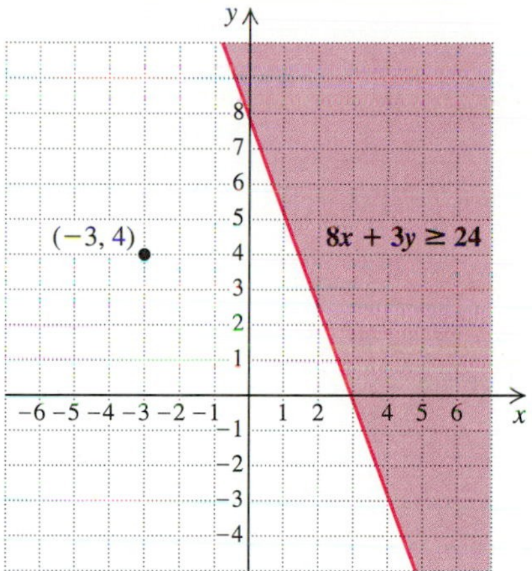

❑

A **linear inequality** is one that we can get from a linear equation by changing the equals sign to an inequality sign. Every linear equation has a graph that is a straight line. The graph of a linear inequality is a half-plane, sometimes including the line along the edge. That line is the graph of what we call a *related equation*. We graph linear inequalities as follows.

To graph an inequality in two variables:

1. Replace the inequality sign with an equals sign and graph this related equation. If the inequality symbol is $<$ or $>$, draw the line dashed. If the inequality symbol is $\leqslant$ or $\geqslant$, draw the line solid.
2. The graph consists of a half-plane, either above or below or to the left or right of the line, and, if the line is solid, the line as well. To determine which half-plane to shade, choose a point not on the line as a test point. Substitute to find whether that point is a solution. If so, shade the half-plane containing that point. If not, shade the other half-plane.

EXAMPLE 4

Graph: $6x - 2y < 12$.

Solution We first graph the line $6x - 2y = 12$. The intercepts are $(0, -6)$ and $(2, 0)$. The point $(3, 3)$ is also on the line. This line forms the boundary of the solutions of the inequality. Since the inequality symbol is $<$, points on the line are not solutions of the inequality and we draw a dashed line with open circles at $(0, -6)$

and (2, 0). To determine which half-plane to shade, we test a point *not* on the line. The point (0, 0) is easy to substitute:

$$\begin{array}{r|l} 6x - 2y & < 12 \\ \hline 6\cdot 0 - 2\cdot 0 & ?\ 12 \\ 0 - 0 & \\ 0 & 12 \quad \text{TRUE} \end{array}$$

Since the inequality $0 < 12$ is *true,* the point (0, 0) is a solution; each point in the half-plane containing (0, 0) is a solution. Thus each point in the other half-plane is *not* a solution. The graph is shown below.

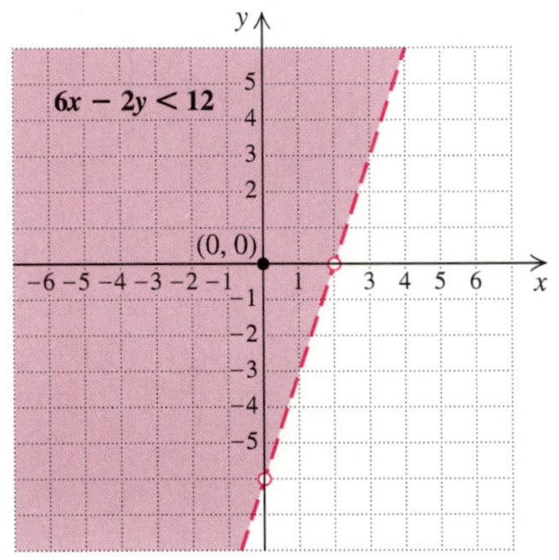

❑

EXAMPLE 5

Graph $x > -3$ on a plane.

Solution There is a missing variable in this inequality. If we graph the inequality on a line, its graph is as follows:

However, we can also write this inequality as $x + 0y > -3$ and consider graphing it in the plane. We use the same technique that we have used with the other examples. We first graph the related equation $x = -3$ in the plane. We then draw the boundary with a dashed line, and use some test point, say, (2, 5):

$$\begin{array}{r|l} x + 0y & > -3 \\ \hline 2 + 0\cdot 5 & ?\ -3 \\ 2 & -3 \quad \text{TRUE} \end{array}$$

Since (2, 5) is a solution, all points in the half-plane containing (2, 5) are solutions. We shade that half-plane. Note that the solutions of $x > -3$ are all pairs with first coordinates greater than -3.

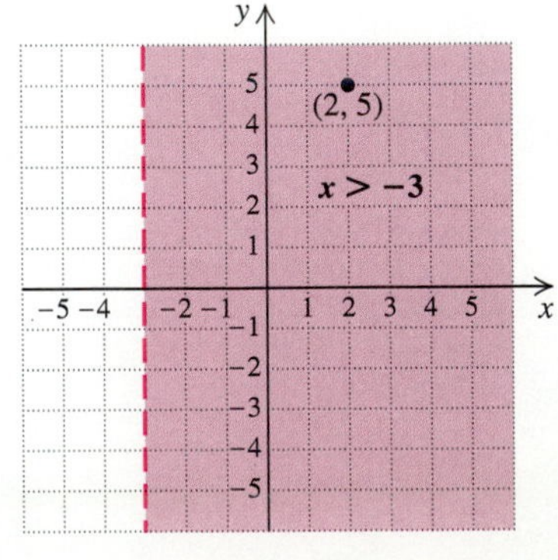

❑

EXAMPLE 6

Graph $y \leq 4$ on a plane.

Solution We first graph $y = 4$ using a solid line to indicate that all points on the line are solutions. We then use $(2, -3)$ as a test point and substitute:

$$\begin{array}{c|c} 0x + y & 4 \\ \hline 0 \cdot 2 + (-3) & ?\ 4 \\ -3 & 4 \end{array} \quad \text{TRUE}$$

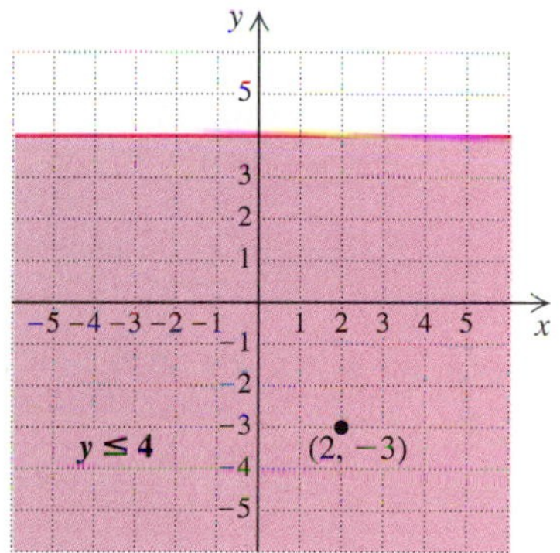

We see that $(2, -3)$ is a solution, so all points in the half-plane containing $(2, -3)$ are solutions. Note that this half-plane consists of all ordered pairs whose y-coordinates are less than or equal to 4. ❑

TECHNOLOGY CONNECTION

We can graph an inequality like $y < 1.2x + 3.49$ on a grapher by using the Shade feature, if it has one. On many calculators, this feature will shade regions only *between* two curves. Thus you may need to enter a second equation, like $y = -100$ or $y = 100$, that you know will bound a shaded region from above or below, out of sight of the $[-10, 10, -10, 10]$ window. To shade $y < 1.2x + 3.49$ in the $[-10, 10, -10, 10]$ window, enter $y_1 = 1.2x + 3.49$ as the upper curve and $y_2 = -100$ as the lower curve. You should see a graph similar to this:

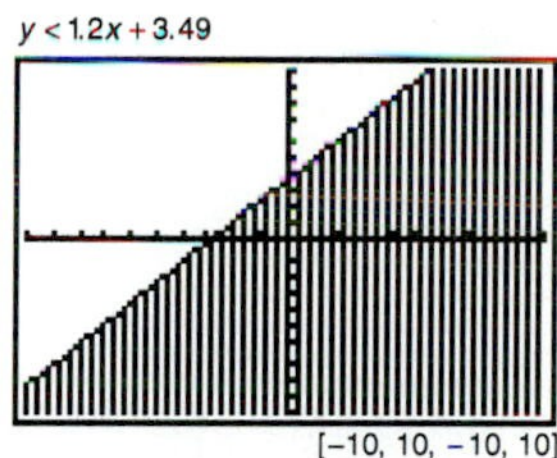

Use a grapher to draw the graphs of the following inequalities. Note that for most graphers we must first solve for y if it is not already isolated. Note too that most graphers draw only solid (not dashed) lines.

TC1. $y > x + 3.5$ **TC2.** $7y \leq 2x + 5$

TC3. $8x - 2y < 11$ **TC4.** $11x + 13y + 4 \geq 0$

Systems of Linear Inequalities

To graph a system of equations, we graph the individual equations and then find the intersection of the individual graphs. We do the same thing for a system of inequalities, that is, we graph each inequality and find the intersection of the individual graphs.

EXAMPLE 7

Graph the system

$$x + y \leq 4,$$
$$x - y < 4.$$

Solution To graph the inequality $x + y \leq 4$, we graph $x + y = 4$ using a solid line. We then consider $(0, 0)$ as a test point and find that it is a solution, so we shade all points in that region red. The arrows near the ends of the line also indicate the half-plane that contains solutions for each inequality.

Next, we graph $x - y < 4$. We graph $x - y = 4$ using a dashed line and consider $(0, 0)$ as a test point. Again, $(0, 0)$ is a solution, so we shade that side of the line using blue shading. The solution set of the system is the region that is shaded purple and part of the line $x + y = 4$.

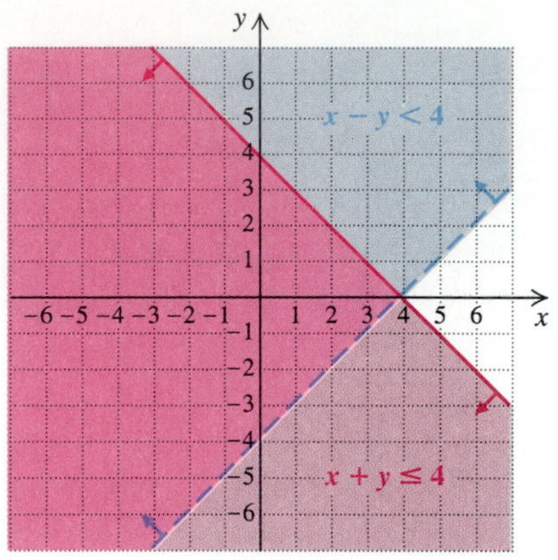

❑

EXAMPLE 8

Graph: $-2 < x \leqslant 3$.

Solution

This is a system of inequalities:

$$-2 < x,$$
$$x \leqslant 3.$$

We graph the equation $-2 = x$, and see that the graph of the first inequality is the half-plane to the right of the line $-2 = x$. It is shaded red.

We graph the second inequality, starting with the line $x = 3$, and find that its graph is the line and also the half-plane to the left of it. It is shaded blue.

The solution set of the system is the region that is the intersection of the individual graphs. Since it is shaded both blue and red, it appears to be purple in the following graph.

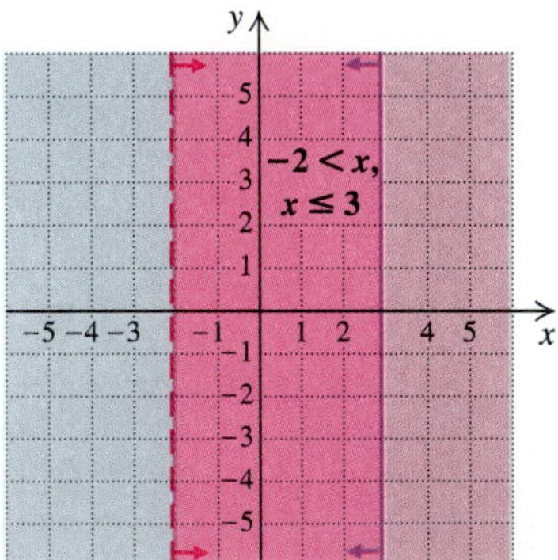

❑

A system of inequalities may have a graph that consists of a polygon and its interior. In the next section, we will need to find the vertices of such a graph.

EXAMPLE 9

Graph the system of inequalities. Find the coordinates of any vertices formed.

$$6x - 2y \leqslant 12, \quad \textbf{(1)}$$
$$y - 3 \leqslant 0, \quad \textbf{(2)}$$
$$x + y \geqslant 0. \quad \textbf{(3)}$$

Solution We graph the lines

$$6x - 2y = 12,$$
$$y - 3 = 0,$$
and $$x + y = 0$$

using solid lines. The regions for each inequality are indicated by the arrows near the ends of the lines. We note where the regions overlap and shade the region of solutions using purple.

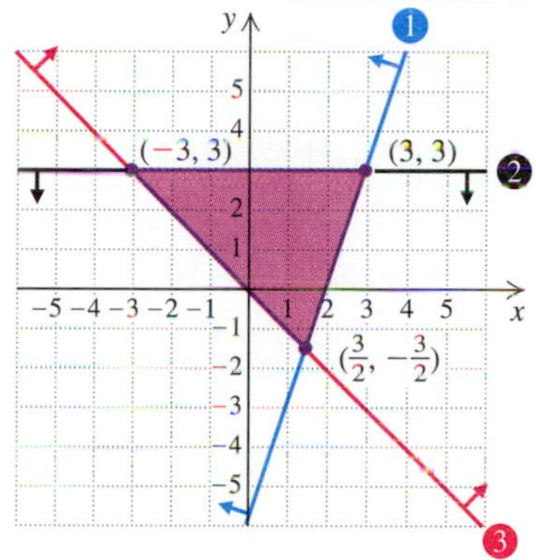

To find the vertices, we solve three different systems of equations. The system of related equations from inequalities (1) and (2) is

$6x - 2y = 12,$
$y - 3 = 0.$ Solving, we obtain the vertex $(3, 3)$.

The system of related equations from inequalities (1) and (3) is

$6x - 2y = 12,$
$x + y = 0.$ Solving, we obtain the vertex $\left(\frac{3}{2}, -\frac{3}{2}\right)$.

The system of related equations from inequalities (2) and (3) is

$y - 3 = 0,$
$x + y = 0.$ Solving, we obtain the vertex $(-3, 3)$. ❑

EXERCISE SET 9.4

Determine whether the ordered pair is a solution of the given inequality.

1. $(-4, 2)$; $2x + y < -5$
2. $(3, -6)$; $4x + 2y \geq 0$
3. $(8, 14)$; $2y - 3x > 5$
4. $(7, 20)$; $3x - y > -1$

Graph on a plane.

5. $y > 2x$
6. $y < 3x$
7. $y < x + 1$
8. $y \leq x - 3$
9. $y > x - 2$
10. $y \geq x + 4$
11. $x + y < 4$
12. $x - y \geq 5$
13. $3x + 4y \leq 12$
14. $2x + 3y < 6$
15. $2y - 3x > 6$
16. $2y - x \leq 4$

17. $3x - 2 \leqslant 5x + y$

18. $2x - 2y \geqslant 8 + 2y$

19. $x < -4$

20. $y \geqslant 2$

21. $y > -2$

22. $x \leqslant 5$

23. $-4 < y < -1$

24. $-2 < y < 3$

25. $-3 \leqslant x \leqslant 3$

26. $-4 \leqslant x \leqslant 4$

27. $0 \leqslant x \leqslant 5$

28. $0 \leqslant y \leqslant 3$

Graph the system of inequalities. Find the coordinates of any vertices formed.

29. $y \leqslant x,$ $y \geqslant -x + 3$

30. $y > x,$ $y < -x + 1$

31. $y \geqslant x,$ $y \leqslant -x + 4$

32. $y \geqslant x,$ $y \leqslant -x + 2$

33. $y \geqslant -2,$ $x \geqslant 1$

34. $y \leqslant -2,$ $x \geqslant 2$

35. $x < 3,$ $y > -3x + 2$

36. $x > -2,$ $y < -2x + 3$

37. $y \geqslant -2,$ $y \geqslant x + 3$

38. $y \leqslant 4,$ $y \geqslant -x + 2$

39. $x + y < 1,$ $x - y < 2$

40. $x + y \leqslant 3,$ $x - y \leqslant 4$

41. $y - 2x \geqslant 1,$ $y - 2x \leqslant 3$

42. $y + 3x > 0,$ $y + 3x < 2$

43. $2y - x \leqslant 2,$ $y - 3x \geqslant -1$

44. $y \leqslant 2x + 1,$ $y \geqslant -2x + 1,$ $x \leqslant 2$

45. $x - y \leqslant 2,$ $x + 2y \geqslant 8,$ $y \leqslant 4$

46. $x + 2y \leqslant 12,$ $2x + y \leqslant 12,$ $x \geqslant 0,$ $y \geqslant 0$

47. $4y - 3x \geqslant -12,$ $4y + 3x \geqslant -36,$ $y \leqslant 0,$ $x \leqslant 0$

48. $8x + 5y \leqslant 40,$ $x + 2y \leqslant 8,$ $x \geqslant 0,$ $y \geqslant 0$

49. $3x + 4y \geqslant 12,$ $5x + 6y \leqslant 30,$ $1 \leqslant x \leqslant 3$

50. $y - x \geqslant 1,$ $y - x \leqslant 3,$ $2 \leqslant x \leqslant 5$

Skill Maintenance

51. One side of a square is 5 less than a side of an equilateral triangle. If the perimeter of the square is the same as the perimeter of the triangle, what is the length of a side of the square? of a side of the triangle?

Solve.

52. $4y - 3x = 8,$ $2x + 5y = -1$

53. $5(3x - 4) = -2(x + 5)$

54. $4(3x + 4) = 2 - x$

Synthesis

55. ◈ Do all systems of linear inequalities have solutions? Why or why not?

56. ◈ Explain how a system of linear inequalities could have a solution set containing exactly one pair.

Graph.

57. $x + y \geqslant 5,$ $x + y \leqslant -3$

58. $x + y \leqslant 8,$ $x + y \leqslant -2$

59. $x - 2y \leqslant 0,$ $-2x + y \leqslant 2,$ $x \leqslant 2,$ $y \leqslant 2,$ $x + y \leqslant 4$

60. $x + y \geqslant 1,$ $-x + y \geqslant 2,$ $x \leqslant 4,$ $y \geqslant 0,$ $y \leqslant 4,$ $x \leqslant 2$

61. *Widths of a basketball floor.* Sizes of basketball floors vary due to building sizes and other constraints such as cost. The length L is to be at most 94 ft and the width W is to be at most 50 ft. Graph a system of inequalities that describes the possible dimensions of a basketball floor.

62. *Hockey wins and losses.* A hockey team determines that it needs at least 60 points for the season in order to make the playoffs. A win w is worth 2 points and a tie t is worth 1 point. Graph a system of inequalities that describes the situation.

63. *Elevators.* Many elevators have a capacity of 1 metric ton (1000 kg). Suppose that c children, each weighing 35 kg, and a adults, each 75 kg, are on an elevator. Graph a system of inequalities that indicates when the elevator is overloaded.

Exercises 64 and 65 can be worked only if your grapher has a Shade feature.

64. Use a grapher to graph the inequality.

a) $3x + 6y > 2$
b) $x - 5y \leqslant 10$
c) $13x - 25y + 10 \leqslant 0$
d) $2x + 5y > 0$

65. Use a grapher to check your answers to Exercises 29–43. If it is available, use the Trace feature to determine the point(s) of intersection.

9.5 Problem Solving Using Linear Programming

Objective Functions and Constraints • Linear Programming

There are many problems in real life in which we want to find a greatest value (a maximum) or a least value (a minimum). For example, if you are in business, you would like to know how to make the *most* profit. Or you might like to know how to make your expenses the *least* possible. Some such problems can be solved using systems of inequalities.

Objective Functions and Constraints

Often a quantity we wish to maximize depends on two or more other quantities. For instance, a gardener's profits P might depend on the number of shrubs s and the number of trees t that are planted. If the gardener makes a \$5 profit from each shrub and a \$9 profit from each tree, the total profit is given by the **objective function**

$$P = 5s + 9t.$$

Thus the gardener might be tempted to simply plant lots of trees since they yield the greater profit. This would be a good idea were it not for the fact that the number of trees and shrubs the gardener plants—and thus the total profit—is subject to the demands, or **constraints,** of the situation. For example, the gardener might be required to plant no more than a total of 10 plants. Thus the objective function would be subject to the *constraint*

$$s + t \leq 10.$$

He or she might also be required to plant at least 3 shrubs. This would subject the objective function to a *second* constraint:

$$s \geq 3.$$

Finally, the gardener might be told to spend no more than \$350 on the plants. If the shrubs cost \$20 each and the trees cost \$50 each, the objective function is subject to a *third* constraint:

The cost of the shrubs plus the cost of the trees cannot exceed \$350.

$$20s \quad + \quad 50t \quad \leq \quad 350$$

In short, the gardener wishes to maximize the objective function

$$P = 5s + 9t$$

subject to the constraints

$$\begin{aligned} s + t &\leq 10, \\ s &\geq 3, \\ 20s + 50t &\leq 350, \\ s &\geq 0, \\ t &\geq 0. \end{aligned}$$

← Because the number of trees and shrubs cannot be negative (applies to $s \geq 0$, $t \geq 0$)

Note that the constraints listed above form a system of linear inequalities that can be graphed.

Linear Programming

The problem facing the gardener is "How many shrubs and trees should be planted, subject to the constraints listed, in order to maximize profit?" To solve such a problem, we use an important result from a branch of mathematics known as **linear programming.**

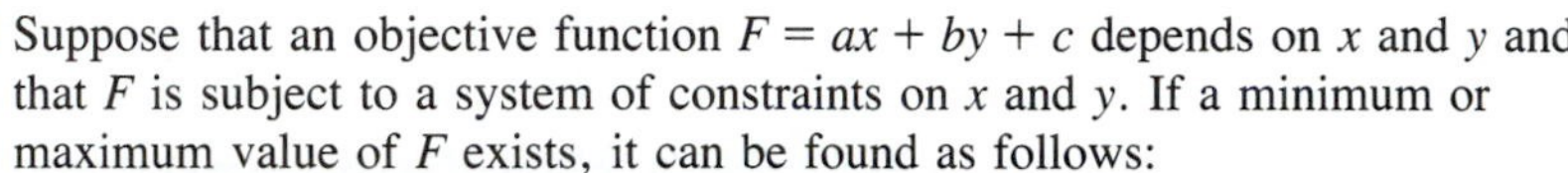

Suppose that an objective function $F = ax + by + c$ depends on x and y and that F is subject to a system of constraints on x and y. If a minimum or maximum value of F exists, it can be found as follows:

1. Graph the system of inequalities and find the vertices.
2. Find the value of the objective function at each vertex. The largest and the smallest of those values are the maximum and the minimum of the function, respectively.

This theorem was proven during World War II, when linear programming was developed to deal with the complicated process of shipping troops and supplies to Europe.

EXAMPLE 1

Solve the gardener's problem discussed above.

Solution We are asked to maximize $P = 5s + 9t$, subject to

$$\begin{aligned} s + t &\leq 10, \\ s &\geq 3, \\ 20s + 50t &\leq 350, \\ s &\geq 0, \\ t &\geq 0. \end{aligned}$$

We graph the system, using the techniques of Section 9.4. The portion of the graph shaded represents all pairs that satisfy the constraints. It is sometimes called the *feasible region*.

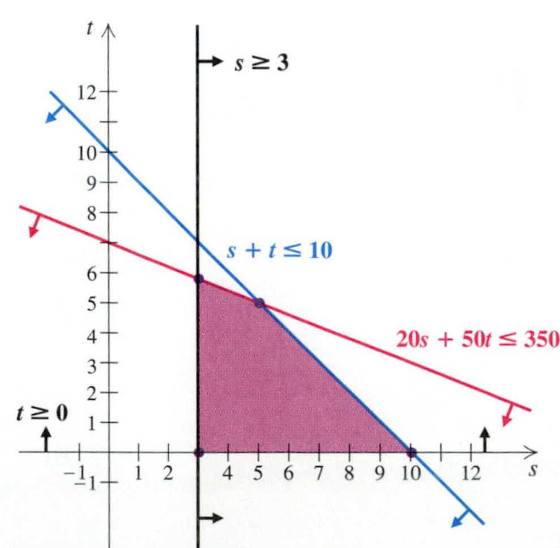

According to the linear programming theorem, P is maximized at one of the vertices of the shaded region. To determine the coordinates of the vertices, we solve the following systems:

$$20s + 50t = 350,$$
$$s = 3;$$

$$s + t = 10,$$
$$20s + 50t = 350;$$

$$s + t = 10,$$
$$t = 0;$$

$$t = 0,$$
$$s = 3.$$

The solutions of the systems are (3, 5.8), (5, 5), (10, 0), and (3, 0), respectively. We now find the value of P at each of these points:

Vertex (s, t)	**Profit** $P = 5s + 9t$	
(3, 5.8)	$5(3) + 9(5.8) = 67.2$	
(5, 5)	$5(5) + 9(5) = 70$	⟵ Maximum
(10,0)	$5(10) + 9(0) = 50$	
(3, 0)	$5(3) + 9(0) = 15$	⟵ Minimum

The largest value of P occurs at (5, 5). Thus profit will be maximized at \$70 if the gardener plants 5 shrubs and 5 trees. Although we were not asked to do so, we have also shown that profit will be minimized at \$15 if 3 shrubs and 0 trees are planted. ❑

EXAMPLE 2

You are taking a test in which multiple-choice questions are worth 10 points each and short-answer questions are worth 15 points each. It takes you 3 min to answer each multiple-choice question and 6 min for each short-answer question. The total time allowed is 60 min, and you are not allowed to answer more than 16 questions. Assuming that all your answers are correct, how many items of each type should you answer in order to get the best score?

Solution

1. FAMILIARIZE. Tabulating information will help us to see the picture.

Type	**Number of Points for Each**	**Time Required for Each**	**Number Answered**
Multiple-choice	10	3 min	x
Short-answer	15	6 min	y
Total time: 60 min			
Total number of items: 16 or fewer			

Note that we have used x to represent the number of multiple-choice questions and y to represent the number of short-answer questions that are answered.

2. TRANSLATE. In this case, it will help to extend the table.

Type	Number of Points for Each	Time Required for Each	Number Answered	Total Time for Type	Total Points for Type
Multiple-choice	10	3 min	x	$3x$	$10x$
Short-answer	15	6 min	y	$6y$	$15y$
Total			$x + y \leq 16$	$3x + 6y \leq 60$	$10x + 15y$
			↑ Because no more than 16 items can be answered	↑ Because the time cannot be more than 60 min	↑ This is the total score on the test.

Suppose that the total score on the test is T. We write T as the objective function in terms of x and y:

$$T = 10x + 15y.$$

We wish to maximize T subject to these facts (constraints) about x and y:

$$x + y \leq 16,$$
$$3x + 6y \leq 60,$$
$$\left.\begin{aligned} x \geq 0, \\ y \geq 0. \end{aligned}\right\} \longleftarrow \text{Because the number of items answered cannot be negative}$$

3. CARRY OUT. The mathematical manipulation consists of graphing the system and evaluating T at each vertex. The graph is as follows:

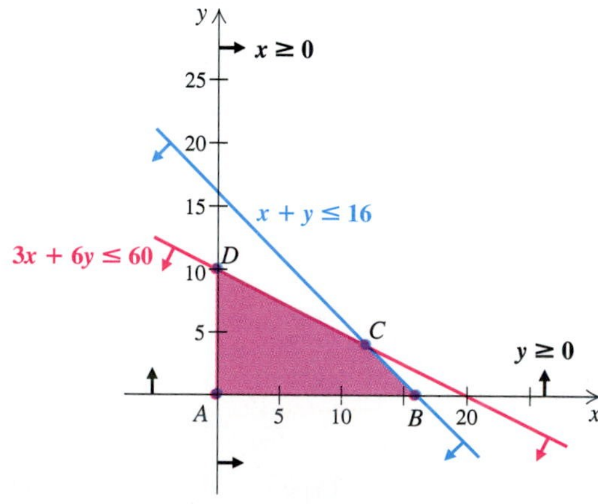

We need to find the coordinates of each vertex. Note that each vertex is the intersection of two lines. Thus we can find each one by solving a system of two linear equations. The coordinates of point A are obviously (0, 0). To find the coordinates of point C, we solve the system

$$3x + 6y = 60, \qquad \textbf{(1)}$$
$$x + y = 16, \qquad \textbf{(2)}$$

as follows:

$$\begin{aligned} 3x + 6y &= 60 \\ -3x - 3y &= -48 \quad \text{Multiplying both sides of Equation (2) by } -3 \\ \hline 3y &= 12 \quad \text{Adding} \\ y &= 4. \end{aligned}$$

Then we find that $x = 12$. Thus the coordinates of vertex C are $(12, 4)$.

Continuing to find the coordinates of the vertices and computing the test score for each ordered pair, we obtain the following:

Vertex (x, y)	**Score** $T = 10x + 15y$
$A\ (0, 0)$	0
$B\ (16, 0)$	160
$C\ (12, 4)$	180
$D\ (0, 10)$	150

The greatest score in the table is 180, obtained when 12 multiple-choice and 4 short-answer questions are answered.

4. **CHECK.** We can go back to the original conditions of the problem and calculate the scores, using the ordered pairs in the table above. We can also check our work, the algebra and the arithmetic. In this case, there is no further checking that we can do without an undue amount of work.

5. **STATE.** The answer is that in order to maximize your score, you should answer 12 multiple-choice questions and 4 short-answer questions. ❑

EXERCISE SET 9.5

Find the maximum and the minimum values of the objective function and the values of x and y for which they occur.

1. $F = 4x + 28y$,
subject to
$5x + 3y \leq 34$,
$3x + 5y \leq 30$,
$x \geq 0$,
$y \geq 0$

2. $G = 14x + 16y$,
subject to
$3x + 2y \leq 12$,
$7x + 5y \leq 29$,
$x \geq 0$,
$y \geq 0$

3. $P = 16x - 2y + 40$,
subject to
$6x + 8y \leq 48$,
$0 \leq y \leq 4$,
$0 \leq x \leq 7$

4. $Q = 24x - 3y + 52$,
subject to
$5x + 4y \leq 20$,
$0 \leq y \leq 4$,
$0 \leq x \leq 3$

5. $F = 5x + 2y + 3$,
subject to
$y \leq 2x + 1$,
$x \leq 5$,
$y \geq 1$

6. $G = 2y - 3x$,
subject to
$y \leq 2x + 1$,
$y \geq -2x + 1$,
$x \leq 2$

Problem Solving

7. The Hockeypuck Biscuit Factory makes two types of biscuits, Biscuit Jumbos and Mitimite Biscuits. The oven can cook at most 200 biscuits per hour. Jumbos each require 2 oz of flour, Mitimites require 1 oz of flour, and there is at most 300 oz of flour available. The income from Jumbos is \$0.10 and from Mitimites is \$0.08. How many of each

type of biscuit should be made in order to maximize income? What is the maximum income?

8. Roschelle owns a car and a moped. She has at most 12 gal of gasoline to be used between the car and the moped. The car's tank holds at most 10 gal and the moped's 3 gal. The mileage for the car is 20 mpg and for the moped is 100 mpg. How many gallons of gasoline should each vehicle use if Roschelle wants to travel as far as possible? What is the maximum number of miles?

9. You are about to take a test that contains matching questions worth 10 points each and essay questions worth 25 points each. You must do at least 3 matching questions, but time restricts doing more than 12. You must do at least 4 essays, but time restricts doing more than 15. You can do no more than 20 questions in total. How many of each type of question must you do to maximize your score? What is this maximum score?

10. You are about to take a test that contains short-answer questions worth 4 points each and word problems worth 7 points each. You must do at least 5 short-answer questions, but time restricts doing more than 10. You must do at least 3 word problems, but time restricts doing more than 10. You can do no more than 18 questions in total. How many of each type of question must you do to maximize your score? What is this maximum score?

11. Yawaka manufactures motorcycles and bicycles. To stay in business, it must produce at least 10 motorcycles each month, but it does not have the facilities to produce more than 60 motorcycles or more than 120 bicycles. The total production of motorcycles and bicycles cannot exceed 160. The profit on a motorcycle is \$134 and on a bicycle, \$20. Find the number of each that should be manufactured in order to maximize profit.

12. Bernie's snack bar sells hamburgers and hot dogs during football games. To stay in business, it must sell at least 10 hamburgers but cannot cook more than 40. It must also sell at least 30 hot dogs but cannot cook more than 70. It cannot cook more than 90 sandwiches altogether. The profit is \$0.33 on a hamburger and \$0.21 on a hot dog. How many of each kind of sandwich should it sell in order to make the maximum profit?

13. Johnson Lumber can convert logs into either lumber or plywood. In a given week, the mill can turn out 400 units of production, of which 100 units of lumber and 150 units of plywood are required by regular customers. The profit on a unit of lumber is \$20 and on a unit of plywood is \$30. How many units of each type should the mill produce in order to maximize the profit?

14. Thano's farm consists of 240 acres of cropland. Thano wishes to plant this acreage in corn or oats. Profit per acre in corn production is \$40 and in oats, \$30. An additional restriction is that the total number of hours of labor during the production period is 320. Each acre of land in corn production uses 2 hr of labor during the production period, while production of oats requires 1 hr per acre. Determine how the land should be divided between corn and oats in order to give maximum profit.

15. Rosa is planning to invest up to \$40,000 in corporate or municipal bonds, or both. The least she is allowed to invest in corporate bonds is \$6000, and she does not want to invest more than \$22,000 in corporate bonds. She also does not want to invest more than \$30,000 in municipal bonds. The interest on corporate bonds is 8% and on municipal bonds is $7\frac{1}{2}\%$. This is simple interest for one year. How much should she invest in each type of bond to earn the most interest? What is the maximum income?

16. Jamaal is planning to invest up to \$22,000 in City Bank or State Bank, or both. He wants to invest at least \$2000 but no more than \$14,000 in City Bank. State Bank does not insure more than a \$15,000 investment, so he will invest no more than that in State Bank. The interest in City Bank is 6% and in State Bank, $6\frac{1}{2}\%$. This is simple interest for one year. How much should he invest in each bank to earn the most interest? What is the maximum income?

17. A pipe tobacco company has 3000 lb of English tobacco, 2000 lb of Virginia tobacco, and 500 lb of Latakia tobacco. To make one batch of Smello tobacco, it takes 12 lb of English tobacco and 4 lb of Latakia. To make one batch of Roppo tobacco, it takes 8 lb of English and 8 lb of Virginia tobacco. The profit is \$10.56 per batch for Smello and \$6.40 for Roppo. How many batches of each kind of tobacco should be made to yield maximum profit? What is the maximum profit? (*Hint:* Organize the information in a table.)

18. It takes a tailoring firm 2 hr of cutting and 4 hr of sewing to make a knit suit. To make a worsted suit, it takes 4 hr of cutting and 2 hr of sewing. At most 20 hr per day are available for cutting and at most 16 hr per day are available for sewing. The profit on a knit suit is $34 and on a worsted suit is $31. How many of each kind of suit should be made to maximize profit?

Synthesis

19. An airline with two types of airplanes, P-1 and P-2, has contracted with a tour group to provide accommodations for a minimum of 2000 first-class, 1500 tourist-class, and 2400 economy-class passengers. Airplane P-1 costs $12,000 per mile to operate and can accommodate 40 first-class, 40 tourist-class, and 120 economy-class passengers, whereas airplane P-2 costs $10,000 per mile to operate and can accommodate 80 first-class, 30 tourist-class, and 40 economy-class passengers. How many of each type of airplane should be used to minimize the operating cost?

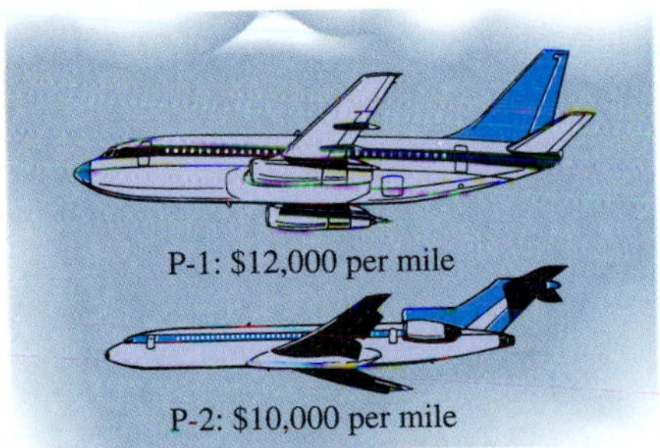

20. A new airplane P-3 becomes available, having an operating cost of $15,000 per mile and accommodating 40 first-class, 40 tourist-class, and 80 economy-class passengers. If airplane P-1 of Exercise 19 were replaced by airplane P-3, how many P-2's and how many P-3's would be needed in order to minimize the operating cost?

21. Guy's Home Furnishings produces chairs and sofas. The chairs require 20 ft of wood, 1 lb of foam rubber, and 2 sq yd of material. The sofas require 100 ft of wood, 50 lb of foam rubber, and 20 sq yd of material. Guy has in stock 1900 ft of wood, 500 lb of foam rubber, and 240 sq yd of material. The chairs can be sold for $20 each and the sofas for $300 each. How many of each should be produced in order to maximize income?

SUMMARY AND REVIEW 9

KEY TERMS

Inequality, p. 440
Solution, p. 440
Solution set, p. 440
Interval notation, p. 442
Open interval, p. 442
Closed interval, p. 442
Half-open interval, p. 442
Compound inequalities, p. 449
Intersection, p. 449
Conjunction, p. 449
Union, p. 452
Disjunction, p. 452
Absolute value, p. 456
Linear inequality, p. 465
Objective function, p. 471
Constraint, p. 471
Linear programming, p. 472

IMPORTANT PROPERTIES AND FORMULAS

Set intersection: $A \cap B = \{x|\ x \text{ is in } A \text{ and } x \text{ is in } B\}$
Set union: $A \cup B = \{x|\ x \text{ is in } A \text{ or in } B, \text{ or both}\}$

For any real numbers a and b with $a < b$, "$a < x$ *and* $x < b$" is equivalent to "$a < x < b$." Intersection corresponds to "and"; union corresponds to "or."

$|a| = a$ if $a \geq 0$; $|a| = -a$ if $a < 0$.

The Absolute-Value Principles for Equations and Inequalities

For any positive number p and any algebraic expression X:

a) The solutions of $|X| = p$ are those numbers that satisfy $X = -p$ or $X = p$.
b) The solutions of $|X| < p$ are those numbers that satisfy $-p < X < p$.
c) The solutions of $|X| > p$ are those numbers that satisfy $X < -p$ or $p < X$.

If $|X| = 0$, then $X = 0$. If p is negative, then $|X| = p$ has no solution.

Suppose that an objective function $F = ax + by + c$ depends on x and y and that F is subject to a system of constraints on x and y. If a minimum or maximum value of F exists, it can be found as follows:

1. Graph the system of inequalities and find the vertices.
2. Find the value of the objective function at each vertex. The largest and the smallest of those values are the maximum and the minimum of the function, respectively.

REVIEW EXERCISES

This chapter's review and test include Skill Maintenance exercises from Sections 2.2, 3.2, 8.2, and 8.3.

Find the solution set. Graph.

1. $x \leq -4$

2. $x + 5 > 6$

3. $a + 7 \leq -14$

4. $y - 5 \geq -12$

5. $4y > -15$

6. $-0.3y < 9$

7. $-6x - 5 < 13$

8. $4y + 3 < -6y - 9$

9. $-\frac{1}{2}x - \frac{1}{4} > \frac{1}{2} - \frac{1}{4}x$

10. $0.3y - 7 < 2.6y + 15$

11. $-2(x - 5) \geq 6(x + 7) - 12$

Solve.

12. You are taking a biology course in which there will be 5 tests. You have scores of 91, 93, 86, and 88 on the first four. You must score a total of 450 in order to get an A. What scores on the last test will give you an A?

13. You are going to invest $30,000, part at 13% and part at 15%. What is the most that can be invested at 13% in order to make at least $4300 interest per year?

14. Find the intersection:

$$\{1, 2, 5, 6, 9\} \cap \{1, 3, 5, 9\}.$$

15. Find the union:

$$\{1, 2, 5, 6, 9\} \cup \{1, 3, 5, 9\}.$$

Solve.

16. $-4 < x + 3 \leq 5$

17. $-15 < -4x - 5 < 0$

18. $3x < -9$ *or* $-5x < -5$

19. $2x + 5 < -17$ *or* $-4x + 10 \leq 34$

20. $x + 5 < -6$ *or* $x + 5 > 6$

21. $2x + 7 \leq -5$ *or* $x + 7 \geq 15$

22. $|x| = 6$

23. $|x| < 0$

24. $|x| \geq 3.5$

25. $|x - 2| = 7$

26. $|2x + 5| < 12$

27. $|3x - 4| \geq 15$

28. $|2x + 5| = |x - 9|$

29. $|5x + 6| = -8$

30. $\left|\frac{x + 4}{8}\right| \leq 1$

31. $2|x - 5| - 7 > 3$

Graph. Find the coordinates of any vertices formed.

32. $y \geq -3,$
$x \geq 2$

33. $x + 3y > -1,$
$x + 3y < 4$

34. $x - 3y \leq 3,$
$x + 3y \geq 9,$
$y \leq 6$

35. Find the maximum and the minimum values of

$$F = 3x + y + 4$$

subject to

$$y \leq 2x + 3,$$
$$x \leq 7,$$
$$y \geq 3.$$

36. LaKenya wants to invest $60,000 in mutual funds and municipal bonds. She does not want to invest more than 50%, or less than 20%, of her money in mutual funds. The minimum investment for municipal bonds is $10,000, and they are guaranteed only up to $40,000, so she will not invest more than $40,000 in municipal bonds. The mutual funds should produce a return of 10%, and the municipal bonds a return of 12%. How much should she invest in each in order to maximize her income? What is her maximum income?

Skill Maintenance

Solve.

37. $5x - 4y = -10,$
$4x + 2y = 5$

38. $3(x + 4) = 2(x - 5)$

39. Graph: $y = -2x - 6$.

40. The perimeter of a rectangular field is 786 ft. The length is 9 ft longer than the width. Find the area of the field.

Synthesis

41. ◈ Explain in your own words why $|x| = p$ has two solutions when p is positive and no solution when p is negative.

42. ◈ Explain why the graph of the solution of a system of linear inequalities is the intersection, not the union, of the individual graphs.

43. Solve: $|2x + 5| \leq |x + 3|$.

44. Determine whether this is true or false: If $x < 3$, then $x^2 < 9$. If false, give an example showing why.

45. Just-For-Fun manufactures marbles with a 1.1-cm diameter and a ±0.03-cm manufacturing tolerance, or allowable variation in diameter. Write the tolerance as an inequality with absolute value.

CHAPTER TEST 9

Solve.

1. $x - 2 < 12$
2. $-0.6y < 30$
3. $-4y - 3 \geq 5$
4. $3a - 5 \leq -2a + 6$
5. $-5y - 1 > -9y + 3$
6. $4(5 - x) < 2x + 5$
7. $-8(2x + 3) + 6(4 - 5x) \geq 2(1 - 7x) - 4(4 + 6x)$
8. You can rent a truck for either \$40 per day with unlimited mileage or \$30 per day with an extra charge of 15¢ a mile. For what numbers of miles traveled would the unlimited mileage plan save you money?
9. On your new job, you can be paid in one of two ways:

   Plan A: A salary of \$30,000 per year;
   Plan B: A salary of \$1600 per month plus a commission of 6% on gross sales.

   For what gross sales is plan A better than plan B?
10. Find the intersection:

    $\{1, 3, 5, 7, 9\} \cap \{3, 5, 11, 13\}$.
11. Find the union:

    $\{1, 3, 5, 7, 9\} \cup \{3, 5, 11, 13\}$.

Solve.

12. $-3 < x - 2 < 4$
13. $-12 \leq -5y - 2 < 0$
14. $-3x > 12$ *or* $4x > -10$
15. $x - 7 \geq -5$ *or* $x - 7 \leq -10$
16. $3x - 2 < 7$ *or* $x - 2 \geq 4$
17. $-\frac{1}{3} \leq \frac{1}{6}x - 1 < \frac{1}{4}$
18. $|x| = 9$
19. $|x| > 3$
20. $|4x - 1| < 4.5$
21. $|-5x - 3| \geq 10$
22. $|x + 10| = |x - 12|$
23. $|2 - 5x| = -10$
24. $\left|\frac{6 - x}{7}\right| \leq 15$

Graph. Find the coordinates of any vertices formed.

25. $x + y \geq 3,$
    $x - y \geq 5$
26. $2y - x \geq -7,$
    $2y + 3x \leq 15,$
    $y \leq 0,$
    $x \leq 0$
27. Find the maximum and the minimum values of

    $$F = 5x + 3y$$

    subject to

    $$x + y \leq 15,$$
    $$1 \leq x \leq 6,$$
    $$0 \leq y \leq 12.$$
28. You are about to take a test that contains questions of type A worth 7 points each and type B worth 12 points each. The total number of questions worked must be at least 8. If you know that type-A questions take 10 min and type-B questions take 8 min and that the maximum time for the test is 80 min, how many of each type of question should you answer in order to maximize your score? What is this maximum score?

Skill Maintenance

Solve.

29. $4(x - 2) - 3(2x + 7) = 4$
30. $3x + 5y = 8,$
    $6x + 9y = -11$
31. Graph: $y = \frac{3}{5}x - 3$.
32. A disc jockey must play 15 commercial spots during one hour of a radio show. Each commercial is either 30 sec or 60 sec long. The total commercial time during that hour is 10 min. How many of each type of commercial were played?

Synthesis

Solve.

33. $|3x - 4| \leq -3$
34. $7x < 8 - 3x < 6 + 7x$

CUMULATIVE REVIEW 1–9

1. Evaluate

$$\frac{2x - y^2}{x + y}$$

for $x = 3$ and $y = -4$.

2. Convert to scientific notation: 5,760,000,000.

3. Determine the slope and the y-intercept for the line given by $7x - 4y = 12$.

4. Find an equation for the line that passes through the points $(-1, 7)$ and $(2, -3)$.

5. Solve the system

$$5x - 2y = -23,$$
$$3x + 4y = 7.$$

6. Solve the system

$$-3x + 4y + z = -5,$$
$$x - 3y - z = 6,$$
$$2x + 3y + 5z = -8.$$

7. Luigi's House of Pizza sold 27 pizzas during one day. Small pizzas sold for \$7.00 and large pizzas sold for \$10.00. The total receipts that day from the pizzas were \$234. How many of each size pizza were sold?

8. The sum of three numbers is 20. The first number is 3 less than twice the third number. The second number minus the third number is -7. What are the numbers?

9. Evaluate: $\begin{vmatrix} 5 & -3 \\ 4 & 6 \end{vmatrix}$.

Solve.

10. $8x = 1 + 16x^2$

11. $625 = 49y^2$

12. $2(x + 5) = 5x - 7$

13. $-0.5y \leqslant 25$

14. $\frac{1}{3}x - \frac{1}{5} \geqslant \frac{1}{5}x - \frac{1}{3}$

15. $-8 < x + 2 < 15$

16. $3x - 2 < -6$ *or* $x + 3 > 9$

17. $|x| > 7$

18. $|3x - 6| = 2$

19. $|4x - 1| \leqslant 14$

20. $\dfrac{2}{n} - \dfrac{7}{n} = 3$

21. $\dfrac{6}{x - 5} = \dfrac{2}{2x}$

22. $\dfrac{3x}{x - 2} - \dfrac{6}{x + 2} = \dfrac{24}{x^2 - 4}$

23. $\dfrac{3x^2}{x + 2} + \dfrac{5x - 22}{x - 2} = \dfrac{-48}{x^2 - 4}$

24. Solve $5m - 3n = 4m + 12$ for n.

25. Solve $P = \dfrac{3a}{a + b}$ for a.

Graph on a plane.

26. $y = \frac{2}{5}x + 1$

27. $3x + 2y = 12$

28. $4x \geqslant 5y + 20$

29. $y < -2$

Perform the indicated operations and simplify.

30. $(2x^2 - 3x + 1) + (6x - 3x^3 + 7x^2 - 4)$

31. $(5x^3y^2)(-3xy^2)$

32. $(3a + b - 2c) - (-4b + 3c - 2a)$

33. $(5x^2 - 2x + 1)(3x^2 + x - 2)$

34. $(2x^2 - 1)^2$

35. $(2x^2 - 1)(2x^2 + 1)$

36. $\dfrac{y^2 - 36}{2y + 8} \cdot \dfrac{y + 4}{y + 6}$

37. $\dfrac{x^4 - 1}{x^2 - x - 2} \div \dfrac{x^2 + 1}{x - 2}$

38. $\dfrac{5ab}{a^2 - b^2} + \dfrac{a + b}{a - b}$

39. $\dfrac{2}{m + 1} + \dfrac{3}{m - 5} - \dfrac{m^2 - 1}{m^2 - 4m - 5}$

40. Simplify: $\dfrac{\frac{1}{x} - \frac{1}{y}}{x + y}$.

41. Divide: $(9x^3 + 5x^2 + 2) \div (x + 2)$.

Factor.

42. $4x^3 + 18x^2$

43. $x^2 + 8x - 84$

44. $16y^2 - 81$

45. $64x^3 + 8$

46. $t^2 - 16t + 64$

47. $x^6 - x^2$

48. $20x^2 + 7x - 3$

49. $x^5 - x^3y + x^2y - y^2$

50. Let $h(x) = x^2 + 1$ and $g(x) = 2x - 5$. Find the following.
a) $h(2)$ **b)** $h(2) + g(2)$ **c)** $(h + g)(x)$

51. If $f(x) = x^2 - 4$ and $g(x) = x^2 - 7x + 10$, find the domain of f/g.

Solve.

52. It takes 45 min for 2 people to shovel a driveway. How long would it take 5 people to shovel the same driveway?

53. Ed's tractor can plow a field in 3 hr. Nell's tractor can plow the field in 1.5 hr. Working together, how long should it take them to plow the field?

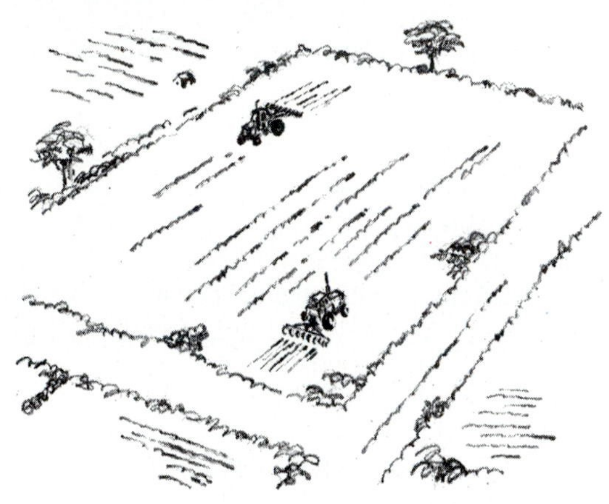

Hours required to mow the field when working together = t

54. The length of a rectangle is 3 ft longer than the width. The area is 54 ft^2. Find the perimeter of the rectangle.

55. The sum of the squares of three consecutive even integers is equal to 8 more than 3 times the square of the second number. Find the integers.

56. Tyler gets a \$54 commission for every water purification system he sells. One week he received \$432 in commissions. How many systems did he sell?

57. Graph the system of inequalities. Find the coordinates of any vertices formed.

$$x + y \leq 2,$$
$$x \geq 0,$$
$$y \geq 1$$

Synthesis

58. Multiply: $(x - 4)^3$.

Solve.

59. $4 \leq |3 - x| \leq 6$

60. $\dfrac{18}{x-9} + \dfrac{10}{x+5} = \dfrac{28x}{x^2 - 4x - 45}$

61. $16x^3 = x$

CHAPTER 10

Rational Exponents and Radicals

AN APPLICATION

Police can estimate the speed at which a car was traveling by measuring its skid marks. The formula

$$r = 2\sqrt{5L}$$

can be used, where r is the speed in miles per hour and L is the length of the skid mark in feet. Estimate the speed of a car that left a skid mark 70 ft long.

This problem appears as Exercise 119 in Section 10.3.

Mark Michael
STATE POLICE OFFICER

"A police officer needs a strong math background. In my work, I use math for everything from daily activity reports to breathalizer tests. Algebra skills are essential for a successful trooper."

In this chapter, we study square roots, cube roots, fourth roots, and so on. We study these in connection with radical expressions that involve these roots, and solve problems involving such roots. We also define and use fractional exponents and complex numbers.

In addition to material from this chapter, the review and test for Chapter 10 include material from Sections 5.7, 5.8, 6.2, and 6.7.

10.1 Radical Expressions

Square Roots • Square-Root Functions • Evaluating Expressions of the Form $\sqrt{a^2}$ • Cube Roots • Odd and Even kth Roots

In this section, we consider roots, such as square roots and cube roots. We look at the symbolism that is used for them and the ways in which we can manipulate symbols to get equivalent expressions. All of this will be important in problem solving.

Square Roots

When a number is raised to the second power, the number is squared. Often we need to know what number was squared in order to produce some value a. If such a number can be found, we call that number a *square root* of a.

Square Root

The number c is a *square root* of a if $c^2 = a$.

For example,

5 is a square root of 25 because $5 \cdot 5 = 25$;

-5 is a square root of 25 because $(-5)(-5) = 25$;

-4 does not have a real-number square root because there is no real number c such that $c^2 = -4$.

Later in this chapter, we will see that there is a number system, different from the real-number system, in which negative numbers do have square roots.

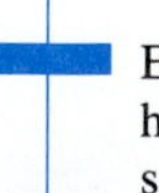

Every positive real number has two real-number square roots. The number 0 has just one square root, 0 itself. Negative numbers do not have real-number square roots.

EXAMPLE 1

Find the two square roots of 64.

Solution The square roots are 8 and -8, because $8^2 = 64$ and $(-8)^2 = 64$. ❑

Principal Square Root

The *principal square root* of a nonnegative number is its nonnegative square root. The symbol $\sqrt{a}$ represents the principal square root of a. To name the negative square root of a, we write $-\sqrt{a}$.

EXAMPLE 2

Simplify each of the following: **(a)** $\sqrt{25}$; **(b)** $\sqrt{\frac{25}{64}}$; **(c)** $-\sqrt{64}$; **(d)** $\sqrt{0.0049}$.

Solution

a) $\sqrt{25} = 5$ — $\sqrt{\ }$ indicates the principal square root.

b) $\sqrt{\frac{25}{64}} = \frac{5}{8}$ — Since $\frac{5}{8} \cdot \frac{5}{8} = \frac{25}{64}$

c) $-\sqrt{64} = -8$ — Since $\sqrt{64} = 8$, $-\sqrt{64} = -8$.

d) $\sqrt{0.0049} = 0.07$ — $(0.07)(0.07) = 0.0049$ ❑

The symbol $\sqrt{\ }$ is called a *radical sign*. An expression written with a radical sign is called a *radical expression*. The expression written under the radical sign is called the *radicand*.

The following are radical expressions:

$$\sqrt{5}, \qquad \sqrt{a}, \qquad -\sqrt{5x}, \qquad \sqrt{\frac{y^2 + 7}{\sqrt{x}}}.$$

An expression like $\sqrt{5}$ can be read as "radical 5," "the square root of 5," or simply "root 5."

Square-Root Functions

For every nonnegative real number x, there is a principal square root $\sqrt{x}$. Thus there is a *square-root function*,

$$f(x) = \sqrt{x}.$$

The domain is the set of all nonnegative real numbers, and the range is the set of all nonnegative real numbers.

x	$\sqrt{x}$	$(x, f(x))$
0	0	(0, 0)
1	1	(1, 1)
4	2	(4, 2)
9	3	(9, 3)

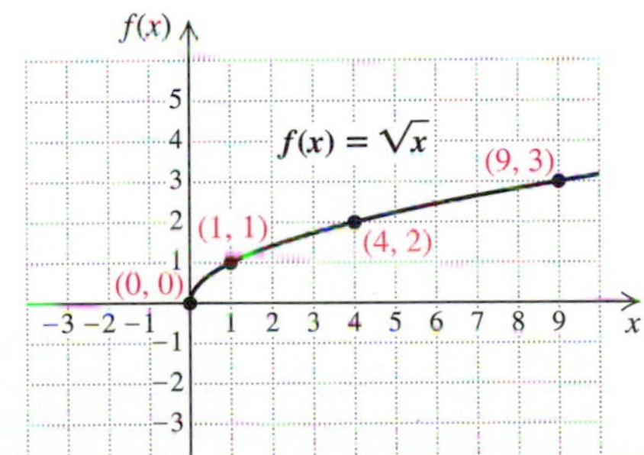

There is another square-root function,

$$g(x) = -\sqrt{x}.$$

The domain of this function is the set of all nonnegative real numbers, and the range is the set of all nonpositive real numbers.

x	$-\sqrt{x}$	$(x, g(x))$
0	0	(0, 0)
1	−1	(1, −1)
4	−2	(4, −2)
9	−3	(9, −3)

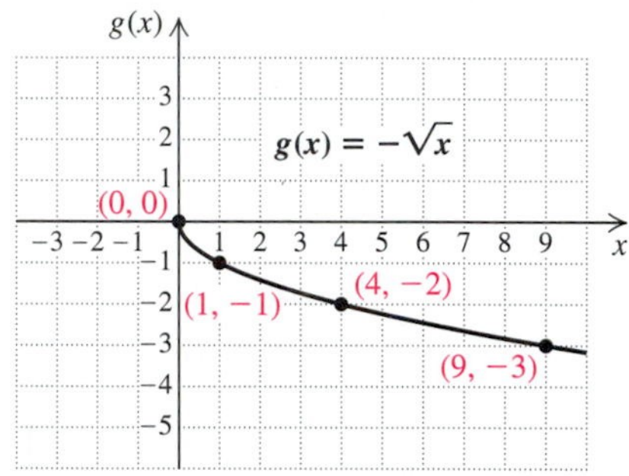

Years ago, the symbol $\sqrt{x}$ was used to represent *all* the square roots of x. That usage has almost completely disappeared, because today we pay a lot more attention to functions. Recall that functions must have exactly one output for each member of the domain.

EXAMPLE 3

For each function, find the indicated function value.

a) $f(x) = \sqrt{3x - 2}$, input 1 **b)** $g(z) = -\sqrt{6z + 4}$, input 2

Solution

a) $f(1) = \sqrt{3 \cdot 1 - 2}$ Substituting

$= \sqrt{1} = 1$ Simplifying and taking the square root

b) $g(2) = -\sqrt{6 \cdot 2 + 4}$ Substituting

$= -\sqrt{16} = -4$ Simplifying, taking the square root, and then taking the opposite ❑

Function Values, Calculators, and Tables. All but the simplest calculators give values for square roots, but they are, for the most part, approximations. For example, if you enter 5 in your calculator and then press the [√‾] key, you will obtain something like

2.23606798,

depending on how your calculator rounds. The exact value is not given by any repeating or terminating decimal. The same will be true of the square root of any whole number that is not a perfect square. We discussed such *irrational numbers* in Chapter 1.

Table 1 at the back of the book contains approximate values of square roots. That table can, of course, be used to find approximate function values for square-root functions. We will have more to say about the table later.

Evaluating Expressions of the Form $\sqrt{a^2}$

We will often have reason to evaluate radical expressions. Note that when 5 or −5 is substituted into the expression $\sqrt{x^2}$, the result is the same.

EXAMPLE 4

Evaluate the expression $\sqrt{x^2}$ for the following values: **(a)** 5; **(b)** 0; **(c)** −5.

Solution

a) $\sqrt{5^2} = \sqrt{25} = 5$

b) $\sqrt{0^2} = \sqrt{0} = 0$

c) $\sqrt{(-5)^2} = \sqrt{25} = 5$

Remember that $\sqrt{m}$ means the nonnegative square root of m. □

We have the following general rule that we can use in simplifying radical expressions.

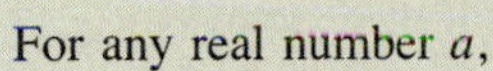

For any real number a,

$$\sqrt{a^2} = |a|.$$

(The principal square root of a^2 is the absolute value of a.)

When an expression contains perfect squares, like $25x^2$ or $(m-3)^2$, in the radicand, we need to use absolute-value signs when simplifying unless we know that the quantities being squared do not represent negative numbers.

EXAMPLE 5

Simplify each expression. Assume that the variable can represent any real number.

a) $\sqrt{(x+1)^2}$ b) $\sqrt{x^2 - 8x + 16}$ c) $\sqrt{(3b)^2}$

Solution

a) $\sqrt{(x+1)^2} = |x+1|$

Since $x + 1$ might be negative (for example, if $x = -3$), absolute-value notation is necessary.

b) $\sqrt{x^2 - 8x + 16} = \sqrt{(x-4)^2} = |x-4|$

Since $x - 4$ might be negative, absolute-value notation is necessary.

c) $\sqrt{(3b)^2} = |3b|$, or $3|b|$

$|3b|$ can be simplified to $3|b|$ because the absolute value of any product is the product of the absolute values. That is, $|a \cdot b| = |a| \cdot |b|$. In this case, $|3b| = |3| \cdot |b|$, or $3|b|$.

□

EXAMPLE 6

Simplify each expression. Assume that no radicands were formed by raising negative quantities to even powers.

a) $\sqrt{y^2}$ b) $\sqrt{(5x+2)^2}$

c) $\sqrt{x^2 - 2x + 1}$ d) $\sqrt{a^6}$

Solution

a) $\sqrt{y^2} = y$ We are assuming that y is nonnegative, so no absolute-value notation is necessary. When y *is* negative, $\sqrt{y^2} \neq y$.

b) $\sqrt{(5x+2)^2} = 5x + 2$ Assuming that $5x + 2$ is nonnegative

c) $\sqrt{x^2 - 2x + 1} = \sqrt{(x-1)^2} = x - 1$ Assuming that $x - 1$ is nonnegative

d) $\sqrt{a^6} = \sqrt{(a^3)^2} = a^3$ Assuming that a^3 is nonnegative □

Cube Roots

We often need to know what number was cubed in order to produce a certain value. When such a number is found, we say that we have found a *cube root*.

Cube Root

The number c is the *cube root* of a if $c^3 = a$. In symbols, we write $\sqrt[3]{a}$ to denote the cube root of a.

For example,

2 is the cube root of 8 because $2^3 = 2 \cdot 2 \cdot 2 = 8$;

-4 is the cube root of -64 because $(-4)^3 = (-4)(-4)(-4) = -64$.

In the real-number system, every number has exactly one cube root. The cube root of a positive number is positive, and the cube root of a negative number is negative. In simplifying expressions involving cube roots, we need not use absolute-value signs.

EXAMPLE 7

For each function, find the indicated function value.

a) $f(y) = \sqrt[3]{y}$, input 125 **b)** $g(x) = \sqrt[3]{x - 3}$, input -24

Solution

a) $f(125) = \sqrt[3]{125} = 5$ Since $5 \cdot 5 \cdot 5 = 125$

b) $g(-24) = \sqrt[3]{-24 - 3}$

$= \sqrt[3]{-27}$

$= -3$ Since $(-3)(-3)(-3) = -27$ ❑

EXAMPLE 8

Simplify: $\sqrt[3]{-8y^3}$.

Solution

$\sqrt[3]{-8y^3} = -2y$ Since $(-2y)(-2y)(-2y) = -8y^3$ ❑

Odd and Even *k*th Roots

The fifth root of a number a is the number c for which $c^5 = a$. There are also 6th roots, 7th roots, and so on. We write $\sqrt[k]{a}$ for the kth root. The number k is called the *index* (plural, indices). When the index is 2, we do not write it.

If k is odd, we say that we are taking an odd root. Every number has just one root for an odd k. If a number is positive, its root is positive. If a number is negative, its root is negative.

EXAMPLE 9

Find each of the following: **(a)** $\sqrt[5]{32}$; **(b)** $\sqrt[5]{-32}$; **(c)** $-\sqrt[5]{32}$; **(d)** $-\sqrt[5]{-32}$; **(e)** $\sqrt[7]{x^7}$; **(f)** $\sqrt[9]{(x-1)^9}$.

Solution

a) $\sqrt[5]{32} = 2$ Since $2^5 = 32$

b) $\sqrt[5]{-32} = -2$ Since $(-2)^5 = -32$

c) $-\sqrt[5]{32} = -2$ Taking the opposite of $\sqrt[5]{32}$

d) $-\sqrt[5]{-32} = -(-2) = 2$ Taking the opposite of $\sqrt[5]{-32}$

e) $\sqrt[7]{x^7} = x$

f) $\sqrt[9]{(x-1)^9} = x - 1$ ❑

Absolute-value signs are never needed when we are finding odd roots.

When the index k in $\sqrt[k]{a}$ is an even number, we say that we are taking an *even* root. Every positive real number has two real kth roots when k is even. One of those roots is positive and one is negative. When k is even, the notation $\sqrt[k]{a}$ indicates the positive kth root. As in the case of square roots, negative numbers do not have real kth roots when k is even. Also, when we are finding even kth roots, absolute-value signs will sometimes be necessary.

EXAMPLE 10

Simplify each expression. Assume that variables can represent any real number.

(a) $\sqrt[4]{16}$; **(b)** $-\sqrt[4]{16}$; **(c)** $\sqrt[4]{-16}$; **(d)** $\sqrt[4]{81x^4}$; **(e)** $\sqrt[6]{(y+7)^6}$

Solution

a) $\sqrt[4]{16} = 2$ Since $2^4 = 16$

b) $-\sqrt[4]{16} = -2$ Taking the opposite of $\sqrt[4]{16}$

c) $\sqrt[4]{-16}$ cannot be simplified. No real-number even root exists.

d) $\sqrt[4]{81x^4} = 3|x|$ Use absolute value since x could represent a negative number.

e) $\sqrt[6]{(y+7)^6} = |y+7|$ Use absolute value since $y + 7$ could be negative. ❑

For any real number a:

a) $\sqrt[k]{a^k} = |a|$ when k is even. We use absolute value when k is even unless a is known to be nonnegative.

b) $\sqrt[k]{a^k} = a$ when k is odd. We do not use absolute value when k is odd.

Observe that for $f(x) = \sqrt{5x - 12}$, we have $f(1) = \sqrt{5 \cdot 1 - 12} = \sqrt{-7}$, which is not a real number. Thus 1 is not in the domain of f. To determine the domain of a function such as $f(x) = \sqrt{5x - 12}$, we find all x-values for which the radicand is nonnegative.

EXAMPLE 11

Determine the domain of $f(x) = \sqrt{5x - 12}$.

Solution We need to find all x-values for which $5x - 12$ is nonnegative. To do so, we solve the inequality $5x - 12 \geq 0$:

$$5x - 12 \geq 0$$
$$5x \geq 12$$
$$x \geq \tfrac{12}{5}.$$

The domain of f is the set of all real numbers greater than or equal to $\frac{12}{5}$—that is,

$$\text{Domain of } f = \{x \mid x \geq \tfrac{12}{5}\}.$$

❑

EXAMPLE 12

Determine the domain of $g(x) = \sqrt[6]{7 - 3x}$.

Solution Since the index is even, the radicand must be nonnegative. We solve the inequality:

$$7 - 3x \geqslant 0$$
$$-3x \geqslant -7$$
$$x \leqslant \tfrac{7}{3}. \quad \text{Multiplying by } -\tfrac{1}{3} \text{ on both sides and reversing the inequality}$$

Thus,

$$\text{Domain of } g = \{x \mid x \leqslant \tfrac{7}{3}\}.$$

EXERCISE SET 10.1

Find the square roots of the number.

1. 16
2. 225
3. 144
4. 9
5. 400
6. 81
7. 49
8. 900

Simplify.

9. $-\sqrt{\frac{49}{36}}$
10. $-\sqrt{\frac{361}{9}}$
11. $\sqrt{196}$
12. $\sqrt{441}$
13. $-\sqrt{\frac{16}{81}}$
14. $-\sqrt{\frac{81}{144}}$
15. $\sqrt{0.09}$
16. $\sqrt{0.36}$
17. $-\sqrt{0.0049}$
18. $\sqrt{0.0144}$

Identify the radicand in the expression.

19. $5\sqrt{p^2 + 4}$
20. $-7\sqrt{y^2 - 8}$
21. $x^2y^2\sqrt[3]{\frac{x}{y + 4}}$
22. $a^2b^3\sqrt[3]{\frac{a}{a^2 - b}}$

For each function, determine whether the specified inputs are members of the domain. If so, find the function values.

23. $f(y) = \sqrt{5y - 10}$;
Inputs: 6, 2, 0, -3
24. $g(x) = \sqrt{x^2 - 25}$;
Inputs: -5, 5, 0, 6, -6
25. $t(x) = -\sqrt{2x + 1}$;
Inputs: 4, -4, 0, 12
26. $p(z) = \sqrt{2z^2 - 20}$;
Inputs: 0, 5, -5, 10, -10
27. $f(t) = \sqrt{t^2 + 1}$;
Inputs: 5, -5, 0, 10, -10
28. $g(x) = -\sqrt{(x + 1)^2}$
Inputs: 1, -1, 3, -3, 5, -5
29. $g(x) = \sqrt{x^3 + 9}$;
Inputs: 2, -2, 3, -3
30. $f(t) = \sqrt{t^3 - 10}$;
Inputs: 2, -2, 3, -3

Simplify. Assume that variables can represent any real number.

31. $\sqrt{16x^2}$
32. $\sqrt{25t^2}$
33. $\sqrt{(-7c)^2}$
34. $\sqrt{(-6b)^2}$
35. $\sqrt{(a + 1)^2}$
36. $\sqrt{(5 - b)^2}$
37. $\sqrt{x^2 - 4x + 4}$
38. $\sqrt{y^2 + 16y + 64}$
39. $\sqrt{4x^2 + 28x + 49}$
40. $\sqrt{9x^2 - 30x + 25}$
41. $\sqrt[4]{625}$
42. $-\sqrt[4]{256}$
43. $\sqrt[5]{-1}$
44. $-\sqrt[5]{-32}$
45. $\sqrt[5]{-\frac{32}{243}}$
46. $\sqrt[5]{-\frac{1}{32}}$
47. $\sqrt[6]{x^6}$
48. $\sqrt[8]{y^8}$
49. $\sqrt[4]{(5a)^4}$
50. $\sqrt[4]{(7b)^4}$
51. $\sqrt[10]{(-6)^{10}}$
52. $\sqrt[12]{(-10)^{12}}$
53. $\sqrt[414]{(a + b)^{414}}$
54. $\sqrt[1976]{(2a + b)^{1976}}$
55. $\sqrt[7]{y^7}$
56. $\sqrt[3]{(-6)^3}$
57. $\sqrt[5]{(x - 2)^5}$
58. $\sqrt[9]{(2xy)^9}$

Simplify. Assume that no radicands were formed by raising negative quantities to even powers.

59. $\sqrt{16x^2}$

60. $\sqrt{25t^2}$

61. $\sqrt{(-6b)^2}$

62. $\sqrt{(-7c)^2}$

63. $\sqrt{(a+1)^2}$

64. $\sqrt{(5+b)^2}$

65. $\sqrt{4x^2+8x+4}$

66. $\sqrt{9x^2+36x+36}$

67. $\sqrt{9t^2-12t+4}$

68. $\sqrt{25t^2-20t+4}$

69. $\sqrt[3]{27}$

70. $-\sqrt[3]{64}$

71. $\sqrt[4]{16x^4}$

72. $\sqrt[4]{(-3x)^4}$

73. $\sqrt[3]{-216}$

74. $-\sqrt[5]{-100,000}$

75. $-\sqrt[3]{-125y^3}$

76. $-\sqrt[3]{-64x^3}$

77. $\sqrt[5]{0.00032(x+1)^5}$

78. $\sqrt[3]{0.000008(y-2)^3}$

79. $\sqrt[6]{64x^{12}}$

80. $\sqrt[4]{81y^{12}}$

For each function, determine whether the specified inputs are members of the domain. If so, find the function values.

81. $f(x) = \sqrt[3]{x+1}$;
Inputs: 7, 26, -9, -65

82. $g(x) = -\sqrt[3]{2x-1}$;
Inputs: 0, -62, -13, 63

83. $g(t) = \sqrt[4]{t-3}$;
Inputs: 19, -13, 1, 84

84. $f(t) = \sqrt[4]{t+1}$;
Inputs: 0, 15, -82, 80

Determine the domain of the function.

85. $f(x) = \sqrt{x+7}$

86. $g(x) = \sqrt{x-10}$

87. $g(t) = \sqrt[4]{2t-9}$

88. $f(x) = \sqrt[4]{x+1}$

89. $g(x) = \sqrt[4]{5-x}$

90. $g(t) = \sqrt[3]{2t-5}$

91. $f(t) = \sqrt[5]{3t+7}$

92. $f(t) = \sqrt[6]{2t+5}$

93. $h(z) = -\sqrt[6]{5z+3}$

94. $d(x) = -\sqrt[4]{7x-5}$

95. $f(t) = 7 + 2\sqrt[8]{3t-5}$

96. $g(t) = 9 - 3\sqrt[8]{5t-4}$

Skill Maintenance

Simplify.

97. $(a^3b^2c^5)^3$

98. $(5a^7b^8)(2a^3b)$

Multiply.

99. $(x-3)(x+3)$

100. $(a+bx)(a-bx)$

Synthesis

101. ◈ If the domain of $f = \{x | x \leq 4\}$, write an equation for $f(x)$ and explain how other equations for $f(x)$ could be formulated.

102. ◈ If the domain of $g = \{x | x \geq 6\}$, write an equation for $g(x)$ and explain how other equations for $g(x)$ could be formulated.

103. *Spaces in a parking lot.* A parking lot has attendants to park the cars. The number N of stalls needed for waiting cars before attendants can get to them is given by the formula $N = 2.5\sqrt{A}$, where A is the number of arrivals in peak hours. Find the number of spaces needed for the given number of arrivals in peak hours: **(a)** 25; **(b)** 36; **(c)** 49; **(d)** 64.

Graph the function.

104. $y = \sqrt{x+3}$

105. $y = \sqrt{x} + 3$

106. $y = \sqrt{x-1}$

107. $y = \sqrt{x} - 2$

10.2 Rational Numbers as Exponents

Rational Exponents • Negative Rational Exponents • Laws of Exponents • Simplifying Radical Expressions

Rational Exponents

Expressions containing rational exponents, like $a^{1/2}$, $5^{-1/4}$, and $(2y)^{4/5}$, have not yet been defined. We will define such expressions so that the usual properties of exponents hold.

Consider $a^{1/2} \cdot a^{1/2}$. If we still want to multiply by adding exponents, it must follow that $a^{1/2} \cdot a^{1/2} = a^{1/2 + 1/2}$, or a^1. Thus we should define $a^{1/2}$ to be a square root of a. Similarly, $a^{1/3} \cdot a^{1/3} \cdot a^{1/3} = a^{1/3 + 1/3 + 1/3}$, or a^1, so $a^{1/3}$ should be defined to mean $\sqrt[3]{a}$.

> For any nonnegative number a and any index n, $a^{1/n}$ means $\sqrt[n]{a}$ (the nonnegative nth root of a). If a is negative, then n must be odd.

EXAMPLE 1

Rewrite without rational exponents: **(a)** $x^{1/2}$; **(b)** $(-27)^{1/3}$; **(c)** $(abc)^{1/5}$.

Solution

a) $x^{1/2} = \sqrt{x}$

b) $(-27)^{1/3} = \sqrt[3]{-27}$
$= -3$

c) $(abc)^{1/5} = \sqrt[5]{abc}$ □

EXAMPLE 2

Rewrite with rational exponents: **(a)** $\sqrt[5]{7xy}$; **(b)** $\sqrt[7]{x^3y/9}$.

Solution Parentheses are required to indicate the base.

a) $\sqrt[5]{7xy} = (7xy)^{1/5}$

b) $\sqrt[7]{\dfrac{x^3y}{9}} = \left(\dfrac{x^3y}{9}\right)^{1/7}$ □

How should we define $a^{2/3}$? If the properties of exponents are to hold, we must have $a^{2/3} = (a^{1/3})^2$ and $a^{2/3} = (a^2)^{1/3}$. This would suggest that $a^{2/3} = (\sqrt[3]{a})^2$ and $a^{2/3} = \sqrt[3]{a^2}$. We make our definition accordingly.

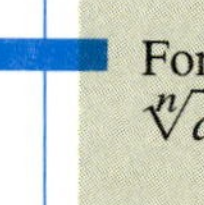

> For any natural numbers m and n ($n \neq 1$) and any real number a for which $\sqrt[n]{a}$ exists,
>
> $a^{m/n}$ means $(\sqrt[n]{a})^m$, or $\sqrt[n]{a^m}$.

EXAMPLE 3

Rewrite without rational exponents: **(a)** $27^{2/3}$; **(b)** $25^{3/2}$.

Solution

a) $27^{2/3} = \sqrt[3]{27^2}$, or $(\sqrt[3]{27})^2$ — This is easier to compute using $(\sqrt[3]{27})^2$.
$= 3^2$, or 9

b) $25^{3/2} = \sqrt[2]{25^3}$, or $(\sqrt[2]{25})^3$ — We normally omit the index 2.
$= 5^3$, or 125 — Taking the square root and cubing □

EXAMPLE 4

Rewrite with rational exponents: **(a)** $\sqrt[3]{9^4}$; **(b)** $(\sqrt[4]{7xy})^5$.

Solution

a) $\sqrt[3]{9^4} = 9^{4/3}$

b) $(\sqrt[4]{7xy})^5 = (7xy)^{5/4}$

The index is the denominator of the rational exponent. ❑

Negative Rational Exponents

Negative rational exponents have a meaning similar to that of negative integer exponents.

For any rational number m/n and any nonzero real number a for which $a^{m/n}$ exists,

$$a^{-m/n} \quad \text{means} \quad \frac{1}{a^{m/n}}.$$

EXAMPLE 5

Rewrite with positive exponents: **(a)** $9^{-1/2}$; **(b)** $(5xy)^{-4/5}$.

Solution

a) $9^{-1/2} = \frac{1}{9^{1/2}}$ $\quad 9^{-1/2}$ is the reciprocal of $9^{1/2}$.

Since $9^{1/2} = \sqrt{9} = 3$, the answer simplifies to $\frac{1}{3}$.

b) $(5xy)^{-4/5} = \frac{1}{(5xy)^{4/5}}$ $\quad (5xy)^{-4/5}$ is the reciprocal of $(5xy)^{4/5}$. ❑

CAUTION! Example 5 shows that a negative exponent does *not* indicate that the expression represents a negative quantity.

Laws of Exponents

The same laws hold for rational-number exponents as for integer exponents. We list them for review.

For any real numbers a and b and any rational exponents m and n for which a^m, a^n, and b^m are defined:

1. $a^m \cdot a^n = a^{m+n}$ — In multiplying, add exponents if the bases are the same.
2. $\frac{a^m}{a^n} = a^{m-n}$ — In dividing, subtract exponents if the bases are the same. (Assume $a \neq 0$.)
3. $(a^m)^n = a^{m \cdot n}$ — To raise a power to a power, multiply the exponents.
4. $(ab)^m = a^m b^m$ — To raise a product to a power, raise each factor to the power and multiply.

EXAMPLE 6

Use the laws of exponents to simplify: **(a)** $3^{1/5} \cdot 3^{3/5}$; **(b)** $7^{1/4}/7^{1/2}$; **(c)** $(7.2^{2/3})^{3/4}$; **(d)** $(a^{-1/3}b^{2/5})^{1/2}$.

Solution

a) $3^{1/5} \cdot 3^{3/5} = 3^{1/5 + 3/5} = 3^{4/5}$ Adding exponents

b) $\dfrac{7^{1/4}}{7^{1/2}} = 7^{1/4 - 1/2} = 7^{1/4 - 2/4} = 7^{-1/4}$ Subtracting exponents after finding a common denominator

c) $(7.2^{2/3})^{3/4} = 7.2^{2/3 \cdot 3/4} = 7.2^{6/12}$ Multiplying exponents

$= 7.2^{1/2}$ Using arithmetic to simplify the exponent

d) $(a^{-1/3}b^{2/5})^{1/2} = a^{-1/3 \cdot 1/2} \cdot b^{2/5 \cdot 1/2}$ Raising a product to a power and multiplying exponents

$= a^{-1/6}b^{1/5}$ ❑

Simplifying Radical Expressions

Rational exponents can be used to simplify some radical expressions. The procedure is as follows.

1. Convert radical expressions to exponential expressions.
2. Use arithmetic and the laws of exponents to simplify.
3. Convert back to radical notation when appropriate.

EXAMPLE 7

Use rational exponents to simplify: **(a)** $\sqrt[6]{x^3}$; **(b)** $\sqrt[3]{8r^{12}}$; **(c)** $\sqrt[8]{a^2b^4}$.

Solution

a) $\sqrt[6]{x^3} = x^{3/6}$ Converting to an exponential expression

$= x^{1/2}$ Using arithmetic to simplify the exponent

$= \sqrt{x}$ Converting back to radical notation

b) $\sqrt[3]{8r^{12}} = (8r^{12})^{1/3}$ Converting to exponential notation

$= 8^{1/3}(r^{12})^{1/3}$ Raising a product to a power

$= 2r^4$ $\sqrt[3]{8} = 2$; $r^{12 \cdot 1/3} = r^4$

c) $\sqrt[8]{a^2b^4} = (a^2b^4)^{1/8}$ Converting to exponential notation

$= a^{2/8} \cdot b^{4/8}$ Raising a product to a power and multiplying exponents

$= a^{1/4} \cdot b^{1/2}$ Using arithmetic to simplify the exponents

$= a^{1/4} \cdot b^{2/4}$ Converting the exponents to fractions that have the least common denominator of the fractions

$= (a^1b^2)^{1/4}$ Multiplying exponents; raising a product to a power (in reverse)

$= \sqrt[4]{ab^2}$ Converting back to radical notation ❑

EXERCISE SET | 10.2

Note: Assume for all exercises that even roots are of nonnegative quantities and that all denominators are nonzero.

Rewrite without fractional exponents.

1. $x^{1/4}$
2. $y^{1/5}$
3. $(8)^{1/3}$
4. $(16)^{1/2}$
5. $81^{1/4}$
6. $64^{1/3}$
7. $9^{1/2}$
8. $25^{1/2}$
9. $(xyz)^{1/3}$
10. $(ab)^{1/4}$
11. $(a^2b^2)^{1/5}$
12. $(x^3y^3)^{1/4}$
13. $a^{2/3}$
14. $b^{3/2}$
15. $16^{3/4}$
16. $4^{7/2}$
17. $49^{3/2}$
18. $27^{4/3}$
19. $9^{5/2}$
20. $81^{3/2}$
21. $(81x)^{3/4}$
22. $(125a)^{2/3}$
23. $(25x^4)^{3/2}$
24. $(9y^6)^{3/2}$
25. $(8a^2b^4)^{2/3}$
26. $(27a^5b)^{2/3}$

Rewrite with fractional exponents.

27. $\sqrt[3]{20}$
28. $\sqrt[3]{19}$
29. $\sqrt{17}$
30. $\sqrt{6}$
31. $\sqrt{x^3}$
32. $\sqrt{a^5}$
33. $\sqrt[5]{m^2}$
34. $\sqrt[5]{n^4}$
35. $\sqrt[4]{cd}$
36. $\sqrt[5]{xy}$
37. $\sqrt[5]{xy^2z}$
38. $\sqrt[7]{x^3y^2z^2}$
39. $(\sqrt{3mn})^3$
40. $(\sqrt[3]{7xy})^4$
41. $(\sqrt[7]{8x^2y})^5$
42. $(\sqrt[6]{2a^5b})^7$

Rewrite with positive exponents.

43. $x^{-1/3}$
44. $y^{-1/4}$
45. $(2rs)^{-3/4}$
46. $(5xy)^{-5/6}$
47. $\left(\frac{1}{10}\right)^{-2/3}$
48. $\left(\frac{1}{8}\right)^{-3/4}$
49. $\frac{1}{x^{-2/3}}$
50. $\frac{1}{x^{-5/6}}$
51. $\frac{5}{\sqrt[4]{x}}$
52. $\frac{8}{\sqrt[3]{a}}$
53. $\frac{3}{5m^{-1/2}}$
54. $\frac{2}{7x^{-1/3}}$

Use the properties of exponents to simplify.

55. $5^{3/4} \cdot 5^{1/8}$
56. $11^{2/3} \cdot 11^{1/2}$
57. $\frac{7^{5/8}}{7^{3/8}}$
58. $\frac{9^{9/11}}{9^{7/11}}$
59. $\frac{8.3^{3/4}}{8.3^{2/5}}$
60. $\frac{3.9^{3/5}}{3.9^{1/4}}$
61. $(10^{3/5})^{2/5}$
62. $(5^{5/4})^{3/7}$
63. $a^{2/3} \cdot a^{5/4}$
64. $x^{3/4} \cdot x^{2/3}$
65. $(x^{2/3})^{3/7}$
66. $(a^{3/2})^{2/5}$
67. $(m^{2/3}n^{1/2})^{1/4}$
68. $(x^{1/3}y^{2/5})^{1/4}$
69. $(a^{-2/3}b^{-1/4})^{-6}$
70. $(m^{-1/5}n^{-5/6})^{-10}$

Use fractional exponents to simplify.

71. $\sqrt[6]{a^4}$
72. $\sqrt[6]{y^2}$
73. $\sqrt[3]{8y^6}$
74. $\sqrt{x^4y^6}$
75. $\sqrt[4]{32}$
76. $\sqrt[8]{81}$
77. $\sqrt[6]{4x^2}$
78. $\sqrt[4]{16x^4y^2}$
79. $\sqrt[5]{32c^{10}d^{15}}$
80. $\sqrt[4]{16x^{12}y^{16}}$
81. $\sqrt[6]{\frac{m^{12}n^{24}}{64}}$
82. $\sqrt[5]{\frac{x^{15}y^{20}}{32}}$
83. $\sqrt[8]{r^4s^2}$
84. $\sqrt[12]{64t^6s^6}$
85. $\sqrt[3]{27a^3b^9}$
86. $\sqrt[4]{81x^8y^8}$

Skill Maintenance

Solve.

87. $x^2 - 1 = 8$
88. $3x - 4 = 5x + 7$
89. $\frac{1}{x} + 2 = 5$
90. For homes under \$100,000, the real-estate transfer tax in Vermont is 0.5% of the selling price. Find the selling price of a home that had a transfer tax of \$467.50.

Synthesis

91. ◈ If $f(x) = (x + 5)^{1/2}(x + 7)^{-1/2}$, find the domain of f. Explain how you found your answer.
92. ◈ How does the graph of $f(x) = x^{1/2}$ compare with the graph of $g(x) = -x^{1/2}$?

Use rational exponents to simplify.

93. $\sqrt[5]{x^2y\sqrt{xy}}$
94. $\sqrt{x^5\sqrt[3]{x^4}}$
95. $\sqrt[4]{\sqrt[3]{8x^3y^6}}$
96. $\sqrt[12]{p^2 + 2pq + q^2}$
97. ▦ *Road pavement messages.* In a psychological study, it was determined that the proper length L of the letters of a word printed on pavement is given by

$$L = \frac{0.000169d^{2.27}}{h},$$

where d is the distance of a car from the lettering and h is the height of the eye above the surface of the road. All units are in meters. This formula says that if a person is h meters above the surface of the road and is to be able to recognize a message d meters away, that message will be the most recognizable if the length of the letters is L. Find L to the nearest tenth of a meter, given d and h.

a) $h = 1$ m, $d = 60$ m
b) $h = 0.9906$ m, $d = 75$ m
c) $h = 2.4$ m, $d = 80$ m
d) $h = 1.1$ m, $d = 100$ m

98. Graph the function $f(x) = x^{3/2}$, for $x \geqslant 0$, and compare it to the graphs of $y = x$ and $y = x^2$.

99. The function $r(t) = 10^{-12}2^{-t/5700}$ expresses the ratio of carbon isotopes to carbon atoms in a fossil that is t years old. What ratio of carbon isotopes to carbon atoms would a 1900-year-old bone have?

10.3 Multiplying and Simplifying with Radical Expressions

Multiplying Radical Expressions • Simplifying by Factoring • Multiplying and Simplifying • Approximating Square Roots

Multiplying Radical Expressions

Note that $\sqrt{4}\sqrt{25} = 2 \cdot 5 = 10$. Also $\sqrt{4 \cdot 25} = \sqrt{100} = 10$. Likewise,

$$\sqrt[3]{27}\sqrt[3]{8} = 3 \cdot 2 = 6 \quad \text{and} \quad \sqrt[3]{27 \cdot 8} = \sqrt[3]{216} = 6.$$

These examples suggest the following.

The Product Rule for Radicals

For any real numbers $\sqrt[k]{a}$ and $\sqrt[k]{b}$,

$$\sqrt[k]{a} \cdot \sqrt[k]{b} = \sqrt[k]{a \cdot b}.$$

(To multiply, multiply the radicands.)

EXAMPLE 1

Multiply: **(a)** $\sqrt{3} \cdot \sqrt{5}$; **(b)** $\sqrt{x+3}\sqrt{x-3}$; **(c)** $\sqrt[3]{4} \cdot \sqrt[3]{5}$; **(d)** $\sqrt[4]{\frac{y}{5}} \cdot \sqrt[4]{\frac{7}{x}}$.

Solution

a) $\sqrt{3} \cdot \sqrt{5} = \sqrt{3 \cdot 5} = \sqrt{15}$

b) $\sqrt{x+3}\sqrt{x-3} = \sqrt{(x+3)(x-3)} = \sqrt{x^2 - 9}$

CAUTION! $\sqrt{x^2 - 9} \neq \sqrt{x^2} - \sqrt{9}$.

c) $\sqrt[3]{4}\,\sqrt[3]{5} = \sqrt[3]{4 \cdot 5} = \sqrt[3]{20}$

d) $\sqrt[4]{\frac{y}{5}} \cdot \sqrt[4]{\frac{7}{x}} = \sqrt[4]{\frac{y}{5} \cdot \frac{7}{x}} = \sqrt[4]{\frac{7y}{5x}}$ ❑

It is important to remember that the product rule for radicals can be used only when radicals have the same index. When indices differ, rational exponents can be useful. Study the steps in the following example carefully.

EXAMPLE 2

Multiply: **(a)** $\sqrt[3]{5} \cdot \sqrt{2}$; **(b)** $\sqrt{x-2} \cdot \sqrt[4]{3y}$.

Solution

a) $\sqrt[3]{5} \cdot \sqrt{2} = 5^{1/3} \cdot 2^{1/2}$ — Converting to exponential notation

$= 5^{2/6} \cdot 2^{3/6}$ — Rewriting so that exponents have a common denominator

$= (5^2 \cdot 2^3)^{1/6}$ — Using the laws of exponents

$= \sqrt[6]{5^2 \cdot 2^3}$ — Converting back to radical notation

$= \sqrt[6]{200}$ — Multiplying under the radical

b) $\sqrt{x-2} \cdot \sqrt[4]{3y} = (x-2)^{1/2}(3y)^{1/4}$ — Converting to exponential notation

$= (x-2)^{2/4}(3y)^{1/4}$ — Writing exponents with a common denominator

$= [(x-2)^2(3y)]^{1/4}$ — Using the laws of exponents

$= \sqrt[4]{(x^2-4x+4) \cdot 3y}$ — Converting back to radical notation

$= \sqrt[4]{3x^2y - 12xy + 12y}$ — Multiplying under the radical ❑

Simplifying by Factoring

Reading the product rule from right to left, we have

$$\sqrt[k]{ab} = \sqrt[k]{a} \cdot \sqrt[k]{b}.$$

This shows a way to factor and thus simplify radical expressions. Consider $\sqrt{20}$. The number 20 has the factor 4, which is a perfect square. Therefore,

$\sqrt{20} = \sqrt{4 \cdot 5}$ — Factoring the radicand (4 is a perfect square)

$= \sqrt{4} \cdot \sqrt{5}$ — Factoring into two radicals

$= 2\sqrt{5}.$ — Taking the square root of 4

To simplify a radical expression by factoring, look for the largest factors of the radicand that are perfect kth powers (where k is the index). Then take the kth root of the resulting factors. A radical expression, with index k, is *simplified* when its radicand has no factors that are perfect kth powers.

EXAMPLE 3

Simplify by factoring: **(a)** $\sqrt{200}$; **(b)** $\sqrt[3]{32}$; **(c)** $\sqrt[4]{48}$.

Solution

a) $\sqrt{200} = \sqrt{100 \cdot 2} = \sqrt{100} \cdot \sqrt{2} = 10\sqrt{2}$ — This is the largest perfect square factor of 200.

b) $\sqrt[3]{32} = \sqrt[3]{8 \cdot 4} = \sqrt[3]{8} \cdot \sqrt[3]{4} = 2\sqrt[3]{4}$ — This is the largest perfect cube (third power) factor of 32.

c) $\sqrt[4]{48} = \sqrt[4]{16 \cdot 3} = \sqrt[4]{16} \cdot \sqrt[4]{3} = 2\sqrt[4]{3}$ — This is the largest fourth-power factor of 48. ❑

In many situations, we can assume that no radicands were formed by raising negative quantities to even powers. We will make this assumption and henceforth refrain from using absolute-value notation when taking even roots.

EXAMPLE 4

Simplify by factoring: **(a)** $\sqrt{5x^2}$; **(b)** $\sqrt{2x^2 - 4x + 2}$; **(c)** $\sqrt{216x^5y^3}$.

Solution

a) $\sqrt{5x^2} = \sqrt{x^2 \cdot 5}$ — Factoring the radicand

$= \sqrt{x^2} \cdot \sqrt{5}$ — Factoring into two radicals

$= x \cdot \sqrt{5}$ — Taking the square root of x^2. We assume $x \geq 0$.

b) $\sqrt{2x^2 - 4x + 2} = \sqrt{2(x-1)^2}$ — Factoring the radicand

$= \sqrt{(x-1)^2} \cdot \sqrt{2}$ — Factoring into two radicals

$= (x-1) \cdot \sqrt{2}$ — Taking the square root of $(x-1)^2$. We assume $x - 1 \geq 0$.

c) $\sqrt{216x^5y^3} = \sqrt{36 \cdot 6 \cdot x^4 \cdot x \cdot y^2 \cdot y}$

$= \sqrt{36 \cdot x^4 \cdot y^2 \cdot 6 \cdot x \cdot y}$ — Factoring the radicand

$= \sqrt{36}\sqrt{x^4}\sqrt{y^2}\sqrt{6xy}$ — Factoring into several radicals

$= 6x^2y\sqrt{6xy}$ — Taking square roots. Note that $36 = 6^2$ and $x^4 = (x^2)^2$. Assume $x, y \geq 0$.

Note: Had we not seen that $216 = 36 \cdot 6$, where 36 is the largest square factor of 216, we could have found the prime factorization

$$2 \cdot 2 \cdot 2 \cdot 3 \cdot 3 \cdot 3.$$

Each pair of identical factors makes a square, so

$$\sqrt{2 \cdot 2 \cdot 2 \cdot 3 \cdot 3 \cdot 3} = \sqrt{2^2 \cdot 3^2 \cdot 2 \cdot 3}$$
$$= 2 \cdot 3\sqrt{2 \cdot 3} = 6\sqrt{6}.$$ ❑

Recall that when a quantity raised to a power is itself raised to a power, the result can be found by multiplying powers. Using this reasoning in reverse, we can simplify by finding factors in the radicand that have powers that are multiples of the index.

EXAMPLE 5

Simplify: $\sqrt{x^7y^{11}z^9}$.

Solution

$$\sqrt{x^7y^{11}z^9} = \sqrt{x^6 \cdot x \cdot y^{10} \cdot y \cdot z^8 \cdot z}$$ Factoring the radicand. Because we're taking the second (square) root, we look for powers that are multiples of 2.

$$= \sqrt{x^6}\,\sqrt{y^{10}}\,\sqrt{z^8}\,\sqrt{xyz}$$ Factoring into several radicals

$$= x^3y^5z^4\,\sqrt{xyz}$$ Note that $x^6 = (x^3)^2$, $y^{10} = (y^5)^2$, and $z^8 = (z^4)^2$. Assume $x, y, z \geq 0$.

Check:
$$(x^3y^5z^4\sqrt{xyz})^2 = (x^3)^2(y^5)^2(z^4)^2(\sqrt{xyz})^2$$
$$= x^6 \cdot y^{10} \cdot z^8 \cdot xyz$$
$$= x^7y^{11}z^9$$

Our check shows that $x^3y^5z^4\sqrt{xyz}$ is the square root of $x^7y^{11}z^9$. ❑

EXAMPLE 6

Simplify: $\sqrt[3]{16a^7b^{11}}$.

Solution

$$\sqrt[3]{16a^7b^{11}} = \sqrt[3]{8 \cdot 2 \cdot a^6 \cdot a \cdot b^9 \cdot b^2}$$ Factoring the radicand. We look for the largest powers that are multiples of 3. Note that $8 = 2^3$.

$$= \sqrt[3]{8} \cdot \sqrt[3]{a^6} \cdot \sqrt[3]{b^9} \cdot \sqrt[3]{2ab^2}$$ Factoring into radicals

$$= 2a^2b^3\sqrt[3]{2ab^2}$$ Taking cube roots

Check:
$$(2a^2b^3\sqrt[3]{2ab^2})^3 = 2^3(a^2)^3(b^3)^3(\sqrt[3]{2ab^2})^3$$
$$= 8 \cdot a^6 \cdot b^9 \cdot 2ab^2$$
$$= 16a^7b^{11}$$

We see that $2a^2b^3\sqrt[3]{2ab^2}$ is the cube root of $16a^7b^{11}$. ❑

Problems like Examples 5 and 6 can also be solved using rational exponents. Let's redo Example 6.

EXAMPLE 6A

Simplify: $\sqrt[3]{16a^7b^{11}}$.

Solution

$$\sqrt[3]{16a^7b^{11}} = (2^4a^7b^{11})^{1/3}$$ Converting to exponential notation

$$= 2^{4/3}a^{7/3}b^{11/3}$$ Multiplying exponents

$$= 2^{1+1/3} \cdot a^{2+1/3} \cdot b^{3+2/3}$$ Rewriting powers as mixed numbers

$$= 2 \cdot 2^{1/3} \cdot a^2 \cdot a^{1/3} \cdot b^3 \cdot b^{2/3}$$ Factoring

$$= 2a^2b^3(2ab^2)^{1/3}$$ Using the laws of exponents

$$= 2a^2b^3\sqrt[3]{2ab^2}$$ ❑

Multiplying and Simplifying

Sometimes after multiplying, we can simplify by factoring.

EXAMPLE 7

Multiply and simplify: **(a)** $\sqrt{15}\,\sqrt{6}$; **(b)** $3\sqrt[3]{25} \cdot 2\sqrt[3]{5}$; **(c)** $\sqrt[4]{8x^3y^5}\,\sqrt[4]{4x^2y^3}$.

Solution

a) $\sqrt{15}\,\sqrt{6} = \sqrt{15 \cdot 6} = \sqrt{90} = \sqrt{9 \cdot 10} = 3\sqrt{10}$

b) $3\sqrt[3]{25} \cdot 2\sqrt[3]{5} = 6 \cdot \sqrt[3]{25 \cdot 5}$
$= 6 \cdot \sqrt[3]{125}$ — Multiplying radicands
$= 6 \cdot 5$, or 30 — Finding the cube root of 125

c) $\sqrt[4]{8x^3y^5}\,\sqrt[4]{4x^2y^3} = \sqrt[4]{8x^3y^5 \cdot 4x^2y^3}$
$= \sqrt[4]{32x^5y^8}$ — Multiplying radicands
$= \sqrt[4]{16x^4y^8 \cdot 2x}$ — Factoring the radicand
$= \sqrt[4]{16}\,\sqrt[4]{x^4}\,\sqrt[4]{y^8}\,\sqrt[4]{2x}$ — Factoring into radicals
$= 2xy^2\,\sqrt[4]{2x}$ — Finding the fourth root

The checks are left for the student. ❑

When radicals have different indices, we multiply as in Example 2 and then simplify.

EXAMPLE 8

Multiply and simplify: $\sqrt{x^3}\,\sqrt[3]{x}$.

Solution

$\sqrt{x^3}\,\sqrt[3]{x} = x^{3/2} \cdot x^{1/3}$ — Converting to exponential notation
$= x^{11/6}$ — Adding exponents: $\frac{3}{2} + \frac{1}{3} = \frac{9}{6} + \frac{2}{6}$
$= x^{1 + 5/6}$ — Writing 11/6 as a mixed number
$= x \cdot x^{5/6}$ — Factoring
$= x\sqrt[6]{x^5}$ — Converting back to radical notation ❑

Approximating Square Roots

We often need to use rational numbers to approximate square roots that are irrational.* Such approximations can be found using a table such as Table 1 on p. 769. They can also be found on a calculator with a square root key. For example, if we were to approximate $\sqrt{37}$ using a calculator, we might get

$\sqrt{37} \approx 6.082762530.$ — Using a calculator with a 10-digit readout

Different calculators give different numbers of digits in their readouts. This may cause some variance in their answers. We might round to the third decimal place. Then

$\sqrt{37} \approx 6.083.$ — This can also be found in Table 1.

Now consider $\sqrt{275}$. To approximate such a root, we can use a calculator, or we can factor and use Table 1. Different procedures can lead to variance in approximations.

*Rational and irrational numbers are discussed in Section 1.4.

EXAMPLE 9

Use a calculator or Table 1 to approximate $\sqrt{275}$. Round to three decimal places.

Solution Using a calculator gives us

$$\sqrt{275} \approx 16.58312395 \approx 16.583.$$

Using factoring and Table 1, we get

$$\begin{aligned}\sqrt{275} &= \sqrt{25 \cdot 11} && \text{Factoring the radicand}\\ &= \sqrt{25} \cdot \sqrt{11} && \text{Factoring into two radicals}\\ &= 5\sqrt{11} && \text{Simplifying}\\ &\approx 5 \times 3.317 && \text{Using Table 1}\\ &= 16.585.\end{aligned}$$

Note the variance in the answers. Because calculators are so much a part of our everyday life, the answers at the back of the book are found using a calculator. If you use a table and get an answer slightly different from the one given at the back, keep in mind that your work may not be wrong. ❑

EXAMPLE 10

Approximate to the nearest thousandth: $\dfrac{16 - \sqrt{640}}{4}$.

Solution

Method 1. Using a calculator, we get

$$\begin{aligned}\frac{16 - \sqrt{640}}{4} &\approx \frac{16 - 25.29822128}{4}\\ &= \frac{-9.298221280}{4}\\ &= -2.324555320\\ &\approx -2.325.\end{aligned}$$

Method 2. Using factoring and Table 1 gives us

$$\begin{aligned}\frac{16 - \sqrt{640}}{4} &= \frac{16 - \sqrt{64 \cdot 10}}{4} && \text{Factoring the radicand}\\ &= \frac{16 - \sqrt{64} \cdot \sqrt{10}}{4} && \text{Factoring into two radicals}\\ &= \frac{16 - 8\sqrt{10}}{4} \\ &= \frac{4(4 - 2\sqrt{10})}{4} = 4 - 2\sqrt{10} && \text{Removing a factor of 1: } \tfrac{4}{4} = 1\\ &\approx 4 - 2(3.162) && \text{Using Table 1}\\ &= 4 - 6.324\\ &= -2.324.\end{aligned}$$

A common error here is to divide 4 into 16 but *not* into 8, and get $4 - 8\sqrt{10}$. Always remember to *remove a factor of* 1.

Note again the variance in answers. ❑

EXAMPLE 11

Use a calculator to approximate $\sqrt{0.000000005768}$.

Solution If the number will not fit into a calculator, we factor:

$$\sqrt{0.000000005768} = \sqrt{57.68 \times 10^{-10}}$$ Factoring the radicand. (Note that the exponent, -10, is divisible by 2.)

$$= \sqrt{57.68} \times \sqrt{10^{-10}}$$ Factoring the expression

$$\approx 7.595 \times 10^{-5}$$ Approximating $\sqrt{57.68}$ with a calculator and finding $\sqrt{10^{-10}}$

$$= 0.00007595.$$

EXERCISE SET 10.3

Note: Assume that no radicands were formed by raising negative quantities to even powers.

Multiply.

1. $\sqrt{3}\,\sqrt{2}$
2. $\sqrt{5}\,\sqrt{7}$
3. $\sqrt[3]{2}\,\sqrt[3]{5}$
4. $\sqrt[3]{7}\,\sqrt[3]{2}$
5. $\sqrt[4]{8}\,\sqrt[4]{9}$
6. $\sqrt[4]{6}\,\sqrt[4]{3}$
7. $\sqrt{3a}\,\sqrt{10b}$
8. $\sqrt{2x}\,\sqrt{13y}$
9. $\sqrt[5]{9t^2}\,\sqrt[5]{2t}$
10. $\sqrt[5]{8y^3}\,\sqrt[5]{10y}$
11. $\sqrt{x-a}\,\sqrt{x+a}$
12. $\sqrt{y-b}\,\sqrt{y+b}$
13. $\sqrt[3]{0.3x}\,\sqrt[3]{0.2x}$
14. $\sqrt[3]{0.7y}\,\sqrt[3]{0.3y}$
15. $\sqrt[4]{x-1}\,\sqrt[4]{x^2+x+1}$
16. $\sqrt[5]{x-2}\,\sqrt[5]{(x-2)^2}$
17. $\sqrt{\frac{6}{x}}\,\sqrt{\frac{y}{5}}$
18. $\sqrt{\frac{7}{t}}\,\sqrt{\frac{s}{11}}$
19. $\sqrt[7]{\frac{x-3}{4}}\,\sqrt[7]{\frac{5}{x+2}}$
20. $\sqrt[6]{\frac{a}{b-2}}\,\sqrt[6]{\frac{3}{b+2}}$

Use rational exponents to write a single radical expression.

21. $\sqrt[3]{7}\cdot\sqrt{2}$
22. $\sqrt[3]{7}\cdot\sqrt[4]{5}$
23. $\sqrt{x}\,\sqrt[3]{2x}$
24. $\sqrt[3]{y}\,\sqrt[5]{3y}$
25. $\sqrt{x}\,\sqrt[3]{x-2}$
26. $\sqrt[4]{3x}\,\sqrt{y+4}$
27. $\sqrt[5]{yx^2}\,\sqrt{xy}$
28. $\sqrt{ab}\,\sqrt[5]{2a^2b^2}$
29. $\sqrt[4]{x(y+1)^2}\,\sqrt[3]{x^2(y+1)}$
30. $\sqrt[5]{2a^2b}\,\sqrt{4ab}$
31. $\sqrt[5]{a^2bc^3}\,\sqrt[3]{ab^2c}$
32. $\sqrt[3]{x^2(y-1)}\,\sqrt[4]{x(y-1)^2}$

Simplify by factoring.

33. $\sqrt{27}$
34. $\sqrt{28}$
35. $\sqrt{45}$
36. $\sqrt{12}$
37. $\sqrt{8}$
38. $\sqrt{18}$
39. $\sqrt{24}$
40. $\sqrt{20}$
41. $\sqrt{180x^4}$
42. $\sqrt{175y^6}$
43. $\sqrt[3]{800}$
44. $\sqrt[3]{270}$
45. $\sqrt[3]{-16x^6}$
46. $\sqrt[3]{-32a^6}$
47. $\sqrt[3]{54x^8}$
48. $\sqrt[3]{40y^3}$
49. $\sqrt[3]{80x^8}$
50. $\sqrt[3]{108m^5}$
51. $\sqrt[4]{32}$
52. $\sqrt[4]{80}$
53. $\sqrt[4]{810}$
54. $\sqrt[4]{160}$
55. $\sqrt[4]{96a^8}$
56. $\sqrt[4]{240x^8}$
57. $\sqrt[4]{162c^4d^6}$
58. $\sqrt[4]{243x^8y^{10}}$
59. $\sqrt[3]{(x+y)^4}$
60. $\sqrt[3]{(a-b)^5}$
61. $\sqrt[3]{8000(m+n)^8}$
62. $\sqrt[3]{-1000(x+y)^{10}}$
63. $\sqrt[5]{-a^6b^{11}c^{17}}$
64. $\sqrt[5]{x^{13}y^8z^{22}}$

Multiply and simplify by factoring.

65. $\sqrt{3}\,\sqrt{6}$
66. $\sqrt{5}\,\sqrt{10}$
67. $\sqrt{15}\,\sqrt{12}$
68. $\sqrt{2}\,\sqrt{32}$
69. $\sqrt{6}\,\sqrt{8}$
70. $\sqrt{18}\,\sqrt{14}$
71. $\sqrt[3]{3}\,\sqrt[3]{18}$
72. $\sqrt{45}\,\sqrt{60}$
73. $\sqrt{5b^3}\,\sqrt{10c^4}$
74. $\sqrt[3]{-6a}\,\sqrt[3]{20a^4}$
75. $\sqrt[3]{10x^5}\,\sqrt[3]{-75x^2}$
76. $\sqrt{2x^3y}\,\sqrt{12xy}$
77. $\sqrt[3]{y^4}\,\sqrt[3]{16y^5}$
78. $\sqrt[3]{5^2t^4}\,\sqrt[3]{5^4t^6}$
79. $\sqrt[3]{(b+3)^4}\,\sqrt[3]{(b+3)^2}$

80. $\sqrt[3]{(x+y)^3}\,\sqrt[3]{(x+y)^5}$

81. $\sqrt{12a^3b}\,\sqrt{8a^4b^2}$

82. $\sqrt{18a^2b^5}\,\sqrt{30a^3b^4}$

83. $\sqrt[5]{a^2(b+c)^4}\,\sqrt[5]{a^4(b+c)^7}$

84. $\sqrt[5]{x^3(y-z)^7}\,\sqrt[5]{x^6(y-z)^9}$

Multiply and simplify. Write the answer in radical notation.

85. $\sqrt[5]{a^3b}\,\sqrt{ab}$

86. $\sqrt{xy^3}\,\sqrt[3]{x^2y}$

87. $\sqrt[3]{4xy^2}\,\sqrt{2x^3y^3}$

88. $\sqrt{3a^4b}\,\sqrt[4]{9ab^3}$

89. $\sqrt[4]{x^3y^5}\,\sqrt{xy}$

90. $\sqrt[5]{a^3b}\,\sqrt{ab}$

91. $\sqrt{a^4b^3c^4}\,\sqrt[3]{ab^2c}$

92. $\sqrt[3]{xy^2z}\,\sqrt{x^3yz^2}$

93. $\sqrt[3]{x^2yz^2}\,\sqrt[4]{xy^3z^3}$

94. $\sqrt[4]{a^2bc^3}\,\sqrt[3]{a^2b^2c^2}$

95. $\sqrt[3]{4a^2(b-5)^2}\,\sqrt{8a(b-5)^3}$

96. $\sqrt{27x^3(y-2)}\,\sqrt[3]{81x(y-2)^5}$

Approximate to the nearest thousandth using a calculator or Table 1.

97. $\sqrt{180}$

98. $\sqrt{124}$

99. $\dfrac{8+\sqrt{480}}{4}$

100. $\dfrac{12-\sqrt{450}}{3}$

101. $\dfrac{16-\sqrt{48}}{20}$

102. $\dfrac{25-\sqrt{250}}{10}$

103. $\dfrac{24+\sqrt{128}}{8}$

104. $\dfrac{96-\sqrt{90}}{12}$

Use a calculator to approximate each of the following.

105. $\sqrt{24{,}500{,}000{,}000}$

106. $\sqrt{16{,}500{,}000{,}000}$

107. $\sqrt{468{,}200{,}000{,}000}$

108. $\sqrt{99{,}400{,}000{,}000}$

109. $\sqrt{0.0000000395}$

110. $\sqrt{0.0000001543}$

111. $\sqrt{0.0000005001}$

112. $\sqrt{0.000010101}$

Skill Maintenance

113. During a one-hour television show, there were 12 commercials. Some of the commercials were 30 sec long and the others were 60 sec long. If the number of 30-sec commercials was 6 less than the total number of minutes of commercial time during the show, how many 60-sec commercials were used during the hour?

114. Multiply: $(2x-3)(2x+3)$.

Factor.

115. $4x^2-49$

116. $2x^2-26x+72$

Synthesis

117. Why are mixed numbers important when multiplying expressions like $\sqrt{ab}\,\sqrt[3]{a^2b^2}$?

118. Is the statement $\sqrt{500}=22.36067977$ true? Why or why not?

119. *Speed of a skidding car.* Police can estimate the speed at which a car was traveling by measuring its skid marks. The formula

$$r=2\sqrt{5L}$$

can be used, where r is the speed in miles per hour and L is the length of a skid mark in feet. Estimate (to the nearest tenth mile per hour) the speed of a car that left skid marks **(a)** 20 ft long; **(b)** 70 ft long; **(c)** 90 ft long.

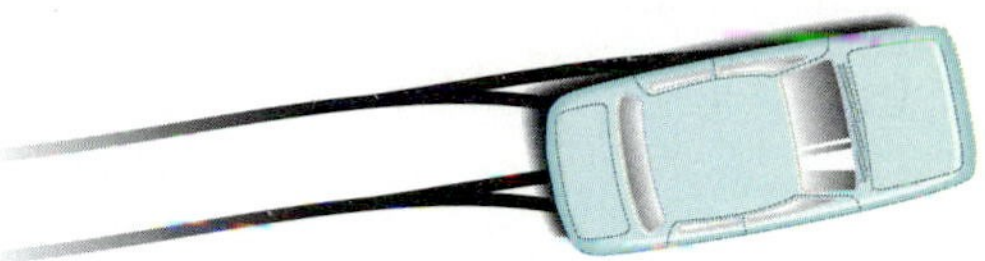

120. *Wind chill temperature.* In cold weather we feel colder if there is wind than if there is not. When the temperature is T degrees Celsius and the wind speed is v meters per second, the *wind chill temperature*, T_w, is the temperature that it feels like. Here is a formula for finding wind chill temperature:

$$T_w=33-\frac{(10.45+10\sqrt{v}-v)(33-T)}{22}.$$

Find the wind chill temperature to the nearest tenth of a degree for the given actual temperatures and wind speeds.

a) $T=7°\text{C},\ v=8$ m/sec
b) $T=0°\text{C},\ v=12$ m/sec
c) $T=-5°\text{C},\ v=14$ m/sec
d) $T=-23°\text{C},\ v=15$ m/sec

121. Solve $\sqrt[3]{5x^{k+1}}\,\sqrt[3]{25x^k}=5x^7$ for k.

122. Solve $\sqrt[5]{4a^{3k+2}}\,\sqrt[5]{8a^{6-k}}=2a^4$ for k.

123. What assumption do we make about x if

$$\sqrt{(2x+3)^2}=2x+3?$$

124. Graph the function given by

$$f(x)=\sqrt{(x-2)^2}.$$

What is the domain of f?

10.4 Dividing and Simplifying Radical Expressions

Dividing Radical Expressions • Roots of Quotients • Powers and Roots Combined

Dividing Radical Expressions

Note that

$$\frac{\sqrt[3]{27}}{\sqrt[3]{8}} = \frac{3}{2} \quad \text{and} \quad \sqrt[3]{\frac{27}{8}} = \frac{3}{2}.$$

This example suggests the following.

The Quotient Rule for Radicals

For any real numbers $\sqrt[k]{a}$ and $\sqrt[k]{b}$, $b \neq 0$,

$$\frac{\sqrt[k]{a}}{\sqrt[k]{b}} = \sqrt[k]{\frac{a}{b}}.$$

(To divide, we divide the radicands. After doing this, we can sometimes simplify by taking roots.)

To help understand the quotient rule for radicals, note that

$$\frac{\sqrt[k]{a}}{\sqrt[k]{b}} = \frac{a^{1/k}}{b^{1/k}} = \left(\frac{a}{b}\right)^{1/k} = \sqrt[k]{\frac{a}{b}}.$$

EXAMPLE 1

Divide and simplify by taking roots, if possible.

a) $\dfrac{\sqrt{80}}{\sqrt{5}}$ b) $\dfrac{3\sqrt{2}}{5\sqrt{3}}$ c) $\dfrac{5\sqrt[3]{32}}{\sqrt[3]{2}}$

d) $\dfrac{\sqrt{72xy}}{2\sqrt{2}}$ e) $\dfrac{\sqrt[4]{33a^9b^3}}{\sqrt[4]{2b^{-1}}}$

Solution

a) $\dfrac{\sqrt{80}}{\sqrt{5}} = \sqrt{\dfrac{80}{5}} = \sqrt{16} = 4$ — We divide the radicands.

b) $\dfrac{3\sqrt{2}}{5\sqrt{3}} = \dfrac{3}{5} \cdot \dfrac{\sqrt{2}}{\sqrt{3}} = \dfrac{3}{5} \cdot \sqrt{\dfrac{2}{3}}$

c) $\dfrac{5\sqrt[3]{32}}{\sqrt[3]{2}} = 5\sqrt[3]{\dfrac{32}{2}} = 5\sqrt[3]{16} = 5\sqrt[3]{8 \cdot 2} = 5\sqrt[3]{8}\sqrt[3]{2} = 5 \cdot 2\sqrt[3]{2} = 10\sqrt[3]{2}$

d) $\dfrac{\sqrt{72xy}}{2\sqrt{2}} = \dfrac{1}{2}\,\dfrac{\sqrt{72xy}}{\sqrt{2}} = \dfrac{1}{2}\sqrt{\dfrac{72xy}{2}} = \dfrac{1}{2}\sqrt{36xy} = \dfrac{1}{2}\sqrt{36}\,\sqrt{xy}$

$= \dfrac{1}{2}\cdot 6\sqrt{xy} = 3\sqrt{xy}$

e) $\dfrac{\sqrt[4]{33a^9b^3}}{\sqrt[4]{2b^{-1}}} = \sqrt[4]{\dfrac{33a^9b^3}{2b^{-1}}} = \sqrt[4]{\dfrac{33a^9b^4}{2}}$

$= \sqrt[4]{a^8b^4}\,\sqrt[4]{\dfrac{33a}{2}} = a^2b\,\sqrt[4]{\dfrac{33a}{2}}$ Note that 8 is the largest power less than 9 that is a multiple of the index 4. Assume that $a \geqslant 0,\ b > 0$. ❑

Remember that when we divide radical expressions by dividing the radicands, both radicals must have the same index. When the indices differ, rational exponents are useful.

EXAMPLE 2

Use rational exponents to write as a single radical expression.

a) $\dfrac{\sqrt[3]{a^2b^4}}{\sqrt{ab}}$ **b)** $\dfrac{\sqrt[4]{x^3y^2}}{\sqrt[3]{x^2y}}$ **c)** $\dfrac{\sqrt[4]{(x+y)^3}}{\sqrt{x+y}}$

Solution

a) $\dfrac{\sqrt[3]{a^2b^4}}{\sqrt{ab}} = \dfrac{(a^2b^4)^{1/3}}{(ab)^{1/2}}$ Converting to exponential notation

$= \dfrac{a^{2/3}b^{4/3}}{a^{1/2}b^{1/2}}$ Using the product and power rules

$= a^{2/3-1/2}b^{4/3-1/2}$ Subtracting exponents

$= a^{1/6}b^{5/6}$

$= (ab^5)^{1/6}$ Converting back to radical notation

$= \sqrt[6]{ab^5}$

b) $\dfrac{\sqrt[4]{x^3y^2}}{\sqrt[3]{x^2y}} = \dfrac{(x^3y^2)^{1/4}}{(x^2y)^{1/3}}$ Converting to exponential notation

$= \dfrac{x^{3/4}y^{2/4}}{x^{2/3}y^{1/3}}$ Using the product and power rules

$= x^{3/4-2/3}y^{2/4-1/3}$ Subtracting exponents

$= x^{1/12}y^{2/12}$

$= (xy^2)^{1/12}$ Converting back to radical notation

$= \sqrt[12]{xy^2}$

c) $\dfrac{\sqrt[4]{(x+y)^3}}{\sqrt{x+y}} = \dfrac{(x+y)^{3/4}}{(x+y)^{1/2}}$ Converting to exponential notation

$= (x+y)^{3/4-1/2}$ Subtracting exponents

$= (x+y)^{1/4}$ Converting back to radical notation

$= \sqrt[4]{x+y}$ ❑

Roots of Quotients

We can reverse the quotient rule to simplify a quotient. We simplify the root of a quotient by taking the roots of the numerator and of the denominator separately.

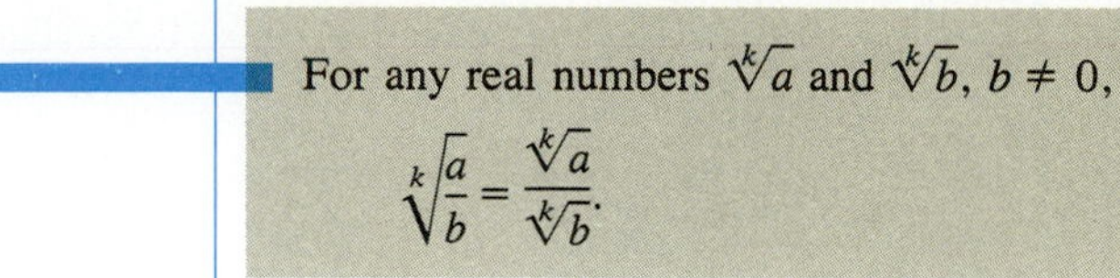

For any real numbers $\sqrt[k]{a}$ and $\sqrt[k]{b}$, $b \neq 0$,

$$\sqrt[k]{\frac{a}{b}} = \frac{\sqrt[k]{a}}{\sqrt[k]{b}}.$$

EXAMPLE 3

Simplify by taking the roots of the numerator and the denominator.

a) $\sqrt[3]{\dfrac{27}{125}}$ **b)** $\sqrt{\dfrac{25}{y^2}}$ **c)** $\sqrt{\dfrac{16x^3}{y^8}}$ **d)** $\sqrt[3]{\dfrac{27y^{14}}{343x^3}}$

Solution

a) $\sqrt[3]{\dfrac{27}{125}} = \dfrac{\sqrt[3]{27}}{\sqrt[3]{125}} = \dfrac{3}{5}$ Taking the cube roots of the numerator and the denominator

b) $\sqrt{\dfrac{25}{y^2}} = \dfrac{\sqrt{25}}{\sqrt{y^2}} = \dfrac{5}{y}$ Taking the square roots of the numerator and the denominator. Assume $y > 0$.

c) $\sqrt{\dfrac{16x^3}{y^8}} = \dfrac{\sqrt{16x^3}}{\sqrt{y^8}} = \dfrac{\sqrt{16x^2 \cdot x}}{\sqrt{y^8}} = \dfrac{4x\sqrt{x}}{y^4}$ Assume $x \geq 0$, $y \neq 0$.

d) $\sqrt[3]{\dfrac{27y^{14}}{343x^3}} = \dfrac{\sqrt[3]{27y^{14}}}{\sqrt[3]{343x^3}} = \dfrac{\sqrt[3]{27y^{12}y^2}}{\sqrt[3]{343x^3}} = \dfrac{\sqrt[3]{27y^{12}}\sqrt[3]{y^2}}{\sqrt[3]{343x^3}} = \dfrac{3y^4\sqrt[3]{y^2}}{7x}$ Assume $x \neq 0$. ❑

Powers and Roots Combined

We saw in Section 10.2 that $a^{m/n}$ means $(\sqrt[n]{a})^m$, or $\sqrt[n]{a^m}$. Thus, $(\sqrt[n]{a})^m$ and $\sqrt[n]{a^m}$ are the same number. This can be illustrated by noting that

$$(\sqrt[3]{8})^2 = 2^2 = 4 \quad \text{and} \quad \sqrt[3]{8^2} = \sqrt[3]{64} = 4.$$

The Power–Root Rule

For any index k for which $\sqrt[k]{a}$ exists, and any integer m for which a^m exists,

$$(\sqrt[k]{a})^m = \sqrt[k]{a^m}.$$

(We can raise to a power and then take a root, or we can take a root and then raise to a power.)

Often, one way of calculating is easier than the other.

EXAMPLE 4

Simplify each expression. Then use the power–root rule to calculate a second way. **(a)** $\sqrt[3]{27^2}$; **(b)** $\sqrt[3]{2^6}$; **(c)** $(\sqrt{5x})^3$; **(d)** $(\sqrt[3]{16x^3y^2})^2$

Solution

a1) $\sqrt[3]{27^2} = \sqrt[3]{729} = 9$ Finding 27^2 and then taking the cube root

a2) $(\sqrt[3]{27})^2 = (3)^2 = 9$ Taking the cube root and then squaring

b1) $\sqrt[3]{2^6} = \sqrt[3]{64} = 4$ Finding 2^6 and then taking the cube root

b2) $(\sqrt[3]{2})^6 = \sqrt[3]{2}\,\sqrt[3]{2}\,\sqrt[3]{2}\cdot\sqrt[3]{2}\,\sqrt[3]{2}\,\sqrt[3]{2} = 2\cdot 2 = 4$

Note that in Example 4(b) we could have easily used fractional exponents: $\sqrt[3]{2^6} = (2^6)^{1/3} = 2^2$ and $(\sqrt[3]{2})^6 = (2^{1/3})^6 = 2^2$.

c1) $(\sqrt{5x})^3 = \sqrt{5x}\,\sqrt{5x}\,\sqrt{5x} = 5x\sqrt{5x}$

c2) $\sqrt{(5x)^3} = \sqrt{5^3x^3} = \sqrt{5^2x^2}\,\sqrt{5x} = 5x\sqrt{5x}$

Assume $x \geq 0$.

d1) $(\sqrt[3]{16x^3y^2})^2 = (\sqrt[3]{8x^3\cdot 2y^2})^2$

$= (2x\sqrt[3]{2y^2})^2$ Simplifying the cube root

$= 2x\sqrt[3]{2y^2}\cdot 2x\sqrt[3]{2y^2}$

$= 4x^2\sqrt[3]{4y^4}$ Multiplying and combining radicals

$= 4x^2\sqrt[3]{y^3\cdot 4y}$

$= 4x^2y\sqrt[3]{4y}$ Simplifying the radical

d2) $\sqrt[3]{(16x^3y^2)^2} = \sqrt[3]{256x^6y^4}$

$= \sqrt[3]{64x^6y^3\cdot 4y}$ Factoring the radicand

$= \sqrt[3]{64x^6y^3}\sqrt[3]{4y}$

$= 4x^2y\sqrt[3]{4y}$ ❑

EXERCISE SET 10.4

Divide. Then simplify by taking roots, if possible. Assume that all radicands represent positive numbers.

1. $\dfrac{\sqrt{21a}}{\sqrt{3a}}$

2. $\dfrac{\sqrt{28y}}{\sqrt{4y}}$

3. $\dfrac{\sqrt[3]{54}}{\sqrt[3]{2}}$

4. $\dfrac{\sqrt[3]{40}}{\sqrt[3]{5}}$

5. $\dfrac{\sqrt{40xy^3}}{\sqrt{8x}}$

6. $\dfrac{\sqrt{56ab^3}}{\sqrt{7a}}$

7. $\dfrac{\sqrt[3]{96a^4b^2}}{\sqrt[3]{12a^2b}}$

8. $\dfrac{\sqrt[3]{189x^5y^7}}{\sqrt[3]{7x^2y^2}}$

9. $\dfrac{\sqrt{144xy}}{2\sqrt{2}}$

10. $\dfrac{\sqrt{75ab}}{3\sqrt{3}}$

11. $\dfrac{\sqrt[4]{48x^9y^{13}}}{\sqrt[4]{3xy^5}}$

12. $\dfrac{\sqrt[5]{64a^{11}b^{28}}}{\sqrt[5]{2ab^2}}$

13. $\dfrac{\sqrt{x^3 - y^3}}{\sqrt{x - y}}$

Hint: Factor and then simplify.

14. $\dfrac{\sqrt{r^3 + s^3}}{\sqrt{r + s}}$

15. $\dfrac{\sqrt[3]{a^2}}{\sqrt[4]{a}}$

16. $\dfrac{\sqrt[3]{x^2}}{\sqrt[5]{x}}$

17. $\dfrac{\sqrt[4]{x^2y^3}}{\sqrt[3]{xy^2}}$

18. $\dfrac{\sqrt[5]{a^3b^4}}{\sqrt[3]{ab^2}}$

19. $\dfrac{\sqrt{ab^3c}}{\sqrt[5]{a^2b^3c}}$

20. $\dfrac{\sqrt[5]{x^3y^4z^9}}{\sqrt{xyz^3}}$

21. $\dfrac{\sqrt[4]{(3x-1)^3}}{\sqrt[5]{(3x-1)^3}}$

22. $\dfrac{\sqrt[4]{(5-3x)^3}}{\sqrt[3]{(5-3x)^2}}$

Simplify by taking the roots of the numerator and the denominator. Assume that all radicands represent positive numbers.

23. $\sqrt{\dfrac{16}{25}}$

24. $\sqrt{\dfrac{100}{81}}$

25. $\sqrt[3]{\dfrac{64}{27}}$

26. $\sqrt[3]{\dfrac{343}{512}}$

27. $\sqrt{\dfrac{49}{y^2}}$

28. $\sqrt{\dfrac{121}{x^2}}$

29. $\sqrt{\dfrac{25y^3}{x^4}}$

30. $\sqrt{\dfrac{36a^5}{b^6}}$

31. $\sqrt[3]{\dfrac{8x^5}{27y^3}}$

32. $\sqrt[3]{\dfrac{64x^7}{216y^6}}$

33. $\sqrt[4]{\dfrac{16a^4}{81}}$

34. $\sqrt[4]{\dfrac{81x^4}{y^8}}$

35. $\sqrt[4]{\dfrac{a^5b^8}{c^{10}}}$

36. $\sqrt[4]{\dfrac{x^9y^{12}}{z^6}}$

37. $\sqrt[5]{\dfrac{32x^6}{y^{11}}}$

38. $\sqrt[5]{\dfrac{243a^9}{b^{13}}}$

39. $\sqrt[6]{\dfrac{x^6y^8}{z^{15}}}$

40. $\sqrt[6]{\dfrac{a^9b^{12}}{c^{13}}}$

Calculate as shown. Then use the power–root rule to calculate another way. Assume that all radicands represent nonnegative numbers.

41. $\sqrt{(6a)^3}$

42. $\sqrt{(7y)^3}$

43. $(\sqrt{16b^2})^3$

44. $(\sqrt{25r^2})^3$

45. $\sqrt{(18a^2b)^3}$

46. $\sqrt{(12x^2y)^3}$

47. $(\sqrt[3]{3c^2d})^4$

48. $(\sqrt[3]{2x^2y})^4$

49. $\sqrt[3]{(5x^2y)^2}$

50. $\sqrt[3]{(6ab^2)^2}$

51. $\sqrt[4]{(x^2y)^3}$

52. $\sqrt[4]{(2a^3)^3}$

53. $(\sqrt[3]{8a^4b})^2$

54. $(\sqrt[3]{27xy^5})^2$

55. $(\sqrt[4]{16x^2y^3})^2$

56. $(\sqrt[4]{16xy^5})^3$

Skill Maintenance

Solve.

57. $\dfrac{12x}{x-4} - \dfrac{3x^2}{x+4} = \dfrac{384}{x^2-16}$

58. $\dfrac{2}{3} + \dfrac{1}{t} = \dfrac{4}{5}$

59. The width of a rectangle is one fourth the length. The area is twice the perimeter. Find the dimensions of the rectangle.

Synthesis

60. ◈ Explain why no assumptions need be made regarding the numbers that x and y represent in Example 4(d).

61. ◈ Explain why $\sqrt[3]{x^6} = x^2$ for any value x, whereas $\sqrt[2]{x^6} = x^3$ only when $x \geq 0$.

62. ▦ *Pendulums.* The *period* of a pendulum is the time it takes to complete one cycle, swinging to and fro. If a pendulum consists of a weight on a string, the period T is given by the formula

$$T = 2\pi\sqrt{\frac{L}{980}},$$

where T is in seconds and L is the length of the pendulum in centimeters. Find to the nearest hundredth of a second the period of a pendulum of length **(a)** 65 cm; **(b)** 98 cm; **(c)** 120 cm. Use 3.14 for π.

Divide and simplify.

63. $\dfrac{7\sqrt{a^2b}\,\sqrt{25xy}}{5\sqrt{a^{-4}b^{-1}}\,\sqrt{49x^{-1}y^{-3}}}$

64. $\dfrac{(\sqrt[3]{81mn^2})^2}{(\sqrt[3]{mn})^2}$

65. $\dfrac{\sqrt{44x^2y^9z}\,\sqrt{22y^9z^6}}{(\sqrt{11xy^8z^2})^2}$

10.5 Addition, Subtraction, and More Multiplication

Addition and Subtraction Involving Radicals • More Multiplication with Radicals

Addition and Subtraction Involving Radicals

Any two real numbers can be added. For instance, the sum of 7 and $\sqrt{3}$ can be expressed as

$$7 + \sqrt{3}.$$

We cannot simplify this name for the sum. **Like radicals,** however, are radicals having the same index and radicand, and they can be simplified using the distributive law.

EXAMPLE 1

Simplify by collecting like radical terms.

a) $6\sqrt{7} + 4\sqrt{7}$

b) $8\sqrt[3]{2} - 7x\sqrt[3]{2} + 5\sqrt[3]{2}$

c) $6\sqrt[5]{4x} + 4\sqrt[5]{4x} - \sqrt[3]{4x}$

Solution

a) $6\sqrt{7} + 4\sqrt{7} = (6 + 4)\sqrt{7}$ Using the distributive law (factoring out $\sqrt{7}$)

$= 10\sqrt{7}$

b) $8\sqrt[3]{2} - 7x\sqrt[3]{2} + 5\sqrt[3]{2} = (8 - 7x + 5)\sqrt[3]{2}$ Factoring out $\sqrt[3]{2}$

$= (13 - 7x)\sqrt[3]{2}$ These parentheses are important!

c) $6\sqrt[5]{4x} + 4\sqrt[5]{4x} - \sqrt[3]{4x} = (6 + 4)\sqrt[5]{4x} - \sqrt[3]{4x}$ Try to do this step mentally.

$= 10\sqrt[5]{4x} - \sqrt[3]{4x}$ ❑

Note that these expressions have the *same* radicand but are *not* like radicals because they have *different* indices!

One way to think of problems like Example 1(a) is as follows: When 4 square roots of 7 are added to 6 square roots of 7, the result is 10 square roots of 7.

Sometimes we need to factor one or more of the radicals in order to have like radical terms.

EXAMPLE 2

Simplify by collecting like radical terms, if possible.

a) $3\sqrt{8} - 5\sqrt{2}$

b) $5\sqrt{2} - 4\sqrt{3}$

c) $5\sqrt[3]{16y^4} + 7\sqrt[3]{2y}$

Solution

a) $3\sqrt{8} - 5\sqrt{2} = 3\sqrt{4 \cdot 2} - 5\sqrt{2}$ Factoring 8

$= 3\sqrt{4} \cdot \sqrt{2} - 5\sqrt{2}$ Factoring $\sqrt{4 \cdot 2}$ into two radicals

$= 3 \cdot 2\sqrt{2} - 5\sqrt{2}$ Taking the square root of 4

$= 6\sqrt{2} - 5\sqrt{2}$

$= (6 - 5)\sqrt{2}$ Factoring out $\sqrt{2}$ to collect like radical terms

$= \sqrt{2}$

b) $5\sqrt{2} - 4\sqrt{3}$ cannot be simplified.

c) $5\sqrt[3]{16y^4} + 7\sqrt[3]{2y} = 5\sqrt[3]{8y^3 \cdot 2y} + 7\sqrt[3]{2y}$

$= 5\sqrt[3]{8y^3} \cdot \sqrt[3]{2y} + 7\sqrt[3]{2y}$ Factoring the first radical

$= 5 \cdot 2y \cdot \sqrt[3]{2y} + 7\sqrt[3]{2y}$ Taking the cube root

$= 10y\sqrt[3]{2y} + 7\sqrt[3]{2y}$

$= (10y + 7)\sqrt[3]{2y}$ Factoring to collect like radical terms ❑

More Multiplication with Radicals

To multiply expressions in which some factors contain more than one term, we use the procedures for multiplying polynomials.

EXAMPLE 3

Multiply.

a) $\sqrt{3}(x - \sqrt{5})$ b) $\sqrt[3]{y}(\sqrt[3]{y^2} + \sqrt[3]{2})$ c) $(4\sqrt{3} + \sqrt{2})(\sqrt{3} - 5\sqrt{2})$

Solution

a) $\sqrt{3}(x - \sqrt{5}) = \sqrt{3} \cdot x - \sqrt{3} \cdot \sqrt{5}$ Using the distributive law

$= x\sqrt{3} - \sqrt{15}$ Multiplying radicals

b) $\sqrt[3]{y}(\sqrt[3]{y^2} + \sqrt[3]{2}) = \sqrt[3]{y} \cdot \sqrt[3]{y^2} + \sqrt[3]{y} \cdot \sqrt[3]{2}$ Using the distributive law

$= \sqrt[3]{y^3} + \sqrt[3]{2y}$ Multiplying radicals

$= y + \sqrt[3]{2y}$ Simplifying $\sqrt[3]{y^3}$

c) F O I L

$(4\sqrt{3} + \sqrt{2})(\sqrt{3} - 5\sqrt{2}) = 4(\sqrt{3})^2 - 20\sqrt{3} \cdot \sqrt{2} + \sqrt{2} \cdot \sqrt{3} - 5(\sqrt{2})^2$

$= 4 \cdot 3 - 20\sqrt{6} + \sqrt{6} - 5 \cdot 2$ Multiplying radicals

$= 12 - 20\sqrt{6} + \sqrt{6} - 10$

$= 2 - 19\sqrt{6}$ Collecting like terms ❑

EXAMPLE 4

Multiply. Assume that all radicands are nonnegative.

a) $(\sqrt{r} + \sqrt{3})(\sqrt{s} + \sqrt{3})$ b) $(\sqrt{3} + x)^2$

c) $(2 - \sqrt{7})(2 + \sqrt{7})$ d) $(\sqrt{a} + \sqrt{b})(\sqrt{a} - \sqrt{b})$

Solution

a) $(\sqrt{r} + \sqrt{3})(\sqrt{s} + \sqrt{3}) = \sqrt{r}\sqrt{s} + \sqrt{r}\sqrt{3} + \sqrt{3}\sqrt{s} + \sqrt{3}\sqrt{3}$ Using FOIL

$= \sqrt{rs} + \sqrt{3r} + \sqrt{3s} + 3$ Multiplying radicals

b) $(\sqrt{3} + x)^2 = (\sqrt{3})^2 + 2x\sqrt{3} + x^2$ Squaring a binomial

$= 3 + 2x\sqrt{3} + x^2$

c) $(2 - \sqrt{7})(2 + \sqrt{7}) = 2^2 - (\sqrt{7})^2$ This is in the same form as a difference of two squares.

$= 4 - 7$

$= -3$

d) $(\sqrt{a} + \sqrt{b})(\sqrt{a} - \sqrt{b}) = (\sqrt{a})^2 - (\sqrt{b})^2$

$= a - b$ ❑

Pairs of expressions in the form $\sqrt{a} + \sqrt{b}$ and $\sqrt{a} - \sqrt{b}$ or $c - \sqrt{b}$ and $c + \sqrt{b}$ are called **conjugates.** Their product is always an expression that has no radicals. Conjugates play an important role in the work we will do in Section 10.6.

As in our earlier work, when different indices appear, rational exponents are useful.

EXAMPLE 5 Multiply: $\sqrt[3]{x^2}(\sqrt{x} + \sqrt[4]{xy^3})$. Assume $x, y \geq 0$.

Solution

$\sqrt[3]{x^2}(\sqrt{x} + \sqrt[4]{xy^3}) = x^{2/3}(x^{1/2} + (xy^3)^{1/4})$ Converting to exponential notation

$= x^{2/3}(x^{1/2} + x^{1/4}y^{3/4})$ Using the laws of exponents

$= x^{2/3} \cdot x^{1/2} + x^{2/3} \cdot x^{1/4}y^{3/4}$ Using the distributive law

$= x^{2/3 + 1/2} + x^{2/3 + 1/4}y^{3/4}$ Adding exponents

$= x^{7/6} + x^{11/12}y^{3/4}$

$= x^{1 + 1/6} + x^{11/12}y^{9/12}$ Writing a mixed number; finding a common denominator

$= x^1x^{1/6} + (x^{11}y^9)^{1/12}$ Using the laws of exponents

$= x\sqrt[6]{x} + \sqrt[12]{x^{11}y^9}$ Converting back to radical notation ❑

EXERCISE SET 10.5

Add or subtract. Simplify by collecting like radical terms, if possible. Assume that all variables and radicands represent nonnegative numbers.

1. $6\sqrt{3} + 2\sqrt{3}$
2. $8\sqrt{5} + 9\sqrt{5}$
3. $9\sqrt[3]{5} - 6\sqrt[3]{5}$
4. $14\sqrt[5]{2} - 6\sqrt[5]{2}$
5. $4\sqrt[3]{y} + 9\sqrt[3]{y}$
6. $6\sqrt[4]{t} - 3\sqrt[4]{t}$
7. $8\sqrt{2} - 6\sqrt{2} + 5\sqrt{2}$
8. $2\sqrt{6} + 8\sqrt{6} - 3\sqrt{6}$
9. $4\sqrt[3]{3} - \sqrt{5} + 2\sqrt[3]{3} + \sqrt{5}$

10. $5\sqrt{7} - 8\sqrt[4]{11} + \sqrt{7} + 9\sqrt[4]{11}$

11. $8\sqrt{27} - 3\sqrt{3}$

12. $9\sqrt{50} - 4\sqrt{2}$

13. $8\sqrt{45} + 7\sqrt{20}$

14. $9\sqrt{12} + 16\sqrt{27}$

15. $18\sqrt{72} + 2\sqrt{98}$

16. $12\sqrt{45} - 8\sqrt{80}$

17. $3\sqrt[3]{16} + \sqrt[3]{54}$

18. $\sqrt[3]{27} - 5\sqrt[3]{8}$

19. $\sqrt{5a} + 2\sqrt{45a^3}$

20. $4\sqrt{3x^3} - \sqrt{12x}$

21. $\sqrt[3]{24x} - \sqrt[3]{3x^4}$

22. $\sqrt[3]{54x} - \sqrt[3]{2x^4}$

23. $\sqrt{8y - 8} + \sqrt{2y - 2}$

24. $\sqrt{12t + 12} + \sqrt{3t + 3}$

25. $\sqrt{x^3 - x^2} + \sqrt{9x - 9}$

26. $\sqrt{4x - 4} - \sqrt{x^3 - x^2}$

27. $5\sqrt[3]{32} - \sqrt[3]{108} + 2\sqrt[3]{256}$

28. $3\sqrt[3]{8x} - 4\sqrt[3]{27x} + 2\sqrt[3]{64x}$

29. $\sqrt{x^3 + x^2} + \sqrt{4x^3 + 4x^2} - \sqrt{9x^3 + 9x^2}$

30. $\sqrt{5x^2 + 4} - 5\sqrt{45x^2 + 36} + 3\sqrt{20x^2 + 16}$

31. $\sqrt[4]{x^5 - x^4} + 3\sqrt[4]{x^9 - x^8}$

32. $\sqrt[4]{16a^4 + 16a^5} - 2\sqrt[4]{a^8 + a^9}$

Multiply. Assume that all variables represent nonnegative real numbers.

33. $\sqrt{6}(2 - 3\sqrt{6})$

34. $\sqrt{3}(4 + \sqrt{3})$

35. $\sqrt{2}(\sqrt{3} - \sqrt{5})$

36. $\sqrt{5}(\sqrt{5} - \sqrt{2})$

37. $\sqrt{3}(2\sqrt{5} - 3\sqrt{4})$

38. $\sqrt{2}(3\sqrt{10} - 2\sqrt{2})$

39. $\sqrt[3]{2}(\sqrt[3]{4} - 2\sqrt[3]{32})$

40. $\sqrt[3]{3}(\sqrt[3]{9} - 4\sqrt[3]{21})$

41. $\sqrt[3]{a}(\sqrt[3]{2a^2} + \sqrt[3]{16a^2})$

42. $\sqrt[3]{x}(\sqrt[3]{3x^2} - \sqrt[3]{81x^2})$

43. $\sqrt[4]{x}(\sqrt[4]{x^7} + \sqrt[4]{3x^2})$

44. $\sqrt[4]{a}(\sqrt[4]{2a} - \sqrt[4]{a^{11}})$

45. $(5 - \sqrt{7})(5 + \sqrt{7})$

46. $(3 + \sqrt{5})(3 - \sqrt{5})$

47. $(\sqrt{5} + \sqrt{8})(\sqrt{5} - \sqrt{8})$

48. $(\sqrt{3} - \sqrt{5})(\sqrt{3} + \sqrt{5})$

49. $(3 - 2\sqrt{7})(3 + 2\sqrt{7})$

50. $(4 - 3\sqrt{2})(4 + 3\sqrt{2})$

51. $(\sqrt{a} + \sqrt{2})(\sqrt{a} + \sqrt{3})$

52. $(2 - \sqrt{x})(1 - \sqrt{x})$

53. $(3 - \sqrt[3]{5})(2 + \sqrt[3]{5})$

54. $(2 + \sqrt[3]{6})(4 - \sqrt[3]{6})$

55. $(2\sqrt{7} - 4\sqrt{2})(3\sqrt{7} + 6\sqrt{2})$

56. $(4\sqrt{5} + 3\sqrt{3})(3\sqrt{5} - 4\sqrt{3})$

57. $(2\sqrt[3]{3} + \sqrt[3]{2})(\sqrt[3]{3} - 2\sqrt[3]{2})$

58. $(3\sqrt[4]{7} + \sqrt[4]{6})(2\sqrt[4]{9} - 3\sqrt[4]{6})$

59. $(1 + \sqrt{3})^2$

60. $(\sqrt{5} + 1)^2$

61. $(a + \sqrt{b})^2$

62. $(x - \sqrt{y})^2$

63. $(2x - \sqrt[4]{y})^2$

64. $(3a + \sqrt[4]{b})^2$

65. $(\sqrt{m} + \sqrt{n})^2$

66. $(\sqrt{r} - \sqrt{s})^2$

67. $\sqrt{a}(\sqrt[3]{a} + \sqrt[4]{ab})$

68. $\sqrt{x}(\sqrt[3]{xy} + \sqrt[4]{x^2y})$

69. $\sqrt[3]{x^2y}(\sqrt{xy} - \sqrt[5]{xy^3})$

70. $\sqrt[4]{a^2b}(\sqrt[3]{a^2b} - \sqrt[5]{a^2b^2})$

71. $(m + \sqrt[3]{n^2})(2m + \sqrt[4]{n})$

72. $(r - \sqrt[4]{s^3})(3r - \sqrt[5]{s})$

73. $(a + \sqrt[4]{b})(a - \sqrt[3]{c})$

74. $(x - \sqrt[3]{y})(x + \sqrt[4]{z})$

75. $(\sqrt{x} - \sqrt[3]{yz})(\sqrt[4]{x} + \sqrt[5]{xz})$

76. $(\sqrt{a} + \sqrt[4]{3b})(\sqrt[3]{2a} - \sqrt[5]{bc})$

Skill Maintenance

77. Solve:

$$\frac{5}{x-1} + \frac{9}{x^2+x+1} = \frac{15}{x^3-1}.$$

78. Divide and simplify:

$$\frac{2x^2 - x - 6}{x^2 + 4x + 3} \div \frac{2x^2 + x - 3}{x^2 - 1}.$$

Synthesis

Perform the indicated operations and simplify. Assume that all variables represent nonnegative real numbers unless otherwise indicated.

79. $\sqrt{432} - \sqrt{6125} + \sqrt{845} - \sqrt{4800}$

80. $\sqrt{1250x^3y} - \sqrt{1800xy^3} - \sqrt{162x^3y^3}$

81. $\frac{1}{2}\sqrt{36a^5bc^4} - \frac{1}{2}\sqrt[3]{64a^4bc^6} + \frac{1}{6}\sqrt{144a^3bc^2}$

82. $7x\sqrt{(x+y)^3} - 5xy\sqrt{x+y} - 2y\sqrt{(x+y)^3}$
(Assume $x + y \geq 0$.)

83. $\sqrt{9 + 3\sqrt{5}}\sqrt{9 - 3\sqrt{5}}$

84. $(\sqrt{x+2} - \sqrt{x-2})^2$

85. $(\sqrt{3} + \sqrt{5} - \sqrt{6})^2$

86. $\sqrt[3]{y}(1 - \sqrt[3]{y})(1 + \sqrt[3]{y})$

87. $(\sqrt[3]{9} - 2)(\sqrt[3]{9} + 4)$

88. $\left[\sqrt{3 + \sqrt{2 + \sqrt{1}}}\right]^4$

10.6 Rationalizing Numerators and Denominators

Rationalizing Denominators • Rationalizing Numerators • Rationalizing When There Are Two Terms

Rationalizing Denominators

Sometimes in mathematics it is useful to find an equivalent expression without a radical in the denominator. This provides a standard notation for expressing results. The procedure for finding such an expression is called **rationalizing a denominator.** We carry this out by multiplying by 1 in either of two ways.

One way is to multiply by 1 *under* the radical to make the denominator of the radicand a perfect power.

EXAMPLE 1

Rationalize the denominator: **(a)** $\sqrt{\frac{7}{3}}$; **(b)** $\sqrt[3]{\frac{5}{16}}$.

Solution

a) We multiply by 1 under the radical, using $\frac{3}{3}$. We do this so that the denominator of the radicand will be a perfect square:

$$\sqrt{\frac{7}{3}} = \sqrt{\frac{7}{3}\cdot\frac{3}{3}} = \sqrt{\frac{21}{9}} = \frac{\sqrt{21}}{\sqrt{9}} = \frac{\sqrt{21}}{3}.$$

b) Note that $16 = 4^2$. Thus, to make the denominator a perfect cube, we multiply under the radical by $\frac{4}{4}$:

$$\sqrt[3]{\frac{5}{16}} = \sqrt[3]{\frac{5}{4\cdot 4}\cdot\frac{4}{4}} = \sqrt[3]{\frac{20}{4^3}} = \frac{\sqrt[3]{20}}{\sqrt[3]{4^3}} = \frac{\sqrt[3]{20}}{4}.$$

□

Another way to rationalize a denominator is to multiply by 1 *outside* the radical in order to eliminate the need for a radical in the denominator.

EXAMPLE 2

Rationalize the denominator.

a) $\sqrt{\dfrac{4}{5b}}$ b) $\dfrac{\sqrt[3]{a}}{\sqrt[3]{9x}}$ c) $\dfrac{3x}{\sqrt[5]{2x^2y^3}}$

Solution

a) We rewrite the expression as a quotient of two radicals. Then we simplify and multiply by 1:

$$\sqrt{\frac{4}{5b}} = \frac{\sqrt{4}}{\sqrt{5b}} = \frac{2}{\sqrt{5b}} \qquad \text{We assume } b > 0.$$

$$= \frac{2}{\sqrt{5b}} \cdot \frac{\sqrt{5b}}{\sqrt{5b}} \qquad \text{Multiplying by 1}$$

$$= \frac{2\sqrt{5b}}{(\sqrt{5b})^2} \qquad \text{Try to do this step mentally.}$$

$$= \frac{2\sqrt{5b}}{5b}.$$

b) To rationalize the denominator $\sqrt[3]{9x}$, we observe that $9x$ is $3 \cdot 3 \cdot x$. To make $9x$ a cube, we need another factor of 3 and two more factors of x. Thus, we multiply by 1, using $\sqrt[3]{3x^2}/\sqrt[3]{3x^2}$:

$$\frac{\sqrt[3]{a}}{\sqrt[3]{9x}} = \frac{\sqrt[3]{a}}{\sqrt[3]{9x}} \cdot \frac{\sqrt[3]{3x^2}}{\sqrt[3]{3x^2}} \qquad \text{Multiplying by 1}$$

$$= \frac{\sqrt[3]{3ax^2}}{\sqrt[3]{27x^3}} \longleftarrow \text{This radicand is a perfect cube.}$$

$$= \frac{\sqrt[3]{3ax^2}}{3x}.$$

c) For the radicand $2x^2y^3$ to be a perfect fifth power, it needs four more factors of 2, three more factors of x, and two more factors of y. Thus we multiply by 1 using $\sqrt[5]{2^4x^3y^2}/\sqrt[5]{2^4x^3y^2}$, or $\sqrt[5]{16x^3y^2}/\sqrt[5]{16x^3y^2}$:

$$\frac{3x}{\sqrt[5]{2x^2y^3}} = \frac{3x}{\sqrt[5]{2x^2y^3}} \cdot \frac{\sqrt[5]{16x^3y^2}}{\sqrt[5]{16x^3y^2}} \qquad \text{Multiplying by 1}$$

$$= \frac{3x\sqrt[5]{16x^3y^2}}{\sqrt[5]{32x^5y^5}} \longleftarrow \text{This radicand is a perfect fifth power.}$$

$$= \frac{3x\sqrt[5]{16x^3y^2}}{2xy} = \frac{3\sqrt[5]{16x^3y^2}}{2y} \qquad \text{Always simplify if possible.}$$

❑

Rationalizing Numerators

Sometimes in calculus it is necessary to rationalize a numerator. To do so, we multiply by 1 to make the radicand in the *numerator* a perfect power.

EXAMPLE 3

Rationalize the numerator.

a) $\sqrt{\dfrac{7}{5}}$

b) $\dfrac{\sqrt[3]{4a^2}}{\sqrt[3]{5b}}$

Solution

a) $\sqrt{\dfrac{7}{5}} = \dfrac{\sqrt{7}}{\sqrt{5}}$

$= \dfrac{\sqrt{7}}{\sqrt{5}} \cdot \dfrac{\sqrt{7}}{\sqrt{7}}$ Multiplying by 1

$= \dfrac{\sqrt{49}}{\sqrt{35}}$ The radicand in the numerator is a perfect square.

$= \dfrac{7}{\sqrt{35}}$

b) $\dfrac{\sqrt[3]{4a^2}}{\sqrt[3]{5b}} = \dfrac{\sqrt[3]{4a^2}}{\sqrt[3]{5b}} \cdot \dfrac{\sqrt[3]{2a}}{\sqrt[3]{2a}}$ Multiplying by 1

$= \dfrac{\sqrt[3]{8a^3}}{\sqrt[3]{10ba}}$ ⟵ This radicand is a perfect cube.

$= \dfrac{2a}{\sqrt[3]{10ab}}$ □

Rationalizing When There Are Two Terms

Recall from Section 10.5 that when a pair of conjugates are multiplied, the product has no radicals in it. Thus when the denominator to be rationalized has two terms, we use the conjugate of the denominator to write a symbol for 1.

EXAMPLE 4

For each expression, write the symbol for 1 that can be used to rationalize the denominator.

a) $\dfrac{3}{x + \sqrt{7}}$

b) $\dfrac{\sqrt{7} + 4}{3 - 2\sqrt{5}}$

Solution

	Expression	*Symbol for 1*
a)	$\dfrac{3}{x + \sqrt{7}}$	$\dfrac{x - \sqrt{7}}{x - \sqrt{7}}$
b)	$\dfrac{\sqrt{7} + 4}{3 - 2\sqrt{5}}$	$\dfrac{3 + 2\sqrt{5}}{3 + 2\sqrt{5}}$

Change the operation sign to obtain the conjugate. Use the conjugate for the numerator and denominator of the symbol for 1.

□

EXAMPLE 5

Rationalize the denominator: $\dfrac{4+\sqrt{2}}{\sqrt{5}-\sqrt{2}}$.

Solution

$$\frac{4+\sqrt{2}}{\sqrt{5}-\sqrt{2}} = \frac{4+\sqrt{2}}{\sqrt{5}-\sqrt{2}} \cdot \frac{\sqrt{5}+\sqrt{2}}{\sqrt{5}+\sqrt{2}}$$

Multiplying by 1, using the conjugate of $\sqrt{5}-\sqrt{2}$, which is $\sqrt{5}+\sqrt{2}$

$$= \frac{(4+\sqrt{2})(\sqrt{5}+\sqrt{2})}{(\sqrt{5}-\sqrt{2})(\sqrt{5}+\sqrt{2})}$$

Multiplying numerators and denominators

$$= \frac{4\sqrt{5}+4\sqrt{2}+\sqrt{2}\sqrt{5}+(\sqrt{2})^2}{(\sqrt{5})^2-(\sqrt{2})^2}$$

Using FOIL

$$= \frac{4\sqrt{5}+4\sqrt{2}+\sqrt{10}+2}{5-2}$$

Squaring in the denominator and the numerator

$$= \frac{4\sqrt{5}+4\sqrt{2}+\sqrt{10}+2}{3}$$

❑

EXAMPLE 6

Rationalize the denominator: $\dfrac{4}{\sqrt{3}+x}$.

Solution

$$\frac{4}{\sqrt{3}+x} = \frac{4}{\sqrt{3}+x} \cdot \frac{\sqrt{3}-x}{\sqrt{3}-x}$$

Multiplying by 1

$$= \frac{4(\sqrt{3}-x)}{(\sqrt{3}+x)(\sqrt{3}-x)}$$

$$= \frac{4(\sqrt{3}-x)}{(\sqrt{3})^2-x^2}$$

$$= \frac{4\sqrt{3}-4x}{3-x^2}$$

❑

To rationalize a numerator with more than one term, we use the conjugate of the numerator.

EXAMPLE 7

Rationalize the numerator: $\dfrac{4+\sqrt{2}}{\sqrt{5}-\sqrt{2}}$.

Solution

$$\frac{4+\sqrt{2}}{\sqrt{5}-\sqrt{2}} = \frac{4+\sqrt{2}}{\sqrt{5}-\sqrt{2}} \cdot \frac{4-\sqrt{2}}{4-\sqrt{2}}$$

Multiplying by 1

$$= \frac{16-(\sqrt{2})^2}{4\sqrt{5}-\sqrt{5}\sqrt{2}-4\sqrt{2}+(\sqrt{2})^2}$$

$$= \frac{14}{4\sqrt{5}-\sqrt{10}-4\sqrt{2}+2}$$

❑

EXERCISE SET 10.6

Rationalize the denominator.

1. $\sqrt{\frac{6}{5}}$
2. $\sqrt{\frac{11}{6}}$
3. $\sqrt{\frac{10}{7}}$
4. $\sqrt{\frac{22}{3}}$
5. $\frac{6\sqrt{5}}{5\sqrt{3}}$
6. $\frac{2\sqrt{3}}{5\sqrt{2}}$
7. $\sqrt[3]{\frac{16}{9}}$
8. $\sqrt[3]{\frac{2}{9}}$
9. $\frac{\sqrt[3]{3a}}{\sqrt[3]{5c}}$
10. $\frac{\sqrt[3]{7x}}{\sqrt[3]{3y}}$
11. $\frac{\sqrt[3]{5y^4}}{\sqrt[3]{6x^4}}$
12. $\frac{\sqrt[3]{3a^4}}{\sqrt[3]{7b^2}}$
13. $\frac{1}{\sqrt[3]{xy}}$
14. $\frac{1}{\sqrt[3]{ab}}$
15. $\sqrt{\frac{7a}{18}}$
16. $\sqrt{\frac{3x}{10}}$
17. $\sqrt{\frac{9}{20x^2y}}$
18. $\sqrt{\frac{5}{32ab^2}}$
19. $\sqrt[3]{\frac{9}{100x^2y^5}}$
20. $\sqrt[3]{\frac{7}{36a^4b}}$

Rationalize the numerator.

21. $\frac{\sqrt{7}}{\sqrt{3x}}$
22. $\frac{\sqrt{6}}{\sqrt{5x}}$
23. $\sqrt{\frac{14}{21}}$
24. $\sqrt{\frac{12}{15}}$
25. $\frac{4\sqrt{13}}{3\sqrt{7}}$
26. $\frac{5\sqrt{21}}{2\sqrt{5}}$
27. $\frac{\sqrt[3]{7}}{\sqrt[3]{2}}$
28. $\frac{\sqrt[3]{5}}{\sqrt[3]{4}}$
29. $\sqrt{\frac{7x}{3y}}$
30. $\sqrt{\frac{6a}{5b}}$
31. $\frac{\sqrt[3]{5y^4}}{\sqrt[3]{6x^5}}$
32. $\frac{\sqrt[3]{3a^5}}{\sqrt[3]{7b^2}}$
33. $\frac{\sqrt{ab}}{3}$
34. $\frac{\sqrt{xy}}{5}$
35. $\sqrt{\frac{x^3y}{2}}$
36. $\sqrt{\frac{ab^5}{3}}$
37. $\frac{\sqrt[3]{a^2b}}{\sqrt[3]{5}}$
38. $\frac{\sqrt[3]{xy^2}}{\sqrt[3]{7}}$
39. $\sqrt[3]{\frac{x^4y^2}{3}}$
40. $\sqrt[3]{\frac{a^5b}{2}}$

Rationalize the denominator. Assume that all variables represent nonnegative numbers and that no denominators are 0.

41. $\frac{5}{8-\sqrt{6}}$
42. $\frac{7}{9+\sqrt{10}}$
43. $\frac{-4\sqrt{7}}{\sqrt{5}-\sqrt{3}}$
44. $\frac{-3\sqrt{2}}{\sqrt{3}-\sqrt{5}}$
45. $\frac{\sqrt{5}-2\sqrt{6}}{\sqrt{3}-4\sqrt{5}}$
46. $\frac{\sqrt{6}-3\sqrt{5}}{\sqrt{3}-2\sqrt{7}}$
47. $\frac{\sqrt{x}-\sqrt{y}}{\sqrt{x}+\sqrt{y}}$
48. $\frac{\sqrt{a}+\sqrt{b}}{\sqrt{a}-\sqrt{b}}$
49. $\frac{5\sqrt{3}-3\sqrt{2}}{3\sqrt{2}-2\sqrt{3}}$
50. $\frac{7\sqrt{2}+4\sqrt{3}}{4\sqrt{3}-3\sqrt{2}}$
51. $\frac{3\sqrt{x}+\sqrt{y}}{2\sqrt{x}+3\sqrt{y}}$
52. $\frac{2\sqrt{a}-\sqrt{b}}{3\sqrt{a}+2\sqrt{b}}$

Rationalize the numerator. Assume that all variables represent nonnegative numbers and that no denominators are 0.

53. $\frac{\sqrt{3}+5}{8}$
54. $\frac{3-\sqrt{2}}{5}$
55. $\frac{\sqrt{3}-5}{\sqrt{2}+5}$
56. $\frac{\sqrt{6}-3}{\sqrt{3}+7}$
57. $\frac{\sqrt{x}-\sqrt{y}}{\sqrt{x}+\sqrt{y}}$
58. $\frac{\sqrt{x}+\sqrt{y}}{\sqrt{x}-\sqrt{y}}$
59. $\frac{4\sqrt{6}-5\sqrt{3}}{2\sqrt{3}+7\sqrt{6}}$
60. $\frac{8\sqrt{2}+5\sqrt{3}}{5\sqrt{3}-7\sqrt{2}}$
61. $\frac{\sqrt{3}+2\sqrt{x}}{\sqrt{3}-\sqrt{x}}$
62. $\frac{\sqrt{5}-3\sqrt{x}}{\sqrt{5}+\sqrt{x}}$
63. $\frac{a+b\sqrt{c}}{a+\sqrt{c}}$
64. $\frac{a\sqrt{b}-c}{\sqrt{b}-c}$

Skill Maintenance

65. Solve:

$$\frac{1}{2}-\frac{1}{3}=\frac{1}{t}.$$

66. Divide and simplify:

$$\frac{1}{x^3 - y^3} \div \frac{1}{(x - y)(x^2 + xy + y^2)}.$$

Synthesis

67. ◈ Explain why it is easier to approximate

$$\frac{\sqrt{2}}{2} \quad \text{than} \quad \frac{1}{\sqrt{2}}$$

if no calculator is available and $\sqrt{2} \approx 1.414213562$.

68. ◈ Use what we know about factoring a difference of two cubes to present a method for rationalizing any denominator of the form $\sqrt[3]{a} - \sqrt[3]{b}$.

For Exercises 69–77, assume that all radicands are positive and that no denominators are 0.

Rationalize the denominator.

69. $\dfrac{a - \sqrt{a + b}}{\sqrt{a + b} - b}$

70. $\dfrac{3\sqrt{y} + 4\sqrt{yz}}{5\sqrt{y} - 2\sqrt{z + y}}$

71. $\dfrac{b + \sqrt{b}}{1 + b + \sqrt{b}}$

Rationalize the numerator.

72. $\dfrac{\sqrt{y + 18} - \sqrt{y}}{18}$

73. $\dfrac{\sqrt{x + 6} - 5}{\sqrt{x + 6} + 5}$

Simplify.

74. $\sqrt{a^2 - 3} - \dfrac{a^2}{\sqrt{a^2 - 3}}$

75. $5\sqrt{\dfrac{x}{y}} + 4\sqrt{\dfrac{y}{x}} - \dfrac{3}{\sqrt{xy}}$

76. $\dfrac{\dfrac{1}{\sqrt{w}} - \sqrt{w}}{\dfrac{\sqrt{w} + 1}{\sqrt{w}}}$

77. $\dfrac{1}{4 + \sqrt{3}} + \dfrac{1}{\sqrt{3}} + \dfrac{1}{\sqrt{3} - 4}$

10.7 Solving Radical Equations

The Principle of Powers • Equations with Two Radical Terms

The Principle of Powers

A **radical equation** has variables in one or more radicands. These are radical equations:

$$\sqrt[3]{2x} + 1 = 5, \qquad \sqrt{x} + \sqrt{4x - 2} = 7.$$

To solve such equations, we need a new principle. Suppose an equation $a = b$ is true. If we square both sides, we get another true equation: $a^2 = b^2$. This can be generalized.

The Principle of Powers

If $a = b$, then $a^n = b^n$ for any natural number n.

Note that the principle of powers is an "if–then" statement. The statement obtained by interchanging the two parts of the sentence—"if $a^n = b^n$ for some natural number n, then $a = b$''—*is not always true*. For example, $3^2 = (-3)^2$ is true, but $3 = -3$ is *not* true. This means that we must always check our solution in the original problem when we use the principle of powers, because $a = b$ and $a^n = b^n$ are not always equivalent equations.

EXAMPLE 1

Solve: $\sqrt{x} - 3 = 4$.

Solution

$$\sqrt{x} - 3 = 4$$
$$\sqrt{x} = 7 \qquad \text{Adding to isolate the radical}$$
$$(\sqrt{x})^2 = 7^2 \qquad \text{Using the principle of powers}$$
$$x = 49$$

Check:

$\sqrt{x} - 3 = 4$	
$\sqrt{49} - 3$?	4
$7 - 3$	
4	4 TRUE

The solution is 49. □

EXAMPLE 2

Solve: $\sqrt{x} - 3 = -5$.

Solution

$$\sqrt{x} - 3 = -5$$
$$\sqrt{x} = -2 \qquad \text{Adding to isolate the radical}$$

The equation $\sqrt{x} = -2$ has no solution because the principal square root of a number is never negative. We continue as in Example 1 for comparison.

$$(\sqrt{x})^2 = (-2)^2 \qquad \text{Using the principle of powers (squaring)}$$
$$x = 4$$

Check:

$\sqrt{x} - 3 = -5$	
$\sqrt{4} - 3$?	-5
$2 - 3$	
-1	-5 FALSE

The number 4 does not check. Thus the equation $\sqrt{x} - 3 = -5$ has no real-number solution. □

The principle of powers does not always give equivalent equations. For this reason, a check is a must!

Note in Example 2 that the equation $x = 4$ has solution 4, but that $\sqrt{x} - 3 = -5$ has *no* solution. Thus the equations $x = 4$ and $\sqrt{x} - 3 = -5$ are *not* equivalent.

EXAMPLE 3

Solve: $x = \sqrt{x + 7} + 5$.

Solution

$$x = \sqrt{x+7} + 5$$

$$x - 5 = \sqrt{x+7}$$ Adding -5 on both sides to isolate the radical term

$$(x-5)^2 = \left(\sqrt{x+7}\right)^2$$ Using the principle of powers; squaring both sides

$$x^2 - 10x + 25 = x + 7$$

$$x^2 - 11x + 18 = 0$$ Adding $-x - 7$ on both sides to write the quadratic equation in standard form

$$(x-9)(x-2) = 0$$ Factoring

$$x = 9 \quad or \quad x = 2$$ Using the principle of zero products

The possible solutions are 9 and 2. Let us check.

Check:

For 9:		For 2:	
$x = \sqrt{x+7} + 5$		$x = \sqrt{x+7} + 5$	
$9 \ ? \ \sqrt{9+7} + 5$		$2 \ ? \ \sqrt{2+7} + 5$	
$9 \mid 9$	TRUE	$2 \mid 8$	FALSE

Since 9 checks but 2 does not, the solution is 9. ❑

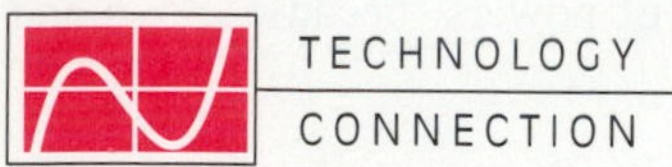

TECHNOLOGY CONNECTION

To solve Example 3 with a grapher, graph the curves $y_1 = x$ and $y_2 = (x+7)^{1/2} + 5$ on the same set of axes.

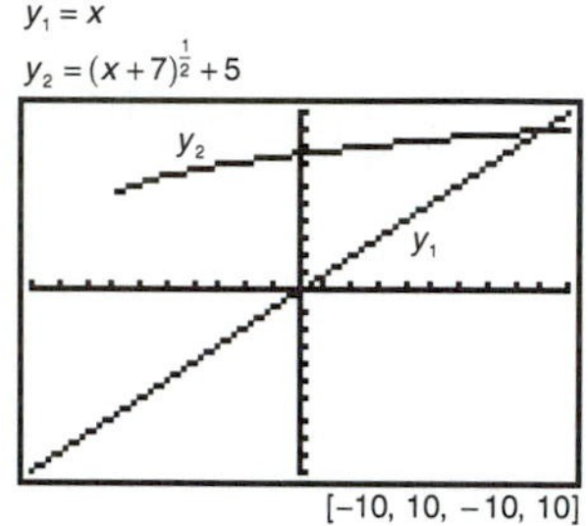

By using the Trace and Zoom features, try to determine the point of intersection. If you lack a Zoom feature, adjust the window with the Range feature. The intersection should occur when $x = 9$.

TC1. Use a grapher to solve Examples 1, 2, 4, 5, and 6. Compare your answers with those found using the algebraic methods shown.

Suppose in Example 3 that we had used the principle of powers *before* we added -5 to each side. We then would have had the expression $\left(\sqrt{x+7} + 5\right)^2$ or $x + 7 + 10\sqrt{x+7} + 25$ on the right side, and the radical would still have been in the problem.

EXAMPLE 4

Solve: $(2x+1)^{1/3} + 5 = 0$.

Solution We rewrite using radical notation and solve as above:

$$(2x+1)^{1/3} + 5 = 0$$

$$\sqrt[3]{2x+1} + 5 = 0$$ Converting to radical notation. This is sometimes performed mentally.

$$\sqrt[3]{2x+1} = -5$$ Adding -5; this isolates the radical term.

$$\left(\sqrt[3]{2x+1}\right)^3 = (-5)^3$$ Using the principle of powers; cubing both sides

$$2x + 1 = -125$$

$$2x = -126$$ Adding -1

$$x = -63$$

Check:

$$\begin{array}{r|l} (2x+1)^{1/3}+5 & = 0 \\ \hline (2(-63)+1)^{1/3}+5 & ?\ 0 \\ (-125)^{1/3}+5 & \\ -5+5 & \\ 0 & 0 \quad \text{TRUE} \end{array}$$

The solution is -63. □

Equations with Two Radical Terms

A general strategy for solving equations with two radical terms is as follows.

1. Isolate one of the radical terms.
2. Use the principle of powers.
3. If a radical remains, perform steps (1) and (2) again.
4. Check possible solutions.

EXAMPLE 5

Solve: $\sqrt{x-3}+\sqrt{x+5}=4$.

Solution

$$\sqrt{x-3}+\sqrt{x+5}=4$$

$$\sqrt{x-3}=4-\sqrt{x+5} \qquad \text{Adding } -\sqrt{x+5}\text{; this isolates one of the radical terms.}$$

$$(\sqrt{x-3})^2=(4-\sqrt{x+5})^2 \qquad \text{Using the principle of powers (squaring both sides)}$$

This is like squaring a binomial. We square 4, then find twice the product of 4 and $-\sqrt{x+5}$, and then the square of $\sqrt{x+5}$.

$$x-3=16-8\sqrt{x+5}+(x+5)$$

$$-3=21-8\sqrt{x+5} \qquad \text{Adding } -x \text{ and collecting like terms}$$

$$-24=-8\sqrt{x+5} \qquad \text{Isolating the remaining radical term}$$

$$3=\sqrt{x+5} \qquad \text{Dividing by } -8$$

$$3^2=(\sqrt{x+5})^2 \qquad \text{Squaring}$$

$$9=x+5$$

$$4=x$$

The number 4 checks and is the solution. □

CAUTION! A common error in solving equations like $\sqrt{x-3}+\sqrt{x+5}=4$ is to obtain $(x-3)+(x+5)$ when squaring the left side. This is wrong because the square of a sum is *not* the sum of the squares. For example, $(\sqrt{9}+\sqrt{16})^2=7^2$, or 49, whereas $(\sqrt{9})^2+(\sqrt{16})^2=9+16$, or 25.

EXAMPLE 6 Solve: $\sqrt{2x-5} = 1 + \sqrt{x-3}$.

Solution

$$\sqrt{2x-5} = 1 + \sqrt{x-3}$$

$$(\sqrt{2x-5})^2 = (1 + \sqrt{x-3})^2$$ One radical is already isolated; we square both sides.

$$2x - 5 = 1 + 2\sqrt{x-3} + (\sqrt{x-3})^2$$

$$2x - 5 = 1 + 2\sqrt{x-3} + (x-3)$$

$$x - 3 = 2\sqrt{x-3}$$ Isolating the remaining radical term

$$(x-3)^2 = (2\sqrt{x-3})^2$$ Squaring both sides

$$x^2 - 6x + 9 = 4(x-3)$$

$$x^2 - 6x + 9 = 4x - 12$$

$$x^2 - 10x + 21 = 0$$

$$(x-7)(x-3) = 0$$ Factoring

$$x = 7 \quad or \quad x = 3$$ Using the principle of zero products

The numbers 7 and 3 check and are the solutions. ❑

EXERCISE SET 10.7

Solve.

1. $\sqrt{2x-3} = 1$
2. $\sqrt{x+3} = 6$
3. $\sqrt{3x} + 1 = 7$
4. $\sqrt{2x} - 1 = 7$
5. $\sqrt{y+1} - 5 = 8$
6. $\sqrt{x-2} - 7 = -4$
7. $\sqrt{y-3} + 4 = 2$
8. $\sqrt{y+4} + 6 = 7$
9. $\sqrt[3]{x+5} = 2$
10. $\sqrt[3]{x-2} = 3$
11. $\sqrt[4]{y-3} = 2$
12. $\sqrt[4]{x+3} = 3$
13. $\sqrt{3y+1} = 9$
14. $\sqrt{2y+1} = 13$
15. $3\sqrt{x} = 6$
16. $8\sqrt{y} = 2$
17. $2y^{1/2} - 7 = 9$
18. $3x^{1/2} + 12 = 9$
19. $\sqrt[3]{x} = -3$
20. $\sqrt[3]{y} = -4$
21. $\sqrt{y+3} - 20 = 0$
22. $\sqrt{x+4} - 11 = 0$
23. $(x+2)^{1/2} = -4$
24. $(y-3)^{1/2} = -2$
25. $\sqrt{2x+3} - 5 = -2$
26. $\sqrt{3x+1} - 4 = -1$
27. $8 = x^{-1/2}$
28. $3 = y^{-1/2}$
29. $\sqrt[3]{6x+9} + 8 = 5$
30. $\sqrt[3]{3y+6} + 2 = 3$
31. $\sqrt{3y+1} = \sqrt{2y+6}$
32. $\sqrt{5x-3} = \sqrt{2x+3}$
33. $2\sqrt{1-x} = \sqrt{5}$
34. $2\sqrt{2y-3} = \sqrt{4y}$
35. $2\sqrt{t-1} = \sqrt{3t-1}$
36. $\sqrt{y+10} = 3\sqrt{2y+3}$
37. $\sqrt{y-5} + \sqrt{y} = 5$
38. $\sqrt{x-9} + \sqrt{x} = 1$
39. $3 + \sqrt{z-6} = \sqrt{z+9}$
40. $\sqrt{4x-3} = 2 + \sqrt{2x-5}$
41. $\sqrt{20-x} + 8 = \sqrt{9-x} + 11$
42. $4 + \sqrt{10-x} = 6 + \sqrt{4-x}$
43. $\sqrt{x+2} + \sqrt{3x+4} = 2$
44. $\sqrt{6x+7} - \sqrt{3x+3} = 1$
45. $\sqrt{4y+1} - \sqrt{y-2} = 3$
46. $\sqrt{y+15} - \sqrt{2y+7} = 1$

47. $\sqrt{3x-5}+\sqrt{2x+3}+1=0$

48. $\sqrt{2m-3}=\sqrt{m+7}-2$

49. $2\sqrt{3x+6}-\sqrt{4x+9}=5$

50. $2\sqrt{x-3}+\sqrt{3x-5}=8$

51. $3\sqrt{t+1}-\sqrt{2t-5}=7$

52. $3\sqrt{7x+1}-\sqrt{12x+21}=9$

Skill Maintenance

53. Solve:

$$\frac{3}{2x}+\frac{1}{x}=\frac{2x+3.5}{3x}.$$

54. The base of a triangle is 2 in. longer than the height. The area is $31\frac{1}{2}$ in^2. Find the height and the base.

Synthesis

55. The principle of powers is an "if–then" statement that becomes false when the sentence parts are interchanged. Give an example of another such if–then statement.

56. Explain a method that could be used to solve

$$\sqrt{x}+\sqrt{2x-1}-\sqrt{3x+2}+\sqrt{2+x}=0.$$

Sighting to the horizon. The function $V(h)=1.2\sqrt{h}$ can be used to approximate the distance V, in miles, that a person can see to the horizon from a height h, in feet.

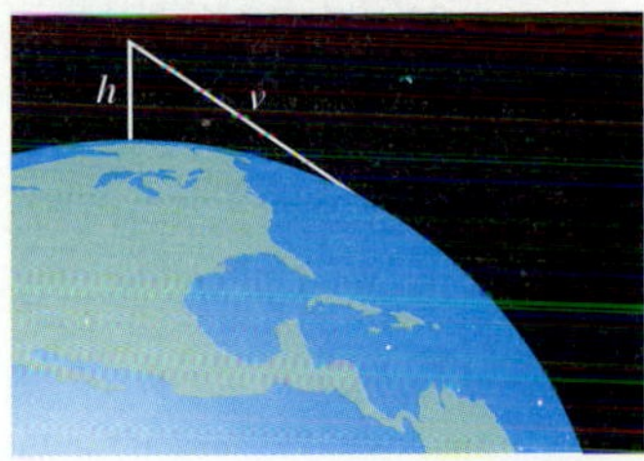

57. How far can you see to the horizon through an airplane window at a height of 30,000 ft?

58. How high above sea level must a sailor climb in order to see 10.2 mi out to sea?

Solve.

59. $\dfrac{x+\sqrt{x+1}}{x-\sqrt{x+1}}=\dfrac{5}{11}$

60. $\left(\dfrac{z}{4}\right)^{1/3}-10=2$

61. $(z^2+17)^{1/4}=3$

62. $\sqrt{\sqrt{y}+49}=7$

63. $\sqrt[3]{x^2+x+15}-3=0$

64. $x^2-5x-\sqrt{x^2-5x-2}=4$

65. $\sqrt{8-b}=b\sqrt{8-b}$

66. $\sqrt{x-2}-\sqrt{x+2}+2=0$

67. $6\sqrt{y}+6y^{-1/2}=37$

68. $\sqrt{a^2+30a}=a+\sqrt{5a}$

10.8 Geometric Applications

Using the Pythagorean Theorem • Two Special Triangles

There are many kinds of problems that involve powers and roots. Many also involve right triangles and the Pythagorean theorem, which we studied in Section 5.8.

EXAMPLE 1

A baseball diamond is actually a square 90 ft on a side. Suppose a catcher fields a bunt along the third-base line 10 ft from home plate. How far would the catcher have to throw the ball to first base? Give an exact answer and an approximation to three decimal places.

Solution We first make a drawing and let d = the distance, in feet, to first base. We see that a right triangle is formed in which the length of the leg from home to first base is 90 ft. The length of the leg from home to where the catcher fields the ball is 10 ft. We substitute these values into the Pythagorean equation to find d:

$$d^2 = 90^2 + 10^2$$
$$d^2 = 8100 + 100$$
$$d^2 = 8200$$
$$d = \sqrt{8200}.$$

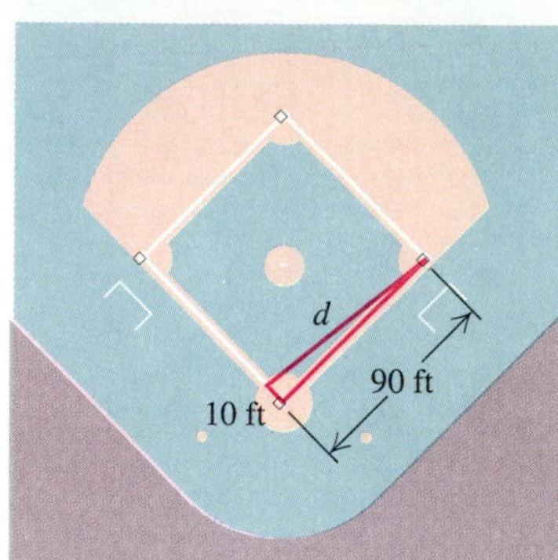

Exact answer: $d = \sqrt{8200}$ ft

Approximation: $d \approx 90.554$ ft

If you use Table 1 to find an approximation, you will need to simplify before finding an approximation in the table:

$$\begin{aligned} d &= \sqrt{8200} \\ &= \sqrt{100 \cdot 82} \\ &= 10\sqrt{82} \\ &\approx 10(9.055) = 90.550 \text{ ft.} \end{aligned}$$

Note that we get a variance in the third decimal place. ❑

EXAMPLE 2

The base of a 40-ft–long guy wire is located 15 ft from the telephone pole that it is anchoring. How high up the pole does the guy wire reach? Give an exact answer and an approximation to three decimal places.

Solution We make a drawing and let h represent the height on the pole that the guy wire reaches. We have a right triangle in which the length of one leg is 15 ft and the length of the hypotenuse is 40 ft. Substituting into the Pythagorean equation, we find h:

$$h^2 + 15^2 = 40^2$$
$$h^2 + 225 = 1600$$
$$h^2 = 1375$$
$$h = \sqrt{1375}.$$

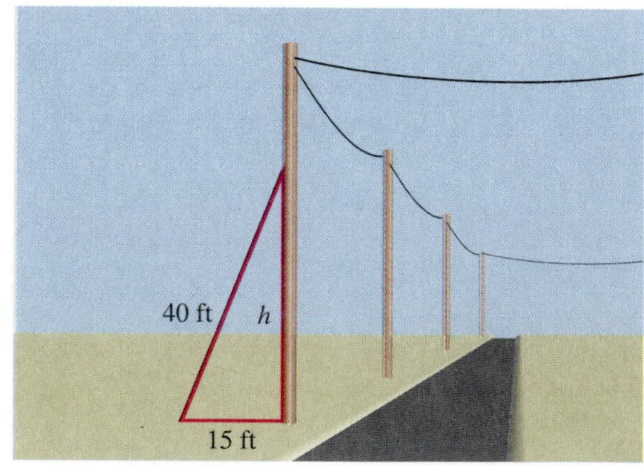

Exact answer:

$h = \sqrt{1375}$ ft

Approximation:

$h \approx 37.081$ ft

Using a calculator ❑

Two Special Triangles

When both legs of a right triangle are the same size, we say that the triangle is an *isosceles right triangle*. If one leg of an isosceles right triangle has length a, we can find the length of the hypotenuse as follows:

$$c^2 = a^2 + b^2$$

$$c^2 = a^2 + a^2 \qquad \text{Because the triangle is isosceles, both legs are the same size.}$$

$$c^2 = 2a^2 \qquad \text{Collecting like terms}$$

$$c = \sqrt{2a^2}$$

$$c = \sqrt{a^2 \cdot 2} = a\sqrt{2}.$$

EXAMPLE 3

One leg of an isosceles right triangle measures 7 cm. Find the length of the hypotenuse. Give an exact answer and an approximation to three decimal places.

Solution We substitute:

$$c = a\sqrt{2} \qquad \text{This equation should be memorized.}$$

$$c = 7\sqrt{2}.$$

Exact answer: $c = 7\sqrt{2}$ cm

Approximation: $c \approx 9.899$ cm ❑

When the hypotenuse of an isosceles right triangle is known, the lengths of the legs can be found.

EXAMPLE 4

The hypotenuse of an isosceles right triangle is 5 ft long. Find the length of a leg. Give an exact answer and an approximation to three decimal places.

Solution We replace c with 5 and solve for a:

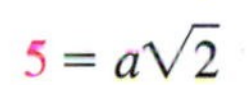

$$5 = a\sqrt{2} \qquad \text{Substituting}$$

$$\frac{5}{\sqrt{2}} = a \qquad \text{Multiplying by } 1/\sqrt{2}$$

$$\frac{5\sqrt{2}}{2} = a. \qquad \text{Rationalizing the denominator}$$

Exact answer: $a = \dfrac{5\sqrt{2}}{2}$ ft

Approximation: $a \approx 3.536$ ft ❑

A second special triangle is known as a 30–60–90 right triangle, so named because of the measures of its angles. Note that in an equilateral triangle, all sides have the same length and all angles are 60°. An altitude, drawn dashed in the figure, bisects, or splits, one angle and one side. Two 30–60–90 right triangles are thus formed.

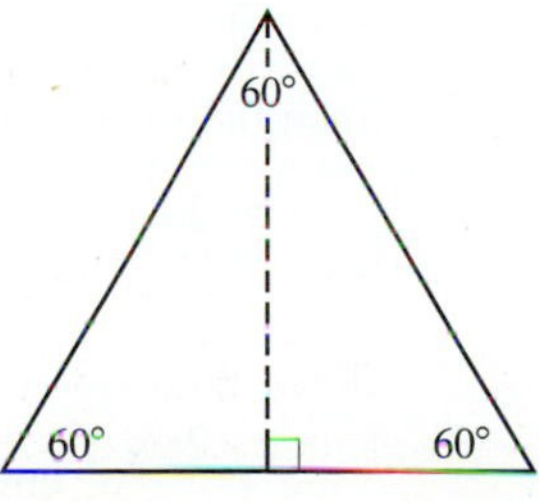

As a result of how the altitude is drawn, if a represents the length of the shorter leg in a 30–60–90 right triangle, then $2a$ represents the length of the hypotenuse. We have

$a^2 + b^2 = (2a)^2$ Using the Pythagorean equation

$a^2 + b^2 = 4a^2$

$b^2 = 3a^2$ Adding $-a^2$ on both sides

$b = \sqrt{3a^2}$

$b = \sqrt{a^2 \cdot 3} = a\sqrt{3}.$

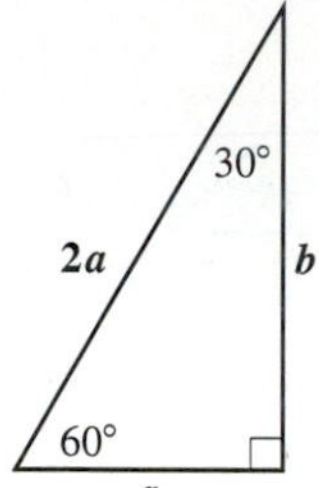

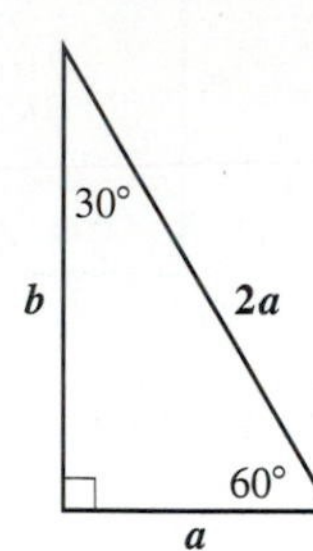

EXAMPLE 5

The shorter leg of a 30–60–90 right triangle measures 8 in. Find the lengths of the other sides. Give exact answers and, where appropriate, approximate to three decimal places.

Solution The hypotenuse is twice as long as the shorter leg, so we have

$c = 2a$ This equation should be memorized.

$c = 2 \cdot 8 = 16$ in.

The length of the longer leg is the length of the shorter leg times $\sqrt{3}$. This gives us

$b = a\sqrt{3}$ This should also be memorized.

$b = 8\sqrt{3}$ in.

Exact answer: $c = 16$ in., $b = 8\sqrt{3}$ in.

Approximation: $b \approx 13.856$ in.

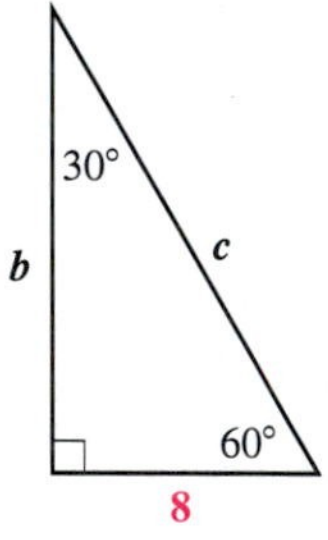

EXAMPLE 6

The length of the longer leg of a 30–60–90 right triangle is 14 cm. Find the length of the hypotenuse. Give an exact answer and an approximation to three decimal places.

Solution The length of the hypotenuse is twice the length of the shorter leg. We first find a, the length of the shorter leg, by using the length of the longer leg:

$14 = a\sqrt{3}$ Substituting 14 for b in $b = a\sqrt{3}$

$\frac{14}{\sqrt{3}} = a$ Multiplying by $1/\sqrt{3}$

$\frac{14\sqrt{3}}{3} = a.$ Rationalizing the denominator

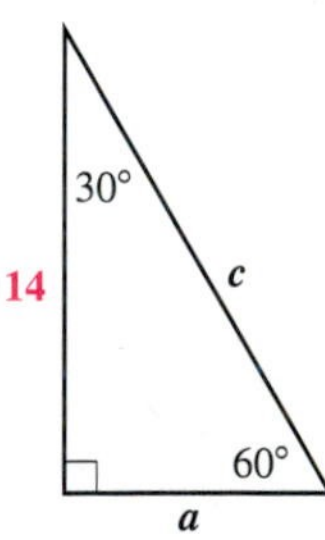

Since the hypotenuse is twice as long as the shorter leg, we have

$c = 2a$

$c = 2 \cdot \frac{14\sqrt{3}}{3}$ Substituting

$c = \frac{28\sqrt{3}}{3}$ cm

Exact answer: $c = \dfrac{28\sqrt{3}}{3}$ cm

Approximation: $c \approx 16.166$ cm ❑

Lengths Within Isosceles and 30–60–90 Right Triangles

The length of the hypotenuse in an isosceles right triangle is the length of a leg times $\sqrt{2}$.

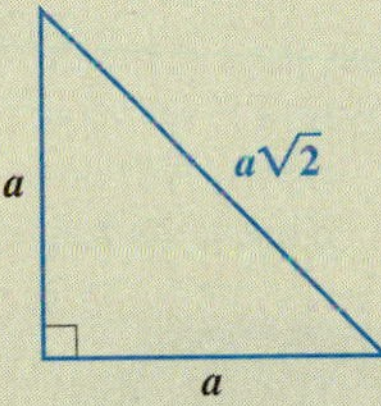

The length of the longer leg in a 30–60–90 right triangle is the length of the shorter leg times $\sqrt{3}$. The hypotenuse is twice as long as the shorter leg.

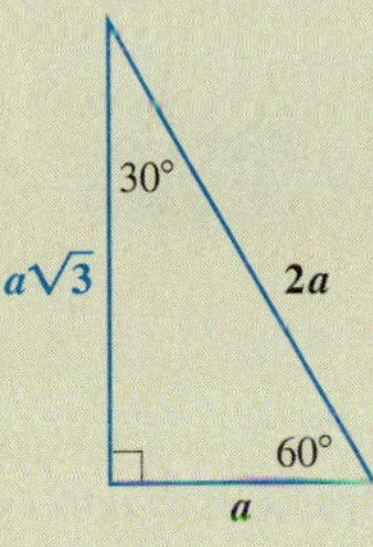

EXERCISE SET 10.8

In a right triangle, find the length of the side not given. Give an exact answer and, where appropriate, an approximation to three decimal places.

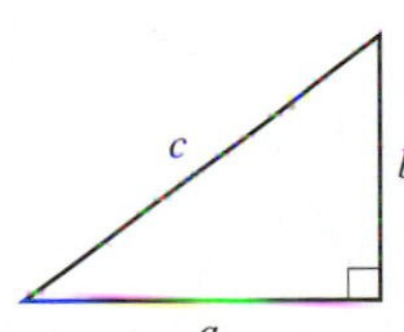

1. $a = 3,\ b = 5$
2. $a = 8,\ b = 10$
3. $a = 12,\ b = 12$
4. $a = 10,\ b = 10$
5. $b = 12,\ c = 13$
6. $a = 5,\ c = 12$
7. $c = 6,\ a = \sqrt{5}$
8. $c = 8,\ a = 4\sqrt{3}$
9. $b = 1,\ c = \sqrt{13}$
10. $a = 1,\ c = \sqrt{20}$
11. $a = 1,\ c = \sqrt{n}$
12. $c = 2,\ a = \sqrt{n}$

In the following problems, give an exact answer and, where appropriate, an approximation to three decimal places.

13. *Guy wire.* How long is a guy wire reaching from the top of a 15-ft pole to a point on the ground 10 ft from the pole?
14. *Softball diamond.* A slow-pitch softball diamond

is actually a square 65 ft on a side. How far is it from home to second base?

15. Suppose the catcher in Example 1 makes a throw to second base. How far is that throw?

16. *Speaker placement.* A stereo receiver is in a corner of a 12-ft by 14-ft room. Speaker wire will run under a rug, diagonally, to a speaker in the far corner. Allowing for 4 ft of slack on both ends, how long a piece of wire should be purchased?

17. *Television sets.* What does it mean to refer to a 20-in. TV set or a 25-in. TV set? Such units refer to the diagonal of the screen. A 20-in. TV set also has a width of 16 in. What is its height?

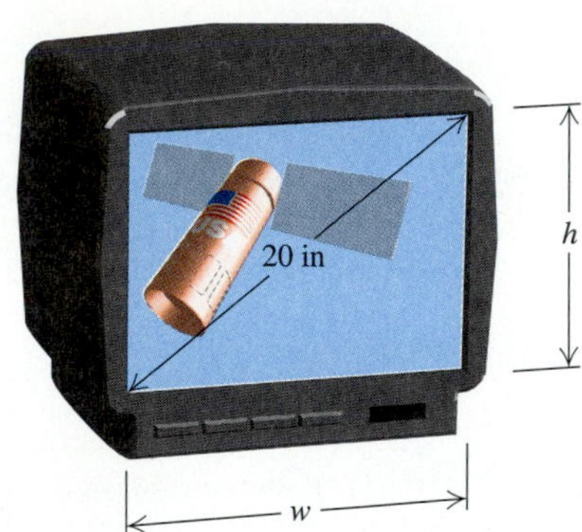

18. *Television sets.* A 25-in. TV set has a screen with a height of 15 in. What is its width?

19. *Vegetable garden.* Benito and Dominique are planting a 30-ft by 40-ft vegetable garden and are laying it out using string. They would like to know the length of a diagonal to make sure that right angles are formed. Find the length of a diagonal.

20. *Distance over water.* To determine the width of a pond, a surveyor locates two stakes at either end of the pond and uses instrumentation to place a third stake so that the distance across the pond is the length of a hypotenuse. If the third stake is 90 m from one stake and 70 m from the other, how wide is the pond?

For each triangle, find the missing length(s). Give an exact answer and, where appropriate, an approximation to three decimal places.

21.

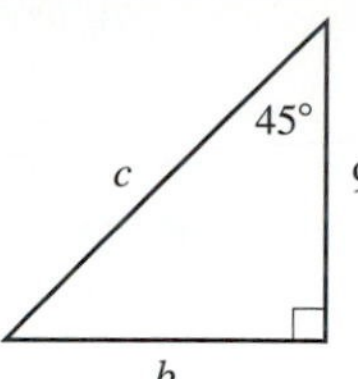

22.

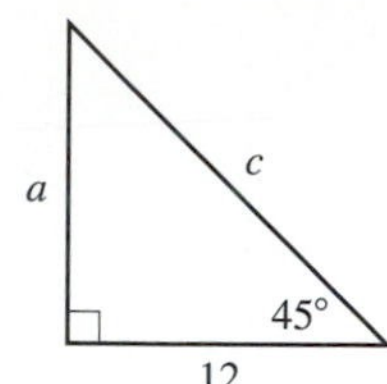

23.

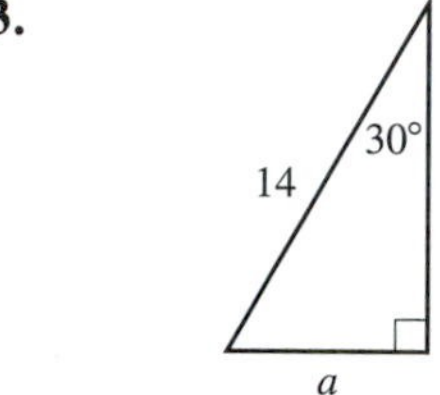

24.

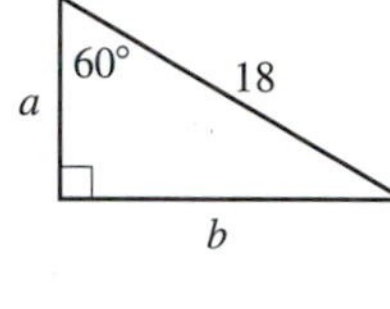

25.

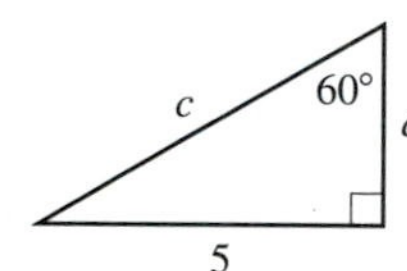

26.

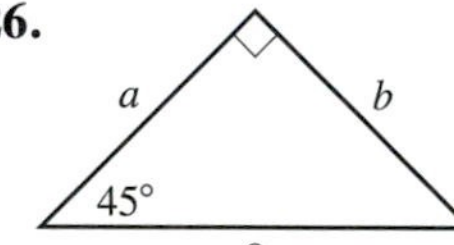

27.

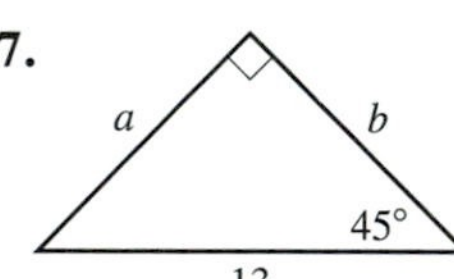

28.

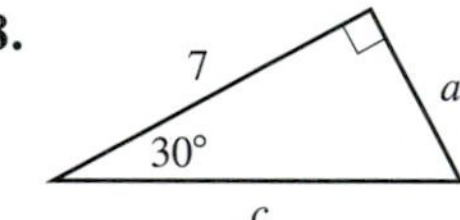

29.

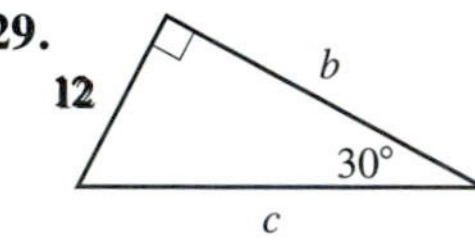

30.

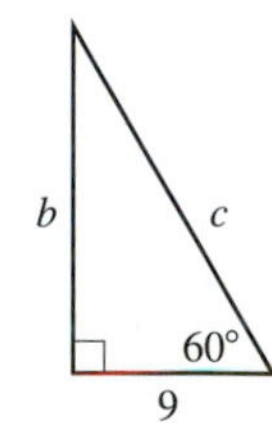

31.

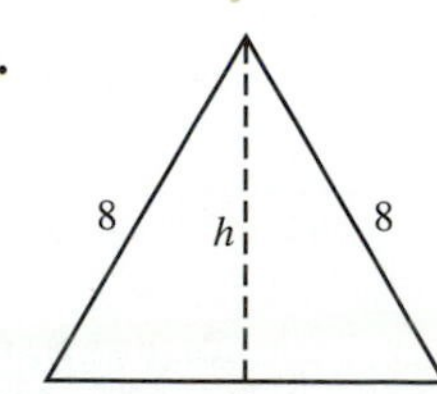

32.

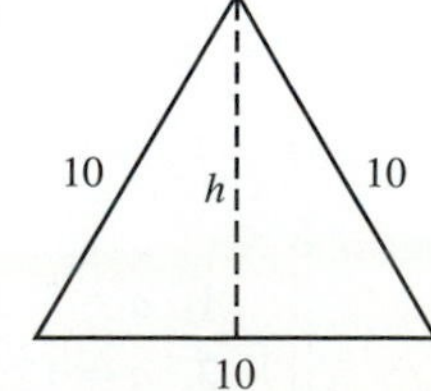

33.

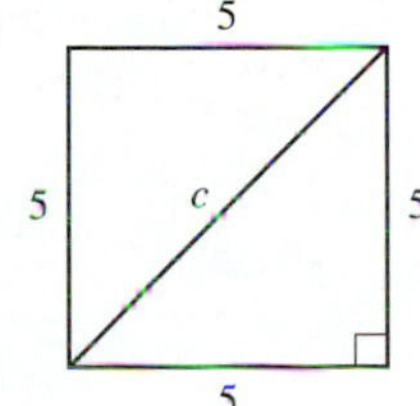

34.

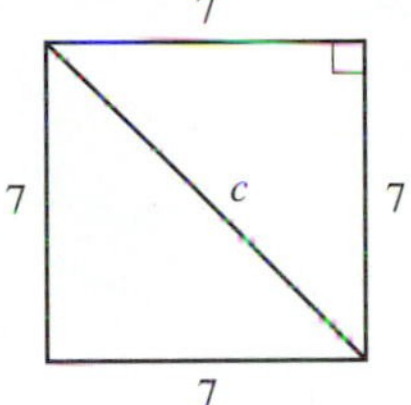

35.

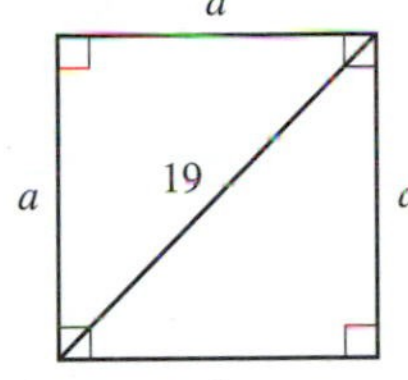

36.

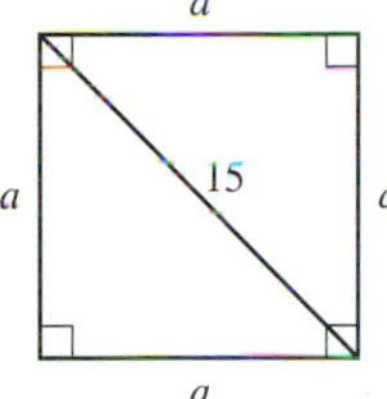

In the following problems, give an exact answer and, where appropriate, an approximation to three decimal places.

37. Triangle ABC has sides of lengths 25 ft, 25 ft, and 30 ft. Triangle PQR has sides of lengths 25 ft, 25 ft, and 40 ft. Which triangle has the greater area and by how much?

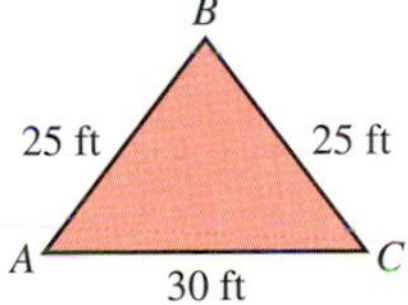

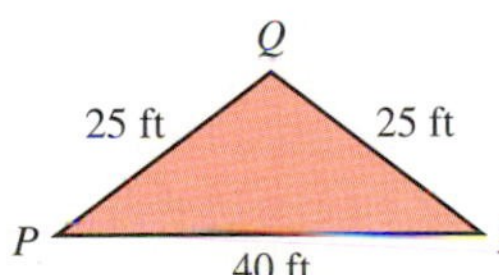

38. *Bridge expansion.* During the summer heat, a 2-mi bridge expands 2 ft in length. Assuming the bulge occurs straight up the middle, how high is the bulge? (The answer may surprise you. In reality, bridges are built with expansion spaces to avoid such buckling.)

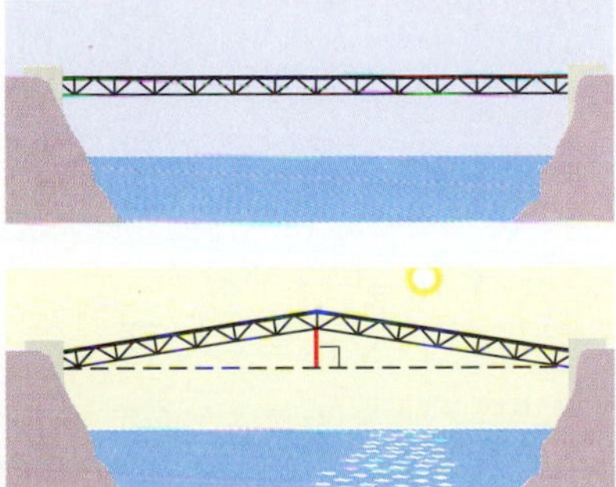

39. *Camping tent.* The entrance to a pup tent is the shape of an equilateral triangle. If the base of the tent is 4 ft wide, how tall is the tent?

40. Each side of a regular octagon has length s. Find a formula for the distance d between the parallel sides of the octagon.

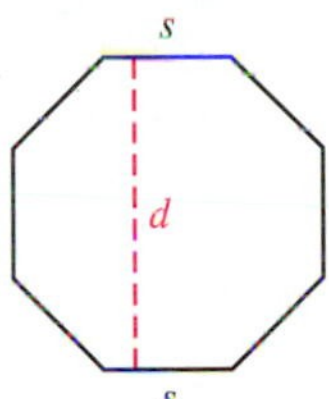

41. The diagonal of a square has length $8\sqrt{2}$ ft. Find the length of a side of the square.

42. The length and the width of a rectangle are given by consecutive integers. The area of the rectangle is 90 cm^2. Find the length of a diagonal of the rectangle.

43. Find all points on the y-axis of a Cartesian coordinate system that are 5 units from the point $(3, 0)$.

44. Find all points on the x-axis of a Cartesian coordinate system that are 5 units from the point $(0, 4)$.

Skill Maintenance

Solve.

45. $x^2 - 11x + 24 = 0$

46. $2x^2 + 11x - 21 = 0$

Synthesis

47. Write a problem for a classmate to solve in which the solution is: "The height of the tepee is $5\sqrt{3}$ ft."

48. Write a problem for a classmate to solve in which the solution is: "The height of the window is $15\sqrt{3}$ ft."

49. A cube measures 5 cm on each side. How long is the diagonal that connects two opposite corners of the cube? Give an exact answer.

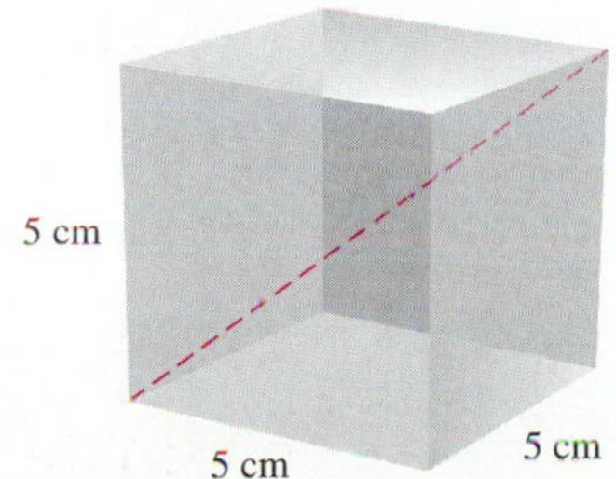

50. Kit's house is 24 ft wide and 32 ft long. The peak of the roof is 6 ft higher than the sides. Kit plans to reshingle the roof. If each packet of shingles covers 100 square feet, how many packets should Kit buy?

10.9 The Complex Numbers

Imaginary and Complex Numbers • Addition and Subtraction • Multiplication • Powers of *i* • Conjugates and Division • Solutions of Equations

Imaginary and Complex Numbers

Negative numbers do not have square roots in the real-number system. However, a larger number system that contains the real-number system is designed so that negative numbers *do* have square roots. That system is called the **complex-number system,** and it makes use of a number that is a square root of -1. We call this new number i.

> We define the number $i = \sqrt{-1}$. That is, $i = \sqrt{-1}$ and $i^2 = -1$.

To express roots of negative numbers in terms of i, we can use the fact that in the complex numbers, $\sqrt{-p} = \sqrt{-1}\sqrt{p}$ when p is a positive real number.

EXAMPLE 1

Express each number in terms of i: **(a)** $\sqrt{-7}$; **(b)** $\sqrt{-16}$; **(c)** $-\sqrt{-13}$; **(d)** $-\sqrt{-64}$; **(e)** $\sqrt{-48}$.

Solution

i is *not* under the radical.

a) $\sqrt{-7} = \sqrt{-1 \cdot 7} = \sqrt{-1} \cdot \sqrt{7} = i\sqrt{7}$, or $\sqrt{7}i$

b) $\sqrt{-16} = \sqrt{-1 \cdot 16} = \sqrt{-1} \cdot \sqrt{16} = i \cdot 4 = 4i$

c) $-\sqrt{-13} = -\sqrt{-1 \cdot 13} = -\sqrt{-1} \cdot \sqrt{13} = -i\sqrt{13}$, or $-\sqrt{13}i$

d) $-\sqrt{-64} = -\sqrt{-1 \cdot 64} = -\sqrt{-1} \cdot \sqrt{64} = -i \cdot 8 = -8i$

e) $\sqrt{-48} = \sqrt{-1 \cdot 48} = \sqrt{-1} \cdot \sqrt{48} = i\sqrt{48} = i \cdot 4\sqrt{3} = 4\sqrt{3}i$, or $4i\sqrt{3}$ ❑

Imaginary Number

An *imaginary number* is a number that can be written $a + bi$, where a and b are real numbers, $b \neq 0$.

Don't let the name "imaginary" fool you. Imaginary numbers appear in such fields as engineering and the physical sciences. The following are examples of imaginary numbers:

$5 + 4i$, $\sqrt{5} - \pi i$, — Here $a \neq 0$, $b \neq 0$.

$17i$. — Here $a = 0$, $b \neq 0$.

When a and b are real numbers and b is allowed to be 0, the number $a + bi$ is said to be **complex.**

Complex Number

A *complex number* is any number that can be written $a + bi$, where a and b are any real numbers. (Note that a and b both can be 0.)

The following are examples of complex numbers:

$7 + 3i$ (here $a \neq 0$, $b \neq 0$); $4i$ (here $a = 0$, $b \neq 0$);

8 (here $a \neq 0$, $b = 0$); 0 (here $a = 0$, $b = 0$).

Complex numbers like $17i$ or $4i$, in which $a = 0$ and $b \neq 0$, are imaginary numbers with no real part. Such numbers are called *pure imaginary* numbers.

Note that when $b = 0$, $a + 0i = a$, so every real number is a complex number. The relationships among various real and complex numbers are shown below.

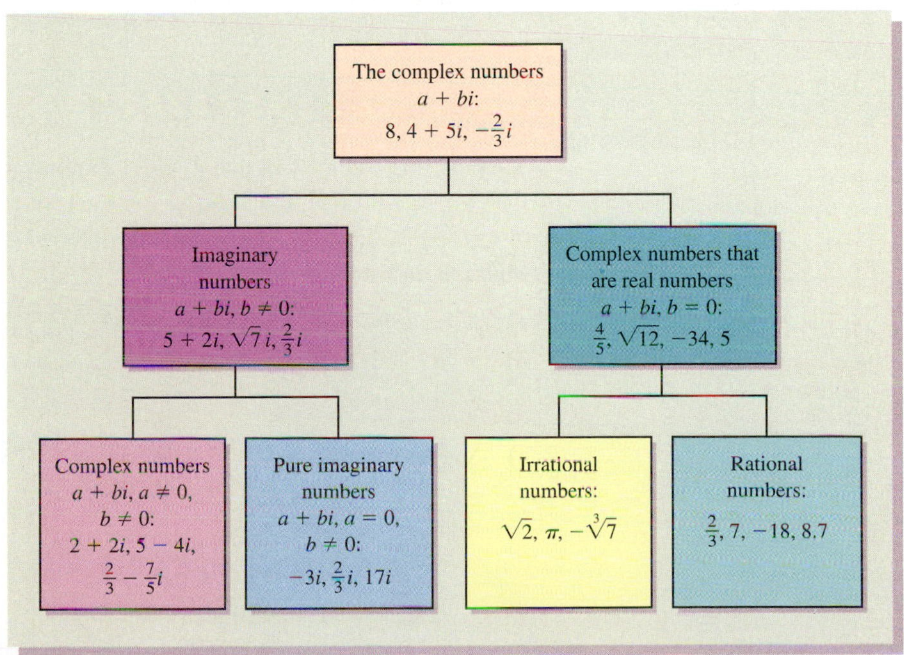

It is important to keep in mind some comparisons between numbers that have real-number roots and those that have complex-number roots that are not real. For example, $\sqrt{-48}$ is a complex number that is not a real number because we are taking the square root of a negative number. *But,* $\sqrt[3]{-48}$ is a real number because each real number has a cube root that is a real number.

Addition and Subtraction

The complex numbers obey the commutative, associative, and distributive laws. Thus we can add and subtract them as we do binomials.

EXAMPLE 2

Add or subtract and simplify.

a) $(8 + 6i) + (3 + 2i)$ **b)** $(4 + 5i) - (6 - 3i)$

Solution

a) $(8 + 6i) + (3 + 2i) = (8 + 3) + (6i + 2i)$ Collecting the real parts and the imaginary parts

$= 11 + (6 + 2)i = 11 + 8i$

b) $(4 + 5i) - (6 - 3i) = (4 - 6) + [5i - (-3i)]$ Note that the 6 and the $-3i$ are both being subtracted.

$= -2 + 8i$ ❑

Multiplication

For complex numbers, the property $\sqrt{a}\sqrt{b} = \sqrt{ab}$ does *not* hold in general, but it does hold when $a = -1$ and b is a nonnegative number. To multiply square roots of negative real numbers, we first express them in terms of i. For example,

$$\sqrt{-2}\cdot\sqrt{-5} = \sqrt{-1}\cdot\sqrt{2}\cdot\sqrt{-1}\cdot\sqrt{5} = i\sqrt{2}\cdot i\sqrt{5}$$
$$= i^2\sqrt{10} = -1\sqrt{10} = -\sqrt{10} \quad \text{is correct!}$$

But

$$\sqrt{-2}\cdot\sqrt{-5} = \sqrt{(-2)(-5)} = \sqrt{10} \quad \text{is wrong!}$$

Keeping this and the fact that $i^2 = -1$ in mind, we multiply in much the same way that we do with real numbers.

EXAMPLE 3

Multiply and simplify.

a) $\sqrt{-16}\cdot\sqrt{-25}$ **b)** $\sqrt{-5}\cdot\sqrt{-7}$ **c)** $-3i\cdot 8i$

d) $-4i(3 - 5i)$ **e)** $(1 + 2i)(1 + 3i)$

Solution

a) $\sqrt{-16}\cdot\sqrt{-25} = \sqrt{-1}\cdot\sqrt{16}\cdot\sqrt{-1}\cdot\sqrt{25}$

$= i\cdot 4\cdot i\cdot 5$

$= i^2\cdot 20$

$= -1\cdot 20$ $i^2 = -1$

$= -20$

b) $\sqrt{-5}\cdot\sqrt{-7} = \sqrt{-1}\cdot\sqrt{5}\cdot\sqrt{-1}\cdot\sqrt{7}$
$= i\cdot\sqrt{5}\cdot i\cdot\sqrt{7}$
$= i^2\cdot\sqrt{35}$
$= -1\cdot\sqrt{35}$ $\quad i^2 = -1$
$= -\sqrt{35}$

c) $-3i\cdot 8i = -24\cdot i^2$
$= -24\cdot(-1)$ $\quad i^2 = -1$
$= 24$

d) $-4i(3 - 5i) = -4i\cdot 3 + (-4i)(-5i)$ — Using the distributive law
$= -12i + 20i^2$
$= -12i - 20$ — $i^2 = -1$
$= -20 - 12i$ — Writing in the form $a + bi$

e) $(1 + 2i)(1 + 3i) = 1 + 3i + 2i + 6i^2$ — Multiplying each term of one number by every term of the other (FOIL)
$= 1 + 3i + 2i - 6$ — $i^2 = -1$
$= -5 + 5i$ — Collecting like terms ❑

Powers of *i*

We now want to simplify certain expressions involving higher powers of i. To do so, we recall that -1 raised to an *even* power is 1, and -1 raised to an *odd* power is -1. Simplifying powers of i can then be done by using the fact that $i^2 = -1$ and expressing the given power of i in terms of i^2. Consider the following:

i, or $\sqrt{-1}$,

$i^2 = -1$,

$i^3 = i^2\cdot i = (-1)i = -i$,

$i^4 = (i^2)^2 = (-1)^2 = 1$,

$i^5 = i^4\cdot i = (i^2)^2\cdot i = (-1)^2\cdot i = i$,

$i^6 = (i^2)^3 = (-1)^3 = -1$.

Note that the powers of i cycle themselves through the values i, -1, $-i$, and 1.

EXAMPLE 4

Simplify: **(a)** i^{37}; **(b)** i^{58}; **(c)** i^{75}; **(d)** i^{80}.

Solution

a) $i^{37} = i^{36}\cdot i = (i^2)^{18}\cdot i = (-1)^{18}\cdot i = 1\cdot i = i$

b) $i^{58} = (i^2)^{29} = (-1)^{29} = -1$

c) $i^{75} = i^{74}\cdot i = (i^2)^{37}\cdot i = (-1)^{37}\cdot i = -1\cdot i = -i$

d) $i^{80} = (i^2)^{40} = (-1)^{40} = 1$ ❑

Now let us simplify other expressions.

EXAMPLE 5

Write the expression in the form $a + bi$: **(a)** $8 - i^2$; **(b)** $17 + 6i^3$; **(c)** $i^{22} - 67i^2$; **(d)** $i^{23} + i^{48}$.

Solution

a) $8 - i^2 = 8 - (-1) = 8 + 1 = 9$

b) $17 + 6i^3 = 17 + 6 \cdot i^2 \cdot i = 17 + 6(-1)i = 17 - 6i$

c) $i^{22} - 67i^2 = (i^2)^{11} - 67(-1) = (-1)^{11} + 67 = -1 + 67 = 66$

d) $i^{23} + i^{48} = (i^{22}) \cdot i + (i^2)^{24}$

$= (i^2)^{11} \cdot i + (-1)^{24} = (-1)^{11} \cdot i + (-1)^{24}$

$= -i + 1 = 1 - i$

Conjugates and Division

Conjugates of complex numbers are defined as follows.

Conjugate of a Complex Number

The *conjugate* of a complex number $a + bi$ is $a - bi$, and the *conjugate* of $a - bi$ is $a + bi$.

EXAMPLE 6

Find the conjugate: **(a)** $5 + 7i$; **(b)** $14 - 3i$; **(c)** $-3 - 9i$; **(d)** $4i$.

Solution

a) $5 + 7i$ The conjugate is $5 - 7i$.

b) $14 - 3i$ The conjugate is $14 + 3i$.

c) $-3 - 9i$ The conjugate is $-3 + 9i$.

d) $4i$ The conjugate is $-4i$. Note that $4i = 0 + 4i$.

When a complex number is multiplied by its conjugate, we get a real number.

EXAMPLE 7

Multiply: **(a)** $(5 + 7i)(5 - 7i)$; **(b)** $(2 - 3i)(2 + 3i)$.

Solution

a) $(5 + 7i)(5 - 7i) = 5^2 - (7i)^2$ Using $(A + B)(A - B) = A^2 - B^2$

$= 25 - 49i^2$

$= 25 - 49(-1)$ $i^2 = -1$

$= 25 + 49$

$= 74$

b) $(2 - 3i)(2 + 3i) = 2^2 - (3i)^2$

$= 4 - 9i^2$

$= 4 - 9(-1)$ $i^2 = -1$

$= 4 + 9$

$= 13$

Conjugates are used when dividing complex numbers.

EXAMPLE 8

Divide and simplify to the form $a + bi$: **(a)** $\dfrac{-5 + 9i}{1 - 2i}$; **(b)** $\dfrac{7 + 3i}{5i}$.

Solution

a) Rewriting i as $\sqrt{-1}$, we can multiply by 1 using the conjugate, as in Section 10.6:

$$\frac{-5+9i}{1-2i} = \frac{-5+9\sqrt{-1}}{1-2\sqrt{-1}} = \frac{-5+9\sqrt{-1}}{1-2\sqrt{-1}} \cdot \frac{1+2\sqrt{-1}}{1+2\sqrt{-1}}.$$ Multiplying by 1

By writing 1 as $(1 + 2i)/(1 + 2i)$, we can use the same method without converting to radical notation:

$$\frac{-5 + 9i}{1 - 2i} = \frac{-5 + 9i}{1 - 2i} \cdot \frac{1 + 2i}{1 + 2i}$$ Multiplying by 1 using the conjugate of the denominator in the symbol for 1

$$= \frac{(-5 + 9i)(1 + 2i)}{(1 - 2i)(1 + 2i)}$$

$$= \frac{-5 - 10i + 9i + 18i^2}{1^2 - 4i^2}$$

$$= \frac{-5 - i - 18}{1 - 4(-1)}$$ $i^2 = -1$

$$= \frac{-23 - i}{5}$$

$$= -\frac{23}{5} - \frac{1}{5}i$$ Writing in the form $a + bi$

b) $$\frac{7 + 3i}{5i} = \frac{7 + 3i}{5i} \cdot \frac{-5i}{-5i}$$ Multiplying by 1 using the conjugate of $0 + 5i$ in the symbol for 1

$$= \frac{-35i - 15i^2}{-25i^2}$$ Multiplying

$$= \frac{-35i - 15(-1)}{-25(-1)}$$ $i^2 = -1$

$$= \frac{15 - 35i}{25}$$

$$= \frac{15}{25} - \frac{35}{25}i$$

$$= \frac{3}{5} - \frac{7}{5}i$$ ❑

Solutions of Equations

The equation $x^2 + 1 = 0$ has no real-number solution, but it has two nonreal complex solutions.

EXAMPLE 9

Determine whether i is a solution of the equation $x^2 + 1 = 0$.

Solution We substitute i for x in the equation.

$$\begin{array}{r|l} x^2 + 1 & 0 \\ \hline i^2 + 1 & ?\ 0 \\ -1 + 1 & \\ 0 & 0 \quad \text{TRUE} \end{array}$$

The number i is a solution. ❑

Any polynomial equation in one variable has complex-number solutions. Sometimes it is not easy to find the solutions, but they always exist.

EXAMPLE 10

Determine whether $1 + i$ is a solution of the equation $x^2 - 2x + 2 = 0$.

Solution We substitute $1 + i$ for x in the equation.

$$\begin{array}{r|l} x^2 - 2x + 2 & 0 \\ \hline (1 + i)^2 - 2(1 + i) + 2 & ?\ 0 \\ 1 + 2i + i^2 - 2 - 2i + 2 & \\ 1 + 2i - 1 - 2 - 2i + 2 & \\ (1 - 1 - 2 + 2) + (2 - 2)i & \\ 0 + 0i & \\ 0 & 0 \quad \text{TRUE} \end{array}$$

The number $1 + i$ is a solution. ❑

EXERCISE SET 10.9

Express in terms of i.

1. $\sqrt{-15}$
2. $\sqrt{-17}$
3. $\sqrt{-16}$
4. $\sqrt{-25}$
5. $-\sqrt{-12}$
6. $-\sqrt{-20}$
7. $\sqrt{-3}$
8. $\sqrt{-4}$
9. $\sqrt{-81}$
10. $\sqrt{-27}$
11. $\sqrt{-98}$
12. $-\sqrt{-18}$
13. $-\sqrt{-49}$
14. $-\sqrt{-125}$
15. $4 - \sqrt{-60}$
16. $6 - \sqrt{-84}$
17. $\sqrt{-4} + \sqrt{-12}$
18. $-\sqrt{-76} + \sqrt{-125}$

Add or subtract and simplify.

19. $(3 + 2i) + (5 - i)$
20. $(-2 + 3i) + (7 + 8i)$
21. $(4 - 3i) + (5 - 2i)$
22. $(-2 - 5i) + (1 - 3i)$
23. $(9 - i) + (-2 + 5i)$
24. $(6 + 4i) + (2 - 3i)$
25. $(3 - i) - (5 + 2i)$
26. $(-2 + 8i) - (7 + 3i)$
27. $(4 - 2i) - (5 - 3i)$
28. $(-2 - 3i) - (1 - 5i)$
29. $(9 + 5i) - (-2 - i)$
30. $(6 - 3i) - (2 + 4i)$

Multiply. Write the answer in the form $a + bi$.

31. $\sqrt{-25}\,\sqrt{-36}$
32. $\sqrt{-81}\,\sqrt{-49}$

33. $\sqrt{-6}\sqrt{-5}$
34. $\sqrt{-7}\sqrt{-10}$
35. $\sqrt{-50}\sqrt{-3}$
36. $\sqrt{-72}\sqrt{-3}$
37. $\sqrt{-48}\sqrt{-6}$
38. $\sqrt{-15}\sqrt{-75}$
39. $5i \cdot 8i$
40. $6i \cdot 9i$
41. $5i \cdot (-7i)$
42. $7i \cdot (-4i)$
43. $5i(3 - 2i)$
44. $4i(5 - 6i)$
45. $-3i(7 - 4i)$
46. $-7i(9 - 3i)$
47. $(3 + 2i)(1 + i)$
48. $(4 + 3i)(2 + 5i)$
49. $(2 + 3i)(6 - 2i)$
50. $(5 + 6i)(2 - i)$
51. $(6 - 5i)(3 + 4i)$
52. $(5 - 6i)(2 + 5i)$
53. $(7 - 2i)(2 - 6i)$
54. $(-4 + 5i)(3 - 4i)$
55. $(5 - 3i)(4 - 5i)$
56. $(7 - 3i)(4 - 7i)$
57. $(-2 + 3i)(-2 + 5i)$
58. $(-3 + 6i)(-3 + 4i)$
59. $(-5 - 4i)(3 + 7i)$
60. $(2 + 9i)(-3 - 5i)$
61. $(3 - 2i)^2$
62. $(5 - 2i)^2$
63. $(2 + 3i)^2$
64. $(4 + 2i)^2$
65. $(-2 + 3i)^2$
66. $(-5 - 2i)^2$

Simplify.

67. i^7
68. i^{11}
69. i^{24}
70. i^{35}
71. i^{42}
72. i^{64}
73. i^9
74. $(-i)^{71}$
75. $(-i)^6$
76. $(-i)^4$
77. $(5i)^3$
78. $(-3i)^5$

Simplify to the form $a + bi$.

79. $7 + i^4$
80. $-18 + i^3$
81. $i^4 - 26i$
82. $i^5 + 37i$
83. $i^2 + i^4$
84. $5i^5 + 4i^3$
85. $i^5 + i^7$
86. $i^{84} - i^{100}$
87. $1 + i + i^2 + i^3 + i^4$
88. $i - i^2 + i^3 - i^4 + i^5$
89. $5 - \sqrt{-64}$
90. $\sqrt{-12} + 36i$

Divide and simplify to the form $a + bi$.

91. $\dfrac{5}{3 - i}$
92. $\dfrac{3}{5 + i}$
93. $\dfrac{2i}{7 + 3i}$
94. $\dfrac{4i}{2 - 5i}$
95. $\dfrac{7}{6i}$
96. $\dfrac{3}{10i}$
97. $\dfrac{8 - 3i}{7i}$
98. $\dfrac{3 + 8i}{5i}$
99. $\dfrac{3 + 2i}{2 + i}$
100. $\dfrac{4 + 5i}{5 - i}$
101. $\dfrac{5 - 2i}{2 + 5i}$
102. $\dfrac{3 - 2i}{4 + 3i}$
103. $\dfrac{3 - 5i}{3 - 2i}$
104. $\dfrac{2 - 7i}{5 - 4i}$

Determine whether the complex number is a solution of the equation.

105. $1 + 2i$; $x^2 - 2x + 5 = 0$
106. $1 - 2i$; $x^2 - 2x + 5 = 0$
107. $1 - i$; $x^2 + 2x + 2 = 0$
108. $2 + i$; $x^2 - 4x - 5 = 0$

Skill Maintenance

Solve.

109. $\dfrac{196}{x^2 - 7x + 49} - \dfrac{2x}{x + 7} = \dfrac{2058}{x^3 + 343}$

110. $\dfrac{5}{t} - \dfrac{3}{2} = \dfrac{4}{7}$

111. $28 = 3x^2 - 17x$

Synthesis

112. A function g is given by

$$g(z) = \frac{z^4 - z^2}{z - 1}.$$

Find $g(2i)$; $g(i + 1)$; $g(2i - 1)$.

113. Evaluate

$$\frac{1}{w - w^2}$$

when

$$w = \frac{1 - i}{10}.$$

Simplify.

114. $\dfrac{i^5 + i^6 + i^7 + i^8}{(1 - i)^4}$

115. $(1 - i)^3(1 + i)^3$

116. $\dfrac{5 - \sqrt{5}i}{\sqrt{5}i}$

117. $\dfrac{6}{1 + \dfrac{3}{i}}$

118. $\left(\dfrac{1}{2} - \dfrac{1}{3}i\right)^2 - \left(\dfrac{1}{2} + \dfrac{1}{3}i\right)^2$

119. $\dfrac{i - i^{38}}{1 + i}$

SUMMARY AND REVIEW 10

KEY TERMS

Square root, p. 484
Principal square root, p. 485
Radical sign, p. 485
Radical expression, p. 485
Radicand, p. 485
Cube root, p. 488
kth root, p. 488
Odd root, p. 488
Index (pl., indices), p. 488
Even root, p. 489
Rational exponent, p. 491
Like radicals, p. 509
Conjugates, p. 511
Rationalizing, p. 513
Radical equation, p. 518
Isosceles right triangle, p. 524
Imaginary number, p. 531
Complex number, p. 531
Conjugate of a complex number, p. 534

IMPORTANT PROPERTIES AND FORMULAS

The number c is a square root of a if $c^2 = a$.
The number c is the cube root of a if $c^3 = a$.

For any real number a:

a) $\sqrt[k]{a^k} = |a|$ when k is even. We use absolute value when k is even unless a is known to be nonnegative.

b) $\sqrt[k]{a^k} = a$ when k is odd. We do not use absolute value when k is odd.

For any nonnegative number a and any index n,

$a^{1/n}$ means $\sqrt[n]{a}$.

If a is negative, then n must be odd.

For any natural numbers m and n ($n \neq 1$), and any real number a for which $\sqrt[n]{a}$ exists,

$a^{m/n}$ means $\left(\sqrt[n]{a}\right)^m$ or $\sqrt[n]{a^m}$.

For any rational number m/n and any nonzero real number a for which $a^{m/n}$ exists,

$a^{-m/n}$ means $\dfrac{1}{a^{m/n}}$.

For any real numbers a and b and any rational exponents m and n for which a^m, a^n, and b^m are defined:

1. $a^m \cdot a^n = a^{m+n}$ In multiplying, add exponents if the bases are the same.
2. $\dfrac{a^m}{a^n} = a^{m-n}$ In dividing, subtract exponents if the bases are the same. (Assume $a \neq 0$.)
3. $(a^m)^n = a^{m \cdot n}$ To raise a power to a power, multiply the exponents.
4. $(ab)^m = a^m b^m$ To raise a product to a power, raise each factor to the power and multiply.

The Product Rule for Radicals

For any real numbers $\sqrt[k]{a}$ and $\sqrt[k]{b}$,

$$\sqrt[k]{a} \cdot \sqrt[k]{b} = \sqrt[k]{a \cdot b}.$$

(To multiply, we multiply the radicands.)

The Quotient Rule for Radicals

For any real numbers $\sqrt[k]{a}$ and $\sqrt[k]{b}$, $b \neq 0$,

$$\frac{\sqrt[k]{a}}{\sqrt[k]{b}} = \sqrt[k]{\frac{a}{b}}.$$

(To divide, we divide the radicands. After doing this, we can sometimes simplify by taking roots.)

The Power–Root Rule

For any index k for which $\sqrt[k]{a}$ exists, and any integer m for which a^m exists,

$$(\sqrt[k]{a})^m = \sqrt[k]{a^m}.$$

(We can raise to a power and then take a root, or we can take a root and then raise to a power.)

Some Ways to Simplify Radical Expressions

1. *Simplifying by factoring.* We factor the radicand, looking for factors raised to powers that are multiples of the index.

   Example: $\sqrt[3]{16} = \sqrt[3]{8}\sqrt[3]{2} = 2\sqrt[3]{2}$

2. *Using rational exponents to simplify.* We convert to exponential notation and then use arithmetic and the laws of exponents to simplify the exponents. Then we convert back to radical notation.

   Example: $\sqrt[3]{p} \cdot \sqrt[4]{q^3} = p^{1/3} \cdot q^{3/4} = p^{4/12} \cdot q^{9/12} = \sqrt[12]{p^4 q^9}$

3. *Collecting like radical terms.*

   Example: $\sqrt{8} + 3\sqrt{2} = \sqrt{4} \cdot \sqrt{2} + 3\sqrt{2} = 2\sqrt{2} + 3\sqrt{2} = 5\sqrt{2}$

4. *Rationalizing denominators*. Radical expressions are usually considered simpler if there are no radicals in the denominator.

Example: $\dfrac{1}{\sqrt{2}} = \dfrac{1}{\sqrt{2}} \cdot \dfrac{\sqrt{2}}{\sqrt{2}} = \dfrac{\sqrt{2}}{2}$

The Principle of Powers

If $a = b$, then $a^n = b^n$ for any natural number n.

A general strategy for solving equations with two radical terms is as follows.

1. Isolate one of the radical terms.
2. Use the principle of powers.
3. If a radical remains, perform steps (1) and (2) again.
4. Check possible solutions.

Special Triangles

The length of the hypotenuse in an isosceles right triangle is the length of a leg times $\sqrt{2}$.

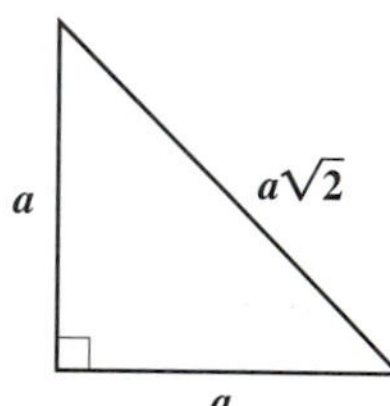

The length of the longer leg in a 30–60–90 right triangle is the length of the shorter leg times $\sqrt{3}$. The hypotenuse is twice as long as the shorter leg.

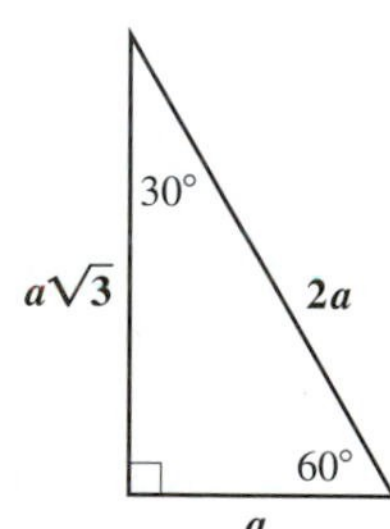

A complex number is any number that can be written $a + bi$, where a and b are any real numbers and $i = \sqrt{-1}$.

REVIEW EXERCISES

This chapter's review and test include Skill Maintenance exercises from Sections 5.7, 5.8, 6.2, and 6.7.

Simplify.

1. $\sqrt{\frac{36}{81}}$

2. $\sqrt{0.0049}$

Let $f(x) = \sqrt{2x - 7}$. Find the following.

3. $f(16)$

4. The domain of f

Simplify. Assume that letters can represent *any* real number.

5. $\sqrt{81a^2}$

6. $\sqrt{(c + 8)^2}$

7. $\sqrt{x^2 - 6x + 9}$

8. $\sqrt{4x^2 + 4x + 1}$

9. $\sqrt[5]{-32}$

10. $\sqrt[3]{-\frac{1}{27}}$

11. $\sqrt[10]{x^{10}}$

12. $-\sqrt[13]{(-3)^{13}}$

13. Rewrite with rational exponents: $(\sqrt[5]{8x^6y^2})^4$.

14. Rewrite without rational exponents: $(5a)^{3/4}$.

Use rational exponents to simplify.

15. $x^{1/3} \cdot y^{1/4}$

16. $\sqrt[3]{\frac{x^9y^{12}}{-8}}$

Approximate to the nearest thousandth using a calculator or Table 1.

17. $\sqrt{245}$

18. $\sqrt{112}$

19. $\frac{14 - \sqrt{245}}{7}$

Multiply and simplify by factoring. Assume that all variables are nonnegative.

20. $\sqrt{3x^2}\,\sqrt{6y^3}$

21. $\sqrt[3]{a^5b}\,\sqrt[3]{27b}$

22. $\sqrt{3x^3y^4}\,\sqrt[3]{9x^2y}$

Divide. Then simplify by taking roots, if possible.

23. $\frac{\sqrt[3]{60xy^3}}{\sqrt[3]{10x}}$

24. $\frac{\sqrt{75x}}{2\sqrt{3}}$ (Assume $x \geq 0$.)

Simplify.

25. $(\sqrt{8xy^2})^2$ (Assume $x, y \geq 0$.)

26. $(\sqrt[3]{4a^2b})^2$

Perform the indicated operation. Simplify by collecting like radical terms, if possible.

27. $12\sqrt[3]{135} - 3\sqrt[3]{40}$

28. $\sqrt{50} + 2\sqrt{18} + \sqrt{32}$

29. $(\sqrt[3]{27} - \sqrt[3]{2})(\sqrt[3]{27} + \sqrt[3]{2})$

30. $(\sqrt{5} - 3\sqrt{8})(\sqrt{5} + 2\sqrt{8})$

31. $(1 - \sqrt{7})^2$

32. Rationalize the denominator. Assume that a and b represent positive numbers.

$$\frac{5\sqrt{12a}}{\sqrt{a} + \sqrt{b}}$$

33. Rationalize the numerator of the expression in Exercise 32.

Solve.

34. $\sqrt{3x - 3} = 1 + \sqrt{x}$

35. $\sqrt[4]{x + 3} = 2$

Solve. Give an exact answer and, where appropriate, an approximation to three decimal places.

36. The diagonal of a square has length $9\sqrt{2}$ cm. Find the length of a side of the square.

37. A bookcase is 1 foot taller than it is wide. A diagonal brace, 1 foot longer than the height of the bookcase, is needed for support. What is the length of the brace?

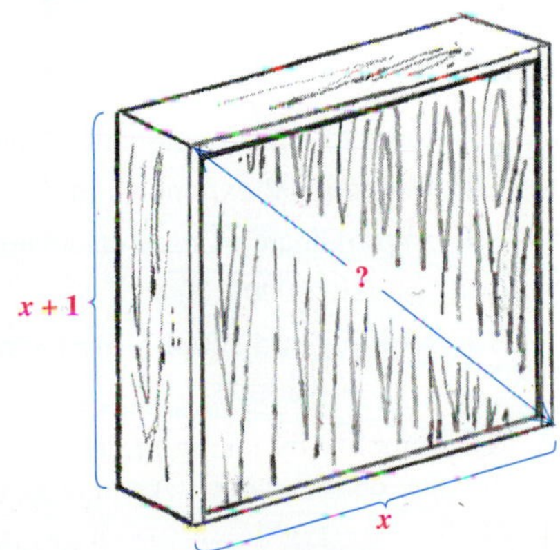

38. Express in terms of i and simplify: $-\sqrt{-8}$.

39. Add: $(-4 + 3i) + (2 - 12i)$.

40. Subtract: $(4 - 7i) - (3 - 8i)$.

Multiply.

41. $(2 + 5i)(2 - 5i)$

42. i^{13}

43. $(6 - 3i)(2 - i)$

Divide and simplify to the form $a + bi$.

44. $\dfrac{-3 + 2i}{5i}$

45. $\dfrac{6 - 3i}{2 + i}$

Skill Maintenance

46. Find three consecutive positive integers such that the product of the first and second integers is 26 less than the product of the second and third integers.

47. Solve:

$$\frac{7}{x + 2} + \frac{5}{x^2 - 2x + 4} = \frac{84}{x^3 + 8}.$$

48. Solve:

$$2x^2 + 3x - 27 = 0.$$

49. Multiply and simplify:

$$\frac{x^2 + 3x}{x^2 - y^2} \cdot \frac{x^2 - xy + 2x - 2y}{x^2 - 9}.$$

Synthesis

50. ◈ Explain why $\sqrt[k]{x^k} = |x|$ when k is even, but $\sqrt[k]{x^k} = x$ when k is odd.

51. ◈ What is the difference between real numbers and complex numbers?

52. Solve:

$$\sqrt{11x + \sqrt{6 + x}} = 6.$$

53. Simplify:

$$\frac{2}{1 - 3i} - \frac{3}{4 + 2i}.$$

CHAPTER TEST 10

1. Simplify: $\sqrt{\dfrac{100}{49}}$.

2. Determine the domain of f if

$$f(t) = \sqrt{2t + 10}.$$

In Questions 3–6, assume that letters can represent *any* real number. Simplify.

3. $\sqrt{36y^2}$

4. $\sqrt{x^2 + 10x + 25}$

5. $\sqrt[3]{-8}$

6. $\sqrt[10]{(-4)^{10}}$

7. Rewrite with fractional exponents: $(\sqrt{5xy^2})^5$.

8. Use rational exponents to simplify: $\sqrt[3]{16a^5b^{10}}$.

9. Approximate using a calculator:

$$\sqrt{0.000001204}.$$

10. Multiply and simplify by factoring:

$$\sqrt[3]{x^4}\sqrt[3]{8x^5}.$$

11. Simplify by taking the roots of the numerator and the denominator. (Assume $x, y > 0$.)

$$\sqrt{\frac{25x^3}{36y^4}}$$

12. Simplify: $(\sqrt[3]{16a^2b})^2$.

13. Simplify and add:

$$3\sqrt{128} + 2\sqrt{18} + 2\sqrt{32}.$$

14. Multiply and simplify:

$$(\sqrt{20} + 2\sqrt{5})(\sqrt{20} - 3\sqrt{5}).$$

15. Multiply and simplify:

$$(2x - \sqrt[3]{y^2})(x + \sqrt{y^3}).$$

16. Rationalize the denominator:

$$\frac{1 + \sqrt{2}}{3 - 5\sqrt{2}}.$$

17. Solve: $\sqrt{y - 6} = \sqrt{y + 9} - 3$.

18. The shorter leg of a 30–60–90 right triangle measures 10 cm. Find the lengths of the other sides. Give exact answers and, where appropriate, approximations to three decimal places.

19. Express in terms of i and simplify: $\sqrt{-18}$.

20. Subtract: $(5 + 8i) - (-2 + 3i)$.

21. Multiply: $\sqrt{-100}\,\sqrt{-9}$.

22. Multiply. Write the answer in the form $a + bi$:

$$(1 - i)^2.$$

23. Divide and simplify to the form $a + bi$:

$$\frac{-7 + 14i}{6 - 8i}.$$

24. Simplify to the form $a + bi$: $5i^{25} - i^{12}$.

Skill Maintenance

25. Solve: $6x^2 = 13x + 5$.

26. Divide and simplify:

$$\frac{x^3 - 27}{x^2 - 16} \div \frac{x^2 + 3x + 9}{x + 4}.$$

27. Solve:

$$\frac{11x}{x + 3} + \frac{33}{x} + 12 = \frac{99}{x^2 + 3x}.$$

28. Find two consecutive even integers whose product is 288.

Synthesis

29. Solve:

$$\sqrt{2x - 2} + \sqrt{7x + 4} = \sqrt{13x + 10}.$$

30. Simplify:

$$\frac{1 - 4i}{4i(1 + 4i)^{-1}}.$$

CHAPTER 11

Quadratic Equations and Functions

AN APPLICATION

A formula relating an athlete's vertical leap V, measured in inches, to hang time T, measured in seconds, is

$$V = 48T^2. \leftarrow \text{This is a quadratic equation.}$$

In Section 11.9, we show how to solve this equation for T:

$$T = \frac{\sqrt{3V}}{12}.$$

Lisa Sonntag, M.S.

EXERCISE PHYSIOLOGIST, NATIONALLY RANKED RACEWALKER

"Math comes up in my daily routine. It is important for interpreting research results and in problem solving. I use math when designing exercise prescriptions and work simulations. Math has also paid off in coaching and in my own racewalking."

Quadratic equations first appeared in Section 5.7. At that time, we used the principle of zero products because all of the equations could be solved by factoring. In this chapter, we will learn methods for solving *any* quadratic equation. These methods are then used to solve applications and to assist us in graphing.

In addition to the material from this chapter, the review and test for Chapter 11 include material from Sections 8.3, 9.1, 10.3, and 10.7.

11.1 Solving Quadratic Equations: The Principle of Square Roots

The Principle of Square Roots • Solving Quadratic Equations of the Type $(x + k)^2 = p$

The following are examples of quadratic equations:

$$x^2 - 7x + 9 = 0, \qquad 5t^2 - 4t = 8, \qquad 6y^2 = -9y, \qquad m^2 = 49.$$

We saw in Chapter 5 that one way to solve an equation like $m^2 = 49$ is to subtract 49 on both sides, factor, and then use the principle of zero products:

$$\begin{aligned} m^2 - 49 &= 0 \\ (m + 7)(m - 7) &= 0 \\ m + 7 = 0 \quad &or \quad m - 7 = 0 \\ m = -7 \quad &or \quad m = 7. \end{aligned}$$

This approach relies on our ability to factor. By using the *principle of square roots*, we can develop a method for solving equations like $m^2 = 49$ that can be used to solve equations for which factoring is impractical.

The Principle of Square Roots

One way to solve an equation like $m^2 = 49$ is to search for a number whose square is 49. There are two such numbers, -7 and 7. Note that -7 and 7 are the square roots of 49.

The Principle of Square Roots

For any real number k, if $x^2 = k$, then $x = \sqrt{k}$ or $x = -\sqrt{k}$.

EXAMPLE 1

Solve: $x^2 = 16$.

Solution We use the principle of square roots:

$$\begin{aligned} x^2 &= 16 \\ x = \sqrt{16} \quad &or \quad x = -\sqrt{16} && \text{Using the principle of square roots} \\ x = 4 \quad &or \quad x = -4. && \text{Simplifying} \end{aligned}$$

We check mentally that $4^2 = 16$ and $(-4)^2 = 16$. The solutions are 4 and -4. □

Unlike the principle of zero products, the principle of square roots can be easily used to solve quadratic equations that have irrational solutions.

EXAMPLE 2

Solve: $5x^2 = 15$.

Solution

$$5x^2 = 15$$

$$x^2 = 3 \qquad \text{Solving for } x^2$$

$$x = \sqrt{3} \quad or \quad x = -\sqrt{3} \qquad \text{Using the principle of square roots}$$

Check:

For $\sqrt{3}$:

$5x^2$	$= 15$
$5(\sqrt{3})^2$	? 15
$5 \cdot 3$	
15	15 TRUE

For $-\sqrt{3}$:

$5x^2$	$= 15$
$5(-\sqrt{3})^2$	? 15
$5 \cdot 3$	
15	15 TRUE

The solutions are $\sqrt{3}$ and $-\sqrt{3}$. ❑

We often use the symbol $\pm\sqrt{k}$ to represent the two numbers $\sqrt{k}$ and $-\sqrt{k}$. The solutions in Example 2 can be written $\pm\sqrt{3}$ (read "plus or minus $\sqrt{3}$").

Sometimes we rationalize denominators to simplify answers.

EXAMPLE 3

Solve: $-3x^2 + 7 = 0$.

Solution

$$-3x^2 + 7 = 0$$

$$x^2 = \frac{7}{3} \qquad \text{Isolating } x^2$$

$$x = \sqrt{\frac{7}{3}} \quad or \quad x = -\sqrt{\frac{7}{3}} \qquad \text{Using the principle of square roots}$$

$$x = \frac{\sqrt{7}}{\sqrt{3}} \cdot \frac{\sqrt{3}}{\sqrt{3}} \quad or \quad x = -\frac{\sqrt{7}}{\sqrt{3}} \cdot \frac{\sqrt{3}}{\sqrt{3}} \qquad \text{Rationalizing the denominators}$$

$$x = \frac{\sqrt{21}}{3} \quad or \quad x = -\frac{\sqrt{21}}{3}$$

The solutions are $\frac{\sqrt{21}}{3}$ and $-\frac{\sqrt{21}}{3}$, or $\pm\frac{\sqrt{21}}{3}$. ❑

Sometimes we get solutions that are imaginary numbers.

EXAMPLE 4

Solve: $4x^2 + 9 = 0$.

Solution

$$4x^2 + 9 = 0$$

$$x^2 = -\tfrac{9}{4} \qquad \text{Isolating } x^2$$

$$x = \sqrt{-\tfrac{9}{4}} \quad or \quad x = -\sqrt{-\tfrac{9}{4}} \qquad \text{Using the principle of square roots}$$

$$x = \tfrac{3}{2}i \quad or \quad x = -\tfrac{3}{2}i \qquad \text{Simplifying}$$

The solutions are $\frac{3}{2}i$ and $-\frac{3}{2}i$, or $\pm\frac{3}{2}i$. We leave the checks to the student. ❑

Solving Quadratic Equations of the Type $(x + k)^2 = p$

Equations like $(x - 5)^2 = 9$ or $(x + 2)^2 = 7$ are of the form $(x + k)^2 = p$. The principle of square roots can be used to solve such equations.

EXAMPLE 5

Solve: **(a)** $(x - 5)^2 = 9$; **(b)** $(x + 2)^2 = 7$.

Solution

a) $(x - 5)^2 = 9$

$x - 5 = 3$ *or* $x - 5 = -3$ Using the principle of square roots

$x = 8$ *or* $x = 2$

The solutions are 8 and 2.

b) $(x + 2)^2 = 7$

$x + 2 = \sqrt{7}$ *or* $x + 2 = -\sqrt{7}$ Using the principle of square roots

$x = -2 + \sqrt{7}$ *or* $x = -2 - \sqrt{7}$

The solutions are $-2 + \sqrt{7}$ and $-2 - \sqrt{7}$, or simply $-2 \pm \sqrt{7}$ (read "-2 plus or minus $\sqrt{7}$"). ❑

In Example 5, the left sides of the equations are squares of binomials. If we can express an equation in such a form, we can proceed as we did in that example.

EXAMPLE 6

Solve by factoring and using the principle of square roots: **(a)** $x^2 + 8x + 16 = 49$; **(b)** $x^2 + 6x + 9 = 10$.

Solution

a) $x^2 + 8x + 16 = 49$ The left side is a trinomial square.

$(x + 4)^2 = 49$ A trinomial square is a binomial squared.

$x + 4 = 7$ *or* $x + 4 = -7$ Using the principle of square roots

$x = 3$ *or* $x = -11$

The solutions are 3 and -11.

b) $x^2 + 6x + 9 = 10$ The left side is a trinomial square.

$(x + 3)^2 = 10$ A trinomial square is a binomial squared.

$x + 3 = \sqrt{10}$ *or* $x + 3 = -\sqrt{10}$ Using the principle of square roots

$x = -3 + \sqrt{10}$ *or* $x = -3 - \sqrt{10}$

The solutions are $-3 + \sqrt{10}$ and $-3 - \sqrt{10}$, or $-3 \pm \sqrt{10}$. ❑

EXERCISE SET 11.1

Solve. Use the principle of square roots.

1. $x^2 = 25$
2. $x^2 = 36$
3. $a^2 = 81$
4. $a^2 = 121$
5. $m^2 = 15$
6. $t^2 = 10$
7. $x^2 = 19$
8. $a^2 = 29$
9. $5a^2 = 35$
10. $3x^2 = 84$
11. $7x^2 = 140$
12. $9m^2 = 72$
13. $4t^2 - 25 = 0$
14. $9a^2 - 4 = 0$

15. $25x^2 + 4 = 0$
16. $9x^2 + 16 = 0$
17. $9y^2 + 2 = 7$
18. $4y^2 - 5 = 16$
19. $3x^2 - 7 = 0$
20. $2x^2 - 3 = 0$
21. $(x - 2)^2 = 49$
22. $(x + 1)^2 = 25$
23. $(x + 3)^2 = 36$
24. $(x - 4)^2 = 81$
25. $(m + 3)^2 = 21$
26. $(m - 3)^2 = 6$
27. $(a + 13)^2 = 8$
28. $(a - 13)^2 = 64$
29. $(x - 7)^2 = 12$
30. $(x + 1)^2 = 14$
31. $(x + 9)^2 = 34$
32. $(t + 2)^2 = 25$
33. $\left(x + \frac{3}{2}\right)^2 = \frac{7}{2}$
34. $\left(y - \frac{3}{4}\right)^2 = \frac{17}{16}$
35. $x^2 - 6x + 9 = 64$
36. $x^2 - 10x + 25 = 100$
37. $y^2 + 14y + 49 = 4$
38. $p^2 + 8p + 16 = 1$
39. $m^2 - 2m + 1 = 5$
40. $t^2 + 6t + 9 = 13$
41. $x^2 + 4x + 4 = 12$
42. $x^2 - 12x + 36 = 18$

Skill Maintenance

Multiply and simplify.

43. $\sqrt{6}\sqrt{10}$
44. $\sqrt{3x}\sqrt{6x^3}$

Solve.

45. $2x + y = 5,$
$y = 4 - x$

46. $2x - y = 5,$
$x + 2y = 3$

Synthesis

47. Explain why 9 is not *the* solution of $x^2 = 81$.

48. Write a quadratic equation that is most easily solved using the principle of square roots. Explain why the principle of zero products would not work as easily on the equation you wrote.

Factor the left side of the equation. Then solve.

49. $x^2 + 5x + \frac{25}{4} = \frac{13}{4}$
50. $x^2 - \frac{7}{3}x + \frac{49}{36} = \frac{7}{36}$
51. $m^2 - \frac{3}{2}m + \frac{9}{16} = \frac{17}{16}$
52. $t^2 + 3t + \frac{9}{4} = \frac{49}{4}$
53. $x^2 + 0.5x + 0.0625 = 13.69$
54. $x^2 + 2.5x + 1.5625 = 9.61$
55. $a^2 - 3.8a + 3.61 = 27.04$
56. $a^2 - 5.2a + 6.76 = 53.29$

11.2 Solving Quadratic Equations: Completing the Square

Completing the Square • Solving by Completing the Square

In Section 11.1, we solved equations like $(x - 5)^2 = 7$ using the principle of square roots. Equations like $x^2 + 8x + 16 = 12$ were also solved using the principle of square roots because the expression on the left side is a trinomial square. We now learn to solve equations like $x^2 + 10x = 4$, in which the left side is not (yet) a trinomial square. The new procedure involves *completing the square* and enables us to solve any quadratic equation.

Completing the Square

Consider the following quadratic equation:

$$x^2 + 10x = 4.$$

We would like to add some constant to both sides of the equation that would make the left side a perfect-square trinomial. To determine what that constant should be, recall that after a binomial of the form $x + a$ is squared, the coefficient of the x-term is $2a$:

$$(x + a)^2 = x^2 + 2ax + a^2.$$

In the trinomial above, a^2 is the square of half the coefficient of x. Returning to the equation $x^2 + 10x = 4$, note that half the coefficient of x is 5, and $5^2 = 25$. This suggests that we should add 25 on both sides:

$$x^2 + 10x = 4$$

$$x^2 + 10x + 25 = 4 + 25 \qquad \text{Adding 25 on both sides}$$

$$(x + 5)^2 = 29. \qquad \text{Factoring the trinomial square}$$

By adding 25 to $x^2 + 10x$, we have *completed the square*. The resulting equation contains the square of a binomial on one side. Thus we can find the solutions by using the principle of square roots, as in Section 11.1:

$$(x + 5)^2 = 29$$

$$x + 5 = \sqrt{29} \quad or \quad x + 5 = -\sqrt{29} \qquad \text{Using the principle of square roots}$$

$$x = -5 + \sqrt{29} \quad or \quad x = -5 - \sqrt{29}.$$

The solutions are $-5 \pm \sqrt{29}$.

The key to solving $x^2 + 10x = 4$ was adding 25 on both sides. Once we did so, the left side became a trinomial square.

Completing the Square

To *complete the square* for an expression like $x^2 + bx$, take half the coefficient of x and square it. Then add that number, which is $(b/2)^2$.

A visual interpretation of completing the square is sometimes helpful. Consider the following figures.

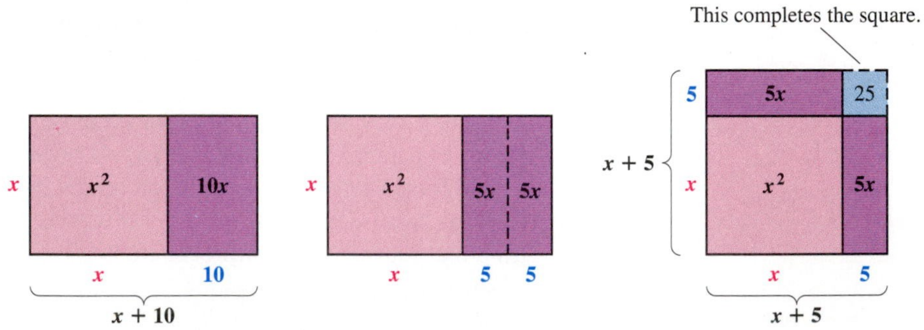

In all figures, the sum of the red and purple areas is $x^2 + 10x$. However, by splitting the purple area in half, we can "complete" a square by adding the blue area. The blue area is $5 \cdot 5$, or 25 square units.

EXAMPLE 1

Complete the square: **(a)** $x^2 - 12x$; **(b)** $x^2 + 5x$.

Solution

a) To complete the square for $x^2 - 12x$, note that the coefficient of x is -12. Half of -12 is -6 and $(-6)^2$ is 36. Thus, $x^2 - 12x$ becomes a trinomial square when

36 is added:

$x^2 - 12x + 36$ is the square of $x - 6$.

That is,

$$x^2 - 12x + 36 = (x - 6)^2.$$

b) To complete the square for $x^2 + 5x$, we take half the coefficient of x and square it:

$$\left(\tfrac{5}{2}\right)^2 = \tfrac{25}{4}.$$

The trinomial $x^2 + 5x + \frac{25}{4}$ is the square of $x + \frac{5}{2}$. That is,

$$x^2 + 5x + \tfrac{25}{4} = \left(x + \tfrac{5}{2}\right)^2.$$

Solving by Completing the Square

The steps that we used to solve $x^2 + 10x = 4$ can be used in a wide variety of problems.

EXAMPLE 2

Solve by completing the square: **(a)** $x^2 + 6x = -8$; **(b)** $x^2 - 10x + 14 = 0$.

Solution

a) To solve $x^2 + 6x = -8$, we take half of 6 and square it, to get 9. Then we add 9 on both sides of the equation. This makes the left side the square of a binomial. Now we can solve:

$$x^2 + 6x + 9 = -8 + 9$$ Adding 9 on both sides to complete the square

$$(x + 3)^2 = 1$$ Writing the trinomial as a binomial squared

$$x + 3 = 1 \quad or \quad x + 3 = -1$$ Using the principle of square roots

$$x = -2 \quad or \quad x = -4.$$

The solutions are -2 and -4.

b) We have

$$x^2 - 10x + 14 = 0$$

$$x^2 - 10x = -14$$ Subtracting 14 on both sides

$$x^2 - 10x + 25 = -14 + 25$$ Adding 25 on both sides to complete the square: $(-10/2)^2 = 25$

$$(x - 5)^2 = 11$$ Factoring

$$x - 5 = \sqrt{11} \quad or \quad x - 5 = -\sqrt{11}$$ Using the principle of square roots

$$x = 5 + \sqrt{11} \quad or \quad x = 5 - \sqrt{11}.$$

The solutions are $5 + \sqrt{11}$ and $5 - \sqrt{11}$, or $5 \pm \sqrt{11}$.

As we saw in Section 5.7, a grapher can be used to find approximate solutions of any quadratic equation that has real-number solutions. To review this, check Example 2(a) by graphing $y = x^2 + 6x + 8$:

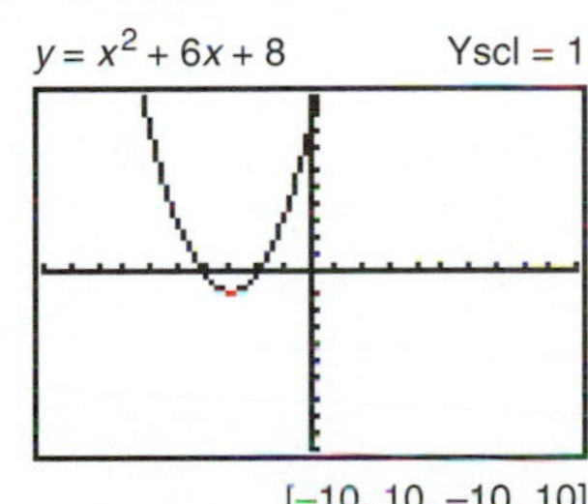

Use the Trace and Zoom features to find the x-intercepts. If you lack a Zoom feature, adjust the window size to magnify the region near each intercept.

TC1. Use a grapher to find the solutions of $x^2 - 8x = 7$. Check the solutions algebraically.

TC2. Can a grapher be used to find *exact* solutions of Example 2(b)? Why or why not?

To complete the square, we must be sure that the coefficient of x^2 is 1. When the x^2-coefficient is not 1, we can multiply or divide on both sides to find an equivalent equation with an x^2-coefficient of 1.

EXAMPLE 3

Solve by completing the square: **(a)** $3x^2 + 6x = -15$; **(b)** $2x^2 - 3x - 1 = 0$.

Solution

a)

$$3x^2 + 6x = -15$$

$$\left.\begin{aligned}\tfrac{1}{3}(3x^2 + 6x) &= \tfrac{1}{3}(-15)\\ x^2 + 2x &= -5\end{aligned}\right\}$$ We multiply by $\frac{1}{3}$ to ensure an x^2-coefficient of 1.

$$x^2 + 2x + 1 = -5 + 1$$ Adding 1 on both sides to complete the square: $\left(\frac{2}{2}\right)^2 = 1$

$$(x + 1)^2 = -4$$ Factoring

$$x + 1 = \sqrt{-4} \quad or \quad x + 1 = -\sqrt{-4}$$ Using the principle of square roots

$$x + 1 = 2i \quad or \quad x + 1 = -2i$$

$$x = -1 + 2i \quad or \quad x = -1 - 2i$$

The solutions are $-1 \pm 2i$.

b)

$$2x^2 - 3x - 1 = 0$$

$$\frac{1}{2}(2x^2 - 3x - 1) = \frac{1}{2} \cdot 0$$ Multiplying by $\frac{1}{2}$ to make the x^2-coefficient 1

$$x^2 - \frac{3}{2}x - \frac{1}{2} = 0$$

$$x^2 - \frac{3}{2}x = \frac{1}{2}$$ Adding $\frac{1}{2}$ on both sides

$$x^2 - \frac{3}{2}x + \frac{9}{16} = \frac{1}{2} + \frac{9}{16}$$ Adding $\frac{9}{16}$ on both sides: $\left[\frac{1}{2}\left(-\frac{3}{2}\right)\right]^2 = \left[-\frac{3}{4}\right]^2 = \frac{9}{16}$. This completes the square on the left side.

$$\left(x - \frac{3}{4}\right)^2 = \frac{8}{16} + \frac{9}{16}$$ Factoring and finding a common denominator

$$\left(x - \frac{3}{4}\right)^2 = \frac{17}{16}$$

$$x - \frac{3}{4} = \frac{\sqrt{17}}{4} \quad or \quad x - \frac{3}{4} = -\frac{\sqrt{17}}{4}$$ Using the principle of square roots

$$x = \frac{3}{4} + \frac{\sqrt{17}}{4} \quad or \quad x = \frac{3}{4} - \frac{\sqrt{17}}{4}$$

The solutions are $\dfrac{3 \pm \sqrt{17}}{4}$. ❑

The steps used in Example 3 can be used to solve any quadratic equation.

Solving by Completing the Square

To solve a quadratic equation $ax^2 + bx + c = 0$ by completing the square:

1. If $a \neq 1$, multiply by $1/a$ on both sides so that the x^2-coefficient is 1.
2. When the x^2-coefficient is 1, rewrite the equation in the form

$$x^2 + bx = -c, \quad \text{or} \quad x^2 + \frac{b}{a}x = -\frac{c}{a} \quad \text{if step (1) has been applied.}$$

3. Take half of the x-coefficient and square it. Add the result on both sides of the equation.
4. Express the side with the variables as the square of a binomial.
5. Use the principle of square roots and complete the solution.

EXERCISE SET 11.2

Complete the square.

1. $x^2 - 2x$
2. $x^2 - 4x$
3. $x^2 + 18x$
4. $x^2 + 22x$
5. $x^2 - x$
6. $x^2 + x$
7. $t^2 + 5t$
8. $y^2 - 9y$
9. $x^2 - \frac{3}{2}x$
10. $x^2 + \frac{4}{3}x$
11. $m^2 + \frac{9}{2}m$
12. $r^2 - \frac{2}{5}r$

Solve by completing the square.

13. $x^2 - 6x - 16 = 0$
14. $x^2 + 8x + 15 = 0$
15. $x^2 + 22x + 21 = 0$
16. $x^2 + 14x - 15 = 0$
17. $3x^2 - 6x - 15 = 0$
18. $3x^2 - 12x - 33 = 0$
19. $x^2 - 10x = 22$
20. $x^2 + 8x = 74$
21. $x^2 + 6x + 13 = 0$
22. $x^2 + 8x + 25 = 0$
23. $x^2 - 7x - 2 = 0$
24. $x^2 + 7x - 2 = 0$
25. $2x^2 + 6x - 56 = 0$
26. $2x^2 - 6x - 56 = 0$
27. $x^2 + \frac{3}{2}x - \frac{1}{2} = 0$
28. $x^2 - \frac{3}{2}x - 2 = 0$
29. $2x^2 - 5x - 3 = 0$
30. $2x^2 + 4x + 1 = 0$
31. $3x^2 + 4x - 1 = 0$
32. $3x^2 - 4x - 3 = 0$
33. $2x^2 = 9x + 5$
34. $2x^2 = 5x + 12$
35. $4x^2 + 12x = 7$
36. $6x^2 + 11x = 10$

Skill Maintenance

37. Twin Cities Roasters has Kenyan coffee worth \$4.50 a pound and Peruvian coffee worth \$7.50 a pound. How much of each kind should be mixed to obtain a 50-lb mixture that is worth \$5.70 a pound?

Solve.

38. $\sqrt{y} = \frac{1}{4}$
39. $\sqrt[5]{x} = -2$
40. $\sqrt[3]{2t + 5} = 2$

Synthesis

41. Explain in your own words how completing the square enables us to solve equations we could not otherwise have solved.

42. A student states that "since solving a quadratic equation by completing the square relies on the principle of square roots, the solutions are always opposites of each other." Is the student correct? Why or why not?

Find b such that the trinomial is a square.

43. $x^2 + bx + 36$
44. $x^2 + bx + 55$
45. $x^2 + bx + 128$
46. $4x^2 + bx + 16$
47. $x^2 + bx + c$
48. $ax^2 + bx + c$

We can use a grapher to solve equations by letting y_1 represent the left side of an equation and y_2 represent the right side, and graphing y_1 and y_2 on the same set of axes. The Trace and Zoom features can then be used to determine the x-coordinate at any point of intersection. Use this approach to find solutions, accurate to two decimal places, of each of the following equations.

49. $(x - 5)^2 = 9$
50. $(x + 3)^2 = 25$
51. $(x + 4)^2 = 13$
52. $(x - 6)^2 = 2$
53. $x^2 - 7x - 2 = 0$ (Exercise 23)
54. $x^2 + 7x - 2 = 0$ (Exercise 24)
55. $2x^2 = 9x + 5$ (Exercise 33)
56. $2x^2 = 5x + 12$ (Exercise 34)

11.3 The Quadratic Formula and Problem Solving

The Quadratic Formula • Problem Solving

We now derive the *quadratic formula*. This formula enables us to solve quadratic equations more quickly than the method of completing the square.

The Quadratic Formula

When mathematicians use a procedure repeatedly on a wide variety of problems, they generally try to find a formula for the procedure. The quadratic formula condenses the many steps used to solve a quadratic equation by completing the square.

Consider a quadratic equation in *standard form*, $ax^2 + bx + c = 0$, with $a > 0$. Our plan is to solve this equation for x by completing the square. As the steps are performed, compare them with those in Example 3(b) in Section 11.2:

$$\frac{1}{a}(ax^2 + bx + c) = \frac{1}{a} \cdot 0$$ Multiplying by $\frac{1}{a}$ to make the x^2-coefficient 1

$$x^2 + \frac{b}{a}x + \frac{c}{a} = 0$$

$$x^2 + \frac{b}{a}x = -\frac{c}{a}.$$ Adding $-\frac{c}{a}$ on both sides

Half of b/a is $b/(2a)$, and the square of $b/(2a)$ is $b^2/(4a^2)$. We add $b^2/(4a^2)$ on both sides:

$$x^2 + \frac{b}{a}x + \frac{b^2}{4a^2} = -\frac{c}{a} + \frac{b^2}{4a^2}$$ Adding $\frac{b^2}{4a^2}$ on both sides. This completes the square on the left side.

$$\left(x + \frac{b}{2a}\right)^2 = -\frac{4ac}{4a^2} + \frac{b^2}{4a^2}$$ Factoring and finding a common denominator

$$\left(x + \frac{b}{2a}\right)^2 = \frac{b^2 - 4ac}{4a^2}$$

$$x + \frac{b}{2a} = \sqrt{\frac{b^2 - 4ac}{4a^2}} \quad or \quad x + \frac{b}{2a} = -\sqrt{\frac{b^2 - 4ac}{4a^2}}.$$ Using the principle of square roots

Since $a > 0$, $\sqrt{4a^2} = 2a$, so we can simplify as follows:

$$x + \frac{b}{2a} = \frac{\sqrt{b^2 - 4ac}}{2a} \quad or \quad x + \frac{b}{2a} = -\frac{\sqrt{b^2 - 4ac}}{2a}.$$

Thus,

$$x = -\frac{b}{2a} + \frac{\sqrt{b^2 - 4ac}}{2a} \quad or \quad x = -\frac{b}{2a} - \frac{\sqrt{b^2 - 4ac}}{2a},$$

so

$$x = -\frac{b}{2a} \pm \frac{\sqrt{b^2 - 4ac}}{2a},$$

or

$$x = \frac{-b \pm \sqrt{b^2 - 4ac}}{2a}.$$

This last equation is the result we were after. It is so useful that it is worth memorizing.

The Quadratic Formula

The solutions of $ax^2 + bx + c = 0$ are given by

$$x = \frac{-b \pm \sqrt{b^2 - 4ac}}{2a}.$$

Note that the formula also holds when $a < 0$. A similar proof would show this, but we will not consider it here.

EXAMPLE 1

Solve using the quadratic formula: **(a)** $4x^2 + 5x - 6 = 0$; **(b)** $x^2 = 4x + 7$; **(c)** $x^2 + x = -1$.

Solution

a) We identify a, b, and c and substitute into the quadratic formula:

$$\underset{a}{4}x^2 + \underset{b}{5}x \underset{c}{-6} = 0$$

$$x = \frac{-b \pm \sqrt{b^2 - 4ac}}{2a}$$

$$x = \frac{-5 \pm \sqrt{5^2 - 4 \cdot 4(-6)}}{2 \cdot 4}$$

Be sure to write the fraction bar all the way across.

$$x = \frac{-5 \pm \sqrt{25 - (-96)}}{8}$$

$$x = \frac{-5 \pm \sqrt{121}}{8}$$

$$x = \frac{-5 \pm 11}{8}$$

$$x = \frac{-5 + 11}{8} \quad or \quad x = \frac{-5 - 11}{8}$$

$$x = \frac{6}{8} \quad or \quad x = \frac{-16}{8}$$

$$x = \frac{3}{4} \qquad or \quad x = -2.$$

The solutions are $\frac{3}{4}$ and -2.

b) We rewrite $x^2 = 4x + 7$ in standard form, identify a, b, and c, and solve using the quadratic formula:

$$\underset{a}{\underset{\uparrow}{1}}x^2 - \underset{b}{\underset{\uparrow}{4}}x - \underset{c}{\underset{\uparrow}{7}} = 0$$

$$x = \frac{-(-4) \pm \sqrt{(-4)^2 - 4(1)(-7)}}{2 \cdot 1} \qquad \text{Substituting into the quadratic formula}$$

$$x = \frac{4 \pm \sqrt{16 + 28}}{2}$$

$$x = \frac{4 \pm \sqrt{44}}{2}.$$

Since $\sqrt{44}$ can be simplified, we have

$$x = \frac{4 \pm \sqrt{4}\sqrt{11}}{2}$$

$$x = \frac{4 \pm 2\sqrt{11}}{2}.$$

Finally, since 2 is a common factor of 4 and $2\sqrt{11}$, we can simplify the fraction by removing a factor of 1:

$$x = \frac{2(2 \pm \sqrt{11})}{2 \cdot 1} \qquad \text{Factoring}$$

$$x = \frac{2}{2} \cdot \frac{2 \pm \sqrt{11}}{1}. \qquad \text{Removing a factor of 1: } \frac{2}{2} = 1$$

The solutions are $2 + \sqrt{11}$ and $2 - \sqrt{11}$, or $2 \pm \sqrt{11}$.

c) We rewrite $x^2 + x = -1$ in standard form and use the quadratic formula:

$$\underset{a}{\underset{\uparrow}{1}}x^2 + \underset{b}{\underset{\uparrow}{1}}x + \underset{c}{\underset{\uparrow}{1}} = 0$$

$$x = \frac{-1 \pm \sqrt{1^2 - 4 \cdot 1 \cdot 1}}{2 \cdot 1} \qquad \text{Substituting into the quadratic formula}$$

$$x = \frac{-1 \pm \sqrt{1 - 4}}{2}$$

$$x = \frac{-1 \pm \sqrt{-3}}{2}$$

$$x = \frac{-1 \pm i\sqrt{3}}{2}.$$

TECHNOLOGY CONNECTION

We saw in Sections 5.7 and 11.2 how graphers can solve quadratic equations. To determine whether quadratic equations are solved more quickly on a grapher or by using the quadratic formula, solve Examples 1(a) and 1(b) both ways. Which method is faster? Which method is more precise? Why?

Try to solve Example 1(c) using a grapher. Can a grapher be used to solve quadratic equations with imaginary-number solutions? Why or why not?

The solutions are $\frac{-1 + i\sqrt{3}}{2}$ and $\frac{-1 - i\sqrt{3}}{2}$. □

The following are general guidelines for solving a quadratic equation.

To solve a quadratic equation:

1. If it is in the form $ax^2 = p$ or $(x + k)^2 = p$, use the principle of square roots as in Section 11.1.
2. When it is not in the form of (1), write it in standard form,

   $$ax^2 + bx + c = 0.$$

3. Try to factor and use the principle of zero products.
4. If it is not possible to factor or if factoring seems difficult, use the quadratic formula.

The solutions of a quadratic equation can always be found using the quadratic formula. They cannot always be found by factoring. When the radicand, $b^2 - 4ac$, is nonnegative, the equation has real-number solutions. When $b^2 - 4ac$ is negative, the equation has imaginary-number solutions.

Problem Solving

EXAMPLE 2

The number of diagonals d of a polygon of n sides is given by the formula

$$d = \frac{n^2 - 3n}{2}.$$

If a polygon has 27 diagonals, how many sides does it have?

Solution

1. **FAMILIARIZE.** A sketch can help us to become familiar with the problem. We draw a hexagon (6 sides) and count the diagonals. As the formula predicts, for $n = 6$, there are

   $$\frac{6^2 - 3 \cdot 6}{2} = \frac{36 - 18}{2} = \frac{18}{2} = 9 \text{ diagonals.}$$

   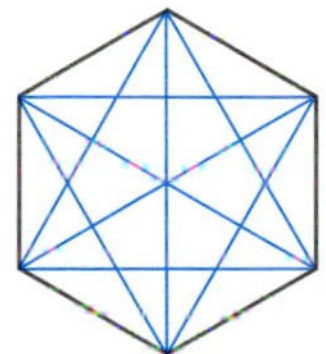

   We might wonder if, when there are three times as many diagonals, there are three times as many sides. Using the above formula, you can confirm that this is *not* the case. Rather than continue guessing, we proceed to a translation.

2. **TRANSLATE.** Since the number of diagonals is 27, we substitute 27 for d:

   $$27 = \frac{n^2 - 3n}{2}.$$

   This gives us a translation.

3. **CARRY OUT.** We solve the equation for n, first reversing the equation for convenience:

$$\frac{n^2 - 3n}{2} = 27$$

$$n^2 - 3n = 54 \quad \text{Multiplying by 2 to clear fractions}$$

$$n^2 - 3n - 54 = 0 \quad \text{Subtracting 54 on both sides}$$

$$(n - 9)(n + 6) = 0 \quad \text{Factoring. There is no need for the quadratic formula here.}$$

$$n - 9 = 0 \quad or \quad n + 6 = 0$$

$$n = 9 \quad or \quad n = -6.$$

4. **CHECK.** Since the number of sides cannot be negative, -6 cannot be a solution. We leave it to the student to show by substitution that 9 checks.

5. **STATE.** The polygon has 9 sides (it is a nonagon). ❑

EXAMPLE 3

The World Trade Center in New York City is 1368 ft tall. How many seconds will it take an object to fall from the top? Round to the nearest hundredth.

Solution

1. **FAMILIARIZE.** If we did not know anything about this problem, we might consider looking up a formula in a mathematics or physics book. A formula that fits this situation is

$$s = 16t^2,$$

where s is the distance, in feet, traveled by a body falling freely from rest in t seconds. This formula is actually an approximation in that it does not account for air resistance. In this problem, we know the distance s to be 1368. We want to determine the time t for the object to reach the ground. If we check a couple of guesses, we can see that the time t must be between 5 and 10 sec.

2. **TRANSLATE.** The distance is 1368 ft and we need to solve for t. We substitute 1368 for s in the formula above to get the following translation:

$$1368 = 16t^2.$$

3. **CARRY OUT.** Because there is no t-term, we can use the principle of square roots to solve:

$$1368 = 16t^2$$

$$\frac{1368}{16} = t^2 \quad \text{Solving for } t^2$$

$$\sqrt{\frac{1368}{16}} = t \quad or \quad -\sqrt{\frac{1368}{16}} = t \quad \text{Using the principle of square roots}$$

$$\frac{\sqrt{1368}}{4} = t \quad or \quad \frac{-\sqrt{1368}}{4} = t$$

$$9.25 \approx t \quad or \quad -9.25 \approx t. \quad \text{Using a calculator or Table 1 and rounding to the nearest hundredth}$$

4. **CHECK.** The number -9.25 cannot be a solution because time cannot be negative in this situation. We substitute 9.25 in the original equation:

$$s = 16(9.25)^2 = 16(85.5625) = 1369.$$

This is close. Remember that we approximated a solution. As we expected in step (1), the solution is between 5 and 10 sec.

5. STATE. It takes about 9.25 sec for the object to fall to the ground from the top of the World Trade Center. ❑

EXAMPLE 4

The hypotenuse of a right triangle is 6 m long. One leg is 1 m longer than the other. Find the lengths of the legs. Round to the nearest hundredth.

Solution

1. FAMILIARIZE. We first make a drawing and label it. We let s = the length of one leg. Then $s + 1$ = the length of the other leg.

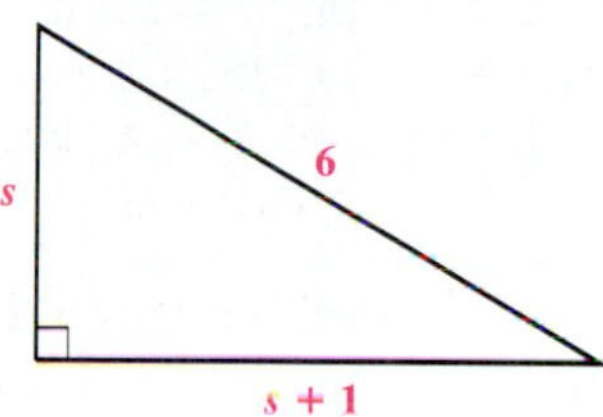

Note that if $s = 3$, then $s + 1 = 4$ and $3^2 + 4^2 = 25 \neq 6^2$. Thus, because of the Pythagorean theorem, we see that $s \neq 3$. Another guess, $s = 4$, is too big since $4^2 + (4 + 1)^2 = 41 \neq 6^2$. Although we have not guessed the solution, we expect the value of s to be between 3 and 4.

2. TRANSLATE. To translate, we use the Pythagorean theorem:

$$s^2 + (s + 1)^2 = 6^2.$$

3. CARRY OUT. We solve the equation:

$$s^2 + (s + 1)^2 = 6^2$$
$$s^2 + s^2 + 2s + 1 = 36$$
$$\underset{a}{\underset{\uparrow}{2}}s^2 + \underset{b}{\underset{\uparrow}{2}}s - \underset{c}{\underset{\uparrow}{35}} = 0$$ This cannot be factored so we use the quadratic formula.

$$s = \frac{-2 \pm \sqrt{2^2 - 4 \cdot 2(-35)}}{2 \cdot 2}$$ Remember: $s = \dfrac{-b \pm \sqrt{b^2 - 4ac}}{2a}$.

$$s = \frac{-2 \pm \sqrt{4 + 280}}{4} = \frac{-2 \pm \sqrt{284}}{4}$$

$s \approx 3.71$ *or* $s \approx -4.71$. Using a calculator or Table 1 and rounding to the nearest hundredth

4. CHECK. Length cannot be negative, so -4.71 does not check. But 3.71 does check. If the smaller leg is 3.71, the other leg is 4.71. Then

$$(3.71)^2 + (4.71)^2 = 13.7641 + 22.1841 = 35.9482$$

and since $35.9482 \approx 6^2$, our approximation checks. Also, note that the value for s, 3.71, is between 3 and 4, as we expected from step (1).

5. STATE. One leg is about 3.71 m long; the other is about 4.71 m long. ❑

EXERCISE SET 11.3

Solve. Try factoring first. If factoring is not posssible or is difficult, use the quadratic formula.

1. $x^2 - 4x = 21$
2. $x^2 + 7x = 18$
3. $x^2 = 6x - 9$
4. $x^2 = 8x - 16$
5. $x^2 + 6x + 4 = 0$
6. $x^2 - 6x - 4 = 0$
7. $y^2 + 16 = 0$
8. $t^2 + 1 = 0$
9. $x^2 - 9 = 0$
10. $x^2 - 4 = 0$
11. $x^2 - 2x - 2 = 0$
12. $x^2 - 4x - 7 = 0$
13. $3u^2 - 5u + 4 = 0$
14. $3p^2 - 2p + 8 = 0$
15. $x^2 + 4x + 4 = 7$
16. $x^2 - 2x + 1 = 5$
17. $3x^2 + 8x + 2 = 0$
18. $3x^2 - 4x - 2 = 0$
19. $x^2 + x + 2 = 0$
20. $x^2 - 2x + 3 = 0$
21. $4y^2 - 4y - 1 = 0$
22. $4y^2 + 4y - 1 = 0$
23. $2t^2 + 6t + 5 = 0$
24. $4y^2 + 3y + 2 = 0$
25. $3x^2 = 5x + 4$
26. $2x^2 + 3x = 1$
27. $25x^2 - 20x + 4 = 0$
28. $36r^2 + 84r + 49 = 0$
29. $6x^2 - 9x = 0$
30. $7x^2 + 2 = 6x$
31. $5t^2 - 7t = -4$
32. $15t^2 + 10t = 0$
33. $5t^2 = 100$
34. $4x^2 = 90$

Solve using the quadratic formula. Use a calculator or Table 1 to approximate the solutions to the nearest thousandth.

35. $x^2 - 4x - 7 = 0$
36. $x^2 + 2x - 2 = 0$
37. $y^2 - 6y - 1 = 0$
38. $y^2 + 10y + 22 = 0$
39. $4x^2 + 4x = 1$
40. $4x^2 = 4x + 1$

Solve. Should irrational answers occur, round to the nearest hundredth.

41. An octagon is a figure with 8 sides. How many diagonals does an octagon have?
42. A decagon is a figure with 10 sides. How many diagonals does a decagon have?
43. A polygon has 14 diagonals. How many sides does it have?
44. A polygon has 2 diagonals. How many sides does it have?
45. The Gateway Arch in St. Louis is 640 ft high. How long would it take an object to fall from the top?
46. Library Square Tower, in Los Angeles, is 1012 ft tall. How long would it take an object to fall from the top?
47. The world record for free-fall to the ground without a parachute by a woman is 175 ft and is held by Kitty O'Neill. Approximately how long did the fall take?

48. The world record for free-fall to the ground without a parachute by a man is 311 ft and is held by Dar Robinson. Approximately how long did the fall take?
49. The hypotenuse of a right triangle is 25 ft long. One leg is 17 ft longer than the other. Find the lengths of the legs.

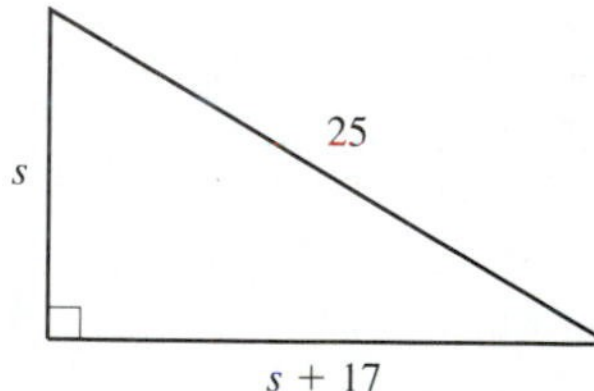

50. The hypotenuse of a right triangle is 26 yd long. One leg is 14 yd longer than the other. Find the lengths of the legs.

51. The length of a rectangle is 2 cm greater than the width. The area is 80 cm^2. Find the length and the width.

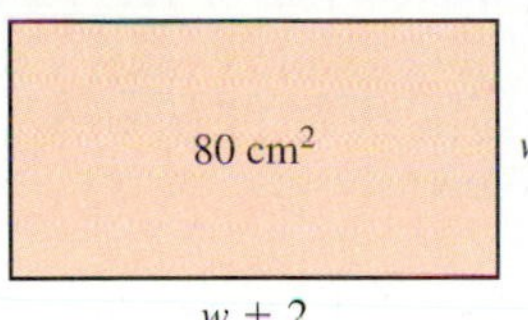

52. The length of a rectangle is 1 m greater than the width. The area is 72 m^2. Find the length and the width.

53. A water pipe runs diagonally under a rectangular yard that is 5 m longer than it is wide. If the pipe is 25 m long, determine the dimensions of the yard.

54. A 26-ft long guy wire is anchored 10 ft from the base of a telephone pole. How far up the pole does the wire reach?

55. The area of a right triangle is 13 m^2. One leg is 2.5 m longer than the other. Find the lengths of the legs.

56. The area of a right triangle is 15.5 cm^2. One leg is 1.2 cm longer than the other. Find the lengths of the legs.

57. The length of a rectangle is 2 in. greater than the width. The area is 20 in^2. Find the length and the width.

58. The length of a rectangle is 3 ft greater than the width. The area is 15 ft^2. Find the length and the width.

59. The length of a rectangle is twice the width. The area is 10 m^2. Find the length and the width.

60. The length of a rectangle is twice the width. The area is 20 cm^2. Find the length and the width.

The formula $A = P(1 + r)^t$ is used to find the value A to which P dollars grows when invested for t years at an annual interest rate r. In Exercises 61–66, find the interest rate for the information provided.

61. \$1000 grows to \$1210 in 2 years

62. \$1000 grows to \$1440 in 2 years

63. \$2560 grows to \$2890 in 2 years

64. \$4000 grows to \$4410 in 2 years

65. \$6250 grows to \$7290 in 2 years

66. \$6250 grows to \$6760 in 2 years

67. Laura has enough mulch to cover 250 ft^2 of garden space. How wide is the largest circular flower garden that Laura can cover with mulch?

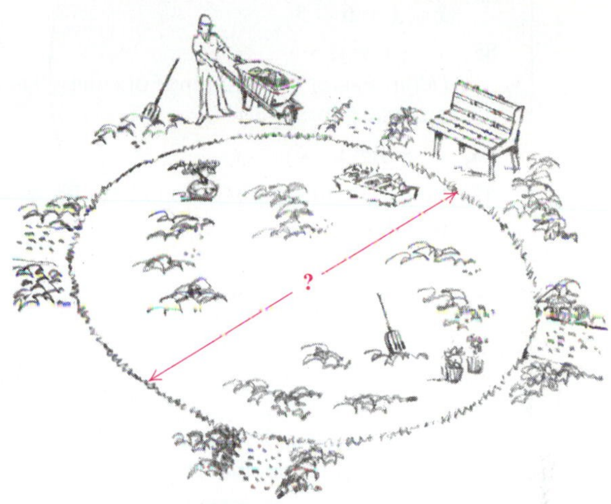

68. A circular oil slick is 20,000 m^2 in area. How wide is the oil slick?

Skill Maintenance

Solve.

69. $\frac{1}{x} + \frac{2}{x+1} = \frac{3}{x}$ **70.** $\frac{3}{x-4} = \frac{2}{x+4}$

Simplify.

71. $\sqrt{40} - 2\sqrt{10} + \sqrt{90}$ **72.** $\sqrt{9000x^{10}}$

Synthesis

73. A student claims to be able to solve any quadratic equation by completing the square. The same student claims to be incapable of understanding why the quadratic formula works. Does this strike you as odd? Why or why not?

74. Write a problem for a classmate to solve. Devise the problem so that a quadratic equation is used to solve it and the solution is an irrational number.

Solve.

75. $5x + x(x - 7) = 0$ **76.** $x(3x + 7) - 3x = 0$

77. $(y + 4)(y + 3) = 15$ **78.** $x^2 + (x + 2)^2 = 7$

79. $(x + 2)^2 + (x + 1)^2 = 0$

80. $(x + 3)^2 + (x + 1)^2 = 0$

81. $x^2 + x - \sqrt{2} = 0$ **82.** $x^2 + \sqrt{5}x - \sqrt{3} = 0$

83. $\dfrac{x^2}{x-4} - \dfrac{7}{x-4} = 0$

84. $\dfrac{1}{x} + \dfrac{1}{x+6} = \dfrac{1}{5}$

85. $x^3 - 8 = 0$
(*Hint:* Factor the difference of cubes. Then use the quadratic formula.)

86. $x^3 + 1 = 0$

87. Find r in this figure. Round to the nearest hundredth.

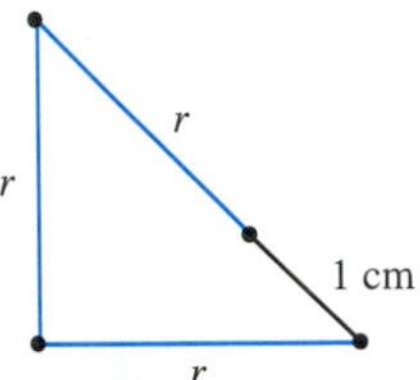

88. Two consecutive integers have squares that differ by 25. Find the integers.

89. Find the area of a square for which the diagonal is one unit longer than the length of the sides.

90. A 20-ft pole is struck by lightning and, while not completely broken, falls over and touches the ground 10 ft from the bottom of the pole. How high up did the pole break?

91. What should the diameter d of a pizza be so that it has the same area as two 10-in. pizzas? Do you get more to eat with a 13-in. pizza or with two 10-in. pizzas?

$d = ?$ = 10 in. + 10 in.

92. How long is the side of a square whose diagonal is 3 cm longer than a side?

93. $4000 is invested at interest rate r. In 2 years, it grows to $5267.03. What is the interest rate?

94. In 2 years, you want to have $3000. How much do you need to invest now if you can get an interest rate of 15.75% compounded annually?

95. In baseball, a batter's strike zone is a rectangular area about 15 in. wide and 40 in. high. Many batters subconsciously enlarge this area by 40% when

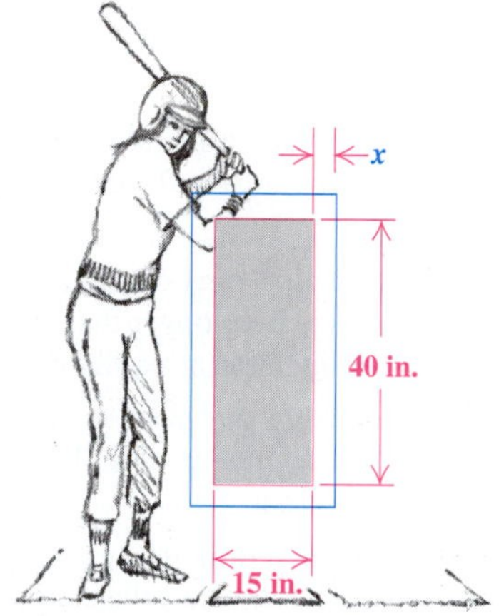

fearful that if they don't swing, the umpire will call the pitch a strike. Assuming that the strike zone is enlarged by an invisible band of uniform width around the actual zone, find the dimensions of the enlarged strike zone.

96. To solve an equation with a grapher, we can let y_1 = the left side of the equation and y_2 = the right side of the equation, and then graph y_1 and y_2 on the same set of axes. The Zoom and Trace features can then be used to find the x-coordinate at any point of intersection. Use a grapher to approximate to the nearest thousandth the solutions of Exercises 35–40. Compare your answers with those found using Table 1 or a calculator.

11.4 Rational Equations and Problem Solving

Rational Equations That Are Quadratic • Solving Problems

When fractions are cleared in a rational equation, we sometimes get a quadratic equation. When we do, we proceed as in solving any quadratic equation, but it is especially important to check all possible solutions (see Section 6.7).

Rational Equations That Are Quadratic

Recall that to solve a rational equation, we multiply on both sides by the LCD.

EXAMPLE 1

Solve: $\dfrac{14}{x+2} - \dfrac{1}{x-4} = 1.$

Solution

$$(x+2)(x-4) \cdot \left[\frac{14}{x+2} - \frac{1}{x-4}\right] = (x+2)(x-4) \cdot 1 \qquad \text{The LCD is } (x+2)(x-4).$$

$$\cancel{(x+2)}(x-4)\frac{14}{\cancel{x+2}} - (x+2)\cancel{(x-4)}\frac{1}{\cancel{x-4}} = (x+2)(x-4)$$

$$14(x-4) - (x+2) = (x+2)(x-4) \qquad \text{Removing factors of 1: } (x+2)/(x+2) \text{ and } (x-4)/(x-4)$$

$$14x - 56 - x - 2 = x^2 - 2x - 8 \qquad \text{Multiplying}$$

$$13x - 58 = x^2 - 2x - 8 \qquad \text{Collecting like terms}$$

$$0 = x^2 - 15x + 50 \qquad \text{Writing standard form}$$

$$0 = (x-5)(x-10) \qquad \text{Factoring}$$

$$x = 5 \quad or \quad x = 10. \qquad \text{The principle of zero products}$$

Since we might have introduced some numbers that are not solutions of the original equation when we cleared fractions, we must check.

Check:

For 5:

$$\frac{14}{x+2} - \frac{1}{x-4} = 1$$

$\dfrac{14}{5+2} - \dfrac{1}{5-4}$?	1
$\dfrac{14}{7} - 1$	
$2 - 1$	
1	1 TRUE

For 10:

$$\frac{14}{x+2} - \frac{1}{x-4} = 1$$

$\dfrac{14}{10+2} - \dfrac{1}{10-4}$?	1
$\dfrac{14}{12} - \dfrac{1}{6}$	
$\dfrac{14}{12} - \dfrac{2}{12}$	
1	1 TRUE

The solutions are 5 and 10. ❑

> If clearing the fractions of a rational equation produces a quadratic equation, solve that resulting equation in the usual way. Remember that possible solutions to rational equations should be checked in the original equation.

Solving Problems

As we saw in Section 6.8, some problems translate to rational equations. Sometimes the rational equation is actually quadratic.

EXAMPLE 2

Josif and Sally work together to type a short story, and it takes them 4 hr. It would take Sally 6 hr more than Josif to type the story alone. How long would each need to type the story if they worked alone?

Solution

1. **FAMILIARIZE.** Let's make a guess. Suppose that it takes Josif 5 hr. Then it would take Sally 6 more hr, or 11 hr. If this is a correct answer, then Josif will type $\frac{4}{5}$ of the story in 4 hr and Sally will type $\frac{4}{11}$ of the story in 4 hr:

$$\frac{4}{5} + \frac{4}{11} = \frac{44}{55} + \frac{20}{55} = \frac{64}{55}.$$

We do not have the correct answer. If we did, they would complete $\frac{55}{55}$ of the story in 4 hr.

Suppose that Josif takes x hr. Then Sally would take 6 hr more, or $(x + 6)$ hr. Josif would type $1/x$ of the story per hour and Sally would type $1/(x + 6)$ of the story per hour. In 4 hr, Josif would type $4(1/x)$ of the story and Sally would type $4[1/(x + 6)]$ of the story.

2. **TRANSLATE.** Since we are told that working together Josif and Sally complete one entire story in 4 hr, we have an equation:

$$\underbrace{4\left(\frac{1}{x}\right)}_{\text{Fraction of the story typed by Josif in 4 hr}} \overset{\text{plus}}{+} \underbrace{4\left(\frac{1}{x+6}\right)}_{\text{fraction of the story typed by Sally in 4 hr}} \overset{\text{equals}}{=} \underbrace{1.}_{\text{one completed story}}$$

3. **CARRY OUT.** We solve the equation:

$$x(x+6)\left[4\left(\frac{1}{x}\right) + 4\left(\frac{1}{x+6}\right)\right] = x(x+6)\cdot 1 \quad \text{Multiplying by the LCD}$$

$$x(x+6)\cdot 4\left(\frac{1}{x}\right) + x(x+6)\cdot 4\left(\frac{1}{x+6}\right) = x(x+6)\cdot 1$$

$$4(x+6) + 4x = x(x+6) \quad \text{Removing factors of 1: } x/x \text{ and } (x+6)/(x+6)$$

$$4x + 24 + 4x = x^2 + 6x$$

$$8x + 24 = x^2 + 6x$$

$$0 = x^2 - 2x - 24 \quad \text{Standard form}$$

$$0 = (x-6)(x+4) \quad \text{Factoring}$$

$$x = 6 \quad or \quad x = -4. \quad \text{Principle of zero products}$$

4. **CHECK.** Since negative time has no meaning in this problem, -4 is not a solution. Let's see if 6 checks. This is the time for Josif to type the story alone. Then Sally would take 12 hr alone. Thus in 4 hr Josif would type $\frac{4}{6}$ of the story, Sally would type $\frac{4}{12}$ of it, and together they would complete $\frac{4}{6} + \frac{4}{12}$, or $\frac{2}{3} + \frac{1}{3}$, of the story. This is all of it, so the numbers check.

5. **STATE.** The answer is that it would take Josif 6 hr alone and Sally 12 hr alone to type the story. ❑

EXAMPLE 3

Makita's motorcycle traveled 300 mi at a certain speed. Had she gone 10 mph faster, the trip would have taken 1 hr less. Find the speed of the motorcycle.

Solution

1. **FAMILIARIZE.** We make a drawing, labeling it with the known and unknown information. We can also organize the information in a table. We let r and t represent the rate, in miles per hour, and time, in hours, respectively.

Distance	Speed	Time
300	r	t
300	$r + 10$	$t - 1$

300 miles

Time t Speed r

300 miles

Time $t - 1$ Speed $r + 10$

Recall that the definition of speed, $r = d/t$, relates the three quantities.

2. **TRANSLATE.** From the first line of the table, we obtain

$$r = \frac{300}{t}.$$

From the second line, we get

$$r + 10 = \frac{300}{t - 1}.$$

3. **CARRY OUT.** We now have a system of equations. We substitute for r from the first equation into the second and solve the resulting equation:

$$\frac{300}{t} + 10 = \frac{300}{t-1} \quad \text{Substituting } 300/t \text{ for } r$$

$$t(t-1)\cdot\left[\frac{300}{t} + 10\right] = t(t-1)\cdot\frac{300}{t-1} \quad \text{Multiplying by the LCD}$$

$$\cancel{t}(t-1)\cdot\frac{300}{\cancel{t}} + t(t-1)\cdot 10 = t\cancel{(t-1)}\cdot\frac{300}{\cancel{t-1}}$$

$$300(t-1) + 10(t^2 - t) = 300t \quad \text{Removing factors of 1: } t/t \text{ and } (t-1)/(t-1)$$

$$10t^2 - 10t - 300 = 0 \quad \text{Standard form}$$

$$t^2 - t - 30 = 0 \quad \text{Multiplying by } \tfrac{1}{10}$$

$$(t-6)(t+5) = 0 \quad \text{Factoring}$$

$$t = 6 \quad or \quad t = -5. \quad \text{Principle of zero products}$$

4. **CHECK.** Note that we have solved for t, not r as required. Since negative time has no meaning in this problem, we disregard the -5 and use 6 hr to find r:

$$r = \frac{300}{6} = 50 \text{ mph.}$$

CAUTION! Always make sure that you find the quantity asked for in the problem.

To see if 50 mph checks, we increase the speed 10 mph to 60 mph and see how long the trip would have taken at that speed:

$$t = \frac{d}{r} = \frac{300}{60} = 5 \text{ hr.}$$

This is 1 hr less than the trip actually took, so we have an answer to the problem.

5. **STATE.** Makita's motorcycle traveled at a speed of 50 mph. ❑

EXERCISE SET 11.4

Solve.

1. $\frac{1}{x} = \frac{x-2}{24}$

2. $\frac{x+3}{14} = \frac{2}{x}$

3. $\frac{1}{2x-1} - \frac{1}{2x+1} = \frac{1}{4}$

4. $\frac{1}{4-x} - \frac{1}{2+x} = \frac{1}{4}$

5. $\frac{50}{x} - \frac{50}{x-5} = -\frac{1}{2}$

6. $3x + \frac{10}{x-1} = \frac{16}{x-1}$

7. $\frac{x+2}{x} = \frac{x-1}{2}$

8. $\frac{x}{3} - \frac{6}{x} = 1$

9. $x - 6 = \frac{1}{x+6}$

10. $x + 7 = \frac{1}{x-7}$

11. $\frac{2}{x} = \frac{x+3}{5}$

12. $\frac{x+3}{x} = \frac{x-4}{3}$

13. $x + 5 = \frac{3}{x-5}$

14. $x - 8 = \frac{1}{x+8}$

15. $\frac{40}{x} - \frac{20}{x-3} = \frac{8}{7}$

16. $\frac{11}{x} + \frac{14}{x+2} = 9$

17. $\frac{5}{x+2} + \frac{3x}{x+6} = 2$

18. $\frac{5}{x+4} + \frac{14}{x+7} = 4$

19. $\frac{3}{3x+1} + \frac{6x}{11x-1} = 1$

20. $\frac{7}{x+2} - \frac{4}{2x+1} = 1$

21. $\frac{16}{5(x-2)(x+2)} + \frac{9}{5(x+2)(x+3)} = \frac{-x}{(x-2)(x+3)}$

22. $\frac{19}{(x-3)(x+3)} - \frac{10}{(x+3)(x-2)} = \frac{x}{(x-3)(x-2)}$

23. $\frac{6}{x^2-4x-5} - \frac{6}{x^2-2x-3} = \frac{x}{x^2-8x+15}$

24. $\frac{4}{x^2-3x+2} - \frac{4}{x^2+2x-3} = \frac{x}{x^2+x-6}$

25. During the first part of a canoe trip, Tim covered 60 km at a certain speed. He then traveled 24 km at a speed that was 4 km/h slower. If the total time for the trip was 8 hr, what was the speed on each part of the trip?

26. During the first part of a trip, Meira's Honda traveled 120 mi at a certain speed. Meira then drove another 100 mi at a speed that was 10 mph slower. If Meira's total trip time was 4 hr, what was her speed on each part of the trip?

27. Sandi's Subaru travels 280 mi at a certain speed. If the car had gone 5 mph faster, the trip would have taken 1 hr less. Find Sandi's speed.

28. Petra's Plymouth travels 200 mi at a certain speed. If the car had gone 10 mph faster, the trip would have taken 1 hr less. Find Petra's speed.

29. A turbo-jet flies 50 mph faster than a super-prop plane. If a turbo-jet goes 2000 mi in 3 hr less time than it takes the super-prop to go 2800 mi, find the speed of each plane.

30. A Cessna flies 600 mi at a certain speed. A Beechcraft flies 1000 mi at a speed that is 50 mph faster, but takes 1 hr longer. Find the speed of each plane.

31. On a sales trip, Gail drives the 600 mi to Richmond at a certain speed. The return trip is made at a speed that is 10 mph slower. Total time for the round trip was 22 hr. How fast did Gail travel on each part of the trip?

32. Naoki travels the 40 mi to Hillsboro at a certain speed. The return trip is made at a speed that is 6 mph slower. Total time for the round trip was 14 hr. How fast did Naoki travel on each part of the trip?

33. A stream flows at 2 mph. A boat travels 24 mi upstream and returns in a total time of 5 hr. What is the speed of the boat in still water?

34. A river flows at 5 mph. A boat travels 60 mi upriver and returns in a total time of 9 hr. What is the speed of the boat in still water?

35. Two pipes are connected to the same tank. When working together, they can fill the tank in 2 hr. The larger pipe, working alone, can fill the tank in 3 hr less time than the smaller one. How long would the smaller one take, working alone, to fill the tank?

36. Two hoses are connected to a swimming pool. When working together, they can fill the pool in 4 hr. The larger hose, working alone, can fill the pool in 6 hr less time than the smaller one. How long would the smaller one take, working alone, to fill the pool?

37. Ellen paddles 1 mi upstream and 1 mi back in a total time of 1 hr. The speed of the river is 2 mph. Find Ellen's speed in still water.

38. Dan rows 10 km upstream and 10 km back in a total time of 3 hr. The speed of the river is 5 km/h. Find Dan's speed in still water.

Skill Maintenance

39. Solve: $\sqrt{3x+1} = \sqrt{2x-1} + 1$.

40. Add: $\frac{1}{x-1} + \frac{1}{x^2-3x+2}$.

41. Multiply and simplify: $\sqrt[3]{18y^3}\,\sqrt[3]{4x^2}$.

Synthesis

42. Write a problem for a classmate to solve. Devise the problem so that **(a)** the solution is found after solving a rational equation, and **(b)** the solution is "The express train travels 90 mph and the local travels 60 mph."

43. Under what circumstances would a negative value for t, time, have meaning?

Solve.

44. $\frac{12}{x^2-9} = 1 + \frac{3}{x-3}$

45. $\frac{x^2}{x-2} - \frac{x+4}{2} + \frac{2-4x}{x-2} + 1 = 0$

46. $\frac{4}{2x+i} - \frac{1}{x-i} = \frac{2}{x+i}$

47. Find a when the reciprocal of $a-1$ is $a+1$.

48. A discount store bought a quantity of beach towels for \$250 and sold all but 15 at a profit of \$3.50 per towel. With the total amount received, the manager could buy 4 more than twice as many as were bought before. Find the cost per towel.

49. Use a grapher to solve Exercises 1, 9, 19, and 39.

11.5 The Discriminant and Solutions to Quadratic Equations

The Discriminant • Writing Equations from Solutions

The Discriminant

From the quadratic formula, we know that the solutions x_1 and x_2 of a quadratic equation are given by

$$x_1 = \frac{-b + \sqrt{b^2 - 4ac}}{2a} \quad \text{and} \quad x_2 = \frac{-b - \sqrt{b^2 - 4ac}}{2a}.$$

The expression $b^2 - 4ac$ shows the nature of the solutions. This expression is called the **discriminant.** If it is 0, then it doesn't matter whether we choose the plus or minus sign in the formula; hence there is just one real solution. If the discriminant is positive, there will be two real solutions. If it is negative, we will be taking the square root of a negative number; hence there will be two imaginary-number solutions, and they will be complex conjugates. We summarize:

Discriminant $b^2 - 4ac$	Nature of Solutions
0	Only one solution; it is a real number
Positive	Two different real-number solutions
Negative	Two different imaginary-number solutions (complex conjugates)

The discriminant also gives information about solving. When the discriminant is a perfect square, the equation can be solved by factoring.

EXAMPLE 1

For each equation, determine the nature of the solutions: **(a)** $9x^2 - 12x + 4 = 0$; **(b)** $x^2 + 5x + 8 = 0$; **(c)** $x^2 + 5x + 6 = 0$.

Solution

a) We have

$$a = 9, \quad b = -12, \quad c = 4.$$

We substitute and compute the discriminant:

$$\begin{aligned} b^2 - 4ac &= (-12)^2 - 4 \cdot 9 \cdot 4 \\ &= 144 - 144 \\ &= 0. \end{aligned}$$

There is just one solution, and it is a real number. Since 0 is a perfect square, the equation can be solved by factoring.

b) We have

$$a = 1, \quad b = 5, \quad c = 8.$$

We substitute and compute the discriminant:

$$\begin{aligned} b^2 - 4ac &= 5^2 - 4 \cdot 1 \cdot 8 \\ &= 25 - 32 \\ &= -7. \end{aligned}$$

Since the discriminant is negative, there are two imaginary-number solutions. The equation cannot be solved by factoring because -7 is not a perfect square.

c) We have

$$a = 1, \quad b = 5, \quad c = 6;$$
$$b^2 - 4ac = 5^2 - 4 \cdot 1 \cdot 6 = 1.$$

Since the discriminant is positive, there are two real-number solutions. The equation can be solved by factoring since the discriminant is a perfect square. ❑

Writing Equations from Solutions

We know by the principle of zero products that $(x - 2)(x + 3) = 0$ has solutions 2 and -3. If we know the solutions of an equation, we can write an equation, using this principle in reverse.

EXAMPLE 2

Find a quadratic equation whose solutions are given.

a) 3 and $-\frac{2}{5}$ **b)** $2i$ and $-2i$ **c)** $\sqrt{3}$ and $-2\sqrt{3}$

Solution

a) We have

$$x = 3 \quad or \quad x = -\tfrac{2}{5}$$

$$x - 3 = 0 \quad or \quad x + \tfrac{2}{5} = 0$$ Getting 0's on one side

$$(x - 3)\left(x + \tfrac{2}{5}\right) = 0$$ Using the principle of zero products (multiplying)

$$x^2 + \tfrac{2}{5}x - 3x - 3 \cdot \tfrac{2}{5} = 0$$ Using FOIL

$$x^2 - \tfrac{13}{5}x - \tfrac{6}{5} = 0,$$ Collecting like terms

or

$$5x^2 - 13x - 6 = 0.$$ Multiplying by 5 on both sides to clear fractions

b) We have

$$x = 2i \quad or \quad x = -2i$$

$$x - 2i = 0 \quad or \quad x + 2i = 0$$ Getting 0's on one side

$$(x - 2i)(x + 2i) = 0$$ Using the principle of zero products (multiplying)

$$x^2 + 2ix - 2ix - (2i)^2 = 0$$ Using FOIL

$$\begin{aligned} x^2 - (2i)^2 &= 0 \\ x^2 - 4i^2 &= 0 \\ x^2 + 4 &= 0. \end{aligned}$$

c) We have

$$x = \sqrt{3} \quad or \quad x = -2\sqrt{3}$$

$$x - \sqrt{3} = 0 \quad or \quad x + 2\sqrt{3} = 0 \qquad \text{Getting 0's on one side}$$

$$(x - \sqrt{3})(x + 2\sqrt{3}) = 0 \qquad \text{Using the principle of zero products}$$

$$x^2 + 2\sqrt{3}x - \sqrt{3}x - 2(\sqrt{3})^2 = 0 \qquad \text{Using FOIL}$$

$$x^2 + \sqrt{3}x - 6 = 0. \qquad \text{Collecting like terms}$$

❑

Note that in Example 2(a) we multiplied on both sides by the LCD, 5. We normally perform this step and thus clear the equation of fractions. Had we preferred, we could have multiplied $x + \frac{2}{5} = 0$ by 5 on both sides, thus clearing fractions *before* using FOIL.

EXERCISE SET 11.5

Determine the nature of the solutions of the equation.

1. $x^2 - 6x + 9 = 0$
2. $x^2 + 10x + 25 = 0$
3. $x^2 + 7 = 0$
4. $x^2 + 2 = 0$
5. $x^2 - 2 = 0$
6. $x^2 - 5 = 0$
7. $4x^2 - 12x + 9 = 0$
8. $4x^2 + 8x - 5 = 0$
9. $x^2 - 2x + 4 = 0$
10. $x^2 + 3x + 4 = 0$
11. $a^2 - 10a + 21 = 0$
12. $t^2 - 8t + 16 = 0$
13. $6x^2 + 5x - 4 = 0$
14. $10x^2 - x - 2 = 0$
15. $9t^2 - 3t = 0$
16. $4m^2 + 7m = 0$
17. $x^2 + 5x = 7$
18. $x^2 + 4x = -6$
19. $y^2 = \frac{1}{2}y - \frac{3}{5}$
20. $y^2 + \frac{9}{4} = 4y$
21. $4x^2 - 4\sqrt{3}x + 3 = 0$
22. $6y^2 - 2\sqrt{3}y - 1 = 0$

Write a quadratic equation having the given numbers as solutions.

23. $-11, 9$
24. $-4, 4$
25. 7, only solution [*Hint:* It must be a "double" solution.]
26. -5, only solution
27. $-3, -5$
28. $-2, -7$
29. $4, \frac{2}{3}$
30. $5, \frac{3}{4}$
31. $\frac{1}{2}, \frac{1}{3}$
32. $-\frac{1}{4}, -\frac{1}{2}$
33. $-\frac{2}{5}, \frac{6}{5}$
34. $\frac{2}{7}, -\frac{3}{7}$
35. $\sqrt{2}, 3\sqrt{2}$
36. $-\sqrt{3}, 2\sqrt{3}$
37. $-\sqrt{5}, -2\sqrt{5}$
38. $-\sqrt{6}, -3\sqrt{6}$
39. $3i, -3i$
40. $4i, -4i$
41. $5 - 2i, 5 + 2i$
42. $2 - 7i, 2 + 7i$
43. $\frac{1 + 3i}{2}, \frac{1 - 3i}{2}$
44. $\frac{2 - i}{3}, \frac{2 + i}{3}$

Skill Maintenance

45. During a one-hour television show, there were 12 commercials. Some of the commercials were 30 sec long and the others were 60 sec long. The amount of time for 30-sec commercials was 6 min less than the total number of minutes of commercial time during the show. How many 30-sec commercials were used? How many 60-sec commercials were used?

Synthesis

46. ◈ Describe a procedure that could be used to write an equation having the first seven natural numbers as solutions.
47. ◈ Explain why each of the following statements is true.

    a) The sum of the solutions of
    $$ax^2 + bx + c = 0$$
    is $-b/a$.

b) The product of the solutions of

$$ax^2 + bx + c = 0$$

is c/a.

48. Find k for which

$$kx^2 - 4x + (2k - 1) = 0$$

and the product of the solutions is 3.

49. The graph of an equation of the form

$$y = ax^2 + bx + c$$

is a curve similar to the one shown below. Determine a, b, and c from the information given.

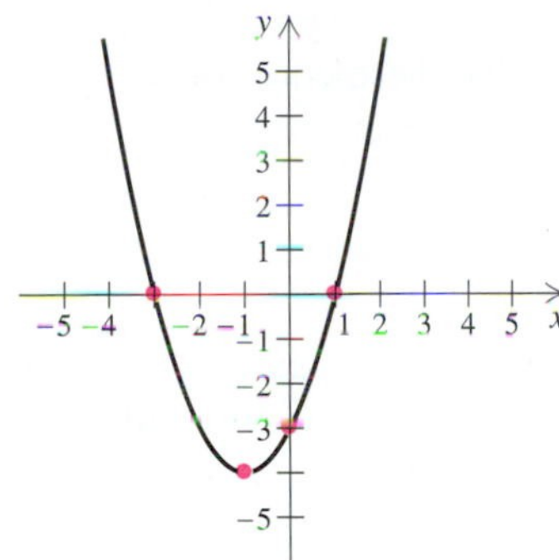

For each equation under the given condition, **(a)** find k, and **(b)** find the other solution.

50. $kx^2 - 17x + 33 = 0$; one solution is 3

51. $kx^2 - 2x + k = 0$; one solution is -3

52. $x^2 - kx + 2 = 0$; one solution is $1 + i$

53. $x^2 - (6 + 3i)x + k = 0$; one solution is 3

54. Find h and k, where $3x^2 - hx + 4k = 0$, the sum of the solutions is -12, and the product of the solutions is 20.

55. Find a quadratic equation for which the sum of the solutions is $\sqrt{3}$ and the product is 8.

56. Let $f(x) = x^2 - 2x - 8$.

a) Find all x such that $f(x) = 0$.
b) Find all x such that $f(x) = 17$.

57. Suppose that $f(x) = ax^2 + bx + c$, with $f(-3) = 0$, $f(\frac{1}{2}) = 0$, and $f(0) = -12$. Find a, b, and c.

58. The sum of the squares of the solutions of

$$x^2 + 2kx - 5 = 0$$

is 26. Find the absolute value of k.

59. When the solutions of each of the equations $x^2 + kx + 8 = 0$ and $x^2 - kx + 8 = 0$ are listed in increasing order, each solution of the second equation is 6 more than the corresponding solution of the first equation. Find k.

11.6

Equations Reducible to Quadratic

Recognizing Equations in Quadratic Form • Using Substitution to Solve

Certain equations that are not really quadratic can be thought of in such a way that they can be solved as quadratic. For example, consider this fourth-degree equation:

$$x^4 - 9x^2 + 8 = 0$$

$$\downarrow \quad \downarrow \quad \downarrow \quad \downarrow$$

$$(x^2)^2 - 9(x^2) + 8 = 0 \qquad \text{Thinking of } x^4 \text{ as } (x^2)^2$$

$$\downarrow \quad \downarrow \quad \downarrow \quad \downarrow$$

$$u^2 - 9u + 8 = 0 \qquad \text{To make this clearer, write } u \text{ instead of } x^2.$$

The equation $u^2 - 9u + 8 = 0$ can be solved by factoring or by the quadratic formula. Then, remembering that $u = x^2$, we can solve for x. Equations that can be solved like this are said to be *reducible to quadratic,* or *in quadratic form.*

EXAMPLE 1

Solve: $x^4 - 9x^2 + 8 = 0$.

Solution Let $u = x^2$. Then we solve the equation by substituting u for x^2:

$$u^2 - 9u + 8 = 0$$

$$(u - 8)(u - 1) = 0 \quad \text{Factoring}$$

$$u - 8 = 0 \quad or \quad u - 1 = 0 \quad \text{Principle of zero products}$$

$$u = 8 \quad or \quad u = 1.$$

CAUTION! A common error is to solve for u and then forget to solve for x. Remember that you must find values for the *original* variable!

We substitute x^2 for u and solve these equations:

$$x^2 = 8 \quad or \quad x^2 = 1$$

$$x = \pm\sqrt{8} \quad or \quad x = \pm 1$$

$$x = \pm 2\sqrt{2} \quad or \quad x = \pm 1.$$

To check, note that when $x = 2\sqrt{2}$, $x^2 = 8$ and $x^4 = 64$. Also, when $x = -2\sqrt{2}$, $x^2 = 8$ and $x^4 = 64$. Similarly, when $x = 1$, $x^2 = 1$ and $x^4 = 1$, and when $x = -1$, $x^2 = 1$ and $x^4 = 1$. Thus instead of making four checks, we need make only two.

Check:

For $\pm 2\sqrt{2}$:

$$\begin{array}{c|l} x^4 - 9x^2 + 8 & 0 \\ \hline (\pm 2\sqrt{2})^4 - 9(\pm 2\sqrt{2})^2 + 8 & ?\ 0 \\ 64 - 9 \cdot 8 + 8 & \\ 0 & 0 \quad \text{TRUE} \end{array}$$

For ± 1:

$$\begin{array}{c|l} x^4 - 9x^2 + 8 & 0 \\ \hline (\pm 1)^4 - 9(\pm 1)^2 + 8 & ?\ 0 \\ 1 - 9 + 8 & \\ 0 & 0 \quad \text{TRUE} \end{array}$$

The solutions are $1, -1, 2\sqrt{2}$, and $-2\sqrt{2}$. ❑

Solutions of equations like $x^4 - 9x^2 + 8 = 0$ will always check unless a mistake has been made in one of the steps. Sometimes, however, rational equations, radical equations, or equations containing fractional exponents are reducible to quadratic. As we saw in Chapters 6 and 10, answers to these equations should be checked in the original equation.

EXAMPLE 2

Solve: $x - 3\sqrt{x} - 4 = 0$.

Solution Let $u = \sqrt{x}$. Then we solve the equation found by substituting u for $\sqrt{x}$ and u^2 for x:

$$u^2 - 3u - 4 = 0$$

$$(u - 4)(u + 1) = 0$$

$$u = 4 \quad or \quad u = -1.$$

Now we substitute $\sqrt{x}$ for u and solve these equations:

$$\sqrt{x} = 4 \quad or \quad \sqrt{x} = -1.$$

Squaring gives us $x = 16$ or $x = 1$.

Check: For 16:

$$\begin{array}{r|l} x - 3\sqrt{x} - 4 = 0 & \\ \hline 16 - 3\sqrt{16} - 4 \ ? & 0 \\ 16 - 3\cdot 4 - 4 & \\ 0 & 0 \quad \text{TRUE} \end{array}$$

For 1:

$$\begin{array}{r|l} x - 3\sqrt{x} - 4 = 0 & \\ \hline 1 - 3\sqrt{1} - 4 \ ? & 0 \\ 1 - 3\cdot 1 - 4 & \\ -6 & 0 \quad \text{FALSE} \end{array}$$

The number 16 checks, but 1 does not. Had we noticed that $\sqrt{x} = -1$ has no solution (since principal roots are never negative), we could have solved only the equation $\sqrt{x} = 4$. The solution is 16. ❑

EXAMPLE 3

Solve: $(x^2 - 1)^2 - (x^2 - 1) - 2 = 0$.

Solution Let $u = x^2 - 1$. Then we solve the equation found by substituting u for $x^2 - 1$:

$$u^2 - u - 2 = 0$$
$$(u - 2)(u + 1) = 0$$
$$u = 2 \quad \text{or} \quad u = -1.$$

Now we substitute $x^2 - 1$ for u and solve these equations:

$$x^2 - 1 = 2 \quad \text{or} \quad x^2 - 1 = -1$$
$$x^2 = 3 \quad \text{or} \quad x^2 = 0$$
$$x = \pm\sqrt{3} \quad \text{or} \quad x = 0.$$

The solutions are $-\sqrt{3}$, $\sqrt{3}$, and 0. ❑

Sometimes great care must be taken in deciding what substitution to make.

EXAMPLE 4

Solve: $y^{-2} - y^{-1} - 2 = 0$.

Solution We rewrite the equation using positive exponents:

$$\frac{1}{y^2} - \frac{1}{y} - 2 = 0.$$

Note that if we let $u = 1/y$, then $u^2 = 1/y^2$. The equation can then be written as a quadratic:

$$u^2 - u - 2 = 0$$
$$(u - 2)(u + 1) = 0$$
$$u = 2 \quad \text{or} \quad u = -1.$$

Now we substitute $1/y$ for u and solve these equations:

$$\frac{1}{y} = 2 \quad \text{or} \quad \frac{1}{y} = -1.$$

Solving gives us

$$y = \frac{1}{2} \quad \text{or} \quad y = \frac{1}{(-1)} = -1.$$

The numbers $\frac{1}{2}$ and -1 both check. They are the solutions. ❑

EXAMPLE 5

Solve: $t^{2/5} - t^{1/5} - 2 = 0$.

Solution Note that $t^{2/5}$ can be rewritten as $(t^{1/5})^2$. The equation can thus be written as $(t^{1/5})^2 - t^{1/5} - 2 = 0$. We let $u = t^{1/5}$ and solve the resulting equation:

$$u^2 - u - 2 = 0$$
$$(u - 2)(u + 1) = 0$$
$$u = 2 \quad or \quad u = -1.$$

Now we substitute $t^{1/5}$ for u and solve:

$$t^{1/5} = 2 \quad or \quad t^{1/5} = -1$$
$$t = 32 \quad or \quad t = -1. \qquad \text{Principle of powers; raising to the 5th power}$$

For 32:

$t^{2/5} - t^{1/5} - 2$	$= 0$
$32^{2/5} - 32^{1/5} - 2$	? 0
$(32^{1/5})^2 - 32^{1/5} - 2$	
$2^2 - 2 - 2$	
0	0 TRUE

For -1:

$t^{2/5} - t^{1/5} - 2$	$= 0$
$(-1)^{2/5} - (-1)^{1/5} - 2$	? 0
$[(-1)^{1/5}]^2 - (-1)^{1/5} - 2$	
$(-1)^2 - (-1) - 2$	
0	0 TRUE

Both numbers check. The solutions are 32 and -1. ❑

EXERCISE SET 11.6

Solve.

1. $x^4 - 10x^2 + 25 = 0$
2. $x^4 - 3x^2 + 2 = 0$
3. $x^4 - 12x^2 + 27 = 0$
4. $x^4 - 9x^2 + 20 = 0$
5. $9x^4 - 14x^2 + 5 = 0$
6. $4x^4 - 19x^2 + 12 = 0$
7. $x - 10\sqrt{x} + 9 = 0$
8. $2x - 9\sqrt{x} + 4 = 0$
9. $3x + 10\sqrt{x} - 8 = 0$
10. $5x + 13\sqrt{x} - 6 = 0$
11. $(x^2 - 9)^2 + 3(x^2 - 9) + 2 = 0$
12. $(x^2 - 4)^2 + 5(x^2 - 4) + 6 = 0$
13. $(x^2 - 6x)^2 - 2(x^2 - 6x) - 35 = 0$
14. $(x^2 - 3x)^2 - 10(x^2 - 3x) + 24 = 0$
15. $(3 + \sqrt{x})^2 - 3(3 + \sqrt{x}) - 10 = 0$
16. $(1 + \sqrt{x})^2 + (1 + \sqrt{x}) - 6 = 0$
17. $x^{-2} - x^{-1} - 6 = 0$
18. $4x^{-2} - x^{-1} - 5 = 0$
19. $2x^{-2} + x^{-1} - 1 = 0$
20. $m^{-2} + 9m^{-1} - 10 = 0$
21. $t^{2/3} + t^{1/3} - 6 = 0$ (*Hint:* Let $u = t^{1/3}$.)
22. $w^{2/3} - 2w^{1/3} - 8 = 0$
23. $z^{1/2} - z^{1/4} - 2 = 0$
24. $m^{1/3} - m^{1/6} - 6 = 0$
25. $x^{2/5} + x^{1/5} - 6 = 0$
26. $x^{1/2} - x^{1/4} - 6 = 0$
27. $t^{1/3} + 2t^{1/6} = 3$
28. $m^{1/2} + 6 = 5m^{1/4}$
29. $\left(\frac{x+3}{x-3}\right)^2 - \left(\frac{x+3}{x-3}\right) - 6 = 0$
30. $\left(\frac{x-4}{x+1}\right)^2 - 2\left(\frac{x-4}{x+1}\right) - 35 = 0$

31. $9\left(\frac{x+2}{x+3}\right)^2 - 6\left(\frac{x+2}{x+3}\right) + 1 = 0$

32. $16\left(\frac{x-1}{x-8}\right)^2 + 8\left(\frac{x-1}{x-8}\right) + 1 = 0$

33. $\left(\frac{y^2-1}{y}\right)^2 - 4\left(\frac{y^2-1}{y}\right) - 12 = 0$

34. $\left(\frac{x^2-2}{x}\right)^2 - 7\left(\frac{x^2-2}{x}\right) - 18 = 0$

Skill Maintenance

35. Multiply and simplify: $\sqrt{3x^2}\,\sqrt{3x^3}$.

36. Solution A is 18% alcohol and solution B is 45% alcohol. How much of each should be mixed together to get 12 L of a solution that is 36% alcohol?

37. Subtract: $\frac{x+1}{x-1} - \frac{x+1}{x^2+x+1}$.

Synthesis

38. ◈ Explain how an equation of the form $ax + b\sqrt{x} + c = 0$ could have no solution.

39. Find all x-intercepts of the graph of the function f given by $f(x) = x^4 - 8x^2 + 7$.

Solve. Check possible solutions by substituting into the original equation.

40. ▦ $6.75x - 35\sqrt{x} - 5.36 = 0$

41. ▦ $\pi x^4 - \pi^2 x^2 - \sqrt{99.3} = 0$

42. $\frac{x}{x-1} - 6\sqrt{\frac{x}{x-1}} - 40 = 0$

43. $\left(\sqrt{\frac{x}{x-3}}\right)^2 - 24 = 10\sqrt{\frac{x}{x-3}}$

44. $\sqrt{x-3} - \sqrt[4]{x-3} = 12$

45. $a^3 - 26a^{3/2} - 27 = 0$

46. $x^6 - 28x^3 + 27 = 0$

47. $x^6 + 7x^3 - 8 = 0$

48. Use a grapher to check your answers to Exercises 1, 3, 41, 46, and 47.

49. Use a grapher to solve $x^4 - x^3 - 13x^2 + x + 12 = 0$.

11.7 Quadratic Functions and Their Graphs

Graphs of $f(x) = ax^2$ • Graphs of $f(x) = a(x-h)^2$ • Graphs of $f(x) = a(x-h)^2 + k$

The following bar graph shows the fall and rise of the rate of unemployment during a recent year. A curve drawn along the graph would approximate the graph of a *quadratic function*. We now consider such graphs.

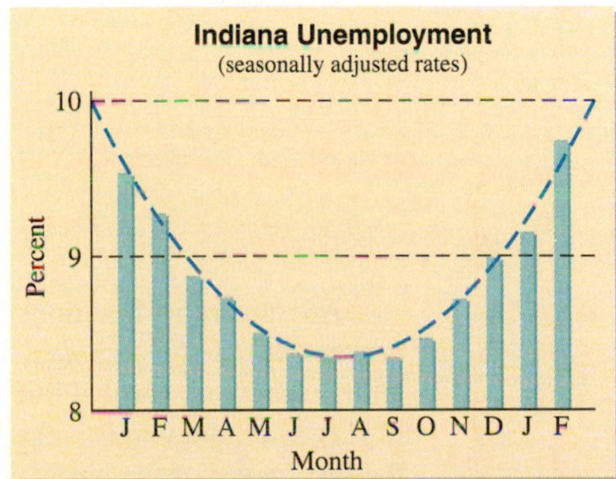

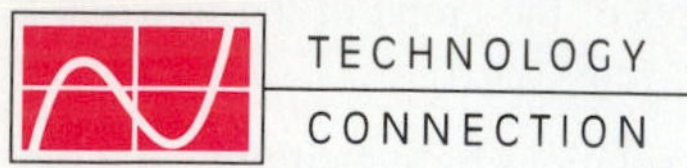

To examine the effect of a when graphing $f(x) = ax^2$, draw the function $y_1 = x^2$ in a $[-5, 5, -10, 10]$ window. Without erasing this curve, graph the function $y_2 = 3x^2$. How do the graphs compare? Now include the graph of $y_3 = \frac{1}{3}x^2$. Find a rule that describes the effect of multiplying x^2 by a, when $a > 1$ and when $0 < a < 1$.

Clear the display and graph $y_1 = x^2$ again. Now include the graph of $y_2 = -x^2$. Notice how it differs from $f(x) = x^2$. Next graph $y_3 = \frac{2}{3}x^2$ and $y_4 = -\frac{2}{3}x^2$ and compare them. Find a rule that describes the effect of multiplying x^2 by a, when $a < -1$ and when $-1 < a < 0$.

Graphs of $f(x) = ax^2$

In Chapter 7, the notion of a function was introduced. We studied equations such as

$$f(x) = 3x + 2 \quad \text{or} \quad y = 3x + 2,$$

whose graphs are straight lines. We now consider equations (or functions) in which the right-hand side is a quadratic polynomial:

$$f(x) = ax^2 + bx + c, \quad a \neq 0.$$

A function of this type is referred to as a **quadratic function.**

EXAMPLE 1

Graph: $f(x) = x^2$.

Solution We choose numbers for x, some positive and some negative, and for each number we compute $f(x)$.

We plot the ordered pairs and connect them with a smooth curve.

x	$f(x) = x^2$	$(x, f(x))$
−3	9	(−3, 9)
−2	4	(−2, 4)
−1	1	(−1, 1)
0	0	(0, 0)
1	1	(1, 1)
2	4	(2, 4)
3	9	(3, 9)

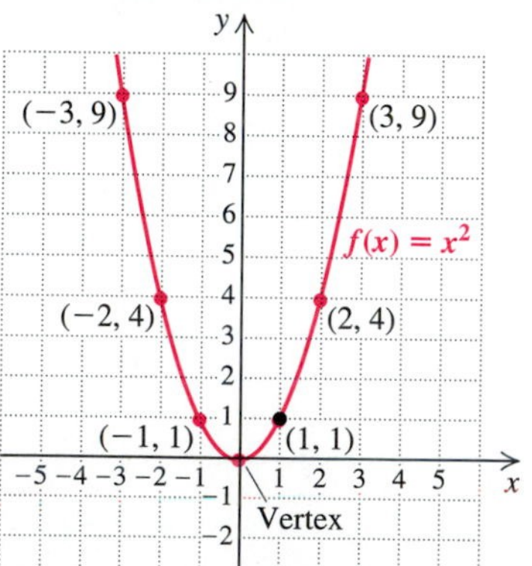

❑

All quadratic functions have graphs similar to the one in Example 1. Such curves are called *parabolas*. They are smooth, cup-shaped curves that are symmetric with respect to a vertical line known as the parabola's *line of symmetry*. In the graph of $f(x) = x^2$, the y-axis is the line of symmetry. If the paper were folded on this line, the two halves of the curve would match. The point (0, 0) is known as the *vertex* of this parabola.

By plotting points, we can see how the graphs of $g(x) = \frac{1}{2}x^2$ and $h(x) = 2x^2$ compare with the graph of $f(x) = x^2$.

x	$h(x) = 2x^2$
-3	18
-2	8
-1	2
0	0
1	2
2	8
3	18

x	$g(x) = \frac{1}{2}x^2$
-3	$\frac{9}{2}$
-2	2
-1	$\frac{1}{2}$
0	0
1	$\frac{1}{2}$
2	2
3	$\frac{9}{2}$

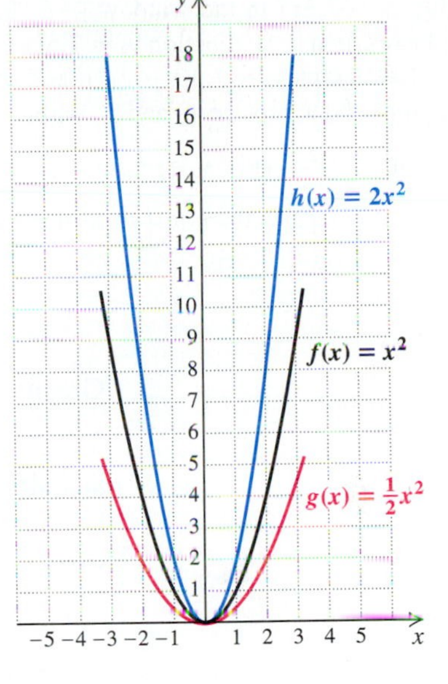

Note that the graph of $g(x) = \frac{1}{2}x^2$ is a flatter parabola than the graph of $f(x) = x^2$, and the graph of $h(x) = 2x^2$ is narrower. The vertex and the line of symmetry, however, have not changed.

When we consider the graph of $k(x) = -\frac{1}{2}x^2$, we see that the parabola opens downward and is the same shape as the graph of $g(x) = \frac{1}{2}x^2$.

x	$k(x) = -\frac{1}{2}x^2$
-3	$-\frac{9}{2}$
-2	-2
-1	$-\frac{1}{2}$
0	0
1	$-\frac{1}{2}$
2	-2
3	$-\frac{9}{2}$

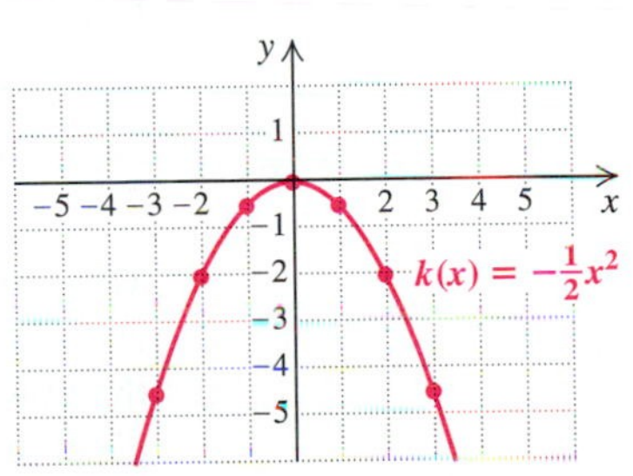

The graph of $g(x) = ax^2$ is a parabola with the vertical axis as its line of symmetry and its vertex at the origin.

If a is positive, the parabola opens upward; if a is negative, the parabola opens downward.

If $|a|$ is greater than 1, the parabola is narrower than $f(x) = x^2$.

If $|a|$ is between 0 and 1, the parabola is flatter than $f(x) = x^2$.

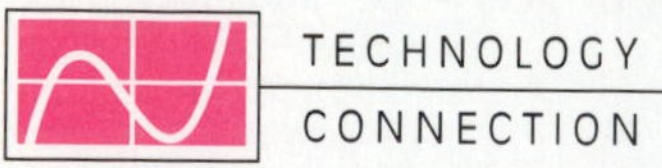

To investigate the effect of h on the graph of $f(x) = a(x - h)^2$, start by graphing $y_1 = 7.3x^2$ in the window $[-5, 5, -5, 5]$. On the same set of axes, graph the function $y_2 = 7.3(x - 1)^2$ and compare the graphs. Next graph $y_3 = 7.3(x - 2.7)^2$ and compare it with the first two graphs. Now replace y_2 and y_3 with

$$y_2 = 7.3(x + 1)^2 \quad \text{and} \quad y_3 = 7.3(x + 2.7)^2.$$

Find a rule that describes the effect of h in the function

$$f(x) = a(x - h)^2.$$

Graphs of $f(x) = a(x - h)^2$

Why not now consider graphs of

$$f(x) = ax^2 + bx + c,$$

where b and c are not both 0? In effect, we will do that, but in a disguised form. It turns out to be convenient to consider functions $f(x) = a(x - h)^2$, that is, where we start with ax^2 but then replace x by $x - h$, where h is some constant.*

EXAMPLE 2

Graph: $f(x) = (x - 3)^2$.

Solution We choose some values for x and compute $f(x)$. Then we plot the points and draw the curve. Comparing these values of x and $f(x)$ with those found in Example 1, we see that when an input here is 3 more than an input for Example 1, the outputs match.

x	$f(x) = (x - 3)^2$	
−1	16	
0	9	
1	4	
2	1	
3	0	⟵ Vertex
4	1	
5	4	
6	9	

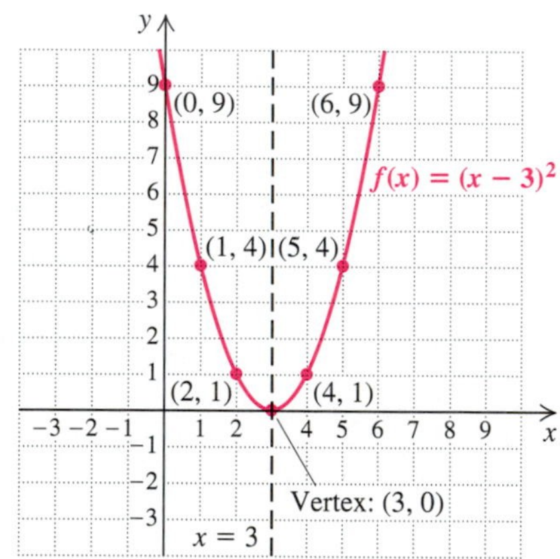

We note that $f(x) = 16$ when $x = -1$ and $f(x)$ gets larger and larger as x gets more and more negative. Thus we go back to positive values to fill out the table. Note that the line $x = 3$ is now the line of symmetry and the point (3, 0) is the vertex. Had we observed that $x = 3$ is the line of symmetry at the outset, then we could have computed some values on one side, such as (4, 1), (5, 4), and (6, 9), and then used symmetry to get their mirror images (2, 1), (1, 4), and (0, 9) without further computation. ❑

The graph of $f(x) = (x - 3)^2$ looks just like the graph of $f(x) = x^2$, except that it is moved, or translated, 3 units to the right.

*The letters h and k are often used to name functions, in which case the notation $h(x)$ and $k(x)$ is used. When h and k appear in expressions like $f(x) = a(x - h)^2 + k$, assume that they represent constants.

The graph of $g(x) = a(x - h)^2$ has the same shape as the graph of $y = ax^2$.
If h is positive, the graph of $y = ax^2$ is shifted h units to the right.
If h is negative, the graph of $y = ax^2$ is shifted $|h|$ units to the left.
The vertex is $(h, 0)$ and the line of symmetry is $x = h$.

EXAMPLE 3 Graph: $g(x) = -2(x + 3)^2$.

Solution We express the equation in the equivalent form $g(x) = -2[x - (-3)]^2$. Then we know that the graph looks like that of $y = 2x^2$ translated 3 units to the left, and it will also open downward since $-2 < 0$. The vertex is $(-3, 0)$, and the line of symmetry is $x = -3$. Plotting points as needed, we obtain the graph shown here.

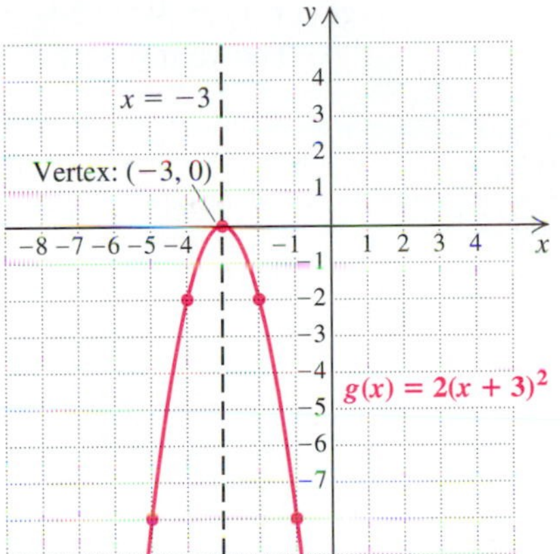

❑

Graphs of $f(x) = a(x - h)^2 + k$

Given a graph of $f(x) = a(x - h)^2$, what happens to it if we add a constant k? Suppose we add 2. This increases each function value $f(x)$ by 2, so the curve is moved up. If k should be -3, the curve is moved down. The vertex of the parabola will be at the point (h, k), and the line of symmetry will be $x = h$.

Note that if a parabola opens upward $(a > 0)$, the function value, or y-value, at the vertex is a least, or *minimum*, value. That is, it is less than the y-value at any other point on the graph. If the parabola opens downward $(a < 0)$, the function value at the vertex will be a greatest, or *maximum*, value.

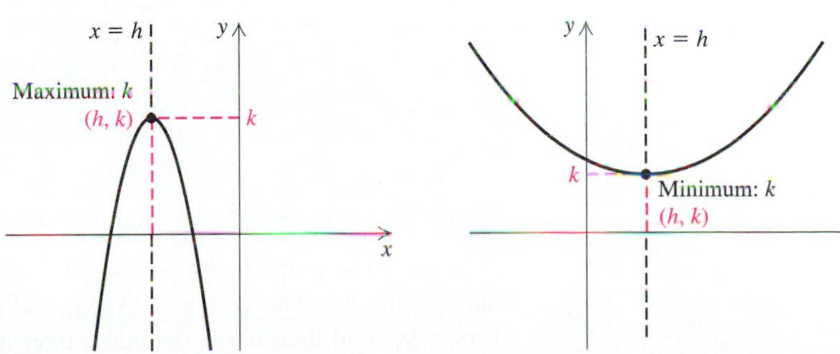

The graph of $f(x) = a(x - h)^2 + k$ looks like that of $y = a(x - h)^2$ translated up or down.

If k is positive, the graph of $y = a(x - h)^2$ is shifted k units up.

If k is negative, the graph of $y = a(x - h)^2$ is shifted $|k|$ units down.

The vertex is (h, k), and the line of symmetry is $x = h$.

For $a > 0$, k is the minimum function value. For $a < 0$, k is the maximum function value.

EXAMPLE 4

Graph $g(x) = (x - 3)^2 - 5$, and find the minimum function value.

Solution We know that the graph looks like that of $f(x) = (x - 3)^2$ (see Example 2) but moved 5 units down. You can confirm this by plotting some points. For instance, $g(4) = (4 - 3)^2 - 5 = -4$, whereas $f(4) = (4 - 3)^2 = 1$.

The vertex is now $(3, -5)$, and the minimum function value is -5.

TECHNOLOGY CONNECTION

To consider the effect of k on the graph of $f(x) = a(x - h)^2 + k$, graph $y_1 = 7.3(x - 1)^2$ in the window $[-5, 5, -5, 5]$. On the same set of axes, graph the function $y_2 = 7.3(x - 1)^2 + 2$ and compare the graphs. Next graph $y_3 = 7.3(x - 1)^2 - 4$ and compare it with the first two graphs. Try other values of k, including fractions and decimals. Find a rule that describes the effect of adding a value k to the function $f(x) = a(x - h)^2$.

x	$g(x) = (x - 3)^2 - 5$	
0	4	
1	−1	
2	−4	
3	−5	⟵ Vertex
4	−4	
5	−1	
6	4	

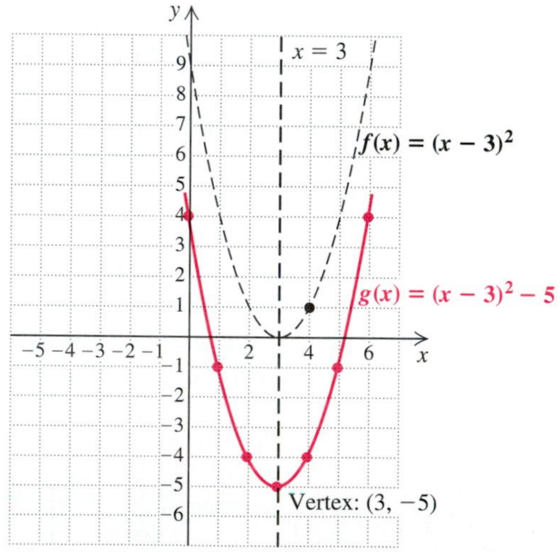

EXAMPLE 5

Graph $h(x) = \frac{1}{2}(x - 3)^2 + 5$, and find the minimum function value.

Solution The graph looks just like that of $f(x) = \frac{1}{2}x^2$ but moved 3 units to the right and 5 units up. The vertex is $(3, 5)$, and the line of symmetry is $x = 3$. We draw $f(x) = \frac{1}{2}x^2$ and then move the curve over and up. We plot a few points as a check. The minimum function value is 5.

x	$h(x) = \frac{1}{2}(x-3)^2 + 5$	
0	$9\frac{1}{2}$	
1	7	
3	5	← Vertex
5	7	
6	$9\frac{1}{2}$	

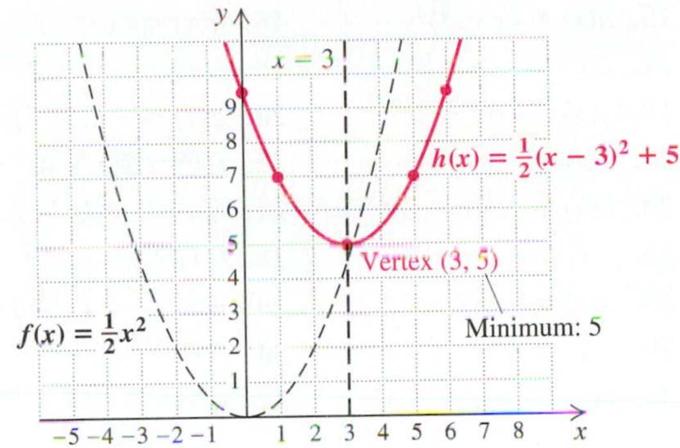

EXAMPLE 6

Graph $y = -2(x+3)^2 + 5$. Find the vertex, the line of symmetry, and the maximum or minimum value.

Solution We first express the equation in the equivalent form

$$y = -2[x - (-3)]^2 + 5.$$

The graph looks like that of $y = -2x^2$ translated 3 units to the left and 5 units up. The vertex is $(-3, 5)$, and the line of symmetry is $x = -3$. Since $-2 < 0$, we know that 5, the second coordinate of the vertex, is the maximum y-value.

We compute a few points as needed. The graph is shown here.

x	$y = -2(x+3)^2 + 5$	
−4	3	
−3	5	← Vertex
−2	3	

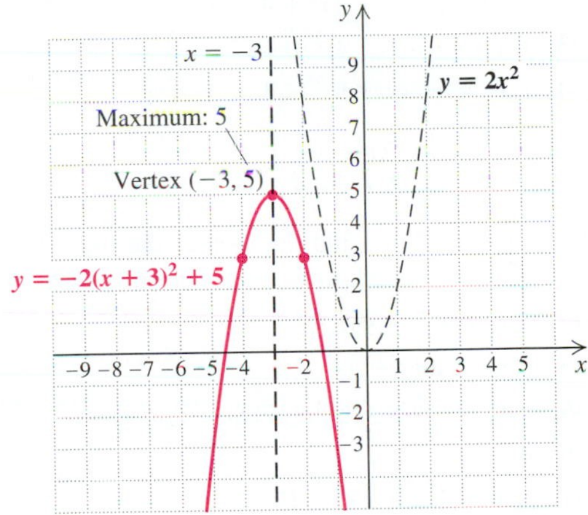

EXERCISE SET 11.7

Graph.

1. $f(x) = x^2$
2. $f(x) = -x^2$
3. $f(x) = -4x^2$
4. $f(x) = -3x^2$
5. $g(x) = \frac{1}{4}x^2$
6. $g(x) = \frac{1}{3}x^2$
7. $h(x) = -\frac{1}{3}x^2$
8. $h(x) = -\frac{1}{4}x^2$
9. $f(x) = \frac{3}{2}x^2$
10. $f(x) = \frac{5}{2}x^2$

For each of the following, graph the function, label the vertex, and draw the line of symmetry.

11. $g(x) = (x+1)^2$
12. $g(x) = (x+4)^2$
13. $f(x) = (x-4)^2$
14. $f(x) = (x-1)^2$

15. $h(x) = (x - 3)^2$
16. $h(x) = (x - 7)^2$
17. $f(x) = -(x + 4)^2$
18. $f(x) = -(x - 2)^2$
19. $g(x) = -(x - 1)^2$
20. $g(x) = -(x + 5)^2$
21. $f(x) = 2(x - 1)^2$
22. $f(x) = 2(x + 4)^2$
23. $h(x) = -\frac{1}{2}(x - 3)^2$
24. $h(x) = -\frac{3}{2}(x - 2)^2$
25. $f(x) = \frac{1}{2}(x + 1)^2$
26. $f(x) = \frac{1}{3}(x + 2)^2$
27. $g(x) = -3(x - 2)^2$
28. $g(x) = -4(x - 7)^2$
29. $f(x) = -2(x + 9)^2$
30. $f(x) = 2(x + 7)^2$
31. $h(x) = -3\left(x - \frac{1}{2}\right)^2$
32. $h(x) = -2\left(x + \frac{1}{2}\right)^2$

For each of the following, graph the function and find the vertex, the line of symmetry, and the maximum value or the minimum value.

33. $f(x) = (x - 3)^2 + 1$
34. $f(x) = (x + 2)^2 - 3$
35. $f(x) = (x + 1)^2 - 2$
36. $f(x) = (x - 1)^2 + 2$
37. $g(x) = (x + 4)^2 + 1$
38. $g(x) = -(x - 2)^2 - 4$
39. $f(x) = \frac{1}{2}(x - 5)^2 + 2$
40. $f(x) = \frac{1}{2}(x + 1)^2 - 2$
41. $h(x) = -2(x - 1)^2 - 3$
42. $h(x) = -2(x + 1)^2 + 4$
43. $f(x) = -3(x + 4)^2 + 1$
44. $f(x) = -2(x - 5)^2 - 3$
45. $g(x) = -\frac{3}{2}(x - 1)^2 + 2$
46. $g(x) = \frac{3}{2}(x + 2)^2 - 1$

Without graphing, find the vertex, the line of symmetry, and the maximum value or the minimum value.

47. $f(x) = 8(x - 9)^2 + 5$
48. $f(x) = 10(x + 5)^2 - 8$
49. $h(x) = -\frac{2}{7}(x + 6)^2 + 11$
50. $h(x) = -\frac{3}{11}(x - 7)^2 - 9$
51. $f(x) = 5\left(x + \frac{1}{4}\right)^2 - 13$
52. $f(x) = 6\left(x - \frac{1}{4}\right)^2 + 19$
53. $f(x) = -7(x - 10)^2 - 20$
54. $f(x) = -9(x + 12)^2 + 23$
55. $f(x) = \sqrt{2}(x + 4.58)^2 + 65\pi$
56. $f(x) = 4\pi(x - 38.2)^2 - \sqrt{34}$

Skill Maintenance

57. Solve the system

$$\begin{aligned} 500 &= 4a + 2b + c, \\ 300 &= a + b + c, \\ 0 &= c. \end{aligned}$$

58. Solve: $6x^2 - 13x + 2 = 0$.

Synthesis

59. ◈ Explain, without plotting points, why the graph of $y = (x + 2)^2$ looks like the graph of $y = x^2$ translated 2 units to the left.

60. ◈ Explain, without plotting points, why the graph of $y = x^2 - 4$ looks like the graph of $y = x^2$ translated 4 units down.

For each of the following, write the equation of the parabola that has the shape of $f(x) = 2x^2$ or $g(x) = -2x^2$ and has a maximum or minimum value at the specified point.

61. Maximum: (0, 4)
62. Minimum: (2, 0)
63. Minimum: (6, 0)
64. Maximum: (0, 3)
65. Maximum: (3, 8)
66. Minimum: (−2, 3)
67. Minimum: (−3, 6)

Write an equation of the parabola that satisfies the following conditions.

68. The parabola has a minimum value at the same point as $f(x) = 3(x - 4)^2$, but for all x, the function values are twice the values obtained from $f(x) = 3(x - 4)^2$.

69. The parabola is the same shape as

$$f(x) = -\tfrac{1}{2}(x - 2)^2 + 4$$

and has a maximum value at the same point as

$$g(x) = -2(x - 1)^2 - 6.$$

Functions other than parabolas can be translated. For a function $f(x)$, if we replace x by $x - h$, where h is a constant, the graph will be moved horizontally. If we add a constant k to a function $f(x)$, the graph will be moved vertically.

Use the graph of the function $y = g(x)$ below for Exercises 70–75.

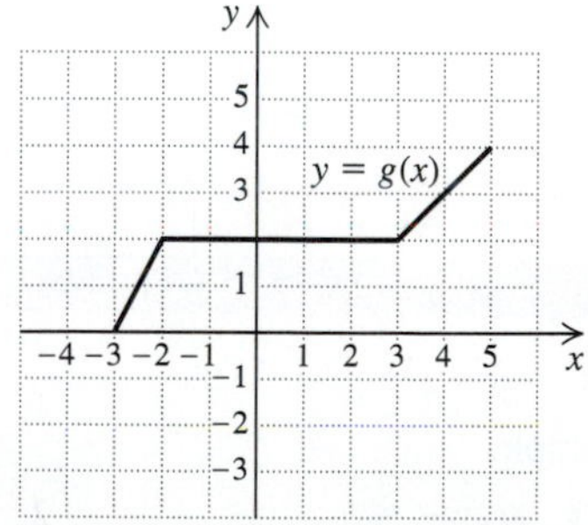

Draw a graph of each of the following.

70. $y = g(x + 2)$
71. $y = g(x - 3)$

72. $y - 3 = g(x)$

73. $y = g(x) + 4$

74. $y + 2 = g(x - 4)$

75. $y = g(x - 2) + 3$

76. Use a grapher to check your graphs for Exercises 7, 25, 39, and 45.

77. Use the Trace feature of a grapher to confirm the maximum and minimum values given as answers to Exercises 35, 43, 51, and 53. Be sure to adjust the window appropriately.

11.8 More About Graphing Quadratic Functions

Completing the Square • Finding x-Intercepts

Completing the Square

By *completing the square* (see Section 11.2), we can always rewrite the polynomial $ax^2 + bx + c$ in the form $a(x - h)^2 + k$. Thus the procedures discussed in Section 11.7 enable us to graph any quadratic function.

EXAMPLE 1

Graph: $g(x) = x^2 - 6x + 4$.

Solution We have

$$g(x) = x^2 - 6x + 4$$
$$= (x^2 - 6x) + 4.$$

To complete the square inside the parentheses, we take half the x-coefficient, $\frac{1}{2}\cdot(-6) = -3$, and square it to get $(-3)^2 = 9$. Then we add $9 - 9$ inside the parentheses:

$$g(x) = (x^2 - 6x + 9 - 9) + 4 \quad \text{The effect is of adding 0.}$$
$$= (x^2 - 6x + 9) + (-9 + 4) \quad \text{Using the associative law of addition to regroup}$$
$$= (x - 3)^2 - 5. \quad \text{Factoring and simplifying}$$

This equation was graphed in Example 4 of Section 11.7. The vertex is $(3, -5)$, and the line of symmetry is $x = 3$.

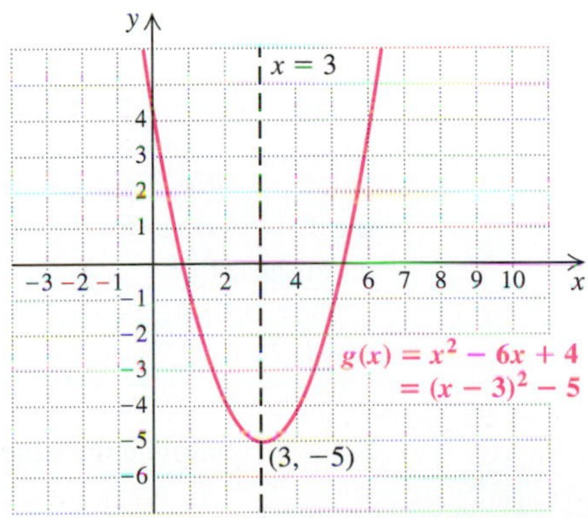

□

When the leading coefficient is not 1, we factor out that number from the first two terms. Then we complete the square.

EXAMPLE 2

Graph: $f(x) = 3x^2 + 12x + 13$.

Solution Since the coefficient of x^2 is not 1, we need to factor out that number—in this case, 3—from the first two terms. Remember that we want the form $f(x) = a(x - h)^2 + k$:

$$\begin{aligned} f(x) &= 3x^2 + 12x + 13 \\ &= 3(x^2 + 4x) + 13. \end{aligned}$$

Now we complete the square as before. We take half of the x-coefficient, $\frac{1}{2} \cdot 4 = 2$, and square it: $2^2 = 4$. Then we add $4 - 4$ inside the parentheses:

$$f(x) = 3(x^2 + 4x + 4 - 4) + 13.$$

This time we must distribute the 3 in order to rearrange terms:

$$\begin{aligned} f(x) &= 3(x^2 + 4x + 4) + 3(-4) + 13 && \text{Distributing to obtain a trinomial square} \\ &= 3(x + 2)^2 + 1. && \text{Factoring and simplifying} \end{aligned}$$

The vertex is $(-2, 1)$, and the line of symmetry is $x = -2$. The coefficient of x^2 is 3, so the graph is narrow and opens upward. We choose a few x-values on either side of the vertex, compute y-values, and then graph the parabola.

x	$f(x) = 3(x + 2)^2 + 1$	
-2	1	← Vertex
-3	4	
-1	4	

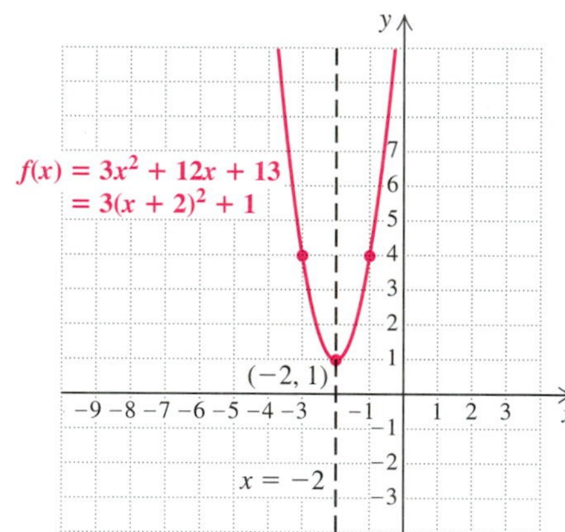

EXAMPLE 3

Graph: $f(x) = -2x^2 + 10x - 7$.

Solution We first find the vertex by completing the square. We factor out -2 from the first two terms of the expression. This makes the coefficient of x^2 inside the parentheses 1:

$$\begin{aligned} f(x) &= -2x^2 + 10x - 7 \\ &= -2(x^2 - 5x) - 7. \end{aligned}$$

Now we complete the square as before. We take half of the x-coefficient and square it to get $\frac{25}{4}$. Then we add $\frac{25}{4} - \frac{25}{4}$ inside the parentheses:

$$f(x) = -2\left(x^2 - 5x + \tfrac{25}{4} - \tfrac{25}{4}\right) - 7$$

$$= -2\left(x^2 - 5x + \tfrac{25}{4}\right) - 2\left(-\tfrac{25}{4}\right) - 7 \qquad \text{Multiplying by } -2 \text{, using the distributive law, and rearranging terms}$$

$$= -2\left(x - \tfrac{5}{2}\right)^2 + \tfrac{11}{2}. \qquad \text{Factoring and simplifying}$$

The vertex is $\left(\frac{5}{2}, \frac{11}{2}\right)$, and the line of symmetry is $x = \frac{5}{2}$. The coefficient of x^2, -2, is negative, so the graph opens downward. We plot a few points on either side of the vertex, including the y-intercept, $f(0)$, and graph the parabola.

x	$f(x)$	
$\frac{5}{2}$	$\frac{11}{2}$	← Vertex
0	-7	← y-intercept
1	1	
4	1	

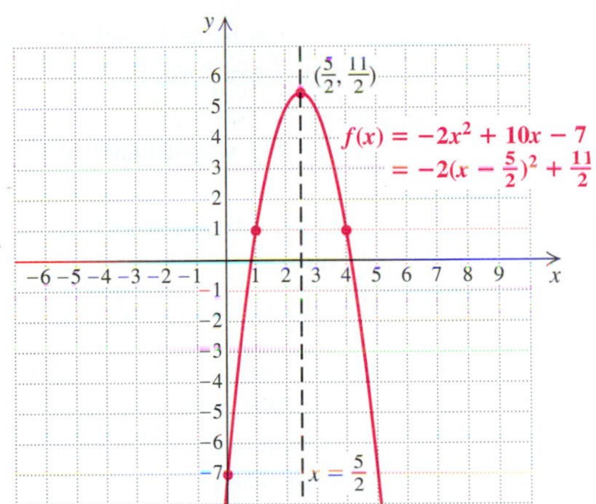

□

We can find a formula for computing the vertex. We do so by completing the square, as in Examples 1–3:

$$f(x) = ax^2 + bx + c$$

$$= a\left(x^2 + \frac{b}{a}x\right) + c. \qquad \text{Factoring } a \text{ out of the first two terms. Check by multiplying.}$$

Half of the x-coefficient, $\frac{b}{a}$, is $\frac{b}{2a}$. We square it to get $\frac{b^2}{4a^2}$ and add $\frac{b^2}{4a^2} - \frac{b^2}{4a^2}$ inside the parentheses. Then we multiply a back through, as follows, and factor:

$$f(x) = a\left(x^2 + \frac{b}{a}x + \frac{b^2}{4a^2} - \frac{b^2}{4a^2}\right) + c$$

$$= a\left(x^2 + \frac{b}{a}x + \frac{b^2}{4a^2}\right) + a\left(-\frac{b^2}{4a^2}\right) + c \qquad \text{Using the distributive law}$$

$$= a\left(x + \frac{b}{2a}\right)^2 + \frac{-b^2}{4a} + \frac{4ac}{4a} \qquad \text{Factoring and finding a common denominator}$$

$$= a\left[x - \left(-\frac{b}{2a}\right)\right]^2 + \frac{4ac - b^2}{4a}.$$

Thus we have the following.

For a parabola given by a quadratic function $f(x) = ax^2 + bx + c$, the vertex is

$$\left(-\frac{b}{2a}, \frac{4ac - b^2}{4a}\right).$$

The x-coordinate of the vertex is $-b/(2a)$. The line of symmetry is $x = -b/(2a)$. The second coordinate of the vertex can be found by substituting into the formula above, but is usually found most easily by evaluating the function at $-b/(2a)$.

Let us look back at Example 3 to see how we can find the vertex directly. From the above formula,

$$\text{the } x\text{-coordinate of the vertex is } -\frac{b}{2a} = -\frac{10}{2(-2)} = \frac{5}{2}.$$

Substituting $\frac{5}{2}$ into $f(x) = -2x^2 + 10x - 7$, we find the second coordinate of the vertex:

$$f\left(\tfrac{5}{2}\right) = -2\left(\tfrac{5}{2}\right)^2 + 10\left(\tfrac{5}{2}\right) - 7 = -2\left(\tfrac{25}{4}\right) + 25 - 7 = \tfrac{11}{2}.$$

The vertex is $\left(\frac{5}{2}, \frac{11}{2}\right)$. The line of symmetry is $x = \frac{5}{2}$.

We have actually developed two methods for finding the vertex. One is by completing the square and the other is by using a formula. You should consult with your instructor about which method to use.

Finding x-Intercepts

The points at which a graph crosses the x-axis are called ***x*-intercepts.** These are, of course, the points at which $y = 0$.

To find the x-intercepts of a quadratic function $f(x) = ax^2 + bx + c$, we solve the equation

$$0 = ax^2 + bx + c.$$

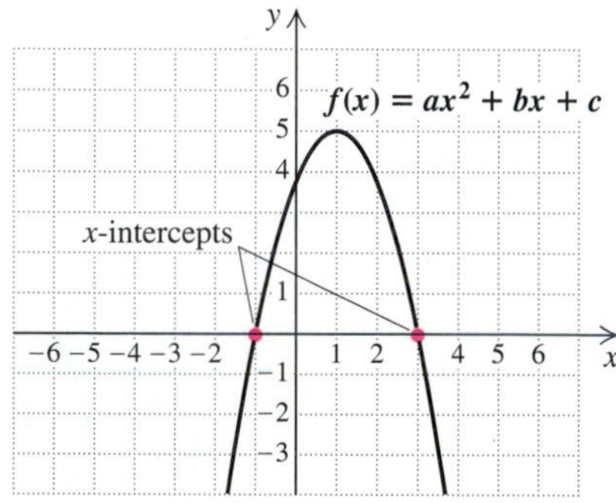

EXAMPLE 4

Find the x-intercepts of the graph of $f(x) = x^2 - 2x - 2$.

Solution We solve the equation

$$0 = x^2 - 2x - 2.$$

The equation is hard to factor, so we use the quadratic formula and get $x = 1 \pm \sqrt{3}$. Thus the x-intercepts are $(1 - \sqrt{3}, 0)$ and $(1 + \sqrt{3}, 0)$.

If graphing, we would approximate, to get $(-0.7, 0)$ and $(2.7, 0)$. □

The discriminant (see Section 11.5), $b^2 - 4ac$, tells us how many real-number solutions the equation $0 = ax^2 + bx + c$ has, so it also indicates how many x-intercepts there are. Compare the following graphs.

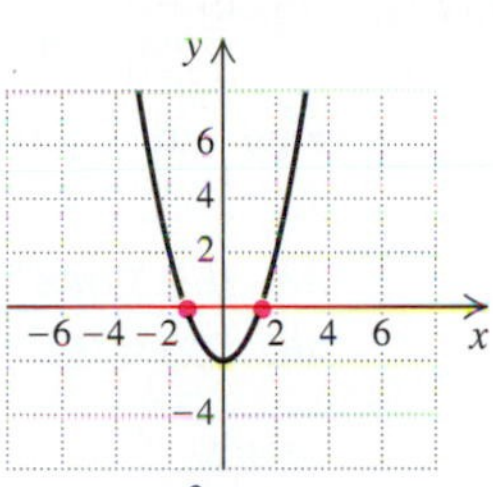

$y = ax^2 + bx + c$
$b^2 - 4ac > 0$
Two real solutions
Two x-intercepts

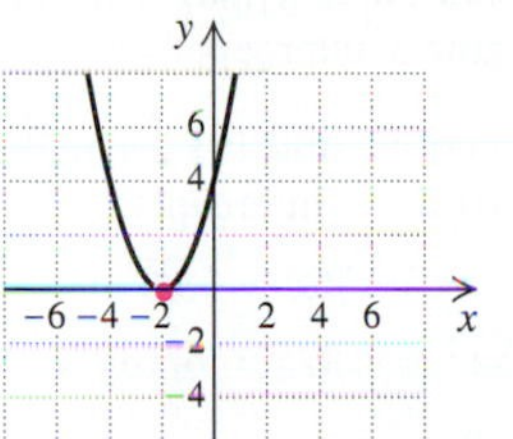

$y = ax^2 + bx + c$
$b^2 - 4ac = 0$
One real solution
One x-intercept

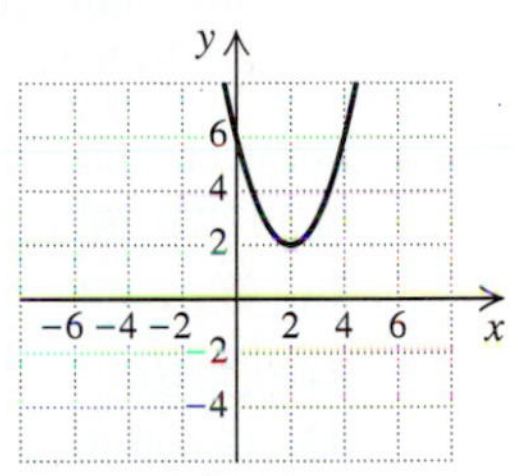

$y = ax^2 + bx + c$
$b^2 - 4ac < 0$
No real solutions
No x-intercepts

EXERCISE SET 11.8

For each quadratic function, **(a)** find the vertex and the line of symmetry and **(b)** graph the function.

1. $f(x) = x^2 - 2x - 3$
2. $f(x) = x^2 + 2x - 5$
3. $g(x) = x^2 + 6x + 13$
4. $g(x) = x^2 - 4x + 5$
5. $f(x) = x^2 + 4x - 1$
6. $f(x) = x^2 - 10x + 21$
7. $h(x) = 2x^2 + 16x + 25$
8. $h(x) = 2x^2 - 16x + 23$
9. $f(x) = -x^2 + 4x + 6$
10. $f(x) = -x^2 - 4x + 3$
11. $g(x) = x^2 + 3x - 10$
12. $g(x) = x^2 + 5x + 4$
13. $f(x) = 3x^2 - 24x + 50$
14. $f(x) = 4x^2 + 8x - 3$
15. $h(x) = x^2 - 9x$
16. $h(x) = x^2 + x$
17. $f(x) = -2x^2 - 4x - 6$
18. $f(x) = -3x^2 + 6x + 2$
19. $g(x) = 2x^2 - 10x + 14$
20. $g(x) = 2x^2 + 6x + 8$
21. $f(x) = -3x^2 - 3x + 1$
22. $f(x) = -2x^2 + 2x + 1$
23. $h(x) = \frac{1}{2}x^2 + 4x + \frac{19}{3}$
24. $h(x) = \frac{1}{2}x^2 - 3x + 2$

Find the x-intercepts. If no x-intercepts exist, state so.

25. $f(x) = x^2 - 4x + 1$
26. $f(x) = x^2 + 6x + 10$
27. $g(x) = -x^2 + 2x + 3$
28. $g(x) = x^2 - 2x - 5$
29. $f(x) = x^2 - 3x - 4$
30. $f(x) = x^2 - 8x + 5$
31. $h(x) = -x^2 + 3x - 2$
32. $h(x) = 2x^2 - 4x + 6$
33. $f(x) = 2x^2 + 4x - 1$
34. $f(x) = x^2 - x + 2$
35. $g(x) = x^2 - x + 1$
36. $g(x) = 4x^2 + 12x + 9$

Skill Maintenance

Solve.

37. $\sqrt{4x - 4} = \sqrt{x + 4} + 1$
38. $\sqrt{5x - 4} + \sqrt{13 - x} = 7$

Synthesis

39. Suppose that the function $f(x) = ax^2 + bx + c$ has x_1 and x_2 as x-intercepts. Explain why the function $g(x) = -ax^2 - bx - c$ will also have x_1 and x_2 as x-intercepts.

40. Compare the graphs of $f(x) = a(x - h)^2 + k$ and $g(x) = -a(x - h)^2 + k$. What requirements, if any, must be placed on a, h, and k if both graphs are to have the same x-intercepts?

For each quadratic function, find **(a)** the maximum or minimum value and **(b)** the x-intercepts.

41. $f(x) = 2.31x^2 - 3.135x - 5.89$

42. $f(x) = -18.8x^2 + 7.92x + 6.18$

43. Graph the function

$$f(x) = x^2 - x - 6.$$

Then use the graph to approximate solutions to the following equations.

a) $x^2 - x - 6 = 2$
b) $x^2 - x - 6 = -3$

44. Graph the function

$$f(x) = \frac{x^2}{8} + \frac{x}{4} - \frac{3}{8}.$$

Then use the graph to approximate solutions to the following equations.

a) $\frac{x^2}{8} + \frac{x}{4} - \frac{3}{8} = 0$

b) $\frac{x^2}{8} + \frac{x}{4} - \frac{3}{8} = 1$

c) $\frac{x^2}{8} + \frac{x}{4} - \frac{3}{8} = 2$

Find an equivalent equation of the type $f(x) = a(x - h)^2 + k$.

45. $f(x) = mx^2 - nx + p$

46. $f(x) = 3x^2 + mx + m^2$

Graph.

47. $f(x) = |x^2 - 1|$

48. $f(x) = |3 - 2x - x^2|$

49. Use a grapher to check your solutions to Exercises 9, 23, 33, 41, 42, 43, and 44.

11.9 Problem Solving and Quadratic Functions

Maximum and Minimum Problems • Fitting Quadratic Functions to Data • Solving Formulas

Let's look now at some of the many situations in which quadratic functions are used for problem solving.

Maximum and Minimum Problems

For a quadratic function, the value $f(x)$ at the vertex will be either greater than any other $f(x)$ or less than any other $f(x)$. If the graph opens upward, f will achieve a minimum. If the graph opens downward, f will achieve a maximum. In certain problems, we want to find a maximum or a minimum. If the problem situation translates to a quadratic function, we can often solve by finding x or $f(x)$ at the vertex.

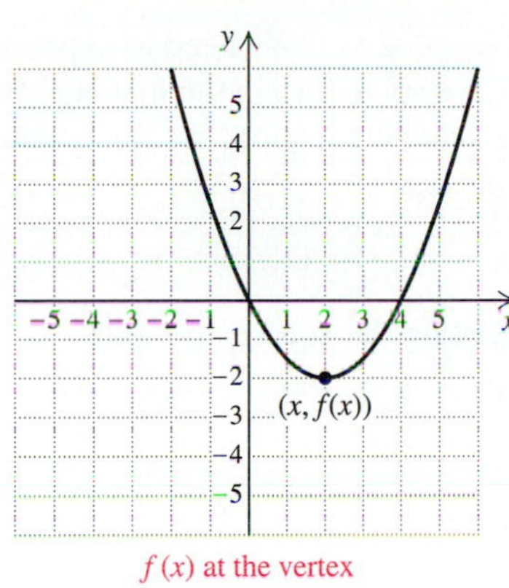

$f(x)$ at the vertex
a minimum

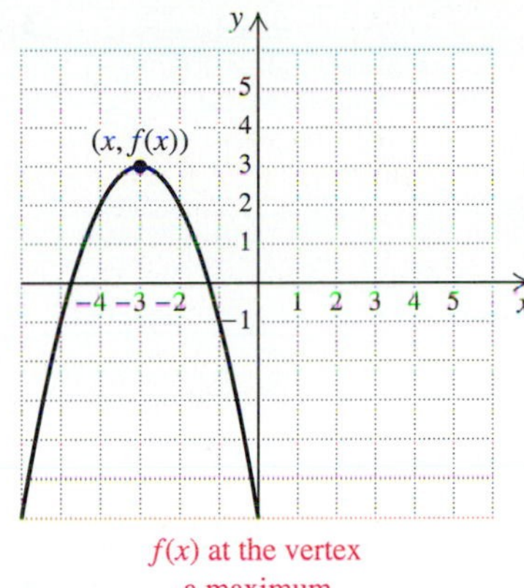

$f(x)$ at the vertex
a maximum

EXAMPLE 1

What are the dimensions of the largest rectangular pen that a farmer can enclose with 64 m of fence?

Solution

1. FAMILIARIZE. We make a drawing and label it.

Perimeter: $2w + 2l = 64$ m

Area: $A = l \cdot w$

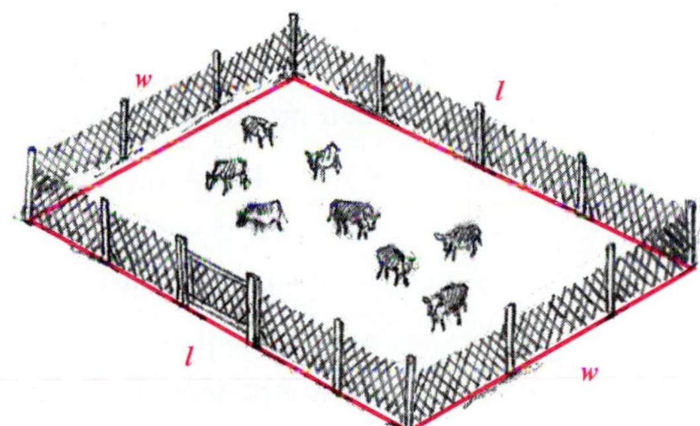

To get a better feel for the problem, we can look at some possible dimensions for a rectangular pen that can be enclosed with 64 m of fence.

l	w	A
22	10	220
20	12	240
18	14	252

2. TRANSLATE. We have two equations, one of which expresses area as a function of length and width:

$$2w + 2l = 64$$

$$A = l \cdot w.$$

3. **CARRY OUT.** We need to express A as a function of l or w but not both. To do so, we solve for l in the first equation to obtain $l = 32 - w$. Substituting for l in the second equation, we get a quadratic function, $A(w)$, or just A:

$$A = (32 - w)w \qquad \text{Substituting for } l$$
$$A = -w^2 + 32w. \qquad \text{This is a parabola opening down, so a maximum exists.}$$

Completing the square, we get

$$A = -(w^2 - 32w + 256 - 256) = -(w - 16)^2 + 256.$$

The maximum function value of 256 occurs when $w = 16$ and $l = 32 - 16$, or 16.

4. **CHECK.** Note that 256 is greater than any of the values for A found in the *Familiarize* step. To be more certain, we could check values other than those used in the *Familiarize* step. For example, if $w = 15$, then $l = 32 - 15 = 17$, and $A = 15 \cdot 17 = 255$. The same area results if $w = 17$ and $l = 15$. Since 256 is greater than 255, it looks as though we have a maximum.

5. **STATE.** For the pen to have maximal area, it should be 16 m by 16 m. ❑

EXAMPLE 2

What is the minimum product of two numbers whose difference is 5? What are the numbers?

Solution

1. **FAMILIARIZE.** We try some pairs of numbers that differ by 5 and compute their products:

$$1 \cdot 6 = 6,$$
$$0 \cdot 5 = 0,$$
$$(-1) \cdot 4 = -4.$$

We suspect that one of the two numbers will be negative and the other positive. Let x represent the larger number and $x - 5$ the other number.

2. **TRANSLATE.** We represent the product of the two numbers by

$$p = x(x - 5), \quad \text{or} \quad p(x) = x^2 - 5x.$$

3. **CARRY OUT.** The function $p(x) = x^2 - 5x$ represents a parabola opening up. Completing the square, we get

$$p(x) = x^2 - 5x + \frac{25}{4} - \frac{25}{4} = \left(x - \frac{5}{2}\right)^2 - \frac{25}{4}.$$

The minimum function value of $-\frac{25}{4}$ occurs when one number is $\frac{5}{2}$ and the other number is $\frac{5}{2} - 5 = -\frac{5}{2}$.

4. **CHECK.** Note that if $x = 2$, then $x - 5 = -3$ and $2(-3) = -6$. Also note that if $x = 3$, then $x - 5 = -2$ and $3(-2) = -6$. Thus, since $-\frac{25}{4} < -6$, it appears that we have a minimum.

5. **STATE.** The numbers $-\frac{5}{2}$ and $\frac{5}{2}$ differ by 5 and yield a minimum product of $-\frac{25}{4}$. ❑

Fitting Quadratic Functions to Data

Whenever we know that a quadratic function fits a certain situation, we can find that function if we know three inputs and their outputs. Each such ordered pair is called a *data point*.

The following example provides a good illustration.

EXAMPLE 3

The instruction booklet for a video cassette recorder (VCR) includes a table relating the time a tape has run to the counter reading.

Counter Reading	Time Tape Has Run (in Hours)
000	0
400	1
700	3
800	4

Find a quadratic function that fits the data. Then predict how long a tape has run when the counter reading is 500.

Solution

1. FAMILIARIZE. The statement of the problem leads us to look for a function of the form

$$T(n) = an^2 + bn + c,$$

where $T(n)$ represents the time, in hours, that the tape has run at counter reading n hundred. We need to determine values of a, b, and c that will fit the given data.

2. TRANSLATE. We substitute some values of n and T:

$$0 = a \cdot 0^2 + b \cdot 0 + c,$$
$$1 = a \cdot 4^2 + b \cdot 4 + c,$$
$$3 = a \cdot 7^2 + b \cdot 7 + c.$$

After simplifying, we see that we need to solve the system

$$0 = c,$$
$$1 = 16a + 4b + c,$$
$$3 = 49a + 7b + c.$$

3. CARRY OUT. Since $c = 0$, it suffices to solve the system

$$1 = 16a + 4b, \qquad \textbf{(1)}$$
$$3 = 49a + 7b. \qquad \textbf{(2)}$$

Multiplying the first equation by -7 and the second equation by 4, we obtain

$$-7 = -112a - 28b,$$
$$12 = 196a + 28b.$$

Thus,

$$5 = 84a \quad \text{Adding equations}$$
$$\tfrac{5}{84} = a.$$

We substitute into one of the original equations to solve for b:

$$1 = 16 \cdot \tfrac{5}{84} + 4b \quad \text{Substituting in Equation (1)}$$
$$1 = \tfrac{80}{84} + 4b$$
$$\tfrac{4}{84} = 4b$$
$$\tfrac{1}{84} = b.$$

Thus the function $T(n) = \frac{5}{84}n^2 + \frac{1}{84}n$ fits the given data. To find how long a tape has been running when the counter reading is 500, we find $T(5)$:

$$T(5) = \tfrac{5}{84} \cdot 5^2 + \tfrac{1}{84} \cdot 5$$
$$= \tfrac{125}{84} + \tfrac{5}{84}$$
$$= \tfrac{130}{84} \approx 1.55 \text{ hr.}$$

4. **CHECK.** Besides rechecking our calculations, we can check to see if the function we obtained will produce other values found in the table. As you should confirm,

$$T(8) \approx 3.9 \text{ hr.}$$

This last value should not bother us since it is fair to assume that the manufacturer of the VCR may have rounded off when making the table or that our model of the situation overlooked some details.

5. **STATE.** The answer is that when the counter reads 500, the tape has been running for just over $1\frac{1}{2}$ hours. ❑

Solving Formulas

Recall that to solve a formula for a certain letter, we use the principles for solving equations to get that letter alone on one side. When square roots appear, we can usually eliminate the radical signs by squaring both sides.

EXAMPLE 4

Period of a pendulum. The time T required for a pendulum of length l to swing back and forth (complete one period) is given by the formula $T = 2\pi\sqrt{l/g}$, where g is the gravitational constant. Solve for l.

Solution We have

$$T = 2\pi \sqrt{\frac{l}{g}}$$
$$T^2 = \left(2\pi \sqrt{\frac{l}{g}}\right)^2 \quad \text{Principle of powers (squaring)}$$
$$T^2 = 2^2\pi^2\frac{l}{g}$$
$$gT^2 = 4\pi^2 l \quad \text{Clearing fractions}$$
$$\frac{gT^2}{4\pi^2} = l. \quad \text{Multiplying by } \frac{1}{4\pi^2}$$

We now have l alone on one side and l does not appear on the other side, so the formula is solved for l. □

In most formulas, the letters represent nonnegative numbers, so we do not need to use absolute-value signs when taking square roots.

EXAMPLE 5

Hang time.* A formula relating an athlete's vertical leap V, in inches, to hang time T, in seconds, is $V = 48T^2$. Solve for T.

Solution

$$48T^2 = V$$

$$T^2 = \frac{V}{48} \qquad \text{Multiplying by } \frac{1}{48} \text{ to get } T^2 \text{ alone}$$

$$T = \sqrt{\frac{V}{48}} \qquad \text{Taking the square root}$$

$$T = \sqrt{\frac{V}{16 \cdot 3} \cdot \frac{3}{3}} \qquad \text{Factoring and multiplying by 1 to rationalize the denominator}$$

$$T = \sqrt{\frac{3V}{144}}$$

$$T = \frac{\sqrt{3V}}{12}$$

□

EXAMPLE 6

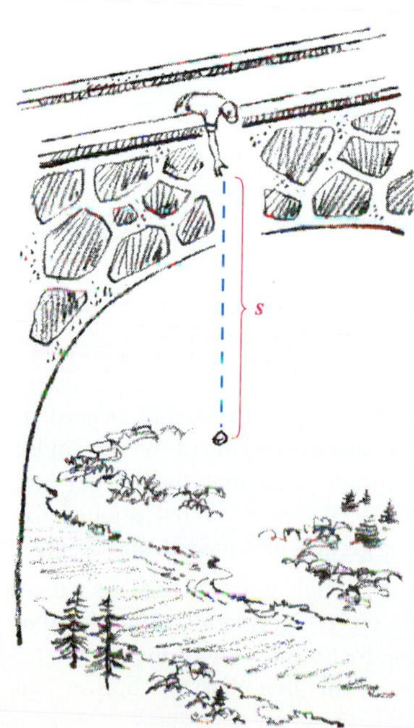

Downward speed. An object tossed downward with an initial speed (velocity) of v_0 will travel a distance of s meters, where $s = 4.9t^2 + v_0t$ and t is measured in seconds. Solve for t.

Solution Since t is squared in one term and raised to the first power in the other term, the equation is quadratic in t. We find standard form and use the quadratic formula:

$$4.9t^2 + v_0t - s = 0 \qquad \text{Writing standard form}$$

$$a = 4.9, \quad b = v_0, \quad c = -s$$

$$t = \frac{-v_0 \pm \sqrt{v_0^2 - 4(4.9)(-s)}}{2(4.9)}.$$

Since the negative square root would yield a negative value for t, we use only the positive root:

$$t = \frac{-v_0 + \sqrt{v_0^2 + 19.6s}}{9.8}.$$

□

*This formula is taken from an article by Peter Brancazio, "The Mechanics of a Slam Dunk," *Popular Mechanics,* November 1991. Courtesy of Professor Peter Brancazio, Brooklyn College.

The following list of steps should help you when solving formulas for a given letter. Try to remember that when solving a formula, you do the same things you would do to solve any equation.

To solve a formula for a letter, say, b:

1. Clear the fractions and use the principle of powers, as needed, until b does not appear in any radicand or denominator. (In some cases, you may clear the fractions first, and in some cases you may use the principle of powers first. Perform these steps until radicals containing b are gone and b is not in any denominator.)
2. Collect all terms with b^2 in them. Also collect all terms with b in them.
3. If b^2 does not appear, you can finish by using just the addition and multiplication principles. Get all terms containing b on one side of the equation; then factor out b. Dividing on both sides will then get b alone.
4. If b^2 appears but b does not appear to the first power, solve the equation for b^2. Then take square roots on both sides.
5. If there are terms containing both b and b^2, put the equation in standard form and use the quadratic formula.

EXERCISE SET 11.9

Solve.

1. A rancher is fencing off a rectangular field with a perimeter of 76 ft. What dimensions will yield the maximum area? What is the maximum area?
2. A carpenter is building a rectangular room with a perimeter of 68 ft. What dimensions will yield the maximum area? What is the maximum area?
3. What is the maximum product of two numbers whose sum is 16? What numbers yield this product?
4. What is the maximum product of two numbers whose sum is 28? What numbers yield this product?
5. What is the maximum product of two numbers whose sum is 22? What numbers yield this product?
6. What is the maximum product of two numbers whose sum is 45? What numbers yield this product?
7. What is the minimum product of two numbers whose difference is 4? What are the numbers?
8. What is the minimum product of two numbers whose difference is 10? What are the numbers?
9. What is the minimum product of two numbers whose difference is 9? What are the numbers?
10. What is the minimum product of two numbers whose difference is 7? What are the numbers?
11. What is the maximum product of two numbers whose sum is -7? What numbers yield this product?
12. What is the maximum product of two numbers whose sum is -9? What numbers yield this product?
13. A farmer decides to enclose a rectangular garden, using the side of a barn as one side of the rectangle. What is the maximum area that the farmer can enclose with 40 ft of fence? What should the dimensions of the garden be to yield this area?
14. A stone mason has enough stones to enclose a rectangular patio with 60 ft of perimeter, assuming that the attached house forms one side of the rectangle. What is the maximum area that the mason can

enclose? What should the dimensions of the patio be to yield this area?

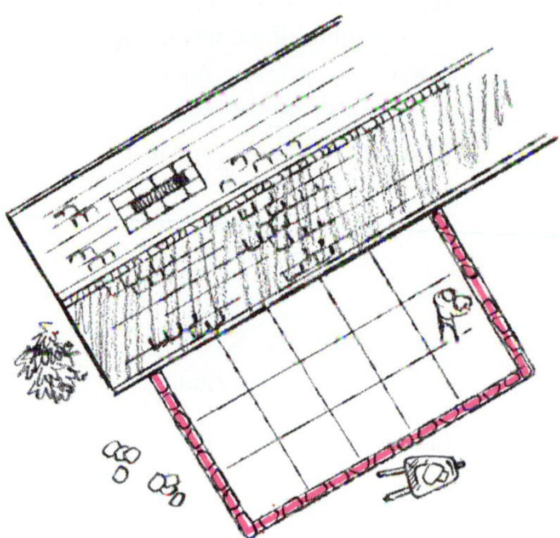

15. A rectangular compost container is to be formed in a corner of a fenced yard, with 8 ft of chicken wire completing the other two sides of the rectangle. If the chicken wire is 3 ft high, what dimensions of the base will maximize the container's volume?

16. A plastics manufacturer plans to produce a one-compartment vertical file by bending the long side of an 8 in. by 14 in. sheet of plastic along two lines to form a U shape. How tall should the file be to maximize the volume that the file can hold?

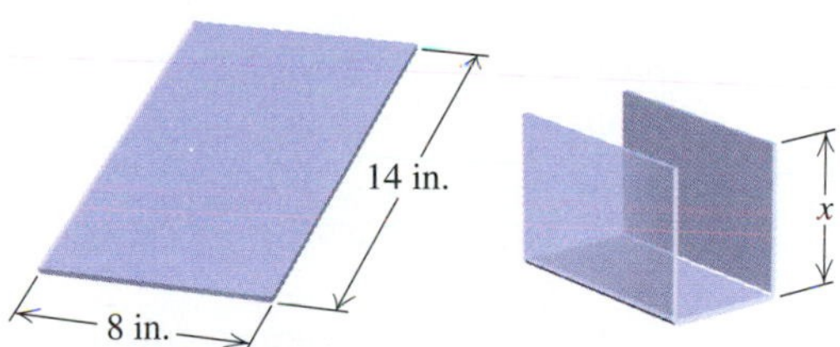

Find the quadratic function that fits the set of data points.

17. (1, 4), (−1, −2), (2, 13)

18. (1, 4), (−1, 6), (−2, 16)

19. (2, 0), (4, 3), (12, −5)

20. (−3, −30), (3, 0), (6, 6)

21. A business earns \$38 in the first week, \$66 in the second week, and \$86 in the third week. The manager graphs the points (1, 38), (2, 66), and (3, 86) and uses a quadratic function to describe the situation.

   a) Find a quadratic function that fits the data.

   b) Using the function, predict the earnings for the fourth week.

22. A business earns \$1000 in its first month, \$2000 in the second month, and \$8000 in the third month. The manager plots the points (1, 1000), (2, 2000), and (3, 8000) and uses a quadratic function to describe the situation.

   a) Find a quadratic function that fits the data.

   b) Using the function, predict the earnings for the fourth month.

23. a) Find a quadratic function that fits the following data.

Travel Speed (in Kilometers per Hour)	Number of Daytime Accidents (for Every 200 Million Kilometers)
60	100
80	130
100	200

   b) Use the function to calculate the number of daytime accidents that occur at 50 km/h.

24. a) Find a quadratic function that fits the following data.

Travel Speed (in Kilometers per Hour)	Number of Nighttime Accidents (for Every 200 Million Kilometers)
60	400
80	250
100	250

   b) Use the function to calculate the number of nighttime accidents that occur at 50 km/h.

25. Pizza Unlimited has the following prices for pizzas.

Diameter	Price
8 in.	\$ 6.00
12 in.	\$ 8.50
16 in.	\$11.50

Is price a quadratic function of diameter? It probably should be, because the price should be proportional to the area, and the area is a quadratic function of the diameter. (The area of a circular region is given by $A = \pi r^2$ or $(\pi/4) \cdot d^2$.)

a) Express price as a quadratic function of diameter using the data points (8, 6), (12, 8.50), and (16, 11.50).

b) Use the function to find the price of a 14-in. pizza.

Recall that total profit P is the difference between total revenue R and total cost C. Given the following total-revenue and total-cost functions, find the total profit, the maximum value of the total profit, and the value of x at which it occurs.

26. $R(x) = 1000x - x^2$,
$C(x) = 3000 + 20x$

27. $R(x) = 200x - x^2$,
$C(x) = 5000 + 8x$

28. $R(x) = 300x - x^2$,
$C(x) = 50 + 80x$

Solve the formula for the indicated letter. Assume that all variables represent nonnegative numbers.

29. $A = 6s^2$, for s
(Surface area of a cube)

30. $A = 4\pi r^2$, for r
(Surface area of a sphere)

31. $F = \dfrac{Gm_1m_2}{r^2}$, for r
(Law of gravity)

32. $N = \dfrac{kQ_1Q_2}{s^2}$, for s
(Number of phone calls between two cities)

33. $E = mc^2$, for c
(Energy–mass relationship)

34. $A = \pi r^2$, for r
(Area of a circle)

35. $a^2 + b^2 = c^2$, for b
(Pythagorean formula in two dimensions)

36. $a^2 + b^2 + c^2 = d^2$, for c
(Pythagorean formula in three dimensions)

37. $N = \dfrac{k^2 - 3k}{2}$, for k
(Number of diagonals of a polygon)

38. $s = v_0t + \dfrac{gt^2}{2}$, for t
(A motion formula)

39. $A = 2\pi r^2 + 2\pi rh$, for r
(Surface area of a right cylindrical solid)

40. $A = \pi r^2 + \pi rs$, for r
(Surface area of a cone)

41. $N = \frac{1}{2}(n^2 - n)$, for n
(Number of games in a league of n teams)

42. $A = A_0(1 - r)^2$, for r
(A business formula)

43. $A = 2w^2 + 4lw$, for w
(Surface area of a rectangular solid)

44. $A = 4\pi r^2 + 2\pi rh$, for r
(Surface area of a capsule)

45. $T = 2\pi\sqrt{\dfrac{l}{g}}$, for g
(A pendulum formula)

46. $W = \sqrt{\dfrac{1}{LC}}$, for L
(An electricity formula)

47. $A = P_1(1 + r)^2 + P_2(1 + r)$, for r
(An investment formula)

48. $A = P_1\left(1 + \dfrac{r}{2}\right)^2 + P_2\left(1 + \dfrac{r}{2}\right)$, for r
(An investment formula)

49. $m = \dfrac{m_0}{\sqrt{1 - \dfrac{v^2}{c^2}}}$, for v
(A relativity formula)

50. Solve the formula given in Exercise 49 for c.

Solve. Examine Exercises 29–50 and Examples 4–6 for the appropriate formula.

51. a) An object is dropped 75 m from an airplane. How long does it take the object to reach the ground?

b) An object is thrown downward 75 m from the plane at an initial velocity of 30 m/sec. How long does it take the object to reach the ground?

c) How far will an object fall in 2 sec, if thrown downward at an initial velocity of 30 m/sec?

52. a) An object is dropped 500 m from an airplane. How long does it take the object to reach the ground?

b) An object is thrown downward 500 m from the plane at an initial velocity of 30 m/sec. How long does it take the object to reach the ground?

c) How far will an object fall in 5 sec, if thrown downward at an initial velocity of 30 m/sec?

53. Michael Jordan of the Chicago Bulls has a vertical leap of about 38 in. What is his hang time?

54. The surface area of a right cylindrical solid of height 3 m is $8\pi\ m^2$. Find the radius of the solid.

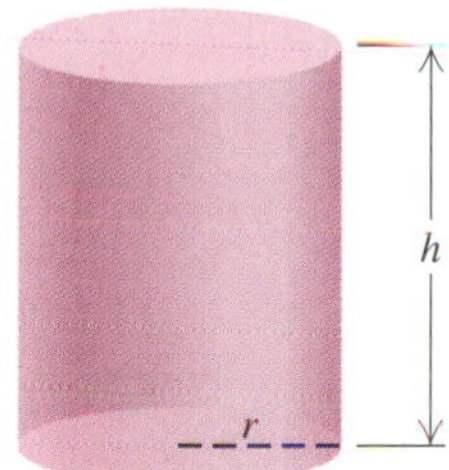

55. In a softball league, each team plays each of the other teams once. If the league plays a total of 91 games, how many teams are in the league?

56. In a volleyball league, each team plays each of the other teams once. If a total of 66 games is played, how many teams are in the league?

57. Jesse is tied to one end of a 40-m elasticized (bungee) cord. The other end of the cord is tied to the middle of a train trestle. If Jesse jumps off the bridge, for how long will he be falling before the cord begins to stretch?

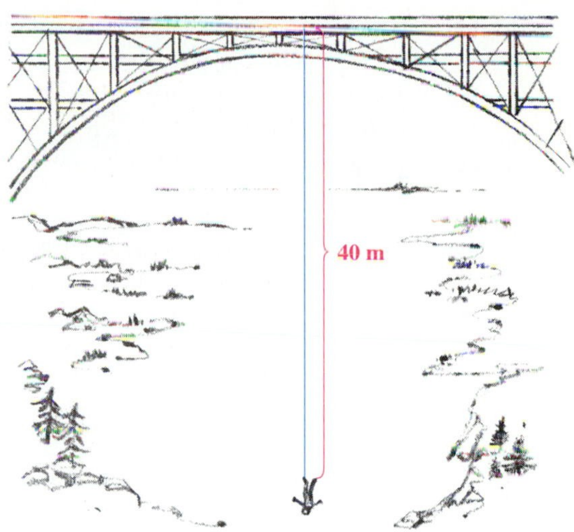

58. Sheila is tied to a bungee cord (see Exercise 57) and falls for 2.5 sec before her cord begins to stretch. How long is the bungee cord?

59. An object thrown downward from a 100-m cliff travels 51.6 m in 3 sec. What was the initial velocity of the object?

60. An object thrown downward from a 200-m cliff travels 91.2 m in 4 sec. What was the initial velocity of the object?

61. A firm invests $3000 in a savings account for two years. At the beginning of the second year, an additional $1700 is invested. If a total of $5253.70 is in the account at the end of the second year, what is the annual interest rate? (*Hint:* See Exercise 47.)

62. A business invests $10,000 in a savings account for two years. At the beginning of the second year, an additional $3500 is invested. If a total of $15,569.75 is in the account at the end of the second year, what is the annual interest rate? (*Hint:* See Exercise 47.)

Skill Maintenance

Solve.

63. $3x + 5 < 5x - 8$

64. $\frac{1}{2}(x - 2) \geq \frac{3}{4}(x + 1)$

Synthesis

65. The sum of the base and the height of a triangle is 38 cm. Find the dimensions for which the area is a maximum, and find the maximum area.

66. The perimeter of a rectangle is 44 ft. Find the least possible length of a diagonal.

67. Solve for n:

$$mn^4 - r^2pm^3 - r^2n^2 + p = 0.$$

68. Solve for t:

$$rt^2 - rt - st^2 + s^2r - st = 0.$$

69. When a theater owner charges \$2 for admission, she averages 100 people attending. For each 10¢ increase in admission price, the average number attending decreases by 1. What should the owner charge in order to make the most money?

70. An orange grower finds that she gets an average yield of 40 bushels (bu) per tree when she plants 20 trees on an acre of ground. Each time she adds a tree to an acre, the yield per tree decreases by 1 bu, due to congestion. How many trees per acre should she plant for maximum yield?

71. The height above the ground of a launched object is a quadratic function of the time that it is in the air. Suppose that a flare is launched from a cliff 64 ft above sea level. If 3 sec after being launched the flare is again level with the cliff, and if 2 sec after that it lands in the sea, what is the maximum height that the flare will reach?

72. The cables supporting a straight-line suspension bridge are parabolic in shape. Suppose that a suspension bridge is being designed to cross a river that is 160 ft wide and that the vertical cables are 30 ft above road level at the bridge's midpoint and are 80 ft above road level at a point 50 ft from the bridge's midpoint. How long are the longest vertical cables?

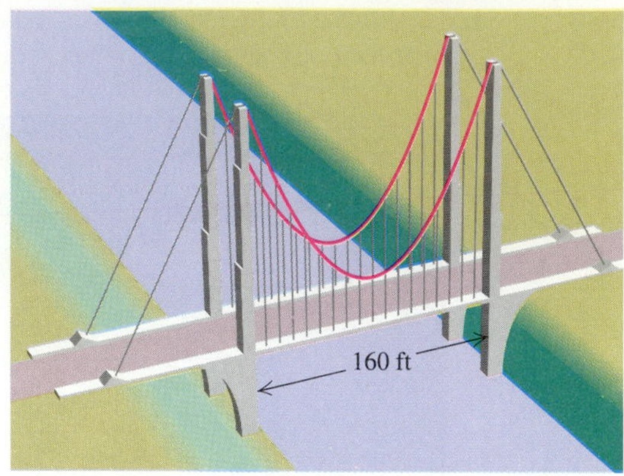

73. Find a formula that expresses the diameter of a right cylindrical solid as a function of its surface area.

74. Find a formula that expresses the length of a cube's three-dimensional diagonal as a function of the cube's volume.

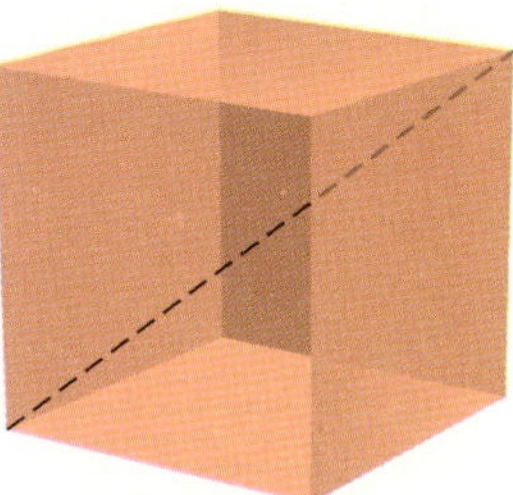

75. Find a formula that expresses the length of a cube's three-dimensional diagonal as a function of the cube's surface area.

11.10 Polynomial and Rational Inequalities

Quadratic and Other Polynomial Inequalities • Rational Inequalities

Quadratic and Other Polynomial Inequalities

Inequalities such as the following are called *polynomial inequalities*:

$$x^3 - 4x^2 + x - 6 > 0, \quad x^2 + 3x - 10 < 0, \quad 5x^2 - 3x + 2 \geqslant 0, \quad 3x - 1 \leqslant 0.$$

Second-degree polynomial inequalities in one variable are called *quadratic inequal-*

ities. We now consider three ways to solve polynomial inequalities. The first two provide understanding, and the last method is the fastest.

The first method for solving polynomial inequalities is to consider the graph of the related function in the plane.

EXAMPLE 1

Solve: $x^2 + 3x - 10 > 0$.

Solution Consider the function $f(x) = x^2 + 3x - 10$ and its graph. Its graph opens upward since the leading coefficient is positive.

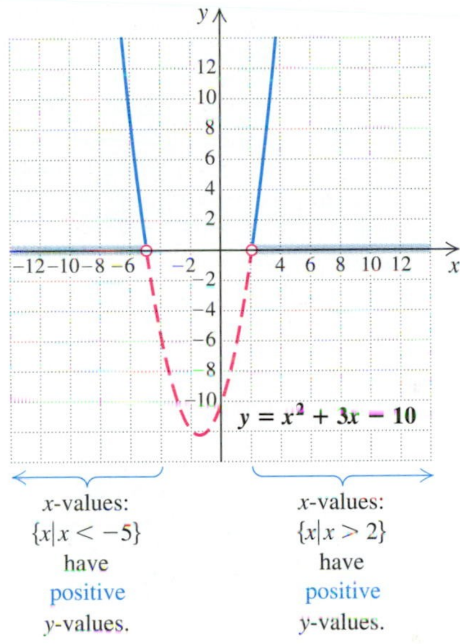

Values of y will be positive to the left and right of the x-intercepts, as shown. We find the intercepts by setting the polynomial equal to 0 and solving:

$$x^2 + 3x - 10 = 0$$
$$(x + 5)(x - 2) = 0$$
$$x + 5 = 0 \quad \text{or} \quad x - 2 = 0$$
$$x = -5 \quad \text{or} \quad x = 2.$$

Thus the solution set of the inequality is

$$\{x \mid x < -5 \text{ or } x > 2\}, \quad \text{or} \quad (-\infty, -5) \cup (2, \infty).$$

We can solve any inequality by considering a graph of the related function and finding intercepts as in Example 1. In some cases, we may need to use the quadratic formula to find the intercepts.

EXAMPLE 2

Solve: $x^2 - 2x \leq 2$.

Solution We first find standard form with 0 on one side:

$$x^2 - 2x - 2 \leq 0.$$

Consider $f(x) = x^2 - 2x - 2$. Its graph opens upward. Function values will be nonpositive between and including its x-intercepts, as shown. We find the intercepts by solving $f(x) = 0$:

$$x = \frac{-b \pm \sqrt{b^2 - 4ac}}{2a} = \frac{-(-2) \pm \sqrt{(-2)^2 - 4 \cdot 1(-2)}}{2 \cdot 1}$$

$$= \frac{2 \pm \sqrt{12}}{2} = \frac{2 \pm 2\sqrt{3}}{2} = 1 \pm \sqrt{3}.$$

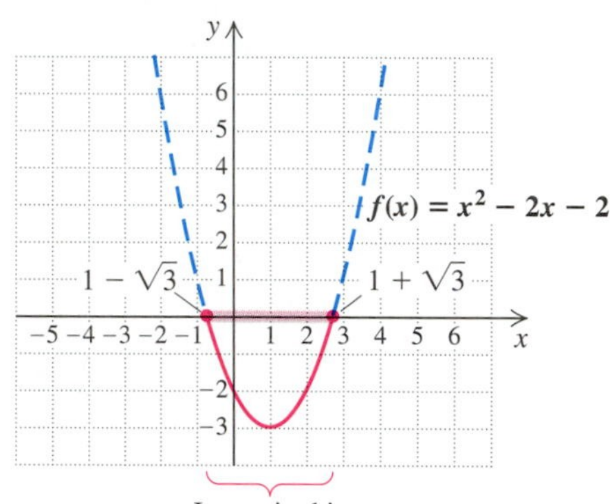

The x-intercepts are $1 + \sqrt{3}$ and $1 - \sqrt{3}$.

The solution set of the inequality is

$$\{x \mid 1 - \sqrt{3} \leq x \leq 1 + \sqrt{3}\}, \quad \text{or} \quad [1 - \sqrt{3}, 1 + \sqrt{3}].$$ ❑

It should be pointed out that we need not actually draw graphs as in the preceding examples. Visualizing the graph will usually suffice.

A second way to solve inequalities works for any polynomial that we can factor into a product of first-degree polynomials.

EXAMPLE 3

Solve: $x^2 + 2x - 3 > 0$.

Solution We factor the inequality, obtaining $(x + 3)(x - 1) > 0$. The solutions of $(x + 3)(x - 1) = 0$ are -3 and 1. They are not solutions of the inequality, but they divide the real-number line in a natural way, as pictured here. The product $(x + 3)(x - 1)$ is positive or negative, for values other than -3 and 1, depending on the signs of the factors $x + 3$ and $x - 1$. We can determine this efficiently with a diagram as follows.

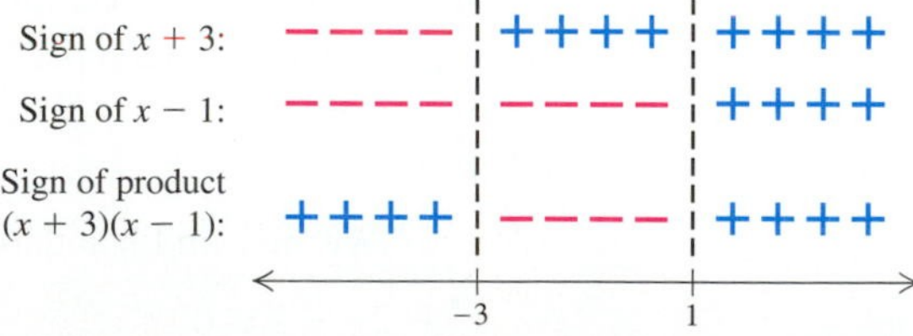

To set up the diagram, solve $x + 3 > 0$. We get $x > -3$. Thus, $x + 3$ is positive for all numbers to the right of -3. We indicate this with + signs. Accordingly, $x + 3$ is negative for all numbers to the left of -3. We indicate that with the − signs.

Similarly, we solve $x - 1 > 0$ and get $x > 1$. Thus, $x - 1$ is positive for all numbers to the right of 1 and negative for all numbers to the left of 1. We indicate this with the + and − signs.

In order for the product $(x + 3)(x - 1)$ to be positive, both factors must be positive or both must be negative. In the diagram, we see that this situation occurs when $x < -3$ and when $x > 1$. The solution set of the inequality is

$$\{x|\ x < -3 \text{ or } x > 1\}, \quad \text{or} \quad (-\infty, -3) \cup (1, \infty).$$ ❑

EXAMPLE 4

Solve: $4x(x + 1)(x - 1) < 0$.

Solution The solutions of $4x(x + 1)(x - 1) = 0$ are -1, 0, and 1. The product $4x(x + 1)(x - 1)$ is positive or negative, depending on the signs of the factors $4x$, $x + 1$, and $x - 1$. We determine this efficiently using a diagram.

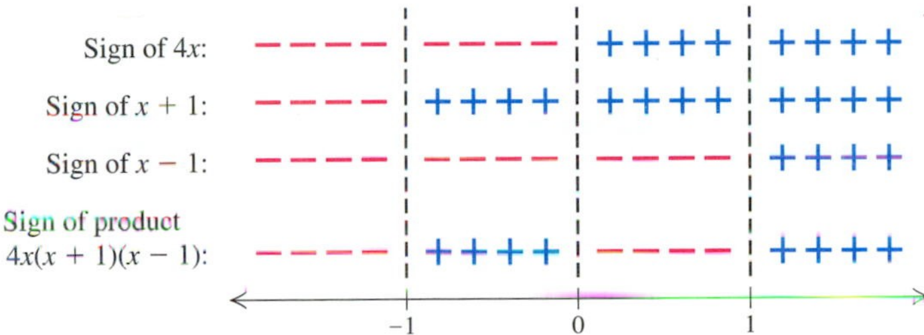

The product of three numbers is negative when it has an odd number of negative factors. We see from the diagram that the solution set is

$$\{x|\ x < -1 \text{ or } 0 < x < 1\}, \quad \text{or} \quad (-\infty, -1) \cup (0, 1).$$ ❑

We now consider our final method for solving polynomial inequalities. In Example 4, we see that the intercepts divide the number line into intervals. If a particular function has a positive output for one number in an interval, it will be positive for *all* the numbers in the interval. Thus we can merely make a test substitution in each interval to solve the inequality.

EXAMPLE 5

Solve: $x^2 + 3x - 10 < 0$.

Solution We set the polynomial equal to 0 and solve. The solutions of $x^2 + 3x - 10 = 0$ or $(x + 5)(x - 2) = 0$ are -5 and 2. We locate them on a number line as follows. Note that the numbers divide the number line into three intervals: A, B, and C.

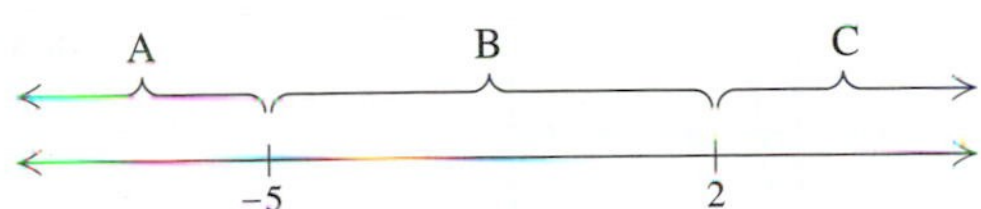

We pick a test number in interval A, say -7, and substitute -7 for x in the function $f(x) = x^2 + 3x - 10$:

$$\begin{aligned} f(-7) &= (-7)^2 + 3(-7) - 10 \\ &= 49 - 21 - 10 = 18. \end{aligned} \qquad f(-7) \text{ is positive.}$$

Since $f(-7) > 0$, the function will be positive for any number in interval A.

Next we try a test number in interval B, say 1, and find $f(1)$:

$$f(1) = 1^2 + 3(1) - 10 = -6. \qquad f(1) \text{ is negative.}$$

Since $f(1) < 0$, the function will be negative for any number in interval B.

Next we try a test number in interval C, say 4, and find $f(4)$:

$$f(4) = (4)^2 + 3(4) - 10$$
$$= 16 + 12 - 10 = 18. \qquad f(4) \text{ is positive.}$$

Since $f(4) > 0$, the function will be positive for any number in interval C. We are looking for numbers x for which $x^2 + 3x - 10 < 0$. Thus any number x in interval B is a solution. If the inequality had been $\leqslant$ or $\geqslant$, we would also have included the intercepts -5 and 2. The solution set is $\{x| -5 < x < 2\}$. ❏

Let us review the last method.

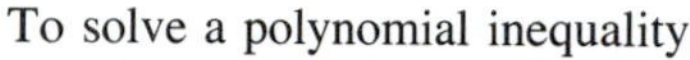

To solve a polynomial inequality:

1. Get 0 on one side, set the polynomial on the other side equal to 0, and solve to find the intercepts.
2. Use the numbers found in step (1) to divide the number line into intervals.
3. Substitute a number from each interval into the related function. If the output is positive, then the function will be positive for all numbers in the interval. If the output is negative, then the function will be negative for all numbers in the interval.
4. Select the intervals for which the inequality is satisfied and write set-builder notation or interval notation for the solution set. Include the intercepts in the solution set if the inequality sign is $\leqslant$ or $\geqslant$.

TECHNOLOGY CONNECTION

To solve the polynomial inequality $2.3x^2 \leqslant 9.11 - 2.94x$, we first rewrite the inequality in the form $2.3x^2 + 2.94x - 9.11 \leqslant 0$ and draw the graph of the function $f(x) = 2.3x^2 + 2.94x - 9.11$.

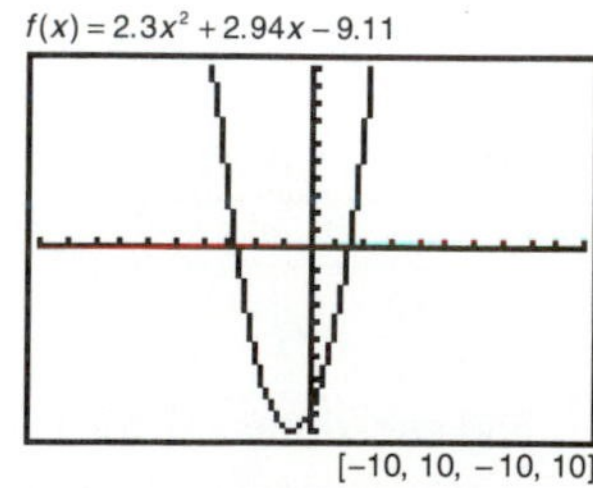

To find the values of x for which $f(x) \leqslant 0$, we focus on the region in which the graph lies *on or below* the x-axis. From this graph, it appears that this region begins somewhere between -3 and -2, and continues to somewhere between 1 and 2. By zooming in on the region between -3 and -2, we can determine the left-hand endpoint. To two decimal places, it is -2.73. Similarly, we can determine that the right-hand endpoint is approximately 1.45. The solution set is approximately $\{x| -2.73 \leqslant x \leqslant 1.45\}$.

Had the inequality been $2.3x^2 > 9.11 - 2.94x$, we would look for portions of the graph that lie *above* the x-axis. An approximate solution set of such an inequality would be $\{x| x < -2.73 \textit{ or } x > 1.45\}$.

Use a grapher to solve the following inequalities to the nearest hundredth.

TC1. $4.32x^2 - 3.54x - 5.34 \leqslant 0$

TC2. $7.34x^2 - 16.55x - 3.89 \geqslant 0$

TC3. $10.85x^2 + 4.28x + 4.44 > 7.91x^2 + 7.43x + 13.03$

TC4. $5.79x^3 - 5.68x^2 + 10.68x - 3.45 > 2.11x^3 + 16.90x - 15.14$

EXAMPLE 6

Solve: $7x(x+3)(x-2) \geqslant 0$.

Solution The solutions of $f(x) = 7x(x+3)(x-2) = 0$ are -3, 0, and 2. They divide the real-number line into four intervals as shown below.

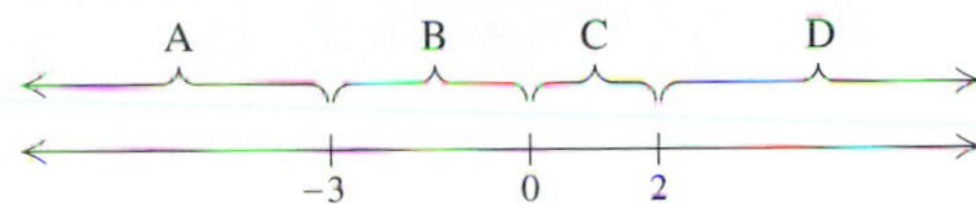

We try test numbers in each interval:

A: Test -5, $f(-5) = 7(-5)(-5+3)(-5-2) = -490$; ← Negative

B: Test -2, $f(-2) = 7(-2)(-2+3)(-2-2) = 56$; ← Positive

C: Test 1, $f(1) = 7(1)(1+3)(1-2) = -28$; ← Negative

D: Test 3, $f(3) = 7(3)(3+3)(3-2) = 126$. ← Positive

Since the inequality symbol is $\geqslant$, we must include the intercepts. The solution set of the inequality is

$$\{x \mid -3 \leqslant x \leqslant 0 \text{ or } 2 \leqslant x\}, \quad \text{or} \quad [-3, 0] \cup [2, \infty).$$

Rational Inequalities

We adapt the preceding method when an inequality involves rational expressions. We call these **rational inequalities.**

EXAMPLE 7

Solve: $\dfrac{x-3}{x+4} \geqslant 2$.

Solution We write the related equation by changing the $\geqslant$ symbol to $=$:

$$\frac{x-3}{x+4} = 2.$$

Then we solve this related equation. We multiply on both sides of the equation by the LCD, which is $x + 4$:

$$(x+4) \cdot \frac{x-3}{x+4} = (x+4) \cdot 2$$

$$x - 3 = 2x + 8$$

$$-11 = x. \quad \text{Solving for } x$$

In the case of rational inequalities, we also need to determine those values that make the denominator 0. We set the denominator equal to 0 and solve:

$$x + 4 = 0$$

$$x = -4.$$

Now we use the numbers -11 and -4 to divide the number line into intervals:

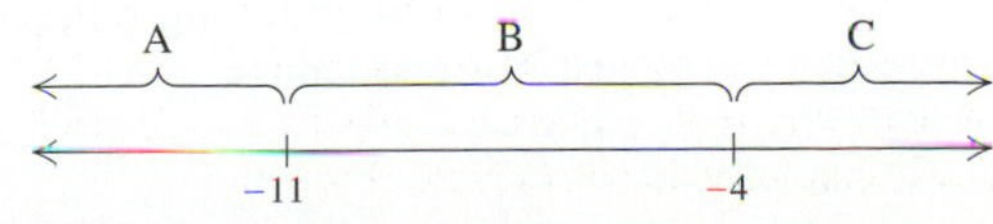

We try test numbers in each interval to see if each satisfies the original inequality

$$\frac{x-3}{x+4} \geqslant 2.$$

A: Test -15, $\dfrac{-15-3}{-15+4} = \dfrac{-18}{-11}$

$= \dfrac{18}{11} \ngeqslant 2$ — -15 *is not* a solution, so interval A is not part of the solution set.

B: Test -8, $\dfrac{-8-3}{-8+4} = \dfrac{-11}{-4}$

$= \dfrac{11}{4} \geqslant 2$ — -8 *is* a solution, so interval B is part of the solution set.

C: Test 1, $\dfrac{1-3}{1+4} = \dfrac{-2}{5}$

$= -\dfrac{2}{5} \ngeqslant 2$ — 1 *is not* a solution, so interval C is not part of the solution set.

The solution set includes the interval B. The number -11 is also included since the inequality symbol is $\geqslant$ and -11 is a solution of the related equation. The number -4 is *not* included since $(x-3)/(x+4)$ is undefined for $x = -4$. Thus the solution set of the original inequality is

$$\{x \mid -11 \leqslant x < -4\}, \quad \text{or} \quad [-11, -4).$$

❑

There is an interesting visual interpretation of Example 7. If we graph the function $f(x) = (x-3)/(x+4)$, we see that the solutions of the inequality $(x-3)/(x+4) \geqslant 2$ can be found by inspection. We simply sketch the line $y = 2$ and locate all x-values for which $f(x) \geqslant 2$.

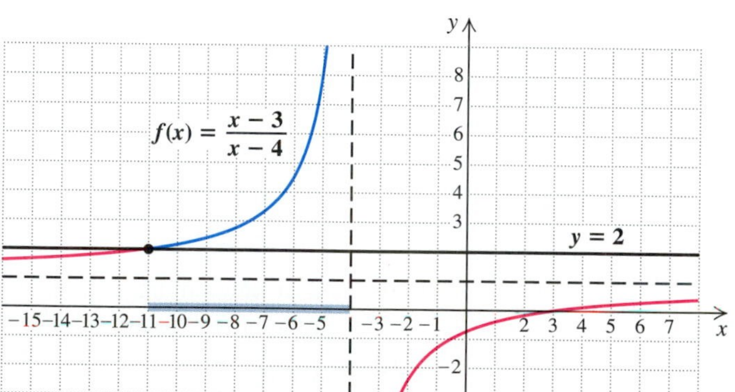

Because graphing rational functions can be very time-consuming, we normally just use test values.

To solve a rational inequality:

1. Change the inequality symbol to an equals sign and solve the related equation.
2. Find any replacements for which the rational expression is undefined.
3. Use the numbers found in steps (1) and (2) to divide the number line into intervals.
4. Substitute a number from each interval into the inequality. If the number is a solution, then the interval to which it belongs is part of the solution set.
5. Select the intervals for which the inequality is satisfied and write set-builder notation or interval notation for the solution set. If the inequality symbol is $\leqslant$ or $\geqslant$, then the solutions to step (1) should also be included in the solution set.

EXERCISE SET 11.10

Solve.

1. $(x-5)(x+3)>0$
2. $(x-4)(x+1)>0$
3. $(x+1)(x-2)\leqslant 0$
4. $(x-5)(x+3)\leqslant 0$
5. $x^2-x-2<0$
6. $x^2+x-2<0$
7. $9-x^2\leqslant 0$
8. $4-x^2\geqslant 0$
9. $x^2-2x+1\geqslant 0$
10. $x^2+6x+9<0$
11. $x^2+8<6x$
12. $x^2-12>4x$
13. $3x(x+2)(x-2)<0$
14. $5x(x+1)(x-1)>0$
15. $(x+3)(x-2)(x+1)>0$
16. $(x-1)(x+2)(x-4)<0$
17. $(x+3)(x+2)(x-1)<0$
18. $(x-2)(x-3)(x+1)<0$
19. $\dfrac{1}{x-4}<0$
20. $\dfrac{1}{x+5}>0$
21. $\dfrac{x+1}{x-3}>0$
22. $\dfrac{x-2}{x+5}<0$
23. $\dfrac{3x+2}{x-3}\leqslant 0$
24. $\dfrac{5-2x}{4x+3}\leqslant 0$
25. $\dfrac{x-1}{x-2}>3$
26. $\dfrac{x+1}{2x-3}<1$
27. $\dfrac{(x-2)(x+1)}{x-5}<0$
28. $\dfrac{(x+4)(x-1)}{x+3}>0$
29. $\dfrac{x}{x-2}\geqslant 0$
30. $\dfrac{x+3}{x}\leqslant 0$
31. $\dfrac{x-5}{x}<1$
32. $\dfrac{x}{x-1}>2$
33. $\dfrac{x-1}{(x-3)(x+4)}<0$
34. $\dfrac{x+2}{(x-2)(x+7)}>0$
35. $2<\dfrac{1}{x}$
36. $\dfrac{1}{x}\leqslant 3$

Skill Maintenance

37. Multiply and simplify: $\sqrt[5]{a^2b}\,\sqrt[3]{ab^2}$.
38. The perimeter of an equilateral triangle is the same as that of a square. If the sides of the triangle are 2 units longer than the sides of the square, how long is each side of the square?

Synthesis

39. ◈ Explain how any quadratic inequality can be solved by examining a parabola that opens upward.

Solve.

40. $x^2+2x>4$
41. $x^4+2x^2\geqslant 0$
42. $x^4+3x^2\leqslant 0$
43. $\left|\dfrac{x+2}{x-1}\right|<3$

44. *Total profit.* A company determines that its total-profit function is given by

$$P(x) = -3x^2 + 630x - 6000.$$

a) A company makes a profit for those nonnegative values of x for which $P(x) > 0$. Find the values of x for which the company makes a profit.

b) A company loses money for those nonnegative values of x for which $P(x) < 0$. Find the values of x for which the company loses money.

45. *Height of a thrown object.* The function

$$S(t) = -16t^2 + 32t + 1920$$

gives the height S, in feet, of an object thrown from a cliff that is 1920 ft high. Here t is the time, in seconds, that the object is in the air.

a) For what times is the height greater than 1920 ft?

b) For what times is the height less than 640 ft?

46. *Number of handshakes.* There are n people in a room. The number N of possible handshakes by the people is given by the function

$$N(n) = \frac{n(n-1)}{2}.$$

For what number of people n is

$$66 \leqslant N \leqslant 300?$$

47. *Number of diagonals.* A polygon with n sides has D diagonals, where D is given by the function

$$D(n) = \frac{n(n-3)}{2}.$$

Find the number of sides n if

$$27 \leqslant D \leqslant 230.$$

Use a grapher to draw the function and find solutions of $f(x) = 0$. Then solve the inequalities $f(x) < 0$ and $f(x) > 0$.

48. $f(x) = x^3 - 2x^2 - 5x + 6$

49. $f(x) = \frac{1}{3}x^3 - x + \frac{2}{3}$

50. $f(x) = x + \frac{1}{x}$

51. $f(x) = x - \sqrt{x},\ x \geqslant 0$

52. $f(x) = x^4 - 4x^3 - x^2 + 16x - 12$

53. $f(x) = \frac{x^3 + x^2 - 2x}{x^2 + x - 6}$

54. Use a grapher to solve Exercises 11, 25, and 35 by drawing two curves, one for each side of the inequality.

SUMMARY AND REVIEW 11

KEY TERMS

Principle of square roots, p. 546
Completing the square, p. 550
The quadratic formula, p. 554
Discriminant, p. 568
Reducible to quadratic, p. 571
Quadratic function, p. 576
Parabola, p. 576
Line of symmetry, p. 576
Vertex, p. 576
Minimum value, p. 579
Maximum value, p. 579
x-intercept, p. 586
Polynomial inequality, p. 598
Quadratic inequality, p. 598
Rational inequality, p. 603

IMPORTANT PROPERTIES AND FORMULAS

The Principle of Square Roots

For any real number k, if $x^2 = k$, then $x = \sqrt{k}$ or $x = -\sqrt{k}$.

To solve a quadratic equation in x by completing the square:

1. If $a \neq 1$, multiply by $1/a$ on both sides so that the x^2-coefficient is 1.
2. When the x^2-coefficient is 1, rewrite the equation in the form $x^2 + bx = -c$, or $x^2 + \frac{b}{a}x = -\frac{c}{a}$ if step (1) has been applied.
3. Complete the square by taking half of the coefficient of x and adding its square on both sides.
4. Express one side as the square of a binomial.
5. Use the principle of square roots and complete the solution.

The Quadratic Formula

The solutions of $ax^2 + bx + c = 0$, $a \neq 0$, are given by

$$x = \frac{-b \pm \sqrt{b^2 - 4ac}}{2a}.$$

To solve a quadratic equation:

1. Check for the form $ax^2 = p$ or $(x + k)^2 = p$. If it is in either of these forms, use the principle of square roots.
2. If it is not in the form of step (1), write it in standard form $ax^2 + bx + c = 0$.
3. Try factoring and using the principle of zero products.
4. If it is not possible to factor or factoring seems difficult, use the quadratic formula.

The solutions of a quadratic equation can always be found using the quadratic formula. They cannot always be found by factoring.

Discriminant $b^2 - 4ac$	Nature of Solutions
0	Only one solution; it is a real number
Positive	Two different real-number solutions
Negative	Two different imaginary-number solutions (complex conjugates)

The graph of $g(x) = ax^2$ is a parabola with the vertical axis as its line of symmetry and its vertex at the origin.

If a is positive, the parabola opens upward; if a is negative, the parabola opens downward.

If $|a|$ is greater than 1, the parabola is narrower than $f(x) = x^2$.

If $|a|$ is between 0 and 1, the parabola is flatter than $f(x) = x^2$.

The graph of $g(x) = a(x - h)^2$ has the same shape as the graph of $y = ax^2$.

If h is positive, the graph of $y = ax^2$ is shifted h units to the right.

If h is negative, the graph of $y = ax^2$ is shifted $|h|$ units to the left.

The vertex is $(h, 0)$, and the line of symmetry is $x = h$.

The graph of $f(x) = a(x - h)^2 + k$ looks like that of $y = a(x - h)^2$ translated up or down.

If k is positive, the graph of $y = a(x - h)^2$ is shifted k units up.

If k is negative, the graph of $y = a(x - h)^2$ is shifted $|k|$ units down.

The vertex is (h, k), and the line of symmetry is $x = h$.

For $a > 0$, k is the minimum function value. For $a < 0$, k is the maximum function value.

For a parabola given by a quadratic function $f(x) = ax^2 + bx + c$, the vertex of the parabola is

$$\left(-\frac{b}{2a}, \frac{4ac - b^2}{4a}\right).$$

The x-coordinate of the vertex is $-b/(2a)$. The line of symmetry is $x = -b/(2a)$. The second coordinate of the vertex can be found by substituting into the formula above, but is usually found most easily by evaluating the function at $-b/(2a)$.

To solve a formula for a letter, say, b:

1. Clear the fractions and use the principle of powers, as needed, until b does not appear in any radicand or denominator. (In some cases you may clear the fractions first, and in some cases you may use the principle of powers first. Perform these steps until radicals containing b are gone and b is not in any denominator.)
2. Collect all terms with b^2 in them. Also collect all terms with b in them.
3. If b^2 does not appear, you can finish by using just the addition and multiplication principles. Get all terms containing b on one side of the equation; then factor out b. Dividing on both sides will then get b alone.
4. If b^2 appears but b does not appear to the first power, solve the equation for b^2. Then take square roots on both sides.
5. If there are terms containing both b and b^2, put the equation in standard form and use the quadratic formula.

To solve a polynomial inequality:

1. Get 0 on one side, set the polynomial on the other side equal to 0, and solve to find the intercepts.
2. Use the numbers found in step (1) to divide the number line into intervals.
3. Substitute a number from each interval into the related function. If the output is positive, then the function will be positive for all numbers in the interval. If the output is negative, then the function will be negative for all numbers in the interval.
4. Select the intervals for which the inequality is satisfied and write set-builder notation or interval notation for the solution set. Include the intercepts in the solution set if the inequality sign is $\leq$ or $\geq$.

To solve a rational inequality:

1. Change the inequality symbol to an equals sign and solve the related equation.
2. Find any replacements for which the rational expression is undefined.
3. Use the numbers found in steps (1) and (2) to divide the number line into intervals.
4. Substitute a number from each interval into the inequality. If the number is a solution, then the interval to which it belongs is part of the solution set.
5. Select the intervals for which the inequality is satisfied and write set-builder notation or interval notation for the solution set. If the inequality symbol is $\leq$ or $\geq$, then the solutions to step (1) should also be included in the solution set.

REVIEW EXERCISES

This chapter's review and test include Skill Maintenance exercises from Sections 8.3, 9.1, 10.3, and 10.7.

Solve.

1. $8x^2 = 24$
2. $14x^2 + 5x = 0$
3. $(x - 6)^2 = 5$
4. $4x^2 + 3x + 1 = 0$
5. $x^2 - 7x + 13 = 0$
6. $20x^2 = 11x + 3$
7. $x^2 + 4x + 1 = 0$. Approximate the solutions to the nearest tenth.

Complete the square. Then write the trinomial square in factored form.

8. $x^2 - 12x$
9. $x^2 + \frac{3}{5}x$

Solve by completing the square. Show your work.

10. $x^2 - 2x - 8 = 0$
11. $x^2 - 6x + 1 = 0$
12. \$2500 grows to \$3025 in 2 years. Use the formula $A = P(1 + r)^t$ to find the interest rate.
13. The Peachtree Center Plaza in Atlanta, Georgia, is 723 ft tall. Use the formula $s = 16t^2$ to approximate how long it would take an object to fall from the top.

Solve.

14. $\frac{x}{x-2} + \frac{4}{x-6} = 0$
15. $\frac{x}{5} = \frac{x+3}{x+7}$
16. $\frac{x}{4} - \frac{4}{x} = 2$
17. $15 + \frac{6}{x-2} = \frac{8}{x+2}$
18. During the first part of a trip, Nina's Nissan traveled 50 mi at a certain speed. Nina drove 80 mi on the second part of the trip at a speed that was 10 mph slower. The total time for the trip was 3 hr. What was the speed on each part of the trip?
19. Working together, Jean and Stacy can cut and split a cord of wood in 4 hr. Working alone, Jean takes 6 hr more than Stacy. How long would it take Stacy to do this job alone?

Determine the nature of the solutions of the equation.

20. $x^2 + 3x - 6 = 0$
21. $x^2 + 2x + 5 = 0$
22. Write a quadratic equation having the solutions $\frac{1}{5}, -\frac{3}{5}$.
23. Write a quadratic equation having -4 as its only solution.

Solve.

24. $x^4 - 13x^2 + 36 = 0$
25. $15x^{-2} - 2x^{-1} - 1 = 0$
26. $2x - 11\sqrt{x} + 12 = 0$
27. $(x^2 - 4)^2 - (x^2 - 4) - 6 = 0$
28. a) Graph: $f(x) = -3(x + 2)^2 + 4$.
    b) Label the vertex.
    c) Draw the line of symmetry.
    d) Find the maximum or the minimum value.
29. For the function $f(x) = 2x^2 - 12x + 23$:
    a) find the vertex and the line of symmetry;
    b) graph the function.
30. Find the x-intercepts: $f(x) = x^2 - 9x + 14$.
31. An object is dropped 120.1 m from an airplane. How long does it take to reach the ground? Round to the nearest second.
32. Solve $N = 3\pi\sqrt{1/p}$ for p.
33. Solve $2A + T = 3T^2$ for T.
34. Little Angels Daycare has 260 yd of fencing available to enclose a rectangular playground. What should the dimensions of the playground be to maximize the area? What is the maximum area?
35. What is the minimum product of two numbers whose difference is 22? What numbers yield this product?
36. Find the quadratic function that fits the data points $(0, -2)$, $(1, 3)$, and $(3, 7)$.

Solve.

37. $x^2 < 6x + 7$
38. $\frac{x-5}{x+3} \leq 0$

Skill Maintenance

39. Metal alloy A is 75% silver. Metal alloy B is 25%

silver. How much of each should be mixed in order to produce 300 kg of an alloy that is 60% silver?

40. Multiply and simplify: $\sqrt[3]{9t^6}\,\sqrt[3]{3s^4t^9}$.

Solve.

41. $4 - 3x \geq x + 5$

42. $\sqrt{x+1} = x - 5$

Synthesis

43. Discuss two ways in which completing the square was used in this chapter.

44. Suppose you know one imaginary-number solution of a quadratic equation with real coefficients. Explain how you can find the other solution.

45. Solve:

$$\frac{26}{x+13} - \frac{14}{x+7} = \frac{12x}{x^2 + 20x + 91}.$$

46. Find h and k if, for $3x^2 - hx + 4k = 0$, the sum of the solutions is 20 and the product is 80.

47. The average of two positive integers is 171. One of the numbers is the square root of the other. Find the integers.

CHAPTER TEST 11

Solve.

1. $3x^2 - 4 = 0$

2. $10x^2 = 101x - 10$

3. $x^2 + x + 1 = 0$

4. $x^2 + 4x = 2$. Approximate the solutions to the nearest tenth, using a calculator or Table 1.

5. $\frac{1}{4-x} + \frac{1}{2+x} = \frac{3}{4}$

6. $x^4 - 5x^2 + 5 = 0$

Complete the square. Then write the trinomial square in factored form.

7. $x^2 + 14x$

8. $x^2 - \frac{2}{7}x$

Solve by completing the square. Show your work.

9. $x^2 + 3x - 18 = 0$

10. $x^2 + 10x + 15 = 0$

Solve.

11. A river flows at a rate of 4 km/hr. A boat travels 60 km upriver and returns in a total time of 8 hr. What is the speed of the boat in still water?

12. Two pipes can fill a tank in $1\frac{1}{2}$ hr. One pipe requires 4 hr longer running alone to fill the tank than the other. How long would it take for the faster pipe, working alone, to fill the tank?

13. Determine the nature of the solutions of the equation $x^2 + 5x + 17 = 0$.

14. Write a quadratic equation having solutions $\sqrt{3}$ and $3\sqrt{3}$.

15. **a)** Graph: $f(x) = 4(x-3)^2 + 5$.
b) Label the vertex.
c) Draw the line of symmetry.
d) Find the maximum or the minimum function value.

16. For the function $f(x) = 2x^2 + 4x - 6$:
a) find the vertex and the line of symmetry;
b) graph the function.

17. Find the x-intercepts of $f(x) = x^2 - x - 6$.

18. Solve $V = \frac{1}{3}\pi(R^2 + r^2)$ for r.

19. What is the maximum product of two numbers having a sum of 8?

20. Find the quadratic function that fits the data points (0, 0), (3, 0), and (5, 2).

Solve.

21. $(x+2)(x-1)(x-2) > 0$

22. $\frac{x(x+1)}{x-3} \geq 0$

Skill Maintenance

Solve.

23. $\sqrt{x + 3} = x - 3$

24. $2(x - 7) < 3(x + 5)$

25. Multiply and simplify: $\sqrt[4]{2a^2b^3}\,\sqrt[4]{a^4b}$.

26. The perimeter of a hexagon with all six sides the same length is the same as the perimeter of a square. One side of the hexagon is 3 less than a side of the square. Find the perimeter of each polygon.

Synthesis

27. One solution of $kx^2 + 3x - k = 0$ is -2. Find the other solution.

28. Solve:

$$\frac{88}{x - 11} - \frac{56}{x - 7} = \frac{32x}{x^2 - 18x + 77}.$$

CHAPTER 12

Exponential and Logarithmic Functions

AN APPLICATION

Because he objected to smoking, and because his first baseball card was issued in cigarette packs, the great shortstop Honus Wagner halted production of his card before many were produced. One of these cards was sold in 1986 for $110,000 and again in 1991 for $451,000.

In Section 12.7, we will use exponential functions to predict the card's value in future years.

Joshua Evans
LELAND'S AUCTION HOUSE

"Sports memorabilia ranging from cards to uniforms to seats in old stadiums often increase in value exponentially. By monitoring the past prices paid for certain items, we can use mathematics to estimate future prices for those same items."

The functions that we consider in this chapter are interesting not only from a purely intellectual point of view, but also for their rich applications to many fields. We will look at such applications as compound interest and population growth, to name just two.

The basis of the theory concerns exponents. We define some functions having variable exponents (*exponential functions*); the rest follows from those functions and their properties.

In addition to material from this chapter, the review and test for Chapter 12 include material from Sections 6.6, 10.9, 11.6, and 11.9.

12.1 Exponential Functions

Graphing Exponential Functions • Equations with x and y Interchanged • Applications of Exponential Functions

The following graph shows the number of cases of AIDS reported and the year in which that number was reported.

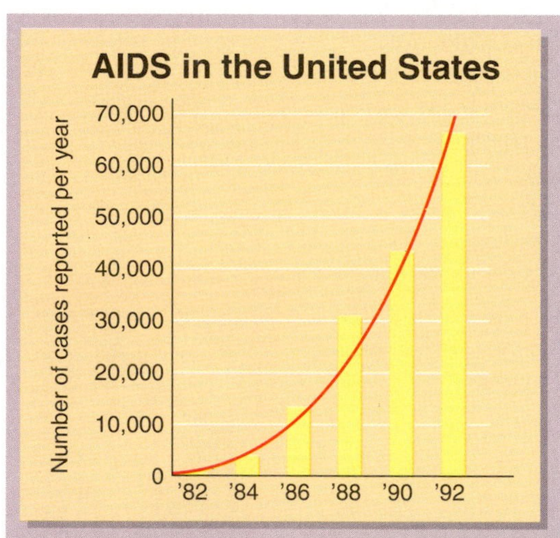

A curve drawn along the graph would approximate the graph of an *exponential function*. We now consider such graphs. We will also study graphs and properties of *logarithmic functions*, which are closely related to the exponential functions. These new functions will appear in a variety of applications.

Graphing Exponential Functions

In Chapter 10, we gave meaning to exponential expressions with rational number exponents, such as

$$5^{1/4}, \quad 3^{-3/4}, \quad 7^{2.34}, \quad 5^{1.73}.$$

For example, $5^{1.73}$, or $5^{173/100}$, represents the 100th root of 5 raised to the 173rd power. We now give meaning to expressions with irrational exponents, such as

$$5^{\sqrt{3}}, \quad 7^{\pi}, \quad 9^{-\sqrt{2}}.$$

Consider $5^{\sqrt{3}}$. Let us think of rational numbers r close to $\sqrt{3}$ and look at 5^r. As r gets closer to $\sqrt{3}$, 5^r gets closer to some real number.

r closes in on $\sqrt{3}$.	5^r closes in on some real number p.
$1 < r < 2$	$5 = 5^1 < p < 5^2 = 25$
$1.7 < r < 1.8$	$15.426 = 5^{1.7} < p < 5^{1.8} = 18.119$
$1.73 < r < 1.74$	$16.189 = 5^{1.73} < p < 5^{1.74} = 16.452$
$1.732 < r < 1.733$	$16.241 = 5^{1.732} < p < 5^{1.733} = 16.267$

As r closes in on $\sqrt{3}$, 5^r closes in on some real number p. We define $5^{\sqrt{3}}$ to be the number p. To seven decimal places,

$$5^{\sqrt{3}} \approx 16.2424508.$$

Any positive irrational exponent can be defined in a similar way. Negative irrational exponents are then defined in the same way as negative integer exponents. Thus the expression a^x has meaning for *any* real number x. The general laws of exponents still hold, but we will not prove that here. We now define exponential functions.

Exponential Function

The function $f(x) = a^x$, where a is a positive constant, $a \neq 1$, is called the *exponential function*, base a.

We require the base a to be positive to avoid the imaginary numbers that would result from taking even roots of negative numbers. The restriction $a \neq 1$ is made to exclude the constant function $f(x) = 1^x$, or $f(x) = 1$.

The following are examples of exponential functions:

$$f(x) = 2^x, \quad f(x) = \left(\tfrac{1}{2}\right)^x, \quad f(x) = (0.4)^x.$$

Note that, in contrast to polynomial functions like $f(x) = x^2$ and $f(x) = x^3$, the variable in an exponential function is in the *exponent*. Let us consider graphs of exponential functions.

EXAMPLE 1

Graph the exponential function $y = f(x) = 2^x$.

Solution We compute some function values, thinking of y as $f(x)$, and list the results in a table. It is a good idea to start by letting $x = 0$.

$f(0) = 2^0 = 1;$ $\quad f(-1) = 2^{-1} = \dfrac{1}{2^1} = \dfrac{1}{2};$

$f(1) = 2^1 = 2;$ $\quad f(-2) = 2^{-2} = \dfrac{1}{2^2} = \dfrac{1}{4};$

$f(2) = 2^2 = 4;$ $\quad f(-3) = 2^{-3} = \dfrac{1}{2^3} = \dfrac{1}{8}.$

$f(3) = 2^3 = 8;$

x	y, or $f(x)$
0	1
1	2
2	4
3	8
-1	$\frac{1}{2}$
-2	$\frac{1}{4}$
-3	$\frac{1}{8}$

Next, we plot these points and connect them with a smooth curve.

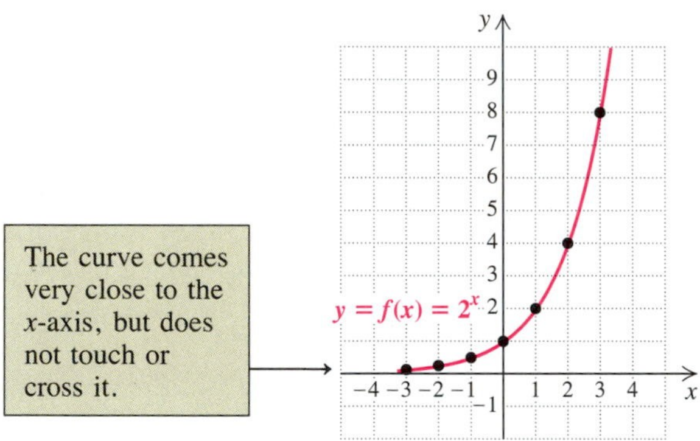

Be sure to plot enough points to determine how steeply the curve rises.

Note that as x increases, the function values increase indefinitely. As x decreases, the function values decrease, getting very close to 0. The x-axis, or the line $y = 0$, is a horizontal *asymptote*, meaning that the curve gets closer and closer to this line the further we move to the left. ❑

EXAMPLE 2 Graph the exponential function $y = f(x) = \left(\frac{1}{2}\right)^x$.

Solution We compute some function values, thinking of y as $f(x)$, and list the results in a table. Before we do this, note that

$$y = f(x) = \left(\tfrac{1}{2}\right)^x = (2^{-1})^x = 2^{-x}.$$

Then we have

$$f(0) = 2^{-0} = 1;$$

$$f(1) = 2^{-1} = \frac{1}{2^1} = \frac{1}{2};$$

$$f(2) = 2^{-2} = \frac{1}{2^2} = \frac{1}{4};$$

$$f(3) = 2^{-3} = \frac{1}{2^3} = \frac{1}{8};$$

$$f(-1) = 2^{-(-1)} = 2^1 = 2;$$

$$f(-2) = 2^{-(-2)} = 2^2 = 4;$$

$$f(-3) = 2^{-(-3)} = 2^3 = 8.$$

x	y, or $f(x)$
0	1
1	$\frac{1}{2}$
2	$\frac{1}{4}$
3	$\frac{1}{8}$
-1	2
-2	4
-3	8

We plot these points and draw the curve. Note that this graph is a reflection across the y-axis of the graph in Example 1. The line $y = 0$ is again an asymptote.

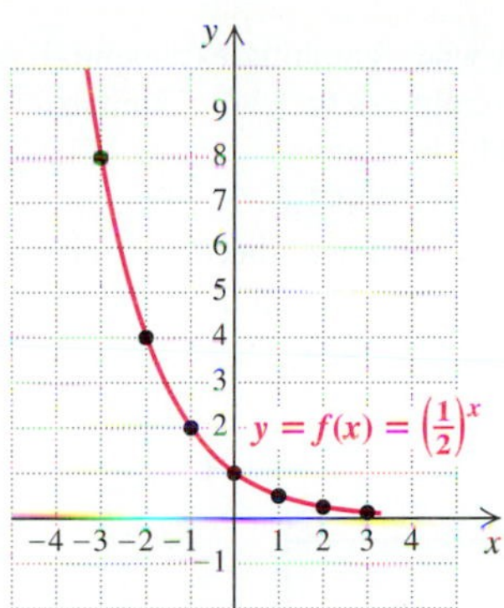

❑

The preceding examples illustrate exponential functions with various bases. Let us list some of their characteristics.

A. When $a > 1$, the function $f(x) = a^x$ increases from left to right. The greater the value of a, the steeper the curve. (See the figure on the left, below.)

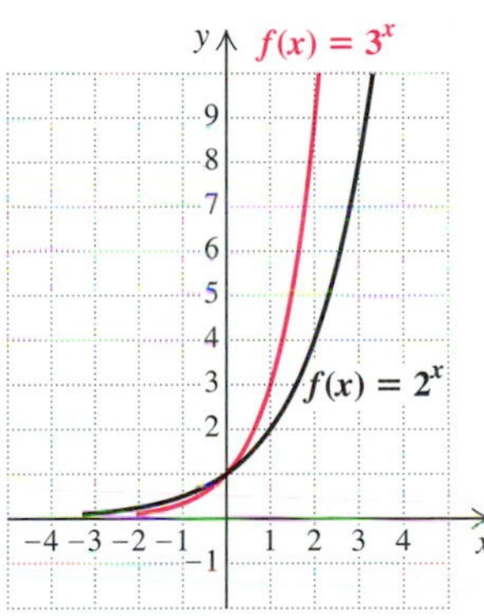

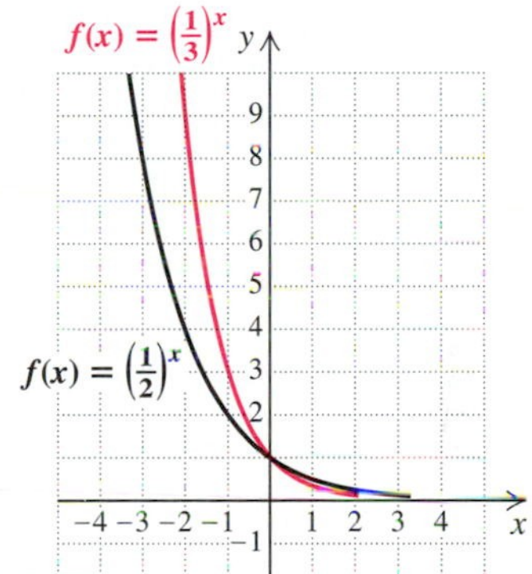

B. When $0 < a < 1$, the function $f(x) = a^x$ decreases from left to right. For smaller values of a, the curve becomes steeper. (See the figure on the right, above.)

C. All exponential functions of the form $f(x) = a^x$ go through the point $(0, 1)$. That is, the y-intercept is $(0, 1)$.

EXAMPLE 3

Graph: $y = f(x) = 2^{x-2}$.

Solution We construct a table of values. Then we plot the points and connect them with a smooth curve. Be sure to note that $x - 2$ is the *exponent*.

$$f(0) = 2^{0-2} = 2^{-2} = \frac{1}{2^2} = \frac{1}{4}$$

$$f(1) = 2^{1-2} = 2^{-1} = \frac{1}{2^1} = \frac{1}{2}$$

$$f(2) = 2^{2-2} = 2^0 = 1$$

$$f(3) = 2^{3-2} = 2^1 = 2$$

$$f(4) = 2^{4-2} = 2^2 = 4$$

$$f(-1) = 2^{-1-2} = 2^{-3} = \frac{1}{2^3} = \frac{1}{8}$$

$$f(-2) = 2^{-2-2} = 2^{-4} = \frac{1}{2^4} = \frac{1}{16}$$

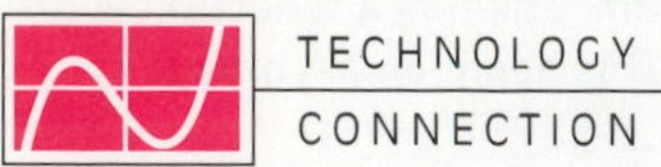

Graphers are especially helpful when graphing exponential functions because bases may not always be whole numbers and function values can quickly become very large. For example, consider the function $y = 5000(1.075)^x$. Because y-values will always be positive and will become quite large, an appropriate window for this function might be $[-10, 10, 0, 15000]$, where the *scale* of the y-axis is 1000.

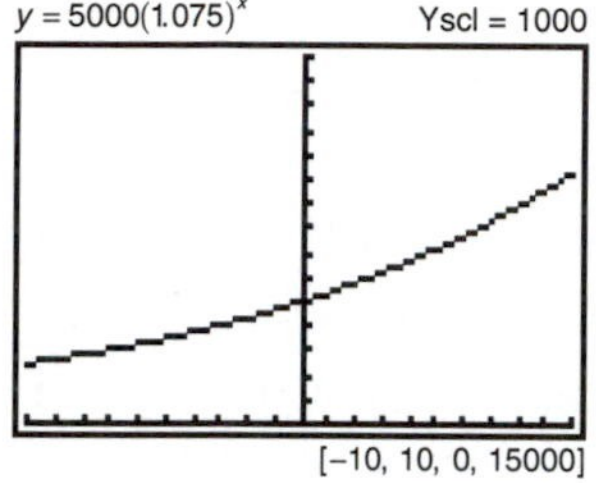

Use a grapher to draw the graph of each function. Select an appropriate viewing box and scale.

TC1. $y = 9.34^x$

TC2. $y = 9.34^{-x}$

TC3. $y = 3.45^{x+5}$

TC4. $y = 8.11^{4.2-x}$

TC5. $y = 20{,}000(1.1225)^x$

TC6. $y = 100{,}000(0.25)^{0.5x}$

x	y, or $f(x)$
0	$\frac{1}{4}$
1	$\frac{1}{2}$
2	1
3	2
4	4
−1	$\frac{1}{8}$
−2	$\frac{1}{16}$

The graph looks just like the graph of $y = 2^x$, but it is translated 2 units to the right. The y-intercept of $y = 2^x$ is $(0, 1)$. The y-intercept of $y = 2^{x-2}$ is $(0, \frac{1}{4})$. The line $y = 0$ is again the asymptote.

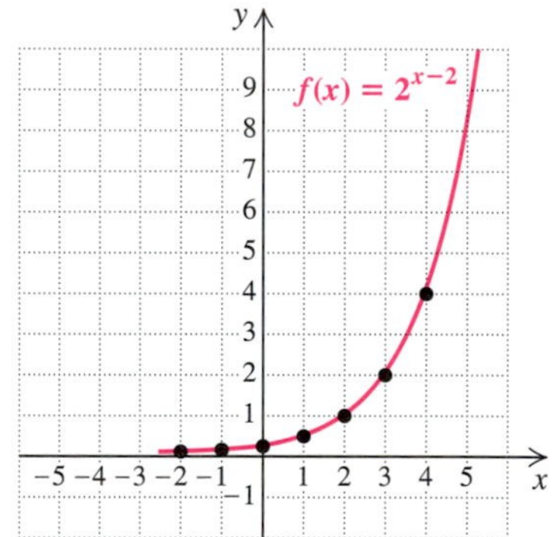

❑

Equations with x and y Interchanged

It will be helpful in later work if we are able to graph an equation in which the x and y in $y = a^x$ have been interchanged.

EXAMPLE 4 Graph: $x = 2^y$.

Solution Note that x is alone on one side of the equation. We can find ordered pairs that are solutions by choosing values for y and then computing values for x.

For $y = 0$, $x = 2^0 = 1$.

For $y = 1$, $x = 2^1 = 2$.

For $y = 2$, $x = 2^2 = 4$.

For $y = 3$, $x = 2^3 = 8$.

For $y = -1$, $x = 2^{-1} = \frac{1}{2^1} = \frac{1}{2}$.

For $y = -2$, $x = 2^{-2} = \frac{1}{2^2} = \frac{1}{4}$.

For $y = -3$, $x = 2^{-3} = \frac{1}{2^3} = \frac{1}{8}$.

x	y
1	0
2	1
4	2
8	3
$\frac{1}{2}$	−1
$\frac{1}{4}$	−2
$\frac{1}{8}$	−3

(1) Choose values for y.
(2) Compute values for x.

We plot the points and connect them with a smooth curve.

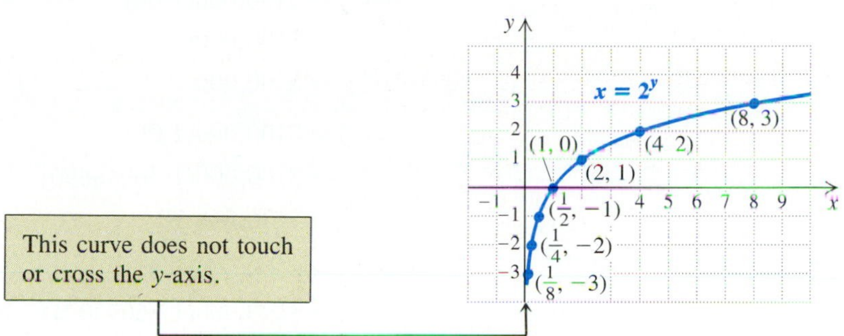

Note too that this curve looks just like the graph of $y = 2^x$, except that it is reflected across the line $y = x$, as shown here.

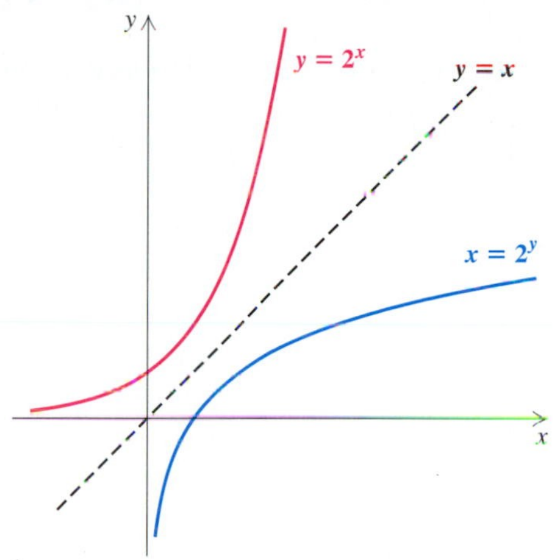

❑

Applications of Exponential Functions

EXAMPLE 5

Interest compounded annually. The amount of money A that a principal P will be worth after t years at interest rate i, compounded annually, is given by the formula

$$A = P(1 + i)^t.$$

Suppose that \$100,000 is invested at 8% interest, compounded annually.

a) Find a function for the amount in the account after t years.
b) Find the amount of money in the account at $t = 0$, $t = 4$, $t = 8$, and $t = 10$.
c) Graph the function.

Solution

a) If $P = \$100{,}000$ and $i = 8\% = 0.08$, we can substitute these values and form the following function:

$$\begin{aligned} A(t) &= \$100{,}000(1 + 0.08)^t \\ &= \$100{,}000(1.08)^t. \end{aligned}$$

b) To find the function values, a calculator with a power key is helpful.

$$A(0) = \$100{,}000(1.08)^0 = \$100{,}000(1) = \$100{,}000$$

$$A(4) = \$100{,}000(1.08)^4 = \$100{,}000(1.36048896) \approx \$136{,}048.90$$

$$A(8) = \$100{,}000(1.08)^8 \approx \$100{,}000(1.85093021) \approx \$185{,}093.02$$

$$A(10) = \$100{,}000(1.08)^{10} \approx \$100{,}000(2.158924997) \approx \$215{,}892.50$$

c) We use the function values computed in part (b), and others if we wish, and draw the graph as follows. Note that the axes are scaled differently because of the large numbers.

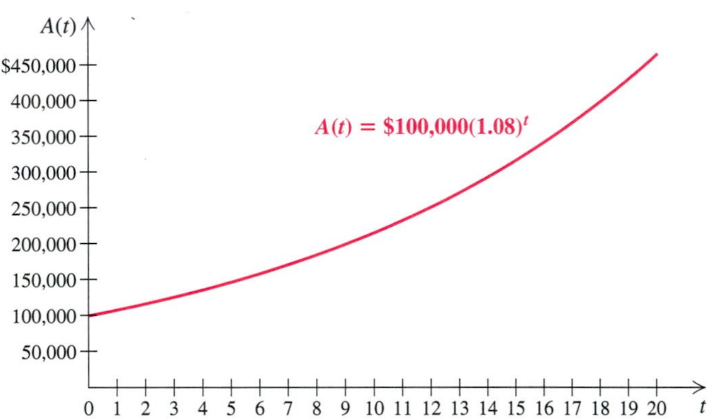

EXERCISE SET 12.1

Graph.

1. $y = f(x) = 2^x$
2. $y = f(x) = 3^x$
3. $y = 5^x$
4. $y = 6^x$
5. $y = 2^{x+1}$
6. $y = 2^{x-1}$
7. $y = 3^{x-2}$
8. $y = 3^{x+2}$
9. $y = 2^x - 3$
10. $y = 2^x + 1$
11. $y = 5^{x+3}$
12. $y = 6^{x-4}$
13. $y = \left(\frac{1}{2}\right)^x$
14. $y = \left(\frac{1}{3}\right)^x$
15. $y = \left(\frac{1}{5}\right)^x$
16. $y = \left(\frac{1}{4}\right)^x$
17. $y = 2^{2x-1}$
18. $y = 3^{4-x}$
19. $y = 2^{x-1} - 3$
20. $y = 2^{x+3} - 4$
21. $x = 3^y$
22. $x = 6^y$
23. $x = \left(\frac{1}{2}\right)^y$
24. $x = \left(\frac{1}{3}\right)^y$
25. $x = 5^y$
26. $x = 4^y$
27. $x = \left(\frac{2}{3}\right)^y$
28. $x = \left(\frac{4}{3}\right)^y$

Graph both equations using the same set of axes.

29. $y = 2^x$, $x = 2^y$

30. $y = 3^x$, $x = 3^y$

31. $y = \left(\frac{1}{2}\right)^x$, $x = \left(\frac{1}{2}\right)^y$

32. $y = \left(\frac{1}{4}\right)^x$, $x = \left(\frac{1}{4}\right)^y$

Solve.

33. *Cases of AIDS.* The total number of Americans who have contracted AIDS is approximated by the exponential function

$$N(t) = 100{,}000(1.4)^t,$$

where $t = 0$ corresponds to 1989.

a) According to the function, how many Americans had been infected as of 1993?
b) Predict the total number of Americans who will have been infected by 1998.
c) Graph the function.

34. *Growth of bacteria.* The bacteria *Escherichi coli* are commonly found in the bladder of human beings. Suppose that 3000 of the bacteria are present at time $t = 0$. Then t minutes later, the number of bacteria present will be

$$N(t) = 3000(2)^{t/20}.$$

a) How many bacteria will be present after 10 min? 20 min? 30 min? 40 min? 60 min?
b) Graph the function.

35. *Recycling aluminum cans.* It is estimated that $\frac{2}{3}$ of all aluminum cans distributed will be recycled each year. A beverage company distributes 250,000 cans. The number still in use after time t, in years, is given by the exponential function

$$N(t) = 250{,}000\left(\tfrac{2}{3}\right)^t.$$

a) How many cans are still in use after 0 years? 1 year? 4 years? 10 years?
b) Graph the function.

36. *Salvage value.* A photocopier is purchased for \$5200. Its value each year is about 80% of the value of the preceding year. Its value, in dollars, after t years is given by the exponential function

$$V(t) = 5200(0.8)^t.$$

a) Find the value of the machine after 0 years, 1 year, 2 years, 5 years, and 10 years.
b) Graph the function.

37. *Compact discs.* The number of compact discs purchased each year is increasing exponentially. The number N, in millions, purchased is given by the exponential function

$$N(t) = 7.5(6)^{0.5t},$$

where t is the number of years after 1985.

a) Find the number of compact discs sold in 1985, 1986, 1988, 1990, 1995, and 2000.
b) Graph the function.

38. *Turkey consumption.* The amount of turkey consumed by each person in this country is increasing exponentially. Assuming $t = 0$ corresponds to 1937, the amount of turkey, in pounds per person, consumed t years after 1937 is given by the exponential function

$$N(t) = 2.3(3)^{0.033t}.$$

a) How much turkey was consumed, per person, in 1940? in 1950? in 1980? in 1988?
b) How much will be consumed, per person, in 2000?
c) Graph the function.

Skill Maintenance

39. Multiply and simplify: $x^{-5} \cdot x^3$.

40. Simplify: $(x^{-3})^4$.

41. Divide and simplify: $\dfrac{x^{-3}}{x^4}$.

42. Simplify: 5^0.

Synthesis

43. Suppose that \$1000 is invested for 5 years at 7% interest, compounded annually. In what year will the most interest be earned? Why?

44. Without using a calculator, explain why 2^π must be greater than 8 but less than 16.

Determine which of the two numbers is larger.

45. $\pi^{1.3}$ or $\pi^{2.4}$

46. $\sqrt{8^3}$ or $8^{\sqrt{3}}$

Graph each of the following. You will find a calculator with a power key most helpful.

47. $f(x) = (2.3)^x$

48. $f(x) = (3.8)^x$

49. $g(x) = (0.125)^x$

50. $g(x) = (0.9)^x$

Graph.

51. $y = 2^x + 2^{-x}$

52. $y = \left|\left(\frac{1}{2}\right)^x - 1\right|$

53. $y = 3^x + 3^{-x}$

54. $y = 2^{-(x-1)^2}$

55. $y = |2^{x^2} - 1|$

56. $y = |2^x - 2|$

Graph both equations using the same set of axes.

57. $y = 3^{-(x-1)}$, $x = 3^{-(y-1)}$

58. $y = 1^x$, $x = 1^y$

59. *Typing speed.* Jim is studying typing. After he has studied for t hours, Jim's speed, in words per minute, is given by the exponential function

$$S(t) = 200[1 - (0.99)^t].$$

a) How fast can Jim type after he has studied for 10 hours? 20 hours? 40 hours? 85 hours?
b) Graph the function.

12.2 Composite and Inverse Functions

Composite Functions • Inverses and One-to-One Functions • Finding Formulas for Inverses • Graphing Functions and Their Inverses • Inverse Functions and Composition

Composite Functions

In the real world, functions frequently occur in which some quantity depends on a variable that, in turn, depends on another variable. For instance, the number of employees hired by a firm may depend on the firm's profits, which may in turn depend on the number of items the firm produces. Functions like this are called **composite functions.**

For example, the function g that gives a correspondence between women's shoe sizes in the United States and those in Italy is given by $g(x) = 2x + 24$, where x is the U.S. size and $g(x)$ is the Italian size. Thus a U.S. size 4 corresponds to a shoe size of $g(4) = 2 \cdot 4 + 24$, or 32, in Italy.

There is also a function that gives a correspondence between women's shoe sizes in Italy and those in Britain. The function is given by $f(x) = \frac{1}{2}x - 14$, where x is the Italian size and $f(x)$ is the corresponding British size. Thus an Italian size 32 corresponds to a British size $f(32) = \frac{1}{2} \cdot 32 - 14$, or 2.

It seems reasonable to conclude that a shoe size of 4 in the United States corresponds to a size of 2 in Britain and that some function h describes this correspondence. Can we find a formula for h? If we look at the following tables, we might guess that such a formula is $h(x) = x - 2$, and that is indeed correct. But, for more complicated formulas, we would need to use algebra.

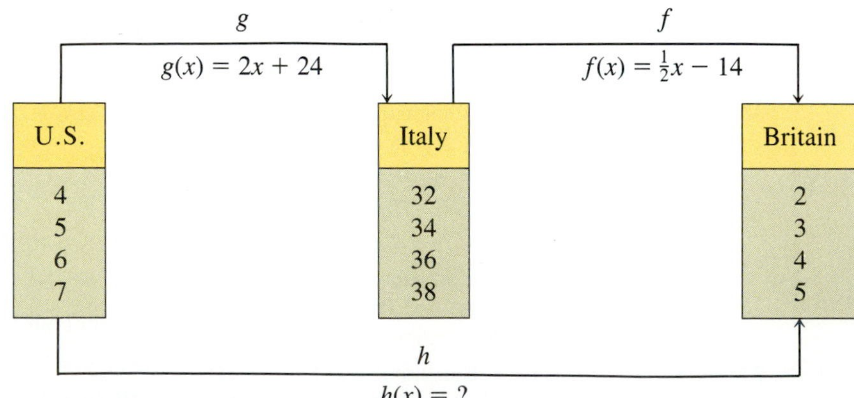

Size x shoes in the United States correspond to size $g(x)$ shoes in Italy, where

$$g(x) = 2x + 24.$$

Size n shoes in Italy correspond to size $f(n)$ shoes in Britain. Thus size $g(x)$ shoes in Italy correspond to size $f(g(x))$ shoes in Britain. Since the x in the expression

$f(g(x))$ represents a U.S. shoe size, we can find the British shoe size that corresponds to a U.S. size x:

$$f(g(x)) = f(2x + 24) = \tfrac{1}{2}\cdot(2x + 24) - 14 \qquad \text{Using } g(x) \text{ as an input}$$
$$= x + 12 - 14 = x - 2.$$

This gives a formula for h: $h(x) = x - 2$. Thus a shoe size of 4 in the United States corresponds to a shoe size of $h(4) = 4 - 2$, or 2, in Britain. The function h is called the *composition* of f and g and is denoted $f \circ g$.

Composition

The *composite function* $f \circ g$, the *composition* of f and g, is defined as

$$f \circ g(x) = f(g(x)).$$

We can visualize the composition of functions as follows.

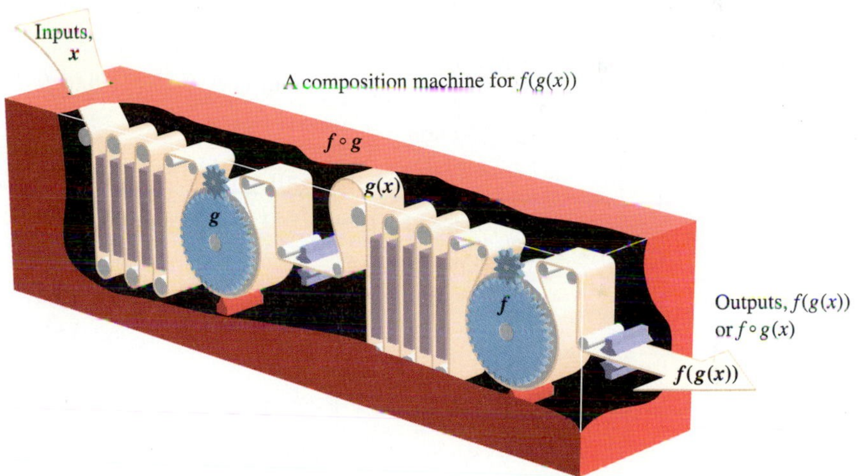

EXAMPLE 1

Given $f(x) = 3x$ and $g(x) = 1 + x^2$:

a) Find $f \circ g(5)$ and $g \circ f(5)$.
b) Find $f \circ g(x)$ and $g \circ f(x)$.

Solution Consider each function separately:

$$f(x) = 3x \qquad \text{This function multiplies each input by 3.}$$

and

$$g(x) = 1 + x^2. \qquad \text{This function adds 1 to the square of each input.}$$

a) To find $f \circ g(5)$, we first find $g(5)$ by substituting in the formula for g: Square 5 and add 1, to get 26. We then use 26 as an input for f:

$$f \circ g(5) = f(g(5)) = f(1 + 5^2)$$
$$= f(26) = 3 \cdot 26 = 78.$$

To find $g \circ f(5)$, we first find $f(5)$ by substituting into the formula for f: Multiply 5 by 3, to get 15. We then use 15 as an input for g:

$$\begin{aligned} g \circ f(5) &= g(f(5)) = g(3 \cdot 5) \\ &= g(15) = 1 + 15^2 = 1 + 225 = 226. \end{aligned}$$

b) We find $f \circ g(x)$ by substituting $g(x)$ for x in the equation for $f(x)$:

$$\begin{aligned} f \circ g(x) &= f(g(x)) = f(1 + x^2) && \text{Substituting } 1 + x^2 \text{ for } g(x) \\ &= 3(1 + x^2) = 3 + 3x^2. && \textit{These} \text{ parentheses indicate multiplication.} \end{aligned}$$

To find $g \circ f(x)$, we substitute $f(x)$ for x in the equation for $g(x)$:

$$\begin{aligned} g \circ f(x) &= g(f(x)) = g(3x) && \text{Substituting } 3x \text{ for } f(x) \\ &= 1 + (3x)^2 \\ &= 1 + 9x^2. \end{aligned}$$

❑

Note in Example 1 that $f \circ g(5) \neq g \circ f(5)$ and that, in general, $f \circ g(x) \neq g \circ f(x)$.

EXAMPLE 2

Given $f(x) = \sqrt{x}$ and $g(x) = x - 1$, find $f \circ g(x)$ and $g \circ f(x)$.

Solution

$$f \circ g(x) = f(g(x)) = f(x - 1) = \sqrt{x - 1}$$
$$g \circ f(x) = g(f(x)) = g(\sqrt{x}) = \sqrt{x} - 1$$

❑

In calculus, one needs to recognize how a function can be expressed as a composition.

EXAMPLE 3

If $h(x) = (7x + 3)^2$, find $f(x)$ and $g(x)$ so that $h(x) = f \circ g(x)$.

Solution To find $h(x)$, we can think of two steps: forming $7x + 3$ and then squaring. This suggests that $g(x) = 7x + 3$ and $f(x) = x^2$. We check by forming the composition:

$$h(x) = f \circ g(x) = f(g(x)) = f(7x + 3) = (7x + 3)^2.$$

This is the most ''obvious'' answer to the question. There can be other less obvious answers. For example, if

$$f(x) = (x - 1)^2 \quad \text{and} \quad g(x) = 7x + 4,$$

then

$$h(x) = f \circ g(x) = f(g(x)) = f(7x + 4) = (7x + 4 - 1)^2 = (7x + 3)^2.$$

❑

Inverses and One-to-One Functions

Let us consider the following two functions. We think of them as relations, or correspondences.

COST OF A 60-SECOND SUPER BOWL COMMERCIAL, BY YEAR

Domain (Set of Inputs)		Range (Set of Outputs)
1967	→	$80,000
1977	→	$324,000
1981	→	$550,000
1983	→	$800,000
1988	→	$1,350,000

U.S. SENATORS AND THEIR STATES

Domain (Set of Inputs)		Range (Set of Outputs)
Wellstone	→	Minnesota
Durenberger	→	Minnesota
Mack	→	Florida
Graham	→	Florida
Bradley	→	New Jersey
Lautenberg	→	New Jersey

Suppose we reverse the arrows. We obtain what is called the **inverse relation.** Are these inverse relations functions?

COST OF A 60-SECOND SUPER BOWL COMMERCIAL, BY YEAR

Range (Set of Outputs)		Domain (Set of Inputs)
1967	←	$80,000
1977	←	$324,000
1981	←	$550,000
1983	←	$800,000
1988	←	$1,350,000

U.S. SENATORS AND THEIR STATES

Range (Set of Outputs)		Domain (Set of Inputs)
Wellstone	←	Minnesota
Durenberger	←	Minnesota
Mack	←	Florida
Graham	←	Florida
Bradley	←	New Jersey
Lautenberg	←	New Jersey

We see that the inverse of the first correspondence is a function, but that the inverse of the second correspondence is not a function.

Recall that for each input, a function provides exactly one output. However, nothing in our definition of function prevents having the same output for two or more different inputs. Thus it is possible for different inputs to correspond to the same output in the range. Only when this possibility is *excluded* is the inverse also a function.

In the Super Bowl function, different inputs have different outputs. It is an example of a **one-to-one function**. In the U.S. Senator function, the inputs *Mack* and *Graham* both have the output *Florida*. Thus, the U.S. Senator function is not one-to-one.

One-to-One Function

A function f is *one-to-one* if different inputs have different outputs. That is, if for any $a \neq b$, we have $f(a) \neq f(b)$, the function f is one-to-one. If a function is one-to-one, then its inverse correspondence is also a function.

How can we tell graphically whether a function is one-to-one and thus has an inverse that is a function?

EXAMPLE 4

Shown here is the graph of a function. Determine whether the function is one-to-one and thus has an inverse that is a function.

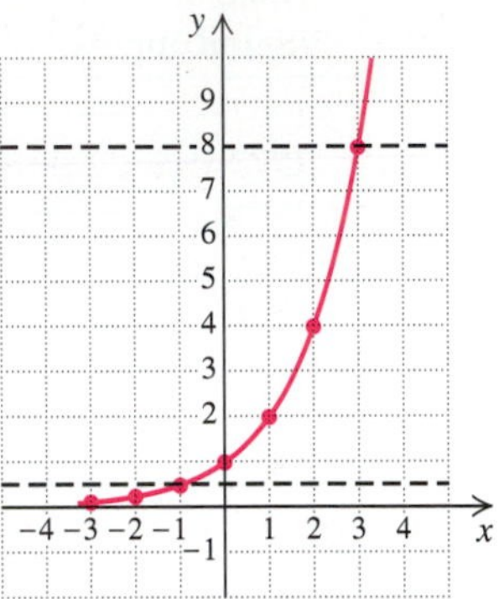

Solution A function is one-to-one if different inputs have different outputs. In other words, no two x-values will have the same y-value. For this function, we cannot find two x-values that have the same y-value. Note also that no horizontal line can be drawn that will cross the graph more than once. The function is one-to-one so its inverse is a function. ❑

The graph of any function must pass the vertical-line test. For a function to have an inverse that is a function, it must pass the *horizontal-line test* as well.

The Horizontal-Line Test

A function is one-to-one and has an inverse that is a function if there is no horizontal line that crosses the graph more than once.

EXAMPLE 5

Determine whether the function $f(x) = x^2$ is one-to-one and has an inverse that is also a function.

Solution The graph of $f(x) = x^2$ is shown here. Many horizontal lines cross the graph more than once, in particular the line $y = 4$.

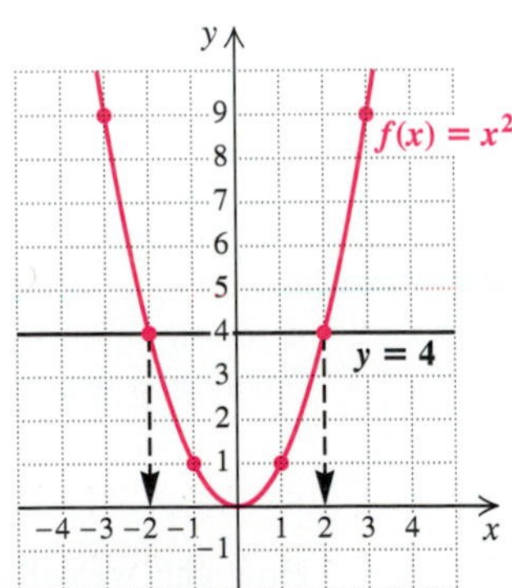

Note that where the line crosses, the first coordinates are -2 and 2. Although these are different inputs, they have the same output. That is, $-2 \neq 2$, but

$$f(-2) = (-2)^2 = 4 = 2^2 = f(2).$$

Thus the function is not one-to-one and no inverse function exists. ❑

Finding Formulas for Inverses

If the inverse of a function f is also a function, it can be named f^{-1} (read "f-inverse").

The -1 in f^{-1} is *not* an exponent!

Suppose a function is described by a formula. If it has an inverse that is a function, how do we find a formula for the inverse? For any equation in two variables, if we interchange the variables, we obtain an equation of the inverse correspondence. If it is a function, we proceed as follows to find a formula for f^{-1}.

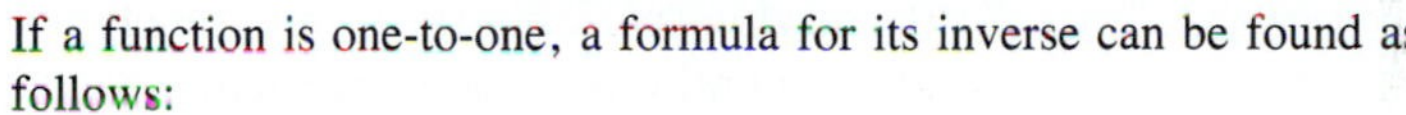

If a function is one-to-one, a formula for its inverse can be found as follows:

1. Replace $f(x)$ by y.
2. Interchange x and y. (This gives the inverse function.)
3. Solve for y.
4. Replace y by $f^{-1}(x)$.

EXAMPLE 6

Determine if the function is one-to-one and if it is, find a formula for $f^{-1}(x)$: **(a)** $f(x) = x + 2$; **(b)** $f(x) = 2x - 3$.

Solution

a) The graph of $f(x) = x + 2$ is shown below. It passes the horizontal-line test, so it is one-to-one. Thus its inverse is a function.

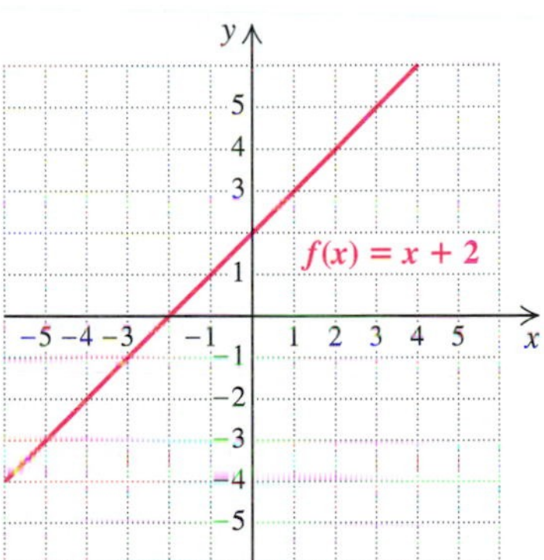

1. Replace $f(x)$ by y: $\quad y = x + 2.$
2. Interchange x and y: $\quad x = y + 2.$ This gives the inverse function.
3. Solve for y: $\quad x - 2 = y.$
4. Replace y by $f^{-1}(x)$: $\quad f^{-1}(x) = x - 2.$ We also "reversed" the equation.

In words, f adds 2 to all inputs. To "undo" f, f^{-1} subtracts 2 from all inputs.

b) The function $f(x) = 2x - 3$ is also linear. Any linear function that is not constant will pass the horizontal-line test. Thus, f is one-to-one.

1. Replace $f(x)$ by y: $\quad y = 2x - 3.$
2. Interchange x and y: $\quad x = 2y - 3.$
3. Solve for y: $\quad x + 3 = 2y$

$$\frac{x+3}{2} = y.$$

4. Replace y by $f^{-1}(x)$: $\quad f^{-1}(x) = \dfrac{x+3}{2}.$ ❑

Let us consider inverses of functions in terms of a function machine. Suppose that a function f programmed into a machine has an inverse that is also a function. Suppose that the function machine has a reverse switch. When the switch is thrown, the machine performs the inverse function f^{-1}. Inputs then enter at the opposite end, and the entire process is reversed.

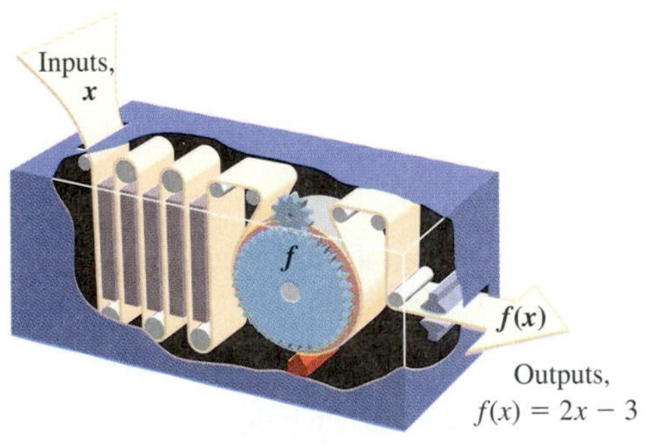

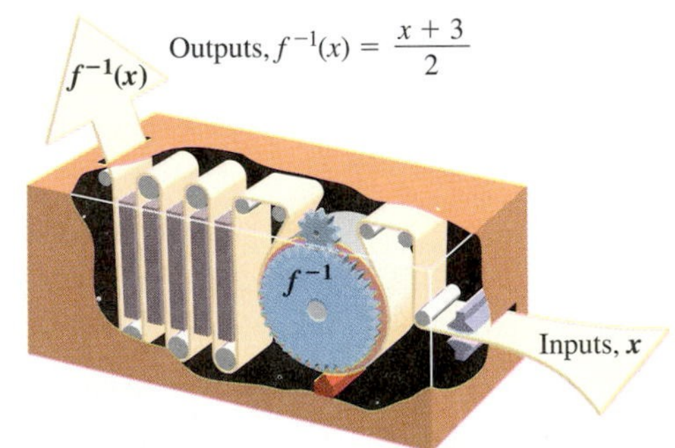

Consider $f(x) = 2x - 3$ and $f^{-1}(x) = \dfrac{x+3}{2}$ from Example 6(b). For the input 5, we have

$$f(5) = 2 \cdot 5 - 3 = 10 - 3 = 7.$$

The output is 7. Now we use 7 for the input in the inverse:

$$f^{-1}(7) = \frac{7+3}{2} = \frac{10}{2} = 5.$$

The function f takes 5 to 7. The inverse function f^{-1} takes the number 7 back to 5.

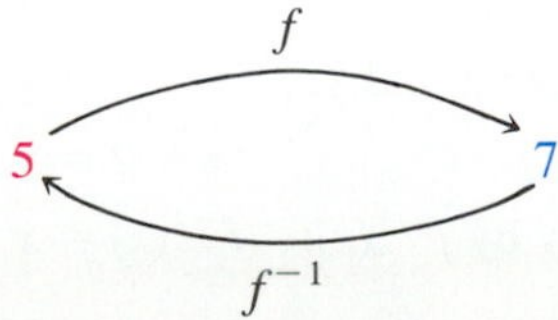

Graphing Functions and Their Inverses

How do the graphs of a function and its inverse compare?

EXAMPLE 7

Graph $f(x) = 2x - 3$ and $f^{-1}(x) = (x + 3)/2$ on the same axes. Then compare.

Solution The graph of each function follows. Note that the graph of f^{-1} can be drawn by reflecting the graph of f across the line $y = x$. That is, if we graph $f(x) = 2x - 3$ in wet ink and fold the paper along the line $y = x$, the graph of $f^{-1}(x) = (x + 3)/2$ will be formed from the impression made by f.

When we interchange y and x in finding a formula for the inverse, we are, in effect, flipping the graph of $f(x) = 2x - 3$ over the line $y = x$. For example, when the coordinates of the y-intercept of the graph of f, $(0, -3)$, are reversed, we get the x-intercept of the graph of f^{-1}, $(-3, 0)$.

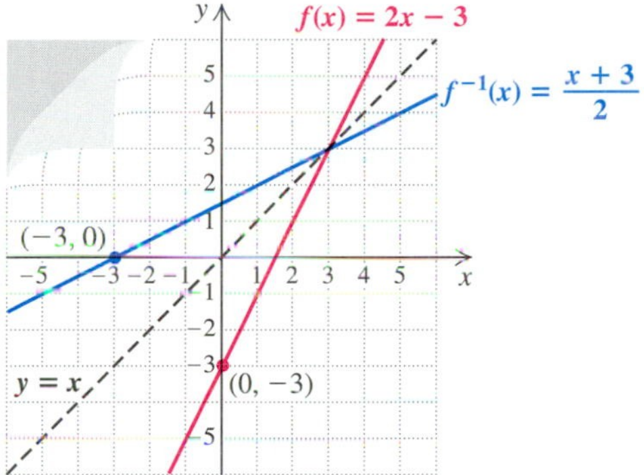

❑

TECHNOLOGY CONNECTION

A grapher can provide a partial check of whether or not two functions are inverses of each other. First we plot the functions believed to be inverses of each other and the line $y = x$, all on the same set of axes. Then, using the fact that a function and its inverse are reflections of each other across the line $y = x$, we can decide if the two functions may be inverses of each other.

Before we do this, however, there is a problem that must be addressed. Most graphers have rectangular, rather than square, displays. This means that a "square" window, like the standard $[-10, 10, -10, 10]$, distorts a graph because one unit in the x-direction is longer than one unit in the y-direction. To correct this, many graphers offer an option to "Square" axes, which guarantees that the unit length is the same in both the x- and y-directions.

To determine whether $y_1 = 2x + 6$ and $y_2 = \frac{1}{2}x - 3$ might be inverses of each other, we have drawn both functions, along with the line $y = x$, on a "squared" set of axes. It appears that y_1 and y_2 are inverses of each other.

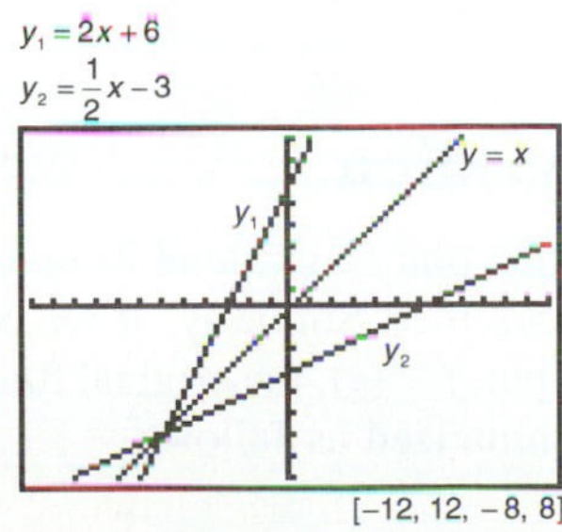

Use a grapher to help match each function in Column A with its inverse from Column B.

Column A	*Column B*
TC1. $y = 5x^3 + 10$	A. $y = \frac{\sqrt[3]{x} - 10}{5}$
TC2. $y = (5x + 10)^3$	B. $y = \sqrt[3]{\frac{x}{5}} - 10$
TC3. $y = 5(x + 10)^3$	C. $y = \sqrt[3]{\frac{x - 10}{5}}$
TC4. $y = (5x)^3 + 10$	D. $y = \frac{\sqrt[3]{x - 10}}{5}$

The graph of f^{-1} is a reflection of the graph of f across the line $y = x$.

EXAMPLE 8

Consider $g(x) = x^3 + 2$.

a) Determine whether the function is one-to-one.
b) If it is one-to-one, find a formula for its inverse.
c) Graph the inverse, if it exists.

Solution

a) The graph of $g(x) = x^3 + 2$ follows. It passes the horizontal-line test and thus has an inverse.

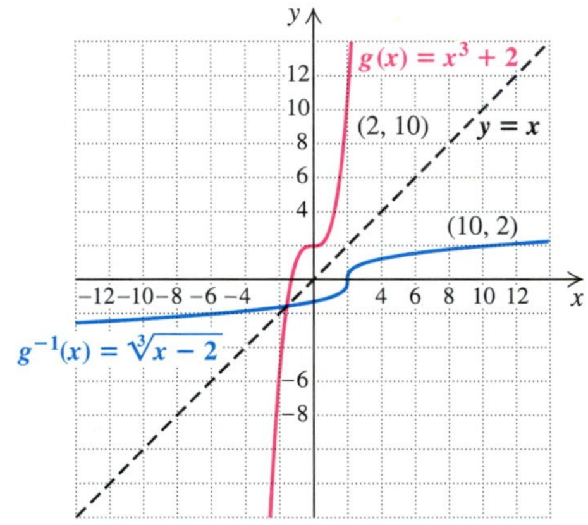

b) **1.** Replace $g(x)$ by y: $\quad y = x^3 + 2.$

2. Interchange x and y: $\quad x = y^3 + 2.$

3. Solve for y: $\quad x - 2 = y^3$

$$\sqrt[3]{x-2} = y.$$

Since a number has only one cube root, we can solve for y.

4. Replace y by $g^{-1}(x)$: $\quad g^{-1}(x) = \sqrt[3]{x-2}.$

c) To find the graph, we reflect the graph of $g(x) = x^3 + 2$ across the line $y = x$, as we did in Example 7. It can also be found by substituting into $g^{-1}(x) = \sqrt[3]{x-2}$ and plotting points. The graphs of g and g^{-1} are shown above using one set of axes. ❑

Inverse Functions and Composition

Suppose that we used some input x for the function f and found its output, $f(x)$. The function f^{-1} would then take that output back to x. Similarly, if we began with an input x for the function f^{-1} and found its output, $f^{-1}(x)$, the original function f would then take that output back to x. This is summarized as follows.

If a function f is one-to-one, then f^{-1} is the unique function such that

$$f^{-1} \circ f(x) = x \quad \text{and} \quad f \circ f^{-1}(x) = x.$$

EXAMPLE 9

Let $f(x) = 2x + 1$. Show that $f^{-1}(x) = (x - 1)/2$.

Solution We find $f^{-1} \circ f(x)$ and $f \circ f^{-1}(x)$ and check to see that each is x.

$$f^{-1} \circ f(x) = f^{-1}(f(x)) = f^{-1}(2x + 1) = \frac{(2x + 1) - 1}{2} = \frac{2x}{2} = x$$

$$f \circ f^{-1}(x) = f(f^{-1}(x)) = f\left(\frac{x - 1}{2}\right) = 2 \cdot \frac{x - 1}{2} + 1 = x - 1 + 1 = x$$ ❑

EXERCISE SET 12.2

Find $f \circ g(x)$ and $g \circ f(x)$.

1. $f(x) = 3x^2 + 2,\ g(x) = 2x - 1$
2. $f(x) = 4x + 3,\ g(x) = 2x^2 - 5$
3. $f(x) = 4x^2 - 1,\ g(x) = 2/x$
4. $f(x) = 3/x,\ g(x) = 2x^2 + 3$
5. $f(x) = x^2 + 1,\ g(x) = x^2 - 1$
6. $f(x) = 1/x^2,\ g(x) = x + 2$

Find $f(x)$ and $g(x)$ such that $h(x) = f \circ g(x)$. Answers may vary.

7. $h(x) = (5 - 3x)^2$
8. $h(x) = 4(3x - 1)^2 + 9$
9. $h(x) = (3x^2 - 7)^5$
10. $h(x) = \sqrt{5x + 2}$
11. $h(x) = \dfrac{1}{x - 1}$
12. $h(x) = \dfrac{3}{x} + 4$
13. $h(x) = \dfrac{1}{\sqrt{7x + 2}}$
14. $h(x) = \sqrt{x - 7} - 3$
15. $h(x) = \dfrac{x^3 + 1}{x^3 - 1}$
16. $h(x) = (\sqrt{x} + 5)^4$

Determine whether the function is one-to-one.

17. $f(x) = 3x - 4$
18. $f(x) = 5 - 2x$
19. $f(x) = x^2 - 3$
20. $f(x) = 1 - x^2$
21. $g(x) = 3^x$
22. $g(x) = (\frac{1}{2})^x$
23. $g(x) = |x|$
24. $h(x) = |x| - 1$

Given the function, **(a)** determine if it is one-to-one; **(b)** if it is one-to-one, find a formula for the inverse.

25. $f(x) = x + 4$
26. $f(x) = x + 7$
27. $f(x) = 5 - x$
28. $f(x) = 9 - x$
29. $g(x) = x - 5$
30. $g(x) = x - 8$
31. $f(x) = 3x$
32. $f(x) = 4x$
33. $g(x) = 3x + 2$
34. $g(x) = 4x + 7$
35. $h(x) = 7$
36. $h(x) = -3$
37. $f(x) = \dfrac{1}{x}$
38. $f(x) = \dfrac{3}{x}$
39. $f(x) = \dfrac{2x + 1}{3}$
40. $f(x) = \dfrac{3x + 2}{5}$
41. $f(x) = x^3 - 1$
42. $f(x) = x^3 + 5$
43. $g(x) = (x - 2)^3$
44. $g(x) = (x + 7)^3$
45. $f(x) = \sqrt{x}$
46. $f(x) = \sqrt{x - 1}$
47. $f(x) = 2x^2 + 3,\ x \geq 0$
48. $f(x) = 3x^2 - 2,\ x \geq 0$

Graph the function and its inverse using the same set of axes.

49. $f(x) = \frac{1}{2}x - 3$
50. $g(x) = x + 4$
51. $f(x) = x^3$
52. $f(x) = x^3 - 1$
53. $y = 2^x$
54. $y = 3^x$

55. $y = \left(\frac{1}{2}\right)^x$

56. $y = \left(\frac{2}{3}\right)^x$

57. $f(x) = 3 - x^2,\ x \geqslant 0$

58. $f(x) = x^2 - 1,\ x \leqslant 0$

59. Let $f(x) = \frac{4}{5}x$. Show that

$$f^{-1}(x) = \tfrac{5}{4}x.$$

60. Let $f(x) = (x + 7)/3$. Show that

$$f^{-1}(x) = 3x - 7.$$

61. Let $f(x) = (1 - x)/x$. Show that

$$f^{-1}(x) = \frac{1}{x + 1}.$$

62. Let $f(x) = x^3 - 5$. Show that

$$f^{-1}(x) = \sqrt[3]{x + 5}.$$

63. *Women's dress sizes in the United States and France.* A size-6 dress in the United States is size 38 in France. A function that will convert dress sizes in the United States to those in France is

$$f(x) = x + 32.$$

a) Find the dress sizes in France that correspond to sizes 8, 10, 14, and 18 in the United States.
b) Determine whether this function has an inverse that is a function. If so, find a formula for the inverse.
c) Use the inverse function to find dress sizes in the United States that correspond to sizes 40, 42, 46, and 50 in France.

64. *Women's dress sizes in the United States and Italy.* A size-6 dress in the United States is size 36 in Italy. A function that will convert dress sizes in the United States to those in Italy is

$$f(x) = 2(x + 12).$$

a) Find the dress sizes in Italy that correspond to sizes 8, 10, 14, and 18 in the United States.
b) Determine whether this function has an inverse that is a function. If so, find a formula for the inverse.
c) Use the inverse function to find dress sizes in the United States that correspond to sizes 40, 44, 52, and 60 in Italy.

Skill Maintenance

65. Find an equation of variation if y varies directly as x and $y = 7.2$ when $x = 0.8$.

66. Find an equation of variation if y varies inversely as x and $y = 3.5$ when $x = 6.1$.

Synthesis

67. ◈ Does the constant function $f(x) = 4$ have an inverse that is a function? If so, find a formula. If not, explain why.

68. ◈ An organization determines that the cost per person of chartering a bus is given by the function

$$C(x) = \frac{100 + 5x}{x},$$

where x = the number of people in the group and $C(x)$ is in dollars. Determine $C^{-1}(x)$ and explain how this inverse function could be used.

69. Use the information in Exercises 63 and 64 to find a function for the French dress size that corresponds to a size x dress in Italy.

In Exercises 70–73, use a grapher to help determine whether or not the given functions are inverses of each other.

70. $f(x) = 0.75x^2 + 2;\ g(x) = \sqrt{\dfrac{4(x - 2)}{3}}$

71. $f(x) = 1.4x^3 + 3.2;\ g(x) = \sqrt[3]{\dfrac{x - 3.2}{1.4}}$

72. $f(x) = \sqrt{2.5x + 9.25};$
$g(x) = 0.4x^2 - 3.7,\ x \geqslant 0$

73. $f(x) = 0.8x^{1/2} + 5.23;$
$g(x) = 1.25(x^2 - 5.23),\ x \geqslant 0$

12.3 Logarithmic Functions

Graphs of Logarithmic Functions • Converting Exponential and Logarithmic Equations • Solving Certain Logarithmic Equations

In this section we consider a type of function called a *logarithm function,* or *logarithmic function*. Such functions have many applications to problem solving.

Graphs of Logarithmic Functions

Consider the exponential function $f(x) = 2^x$. Does this function have an inverse that is a function? We see from the graph that this function is one-to-one so f^{-1} *is* a function.

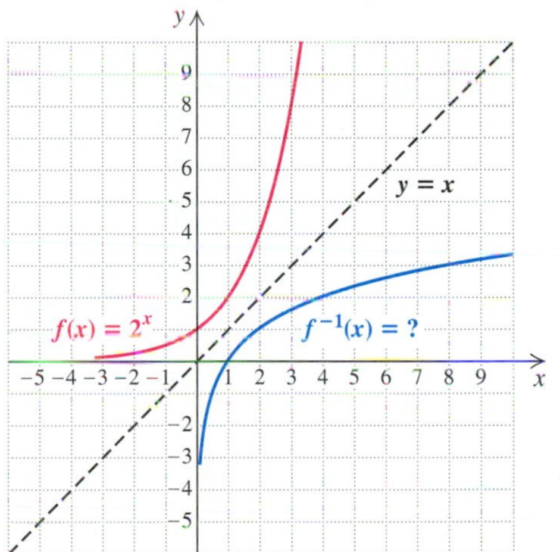

To find a formula for f^{-1}, we use the method of Section 12.2:

$$f(x) = 2^x.$$

1. Replace $f(x)$ by y: $\quad y = 2^x$.
2. Interchange x and y: $\quad x = 2^y$.
3. Solve for y: $\quad y =$ the power to which we raise 2 to get x.
4. Replace y by $f^{-1}(x)$: $\quad f^{-1}(x) =$ the power to which we raise 2 to get x.

We now define a new symbol to replace the words "the power to which we raise 2 to get x":

> **$\log_2 x$, read "the logarithm, base 2, of x", means "the power to which we raise 2 to get x."**

Thus if $f(x) = 2^x$, then $f^{-1}(x) = \log_2 x$. Note that $f^{-1}(8) = \log_2 8 = 3$, because 3 is *the power to which we raise 2 to get 8.*

Although expressions like $\log_2 13$ cannot be simplified, we must remember that $\log_2 13$ represents *the power to which we raise* 2 *to get* 13.

For any exponential function $f(x) = a^x$, the inverse is called a **logarithmic function, base *a*.** The graph of the inverse can, of course, be drawn by reflecting the graph of $f(x) = a^x$ across the line $y = x$. It will be helpful to remember that the inverse of $f(x) = a^x$ is given by $f^{-1}(x) = \log_a x$. Normally, we use a number a that is greater than 1 for the logarithm base.

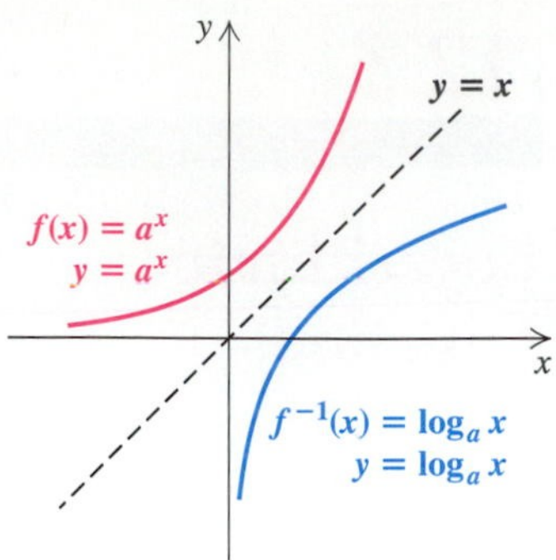

We define $y = \log_a x$ as that number y such that $a^y = x$, where $x > 0$ and a is a positive constant other than 1.

It is helpful in dealing with logarithmic functions to remember that the logarithm of a number is an *exponent*. It is the exponent y in a^y. You might also think to yourself, "The logarithm, base a, of a number x is the power to which a must be raised in order to get x."

A logarithm is an exponent.

The following is a comparison of exponential and logarithmic functions.

Exponential Function	**Logarithmic Function**
$y = a^x$	$x = a^y$
$f(x) = a^x$	$f(x) = \log_a x$
$a > 0,\ a \neq 1$	$a > 0,\ a \neq 1$
The input x can be any real number.	The output y can be any real number.
$y > 0$	$x > 0$
Outputs are positive.	Inputs are positive.

EXAMPLE 1 Graph: $y = f(x) = \log_5 x$.

Solution The equation $y = \log_5 x$ is equivalent to $5^y = x$. We can find ordered pairs that are solutions by choosing values for y and computing the x-values.

For $y = 0$, $x = 5^0 = 1$.
For $y = 1$, $x = 5^1 = 5$.
For $y = 2$, $x = 5^2 = 25$.
For $y = -1$, $x = 5^{-1} = \frac{1}{5}$.
For $y = -2$, $x = 5^{-2} = \frac{1}{25}$.

x, or 5^y	y
1	0
5	1
25	2
$\frac{1}{5}$	-1
$\frac{1}{25}$	-2

(1) Select y.
(2) Compute x.

We plot the set of ordered pairs and connect the points with a smooth curve. The graph of $y = 5^x$ has been shown only for reference.

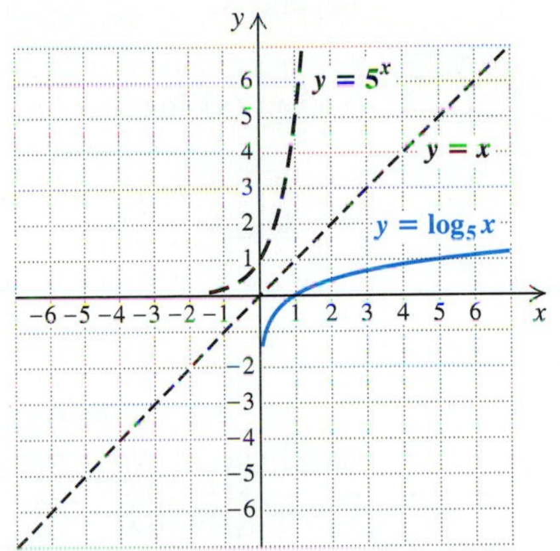

❑

Converting Exponential and Logarithmic Equations

We use the definition of logarithm to convert from *exponential equations* to *logarithmic equations:*

$$y = \log_a x \quad \text{is equivalent to} \quad a^y = x.$$

> CAUTION! **Be sure to memorize this relationship!** It is probably the most important definition in the chapter. Many times this definition will serve as a justification for a step that we are considering.

EXAMPLE 2

Convert to logarithmic equations: **(a)** $8 = 2^x$; **(b)** $y^{-1} = 4$; **(c)** $a^b = c$.

Solution

a) $8 = 2^x$ is equivalent to $x = \log_2 8$ — The exponent is the logarithm. The base remains the same.

b) $y^{-1} = 4$ is equivalent to $-1 = \log_y 4$

c) $a^b = c$ is equivalent to $b = \log_a c$ ❑

We also use the definition of logarithm to convert from logarithmic equations to exponential equations.

EXAMPLE 3

Convert to exponential equations: **(a)** $y = \log_3 5$; **(b)** $-2 = \log_a 7$; **(c)** $a = \log_b d$.

Solution

a) $y = \log_3 5$ is equivalent to $3^y = 5$ — The logarithm is the exponent. The base remains the same.

b) $-2 = \log_a 7$ is equivalent to $a^{-2} = 7$

c) $a = \log_b d$ is equivalent to $b^a = d$ ❑

Solving Certain Logarithmic Equations

Some logarithmic equations can be solved by first converting to exponential equations.

EXAMPLE 4

Solve: **(a)** $\log_2 x = -3$; **(b)** $\log_x 16 = 2$.

Solution

a) $\log_2 x = -3$

$2^{-3} = x$ Converting to an exponential equation

$\frac{1}{8} = x$ Computing 2^{-3}

Check: For $\frac{1}{8}$ to be the solution, $\log_2 \frac{1}{8}$ should equal -3. Since $2^{-3} = \frac{1}{8}$, we know that $\frac{1}{8}$ checks and is the solution.

b) $\log_x 16 = 2$

$x^2 = 16$ Converting to an exponential equation

$x = 4$ *or* $x = -4$ Principle of square roots

Check: $\log_4 16 = 2$ because $4^2 = 16$. Thus, 4 is a solution. Because all logarithm bases must be positive, $\log_{-4} 16$ is not defined. Therefore, -4 is not a solution. Logarithm bases must be positive because logarithms are defined using exponential functions that are defined only for positive bases. ❑

Solving for x in $\log_b a = x$ amounts to finding the logarithm, base b, of the number a. To think of finding logarithms as solving equations may help.

EXAMPLE 5

Simplify: **(a)** $\log_{10} 1000$; **(b)** $\log_{10} 0.01$; **(c)** $\log_5 1$.

Solution

a) Method 1. Let $\log_{10} 1000 = x$. Then

$10^x = 1000$ Converting to an exponential equation

$10^x = 10^3$ Writing 1000 as a power of 10

$x = 3.$ The exponents must be the same.

Therefore, $\log_{10} 1000 = 3$.

Method 2. Think of the meaning of $\log_{10} 1000$. It is the exponent to which we raise 10 to get 1000. That exponent is 3. Therefore, $\log_{10} 1000 = 3$.

b) Method 1. Let $\log_{10} 0.01 = x$. Then

$10^x = 0.01$ Converting to an exponential equation

$10^x = \frac{1}{100}$

$10^x = 10^{-2}$ Writing $\frac{1}{100}$ as a power of 10

$x = -2.$ Equating exponents

Therefore, $\log_{10} 0.01 = -2$.

Method 2. $\log_{10} 0.01$ is the exponent to which we raise 10 to get 0.01. If we note that $0.01 = 1/100 = 1/10^2$, it follows that the exponent is -2. Therefore, $\log_{10} 0.01 = -2$.

c) **Method 1.** Let $\log_5 1 = x$. Then

$5^x = 1$ Converting to an exponential equation

$5^x = 5^0$ Writing 1 as a power of 5

$x = 0.$

Therefore, $\log_5 1 = 0$.

Method 2. $\log_5 1$ is the exponent to which we raise 5 to get 1. That exponent is 0. Therefore, $\log_5 1 = 0$. ❑

Example 5(c) illustrates an important property of logarithms.

The logarithm, base a, of 1 is always 0: $\log_a 1 = 0$.

This follows from the fact that $a^0 = 1$ is equivalent to the logarithmic equation $\log_a 1 = 0$.

Another property results from the fact that $a^1 = a$. This is equivalent to the equation $\log_a a = 1$.

The logarithm, base a, of a is always 1: $\log_a a = 1$.

Thus, $\log_{10} 10 = 1$, $\log_8 8 = 1$, and so on.

EXERCISE SET 12.3

Graph.

1. $y = \log_2 x$
2. $y = \log_{10} x$
3. $y = \log_6 x$
4. $y = \log_3 x$
5. $f(x) = \log_4 x$
6. $f(x) = \log_5 x$
7. $f(x) = \log_{1/2} x$
8. $f(x) = \log_{2.5} x$

Graph both functions using the same set of axes.

9. $f(x) = 3^x$, $f^{-1}(x) = \log_3 x$
10. $f(x) = 4^x$, $f^{-1}(x) = \log_4 x$

Convert to logarithmic equations.

11. $10^3 = 1000$
12. $10^2 = 100$
13. $5^{-3} = \frac{1}{125}$
14. $4^{-5} = \frac{1}{1024}$
15. $8^{1/3} = 2$
16. $16^{1/4} = 2$
17. $10^{0.3010} = 2$
18. $10^{0.4771} = 3$
19. $e^2 = t$
20. $p^k = 3$
21. $Q^t = x$
22. $p^m = V$
23. $e^2 = 7.3891$
24. $e^3 = 20.0855$
25. $e^{-2} = 0.1353$
26. $e^{-4} = 0.0183$

Convert to exponential equations.

27. $t = \log_3 8$
28. $h = \log_7 10$
29. $\log_5 25 = 2$
30. $\log_6 6 = 1$
31. $\log_{10} 0.1 = -1$
32. $\log_{10} 0.01 = -2$
33. $\log_{10} 7 = 0.845$
34. $\log_{10} 3 = 0.4771$
35. $\log_e 20 = 2.9957$
36. $\log_e 10 = 2.3026$
37. $\log_t Q = k$
38. $\log_m P = a$
39. $\log_e 0.25 = -1.3863$
40. $\log_e 0.989 = -0.0111$
41. $\log_r T = -x$
42. $\log_c M = -w$

Solve.

43. $\log_3 x = 2$
44. $\log_4 x = 3$
45. $\log_x 125 = 3$
46. $\log_x 64 = 3$
47. $\log_2 16 = x$
48. $\log_5 25 = x$
49. $\log_3 27 = x$
50. $\log_4 16 = x$
51. $\log_x 13 = 1$
52. $\log_x 23 = 1$
53. $\log_6 x = 0$
54. $\log_9 x = 1$
55. $\log_2 x = -1$
56. $\log_3 x = -2$
57. $\log_8 x = \frac{1}{3}$
58. $\log_{32} x = \frac{1}{5}$

Find each of the following.

59. $\log_{10} 100$
60. $\log_{10} 100{,}000$
61. $\log_{10} 0.1$
62. $\log_{10} 0.001$
63. $\log_{10} 1$
64. $\log_{10} 10$
65. $\log_5 625$
66. $\log_2 64$
67. $\log_5 \frac{1}{25}$
68. $\log_2 \frac{1}{16}$
69. $\log_3 1$
70. $\log_4 4$
71. $\log_e e$
72. $\log_e 1$
73. $\log_{27} 9$
74. $\log_8 2$
75. $\log_e e^3$
76. $\log_e e^{-4}$
77. $\log_{10} 10^t$
78. $\log_3 3^p$

Skill Maintenance

Simplify.

79. $\dfrac{\frac{3}{x} - \frac{2}{xy}}{\frac{2}{x^2} + \frac{1}{xy}}$
80. $\dfrac{\frac{4+x}{x^2+2x+1}}{\frac{3}{x+1} - \frac{2}{x+2}}$

Rename.

81. 8^{-4}
82. $x^{4/5}$
83. $t^{-2/3}$
84. 5^1

Synthesis

85. ◈ Explain why 1 is excluded from being a logarithmic base.
86. ◈ Explain why the number $\log_2 13$ is between 3 and 4.
87. Graph both equations using the same set of axes: $y = \left(\frac{3}{2}\right)^x$, $y = \log_{3/2} x$.

Graph.

88. $y = \log_2 (x - 1)$
89. $y = \log_3 |x + 1|$

Solve.

90. $|\log_3 x| = 3$
91. $\log_{125} x = \frac{2}{3}$
92. $\log_4 (3x - 2) = 2$
93. $\log_8 (2x + 1) = -1$
94. $\log_{10} (x^2 + 21x) = 2$

Simplify.

95. $\log_{1/4} \frac{1}{64}$
96. $\log_{81} 3 \cdot \log_3 81$
97. $\log_{10} (\log_4 (\log_3 81))$
98. $\log_2 (\log_2 (\log_4 256))$
99. $\log_{1/5} 25$

12.4 Properties of Logarithmic Functions

Logarithms of Products • Logarithms of Powers • Logarithms of Quotients • Using the Properties Together

Logarithmic functions are important in many applications and in more advanced mathematics. We now establish some basic properties that are useful in manipulating expressions involving logarithms. As the proofs of these properties reveal, the properties of logarithms are related to the properties of exponents.

Logarithms of Products

The first property we discuss is reminiscent of the property $a^m \cdot a^n = a^{m+n}$.

The Product Rule for Logarithms

For any positive numbers M, N, and a $(a \neq 1)$,

$$\log_a MN = \log_a M + \log_a N.$$

(The logarithm of a product is the sum of the logarithms of the factors.)

EXAMPLE 1 Express as a sum of logarithms: $\log_2 (4 \cdot 16)$.

Solution We have

$$\log_2 (4 \cdot 16) = \log_2 4 + \log_2 16. \qquad \text{Using the product rule}$$

As a check, note that

$$\log_2 (4 \cdot 16) = \log_2 64 = 6$$

and that

$$\log_2 4 + \log_2 16 = 2 + 4 = 6.$$ ❑

EXAMPLE 2 Express as a single logarithm: $\log_{10} 0.01 + \log_{10} 1000$.

Solution We have

$$\begin{aligned} \log_{10} 0.01 + \log_{10} 1000 &= \log_{10} (0.01 \times 1000) \qquad \text{Using the product rule} \\ &= \log_{10} 10. \end{aligned}$$ ❑

A Proof of the Product Rule. Let $\log_a M = x$ and $\log_a N = y$. Converting to exponential equations, we have $a^x = M$ and $a^y = N$.

Now we multiply the latter two equations, to obtain

$$MN = a^x \cdot a^y = a^{x+y}.$$

Converting back to a logarithmic equation, we get

$$\log_a MN = x + y.$$

Recalling what x and y represent, we get

$$\log_a MN = \log_a M + \log_a N.$$ ❑

Logarithms of Powers

The second basic property is related to the property $(a^m)^n = a^{mn}$.

The Power Rule for Logarithms

For any positive numbers M and a $(a \neq 1)$, and any real number p,

$$\log_a M^p = p \cdot \log_a M.$$

(The logarithm of a power of M is the exponent times the logarithm of M.)

EXAMPLE 3

Express as a product: **(a)** $\log_a 9^{-5}$; **(b)** $\log_a \sqrt[4]{5}$.

Solution

a) $\log_a 9^{-5} = -5 \log_a 9$ Using the power rule

b) $\log_a \sqrt[4]{5} = \log_a 5^{1/4}$ Writing exponential notation

$= \frac{1}{4} \log_a 5$ Using the power rule ❑

A Proof of the Power Rule. Let $x = \log_a M$. We then convert to an exponential equation, to get $a^x = M$. Raising both sides to the pth power, we obtain

$$(a^x)^p = M^p, \quad \text{or} \quad a^{xp} = M^p.$$

Converting back to a logarithmic equation gives us

$$\log_a M^p = xp.$$

But $x = \log_a M$, so substituting, we have

$$\log_a M^p = (\log_a M)p = p \cdot \log_a M.$$

❑

Logarithms of Quotients

The third property that we study is similar to the property $a^m/a^n = a^{m-n}$.

The Quotient Rule for Logarithms

For any positive numbers M, N, and a ($a \neq 1$),

$$\log_a \frac{M}{N} = \log_a M - \log_a N.$$

(The logarithm of a quotient is the logarithm of the dividend minus the logarithm of the divisor.)

EXAMPLE 4

Express as a difference of logarithms: $\log_t (6/U)$.

Solution

$\log_t \frac{6}{U} = \log_t 6 - \log_t U$ Using the quotient rule ❑

EXAMPLE 5

Express as a single logarithm: $\log_b 17 - \log_b 27$.

Solution

$\log_b 17 - \log_b 27 = \log_b \frac{17}{27}$ Using the quotient rule "in reverse" ❑

A Proof of the Quotient Rule. Our proof uses both the product and power rules:

$\log_a \frac{M}{N} = \log_a MN^{-1}$

$= \log_a M + \log_a N^{-1}$ Using the product rule

$= \log_a M + (-1) \log_a N$ Using the power rule

$= \log_a M - \log_a N.$ ❑

Using the Properties Together

EXAMPLE 6

Express in terms of logarithms of x, y, z, m, and n.

a) $\log_a \dfrac{x^2y^3}{z^4}$ b) $\log_a \sqrt[4]{\dfrac{xy}{z^3}}$ c) $\log_b \dfrac{xy}{m^3n^4}$

Solution

a) $\log_a \dfrac{x^2y^3}{z^4} = \log_a (x^2y^3) - \log_a z^4$ Using the quotient rule

$= \log_a x^2 + \log_a y^3 - \log_a z^4$ Using the product rule

$= 2 \log_a x + 3 \log_a y - 4 \log_a z$ Using the power rule

b) $\log_a \sqrt[4]{\dfrac{xy}{z^3}} = \log_a \left(\dfrac{xy}{z^3}\right)^{1/4}$ Writing exponential notation

$= \dfrac{1}{4} \cdot \log_a \dfrac{xy}{z^3}$ Using the power rule

$= \dfrac{1}{4} (\log_a xy - \log_a z^3)$ Using the quotient rule. Parentheses are important.

$= \dfrac{1}{4} (\log_a x + \log_a y - 3 \log_a z)$ Using the product and power rules

c) $\log_b \dfrac{xy}{m^3n^4} = \log_b xy - \log_b m^3n^4$ Using the quotient rule

$= (\log_b x + \log_b y) - (\log_b m^3 + \log_b n^4)$ Using the product rule

$= \log_b x + \log_b y - \log_b m^3 - \log_b n^4$ Removing parentheses

$= \log_b x + \log_b y - 3 \log_b m - 4 \log_b n$ Using the power rule

❑

EXAMPLE 7

Express as a single logarithm.

a) $\dfrac{1}{2} \log_a x - 7 \log_a y + \log_a z$

b) $\log_a \dfrac{b}{\sqrt{x}} + \log_a \sqrt{bx}$

Solution

a) $\dfrac{1}{2} \log_a x - 7 \log_a y + \log_a z$

$= \log_a x^{1/2} - \log_a y^7 + \log_a z$ Using the power rule

$= (\log_a \sqrt{x} - \log_a y^7) + \log_a z$ Using parentheses to emphasize the order of operations; $x^{1/2} = \sqrt{x}$

$= \log_a \dfrac{\sqrt{x}}{y^7} + \log_a z$ Using the quotient rule

$= \log_a \dfrac{z\sqrt{x}}{y^7}$ Using the product rule

b) $\log_a \dfrac{b}{\sqrt{x}} + \log_a \sqrt{bx} = \log_a \dfrac{b}{\sqrt{x}} \sqrt{bx}$ — Using the product rule

$= \log_a b\sqrt{b}$ — Removing a factor of 1: $\sqrt{x}/\sqrt{x} = 1$

$= \log_a b^{3/2}$, or $\dfrac{3}{2} \log_a b$ — Since $b\sqrt{b} = b^1 \cdot b^{1/2}$ ❑

EXAMPLE 8

Given $\log_a 2 = 0.301$ and $\log_a 3 = 0.477$, find each of the following: **(a)** $\log_a 6$; **(b)** $\log_a \frac{2}{3}$; **(c)** $\log_a 81$; **(d)** $\log_a \frac{1}{3}$; **(e)** $\log_a 2a$; **(f)** $\log_a 5$.

Solution

a) $\log_a 6 = \log_a (2 \cdot 3) = \log_a 2 + \log_a 3$ — Using the product rule

$= 0.301 + 0.477 = 0.778$

b) $\log_a \frac{2}{3} = \log_a 2 - \log_a 3$ — Using the quotient rule

$= 0.301 - 0.477 = -0.176$

c) $\log_a 81 = \log_a 3^4 = 4 \log_a 3$ — Using the power rule

$= 4(0.477) = 1.908$

d) $\log_a \frac{1}{3} = \log_a 1 - \log_a 3$ — Using the quotient rule

$= 0 - 0.477 = -0.477$

e) $\log_a 2a = \log_a 2 + \log_a a$ — Using the product rule

$= 0.301 + 1 = 1.301$

f) $\log_a 5$ *cannot be found using these properties.*
($\log_a 5 \neq \log_a 2 + \log_a 3$) ❑

A final property follows from the product rule: Since $\log_a a^k = k \log_a a$, and $\log_a a = 1$, we have $\log_a a^k = k$.

The Logarithm of the Base to a Power

For any base a,

$$\log_a a^k = k.$$

(The logarithm, base a, of a to a power is the power.)

This property also follows from the definition of logarithm: k is the power to which you raise a in order to get a^k.

EXAMPLE 9

Simplify: **(a)** $\log_3 3^7$; **(b)** $\log_{10} 10^{5.6}$; **(c)** $\log_e e^{-t}$.

Solution

a) $\log_3 3^7 = 7$ — 7 is the power to which you raise 3 in order to get 3^7.

b) $\log_{10} 10^{5.6} = 5.6$

c) $\log_e e^{-t} = -t$ ❑

CAUTION! Keep in mind that, in general,

$$\log_a (M + N) \neq \log_a M + \log_a N,$$
$$\log_a (M - N) \neq \log_a M - \log_a N,$$
$$\log_a MN \neq (\log_a M)(\log_a N), \text{ and}$$
$$\log_a (M/N) \neq (\log_a M) \div (\log_a N).$$

EXERCISE SET 12.4

Express as a sum of logarithms.

1. $\log_2 (32 \cdot 8)$
2. $\log_3 (27 \cdot 81)$
3. $\log_4 (64 \cdot 16)$
4. $\log_5 (25 \cdot 125)$
5. $\log_c Bx$
6. $\log_t 5Y$

Express as a single logarithm.

7. $\log_a 6 + \log_a 70$
8. $\log_b 65 + \log_b 2$
9. $\log_c K + \log_c y$
10. $\log_t H + \log_t M$

Express as a product.

11. $\log_a x^3$
12. $\log_b t^5$
13. $\log_c y^6$
14. $\log_{10} y^7$
15. $\log_b C^{-3}$
16. $\log_c M^{-5}$

Express as a difference of logarithms.

17. $\log_a \frac{67}{5}$
18. $\log_t \frac{T}{7}$
19. $\log_b \frac{3}{4}$
20. $\log_a \frac{y}{x}$

Express as a single logarithm.

21. $\log_a 15 - \log_a 7$
22. $\log_b 42 - \log_b 7$

Express in terms of logarithms of w, x, y, and z.

23. $\log_a x^2y^3z$
24. $\log_a xy^4z^3$
25. $\log_b \frac{xy^2}{z^3}$
26. $\log_b \frac{x^2y^5}{w^4z^7}$
27. $\log_c \sqrt[3]{\frac{x^4}{y^3z^2}}$
28. $\log_a \sqrt{\frac{x^6}{y^5z^8}}$
29. $\log_a \sqrt[4]{\frac{x^8y^{12}}{a^3z^5}}$
30. $\log_a \sqrt[3]{\frac{x^6y^3}{a^2z^7}}$

Express as a single logarithm and simplify, if possible.

31. $\frac{2}{3} \log_a x - \frac{1}{2} \log_a y$
32. $\frac{1}{2} \log_a x + 3 \log_a y - 2 \log_a x$
33. $\log_a 2x + 3(\log_a x - \log_a y)$
34. $\log_a x^2 - 2 \log_a \sqrt{x}$
35. $\log_a \frac{a}{\sqrt{x}} - \log_a \sqrt{ax}$
36. $\log_a (x^2 - 4) - \log_a (x - 2)$

Given $\log_b 3 = 1.099$ and $\log_b 5 = 1.609$, find each of the following.

37. $\log_b 15$
38. $\log_b \frac{3}{5}$
39. $\log_b \frac{5}{3}$
40. $\log_b \frac{1}{3}$
41. $\log_b \frac{1}{5}$
42. $\log_b \sqrt{b}$
43. $\log_b \sqrt{b^3}$
44. $\log_b 3b$
45. $\log_b 5b$
46. $\log_b 9$
47. $\log_b 25$
48. $\log_b 75$

Simplify.

49. $\log_t t^9$
50. $\log_p p^4$
51. $\log_e e^m$
52. $\log_Q Q^{-2}$

Solve for x.

53. $\log_3 3^4 = x$
54. $\log_5 5^7 = x$
55. $\log_e e^x = -7$
56. $\log_a a^x = 2.7$

Skill Maintenance

Compute and simplify. Express answers in the form $a + bi$, where $i^2 = -1$.

57. i^{29}
58. $(2 + i)(2 - i)$
59. $\frac{2 + i}{2 - i}$
60. $(7 - 8i) - (-16 + 10i)$

Synthesis

61. ◈ Explain why $a^{\log_a 5} = 5$.

62. ◈ The product, power, and quotient rules enable us to simplify expressions like $\log_a (rs/pv)$. Explain why such expressions can always be simplified without ever using the quotient rule.

Express as a single logarithm and simplify, if possible.

63. $\log_a (x^8 - y^8) - \log_a (x^2 + y^2)$

64. $\log_a (x + y) + \log_a (x^2 - xy + y^2)$

Express as a sum or difference of logarithms.

65. $\log_a \sqrt{1 - s^2}$

66. $\log_a \dfrac{c - d}{\sqrt{c^2 - d^2}}$

67. If $\log_a x = 2$, $\log_a y = 3$, and $\log_a z = 4$, what is

$$\log_a \frac{\sqrt[3]{x^2 z}}{\sqrt[3]{y^2 z^{-2}}}?$$

68. If $\log_a x = 2$, what is $\log_a (1/x)$?

69. If $\log_a x = 2$, what is $\log_{1/a} x$?

Determine whether each of the following is true. Assume a, x, P, and $Q > 0$.

70. $\dfrac{\log_a P}{\log_a Q} = \log_a \dfrac{P}{Q}$

71. $\dfrac{\log_a P}{\log_a Q} = \log_a P - \log_a Q$

72. $\log_a 3x = \log_a 3 + \log_a x$

73. $\log_a 3x = 3 \log_a x$

74. $\log_a (P + Q) = \log_a P + \log_a Q$

75. $\log_a x^2 = 2 \log_a x$

76. Prove that for $a > 0$ and $x \geq \sqrt{3}$,

$$\log_a \frac{x + \sqrt{x^2 - 3}}{3} = -\log_a \left(x - \sqrt{x^2 - 3}\right).$$

12.5 Common and Natural Logarithms

Common Logarithms on a Calculator • The Base *e* and Natural Logarithms on a Calculator • Changing Logarithmic Bases • Graphs of Exponential and Logarithmic Functions, Base *e*

Any positive number different from 1 can be used as the base of a logarithmic function. However, some numbers are easier to use than others, and there are logarithm bases that fit into certain applications more naturally than others. Base-10 logarithms, called **common logarithms,** are useful because they are the same base as our "commonly" used decimal system for naming numbers.

Before calculators became so widely available, common logarithms were extensively used in calculations. In fact, that is why logarithms were invented. Another logarithm base widely used today is an irrational number named e. We will consider e and natural logarithms later in this section. We first consider common logarithms.

Common Logarithms on a Calculator

Before the invention of calculators, tables were developed to list common logarithms. Today we find common logarithms using calculators.

The abbreviation **log,** with no base written, is understood to mean logarithm base 10, or a common logarithm. Thus

$$\log 17 \quad \text{means} \quad \log_{10} 17.$$

On scientific calculators, the key for common logarithms is usually marked [LOG]. To find the common logarithm of a number, we enter that number and press the [LOG] key. Table 2 at the back of this text can also be used.

EXAMPLE 1

Use a calculator to find each number: **(a)** log 53,128; **(b)** log 0.000128.

Solution

a) We enter 53,128 and then press the [log] key. We find that

$$\log 53{,}128 \approx 4.7253. \quad \text{Rounded to four decimal places}$$

b) We enter 0.000128 and then press the [log] key. We find that

$$\log 0.000128 \approx -3.8928. \quad \text{Rounded to four decimal places}$$

❑

The inverse of a logarithmic function is, of course, an exponential function. The inverse of finding a logarithm is called finding an *antilogarithm* or an *inverse logarithm*. To find an antilogarithm, we evaluate a power of the base:

$$\text{for} \quad f(x) = \log x, \quad f^{-1}(x) = \text{antilog } x = 10^x.$$

Generally, there is no key on a calculator marked "antilog." It is up to you to know that to find the inverse, or antilogarithm, you must use the [10^x] key, if there is one. If there is no such key, then you must raise 10 to the x power using a key marked [x^y]. Often the [log] key serves as the [10^x] key after a "shift" or "inverse" key is pushed.

EXAMPLE 2

Use a calculator to find each number.

a) antilog 2.1792

b) antilog (−4.678834)

Solution

a) We enter 2.1792 and then press the [10^x] key. We find that

$$\text{antilog } 2.1792 = 10^{2.1792} \approx 151.078.$$

b) $\text{antilog } (-4.678834) = 10^{-4.678834} \approx 0.00002095$ ❑

The Base *e* and Natural Logarithms on a Calculator

When interest is computed n times a year, the compound interest formula is

$$A = P\left(1 + \frac{i}{n}\right)^{nt},$$

where A is the amount that an initial investment P will be worth after t years at interest rate i. Suppose that \$1 is an initial investment at 100% interest for 1 year (no bank would pay this). The preceding formula becomes a function A defined in terms of the number of compounding periods n:

$$A(n) = \left(1 + \frac{1}{n}\right)^n.$$

Let us find some function values. We round to six decimal places, using a calculator with a power key [x^y].

n	$A(n) = \left(1 + \frac{1}{n}\right)^n$
1 (compounded annually)	\$2.00
2 (compounded semiannually)	\$2.25
3	\$2.370370
4 (compounded quarterly)	\$2.441406
5	\$2.488320
100	\$2.704814
365 (compounded daily)	\$2.714567
8760 (compounded hourly)	\$2.718127

The numbers in this table get closer and closer to a very important number in mathematics, called e. Being irrational, its decimal representation does not terminate or repeat.

$$e \approx 2.7182818284\ldots$$

Logarithms base e are called **natural logarithms,** or **Napierian logarithms,** in honor of John Napier (1550–1617), who first "discovered" logarithms.

The abbreviation "ln" is generally used with natural logarithms. Thus,

$\ln 53$ means $\log_e 53$.

The calculator key [ln] is used for natural logarithms.

EXAMPLE 3

Find ln 4568.

Solution We enter 4568 and then press the [ln] key. We find that

$\ln 4568 \approx 8.4268.$ Rounded to four decimal places ❑

To find the antilogarithm, base e, we use the [e^x] key, if there is one. If not, we can use a power key [x^y] and an approximation for e, say, 2.71828. Often the [ln] key serves as the [e^x] key after a "shift" or "inverse" key is pushed.

EXAMPLE 4

Find $\text{antilog}_e\,(-5.6734)$.

Solution The problem gives the exponent. We enter -5.6734 and then press the [e^x] key. We find that

$$\text{antilog}_e\,(-5.6734) = e^{-5.6734} \approx 0.003436.$$ ❑

Changing Logarithmic Bases

Most calculators give the values of both common logarithms and natural logarithms. To find a logarithm with some other base, we can use the following conversion formula.

The Change-of-Base Formula

For any logarithmic bases a and b, and any positive number M,

$$\log_b M = \frac{\log_a M}{\log_a b}.$$

Proof. Let $x = \log_b M$. Then, writing an equivalent exponential equation, we have $b^x = M$. Next we take the logarithm base a on both sides. This gives us

$$\log_a b^x = \log_a M.$$

By the power rule for logarithms,

$$x \log_a b = \log_a M,$$

and solving for x, we obtain

$$x = \frac{\log_a M}{\log_a b}.$$

But $x = \log_b M$, so we have

$$\log_b M = \frac{\log_a M}{\log_a b},$$

which is the change-of-base formula. ❑

EXAMPLE 5

Find $\log_5 8$ using common logarithms.

Solution Let $a = 10$, $b = 5$, and $M = 8$. Then we substitute into the change-of-base formula:

$$\log_5 8 = \frac{\log_{10} 8}{\log_{10} 5} \qquad \text{Substituting into } \log_b M = \frac{\log_a M}{\log_a b}$$

$$\approx \frac{0.9031}{0.6990} \qquad \text{Using the LOG key twice}$$

$$\approx 1.2920. \qquad \text{When using a calculator, you need not round before dividing.}$$

To check, we use a calculator with an x^y key to verify that

$$5^{1.2920} \approx 8.$$ ❑

We can also use base e for a conversion.

EXAMPLE 6

Find $\log_4 31$ using natural logarithms.

Solution Substituting e for a, 4 for b, and 31 for M, we have

$$\log_4 31 = \frac{\log_e 31}{\log_e 4} \qquad \text{Substituting into } \log_b M = \frac{\log_a M}{\log_a b}$$

$$= \frac{\ln 31}{\ln 4}$$

or

$$\log_4 31 \approx \frac{3.4340}{1.3863} \quad \text{Using the } \boxed{\text{LN}} \text{ key twice}$$
$$\approx 2.4771.$$

❑

Graphs of Exponential and Logarithmic Functions, Base *e*

EXAMPLE 7

Graph $f(x) = e^x$ and $g(x) = e^{-x}$.

Solution We use a calculator with an $\boxed{e^x}$ key to find approximate values of e^x and e^{-x}. Using these values, we can draw the graphs of the functions.

x	e^x	e^{-x}
0	1	1
1	2.7	0.4
2	7.4	0.1
−1	0.4	2.7
−2	0.1	7.4

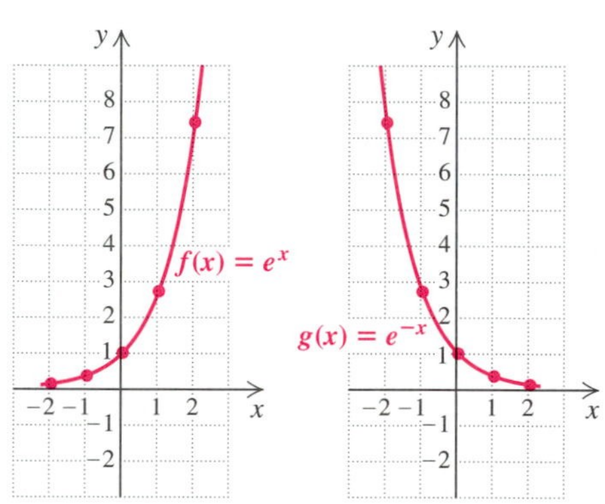

❑

EXAMPLE 8

Graph: $f(x) = e^{-0.5x}$.

Solution We find some solutions with a calculator, plot them, and then draw the graph. For example, $f(2) = e^{-0.5(2)} = e^{-1} \approx 0.4$.

x	$e^{-0.5x}$
0	1
1	0.6
2	0.4
3	0.2
−1	1.6
−2	2.7
−3	4.5

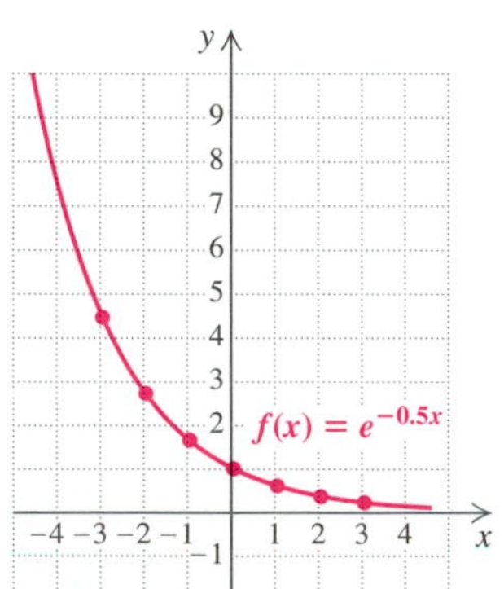

❑

EXAMPLE 9

Graph: **(a)** $g(x) = \ln x$; **(b)** $f(x) = \ln(x + 3)$.

Solution

a) We find some solutions with a calculator and then draw the graph. As expected, the graph is a reflection across the line $y = x$ of the graph of $y = e^x$.

x	$\ln x$
1	0
4	1.4
7	1.9
0.5	−0.7

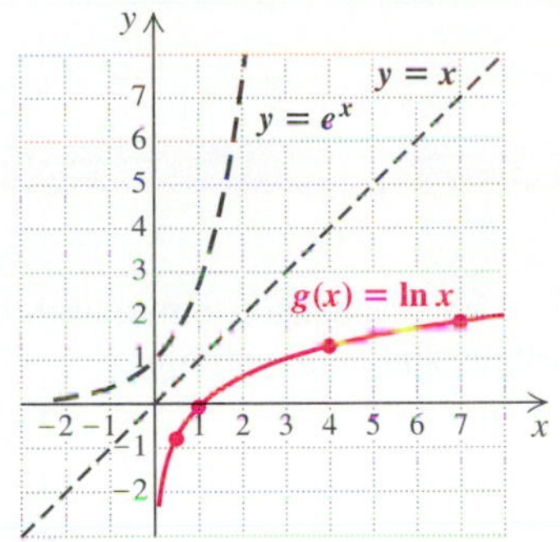

b) We find some solutions with a calculator, plot them, and then draw the graph.

x	$\ln(x+3)$
0	1.1
1	1.4
2	1.6
3	1.8
4	1.9
−1	0.7
−2	0
−2.5	−0.7

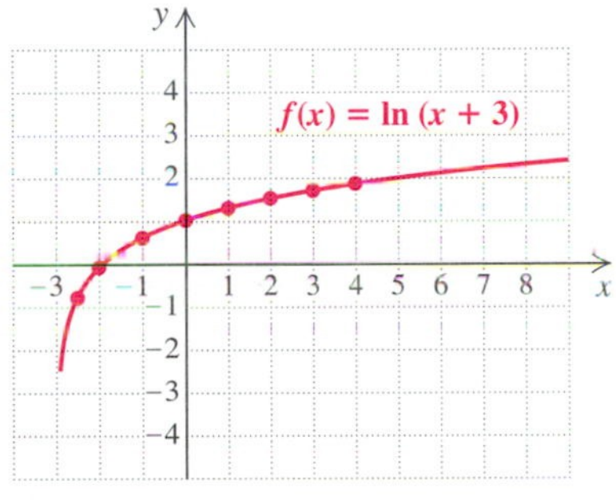

Note that the graph of $y = \ln(x+3)$ is the graph of $y = \ln x$ translated 3 units to the left. ❑

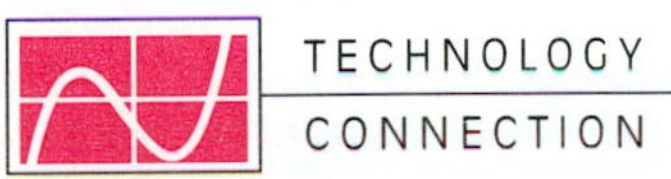

Graphs that involve e^x and/or $\ln x$ are easily drawn on a grapher because there are usually keys for these functions. Since these functions are inverses of each other, most graphers use the same key for both.

To graph a function like $y_1 = 3.4 \ln x - 0.25e^x$, check the domain and set the window accordingly. Since $\ln x$ is not defined for $x \leq 0$, our window need not include any negative values of x. The figure at right shows the function drawn in the window $[0, 5, -10, 10]$.

$y_1 = 3.4 \ln x - 0.25e^x$

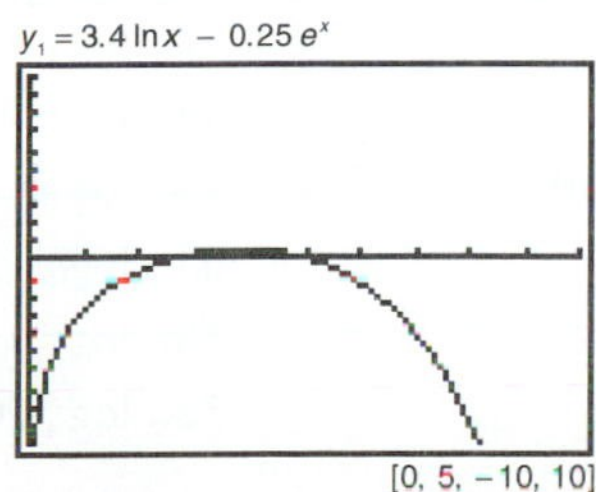

$[0, 5, -10, 10]$

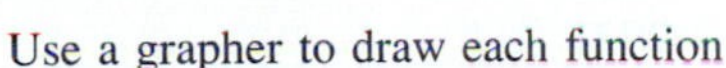
Use a grapher to draw each function.

TC1. $f(x) = 2.3e^x - 3.8$

TC2. $f(x) = \dfrac{3e^x - 6.3}{e^x + 2.1}$

TC3. $f(x) = xe^{-2.3x} + 3x^2$

TC4. $f(x) = 7.4e^x \ln x$

TC5. $f(x) = 5.3 \ln(x - 2.1)$

TC6. $f(x) = 2x^3 \ln x$

EXERCISE SET 12.5

Use a calculator to find each of the following logarithms and antilogarithms.

1. log 2
2. log 5
3. log 6.34
4. log 5.02
5. log 45
6. log 74
7. log 437
8. log 295
9. log 13,400
10. log 93,100
11. log 0.052
12. log 0.387
13. antilog 3
14. antilog 5
15. antilog 2.7
16. antilog 14.8
17. antilog 0.477133
18. antilog 0.06532
19. antilog (−0.5465)
20. antilog (−0.3404)
21. $10^{-2.9523}$
22. $10^{4.8982}$
23. ln 2
24. ln 3
25. ln 62
26. ln 30
27. ln 4365
28. ln 901.2
29. ln 0.0062
30. ln 0.00073
31. antilog_e 3.6052
32. antilog_e 4.9312
33. antilog_e (−6.0751)
34. antilog_e (−2.3001)
35. antilog_e 0.00567
36. antilog_e 0.01111
37. antilog_e 34
38. antilog_e 56
39. $e^{2.0325}$
40. $e^{-1.3783}$

Find each of the following logarithms using the change-of-base formula.

41. $\log_6 100$
42. $\log_3 18$
43. $\log_2 10$
44. $\log_7 50$
45. $\log_{200} 30$
46. $\log_{100} 30$
47. $\log_{0.5} 5$
48. $\log_{0.1} 3$
49. $\log_2 0.2$
50. $\log_2 0.08$
51. $\log_\pi 58$
52. $\log_\pi 200$

Graph.

53. $f(x) = e^x$
54. $f(x) = e^{3x}$
55. $f(x) = e^{-3x}$
56. $f(x) = e^{-x}$
57. $f(x) = e^{x-1}$
58. $f(x) = e^{x+2}$
59. $f(x) = e^{-x} - 3$
60. $f(x) = e^x + 3$
61. $f(x) = 5e^{0.2x}$
62. $f(x) = 8e^{0.6x}$
63. $f(x) = 20e^{-0.5x}$
64. $f(x) = 10e^{-0.4x}$
65. $f(x) = \ln (x + 4)$
66. $f(x) = \ln (x + 1)$
67. $f(x) = 3 \ln x$
68. $f(x) = 2 \ln x$
69. $f(x) = \ln (x - 2)$
70. $f(x) = \ln (x - 3)$
71. $f(x) = \ln x - 3$
72. $f(x) = \ln x + 2$

Skill Maintenance

Solve for x.

73. $4x^2 - 25 = 0$
74. $5x^2 - 7x = 0$

Solve.

75. $x^{1/2} - 6x^{1/4} + 8 = 0$
76. $2y - 7\sqrt{y} + 3 = 0$

Synthesis

77. ◈ Explain how the graph of $f(x) = e^x$ could be used to graph the function given by $g(x) = 1 + \ln x$.
78. ◈ Explain how the graph of $f(x) = \ln x$ could be used to graph the function given by $g(x) = e^{x-1}$.
79. Find a formula for converting common logarithms to natural logarithms.
80. Find a formula for converting natural logarithms to common logarithms.

Solve for x.

81. $\log 374x = 4.2931$
82. $\log 95x^2 = 3.0177$
83. $\log 692 + \log x = \log 3450$
84. $\dfrac{4.31}{\ln x} = \dfrac{28}{3.01}$

12.6 Solving Exponential and Logarithmic Equations

Solving Exponential Equations • Solving Logarithmic Equations

Solving Exponential Equations

Equations with variables in exponents, such as $5^x = 12$ and $2^{7x} = 64$, are called **exponential equations.** Sometimes, as in the case of $2^{7x} = 64$, we can write each side as a power of the same number:

$$2^{7x} = 2^6.$$

Since the base is the same, 2, the exponents are the same. We can equate exponents and then solve:

$$7x = 6$$
$$x = \tfrac{6}{7}.$$

We use the following property.

> For any $a > 0$, $a \neq 1$,
>
> $a^x = a^y$ is equivalent to $x = y$.

This follows from the fact that $f(x) = a^x$ is a one-to-one function, so that if two outputs a^x and a^y are equal, their inputs are equal.

EXAMPLE 1

Solve: $2^{3x-5} = 16$.

Solution Note that $16 = 2^4$. Thus we can write each side as a power of the same number:

$$2^{3x-5} = 2^4.$$

Since the base is the same, 2, the exponents must be the same. Thus,

$$3x - 5 = 4 \quad \text{Equating exponents}$$
$$3x = 9$$
$$x = 3.$$

Check:

2^{3x-5}	$= 16$
$2^{3\cdot 3-5}$	? 16
2^{9-5}	
2^4	
16	16 TRUE

The solution is 3. ❑

When it does not seem possible to write each side as a power of the same base, we can take the common or natural logarithm on each side and then use the power rule for logarithms.

EXAMPLE 2

Solve: $5^x = 12$.

Solution

$$5^x = 12$$

$$\log 5^x = \log 12 \quad \text{Taking the common logarithm on both sides}$$

$$x \log 5 = \log 12 \quad \text{Using the power rule}$$

$$x = \frac{\log 12}{\log 5} \quad \leftarrow$$

CAUTION! This is not $\log 12 - \log 5$!

This is an exact answer. We cannot simplify further, but we can approximate using a calculator or a table:

$$x = \frac{\log 12}{\log 5} \approx \frac{1.0792}{0.6990} \approx 1.544.$$

You can check this by finding $5^{1.544}$ using the $\boxed{x^y}$ key on a calculator. ❑

If we prefer, we can take the logarithm with e as the base. This will often ease our work.

EXAMPLE 3

Solve: $e^{0.06t} = 1500$.

Solution We take the natural logarithm on both sides:

$$\ln e^{0.06t} = \ln 1500 \quad \text{Taking the natural logarithm on both sides}$$

$$0.06t = \ln 1500 \quad \text{Finding the logarithm of the base to a power: } \log_a a^k = k$$

$$0.06t \approx 7.3132 \quad \text{Using a calculator}$$

$$t \approx 121.89.$$

❑

Solving Logarithmic Equations

Equations containing logarithmic expressions are called **logarithmic equations.** We solved some such equations in Section 12.3. We did so by converting to an equivalent exponential equation.

EXAMPLE 4

Solve: $\log_4 (8x - 6) = 3$.

Solution We write an equivalent exponential equation:

$$8x - 6 = 4^3$$

$$8x = 70 \quad 4^3 = 64 \text{ and } 64 + 6 = 70$$

$$x = \tfrac{70}{8}, \quad \text{or } \tfrac{35}{4}.$$

The check is left for the student. The solution is $\frac{35}{4}$. ❑

To solve logarithmic equations, first try to obtain a single logarithmic expression on one side and then write an equivalent exponential equation.

EXAMPLE 5 Solve: $\log x + \log (x - 3) = 1$.

Solution We have common logarithms here. It will help to write in the base, 10.

$$\log_{10} x + \log_{10} (x - 3) = 1$$

$$\log_{10} [x(x - 3)] = 1 \quad \text{Using the product rule for logarithms to obtain a single logarithm}$$

$$x(x - 3) = 10^1 \quad \text{Writing an equivalent exponential equation}$$

$$x^2 - 3x = 10$$

$$x^2 - 3x - 10 = 0$$

$$(x + 2)(x - 5) = 0 \quad \text{Factoring}$$

$$x + 2 = 0 \quad \text{or} \quad x - 5 = 0 \quad \text{Principle of zero products}$$

$$x = -2 \quad \text{or} \quad x = 5$$

Check: For −2:

$\log x + \log (x - 3) = 1$	
$\log (-2) + \log (-2 - 3)$	? 1

The number −2 *does not check* because negative numbers do not have logarithms.

For 5:

$\log x + \log (x - 3) = 1$	
$\log 5 + \log (5 - 3)$	? 1
$\log 5 + \log 2$	
$\log 10$	
1	1 TRUE

The solution is 5. ❑

TECHNOLOGY CONNECTION

Exponential and logarithmic equations can be solved using a grapher by graphing the function on each side of the equals sign, making sure that the window is large enough to show all points of intersection. Then zoom in on any such points and use the Trace feature to determine their x-coordinates.

For example, to solve $e^{0.5x} - 7 = 2x + 6$, we graph the curves $y_1 = e^{0.5x} - 7$ and $y_2 = 2x + 6$ as shown in the figure at right. We then adjust the window and use the Trace feature to pinpoint any intersections. The x-coordinates at the intersections are approximately −6.48 and 6.52.

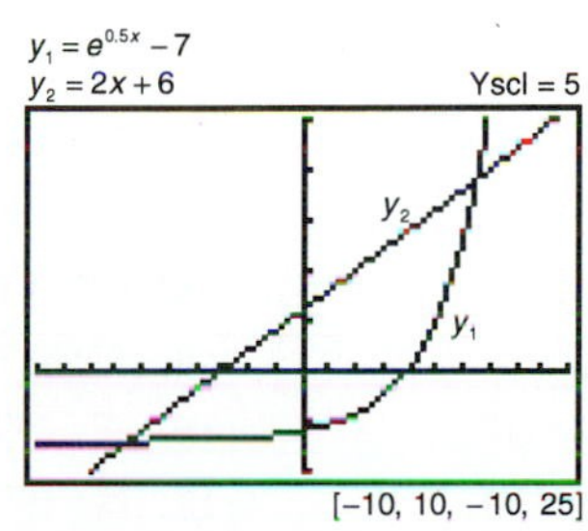

Use a grapher to find solutions, accurate to the nearest hundredth, for each of the following equations.

TC1. $e^{7x} = 14$

TC2. $8e^{0.5x} = 3$

TC3. $xe^{3x - 1} = 5$

TC4. $4 \ln (x + 3.4) = 2.5$

TC5. $\ln 3x = 3x - 8$

TC6. $\ln x^2 = -x^2$

EXAMPLE 6 Solve: $\log_2 (x + 7) - \log_2 (x - 7) = 3$.

Solution

$$\log_2 (x + 7) - \log_2 (x - 7) = 3$$

$$\log_2 \frac{x+7}{x-7} = 3$$ Using the quotient rule for logarithms to obtain a single logarithm

$$\frac{x+7}{x-7} = 2^3$$ Writing an equivalent exponential expression

$$\frac{x+7}{x-7} = 8$$

$$x + 7 = 8(x - 7)$$ Multiplying by the LCD, $x - 7$

$$x + 7 = 8x - 56$$ Using the distributive law

$$63 = 7x$$

$$\frac{63}{7} = x$$

$$9 = x$$

Check:

$\log_2 (x + 7) - \log_2 (x - 7) = 3$	
$\log_2 (9 + 7) - \log_2 (9 - 7)$?	3
$\log_2 16 - \log_2 2$	
$4 - 1$	
3	3 TRUE

The solution is 9. ❑

EXERCISE SET 12.6

Solve.

1. $2^x = 8$
2. $3^x = 81$
3. $4^x = 256$
4. $5^x = 125$
5. $2^{2x} = 32$
6. $4^{3x} = 64$
7. $3^{5x} = 27$
8. $5^{7x} = 625$
9. 🖩 $2^x = 9$
10. 🖩 $2^x = 30$
11. 🖩 $2^x = 10$
12. 🖩 $2^x = 33$
13. $5^{4x-7} = 125$
14. $4^{3x+5} = 16$
15. $3^{x^2} \cdot 3^{4x} = \frac{1}{27}$
16. $3^{5x} \cdot 3^{2x^2} = 27$
17. 🖩 $4^x = 7$
18. 🖩 $8^x = 10$
19. 🖩 $e^t = 100$
20. 🖩 $e^t = 1000$
21. 🖩 $e^{-t} = 0.1$
22. 🖩 $e^{-t} = 0.01$
23. 🖩 $e^{-0.02t} = 0.06$
24. 🖩 $e^{0.07t} = 2$
25. 🖩 $2^x = 3^{x-1}$
26. 🖩 $3^{x+2} = 5^{x-1}$
27. 🖩 $(2.8)^x = 41$
28. 🖩 $(3.4)^x = 80$
29. 🖩 $20 - (1.7)^x = 0$
30. 🖩 $125 - (4.5)^y = 0$
31. $\log_3 x = 3$
32. $\log_5 x = 4$
33. $\log_2 x = -3$
34. $\log_4 x = \frac{1}{2}$
35. $\log x = 1$
36. $\log x = 3$
37. $\log x = -2$
38. $\log x = -3$
39. $\ln x = 2$
40. $\ln x = 1$
41. $\ln x = -1$
42. $\ln x = -3$
43. $\log_5 (2x - 7) = 3$
44. $\log_2 (7 - 6x) = 5$
45. $\log x + \log (x - 9) = 1$
46. $\log x + \log (x + 9) = 1$

47. $\log x - \log (x + 3) = -1$

48. $\log (x + 9) - \log x = 1$

49. $\log_2 (x + 1) + \log_2 (x - 1) = 3$

50. $\log_4 (x + 3) - \log_4 (x - 5) = 2$

51. $\log_4 (x + 6) - \log_4 x = 2$

52. $\log_2 x + \log_2 (x - 2) = 3$

53. $\log_4 (x + 3) + \log_4 (x - 3) = 2$

54. $\log_5 (x + 4) + \log_5 (x - 4) = 2$

Skill Maintenance

Simplify.

55. $(125x^7y^{-2}z^6)^{-2/3}$

56. i^{79}

Solve.

57. $E = mc^2$, for c (Assume $E, m, c > 0$.)

58. $x^4 + 400 = 104x^2$

Synthesis

59. ◈ Explain how Exercises 35–38 could be solved using the graph of $f(x) = \log x$.

60. ◈ In Example 2, we took the common logarithm on both sides. What would have happened had we taken the natural logarithm on both sides?

Solve.

61. $8^x = 16^{3x + 9}$

62. $27^x = 81^{2x - 3}$

63. $\log_6 (\log_2 x) = 0$

64. $\log_x (\log_3 27) = 3$

65. $\log_5 \sqrt{x^2 - 9} = 1$

66. $x \log \frac{1}{8} = \log 8$

67. $\log (\log x) = 5$

68. $2^{x^2 + 4x} = \frac{1}{8}$

69. $\log x^2 = (\log x)^2$

70. $\log_5 |x| = 4$

71. $\log x^{\log x} = 25$

72. $\log \sqrt{2x} = \sqrt{\log 2x}$

73. $(81^{x - 2})(27^{x + 1}) = 9^{2x - 3}$

74. $3^{2x} - 8 \cdot 3^x + 15 = 0$

75. $3^{2x} - 3^{2x - 1} = 18$

76. Given that $2^y = 16^{x - 3}$ and $3^{y + 2} = 27^x$, find the value of $x + y$.

77. If $x = (\log_{125} 5)^{\log_5 125}$, what is the value of $\log_3 x$?

78. Use a grapher to check the solutions of Exercises 11, 19, 25, 35, and 47.

12.7 Applications of Exponential and Logarithmic Functions

Exponential Growth • Exponential Decay

We now consider applications of exponential and logarithmic functions. A calculator with logarithmic and power keys would be most helpful.

EXAMPLE 1

Chemistry: pH of substances. In chemistry the pH of a substance is defined as follows:

$$\text{pH} = -\log [\text{H}^+],$$

where $[\text{H}^+]$ is the hydrogen ion concentration in moles per liter.

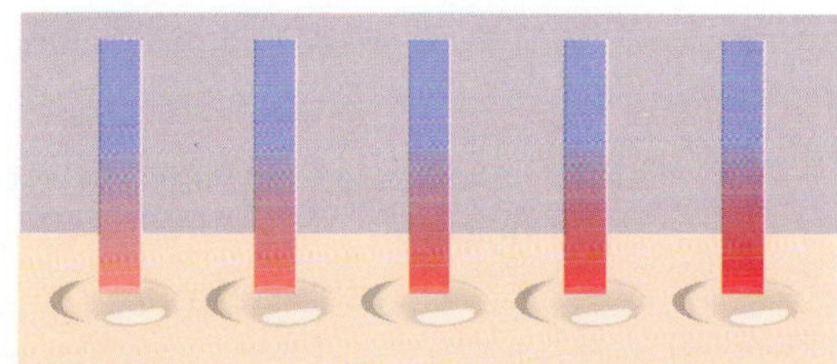

a) The hydrogen ion concentration of pineapple juice is 1.6×10^{-4} moles per liter. Find the pH.

b) The pH of a common hair rinse is 2.9. Find the hydrogen ion concentration.

Solution

a) To find the pH of pineapple juice, we substitute 1.6×10^{-4} for $[H^+]$ in the formula for pH:

$$\begin{aligned} pH = -\log [H^+] &= -\log [1.6 \times 10^{-4}] \\ &= -[\log 1.6 + \log 10^{-4}] \\ &= -[\log 1.6 + (-4)] \quad &&\text{log 1.6 is in Table 2.} \\ &\approx -[0.2041 + (-4)] \\ &\approx -(-3.7959) \approx 3.8 \quad &&\text{This can be found directly using a calculator.} \end{aligned}$$

The pH of pineapple juice is about 3.8.

b) To find the hydrogen ion concentration of the hair rinse, we substitute 2.9 for pH in the formula and solve for $[H^+]$:

$$\begin{aligned} 2.9 &= -\log [H^+] \\ -2.9 &= \log [H^+] \quad &&\text{Multiplying by } -1 \text{ on both sides} \\ 10^{-2.9} &= [H^+] \quad &&\text{Writing an equivalent exponential equation} \\ 0.0013 &\approx [H^+] \\ [H^+] &\approx 1.3 \times 10^{-3} \text{ moles per liter.} \quad &&\text{Writing scientific notation: a number between 1 and 10, times a power of 10} \end{aligned}$$

❑

EXAMPLE 2

Earthquake magnitude. The magnitude R (measured on the Richter scale) of an earthquake of intensity I is defined as

$$R = \log \frac{I}{I_0},$$

where I_0 is a minimum intensity used for comparison. We can regard I_0 as the intensity of the weakest earthquake that can be recorded on a seismograph. When one earthquake is 10 times as intense as another, its magnitude on the Richter scale is 1 higher. If an earthquake is 100 times as intense as another, its magnitude on the Richter scale is 2 higher, and so on. Thus an earthquake whose magnitude is 7 on the Richter scale is 10 times as intense as an earthquake whose magnitude is 6. The San Francisco (Loma Prieta) earthquake of 1989 had an intensity of $10^{7.2}I_0$. What was its magnitude on the Richter scale?

Solution We substitute into the formula:

$$R = \log \frac{10^{7.2}I_0}{I_0} = \log 10^{7.2} = 7.2.$$

The magnitude on the Richter scale was 7.2. ❑

EXAMPLE 3

Interest compounded annually. Suppose that \$30,000 is invested at 8% interest, compounded annually. In t years, it will grow to the amount A given by the function

$$A(t) = 30{,}000(1.08)^t.$$

(See Example 5 in Section 12.1.)

a) How long will it take until there is \$150,000 in the account?
b) Let $T =$ the amount of time it takes for the \$30,000 to double itself. Find the *doubling time, T.*

Solution

a) We set $A(t) = 150{,}000$ and solve for t:

$$150{,}000 = 30{,}000(1.08)^t$$

$$\frac{150{,}000}{30{,}000} = 1.08^t \qquad \text{Dividing by 30,000 on both sides}$$

$$5 = 1.08^t$$

$$\log 5 = \log 1.08^t \qquad \text{Taking the common logarithm on both sides}$$

$$\log 5 = t \log 1.08 \qquad \text{Using the power rule for logarithms}$$

$$\frac{\log 5}{\log 1.08} = t.$$

We simplify further by using a calculator or Table 2 and approximating:

$$t = \frac{\log 5}{\log 1.08} \approx \frac{0.69897}{0.03342} \approx 20.9.$$

It will take about 20.9 years for the \$30,000 to grow to \$150,000.

Calculator Note: When doing a calculation like this on your calculator, it is best not to stop and round the approximate values of the logarithms. Just find and divide. Answers will be found that way in the exercises. You may notice some variation in the last one or two decimal places if you round as you go.

b) To find the doubling time T, we set $A(t) = 60{,}000$ and $t = T$ and solve for T:

$$60{,}000 = 30{,}000(1.08)^T$$

$$2 = (1.08)^T \qquad \text{Dividing by 30,000 on both sides}$$

$$\log 2 = \log (1.08)^T \qquad \text{Taking the common logarithm on both sides}$$

$$\log 2 = T \log 1.08 \qquad \text{Using the power rule for logarithms}$$

$$T = \frac{\log 2}{\log 1.08} \approx \frac{0.30103}{0.03342} \approx 9.0.$$

The doubling time is about 9 years. ❑

Exponential Growth

An equation of the form

$$P(t) = P_0 e^{kt}$$

can model the growth of many things, ranging from growing populations to investments that are increasing in value. In this equation, P_0 is the population at time 0, $P(t)$ is the population at time t, and k is a positive constant that depends on the situation. The constant k is often called the **exponential growth rate.** You should regard the exponential growth rate as a population's rate of growth at any *instant* in time. Since the population is continually growing, the percent of total growth after one year will exceed the exponential growth rate.

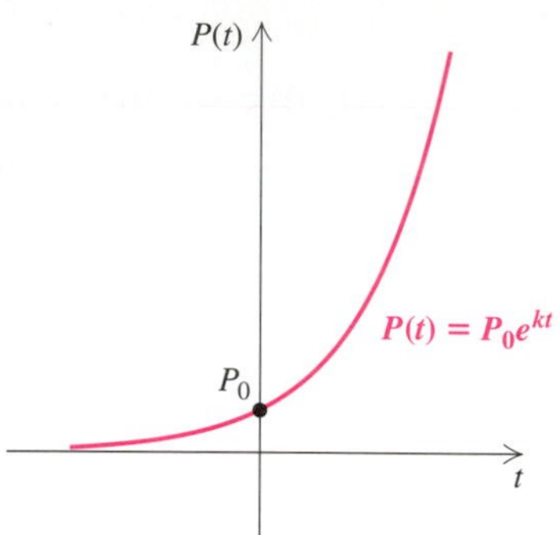

EXAMPLE 4

Growth of the United States. In 1992, the population of the United States was 249 million and the exponential growth rate was 0.9% per year.

a) Find the exponential growth function.
b) What would you expect the population to be in the year 2000?

Solution

a) In 1992, at $t = 0$, the population was 249 million. We substitute 249 for P_0 and 0.9%, or 0.009, for k to obtain the exponential growth function:

$$P(t) = 249e^{0.009t}.$$

b) In 2000, we have $t = 8$. That is, 8 years have passed. To find the population in 2000, we substitute 8 for t:

$$P(8) = 249e^{0.009(8)} \quad \text{Substituting 8 for } t$$
$$= 249e^{0.072}$$
$$\approx 249(1.0747) \quad \text{Finding } e^{0.072} \text{ using a calculator}$$
$$\approx 267.6 \text{ million.}$$

The population of the United States in 2000 will be about 267.6 million. ❑

EXAMPLE 5

Spread of AIDS. The number of people $N(t)$ infected with a contagious disease at time t usually increases exponentially. Through August 1989, 100,000 cases of AIDS had been reported in the United States. By the end of December 1991, the number had grown to 200,000.

a) Find the exponential growth rate and the exponential growth function.
b) Predict the year in which the 500,000th case will occur.

Solution

a) We use the equation $N(t) = N_0e^{rt}$, where t is the number of months since August 1989. In August 1989, at $t = 0$, a total of 100,000 cases had been reported. We substitute 100,000 for N_0:

$$N(t) = 100{,}000e^{rt}.$$

To find the exponential growth rate r, observe that the 200,000th case was reported in December 1991, or 28 months after August 1989. Substituting, we can solve for r:

$$N(28) = 100{,}000e^{r \cdot 28}$$

$$200{,}000 = 100{,}000e^{28r}$$

$$2 = e^{28r} \qquad \text{Dividing by 100,000 on both sides}$$

$$\ln 2 = \ln e^{28r} \qquad \text{Taking the natural logarithm on both sides}$$

$$\ln 2 = 28r \qquad \text{Finding the logarithm of the base to a power}$$

$$\frac{\ln 2}{28} = r \qquad \text{Dividing by 28 on both sides}$$

$$0.025 \approx r. \qquad \text{Using a calculator and rounding}$$

The exponential function is $N(t) = 100{,}000e^{0.025t}$, where t is measured in months since August 1989.

b) To predict when the 500,000th case will occur, we replace $N(t)$ with 500,000 and solve for t:

$$500{,}000 = 100{,}000e^{0.025t}$$

$$5 = e^{0.025t} \qquad \text{Dividing by 100,000 on both sides}$$

$$\ln 5 = \ln e^{0.025t} \qquad \text{Taking the natural logarithm on both sides}$$

$$\ln 5 = 0.025t$$

$$\frac{\ln 5}{0.025} = t$$

$$64 \approx t. \qquad \text{Using a calculator}$$

Since 64 months is 5 years and 4 months, we predict that the 500,000th case will occur 5 years and 4 months from August 1989, or in December 1994. ❑

EXAMPLE 6

Interest compounded continuously. Suppose an amount of money P_0 is invested in a savings account at interest rate k, compounded continuously. That is, suppose that interest is computed every "instant" and added to the amount in the account. The balance $P(t)$, after t years, is given by the exponential function

$$P(t) = P_0e^{kt}.$$

a) Suppose that \$30,000 is invested and grows to \$44,754.75 in 5 years. Find the exponential growth function.

b) After what amount of time will the \$30,000 double itself?

Solution

a) At $t = 0$, $P(0) = 30{,}000$. Thus the exponential growth function is

$$P(t) = 30{,}000e^{kt}, \qquad \text{where } k \text{ must still be determined.}$$

We know that at $t = 5$, $P(5) = 44{,}754.75$. We substitute and solve for k:

$$44{,}754.75 = 30{,}000e^{k(5)} = 30{,}000e^{5k}$$

$$\frac{44{,}754.75}{30{,}000} = e^{5k} \qquad \text{Dividing on both sides by 30,000}$$

$$1.491825 = e^{5k}$$

$$\ln 1.491825 = \ln e^{5k} \qquad \text{Taking the natural logarithm on both sides}$$

$$0.4 \approx 5k \qquad \text{Finding ln 1.491825 on a calculator and simplifying } \ln e^{5k}$$

$$\frac{0.4}{5} = 0.08 \approx k.$$

The interest rate is about 0.08, or 8%, compounded continuously. Note that since interest is being compounded continuously, the interest earned each year is more than 8%. The exponential growth function is

$$P(t) = 30{,}000e^{0.08t}.$$

b) To find the doubling time T, we replace $P(T)$ with 60,000 and solve for T:

$$60{,}000 = 30{,}000e^{0.08T}$$

$$2 = e^{0.08T}$$

$$\ln 2 = \ln e^{0.08T} \qquad \text{Taking the natural logarithm on both sides}$$

$$\ln 2 = 0.08T$$

$$\frac{\ln 2}{0.08} = T \qquad \text{Dividing}$$

$$\frac{0.693147}{0.08} \approx T$$

$$8.7 \approx T.$$

Thus the original investment of \$30,000 will double in about 8.7 years. ❑

Comparing Examples 3(b) and 6(b), we see that for any specified interest rate, continuous compounding gives the highest yield and the shortest doubling time.

Exponential Decay

The function

$$P(t) = P_0e^{-kt}, \qquad k > 0,$$

can model the decline, or decay, of a population or quantity. An example is the decay of a radioactive substance. Here P_0 is the amount of the substance at time $t = 0$, $P(t)$ is the amount of the substance remaining at time t, and k is a positive constant that depends on the situation. The constant k is called the **decay rate.** The **half-life** of a substance is the amount of time necessary for half of the substance to decay.

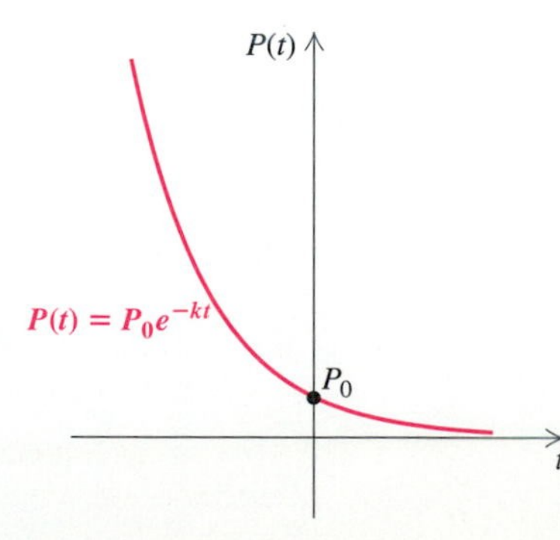

EXAMPLE 7

Carbon dating. The radioactive element carbon-14 has a half-life of 5750 years. The percentage of carbon-14 present in the remains of animal bones can be used to determine age. How old is an animal bone that has lost 40% of its carbon-14?

Solution We first find k. To do so, we use the concept of half-life. When $t = 5750$ (the half-life), $P(t)$ will be half of P_0. Then

$$0.5P_0 = P_0e^{-k(5750)}$$

$$0.5 = e^{-5750k} \quad \text{Dividing by } P_0 \text{ on both sides}$$

$$\ln 0.5 = \ln e^{-5750k} \quad \text{Taking the natural logarithm on both sides}$$

$$\ln 0.5 = -5750k$$

$$\frac{\ln 0.5}{-5750} = k$$

$$0.00012 \approx k.$$

Now we have the function

$$P(t) = P_0e^{-0.00012t}.$$

(*Note:* This equation can be used for any subsequent carbon-dating problem.) If an animal bone has lost 40% of its carbon-14 from an initial amount P_0, then $60\%(P_0)$ is the amount present. To find the age t of the bone, we solve this equation for t:

$$0.6P_0 = P_0e^{-0.00012t} \quad \text{We want to find } t \text{ for which } P(t) = 0.6P_0.$$

$$0.6 = e^{-0.00012t}$$

$$\ln 0.6 = \ln e^{-0.00012t}$$

$$-0.5108 \approx -0.00012t$$

$$t \approx \frac{-0.5108}{-0.00012}$$

$$t \approx 4257.$$

The animal bone is about 4257 years old. ❑

EXERCISE SET 12.7

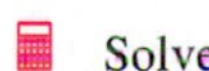 Solve.

1. *Compact discs.* The number of compact discs purchased each year is increasing exponentially. The number N, in millions, purchased is given by

$$N(t) = 7.5(6)^{0.5t},$$

where $t = 0$ corresponds to 1985, $t = 1$ corresponds to 1986, and so on, t being the number of years after 1985.

a) After what amount of time will one billion compact discs be sold in a year?

b) What is the doubling time on the sale of compact discs?

2. *Spread of a rumor.* The number of people who have heard a rumor increases exponentially. If all who hear a rumor repeat it to two people a day, and if 20 people start the rumor, the number of people N who have heard the rumor after t days is given by

$$N(t) = 20(3)^t.$$

a) After what amount of time will 1000 people have heard the rumor?

b) What is the doubling time on the number of people who have heard the rumor?

3. *Interest compounded annually.* Suppose that \$50,000 is invested at 6% interest, compounded annually. After t years, it grows to the amount A given by the function

$$A(t) = 50{,}000(1.06)^t.$$

a) After what amount of time will there be \$450,000 in the account?

b) Find the doubling time.

4. *Turkey consumption.* The amount of turkey consumed by each person in one year in the United States is increasing exponentially. Assume that $t = 0$ corresponds to 1937. The amount of turkey, in pounds per person, consumed t years after 1937 is given by the function

$$N(t) = 2.3(3)^{0.033t}.$$

a) After what amount of time will the consumption rate be 20 lb of turkey per year?

b) What is the doubling time on the consumption of turkey?

5. *Recycling aluminum cans.* Approximately two thirds of all aluminum cans distributed will be recycled each year. A beverage company distributes 250,000 cans. The number still in use after t years is given by the function

$$N(t) = 250{,}000\left(\tfrac{2}{3}\right)^t.$$

a) After how many years will 60,000 cans still be in use?

b) After what amount of time will only 1000 cans still be in use?

6. *Salvage value.* An office machine is purchased for \$5200. Its value each year is about 80% of the preceding year. Its value in dollars after t years is given by the exponential function

$$V(t) = 5200(0.8)^t.$$

a) After what amount of time will the salvage value be \$1200?

b) After what amount of time will the salvage value be half the original value?

7. *Interest compounded continuously.* Suppose that P_0 is invested in a savings account where interest is compounded continuously at 9% per year. That is, the balance $P(t)$ after time t, in years, is an exponential function of the form

$$P(t) = P_0e^{kt}.$$

a) Express $P(t)$ in terms of P_0 and 0.09.

b) Suppose that \$1000 is invested. What is the balance after 1 year? after 2 years?

c) When will an investment of \$1000 double itself?

8. *Interest compounded continuously.* Suppose that P_0 is invested in a savings account where interest is compounded continuously at 10% per year. That is, the balance $P(t)$ after time t, in years, is an exponential function of the form

$$P(t) = P_0e^{kt}.$$

a) Express $P(t)$ in terms of P_0 and 0.10.

b) Suppose that \$20,000 is invested. What is the balance after 1 year? after 2 years?

c) When will an investment of \$20,000 double itself?

9. *Population growth.* The exponential growth rate of the population of Europe west of Russia is 1% per year. What is the doubling time?

10. *Population growth.* The exponential growth rate of the population of Central America is 3.5% per year (one of the highest in the world). What is the doubling time?

11. *World population growth.* The population of the world was 5.2 billion in 1990. The exponential growth rate was 1.6% per year.

a) Find the exponential growth function.

b) Estimate the population of the world in 2000.

c) When will the world population be 8.0 billion?

12. *Growth of bacteria.* The bacteria *Escherichi coli* are commonly found in the bladder of human beings. Suppose that 3000 of the bacteria are present at time $t = 0$. Then t minutes later, the number of bacteria present will be

$$N(t) = 3000(2)^{t/20}.$$

a) After what amount of time will there be 60,000 bacteria?

b) If 100,000,000 bacteria accumulate, a bladder infection can occur. What amount of time would have to pass in order for a possible infection to occur?

c) What is the doubling time?

13. *Forgetting.* Students in an English class took a final exam. They took equivalent forms of the exam in monthly intervals thereafter. The average score $S(t)$, in percent, after t months was found to be given by

$$S(t) = 68 - 20 \log (t + 1), \quad t \geq 0.$$

a) What was the average score when they initially took the test, $t = 0$?

b) What was the average score after 4 months? after 24 months?

c) Graph the function.

d) After what time t was the average score 50?

14. *Advertising.* A model for advertising response is given by

$$N(a) = 2000 + 500 \log a, \quad a \geq 1,$$

where $N(a)$ = the number of units sold and a = the amount spent on advertising, in thousands of dollars.

a) How many units were sold after spending \$1000 ($a = 1$) on advertising?
b) How many units were sold after spending \$8000?
c) Graph the function.
d) How much would have to be spent in order to sell 5000 units?

Consider the pH formula of Example 1 for Exercises 15–22.

Find the pH, given the hydrogen ion concentration.

15. A common brand of mouthwash:

$$[H^+] = 6.3 \times 10^{-7} \text{ moles per liter}$$

16. A common brand of insect repellent:

$$[H^+] = 4.0 \times 10^{-8} \text{ moles per liter}$$

17. Eggs:

$$[H^+] = 1.6 \times 10^{-8} \text{ moles per liter}$$

18. Tomatoes:

$$[H^+] = 6.3 \times 10^{-5} \text{ moles per liter}$$

Find the hydrogen ion concentration of each substance, given the pH.

19. Tap water: pH = 7

20. Rainwater: pH = 5.4

21. Orange juice: pH = 3.2

22. Wine: pH = 4.8

23. The San Francisco earthquake of 1906 had an intensity of $10^{8.25}$ times I_0. What was its magnitude on the Richter scale?

24. In 1986, there was an earthquake near Cleveland, Ohio. It had an intensity of 10^5 times I_0. What was its magnitude on the Richter scale?

25. *Oil demand.* The exponential growth rate of the demand for oil in the United States is 10% per year. When will the demand be double that of 1993?

26. *Coal demand.* The exponential growth rate of the demand for coal in the world is 4% per year. When will the demand be double that of 1993?

27. *Heart transplants.* In 1967, Dr. Christian Barnard of South Africa stunned the world by performing the first heart transplant. There was 1 transplant in 1967. In 1987, there were 1418 such transplants.

a) Find an exponential growth function that fits the data.
b) Use the function to predict the number of heart transplants in 1998.

28. *The cost of a first-class postage stamp.* The cost of a first-class postage stamp became 3¢ in 1932 and the exponential growth rate was 3.8% per year. The exponential growth function was

$$P(t) = 3e^{0.038t}.$$

a) The cost of first-class postage increased to 29¢ in 1991. Use the given function to see what the predicted cost was for 1991 and compare with the actual cost.
b) What will the cost of a first-class postage stamp be in 2000?
c) When will the cost of a first-class postage stamp be \$1.00?

29. An ivory tusk has lost 20% of its carbon-14. How old is the tusk?

30. A piece of wood has lost 10% of its carbon-14. How old is the piece of wood?

31. The decay rate of iodine-131 is 9.6% per day. What is its half-life?

32. The decay rate of krypton-85 is 6.3% per year. What is its half-life?

33. The half-life of polonium is 3 minutes. What is its decay rate?

34. The half-life of lead is 22 years. What is its decay rate?

35. *Value of a sports card.* Because he objected to smoking, and because his first baseball card was issued in cigarette packs, the great shortstop Honus Wagner halted production of his card before many were produced. One of these cards was sold in 1986 for \$110,000 and again in 1991 for \$451,000. For the following questions, assume that the card's value increases exponentially.

a) Find an exponential function $V(t)$, if $V_0 = 110{,}000$.
b) Predict the card's value in 1998.
c) What is the doubling time for the value of the card?
d) In what year will the value of the card be \$9,000,000?

36. *Value of a Van Gogh painting.* The Van Gogh painting *Irises,* shown on the following page, sold for \$84,000 in 1947 and was sold again for \$53,900,000 in 1987. Assume that the growth in the value V of the painting is exponential.

a) Find the exponential growth rate k and determine the exponential growth function, assuming $V_0 = 84{,}000$.

b) Estimate the value of the painting in 1997.
c) What is the doubling time for the value of the painting?
d) How long after 1947 will the value of the painting be $1 billion?

Van Gogh's *Irises*, a 28-by-32-inch oil on canvas.

Skill Maintenance

Simplify.

37. $\dfrac{\dfrac{x-5}{x+3}}{\dfrac{x}{x-3}+\dfrac{2}{x+3}}$

38. $\dfrac{\dfrac{3}{a}+\dfrac{5}{b}}{\dfrac{2}{a^2}-\dfrac{4}{b^2}}$

Synthesis

39. *Atmospheric pressure.* Atmospheric pressure P at altitude a is given by

$$P = P_0 e^{-0.00005a},$$

where P_0 = the pressure at sea level ≈ 14.7 lb/in^2 (pounds per square inch). Explain how a barometer, or some other device for measuring atmospheric pressure, can be used to find the height of a skyscraper.

40. Examine the restriction on t in Exercise 13.
a) What upper limit might be placed on t?
b) In practice, would this upper limit ever be enforced? Why or why not?

41. Write a problem for a classmate to solve in which information is provided and the classmate is asked to find an exponential growth function. Make the problem as realistic as possible.

42. *Supply and demand.* The supply and demand for the sale of stereos by a sound company are given by

$$S(x) = e^x \quad \text{and} \quad D(x) = 162{,}755e^{-x},$$

where $S(x)$ = the price at which the company is willing to supply x stereos and $D(x)$ = the demand price for a quantity of x stereos. Find the equilibrium point. (For reference, see Section 8.8.)

SUMMARY AND REVIEW 12

KEY TERMS

Exponential function, p. 615
Asymptote, p. 616
Composite functions, p. 622
Inverse relation, p. 625
One-to-one function, p. 625
Horizontal-line test, p. 626
Logarithmic function, p. 633
Exponential equation, p. 635
Logarithmic equation, p. 635
Common logarithm, p. 644
Antilogarithm, p. 645
Natural logarithm, p. 646
Exponential growth rate, p. 657
Decay rate, p. 660
Half-life, p. 660

IMPORTANT PROPERTIES AND FORMULAS

Exponential function:	$f(x) = a^x$
Interest compounded annually:	$A = P(1 + i)^t$
Composition of f and g:	$f \circ g(x) = f(g(x))$

To Find a Formula for the Inverse of a Function

If a function is one-to-one, a formula for its inverse can be found as follows:

1. Replace $f(x)$ by y.
2. Interchange x and y.
3. Solve for y.
4. Replace y by $f^{-1}(x)$.

Definition of logarithm: $y = \log_a x$ is that number y such that $a^y = x$, where $x > 0$ and a is a positive constant other than 1

Properties of logarithms:

$\log_a MN = \log_a M + \log_a N$, $\quad \log_a \frac{M}{N} = \log_a M - \log_a N$,

$\log_a M^p = p \cdot \log_a M$, $\quad \log_a 1 = 0$,

$\log_a a = 1$, $\quad \log_a a^k = k$,

$\log M = \log_{10} M$, $\quad e \approx 2.7182818284\ldots$,

$\ln M = \log_e M$, $\quad \log_b M = \frac{\log_a M}{\log_a b}$

Exponential growth:	$P(t) = P_0e^{kt}$
Exponential decay:	$P(t) = P_0e^{-kt}$
Interest compounded continuously:	$P(t) = P_0e^{kt}$, where P_0 is the principal invested for t years at interest rate k
Carbon dating:	$P(t) = P_0e^{-0.00012t}$

REVIEW EXERCISES

This chapter's review and test include Skill Maintenance exercises from Sections 6.6, 10.9, 11.6, and 11.9.

Graph.

1. $f(x) = 3^{x-2}$

2. $x = 3^{y-2}$

3. $y = \log_3 x$

4. $f(x) = e^{x+1}$

5. Find $f \circ g(x)$ and $g \circ f(x)$ if $f(x) = x^2$ and $g(x) = 3x - 5$.

6. Determine whether $f(x) = 4 - x^2$ is one-to-one.

Find a formula for the inverse.

7. $f(x) = x + 2$

8. $g(x) = \frac{2x - 3}{7}$

9. $f(x) = \frac{2}{x + 5}$

10. $g(x) = 8x^3$

Convert to an exponential equation.

11. $\log_4 16 = x$

12. $\log_{10} 2 = 0.3010$

13. $\log_{1/2} 8 = -3$

14. $\log_{16} 8 = \frac{3}{4}$

Convert to a logarithmic equation.

15. $10^4 = 10{,}000$

16. $25^{1/2} = 5$

17. $7^{-2} = \frac{1}{49}$

18. $(2.718)^3 = 20.1$

Express in terms of logarithms of x, y, and z.

19. $\log_a x^4y^2z^3$

20. $\log_a \frac{xy}{z^2}$

21. $\log \sqrt[4]{\frac{z^2}{x^3y}}$

22. $\log_q \left(\frac{x^2y^{1/3}}{z^4}\right)$

Express as a single logarithm.

23. $\log_a 8 + \log_a 15$

24. $\log_a 72 - \log_a 12$

25. $\frac{1}{2} \log a - \log b - 2 \log c$

26. $\frac{1}{3}[\log_a x - 2 \log_a y]$

Simplify.

27. $\log_m m$

28. $\log_m 1$

29. $\log_m m^{17}$

30. $\log_m m^{-7}$

Given $\log_a 2 = 1.8301$ and $\log_a 7 = 5.0999$, find each of the following.

31. $\log_a 14$

32. $\log_a \frac{2}{7}$

33. $\log_a 28$

34. $\log_a 3.5$

35. $\log_a \sqrt{7}$

36. $\log_a \frac{1}{4}$

Find each of the following using a calculator.

37. log 0.00627

38. log 72,800,000

39. antilog 4.4742

40. antilog (−1.4425)

41. antilog 2.3294

42. log 0.004937

43. log 394,900

44. antilog (−6.7889)

45. ln 23,912.2

46. ln 0.06774

47. $\text{antilog}_e (-10.56)$

48. $\text{antilog}_e 45$

Find each of the following logarithms using the change-of-base formula and a calculator or Table 2.

49. $\log_5 2$

50. $\log_{12} 70$

Solve.

51. $\log_3 x = -2$

52. $\log_x 32 = 5$

53. $\log x = -4$

54. $\ln x = 2$

55. $4^{2x-5} = 16$

56. $4^x = 8.3$

57. $\log_4 16 = x$

58. $\log (x^2 - 9) - \log (x - 3) = 1$

59. $\log_4 x + \log_4 (x - 6) = 2$

60. $\log x + \log (x - 15) = 2$

61. $\log_3 (x - 4) = 3 - \log_3 (x + 4)$

62. *Forgetting.* In a business class, students were tested at the end of the course on a final exam. They were tested again after 6 months. The forgetting formula was determined to be

$$S(t) = 62 - 18 \log (t + 1),$$

where t is the time, in months, after taking the first test.

a) Determine the average score when they first took the test (when $t = 0$).
b) What was the average score after 6 months?
c) After what time was the average score 34?

63. *Cost of a prime-rib dinner.* The average cost C of a prime-rib dinner was \$4.65 in 1962. In 1986, it was \$15.81. Assume that the cost increases exponentially.

a) Find k and write the exponential growth function.
b) Predict the cost of a prime-rib dinner in 2002.
c) When will the average cost of a prime-rib dinner be \$30?
d) What is the doubling time?

64. The population of Riverton doubled in 16 years. What was the exponential growth rate?

65. How long will it take \$7600 to double itself if it is invested at 8.4%, compounded continuously?

66. How old is a skeleton that has lost 34% of its carbon-14?

67. What is the pH of a substance whose hydrogen ion concentration is 2.3×10^{-7} moles per liter?

68. An earthquake has an intensity of $10^{8.3}$ times I_0. What is its amplitude on the Richter scale?

Skill Maintenance

69. Solve $aT^2 + bT = Q$ for T.

70. Solve: $x^4 + 80 = 21x^2$.

71. Divide: $\frac{4 - 5i}{1 + 3i}$.

72. Simplify:

$$\frac{\frac{1}{ab} - \frac{2}{bc}}{\frac{2}{ac} + \frac{3}{ab}}.$$

Synthesis

73. Explain why negative numbers do not have logarithms.

74. Explain why taking the natural or common logarithm on each side of an equation produces an equivalent equation.

Solve.

75. $\ln(\ln x) = 3$

76. $2^{x^2+4x} = \frac{1}{8}$

77. $5^{x+y} = 25,$
$2^{2x-y} = 64$

CHAPTER TEST 12

Graph.

1. $f(x) = 2^{x+3}$

2. $f(x) = \log_7 x$

3. Find $f \circ g(x)$ and $g \circ f(x)$ if $f(x) = x + x^2$ and $g(x) = 5x - 2$.

4. Determine whether $f(x) = 2 - |x|$ is one-to-one.

Find a formula for the inverse.

5. $f(x) = 4x - 3$

6. $g(x) = \dfrac{x-2}{4}$

7. $f(x) = \dfrac{x+1}{x-2}$

Convert to a logarithmic equation.

8. $4^{-3} = x$

9. $256^{1/2} = 16$

Convert to an exponential equation.

10. $\log_4 16 = 2$

11. $m = \log_7 49$

12. Express in terms of logarithms of a, b, and c:
$$\log \frac{a^3 b^{1/2}}{c^2}.$$

13. Express as a single logarithm:
$$\tfrac{1}{3}\log_a x - 3\log_a y + 2\log_a z.$$

Simplify.

14. $\log_t t^{23}$

15. $\log_p p$

16. $\log_c 1$

Given $\log_a 2 = 0.301$, $\log_a 6 = 0.778$, and $\log_a 7 = 0.845$, find each of the following.

17. $\log_a \frac{2}{7}$

18. $\log_a \sqrt{24}$

19. $\log_a 21$

Find each of the following using a calculator.

20. $\log 0.0123$

21. antilog 5.6484

22. antilog (-7.2614)

23. $\log 12{,}340$

24. $\ln 0.01234$

25. $\text{antilog}_e\,(5.6774)$

26. Find $\log_{18} 31$ using the change-of-base formula and a calculator.

Solve.

27. $\log_x 25 = 2$

28. $\log_4 x = \frac{1}{2}$

29. $\log x = 4$

30. $5^{4-3x} = 125$

31. $7^x = 1.2$

32. $\ln x = \frac{1}{4}$

33. $\log(x^2 - 1) - \log(x - 1) = 1$

34. *Walking speed.* The average walking speed R of people living in a city of population P, in thousands, is given by
$$R = 0.37 \ln P + 0.05,$$
where R is in feet per second.
a) The population of Akron, Ohio, is 660,000. Find the average walking speed.
b) A city has an average walking speed of 2.6 ft/sec. Find the population.

35. *Population of Canada.* The population of Canada was 24 million in 1981, and the exponential growth rate was 1.2% per year.
a) Write an exponential function describing the population of Canada.
b) What will the population be in 1998? in 2010?
c) When will the population be 30 million?
d) What is the doubling time?

36. The population of Clay County doubled in 20 years. What was the exponential growth rate?

37. How long will it take an investment to double itself if it is invested at 7.6%, compounded continuously?

38. How old is an animal bone that has lost 43% of its carbon-14?

39. An earthquake has an intensity of $10^{8.34}$ times I_0. What is its magnitude on the Richter scale?

40. The hydrogen ion concentration of water is 1.0×10^{-7} moles per liter. What is the pH?

Skill Maintenance

41. Solve: $y - 9\sqrt{y} + 8 = 0$.

42. Solve $S = at^2 - bt$ for t.

43. Multiply: $(2 + 5i)(2 - 5i)$.

44. Simplify:

$$\frac{\dfrac{1}{x^2 - 4}}{\dfrac{1}{x + 2} + \dfrac{1}{x - 2}}.$$

Synthesis

45. Solve: $\log_5 |2x - 7| = 4$.

46. If $\log_a x = 2$, $\log_a y = 3$, and $\log_a z = 4$, find

$$\log_a \frac{\sqrt[3]{x^2 z}}{\sqrt[3]{y^2 z^{-1}}}.$$

CUMULATIVE REVIEW 1–12

1. Evaluate $\frac{x^0 + y}{-z}$ when $x = 6$, $y = 9$, and $z = -5$.

Simplify.

2. $\left|-\frac{5}{2} + \left(-\frac{7}{2}\right)\right|$

3. $(-2x^2y^{-3})^{-4}$

4. $(-5x^4y^{-3}z^2)(-4x^2y^2)$

5. $\frac{3x^4y^6z^{-2}}{-9x^4y^2z^3}$

6. $2x - 3 - 2[5 - 3(2 - x)]$

7. $3^3 + 2^2 - (32 \div 4 - 16 \div 8)$

Solve.

8. $8(2x - 3) = 6 - 4(2 - 3x)$

9. $(5x - 2)(4x + 20) = 0$

10. $4x - 3y = 15,$
$3x + 5y = 4$

11. $x + y - 3z = -1,$
$2x - y + z = 4,$
$-x - y + z = 1$

12. $5 = x^2 + 6x$

13. $x(x - 3) = 10$

14. $\frac{7}{x^2 - 5x} - \frac{2}{x - 5} = \frac{4}{x}$

15. $\frac{8}{x + 1} + \frac{11}{x^2 - x + 1} = \frac{24}{x^3 + 1}$

16. $\sqrt{x - 1} = \sqrt{x + 4} - 1$

17. $\sqrt[3]{2x} = 1$

18. $3x^2 + 75 = 0$

19. $x - 8\sqrt{x} + 15 = 0$

20. $x^4 - 13x^2 + 36 = 0$

21. $\log_8 x = 1$

22. $\log_x 49 = 2$

23. $9^x = 27$

24. $\log x - \log (x - 8) = 1$

25. $x^2 + 4x > 5$

26. $|2x - 3| \geq 9$

Solve.

27. $D = \frac{ab}{b + a}$, for a

28. $\frac{1}{p} + \frac{1}{q} = \frac{1}{f}$, for q

29. $M = \frac{2}{3}(A + B)$, for B

Evaluate.

30. $\begin{vmatrix} 6 & -5 \\ 4 & -3 \end{vmatrix}$

31. $\begin{vmatrix} 7 & -6 & 0 \\ -2 & 1 & 2 \\ -1 & 1 & -1 \end{vmatrix}$

Solve.

32. Twenty-four plus five times a number is eight times the number. Find the number.

33. The perimeter of a rectangular garden is 112 m. The length is 16 m more than the width. Find the length and the width.

34. In triangle ABC, the measure of angle B is three times the measure of angle A. The measure of angle C is 105° greater than the measure of angle A. Find the angle measures.

35. Phil can build a shed from a lumber kit in 10 hr. Jenny can build the same shed in 12 hr. How long would it take Phil and Jenny, working together, to build the shed?

36. Swim Clean is 30% muriatic acid. Pure Swim is 80% muriatic acid. How many liters of each should be mixed together in order to get 100 L of a solution that is 50% muriatic acid?

37. A boat can move at a speed of 5 km/h in still water. The boat travels 42 km downstream in the same time that it takes to travel 12 km upstream. What is the speed of the stream?

38. What is the minimum product of two numbers whose difference is 14? What are the numbers that yield this product?

39. The speed of a passenger train is 13 mph faster than that of a freight train. The passenger train travels 160 mi in the same time it takes the freight train to travel 108 mi. Find the speed of each train.

Forgetting. Students in a biology class took a final exam. A forgetting formula for the average exam grade was determined to be

$$S(t) = 78 - 15 \log (t + 1),$$

where t is the number of months after the final was taken.

40. The average score when the students first took the test is when $t = 0$. Find the students' average score on the final exam.

41. What would the average score be on a retest after 4 months?

Population growth. The population of Europe west of Russia was 430 million in 1961, and the exponential growth rate was 1% per year.

42. Write an exponential function describing the growth of the population of Europe.

43. Predict what the population will be in 1997; in 2001.

44. y varies directly as the square of x and inversely as z, and $y = 2$ when $x = 5$ and $z = 100$. What is y when $x = 3$ and $z = 4$?

Perform the indicated operations and simplify.

45. $(5p^2q^3 - 4p^3q + 6pq - p^2 + 3) + (2p^2q^3 + 2p^3q + p^2 - 5pq - 9)$

46. $(11x^2 - 6x - 3) - (3x^2 + 5x - 2)$

47. $(3x^2 - 2y)^2$

48. $(5a + 3b)(2a - 3b)$

49. $\dfrac{x^2 + 8x + 16}{2x + 6} \div \dfrac{x^2 + 3x - 4}{x^2 - 9}$

50. $\dfrac{1 + \dfrac{3}{x}}{x - 1 - \dfrac{12}{x}}$

51. $\dfrac{a^2 - a - 6}{a^3 - 27} \cdot \dfrac{a^2 + 3a + 9}{6}$

52. $\dfrac{3}{x + 6} - \dfrac{2}{x^2 - 36} + \dfrac{4}{x - 6}$

Factor.

53. $xy - 2xz + xw$

54. $1 - 125x^3$

55. $6x^2 + 8xy - 8y^2$

56. $x^4 - 4x^3 + 7x - 28$

57. $a^2 - 10a + 25 - 81b^2$

58. $2m^2 + 12mn + 18n^2$

59. $x^4 - 16y^4$

60. For the function described by

$$h(x) = -3x^2 + 4x + 8,$$

find $h(-2)$.

61. Divide: $(x^4 - 5x^3 + 2x^2 - 6) \div (x - 3)$.

62. Multiply $(5.2 \times 10^4)(3.5 \times 10^{-6})$. Write scientific notation for the answer.

63. Divide: $\dfrac{3.4 \times 10^5}{6.8 \times 10^{-9}}$.

Write scientific notation for the answer.

For the radical expressions that follow, assume that all variables represent positive numbers.

64. Divide and simplify: $\dfrac{\sqrt[3]{40xy^8}}{\sqrt[3]{5xy}}$.

65. Multiply and simplify: $\sqrt{7xy^3} \cdot \sqrt{28x^2y}$.

66. Rewrite without fractional exponents: $(27a^6b)^{4/3}$.

67. Rationalize the denominator: $\dfrac{3 - \sqrt{y}}{2 - \sqrt{y}}$.

68. Divide and simplify: $\dfrac{\sqrt[5]{x + 5}}{\sqrt{x + 5}}$.

69. Multiply these complex numbers:

$$(1 + i\sqrt{3})(6 - 2i\sqrt{3}).$$

70. Divide these complex numbers: $\dfrac{3 - 2i}{4 - 3i}$.

71. Find the inverse of f if $f(x) = 7 - 2x$.

72. Find an equation of the line containing the points $(0, -3)$ and $(-1, 2)$.

73. Find an equation of the line containing the point $(-3, 5)$ and perpendicular to the line whose equation is $2x + y = 6$.

Graph.

74. $5x = 15 + 3y$

75. $y = 2x^2 - 4x - 1$

76. $y = \log_3 x$

77. $y = 3^x$

78. $-2x - 3y \leq 6$

79. Graph: $f(x) = 2(x + 3)^2 + 1$.

a) Label the vertex.

b) Draw the line of symmetry.

c) Find the maximum or minimum value.

80. Express in terms of logarithms of a, b, and c:

$$\log\left(\frac{a^2c^3}{b}\right).$$

81. Express as a single logarithm:

$$3 \log x - \frac{1}{2} \log y - 2 \log z.$$

82. Convert to an exponential equation: $\log_a 5 = x$.

83. Convert to a logarithmic equation: $x^3 = t$.

Find each of the following using a calculator.

84. log 0.05566

85. antilog 5.4453

86. ln 12.78

87. $\text{antilog}_e\ (-3.6762)$

88. Solve: $3^{5x} = 7$.

Synthesis

89. Solve: $\dfrac{5}{3x - 3} + \dfrac{10}{3x + 6} = \dfrac{5x}{x^2 + x - 2}$.

90. Solve: $\log \sqrt{3x} = \sqrt{\log 3x}$.

91. A train travels 280 mi at a certain speed. If the speed had been increased 5 mph, the trip could have been made in 1 hr less time. Find the actual speed.

CHAPTER 13

Conic Sections

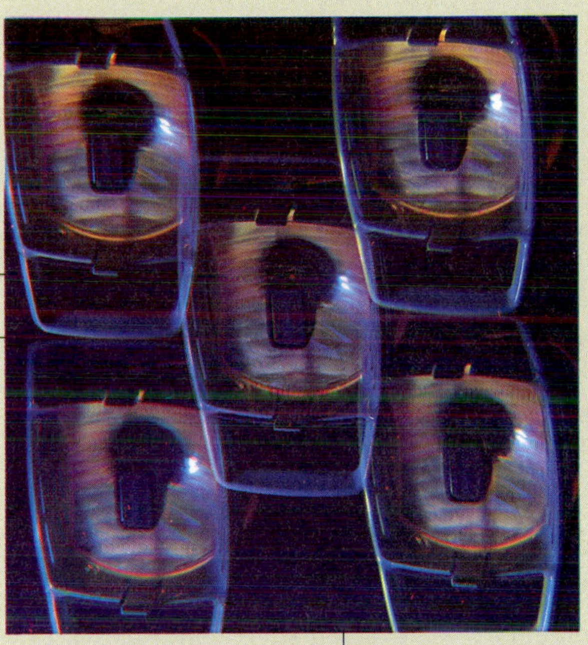

AN APPLICATION

The light source in a dental lamp shines against a reflector that is shaped like a portion of an *ellipse* in which the light source is one *focus* of the ellipse. Reflected light enters a patient's mouth at the other focus of the ellipse. If the ellipse from which the reflector was formed is 2 ft wide and 6 ft long, how far should the light source be from a patient's mouth?

This problem appears as Exercise 57 in Section 13.3.

Dr. Roschelle Major-Banks
DENTIST

"A good math background is critical for those working in dentistry: Dental assistants need math for their training, orthodontists use physics extensively in shaping braces, and dental technologists work with precise measurements when making dental appliances."

The ellipse described in the chapter opening is a curve known as a *conic section*, meaning that the curve is formed as a cross section of a cone. In this chapter, we will study equations whose graphs are conic sections. We have already studied two conic sections, *lines* and *parabolas*, in some detail in Chapters 3, 7, and 11. There are many applications involving conics, and we will consider some of them in this chapter.

In addition to material from this chapter, the review and test for Chapter 13 include material from Sections 6.8, 10.3, 10.6, and 11.3.

13.1 The Distance and Midpoint Formulas

The Distance Formula • Midpoints of Segments

The Distance Formula

For our work with conic sections in this chapter, we now develop a formula for finding the distance between two points whose coordinates are known.

Suppose that two points are on a horizontal line, and thus have the same second coordinate. We can find the distance between them by subtracting their first coordinates. This difference may be negative, depending on the order in which we subtract. So to make sure we get a positive number, we take the absolute value of this difference. The distance between two points on a horizontal line (x_1, y_1) and (x_2, y_1) is thus $|x_2 - x_1|$. Similarly, the distance between two points on a vertical line (x_2, y_1) and (x_2, y_2) is $|y_2 - y_1|$.

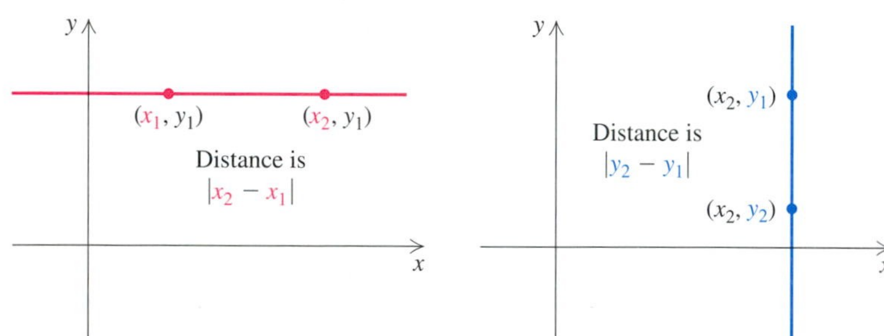

EXAMPLE 1

Find the distance between each pair of points:
(a) $(-2, 13)$ and $(-2, -1)$; **(b)** $(-1, 5)$ and $(-6, 5)$.

Solution

a) Since the first coordinates are the same, the points $(-2, 13)$ and $(-2, -1)$

would lie on a vertical line. The distance between them is the absolute value of the difference of the second coordinates. The distance is

$$|13 - (-1)| = |14| = 14.$$

If we subtract in a different order, the distance is still

$$|-1 - 13| = |-14| = 14.$$

b) The points $(-1, 5)$ and $(-6, 5)$ would lie on a horizontal line. The distance between them is

$$|-1 - (-6)| = |5| = 5.$$ ❑

Now consider two points (x_1, y_1) and (x_2, y_2), as shown. Since $x_1 \neq x_2$ and $y_1 \neq y_2$, these points are vertices of a right triangle. The other vertex is then (x_2, y_1). The lengths of the legs are $|x_2 - x_1|$ and $|y_2 - y_1|$. We find d, the length of the hypotenuse, by using the Pythagorean theorem:

$$d^2 = |x_2 - x_1|^2 + |y_2 - y_1|^2.$$

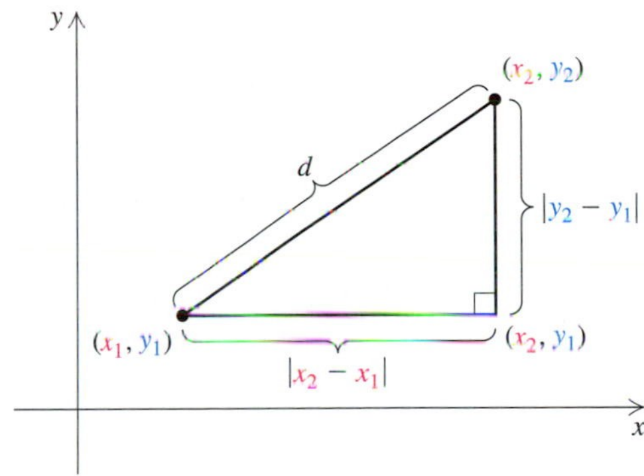

Since the square of a number is the same as the square of its opposite, we don't really need the absolute-value signs. Thus,

$$d^2 = (x_2 - x_1)^2 + (y_2 - y_1)^2.$$

Taking the principal square root, we obtain the distance between two points.

The Distance Formula

The distance d between any two points (x_1, y_1) and (x_2, y_2) is given by

$$d = \sqrt{(x_2 - x_1)^2 + (y_2 - y_1)^2}.$$

As in finding distances between points on vertical or horizontal lines, it does not matter which point is chosen to be (x_1, y_1) or (x_2, y_2) in the formula.

EXAMPLE 2

Find the distance between $(5, -1)$ and $(-4, 6)$. Find an exact answer and an approximation to three decimal places.

Solution We substitute into the distance formula:

$$\begin{aligned} d &= \sqrt{(-4-5)^2 + [6-(-1)]^2} && \text{Substituting} \\ &= \sqrt{(-9)^2 + 7^2} \\ &= \sqrt{130} \approx 11.402. && \text{Using a calculator} \end{aligned}$$

❑

We can use the distance formula to determine whether three points are vertices of a right triangle.

EXAMPLE 3

Determine whether the points $A(-2, 2)$, $B(4, -3)$, and $C(-2, -3)$ are vertices of a right triangle.

Solution First we find the squares of the distances between the points:

B to A

$$\begin{aligned} (d_1)^2 &= [4-(-2)]^2 + (-3-2)^2 \\ &= (6)^2 + (-5)^2 = 61, && \text{The distance between } A \text{ and } B \text{ is } \sqrt{61}. \end{aligned}$$

C to A

$$\begin{aligned} (d_2)^2 &= [-2-(-2)]^2 + (-3-2)^2 \\ &= (0)^2 + (-5)^2 = 25, && \text{The distance between } A \text{ and } C \text{ is } \sqrt{25}, \text{ or } 5. \end{aligned}$$

C to B

$$\begin{aligned} (d_3)^2 &= (-2-4)^2 + [-3-(-3)]^2 \\ &= (-6)^2 + (0)^2 = 36. && \text{The distance between } B \text{ and } C \text{ is } \sqrt{36}, \text{ or } 6. \end{aligned}$$

Since $(d_2)^2 + (d_3)^2 = (d_1)^2$, the points *are* vertices of a right triangle. ❑

Midpoints of Segments

The distance formula can be used to verify or derive a formula for finding the coordinates of the *midpoint* of a segment when the coordinates of the endpoints are known. We state the formula and leave its proof to the exercises. Note that although the distance formula involves both subtraction and addition, the midpoint formula uses only addition.

The Midpoint Formula

If the endpoints of a segment are (x_1, y_1) and (x_2, y_2), then the coordinates of the midpoint are

$$\left(\frac{x_1 + x_2}{2}, \frac{y_1 + y_2}{2}\right).$$

(Average the coordinates of the endpoints to find the coordinates of the midpoint.)

EXAMPLE 4

Find the midpoint of the segment with endpoints $(-2, 3)$ and $(4, -6)$.

Solution Using the midpoint formula, we obtain

$$\left(\frac{-2+4}{2}, \frac{3+(-6)}{2}\right), \quad \text{or} \quad \left(\frac{2}{2}, \frac{-3}{2}\right), \quad \text{or} \quad \left(1, -\frac{3}{2}\right).$$

❑

EXERCISE SET 13.1

Find the distance between the pair of points. Where appropriate, find an approximation to three decimal places.

1. (9, 5) and (6, 1)
2. (1, 10) and (7, 2)
3. (0, −7) and (3, −4)
4. (6, 2) and (6, −8)
5. (2, 2) and (−2, −2)
6. (5, 21) and (−3, 1)
7. (8.6, −3.4) and (−9.2, −3.4)
8. (5.9, 2) and (3.7, −7.7)
9. $\left(\frac{5}{7}, \frac{1}{14}\right)$ and $\left(\frac{1}{7}, \frac{11}{14}\right)$
10. $\left(0, \sqrt{7}\right)$ and $\left(\sqrt{6}, 0\right)$
11. (−23, 10) and (56, −17)
12. (34, −18) and (−46, −38)
13. (a, b) and (0, 0)
14. (0, 0) and (p, q)
15. $(-1, 3k)$ and $(6, 2k)$
16. $(a, -3)$ and $(2a, 5)$
17. $\left(\sqrt{2}, -\sqrt{3}\right)$ and $\left(-\sqrt{7}, \sqrt{5}\right)$
18. $\left(\sqrt{8}, \sqrt{3}\right)$ and $\left(-\sqrt{5}, -\sqrt{6}\right)$
19. (1000, −240) and (−2000, 580)
20. (−3000, 560) and (−430, −640)

Determine whether the points are vertices of a right triangle.

21. (9, 6), (−1, 2), and (1, −3)
22. (3, 0), (−1, −3), and (−1, 8)
23. (−5, 1), (−1, −2), and (4, 10)
24. (−8, −5), (6, 1), and (−4, 5)

Find the midpoint of the segment with the given endpoints.

25. (−3, 6) and (2, −8)
26. (6, 7) and (7, −9)
27. (8, 5) and (−1, 2)
28. (−1, 2) and (1, −3)
29. (−8, −5) and (6, −1)
30. (8, −2) and (−3, 4)
31. (−3.4, 8.1) and (2.9, −8.7)
32. (4.1, 6.9) and (5.2, −6.9)
33. $(-a, b)$ and (a, b)
34. $(-c, d)$ and $(c, -d)$
35. $\left(\frac{1}{6}, -\frac{3}{4}\right)$ and $\left(-\frac{1}{3}, \frac{5}{6}\right)$
36. $\left(-\frac{4}{5}, -\frac{2}{3}\right)$ and $\left(\frac{1}{8}, \frac{3}{4}\right)$
37. $\left(\sqrt{2}, -1\right)$ and $\left(\sqrt{3}, 4\right)$
38. $\left(9, 2\sqrt{3}\right)$ and $\left(-4, 5\sqrt{3}\right)$

Skill Maintenance

Solve.

39. $x^2 + 2x - 1 = 0$
40. $x^2 + x + 5 = 0$
41. $2a^2 - a - 3 = 0$
42. $y^2 + y = 10$

Synthesis

43. ◈ Outline a procedure that would use the distance formula to determine whether three points, (x_1, y_1), (x_2, y_2), and (x_3, y_3), are collinear (lie on the same line).
44. ◈ Explain why the distance formula holds for two points on a horizontal line.
45. Find the point on the y-axis that is equidistant from (2, 10) and (6, 2).
46. Find the point on the x-axis that is equidistant from (−1, 3) and (−8, −4).
47. Find the midpoint of the segment with the endpoints $\left(2 - \sqrt{3}, 5\sqrt{2}\right)$ and $\left(2 + \sqrt{3}, 3\sqrt{2}\right)$.

Find the distance between the pair of points.

48. $(6m, -7n)$ and $(-2m, n)$
49. $\left(\sqrt{d}, -\sqrt{3c}\right)$ and $\left(\sqrt{d}, \sqrt{3c}\right)$
50. $\left(-3\sqrt{3}, 1 - \sqrt{6}\right)$ and $\left(\sqrt{3}, 1 + \sqrt{6}\right)$
51. (5.989, 2.001) and (3.712, −7.784)

52. Prove the midpoint formula by showing that

i) the distance from (x_1, y_1) to

$$\left(\frac{x_1 + x_2}{2}, \frac{y_1 + y_2}{2}\right)$$

equals the distance from (x_2, y_2) to

$$\left(\frac{x_1 + x_2}{2}, \frac{y_1 + y_2}{2}\right); \text{ and}$$

ii) the points

$$(x_1, y_1), \left(\frac{x_1 + x_2}{2}, \frac{y_1 + y_2}{2}\right),$$

and

$$(x_2, y_2)$$

lie on the same line (see Exercise 43).

13.2 Conic Sections: Parabolas and Circles

Parabolas • Circles

In this section and the next, we study curves formed by cross sections of cones. These curves are graphs of second-degree equations in two variables. Some are shown below:

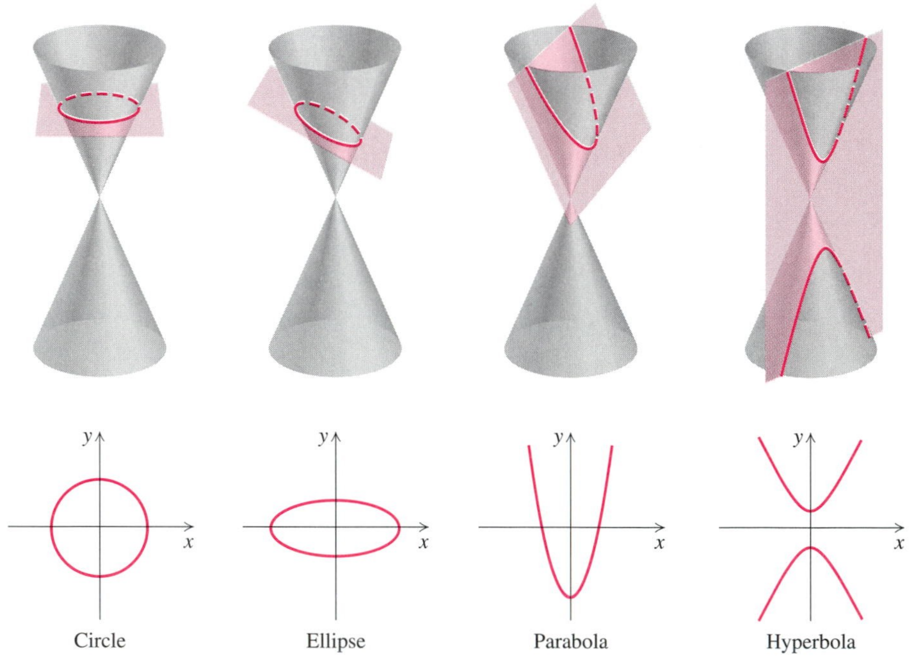

Parabolas

When a cone is cut as shown in the third figure above, the conic section formed is a **parabola.** Parabolas have many applications in electricity, mechanics, and optics. The cross section of a satellite dish is a parabola. Cables that support bridges that are straight lines are shaped like parabolas. (Free-hanging cables have a different shape, called a ‘‘catenary.’’)

Equation of a Parabola

Parabolas have equations as follows:

$y = ax^2 + bx + c$ (Line of symmetry parallel to the y-axis);

$x = ay^2 + by + c$ (Line of symmetry parallel to the x-axis).

We found in Chapter 11 that the graphs of quadratic functions of the form $f(x) = ax^2 + bx + c$ are parabolas. Our goal here is to review those graphs and extend those ideas to equations of the type $x = ay^2 + by + c$. These parabolas have lines of symmetry parallel to the x-axis and open to the right or to the left.

EXAMPLE 1

Graph: $y = x^2 - 4x + 8$.

Solution We first find the vertex. We can do so in either of two ways. The first way is by completing the square:

$$\begin{aligned} y &= (x^2 - 4x) + 8 \\ &= (x^2 - 4x + 4 - 4) + 8 && \tfrac{1}{2}(-4) = -2;\ (-2)^2 = 4 \\ &= (x^2 - 4x + 4) + (-4 + 8) && \text{Regrouping} \\ &= (x - 2)^2 + 4. && \text{Factoring and simplifying} \end{aligned}$$

The vertex is (2, 4).

A second way to find the vertex is to recall from Section 11.8 that *the x-coordinate of the vertex of the parabola given by $y = ax^2 + bx + c$ is $-b/(2a)$.* Thus, instead of completing the square as above, we could have evaluated the formula

$$x = -\frac{b}{2a} = -\frac{-4}{2(1)} = 2,$$

and found the y-coordinate of the vertex by substituting 2 for x:

$$y = x^2 - 4x + 8 = 2^2 - 4(2) + 8 = 4.$$

Either way we know that the vertex is (2, 4). We choose some x-values on both sides of the vertex and compute the corresponding y-values. Then we plot the points and graph the parabola. Since the coefficient of x^2, 1, is positive, we know that the graph opens up. It is easy to find the y-intercept by finding y when $x = 0$.

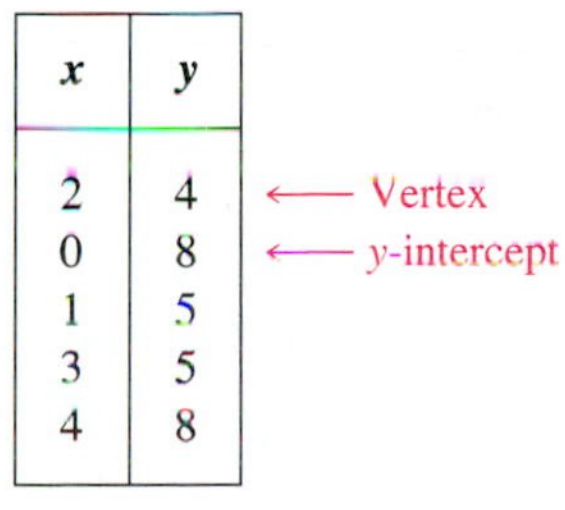

x	y	
2	4	← Vertex
0	8	← y-intercept
1	5	
3	5	
4	8	

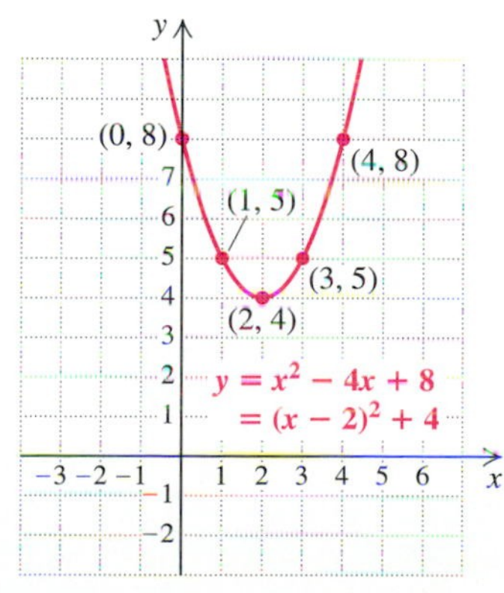

□

To graph an equation in the form $y = ax^2 + bx + c$:

1. Find the vertex (h, k) either by completing the square to find an equivalent equation $y = a(x - h)^2 + k$, or by using $-b/(2a)$ to find the x-coordinate and substituting to find the y-coordinate.
2. Choose other values for x on both sides of the vertex, and compute the corresponding y-values.
3. The graph opens upward for $a > 0$ and downward for $a < 0$.

EXAMPLE 2

Graph: $x = y^2 - 4y + 8$.

Solution This equation is like that in Example 1 except that x and y are interchanged. The vertex is (4, 2) instead of (2, 4). To find ordered pairs, we first choose values for y on each side of the vertex. Then we compute values for x. A table is shown, together with the graph. Note that in this table the x- and y-values of the table in Example 1 are interchanged.

x	y	
4	2	← Vertex
8	0	← x-intercept
5	1	
5	3	
8	4	

(1) Choose these values for y.

(2) Compute these values for x.

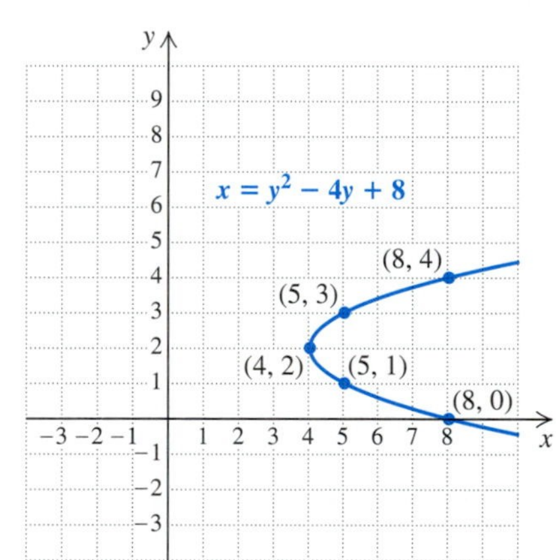

To graph an equation in the form $x = ay^2 + by + c$:

1. Find the vertex (h, k) either by completing the square to find an equivalent equation $x = a(y - k)^2 + h$ or by using $-b/(2a)$ to find the y-coordinate and substituting to find the x-coordinate.
2. Choose other values for y that are above and below the vertex, and compute the corresponding x-values.
3. The graph opens to the right if $a > 0$ and to the left if $a < 0$.

EXAMPLE 3

Graph: $x = -2y^2 + 10y - 7$.

Solution Completing the square, we have

$$x = -2y^2 + 10y - 7 = -2(y^2 - 5y) - 7$$

$$= -2\left(y^2 - 5y + \tfrac{25}{4}\right) - 7 - (-2)\tfrac{25}{4}$$ $\frac{1}{2}(-5) = \frac{-5}{2}$; $\left(\frac{-5}{2}\right)^2 = \frac{25}{4}$; we add and subtract $(-2)\frac{25}{4}$

$$= -2\left(y - \tfrac{5}{2}\right)^2 + \tfrac{11}{2}.$$ Factoring and simplifying

The vertex is $\left(\frac{11}{2}, \frac{5}{2}\right)$.

For practice, we also find the vertex by first computing the second coordinate, $y = -b/(2a)$, and then substituting to find the first coordinate:

$$y = -\frac{b}{2a} = -\frac{10}{2(-2)} = \frac{5}{2}$$

$$x = -2y^2 + 10y - 7 = -2\left(\tfrac{5}{2}\right)^2 + 10\left(\tfrac{5}{2}\right) - 7 = \tfrac{11}{2}.$$

To find ordered pairs, we first choose values for y and then compute values for x. A table is given, together with the graph. The graph opens to the left because the coefficient of y^2, -2, is negative.

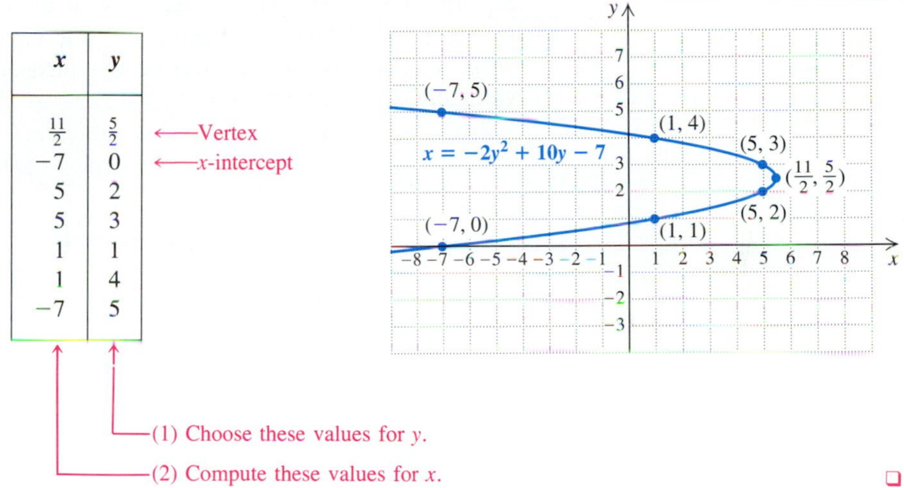

x	y	
$\frac{11}{2}$	$\frac{5}{2}$	← Vertex
-7	0	← x-intercept
5	2	
5	3	
1	1	
1	4	
-7	5	

Circles

The distance formula that we developed in Section 13.1 can be used to develop an equation for another conic section, shown in the figure at the beginning of this section, the circle. A **circle** is a set of points in a plane that are a fixed distance r, called the **radius,** from a fixed point (h, k), called the **center.** If a point (x, y) is on the circle, then by the definition of a circle and the distance formula, it must follow that

$$r = \sqrt{(x-h)^2 + (y-k)^2}.$$

Squaring both sides gives an equation of the circle in standard form: $(x-h)^2 + (y-k)^2 = r^2$. When $h = 0$ and $k = 0$, the circle is centered at the origin. Otherwise, we can think of that circle being translated $|h|$ units horizontally and $|k|$ units vertically.

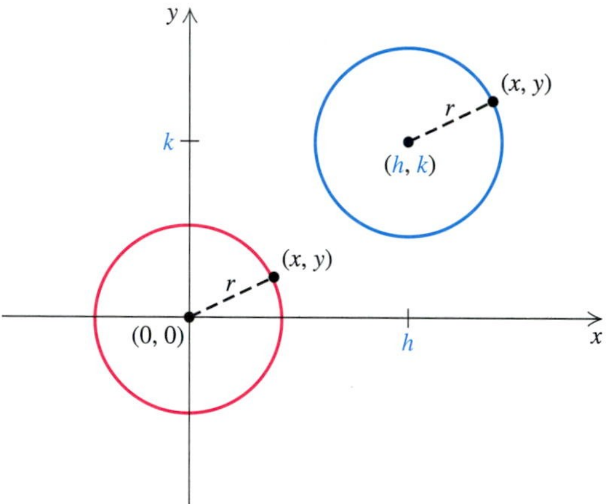

EXAMPLE 4

Find an equation of the circle having center $(4, -5)$ and radius 6.

Solution Using the standard form, we obtain

$$(x - 4)^2 + [y - (-5)]^2 = 6^2, \qquad \text{Using } (x-h)^2 + (y-k)^2 = r^2$$

or

$$(x - 4)^2 + (y + 5)^2 = 36.$$

❑

EXAMPLE 5

Find the center and the radius and then graph the circle.

a) $(x - 2)^2 + (y + 3)^2 = 4^2$
b) $x^2 + y^2 + 8x - 2y + 15 = 0$

Solution

a) We write standard form:

$$(x - 2)^2 + [y - (-3)]^2 = 4^2.$$

The center is $(2, -3)$ and the radius is 4. Now the graph is easy to draw using a compass.

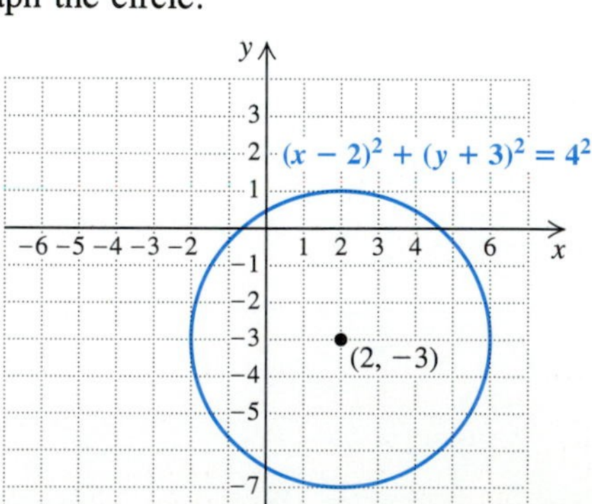

b) To write the equation $x^2 + y^2 + 8x - 2y + 15 = 0$ in standard form, we complete the square twice, once with $x^2 + 8x$ and once with $y^2 - 2y$:

$$x^2 + y^2 + 8x - 2y + 15 = 0$$

$$x^2 + 8x + y^2 - 2y + 15 = 0 \quad \text{Using the commutative law}$$

$$x^2 + 8x \quad + y^2 - 2y \quad = -15 \quad \text{Adding } -15 \text{ on both sides}$$

$$x^2 + 8x + 16 + y^2 - 2y + 1 = -15 + 16 + 1 \quad \text{Adding } \left(\tfrac{8}{2}\right)^2\text{, or 16, and } \left(-\tfrac{2}{2}\right)^2\text{, or 1, on both sides}$$

$$(x + 4)^2 + (y - 1)^2 = 2 \quad \text{Factoring}$$

$$[x - (-4)]^2 + (y - 1)^2 = (\sqrt{2})^2. \quad \text{Writing standard form}$$

The center is $(-4, 1)$ and the radius is $\sqrt{2}$.

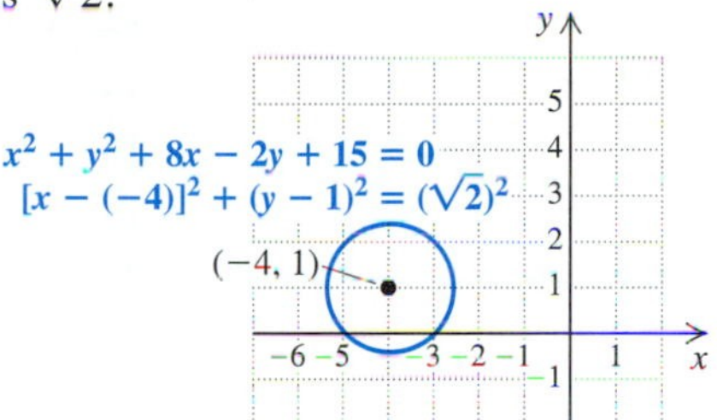

Graphing an equation of a circle on a grapher involves two steps:

1. Solve the equation for y. The result will include a $\pm$ sign in front of a radical.
2. Graph two functions, one including the + sign and the other including the − sign, on the same set of axes. (Because a grapher can graph only functions and a circle is not a function, we must divide it into two functions and graph both parts.)

For example, to graph the circle $x^2 + y^2 - 6x + 2y - 6 = 0$, we rewrite it as a quadratic equation in y, that is, $y^2 + 2y + (x^2 - 6x - 6) = 0$. The quadratic formula then gives us

$$y = \frac{-2 \pm \sqrt{4 - 4(x^2 - 6x - 6)}}{2},$$

which simplifies to

$$y = -1 \pm \sqrt{-x^2 + 6x + 7}$$

or

$$y_1 = -1 + \sqrt{-x^2 + 6x + 7} \quad \text{and}$$
$$y_2 = -1 - \sqrt{-x^2 + 6x + 7}.$$

When both functions are graphed (in a "squared" window, to eliminate distortion), the result is the graph shown here:

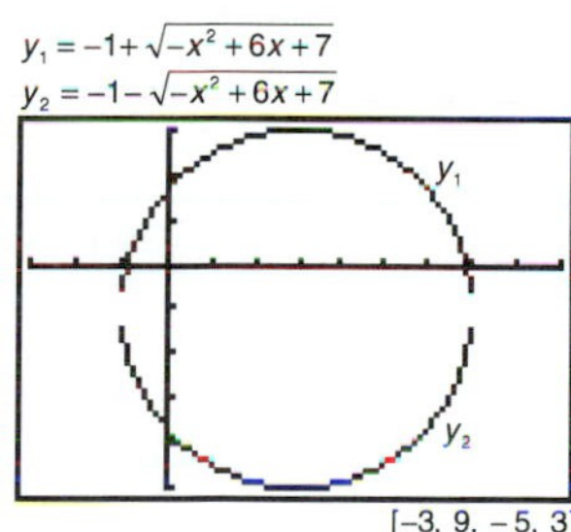

Most graphers have trouble connecting the two pieces because the segments become nearly vertical in this region.

Use a grapher to draw the graph of each of the following circles.

TC1. $x^2 + y^2 - 16 = 0$

TC2. $4x^2 + 4y^2 = 100$

TC3. $x^2 + y^2 + 14x - 16y + 54 = 0$

TC4. $x^2 + y^2 - 10x - 11 = 0$

EXERCISE SET 13.2

Graph.

1. $y = x^2$
2. $x = y^2$
3. $x = y^2 + 4y + 1$
4. $y = x^2 - 2x + 3$
5. $y = -x^2 + 4x - 5$
6. $x = 4 - 3y - y^2$
7. $x = y^2 + 1$
8. $x = 2y^2$
9. $x = -1 \cdot y^2$
10. $x = y^2 - 1$
11. $x = -y^2 + 2y$
12. $x = y^2 + y - 6$
13. $x = 8 - y - y^2$
14. $y = x^2 + 2x + 1$
15. $y = x^2 - 2x + 1$
16. $y = -\frac{1}{2}x^2$
17. $x = -y^2 + 2y + 3$
18. $x = -y^2 - 2y + 3$
19. $x = -2y^2 - 4y + 1$
20. $x = 2y^2 + 4y - 1$

Find an equation of the circle satisfying the given conditions.

21. Center (0, 0), radius 7
22. Center (0, 0), radius 4
23. Center (−2, 7), radius 3
24. Center (5, 6), radius 1
25. Center (−4, 3), radius $\sqrt{5}$
26. Center (−2, 7), radius $\sqrt{3}$
27. Center (−7, −2), radius $5\sqrt{2}$
28. Center (−5, −8), radius $2\sqrt{5}$
29. Center (0, 0), passing through (−3, 4)
30. Center (3, −2), passing through (11, −2)
31. Center (−4, 1), passing through (−2, 5)
32. Center (0, 0), passing through (1.8, 2.6)

Find the center and the radius of the circle. Then graph the circle.

33. $x^2 + y^2 = 36$
34. $x^2 + y^2 = 25$
35. $(x + 1)^2 + (y + 3)^2 = 4$
36. $(x - 2)^2 + (y + 3)^2 = 1$
37. $(x - 8)^2 + (y + 3)^2 = 40$
38. $(x + 5)^2 + (y - 1)^2 = 75$
39. $x^2 + y^2 = 2$
40. $x^2 + y^2 = 3$
41. $(x - 5)^2 + y^2 = \frac{1}{4}$
42. $x^2 + (y - 1)^2 = \frac{1}{25}$
43. $x^2 + y^2 + 8x - 6y - 15 = 0$
44. $x^2 + y^2 + 6x - 4y - 15 = 0$
45. $x^2 + y^2 - 8x + 2y + 13 = 0$
46. $x^2 + y^2 + 6x + 4y + 12 = 0$
47. $x^2 + y^2 - 4x = 0$
48. $x^2 + y^2 + 6x = 0$
49. $x^2 + y^2 + 10y - 75 = 0$
50. $x^2 + y^2 - 8x - 84 = 0$
51. $x^2 + y^2 + 7x - 3y - 10 = 0$
52. $x^2 + y^2 - 21x - 33y + 17 = 0$
53. $4x^2 + 4y^2 = 1$
54. $25x^2 + 25y^2 = 1$

Skill Maintenance

55. A rectangle 10 in. long and 6 in. wide is bordered by a strip of uniform width. If the perimeter of the larger rectangle is twice that of the smaller rectangle, what is the width of the border?
56. One airplane flies 60 mph faster than another. To fly a certain distance, the faster plane takes 4 hr and the slower plane takes 4 hr and 24 min. What is the distance?

Synthesis

57. ◈ Outline a procedure for finding the equation of a circle if the center and one point on the circle are known.
58. ◈ Describe the graph of an equation of the form $(x - h)^2 + (y - k)^2 = 0$.
59. Use a graph of the equation $x = y^2 - y - 6$ to approximate the solutions of each of the following equations.
    a) $y^2 - y - 6 = 2$ (*Hint:* Graph $x = 2$ on the same set of axes as the graph of $x = y^2 - y - 6$.)
    b) $y^2 - y - 6 = -3$
60. The horsepower of a certain kind of engine is given by the fomula

$$H = \frac{D^2N}{2.5},$$

where N is the number of cylinders and D is the diameter, in inches, of each piston. Graph this equation, assuming that $N = 6$ (a six-cylinder engine). Let D run from 2.5 to 8.

Using the same set of axes, graph the pair of equations. Try to discover a way to obtain one graph from the other without computing points for the second graph. (*Hint:* Review Section 12.2.)

61. $y = -2x^2 + 3, \quad x = -2y^2 + 3$

62. $y = x^2 + 2x - 3, \quad x = y^2 + 2y - 3$

Find the center and the radius of the circle.

63. $4x^2 + 4y^2 + 4x - 8y + 1 = 0$

64. $9x^2 + 9y^2 + 18x + 6y = 26$

65. Ace Carpentry needs to cut an arch for the top of an entranceway. The arch needs to be 8 ft wide and 2 ft high. To draw the arch, the carpenters will use a stretched string with chalk attached at an end as a compass.

a) Using a coordinate system, locate the center of the circle.

b) What radius should the carpenters use to draw the arch?

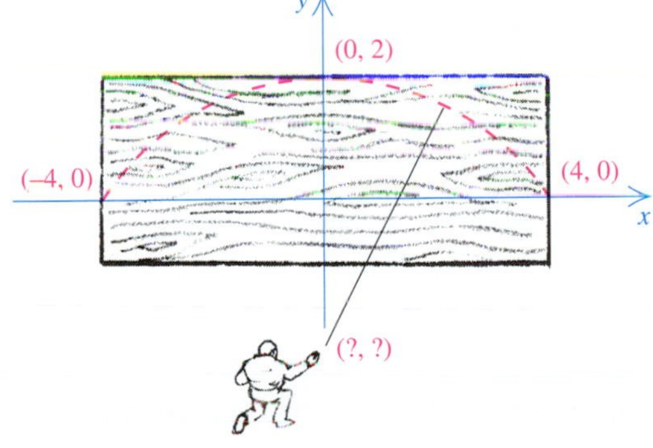

Find an equation of a circle satisfying the given conditions.

66. Center $(3, -5)$ and tangent to (touching at one point) the y-axis

67. Center $(-7, -4)$ and tangent to the x-axis

68. The endpoints of a diameter are $(7, 3)$ and $(-1, -3)$.

69. Center $(-3, 5)$ with a circumference of 8π units

70. A ferris wheel has a radius of 24.3 ft. Assuming that the center is 30.6 ft off the ground and that the origin is below the center, as in the following figure, find an equation of the circle.

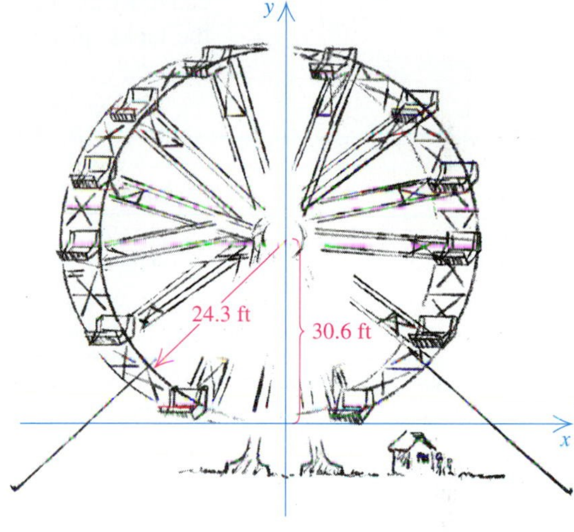

13.3

Conic Sections: Ellipses and Hyperbolas

Ellipses • Hyperbolas • Hyperbolas (Nonstandard Form) • Classifying Graphs of Equations

Ellipses

When a cone is cut at an angle, as shown, the conic section formed is an *ellipse*. You can draw an ellipse by sticking two tacks in a piece of cardboard. Then tie a string to the tacks, place a pencil as shown, and draw.

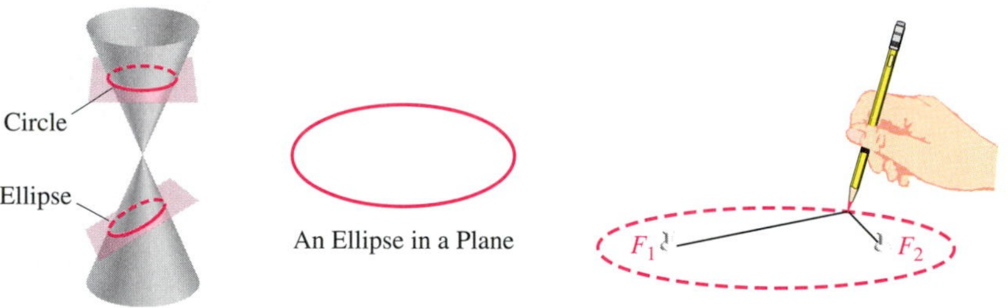

The formal mathematical definition is related to this method of drawing an ellipse. An **ellipse** is defined as the set of all points in a plane such that the *sum* of the distances from two fixed points F_1 and F_2 (called **foci;** singular, **focus**) is constant. In the figure shown, the tacks are at the foci. The midpoint of the segment F_1F_2 is the **center.** Ellipses have equations as follows. The proof is left to the exercises.

Equation of an Ellipse

The equation of an ellipse centered at the origin and parallel to an axis is

$$\frac{x^2}{a^2} + \frac{y^2}{b^2} = 1, \quad a, b > 0, \quad a \neq b. \qquad \text{(Standard form)}$$

We can think of a circle as a special kind of ellipse. A circle is formed when $a = b$ and the angle at which the cutting plane cuts the circle is 90°. If $a = b$, the equation of an ellipse becomes

$$\frac{x^2}{a^2} + \frac{y^2}{a^2} = 1 \qquad a = b$$

$$x^2 + y^2 = a^2. \qquad \text{Multiplying by } a^2$$

This is the equation of a circle centered at the origin with radius a. The foci of a circle, F_1 and F_2, are the same point, the center. An ellipse with its foci close together is very nearly a circle.

Ellipses with centers other than the origin are discussed in Exercises 59 and 60.

When graphing ellipses, it helps to first find the intercepts. If we replace x by 0, we can find the y-intercepts:

$$\frac{0^2}{a^2} + \frac{y^2}{b^2} = 1$$
$$\frac{y^2}{b^2} = 1$$
$$y^2 = b^2$$
$$y = \pm b.$$

Thus the y-intercepts are $(0, b)$ and $(0, -b)$. Similarly, the x-intercepts are $(a, 0)$ and $(-a, 0)$.

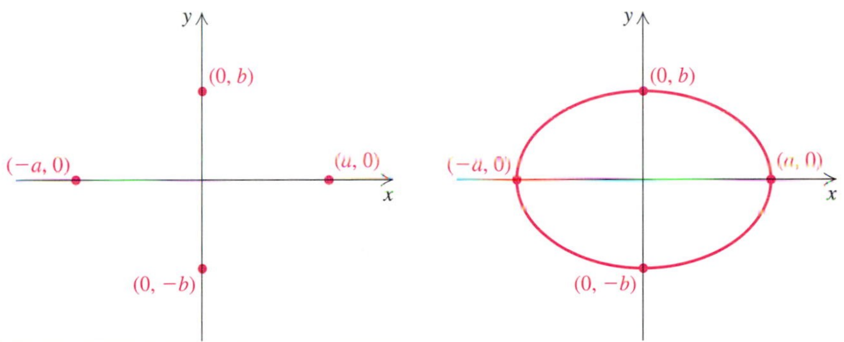

Plotting these points and filling in an oval-shaped curve, we get a graph of the ellipse. If a more precise graph is desired, we can plot more points.

For the ellipse

$$\frac{x^2}{a^2} + \frac{y^2}{b^2} = 1,$$

the x-intercepts are $(-a, 0)$ and $(a, 0)$. The y-intercepts are $(0, -b)$ and $(0, b)$.

EXAMPLE 1

Graph the ellipse $\dfrac{x^2}{4} + \dfrac{y^2}{9} = 1$.

Solution Note that

$$\frac{x^2}{4} + \frac{y^2}{9} = \frac{x^2}{2^2} + \frac{y^2}{3^2}.$$

Thus the x-intercepts are $(-2, 0)$ and $(2, 0)$, and the y-intercepts are $(0, 3)$ and $(0, -3)$. We plot these points and connect them with an oval-shaped curve. To plot some other points, we let $x = 1$ and solve for y:

$$\frac{1^2}{4} + \frac{y^2}{9} = 1$$

$$36\left(\frac{1}{4} + \frac{y^2}{9}\right) = 36 \cdot 1$$

$$36 \cdot \frac{1}{4} + 36 \cdot \frac{y^2}{9} = 36$$

$$9 + 4y^2 = 36$$

$$4y^2 = 27$$

$$y^2 = \frac{27}{4}$$

$$y = \pm\sqrt{\frac{27}{4}}$$

$$y \approx \pm 2.6.$$

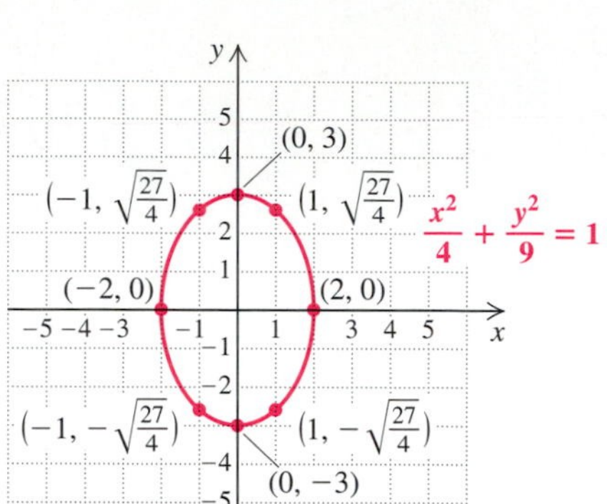

Thus, $(1, 2.6)$ and $(1, -2.6)$ can also be used to draw the graph. Similarly, the points $(-1, 2.6)$ and $(-1, -2.6)$ can also be computed and plotted. ❑

EXAMPLE 2

Graph: $4x^2 + 25y^2 = 100$.

Solution We write the equation in standard form by multiplying on both sides by $\frac{1}{100}$:

$$\frac{1}{100}(4x^2 + 25y^2) = \frac{1}{100}(100)$$

$$\frac{1}{100}(4x^2) + \frac{1}{100}(25y^2) = 1$$

$$\frac{x^2}{25} + \frac{y^2}{4} = 1$$

$$\frac{x^2}{5^2} + \frac{y^2}{2^2} = 1.$$

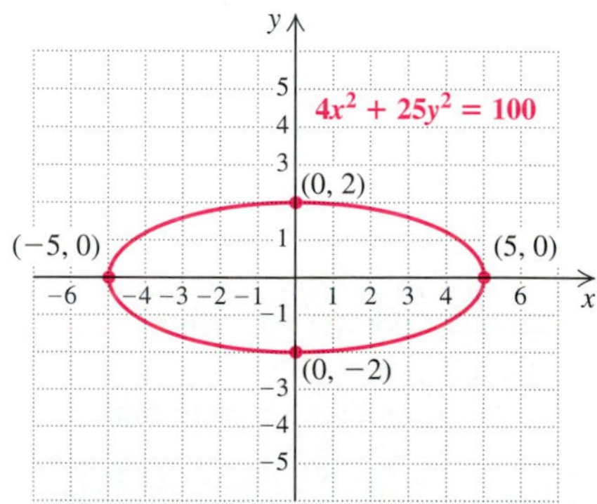

The x-intercepts are $(-5, 0)$ and $(5, 0)$, and the y-intercepts are $(0, -2)$ and $(0, 2)$. We plot the intercepts and connect them with an oval-shaped curve. Other points can also be computed and plotted. ❑

Ellipses have many applications. Earth satellites travel in elliptical orbits, and planets travel around the sun in elliptical orbits with the sun at one focus.

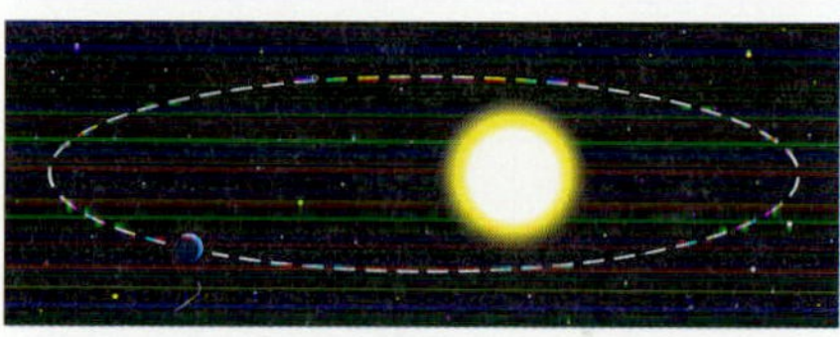

Planetary orbit

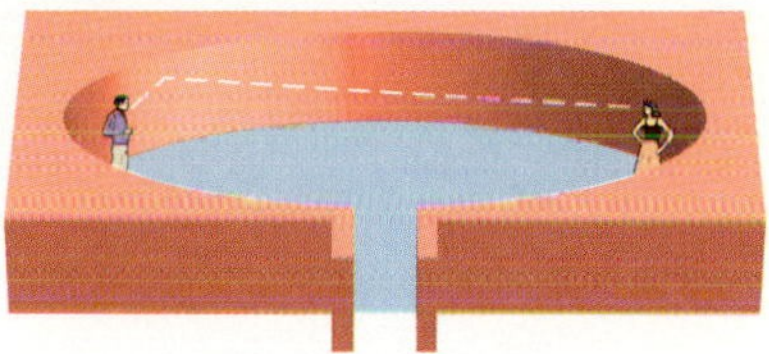

Whispering gallery

An interesting application found in some buildings is a *whispering gallery,* which is elliptical. Persons with their heads at the foci can whisper and hear each other clearly, while persons at other positions cannot hear them. This happens when sound waves emanating at one focus are reflected to the other focus, being concentrated there.

A dentist often uses a reflector light. So that the light does not hit you in the eyes, it is covered and reflected. The reflection is directed toward your mouth. One focus is at the light source and the other is at your mouth.

Hyperbolas

A **hyperbola** looks like a pair of parabolas, but the actual shapes are different. A hyperbola has two **vertices** (singular, **vertex**) and the line through the vertices is known as an **axis.** The point halfway between the vertices is called the **center.**

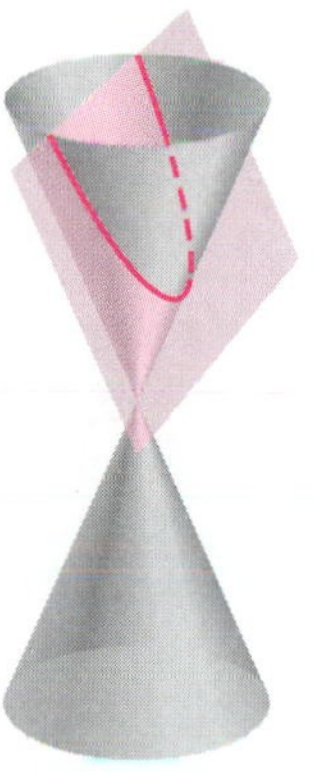

Parabola

Hyperbola in three dimensions

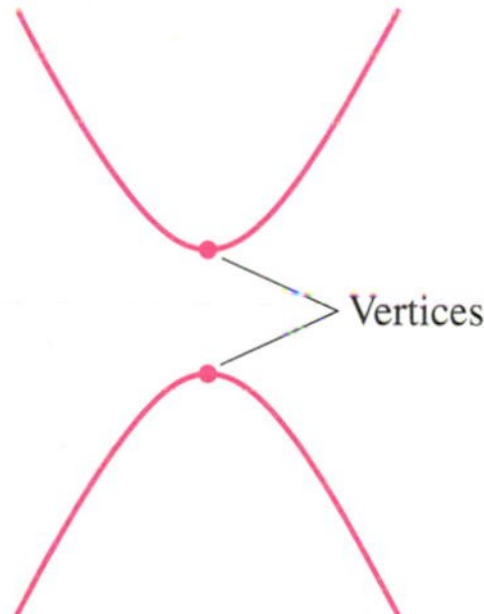

Hyperbola in a plane

Equation of a Hyperbola

Hyperbolas with their centers at the origin and vertical or horizontal axes have equations as follows:

$$\frac{x^2}{a^2} - \frac{y^2}{b^2} = 1 \qquad \text{(Axis horizontal);}$$

$$\frac{y^2}{b^2} - \frac{x^2}{a^2} = 1 \qquad \text{(Axis vertical).}$$

Note carefully that these equations have a 1 on the right and a minus sign between the terms.

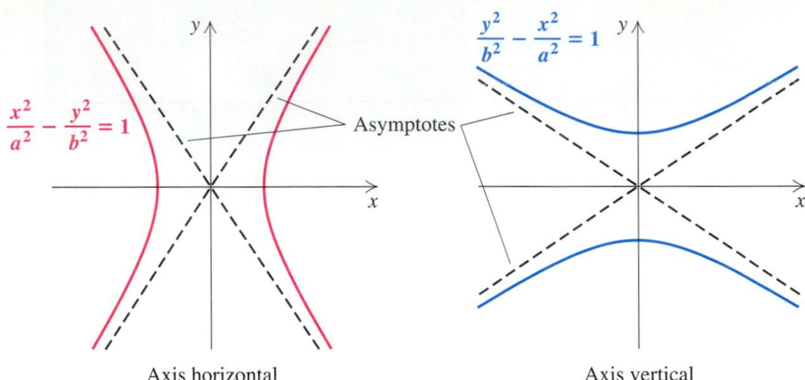

Hyperbolas with horizontal or vertical axes and centers *not* at the origin are discussed in the exercise set.

To graph a hyperbola, it helps to begin by graphing the lines called **asymptotes.**

Asymptotes of a Hyperbola

For hyperbolas with equations as given in the preceding box, the asymptotes are the lines

$$y = \frac{b}{a}x \quad \text{and} \quad y = -\frac{b}{a}x.$$

As a hyperbola gets farther away from the origin, it gets closer and closer to its asymptotes. The larger $|x|$ gets, the closer the graph gets to an asymptote. The asymptotes act to "constrain" the graph of a hyperbola. Parabolas are *not* constrained by any asymptotes.

The next thing to do after sketching asymptotes is to plot vertices. Then it is easy to sketch the curve.

EXAMPLE 3

Graph: $\frac{x^2}{4} - \frac{y^2}{9} = 1$.

Solution Note that

$$\frac{x^2}{4} - \frac{y^2}{9} = \frac{x^2}{2^2} - \frac{y^2}{3^2},$$

so $a = 2$ and $b = 3$. The asymptotes are thus

$$y = \frac{3}{2}x \quad \text{and} \quad y = -\frac{3}{2}x.$$

We sketch them, as shown on the left in the figure on the following page.

For horizontal or vertical hyperbolas centered at the origin, the vertices will also be intercepts. Since this hyperbola is horizontal, we replace y with 0 and solve for x. We see that $x^2/2^2 = 1$ when $x = \pm 2$. The intercepts are $(2, 0)$ and $(-2, 0)$. There are intercepts on only one axis. If we replace x with 0, we see that $y^2/9 = -1$ has no real-number solution.

Finally, we plot the intercepts and sketch the graph. Through each intercept, we draw a smooth curve that approaches the asymptotes closely, as shown on the right below.

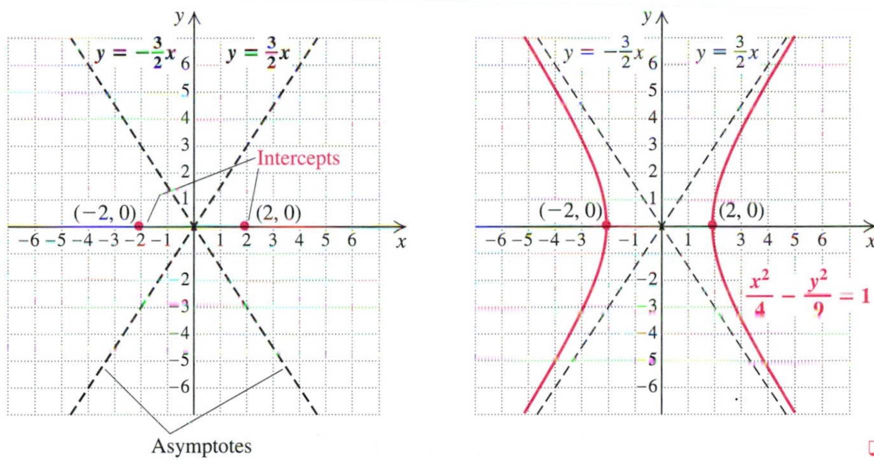

EXAMPLE 4

Graph: $\dfrac{y^2}{36} - \dfrac{x^2}{4} = 1.$

Solution Note that

$$\frac{y^2}{36} - \frac{x^2}{4} = \frac{y^2}{6^2} - \frac{x^2}{2^2} = 1.$$

The intercept distance is found in the term without the minus sign. Here there is a y in this term, so the intercepts are on the y-axis.

The asymptotes are thus $y = \frac{6}{2}x$ and $y = -\frac{6}{2}x$, or $y = 3x$ and $y = -3x$.

With the numbers 6 and 2, we can quickly sketch a rectangle to use as a guide. Thinking of ± 2 as x coordinates and ± 6 as y-coordinates, we form all possible ordered pairs: $(2, 6)$, $(2, -6)$, $(-2, 6)$, and $(-2, -6)$. We plot these pairs and lightly sketch a rectangle through them. The asymptotes pass through the corners (see the figure on the left on page 690). Since the hyperbola is vertical, we graph the y-intercepts. They are $(0, 6)$ and $(0, -6)$. We now draw curves through the intercepts toward the asymptotes, as shown on the right in the figure on the following page.

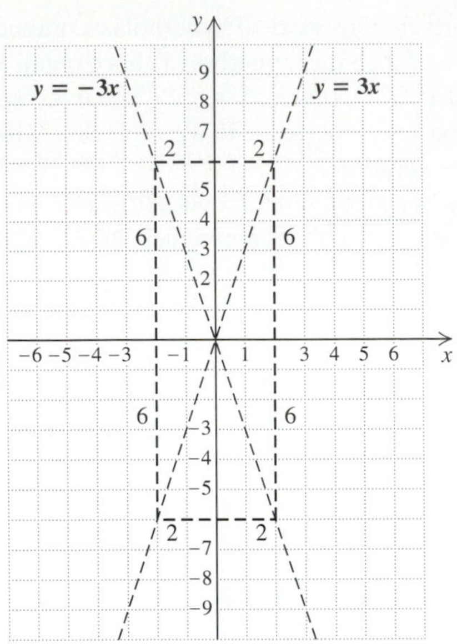

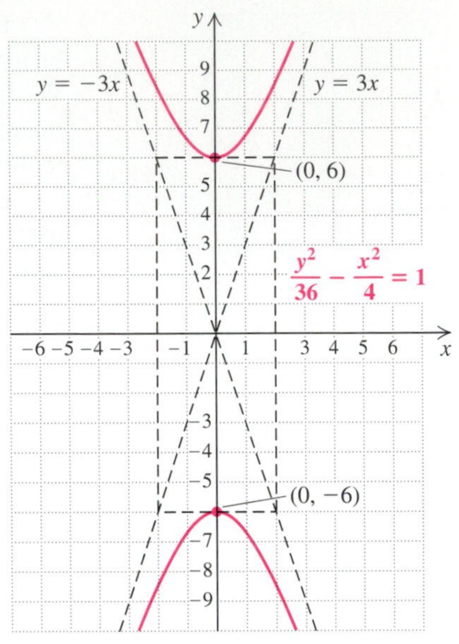

❑

Hyperbolas (Nonstandard Form)

The equations that we have just seen for hyperbolas are the standard ones, but there are other hyperbolas. We consider some of them.

> Hyperbolas having the x- and y-axes as asymptotes have equations as follows:
>
> $xy = c$, where c is a nonzero constant.

EXAMPLE 5

Graph: $xy = -8$.

Solution We first solve for y:

$$y = -\frac{8}{x}.$$

Next, we find some solutions, keeping the results in a table. Note that we cannot use 0 for x and that for large values of $|x|$, y will be close to 0. We plot the points and connect the points in the second quadrant with a smooth curve. Similarly, we connect the points in the fourth quadrant with a smooth curve. Remember that the axes serve as asymptotes.

x	y
2	−4
−2	4
4	−2
−4	2
1	−8
−1	8
8	−1
−8	1

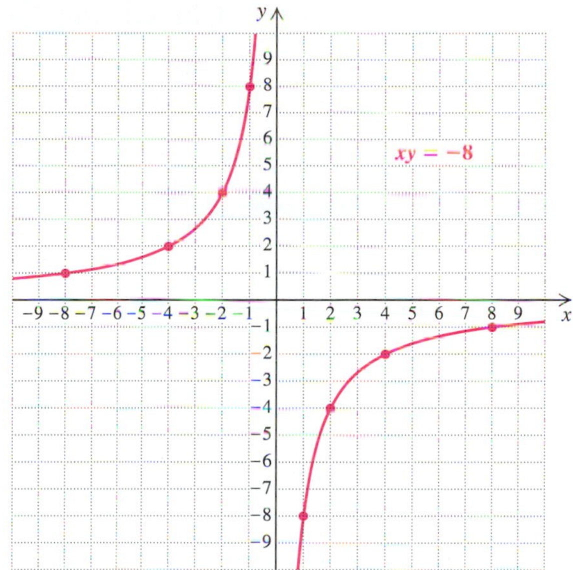

TECHNOLOGY CONNECTION

The procedure used to draw the graph of an ellipse or a hyperbola on a grapher is similar to that used to draw a circle. For example, let's draw the graph of the hyperbola given by the equation

$$\frac{x^2}{25} - \frac{y^2}{49} = 1.$$

Solving for y gives us

$$y_1 = \frac{\sqrt{49x^2 - 1225}}{5} = \frac{7}{5}\sqrt{x^2 - 25}$$

and

$$y_2 = \frac{-\sqrt{49x^2 - 1225}}{5} = -\frac{7}{5}\sqrt{x^2 - 25}.$$

When the two pieces are drawn on the same squared axes, the result is the following:

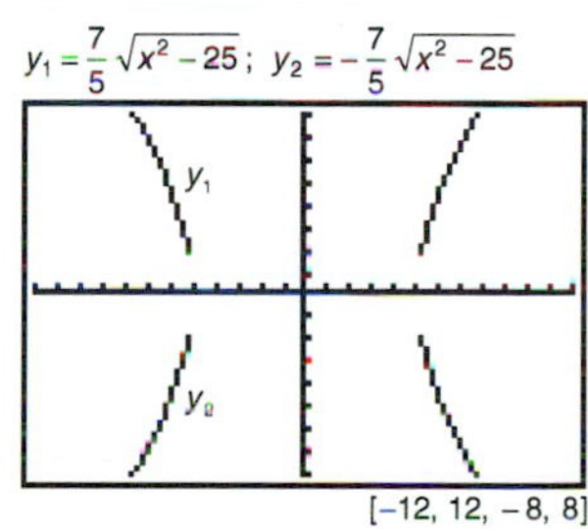

Again, note the problem that the grapher has where the graph is nearly vertical.

Use a grapher to draw the graph of each of the following ellipses and hyperbolas. Use squared axes so that the shapes are not distorted.

TC1. $\frac{x^2}{16} + \frac{y^2}{60} = 1$

TC2. $16x^2 + 3y^2 = 64$

TC3. $\frac{y^2}{20} - \frac{x^2}{64} = 1$

TC4. $9x^2 - 45y^2 = 441$

Hyperbolas have many applications. A jet breaking the sound barrier creates a sonic boom whose wave front has the shape of a cone. The cone intersects the ground in one branch of a hyperbola. Some comets travel in hyperbolic orbits, and a cross section of an amphitheater may be hyperbolic in shape.

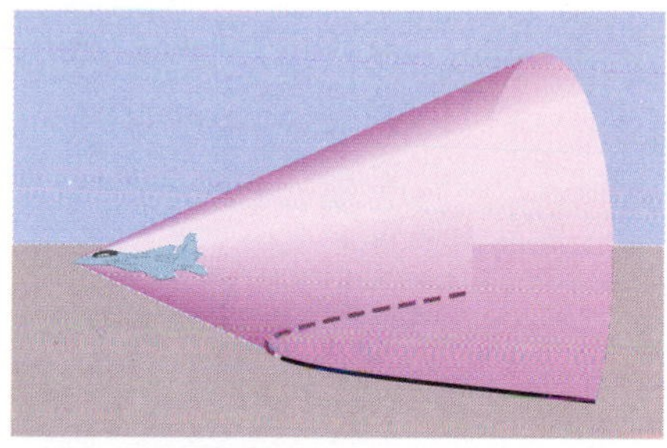

Classifying Graphs of Equations

The following is a summary of the equations and graphs of conic sections studied in this chapter.

Parabola

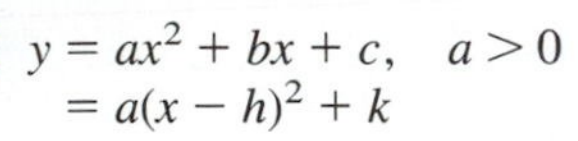

$$\begin{aligned} y &= ax^2 + bx + c, \quad a > 0 \\ &= a(x - h)^2 + k \end{aligned}$$

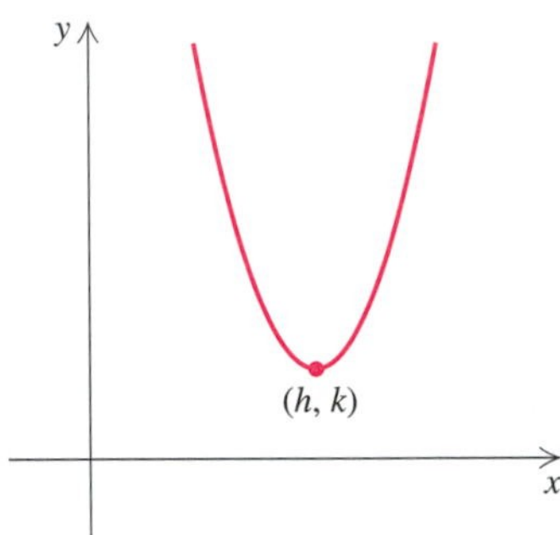

$$\begin{aligned} y &= ax^2 + bx + c, \quad a < 0 \\ &= a(x - h)^2 + k \end{aligned}$$

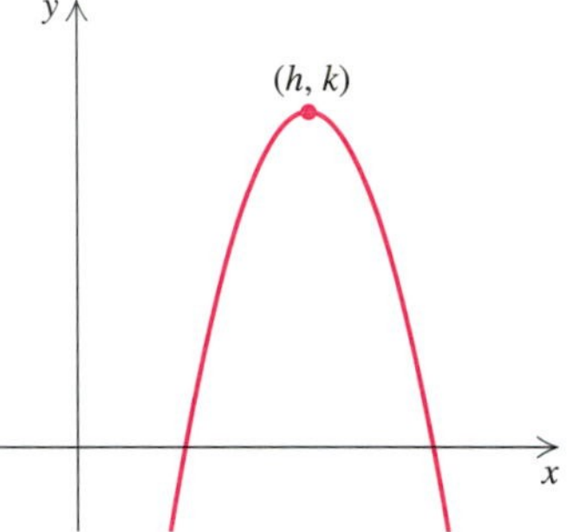

$$\begin{aligned} x &= ay^2 + by + c, \quad a > 0 \\ &= a(y - k)^2 + h \end{aligned}$$

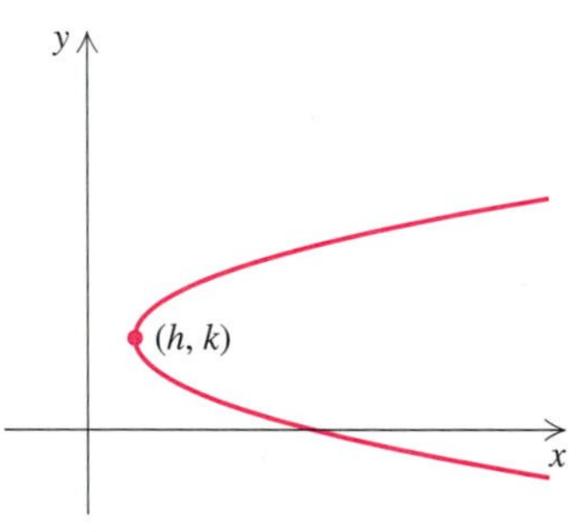

$$\begin{aligned} x &= ay^2 + by + c, \quad a < 0 \\ &= a(y - k)^2 + h \end{aligned}$$

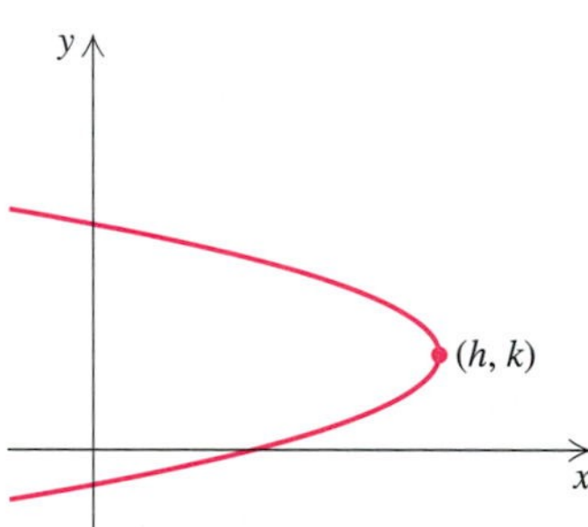

Circle

Center at the origin:

$$x^2 + y^2 = r^2$$

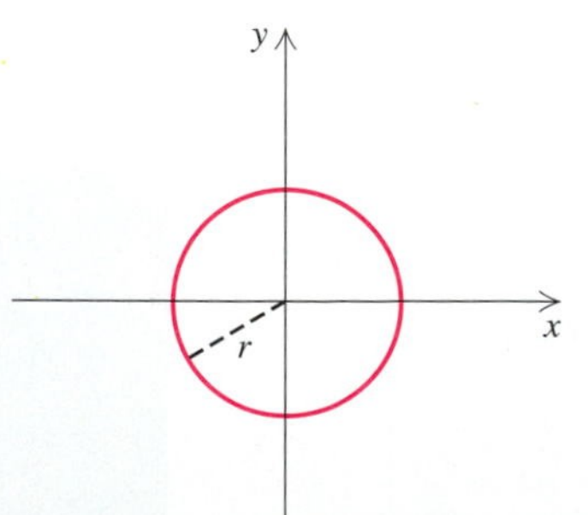

Center at (h, k):

$$(x - h)^2 + (y - k)^2 = r^2$$

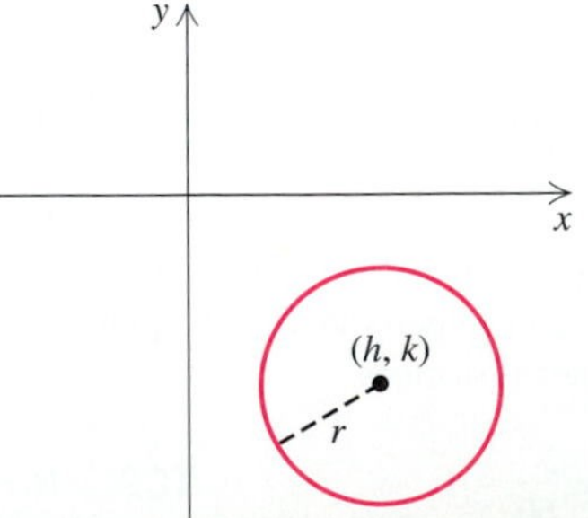

Ellipse

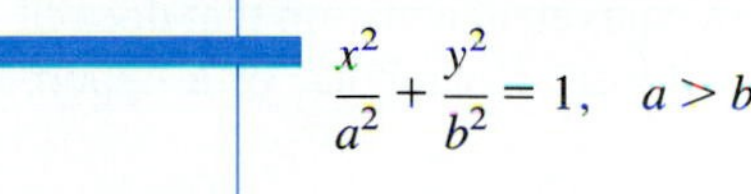

$\dfrac{x^2}{a^2} + \dfrac{y^2}{b^2} = 1, \quad a > b$

$\dfrac{x^2}{a^2} + \dfrac{y^2}{b^2} = 1, \quad b > a$

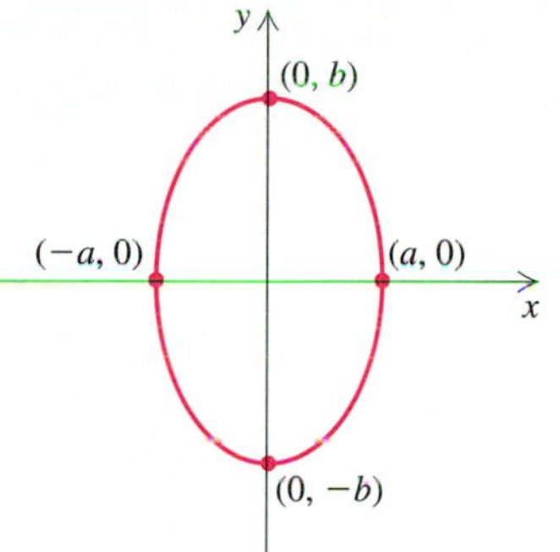

Hyperbola

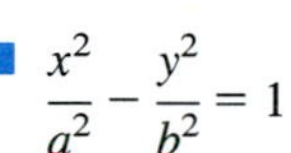

$\dfrac{x^2}{a^2} - \dfrac{y^2}{b^2} = 1$

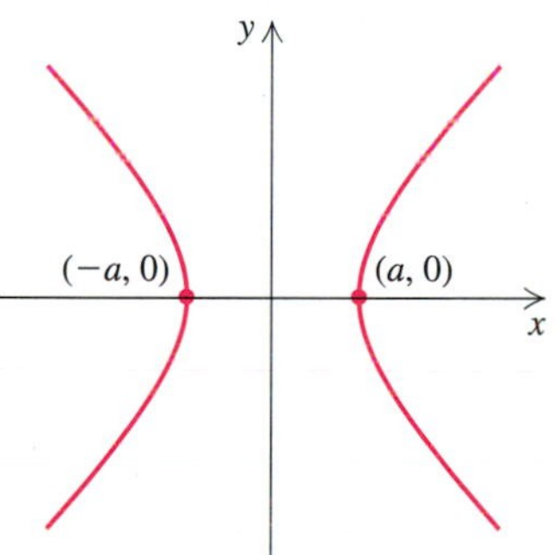

$\dfrac{y^2}{b^2} - \dfrac{x^2}{a^2} = 1$

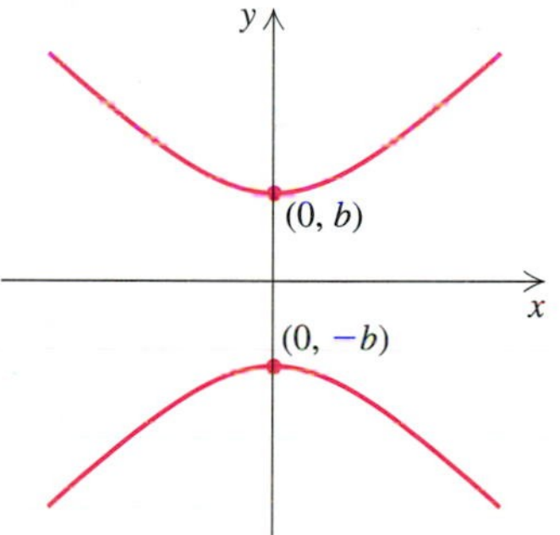

$xy = c, \quad c > 0$

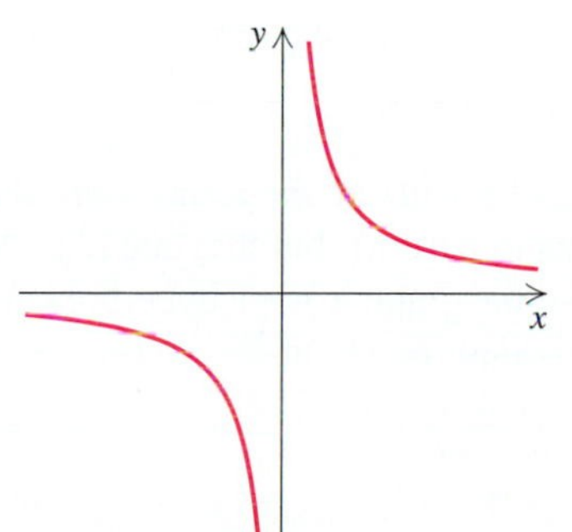

$xy = c, \quad c < 0$

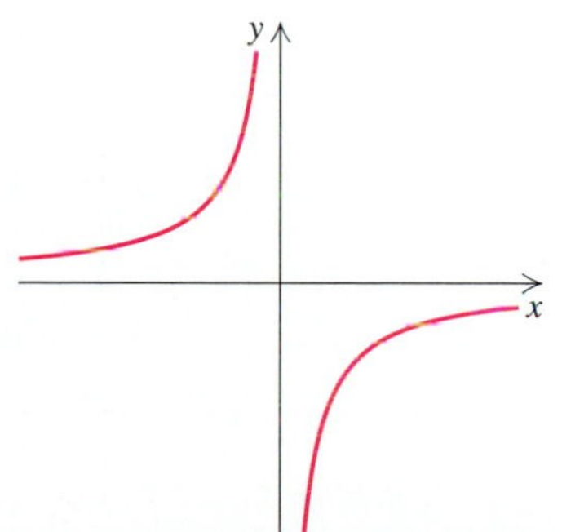

Suppose we encounter an equation that is not in one of the preceding forms. Sometimes we can find an equivalent equation that does fit one of the forms and then classify it as a circle, an ellipse, a parabola, or a hyperbola.

EXAMPLE 6

Classify the graph of the equation as a circle, an ellipse, a parabola, or a hyperbola.

a) $5x^2 = 20 - 5y^2$ **b)** $x + 3 + 8y = y^2$
c) $x^2 = y^2 + 4$ **d)** $x^2 = 16 - 4y^2$

Solution

a) We get the terms with variables on one side by adding $5y^2$:

$$5x^2 + 5y^2 = 20.$$

The fact that x and y are *both* squared tells us that we do not have a parabola. The fact that the squared terms are *added* tells us that we do not have a hyperbola. Do we have a circle? To find out, we need to get $x^2 + y^2$ by itself. We can do that by factoring the 5 out of both terms on the left and then dividing by 5 on both sides:

$$5(x^2 + y^2) = 20$$
$$x^2 + y^2 = 4$$
$$x^2 + y^2 = 2^2.$$

We can see that the graph is a circle with center at the origin and radius 2.

b) Since only one of the variables is squared, we can find the following equivalent equation:

$$x = y^2 - 8y - 3.$$

Thus the graph is a horizontal parabola that opens to the right since the coefficient of y^2, 1, is positive.

c) Both variables are squared, so we do not have a parabola. We can subtract y^2 on both sides and divide by 4 to obtain

$$\frac{x^2}{2^2} - \frac{y^2}{2^2} = 1.$$

The minus sign in this form tells us that the graph of the equation is a hyperbola.

d) Both variables are squared, so the graph is not a parabola. We obtain the following equivalent equation:

$$x^2 + 4y^2 = 16.$$

If the coefficients of the terms were the same, we would have the graph of a circle, as in part (a), but they are not. We have a plus sign between the squared terms, so the graph is not a hyperbola. We divide by 16 on both sides and obtain the equivalent equation

$$\frac{x^2}{16} + \frac{y^2}{4} = 1$$

whose graph is an ellipse. ❑

EXERCISE SET 13.3

Graph the ellipse.

1. $\frac{x^2}{4} + \frac{y^2}{1} = 1$
2. $\frac{x^2}{1} + \frac{y^2}{4} = 1$
3. $\frac{x^2}{16} + \frac{y^2}{25} = 1$
4. $\frac{x^2}{9} + \frac{y^2}{25} = 1$
5. $4x^2 + 9y^2 = 36$
6. $9x^2 + 4y^2 = 36$
7. $16x^2 + 9y^2 = 144$
8. $9x^2 + 16y^2 = 144$
9. $2x^2 + 3y^2 = 6$
10. $5x^2 + 7y^2 = 35$
11. $4x^2 + 9y^2 = 1$
12. $25x^2 + 16y^2 = 1$
13. $5x^2 + 12y^2 = 60$
14. $8x^2 + 3y^2 = 24$

Graph the hyperbola.

15. $\frac{x^2}{16} - \frac{y^2}{16} = 1$
16. $\frac{y^2}{9} - \frac{x^2}{9} = 1$
17. $\frac{y^2}{16} - \frac{x^2}{9} = 1$
18. $\frac{x^2}{9} - \frac{y^2}{4} = 1$
19. $\frac{x^2}{25} - \frac{y^2}{36} = 1$
20. $\frac{y^2}{9} - \frac{x^2}{25} = 1$
21. $x^2 - y^2 = 4$
22. $y^2 - x^2 = 25$
23. $4y^2 - 9x^2 = 36$
24. $25x^2 - 16y^2 = 400$
25. $xy = 6$
26. $xy = -4$
27. $xy = -9$
28. $xy = 3$
29. $xy = -1$
30. $xy = -2$
31. $xy = 2$
32. $xy = 1$

Classify each of the following as the equation of a circle, an ellipse, a parabola, or a hyperbola.

33. $x^2 + y^2 - 10x + 8y - 40 = 0$
34. $y + 1 = 2x^2$
35. $9x^2 - 4y^2 - 36 = 0$
36. $1 - 3y = 2y^2 - x$
37. $4x^2 + 25y^2 - 100 = 0$
38. $y^2 + x^2 = 7$
39. $x^2 + y^2 = 2x + 4y + 4$
40. $2y + 13 + x^2 = 8x - y^2$
41. $4x^2 = 64 - y^2$
42. $y = \frac{1}{x}$
43. $x - \frac{3}{y} = 0$
44. $x - 4 = y^2 + 5$
45. $y + 6x = x^2 + 6$
46. $x^2 = 16 + y^2$
47. $9y^2 = 36 + 4x^2$
48. $3x^2 + 5y^2 + x^2 = y^2 + 49$

Skill Maintenance

49. Simplify: $\sqrt[3]{125t^{15}}$.
50. Solve: $2x^2 + 10 = 0$.
51. Rationalize the denominator:
$$\frac{4\sqrt{2} - 5\sqrt{3}}{6\sqrt{3} - 8\sqrt{2}}.$$
52. An airplane travels 500 mi at a certain speed. A larger plane travels 1620 mi at a speed that is 320 mph faster, but takes 1 hr longer. Find the speed of each plane.

Synthesis

53. ◈ An eccentric person builds a pool table in the shape of an ellipse with a hole at one focus and a tiny dot at the other. Guests are amazed at how many bank shots the owner of the pool table makes. Explain.
54. The maximum distance of the planet Mars from the sun is 2.48×10^8 miles. The minimum distance is 3.46×10^7 miles. The sun is at one focus of the elliptical orbit. Find the distance from the sun to the other focus.
55. Let $(-c, 0)$ and $(c, 0)$ be the foci of an ellipse. Any point $P(x, y)$ is on the ellipse if the sum of the distances from the foci to P is some constant. Use $2a$ to represent this constant.
    a) Show that an equation for the ellipse is given by
    $$\frac{x^2}{a^2} + \frac{y^2}{a^2 - c^2} = 1.$$
    b) Substitute b^2 for $a^2 - c^2$ to get standard form.

56. The Oval Office of the President of the United States is an ellipse 31 ft wide and 38 ft long. Show in a sketch precisely where the President and an adviser could sit to best use the room's acoustics. (*Hint:* See Exercise 55(b) and the discussion following Example 2.)

57. The light source in a dental lamp shines against a reflector that is shaped like a portion of an ellipse in which the light source is one focus of the ellipse. Reflected light enters a patient's mouth at the other focus of the ellipse. If the ellipse from which the reflector was formed is 2 ft wide and 6 ft long, how far should the patient's mouth be from the light source? (*Hint:* See Exercise 55(b).)

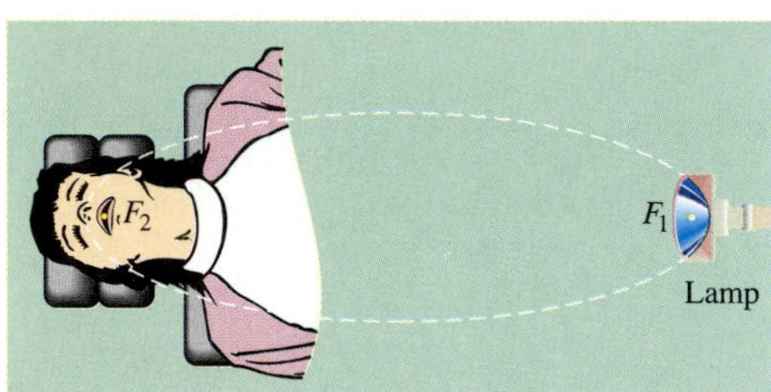

58. Find an equation of an ellipse that has x-intercepts $(-9, 0)$ and $(9, 0)$ and y-intercepts $(0, -11)$ and $(0, 11)$.

The standard form of an ellipse parallel to an axis and centered at (h, k) is

$$\frac{(x-h)^2}{a^2} + \frac{(y-k)^2}{b^2} = 1.$$

The vertices are $(a + h, k)$, $(-a + h, k)$, $(h, b + k)$, and $(h, -b + k)$. For each of the following equations of ellipses, complete the square, if necessary, and find an equivalent equation in standard form. Find the center and the vertices. Then graph the ellipse.

59. $16x^2 + y^2 + 96x - 8y + 144 = 0$

60. $4x^2 + 25y^2 - 8x + 50y = 71$

Find an equation of a hyperbola satisfying the given conditions.

61. Having intercepts $(0, 8)$ and $(0, -8)$ and asymptotes $y = 4x$ and $y = -4x$

62. Having intercepts $(8, 0)$ and $(-8, 0)$ and asymptotes $y = 4x$ and $y = -4x$

Hyperbolas centered at (h, k), with horizontal or vertical axes, have standard equations as follows, where the asymptotes are

$$y - k = \pm\frac{b}{a}(x - h).$$

The vertices are labeled in the figures:

$$\frac{(x-h)^2}{a^2} - \frac{(y-k)^2}{b^2} = 1$$

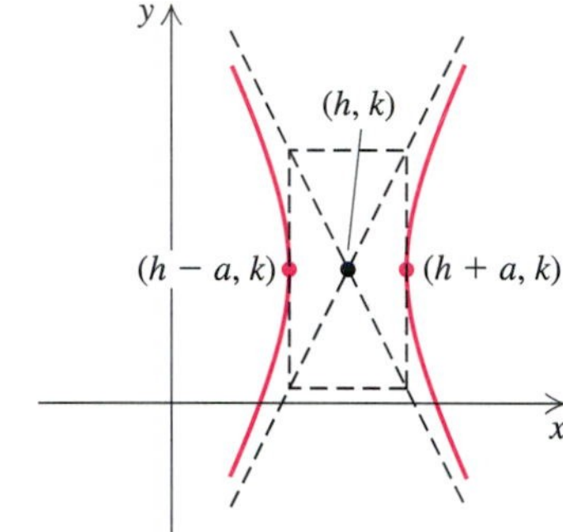

$$\frac{(y-k)^2}{b^2} - \frac{(x-h)^2}{a^2} = 1$$

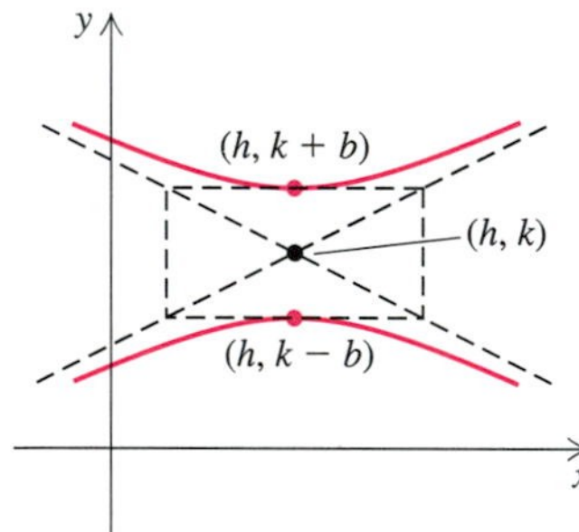

For each of the following equations of hyperbolas, complete the square, if necessary, and write in standard form. Find the center, the vertices, and the asymptotes. Then graph the hyperbola.

63. $\dfrac{(x-2)^2}{9} - \dfrac{(y+1)^2}{16} = 1$

64. $4x^2 - y^2 + 24x + 4y + 28 = 0$

65. $4y^2 - 25x^2 - 8y - 100x - 196 = 0$

66. $x^2 - y^2 - 2y - 4x = 6$

13.4 Nonlinear Systems of Equations

Systems Involving One Nonlinear Equation • Systems of Two Nonlinear Equations • Problem Solving

The equations that have appeared in systems of two equations have thus far all been linear. We now consider systems of two equations in which at least one equation is not linear.

Systems Involving One Nonlinear Equation

We consider a system involving an equation of a circle and an equation of a line. Let's think about the possible ways in which a circle and a line can intersect. The three possibilities are shown in the figure below. For L_1 there is no point of intersection; hence the system of equations has no real solution. For L_2 there is one point of intersection, hence one real solution. For L_3 there are two points of intersection, hence two real solutions.

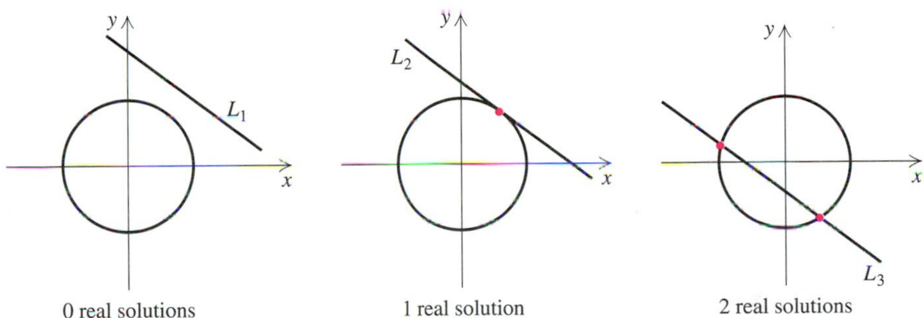

Remember that we used both *elimination* and *substitution* to solve systems of linear equations. When solving systems in which one equation is of first degree and one is of second degree, it is preferable to use the *substitution* method.

EXAMPLE 1

Solve the system

$$x^2 + y^2 = 25, \quad (1) \quad \text{(The graph is a circle.)}$$
$$3x - 4y = 0. \quad (2) \quad \text{(The graph is a line.)}$$

Solution First, we solve the linear equation (2) for x:

$$x = \tfrac{4}{3}y. \quad (3)$$

Then we substitute $\frac{4}{3}y$ for x in Equation (1) and solve for y:

$$\left(\tfrac{4}{3}y\right)^2 + y^2 = 25$$
$$\tfrac{16}{9}y^2 + y^2 = 25$$
$$\tfrac{25}{9}y^2 = 25$$
$$y^2 = 9 \quad \text{Multiplying by } \tfrac{9}{25} \text{ on both sides}$$
$$y = \pm 3.$$

Now we substitute these numbers for y in Equation (3) and solve for x:

for $y = 3$, $x = \frac{4}{3}(3) = 4$; for $y = -3$, $x = \frac{4}{3}(-3) = -4$.

Check: For (4, 3):

$x^2 + y^2 = 25$		
$4^2 + 3^2$	?	25
$16 + 9$		
25		25 TRUE

$3x - 4y = 0$		
$3(4) - 4(3)$	?	0
$12 - 12$		
0		0 TRUE

It is left to the student to confirm that $(-4, -3)$ also checks in both equations.

The pairs (4, 3) and $(-4, -3)$ check, so they are solutions. We can see the solutions in the graph. The graph of Equation (1) is a circle, and the graph of Equation (2) is a line. The graphs intersect at the points (4, 3) and $(-4, -3)$.

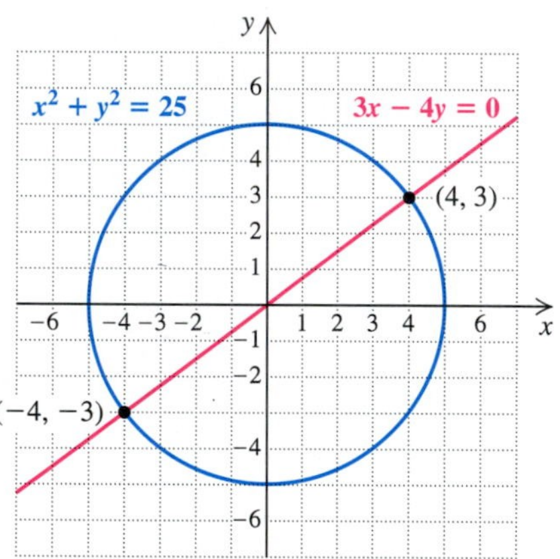

❑

EXAMPLE 2

Solve the system

$$y + 3 = 2x, \qquad (1)$$
$$x^2 + 2xy = -1. \qquad (2)$$

Solution First we solve the linear equation (1) for y:

$$y = 2x - 3. \qquad (3)$$

Then we substitute $2x - 3$ for y in Equation (2) and solve for x:

$$x^2 + 2x(2x - 3) = -1$$
$$x^2 + 4x^2 - 6x = -1$$
$$5x^2 - 6x + 1 = 0$$
$$(5x - 1)(x - 1) = 0 \qquad \text{Factoring}$$
$$5x - 1 = 0 \quad or \quad x - 1 = 0 \qquad \text{Using the principle of zero products}$$
$$x = \tfrac{1}{5} \quad or \quad x = 1.$$

Now we substitute these numbers for x in Equation (3) and solve for y:

for $x = \frac{1}{5}$, $y = 2(\frac{1}{5}) - 3 = -\frac{13}{5}$; for $x = 1$, $y = 2(1) - 3 = -1$.

The pairs $(\frac{1}{5}, -\frac{13}{5})$ and $(1, -1)$ check, so they are solutions. ❑

EXAMPLE 3 Solve the system

$$x + y = 5, \quad (1) \quad \text{(The graph is a line.)}$$
$$y = 3 - x^2. \quad (2) \quad \text{(The graph is a parabola.)}$$

Solution We substitute $3 - x^2$ for y in the first equation:

$$x + 3 - x^2 = 5$$
$$-x^2 + x - 2 = 0 \qquad \text{Adding } -5 \text{ on both sides and rearranging}$$
$$x^2 - x + 2 = 0. \qquad \text{Multiplying by } -1 \text{ on both sides}$$

To solve this equation, we need the quadratic formula:

$$x = \frac{-b \pm \sqrt{b^2 - 4ac}}{2a}$$
$$= \frac{-(-1) \pm \sqrt{(-1)^2 - 4 \cdot 1 \cdot 2}}{2(1)}$$
$$= \frac{1 \pm \sqrt{1 - 8}}{2}$$
$$= \frac{1 \pm \sqrt{-7}}{2}$$
$$= \frac{1}{2} \pm \frac{\sqrt{7}}{2}i.$$

Solving Equation (1) for y gives us $y = 5 - x$. Substituting values for x gives

$$y = 5 - \left(\frac{1}{2} + \frac{\sqrt{7}}{2}i\right) = \frac{9}{2} - \frac{\sqrt{7}}{2}i \quad \text{and} \quad y = 5 - \left(\frac{1}{2} - \frac{\sqrt{7}}{2}i\right) = \frac{9}{2} + \frac{\sqrt{7}}{2}i.$$

The solutions are

$$\left(\frac{1}{2} + \frac{\sqrt{7}}{2}i, \frac{9}{2} - \frac{\sqrt{7}}{2}i\right) \quad \text{and} \quad \left(\frac{1}{2} - \frac{\sqrt{7}}{2}i, \frac{9}{2} + \frac{\sqrt{7}}{2}i\right).$$

There are no real-number solutions. Note in the figure below that the graphs do not intersect. Getting only nonreal solutions tells us that the graphs do not intersect.

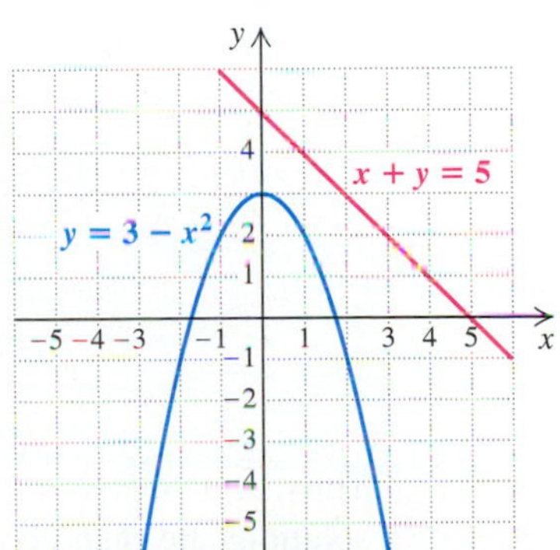

❑

Systems of Two Nonlinear Equations

We now consider systems of two second-degree equations. Graphs of such systems can involve any two conic sections. The following figure shows some ways in which a circle and a hyperbola can intersect.

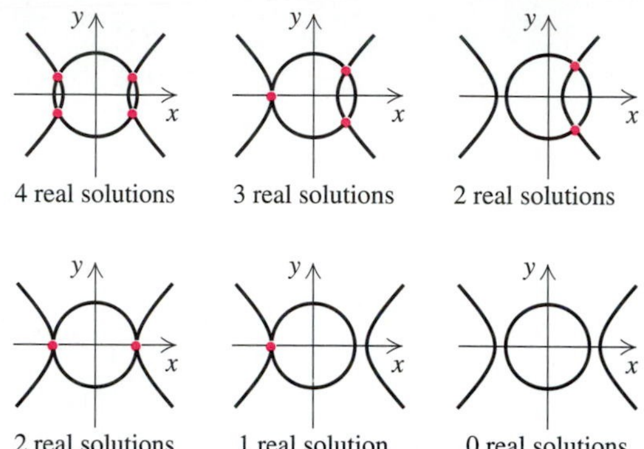

To solve systems of two second-degree equations, we can use either the substitution method or the elimination method. The elimination method is generally better when each equation is of the form $Ax^2 + By^2 = C$. Then we can eliminate an x^2- or a y^2-term in a manner similar to the procedure we used for systems of linear equations in Chapter 8.

EXAMPLE 4

Solve the system

$$2x^2 + 5y^2 = 22, \qquad (1)$$
$$3x^2 - y^2 = -1. \qquad (2)$$

Solution Here we multiply Equation (2) by 5 and then add the equations:

$$
\begin{aligned}
2x^2 + 5y^2 &= 22 \\
15x^2 - 5y^2 &= -5 && \text{Multiplying by 5 on both sides of Equation (2)} \\
\hline
17x^2 \qquad &= 17 && \text{Adding} \\
x^2 &= 1 \\
x &= \pm 1.
\end{aligned}
$$

If $x = 1$, $x^2 = 1$, and if $x = -1$, $x^2 = 1$, so substituting 1 or -1 for x in Equation (2), we have

$$
\begin{aligned}
3 \cdot (\pm 1)^2 - y^2 &= -1 \\
3 - y^2 &= -1 \\
-y^2 &= -4 \\
y^2 &= 4 \\
y &= \pm 2.
\end{aligned}
$$

Thus, if $x = 1$, $y = 2$ or $y = -2$; and if $x = -1$, $y = 2$ or $y = -2$. The possible solutions are then $(1, 2)$, $(1, -2)$, $(-1, 2)$, and $(-1, -2)$.

Check: Since $(2)^2 = 4$, $(-2)^2 = 4$, $(1)^2 = 1$, and $(-1)^2 = 1$, we can check all four pairs at once.

$2x^2 + 5y^2 = 22$	
$2(\pm 1)^2 + 5(\pm 2)^2$?	22
$2 + 20$	
22	22 TRUE

$3x^2 - y^2 = -1$	
$3(\pm 1)^2 - (\pm 2)^2$?	-1
$3 - 4$	
-1	-1 TRUE

The solutions are $(1, 2)$, $(1, -2)$, $(-1, 2)$, and $(-1, -2)$. ❑

When a product of variables is in one equation and the other is of the form $Ax^2 + By^2 = C$, we often solve for a variable in the equation with the product and then use substitution.

EXAMPLE 5

Solve the system

$$x^2 + 4y^2 = 20, \qquad (1)$$
$$xy = 4. \qquad (2)$$

Solution First we solve Equation (2) for y:

$$y = \frac{4}{x}.$$

TECHNOLOGY CONNECTION

Because the algebra is often difficult, finding the solution(s) to systems of nonlinear equations provides an excellent opportunity to use a grapher. As with systems of linear equations, we simply zoom in as often as necessary and then use the Trace feature to read the coordinates of the points of intersection. Using a grapher restricts solutions to real numbers. Few graphers have the ability to find imaginary solutions.

For example, to solve the system

$$x + 2y = 10,$$
$$x^2 - 5y = 25,$$

we first solve each equation for y:

$$y_1 = \frac{10 - x}{2},$$

$$y_2 = \frac{x^2 - 25}{5}.$$

Both equations are then graphed.

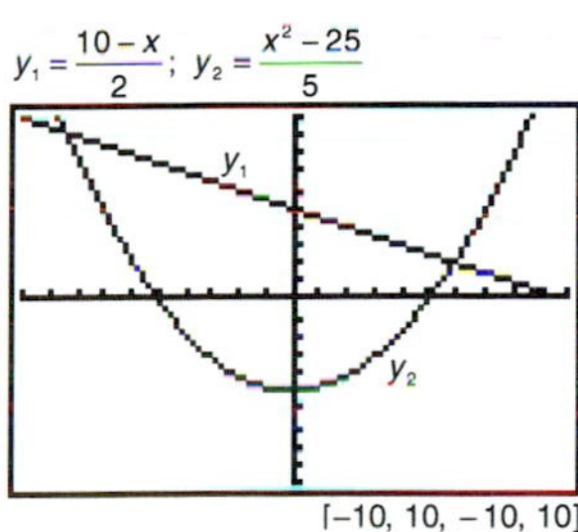

Zooming in on the points of intersection, we determine coordinates, to two decimal places, to be $(-8.43, 9.22)$ and $(5.93, 2.03)$.

Use a grapher to solve each system. Round all values to two decimal places.

TC1. $4xy - 7 = 0$,
$x - 3y - 2 = 0$

TC2. $x^2 + y^2 = 14$,
$16x + 7y^2 = 0$

Then we substitute $4/x$ for y in Equation (1) and solve for x:

$$x^2 + 4\left(\frac{4}{x}\right)^2 = 20$$

$$x^2 + \frac{64}{x^2} = 20$$

$$x^4 + 64 = 20x^2 \qquad \text{Multiplying by } x^2$$

$$x^4 - 20x^2 + 64 = 0 \qquad \text{Obtaining standard form. This equation is reducible to quadratic.}$$

$$u^2 - 20u + 64 = 0 \qquad \text{Letting } u = x^2$$

$$(u - 16)(u - 4) = 0 \qquad \text{Factoring}$$

$$u = 16 \quad or \quad u = 4. \qquad \text{Using the principle of zero products}$$

Now we substitute x^2 for u and solve these equations:

$$x^2 = 16 \quad or \quad x^2 = 4$$

$$x = \pm 4 \quad or \quad x = \pm 2.$$

Then $x = 4$ or $x = -4$ or $x = 2$ or $x = -2$. Since $y = 4/x$, if $x = 4$, $y = 1$; if $x = -4$, $y = -1$; if $x = 2$, $y = 2$; and if $x = -2$, $y = -2$. The ordered pairs $(4, 1)$, $(-4, -1)$, $(2, 2)$, and $(-2, -2)$ check. They are the solutions. ❑

Problem Solving

We now consider solving problems in which the translation is to a system of equations in which at least one is not linear.

EXAMPLE 6

For a building at a community college, an architect wants to lay out a rectangular piece of ground that has a perimeter of 204 m and an area of 2565 m^2. Find the dimensions of the piece of ground.

Solution

1. **FAMILIARIZE.** We draw a picture of the field, labeling the drawing. We let l = the length and w = the width, both in meters.

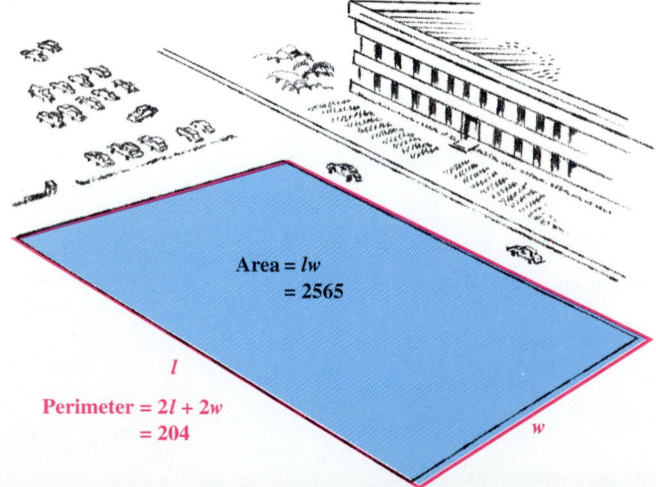

2. **TRANSLATE.** We then have the following translation:

$$\text{Perimeter:} \quad 2w + 2l = 204;$$
$$\text{Area:} \quad lw = 2565.$$

3. **CARRY OUT.** We solve the system

$$2w + 2l = 204,$$
$$lw = 2565.$$

We solve the second equation for l and get $l = 2565/w$. Then we substitute $2565/w$ for l in the first equation and solve for w:

$$2w + 2\left(\frac{2565}{w}\right) = 204$$
$$2w^2 + 2(2565) = 204w \quad \text{Multiplying by } w$$
$$2w^2 - 204w + 2(2565) = 0 \quad \text{Standard form}$$
$$w^2 - 102w + 2565 = 0 \quad \text{Multiplying by } \tfrac{1}{2}$$

Factoring could be used instead of the quadratic formula, but the numbers are quite large.

$$w = \frac{-(-102) \pm \sqrt{(-102)^2 - 4 \cdot 1 \cdot 2565}}{2 \cdot 1}$$
$$w = \frac{102 \pm \sqrt{144}}{2} = \frac{102 \pm 12}{2}$$
$$w = 57 \quad or \quad w = 45.$$

If $w = 57$, then $l = 2565/w = 2565/57 = 45$. If $w = 45$, then $l = 2565/w = 2565/45 = 57$. Since length is usually considered to be longer than width, we have the solution $l = 57$ and $w = 45$, or $(57, 45)$.

4. **CHECK.** If $l = 57$ and $w = 45$, the perimeter is $2 \cdot 57 + 2 \cdot 45$, or 204. The area is $57 \cdot 45$, or 2565. The numbers check.

5. **STATE.** The length is 57 m and the width is 45 m. ❑

EXAMPLE 7

The area of a rectangular Oriental rug is 300 ft^2, and the length of a diagonal is 25 ft. Find the dimensions of the rug.

Solution

1. **FAMILIARIZE.** We draw a picture and label it. Note that there is a right triangle in the figure. We let l = the length and w = the width, both in feet.

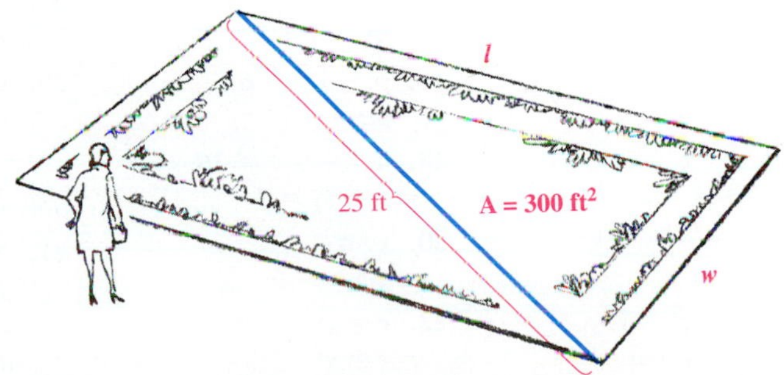

2. **TRANSLATE.** We translate to a system of equations:

$l^2 + w^2 = 25^2,$ Using the Pythagorean theorem

$lw = 300.$ Using the formula for the area of a rectangle

3. **CARRY OUT.** We solve the system

$$l^2 + w^2 = 625,$$
$$lw = 300$$

to get (20, 15), (15, 20), (−20, −15), and (−15, −20).

4. **CHECK.** Measurements cannot be negative and length is usually greater than width, so we check only (20, 15). In the right triangle, $15^2 + 20^2 = 225 + 400 = 625$, which is 25^2. The area is $15 \cdot 20$, or 300, so we have a solution.

5. **STATE.** The length is 20 ft and the width is 15 ft. ❑

EXERCISE SET 13.4

Solve. In most cases, graphs can be used to confirm the solutions.

1. $x^2 + y^2 = 25,$ $y - x = 1$
2. $x^2 + y^2 = 100,$ $y - x = 2$
3. $4x^2 + 9y^2 = 36,$ $3y + 2x = 6$
4. $9x^2 + 4y^2 = 36,$ $3x + 2y = 6$
5. $y^2 = x + 3,$ $2y = x + 4$
6. $y = x^2,$ $3x = y + 2$
7. $x^2 - xy + 3y^2 = 27,$ $x - y = 2$
8. $2y^2 + xy + x^2 = 7,$ $x - 2y = 5$
9. $x^2 + 4y^2 = 25,$ $x + 2y = 7$
10. $y^2 - x^2 = 16,$ $2x - y = 1$
11. $x^2 - xy + 3y^2 = 5,$ $x - y = 2$
12. $m^2 + 3n^2 = 10,$ $m - n = 2$
13. $3x + y = 7,$ $4x^2 + 5y = 24$
14. $2y^2 + xy = 5,$ $4y + x = 7$
15. $a + b = 7,$ $ab = 4$
16. $p + q = -6,$ $pq = -7$
17. $2a + b = 1,$ $b = 4 - a^2$
18. $4x^2 + 9y^2 = 36,$ $x + 3y = 3$
19. $a^2 + b^2 = 89,$ $a - b = 3$
20. $xy = 4,$ $x + y = 5$
21. $x^2 + y^2 = 25,$ $y^2 = x + 5$
22. $y = x^2,$ $x = y^2$
23. $x^2 + y^2 = 9,$ $x^2 - y^2 = 9$
24. $y^2 - 4x^2 = 4,$ $4x^2 + y^2 = 4$
25. $x^2 + y^2 = 25,$ $xy = 12$
26. $x^2 - y^2 = 16,$ $x + y^2 = 4$
27. $x^2 + y^2 = 4,$ $16x^2 + 9y^2 = 144$
28. $x^2 + y^2 = 25,$ $25x^2 + 16y^2 = 400$
29. $x^2 + y^2 = 16,$ $y^2 - 2x^2 = 10$
30. $x^2 + y^2 = 14,$ $x^2 - y^2 = 4$
31. $x^2 + y^2 = 5,$ $xy = 2$
32. $x^2 + y^2 = 20$ $xy = 8$
33. $x^2 + y^2 = 13,$ $xy = 6$
34. $x^2 + 4y^2 = 20,$ $xy = 4$
35. $3xy + x^2 = 34,$ $2xy - 3x^2 = 8$
36. $2xy + 3y^2 = 7,$ $3xy - 2y^2 = 4$
37. $xy - y^2 = 2,$ $2xy - 3y^2 = 0$
38. $4a^2 - 25b^2 = 0,$ $2a^2 - 10b^2 = 3b + 4$
39. $x^2 - y = 5,$ $x^2 + y^2 = 25$
40. $ab - b^2 = -4,$ $ab - 2b^2 = -6$

Solve.

41. A computer parts company wants to make a rectangular memory board that has a perimeter of 28 cm and a diagonal of length 10 cm. What are the dimensions of the board?

42. A bathroom tile company wants to make a new rectangular tile that has a perimeter of 6 in. and a diagonal of length $\sqrt{5}$ in. What are the dimensions of the tile?

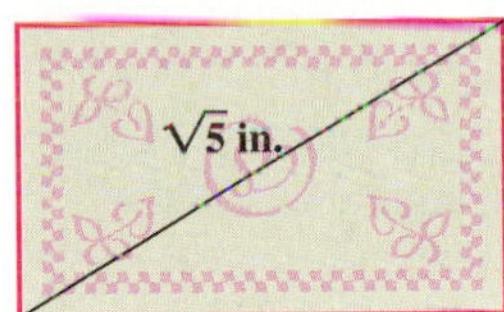

43. A rectangle has an area of 20 in^2 and a perimeter of 18 in. Find its dimensions.

44. A rectangle has an area of 2 yd^2 and a perimeter of 6 yd. Find its dimensions.

45. It will take 210 yd of fencing to enclose a rectangular field. The area of the field is 2250 yd^2. What are the dimensions of the field?

46. The diagonal of a rectangle is 1 ft longer than the length of the rectangle and 3 ft longer than twice the width. Find the dimensions of the rectangle.

47. The product of the lengths of the legs of a right triangle is 156. The hypotenuse has length $\sqrt{313}$. Find the lengths of the legs.

48. The product of two numbers is 60. The sum of their squares is 136. Find the numbers.

49. A garden contains two square peanut beds. Find the length of each bed if the sum of their areas is 832 ft^2 and the difference of their areas is 320 ft^2.

50. A certain amount of money saved for 1 yr at a certain interest rate yielded \$7.50. If the principal had been \$25 more and the interest rate 1% less, the interest would have been the same. Find the principal and the rate.

51. The area of a rectangle is $\sqrt{3}$ m^2, and the length of a diagonal is 2 m. Find the dimensions.

52. The area of a rectangle is $\sqrt{2}$ m^2, and the length of a diagonal is $\sqrt{3}$ m. Find the dimensions.

Skill Maintenance

Simplify.

53. $\sqrt{48}$

54. $\sqrt[4]{32a^{24}d^9}$

55. A boat travels 4 mi upstream and 4 mi back downstream. The total time for the trip is 3 hr. The speed of the stream is 2 mph. Find the speed of the boat in still water.

56. Rationalize the denominator: $\dfrac{\sqrt{x}-\sqrt{h}}{\sqrt{x}+\sqrt{h}}$.

Synthesis

57. ◈ Write a problem for a classmate to solve. Devise the problem so that a system of two nonlinear equations must be solved.

58. A piece of wire 100 cm long is to be cut into two pieces and those pieces are each to be bent to make a square. The area of one square is to be 144 cm^2 greater than that of the other. How should the wire be cut?

59. Find the equation of a circle that passes through $(-2, 3)$ and $(-4, 1)$ and whose center is on the line $5x + 8y = -2$.

60. Find the equation of an ellipse centered at the origin that passes through the points $(2, -3)$ and $(1, \sqrt{13})$.

61. Four squares with sides 5 in. long are cut from the corners of a rectangular metal sheet that has an area of 340 in^2. The edges are bent up to form an open box with a volume of 350 in^3. Find the dimensions of the box.

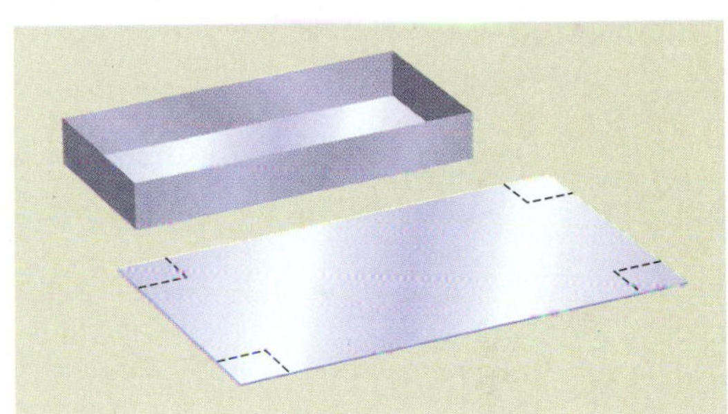

62. A company keeps records of the total revenue (money taken in) from the sale of x units of a product. It determines that total revenue R is given by

$$R = 100x - x^2.$$

The company also keeps records of the total cost of producing x units of the same product. It determines that the total cost C is given by

$$C = 20x + 1500.$$

A break-even point is a value of x for which total revenue is the same as total cost; that is, $R = C$. Find the break-even points.

Solve.

63. $p^2 + q^2 = 13,$

$\dfrac{1}{pq} = -\dfrac{1}{6}$

64. $a + b = \dfrac{5}{6},$

$\dfrac{a}{b} + \dfrac{b}{a} = \dfrac{13}{6}$

SUMMARY AND REVIEW 13

KEY TERMS

Conic section, p. 672
Midpoint, p. 674
Parabola, p. 676
Circle, p. 680
Radius, p. 680
Center, p. 680
Ellipse, p. 684
Foci (singular, focus) p. 684
Hyperbola, p. 687
Vertices (singular, vertex), p. 687
Asymptote, p. 688

IMPORTANT PROPERTIES AND FORMULAS

The Distance Formula

The distance d between any two points (x_1, y_1) and (x_2, y_2) is given by

$$d = \sqrt{(x_2 - x_1)^2 + (y_2 - y_1)^2}.$$

The Midpoint Formula

If the endpoints of a segment are (x_1, y_1) and (x_2, y_2), then the coordinates of the midpoint are

$$\left(\frac{x_1 + x_2}{2}, \frac{y_1 + y_2}{2}\right).$$

(See the summary of graphs at the end of Section 13.3.)

Parabola

Vertical with vertex at (h, k):

$$y = ax^2 + bx + c$$
$$= a(x - h)^2 + k$$

Horizontal with vertex at (h, k):

$$x = ay^2 + by + c$$
$$= a(y - k)^2 + h$$

Circle

Center at the origin:

$$x^2 + y^2 = r^2$$

Center at (h, k):

$$(x - h)^2 + (y - k)^2 = r^2$$

Ellipse

Center at the origin

Foci on x-axis: $\frac{x^2}{a^2} + \frac{y^2}{b^2} = 1, \quad a > b$ Foci on y-axis: $\frac{x^2}{a^2} + \frac{y^2}{b^2} = 1, \quad b > a$

Hyperbola

Center at the origin

Axis horizontal: $\frac{x^2}{a^2} - \frac{y^2}{b^2} = 1$ Axis vertical: $\frac{y^2}{b^2} - \frac{x^2}{a^2} = 1$

x- and y-axes asymptotes: $xy = c$

REVIEW EXERCISES

This chapter's review and test include Skill Maintenance exercises from Sections 6.8, 10.3, 10.6, and 11.3.

Find the distance between the pair of points. Where appropriate, find an approximation to three decimal places.

1. (2, 6) and (6, 6) **2.** (−1, 1) and (−5, 4)

3. (4, 7) and (−3, −2) **4.** (2, 3a) and (−1, a)

Find the midpoint of the segment with the given endpoints.

5. (1, 6) and (7, 6) **6.** (−1, 1) and (−5, 4)

7. (4, 7) and (−3, −2) **8.** (2, 3a) and (−1, a)

Find the center and the radius.

9. $(x + 2)^2 + (y - 3)^2 = 2$

10. $(x - 5)^2 + y^2 = 49$

11. $x^2 + y^2 - 6x - 2y + 1 = 0$

12. $x^2 + y^2 + 8x - 6y - 10 = 0$

13. Find an equation of the circle with center (−4, 3) and radius $4\sqrt{3}$.

14. Find an equation of the circle with center (7, −2) and radius $2\sqrt{5}$.

Classify the equation as a circle, an ellipse, a parabola, or a hyperbola. Then graph.

15. $4x^2 + 4y^2 = 100$ **16.** $9x^2 + 2y^2 = 18$

17. $y = -x^2 + 2x - 3$ **18.** $\frac{y^2}{9} - \frac{x^2}{4} = 1$

19. $xy = 9$ **20.** $x = y^2 + 2y - 2$

21. $xy = -3$

22. $x^2 + y^2 + 6x - 8y - 39 = 0$

Solve.

23. $x^2 - y^2 = 33,$ $x + y = 11$ **24.** $x^2 - 2x + 2y^2 = 8,$ $2x + y = 6$

25. $x^2 - y = 3,$ $2x - y = 3$ **26.** $x^2 + y^2 = 25,$ $x^2 - y^2 = 7$

27. $x^2 - y^2 = 3,$ $y = x^2 - 3$ **28.** $x^2 + y^2 = 18,$ $2x + y = 3$

29. $x^2 + y^2 = 100,$ $2x^2 - 3y^2 = -120$ **30.** $x^2 + 2y^2 = 12,$ $xy = 4$

31. A rectangle has a perimeter of 38 m and an area of 84 m^2. What are its dimensions?

32. Find two positive integers whose sum is 12 and the sum of whose reciprocals is $\frac{3}{8}$.

33. The perimeter of a square is 12 cm more than the perimeter of another square. Its area exceeds the area of the other by 39 cm^2. Find the perimeter of each square.

34. The sum of the areas of two circles is 130π ft^2. The difference of the circumferences is 16π ft. Find the radius of each circle.

Skill Maintenance

35. Simplify: $\sqrt[3]{81a^8b^{10}}$.

36. Solve: $x^2 + 2x + 5 = 0$.

37. Rationalize the numerator: $\dfrac{4-\sqrt{a}}{2+\sqrt{a}}$.

38. The speed of a moving sidewalk at an airport is 5 ft/sec. A person can walk 55 ft forward on the moving sidewalk in the same time it takes to walk 5 ft in the opposite direction. At what rate would the person walk on a nonmoving sidewalk?

Synthesis

39. ◈ Is a circle a special type of ellipse? Why or why not?

40. ◈ Explain why function notation is not used in this chapter, and list the graphs discussed for which function notation could be used.

41. Solve:

$$3x^2 + 4y^2 = 8,$$
$$x^2 - y^2 = 5.$$

42. Find the points whose distance from (8, 0) and from (−8, 0) is 10.

43. Find an equation of the circle that passes through (−2, −4), (5, −5), and (6, 2).

44. Find an equation of the ellipse with the following vertices: (−7, 0), (7, 0), (0, −3), and (0, 3).

45. Find the point on the x-axis that is equidistant from (−3, 4) and (5, 6).

CHAPTER TEST 13

Find the distance between the pair of points. If appropriate, find an approximation to three decimal places.

1. (4, −1) and (−5, 8)
2. (3, $-a$) and (−3, a)

Find the midpoint of the segment with the given endpoints.

3. (4, −1) and (−5, 8)
4. (3, $-a$) and (−3, a)

Find the center and the radius of the circle.

5. $(x+2)^2 + (y-3)^2 = 64$
6. $x^2 + y^2 + 4x - 6y + 4 = 0$

Classify the equation as a circle, an ellipse, a parabola, or a hyperbola. Then graph.

7. $y = x^2 - 4x - 1$
8. $x^2 + y^2 + 2x + 6y + 6 = 0$
9. $\dfrac{x^2}{9} - \dfrac{y^2}{4} = 1$
10. $16x^2 + 4y^2 = 64$
11. $xy = -5$
12. $x = -y^2 + 4y$

Solve.

13. $\dfrac{x^2}{16} + \dfrac{y^2}{9} = 1,$
$3x + 4y = 12$

14. $x^2 + y^2 = 16,$
$\dfrac{x^2}{16} - \dfrac{y^2}{9} = 1$

15. In a rational expression, the sum of the values of the numerator and the denominator is 23. The product of their values is 120. Find the values of the numerator and the denominator.

16. A rectangle with diagonal of length $5\sqrt{5}$ has an area of 22. Find the dimensions of the rectangle.

17. Two squares are such that the sum of their areas is 8 m^2 and the difference of their areas is 2 m^2. Find the length of a side of each square.

18. A rectangle has a diagonal of length 20 ft and a perimeter of 56 ft. Find the dimensions of the rectangle.

Skill Maintenance

19. Solve: $x^2 + 2x = 5$.

20. Simplify: $\sqrt[3]{48a^5b^{18}}$.

21. Rationalize the denominator: $\dfrac{4-\sqrt{a}}{2+\sqrt{a}}$.

22. A boat travels 6 mi upstream in the same time it takes to travel 30 mi downstream. The speed of the stream is 4 mph. Find the speed of the boat in still water.

Synthesis

23. Find an equation of the ellipse with the following vertices: (1, 3), (6, 6), (11, 3), and (6, 0).

24. Find the point on the y-axis that is equidistant from (−3, −5) and (4, −7).

25. The sum of two numbers is 36, and the product is 4. Find the sum of the reciprocals of the numbers.

CHAPTER 14

Sequences, Series, and Combinatorics

AN APPLICATION

A student loan is in the amount of $6000. Interest is 9%, compounded annually, and the entire amount is to be repaid after 10 years. How much is to be paid back?

In Section 14.3, we will find that the amount due can be determined using the 11th term of a sequence:

$$6000,\quad 1.09 \cdot 6000,\quad (1.09)^2 6000,\quad (1.09)^3 6000,\ \ldots .$$

Dr. Jesse G. Jackson
CHIEF FINANCIAL AID OFFICER

"I am constantly amazed at how important a role mathematics plays not just in the day-to-day operations of our office, but in the lives of our graduating students. I really never expected to use math nearly as much as I now do."

The first three sections of this chapter are devoted to *sequences* and *series*. A sequence is simply an ordered list. For example, when a baseball coach writes a batting order, a sequence is being formed. When the members of a sequence are numbers, we can find their sum. Such a sum is called a *series*.

In the last part of this chapter, we begin a brief study of *probability*. The theory of probability has important applications to business, medicine, sociology, psychology, science, and to games of chance.

In addition to the material in this chapter, the review and test for Chapter 14 include material from Sections 8.2, 12.4, 12.6, and 13.2.

14.1 Sequences and Series

Sequences • Finding the General Term • Sums and Series • Sigma Notation

Sequences

Suppose that \$1000 is invested at 8%, compounded annually. The amounts to which the account will grow after 1 year, 2 years, 3 years, 4 years, and so on, are as follows:

(1) ↓ \$1080.00, (2) ↓ \$1166.40, (3) ↓ \$1259.71, (4) ↓ \$1360.49, ...

Note that we can think of this as a function that pairs 1 with the number \$1080.00, 2 with the number \$1166.40, 3 with the number \$1259.71, 4 with the number \$1360.49, and so on. A sequence is thus a *function,* where the domain is a set of consecutive positive integers beginning with 1.

If we continue computing the amounts in the account forever, we obtain an **infinite sequence,** with function values

\$1080.00, \$1166.40, \$1259.71, \$1360.49, \$1469.33, \$1586.87, ...

The three dots at the end indicate that the sequence goes on without stopping. If we stop after a certain number of years, we obtain a **finite sequence:**

\$1080.00, \$1166.40, \$1259.71, \$1360.49.

Sequences

An *infinite sequence* is a function having for its domain the set of positive integers: $\{1, 2, 3, 4, 5, \ldots\}$.

A *finite sequence* is a function having for its domain a set of positive integers $\{1, 2, 3, 4, 5, \ldots, n\}$, for some positive integer n.

As another example, consider the sequence given by

$$a(n) = 2^n, \quad \text{or} \quad a_n = 2^n.$$

The notation a_n means the same as $a(n)$ but is used more commonly with sequences. Some of the function values (also known as *terms* of the sequence) are as follows:

$$a_1 = 2^1 = 2,$$
$$a_2 = 2^2 = 4,$$
$$a_3 = 2^3 = 8,$$
$$a_6 = 2^6 = 64.$$

The first term of the sequence is a_1, the fifth term is a_5, and the nth term, or **general term,** is a_n. This sequence can also be denoted in the following ways:

2, 4, 8, . . . ; or

2, 4, 8, . . . , 2^n, ...

The 2^n emphasizes that the nth term of this sequence is found by raising 2 to the nth power.

EXAMPLE 1

Find the first 4 terms and the 57th term of the sequence whose general term is given by $a_n = (-1)^n/(n + 1)$.

Solution

$$a_1 = \frac{(-1)^1}{1+1} = -\frac{1}{2}$$
$$a_2 = \frac{(-1)^2}{2+1} = \frac{1}{3}$$
$$a_3 = \frac{(-1)^3}{3+1} = -\frac{1}{4}$$
$$a_4 = \frac{(-1)^4}{4+1} = \frac{1}{5}$$
$$a_{57} = \frac{(-1)^{57}}{57+1} = -\frac{1}{58}$$

Note that the expression $(-1)^n$ causes the signs of the terms to alternate between positive and negative, depending on whether n is even or odd. ❑

Finding the General Term

When only the first few terms of a sequence are known, we do not know for sure what the general term is, but we can make a prediction by looking for a pattern.

EXAMPLE 2

For each sequence, predict the general term.

a) 1, 4, 9, 16, 25, . . .

b) $\sqrt{1}, \sqrt{2}, \sqrt{3}, \sqrt{4}, \ldots$

c) $-1, 2, -4, 8, -16, \ldots$

d) 2, 4, 8, . . .

Solution

a) 1, 4, 9, 16, 25, . . .
These are squares of consecutive positive integers, so the general term may be n^2.

b) $\sqrt{1}, \sqrt{2}, \sqrt{3}, \sqrt{4}, \ldots$
These are square roots of consecutive positive integers, so the general term may be $\sqrt{n}$.

c) $-1, 2, -4, 8, -16, \ldots$
These are powers of 2 with alternating signs, so the general term may be $(-1)^n[2^{n-1}]$.

d) $2, 4, 8, \ldots$
If we see the pattern of powers of 2, we will see 16 as the next term and guess 2^n for the general term. Then the sequence could be written with more terms as

$$2, 4, 8, 16, 32, 64, 128, \ldots .$$

If we see that we can get the second term by adding 2, the third term by adding 4, the next term by adding 6, and so on, we will see 14 as the next term. A general term for the sequence is then $n^2 - n + 2$, and the sequence can be written with more terms as

$$2, 4, 8, 14, 22, 32, 44, 58, \ldots .$$

❑

Example 2(d) illustrates that with few given terms, the uncertainty about the nth term is greater.

Sums and Series

Series

Given the infinite sequence

$$a_1, a_2, a_3, a_4, \ldots, a_n, \ldots,$$

the sum of the terms

$$a_1 + a_2 + a_3 + \cdots + a_n + \cdots$$

is called an *infinite series*. A *partial sum* is the sum of the first n terms

$$a_1 + a_2 + a_3 + \cdots + a_n.$$

A partial sum is also called a *finite series* and is denoted S_n.

For instance, the sequence

$$3, 5, 7, 9, \ldots, 2n + 1, \ldots$$

has the following partial sums:

$S_1 = 3,$ — This is the first term of the given sequence.
$S_2 = 3 + 5 = 8,$ — This is the sum of the first two terms.
$S_3 = 3 + 5 + 7 = 15,$ — The sum of the first three terms
$S_4 = 3 + 5 + 7 + 9 = 24.$ — The sum of the first four terms

EXAMPLE 3 For the sequence $-2, 4, -6, 8, -10, 12, -14$, find: **(a)** S_3; **(b)** S_5.

Solution

a) $S_3 = -2 + 4 + (-6) = -4$

b) $S_5 = -2 + 4 + (-6) + 8 + (-10) = -6$ □

Sigma Notation

When the general term of a sequence is known, the Greek letter $\sum$ (sigma) can be used to write a series. For example, the sum of the first four terms of the sequence 3, 5, 7, 9, . . . , $2k + 1$, . . . can be named as follows, using *sigma notation*, or *summation notation:*

$$\sum_{k=1}^{4} (2k + 1).$$

This is read "the sum as k goes from 1 to 4 of $(2k + 1)$." The letter k is called the *index of summation*. Sometimes the index of summation starts at a number other than 1.

EXAMPLE 4

Find and evaluate the sum.

a) $\sum_{k=1}^{5} k^2$ **b)** $\sum_{k=1}^{4} (-1)^k(2k)$ **c)** $\sum_{k=0}^{3} (2^k + 5)$

Solution

a) $\sum_{k=1}^{5} k^2 = 1^2 + 2^2 + 3^2 + 4^2 + 5^2 = 1 + 4 + 9 + 16 + 25 = 55$

Evaluate k^2 for all integers from 1 through 5. Then add.

b) $\sum_{k=1}^{4} (-1)^k(2k) = (-1)^1(2 \cdot 1) + (-1)^2(2 \cdot 2) + (-1)^3(2 \cdot 3) + (-1)^4(2 \cdot 4)$

$= -2 + 4 - 6 + 8 = 4$

c) $\sum_{k=0}^{3} (2^k + 5) = (2^0 + 5) + (2^1 + 5) + (2^2 + 5) + (2^3 + 5)$

$= 6 + 7 + 9 + 13 = 35$ □

EXAMPLE 5

Write sigma notation for the sum.

a) $1 + 4 + 9 + 16 + 25$ **b)** $-1 + 3 - 5 + 7$ **c)** $3 + 9 + 27 + 81 + \cdots$

Solution

a) $1 + 4 + 9 + 16 + 25$

This is a sum of squares, $1^2 + 2^2 + 3^2 + 4^2 + 5^2$, so the general term is k^2. Sigma notation is

$$\sum_{k=1}^{5} k^2.$$

b) $-1 + 3 - 5 + 7$
Except for the alternating signs, this is the sum of the first four positive odd numbers. Note that $2k - 1$ is a formula for the kth positive odd number, and $(-1)^k = 1$ when k is even and $(-1)^k = -1$ when k is odd. The general term is thus $(-1)^k(2k - 1)$, beginning with $k = 1$. Sigma notation is

$$\sum_{k=1}^{4} (-1)^k(2k - 1).$$

To check, we can evaluate $(-1)^k(2k - 1)$ using 1, 2, 3, and 4. Then we can write the sum of the four terms.

c) $3 + 9 + 27 + 81 + \cdots$
This is a sum of powers of 3, and it is also an infinite series. We use the symbol ∞ to represent infinity and name the infinite series using sigma notation as follows:

$$\sum_{k=1}^{\infty} 3^k.$$

❑

EXERCISE SET 14.1

In each of the following, the nth term of a sequence is given. In each case find the first 4 terms; the 10th term, a_{10}; and the 15th term, a_{15}.

1. $a_n = 3n + 1$

2. $a_n = 3n - 1$

3. $a_n = \dfrac{n}{n + 1}$

4. $a_n = n^2 + 1$

5. $a_n = n^2 - 2n$

6. $a_n = \dfrac{n^2 - 1}{n^2 + 1}$

7. $a_n = n + \dfrac{1}{n}$

8. $a_n = \left(-\dfrac{1}{2}\right)^{n-1}$

9. $a_n = (-1)^n n^2$

10. $a_n = (-1)^n(n + 3)$

11. $a_n = (-1)^{n+1}(3n - 5)$

12. $a_n = (-1)^n(n^3 - 1)$

Find the indicated term of the sequence.

13. $a_n = 4n - 7$; a_8

14. $a_n = 5n + 11$; a_9

15. $a_n = (3n + 4)(2n - 5)$; a_7

16. $a_n = (3n + 2)^2$; a_6

17. $a_n = (-1)^{n-1}(3.4n - 17.3)$; a_{12}

18. $a_n = (-2)^{n-2}(45.68 - 1.2n)$; a_{23}

19. $a_n = 5n^2(4n - 100)$; a_{11}

20. $a_n = 4n^2(11n + 31)$; a_{22}

21. $a_n = \left(1 + \dfrac{1}{n}\right)^2$; a_{20}

22. $a_n = \left(1 - \dfrac{1}{n}\right)^3$; a_{15}

23. $a_n = \log 10^n$; a_{43}

24. $a_n = \ln e^n$; a_{67}

Predict the general term, or nth term, a_n, of the sequence. Answers may vary.

25. 1, 3, 5, 7, 9, . . .

26. 3, 9, 27, 81, 243, . . .

27. −2, 6, −18, 54, . . .

28. −2, 3, 8, 13, 18, . . .

29. $\frac{2}{3}, \frac{3}{4}, \frac{4}{5}, \frac{5}{6}, \frac{6}{7}, \ldots$

30. $\sqrt{2}, \sqrt{4}, \sqrt{6}, \sqrt{8}, \sqrt{10}, \ldots$

31. $\sqrt{3}, 3, 3\sqrt{3}, 9, 9\sqrt{3}, \ldots$

32. $1 \cdot 2, 2 \cdot 3, 3 \cdot 4, 4 \cdot 5, \ldots$

33. −1, −4, −7, −10, −13, . . .

34. log 1, log 10, log 100, log 1000, . . .

Find the indicated partial sum for the sequence.

35. 1, 2, 3, 4, 5, 6, 7, . . . ; S_7

36. 1, −3, 5, −7, 9, −11, . . . ; S_8

37. 2, 4, 6, 8, . . . ; S_5

38. 1, $\frac{1}{4}$, $\frac{1}{9}$, $\frac{1}{16}$, $\frac{1}{25}$, . . . ; S_5

Rename and evaluate the sum.

39. $\sum_{k=1}^{5} \frac{1}{2k}$

40. $\sum_{k=1}^{6} \frac{1}{2k+1}$

41. $\sum_{k=0}^{5} 2^k$

42. $\sum_{k=4}^{7} \sqrt{2k-1}$

43. $\sum_{k=7}^{10} \log k$

44. $\sum_{k=0}^{4} \pi k$

45. $\sum_{k=1}^{8} \frac{k}{k+1}$

46. $\sum_{k=1}^{4} \frac{k-2}{k+3}$

47. $\sum_{k=1}^{5} (-1)^k$

48. $\sum_{k=1}^{5} (-1)^{k+1}$

49. $\sum_{k=1}^{8} (-1)^{k+1} 3^k$

50. $\sum_{k=1}^{7} (-1)^k 4^{k+1}$

51. $\sum_{k=0}^{5} (k^2 - 2k + 3)$

52. $\sum_{k=0}^{5} (k^2 - 3k + 4)$

53. $\sum_{k=1}^{10} \frac{1}{k(k+1)}$

54. $\sum_{k=1}^{10} \frac{2^k}{2^k+1}$

Rewrite the sum using sigma notation.

55. $\frac{1}{2} + \frac{2}{3} + \frac{3}{4} + \frac{4}{5} + \frac{5}{6} + \frac{6}{7}$

56. $3 + 6 + 9 + 12 + 15$

57. $-2 + 4 - 8 + 16 - 32 + 64$

58. $\frac{1}{1^2} + \frac{1}{2^2} + \frac{1}{3^2} + \frac{1}{4^2} + \frac{1}{5^2}$

59. $4 - 9 + 16 - 25 + \cdots + (-1)^n n^2$

60. $9 - 16 + 25 + \cdots + (-1)^{n+1} n^2$

61. $5 + 10 + 15 + 20 + 25 + \cdots$

62. $7 + 14 + 21 + 28 + 35 + \cdots$

63. $\frac{1}{1 \cdot 2} + \frac{1}{2 \cdot 3} + \frac{1}{3 \cdot 4} + \frac{1}{4 \cdot 5} + \cdots$

64. $\frac{1}{1 \cdot 2^2} + \frac{1}{2 \cdot 3^2} + \frac{1}{3 \cdot 4^2} + \frac{1}{4 \cdot 5^2} + \cdots$

Skill Maintenance

Simplify.

65. $\log_3 3$

66. $\log_3 1$

67. $\log_3 3^7$

68. $\text{loc}_c\, c$

Synthesis

69. Explain why the equation

$$\sum_{k=1}^{n} (a_k + b_k) = \sum_{k=1}^{n} a_k + \sum_{k=1}^{n} b_k$$

is true for any positive integer n. What laws are used to justify this result?

70. **a)** Find the first few terms of the sequence $a_n = n^2 - n + 41$.

b) What pattern do you observe?

c) Find the 41st term. Does the pattern you found in part (b) still hold?

Find the first five terms of the sequence; then find S_5.

71. $a_n = \frac{1}{2^n} \log 1000^n$

72. $a_n = i^n,\ i = \sqrt{-1}$

73. $a_n = \ln (1 \cdot 2 \cdot 3 \cdots n)$

Find decimal notation, rounded to six decimal places, for the first six terms of the sequence.

74. $a_n = \sqrt{n+1} - \sqrt{n}$

75. $a_n = \left(1 + \frac{1}{n}\right)^n$

Some sequences are given by a *recursive definition*. The value of the first term, a_1, is given, and then we are told how to find each subsequent term from the term preceding it in the sequence. Find the first six terms of each of the following recursively defined sequences.

76. $a_1 = 1, \quad a_{n+1} = 3a_n - 2$

77. $a_1 = 0, \quad a_{n+1} = a_n^2 + 4$

78. A single cell of bacterium divides into two every 15 min. Suppose the same rate of division is maintained for 4 hr. Give a sequence that lists the number of cells after successive 15-min periods.

79. The value of an office machine is \$5200. Its scrap value each year is 75% of its value the year before. Give a sequence that lists the scrap value of the machine at the start of each year for a 10-year period.

80. Katrina gets \$6.20 for working in a warehouse for a publishing company. Each year she gets a \$0.40 hourly raise. Give a sequence that lists Katrina's hourly salary over a 10-year period.

14.2 Arithmetic Sequences and Series

Arithmetic Sequences • Sum of the First *n* Terms of an Arithmetic Sequence • Problem Solving

In this section, we concentrate on sequences and series that are said to be arithmetic (pronounced ăr′ĭth-mĕt′-ĭk).

Arithmetic Sequences

In an **arithmetic sequence,** all terms (other than the first) can be found by adding the same number to the preceding term. For example, the sequence 2, 5, 8, 11, 14, 17, . . . is arithmetic because adding 3 to any term produces the next term. In other words, the difference between any term and the preceding one is 3. Arithmetic sequences are also called *arithmetic progressions*.

Arithmetic Sequence

A sequence is *arithmetic* if there exists a number d, called the *common difference,* such that $a_{n+1} = a_n + d$ for any integer $n \geq 1$.

EXAMPLE 1

For each arithmetic sequence, identify the first term, a_1, and the common difference, d.

a) 4, 9, 14, 19, 24, . . .
b) 27, 20, 13, 6, −1, −8, . . .
c) 2, $2\frac{1}{2}$, 3, $3\frac{1}{2}$, . . .

Solution To find a_1, we simply use the first term listed. To find d, we choose any term beyond the first and subtract the preceding term from it.

	Sequence	*First Term,* a_1	*Common Difference,* d
a)	4, 9, 14, 19, 24, . . .	4	5 ← 9 − 4 = 5
b)	27, 20, 13, 6, −1, −8, . . .	27	−7 ← 20 − 27 = −7
c)	2, $2\frac{1}{2}$, 3, $3\frac{1}{2}$, . . .	2	$\frac{1}{2} \leftarrow 2\frac{1}{2} - 2 = \frac{1}{2}$

Here we found the common difference by subtracting a_1 from a_2. Had we subtracted a_2 from a_3 or a_3 from a_4 we would have obtained the same values for d. Thus we can check by adding d to each term in a sequence to see if we progress to the next term.

Check: **a)** $4 + 5 = 9$, $9 + 5 = 14$, $14 + 5 = 19$, $19 + 5 = 24$
b) $27 + (-7) = 20$, $20 + (-7) = 13$, $13 + (-7) = 6$, $6 + (-7) = -1$, $-1 + (-7) = -8$
c) $2 + \frac{1}{2} = 2\frac{1}{2}$, $2\frac{1}{2} + \frac{1}{2} = 3$, $3 + \frac{1}{2} = 3\frac{1}{2}$ ❑

To find a formula for the general, or nth, term of any arithmetic sequence, we denote the common difference by d and write out the first few terms:

$$a_1,$$
$$a_2 = a_1 + d,$$
$$a_3 = a_2 + d = (a_1 + d) + d = a_1 + 2d, \qquad \text{Substituting for } a_2$$
$$a_4 = a_3 + d = (a_1 + 2d) + d = a_1 + 3d. \qquad \text{Substituting for } a_3$$

Note that the coefficient of d in each case is 1 less than the subscript.

Generalizing, we obtain the following formula.

Formula 1

The nth term of an arithmetic sequence is given by

$$a_n = a_1 + (n - 1)d, \quad \text{for any integer } n \geq 1.$$

EXAMPLE 2

Find the 14th term of the arithmetic sequence 4, 7, 10, 13, ...

Solution First we note that $a_1 = 4$, $d = 3$, and $n = 14$. Then using Formula 1, we obtain

$$a_n = a_1 + (n - 1)d$$
$$a_{14} = 4 + (14 - 1) \cdot 3 = 4 + 13 \cdot 3 = 4 + 39 = 43.$$

The 14th term is 43. ❑

EXAMPLE 3

In the sequence in Example 2, which term is 301? That is, find n if $a_n = 301$.

Solution We substitute into Formula 1 and solve for n:

$$a_n = a_1 + (n - 1)d$$
$$301 = 4 + (n - 1) \cdot 3$$
$$301 = 4 + 3n - 3$$
$$300 = 3n$$
$$100 = n.$$

The term 301 is the 100th term of the sequence. ❑

Given two terms and their places in an arithmetic sequence, we can construct the sequence.

EXAMPLE 4

The third term of an arithmetic sequence is 8, and the sixteenth term is 47. Find a_1 and d and construct the sequence.

Solution We know that $a_3 = 8$ and $a_{16} = 47$. Thus we would have to add d thirteen times to get from 8 to 47. That is,

$$8 + 13d = 47. \qquad a_3 \text{ and } a_{16} \text{ are 13 terms apart.}$$

Solving $8 + 13d = 47$, we obtain

$$13d = 39$$
$$d = 3.$$

We subtract d twice from a_3 to get to a_1. Thus,

$$a_1 = 8 - 2 \cdot 3 = 2. \qquad a_1 \text{ and } a_3 \text{ are 2 terms apart.}$$

The sequence is 2, 5, 8, 11, ... Note that we could have subtracted d 15 times from a_{16} in order to find a_1. ❑

In general, d should be subtracted $(n - 1)$ times from a_n in order to find a_1.

Sum of the First n Terms of an Arithmetic Sequence

When we add the terms of an arithmetic sequence, we form an **arithmetic series.** To find a formula for computing S_n when the series is arithmetic, we denote the first n terms as follows:

This is the next-to-last-term. If you add d to this term, the result is a_n.

$$a_1,\ (a_1 + d),\ (a_1 + 2d),\ .\ .\ .\ ,(a_n - 2d),\ (a_n - d),\ a_n$$

This term is two terms back from the last. If you add d to this term, you get the next-to-last term, $a_n - d$.

Then S_n is given by

$$S_n = a_1 + (a_1 + d) + (a_1 + 2d) + \cdots + (a_n - 2d) + (a_n - d) + a_n. \quad \textbf{(1)}$$

Reversing the order of addition, we have

$$S_n = a_n + (a_n - d) + (a_n - 2d) + \cdots + (a_1 + 2d) + (a_1 + d) + a_1. \quad \textbf{(2)}$$

If we add corresponding terms of each side of Equations (1) and (2), we get

$$2S_n = [a_1 + a_n] + [(a_1 + d) + (a_n - d)] + [(a_1 + 2d) + (a_n - 2d)]$$
$$+ \cdots + [(a_n - 2d) + (a_1 + 2d)] + [(a_n - d) + (a_1 + d)] + [a_n + a_1].$$

This simplifies to

$$2S_n = [a_1 + a_n] + [a_1 + a_n] + [a_1 + a_n]$$
$$+ \cdots + [a_n + a_1] + [a_n + a_1] + [a_n + a_1].$$

Since $(a_1 + a_n)$ is being added n times, it follows that

$$2S_n = n(a_1 + a_n),$$

from which we obtain the following formula.

Formula 2

The sum of the first n terms of an arithmetic sequence is given by

$$S_n = \frac{n}{2}(a_1 + a_n).$$

EXAMPLE 5

Find the sum of the first 100 positive even numbers.

Solution The sum is

$$2 + 4 + 6 + \cdot \cdot \cdot + 198 + 200.$$

This is the sum of the first 100 terms of the arithmetic sequence for which

$$a_1 = 2, \qquad a_n = 200, \quad \text{and} \quad n = 100.$$

Substituting in the formula

$$S_n = \frac{n}{2}(a_1 + a_n),$$

we get

$$S_{100} = \frac{100}{2}(2 + 200) = 50(202) = 10{,}100.$$ ❑

Formula 2 is useful when we know a_1 and a_n, the first and last terms. When a_n is unknown, but a_1, n, and d are known, we can find S_n by using Formulas 1 and 2 together.

EXAMPLE 6

Find the sum of the first 15 terms of the arithmetic sequence 4, 7, 10, 13, ...

Solution Note that

$$a_1 = 4, \qquad d = 3, \quad \text{and} \quad n = 15.$$

Before using Formula 2, we find a_{15}:

$$a_{15} = 4 + (15 - 1)3 \qquad \text{Substituting into Formula 1}$$
$$= 4 + 14 \cdot 3 = 46.$$

Thus,

$$S_{15} = \tfrac{15}{2}(4 + 46) \qquad \text{Using Formula 2}$$
$$= \tfrac{15}{2}(50) = 375.$$ ❑

Problem Solving

For some problem-solving situations, the translation may involve sequences or series. We look at some examples.

EXAMPLE 7

Chris takes a job, starting with an hourly wage of \$14.25, and is promised a raise of 15¢ per hour every 2 months for 5 years. At the end of 5 years, what will be Chris's hourly wage?

Solution

1. **FAMILIARIZE.** It helps to write down the hourly wage for several two-month time periods.

Beginning:	14.25,
After two months:	14.40,
After four months:	14.55,
and so on.	

What appears is a sequence of numbers: 14.25, 14.40, 14.55, ... Is it an arithmetic sequence? Yes, because we add 0.15 each time to get the next term.

We ask ourselves what we know about arithmetic sequences. The pertinent formulas are

$$a_n = a_1 + (n - 1)d$$

and

$$S_n = \frac{n}{2}(a_1 + a_n).$$

In this case, we are not looking for a sum, so it is probably the first formula that will give us our answer. We want to know the last term in a sequence. We will need to know a_1, n, and d. From our list above, we see that

$$a_1 = 14.25 \quad \text{and} \quad d = 0.15.$$

What is n? That is, how many terms are in the sequence? Each year there are 6 raises, since Chris gets a raise every 2 months. There are 5 years, so the total number of raises will be $5 \cdot 6$, or 30. There will be 31 terms: the original wage and 30 increased rates.

2. **TRANSLATE.** We want to find a_n for the arithmetic sequence in which $a_1 = 14.25$, $d = 0.15$, and $n = 31$.

3. **CARRY OUT.** Substituting in Formula 1 gives us

$$\begin{aligned} a_{31} &= 14.25 + (31 - 1) \cdot 0.15 \\ &= 18.75. \end{aligned}$$

4. **CHECK.** We can check the calculations. We can also calculate in a slightly different way for another check. For example, at the end of a year, there will be 6 raises, for a total raise of \$0.90. At the end of 5 years, the total raise will be $5 \times \$0.90$, or \$4.50. If we add that to the original wage of \$14.25, we obtain \$18.75. The answer checks.

5. **STATE.** At the end of 5 years, Chris's hourly wage will be \$18.75. ❑

Example 7 is one in which the calculations or the translation could be done in a number of ways. There is often a variety of ways in which a problem can be solved. You should use the one that is best or easiest for you. In this chapter, however, we will concentrate on the use of sequences and series and their related formulas in problem solving.

EXAMPLE 8

A stack of telephone poles has 30 poles in the bottom row. There are 29 poles in the second row, 28 in the next row, and so on. How many poles are in the stack if there are 5 poles in the top row?

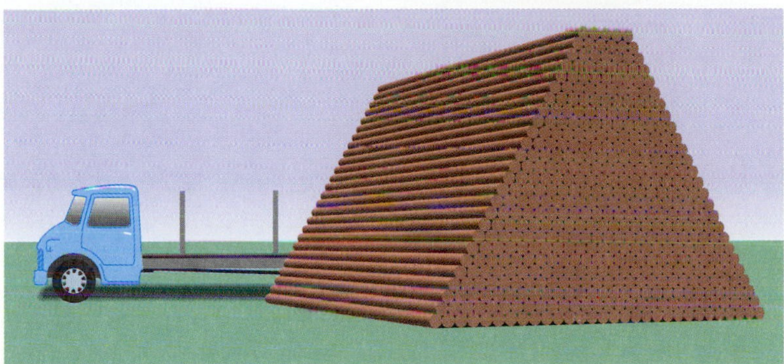

Solution

1. FAMILIARIZE. A picture will help in this case. The following figure shows the ends of the poles and the way in which they stack. There are 30 poles on the bottom, and we see that there will be one fewer in each succeeding row. How many rows will there be?

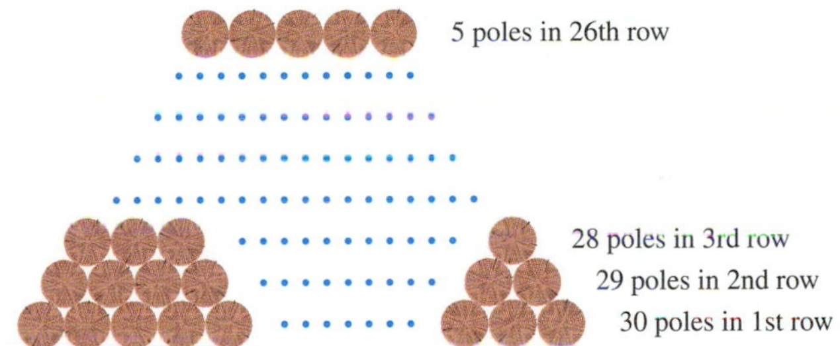

We go from 30 poles in a row, down to 5 poles in the top row, so there must be 26 rows.

We want the sum

$$30 + 29 + 28 + \cdots + 5.$$

Thus we have an arithmetic series. We recall, or look up if necessary, the formula

$$S_n = \frac{n}{2}(a_1 + a_n).$$

2. TRANSLATE. We want to find the sum of the first 26 terms of an arithmetic sequence in which $a_1 = 30$ and $a_{26} = 5$.

3. CARRY OUT. Substituting into Formula 2, we have

$$S_{26} = \tfrac{26}{2}(30 + 5)$$
$$= 13 \cdot 35 = 455.$$

4. **CHECK.** In this case, we can check the calculations by doing them again. A longer, harder way would be to do the entire addition:

$$30 + 29 + 28 + \cdots + 5.$$

5. **STATE.** There are 455 poles in the stack. ❑

EXERCISE SET 14.2

Find the first term and the common difference.

1. 2, 7, 12, 17, . . .
2. 1.06, 1.12, 1.18, 1.24, . . .
3. 7, 3, −1, −5, . . .
4. −9, −6, −3, 0, . . .
5. $\frac{3}{2}, \frac{9}{4}, 3, \frac{15}{4}, \ldots$
6. $\frac{3}{5}, \frac{1}{10}, -\frac{2}{5}, \ldots$
7. \$2.12, \$2.24, \$2.36, \$2.48, . . .
8. \$214, \$211, \$208, \$205, . . .
9. Find the 12th term of the arithmetic sequence 2, 6, 10, ...
10. Find the 11th term of the arithmetic sequence 0.07, 0.12, 0.17, ...
11. Find the 17th term of the arithmetic sequence 7, 4, 1, ...
12. Find the 14th term of the arithmetic sequence $3, \frac{7}{3}, \frac{5}{3}, \ldots$.
13. Find the 13th term of the arithmetic sequence \$1200, \$964.32, \$728.64, ...
14. Find the 10th term of the arithmetic sequence \$2345.78, \$2967.54, \$3589.30, ...
15. In the sequence of Exercise 9, what term is 106?
16. In the sequence of Exercise 10, what term is 1.67?
17. In the sequence of Exercise 11, what term is −296?
18. In the sequence of Exercise 12, what term is −27?
19. Find a_{17} when $a_1 = 5$ and $d = 6$.
20. Find a_{20} when $a_1 = 14$ and $d = -3$.
21. Find a_1 when $d = 4$ and $a_8 = 33$.
22. Find a_1 when $d = 8$ and $a_{11} = 26$.
23. Find n when $a_1 = 5$, $d = -3$, and $a_n = -76$.
24. Find n when $a_1 = 25$, $d = -14$, and $a_n = -507$.
25. For an arithmetic sequence in which $a_{17} = -40$ and $a_{28} = -73$, find a_1 and d. Write the first 5 terms of the sequence.
26. In an arithmetic sequence, $a_{17} = \frac{25}{3}$ and $a_{32} = \frac{95}{6}$. Find a_1 and d. Write the first 5 terms of the sequence.
27. Find the sum of the first 20 terms of the arithmetic series $5 + 8 + 11 + 14 + \cdots$.
28. Find the sum of the first 14 terms of the arithmetic series $11 + 7 + 3 + \cdots$.
29. Find the sum of the first 300 natural numbers.
30. Find the sum of the first 400 natural numbers.
31. Find the sum of the even numbers from 2 to 100, inclusive.
32. Find the sum of the odd numbers from 1 to 99, inclusive.
33. Find the sum of all multiples of 7 from 7 to 98, inclusive.
34. Find the sum of all multiples of 4 that are between 14 and 523.
35. If an arithmetic series has $a_1 = 2$ and $d = 5$, find S_{20}.
36. If an arithmetic series has $a_1 = 7$ and $d = -3$, find S_{32}.

Problem Solving

37. A gardener is making a triangular planting, with 35 plants in the front row, 31 in the second row, 27 in the third row, and so on. If the pattern is consistent, how many plants will there be in the last row? How many plants are there altogether?

38. A formation of a marching band has 14 marchers in the front row, 16 in the second row, 18 in the third row, and so on, for 25 rows. How many marchers are in the last row? How many marchers are there altogether?

39. How many poles will be in a pile of telephone poles if there are 50 in the first layer, 49 in the second, and so on, until there are 6 in the last layer?

40. If 10¢ is saved on October 1, 20¢ on October 2, 30¢ on October 3, and so on, how much is saved during October? (October has 31 days.)

41. A family saves money in an arithmetic sequence: \$600 the first year, \$700 the second, and so on, for 20 years. How much do they save in all (disregarding interest)?

42. Jacob saves \$30 on August 1, \$50 on August 2, \$70 on August 3, and so on. How much will he have saved in August? (August has 31 days.)

43. Theaters are often built with more seats per row as the rows move toward the back. Suppose the main floor of a theater has 28 seats in the first row, 32 in the second, 36 in the third, and so on, for 50 rows. How many seats are on the main floor?

44. Shirley sets up an investment such that it will return \$5000 the first year, \$6125 the second year, \$7250 the third year, and so on, for 25 years. How much in all is received from the investment?

Skill Maintenance

Convert to an exponential equation.

45. $\log_a P = k$

46. $\ln t = a$

Find an equation of the circle satisfying the given conditions.

47. Center $(0, 0)$, radius 9

48. Center $(-2, 5)$, radius $3\sqrt{2}$

Synthesis

49. ◈ The sum of the first n terms of an arithmetic sequence is given by

$$S_n = \frac{n}{2}[2a_1 + (n-1)d].$$

Use Formulas 1 and 2 to explain how this equation is derived.

50. ◈ It is said that as a young child, the mathematician Karl F. Gauss (1777–1855) was able to compute the sum $1 + 2 + 3 + \cdots + 100$ very quickly in his head. Explain how Gauss might have done this and present a formula for the sum of the first n natural numbers.

51. Find three numbers in an arithmetic sequence such that the sum of the first and third is 10 and the product of the first and second is 15.

52. Find a formula for the sum of the first n consecutive odd numbers starting with 1:

$$1 + 3 + 5 + \cdots + (2n - 1).$$

53. ▦ In an arithmetic sequence, $a_1 = \$8760$ and $d = -\$798.23$. Find the first 10 terms of the sequence.

54. ▦ Find the sum of the first 10 terms of the sequence given in Exercise 53.

55. Prove that if p, m, and q are consecutive terms in an arithmetic sequence, then

$$m = \frac{p+q}{2}.$$

56. *Business: Straight-line depreciation.* A company buys an office machine for \$5200 on January 1 of a given year. The machine is expected to last for 8 years, at the end of which time its *trade-in,* or *salvage, value* will be \$1100. If the company figures the decline in value to be the same each year, then the trade-in values, after t years, $0 \leq t \leq 8$, form an arithmetic sequence given by

$$a_t = C - t\left(\frac{C - S}{N}\right),$$

where C = the original cost of the item (\$5200), N = the years of expected life (8), and S = the salvage value (\$1100).

a) Find the formula for a_t for the straight-line depreciation of the office machine.

b) Find the salvage value after 0 years, 1 year, 2 years, 3 years, 4 years, 7 years, and 8 years.

14.3 Geometric Sequences and Series

Geometric Sequences • Sum of the First n Terms of a Geometric Sequence • Infinite Geometric Series • Problem Solving

In an arithmetic sequence, we added a certain number to each term to get the next term. With the kind of sequence we consider now, each term is *multiplied* by a certain number to get the next term. These are called *geometric sequences* or *geometric progressions*. We also consider *geometric series*.

Geometric Sequences

Consider the sequence

$$2, 6, 18, 54, 162, \ldots .$$

If we multiply each term by 3, we obtain the next term. Sequences in which each term can be multiplied by a certain number in order to get the next term are called **geometric.** We call this multiplier the *common ratio* because it is found by dividing any term by the preceding term.

Geometric Sequence

A sequence is *geometric* if there exists a number r, called the *common ratio,* such that

$$\frac{a_{n+1}}{a_n} = r, \quad \text{or} \quad a_{n+1} = a_n \cdot r \quad \text{for any integer } n \geq 1.$$

EXAMPLE 1

For each geometric sequence, identify the common ratio.

a) 3, 6, 12, 24, 48, . . .
b) 3, −6, 12, −24, 48, −96, . . .
c) \$5200, \$3900, \$2925, \$2193.75, . . .
d) \$1000, \$1080, \$1166.40, . . .

Solution

	Sequence	*Common Ratio*	
a)	3, 6, 12, 24, 48, . . .	2	$\frac{6}{3} = 2$, $\frac{12}{6} = 2$, and so on
b)	3, −6, 12, −24, 48, −96, . . .	−2	$\frac{-6}{3} = -2$, $\frac{12}{-6} = -2$, and so on
c)	\$5200, \$3900, \$2925, \$2193.75, . . .	0.75	$\frac{\$3900}{\$5200} = 0.75$, $\frac{\$2925}{\$3900} = 0.75$
d)	\$1000, \$1080, \$1166.40, . . .	1.08	$\frac{\$1080}{\$1000} = 1.08$

❑

We now find a formula for the general, or nth, term of a geometric sequence. Let a_1 be the first term and let r be the common ratio. We write out the first few terms as follows:

$$
\begin{aligned}
&a_1, \\
&a_2 = a_1 r, \\
&a_3 = a_2 r = (a_1 r) r = a_1 r^2, \qquad \text{Substituting } a_1 r \text{ for } a_2 \\
&a_4 = a_3 r = (a_1 r^2) r = a_1 r^3. \qquad \text{Substituting } a_1 r^2 \text{ for } a_3
\end{aligned}
$$

Note that the exponent is 1 less than the subscript.

Generalizing, we obtain the following.

Formula 3

The nth term of a geometric sequence is given by

$$a_n = a_1 r^{n-1}, \quad \text{for any integer } n \geq 1.$$

EXAMPLE 2

Find the 7th term of the geometric sequence 4, 20, 100, ...

Solution First we note that

$$a_1 = 4 \quad \text{and} \quad n = 7.$$

To find the common ratio, we can divide any term by its predecessor, provided it has one. Since the second term is 20 and the first is 4, we get

$$r = \frac{20}{4}, \quad \text{or } 5.$$

The formula

$$a_n = a_1 r^{n-1}$$

gives us

$$a_7 = 4 \cdot 5^{7-1} = 4 \cdot 5^6 = 4 \cdot 15{,}625 = 62{,}500.$$ ❑

EXAMPLE 3

Find the 10th term of the geometric sequence

$$64, -32, 16, -8, \ldots .$$

Solution First we note that

$$a_1 = 64, \qquad n = 10, \quad \text{and} \quad r = \frac{-32}{64}, \quad \text{or } -\frac{1}{2}.$$

Then, using Formula 3, we have

$$a_{10} = 64 \cdot \left(-\frac{1}{2}\right)^{10-1} = 64 \cdot \left(-\frac{1}{2}\right)^9 = 2^6 \cdot \left(-\frac{1}{2^9}\right) = -\frac{1}{2^3} = -\frac{1}{8}.$$ ❑

Sum of the First n Terms of a Geometric Sequence

We want to find a formula for S_n when a sequence is geometric:

$$a_1, a_1 r, a_1 r^2, a_1 r^3, \ldots, a_1 r^{n-1}, \ldots .$$

The **geometric series** S_n is given by

$$S_n = a_1 + a_1r + a_1r^2 + \cdots + a_1r^{n-2} + a_1r^{n-1}. \quad \textbf{(1)}$$

We want to develop a formula that allows us to find this sum without a great deal of adding. If we multiply on both sides of Equation (1) by r, we have

$$rS_n = a_1r + a_1r^2 + a_1r^3 + \cdots + a_1r^{n-1} + a_1r^n. \quad \textbf{(2)}$$

Subtracting corresponding sides of Equation (2) from Equation (1), we see that the red terms drop out, leaving

$$S_n - rS_n = a_1 - a_1r^n,$$

or

$$S_n(1 - r) = a_1(1 - r^n). \qquad \text{Factoring}$$

Dividing on both sides by $1 - r$ gives us the following formula.

Formula 4

The sum of the first n terms of a geometric sequence is given by

$$S_n = \frac{a_1(1 - r^n)}{1 - r}, \quad \text{for any } r \neq 1.$$

EXAMPLE 4

Find the sum of the first 7 terms of the geometric sequence 3, 15, 75, 375, ...

Solution First we note that

$$a_1 = 3, \; n = 7, \quad \text{and} \quad r = \frac{15}{3}, \quad \text{or } 5.$$

Then, using Formula 4, we have

$$\begin{aligned} S_7 &= \frac{3(1 - 5^7)}{1 - 5} = \frac{3(1 - 78{,}125)}{-4} \\ &= \frac{3(-78{,}124)}{-4} \\ &= 58{,}593. \end{aligned}$$

❑

Infinite Geometric Series

Suppose we consider the sum of the terms of an infinite geometric sequence, such as 2, 4, 8, 16, 32, ... We get what is called an **infinite geometric series:**

$$2 + 4 + 8 + 16 + 32 + \cdots.$$

Here, as n grows larger and larger, the sum of the first n terms, S_n, becomes larger and larger without bound. There are also infinite series that get closer and closer to some specific number. Here is an example:

$$\frac{1}{2} + \frac{1}{4} + \frac{1}{8} + \frac{1}{16} + \cdots + \frac{1}{2^n} + \cdots.$$

Let's consider S_n for the first five values of n:

$$S_1 = \tfrac{1}{2} = \tfrac{1}{2} = 0.5,$$
$$S_2 = \tfrac{1}{2} + \tfrac{1}{4} = \tfrac{3}{4} = 0.75,$$
$$S_3 = \tfrac{1}{2} + \tfrac{1}{4} + \tfrac{1}{8} = \tfrac{7}{8} = 0.875,$$
$$S_4 = \tfrac{1}{2} + \tfrac{1}{4} + \tfrac{1}{8} + \tfrac{1}{16} = \tfrac{15}{16} = 0.9375,$$
$$S_5 = \tfrac{1}{2} + \tfrac{1}{4} + \tfrac{1}{8} + \tfrac{1}{16} + \tfrac{1}{32} = \tfrac{31}{32} = 0.96875.$$

The denominator of the sum is 2^n, where n is the subscript of S. The numerator is $2^n - 1$.

Thus, for this particular series, we have

$$S_n = \frac{2^n - 1}{2^n} = \frac{2^n}{2^n} - \frac{1}{2^n} = 1 - \frac{1}{2^n}.$$

Note that the value of S_n is less than 1 for any value of n, but as n gets larger and larger, the values of S_n get closer and closer to 1. We say that 1 is the *limit* of S_n and that 1 is the sum of this infinite geometric sequence. An infinite geometric series is denoted S_∞. It can be shown (but we will not do it here) that the sum of the terms of an infinite geometric sequence exists if and only if $|r| < 1$ (that is, the absolute value of the common ratio is less than 1).

To find a formula for the sum of an infinite geometric sequence, we first consider the sum of the first n terms:

$$S_n = \frac{a_1(1 - r^n)}{1 - r} = \frac{a_1 - a_1 r^n}{1 - r}. \qquad \text{Using the distributive law}$$

For $|r| < 1$, it follows that values of r^n get closer and closer to 0 as n gets larger. (Choose a number between -1 and 1 and check this by finding larger and larger powers on your calculator.) As r^n gets closer and closer to 0, so does $a_1 r^n$. Thus S_n gets closer and closer to $a_1/(1 - r)$.

Formula 5

When $|r| < 1$, the limit of an infinite geometric series is given by

$$S_\infty = \frac{a_1}{1 - r}.$$

EXAMPLE 5

Determine whether each infinite geometric series has a limit. If a limit exists, find it.

a) $1 + 3 + 9 + 27 + \cdots$ **b)** $-2 + 1 - \tfrac{1}{2} + \tfrac{1}{4} - \tfrac{1}{8} + \cdots$

Solution

a) Here $r = 3$, so $|r| = |3| = 3$. Since $|r| \not< 1$, the series does *not* have a limit.

b) Here $r = -\tfrac{1}{2}$, so $|r| = |-\tfrac{1}{2}| = \tfrac{1}{2}$. Since $|r| < 1$, the series *does* have a limit. We find the limit by substituting into Formula 5:

$$S_\infty = \frac{-2}{1-\left(-\frac{1}{2}\right)} = \frac{-2}{\frac{3}{2}} = -\frac{4}{3}.$$

❑

EXAMPLE 6

Find fractional notation for 0.63636363 . . . .

Solution We can express this as

$$0.63 + 0.0063 + 0.000063 + \cdots.$$

This is an infinite geometric series, where $a_1 = 0.63$ and $r = 0.01$. Since $|r| < 1$, this series has a limit:

$$S_\infty = \frac{a_1}{1-r} = \frac{0.63}{1-0.01} = \frac{0.63}{0.99} = \frac{63}{99}.$$

Thus fractional notation for 0.63636363 . . . is $\frac{63}{99}$, or $\frac{7}{11}$. ❑

Problem Solving

For some problem-solving situations, the translation may involve geometric sequences or series.

EXAMPLE 7

Suppose someone offered you a job for the month of September (30 days) under the following conditions. You will be paid \$0.01 for the first day, \$0.02 for the second, \$0.04 for the third, and so on, doubling your previous day's salary each day. How much would you earn? (Would you take the job? Make a guess before reading further.)

Solution

1. FAMILIARIZE. You earn \$0.01 the first day, \$0.01(2) the second day, \$0.01(2)(2) the third day, and so on. The amounts form a geometric sequence with $a_1 = \$0.01$, $r = 2$, and $n = 30$.

2. TRANSLATE. The amount earned is the geometric series

$$\$0.01 + \$0.01(2) + \$0.01(2^2) + \$0.01(2^3) + \cdots + \$0.01(2^{29}),$$

where

$$a_1 = \$0.01, \quad n = 30, \quad \text{and} \quad r = 2.$$

3. CARRY OUT. Using the formula

$$S_n = \frac{a_1(1-r^n)}{1-r},$$

we have

$$S_{30} = \frac{\$0.01(1-2^{30})}{1-2}$$
$$= \frac{\$0.01(-1{,}073{,}741{,}823)}{-1} \qquad \text{Using a calculator}$$
$$= \$10{,}737{,}418.23.$$

4. **CHECK.** The calculations can be repeated as a check.

5. **STATE.** The pay exceeds \$10.7 million for the month. Most people would probably take the job! ❑

EXAMPLE 8

A student loan is in the amount of \$6000. Interest is to be 9% compounded annually, and the entire amount is to be paid after 10 years. How much is to be paid back?

Solution

1. **FAMILIARIZE.** Suppose we let P represent any principal amount. At the end of one year, the amount owed will be $P + 0.09P$, or $1.09P$. That amount will be the principal for the second year. The amount owed at the end of the second year will be $1.09 \times$ New principal $= 1.09(1.09P)$, or 1.09^2P. Thus the amount owed at the beginning of successive years is as follows:

$$\underset{P,}{\overset{(1)}{\downarrow}} \quad \underset{1.09P,}{\overset{(2)}{\downarrow}} \quad \underset{1.09^2P,}{\overset{(3)}{\downarrow}} \quad \underset{1.09^3P,}{\overset{(4)}{\downarrow}} \quad \text{and so on.}$$

We have a geometric sequence. The amount owed at the beginning of the 11th year will be the amount owed at the end of the 10th year.

2. **TRANSLATE.** We have a geometric sequence with $a_1 = 6000$, $r = 1.09$, and $n = 11$. The appropriate formula is

$$a_n = a_1r^{n-1}.$$

3. **CARRY OUT.** We substitute and calculate:

$$\begin{aligned} a_{11} &= \$6000(1.09)^{11-1} = \$6000(1.09)^{10} \\ &\approx \$6000(2.3673637) \qquad \text{Using a calculator to approximate } 1.09^{10} \\ &\approx \$14{,}204.18. \qquad \text{Rounded to the nearest hundredth} \end{aligned}$$

4. **CHECK.** A check, by repeating the calculations, is left to the student.

5. **STATE.** A total of \$14,204.18 is to be paid back at the end of 10 years. ❑

EXAMPLE 9

A bungee jumper rebounds 60% of the height jumped. A bungee jump is made using a cord that stretches to 200 ft.

a) After jumping and then rebounding 9 times, how far has a bungee jumper traveled upward (the total rebound distance)?

b) Approximately how far will a jumper have traveled upward (bounced) before coming to rest?

Solution

1. **FAMILIARIZE.** Let's do some calculations and look for a pattern.

First fall:	200 ft
First rebound:	0.6×200, or 120 ft
Second fall:	120 ft, or 0.6×200
Second rebound:	0.6×120, or $0.6(0.6 \times 200)$, which is 72 ft
Third fall:	72 ft, or $0.6(0.6 \times 200)$
Third rebound:	0.6×72, or $0.6(0.6(0.6 \times 200))$, which is 43.2 ft

The rebound distances form a geometric sequence:

$$\overset{①}{0.6 \times 200}, \quad \overset{②}{0.6^2 \times 200}, \quad \overset{③}{0.6^3 \times 200}, \quad \overset{④}{0.6^4 \times 200}, \; . \; . \; . \; ,$$

or

$$120, \quad 0.6 \times 120, \quad 0.6^2 \times 120, \quad 0.6^3 \times 120, \; . \; . \; . \; .$$

2. TRANSLATE.

a) The total rebound distance after 9 bounces is the sum of a geometric sequence. The first term is 120 and the common ratio is 0.6. There will be 9 terms, so we can use Formula 4:

$$S_n = \frac{a_1(1 - r^n)}{1 - r}.$$

b) Theoretically, the jumper will never stop bouncing. Realistically, the bouncing will eventually stop. We can approximate the actual distance bounced by considering an infinite number of bounces. We use Formula 5:

$$S_\infty = \frac{a_1}{1 - r}.$$

3. CARRY OUT.

a) We substitute into the formula and calculate:

$$S_9 = \frac{120[1 - (0.6)^9]}{1 - 0.6}$$

$$\approx 297. \qquad \text{Using a calculator}$$

b) We substitute and calculate:

$$S_\infty = \frac{120}{1 - 0.6} = 300.$$

4. CHECK. We can do the calculations again.

5. STATE.

a) In 9 bounces, the bungee jumper will have traveled upward a total distance of about 297 ft.

b) The jumper will travel upward about 300 ft before coming to rest. □

EXERCISE SET 14.3

Find the common ratio for the geometric sequence.

1. 2, 4, 8, 16, . . .

2. 12, −4, $\frac{4}{3}$, $-\frac{4}{9}$, . . .

3. 1, −1, 1, −1, . . .

4. −5, −0.5, −0.05, −0.005, . . .

5. $\frac{1}{2}$, $-\frac{1}{4}$, $\frac{1}{8}$, $-\frac{1}{16}$, . . .

6. $\frac{2}{3}$, $-\frac{4}{3}$, $\frac{8}{3}$, $-\frac{16}{3}$, . . .

7. 75, 15, 3, $\frac{3}{5}$, . . .

8. 6.275, 0.6275, 0.06275, . . .

9. $\frac{1}{x}, \frac{1}{x^2}, \frac{1}{x^3}, \ldots$

10. $5, \frac{5m}{2}, \frac{5m^2}{4}, \frac{5m^3}{8}, \ldots$

11. \$780, \$858, \$943.80, \$1038.18, . . .

12. \$5600, \$5320, \$5054, \$4801.30, . . .

Find the indicated term for the geometric sequence.

13. 2, 4, 8, 16, . . . ; the 6th term

14. 2, −10, 50, −250, . . . ; the 9th term

15. $2, 2\sqrt{3}, 6, \ldots$; the 9th term

16. 1, −1, 1, −1, . . . ; the 57th term

17. $\frac{8}{243}, \frac{8}{81}, \frac{8}{27}, \ldots$; the 10th term

18. $\frac{7}{625}, \frac{-7}{125}, \frac{7}{25}, \ldots$; the 13th term

19. \$1000, \$1080, \$1166.40, . . . ; the 12th term

20. \$1000, \$1070, \$1144.90, . . . ; the 11th term

Find the *n*th, or general, term for the geometric sequence.

21. 1, 3, 9, . . .

22. 25, 5, 1, . . .

23. 1, −1, 1, −1, . . .

24. 2, 4, 8, . . .

25. $\frac{1}{x}, \frac{1}{x^2}, \frac{1}{x^3}, \ldots$

26. $5, \frac{5m}{2}, \frac{5m^2}{4}, \ldots$

For Exercises 27–34, use Formula 4 to find the indicated sum.

27. S_7 for the geometric series $6 + 12 + 24 + \cdots$

28. S_6 for the geometric series $16 - 8 + 4 - \cdots$

29. S_7 for the geometric series $\frac{1}{18} - \frac{1}{6} + \frac{1}{2} - \cdots$

30. S_5 for the geometric series $6 + 0.6 + 0.06 + \cdots$

31. S_8 for the series $1 + x + x^2 + x^3 + \cdots$

32. S_{10} for the series $1 + x^2 + x^4 + x^6 + \cdots$

33. S_{16} for the geometric sequence \$200, \$200(1.06), \$200(1.06)^2, . . .

34. S_{23} for the geometric sequence \$1000, \$1000(1.08), \$1000(1.08)^2, . . .

Determine whether the infinite geometric series has a limit. If a limit exists, find it.

35. $4 + 2 + 1 + \cdots$

36. $7 + 3 + \frac{9}{7} + \cdots$

37. $25 + 20 + 16 + \cdots$

38. $12 + 9 + \frac{27}{4} + \cdots$

39. $100 - 10 + 1 - \frac{1}{10} + \cdots$

40. $-6 + 18 - 54 + 162 - \cdots$

41. $8 + 40 + 200 + \cdots$

42. $-6 + 3 - \frac{3}{2} + \frac{3}{4} - \cdots$

43. $0.3 + 0.03 + 0.003 + \cdots$

44. $0.37 + 0.0037 + 0.000037 + \cdots$

45. $\$500(1.02)^{-1} + \$500(1.02)^{-2} + \$500(1.02)^{-3} + \cdots$

46. $\$1000(1.08)^{-1} + \$1000(1.08)^{-2} + \$1000(1.08)^{-3} + \cdots$

Find fractional notation for the infinite sum. (These are geometric series.)

47. 0.4444 . . .

48. 9.999999 . . .

49. 0.55555 . . .

50. 0.66666 . . .

51. 0.15151515 . . .

52. 0.12121212 . . .

Solve. Use a calculator as needed for evaluating formulas.

53. A ping-pong ball is dropped from a height of 16 ft and always rebounds one fourth of the distance fallen. How high does it rebound the 6th time?

54. Approximate the total of the rebound heights of the ball in Exercise 53.

55. Yorktown has a current population of 100,000, and the population is increasing by 3% each year. What will the population be in 15 years?

56. How long will it take for the population of Yorktown to double? (See Exercise 55.)

57. A student borrows \$1200. The loan is to be repaid in 13 years at 12% interest, compounded annually. How much will be repaid at the end of 13 years?

58. A piece of paper is 0.01 in. thick. It is folded repeatedly in such a way that its thickness is doubled each time for 20 times. How thick is the result?

59. A superball dropped from the top of the Washington Monument (556 ft high) always rebounds three fourths of the distance fallen. How far (up and down) will the ball have traveled when it hits the ground for the 6th time?

60. Approximate the total distance that the ball of Exercise 59 will have traveled when it comes to rest.

61. Suppose you accepted a job for the month of February (28 days) under the following conditions. You will be paid \$0.01 the 1st day, \$0.02 the 2nd, \$0.04 the 3rd, and so on, doubling your previous day's salary each day. How much would you earn?

62. Leslie is saving money in a savings account for retirement. At the beginning of each year, \$1000 is invested at 11%, compounded annually. How much will be in the retirement fund at the end of 40 years?

Skill Maintenance

Solve the system.

63. $5x - 2y = -3,$
$2x + 5y = -24$

64. $x - 2y + 3z = 4,$
$2x - y + z = -1,$
$4x + y + z = 1$

Synthesis

65. ◈ Write a problem for a classmate to solve. Devise the problem so that a geometric series is involved and the solution is "The total amount in the bank is

$$\$900(1.08)^{40},$$

or about \$19,550."

66. ◈ The infinite series

$$S_\infty = 2 + \frac{1}{2} + \frac{1}{2\cdot3} + \frac{1}{2\cdot3\cdot4} + \frac{1}{2\cdot3\cdot4\cdot5} + \frac{1}{2\cdot3\cdot4\cdot5\cdot6} + \cdots$$

is not geometric, but it does have a sum. Using S_1, S_2, S_3, S_4, S_5, and S_6, make a conjecture about the value of S_∞ and explain your reasoning.

67. Find the sum of the first n terms of

$$1 + x + x^2 + \cdots.$$

68. Find the sum of the first n terms of

$$x^2 - x^3 + x^4 - x^5 + \cdots.$$

69. The sides of a square are each 16 cm long. A second square is inscribed by joining the midpoints of the sides, successively. In the second square we repeat the process, inscribing a third square. If this process is continued indefinitely, what is the sum of all of the areas of all the squares? (*Hint:* Use an infinite geometric series.)

14.4 Combinatorics: Permutations

Fundamental Counting Principle • Permutations • Factorial Notation • Permutations of n Objects Taken r at a Time • Circular Permutations

To study probability, it is first necessary to be able to determine the number of ways in which a set can be arranged or combined, certain objects can be chosen, or a succession of events can occur. This study of the theory of counting is called **combinatorics.**

Fundamental Counting Principle

EXAMPLE 1 How many 3-letter code symbols can be formed with the letters A, B, C without repetition?

Solution Examples of such symbols are ABC, CBA, ACB, and so on. Consider placing the letters in these frames.

We can select any of the 3 letters for the first letter in the symbol. Once this letter has been selected, the second must be selected from the 2 remaining letters. The third letter is already determined, since only 1 possibility is left. The possibilities can be arrived at with a **tree diagram.**

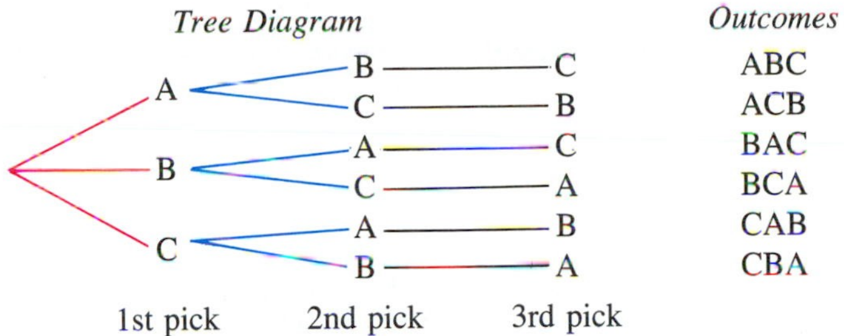

There are $3 \cdot 2 \cdot 1$, or 6, possibilities. The set of all of them is as follows:

{ABC, ACB, BAC, BCA, CAB, CBA}. ❑

Suppose we perform an experiment such as selecting letters (as in the preceding example), flipping a coin, or drawing a card. The results are called **outcomes.** An **event** is a set of outcomes. The following principle concerns events that occur together, or are combined.

Fundamental Counting Principle

Given a combined action, or event, in which the first action can be performed in n_1 ways, the second action can be performed in n_2 ways, and so on, the total number of ways in which the combined action can be performed is the product

$$n_1 \cdot n_2 \cdot n_3 \cdot \cdots \cdot n_k.$$

In Example 1, the first letter can be chosen in 3 ways, the second in 2 ways, and the third in 1 way. So, by the fundamental counting principle, the total number of ways in which all 3 letters can be selected is $3 \cdot 2 \cdot 1 = 6$.

EXAMPLE 2

How many 3-letter code symbols can be formed with the letters A, B, and C with repetition allowed?

Solution There are 3 choices for the first letter and, since we allow repetition, 3 choices for the second and 3 for the third. Thus by the fundamental counting principle, there are $3 \cdot 3 \cdot 3$, or 27, 3-letter codes. ❑

EXAMPLE 3

Uncle Ed's Cafeteria offers a special dinner platter. A customer may choose one entree, one side dish, and one dessert. The cafeteria prepares 4 entrees, 8 side dishes, and 5 desserts. How many different dinner platters are possible?

Solution There are 4 choices for the entree, 8 for the side dish, and 5 for the dessert. Thus by the fundamental counting principle, there are $4 \cdot 8 \cdot 5$, or 160 different dinner platters. ❑

Permutations

We now turn our attention to the part of combinatorics that deals with the study of *permutations*. The study of permutations involves *order* and *arrangement*.

A *permutation* of a set of n objects is an ordered arrangement of all n objects.

Consider, for example, a set of 4 objects:

{A, B, C, D}.

If we place the letters in a particular order, we say that we have made an ordered arrangement of the set. To find the number of ordered arrangements of the set, we select a first letter: There are 4 choices. Then we select a second letter: There are 3 choices. Then we select a third letter: There are 2 choices. Finally, there is 1 choice for the last selection. Thus by the fundamental counting principle, there are $4 \cdot 3 \cdot 2 \cdot 1$, or 24, permutations of a set of 4 objects.

We can find a formula for the total number of permutations of all objects in a set of n objects. We have n choices for the first selection, $n - 1$ for the second, $n - 2$ for the third, and so on. For the nth selection, there is only 1 choice.

Permutations of n objects

The total number of permutations of a set of n objects, denoted ${}_nP_n$, is given by

$${}_nP_n = n(n-1)(n-2) \cdots (3)(2)(1).$$

EXAMPLE 4

Find **(a)** ${}_4P_4$ and **(b)** ${}_7P_7$.

Solution

a) ${}_4P_4 = 4 \cdot 3 \cdot 2 \cdot 1 = 24$

b) ${}_7P_7 = 7 \cdot 6 \cdot 5 \cdot 4 \cdot 3 \cdot 2 \cdot 1 = 5040$ ❑

EXAMPLE 5

In how many different ways can 9 different envelopes be placed in 9 mailboxes, one envelope to a box?

Solution

$${}_9P_9 = 9 \cdot 8 \cdot 7 \cdot 6 \cdot 5 \cdot 4 \cdot 3 \cdot 2 \cdot 1 = 362{,}880$$ ❑

Factorial Notation

Products of successive natural numbers, such as $7 \cdot 6 \cdot 5 \cdot 4 \cdot 3 \cdot 2 \cdot 1$ and $9 \cdot 8 \cdot 7 \cdot 6 \cdot 5 \cdot 4 \cdot 3 \cdot 2 \cdot 1$, are used so often that it is convenient to adopt a notation for them.

For the product $7 \cdot 6 \cdot 5 \cdot 4 \cdot 3 \cdot 2 \cdot 1$, we write 7!, read "7-factorial."

Factorial Notation

For any natural number n,

$$n! = n(n-1)(n-2) \cdots (3)(2)(1).$$

Here are some examples.

$$\begin{aligned}
7! &= 7 \cdot 6 \cdot 5 \cdot 4 \cdot 3 \cdot 2 \cdot 1 = 5040\\
6! &= 6 \cdot 5 \cdot 4 \cdot 3 \cdot 2 \cdot 1 = 720\\
5! &= 5 \cdot 4 \cdot 3 \cdot 2 \cdot 1 = 120\\
4! &= 4 \cdot 3 \cdot 2 \cdot 1 = 24\\
3! &= 3 \cdot 2 \cdot 1 = 6\\
2! &= 2 \cdot 1 = 2\\
1! &= 1 = 1
\end{aligned}$$

We also define 0! to be 1. This is so certain formulas and theorems can be stated concisely and with a consistent pattern.

We can now simplify the permutation formula as follows:

$${}_nP_n = n!$$

Note that $8! = 8 \cdot 7!$. To see this, note that

$$\begin{aligned}
8! &= 8 \cdot 7 \cdot 6 \cdot 5 \cdot 4 \cdot 3 \cdot 2 \cdot 1\\
&= 8 \cdot (7 \cdot 6 \cdot 5 \cdot 4 \cdot 3 \cdot 2 \cdot 1)\\
&= 8 \cdot 7!.
\end{aligned}$$

Generalizing, we get the following.

For any natural number n, $n! = n(n-1)!$.

By using this result repeatedly, we can further manipulate factorial notation.

EXAMPLE 6 Rewrite 7! with a factor of 5!.

Solution

$$7! = 7 \cdot 6 \cdot 5!$$

□

Permutations of n Objects Taken r at a Time

Consider a set of 6 objects, say {A, B, C, D, E, F}. How many ordered arrangements are there having 3 members without repetition? We can select the first object in 6 ways. There are then 5 choices for the second and then 4 choices for the third. By the fundamental counting principle, there are then $6 \cdot 5 \cdot 4$ ways to construct the ordered arrangement. In other words, there are $6 \cdot 5 \cdot 4$ permutations of a set of 6 objects taken 3 at a time. Note that if we multiply by 1 we have

$$6 \cdot 5 \cdot 4 = \frac{6 \cdot 5 \cdot 4 \cdot 3 \cdot 2 \cdot 1}{3 \cdot 2 \cdot 1}, \quad \text{or} \quad \frac{6!}{3!}.$$

A *permutation* of a set of n objects taken r at a time is an ordered arrangement of r objects taken from the set.

Consider a set of n objects from which an ordered arrangement of r objects is selected. The first object can be selected in n ways. The second can be selected in $n - 1$ ways, and so on. The rth can be selected in $n - (r - 1)$ ways. By the fundamental counting principle, the total number of permutations is

$$n(n-1)(n-2) \cdots [n-(r-1)].$$

We now multiply by 1:

$$n(n-1)(n-2) \cdots [n-(r-1)]\frac{(n-r)!}{(n-r)!}$$
$$= \frac{n(n-1)(n-2)(n-3) \cdots [n-(r-1)](n-r)!}{(n-r)!}.$$

The numerator is the product of all natural numbers from n to 1, or $n!$. Thus the total number of permutations is

$$\frac{n!}{(n-r)!}.$$

This gives us the following result.

Permutations of n Objects Taken r at a Time

The number of permutations of a set of n objects taken r at a time, denoted ${}_nP_r$, is given by

$${}_nP_r = n(n-1)(n-2) \cdots [n-(r-1)] \quad (1)$$

or

$${}_nP_r = \frac{n!}{(n-r)!}. \quad (2)$$

Formula (1) is most useful in application, but formula (2) will be important in a later development.

EXAMPLE 7

Compute ${}_6P_4$ using both of the above formulas.

Solution Using formula (1), we have

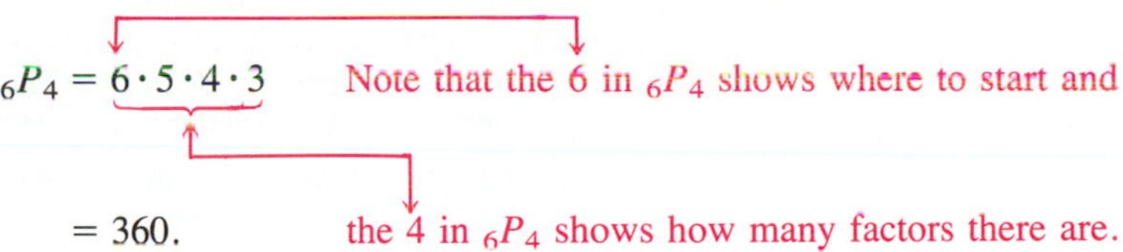

$$ {}_6P_4 = 6 \cdot 5 \cdot 4 \cdot 3 \qquad \text{Note that the 6 in } {}_6P_4 \text{ shows where to start and} $$
$$ = 360. \qquad \text{the 4 in } {}_6P_4 \text{ shows how many factors there are.} $$

Using formula (2), we have

$$ {}_6P_4 = \frac{6!}{(6-4)!} = \frac{6!}{2!} = \frac{6 \cdot 5 \cdot 4 \cdot 3 \cdot 2 \cdot 1}{2 \cdot 1} = 6 \cdot 5 \cdot 4 \cdot 3 = 360. $$ ❑

EXAMPLE 8

In how many ways can the letters of the set {A, B, C, D, E, F, G} be arranged without repetition to form code words of (a) 5 letters? (b) 2 letters?

Solution

a) ${}_7P_5 = 7 \cdot 6 \cdot 5 \cdot 4 \cdot 3 = 2520$
b) ${}_7P_2 = 7 \cdot 6 = 42$ ❑

EXAMPLE 9

A baseball manager arranges the batting order as follows: The 4 infielders will bat first, then the other 5 players will follow. How many different batting orders are possible?

Solution The infielders can bat in 4! different ways; the rest in 5! different ways. Then by the fundamental counting principle, we have ${}_4P_4 \cdot {}_5P_5 = 4! \cdot 5!$, or 2880, possible batting orders. ❑

Circular Permutations

Suppose that 5 friends are to be seated around a circular table. How many ways can they be arranged? The arrangements, or permutations, will be different only if the friends are not seated by the same people at both their right and left. Thus the two arrangements shown here are actually the same permutation.

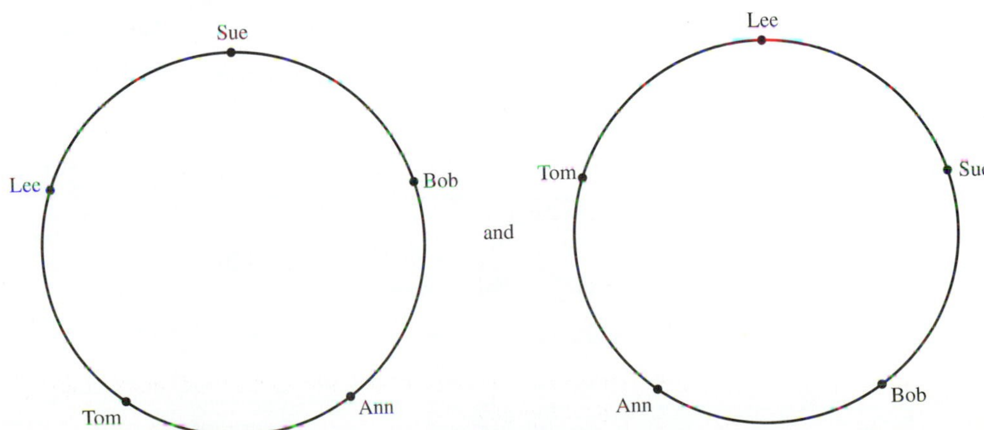

These arrangements would be considered different if all the friends were seated in a straight line. In fact, all 5 of the different permutations shown in the following figure would be equivalent if each line were curled into a circle.

Sue	Bob	Ann	Tom	Lee
Lee	Sue	Bob	Ann	Tom
Tom	Lee	Sue	Bob	Ann
Ann	Tom	Lee	Sue	Bob
Bob	Ann	Tom	Lee	Sue

For each circular permutation of these 5 names, there are 5 distinguishable ordered arrangements on a line. We know that the number of arrangements on the line is ${}_5P_5$, or 5! Therefore, the number of circular permutations is

$$\frac{{}_5P_5}{5} = \frac{5!}{5} = 4! = 24.$$

Circular Permutations

The number of distinct (different) circular arrangements of n objects is $(n - 1)!$

EXAMPLE 10

In how many ways can 6 different foods be arranged in 6 dishes around a lazy Susan?

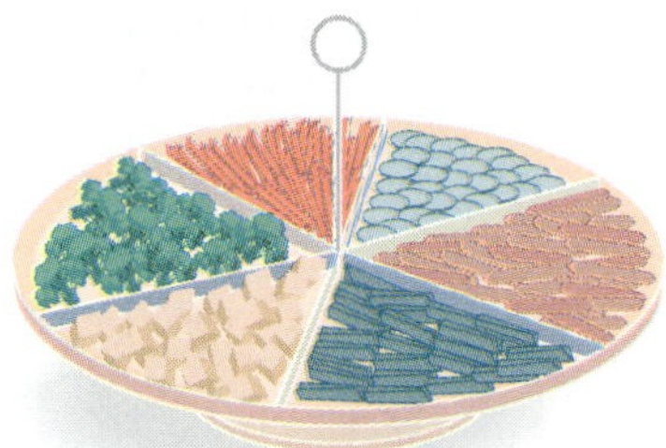

Solution

$$(6 - 1)! = 5 \cdot 4 \cdot 3 \cdot 2 \cdot 1 = 120$$

❑

EXERCISE SET | 14.4

Evaluate.

1. 9!
2. 10!
3. 11!
4. 12!
5. 0!
6. 1!
7. $\frac{7!}{4!}$
8. $\frac{8!}{6!}$
9. $\frac{9!}{5!}$
10. $\frac{10!}{7!}$
11. $(8 - 3)!$
12. $(9 - 5)!$
13. $8! - 3!$
14. $9! - 5!$
15. ${}_6P_6$
16. ${}_5P_5$
17. ${}_4P_3$
18. ${}_7P_5$
19. ${}_{10}P_7$
20. ${}_{10}P_3$
21. ${}_6P_1$
22. ${}_{12}P_1$
23. ${}_6P_5$
24. ${}_{12}P_{11}$

Answer each of the following exercises using permutation notation, factorial notation, or other products. Then evaluate.

25. Hercules Transportation ships goods from New York City to San Francisco. It can use 3 routes from New York to Indianapolis, 2 routes from Indianapolis to Denver, and 2 routes from Denver to San Francisco. How many different routes can it use between New York and San Francisco?

26. Jeff is going to order a new pickup truck. He must choose 1 of 5 wheel styles, 1 of 10 paint colors, 1 of 4 interior colors, and 1 of 4 side mirror styles. How many different ways can he select the truck?

How many permutations are there of the letters in each of the following words, if all of the letters are used?

27. OWL
28. WE
29. QUALIFY
30. TIMES

31. How many permutations are there of the letters of the word QUALIFY if the letters are taken 4 at a time?

32. How many permutations are there of the letters of the word TIMES if the letters are taken 3 at a time?

33. How many 5-digit numbers can be named using the digits 5, 6, 7, 8, and 9 without repetition? with repetition?

34. How many 4-digit numbers can be named using the digits 2, 3, 4, and 5 without repetition? with repetition?

35. In how many ways can 5 students be arranged in a straight line?

36. In how many ways can 7 athletes be arranged in a straight line?

37. How many 7-digit phone numbers can be formed with the digits 0, 1, 2, 3, 4, 5, 6, 7, 8, and 9, assuming that no digit is used more than once and the first digit is not 0?

38. How many ways can a president, vice president, secretary, and treasurer be chosen from a committee of 12 people?

39. A penny, nickel, dime, quarter, and half-dollar are arranged in a straight line.
    a) Considering just the coins, in how many ways can they be lined up?
    b) Considering the coins and heads and tails, in how many ways can they be lined up?

40. A penny, nickel, dime, and quarter are arranged in a straight line.
    a) Considering just the coins, in how many ways can they be lined up?
    b) Considering the coins and heads and tails, in how many ways can they be lined up?

41. Compute ${}_{52}P_4$.

42. Compute ${}_{50}P_5$.

43. A state forms its license plates by first listing a number that corresponds to the county in which the car owner lives (the names of the counties are alphabetized and the number is its location in that order). Then the plate lists a letter of the alphabet, and this is followed by a number from 1 to 9999. How many such plates are possible if there are 80 counties?

44. *Zip codes.* Zip codes in Canada are a series of numerals and letters. A zip code for Montreal, Quebec, is H2N 1M5. It consists of a letter for the first, third, and fifth places, and a number from 0 to 9 in the second, fourth, and sixth places.
    a) How many such zip codes are possible?
    b) There are 27 million people in Canada. Can each person have his or her own zip code?

45. *Zip codes.* Zip codes in the United States are 5-digit numbers. A zip code in Dallas, Texas, is 75247.
    a) How many zip codes are possible if any of the digits 0 to 9 can be used?
    b) If each post office has its own zip code, how many possible post offices can there be?

46. *Zip codes.* Zip codes are sometimes given using a 9-digit number like 75247-5456, where the last 4 digits represent a post office box number.

a) How many 9-digit zip codes are possible?

b) There are 261 million people in the United States. If each person had a private zip code and post office box, would there be enough zip codes?

47. *Social security numbers.* A social security number is a 9-digit number like 293-36-0391.

a) How many social security numbers can there be?

b) There are 261 million people in the United States. Can each person have a social security number?

48. How "long" is 15!? You own 15 different books and decide to actually make up all possible ordered arrangements of the books on a shelf. About how long, in years, would it take if you can make one arrangement per second?

49. In how many ways can the numbers on a clock face be arranged?

50. In how many ways can King Arthur and his 12 knights sit at his Round Table?

51. In how many ways can 6 students be arranged in a straight line? in a circle?

52. In how many ways can 7 candidates be arranged in a straight line? in a circle?

Skill Maintenance

Given $\log_b 7 = 1.946$ and $\log_b 5 = 1.609$, find each of the following.

53. $\log_b 35$ **54.** $\log_b \frac{5}{7}$ **55.** $\log_b 49$

Simplify.

56. $\log_p p^{10}$ **57.** $\log_b b^{-5}$ **58.** $\log_Q Q^m$

Synthesis

59. Write a problem for which the solution requires the use of both the fundamental counting principle and permutations.

60. Explain why more code words can be formed from a set of letters if repetition is allowed than if it is not.

Solve for n.

61. ${}_nP_5 = 7 \cdot {}_nP_4$ **62.** ${}_nP_4 = 8 \cdot {}_{n-1}P_3$

63. ${}_nP_5 = 9 \cdot {}_{n-1}P_4$ **64.** ${}_nP_4 = 8 \cdot {}_nP_3$

65. In how many ways can 3 men and 3 women be seated in a row of 6 seats:

a) with no seating restrictions?

b) if men and women must alternate?

c) if a particular man and woman must sit together?

d) if a particular man and woman must not sit together?

66. A car holds 3 people in the front and 3 people in the back. In how many ways can 6 people be seated in the car:

a) with no seating restrictions?

b) if two people must sit together?

c) if two particular couples must each sit together?

67. One method for factoring a trinomial $ax^2 + bx + c$ is the trial-and-error method. Factorizations of the form $(px + q)(rx + s)$ are formed, where $pr = a$ and $qs = c$. The number of possible factorizations thus depends on the number of ways a and c can be factored. How many possible trial factorizations are there of $6x^2 + 73x + 12$?

68. With reference to Exercise 67, how many possible trial factorizations are there of $5x^2 + 12x + 7$?

14.5 Combinatorics: Combinations

Combinations • Combinations of n Objects Taken r at a Time

If you play cards, you know that in most situations the *order* in which you hold cards *is not important*! It is just the contents of the hand, or set, of cards. We may sometimes make selections from a set *without regard to order*. Such selections are called **combinations.**

A → B → C　　C A B

Permutation: Order considered!　　*Combination:* Order *not* considered!

Combinations

EXAMPLE 1

Find all the combinations of taking 3 elements from the set of 5 elements {A, B, C, D, E}. How many are there?

Solution The combinations are

{A, B, C}, {A, B, D}, {A, B, E}, {A, C, D,}, {A, C, E},
{A, D, E}, {B, C, D}, {B, C, E}, {B, D, E,}, {C, D, E}.

There are 10 combinations of 5 objects taken 3 at a time. ❑

When we find all the combinations of 5 objects taken 3 at a time, we are finding all the 3-element subsets. When we are naming a set, the order of the listing is *not* important. Thus,

{A, C, B} names the same set as {A, B, C}.

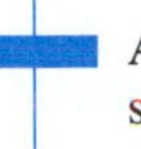

A *combination* of r objects chosen from a set of n objects is a subset of the set of n objects.

Because the elements of a subset may be listed in any order, it is important to remember that *when thinking of combinations we do not think about order.*

EXAMPLE 2

Find all the subsets of the set {A, B, C}. Identify these as combinations. How many subsets are there in all?

Solution

a) The empty set has 0 elements in it. It is denoted ∅. The empty set is a subset of every set. In this case, it is the combination of 3 objects taken 0 at a time. There is 1 such combination, ∅.

b) The following are all the one-element subsets of {A, B, C}:

{A}, {B}, {C}.

These are the combinations of 3 objects taken 1 at a time. There are 3 such combinations.

c) The following are all the two-element subsets of {A, B, C}:

{A, B}, {A, C}, {B, C}.

These are the combinations of 3 objects taken 2 at a time. There are 3 such combinations.

d) The following are all the three-element subsets of {A, B, C}:

{A, B, C}.

These are the combinations of 3 objects taken 3 at a time. There is only 1 such combination. A set is always a subset of itself.

The total number of subsets is 1 + 3 + 3 + 1, or 8. ❑

Combinations of *n* Objects Taken *r* at a Time

We want to develop a formula for computing the number of combinations of n objects taken r at a time without actually listing the combinations, or subsets.

The notation for the number of combinations taken r at a time from a set of n objects is denoted ${}_nC_r$.

We call ${}_nC_r$ **combination notation.** In Example 2 we see that

$${}_3C_0 = 1, \quad {}_3C_1 = 3, \quad {}_3C_2 = 3, \quad \text{and} \quad {}_3C_3 = 1.$$

We want to derive a general formula for ${}_nC_r$, for any whole number $r \leq n$. Let us return to Example 1 and compare the number of combinations with the number of permutations.

Combinations		*Permutations*					
{A, B, C}	⟶	ABC	BCA	CAB	CBA	BAC	ACB
{A, B, D}	⟶	ABD	BDA	DAB	DBA	BAD	ADB
{A, B, E}	⟶	ABE	BEA	EAB	EBA	BAE	AEB
{A, C, D}	⟶	ACD	CDA	DAC	DCA	CAD	ADC
{A, C, E}	⟶	ACE	CEA	EAC	ECA	CAE	AEC
{A, D, E}	⟶	ADE	DEA	EAD	EDA	DAE	AED
{B, C, D}	⟶	BCD	CDB	DBC	DCB	CBD	BDC
{B, C, E}	⟶	BCE	CEB	EBC	ECB	CBE	BEC
{B, D, E}	⟶	BDE	DEB	EBD	EDB	DBE	BED
{C, D, E}	⟶	CDE	DEC	ECD	EDC	DCE	CED

Note that each combination of 3 objects, say {A, C, E}, yields 3!, or 6, permutations, as shown. It follows that

$$3! \cdot {}_5C_3 = {}_5P_3 = 5 \cdot 4 \cdot 3 = 60,$$

so

$$_5C_3 = \frac{_5P_3}{3!} = \frac{5 \cdot 4 \cdot 3}{3 \cdot 2 \cdot 1} = 10. \qquad \text{Dividing by } 3!$$

In general, the number of combinations of n objects taken r at a time, $_nC_r$, times the number of permutations of these r objects, $r!$, must equal the number of permutations of n objects taken r at a time:

$$r! \cdot {_nC_r} = {_nP_r}.$$

Dividing on both sides by $r!$, we have

$$_nC_r = \frac{_nP_r}{r!} = {_nP_r} \cdot \frac{1}{r!} = \frac{n!}{(n-r)!} \cdot \frac{1}{r!} = \frac{n!}{(n-r)!r!}.$$

Combinations of n Objects Taken r at a Time

The total number of combinations of n objects taken r at a time, denoted $_nC_r$, is given by

$$_nC_r = \frac{n!}{(n-r)!r!}.$$

We can make some general observations. First, it is always true that $_nC_n = 1$ because a set with n objects has only 1 subset with n objects, the set itself. Second, $_nC_1 = n$ because a set with n objects has n subsets with 1 element each. Finally, $_nC_0 = 1$ because a set with n objects has only one subset with 0 elements, namely, the empty set $\varnothing$.

There is another kind of notation that is also used for $_nC_r$. It is called **binomial coefficient** notation. The reason for such terminology will be seen later.

Binomial Coefficient

$$\binom{n}{r} = {_nC_r}$$

It is important to remember that $\binom{n}{r}$ does not mean $n \div r$, or $\frac{n}{r}$. The notation $\binom{n}{r}$ is often read "n choose r."

EXAMPLE 3

Simplify: **(a)** $\binom{7}{5}$; **(b)** $\binom{7}{2}$; **(c)** $\binom{6}{6}$.

Solution

a) $$\binom{7}{5} = \frac{7!}{(7-5)!5!}$$

$$= \frac{7!}{2!5!} = \frac{7 \cdot 6 \cdot 5!}{2 \cdot 1 \cdot 5!} = \frac{7 \cdot 6}{2 \cdot 1} = \frac{7 \cdot 3 \cdot 2}{2} = 21$$

b) $\binom{7}{2} = \frac{7!}{5!2!} = \frac{7 \cdot 6 \cdot 5!}{5! \cdot 2 \cdot 1} = \frac{7 \cdot 6}{2} = 7 \cdot 3 = 21$

c) $\binom{6}{6} = \frac{6!}{0!6!} = \frac{6!}{1 \cdot 6!}$ Since $0! = 1$

$= \frac{6!}{6!} = 1$ ❑

In Example 3, we saw that

$$\binom{7}{5} = \binom{7}{2}.$$

This says that the number of 5-element subsets of a set of 7 objects is the same as the number of 2-element subsets of a set of 7 objects. This pattern is true in general:

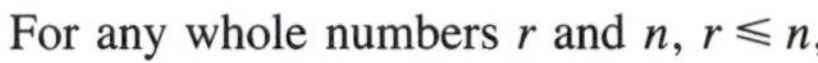

For any whole numbers r and n, $r \leqslant n$,

$$\binom{n}{r} = \binom{n}{n-r} \quad \text{and} \quad {}_nC_r = {}_nC_{n-r}.$$

The number of subsets of size r of a set with n objects is the same as the number of subsets of size $n - r$. The number of combinations of n objects taken r at a time is the same as the number of combinations of n objects taken $n - r$ at a time.

We can use combinations to determine the number of ways subsets can be chosen.

EXAMPLE 4

State lotto. A state runs a 6-out-of-44-number lotto twice a week that pays at least \$1.5 million. You purchase a card for \$1 and pick any 6 numbers from 1 to 44. How many possible 6-number combinations are there for drawing?

Solution No order is implied here. You pick any 6 numbers from 1 to 44. Thus the number of combinations is

$$_{44}C_6 = \binom{44}{6} = \frac{44!}{38!6!} = \frac{44 \cdot 43 \cdot 42 \cdot 41 \cdot 40 \cdot 39 \cdot 38!}{38! \cdot 6 \cdot 5 \cdot 4 \cdot 3 \cdot 2 \cdot 1}$$
$$= 7{,}059{,}052.$$ ❑

EXAMPLE 5

A company employs 5 analysts and 7 accountants. Management is forming a search committee containing 3 analysts and 4 accountants. How many different committees can be formed?

Solution First determine the number of ways that 3 analysts can be chosen from the group of 5. Since the order in which they are chosen is not important, the analysts can be selected in $_5C_3$ ways.

Similarly, the 4 accountants can be chosen in $_7C_4$ ways.

If we use the fundamental counting principle, it follows that the number of possible committees is

$$_5C_3 \cdot {}_7C_4 = \frac{5!}{2!3!} \cdot \frac{7!}{3!4!} = 10 \cdot 35 = 350.$$ ❑

Compare the solutions of Example 9 in Section 14.4 and Example 5 in this section. In Example 9 of Section 14.4, order was important, so permutations were used. In Example 5 of this section, the groups chosen were subsets, in which order was not important, so combinations were used.

EXERCISE SET 14.5

Evaluate.

1. $_{13}C_2$ **2.** $_9C_6$ **3.** $\binom{13}{11}$ **4.** $\binom{9}{3}$

5. $\binom{7}{1}$ **6.** $\binom{8}{8}$ **7.** $\frac{_5P_3}{3!}$ **8.** $\frac{_{10}P_5}{5!}$

9. $\binom{6}{0}$ **10.** $\binom{6}{3}$ **11.** $_{12}C_{11}$ **12.** $_{12}C_{10}$

13. $_{20}C_{18}$ **14.** $_{30}C_3$

15. $\binom{35}{2}$ **16.** $\binom{40}{38}$

17. $_{10}C_5$ **18.** $_{15}C_{11}$

Answer each of the following exercises using permutation notation, combination notation, factorial notation, or other products. Then evaluate.

19. There are 23 students in a business club. How many sets of 4 officers can be selected?

20. How many basketball games can be played in a 9-team league if each team plays all other teams once? twice?

21. On a test a student is to select 6 out of 10 questions. In how many ways can the student do this?

22. On a test, a student is to select 7 out of 11 questions. In how many ways can the student do this?

23. How many lines are determined by 8 points, no 3 of which are collinear? How many triangles are determined by the same points?

24. How many lines are determined by 7 points, no 3 of which are collinear? How many triangles are determined by the same points?

25. Of the first 10 questions on a test, a student must answer 7. On the next 5 questions, the student must answer 3. In how many ways can this be done?

26. Of the first 8 questions on a test, a student must answer 6. On the next 4 questions, the student must answer 3. In how many ways can this be done?

27. Suppose the Senate of the United States consists of 58 Democrats and 42 Republicans. How many committees made up of 6 Democrats and 4 Republicans can be formed? You need not simplify the expression.

28. Suppose the Senate of the United States consists of 63 Republicans and 37 Democrats. How many committees made up of 8 Republicans and 12 Democrats can be formed? You need not simplify the expression.

29. A restaurant's Sunday buffet offers 6 seafood entrees, 9 vegetables, and 7 sauces. How many possible combinations of seafood, vegetable, and sauce are there?

30. Two 6-sided dice are rolled, one a blue die and one red. How many different ways can the two fall?

31. How many different 5-card poker hands are possible with a 52-card deck? (See Section 14.7 for a description of a 52-card deck.) You need not simplify the expression.

32. How many different 13-card bridge hands can be dealt from a standard deck of 52 cards? You need not simplify the expression.

33. A university offers 5 science courses, 6 humanity courses, and 3 literature courses. In how many ways can a student choose 2 science courses, 3 humanity courses, and 1 literature course?

34. In how many ways can a committee of 4 students and 3 professors be chosen out of a group of 7 students and 5 professors?

35. In how many ways can 9 different books be distributed among 3 children so that the oldest gets 4, the middle gets 3, and the youngest gets 2?

36. In how many ways can 8 different compact discs be distributed among 3 students if the first gets 2, the second gets 5, and the third gets 1?

37. Pizza Shack has the following toppings for pizzas:

> extra cheese, pepperoni, sausage, mushroom, onion, green pepper, beef, Canadian bacon, black olives, ham.

How many different kinds of pizza can Pizza Shack serve (excluding size and thickness of pizza)?

38. Pizza Shack serves round pizzas in sizes 10-in. and 16-in. and three thicknesses, original, thin, and pan. Using the toppings listed in Exercise 37 and considering size and thickness, how many different kinds of pizza can Pizza Shack serve?

39. How many 5-card poker hands consisting of 3 aces and 2 cards that are not aces are possible with a 52-card deck? (See Section 14.7 for a description of a 52-card deck.)

40. How many 5-card poker hands consisting of 2 kings and 3 cards that are not kings are possible with a 52-card deck?

41. Bresler's Ice Cream, a national firm, sells ice cream in 33 flavors.

a) How many 3-dip cones are possible if order of flavors is to be considered and no flavor is repeated?

b) How many 3-dip cones are possible if order is to be considered and flavors can be repeated?

c) How many 3-dip cones are possible if order is not considered and no flavor is repeated?

42. Baskin-Robbins Ice Cream, a national firm, sells ice cream in 31 flavors.

a) How many 2-dip cones are possible if order of flavors is to be considered and no flavor is repeated?

b) How many 2-dip cones are possible if order is to be considered and flavors can be repeated?

c) How many 2-dip cones are possible if order is not considered and no flavor is repeated?

Skill Maintenance

Solve.

43. $2^x = \frac{1}{4}$

44. $3^{2x} = 27$

45. $\log_5 (x + 1) = 2$

46. $\log x + \log (x - 3) = 1$

Synthesis

47. ◈ Explain the difference between permutations and combinations.

48. ◈ A restaurant that allows its customers to top their own hamburgers advertises that with 28 toppings to choose from, there are 268,435,456 combinations. Describe a procedure for determining this number using the formulas developed in this section.

Simplify.

49. $\binom{m}{1}$ **50.** $\binom{m}{m-1}$ **51.** $\binom{m}{0}$ **52.** $\binom{m}{m-2}$

Solve for n.

53. $\binom{n+1}{3} = 2 \cdot \binom{n}{2}$

54. $\binom{n}{n-2} = 6$

55. $\binom{n+2}{4} = 6 \cdot \binom{n}{2}$

56. $\binom{n}{3} = 2 \cdot \binom{n-1}{2}$

57. In a single-elimination sports tournament consisting of n teams, a team is eliminated when it loses one game. How many games are required to complete the tournament?

58. In a double-elimination softball tournament consisting of n teams, a team is eliminated when it loses two games. At most, how many games are required to complete the tournament?

59. There are m points on a circle. How many triangles can be inscribed with these points as vertices?

60. A set of m parallel lines crosses another set of n parallel lines. How many parallelograms are formed?

61. Prove that

$$\binom{n}{r} = \binom{n}{n-r}$$

for any whole numbers n and r.

14.6 The Binomial Theorem

Binomial Expansion Using Pascal's Triangle • Binomial Expansion Using Factorial Notation

A binomial, defined in Chapter 4, is a polynomial with two terms, such as $3x + 5$ or $x^2 + 2xy^3$.

In Section 4.5, we developed a method for computing quickly the square of a binomial:

$$(A + B)^2 = A^2 + 2AB + B^2.$$

The expression on the right in the equation above is called the *expansion* of $(A + B)^2$. In this section, we will consider two methods for determining expanded forms of binomials raised to powers greater than 2.

Binomial Expansion Using Pascal's Triangle

Consider the following expanded powers of $(a + b)^n$:

$$(a + b)^0 = 1$$
$$(a + b)^1 = a + b$$
$$(a + b)^2 = a^2 + 2a^1b^1 + b^2$$
$$(a + b)^3 = a^3 + 3a^2b^1 + 3a^1b^2 + b^3$$
$$(a + b)^4 = a^4 + 4a^3b^1 + 6a^2b^2 + 4a^1b^3 + b^4$$
$$(a + b)^5 = a^5 + 5a^4b^1 + 10a^3b^2 + 10a^2b^3 + 5a^1b^4 + b^5.$$

Each expansion is a polynomial. There are some patterns to be noted:

1. There is one more term than the power of the binomial, n. That is, there are $n + 1$ terms in the expansion of $(a + b)^n$.
2. In each term, the sum of the exponents is the power to which the binomial is raised.
3. The exponents of a start with n, the power of the binomial, and decrease to 0. The last term has no factor of a. The first term has no factor of b, so powers of b start with 0 and increase to n.
4. The coefficients start at 1 and increase through certain values about "half"-way and then decrease through these same values back to 1. Let's study the coefficients further.

Suppose we want to find an expansion of $(a + b)^8$. The patterns we noticed above indicate 9 terms in the expansion:

$$a^8 + c_1a^7b + c_2a^6b^2 + c_3a^5b^3 + c_4a^4b^4 + c_5a^3b^5 + c_6a^2b^6 + c_7ab^7 + b^8.$$

How can we determine the values for the c's? We can answer this question in two different ways. The first method seems to be the easiest, but is not always. It involves writing down the coefficients in a triangular array as follows. We form what is known as **Pascal's triangle:**

$(a + b)^0$: 1

$(a + b)^1$: 1 1

$(a + b)^2$: 1 2 1

$(a + b)^3$: 1 3 3 1

$(a + b)^4$: 1 4 6 4 1

$(a + b)^5$: 1 5 10 10 5 1

There are many patterns in the triangle. Find as many as you can.

Perhaps you discovered a way to write the next row of numbers, given the numbers in the row above it. There are always 1's on the outside. Each remaining number is the sum of the two numbers above:

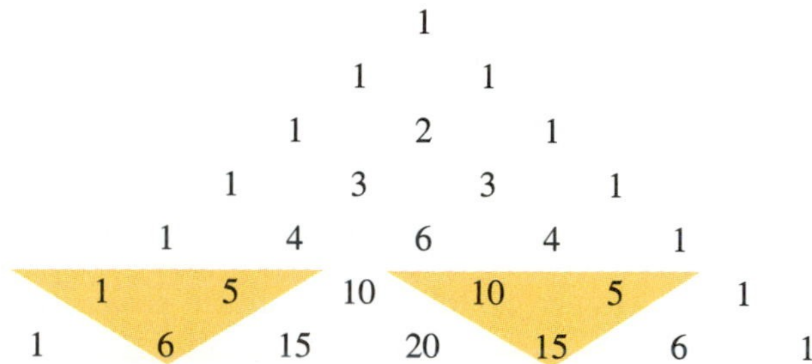

We see that in the last row

the 1st and last numbers are 1;
the 2nd number is 1 + 5, or 6;
the 3rd number is 5 + 10, or 15;
the 4th number is 10 + 10, or 20;
the 5th number is 10 + 5, or 15; and
the 6th number is 5 + 1, or 6.

Thus the expansion of $(a+b)^6$ is

$$(a+b)^6 = 1a^6 + 6a^5b + 15a^4b^2 + 20a^3b^3 + 15a^2b^4 + 6ab^5 + 1b^6.$$

To find the expansion for $(a+b)^8$, we complete two more rows of Pascal's triangle:

$$\begin{array}{ccccccccccccccccc}
&&&&&&&&1&&&&&&&&\\
&&&&&&&1&&1&&&&&&&\\
&&&&&&1&&2&&1&&&&&&\\
&&&&&1&&3&&3&&1&&&&&\\
&&&&1&&4&&6&&4&&1&&&&\\
&&&1&&5&&10&&10&&5&&1&&&\\
&&1&&6&&15&&20&&15&&6&&1&&\\
&1&&7&&21&&35&&35&&21&&7&&1&\\
1&&8&&28&&56&&70&&56&&28&&8&&1
\end{array}$$

Thus the expansion of $(a+b)^8$ is

$$(a+b)^8 = 1a^8 + 8a^7b + 28a^6b^2 + 56a^5b^3 + 70a^4b^4 + 56a^3b^5 + 28a^2b^6 + 8ab^7 + 1b^8.$$

We can generalize our results as follows:

The Binomial Theorem (Form 1)

For any binomial $a+b$ and any natural number n,

$$(a+b)^n = c_0a^nb^0 + c_1a^{n-1}b^1 + c_2a^{n-2}b^2 + \cdots + c_{n-1}a^1b^{n-1} + c_na^0b^n,$$

where the numbers $c_0, c_1, c_2, \ldots, c_n$ are from the $(n+1)$st row of Pascal's triangle.

EXAMPLE 1

Expand: $(u-v)^5$.

Solution We note that $a = u$, $b = -v$, and $n = 5$. We use the 6th row of Pascal's triangle: 1 5 10 10 5 1.

Then we have

$$\begin{aligned}(u-v)^5 &= [u+(-v)]^5\\ &= 1(u)^5 + 5(u)^4(-v)^1 + 10(u)^3(-v)^2 + 10(u)^2(-v)^3 + 5(u)^1(-v)^4 + 1(-v)^5\\ &= u^5 - 5u^4v + 10u^3v^2 - 10u^2v^3 + 5uv^4 - v^5.\end{aligned}$$

Note that the signs of the terms alternate between + and −. When $-v$ is raised to an odd power, the sign is −. ❑

EXAMPLE 2

Expand: $\left(2t + \dfrac{3}{t}\right)^6$.

Solution We note that $a = 2t$, $b = 3/t$, and $n = 6$. We use the 7th row of Pascal's triangle: 1 6 15 20 15 6 1.

Then we have

$$\left(2t + \frac{3}{t}\right)^6 = 1(2t)^6 + 6(2t)^5\left(\frac{3}{t}\right)^1 + 15(2t)^4\left(\frac{3}{t}\right)^2 + 20(2t)^3\left(\frac{3}{t}\right)^3$$
$$+ 15(2t)^2\left(\frac{3}{t}\right)^4 + 6(2t)^1\left(\frac{3}{t}\right)^5 + 1\left(\frac{3}{t}\right)^6$$
$$= 64t^6 + 6(32t^5)\left(\frac{3}{t}\right) + 15(16t^4)\left(\frac{9}{t^2}\right) + 20(8t^3)\left(\frac{27}{t^3}\right)$$
$$+ 15(4t^2)\left(\frac{81}{t^4}\right) + 6(2t)\left(\frac{243}{t^5}\right) + \frac{729}{t^6}$$
$$= 64t^6 + 576t^4 + 2160t^2 + 4320 + 4860t^{-2} + 2916t^{-4} + 729t^{-6}.$$

❑

Binomial Expansion Using Factorial Notation

The disadvantage in using Pascal's triangle is that we must compute all the preceding rows in the table to obtain the row needed for the expansion. The following method avoids this difficulty. It will also enable us to find a specific term—say, the 8th term—without computing all the other terms in the expansion. This method is useful in such courses as finite mathematics, calculus, and statistics, and uses the *binomial coefficient* notation

$$\binom{n}{r}$$

developed in Section 14.5.

The Binomial Theorem (Form 2)

For any binomial $a + b$ and any natural number n,

$$(a + b)^n = \binom{n}{0}a^n + \binom{n}{1}a^{n-1}b + \binom{n}{2}a^{n-2}b^2 + \cdots + \binom{n}{n}b^n.$$

EXAMPLE 3

Expand: $(3x + y)^4$.

Solution We use the binomial theorem (Form 2) with $a = 3x$, $b = y$, and $n = 4$:

$$(3x + y)^4 = \binom{4}{0}(3x)^4 + \binom{4}{1}(3x)^3(y) + \binom{4}{2}(3x)^2(y)^2 + \binom{4}{3}(3x)(y)^3 + \binom{4}{4}(y)^4$$
$$= \frac{4!}{4!0!}3^4x^4 + \frac{4!}{3!1!}3^3x^3y + \frac{4!}{2!2!}3^2x^2y^2 + \frac{4!}{1!3!}3xy^3 + \frac{4!}{0!4!}y^4$$
$$= 81x^4 + 108x^3y + 54x^2y^2 + 12xy^3 + y^4. \quad \text{Simplifying}$$

❑

EXAMPLE 4

Expand: $(x^2 - 2y)^5$.

Solution We have $a = x^2$, $b = -2y$, and $n = 5$:

$$\begin{aligned}(x^2 - 2y)^5 &= \binom{5}{0}(x^2)^5 + \binom{5}{1}(x^2)^4(-2y) + \binom{5}{2}(x^2)^3(-2y)^2 \\ &\quad + \binom{5}{3}(x^2)^2(-2y)^3 + \binom{5}{4}(x^2)(-2y)^4 + \binom{5}{5}(-2y)^5 \\ &= \frac{5!}{5!0!}x^{10} + \frac{5!}{4!1!}x^8(-2y) + \frac{5!}{3!2!}x^6(-2y)^2 + \frac{5!}{2!3!}x^4(-2y)^3 \\ &\quad + \frac{5!}{1!4!}x^2(-2y)^4 + \frac{5!}{0!5!}(-2y)^5 \\ &= x^{10} - 10x^8y + 40x^6y^2 - 80x^4y^3 + 80x^2y^4 - 32y^5.\end{aligned}$$

❑

Note that in the binomial theorem (Form 2), $\binom{n}{0}a^nb^0$ gives us the first term, $\binom{n}{1}a^{n-1}b^1$ gives us the second term, $\binom{n}{2}a^{n-2}b^2$ gives us the third term, and so on. This can be generalized to give a method for finding a specific term without writing the entire expansion.

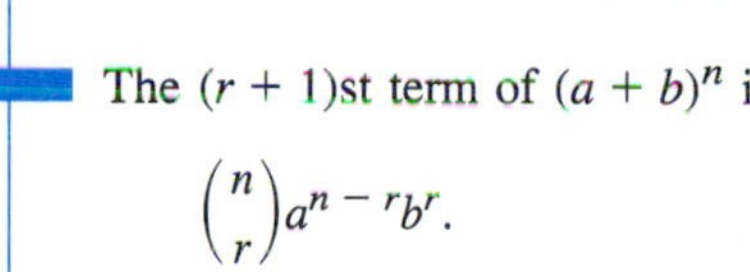

The $(r + 1)$st term of $(a + b)^n$ is

$$\binom{n}{r}a^{n-r}b^r.$$

EXAMPLE 5

Find the 5th term in the expansion of $(2x - 3y)^7$.

Solution First, we note that $5 = 4 + 1$. Thus, $r = 4$, $a = 2x$, $b = -3y$, and $n = 7$. Then the 5th term of the expansion is

$$\binom{7}{4}(2x)^{7-4}(-3y)^4, \quad \text{or} \quad \frac{7!}{3!4!}(2x)^3(-3y)^4, \quad \text{or} \quad 22{,}680x^3y^4.$$

❑

It is because of the binomial theorem that $\binom{n}{r}$ is called a *binomial coefficient*. We can now explain why 0! is defined to be 1. In the binomial expansion, we want $\binom{n}{0}$ to equal 1 and we also want the definition

$$\binom{n}{r} = \frac{n!}{(n-r)!r!}$$

to hold for all whole numbers n and r. Thus we must have

$$\binom{n}{0} = \frac{n!}{(n-0)!0!} = \frac{n!}{n!0!} = 1.$$

This will be satisfied if 0! is defined to be 1.

EXERCISE SET 14.6

Expand. Use both of the methods shown in this section.

1. $(m + n)^5$
2. $(a - b)^3$
3. $(x - y)^6$
4. $(p + q)^4$
5. $(x^2 - 3y)^5$
6. $(3c - d)^7$
7. $(3c - d)^6$
8. $(t^{-2} + 2)^6$
9. $(x - y)^7$
10. $(x - y)^5$
11. $\left(\frac{1}{x} + y\right)^7$
12. $(2s - 3t^2)^3$
13. $\left(a - \frac{2}{a}\right)^9$
14. $\left(2x + \frac{1}{x}\right)^9$
15. $(a^2 + 2b^3)^4$
16. $(x^3 + 2)^6$
17. $(\sqrt{3} - t)^5$
18. $(\sqrt{5} + t)^6$
19. $(x^{-2} + x^2)^4$
20. $\left(\frac{1}{\sqrt{x}} - \sqrt{x}\right)^6$

Find the indicated term of the binomial expression.

21. 3rd, $(a + b)^6$
22. 6th, $(x + y)^7$
23. 12th, $(a - 2)^{14}$
24. 11th, $(x - 3)^{12}$
25. 5th, $(2x^3 - \sqrt{y})^8$
26. 4th, $\left(\frac{1}{b^2} + \frac{b}{3}\right)^7$
27. Middle, $(2u - 3v^2)^{10}$
28. Middle two, $(\sqrt{x} + \sqrt{3})^5$

Skill Maintenance

Find the center and the radius of each circle.

29. $x^2 + (y - 1)^2 = 49$
30. $x^2 + y^2 + 6x - 3 = 0$

Express as a single logarithm and simplify, if possible.

31. $2 \log_a x - 3 \log_a y - \frac{1}{2} \log_a x$
32. $\log_a x^3 - \log_a 2x$

Synthesis

33. Compare the two forms of the binomial theorem given in this section. Under what circumstances would one be better to use than the other?
34. A student asserts that the first and last coefficients of the expansion of any binomial are always 1. Is the student correct? Why or why not?
35. At one point in a recent season, Barry Bonds of the San Francisco Giants had a batting average of 0.313. Suppose he came to bat 5 times in a game. The probability of his getting exactly 3 hits is the 3rd term of the binomial expansion of $(0.313 + 0.687)^5$. Find that term and use your calculator to estimate the probability.
36. The probability that a woman will be either widowed or divorced is 85%. Suppose 8 women are interviewed. The probability that exactly 5 of them will be either widowed or divorced in their lifetime is the 6th term of the binomial expansion of $(0.15 + 0.85)^8$. Find that term and use your calculator to estimate the probability.
37. In reference to Exercise 35, the probability that Bonds will get *at most* 3 hits is found by adding the last 4 terms of the binomial expansion of $(0.313 + 0.687)^5$. Find these terms and use your calculator to estimate the probability.
38. In reference to Exercise 36, the probability that *at least* 6 of the women will be widowed or divorced is found by adding the last three terms of the binomial expansion of $(0.15 + 0.85)^8$. Find these terms and use your calculator to estimate the probability.
39. Expand $(1 + i)^4$, where $i^2 = -1$.
40. Expand $(1 - i)^5$, where $i^2 = -1$.
41. Find the term of
$$\left(\frac{3x^2}{2} - \frac{1}{3x}\right)^{12}$$
that does not contain x.
42. Find the middle term of $(x^2 - 6y^{3/2})^6$.
43. Find the ratio of the 4th term of
$$\left(p^2 - \frac{1}{2}p\sqrt[3]{q}\right)^5$$
to the 3rd term.
44. Find the term containing $\frac{1}{x^{1/6}}$ of
$$\left(\sqrt[3]{x} - \frac{1}{\sqrt{x}}\right)^7.$$
45. What is the degree of $(x^2 + 3)^4$?

14.7 Probability

Experimental and Theoretical Probability • Experimental Probabilities • Theoretical Probabilities • Origin and Use of Probability

We say that when a coin is tossed, the chances that it will fall heads are 1 out of 2, or the **probability** that it will fall heads is $\frac{1}{2}$. Of course this does not mean that if a coin is tossed ten times, it will necessarily fall heads exactly five times. If the coin is tossed a great number of times, however, it will fall heads very nearly half of them.

Experimental and Theoretical Probability

If we toss a coin a great number of times, say 1000, and count the number of heads, we can determine the probability of getting a head. If there are 503 heads, we would calculate the probability of getting a head to be

$$\frac{503}{1000}, \quad \text{or} \quad 0.503.$$

This is an **experimental** determination of probability. Such a determination of probability is quite common. Here, for example, are some probabilities that have been determined *experimentally*:

1. If you kiss someone who has a cold, the probability of your catching a cold is 0.07.
2. A person just released from prison has an 80% probability of returning.

If we consider a coin and *reason* that it is just as likely to fall heads as tails, we would calculate the probability to be $\frac{1}{2}$. This is a **theoretical** determination of probability. Here, for example, are some probabilities that have been determined *theoretically*:

1. If there are 30 people in a room, the probability that two of them have the same birthday (excluding year of birth) is 0.706.
2. If a deck of 52 playing cards is thoroughly shuffled and a card is selected, the probability that the card is a jack is $\frac{1}{13}$, or about 0.077.

Experimental Probabilities

We first consider experimental determination of probability. The basic principle we use in computing such probabilities is as follows.

Principle *P* (Experimental)

An experiment is performed in which n observations are made. If a situation E, or event, occurs m times out of the n observations, then we say that the *experimental probability* of that event is given by

$$P(E) = \frac{m}{n}.$$

EXAMPLE 1

Sociological survey. An actual experiment was conducted to determine the number of people who are left-handed, right-handed, or both. The results are shown in the graph.

a) Determine the probability that a person is left-handed.
b) Determine the probability that a person is ambidextrous (uses both hands equally well).

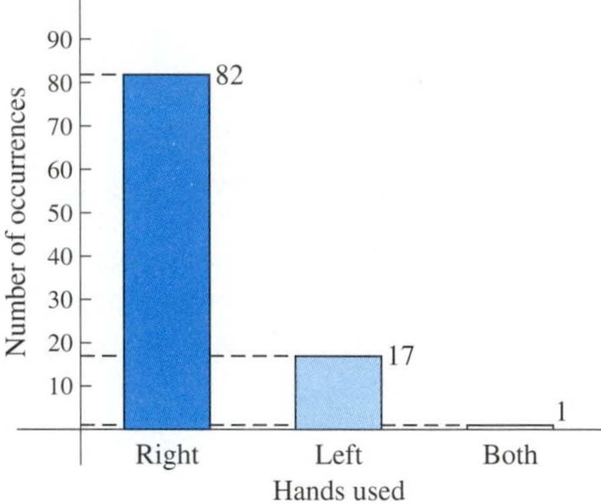

Solution

a) The number of people who were right-handed was 82, the number who were left-handed was 17, and there was 1 person who was ambidextrous. The total number of observations was 82 + 17 + 1, or 100. Thus the probability that a person is left-handed is P, where

$$P = \frac{17}{100}.$$

b) The probability that a person is ambidextrous is P, where

$$P = \frac{1}{100}.$$

❑

EXAMPLE 2

TV ratings. Television networks are always concerned about the percentage of TV sets that are tuned to their programming. Because contacting every home in the country is not practical, a sample, or portion, of the homes are contacted. This is done by attaching an electronic device to the TVs of about 1400 homes across the country. Viewing information is then fed into a computer.

In 1993, CBS premiered a new late night show, competing with shows on ABC and NBC. The number of homes tuned to each show on the first night of the new show are shown in the following table.

Show	*Late Show with David Letterman* (CBS)	*Nightline* (ABC)	*The Tonight Show with Jay Leno* (NBC)	Other
Number of TV Sets Tuned to Show	518	210	196	476

What is the probability that a TV set was tuned to the new *Late Show with David Letterman*?

Solution The total number of observations was 1400. The situation of being tuned to CBS occurred 518 times. Thus the probability that a set was tuned to the new show is P, where

$$P = \frac{518}{1400} = 0.37 = 37\%.$$

❑

Theoretical Probabilities

We need some terminology before we can continue. Suppose we perform an experiment such as flipping a coin, throwing a dart, drawing a card from a deck, or checking an item off an assembly line for quality. The results of an experiment are called **outcomes.** The set of all possible outcomes is called the **sample space.** An **event** is a set of outcomes, that is, a subset of the sample space. For example, for the experiment ''throwing a dart,'' suppose the dartboard is as shown.

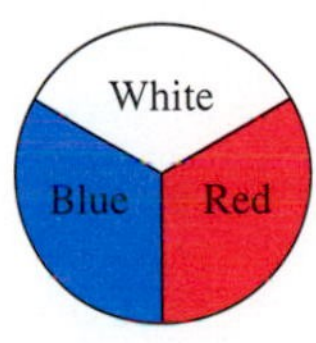

Then one event is

{blue}, (the outcome is ''hitting blue'')

which is a subset of the sample space

{blue, white, red}, (sample space)

assuming that the dart must hit the target somewhere.

We denote the probability that an event E occurs as $P(E)$. For example, when flipping a coin, ''getting a head'' may be denoted by H. Then $P(H)$ represents the probability of getting a head. When all the outcomes of an experiment have the same probability of occurring, we say that they are *equally likely*. To see the distinction between events that are equally likely and those that are not, consider the dartboards shown on the next page.

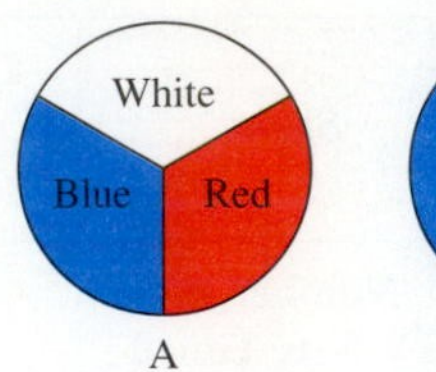

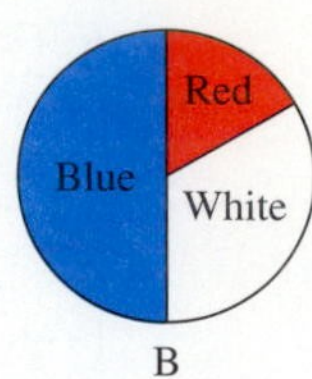

For dartboard A, the events hitting *blue, white,* and *red* are equally likely, but for board B they are not. When a sample space is comprised of equally likely events, it is easy to calculate certain probabilities.

Principle P (Theoretical)

If an event E can occur m ways out of n possible equally likely outcomes of a sample space S, then the *theoretical probability* of that event is given by

$$P(E) = \frac{m}{n}.$$

A die (pl., dice) is a cube, with six faces, each containing a number of dots from 1 to 6.

EXAMPLE 3

What is the probability of rolling a 3 on a die?

Solution On a fair die, there are 6 equally likely outcomes and there is 1 way to get a 3. By Principle P, $P(3) = \frac{1}{6}$. ❑

EXAMPLE 4

What is the probability of rolling an even number on a die?

Solution The event is getting an *even* number. It can occur in 3 ways (getting 2, 4, or 6). The number of equally likely outcomes is 6. By Principle P, $P(\text{even}) = \frac{3}{6}$, or $\frac{1}{2}$. ❑

EXAMPLE 5

Suppose we select, without looking, one marble from a bag containing 3 red marbles and 4 green marbles. What is the probability of selecting a red marble?

Solution There are 7 equally likely ways of selecting any marble, and since the number of ways of getting a red marble is 3,

$$P(\text{selecting a red marble}) = \tfrac{3}{7}.$$ ❑

We now use a number of examples related to a standard deck of 52 playing cards. Such a deck is made up as shown in the following figure.

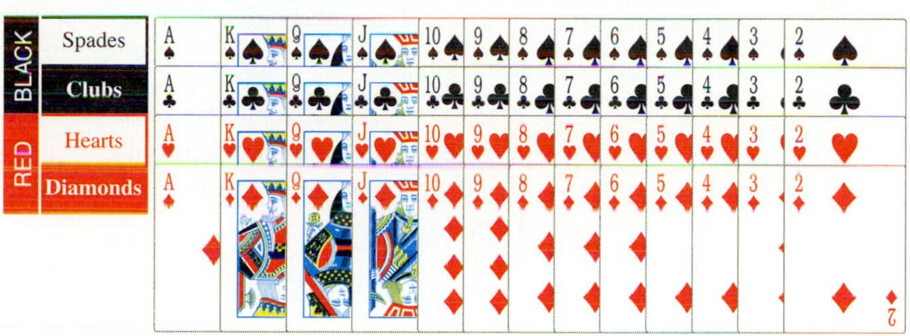

EXAMPLE 6

What is the probability of drawing an ace from a well-shuffled deck of 52 cards?

Solution Since there are 52 outcomes (cards in the deck) and they are equally likely (from a well-shuffled deck) and there are 4 ways to obtain an ace, by Principle P we have

$$P(\text{drawing an ace}) = \frac{4}{52}, \quad \text{or} \quad \frac{1}{13}. \qquad \square$$

The following are some results that follow from Principle P.

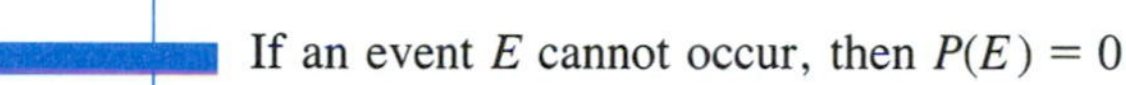

If an event E cannot occur, then $P(E) = 0$.

For example, in coin tossing, the event that a coin would land on its edge has probability 0.

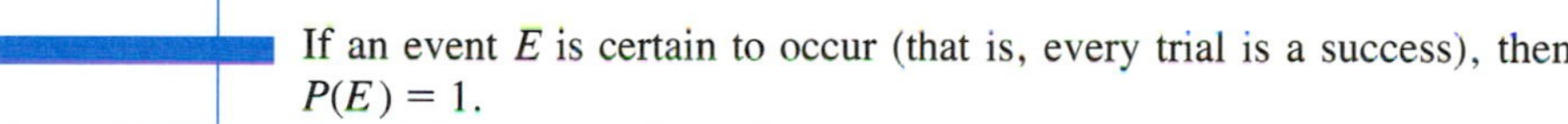

If an event E is certain to occur (that is, every trial is a success), then $P(E) = 1$.

For example, in coin tossing, the event that a coin falls either heads or tails has probability 1.

In general:

The probability that an event E will occur is a number from 0 to 1:

$$0 \leq P(E) \leq 1.$$

In the following examples we use the combinations that we studied in Section 14.5 to calculate theoretical probabilities.

EXAMPLE 7

Suppose 2 cards are drawn from a well-shuffled deck of 52 cards. What is the probability that both of them are spades?

Solution The probability will be

$$P(\text{getting 2 spades}) = \frac{m}{n} = \frac{\text{The number of ways of drawing 2 spades}}{\text{The number of ways of drawing any 2 cards}}.$$

The number of ways n of drawing 2 cards from a deck of 52 is ${}_{52}C_2$. Now 13 of the 52 cards are spades, so the number of ways m of drawing 2 *spades* is ${}_{13}C_2$. Thus,

$$P(\text{getting 2 spades}) = \frac{m}{n} = \frac{{}_{13}C_2}{{}_{52}C_2} = \frac{78}{1326} = \frac{1}{17}.$$ ❑

EXAMPLE 8

Suppose 2 people are selected at random from a group that consists of 6 men and 4 women. What is the probability that both of them are women?

Solution We need to know how many ways n any 2 people can be selected, and also how many ways m 2 women can be selected. The number of ways of selecting 2 people from a group of 10 is ${}_{10}C_2$. The number of ways of selecting 2 women from a group of 4 is ${}_4C_2$. Thus the probability of selecting 2 women from the group of 10 is P, where

$$P = \frac{{}_4C_2}{{}_{10}C_2} = \frac{6}{45} = \frac{2}{15}.$$ ❑

EXAMPLE 9

Suppose 3 people are selected at random from a group that consists of 6 men and 4 women. What is the probability that 1 man and 2 women are selected?

Solution The number of ways of selecting 3 people from a group of 10 is ${}_{10}C_3$. One man can be selected in ${}_6C_1$ ways, and 2 women can be selected in ${}_4C_2$ ways. By the fundamental counting principle, the number of ways of selecting 1 man and 2 women is ${}_6C_1 \cdot {}_4C_2$. Thus the probability is

$$P = \frac{{}_6C_1 \cdot {}_4C_2}{{}_{10}C_3}, \quad \text{or} \quad \frac{3}{10}.$$ ❑

EXAMPLE 10

What is the probability of getting a total of 8 on a roll of a pair of dice? (Assume that the dice are different, say one blue and one black.)

Solution On each die, there are 6 possible outcomes. The outcomes are paired so there are $6 \cdot 6$, or 36, possible ways in which the two can fall.

Blue die						
6	(1, 6)	(2, 6)	(3, 6)	(4, 6)	(5, 6)	(6, 6)
5	(1, 5)	(2, 5)	(3, 5)	(4, 5)	(5, 5)	(6, 5)
4	(1, 4)	(2, 4)	(3, 4)	(4, 4)	(5, 4)	(6, 4)
3	(1, 3)	(2, 3)	(3, 3)	(4, 3)	(5, 3)	(6, 3)
2	(1, 2)	(2, 2)	(3, 2)	(4, 2)	(5, 2)	(6, 2)
1	(1, 1)	(2, 1)	(3, 1)	(4, 1)	(5, 1)	(6, 1)
	1	2	3	4	5	6 Black die

The pairs that total 8 are as shown. Thus there are 5 possible ways of getting a total of 8, so the probability is $\frac{5}{36}$. ❑

Origin and Use of Probability

A desire to calculate odds in games of chance gave rise to the theory of probability. Today the theory of probability and its closely related field, mathematical statistics, have many applications, most of them not related to games of chance. Opinion polls, with such uses as predicting elections, are a familiar example. Quality control, in which a prediction about the percentage of faulty items manufactured is made, is an important application, among many, in business. Still other applications are in the areas of genetics, medicine, and the kinetic theory of gases.

EXERCISE SET 14.7

1. In an actual survey, 100 people were polled to determine the probability of a person wearing either glasses or contact lenses. Of those polled, 57 wore either glasses or contacts. What is the probability that a person wears either glasses or contacts? What is the probability that a person wears neither?

2. In another survey 100 people were polled and asked to select a number from 1 to 5. The results are shown in the following table.

Number of Choices	1	2	3	4	5
Number of People Who Selected That Number	18	24	23	23	12

What is the probability that the number selected is 1? 2? 3? 4? 5? What general conclusion might a psychologist make from this experiment?

Linguistics. An experiment was conducted to determine the relative occurrence of various letters of the English alphabet. A paragraph from a newspaper, one from a textbook, and one from a magazine were considered. In all, there were 1044 letters. The number of occurrences of each letter of the alphabet is listed in the following table.

Letter	A	B	C	D	E	F	G
Number of Occurrences	78	22	33	33	140	24	22

Letter	H	I	J	K	L	M
Number of Occurrences	63	60	2	9	35	30

Letter	N	O	P	Q	R	S	T
Number of Occurrences	74	74	27	4	67	67	95

Letter	U	V	W	X	Y	Z
Number of Occurrences	31	10	22	8	13	1

Round answers to Exercises 3–6 to three decimal places.

3. What is the probability of the occurrence of the letter A? E? I? O? U?

4. What is the probability of a vowel occurring?

5. What is the probability of a consonant occurring?

6. Which letter has the least probability of occurring? What is the probability of this letter not occurring?

Suppose we draw a card from a well-shuffled deck of 52 cards.

7. How many equally likely outcomes are there?

8. What is the probability of drawing a queen?

9. What is the probability of drawing a heart?

10. What is the probability of drawing a club?

11. What is the probability of drawing a 4?

12. What is the probability of drawing a red card?

13. What is the probability of drawing a black card?

14. What is the probability of drawing an ace or a two?

15. What is the probability of drawing a 9 or a king?

Suppose we select, without looking, one marble from a bag containing 4 red marbles and 10 green marbles.

16. What is the probability of selecting a red marble?

17. What is the probability of selecting a green marble?

18. What is the probability of selecting a purple marble?

19. What is the probability of selecting a white marble?

Suppose 4 cards are drawn from a well-shuffled deck of 52 cards.

20. What is the probability that all 4 are spades?

21. What is the probability that all 4 are hearts?

22. If 4 marbles are drawn at random all at once from a bag containing 8 white marbles and 6 black marbles, what is the probability that 2 will be white and 2 will be black?

23. From a group of 8 men and 7 women, a committee of 4 is chosen. What is the probability that 2 men and 2 women will be chosen?

24. What is the probability of getting a total of 6 on a roll of a pair of dice?

25. What is the probability of getting a total of 3 on a roll of a pair of dice?

26. What is the probability of getting snake eyes (a total of 2) on a roll of a pair of dice?

27. What is the probability of getting box-cars (a total of 12) on a roll of a pair of dice?

28. From a bag containing 5 nickels, 8 dimes, and 7 quarters, 5 coins are drawn at random, all at once. What is the probability of getting 2 nickels, 2 dimes, and 1 quarter?

29. From a bag containing 6 nickels, 10 dimes, and 4 quarters, 6 coins are drawn at random, all at once. What is the probability of getting 3 nickels, 2 dimes, and 1 quarter?

30. Tracy takes 1 each of vitamins A, C, E, and B-12 a day. He fills a bottle with 7 days worth of all the vitamins needed. The next day, Tracy takes 4 vitamins at random, all at once, from the bottle. What is the probability that the vitamins will be the 4 desired?

31. Toni is given a box of canned goods and told that it contains 4 cans of meat, 10 cans of vegetables, 6 cans of fruit, and 5 cans of soup. However, the labels have been removed from the cans and they all appear to be identical. Toni selects 4 cans at random, all at once, and opens them for dinner. What is the probability that she will open 1 can each of meat, fruit, vegetables, and soup?

Roulette. A roulette wheel contains slots numbered 00, 0, 1, 2, 3, . . . , 35, 36. Eighteen of the slots numbered 1 through 36 are colored red and eighteen are colored black. The 00 and 0 slots are uncolored. The wheel is spun, and a ball is rolled around the rim until it falls into a slot. What is the probability that the ball falls in:

32. a black slot?

33. a red slot?

34. a red or black slot?

35. the 00 slot?

36. the 0 slot?

37. either the 00 or 0 slot? (Here the house always wins.)

38. an odd-numbered slot?

Skill Maintenance

Solve.

39. $2x + 5y = 7,$
$3x + 2y = 16$

40. $2^{x^2 + 2x} = \frac{1}{2}$

41. Express in terms of logarithms of x, y, and z:

$$\log_a \frac{x^2y}{z^3}.$$

42. Find an equation of the circle with center $(0, -3)$ and radius $2\sqrt{3}$.

Synthesis

43. ◈ Use Principle P to explain why the probability of an event cannot be greater than 1.

44. ◈ Explain the difference between experimental and theoretical probability.

Five-card poker hands and probabilities. In part (a) of each problem, give a reasoned expression as well as the answer. Read all the problems before beginning.

45. ▦ How many 5-card poker hands can be dealt from a standard 52-card deck?

46. ▦ A *royal flush* consists of a 5-card hand with A-K-Q-J-10 of the same suit.

a) How many royal flushes are there?

b) What is the probability of getting a royal flush?

47. A *straight flush* consists of 5 cards in sequence in the same suit, but excludes royal flushes. An ace can be used low, before a two.
   a) How many straight flushes are there?
   b) What is the probability of getting a straight flush?

48. *Four of a kind* is a 5-card hand in which exactly 4 of the cards are of the same denomination, such as J-J-J-J-6, 7-7-7-7-A, or 2-2-2-2-5.
   a) How many are there?
   b) What is the probability of getting four of a kind?

49. A *full house* consists of a pair and three of a kind, such as Q-Q-Q-4-4.
   a) How many are there?
   b) What is the probability of getting a full house?

50. A *pair* is a 5-card hand in which just 2 of the cards are of the same denomination, such as Q-Q-8-A-3.
   a) How many are there?
   b) What is the probability of getting a pair?

51. *Three of a kind* is a 5-card hand in which exactly 3 of the cards are of the same denomination and the other 2 are *not* of the same denomination, such as Q-Q-Q-10-7.
   a) How many are there?
   b) What is the probability of getting three of a kind?

52. A *flush* is a 5-card hand in which all the cards are of the same suit, but not all in sequence (not a straight flush or royal flush).
   a) How many are there?
   b) What is the probability of getting a flush?

53. *Two pairs* is a hand like Q-Q-3-3-A.
   a) How many are there?
   b) What is the probability of getting two pairs?

54. A *straight* is any 5 cards in sequence, but not of the same suit—for example, 4 of spades, 5 of spades, 6 of diamonds, 7 of hearts, and 8 of clubs.
   a) How many are there?
   b) What is the probability of getting a straight?

SUMMARY AND REVIEW 14

KEY TERMS

Infinite sequence, p. 710
Finite sequence, p. 710
General term, p. 711
Infinite series, p. 712
Partial sum, p. 712
Finite series, p. 712
Sigma notation, p. 713
Summation notation, p. 713
Index of summation, p. 713
Arithmetic sequence, p. 716
Arithmetic progression, p. 716
Common difference, p. 716
Arithmetic series, p. 718
Geometric sequence, p. 724
Geometric progression, p. 724
Common ratio, p. 724
Geometric series, p. 726
Infinite geometric series, p. 726
Combinatorics, p. 732
Tree diagram, p. 733
Outcome, p. 733
Event, p. 733
Permutation, p. 734
Factorial notation, p. 735
Combination, p. 741
Binomial coefficient, p. 743
Pascal's triangle, p. 748
Binomial theorem, p. 749
Probability, p. 753
Experimental probability, p. 753
Theoretical probability, p. 753
Sample space, p. 755

IMPORTANT PROPERTIES AND FORMULAS

Arithmetic sequence: $a_{n+1} = a_n + d$

nth term of an arithmetic sequence: $a_n = a_1 + (n-1)d$

Sum of the first n terms of an arithmetic sequence: $S_n = \frac{n}{2}(a_1 + a_n)$

Geometric sequence: $a_{n+1} = a_n \cdot r$

nth term of a geometric sequence: $a_n = a_1 r^{n-1}$

Sum of the first n terms of a geometric sequence: $S_n = \frac{a_1(1-r^n)}{1-r}$

Limit of an infinite geometric series: $S_\infty = \frac{a_1}{1-r}, \quad |r| < 1$

Factorial notation: $n! = n(n-1)(n-2) \cdot \cdots \cdot 3 \cdot 2 \cdot 1$

Binomial coefficient: $\binom{n}{r} = \frac{n!}{(n-r)!r!}$

Binomial theorem:
$$(a+b)^n = \binom{n}{0}a^n + \binom{n}{1}a^{n-1}b + \binom{n}{2}a^{n-2}b^2 + \cdots + \binom{n}{n}b^n$$

Fundamental Counting Principle

Given a combined action, or event, in which the first action can be performed in n_1 ways, the second action can be performed in n_2 ways, and so on, then the total number of ways the combined action can be performed is the product

$$n_1 \cdot n_2 \cdot n_3 \cdot \cdots \cdot n_k.$$

Permutations of n Objects Taken r at a Time

The number of permutations of a set of n objects taken r at a time, denoted $_nP_r$, is given by

$$_nP_r = n(n-1)(n-2) \cdot \cdots \cdot [n-(r-1)],$$

or

$$_nP_r = \frac{n!}{(n-r)!}.$$

Combinations of n Objects Taken r at a Time

The total number of combinations of n objects taken r at a time, denoted $_nC_r$, is given by

$$_nC_r = \frac{n!}{r!(n-r)!}.$$

Principle P (Experimental)

An experiment is performed in which n observations are made. If a situation E, or event, occurs m times out of the n observations, then we say that the *experimental probability* of that event is given by

$$P(E) = \frac{m}{n}.$$

Principle P (Theoretical)

If an event E can occur m ways out of n possible equally likely outcomes of a sample space S, then the theoretical probability of that event is given by

$$P(E) = \frac{m}{n}.$$

If an event E cannot occur, then $P(E) = 0$.
If an event E is certain to occur, then $P(E) = 1$.
The probability that an event E will occur is a number from 0 to 1:

$$0 \leq P(E) \leq 1.$$

REVIEW EXERCISES

This chapter's review and test include Skill Maintenance exercises from Sections 8.2, 12.4, 12.6, and 13.2.

Find the first 4 terms; the 8th term, a_8; and the 12th term, a_{12}.

1. $a_n = 4n - 3$

2. $a_n = \dfrac{n-1}{n^2+1}$

Predict the general term. Answers may vary.

3. $-2, -4, -6, -8, -10, \ldots$

4. $1, 4, 9, 16, 25, \ldots$

Rename and evaluate the sum.

5. $\sum_{k=1}^{5} (-2)^k$

6. $\sum_{k=2}^{7} (1-2k)$

Rewrite using sigma notation.

7. $4 + 8 + 12 + 16 + 20$

8. $\dfrac{-1}{2} + \dfrac{1}{4} + \dfrac{-1}{8} + \dfrac{1}{16} + \dfrac{-1}{32}$

9. Find the 14th term of the arithmetic sequence $-6, 1, 8, \ldots$.

10. Find d when $a_1 = 11$ and $a_{10} = 35$. Assume an arithmetic sequence.

11. Find a_1 and d when $a_{12} = 25$ and $a_{24} = 40$. Assume an arithmetic sequence.

12. Find the sum of the first 17 terms of the arithmetic series $-8 + (-11) + (-14) + \cdots$.

13. Find the sum of all the multiples of 6 from 12 to 318 inclusive.

Solve.

14. An auditorium has 31 seats in the first row, 33 seats in the second row, 35 seats in the third row, and so on, for 18 rows. How many seats are there in the 17th row?

15. In Exercise 14, how many seats are there in all 18 rows of the auditorium?

16. Find the 20th term of the geometric sequence $2, 2\sqrt{2}, 4, \ldots$.

17. Find the common ratio of the geometric sequence $2, \frac{4}{3}, \frac{8}{9}, \ldots$.

18. Find the nth term of the geometric sequence $-2, 2, -2, \ldots$.

19. Find the nth term of the geometric sequence $3, \frac{3}{4}x, \frac{3}{16}x^2, \ldots$.

20. Find S_6 for the geometric series

$$3 + 12 + 48 + \cdots.$$

21. Find S_{12} for the geometric series

$$3x - 6x + 12x - \cdots.$$

Determine whether the infinite geometric series has a limit. If a limit exists, find it.

22. $6 + 3 + 1.5 + 0.75 + \cdots$

23. $0.04 + 0.008 + 0.0016 + \cdots$

24. $2 + (-2) + 2 + (-2) + \cdots$

25. $0.04 + 0.08 + 0.16 + 0.32 + \cdots$

26. $\$2000 + \$1900 + \$1805 + \$1714.75 + \cdots$

27. Find fractional notation for $0.555555\ldots$.

28. Find fractional notation for $0.39393939\ldots$.

Solve.

29. You take a job, starting with an hourly wage of \$11.40. You are promised a raise of 20¢ per hour every 3 months for 8 years. At the end of 8 years, what will be your hourly wage?

30. A stack of logs has 42 poles in the bottom row. There are 41 poles in the second row, 40 poles in the third row, and so on. How many poles are in the stack?

31. A student loan is in the amount of \$10,000. Interest is 7%, compounded annually, and the amount is to be paid off in 12 years. How much is to be paid back?

32. Find the total rebound distance of a ball, given that it is dropped from a height of 12 m and each rebound is one third of the preceding one.

Simplify.

33. $8!$

34. ${}_6P_2$

35. $\dfrac{9!}{3!}$

36. $\dbinom{8}{3}$

37. ${}_7C_4$

38. If 9 different signal flags are available, how many different displays are possible using 4 flags in a row?

39. The Greek alphabet contains 24 letters. How many fraternity or sorority names can be formed using 3 different letters?

40. In how many different ways can 6 books be arranged on a shelf?

41. The winner of a contest can choose any 8 of 15 prizes. How many different selections can be made?

42. A manufacturer of houses has one floor plan but achieves variety by having 3 different colored roofs, 4 different ways of attaching the garage, and 3 different types of entrance. Find the number of different houses that can be produced.

43. Find the 3rd term of $(a + b)^{20}$.

44. Expand: $(x - 2y)^4$.

45. Before an election, a poll was conducted to see which candidate was favored. Three people were running for a particular office. During the polling, 86 favored A, 97 favored B, and 23 favored C. Assuming that the poll is a valid indicator of the election, what is the probability that a person will vote for A? for B? for C?

46. What is the probability of rolling a 10 on a roll of a pair of dice? on a roll of one die?

47. From a deck of 52 cards, 1 card is drawn. What is the probability that it is a club?

48. From a deck of 52 cards, 3 are drawn at random without replacement. What is the probability that 2 are aces and 1 is a king?

Skill Maintenance

Solve.

49. $3x - y = 7,$
$2x + 3y = 5$

50. $\log (x + 5) - \log x = 1$

51. Express in terms of logarithms of a, b, c, and d:

$$\log \sqrt{\frac{a^6b^8}{c^2d^4}}.$$

52. Find the center and radius of the circle given by
$$(x - 3)^2 + (y + 4)^2 = 75.$$

Synthesis

53. ◈ Explain what happens to the terms of a geometric sequence with $|r| < 1$ as n gets larger.

54. ◈ Explain why a sequence whose terms have alternating signs cannot be an arithmetic sequence.

55. Find the sum of the first n terms of the geometric series $1 - x + x^2 - x^3 + \cdots$.

56. Expand: $(x^{-3} + x^3)^5$.

CHAPTER TEST 14

1. Find the first 5 terms and the 16th term of a sequence with general term $a_n = 6n - 5$.

2. Predict the general term of the sequence
$\frac{4}{3}, \frac{4}{9}, \frac{4}{27}, \ldots$.

3. Rename and evaluate:
$$\sum_{k=1}^{5} (3 - 2^k).$$

4. Rewrite using sigma notation:
$$1 + 8 + 27 + 64 + 125.$$

5. Find the 12th term, a_{12}, of the arithmetic sequence $9, 4, -1, \ldots$.

Assume arithmetic sequences for Questions 6 and 7.

6. Find the common difference d when $a_1 = 9$ and $a_7 = 11\frac{1}{4}$.

7. Find a_1 and d when $a_5 = 16$ and $a_{10} = -3$.

8. Find the sum of all the multiples of 12 from 24 to 240 inclusive.

9. Find the 6th term of the geometric sequence $72, 18, 4\frac{1}{2}, \ldots$.

10. Find the common ratio of the geometric sequence $22\frac{1}{2}, 15, 10, \ldots$.

11. Find the nth term of the geometric sequence $3, -9, 27, \ldots$.

12. Find the sum of the first 9 terms of the geometric series
$$(1 + x) + (2 + 2x) + (4 + 4x) + \cdots.$$

Determine whether the infinite geometric series has a limit. If a limit exists, find it.

13. $0.5 + 0.25 + 0.125 + \cdots$

14. $0.5 + 1 + 2 + 4 + \cdots$

15. $\$1000 + \$80 + \$6.40 + \cdots$

16. Find fractional notation for $0.85858585\ldots$.

17. You take a job, starting with an hourly wage of \$12.80. You are promised a raise of 20¢ per hour every 4 months for 8 years. At the end of 8 years, what will be your hourly wage?

18. A stack of poles has 52 poles in the bottom row. There are 51 poles in the second row, 50 poles in the third row, and so on. How many poles are in the stack?

19. A student loan is in the amount of \$20,000. Interest is to be 8%, compounded annually, and the amount is to be paid off in 10 years. How much is to be paid back?

20. Find the total rebound distance of a ball that is dropped from a height of 18 m, with each rebound two thirds of the preceding one.

21. Simplify: $\binom{13}{11}$.

22. In how many different ways can 6 people be seated in a row?

23. On a test, a student must answer 4 out of 7 questions. In how many ways can this be done?

24. From a group of 20 seniors and 14 juniors, how many committees consisting of 3 seniors and 2 juniors are possible?

25. Ben and Jerry's Homemade Inc., an international firm, sells ice cream in 20 flavors. How many 3-dip cones are possible if order of flavors is to be considered and no flavor is repeated?

26. Expand: $(x^2 - 3y)^5$.

27. Find the 4th term in the expansion of $(a + x)^{12}$.

28. What is the probability of getting a total of 7 on a roll of a pair of dice?

29. From a deck of 52 cards, 1 card is drawn. What is the probability of drawing an ace?

30. If 3 marbles are drawn at random all at once from a bag containing 5 green marbles, 7 red marbles, and 4 white marbles, what is the probability that 2 will be green and 1 will be white?

Skill Maintenance

31. Solve: $4^{2x-3} = 64$.

32. Find an equation of the circle with center $(1, -2)$ and radius $3\sqrt{3}$.

33. Solve:

$$y = 3x + 5,$$
$$2x + 5y = 8.$$

34. Express as a single logarithm:

$$\log_b 7 - \log_b 5.$$

Synthesis

35. Find a formula for the sum of the first n even natural numbers:

$$2 + 4 + 6 + \cdots + 2n.$$

36. Find the sum of the first n terms of

$$1 + \frac{1}{x} + \frac{1}{x^2} + \frac{1}{x^3} + \cdots.$$

CUMULATIVE REVIEW 1–14

Simplify.

1. $(-9x^2y^3)(5x^4y^{-7})$

2. $|-3.5 + 9.8|$

3. $2y - [3 - 4(5 - 2y) - 3y]$

4. $(10 \cdot 8 - 9 \cdot 7)^2 - 54 \div 9 - 3$

5. Evaluate

$$\frac{ab - ac}{bc}$$

when $a = -2$, $b = 3$, and $c = -4$.

Perform the indicated operations and simplify.

6. $(5a^2 - 3ab - 7b^2) - (2a^2 + 5ab + 8b^2)$

7. $(-3x^2 + 4x^3 - 5x - 1) + (9x^3 - 4x^2 + 7 - x)$

8. $(2a - 1)(3a + 5)$

9. $(3a^2 - 5y)^2$

10. $\dfrac{1}{x-2} - \dfrac{4}{x^2-4} + \dfrac{3}{x+2}$

11. $\dfrac{x^2 - 6x + 8}{3x + 9} \cdot \dfrac{x + 3}{x^2 - 4}$

12. $\dfrac{3x + 3y}{5x - 5y} \div \dfrac{3x^2 + 3y^2}{5x^3 - 5y^3}$

13. $\dfrac{x - \dfrac{a^2}{x}}{1 + \dfrac{a}{x}}$

Factor.

14. $4x^2 - 12x + 9$

15. $27a^3 - 8$

16. $a^3 + 3a^2 - ab - 3b$

17. $15y^4 + 33y^2 - 36$

18. For the function described by

$$f(x) = 3x^2 - 4x,$$

find $f(-2)$.

19. Divide:

$$(7x^4 - 5x^3 + x^2 - 4) \div (x - 2).$$

Solve.

20. $9(x - 1) - 3(x - 2) = 1$

21. $2x + 1 > 5$ *or* $x - 7 \leq 3$

22. $x^2 - 2x = 48$

23. $\dfrac{6}{x} + \dfrac{6}{x+2} = \dfrac{5}{2}$

24. $5x + 3y = 2,$
$3x + 5y = -2$

25. $x + y - z = 0,$
$3x + y + z = 6,$
$x - y + 2z = 5$

26. $\sqrt{x - 5} = 5 - \sqrt{x}$

27. $x^4 - 29x^2 + 100 = 0$

28. $x^2 + y^2 = 8,$
$x^2 - y^2 = 2$

29. $5^x = 8$

30. $\log(x^2 - 25) - \log(x + 5) = 3$

31. $\log_4 x = -2$

32. $7^{2x+3} = 49$

33. $|2x - 1| \leq 5$

34. $7x^2 + 14 = 0$

35. $x^2 + 4x = 3$

36. $y^2 + 3y > 10$

Solve.

37. The perimeter of a rectangle is 34 ft. The length of a diagonal is 13 ft. Find the dimensions of the rectangle.

38. A music club offers two types of memberships. Limited members pay a fee of $10 a year and can buy CD's for $10 each. Preferred members pay $20 a year and can buy CD's for $7.50 each. For what numbers of annual CD purchases would it be less expensive to be a preferred member?

39. Find three consecutive integers whose sum is 198.

40. A pentagon with all five sides the same size has a perimeter equal to that of an octagon in which all eight sides are the same size. One side of the pentagon is 2 less than 3 times one side of the octagon. What is the perimeter of each figure?

41. Swim Clean is 30% muriatic acid. Pure Swim is 80% muriatic acid. How many liters of each should be mixed together in order to get 100 L of a solution that is 50% muriatic acid?

42. An airplane can fly 190 mi with the wind in the same time it takes to fly 160 mi against the wind. The speed of the wind is 30 mph. How fast can the plane fly in still air?

43. Bianca can do a certain job in 21 min. Dahlia can do the same job in 14 min. How long would it take to do the job if the two worked together?

44. The centripetal force F of an object moving in a circle varies directly as the square of the velocity v and inversely as the radius r of the circle. If $F = 8$ when $v = 1$ and $r = 10$, what is F when $v = 2$ and $r = 16$?

45. A farmer wants to fence in a rectangular area next to a river. (Note that no fence will be needed along the river.) What is the area of the largest region that can be fenced in with 100 ft of fencing?

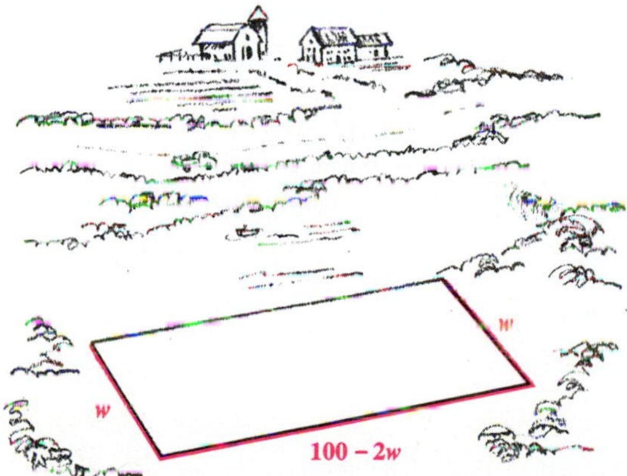

Graph.

46. $3x - y = 6$

47. $\dfrac{x^2}{25} + \dfrac{y^2}{4} = 1$

48. $y = \log_2 x$

49. $2x - 3y < -6$

50. Graph: $f(x) = -2(x - 3)^2 + 1$.

a) Label the vertex.

b) Draw the line of symmetry.

c) Find the maximum or minimum value.

51. Solve $V = P - Prt$ for r.

52. Solve $I = \dfrac{R}{R + r}$ for R.

53. Find an equation of the line containing the point $(-1, 4)$ and perpendicular to the line whose equation is $3x - y = 6$.

Evaluate.

54. $\begin{vmatrix} -5 & -7 \\ 4 & 6 \end{vmatrix}$

55. $\begin{vmatrix} 2 & -1 & 1 \\ 1 & 2 & 0 \\ 3 & -1 & 1 \end{vmatrix}$

56. Multiply $(8.9 \times 10^{-17})(7.6 \times 10^4)$. Write scientific notation for the answer.

57. Multiply and simplify: $\sqrt{8x}\,\sqrt{8x^3y}$.

58. Simplify: $(25x^{4/3}y^{1/2})^{3/2}$.

59. Divide and simplify:
$$\frac{\sqrt[3]{15x}}{\sqrt[3]{3y^2}}.$$

60. Rationalize the denominator:
$$\frac{1 - \sqrt{x}}{1 + \sqrt{x}}.$$

61. Write a single radical expression:
$$\frac{\sqrt[3]{(x + 1)^5}}{\sqrt{(x + 1)^3}}.$$

62. Multiply these complex numbers:
$$(3 + 2i)(4 - 7i).$$

63. Write a quadratic equation whose solutions are $\frac{1}{2}$ and -1.

64. Find the center and the radius of the circle
$$x^2 + y^2 + 6y - 13 = 0.$$

65. Express as a single logarithm:
$$\tfrac{2}{3} \log_a x - \tfrac{1}{2} \log_a y + 5 \log_a z.$$

66. Convert to an exponential equation: $\log_a c = 5$.

Find each of the following using a calculator or table.

67. $\log 128.3$

68. antilog (-1.4507)

69. ln 128.3

70. $\text{antilog}_e\,(-1.4507)$

Population growth. The Virgin Islands has an exponential growth rate of −1.0%. In 1992, the population was 98,942.

71. Write an exponential function describing the growth of the population of the Virgin Islands. (Note that the "growth" rate is negative; the population is actually decreasing.)

72. Predict the population for the year 2005.

73. Find the distance between the points $(-1, -5)$ and $(2, -1)$.

74. Find the 21st term of the arithmetic sequence 19, 12, 5, ...

75. Find the sum of the first 25 terms of the arithmetic series $-1 + 2 + 5 + \cdots$.

76. Find the general term of the geometric sequence 16, 4, 1, ...

77. Find the 7th term of $(a - 2b)^{10}$.

78. Find the sum of the first 9 terms of the geometric series $1 + 1.5 + 2.25 + \cdots$.

79. On Mark's 9th birthday, his grandmother opened a savings account for him with $100. The account draws 6% interest, compounded annually. If Mark neither adds to nor withdraws any money from the bank, how much will be in the account on his 18th birthday?

80. How many code words can be formed using 4 out of the 5 letters of the word "wheat" if the letters are not repeated?

81. What is the probability of drawing a heart from a well-shuffled deck of 52 cards?

Synthesis

Solve.

82. $\dfrac{9}{x} - \dfrac{9}{x + 12} = \dfrac{108}{x^2 + 12x}$

83. $\log_2 (\log_3 x) = 2$

84. y varies directly as the cube of x and x is multiplied by 0.5. What is the effect on y?

85. Divide these complex numbers:

$$\frac{2\sqrt{6} + 4\sqrt{5}i}{2\sqrt{6} - 4\sqrt{5}i}$$

86. Diaphantos, a famous mathematician, spent $\frac{1}{6}$ of his life as a child, $\frac{1}{12}$ as an adolescent, and $\frac{1}{7}$ as a bachelor. Five years after he was married, he had a son who died 4 years before his father at half his father's final age. How long did Diaphantos live?

TABLES

TABLE 1 Powers, Roots, and Reciprocals

n	n^2	n^3	$\sqrt{n}$	$\sqrt[3]{n}$	$\sqrt{10n}$	$\frac{1}{n}$	n	n^2	n^3	$\sqrt{n}$	$\sqrt[3]{n}$	$\sqrt{10n}$	$\frac{1}{n}$
1	1	1	1.000	1.000	3.162	1.0000	**51**	2,601	132,651	7.141	3.708	22.583	.0196
2	4	8	1.414	1.260	4.472	.5000	**52**	2,704	140,608	7.211	3.733	22.804	.0192
3	9	27	1.732	1.442	5.477	.3333	**53**	2,809	148,877	7.280	3.756	23.022	.0189
4	16	64	2.000	1.587	6.325	.2500	**54**	2,916	157,464	7.348	3.780	23.238	.0185
5	25	125	2.236	1.710	7.071	.2000	**55**	3,025	166,375	7.416	3.803	23.452	.0182
6	36	216	2.449	1.817	7.746	.1667	**56**	3,136	175,616	7.483	3.826	23.664	.0179
7	49	343	2.646	1.913	8.367	.1429	**57**	3,249	185,193	7.550	3.849	23.875	.0175
8	64	512	2.828	2.000	8.944	.1250	**58**	3,364	195,112	7.616	3.871	24.083	.0172
9	81	729	3.000	2.080	9.487	.1111	**59**	3,481	205,379	7.681	3.893	24.290	.0169
10	100	1,000	3.162	2.154	10.000	.1000	**60**	3,600	216,000	7.746	3.915	24.495	.0167
11	121	1,331	3.317	2.224	10.488	.0909	**61**	3,721	226,981	7.810	3.936	24.698	.0164
12	144	1,728	3.464	2.289	10.954	.0833	**62**	3,844	238,328	7.874	3.958	24.900	.0161
13	169	2,197	3.606	2.351	11.402	.0769	**63**	3,969	250,047	7.937	3.979	25.100	.0159
14	196	2,744	3.742	2.410	11.832	.0714	**64**	4,096	262,144	8.000	4.000	25.298	.0156
15	225	3,375	3.873	2.466	12.247	.0667	**65**	4,225	274,625	8.062	4.021	25.495	.0154
16	256	4,096	4.000	2.520	12.648	.0625	**66**	4,356	287,496	8.124	4.041	25.690	.0152
17	289	4,913	4.123	2.571	13.038	.0588	**67**	4,489	300,763	8.185	4.062	25.884	.0149
18	324	5,832	4.243	2.621	13.416	.0556	**68**	4,624	314,432	8.246	4.082	26.077	.0147
19	361	6,859	4.359	2.668	13.784	.0526	**69**	4,761	328,509	8.307	4.102	26.268	.0145
20	400	8,000	4.472	2.714	14.142	.0500	**70**	4,900	343,000	8.367	4.121	26.458	.0143
21	441	9,261	4.583	2.759	14.491	.0476	**71**	5,041	357,911	8.426	4.141	26.646	.0141
22	484	10,648	4.690	2.802	14.832	.0455	**72**	5,184	373,248	8.485	4.160	26.833	.0139
23	529	12,167	4.796	2.844	15.166	.0435	**73**	5,329	389,017	8.544	4.179	27.019	.0137
24	576	13,824	4.899	2.884	15.492	.0417	**74**	5,476	405,224	8.602	4.198	27.203	.0135
25	625	15,625	5.000	2.924	15.811	.0400	**75**	5,625	421,875	8.660	4.217	27.386	.0133
26	676	17,576	5.099	2.962	16.125	.0385	**76**	5,776	438,976	8.718	4.236	27.568	.0132
27	729	19,683	5.196	3.000	16.432	.0370	**77**	5,929	456,533	8.775	4.254	27.749	.0130
28	784	21,952	5.292	3.037	16.733	.0357	**78**	6,084	474,552	8.832	4.273	27.928	.0128
29	841	24,389	5.385	3.072	17.029	.0345	**79**	6,241	493,039	8.888	4.291	28.107	.0127
30	900	27,000	5.477	3.107	17.321	.0333	**80**	6,400	512,000	8.944	4.309	28.284	.0125
31	961	29,791	5.568	3.141	17.607	.0323	**81**	6,561	531,441	9.000	4.327	28.460	.0123
32	1,024	32,768	5.657	3.175	17.889	.0312	**82**	6,724	551,368	9.055	4.344	28.636	.0122
33	1,089	35,937	5.745	3.208	18.166	.0303	**83**	6,889	571,787	9.110	4.362	28.810	.0120
34	1,156	39,304	5.831	3.240	18.439	.0294	**84**	7,056	592,704	9.165	4.380	28.983	.0119
35	1,225	42,875	5.916	3.271	18.708	.0286	**85**	7,225	614,125	9.220	4.397	29.155	.0118
36	1,296	46,656	6.000	3.302	18.974	.0278	**86**	7,396	636,056	9.274	4.414	29.326	.0116
37	1,369	50,653	6.083	3.332	19.235	.0270	**87**	7,569	658,503	9.327	4.431	29.496	.0115
38	1,444	54,872	6.164	3.362	19.494	.0263	**88**	7,744	681,472	9.381	4.448	29.665	.0114
39	1,521	59,319	6.245	3.391	19.748	.0256	**89**	7,921	704,969	9.434	4.465	29.833	.0112
40	1,600	64,000	6.325	3.420	20.000	.0250	**90**	8,100	729,000	9.487	4.481	30.000	.0111
41	1,681	68,921	6.403	3.448	20.248	.0244	**91**	8,281	753,571	9.539	4.498	30.166	.0110
42	1,764	74,088	6.481	3.476	20.494	.0238	**92**	8,464	778,688	9.592	4.514	30.332	.0109
43	1,849	79,507	6.557	3.503	20.736	.0233	**93**	8,649	804,357	9.644	4.531	30.496	.0108
44	1,936	85,184	6.633	3.530	20.976	.0227	**94**	8,836	830,584	9.695	4.547	30.659	.0106
45	2,025	91,125	6.708	3.557	21.213	.0222	**95**	9,025	857,375	9.747	4.563	30.822	.0105
46	2,116	97,336	6.782	3.583	21.448	.0217	**96**	9,216	884,736	9.798	4.579	30.984	.0104
47	2,209	103,823	6.856	3.609	21.679	.0213	**97**	9,409	912,673	9.849	4.595	31.145	.0103
48	2,304	110,592	6.928	3.634	21.909	.0208	**98**	9,604	941,192	9.899	4.610	31.305	.0102
49	2,401	117,649	7.000	3.659	22.136	.0204	**99**	9,801	970,299	9.950	4.626	31.464	.0101
50	2,500	125,000	7.071	3.684	22.361	.0200	**100**	10,000	1,000,000	10.000	4.642	31.623	.0100

TABLE 2 Common Logarithms

x	0	1	2	3	4	5	6	7	8	9
1.0	.0000	.0043	.0086	.0128	.0170	.0212	.0253	.0294	.0334	.0374
1.1	.0414	.0453	.0492	.0531	.0569	.0607	.0645	.0682	.0719	.0755
1.2	.0792	.0828	.0864	.0899	.0934	.0969	.1004	.1038	.1072	.1106
1.3	.1139	.1173	.1206	.1239	.1271	.1303	.1335	.1367	.1399	.1430
1.4	.1461	.1492	.1523	.1553	.1584	.1614	.1644	.1673	.1703	.1732
1.5	.1761	.1790	.1818	.1847	.1875	.1903	.1931	.1959	.1987	.2014
1.6	.2041	.2068	.2095	.2122	.2148	.2175	.2201	.2227	.2253	.2279
1.7	.2304	.2330	.2355	.2380	.2405	.2430	.2455	.2480	.2504	.2529
1.8	.2553	.2577	.2601	.2625	.2648	.2672	.2695	.2718	.2742	.2765
1.9	.2788	.2810	.2833	.2856	.2878	.2900	.2923	.2945	.2967	.2989
2.0	.3010	.3032	.3054	.3075	.3096	.3118	.3139	.3160	.3181	.3201
2.1	.3222	.3243	.3263	.3284	.3304	.3324	.3345	.3365	.3385	.3404
2.2	.3424	.3444	.3464	.3483	.3502	.3522	.3541	.3560	.3579	.3598
2.3	.3617	.3636	.3655	.3674	.3692	.3711	.3729	.3747	.3766	.3784
2.4	.3802	.3820	.3838	.3856	.3874	.3892	.3909	.3927	.3945	.3962
2.5	.3979	.3997	.4014	.4031	.4048	.4065	.4082	.4099	.4116	.4133
2.6	.4150	.4166	.4183	.4200	.4216	.4232	.4249	.4265	.4281	.4298
2.7	.4314	.4330	.4346	.4362	.4378	.4393	.4409	.4425	.4440	.4456
2.8	.4472	.4487	.4502	.4518	.4533	.4548	.4564	.4579	.4594	.4609
2.9	.4624	.4639	.4654	.4669	.4683	.4698	.4713	.4728	.4742	.4757
3.0	.4771	.4786	.4800	.4814	.4829	.4843	.4857	.4871	.4886	.4900
3.1	.4914	.4928	.4942	.4955	.4969	.4983	.4997	.5011	.5024	.5038
3.2	.5051	.5065	.5079	.5092	.5105	.5119	.5132	.5145	.5159	.5172
3.3	.5185	.5198	.5211	.5224	.5237	.5250	.5263	.5276	.5289	.5307
3.4	.5315	.5328	.5340	.5353	.5366	.5378	.5391	.5403	.5416	.5428
3.5	.5441	.5453	.5465	.5478	.5490	.5502	.5514	.5527	.5539	.5551
3.6	.5563	.5575	.5587	.5599	.5611	.5623	.5635	.5647	.5658	.5670
3.7	.5682	.5694	.5705	.5717	.5729	.5740	.5752	.5763	.5775	.5786
3.8	.5798	.5809	.5821	.5832	.5843	.5855	.5866	.5877	.5888	.5899
3.9	.5911	.5922	.5933	.5944	.5955	.5966	.5977	.5988	.5999	.6010
4.0	.6021	.6031	.6042	.6053	.6064	.6075	.6085	.6096	.6107	.6117
4.1	.6128	.6138	.6149	.6160	.6170	.6180	.6191	.6201	.6212	.6222
4.2	.6232	.6243	.6253	.6263	.6274	.6284	.6294	.6304	.6314	.6325
4.3	.6335	.6345	.6355	.6365	.6375	.6385	.6395	.6405	.6415	.6425
4.4	.6435	.6444	.6454	.6464	.6474	.6484	.6493	.6503	.6513	.6522
4.5	.6532	.6542	.6551	.6561	.6571	.6580	.6590	.6599	.6609	.6618
4.6	.6628	.6637	.6646	.6656	.6665	.6675	.6684	.6693	.6702	.6712
4.7	.6721	.6730	.6739	.6749	.6758	.6767	.6776	.6785	.6794	.6803
4.8	.6812	.6821	.6830	.6839	.6848	.6857	.6866	.6875	.6884	.6893
4.9	.6902	.6911	.6920	.6928	.6937	.6946	.6955	.6964	.6972	.6981
5.0	.6990	.6998	.7007	.7016	.7024	.7033	.7042	.7050	.7059	.7067
5.1	.7076	.7084	.7093	.7101	.7110	.7118	.7126	.7135	.7143	.7152
5.2	.7160	.7168	.7177	.7185	.7193	.7202	.7210	.7218	.7226	.7235
5.3	.7243	.7251	.7259	.7267	.7275	.7284	.7292	.7300	.7308	.7316
5.4	.7324	.7332	.7340	.7348	.7356	.7364	.7372	.7380	.7388	.7396
x	0	1	2	3	4	5	6	7	8	9

TABLE 2 (continued)

x	0	1	2	3	4	5	6	7	8	9
5.5	.7404	.7412	.7419	.7427	.7435	.7443	.7451	.7459	.7466	.7474
5.6	.7482	.7490	.7497	.7505	.7513	.7520	.7528	.7536	.7543	.7551
5.7	.7559	.7566	.7574	.7582	.7589	.7597	.7604	.7612	.7619	.7627
5.8	.7634	.7642	.7649	.7657	.7664	.7672	.7679	.7686	.7694	.7701
5.9	.7709	.7716	.7723	.7731	.7738	.7745	.7752	.7760	.7767	.7774
6.0	.7782	.7789	.7796	.7803	.7810	.7818	.7825	.7832	.7839	.7846
6.1	.7853	.7860	.7868	.7875	.7882	.7889	.7896	.7903	.7910	.7917
6.2	.7924	.7931	.7938	.7945	.7952	.7959	.7966	.7973	.7980	.7987
6.3	.7993	.8000	.8007	.8014	.8021	.8028	.8035	.8041	.8048	.8055
6.4	.8062	.8069	.8075	.8082	.8089	.8096	.8102	.8109	.8116	.8122
6.5	.8129	.8136	.8142	.8149	.8156	.8162	.8169	.8176	.8182	.8189
6.6	.8195	.8202	.8209	.8215	.8222	.8228	.8235	.8241	.8248	.8254
6.7	.8261	.8267	.8274	.8280	.8287	.8293	.8299	.8306	.8312	.8319
6.8	.8325	.8331	.8338	.8344	.8351	.8357	.8363	.8370	.8376	.8382
6.9	.8388	.8395	.8401	.8407	.8414	.8420	.8426	.8432	.8439	.8445
7.0	.8451	.8457	.8463	.8470	.8476	.8482	.8488	.8494	.8500	.8506
7.1	.8513	.8519	.8525	.8531	.8537	.8543	.8549	.8555	.8561	.8567
7.2	.8573	.8579	.8585	.8591	.8597	.8603	.8609	.8615	.8621	.8627
7.3	.8633	.8639	.8645	.8651	.8657	.8663	.8669	.8675	.8681	.8686
7.4	.8692	.8698	.8704	.8710	.8716	.8722	.8727	.8733	.8739	.8745
7.5	.8751	.8756	.8762	.8768	.8774	.8779	.8785	.8791	.8797	.8802
7.6	.8808	.8814	.8820	.8825	.8831	.8837	.8842	.8848	.8854	.8859
7.7	.8865	.8871	.8876	.8882	.8887	.8893	.8899	.8904	.8910	.8915
7.8	.8921	.8927	.8932	.8938	.8943	.8949	.8954	.8960	.8965	.8971
7.9	.8976	.8982	.8987	.8993	.8998	.9004	.9009	.9015	.9020	.9025
8.0	.9031	.9036	.9042	.9047	.9053	.9058	.9063	.9069	.9074	.9079
8.1	.9085	.9090	.9096	.9101	.9106	.9112	.9117	.9122	.9128	.9133
8.2	.9138	.9143	.9149	.9154	.9159	.9165	.9170	.9175	.9180	.9186
8.3	.9191	.9196	.9201	.9206	.9212	.9217	.9222	.9227	.9232	.9238
8.4	.9243	.9248	.9253	.9258	.9263	.9269	.9274	.9279	.9284	.9289
8.5	.9294	.9299	.9304	.9309	.9315	.9320	.9325	.9330	.9335	.9340
8.6	.9345	.9350	.9355	.9360	.9365	.9370	.9375	.9380	.9385	.9390
8.7	.9395	.9400	.9405	.9410	.9415	.9420	.9425	.9430	.9435	.9440
8.8	.9445	.9450	.9455	.9460	.9465	.9469	.9474	.9479	.9484	.9489
8.9	.9494	.9499	.9504	.9509	.9513	.9518	.9523	.9528	.9533	.9538
9.0	.9542	.9547	.9552	.9557	.9562	.9566	.9571	.9576	.9581	.9586
9.1	.9590	.9595	.9600	.9605	.9609	.9614	.9619	.9624	.9628	.9633
9.2	.9638	.9643	.9647	.9652	.9657	.9661	.9666	.9671	.9675	.9680
9.3	.9685	.9689	.9694	.9699	.9703	.9708	.9713	.9717	.9722	.9727
9.4	.9731	.9736	.9741	.9745	.9750	.9754	.9759	.9763	.9768	.9773
9.5	.9777	.9782	.9786	.9791	.9795	.9800	.9805	.9809	.9814	.9818
9.6	.9823	.9827	.9832	.9836	.9841	.9845	.9850	.9854	.9859	.9863
9.7	.9868	.9872	.9877	.9881	.9886	.9890	.9894	.9899	.9903	.9908
9.8	.9912	.9917	.9921	.9926	.9930	.9934	.9939	.9943	.9948	.9952
9.9	.9956	.9961	.9965	.9969	.9974	.9978	.9983	.9987	.9991	.9996
x	0	1	2	3	4	5	6	7	8	9

TABLE 3 Geometric Formulas

Plane Geometry:

Rectangle
Area: $A = lw$
Perimeter: $P = 2l + 2w$

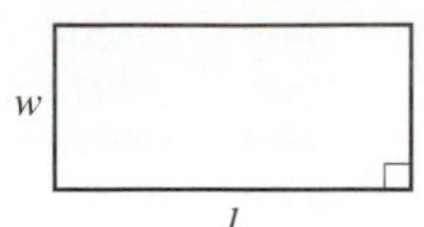

Square
Area: $A = s^2$
Perimeter: $P = 4s$

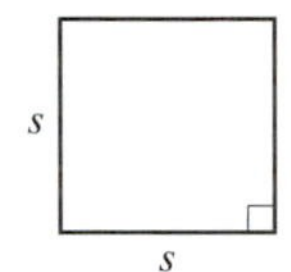

Triangle
Area: $A = \frac{1}{2}bh$

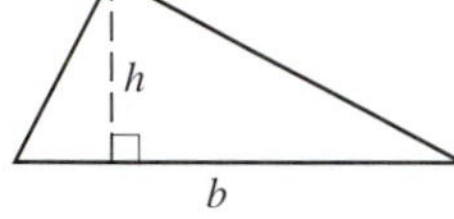

Triangle
Sum of Angle Measures:
$A + B + C = 180°$

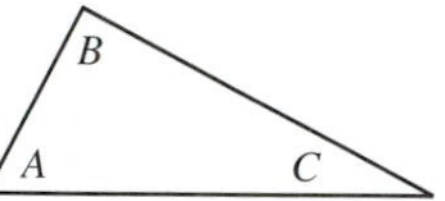

Right Triangle
Pythagorean Theorem (Equation):
$a^2 + b^2 = c^2$

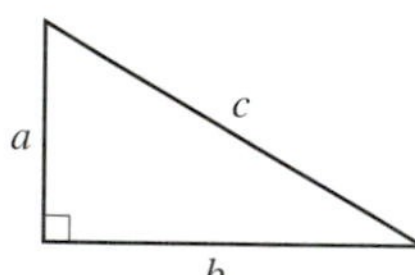

Parallelogram
Area: $A = bh$

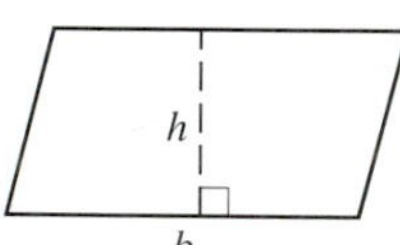

Trapezoid
Area: $A = \frac{1}{2}h(a + b)$

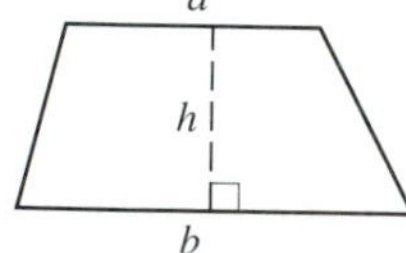

Circle
Area: $A = \pi r^2$
Circumference:
$C = \pi D = 2\pi r$
($\frac{22}{7}$ and 3.14 are different approximations for π)

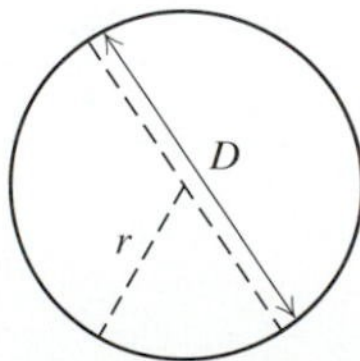

Solid Geometry:

Rectangular Solid
Volume: $V = lwh$

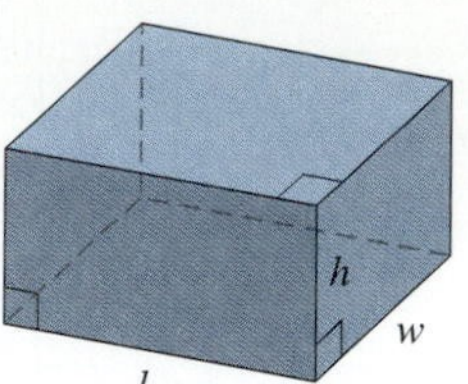

Cube
Volume: $V = s^3$

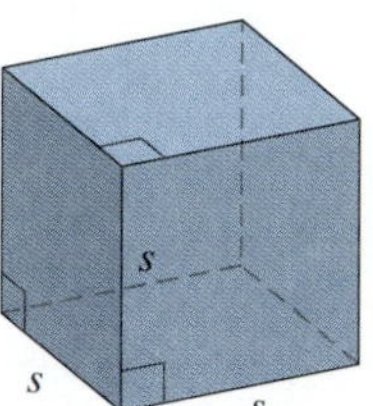

Right Circular Cylinder
Volume: $V = \pi r^2 h$
Total Surface Area:
$S = 2\pi rh + 2\pi r^2$

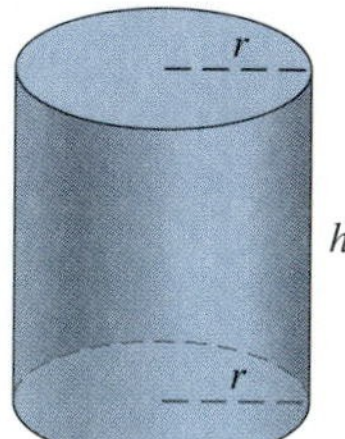
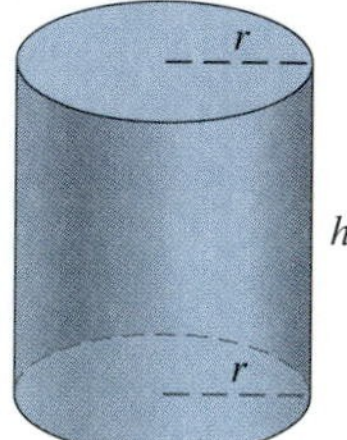

Right Circular Cone
Volume: $V = \frac{1}{3}\pi r^2 h$
Total Surface Area:
$S = \pi r^2 + \pi rs$
Slant Height:
$s = \sqrt{r^2 + h^2}$

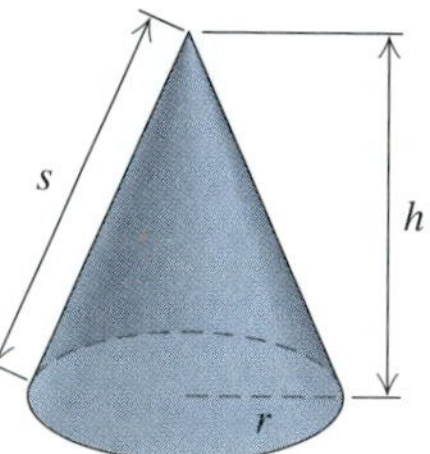

Sphere
Volume: $V = \frac{4}{3}\pi r^3$
Surface Area: $S = 4\pi r^2$

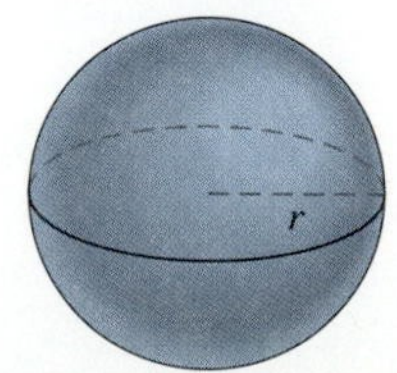

TABLE 4 Fractional and Decimal Equivalents

Fractional Notation	$\frac{1}{10}$	$\frac{1}{8}$	$\frac{1}{6}$	$\frac{1}{5}$	$\frac{1}{4}$	$\frac{3}{10}$	$\frac{1}{3}$	$\frac{3}{8}$	$\frac{2}{5}$	$\frac{1}{2}$	$\frac{3}{5}$	$\frac{5}{8}$	$\frac{2}{3}$	$\frac{7}{10}$	$\frac{3}{4}$	$\frac{4}{5}$	$\frac{5}{6}$	$\frac{7}{8}$	$\frac{9}{10}$	$\frac{1}{1}$
Decimal Notation	0.1	0.125	$0.16\overline{6}$	0.2	0.25	0.3	$0.333\overline{3}$	0.375	0.4	0.5	0.6	0.625	$0.666\overline{6}$	0.7	0.75	0.8	$0.83\overline{3}$	0.875	0.9	1
Percent Notation	10%	12.5% or $12\frac{1}{2}\%$	$16.6\overline{6}\%$ or $16\frac{2}{3}\%$	20%	25%	30%	$33.3\overline{3}\%$ or $33\frac{1}{3}\%$	37.5% or $37\frac{1}{2}\%$	40%	50%	60%	62.5% or $62\frac{1}{2}\%$	$66.6\overline{6}\%$ or $66\frac{2}{3}\%$	70%	75%	80%	$83.3\overline{3}\%$ or $83\frac{1}{3}\%$	87.5% or $87\frac{1}{2}\%$	90%	100%

ANSWERS

CHAPTER 1

Exercise Set 1.1, pp. 6–8

1. 42 **3.** 16 **5.** 1 **7.** 6 **9.** 2 **11.** $\frac{1}{2}$ **13.** 20 **15.** 220 mi **17.** 100.1 sq cm **19.** 150 sec, 450 sec, 10 min **21.** $b + 6$, or $6 + b$ **23.** $c - 9$ **25.** $6 + q$, or $q + 6$ **27.** $b + a$, or $a + b$ **29.** $y - x$ **31.** $x \div w$, or $\dfrac{x}{w}$ **33.** $n - m$ **35.** $r + s$, or $s + r$ **37.** $2x$ **39.** $\dfrac{1}{3}t$, or $\dfrac{t}{3}$ **41.** $97\%n$, or $0.97n$ **43.** $\$d - \$29.95$ **45.** Yes **47.** No **49.** No **51.** Yes **53.** No **55.** $x + 60 = 112$ **57.** $42y = 2352$ **59.** $s + 35 = 64$ **61.** $4c = \$7.96$ **63.** ◈ **65.** $y + 2x$ **67.** $2x - 3$ **69.** $b - 2$ **71.** $s + s + s + s$, or $4s$ **73.** 6 **75.** 6 **77.** $w + 4$ **79.** $t - 3$, $t + 3$

Exercise Set 1.2, pp. 13–14

1. $5 + y$ **3.** $ab + 5$ **5.** $3y + 9x$ **7.** $2(3 + a)$ **9.** tr **11.** $a5$ **13.** $5 + ba$ **15.** $(a + 3)2$ **17.** $(x + y) + 2$ **19.** $9 + (m + 2)$ **21.** $ab + (c + d)$ **23.** $(5a)b$ **25.** $6(mn)$ **27.** $(3 \cdot 2)(a + b)$ **29.** $2 + (b + a)$, $(2 + a) + b$; answers may vary **31.** $(ba)7$, $a(7b)$; answers may vary

33.

$(3a)4 = 4(3a)$	Commutative law
$= (4 \cdot 3)a$	Associative law
$= 12a$	Simplifying

35.

$5 + (2 + x) = (5 + 2) + x$	Associative law
$= x + (5 + 2)$	Commutative law
$= x + 7$	Simplifying

37. $2b + 10$ **39.** $7 + 7t$ **41.** $3x + 3$ **43.** $4 + 4y$ **45.** $30x + 12$ **47.** $7x + 28 + 42y$ **49.** $2a + 2b$ **51.** $5x + 5y + 10$ **53.** $2(x + y)$ **55.** $5(1 + y)$ **57.** $3(x + 4y)$ **59.** $5(x + 2 + 3y)$ **61.** $9(x + 1)$ **63.** $3(3x + y)$ **65.** $2(a + 8b + 32)$ **67.** $11(x + 4y + 11)$ **69.** $t - 9$ **71.** ◈ **73.** Yes; commutative law of addition **75.** Yes; distributive law and commutative laws of addition and multiplication **77.** ◈ **79.** ◈

Exercise Set 1.3, pp. 20–22

1. $4 \cdot 14$, $7 \cdot 8$ **3.** $1 \cdot 93$, $3 \cdot 31$ **5.** $2 \cdot 7$ **7.** $3 \cdot 11$ **9.** $3 \cdot 3$ **11.** $7 \cdot 7$ **13.** $2 \cdot 3 \cdot 3$ **15.** $2 \cdot 2 \cdot 2 \cdot 5$ **17.** $2 \cdot 3 \cdot 3 \cdot 5$ **19.** $2 \cdot 3 \cdot 5 \cdot 7$ **21.** Prime **23.** $7 \cdot 17$ **25.** $\frac{2}{5}$ **27.** $\frac{7}{2}$ **29.** $\frac{1}{7}$ **31.** 8 **33.** $\frac{1}{4}$ **35.** 5 **37.** $\frac{17}{21}$ **39.** $\frac{13}{7}$ **41.** $\frac{4}{3}$ **43.** $\frac{1}{8}$ **45.** $\frac{51}{8}$ **47.** 1 **49.** $\frac{7}{6}$ **51.** $\dfrac{3b}{7a}$ **53.** $\dfrac{5}{x}$ **55.** $\frac{5}{6}$ **57.** $\frac{1}{2}$ **59.** $\frac{5}{18}$ **61.** $\frac{31}{60}$ **63.** $\frac{35}{18}$ **65.** $\frac{10}{3}$ **67.** $\frac{1}{2}$ **69.** $\frac{5}{36}$ **71.** 500 **73.** $\frac{3}{40}$ **75.** $\dfrac{5b}{3a}$ **77.** $\dfrac{x - 2}{6}$ **79.** $5(3 + x)$; answers may vary **81.** ◈ **83.** $\frac{2}{3}$ **85.** $\dfrac{3sb}{2}$ **87.** $\dfrac{r}{g}$ **89.** 24 in. **91.** $\frac{28}{45}$ m^2 **93.** $\frac{20}{9}$ m **95.** ◈

Exercise Set 1.4, pp. 29–30

1. 5, -12 **3.** -170, 950 **5.** -1286, 29,028 **7.** 750, -125 **9.** 20, -150, 300 **11.** **13.**

15.

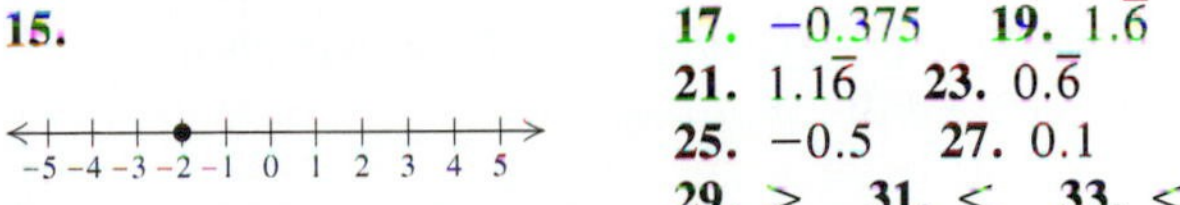

17. -0.375 **19.** $1.\overline{6}$ **21.** $1.1\overline{6}$ **23.** $0.\overline{6}$ **25.** -0.5 **27.** 0.1 **29.** $>$ **31.** $<$ **33.** $<$ **35.** $<$ **37.** $>$ **39.** $<$ **41.** $>$ **43.** $<$ **45.** $x < -6$ **47.** $y \geq -10$ **49.** True **51.** False **53.** True **55.** $-5 > x$ **57.** $120 > -20$

59. $-500{,}000 < 1{,}000{,}000$ **61.** $s \leq 95$
63. $p \leq 15{,}000$ **65.** 3 **67.** 10 **69.** 0 **71.** 24
73. $\frac{2}{3}$ **75.** 43.9 **77.** 5
79. Answers may vary. $-\frac{9}{7}$, 0, $4\frac{1}{2}$, -1.97, -491, 128, $\frac{3}{11}$, $-\frac{1}{7}$, 0.000011, $-26\frac{1}{3}$
81. Answers may vary. $-\pi$, $\sqrt{42}$, 8.4262262226 . . .
83. $\frac{3}{5}$ **85.** $5 + ab$, $ba + 5$, or $5 + ba$ **87.** ◈
89. -17, -12, 5, 13 **91.** $-\frac{4}{3}, \frac{4}{9}, \frac{4}{8}, \frac{4}{6}, \frac{4}{5}, \frac{4}{3}, \frac{4}{2}$ **93.** $>$
95. $=$ **97.** $<$ **99.** 7, -7
101. **(a)** $\frac{5}{9}$; **(b)** $\frac{1}{9}$; **(c)** $\frac{2}{9}$; **(d)** 1

Exercise Set 1.5, pp. 34–35

1. -7 **3.** -4 **5.** 0 **7.** -8 **9.** -7 **11.** -27
13. 0 **15.** -42 **17.** 0 **19.** 3 **21.** -9 **23.** 7
25. 2 **27.** -26 **29.** -22 **31.** 32 **33.** 0
35. 45 **37.** -1.8 **39.** -8.1 **41.** $-\frac{1}{5}$ **43.** $-\frac{8}{7}$
45. $-\frac{3}{8}$ **47.** $-\frac{29}{35}$ **49.** 39 **51.** 50 **53.** -1093
55. 8-yd gain **57.** 13-mb drop **59.** She owes \$85.
61. $11a$ **63.** $13x$ **65.** $11x$ **67.** $-2m$ **69.** $4a$
71. $1 - 2x$ **73.** $16 + 11x$ **75.** $16 + 9m$
77. $21z + 7y + 14$ **79.** ◈ **81.** $\$65\frac{1}{4}$ **83.** $-5y$
85. $-7m$ **87.** $3x$

Exercise Set 1.6, pp. 40–42

1. -24 **3.** 9 **5.** 26.9 **7.** -9 **9.** $\frac{14}{3}$
11. -0.101 **13.** -65 **15.** $\frac{5}{3}$ **17.** 1 **19.** -7
21. -4 **23.** -7 **25.** -6 **27.** 0 **29.** -4
31. -7 **33.** -6 **35.** 0 **37.** 0 **39.** 14 **41.** 11
43. -14 **45.** 5 **47.** -1 **49.** 18 **51.** -5
53. -3 **55.** -21 **57.** 5 **59.** -8 **61.** 12
63. -23 **65.** -68 **67.** -73 **69.** 116 **71.** 0
73. $-\frac{1}{4}$ **75.** $\frac{1}{12}$ **77.** $-\frac{17}{12}$ **79.** -2.8 **81.** -0.01
83. $-\frac{1}{2}$ **85.** $\frac{6}{7}$ **87.** $1.5 - (-3.5)$; 5
89. $79 - (114)$; 193 **91.** -54 **93.** 34
95. Negative three point two minus five point eight; -9
97. Negative two hundred thirty minus negative five hundred; 270 **99.** 37 **101.** -62 **103.** -139
105. 6 **107.** $3x$, $-2y$ **109.** -5, $3m$, $-6mn$
111. 5, $-a$, $-6b$, 2 **113.** $-5a$ **115.** $-2m - 5$
117. $-6x + 5$ **119.** $-7 - 8t$ **121.** $10x + 7$
123. $15x + 66$ **125.** $-\$330.54$ **127.** 50°C
129. 116 m **131.** 432 ft^2 **133.** ◈ **135.** True. For example, for $m = 5$ and $n = 3$, $5 > 3$ and $5 - 3 = 2 > 0$; for $m = -4$ and $n = -9$, $-4 > -9$ and $-4 - (-9) = 5 > 0$. **137.** False. For example, let $m = 2$ and $n = -2$. Then 2 and -2 are opposites, but $2 - (-2) = 4 \neq 0$.
139. ◈

Exercise Set 1.7, pp. 47–48

1. -16 **3.** -42 **5.** -24 **7.** -72 **9.** 16
11. 42 **13.** -120 **15.** -238 **17.** 1200 **19.** 98
21. -72 **23.** 21.7 **25.** $-\frac{2}{5}$ **27.** $\frac{1}{12}$ **29.** -17.01
31. $-\frac{5}{12}$ **33.** 420 **35.** $\frac{2}{7}$ **37.** -60 **39.** 150
41. 0 **43.** -720 **45.** $-30{,}240$ **47.** -6
49. -13 **51.** -2 **53.** 4 **55.** -8 **57.** 2
59. -12 **61.** -8 **63.** Undefined **65.** $-\frac{88}{9}$ **67.** 0
69. Indeterminate **71.** $\dfrac{-9}{5}, -\dfrac{9}{5}$ **73.** $\dfrac{36}{-11}, -\dfrac{36}{11}$
75. $\dfrac{-7}{3}, \dfrac{7}{-3}$ **77.** $\dfrac{x}{-2}, -\dfrac{x}{2}$ **79.** $-\dfrac{7}{3}$ **81.** $-\dfrac{13}{47}$
83. $-\dfrac{1}{10}$ **85.** $\dfrac{1}{4.3}$ **87.** $-\frac{3}{5}$ **89.** $-\frac{1}{1}$, or -1 **91.** $\frac{6}{35}$
93. $\frac{35}{12}$ **95.** $-\frac{11}{5}$ **97.** $\frac{5}{28}$ **99.** $-\frac{13}{7}$ **101.** $-\frac{9}{8}$
103. $-\frac{7}{36}$ **105.** -3 **107.** $\frac{5}{3}$ **109.** -2 **111.** $-\frac{5}{7}$
113. $-\frac{11}{9}$ **115.** $-\frac{3}{2}$ **117.** $\frac{5}{9}$ **119.** $-\frac{1}{2}$ **121.** $-\frac{1}{2}$
123. $\frac{22}{39}$ **125.** ◈ **127.** There are none.
129. Negative **131.** Positive **133.** Negative
135. Distributive law; law of opposites; multiplicative property of 0; law of opposites

Exercise Set 1.8, pp. 54–56

1. 10^3 **3.** x^7 **5.** $(3y)^4$ **7.** 16 **9.** 9 **11.** 1
13. 64 **15.** -64 **17.** 7 **19.** $16a^2$ **21.** $-343x^3$
23. 19 **25.** 86 **27.** 7 **29.** 5 **31.** 12 **33.** -8
35. -7 **37.** -7 **39.** -4 **41.** -334 **43.** 14
45. 1880 **47.** 16 **49.** 1 **51.** -26 **53.** 37
55. -6 **57.** 4 **59.** 144 **61.** 2 **63.** $-\frac{21}{38}$
65. $-\frac{4}{3}$ **67.** -8 **69.** 4 **71.** $\frac{55}{4}$ **73.** 6
75. $-2x - 7$ **77.** $-5x + 8$ **79.** $-4a + 3b - 7c$
81. $-3x^2 - 5x + 1$ **83.** $5x - 3$ **85.** $-3a + 9$
87. $5x - 6$ **89.** $-19x + 2y$ **91.** $9y - 25z$
93. $x^2 + 2$ **95.** $7x^3 - 5x$ **97.** $37a^2 - 23ab + 35b^2$
99. $-22t^3 - t^2 + 9t$ **101.** $12x + 30$ **103.** $3x^2 + 30$
105. $9x - 18$ **107.** $-4x^3 - 64$ **109.** $2x + 9$
111. ◈ **113.** $-4z$ **115.** $x - 3$ **117.** ◈
119. False **121.** False **123.** False **125.** True

Review Exercises: Chapter 1, pp. 57–58

1. [1.1] 15 **2.** [1.1] 6 **3.** [1.1] 5 **4.** [1.1] 4
5. [1.8] -15 **6.** [1.8] -5 **7.** [1.1] $z - 8$
8. [1.1] $3x$ **9.** [1.1] $\frac{1}{3}y$ **10.** [1.1] No
11. [1.1] $6x = 6768$ **12.** [1.2] $y + 2x$
13. [1.2] $x \cdot 2 + y$ **14.** [1.2] $2x + (y + z)$
15. [1.2] $4(yx)$, $(4y)x$, $y(4x)$ **16.** [1.2] $18x + 30y$
17. [1.2] $40x + 24y + 16$ **18.** [1.2] $7(3x + y)$
19. [1.2] $7(5x + 2 + y)$ **20.** [1.3] $\frac{5}{12}$ **21.** [1.3] 10
22. [1.3] $\frac{31}{36}$ **23.** [1.3] $\frac{1}{4}$ **24.** [1.3] $\frac{3}{5}$ **25.** [1.3] $\frac{72}{25}$
26. [1.4] -45, 72 **27.** [1.4] $x \leq 300$
28. [1.4]

$\frac{-1}{3}$
−5 −4 −3 −2 −1 0 1 2 3 4 5

29. [1.4] $x > -3$ **30.** [1.4] False **31.** [1.4] -0.875
32. [1.4] 1 **33.** [1.6] -5 **34.** [1.5] -3

35. [1.5] $-\frac{7}{12}$ **36.** [1.5] -4 **37.** [1.5] -5
38. [1.6] 4 **39.** [1.6] $-\frac{7}{5}$ **40.** [1.6] -7.9
41. [1.7] 54 **42.** [1.7] -9.18 **43.** [1.7] $-\frac{2}{7}$
44. [1.7] -210 **45.** [1.7] -7 **46.** [1.7] -3
47. [1.7] $\frac{3}{4}$ **48.** [1.8] 92 **49.** [1.8] 62 **50.** [1.8] 48
51. [1.8] 168 **52.** [1.8] $\frac{21}{8}$ **53.** [1.8] $\frac{103}{17}$
54. [1.5] $7a - 3b$ **55.** [1.6] $-2x + 5y$ **56.** [1.6] 7
57. [1.7] $-\frac{1}{7}$ **58.** [1.8] $(2x)^4$ **59.** [1.8] $-27y^3$
60. [1.8] $-3a + 9$ **61.** [1.8] $-2b + 21$
62. [1.8] $-3x + 9$ **63.** [1.8] $12y - 34$
64. [1.8] $5x + 24$ **65.** [1.8] $-15x + 25$
66. [1.2], [1.5], [1.8] ◈ The distributive law is used in factoring algebraic expressions, multiplying algebraic expressions, collecting like terms, finding the opposite of a sum, and subtracting algebraic expressions.
67. [1.8] ◈ A negative quantity raised to an even power is positive; a negative quantity raised to an odd power is negative.
68. [1.8] 25,281 **69.** [1.4] **(a)** $\frac{3}{11}$; **(b)** $\frac{10}{11}$
70. [1.8] $-\frac{5}{8}$ **71.** [1.8] -2.1

Test: Chapter 1, pp. 58–59

1. [1.1] 6 **2.** [1.1] $x - 9$ **3.** [1.1] 240 ft^2
4. [1.2] $q + 3p$ **5.** [1.2] $(x \cdot 4) \cdot y$ **6.** [1.2] $18 - 3x$
7. [1.2] $-5y + 5$ **8.** [1.2] $11(1 - 4x)$
9. [1.2] $7(x + 3 + 2y)$ **10.** [1.3] $2 \cdot 2 \cdot 3 \cdot 5 \cdot 5$
11. [1.4] $<$ **12.** [1.4] $>$ **13.** [1.4] $>$ **14.** [1.4] $<$
15. [1.4] 7 **16.** [1.4] $\frac{9}{4}$ **17.** [1.4] 2.7 **18.** [1.6] $-\frac{2}{3}$
19. [1.7] $-\frac{7}{4}$ **20.** [1.6] 8 **21.** [1.4] $-2 \geq x$
22. [1.6] 7.8 **23.** [1.5] -8 **24.** [1.5] $\frac{7}{40}$
25. [1.6] 10 **26.** [1.6] -2.5 **27.** [1.6] $\frac{7}{8}$
28. [1.7] -48 **29.** [1.7] $\frac{3}{16}$ **30.** [1.7] -9
31. [1.7] $\frac{3}{4}$ **32.** [1.7] -9.728 **33.** [1.8] -173
34. [1.6] 12 **35.** [1.8] -4 **36.** [1.8] 448
37. [1.6] $22y + 21a$ **38.** [1.8] $16x^4$ **39.** [1.8] $2x + 7$
40. [1.8] $9a - 12b - 7$ **41.** [1.8] $68y - 8$
42. [1.1] 15 **43.** [1.3] $\frac{23}{70}$ **44.** [1.8] 15 **45.** [1.8] $4a$

CHAPTER 2

Exercise Set 2.1, pp. 67–68

1. 4 **3.** -20 **5.** -14 **7.** -18 **9.** 15 **11.** -14
13. 2 **15.** 20 **17.** -6 **19.** $\frac{7}{3}$ **21.** $-\frac{7}{4}$ **23.** $\frac{41}{24}$
25. $-\frac{1}{20}$ **27.** 5.1 **29.** -5 **31.** 6 **33.** 9 **35.** 12
37. -40 **39.** 1 **41.** -7 **43.** -6 **45.** 6
47. -63 **49.** 36 **51.** -21 **53.** $-\frac{3}{5}$ **55.** $\frac{3}{2}$
57. $\frac{9}{2}$ **59.** 7 **61.** 4.5 **63.** -27 **65.** -12 **67.** $\frac{1}{2}$
69. -15 **71.** -2.5 **73.** $7x$ **75.** $x - 4$ **77.** ◈
79. 342.246 **81.** No solution **83.** 12, -12
85. All real numbers **87.** No solution **89.** 5
91. $\frac{b}{3a}$ **93.** $1 - c - a$ **95.** 11,074 **97.** ◈

Exercise Set 2.2, pp. 73–75

1. 5 **3.** 8 **5.** 10 **7.** 14 **9.** -8 **11.** -8
13. -7 **15.** 15 **17.** 6 **19.** 4 **21.** 6 **23.** -3
25. 1 **27.** -20 **29.** 6 **31.** 7 **33.** 2 **35.** 5
37. 2 **39.** 10 **41.** 4 **43.** 0 **45.** -1 **47.** $\frac{64}{3}$
49. $\frac{2}{5}$ **51.** -2 **53.** -4 **55.** 0.8 **57.** $-\frac{28}{27}$ **59.** 6
61. 2 **63.** 6 **65.** 8 **67.** 1 **69.** 17 **71.** $-\frac{5}{3}$
73. -3 **75.** 2 **77.** 6 **79.** $\frac{4}{7}$ **81.** 8 **83.** $\frac{11}{18}$
85. $-\frac{51}{31}$ **87.** 2 **89.** -6.5 **91.** $<$ **93.** ◈
95. 4.4233464 **97.** $-\frac{7}{2}$ **99.** $\frac{837{,}353}{1929}$ **101.** -2
103. 0 **105.** -2

Exercise Set 2.3, pp. 78–79

1. $b = \frac{A}{h}$ **3.** $r = \frac{d}{t}$ **5.** $P = \frac{I}{rt}$ **7.** $a = \frac{F}{m}$
9. $w = \frac{P - 2l}{2}$ **11.** $r^2 = \frac{A}{\pi}$ **13.** $b = \frac{2A}{h}$ **15.** $m = \frac{E}{c^2}$
17. $d = 2Q - c$ **19.** $b = 3A - a - c$ **21.** $t = \frac{3k}{v}$
23. $y = \frac{C - Ax}{B}$ **25.** $b = \frac{2A - ah}{h}$ **27.** $a = \frac{Q}{3 + 5c}$
29. $P = \frac{A}{1 + rt}$ **31.** $S = \frac{360A}{\pi r^2}$ **33.** $t = \frac{R - 3.85}{-0.0075}$
35. -42 **37.** -29 **39.** ◈ **41.** $y = \frac{z^2}{t}$
43. $t = \frac{q - rs}{r}$, or $t = \frac{q}{r} - s$ **45.** $x = \frac{a - cy}{c + b}$ **47.** ◈

Exercise Set 2.4, pp. 83–84

1. 0.76 **3.** 0.547 **5.** 1 **7.** 0.0061 **9.** 2.4
11. 0.0325 **13.** 454% **15.** 99.8% **17.** 200%
19. 7.2% **21.** 920% **23.** 0.68% **25.** 12.5%
27. 68% **29.** 75% **31.** 70% **33.** 60% **35.** $66\frac{2}{3}$%
37. 25% **39.** 24% **41.** 150 **43.** 2.5 **45.** 546
47. 125% **49.** 0.8 **51.** 5% **53.** 3.75% **55.** 27
57. 86.4% **59.** \$36 **61.** 0.92 **63.** -13.2 **65.** ◈
67. Rollie's Music: \$12.83; Sound Warp: \$12.97
69. 62.5% **71.** $c = \frac{T}{1 + r}$

Exercise Set 2.5, pp. 92–95

1. $2x - 3$ **3.** $\frac{1}{2} \cdot 7x$ **5.** $5(a + 3)$ **7.** $L + 2$
9. $b - 30\%b$, or $b - 0.3b$, or $0.7b$
11. $x + (x + 2) + (x + 4)$ **13.** $34.95 + 0.27m$
15. **(a)** $2w$; **(b)** $\frac{1}{2}l$ **17.** Second: $3x$; third: $x + 30$
19. 14 **21.** 11 **23.** 19 **25.** -10 **27.** 40
29. 20 m, 40 m, 120 m **31.** 136 and 137
33. 56, 58 **35.** 35, 36, 37 **37.** 61, 63, 65

39. $l = 160$ ft, $w = 100$ ft; 16,000 ft^2
41. $l = 27.9$ cm; $w = 21.6$ cm **43.** 22.5° **45.** \$16
47. \$4400 **49.** 450.5 mi **51.** 28°, 84°, 68°
53. 2020 **55.** $3(x - 4y + 20)$ **57.** ◈ **59.** 20
61. 19¢ **63.** 12 cm, 9 cm **65.** 30
67. 0.726175 in. **69.** ◈

Exercise Set 2.6, pp. 101–102

1. **(a)** Yes; **(b)** yes; **(c)** no; **(d)** yes; **(e)** yes
3. **(a)** No; **(b)** no; **(c)** yes; **(d)** yes; **(e)** no
5.

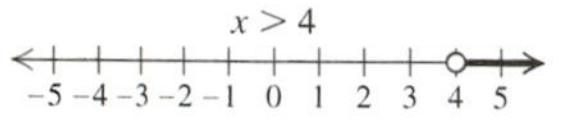

7.

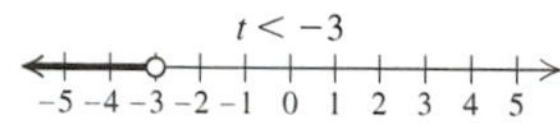

9. $m \geq -1$

11. $-3 < x \leq 4$

13. $0 < x < 3$

15. $\{x|x > -1\}$
17. $\{x|x \leq 2\}$
19. $\{x|x < -2\}$
21. $\{x|x \geq 0\}$

23. $\{y|y > 3\}$,

25. $\{x|x \leq -18\}$,

27. $\{x|x < 16\}$,

29. $\{x|x \geq 8\}$,

31. $\{y|y > -5\}$,

33. $\{x|x \leq 2\}$,

35. $\{x|x \geq 13\}$ **37.** $\{x|x < 4\}$ **39.** $\{c|c > 0\}$
41. $\left\{y|y \leq \frac{1}{4}\right\}$ **43.** $\left\{x|x > \frac{7}{12}\right\}$ **45.** $\{x|x > 0\}$

47. $\{x|x < 7\}$,

49. $\{y|y \leq 9\}$,

51. $\left\{x|x < \frac{13}{7}\right\}$,

53. $\{x|x > -3\}$,

55. $\left\{y|y \geq -\frac{2}{5}\right\}$ **57.** $\{x|x \geq -6\}$ **59.** $\{y|y \leq 4\}$
61. $\left\{x|x > \frac{17}{3}\right\}$ **63.** $\left\{y|y < -\frac{1}{14}\right\}$ **65.** $\left\{x|\frac{3}{10} \geq x\right\}$
67. $\{x|x < 8\}$ **69.** $\{y|y \geq 6\}$ **71.** $\{x|x \leq 6\}$
73. $\{x|x < -3\}$ **75.** $\{x|x \geq -2\}$ **77.** $\{y|y < -3\}$
79. $\{x|x > -3\}$ **81.** $\left\{y|y < -\frac{10}{3}\right\}$ **83.** $\{x|x > -10\}$

85. $\{y|y < 2\}$ **87.** $\{y|y \geq 3\}$ **89.** $\{y|y > -2\}$
91. $\left\{x|x < \frac{9}{5}\right\}$ **93.** $\{x|x > -4\}$ **95.** $\{n|n \geq 70\}$
97. $\{x|x \leq 9\}$ **99.** $\{y|y \leq -3\}$ **101.** $\{y|y < 6\}$
103. $\{d|d \geq 17\}$ **105.** $\left\{t|t < -\frac{5}{3}\right\}$ **107.** $\{r|r > -3\}$
109. $\{x|x \geq 8\}$ **111.** $\left\{x|x < \frac{11}{18}\right\}$ **113.** 32 **115.** ◈
117. $\left\{x|x \leq \frac{4}{7}\right\}$ **119.** $\{x|x \leq -4a\}$ **121.** $\left\{x|x > \frac{y-b}{a}\right\}$
123. **(a)** Yes; **(b)** yes; **(c)** no; **(d)** no; **(e)** no; **(f)** yes; **(g)** yes

Exercise Set 2.7, pp. 105–107

1. $x > 4$ **3.** $x \leq -6$ **5.** $t \leq 80$ **7.** $75 < a < 100$
9. $p \geq 1200$ **11.** $y \leq 500$ **13.** $3x + 2 < 13$
15. $\{s|s \geq 97\}$ **17.** $\{m|m \leq 341.4$ mi$\}$
19. $\{l|l < 21.5$ cm$\}$ **21.** $\{t|t \geq 3.5$ hr$\}$
23. $\{t|t > 1934\}$ **25.** $\{w|w > 18$ wk$\}$
27. $\{C|C < 31.1°\}$ **29.** $\{n|n > 5\}$ **31.** $\{w|w > 14$ cm$\}$
33. $\{b|b > 6$ cm$\}$ **35.** $\{c|c \geq 21\}$
37. $\{s|s \leq \$49.02\}$. The most Angelo can spend for each sweater is \$49.02. **39.** $\{l|l \geq 64$ km$\}$
41. $\{Y|Y \geq 2001\}$ **43.** -160 **45.** $4x^2 - 4x - 7$
47. ◈ **49.** $\{s|s \leq 8$ cm and s is positive$\}$
51. More than 6 hr **53.** Between 5 and 9 hr **55.** ◈

Review Exercises: Chapter 2, pp. 109–110

1. [2.1] -22 **2.** [2.1] 7 **3.** [2.1] -192 **4.** [2.1] 1
5. [2.1] $-\frac{7}{3}$ **6.** [2.1] 25 **7.** [2.1] $\frac{1}{2}$ **8.** [2.1] $-\frac{15}{64}$
9. [2.1] 9.99 **10.** [2.1] -8 **11.** [2.2] -5
12. [2.2] $-\frac{1}{3}$ **13.** [2.2] 4 **14.** [2.2] 3 **15.** [2.2] 4
16. [2.2] 16 **17.** [2.2] 6 **18.** [2.2] -3
19. [2.2] 12 **20.** [2.2] 4 **21.** [2.3] $d = \frac{C}{\pi}$
22. [2.3] $B = \frac{3V}{h}$ **23.** [2.3] $a = 2A - b$ **24.** [2.4] 0.007
25. [2.4] 44% **26.** [2.4] 20% **27.** [2.4] 360
28. [2.6] Yes **29.** [2.6] No **30.** [2.6] Yes
31. [2.6] $4x - 6 < x + 3$

32. [2.6] $-2 < x \leq 5$

33. [2.6] $y > 0$

34. [2.6] $\left\{y|y \geq -\frac{1}{2}\right\}$
35. [2.6] $\{x|x \geq 7\}$
36. [2.6] $\{y|y > 2\}$
37. [2.6] $\{y|y \leq -4\}$ **38.** [2.6] $\{x|x < -11\}$
39. [2.6] $\{y|y > -7\}$ **40.** [2.6] $\{x|x > -6\}$
41. [2.6] $\left\{x|x > -\frac{9}{11}\right\}$ **42.** [2.6] $\{y|y \leq 7\}$
43. [2.6] $\left\{x|x \geq -\frac{1}{12}\right\}$ **44.** [2.5] \$591 **45.** [2.5] 27

46. [2.5] 3 m, 5 m **47.** [2.5] 9 **48.** [2.5] 57, 59
49. [2.5] Width: 11 cm; length: 17 cm **50.** [2.5] $220
51. [2.5] $26,087 **52.** [2.5] 35°, 85°, 60°
53. [2.7] 86 **54.** [2.7] $\{w|w > 17 \text{ cm}\}$ **55.** [1.1] 5
56. [1.2] $12t + 8 + 4s$ **57.** [1.7] -2.3
58. [1.8] $-43x + 8y$
59. [2.1], [2.6] ◈ Multiplying on both sides of an equation by *any* number results in an equivalent equation. When multiplying on both sides of an inequality, the sign of the number being multiplied by must be considered. If the number is 0, there is no equivalent inequality; if the number is positive, the direction of the inequality symbol remains unchanged; if the number is negative, the direction of the inequality symbol must be reversed to produce an equivalent inequality.
60. [2.1], [2,6] ◈ The solutions to an equation can each be checked. The solutions to an inequality are too numerous to check. Checking a few numbers from the solution set found cannot guarantee the answer is correct, although if any number does not check, the answer found is incorrect.
61. [2.5] Amazon: 6437 km; Nile: 6671 km
62. [2.5] $14,150 **63.** [1.4], [2.2] 23, -23
64. [1.4], [2.1] 20, -20 **65.** [2.3] $a = \dfrac{y-3}{2-b}$

Test: Chapter 2, pp. 110–111

1. [2.1] 8 **2.** [2.1] 26 **3.** [2.1] -6 **4.** [2.1] 49
5. [2.2] -12 **6.** [2.2] 2 **7.** [2.1] -8 **8.** [2.1] $-\frac{7}{20}$
9. [2.2] 7 **10.** [2.2] -5 **11.** [2.2] 2.5
12. [2.6] $\{x|x \leqslant -4\}$ **13.** [2.6] $\{x|x > -13\}$
14. [2.6] $\{x|x \leqslant 5\}$ **15.** [2.6] $\{y|y \leqslant -13\}$
16. [2.6] $\{y|y \geqslant 8\}$ **17.** [2.6] $\left\{x|x \leqslant -\frac{1}{20}\right\}$
18. [2.6] $\{x|x < -6\}$ **19.** [2.6] $\{x|x \leqslant -1\}$
20. [2.3] $r = \dfrac{A}{2\pi h}$ **21.** [2.3] $l = \dfrac{2w - P}{-2}$ **22.** [2.4] 2
23. [2.4] 5.4% **24.** [2.4] 21 **25.** [2.4] 44%

26. [2.6]
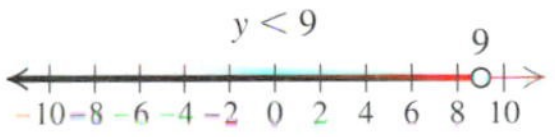

27. [2.6]
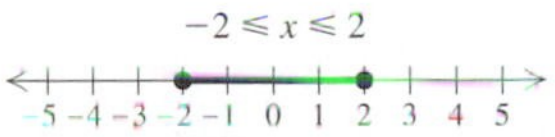

28. [2.5] Width: 7 cm; length: 11 cm **29.** [2.5] 6
30. [2.5] 81, 83, 85 **31.** [2.5] $2500 **32.** [2.7] $\{x|x > 6\}$ **33.** [2.7] $\{l|l \geqslant 174 \text{ yd}\}$ **34.** [1.1] $x - 10$
35. [1.2] $3(a + 8b + 4)$ **36.** [1.7] $\frac{3}{10}$ **37.** [1.8] 5
38. [2.3] $d = \dfrac{ca - 1}{c}$ **39.** [1.4], [2.2] 15, -15
40. [2.5] 60

CHAPTER 3

Exercise Set 3.1, pp. 119–121

1. 3 **3.** 120 lb **5.** Boredom **7.** Approximately 5%
9. 1 **11.** The second month **13.** 12% **15.** $672
17. 4% **19.** 1982 **21.** 1953, 1963, 1970 **23.** 1955
25. Approximately 5% **27.** 3.7% **29.** 70.1%
31. 1743; 360; 270

33.
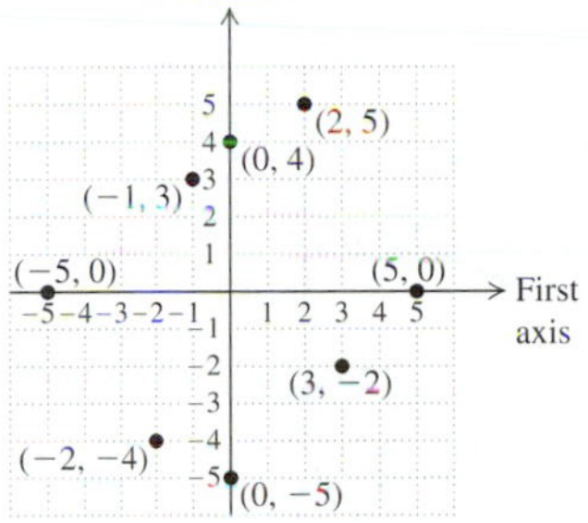

35. II **37.** IV **39.** III **41.** I
43. Negative, negative
45. *A*: (3, 3), *B*: (0, −4), *C*: (−5, 0), *D*: (−1, −1), *E*: (2, 0) **47.** $\frac{2}{3}$ **49.** $-\frac{13}{35}$ **51.** ◈ **53.** I or IV
55. I or III **57.** (−1, −5)

59.
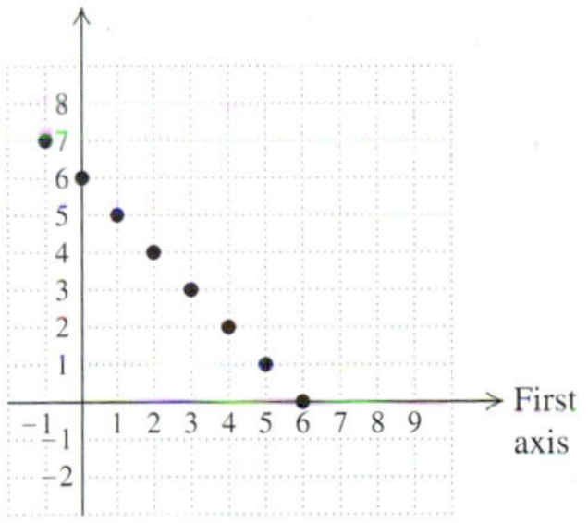

Answers may vary

61. 26
63. Latitude 32.5° North, longitude 64.5° West

Technology Connection, Section 3.2

TC1.
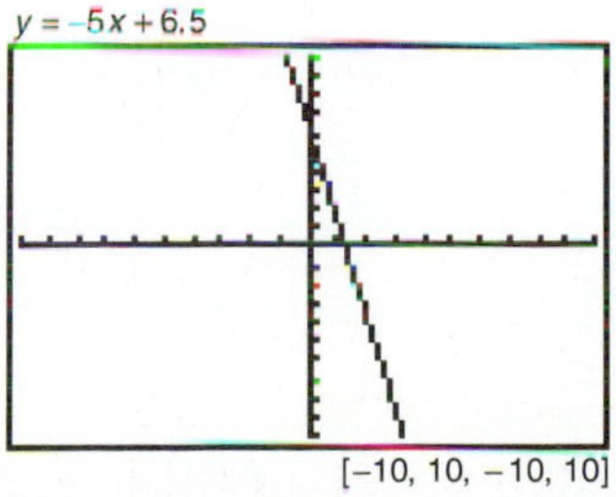

TC2. $y = 3x - 4.5$

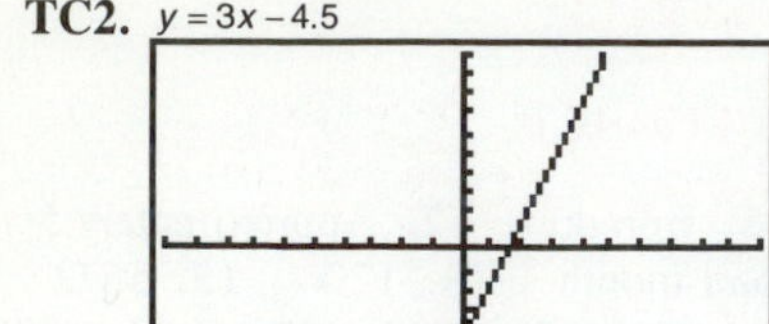

[−10, 10, −10, 10]

TC3. 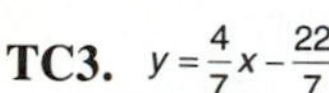 $y = \frac{4}{7}x - \frac{22}{7}$

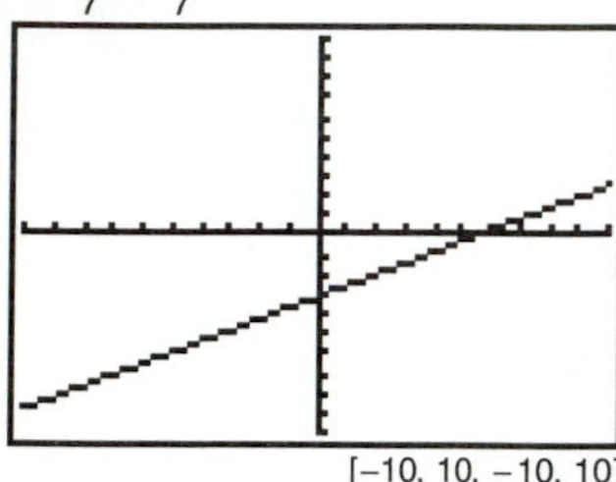

[−10, 10, −10, 10]

TC4. $y = -\frac{11}{5}x - 4$

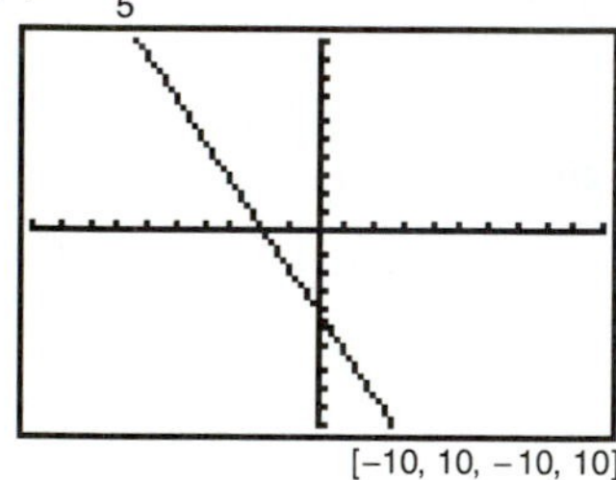

[−10, 10, −10, 10]

Exercise Set 3.2, p. 128–129

1. Yes **3.** No **5.** No

7.

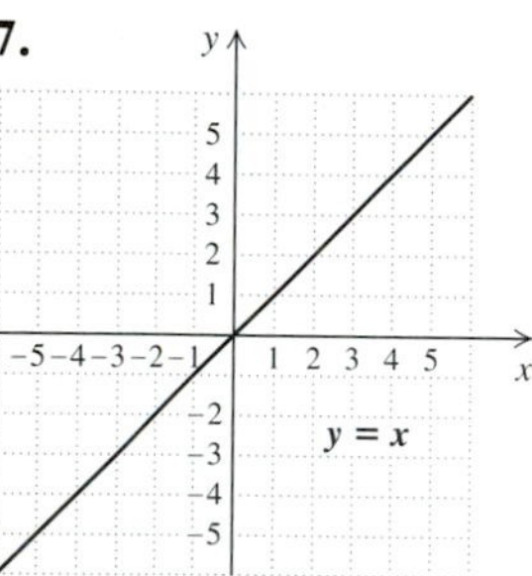

9.

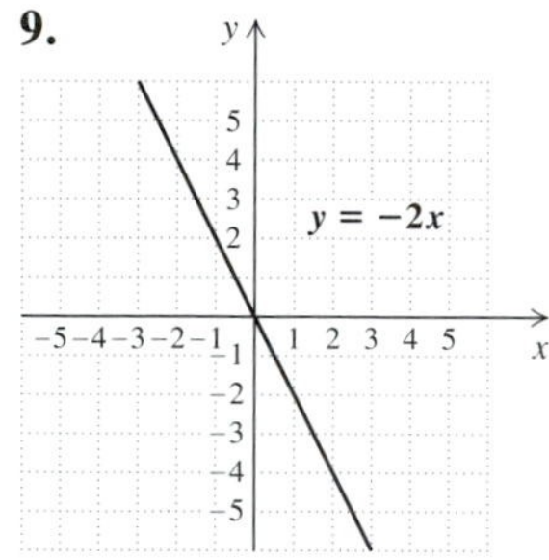

11.

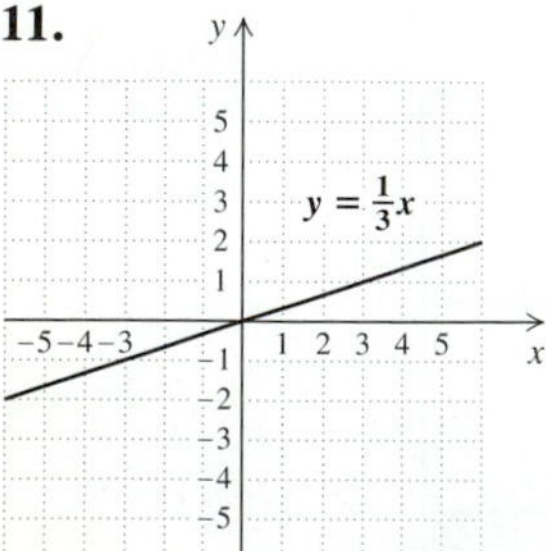

13.

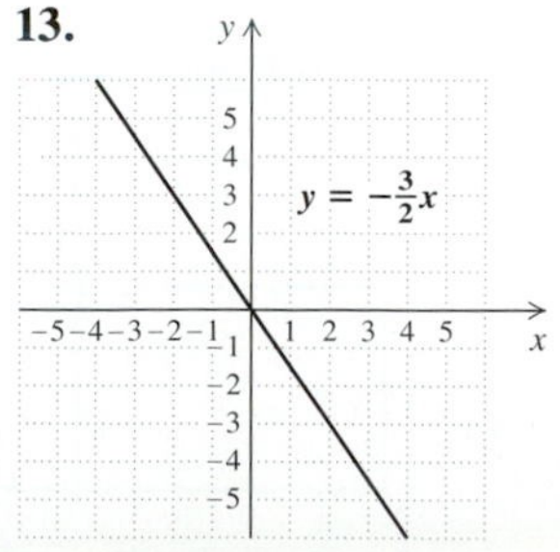

15.

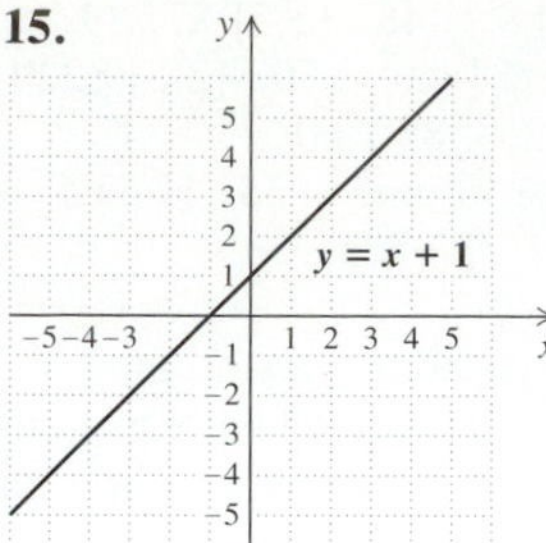

17.

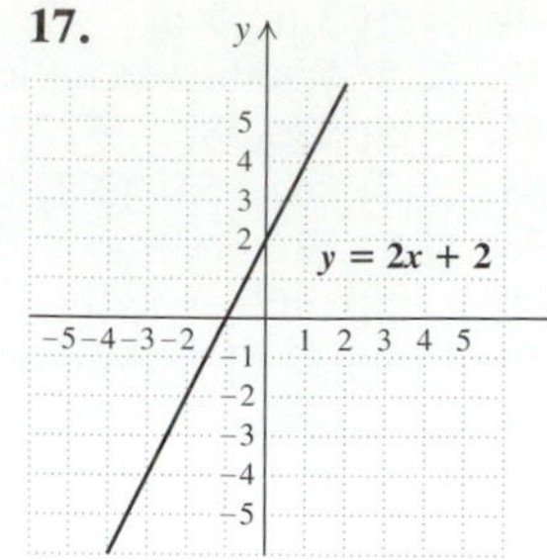

19.

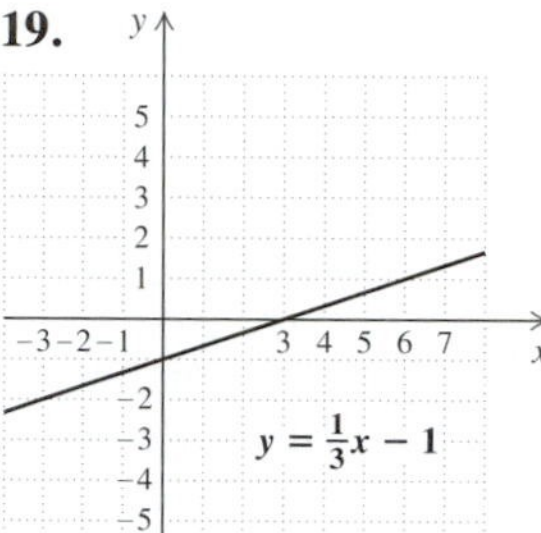

21.

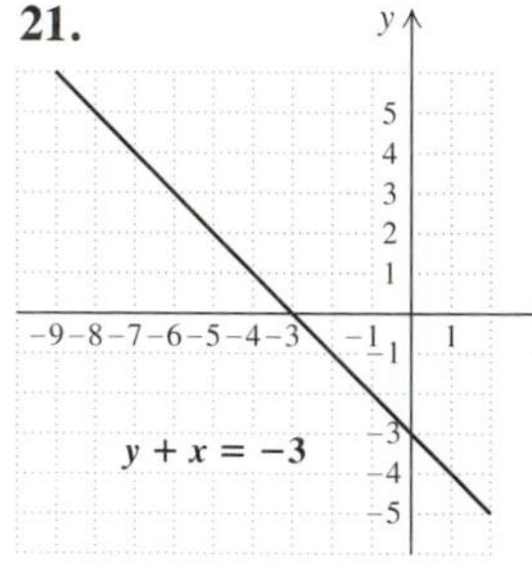

23.

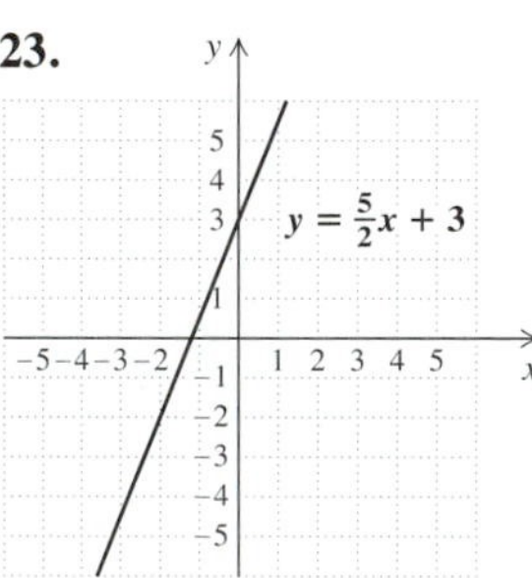

25.

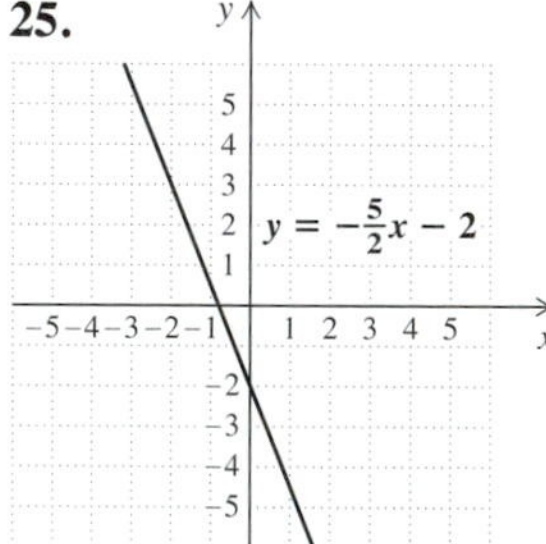

27.

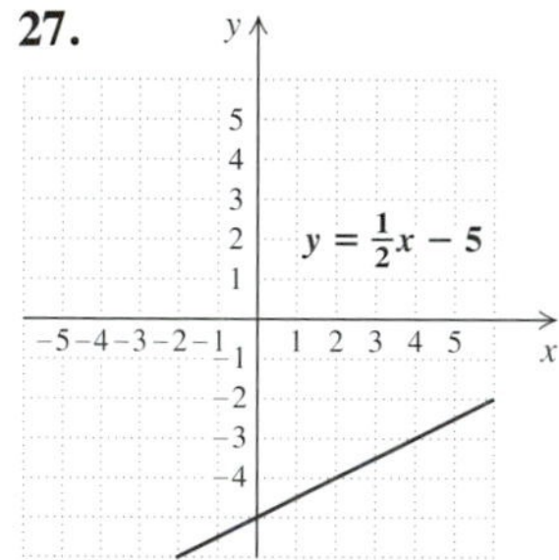

29.

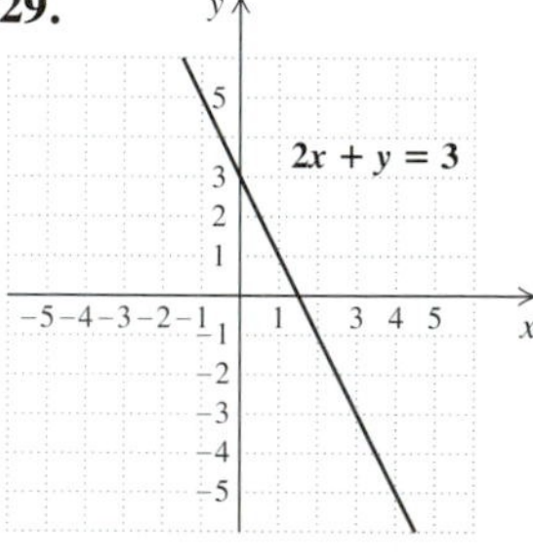

31.

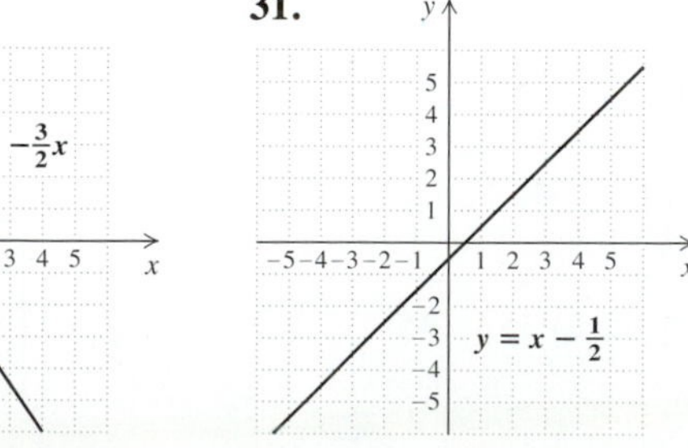

33.

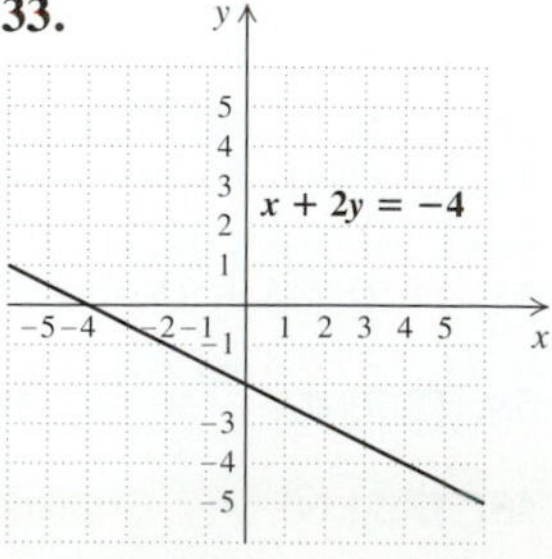

35.

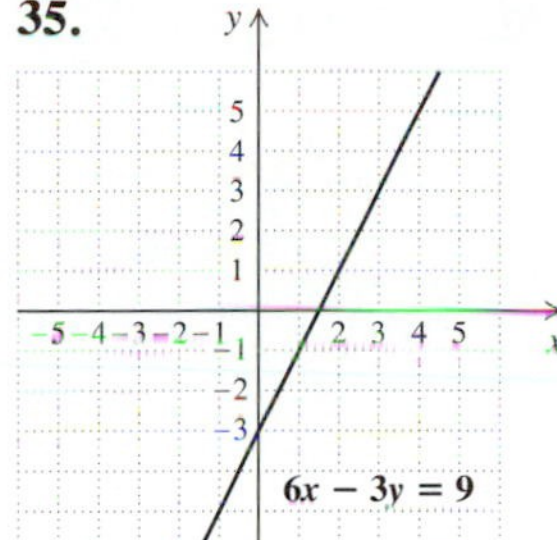

37.

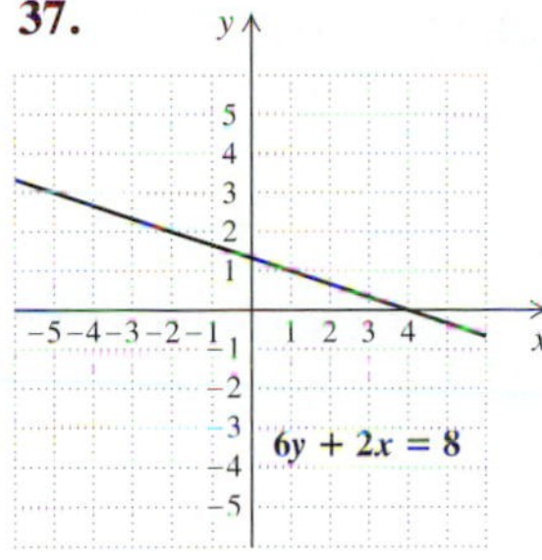

39. −9 **41.** $y = \dfrac{C - Ax}{B}$ **43.** ◈

45.

x	0	−1	1	−2	2	−3	3
y	1	2	2	5	5	10	10

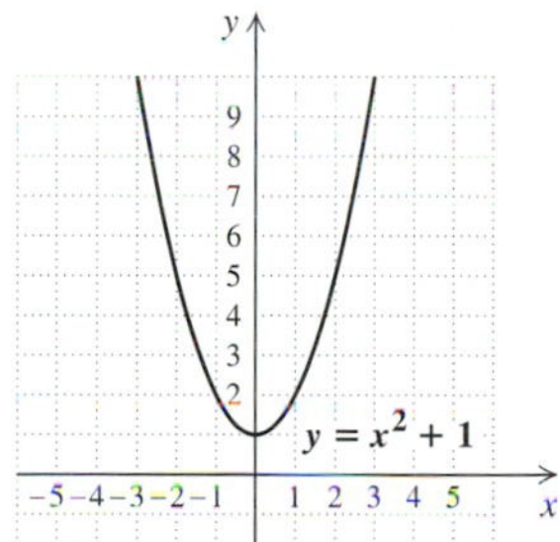

47. (15, 0), (12, 1), (9, 2), (6, 3), (3, 4), (0, 5)
49. $5n + 25q = 235$; (27, 4), (7, 8), (47, 0), answers may vary

51. $y = -2.8x + 3.5$

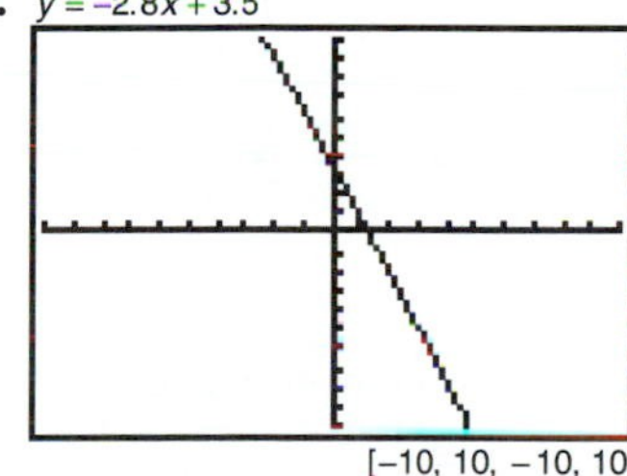

[−10, 10, −10, 10]

53. $y = \frac{2}{7}x - \frac{24}{5}$

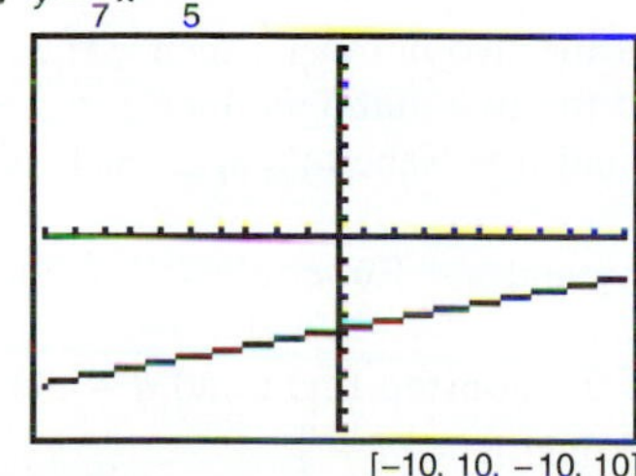

[−10, 10, −10, 10]

55. ◈

Exercise Set 3.3, p. 134–135

1. **(a)** (0, 3); **(b)** (4, 0) **3.** **(a)** (0, 5); **(b)** (−3, 0)
5. **(a)** (0, 4); **(b)** (10, 0) **7.** **(a)** (0, −8); **(b)** (6, 0)
9. **(a)** (0, 8); **(b)** $(-\frac{4}{3}, 0)$ **11.** **(a)** (0, 2); **(b)** $(-\frac{2}{3}, 0)$

13.

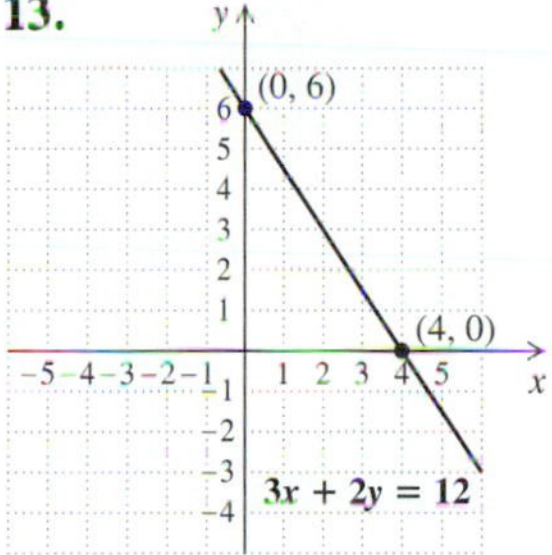

15.

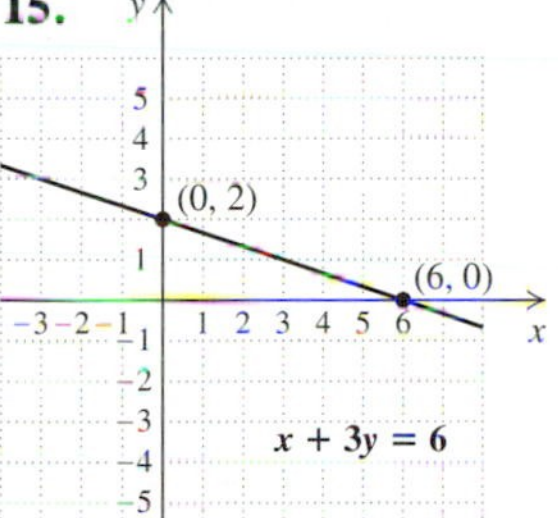

17.

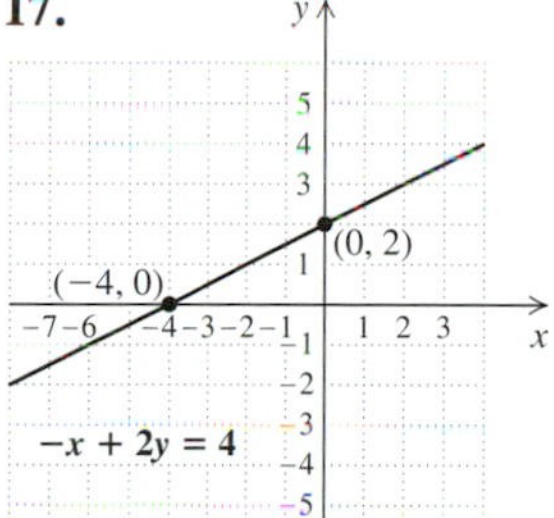

19.

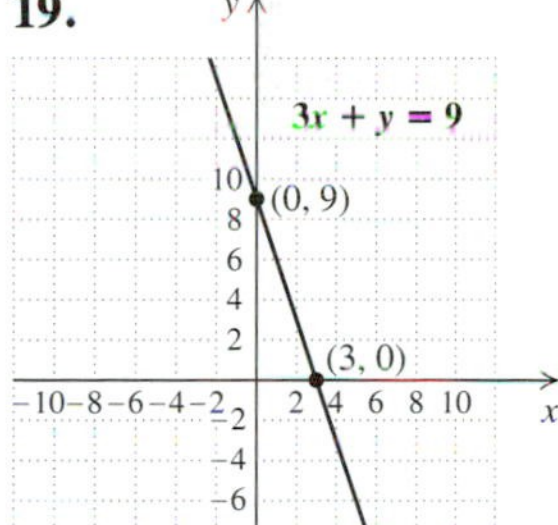

21.

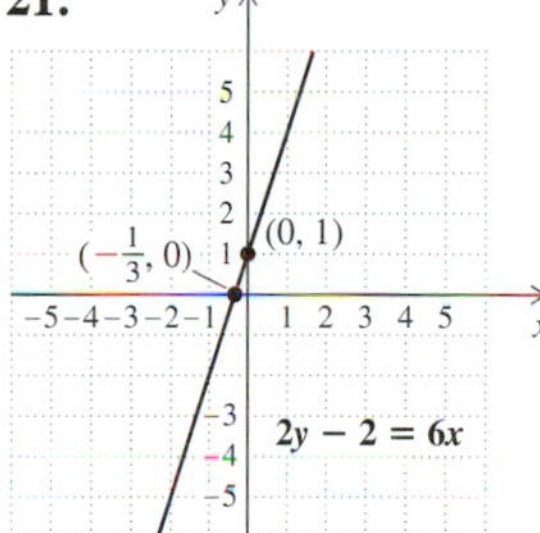

23.

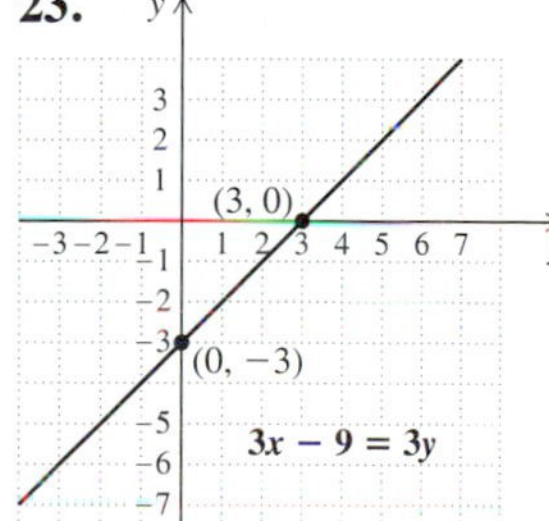

25.

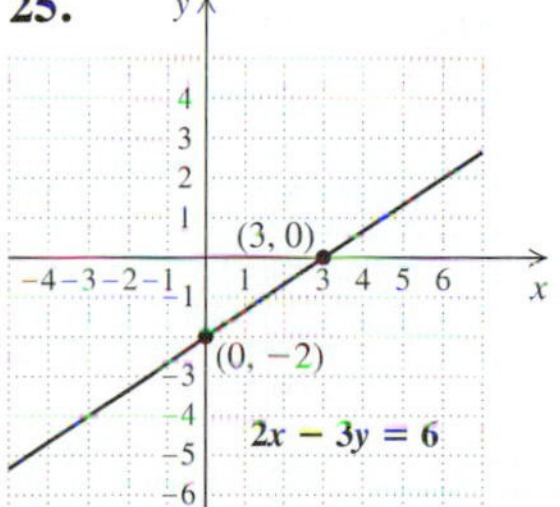

27.

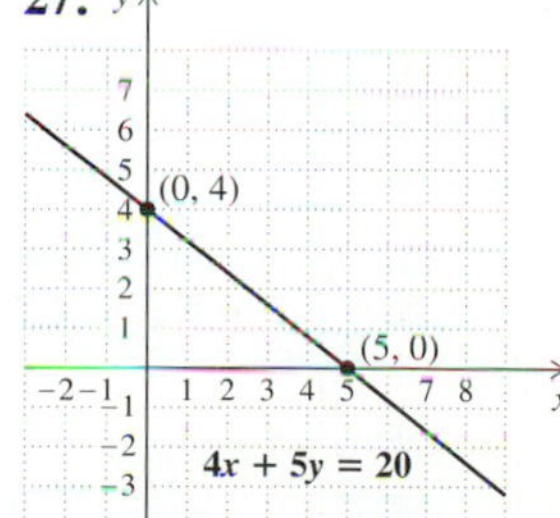

29.

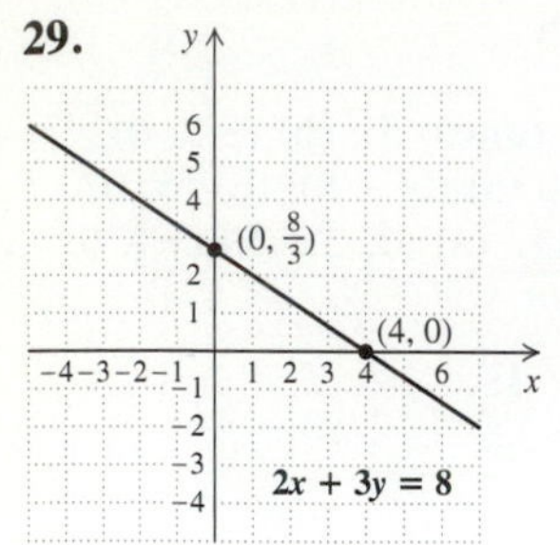

31.

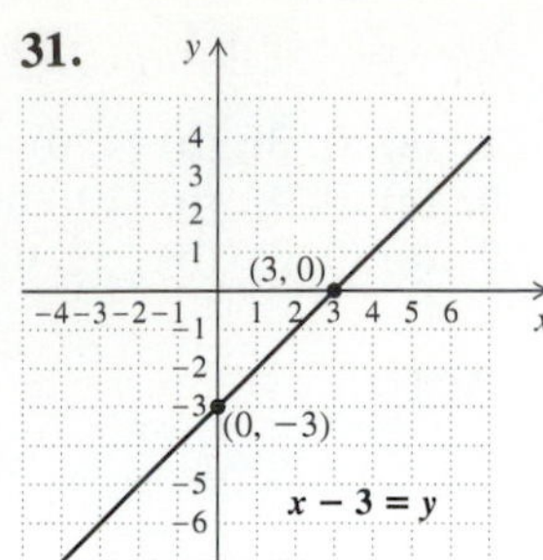

49.

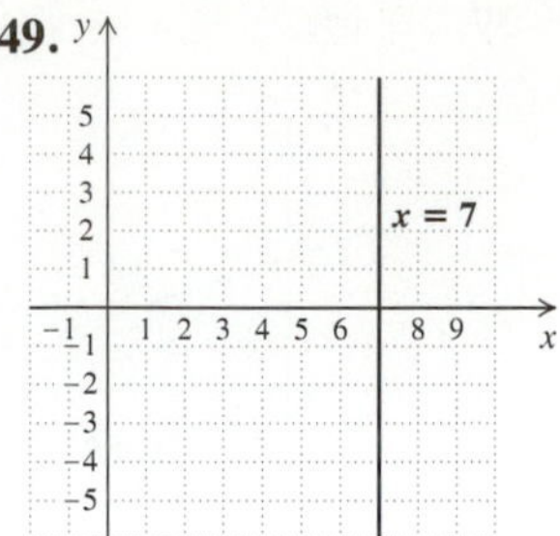

51.

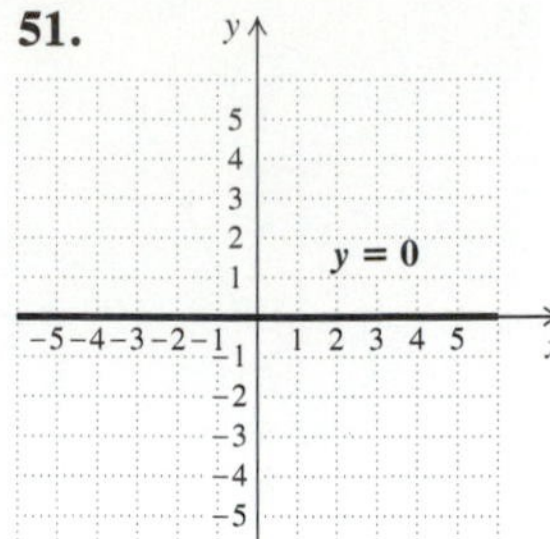

33.

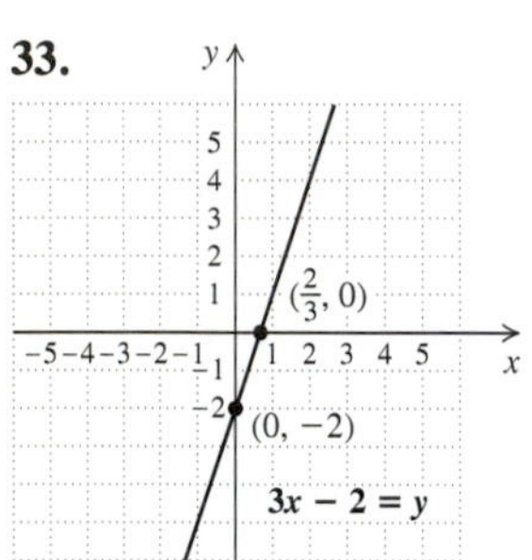

35.

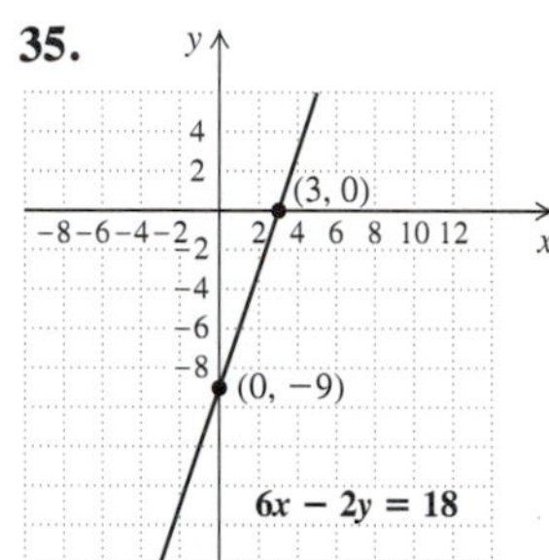

53.

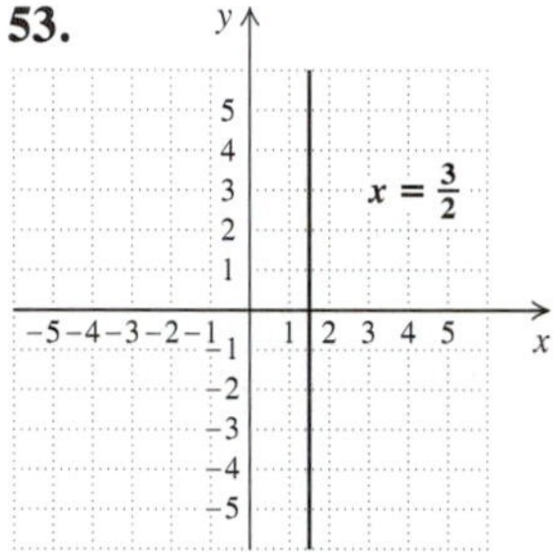

55.

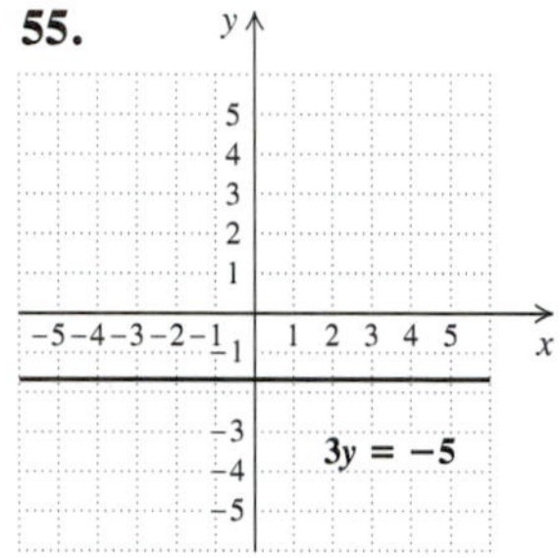

37.

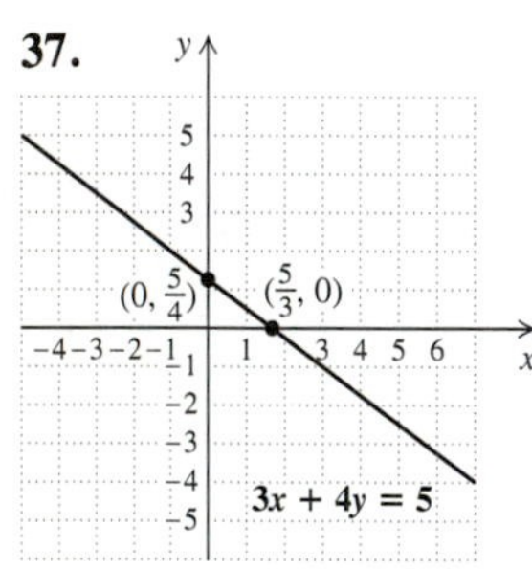

39.

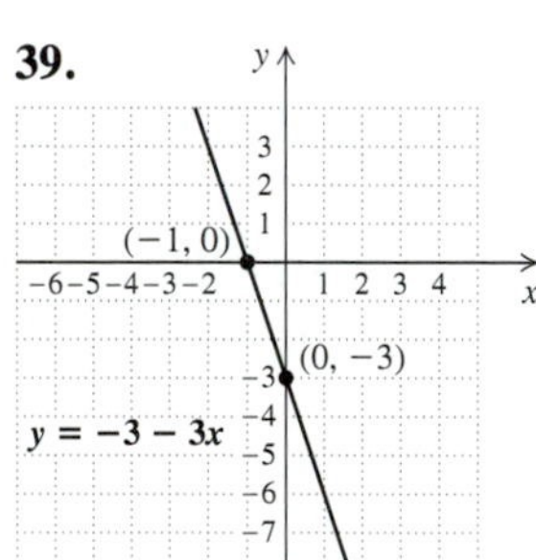

57.

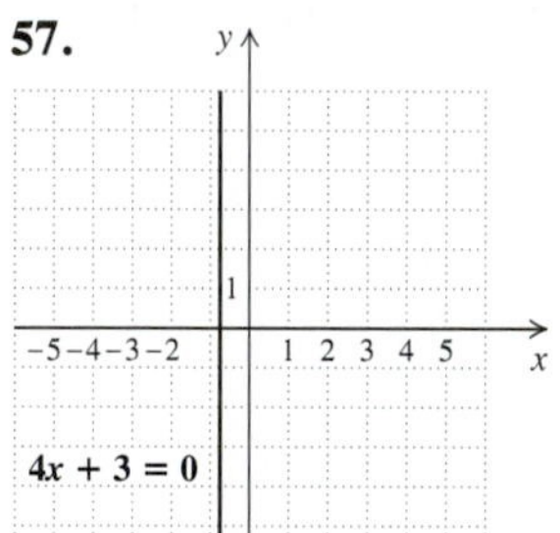

59.

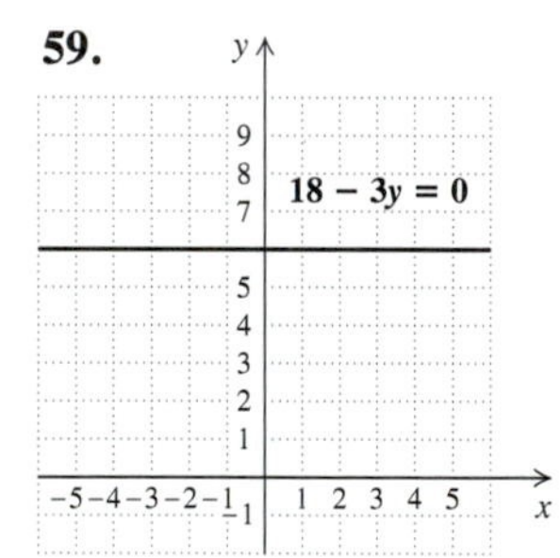

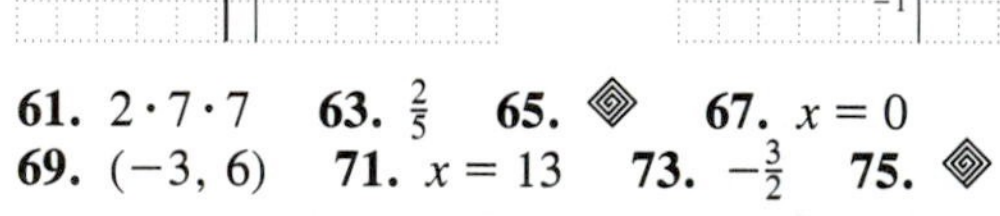

61. $2 \cdot 7 \cdot 7$ **63.** $\frac{2}{5}$ **65.** ◈ **67.** $x = 0$
69. $(-3, 6)$ **71.** $x = 13$ **73.** $-\frac{3}{2}$ **75.** ◈

41.

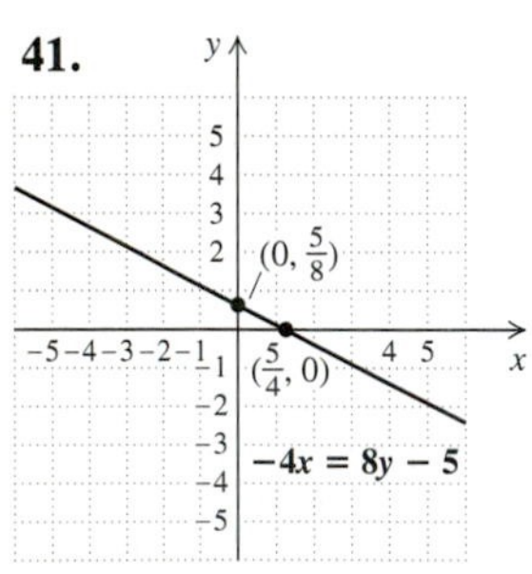

43.

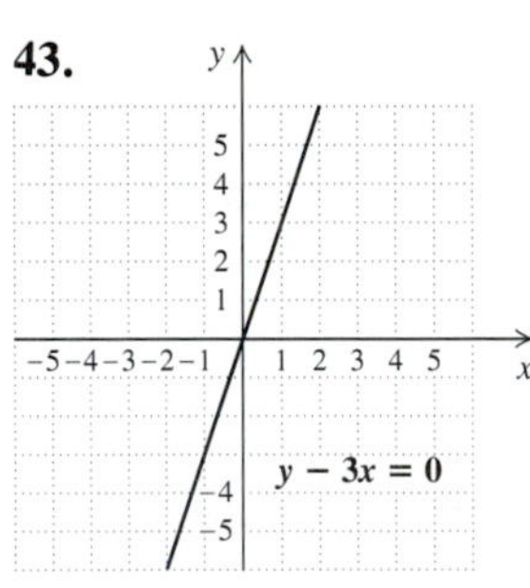

45.

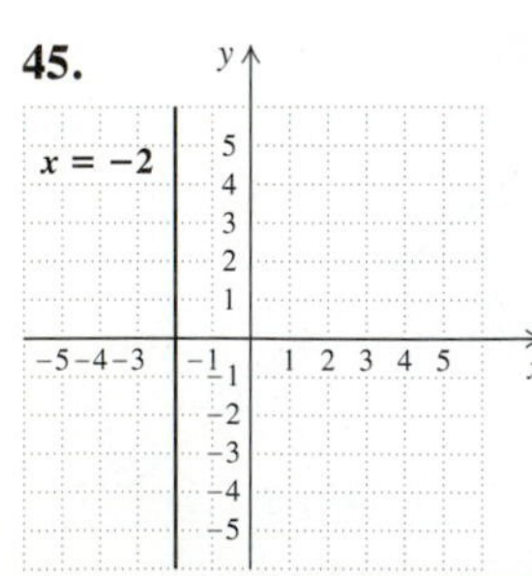

47.

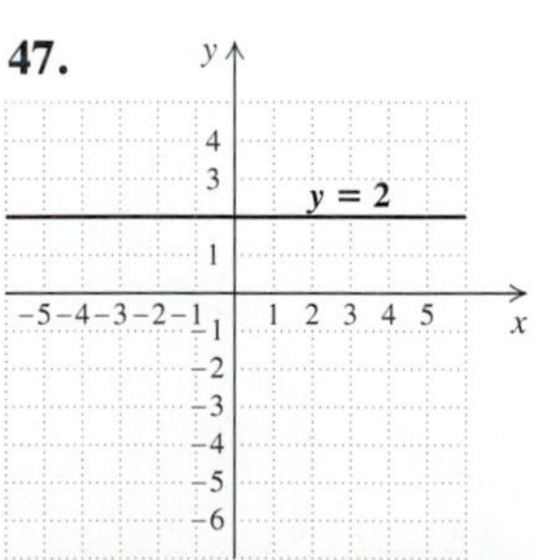

Technology Connection, Section 3.4

TC1. \$116.45 **TC2.** 17,750 ft **TC3.** \$327.50

Exercise Set 3.4, pp. 139–142

1. Let x and y represent the two numbers; then $x + y = 27$.
3. Let x and y represent the two numbers; then $x + 2y = 65$.
5. Let x and y represent the two numbers; then $x = 3y$.
7. Let x and y represent the two numbers; then $x = y + 5$.
9. Let h = Hank's age and n = Nanette's age; then $h + 7 = 2n$.
11. Let x = Lois's salary and y = Roberta's salary; then $x = 3y + 170$.
13. Let n = the time of the nonstop flight and d = the time of the direct flight; then $n = \frac{1}{2}d + \frac{3}{4}$.
15. Let p = the cost of a pizza and s = the cost of a sandwich; then $3p + 2s = 37$.

17. **(a)** 55 mi, 110 mi, 275 mi, 550 mi; **(b)**

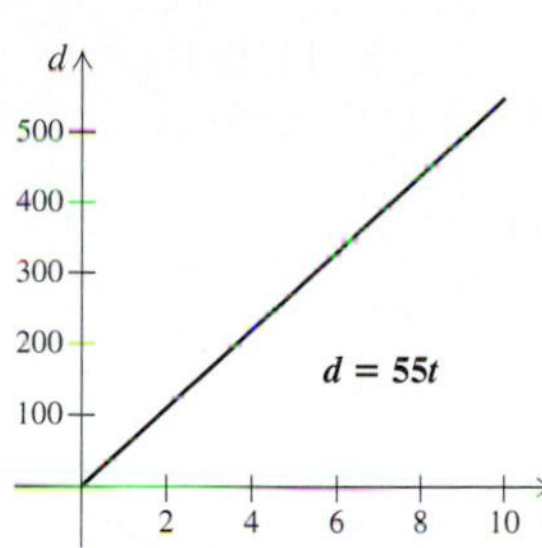

19. **(a)** 2, 3, 4, 5, 6; **(b)**

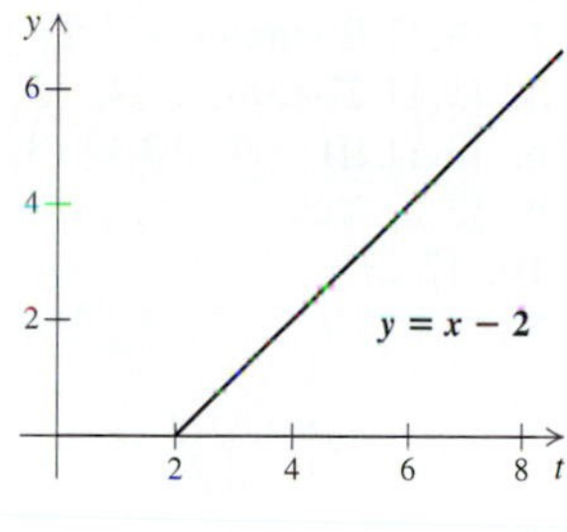

21. **(a)** $12\frac{1}{2}$ times; **(b)**

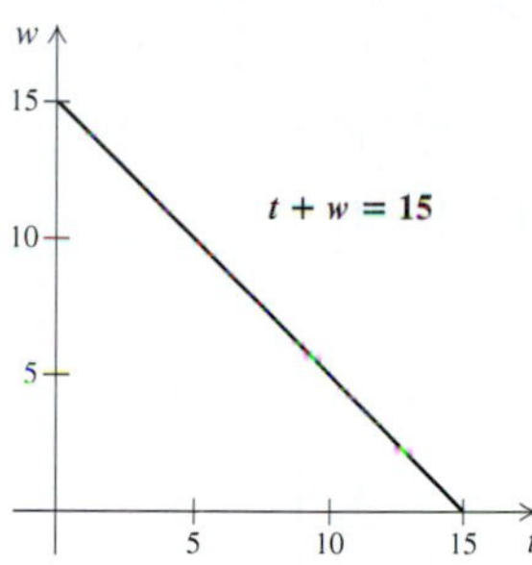

23.

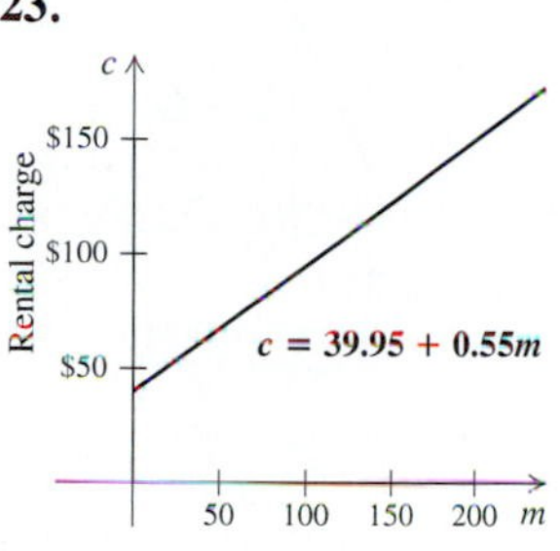

Approximately \$140

25.

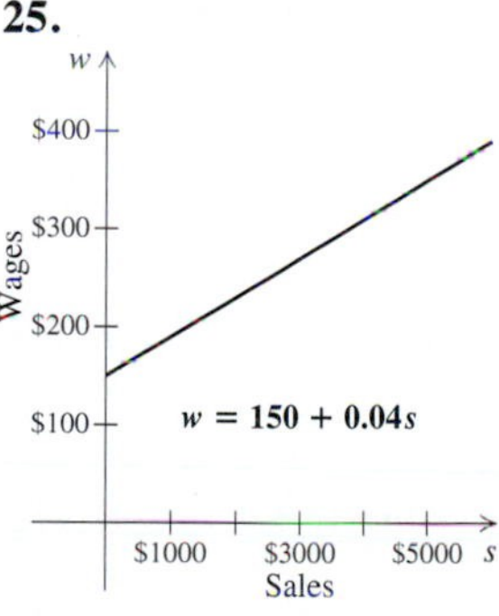

\$330

27.

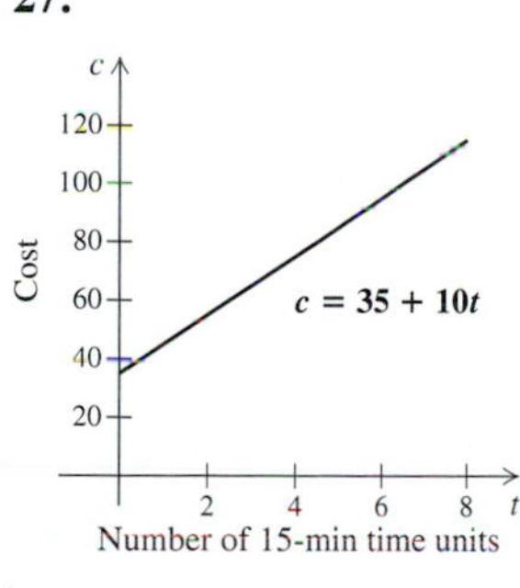

\$95

29.

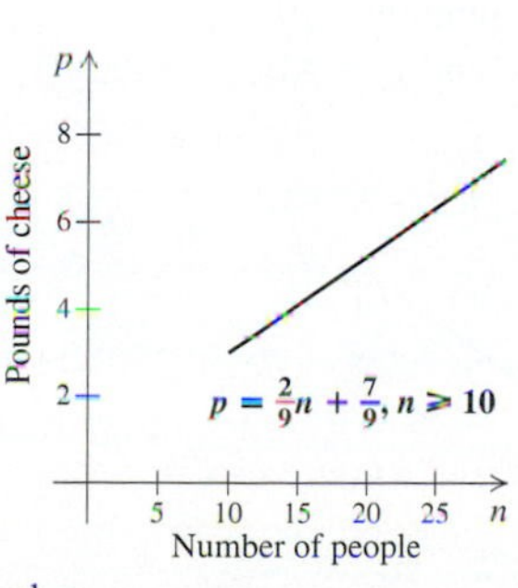

$5\frac{1}{2}$ lb

31.

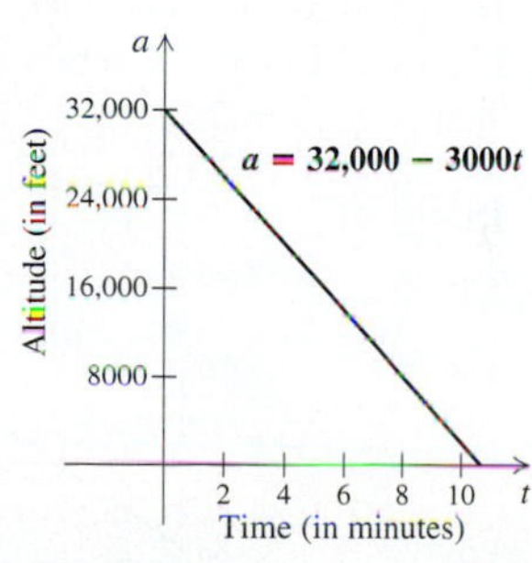

8000 ft

33. $t = \dfrac{s - d}{v}$ **35.** ◈

37.

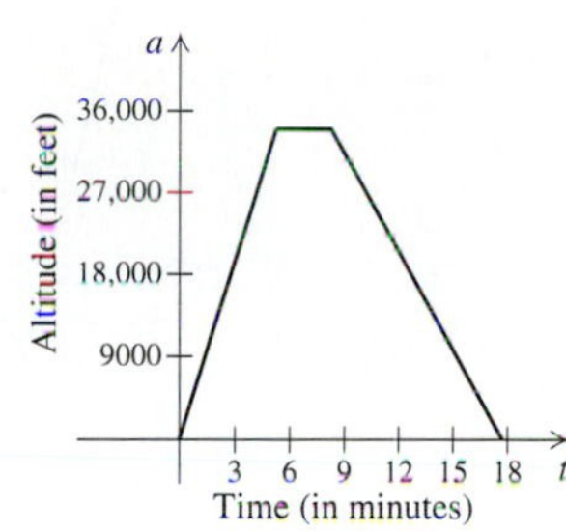

39.

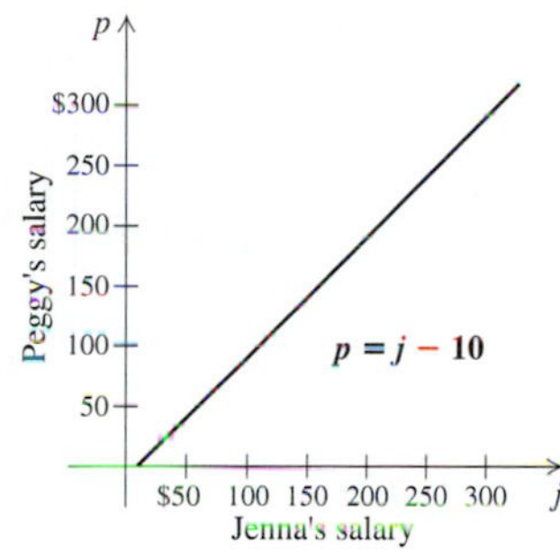

41. \$280.13

Review Exercises: Chapter 3, p. 143

1. [3.1] 11% **2.** [3.1] Years 3 and 4

3.–5. [3.1]

(2, 5)
(−4, −2)
(0, −3)

6. [3.1] IV **7.** [3.1] III **8.** [3.1] I
9. [3.1] (−5, −1) **10.** [3.1] (−2, 5)
11. [3.1] (3, 0) **12.** [3.2] No **13.** [3.2] Yes
14. [3.2] **15.** [3.2]

$y = 2x - 5$

$y = -\frac{3}{4}x$

16. [3.2]

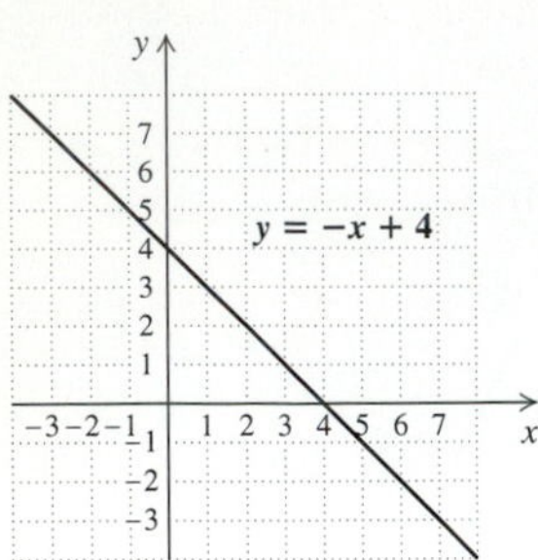

17. [3.2]

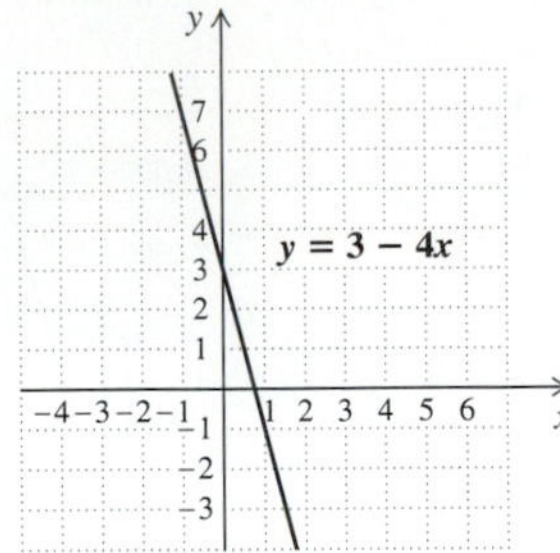

18. [3.3]

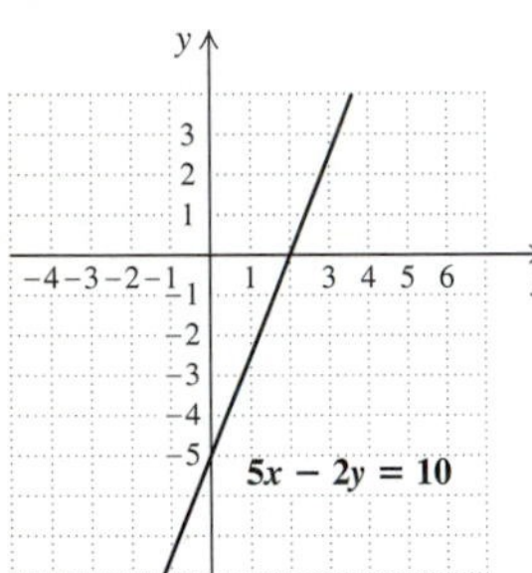

19. [3.3]

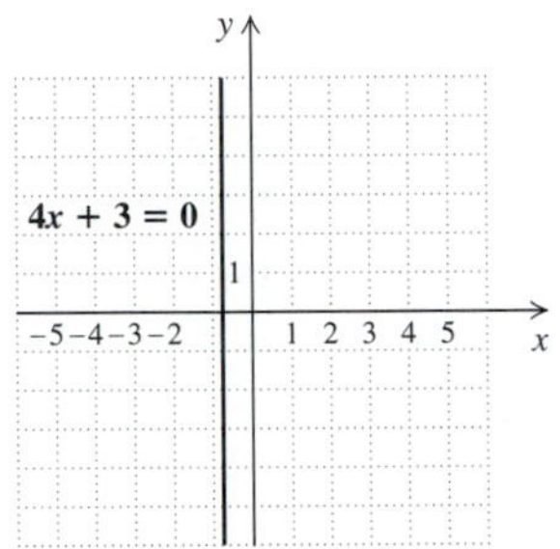

20. [3.3] x-intercept: (6, 0); y-intercept: (0, −3)
21. [3.3] x-intercept: (13.5, 0); y-intercept: (0, 9)
22. [3.4] Let $x =$ the first number and $y =$ the second; then $x = 2y - 3$.
23. [3.4] **(a)** 2.93 ft, 117.2 ft, 293 ft; **(b)**

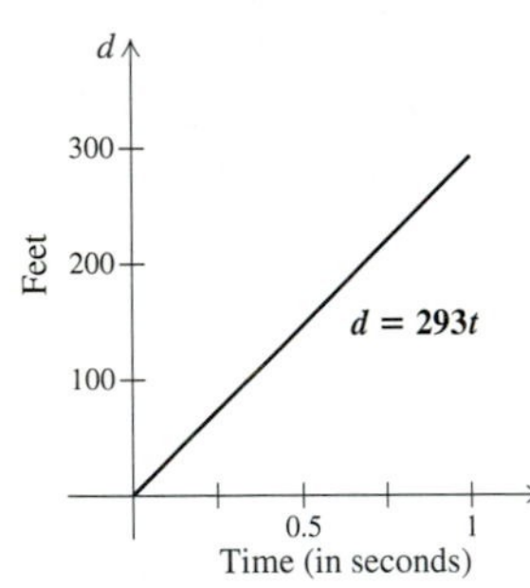

24. [3.4], Approx. $15

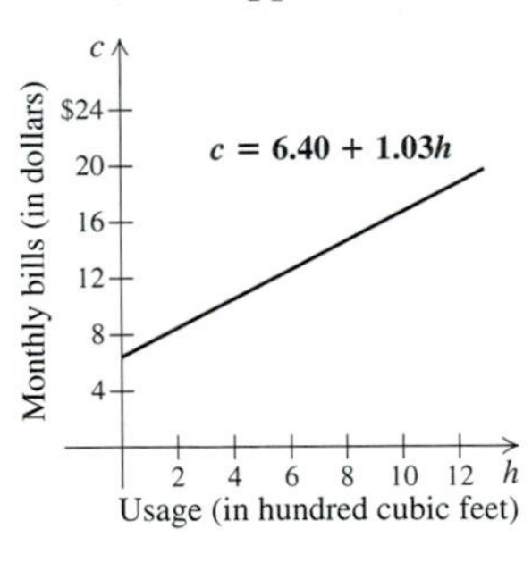

25. [1.3] $\frac{19}{24}$ **26.** [1.4] −0.875 **27.** [2.2] 8
28. [2.3] $m = 2A - n$
29. [3.4] ◈ A business might use a graph to quickly look up prices (as in the rental truck example), or to plot how total sales change from year to year. Many other applications exist.
30. [3.2] ◈ The y-intercept is the point at which the graph crosses the y-axis. Since a point on the y-axis is neither left nor right of the origin, the first or x-coordinate of the point is 0.
31. [3.2] −1 **32.** [3.2] 19
33. [3.1] Area = 45, perimeter = 28
34. [3.2] (0, 4), (1, 3), (−1, 3); answers may vary

Test: Chapter 3, p. 144

1. [3.1] Bachelor's **2.** [3.1] 581,991
3. [3.1] 274,901 **4.** [3.1] 417,197 **5.** [3.1] II
6. [3.1] III **7.** [3.1] (3, 4) **8.** [3.1] (0, −4)
9. [3.2] Yes
10. [3.2]

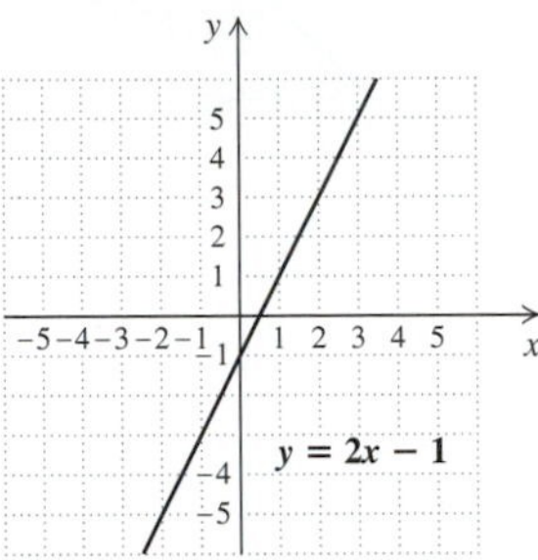

11. [3.3]

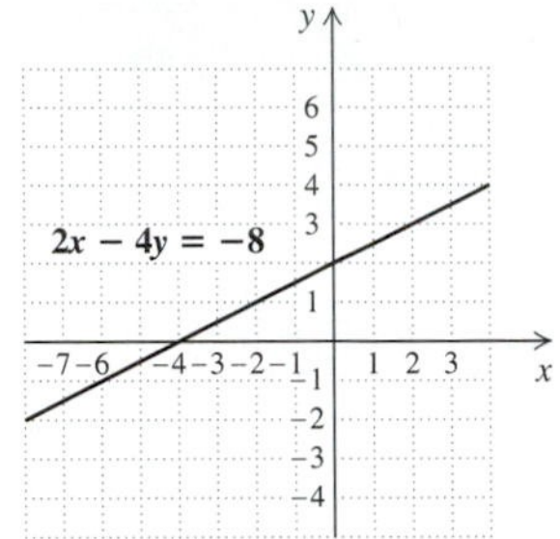

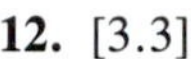

12. [3.3]

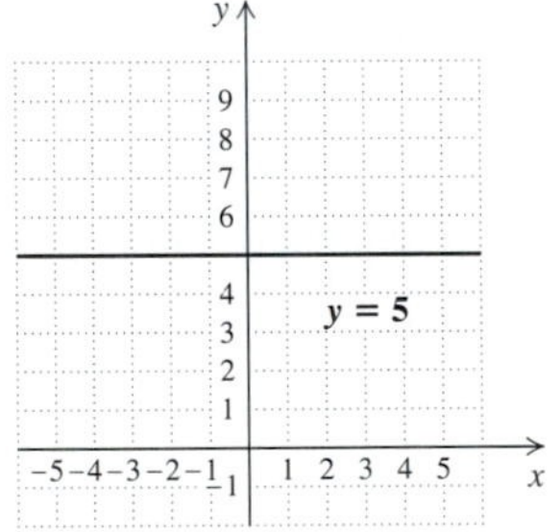

13. [3.2]

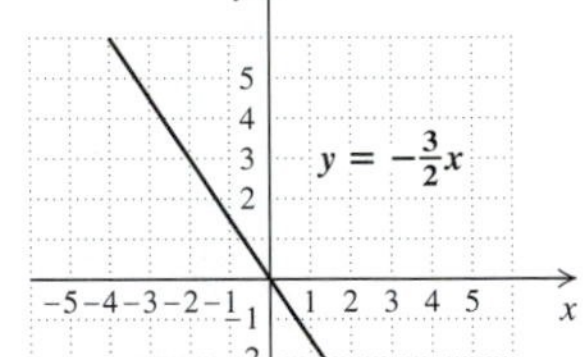

14. [3.3]

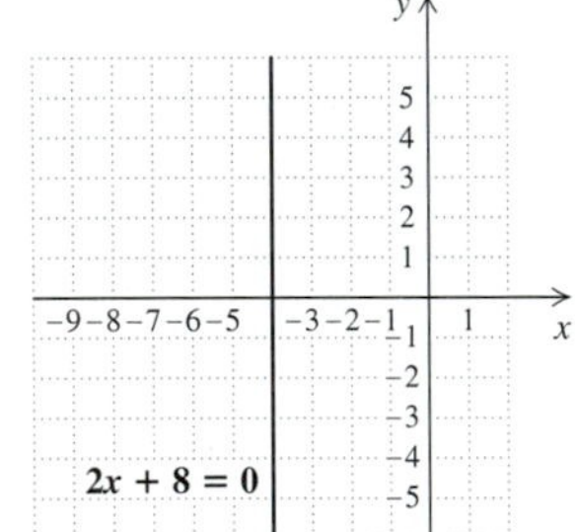

15. [3.3] x-intercept: (9, 0); y-intercept: (0, −15)
16. [3.3] x-intercept: (10, 0); y-intercept: (0, 2.5)
17. [3.4] Let $g =$ Greta's salary and $a =$ Alice's salary; then $g = 50 + 2a$.

18. [3.4]

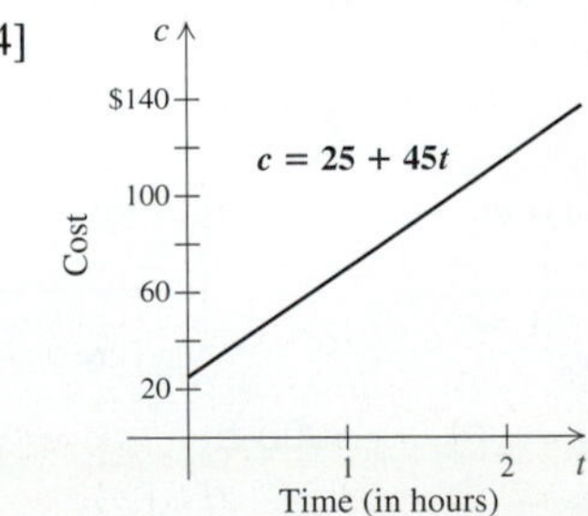

Approximately $50

19. [1.3] $\frac{11}{5}$ **20.** [1.4] $>$ **21.** [2.2] $\frac{4}{15}$

22. [2.3] $x = \dfrac{b}{m + n}$

23. [3.1] Area: 25, perimeter: 20 **24.** [3.3] $y = 3$

CUMULATIVE REVIEW: 1–3

1. [1.1] $\frac{5}{2}$ **2.** [1.2] $12x - 15y + 21$
3. [1.2] $3(x + 3y + 5)$ **4.** [1.3] $2 \cdot 3 \cdot 7$ **5.** [1.4] 0.45
6. [1.4] 4 **7.** [1.6] -5 **8.** [1.7] $\frac{1}{5}$ **9.** [1.6] $-x - y$
10. [2.4] 0.785 **11.** [1.3] $\frac{11}{60}$ **12.** [1.5] 2.6
13. [1.7] 7.28 **14.** [1.7] $-\frac{5}{12}$ **15.** [1.8] -2
16. [1.8] 27 **17.** [1.8] $-2y - 7$ **18.** [1.8] $5x + 11$
19. [2.1] -1.2 **20.** [2.1] -21 **21.** [2.2] 9
22. [2.1] $-\frac{20}{3}$ **23.** [2.2] 2 **24.** [2.1] $\frac{13}{8}$
25. [2.2] $-\frac{17}{21}$ **26.** [2.2] -17 **27.** [2.2] 2
28. [2.6] $\{x \mid x < 16\}$ **29.** [2.6] $\{x \mid x \leq -\frac{11}{8}\}$

30. [2.3] $h = \dfrac{A - \pi r^2}{2\pi r}$ **31.** [3.1] IV

32. [2.6]

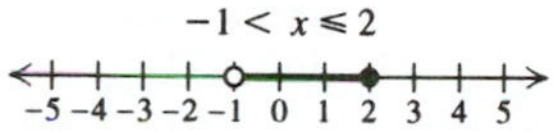

33. [3.3]

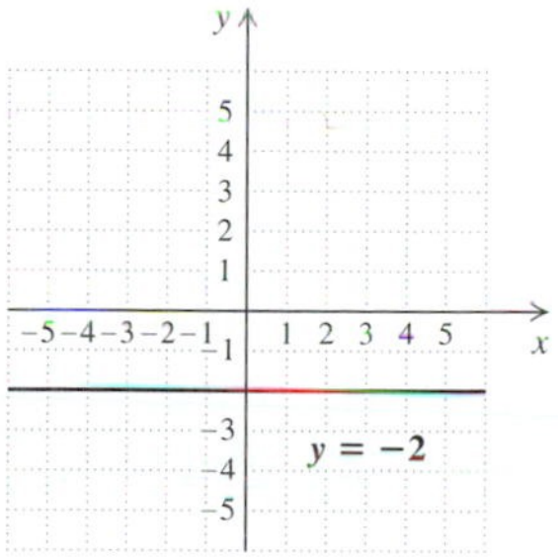

34. [3.3]

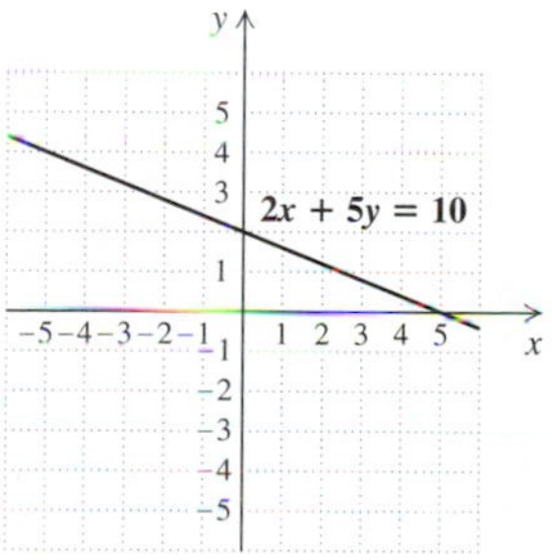

35. [3.2]

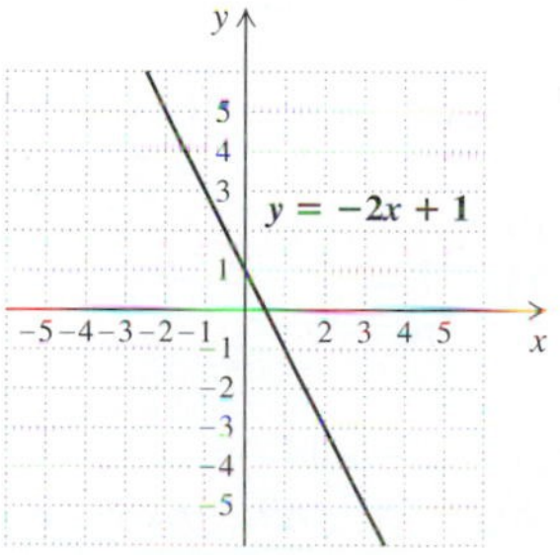

36. [3.2]

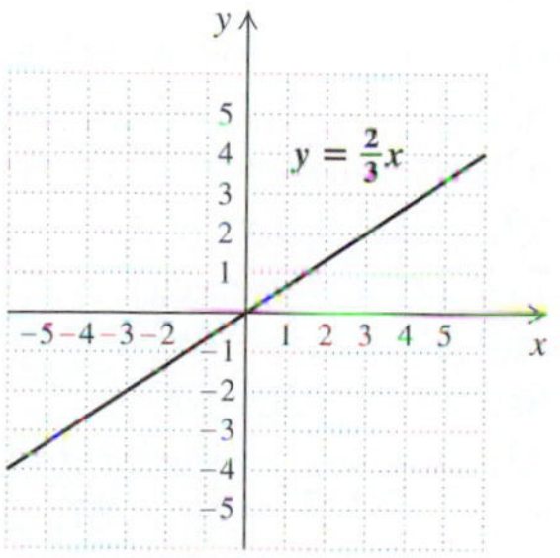

37. [3.3] x-intercept: (10.5, 0); y-intercept: (0, -3)
38. [3.3] x-intercept: $(-\frac{5}{4}, 0)$; y-intercept: (0, 5)
39. [2.4] \$13.60 **40.** [2.5] 154 **41.** [2.5] \$45
42. [2.5] 50 m, 53 m, 40 m **43.** [2.5] \$1500

44. [2.7] $x \geq 78$
45. [3.4] Let s = the cost of steak and c = the cost of chicken; then $s = 5 + 3c$.

46. [3.4]

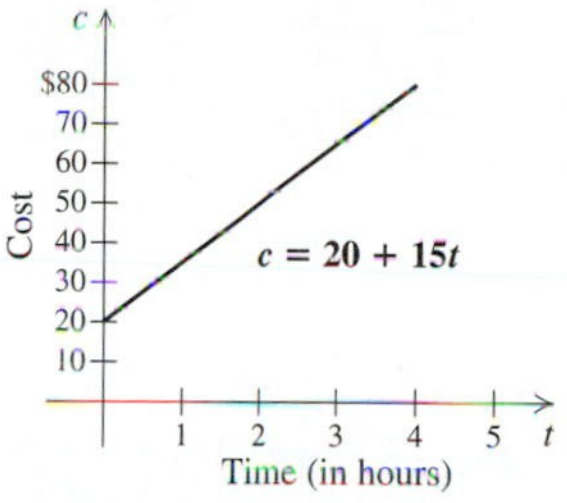

About \$55 for a $2\frac{1}{2}$ hr job

47. [2.4], [2.5] \$25,000 **48.** [1.4], [2.2] 4, -4
49. [2.2] All real numbers **50.** [2.2] No solution
51. [2.2] 3 **52.** [2.2] All real numbers

53. [2.3] $Q = \dfrac{2 - pm}{p}$

CHAPTER 4

Exercise Set 4.1, pp. 153–154

1. 2^7 **3.** 8^{14} **5.** x^7 **7.** 9^{38} **9.** $(3y)^{12}$
11. $(7y)^{17}$ **13.** a^5b^9 **15.** x^4y^{14} **17.** r^{12} **19.** x^5y^5
21. 7^3 **23.** 8^6 **25.** y^4 **27.** $5a$ **29.** 6^3x^5
31. $3m^3$ **33.** a^7b^6 **35.** m^9n^4 **37.** 1 **39.** 5
41. 1 **43.** 1 **45.** 4 **47.** x^{12} **49.** 2^{24} **51.** m^{35}
53. a^{75} **55.** $9x^2$ **57.** $-8a^3$ **59.** $16m^6$
61. $27a^6b^3$ **63.** $a^{15}b^{10}$ **65.** $25x^8y^{10}$

67. $\dfrac{a^3}{64}$ **69.** $\dfrac{49}{25a^2}$ **71.** $\dfrac{a^8}{b^{12}}$ **73.** $\dfrac{y^6}{4}$

75. $\dfrac{125x^6}{y^9}$ **77.** $\dfrac{a^{12}}{16b^{20}}$ **79.** $\dfrac{8a^6}{27b^{12}}$ **81.** $\dfrac{16x^6y^{10}}{9z^{14}}$

83. $3(s + t + 8)$ **85.** $5x$ **87.** ◈ **89.** 25 **91.** a^{2k}
93. 2 **95.** $>$ **97.** $<$
99. Let $x = 2$; then $3x^2 = 12$, and $(3x)^2 = 36$.

101. Let $x = 1$; then $\dfrac{x + 2}{2} = \dfrac{3}{2}$, and $x = 1$.

103. \$15,638.03 **105.** $25x^2$

Exercise Set 4.2, pp. 159–161

1. Trinomial **3.** None of these **5.** Binomial
7. Monomial **9.** 2, $-3x$, x^2 **11.** $6x^2$ and $-3x^2$
13. $2x^4$ and $-3x^4$; $5x$ and $-7x$ **15.** -3, 6
17. 5, 3, 3 **19.** -7, 6, 3, 7 **21.** -5, 6, -3, 8, -2
23. 1, 0; 1 **25.** 2, 1, 0; 2 **27.** 3, 2, 1, 0; 3
29. 2, 1, 6, 4; 6

31.

Term	Coefficient	Degree of Term	Degree of Polynomial
$-7x^4$	-7	4	4
$6x^3$	6	3	
$-3x^2$	-3	2	
$8x$	8	1	
-2	-2	0	

33. $-3x$ **35.** $-8x$ **37.** $11x^3 + 4$ **39.** $x^3 - x$
41. $4b^5$ **43.** $\frac{3}{4}x^5 - 2x - 42$ **45.** x^4 **47.** $\frac{15}{16}x^3 - \frac{7}{6}x^2$
49. $x^6 + x^4$ **51.** $13x^3 - 9x + 8$ **53.** $-5x^2 + 9x$
55. $12x^4 - 2x + \frac{1}{4}$ **57.** -18 **59.** 19 **61.** -12
63. 2 **65.** 4 **67.** 11 **69.** Approximately 449
71. 1112 ft **73.** \$18,750 **75.** \$155,000
77. 62.8 cm **79.** 78.5 m^2

81.

t	$-t^2 + 6t - 4$
1	1
2	4
3	5
4	4
5	1

83. 274 and 275 **85.** ◈ **87.** $5x^9 + 4x^8 + x^2 + 5x$
89. 99, -99; 50, -50; 99, -99
91. $-6x^5 + 14x^4 - x^2 + 11$; answers may vary **93.** 10

95.

d	$-0.0064d^2 + 0.8d + 2$
0	2
30	20.24
60	26.96
90	22.16
120	5.84

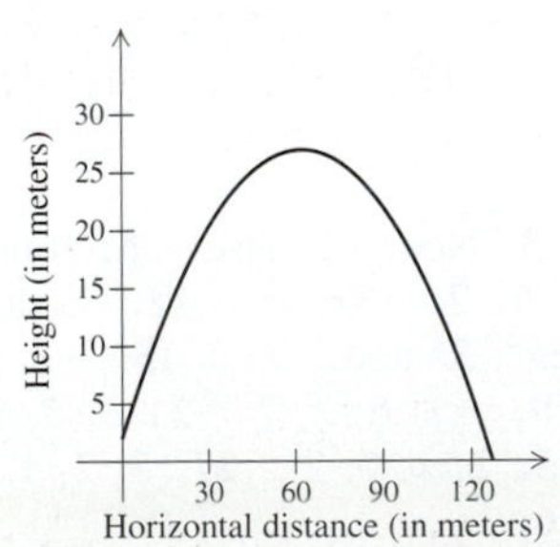

Exercise Set 4.3, pp. 166–169

1. $-x + 5$ **3.** $x^2 - 5x - 1$ **5.** $2x^2$
7. $5x^2 + 3x - 30$ **9.** $-2.2x^3 - 0.2x^2 - 3.8x + 23$
11. $12x^2 + 6$ **13.** $9x^8 + 8x^7 - 3x^4 + 2x^2 - 2x + 5$
15. $-\frac{1}{2}x^4 + \frac{2}{3}x^3 + x^2$
17. $0.01x^5 + x^4 - 0.2x^3 + 0.2x + 0.06$
19. $-3x^4 + 3x^2 + 4x$
21. $1.05x^4 + 0.36x^3 + 14.22x^2 + x + 0.97$
23. $-(-5x)$, $5x$ **25.** $-(-x^2 + 10x - 2)$, $x^2 - 10x + 2$
27. $-(12x^4 - 3x^3 + 3)$, $-12x^4 + 3x^3 - 3$
29. $-3x + 7$ **31.** $-4x^2 + 3x - 2$
33. $4x^4 - 6x^2 - \frac{3}{4}x + 8$ **35.** $7x - 1$
37. $-x^2 - 7x + 5$ **39.** -18
41. $6x^4 + 3x^3 - 4x^2 + 3x - 4$
43. $4.6x^3 + 9.2x^2 - 3.8x - 23$ **45.** $2x^2 + 14$
47. $-2x^5 - 6x^4 + x + 2$ **49.** $9x^2 + 9x - 8$
51. $\frac{3}{4}x^3 - \frac{1}{2}x$ **53.** $0.06x^3 - 0.05x^2 + 0.01x + 1$
55. $3x + 6$ **57.** $11x^4 + 12x^3 - 9x^2 - 8x - 9$
59. $-4x^5 + 9x^4 + 6x^2 + 16x + 6$ **61.** $x^4 - x^3 + x^2 - x$
63. **(a)** $5x^2 + 4x$; **(b)** 57, 352 **65.** $14y + 17$
67. $(r + 9)(r + 11)$; $9r + 99 + r^2 + 11r$ **69.** $\pi r^2 - 9\pi$
71. $z^2 - 27z + 72$ **73.** $144 - 4x^2$ **75.** $\frac{115}{22}$ **77.** 4
79. ◈ **81.** $11a^2 - 18a - 4$ **83.** $-10y^2 - 2y - 10$
85. $-3y^4 - y^3 + 5y - 2$ **87.** $569.607x^3 - 15.168x$
89. $28a + 90$ **91.** **(a)** $P = -x^2 + 280x - 5000$;
(b) \$10,375; **(c)** \$13,000

Exercise Set 4.4, pp. 173–175

1. $42x^2$ **3.** x^4 **5.** $-x^8$ **7.** $6x^6$ **9.** $28t^8$
11. $-0.02x^{10}$ **13.** $\frac{1}{15}x^4$ **15.** 0 **17.** $-24x^{11}$
19. $-3x^2 + 15x$ **21.** $4x^2 + 4x$ **23.** $5x^2 + 35x$
25. $x^5 + x^2$ **27.** $6x^3 - 18x^2 + 3x$ **29.** $12x^3 + 24x^2$
31. $-6x^4 - 6x^3$ **33.** $18y^6 + 24y^5$
35. $42x^{54} + 60x^{15} + 18x^{61} + 180x^{19}$ **37.** $x^2 + 9x + 18$
39. $x^2 + 3x - 10$ **41.** $x^2 - 7x + 12$ **43.** $x^2 - 9$
45. $25 - 15x + 2x^2$ **47.** $4x^2 + 20x + 25$
49. $9y^2 - 16$ **51.** $x^2 - \frac{21}{10}x - 1$ **53.** $x^3 - 1$
55. $4x^3 + 14x^2 + 8x + 1$
57. $3y^4 - 6y^3 - 7y^2 + 18y - 6$ **59.** $x^6 + 2x^5 - x^3$
61. $-10x^5 - 9x^4 + 7x^3 + 2x^2 - x$
63. $x^4 - x^2 - 2x - 1$ **65.** $4x^4 + 8x^3 - 9x^2 - 10x + 8$
67. $x^4 + 8x^3 + 12x^2 + 9x + 4$
69. $-4x^4 + 6x^3 - 2x^2 - 13x + 10$
71. $2x^4 - 7x^3 + 2x^2 - x - 2$ **73.** $x^4 - 1$
75. $x^4 - 2x^3 - 4x^2 + 9$ **77.** $-\frac{3}{4}$ **79.** ◈
81. $84y^2 - 30y$
83. $V = 4x^3 - 48x^2 + 144x$; $S = -4x^2 + 144$
85. $A = \frac{1}{2}b^2 + 2b$ **87.** $2x^2 + 18x + 36$ **89.** $16x + 16$

Exercise Set 4.5, pp. 181–183

1. $x^3 + x^2 + 3x + 3$ **3.** $x^4 + x^3 + 2x + 2$
5. $y^2 - y - 6$ **7.** $9x^2 + 15x + 6$ **9.** $5x^2 + 4x - 12$
11. $9t^2 - 1$ **13.** $4x^2 - 6x + 2$ **15.** $p^2 - \frac{1}{16}$

17. $x^2 - 0.01$ **19.** $2x^3 + 2x^2 + 6x + 6$
21. $-2x^2 - 11x + 6$ **23.** $a^2 + 14a + 49$
25. $1 - x - 6x^2$ **27.** $x^5 + 3x^3 - x^2 - 3$
29. $3x^6 - 2x^4 - 6x^2 + 4$ **31.** $6x^7 + 18x^5 + 4x^2 + 12$
33. $8x^6 + 65x^3 + 8$ **35.** $4x^3 - 12x^2 + 3x - 9$
37. $4y^6 + 4y^5 + y^4 + y^3$ **39.** $x^2 - 16$ **41.** $4x^2 - 1$
43. $25m^2 - 4$ **45.** $4x^4 - 9$ **47.** $9x^8 - 16$
49. $x^{12} - x^4$ **51.** $x^8 - 9x^2$ **53.** $x^{24} - 9$
55. $4y^{16} - 9$ **57.** $x^2 + 4x + 4$ **59.** $9x^4 + 6x^2 + 1$
61. $a^2 - a + \frac{1}{4}$ **63.** $9 + 6x + x^2$ **65.** $x^4 + 2x^2 + 1$
67. $4 - 12x^4 + 9x^8$ **69.** $25 + 60t^2 + 36t^4$
71. $49x^2 - 4.2x + 0.09$ **73.** $10a^5 - 5a^3$
75. $x^4 + x^3 - 6x^2 - 5x + 5$ **77.** $9 - 12x^3 + 4x^6$
79. $4x^3 + 24x^2 - 12x$ **81.** $4x^4 - 2x^2 + \frac{1}{4}$
83. $-1 + 9p^2$ **85.** $15t^5 - 3t^4 + 3t^3$
87. $36x^8 + 48x^4 + 16$ **89.** $12x^3 + 8x^2 + 15x + 10$
91. $64 - 96x^4 + 36x^8$ **93.** $t^3 - 1$ **95.** 25; 49
97. 56; 16 **99.** $x^2 + 6x + 9$ **101.** $t^2 + 7t + 12$
103. $a^2 + 8a + 7$
105. Television: 50 watts; lamps: 500 watts; air conditioner: 2000 watts
107. ◈ **109.** $8y^3 + 72y^2 + 160y$ **111.** $81x^4 - 16$
113. $625t^{12} - 450t^6 + 81$
115. $4567.0564x^2 + 435.891x + 10.400625$
117. $400 - 4 = 396$ **119.** -7
121. $V = w^3 + 3w^2 + 2w$ **123.** $V = h^3 - 3h^2 + 2h$
125. $F^2 - (F - 17)(F - 7)$; $24F - 119$
127. (a) $A^2 + AB$; (b) $AB + B^2$; (c) $A^2 - B^2$; (d) $(A + B)(A - B) = A^2 - B^2$
129. $100x^2 + 100x + 25$, or $100(x^2 + x) + 25$. Add the first digit to its square, multiply by 100, and add 25.

Exercise Set 4.6, pp. 187–190

1. -1 **3.** -7 **5.** \$11,664 **7.** \$12,597.12
9. 20.60625 in^2
11. Coefficients: 1, −2, 3, −5; degrees: 4, 2, 2, 0; 4
13. Coefficients: 17, −3, −7; degrees: 5, 5, 0; 5
15. $-a - 2b$ **17.** $3x^2y - 2xy^2 + x^2$ **19.** $8u^2v - 5uv^2$
21. $20au + 10av$ **23.** $x^2 - 4xy + 3y^2$
25. $-4r^3 + 2rs - 9s^2$ **27.** $3r + s - 4$
29. $-x^2 - 8xy - y^2$ **31.** $2ab$
33. $-2a + 4b + 3c - 8d$ **35.** $-8x + 8y$
37. $6z^2 + 7zu - 3u^2$ **39.** $a^4b^2 - 7a^2b + 10$
41. $a^6 - b^2c^2$ **43.** $y^6x + y^4x + y^4 + 2y^2 + 1$
45. $12x^2y^2 + 2xy - 2$ **47.** $12 - c^2d^2 - c^4d^4$
49. $m^3 + m^2n - mn^2 - n^3$
51. $x^9y^9 - x^6y^6 + x^5y^5 - x^2y^2$ **53.** $x^2 + 2xh + h^2$
55. $r^6t^4 - 8r^3t^2 + 16$ **57.** $p^8 + 2m^2n^2p^4 + m^4n^4$
59. $4a^2 - b^2$ **61.** $c^4 - d^2$ **63.** $a^2b^2 - c^2d^4$
65. $x^2 + 2xy + y^2 - 9$ **67.** $x^2 - y^2 - 2yz - z^2$
69. $a^2 - b^2 - 2bc - c^2$ **71.** $x^2 + 2xy + y^2$
73. $x^2 - z^2$ **75.** \$60 per ton **77.** December 1987
79. ◈ **81.** $4xy - 4y^2$ **83.** $2xy + \pi x^2$
85. 3.535 L **87.** ◈

Exercise Set 4.7, pp. 194–195

1. $3x^4 - \frac{1}{2}x^3$ **3.** $1 - 2u - u^4$ **5.** $5t^2 + 8t - 2$
7. $-4x^4 + 4x^2 + 1$ **9.** $6x^2 - 10x + \frac{3}{2}$
11. $4x^2 - \frac{3}{2}x + \frac{1}{2}$ **13.** $x^2 + 3x + 2$ **15.** $-3rs - r + 2s$
17. $x + 2$ **19.** $x - 5 - \dfrac{50}{x - 5}$ **21.** $x - 2 - \dfrac{2}{x + 6}$
23. $x - 3$ **25.** $x^4 - x^3 + x^2 - x + 1$
27. $2x^2 - 7x + 4$ **29.** $x^3 - 6$ **31.** $x^3 + 2x^2 + 4x + 8$
33. $t^2 + 1$ **35.** 25,543.75 ft^2

37.

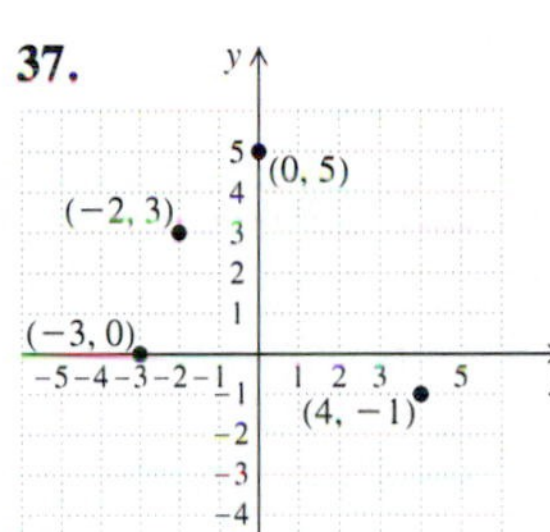

39. ◈ **41.** $x^2 + 5$
43. $a + 3 + \dfrac{5}{5a^2 - 7a - 2}$
45. $2x^2 + x - 3$
47. $3a^{2h} + 2a^h - 5$
49. 2

Exercise Set 4.8, pp. 198–199

1. $x^2 - x + 1$, R -4, or $x^2 - x + 1 + \dfrac{-4}{x - 1}$

3. $a + 7$, R -47, or $a + 7 + \dfrac{-47}{a + 4}$

5. $x^2 - 5x - 23$, R -43, or $x^2 - 5x - 23 + \dfrac{-43}{x - 2}$

7. $3x^2 - 2x + 2$, R -3, or $3x^2 - 2x + 2 + \dfrac{-3}{x + 3}$

9. $y^2 + 2y + 1$, R 12, or $y^2 + 2y + 1 + \dfrac{12}{y - 2}$

11. $3x^3 + 9x^2 + 2x + 6$ **13.** $x^2 + 3x + 9$
15. $y^4 + y^3 + y^2 + y + 1$ **17.** $3x^3 + 2x^2 - 2x - 3$, R 2, or $3x^3 + 2x^2 - 2x - 3 + \dfrac{2}{x + 2}$

19. $3x^2 + 6x - 3$, R 2, or $3x^2 + 6x - 3 + \dfrac{2}{x + \frac{1}{3}}$

21. $-7y$ **23.** $11m + 6$ **25.** ◈
27. (a) Remainder is 0; (b) If $f(x) = (x - 4) \cdot p(x)$ for some polynomial $p(x)$, then $f(4) = (4 - 4)\ p(4) = 0$; (c) $4^3 - 5(4)^2 + 5(4) - 4 = 0$

Exercise Set 4.9, pp. 205–207

1. $\dfrac{1}{3^2} = \dfrac{1}{9}$ **3.** $\dfrac{1}{10^4} = \dfrac{1}{10{,}000}$ **5.** $\dfrac{1}{7^3} = \dfrac{1}{343}$ **7.** $\dfrac{1}{a^3}$

9. y^4 **11.** z^n **13.** $\dfrac{1}{2}$ **15.** 16 **17.** 4^{-3} **19.** x^{-3}

21. a^{-4} **23.** p^{-n} **25.** 5^{-1} **27.** t^{-1} **29.** 3^3

31. x^{-1}, or $\frac{1}{x}$ **33.** x^{-13}, or $\frac{1}{x^{13}}$ **35.** m^{-6}, or $\frac{1}{m^6}$

37. $(8x)^{-4}$, or $\frac{1}{(8x)^4}$ **39.** 1 **41.** $a^{-7}b^{-11}$, or $\frac{1}{a^7b^{11}}$

43. x^9 **45.** z^{-4}, or $\frac{1}{z^4}$ **47.** x^3 **49.** x^2

51. a^{-15}, or $\frac{1}{a^{15}}$ **53.** 5^{-6}, or $\frac{1}{5^6}$ **55.** x^{12}

57. m^{-21}, or $\frac{1}{m^{21}}$ **59.** $a^{-3}b^{-3}$, or $\frac{1}{a^3b^3}$

61. $5^{-2}a^{-2}b^{-2}$, or $\frac{1}{25a^2b^2}$ **63.** $36x^{-10}$, or $\frac{36}{x^{10}}$

65. $x^{-12}y^{-15}$, or $\frac{1}{x^{12}y^{15}}$ **67.** $x^{24}y^8$

69. $9x^6y^{-16}z^{-6}$, or $\frac{9x^6}{y^{16}z^6}$ **71.** $x^{-1}y^{-6}z^4$, or $\frac{z^4}{xy^6}$

73. $m^5n^5p^{-7}$, or $\frac{m^5n^5}{p^7}$ **75.** $\frac{y^{-6}}{2^{-3}}$, or $\frac{8}{y^6}$ **77.** $\frac{27}{a^6}$

79. $\frac{x^6y^3}{z^3}$ **81.** $\frac{a^{-4}b^{-2}}{c^{-2}d^{-6}}$, or $\frac{c^2d^6}{a^4b^2}$ **83.** 2140

85. 0.00692 **87.** 784,000,000 **89.** 0.0000000008764
91. 100,000,000 **93.** 0.0001 **95.** 2.5×10^4
97. 3.71×10^{-3} **99.** 7.8×10^{10} **101.** 9.07×10^{17}
103. 3.74×10^{-6} **105.** 1.8×10^{-8} **107.** 10^7
109. 10^{-9} **111.** 6×10^9 **113.** 3.38×10^4
115. 8.1477×10^{-13} **117.** 2.5×10^{13}
119. 5×10^{-4} **121.** 3×10^{-21}
123. Approximately 1.231×10^{-5}
125. 2.3725×10^9 gal
127. 1.512×10^{10} cu ft, 1.324512×10^{14} cu ft
129. $8a$

131.

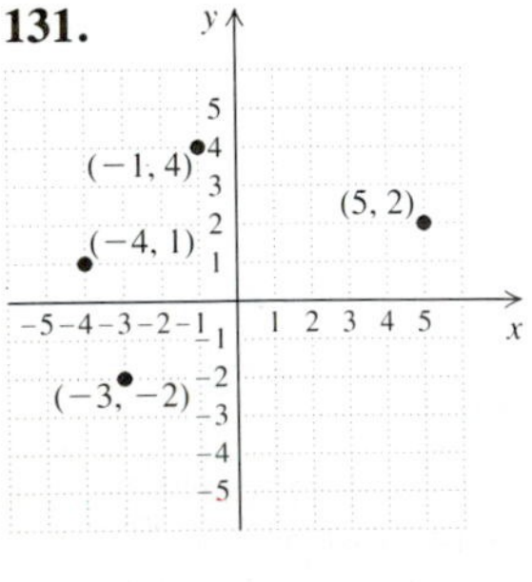

(−1, 4) (5, 2) (−4, 1) (−3, −2)

133. 2.478125×10^{-1}
135. 3.5×10^{-10}
137. 2 **139.** 5
141. a^n **143.** False
145. False

Review Exercises: Chapter 4, pp. 208–210

1. [4.1] y^{11} **2.** [4.1] $(3x)^{14}$ **3.** [4.1] t^8 **4.** [4.1] 4^3

5. [4.1] 1 **6.** [4.1] $\frac{9t^8}{4s^6}$ **7.** [4.1] $-8x^3y^6$

8. [4.1] $36x^8$ **9.** [4.1] $18x^8$ **10.** [4.2] $3x^2$, $6x$, $\frac{1}{2}$

11. [4.2] $-4y^5$, $7y^2$, $-3y$, -2 **12.** [4.2] 6, 17

13. [4.2] 4, 6, −5, $\frac{5}{3}$ **14.** [4.2] 3, 1, 0; 3

15. [4.2] 0, 4, 9, 6, 3; 9 **16.** [4.2] Binomial
17. [4.2] None of these **18.** [4.2] Monomial

19. [4.2] $-x^2 + 9x$ **20.** [4.2] $-\frac{1}{4}x^3 + 4x^2 + 7$

21. [4.2] $-3x^5 + 25$ **22.** [4.2] $-2x^2 - 3x + 2$

23. [4.2] $10x^4 - 7x^2 - x - \frac{1}{2}$ **24.** [4.2] −17

25. [4.2] 10 **26.** [4.3] $x^5 - 2x^4 + 6x^3 + 3x^2 - 9$
27. [4.3] $2x^5 - 6x^4 + 2x^3 - 2x^2 + 2$
28. [4.3] $2x^2 - 4x - 6$ **29.** [4.3] $x^5 - 3x^3 - 2x^2 + 8$

30. [4.3] $\frac{3}{4}x^4 + \frac{1}{4}x^3 - \frac{1}{3}x^2 - \frac{7}{4}x + \frac{3}{8}$

31. [4.3] $-x^5 + x^4 - 5x^3 - 2x^2 + 2x$
32. **(a)** [4.3] $4w + 8$; **(b)** [4.4] $w^2 + 4w$
33. [4.4] $-12x^3$ **34.** [4.5] $49x^2 + 14x + 1$

35. [4.5] $x^2 + \frac{7}{6}x + \frac{1}{3}$ **36.** [4.5] $0.3x^2 + 0.65x - 8.45$

37. [4.4] $12x^3 - 23x^2 + 13x - 2$
38. [4.5] $x^2 - 18x + 81$
39. [4.4] $15x^7 - 40x^6 + 50x^5 + 10x^4$
40. [4.5] $x^2 - 3x - 28$
41. [4.5] $x^2 - 1.05x + 0.225$
42. [4.4] $x^7 + x^5 - 3x^4 + 3x^3 - 2x^2 + 5x - 3$
43. [4.5] $9y^4 - 12y^3 + 4y^2$ **44.** [4.5] $2t^4 - 11t^2 - 21$
45. [4.4] $4x^5 - 5x^4 - 8x^3 + 22x^2 - 15x$
46. [4.5] $9x^4 - 16$ **47.** [4.5] $4 - x^2$
48. [4.5] $13x^2 - 172x + 39$ **49.** [4.6] 49
50. [4.6] Coefficients: 1, −7, 9, −8; degrees: 6, 2, 2, 0; 6
51. [4.6] Coefficients: 1, −1, 1; degrees: 16, 40, 23; 40
52. [4.6] $9w - y - 5$
53. [4.6] $m^6 - 2m^2n + 2m^2n^2 + 8n^2m - 6m^3$
54. [4.6] $-9xy - 2y^2$
55. [4.6] $11x^3y^2 - 8x^2y - 6x^2 - 6x + 6$ **56.** [4.6] $p^3 - q^3$

57. [4.6] $9a^8 - 2a^4b^3 + \frac{1}{9}b^6$ **58.** [4.7] $5x^2 - \frac{1}{2}x + 3$

59. [4.7] $3x^2 - 7x + 4 + \frac{1}{2x + 3}$ **60.** [4.7] $t^3 + 2t - 3$

61. [4.7] $2x^2 + 1 + \frac{x^2 + 2}{x^3 - 1}$

62. [4.8] $x^2 + 6x + 20 + \frac{54}{x - 3}$

63. [4.8] $4x^2 - 6x + 18 + \frac{-59}{x + 3}$ **64.** [4.9] $\frac{1}{y^4}$

65. [4.9] t^{-5} **66.** [4.9] $\frac{1}{7^2}$ **67.** [4.9] $\frac{1}{a^{13}b^7}$

68. [4.9] $\frac{1}{x^{12}}$ **69.** [4.9] $\frac{x^6}{4y^2}$ **70.** [4.9] $\frac{y^3}{8x^3}$
71. [4.9] 8,300,000 **72.** [4.9] 3.28×10^{-5}
73. [4.9] 2.09×10^4 **74.** [4.9] 5.12×10^{-5}
75. [4.9] 6.205×10^{10} **76.** [1.5] $13x + \frac{22}{15}$
77. [2.5] $w = 125.5$ m, $l = 144.5$ m
78. [2.6] $\{x|x \geq -2\}$ **79.** [3.1] IV
80. [4.1] ◈ In the expression $5x^3$, the exponent refers only to the x. In the expression $(5x)^3$, the entire expression within the parentheses is cubed.
81. [4.3] ◈ The sum of two polynomials of degree n will also have degree n, since only the coefficients are added and the variables remain unchanged. An exception to this occurs when the leading terms of the two polynomials are opposites. The sum of those terms is then zero and the sum of the polynomials will have a degree less than n.
82. [4.2], [4.5] **(a)** 3; **(b)** 2 **83.** [4.1], [4.2] $-28x^8$
84. [4.2] $8x^4 + 4x^3 + 5x - 2$
85. [4.5] $-4x^6 + 3x^4 - 20x^3 + x^2 - 16$
86. [2.2], [4.5] $\frac{94}{13}$

Test: Chapter 4, pp. 210–211

1. [4.1] x^9 **2.** [4.1] $(4a)^{11}$ **3.** [4.1] 3^3 **4.** [4.1] 1
5. [4.1] x^6 **6.** [4.1] $-27y^6$ **7.** [4.1] $-216x^{21}$
8. [4.1] $-24x^{21}$ **9.** [4.2] Binomial
10. [4.2] $\frac{1}{3}, -1, 7$ **11.** [4.2] 3, 0, 1, 6; 6
12. [4.2] -7 **13.** [4.2] $5a^2 - 6$ **14.** [4.2] $\frac{7}{4}y^2 - 4y$
15. [4.2] $x^5 + 2x^3 + 4x^2 - 8x + 3$
16. [4.3] $4x^5 + x^4 + 2x^3 - 8x^2 + 2x - 7$
17. [4.3] $5x^4 + 5x^2 + x + 5$
18. [4.3] $-4x^4 + x^3 - 8x - 3$
19. [4.3] $-x^5 + 0.7x^3 - 0.8x^2 - 21$
20. [4.4] $-12x^4 + 9x^3 + 15x^2$ **21.** [4.5] $x^2 - \frac{2}{3}x + \frac{1}{9}$
22. [4.5] $9x^2 - 100$ **23.** [4.5] $3b^2 - 4b - 15$
24. [4.5] $x^{14} - 4x^8 + 4x^6 - 16$
25. [4.5] $48 + 34y - 5y^2$
26. [4.4] $6x^3 - 7x^2 - 11x - 3$
27. [4.5] $25t^2 + 20t + 4$
28. [4.6] $-5x^3y - x^2y^2 + xy^3 - y^3 + 19$
29. [4.6] $8a^2b^2 + 6ab - 4b^3 + 6ab^2 + ab^3$
30. [4.6] $9x^{10} - 16y^{10}$ **31.** [4.7] $4x^2 + 3x - 5$
32. [4.7] $2x^2 - 4x - 2 + \frac{17}{3x + 2}$
33. [4.8] $x^2 + 9x + 40 + \frac{153}{x - 4}$ **34.** [4.9] $\frac{1}{5^3}$
35. [4.9] y^{-8} **36.** [4.9] $\frac{1}{6^5}$ **37.** [4.9] $\frac{y^5}{x^5}$
38. [4.9] $\frac{b^4}{16a^{12}}$ **39.** [4.9] $\frac{c^3}{a^3b^3}$ **40.** [4.9] 3.9×10^9
41. [4.9] 0.00000005 **42.** [4.9] 1.75×10^{17}
43. [4.9] 1.296×10^{22}
44. [4.9] Approximately 2.49×10^2
45. [2.6] $\{x|x > 13\}$
46. [3.1]

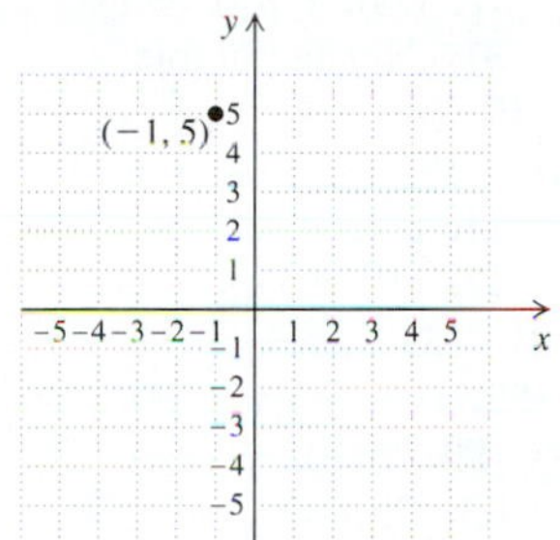

47. [1.5] $-\frac{7}{20}$ **48.** [2.5] 100°, 25°, 55°
49. [4.5] $V = l(l - 2)(l - 1) = l^3 - 3l^2 + 2l$
50. [2.2], [4.5] $\frac{100}{21}$

CHAPTER 5

Exercise Set 5.1, pp. 217–218

1. Answers may vary. $(6x)(x^2)$, $(3x^2)(2x)$, $(2x^2)(3x)$
3. Answers may vary. $(-3x^2)(3x^3)$, $(-x)(9x^4)$, $(3x^2)(-3x^3)$
5. Answers may vary. $(6x)(4x^3)$, $(-3x^2)(-8x^2)$, $(2x^3)(12x)$
7. $x(x - 4)$ **9.** $2x(x + 3)$ **11.** $x^2(x + 6)$
13. $8x^2(x^2 - 3)$ **15.** $2(x^2 + x - 4)$
17. $17xy(x^4y^2 + 2x^2y + 3)$ **19.** $x^2(6x^2 - 10x + 3)$
21. $x^2y^2(x^3y^3 + x^2y + xy - 1)$
23. $2x^3(x^4 - x^3 - 32x^2 + 2)$
25. $0.8x(2x^3 - 3x^2 + 4x + 8)$
27. $\frac{1}{3}x^3(5x^3 + 4x^2 + x + 1)$ **29.** $(y + 3)(y + 4)$
31. $(x + 3)(x^2 + 2)$ **33.** $(y + 8)(y^2 + 1)$
35. $(x + 3)(x^2 + 2)$ **37.** $(x + 3)(2x^2 + 1)$
39. $(2x - 3)(4x^2 + 3)$ **41.** $(3x - 4)(4x^2 + 1)$
43. $(x + 8)(x^2 - 3)$ **45.** $(w - 7)(w^2 + 4)$
47. Not factorable by grouping **49.** $(x - 4)(2x^2 - 9)$
51.

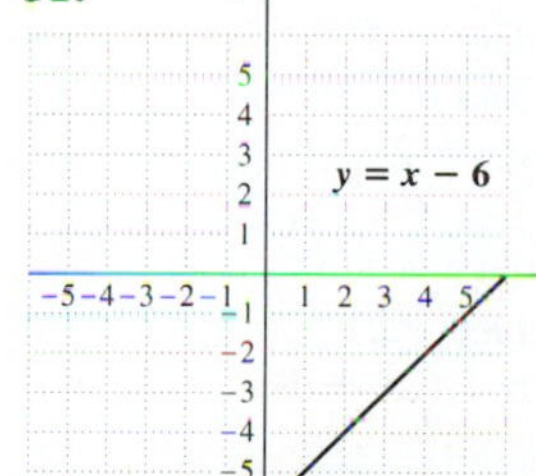

53. 12
55. $y^2 + 12y + 35$
57. $y^2 - 49$ **59.** ◈
61. $(2x^2 + 3)(2x^3 + 3)$
63. $(x^5 + 1)(x^7 + 1)$
65. Not factorable
67. ◈

Exercise Set 5.2, pp. 223–224

1. $(x + 5)(x + 3)$ **3.** $(x + 4)(x + 3)$ **5.** $(x - 3)^2$
7. $(x + 7)(x + 2)$ **9.** $(b + 4)(b + 1)$ **11.** $\left(x + \frac{1}{3}\right)^2$
13. $(d - 5)(d - 2)$ **15.** $(y - 10)(y - 1)$
17. $(x + 7)(x - 6)$ **19.** $2(x + 2)(x - 9)$
21. $x(x - 8)(x + 2)$ **23.** $(y + 5)(y - 9)$
25. $(x + 9)(x - 11)$ **27.** $c^2(c + 8)(c - 7)$
29. $2(a + 7)(a - 5)$ **31.** Not factorable
33. Not factorable **35.** $(x + 10)^2$
37. $3x(x - 25)(x + 4)$ **39.** $(x - 24)(x + 3)$
41. $(x - 16)(x - 9)$ **43.** $a^2(a + 12)(a - 11)$
45. $(x - 15)(x - 8)$ **47.** $(12 + x)(9 - x)$
49. $(y - 0.4)(y + 0.2)$ **51.** $(p + 5q)(p - 2q)$
53. Not factorable **55.** $(s - 5t)(s + 3t)$
57. $2x(x - 3)(x - 2)$ **59.** $7a^7(a + 1)(a - 5)$
61. $3x^2 + 22x + 24$ **63.** 29,443 **65.** ◈
67. 15, −15, 27, −27, 51, −51 **69.** $\left(x + \frac{1}{4}\right)\left(x - \frac{3}{4}\right)$
71. $(x + 5)\left(x - \frac{5}{7}\right)$ **73.** $(b^n + 5)(b^n + 2)$
75. $(x + 1)(a + 2)(a + 1)$ **77.** $2x^2(4 - \pi)$

Exercise Set 5.3, pp. 230–231

1. $(2x + 1)(x - 4)$ **3.** $(5x - 9)(x + 2)$
5. $(2x + 7)(3x + 1)$ **7.** $(3x + 1)(x + 1)$
9. $(2x + 5)(2x - 3)$ **11.** $(2x + 1)(x - 1)$
13. $(3x + 8)(3x - 2)$ **15.** $(3x + 1)(x - 2)$
17. $(3x + 4)(4x + 5)$ **19.** $(7x - 1)(2x + 3)$
21. $(3x + 4)(3x + 2)$ **23.** $(7 - 3x)^2$
25. $(x + 2)(24x - 1)$ **27.** $(7x + 4)(5x - 11)$
29. Prime **31.** $4(3x - 2)(x + 3)$
33. $6(5x - 9)(x + 1)$ **35.** $2(3x + 5)(x - 1)$
37. $(3x - 1)(x - 1)$ **39.** $4(3x + 2)(x - 3)$
41. $(2x + 1)(x - 1)$ **43.** $(3x - 8)(3x + 2)$
45. $5(3x + 1)(x - 2)$ **47.** $x(3x + 4)(4x + 5)$
49. $x^2(2x + 3)(7x - 1)$ **51.** $3x(8x - 1)(7x - 1)$
53. $(5x - 3)(3x - 2)$ **55.** $(5t + 8)^2$
57. $2x(3x + 5)(x - 1)$ **59.** $(25x + 64)(x + 1)$
61. Prime **63.** $(4m - 5n)(3m + 4n)$
65. $(3a - 5b)(2a + 3b)$ **67.** $(3a + 2b)(3a + 4b)$
69. $(5p + 2q)(7p + 4q)$ **71.** $6(3x - 4y)(x + y)$
73. $(y + 4)(y + 1)$ **75.** $(x - 4)(x - 1)$
77. $(3x + 2)(2x + 3)$ **79.** $(3x - 4)(x - 4)$
81. $(7x - 8)(5x + 3)$ **83.** $(2x + 3)(2x - 3)$
85. $(2x + 1)(x - 4)$ **87.** $(5x - 9)(x + 2)$
89. $(3x + 1)(2x + 7)$ **91.** $(3x + 1)(x + 1)$
93. $(2x + 5)(2x - 3)$ **95.** $(2x + 1)(x - 1)$
97. $(3x + 8)(3x - 2)$ **99.** $(3x + 1)(x - 2)$
101. $(4x + 5)(3x + 4)$ **103.** $(7x - 1)(2x + 3)$
105. $(3x + 2)(3x + 4)$ **107.** $(3x - 7)^2$
109. 6369 km, 3949 mi

111.

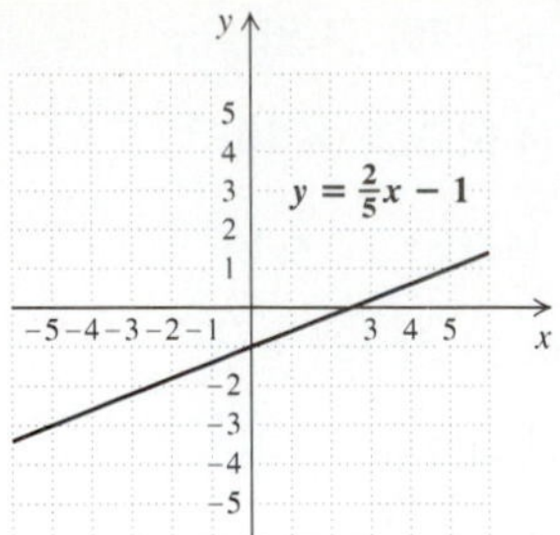

113. ◈ **115.** $(3x^5 - 2)^2$ **117.** $(10x^n + 3)(2x^n + 1)$
119. $(x^{3a} - 1)(3x^{3a} + 1)$
121. $-2(a + 1)^n(a + 3)^2(a + 6)$

Exercise Set 5.4, pp. 238–239

1. Yes **3.** No **5.** No **7.** No **9.** $(x - 7)^2$
11. $(x + 8)^2$ **13.** $(x - 1)^2$ **15.** $(x + 2)^2$
17. $(3x + 1)^2$ **19.** $(4y - 7)^2$, or $(7 - 4y)^2$
21. $2(x - 1)^2$ **23.** $x(x - 9)^2$ **25.** $5(2x + 5)^2$
27. $(7 - 3x)^2$ **29.** $5(y + 1)^2$ **31.** $2(1 + 5x)^2$
33. $(2p + 3q)^2$ **35.** $(a - 7b)^2$ **37.** $(8m + n)^2$
39. $(4s - 5t)^2$ **41.** Yes **43.** No **45.** No
47. Yes **49.** $(y + 2)(y - 2)$ **51.** $(p + 3)(p - 3)$
53. $(t + 7)(t - 7)$ **55.** $(a + b)(a - b)$
57. $(5t + m)(5t - m)$ **59.** $(10 + k)(10 - k)$
61. $(4a + 3)(4a - 3)$ **63.** $(2x + 5y)(2x - 5y)$
65. $2(2x + 7)(2x - 7)$ **67.** $x(6 + 7x)(6 - 7x)$
69. $(7a^2 + 9)(7a^2 - 9)$ **71.** $(x^2 + 1)(x + 1)(x - 1)$
73. $4(x^2 + 4)(x + 2)(x - 2)$
75. $(y^4 + 1)(y^2 + 1)(1 + y)(1 - y)$ **77.** $3x(x - 4)^2$
79. $(x^6 + 4)(x^3 + 2)(x^3 - 2)$ **81.** $\left(y + \frac{1}{4}\right)\left(y - \frac{1}{4}\right)$
83. $a^6(a - 1)^2$ **85.** $\left(5 + \frac{1}{7}x\right)\left(5 - \frac{1}{7}x\right)$
87. $(4m^2 + t^2)(2m + t)(2m - t)$
89. $(m - 7)(m + 2)(m - 2)$ **91.** $(a - 2)(a + b)(a - b)$
93. $(a + b + 10)(a + b - 10)$
95. $(a + b + 3)(a + b - 3)$
97. $(r - 1 - 2s)(r - 1 + 2s)$
99. $2(5a + m + n)(5a - m - n)$
101. $(3 - a + b)(3 + a - b)$ **103.** $s \geqslant 77$
105. $x^{12}y^{12}$ **107.** ◈ **109.** Prime **111.** $2x(3x + 1)^2$
113. $(x^4 + 2^4)(x^2 + 2^2)(x + 2)(x - 2)$
115. $3x^3(x + 2)(x - 2)$ **117.** $2x\left(3x + \frac{2}{5}\right)\left(3x - \frac{2}{5}\right)$
119. $p(0.7 + p)(0.7 - p)$ **121.** $x(x + 6)$
123. $\left(x + \dfrac{1}{x}\right)\left(x - \dfrac{1}{x}\right)$
125. $(9 + b^{2k})(3 + b^k)(3 - b^k)$ **127.** $(3b^n + 2)^2$
129. $(y + 4)^2$ **131.** $(3x + 7)(3x - 7)^2$ **133.** 9
135. $(x + 1)^2 - x^2 = ((x + 1) + x)((x + 1) - x)$
$= ((x + 1) + x)(1) = (x + 1) + x$

Exercise Set 5.5, pp. 241–242

1. $(x + 2)(x^2 - 2x + 4)$ **3.** $(y - 4)(y^2 + 4y + 16)$
5. $(w + 1)(w^2 - w + 1)$ **7.** $(2a + 1)(4a^2 - 2a + 1)$
9. $(y - 2)(y^2 + 2y + 4)$ **11.** $(2 - 3b)(4 + 6b + 9b^2)$
13. $(4y + 1)(16y^2 - 4y + 1)$
15. $(2x + 3)(4x^2 - 6x + 9)$ **17.** $(a - b)(a^2 + ab + b^2)$
19. $\left(a + \frac{1}{2}\right)\left(a^2 - \frac{1}{2}a + \frac{1}{4}\right)$ **21.** $2(y - 4)(y^2 + 4y + 16)$
23. $3(2a + 1)(4a^2 - 2a + 1)$
25. $r(s + 4)(s^2 - 4s + 16)$
27. $5(x - 2z)(x^2 + 2xz + 4z^2)$
29. $(x + 0.1)(x^2 - 0.1x + 0.01)$
31. $8(2x^2 - t^2)(4x^4 + 2x^2t^2 + t^4)$
33. $2y(y - 4)(y^2 + 4y + 16)$
35. $(z + 1)(z^2 - z + 1)(z - 1)(z^2 + z + 1)$
37. $(t^2 + 4y^2)(t^4 - 4t^2y^2 + 16y^4)$ **39.** $s \geq 71$
41. $-\frac{11}{12}$ **43.** ◈ **45.** $(x^{2a} + y^b)(x^{4a} - x^{2a}y^b + y^{2b})$
47. $3(x^a + 2y^b)(x^{2a} - 2x^ay^b + 4y^{2b})$
49. $\frac{1}{3}\left(\frac{1}{2}xy + z\right)\left(\frac{1}{4}x^2y^2 - \frac{1}{2}xyz + z^2\right)$
51. $7\left(x + \frac{1}{2}\right)\left(x^2 - \frac{1}{2}x + \frac{1}{4}\right)$ **53.** $y(3x^2 + 3xy + y^2)$
55. $4(3a^2 + 4)$

Exercise Set 5.6, pp. 246–247

1. $(x + 12)(x - 12)$ **3.** $(p + 8)^2$
5. $(2x - 3)(x - 4)$ **7.** $x(x + 12)^2$
9. $(x + 3)(x - 2)(x + 2)$ **11.** $(3x + 5y)(3x - 5y)$
13. $4x(x - 2)(5x + 9)$ **15.** $b(a - 2)(a^2 + 2a + 4)$
17. $x(x^2 + 7)(x - 3)$ **19.** $x^3(x - 7)^2$
21. $(m + 1)(m^2 - m + 1)(m - 1)(m^2 + m + 1)$
23. Prime **25.** $4(x^2 + 4)(x + 2)(x - 2)$ **27.** Prime
29. $x^3(x - 3)(x - 1)$ **31.** $(x + 3 + 4y)(x + 3 - 4y)$
33. $12n^2(1 + 2n)$ **35.** $9xy(xy - 4)$ **37.** $2\pi r(h + r)$
39. $(a + b)(2x + 1)$ **41.** $2(2x + 3y)(4x^2 - 6xy + 9y^2)$
43. $(n + 2)(n + p)$ **45.** $(a + d)(c - b)$ **47.** $(x - y)^2$
49. $(3c + d)^2$ **51.** $7(p^2 + q^2)(p + q)(p - q)$
53. $(5z + y)^2$ **55.** $a^3(a - b)(a + 5b)$
57. $(a + b)(a - 2b)$ **59.** $(m + 20n)(m - 18n)$
61. $(mn - 8)(mn + 4)$ **63.** $a^3(ab + 5)(ab - 2)$
65. $(7m - 8n)^2$ **67.** $x^4(x + 2y)(x - y)$ **69.** $\left(6a - \frac{5}{4}\right)^2$
71. $\left(\frac{1}{2}a + \frac{1}{3}b\right)^2$ **73.** $(9a^2 + b^2)(3a - b)(3a + b)$
75. $(w - 7)(w + 2)(w - 2)$

77.

$y = -4x + 7$		
11 ?	$-4(-1) + 7$	
	$4 + 7$	
11	11	TRUE

$y = -4x + 7$		
7 ?	$-4 \cdot 0 + 7$	
	$0 + 7$	
7	7	TRUE

$y = -4x + 7$		
-5 ?	$-4 \cdot 3 + 7$	
	$-12 + 7$	
-5	-5	TRUE

79. $X = \dfrac{A + 7}{a + b}$

81. ◈ **83.** $-10 = -10$; probably correct
85. $(y - 2)(y + 3)(y - 3)$ **87.** $(a + 4)(a^2 + 1)$
89. $(x + 3)(x - 3)(x^2 + 2)$ **91.** $(x - 1)(x + 2)(x - 2)$
93. $(y - 1)^3$ **95.** $[(y + 4) + x]^2$
97. $(x^4 + 16)(x^2 + 4)(x + 2)(x - 2)$

Technology Connection, Section 5.7

TC1. $-4.65, 0.65$ **TC2.** $-0.37, 5.37$
TC3. $-4.56, -8.98$ **TC4.** No solution

Exercise Set 5.7, pp. 252–253

1. $-8, -6$ **3.** $3, -5$ **5.** $-12, 11$ **7.** $0, -5$
9. $0, -10$ **11.** $-\frac{5}{2}, -4$ **13.** $-\frac{1}{5}, 3$ **15.** $4, \frac{1}{4}$
17. $0, \frac{2}{3}$ **19.** $0, 18$ **21.** $-\frac{1}{10}, \frac{1}{27}$ **23.** $\frac{1}{3}, 20$
25. $0, \frac{2}{3}, \frac{1}{2}$ **27.** $-1, -5$ **29.** $-9, 2$ **31.** $3, 5$
33. $0, 8$ **35.** $0, -19$ **37.** $4, -4$ **39.** $\frac{2}{3}, -\frac{2}{3}$
41. -3 **43.** 4 **45.** $0, \frac{6}{5}$ **47.** $\frac{5}{3}, -1$ **49.** $\frac{2}{3}, -\frac{1}{4}$
51. $7, -2$ **53.** $\frac{9}{8}, -\frac{9}{8}$ **55.** $-3, 1$ **57.** $-\frac{2}{3}, -4$
59. $(3, 0), (-2, 0)$ **61.** $(2, 0), (-4, 0)$
63. $\left(\frac{3}{2}, 0\right), (-3, 0)$ **65.** $(a + b)^2$ **67.** $2x + 5 < 19$
69. ◈ **71.** ◈
73. **(a)** $x^2 - x - 12 = 0$; **(b)** $x^2 + 7x + 12 = 0$;
(c) $4x^2 - 4x + 1 = 0$; **(d)** $x^2 - 25 = 0$;
(e) $40x^3 - 14x^2 + x = 0$
75. 4 **77.** $\frac{1}{8}, -\frac{1}{8}$ **79.** $4, -4$
81. **(a)** $9x^2 - 12x + 24 = 0$; **(b)** $x^2 - 3x - 18 = 0$;
(c) $4x^2 + 8x + 36 = 0$; **(d)** $(2x + 8)(2x - 5) = 0$;
(e) $(x + 1)(5x - 5) = 0$; **(f)** $2x^2 + 20x - 4 = 0$
83. ◈ **85.** $-2.33, -6.77$ **87.** $-4.59, -9.15$
89. $-3.25, -6.75$

Exercise Set 5.8, pp. 258–261

1. $-\frac{3}{4}, 1$ **3.** $2, 4$ **5.** 14 and 15
7. 12 and 14, -14 and -12
9. 15 and 17, -17 and -15
11. Length: 12 m; width: 8 m **13.** 5
15. Height: 4 cm; base: 14 cm **17.** 6 m **19.** 5 and 7
21. 506 **23.** 12 **25.** 780 **27.** 20
29. Hypotenuse: 17 ft; leg: 15 ft
31. Dining room: 12 ft by 12 ft; kitchen: 10 ft by 12 ft
33.

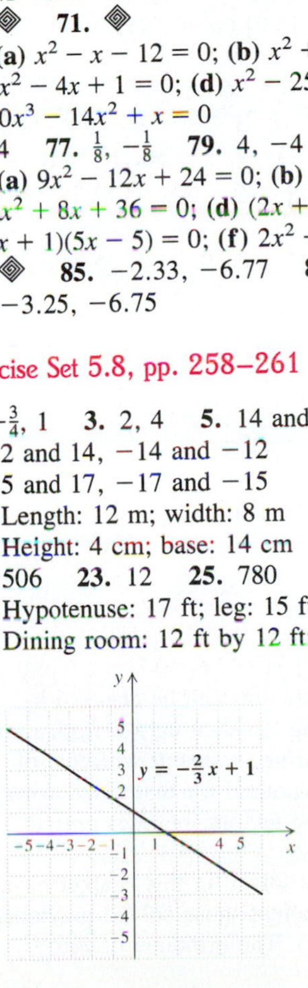

$y = -\frac{2}{3}x + 1$

35. 7 **37.** ◈ **39.** 5 ft **41.** 37
43. 30 cm by 15 cm **45.** 100 cm^2; 225 cm^2

Review Exercises: Chapter 5, pp. 262–263

1. [5.1] Answers may vary. $(-5x)(2x)$, $(-x)(10x)$, $(-2x)(5x)$
2. [5.1] Answers may vary. $(4x^2)(9x^3)$, $(18x)(2x^4)$, $(-6x^3)(-6x^2)$
3. [5.1] $x(x-5)$ **4.** [5.4] $3y^2(2y+1)(2y-1)$
5. [5.4] $(3x+2)(3x-2)$ **6.** [5.2] $(x+6)(x-2)$
7. [5.4] $(x+7)^2$ **8.** [5.4] $(x+2)(x+3)(x-3)$
9. [5.1] $(x^2+3)(x+1)$ **10.** [5.3] $(3x-1)(2x-1)$
11. [5.4] $(x^2+9)(x+3)(x-3)$
12. [5.3] $3x(3x-5)(x+3)$ **13.** [5.4] $2(x+5)(x-5)$
14. [5.1] $(x^3-2)(x+4)$
15. [5.4] $(4x^2+1)(2x+1)(2x-1)$
16. [5.1] $4x^4(2x^2-8x+1)$ **17.** [5.4] $3(2x+5)^2$
18. [5.4] Prime **19.** [5.2] $x(x-6)(x+5)$
20. [5.5] $(1+a)(1-a+a^2)$ **21.** [5.4] $(3x-5)^2$
22. [5.3] $2(3x+4)(x-6)$ **23.** [5.4] $(x-3)^2$
24. [5.3] $(2x+1)(x-4)$ **25.** [5.4] $2(3x-1)^2$
26. [5.5] $(3x-2)(9x^2+6x+4)$
27. [5.5] $2y(3x^2-1)(9x^4+3x^2+1)$
28. [5.4] $(5x-2)^2$ **29.** [5.2] $(xy+4)(xy-3)$
30. [5.4] $3(2a+7b)^2$
31. [5.4] $(a-b+2t)(a-b-2t)$
32. [5.4] $32(x^2-2y^2z^2)(x^2+2y^2z^2)$ **33.** [5.7] 1, −3
34. [5.7] −7, 5 **35.** [5.7] −4, 3 **36.** [5.7] $\frac{2}{3}$, $\frac{3}{2}$
37. [5.7] $\frac{1}{2}$, $\frac{5}{4}$ **38.** [5.7] 8, −2 **39.** [5.8] 3, −2
40. [5.7] $(-1, 0)$, $(\frac{5}{2}, 0)$
41. [5.8] −19 and −17, 17 and 19 **42.** [5.8] 5
43. [1.6] $\frac{1}{2}$ **44.** [3.2]

44. [3.2]

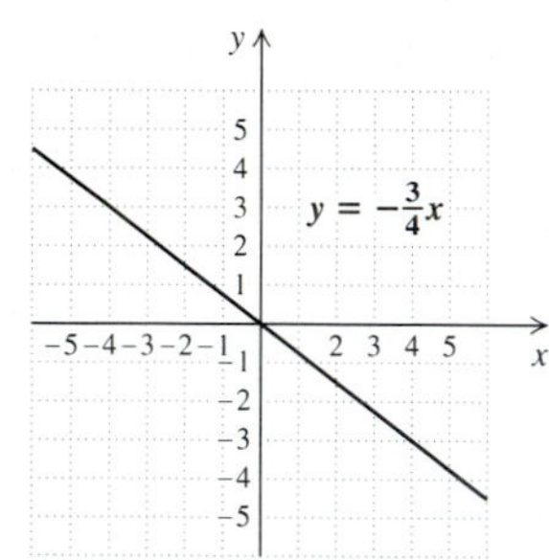

45. [4.1] m^2n^5 **46.** [2.7] $\frac{1}{2}x - 2 \leq 10$
47. [5.1], [5.6] ◈ Factorizations can be checked by multiplying or by evaluating. Evaluating a factorization for one number and comparing the result with that of evaluating the original polynomial for that number may be quicker than multiplying the factors, but it is only a partial check.
48. [5.7] ◈ The equations solved in this chapter have an x^2-term (are quadratic), whereas those solved previously have no x^2-term (are linear). The principle of zero products is used to solve quadratic equations and is not used to solve linear equations.
49. [5.8] $2\frac{1}{2}$ cm **50.** [5.8] 0, 2
51. [5.8] $l = 12$, $w = 6$ **52.** [5.7] No real solution
53. [5.7] 2, −3, $\frac{5}{2}$

Test: Chapter 5, p. 263

1. [5.1] Answers may vary. $4x \cdot x^2$, $2x^2 \cdot 2x$, $2 \cdot 2x^3$
2. [5.2] $(x-5)(x-2)$ **3.** [5.4] $(x-5)^2$
4. [5.1] $2y^2(3-4y+2y^2)$ **5.** [5.1] $(x^2+2)(x+1)$
6. [5.1] $x(x-5)$ **7.** [5.2] $x(x+3)(x-1)$
8. [5.3] $2(5x-6)(x+4)$ **9.** [5.4] $(2x+5)(2x-5)$
10. [5.2] $(x-4)(x+3)$ **11.** [5.1] $m(24m^2+40m+3)$
12. [5.4] $3(w+5)(w-5)$
13. [5.5] $2ab(2a^2+3b^2)(4a^4-6a^2b^2+9b^4)$
14. [5.4] $3(x^2+4)(x+2)(x-2)$ **15.** [5.4] $(7x-6)^2$
16. [5.3] $(5x-1)(x-5)$
17. [5.4] $(y+4+10t)(y+4-10t)$
18. [5.5] $3(r-1)(r^2+r+1)$
19. [5.4] $(x+5)(x+2)(x-2)$
20. [5.3] $3t(2t+5)(t-1)$
21. [5.2] $3(m+2n)(m-5n)$ **22.** [5.7] 5, −4
23. [5.7] −5, $\frac{3}{2}$ **24.** [5.7] −4, 7 **25.** [5.8] 8, −3
26. [5.8] $l = 8$ cm, $w = 5$ cm **27.** [1.6] −0.4
28. [2.7] $\{l \mid l < 13\}$

29. [3.2]

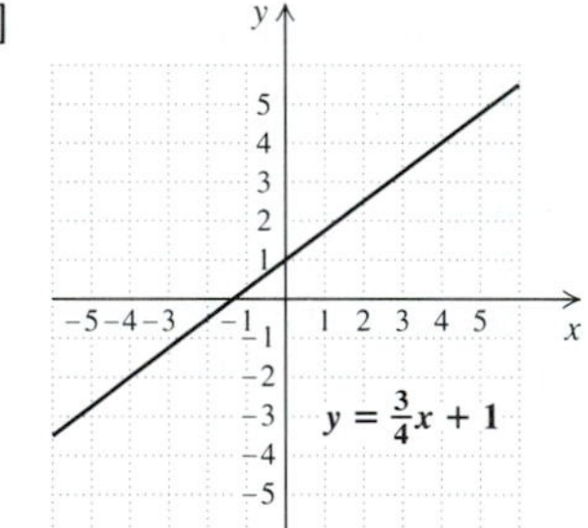

30. [4.1] $-27x^3y^{15}$ **31.** [5.8] $l = 15$, $w = 3$
32. [5.2] $(a-4)(a+8)$

CHAPTER 6

Exercise Set 6.1, pp. 270–271

1. 0 **3.** 8 **5.** $-\frac{5}{2}$ **7.** 7, −4 **9.** −5, 5 **11.** $\frac{a^2}{3b}$

13. $\frac{5}{2xy^4}$ **15.** $\frac{3}{2}$ **17.** $\frac{a-5}{a+1}$ **19.** $\frac{8}{3x^2}$ **21.** $\frac{x-3}{x}$

23. $\frac{m+1}{2m+3}$ **25.** $\frac{a-3}{a+2}$ **27.** $\frac{t+2}{2(t-4)}$ **29.** $\frac{x+5}{x-5}$

31. $a+1$ **33.** $\frac{x^2+1}{x+1}$ **35.** $\frac{3}{2}$ **37.** $\frac{6}{t-3}$ **39.** $\frac{a-3}{a-4}$

41. $\frac{t-2}{t+2}$ **43.** -1 **45.** -1 **47.** -6 **49.** $-a-1$
51. $(x+7)(x+1)$
53.

5x + 2y = 20
(0, 10)
(4, 0)

55. ◈ **57.** $x+2y$ **59.** $\frac{(t-1)(t-9)^2}{(t^2+9)(t+1)}$ **61.** $\frac{x-y}{x-5y}$ **63.** ◈

Exercise Set 6.2, pp. 274–276

1. $\frac{3x(x+4)}{2(x-1)}$ **3.** $\frac{(x-1)(x+1)}{(x+2)(x+2)}$ **5.** $\frac{(2x+3)(x+1)}{4(x-5)}$
7. $\frac{(a-5)(a+2)}{(a^2+1)(a^2-1)}$ **9.** $\frac{(x+1)(x-1)}{(2+x)(x+1)}$ **11.** $\frac{56x}{3}$
13. $\frac{2}{dc^2}$ **15.** $\frac{x+2}{x-2}$ **17.** $\frac{(a+5)(a-5)(2a-5)}{(a-3)(a-1)(2a+5)}$
19. $\frac{(a+3)(a-3)}{a(a+4)}$ **21.** $\frac{2a}{a-2}$ **23.** $\frac{t-5}{t+5}$
25. $\frac{5(a+6)}{a-1}$ **27.** $\frac{(x-1)(x-3)^3}{(x+3)(x+1)}$
29. $\frac{(a+2)(a-2)(a-1)}{(a+1)^2(a^4+1)}$ **31.** $\frac{t-2}{t-1}$ **33.** $\frac{x}{4}$
35. $\frac{1}{x^2-y^2}$ **37.** $\frac{x^2-4x+7}{x^2+2x-5}$ **39.** $\frac{3}{10}$ **41.** $\frac{1}{4}$ **43.** $\frac{y^2}{x}$
45. $\frac{(a+2)(a+3)}{(a-3)(a-1)}$ **47.** $\frac{(x-1)^2}{x}$ **49.** $\frac{1}{2}$
51. $\frac{(y+3)(y^2+1)}{y+1}$ **53.** $\frac{15}{8}$ **55.** $\frac{15}{4}$ **57.** $\frac{a-5}{3(a-1)}$
59. $2x+1$ **61.** $\frac{(x+2)^2}{x}$ **63.** $\frac{3}{2}$ **65.** $\frac{c+1}{c-1}$
67. $\frac{y-3}{2y-1}$ **69.** $\frac{1}{(c-5)^2}$ **71.** $\frac{t+5}{t-5}$ **73.** 4
75. $8x^3-11x^2-3x+12$ **77.** ◈
79. $\frac{a}{(c-3d)(2a+5b)}$ **81.** $-\frac{1}{b^2}$ **83.** x **85.** $\frac{4}{x+7}$
87. $\frac{(t-1)(t-9)(t-9)}{(t^2+9)(t+1)}$ **89.** $\frac{3(y+2)^3}{y(y-1)}$

Exercise Set 6.3, pp. 280–281

1. $\frac{8}{x}$ **3.** $\frac{3x+1}{15}$ **5.** $\frac{6}{a+3}$ **7.** $\frac{4}{a+6}$ **9.** $\frac{y+4}{y}$
11. $\frac{11x+8}{x+1}$ **13.** $\frac{7x+2}{x+1}$ **15.** $a+5$ **17.** $x-4$
19. $t+7$ **21.** $\frac{1}{x+2}$ **23.** $\frac{a+1}{a+6}$ **25.** $\frac{t-4}{t+3}$
27. $\frac{x+6}{x-5}$ **29.** $\frac{x}{4}$ **31.** $-\frac{1}{t}$ **33.** $\frac{-x+7}{x-6}$ **35.** $\frac{4a}{3}$
37. $\frac{13}{a}$ **39.** $\frac{4x-5}{4}$ **41.** $y+3$ **43.** $\frac{2b-14}{b^2-16}$
45. $\frac{-5}{t-4}$ **47.** $\frac{x-2}{x-7}$ **49.** $\frac{2x-16}{x^2-16}$ **51.** $\frac{2x-4}{x-9}$
53. $\frac{-4}{x-1}$
55.

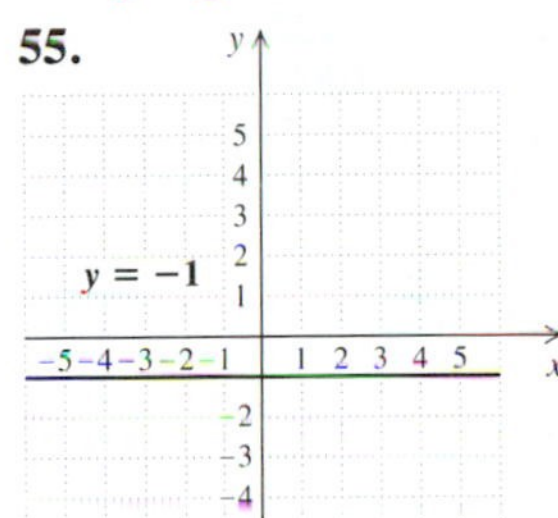

57.

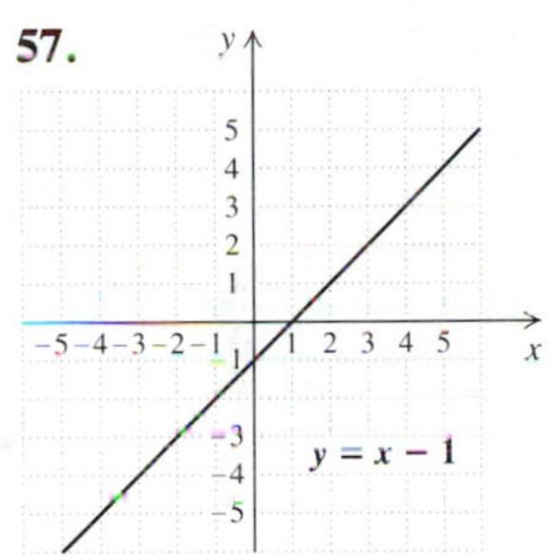

59. ◈ **61.** $\frac{18x+5}{x-1}$ **63.** 0 **65.** $\frac{20}{2y-1}$ **67.** 0
69. $\frac{x}{3x+1}$ **71.** ◈

Exercise Set 6.4, pp. 285–286

1. 108 **3.** 72 **5.** 126 **7.** 360 **9.** 420 **11.** $\frac{65}{72}$
13. $\frac{161}{600}$ **15.** $\frac{151}{180}$ **17.** $12x^3$ **19.** $18x^2y^2$
21. $6(y-3)$ **23.** $t(t+2)(t-2)$
25. $(x+2)(x-2)(x+3)$ **27.** $t(t+2)^2(t-4)$
29. $18a^5b^6$ **31.** $30x^2y^2z^3$ **33.** $(a-1)^2(a+1)$
35. $(m-2)^2(m-3)$ **37.** $(2+3x)(2-3x)$
39. $10v(v+3)(v+4)$ **41.** $18x^3(x-2)^2(x+1)$
43. $6x^3(x+2)^2(x-2)$ **45.** $\frac{14}{12x^5}, \frac{x^2y}{12x^5}$
47. $\frac{12b}{8a^2b^2}, \frac{5a}{8a^2b^2}$
49. $\frac{(x+3)(x+1)}{(x+3)(x+2)(x-2)}, \frac{(x-2)^2}{(x+3)(x+2)(x-2)}$
51. $\frac{3(t+2)(t-2)}{t(t+2)(t-2)}, \frac{4t(t-2)}{t(t+2)(t-2)}, \frac{t^2(t+2)}{t(t+2)(t-2)}$
53. $\frac{(x+1)(2x+3)}{(2x-3)(2x+3)}, \frac{x-2}{(2x+3)(2x-3)}, \frac{(x+1)(2x-3)}{(2x+3)(2x-3)}$
55. $(x-4)(x-15)$ **57.** $x^2-9x+18$ **59.** ◈
61. 1440 **63.** 24 min

Exercise Set 6.5, pp. 291–293

1. $\frac{2x+5}{x^2}$ **3.** $\frac{-1}{24r}$ **5.** $\frac{4x+6y}{x^2y^2}$ **7.** $\frac{4-3t}{18t^3}$
9. $\frac{5x+9}{24}$ **11.** $\frac{-x-4}{6}$ **13.** $\frac{a^2+16a+16}{16a^2}$
15. $\frac{7z-12}{12z}$ **17.** $\frac{x^2+4xy+y^2}{x^2y^2}$ **19.** $\frac{4x^2-13xt+9t^2}{3x^2t^2}$
21. $\frac{6x}{(x-2)(x+2)}$ **23.** $\frac{2x-40}{(x+5)(x-5)}$ **25.** $\frac{11x+2}{3x(x+1)}$
27. $\frac{3-5t}{2t(t-1)}$ **29.** $\frac{x^2+6x}{(x-4)(x+4)}$ **31.** $\frac{16}{3(z+4)}$
33. $\frac{3x-1}{(x-1)^2}$ **35.** $\frac{-t-9}{(t+3)(t-3)}$ **37.** $\frac{11a}{10(a-2)}$
39. $\frac{-2a^2}{(x+a)(x-a)}$ **41.** $\frac{2x^2+8x+16}{x(x+4)}$
43. $\frac{x-3}{(x+1)(x+3)}$ **45.** $\frac{x^2+5x+1}{(x+1)^2(x+4)}$
47. $\frac{x^2-48}{(x+7)(x+8)(x+6)}$ **49.** $\frac{3x^2+19x-20}{(x+3)(x-2)^2}$
51. $\frac{y^2+10y+11}{(y+7)(y-7)}$ **53.** $\frac{13x+20}{(4-x)(4+x)}$
55. $\frac{-a-2}{(a+1)(a-1)}$ **57.** $\frac{10x+6y}{(x+y)(x-y)}$ **59.** $\frac{2}{y(y-1)}$
61. $\frac{z-3}{2z-1}$ **63.** $\frac{-3x+1}{(2x-3)(x+1)}$ **65.** $\frac{1}{2c-1}$
67. $\frac{2}{x+y}$

69.

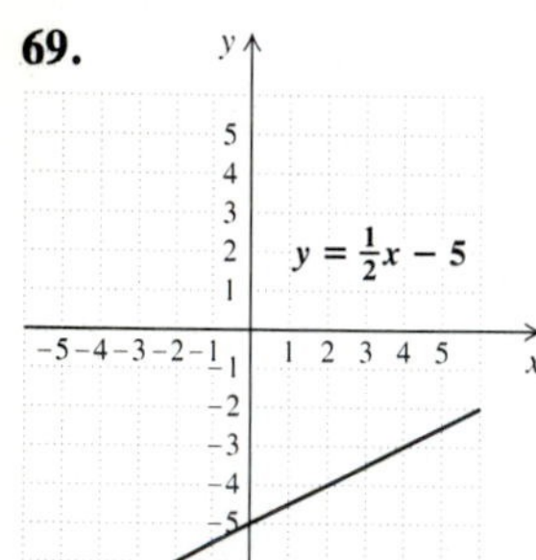

71.

$y = 3$

73. ◈ **75.** Perimeter: $\frac{16y+28}{15}$; area: $\frac{y^2+2y-8}{15}$
77. $\frac{(z+6)(2z-3)}{(z+2)(z-2)}$ **79.** $\frac{-3xy-3a+6x}{(a+2x)(a-2x)(y-3)^2}$
81. $\frac{a}{a-b}+\frac{3b}{b-a}$; answers may vary

Exercise Set 6.6, pp. 299–300

1. $\frac{25}{4}$ **3.** $\frac{1}{3}$ **5.** $\frac{1+4x}{1-3x}$ **7.** -6 **9.** $\frac{5}{3y^2}$ **11.** 8
13. $x-8$ **15.** $\frac{y}{y-1}$ **17.** $\frac{5-a}{a}$ **19.** $\frac{x+y}{x}$
21. $\frac{3m^2+2}{4m-3m^2}$ **23.** $\frac{10-3x^2}{12x^2+6}$ **25.** $\frac{3a+8b}{15b^2-2}$
27. $\frac{2y^2+3xy}{2x+y^2}$ **29.** $\frac{3a^4-2}{2a^4+3a}$ **31.** $\frac{-1}{x(x+h)}$
33. $\frac{y-3}{y+5}$ **35.** $\frac{x+5}{2x-3}$ **37.** $\frac{x-2}{x-3}$
39. $\frac{4x-7}{7x-9}$ **41.** $\frac{2a-3}{a+1}$ **43.** $\frac{2(x+1)(x-2)}{(x+3)(2x-1)}$
45. $23x^4+50x^3+23x^2-163x+41$ **47.** ◈
49. $\frac{(x-1)(3x-2)}{5x-3}$ **51.** $-\frac{ac}{bd}$ **53.** $\frac{3x+2}{2x+1}$ **55.** $\frac{1}{a-b}$

Exercise Set 6.7, p. 305

1. $\frac{47}{2}$ **3.** -6 **5.** $\frac{24}{7}$ **7.** $-4, -1$ **9.** $4, -4$
11. 3 **13.** $\frac{14}{3}$ **15.** 10 **17.** 5 **19.** $\frac{5}{2}$ **21.** -1
23. $\frac{17}{2}$ **25.** No solution **27.** -5 **29.** $\frac{5}{3}$ **31.** $\frac{1}{2}$
33. No solution **35.** -13
37. $a^{-6}b^{-15}$, or $\frac{1}{a^6b^{15}}$ **39.** $\frac{16x^4}{t^8}$ **41.** ◈ **43.** 7
45. 3 **47.** 2, -2 **49.** 4 **51.**

Exercise Set 6.8, pp. 313–316

1. $-1, 2$ **3.** 1 **5.** $2\frac{2}{9}$ hr **7.** $25\frac{5}{7}$ min **9.** $5\frac{1}{7}$ hr
11. $3\frac{3}{7}$ hr
13. 30 km/h, 70 km/h

Speed	Time
r	t
$r + 40$	t

Speed	Time
r	$\frac{150}{r}$
$r + 40$	$\frac{350}{r+40}$

15. Passenger: 80 km/h; freight: 66 km/h

Speed	Time
$r - 14$	t
r	t

Speed	Time
$r - 14$	$\frac{330}{r-14}$
r	$\frac{400}{r}$

17. $1\frac{7}{17}$ hr **19.** 3 hr **21.** 9 **23.** 2.3 km/h
25. 582 **27.** 702 km **29.** 1.92 g **31.** 10.5 **33.** $\frac{8}{3}$
35. 6.25 **37.** 287 **39.** 2074 **41.** 20
43. **(a)** 1.92 tons; **(b)** 28.8 lb **45.** $\frac{36}{68}$ **47.** 1
49. $13y^3 - 14y^2 + 12y - 73$ **51.** ◈ **53.** $\frac{3}{4}$
55. 2 mph **57.** $\frac{A}{C} = \frac{B}{D}$; $\frac{D}{B} = \frac{C}{A}$; $\frac{D}{C} = \frac{B}{A}$ **59.** $9\frac{3}{13}$ days
61. 45 mph **63.** $66\frac{2}{3}$ ft

Exercise Set 6.9, pp. 320–322

1. $r = \frac{S}{2\pi h}$ **3.** $b = \frac{2A}{h}$ **5.** $n = \frac{s}{180} + 2$, or $n = \frac{s + 360}{180}$
7. $b = \frac{3V - kB - 4kM}{k}$ **9.** $r = \frac{L}{l - S}$ **11.** $h = \frac{2A}{b_1 + b_2}$
13. $a = \frac{d}{b - c}$ **15.** $p = \frac{r}{q}$ **17.** $d = \frac{c}{a + b}$
19. $z = \frac{x - y}{p + q}$ **21.** $f = \frac{pq}{q + p}$ **23.** $p = \frac{ar}{v^2L}$
25. $n = \frac{a}{c(1 + b)}$ **27.** $b = \frac{a}{3S - 2}$ **29.** $F = \frac{9C + 160}{5}$
31. $d_2 = \frac{d_1 W_2}{W_1}$ **33.** $a = \frac{S(1 - r)}{1 - r^n}$
35. $m = \frac{-t}{f - g}$, or $m = \frac{t}{g - f}$ **37.** 12 **39.** $5\frac{5}{9}$ ohms
41. $r = \frac{A - P}{P}$ **43.** $-\frac{3}{4}$ **45.** $(x + 2)(x - 15)$ **47.** ◈
49. $T = \frac{FP}{u + EF}$ **51.** $-40°$ **53.** $T = -\frac{I_f}{I_t} + 1$

Review Exercises: Chapter 6, pp. 324–325

1. [6.1] 0 **2.** [6.1] 6 **3.** [6.1] −6, 6 **4.** [6.1] −6, 5
5. [6.1] −2 **6.** [6.1] 0, 3, 5 **7.** [6.1] $\frac{x - 2}{x + 1}$
8. [6.1] $\frac{7x + 3}{x - 3}$ **9.** [6.1] $\frac{y - 5}{y + 5}$ **10.** [6.2] $\frac{a - 6}{5}$
11. [6.2] $\frac{6}{2t - 1}$ **12.** [6.2] $-20t$ **13.** [6.2] $\frac{2x^2 - 2x}{x + 1}$
14. [6.2] $\frac{(x^2 + 1)(2x + 1)}{(x - 2)(x + 1)}$ **15.** [6.2] $\frac{(t + 4)^2}{t + 1}$
16. [6.4] $30x^2y^2$ **17.** [6.4] $x^4(x + 1)(x - 1)$
18. [6.4] $(y - 2)(y + 2)(y + 1)$ **19.** [6.3] $\frac{-3x + 18}{x + 7}$
20. [6.5] −1 **21.** [6.3] $\frac{4}{x - 4}$ **22.** [6.5] $\frac{x + 5}{2x}$
23. [6.3] $\frac{2x + 3}{x - 2}$ **24.** [6.5] $\frac{2a}{a - 1}$ **25.** [6.3] $d + c$
26. [6.5] $\frac{-x^2 + x + 26}{(x - 5)(x + 5)(x + 1)}$ **27.** [6.5] $\frac{2(x - 2)}{x + 2}$
28. [6.5] $\frac{8x + 3}{2x(2x + 1)}$ **29.** [6.6] $\frac{z}{1 - z}$
30. [6.6] $\frac{2x^4y^2 + x^3}{y + xy}$ **31.** [6.6] $c - d$ **32.** [6.7] 8
33. [6.7] $-\frac{1}{2}$ **34.** [6.7] 3, −5 **35.** [6.8] $5\frac{1}{7}$ hr
36. [6.8] 240 km/h, 280 km/h **37.** [6.8] −2
38. [6.8] 160 **39.** [6.8] $x = 6$ **40.** [6.9] $s = \frac{rt}{r - t}$
41. [6.9] $C = \frac{5}{9}(F - 32)$, or $C = \frac{5}{9}F - \frac{160}{9}$ **42.** [2.1] 11
43. [3.3] (3, 0), (0, −6)

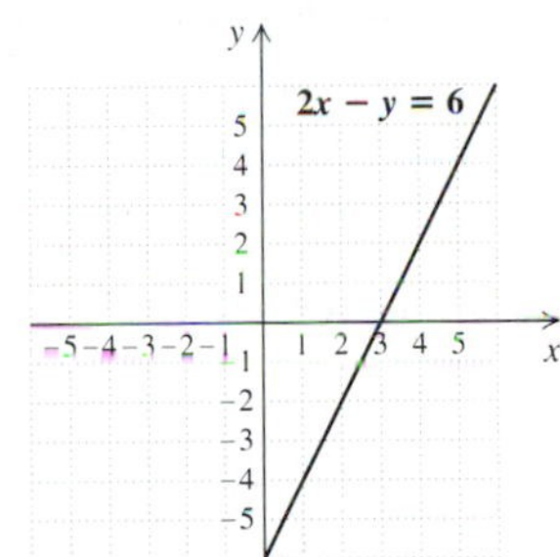

44. [5.2] $(x + 12)(x - 4)$
45. [4.3] $-2x^3 + 3x^2 + 12x - 18$
46. [6.5], [6.6], [6.7] ◈ A student should master factoring before beginning a study of rational equations because it is necessary to factor when finding the LCD of the rational expressions. It may also be necessary to factor to use the principle of zero products after fractions have been cleared.
47. [6.5] ◈ Although multiplying the denominators of the expressions being added results in a common denominator, it is often not the *least* common denominator. Using a common denominator other than the LCD makes the expressions more complicated, requires additional simplifying after the addition has been performed, and leaves more room for error.
48. [6.2] $\frac{5(a + 3)^2}{a}$ **49.** [6.3] $\frac{10a}{(a - b)(b - c)}$

Test: Chapter 6, pp. 325–326

1. [6.1] 0 **2.** [6.1] −8 **3.** [6.1] −7, 7
4. [6.1] 1, 2 **5.** [6.1] 1 **6.** [6.1] 0, −3, −5
7. [6.1] $\frac{3x + 7}{x + 3}$ **8.** [6.2] $\frac{a + 5}{2}$ **9.** [6.2] $\frac{(5x + 1)(x + 1)}{3x(x + 2)}$
10. [6.4] $(y - 3)(y + 3)(y + 7)$ **11.** [6.3] $\frac{23 - 3x}{x^3}$

12. [6.3] $\frac{8-2t}{t^2+1}$ **13.** [6.3] $\frac{-3}{x-3}$ **14.** [6.3] $\frac{2x-5}{x-3}$

15. [6.5] $\frac{8t-3}{t(t-1)}$

16. [6.5] $\frac{-x^2-7x-15}{(x+4)(x-4)(x+1)}$

17. [6.5] $\frac{x^2+2x-7}{(x-1)^2(x+1)}$

18. [6.6] $\frac{3y+1}{y}$

19. [6.6] $\frac{3a^2b^2-2a^3}{a^3b^2+2b^2}$ **20.** [6.7] 12 **21.** [6.7] −3, 5

22. [6.8] 4 **23.** [6.8] 16 **24.** [6.8] 45 km/h, 65 km/h

25. [6.9] $t=\frac{d}{r+w}$ **26.** [2.1] $-\frac{3}{7}$

27. [3.3] (10, 0), (0, 4)

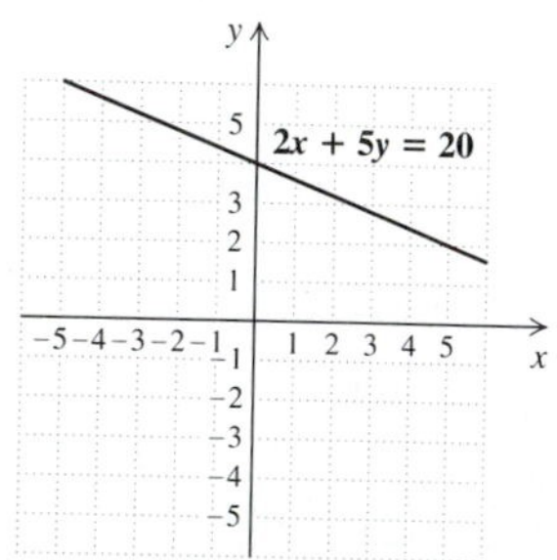

28. [5.2] $(x-9)(x+5)$ **29.** [4.3] $13x^2-29x+76$

30. [6.8] Reggie: 10 hr; Rema: 4 hr **31.** [6.6] $\frac{3a+2}{2a+1}$

CUMULATIVE REVIEW: 1–6

1. [1.2] $2b+a$ **2.** [1.4] $-3.1>-3.15$ **3.** [1.8] 49
4. [1.8] $-8x+28$ **5.** [1.5] $-\frac{43}{8}$ **6.** [1.7] 1
7. [1.7] −6.2 **8.** [1.8] 8 **9.** [2.2] 10
10. [2.2] −3 **11.** [2.1] $\frac{9}{2}$ **12.** [2.2] −2
13. [2.2] $\frac{8}{3}$ **14.** [2.2] $-\frac{1}{4}$ **15.** [2.2] −8
16. [2.1] $-\frac{1}{2}$ **17.** [2.6] $\{y|y\leq-\frac{2}{3}\}$ **18.** [5.7] $\frac{4}{3}, -\frac{5}{2}$
19. [5.7] $\frac{1}{2}$, −4 **20.** [5.7] 4, −4 **21.** [6.7] 2

22. [6.7] 1 **23.** [6.7] −13 **24.** [6.9] $t=\frac{4b}{A}$

25. [6.9] $n=\frac{tm}{t-m}$ **26.** [6.9] $c=\frac{a-b}{r}$

27. [1.6] $\frac{3}{2}x+2y-3z$ **28.** [4.2] $-4x^3-\frac{1}{7}x^2-2$

29. [3.2]

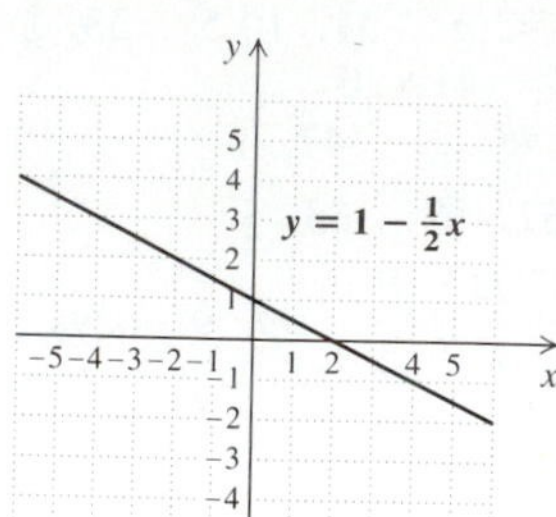

30. [3.3]

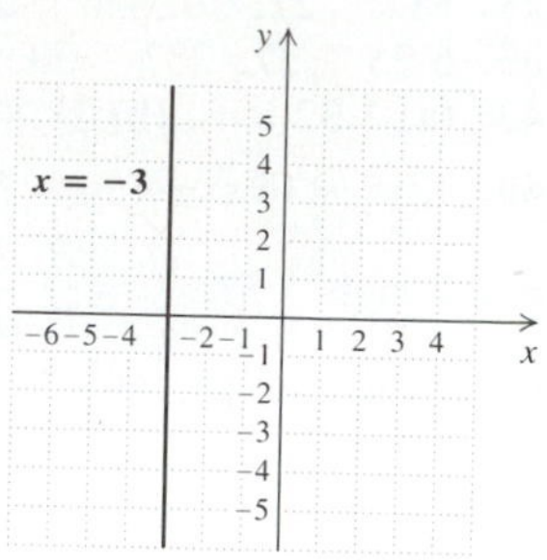

31. [3.3]

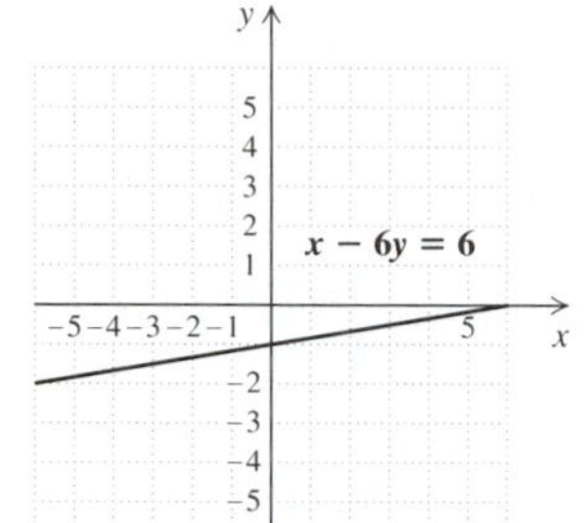

32. [4.1] x^{10} **33.** [4.9] z^{11} **34.** [4.1] $-27x^6y^3$
35. [4.3] $-y^3-2y^2-2y+7$
36. [1.2] $12x+16y+4z$ **37.** [4.5] a^2-9
38. [4.4] $2x^5+x^3-6x^2-x+3$
39. [4.5] $36x^2-60xy+25y^2$
40. [4.5] $6x^7-12x^5+9x^2-18$ **41.** [4.5] $4x^6-1$
42. [5.1] $2x(3-x-12x^3)$ **43.** [5.4] $(4x+9)(4x-9)$
44. [5.2] $(x-6)(x-4)$ **45.** [5.3] $(2x+1)(4x+3)$
46. [5.3] $2(3x-2)(x-4)$ **47.** [5.4] $2(x+3)(x-3)$
48. [5.4] $(4x+5)^2$ **49.** [5.3] $(3x-2)(x+4)$

50. [5.1] $(x^3-3)(x+2)$ **51.** [6.2] $\frac{y-6}{2}$

52. [6.2] 1 **53.** [6.5] $\frac{a^2+7ab+b^2}{a^2-b^2}$ **54.** [6.3] $\frac{-2x-5}{x-4}$

55. [6.6] $\frac{x}{x-2}$ **56.** [6.6] $\frac{t+2t^3}{t^3-2}$

57. [4.7] $5x^2-4x+2+\frac{2}{3x}+\frac{6}{x^2}$

58. [4.7] $15x^3-57x^2+177x-529+\frac{1605}{x+3}$

59. [2.5] −278 and −276 **60.** [2.4] 544.32
61. [2.5] −15 **62.** [2.7] $\{s|s\leq225\}$ **63.** [5.8] 14 ft
64. [6.8] 40 km/h, 50 km/h **65.** [6.8] $3\frac{3}{7}$ hr
66. [4.3], [4.5] 12 **67.** [1.4], [2.2] −144, 144
68. [4.5] $16y^6-y^4+6y^2-9$
69. [5.4] $2(a^{16}+81b^{20})(a^8+9b^{10})(a^4+3b^5)(a^4-3b^5)$
70. [5.7] 4, −7, 12 **71.** [1.4], [1.6] −7

CHAPTER 7

Exercise Set 7.1, pp. 336–338

1. $\frac{2}{3}$ **3.** 1 **5.** $\frac{1}{3}$ **7.** 3 **9.** $-\frac{1}{2}$ **11.** $-\frac{3}{2}$ **13.** -2
15. Undefined **17.** $-\frac{3}{4}$ **19.** $-\frac{4}{5}$ **21.** 7 **23.** $-\frac{2}{3}$
25. -2 **27.** 0 **29.** Undefined **31.** 0
33. Undefined **35.** 0 **37.** Undefined **39.** 0
41. 6.7% **43.** 0.08 or 8% **45.** 3% **47.** 0.6 ft
49. $45x^2 - 15x$ **51.** $x^2 - 49$ **53.** ◈
55. $\{m|m \geqslant \frac{4}{3}\}$ **57.** 27 candles per hour **59.** ◈

Technology Connection, Section 7.2

TC1. $y_1 = -\frac{3}{4}x - 2$ $y_2 = -\frac{1}{5}x - 2$
$y_3 = -\frac{3}{4}x - 5$ $y_4 = -\frac{1}{5}x - 5$

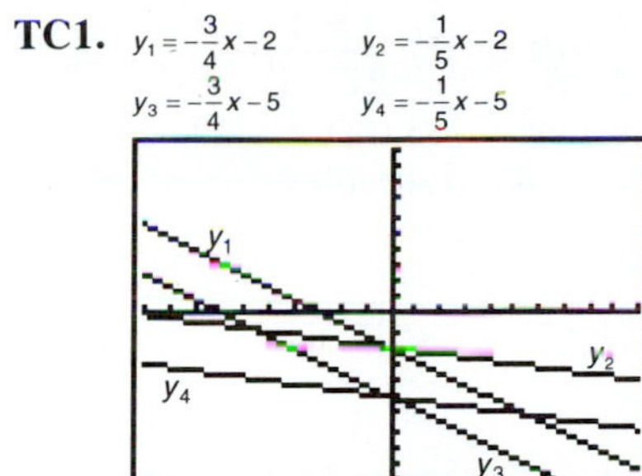

TC2. $y = 2x - 1$, $y = 2x + 4$, $y = -x + 3$, $y = 4x + 3$; Answers may vary.

Exercise Set 7.2, pp. 343–344

1.

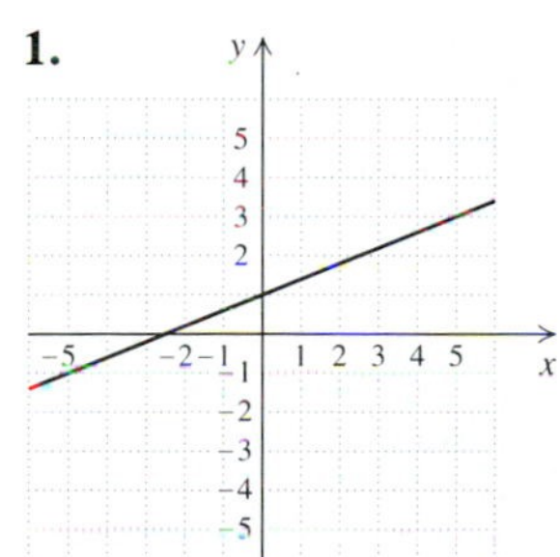

3.

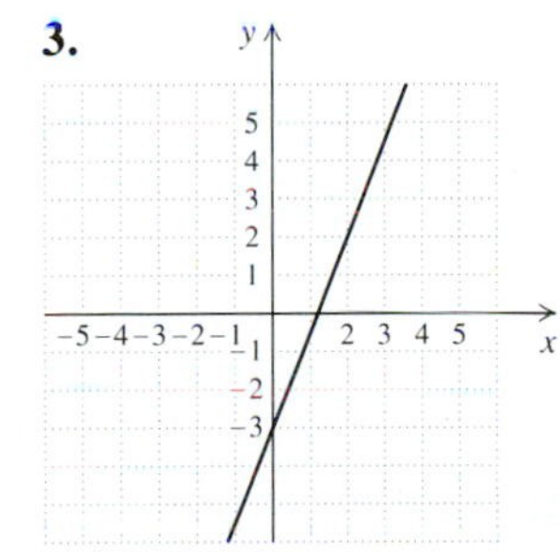

5.

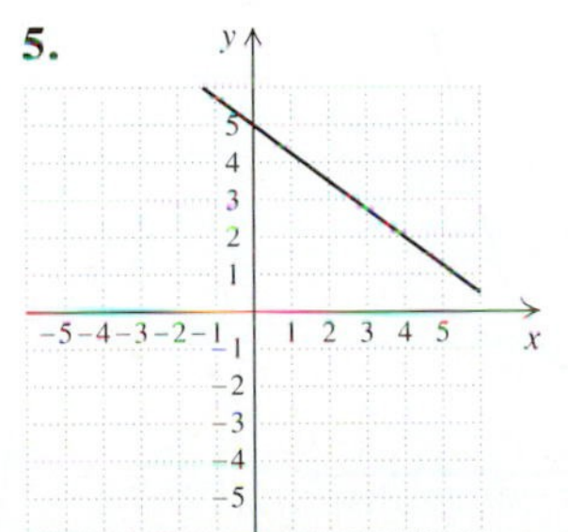

7.

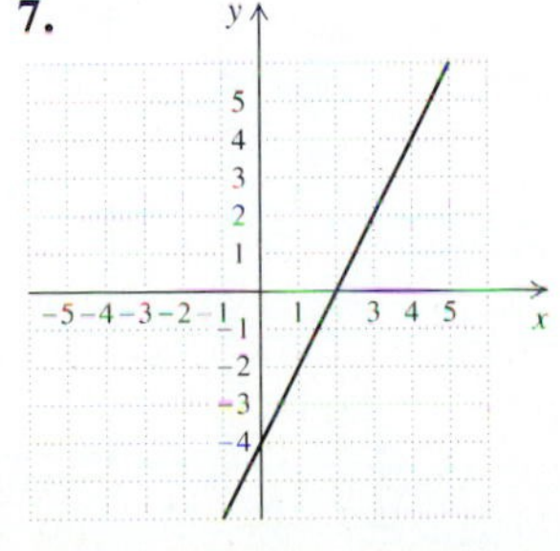

9.

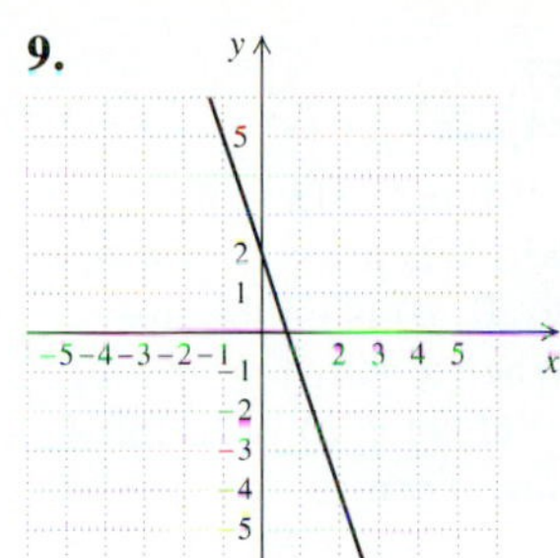

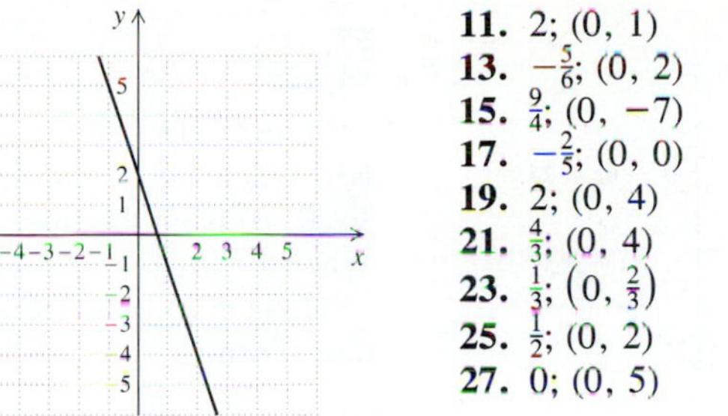

11. 2; (0, 1)
13. $-\frac{5}{6}$; (0, 2)
15. $\frac{9}{4}$; (0, −7)
17. $-\frac{2}{5}$; (0, 0)
19. 2; (0, 4)
21. $\frac{4}{3}$; (0, 4)
23. $\frac{1}{3}$; $\left(0, \frac{2}{3}\right)$
25. $\frac{1}{2}$; (0, 2)
27. 0; (0, 5)

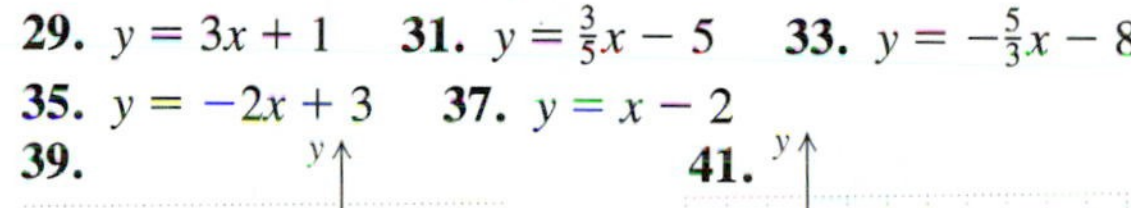

29. $y = 3x + 1$ **31.** $y = \frac{3}{5}x - 5$ **33.** $y = -\frac{5}{3}x - 8$
35. $y = -2x + 3$ **37.** $y = x - 2$

39.

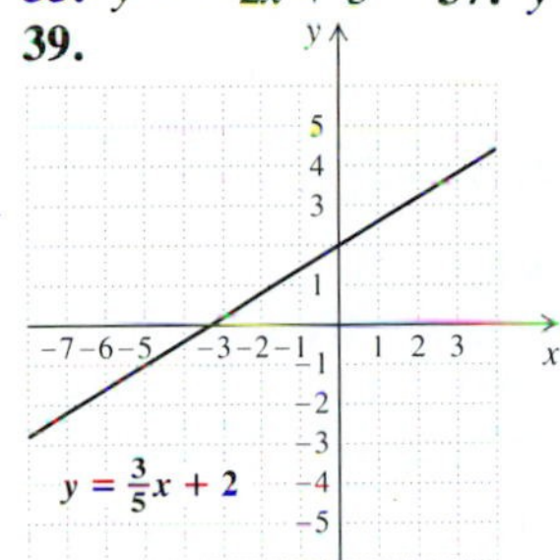

41.

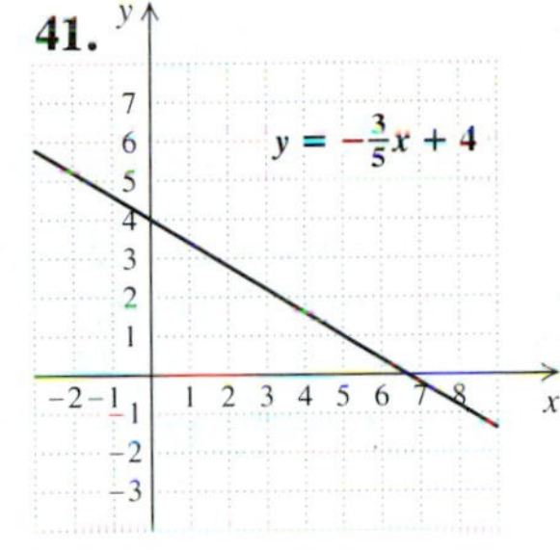

43.

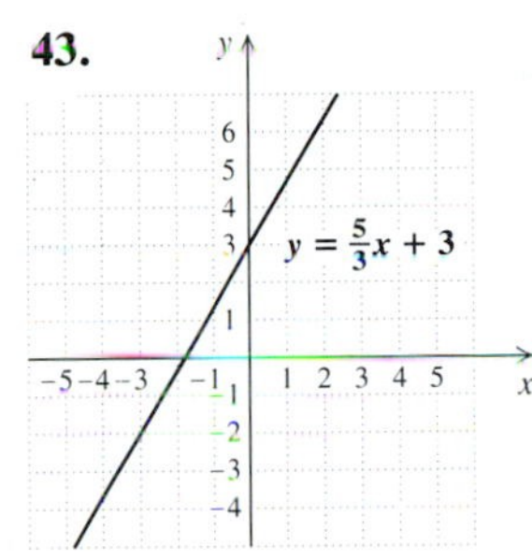

45.

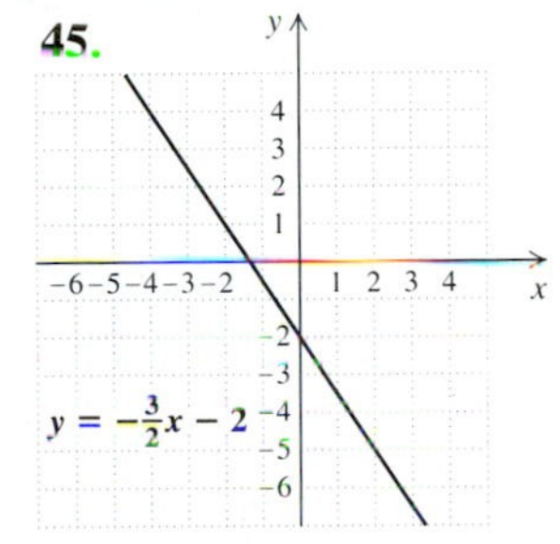

47.

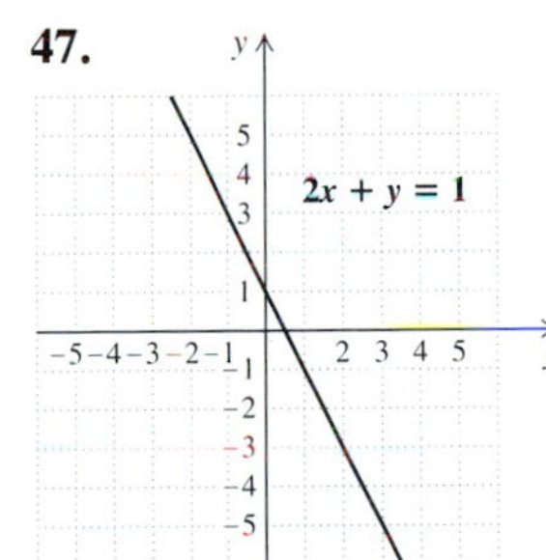

49.

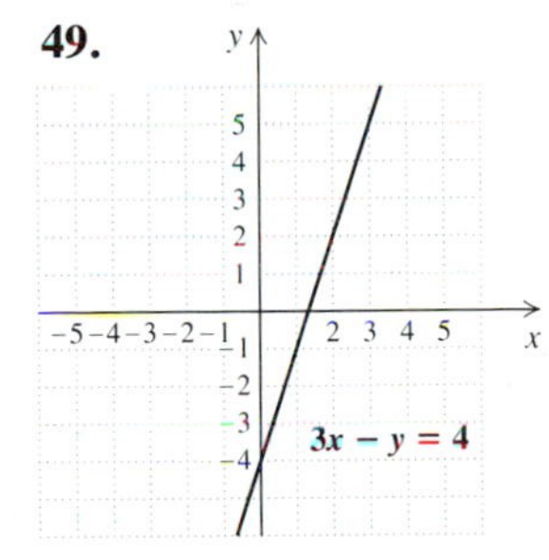

51.

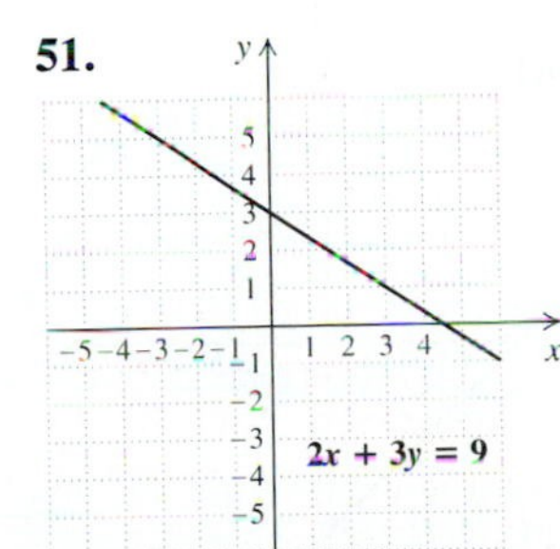

53.

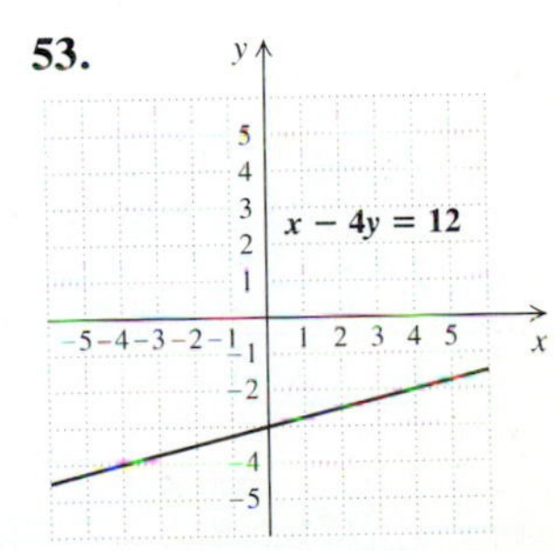

55.

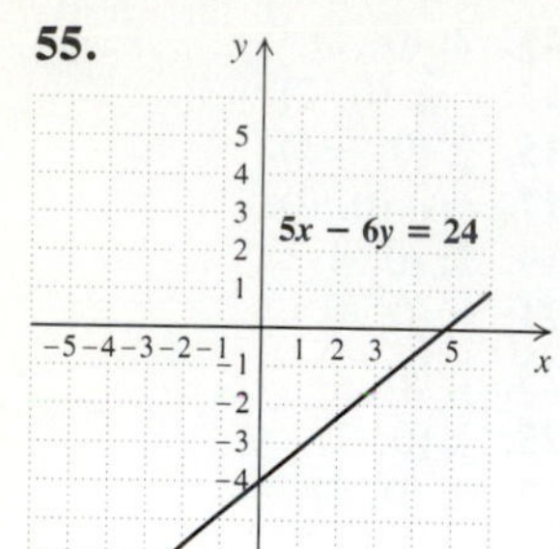

57. 0, 3
59. 13 and 15, −15 and −13
61. ◈
63. When $x = 0$, $y = b$, so $(0, b)$ is on the line. When $x = 1$, $y = m + b$, so $(1, m + b)$ is on the line. Then

$$\text{slope} = \frac{(m + b) - b}{1 - 0} = m.$$

65. $y = \frac{3}{2}x - 2$

Exercise Set 7.3, pp. 349–350

1. $y - 5 = 5(x - 2)$ **3.** $y - 4 = \frac{3}{4}(x - 2)$ **5.** $y - (-6) = 1 \cdot (x - 2)$ **7.** $y - 0 = -3(x - (-7))$ **9.** $y - 6 = \frac{2}{3}(x - 5)$ **11.** $y = 2x + 1$ **13.** $y = -x + 9$ **15.** $y = \frac{1}{2}x + 4$ **17.** $y = -\frac{1}{3}x - 7$ **19.** $y = \frac{5}{4}x - 5$ **21.** $y = \frac{1}{4}x + \frac{5}{2}$ **23.** $y = -\frac{1}{2}x + 4$ **25.** $y = -\frac{3}{2}x + \frac{13}{2}$ **27.** $y = \frac{2}{5}x - 2$ **29.** $y = \frac{3}{4}x - \frac{5}{2}$

31.

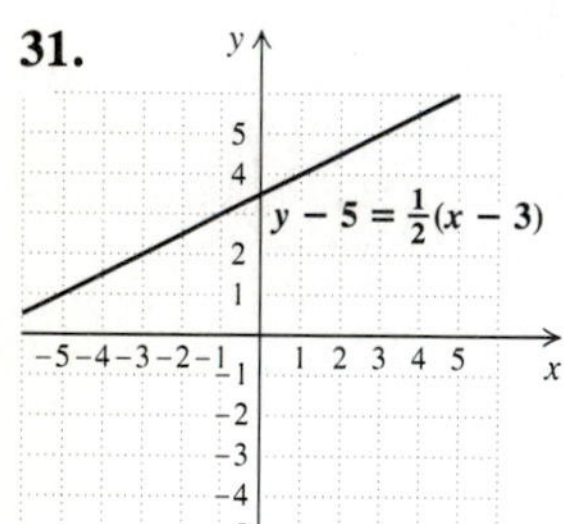

33.

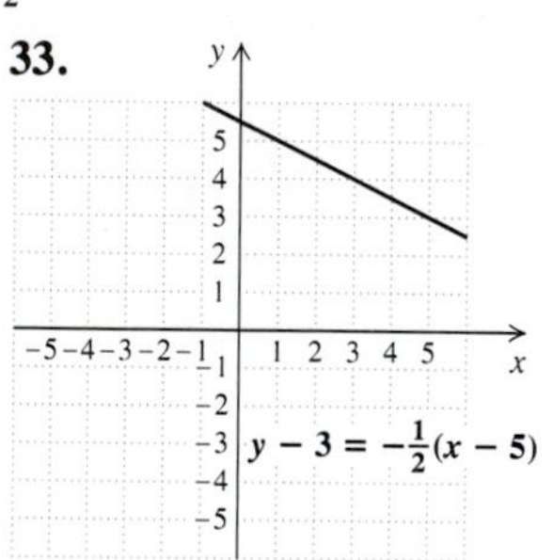

35.

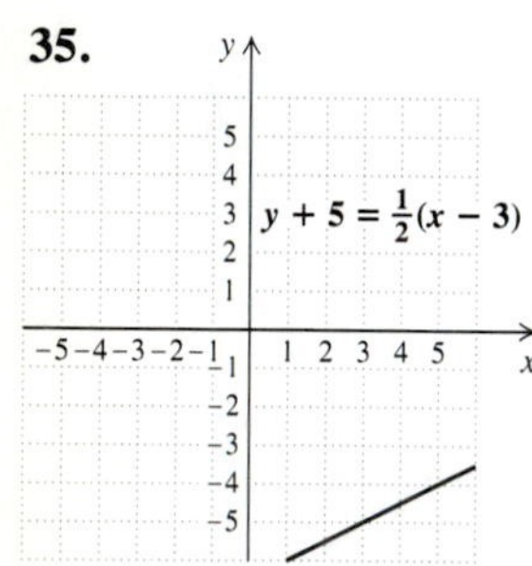

37.

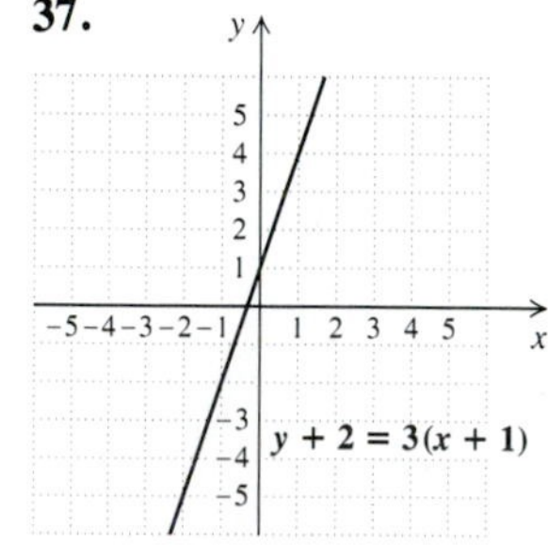

39.

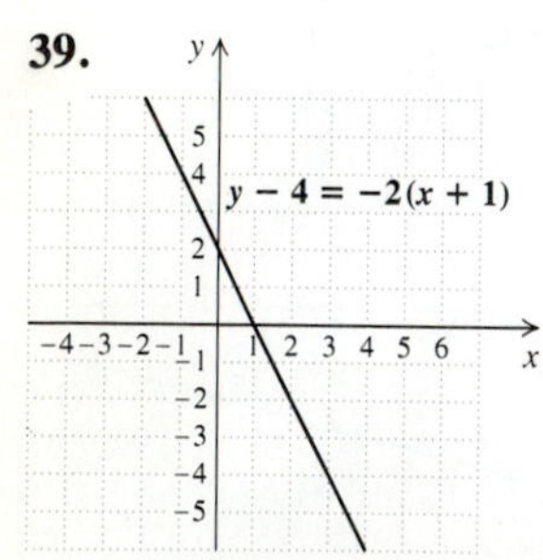

41.

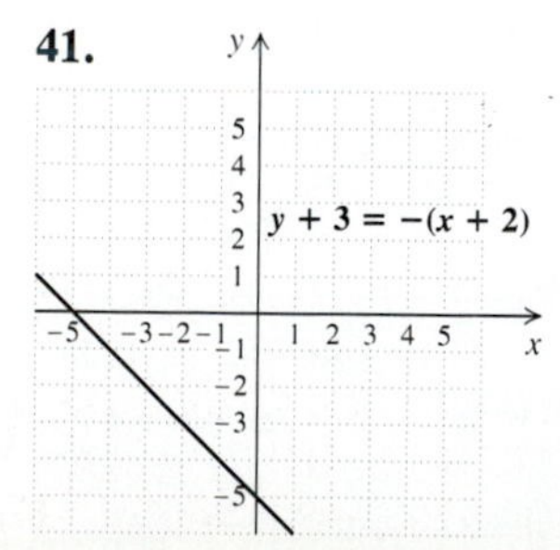

43. Yes **45.** No **47.** Yes **49.** $y = -\frac{1}{2}x + \frac{17}{2}$ **51.** $y = \frac{5}{7}x - \frac{17}{7}$ **53.** $y = \frac{1}{3}x + 4$ **55.** Yes **57.** No **59.** Yes **61.** $y = \frac{1}{2}x + 4$ **63.** $y = \frac{4}{3}x - 6$ **65.** $y = \frac{5}{2}x + 9$ **67.** $7x^2y^2(x + 5y^4)$ **69.** $\frac{x + 1}{2x(x - 1)}$ **71.** ◈ **73.** $y = 3x - 9$ **75.** $y = \frac{1}{2}x + 1$

Exercise Set 7.4, pp. 355–359

1. No **3.** Yes **5.** Yes **7.** No **9.** Yes **11.** Function **13.** Function **15.** A relation, but not a function. **17.** **(a)** 1; **(b)** −3; **(c)** −6; **(d)** 9; **(e)** a + 3 **19.** **(a)** 4; **(b)** 9; **(c)** 49; **(d)** $5t^2 + 4$; **(e)** $20a^2 + 4$ **21.** **(a)** 15; **(b)** 32; **(c)** 20; **(d)** 4; **(e)** $27r^2 + 6r - 1$ **23.** **(a)** $\frac{3}{5}$; **(b)** $\frac{1}{3}$; **(c)** $\frac{4}{7}$; **(d)** 0; **(e)** $\frac{x - 1}{2x - 1}$ **25.** $4\sqrt{3}$ cm^2 **27.** 36π in$^2 \approx 113.04$ in^2 **29.** 14°F **31.** 159.48 cm **33.** 75 **35.** 1.4 million **37.** 3.5 drinks

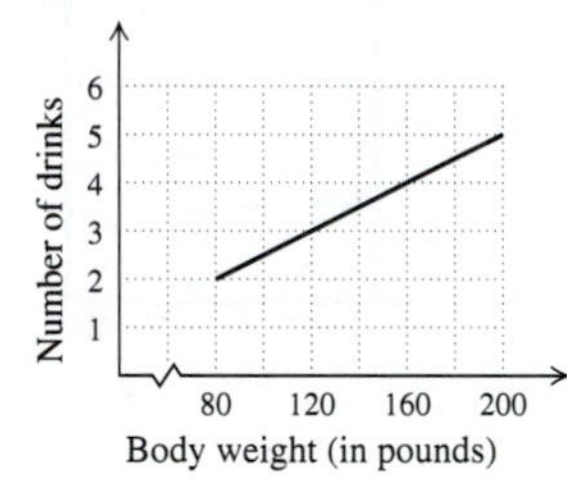

39. About 2.5 hr

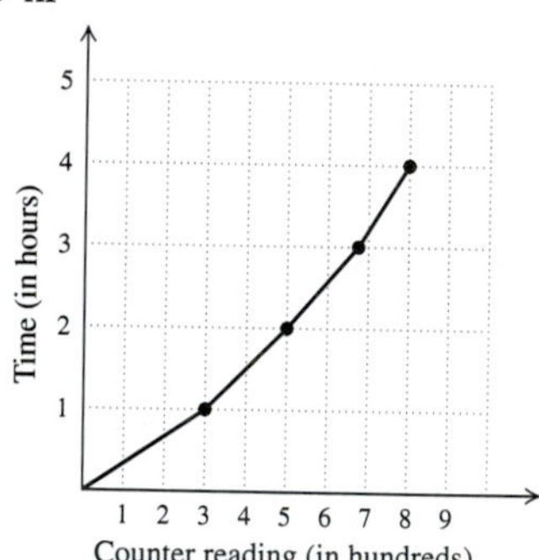

41. 64,000

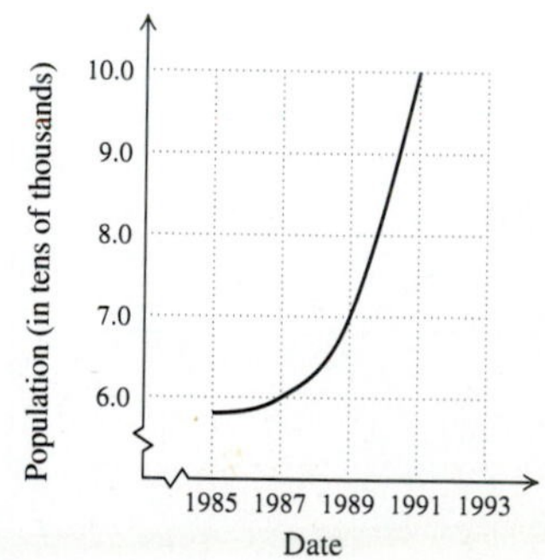

43. About \$310,000

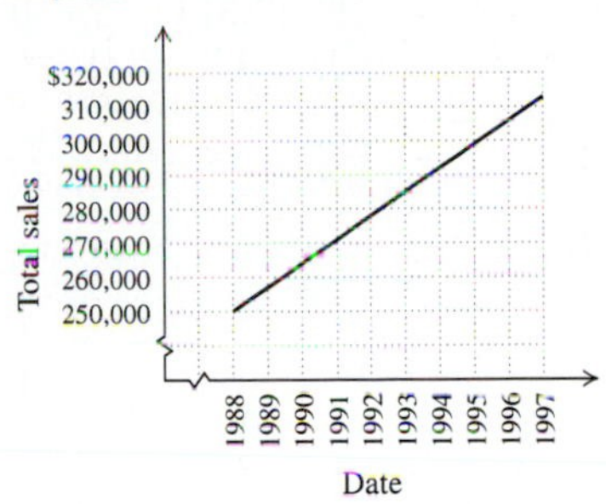

45. Yes **47.** Yes **49.** No **51.** No **53.** $3x^3(x-3)$ **55.** $(x-7)(x^2+3)$ **57.** ◈ **59.** **(a)** 3.1497708; **(b)** 55.7314683; **(c)** 3178.20675; **(d)** 1166.70323 **61.** $g(x) = \frac{15}{4}x - \frac{13}{4}$ **63.** ◈ **65.** ◈ **67.** At 2 min, 40 sec and at 5 min, 40 sec **69.** 1 every 3 min

Exercise Set 7.5, pp. 365–367

1. 1 **3.** 7 **5.** −29 **7.** −41 **9.** −30 **11.** 110 **13.** $\frac{1}{2}$ **15.** $\frac{10}{11}$ **17.** 13 **19.** $x^2 - x + 1$ **21.** 5 **23.** 2 **25.** 42 **27.** $-\frac{3}{4}$ **29.** $\frac{1}{6}$ **31.** $\{x|x$ is a real number and $x \neq 3\}$ **33.** $\{x|x$ is a real number and $x \neq 0$ and $x \neq -4\}$ **35.** $\{x|x$ is a real number$\}$ **37.** $\{x|x$ is a real number and $x \neq 2$ and $x \neq 6\}$ **39.** $\{x|x$ is a real number$\}$ **41.** $\{x|x$ is a real number and $x \neq 2\}$ **43.** $\{x|x$ is a real number and $x \neq 1\}$ **45.** $\{x|x$ is a real number and $x \neq 2$ and $x \neq 4\}$ **47.** $\{x|x$ is a real number and $x \neq -2$ and $x \neq \frac{4}{3}\}$ **49.** $\{x|x$ is a real number and $x \neq 3\}$ **51.** $\{x|x$ is a real number and $x \neq 4\}$ **53.** $\{x|x$ is a real number and $x \neq 4$ and $x \neq 5\}$ **55.** $\{x|x$ is a real number and $x \neq 0$ and $x \neq 4\}$ **57.** $\{x|x$ is a real number and $x \neq 1$ and $x \neq \frac{4}{3}\}$ **59.** $\{x|x$ is a real number and $x \neq 2$, $x \neq 3$, and $x \neq 4\}$ **61.** $\{x|0 \leq x \leq 9\}$; $\{x|3 \leq x \leq 10\}$; $\{x|3 \leq x \leq 9\}$; $\{x|3 \leq x \leq 9\}$

63.

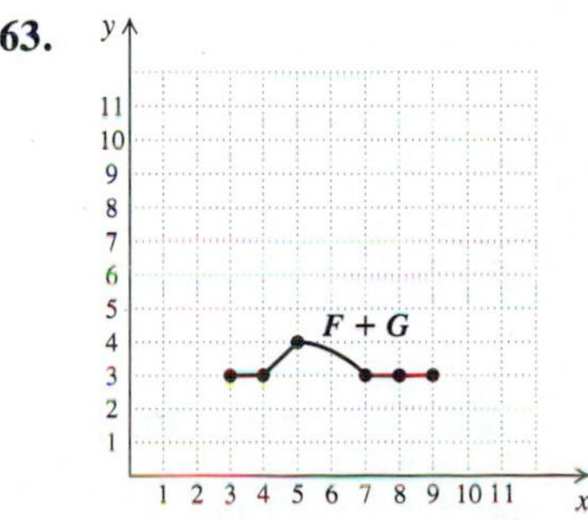

65. $2x^3 + 5x^2 - 4x - 3$ **67.** $(a+6)(3a^2-4)$ **69.** ◈ **71.** $\{x|x$ is a real number and $x \neq 2$, $x \neq 3$, and $x \neq 1\}$ **73.** For $f+g$, $f-g$, and $f \cdot g$, domain $= \{-2, -1, 0, 1\}$; for f/g, domain $= \{-2, 0, 1\}$ **75.** $\{x|x$ is a real number and $x \neq 1$, $x \neq -1$, $x \neq 2$, $x \neq -2$, and $x \neq 3\}$ **77.** Answers may vary. $f(x) = \dfrac{1}{x+2}$, $g(x) = \dfrac{1}{x-5}$ **79.** **(a)** 67 million; **(b)** 50 million; **(c)** 50 million

Exercise Set 7.6, pp. 373–376

1. $k = 8$; $y = 8x$ **3.** $k = 3.6$; $y = 3.6x$ **5.** $k = \frac{15}{4}$; $y = \frac{15}{4}x$ **7.** $k = 1.6$; $y = 1.6x$ **9.** 6 amperes **11.** \$4.29 **13.** 50 kg **15.** 3.36 **17.** $k = 60$; $y = \frac{60}{x}$ **19.** $k = 12$; $y = \frac{12}{x}$ **21.** $k = 36$; $y = \frac{36}{x}$ **23.** $k = 9$; $y = \frac{9}{x}$ **25.** $\frac{32}{9}$ amperes **27.** 160 cm^3 **29.** $6\frac{2}{3}$ hours **31.** $y = 15x^2$ **33.** $y = \dfrac{0.0015}{x^2}$ **35.** $y = xz$ **37.** $y = 0.3xz^2$ **39.** $y = \dfrac{4wx^2}{z}$ **41.** $y = \dfrac{xz}{5wp}$ **43.** 36 mph **45.** 6.25 km **47.** 1600 km **49.** 2 mm **51.** $(a+b)(c-1)$ **53.** $4x^2 + 0.4x + 0.01$ **55.** ◈ **57.** ◈ **59.** $P = kS$, $k = 8$

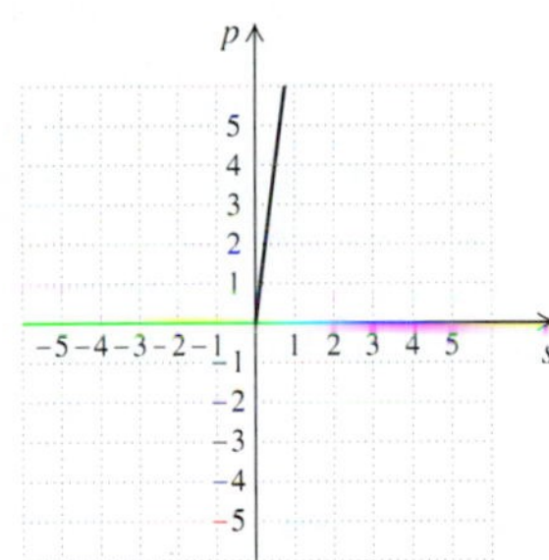

61. $B = kN$ **63.** **a)** $k \approx 0.001$; $N = \dfrac{0.001\,P_1P_2}{d^2}$ **b)** 1173 km **65.** Q varies directly as the square of p and inversely as the cube of q. **67.** If $p = kq$, then $q = \frac{1}{k}p$. Since k is a constant, so is $\frac{1}{k}$, and q varies directly as p.

Review Exercises: Chapter 7, pp. 378–379

1. [7.1] 0 **2.** [7.1] $\frac{7}{3}$ **3.** [7.1] $-\frac{3}{7}$ **4.** [7.1] $\frac{3}{2}$ **5.** [7.1] 0 **6.** [7.1] Undefined **7.** [7.1] 2 **8.** [7.1] 7% **9.** [7.1] 0 **10.** [7.2] $\frac{3}{5}$ **11.** [7.2] −2 **12.** [7.1] Undefined **13.** [7.2] −9, (0, 46) **14.** [7.2] −1, (0, 9) **15.** [7.2] $\frac{1}{3}$, $(0, -\frac{2}{3})$ **16.** [7.2] $y = -2x - 4$ **17.** [7.2] $y = 1.5x + 1$

18. [7.2]

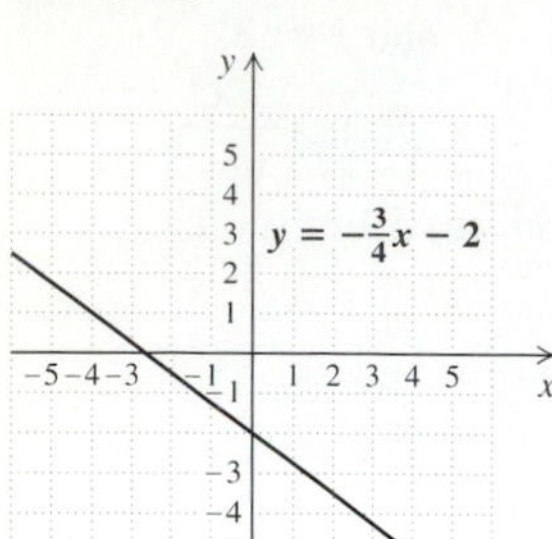

19. [7.2]

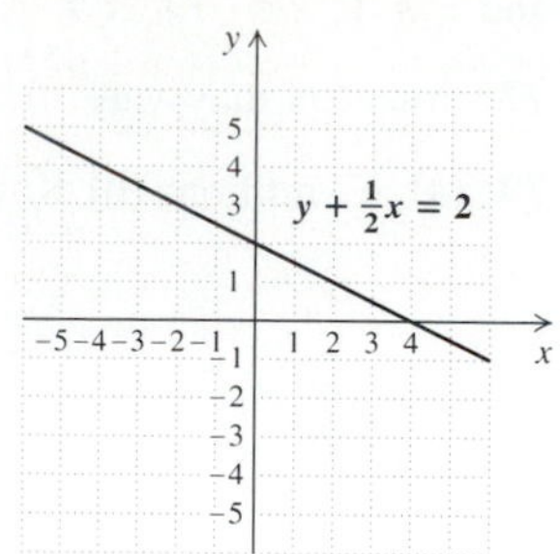

20. [7.3]

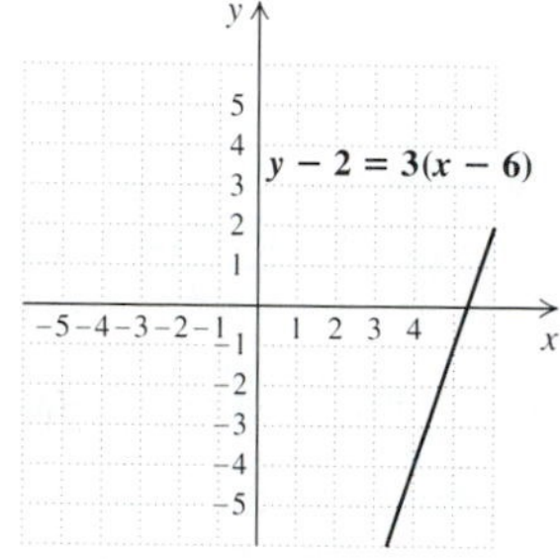

21. [7.3] $y - 2 = 3(x - 1)$ **22.** [7.3] $y - (-5) = \frac{2}{3}(x - (-2))$ **23.** [7.3] $y = x + 2$ **24.** [7.3] $y = \frac{1}{2}x - 1$ **25.** [7.3] Parallel **26.** [7.3] Not Parallel **27.** [7.3] Not perpendicular **28.** [7.3] Perpendicular **29.** [7.3] $y = 2x - 3$ **30.** [7.3] $y = -\frac{1}{2}x - \frac{1}{2}$ **31.** [7.4] -5 **32.** [7.4] -8 **33.** [7.5] 57 **34.** [7.5] -10 **35.** [7.5] $-\frac{7}{4}$ **36.** [7.4] $2a + 2b - 5$ **37.** [7.5] $\{x|x$ is a real number$\}$ **38.** [7.5] $\{x|x$ is a real number and $x \neq \frac{5}{2}\}$ **39.** [7.4] Yes **40.** [7.6] $9\frac{3}{5}$ lb **41.** [7.6] 8.25 cm **42.** [4.4] $3x^3 - 8x^2 + 5x - 2$ **43.** [4.5] $\frac{1}{4}y^2 + \frac{1}{4}y + \frac{1}{16}$ **44.** [5.1] $(x^2 + 2)(x - 1)$

45. [6.1] $\dfrac{a + 2}{2a + 1}$

46. [7.3] ◈ Point–slope form would be more useful than slope–intercept form if we were asked to find an equation for a line with a specified slope that passes through a specified point that is not the y-intercept.

47. [7.1] ◈ The slope of a line is the rise between two points on the line divided by the run between those points. For a vertical line, there is no run between any two points, and division by 0 is undefined; therefore, the slope is undefined. For a horizontal line, there is no rise between any two points, so the slope is 0/run, or 0.

48. [7.2] $y = -2x - 3$ **49.** [7.3] $-\frac{9}{2}$ **50.** [7.2] $y = -x$ **51.** [7.2] $-\frac{b}{a}$; $(0, b)$, $(a, 0)$

Test: Chapter 7, pp. 379–380

1. [7.1] Undefined **2.** [7.1] $\frac{7}{12}$ **3.** [7.2] -2 **4.** [7.1] 0 **5.** [7.1] Undefined **6.** [7.2] 2, $\left(0, -\frac{1}{4}\right)$ **7.** [7.2] $\frac{4}{3}$, $(0, -2)$ **8.** [7.2] $y = \frac{1}{2}x - 7$ **9.** [7.2] $y = -4x + 3$ **10.** [7.3] $y - 5 = 1(x - 3)$ **11.** [7.3] $y = -3(x - (-2))$ **12.** [7.3] $y = -3x + 4$ **13.** [7.3] $y = \frac{1}{4}x - 2$

14. [7.2]

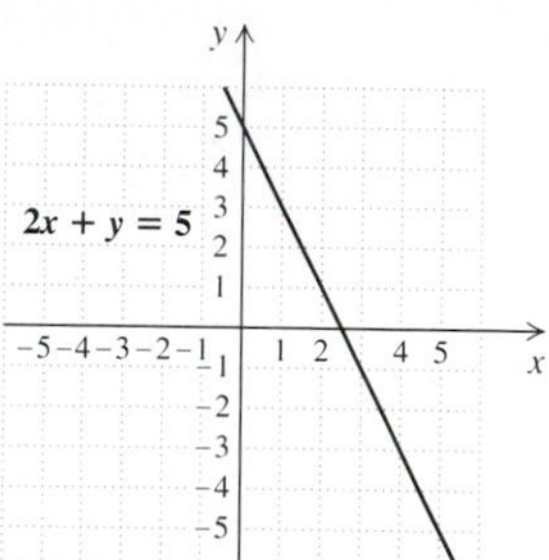

15. [7.3]

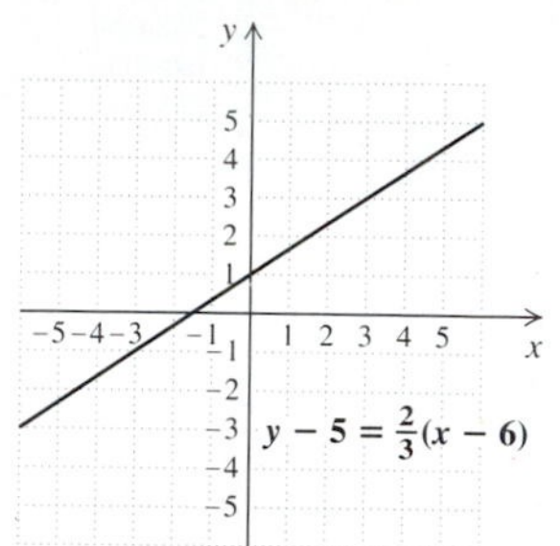

16. [7.3] Parallel **17.** [7.3] Not perpendicular **18.** [7.4], [7.5] **(a)** -4; **(b)** 5; **(c)** -2; **(d)** -5 **19.** [7.5] **(a)** $\{x|x$ is a real number and $x \neq 0\}$; **(b)** $\{x|x$ is a real number$\}$; **(c)** $\{x|x$ is a real number and $x \neq 0\}$; **(d)** $\{x|x$ is a real number and $x \neq 0$ and $x \neq -5\}$ **20.** [7.6] $y = \dfrac{243}{x}$ **21.** [7.6] $y = 0.136r^2$ **22.** [4.4] $-3y^7 + 9y^5 - 21y^4$ **23.** [4.5] $x^2 - 0.01$ **24.** [5.1] $3x(2x^2 + x - 1)$ **25.** [6.1] $\dfrac{-x + 4}{x}$ **26.** [7.3] Answers may vary; $(25, -1)$, $(-25, 1)$, $(1, -\frac{1}{25})$, $(-1, \frac{1}{25})$ **27.** [7.4], [7.5] **a)** 30 miles; **b)** 15 mph

CHAPTER 8

Technology Connection, Section 8.1

TC1. $x = 1.59$, $y = 5.72$ **TC2.** $x = -0.26$, $y = 57.06$ **TC3.** $x = 2.08$, $y = 1.04$ **TC4.** $x = 0.87$, $y = -0.32$

Exercise Set 8.1, pp. 388–390

1. $x + y = -42$, $x - y = 52$ **3.** $x + y = 40$, $4.95x + 7.95y = 282$ **5.** $x + y = 90$, $x + \frac{1}{2}y = 64$ **7.** $x + y = 250$, $3.50x + 7y = 1347.50$ **9.** $2w + 2l = 228$, $w = l - 42$ **11.** $x + y = 40$, $2x + 3y = 89$ **13.** $x + y = 400$, $20x + 30y = 11{,}000$ **15.** $x + y = 12$, $30x + 60y = 600$ **17.** Yes **19.** No **21.** Yes **23.** No **25.** Yes **27.** $(3, 1)$ **29.** $(3, 2)$ **31.** $(1, -5)$ **33.** $(2, 1)$ **35.** $\left(\frac{5}{2}, -2\right)$ **37.** $(3, -2)$ **39.** $\varnothing$ **41.** $(4, -5)$ **43.** $(3, -4)$ **45.** $\{(x, y)|2x - 3y = 6\}$ **47.** All except 39 **49.** 45 **51.** -3 **53.** $\frac{9}{20}$ **55.** ◈ **57.** 1983 and 1990

59. 1987 **61.** Answers may vary.
(a) $x + y = 6$, $x - y = 4$; **(b)** $x + y = 1$, $2x + 2y = 3$;
(c) $x + y = 1$, $2x + 2y = 2$
63. $b = 2s$, $b - 10 = 3(s - 10)$
65. $2l + 2w = 156$, $l = 4(w - 6)$ **67.** $(3, 3)$, $(-5, 5)$
69. $(-0.39, -1.10)$ **71.** $(-0.13, 0.67)$

Exercise Set 8.2, pp. 396–397

1. $(-4, 3)$ **3.** $(-3, -15)$ **5.** $(2, -2)$ **7.** $(-2, 1)$
9. $\left(\frac{1}{2}, \frac{1}{2}\right)$ **11.** $\left(\frac{19}{8}, \frac{1}{8}\right)$ **13.** $\varnothing$ **15.** $\{(x, y)|x - 3 = y\}$
17. $(1, 2)$ **19.** $(3, 0)$ **21.** $(-1, 2)$ **23.** $\left(\frac{128}{31}, -\frac{17}{31}\right)$
25. $(6, 2)$ **27.** $\left(\frac{140}{13}, -\frac{50}{13}\right)$ **29.** $(4, 6)$
31. $\left(\frac{110}{19}, -\frac{12}{19}\right)$ **33.** $\{(x, y)|2x + 3y = 1\}$ **35.** $\varnothing$
37. $\left(\frac{1}{2}, -\frac{1}{2}\right)$ **39.** $\left(-\frac{4}{3}, -\frac{19}{3}\right)$ **41.** $\left(\frac{1000}{11}, -\frac{1000}{11}\right)$
43. $(140, 60)$ **45.** $18a^5b^6$ **47.** ◈
49. $(23.118879, -12.039964)$ **51.** $\left(\dfrac{a + 2b}{7}, \dfrac{a - 5b}{7}\right)$
53. $p = 2$, $q = -\frac{1}{3}$ **55.** $\left(-\frac{1}{4}, -\frac{1}{2}\right)$

Exercise Set 8.3, pp. 405–407

1. 5, −47 **3.** 12 white, 28 printed **5.** 38° and 52°
7. 115 children's plates, 135 adults' plates
9. Width = 36 ft, length = 78 ft
11. 31 2-point, 9 3-point
13. Lumber: 100; plywood: 300
15. 4 30-sec spots; 8 60-sec spots
17. 150 lb soybean meal, 200 lb corn meal
19. $12\frac{1}{2}$ L of Arctic Antifreeze, $7\frac{1}{2}$ L of Frost-No-More
21. $6800 at 9%, $8200 at 10%
23. $12,500 at 10%, $14,500 at 12%
25. 84 adult, 33 children **27.** Paula is 32; Bob is 20
29. Length = 76 m, width = 19 m
31. 7 $5 bills, 15 $1 bills **33.** 375 km **35.** $1\frac{3}{4}$ hr
37. 24 mph **39.** $(x + 1)(x + 5)$ **41.** $y = -\frac{3}{4}x - \frac{7}{2}$
43. Burl 40, son 20
45. Width = $\frac{102}{5}$ in., length = $\frac{288}{5}$ in. **47.** $4\frac{4}{7}$ L
49. 82 **51.** First train: 36 km/h; second train: 54 km/h
53. 3 girls, 4 boys

Exercise Set 8.4, pp. 413–414

1. Yes **3.** $(1, 2, 3)$ **5.** $(-1, 5, -2)$ **7.** $(3, 1, 2)$
9. $(-3, -4, 2)$ **11.** $(2, 4, 1)$ **13.** $(-3, 0, 4)$
15. Dependent **17.** $\left(\frac{1}{2}, 4, -6\right)$ **19.** $\left(\frac{1}{2}, \frac{1}{3}, \frac{1}{6}\right)$
21. $\left(\frac{1}{2}, \frac{2}{3}, -\frac{5}{6}\right)$ **23.** $(15, 33, 9)$ **25.** $\left(\frac{1}{4}, -\frac{1}{2}, -\frac{1}{4}\right)$
27. $\left(\frac{98}{5}, \frac{304}{5}, \frac{498}{5}\right)$ **29.** $\varnothing$ **31.** Dependent
33. −9 **35.** ◈ **37.** $(1, -1, 2)$
39. $(1, -2, 4, -1)$ **41.** $\left(-1, \frac{1}{5}, -\frac{1}{2}\right)$ **43.** 12
45. $3x + 4y + 2z = 12$

Exercise Set 8.5, pp. 417–419

1. 17, 9, 79 **3.** 4, 2, −1
5. $A = 34°$, $B = 104°$, $C = 42°$
7. $A = 25°$, $B = 50°$, $C = 105°$
9. $41.1 billion on newspaper, $36 billion on television, $7.7 billion on radio
11. Steak: 2; baked potato: 1; broccoli: 2
13. Oriental: 385; African-American: 200; Caucasian: 154
15. A: 1500; B: 1900; C: 2300
17. A: 900 gal/hr; B: 1300 gal/hr; C: 1500 gal/hr
19. $45,000 at 8%; $15,000 at 6%; $20,000 at 9%
21. $\dfrac{2 + t}{t + 1}$ **23.** $\dfrac{t^2 + 2t + 2}{t(t + 1)}$
25. 20 **27.** 180° **29.** 35

Exercise Set 8.6, pp. 422–423

1. $\left(\frac{3}{2}, \frac{5}{2}\right)$ **3.** $(-4, 3)$ **5.** $\left(\frac{1}{2}, \frac{3}{2}\right)$ **7.** $\left(\frac{3}{2}, -4, 3\right)$
9. $(2, -2, 1)$ **11.** $\left(4, \frac{1}{2}, -\frac{1}{2}\right)$ **13.** $(1, -3, -2, -1)$
15. 4 dimes, 30 nickels **17.** Mix 5 lb of the $4.05-per-pound granola with 10 lb of the $2.70-per-pound granola. **19.** $400 at 7%, $500 at 8%, $1600 at 9% **21.** $(x + 3)(x - 2)$ **23.** $\left(x + \frac{1}{2}\right)\left(x + \frac{1}{4}\right)$
25. ◈ **27.** 1324

Exercise Set 8.7, p. 427

1. 3 **3.** 36 **5.** −10 **7.** −3 **9.** 5 **11.** $(2, 0)$
13. $\left(-\frac{25}{2}, -\frac{11}{2}\right)$ **15.** $\left(\frac{3}{2}, \frac{13}{14}, \frac{33}{14}\right)$ **17.** $(2, -1, 4)$
19. $(1, 2, 3)$ **21.** $x^2 + 2x + 5$ **23.** $\left\{x|x \text{ is a real number and } x \neq -\frac{5}{2}\right\}$ **25.** 3 **27.** An equation of the line through (x_1, y_1) and (x_2, y_2) is

$$y - y_1 = \frac{y_2 - y_1}{x_2 - x_1}(x - x_1),$$

which is equivalent to

$$yx_2 - yx_1 - y_1x_2 + y_1x_1 = y_2x - y_2x_1 - y_1x + y_1x_1$$

or

$$y_2x_1 + y_1x - y_2x + yx_2 - yx_1 - y_1x_2 = 0. \quad (1)$$

$$\begin{vmatrix} x & y & 1 \\ x_1 & y_1 & 1 \\ x_2 & y_2 & 1 \end{vmatrix} = 0$$

is equivalent to

$$x(y_1 - y_2) - x_1(y - y_2) + x_2(y - y_1) = 0$$

or

$$xy_1 - xy_2 - x_1y + x_1y_2 + x_2y - x_2y_1 = 0. \quad (2)$$

Equations (1) and (2) are equivalent.

Exercise Set 8.8, pp. 431–433

1. **(a)** $P(x) = 20x - 600{,}000$; **(b)** 30,000 units
3. **(a)** $P(x) = 50x - 120{,}000$; **(b)** 2400 units
5. **(a)** $P(x) = 80x - 10{,}000$; **(b)** 125 units
7. **(a)** $P(x) = 40x - 75{,}000$; **(b)** 1875 units
9. **(a)** $P(x) = 75x - 195{,}000$; **(b)** 2600 units

11. ($10, 1400) **13.** ($22, 474) **15.** ($50, 6250) **17.** ($10, 1070) **19.** **(a)** $C(x) = 750x + 125{,}000$; **(b)** $R(x) = 1050x$; **(c)** $P(x) = 300x - 125{,}000$; **(d)** $5000 loss, $85,000 profit; **(e)** 417 units **21.** **(a)** $C(x) = 20x + 10{,}000$; **(b)** $R(x) = 100x$; **(c)** $P(x) = 80x - 10{,}000$; **(d)** $150,000 profit, $6000 loss; **(e)** 125 units **23.** $(x + 3)(x + 4)$ **25.** $\frac{3x + 1}{x^2}$ **27.** ($5, 300) **29.** ◈ **31.** **(a)** 4526; **(b)** $870

Review Exercises: Chapter 8, pp. 435–436

1. [8.1] (3, 2) **2.** [8.1] (−2, 1) **3.** [8.2] $\left(\frac{2}{5}, -\frac{4}{5}\right)$ **4.** [8.2] ∅ **5.** [8.2] $\left(-\frac{11}{15}, -\frac{43}{30}\right)$ **6.** [8.2] $\left(\frac{37}{19}, \frac{53}{19}\right)$ **7.** [8.2] $\left(\frac{76}{17}, -\frac{2}{119}\right)$ **8.** [8.2] (2, 2) **9.** [8.3] CD, $14; cassette, $9 **10.** [8.3] 4 hours **11.** [8.3] 10 liters of Cleanse-O, 30 liters of Tingle **12.** [8.4] (10, 4, −8) **13.** [8.4] $\left(-\frac{7}{3}, \frac{125}{27}, \frac{20}{27}\right)$ **14.** [8.4] (2, 0, 4) **15.** [8.2] ∅ **16.** [8.4] $\left(\frac{8}{9}, -\frac{2}{3}, \frac{10}{9}\right)$ **17.** [8.2] $\{(x, y) \mid 3x + 4y = 6\}$ **18.** [8.4] $\left(2, \frac{1}{3}, -\frac{2}{3}\right)$ **19.** [8.5] $A = 90°$, $B = 67\frac{1}{2}°$, $C = 22\frac{1}{2}°$ **20.** [8.5] 641 **21.** [8.5] $20 bills: 5; $5 bills: 15; $1 bills: 19 **22.** [8.6] $\left(55, -\frac{89}{2}\right)$ **23.** [8.6] (−1, 1, 3) **24.** [8.7] 2 **25.** [8.7] 9 **26.** [8.7] (6, −2) **27.** [8.7] (−3, 0, 4) **28.** [8.8] ($3, 81) **29.** [8.8] **(a)** $C(x) = 175x + 35{,}000$; **(b)** $R(x) = 225x$; **(c)** $P(x) = 50x - 35{,}000$; **(d)** $25,000 profit, $10,000 loss; **(e)** 700 beds **30.** [5.2] $(x + 11)(x + 1)$ **31.** [5.3] $(11x + 1)(x + 11)$ **32.** [6.5] $\frac{x^2 - 2x - 1}{(x + 1)(x - 1)}$ **33.** [7.5] $-x^2 + x + 4$ **34.** [8.1] ◈ Every point on the line that is the graph of a linear equation is a solution of that equation. If two lines intersect, the point of intersection is on both lines, and is thus a solution of both equations graphed. A solution of both equations in a system of two equations is a solution of the system. **35.** [8.4] ◈ A system of equations can be both dependent and inconsistent if it is equivalent to a system with fewer equations that has no solution. An example is a system of three equations in three unknowns in which two of the equations represent the same plane, and the third represents a parallel plane. **36.** [8.2] (2, 1, −2) **37.** [8.5] $a = -\frac{2}{3}$, $b = -\frac{4}{3}$, $c = 3$; $f(x) = -\frac{2}{3}x^2 - \frac{4}{3}x + 3$

Test: Chapter 8, pp. 436–437

1. [8.2] $\left(3, -\frac{11}{3}\right)$ **2.** [8.2] $\left(\frac{15}{7}, -\frac{18}{7}\right)$ **3.** [8.2] $\left(\frac{1}{2}, -\frac{1}{6}\right)$ **4.** [8.2] ∅ **5.** [8.3] $l = 30$, $w = 18$ **6.** [8.3] 15 sheets of oak; 3 sheets of pine **7.** [8.4] Dependent **8.** [8.4] $\left(2, -\frac{1}{2}, -1\right)$ **9.** [8.4] ∅ **10.** [8.4] (0, 1, 0) **11.** [8.6] $\left(\frac{34}{107}, -\frac{104}{107}\right)$ **12.** [8.6] (3, 1, −2) **13.** [8.7] 34 **14.** [8.7] 55 **15.** [8.7] $\left(\frac{13}{18}, \frac{7}{27}\right)$ **16.** [8.5] 3.5 hr **17.** [8.8] ($3, 55) **18.** [8.8] **(a)** $C(x) = 40{,}000 + 45x$; **(b)** $R(x) = 80x$; **(c)** $P(x) = 35x - 40{,}000$; **(d)** $12,500 profit; **(e)** 1143 rackets **19.** [5.3] $(3x + 5)(2x - 5)$ **20.** [5.2] $(x - 1)(x - 13)$ **21.** [6.5] $\frac{3t + 8}{(t + 2)(t - 2)}$ **22.** [7.5] $\{x \mid x$ is a real number and $x \neq 1\}$ **23.** [7.4], [8.3] $m = 7$, $b = 10$ **24.** [8.5] Adults: 1651; senior citizens: 1346; children: 335

CHAPTER 9

Exercise Set 9.1, pp. 446–448

1. $\{x \mid x > 4\}$, $(4, \infty)$

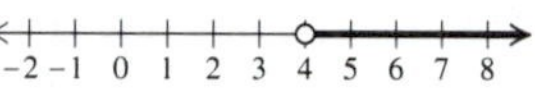

3. $\{t \mid t \leq 6\}$, $(-\infty, 6]$

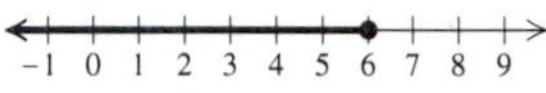

5. $\{y \mid y < -3\}$, $(-\infty, -3)$

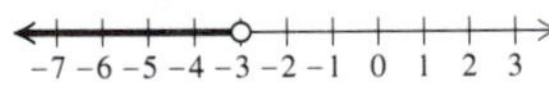

7. $\{x \mid x \geq -6\}$, $[-6, \infty)$

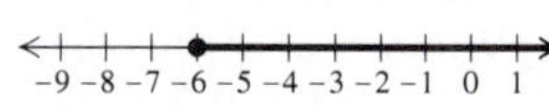

9. $\{x \mid x > -5\}$, or $(-5, \infty)$

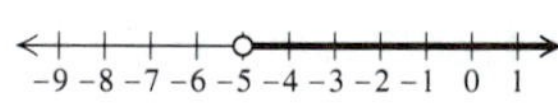

11. $\{a \mid a \leq -21\}$, or $(-\infty, -21]$

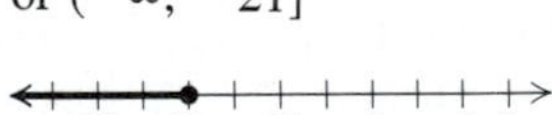

13. $\{x \mid x \geq 3\}$ or $[3, \infty)$

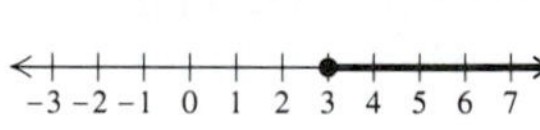

15. $\{x \mid x \leq 0.9\}$, or $(-\infty, 0.9]$

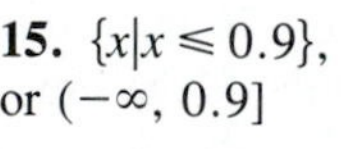

17. $\left\{x \mid x \leq \frac{5}{6}\right\}$, or $\left(-\infty, \frac{5}{6}\right]$

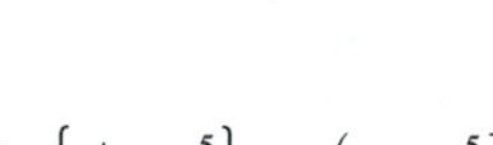

19. $\{x \mid x < 6\}$, or $(-\infty, 6)$

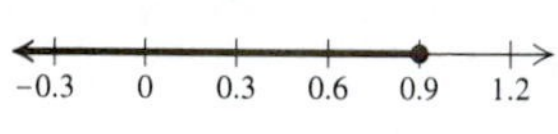

21. $\{y \mid y \leq -3\}$, or $(-\infty, -3]$

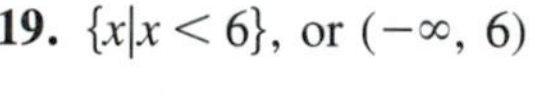

23. $\left\{x \mid x \leq \frac{1}{2}\right\}$, or $\left(-\infty, \frac{1}{2}\right]$

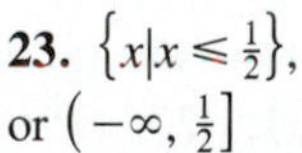

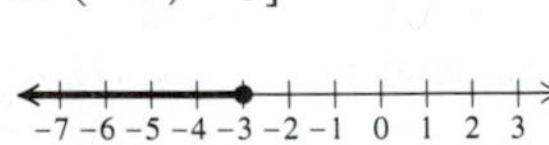

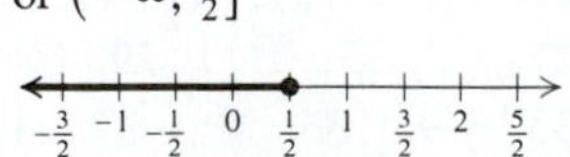

25. $\left\{x \mid x \leq \frac{11}{6}\right\}$, or $\left(-\infty, \frac{11}{6}\right]$ **27.** $\left\{m \mid m > \frac{7}{3}\right\}$, or $\left(\frac{7}{3}, \infty\right)$ **29.** $\left\{x \mid x \leq \frac{4}{7}\right\}$, or $\left(-\infty, \frac{4}{7}\right]$

31. Mileage < 69.7
33. All calls greater than $1\frac{1}{2}$ min
35. Gross sales greater than $15,000
37. $\{n \mid n < 100 \text{ hr}\}$ **39.** $20,000 **41.** $1.99
43. $\{b \mid b > \$1450\}$ **45.** **(a)** $\{t \mid t < 1945.4°\text{F}\}$;
(b) $\{t \mid t < 1761.44°\text{F}\}$ **47.** **(a)** $\{x \mid x \leq 8181\}$;
(b) $\{x \mid x > 8181\}$
49.

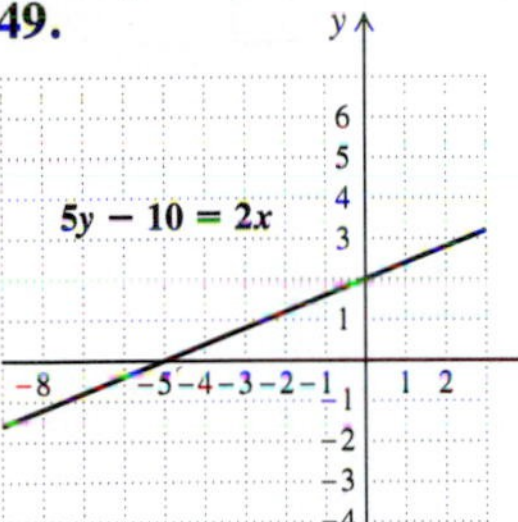

51. 7 **53.** ◈
55. $\left\{x \,\middle|\, x \leq \dfrac{2}{a-1}\right\}$
57. $\left\{y \,\middle|\, y \geq \dfrac{5b+2a}{b(a-2)}\right\}$
59. $\left\{x \,\middle|\, x > \dfrac{4m-2c}{d-(5c+2m)}\right\}$

61. False; $a = 2$, $b = 3$, $c = 4$, $d = 5$; $2 - 4 = 3 - 5$
63. ◈ **65.** All real numbers
67. All real numbers except 0

Exercise Set 9.2, pp. 454–456

1. $\{6, 8\}$ **3.** $\varnothing$ **5.** $\{1, 2, 3, 4\}$
7. $1 < x < 6$; $(1, 6)$ **9.** $-7 \leq y \leq -3$; $[-7, -3]$
11. $-3 < x \leq 4$; $(-3, 4]$ **13.** $-6 < x \leq 2$; $(-6, 2]$
15. $-2 \leq x < 5$; $[-2, 5)$ **17.** $1 \leq x < 5$; $[1, 5)$
19. $\{x \mid -4 < x < 6\}$, or $(-4, 6)$ **21.** $\{y \mid -2 < y \leq 2\}$, or $(-2, 2]$ **23.** $\left\{x \mid -\frac{5}{3} \leq x \leq \frac{4}{3}\right\}$, or $\left[-\frac{5}{3}, \frac{4}{3}\right]$
25. $\{x \mid -1 < x \leq 6\}$, or $(-1, 6]$ **27.** $\left\{x \mid -\frac{3}{2} \leq x < \frac{9}{2}\right\}$, or $\left[-\frac{3}{2}, \frac{9}{2}\right)$ **29.** $\{x \mid 10 < x \leq 14\}$, or $(10, 14]$
31. $\{1, 4, 5, 6, 7, 8, 11\}$ **33.** $\{1, 2, 3, 4, 5, 6, 8\}$
35. $\{4, 8, 11\}$
37.

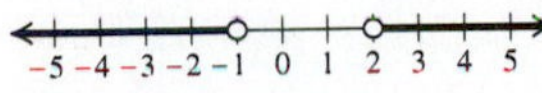

39.

41. $\{x \mid x < -9 \text{ or } x > -5\}$, or $(-\infty, -9) \cup (-5, \infty)$
43. $\left\{x \mid x \leq \frac{5}{2} \text{ or } x \geq 11\right\}$, or $\left(-\infty, \frac{5}{2}\right] \cup [11, \infty)$
45. $\{x \mid x \geq -3\}$, or $[-3, \infty)$
47. $\left\{x \mid x \leq -\frac{5}{4} \text{ or } x > -\frac{1}{2}\right\}$, or $\left(-\infty, -\frac{5}{4}\right] \cup \left(-\frac{1}{2}, \infty\right)$
49. $\{x \mid x \text{ is a real number}\}$, or $\mathbb{R}$, or $(-\infty, \infty)$
51. $\{x \mid x < -4 \text{ or } x > 2\}$, or $(-\infty, -4) \cup (2, \infty)$
53. $\left\{x \mid x < \frac{79}{4} \text{ or } x > \frac{89}{4}\right\}$, or $\left(-\infty, \frac{79}{4}\right) \cup \left(\frac{89}{4}, \infty\right)$
55. $\left\{x \mid x \leq -\frac{13}{2} \text{ or } x \geq \frac{29}{2}\right\}$, or $\left(-\infty, -\frac{13}{2}\right] \cup \left[\frac{29}{2}, \infty\right)$
57. $\{x \mid x \text{ is a real number}\}$, or $\mathbb{R}$, or $(-\infty, \infty)$
59. $\left(-\frac{16}{13}, -\frac{41}{13}\right)$ **61.** $-\frac{27}{7}$ **63.** ◈
65. **(a)** $1945.4° \leq F < 4820°$; **(b)** $1761.44° \leq F < 3956°$
67. Sizes between 6 and 13 **69.** From 1973 to 1987
71. $\left\{a \mid -\frac{3}{2} \leq a \leq 1\right\}$, or $\left[-\frac{3}{2}, 1\right]$; **73.** $\{x \mid -4 < x \leq 1\}$, or $(-4, 1]$;
75. $\left\{x \mid -\frac{1}{8} < x < \frac{1}{2}\right\}$, or $\left(-\frac{1}{8}, \frac{1}{2}\right)$;
77. True **79.** False **81.** False **83.** $\left\{x \mid -\frac{2}{5} \leq x \leq 2\right\}$, or $\left[-\frac{2}{5}, 2\right]$

Exercise Set 9.3, pp. 461–463

1. $\{-3, 3\}$ **3.** $\varnothing$ **5.** $\{0\}$ **7.** $\{-5.5, 5.5\}$
9. $\{-9, 15\}$ **11.** $\left\{-\frac{1}{2}, \frac{7}{2}\right\}$ **13.** $\left\{-\frac{3}{2}, \frac{17}{2}\right\}$ **15.** $\varnothing$
17. $\{-11, 11\}$ **19.** $\{-8, 8\}$ **21.** $\left\{-\frac{11}{5}, \frac{11}{5}\right\}$
23. $\{-7, 8\}$ **25.** $\{-12, 2\}$ **27.** $\{-1, 2\}$
29. $\left\{-\frac{1}{3}, 3\right\}$ **31.** 4 **33.** 11 **35.** 33 **37.** 34
39. $\left\{-\frac{11}{2}, \frac{3}{4}\right\}$ **41.** $\left\{-\frac{3}{2}\right\}$ **43.** $\left\{-\frac{3}{5}, 5\right\}$
45. $\mathbb{R}$ **47.** $\left\{-\frac{3}{2}\right\}$ **49.** $\left\{0, \frac{24}{23}\right\}$ **51.** $\left\{\frac{8}{3}, 32\right\}$
53. $\{x \mid -3 < x < 3\}$, or $(-3, 3)$; **55.** $\{x \mid x \leq -2 \text{ or } x \geq 2\}$, or $(-\infty, -2] \cup [2, \infty)$;
57. $\{t \mid t \leq -5.5 \text{ or } t \geq 5.5\}$, or $(-\infty, -5.5] \cup [5.5, \infty)$; **59.** $\{x \mid 2 < x < 4\}$, or $(2, 4)$;
61. $\{x \mid -7 \leq x \leq 3\}$, or $[-7, 3]$;
63. $\{x \mid x < 2 \text{ or } x > 4\}$, or $(-\infty, 2) \cup (4, \infty)$; **65.** $\left\{x \mid -\frac{1}{2} \leq x \leq \frac{7}{2}\right\}$, or $\left[-\frac{1}{2}, \frac{7}{2}\right]$;

67. $\{y|y \text{ is a real number}\}$, or $\mathbb{R}$, or $(-\infty, \infty)$;

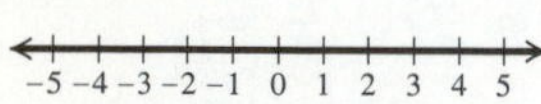

69. $\left\{x|x \leqslant -\frac{5}{4} \text{ or } x \geqslant \frac{23}{4}\right\}$, or $\left(-\infty, -\frac{5}{4}\right] \cup \left[\frac{23}{4}, \infty\right)$;

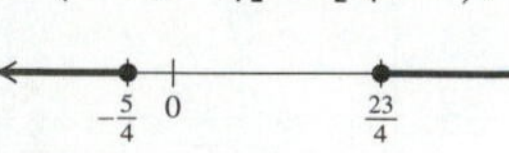

71. $\{y|-9 < y < 15\}$, or $(-9, 15)$;

73. $\left\{x|-\frac{7}{2} \leqslant x \leqslant \frac{1}{2}\right\}$, or $\left[-\frac{7}{2}, \frac{1}{2}\right]$;

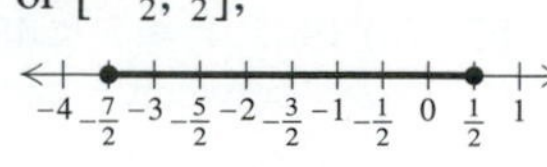

75. $\left\{y|y < -\frac{4}{3} \text{ or } y > 4\right\}$, or $\left(-\infty, -\frac{4}{3}\right) \cup (4, \infty)$;

77. $\left\{x|-\frac{5}{4} \leqslant x \leqslant \frac{23}{4}\right\}$, or $\left[-\frac{5}{4}, \frac{23}{4}\right]$

79. $\varnothing$

81. $\{x|x < -4 \text{ or } x > 5\}$, or $(-\infty, -4) \cup (5, \infty)$;

83. $\{x|-5 < x < 19\}$, or $(-5, 19)$;

85. $\left\{x|x \leqslant -\frac{2}{15} \text{ or } x \geqslant \frac{14}{15}\right\}$, or $\left(-\infty, -\frac{2}{15}\right] \cup \left[\frac{14}{15}, \infty\right)$;

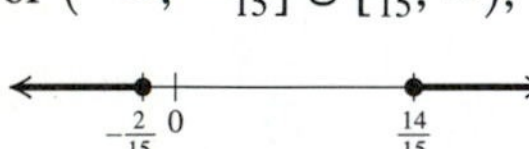

87. $\{m|-12 \leqslant m \leqslant 2\}$, or $[-12, 2]$;

89. $\{g|g \text{ is a real number}\}$, or $\mathbb{R}$, or $(-\infty, \infty)$;

91. $\{x|-1 \leqslant x \leqslant 2\}$, or $[-1, 2]$;

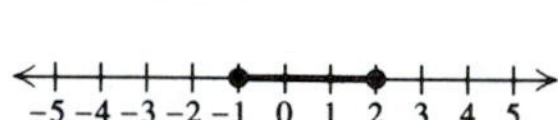

93. 24,640 m^2 **95.** ◈ **97.** $\left\{x|x \geqslant \frac{5}{2}\right\}$

99. $\{x|x \geqslant -5\}$ **101.** $\left\{1, -\frac{1}{4}\right\}$ **103.** $\varnothing$

105. $\mathbb{R}$ **107.** $|x| < 3$ **109.** $|x| \geqslant 6$

111. $|x + 3| > 5$ **113.** $|p - 5| \leqslant \frac{1}{96}$, or $|5 - p| \leqslant \frac{1}{96}$, or $|p - 60| \leqslant \frac{1}{8}$, or $|60 - p| \leqslant \frac{1}{8}$ **115.** ▣

Technology Connection, Section 9.4

TC1.

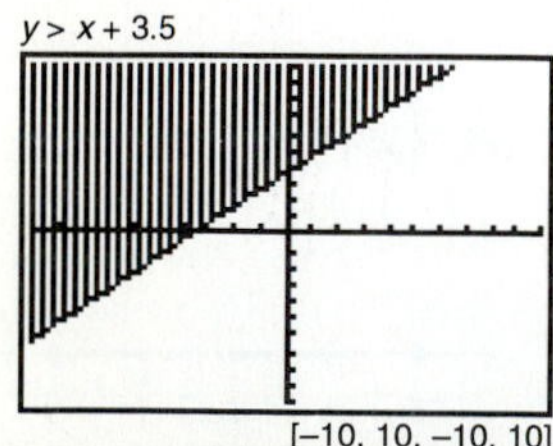

TC2.

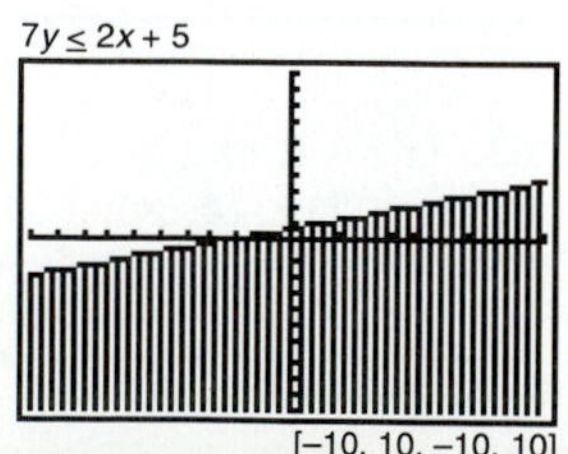

TC3.

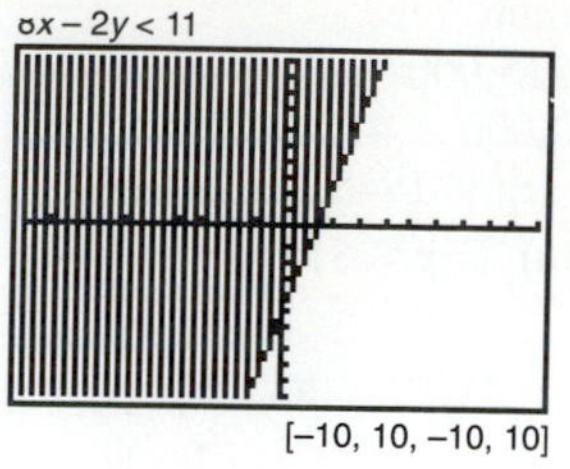

TC4.

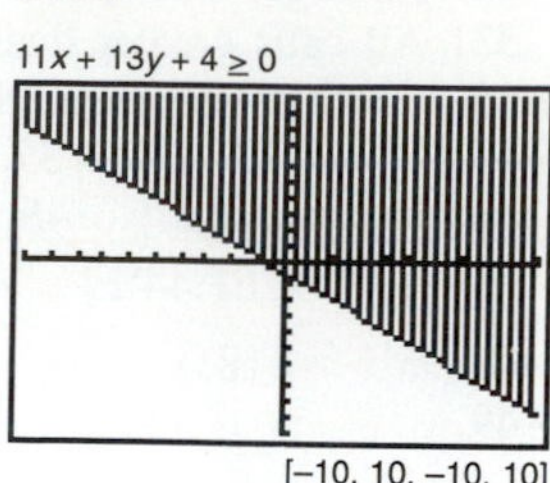

Exercise Set 9.4, pp. 469–470

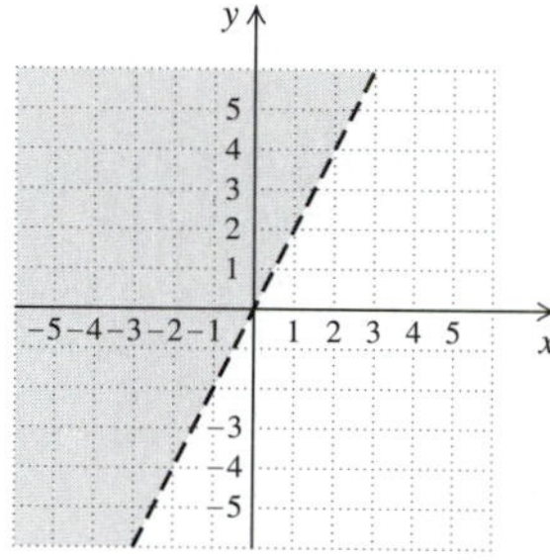

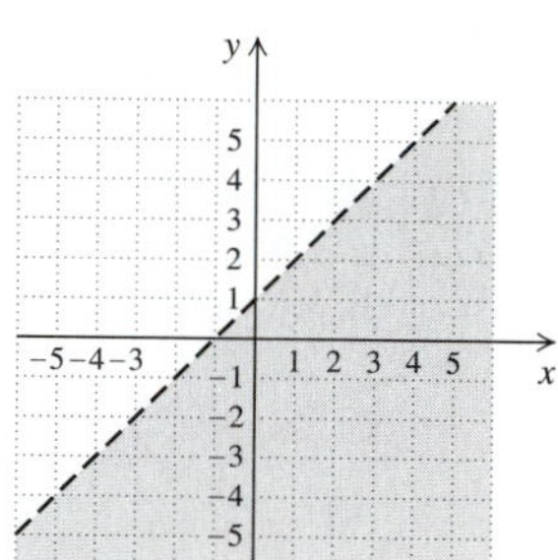

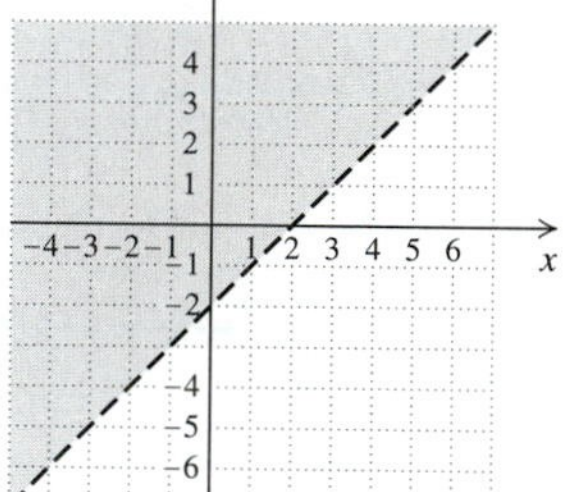

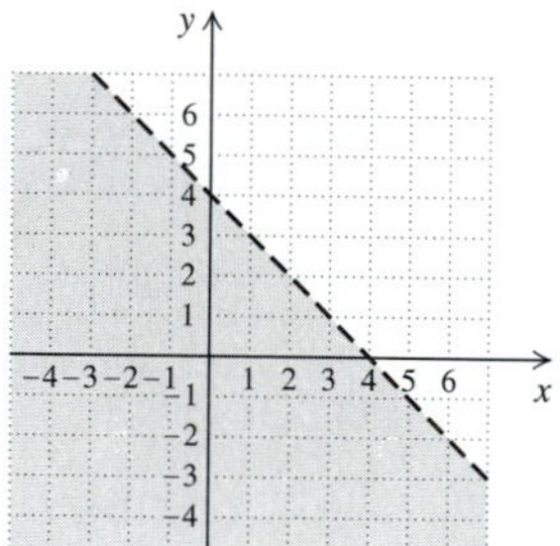

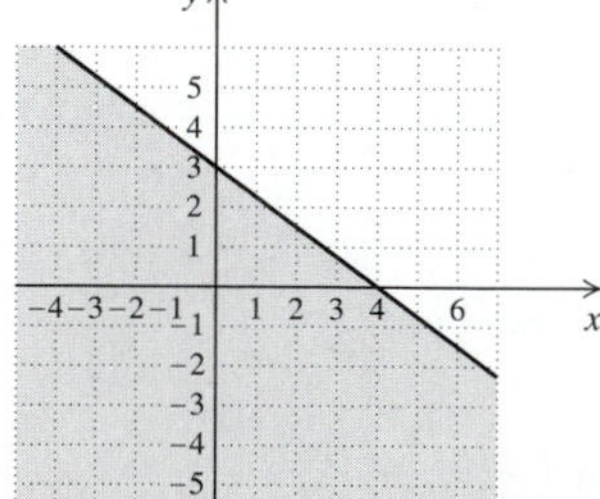

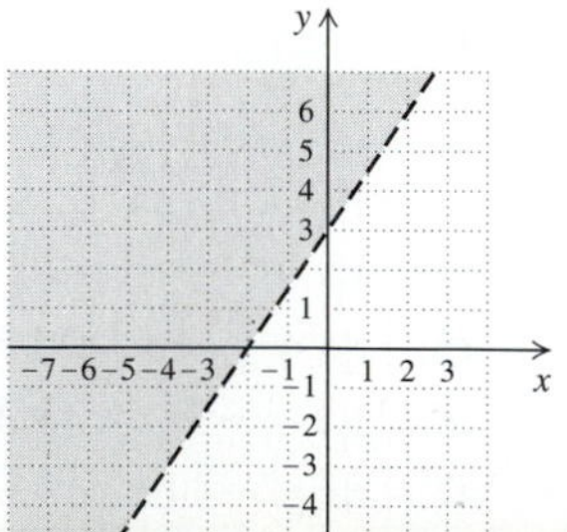

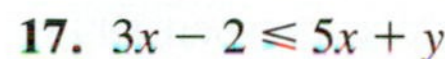

17. $3x - 2 \leq 5x + y$

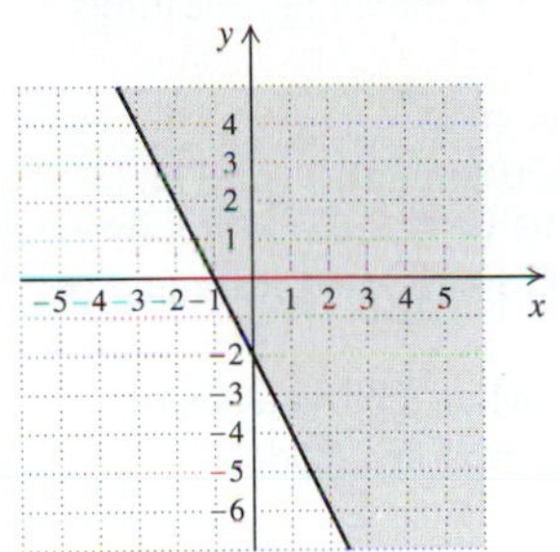

19. $x < -4$

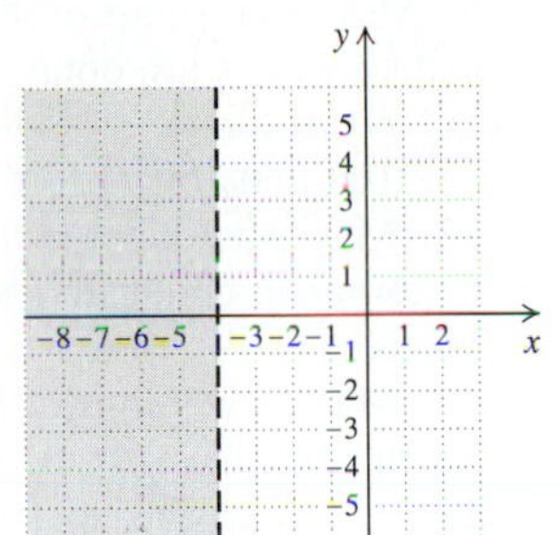

33.

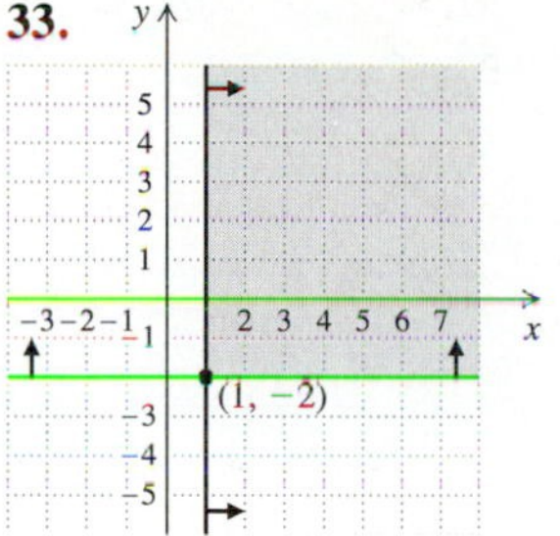

35.

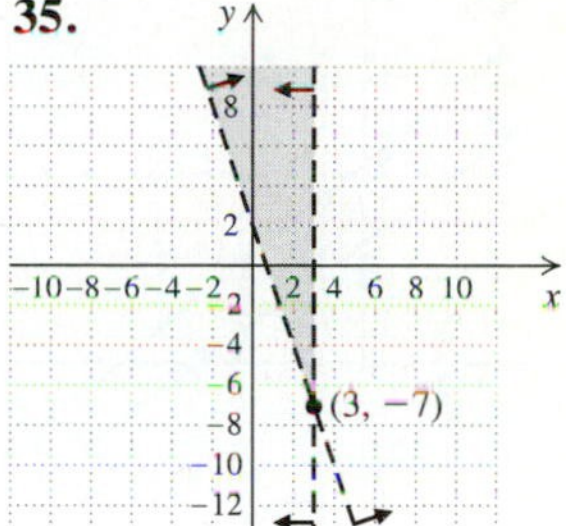

21. $y > -2$

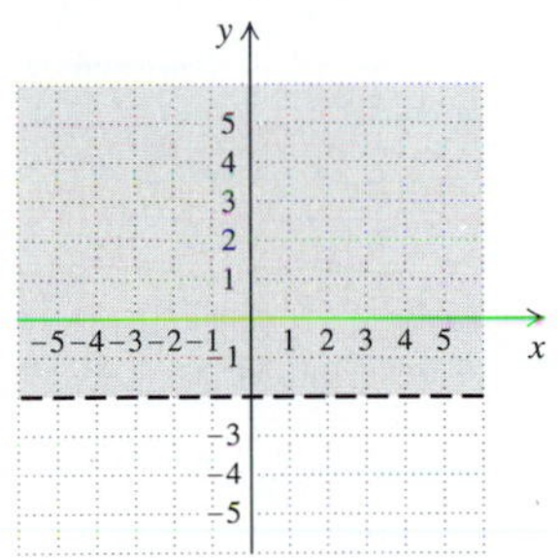

23. $-4 < y < -1$

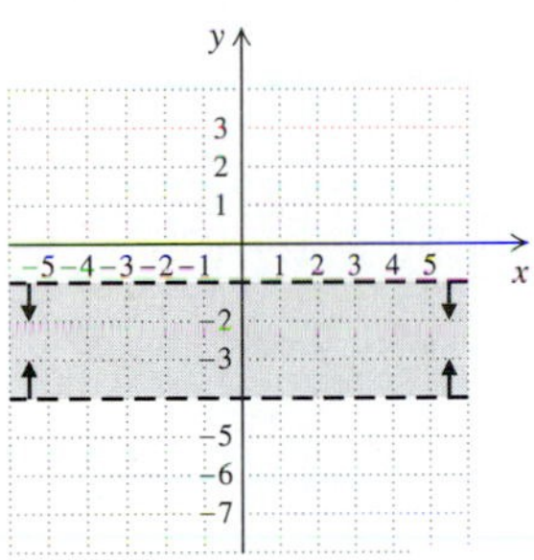

37.

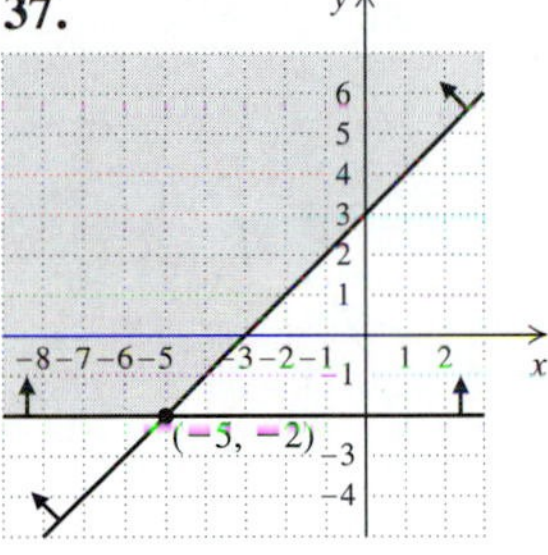

39.

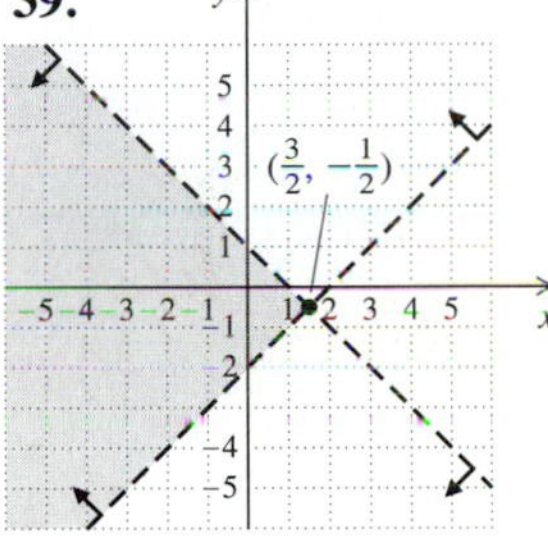

25. $-3 \leq x \leq 3$

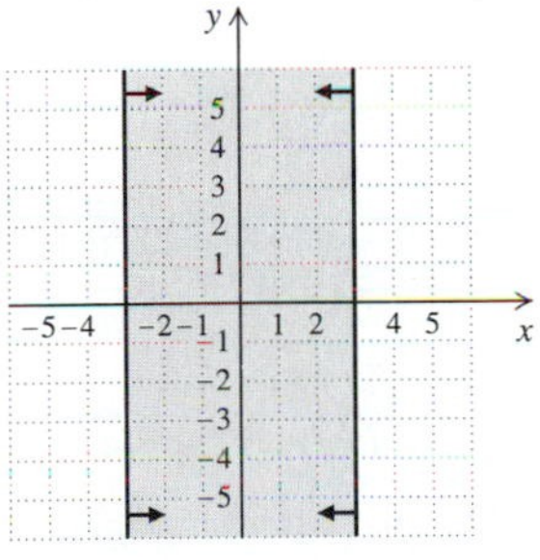

27. $0 \leq x \leq 5$

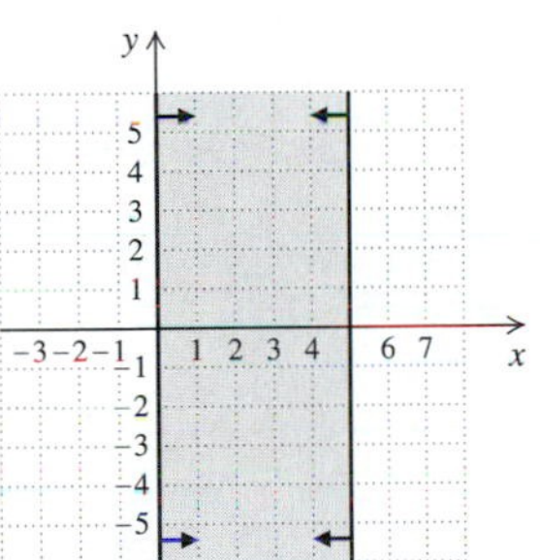

41.

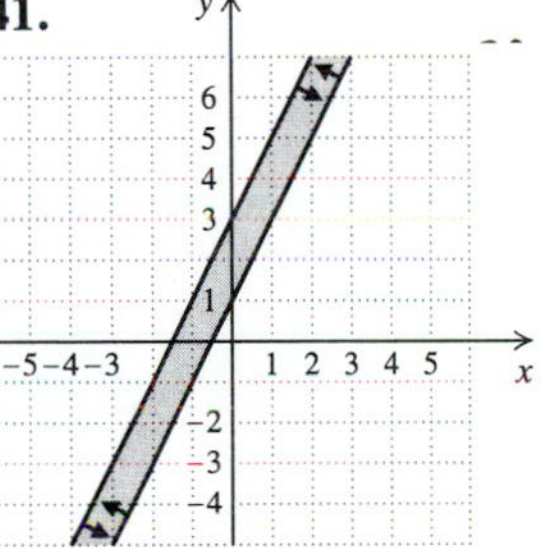

43.

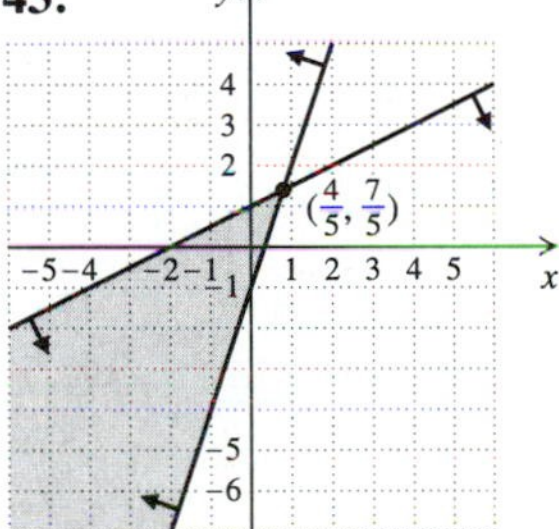

29.

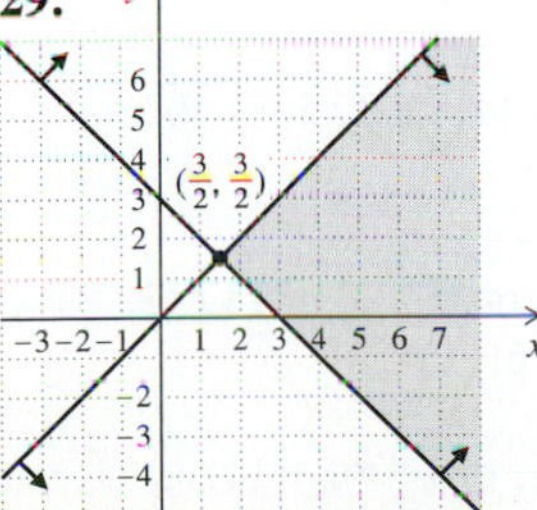

31.

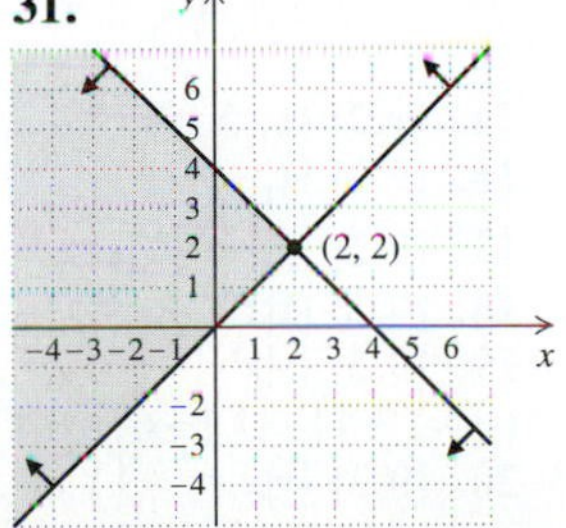

45.

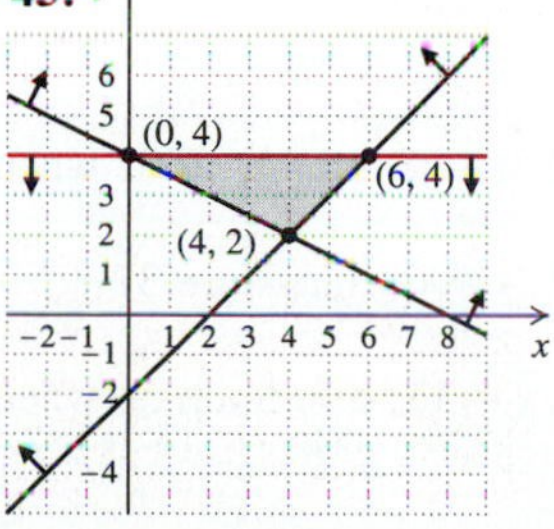

47.

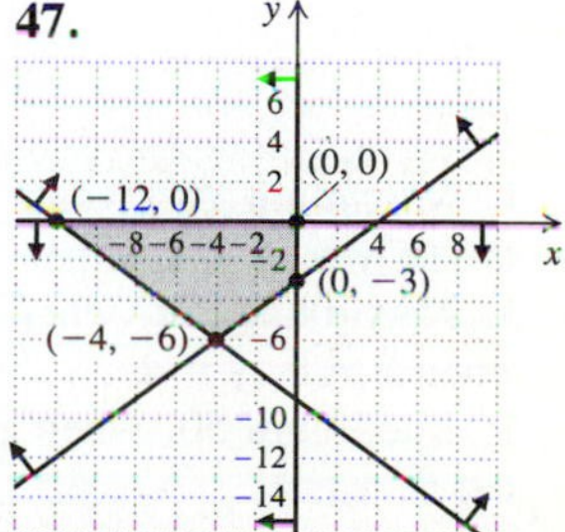

49.

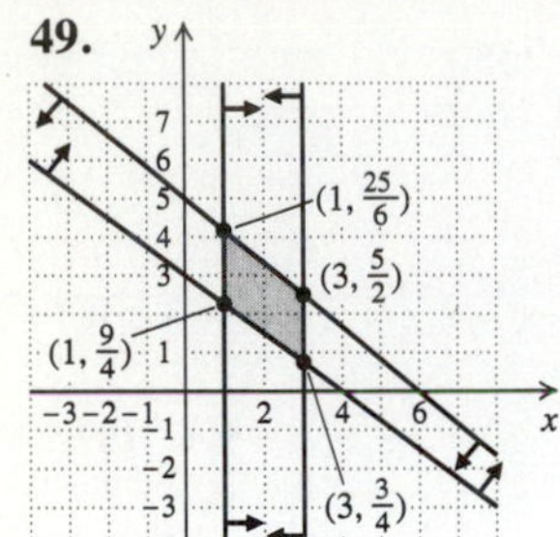

51. 15; 20 **53.** $\frac{10}{17}$

55.

57.

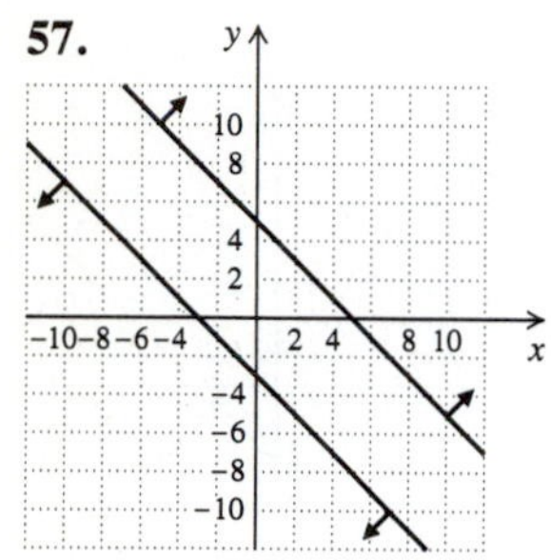

59.

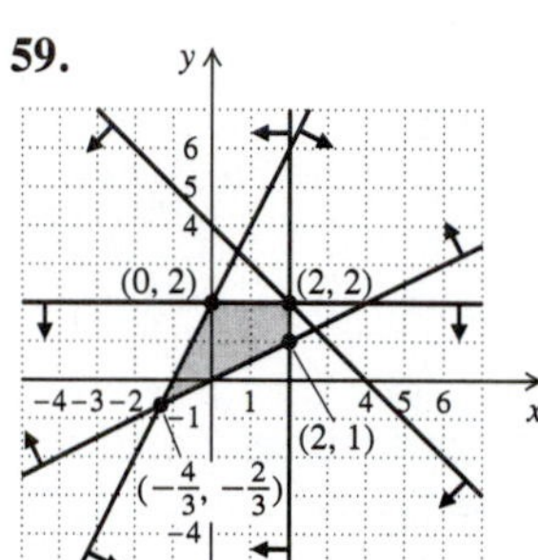

61. $0 \leq W \leq 50,$
$0 \leq L \leq 94$

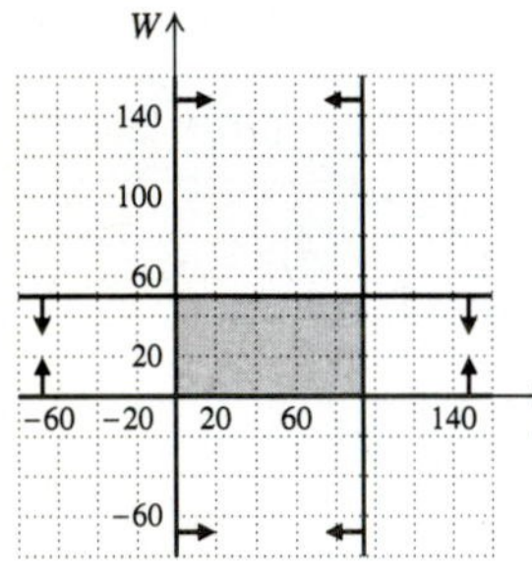

63. $35c + 75a > 1000,$
$a \geq 0,$
$c \geq 0$

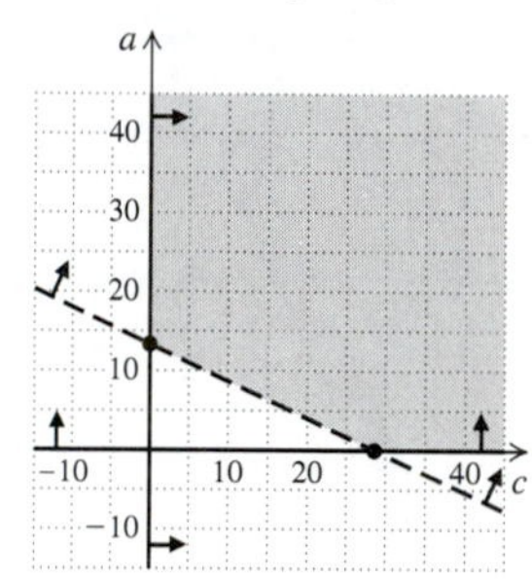

65.

Exercise Set 9.5, pp. 475–477

1. Maximum 168, when $x = 0$, $y = 6$; minimum 0, when $x = 0$, $y = 0$
3. Maximum 152, when $x = 7$, $y = 0$; minimum 32, when $x = 0$, $y = 4$
5. Maximum 50, when $x = 5$, $y = 11$; minimum 5, when $x = 0$, $y = 1$
7. Maximum income of \$18 an hour occurs when 100 of each type of biscuit is made
9. Maximum score of 425 when 5 matching questions and 15 essays are done
11. 60 motorcycles and 100 bicycles
13. 100 units of lumber and 300 units of plywood
15. Maximum income of \$3110 when \$22,000 is invested in corporate bonds and \$18,000 is invested in municipal bonds
17. Maximum profit of \$2520 when 125 batches of Smello and 187.5 batches of Roppo are made
19. 30 P-1 airplanes, 10 P-2 airplanes
21. 25 chairs, 9 sofas

Review Exercises: Chapter 9, p. 479

1. [9.1] $\{x|x \leq -4\}$, or $(-\infty, -4]$;

2. [9.1] $\{x|x > 1\}$, or $(1, \infty)$;

3. [9.1] $\{a|a \leq -21\}$, or $(-\infty, -21]$;

4. [9.1] $\{y|y \geq -7\}$, or $[-7, \infty)$;

5. [9.1] $\{y|y > -\frac{15}{4}\}$, or $(-\frac{15}{4}, \infty)$;

6. [9.1] $\{y|y > -30\}$, or $(-30, \infty)$;

7. [9.1] $\{x|x > -3\}$, or $(-3, \infty)$;

8. [9.1] $\{y|y < -\frac{6}{5}\}$, or $(-\infty, -\frac{6}{5})$;

9. [9.1] $\{x|x < -3\}$, or $(-\infty, -3)$;

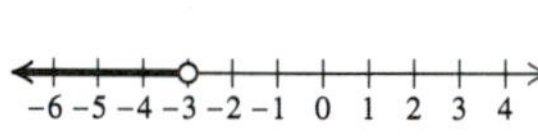

10. [9.1] $\{y|y > -\frac{220}{23}\}$, or $(-\frac{220}{23}, \infty)$;

11. [9.1] $\{x|x \leq -\frac{5}{2}\}$, or $(-\infty, -\frac{5}{2}]$;

12. [9.1] Scores higher than or equal to 92
13. [9.1] \$10,000
14. [9.2] $\{1, 5, 9\}$ **15.** [9.2] $\{1, 2, 3, 5, 6, 9\}$
16. [9.2] $\{x|-7 < x \leq 2\}$, or $(-7, 2]$
17. [9.2] $\{x|-\frac{5}{4} < x < \frac{5}{2}\}$, or $(-\frac{5}{4}, \frac{5}{2})$
18. [9.2] $\{x|x < -3 \text{ or } x > 1\}$, or $(-\infty, -3) \cup (1, \infty)$
19. [9.2] $\{x|x < -11 \text{ or } x \geq -6\}$, or $(-\infty, -11) \cup [-6, \infty)$
20. [9.2] $\{x|x < -11 \text{ or } x > 1\}$, or $(-\infty, -11) \cup (1, \infty)$

21. [9.2] $\{x|x \leq -6 \text{ or } x \geq 8\}$, or $(-\infty, -6] \cup [8, \infty)$
22. [9.3] $\{6, -6\}$ **23.** [9.3] $\varnothing$ **24.** [9.3] $\{x|x \leq -3.5 \text{ or } x \geq 3.5\}$, or $(-\infty, -3.5] \cup [3.5, \infty)$
25. [9.3] $\{-5, 9\}$ **26.** [9.3] $\left\{x\middle|-\frac{17}{2} < x < \frac{7}{2}\right\}$, or $\left(-\frac{17}{2}, \frac{7}{2}\right)$
27. [9.3] $\left\{x\middle|x \leq -\frac{11}{3} \text{ or } x \geq \frac{19}{3}\right\}$, or $\left(-\infty, -\frac{11}{3}\right] \cup \left[\frac{19}{3}, \infty\right)$
28. [9.3] $\left\{-14, \frac{4}{3}\right\}$ **29.** [9.3] $\varnothing$
30. [9.3] $\{x|-12 \leq x \leq 4\}$, or $[-12, 4]$
31. [9.3] $\{x|x < 0 \text{ or } x > 10\}$, or $(-\infty, 0) \cup (10, \infty)$
32. [9.4] $(2, -3)$

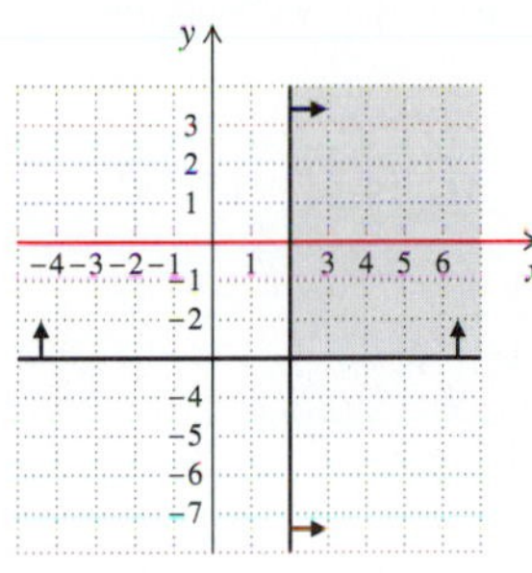

33. [9.4]

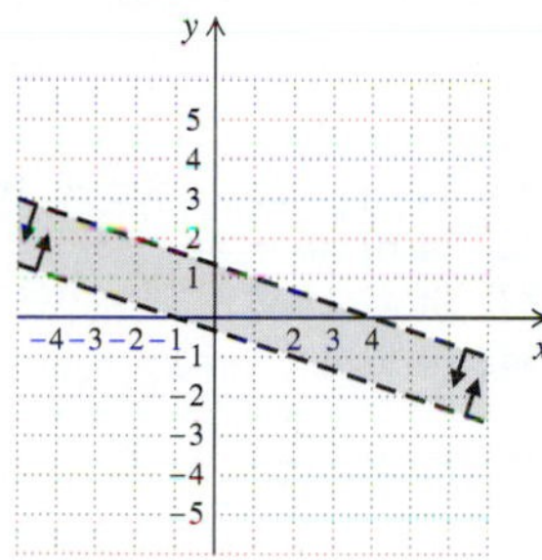

34. [9.4]

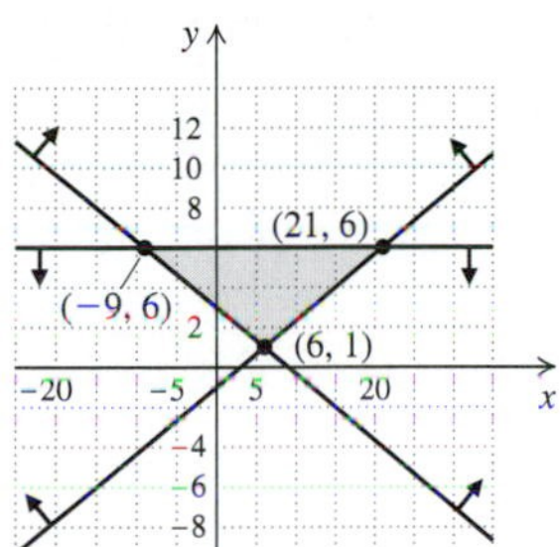

35. [9.5] Maximum 42 when $x = 7$, $y = 17$; minimum 7 when $x = 0$, $y = 3$
36. [9.5] $40,000 in municipal bonds, $20,000 in mutual funds; maximum income $6800 **37.** [8.2] $\left(0, \frac{5}{2}\right)$
38. [2.2] -22

39. [3.2]

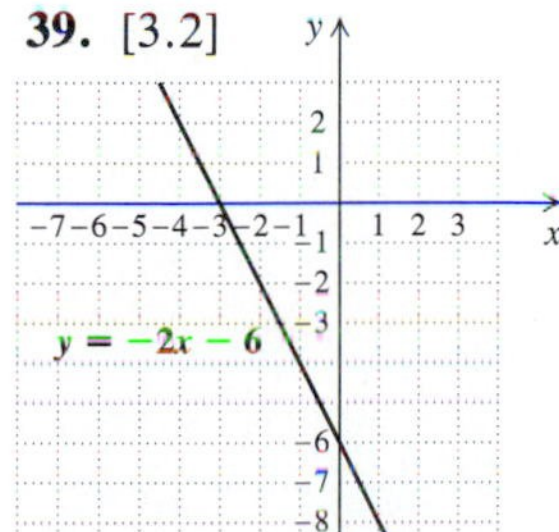

40. [8.3] 38,592 ft^2

41. [9.3] ◈ The equation $|x| = p$ has two solutions when p is positive because x can be either p or $-p$. The same equation has no solution when p is negative because no number has a negative absolute value.
42. [9.4] ◈ The solution set of a system of inequalities is all numbers that make *all* the individual inequalities true. This consists of numbers that are common to all the individual solution sets, or the intersection of the graphs.
43. [9.3] $\left\{x\middle|-\frac{8}{3} \leq x \leq -2\right\}$, or $\left[-\frac{8}{3}, -2\right]$
44. [9.1] False; $-4 < 3$ is true, but $(-4)^2 < 9$ is false.
45. [9.3] $|x - 1.1| \leq 0.03$

Test: Chapter 9, p. 480

1. [9.1] $\{x|x < 14\}$, or $(-\infty, 14)$
2. [9.1] $\{y|y > -50\}$, or $(-50, \infty)$
3. [9.1] $\{y|y \leq -2\}$, or $(-\infty, -2]$
4. [9.1] $\left\{a\middle|a \leq \frac{11}{5}\right\}$, or $\left(-\infty, \frac{11}{5}\right]$
5. [9.1] $\{y|y > 1\}$, or $(1, \infty)$
6. [9.1] $\left\{x\middle|x > \frac{5}{2}\right\}$, or $\left(\frac{5}{2}, \infty\right)$
7. [9.1] $\left\{x\middle|x \leq \frac{7}{4}\right\}$, or $\left(-\infty, \frac{7}{4}\right]$
8. [9.1] More than 66.7 miles
9. [9.1] Sales less than $180,000
10. [9.2] $\{3, 5\}$
11. [9.2] $\{1, 3, 5, 7, 9, 11, 13\}$
12. [9.2] $\{x|-1 < x < 6\}$, or $(-1, 6)$
13. [9.2] $\left\{y\middle|-\frac{2}{5} < y \leq 2\right\}$, or $\left(-\frac{2}{5}, 2\right]$
14. [9.2] $\left\{x\middle|x < -4 \text{ or } x > -\frac{5}{2}\right\}$, or $(-\infty, -4) \cup \left(-\frac{5}{2}, \infty\right)$
15. [9.2] $\{x|x \leq -3 \text{ or } x \geq 2\}$, or $(-\infty, -3] \cup [2, \infty)$
16. [9.2] $\{x|x < 3 \text{ or } x \geq 6\}$, or $(-\infty, 3) \cup [6, \infty)$
17. [9.2] $\left\{x\middle|4 \leq x < \frac{15}{2}\right\}$, or $\left[4, \frac{15}{2}\right)$
18. [9.3] $\{9, -9\}$
19. [9.3] $\{x|x < -3 \text{ or } x > 3\}$, or $(-\infty, -3) \cup (3, \infty)$
20. [9.3] $\left\{x\middle|-\frac{7}{8} < x < \frac{11}{8}\right\}$, or $\left(-\frac{7}{8}, \frac{11}{8}\right)$
21. [9.3] $\left\{x\middle|x \leq -\frac{13}{5} \text{ or } x \geq \frac{7}{5}\right\}$, or $\left(-\infty, -\frac{13}{5}\right] \cup \left[\frac{7}{5}, \infty\right)$
22. [9.3] $\{1\}$ **23.** [9.3] $\varnothing$
24. [9.3] $\{x|-99 \leq x \leq 111\}$, or $[-99, 111]$
25. [9.4]

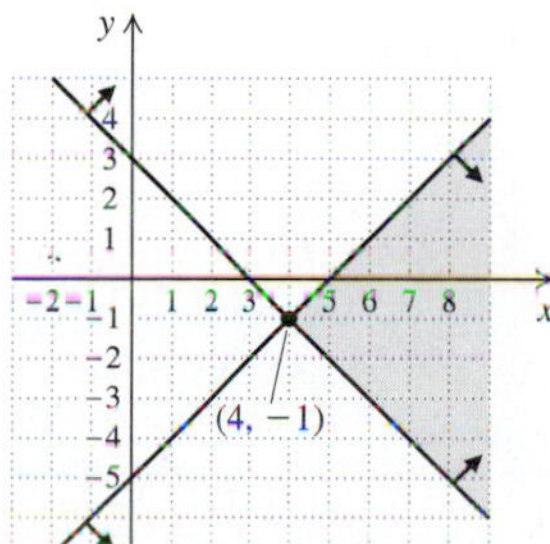

26. [9.4]

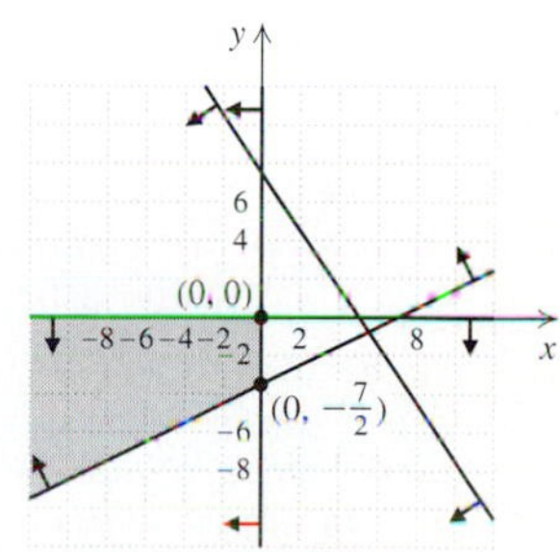

27. [9.5] Maximum 57 when $x = 6$, $y = 9$; minimum 5 when $x = 1$, $y = 0$ **28.** [9.5] 0 of A, 10 of B; 120

29. [2.2] $-\frac{33}{2}$ **30.** [8.2] $\left(-\frac{127}{3}, 27\right)$
31. [3.2]

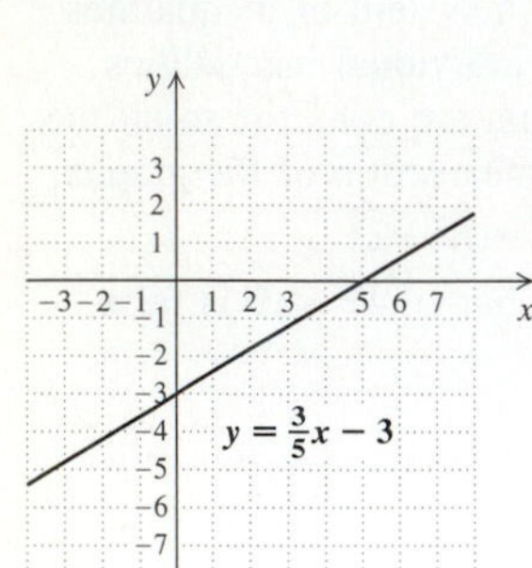

32. [8.3] 30-sec: 10; 60-sec: 5
33. [9.3] $\varnothing$
34. [9.2] $\{x|\frac{1}{5} < x < \frac{4}{5}\}$, or $\left(\frac{1}{5}, \frac{4}{5}\right)$

28. [9.4]

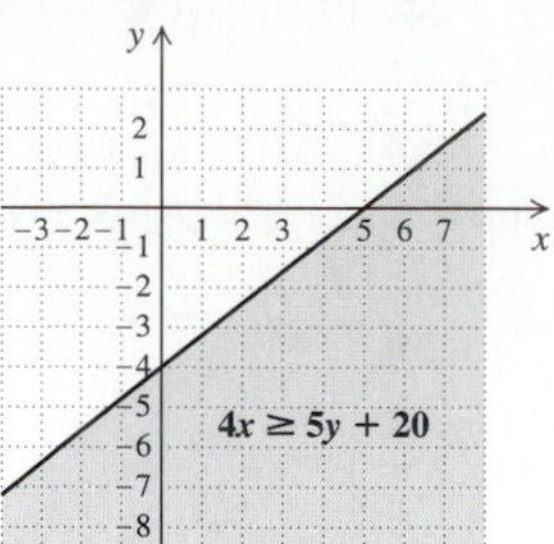

29. [9.4]

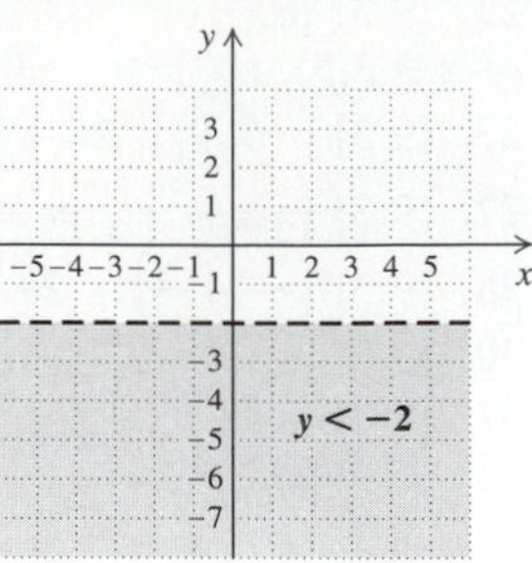

30. [4.3] $-3x^3 + 9x^2 + 3x - 3$ **31.** [4.6] $-15x^4y^4$
32. [4.6] $5a + 5b - 5c$
33. [4.4] $15x^4 - x^3 - 9x^2 + 5x - 2$
34. [4.5] $4x^4 - 4x^2 + 1$ **35.** [4.5] $4x^4 - 1$
36. [6.2] $\frac{y-6}{2}$ **37.** [6.2] $x - 1$
38. [6.3] $\frac{a^2 + 7ab + b^2}{(a-b)(a+b)}$ **39.** [6.3] $\frac{-m^2 + 5m - 6}{(m+1)(m-5)}$
40. [6.6] $\frac{y-x}{xy(x+y)}$ **41.** [4.7] $9x^2 - 13x + 26 + \frac{-50}{x+2}$
42. [5.1] $2x^2(2x+9)$ **43.** [5.2] $(x-6)(x+14)$
44. [5.4] $(4y-9)(4y+9)$
45. [5.5] $8(2x+1)(4x^2-2x+1)$ **46.** [5.4] $(t-8)^2$
47. [5.6] $x^2(x-1)(x+1)(x^2+1)$
48. [5.3] $(4x-1)(5x+3)$ **49.** [5.1] $(x^3+y)(x^2-y)$
50. [7.4], [7.5] **(a)** 5; **(b)** 4; **(c)** $x^2 + 2x - 4$
51. [7.5] $\{x|x$ is a real number and $x \neq 2$ and $x \neq 5\}$
52. [7.6] 18 min
53. [6.8] 1 hr **54.** [5.8] 30 ft
55. [5.8] All such sets of even integers satisfy this condition.
56. [2.5] 8
57. [9.4]

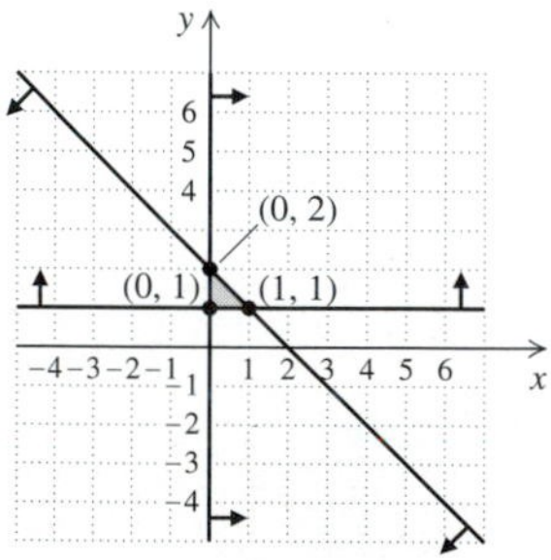

58. [4.4] $x^3 - 12x^2 + 48x - 64$
59. [9.3] $\{x|-3 \leq x \leq -1$ *or* $7 \leq x \leq 9\}$; or $[-3, -1] \cup [7, 9]$
60. [6.7] All real numbers except 9 and -5
61. [5.7] $0, \frac{1}{4}, -\frac{1}{4}$

CUMULATIVE REVIEW: 1–9

1. [1.8], [4.6] 10 **2.** [4.9] 5.76×10^9
3. [7.2] Slope is $\frac{7}{4}$; y-intercept is $(0, -3)$
4. [7.3] $y = -\frac{10}{3}x + \frac{11}{3}$ **5.** [8.2] $(-3, 4)$
6. [8.4] $(-2, -3, 1)$ **7.** [8.3] 12 small, 15 large
8. [8.5] First number is 12, second number is $\frac{1}{2}$, third number is $7\frac{1}{2}$.
9. [8.7] 42 **10.** [5.7] $\frac{1}{4}$ **11.** [5.7] $-\frac{25}{7}, \frac{25}{7}$
12. [2.2] $\frac{17}{3}$
13. [9.1] $\{y|y \geq -50\}$; or $[-50, \infty)$
14. [9.1] $\{x|x \geq -1\}$; or $[-1, \infty)$
15. [9.2] $\{x|-10 < x < 13\}$; or $(-10, 13)$
16. [9.2] $\{x|x < -\frac{4}{3}$ *or* $x > 6\}$; or $\left(-\infty, -\frac{4}{3}\right) \cup (6, \infty)$
17. [9.3] $\{x|x < -7$ *or* $x > 7\}$; or $(-\infty, -7) \cup (7, \infty)$
18. [9.3] $\left\{\frac{4}{3}, \frac{8}{3}\right\}$
19. [9.3] $\left\{x|-\frac{13}{4} \leq x \leq \frac{15}{4}\right\}$; or $\left[-\frac{13}{4}, \frac{15}{4}\right]$
20. [6.7] $-\frac{5}{3}$ **21.** [6.7] -1 **22.** [6.7] No solution
23. [6.7] $\frac{1}{3}$ **24.** [2.3] $n = \frac{m-12}{3}$ **25.** [6.9] $a = \frac{Pb}{3-P}$
26. [3.2]

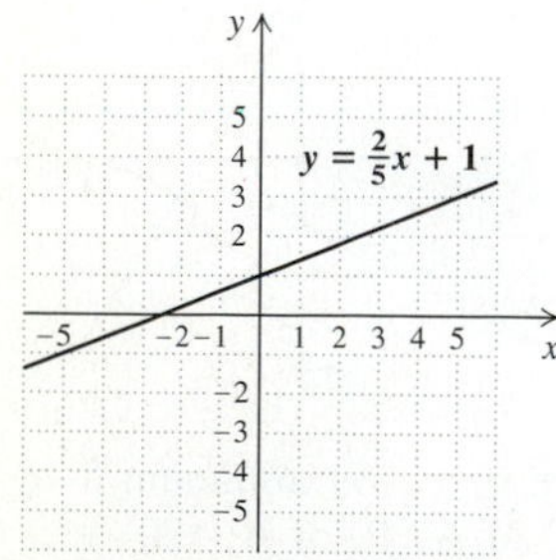

27. [3.3]

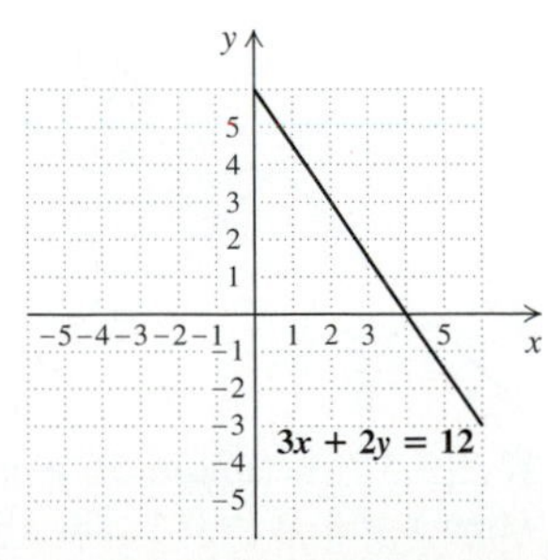

CHAPTER 10

Exercise Set 10.1, pp. 490–491

1. 4, −4 **3.** 12, −12 **5.** 20, −20 **7.** 7, −7 **9.** $-\frac{7}{6}$ **11.** 14 **13.** $-\frac{4}{9}$ **15.** 0.3 **17.** −0.07 **19.** p^2+4 **21.** $\frac{x}{y+4}$ **23.** $f(6)=\sqrt{20}$, $f(2)=0$, 0 and −3 are not in the domain **25.** $t(4)=-3$, −4 is not in the domain, $t(0)=-1$, $t(12)=-5$ **27.** $f(5)=\sqrt{26}$, $f(-5)=\sqrt{26}$, $f(0)=1$, $f(10)=\sqrt{101}$, $f(-10)=\sqrt{101}$ **29.** $g(2)=\sqrt{17}$, $g(-2)=1$, $g(3)=6$, −3 is not in the domain **31.** $4|x|$ **33.** $7|c|$ **35.** $|a+1|$ **37.** $|x-2|$ **39.** $|2x+7|$ **41.** 5 **43.** −1 **45.** $-\frac{2}{3}$ **47.** $|x|$ **49.** $5|a|$ **51.** 6 **53.** $|a+b|$ **55.** y **57.** $x-2$ **59.** $4x$ **61.** $6b$ **63.** $a+1$ **65.** $2(x+1)$, or $2x+2$ **67.** $3t-2$ **69.** 3 **71.** $2x$ **73.** −6 **75.** $5y$ **77.** $0.2(x+1)$ **79.** $2x^2$ **81.** $f(7)=2$, $f(26)=3$, $f(-9)=-2$, $f(-65)=-4$ **83.** $g(19)=2$, −13 and 1 are not in the domain, $g(84)=3$ **85.** $\{x|x\geq -7\}$ **87.** $\{t|t\geq \frac{9}{2}\}$ **89.** $\{x|x\leq 5\}$ **91.** $\mathbb{R}$ **93.** $\{z|z\geq -\frac{3}{5}\}$ **95.** $\{t|t\geq \frac{5}{3}\}$ **97.** $a^9b^6c^{15}$ **99.** x^2-9 **101.** ◈ **103.** (a) 13; (b) 15; (c) 18; (d) 20

105.

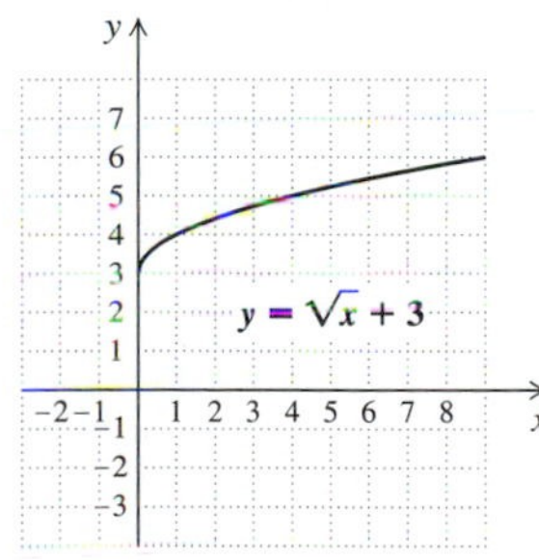

107.

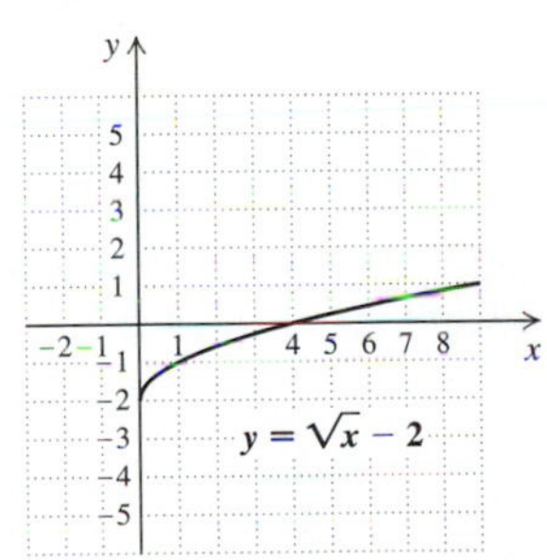

Exercise Set 10.2, pp. 495–496

1. $\sqrt[4]{x}$ **3.** 2 **5.** 3 **7.** 3 **9.** $\sqrt[3]{xyz}$ **11.** $\sqrt[5]{a^2b^2}$ **13.** $\sqrt[3]{a^2}$ **15.** 8 **17.** 343 **19.** 243 **21.** $\sqrt[4]{81^3x^3}$, or $27\sqrt[4]{x^3}$ **23.** $125x^6$ **25.** $\sqrt[3]{64a^4b^8}$, or $4ab^2\sqrt[3]{ab^2}$ **27.** $20^{1/3}$ **29.** $17^{1/2}$ **31.** $x^{3/2}$ **33.** $m^{2/5}$ **35.** $(cd)^{1/4}$ **37.** $(xy^2z)^{1/5}$ **39.** $(3mn)^{3/2}$ **41.** $(8x^2y)^{5/7}$ **43.** $\frac{1}{x^{1/3}}$ **45.** $\frac{1}{(2rs)^{3/4}}$ **47.** $10^{2/3}$ **49.** $x^{2/3}$ **51.** $\frac{5}{x^{1/4}}$ **53.** $\frac{3m^{1/2}}{5}$ **55.** $5^{7/8}$ **57.** $7^{1/4}$ **59.** $8.3^{7/20}$ **61.** $10^{6/25}$ **63.** $a^{23/12}$ **65.** $x^{2/7}$ **67.** $m^{1/6}n^{1/8}$ **69.** $a^4b^{3/2}$ **71.** $\sqrt[3]{a^2}$ **73.** $2y^2$ **75.** $2\sqrt[4]{2}$ **77.** $\sqrt[3]{2x}$ **79.** $2c^2d^3$ **81.** $\frac{m^2n^4}{2}$ **83.** $\sqrt[4]{r^2s}$ **85.** $3ab^3$ **87.** −3, 3 **89.** $\frac{1}{3}$ **91.** ◈ **93.** $\sqrt[10]{x^5y^3}$ **95.** $\sqrt[4]{2xy^2}$ **97.** (a) 1.8 m; (b) 3.1 m; (c) 1.5 m; (d) 5.3 m **99.** 7.937×10^{-13} to 1

Exercise Set 10.3, pp. 502–503

1. $\sqrt{6}$ **3.** $\sqrt[3]{10}$ **5.** $\sqrt[4]{72}$ **7.** $\sqrt{30ab}$ **9.** $\sqrt[5]{18t^3}$ **11.** $\sqrt{x^2-a^2}$ **13.** $\sqrt[3]{0.06x^2}$ **15.** $\sqrt[4]{x^3-1}$ **17.** $\sqrt{\frac{6y}{5x}}$ **19.** $\sqrt[7]{\frac{5x-15}{4x+8}}$ **21.** $\sqrt[6]{392}$ **23.** $\sqrt[6]{4x^5}$ **25.** $\sqrt[6]{x^5-4x^4+4x^3}$ **27.** $\sqrt[10]{x^9y^7}$ **29.** $\sqrt[12]{x^{11}(y+1)^{10}}$ **31.** $\sqrt[15]{a^{11}b^{13}c^{14}}$ **33.** $3\sqrt{3}$ **35.** $3\sqrt{5}$ **37.** $2\sqrt{2}$ **39.** $2\sqrt{6}$ **41.** $6x^2\sqrt{5}$ **43.** $2\sqrt[3]{100}$ **45.** $-2x^2\sqrt[3]{2}$ **47.** $3x^2\sqrt[3]{2x^2}$ **49.** $2x^2\sqrt[3]{10x^2}$ **51.** $2\sqrt[4]{2}$ **53.** $3\sqrt[4]{10}$ **55.** $2a^2\sqrt[4]{6}$ **57.** $3cd\sqrt[4]{2d^2}$ **59.** $(x+y)\sqrt[3]{x+y}$ **61.** $20(m+n)^2\sqrt[3]{(m+n)^2}$ **63.** $-ab^2c^3\sqrt[5]{abc^2}$ **65.** $3\sqrt{2}$ **67.** $6\sqrt{5}$ **69.** $4\sqrt{3}$ **71.** $3\sqrt[3]{2}$ **73.** $5bc^2\sqrt{2b}$ **75.** $-5x^2\sqrt[3]{6x}$ **77.** $2y^3\sqrt[3]{2}$ **79.** $(b+3)^2$ **81.** $4a^3b\sqrt{6ab}$ **83.** $a(b+c)^2\sqrt[5]{a(b+c)}$ **85.** $a\sqrt[10]{ab^7}$ **87.** $2xy^2\sqrt[6]{2x^5y}$ **89.** $xy\sqrt[4]{xy^3}$ **91.** $a^2b^2c^2\sqrt[6]{a^2bc^2}$ **93.** $yz\sqrt[12]{x^{11}yz^5}$ **95.** $2^2a(b-5)^2\sqrt[6]{2a(b-5)}$ **97.** 13.416 **99.** 7.477 **101.** 0.454 **103.** 4.414 **105.** 156,524.7584 **107.** 684,251.4158 **109.** 0.0001987460691 **111.** 0.0007071774883 **113.** 8 **115.** $(2x-7)(2x+7)$ **117.** ◈ **119.** (a) 20 mph; (b) 37.4 mph; (c) 42.4 mph **121.** 10 **123.** $x\geq -\frac{3}{2}$

Exercise Set 10.4, pp. 507–508

1. $\sqrt{7}$ **3.** 3 **5.** $y\sqrt{5y}$ **7.** $2\sqrt[3]{a^2b}$ **9.** $3\sqrt{2xy}$ **11.** $2x^2y^2$ **13.** $\sqrt{x^2+xy+y^2}$ **15.** $\sqrt[12]{a^5}$ **17.** $\sqrt[12]{x^2y}$ **19.** $\sqrt[10]{ab^9c^3}$ **21.** $\sqrt[20]{(3x-1)^3}$ **23.** $\frac{4}{5}$ **25.** $\frac{4}{3}$ **27.** $\frac{7}{y}$ **29.** $\frac{5y\sqrt{y}}{x^2}$ **31.** $\frac{2x\sqrt[3]{x^2}}{3y}$ **33.** $\frac{2a}{3}$ **35.** $\frac{ab^2}{c^2}\sqrt[4]{\frac{a}{c^2}}$ **37.** $\frac{2x}{y^2}\sqrt[5]{\frac{x}{y}}$ **39.** $\frac{xy}{z^2}\sqrt[6]{\frac{y^2}{z^3}}$ **41.** $6a\sqrt{6a}$ **43.** $64b^3$ **45.** $54a^3b\sqrt{2b}$ **47.** $3c^2d\sqrt[3]{3c^2d}$ **49.** $x\sqrt[3]{25xy^2}$ **51.** $x\sqrt[4]{x^2y^3}$ **53.** $4a^2\sqrt[3]{a^2b^2}$ **55.** $4xy\sqrt[4]{y^2}$, or $4xy\sqrt{y}$ **57.** 8 **59.** Length is 20; width is 5 **61.** ◈ **63.** a^3bxy^2 **65.** $2yz\sqrt{2z}$

Exercise Set 10.5, pp. 511–512

1. $8\sqrt{3}$ **3.** $3\sqrt[3]{5}$ **5.** $13\sqrt[3]{y}$ **7.** $7\sqrt{2}$ **9.** $6\sqrt[3]{3}$
11. $21\sqrt{3}$ **13.** $38\sqrt{5}$ **15.** $122\sqrt{2}$ **17.** $9\sqrt[3]{2}$
19. $(1+6a)\sqrt{5a}$ **21.** $(2-x)\sqrt[3]{3x}$ **23.** $3\sqrt{2y-2}$
25. $(x+3)\sqrt{x-1}$ **27.** $15\sqrt[3]{4}$ **29.** 0
31. $(3x^2+x)\sqrt[4]{x-1}$ **33.** $2\sqrt{6}-18$
35. $\sqrt{6}-\sqrt{10}$ **37.** $2\sqrt{15}-6\sqrt{3}$ **39.** -6
41. $3a\sqrt[3]{2}$ **43.** $x^2+\sqrt[4]{3x^3}$ **45.** 18 **47.** -3
49. -19 **51.** $a+\sqrt{3a}+\sqrt{2a}+\sqrt{6}$
53. $6+\sqrt[3]{5}-\sqrt[3]{25}$ **55.** -6
57. $2\sqrt[3]{9}-3\sqrt[3]{6}-2\sqrt[3]{4}$ **59.** $4+2\sqrt{3}$
61. $a^2+2a\sqrt{b}+b$ **63.** $4x^2-4x\sqrt[4]{y}+\sqrt{y}$
65. $m+2\sqrt{mn}+n$ **67.** $\sqrt[6]{a^5}+\sqrt[4]{a^3b}$
69. $x\sqrt[6]{xy^5}-\sqrt[15]{x^{13}y^{14}}$
71. $2m^2+m\sqrt[4]{n}+2m\sqrt[3]{n^2}+\sqrt[12]{n^{11}}$
73. $a^2-a\sqrt[3]{c}+a\sqrt[4]{b}-\sqrt[12]{b^3c^4}$
75. $\sqrt[4]{x^3}+\sqrt[10]{x^7z^2}-\sqrt[12]{x^3y^4z^4}-\sqrt[15]{x^3y^5z^8}$
77. $-\frac{19}{5}$ **79.** $-28\sqrt{3}-22\sqrt{5}$
81. $ac\left[(3ac+2)\sqrt{ab}-2c\sqrt[3]{ab}\right]$ **83.** 6
85. $14+2\sqrt{15}-6\sqrt{2}-2\sqrt{30}$ **87.** $3\sqrt[3]{3}+2\sqrt[3]{9}-8$

Exercise Set 10.6, pp. 517–518

1. $\frac{\sqrt{30}}{5}$ **3.** $\frac{\sqrt{70}}{7}$ **5.** $\frac{2\sqrt{15}}{5}$ **7.** $\frac{2\sqrt[3]{6}}{3}$
9. $\frac{\sqrt[3]{75ac^2}}{5c}$ **11.** $\frac{y\sqrt[3]{180x^2y}}{6x^2}$ **13.** $\frac{\sqrt[3]{x^2y^2}}{xy}$
15. $\frac{\sqrt{14a}}{6}$ **17.** $\frac{3\sqrt{5y}}{10xy}$ **19.** $\frac{\sqrt[3]{90xy}}{10xy^2}$ **21.** $\frac{7}{\sqrt{21x}}$
23. $\frac{2}{\sqrt{6}}$ **25.** $\frac{52}{3\sqrt{91}}$ **27.** $\frac{7}{\sqrt[3]{98}}$ **29.** $\frac{7x}{\sqrt{21xy}}$
31. $\frac{5y^2}{x\sqrt[3]{150x^2y^2}}$ **33.** $\frac{ab}{3\sqrt{ab}}$ **35.** $\frac{x^2y}{\sqrt{2xy}}$
37. $\frac{ab}{\sqrt[3]{5ab^2}}$ **39.** $\frac{x^2y}{\sqrt[3]{3x^2y}}$ **41.** $\frac{5(8+\sqrt{6})}{58}$
43. $-2\sqrt{7}(\sqrt{5}+\sqrt{3})$ **45.** $\frac{\sqrt{15}+20-6\sqrt{2}-8\sqrt{30}}{-77}$
47. $\frac{x-2\sqrt{xy}+y}{x-y}$ **49.** $\frac{4+3\sqrt{6}}{2}$
51. $\frac{6x-7\sqrt{xy}-3y}{4x-9y}$ **53.** $\frac{-11}{4(\sqrt{3}-5)}$
55. $\frac{-22}{\sqrt{6}+5\sqrt{2}+5\sqrt{3}+25}$ **57.** $\frac{x-y}{x+2\sqrt{xy}+y}$
59. $\frac{7}{43\sqrt{2}+66}$ **61.** $\frac{3-4x}{3-3\sqrt{3x}+2x}$
63. $\frac{a^2-b^2c}{a^2+(a-ab)\sqrt{c}-bc}$ **65.** 6 **67.** ◈
69. $\frac{ab+(a-b)\sqrt{a+b}-a-b}{a+b-b^2}$
71. $\frac{b^2+\sqrt{b}}{b^2+b+1}$ **73.** $\frac{x-19}{x+10\sqrt{x+6}+31}$
75. $\left(\frac{5x+4y-3}{xy}\right)\sqrt{xy}$ **77.** $\frac{7\sqrt{3}}{39}$

Technology Connection, Section 10.7

TC1. x-coordinates of points of intersection should approximate the solutions of the examples.

Exercise Set 10.7, pp. 522–523

1. 2 **3.** 12 **5.** 168 **7.** No solution **9.** 3
11. 19 **13.** $\frac{80}{3}$ **15.** 4 **17.** 64 **19.** -27
21. 397 **23.** No solution **25.** 3 **27.** $\frac{1}{64}$ **29.** -6
31. 5 **33.** $-\frac{1}{4}$ **35.** 3 **37.** 9 **39.** 7 **41.** $\frac{80}{9}$
43. -1 **45.** 6, 2 **47.** No solution **49.** 10
51. 15 **53.** 2 **55.** ◈ **57.** About 208 miles
59. $-\frac{8}{9}$ **61.** -8, 8 **63.** -4, 3 **65.** 1, 8
67. $\frac{1}{36}$, 36

Exercise Set 10.8, pp. 527–530

1. $\sqrt{34}$; 5.831 **3.** $\sqrt{288}$; 16.971 **5.** 5
7. $\sqrt{31}$; 5.568 **9.** $\sqrt{12}$; 3.464 **11.** $\sqrt{n-1}$
13. $\sqrt{325}$; 18.028 ft **15.** $\sqrt{14{,}500}$; 120.416 ft
17. 12 in. **19.** 50 ft **21.** $b=9$; $c=9\sqrt{2}\approx 12.728$
23. $a=7$; $b=7\sqrt{3}\approx 12.124$
25. $a=\frac{5\sqrt{3}}{3}\approx 2.887$; $c=\frac{10\sqrt{3}}{3}\approx 5.774$
27. $a=\frac{13\sqrt{2}}{2}\approx 9.192$; $b=\frac{13\sqrt{2}}{2}\approx 9.192$
29. $c=24$; $b=12\sqrt{3}\approx 20.785$ **31.** $h=4\sqrt{3}\approx 6.928$
33. $c=5\sqrt{2}\approx 7.071$ **35.** $\frac{19\sqrt{2}}{2}\approx 13.435$
37. Neither; both have the same area, 300 ft^2
39. $h=2\sqrt{3}\approx 3.464$ ft **41.** 8 ft **43.** $(0,-4)$, $(0,4)$
45. 3, 8 **47.** ◈ **49.** $\sqrt{75}$ cm

Exercise Set 10.9, pp. 536–537

1. $i\sqrt{15}$, or $\sqrt{15}i$ **3.** $4i$ **5.** $-2i\sqrt{3}$, or $-2\sqrt{3}i$
7. $i\sqrt{3}$, or $\sqrt{3}i$ **9.** $9i$ **11.** $7i\sqrt{2}$, or $7\sqrt{2}i$
13. $-7i$ **15.** $4-2\sqrt{15}i$ **17.** $(2+2\sqrt{3})i$

19. $8 + i$ **21.** $9 - 5i$ **23.** $7 + 4i$ **25.** $-2 - 3i$
27. $-1 + i$ **29.** $11 + 6i$ **31.** -30 **33.** $-\sqrt{30}$
35. $-5\sqrt{6}$ **37.** $-12\sqrt{2}$ **39.** -40 **41.** 35
43. $10 + 15i$ **45.** $-12 - 21i$ **47.** $1 + 5i$
49. $18 + 14i$ **51.** $38 + 9i$ **53.** $2 - 46i$ **55.** $5 - 37i$
57. $-11 - 16i$ **59.** $13 - 47i$ **61.** $5 - 12i$
63. $-5 + 12i$ **65.** $-5 - 12i$ **67.** $-i$ **69.** 1
71. -1 **73.** i **75.** -1 **77.** $-125i$ **79.** 8
81. $1 - 26i$ **83.** 0 **85.** 0 **87.** 1 **89.** $5 - 8i$
91. $\frac{3}{2} + \frac{1}{2}i$ **93.** $\frac{3}{29} + \frac{7}{29}i$ **95.** $-\frac{7}{6}i$ **97.** $-\frac{3}{7} - \frac{8}{7}i$
99. $\frac{8}{5} + \frac{1}{5}i$ **101.** $-i$ **103.** $\frac{19}{13} - \frac{9}{13}i$ **105.** Yes
107. No **109.** 7 **111.** $-\frac{4}{3}, 7$ **113.** $\frac{250}{41} + \frac{200}{41}i$
115. 8 **117.** $\frac{3}{5} + \frac{9}{5}i$ **119.** 1

Review Exercises: Chapter 10, pp. 541–542

1. [10.1] $\frac{2}{3}$ **2.** [10.1] 0.07 **3.** [10.1] 5
4. [10.1] $\{x \mid x \geq \frac{7}{2}\}$, or $[\frac{7}{2}, \infty)$ **5.** [10.1] $9|a|$
6. [10.1] $|c + 8|$ **7.** [10.1] $|x - 3|$ **8.** [10.1] $|2x + 1|$
9. [10.1] -2 **10.** [10.1] $-\frac{1}{3}$ **11.** [10.1] $|x|$
12. [10.1] 3 **13.** [10.2] $(8x^6y^2)^{4/5}$, or $8^{4/5}x^{24/5}y^{8/5}$
14. [10.2] $\sqrt[4]{(5a)^3}$ **15.** [10.2] $\sqrt[12]{x^4y^3}$ **16.** [10.2] $-\dfrac{x^3y^4}{2}$
17. [10.3] 15.652 **18.** [10.3] 10.583
19. [10.3] -0.236 **20.** [10.3] $3xy\sqrt{2y}$
21. [10.3] $3a\sqrt[3]{a^2b^2}$ **22.** [10.3] $3x^2y^2\sqrt[6]{3xy^2}$
23. [10.4] $y\sqrt[3]{6}$ **24.** [10.4] $\dfrac{5\sqrt{x}}{2}$
25. [10.4] $8xy^2$ **26.** [10.4] $2a\sqrt[3]{2ab^2}$
27. [10.5] $30\sqrt[3]{5}$ **28.** [10.5] $15\sqrt{2}$
29. [10.5] $9 - \sqrt[3]{4}$ **30.** [10.5] $-43 - 2\sqrt{10}$
31. [10.5] $8 - 2\sqrt{7}$ **32.** [10.6] $\dfrac{10a\sqrt{3} - 10\sqrt{3ab}}{a - b}$
33. [10.6] $\dfrac{30a}{\sqrt{3a}(\sqrt{a} + \sqrt{b})}$, or $\dfrac{30a}{\sqrt{3}(a + \sqrt{ab})}$
34. [10.7] 4 **35.** [10.7] 13 **36.** [10.8] 9 cm
37. [10.8] 5 ft **38.** [10.9] $-2\sqrt{2}i$
39. [10.9] $-2 - 9i$ **40.** [10.9] $1 + i$ **41.** [10.9] 29
42. [10.9] i **43.** [10.9] $9 - 12i$ **44.** [10.9] $\frac{2}{5} + \frac{3}{5}i$
45. [10.9] $\frac{9}{5} - \frac{12}{5}i$ **46.** [5.8] 12, 13, 14
47. [6.7] $\frac{23}{7}$ **48.** [5.7] $-\frac{9}{2}, 3$
49. [6.2] $\dfrac{x(x + 2)}{(x + y)(x - 3)}$
50. [10.1] ◈ An absolute-value sign must be used to simplify $\sqrt[k]{x^k}$ when k is even, since x may be negative. If x is negative while k is even, the radical expression cannot be simplified to x, since $\sqrt[k]{x^k}$ represents the principal, or positive, root. When k is odd, there is only one root, and it will be positive or negative depending on the sign of x. Thus there is no absolute-value sign when k is odd.
51. [10.9] ◈ Every real number is a complex number, but there are complex numbers that are not real. A complex number $a + bi$ is not real if $b \neq 0$.
52. [10.7] 3 **53.** [10.9] $-\frac{2}{5} + \frac{9}{10}i$

Test: Chapter 10, pp. 542–543

1. [10.1] $\frac{10}{7}$ **2.** [10.1] $\{t \mid t \geq -5\}$, or $[-5, \infty)$
3. [10.1] $6|y|$ **4.** [10.1] $|x + 5|$ **5.** [10.1] -2
6. [10.1] 4 **7.** [10.2] $(5xy^2)^{5/2}$
8. [10.2] $2ab^3\sqrt[3]{2a^2b}$ **9.** [10.3] 0.00109727
10. [10.3] $2x^3$ **11.** [10.4] $\dfrac{5x\sqrt{x}}{6y^2}$
12. [10.4] $4a\sqrt[3]{4ab^2}$ **13.** [10.5] $38\sqrt{2}$
14. [10.5] -20
15. [10.5] $2x^2 - x\sqrt[3]{y^2} + 2xy\sqrt{y} - y^2\sqrt[6]{y}$
16. [10.6] $-\dfrac{13 + 8\sqrt{2}}{41}$ **17.** [10.7] 7
18. [10.8] Longer leg: $10\sqrt{3} \approx 17.321$ cm; hypotenuse: 20 cm **19.** [10.9] $3\sqrt{2}i$ **20.** [10.9] $7 + 5i$
21. [10.9] -30 **22.** [10.9] $-2i$ **23.** [10.9] $-\frac{77}{50} + \frac{7}{25}i$
24. [10.9] $-1 + 5i$ **25.** [5.7] $-\dfrac{1}{3}, \dfrac{5}{2}$ **26.** [6.2] $\dfrac{x - 3}{x - 4}$
27. [6.7] No solution **28.** [5.8] 16, 18 and -18, -16
29. [10.7] 3 **30.** [10.9] $-\frac{17}{4}i$

CHAPTER 11

Exercise Set 11.1, pp. 548–549

1. $5, -5$ **3.** $9, -9$ **5.** $\sqrt{15}, -\sqrt{15}$
7. $\sqrt{19}, -\sqrt{19}$ **9.** $\sqrt{7}, -\sqrt{7}$ **11.** $2\sqrt{5}, -2\sqrt{5}$
13. $\dfrac{5}{2}, -\dfrac{5}{2}$ **15.** $\dfrac{2}{5}i, -\dfrac{2}{5}i$ **17.** $\dfrac{\sqrt{5}}{3}, -\dfrac{\sqrt{5}}{3}$
19. $\dfrac{\sqrt{21}}{3}, -\dfrac{\sqrt{21}}{3}$ **21.** $9, -5$ **23.** $3, -9$
25. $-3 \pm \sqrt{21}$ **27.** $-13 \pm 2\sqrt{2}$ **29.** $7 \pm 2\sqrt{3}$
31. $-9 \pm \sqrt{34}$ **33.** $\dfrac{-3 \pm \sqrt{14}}{2}$ **35.** $11, -5$
37. $-9, -5$ **39.** $1 \pm \sqrt{5}$ **41.** $-2 \pm 2\sqrt{3}$
43. $2\sqrt{15}$ **45.** $(1, 3)$ **47.** ◈ **49.** $\dfrac{-5 \pm \sqrt{13}}{2}$

51. $\dfrac{3 \pm \sqrt{17}}{4}$ **53.** 3.45, −3.95 **55.** 7.1, −3.3

Technology Connection, Section 11.2

TC1. x-intercepts should be approximations of $4 + \sqrt{23}$ and $4 - \sqrt{23}$. **TC2.** A grapher can only give rational-number approximations of the two irrational solutions.

Exercise Set 11.2, pp. 553

1. $x^2 - 2x + 1$ **3.** $x^2 + 18x + 81$ **5.** $x^2 - x + \frac{1}{4}$
7. $t^2 + 5t + \frac{25}{4}$ **9.** $x^2 - \frac{3}{2}x + \frac{9}{16}$ **11.** $m^2 + \frac{9}{2}m + \frac{81}{16}$
13. −2, 8 **15.** −21, −1 **17.** $1 \pm \sqrt{6}$
19. $5 \pm \sqrt{47}$ **21.** $-3 \pm 2i$ **23.** $\dfrac{7 \pm \sqrt{57}}{2}$
25. −7, 4 **27.** $\dfrac{-3 \pm \sqrt{17}}{4}$ **29.** $-\dfrac{1}{2}$, 3
31. $\dfrac{-2 \pm \sqrt{7}}{3}$ **33.** $-\frac{1}{2}$, 5 **35.** $-\frac{7}{2}, \frac{1}{2}$
37. 30 lb Kenyan; 20 lb Peruvian **39.** −32
41.
43. 12, −12
45. $\pm 16\sqrt{2}$
47. $\pm 2\sqrt{c}$
49. 8.00, 2.00
51. −0.39, −7.61
53. 7.27, −0.27
55. −0.50, 5.00

Exercise Set 11.3, pp. 560–562

1. 7, −3 **3.** 3 **5.** $-3 \pm \sqrt{5}$ **7.** $\pm 4i$ **9.** −3, 3
11. $1 \pm \sqrt{3}$ **13.** $\dfrac{5 \pm i\sqrt{23}}{6}$ **15.** $-2 \pm \sqrt{7}$
17. $\dfrac{-4 \pm \sqrt{10}}{3}$ **19.** $\dfrac{-1 \pm i\sqrt{7}}{2}$ **21.** $\dfrac{1 \pm \sqrt{2}}{2}$
23. $\dfrac{-3 \pm i}{2}$ **25.** $\dfrac{5 \pm \sqrt{73}}{6}$ **27.** $\dfrac{2}{5}$ **29.** 0, $\dfrac{3}{2}$
31. $\dfrac{7 \pm i\sqrt{31}}{10}$ **33.** $\pm 2\sqrt{5}$
35. 5.317, −1.317 **37.** 6.162, −0.162
39. 0.207, −1.207 **41.** 20 **43.** 7
45. About 6.32 sec **47.** About 3.31 sec
49. 7 ft, 24 ft **51.** Width: 8 cm; length: 10 cm
53. 15 m by 20 m **55.** 4 m, 6.5 m
57. Width: 3.58 in.; length: 5.58 in.
59. Width: 2.24 m; length: 4.47 m **61.** 10%
63. 6.25% **65.** 8% **67.** 17.84 ft **69.** No solution
71. $3\sqrt{10}$ **73.** **75.** 0, 2 **77.** $\dfrac{-7 \pm \sqrt{61}}{2}$
79. $\dfrac{-3 \pm i}{2}$ **81.** $\dfrac{-1 \pm \sqrt{1 + 4\sqrt{2}}}{2}$ **83.** $\pm \sqrt{7}$
85. 2, $-1 \pm i\sqrt{3}$ **87.** $1 + \sqrt{2} \approx 2.41$ cm
89. $3 + 2\sqrt{2} \approx 5.828$
91. $10\sqrt{2} \approx 14.14$ in.; two 10-in. pizzas
93. 14.75% **95.** About 19 in. by 44 in.

Exercise Set 11.4, pp. 566–567

1. 6, −4 **3.** $\pm \frac{3}{2}$ **5.** −20, 25 **7.** −1, 4
9. $\pm \sqrt{37}$ **11.** −5, 2 **13.** $\pm 2\sqrt{7}$ **15.** 10, $\frac{21}{2}$
17. 2, 3 **19.** $\frac{1}{15}$, 2 **21.** −1, −6 **23.** −4
25. First part, 12 km/h; second part, 8 km/h **27.** 35 mph **29.** Super-prop: 350 mph; turbo-jet: 400 mph
31. Speed out: 60 mph; speed back: 50 mph
33. 10 mph **35.** 6 hr **37.** 3.24 mph **39.** 1, 5
41. $2y\sqrt[3]{9x^2}$ **43.** **45.** $4 \pm 2\sqrt{2}$ **47.** $\pm \sqrt{2}$
49.

Exercise Set 11.5, pp. 570–571

1. One real **3.** Two imaginary **5.** Two real
7. One real **9.** Two imaginary **11.** Two real
13. Two real **15.** Two real **17.** Two real
19. Two imaginary **21.** One real
23. $x^2 + 2x - 99 = 0$ **25.** $x^2 - 14x + 49 = 0$
27. $x^2 + 8x + 15 = 0$ **29.** $3x^2 - 14x + 8 = 0$
31. $6x^2 - 5x + 1 = 0$ **33.** $25x^2 - 20x - 12 = 0$
35. $x^2 - 4\sqrt{2}x + 6 = 0$ **37.** $x^2 + 3\sqrt{5}x + 10 = 0$
39. $x^2 + 9 = 0$ **41.** $x^2 - 10x + 29 = 0$
43. $2x^2 - 2x + 5 = 0$ **45.** Six 30-second commercials; six 60-second commercials **47.** **49.** $a = 1, b = 2, c = -3$ **51.** **(a)** $k = -\frac{3}{5}$; **(b)** $-\frac{1}{3}$ **53.** **(a)** $k = 9 + 9i$; **(b)** $3 + 3i$ **55.** $x^2 - \sqrt{3}x + 8 = 0$ **57.** $a = 8$, $b = 20$, $c = -12$ **59.** $k = 6$

Exercise Set 11.6, pp. 574–575

1. $\pm \sqrt{5}$ **3.** $\pm \sqrt{3}, \pm 3$ **5.** $\pm 1, \pm \dfrac{\sqrt{5}}{3}$ **7.** 81, 1
9. $\frac{4}{9}$ **11.** $\pm \sqrt{7}, \pm 2\sqrt{2}$ **13.** 7, −1, 5, 1 **15.** 4
17. $-\frac{1}{2}, \frac{1}{3}$ **19.** −1, 2 **21.** −27, 8 **23.** 16
25. 32, −243 **27.** 1 **29.** 6, 1 **31.** $-\frac{3}{2}$
33. $3 \pm \sqrt{10}$, $-1 \pm \sqrt{2}$ **35.** $3x^2\sqrt{x}$
37. $\dfrac{x^3 + x^2 + 2x + 2}{x^3 - 1}$
39. $(\sqrt{7}, 0)$, $(-\sqrt{7}, 0)$, $(1, 0)$, $(-1, 0)$ **41.** ± 1.99
43. $\frac{432}{143}$ **45.** 9 **47.** −2, 1 **49.** −3, −1, 1, 4

Exercise Set 11.7, pp. 581–583

1.

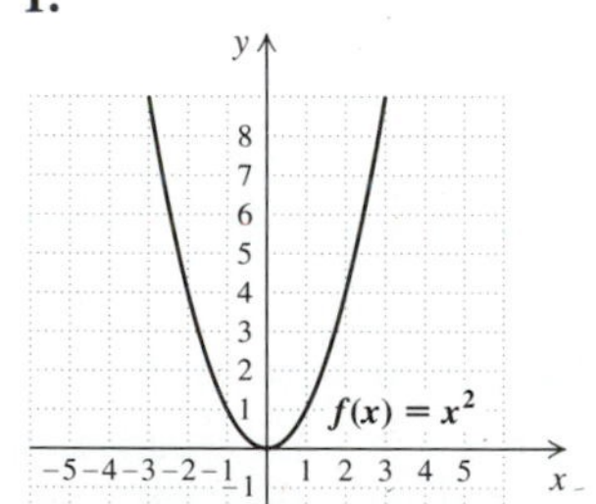

3.

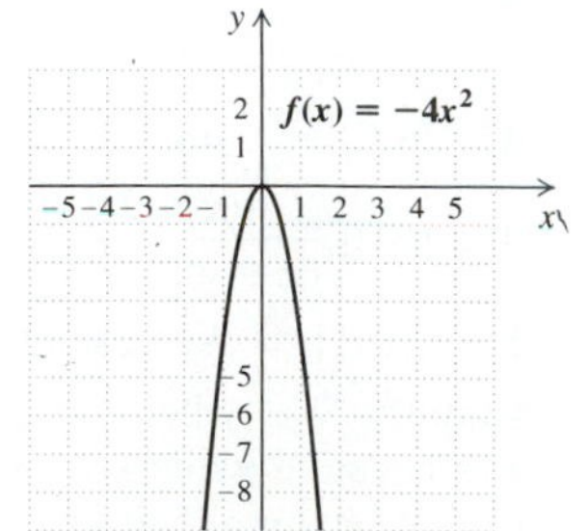

5.

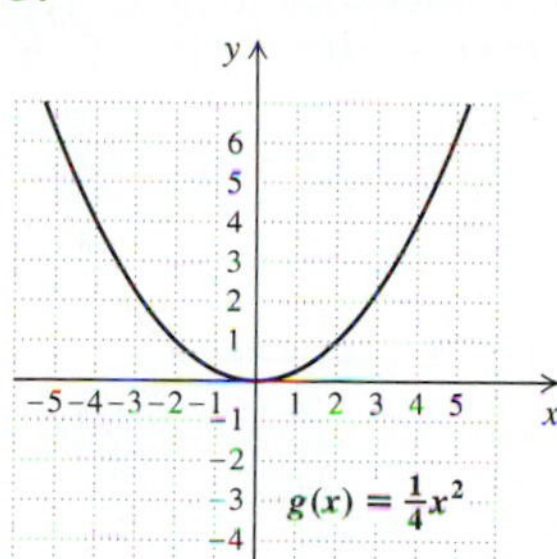

7.

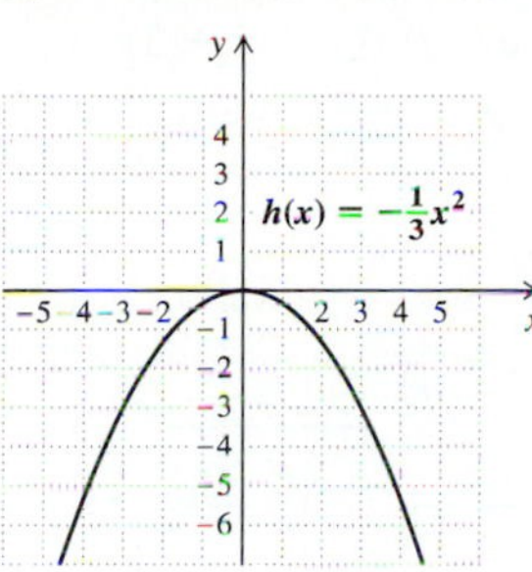

9.

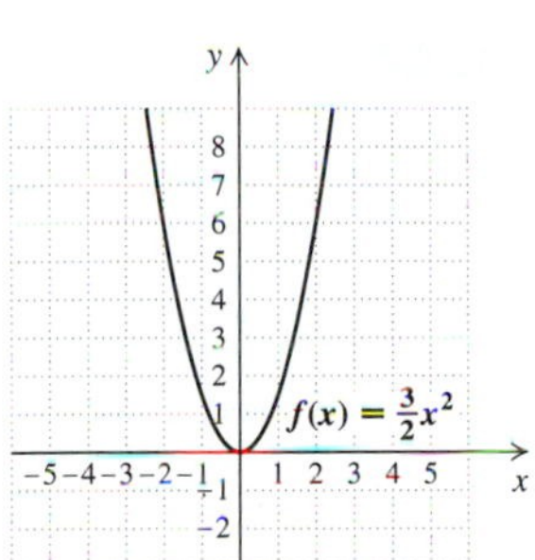

11. Vertex: $(-1, 0)$; line of symmetry: $x = -1$

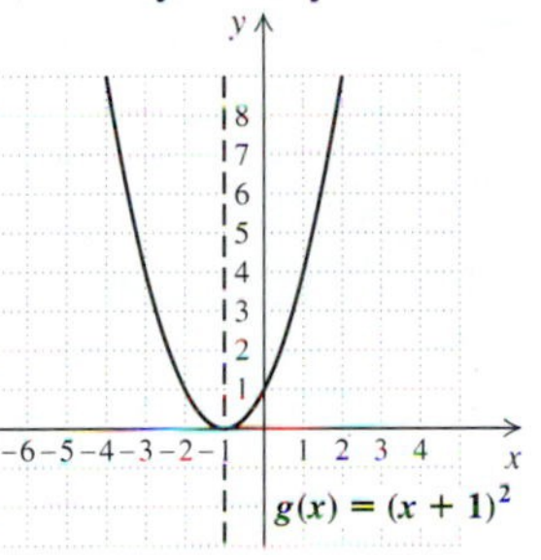

13. Vertex: $(4, 0)$; line of symmetry: $x = 4$

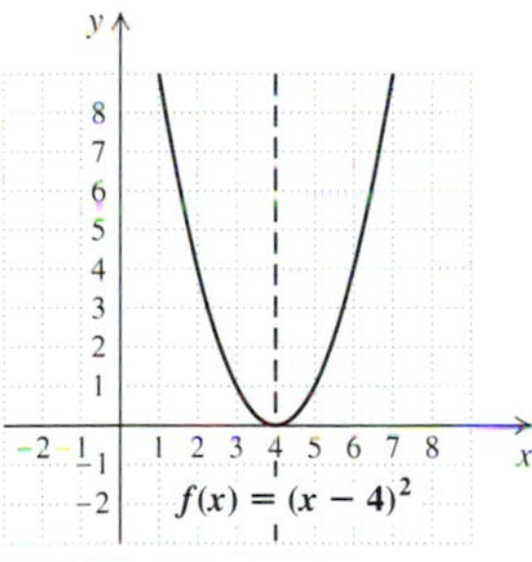

15. Vertex: $(3, 0)$; line of symmetry: $x = 3$

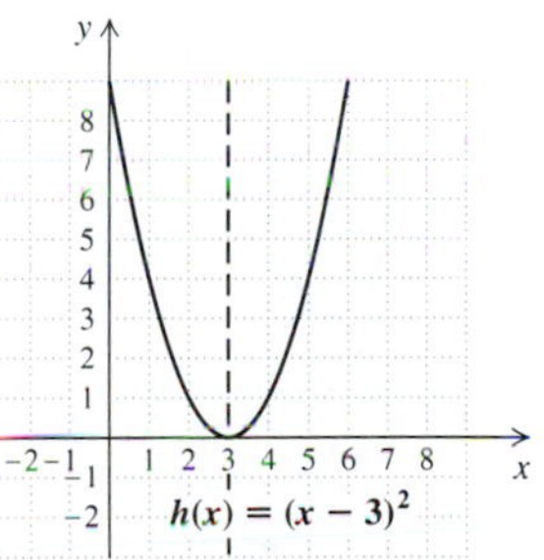

17. Vertex: $(-4, 0)$; line of symmetry: $x = -4$

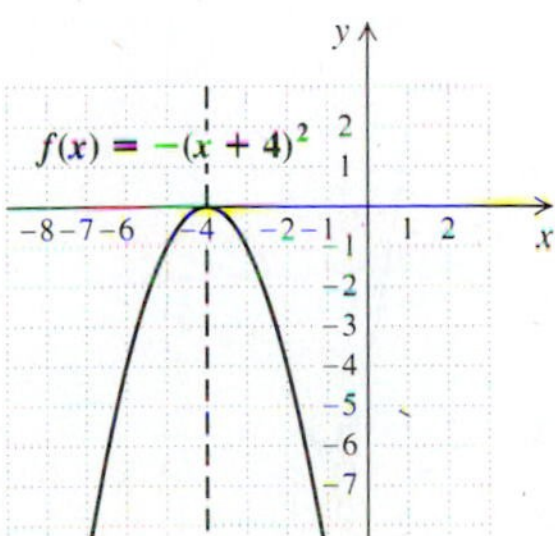

19. Vertex: $(1, 0)$; line of symmetry: $x = 1$

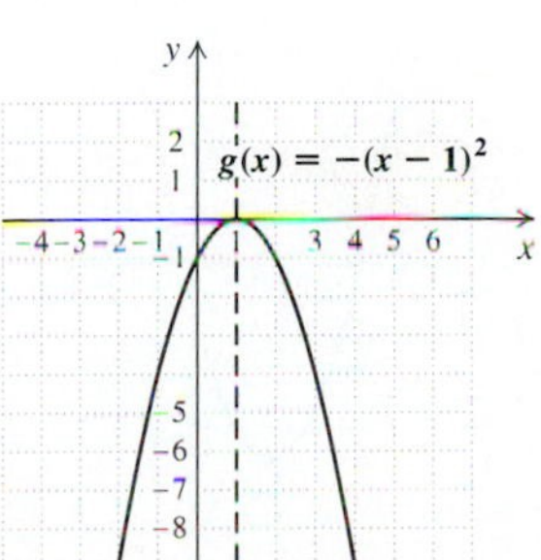

21. Vertex: $(1, 0)$; line of symmetry: $x = 1$

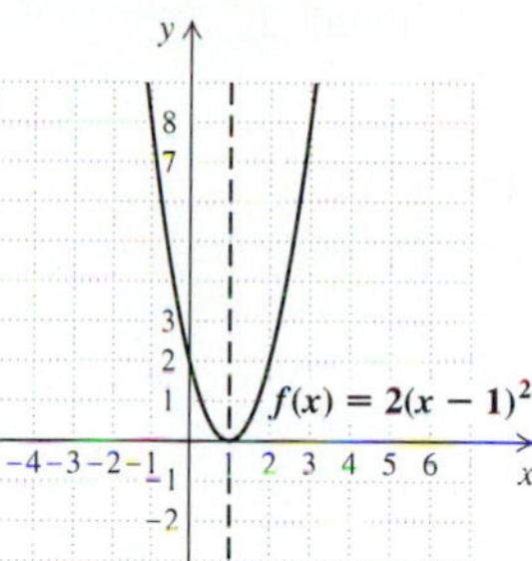

23. Vertex: $(3, 0)$; line of symmetry: $x = 3$

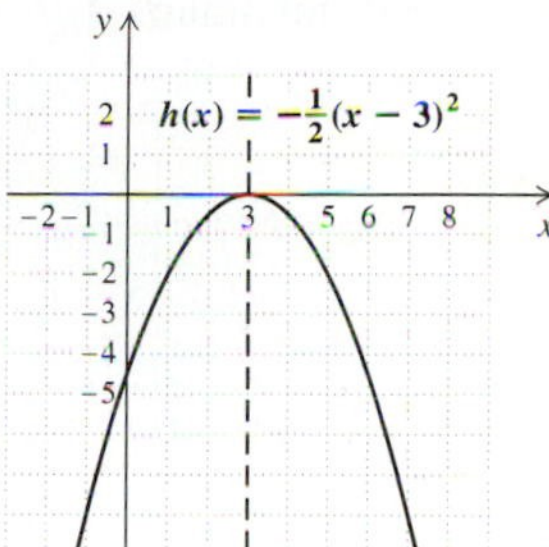

25. Vertex: $(-1, 0)$; line of symmetry: $x = -1$

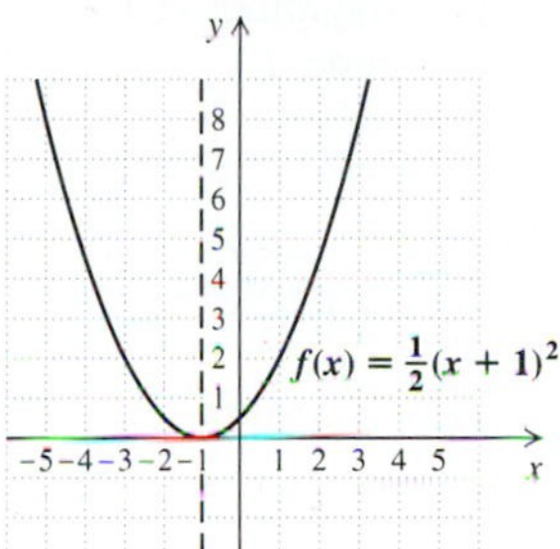

27. Vertex: $(2, 0)$; line of symmetry: $x = 2$

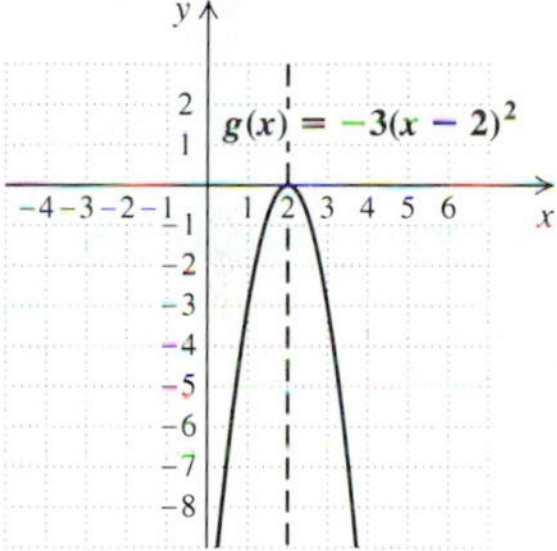

29. Vertex: $(-9, 0)$; line of symmetry: $x = -9$

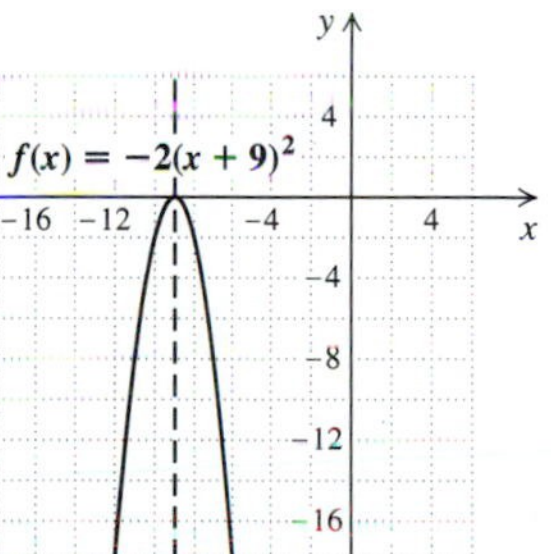

31. Vertex: $\left(\frac{1}{2}, 0\right)$; line of symmetry: $x = \frac{1}{2}$

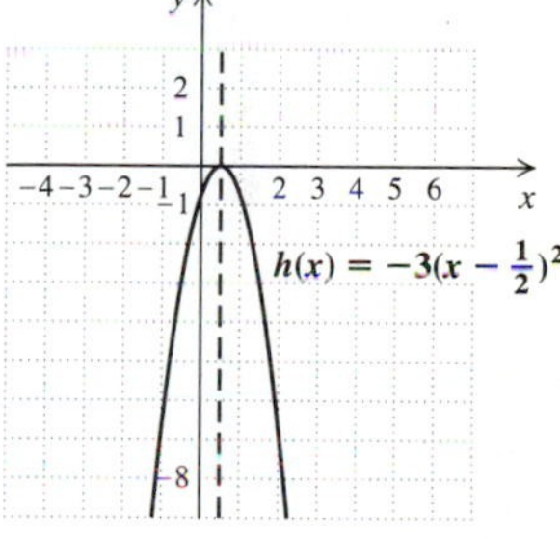

33. Vertex: $(3, 1)$; line of symmetry: $x = 3$; minimum: 1

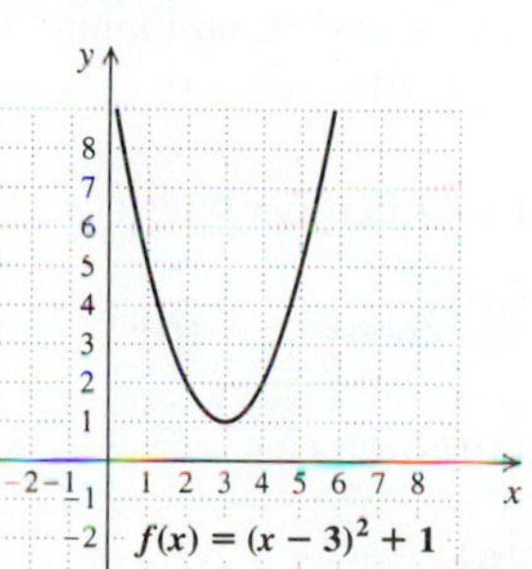

35. Vertex: $(-1, -2)$; line of symmetry: $x = -1$; minimum: -2

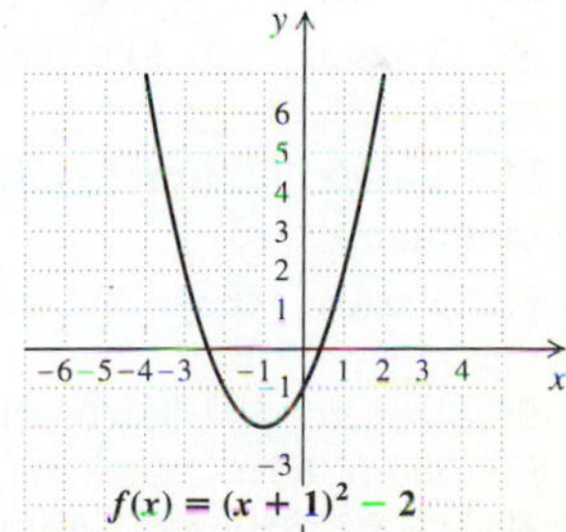

37. Vertex: $(-4, 1)$; line of symmetry: $x = -4$; minimum: 1

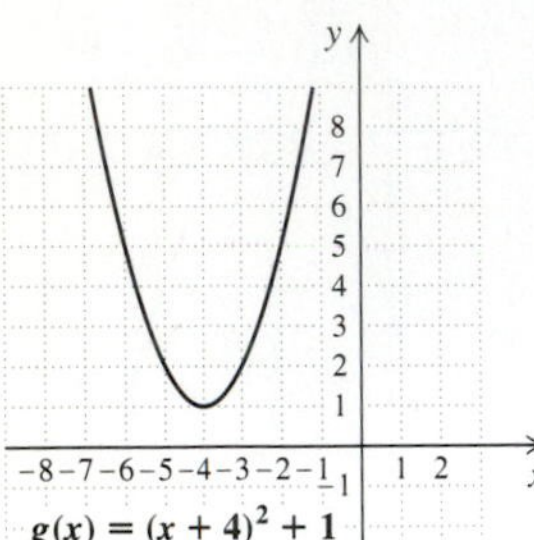

39. Vertex: $(5, 2)$; line of symmetry: $x = 5$; minimum: 2

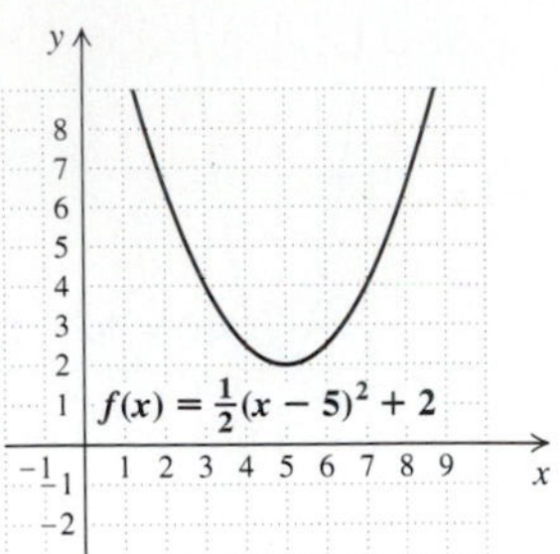

41. Vertex: $(1, -3)$; line of symmetry: $x = 1$; maximum: -3

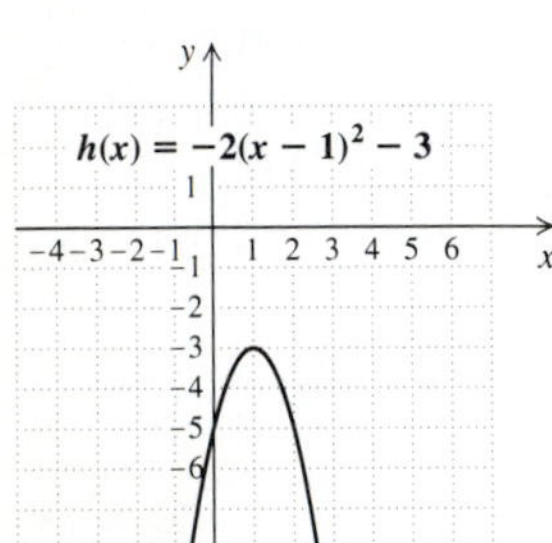

43. Vertex: $(-4, 1)$; line of symmetry: $x = -4$; maximum: 1

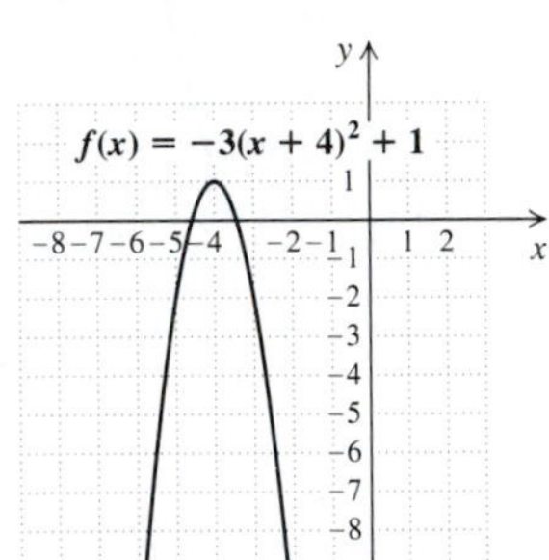

45. Vertex: $(1, 2)$; line of symmetry: $x = 1$; maximum: 2

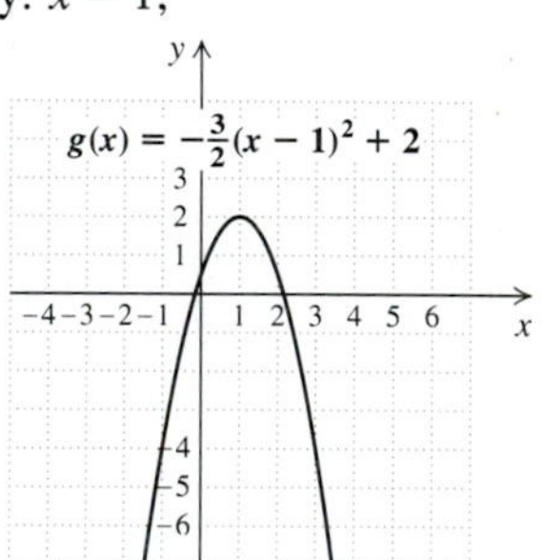

47. Vertex: $(9, 5)$; line of symmetry: $x = 9$; minimum: 5 **49.** Vertex: $(-6, 11)$; line of symmetry: $x = -6$; maximum: 11 **51.** Vertex: $\left(-\frac{1}{4}, -13\right)$; line of symmetry: $x = -\frac{1}{4}$; minimum: -13 **53.** Vertex: $(10, -20)$; line of symmetry: $x = 10$; maximum: -20 **55.** Vertex: $(-4.58, 65\pi)$; line of symmetry: $x = -4.58$; minimum: 65π **57.** $(-50, 350, 0)$ **59.** ◈ **61.** $f(x) = -2x^2 + 4$

63. $f(x) = 2(x - 6)^2$ **65.** $f(x) = -2(x - 3)^2 + 8$ **67.** $f(x) = 2(x + 3)^2 + 6$ **69.** $g(x) = -\frac{1}{2}(x - 1)^2 - 6$

71.

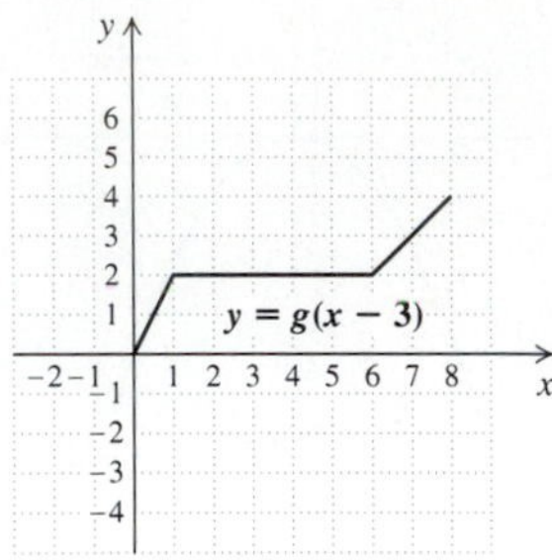

73.

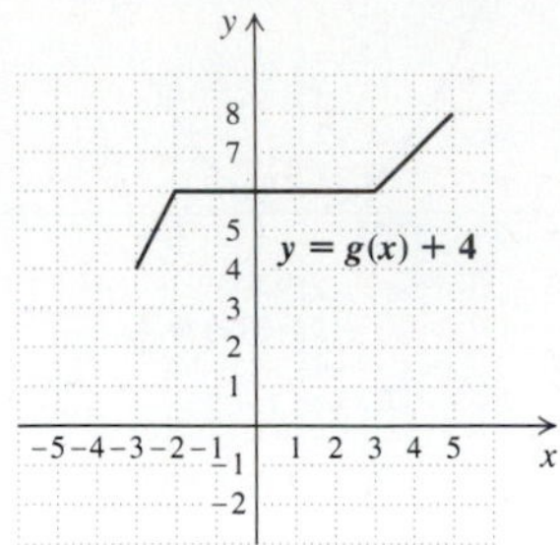

75.

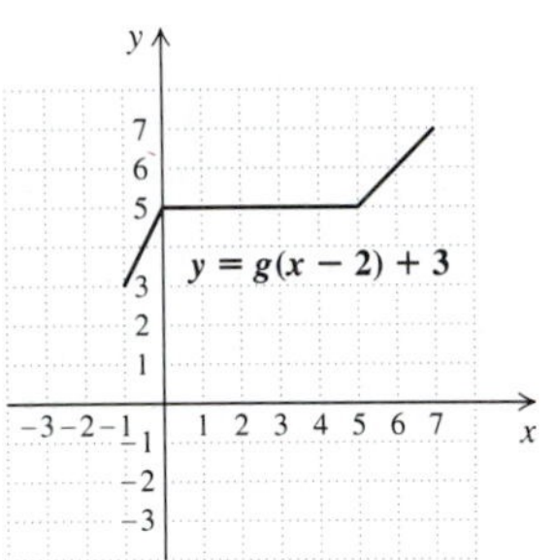

77. ▣

Exercise Set 11.8, pp. 587–588

1. $f(x) = (x - 1)^2 - 4$

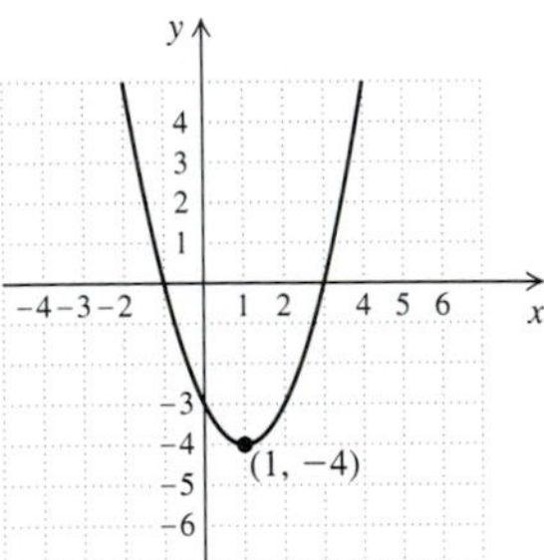

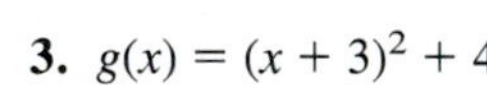
3. $g(x) = (x + 3)^2 + 4$

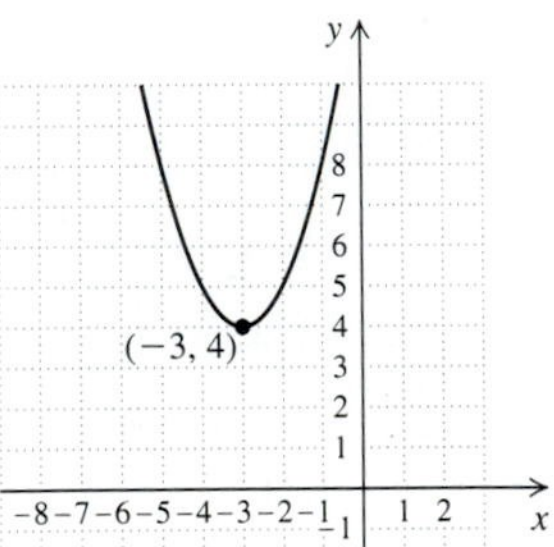

5. $f(x) = (x + 2)^2 - 5$

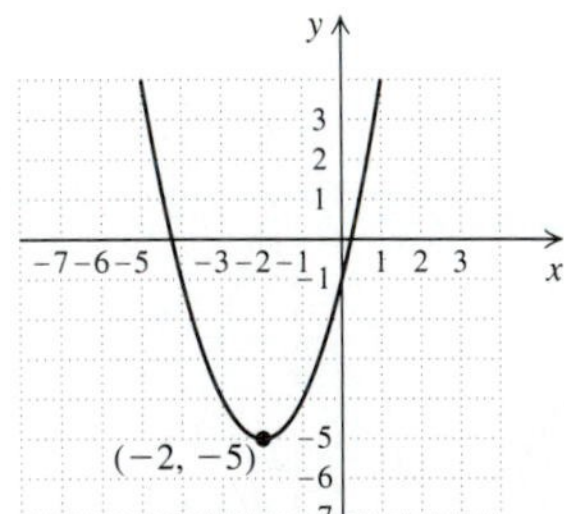

7. $h(x) = 2(x + 4)^2 - 7$

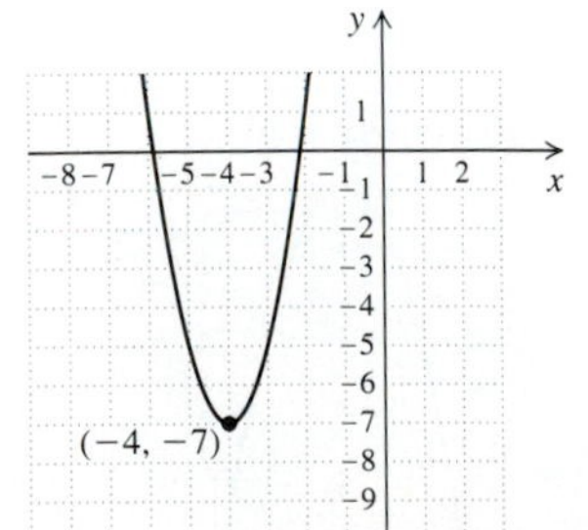

9. $f(x) = -(x - 2)^2 + 10$

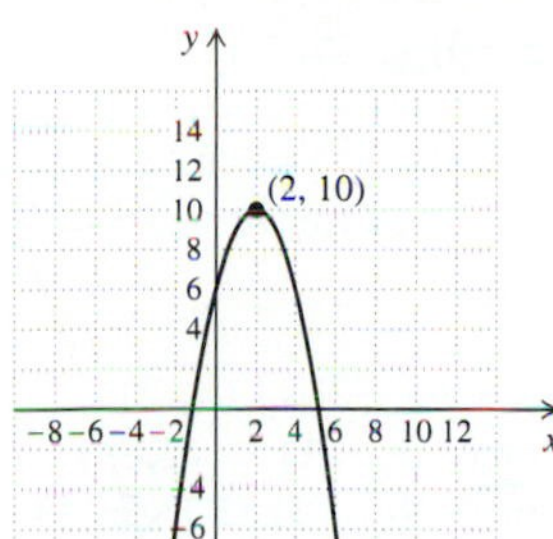

11. $g(x) = \left(x + \frac{3}{2}\right)^2 - \frac{49}{4}$

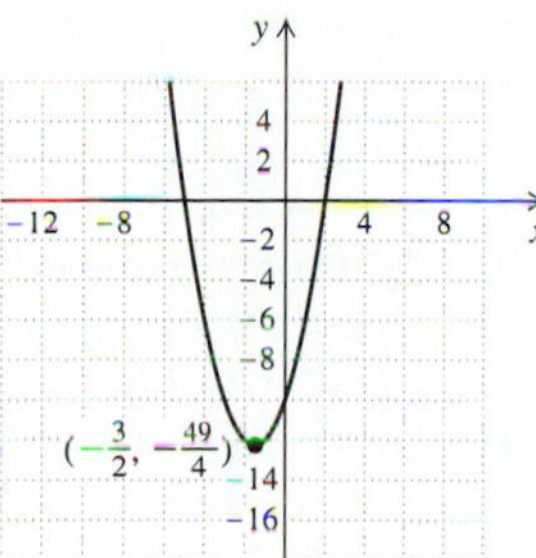

13. $f(x) = 3(x - 4)^2 + 2$

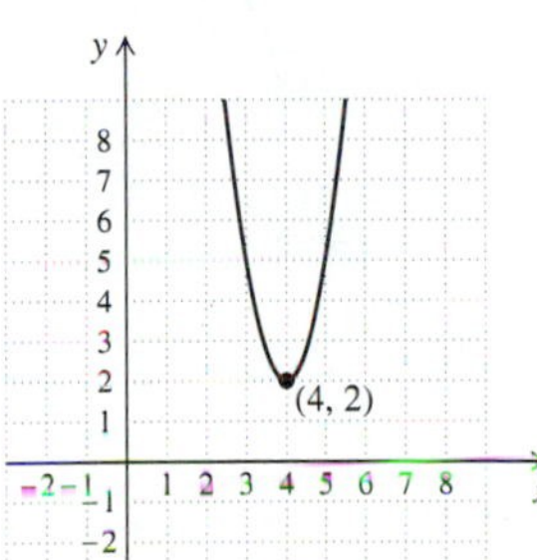

15. $h(x) = \left(x - \frac{9}{2}\right)^2 - \frac{81}{4}$

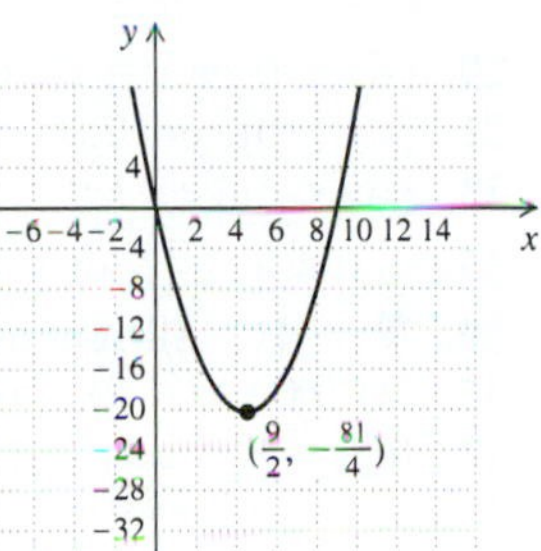

17. $f(x) = -2(x + 1)^2 - 4$

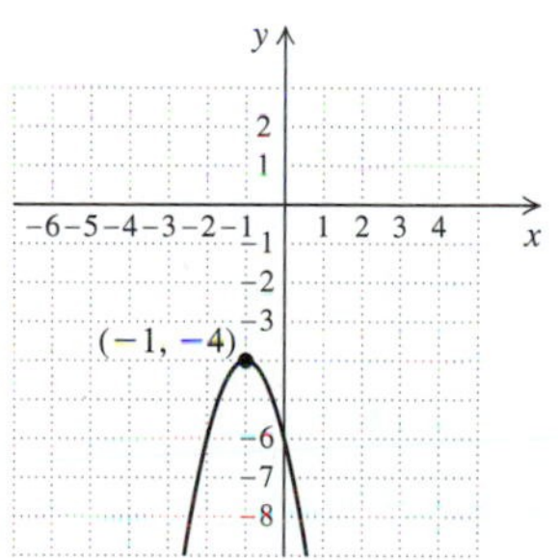

19. $g(x) = 2\left(x - \frac{5}{2}\right)^2 + \frac{3}{2}$

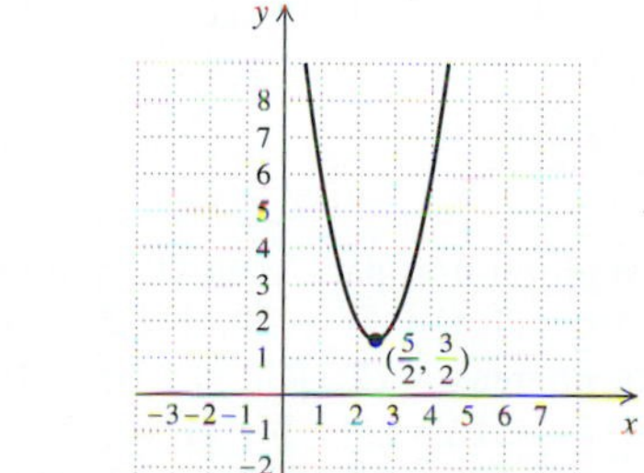

21. $f(x) = -3\left(x + \frac{1}{2}\right)^2 + \frac{7}{4}$

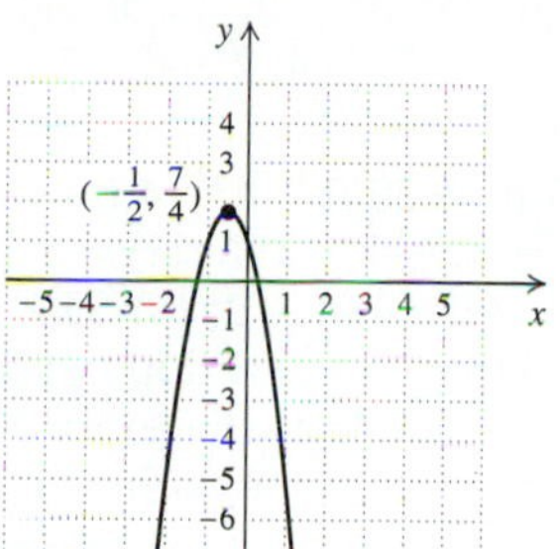

23. $h(x) = \frac{1}{2}(x + 4)^2 - \frac{5}{3}$

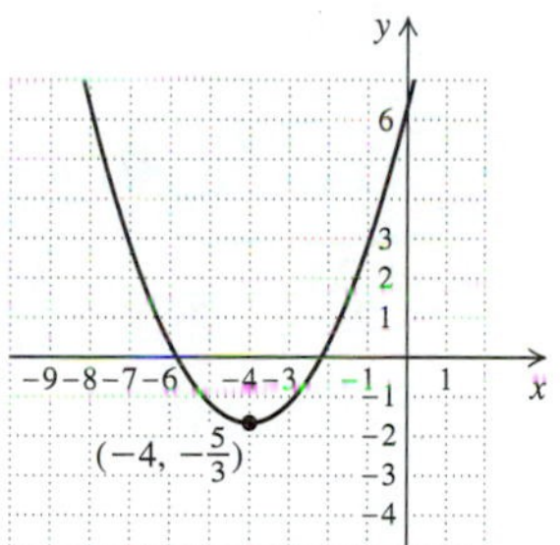

25. $(2 + \sqrt{3}, 0), (2 - \sqrt{3}, 0)$ **27.** $(3, 0), (-1, 0)$
29. $(4, 0), (-1, 0)$ **31.** $(2, 0), (1, 0)$
33. $\left(\dfrac{-2 + \sqrt{6}}{2}, 0\right), \left(\dfrac{-2 - \sqrt{6}}{2}, 0\right)$
35. None exists. **37.** 5 **39.** ◈
41. **(a)** Minimum: -6.953660714;
(b) $(-1.056433682, 0), (2.413576539, 0)$
43. **(a)** $3.4, -2.4$; **(b)** $2.3, -1.3$
45. $f(x) = m\left(x - \dfrac{n}{2m}\right)^2 + \dfrac{4mp - n^2}{4m}$
47. **49.**

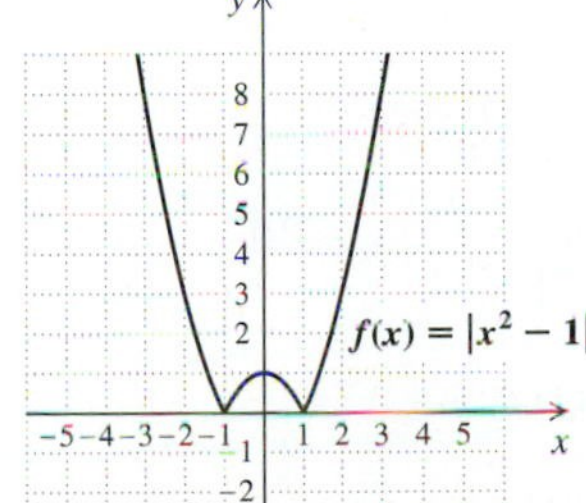

Exercise Set 11.9, pp. 594–598

1. 19 ft by 19 ft; 361 ft^2 **3.** 64; 8 and 8
5. 121; 11 and 11 **7.** -4; 2 and -2
9. $-\frac{81}{4}$; $-\frac{9}{2}$ and $\frac{9}{2}$ **11.** $\frac{49}{4}$; $-\frac{7}{2}$ and $-\frac{7}{2}$
13. 200 ft^2; 10 ft by 20 ft **15.** 4 ft by 4 ft
17. $f(x) = 2x^2 + 3x - 1$ **19.** $f(x) = -\frac{1}{4}x^2 + 3x - 5$
21. **(a)** $f(x) = -4x^2 + 40x + 2$; **(b)** $98
23. **(a)** $A(s) = 0.05s^2 - 5.5s + 250$, where $A(s)$ is the number of daytime accidents (for every 200 million km) and s is the speed of travel in km/h; **(b)** 100
25. **(a)** $P(d) = \frac{1}{64}d^2 + \frac{5}{16}d + \frac{5}{2}$; **(b)** $9.94
27. $P(x) = -x^2 + 192x - 5000$; a maximum profit of $4216 occurs when 96 units are produced and sold.
29. $s = \sqrt{\frac{A}{6}}$ **31.** $r = \sqrt{\frac{Gm_1m_2}{F}}$ **33.** $c = \sqrt{\frac{E}{m}}$
35. $b = \sqrt{c^2 - a^2}$ **37.** $k = \frac{3 + \sqrt{9 + 8N}}{2}$
39. $r = \frac{-\pi h + \sqrt{\pi^2h^2 + 2\pi A}}{2\pi}$ **41.** $n = \frac{1 + \sqrt{1 + 8N}}{2}$
43. $w = \frac{-2l + \sqrt{4l^2 + 2A}}{2}$ **45.** $g = \frac{4\pi^2 l}{T^2}$
47. $r = -1 + \frac{-P_2 + \sqrt{P_2^2 + 4AP_1}}{2P_1}$
49. $v = \frac{c}{m}\sqrt{m^2 - m_0^2}$
51. **(a)** 3.9 sec; **(b)** 1.9 sec; **(c)** 79.6 m **53.** 0.89 sec
55. 14 **57.** 2.9 sec **59.** 2.5 m/sec **61.** 7%
63. $\{x|x > \frac{13}{2}\}$, or $(\frac{13}{2}, \infty)$
65. Base: 19 cm, height: 19 cm; 180.5 cm^2
67. $n = \pm\sqrt{\frac{r^2 \pm \sqrt{r^4 + 4m^4r^2p - 4mp}}{2m}}$ **69.** $6
71. 78.4 ft **73.** $d = \frac{-\pi h + \sqrt{\pi^2h^2 + 2\pi A}}{\pi}$
75. $L = \sqrt{\frac{A}{2}}$

Technology Connection, Section 11.10

TC1. $\{x|-0.78 \leq x \leq 1.59\}$, or $[-0.78, 1.59]$
TC2. $\{x|x \leq -0.21 \text{ or } x \geq 2.47\}$, or $(-\infty, -0.21] \cup [2.47, \infty)$
TC3. $\{x|x < -1.26 \text{ or } x > 2.33\}$, or $(-\infty, -1.26) \cup (2.33, \infty)$
TC4. $\{x|x > -1.37\}$, or $(-1.37, \infty)$

Exercise Set 11.10, pp. 605–606

1. $\{x|x < -3 \text{ or } x > 5\}$, or $(-\infty, -3) \cup (5, \infty)$
3. $\{x|-1 \leq x \leq 2\}$, or $[-1, 2]$
5. $\{x|-1 < x < 2\}$, or $(-1, 2)$
7. $\{x|x \leq -3 \text{ or } x \geq 3\}$, or $(-\infty, -3] \cup [3, \infty)$
9. $\{x|x$ is a real number$\}$, or $(-\infty, \infty)$
11. $\{x|2 < x < 4\}$, or $(2, 4)$
13. $\{x|x < -2 \text{ or } 0 < x < 2\}$, or $(-\infty, -2) \cup (0, 2)$
15. $\{x|-3 < x < -1 \text{ or } x > 2\}$, or $(-3, -1) \cup (2, \infty)$
17. $\{x|x < -3 \text{ or } -2 < x < 1\}$, or $(-\infty, -3) \cup (-2, 1)$
19. $\{x|x < 4\}$, or $(-\infty, 4)$
21. $\{x|x < -1 \text{ or } x > 3\}$, or $(-\infty, -1) \cup (3, \infty)$
23. $\{x|-\frac{2}{3} \leq x < 3\}$, or $[-\frac{2}{3}, 3)$
25. $\{x|2 < x < \frac{5}{2}\}$, or $(2, \frac{5}{2})$
27. $\{x|x < -1 \text{ or } 2 < x < 5\}$, or $(-\infty, -1) \cup (2, 5)$
29. $\{x|x \leq 0 \text{ or } x > 2\}$, or $(-\infty, 0] \cup (2, \infty)$
31. $\{x|x > 0\}$, or $(0, \infty)$
33. $\{x|x < -4 \text{ or } 1 < x < 3\}$, or $(-\infty, -4) \cup (1, 3)$
35. $\{x|0 < x < \frac{1}{2}\}$, or $(0, \frac{1}{2})$ **37.** $\sqrt[15]{a^{11}b^{13}}$
39. ◈ **41.** All real numbers **43.** $\{x|x < \frac{1}{4} \text{ or } x > \frac{5}{2}\}$
45. **(a)** $\{t|0 < t < 2\}$; **(b)** $\{t|t > 10\}$
47. $\{n|9 \leq n \leq 23\}$, or $[9, 23]$
49. $f(x) < 0$ for $\{x|x < -2\}$, $f(x) > 0$ for $\{x|-2 < x < 1 \text{ or } x > 1\}$
51. $f(x) < 0$ for $\{x|0 < x < 1\}$, $f(x) > 0$ for $\{x|x > 1\}$
53. $f(x) < 0$ for $\{x|x < -3 \text{ or } -2 < x < 0 \text{ or } 1 < x < 2\}$, $f(x) > 0$ for $\{x|-3 < x < -2 \text{ or } 0 < x < 1 \text{ or } x > 2\}$

Review Exercises: Chapter 11, pp. 609–610

1. [11.1] $\pm\sqrt{3}$ **2.** [11.2] $0, -\frac{5}{14}$
3. [11.1] $6 \pm \sqrt{5}$ **4.** [11.3] $\frac{-3 \pm i\sqrt{7}}{8}$
5. [11.3] $\frac{7 \pm i\sqrt{3}}{2}$ **6.** [11.3] $\frac{3}{4}, -\frac{1}{5}$
7. [11.3] $-0.3, -3.7$
8. [11.2] $x^2 - 12x + 36$; $(x - 6)^2$
9. [11.2] $x^2 + \frac{3}{5}x + \frac{9}{100}$; $(x + \frac{3}{10})^2$
10. [11.2] 4, -2 **11.** [11.2] $3 \pm 2\sqrt{2}$
12. [11.3] 10% **13.** [11.3] 6.7 sec **14.** [11.4] 4, -2
15. [11.4] -5, 3 **16.** [11.4] $4 \pm 4\sqrt{2}$
17. [11.4] $\frac{1 \pm \sqrt{481}}{15}$
18. [11.4] 50 mph on first part, then 40 mph on second part **19.** [11.4] 6 hr **20.** [11.5] Two real
21. [11.5] Two nonreal **22.** [11.5] $25x^2 + 10x - 3 = 0$
23. [11.5] $x^2 + 8x + 16 = 0$ **24.** [11.6] 2, -2, 3, -3
25. [11.6] 3, -5 **26.** [11.6] $\frac{9}{4}$, 16 **27.** [11.6] $\pm\sqrt{7}, \pm\sqrt{2}$

28. [11.7]

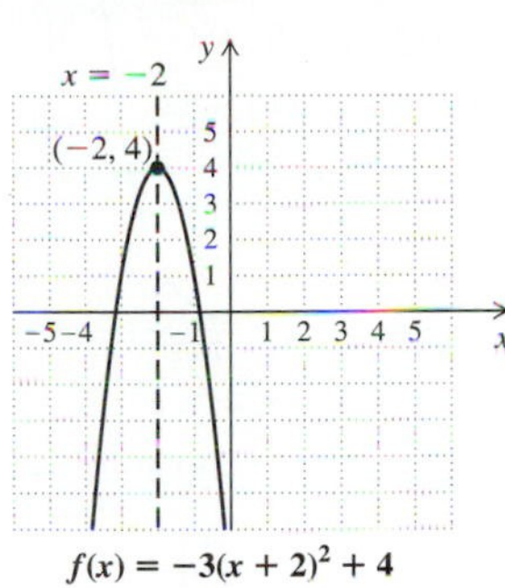

29. [11.8] **(a)** Vertex: (3, 5), line of symmetry: $x = 3$; **(b)**

30. [11.8] (7, 0), (2, 0) **31.** [11.9] 5 sec

32. [11.9] $p = \dfrac{9\pi^2}{N^2}$ **33.** [11.9] $T = \dfrac{1 \pm \sqrt{1 + 24A}}{6}$

34. [11.9] 65 yd by 65 yd; 4225 yd^2
35. [11.9] −121; 11 and −11
36. [11.9] $f(x) = -x^2 + 6x - 2$
37. [11.10] $\{x \mid -1 < x < 7\}$, or $(-1, 7)$
38. [11.10] $\{x \mid -3 < x \leq 5\}$, or $(-3, 5]$
39. [8.3] 210 kg of A; 90 kg of B
40. [10.3] $3t^5s\sqrt[3]{s}$ **41.** [9.1] $\{x \mid x \leq -\frac{1}{4}\}$, or $(-\infty, -\frac{1}{4}]$
42. [10.7] 8 **43.** [11.2], [11.3], [11.8] ◈ Completing the square was used to solve quadratic equations and to graph quadratic functions by rewriting the function in the form $f(x) = a(x - h)^2 + k$.
44. [11.5] ◈ If you know one imaginary-number solution of a quadratic equation with real coefficients, the other solution is the complex conjugate of the first solution.
45. [11.4] All real numbers except −13 and −7
46. [11.5] $h = 60$, $k = 60$ **47.** [11.5] 18 and 324

Test: Chapter 11, pp. 610–611

1. [11.1] $\pm\dfrac{2\sqrt{3}}{3}$ **2.** [11.2] $\dfrac{1}{10}$, 10

3. [11.3] $\dfrac{-1 \pm i\sqrt{3}}{2}$

4. [11.3] 0.4, −4.4 **5.** [11.4] 0, 2

6. [11.6] $\pm\sqrt{\dfrac{5 + \sqrt{5}}{2}}, \pm\sqrt{\dfrac{5 - \sqrt{5}}{2}}$

7. [11.2] $x^2 + 14x + 49$; $(x + 7)^2$
8. [11.2] $x^2 - \frac{2}{7}x + \frac{1}{49}$; $\left(x - \frac{1}{7}\right)^2$
9. [11.2] −6, 3 **10.** [11.2] $-5 \pm \sqrt{10}$
11. [11.4] 16 km/h **12.** [11.4] 2 hr

13. [11.5] Two nonreal **14.** [11.5] $x^2 - 4\sqrt{3}x + 9 = 0$
15. [11.7]

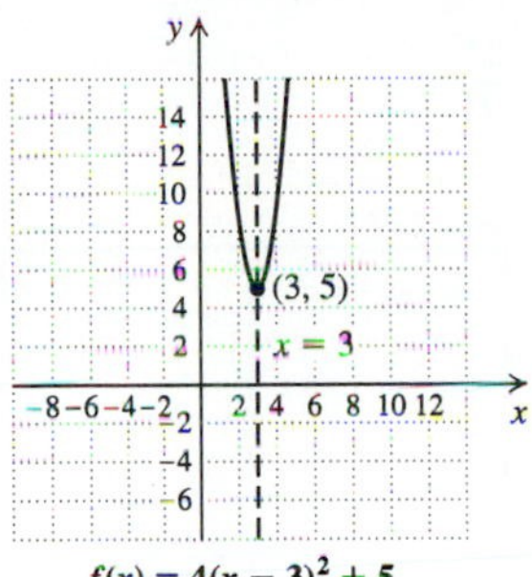

16. [11.8] **(a)** (−1, −8), $x = -1$;
(b)

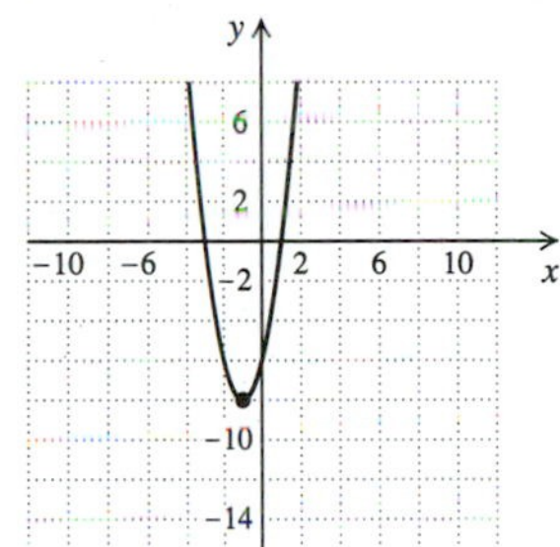

17. [11.8] (3, 0) and (−2, 0)

18. [11.9] $r = \sqrt{\dfrac{3V}{\pi} - R^2}$ **19.** [11.9] 16

20. [11.9] $f(x) = \frac{1}{5}x^2 - \frac{3}{5}x$
21. [11.10] $\{x \mid -2 < x < 1 \text{ or } x > 2\}$, or $(-2, 1) \cup (2, \infty)$
22. [11.10] $\{x \mid -1 \leq x \leq 0 \text{ or } x > 3\}$, or $[-1, 0] \cup (3, \infty)$
23. [10.7] 6
24. [9.1] $\{x \mid x > -29\}$ or $(-29, \infty)$
25. [10.3] $ab\sqrt[4]{2a^2}$
26. [8.3] Each is 36.
27. [11.5] $\frac{1}{2}$
28. [11.4] All real numbers except 11 and 7

CHAPTER 12

Technology Connection, Section 12.1

TC1.

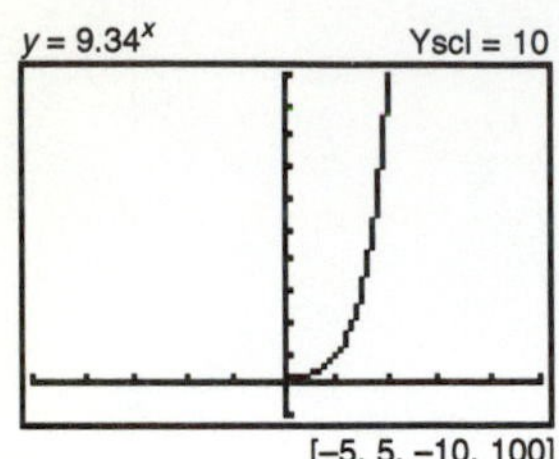

TC2.

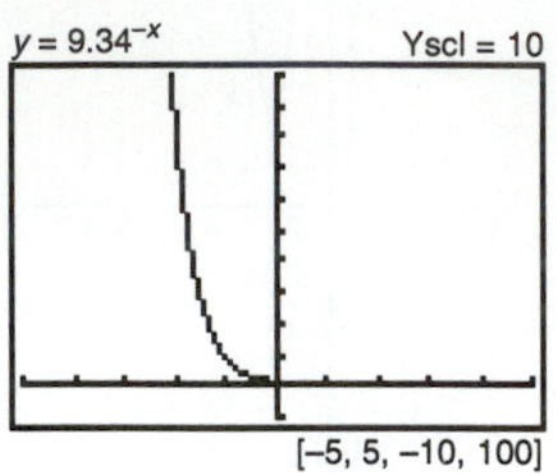

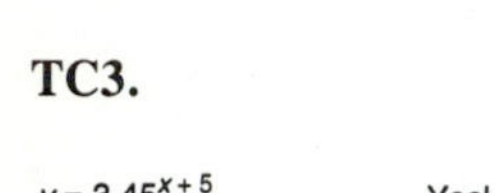

TC3.

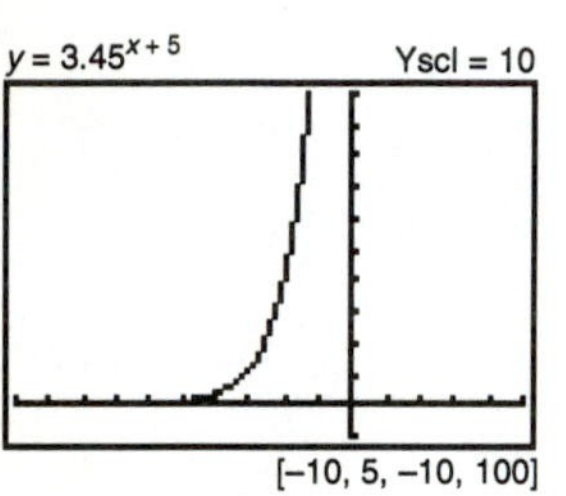

TC4.

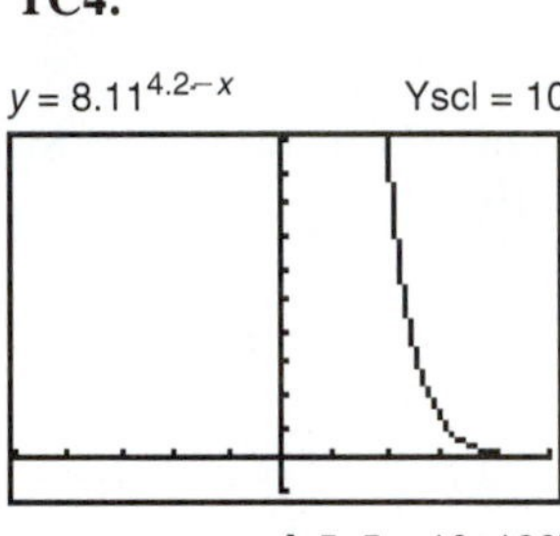

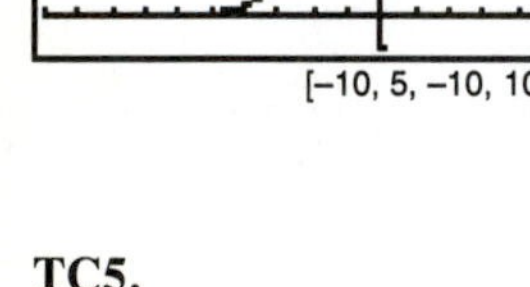

TC5.

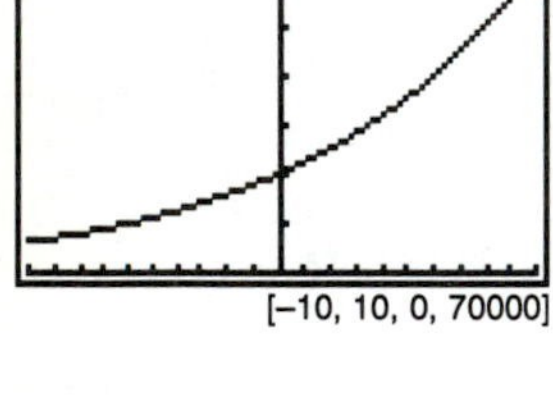

TC6.

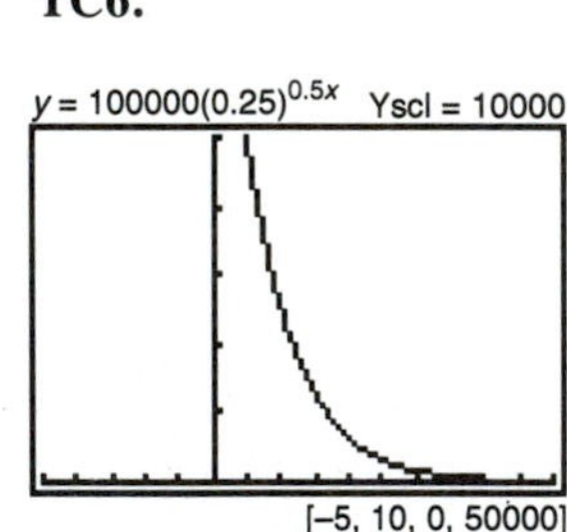

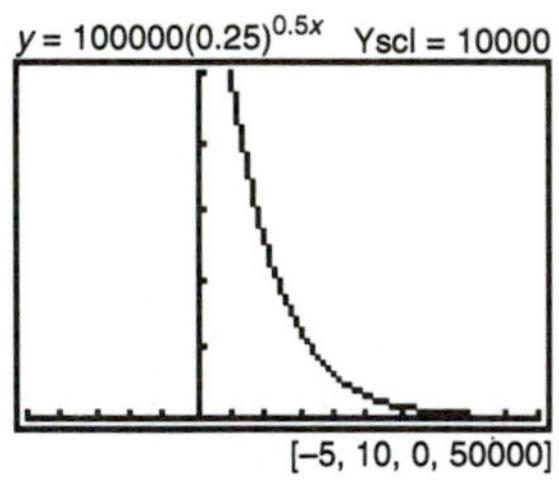

Exercise Set 12.1, pp. 620–621

1.

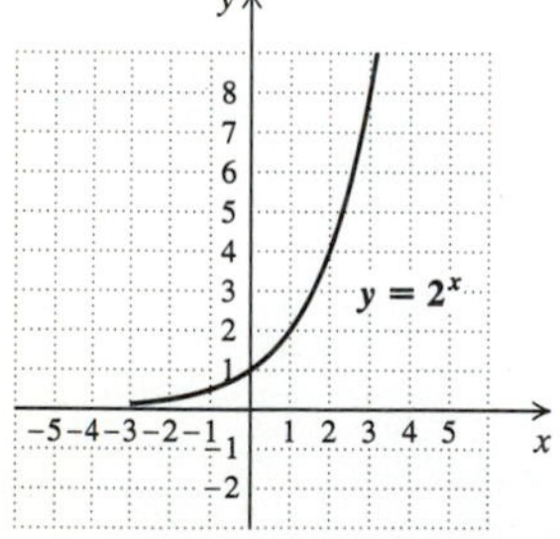

3.

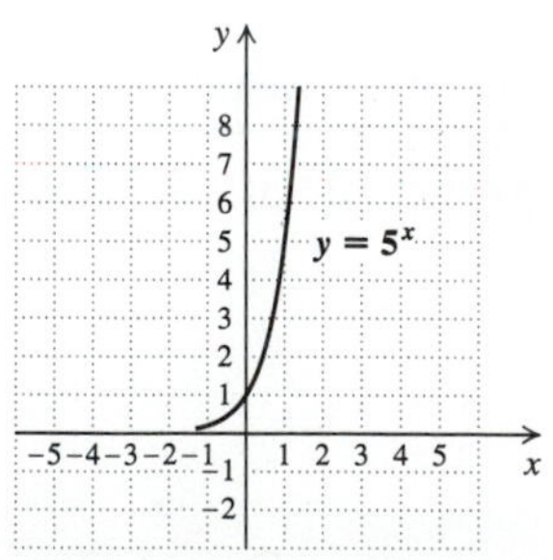

5.

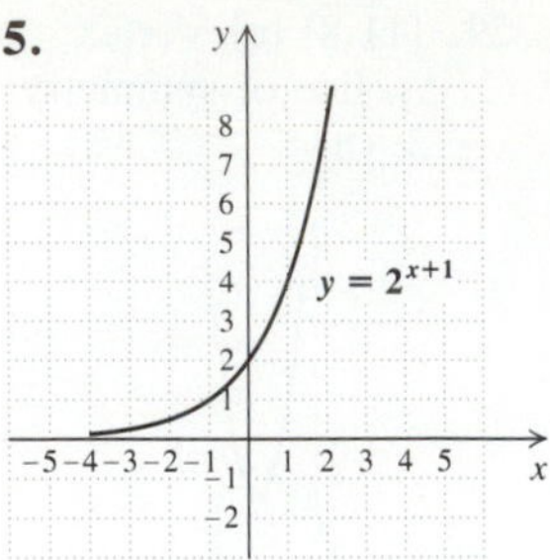

7.

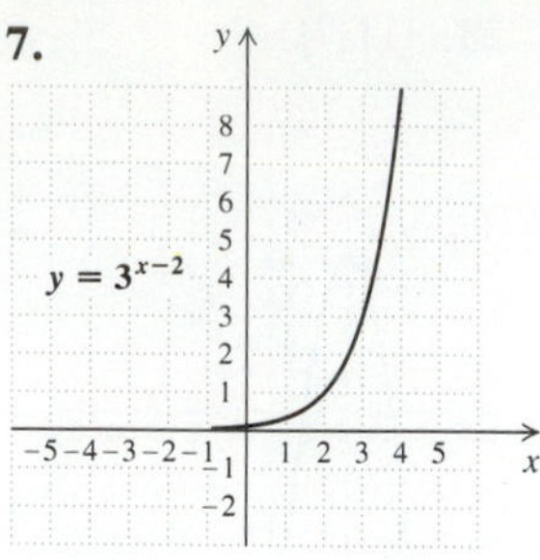

9.

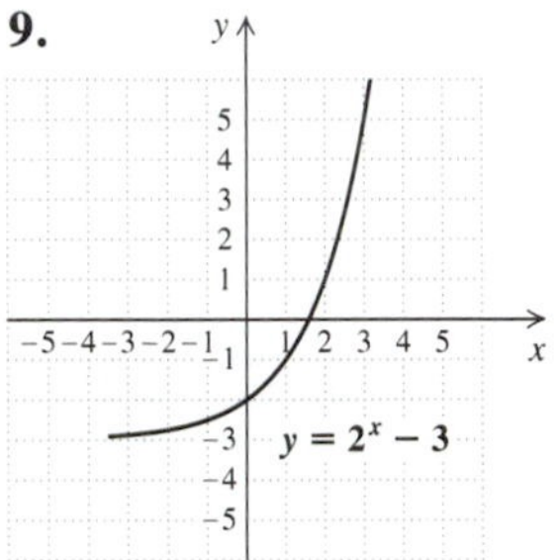

11.

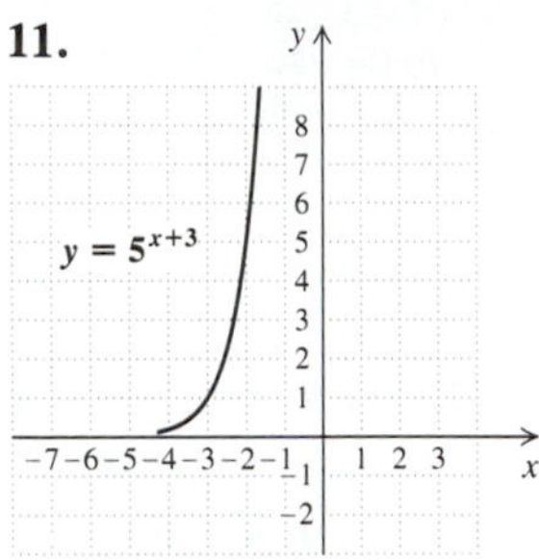

13.

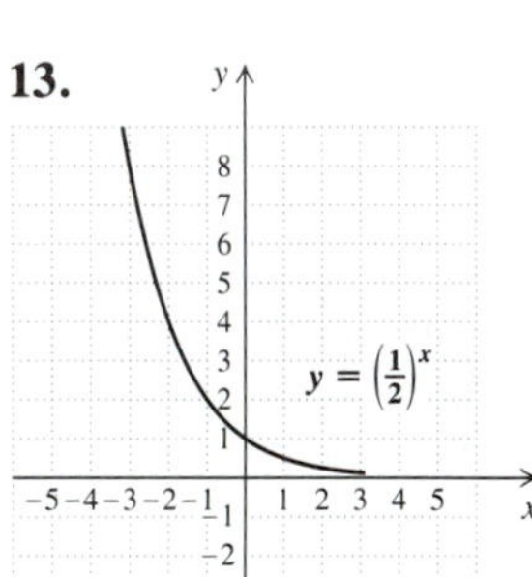

15.

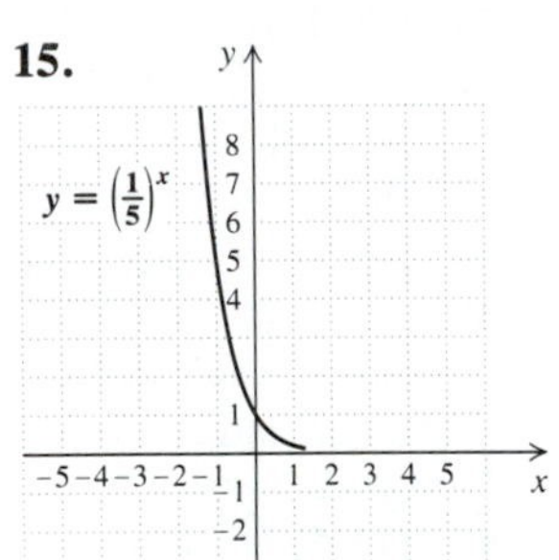

17.

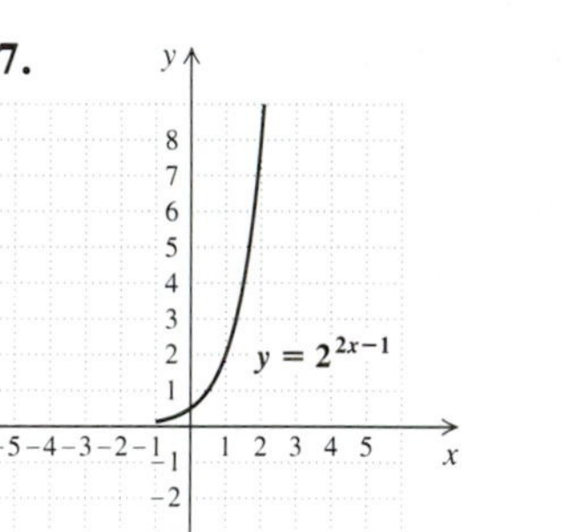

19.

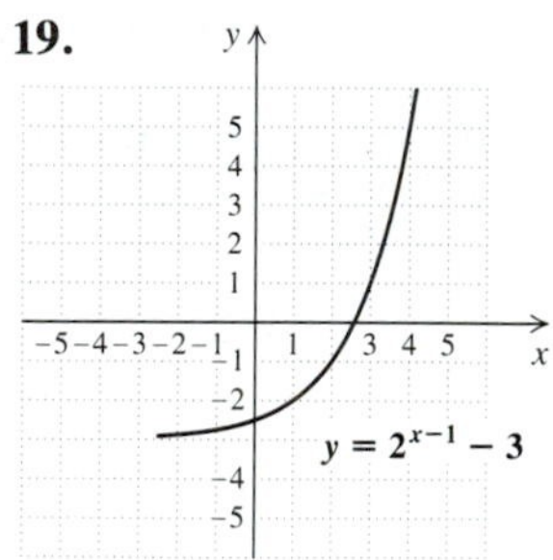

21.

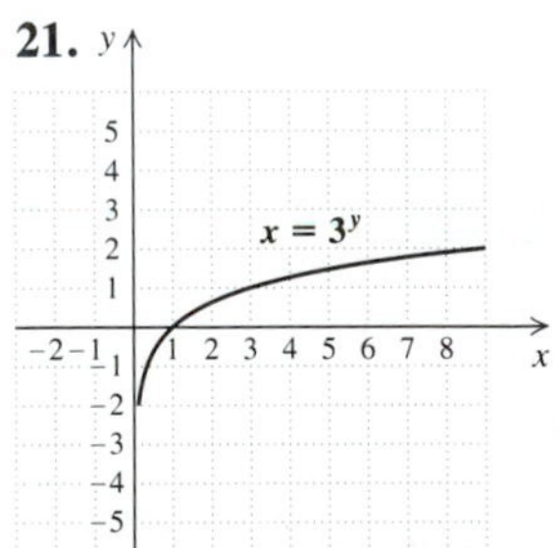

23.

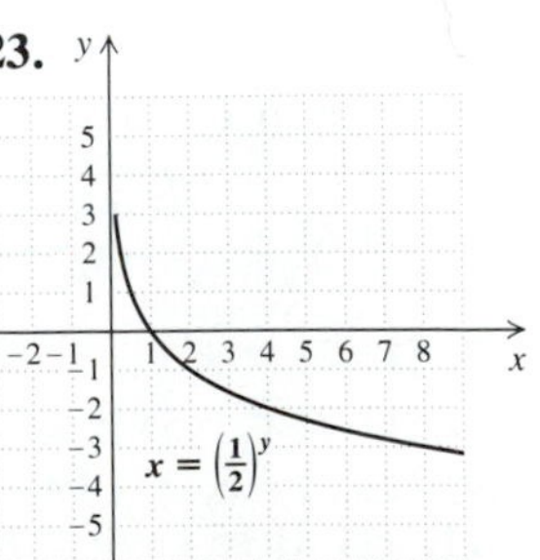

25.

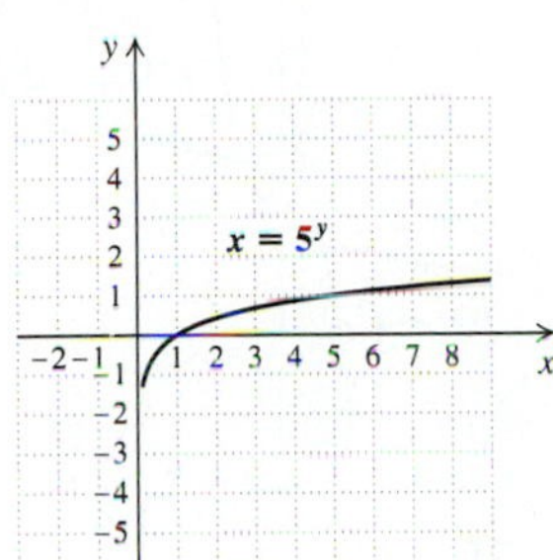

27.

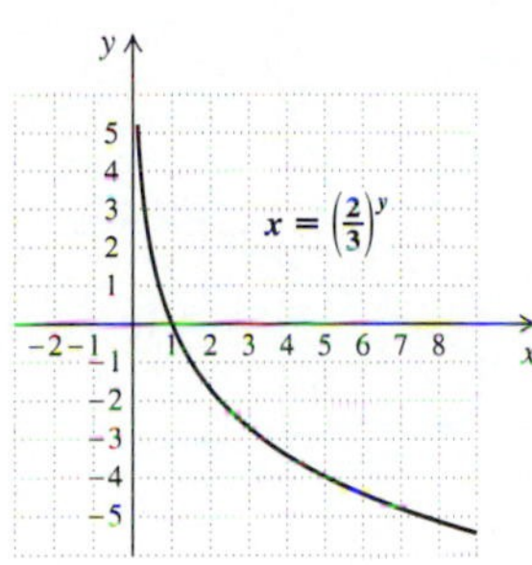

29.

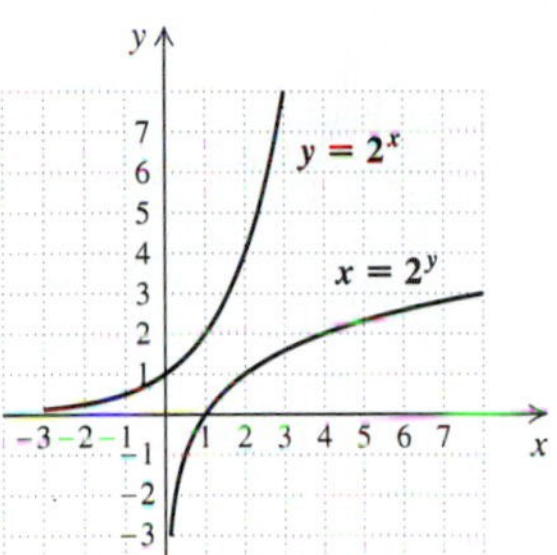

31.

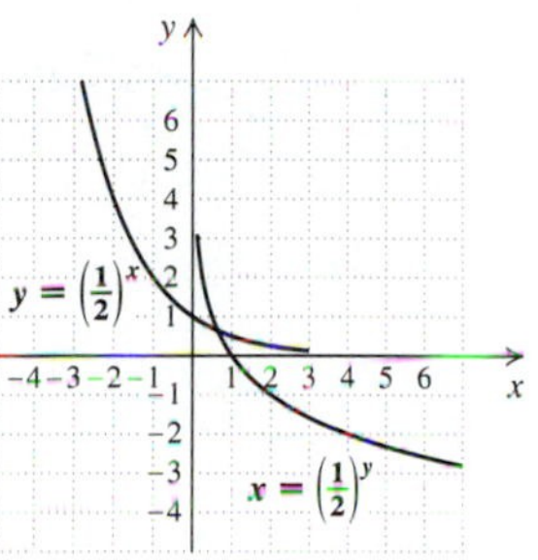

33. **(a)** 384,160; **(b)** 2,066,105;
(c)

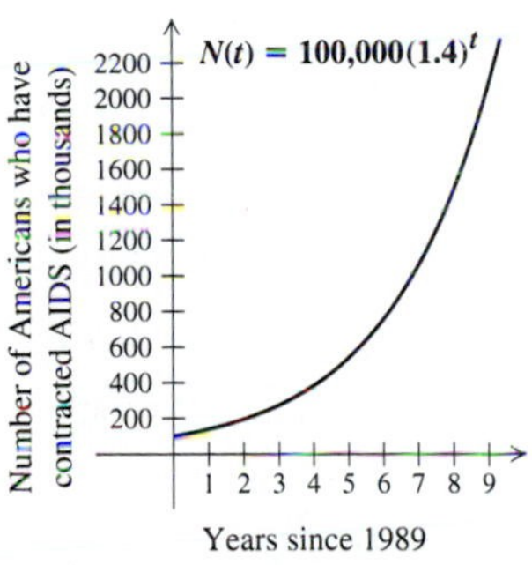

35. **(a)** 250,000; 166,667; 49,383; 4335;
(b)

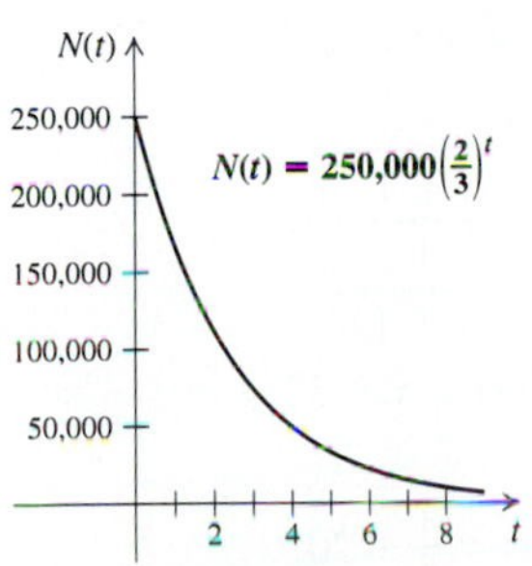

37. **(a)** 7.5 million, 18.4 million, 110.2 million, 661.4 million, 58,320 million, 5,142,752.7 million;
(b)

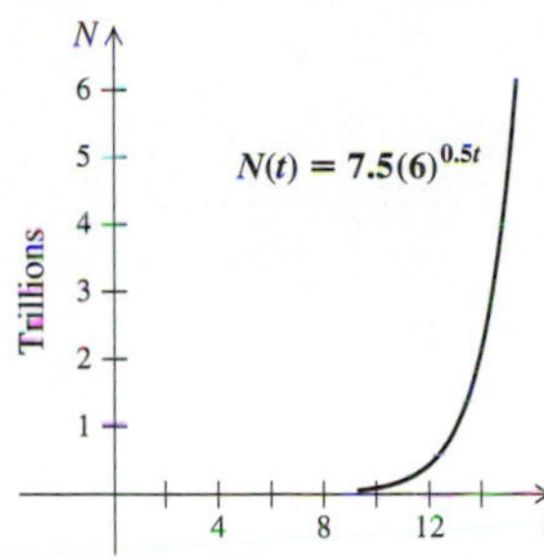

39. x^{-2}, or $\frac{1}{x^2}$ **41.** x^{-7}, or $\frac{1}{x^7}$ **43.** 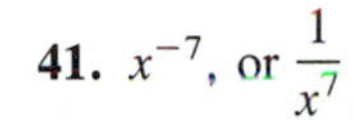 **45.** $\pi^{2.4}$

47.

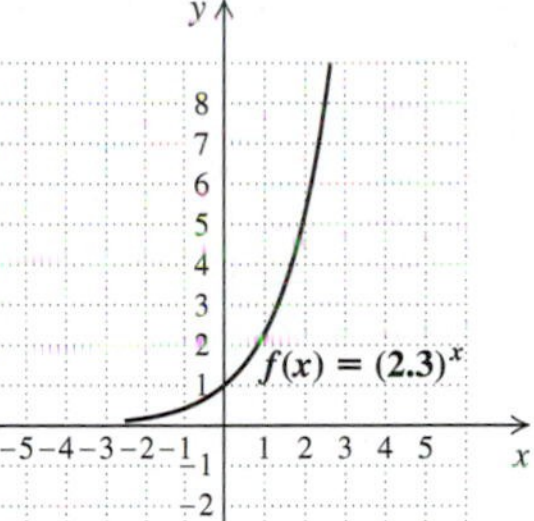

49.

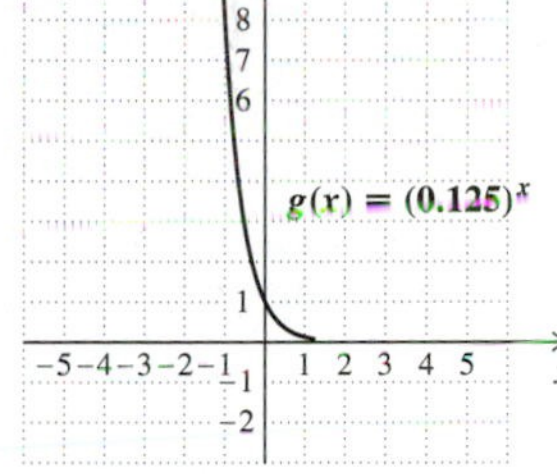

51.

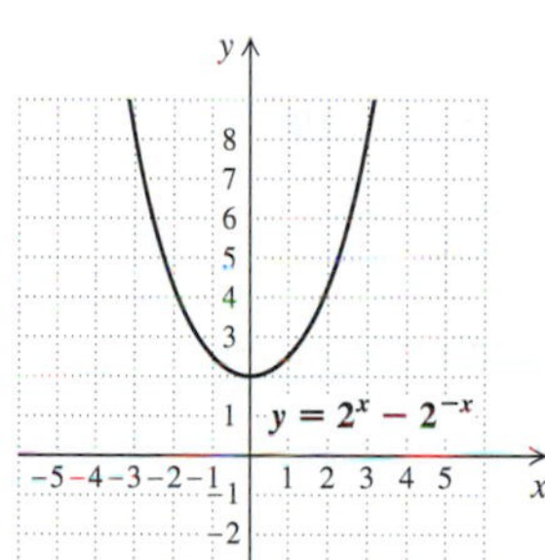

53.

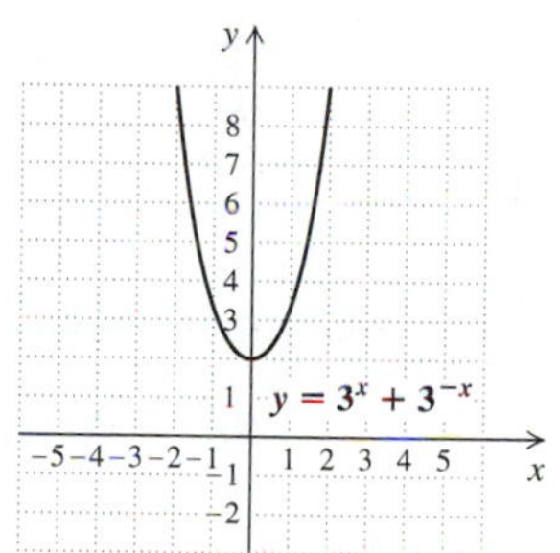

55.

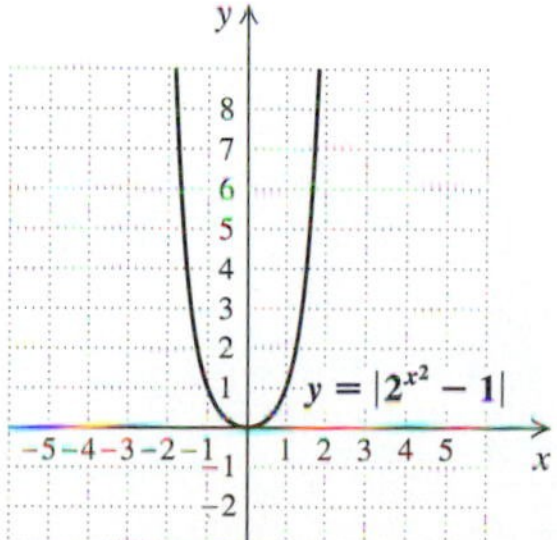

57.

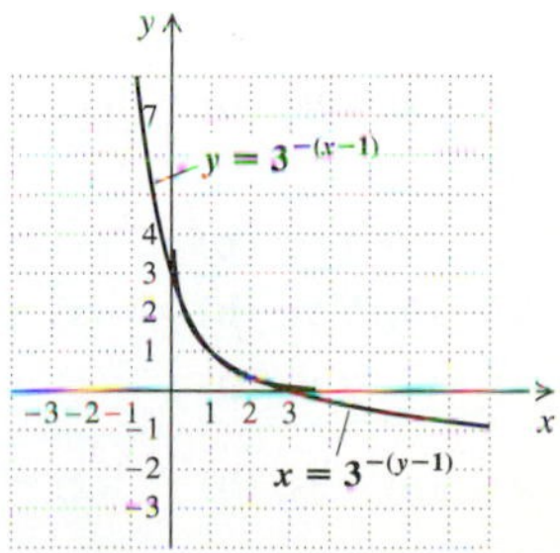

59. **(a)** 19 wpm, 36 wpm, 66 wpm, 115 wpm;
(b)

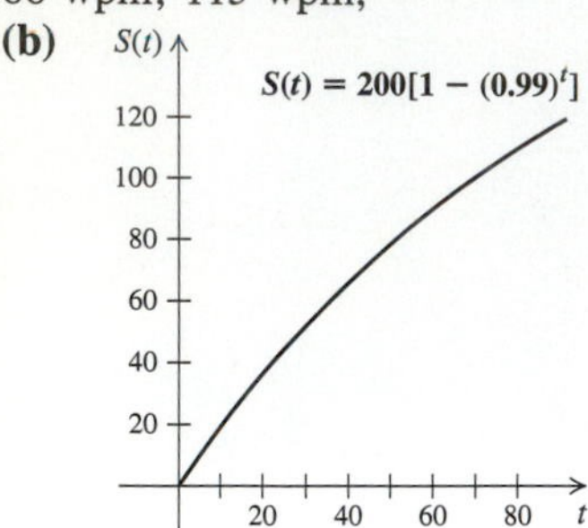

Technology Connection, Section 12.2

TC1. C **TC2.** A **TC3.** B **TC4.** D

Exercise Set 12.2, pp. 631–632

1. $f \circ g(x) = 12x^2 - 12x + 5$; $g \circ f(x) = 6x^2 + 3$

3. $f \circ g(x) = \dfrac{16}{x^2} - 1$; $g \circ f(x) = \dfrac{2}{4x^2 - 1}$

5. $f \circ g(x) = x^4 - 2x^2 + 2$; $g \circ f(x) = x^4 + 2x^2$

7. $f(x) = x^2$; $g(x) = 5 - 3x$

9. $f(x) = x^5$; $g(x) = 3x^2 - 7$

11. $f(x) = \dfrac{1}{x}$; $g(x) = x - 1$

13. $f(x) = \dfrac{1}{\sqrt{x}}$; $g(x) = 7x + 2$

15. $f(x) = \dfrac{x+1}{x-1}$; $g(x) = x^3$ **17.** Yes **19.** No

21. Yes **23.** No **25.** **(a)** Yes; **(b)** $f^{-1}(x) = x - 4$

27. **(a)** Yes; **(b)** $f^{-1}(x) = 5 - x$

29. **(a)** Yes; **(b)** $g^{-1}(x) = x + 5$

31. **(a)** Yes; **(b)** $f^{-1}(x) = \dfrac{x}{3}$

33. **(a)** Yes; **(b)** $g^{-1}(x) = \dfrac{x-2}{3}$

35. No **37.** **(a)** Yes; **(b)** $f^{-1}(x) = \dfrac{1}{x}$

39. **(a)** Yes; **(b)** $f^{-1}(x) = \dfrac{3x-1}{2}$

41. **(a)** Yes; **(b)** $f^{-1}(x) = \sqrt[3]{x+1}$

43. **(a)** Yes; **(b)** $g^{-1}(x) = \sqrt[3]{x} + 2$

45. **(a)** Yes; **(b)** $f^{-1}(x) = x^2,\ x \geq 0$

47. **(a)** Yes; **(b)** $f^{-1}(x) = \sqrt{\dfrac{x-3}{2}}$

49.

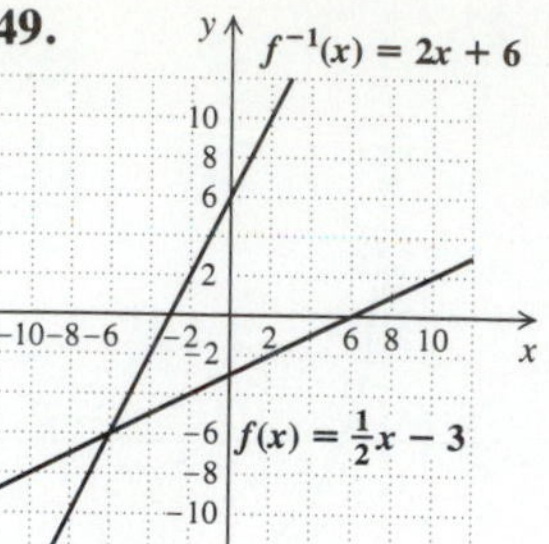

51.

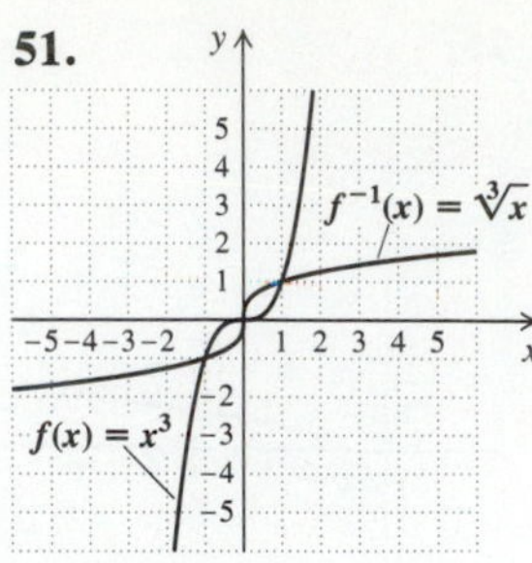

53.

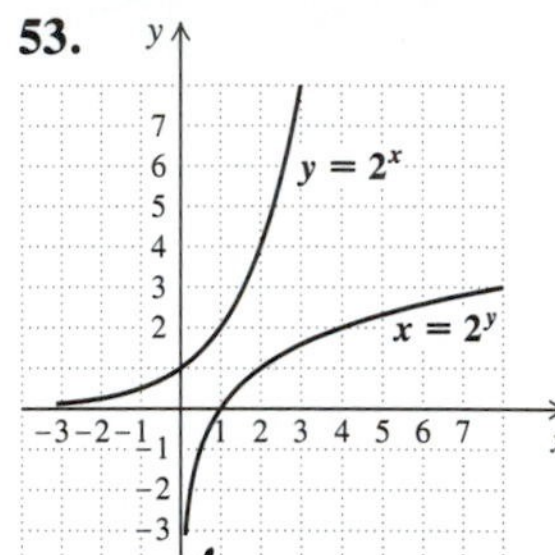

55.

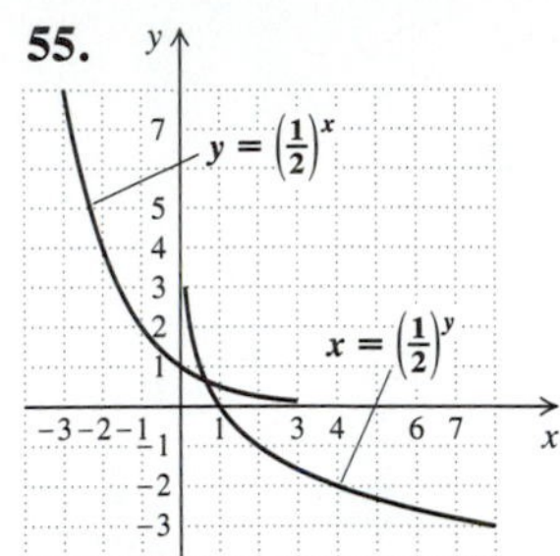

57.

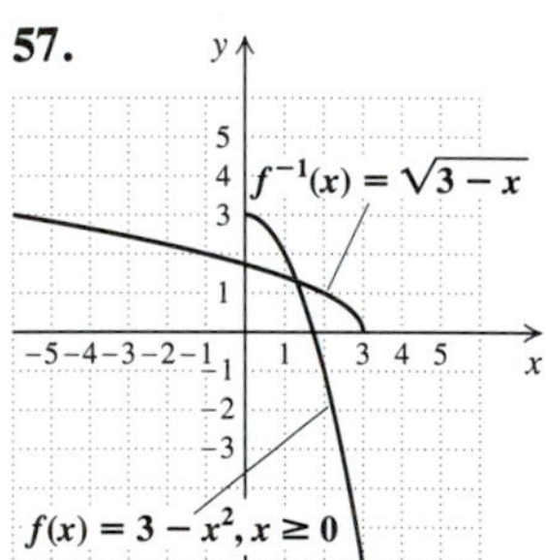

59. **(1)** $f^{-1} \circ f(x) = f^{-1}(f(x)) = f^{-1}\left(\frac{4}{5}x\right) = \frac{5}{4}\left(\frac{4}{5}x\right) = x$;
(2) $f \circ f^{-1}(x) = f(f^{-1}(x)) = f\left(\frac{5}{4}x\right) = \frac{4}{5}\left(\frac{5}{4}x\right) = x$

61. **(1)** $f^{-1} \circ f(x) = f^{-1}(f(x)) = f^{-1}\left(\dfrac{1-x}{x}\right)$

$$= \frac{1}{\left(\dfrac{1-x}{x}\right) + 1} = \frac{1}{\dfrac{1-x+x}{x}}$$

$$= x;$$

(2) $f \circ f^{-1}(x) = f(f^{-1}(x)) = f\left(\dfrac{1}{x+1}\right)$

$$= \frac{1 - \left(\dfrac{1}{x+1}\right)}{\left(\dfrac{1}{x+1}\right)} = \frac{\dfrac{x+1-1}{x+1}}{\dfrac{1}{x+1}} = x$$

63. **(a)** 40, 42, 46, 50; **(b)** $f^{-1}(x) = x - 32$;
(c) 8, 10, 14, 18

65. $y = 9x$ **67.** ◈ **69.** $g(x) = \dfrac{x}{2} + 20$

71. Inverses **73.** Not inverses

Exercise Set 12.3, pp. 637–638

1.

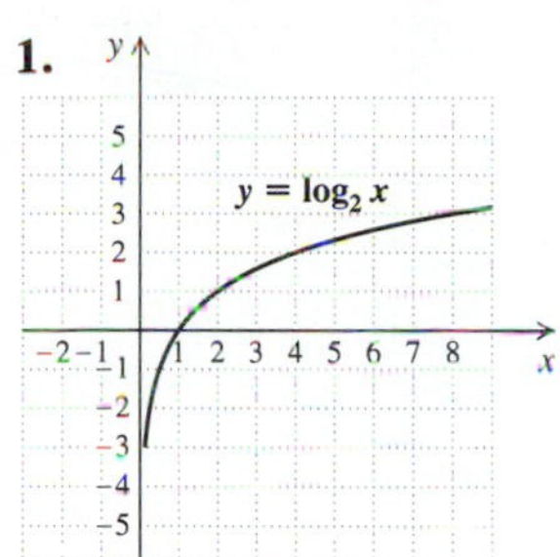

3.

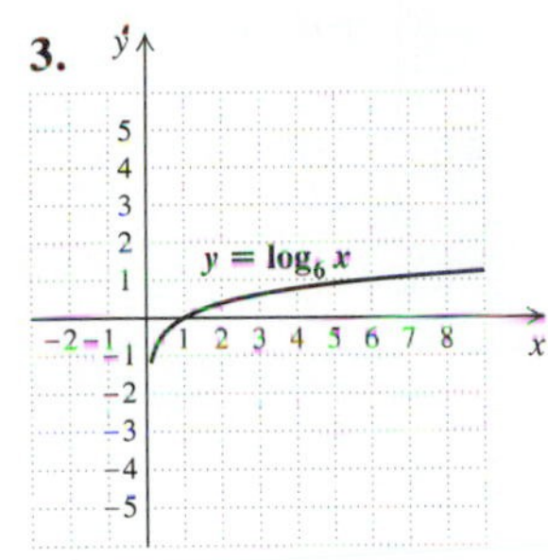

5.

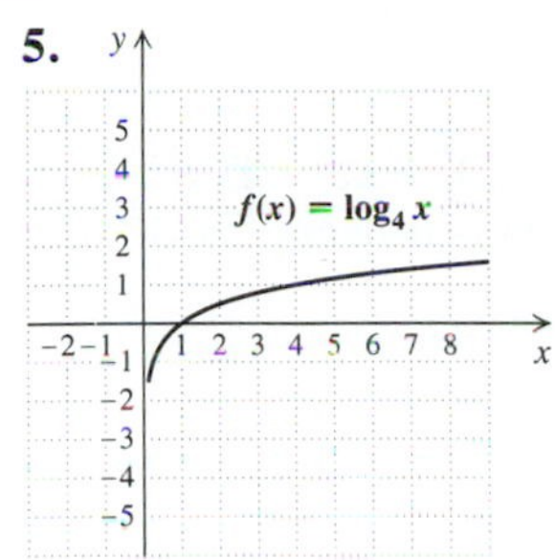

7.

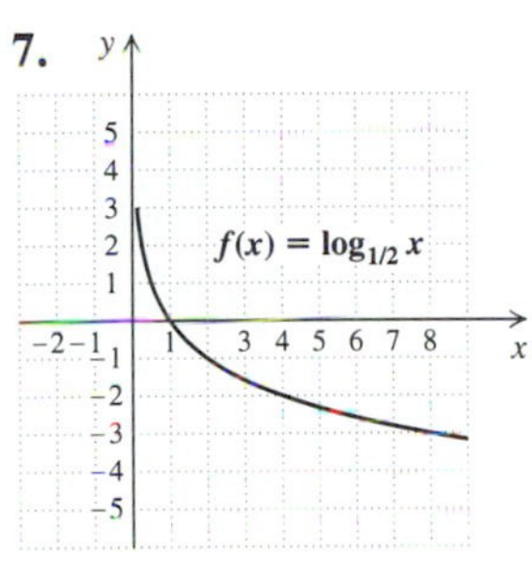

9.

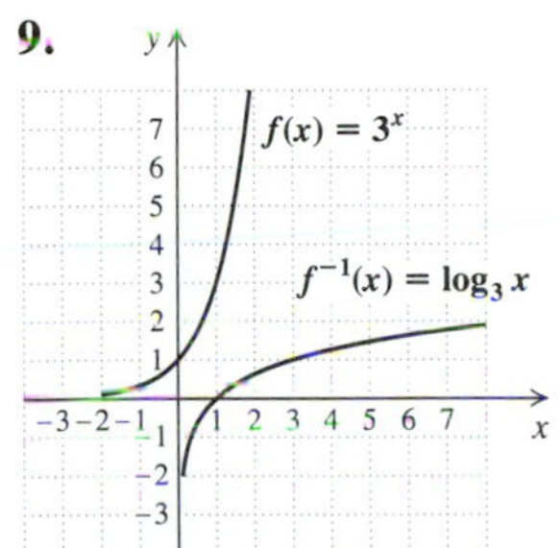

11. $3 = \log_{10} 1000$
13. $-3 = \log_5 \frac{1}{125}$
15. $\frac{1}{3} = \log_8 2$
17. $0.3010 = \log_{10} 2$
19. $2 = \log_e t$
21. $t = \log_Q x$
23. $2 = \log_e 7.3891$
25. $-2 = \log_e 0.1353$
27. $3^t = 8$ **29.** $5^2 = 25$
31. $10^{-1} = 0.1$

33. $10^{0.845} = 7$ **35.** $e^{2.9957} = 20$ **37.** $t^k = Q$
39. $e^{-1.3863} = 0.25$ **41.** $r^{-x} = T$ **43.** 9 **45.** 5
47. 4 **49.** 3 **51.** 13 **53.** 1 **55.** $\frac{1}{2}$ **57.** 2
59. 2 **61.** −1 **63.** 0 **65.** 4 **67.** −2 **69.** 0
71. 1 **73.** $\frac{2}{3}$ **75.** 3 **77.** t

79. $\frac{x(3y-2)}{2y+x}$ **81.** $\frac{1}{4096}$ **83.** $\frac{1}{\sqrt[3]{t^2}}$ **85.** ◈

87.

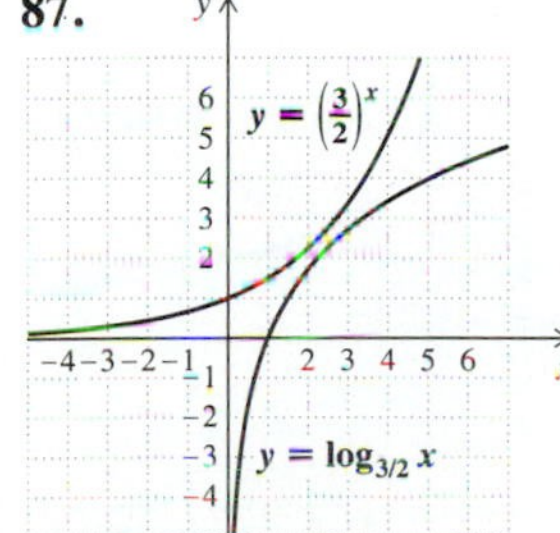

89.

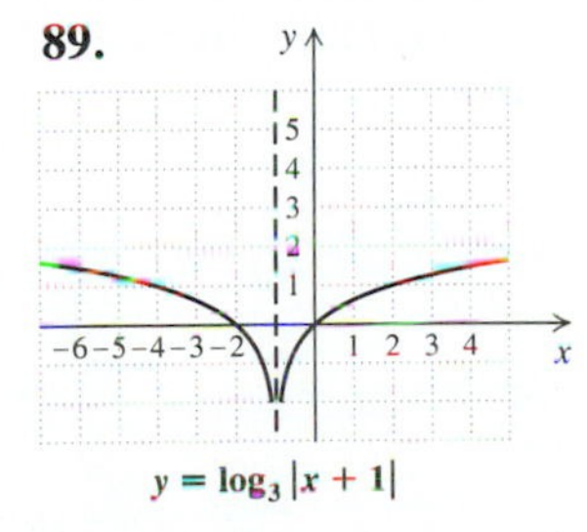

91. 25 **93.** $-\frac{7}{16}$ **95.** 3 **97.** 0 **99.** −2

Exercise Set 12.4, pp. 643–644

1. $\log_2 32 + \log_2 8$ **3.** $\log_4 64 + \log_4 16$
5. $\log_c B + \log_c x$
7. $\log_a (6 \cdot 70)$ **9.** $\log_c (K \cdot y)$ **11.** $3 \log_a x$
13. $6 \log_c y$ **15.** $-3 \log_b C$ **17.** $\log_a 67 - \log_a 5$
19. $\log_b 3 - \log_b 4$ **21.** $\log_a \frac{15}{7}$
23. $2 \log_a x + 3 \log_a y + \log_a z$
25. $\log_b x + 2 \log_b y - 3 \log_b z$
27. $\frac{1}{3}(4 \log_c x - 3 \log_c y - 2 \log_c z)$
29. $\frac{1}{4}(8 \log_a x + 12 \log_a y - 3 - 5 \log_a z)$
31. $\log_a \frac{\sqrt[3]{x^2}\sqrt{y}}{y}$ **33.** $\log_a \frac{2x^4}{y^3}$ **35.** $\log_a \frac{\sqrt{a}}{x}$
37. 2.708 **39.** 0.51 **41.** −1.609 **43.** $\frac{3}{2}$
45. 2.609 **47.** 3.218 **49.** 9 **51.** m **53.** 4
55. −7 **57.** i **59.** $\frac{3}{5} + \frac{4}{5}i$ **61.** ◈
63. $\log_a (x^6 - x^4y^2 + x^2y^4 - y^6)$
65. $\frac{1}{2} \log_a (1 - s) + \frac{1}{2} \log_a (1 + s)$ **67.** $\frac{10}{3}$ **69.** −2
71. False **73.** False **75.** True

Technology Connection, Section 12.5

TC1.

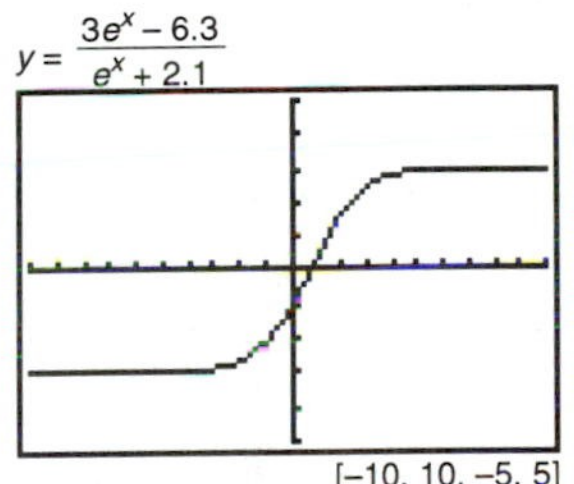

TC2.

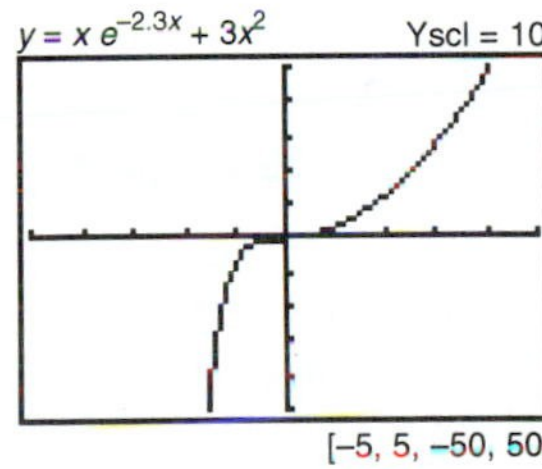

TC3.

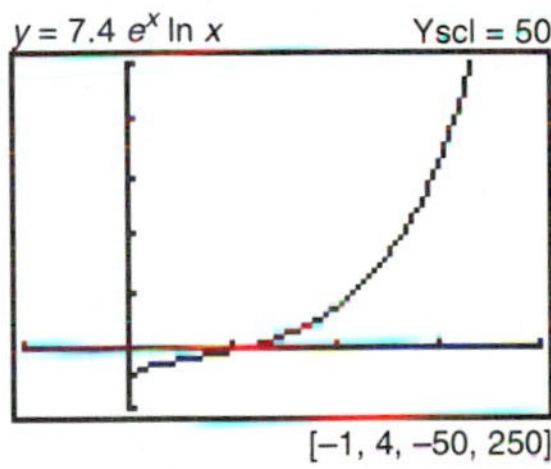

TC4.

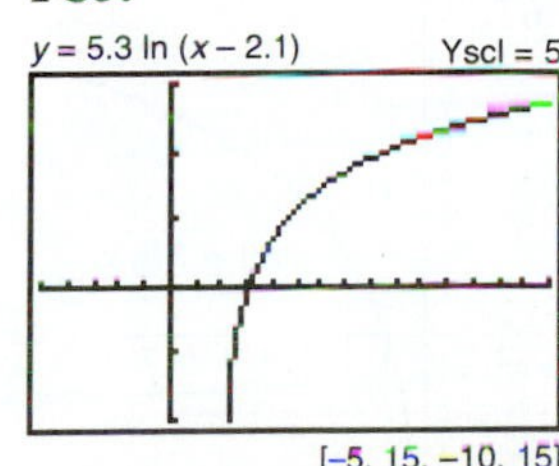

TC5.

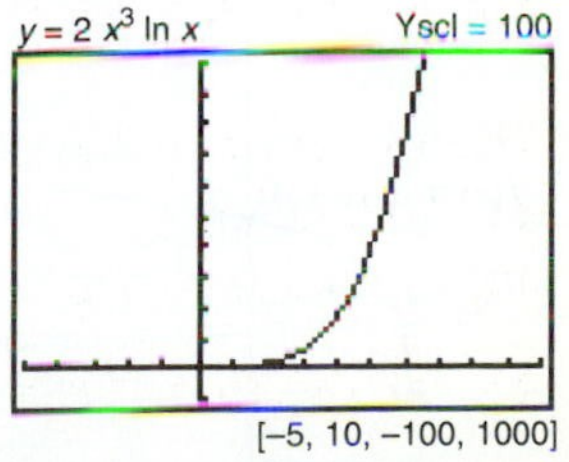

TC6.

Exercise Set 12.5, p. 650

1. 0.3010 **3.** 0.8021 **5.** 1.6532 **7.** 2.6405 **9.** 4.1271 **11.** −1.2840 **13.** 1000 **15.** 501.1872 **17.** 3.0001 **19.** 0.2841 **21.** 0.0011 **23.** 0.6931 **25.** 4.1271 **27.** 8.3814 **29.** −5.0832 **31.** 36.7890 **33.** 0.0023 **35.** 1.0057 **37.** 5.8346×10^{14} **39.** 7.6331 **41.** 2.5702 **43.** 3.3219 **45.** 0.6419 **47.** −2.3219 **49.** −2.3219 **51.** 3.5471

53.

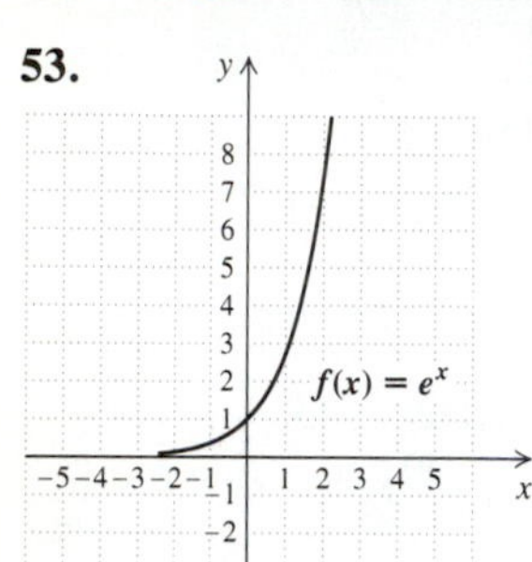

55.

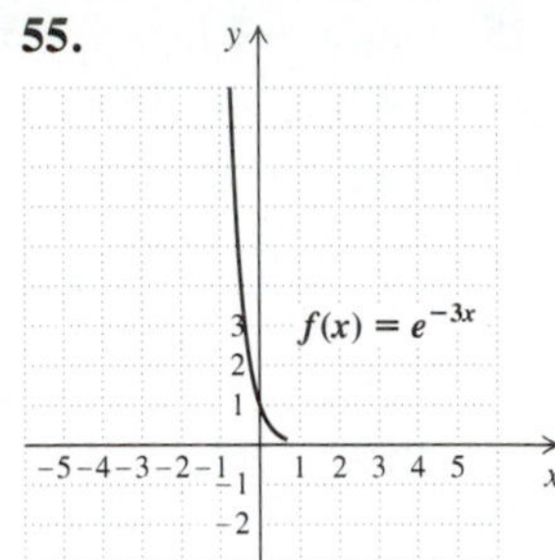

57.

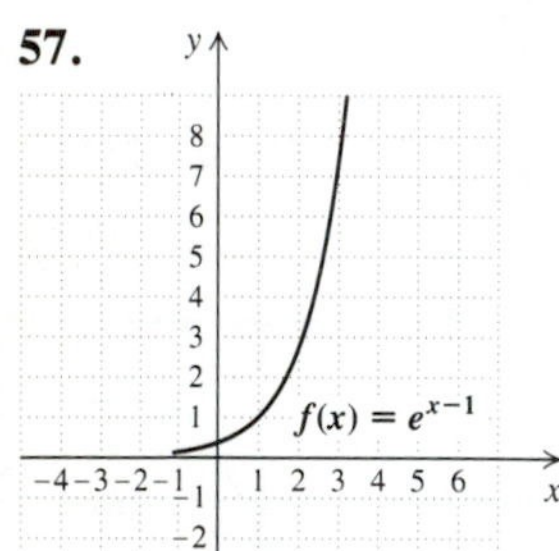

59.

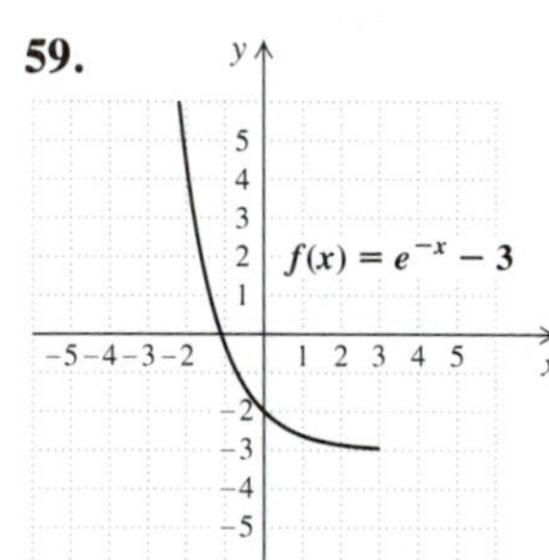

61.

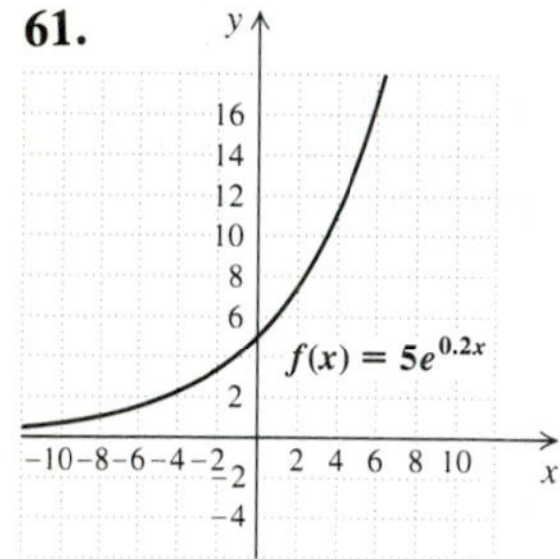

63.

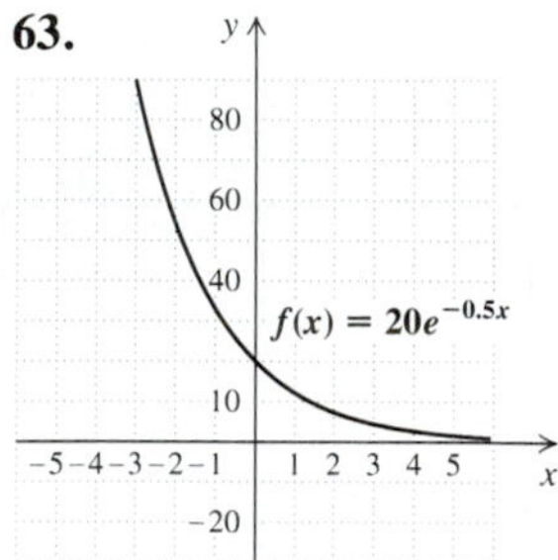

65.

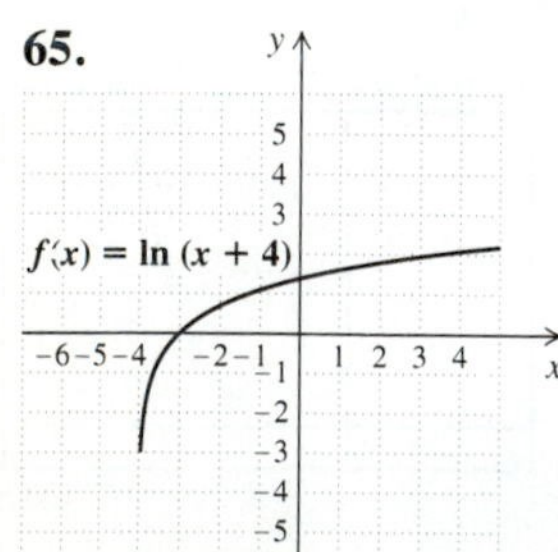

67.

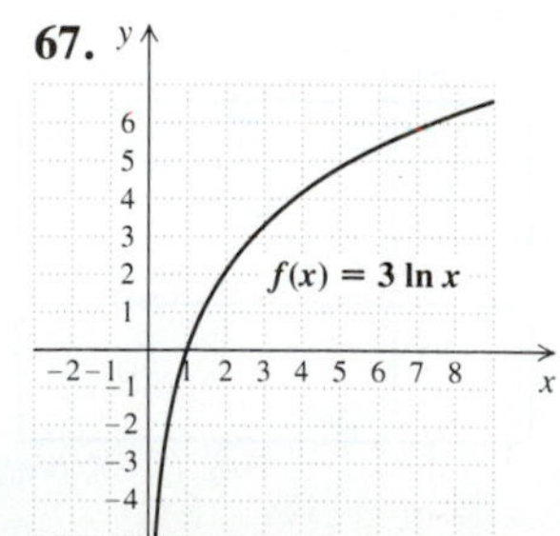

69.

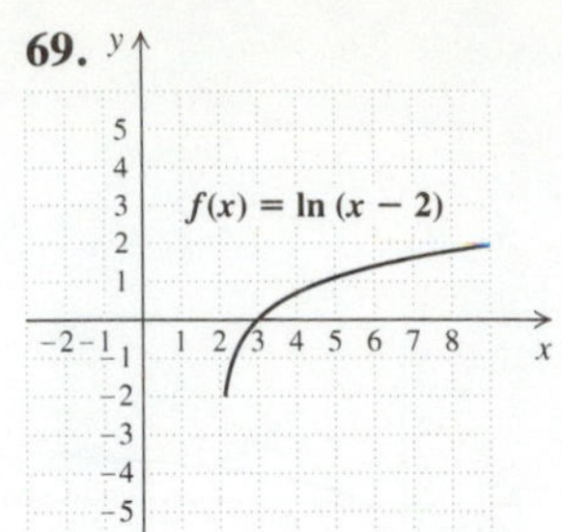

71.

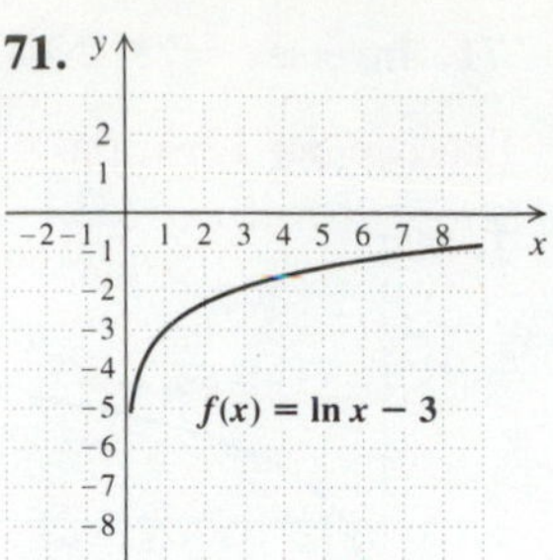

73. $\frac{5}{2}, -\frac{5}{2}$ **75.** 16, 256 **77.** ◈

79. $\ln M = \dfrac{\log M}{\log e}$ **81.** 52.5084 **83.** 4.9855

Technology Connection, Section 12.6

TC1. 0.38 **TC2.** −1.96 **TC3.** 0.90 **TC4.** −1.53 **TC5.** 3.45, 0.0001 (the function is undefined at zero) **TC6.** −0.75, 0.75

Exercise Set 12.6, pp. 654–655

1. 3 **3.** 4 **5.** $\frac{5}{2}$ **7.** $\frac{3}{5}$ **9.** 3.170 **11.** 3.322 **13.** $\frac{5}{2}$ **15.** −3, −1 **17.** 1.404 **19.** 4.605 **21.** 2.303 **23.** 140.671 **25.** 2.710 **27.** 3.607 **29.** 5.646 **31.** 27 **33.** $\frac{1}{8}$ **35.** 10 **37.** $\frac{1}{100}$

39. $e^2 \approx 7.389$ **41.** $\dfrac{1}{e} \approx 0.368$ **43.** 66 **45.** 10

47. $\frac{1}{3}$ **49.** 3 **51.** $\frac{2}{5}$ **53.** 5

55. $\dfrac{1}{25}x^{-14/3}y^{4/3}z^{-4}$, or $\dfrac{y^{4/3}}{25x^{14/3}z^4}$ **57.** $c = \sqrt{\dfrac{E}{m}}$

59. ◈ **61.** −4 **63.** 2 **65.** $\pm\sqrt{34}$ **67.** $10^{100{,}000}$ **69.** 1, 100 **71.** $\frac{1}{100{,}000}$, 100,000 **73.** $-\frac{1}{3}$ **75.** $\frac{3}{2}$ **77.** −3

Exercise Set 12.7, pp. 661–664

1. (a) 5.5 years; (b) 0.8 year **3.** (a) 37.7 years; (b) 11.9 years **5.** (a) 3.5 years; (b) 13.6 years **7.** (a) $P(t) = P_0e^{0.09t}$; (b) \$1094.17, \$1197.22; (c) 7.7 years **9.** 69.3 years **11.** (a) $P(t) = 5.2e^{0.016t}$, where t is the number of years after 1990; (b) 6.1 billion; (c) 2017 **13.** (a) 68%; (b) 54%, 40%; (c)

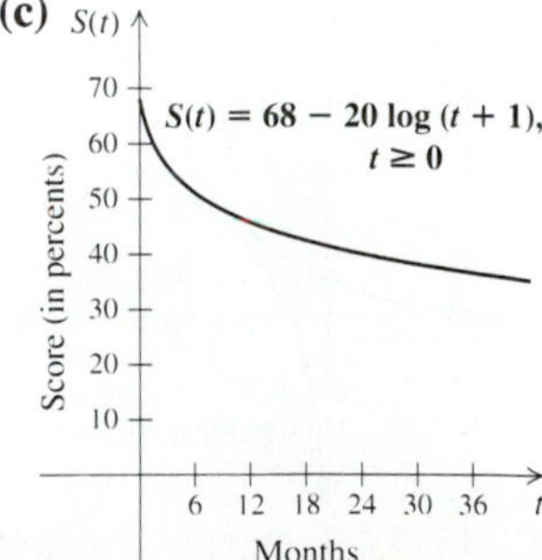

(d) 6.9 months

15. 6.2 **17.** 7.8 **19.** 10^{-7} moles per liter **21.** 6.3×10^{-4} moles per liter **23.** 8.25 **25.** 2000 **27.** **(a)** $N(t) = e^{0.363t}$, where t is the number of years after 1967; **(b)** 77,111 **29.** 1860 years **31.** 7.2 days **33.** 23% per minute **35.** **(a)** $V(t) = 110{,}000e^{0.2822t}$, where t is the number of years after 1986; **(b)** \$3,251,528; **(c)** 2.5 years; **(d)** 2002

37. $\dfrac{(x-5)(x-3)}{x^2+5x-6}$ **39.** ◈ **41.** ◈

Review Exercises: Chapter 12, pp. 665–667

1. [12.1]

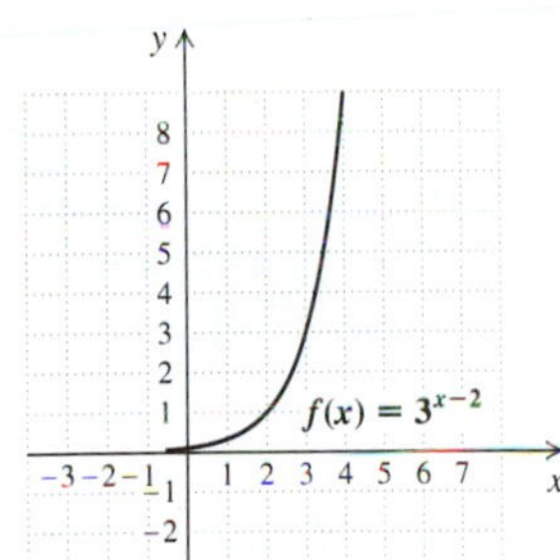

2. [12.1]

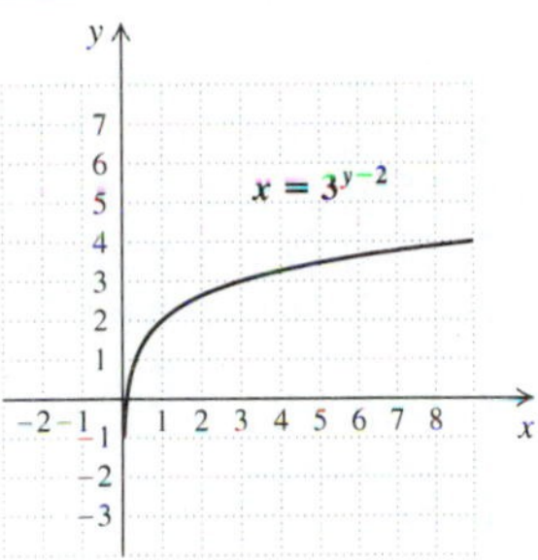

3. [12.3]

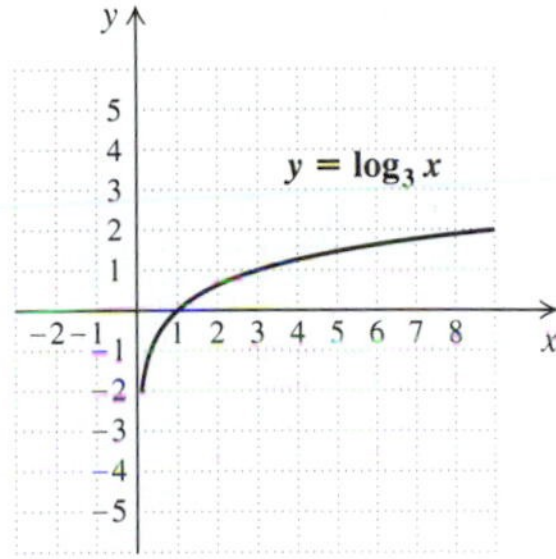

4. [12.5]

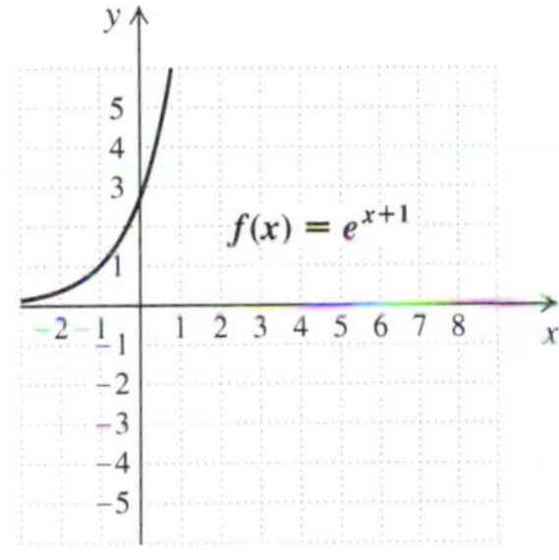

5. [12.2] $f \circ g(x) = 9x^2 - 30x + 25$, $g \circ f(x) = 3x^2 - 5$
6. [12.2] No **7.** [12.2] $f^{-1}(x) = x - 2$

8. [12.2] $g^{-1}(x) = \dfrac{7x+3}{2}$

9. [12.2] $f^{-1}(x) = \dfrac{2-5x}{x}$, or $\dfrac{2}{x} - 5$

10. [12.2] $g^{-1}(x) = \dfrac{\sqrt[3]{x}}{2}$ **11.** [12.3] $4^x = 16$

12. [12.3] $10^{0.3010} = 2$ **13.** [12.3] $\left(\frac{1}{2}\right)^{-3} = 8$
14. [12.3] $16^{3/4} = 8$ **15.** [12.3] $\log_{10} 10{,}000 = 4$
16. [12.3] $\log_{25} 5 = \frac{1}{2}$ **17.** [12.3] $\log_7 \frac{1}{49} = -2$
18. [12.3] $\log_{2.718} 20.1 = 3$
19. [12.4] $4 \log_a x + 2 \log_a y + 3 \log_a z$
20. [12.4] $\log_a x + \log_a y - 2 \log_a z$
21. [12.4] $\frac{1}{4}(2 \log z - 3 \log x - \log y)$
22. [12.4] $2 \log_q x + \frac{1}{3} \log_q y - 4 \log_q z$
23. [12.4] $\log_a (8 \cdot 15)$, or $\log_a 120$

24. [12.4] $\log_a \frac{72}{12}$, or $\log_a 6$ **25.** [12.4] $\log \dfrac{a^{1/2}}{bc^2}$

26. [12.4] $\log_a \sqrt[3]{\dfrac{x}{y^2}}$ **27.** [12.3] 1 **28.** [12.3] 0

29. [12.4] 17 **30.** [12.4] −7 **31.** [12.4] 6.93
32. [12.4] −3.2698 **33.** [12.4] 8.7601
34. [12.4] 3.2698 **35.** [12.4] 2.54995
36. [12.4] −3.6602 **37.** [12.5] −2.2027
38. [12.5] 7.8621 **39.** [12.5] 29,798.88
40. [12.5] 0.0361 **41.** [12.5] 213.50
42. [12.5] −2.3065 **43.** [12.5] 5.5965
44. [12.5] 0.000000163 **45.** [12.5] 10.0821
46. [12.5] −2.6921 **47.** [12.5] 0.00002593
48. [12.5] 3.4934×10^{19} **49.** [12.5] 0.4307
50. [12.5] 1.7097 **51.** [12.6] $\frac{1}{9}$ **52.** [12.6] 2
53. [12.6] $\frac{1}{10{,}000}$ **54.** [12.6] $e^2 \approx 7.3891$ **55.** [12.6] $\frac{7}{2}$
56. [12.6] 1.5266 **57.** [12.6] 2 **58.** [12.6] 7
59. [12.6] 8 **60.** [12.6] 20 **61.** [12.6] $\sqrt{43}$
62. [12.7] **(a)** 62; **(b)** 46.8; **(c)** 35 months
63. [12.7] **(a)** $k = 0.05$, $C(t) = \$4.65e^{0.05t}$; **(b)** \$34.36; **(c)** 1999; **(d)** 13.9 years
64. [12.7] 4.3% **65.** [12.7] 8.25 years
66. [12.7] 3463 years **67.** [12.7] 6.6 **68.** [12.7] 8.3

69. [11.9] $T = \dfrac{-b \pm \sqrt{b^2 + 4aQ}}{2a}$ **70.** [11.6] ± 4, $\pm\sqrt{5}$

71. [10.9] $\dfrac{-11}{10} - \dfrac{17}{10}i$ **72.** [6.6] $\dfrac{c-2a}{2b+3c}$

73. [12.3] ◈ Negative numbers do not have logarithms because logarithm bases are positive, and there is no power to which a positive number can be raised to yield a negative number. **74.** [12.6] ◈ Taking the logarithm on each side of an equation produces an equivalent equation because the logarithm function is one-to-one. If two quantities are equal, their logarithms must be equal, and if the logarithms of two quantities are equal, the quantities must be the same.
75. [12.6] e^{e^3} **76.** [12.6] −3, −1 **77.** [12.6] $\left(\frac{8}{3}, -\frac{2}{3}\right)$

Test: Chapter 12, pp. 667–668

1. [12.1]

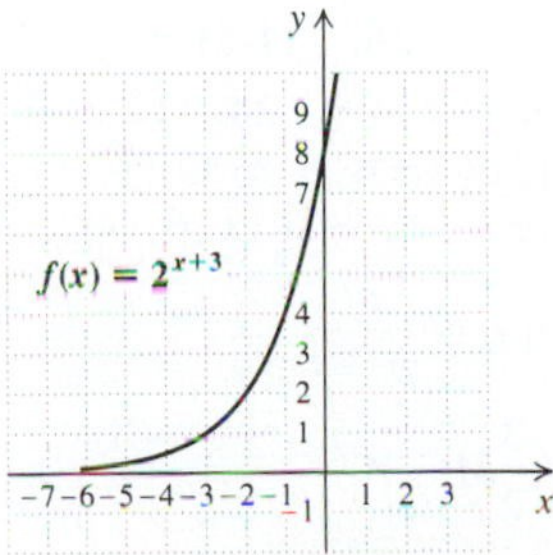

2. [12.3]

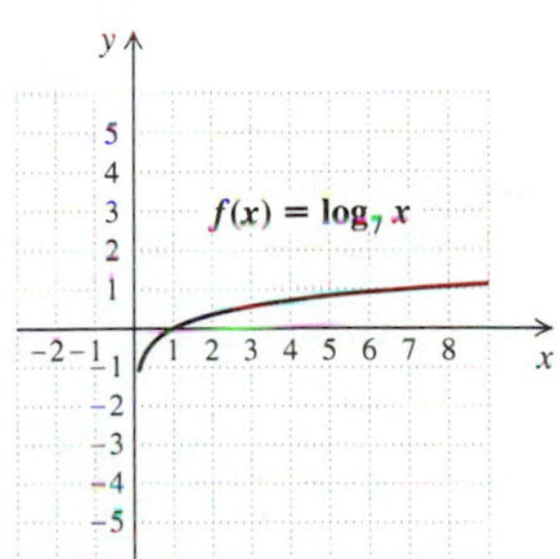

3. [12.2] $f \circ g(x) = 25x^2 - 15x + 2$, $g \circ f(x) = 5x^2 + 5x - 2$

4. [12.2] No **5.** [12.2] $f^{-1}(x) = \frac{x+3}{4}$

6. [12.2] $g^{-1}(x) = 4x + 2$ **7.** [12.2] $f^{-1}(x) = \frac{1+2x}{x-1}$

8. [12.3] $\log_4 x = -3$ **9.** [12.3] $\log_{256} 16 = \frac{1}{2}$

10. [12.3] $4^2 = 16$ **11.** [12.3] $7^m = 49$

12. [12.4] $3 \log a + \frac{1}{2} \log b - 2 \log c$

13. [12.4] $\log_a \frac{x^{1/3}z^2}{y^3}$ **14.** [12.4] 23 **15.** [12.3] 1

16. [12.3] 0 **17.** [12.4] −0.544 **18.** [12.4] 0.69
19. [12.4] 1.322 **20.** [12.5] −1.9101
21. [12.5] 445,040.98 **22.** [12.5] 0.000000054777
23. [12.5] 4.0913 **24.** [12.5] −4.3949
25. [12.5] 292.19 **26.** [12.5] 1.1881 **27.** [12.6] 5
28. [12.6] 2 **29.** [12.6] 10,000 **30.** [12.6] $\frac{1}{3}$
31. [12.6] 0.0937 **32.** [12.6] $e^{1/4} \approx 1.2840$
33. [12.6] 9 **34.** [12.7] **(a)** 2.45 ft/sec; **(b)** 984,262
35. [12.7] **(a)** $P(t) = 24e^{0.012t}$, where $P(t)$ is in millions and t is in years; **(b)** 29 million, 34 million; **(c)** 2000; **(d)** 57.8 years **36.** [12.7] 3.5% **37.** [12.7] 9.1 years
38. [12.7] 4684 years **39.** [12.7] 8.34 **40.** [12.7] 7.0

41. [11.6] 1, 64 **42.** [11.9] $t = \frac{b \pm \sqrt{b^2 + 4aS}}{2a}$

43. [10.9] 29 **44.** [6.6] $\frac{1}{2x}$ **45.** [12.6] 316, −309

46. [12.4] 2

CUMULATIVE REVIEW: 1–12

1. [1.1], [4.1] 2 **2.** [1.4], [1.5] 6 **3.** [4.9] $\frac{y^{12}}{16x^8}$

4. [4.9] $\frac{20x^6z^2}{y}$ **5.** [4.9] $\frac{-y^4}{3z^5}$ **6.** [1.8] $-4x - 1$

7. [1.8] 25 **8.** [2.2] $\frac{11}{2}$ **9.** [5.7] $\frac{2}{5}$, −5
10. [8.2] (3, −1) **11.** [8.4] (1, −2, 0)
12. [11.3] $-3 \pm \sqrt{14}$ **13.** [5.7] 5, −2 **14.** [6.7] $\frac{9}{2}$
15. [11.4] $\frac{5}{8}$ **16.** [10.7] 5 **17.** [10.7] $\frac{1}{2}$
18. [11.1] $\pm 5i$ **19.** [11.6] 9, 25 **20.** [11.6] ±3, ±2
21. [12.3] 8 **22.** [12.3] 7 **23.** [12.6] $\frac{3}{2}$ **24.** [12.6] $\frac{80}{9}$
25. [11.10] $\{x|x < -5 \text{ or } x > 1\}$, or $(-\infty, -5) \cup (1, \infty)$
26. [9.3] $\{x|x \leq -3 \text{ or } x \geq 6\}$, or $(-\infty, -3] \cup [6, \infty)$

27. [6.9] $a = \frac{Db}{b - D}$ **28.** [6.9] $q = \frac{pf}{p - f}$

29. [2.3] $B = \frac{3M - 2A}{2}$, or $B = \frac{3}{2}M - A$

30. [8.7] 2 **31.** [8.7] 3 **32.** [2.5] 8
33. [2.5] 36 m, 20 m

34. [8.5] A: 15°; B: 45°; C: 120° **35.** [6.8] $5\frac{5}{11}$ hr
36. [8.3] 60 L of Swim Clean; 40 L of Pure Swim
37. [6.8] $2\frac{7}{9}$ km/h **38.** [11.9] −49, −7 and 7
39. [6.8] Freight: 27 mph; passenger: 40 mph
40. [12.7] 78 **41.** [12.7] 67.5 **42.** [12.7] $P(t) = 430e^{0.01t}$, where P is in millions and t is in years since 1961.
43. [12.7] 616 million, 641 million **44.** [7.6] 18
45. [4.6] $7p^2q^3 - 2p^3q + pq - 6$
46. [4.3] $8x^2 - 11x - 1$ **47.** [4.5] $9x^4 - 12x^2y + 4y^2$
48. [4.5] $10a^2 - 9ab - 9b^2$

49. [6.2] $\frac{(x+4)(x-3)}{2(x-1)}$ **50.** [6.6] $\frac{1}{x-4}$

51. [6.2] $\frac{a+2}{6}$ **52.** [6.5] $\frac{7x+4}{(x+6)(x-6)}$

53. [5.1] $x(y - 2z + w)$
54. [5.5] $(1 - 5x)(1 + 5x + 25x^2)$
55. [5.3] $2(3x - 2y)(x + 2y)$ **56.** [5.1] $(x^3 + 7)(x - 4)$
57. [5.4] $(a - 5 + 9b)(a - 5 - 9b)$ **58.** [5.4] $2(m + 3n)^2$
59. [5.4] $(x - 2y)(x + 2y)(x^2 + 4y^2)$ **60.** [7.4] −12

61. [4.7] $x^3 - 2x^2 - 4x - 12 + \frac{-42}{x-3}$

62. [4.9] 1.8×10^{-1} **63.** [4.9] 5.0×10^{13}
64. [10.4] $2y^2 \sqrt[3]{y}$ **65.** [10.3] $14xy^2 \sqrt{x}$

66. [10.2] $81a^8b \sqrt[3]{b}$ **67.** [10.6] $\frac{6 + \sqrt{y} - y}{4 - y}$

68. [10.4] $\sqrt[10]{\frac{1}{(x+5)^3}}$ **69.** [10.9] $12 + 4\sqrt{3}i$

70. [10.9] $\frac{18}{25} + \frac{1}{25}i$ **71.** [12.2] $f^{-1}(x) = \frac{x-7}{-2}$

72. [7.3] $y = -5x - 3$ **73.** [7.3] $y = \frac{1}{2}x + \frac{13}{2}$

74. [3.3]

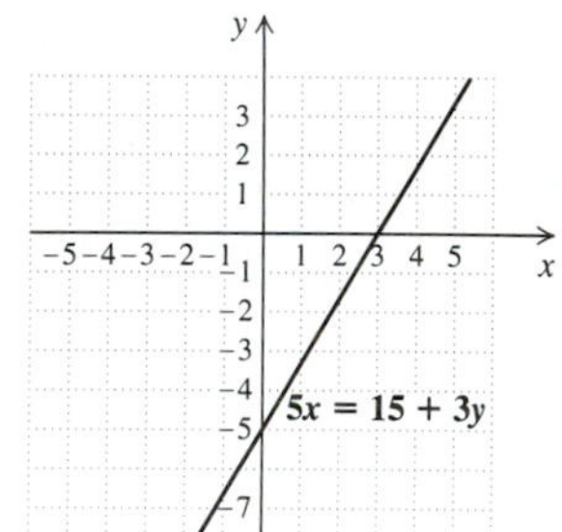

75. [11.8]

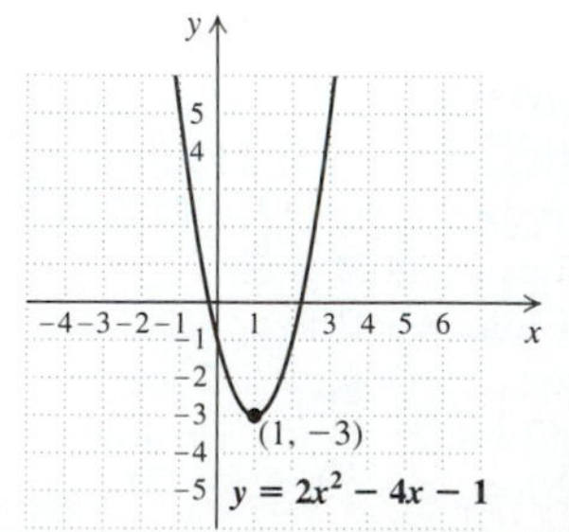

76. [12.3]

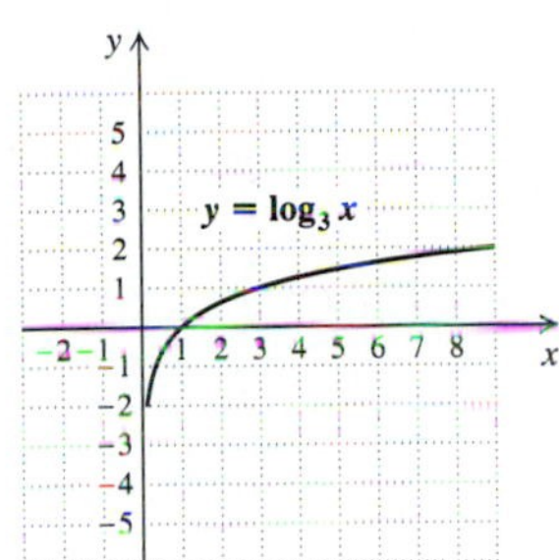

77. [12.1]

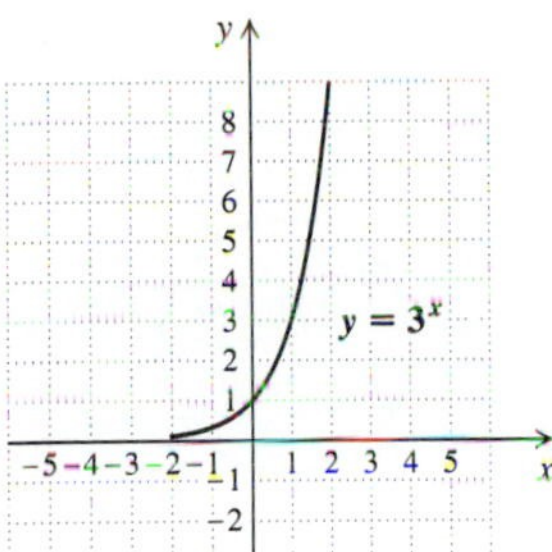

78. [9.4]

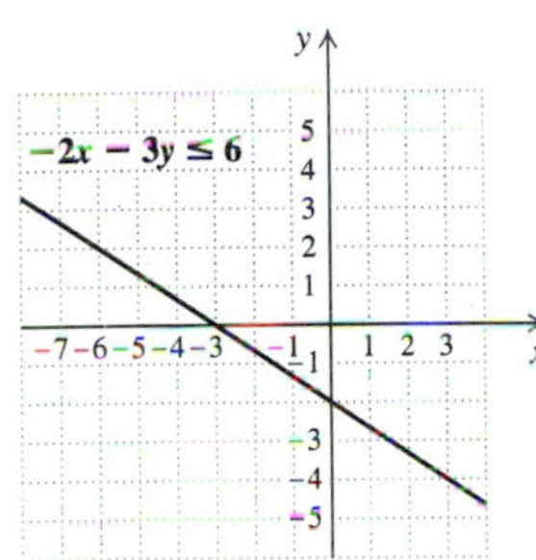

79. [11.7]

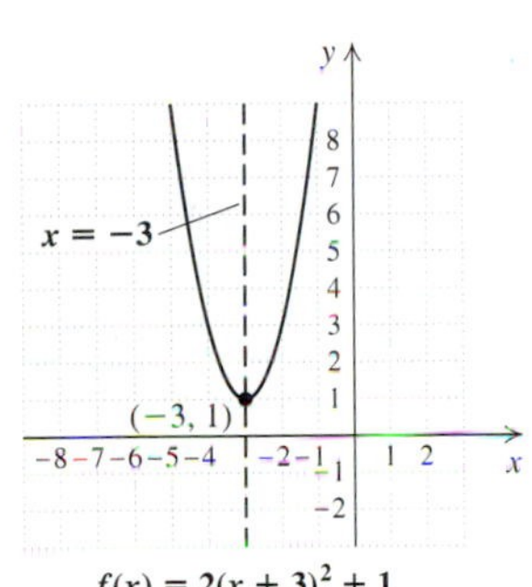

$f(x) = 2(x + 3)^2 + 1$
Minimum: 1

80. [12.4] $2 \log a + 3 \log c - \log b$

81. [12.4] $\log \left(\dfrac{x^3}{y^{1/2} z^2} \right)$ **82.** [12.3] $a^x = 5$

83. [12.3] $\log_x t = 3$ **84.** [12.5] -1.2545

85. [12.5] 278,804.64 **86.** [12.5] 2.5479

87. [12.5] 0.0253 **88.** [12.6] 0.354

89. [6.7] All real numbers except 1 and -2

90. [12.6] $\frac{1}{3}$, $\frac{10{,}000}{3}$ **91.** [11.4] 35 mph

CHAPTER 13

Exercise Set 13.1, pp. 675–676

1. 5 **3.** $\sqrt{18} \approx 4.243$ **5.** $\sqrt{32} \approx 5.657$

7. 17.8 **9.** $\dfrac{\sqrt{41}}{7} \approx 0.915$ **11.** $\sqrt{6970} \approx 83.487$

13. $\sqrt{a^2 + b^2}$ **15.** $\sqrt{49 + k^2}$

17. $\sqrt{17 + 2\sqrt{14} + 2\sqrt{15}} \approx 5.677$

19. $\sqrt{9{,}672{,}400} \approx 3110.048$ **21.** Yes **23.** No

25. $\left(-\dfrac{1}{2}, -1 \right)$ **27.** $\left(\dfrac{7}{2}, \dfrac{7}{2} \right)$ **29.** $(-1, -3)$

31. $(-0.25, -0.3)$ **33.** $(0, b)$ **35.** $\left(-\dfrac{1}{12}, \dfrac{1}{24} \right)$

37. $\left(\dfrac{\sqrt{2} + \sqrt{3}}{2}, \dfrac{3}{2} \right)$ **39.** $-1 \pm \sqrt{2}$ **41.** $\frac{3}{2}$, -1

43.  **45.** $(0, 4)$ **47.** $(2, 4\sqrt{2})$ **49.** $2\sqrt{3c}$

51. Approximately 10.046

Technology Connection, Section 13.2

TC1.

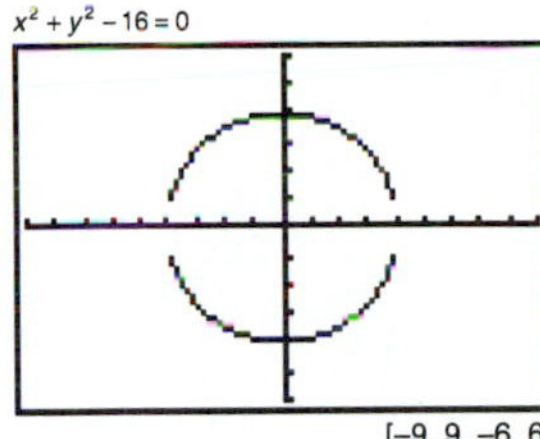

TC2.

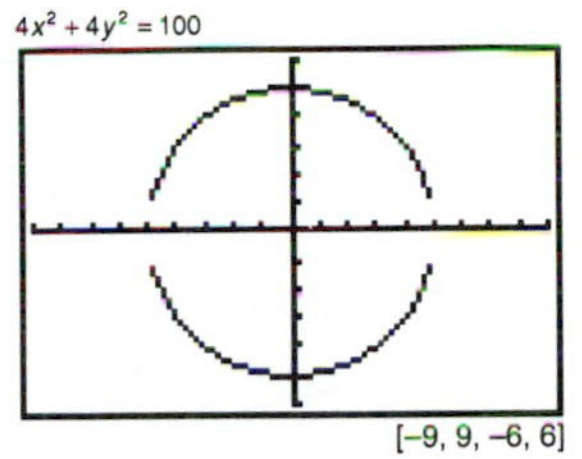

TC3.

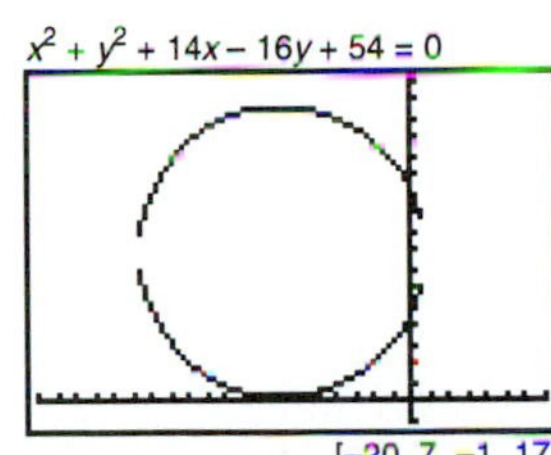

TC4.

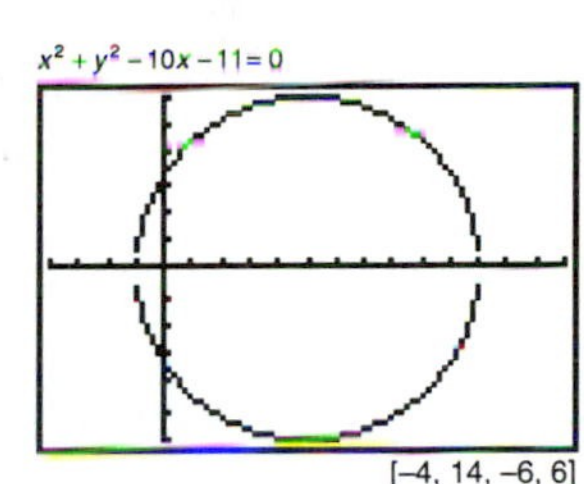

Exercise Set 13.2, pp. 682–683

1.

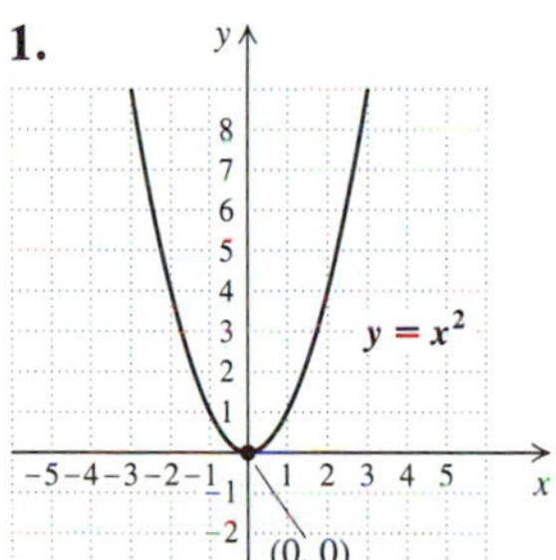

3.

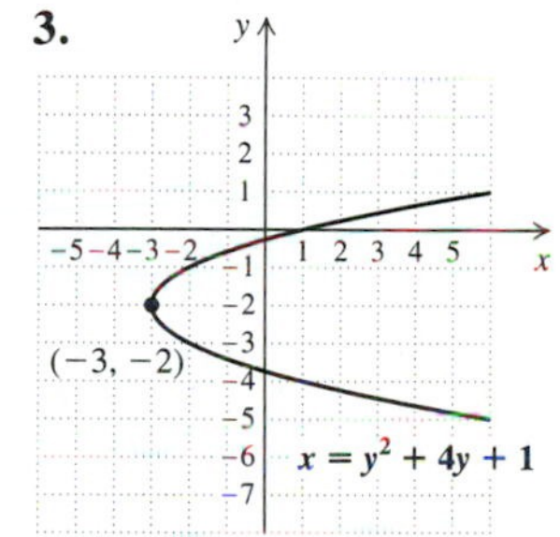

5.

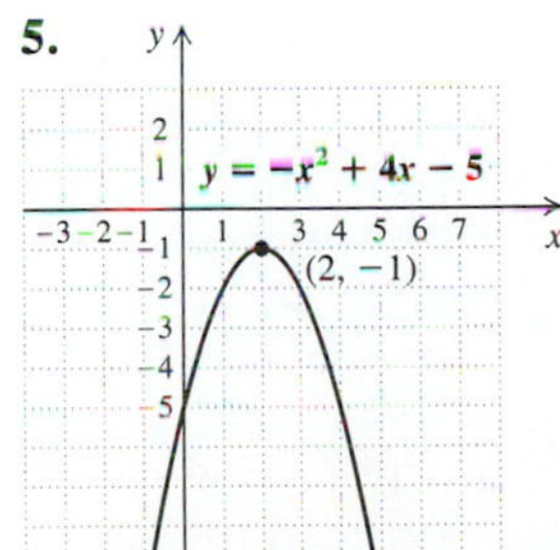

7.

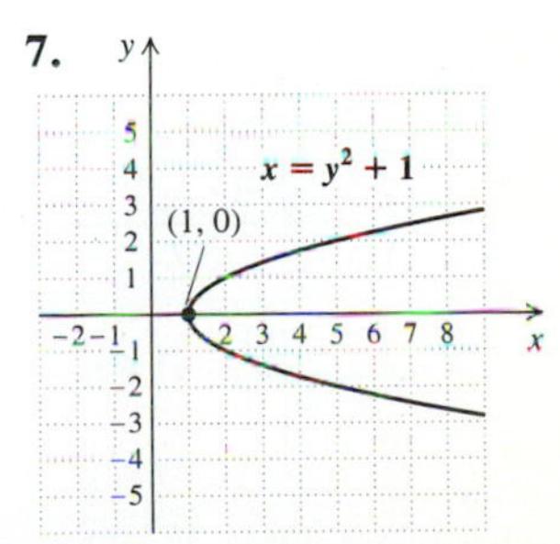

9.

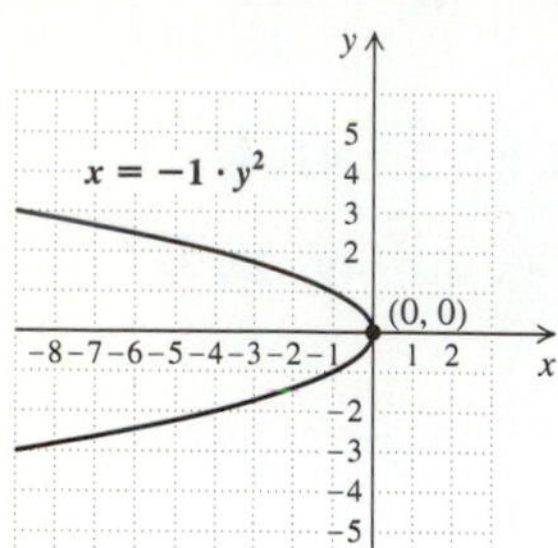

11.

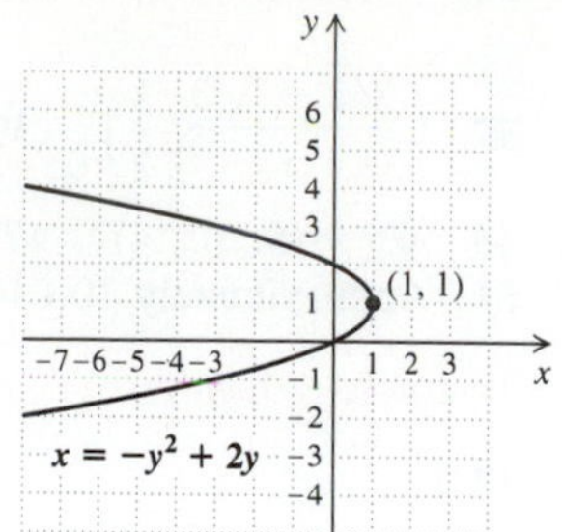

13.

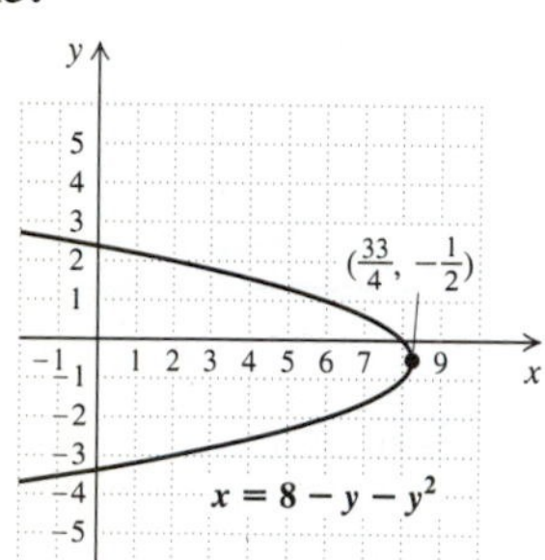

15.

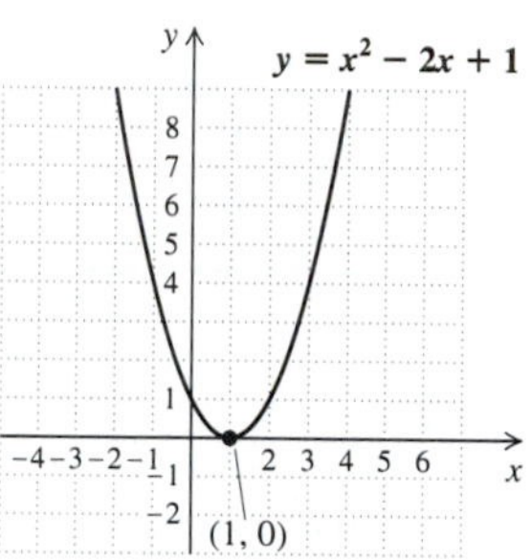

17.

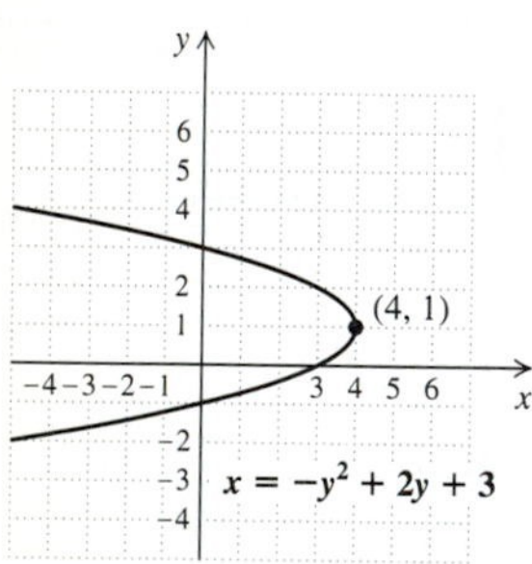

19.

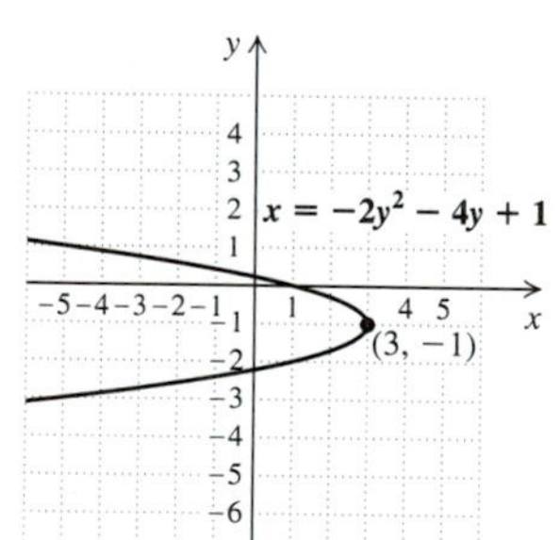

21. $x^2 + y^2 = 49$ **23.** $(x + 2)^2 + (y - 7)^2 = 9$
25. $(x + 4)^2 + (y - 3)^2 = 5$
27. $(x + 7)^2 + (y + 2)^2 = 50$ **29.** $x^2 + y^2 = 25$
31. $(x + 4)^2 + (y - 1)^2 = 20$
33. (0, 0), 6 **35.** (−1, −3), 2

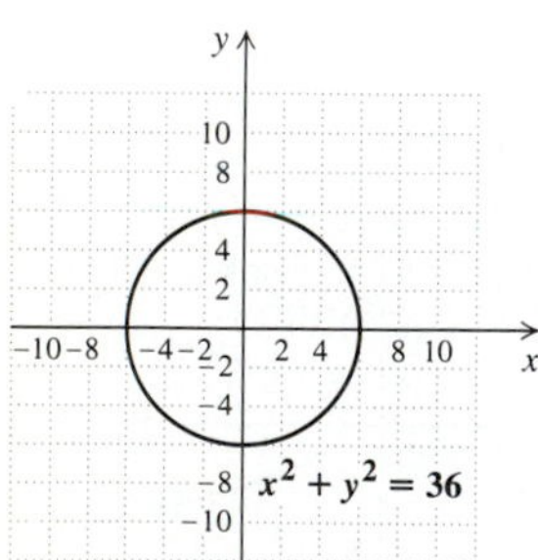

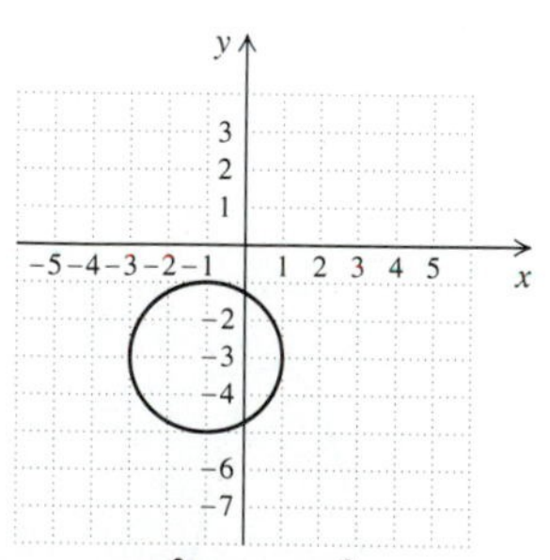

$(x + 1)^2 + (y + 3)^2 = 4$

37. (8, −3), $2\sqrt{10}$

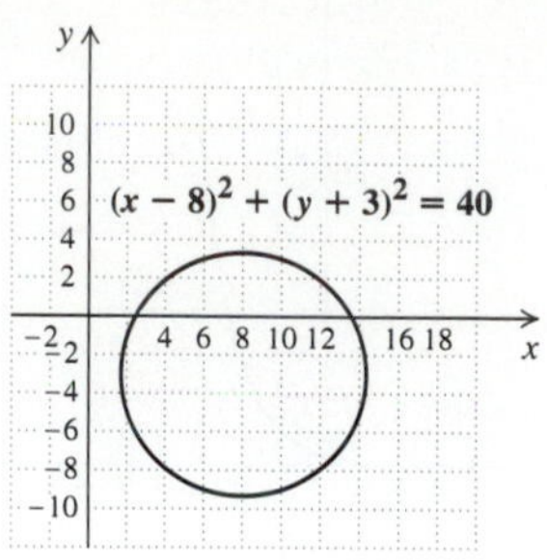

39. (0, 0), $\sqrt{2}$

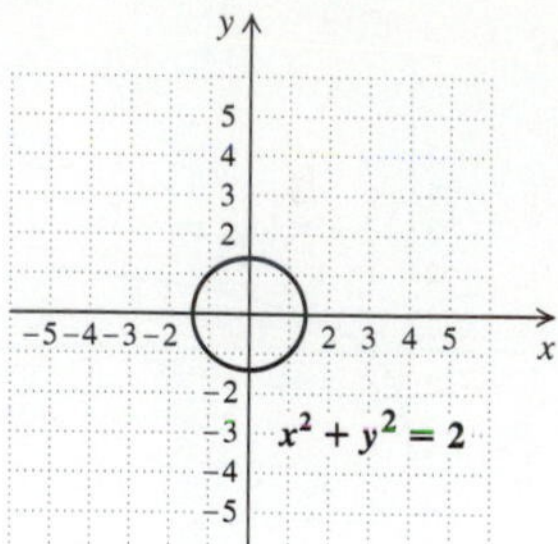

41. (5, 0), $\frac{1}{2}$

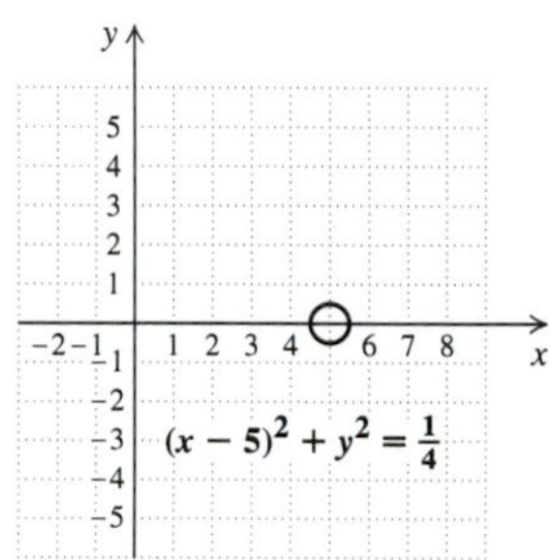

43. (−4, 3), $2\sqrt{10}$

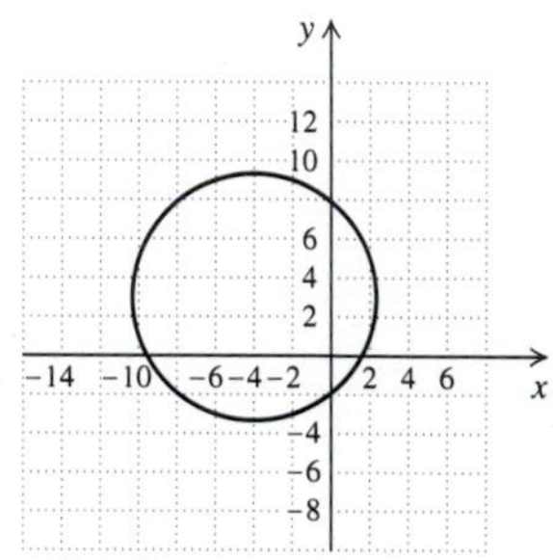

$x^2 + y^2 + 8x - 6y - 15 = 0$

45. (4, −1), 2

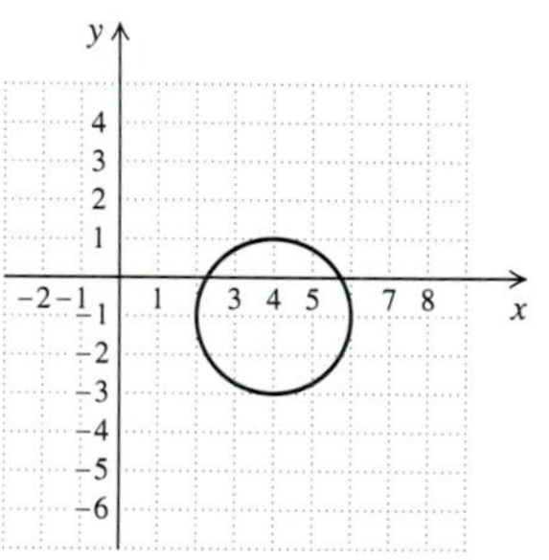

$x^2 + y^2 - 8x + 2y + 13 = 0$

47. (2, 0), 2

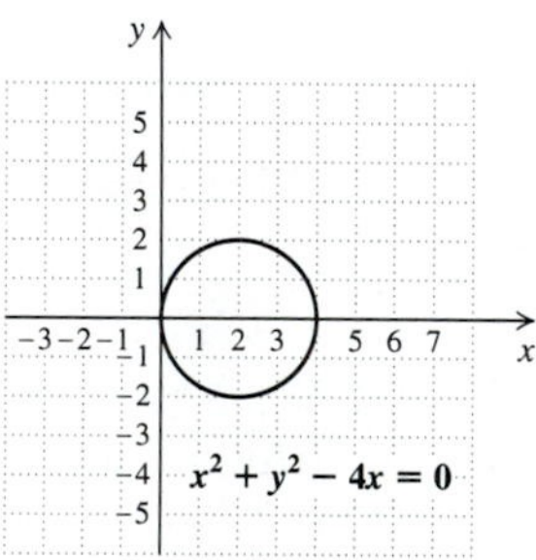

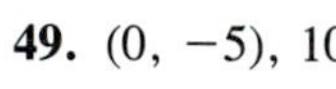
49. (0, −5), 10

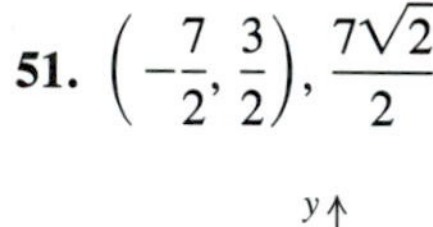
51. $\left(-\frac{7}{2}, \frac{3}{2}\right), \frac{7\sqrt{2}}{2}$

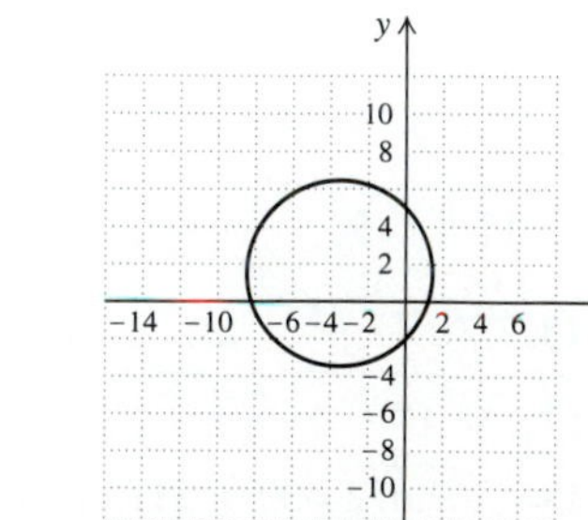

$x^2 + y^2 + 10y - 75 = 0$

$x^2 + y^2 + 7x - 3y - 10 = 0$

53. $(0, 0), \frac{1}{2}$

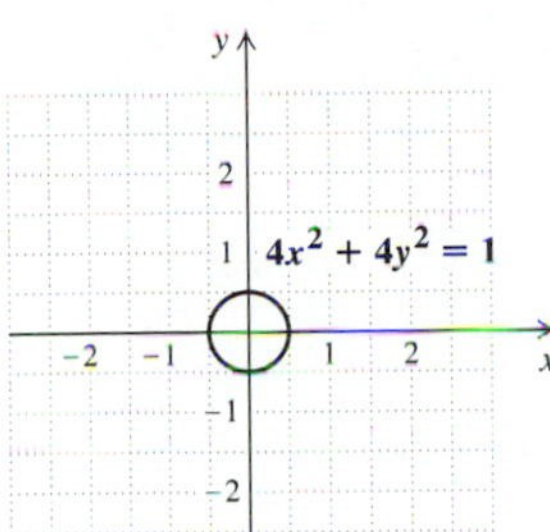

55. 4 in. **57.** ◈
59. **(a)** 3.4, −2.4; **(b)** 2.3, −1.3

61. Reflect one graph across the line $y = x$ to obtain the other.

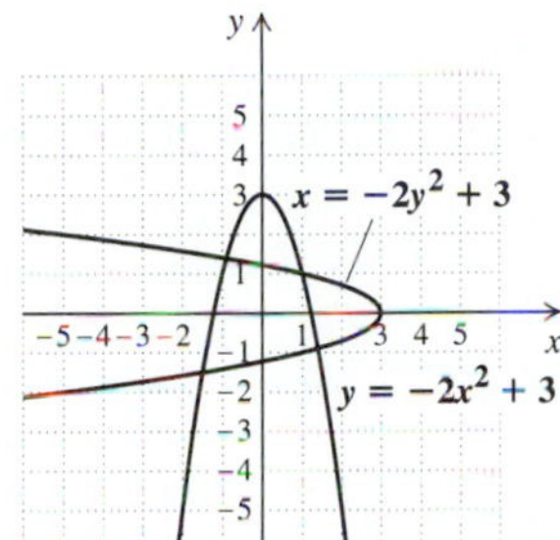

63. $\left(-\frac{1}{2}, 1\right), 1$ **65.** **(a)** $(0, -3)$; **(b)** 5 ft
67. $(x + 7)^2 + (y + 4)^2 = 16$
69. $(x + 3)^2 + (y - 5)^2 = 16$

Technology Connection, Section 13.3

TC1.

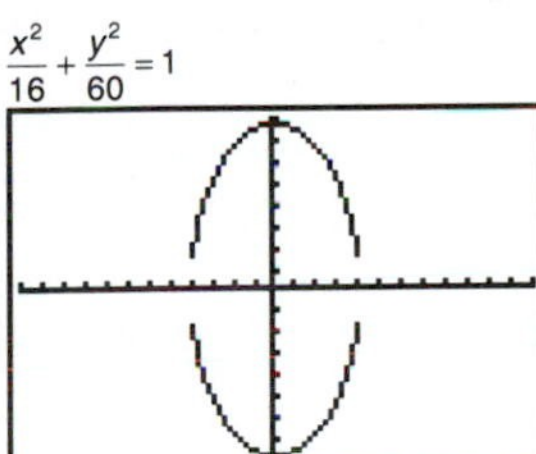

TC2.

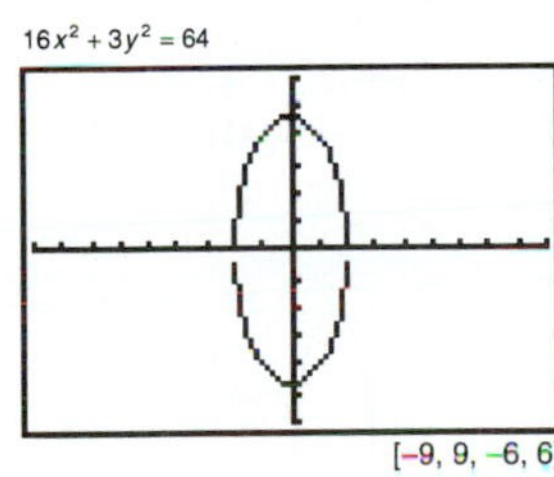

TC3.

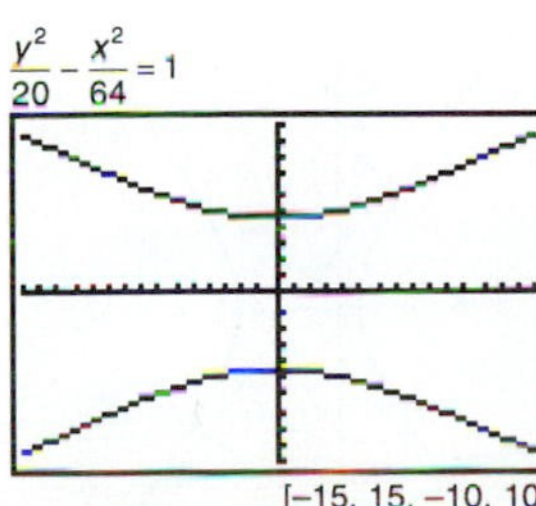

TC4.

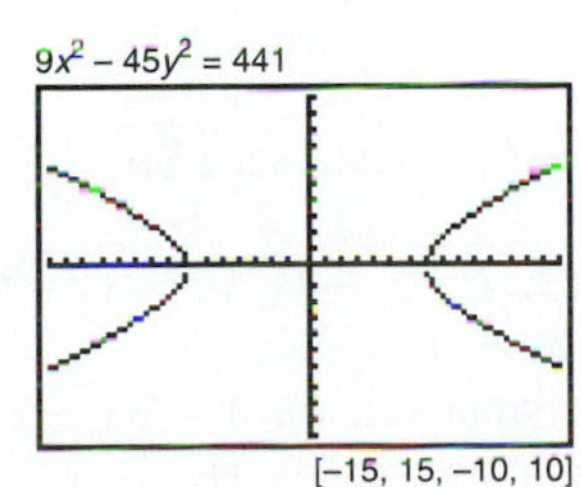

Exercise Set 13.3, pp. 695–696

1.

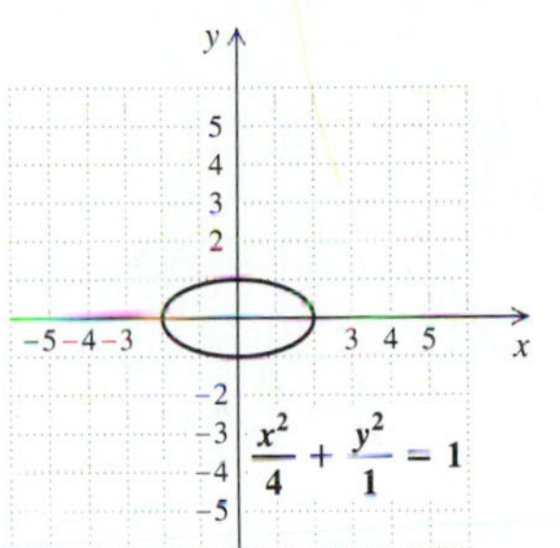

3.

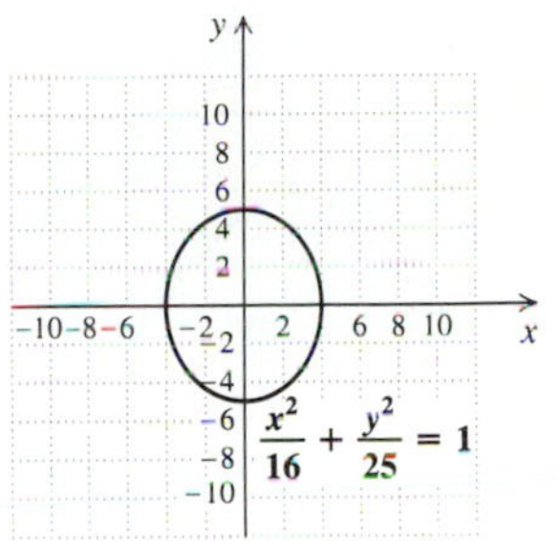

5.

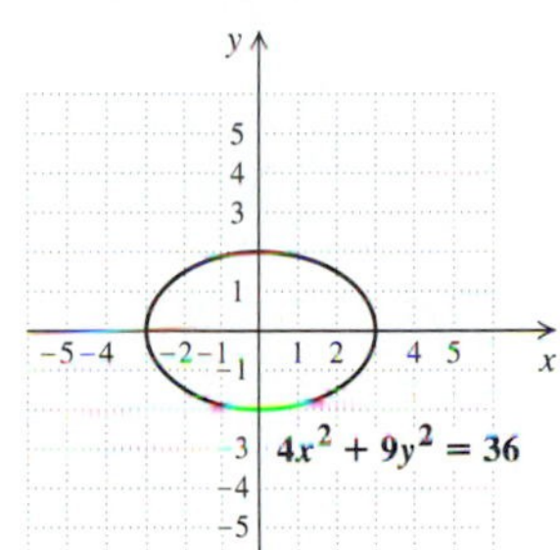

7.

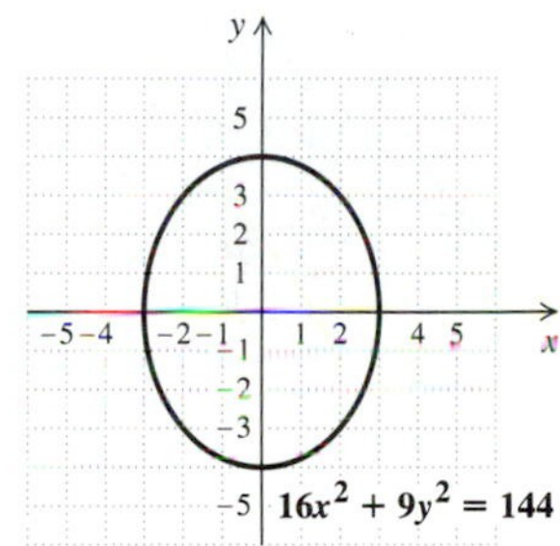

9.

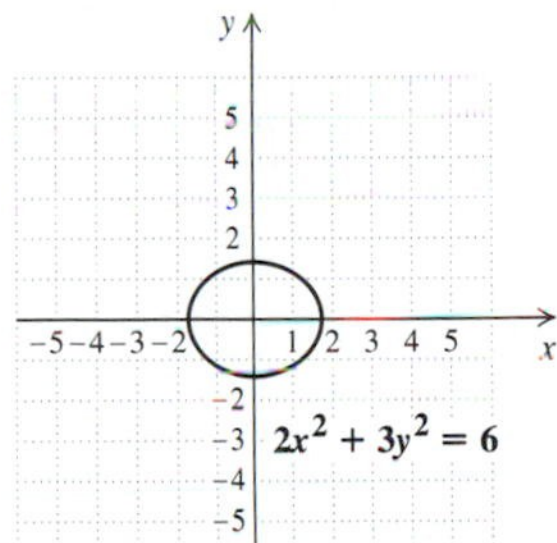

11.

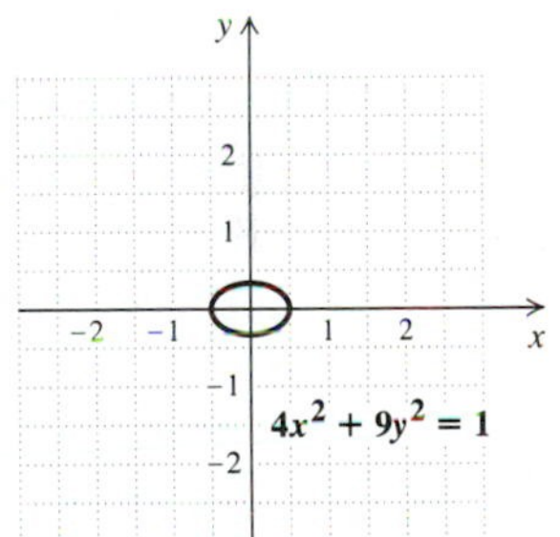

13.

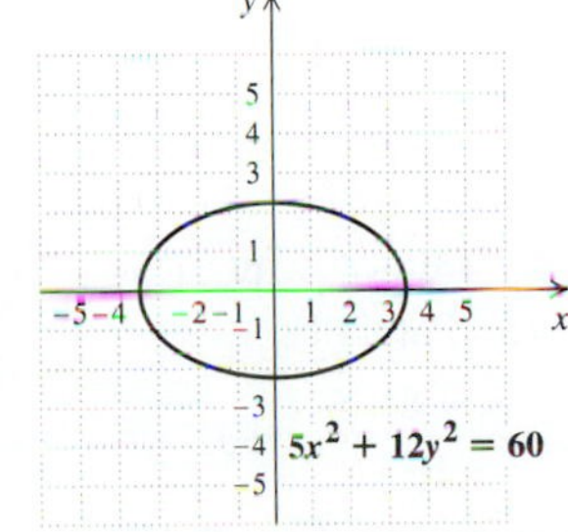

15.

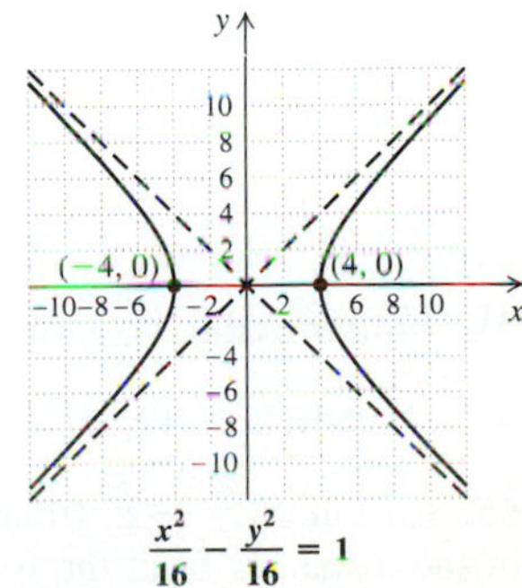

17.

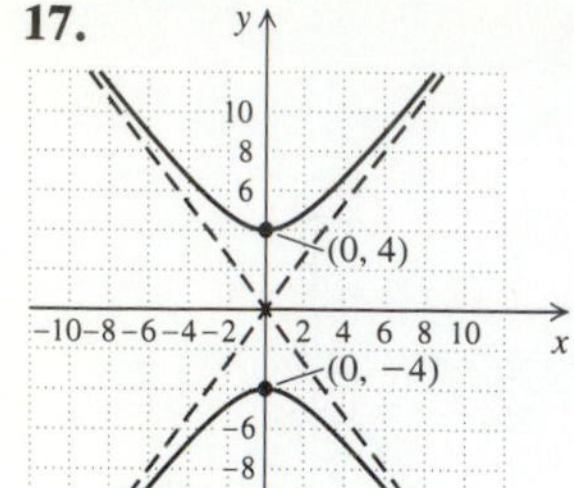

$\frac{y^2}{16} - \frac{x^2}{9} = 1$

19.

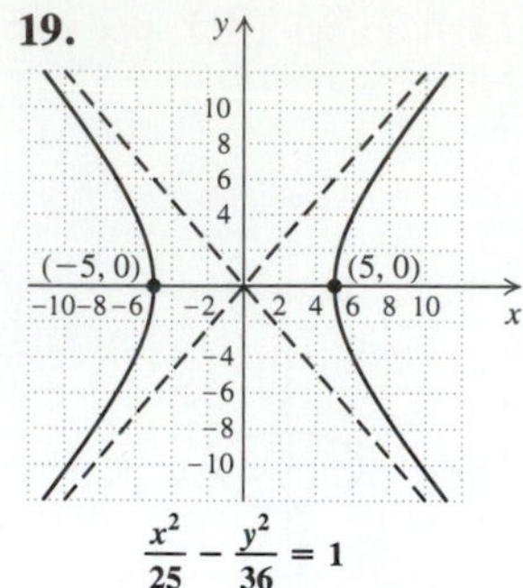

$\frac{x^2}{25} - \frac{y^2}{36} = 1$

21.

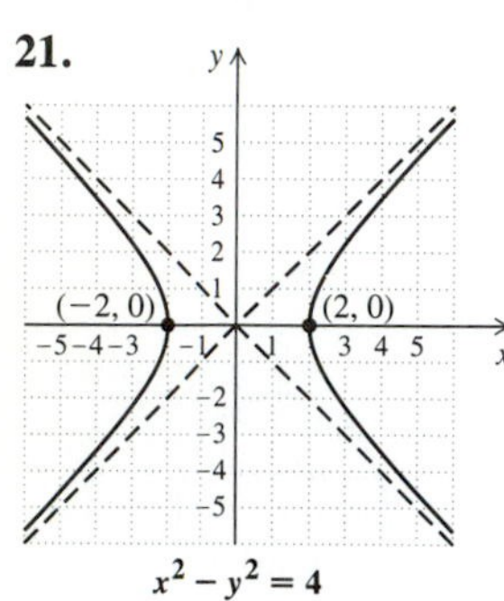

$x^2 - y^2 = 4$

23.

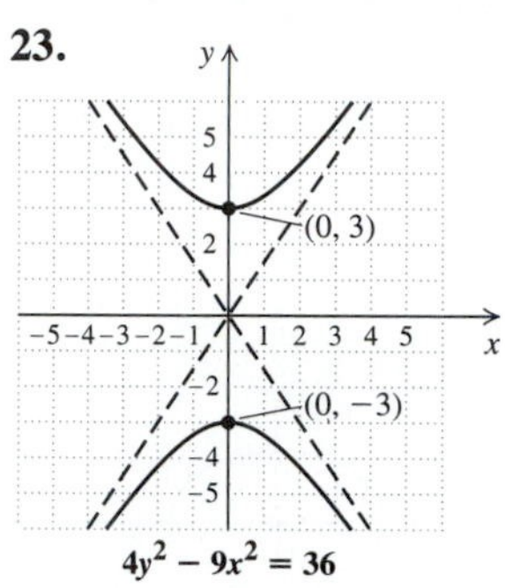

$4y^2 - 9x^2 = 36$

25.

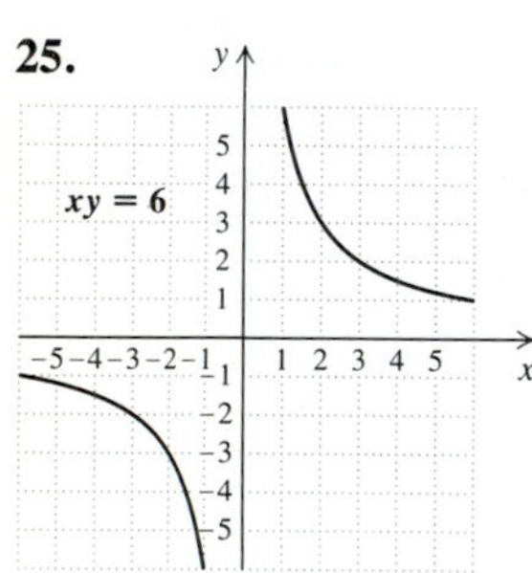

27.

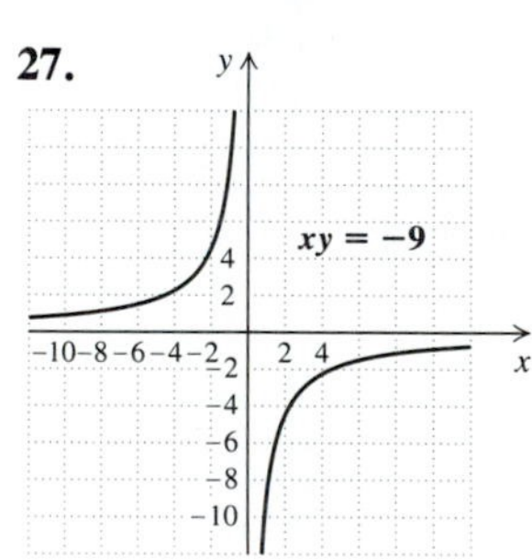

29.

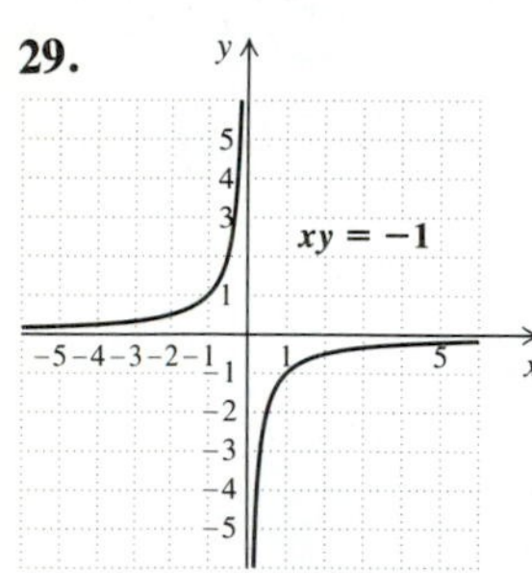

31.

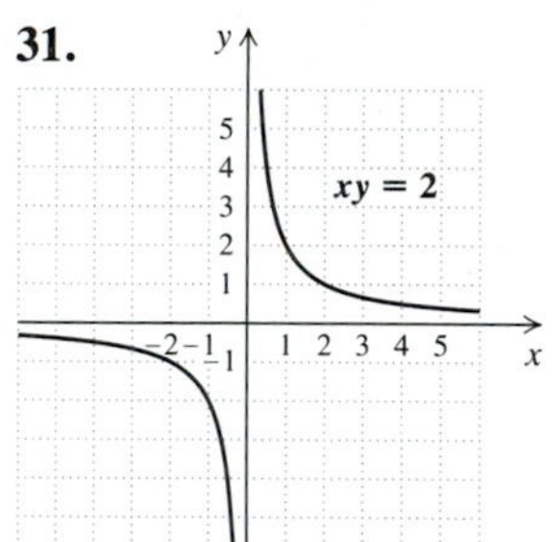

33. Circle **35.** Hyperbola **37.** Ellipse **39.** Circle **41.** Ellipse **43.** Hyperbola **45.** Parabola

47. Hyperbola **49.** $5t^5$ **51.** $\frac{13 + 8\sqrt{6}}{10}$ **53.** ◈

55. **(a)** Let $F_1 = (-c, 0)$ and $F_2 = (c, 0)$. Then the sum of the distances from the foci to P is $2a$. By the distance formula,

$$\sqrt{(x+c)^2 + y^2} + \sqrt{(x-c)^2 + y^2} = 2a, \quad \text{or}$$
$$\sqrt{(x+c)^2 + y^2} = 2a - \sqrt{(x-c)^2 + y^2}.$$

Squaring, we get

$$(x+c)^2 + y^2 = 4a^2 - 4a\sqrt{(x-c)^2 + y^2} + (x-c)^2 + y^2,$$

or $x^2 + 2cx + c^2 + y^2$

$$= 4a^2 - 4a\sqrt{(x-c)^2 + y^2} + x^2 - 2cx + c^2 + y^2.$$

Thus

$$-4a^2 + 4cx = -4a\sqrt{(x-c)^2 + y^2}$$
$$a^2 - cx = a\sqrt{(x-c)^2 + y^2}.$$

Squaring again, we get

$$a^4 - 2a^2cx + c^2x^2 = a^2(x^2 - 2cx + c^2 + y^2)$$
$$a^4 - 2a^2cx + c^2x^2 = a^2x^2 - 2a^2cx + a^2c^2 + a^2y^2,$$

or

$$x^2(a^2 - c^2) + a^2y^2 = a^2(a^2 - c^2)$$
$$\frac{x^2}{a^2} + \frac{y^2}{a^2 - c^2} = 1.$$

(b) When P is at $(0, b)$, it follows that $b^2 = a^2 - c^2$. Substituting, we have

$$\frac{x^2}{a^2} + \frac{y^2}{b^2} = 1.$$

57. 5.66 ft

59. $\frac{(x+3)^2}{1} + \frac{(y-4)^2}{16} = 1$; **61.** $\frac{y^2}{64} - \frac{x^2}{4} = 1$

C: $(-3, 4)$;
V: $(-2, 4)$, $(-4, 4)$, $(-3, 8)$, $(-3, 0)$

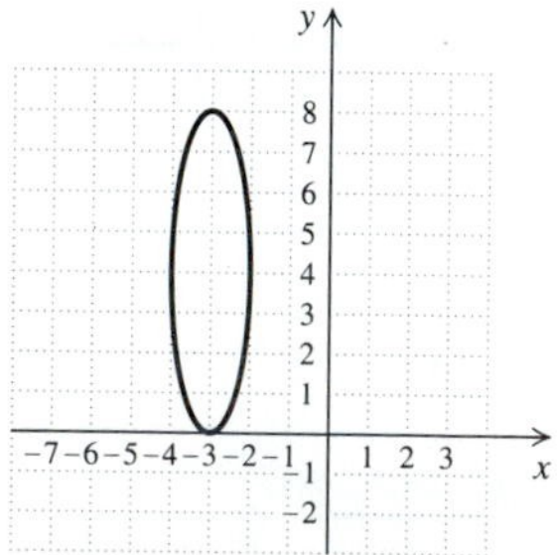

$16x^2 + y^2 + 96x - 8y + 144 = 0$

63. C: $(2, -1)$; V: $(-1, -1)$, $(5, -1)$;
asymptotes: $y + 1 = \frac{4}{3}(x - 2)$, $y + 1 = -\frac{4}{3}(x - 2)$;

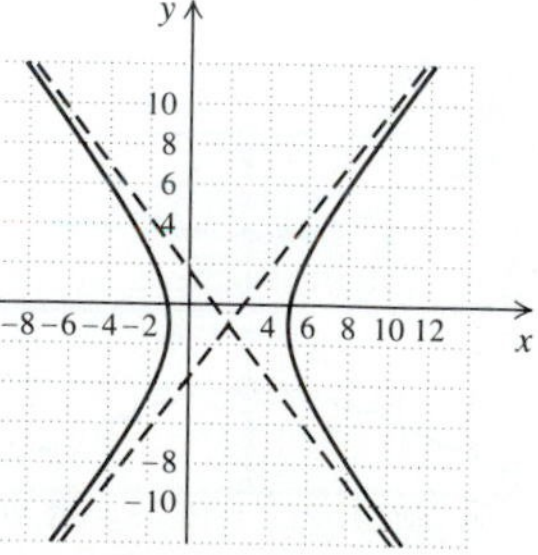

65. $\frac{(y-1)^2}{25} - \frac{(x+2)^2}{4} = 1$; C: $(-2, 1)$; V: $(-2, 6)$, $(-2, -4)$; asymptotes: $y - 1 = \frac{5}{2}(x + 2)$, $y - 1 = -\frac{5}{2}(x + 2)$;

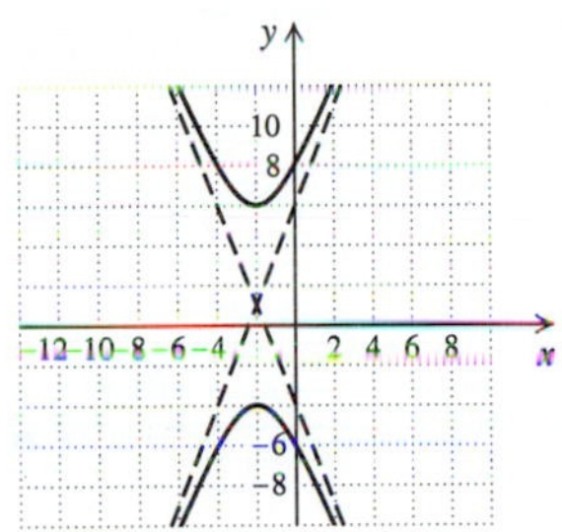

Technology Connection, Section 13.4

TC1. $(-1.50, -1.17)$; $(3.5, 0.5)$
TC2. $(-2.77, 2.52)$; $(-2.77, -2.52)$

Exercise Set 13.4, pp. 704–705

1. $(-4, -3)$, $(3, 4)$ **3.** $(0, 2)$, $(3, 0)$ **5.** $(-2, 1)$
7. $\left(\frac{5+\sqrt{70}}{3}, \frac{-1+\sqrt{70}}{3}\right)$, $\left(\frac{5-\sqrt{70}}{3}, \frac{-1-\sqrt{70}}{3}\right)$
9. $(3, 2)$, $\left(4, \frac{3}{2}\right)$ **11.** $\left(\frac{7}{3}, \frac{1}{3}\right)$, $(1, -1)$
13. $(1, 4)$, $\left(\frac{11}{4}, -\frac{5}{4}\right)$
15. $\left(\frac{7+\sqrt{33}}{2}, \frac{7-\sqrt{33}}{2}\right)$, $\left(\frac{7-\sqrt{33}}{2}, \frac{7+\sqrt{33}}{2}\right)$
17. $(3, -5)$, $(-1, 3)$ **19.** $(8, 5)$, $(-5, -8)$
21. $(4, -3)$, $(4, 3)$, $(-5, 0)$ **23.** $(-3, 0)$, $(3, 0)$
25. $(-4, -3)$, $(-3, -4)$, $(3, 4)$, $(4, 3)$
27. $\left(\frac{6\sqrt{21}}{7}, \frac{4i\sqrt{35}}{7}\right)$, $\left(\frac{6\sqrt{21}}{7}, -\frac{4i\sqrt{35}}{7}\right)$, $\left(-\frac{6\sqrt{21}}{7}, \frac{4i\sqrt{35}}{7}\right)$, $\left(-\frac{6\sqrt{21}}{7}, -\frac{4i\sqrt{35}}{7}\right)$
29. $(-\sqrt{2}, -\sqrt{14})$, $(-\sqrt{2}, \sqrt{14})$, $(\sqrt{2}, -\sqrt{14})$, $(\sqrt{2}, \sqrt{14})$
31. $(-2, -1)$, $(-1, -2)$, $(1, 2)$, $(2, 1)$
33. $(-3, -2)$, $(-2, -3)$, $(2, 3)$, $(3, 2)$
35. $(2, 5)$, $(-2, -5)$ **37.** $(3, 2)$, $(-3, -2)$
39. $(-3, 4)$, $(3, 4)$, $(0, -5)$
41. Length: 8 cm; width: 6 cm
43. Length: 5 in.; width: 4 in.
45. Length: 75 yd; width: 30 yd **47.** 13 and 12
49. 24 ft, 16 ft **51.** Length: $\sqrt{3}$ m; width: 1 m
53. $4\sqrt{3}$ **55.** 3.7 mph **57.** ◈
59. $(x + 2)^2 + (y - 1)^2 = 4$
61. 10 in. by 7 in. by 5 in.
63. $(-2, 3)$, $(2, -3)$, $(-3, 2)$, $(3, -2)$

Review Exercises: Chapter 13, pp. 707–708

1. [13.1] 4 **2.** [13.1] 5 **3.** [13.1] $\sqrt{130} \approx 11.402$
4. [13.1] $\sqrt{9 + 4a^2}$ **5.** [13.1] $(4, 6)$
6. [13.1] $\left(-3, \frac{5}{2}\right)$ **7.** [13.1] $\left(\frac{1}{2}, \frac{5}{2}\right)$
8. [13.1] $\left(\frac{1}{2}, 2a\right)$ **9.** [13.2] $(-2, 3)$, $\sqrt{2}$
10. [13.2] $(5, 0)$, 7 **11.** [13.2] $(3, 1)$, 3
12. [13.2] $(-4, 3)$, $\sqrt{35}$
13. [13.2] $(x + 4)^2 + (y - 3)^2 = 48$
14. [13.2] $(x - 7)^2 + (y + 2)^2 = 20$
15. [13.3], [13.2] Circle

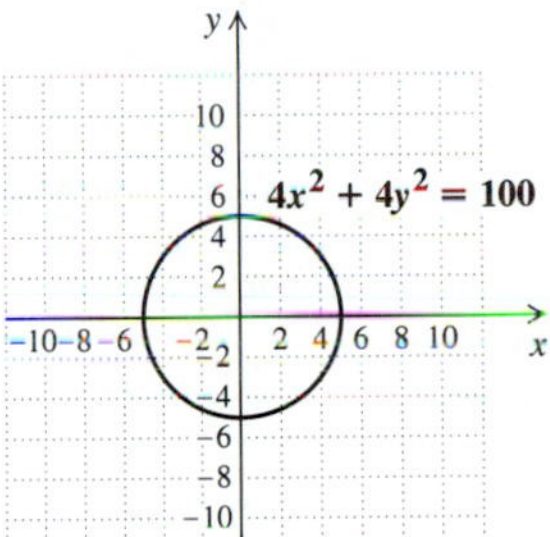

16. [13.3] Ellipse

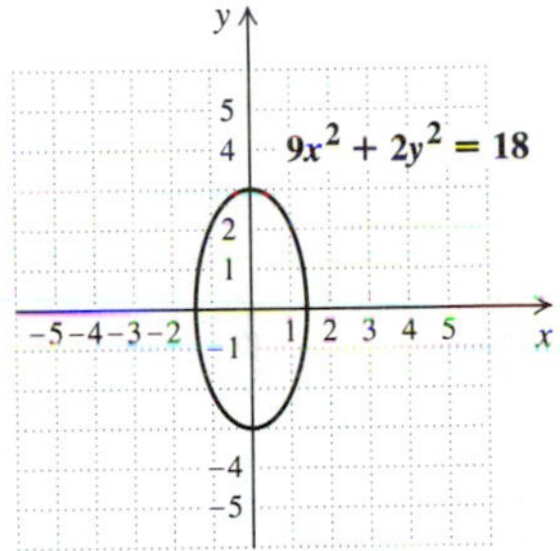

17. [13.3], [13.2] Parabola

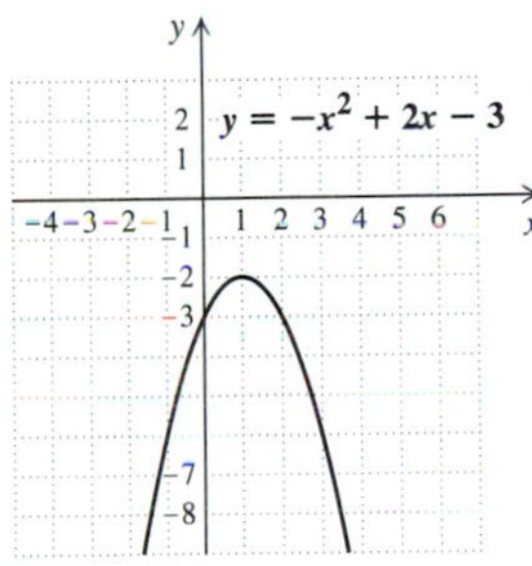

18. [13.3] Hyperbola

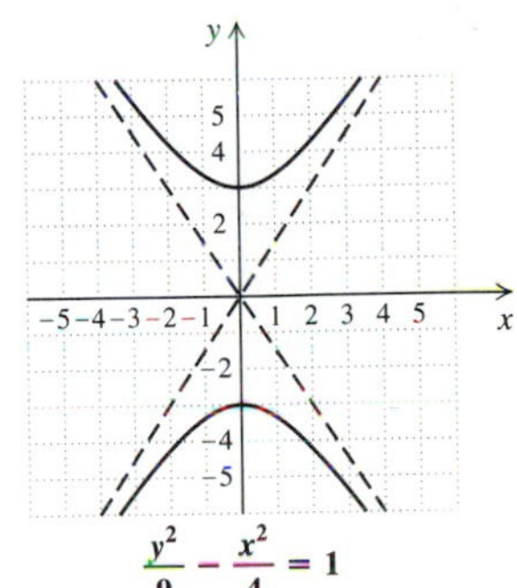

19. [13.3] Hyperbola

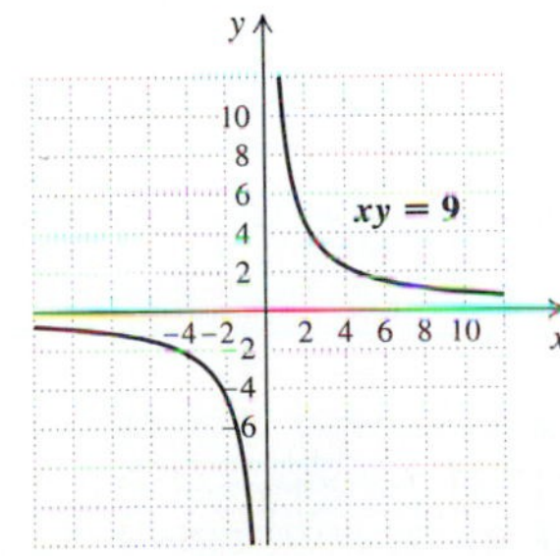

20. [13.3], [13.2] Parabola

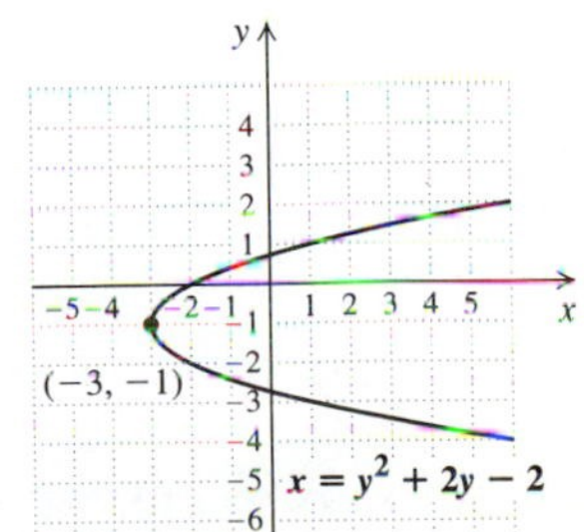

21. [13.3] Hyperbola

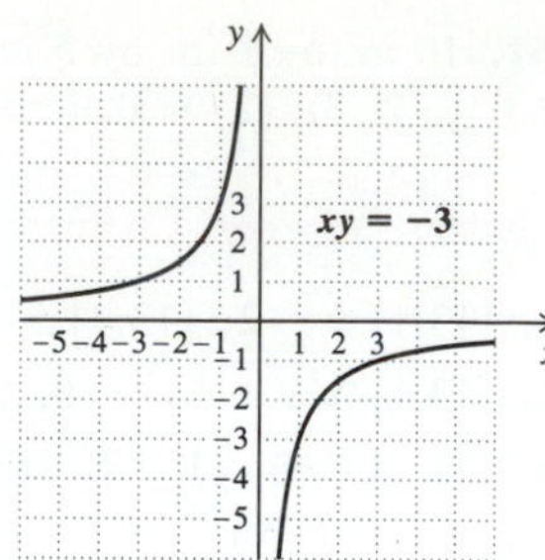

22. [13.3], [13.2] Circle

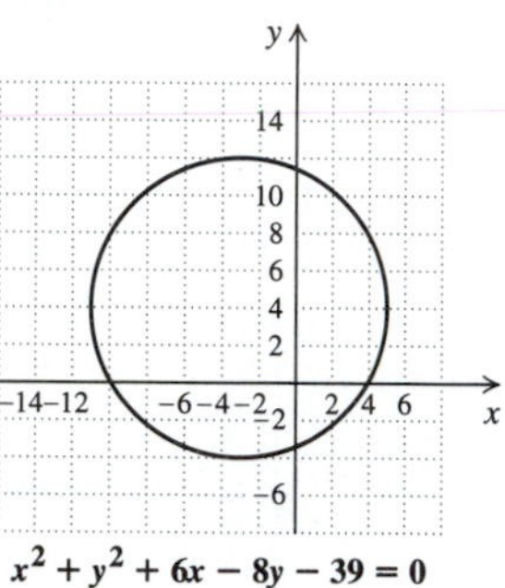

23. [13.4] (7, 4) **24.** [13.4] (2, 2), $\left(\frac{32}{9}, -\frac{10}{9}\right)$
25. [13.4] (0, −3), (2, 1)
26. [13.4] (4, 3), (4, −3), (−4, 3), (−4, −3)
27. [13.4] (2, 1), $(\sqrt{3}, 0)$, (−2, 1), $(-\sqrt{3}, 0)$
28. [13.4] (3, −3), $\left(-\frac{3}{5}, \frac{21}{5}\right)$
29. [13.4] (6, 8), (6, −8), (−6, 8), (−6, −8)
30. [13.4] (2, 2), (−2, −2), $(2\sqrt{2}, \sqrt{2})$, $(-2\sqrt{2}, -\sqrt{2})$
31. [13.4] 12 m by 7 m **32.** [13.4] 4 and 8
33. [13.4] 32 cm, 20 cm **34.** [13.4] 3 ft, 11 ft
35. [10.3] $3a^2b^3\sqrt[3]{3a^2b}$ **36.** [11.3] $-1 \pm 2i$
37. [10.6] $\dfrac{16 - a}{8 + 6\sqrt{a} + a}$ **38.** [6.8] 6 ft/sec
39. [13.2], [13.3] ◈ A circle is a special type of ellipse, where $a = b$.
40. [13.2], [13.3] ◈ Function notation is not used in this chapter because many of the relations are not functions. Function notation could be used for vertical parabolas and for hyperbolas that have the axes as asymptotes.
41. [13.4] (2, i), (2, $-i$), (−2, i), (−2, $-i$)
42. [13.1], [13.4] (0, 6), (0, −6)
43. [13.2], [13.4] $(x - 2)^2 + (y + 1)^2 = 25$
44. [13.3] $\dfrac{x^2}{49} + \dfrac{y^2}{9} = 1$ **45.** [13.1] $\left(\dfrac{9}{4}, 0\right)$

Test: Chapter 13, p. 708

1. [13.1] $9\sqrt{2} \approx 12.728$ **2.** [13.1] $2\sqrt{9 + a^2}$
3. [13.1] $\left(-\frac{1}{2}, \frac{7}{2}\right)$ **4.** [13.1] (0, 0)
5. [13.2] (−2, 3), 8 **6.** [13.2] (−2, 3), 3

7. [13.2], [13.3] Parabola

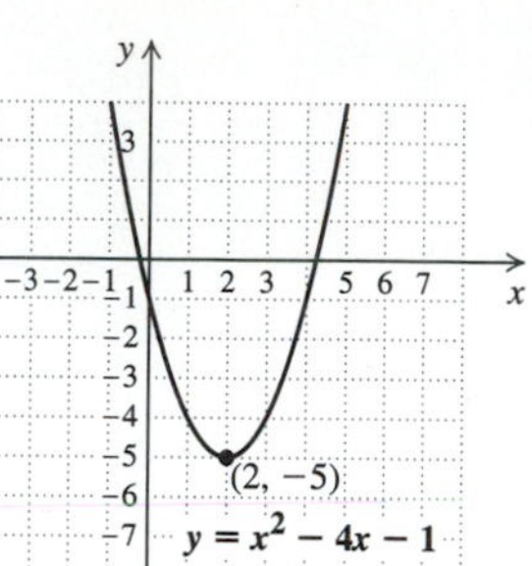

8. [13.2], [13.3] Circle

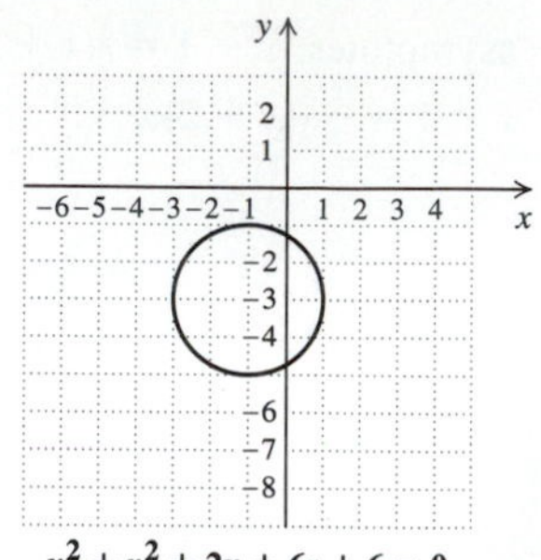

9. [13.3] Hyperbola

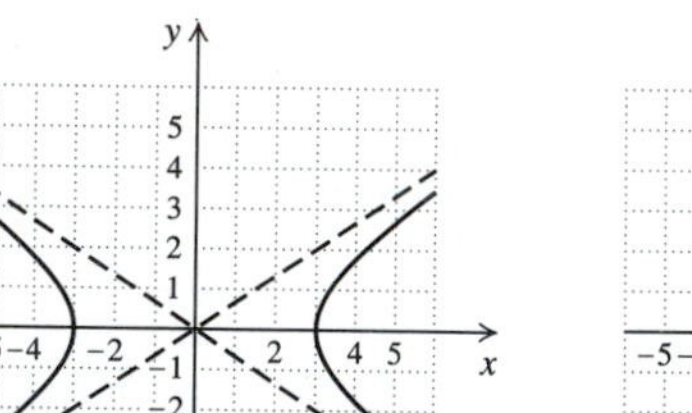

10. [13.3] Ellipse

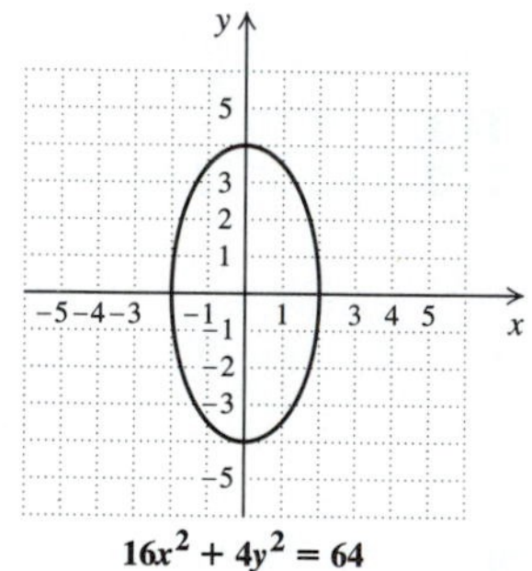

11. [13.3] Hyperbola

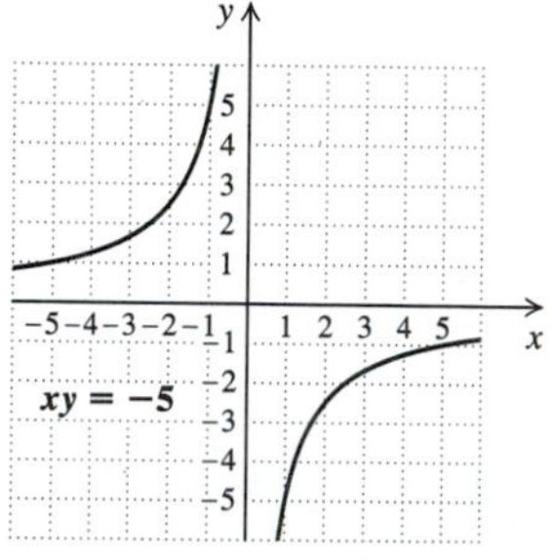

12. [13.2], [13.3] Parabola

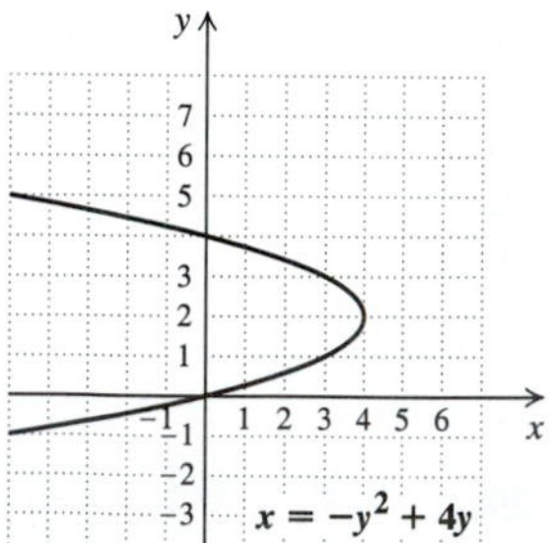

13. [13.4] (0, 3), (4, 0) **14.** [13.4] (4, 0), (−4, 0)
15. [13.4] 15 and 8 or 8 and 15 **16.** [13.4] 2 by 11
17. [13.4] $\sqrt{5}$ m, $\sqrt{3}$ m **18.** [13.4] 16 ft by 12 ft
19. [11.3] $-1 \pm \sqrt{6}$ **20.** [10.3] $2ab^6\sqrt[3]{6a^2}$
21. [10.6] $\dfrac{8 - 6\sqrt{a} + a}{4 - a}$ **22.** [6.8] 6 mph
23. [13.3] $\dfrac{(x - 6)^2}{25} + \dfrac{(y - 3)^2}{9} = 1$
24. [13.1] $\left(0, -\frac{31}{4}\right)$ **25.** [13.4] 9

CHAPTER 14

Exercise Set 14.1, pp. 714–715

1. 4, 7, 10, 13; 31; 46 **3.** $\frac{1}{2}, \frac{2}{3}, \frac{3}{4}, \frac{4}{5}; \frac{10}{11}; \frac{15}{16}$
5. −1, 0, 3, 8; 80; 195 **7.** 2, $2\frac{1}{2}$, $3\frac{1}{3}$, $4\frac{1}{4}$; $10\frac{1}{10}$; $15\frac{1}{15}$
9. −1, 4, −9, 16; 100; −225
11. −2, −1, 4, −7; −25; 40 **13.** 25 **15.** 225
17. −23.5 **19.** −33,880 **21.** $\frac{441}{400}$ **23.** 43
25. $2n - 1$ **27.** $(-1)^n 2(3)^{n-1}$ **29.** $\dfrac{n+1}{n+2}$ **31.** $3^{n/2}$
33. $-(3n - 2)$ **35.** 28 **37.** 30
39. $\frac{1}{2} + \frac{1}{4} + \frac{1}{6} + \frac{1}{8} + \frac{1}{10} = \frac{137}{120}$
41. $2^0 + 2^1 + 2^2 + 2^3 + 2^4 + 2^5 = 63$
43. $\log 7 + \log 8 + \log 9 + \log 10 \approx 3.7024$
45. $\frac{1}{2} + \frac{2}{3} + \frac{3}{4} + \frac{4}{5} + \frac{5}{6} + \frac{6}{7} + \frac{7}{8} + \frac{8}{9} = \frac{15,551}{2520}$
47. $(-1)^1 + (-1)^2 + (-1)^3 + (-1)^4 + (-1)^5 = -1$
49. $(-1)^2 3^1 + (-1)^3 3^2 + (-1)^4 3^3 + (-1)^5 3^4 + (-1)^6 3^5 + (-1)^7 3^6 + (-1)^8 3^7 + (-1)^9 3^8 = -4920$
51. $3 + 2 + 3 + 6 + 11 + 18 = 43$
53. $\frac{1}{1\cdot 2} + \frac{1}{2\cdot 3} + \frac{1}{3\cdot 4} + \frac{1}{4\cdot 5} + \frac{1}{5\cdot 6} + \frac{1}{6\cdot 7} + \frac{1}{7\cdot 8} + \frac{1}{8\cdot 9} + \frac{1}{9\cdot 10} + \frac{1}{10\cdot 11} = \frac{10}{11}$
55. $\sum_{k=1}^{6} \frac{k}{k+1}$ **57.** $\sum_{k=1}^{6} (-2)^k$ **59.** $\sum_{k=2}^{n} (-1)^k k^2$
61. $\sum_{k=1}^{\infty} 5k$ **63.** $\sum_{k=1}^{\infty} \frac{1}{k(k+1)}$ **65.** 1 **67.** 7
69. ◈ **71.** $\frac{3}{2}, \frac{3}{2}, \frac{9}{8}, \frac{3}{4}, \frac{15}{32}; \frac{171}{32}$
73. 0, ln 2, ln 6, ln 24, ln 120; ln 34,560
75. 2, 2.25, 2.370370, 2.441406, 2.488320, 2.521626
77. 0, 4, 20, 404, 163,220, 26,640,768,404
79. \$5200, \$3900, \$2925, \$2193.75, \$1645.31, \$1233.98, \$925.49, \$694.12, \$520.59, \$390.44

Exercise Set 14.2, pp. 722–723

1. $a_1 = 2, d = 5$ **3.** $a_1 = 7, d = -4$ **5.** $a_1 = \frac{3}{2}, d = \frac{3}{4}$
7. $a_1 = \$2.12, d = \$0.12$ **9.** 46 **11.** −41
13. −\$1628.16 **15.** 27th **17.** 102nd **19.** 101
21. 5 **23.** 28 **25.** $a_1 = 8; d = -3$; 8, 5, 2, −1, −4
27. 670 **29.** 45,150 **31.** 2550 **33.** 735 **35.** 990
37. 3; 171 **39.** 1260 **41.** \$31,000 **43.** 6300
45. $a^k = P$ **47.** $x^2 + y^2 = 81$ **49.** ◈ **51.** 3, 5, 7
53. \$8760, \$7961.77, \$7163.54, \$6365.31, \$5567.08, \$4768.85, \$3970.62, \$3172.39, \$2374.16, \$1575.93
55. Let d = the common difference. Since p, m, and q form an arithmetic sequence, $m = p + d$ and $q = p + 2d$. Then $\frac{p+q}{2} = \frac{p + (p + 2d)}{2} = p + d = m$.

Exercise Set 14.3, pp. 730–732

1. 2 **3.** −1 **5.** $-\frac{1}{2}$ **7.** $\frac{1}{5}$ **9.** $\frac{1}{x}$ **11.** 1.1
13. 64 **15.** 162 **17.** 648 **19.** \$2331.64
21. $a_n = 3^{n-1}$ **23.** $a_n = (-1)^{n-1}$ **25.** $a_n = \frac{1}{x^n}$
27. 762 **29.** $\frac{547}{18}$
31. $\frac{1 - x^8}{1 - x}$, or $(1 + x)(1 + x^2)(1 + x^4)$ **33.** \$5134.51
35. 8 **37.** 125 **39.** $\frac{1000}{11}$ **41.** No **43.** $\frac{1}{3}$
45. \$25,000 **47.** $\frac{4}{9}$ **49.** $\frac{5}{9}$ **51.** $\frac{5}{33}$ **53.** $\frac{1}{256}$ ft
55. 155,797 **57.** \$5236.19 **59.** 3100.35 ft
61. \$2,684,354.55 **63.** $\left(-\frac{63}{29}, -\frac{114}{29}\right)$ **65.** ◈
67. $\frac{1 - x^n}{1 - x}$ **69.** 512 cm^2

Exercise Set 14.4, pp. 739–740

1. 362,880 **3.** 39,916,800 **5.** 1 **7.** 210 **9.** 3024
11. 120 **13.** 40,314 **15.** 720 **17.** 24
19. 604,800 **21.** 6 **23.** 720 **25.** 3 · 2 · 2, or 12
27. 3!; 6 **29.** 7!, or 5040 **31.** $_7P_4$, or 840 **33.** 5!, or 120; 5^5, or 3125 **35.** 5!, or 120
37. 9 · 9 · 8 · 7 · 6 · 5 · 4, or 544,320 **39.** **(a)** 5!, or 120; **(b)** 10 · 8 · 6 · 4 · 2, or 3840 **41.** 52 · 51 · 50 · 49, or 6,497,400 **43.** 80 · 26 · 9999, or 20,797,920
45. **(a)** 10^5 or 100,000; **(b)** 100,000 **47.** **(a)** 10^9, or 1,000,000,000; **(b)** yes **49.** (12 − 1)!, or 39,916,800
51. 6!, or 720; (6 − 1)!, or 120 **53.** 3.555
55. 3.892 **57.** −5 **59.** ◈ **61.** 11 **63.** 9
65. **(a)** 6!, or 720; **(b)** 3! · 3! · 2, or 72; **(c)** 2! · 5!, or 240; **(d)** 720 − 240, or 480 **67.** 4 · 3, or 12

Exercise Set 14.5, pp. 745–747

1. 78 **3.** 78 **5.** 7 **7.** 10 **9.** 1 **11.** 12
13. 190 **15.** 595 **17.** 252
19. $\binom{23}{4}$, or 8855 **21.** $\binom{10}{6}$, or 210
23. $\binom{8}{2}$, or 28; $\binom{8}{3}$, or 56
25. $\binom{10}{7} \cdot \binom{5}{3}$, or 1200 **27.** $\binom{58}{6} \cdot \binom{42}{4}$
29. 6 · 9 · 7, or 378 **31.** $\binom{52}{5}$
33. $\binom{5}{2} \cdot \binom{6}{3} \cdot \binom{3}{1}$, or 600
35. $\binom{9}{4} \cdot \binom{5}{3} \cdot \binom{2}{2}$, or 1260

37. $\binom{10}{0} + \binom{10}{1} + \binom{10}{2} + \binom{10}{3} + \binom{10}{4} + \binom{10}{5} + \binom{10}{6} + \binom{10}{7} + \binom{10}{8} + \binom{10}{9} + \binom{10}{10}$, or 1024

39. $\binom{4}{3} \cdot \binom{48}{2}$, or 4512

41. **(a)** ${}_{33}P_3$, or 32,736; **(b)** 33^3, or 35,937; **(c)** ${}_{33}C_3$, or 5456 **43.** -2 **45.** 24 **47.** ◈ **49.** m **51.** 1 **53.** 5 **55.** 7 **57.** $n - 1$

59. $\binom{m}{3}$, or $\dfrac{m(m-1)(m-2)}{6}$

61. $$\binom{n}{n-r} = \frac{n!}{[n-(n-r)]!\,(n-r)!} = \frac{n!}{r!\,(n-r)!} = \binom{n}{r}$$

Exercise Set 14.6, p. 752

1. $m^5 + 5m^4n + 10m^3n^2 + 10m^2n^3 + 5mn^4 + n^5$
3. $x^6 - 6x^5y + 15x^4y^2 - 20x^3y^3 + 15x^2y^4 - 6xy^5 + y^6$
5. $x^{10} - 15x^8y + 90x^6y^2 - 270x^4y^3 + 405x^2y^4 - 243y^5$
7. $729c^6 - 1458c^5d + 1215c^4d^2 - 540c^3d^3 + 135c^2d^4 - 18cd^5 + d^6$ **9.** $x^7 - 7x^6y + 21x^5y^2 - 35x^4y^3 + 35x^3y^4 - 21x^2y^5 + 7xy^6 - y^7$
11. $x^{-7} + 7x^{-6}y + 21x^{-5}y^2 + 35x^{-4}y^3 + 35x^{-3}y^4 + 21x^{-2}y^5 + 7x^{-1}y^6 + y^7$
13. $a^9 - 18a^7 + 144a^5 - 672a^3 + 2016a - 4032a^{-1} + 5376a^{-3} - 4608a^{-5} + 2304a^{-7} - 512a^{-9}$
15. $a^8 + 8a^6b^3 + 24a^4b^6 + 32a^2b^9 + 16b^{12}$
17. $9\sqrt{3} - 45t + 30\sqrt{3}t^2 - 30t^3 + 5\sqrt{3}t^4 - t^5$
19. $x^{-8} + 4x^{-4} + 6 + 4x^4 + x^8$ **21.** $15a^4b^2$
23. $-745{,}472a^3$ **25.** $1120x^{12}y^2$

27. $-1{,}959{,}552u^5v^{10}$ **29.** $(0, 1)$, 7 **31.** $\log_a \dfrac{x^{3/2}}{y^3}$

33. ◈ **35.** $\binom{5}{2}(0.313)^3(0.687)^2 \approx 0.145$

37. $\binom{5}{2}(0.313)^3(0.687)^2 + \binom{5}{3}(0.313)^2(0.687)^3 + \binom{5}{4}(0.313)(0.687)^4 + \binom{5}{5}(0.687)^5 \approx 0.964$

39. $1 + 4i - 6 - 4i + 1 = -4$

41. 9th term: $\dfrac{55}{144}$ **43.** $-\dfrac{\sqrt[3]{q}}{2p}$ **45.** 8

Exercise Set 14.7, pp. 759–761

1. 0.57, 0.43 **3.** 0.075, 0.134, 0.057, 0.071, 0.030
5. 0.633 **7.** 52 **9.** $\frac{1}{4}$ **11.** $\frac{1}{13}$ **13.** $\frac{1}{2}$ **15.** $\frac{2}{13}$
17. $\frac{5}{7}$ **19.** 0 **21.** $\frac{11}{4165}$ **23.** $\frac{28}{65}$ **25.** $\frac{1}{18}$ **27.** $\frac{1}{36}$
29. $\frac{30}{323}$ **31.** $\frac{24}{253}$ **33.** $\frac{9}{19}$ **35.** $\frac{1}{38}$ **37.** $\frac{1}{19}$

39. $(6, -1)$ **41.** $2 \log_a x + \log_a y - 3 \log_a z$
43. ◈ **45.** 2,598,960 **47.** **(a)** 36; **(b)** 0.0000139
49. **(a)** $(13 \cdot {}_4C_3) \cdot (12 \cdot {}_4C_2) = 3744$;
(b) $\dfrac{3744}{{}_{52}C_5} = \dfrac{3744}{2{,}598{,}960} \approx 0.00144$

51. **(a)** $13 \cdot \binom{4}{3} \cdot \binom{48}{2} - 3744 = 54{,}912$;
(b) $\dfrac{54{,}912}{{}_{52}C_5} = \dfrac{54{,}912}{2{,}598{,}960} \approx 0.0211$

53. **(a)** $\binom{13}{2}\binom{4}{2}\binom{4}{2}\binom{44}{1} = 123{,}552$;
(b) $\dfrac{123{,}552}{{}_{52}C_5} = \dfrac{123{,}552}{2{,}598{,}960} \approx 0.0475$

Review Exercises: Chapter 14, pp. 763–765

1. [14.1] 1, 5, 9, 13; 29; 45
2. [14.1] 0, $\frac{1}{5}$, $\frac{1}{5}$, $\frac{3}{17}$; $\frac{7}{65}$; $\frac{11}{145}$ **3.** [14.1] $a_n = -2n$
4. [14.1] $a_n = n^2$
5. [14.1] $-2 + 4 + (-8) + 16 + (-32) = -22$
6. [14.1] $-3 + (-5) + (-7) + (-9) + (-11) + (-13) = -48$

7. [14.1] $\sum_{k=1}^{5} 4k$ **8.** [14.1] $\sum_{k=1}^{5} \dfrac{1}{(-2)^k}$

9. [14.2] 85 **10.** [14.2] $\frac{8}{3}$
11. [14.2] $d = 1.25$, $a_1 = 11.25$ **12.** [14.2] -544
13. [14.2] 8580 **14.** [14.2] 63 **15.** [14.2] 864
16. [14.3] $1024\sqrt{2}$ **17.** [14.3] $\frac{2}{3}$

18. [14.3] $a_n = 2(-1)^n$ **19.** [14.3] $a_n = 3\left(\dfrac{x}{4}\right)^{n-1}$

20. [14.3] 4095 **21.** [14.3] $-4095x$ **22.** [14.3] 12
23. [14.3] 0.05 **24.** [14.3] No **25.** [14.3] No
26. [14.3] \$40,000 **27.** [14.3] $\frac{5}{9}$ **28.** [14.3] $\frac{13}{33}$
29. [14.2] \$17.80 **30.** [14.2] 903
31. [14.3] \$22,521.92 **32.** [14.3] 6 m
33. [14.4] 40,320 **34.** [14.4] 30 **35.** [14.4] 60,480
36. [14.5] 56 **37.** [14.5] 35 **38.** [14.4] 3024
39. [14.4] 12,144 **40.** [14.4] 720 **41.** [14.5] 6435
42. [14.5] 36 **43.** [14.6] 190 $a^{18}b^2$
44. [14.6] $x^4 - 8x^3y + 24x^2y^2 - 32xy^3 + 16y^4$
45. [14.7] $\frac{86}{206} \approx 0.42$, $\frac{97}{206} \approx 0.47$, $\frac{23}{206} \approx 0.11$
46. [14.7] $\frac{1}{12}$, 0 **47.** [14.7] $\frac{1}{4}$ **48.** [14.7] $\frac{6}{5525}$
49. [8.2] $\left(\frac{26}{11}, \frac{1}{11}\right)$ **50.** [12.6] $\frac{5}{9}$
51. [12.4] $3 \log a + 4 \log b - \log c - 2 \log d$
52. [13.2] $(3, -4)$, $5\sqrt{3}$
53. [14.3] ◈ For a geometric sequence with $|r| < 1$, as n gets larger, the absolute value of the terms gets smaller, since $|r^n|$ gets smaller. **54.** [14.2] ◈ A sequence whose terms have alternating signs does not have a

common difference, and therefore cannot be an arithmetic sequence. **55.** [14.3] $\frac{1-(-x)^n}{x+1}$

56. [14.6] $x^{-15} + 5x^{-9} + 10x^{-3} + 10x^3 + 5x^9 + x^{15}$

Test: Chapter 14, pp. 765–766

1. [14.1] 1, 7, 13, 19, 25; 91 **2.** [14.1] $a_n = 4\left(\frac{1}{3}\right)^n$

3. [14.1] $1 - 1 - 5 - 13 - 29 = -47$

4. [14.1] $\sum_{k=1}^{5} k^3$ **5.** [14.2] -46 **6.** [14.2] $\frac{3}{8}$

7. [14.2] $a_1 = 31.2$; $d = -3.8$ **8.** [14.2] 2508

9. [14.3] $\frac{9}{128}$ **10.** [14.3] $\frac{2}{3}$ **11.** [14.3] $(-1)^{n+1}3^n$

12. [14.3] $511 + 511x$ **13.** [14.3] 1

14. [14.3] No **15.** [14.3] $\frac{\$25{,}000}{23} \approx \$1086.96$

16. [14.3] $\frac{85}{99}$ **17.** [14.2] $17.60 **18.** [14.2] 1378

19. [14.3] $43,178.50 **20.** [14.3] 36 m **21.** [14.5] 78

22. [14.4] 720 **23.** [14.5] 35 **24.** [14.5] 103,740

25. [14.4] 6840 **26.** [14.6] $x^{10} - 15x^8y + 90x^6y^2 - 270x^4y^3 + 405x^2y^4 - 243y^5$ **27.** [14.6] $220a^9x^3$

28. [14.7] $\frac{1}{6}$ **29.** [14.7] $\frac{1}{13}$ **30.** [14.7] $\frac{1}{14}$

31. [12.6] 3 **32.** [13.2] $(x-1)^2 + (y+2)^2 = 27$

33. [8.2] $(-1, 2)$ **34.** [12.4] $\log_b \frac{7}{5}$ **35.** [14.2] $n(n+1)$

36. [14.3] $\frac{1-\left(\frac{1}{x}\right)^n}{1-\frac{1}{x}}$, or $\frac{x^n - 1}{x^{n-1}(x-1)}$

CUMULATIVE REVIEW: 1–14

1. [4.9] $-45x^6y^{-4}$, or $\frac{-45x^6}{y^4}$ **2.** [1.4], [1.5] 6.3

3. [1.8] $-3y + 17$ **4.** [1.8] 280 **5.** [1.1], [4.6] $\frac{7}{6}$

6. [4.6] $3a^2 - 8ab - 15b^2$

7. [4.3] $13x^3 - 7x^2 - 6x + 6$ **8.** [4.5] $6a^2 + 7a - 5$

9. [4.5] $9a^4 - 30a^2y + 25y^2$

10. [6.5] $\frac{4}{x+2}$ **11.** [6.2] $\frac{x-4}{3(x+2)}$

12. [6.2] $\frac{(x+y)(x^2+xy+y^2)}{x^2+y^2}$

13. [6.6] $x - a$ **14.** [5.4] $(2x-3)^2$

15. [5.5] $(3a-2)(9a^2+6a+4)$

16. [5.1] $(a^2-b)(a+3)$ **17.** [5.3] $3(y^2+3)(5y^2-4)$

18. [7.4] 20 **19.** [4.7] $7x^3 + 9x^2 + 19x + 38 + \frac{72}{x-2}$

20. [2.2] $\frac{2}{3}$ **21.** [9.2] $\{x|x \text{ is a real number}\}$ or $(-\infty, \infty)$

22. [5.7] 8, -6 **23.** [6.7] $-\frac{6}{5}$, 4 **24.** [8.2] $(1, -1)$

25. [8.4] $(2, -1, 1)$ **26.** [10.7] 9 **27.** [11.6] ± 5, ± 2 **28.** [13.4] $(\sqrt{5}, \sqrt{3})$, $(\sqrt{5}, -\sqrt{3})$, $(-\sqrt{5}, \sqrt{3})$, $(-\sqrt{5}, -\sqrt{3})$ **29.** [12.6] $\ln 8/\ln 5 \approx 1.2920$

30. [12.6] 1005 **31.** [12.3] $\frac{1}{16}$ **32.** [12.6] $-\frac{1}{2}$

33. [9.3] $\{x|-2 \leq x \leq 3\}$ **34.** [11.1] $\pm i\sqrt{2}$

35. [11.3] $-2 \pm \sqrt{7}$ **36.** [11.10] $\{y|y < -5 \text{ or } y > 2\}$

37. [13.4] 5 ft by 12 ft **38.** [2.7] More than 4 CDs

39. [2.5] 65, 66, 67 **40.** [8.3] $11\frac{3}{7}$ **41.** [8.3] 60 L Swim Clean, 40 L of Pure Swim **42.** [6.8] 350 mph

43. [6.8] $8\frac{2}{5}$ min or 8 min, 24 sec **44.** [7.6] 20

45. [11.9] 1250 ft²

46. [3.3]

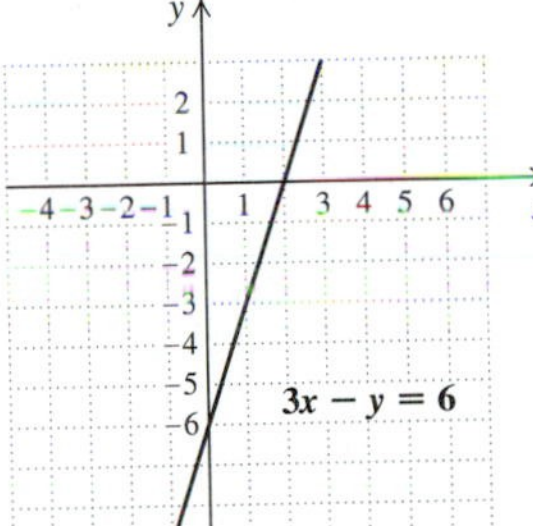

47. [13.3]

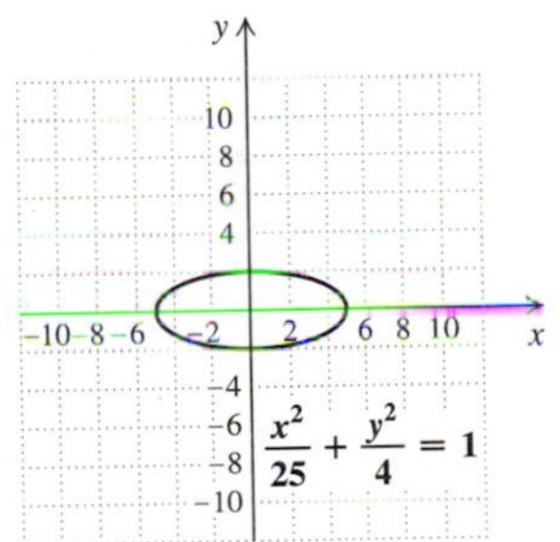

48. [12.3]

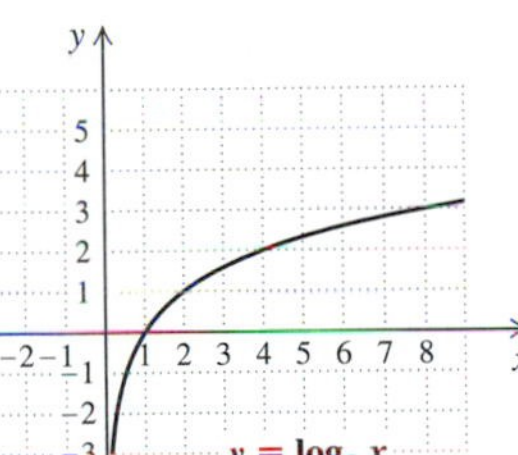

49. [9.4]

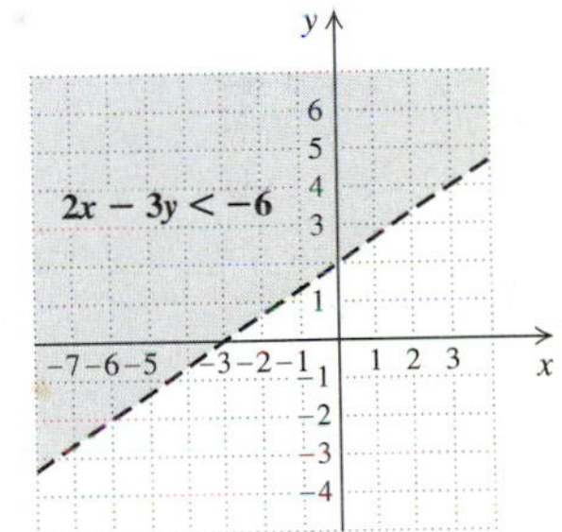

50. [11.7]

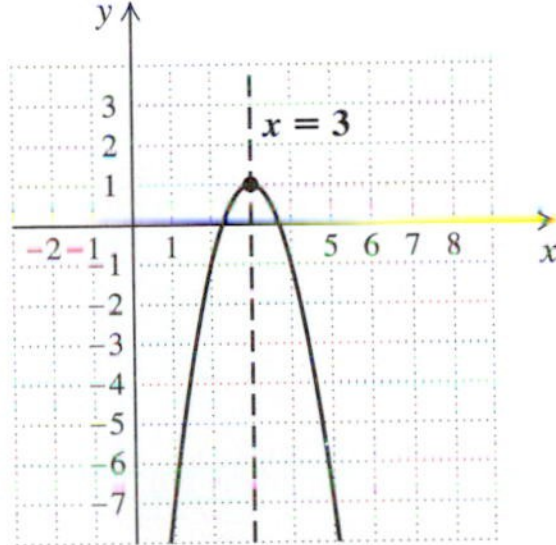

$f(x) = -2(x-3)^2 + 1$
Vertex (3, 1)
Maximum: 1

51. [2.3] $r = \frac{V - P}{-Pt}$ **52.** [6.9] $R = \frac{Ir}{1 - I}$
53. [7.3] $y = -\frac{1}{3}x + \frac{11}{3}$ **54.** [8.7] -2 **55.** [8.7] -2
56. [4.9] 6.8×10^{-12} **57.** [10.3] $8x^2\sqrt{y}$
58. [10.2] $125x^2y^{3/4}$ **59.** [10.4] $\frac{\sqrt[3]{5xy}}{y}$
60. [10.6] $\frac{1 - 2\sqrt{x} + x}{1 - x}$ **61.** [10.4] $\sqrt[6]{x + 1}$
62. [10.9] $26 - 13i$ **63.** [11.5] $2x^2 + x - 1$
64. [13.2] Center: $(0, -3)$; radius $= \sqrt{22}$
65. [12.4] $\log_a \frac{\sqrt[3]{x^2} \cdot z^5}{\sqrt{y}}$ **66.** [12.3] $a^5 = c$
67. [12.5] 2.1082 **68.** [12.5] 0.0354
69. [12.5] 4.8544 **70.** [12.5] 0.2344
71. [12.7] $P(t) = 98{,}942e^{-0.01t}$ **72.** [12.7] 86,881
73. [13.1] 5 **74.** [14.2] -121 **75.** [14.2] 875
76. [14.3] $16(\frac{1}{4})^{n-1}$ **77.** [14.6] $13{,}440a^4b^6$
78. [14.3] 74.88671875 **79.** [14.3] \$168.95
80. [14.4] 120 **81.** [14.7] $\frac{1}{4}$
82. [6.7] All real numbers except 0 and -12
83. [12.6] 81 **84.** [7.6] y gets divided by 8
85. [10.9] $-\frac{7}{13} + \frac{2\sqrt{30}}{13}i$ **86.** [8.5] 84 yr

INDEX

A

Absolute value, 28, 456, 478
  equations with, 457
  inequalities with, 459
  of a product, 487
  and roots, 487, 488, 489, 538
Addition
  associative law of, 10, 57
  commutative law of, 9, 56
  of complex numbers, 532
  of exponents, 148, 493, 539
  of fractional expressions, 18, 19, 276, 278, 287, 322, 323
  of functions, 360, 377
  of logarithms, 639, 665
  of polynomials, 162, 185
  of radical expressions, 509, 539
  of rational expressions, 276, 278, 287, 322, 323
  of real numbers, 30, 32
Addition principle
  for equations, 63, 108
  for inequalities, 97, 108, 440, 478
Additive inverse, 36. *See also* Opposite.
Algebra of functions, 360, 377
Algebraic expressions, 3
  evaluating, 3
  terms, 39
  translating to, 3
Algebraic fractions, 266
Angles
  complementary, 382
  supplementary, 388
Applications, *see* Applied problems; Index of Applications
Applied problems, 3, 7, 14, 23, 29, 30, 33–35, 39, 41, 42, 57, 75–79, 82–95, 103–107, 109–111, 114–117, 119–121, 136–145, 157, 158, 160, 161, 164, 165, 167, 168, 174, 175, 182, 183, 187–190, 195, 205, 206, 207, 209, 211, 223, 224, 231, 238, 242, 254–263, 276, 286, 306–316, 324–327, 336–338, 344, 353, 356–359, 366, 368, 371–376, 378–380, 397–407, 414–419, 422, 423, 428–437, 444, 448, 455, 456, 462, 463, 472–477, 479–482, 491, 495, 496, 503, 508, 523, 524, 527–530, 541, 542, 553, 557–562, 564, 565, 567, 570, 575, 589, 590, 594–598, 605, 606, 609–611, 619, 632, 655–664, 666–670, 682, 683, 695, 702–705, 707, 708, 715, 719–723, 728, 729, 731, 732–734, 737–740, 744–747, 752, 754, 756–761, 764–768. *See also* Index of Applications.
Antilogarithm, 645
Approximately equal to (≈), 27
Approximating square roots, 26, 486, 500
Area, *see* Index of Applications: Geometric Applications
Arithmetic progression, *see* Arithmetic sequence
Arithmetic sequence, 716, 762
  common difference, 716
  nth term, 717, 762
  sum of first n terms, 719, 762
Arithmetic series, 718
Associative laws, 10, 57
Asymptote
  of an exponential function, 616
  of a hyperbola, 688
Axes of graphs, 117
Axis
  on a coordinate plane, 117
  of a hyperbola, 687

B

Back-substitution, 420
Bar graph, 114
Base, 49
Binomial coefficient notation, 743, 750, 762
Binomial expansion, 747
  using factorial notation, 750, 762
  using Pascal's triangle, 749
  $(r + 1)$st term, 751

Binomial theorem, 749, 750, 762
Binomials, 155
as divisors, 192
product of, 171
FOIL method, 175
sum and difference of expressions, 177, 208
squares of, 178, 208
Braces, 50
Brackets, 50
Break-even analysis, 428
Break-even point, 429

C

Calculator
antilogarithms on, 645
approximating square roots on, 26, 486, 500
common logarithms on, 644
Canceling, 17, 268
Carbon dating, 661, 665
Carry out, 85, 109
Catenary, 676
Center
of a circle, 680
of an ellipse, 684
of a hyperbola, 687
Change-of-base formula, 647
Changing the sign, 37, 163
Checking
by evaluating, 246, 269
in problem-solving process, 85, 109
solutions of equations, 64, 249, 301, 457, 519, 563, 572, 573, 636, 651, 653, 698, 701
solutions of inequalities, 97, 463
solutions of systems of equations, 385, 391, 393, 407
Circle, 680, 692, 706
Circle graph, 117
Circular permutations, 738
Circumference, *see* Index of Applications: Geometric Applications
Clearing decimals, 71, 395, 409
Clearing fractions, 71, 294, 301, 395, 409, 563
Closed interval, 442
Coefficients, 66, 155, 184
leading, 218
Collecting like terms, 33, 39, 52, 156, 185
in equation solving, 70
Columns of a matrix, 419
Combinations, 741, 743, 744, 763
notation, 742, 743
Combinatorics, 732
Combining like terms, *see* Collecting like terms
Common denominators, 19. *See also* Least common denominator.
Common difference, 716
Common factor, 12, 215
Common logarithms, 644
Common ratio, 724
Commutative laws, 9, 56
Complementary angles, 382
Completing the square, 549
and circle equations, 681
and ellipse equations, 696
and graphing, 583, 677, 678
and hyperbola equations, 696
solving equations by, 551, 607
Complex numbers, 530, 531, 540
addition of, 532
conjugates of 534, 535
division of, 535
multiplication of, 532
as solutions of equations, 535
subtraction of, 532
Complex rational, or fractional, expressions, 293
simplifying, 293, 296, 323
Composite functions, 622, 623, 665
and inverses, 631
Composite number, 15
Compound interest, 154, 645, 659, 665. *See also* Index of Applications: Business and Economics.
Conic sections, 672
Conjugates, 511, 515, 534, 535
Conjunction, 449–451
Consecutive integers, 86, 87
Consistent system of equations, 387, 395, 411, 412
Constant, 2
function, 353, 376
of proportionality, 368, 370
variation, 368, 370
Constraints, 471
Coordinates, 117
finding, 119
on the globe, 121
Correspondence, 350. *See also* Function.
Cost, total, 160, 428
Cramer's rule, 424, 426, 434
Cross-multiplying, 311
Cross-products, 311
Cube root, 488, 538
Cubes, factoring sums or differences, 239–241, 261

D

Data point, 591
Decay rate, 660
Decimal notation
converting from/to percent notation, 81
converting from/to scientific notation, 203
for irrational numbers, 26
for rational numbers, 25
repeating, 25
terminating, 25
Decimals, clearing, 71, 395, 409
Degree
of a polynomial, 156, 184
of term, 155, 184
of zero, 156
Demand, 430
Denominator, 15
common, 19
least common, 71, 281, 282, 323
opposite, 278, 290
rationalizing, 513, 516, 540
Dependent system of equations, 387, 395, 412, 424
Dependent variable, 352
Depreciation, 141, 723
Descending order, 157
Determinant, 423, 425, 434
Diagonals, number of, 557. *See also* Index of Applications.
Difference, *see* Subtraction
Difference of logarithms, 640, 665
Differences of cubes, factoring, 239–241, 261
Differences of squares, 234
factoring, 235, 237, 241, 261
Dimensions of a matrix, 423
Direct variation, 368, 377
Discriminant, 568, 587, 607
Disjunction, 452, 453
Distance, 7
Distance formula
for distance between two points, 673, 706
for distance traveled, 309
Distributive law, 11, 57
and collecting like terms, 33
and factoring, 12

Division
  of complex numbers, 535
  using exponents, 149, 493, 539
  of fractional expressions, 19, 273, 322
  of functions, 360, 374
  of polynomials, 191, 192
  of radical expressions, 504, 539
  of rational expressions, 273, 322
  of real numbers, 44, 45
  and reciprocals, 19, 273
  using scientific notation, 204
  synthetic, 197
  by zero, 44, 46, 47
Domain, 350, 376
  and graphs, 365
  of a square root function, 489
  of a sum, difference, product, or quotient of functions, 362, 377
Doubling time, 657
Dummy variable, 353

E

e, 646
Element of a matrix, 420
Elimination method, solving systems of equations, 393, 395, 408, 409, 434, 700
Ellipse, 684, 693, 696, 707
  applications, 687
  intercepts, 685
Empty set, 451
Entries of a matrix, 420
Equally likely outcomes, 755
Equations, 5
  absolute value, 457
  of a circle, 680, 692, 706
  of direct variation, 368
  of an ellipse, 684, 693, 696, 707
  equivalent, 63
  exponential, 635
  fractional, 301
  graphs of, *see* Graphing
  of a hyperbola, 687, 690, 693, 696, 707
  of inverse variation, 370
  linear, 122, 407. *See also* Graphing.
  logarithmic, 635
  of a parabola, 677, 706
  containing parentheses, 72
  point-slope, 345, 377
  quadratic, 247, 248
  in quadratic form, 571
  radical, 518
  rational, 301
  reducible to linear, 397
  reducible to quadratic, 571
  related, 465
  reversing, 64
  slope-intercept, 340, 377
  solutions, 5, 62, 122, 535
  solving, *see* Solving equations
  systems of, 383
  translating to, 6, 136
  of variation, 368, 370
  writing from solutions, 569
  See also Formulas.
Equilateral triangle, 525
Equilibrium point, 430
Equivalent equations, 63
Equivalent expressions, 9, 285
Evaluating expressions, 2, 3
  and checking, 246
  polynomials, 157, 184
Even roots, 489, 538
Event, 755
Expansion, binomial, 747, 749
Experimental probability, 753, 754, 763
Exponential decay, 660, 665
Exponential equations, 635, 651
Exponential function, 614, 615, 634, 665
Exponential growth, 657, 665
Exponential notation, 49. *See also* Exponents.
Exponents, 49
  definitions for, 152, 198,
  dividing using, 149, 493, 539
  multiplying using, 148, 493, 539
  negative, 200, 202, 208
  one as, 50, 152, 202, 208
  raising a power to a power, 150, 493, 539
  raising a product to a power, 151, 493, 539
  raising a quotient to a power, 152, 202, 208
  rational, 492, 493, 538
  rules for, 152, 202, 208, 493, 539
  zero as, 150, 152, 202, 208
Expressions
  algebraic, 3
  equivalent, 9, 285
  evaluating, 2
  fractional, 266
  radical, 485
  rational, 266
  simplifying, *see* Simplifying
  terms of, 12
  value of, 3
  variable, 2

F

Factorial notation, 735, 763
  and binomial expansion, 750, 762
Factoring, 12
  common factor, 12, 215
  numbers, 14
  polynomials, 214, 242, 262
    checking by evaluating, 246
    with a common factor, 215
    completely, 236
    differences of cubes, 239–241, 261
    differences of squares, 235, 237, 261
    by grouping, 216, 217, 229, 236
    monomials, 214
    squares of binomials, 233, 261
    sums of cubes, 239–241
    trinomial squares, 233, 261
    trinomials, 219–230
  radical expressions, 497, 499, 539
  solving equations by, 249, 250
Factorizations, 15. *See also* Factoring.
Factors, 12, 14, 195, 214. *See also* Factoring.
  common, 12, 215
Falling object
  distance traveled, 160, 320, 375, 593, 596, 597
  time of fall, 558, 560, 596, 597, 609
Familiarize, 85, 109
Feasible region, 472
Finite sequence, 710
Finite series, 712
First coordinate, 117
Fitting quadratic functions to data, 591
Five-step process for problem solving, 85, 109
Fixed costs, 428
Focus (foci) of an ellipse, 684
FOIL method, 175
  and factoring, 224
Formulas, 75
  change-of-base, 647
  compound-interest, *see* Index of Applications: Business and Economics
  diagonals of a polygon, 557
  distance between two points, 673, 706
  distance traveled, 309
  for inverse of a function, 627, 665
  midpoint, 674, 706

quadratic, 555, 607
solving for given letter, 77, 317, 592, 594, 608
See also Equations.
Fractional equations, 301
Fractional expressions, 266
addition of, 18, 19
complex, 293
division of, 19
multiplication of, 16
simplifying, 17
subtraction of, 18, 19
See also Fractional notation; Rational expressions.
Fractional notation, 15
converting to percent notation, 81
for one, 16
for rational numbers, 25
simplifying, 17
See also Fractional expressions; Rational expressions.
Fractions
algebraic, 266
clearing, 71, 294, 301, 395, 409
improper, 18
See also Fractional expressions; Fractional notation; Rational expressions
Function, 350, 351
composite, 622, 623, 665
constant, 353
domain, 350, 362
exponential, 614, 615, 634, 665
graphs of, 353, 354, 361, 362, 377, 629, 630
greatest integer, 359
inputs, 352
linear, 353
logarithmic, 614, 634, 665
"machine," 352, 628
notation, 352
objective, 471
one-to-one, 625
outputs, 352
quadratic, 576
range, 350
sequence as, 710
square root, 485, 486
vertical-line test, 354, 377
Fundamental counting principle, 733, 763

G
Games played, 158
General term of a sequence, 711
Geometric progression, *see* Geometric series
Geometric sequence, 724, 762
common ratio, 724
*n*th term, 725, 762
sum of first *n* terms, 726
Geometric series, 726
infinite, 726
limit, 727, 762
Globe, coordinates on, 121
Grade, 329, 335
Graph
bar, 114
circle, 117
line, 116
scale of, 137, 138
See also Graphing.
Grapher, 127
Graphing
circles, 680, 692
ellipses, 685, 693
equations, 123
of direct variation, 368
horizontal line, 131
using intercepts, 129
of inverse variation, 370
and point-slope form, 346
and problem solving, 136
quadratic, 251, 677, 678, 692, 706
slope, 333
vertical line, 131
x-intercept, 129, 251, 252
$y=mx$, 123
$y=mx+b$, 125
y-intercept, 127, 129
using *y*-intercept and slope, 339, 341
functions, 353, 354, 361, 362, 365, 377, 607, 608
exponential, 615–618, 648
and their inverses, 629, 630
logarithmic, 633, 634, 648
quadratic, 576–581, 583–586
hyperbolas, 688, 691, 693
inequalities, 95
in one variable, 95, 441, 465–467
systems of, 467
in two variables, 464, 465
numbers, 24, 27
parabolas, 583–587, 677, 678, 692, 706
points on a plane, 117
solving equations by, 304, 385, 395, 408
systems of equations, 385, 395, 408, 411, 412
Greater than (>), 27
Greater than or equal to (≥), 27
Greatest integer function, 359
Grouping
in addition, 10, 57
factoring by, 216, 217, 229, 236
in multiplication, 10, 57
symbols, 50
Growth rate, exponential, 657

H
Half-life, 660
Half-open intervals, 442
Horizon, distance to, 523
Horizontal-line test, 626
Horizontal lines, 131
slope, 335, 377
Hyperbola, 687, 690, 693, 696, 707
applications, 691
asymptotes, 688
Hypotenuse, 257

I
i, 530
powers of, 533
Identity property
of 1, 16, 57
of 0, 32
Imaginary number, 531
Improper fraction, 18
Inconsistent system of equations, 387, 395, 411, 412, 424
Independent system of equations, 395, 412
Independent variable, 352
Indeterminate, 46, 47
Index of a radical expression, 488
Index of summation, 713
Indices, 488
Inequalities, 27, 95, 440
with absolute value, 459
addition principle for, 97, 108
conjunction, 449–451
disjunction, 452, 453
graphs of, 95
linear, 465
in one variable, 95, 441, 465–467
system of, 467
in two variables, 464, 465
linear, 465

system of, 467
multiplication principle for, 98, 108
polynomial, 598
quadratic, 598
rational, 603
solution set, 97, 440
solutions of, 95, 440, 463
solving, *see* Solving inequalities
translating to, 28, 103
Infinite sequence, 710
Infinite series, 712
geometric, 726
limit, 727, 762
Infinity, 442
Inputs, 352
Integers, 22, 23
Intercepts, 127, 129, 251, 252, 586, 587, 685
graphs using, 129, 340
Interest, *see* Index of Applications: Business and Economics
Intersection, 449, 478
Interval notation, 442
Inverse
additive, 36. *See also* Opposite of a number.
of a function, 625, 627, 631, 665
multiplicative, 19
relation, 625
variation, 370, 377
Irrational numbers, 25
decimal notation, 26
graphing, 27
Isosceles right triangle, 524, 527, 540

J
Joint variation, 371, 377

K
*k*th root, 488, 489, 538

L
Largest common factor, 215
Law of opposites, 37, 57
Laws of exponents, 493, 539
LCD, *see* Least common denominator
LCM, *see* Least common multiple
Leading coefficient, 218
Least common denominator, 71, 281, 282, 323
Least common multiple, 281–283
Legs of a right triangle, 257
Less than (<), 27
Less than or equal to (≤), 27
Like radicals, 509, 539
Like terms, 33, 52, 156, 185
collecting (or combining), 33, 39, 52, 156, 185
Limit, infinite geometric series, 727
Line
horizontal, 131, 335, 377
point-slope equation, 345, 377
slope, 333, 377
slope-intercept equation, 340, 377
vertical, 131, 335, 377
See also Lines.
Line graph, 116
Line of symmetry, 576, 586, 608, 677
Linear equations, 122
graphing, 123, 129, 131, 133, 142
in three variables, 407
Linear function, 353
Linear inequalities, 465
systems of, 467
Linear programming, 472, 478
Lines
parallel, 347, 377
perpendicular, 348, 377
Log, 644
Logarithm
common, 644
changing bases, 647
definition, 634, 665
inverse, 645
natural, 646
notation, 633
of powers, 639, 665
of a product, 639, 665
properties, 637, 639, 640, 642, 665
of a quotient, 640, 665
Logarithm function, *see* Logarithmic function
Logarithmic equations, 635, 636, 652
Logarithmic function, 614, 633, 634, 665

M
Magic number, 190
Matrices, *see* Matrix
Matrix, 419
columns, 419
determinant of, 423, 425, 434
dimensions, 423
elements, 420
entries, 420
row-equivalent, 421
rows, 419
and solving systems of equations, 419
square, 423
Maximum value
of an objective function, 472
of a quadratic function, 579, 580, 588
Midpoint formula, 674, 706
Minimum value
of an objective function, 472
of a quadratic function, 579, 580, 588
Mixture problems, *see* Index of Applications: General Interest
Monomials, 154
as divisors, 191
factoring, 214
and multiplying, 169
Motion problems, *see* Index of Applications: General Interest
Multiplication
associative law of, 10, 57
commutative law of, 9, 56
of complex numbers, 532
using exponents, 148, 493, 539
of fractional expressions, 16, 271, 322
of functions, 360, 377
of polynomials, 169–173, 175–181, 186
of radical expressions, 496, 499, 510, 539
of rational expressions, 271, 322
of real numbers, 42–45
using scientific notation, 204
by zero, 43, 57
See also Multiplying.
Multiplication principle
for equations, 65, 108
for inequalities, 98, 108, 440, 478
Multiplicative inverse, 19
Multiplicative property of zero, 43, 57
Multiplying
exponents, 150, 493, 539
by 1, 16, 57, 282, 283, 287, 294
by -1, 53, 57
See also Multiplication.

N
*n*th term
of an arithmetic sequence, 717
of a geometric sequence, 725, 762
Naperian logarithms, 646
Natural logarithms, 646
Natural numbers, 14
Negative exponents, 200, 202, 208, 493, 538
Negative integers, 23

Negative one, property of, 53, 57
Negative square root, 485
Nonlinear systems of equations, 697, 700
Nonnegative integers, 23
Notation
  binomial coefficient, 743, 762
  combination, 742, 743
  exponential, 49
  factorial, 735, 763
  fractional, 15
  function, 352
  inverse function, 627
  logarithm, 633
  percent, 80, 108
  permutation, 742
  scientific, 202, 203, 208
  set, 97
  sigma, 713
  summation, 713
Number line
  addition on, 30
  and graphing, 24, 27
  order on, 27
Numbers
  complex, 530, 531, 540
  composite, 15
  factoring, 14
  imaginary, 531
  integers, 22
  irrational, 25
  natural, 14
  order of, 27
  prime, 15
  rational, 24
  real, 26
  signs of, 37
  whole, 14
Numerator, 15
  rationalizing, 514, 516

O

Objective function, 471
Odd roots, 488, 538
One
  as exponent, 50, 152, 202, 208
  fractional notation for, 16
  identity property of, 16, 57
  removing a factor of, 17
One-to-one function, 625
Open interval, 442
Operations
  order of, 50, 57
  row-equivalent, 421, 422, 434
Opposite
  and changing the sign, 37
  factors, 270
  of a number, 22, 23, 36
  of a polynomial, 163
  in subtraction, 38
  of a sum, 53, 57
  See also Additive inverse.
Opposites, law of, 37, 57
Order
  descending, 157
  on number line, 27
  of operations, 50, 57
Ordered pairs, 118
Origin, 117
Outcomes, 755
Outputs, 352

P

Pairs, ordered, 118
Parabolas, 576, 676, 692, 706
  line of symmetry, 576, 586, 608, 677
  vertex, 586, 608, 677
  x-intercepts, 586, 587
Parallel lines, 347, 377
Parallelogram, *see* Index of Applications: Geometric Applications
Parentheses, 50
  in equations, 72
  removing, 12, 52, 53
Parking-lot arrival spaces, 491
Partial sum, 712
Pascal's triangle, 748
  and binomial expansion, 749
Pendulum, period of, 508, 592
Percent, problems involving, *see* Index of Applications: General Interest
Percent notation, 80, 108
  converting to/from decimal notation, 81
Perimeter, *see* Index of Applications: Geometric Applications
Period of a pendulum, 508, 592
Permutations, 734, 735, 736, 763
  circular, 738
Perpendicular lines, 348, 377
Pi (π), 25
Plotting points, 117
Point-slope equation, 345, 377
Points, coordinates of, 117
Polygon, number of diagonals, 577. *See also* Index of Applications.
Polynomial inequalities, 598
  solving, 599–603, 608
Polynomials, 154
  addition of, 162, 185
  additive inverse of, 163
  binomials, 155
  coefficients in, 155, 184
  collecting like terms (or combining similar terms), 156
  degree of, 156, 184
  in descending order, 157
  division of, 191, 192
  evaluating, 157, 184
  factoring, *see* Factoring, polynomials
  least common multiple of, 283
  monomials, 154
  multiplication of, 169–173, 175–181, 186
  opposite of, 163
  prime, 222
  in several variables, 183
  subtraction of, 163, 185
  terms of, 155, 184
  trinomials, 155
  value of, 157
Population decay, *see* Exponential decay
Population growth, *see* Exponential growth
Positive integers, 23
Positive square root, 485
Power, 49
  of *i*, 533
  powers, principle of, 518, 540
  raising to a power, 150, 493, 539
  roots of, 506, 539
  See also Exponents.
Power rule
  for exponents, 151, 152, 202, 208
  for logarithms, 639, 665
Prime factorization, 15
  and LCM, 282
Prime numbers, 15
Prime polynomial, 222
Principal square root, 485
Principle of square roots, 546, 607
Principle of powers, 518, 540
Principle of zero products, 248, 262
Probability, 753
  event, 755
  experimental, 753, 754, 763
  origin and use, 759
  outcomes, 755
  sample space, 755
  theoretical, 753, 756, 763
Problem solving
  five-step process, 85, 109

and graphing, 136
other tips, 92
See also Applied problems.
Problems, applied, *see* Applied Problems; Index of Applications
Product
logarithm of, 639, 665
raising to a power, 151, 152, 202, 208, 493, 539
of sums and differences, 177, 208
of two binomials, 171, 175–181
See also Multiplication; Multiplying.
Product rule
for exponential notation, 148, 152, 202, 208
for logarithms, 639
for radicals, 496, 539
Profit, total, 168
Progression
arithmetic, 716
geometric, 724
Projections, 365
Properties of logarithms, 637, 639, 640, 642, 665
Property of -1, 53, 57
Proportion, 311
Proportional, 311, 368, 370
Proportionality, constant of, 368, 370
Pythagorean theorem, 258, 262, 596

Q

Quadrants, 118
Quadratic equations, 247, 248, 546
discriminant, 568, 587, 607
graphs of, 251, 677, 678, 692, 706
solutions, nature of, 568, 607
solving, 557, 607
by completing the square, 551, 553, 607
by factoring, 249, 250
using quadratic formula, 555
using principle of square roots, 546, 548
using principle of zero products, 248–250, 262
in standard form, 554
writing from solutions, 569
Quadratic form, equations in, 571
Quadratic formula, 555, 607
Quadratic function, 576
fitting to data, 591
graphs of, 576–581, 583–586, 607, 608
maximum/minimum value, 579, 588
Quadratic inequalities, 598
solving, 599–603, 608
Quotient, raising to a power, 152, 202, 208
Quotient, root of, 506
Quotient rule
for exponential notation, 149, 152, 202, 208
for logarithms, 640, 665
for radicals, 504, 539

R

Radical equations, 518
Radical expressions, 485
adding, 509, 539
conjugates, 511, 515
dividing, 504, 539
in equations, 518
factoring, 497, 499, 539
index (indices), 488
multiplying, 496, 499, 510, 539
product rule, 496, 539
quotient rule, 504, 539
radicand, 485
rationalizing denominators, 513, 516, 540
rationalizing numerators, 514, 516
simplifying, 494, 497, 499, 539, 540
subtracting, 509, 539
See also Square roots.
Radical sign, 485
Radicals, like, 509, 539
Radicand, 485
Radius, 680
Raising a power to a power, 150, 493, 539
Raising a product to a power, 151, 202, 208, 493, 539
Raising a quotient to a power, 152, 202, 208
Range
of a function, 350
on a grapher, 127, 138
of a relation, 351
Rate, 331
Ratio, 311
common, 724
and slope, 333
See also Proportion.
Rational equations, 301, 563
Rational exponents, 492, 493, 538
Rational expressions, 266
addition, 276, 278, 287, 322, 323
complex, 293
division, 273, 322
multiplying, 271, 322
reciprocals, 273
simplifying, 267–270
subtraction, 277, 278, 287, 322, 323
undefined, 266
See also Fractional expressions; Rational numbers.
Rational inequalities, 603, 608
Rational numbers, 24, 25. *See also* Fractional expressions; Rational expressions.
Rationalizing denominators, 513, 516, 540
Rationalizing numerators, 514, 516
Real-number system, 26
Real numbers, 26
addition, 30, 32
division, 44
graphing, 27
multiplication, 42–45
order, 27
subtraction, 38
Reciprocals, 19, 200, 273
Recursively defined sequence, 715
Reducible to linear equations, 397
Reducible to quadratic equations, 571
Region, feasible, 472
Related equation, 465
Relation, 351
inverse, 625
Remainder, 193
Removing a factor of 1, 17, 267, 269
Removing parentheses, 12, 52, 53
Repeating decimals, 25
Revenue, total, 160, 428
Reversing equations, 64
Reversing the sign, 37
Richter scale, 656
Right triangle, 257
isosceles, 524, 527, 540
30–60–90, 525, 527, 540
Rise, 333
Roots
cube, 488, 538
even, 489, 538
*k*th, 488, 489, 538
odd, 488, 538
and powers, 506, 539
of quotients, 506
and rational exponents, 492, 538
square, *see* Square roots; Radical expressions
Row-equivalent matrices, 421

Row-equivalent operations, 421, 422, 434
Rows of a matrix, 419
Rules for exponents, 152, 202, 208, 493, 539
Run, 333

S
Sample space, 755
Scale of a graph, 137, 138
Scientific notation, 202, 203, 208
  converting from/to decimal notation, 203
  dividing using, 204
  multiplying using, 204
  and problem solving, 205
Second coordinate, 117
Sequence mode, 342
Set-builder notation, 97, 440
Sequence, 710
  arithmetic, 716, 762
  finite, 710
  general term, 711
  geometric, 724, 762
  infinite, 710
  recursive, 715
  terms, 711
Series
  arithmetic, 718
  finite, 711
  geometric, 726
  infinite, 711
Sets, 22
  empty, 451
  intersection of, 449, 478
  notation, 97
  solution, 97
  subset, 22
  union of, 452, 478
Shade, on a grapher, 467
Sigma notation, 713
Signs of numbers, 37
Similar terms, 33, 52, 156, 185. *See also* Like terms.
Similar triangles, 311
Simple interest, *see* Index of Applications: Business and Economics
Simplifying
  complex rational expressions, 293, 296, 323
  fractional expressions, 17
  fractional notation, 17
  radical expressions, 494, 497, 499, 539, 540
  rational expressions, 267–270
  removing parentheses, 12, 52, 53
Simultaneous mode, 342
Slant, *see* Slope
Slope, 333, 377
  applications, 335
  and graphing, 339, 341, 346
  of horizontal line, 335, 377
  of parallel lines, 347, 377
  of perpendicular lines, 348, 377
  point-slope equation, 345, 377
  slope-intercept equation, 340, 377
  of vertical line, 335, 377
Slope-intercept equation, 340, 377
Solution set, 97, 440
Solutions
  of equations, 5, 62, 122
    complex-number, 535
    quadratic, nature of, 568, 607
  of inequalities, 95, 440, 463
  of systems of equations, 384, 407
Solving equations, 5, 62
  absolute value, 457, 478
  using the addition principle, 63, 108
  clearing decimals, 71, 395, 409
  clearing fractions, 71, 395, 409, 563
  collecting like terms, 70
  containing parentheses, 72
  exponential, 651
  by factoring, 249, 250
  fractional, 301, 323
  by graphing, 304
  logarithmic, 636, 652
  using the multiplication principle, 65, 108
  with parentheses, 72
  using principle of powers, 518, 521, 540
  using principle of zero products, 248, 262
  using principles together, 68
  procedure, 73, 108
  quadratic, 557, 607
    by completing the square, 551, 553, 607
    by factoring, 249, 250
    using quadratic formula, 555
    using principle of square roots, 546, 548
    using principle of zero products, 248–250
  radical, 518, 521, 540
  rational, 301, 323, 563
  reducible to quadratic, 572
  systems of
    using Cramer's rule, 424, 426, 434
    by elimination method, 393, 395, 408, 409, 434, 700
    by graphing, 385, 395
    using matrices, 419
    nonlinear, 697, 700
    by substitution method, 391, 395, 697, 700
  See also Solving formulas.
Solving formulas, 77, 317, 592, 594, 608
Solving inequalities
  with absolute value, 459, 460, 478
  using addition principle, 97, 108, 440, 478
  conjunctions, 450
  disjunctions, 453
  using multiplication principle, 98, 108, 440, 478
  polynomial, 599–603
  using principles together, 99
  rational, 603–605, 608
  solution set, 97
  systems of, 467
Square of a binomial, 178, 208, 232
  factoring, 233, 261
Square matrix, 423
Square root function, 485, 486
Square roots, 26, 484, 538
  and absolute value, 487, 538
  approximating, 26, 486, 500
  on a calculator, 26, 486, 500
  negative, 485
  of negative numbers, 484, 530
  positive, 485
  principal, 485
  principle of, 546, 607
  and tables, 486, 500
  See also Radical expressions.
Squares
  differences of, 234
    factoring, 235, 237, 261
  sum of, 236, 241
Standard form
  of a linear equation in three variables, 407
  of a quadratic equation, 554
State answer, problem solving process, 85, 109
Straight-line depreciation, 723

Subset, 22
Substituting, 3. *See also* Evaluating expressions.
Substitution method for solving equations, 391, 395, 408, 697, 700
Subtraction
by adding the opposite, 38
of complex numbers, 532
of exponents, 149, 493, 539
of fractional expressions, 18, 19, 277, 278, 287, 322, 323
of functions, 360, 377
of logarithms, 640, 665
of polynomials, 163, 185
of radical expressions, 509, 539
of rational expressions, 277, 278, 287, 322, 323
of real numbers, 38
Sum
of cubes, factoring, 239–241, 261
of first *n* terms, arithmetic sequence, 719, 762
of first *n* terms, geometric sequence, 726, 762
of logarithms, 639, 665
opposite of, 53, 57
of squares, 236, 241
Sum and difference of two terms, product of, 177
Summation notation, 713
Supplementary angles, 388
Supply, 430
Surface area, *see* Index of Applications: Geometric Applications
Symmetry, line of, 576, 586, 608, 677
Systems of equations, 383, 407
consistent, 387, 395, 411, 412
dependent, 387, 395, 412, 424
inconsistent, 387, 395, 411, 412, 424
independent, 395, 412
nonlinear, 697, 700
reducible to linear, 397
solutions of, 384, 407
solving
using Cramer's rule, 424, 426
by elimination method, 393, 395, 408, 409, 700
by graphing, 385, 395, 408, 411, 412
using matrices, 419
by substitution method, 391, 395, 408, 697, 700
Systems of linear inequalities, 467

T
Table of square roots, using, 486, 500
Technology Connection, 127, 138, 251, 304, 342, 387, 461, 467, 520, 551, 556, 576, 578, 580, 602, 618, 629, 649, 653, 681, 691, 701
Temperature conversion, 79, 106, 107, 321, 322, 356, 447, 455
Terminating decimal, 25
Terms, 12, 39
coefficients of, 155
collecting (or combining) like, 33, 39, 185
degrees of, 155, 184
like, 33, 52, 156, 185, 509, 539
of polynomials, 155
of a sequence, 711
similar, 33, 52, 156, 185
Theoretical probability, 753, 756, 763
30–60–90 triangle, 525, 527, 540
Total cost, 428
Total profit, 168, 428
Total revenue, 160, 428
Trace, 138
Translating
to algebraic expressions, 3
to equations, 6, 136
to inequalities, 28, 103
in problem-solving process, 85, 109, 382
Tree diagram, 733
Trial-and-error factoring, 219–228
Triangle
area, *see* Index of Applications: Geometric Applications
equilateral, 525
isosceles, 524, 527, 540
Pascal's, 748
right, 257, 524, 525, 527, 540
similar, 311
30–60–90, 525, 527, 540
Trinomial squares, 232
factoring, 233, 261
Trinomials, 155
factoring, 219–290, 233, 261

U
Union, 452, 478

V
Value
of an expression, 3
of a polynomial, 157
Variable, 2
dependent, 352
dummy, 353
independent, 352
Variable costs, 428
Variable expression, 2
Variation
constant, 368, 370
direct, 368, 376, 377
inverse, 370, 377
joint, 371, 377
Vertex
of an ellipse, 696
of a parabola, 576, 586, 608, 677
of a hyperbola, 687, 696
Vertical lines, 131
slope, 335, 377
Vertical-line test, 354, 377
Volume, *see* Index of Applications: Geometric Applications

W
Whispering gallery, 687
Whole numbers, 14
Wildlife populations, estimating, 312, 315
Wind chill temperature, 503
Window, 127
Work principle, 309, 323
Writing equations from solutions, 569

X
x-intercept, 129, 251, 252
of an ellipse, 685
of a parabola, 586, 587

Y
y-intercept, 127, 129
of an ellipse, 685
and graphing, 340

Z
Zero
degree of, 156
division by, 44, 46, 47
as exponent, 150, 152, 202, 208
identity property, 32
multiplicative property, 43, 57
Zero products, principle of, 248, 262
Zoom, 138

1. 85
2. 112 } 80.5 %

CALCULATOR KEYS
RELEVANT TO INTERMEDIATE ALGEBRA

SELECTED KEYS OF THE GRAPHING CALCULATOR

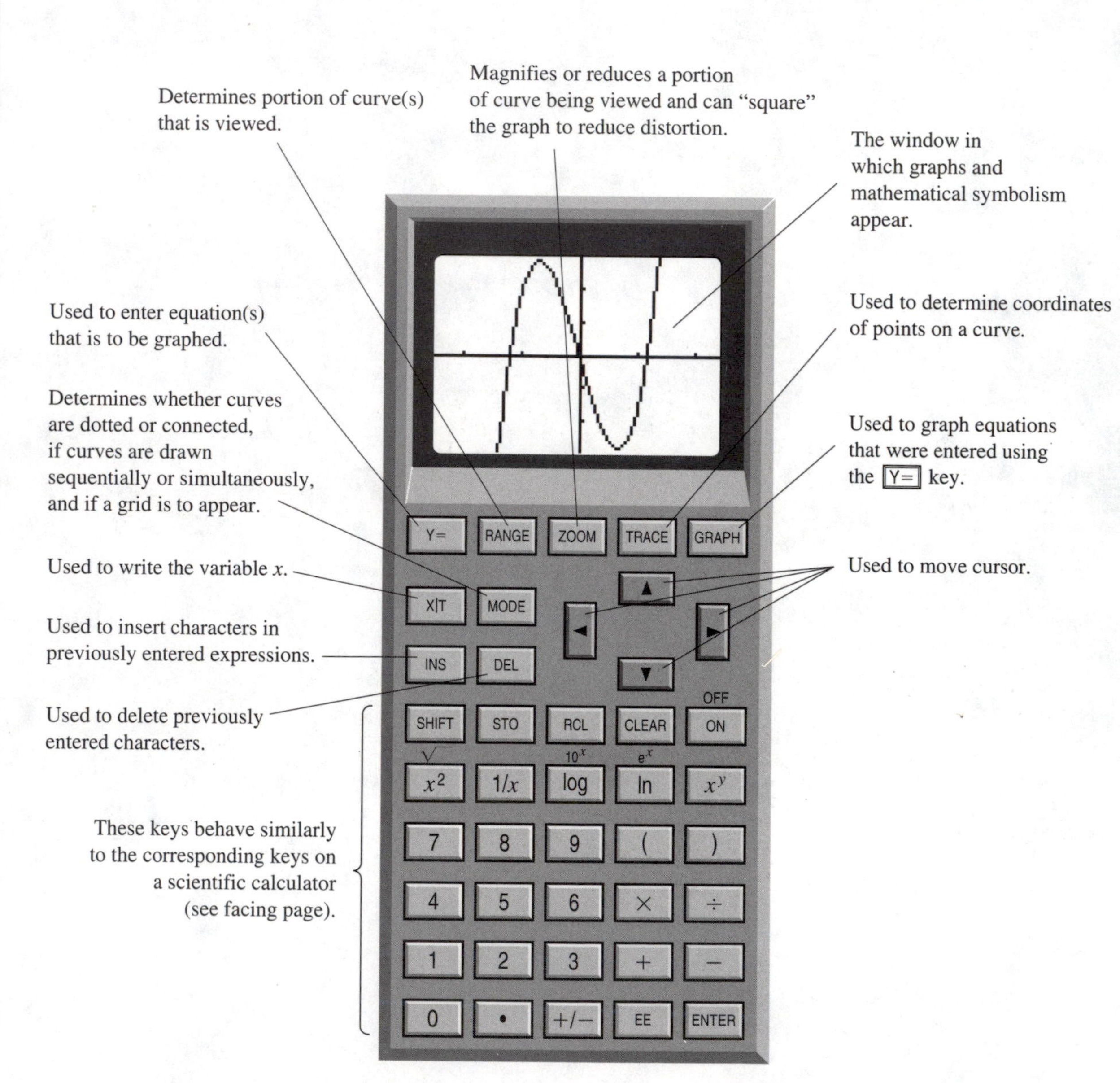

The use of a graphing calculator is optional in this text.